DYMAT 2015 – 11th International Conference on the Mechanical and Physical Behaviour of Materials under Dynamic Loading

EPJ Web of Conferences Volume 94 (2015)

Lugano, Switzerland
7 - 11 September 2015

Editors:

Ezio Cadoni

ISBN: 978-1-5108-2330-3

Printed from e-media with permission by:

Curran Associates, Inc.
57 Morehouse Lane
Red Hook, NY 12571

proceedings.com

Some format issues inherent in the e-media version may also appear in this print version.

Printed by Curran Associates, Inc. (2016)

For additional information, please contact EDP Sciences – Web of Conferences
at the address below.

EDP Sciences – Web of Conferences
17, Avenue du Hoggar
Parc d'Activité de Courtabœuf
BP 112
F-91944 Les Ulis Cedex A
France

Phone: +33 (0) 1 69 18 75 75
Fax: +33 (0) 1 69 28 84 91

contact-edps@webofconferences.org

Additional copies of this publication are available from:

Curran Associates, Inc.
57 Morehouse Lane
Red Hook, NY 12571 USA
Phone: 845-758-0400
Fax: 845-758-2633
Email: curran@proceedings.com
Web: www.proceedings.com

TABLE OF CONTENTS

EPJ Web of Conferences **94**, 00001 (2015)
DOI: 10.1051/epjconf/20159400001

Foreword

DYMAT, the European Association for research into the dynamic behaviour of materials and its applications, has been organizing a five days single sessions conference of oral presentations and poster exhibitions since 1983. The triennially DYMAT International Conferences are the platform to present the most recent scientific achievements on dynamic behaviour of materials.

Being an association of leading research institutions in Europe and well linked to the relevant industry sectors, DYMAT can claim excellence in science along with industry oriented research and a variety of platforms to publish. Where the spectrum of events to gather and to exchange is ranging from the International Conference over Technical Meetings, Working Groups to Student Camps. The success of the institution also became visible at the exceptional meeting of former presidents and governing board members 2013 in Paris when DYMAT celebrated its 30 anniversary.

2014 was the DYMAT year to celebrate the centenary of the epoch making invention by Bertram Hopkinson (1894–1918). With his 1914 seminal paper, "A method of measuring the pressure produced in the detonation of high explosives or by the impact of bullets" Phil. Trans. R. Soc. Lond. A213 437-456, Bertram Hopkinson pioneered the development of an experimental set-up which even nowadays seems to be in greater demand than ever.

The Hopkinson Centenary Conference was intended to commemorate this ground-breaking invention and to reflect the most recent scientific developments in the area of Hopkinson bar type tests for dynamic material testing. With Cambridge University as the place where Bertram Hopkinson studied at Trinity College and where he was Professor for Mechanisms and Applied Mechanics, a perfect location could be found for the Centenary Conference.

DYMAT 2015 will have the honour of being the third DYMAT International Conference to give the John S. Rinehart's Award. This distinction, established in 1990 to recognize outstanding effort and creative work in the science and technology of dynamic processes in materials, has been awarded every five years on the occasion of several EXPLOMET International Conferences. Since 2009, DYMAT is the Custodian of the Award. A special jury, composed of the members of the DYMAT International Advisory Committee and members of the DYMAT Governing Board, voted and selected two winners. Our sincere congratulations go to Carlo Albertini and to Ron Armstrong, the two recipients of the John S. Rinehart Award 2015.

My sincere gratefulness goes to Professor Ezio Cadoni, chairman of the Conference and co-editor of the 2015 EPJ-ST issue. Ezio has completed a remarkable work in the organization and the success of this conference. My thanks go to him and his colleagues, who greatly contributed to the success of DYMAT 2015.

Finally, I would also like to thank the members of the International Advisory Committee and the Governing Board for their commitment in ensuring the high scientific level of the conference.

Freiburg, April 12th 2015
Stefan Hiermaier
DYMAT President

Article available at http://www.epj-conferences.org or http://dx.doi.org/10.1051/epjconf/20159400001

EPJ Web of Conferences **94**, 00002 (2015)
DOI: 10.1051/epjconf/20159400002

Preface

In this book are published the papers of the XI International Conference on the Mechanical and Physical Behaviour of Materials under Dynamic Loading – DYMAT2015 held in Lugano (Switzerland) September 7–11, 2015. After 30 years from the first edition for the first time this important conference arrives in Switzerland.

In continuity with previous DYMAT conferences [Paris (1985), Ajaccio (1988), Strasbourg (1991), Oxford (1994), Toledo (1997), Krakow (2000), Porto (2003), Dijon (2006), Bruxelles (2009), Freiburg (2012)], this book contains 186 scientific papers from 32 countries from 5 continents that collectively provide an excellent snapshot of the state-of-the-arts in the field of the mechanical properties of materials at high rates of strain and highlights some remarkable modern developments and industrial applications. This volume covers five topics such as Experimental Techniques (80 papers), Microstructural Effects (27 papers), Biomechanics (6 papers), Modelling and Numerical simulation (62 papers) and Industrial Application (11 papers).

The conference was organized by the DynaMat Laboratory of the University of Applied Sciences and Arts of Southern Switzerland (SUPSI) engaged in research in these particular areas since 2006.

I would like to take this opportunity to thank the authors, all reviewers for their generous efforts.

A sincere thanks to all participants, sponsors and institutions, for their great support towards a successful DYMAT2015.

Finally, my acknowledgement to DYMAT Governing Board and International Advisory Committee members for their enthusiastic support in all phases of the DYMAT2015 organization, and for the very important management help.

Last but not least I feel indebted to my family for all the time dedicated to the organization of this event.

Ezio Cadoni
DYMAT2015 Chair
Vice-President of DYMAT

Article available at http://www.epj-conferences.org or http://dx.doi.org/10.1051/epjconf/20159400002

edp sciences

EPJ Web of Conferences

Statement of Peer review

In submitting conference proceedings to EPJ Web of Conferences, I certify to the Publisher that I adhere to the **Policy on Publishing Integrity** of the journal in order to safeguard good scientific practice in publishing.

1. All articles have been subjected to peer review administered by the proceedings editors.
2. Reviews have been conducted by expert referees, who have been requested to provide unbiased and constructive comments aimed, whenever possible, at improving the work.
3. Proceedings editors have taken all reasonable steps to ensure the quality of the materials they publish and their decision to accept or reject a paper for publication has been based only on the merits of the work and the relevance to the journal.

Title, date and place of the conference

DYMAT2015
11th International Conference on the Mechanical and Physical Behaviour of Materials under Dynamic Loading
September 7-11, 2015
Lugano (Switzerland)

Proceedings editor(s):

Prof. Dr. Ezio Cadoni

Lugano
April 14, 2015

Date and editor's signature

EPJ Web of Conferences **94**, 01001 (2015)
DOI: 10.1051/epjconf/20159401001

Dynamic compressive response of wrought and additive manufactured 304L stainless steels

Erik Nishida[1,a], Bo Song[1], Michael Maguire[1], David Adams[1], Jay Carroll[1], Jack Wise[1], Joseph Bishop[1], and Todd Palmer[2]

[1] Sandia National Laboratories, 1515 Eubank Blvd. SE, New Mexico, USA
[2] Pennsylvania State University, MatSE Department, 301 Applied Science Building, Pennsylvania, USA

Abstract. Additive manufacturing (AM) technology has been developed to fabricate metal components that include complex prototype fabrication, small lot production, precision repair or feature addition, and tooling. However, the mechanical response of the AM materials is a concern to meet requirements for specific applications. Differences between AM materials as compared to wrought materials might be expected, due to possible differences in porosity (voids), grain size, and residual stress levels. When the AM materials are designed for impact applications, the dynamic mechanical properties in both compression and tension need to be fully characterized and understood for reliable designs. In this study, a 304L stainless steel was manufactured with AM technology. For comparison purposes, both the AM and wrought 304L stainless steels were dynamically characterized in compression Kolsky bar techniques. They dynamic compressive stress-strain curves were obtained and the strain rate effects were determined for both the AM and wrought 304L stainless steels. A comprehensive comparison of dynamic compressive response between the AM and wrought 304L stainless steels was performed. SAND2015-0993 C.

1. Introduction

Additive manufacturing (AM), also commonly known as 3D printing, has been increasingly utilized in many applications that include rapid prototyping and manufacturing, as well as mass customization for automobile, aerospace, construction, and defense industries. For example, Sandia National Laboratories developed a Laser Engineered Net Shaping (LENS) technology, a type of AM technology which utilizes a combination of metal powders and a laser to create metallic components designed by computer aided design (CAD) models [1]. New AM processes that are LENS-like have been recently developed to create components at much higher deposition rates. Often as AM materials are created, residual stresses may occur and grain structures may change [2], which can lead to uncertainties in the mechanical behaviour of the material. Particularly, the AM materials may be utilized in abnormal mechanical environments where the materials are subjected to impact loading. Understanding the dynamic response of the AM materials is critical for applications in terms of reliability, particularly when compared to wrought materials.

In this study, a wrought and AM processed 304L stainless steel (SS) were dynamically characterized to compare compressive stress-strain curves at various strain rates with a split Hopkinson pressure bar (SHPB) or Kolsky compression bar.

2. Materials & experimental procedure

The 304L SS used in this paper has been formed using a high-powered (3.8 kW) LENS-like AM process developed at Pennsylvania State University. The chemical composition of the AM 304L SS was exactly the same as that of the wrought 304L SS. The compressive specimens for the wrought 304L SS were machined in a disc shape with a diameter of 6.35 mm and a thickness of 3.175 mm along the longitudinal and transverse directions, respectively. The compressive specimens for the AM 304L had the same dimensions, but were machined in an X, Y, and Z direction, respectively. It is noted that the AM 304L specimens were treated as delivered without any additional heat treatment or annealing process. The purpose of making the compressive specimens in different directions for both wrought and AM 304L SS is to examine and compare the isotropic characteristic for both wrought and AM 304L stainless steels. In this study, we characterized the wrought 304L SS in both longitudinal and transverse directions, and the AM 304L SS in the X direction. The X direction refers to the travel direction of the laser during the LENS process.

The dynamic compression experiments were conducted with a Kolsky compression bar at Sandia National Laboratories. The Kolsky compression bar, also called the SHPB, was originally developed in the 1940s [3]. As shown in Fig. 1, a general Kolsky compression bar set-up consists of a striker, incident bar, transmission bar, and momentum bar.

The striker is usually launched with a gas gun to impact the incident bar, generating a compressive stress wave (incident wave) that is recorded with the strain gages on the incident bar. When the incident wave propagates to the specimen, part of the stress wave is reflected back into the incident bar and recorded with the same strain gages due to the mismatch of mechanical impedance between the

[a] Corresponding author: eenishi@sandia.gov

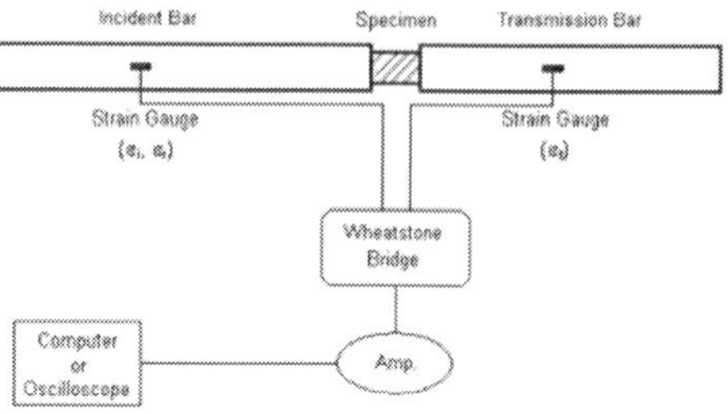

Figure 1. General Schematic of a Kolsky Compression Bar Set-Up.

pressure bars and the specimen. The rest of the stress wave transmits into the transmission bar though the specimen and is recorded with strain gages on the transmission bar. Using one-dimensional, wave propagation analysis, the measured strain signals from the strain gages can be used to determine the dynamic behaviour of the sample,

$$\dot{\varepsilon} = -\frac{2c_0}{L}\varepsilon_r \tag{1}$$

$$\sigma = \frac{A_0}{A_s}E\varepsilon_t \tag{2}$$

where L and $A_s$are the original sample length and cross-sectional area; c_0, E, and A_0 are the elastic bar-wave velocity, Young's modulus, and cross-sectional area of the bars; ε_r and ε_t are the measured reflected and transmitted strain signals.

It is noted that Eqs. ((1) and (2)) are based on the assumption of stress equilibrium during dynamic loading. However, this assumption may not be automatically satisfied in conventional Kolsky bar experiments. In addition, the strain rate in the specimen may not be constant, particularly when the specimen exhibits significant work hardening behavior, which may further generate uncertainties in determining strain rate sensitivity of the material under investigation.

Pulse shaping techniques have been developed and extensively used in Kolsky bar experiments in order to minimize stress wave dispersion and create a desired loading pulse, allowing for constant strain rate deformation and dynamic stress equilibrium [4]. In this study, a double pulse shaping technique that consists of a small copper disc and a steel disc was employed.

The pressure bars used in this study were made of C300 Maraging steel and had a common diameter of 19.05 mm. The incident and transmission bars were 3658 mm and 1829 mm long, respectively. The length of the striker was varied for compression tests at various strain rates.

3. Experimental results and discussion

Dynamic compression tests were conducted at 2500 s^{-1}, 1500 s^{-1}, and 500 s^{-1} in the longitudinal and transverse directions for the wrought 304L SS and X direction for the AM 304L SS.

Figure 2 shows a typical set of oscilloscope records of strain gage signals on the incident bar (solid line) and

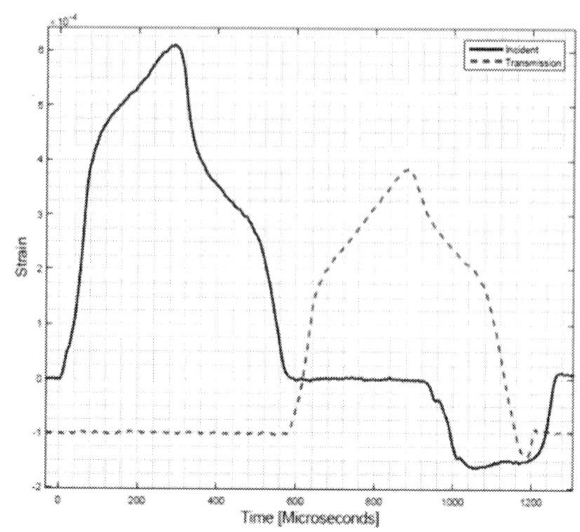

Figure 2. Typical Pulse Shaped Compression Oscilloscope Record for the Wrought 304L SS Specimens.

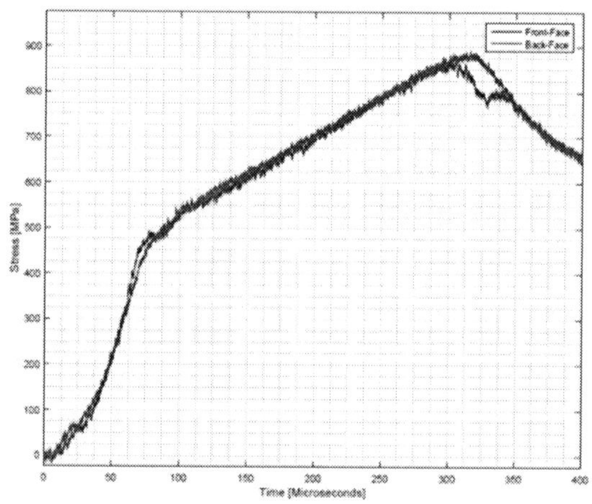

Figure 3. Sample Stress Equilibrium Plot on the Front-Face and Back-Face of the Specimen.

transmission bar (dotted line) in a compression Kolsky bar experiment for the wrought 304L SS specimens.

Figure 2 clearly shows a significantly different profile in the incident wave from the conventional trapezoidal incident wave in Kolsky bar experiments, due to the use of the pulse shaping technique. This modified incident pulse generated a nearly flat reflected pulse, which represents a constant strain rate when the specimen is equilibrated.

Figure 3 shows the stress equilibrium process in the specimen.

The specimen stress at the transmission bar end (back-face) was calculated with Eq. (2). The specimen stress at the incident bar end (front-face) was calculated with the 2-wave method,

$$\sigma_1 = \frac{A_0}{A_s}E(\varepsilon_i + \varepsilon_r) \tag{3}$$

where is the measured incident strain signal. The comparison of stresses at both ends of the specimen in Fig. 3 indicates that the specimen was in stress

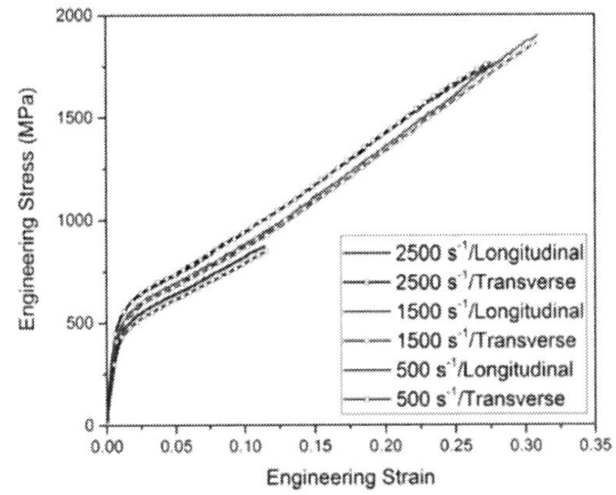

Figure 4. Compression Engineering Stress-Strain Curves (Longitudinal and Transverse Directions) of Wrought 304L SS at Various Strain Rates.

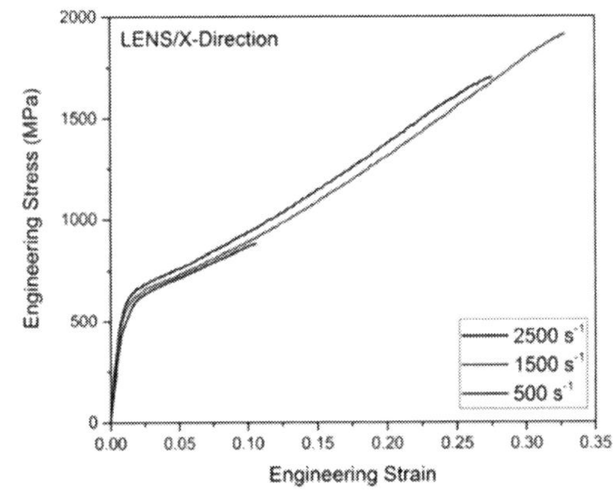

Figure 5. Compression Engineering Stress-Strain Curves (X Direction) of AM 304L SS at Various Strain Rates.

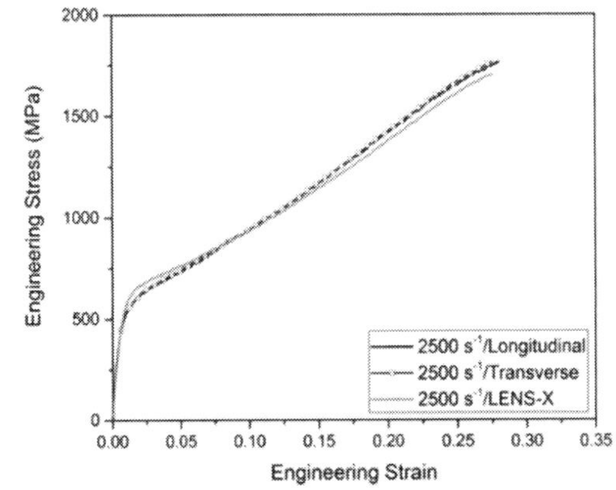

Figure 6. Compression Engineering Stress-Strain Comparison between Wrought and AM 304L SS at $2500\,s^{-1}$.

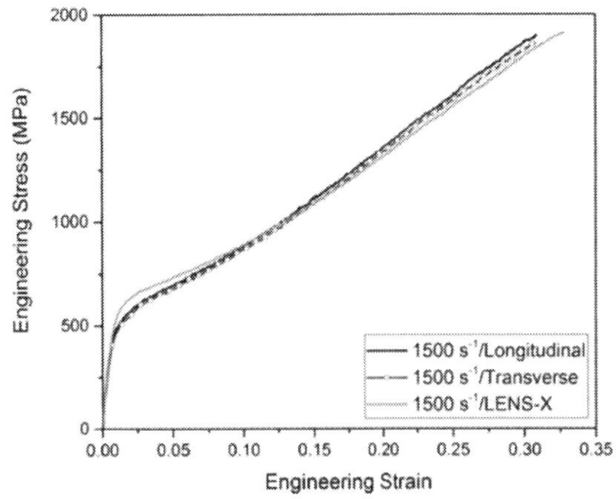

Figure 7. Compression Engineering Stress-Strain Comparison between Wrought and AM 304L SS at $1500\,s^{-1}$.

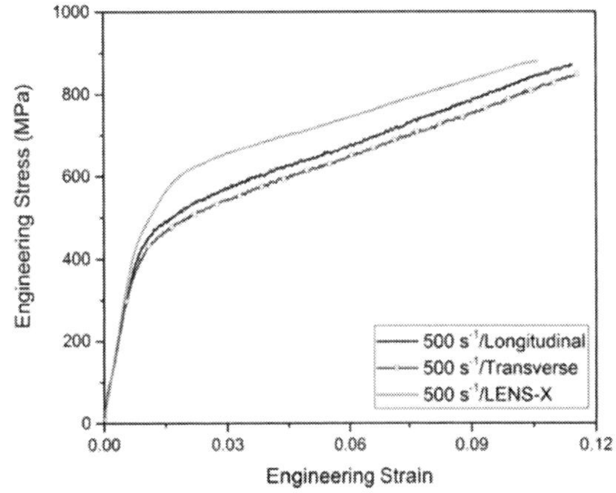

Figure 8. Compression Engineering Stress-Strain Comparison between Wrought and AM 304L SS at $500\,s^{-1}$.

Figure 9. Wrought (Left) and AM (Right) 304L SS Specimens After Testing.

equilibrium nearly over the entire duration of loading. Since the conditions of stress equilibrium was satisfied, the engineering stress-strain curve could be calculated with Eqs. (1) and (2).

Following the same procedure, the compressive stress-strain curves of the wrought and AM 304L stainless steels in different directions were obtained at different strain rates, which are plotted in Figs. 4 and 5, respectively.

Figure 4 shows a nearly isotropic work hardening characteristic in the dynamic compressive stress-strain response for the wrought 304L SS. In addition, the stress-strain curves for the wrought 304L SS show moderate sensitivity to strain rate in both longitudinal and transverse directions. The yield strength and flow stress increase with increasing strain rate.

As shown in Fig. 5, the dynamic compressive response of the AM 304L SS in the X direction had very similar work hardening characteristics to the wrought 304L SS. However, the strain rate dependence seems less significant than the wrought 304L SS.

For comparison purposes, the compressive stress-strain curves of the AM 304L SS are plotted with the wrought 304L SS at each strain rate in Fig. 8, Fig. 9, and Fig. 10, respectively.

It is apparent that the AM 304L SS in the X direction has higher yield strengths than the wrought 304L SS in both the longitudinal and transverse directions when the strain is smaller than 10%. As strain increases though (above 15%), the stress in the AM 304L SS approaches the wrought 304L SS and eventually becomes lower than the wrought 304L SS.

Although minimal differences are seen when comparing stress-strain curves, a visual difference is apparent in the specimens after testing, as shown in Fig. 9.

The specimen on the left represents the plastically deformed wrought 304L SS specimen and the AM 304L SS specimen after dynamic loading is shown in the right. It is observed that the wrought 304L SS specimen had a smooth surface with a slight barrel shape on the side due to friction constriction between the specimen and the bar ends. However, the surface of the AM 304L SS specimen shows much rougher ends and side surfaces, which possibly indicates some microstructural changes. The microscopic grain-scale morphology on the AM 304L SS is currently under investigation.

4. Conclusions

A pulse-shaped Kolsky compression bar has been employed to characterize both wrought and AM 304L stainless steels at various high strain rates from $500\,s^{-1}$ to $2500\,s^{-1}$. The wrought 304L SS was characterized in both longitudinal and transverse directions; whereas the AM 304L SS was characterized in the X direction (the travel direction of laser during AM process) at this moment. The pulse shaper material was carefully designed to obtain nearly constant strain rate in the specimen under dynamic stress equilibrium. The dynamic compressive stress-strain curves were obtained for both wrought and AM 304L stainless steels. Both materials showed very similar work hardening characteristics. However, in comparison to the wrought 304L SS, the yield strength and flow stress for the AM 304L SS were higher at small strains (below 10%) but dropped lower when the strain was larger than 15%, indicating a milder work hardening behaviour in the AM 304L SS along the X direction. Within the dynamic strain rate range investigated in this study, the AM 304L SS also exhibited less strain rate sensitivity than the wrought 304L SS. The surface morphologies for the wrought and AM 304L SS specimens after dynamic tests were significantly different. While still under investigation, speculation can be made that the bands of small grains separated by larger grains is likely responsible for the surface roughness of the AM 304L SS specimens

The authors would like to acknowledge Professor Palmer at Pennsylvania State University for preparing the AM 304L SS specimen.

Sandia National Laboratories is a multi-program laboratory managed and operated by Sandia Corporation, a wholly owned subsidiary of Lockheed Martin Corporation, for the U.S. Department of Energy's National Nuclear Security Administration under contract DE-AC04-94AL85000.

References

[1] Laser Engineered Net Shaping, *SAND2002-3539W*

[2] E.R. Denlinger, J.C. Heigel, P. Michaleris, T.A. Palmer, J. Mat. Processing Tech., **215**, p.123-131 (2014)

[3] B. Song, B.R. Antoun, K Connelly, J. Korellis, W-Y Lu, Meas. Sci. Tech., **22**, 045704 (2011)

[4] E.E. Nishida, *Master's Thesis*, Purdue University (2010)

EPJ Web of Conferences **94**, 01002 (2015)
DOI: 10.1051/epjconf/20159401002

Experimental issues and interpretation of the necking phenomena in the dynamic characterization via Hopkinson bar

Giuseppe Mirone[a] and Domenico Corallo

University of Catania, Dept. of Industrial Engineering, Viale A. Doria 6, 95125 Catania, Italy

Abstract. The usual procedures for processing experimental data from Hopkinson bars (HB) are based on modeling hypotheses which, in some cases, may lead to considerable approximations. The effects of the specimen geometry and of the necking are analysed, both experimentally and by FE analyses, for assessing how much the conventional values of strain rate, strain and stress, based on strain gauges readings along the HB, are affected by approximations in different testing conditions. Also, from the way the necking affects the load/area ratio, useful considerations arise about the hardening response of metal alloys under high strain rates. The sensitivity of stress-strain curves to the time-translations of strain waves along the bars is assessed by comparing the above curves to others where the strain, the strain rate and the current cross section come from speed camera image analyses. This clearly shows that the effective strain rate in the most strained areas of the specimens is many times higher than the nominal value, and also allows to explore how much the engineering curve is poorly representative of the material response, due to strain localization in the post-necking phase. The necking-affected length compared to the total gauge length is also found to significantly influence the strain response of the specimens, in a way that can easily mislead the experimenters in the derivation of the material ductility. An empirical relationship, already found to be valid in the post-necking quasi-static regime for transforming the true stress (load / current area) into an estimation of the cross section-averaged equivalent stress, is also found to apply at high strain rates if only the appropriate considerations are done on the way the equivalent stress is considered.

1. Introduction

The tensile tests performed at high strain rates by way of the Split Hopkinson Tension Bar (SHTB) are subjected to the same necking-induced stress-strain uniformity issues typical of the static tension tests, together with other possible inaccuracy causes, more directly inherent the strain-rate-dependent nature of the SHTB tests.

The reflected wave εr obtained from strain gauge readings on the input bar only allows to derive the gross elongation rate of the entire specimen, including the contribution of fillets and shoulders always present in tensile specimens. Then the standard SHTB formulas are approximate due to the stress, the strain and the strain rate non-uniformities along the specimen length, up to very large extents depending on the geometry of the specimens.

Verleysen et al. [1,2] and Sato et al. [3] addressed this topic by providing a quantitative estimation of how the elongation rate and the engineering deformation, based on classical SHTB strain gauge readings on the bars, differ from the effective strain of the gage length of the specimen, based on image-analysis measurements.

Sato et al. [3] also evidenced the necking-induced progressive departure of the local true strain from the engineering strain, as it was also shown in Mirone [4] in terms of strain rate. When the strain measurements in SHTB experiments are only based on strain gauge readings, without speed camera image analysis, also the length of the specimen may greatly affect the calculation of the engineering strain, as also Rusinek et al. [5], Osovski et al. [6] and Rodriguez-Martinez et al. [7] pointed out. For this reason in [6] is proposes a "structural strain" for taking into account the strain localization induced by the necking.

Also the non-uniformity of stress distributions and its deviation from uniaxiality are directly related to the specimen shape, length and to the necking-induced perturbations, but the strain-rate related aspects of necking are somehow less investigated and still deserve further attention. In [4], starting from FE analyses of SHTB tests validated by literature experiments [17], was shown that the necking-affected experimental true stress, divided by the flow stress, obeyed to the same material-independent *MLR* function suitable for the static case [18–21], if only the flow stress corresponding to the nominal engineering strain rate was considered through the adopted material model and the known material constants.

2. Specimen geometry, necking and engineering vs. true curve approximation

The most classical equations used in SHTB experiments for the stress / strain / strain rate characterization of materials are based on the elastic strain waves measured on the input bar (incident ε_i, reflected ε_r) and the output bar (transmitted ε_t):

$$\dot{\varepsilon}_{Eng}(t) = 2 \cdot \frac{c_0}{L_0} \cdot \varepsilon_r(t) \quad (1)$$

[a] Corresponding author: gmirone@dii.unict.it

$$\varepsilon_{Eng}(t) = -2 \cdot \frac{c_0}{L_0} \cdot \int_0^t \varepsilon_r(t) \cdot dt \tag{2}$$

$$\sigma_{Eng}(t) = \frac{E_{bar} \cdot A_{bar}}{A_{0-Sp}} \cdot \varepsilon_t(t) \tag{3}$$

with L_0 reference gauge length of the smooth, unnotched specimen, c_0 the nominal sound speed in the bars E, A_{bar} and A_{0-Sp} *the* elastic modulus, the cross section of the bars and that of the undeformed smooth specimen, respectively. For a given SHTB system, the maximum achievable strain rate is inversely proportional to L_0, leading many researchers toward the adoption of specimens as short as possible, for increasing the achievable strain rate. This tendency is directly opposite to the need of long specimens arising from the hypothesis, intrinsic in Eqs. (1) and (2), that all the elongation of the specimen is concentrated within the gauge length L_0. In fact, $2 \cdot c_0 \cdot \varepsilon_r(t)$ in Eq. (1) expresses the time derivative of the total specimen length L_{tot}, including the length of shoulders and fillets always present in tension samples. So, if the ratio L_0/A_{Sp} is significant (slender, "long" specimens), then the elongation of the specimen shoulders is much smaller than that of the gauge length and the former can be neglected: only in this case Eq. (1) gives a good approximation of the engineering strain.

On the contrary, if the ratio L_0/A_{Sp} is rather small (stub, "short" specimens), then the elongation of shoulders and fillets is comparable to that of the gage length and cannot be neglected, so Eq. (1) gives a poor approximation of the engineering strain.

This first requirement for long specimens in SHTB tests can only be overcome if the elongation of the gage length is recorded during a test, without including also the elongation of specimen fillets and shoulders. Given the speeds typical of SHTB, this can only be done by fast video acquisitions and image analysis, so that $\Delta L_0/L_0$ and its time derivative are determined at known time intervals. Then, short specimens for increasing the achievable strain rate of SHTB tests can only be adopted if speed camera images are available, and in such a case Eqs. (1) and (2) should not be used.

Another aspect pushing toward the undesirable choice of "long" specimens is the necking-induced perturbation of the strain uniformity; the necking always affects a finite volume of the specimen roughly extending 2–3 times the undeformed diameter along the specimens axis. And, clearly, such necking-affected zone of the specimens deforms and elongates much more than the remaining segments of the gage length, if any. For whatever arbitrary given necking-affected segment L_N of a specimen with its own local "true" strain $Log(A_{Sp-0}/A_{Neck})$, the longer is the total gage length L_0, the smaller is the weight of the necking-elongated segment over the total gage length, the smaller is the engineering strain $\Delta L_0/L_0$.

This is why shorter specimens apparently exhibit larger strains than longer ones; such difference is only apparent as local measurements of more pertinent true-strain ε_{True}, would show identical values at identical area reductions:

$$\varepsilon_{True}(t) = Log \frac{A_{0-Neck}}{A_{Neck}}. \tag{4}$$

Such second reason for adopting longer specimens is intrinsic to the necking phenomenon, it is also common to the static testing, and cannot be overcome in any way because, in principle, only infinitely long specimens would elongate with a negligible influence of the necking. The only possible countermeasure for this problem is that of adopting the true strain instead of the engineering strain; which, as before, can only be done by image analysis and fast video acquisition. The variability of the resisting cross section can be easily sorted out by fast video capture of the evolving neck section during experiments and successive image analysis:

$$\sigma_{True}(t) = \frac{E_{bar} \cdot A_{bar}}{A_{Sp}} \cdot \varepsilon_t(t). \tag{5}$$

The "true" stress σ_{True} correctly describes the hardening only up to the inception of necking, after which the increasing stress triaxiality causes the true stress to considerably depart from the equivalent flow stress necessary for the material characterization.

The ratio $\sigma_{Eq}/\sigma_{True}$ starts from 1 when the necking begins, and evolves by decreasing as the plastic strain flows, until fracture occurs. It has been shown that during the post-necking plastic straining up to failure, for the majority of structural metals such a ratio has always evolved in the range 0.75–1:

$$\begin{aligned} &\sigma_{Eq}(t) < \sigma_{True}(t) \\ &0.75 < \tfrac{\sigma_{Eq}(t)}{\sigma_{Eq}(t)} < 1. \end{aligned} \tag{6}$$

The above ratio was found to be a material-independent function of the difference $\varepsilon_{True} - \varepsilon_N$, between the current true strain and the Considére strain (which, of course, is material-dependent), at which necking initiates.

The MLR polynomial in [18–21] approximates such a function within an error below 5%, and then constitutes a tool for the necking correction of static true curves, which also resulted suitable for correcting dynamic true curves from SHTB as reported in [4,19,20].

3. Quantitative evaluation of length-induced and necking-induced approximations in classical SHTB tests

Round bar specimens are machined with different lengths from 5 different metal alloys (Aluminium 2011, *AL*, Silicon Bronze *BR*, inox steel AISI 304, *X*, mild steel FE370, *FE*), according to the shape depicted in Fig. 1 and the nominal dimensions in Table 1.

The direct tension SHTB used consists of two Al 7075 bars with a 16 mm diameter, 3 m long the output bar, 4.5 m the input bar. A preload up to 100 kN is imposed to the initial 1.5 m segment of the input bar and is released by fracturing a fragile sacrificial element which maintains closed a two-jaws gripping mechanism. A rise time of about 100 μs is achieved. Signals acquisition is done by a 250 kHz 8 channels strain gauge station and a3 MHz digitalizing and sampling board.

Tests at different strain rates have been carried out either by imposing different incident waves on identical

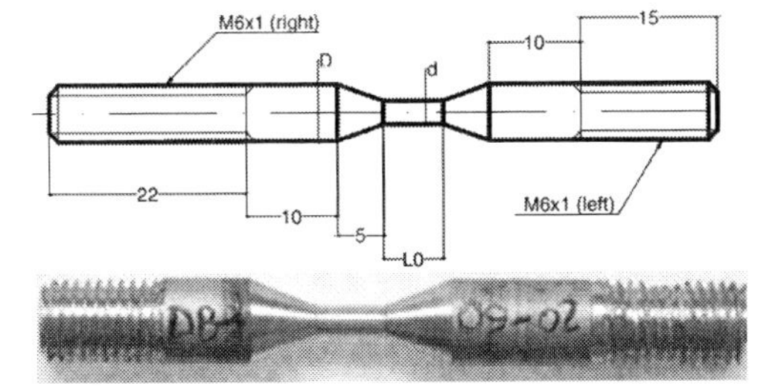

Figure 1. Specimens shape for multi-length, conventional SHTB tests.

Table 1. Specimens data for multi-length, conventional SHTB tests.

Material	*Spec. name*	*d*	*D*	*L0*
Alum. 2011	AL-03-03	3	8	3
"	AL-03-09	3	8	9
"	ALs20-i	2.5	6	2.5/ 6/ 10
Bronze	BR-03-03	3	8	3
"	BR-03-09	3	8	9
AISI 304	X-03-03	3	8	3
"	X-03-09	3	8	9
FE370	FE-03-03	3	8	3
"	FE-03-09	3	8	9

specimens, or also by imposing the same incident wave to similar specimens of the same material, where only the gage length is varied. In this way the interacting effects of the elongation rate and of the specimen shape are investigated. This first series of standard tests is based on the classical SHTB theory, so only the engineering stress, strain and strain rates are evaluated according to Eqs. (1–3). Figures 2 and 3 show the results for the AL2011 alloy in terms of engineering stress-strain curves and of engineering strain rate vs. time, respectively. Short specimens ($L/d = 1$) only have a moderate dependency on the engineering strain rate; in fact, when ε_{Eng} varies between 1000 and 7000 s^{-1}, only a small variability within 20% in the ultimate stress and 15% in the failure strain occurs. A similar evidence is reported in Figs. 2b) and 3b) for the "long" specimens ($L/d = 3$). Instead the comparison between Figs. 2a) and 2b) suggests that the effect of L/d is much more evident than that of the elongation rate; in fact, the static and dynamic failure strains of the short specimens in Fig. 2a) are almost identical each other, while in Fig. 2b), on the contrary, the dynamic failure strain is about 1.5 times greater than its static counterpart. In Figs. 2c) and 3c) three series of tests with the same incident wave (same preload of the input bar) and different L/d are compared; given that the effect of the engineering strain rate on specimens with the same shape is limited to within 20% as shown in Figs. 2a) and 2b), then the large differences between the three dynamic series in Fig. 2c) are only due to the different length/diameter ratio of the specimens.

It is clear that the material ductility at the local scale is not really increasing from about 0.2 to 0.7 when passing from long specimens to short ones, as suggested by Fig. 2c); such a wrong appearance, spoiling both static and

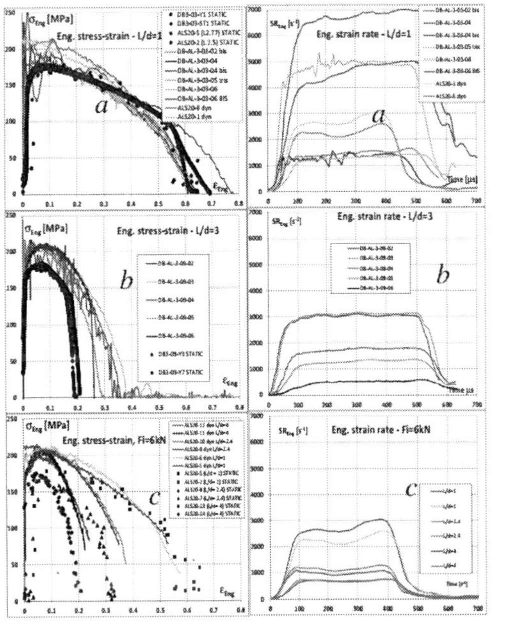

Figures 2-3. Engineering curves for AL.

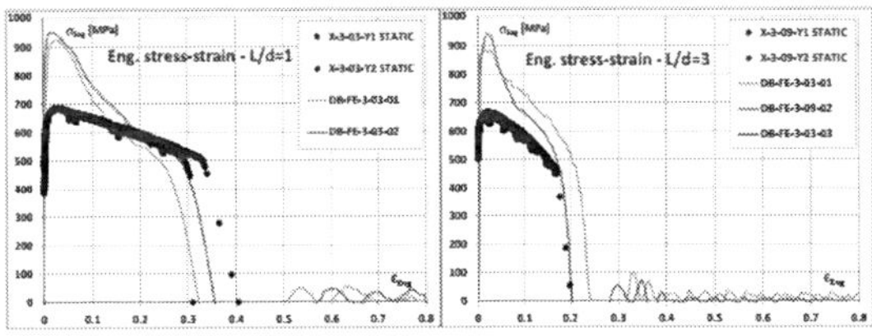

Figure 4. Engineering stress-strain curves for FE370.

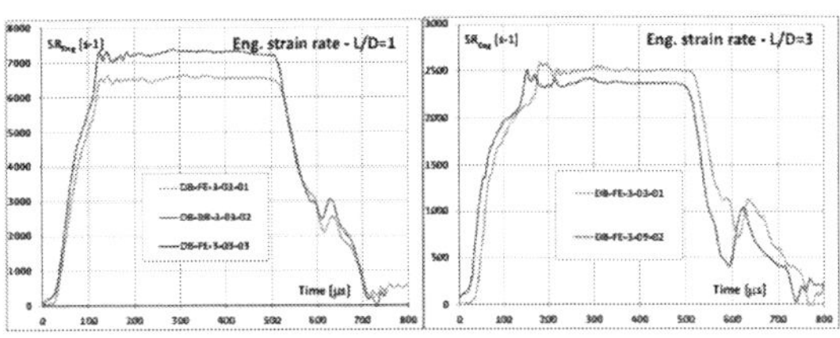

Figure 5. Engineering strain rate curves for FE370.

dynamic curves, is only a specimen-shape-dependent issue intrinsic in the engineering stress-strain curves.

Similar experiments are carried out with Bronze, AISI 304 and steel FE370, but only part of the respective results is reported below due to pages limit requirements.

The dependence of the hardening stress on the strain rate follows the usual predictable response as the former increases when the latter does, to different extents depending on the material sensitivity: the bronze hardening is almost strain-rate insensitive, while the Al alloy, the FE 37 and the AISI steels show strain-rate-promoted amplifications of the ultimate stress ranging from 12% to 28%. The apparent increase of the engineering strain in too short specimens reflects how the elongation of the necked zone affects the elongation of the entire gage length, and its occurrence is evident for all materials at both static

Table 2. Dynamic/static engineering ductility.

	L/d=1	L/d=3
Al 2011	1	1.5
Bronze	1	1.4
AISI304	0.65	0.85
FE370	0.95	1.05

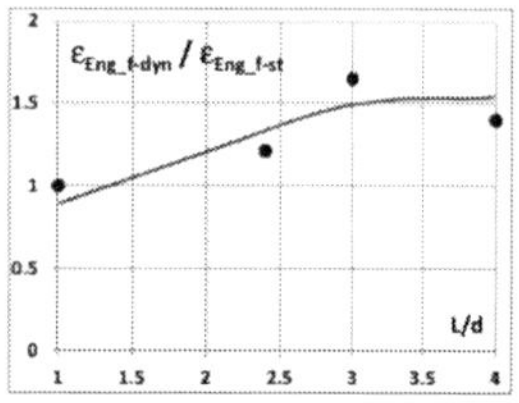

Figure 6. Specimen slenderness and strain increase.

and dynamics strain rates; instead, the magnitude of such apparent strain increase is related to the strain rate in a way that also depends on the material considered.

In fact, the ratio $\varepsilon_{Eng\_f\_Dyn}/\varepsilon_{Eng\_f\_St}$ between the average dynamic and static engineering strains at failure, reported in Table 2 against the L/D ratio for all materials, shows a similar trend for the non-ferrous alloys tested (in both *Al* and *Br* tests the failure strain has null increase for short specimens and a 50% increase for longer specimens), while it is rather different for the steels responses (ductility almost insensitive to the strain rate for the FE, regardless of the L/d ratio, and strain-rate-promoted ductility reduction for the AISI steel). The ratio $\varepsilon_{Eng\_f\_Dyn}/\varepsilon_{Eng\_f\_St}$ is plotted in Fig. 6 against the slenderness ratio L/d for the Al alloy, the only one for which four different gauge lengths have been tested. More tests with many different L/d ratios would be useful for a detailed identification of the above mechanism, but Fig. 6 already indicates that, for the alloy at hand, the effect of the specimen length on the engineering strain tends to become somehow stabilized at L/d of about 3; such a feature is not a real complete saturation and may either happen for other materials at different values of the slenderness ratio. Such experimental evidence is related to the considerations that the necked zone may largely affect very short specimens where the necking almost extends all over the entire gage length, while produces a less pronounced effect if specimens are longer and the necked zone only constitutes a partial or even a small segment of the entire gage length.

In Fig. 7 are reported two Al specimens with L/d = 1 and 3 respectively, before tests and after failure. It is clearly visible that the length of the neck-affected zone is similar in both cases, but for shorter specimens it covers the entire gage length while for longer specimens it just covers about 50% of the total gage length. Then it follows that the necking in shorter specimens with L/d = 1 affects the engineering strain to a maximum extent and, also, the interaction of the specimen shoulders with the gage length makes very likely that the uniaxiality of stress and the true stress-true strain curve are both spoiled well before of necking. Such short specimens behave more like notched bars than they do as smooth ones. The only way

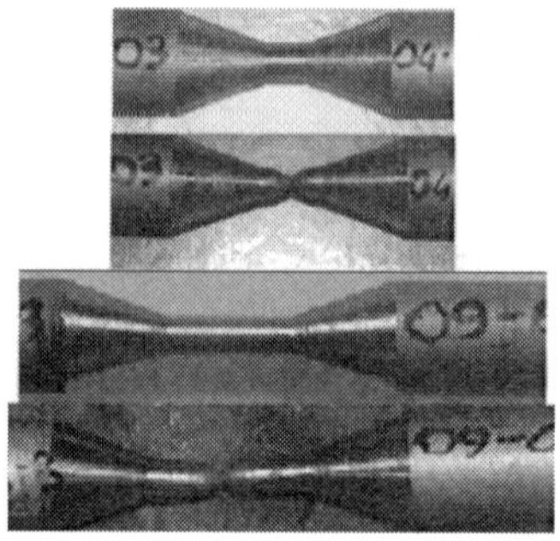

Figure 7. Neck-affected zone in short and long specimens.

for getting rid of these length-related issues is to rely on true stress and true strain measurements, provided that the gage lenght is long enough to prevent interaction of the specimen shoulders and fillets with the necked zone. Similar investigations conducted with the other alloys tested shown that a gage length two-three times longer than the diameter is essential for ensuring that a fraction of the gage length remains necking-free until failure, so that standard stress state of smooth unnotched specimen is achieved and virtually flawless true stress –true strain measurements can be performed.

The main results of the present section can then be summarized as follows:

- The gage length of specimens greatly affects the accuracy of engineering strain and strain rate measurements;
- The apparent increase in the engineering ductility of shorter specimens is only due to the amplified elongation of the necking zone, which in these cases either covers large amounts of the gage length or the entire gage length itself, so largely affecting the whole definition of engineering strain.
- When the necking covers the entire gage length it is very likely that the specimens behave as notched bars; then also the stress state at the local scale and the true stress-true strain data are affected by error.
- The length-induced spoiling of engineering strain measurements decreases and tends to saturate as the gage length increases, but such a saturation may be largely material-dependent and no real complete saturation is ensured at any length.
- On the contrary, the length-induced spoiling of true stress-true strain data only occurs when the entire gage length is affected by the necking; such a negative occurrence can be prevented by simply adopting L/d ratios greater of 2 or three, depending on the necking attitude of the given materials. It is understood that such a countermeasure poses a limitation on the maximum strain rate achievable on a given SHTB equipment for a given specimen diameter.

The analysis of the true stress-true strain characterization at high strain rates is the subject of investigation in the successive section.

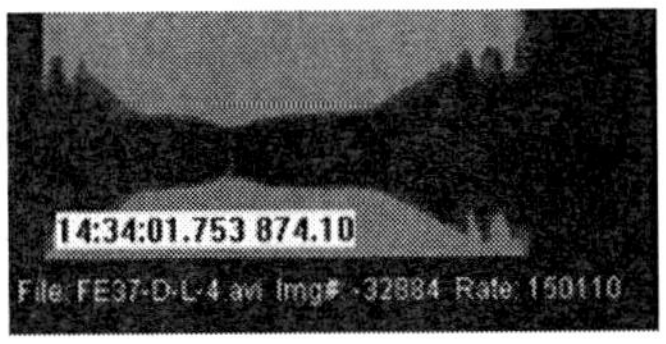

Figure 8. FEN specimen at incipient failure.

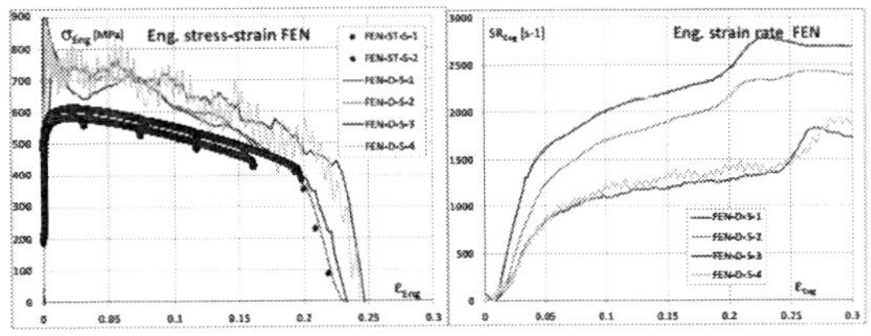

Figure 9. Engineering curves @ different strain rates.

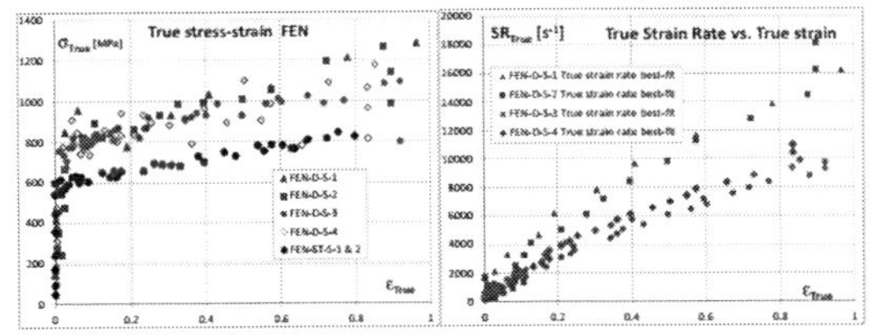

Figure 10. True curves @ different strain rates.

4. Evidences of interaction between true stress, true strain and engineering strain rate

A new series of tests has been carried out with an FE370 steel nominally identical to the previous one; however the material is obtained from a different supplier at a different time so that a slightly different response is found. The new FE370 steel is called FeN and the specimens have a ratio L/d = 2.5. Two series of dynamic tests have been performed with the direct-tension SHTB apparatus, the "faster" series with an input bar preload of 50kN and a nominal $\dot{\varepsilon}_{Eng} = 2500\,s^{-1}$, the "slower" series with a bar preload of about 35kN generating a nominal $\dot{\varepsilon}_{Eng} = 1800\,s^{-1}$. Fast image recordings were taken via a Phantom V7 speed camera at rates between 70 kfps and 150 kfps, (Fig. 8), so that synchronization of frames with the load history from the output bar and successive image analysis allowed to calculate the engineering and the true stresses, strains and strain rates.

Huge difference between engineering curves and true curves are reported in Figs. 9 and 10 showing the results from the "camera-assisted" experiments. The most visible evidence conveyed by Figs. 9 and 10 is that the engineering curves hide and distort data over most of the material straining life; in fact, the apparent softening and the very low ductilities suggested by the engineering curves $\sigma_{Eng}(\varepsilon_{Eng})$ are not real material features, but only inaccuracies intrinsic in the engineering approach, already known to afflict also the static elastic-plastic characterization. The failure strain form the engineering curves is close to 0.2–0.23 while that from the true curves is about 5 times greater (0.9–1); also the stress is largely underestimated by the engineering curves, reporting stresses at failure of about 500 MPa instead of the 800–1200 MPa of the true stress range really exhibited by the material at incipient failure. Finally, the engineering strain rate is found to provide a huge underestimation of the true-strain rate, of a factor between 6 and 7; such an underestimation may radically change the outcome of a dynamic material characterization.

These considerations make clear why the knowledge of the true stress-true strain curves is an essential prerequisite for a reasonable material characterization, at both static and dynamic strain rates.

A commonly accepted hypothesis for the dynamic material characterization is that of assuming that the effects of strain and strain rate on the equivalent stress are uncoupled each other, in general:

$$\sigma_{Eq}\left(\varepsilon_{Eq}, \dot{\varepsilon}_{Eq}\right) = \sigma_{Eq\_S}\left(\varepsilon_{Eq}\right) \cdot R\left(\dot{\varepsilon}_{Eq}\right) \qquad (7)$$

where $\sigma_{Eq\_S}(\varepsilon_{Eq})$ is the static material curve and $R\left(\dot{\varepsilon}_{Eq}\right)$ is the strain-rate-promoted amplification of the static hardening. Apart from the temperature effect which is not included here, widely known material models like those by Johnson-Cook or by Cowper-Symonds, are special cases of Eq. (1). Then the dynamic characterization consists of determining both right-hand terms of Eq. (7). The experimental measurement of quasi-static σ_{Eq_S} and ε_{Eq} is only possible from smooth tensile specimens before the necking initiates, when $\sigma_{Eq} = \sigma_{True}$ and $\varepsilon_{Eq} = \varepsilon_{True}$, respectively. After the necking starts ε_{Eq} remains quite close to ε_{True}, but σ_{Eq} progressively departs from σ_{True}.

The FEN steel exhibits early necking ($\varepsilon_N \sim 0.02 \div 0.04$) and rather long post-necking range (true strain at failure $\varepsilon_{True_F} \sim 0.9 \div 1$ at all the strain rates tested), so the departure between σ_{Eq} and σ_{True} is expected to be considerable. It was demonstrated in [18, 19] that the true stress and the equivalent stress at quasi-static rates are connected by the material-independent function *MLR*, according to Eq. (8):

$$\sigma_{Eq\_S}\left(\varepsilon_{Eq}\right) = \sigma_{True\_S}\left(\varepsilon_{Eq}\right) \cdot MLR\left(\varepsilon_{Eq} - \varepsilon_N\right). \qquad (8)$$

The ratio of dynamic true stress to static true stress has been plotted against the true strain rate for all the tests, as visible in the solid-line curves of Fig. 11a. Most of the literature, together with the material models mentioned before, assume that the stress amplification, expressed here by the ratio R, is a material characteristic function which, then, should not change for whatever strain-strain rate history occurring to the same metal; instead the experimental solid-line curves in Fig. 19a show two rather different trends of the ratio R for the two series of tests, with a scatter of about 5% around the mean value. The reason of this difference cannot be explained at the moment, and it is not even clear whether or not it is physically sounding; however it is in contrast to the hypothesis expressed by Eq. (1), and suggests that different functions R may exist or, which is more acceptable, that the function R should also depend on ε_{Eq}, so that the effects of the strain and of the strain rate is not uncoupled anymore.

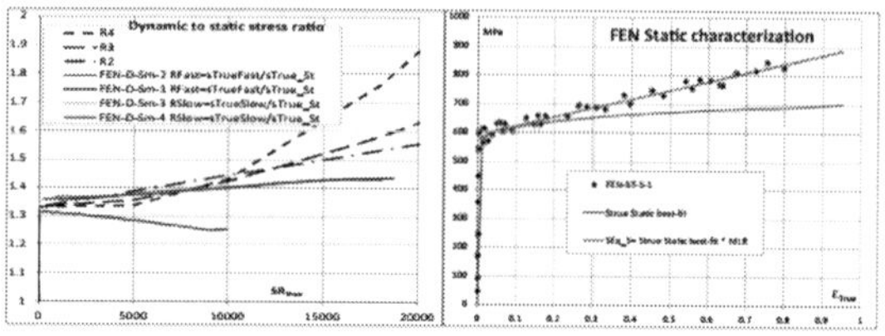

Figure 11. Static and dynamic stress-strain data.

Figure 12. Deformed model of the SHTB specimen.

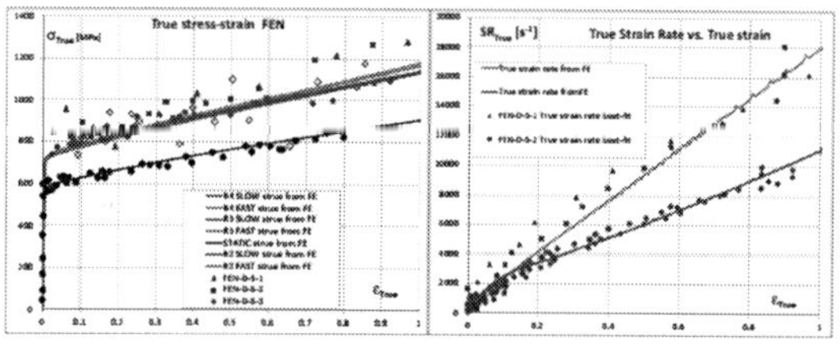

Figure 13. True stress, true strain and true strain rate from finite elements and from experiments.

Further detailed investigation on this subject is certainly deserved but lies beyond the timing of this work.

Then the dynamic characterization of the FEN alloy is attempted here by an inverse calibration procedure based on (1); the function R, adjusted iteratively, is joined to the Static flow curve from Eq. (8) until two FE analyses, driven by the same material parameters, are able to correctly simulate both the "slow" and the "fast" experiments. The three final choices of the R ratio and the static flow curve are plotted as the black dashed curves in Fig. 11a and the lower curve of Fig. 11b, respectively.

The finite element model is made of 15900 axysimmetric quad-4 full integration elements for the input / output bars and the specimen. The material is described by Eq. (1) via the fortran user subroutine Yiel available within the MSC-MARC finite elements code. The output produced by the FE runs driven by the above material data is compared to the experimental data in Fig. 13.

While the true strain vs. true strain rate behavior is modeled correctly by finite elements for both dynamic tests, the simulated "fast" true curve from forces and displacements readings at the proper mesh nodes is rather underestimated. Only the simulated true curves for the static and the "slow" dynamic tests are satisfactory; and a very low sensitivity to the choice of $R$is also found.

5. Conclusions

An experimental campaign based on static and dynamic SHTB tests has been carried out for investigating various aspects of the stress-strain characterization of metals at strain rates ranging from quasistatic to 20000 s-1.

A first series of conventional SHTB tests has shown that the initial gage length of the specimen can significantly affect the test if the ratio L/d is lower than about 2.5. These tests also highlighted the uncertainties intrinsic in the engineering definition of the stress-strain curves.

A second series of SHTB tests, including the acquisition of speed-camera images, allowed to determine the true stress-true strain curves and to calculate the true-strain rate at the neck section. The experimental true stress-true strain curves fully highlighted the approximations and the lack of information conveyed by the engineering stress-strain data: for the FEN alloy, the standard strain-gauge-based SHTB procedures led to stress, strain and strain rate values about 2.5, 5 and 7 times smaller than their real values, respectively. The true strain rate vs. true strain curves obtained from finite elements are accurate, but the predicted true stress was accurate only for static and for slow dynamic tests; a visible true stress underestimation resulted from the simulation of fast dynamic tests.

References

[1] Verleysen P., Degrieck J., Int. J. Impact Engg, 2004. **30**: 239–253

[2] Sato K., Yu Q., Hiramoto J., Urabe T., Yoshitake A., Int. J Impact Engg 2015; **75**: 11–26

[3] Mirone G. The dynamic effect of necking in Hopkinson bar tension tests. Mech. Mater. 2013. **58** 84–96.

[4] A. Rusinek, R. Zaera, J.R. Klepaczko, R. Cheriguene, Acta Materialia 2005; **53**: 5387–5400

[5] S. Osovski, D. Rittel, J.A. Rodríguez-Martínez, R. Zaera, Mechanics of Materials. 2013; **62** : 1–13

[6] J.A. Rodrígue z-Martínez, D. Rittel, R. Zaera, S. Osovski. 2013 ; **54**:206–216

[7] Yang L.M., Shim V.P., Int J Impact Engg. 2005; **31**:129

[8] K. Nilsson, Int. J Impact Engg. 2004; **30**: 1353–1367

[9] Besnard G., Hild F., Lagrange J. M., Martinuzzi P., Roux S., International J Impact Engg. 2012; **49**: 1353–1367

[10] R. Zaera, J.A. Rodríguez-Martínez, G. Vadillo, J. Fernández-Sáez, International Journal of Mechanics and Physics of Solids. 2014; **64**: 316–337.

[11] Guan Zhiping. Materials and Design. 2014; 56: 209–218

[12] Z. Xue, A. Vaziri, J.W. Hutchinson. Journal of the Mechanics and Physics of Solids. 2008; **56**: 93–113

[13] V. Tarigopula, O.S. Hopperstad, M. Langseth, A.H. Clausen, F. Hild, Int. J. Solids Struct. **45** (2008) 601–619

[14] D. Anderson, S. Winkler, A. Bardelcik, M.J. Worswick, Materials and Design, 2014; **60**:198–207

[15] C. Roth, Dirk Mohr, Int. J Plasticity. 2014; **56**: 19-44

[16] G. Mirone, Int. J Sol. Struct., 2004; **41**: 3545–3564

[17] G. Mirone, D. Corallo, Engg Fract. Mech., 2013; **102**:118–145

[18] G. Mirone, Mech. Mat., 2008; **40**: 685–694

[19] G. Mirone, Int. J. Damage Mech, 2004; **13**: 241–261

[20] B. Erice, F. Gálvez, D.A. Cendón, V. Sánchez-Gálvez, Engg Fract. Mech., 2012; **79**: 1–17

EPJ Web of Conferences **94**, 01003 (2015)
DOI: 10.1051/epjconf/20159401003

Digital image analysis of ASB-assisted failure of impacted structural materials

Patrice Longère[1,a], Emile Roux[1], Olivier Cherrier[1], Thierry Millot[2], Didier Capdeville[3], and Jacques Petit[3]

[1] Université Fédérale Toulouse Midi-Pyrénées, ISAE-SUPAERO, ICA (FRE CNRS 3687), Toulouse, France
[2] DCNS Research, Indret, France
[3] CEA-Centre de Gramat, Gramat, France

Abstract. Experimental analysis of fracture mechanisms is a key point to understand and further reproduce physical phenomena involved in structural material failure. Actually, crack propagation under high loading rate is a strongly coupled thermo-mechanical problem involving large deformation, high strain rate and (quasi) adiabatic conditions. The work presented herein aims at analyzing the successive steps of the dynamic deterioration of a high strength structural material, namely the ARMOX500T armor steel, leading to the ultimate fracture. The Kalthoff and Winkler impact test, consisting in impacting the edge of a double notched plate, was retained for that purpose. Images are captured at high frame rate (1M frame/s) during the impact loading then analyzed according to a line tracking method. Three stages are accordingly observed: first the progressive development of a weak localization of the deformation within a wide zone behind the notch tips, second the propagation of a narrow band of strong localization, and finally the propagation of a crack. The propagation of the white band is associated to the development of an adiabatic shear band (ASB), as a precursor of the crack. Corresponding displacement fields are identified thanks to the adopted line tracking method, allowing for determining the evolution of mechanical quantities, including notably notch tip shear displacement and velocity, ASB-related white band velocity, global and local shear strain and shear strain rate magnitudes, in view of being implemented into numerical models.

1. Introduction

Analysis of crack propagation in high strength steel under impact loading is the purpose of the present work. With this aim in view, Kalthoff & Winkler [1] impact tests were carried out at various impact velocities. Depending notably on the impact velocity, material grade and temperature, different phenomena can be observed: failure under crack propagation (Mode I and/or II) and/or adiabatic shear banding (ASB). The knowledge of the fracture mechanisms involved in structures during impact loading and of their chronology is a key point to understand and further prevent failure process.

Kalthoff & Winkler impact test consist in impacting the edge of a double notched plate with a cylindrical projectile (see Fig. 1). In the present work, a series of images was captured at very high frame rate (1 Mfps) and high spatial resolution (312 × 260 pixel^2) during the impact loading.

A numerical method of image analysis is presented. This method aims at tracking lines that have been beforehand engraved on the sample. The goal is on one hand to identify the succession of kinematic displacement fields that leads to the ultimate dynamic failure of the structure, and on the other hand to quantify the velocity of the physical phenomena involved in the dynamic failure at stake (adiabatic shear band tip velocity, crack tip velocity).

Under the condition of the reported test (impact at 156 m/s, room temperature) an ASB-assisted failure is observed in plates consisted of the high strength ARMOX500T steel as a candidate for armor application. Tow phenomena are clearly observed, first the ASB development and then the crack propagation in the wake of the ASB. This two mechanisms leading ultimately to the failure of the structure are analyzed thanks to the image analysis outlined in the present paper. The final aim of this digital image analysis is to give qualitative and quantitative information for further model development and enhancement [3,4].

[a] Corresponding author: Patrice.longere@isae.fr

2. Experimental procedure

Impact tests were carried out using a 3 m length- and 25 mm inner diameter-gas launcher. The projectile is a 25 mm length- and 20 mm diameter-steel cylinder settled in a 25 mm outer diameter foam cylinder. Double notched plate dimensions are given in Fig. 2. The 20 mm length- and 300 μm height-notches were machined by electrical discharge machining (EDM) and 2 mm spaced out-vertical straight lines were engraved on the plate, see Fig. 3. Optical sensors allow for measuring the projectile velocity just before the impact. A Shimadzu HPV-1 ultra-high speed camera was used to observe the projectile/plate interaction at 1 Mfps with a 312 × 260 pixel^2 spatial resolution.

3. Digital image analysis

The lines engraved on a side of the plate are used to track the horizontal x-displacement during the impact loading.

Article available at http://www.epj-conferences.org or http://dx.doi.org/10.1051/epjconf/20159401003

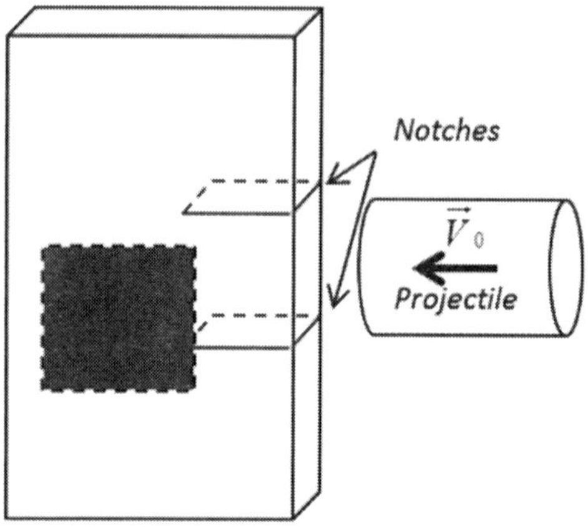

Figure 1. Schematic view of the Kalthoff and Winkler like test with the zone of interest for ultra-high speed frame rate recording.

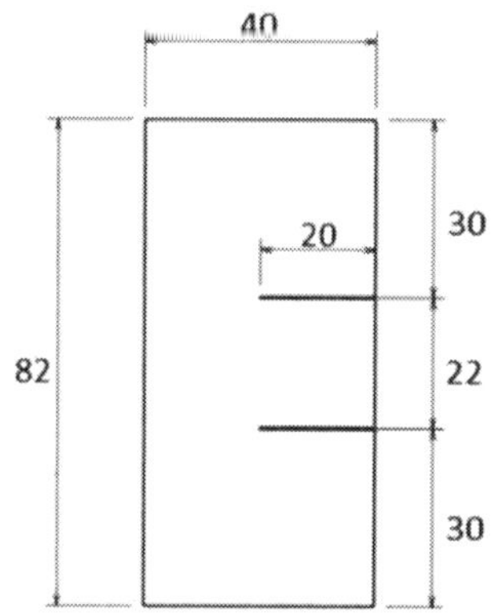

Figure 2. Dimensions of the 6mm-thick double notched Kalthoff and Winkler type plate used in the present work.

Neglecting the vertical y-displacement, the curvature of the lines can be directly linked to shear strain. The goal is therefore to describe the evolution of the engraved lines with analytical functions. These functions are then used to quantify the displacement and strain fields along the lines.

An incremental approach (image by image) is developed within Matlab, the key steps of the algorithm are:

- Isolation of each line within a sub-image (Fig. 4a).
- Detection of the contour of the line using a Sobel filtering method (Fig. 4b). At this stage the line is described by a cloud of points.
- Identification of the parameters of the analytical function by linear regression based on the cloud of points.

The displacement of each point of the line is described by the following expression:

$$U_x(y,t) = U_x^{MM}(t) + U_x^{WHS}(y,t) + H(\gamma_{c1})\,U_x^{ASB}(y,t) + H(\gamma_{c2})\,U_x^{crack}(y,t). \quad (1)$$

Each components of the function in Eq. (1) is linked to a specific area or stage of the deformation process

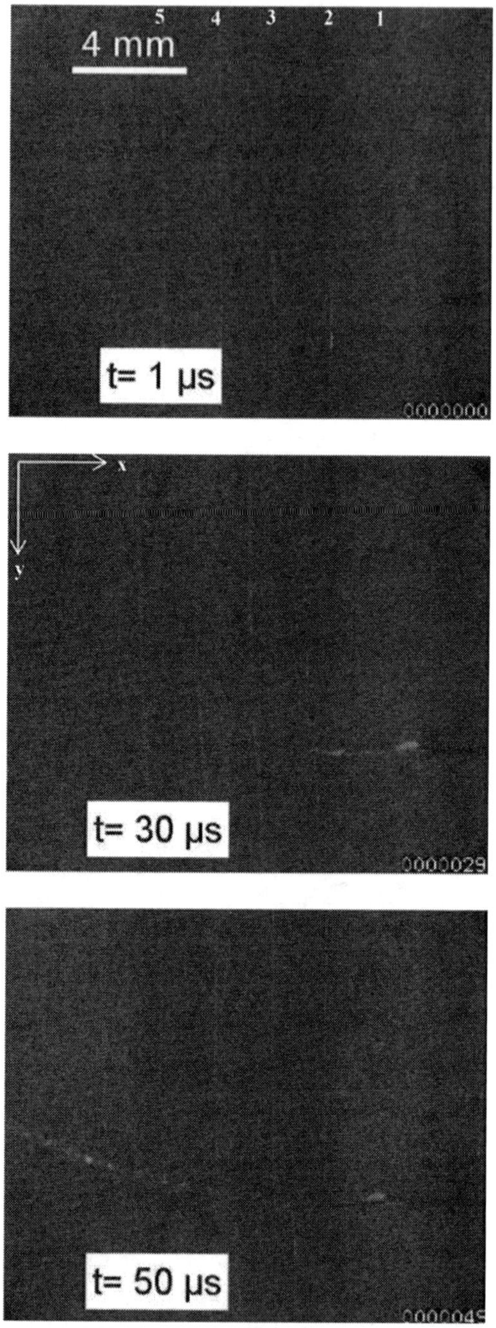

Figure 3. Series of images recorded at 1 Mfps. Vertical straight lines were beforehand engraved on the plate for line tracking. Impact at 156 m/s on ARMOX500T plate.

(see Fig. 5), and each of them involved specific parameters:

- The mater motion : $U_x^{MM}(t)$
- The displacement field within the weakly heterogeneous strain area assumed in the form:

$$U_x^{WHS}(y,t) = \bar{U}_x^{WHS}(t).\sum_{i=1}^{2}\tanh\left[d_i(t).\left(y - y_{offset}\right)\right].$$

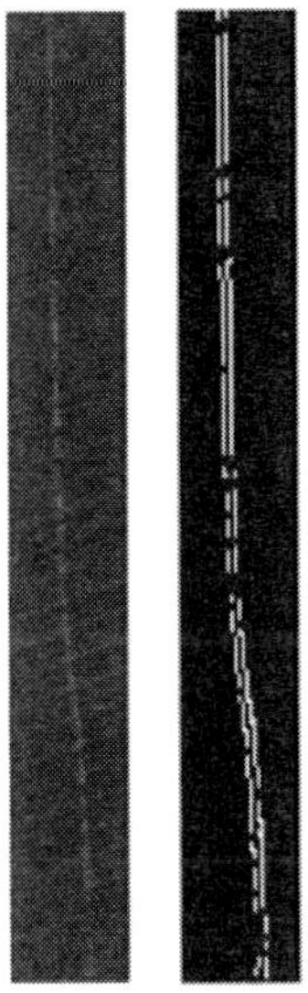
(a) (b)

Figure 4. Principle of line digitalization method. a) Sub-image containing one line, b) results of the Sobel filtering method for edge detection.

- The displacement field within the ASB-related strongly heterogeneous strain area assumed in the form:

$$U_x^{ASB}(y,t) = \begin{cases} \frac{\bar{\gamma}^{ASB}(t).w}{\pi}\sin\left(\pi\frac{y-y_{offset}}{w}\right) \\ \qquad if \quad y - y_{offset} \in \left[-\frac{w}{2}, \frac{w}{2}\right] \\ +\frac{\bar{\gamma}^{ASB}(t).w}{\pi} \quad if \quad y - y_{offset} > \frac{w}{2} \\ -\frac{\bar{\gamma}^{ASB}(t).w}{\pi} \quad if \quad y - y_{offset} < -\frac{w}{2}. \end{cases}$$

This component of the function is activated as soon as the critical shear strain $\gamma_{c1}$is reached (by the mean of the Heaviside function$H(\gamma_{c1})$).

- The discontinuous displacement field induced by the crack and assumed in the form:

$$U_x^{crack}(y,t) = \bar{U}_x^{crack}(t)\,x \begin{cases} -1/2 & if \quad y - y_{offset} > 0 \\ +1/2 & if \quad y - y_{offset} < 0. \end{cases}$$

This component of the function is activated as soon as the critical shear strain $\gamma_{c2}$is reached.

Nine coefficients are needed to described the displacement field (equation 1): $\bar{U}_x^{WHS}(t)$, $d_1(t)$, $d_2(t)$,y_{offset}, γ_{c1}, $\bar{\gamma}^{ASB}(t)$, w, γ_{c2}, $\bar{U}_x^{crack}(t)$.

The values of these coefficients are identified using a weighted linear regression method. The main advantage of this weighted method is to overcome the non-regular sampling issues coming from the edge detection method. Moreover a constraint is added on the parameters $\bar{U}_x^{WHS}(t)$, $d_1(t)$, $d_2(t)$, $\bar{\gamma}^{ASB}(t)$ et $\bar{U}_x^{crack}(t)$: these parameters must increase with time. These extra constraints disable the capacity of the relaxation phenomena detection but increase the stability of the method.

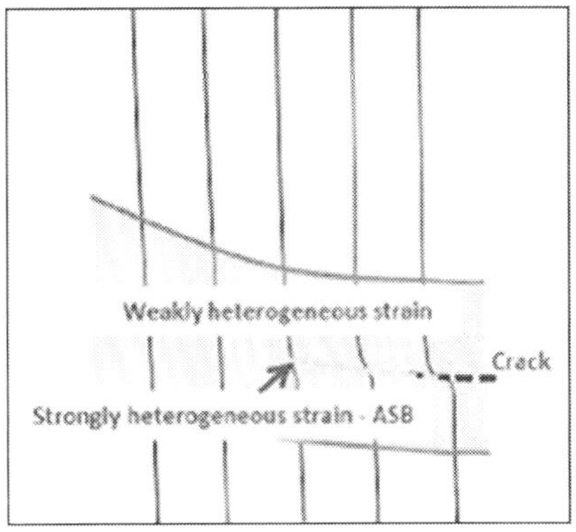

Figure 5. Schematic description of the areas and phenomena involved in the ASB-assisted failure for displacement field identification.

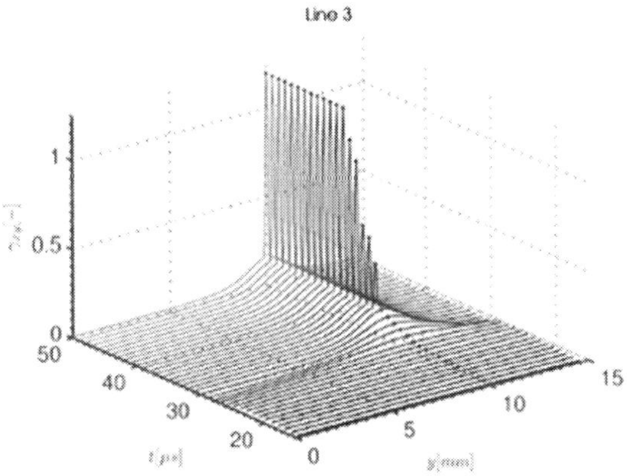

Figure 6. Time history of the shear strain along the median line.

Once the displacement field is known, the function (Eq. (1)) can be differentiated with respect to y to compute the shear strain field along the line:

$$\gamma_{xy}(y,t) = \frac{\partial U_x(y,t)}{\partial y}. \tag{2}$$

4. Results

4.1. Displacement and shear strain-shear strain rate fields

Examples of shear strain fields (Eq. (2)) resulting from the digital image analysis are presented in Figs. 6–8 for the median line (line 3). Shear strain fields are then derived with respect to time to evaluate the shear strain rate fields during ASB development. According to Fig. 9, the shear strain rate may reach values greater than 10^5 s^{-1} within the ASB. Those values must be considered as lower bounds for the true shear strain rate, and may locally be higher.

4.2. Velocity measurements

Analyzing the recorded images also allows measuring the velocities of different phenomena involved in the ASB-assisted failure of the plate. In particular, for the test

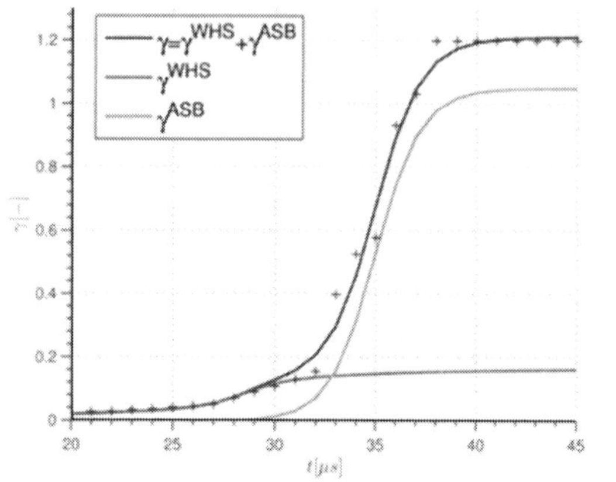

Figure 7. Decomposition of the shear strain into a contribution due to the weak localization (WHS) and a contribution due to strong localization (ASB).

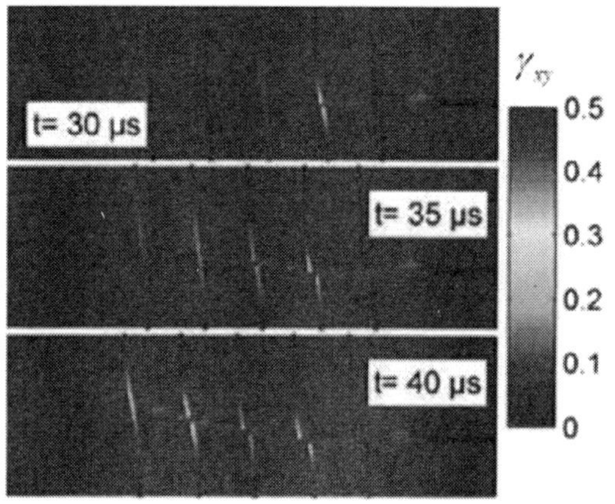

Figure 8. Shear strain field representation as determined from the line tracking method.

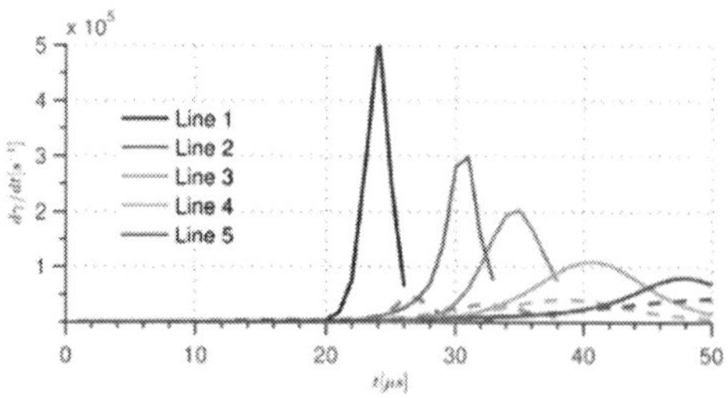

Figure 9. Time history of the shear strain rate in the ASB middle of each line.

presented in the present paper the key values summarized in Fig. 10 are:

- projectile velocity: 156 m/s.
- pre-notch shear velocity (V_s): 40 m/s.
- ASB tip maximal velocity: 650 m/s (lower bound).
- crack tip velocity in the wake of the ASB: 260 m/s (lower bound).

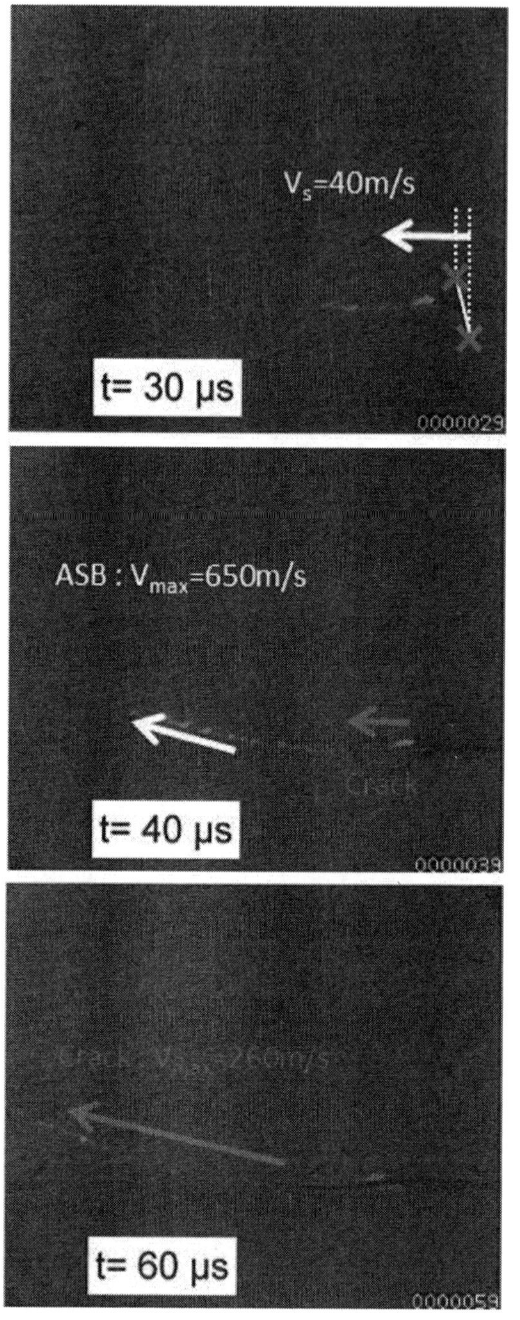

Figure 10. Shear velocity at the notch tip, ASB velocity and crack tip velocity.

5. Conclusions

The work presented herein aims at analyzing the failure mechanisms involved in an ARMOX500T steel plate when submitted to impact loading. Images were captured at high frame rate (1M frame/s) during the impact loading then analyzed according to a line tracking method. Three stages are accordingly observed: first the progressive development of a weak localization of the deformation within a wide zone behind the notch tips, second the propagation of a narrow band of strong localization (adiabatic shear band), and finally the propagation of a crack in the wake of the adiabatic shear band.

Displacement fields are identified thanks to the adopted line tracking method, allowing for determining the evolution of mechanical quantities, including notably pre-notch tip shear displacement and velocity, ASB-tip velocity, global and local shear strain and shear strain rate magnitudes.

Further works are in progress to go deeper in the understanding of ASB-assisted failure mechanism. The final aim of this work is to give qualitative and quantitative information in view of developing, enhancing and calibrating failure models for structural applications.

The authors would like to thank DGA/MRIS (French Ministry of Defence) for its financial support.

References

[1] J.F. Kalthoff, S. Winkler, Failure mode transition at high rates of shear loading, Proc Int Conf Impact Load. Dyn. Behav. Mater. (1987) 185–195.

[2] J. Duffy, Y.C. Chi, On the measurement of local strain and temperature during the formation of adiabatic shear bands, Mater. Sci. Eng. A. 157 (1992) 195–210.

[3] P. Longère, A. Dragon, X. Deprince, Numerical Study of Impact Penetration Shearing Employing Finite Strain Viscoplasticity Model Incorporating Adiabatic Shear Banding, J. Eng. Mater. Technol., ASME, 131 (2009) 1–14.

[4] S. Su, L. Stainier, S. Mercier, Energy-based variational modeling of fully formed adiabatic shear bands, Eur. J. Mech. – A Solids. 47 (2014) 1–13.

EPJ Web of Conferences **94**, 01004 (2015)
DOI: 10.1051/epjconf/20159401004

Strain rate effects on reinforcing steels in tension

Ezio Cadoni[1] and Daniele Forni[1,2,a]

[1] DynaMat Laboratory, University of Applied Sciences of Southern Switzerland, Switzerland
[2] Department of Structural, Geotechnical and Building Engineering, Politecnico di Torino, Italy

Abstract. It is unquestionable the fact that a structural system should be able to fulfil the function for which it was created, without being damaged to an extent disproportionate to the cause of damage. In addition, it is an undeniable fact that in reinforced concrete structures under severe dynamic loadings, both concrete and reinforcing bars are subjected to high strain-rates. Although the behavior of the reinforcing steel under high strain rates is of capital importance in the structural assessment under the abovementioned conditions, only the behaviour of concrete has been widely studied. Due to this lack of data on the reinforcing steel under high strain rates, an experimental program on rebar reinforcing steels under high strain rates in tension is running at the DynaMat Laboratory. In this paper a comparison of the behaviour in a wide range of strain-rates of several types of reinforcing steel in tension is presented. Three reinforcing steels, commonly proposed by the European Standards, are compared: B500A, B500B and B500C. Lastly, an evaluation of the most common constitutive laws is performed.

1. Introduction

Failure in engineering is understandably considered a negative occurrence, something that should be avoided at almost any cost. But sometimes a component must fail so that the larger system to succeed. This is a challenge for engineers, because they have to design systems and devices having well defined and predictable breaking points. On the other hand, the structural design should avoid that the presence of a human error or in case of deliberate act of terrorism can be the cause of a catastrophic collapse.

For these reasons, in a safe design is of extreme importance to consider all the possible loadings that could lead to failures, which could be caused by dynamic loads due to natural causes, like strong earthquakes, or accidentally caused by impacts or blasts.

Historically, the behaviour of structures under severe dynamic loading has become an important issue in civil engineering during the last decades because of the increasing number of deliberate terroristic attacks. Worldwide interests, renewed after the 9/11 World Trade Centre disaster, and because of the fresh outbreak of terroristic threats on civilian targets, have further increased the need for knowledge in this field [1]. Consequently, the knowledge of the mechanical properties at high strain rates are needed for the assessment of structures against severe dynamic loads.

In particular, for reinforced concrete structures both concrete and rebar properties have to be considered. This because these two materials have a completely different behavior: the concrete is strong enough in compression, but has a very limited strength in the tensile part of the structure, where the reinforcing steel bars are properly used.

In spite of this, the attention of many researchers was focused on this problem, only the dynamic behavior of concrete [2–4] was deeply studied. Consequently, the dynamic behavior of the reinforcing steels remained quite neglected and only few and/or relatively old data were available [5–8].

Malvar [9], considering strain rate tests based on a literature review, proposed a strength enhancement of steel rebar in terms of the dynamic increase factor (DIF). Recently, the new version of the Model Code 2010 [10], has re-adopted the strain rate effect on the reinforcing steel properties based on the CEB Bulletin 187 [4].

In case of high dynamic loadings, high strain rates in the range of 10^1 to $10^3\,s^{-1}$ are generated [12–16]. For this reason, new information on the high strain rate behavior of modern rebars is extremely necessary [13,14]. According to the current European standards [11], there are three classes of reinforcement, A, B and C, which indicate increasing ductility. In this paper a discussion of the strain rate sensitivity of the main reinforcing steel properties for all three classes B500A, B500B and B500C steels is reported. In the case of B500A and of B500B steel, were used specimens coming from bars having 12 mm in diameter [14]. While for the B500C steel [17], considered as a composite bar made of a hardened outer cortical layer, an intermediate hardened layer and a soft inner core, the effects of the strain rate on the tensile properties of these three layers were evaluated on specimens coming from a bar having 40 mm in diameter.

Finally, an evaluation of the most common constitutive laws, such as Johnson-Cook [18] and Cowper and Symonds [19] relationships, were performed.

2. Experimental procedure

The dynamic behavior of reinforcing steels was analysed in a wide range of strain rates: quasi-static, intermediate and high strain rates.

[a] Corresponding author: daniele.forni@supsi.ch

Article available at http://www.epj-conferences.org or http://dx.doi.org/10.1051/epjconf/20159401004

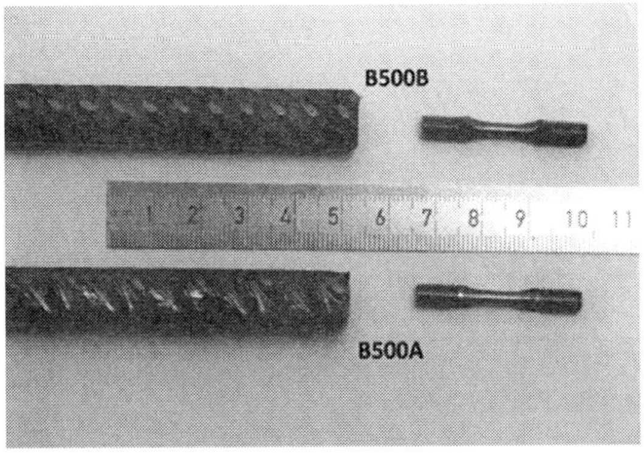

Figure 1. Photo of the B500A and B500B rebar and specimens.

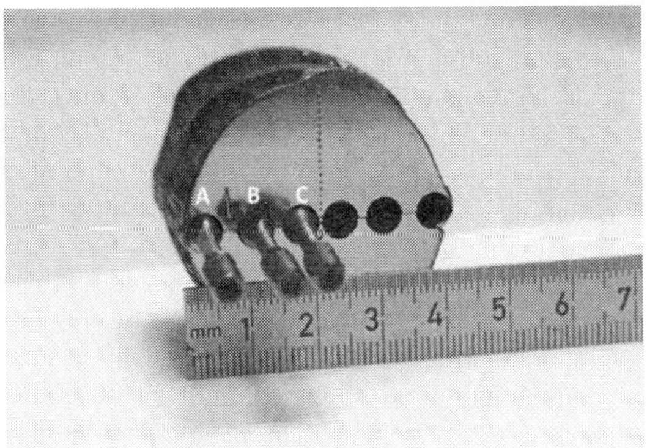

Figure 2. Photo of the B500C rebar and specimens.

The specimens used were cylindrical samples having 5 mm of gauge length and a diameter of 3 mm [12–14]. In case of B500A, B500B samples were obtained, from rebars (having a diameter of 12 mm), by turning, while for B500C material, samples were firstly extracted from a bar of 40 mm by means of a wire electrical discharge machine as described in [17] as shown in Fig. 2.

The quasi-static tests were performed at $10^{-3}\,s^{-1}$ by means of a universal electro-mechanical testing machine Zwick/Roell-Z50 with a maximum load bearing capacity of 50 kN. The intermediate tensile tests were performed at 5 and $25\,s^{-1}$ by means of a Hydro-pneumatic machine (HPM), whose functioning was widely explained in [20–23]. Three high strain rates 250, 500, and $1000\,s^{-1}$ were carried out by means of a Split Hopkinson Tensile Bar (SHTB) [12–14].

3. Results and discussion

The reinforcing steels were investigated in a wide range of strain rates. The results, in terms of stress versus strain curves, of the tensile tests at the different strain rate are shown in Figs. 5–8. In particular in Fig. 5 are depicted the curves for the B500A steels, while in Fig. 6 the curves for B500B steel are shown. These figures shown how these steels have very high ductility in dynamic regime.

In case of B500C the represented data showed in Figs. 6–8 were obtained by means of the weighted strength

Figure 3. Hydro-Pneumatic Machine for intermediate strain rates tests.

Figure 4. Split Hopkinson Tensile Bar device for high strain rates tests.

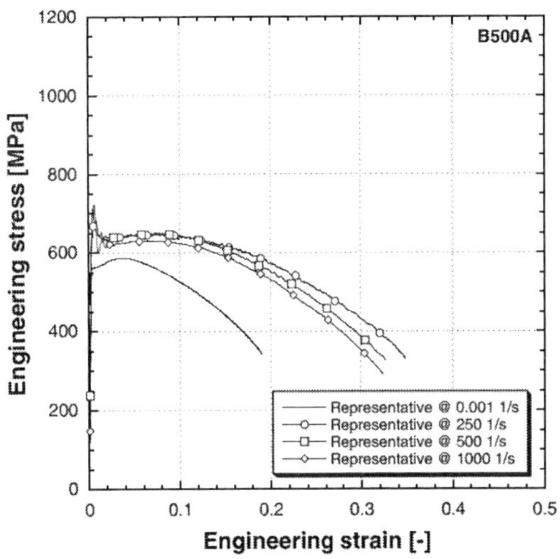

Figure 5. B500A mechanical properties trend versus strain-rate.

values of the three different zones [14] in function of the respective area of the cross section in the rebar. Also in this case the mechanical characteristics increase with the increasing of the strain rate (i.e. ultimate tensile strength as well as the uniform strain). Observing the fracture strain at different strain rate can be noted how this increment

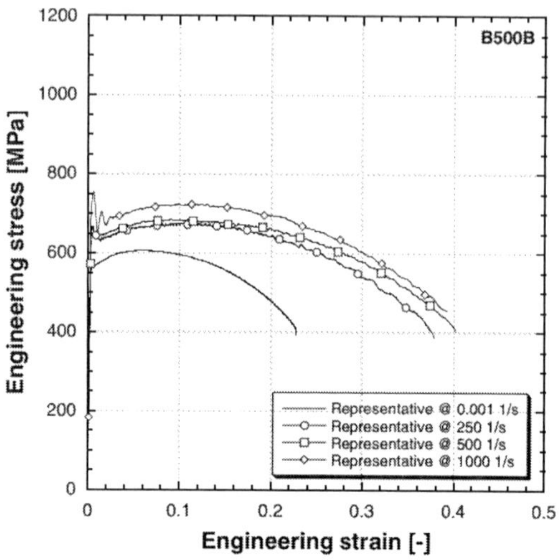

Figure 6. B500B mechanical properties trend versus strain-rate.

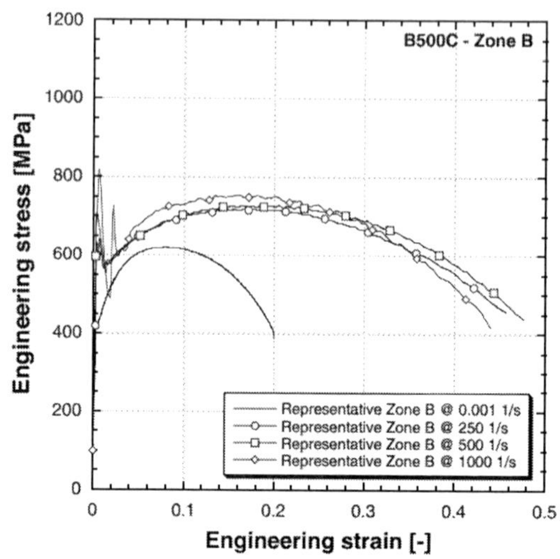

Figure 8. B500C mechanical properties trend versus strain-rate.

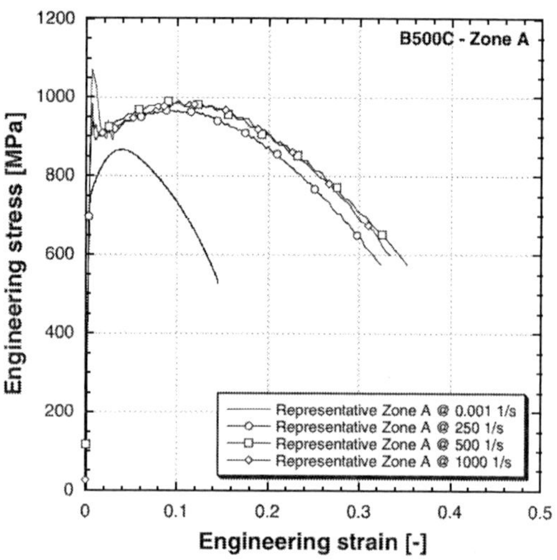

Figure 7. B500C mechanical properties trend versus strain-rate.

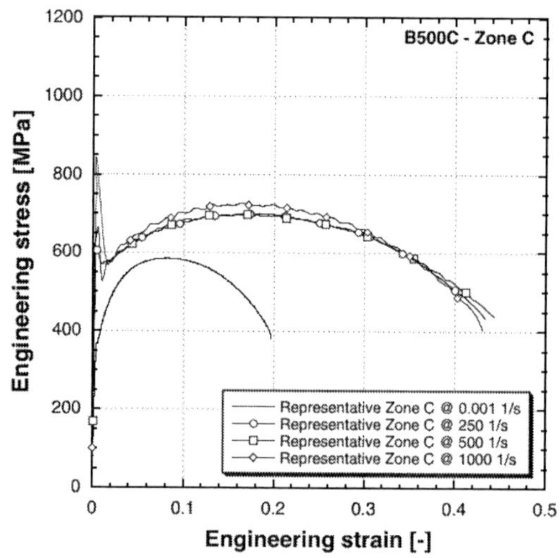

Figure 9. B500C mechanical properties trend versus strain-rate.

becomes larger for the softer material such as the core material in B500C [17].

More information about the evolution of the mechanical parameter of these reinforcing steel can be obtained looking at the Figs. 9–11.

In Fig. 9 a representation of the strain-rate sensitivity of this B500A reinforcing steel is given, with the tendency line of the yield strength and the Ultimate Tensile Strength (UTS) as a function of strain rate and tendency line of the uniform and fracture strain, and the reduction in cross-sectional area. For each of these trends, a visual trend line is included to help distinguish the data sets.

The same data were analysed for the B500B reinforcing steel (Fig. 10) and for B500C, as average weighted values (Fig. 11). It is possible to point out that even if the B500C steel is more ductile with respect to the B500A steel in case of static conditions, at higher strain rates this difference is lower. The B500C steel has a pronounced decrease of ductility at high strain rates.

A fundamental property for rebar is the ductility that can be measured with the ratio between the UTS and yield stress. This parameter is represented in Fig. 12 for B500A and B500B reinforcing steels.

The B500B reinforcing steel shows a constant decrease of the ductility parameter while the B500A reinforcing steel is quite constant for the higher velocities.

4. Constitutive laws

The experimental results were used to calibrate two well-known material constitutive relationships such as Johnson-Cook [19] and Cowper-Symonds [20]. These are widely used models. The first is based on three independent phenomena, the isotropic hardening, the strain-rate hardening and the thermal softening.

$$\sigma_{JC} = (A + B \cdot \epsilon_p^n) \cdot \left(1 + C \cdot \ln \dot{\epsilon}_p/\dot{\epsilon}_0\right) \cdot (1 - \hat{T}^m). \quad (1)$$

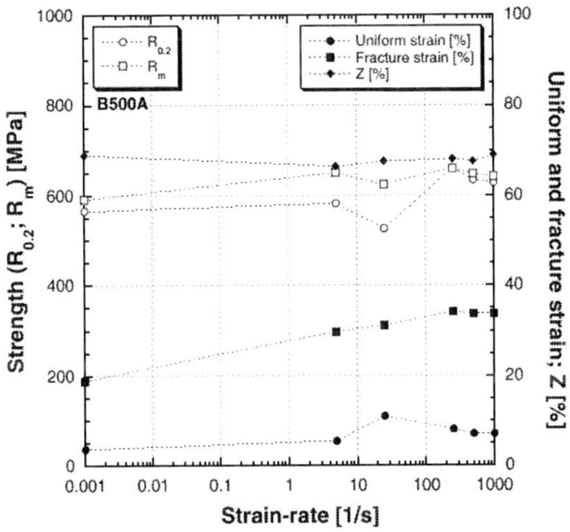

Figure 10. B500A mechanical properties trend versus strain-rate.

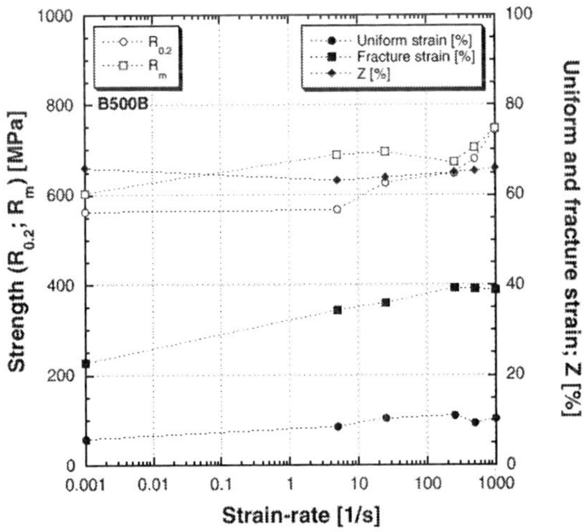

Figure 11. B500B mechanical properties trend versus strain-rate.

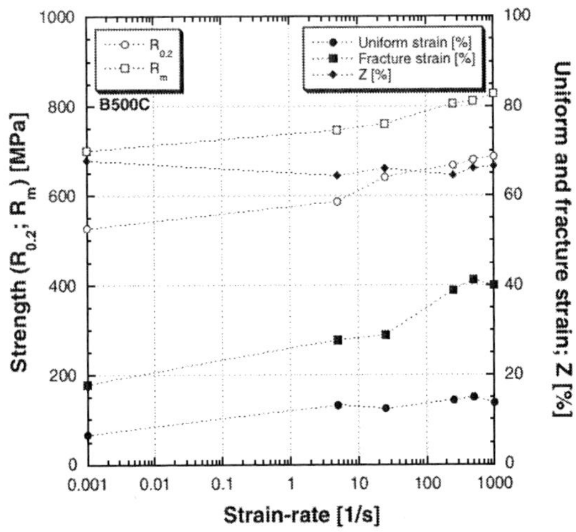

Figure 12. B500C mechanical properties trend versus strain-rate.

Table 1. Johnson-Cook parameters.

	A	B	n	C
	[MPa]	[MPa]	[-]	[-]
B500A	564	946	0.875	0.01769
B500B	571	643	0.720	0.02139

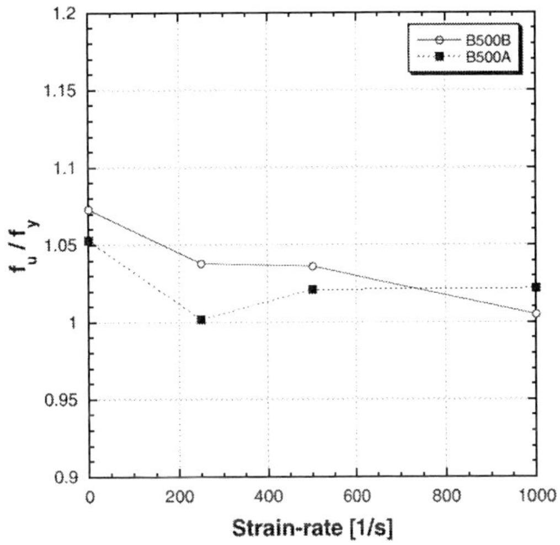

Figure 13. B500C mechanical properties trend versus strain-rate.

Table 2. Cowper-Johnson-Cook parameters.

Material	D	q
B500A	141917	2.816
B500B	599	0.753

The parameters to be determined from the experimental results are A, B, and n in order to take into account the strain-hardening, as well as C and m that represent the strain-rate sensitivity and the thermal softening respectively. In this case, the thermal softening parameter is set equal to one, due to the fact the temperature effect at high strain rates is actually under investigation, and only room temperature tests have been considered. The previously mentioned material parameters are reported in Table 1.

In Fig. 13 the comparison between Johnson-Cook model and experiments are depicted.

The second model is a strain rate dependent elastic–plastic model and is able to relate the dynamic stress with the strain rate by means of two constants D and q:

$$\frac{f_d}{f_0} = 1 + \left(\dot{\epsilon}_p/D\right)^{1/q} \qquad (2)$$

where f_d is the dynamic yield true stress, f_0 is the static yield true stress, $\dot{\epsilon}p$ is the strain rate, D and q are material parameters.

In Table 2 the D and q parameters of two reinforcing steel are reported while in Fig. 14 the true yield stresses versus strain rate and the model prevision are shown.

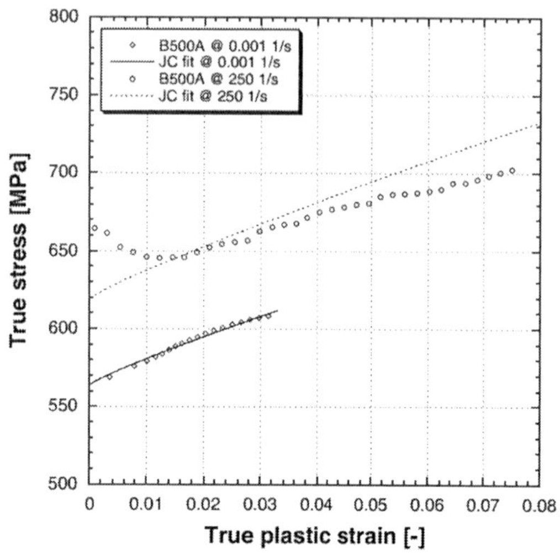

Figure 14. Jonhson Cook model for B500A steel.

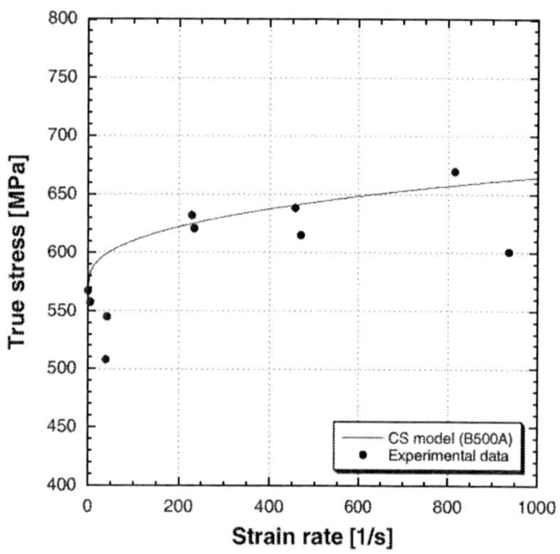

Figure 15. Cowper-Symonds model for B500A steel.

5. Conclusions

The strain rate effects on the behaviour of different reinforcing steels in tension were investigated.

The mechanical behaviour of B500A, B500B and B500C reinforcing steel was analysed by two experimental set-ups such as Hydro-Pneumatic Machine and Split Hopkinson Tensile Bar device used respectively for intermediate and high strain rate.

The examined reinforcing steels show high strain rate dependency.

Results on different layers of B500C confirm the fact that this quenched and self-tempered reinforced steel has to be considered as a composite structure. For this reason a special attention should be attended in order to simulate the structural performance of a real rebar subjected to high strain rates [16].

Two widely used material models were calibrated on the basis of the experimental results obtained.

The predicted flow stress by Johnson-Cook model has good agreement with the experimental results.

The parameters search will be concluded with the high strain rate tests in temperature, providing the parameter m and calibrating other constitutive relationship as Zerilli-Armstrong and others.

The authors wish to thank Matteo Dotta and Nicoletta Tesio of the DynaMat Laboratory, University of Applied Sciences of Southern Switzerland, for their help in the experiments. The research was financially supported by the project COST C12.0051 entitled "Behaviour of structural steels under fire in a wide range of strain rates" funded by Swiss State Secretariat of Education, Research and Innovation of the Swiss Confederation in the frame of the COST Action TU0904 "Integrated Fire Engineering and Response".

References

[1] COST TU0601 - Robustness of Structures: `http://www.cost.eu/COST_Actions/tud/Actions/TU0601`
[2] P.H. Bishoffs, S.H. Perry, Mat. Struct. **24**, 425–450 (1991)
[3] L.J. Malvar, C.A. Ross, ACI Mat. J. **95(6)**, 735–739 (1998)
[4] Comité Euro-International du Béton, Bulletin **187**, (1988)
[5] J. Seabold, Tech. Rep. R695, (1970)
[6] W.L. Cowell, Tech. Rep. R394, (1965)
[7] W. Keenan, J. Tancreto, G. Meyers, F. Johnson, J. Hopkins, H. Nickerson, W. Armstrong, Tech. Mem. 51-83-19 (prev. 2591TM), (1983)
[8] R.G. Crum, ACI J. Proc. **56(1)**, 59–61 (1959)
[9] L.J. Malvar, ACI Mat. J. **95**, 609–616 (1998)
[10] FIB-International Federation for Structural Concrete, Model Code 2010, **Bulletin 66**, (2012)
[11] European Standard, EN 10080, (2005)
[12] E. Cadoni, L. Fenu, D. Forni, *Constr. Build. Mat.* **35**, 399–407 (2012)
[13] E. Cadoni, M. Dotta, D. Forni, N. Tesio, *Appl. Mech. Mat.* **82**, 86–91 (2011)
[14] E. Cadoni, M. Dotta, D. Forni, N. Tesio, *Mater. Struct.* 1–11 (2014)
[15] N.K. Singh, E. Cadoni, M.K. Singha, N.K. Gupta, J. *Eng. Mech-ASCE* **139**(9), 1197–1206 (2013)
[16] C. Albertini, E. Cadoni, G. Solomos, *Philos. Trans. A Math. Phys. Eng. Sci.* **372** (2015)
[17] E. Cadoni, M. Dotta, D. Forni, N. Tesio, C. Albertini, *Mat. Des.* **49**, 657–666 (2013)
[18] G.J. Johnson, W.J. Cook, *Eng. Fract. Mech.* 21–31 (1985)
[19] G.R. Cowper, P.S. Symonds, *Div. Appl. Math. Rep.* (1957)
[20] D. Asprone, E. Cadoni, A. Prota, *ACI Struct. J.* **106**, 523–529 (2009)
[21] E. Cadoni, M. Dotta, D. Forni, P. Spaetig, *J. Nucl. Mat.* **414(3)**, 360–366 (2011)
[22] A. Caverzan, E. Cadoni, M. di Prisco, *Int. J. Imp. Engng.* **45**, 28–38 (2012)
[23] A. Caverzan, E. Cadoni, M. di Prisco, *Mech. Mat.* **59**, 87–109 (2013)

EPJ Web of Conferences **94**, 01005 (2015)
DOI: 10.1051/epjconf/20159401005

Recent developments in dynamic testing of materials

Amos Gilat[a] and Jeremy D. Seidt

The Ohio State University, Department of Mechanical and Aerospace Engineering 201 W. 19th Ave., Columbus, OH 43210, USA

Abstract. New techniques for dynamic characterization of materials that have been developed in the last three years (since the last DYMAT conference in 2012), and results from recent dynamic testing of Inconel 718 are presented. The first development is a dynamic punch test in which three dimensional Digital Image Correlation (DIC) is used to measure the deformation of the rear surface of a specimen as it being penetrated. The second experimental technique that is under development is a dynamic tension experiment in which full-field strain measurement with DIC and full-field temperature measurement are done simultaneously during the test.

1. Introduction

Numerical simulation of the response of materials to loads has reached a level of maturity at which it can be used with confidence for design. Numerical codes like LS-DYNA and Abaqus include many material models for plastic deformation and failure (constitutive relations) that can be selected for specific applications. The various models require input parameters that are specific to the material that is being simulated. Accurate experimental results from tests under well-controlled conditions are needed for the determinations of the parameters in the material models as well as for the validation.

The current paper presents two new testing configurations that have been developed recently for the purpose of providing fundamental data for determining the parameters in material models for dynamic plastic deformation and failure. In addition, results from testing Inconel 718 over a wide range of strain rates and temperatures are presented.

2. Dynamic punch experiment

A dynamic punch test in which three-dimensional Digital Image Correlation (DIC) is used to measure the deformation of the rear surface of a specimen as it being penetrated is carried out by using a large diameter (50 mm) compression Split Hopkinson Bar (SHB) apparatus, Fig. 1. Flat, round Ti-6Al-4V disk specimens are attached to the transmitter bar of a compression SHB apparatus and a tungsten carbide punch is attached to the incident bar and positioned such that it is in contact with the disk, Fig. 2. During a test, a compression wave is introduced into the incident bar which causes the punch to penetrate into the specimen. The specimen is mounted on a die fixture that is slotted on both sides such that the rear surface of the disk specimen is visible to two high speed cameras. This

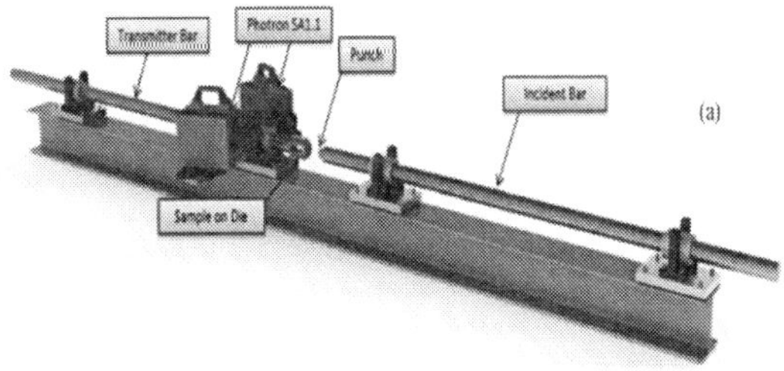

Figure 1. Dynamic punch test in a compression SHB apparatus.

provides a stereographic view of the specimen that is used to measure full-field displacement directly on the specimen using three-dimensional DIC. The contact force between the punch and the disk is determined from the wave in the transmitter bar.

Two Photron SA1.1 cameras running at 100,000 frames per second (10 μs interval) at 192 pixel by 192 pixel resolution record the rear surface deformation of the disk specimen. The images are processed by commercial Digital Image Correlation (DIC) software (VIC-3D).

Typical wave profiles recorded in a test are shown in Fig. 3. The amplitude of the incident wave in this experiment is 400 kN. Figure 4 shows the history of the strains at two points (the centre point and the point that failure first occurs) at the back surface of the specimen. The last DIC image recorded before failure is shown in Fig. 5.

Results from tests with punches of various geometries show that the punch geometry greatly influences the punching force and the failure mode. The data is used to construct and validate deformation and failure models.

3. Dynamic response of Inconel

Inconel 718 has been tested in compression, tension and shear at various strain rates ranging from 0.0001 s^{-1} up

[a] Corresponding author: gilat.1@osu.edu

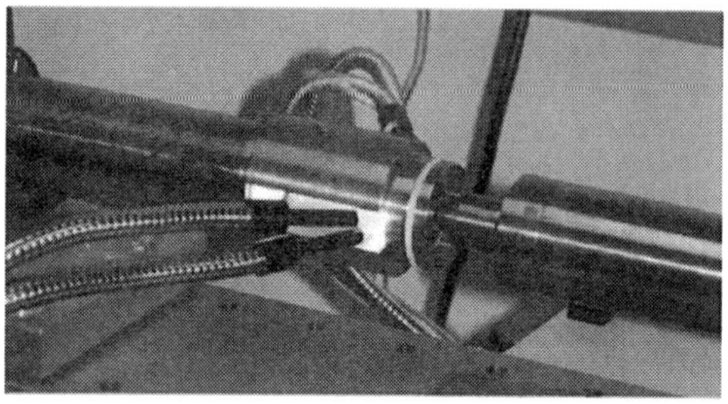

Figure 2. Close-up of the dynamic punch experiment.

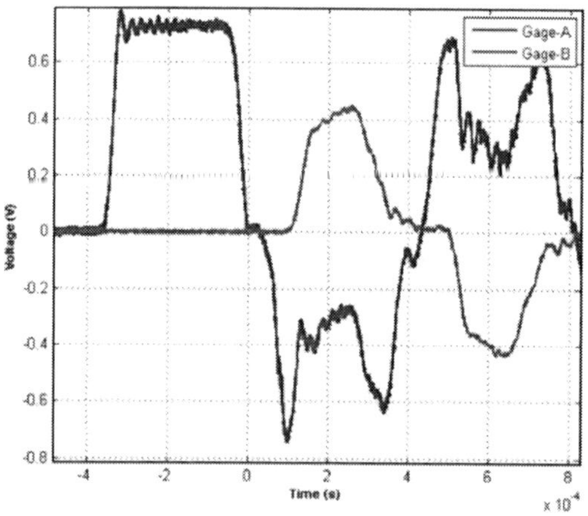

Figure 3. Waves recorded in a dynamic punch experiment.

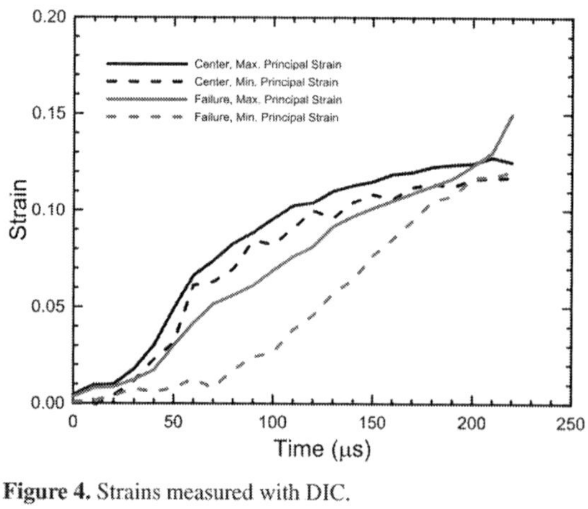

Figure 4. Strains measured with DIC.

to $5.200\,s^{-1}$. A hydraulic frame is used for the quasi-static tests and the SHB technique (with DIC used to measure strains directly on the specimen) is used for the dynamics tests. Results from tensile and compression tests with specimens initially at room temperature are shown in Figs. 6 and 7, respectively.

The results show a significant effect of the strain rate on the stress. There is also a noticeable difference in

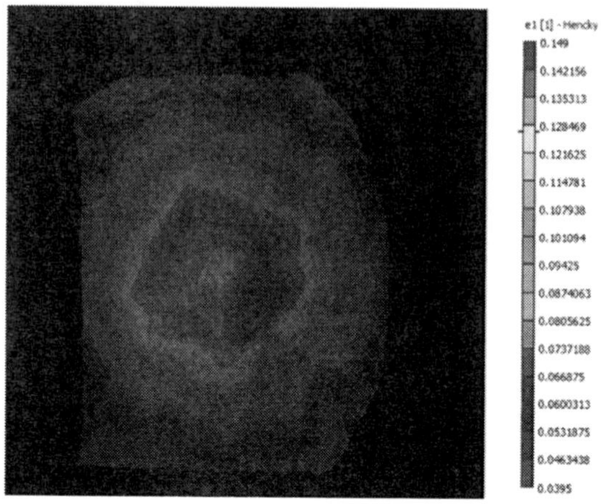

Figure 5. Last DIC image prior to failure (red dot is the failure point).

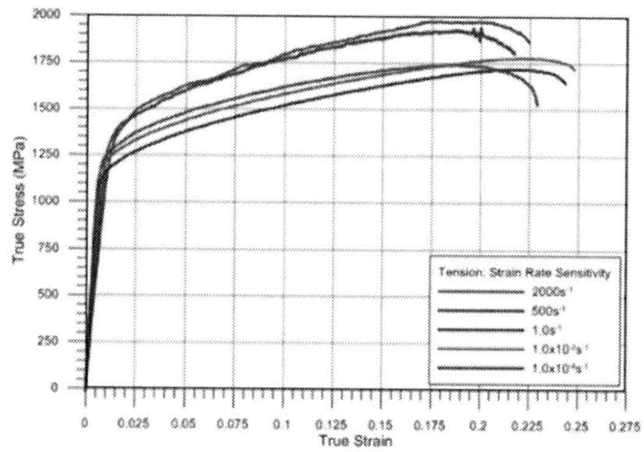

Figure 6. Tensile stress strain curves for Inconel 718 at various strain rates.

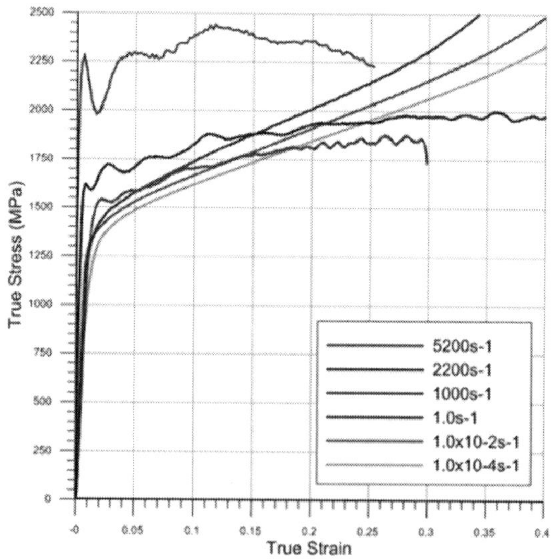

Figure 7. Compression stress strain curves for Inconel 718 at various strain rates.

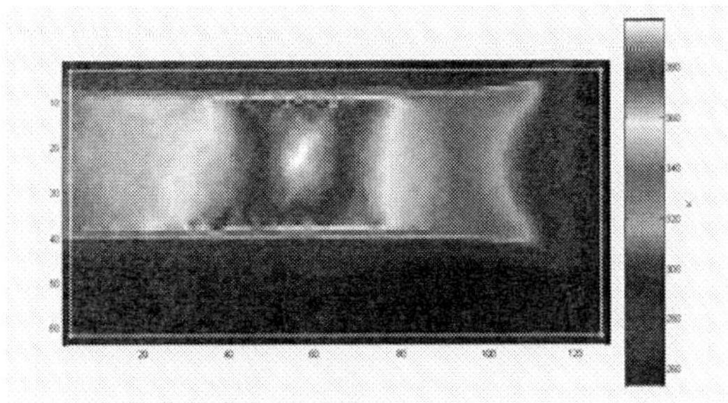

Figure 8. IR measurement in a dynamic tensile test.

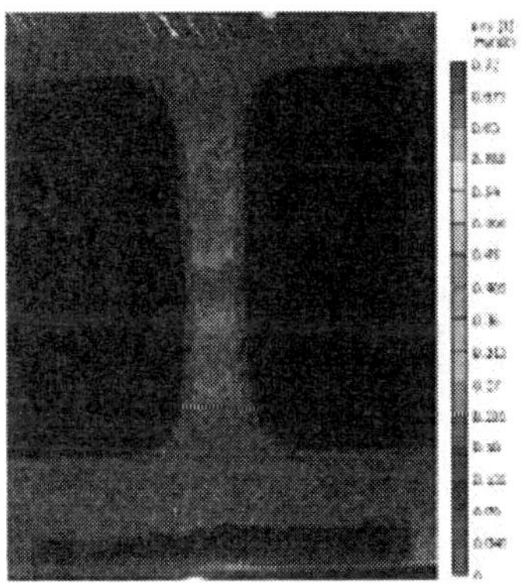

Figure 9. DIC strain measured in a dynamic tensile test.

the response between tension and compression. Higher strain rates are achieved in the compression tests which show an increase in strain rate sensitivity at strain rate of about $5.000\,s^{-1}$.

4. Full-field strain and temperature measurements during dynamic tensile test

Under development is a dynamic tensile experiment that features a full-field strain measurement, and a full-field temperature measurement at the same time (on opposite sides of the specimen). The strains are measured with DIC and the temperature is measured with a high speed IR camera. Pilot experiments have been conducted at strain rate of $150\,s^{-1}$ and $500\,s^{-1}$. Figures 8 and 9 present results from tensile testing of specimen made of 304 stainless steel at strain rate of approximately $150\,s^{-1}$. Figure 8 shows the last frame before failure that was recorded with an IR camera (Telops Fast MWIR 1500, resolution 128×64, frame rate 10.000 fps). The maximum temperature observed is about 270 °C. The full-field strain measured with DIC at the same time is shown in Fig. 9 (Photron SA1.1, resolution 448×592, 20.000 fps). The maximum strain the is measured in the neck is about 0.8.

The research was supported by the USA Federal Aviation Administration, Grant No. 11-G-003. The authors are grateful to Mr. Bill Emmerling and Dr. Chip Queitzsch of the FAA. Thanks also to Professor Veli-Tapani Kuokkala from Tampere University of Technology for his contribution to the development of the full-field temperature measurements.

EPJ Web of Conferences **94**, 01006 (2015)
DOI: 10.1051/epjconf/20159401006

Estimation of the restitution coefficient by strain measurement

Naoya Nishimura[1,a], Katsuhiko Murase[1], Takeru Watanabe[2], Keisuke Niimi[1], and Makoto Fukuhara[2]

[1] Meijo University, Department of Vehicle and Mechanical Engineering, 1-501 Shiogamaguchi, Tempaku-ku, Nagoya 468-8502, Japan
[2] Meijo University, Graduate School of Science and Technology, 1-501 Shiogamaguchi, Tempaku-ku, Nagoya 468-8502, Japan

Abstract. In order to clarify the collision phenomenon of the traffic machine in motion, such as automobile and aircraft, it is necessary to examine not only the deformation and the behavior of composed structural member but also the dynamic characteristic of friction coefficient and restitution coefficient in the collided contact surface. In this study, the restitution coefficient was estimated from the strain waveform measured by collision test using the Air Gun. The velocity and deformation dependence of the restitution coefficient was examined by carrying out the collision test at the different speed. As a result, in the metal test piece, restitution coefficient decreased by the plastic deformation with the increasing of the collision speed, and the velocity dependence was confirmed.

1. Introduction

It is known that the restitution coefficient between two objects differs by the shape, dimension and material distribution (density distribution, each coefficient distribution of elasticity, plasticity and viscosity, Poisson ratio distribution, etc.) of each object and also the deformation characteristic of contact point [1–3]. However, role and application of the restitution coefficient as basic data which macroscopically and approximately estimates collision behavior between two objects are wide, and it is broadly used in various fields. As rebound behavior evaluation of two objects in the elastic region, the basic research on the estimation of the rebounding characteristics by the numerical solution which used restitution coefficient, mass and moment of inertia of each objects as a parameter is reported [4]. And, in order to generally understand the collision collapse behavior of the whole structure, the method which calculates the restitution coefficient as near zero value is also used by temporarily assuming composing material and member the rigid-plasticity [1,5]. The experimental technique, which evaluates the restitution coefficient in the collision including plastic collapse phenomenon like automobile collision, has not yet been established.

The restitution coefficient is derived from momentum conservation law of the collision object, and it is defined by the speed of the object before and after the collision [6]. Generally, in the case of simple shape object, the restitution coefficient is often calculated from falling height and rebounding height of the object to the other stationary object by the drop type. However, accuracy and reproducibility for this method are low. On the other hand, it is reported that an accurate restitution can get from the approach and restitution impulse by measuring the contact force at the collision contact surface of two objects [7], but it has not measured and evaluated by this technique yet. It is necessary to establish the technique which estimates experimentally the restitution coefficient including plastic collapse phenomenon, such as automobile collision.

In this study, the aim is to establish a technique that measure and estimate standardly the restitution coefficient in the collision which include plastic collapse phenomenon. As a first step, an impulse in the two object collision was estimated from the strain waveform which generated in the stationary object before the collision, and the validity of evaluating method of the restitution coefficient was examined. The velocity and deformation dependence of the restitution coefficient was evaluated from the approach and restitution impulse in the collision test at the different speed by using the Air Gun.

2. Restitution coefficient and impulse

The velocity change of two objects (mass m1 and m2) in one dimensional collision is considered, as shown in Fig. 1. The velocity of two objects before and after the collision are v_1, v_2 and v_1', v_2', respectively. The restitution coefficient e is defined by conservation of momentum as following [6].

$$e = \frac{v_2' - v_1'}{v_1 - v_2}. \tag{1}$$

The whole momentum is conserved before and after the collision because the objects does not forced except for the collision force (in case of Fig. 1). Therefore, the momentum is given by following equation, where v_c is the velocity when two objects becomes the same velocity.

$$m_1 v_1 + m_2 v_2 = m_1 v_1' + m_2 v_2' = (m_1 + m_2)\, v_c. \tag{2}$$

[a] Corresponding author: nisimura@meijo-u.ac.jp

Article available at http://www.epj-conferences.org or http://dx.doi.org/10.1051/epjconf/20159401006

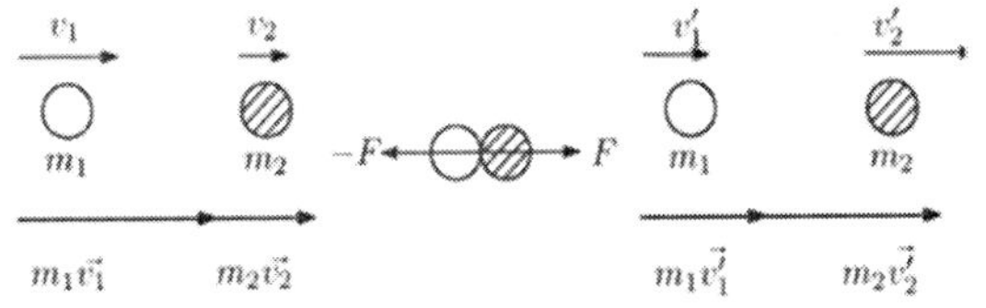

Figure 1. The collision of two objects.

The impulse during the collision is calculated from Eq. (2).

$$P_1 = m_1(v_1 - v_c) = m_2(v_c - v_2)$$
$$P_2 = m_1(v_c - v_1') = m_2(v_2' - v_c) \tag{3}$$

where P_1 and P_2 are the impulse when the two objects are approaching and restituting, respectively. The restitution coefficient is shown in the ratio of the restitution impulse to the approach impulse from Eq. (3), as following equation.

$$\frac{P_2}{P_1} = \frac{v_2' - v_1'}{v_1 - v_2} = e. \tag{4}$$

The force F generated in the object is given by following relation.

$$F = \sigma A = \varepsilon E A \tag{5}$$

where A, E, σ and ε are the cross-sectional area, longitudinal modulus of elasticity, stress and strain of the object, respectively. An impulse P is defined in the product of the force and the action time of that force, and it given by following equation from Eq. (5).

$$P = \int F dt = \int \varepsilon E A dt. \tag{6}$$

The approach and restitution period from the contact to the separation of the objects in the collision, and the strain generated in the object during that period are considered from the strain waveform, as shown in Fig. 2. When the object is collided by other object (t = 0), the strain wave generates in object. The strain takes maximum value at time t = t_1 and the objects approach extremely. The restitution of objects starts at t = t_2 after the duration of the maximum strain. The contact of objects finishes at t = t_3 (objects are separated and strain takes zero). The approach impulse P_1 (period from 0 to t_1) and restitution impulse P_2 (period from t_2 to t_3) are calculated by following equation from Eq. (6).

$$P_1 = \int_0^{t_1} \varepsilon_1(t) E A dt = E A \int_0^{t_1} \varepsilon_1(t) dt$$
$$P_2 = \int_{t_2}^{t_3} \varepsilon_2(t) E A dt = E A \int_{t_2}^{t_3} \varepsilon_2(t) dt. \tag{7}$$

Consequently, the restitution coefficient e is estimated by following equation from Eqs. (4) and (7).

$$e = \frac{P_2}{P_1} = \frac{E A \int_{t_2}^{t_3} \varepsilon_2(t) dt}{E A \int_0^{t_1} \varepsilon_1(t) dt} = \frac{\int_{t_2}^{t_3} \varepsilon_2(t) dt}{\int_0^{t_1} \varepsilon_1(t) dt}. \tag{8}$$

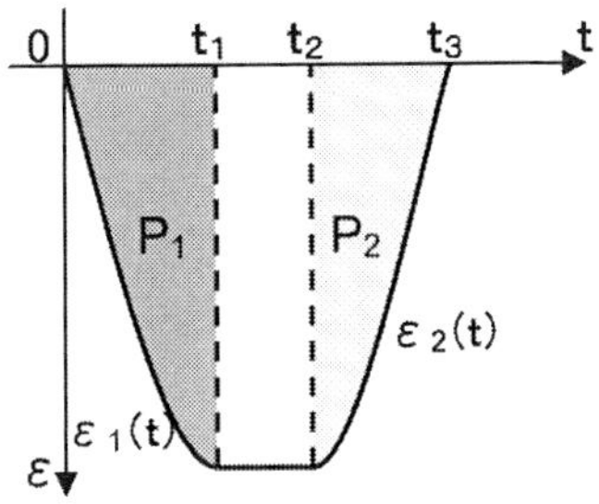

Figure 2. Strain waveform and impulse.

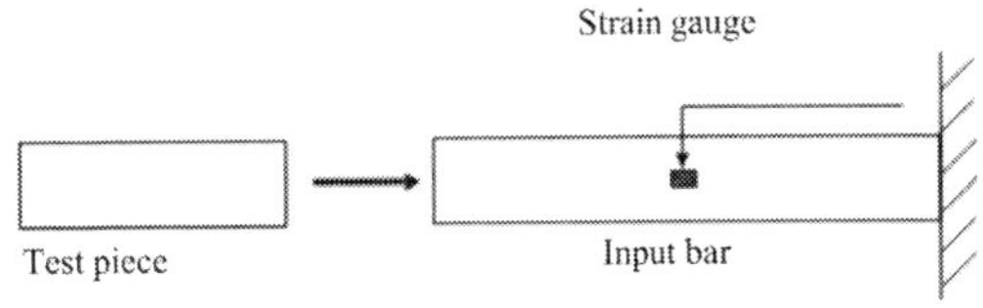

Figure 3. Schematic of collision test.

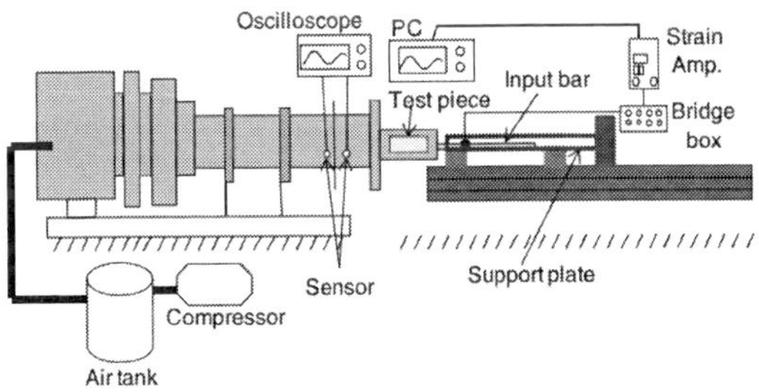

Figure 4. Schematic of the experimental equipments.

By measuring the strain wave generated by collision, the evaluation of the restitution coefficient is possible from the strain change and time during the approach and restitution period.

3. Collision test

3.1. Test piece

The restitution coefficient of test piece is estimated by measuring the deformation of elastic body from one dimensional collision test (wave propagation), as shown in Fig. 3. The test piece made of aluminum alloy (A5052) is carried out the collision test. The dimensions of test piece are 29.5 mm diameter and 100 mm length. The collision side of test piece is machined in R160 arc shape. Stainless steel (SUS304) of 20 mm diameter and 500 mm length is used for the input bar.

3.2. Test procedure

In the collision test used Air Gun (Fig. 4), the test piece which is accelerated by compression air is collided with the input bar. The strain gauge was bonded to the input bar (100 mm away from collision side). The stress wave which propagated in the input bar was measured by the oscilloscope in of 100 MHz sampling [8].

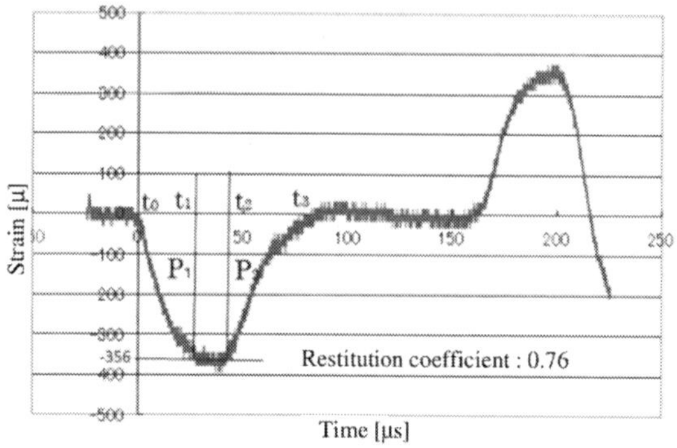

Figure 5. Example of strain waveform. (collision velocity 4.5 m/s).

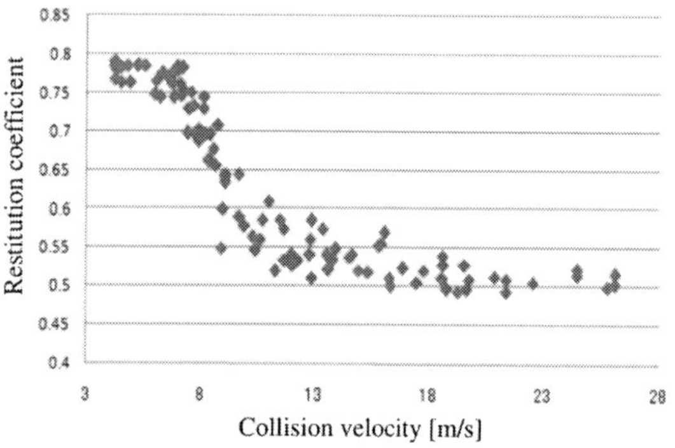

Figure 6. Dependence of restitution coefficient on collision velocity (Aluminum alloy).

The approach impulse P_1 and restitution impulse P_2 were estimated from the before and after maximum strain value in the measured strain waveform, and the restitution coefficient was obtained from Eq. (8). The collision test was carried out in 4–26 m/s velocity range, and the effectiveness of this technique which estimated the restitution coefficient by the strain measurement was examined.

4. Results and discussions

An example of the strain waveform measured at the collision test is shown in Fig. 5. The moving average processing was conducted to the measurement waveform, since the error by some noise and digitizations is contained in the signal. Maximum strain value, the closest approach time t_1, the restitution start time t_2 and the contact finish time t_3 was estimated from measured waveform, and then the restitution coefficient was calculated from the approach impulse P_1 and restitution impulse P_2.

The restitution coefficient of aluminum alloy for stainless steel at each collision velocity is shown in Fig. 6. The value of the restitution coefficient calculated by strain measurement in the elastic region (low collision velocity) was equivalent to the value shown in literature [9].

The restitution coefficient rapidly decreases after approximately 7 m/s collision velocities, though it takes constant value until that velocity. R160 arc shape which was the impact side of test piece was partially deformed plastically at the collision velocity 7 m/s to 10 m/s. The plastic deformation increased with the increase in collision velocity in this velocity range. It is considered that the kinetic energy of test piece was spent in the energy of plastic deformation and the restitution coefficient decreased with that plastic deformation. And also, in the collision over 10 m/s velocity, the whole arc part of the test piece perfectly deforms plastically. Further deformation is accumulated over the rod cross section and more large collision force is required for the plastic deformation. Therefore, the drop in the restitution coefficient seems to be slowly.

Measured strain also considered the plastic deformation, and it was confirmed that this technique was effective for the estimation of restitution coefficient containing the plastic collapse phenomenon.

5. Conclusions

In this study, the restitution coefficient was estimated from strain and impulse in the collision object measured by collision test at the different velocity, and the velocity and deformation dependence of the restitution coefficient were investigated. Consequently, in metal test piece, the restitution coefficient decreased by the plastic deformation with the increase in collision velocity, and the velocity dependence was confirmed. It was validated that the estimation of restitution coefficient by this technique was effective for the measurement in both of the elastic and plastic region. As the future work, the effect of shape and material of test piece on restitution coefficient and accuracy improvement of this technique will be attempted by the collision test at the wide velocity range.

References

[1] Editorial committee of automotive technology handbook, *Automotive technology handbook* (Vol.2 environment and safety), Society of automotive engineers of Japan, pp. 85–114 (2005).
[2] I. Emori, *Automobile accident engineering*, p. 1–15, Gijutsu Syoin (1974).
[3] R. C. Hibbeler, *Engineering Mechanics : Dynamics*, 11th edition, Pearson Prentice Hall, pp. 242–245 (2007).
[4] S. Yoshimura and K. Kimura, Use of Impact Vibration for Separation of Objects with Different Profiles or Restitution, Transaction of JSME-C, Vol. 65, No. 630, pp. 449–456 (1999).
[5] N. Jones, *Structural Impact*, Cambridge University Press, pp. 415–420 (1989).
[6] N. Ishikawa, et al., *Fundamental Impact engineering*, Morikita Publishing Co., Ltd., p. 33–36 (2008).
[7] W. Goldsmith, *Impact*, Edward Arnold Ltd, pp. 4–6 (1960).
[8] T. Hayashi and Y. Tanaka, *Impact engineering*, Nikkan Kogyo Shinbun Ltd., p. 169–183 (1986).
[9] W. Goldsmith, *Impact*, Edward Arnold Ltd, pp. 258–259 (1960).

EPJ Web of Conferences **94**, 01007 (2015)
DOI: 10.1051/epjconf/20159401007

Application backwards characteristics analysis method to dynamic response of metals under high pressure

Hao Pan[a], Xiaomian Hu, and Zihui Wu

National Key Laboratory of Computational Physics, Institute of Applied Physics and Computational Mathematics, 100088 Beijing, China

Abstract. Dynamic yield strength of metals/alloys depends on loading pressure and rates sensitively. With the development of laser interferometer measurement system, extracting strength information from window/free surface velocity profiles in shock and ramp loading experiments is becoming an important method to investigate materials' dynamic response under high pressure and high strain rates. Backwards characteristics analysis method (BCAM) can analyze the velocity profiles more reasonable because it accounts for bending of the incoming characteristics due to impedance mismatch between the sample and window. Synthetic analyses of reverse impact experiment and graded-density impactor loading-releasing experiment suggest that BCAM can give more accurate results including sound speed-particle velocity and yield strength at high pressure than incremental impedance matching method. We use BCAM to analyze velocity profiles of Sn in shock-release experiments and obtain its shear modulus and yield strength at different shock pressure and investigate its phase transition and dynamic unloading response.

1. Introduction

Metals /alloys under extreme loading such as impact, explosion, penetration, etc. have attracted much attention because of interesting phenomena and great challenges to experiment and simulation. A number of experiments have shown that the dynamic yield strength of metals depends on loading pressure and rates very sensitively [1–3]. Techniques for strength measurement at high pressures include comparison to hydrostatic response, lateral stress gauges, growth of Rayleigh-Taylor instabilities [4,5], etc.. In practice, with the development of laser interferometer measurement system, self-consistent method [6] is becoming an important method to investigate materials' dynamic response under high pressure and high strain rates. The key problem of this method is to find a valid analysis technique to extract strength information from window/free surface velocity profiles in shock or ramp compression and unload experiments. Common used Lagrangian analysis methods include backward integration method (BIM) [7], incremental impedance matching method (IIMM) [8], transfer function method (TFM) [9] and backwards characteristics analysis method (BCAM) [10,11]. Compared to other methods, the BCAM can account for bending of the incoming characteristics due to impedance differences between the sample and window, moreover, this method also do not need to know the form of equations of state (EOS) and constitutive model. However, this method has been limited to analyzing only the loading portion of a ramp experiment and cannot account for elastic-plastic unloading due to an assumption of a unique one-to-one sound speed for a given pressure or particle velocity in this method.

In this work, we attempt to extend the BCAM to analyze the reverse impact experiment and graded-density impactor (GDI) loading-releasing experiment [12]. Synthetic analyses suggest that the BCAM can give more accurate results including sound speed-particle velocity and yield strength at high pressure than the IIMM. We also use the BCAM to analyze velocity profiles of Sn in shock-release experiments and obtain its shear modulus and yield strength at different shock pressures and investigate its phase transition and dynamic unloading response.

2. Backwards characteristics analysis method

The BCAM was originally proposed as a way to extract stress-strain response of a ramp-wave loaded material described by Ref. [10,11]. Once the material contains a shock wave, the wave characteristics in the material require accounting for the shock wave and its relevant rarefaction waves. The sketch map of the characteristics with a shock wave is given in Fig. 1. Figure 2 shows the idealized interface velocity and in-situ velocity profiles of two thickness samples measured from the shock-release experiment.

Since the BCAM requires an iterative approach, an initial guess at the in-situ particle velocity profiles of two thickness samples should make. Then the longitudinal Lagrangian wave speed can be defined as $C_L = (\partial X/\partial t)u$. These velocity profiles are used to compute material response by

$$d\sigma = \rho_0 C_L du^* \quad (1)$$

$$d\varepsilon = du^*/C_L\left(u^*\right). \quad (2)$$

[a] Corresponding author: panhao1979@sina.com

Article available at http://www.epj-conferences.org or http://dx.doi.org/10.1051/epjconf/20159401007

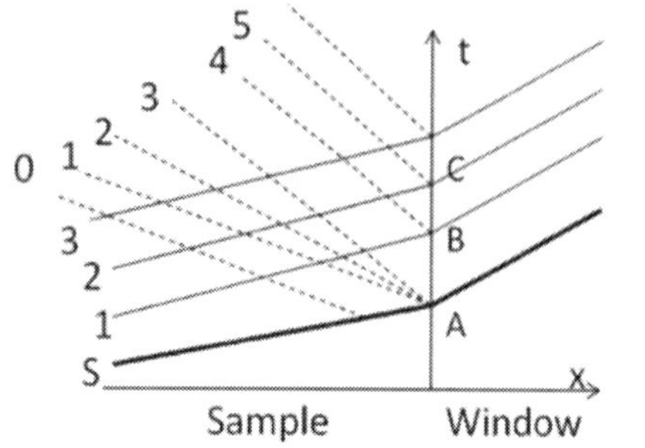

Figure 1. Schematic x-t diagram of BCAM with shock wave.

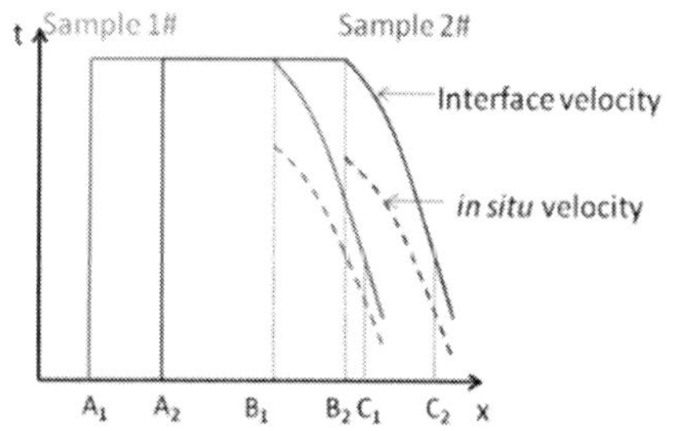

Figure 2. The idealized interface velocity and in-situ velocity profiles of two thickness samples measured from the shock-release experiment.

Where σ is stress, ρ_0 is initial density, ε is strain, μ^* is in-situ velocity. The Riemann invariant R^+ alone positive characteristics and R^- along negative characteristics for isentropic flow are

$$R^{\pm} = u_p + F \tag{3}$$

$$F = \int \frac{1}{\rho_0 C_L} = u^*. \tag{4}$$

The particle velocity u_p and in-situ velocity at the point where the negative and positive characteristics intersect are

$$u_p = \frac{R^+ + R^-}{2} \tag{5}$$

$$F = u* = \frac{R^+ - R^-}{2}. \tag{6}$$

The in-situ velocity and stress behind the shock wave front can be determined using the sample's Hugoniot. The particle velocities on the boundary are the experimentally measured interface velocity. The relevant stress along the interface can be calculated from the known window's Hugoniot and EOS. Because 0th negative characteristic is emanating from the initial state of the shock wave, along the negative characteristic the new in-situ velocity profiles $u^*(t)$ can be obtained by [13]

$$\left(t_{ipos}\right)_R = \left(t_{0,ipos}\right) + \frac{X_R - X_{(0,ipos)}}{C_{L_{(0,ipos)}}} \tag{7}$$

$$\left(u^*_{ipos}\right)_R = u^*_{(0,ipos)}. \tag{8}$$

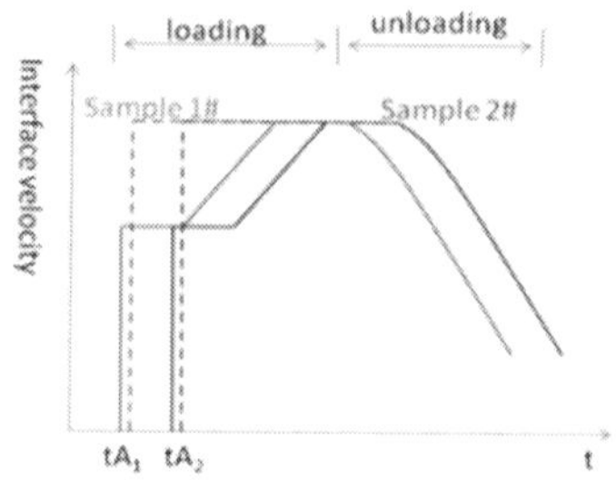

Figure 3. The idealized interface velocity and in-situ velocity profiles of two thickness samples measured from the GDI loading-release experiment. Dash curves are the virtual shock waves. t_A is the arriving time of the shock wave.

In practice, the in-situ velocity profiles achieve convergence with fewer than 10 backward calculations and the shear modulus G and yield strength Y can be obtained from $C_L(u^*)$ using the self-consistent method which are [6]

$$G = \frac{3}{4}\frac{\rho_0^2}{\rho}\left(C_L^2 - C_B^2\right) \tag{9}$$

$$Y = -\frac{3}{4}\rho_0 \int \frac{C_L^2 - C_B^2}{C_L} du. \tag{10}$$

The above analysis method can be directly applied to the shock-release experiment. But for the GDI loading-releasing experiment, the measured interface velocity profiles require to be divided into loading portion and unloading portion, as shown in Fig. 3. Then we use the BCAM to analyze the loading portion and obtain the material's response under loading. However, there is some difference to the unloading portion. Since the GDI experiment has a steady peak state in Fig. 3, this provides a possibility to replace the loading portion with a virtual shock wave. The relevant pressure and particle velocity of the virtual shock wave is defined as the peak pressure and in-situ velocity of the BCAM analysis for the loading portion. Then the BCAM is used to analyze the unloading portion of the profiles in Fig. 3. But the arriving time t_A of the virtual shock wave cannot be defined very properly. So we adjust the arriving time difference between the two shock waves $t_c = t_{A1} - t_{A2}$ to make the bulk sound speed C_B approach to the sound speed calculated by EOS, as shown in Fig. 4. Finally the whole loading and unloading response of material can be obtained from the interface velocity profiles by the BCAM.

3. Results and discussion

3.1. Application to synthetic experiments

First we use the BCAM to analyze a synthetic reverse impact experiment. The sketch map of this experiment is given in Fig. 5. 2 mm Ta is impacted against LiF window at 3 km/s. Ta is modeled with a Mie-Grüneisen EOS and a modified Steinberg-Coharn-Gunian (SCG) strength model [14]. The modified SCG uses an effect shear modulus to describe the Bauschinger effect which can simulate a more realistic quasi-elastic unloading behaviour [6]. LiF is modeled with a Mie-Grüneisen

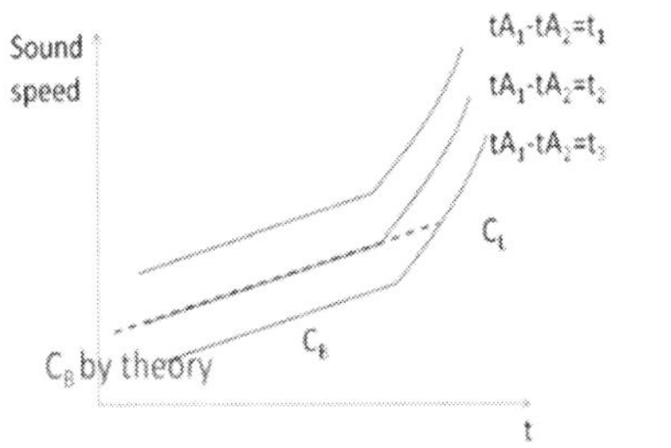

Figure 4. Lagrangian sound speed vs. particle velocity with different $t_c = t_{A1}$-t_{A2}. Adjusting t_c to make the bulk sound speed $C_B$approach to the sound speed calculated by EOS.

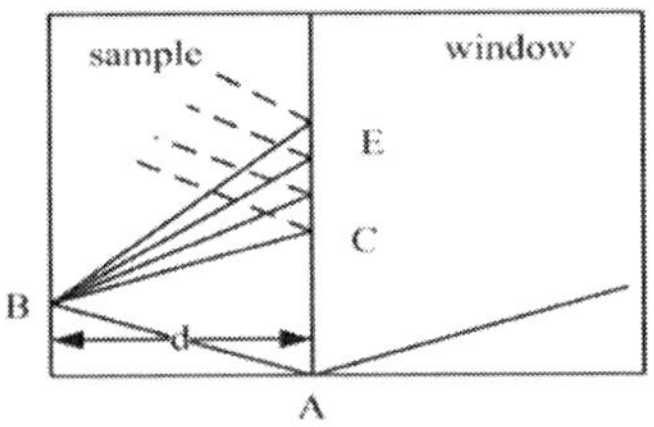

Figure 5. Sketch map of reverse impact experiment.

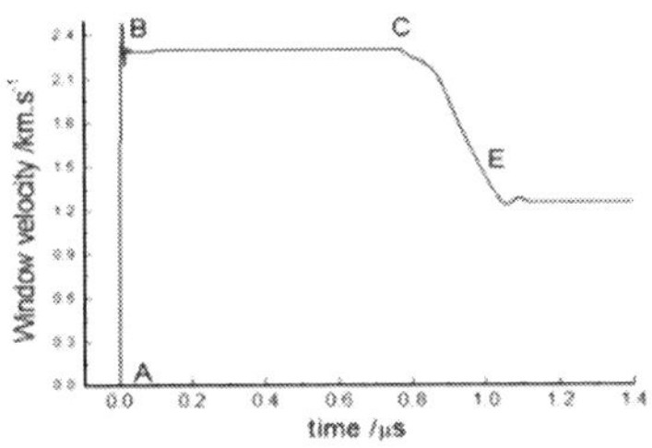

Figure 6. Simulated Ta/LiF interface velocity profile of the reverse impact experiment.

EOS and an elastic-perfectly plastic strength model with constant yield strength. The simulation is performed using a 1-D hydrodynamics code and the spatial resolution is 10 μm. The peak pressure in Ta is 30GPa. Figure 6 shows the simulated interface velocity profile of this experiment.

The IIMM and BCAM are used to analyze the interface velocity profile shown in Fig. 6. The releasing sound speed vs. in-situ particle velocity of the two methods is given in Fig. 7. The subsequent yield strength and shear modulus calculated by Eq. (9) and Eq. (10) are given in Table 1.

As shown in Fig. 7, though the initial unloading path by the two methods is almost same, the bulk sound speeds are not in agreement with each other. Further, difference of yield strength and shear modulus between the IIMM and BCAM is 5.5% and 1.7% separate and the strength and shear modulus of BCAM are in better agreement with the input data, as shown in Table 1.

Another application is to the synthetic GDI loading-releasing experiment. The sketch map of the experiment is given in Fig. 8. A thirteen layers graded-density impactor is impacted against two thickness steel samples. Steel is also modeled with a Mie-Grüneisen EOS and a modified SCG strength model. Figure 9 shows the simulated steel/LiF velocity profiles of two samples. Table 2 shows

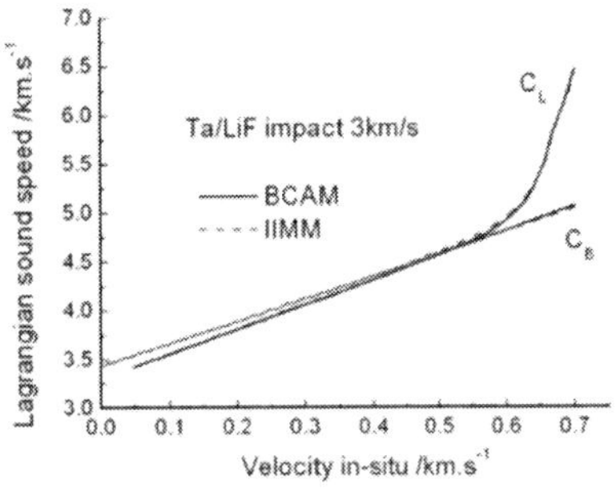

Figure 7. Releasing sound speed vs. in-situ particle velocity analyzed by the IIMM and BCAM for the synthetic profile in Fig. 6.

Table 1. Y & G analyzed by the IIMM and BCAM for the synthetic reverse impact experiment.

	Y/GPa	G/GPa
BCAM	1.44	113
IIMM	1.52	115
INPUT	1.45	113

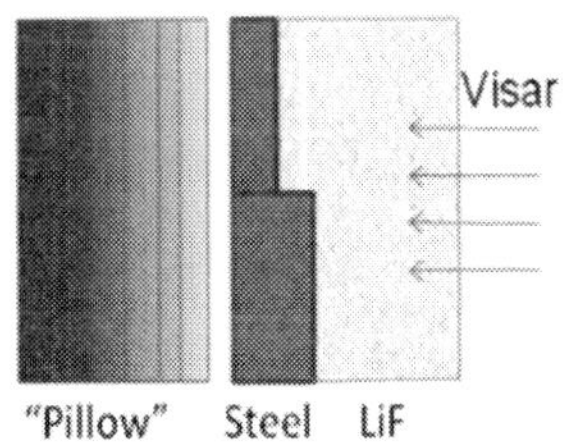

Figure 8. Sketch map of the GDI loading- releasing experiment.

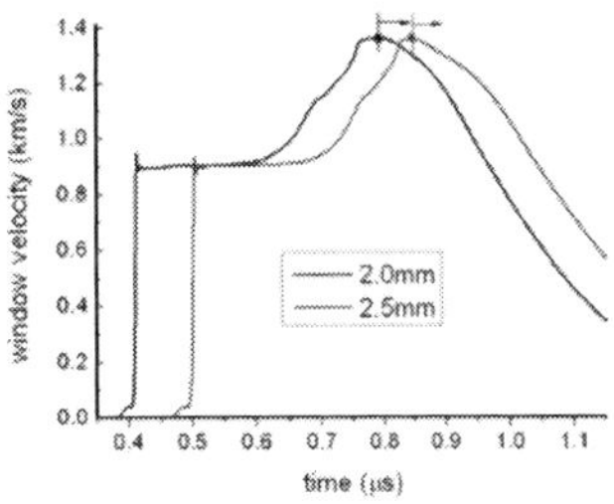

Figure 9. Simulated steel/LiF interface velocity profiles of the GDI experiment.

Table 2. Y & G analyzed by the IIMM and BCAM for the profiles in Fig. 9.

	Y/GPa	G/GPa
BCAM	1.68	149
IIMM	1.90	156
INPUT	1.73	145

the results of Y and G analysis using the IIMM and BCAM for the profiles shown in Fig. 9.

As shown in Table 2, as well as to the reverse impact experiment, compared to the IIMM, the BCAM yield strength and shear modulus is also in better agreement with the input data of the GDI experiment.

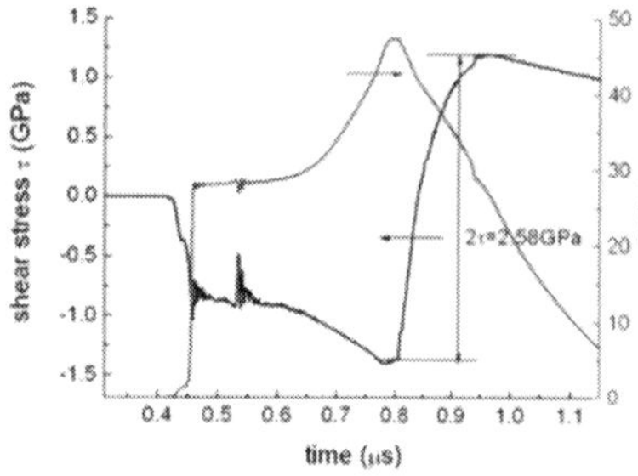

Figure 10. Simulated in-situ shear stress and pressure profiles at 2.0 mm into the steel which gives a mean shear stress value of 2.58 GPa.

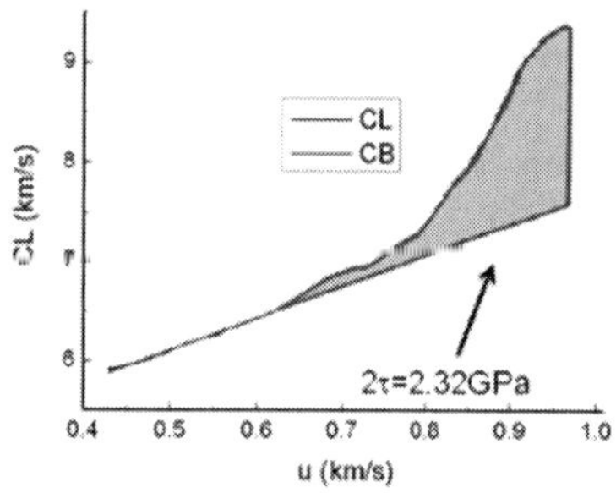

Figure 11. Results of the releasing sound speed vs. particle velocity analysis using BCAM for the velocity profiles simulated by the rate-dependent response. The analyzed mean shear stress is 2.32 GPa.

3.2. Rate-dependent response considerations

As we know, strength of metals commonly depends on the strain rate sensitively, but one assumption in the BCAM is that of a rate-independent response. In order to validate the use of the BCAM analysis for a rate-dependent material, the synthetic data analysis of GDI experiment is repeated with a rate-dependent response. So a viscoplastic deformation constitutive model is used in the simulation, which is [15]

$$\dot{\varepsilon} = \dot{\varepsilon}_e + \dot{\varepsilon}_p \quad (11)$$

$$\dot{\bar{\varepsilon}}_p = \frac{2\sqrt{2}}{9} A\left(\bar{\sigma} - Y_{SG}\right)^2. \quad (12)$$

Where ε_e is elastic strain, $\varepsilon_p$is plastic strain and $\bar{\sigma}$ is equivalent stress and A is constant, $A = 3 \times 10^7/\mathrm{GPa}^{-2}.\mathrm{s}^{-1}$ for steel [16].

The simulated in-situ shear stress and pressure profiles at 2.0 mm into the steel are given in Fig. 10, where τ is equivalent shear stress. In the context of a yield surface, 2τ is equivalent to the yield strength. The BCAM analysis of releasing sound speed vs. particle velocity for the velocity profiles simulated by the viscoplastic model is given in Fig. 11.

The BCAM shear modulus for the profiles simulated by the viscoplastic model is according with the input data. This suggest that the rate-dependent response does not have obviously influence on the shear modulus analysis. However, as shown in Fig. 9 and Fig. 10, the rate-dependent response will produce a larger error of ~10% in the BCAM mean shear stress or yield strength. That may be caused by the reflecting waves from the interface which will perturb the strain rate state of the sample, but we still think this error is acceptable in measuring dynamic strength of metals under high pressure.

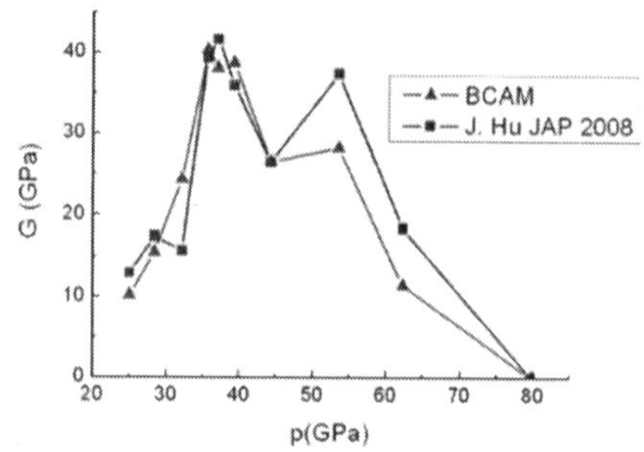

Figure 12. Shear modulus G vs. shock pressure p.

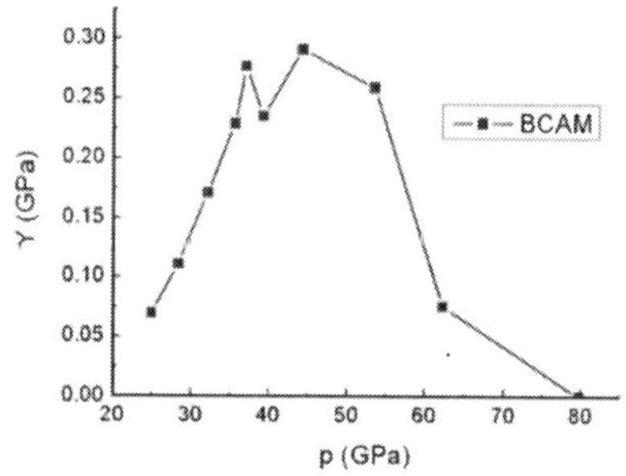

Figure 13. Yield strength Y vs. shock pressure p.

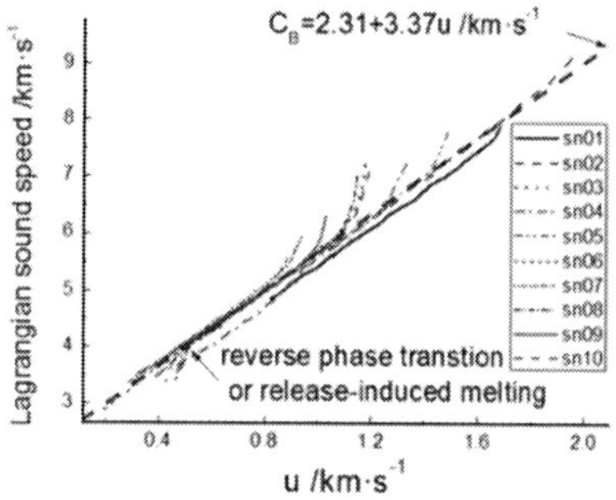

Figure 14. Releasing $C_L$vs. u of Sn at different shock pressure by BCAM.

3.3. Analysis of Sn

Sn is a kind of complex-structure metal with polymorphic transitions at high pressure and temperature. We use the BCAM to re-analyze the velocity profiles of plate impact experiments of Sn [17]. Figures 12 and 13 shows the analyses of G and Y at different high pressure.

As shown in Figs. 12 and 13, the same as the Ref. [17], above ~39GPa, shear modulus and yield strength gradually reduce to zero resulting from shock-induced melting. However, unlike to the Ref., results do not show an obvious BCT to BCC transition at ~34 GPa. Figure 14 shows the releasing $C_L(u)$ at different shock pressure analyzed by the BCAM.

As shown in Fig. 14, at $u \approx 0.5\,\text{km.s}^{-1}$ or $p \approx 12\,\text{GPa}$, results of Sn03 and Sn10 show a reverse phase transition or release-induced melting. Except Sn01 and Sn04, unloading paths of all experiments are very similar. Bulk sound speed is nearly linear with respect to particle velocity and the equation is $C_B = 2.31 + 3.37\text{u/km.s}^{-1}$.

4. Conclusions

We extend the BCAM to analyze the synthetic velocity profiles of the reverse impact and GDI loading-releasing experiments and extract dynamic strength of metals at high pressures. Synthetic analyses suggest that the BCAM can give more accurate results including sound speed-particle velocity and yield strength at high pressure than the IIMM. The synthetic GDI experiment simulated by a viscoplastic deformation constitutive model is used to validate the use of the BCAM analysis for a rate-dependent material. Results reveal that the rate-dependent response can produce an error of ∼10% in the analysis of the yield strength. We also use the BCAM to obtain shear modulus and yield strength of Sn at different shock pressure. The results do not support the shock-induced BCT to BCC transition.

References

[1] T. J. Vogler, T. Ao, J. R. Asay, Int. J. Plast. **25**, 671 (2009)

[2] E. B. Zaretsky, G. I. Kanel, J. Appl. Phys. **114**, 083511 (2013)

[3] J. L. Brown, C. S. Alexander, J. R. Asay, T. J. Vogler, D. H. Dolan, J. L. Belof, J. Appl. Phys. **115**, 043530 (2014)

[4] J. F. Barnes, P. J. Blewett, R. G. McQueen, K. A. Meyer, D. Venable, J. Appl. Phys. **45**, 727 (1974)

[5] Hye-Sook Park, K. T. Lorenz, R. M. Cavallo, S. M. Pollaine, S. T. Prisbrey, R. E. Rudd, R. C. Becker, J. V. Bernier, B. A. Remington, Phys. Rev. Lett. **104**,135504 (2010)

[6] J. R. Asay, J. Lipkin, J. Appl. Phys. **49**, 4242 (1978).

[7] D. B. Hayes, " Backward Integration of the Equations of Motion to Correct for Free Surface Perturbations", Sandia National Laboratories Report SAND2001-1440 (2001)

[8] J. Lipkin, J. R. Asay, J. Appl. Phys. **48**, 182 (1977)

[9] J. L. Brown, C. S. Alexander, J. R. Asay, T. J. Vogler, J. L. Ding, J. Appl. Phys. **114**, 223518 (2013)

[10] S. D. Rothman, J. Maw, J. Phys. **134**, 745 (2006)

[11] S. D. Rothman, J. P. Davis, J. Maw, C. M. Robinson, K. Parker, J. Palmer, J. Phys. D **38**, 733 (2005)

[12] L. P. Martin, J. R. Patterson, D. Orlikowski, J. H. Nguyen, J. Appl. Phys. **102**, 023507 (2007)

[13] J. P. Davis, "User Manual for CHARICE 1.0:", Sandia National Laboratories Report SAND2007-4984 (2007)

[14] D. J. Steinberg, S. G. Cochran, M. W. Guinan, J. Appl. Phys. **51**, 1498 (1980)

[15] J. L. Ding, J. R. Asay, T. Ao, J. Appl. Phys. **107**, 083508 (2010)

[16] J. W. Swegle, D. E. Grady, J. Appl. Phys. **58**, 692 (1985)

[17] J. Hu, X. Zhou, C. Dai, H. Tan, J. Li, J. Appl. Phys. **104**, 083520 (2008)

EPJ Web of Conferences **94**, 01008 (2015)
DOI: 10.1051/epjconf/20159401008

Experimental study on dynamic splitting of recycled concrete using SHPB

Yubin Lu[1,a], Shuisheng Yu[2], and Yong Cai[1]

[1] Ministry of Education Key Laboratory of Technology for Manufacturing Process, Southwest University of Science and Technology, 621010 Mianyang, Sichuan, China
[2] College of Resources and Civil Engineering, Northeastern University, Shenyang 110819, PR China

Abstract. To study the recycled concrete splitting tensile properties and fracture state with various recycled coarse aggregate replacement percentage (i.e. 0%, 25%, 50%, 75% and 100%), the dynamic splitting test of recycled concrete was carried out using large diameter (75 mm) split Hopkinson pressure bar (SHPB). The results show that the recycled concrete splitting tensile strength increases with the increase of loading rate, and the loading rate also affects the recycled concrete fracture state, which indicates that the recycled concrete has obvious rate sensitivity. The damage state of the recycled concrete is not only the destruction of the interface between coarse aggregate and cement mortar, but also associates with the fracture damage of aggregates. Under the same water cement ratio, when the replacement percentage of coarse aggregates is around 50%–75%, the gradation of natural and recycled coarse aggregate is optimal, and thus the splitting tensile strength is the largest. This study offers theoretical basis for the engineering applications of recycled concrete.

1. Introduction

Recycled concrete is a new type of concrete that made from the disuse concrete after a series of courses of disposal, such as breaking up, washing, filtration, classification, combination on a certain proportion and so on [1]. Recycled concrete can realize the effective recycling of waste concrete, for protecting environment, saving resources, developing ecological architecture are of great significance, which is often considered as one of the main measures of developing green ecological concrete [2]. So far, in the tensile test of concrete studies, scholars [3–14] studied the tensile properties of concrete under quasi static or dynamic loading from different perspectives, and many different results were obtained. The results related with the concrete strength grade, water cement ratio, aggregate type, particle size distribution, testing methods and so on. But the research of dynamic splitting tensile properties of recycled concrete is less. Based on this, the dynamic splitting tensile test was carried out using 75 mm diameter split Hopkinson pressure bar (SHPB) in this study, researching the recycled concrete dynamic splitting mechanics performance and fracturing situation, discussing the influence of recycled coarse aggregate replacement percentage to recycled concrete splitting strength, accumulating the data of dynamic splitting tensile properties of recycled concrete, and providing reference for the engineering application of recycled concrete.

2. SHPB test principles, test materials and combination ratio

2.1. Test principles

The dynamic splitting tests were carried out using the 75 mm diameter split Hopkinson pressure bar with the Brazilian disc tensile test method, which is also known as the splitting test and is widely used in tensile strength tests. The splitting test is adopted as the standard method for determination of the tensile strength of concrete by many international concrete specifications [e.g. 14–17]. The test schematic is shown in Fig. 1.

Based on the analysis of dynamic splitting tensile experimental data, Li et al. [18] pointed out that the calculation formula with modifying the nonuniform effect should be adopted. The time starting point is when the left end of specimen is loaded, the transmitted waveform is translated to the left with the time of t_0 and $\varepsilon_t^*(t)$ is obtained, namely $\varepsilon_t^*(t) = \varepsilon_t(t + t_0)$, which can eliminate the influence of the time nonuniformity on the test results. At the same time, due to using the three measured waves to calculate, the influence of the spatial inhomogeneity can also be eliminated. So the loading force on the specimen ends in the splitting test is,

$$P(t) = \frac{\pi E R^2}{2}\left(\varepsilon_i(t) + \varepsilon_r(t) + \varepsilon_t^*(t)\right) \tag{1}$$

where E is the elastic modulus of SHPB bars, R is the radius of SHPB bars, $\varepsilon_i(t)$ and $\varepsilon_r(t)$ is the incident wave and reflected wave, respectively. So the maximum value of $P(t)$, $P_{\max}$, can be got, and then the dynamic splitting

[a] Corresponding author: yubinluzju@hotmail.com

Article available at http://www.epj-conferences.org or http://dx.doi.org/10.1051/epjconf/20159401008

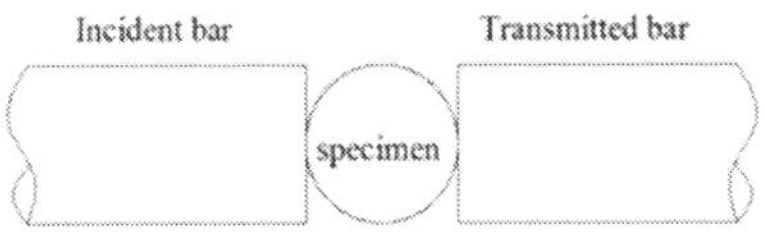

Figure 1. Sketch of dynamic splitting test.

Table 1. The combination ratio of recycled concrete in test.

Specimen type	PC-0	RC-25	RC-50	RC-75	RC-100
Cement	1	1	1	1	1
Water	0.477	0.477	0.477	0.477	0.477
Sand	1.465	1.465	1.465	1.465	1.465
Natural coarse aggregate	3.418	2.564	1.709	0.854	0
Recycled coarse aggregate	0	0.854	1.709	2.564	3.418
Replacement percent-age/%	0	25	50	75	100

tensile strength, f_{td}, can be calculated,

$$f_{td} = \frac{2P_{\max}}{\pi LD} \quad (2)$$

in which L and D is the thickness and diameter of specimens, respectively.

Additionally, the loading rate in the specimen can be estimated by f_{td}/τ, where τ is the time lag between the start of the transmitted stress wave and the maximum transmitted stress wave.

2.2. Test materials and combination ratio

Raw materials used for recycled concrete specimens are (1)recycled coarse aggregates taken from waste concrete of a Mianyang road, the design strength of this waste concrete is unclear, and the waste concrete is broken by a jaw crusher, graded and cleaned to recycled coarse aggregates of diameters from 10 to 20 mm; (2)natural coarse aggregates are natural cobbles of diameters from 10 to 20 mm; (3)ordinary Portland cement of 42.5 grade is used; (4)fine aggregates are natural river sand of continuous grading with fineness modulus of 2.28; and (5)Mianyang tap water is used.

Concrete mix proportions largely influence the mechanical properties of concrete. To ensure the flowability during concreting and the strength after molding of recycled concrete, the mix proportions of recycled concrete specimens used in this study are cement: water: fine aggregate: coarse aggregate = 1:0.477:1.467:3.422, where coarse aggregates include natural coarse aggregates and recycled coarse aggregates. According to the difference of replacement ratios of recycled coarse aggregates, there are five different types of specimens, as shown in Table 1. PC-0 stands normal concrete specimens, and RC-25, RC-50, RC-75 and RC-100 stand recycled concrete specimens with recycled coarse aggregate replacement ratio of 25%, 50%, 75% and 100%,

Table 2. The results of quasi-static splitting tests.

Replacement percentage	Number	Splitting strength /MPa	Average of splitting strength /MPa
0	A-1	3.27	3.07
	A-2	3.04	
	A-3	2.89	
25%	B-1	2.81	2.68
	B-2	2.48	
	B-3	2.74	
50%	C-1	3.34	3.24
	C-2	3.38	
	C-3	3.00	
75%	D-1	3.17	3.06
	D-2	2.98	
	D-3	3.05	
100%	E-1	2.39	2.31
	E-2	2.12	
	E-3	2.41	

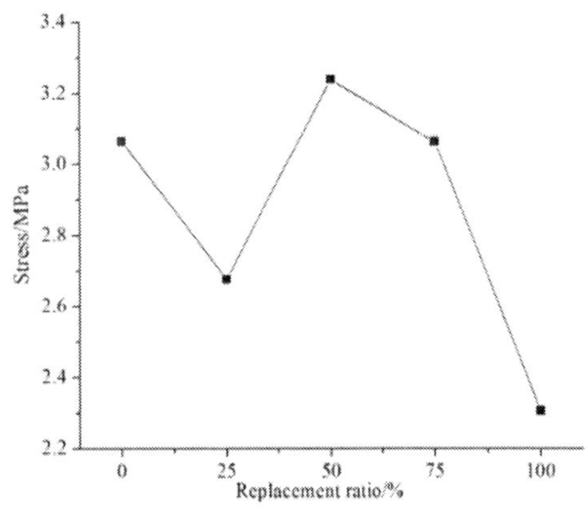

Figure 2. Relation of average quasi-static splitting strength of recycled concrete and replacement ratio of coarse aggregates.

respectively. The diameter and thickness of cylindrical specimens is 74 and 37 mm, respectively. Specimens with the same dimensions are used in the quasi-static splitting tests.

3. The test results and discussion

In the dynamic splitting tests, to ensure that uniform stress state can be achieved as early as possible in the reclycled concrete samples, the amplitude of pulse oscillation can be reduced, and correct pulse shape is obtained for approximate constant strain-rate loading, soft rubber (bicycle inner tube) with the diameter of 20 mm and the thickness of 1.5 mm is employed as pulse shapers through multiple trial tests.

3.1. The test results

The quasi-static splitting tests were carried out with the MTS-815 Flex Test GT rock mechanics test system, and the test results are summarized in Table 2.

The quasi-static splitting test results are shown in Fig. 2. From Fig. 2, it is observed that, in the quasi-static

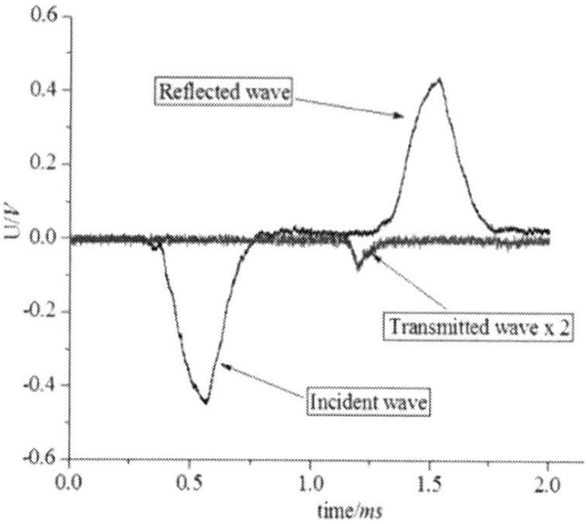

Figure 3. Typical waveforms of dynamic splitting tests.

splitting tests, the relationship between the splitting tensile strength and the recycled coarse aggregate replacement ratio has no obvious regularity. When the replacement percentage is 75%, the splitting tensile strength of recycled concrete approaches that of normal concrete, while the replacement percentage is 25% and 100%, the splitting tensile strength of recycled concrete is obviously lower than that of normal concrete with 12.7% and 24.8%, respectively, and when the replacement percentage is 50%, the tensile splitting strength is 5.7% higher than that of normal concrete. These results does not agree with those of Yang et al. [19], who thought that the recycled concrete tensile strength is decreased with the increase of recycled coarse aggregate replacement percentage. However, these results are consistent with the quasi static compressive test results of Xiao et al. [20], who found that when the recycled coarse aggregate replacement percentage was 50%, the compressive strength of recycled concrete was higher than that of normal concrete. They suggested that the reason may be the gradation of recycled coarse aggregate and natural aggregate being best under this situation, or may be due to the randomicity of the test results.

The typical waveform of recycled concrete specimens in SHPB tests is shown in Fig. 3. It is seen from Fig. 3 that the transmission wave is far less than the incident wave, which is because that the contact face between specimen and SHPB bars is tangent, the contact area is small, the density and wave velocity of the recycled concrete specimen are smaller than those of the SHPB bar material, making its wave impedance less compared with that of SHPB bars, and leading to the signal on the transmission bar being smaller [21]. The dynamic splitting tensile strength was obtained at different impact air pressures (i.e. 0.4, 0.5 and 0.6 MPa) according to the transmission wave signals of recycled concrete specimens, and the experimental data are summarized in Table 3.

To describe the enhancement of dynamic splitting tensile strength of recycled concrete specimens, the dynamic splitting tensile strength increase factor is defined as,

$$DIF = f_{td}/f_{ts} \quad (3)$$

Table 3. The results of dynamic splitting tests.

Replacement percentage	Air pressure/ MPa	Number	Splitting strength /MPa	Average of splitting strength /MPa
0%	0.4	A-1	3.10	3.19
		A-2	3.28	
	0.5	A-5	5.90	7.56
		A-6	9.22	
	0.6	A-8	9.50	10.26
		A-9	11.02	
25%	0.4	B-1	3.15	3.03
		B-2	2.91	
	0.5	B-4	7.76	7.47
		B-5	5.23	
		B-6	9.41	
	0.6	B-7	11.60	11.47
		B-8	11.33	
50%	0.4	C-1	3.17	3.61
		C-2	4.04	
	0.5	C-4	8.30	9.09
		C-5	8.00	
		C-6	10.98	
	0.6	C-7	14.72	14.17
		C-8	12.96	
		C-9	14.82	
75%	0.4	D-1	4.79	4.70
		D-2	3.98	
		D-3	5.32	
	0.5	D-4	9.68	9.13
		D-5	9.66	
		D-6	8.05	
	0.6	D-7	11.15	10.28
		D-9	9.40	
100%	0.4	E-1	2.99	3.46
		E-2	3.93	
	0.5	E-4	6.91	7.69
		E-5	10.40	
		E-6	5.75	
	0.6	E-7	10.80	10.45
		E-8	9.13	
		E-9	11.41	

where f_{ts} is the quasi-static splitting tensile strength of recycled concrete.

The relationship of the recycled concrete dynamic splitting tensile strength increase factor (DIF) and impact air pressure or loading rate is shown in Fig. 4. It is found that the DIF increases with the increase of impact air pressure or loading rate, which indicates that the dynamic splitting tensile strength of recycled concrete has obvious rate sensitivity, viz. the DIF largely increases with the increase of loading rate. Lu et al. [22] explained this kind of phenomenon, who believed that, in the macroscale level, the dynamic tensile strength enhancement obtained from dynamic tensile tests on concrete-like materials is a kind of material intrinsic property rather than caused by the structure effect. A micromechanics model was proposed by Lu et al. [22] based on the dynamic fracture mechanics from the analysis results, and the phenomenon was pointed

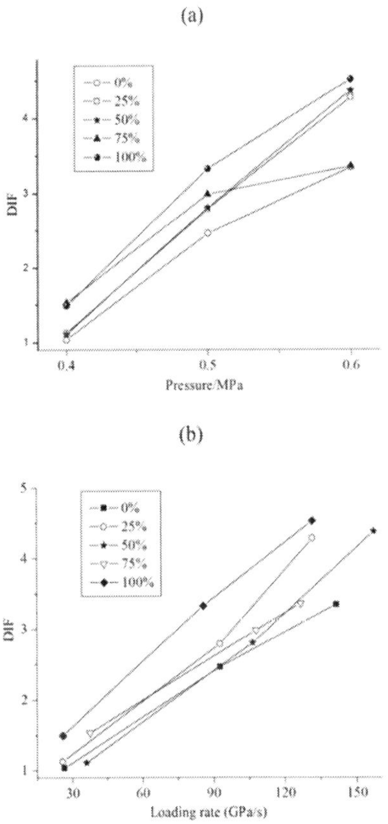

Figure 4. Relation of DIF with (a) impact gas pressure and (b) loading rate.

out to be largely caused by the inertia of micro cracks. Among the five groups of tests, when the replacement percentage is 0%, the DIF is lower than that under other replacement percentages, which reveals that strain-rate sensitivity of the dynamic splitting tensile strength of normal concrete is relatively low.

As shown in Fig. 5, the dynamic splitting failure pattern of recycled concrete specimens is mainly breaking into two parts symmetrically along the loading axis, meeting the test validity condition of conventional Brazilian disc tests [23]. When the impact air pressure is small, namely the impact velocity is small, and the dynamic splitting fracture form of recycled concrete specimens is appearing crushed zones along the radial loading end surfaces and splitting into two parts with large blocks. With the increase of loading velocity, the pressure along the radial direction increases, the crushing area also increases, the critical crack initial length of specimen decreases and the range of crack angles that can be extended increases. Thus, the number of extending cracks increases, the size of fragments decreases, the broken degree of specimen increases, and the specimen local failure gradually becomes obvious. This reflects the rate sensitivity of recycled concrete on its splitting tensile failure mode, which indicates the strain-rate effect of recycled concrete from another view.

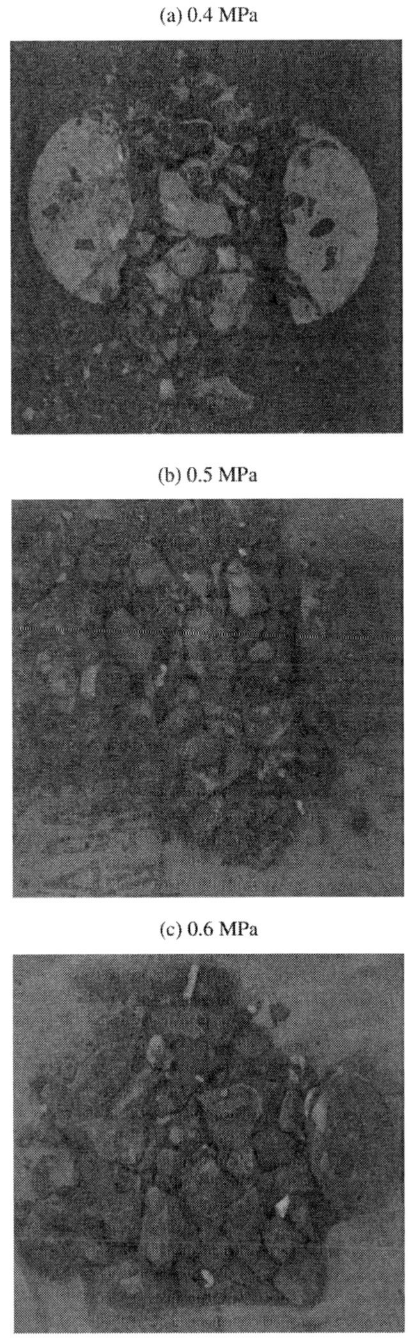

Figure 5. The destruction state of recycled concrete under different impact air pressures.

3.2. The relationship between recycled concrete dynamic splitting tensile strength and recycled coarse aggregate replacement percentage

The relationship between the average dynamic splitting tensile strength of recycled concrete under different impact air pressures and recycled coarse aggregate replacement percentage is shown in Fig. 6.

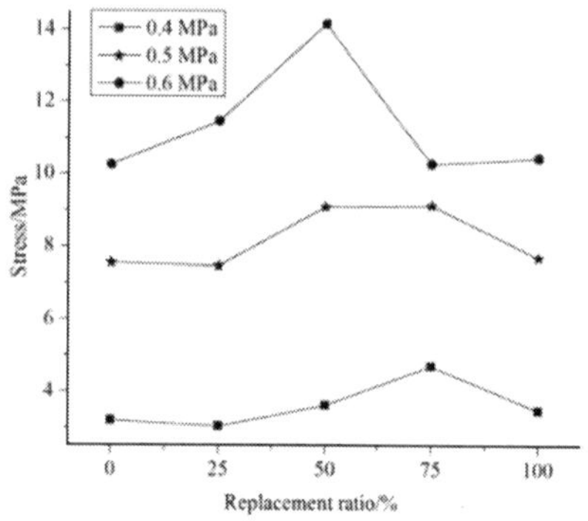

Figure 6. Relation between the average dynamic splitting tensile strength of recycled concrete and replacement ratio of coarse aggregates under different gas pressures.

It shows that the dynamic splitting tensile strength under different recycled coarse aggregate replacement percentages is basically greater than that of normal concrete. Under the impact air pressure of 0.4 MPa, the average dynamic splitting tensile strength of recycled concrete increases with the range from −5% to 47.3% compared with that of normal concrete; when the replacement percentage is 75%, the increase is maximum with 47.3%; and it is 25%, the average dynamic splitting tensile strength decreases by 5%. Under the impact air pressure of 0.5 MPa, when the replacement percentage is 75%, the increase is largest with 20.8%; the strength of RAC with the replacement percentage of 50% is basically equal with that of 75%; and it is 25%, the strength decreases by 1.2%. Under the impact air pressure of 0.6 MPa, when the replacement percentage is 50%, the increase is also largest with 38.1%; and it is 75%, the strength of RAC is basically equal with that of normal concrete. Accordingly, under 0.4, 0.5 and 0.6 MPa, when the recycled coarse aggregate replacement percentage are 50% and 75%, the dynamic tensile splitting strength of recycled concrete is the largest, which is consistent with the quasi-static splitting test results. Therefore, experimental results in this study indicate that when the replacement percentage of recycled coarse aggregate is around 50%–75%, its gradation is best with natural coarse aggregate.

From the increase value of recycled concrete dynamic tensile splitting strength under different air pressure, it can be seen that with the increase of recycled coarse aggregate replacement percentage, the dynamic splitting tensile strength appears the phenomenon of first increase and then decrease. The reasons for this phenomenon can be approximately explained by the following points. (1) In dynamic splitting tests, under the same recycled coarse aggregate replacement percentage, because of the micro crack inertia effect the aggregate crack extension requires higher loading to make aggregate fracture (as shown in Fig. 7), leading concrete specimen damage, thus the recycled concrete dynamic splitting tensile strength increases. (2) Due to the production of new micro cracks and fracture on the recycled coarse aggregate, the mechanical properties of recycled coarse aggregates

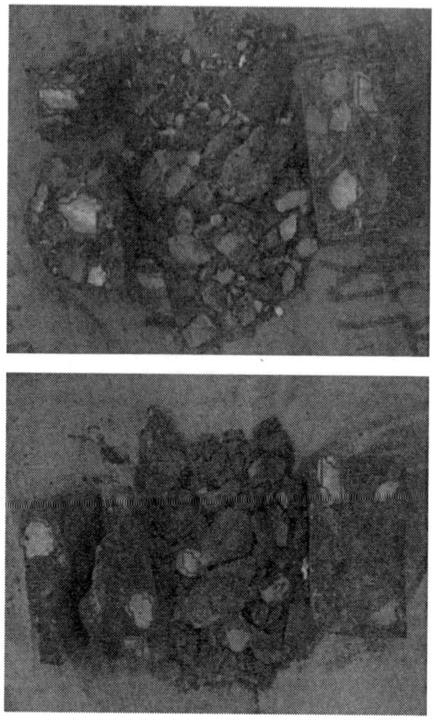

Figure 7. Fracture state of coarse aggregates.

are lowered. On the other hand, because recycled coarse aggregates are wrapped by cement mortar, the hydrophilic of recycled coarse aggregates is relatively strong, and many micro cracks on the recycled coarse aggregate surfaces take in new cement particles, making the hydration of contact areas between the recycled coarse aggregate and cement mortar more completely, forming dense interface structures, and thus improving the chemical bonding strength of interfaces. In a word, the mechanical properties of recycled concrete are influenced by recycled coarse aggregate and the chemical bonding strength of interfaces. (3) When the recycled coarse aggregate replacement percentage is around 50%–75%, the gradation between the recycled and natural coarse aggregates is best, so the dynamic splitting tensile strength is relatively lager. With the increase of replacement percentage, the influence of recycled coarse aggregates is greater than the chemical bonding strength of interfaces. Since the mechanical properties of recycled coarse aggregates is reduced, the dynamic splitting tensile strength of recycled concrete decreases.

4. Conclusions

In this study, the dynamic splitting test of recycled concrete (recycled coarse aggregate replacement percentage is 0%, 25%, 50%, 25% and 100%) was carried out using large diameter (75mm) SHPB, meanwhile the quasi-static splitting test was carried out, the following conclusions are drawn.

(1) In dynamic splitting tests, when the recycled coarse aggregate replacement percentage is around 50%–75%, the

dynamic splitting tensile strength of recycled concrete is largest, this is consistent with the results in quasi-static splitting tests, which further validates that at this situation the gradation formed between recycled and natural coarse aggregates is best.

(2) Because of the micro crack inertia effect, the aggregate crack extension requires higher loading to make aggregate fracture, leading concrete specimen damage, thus the dynamic splitting tensile strength of recycled concrete increases. The strain-rate sensitivity of recycled concrete is obvious, and the fragment degree of recycled concrete is influenced by the strain-rate.

(3) Due to the appearance of new micro cracks and fracture on the recycled coarse aggregate, the recycled coarse aggregate mechanical properties decrease. As recycled coarse aggregates are wrapped by cement mortar, the hydrophilic is relatively strong, and many micro cracks on the recycled coarse aggregate surfaces intake new cement particles, making the hydration of contact areas between recycled coarse aggregates and cement mortar more prefect, forming dense interface structures, and thus improving the chemical bonding strength of interfaces. In a word, the mechanical properties of recycled concrete are influenced by both coarse aggregates and the chemical bonding strength of interfaces.

The authors would like to acknowledge the sponsor by the National Natural Science Foundation of China (51308480).

References

[1] J. Z. Xiao, J. B. Li, Y. Lan. Concrete, **25**, 17–20 (2003) (in Chinese)

[2] Z. X. Xing, Y. N. Zhou. J. North China Institute Water Conservancy Hydroelectric Power, **19**:30–32 (1998) (in Chinese)

[3] J. W. Tedsco, C. A. Ross, S. T. Kuennen. ACI Mater. J., **90**, 162–169 (1993)

[4] B. L. Atchley, H. L. Furr. ACI Mater. J., **64**, 745–756 (1967)

[5] M. L. Hughes, J. W. Tedesco, C. A. Ross. Comput. Struct., **47**, 653–671 (1993)

[6] D. E. Lambert, C. A. Ross. Int J Impact Eng., **21**(24): 985–998 (2000).

[7] J. T. Gomez, A. Shukla, A. Sharma. Theo. Appl. Fract. Mech., **25**, 37–49 (2001)

[8] C. J. Jiao, G. P. Jiang, L. Gao. Acta Armament., **31**, 469–472 (2010) (in Chinese)

[9] J. Hu, X. T. Wu. Explosive Shock Waves, **31**, 402–406 (2011) (in Chinese)

[10] W. Sun, C. J. Jiao. J. Guangzhou Uni., **10**, 42–47 (2011) (in Chinese)

[11] A. Behnood, M. Ghandehari. Fire Safe. J., **44**, 1015–1022 (2009)

[12] Y. Choi, R. L. Yuan. Cem. Concr. Res., **35**, 1587–1591 (2005)

[13] C. Parra, M. Valcuende, F. Gomez. Construct. Build. Mater., **25**, 201–207 (2011)

[14] H. Tanyildizi, A. Coskun. Construct. Build. Mater., **22**, 2269–2275 (2008)

[15] American Society for Testing and Materials (ASTM). *Standard test method for spitting tensile strength of cylindrical of foam concrete.* ASTM C496-90, 4:266-269 (1991)

[16] British Standards Institution. BS1881:Part117: *Testing concrete method for the determination of tensile strength.* London: British Standards Institution (1983)

[17] ISO4108 (E) Concrete: *Determination of tensile splitting strength of test specimens.* International Standard, 1–3 (1980)

[18] W. Li, H. P. Xie, Q. Z. Wang. Explosive Shock Waves, **26**, 12–20 (2006) (in Chinese)

[19] F. Yang, A. J. Chen, J. Wang, X. P. Sun. Concrete, **12**, 11–14 (2012) (in Chinese)

[20] J. Z. Xiao, J. B. Li, Z. P. Sun, X. M. Hao. J. Tongji. Uni., **32**, 1558–1561 (2004) (in Chinese)

[21] Q. Ping, Q. Y. Ma, J. S. Zhang, P Yuan, B. S. Chen. Chin J Rock Mech. Eng., **31**, 3363–3369 (2012) (in Chinese)

[22] Y. B. Lu, H. J. Wu, L. M. Zhao. Explosive Shock Waves, **33**, 275–281 (2013) (in Chinese)

[23] F. Q. Gong,X. B. Li, J. Zhao. Chin J Rock Mech. Eng., **29**, 881–891 (2010) (in Chinese)

EPJ Web of Conferences **94**, 01009 (2015)
DOI: 10.1051/epjconf/20159401009

Damage characterization for particles filled semi-crystalline polymer

Franck Lauro[1,a], Romain Balieu[2], Bruno Bennani[1], Gregory Haugou[1], Benjamin Bourel[1], Fahmi Chaari[1], Tsukatada Matsumoto[3], and Ernesto Mottola[3]

[1] University of Valenciennes and Hainaut Cambrésis LAMIH, UMR CNRS 8201, 59313 Valenciennes, France
[2] KTH Royal Institute of Technology, Division of Highway and Railway Engineering, 10044 Stockholm, Sweden
[3] TOYOTA MOTOR EUROPE, 1140 Brussels, Belgium

Abstract. Damage evolution and characterization in semi-crystalline polymer filled with particles under various loadings is still a challenge. A specific damage characterization method using Digital Image Correlation is proposed for a wide range of strain rates considering tensile tests with hydraulic jacks as well as Hopkinson's bars. This damage measurement is obtained by using and adapting the SEE method [1] which was developed to characterize the behaviour laws at constant strain rates of polymeric materials in dynamic. To validate the characterization process, various damage measurement techniques are used under quasi-static conditions before to apply the procedure in dynamic. So, the well-known damage characterization by loss of stiffness technique under quasi-static loading is applied to a polypropylene. In addition, an in-situ tensile test, carried out in a microtomograph, is used to observe the cavitation phenomenon in real time. A good correlation is obtained between all these techniques and consequently the proposed technique is supposed suitable for measuring the ductile damage observed in semi-crystalline polymers under dynamic loading. By applying it to the semi-crystalline polymer at moderate and high speed loadings, the damage evolution is measured and it is observed that the damage evolution is not strain rate dependent but the failure strain on the contrary is strain rate dependent.

1. Introduction

Damage characterization in semi-crystalline polymers is still a real challenge. Due to the deformation process, an early necking occurs and leads to a non-homogeneous strain and strain rate field into tensile specimen. This inhomogeneity state is then a problem to determine constant strain rate behaviour law to be used in finite element simulations and consequently to obtain damage evolution as well. Mostly damage studies based on volume variation have been conducted under static conditions and by controlling the in real time the tensile displacement in order to obtain a constant strain rate. A video controlled-system is generally used to measure the volume variation with seven spots painted on the tensile specimen [1,2]. It is also possible to use inverse technique by identifying the parameters of a damage model by comparing and correlating same measurements obtained from experiments and finite element simulations and also taking into account the initial stress triaxiality ratio state [3–5]. Nevertheless, if you want to obtain the damage evolution without numerical simulations and for dynamic loadings, all these methods are not adapted for this particular condition. In consequence, another approach is needed. In this paper, the $SE\dot{E}$ method [6], which was first developed for the polymer behaviour identification on a large strain rate range, is extended to the damage measurement on a large strain rate range. This method is applied to a polypropylene filled with 20% wt of mainly talc particles called Hostacom by LyondellBasell, for various strain rates and on small volumes all along the tensile test specimen. To validate this approach, the method is first applied under quasi static loadings and compared with the loss of stiffness technique. Tensile tests inside a microtomograph are also performed to complete the validation as it is a very powerful technique to visualize the damage inside the specimen [7,8]. As good results are obtained by comparing these various techniques, the $SE\dot{E}$ method for damage measurement is then applied to dynamic tensile tests carried out on a hydraulic jack and Hopkinson bars'. The results obtained for this semi-crystalline polymer filled with mineral particles highlight the independence of the damage evolution to the strain rates.

2. The loss of stiffness technique versus the SEE method for damage

The damage evolution is given by the loss of stiffness technique by measuring the evolution of the Young modulus during a cyclic tensile loading [9] as illustrated Fig. 1 and by calculating the damage variable D by:

$$D = 1 - \frac{\tilde{E}}{E} \tag{1}$$

with E the initial Young modulus and $\tilde{E}$ the damaged Young modulus.

The corresponding behaviour obtained for the semi-crystalline polymer is presented Fig. 2. This result is obtained with a notched specimen loaded with a

[a] Corresponding author: franck.lauro@univ-valenciennes.fr

Article available at http://www.epj-conferences.org or http://dx.doi.org/10.1051/epjconf/20159401009

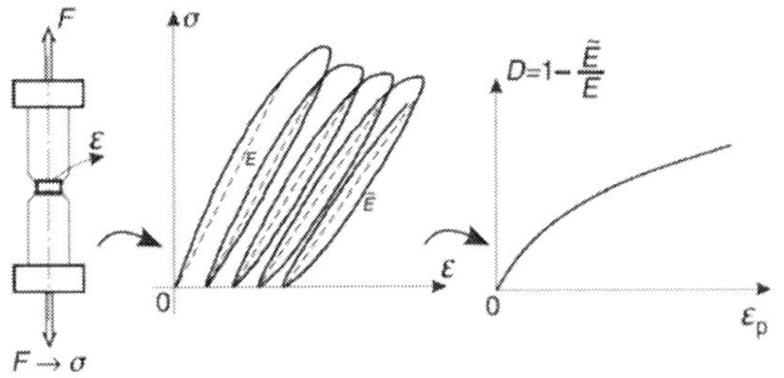

Figure 1. Procedure for damage measurement by loss of stiffness technique: Step1: Longitudinal strain measurement on lighted notched tensile specimen, Step2: measurement of an equivalent Young modulus for each hysteresis, Step 3: calculation of the damage evolution.

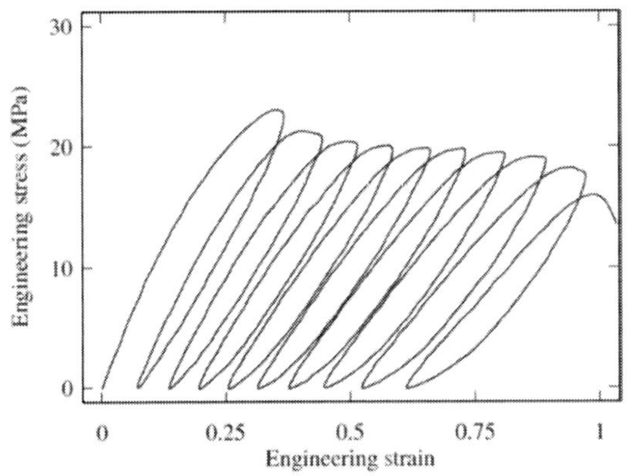

Figure 2. Engineering stress-strain curve obtained by the tensile test with repeated unloadings.

prescribed displacement of 1mm/min for both loadings and unloadings. Viscoelastic behaviour is observed with some hysteresis in the loading-unloading phases. In the classical loss of stiffness technique, the Young modulus evolution is captured by measuring with the unloading slope. Due to the non-linearity of the results for this material, the damage is then calculated with an apparent Young modulus obtained by taking the upper point, at the crossing between loading and unloading curves, and the lower point, at zero stress.

For the $SE\dot{E}$method, the damage evolution is obtained at constant strain rate by comparing the behaviour laws under the assumption of incompressibility and transverse isotropy. The $SE\dot{E}$ method used the digital image correlation to obtain the strains and strain rates in different Zone Of Interest (ZOI) on the tensile specimen during a tensile test. The true stress which takes the reduction of the real cross-sectional area into account in each ZOI, is therefore calculated assuming the compressibility (transverse isotropy) $\sigma_{yy_i}^{com}$ or the incompressibility $\sigma_{yy_i}^{inc}$ hypothesis. The damage evolution is therefore determined by the following relation:

$$D = 1 - \frac{\sigma_{yy_i}^{com}}{\sigma_{yy_i}^{inc}}. \quad (2)$$

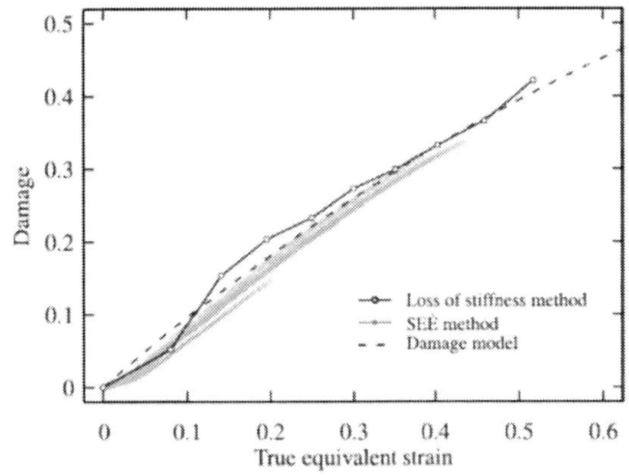

Figure 3. Damage evolution obtained with the loss of stiffness technique and the SEE method and the corresponding damage model.

With

$$\sigma_{yy_i}^{inc} = \frac{F_i}{S_{o_i}} \exp(\varepsilon_{yy_i}); \sigma_{yy_i}^{com} = \frac{F_i}{S_{o_i}} \exp(-2\varepsilon_{xx_i}) \quad (3)$$

where F_i, S_{o_i}, ε_{xx_i} and ε_{yy_i} are the force, the initial cross section, the transversal and longitudinal strains of the ZOI subscribed i. A damage model (4) is also identified based on the work of Balieu [10] and presented in Fig. 3 with both the damage from the loss of stiffness technique and the $SE\dot{E}$ method.

$$D = 1 - \exp\left(\frac{k}{k_c}\right) \quad (4)$$

with k the equivalent plastic strain and k_c a material parameter.

Results obtained by the both techniques are close and validate the approach for measuring the damage evolution by the $SE\dot{E}$ method. To conclude this static part of the damage measurement, an in situ tensile test on a small specimen is performed inside a microtromograph (μcT) at three different deformations to observe and quantify the porosity. μCT imaging is performed thanks to a 1172 Skyscan system, with 40 kV and 100 mA settings. The voxel size for all μCT data is adjusted to 3.97 mm. This choice ensures capturing small voids size (around 8 mm) with a reasonable imaging duration. Under these conditions, the acquisition of the sample geometry in each mechanical state takes about 8 h. A first μCT-scan is carried out at the initial state (i.e. undamaged material) and two other μCT-scans are performed to follow the evolution of the specimen microstructure at different deformed states. The reconstruction of tomography data leads to a set of grayscale images (Fig. 4) obtained along the principal direction of the sample.

In order to correctly estimate the damage, an accurate measurement of the cross section of the specimen is necessary. The first use of μCT is to analyse the outer shape of the specimen and how it changed during deformation. The edges of the cross section are detected and the sum of all the pixels leads to the effective cross sectional area S. The real resisting cross sectional area

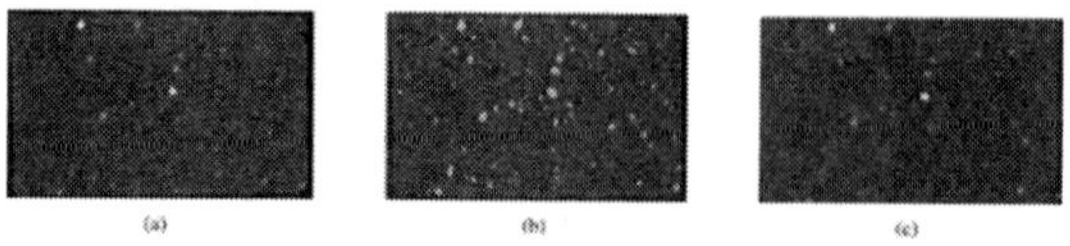

Figure 4. Initial cross section visualization, a) Grayscale image, b) particles distribution in green, c) voids distribution in red.

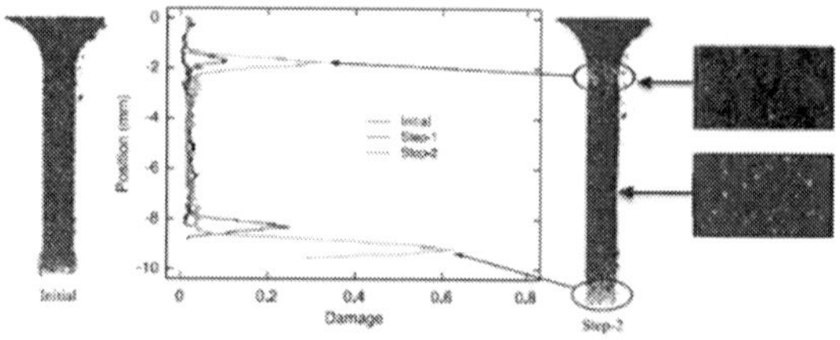

Figure 5. Evolution of the damage measured by μCT.

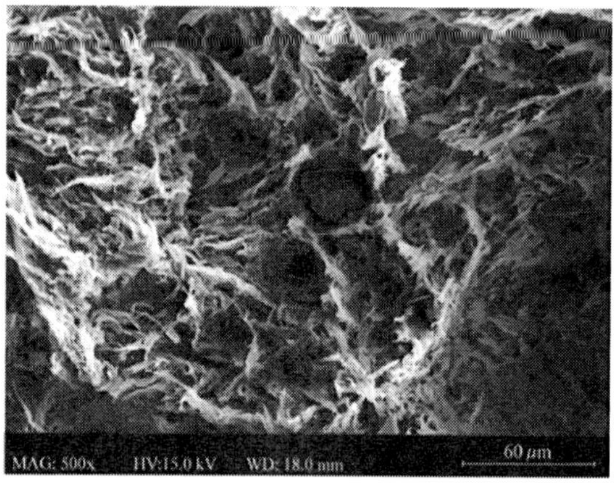

Figure 6. High magnification SEM micrograph of the fracture surface of particles filled semi-crystalline polymer under tensile loading at 1 mm/min.

can also be computed by retrieving the void surface (S_D) from the effective cross section area. The effective cross sectional area along the length gauge of the specimen is presented in Fig. 5 for the initial and two deformed states. In this test, the deformation occurs mainly at the extremities of the length gauge. At the bottom of the specimen, where the displacement is applied, effective cross-sectional area decreases strongly.

The tensile test in-situ μCT is an interesting technical solution to visualize the evolution of the voids during deformation. Unfortunately, the localization of the deformations in two very small parts of the specimen makes the deformation measurement difficult. The calculated damage value is therefore, in this case, not accurately correlated to the deformation. Nevertheless, the three techniques converge to an equivalent measurement of the damage evolution with important volume of porosities at the last stage before as highlighted in the image of the fracture surface in Fig. 6.

After validation of the damage measurement by the $SE\dot{E}$ method in static, this method is then applied on dynamic tensile tests at various speed loadings to evaluate the influence of the strain rate on the damage evolution.

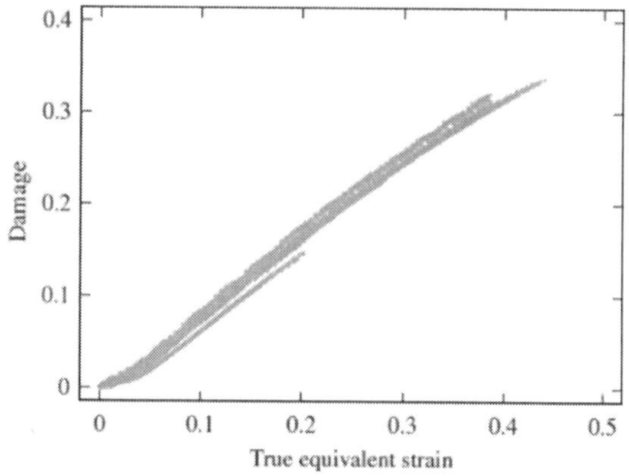

Figure 7. Damage evolution obtained with the SEE method from low to high strain rates.

3. Application to dynamic tensile tests

A high speed hydraulic machine (Instron 65 kN-20 m/s) is used for the dynamic loadings. A piezoelectric load cell-calibrated in the range [0;5] kN with a precision of 25 N is fixed in the upper part of the machine and a LVDT sensor is used to measure the displacement of the jack during the test. The lower part is composed of the clamping system used to load the specimen once the velocity is constant. The strain rate effects are studied at 0.08, 0.8 and 4 m/s which correspond to average equivalent strain rates of 2.6, 26 and 133 s^{-1}, respectively. Due to ringing phenomenon of the assembly (6 kHz), tensile tests are also performed on direct tensile Hopkinson bars over 200 s^{-1}. This equipment is composed of two aluminium bars made of 6060 aluminium alloy with 20 mm in diameter and a hollow striker made of Polyamide 66 with respect to impedance matching issues. The length of the input, output bar and the striker are respectively 5.63, 4 and 1 m. For these tests, two input pressures are used, 9 and 17 bars, which correspond in terms of velocities of the striker to 6.7 and 11.5 m/s. Two approximate strain rates of 350 and 600 s^{-1} can be obtained during the tests with these velocities. High speed imaging is used with digital image correlation in order to obtain the behaviour laws at different strain rates and then calculate the damage corresponding.

The results obtained by using the $SE\dot{E}$ method for damage measurement with all these static and dynamic tests are presented Fig. 7. All the points plotted in this figure are extracted from all the ZOI on the tensile specimens. As consequence, the damage evolution for this mineral filled polypropylene is always the same whatever the strain rate imposed by the tensile conditions. The damage is then not strain rate sensitive. Nevertheless, fracture of the specimens occur at different fracture strain according the speed loading conditions which means that the fracture strain is strain rate dependent.

Average values of the fracture strains according the initial strain rates are plotted Fig. 8. For low strain rates, the fracture strain first increase which could link to the small increase of the temperature due to deformation which helps the molecular chain motion. When the strain rates increase,

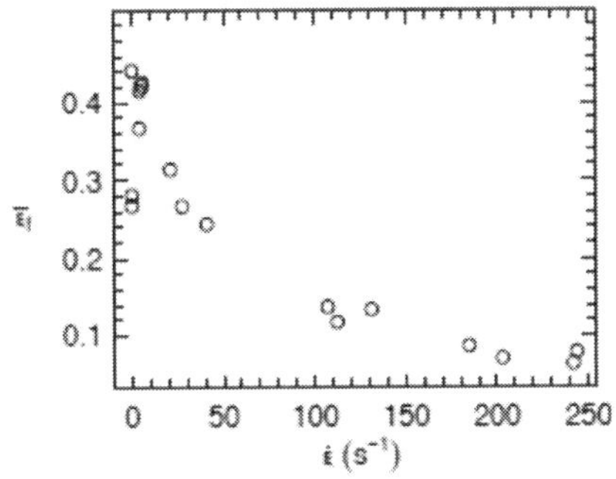

Figure 8. Fracture strain evolution for a tensile test with different strain rates.

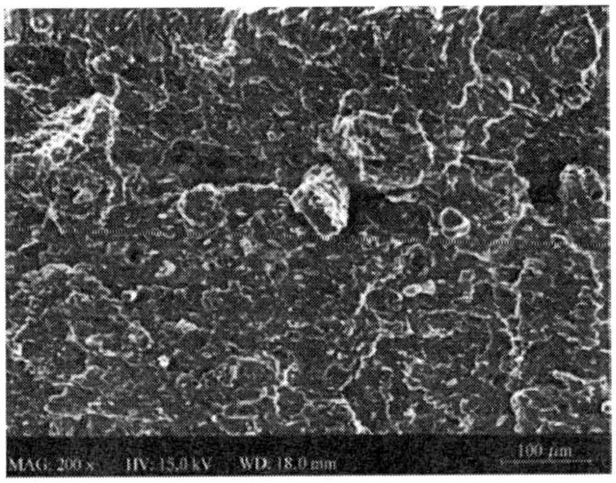

Figure 9. Fracture surface obtained by magnification SEM micrograph under tensile loading at 600 s^{-1}.

this temperature effect disappears and a ductile to fragile transition is observed as the pictures of the fracture surface highlighted it (Fig. 6 and Fig. 9).

4. Conclusions

A method is proposed to determine the damage evolution on a large strain rate range for particles filled semi-crystalline polymers. This method used information obtained by digital image correlation on tensile specimens loaded at different speeds. The damage is then calculated on the basis of the $SE\dot{E}$ method. To validate the damage measurement with this method, results are compared with thus obtained by the loss of stiffness technique in static and correlated to information obtained with an in-situ μCT tensile test. As the results are in agreement, this method is applied to dynamic tensile tests on hydraulic jack and Hopkinson bars. The damage measurement obtained highlights that the damage evolution is independent of the strain rates. Nevertheless, the fracture strains are strain rate dependant with a ductile to fragile transition at high speed loadings.

The present research work has been supported by International Campus on Safety and Intermodality in Transportation, the Région Nord Pas de Calais, the European Community, the Délégation Régionale à la Recherche et à la Technologie, the Ministère de l'Enseignement Supérieur et de la Recherche, and the Centre National de la Recherche Scientifique: the authors gratefully acknowledge the support of these institutions.

References

[1] C. G'Sell, J. Jonas, J Mater Sci, **14**, 583–91 (1979)

[2] F. Zaïri, M. Naït-Abdelaziz, J. Gloaguen, J. Lefebvre, Int. J. Plasticity, **24**(6), 945–65 (2008)

[3] G. Boisot, L. Laiarinandrasana, J. Besson, C. Fond, G. Hochstetter, Int. J. of Solids and Structures **48**, 2642–2654 (2011)

[4] F. Lauro, B. Bennani, P. Croix, J. Oudin, J. Mater. Process Technol., **118** (13), 472–477 (2001)

[5] F. Lauro, B. Bennani, P. Drazetic, J. Oudin, X. Ni, Commun. Numer. Methods Eng., **13** (7), 113–126 (1997)

[6] F. Lauro, B. Bennani, D. Morin, A Epee, Int. J. Impact Eng., **37**(6),715–722 (2010)

[7] L. Laiarinandrasana, T.F. Morgeneyer, H. Proudhon, C. Regrain, J. Polym. Sci. B. Polym. Phys. **48**, 1516–1525 (2010)

[8] R. Balieu, F. Lauro, B. Bennani, G. Haugou, F. Chaari, T. Matsumoto, E. Mottola, Int. J. Impact Eng., **76**, 1–8 (2015)

[9] J. Lemaitre, J. Eng. Mater. Technol., **107**(1), 83–89 (1985)

[10] R. Balieu, F. Lauro, B. Bennani, R. Delille, T. Matsumoto, E. Mottola, Int. J. Plasticity, **51**, 241–270 (2013)

EPJ Web of Conferences **94**, 01010 (2015)
DOI: 10.1051/epjconf/20159401010

Assessment of dynamic mechanical behaviour of reinforced concrete beams using a blast simulator

Marco Peroni[a], George Solomos, Alessio Caverzan, Martin Larcher, and Georgios Valsamos

European Commission Joint Research Centre (JRC), Institute for the Protection and Security of the Citizen (IPSC), European Laboratory for Structural Assessment (ELSA), via E. Fermi 2749, 21027 Ispra (VA), Italy

Abstract. Critical infrastructures may become the target of terrorist bombing attacks or may have to withstand explosive loads due to accidents. The impulsive load connected to explosions is delivered to the structure in a few milliseconds forcing it to respond or fail in a peculiar mode. With reference to the above scientific framework this work presents an innovative apparatus designed and developed at the European Laboratory for Structural Assessment to reproduce a blast pressure history without using explosives. This apparatus is practically a hybrid nitrogen-spring-driven actuator that accelerates masses of up to 100 kg to a maximum velocity of about 25 m/s that impact against the tested structure. The pressure-load history applied to the structure is modulated and reshaped using appropriate layers of elastic soft materials (such as polymeric foams) placed between the specimen and the impacting masses. Specific instrumentation has extensively been utilised to investigate the blast simulator performance and to precisely measure the pressure loads applied to the specimen. A series of tests on real scale reinforced concrete beams/columns (250 × 250 × 2200 mm) has been performed to efficiently assess the performance and potentiality of the new blast simulator. Results are under evaluation. In addition to the experimental work, a series of numerical simulations by means of the explicit FEM code EUROPLEXUS have been carried out to support and improve the equipment design.

1. Introduction

Critical infrastructures in fields such as energy, health, communication, government, transport etc. are made of physical structures, or are housed in physical structures. Such structures may naturally become the target of terrorist bombing attacks. Measures to protect them will certainly be taken, involving prevention, intelligence, detection, deterrence etc., but if everything fails, it is very important that the mechanical structure itself mitigates some effects of the explosion and maintains certain functionalities.

A typical pressure wave curve (which eventually will load a structure) at some distance from an explosion is shown in Fig. 1. Its main characteristics concerning damaging effects on structures are the magnitude of the overpressure, the duration of the positive phase and especially its positive impulse, i.e., the area under the curve over the positive phase. This impulsive load will be delivered to a structure in a few milliseconds forcing it to respond or fail in a peculiar mode. This necessitates that models and design techniques for blast resistant structures be thoroughly validated with reliable data from field tests. However, such tests with actual explosions are expensive, they are usually performed within military grounds, and are often not easily reproducible. Thus, alternative testing methods are desirable, and this has been the case at the University of California in San Diego, where the first blast simulator facility has been built (2006).

As claimed, the effects of explosions are generated without the use of explosive materials. The facility produces repeatable, controlled blast load simulations on full-scale columns and other structural components [1]. The simulator recreates the speed and force of explosive shock waves through servo-controlled hydraulic actuators that punch properly the test specimens.

With the ongoing work a similar blast simulation capability is under development within the EU by the JRC-ELSA, where long and strong experience with servo-controlled actuators resides. Concerning the currently required fast actuators, an alternative design concept has been implemented, which is believed to be capable of generating impacting loads resembling closer those of the real explosions of Fig. 1. This could allow the realistic testing of components to "simulated" explosions and would provide the necessary data for the validation of the computer tools.

The development of this technology will be important both for the research and the practicing engineers and architects who need design rules and guidelines. Besides characterizing blast effects on structural systems, the methodology will contribute to evaluating technologies for hardening and retrofitting buildings and bridges against terrorist bomb attacks. Further, it will help in the investigation of the problem of progressive collapse, i.e. the phenomenon where a local failure propagates in a disproportionate manner to lead to global failure, like the building collapse in the Oklahoma City bombing.

2. Experimental setup – instrumentation

The blast simulator is presently foreseen to be a small pneumatic/hydraulic facility and the sketch below summarizes the main parts of the apparatus (Fig. 2).

[a] Corresponding author: marco.peroni@jrc.ec.europa.eu

Article available at http://www.epj-conferences.org or http://dx.doi.org/10.1051/epjconf/20159401010

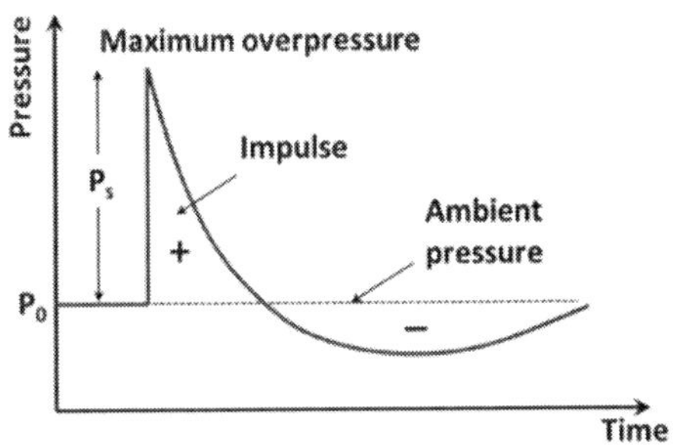

Figure 1. Pressure curve characteristics in free-air explosions.

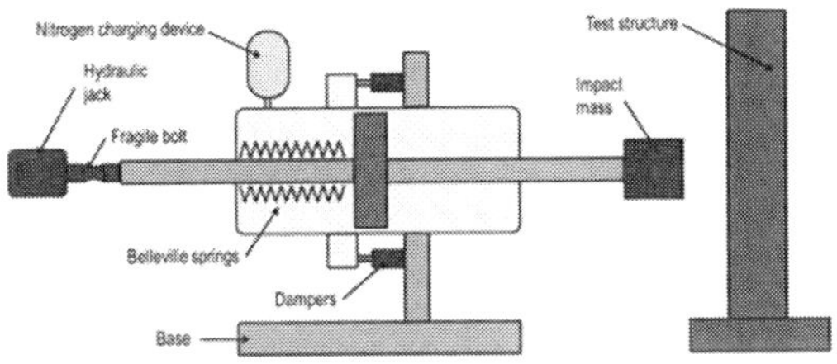

Figure 2. Sketch of blast simulator.

The operating principle of the testing rig is quite simple: the shaft is pulled through a fragile bolt by a hydraulic jack and this action compresses a series of Belleville springs inside the blast actuator. At this point there is the option to also raise the pressure inside the active chamber of the blast actuator by charging it with nitrogen (maximum pressure 100 bar). By continuing the pulling of the shaft (or, by increasing the nitrogen pressure) the bolt suddenly breaks and the piston and shaft rapidly accelerate pushing the impacting mass, which is attached at the other extreme of the shaft. When the piston has done most of its stroke, it starts to decelerate and the impacting mass is detached from it and collides with the tested structure reproducing local pressures similar to those of a blast wave. The implementation of this operating principle and details of the equipment can be found in [2] together with some preliminary tests. The structural elements tested in this work are two horizontally placed, simply supported beams of reinforced concrete.

The setup is similar to a dynamic three point bending, where the dynamic loading is applied over an area on the central portion of the beam, as shown in Figs. 3 and 4.

The two horizontally reacting aluminium supports have a semi-cylindrical geometry and are connected through a bar to the Reaction Wall (a very stiff concrete structure). The beam weight is vertically supported at two points by means of two Teflon sliders, which reduce horizontal friction during the test. The frame that supports the beam and the sliders is a modular structure made of aluminium profiles as is also the structure that supports the two linear bearing guides of the impacting mass. This frame allows a complete optical accessibility to the specimen from the top, where a high speed camera is placed.

The impact mass, composed of several parts, is designed to apply a pressure history similar to that of Fig. 1 and at the same time to effectively measure

Figure 3. Final setup for blast experiments on RC beams.

Figure 4. Blast simulator: specimen supports.

the contact forces between mass and specimen. This is essentially achieved by combining an instrumented mass composed of some rigid, light plates, in the front, connected through some load cells to the heavy, main mass behind (stainless steel). In this concept the plate masses (negligible compared to the stainless steel mass) do not affect the load cells results and the force signals are directly measured and not derived from calculations.

Specifically, the configuration tested in this campaign is shown in Fig. 5. The block of the impacting mass is composed of three aluminium plates ($290 \times 250 \times 10$ mm) connected with three independent piezoelectric load cells to a heavy stainless steel prismatic mass ($75 \times 75 \times 850$ mm). In front of each plate a layer of polyurethane foam has been placed to smooth the pressure pulse and reproduce closer the blast pressure history. The mass has been designed to slide on two linear bearing carriages and to rotate around two bearings rigidly connected to the carriages. These degrees of freedom reduce drastically the forces transmitted to the guiding rails, improving the safety and the lifetime of the equipment.

After the first test, two additional springs have been added (Fig. 5), in order to control rotations and avoid possible misalignment of the impacting mass during its free movement before the impact. The impactor, in its current design, can reliably measure a maximum load of 1980 kN, a suitable value also for impacting velocity greater than 15 m/s.

Figure 5. Blast simulator: impacting mass.

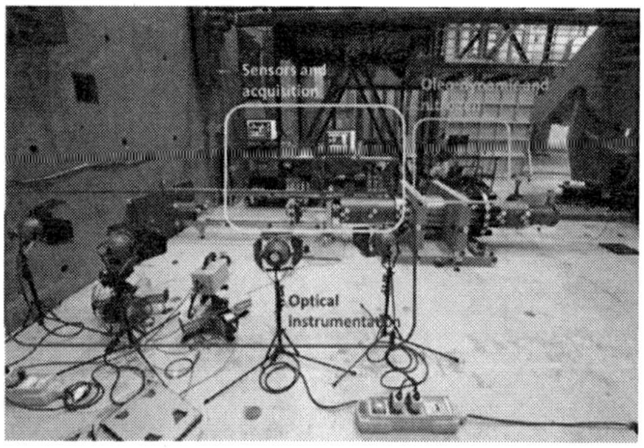

Figure 6. Blast simulator: instrumentation.

The instrumentation (electronic components and transducers) and the adopted devices are described below [2]. Figure 6 shows an example of typical instrumentation adopted in a blast simulator test.

The actuator is a servo-hydraulic axis electronically controlled with an in-house control system based on EtherCAT technology. In the control zone all components concerning the nitrogen booster have been placed to have a complete overview of actuator power devices. For what concerns the test instrumentation all sensors and transducers are acquired with a high speed transient recorder. The transducers typically adopted are piezoelectric accelerometers, load cells and pressure sensors and a series of photoelectric cells. In addition, two high speed cameras are normally used, synchronized to the transient recorder. The detailed list of instrumentation adopted and deployed during the experiments carried out is given below.

- 2 acquisition boards GAGE Octopus of 8 channels each with 20 Msample/s per channel. Considering the test duration, a sampling frequency of 200 kHz has been applied (pre-trigger 10 000 points, post-trigger 100 000 points).
- 1 High Speed camera IDT Y4 with 14 mm Nikkor lens. It films laterally the evolution of the whole experiment at a frequency of 800 fps (pre-trigger 800 frames, post-trigger 800 frames).
- 1 High speed camera Photron SA1 with 21 mm Zeiss Distagon lens. It films the upper face of the specimen and impactor, for optical/ photogrammetric analysis purposes, at a frequency of 5000 fps (pre-trigger 0 frames, post-trigger 5000 frames).
- 9 Charge Amplifiers Kistler 5015 for the conditioning of piezoelectric sensors.
- 2 Piezoelectric load cells Kistler 9106A (full scale 330 kN) placed on the Reaction Wall supports.
- 3 Piezoelectric load cells Kistler 9106A (full scale 330 kN) interposed between the three aluminium plates and the main steel mass of the impactor.
- 2 Piezoelectric accelerometers Kistler 8202A (full scale 2000 g) mounted on the steel mass of the impactor to evaluate its acceleration and its possible rotation.
- 1 Piezoelectric accelerometer Kistler 8202A (full scale 2000 g) mounted on the central aluminium plate of the impactor to evaluate possible inaccuracies in the force measurement due to the plate inertia.
- 1 Piezoelectric pressure sensor Kistler 601H on the pressurized chamber of the blast actuator.
- 3 Photoelectric cells Pepperl Fuchs for triggering the data acquisition and for measuring locally the impactor velocity

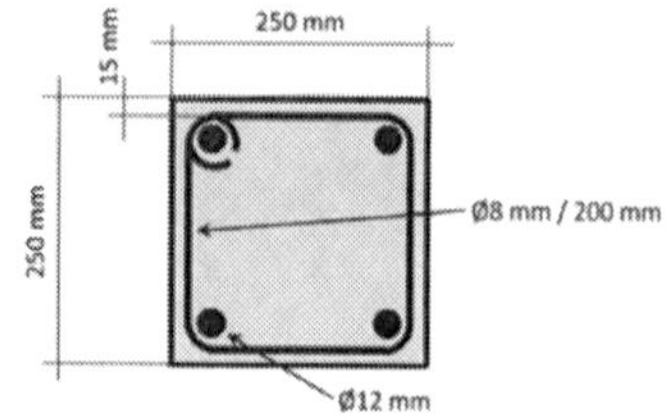

Figure 7. Reinforcement of the RC beams.

3. Experimental tests

This section summarizes some results of the tests performed with the blast actuator on two reinforced concrete (RC) beams in order to evaluate the performance of this equipment and its capability to simulate the effects of a blast explosion without using explosives.

The tests have been conducted on full scale RC beams of dimensions 250 × 250 × 2200 mm, manufactured with standard materials and reinforcement. The designed compressive strength of the concrete is around f_{ck} = 20 MPa with a maximum aggregate size of 20 mm. The reinforcement consists of 4 longitudinal deformed bars of 12 mm diameter and stirrups of 8 mm diameter spaced every 200 mm, as shown in Fig. 7.

The test campaign has been conducted taking into account safety issues due to the elevated levels of energy stored and quickly released during the experiments. For this reason and taking into account the high complexity of the equipment under operation, the energy involved during the experiments has been increased gradually. The set-up is also covered by a safety box (a mixed aluminium and polycarbonate panel structure) that protects the operators conducting the experiment from accidental debris.

The first three tests have been conducted on the same specimen with an increasing impacting velocity, and thus several of the testing parameters (especially the specimen damage state) are not well defined. Although these tests

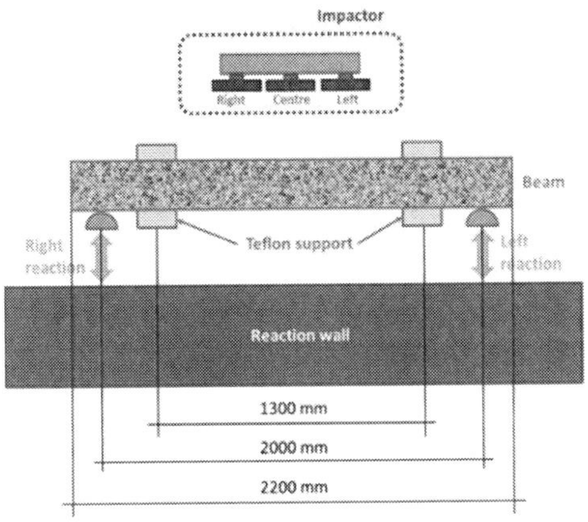

Figure 8. Schematic top view of the geometry of the experiments.

do not have a high scientific value, they have allowed a better knowledge of the equipment performance and of the test evolution. The tests have been performed by storing energy only in the actuator spring and a maximum velocity of about 12 m/s has been reached. Following this test, the column has been judged to be overly damaged to continue with additional tests on it.

On the contrary, the last test presented in this paper can be considered to have a greater scientific value even though a series of experimental setbacks have occurred (the acquisition of piezoelectric sensors failed for a bug of the transient recorder). The test has been performed by employing the energy of the springs and of pressurized nitrogen in order to increase the impact velocity of the mass (about 17 m/s). The beam has been seriously damaged and patterns of a shear failure mode have appeared (also typical of blast load failure).

A first analysis of the recorded data has been made and some results in the form of images and diagrams are presented in the following. Figure 8 provides a general picture of the geometry of the impactor, the beam specimen and its supports, which helps in understanding the diagrams produced below for each one of the tests. Figure 9 presents a selection of representative frames of the high-speed photo-sequence captured during the last test with the high speed camera placed perpendicularly above the specimen.

The impacting mass hits normally the RC beam (frames 1 and 2) and forces it to bend, as shown in the frames 3 and 4. For the specific test, due the non-symmetric behaviour of the specimen, the beam mainly fails in the left part as underlined by the rotation of the impacting mass (frame 4). In addition to the qualitative information that can be deduced by the high speed camera photo sequence (type of damage, possible rebounds, impact misalignments, etc.), special analyses have been employed for these optical data to obtain quantitative measures.

Specifically, two different algorithms have been applied:

- a fast tracking algorithm [3] that returns an accurate measure (in the order of 0.01 mm) of the position of the particular Gaussian grayscale targets applied to the column and to the impacting mass (Fig. 9);
- a full-field technique, the so called optical flow algorithm [4], that returns the displacement of each image pixel (with less precision compared with the previous algorithm), useful to study the evolution of the fracture paths during the test.

Figure 9. High speed photo sequence; time = 14 ms, 24 ms, 34 ms and 40 ms.

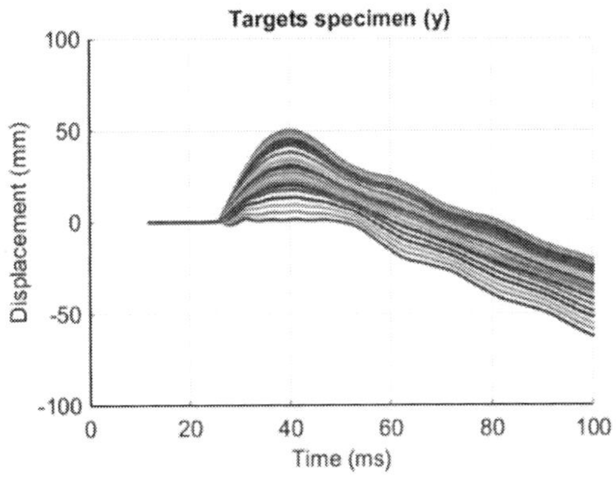

Figure 10. Beam target displacements (along impact direction).

Figure 10 shows an example of quantitative data extracted with the first of the algorithm described above. The beam target displacements (along the impact direction) have

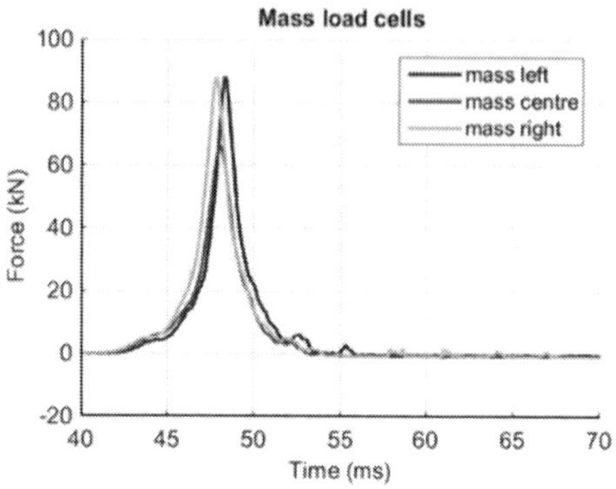

Figure 11. Mass load cell signals.

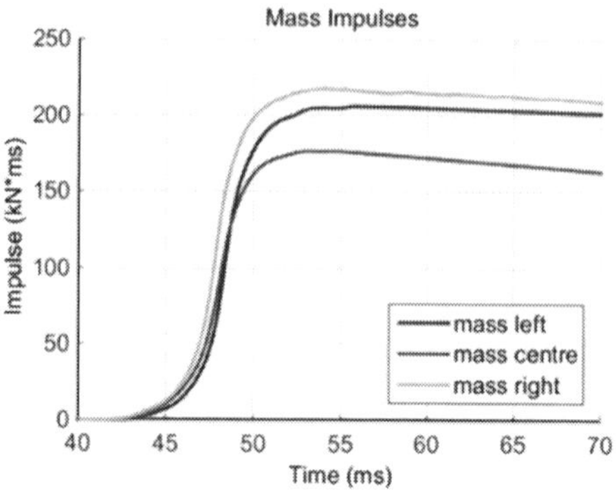

Figure 12. Impulses applied by three plates.

been presented. The impact starts approximatively at 22 ms and the maximum deflection occurs at 40 ms as underlined by the graph (the contact between the mass and the specimen lasts less than 8 ms). Obviously the curves that present the greater deflections are related to targets located in the central part of the column. After 40 ms the elastic unloading of the damaged column starts, which causes the rebound of the specimen in the direction opposite to the impact. At 100 ms the bending oscillations due to the impact are substantially finished and the displacements of the targets are representative of the damaged shape of the beam. A permanent deflection of about 50 mm has been reached in the central part of the beam.

For what concerns the piezoelectric sensors placed on the impacting mass, Fig. 11 shows the signals recorded during the last experiment conducted on the first beam (the specimen was already damaged by previous experiments performed at low impact velocities). The direct measures of the impacting forces are totally noiseless and no ringing phenomena of the transducers occur in the duration of the test. In addition it is possible to observe that the force transmitted by the central plate is lower compared with those of the two lateral plates due to the pre-existing damage in the specimen. In any case the peak load and the duration of the pulses generated by the equipment are comparable with the experimental data obtained from experiments conducted with explosives.

Figure 12 shows the impulses related to Fig. 11. As expected, the signal derived by the central plate is lower compared with the two lateral ones.

4. Numerical simulations

Finite element analysis has been performed, in order to verify the experimental results and to ameliorate the final configuration of the apparatus. The geometrical model presented in Fig. 8 has been discretised, as depicted in Fig. 13, with finite elements via the CAST3M code [5]. The numerical calculations have been performed with the EUROPLEXUS explicit finite element code [6]. Hexahedral solid elements with eight nodes, one integration point and linear displacement interpolation have overall been preferred.

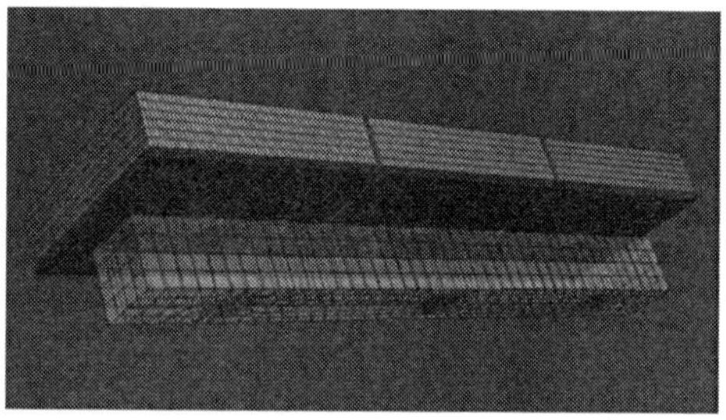

Figure 13. Finite element model of the impacting mass.

Different element types and sizes are employed in the representation of the impacting mass. For the massive steel part a relatively coarse mesh has been used since it will not undergo high strains (cubic edge size 20 mm). For the load cells, the aluminium plates and the polyurethane foam finer meshes have been used with an average cubic edge of 5 mm.

The concrete beam has been discretised with cubic finite elements of 20 mm edge size, while the reinforcement has been modelled with 1-D bar elements. The bonding between the concrete and the reinforcement has been modelled with a procedure that links the displacement of the nodes belonging to the continuum-like elements of concrete, with those of the bar-like elements of steel.

For the massive steel part of the impactor, the load cells and the beam reinforcement a Von Mises material law has been employed. Elasto-plasticity is implemented via a radial return algorithm and isotropic hardening is included. For the representation of the aluminium, a perfectly plastic Von Mises material law has been used. The concrete part has been introduced via a three-invariant cap model with mixed hardening and isotropic damage that contains also the strain-rate effect [6]. Finally, for the elastic foam a hyperelastic material law has been used, based on a strain energy equation [7]. The energy equation represents the stress-strain curve, with several parameters derived from material tests. The parameters have been defined after a best-fit procedure of the given stress-strain curve. Table 1 presents the mechanical properties (density, Young modulus, Poisson ratio) of the metallic and concrete parts

Table 1. Mechanical properties used for the numerical model.

	ρ[kg/m^3]	E [GPa]	ν
Steel	7800	210	0.30
Aluminium	2800	70	0.35
Load cell	6488	131	0.30
Concrete	2400	30	0.21

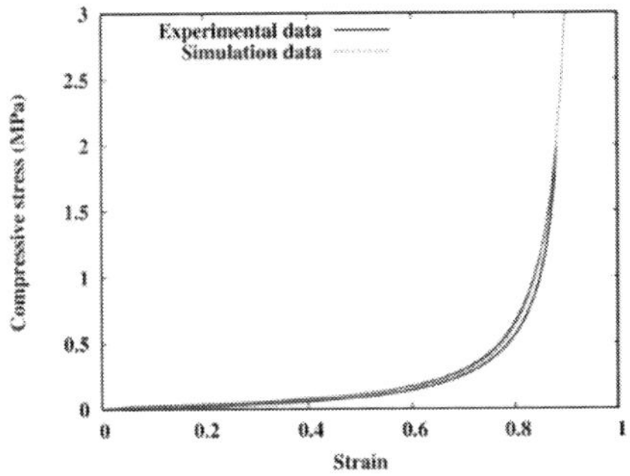

Figure 14. Stress-strain curve for the polyurethane foam.

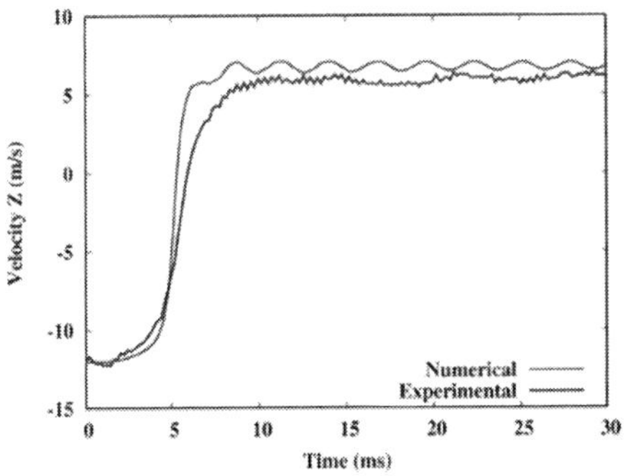

Figure 15. Velocity of the impactor.

of the structure, while Fig. 14 presents the stress-strain curve for the polyurethane foam (both experimental and numerical).

The impactor is moving with an initial velocity until it collides with the specimen. The pinball method [8,9] has been used for the contact between the impacting mass and the specimen. The RC beam has been blocked in the vertical direction for all the points lying on the support lines.

Figures 15 and 16 present the comparison between the experimental and numerical results concerning the velocity and the displacement of the impactor in one of the initial tests. As seen, there is a relatively good agreement between the two approaches.

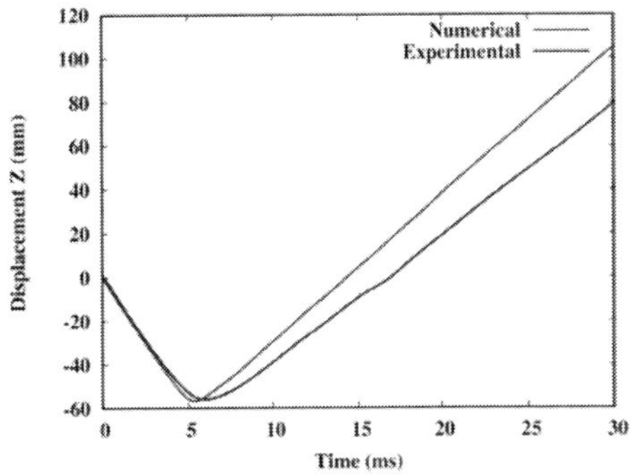

Figure 16. Displacement of the impactor.

5. Conclusions

Towards developing a blast simulator, a fast, spring-gas driven actuator and an innovative impactor have been constructed. Their performance has been assessed through appropriately instrumented tests. The results obtained from these tests demonstrate that the developed blast actuator is capable of reproducing through impact the required pressure levels to bring real structural members (reinforced concrete beams) to failure. Numerical simulations are also conducted, successfully modelling the experiments and pointing to improvements in the performance of the apparatus.

References

[1] G.A. Hegemier, F. Seible, T. Rodriguez-Nikl, K. Arnett, Blast Mitigation of Critical Infrastructure Components and Systems, FIB Proceedings of the 2nd International Congress, June 5–8 2006, Naples, Italy, paper ID: 11–2

[2] M. Peroni, G. Solomos, B. Viaccoz, Blast Simulator project: Performance assessment and preliminary tests, Report JRC 86864

[3] P. Caperan, M. Poljansek, E. Gutierrez, S. Primi, C. Paulotto, Comp. Struct. **94**, 2012

[4] C. Zach, T. Pock, H. Bischof, A Duality Based Approach for Realtime TV-L^1 Optical Flow, Pattern Recognition, Springer, 2007

[5] `http://www-cast3m.cea.fr`

[6] EUROPLEXUS User's manual, 2015; `http://www-epx.cea.fr`

[7] S. Kolling. P.A. Du Bois. D.J. Benson and W.W. Feng. Comp. Mech., **40**:885–899, 2007

[8] T. Belytschko, M.O. Neal, Int. J. Num. Meth. Engng., 31:547–572, 1991

[9] F. Casadei, A general Impact-Contact method based on hierarchic pinballs for the Europlexus software system, Report JRC I.03.176

EPJ Web of Conferences **94**, 01011 (2015)
DOI: 10.1051/epjconf/20159401011

Strain localization during tensile Hopkinson bar testing of commercially pure titanium and Ti6Al4V titanium alloy

Wojciech Moćko[1,a], Leopold Kruszka[2], and Adam Brodecki[1]

[1]Motor Transport Institute, 80, Jagiellońska st., 03-301 Warsaw, Poland
[2]Gen. Jaroslaw Dabrowski Military University of Technology, 2, Gen. Sylwester Kaliski st., 00-908 Warsaw, Poland

Abstract. The goal of the analysis was to determine the strain localization for various specimen shapes (type A and type B according to PN-EN ISO 26203-1 standard) and different loading conditions, i.e. quasi- static and dynamic. Commercially pure titanium (Grade 2) and titanium alloy Ti6Al4V (Grade 5) were selected for the tests. Tensile loadings were applied out using servo-hydraulic testing machine and tensile Hopkinson bar with pre-tension. The results were recorded using ARAMIS system cameras and fast camera Phantom V1210, respectively at quasi-static and dynamic loading conditions. Further, specimens outline was determined on the basis of video data using TEMA MOTION software. The strain distribution on the specimen surface was estimated using digital image correlation method. The larger radius present in the specimen of type B in comparison to specimen of type A, results in slight increase of the elongation for commercially pure titanium at both quasi-static and dynamic loading conditions. However this effect disappears for Ti6Al4V alloy. The increase of the elongation corresponds to the stronger necking effect. Material softening due to increase of temperature induced by plastic work was observed at dynamic loading conditions. Moreover lower elongation at fracture point was found at high strain rates for both materials.

1. Introduction

A normalized specimens are usually applied for the determining of mechanical properties of titanium and its alloys using tensile Hopkinson bar. The small ratio between length and width of specimen (in comparison to typical tensile tests at quasi-static loading conditions) together with limited work hardening leads to necking at low strain values. As a consequence determination of the true stress-strain curve is complicated. Nonhomogeneous deformation of the specimens may be determined using digital speckle photography to obtain in-plane point-wise displacement and strain field [1]. Subsequently, results of strain distribution measurement were used to develop methodology based on inverse modelling for estimating visco-plastic material parameters at high strain rate conditions. Test were carried out using Hopkinson bar under compressive loading conditions for a mild steel. Necking process was also investigated using digital image correlation method during tensile tests of dual phase steel [2,3]. It was found that strain localization started even before the maximum load was attained in the specimen. An elasto-viscoplastic constitutive model was used to predict the observed stress-strain behaviour and strain localization. A good agreement between FEM simulation and DIC result was obtained. Wide range of tensile tests for a set of advanced high strength steels was carried out by Sato [4,5]. Analysis were performed at wide range of strain rates using a newly developed strain analysis system.

Digital image correlation method was also applied in order to study polymeric foam under impact loading using large diameter nylon Hopkinson bar [6]. Application of optical field measurement methods enables to determine non-homogenous strain and stress field. Similarly, fast camera videos recorded during Hopkinson bar tests of carbon-epoxy composites were used to analyse deformation mechanisms under high strain rate deformation regime [7]. Strain rate effect on modulus, yield, ultimate strength, strain to failure and on the in-plane shear properties was studied. Results of tensile tests obtained using digital image correlation method may be used to determine stress-strain curve including post-necking strain [8]. Hourglass type specimens were used for tensile tests, and the stress-strain curves were identified through an iteration process using finite element analysis. The strain at the position of minimum diameter was measured by DIC method.

Hopkinson bar test are carried out under compressive, shear and tensile loading conditions. Combined experimental-numerical approach was applied to extract the strain rate and temperature dependent mechanical behaviour from high strain rate experiments. The method [9–11] involves the identification of the material model parameters used for the finite element simulation. The technique was applied to determine stress-strain behaviour of Ti6Al4V titanium alloy using both high strain rate in-plane shear and tensile tests.

Fracture initiation is also studied using optical field measurement systems. The tests including low, intermediate and high strain rate tensile experiments were carried out on flat smooth, notched and central-hole tensile specimens extracted from AHSS sheet. A split Hopkinson bar was used to perform the high strain rate tension experiments. Selected surface strains, as well as local displacements were measured using high speed photography in conjunction with planar digital image correlation [12].

[a] Corresponding author: wojciech.mocko@its.waw.pl

Article available at http://www.epj-conferences.org or http://dx.doi.org/10.1051/epjconf/20159401011

Table 1. List of specimen used in experiments.

Spec No.	Material	Shape	Strain rate $[s^{-1}]$	Yield point [MPa]	Tensile strength [MPa]	Elongation
3	Grade 2	type A	1.0×10^{-2}	470	510	0.42
4	Grade 2	type A	1.0×10^{-2}	480	517	0.42
9	Grade 2	type B	1.0×10^{-2}	478	514	0.43
10	Grade 2	type B	1.0×10^{-2}	476	518	0.44
11	Grade 2	type A	5.8×10^{2}	623	610	0.29
12	Grade 2	type A	5.8×10^{2}	628	618	0.29
13	Grade 2	type B	6.0×10^{2}	612	606	0.30
14	Grade 2	type B	6.2×10^{2}	626	615	0.32
17	Grade 5	type A	1.0×10^{-2}	1003	1035	0.26
18	Grade 5	type A	1.0×10^{-2}	1007	1038	0.26
23	Grade 5	type B	1.0×10^{-2}	1007	1043	0.26
24	Grade 5	type B	1.0×10^{-2}	1017	1045	0.26
25	Grade 5	type A	4.5×10^{2}	1115	1086	0.19
26	Grade 5	type A	5.0×10^{2}	1115	1094	0.19
27	Grade 5	type B	4.5×10^{2}	1113	1092	0.19
28	Grade 5	type B	4.5×10^{2}	1126	1107	0.21

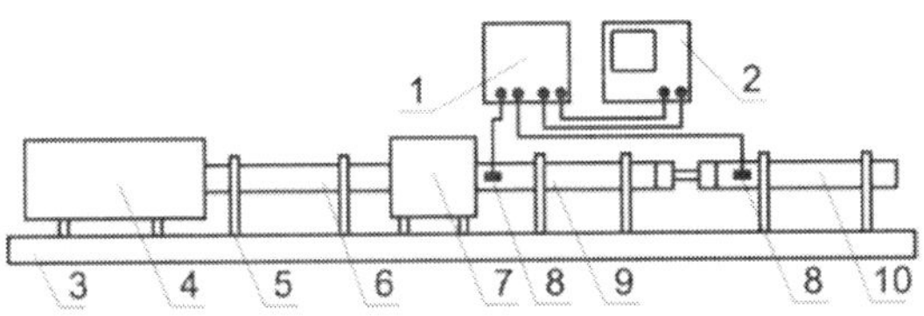

Figure 1. Scheme of the tensile Hopkinson bar with pre-tension. 1 – Wheatston's bridge, 2 – digital oscilloscope, 3 – base, 4 – pre-strain hydraulic actuator, 5 – support, 6 – pre-strained part of incident bar, 7 – hydraulic clamp, 8 – tensometers, 9 – free part of incident bar, 10 – transmitter bar.

This paper presents analysis of the necking phenomenon during quasi-static and dynamic tensile tests of commercially pure titanium (CPT) and Ti6Al4V titanium alloy. Application of digital image correlation method enables to investigate differences between low and high strain rate deformation of those materials.

2. Experimental method

2.1. Specimens

At both, quasi-static and dynamic deformation rates the specimens of type A according to standard PN-EN ISO 26203-1 were used. The length and width of gauge were equal to 7 mm and 4 mm, respectively. The specimens were cut using electro discharge machining from the 3 mm thick sheet of commercially pure titanium (CPT) and TiAl6V4 titanium alloy. List of the specimens and loading parameters are shown in Table 1.

2.2. Tensile test

The tensile tests were carried out using the servo-hydraulic testing machine manufactured by Instron company and pre-tension Hopkinson bar [13,14] at quasi-static and dynamic range of strain rates. The testing stand at Motor Transport Institute laboratory [15,16] is equipped with the bars of 20 mm in diameter made of 7075-T6 aluminium alloy (Fig. 1). The incident bar of 3600 mm in length is divided into a pre-tension section with a length of 1600 mm and a free end. The clamp which confines the bar during initial loading with the use of a hydraulic actuator is placed between the pre-tension and free section of the incident bar. The transmission bar length is equal to 1800 mm. The history of elastic wave in the bars is determined using tensometers, than amplified at the broad-band bridge and finally recorded by a digital oscilloscope.

2.3. Digital image correlation system

The test was recorded using Phantom V1210 fast camera at a resolution of 384 × 128 and frame rate equal to 150 000 at dynamic range of loading. In order to obtain clear view without blurring of measurement grid required in DIC analysis the shutter time was set to 2 μs. Very short time of frame acquisition requires very strong lighting, therefore two COOLH illuminators emitting very focused light beam were applied. The arrangement including Hopkinson bar, fast camera and lighting system is presented in Fig. 2. At the quasi-static deformation rates ARAMIS 4M video system with resolution equal to 2400 × 1728 pixels was used. The videos recorded during experiments were subsequently analysed by DIC method implemented in the ARAMIS software to determine 2D field of displacement.

3. Experimental results

3.1. Quasi-static and dynamic tensile curves

Stress-strain curves of CPT and TiAl6V4 alloy are shown in Fig. 3 and Fig. 4 respectively. Small, but clearly to observe influence of the specimen shape on the elongation and necking at fracture point were found for CPT. The larger radius present in the specimen of type B in comparison to specimen of type A, results in slight increase of the elongation for both quasi-static and dynamic loading conditions. Influence of the specimen

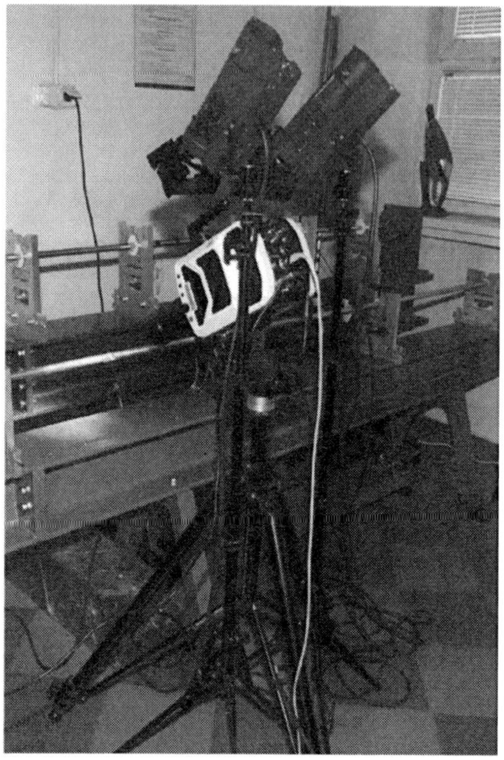

Figure 2. TSHPB stand with high speed camera.

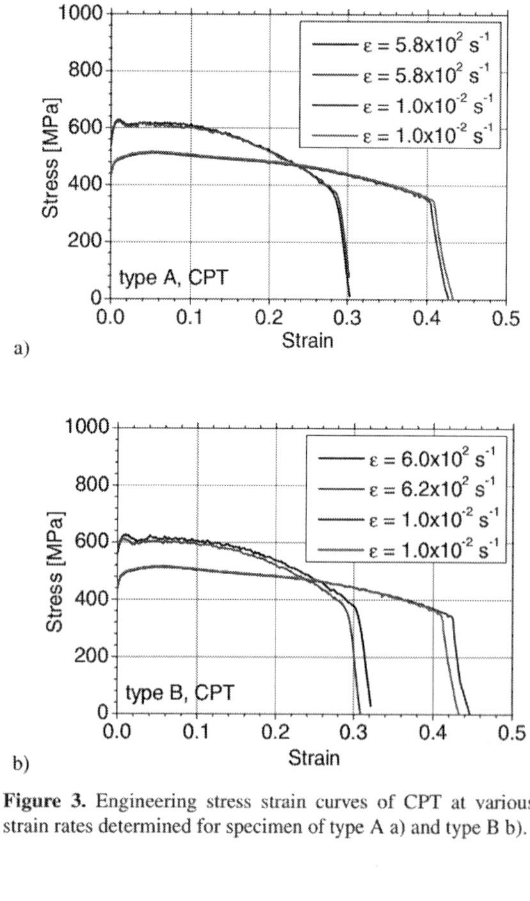

Figure 3. Engineering stress strain curves of CPT at various strain rates determined for specimen of type A a) and type B b).

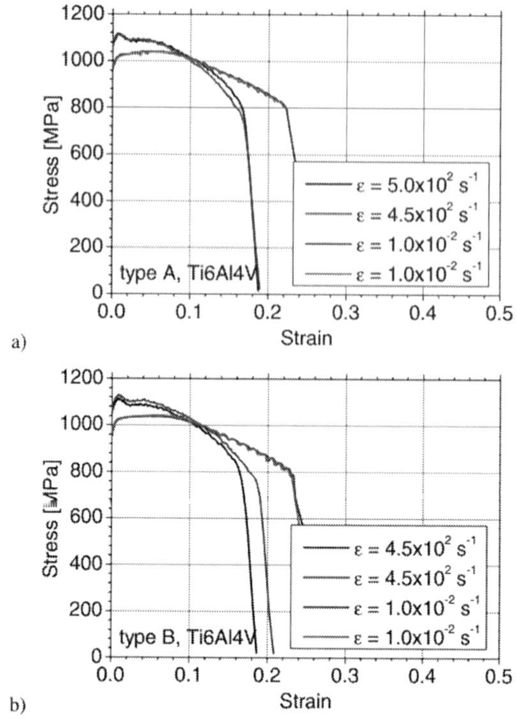

Figure 4. Engineering stress strain curves of Ti6Al4V titanium alloy at various strain rates determined for specimen of type A a) and type B b).

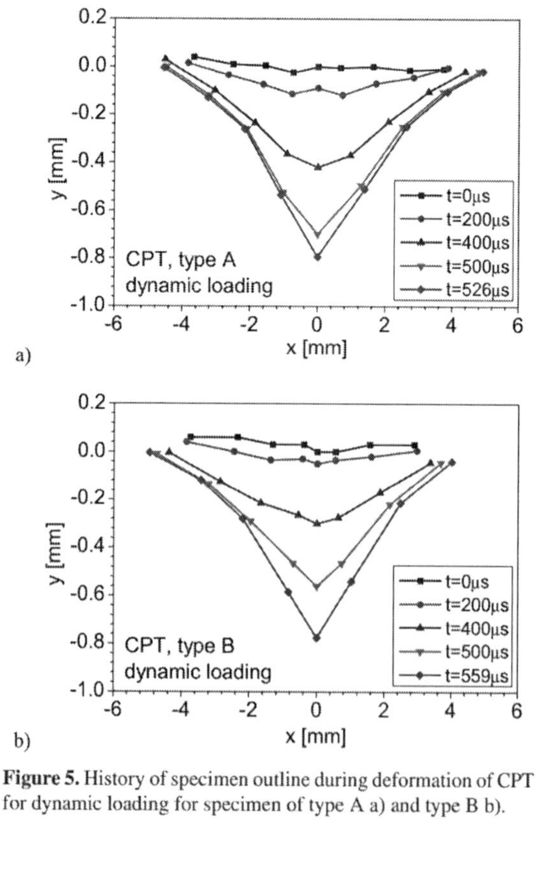

Figure 5. History of specimen outline during deformation of CPT for dynamic loading for specimen of type A a) and type B b).

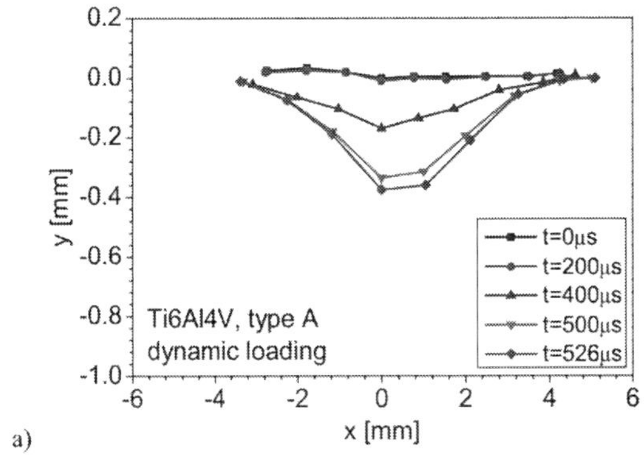

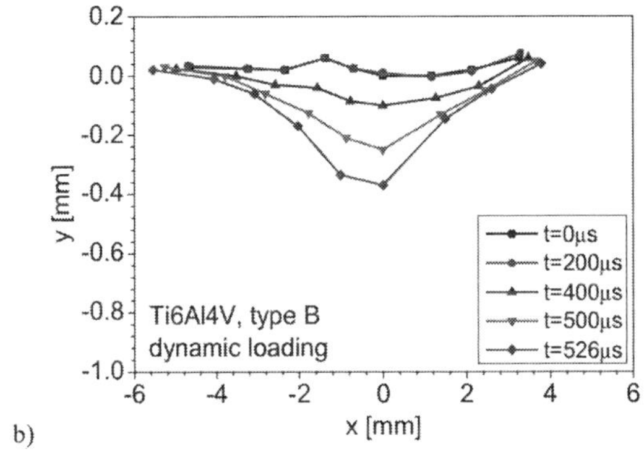

Figure 6. History of specimen outline during deformation of Ti6l4V titanium alloy for dynamic loading for specimen of type A a) and type B b).

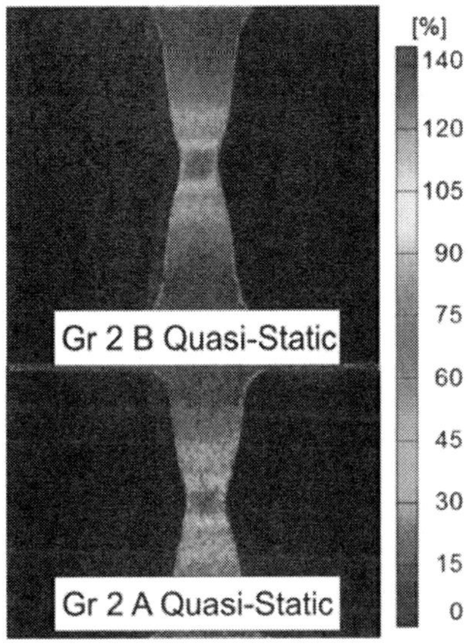

Figure 7. Strain distribution determined during quasi-static tensile test of CPT.

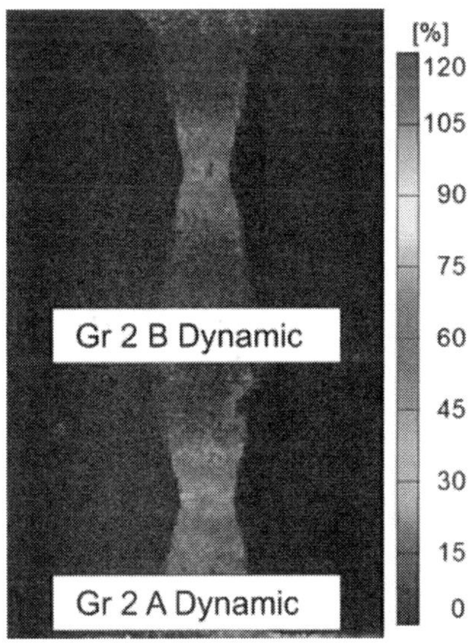

Figure 8. Strain distribution determined dynamic tensile test of CPT.

shape on elongation for TiAl6V4 may be neglected. The increase of the elongation corresponds to the stronger necking effect (if the necking is stronger the width of specimen in the necking point is lower). Exact elongation values are shown in Table 1.

The influence of the strain rate on the tensile curve is clear to observe. Both materials shows strain rate hardening effect. Yield stress at dynamic loading condition is about 100 MPa higher than for quasi-static loading conditions. Material softening due to increase of temperature induced by plastic work was observed at dynamic loading conditions. Moreover lower elongation at fracture point was found at high strain rates for both materials. This behaviour may be related to temperature increase and plastic deformation mechanisms as well.

3.2. Evolution of specimen thickness at necking point

Evolution of the specimen outline during tensile tests was determined at dynamic loading conditions using TEMA MOTION software on the basis of recorded videos. The charts shows average value determined on the basis of measurement of both edges of specimen. The change of specimen shape at subsequent step of deformation is show in Fig. 5 and Fig. 6 for CPT and TiAl6V4 alloy respectively. It may be observed that necking effect is stronger for the CPT independently on the shape of specimen. Moreover it may be found that for the specimens of shape of type A for both materials necking grows quicker than for the type B. Value of reduction of the specimen width at necking point is equal 0.8 mm and 0.4 mm for the CPT and TiAl6V4 alloy respectively.

3.3. Strain distribution at specimen surface

Results of analysis conducted with the use of digital correlation method are shown in Figs. 7–9. Presented frames were taken form videos just before macroscopically observed fracture. In all cases strain localization is ease to observe. It is located in the middle of gauge. The strain

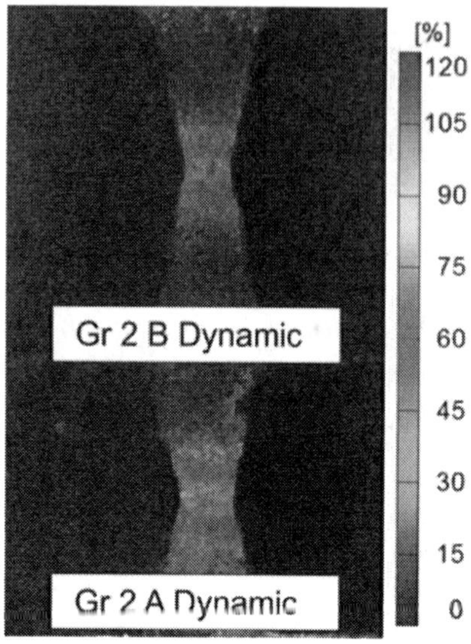

Figure 9. Strain distribution determined quasi-static test of Ti6Al4V alloy.

field is not uniform across specimen with. It may be found that strain value at necking region is highest near specimen longitudinal axis. Comparison between quasi-static and dynamic deformation mechanism of CPT is presented in Fig. 7 and Fig. 8 respectively. Maximum values of strain determined locally using DIC method are strongly dependent on strain rate. Local strain before fracture is equal to 1.4 (Fig. 7) at static loading conditions and equal to 1.0 (Fig. 8) at dynamic loadings. It may be stated that locally observed behaviour is in coincidence with macroscopically determined strain. Strain estimated using Hopkinson bar methodology was equal to 0.3 whereas strain measured with extensometer during quasi-static tensile test was equal to 0.42. Taking into account both method of strain determining (local and macroscopic) it may be found that elongation at quasi-static loading conditions is 330% higher in comparison to the dynamic one.

Locally observed strain of Ti6Al4V is equal to 0.6 at quasi-static loading conditions whereas macroscopically observed strain is equal to 0.25. It means that local strain is 240% higher than macroscopically determined one. It may be stated that necking phenomenon is stronger for the CPT than for Ti6Al4V alloy.

4. Summary

Tensile tests carried out using Hopkinson bar with pre-tension were recorded using high speed camera. Subsequently, the videos were analysed using digital image correlation method. On the basis of determined strain distribution, it was found that locally observed strain is 330% and 240% higher than macroscopic one, respectively, for the CPT and Ti6Al4V alloy. Moreover, elongation value depends on strain rate. Locally determined strain at quasi-static tensile tests of CPT was equal to 1.4, whereas at dynamic loading conditions was equal to 1.0. Macroscopic strain shows similar behaviour, i.e. strain was equal to 0.42 and 0.3 respectively at quasi-static and dynamic loading conditions.

This study was supported by the Polish National Centre for Research and Development (GRAF-TECH/NCBR/14/26/2013)

References

[1] J. Kajberg, B. Wikman, Viscoplastic parameter estimation by high strain-rate experiments and inverse modelling – Speckle measurements and high-speed photography, Int. J. Solids Struct. **44**, 145–164 (2007)

[2] V. Tarigopula, O.S. Hopperstad, M. Langseth, A.H. Clausen, F. Hild, A study of localisation in dual-phase high-strength steels under dynamic loading using digital image correlation and FE analysis, Int. J. Solids Struct. **45**, 601–619 (2008)

[3] J. Qin, R. Chen, X. Wen, Y. Lin, M. Liang, F. Lu, Mechanical behavior of dual-phase high-strength steel under high strain rate tensile loading, Mat. Sci. Eng. A-Struct. **586**, 62–70 (2013)

[4] K. Sato, Q. Yu, J. Hiramoto, T. Urabe, A. Yoshitake, A method to investigate strain rate effects on necking and fracture behaviors of advanced high-strength steels using digital imaging strain analysis, Int. J. Impact Eng. **75**, 11–26 (2015)

[5] W. Moćko, Comparison of energy absorption properties of high nitrogen austenitic steel and cast alloy determined using low velocity perforation test. Arch. Metall. Mater. **59**, 65–69 (2014)

[6] J. Liu, D. Saletti, S. Pattofatto, H. Zhao, Impact testing of polymeric foam using Hopkinson bars and digital image analysis, Polym. Test. **36**, 101–109 (2014)

[7] H. Koerber, J. Xavier, P.P. Camanho, High strain rate characterisation of unidirectional carbon-epoxy IM7-8552 in transverse compression and in-plane shear using digital image correlation, Mech. Mater. **42**, 1004–1019 (2010)

[8] M. Kamaya, M. Kawakubo, A procedure for determining the true stress–strain curve over a large range of strains using digital image correlation and finite element analysis. Mech. Mater. **43**, 243–253 (2011)

[9] J. Peirs, P. Verleysen, W. Van Paepegem, J. Degrieck, Determining the stressestrain behaviour at large strains from high strain rate tensile and shear experiments, Int. J. Impact Eng. **38**, 406-415 (2011)

[10] W. Moćko, J. Janiszewski, M. Grązka, Application of an extended Rusinek-Klepaczko constitutive model to predict the mechanical behavior of 6082-T6 aluminum under Taylor impact test conditions. J Strain Analysis **48**, 364–375 (2013)

[11] W. Moćko, J. Janiszewski, J. Radziejewska, M. Grązka, Analysis of deformation history and damage initiation for 6082-T6 aluminium alloy loaded at

classic and symmetric Taylor impact test conditions. Int. J. Impact Eng. **75**, 203–213 (2015)

[12] C. C. Roth, D. Mohr, Effect of strain rate on ductile fracture initiation in advanced high strength steel sheets: Experiments and modeling. Int. J. Plasticity **56**, 19–44 (2014)

[13] G.H. Staab, A. Gilat, A Direct Tension SHB for high strain rate testing, Exp. Mech. **31**, 232–235 (1991)

[14] H. Kolsky, An Investigation of the mechanical properties of materials at very hign rates of deformation of loading. Proc. Phys. Soc. **62B**, 647–700 (1949)

[15] W. Moćko, Analysis of the impact of the frequency range of the tensometer bridge and projectile geometry on the results of the measurement by the split Hopkinson pressure bar method. Metrol. Meas. Syst. **20**, 555–564 (2013)

[16] W. Moćko, J.A. Rodríguez-Martínez, Z.L. Kowalewski, A. Rusinek, Compressive viscoplastic response of 6082-T6 and 7075-T6 aluminium alloys under wide range of strain rate at room temperature: Experiments and modelling. Strain **48**, 498–509 (2012)

EPJ Web of Conferences **94**, 01012 (2015)
DOI: 10.1051/epjconf/20159401012

Determination of dynamic fracture toughness using a new experimental technique

Carl M. Cady[a], Cheng Liu, and Manuel L. Lovato

MST-8, MS-G755, Los Alamos National Laboratory, Los Alamos NM 87545, USA

Abstract. In other studies dynamic fracture toughness has been measured using Charpy impact and modified Hopkinson Bar techniques. In this paper results will be shown for the measurement of fracture toughness using a new test geometry. The crack propagation velocities range from ~0.15 mm/s to 2.5 m/s. Digital image correlation (DIC) will be the technique used to measure both the strain and the crack growth rates. The boundary of the crack is determined using the correlation coefficient generated during image analysis and with interframe timing the crack growth rate and crack opening can be determined. A comparison of static and dynamic loading experiments will be made for brittle polymeric materials. The analysis technique presented by Sammis *et al.* [1] is a semi-empirical solution, however, additional Linear Elastic Fracture Mechanics analysis of the strain fields generated as part of the DIC analysis allow for the more commonly used method resembling the crack tip opening displacement (CTOD) experiment. It should be noted that this technique was developed because limited amounts of material were available and crack growth rates were to fast for a standard CTOD method.

1. Introduction

The primary reason for this investigation was to determine the most accurate means of measuring the effect of aging in polymeric materials and polymer bonded composites. It is strongly believed that fracture toughness is the property that is most sensitive to chemical and microstructural changes that occur in these materials. Initial characterization was pursued using several common techniques including 3 and 4 point bend tests, Brazil tests, and Brazil tests with initial damage [2–6]. Each of these test geometries were non-ideal because the crack growth rates were not easily controlled and there was significant variability in test results. Finally, the "compression-fracture" test geometry introduced by Sammis *et al.* [1] was settled on because the crack growth rate can be easily controlled using the loading rate and repeat tests show minimal sample to sample variation. In addition to the fracture toughness tests, simple tensile tests were performed to find Poisson's ratio and the stiffness modulus. Initially, the semi-empirical equation introduced by Sammis *et al.* [1] for fracture toughness was used to generate a curve for the fracture toughness. Evaluation of the deformation data indicated that a linear elastic fracture mechanics (LEFM) solution could be used to find the unique solution for fracture toughness. The materials studied for this research were Diallyl Phthalate (DAP), a rigid polymer having good dimensional stability and impact resistance, and an epoxy bonded ceramic (EBC). The binder content of the EBC is 10 wt% (~22 vol.%) with a ceramic size distribution varying from 10 to 350 μm. The materials have little strain to failure; therefore, lubrication was assumed to be unnecessary.

[a] Corresponding author: cady@lanl.gov

2. Experimental methods

Mechanical testing was conducted on an Instron 1125 electro-mechanical load frame with a MTS "renew" control package for slow loading rate tests and on an MTS 810 servo-hydraulic system with a cross-head speed up to 4.5 m/s for the dynamic experiments. For the low rate tests a sub press was used to ensure axial loading and minimize the risk of off axis deflection. No lubrication was used at the loading surfaces between the frame and the sample but optically flat platens were used as the contact surface. For the high rate tests the sample was bonded to the fixed side of the load chain using vacuum grease. Because failure was seen to occur at small displacements and is caused by tensile crack growth, the increased friction was assumed to have little influence on the test results. As in the low rate tests the unconstrained end of the sample was unlubricated. Experimental data are captured simultaneously on the load frame controller and on an image capture system for the low rate tests. At high rates, contact and failure images were synchronized with the load-displacement data so that the images could be matched to corresponding loads.

The characterization technique used to measure displacement and strain in these experiments is the digital image correlation (DIC). DIC has been demonstrated to be robust, flexible, applicable to large deformation and over a wide range of size scales and very affordable (as sophisticated optics instrumentation is usually not required). The underlying principle of digital image correlation as a deformation measurement technique is rather simple. It relies on the computer vision approach to extract the whole-field displacement data, that is, by comparing the features in a pair of digital images of a specimen surface before and after deformation. Since the displacements are calculated directly by correlating

Article available at http://www.epj-conferences.org or http://dx.doi.org/10.1051/epjconf/20159401012

the two digital images, processing of fringe patterns or other secondary phenomena to yield deformations is not required. Details of standard DIC analysis are described elsewhere [7].

The advantage to using DIC is that the images can be used to look at both the global strain and local strain, and it is possible to reanalyze the data in many different ways. This means that it is possible to get a large amount of information from a single experiment. The DIC technique can detect motion as small as 0.02 pixels. As a result, if the gage section is 100 pixels across in the image, the accuracy of strain measurement will be 0.02%, assuming the deformation in the gage section is uniform. The larger the image size the higher accuracy of measurement can achieve. Since the strain from the DIC technique does not have any machine compliance, it very accurately shows the early loading stages. It is essentially like putting a strain gage of variable size or multiple strain gages on the test specimen. Further details on the stress analysis for the "compression-fracture" geometry will be described in the section on sample geometry.

Another key characteristic of the DIC technique is the ability to detect a breakdown in image quality. For the analysis conducted in this study this is a critical capability of the software provided. This parameter is known as the correlation coefficient. The definition of a crack in this study is defined as the value when "image correlation" begins to break down. It is through this correlation breakdown that we can measure both the crack length and the crack growth rates. Both measurements are useful and necessary in in order to calculate fracture toughness.

3. Sample geometry

One of the principle drivers of this work was the limited availability of the materials used in this study. Initial tensile samples were machined having a gage dimension of 14 mm length × 3 mm width and 3 mm thickness. It was possible to measure the Poisson's ratio of these tensile samples using the DIC technique described above. As a result of preliminary tests it was determined that a test geometry that would slow the crack growth rate was necessary. The sample geometry was chosen based on work of Sammis *et al.* [1] and Plaisted, *et al.* [8]. In this case a rectangular sample with a center hole was used as the new test sample. It will be referred to as the "compression-fracture" (CF) sample throughout this paper. The nominal dimensions of the test sample are 25 mm in length or height (H) by 9 mm width (W) by the available material thickness (~4 mm for the DAP samples and 9 mm for the EBC material). A 3 mm diameter hole was machined in the sample centered on the length and width as shown in Fig. 1.

4. Experimental evaluation

Initially, the intention of this geometry was to generate crack growth in the test sample that is slow enough to capture images during deformation in order to measure crack growth rates. Early experiments showed that the load displacement curves do not provide any indication

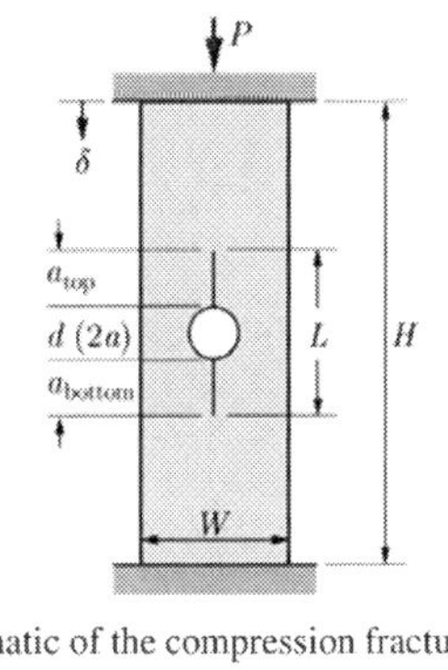

Figure 1. Schematic of the compression fracture test geometry.

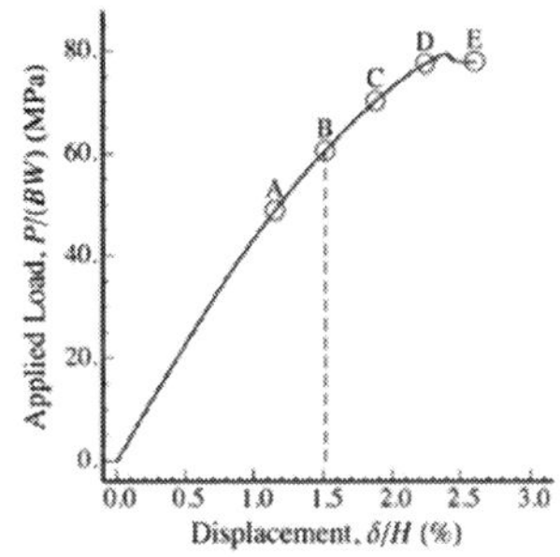

Figure 2. Example load displacement curve for DAP with indicators that correspond to sequential strain and correlation maps in the next figure. Location B is the moment when the correlation value begins to change rapidly.

of damage initiation. This is actually a positive as it means that the onset of cracking and crack growth is a graceful process for this geometry. However, it also makes it necessary to compare the DIC image evaluation with the load displacement data. An example plot of the load displacement curve (Fig. 2) for DAP is shown with indicators showing times when corresponding strain field maps are presented (Fig. 3). Using the technique to determine the correlation coefficient it is possible to define the location and time (displacement) when damage initiates. Images corresponding to point "B" give an indication of the onset of damage (as seen in the correlation maps) with further crack growth progressing to and beyond position "E". By analyzing all images using this technique it is possible to observe that there are two dominant cracks present, one propagating upward and the other down. This methodology is used to generate unique correlation and strain maps for every frame of the video that captured the test, thus, providing information about crack location, length, and growth rate.

It is necessary to elaborate on the technique used to determine the initiation and boundary of the cracks. This is done first by analyzing the maximum correlation value as it evolves during a test (Fig. 4). This plot shows a relatively flat, unchanging, value during the early part of the test, typically associated with the elastic loading of the sample. At the time when a crack initiates and begins to grow the correlation value also increases rapidly. The correlation value where this transition occurs is called the critical correlation coefficient. This coefficient will be used

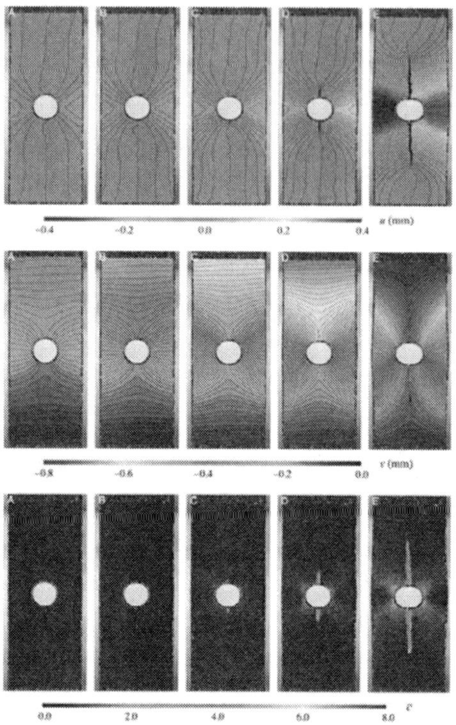

Figure 3. Images, marked with corresponding letters, associated with the moments in time indicated in Fig. 3. The top row of images is for the horizontal displacement fields, u. The middle row is for vertical displacement fields, v, and the bottom row are maps of the correlation coefficient.

to track the boundary of the "damaged" region showing where the crack is located and the rate at which it grows for the entire duration of the test. The maximum correlation value is not specifically related to a single point but once a crack is initiated the correlation value of that point will increase rapidly and this curve can be thought of as an accurate representation of the correlation coefficient at the damage initiation site. Typically, the maximum correlation value is associated with the region, if not exact position, of the onset of damage and the perimeter of the damage region can be traced when the value of correlation exceeds the critical correlation value.

The crack does not grow uniformly in both directions during the test (Fig. 5). The load versus displacement and the crack length versus displacement as analyzed from the correlation map (Fig. 3) shows this non-uniform crack growth.

5. Fracture toughness calculation

5.1. Semi-empirical solution

The fracture toughness, or critical stress intensity factor for the compression-fracture specimen is derived from the following equation. It has been modified to reflect the fact that the crack lengths may be different as they propagate from the center hole toward the top and bottom of the test sample. In this case the initial hole diameter is

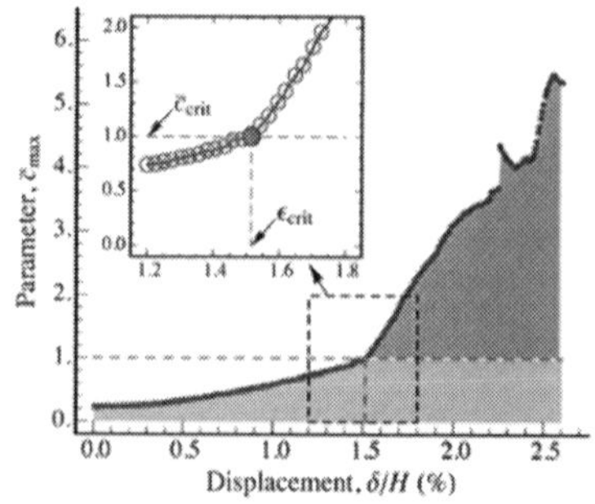

Figure 4. Plot of the correlation coefficient versus time. The inset shows how the critical correlation coefficient is determined. This plot is unique for each sample tested.

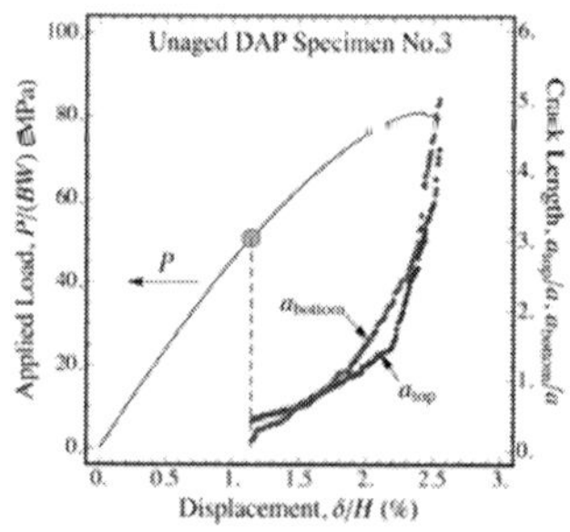

Figure 5. Plot of the applied load and crack length versus displacement with the upper and lower crack being individually represented.

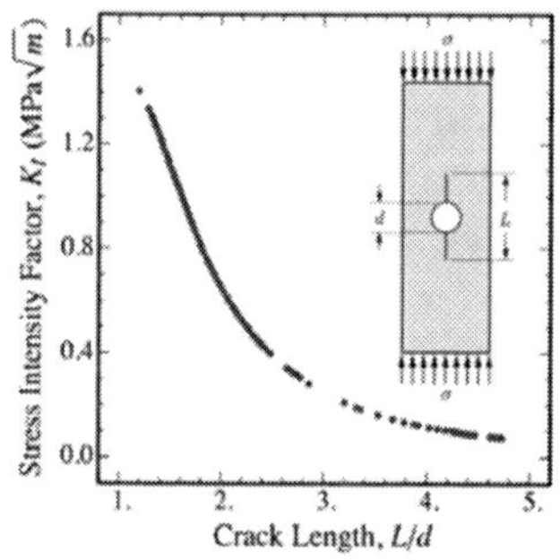

Figure 6. Example of the critical stress intensity factor for DAP using the compression fracture geometry.

a, the stress is the far field stress and is found by dividing the applied load by the loading area (width x thickness). a_{top} and a_{bottom} are the crack length propagating from the top and bottom of the hole respectively.

$$K_{IC} = -\left(\frac{a_{top} + a_{bottom}}{2a}\right)^{1/2} \left\{\frac{1.1}{\left(1 + \frac{a_{top}+a_{bottom}}{2a}^{3.3}\right)}\right\} \sigma\sqrt{\pi a}. \tag{1}$$

Using the equation above a calculation of the stress intensity factor as a function of crack length was calculated assuming that the stress applied is equal to the applied load divided by the width and thickness of the sample at the contact surface (Fig. 6). The solution for the stress intensity

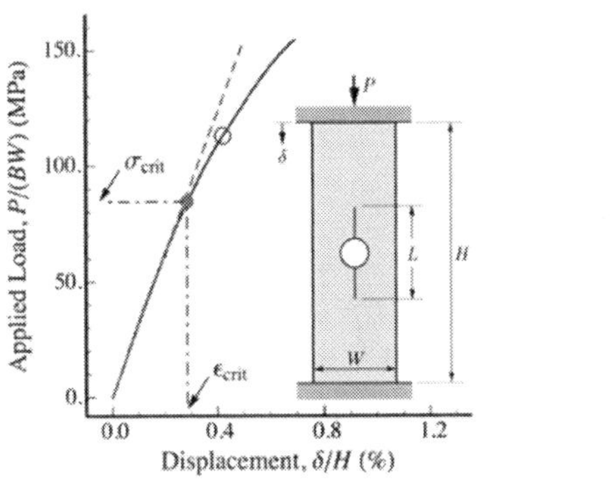

Figure 7. Comparison plot of the critical stress intensity for several experiments for un-aged and aged DAP.

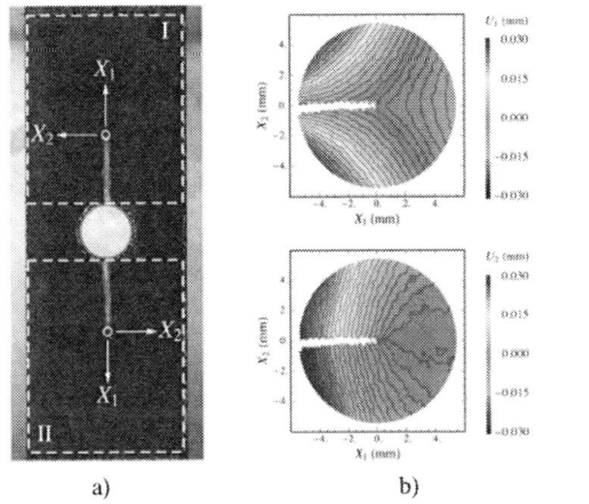

Figure 8. a) Partitioning of the cracks propagating from the center hole, and b) strain maps in the "new" coordinate system for one end of the test sample.

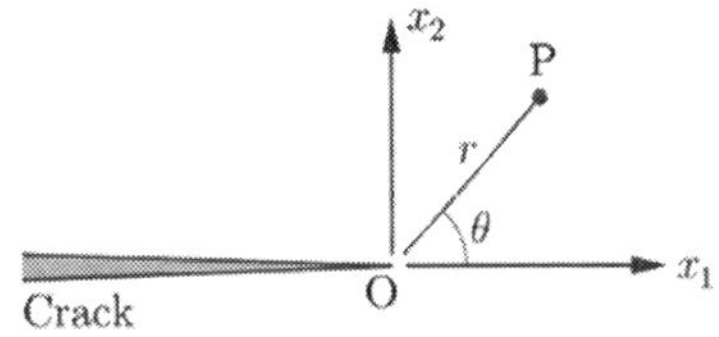

Figure 9. Crack-tip coordinate system.

factor is a semi-empirical in formula used to calculate the fracture toughness on a PMMA plate[1]. We have applied the technique as best we could to analyze our test results.

It should be noted that this experimental technique does not give a single value for the stress intensity factor. It does, however, show how crack growth and mechanical response to loading are correlated. Also, because the stress intensity factor is so closely tied to crack growth our hypothesis that this property is most sensitive to material change is likely correct.

It is important to remember that there are contributing factors that influence the initiation of fracture in all materials like those presented here. The factor that most greatly influences crack initiation is the finish quality of the hole. A very fine finish will likely cause initiation to be very high, whereas a rough finish is likely to improve the repeatability of the experiments. In the experiments presented here there was no specification made on the hole finish quality. This may be a contributing factor in the magnitude of the initial fracture toughness value.

5.2. Linear elastic fracture mechanics solution

In this section, we will apply Linearly Elastic Fracture Mechanics (LEFM) to further analyze the experimental data and extract the unique solution of fracture toughness for a given material. The definition of critical stress, σ_{crit} and critical deformation, ε_{crit}, when crack initiation occurs, is shown using solid symbol in Fig. 7. We see that prior to the critical moment, the applied load, P, varies linearly as a function of the deformation, δ, as indicated by the dashed line. The deviation from the linearity starts near the critical moment, which suggests that the apparent nonlinearity between applied load, P, and the deformation, δ, is the result of the changing boundary or the motion of the growing crack. Therefore, all material elements within the specimen, except those very close to the crack tip, only experience elastic deformation. As a result, Linearly Elastic Fracture Mechanics (LEFM) theory can be applied to analyze the experimental data.

The displacement field measurement, obtained using DIC, over the surface of the specimen forms the basis of our data analysis. Consider a particular moment during the compression of a specimen, as shown in Fig. 8a, where two tensile cracks have initiated and extended. For each tensile crack, we choose a coordinate system (X_1, X_2) originating at the crack tip, determined according to our experimental scheme, the X_1-axis is parallel to the loading axis, or in the crack growth direction, and the X_2-axis is perpendicular to the loading axis. A rectangular region that surrounds each tensile crack is identified with (I and II). The displacement field surrounding the tensile crack at one end of the sample is shown in Fig. 8b, where U_1 is the component of strain (displacement) along the X_1-direction and U_2 is the component of strain (displacement) along the X_2-direction. We also assign that (wording) ($U_1 = U_2 = 0$) at the origin of the (X_1, X_2) coordinate system.

For extracting crack tip parameters, a mathematical representation of the displacement field surrounding the crack is needed, and such a representation is usually expressed in the crack-tip coordinate system, as shown in Fig. 9. The (x_1, x_2) system is originated at the crack tip and the x_1-axis is tangent to the crack at the crack tip and x_2-axis is perpendicular to the crack.

According to the asymptotic analysis of the elastic fracture mechanics [9], the displacement field can be represented in the asymptotic form,

$$\begin{aligned} u(r,\theta) = \Sigma_{n=1}^{\infty} \frac{A_n r^{n/2}}{2\mu\sqrt{2\pi}} \left\{ \kappa \cos\frac{n\theta}{2} - \frac{n}{2}\cos\left(\frac{n}{2}-2\right)\theta \right. \\ \left. + \left[\frac{n}{2} + (-1)^n\right]\cos\frac{n\theta}{2} \right\} + \Sigma_{n=1}^{\infty} \frac{B_n r^{n/2}}{2\mu\sqrt{2\pi}} \\ \times \left\{ \kappa \sin\frac{n\theta}{2} - \frac{n}{2}\sin\left(\frac{n}{2}-2\right)\theta + \left[\frac{n}{2} - (-1)^n\right]\sin\frac{n\theta}{2} \right\}, \end{aligned} \quad (2)$$

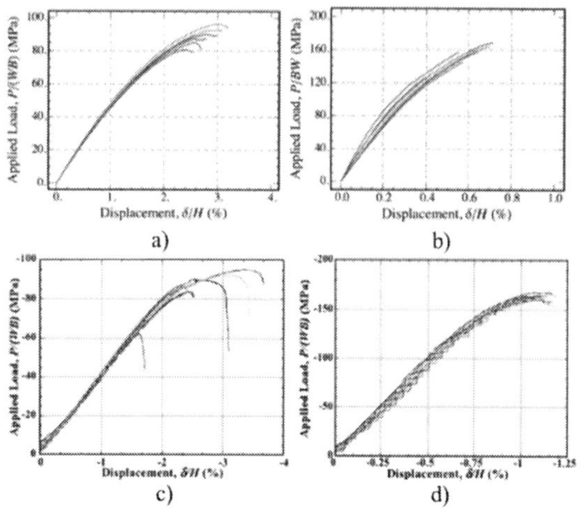

Figure 10. Typical load displacement curves for a) DAP, low rate, b) EBC, low rate, c) DAP high rate and d) EBC high rate.

$$v(r,\theta) = \Sigma_{n=1}^{\infty} \frac{A_n r^{n/2}}{2\mu\sqrt{2\pi}} \left\{ \kappa \sin\frac{n\theta}{2} - \frac{n}{2}\sin\left(\frac{n}{2}-2\right)\theta \right.$$
$$\left. - \left[\frac{n}{2} + (-1)^n\right]\sin\frac{n\theta}{2} \right\} + \Sigma_{n=1}^{\infty} \frac{B_n r^{n/2}}{2\mu\sqrt{2\pi}}$$
$$\times \left\{ -\kappa\cos\frac{n\theta}{2} - \frac{n}{2}\cos\left(\frac{n}{2}-2\right)\theta + \left[\frac{n}{2} - (-1)^n\right]\cos\frac{n\theta}{2} \right\}. \tag{3}$$

Where (r,θ) is the polar coordinates of a point, μ is the shear modulus of the material, and parameter κ is related to the Poisson's ratio v through

$$\kappa = \begin{cases} \dfrac{3-v}{1+v}, & \text{for plane stress,} \\ 3-4v, & \text{for plane strain.} \end{cases}$$

A_n and B_n are the coefficients of the asymptotic expansion and they cannot be determined by the asymptotic analysis itself but by the far-field conditions. In particular, for the leading term, $n = 1$, we have $A_I = K_I$ and $B_I = K_{II}$, where K_I and K_{II} are the stress intensity factors at the crack tip.

The asymptotic representation for a dynamically growing crack will be used to extract the dynamic stress intensity factor, or the dynamic fracture toughness, at the crack tip. Such an asymptotic representation can be found in [10,11].

6. Experimental results

Typical load displacement responses for the two materials in this study are shown Fig. 10. From these plots it can be concluded that the test results are reproducible with acceptable differences in their loading behavior for all loading rates. These results also indicate that failure occurs at nominally the same load level independent of rate (with one exception). The principle difference between the low rate and high rate tests is seen in the strain to failure. In general there is an increase in the strain to failure as the loading rate increased. This is consistent with other high stiffness, low ductility materials.

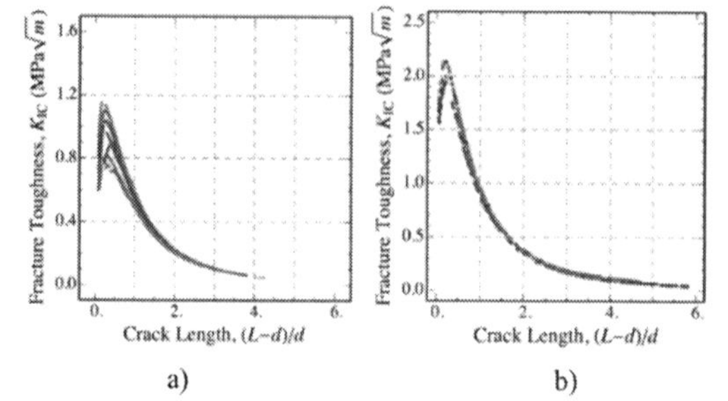

Figure 11. Plots of the semi-empirical solution for fracture toughness for a) DAP and b) EBC.

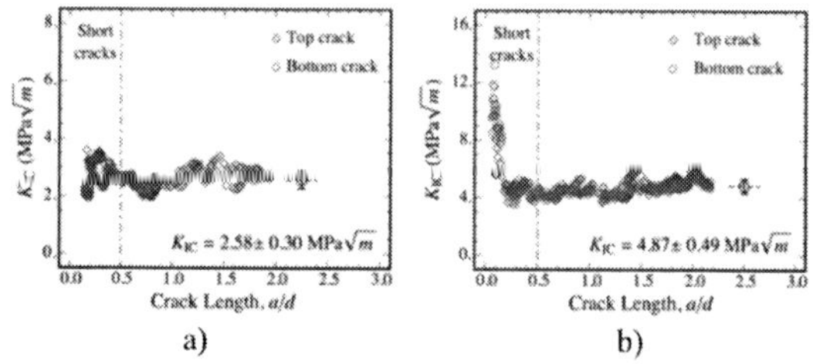

Figure 12. Example of LEFM solution for slow crack growth (K_{IC}) in a) DAP and b) EBC.

The crack velocity was measured for the different materials over the loading rates and was found to be somewhat variable. For the EBC material, the crack velocity was fairly constant through most of the deformation but slowed during the last ~15% of each test. For the low loading rate test the crack velocity averaged 0.18 mm/s and at the higher loading rate it was ~2.5 m/s. The DAP material has a crack growth behavior that increases with displacement for the duration of the test (see Fig. 5).

A plot showing with the results of several tests using the semi-empirical solution for both materials shows reasonably good reproducibility (Fig. 11), if not the well understood material property definition for fracture toughness.

Typical plots for the LEFM solution for the low loading rate tests are presented below (Fig. 12). These plots provide insight into the development of fracture toughness. If the crack is arrested the value of K_{IC} increases, when it is growing quickly, K_{IC} drops. Uniform crack growth reduces the magnitude of the scatter and there is a minimum crack length for uniform crack growth on the order of the initial hole radius.

Uncertainties in the value for fracture toughness calculated using LEFM come from two sources. First, the scatter found within the analysis of a single test, and second, sample to sample variation. It is thought that, like in metals, a representative sample size is necessary to randomize the material being tested that will produce more uniform test results. At this point this is conjecture with out experimental support.

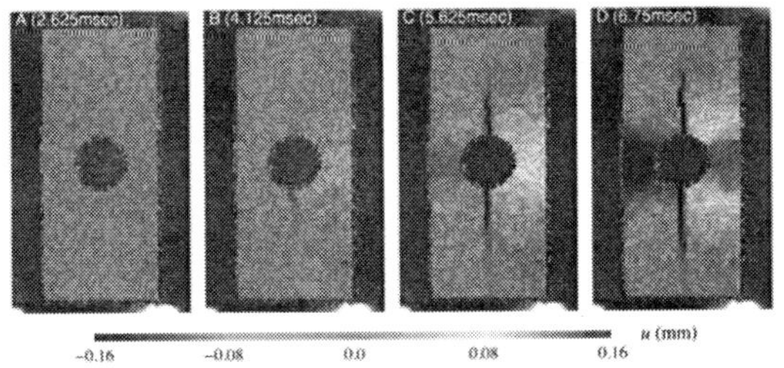

Figure 13. Displacement field map for EBC showing crack position as a function of time.

The challenge of the high strain rate testing is the resolution of the high speed camera system used. The system available for these experiments was only 320 pixel by 260 pixel. Unfortunately this made the determination of the crack location somewhat speculative and for the experiments conducted here the location of the crack was best determined from the u displacement field (Fig. 13) instead of the correlation field that was used in the earlier analysis.

The image quality made it challenging to generate the correlation filed maps that are essential in tracking precisely the location of the crack tip. Since this is critical information in determining the fracture toughness the calculation for the fracture toughness using this technique will be postponed until a camera with a higher resolution can be found for these experiments.

7. Conclusions

The compression-fracture test geometry is suitable for measuring the fracture toughness of the materials in question and coordinated testing and evaluation techniques using high speed imaging, digital image correlation, and standard data acquisition from the load frames allowed for a calculation of fracture toughness using the semi-empirical formula and LEFM with the only limitation of the analysis being subject to the capabilities of the equipment available.

The analysis technique to understand the fracture toughness of brittle and quasi-brittle materials was developed in this work. One of the benefits of using image capture and DIC is that it allows for re-evaluation of the test at any time in the future since the entire test is recorded for evaluation. A scheme for quantifying the extent of damage and cracking was also employed to determine the moment of crack initiation and the history of crack growth. This information was used to calculate the fracture toughness once the images were fully characterized.

The both materials exhibited easily traceable cracks and easily measureable crack growth rates that allowed for consistent calculation of the fracture toughness using the compression fracture test. A comparison fracture processes between the two materials studied indicates that the crack growth rate in the DAP material gradually increases, whereas, in the EBC material the growth rate is fairly uniform at the beginning of the test and gradually decreases near the end of the test due to loading constraints inherent in the test geometry.

The results for the stress intensity factor for the semi-empirical solution has many constraints and assumptions which may not be correct, however, it does provide a straight forward comparison tool that will likely show material changes if they are present and significant. The mathematics and analysis scheme for LEFM solution has just recently been developed and the results shown here are for the first set of samples analyzed using this technique. One of the benefits of this technique is that the development of the crack and calculation of fracture toughness are tied together during this analysis. Therefore the value determined for K_{IC} can be averaged over a test with non-uniform crack extension at the beginning or end of the test removed from the solution.

The solution for the dynamic fracture toughness using LEFM will require using a modification to Eqs. (2) and (3) that are described by Freund [11]. It was possible to predict the approximate location of the crack from the displacement fields and therefor an approximation of the crack velocity. The test geometry is suitable for high rate tests, but a system with high resolution at high framing rates is necessary to determine the crack location.

Los Alamos National Laboratory is operated by Los Alamos National Security, LLC, for the National Nuclear Security Administration of the U.S. Department of Energy under contract DE-AC52-06NA25396.

References

[1] C. G. Sammis and M.F. Ashby, *Acta Metall.*, **34** (3) pp 511–526 (1986)

[2] Y. Huang, C. Liu, M.G. Stout *Acta mater* **44** (3) pp 1223–1232 (1996)

[3] C. Liu, Y. Huang, M.L. Lovato, M.G. Stout, *I J Fract*, **87** (3) pp 241–263 (1997)

[4] R. Bedsole, H.V. Tippur. *J Eng Mat and Tech Conference Proceedings* **135** (3) pp 1–11 (2013)

[5] K.C. Jajam, H.V. Tippur, *Composites Part B, Eng* **43** (8) pp 3467–3481 (2012)

[6] K.C. Jajam, M.M. Rahman, M.V. Hosur, H.V. Tippur, *Composites Part A, Appl Sci & Mfg* **59**, pp 57–69 (2014)

[7] M.A. Sutton, S.R. McNeill, J.D. Helm, Y.L. Chao, (2000) Advances in Two-Dimensional and Three-Dimensional Computer Vision, *Topics Appl. Phys.* **77**, pp 323–372 (2000)

[8] T.A. Plaisted, A.V. Amirkhizi, S. Nemat-Nasser, *I J Fract*, **141**, pp 447–457 (2006)

[9] J.R. Rice, *Fracture – An Advanced Treatise*, Vol. II, H. Liebowitz (ed.), Academic, New York, pp. 191–308, 1968

[10] L.B. Freund, *Dynamic Fracture Mechanics*, Cambridge University Press, New York, 1990

[11] C. Liu and A.J. Rosakis, *Journal of Elasticity*, **35**, pp. 27–60, 1994

EPJ Web of Conferences **94**, 01013 (2015)
DOI: 10.1051/epjconf/20159401013

Why does necking ignore notches in dynamic tension?

Y. Rotbaum, S. Osovski, and D. Rittel[a]

Faculty of Mechanical Engineering, Technion, 32000 Haifa, Israel

Abstract. Recent experimental work has revealed that necking of tensile specimens, subjected to dynamic loading, is a deterministic phenomenon, governed by the applied boundary conditions. Furthermore it was shown that the potential sited, dictated by the boundary conditions, may prevail even in the presence of a notch, thus necking may occur away of the notched region. The present paper combines experimental and numerical work to address this issue. Specifically, it is shown that the dynamic tensile failure locus is dictated by both the applied velocity boundary condition and the material mechanical properties, specifically strain-rate sensitivity and strain-rate hardening. It is shown that at sufficiently high impact velocities, the flows stress in the notch vicinity becomes quite higher than in the rest of the specimen, so that while the former resists deformation, it transfers the load to the latter, resulting in the formation of a local neck and failure away from the notch. Small local perturbations in the material properties are shown to be sufficient to stabilize the structure under local failure until a neck forms elsewhere. While the physical observations are quite counterintuitive with respect to the engineering views of stress concentrator's effect, the present work rationalizes those observations and also provides information for the designers of dynamically tensioned structures that may contain notches or similar flaws.

1. Necking

The mechanical response of a structural element under external loads may be strongly influenced by the presence of geometrical perturbations, amounting to a local increase in the stress field surrounding it. The importance of notched members under tension has led to a vast body of works focusing on evaluating the stress intensity factors around geometrical discontinuities [1–4], and specifically the differences between quasi-static and dynamic loading scenarios [5–7]. An underlying assumption in all of the above-mentioned studies is that the fracture locus will be that of the geometrical imperfection, or any other given flaw, even in situations where inertia plays an important role. Material heterogeneities related to the manufacturing process are vastly ignored or considered secondary, with respect to the presence of the dominant flaw. It was recently shown that under dynamic tensile loading, the necking locations is a deterministic event resulting from the applied boundary conditions [8]. Furthermore, it was shown experimentally that the necking location, as dictated by the applied boundary conditions, may prevail even in the presence of a geometrical perturbation introduced as a notch [9]. The results of presented in [9] imply that *the presence of a geometrical imperfection will not necessarily dictate the dynamic failure locus, as commonly assumed in the literature.*

Here we present a hybrid experimental-numerical study, examining several potent physical factors responsible for the selection of the dynamic failure locus.

Experimental results for 90 dynamic tensile notched 15-5 PH (annealed) and 4340 steel specimens are presented, and are shown to follow the trends found in [9],

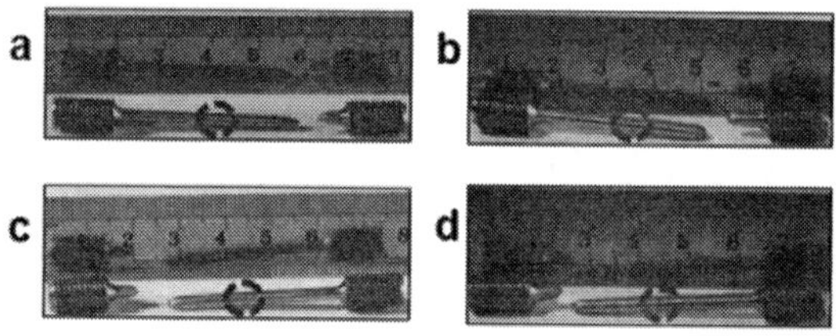

Figure 1. Dynamically loaded notched tensile specimens showing failure outside of the notch. Specimens shown in (a) and (b) were loaded at 15 m/s and failed on the impacted side. Specimens shown in (c) and (d) were loaded at 16 m/s and 18 m/s respectively and failed at the opposite side.

as shown in Fig. 1. The notch depth was chosen as such that will ensure failure within the notch under quasi-static loading.

The dynamic tensile tests were modeled using the commercial finite element package Abaqus explicit [10]. The boundary conditions used in the simulations, were set as a symmetrical trapezoidal velocity profile, applied on the free surface of the incident bar, with maximum amplitude in the range of 7–21 m/s. The pulse duration was taken to be the same as in the experiments. The response of the elastic viscoplastic material is calculated within the framework of J_2 flow theory. The viscoplastic response was taken as such that will account for increased rate sensitivity at high strain rates, following Zhou et al. [11]. The viscoplastic response is described by a power law up to a transition strain rate $\dot{\varepsilon}_m$ from which it becomes exponentials. For a detailed description of the numerical mode, the reader is referred to [12]. The material properties used in the simulations are of a generic character and do not represent accurately the experimental materials of

[a] Corresponding author: meritte1@technion.ac.il

Article available at http://www.epj-conferences.org or http://dx.doi.org/10.1051/epjconf/20159401013

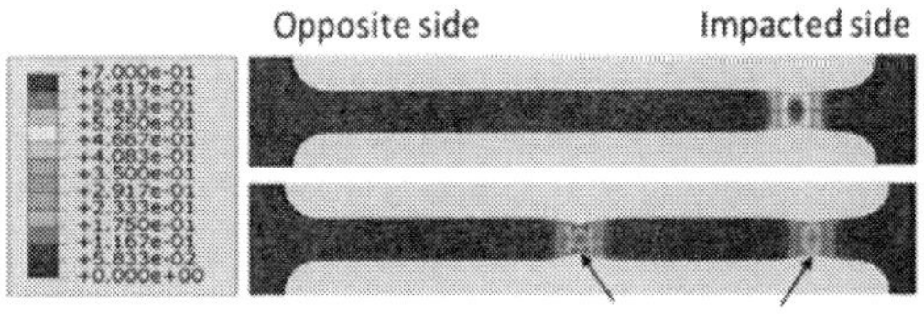

Figure 2. Numerical simulation of dynamically deformed representative steel specimens stretched at a peak velocity of 17.5 [m/s]. The upper specimen is smooth while the lower contains a 10% notch. The color map represents the equivalent plastic strain values.

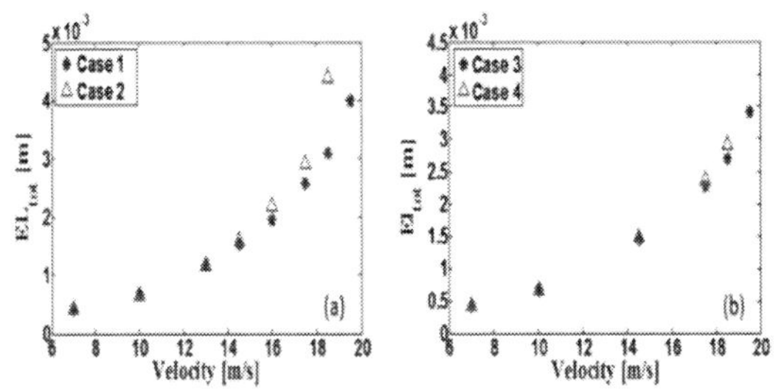

Figure 3. Macroscopic elongation at equal radial displacement values (meeting time). (A) Cases 1 and 2. (b) Cases 3 and 4.

the study. As such, they are used to explore trends and not reproduce experimental results. Representative results of the numerical simulations of smooth and 10% center-notched samples, stretched at a peak velocity of 17.5 m/s, are shown in Fig. 2. For a smooth specimen, the failure location is on the impacted side, in correspondence with [8]. For the notched specimen, one observes a strong competition between the neck on the impacted side and the imperfection in the middle of the gauge. This is the outcome of the joint influence of inertia and stress waves against local the stress concentration.

In the following, the competition between potential necking sites is described by comparing the evolution of the local radial displacement with respect to the total elongations. The overall elongation at which the radial displacement in the two potential sites is equal, provides an estimation as to the prevalence of the notched site over the one dictated by the boundary conditions. Once the two curves meet, the displacement under one of them starts to grow rapidly while the other is unloaded. The role of the viscoplastic behavior was studied by varying the transition strain rate $\dot{\varepsilon}_m$ from 8000 s^{-1} (cases 1&2) to from 16000 s^{-1} (cases 3&4). The exponent of the rate dependency at high strain rate of cases 1&3 is identical and lower than the one used in cases 2&4. In Fig. 3 the effect of the viscoplastic behavior on the necking location is presented. One should emphasize that the case with a higher elongation prior to the prevalence of one of the potential sites, is more stable in the sense that larger total elongation to fracture can be reached. Thus, specimens at whom the competition between the two localization sites lasts longer, promotes a higher energy absorption capability.

An important issue, which is generally overlooked, concerns the machining of the notch. The latter

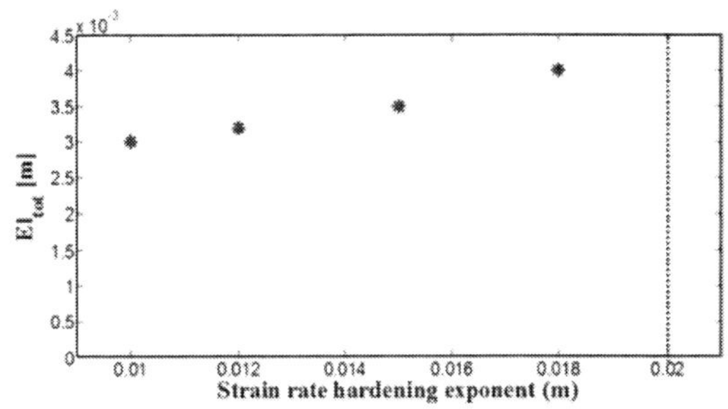

Figure 4. The nominal elongation of the specimen at the meeting time for $V_{max} = 18.5$ m/s, for different values of m_{notch}. Note that for m = 0.02, there is no meeting point as the neck dominates throughout the process.

has a definite local hardening effect, which can be somewhat mitigated by annealing treatments. An obvious consequence of the machining process is an increase in the yield strength due to the storage of cold work. The effect of small variations in the strain rate sensitivity was studied for case 1 with a fixed maximal velocity of 18.5 m/s. For that purpose an area with a radius of 130 μm around the notch root was selected to have slightly different material properties than the rest of the specimen. The strain rate hardening exponent in the vicinity of the notch-tip was varied in the range $m_{notch} = 0.012$, 0.015, 0.017, 0.020. The maximum difference in stress between the notch and the rest of the specimen at a plastic strain of 40% for $m_{notch} = 0.020$ is 48 MPa which corresponds to ~4.65%. In Fig. 4 the total elongation up to the meeting time of the radial displacements in the two potential sites is plotted with respect to m_{notch}. As evident from Fig. 4, when the notch area is more hardened, the sample elongates more before reaching to its final failure state, indicating that the notch is less dominant in the competition between the potential sites. For $m_{notch} = 0.02$, the necking site dictated by the boundary conditions prevailed throughout the deformation process. Therefore one can suggest that minor perturbations in the material flow stress can cause a significant change in the overall structural response.

The main conclusions arising from the work presented here are:

1. The locus of dynamic tensile failure of a flawed (notched) structure is dictated by the applied velocity as a boundary condition, strain rate sensitivity and rate-hardening of the material.
2. Small perturbations in the material's constitutive behavior, resulting from the machining process, can promote a significant change in the structural response under dynamic tensile loads.
3. At sufficiently high velocities, the very high strain rates in the notch stabilize the deformation and transfer the load to another part of the specimen which ultimately grows a neck there.
4. When designing a tensile notched structural component, the dynamic mechanical properties and boundary conditions should both be carefully taken into account to predict the failure locus.

References

[1] E. Coker, K. Chakko, Y. Satake, Proc Inst Eng Shipbuilding Scotland, **63**, 34–94 (1919)

[2] R. Howland, Phil. Trans. Roy. Soc. Series A, Containing Papers of a Mathematical or Physical Character: 49–86 (1930)

[3] M. Strandberg, Eng. Fract. Mech., **68**, 577–585 (2001)

[4] M. Zappalorto, P. Lazzarin, Eng. Fract. Mech., **78**, 2691–2706 (2011)

[5] W. James, W. North, J Strain Anal, **4**, 261–266 (1969)

[6] N. Nakayama, M. Ohashi, H. Takeishi, JSME Int. J Series A, Solid Mech. Mat. Eng, **41**, 326–331 (1998)

[7] H. Matsumoto, T. Adachi, Y. Kakuhama, K. Fukuzawa, JSME Int. J. Ser 1, Solid Mech., Strength of Materials, **33**, 37–43 (1990)

[8] S. Osovski, D. Rittel, J.A. Rodríguez-Martínez, R. Zaera, Dynamic tensile necking: influence of specimen geometry and boundary conditions. Mech. Mat. **62**, 1–13 (2013)

[9] D. Rittel, Y. Rotbaum, J.A. Rodríguez-Martínez, D. Sory, R. Zaera, Exp. Mech., **54**(6), 1099–1109 (2014)

[10] Hibbett, Karlsson, Sorensen, Hibbitt. (1998) ABAQUS/standard: User's Manual, Hibbitt, Karlsson & Sorensen

[11] M. Zhou, A. Needleman, R.J. Clifton, J. Mech. Phys. Sol., **42**, 423–458 (1994)

[12] Y. Rotbaum, S. Osovski, D. Rittel, J. Mech. Phys. Sol., **78**, 173–185 (2015)

EPJ Web of Conferences **94**, 01014 (2015)
DOI: 10.1051/epjconf/20159401014

Experimental studies on the deformation and rupture of thin metal plates subject to underwater shock wave loading

Pengwan Chen[a], Han Liu, Shaolong Zhang, Haibo Liu, Ang Chen, and Baoqiao Guo

State Key Laboratory of Explosion Science and Technology, Beijing Institute of Technology, Beijing 100081, China

Abstract. In this paper, the dynamic deformation and rupture of thin metal plates subject to underwater shock wave loading are studied by using high-speed 3D digital image correlation (3D-DIC). An equivalent device consist of a gas gun and a water anvil tube was used to supplying an exponentially decaying pressure in lieu of explosive detonation which acted on the panel specimen. The thin metal plate is clamped on the end of the shock tube by a flange. The deformation and rupture process of the metal plates subject to underwater shock waves are recorded by two high-speed cameras. The shape, displacement fields and strain fields of the metal plates under dynamic loading are obtained by using VIC-3D digital image correlation software. The strain gauges also were used to monitor the structural response on the selected position for comparison. The DIC data and the strain gauges results show a high level of correlation, and 3D-DIC is proven to be an effective method to measure 3D full-field dynamic response of structures under underwater impact loading. The effects of pre-notches on the failure modes of thin circular plate were also discussed.

1. Introduction

Plated structures are a kind of important basic element of marine structures and battleships, which may be subjected to an underwater explosion by the attack of a torpedo or a depth charge [1]. In ship design application, the ship plate shell with a small curvature is supported by welded stiffeners at its edges, so the shell between the stiffeners can be considered as a flat panel [2]. Non-contact underwater explosion is the important source of threat on the ship structure, and Understanding the response include deformation and failure modes subjected can supply an important reference for designing.

The response of clamped plate structures subjected to air and underwater blast loading has been studied for many years. In the underwater impact area, Ramajeyathilagam et al. [3–5] carried out many experimental and numerical investigations on the deformation and tensile tearing of air-backed metal plates and shell panels. The results indicated that the shock impulsive can be estimated in accordance with the Cole's empirical formula [6] and Taylor's plate theory [7]. Hung et al. [8] also presented experimental studies on the dynamic elastic response of the aluminium alloy panel subjected blast loading in the water tank. The research reports the underwater pressure history, the moving acceleration history and strain history of the plate surface.

For the underwater blast experimental study in laboratory scale, the reappear of underwater environment, production of the ideal blast impulsive loading and safety protection are all difficult and limited by the technology. With the development of technology, an experimental apparatus incorporating fluid-structure interaction (FSI) effects was recently developed to test scaled structures by Espinosa et al. [9]. The set-up allows characterization of the response of solid and sandwich structures subjected to underwater blast impulsive loading. The calibration plate impact experiments confirmed that the FSI setup can generate an exponentially decaying pressure history. The full field out-of-plane deformation profile of annealed Steel plates was recorded in real time by Shadow Moiré and high speed photography. McShane et al. [10] also designed a similar equivalent device based on the impact produced the underwater shock wave loading. The potentiation of polymer coating for the resistence performance of the pure copper plates subjected underwater impact loading was analyzed by used this set-up, and four typical failure models are recorded by high-speed photography.

However the level of understanding of the response of these structures at these high loading rates is not as established as that under static conditions. In this paper, A FSI apparatus was developed to generate an exponentially decaying pressure in lieu of explosive detonation by Xiang [11] in reference of Espinosa's works [9]. Combined with this equivalent device, 3D DIC technology and high-speed photography, the dynamic response of air-backed copper circular plates with pre-notches subjected to underwater impulsive loading was recorded in real-time. Compared with the strain gauge results and DIC results, the accuracy and advantage of DIC technology was proven obviously. The final failure modes for specimens with different pre-notches are evaluated by recycling the specimens. Moreover, the effects of pre-notches on the failure modes of thin circular plate were discussed. This work will be useful to those involved in research into the response of structures to explosive loading, a subject that has become increasingly important with heightened public awareness of potential explosive threats to civilian safety.

[a] Corresponding author: pwchen@bit.edu.cn

Article available at http://www.epj-conferences.org or http://dx.doi.org/10.1051/epjconf/20159401014

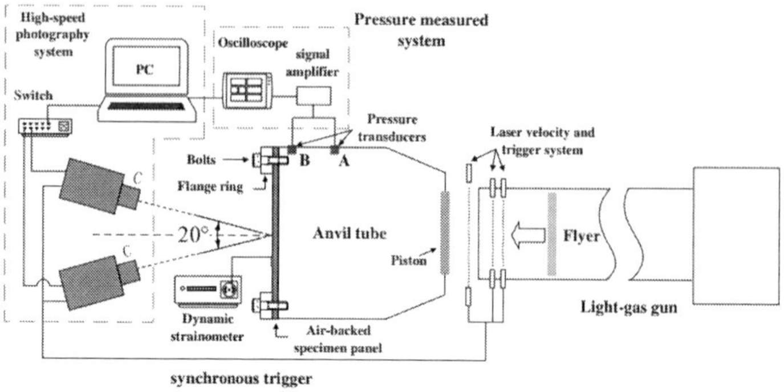

Figure 1. Principle diagram of the experiment setup.

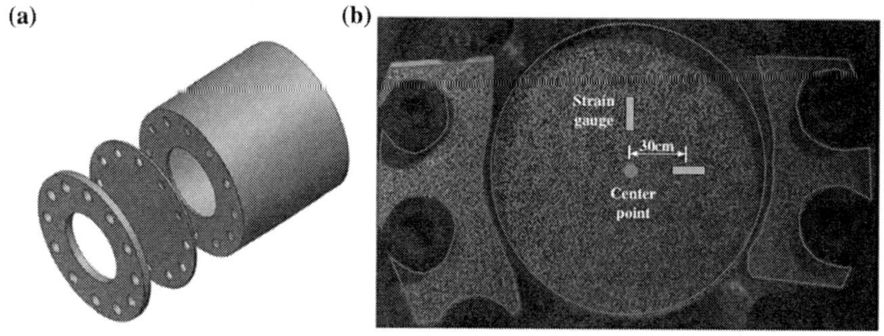

Figure 2. Principle diagram of the specimens: (a) the schematic of clamped support condition, (b) the schematic of selected computing area.

2. Experimental details

The simplified FSI equipment was developed by Xiang et al. (From School of Aerospace Engineering in Beijing Institute of Technology). The diagram of experimental setup is shown in Fig. 1. As seen from this figure, the simplified FSI equipment consists of two parts: a gas gun and a conical anvil tube filling up with water. A piston was sealing the conical anvil tube in the inlet of conical anvil tube opposite to the gas gun, and the disc specimens were clamped by the flange on the other bottom as shown in Fig. 2a. The flyer is driving by the gas gun and impact the piston, and the flyer speed and the piston thickness determine the peak pressure and decay time in the tube respectively [11]. In this paper, the strength of shock wave loading is control by changing the speed of the flyer (depended on the pressure of the gas gun). The speed of the flyer can be measured by two pairs of laser gauges position at the outlet of gas gun, and then the flyer fight through the laser trigger that synchronizer trigger the cameras. The pressure gauges respectively near the plate and in the center of conical anvil tube are used to measure the pressure of water shock wave during the loading event. So the peak pressure values respectively at a certain position A (in the anvil) can be predicted by Eq. (1), namely:

$$P_A = \frac{Z_w Z_p}{Z_w + Z_p} V_0 \left(\frac{D}{D_A}\right)^2, \tag{1}$$

$$Z_w = 40.82 \times 10^6, \quad Z_p = 1.46 \times 10^6.$$

Where V_0 is the the flyer impact velocity, Z_p and Z_w are the acoustic impedances of the piston of the piston and the water, where D and D_A are the diameters of the tube at the position A and impact locations.

Two Fastcam SA5 (produced by Photron Corporation in Japan) high-speed digital cameras were mounted in a stereo configuration (the angle is about 20°) to record the synchronized images of the deformation and rupture process of the sample specimens during the blast loading. The cameras are far from the explosive vessel and behind the observing windows in order to prevent vibration interference and protect the cameras during testing. In order to maximize the common field of view and the level of correlation, the two cameras were rotated and focused. The captured rate will be 50,000, and 75,000 frames per second. Three halogen lamps with a power of 1 kilowatt were used as the lighting source. The strain gauges were used to measurement the in-plane strain along radial and tangential directions shown in Fig. 2b.

In this paper, two circular plates made of red copper with a diameter of 292 mm and same thickness of 1 mm were preparation before the testes subject to underwater shock wave loading. Through changing the types of the specimen surface (no pre-notch or pre-notch), the effect of pre-crack on the performance of specimens under underwater shock loading is understood with different pressure strength. The shapes of pre-crack is cross, and the position of pre-crack are in the center of plates. The length of the cross crack is 30 mm, the width is 1 mm, and the depth is 0.5 mm. For comparison, a circular plate

Table 1. Summary of experimental conditions of all specimens.

No.	Clamped condition	Pre-notch	V_0(m/s)	P_0(MPa)	P_B(MPa)
I	fully	no	152.8	83.5	91.1
II	fully	cross	168.9	92.3	102.1
III	un-fully	cross	101.9	55.7	61.2

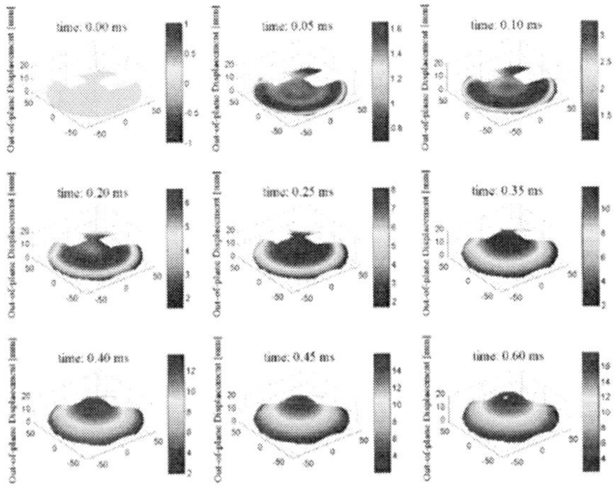

Figure 3. Principle diagram of the experiment setup.

with a diameter of 200 mm and same size cross pre-notch subjected a lower loading are also tested, and the specmen was only fixed while bolts not through the plate, hereinafter referred to as "un-fully clamped". Table 1 is a summary of experimental conditions of all specimens, including the support condition, the pre-notch of the metal plates, the impact velocity V_0, the theoretical peak pressure P_0 and the measured peak pressure at position A P_A.

Before prepared the random speckle of samples, the surface must be cleaned and polished. Then, the high contrast speckle pattern is prepared on the plate by spraying white paint and marking the random points by a black marker pen on the plate surface, and the black speckle dots with a diameter of about 5 pixels when seen by two cameras are thought perfect for the random high-contrast pattern. As shown in Fig. 2b, three areas on the specimen surface and flange surface were selected as the AOI (area of interest) which is used to analyze the deformation fields by comparing the difference before and after deformation. During the shock, the shock absorbers cannot completely prevented the motion of the anvil tube. So the DIC measured motion on the specimen surface is absolute displacement include the plate deflection and the tube rigid motion. The areas on the flange surface were used to calculate the rigid displacement of the anvil tube.

3. Results and discussion

3.1. DIC results of no pre-notched plate

The evolution of displacement (W) fields on the surface of no pre-notched specimen at different moments are

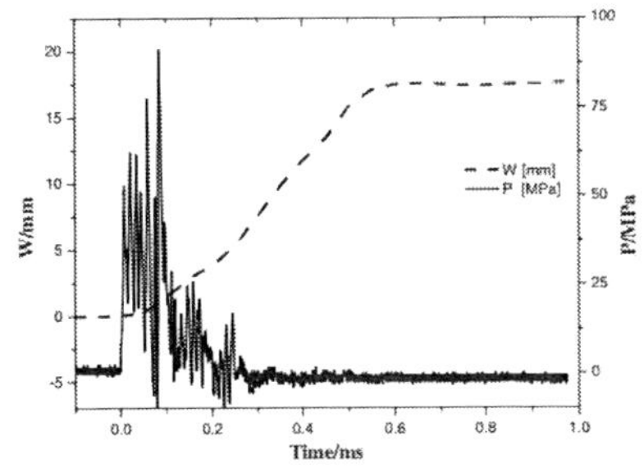

Figure 4. Diagram of the deflection history of point O and the pressure history of position B.

shown in Fig. 3. The major component is the out-of-plane displacement (absolute displacement). At 0.05 ms, a visible symmetric deformation ring can be found in the boundary of AOI, and then the deformation region rapidly spreads to the center along the radial direction until 0.25 ms. It is worth noticed that the shape of the deformation area is flat nearly except the deformation ring in the first, and evolves into a symmetric curved shape after 0.35 ms with the propagation of deformation ring appeared in the boundary. The displacement ring can be seen as plastic yield line of the first bending modal mode shape of the plate.

Figure 4 shows the comparison between the deflection history of centeral point O and the pressure history of position B for the same specimen. The initial moment of pressure history of position B is unified with the initial moment of the pressure history loading on the wet side of the specimen. At 0.1 ms, the first loading from the incident shock wave has vanished, the plate continuous to deform due to multiple followed incident and reflected shock waves and the inertial effects of the plates. It is indicated that the resultant plastic deformation is caused by the combined effect of the reflected shock, the incident shock and the support condition of the structure.

Strain is used to describe the relative change of the material's shape as a normalization of the deformation [35]. The evolution of the in-plane principal strain fields of the no pre-notched specimen at different moments is shown in Fig. 5. It can be seen that until 0.15 ms the higher principal strains occur close to the boundaries. During the loading, the incident underwater impulse forces all points of the panel to move out. But points close the clamping area are constrained and higher strains occur in the boundary areas. Further, when inertia forces and the reflection waves are taking over, strains are developing in the centre of the plate reaching first peak values 5 at 0.55 ms in Fig. 5. Somewhat counter-intuitively, the entire central and boundary region of the field of view is at a higher strain than the surrounding plastically deformed circular region in follow.

Figure 6 shows the comparison of the normal strains ε_{xx} and ε_{yy} between the DIC results and the strain gauge results on the strain gauge locations as shown in Fig. 2. The DIC results are acquired from the strain date of symmetric points to the center. From this figure, the strain histories of ε_{xx} and ε_{yy} all correlate well until 0.6 ms between the DIC and the strain gauge. After this moment, the

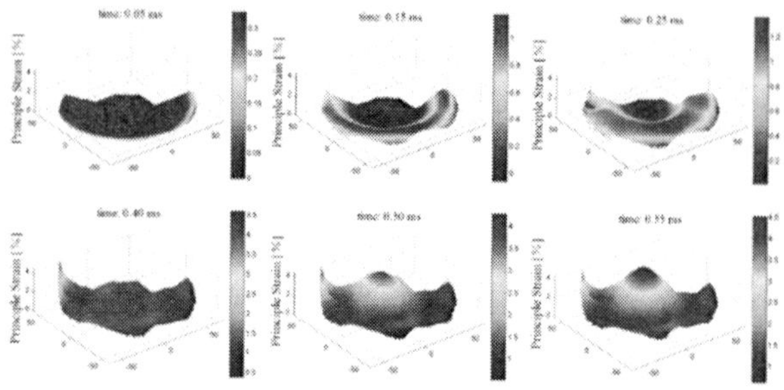

Figure 5. Full-field resultant in-plane principal strain at different time steps.

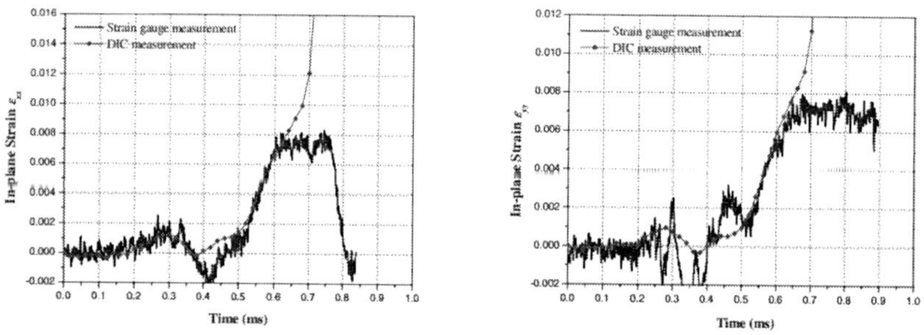

(a) histories of ε_{xx} (b) histories of ε_{yy}

Figure 6. Comparison of the normal strains ε_{xx} and ε_{yy} between the DIC results and the strain gauge results.

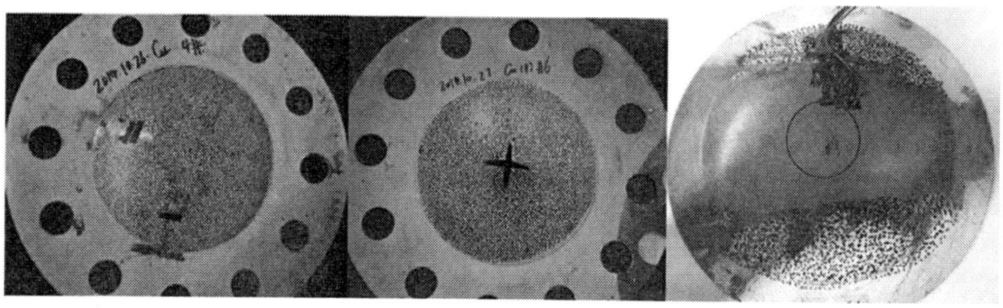

(a) specimen I b) specimen II (c) specimen III

Figure 7. Failure modes observed for different specimens.

strain history curves stop increasing and even decrease, possibility caused by the strain gauges falling off from the specimen surface. The comparison indicates that DIC technology is a dependable measured method.

3.2. Recovering and discussion

After the blast tests, all specimens were collected for further analysis. Figure 7 shows the photographs of four recovered specimens after tests. Three types of failure modes can be identified. For the specimen without pre-notch, only large ductile deformation can be observed in Fig. 7a; for the pre-notched specimen, large ductile deformation and local petalling at the location of pre-notch can be observed in Fig. 7b; for the un-fully clamped specimen, in addition to large ductile deformation, local necking in the center can also be observed in Fig. 7c, which is confined to the location of the pre-notch.

4. Conclusions

This paper focus on the dynamic deformation and rupture of pre-notched circular copper plates subjected to underwater impulse loading. High-speed photography, combined with 3D-DIC has been successfully used to acquire the out-of-plane displacement and in-plane principal strain on the surface of the circular plates during the blast process. Compared with the strain gauge results and DIC results, the accuracy and advantage of DIC technology was proven obviously. Through change the clamped condition and the loaded intensity, the evolution of the failure modes has been observed and discussed. Moreover, the effects of pre-notches on the failure modes of thin circular plate were discussed. 3D-DIC is also proven to be an effective method to measure 3D full-field dynamic response of structures under underwater impact loading.

References

[1] Rajendran, R.L.J. Marine Structures, **22**, 2 (2009)
[2] EN, F.. ONR, 1 (1947)
[3] Ramajeyathilagam K, V.C.P., International Journal of Impact Engineering (2004)
[4] Ramajeyathilagam, K., P.V. C and B.R. V.. International Journal of Impact Engineering, **24**, 10 (2000)
[5] Ramajeyathilagam, K., C.P.. Shock and Vibration, **8**, 5 (2001)
[6] RH, C., *Underwater explosions* (Dover Publications Inc, New York, USA, 1948)
[7] GI, T.. Compendium Underwater Explosion Res ONR, 1: p. 1155–74 (1950)
[8] Hung, C.F., P.Y. Hsu and J.J. Hwang-Fuu. International Journal of Impact Engineering, **31**, 2 (2005)
[9] Espinosa, H.D., S. Lee and N. Moldovan, Experimental Mechanics, **46**, 6 (2006)
[10] McShane, G.J., et al., International Journal of Solids and Structures, **45**, 16 (2008)
[11] Xiang Da-lin, Rong Ji-li, He Xuan, Hu Chang-hua, Li Jian, Zhang Wei, Ren Peng, Acta Armamentarii (in CHINA), **39**, 6 (2014)

EPJ Web of Conferences **94**, 01015 (2015)
DOI: 10.1051/epjconf/20159401015

Behaviour of plated structures subjected to blast loading

Vegard Aune[a], Tore Børvik, and Magnus Langseth

Structural Impact Laboratory (SIMLab), Centre for Research-based Innovation (CRI), Department of Structural Engineering, Norwegian University of Science and Technology (NTNU), NO-7491 Trondheim, Norway

Abstract. An experimental investigation using a new shock tube facility to study blast-load effects on thin aluminium plates is presented. The shock tube is designed to expose materials and structures to extreme loading conditions, such as accidental explosions or terrorist attacks. The intensity of the loading in the present study was determined by the initial conditions of the compressed gas, i.e. volume and pressure, and the resulting loading on the target plate was compared to experimental data from explosive detonations found in the literature. The square plates were manufactured from a low-strength aluminium alloy and had an exposed area of $0.3 \times 0.3\,m^2$. Piezoelectric pressure sensors were used for pressure recordings and synchronized with two high-speed cameras operating at a frame rate of 21,000 fps in a stereoscopic setup to capture the dynamic response using a three-dimensional digital image correlation (3D-DIC) technique. The experiment showed that the shock tube is capable of recreating a loading similar to that of an unconfined far-field airblast, and worked as an easily controllable alternative to explosive detonations when studying the dynamic response of structures subjected to blast loading.

1. Introduction

Protection of engineering structures against blast loading has received a lot of attention in recent years [1–3]. Depending on the blast intensity the material and structural behaviour may become significantly different from their quasi-static behaviour if inertia and strain-rate effects become dominant. This results in the need to consider these effects in the design of protective structures involving blast events.

Due to the complexity in both the explosive loading and the resulting structural response, advanced numerical techniques are necessary for sufficient insight in the structural component optimization. However, before using such computational methods their performance should be validated in terms of reliability, robustness and effectiveness in predicting both blast loading and structural response. By its very nature experimental validation is ideal as it represents the actual physics of the problem, and controllable small-scale experiments could therefore be used to validate the computational methods and improve the understanding of the response of structures and materials under extreme loading situations (such as explosions). It may seem inviting to use scaled explosive detonations in such small-scale experiments. However, experimentation involving high-explosives may be hazardous and involves significant legal responsibilities which makes high-explosives less ideal for research purposes. An alternative to explosive detonations is the shock tube technique where pressure loading is generated by a rapid release of high-pressurized gas which creates a uniform shock wave propagating down a tube towards a test specimen. The loading is well defined by the initial conditions in the pressure chambers, and this technique serves as an easily controllable and repeatable alternative to explosive detonations. Shock tube experiments also avoid the complexity of ground reflections, light flashes, fireballs and other potential challenges related to high-explosive detonations.

The objective of this study is to investigate the performance of a new shock tube facility in recreating a loading similar to that from a spherical airblast, so that the shock tube can be used as an alternative to explosive detonations when studying the dynamic response of aluminium plates subjected to blast loading. This is done by comparing the pressure time-history from the shock tube experiment to experimental data from explosive detonations found in the literature [4,5]. The dynamic response of the plate was studied using pressure measurements and two high-speed cameras in a stereoscopic setup, and the deformation field was found using three-dimensional digital image correlation (3D-DIC).

[a] Corresponding author: `vegard.aune@ntnu.no`

2. Experimental setup

2.1. Material

The 0.8 mm thick aluminium plate was manufactured from a low-strength, strain-hardened and cold-rolled sheet of the alloy EN AW 1050A-H14 produced by Hindalco Industries Ltd. This is 99.6% pure aluminium subjected to annealing before work hardened by rolling. The material is often used for sheet work where high mechanical properties are not required. The nominal chemical composition is provided in Table 1, while the nominal yield and tensile strengths were given by the producer to be 110 MPa and 116 MPa, respectively.

Article available at http://www.epj-conferences.org or http://dx.doi.org/10.1051/epjconf/20159401015

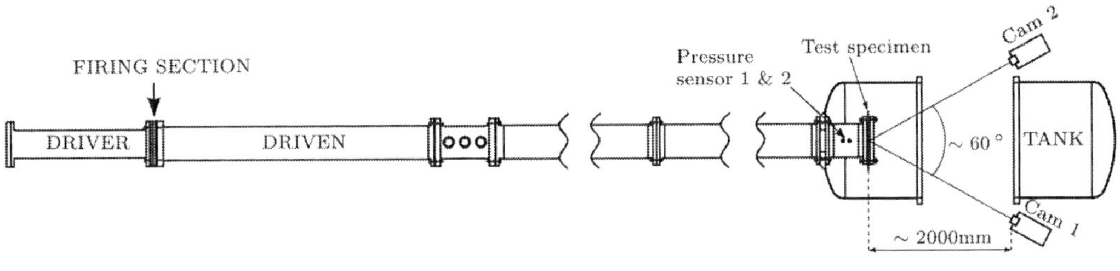

(a) Sketch of experimental setup (seen from above)

(b) Picture of the shock tube (seen from the driver)

(c) Picture of the firing section

(d) Picture of the window section

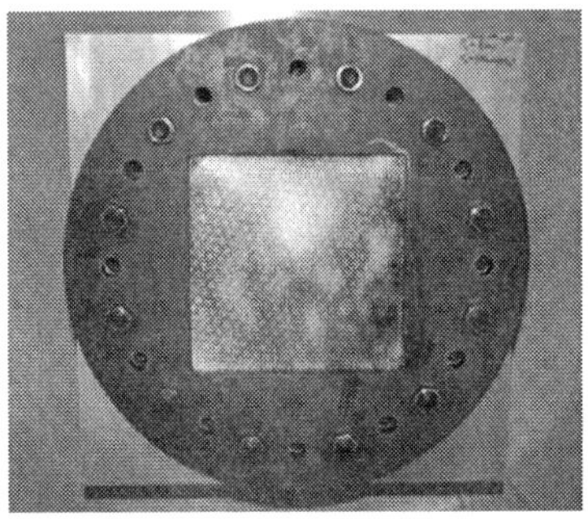

(e) Picture of clamping and DIC speckle pattern

Figure 1. Experimental setup in the SIMLab Shock Tube Facility. The speckle pattern is seen from the cameras.

Table 1. Chemical composition of EN AW 1050A-H14 (in wt.%).

Si	Fe	Cu	Mn	Mg
0.100	0.280	0.003	0.010	0.002
Zn	Ti	Cr	Al	
0.002	0.009	0.002	Rest	

2.2. Shock tube experiment

The shock tube experiment was performed in the new shock tube facility at the Structural Impact Laboratory, NTNU (Fig. 1). The overall length of the tube is 18.275 m and is made from P355NH which is a stainless steel designed for pressure purposes according to EN 13445-1:2009. The high-pressure chamber of the shock tube (called *driver* in Fig. 1a and shown in Fig. 1b) is manufactured with a total length of 2.02 m and an inner diameter of 0.331 m where the internal wall is dull polished to obtain a smooth surface. The driver is followed by the firing section which consists of two intermediate pressure chambers that can be separated by membranes (Fig. 1c). This enables the total pressure difference to be achieved stepwise. The experiment starts by filling the driver and firing section with pressurized air, and the pressure in the intermediate chambers is operated such that the desired pressure is obtained in the driver. A LABVIEW-based program has been developed and solenoid valves (ASCO Series 223) have been installed on the gas-filling lines to control the filling procedure, making the process fully automated based on the signals given by the pressure sensors (BAUMER PBMN-24B31) mounted to the driver and intermediate pressure chambers. The rupture of the membranes is initiated by controlled venting of the intermediate pressure closest to the driver, using two solenoid valves (ASCO Series 223). This ensures a controlled rupture of the membranes.

The sudden release of the high pressure generates a shock wave propagating down the low-pressure chamber (called *driven* in Fig. 1a). The inner cross-section in the driven section starts with a transition from circular to a square cross-section, where an epoxy material is used to obtain a smooth surface and square cross-section of 0.3 × 0.3 m^2 inside the surrounding tube. The epoxy material works as a practically incompressible material while the surrounding tube ensures the structural strength. The square cross-section downstream the firing section was chosen to accommodate plane parallel windows (Fig. 1d) which simplifies the use of optical techniques, and enables the installation of test objects in threaded holes in the tube floor.

The maximum working pressure of the driver section is limited to 17 MPa while the driven section, window section and dump tank are limited to 10 MPa, 5 MPa and 1.4 MPa, respectively. All respective parts of the facility were tested at a pressure 45% higher than the working pressure for a few minutes to ensure sufficient strength for routine use.

The experiment in the present work used aluminium inserts of diameter 0.33 m to reduce the initial volume in the driver to 23.2 dm^3, and it was sufficient with two Mellinex membranes of 0.5 mm separated by only one intermediate pressure in the firing section. The high pressure was released when the overpressure in the driver section was 802.6 kPa, while the driven section was operated at ambient pressure (100.13 kPa) and temperature (20.5°C) before releasing the shock wave. The square plate specimen with dimensions of 0.625 × 0.625 × 0.0008 m^3 had an exposed area of 0.3 × 0.3 m^2, and was clamped to the end flange of the tube using M24 bolts and a clamping frame in an attempt to achieve fixed boundary conditions (Fig. 1e).

Piezoelectric pressure sensors (Kistler 603B), corresponding charge amplifiers (Kistler 5064) and data acquisition system from National Instruments (NI USB-6356) were used to measure the pressure downstream the firing section with a samling frequency of 500 kHz during the experiments. The pressure transducers were placed in threaded adapters (Kistler 6501) and flush mounted in the tube wall, measuring the pressure behind the incident and reflected shock wave 29.5 cm (sensor 2) and 39.5 cm (sensor 1) upstream of the test specimen (see Fig. 1a). The delay in time of arrival of the shock wave at the pressure sensors was used to determine shock wave velocity and Mach number. A thin layer of insulating silicone (Kistler 1051) was used to shield the pressure transducers against heat transfer from the shock wave since the sensors are only designed for temperatures up to 200°C.

2.3. DIC measurements

The plate was painted with a speckle pattern (Fig. 1e) and a three-dimensional digital image correlation (3D-DIC) analysis was conducted using a stereovision setup with two Phantom v1610 high-speed cameras. The separation angle between the optical axes of the cameras was approximately 60° (see Fig. 1a). The recording rate was chosen to 21,000 fps (21 kHz) with an image resolution of 896 × 800 pixels and 12-bit grey level digitization. An independent data acquisition system from National Instruments (NI USB-6356) was used to synchronize the pressure recordings with the images from the high-speed cameras. The camera calibration and image analysis were carried out in a post-processing phase using an in-house DIC code. The resulting residual root-mean-square (RMS) from the calibration was calculated to be approximately 0.22 mm (~0.5 pixels) for both cameras.

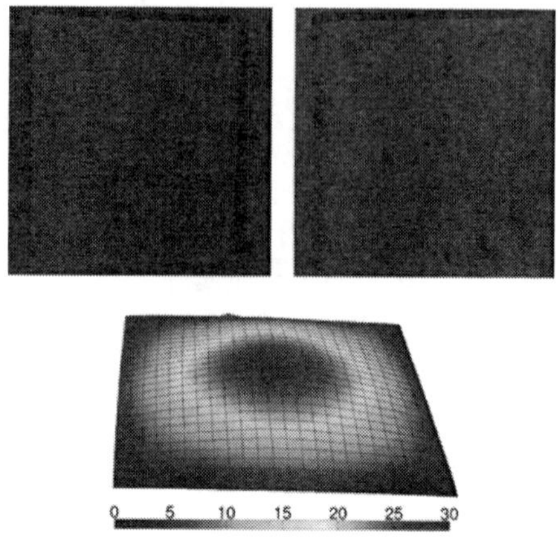

Figure 2. Results from 3D-DIC for the shock tube experiment. The colour scaling on the 3D-model indicates out-of-plane displacement (in mm)

Figure 2 shows an example of recorded images using the two synchronized high-speed cameras with the resulting DIC meshes plotted on top. The top images show two matching images from camera 1 (top-left) and camera 2 (top-right). The corresponding 3D model from DIC is presented in the lower image.

3. Experimental results

3.1. Material tests

Tensile tests on dog-bone specimens from the base material with clamped ends were performed to determine the quasi-static behaviour of the material while the dynamic material behaviour may be found in the literature [6,7]. The tests were performed in a Zwick/Roell Z030 testing machine at a constant deformation rate of 2.1 mm/min. This corresponds to an initial strain rate of $\dot{\varepsilon} = 5 \times 10^{-4}\,s^{-1}$ for a gauge length of 70 mm. Three parallel tests were performed in three different directions (0°, 45° and 90°) with respect to the rolling direction of the plate. Two-dimensional DIC was used to measure the displacement field and synchronized with the force measured by the hydraulic test machine at a sampling rate of 4 Hz.

The results are plotted as nominal stress-strain curves in Fig. 3. It is seen that the material is slightly anisotropic both in flow stress and elongation to failure.

3.2. Shock tube performance

Figure 4 shows the measured pressure time-history at the respective sensors, while Table 2 gives the characteristic loading parameters for the sensor closest to the target plate (sensor 2) and the maximum out-of-plane displacement of the centre point $d_{z,max}$. The reported blast wave parameters are limited to those related to the positive phase, i.e. the peak reflected overpressure $p_{r,max}$, positive duration t_+

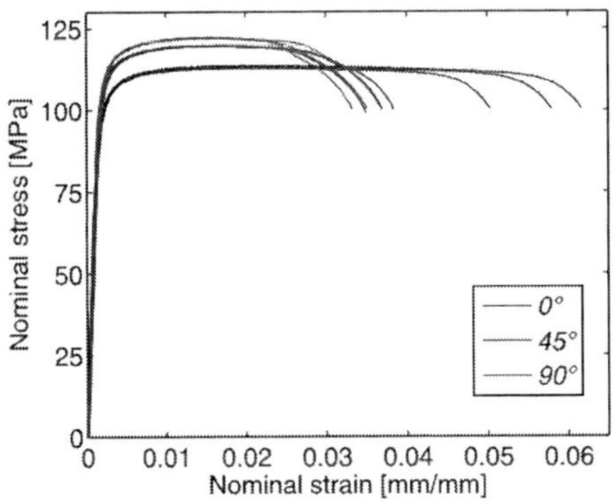

Figure 3. Nominal stress-strain curves from uniaxial tensile tests at three different loading directions.

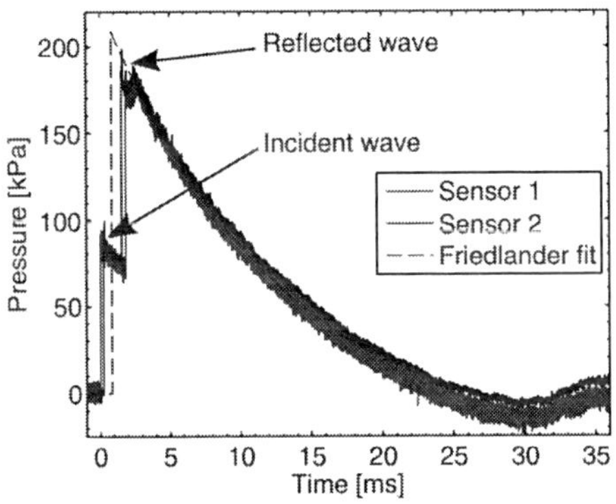

Figure 4. Pressure time-history from the experiment.

and impulse I_+. The impulse was found by numerical integration of the pressure curve during the positive duration at the selected sensor. The time of arrival of the incident shock wave at the first pressure transducer (sensor 1) was used as the starting point for the recordings in the experiment ($t = 0$).

A commonly used approximation to describe the pressure time-history for blast loading on a continuous format is the Friedlander equation [8], i.e.

$$p(t) = p_r\left(1 - \frac{t}{t_+}\right)\exp\left(\frac{-bt}{t_+}\right). \tag{1}$$

Curve fitting using the Levenberg-Marquardt algorithm in MATLAB gives the corresponding parameters of the Friedlander equation, and the reflected overpressure $p_{r,max}$ at the plate was estimated to be 208.8 kPa and the decay coefficient b was determined to be 1.26. This is plotted as a dotted line in Fig. 4 indicating the actual loading on the plate, and that the pressure loading in the shock tube follows the same exponential decay as an explosive detonation.

The difference in time of arrival at sensor 1 and 2 was used to determine the velocity U_s of the incident shock wave to be 446.4 m/s, which corresponds to a Mach number M_s of 1.3. Knowing the shock velocity the ideal gas theory may be used to calculate the incident and reflected pressures. The derivation of these expressions is found in the literature (see e.g. [9]) and is therefore only

Table 2. Experimental results.

Test	$p_{r,max}$ [kPa]	t_+ [ms]	I_+ [kPa · ms]	t_+/T_n [-]	$d_{z,max}$ [mm]
Shock tube	208.8	25.4	1816.9	1.9	29.5

briefly repeated below. The pressure behind the incident shock wave p_2 is given as

$$p_2 = p_1\left(1 + \frac{2\gamma_1}{\gamma_1 + 1}(M_s^2 - 1)\right) \tag{2}$$

where p_1 is the initial pressure in the driven section and γ_1 is the corresponding ratio of specific heat at constant pressure to that at constant volume for the low-pressure gas. The Mach number M_s is the ratio of the incident shock wave velocity U_s to the speed of sound $a_1 = \sqrt{\gamma_1 R T_1/m_1}$ in the undisturbed low-pressure gas, where T_1 is the measured initial temperature (in Kelvin), m_1 is the molecular weight and R is the universal gas constant.

The pressure p_5 behind the reflected shock wave from a rigid wall can be calculated as

$$p_5 = p_2\left(\frac{(\alpha_1 + 2)(p_2/p_1) - 1}{(p_2/p_1) + \alpha_1}\right) \tag{3}$$

where the relation $\alpha_1 = (\gamma_1 + 1)/(\gamma_1 - 1)$ is introduced for simplicity. Thus, using the relations from the ideal gas theory and the measured Mach number M_s of 1.3, the incident overpressure $p_2 - p_1$ and reflected overpressure $p_5 - p_1$ are given respectively as 80.6 kPa and 210.9 kPa, which is in good agreement with the corresponding pressures in Fig. 4 and indicates that the pressure recordings in this study are reliable.

It is well known that the shock tube technique often replicates some of the characteristics of a pressure loading from an unconfined far-field airblast detonation, i.e. steep rise time and a relatively long decay. This is confirmed by the theory and experiments by Kingery and Bulmash [5], where the reflected overpressure and impulse generated in the shock tube experiment are found to be equivalent to a spherical airburst of $W = 1967$ kg TNT at a stand-off distance of $R = 38$ m. Using the Hopkinson-Cranz scaling law ($Z = R/W^{1/3}$) the blast loading can be categorized with respect to the scaled distance, which is $Z = 3.0\,\mathrm{m/kg^{1/3}}$ in this experiment. The experiment may therefore be defined as a far-field airblast [10].

Figure 5 shows an example of synchronized loading and response histories during the experiment. The figure also include a selection of corresponding DIC images in terms of 3D topography maps, contours of the transverse displacement field and deformation profiles at characteristic times. It is seen that the response is typical for blast loaded plates, i.e. the plate experience large permanent deformations before it vibrates at its new equilibrium position. The DIC images also show plastic hinges traveling from the supports toward the centre of the plate, determining the deformed shape of the plate which is similar to a global dome (Figs. 5c–e). Moreover, the ratio of the positive duration t_+ of the pressure pulse and the natural period of vibration T_n for the target plate may be used to characterize the loading (see e.g. [11]) as one

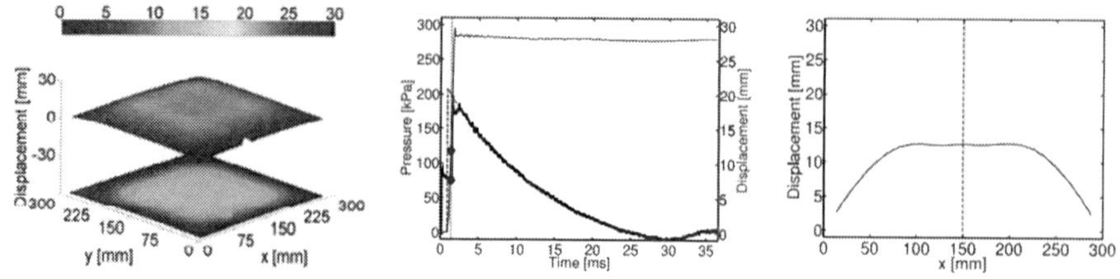

(a) t = 1.38 ms

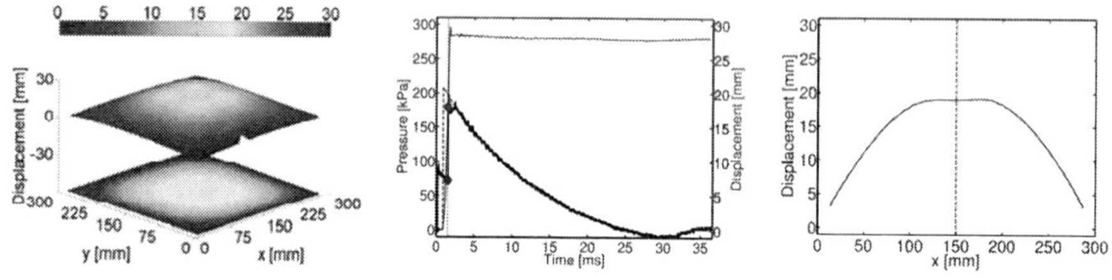

(b) t = 1.52 ms

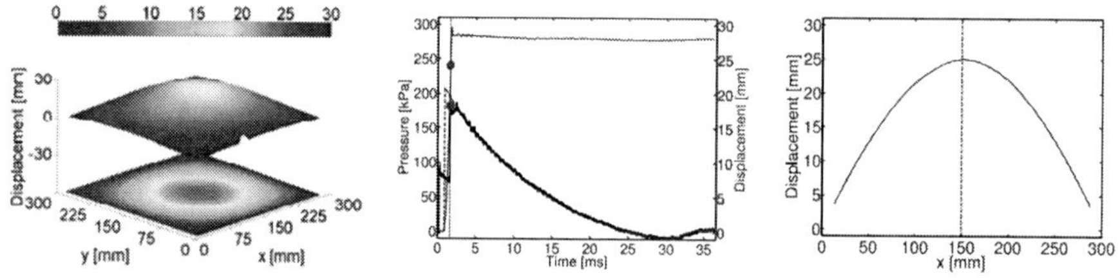

(c) t = 1.62 ms

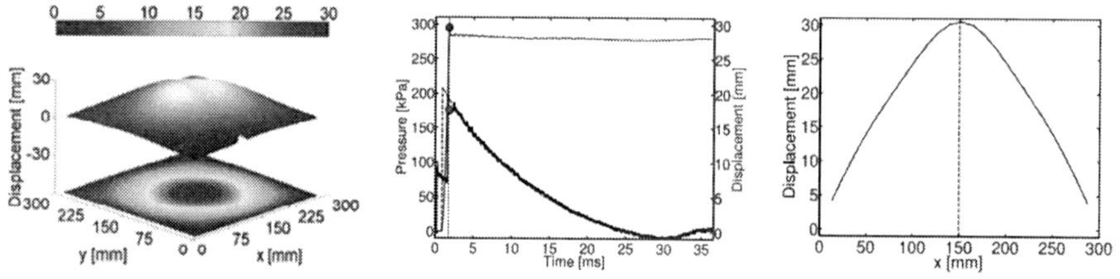

(d) t = 1.76 ms

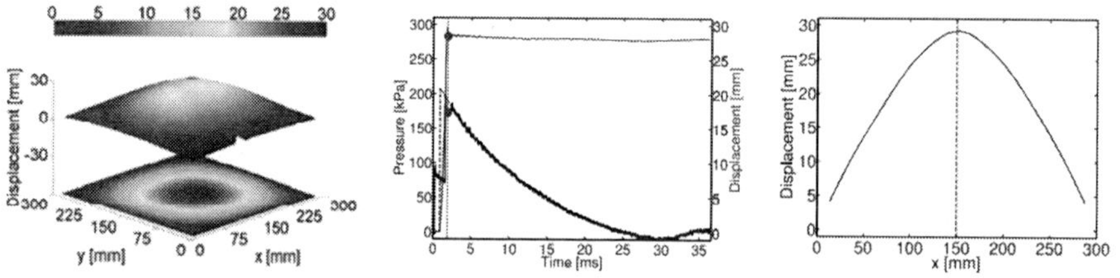

(e) t = 1.90 ms

Figure 5. A selection of the recorded DIC images and synchronized data from the shock tube experiment.

of the following domains, i.e.

$$\begin{aligned} \frac{t_+}{T_n} &< 0.064 && \text{(Impulsive domain)} \\ 0.064 < \frac{t_+}{T_n} &< 6.4 && \text{(Dynamic domain)} \qquad (4) \\ 6.4 < \frac{t_+}{T_n} & && \text{(Quasi-static domain)} \end{aligned}$$

Using the properties of an equivalent SDOF model of a square plate with fixed boundary conditions as given by e.g. Biggs [12], the natural period of vibration is found to be 13.41 ms for the plate used in this study. Thus, based on the ratio in Table 2, the loading can be classified as dynamic which indicates that the response is significantly influenced by the profile of the load history.

4. Concluding remarks

The experiment showed that the shock tube is capable of recreating a loading similar to that of an unconfined far-field airblast, and the recorded incident Mach number M_s shows that the pressure measurements were in good agreement with the idealized gas theory.

This indicates that the shock tube can be used to investigate the material effects, dynamic response, fluid-structure interaction problems, and typical failure modes of structures exposed to blast loading. Suggesting that blast loading may be generated in the shock tube as a cost-effective, easily controllable alternative to explosive detonations. Such experiments are useful for improving the understanding of the response of structures and materials subjected to unintentional external loads, and can be used to compare materials and structures loaded under similar conditions.

Financial support from the Structural Impact Laboratory (SIMLab), Centre for Research-based Innovation (CRI) at the Norwegian University of Science and Technology, is gratefully appreciated. The authors would also like to express their gratitude to Mr. Trond Aucstad and Mr. Tore Wisth from CRI-SIMLab for their contributions during the experimental work.

References

[1] A. G. Hanssen, L. Enstock and M. Langseth, International Journal of Impact Engineering **27**, page 593–618 (2002)

[2] T. Børvik, A. G. Hanssen, S. Dey, H. Langberg and M. Langseth, Engineering Structures **30**, page 1605–1620 (2008)

[3] T. Børvik, A. Burbach, H. Langberg and M. Langseth, Engineering Structures **30**, page 1621–1631 (2008)

[4] Department of Defense (DoD), *Structures to Resist the Effects of Accidental Explosions* (Report No. UFC 3-340-02, Washington, D.C.: Department of Defense, 2008)

[5] C. N. Kingery, G. Bulmash, Defence Technical Information Center, Ballistic Research Laboratory, Aberdeen Proving Ground, Maryland (1984)

[6] J. A. Zukas, T. Nicholas, H. F. Swift, L. B. Greszczuk and D. R. Curran *Impact Dynamics* (Wiley & Sons, New York, USA, 1982) page 316–320

[7] D. K. Christoulis, S. Guetta, V. Guipont and M. Jeandin, Journal of Thermal Spray Technology **20**, page 523–533 (2011)

[8] D. Cormie, G. Mays and P. Smith, *Blast Effects on Buildings, 2nd edn.* (Thomas Telford, London, UK, 2009) page 39–41

[9] A. G. Gaydon and I. R. Hurle, *The Shock Tube in High-Temperature Chemical Physics* (Chapman and Hall Ltd., London, UK, 1963) page 9–28

[10] D. O. Dusenberry, *Handbook for Blast Resistant Design of Buildings* (Wiley & Sons Inc., New Jersey, USA, 2010) page 246–248

[11] W. E. Baker, P. A. Cox, P. S. Westine, J. J. Kulesz and R. A. Strehlow *Explosion Hazards and Evaluation* (Elsevier, Amsterdam, 1983) page 273–281

[12] J. M. Biggs, *Introduction to Structural Dynamics* (McGraw-Hill, New York, 1964) page 214

EPJ Web of Conferences **94**, 01016 (2015)
DOI: 10.1051/epjconf/20159401016

Dynamic behaviour of "Collapsible" concrete

Alessio Caverzan[a], Marco G.L. Lamperti Tornaghi[b], Marco Peroni[c], and George Solomos[d]

European Commission Joint Research Centre (JRC), Institute for the Protection and Security of the Citizen (IPSC), European Laboratory for Structural Assessment (ELSA), via E. Fermi 2749, Ispra (VA), 21027 Italy

Abstract. In this work a particular cement composite material for protection of structures and infrastructures against accidental actions, such as blast or impact, has been investigated. An experimental procedure has been developed in order to assess static and dynamic behaviour of energy absorbing cementitious composites. The granular cementitious composite has been studied focusing attention to compressive strength, high deformation and energy dissipation capacity which are important characteristics for an absorber material. An experimental characterization of the material behaviour under compressive static and dynamic loadings has been carried out. Different deformation velocities have been studied in order to define the material behaviour in a wide range of strain rates. The velocity range up to 0.1 m/s is investigated by means of a universal servo-hydraulic MTS 50 kN testing machine. Some preliminary results have been reported and discussed in the present work.

1. Introduction

One of the major concerns of our society focuses on the development of strategies to provide structures and infrastructures, especially if they are critical, with adequate measures of either active or passive protection against accidental actions, such as impact and blast. During the last years, researchers developed several high performance materials in order to increase the blast resistance performance of structures, focusing their attention mainly on strength, stiffness and resilience. For example the use of High Performance Fiber Reinforced Cementitious Composites (HPFRCCs) has been intensively studied for repairing, retrofitting, upgrading and/or protecting existing structures. This category of materials is able to sustain large amount of strain with not negligible stress. Furthermore, it has been recently shown ([1,2]) that HPFRCCs are able to retain a significant level of their original performance, even when exposed to conditions such as high temperature and high strain rates. However, as the American Concrete Institute underlines, a high-performance concrete is a "concrete meeting special combinations of performance and uniformity requirements that cannot always be achieved routinely using conventional constituents and normal mixing, placing, and curing practices" [3].

Starting from this observation, the search of a different class of materials, meeting the needs and performance requirements for such special conditions, has been initiated. O'Neil et al. [4] have developed a concrete called "frangible concrete", due to its property of breaking into small fragments, which are less harmful as well as easily lost in the environment. The material is thus able to absorb and dissipate, through the diffused fracture process, a high amount of energy. A similar concept has been developed by Caverzan and Ferrara [5] in tailoring a "collapsible concrete". This material has shown merits in dissipating a significant amount of the energy input generated by either a far or a near field explosion and thus prevent it from being transferred to critical load bearing elements. A possible application of the "collapsible concrete" for protection could be found in protective sacrificial screeds. The use of protective screeds for reinforced concrete structural elements has already been investigated, metallic foams have been especially used in such studies [6–9]. However, real applications on existing or new structures have never developed due to their high cost. The key idea highlighted by Ferrara et al. [10,11] is to have a cementitious composite featuring a significant energy dissipation at a quite low strength level and low cost.

The relevant deformation capacity is mainly dependent on a rearrangement of the aggregate granular skeleton obtained after the failure of weak inter-granular bonds. Three peculiar concepts have been applied: the use of cement paste characterised by a low strength and a not negligible fracture toughness (obtained also by using air entraining admixture AEA); an aggregate granular skeleton with a grain size distribution as narrow as possible which can lead to a high capacity of the material to undergo significant deformation and finally a low dosage of cement paste in order to assure just a minimum inter-granular bonding without filling the voids between the aggregate.

In order to study in depth the potentiality of "collapsible concrete", by investigating in particular its dynamic behaviour at different strain rates, an extended experimental campaign has been planned. In the present work preliminary results have been reported and some strategies are highlighted in order to improve the material performance.

[a] e-mail: alessio.caverzan@jrc.ec.europa.eu
[b] e-mail: marco.lamperti@jrc.ec.europa.eu
[c] e-mail: marco.peroni@jrc.ec.europa.eu
[d] e-mail: george.solomos@jrc.ec.europa.eu

Article available at http://www.epj-conferences.org or http://dx.doi.org/10.1051/epjconf/20159401016

Table 1. Mix Design.

Constituent	dosage
Cement type I 32.5	195.12 kg/m^3
LWA	376.02 kg/m^3
Water	117.12 l/m^3
AEA	2.34 l/m^3

2. Material and experimental test description

The "collapsible" concrete studied in the present work has been cast following the mix design reported in Table 1. The material obtained is characterised by a water-cement ratio (w/c) equal to 0.6 and a percentage of mortar in volume equal to around 26% which correspond to the minimum theoretical void ratio of a densely packed mono-size arrangement of spheres. Expanded clay Light Weight Aggregate (LWA) with a nominal grain size range equal to 3–8 mm was chosen as aggregates. Sixteen cylindrical samples of diameter and length equal to 80 mm have been cast. Differently from what it was done previously by Ferrara et al. [10, 11] the ratio between diameter and length (d/l) has been limited to 1 in order to reduce the influence of friction even if this contribution is important when practical applications on real structural elements are considered.

An experimental program has been carried out to properly identify the mechanical properties of "collapsible" concrete under static and dynamic conditions. Uniaxial compression tests on the cylindrical samples have been performed by means of a universal servo-hydraulic MTS testing machine (Fig. 1a) in the European Laboratory for Structural Assessment (ELSA). The machine has a maximum load capacity of 50 kN and it can impose a maximum displacement rate equal to 300 mm/s. A Kistler 9341 piezoelectric load cell with a maximum capacity of 30 kN has been installed in order to improve the quality of the signal recorded during the tests (Fig. 1b).

Signals have been acquired by employing a National Instrument PXIe-6366 Simultaneous Data Acquisition recorder with a maximum sampling-rate of 2 MHz. In addition a high-speed camera IDT-Y4 has been intensively used. It is able to perform at 4000 fps with maximum resolution, however the acquired frame-rate can be increased to 200 000 fps by reducing substantially one dimension of the recorded image.

(a)

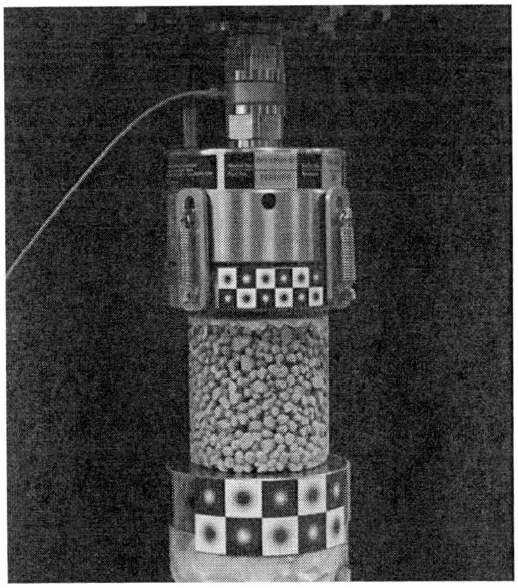

(b)

Figure 1. Test set-up: press device and high-speed camera (a); detail of the sample, piezoelectric loading-cell and optical targets (b).

3. Test results and discussion

Uniaxial compression tests have been performed in order to investigate the influence of strain rate on the material behaviour. Four different displacement rates have been applied (0.1, 1, 10, 100 mm/s) and some reference quasi-static tests have been performed exploiting the same test set-up configuration. The results of compressive tests are shown in Figs. 2–6 in terms of nominal longitudinal stress (σ_v) versus longitudinal strain (ε_v). The behaviour exhibited is different from what has been reported by Ferrara et al. [11], the stress plateau observed after the initial linear elastic range has not been clearly identified in this series of tests, only few samples have shown a limited plateau up to 0.05 strain. It is thought that the main reason for that should be searched in the different specimen geometry and test set-up. In fact, previous experimental campaigns have been conducted on cylindrical samples of diameter equal to 200 mm and length equal to 50 mm, and consequently the test set-up was characterized by particular boundary conditions that accentuate the contribution of friction. The boundary conditions and sample geometries selected by Ferrara

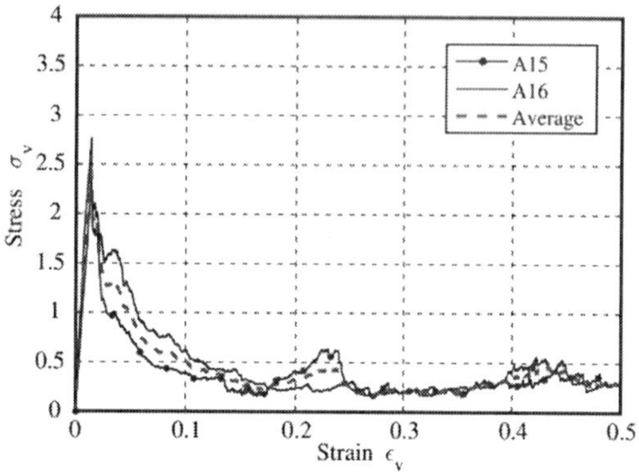

Figure 2. Stress vs. strain curves for the static tests.

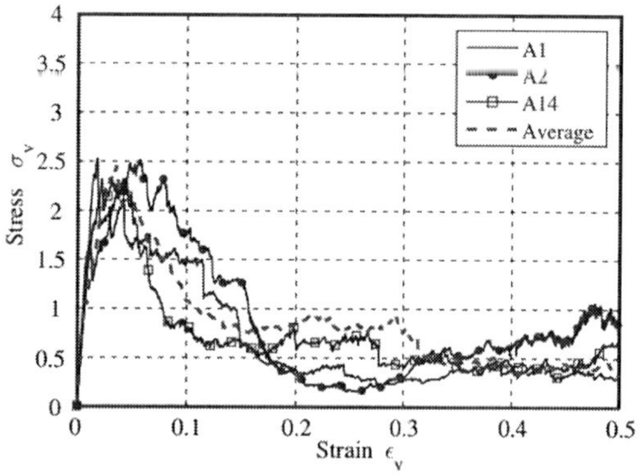

Figure 3. Stress vs. strain curves for the dynamic tests with a displacement rate $\dot{\delta}$ equal to 0.1 mm/s.

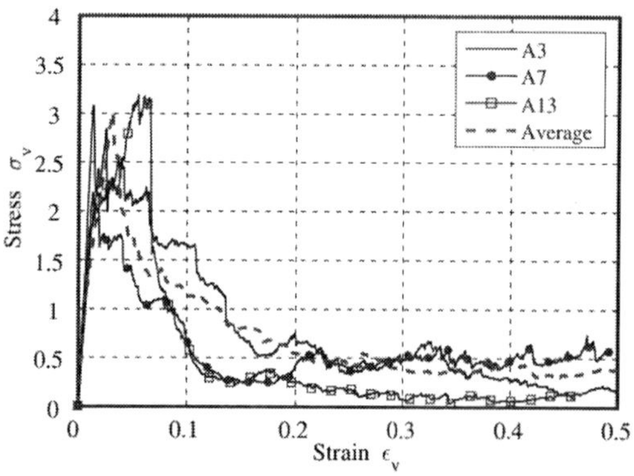

Figure 4. Stress vs. strain curves for the dynamic tests with a displacement rate $\dot{\delta}$ equal to 1 mm/s.

et al. [10] were justified by the needs to focus on some particular engineering application as explained by the authors. Moreover, the decay in the material performance, linked to the reduction of friction contribution, has been emphasised by the higher compressive peak strength obtained in these new experimental tests.

The use of a mortar with w/c ratio relatively low has led to compressive strength of matrix higher than

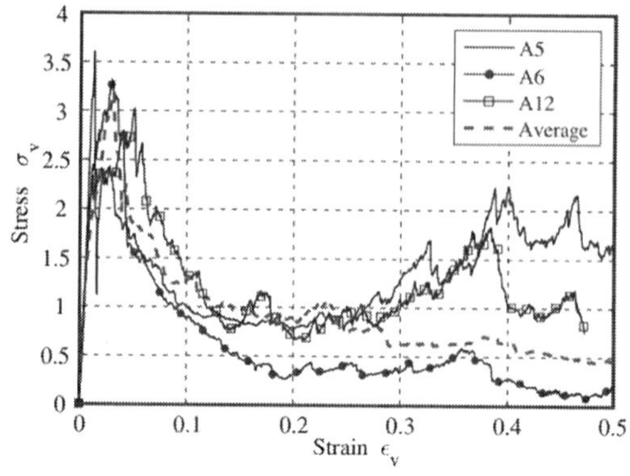

Figure 5. Stress vs. strain curves for the dynamic tests with a displacement rate $\dot{\delta}$ equal to 10 mm/s.

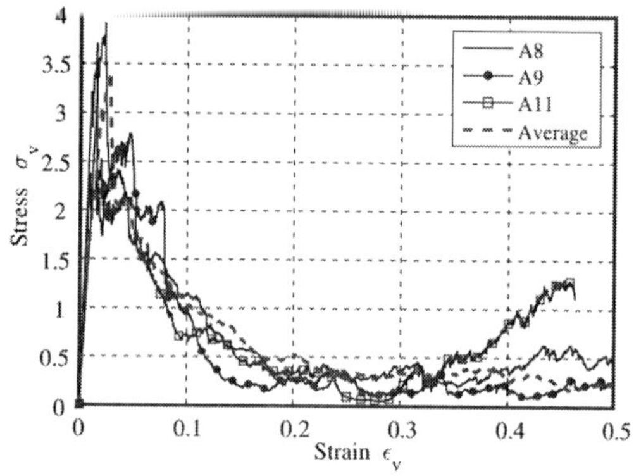

Figure 6. Stress vs. strain curves for the dynamic tests with a displacement rate $\dot{\delta}$ equal to 100 mm/s.

Figure 7. Pieces of failed samples. Intra-granular failure can be clearly observed.

the light weight aggregates strength. As Figs. 2–6 show, when the peak strength is reached a sudden release of elastic energy generates a drop in the stress level associated to aggregate failures (Figs. 7 and 9). The intra-granular failure does not allow to develop a proper

Table 2. Peak strengths and energies absorbed.

Sample code	$\dot{\delta}$ [mm/s]	σ_{Peak} [MPa]	$\sigma_{Peak,av}$ [MPa] (std)	G_{tot} [J/dm^3]	$G_{tot,av}$ [J/dm^3] (std)	G_{10} [J/dm^3]	$G_{10,av}$ [J/dm^3] (std)	G_{20} [J/dm^3]	$G_{20,av}$ [J/dm^3] (std)
A15	0.01	2.77	2.60	245.34	225.69	118.65	100.87	154.76	133.07
A16	0.01	2.44	(0.23)	206.04	(27.80)	83.09	(25.14)	111.38	(30.67)
A1	0.1	2.53	2.46	349.64	393.44	164.50	169.46	247.80	253.37
A2	0.1	2.52	0.11	459.23	(58.01)	192.90	(21.39)	296.50	(40.63)
A14	0.1	2.34		371.44		150.99		215.82	
A3	1	2.84	3.04	391.73	318.82	189.43	171.57	278.88	222.92
A7	1	3.08	(0.18)	310.07	(68.95)	134.25	(32.33)	167.05	(55.92)
A13	1	3.19		254.66		191.05		222.83	
A5	10	3.61	3.31	683.41	534.43	164.64	181.26	251.55	260.59
A6	10	3.32	(0.30)	323.09	(188.08)	175.32	(20.26)	228.37	(37.57)
A12	10	3.02		596.78		203.83		301.86	
A8	100	3.92	3.47	358.82	329.85	185.59	178.09	247.68	227.75
A9	100	2.79	(0.60)	286.23	(38.45)	190.30	(17.23)	224.43	(18.50)
A11	100	3.71		344.49		158.39		211.13	

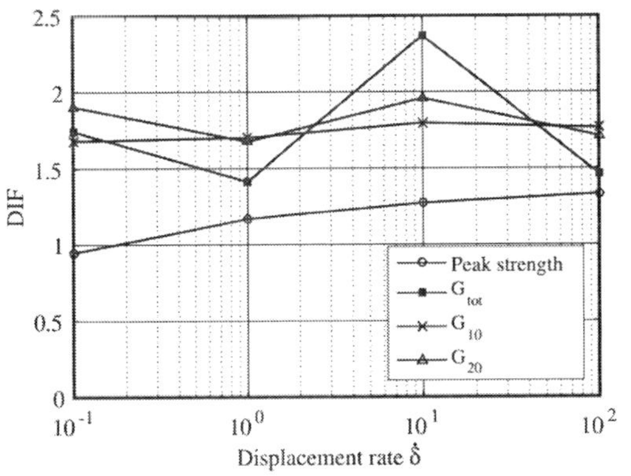

Figure 8. Dynamic increase factor trends for peak strength and energies.

grain redistribution which is the principal resource of energy dissipation. For such reason relative low values of dissipated energy can be observed in Table 2, where peak strength and energies absorbed up to different strain levels (0.1 → G_{10}, 0.2 → G_{20} and 0.5 → G_{tot}) are listed. In order to highlight the influence of strain rate on the material behaviour the Dynamic Increase Factors (DIF), computed as the ratio between dynamic and static material characteristics are listed in Table 3. While the DIF of peak strength shows an increasing linear trend from 1 to around 1.33 for displacement rate equal to 100 mm/s, the DIF trends related to energies are not clearly defined especially when the total energy G is considered. The DIF for the energy G_{10} related to a strain level equal to 0.1 exhibits a stable value close to 1.7. This would imply that in the first part of the fracture process the material is able to absorb 70% more energy in dynamic conditions.

Table 3. Dynamic Increase Factors of peak strength and energies.

$\dot{\delta}$	DIF_{Peak}	DIF_{Gtot}	DIF_{G10}	DIF_{G20}
0.1	0.95	1.74	1.68	1.90
1	1.17	1.41	1.70	1.68
10	1.27	2.37	1.80	1.96
100	1.33	1.46	1.77	1.71

4. Conclusions

The behaviour of "collapsible" concrete when subjected to different displacement rates has been investigated. A preliminary experimental campaign has been carried out by means of uniaxial compression tests. Tests have been carried out by means of a universal servo-hydraulic MTS testing machine, which is able to impose a maximum displacement rate equal to 300 mm/s. From the experimental results presented in this research work some important conclusion can be drawn for the specific concrete mix under consideration:

- the sample geometry and the boundary conditions adopted have limited the influence of the friction, highlighting a different behaviour from that reported in previous studies. However, the friction and its passive confinement effect play an important role when real applications are considered and can not always be neglected;
- the change in material behaviour has been emphasized by the relatively high compressive strength of the matrix with respect to the LWA strength, which has led to an intra-granular failure. This type of rupture has not allowed the development of a proper granular compaction which is the main resource of energy absorption;
- "collapsible" concrete has exhibited a strain rate sensitivity in terms of compressive strength, the DIF

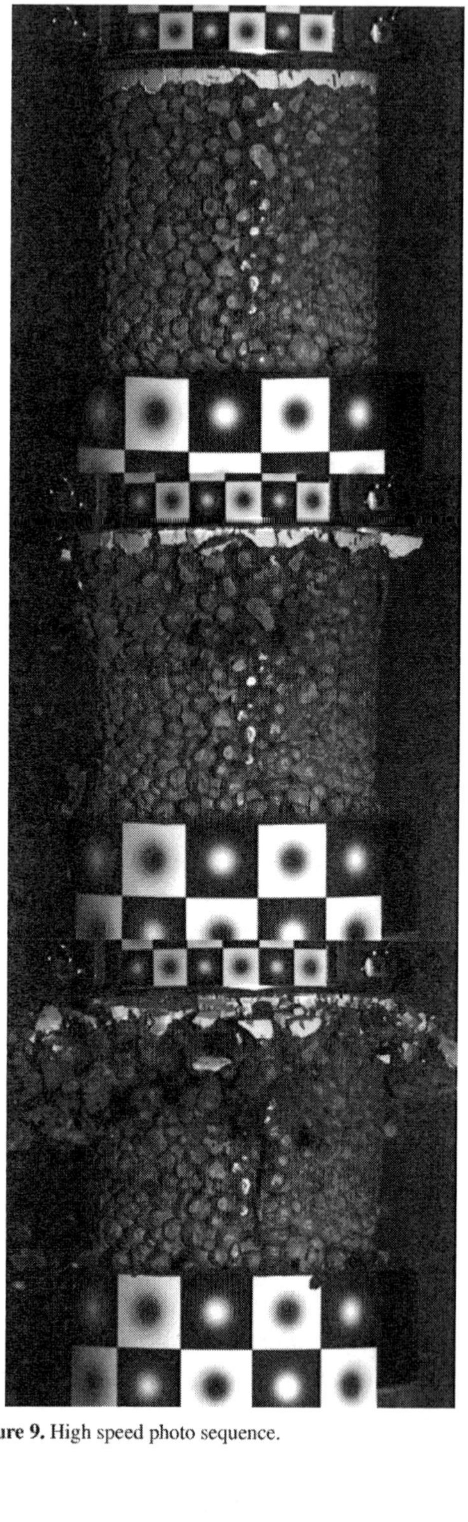

Figure 9. High speed photo sequence.

has increased up to 1.33 when passing from quasi static to dynamic tests at 100 mm/s;

- even if the behaviour underlined is slightly different from previous research in dynamic regime, the energy absorbed in the first stage (up to 0.1 strain) is about 70% higher than what has been recorded in quasi-static tests.

The adopted experimental procedure is going to be used as an efficient method to evaluate similar type of materials; research work is continuing using different aggregates and changing matrix properties in order to improve the material dynamic performance. For instance, it is expected that a combined use of micro-fibres and matrix characterised by a lower compressive strength could improve significantly the material behaviour observed in the tests. The effects of passive confinement due to friction could be replaced by an internal confinement, obtained by introducing an appropriate amount of micro-fibre; this aspect is part of the research work in progress at the ELSA laboratories.

References

[1] A. Caverzan, E. Cadoni, M. di Prisco, Mechanics of Materials **59**, 87 (2013)

[2] A. Caverzan, M. Colombo, M. di Prisco, B. Rivolta, Materials and Structures pp. 1–13 (2014)

[3] A.C.I. committee 116, *"Cement and Concrete Terminology", ACI Manual of Concrete Practice, Part 1.*, American Concrete Institute, (2012)

[4] E.F. O' Neil, W. Shen, H.M. Jennings, J.J. Thomas, T. Cummins, ACI Materials Journal **109** (2012)

[5] A. Caverzan, L. Ferrara, *"Collapsible" concrete as a blast and impact absorber: From material concept to static characterization* (2012), pp. 367–370, `http://www.scopus.com/inward/record.url?eid=2-s2.0-84883345873&partnerID=40&md5=a29dc320fce14b41d27208594ed8fa61`

[6] A. Hanssen, L. Enstock, M. Langseth, International Journal of Impact Engineering **27**, 593 (2002)

[7] A. Schenker, I. Anteby, E. Nizri, B. Ostraich, Y. Kivity, O. Sadot, O. Haham, R. Michaelis, E. Gal, G. Ben-Dor, J. Struct. Eng. **131**, 1233 (2005)

[8] C. Cady, G. Gray III, C. Liu, M. Lovato, T. Mukai, Materials Science and Engineering: A **525**, 1 (2009)

[9] O. Sadot, I. Anteby, S. Harush, O. Levintant, E. Nizri, B. Ostraich, A. Schenker, E. Gal, Y. Kivity, G. Ben-Dor, J. Struct. Eng. **131**, 1226 (2005)

[10] L. Ferrara, A. Caverzan, A. Peled, Materials and Structures, Submitted (2015)

[11] L. Ferrara, A. Caverzan, L. Nahum, A. Peled, Materials and Structures, Submitted (2015)

EPJ Web of Conferences **94**, 01017 (2015)
DOI: 10.1051/epjconf/20159401017

The dynamic Virtual Fields Method on rubbers at medium and high strain rates

Sung-Ho Yoon[a] and Clive R. Siviour

Department of Engineering Science, University of Oxford, Parks Road, Oxford, UK

Abstract. Elastomeric materials are widely used for energy absorption applications, often experiencing high strain rate deformations. The mechanical characterization of rubbers at high strain rates presents several experimental difficulties, especially associated with achieving adequate signal to noise ratio and static stress equilibrium, when using a conventional technique such as the split Hopkinson pressure bar. In the present study, these problems are avoided by using the dynamic Virtual Fields Method (VFM) in which acceleration fields, clearly generated by the non-equilibrium state, are utilized as a force measurement with in the frame work of the principle of virtual work equation. In this paper, two dynamic VFM based techniques are used to characterise an EPDM rubber. These are denoted as the linear and nonlinear VFM and are developed for (respectively) medium (drop-weight) and high (gas-gun) strain-rate experiments. The use of the two VFMs combined with high-speed imaging analysed by digital imaging correlation allows the identification of the parameters of a given rubber mechanical model; in this case the Ogden model is used.

1. Introduction

Rubbers are often used for load mitigation in components that may experience dynamic loading. Most rubbers are observed to exhibit rate dependent mechanical behaviour, and are usually stiffer at high strain rates due to their viscoelastic characteristics [1]. Hence, the effective use of a rubber can only be achieved when its dynamic mechanical properties are well characterized. Several studies have included high speed experiments on rubbers in compression and tension using the split Hopkinson bars [2–4] and newly developed techniques [5,6]. Although many improvements have been made to this traditional technique, the inherent softness of rubber produces experimental difficulties in achieving precise force measurements and a static stress equilibrium state. These difficulties become more magnified as the strain rate increases.

Recently, a new experimental technique has been developed for performing dynamic tests [7–9] without a traditional force measurement by means of high speed imaging and the dynamic Virtual Fields Method (VFM) [10]. This method can be used to identify material parameters of a mechanical model for rubbers (hyperelastic model) using acceleration fields to identify the stress in the specimen. Removing the requirement for a further force measurement is advantageous for dynamic tests on rubbers with regard to the aforementioned experimental difficulties.

In the present study, the dynamic VFM was used in order to characterize mechanical properties of EPDM rubbers. Two VFM schemes are introduced: the linear and nonlinear VFMs. The linear VFM, involving the isotropic linear elastic model, is applied to a drop-weight test, able to produce medium strain rates and small amplitude dynamic loading. A series of effective Young's moduli is obtained after different static pre-stretches. This set of moduli is then used in an optimization procedure to calculate the material parameters of a hyperelastic model. On the other hand, the nonlinear VFM uses a hyperelastic model within its calculation procedure. This method directly produces material parameters from a continuous, large amplitude, high strain-rate dynamic deformation, in this case generated by a gas gun experiment. The two VFMs are described in detail, before being applied to data from FEM simulations and experiments on EPDM rubber.

2. Virtual Fields Method

2.1. Linear VFM

The VFM [11] is an inverse method that can be used to characterize parameters of a given mechanical constitutive model by applying the following principle of virtual work (PVW) equation to experimental data:

$$-\int_V \sigma : \boldsymbol{\varepsilon}^* dV + \int_{S_f} \boldsymbol{T} \cdot \boldsymbol{u}^* dS = \int_V \rho \boldsymbol{a} \cdot \boldsymbol{u}^* dV \quad (1)$$

where

σ	*actual Cauchy stress tensor*
$\boldsymbol{T}$	*actual loading vector*
$\boldsymbol{u}^*$	*virtual displacement vector*
ε^*	*virtual strain tensor*
	(the spatial derivative of $\boldsymbol{u}^*$)
V	*current volume*
S_f	*loaded surface(s)*
ρ	*density*
':' and '·'	*the dot product for matrices and vectors.*

[a] Corresponding author: sung-ho.yoon@eng.ox.ac.uk

Article available at http://www.epj-conferences.org or http://dx.doi.org/10.1051/epjconf/20159401017

The integration domain and strain term of Eq. (1) are defined with respect to the current coordinates, $x(t)$ and $y(t)$. The PVW for the dynamic VFM can be constructed for a two dimensional case using the following special virtual fields [8] written as (L: specimen length)

$$\begin{cases} u_x^{*(1)} = x(x-L) \\ u_y^{*(1)} = 0 \end{cases} \rightarrow \begin{cases} \varepsilon_x^{*(1)} = 2x - L \\ \varepsilon_y^{*(1)} = 0 \\ \varepsilon_{xy}^{*(1)} = 0 \end{cases}$$
$$\begin{cases} u_x^{*(2)} = 0 \\ u_y^{*(2)} = x(x-L)y \end{cases} \rightarrow \begin{cases} \varepsilon_x^{*(2)} = 0 \\ \varepsilon_y^{*(2)} = x(x-L) \\ \varepsilon_{xy}^{*(2)} = (2x-L)y \end{cases} \tag{2}$$

and assuming constant stress in the thickness direction of a thin specimen:

$$-\int_S \boldsymbol{\sigma} : \boldsymbol{\varepsilon}^* dS = \int_S \rho \boldsymbol{a} \cdot \boldsymbol{u}^* dS \tag{3}$$

where S is a surface area in the current configuration. In particular, the use of these virtual fields removes the traction term, so that force measurements are not required. In the stress term σ, the (two dimensional) isotropic linear elastic constitutive relation is used with two stiffness parameters Q_{xx} and Q_{xy} in order to derive the following system of a linear equation.

$$\begin{bmatrix} M_{11} & M_{12} \\ M_{21} & M_{22} \end{bmatrix} \begin{pmatrix} Q_{xx} \\ Q_{xy} \end{pmatrix} = \begin{pmatrix} N_1 \\ N_2 \end{pmatrix}$$
$$M_{11} = \int_S \varepsilon_x \varepsilon_x^{*(1)} + \varepsilon_y \varepsilon_y^{*(1)} + \frac{1}{2} \varepsilon_{xy} \varepsilon_{xy}^{*(1)} dS$$
$$M_{12} = \int_S \varepsilon_x \varepsilon_y^{*(1)} + \varepsilon_y \varepsilon_x^{*(1)} - \frac{1}{2} \varepsilon_{xy} \varepsilon_{xy}^{*(1)} dS$$
$$M_{21} = \int_S \varepsilon_x \varepsilon_x^{*(2)} + \varepsilon_y \varepsilon_y^{*(2)} + \frac{1}{2} \varepsilon_{xy} \varepsilon_{xy}^{*(2)} dS \tag{4}$$
$$M_{22} = \int_S \varepsilon_x \varepsilon_y^{*(2)} + \varepsilon_y \varepsilon_x^{*(2)} - \frac{1}{2} \varepsilon_{xy} \varepsilon_{xy}^{*(2)} dS$$
$$N_1 = -\int_S \rho (a_x u_x^{*(1)} + a_y u_y^{*(1)}) dS$$
$$N_2 = -\int_S \rho (a_x u_x^{*(2)} + a_y u_y^{*(2)}) dS.$$

The material is assumed to be incompressible, so that the density ρ is constant during the loading. Using experimental $\boldsymbol{\varepsilon}(t)$ and $\boldsymbol{a}(t)$ data, Eq. (4) is linearly solved to obtain $Q_{xx}(t)$ and $Q_{xy}(t)$. This procedure is used on the data from the drop-weight experiment without pre-stretching.

For the pre-stretching case, it is necessary to consider the static stress field from the pre deformation. Assuming that the pre-existing stress $\boldsymbol{\sigma}^{pre}$ and the additional dynamic stress $d\boldsymbol{\sigma}$ fields are independent, Eq. (3) becomes

$$\int_S (\boldsymbol{\sigma}^{pre} + d\boldsymbol{\sigma}) : \boldsymbol{\varepsilon}^* dS = -\int_S \rho \boldsymbol{a} \cdot \boldsymbol{u}^* dS \tag{5-1}$$

$$\int_S d\boldsymbol{\sigma} : \boldsymbol{\varepsilon}^* dS = -\int_S \rho \boldsymbol{a} \cdot \boldsymbol{u}^* dS - \int_S \boldsymbol{\sigma}^{pre} : \boldsymbol{\varepsilon}^* dS. \tag{5-2}$$

The dynamic stress term $d\boldsymbol{\sigma}$ is a function of an incremental dynamic true strain $d\boldsymbol{\varepsilon}$ that does not include the strain from pre-deformations. Eqs. ((5)-(2)) can be expanded to the components form similar to Eq. (4). The last term of Eqs. ((5)-(2)) has only an axial component (longitudinal direction) with an assumption that the transverse pre-stress is very small compared to the axial in the specimens used in this study.

The use of Eqs. ((5)-(2)) on several pre-stretched states produces a set of effective Young's modulus values, including one from Eq. (4). These moduli are then used in an optimization scheme to obtain material parameters, assuming that they are each slope of the secant line between their pre-strain $\varepsilon_{x,p}$ and final strain $\varepsilon_{x,f}$, each of which is calculated from the length at the relevant condition with respect to the initial length.

$$\begin{cases} \text{mean}\left(\frac{\partial \sigma_x}{\partial \varepsilon_x}\left(A_1, \ldots A_i, \varepsilon_{x,p}^1\right) + \frac{\partial \sigma_x}{\partial \varepsilon_x}\left(A_1, \ldots A_i, \varepsilon_{x,f}^1\right)\right) \\ \qquad -E_1 = 0 \\ \qquad \vdots \\ \text{mean}\left(\frac{\partial \sigma_x}{\partial \varepsilon_x}\left(A_1, \ldots A_i, \varepsilon_{x,p}^n\right) + \frac{\partial \sigma_x}{\partial \varepsilon_x}\left(A_1, \ldots A_i, \varepsilon_{x,f}^n\right)\right). \\ \qquad -E_n = 0 \end{cases} \tag{6}$$

Here, the Es are the identified Young's moduli, from Eq. (4) for the non-pre-stretching case ($\varepsilon_{x,p}^1 = 0$) and Eqs ((5)-(2)) for the pre-stretching cases at their pre-true axial strains $\varepsilon_{x,p}^n$. The axial true stress term σ_x in Eq. (6) can be defined by any material model suitable for rubbers with parameters A_i, and it is the A_i which are identified by the optimisation procedure. In this study, the one-term Ogden model [12] is adopted because it is found that this model was able to describe the static stress-strain curve of the EPDM rubbers tested in the present study. The uniaxial form of the incompressible one-term Ogden model used in ABAQUS [13] is written as

$$\sigma_x = 2\frac{\mu}{\alpha}\left(\lambda_x^{\alpha-1} - \lambda_x^{-0.5\alpha-1}\right) \tag{7}$$

where μ and α are the material parameters; λ_x is an axial stretch ratio ($\lambda_x = \exp(\varepsilon_x)$). Applying Eq. (7) to Eq. (6), the optimization of Eq. (6) leads to the identification of μ and α. This optimization procedure was conducted by a built-in nonlinear equation solver (fsolve) in Matlab®.

2.2. Nonlinear VFM

In the nonlinear VFM, the PVW equation contains the one-term Ogden model in the stress term. This stress term is expressed by the first Piola-Kirchoff stress and thus the integration domain of the PVW is based on the initial surface area of the specimen S_0. A similar VFM has been used for static tests on rubbers [14]. The PVW of the nonlinear VFM with one of the virtual fields from Eq. (2) is written as

$$-\int_{S_0} \boldsymbol{\Pi} : \frac{\partial \mathbf{u}_0^*}{\partial \mathbf{x}_0} dS_0 = \int_{S_0} \rho \boldsymbol{a} \cdot \boldsymbol{u}_0^* dS_0. \tag{8}$$

Where

Π *the first Piola-Kirchoff stress tensor*
u_0^* *u^* defined on the initial configuration,* $u_0^* = u^*(x_0, y_0)$
S_0 *initial surface area*
x_0 *initial coordinate, x_0 and y_0.*

Unlike the linear VFM, only the first set of the virtual fields of Eq. (2) can be implemented because the use of the experiment data from the entire long loading history generates many PVW equations. Using these PVW equations and with the first virtual fields of Eq. (2), which is based on the initial coordinates, the following cost function is constructed

$$\Phi = \sum_{k=1}^{N} \left[\int_{S_0} \Pi_x (A_1, \ldots, A_i, t_k) \frac{\partial u^*_{x_0}}{\partial x_0} dS_0 + \int_{S_0} \rho a_x(t_k) u^*_{x_0} dS_0 \right]^2. \quad (9)$$

where N indicates the total number of data from the simulation or experiment. By minimizing this cost function using the Matlab® *fmincon* function, the material parameters A_i determining Π_x are directly obtained. As used in the linear VFM, the one-term Ogden model can be used to express the stress term $\mathbf{\Pi}$. It is convenient to define $\mathbf{\Pi}$ with respect to principal stretch ratio λ_i as the Ogden model is defined in the principal directions. The Piola-Kirchoff stress under plan stress $\Pi_3 = 0$ is

$$\Pi_1 = -\frac{1}{\lambda_1}\frac{2\mu}{\alpha^2}\left(\alpha\lambda_3^{\alpha}\right) + \frac{2\mu}{\alpha^2}\left(\alpha\lambda_1^{\alpha-1}\right)$$
$$\Pi_2 = -\frac{1}{\lambda_2}\frac{2\mu}{\alpha^2}\left(\alpha\lambda_3^{\alpha}\right) + \frac{2\mu}{\alpha^2}\left(\alpha\lambda_2^{\alpha-1}\right). \quad (10)$$

These principal stresses are transformed by an eigen-vector matrix of the plane strain tensor in order to obtain Π_x in Eq. (9). A similar procedure was used in [15].

3. FEM simulations

3.1. Linear VFM

A finite element simulation was conducted using ABAQUS/Explicit. A rectangular geometry similar to the actual drop-weight test specimen was made with length $L = 38$ mm and width $W = 8$ mm. The length and width are aligned along x and y directions, respectively. CPS4R (plane stress quadrilateral element) with 0.2 mm element size was chosen; a velocity boundary condition of 5 m s^{-1} in the x direction was applied at the right end of the geometry with a rise time of 0.2 ms. Every 20 μs, the displacement field $\boldsymbol{u}$ was extracted, in order to resemble the actual imaging speed of 50,000 fps. Each displacement field was polluted by white Gaussian noise with a standard deviation of 4×10^{-4} mm; this noise level is similar to that of the displacement fields obtained in the actual imaging of the drop-weight and gas-gun experiments. This polluted $\boldsymbol{u}$ was used to numerically calculate $\boldsymbol{\varepsilon}$ and $\boldsymbol{a}$, by respectively using the strain-displacement matrix of an isoparametric quadrilateral element and a simple finite second-order central approximation. For the pre-stretching cases, a general static step was introduced before applying the dynamic loading in the explicit simulation. There were 7 different pre-deformations in order to produce engineering pre-strains of 0.1 to 0.7 with an interval of 0.1. After this pre-stretching, the same velocity boundary condition was used but applied along a plane 38 mm away from the fixed end. The one-term Ogden material model was adopted; for its parameters μ and α, 1, 2, 3 and 4 were given. Thus, there were 16 simulations in total for each parameter case.

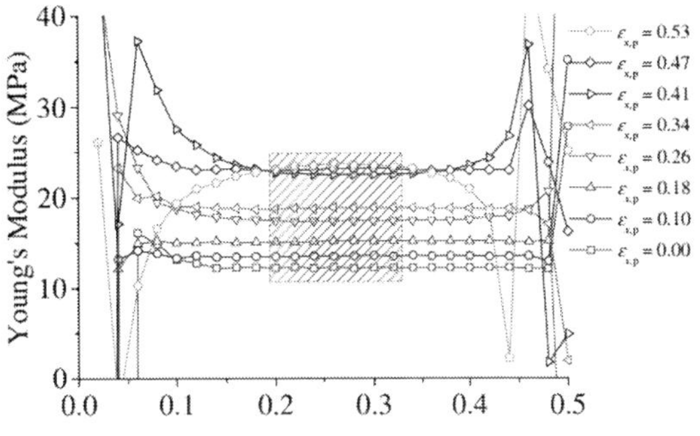

Figure 1. A series of the Young's modulus identifications for the simulation with $\mu = 4$ MPa and $\alpha = 2$.

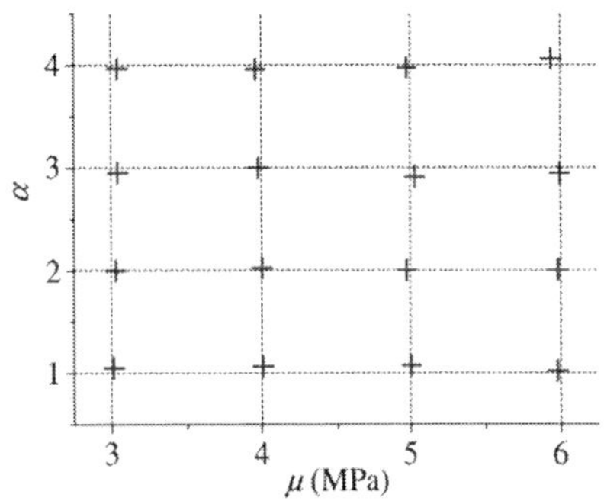

Figure 2. All identified parameters from the liner VFM on the drop-weight simulations with respect to the given parameters.

As an example, Fig. 1 shows the histories of the Young's modulus identification for the simulation with parameters $\mu = 4$ MPa and $\alpha = 2$. Next to this figure, the true pre-strain $\varepsilon_{x,p}$ for each identification is shown. The Young's moduli E_n used in Eq. (6) are averaged values over the period (0.20–0.34 ms) as indicated by the hatched box in Fig. 1. This period is chosen so that the estimation is stable and the identified Poisson's ratio is bounded between 0.45–0.50, as expected from material incompressibility. Applying E_n to Eq. (6), the Ogden parameters are identified as $\mu = 4.03$ MPa and $\alpha = 2.00$. Using the same method, all other parameters were identified. As shown in Fig. 2, the identified parameters (cross symbol) are well matched with the given values (intersected points of dashed lines). There is a slight scatter of the identified parameters. This deviation is caused by the added noise and the averaging procedure.

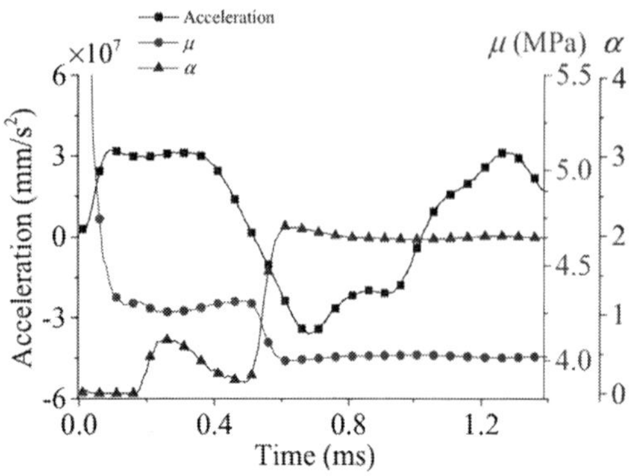

Figure 3. Averaged acceleration profile and the history of μ and α identifications.

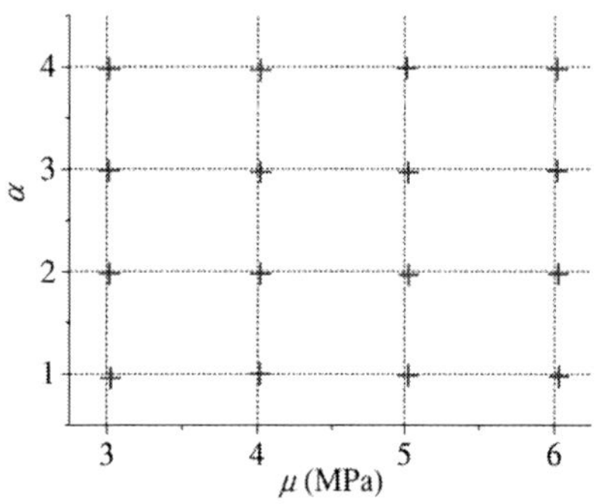

Figure 4. All identified parameters from the nonlinear VFM on the gas-gun simulations with respect to the given parameters.

3.2. Nonlinear VFM

A similar FEM was used for simulating a gas gun experiment and the nonlinear VFM procedure. The geometry was given as $L = 40$ mm and $W = 11$ mm. A faster velocity boundary condition was applied, which peaked at 14 m s^{-1} with a rise time of 0.5 ms. This velocity condition is similar to that produced in the actual gas-gun experiment. The output spacing was 10 μs, i.e. an imaging speed of 100,000 fps, with a total loading duration of 1.4 ms. White Gaussian noise was added to $\boldsymbol{u}$ as in the linear VFM and the calculation of $\boldsymbol{\varepsilon}$ and $\boldsymbol{a}$ was performed.

These field data were then applied to Eq. (9) to identify the material parameters. Figure 3 shows the history of the parameter identifications on the simulation with $\mu = 4$ MPa and $\alpha = 2$. The acceleration profile is an averaged value over the specimen surface at each time; the sign change of the acceleration indicates the moment of the wave reflection off the fixed end of the specimen. Each predicted parameter at time t_k is obtained by solving Eq. (9) with $\varepsilon(t)$ and $\mathbf{a}(t)$ from $t = 0$ to $t = t_k$. As can be seen in Fig. 3, the identifications of μ and α do not reach the given values during the incident wave period (0.0–0.6 ms). The identifications approach the given parameters after the first wave reflection (after around 0.6 ms) in a stepwise manner. The final identified parameters are taken at the end of the time step. Using the same procedure, all other identified parameters are presented in Fig. 4 showing a good match with the given values.

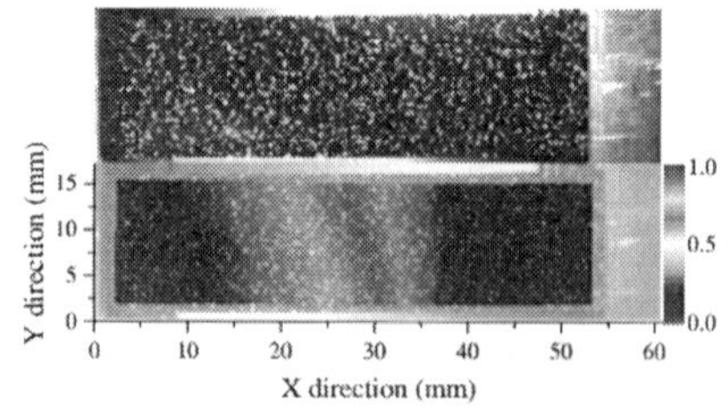

Figure 5. Pictures of the drop-weight test before (upper) and during (bottom) loading.

4. Experiment

A sheet of a commercial EPDM rubber (Amarin Rubber & Plastic Ltd) was used for the experimental study. Three uniaxial specimens were prepared for static ($L = 60$ & $W = 20$ mm), drop-weight and gas-gun ($L = 40$ & $W = 11$ mm) tests. The thickness is uniform at 2 mm. The uniaxial static test was conducted at three true strain rates: 0.01, 0.001 and 0.0001 s^{-1}. Their true stress-strain curves are presented in Fig. 8. The strain data was measured by means of digital image correlation (DIC) software (Davis 8.2.0, LaVision) applied on test images taken by a USB camera (USB 3.0 Digital Camera, Point Grey Research, Inc.). For the DIC analysis, a white speckle pattern was made on the specimens by spray paint. The same speckle pattern was applied to the specimens for the drop-weight and gas-gun experiments. Images of the drop-weight test with this pattern is shown in Fig. 5. The drop-weight applies a displacement in the x direction on the right hand side of the specimen. The colour map represents $\boldsymbol{u}_x$ field obtained from DIC analysis.

4.1. Linear VFM and drop-weight tests

In the drop-weight experiments, the EPDM specimens were located on the top of a (vertically set) long loading bar. After clamping, the effective specimen length is about 37 mm. The bar is loaded by means of a drop-weight which introduces a tensile stress wave which dynamically stretches the bottom of the specimen. A high-speed camera (FASTCAM SA5, Photron) is used to take images of the specimen at a speed of 50,000 fps. For the pre-stretching test, manual stretching was applied prior to the dynamic loading. The upper clamp was fixed and the pre-force measured with a load cell installed above. The pre-strains were calculated by comparing the distances of two pre-marked points on the specimen images.

The imaging data from one non-pre-stretching and several pre-stretching tests were analysed using DIC software to produce their displacement and strain fields. Each displacement data point was temporally fitted to a 9 degree polynomial, which was double differentiated to calculate acceleration. The test data were applied to the linear VFM procedure to obtain a series of Young's modulus identifications. This result from the first experiment set (*SET 1*) is shown in Fig. 6. The hatched area indicates the period where the Young's moduli were averaged. The averaged moduli are listed in Table 1 for *SET 1* and 2 tests. These two sets of all moduli were applied to the optimization procedure of Eq. (6); the

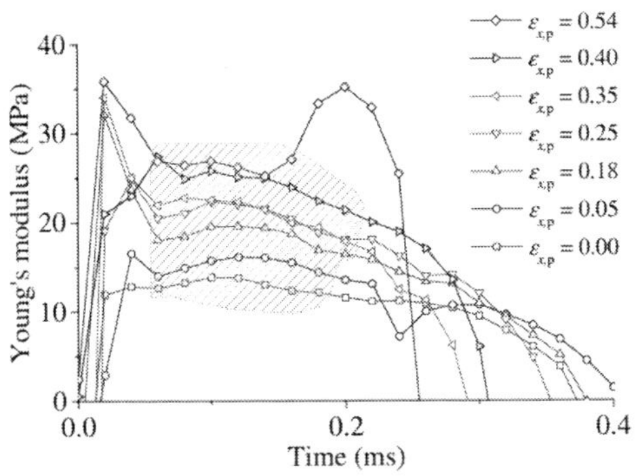

Figure 6. Identified Young's moduli from one non-pre-stretching and six pre-stretching drop-weight tests and the application of the linear VFM.

Table 1. Averaged Young's moduli, strain rates of the two drop-weight test sets and the identified Ogden parameters.

SET 1			*SET 2*		
E(MPa)	$\dot{\varepsilon}_x(s^{-1})$	$\varepsilon_{x,p}$	E(MPa)	$\dot{\varepsilon}_x(s^{-1})$	$\varepsilon_{x,p}$
11.8	80	0.00	12.3	103	0.00
14.9	78	0.05	12.6	96	0.08
17.1	86	0.18	13.0	108	0.14
20.6	90	0.25	14.6	102	0.15
21.0	96	0.35	17.4	122	0.23
22.2	98	0.40	20.4	118	0.31
26.6	120	0.54	23.4	109	0.34
			27.7	125	0.63
Averaged $\dot{\varepsilon}_x = 102\,s^{-1}$					
Optimized Ogden parameters: $\mu = 4.39$ MPa, $\alpha = 1.71$					

identified parameter are $\mu = 4.39$ MPa and $\alpha = 1.71$. The strain rate value for each test in Table 1 were obtained by averaging the field data only over the deformed area.

4.2. Nonlinear VFM and gas-gun tests

The gas-gun experiment in the present study is based on the high strain rate experiment on a yarn of polymeric fibres conducted by Russell *et al.* [16]. For the present tests, an aluminium block (80 × 80 × 30 mm) was prepared, on which one end of two EPDM specimens was fixed. The other ends were clamped to a rigid frame attached to the barrel of the gas gun. An aluminium projectile was accelerated by the gas gun to apply impact loading on the block at speeds between 50 and 80 m s^{-1}. After this impact, the high-speed imaging was triggered at an imaging speed of 100,000 fps. The high-speed imaging was conducted on one of the two specimens. The captured images were analysed in the same manner as the drop-weigh test in order to produce $\boldsymbol{u}$, $\boldsymbol{\varepsilon}$ and $\boldsymbol{a}$. These field data were then applied to Eq. (9) to identify the Ogden parameters. There were four gas-gun tests with different impacting velocities.

Figure 7 shows the histories of the identified parameters and the averaged acceleration and strain rate profiles from one of the tests. Similarly as shown in the simulation, the identification of μ and α becomes more stable and converges in a stepwise manner as the range of the data period included in Eq. (9) increases. The stable

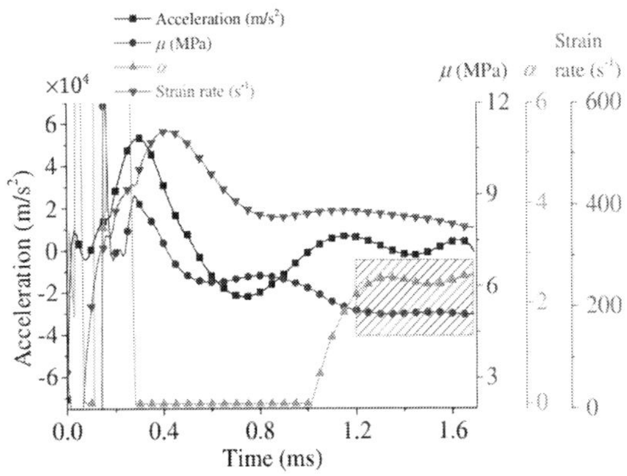

Figure 7. Averaged acceleration & strain rate profiles and the history of the μ and α identifications of the EPDM specimens tested by the gas-gun experiment (Test 3).

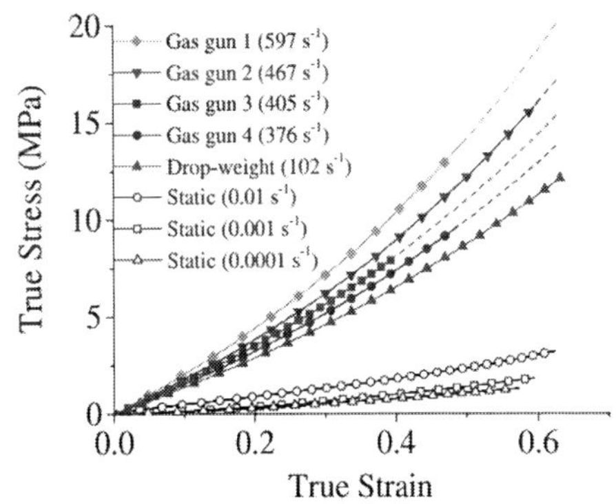

Figure 8. Quasi-static stress strain curves and the reconstructed Ogden curves using the parameters from Tables 1 and 2.

Table 2. Identified Ogden parameters of the EPDM specimens from the gas-gun experiment and the nonlinear VFM.

Test	$\dot{\varepsilon}_x(s^{-1})$	μ(MPa)	α
1	597	6.52	2.12
2	467	5.70	2.05
3	405	5.26	1.95
4	376	4.87	1.85

identification period is indicated by the hatched box in Fig. 7, in which the μ and α values are averaged to obtain the final identified parameters. To calculate an average strain rate, the strain rate profile from 0.4 to 1.7 ms was chosen to be averaged, as an example. This procedure was identically repeated for all other gas-gun experiments. The identified parameters are given in Table 2.

5. Discussion and conclusions

The Ogden uniaxial curves corresponding to the identified parameters from the drop-weight and gas-gun experiments are shown in Fig. 8. This figure shows a clear rate dependency of the EPDM rubber captured by the two dynamic experiments and their linear and nonlinear VFMs.

One question might arise regarding the possibility that the gas-gun and nonlinear VFM replace the drop-weight and linear VFM which need the relatively complicated pre-stretching procedures. However, it is experimentally difficult for the gas-gun experiment to produce a single loading pulse, which can generate a medium strain rate (e.g., $100\,s^{-1}$), and, at the same time, a large strain with a constant strain rate. Thus, the drop-weight experiment with the pre-stretching procedure is still useful technique to fill the gap between high and quasi-static strain rates. The drop-weight test is also advantageous for a high-speed imaging condition with a small field of view as its required deformation amplitude is small. The usefulness of the gas-gun test and its VFM procedure lies in the capability of introducing high strain rates and the experimental expediency of using single loading pulse so that any unexpected material behaviour from pre-stretching can be avoided.

Effort sponsored by the Air Force Office of Scientific Research, Air Force Material Command, USAF, under grant number FA8655-12-1-2015. The U.S Government is authorized to reproduce and distribute reprints for Governmental purpose notwithstanding any copyright notation thereon. The authors thank S Fuller of AFOSR and M Snyder and R Pollak of EOARD for their support. Finally we thank Professor F Pierron for his invaluable help with the Virtual Fields Method.

References

[1] Sarva S.S., Deschanel S., Boyce M.C., Chen W. (2007) Stress–strain behavior of a polyurea and a polyurethane from low to high strain rates. Polymer (Guildf) 48:2208–2213.

[2] Chen W., Zhang B., Forrestal M.J. (1999) A split Hopkinson bar technique for low-impedance materials. Exp Mech 39:81–85.

[3] Song B., Chen W. (2003) One-Dimensional Dynamic Compressive Behavior of EPDM Rubber. J Eng Mater Technol 125:294.

[4] Shergold O.A., Fleck N.A., Radford D. (2006) The uniaxial stress versus strain response of pig skin and silicone rubber at low and high strain rates. Int. J. Impact Eng. 32:1384–1402.

[5] Roland C.M., Twigg J.N., Vu Y., Mott P.H. (2007) High strain rate mechanical behavior of polyurea. Polymer (Guildf) 48:574–578.

[6] Nie X., Song B., Ge Y., et al. (2008) Dynamic Tensile Testing of Soft Materials. Exp Mech 49:451–458.

[7] Pierron F., Sutton M.a., Tiwari V. (2010) Ultra High Speed DIC and Virtual Fields Method Analysis of a Three Point Bending Impact Test on an Aluminium Bar. Exp Mech 51:537–563.

[8] Moulart R., Pierron F., Hallett S., Wisnom M. (2011) Full-Field Strain Measurement and Identification of Composites Moduli at High Strain Rate with the Virtual Fields Method. Exp Mech 51:509–536.

[9] Pierron F., Forquin P. (2012) Ultra-High-Speed Full-Field Deformation Measurements on Concrete Spalling Specimens and Stiffness Identification with the Virtual Fields Method. Strain 48:388–405.

[10] Pierron F., Zhu H., Siviour C. (2014) Beyond Hopkinson's bar. Philos Trans R Soc A Math Phys Eng Sci 372:20130195.

[11] Pierron F., Grédiac M. (2012) The virtual fields method?: extracting constitutive mechanical parameters from full-field deformation measurements (Springer, New York)

[12] Ogden R.W. (1972) Large Deformation Isotropic Elasticity – On the Correlation of Theory and Experiment for Incompressible Rubberlike Solids. Proc R Soc London A Math Phys Sci 326:565–584.

[13] ABAQUS (2011) ABAQUS 6.11 analysis user's manual. Abaqus 6.11 Doc.

[14] Promma N., Raka B., Grédiac M., et al. (2009) Application of the virtual fields method to mechanical characterization of elastomeric materials. Int J Solids Struct 46:698–715.

[15] Palmieri G., Sasso M., Chiappini G., Amodio D. (2011) Virtual Fields Method on Planar Tension Tests for Hyperelastic Materials Characterisation. Strain 47:196–209.

[16] Russell B.P., Karthikeyan K., Deshpande V.S., Fleck N. a. (2013) The high strain rate response of Ultra High Molecular-weight Polyethylene: From fibre to laminate. Int J Impact Eng 60:1–9.

EPJ Web of Conferences **94**, 01018 (2015)
DOI: 10.1051/epjconf/20159401018

Development of a finite element model for the simulation of parabolic impact of sandwich panels

Karthik Ram Ramakrishnan[1,2,a], Sandra Guérard[1,b], Laurent Mahéo[3], Krishna Shankar[2], and Philippe Viot[1]

[1] Arts et Métiers ParisTech, I2M, UMR 5295, 33400 Talence, France
[2] School of Engineering and Information Technology, UNSW, Canberra 2600, Australia
[3] LIMATB, Université de Bretagne Sud, 56321 Lorient Cedex, France

Abstract. Sandwich panels are lightweight structures of two thin high strength facesheets bonded to either side of a thick low density core such as foams and honeycombs. It is necessary to study the impact response of sandwich structures in order to ensure the reliability and safety of these structures. The response of sandwich panels to impact loading is usually studied for impact at normal angle of incidence. In real engineering situations, the structures are more frequently loaded at some oblique angle or with a complex trajectory. It is easy to carry out normal impact tests using devices like the drop tower, but impacts at oblique angles are difficult to characterise experimentally. A tri-dimensional impact device called Hexapod has been developed to experimentally study the impact loading of sandwich plates with a parabolic trajectory. The Hexapod is a modified Gough-Stewart platform that can be moved independently in the six degrees of freedom, corresponding to three translation axes and three rotation axes. In this paper, an approach for modelling the parabolic impact of sandwich structures with thin metallic facesheets and polymer foam core using commercial finite element code LS-DYNA software is presented. The results of the FE model of sandwich panels are compared with experimental data in terms of the time history of vertical and horizontal components of force. A comparison of the strain history obtained from Digital Image Correlation and LS-Dyna model are also presented.

1. Introduction

Sandwich panels are a special form of laminated composites comprising of two thin high strength facesheets bonded to either side of a thick low density core such as foams and honeycombs. This provides a lightweight structure with high bending stiffness. The skins are designed to resist tensile and compressive stresses and are usually made of aluminium, polymers or fibre reinforced polymers. The core is designed to resist compression and shear stresses and a range of materials including end grain Balsa wood, rigid polymer foams, aluminium or Nomex honeycombs are used as cores. The standard test method for the study of low velocity impact response of sandwich panels are for normal incidence impact, i.e. the impact loading is applied along one fixed direction that is perpendicular to the plane of the target plate. Test devices like drop tower are designed for normal impact condition and the response of the panels are analysed in terms of peak load, absorbed energy and deflection at peak load. However, normal impacts rarely occur in real engineering situations and structures are more frequently loaded at some oblique angle or complex trajectory. While normal incidence impact tests are easy to carry out using devices such as the drop tower, oblique tests are difficult to undertake experimentally, given the need to guide the projectile, coupled with the presence of a horizontal force component that applies a load to the required guide rails. Zhou et al. [1] found that the impact energy for an oblique impact of sandwich composite is dissipated in shearing both the composite and the foam around the perimeter of the projectile, as well as crushing the foam ahead of the impactor. The perforation energy was found to increase with increasing angle of incidence. Goswami et al. [2] found that the debonding along the facesheets and the core interface of sandwich panels is predominated by a sliding mode and observed that low relative density of the core will result in more critical debonding [2]. Zhou et al. [1] also observed that the shear fracture properties of the foams had a significant effect on the perforation resistance of the sandwich structures, suggesting that this mode of failure is important in determining the impact resistance of sandwich structures. Chai and Zhu [3] reviewed the existing literature on the low velocity impact of sandwich structures and found that studies on sandwich panel subjected to low velocity impact at angles other than perpendicular to the plane of the sandwich were limited. Guérard et al. [4] noted that normal impacts do not correspond to the complex loading that real structures encounter and developed a tri-dimensional impact device called Hexapod to experimentally study the impact loading of sandwich plates with a parabolic trajectory. Sandwich plates with metallic facesheets and polymeric foam core were investigated.

Experiments are time consuming and expensive, and predicting the response of sandwich plates is complicated due to effects such as material and geometric nonlinearities, transverse shear effects and multiple and coupled damage modes [5]. These problems and cost of experimentation can be avoided by using numerical methods,

[a] e-mail: karthik.ramakrishnan@adfa.edu.au
[b] e-mail: sandra.guerard@ensam.eu

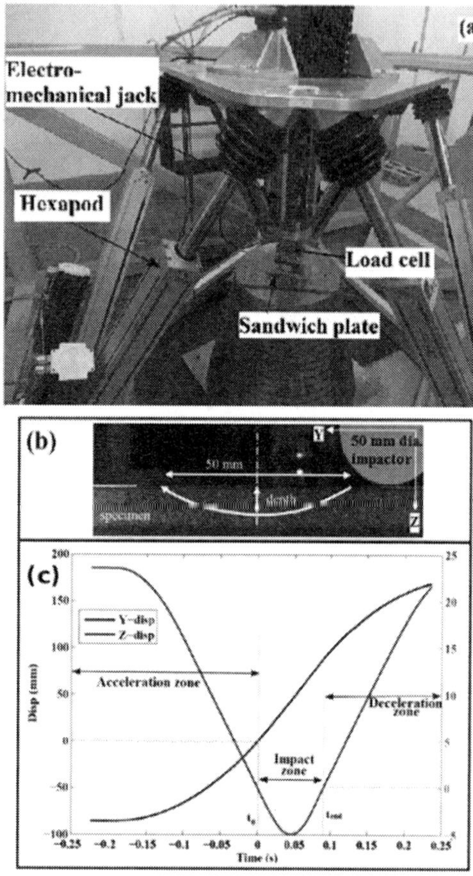

Figure 1. (a) Hexapod setup for tri-dimensional impact, (b) Parabolic trajectory and (c) Displacement input for the Hexapod.

which can help in predicting the energy absorption and peak loads by considering competing mechanisms [6–8]. The combination of improved constitutive modelling and decreased computational costs make it possible to utilise continuum modelling of sandwich composites to numerically simulate impact response for a much wider range of impactor geometries and velocities [9]. In this paper, a Finite Element model is presented for the simulation of the parabolic impact of sandwich plates.

2. Experimental setup

Low-velocity impact tests were performed using the Hexapod facility available in the Dynamics platform of DuMAS Department of I2M laboratory in Arts et Métiers Paristech, Bordeaux. The Hexapod, which is a modified Gough-Stewart platform, a type of parallel robot constituted of a fixed and a moving platform is shown in Fig. 1. The top plate can be moved independently in the six degrees of freedom, corresponding to three translation axes and three rotation axes. Its horizontal velocity can reach 1.4 m/s and the maximal vertical velocity is 1 m/s. A seventh electromechanical jack (of maximum velocity 3.5 m/s) is implemented on the top plate to increase the vertical speed. A cylindrical impactor is attached to the seventh jack and the sandwich sample is placed on a rigid platform. The velocity achieved with this configuration of Hexapod is equivalent to the drop tower but with the advantage of imposing a 3D trajectory to the projectile [4]. The Hexapod is setup in such a way that it is possible to change the impactor easily with other geometries such as wedge-shaped impactor.

The Hexapod is instrumented with a tri-axial load cell (Kistler, 9377C, Fx and Fy range: 3000 N, Fz range: 5000N) which is placed between the cylindrical impactor and the bottom end of the seventh jack. Load history of each impact event was measured by the 3D load cell located between the impactor and the jack extremity. A high-speed camera (Photron Fastcam APX RS, 5000 frames per second) was used to film the impact. The side of the tested sample was coated with a fine spray of paint to enhance the random texture of the foam and allow for Digital Image Correlation (DIC) calculation (VIC-2D, Correlated Solution).

The Hexapod setup was used for three dimensional impact testing of sandwich samples made of Aluminium facesheets and PVC closed-cell foam (Divinycell) core by Guérard et al. [4]. Sandwich samples with PVC foam core of thickness 20 mm and 2024 aluminium facesheets of two different thicknesses: 0.5 mm and 1 mm were considered. The sandwich panels were assembled using a two-component epoxy adhesive (Araldite AW 106R/hardener HV 953U) which was chosen for their high shear and peel strength. The dimensions of the sandwich samples were 200 mm long and 120 mm wide.

In the case of a parabolic impact, a trajectory is imposed to the impactor in terms of displacement curves in the Y- and Z-axis, corresponding to the horizontal and vertical directions. This is in contrast to the normal impact testing using a drop tower, where the impactor is raised to a particular height and released and the potential energy of the impactor is converted to kinetic energy. The parameters defining the parabolic impact are the depth of penetration into the impacted specimen and the width or opening of the parabola. A parabolic displacement shown in Fig. 1b was imposed to the cylindrical impactor with a depth of penetration of 5 mm into the impacted specimen and the width of the parabola was 50 mm. Figure 1c also shows the translation displacement in the Y-direction and Z-direction, which is input to the hexapod. This represents the entire trajectory input to the impactor with the beginning of the contact with the sandwich target represented by t_0 and the end of contact denoted by t_{end}. The horizontal velocity during impact (along Y direction) was constant and equal to 1 m/s and the vertical acceleration was equal to 4 m/s^2.

3. Finite element model of sandwich plate

A numerical model was developed using the explicit finite element software, LS-Dyna. An LS-Dyna model, shown in the Fig. 2, was created to simulate the parabolic impact of the aluminium – Divinycell sandwich. The aluminium

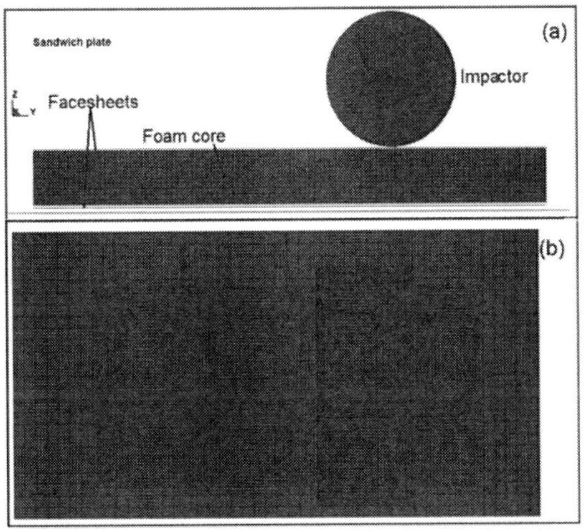

Figure 2. LS-Dyna model of Aluminium – Divinycell sandwich and cylindrical impactor.

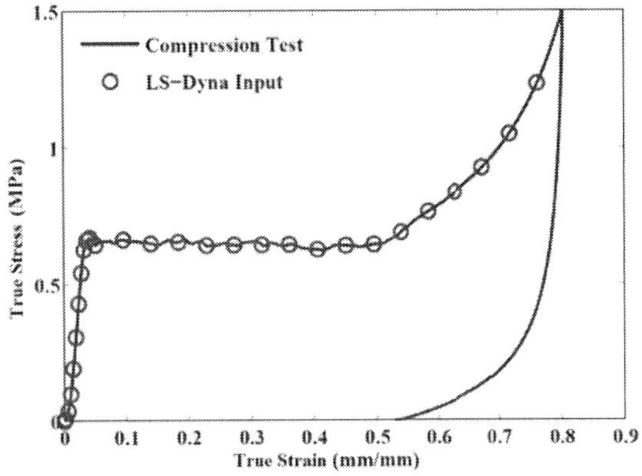

Figure 3. Material input for Divinycell foam from uniaxial compression test.

– Divinycell sandwich plate was modelled with solid elements for the core and shell elements for the thin facesheets. Two types of sandwich plates were modelled, with 0.5 mm thick facesheet and 1 mm thick facesheet. The aluminium facesheets were modelled using Belytschko-Tsai shell elements and the foam core was modelled using constant stress brick elements.

Foams are a type of cellular solids which can be described as a collection of regular or irregular thin-walled cells. The macroscopic compressive behaviour is determined by the compressive behaviour of individual cells and the collective interactions between neighbouring cells [10]. The simulation of the impact behaviour of sandwich composites is usually more complex than that of monolithic laminates. Feng and Aymerich [11] found that the mechanical properties of the core material has a strong influence on the response of the the facesheets. The failure mechanisms may also be significantly different from those developing in conventional laminates. Therefore, an accurate FE analysis of a sandwich structures requires including complex material models capable of capturing the mechanical behaviour of the core as well as the facesheets.

LS-Dyna offers a variety of material models, each with capabilities designed to capture the unique behaviour of different types of foam, for instance, open cell foams and closed cell foams. The material model chosen for the PVC foam core was MAT CRUSHABLE FOAM available in the material library of LS-Dyna as MAT63. The modelling of the foam core behaviour requires the use of models to reproduce the crushing behaviour of these cellular materials. Croop and Lobo [12] described that irrespective of the type of the foam, there are generally three zones in the compressive stress-strain relationship of foam, namely, initial elastic region up to yield, plateau compaction region and a densification region. The stress-strain response of the Divinycell foam obtained from uniaxial compression tests using Zwick electromechanical testing machine was input to the material model as shown in the Fig. 3. The aluminium alloy facesheets were modelled with a PLASTIC KINEMATIC material law and the steel impactor was modelled with MAT RIGID. The input parameters of these materials were obtained from manufacturer datasheets.

The steel cylindrical impactor was given displacement input in the Y and Z direction by defining a curve to BOUNDARY PRESCRIBED MOTION keyword. This is in contrast to the normal impact simulation where an initial velocity input is given to the impactor. The rotation motion of the cylindrical impactor was constrained. The simulation was initiated just before the contact to reduce the run-time, with the impactor placed 0.1 mm above the facesheet. The total duration of the simulation is 90 milliseconds corresponding to the impact duration observed in the experiment.

The progression of core effective strain for parabolic impact with cylindrical impactor is shown in Fig. 4. It can be seen that initially when the impactor comes into contact with the sandwich plate, the strain is limited to a small region of the sandwich plate at Time = 0.01 s and this region grows with increasing contact duration. The strain contour for contact time of 26, 45, and 86 milliseconds are also shown in the figure. These times correspond to the point of maximum force, maximum displacement and end of contact, respectively. It can be seen that there is a strain component that is ahead of the impactor as the core elements undergo deformation.

The LS-Dyna model is validated by comparing the force-time histories with experimental data presented by Guérard et al. [4]. A comparison of the vertical and horizontal component of forces obtained from the LS-Dyna model and the experiments for the sample with 0.5-mm-thick facesheet are shown in Fig. 5a. The experimental force values are in solid lines and the FE results are in dash lines. It can be seen that the maximum vertical force (Z-force) from the experiment compares well with the simulation but the maximum horizontal force (Y-force) obtained from the simulation is much lower than the experiment. Guérard et al. [4] reported that the peak of the Z-force (2544 N) occurs before the maximum penetration depth whereas the peak of Y-force (1850 N)

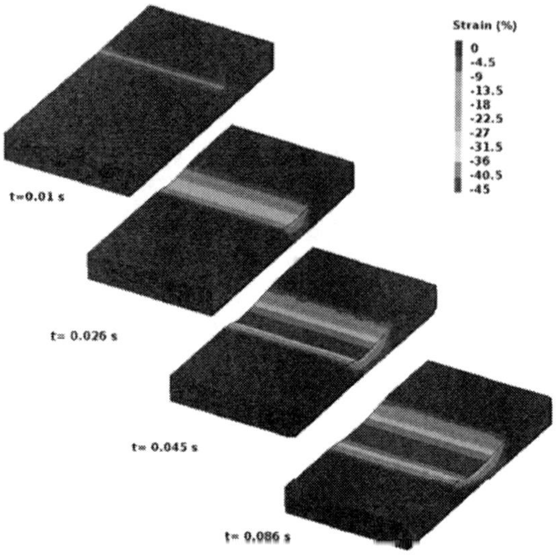

Figure 4. Progression of the impact on sandwich sample with 1 mm thick Al facesheet.

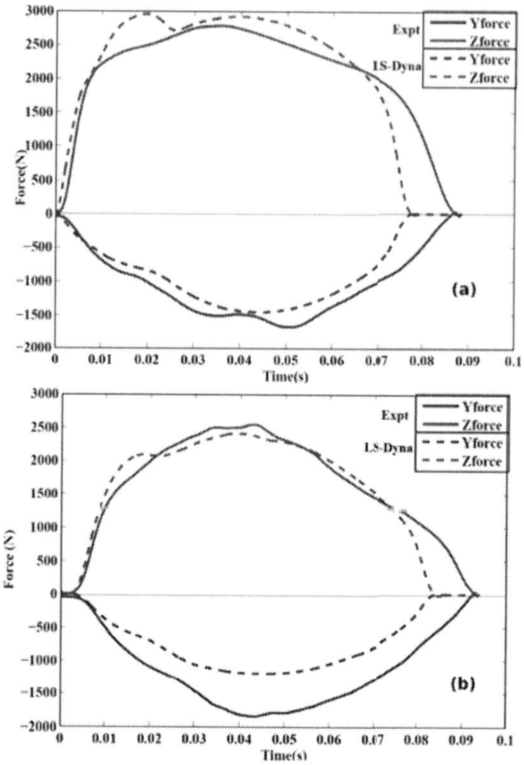

Figure 5. Force-time results obtained for parabolic impact of Aluminium Divinycell sandwich plates (a) 0.5 mm thick facesheet and (b) 1 mm thick facesheet.

occurs slightly after which was attributed to the formation of a small wrinkle in front of the impactor during the test which generates a resistant horizontal force. The Y-force is higher in the experiment compared to the LS-Dyna model for the same reason. There is no wrinkling in the simulation and hence the value of Y-force is lower.

For the sample with 1 mm thick facesheet, shown in Fig. 5b, the maximum vertical force obtained from the simulation is comparable to the experiment. It can be observed that the Z-force is slightly higher for thicker aluminium facesheet with a peak of 2780 N. The Y-force, with a peak of 1670 N is lower for the 1 mm thick facesheet compared to the 0.5 mm thick facesheet. This is explained by the fact that there is no observable wrinkling with the 1 mm thick facesheet. The horizontal component of force is under-estimated which suggests that the modelling of shear response can be improved by using a foam material model which is capable of simulating multi-axial loading.

It is also observed that the contact duration is shorter for the LS-Dyna model in both cases. One of the possible causes for this phenomenon is the permanent deformation observed in the foam material ahead of the point of contact. However, the overall characteristics of the response of the sandwich plate to parabolic impact was simulated and it can be concluded from these simulations that an LS-Dyna model can be used for the numerical modelling of tri-dimensional impacts.

A comparison of the strain contour of the sandwich panel is shown in Fig. 6. The effective strain obtained from Digital Image Correlation (DIC) of the high speed camera images using Vic2D software is compared with the strain contour obtained from LS-Dyna simulation for the case of the sandwich plate with 1 mm thick aluminium facesheets. The maximum strain at the point of maximum penetration, i.e. at 45 milliseconds, is found to be 37.8%, while the corresponding strain obtained from LS-Dyna simulation is 32.98%. There is large localised strain immediately below the impactor and there is permanent plastic deformation in both the top facesheet and the core. This is different to the normal impact case, where there is elastic recovery of the top facesheet and core crushing in the foam, which results in debonding between the core and facesheet.

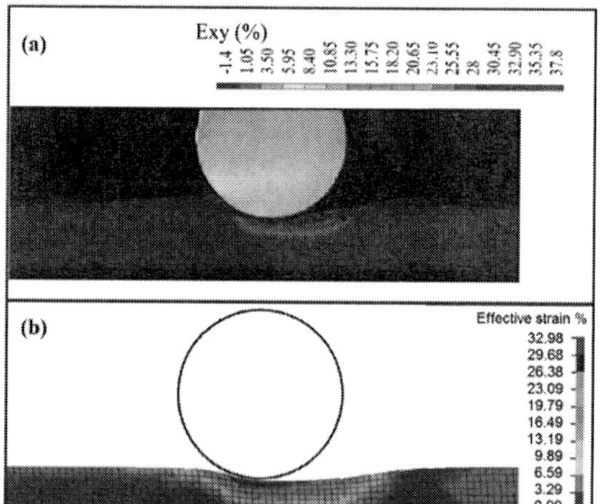

Figure 6. Effective strain of Aluminium Divinycell sandwich plates (a) DIC and (b) LS-Dyna.

4. Conclusion

A new method to conduct low velocity tri-dimensional impact tests on sandwich panels using a Hexapod with parabolic trajectory of the impactor was presented. An LS-Dyna model was developed to reproduce the response of the sandwich panels made with Aluminium facesheets of two different thickness and PVC foam core. A crushable foam material model was used for the simulation of the Divinycell foam and bilinear kinematic model was used for the aluminium facesheets. The results of the numerical model were compared with experimental data presented by Guérard et al. [4] in terms of macroscopic force response and strain contour from digital image correlation. The peak force and maximum strain from the FE model show good correlation with experimental data. The LS-Dyna model provides reasonable results for the parabolic impact but more work is required to improve the model, particularly shear behaviour of foam and failure modes like debonding.

References

[1] J. Zhou, M.Z. Hassan, Z. Guan, and W.J. Cantwell. The low velocity impact response of foam-based sandwich panels. *Composites Science and Technology*, 72(14):1781–1790, 2012.

[2] S. Goswami and W. Becker. The effect of facesheet/core delamination in sandwich structures under transverse loading. *Composite Structures*, 54(4):515–521, 2001.

[3] G.B. Chai and S. Zhu. A review of low- velocity impact on sandwich structures. *Proceedings of the Institution of Mechanical Engineers, Part L: Journal of Materials Design and Applications*, 225(4): 207–230, 2011.

[4] S. Guérard, J-L. Barou, L. Mahéo, and P. Viot. Development of a new experimental device for tridimensional impacts – results on foam-core sandwich panels. In *4th International conference on impact loading of lightweight structures*, 2014.

[5] J. Mackerle. Finite element analyses of sandwich structures: a bibliography 1980–2001. *Engineering Computations*, 19(2):206–245, 2002.

[6] R.C. Batra, G. Gopinath, and J.Q. Zheng. Damage and failure in low energy impact of fiber-reinforced polymeric composite laminates. *Composite Structures*, 94(2):540–547, 2012.

[7] C. Bouvet, S. Rivallant, and J.J. Barrau. Low velocity impact modeling in composite laminates capturing permanent indentation. *Composites Science and Technology*, 72(16):1977–1988, 2012.

[8] A. Rajaneesh, I. Sridhar, and S. Rajendran. Impact modeling of foam cored sandwich plates with ductile or brittle faceplates. *Composite Structures*, 94(5):1745–1754, 2012.

[9] J.A. Nemes and K.E. Simmonds. Low-velocity impact response of foam-core sandwich composites. *Journal of Composite Materials*, 26(4):500–519, 1992.

[10] E.A. Flores-Johnson and Q.M. Li. Experimental study of the indentation of sandwich panels with carbon fibre-reinforced polymer face sheets and polymeric foam core. *Composites Part B: Engineering*, 42(5):1212–1219, 2011.

[11] D. Feng and F. Aymerich. Damage prediction in composite sandwich panels subjected to low-velocity impact. *Composites Part A: Applied Science and Manufacturing*, 52(0):12–22, 2013.

[12] B. Croop and H. Lobo. Selecting material models for the simulation of foams in ls-dyna. In *7th European LS-DYNA Conference*, 2009.

EPJ Web of Conferences **94**, 01019 (2015)
DOI: 10.1051/epjconf/20159401019

Computational modeling of dynamic mechanical properties of pure polycrystalline magnesium under high loading strain rates

Qizhen Li[1,2,a]

[1]School of Mechanical and Materials Engineering, Washington State University, Pullman, WA 99164, USA
[2]Department of Chemical and Materials Engineering, University of Nevada, Reno, Reno, Nevada 89557, USA

Abstract. Computational simulations were performed to investigate the dynamic mechanical behavior of pure polycrystalline magnesium under different high loading strain rates with the values of 800, 1000, 2000, and 3600 s^{-1}. The Johnson-Cook model was utilized in the simulations based on finite element modeling. The results showed that the simulations provided well-matched predictions of the material behavior such as the strain rate-time history, the stress-strain curve, and the temperature increase. Under high loading strain rates, the tested material experienced linear strain hardening at the early stage of plastic deformation, increased strain hardening at the intermediate plastic deformation region, and decreased strain hardening at the region before fracture. The strain hardening rates for the studied high loading strain rate cases do not vary much with the change of strain rates.

1. Introduction

As the lightest structural metal, magnesium is an attractive material candidate for various structural components in automobiles, aircrafts, etc.. These components often experience high strain rate dynamic loadings during their actual service. Thus, research efforts were devoted to study dynamic mechanical properties of magnesium and its alloys such as ZK60, Mg-Gd-Y, AZ80, AZ31, magnesium single crystal, and pure polycrystalline magnesium [1–6].

The experimental investigations require a split Hopkinson pressure bar (SHPB) system and it can be costly. Therefore, it is often appealing to investigate the dynamic mechanical properties through computational simulations. To achieve meaningful simulations, appropriate theoretical models are needed. Among the theoretical models for high strain rate loading conditions, the Zerilli-Armstrong model [7–10] and the Johnson-Cook model [11, 12] are the most commonly used because of their simple forms. The model's simplicity can result in the save of computational resources.

This study performed computational simulations of mechanical behavior of pure polycrystalline magnesium under a range of high loading strain rates using finite element method. The loading strain rates are chosen to be the same as those experimental strain rates in the reference [6]. The simulation results were compared with the experimental data for validation.

2. Computational method and data analysis

ABAQUS\Explicit is utilized for three-dimensional modeling of high strain rate tests. To be the same as the experimental SHPB set-up as shown in the schematic

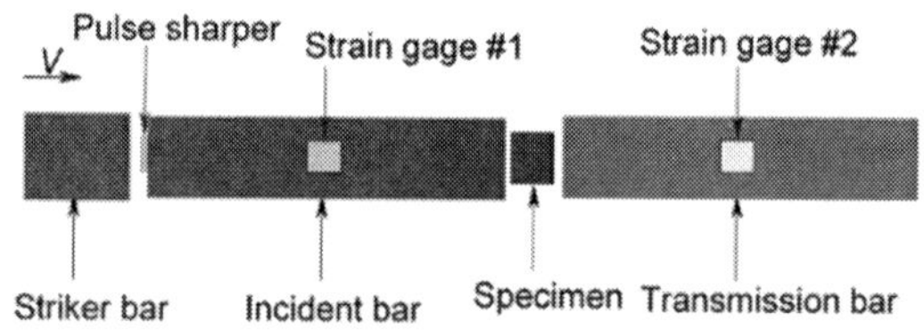

Figure 1. Schematic drawing of a SHPB system. The striker bar applies the loading stress wave, the specimen is held between the incident bar and the transmission bar, and strains are measured using strain gauges #1 and #2.

drawing in Fig. 1, aluminum is chosen to be the material used for the striker, incident, and transmission bars. Eight-node solid elements are used for the bars and samples in the simulations. The bars are assumed to behave elastically. The Johnson-Cook model was implemented in the modeling. Simulations are performed for the same strain rates as the reference [6]: 800, 1000, 2000, and 3600 s^{-1}.

When performing a high strain rate testing with a SHPB system, the striker bar is first accelerated to a certain velocity V, then it hits the incident bar to generate a incident stress wave in this incident bar. This stress wave has $\sigma = \rho CV/2$, where C is the wave speed and $C = \sqrt{E/\rho}$. ρ and E are the density and Young's modulus of the bar. Since aluminum is used for the bars, the Young's modulus E is 75 GPa, the density ρ is 2700 kg/m^3, and the speed$C$ of the stress wave is ~5270 m/s. The collected data are the incident strain $\varepsilon_I(t)$, the reflected strain $\varepsilon_R(t)$, and the transmission strain $\varepsilon_T(t)$. These three strains are functions of testing time, and the summation of the absolute reflected strain and the absolute transmission strain equals to the absolute incident strain, i.e. $|\varepsilon_I(t)| = |\varepsilon_R(t)| + |\varepsilon_T(t)|$. These collected strains are employed to compute the high strain rate dynamic stress σ(t), strain ε(t),

[a] Corresponding author: Qizhen.li@wsu.edu

Article available at http://www.epj-conferences.org or http://dx.doi.org/10.1051/epjconf/20159401019

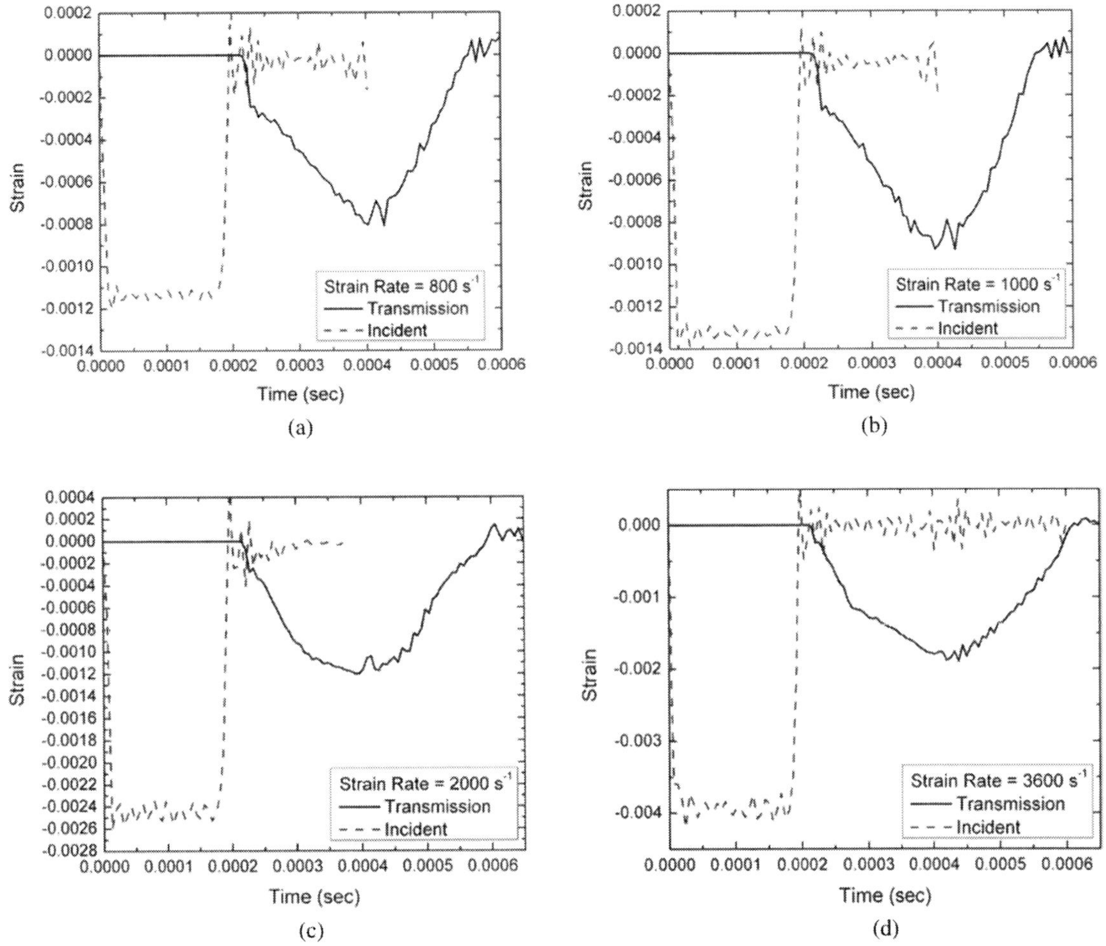

Figure 2. Strain-time histories of the incident strain and the transmission strain for different high strain rates. (a) $800\,s^{-1}$; (b) $1000\,s^{-1}$; (c) $2000\,s^{-1}$; (d) $3600\,s^{-1}$. Note: the incident strain is the dashed curve and the transmission strain is the solid curve for each strain rate case.

and strain rate $\dot{\varepsilon}(t)$ based on one-dimensional wave theory [13] as follows.

$$\sigma(t) = \frac{EA_0}{A}\varepsilon_T(t) \quad (1)$$

$$\varepsilon(t) = -\frac{2C}{L}\int_0^t \varepsilon_R(t)dt \quad (2)$$

$$\dot{\varepsilon}(t) = \frac{d\varepsilon}{dt} = -\frac{2C}{L}\varepsilon_R(t). \quad (3)$$

Where A_0, A, and L are the cross-section area of the bar, the cross-section of the specimen, and the original length of the specimen.

3. Results and discussions

Simulations were performed for the four loading strain rate cases: 800, 1000, 2000, and $3600\,s^{-1}$. These four strain rates were chosen because they were employed in the experimental testing reported in the reference [6]. For each of the four strain rates simulated in this work, the incident strain $\varepsilon_I(t)$ and the transmission strain $\varepsilon_T(t)$ were extracted from the incident and transmission bars and reported in Fig. 2.

Figure 3 reports the strain rate versus time curves from experimental testing and finite element modeling for different strain rates (800, 1000, 2000, and $3600\,s^{-1}$). The experimental curves show (a) the strain rate fluctuation with time; (b) the higher the strain rate is, the shorter the testing time is; and (c) the strain rate reaches a peak value early in a test and then diminishes slightly as the test continues. The computational curves show that the strain rate reaches the nominal value faster than the experimental curves. This may be due to the inertia of deformation process corresponding to the real experiments. For all the tested strain rates, the strain rate drops after reaching a maximum value. This relates to the strain waves prorogating in the incident and transmission bars. As shown in Fig. 2, a constant-amplitude incident strain or stress pulse is generated by the striking bar;

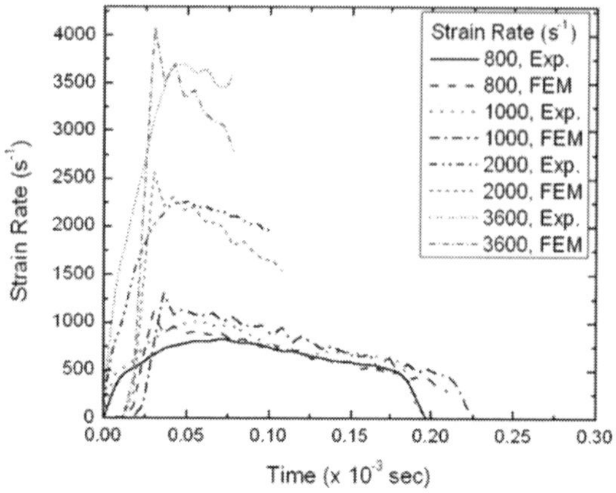

Figure 3. Strain rate versus time for the studied dynamic tests with strain rates of 800, 1000, 2000, and $3600\,s^{-1}$, at room temperature from experimental testing and finite element modeling.

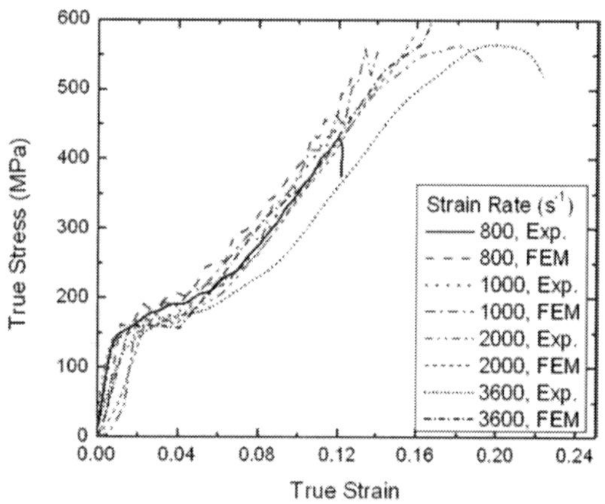

Figure 4. True stress-true strain curves for the stain rates of 800, 1000, 2000, and $3600\,s^{-1}$ from experimental testing and finite element modeling.

this pulse is partially transmitted and partially reflected. The summation of the transmitted and reflected pulses equals to the incident pulse. The figures show that the transmitted pulse is not constant, thus the reflected pulse is also not constant. According to Eq. (3), the reflected pulse is proportional to the strain rate. Thus, the strain rate is not constant, which is true to both experimental and simulated data. These data reported in Fig. 2 are employed to calculate the stress-strain curves for all the cases. The computed stress-strain curves are reported in Fig. 4. To make a comparison with the experimental testing, the experimental stress-strain curves for all high strain rate cases are also included in Fig. 4. The results show that the stress-strain curves of the material from the modeling match the experimental curves.

The temperature increase (ΔT) for each dynamic strain rate can be computed. It is known that most of the work during the high strain rate testing is converted to heat. In the simulations, the fraction of energy converted to heat is assumed to be 90%. The temperature increases can also be calculated from the experimental stress-strain curves and the values are 15.0, 19.0, 32.4, and 38.2 K for the strain rates of 800, 1000, 2000, and $3600\,s^{-1}$ respectively. The temperature changes are 15.0, 19.0, 32.4, and 38.2 K from the modeling for the strain rate of 800, 1000, 2000, and $3600\,s^{-1}$ respectively. The predictions of ΔT by finite element modeling match those from experimental data.

4. Conclusion

Dynamic mechanical behaviour of pure polycrystalline magnesium samples were simulated using finite element method. The loading strain rates were chosen to be 800, 1000, 2000, and $3600\,s^{-1}$ to compare with the reported experimental data [6]. Overall, the simulations provided well-matched predictions of the material behavior such as the strain rate-time history, the stress-strain curve, and the temperature increase. The high strain rate responding features were captured by the computational modeling. The dynamic stress-strain curves show linear strain hardening at the early stage of plastic deformation, increased strain hardening at the intermediate plastic deformation region, and decreased strain hardening at the region before fracture. The strain hardening rates for the cases with the studied high loading strain rates have negligible variation when the strain rates vary in the range of 800–$3600\,s^{-1}$. Computational modeling is very useful for obtaining material properties under different loading strain rates while reducing the cost for performing actual experimental tests.

The support for the research from the national science foundation under DMR-1449607 and the US Department of Energy, Office of Basic Energy Sciences under Grant No. DE-SC0002144 is greatly appreciated.

References

[1] Y. Fan, Q. Wang, J. Ning, J. Chen, W. Ji, Journal of Applied Mechanics **77**, 051902, 77 (2010).
[2] E. El-Magd, and M. Abouridouane, International Journal of Impact Engineering **32**, 741 (2006).
[3] Q.Z. Li, Journal of Applied Physics **109**, 103514 (2011).
[4] Q.Z. Li, Materials Science and Engineering A **568**, 96 (2013).
[5] M.T. Tucker, M.F. Horstemeyer, P.M. Gullett, H.E. Kadiri, W.R. Whittington, Scripta Materialia **60**, 182 (2009).
[6] Q.Z. Li, Materials Science and Engineering A **540**, 130 (2012).
[7] F.J. Zerilli, R.W. Armstrong, Acta Metallurgica et Materialia **40(8)**, 1803 (1992).
[8] F.J. Zerilli, R.W. Armstrong, Journal of Applied Physics **61(5)**, 1816 (1987).
[9] F.J. Zerilli, R.W. Armstrong, Journal de Physique IV **49(C3)**, 529 (1988).

[10] R.W. Armstrong, Q.Z. Li, Metallurgical and Materials Transactions A, 2015, DOI: 10.1007/s11661-015-2779-6
[11] G.R. Johnson, W.H. Cook, 7th international Symposium on Ballistics, Hague, Netherlands 541 (1983).
[12] G.R. Johnson, W.H. Cook, Engineering Fracture Mechanics **21**, 31 (1985).
[13] H.E. Boyer, and T.L. Gall, Metals handbook (9th ed.), American Society for Metals, Materials Park (OH) 190 (1985).

EPJ Web of Conferences 94, 01020 (2015)
DOI: 10.1051/epjconf/20159401020

High strain rate response of UHP(FR)C in compression

Anatoly Bragov[1], Alexandr Konstantinov[1], Andrey Lomunov[1], Daniele Forni[2,3], Gianmario Riganti[3], and Ezio Cadoni[3,a]

[1]Research Institute of Mechanics, Lobachevsky State University of Nizhny Novgorod, Nizhny Novgorod, Russia
[2]Department of Structural, Geotechnical and Building Engineering, Politecnico di Torino, 10129 Torino, Italy
[3]DynaMat Laboratory, University of Applied Sciences of Southern Switzerland, 6952 Canobbio, Switzerland

Abstract. The objective of this study was to investigate the compression behaviour of the UHPFRC and its matrix (UHPC) under high strain rate. Two experimental set-ups were used for compression testing: a traditional Split Hopkinson Pressure Bars and a compression version of the Modified Hopkinson Bar. The tests were conducted, in the range of 100–500 s^{-1} on cylindrical specimens with both diameter and height of 20 mm. Results show significant increases in peak strength and dissipated energy.

1. Introduction

Nowadays Ultra-High Performance (Fibre-Reinforced) Concretes – UHP(FR)Cs are principally used for rehabilitation and strengthening of structures. While many authors have examined the quasi-static behaviour (for example [1–6]), there is still a lack of knowledge about the material behaviour of UHP(FR)C under high strain rates above 100 s^{-1}. Recently, a few works about the tensile behaviour have been published as [7–11].

The objective of this study was to investigate the compression behaviour of the UHPFRC and its matrix under high strain rate. Two experimental set-ups were used for compression testing: a traditional Split Hopkinson Pressure Bars [12] and a compression version of the Modified Hopkinson Bar. The tests were conducted in the range of 100–600 s^{-1} on cylindrical specimens with both diameter and height of 20 mm.

2. Materials

The materials here analysed were commercial products characterized by high strength and durability, thanks to the exceptional properties of the matrix (UHPC). UHPFRC was obtained adding to the UHPC an elevate percentage of high strength steel fibres reinforcement (3% in volume) having a diameter of 0.16 mm and a length of 13 mm. The UHP(FR)C specimens had a cylindrical shape with diameter and height of 20 mm. All specimens were drilled out from standard cube (150 mm side). The results of the quasi-static compressive tests were $f_C = 104 \pm 22$ MPa for UHPC and $f_C = 127 \pm 21$ MPa for UHPFRC.

The Young's modulus was 51 GPa. Moreover UHPFRC had high durability guaranteed by high water and low gas permeability, very high resistance to chloride penetration, carbonation, acid attack and abrasion.

[a] Corresponding author: ezio.cadoni@supsi.ch

3. Experimental set-ups

The high strain rate compressive behaviour was examined by means of two set-ups.

3.1. Split Hopkinson Pressure Bar

The dynamic compression tests were performed using a Split Hopkinson Pressure Bar (SHPB), installed at the Dynamic Testing of Materials Laboratory of the Research Institute of Mechanics – Lobachevsky State University of Nizhny Novgorod (Russia), which consists of an input (2) and an output (5) bars with the specimen (4) sandwiched between them as schematically shown in Fig. 1. When the strike bar (1) impacts onto the input bar a compressive incident wave ε_I(t) travels along the input bar. Once it reaches the specimen, a reflected wave ε_R (t) and a transmitted wave ε_T(t) are generated, propagating along the input and output bar respectively. According to the one-dimension wave propagation theory the forces and particle velocities/displacements at the two faces of specimen can be determined by those three waves recorded. In Fig. 1a is also shown the Lagrangian graph describing the strain history of the two bars.

In Fig. 2 the incident, reflected and transmitted pulses measured on input and output bars in a dynamic compression test of UHPC specimen are shown. It can be observed as at high strain rate the specimen failure is reached just in the first cycle of loading. The time necessary to bring the specimen at failure in compression is about 100 μs.

With this set-up were carried out the test with higher strain rates.

3.2. Modified Hopkinson Bar in Compression

The Modified Hopkinson Bar for compression test installed at the DynaMat Laboratory (Fig. 3) consists of a hydraulic actuator (1) that put in tension a pre-stressed loading bar (2) thanks to a blocking ring (3) placed at the other extremity of this bar (see Fig. 4a). The pre-stressed

Article available at http://www.epj-conferences.org or http://dx.doi.org/10.1051/epjconf/20159401020

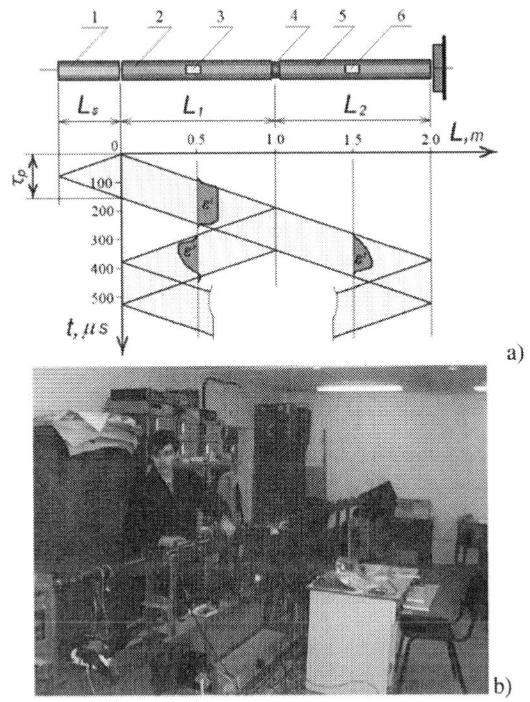

Figure 1. Dynamic compression testing set-up: a) scheme and Lagrangian graph; b) view.

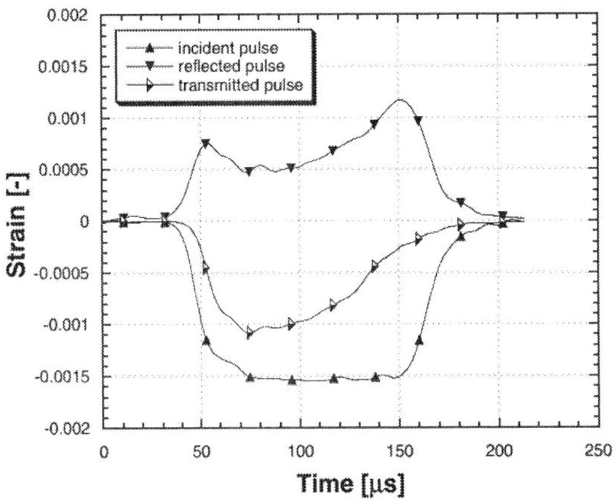

Figure 2. Signals measured on the input and output of UHPC.

bar was a high strength steel bar having a diameter of 12 mm and a length of 6 m. This pre-stressed bar is directly connected to an aluminium bar with a diameter of 30 mm and a length of 3 m working as input bar (4). The specimen (5) was sandwiched between the input bar and another identical bar used as output bar (6). Pulling the pre-stressed bar it is possible to drive the test by the energy stored in it. The test starts when a fragile bolt (see Fig. 4b) positioned between the pre-stressed bar and the hydraulic actuator (1) by suddenly breaks. Consequently a rectangular stress wave pulse is generated and propagates through the input bar, the specimen (5) and the output bar (6), causing a

Figure 3. Modified Hokinson Bar set-up in compression.

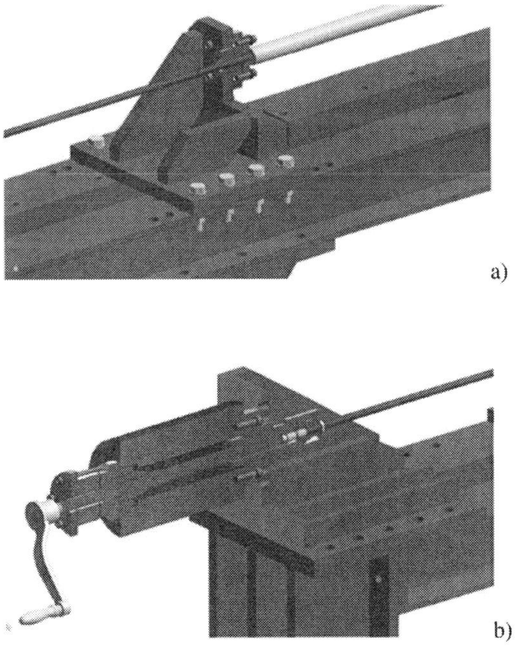

Figure 4. Details of the MHB in compression: a) blocking ring; b) fragile bolt.

state of compression stress in the specimen. By the strain gauges placed on the input and output bars the test signals were obtained and in Fig. 5 the input and output signals (Fig. 5a) as well as the incident, reflected and transmitted signals (Fig. 5b) are shown.

4. Results

The tests performed by means of SHPB were carried out using two (aluminium and steel) strikers having a diameter of 20 mm and a length of 300 mm. To obtain the strain rates three different pressure values were used (1, 3 and 6 bars). In the MHB only one velocity were used in order to compare the UHPFRC and UHPC (the preload was 50 kN).

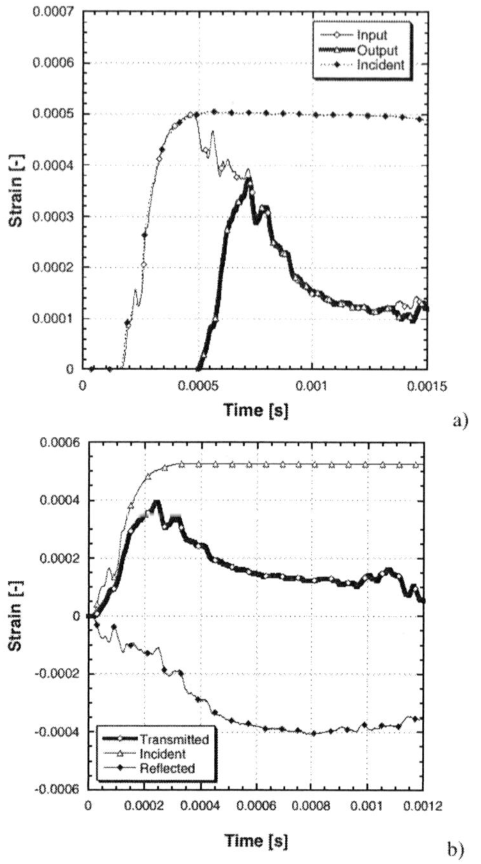

Figure 5. Compression test: a) input and output signals; b) incident, reflected and transmitted signals.

4.1. Ultra High Performance Concrete

4.1.1. SHPB results

The tests on UHPC specimen were performed using the aluminium striker at 1 and 3 bars as pressure in the gas gun obtaining an impact velocity equal to 15.4 and 27.2 m/s respectively. The strain rate obtained was in average 181 and 462 s^{-1}. As well known also in this case it can be observed as the compressive strength increases with increasing strain rate.

In Tables 1 and 2 are collected the results at the two strain rates while in Figs. 6 and 7 the stress versus time curves are shown.

4.1.2. MHB results

The compression test using the MHB was performed imposing a 50 kN of preload that produce an equivalent impact of 11.3 m/s. In Table 3 and Fig. 8 the results are shown.

Table 1. Results on UHPC_pressure 1 bar. v_{impact} = 15.44 m/s.

Specimen	Stress Rate [GPa/s]	Max Stress [MPa]	Fract. strain [‰]	Fract. time [μs]	Strain rate [s^{-1}]	Total energy [kJ/m^2]
UHPC_29	5'311	96.22	9	65	200	37.51
UHPC_30	5'509	100.28	11	62	214	39.13
UHPC_31	5'995	91.22	5	40	148	27.08
UHPC_34	6'820	105.95	6	40	160	29.57
Average	5'909 (672)	98.42 (6.24)	*7.97* *(2.84)*	*52* *(14)*	*181* *(32)*	*33.32* *(5.89)*

Table 2. Results on UHPC_pressure 3 bar. v_{impact} =27.2 m/s.

Specimen	Stress Rate [GPa/s]	Max Stress [MPa]	Fract. strain [‰]	Fract. time [μs]	Strain rate [s^{-1}]	Total energy [kJ/m^2]
UHPC_35	12'990	167.6	6.28	24	465	65.21
UHPC_36	9'364	126.7	7.83	23	588	67.17
UHPC_37	14'783	197.9	6.17	22	409	73.33
UHPC_38	17'010	198.7	6.16	19	385	63.32
Average	*13'537* *(3'232)*	*172.7* *(33.9)*	*6.61* *(0.81)*	*22* *(2)*	*462* *(91)*	*67.26* *(4.34)*

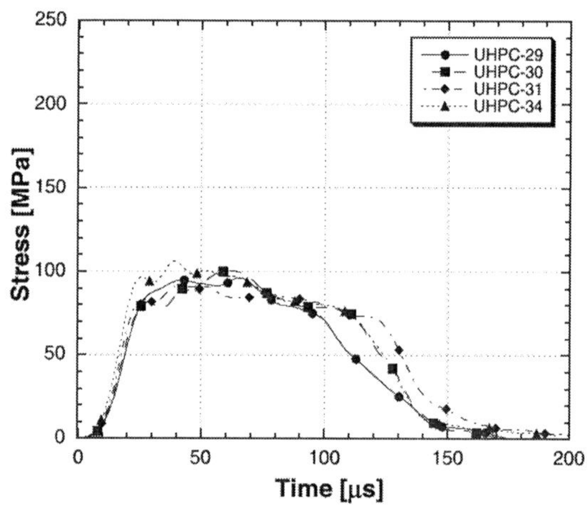

Figure 6. Results on UHPC at 181 s^{-1}.

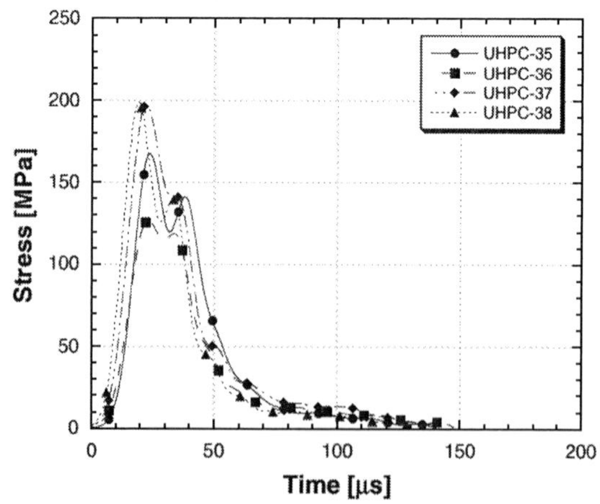

Figure 7. Results on UHPC at 462 s^{-1}.

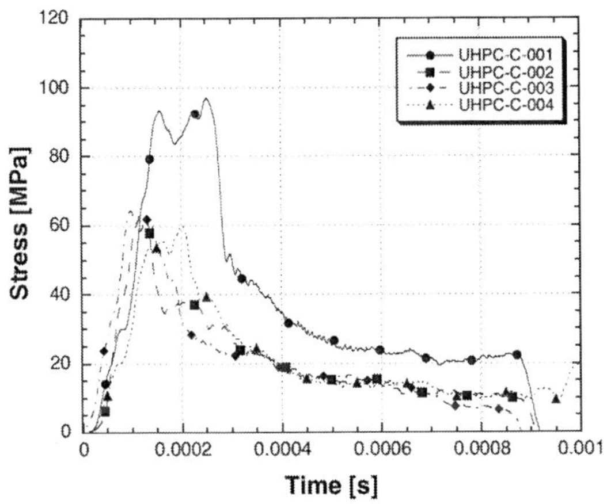

Figure 8. Results on UHPC at 136 s^{-1}.

Table 3. Results on UHPC with MHB.

Specimen	Stress Rate [GPa/s]	Stress [MPa]	Fract. strain [‰]	Fract. time μs]	Strain rate [s^{-1}]	Tot. energy [kJ/m^2]
UHPC_01	1176	97.06	22.19	252	141	133.7
UHPC_02	934	64.1	8.51	121	153	61.65
UHPC_03	722	64.23	5.58	98	113	54.94
UHPC_04	340	60.15	13.18	201	137	70.87
Average	*793* (354)	*71.39* (17.22)	*12.37* (7.26)	*168* (71)	136 (17)	*80.30* (36.22)

Table 4. Results on UHPFRC_pressure 3 bar. $v_{impact} = 27.2$ m/s.

Specimen	Stress Rate [GPa/s]	Max Stress [MPa]	Fract. strain [‰]	Fract. time μs]	Strain rate [s^{-1}]	Tot. energy [kJ/m^2]
UHPFRC_37	13'676	179.64	6.82	23	428	101.33
UHPFRC_38	14'463	183.45	13.04	36	392	117.17
UHPFRC_39	13'811	179.29	7.27	22	460	107.11
UHPFRC_40	11'258	153.96	16.89	36	585	119.65
Average	*13'302* (1'405)	*174.09* (13.55)	*11.00* (4.84)	*29* (8)	*466* (84)	*111.32* (8.58)

4.2. Ultra High Performance Fibre Reinforced Concrete

4.2.1. SHPB results

For the tests on UHPFRC the pressure in the gas gun was incremented at 3 and 6 bars so impact velocities equal to 27.2 m/s (for aluminium striker) and 20.8 m/s (for steel striker) were obtained. The resulting strain rates were respectively. 466 and 688 s^{-1}.

The results on UHPFRC are resumed in Tables 4 and 5 and in Figs. 9 and 10.

4.2.2. MHB results

The tests performed by means of MHB had exactly the same condition used for UHPC.

The strain rate obtained was in average equal to 111s^{-1}.

Table 5. Results on UHPFRC_pressure 3 bar. $v_{impact} = 20.8$ m/s.

Specimen	Stress Rate [GPa/s]	Max Stress [MPa]	Fract. strain [‰]	Fract. time μs]	Strain rate [s^{-1}]	Tot. energy [kJ/m^2]
UHPFRC_44	15'622	178.74	10.53	21	754	108.51
UHPFRC_45	15'589	176.10	8.83	19	777	107.38
UHPFRC_46	19'445	249.85	8.45	22	537	81.91
UHPFRC_47	19'641	229.10	9.83	21	684	95.64
Average	*17'574* (2'275)	*208.45* (36.83)	*9.41* (0.94)	*21* (1)	*688* (108)	*98.36* (12.41)

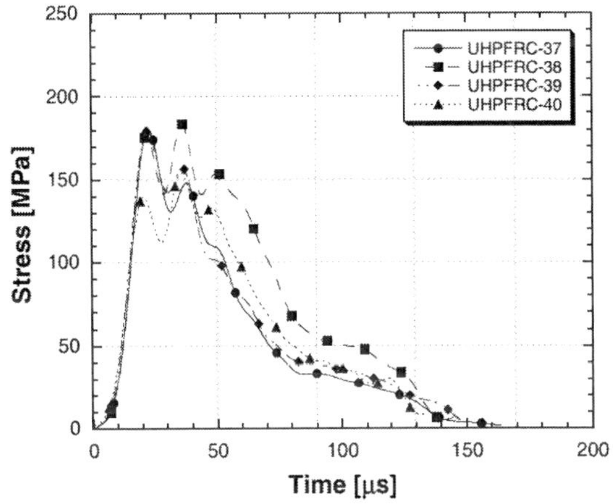

Figure 9. Results on UHPFRC at 466 s^{-1}.

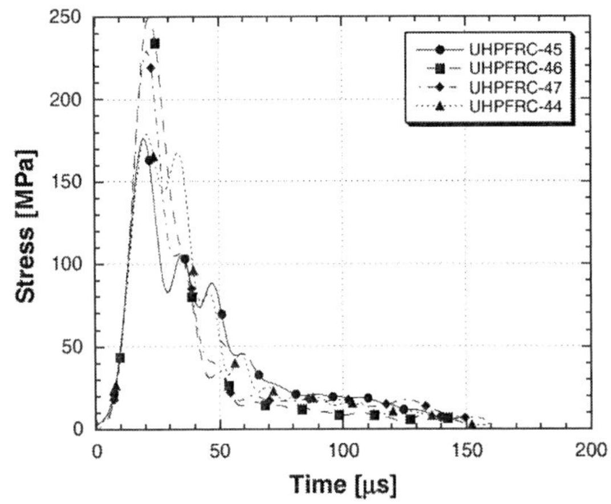

Figure 10. Results on UHPFRC at 688 s^{-1}.

5. Discussion

UHPC is a brittle material while UHPFRC can reduce the brittleness thanks to the addition of fibers in the mix. In particular the presence of the fibres enhances the post-peak behaviour. In Figs. 12 and 13 the stress versus displacement curves of UHPC and UHPFRC specimens tested in tension at about 450 and 700 s^{-1} are shown.

From these results it can be easily recognized the fibres role in the post-peak behaviour. The energy dissipation capacity is enhanced because of higher stress level for the same strain level. The peak strength seems not be

Table 6. Results on UHPFRC with MHB.

Specimen	Stress Rate [GPa/s]	Max Stress [MPa]	Fract. strain [‰]	Fract. time [μs]	Strain rate [s^{-1}]	Tot. energy [kJ/m^2]
UHPFRC_002	677	100.5	16	279	78	93.70
UHPFRC_003	955	92.53	12	211	108	113.19
UHPFRC_004	716	88.92	24	255	135	109.70
UHPFRC_005	859	91.68	17	266	124	107.57
Average	*802*	*93.41*	*17*	*255*	111	*106.06*
	(127)	*(4.9)*	*(5)*	*(29)*	(25)	*(8.5)*

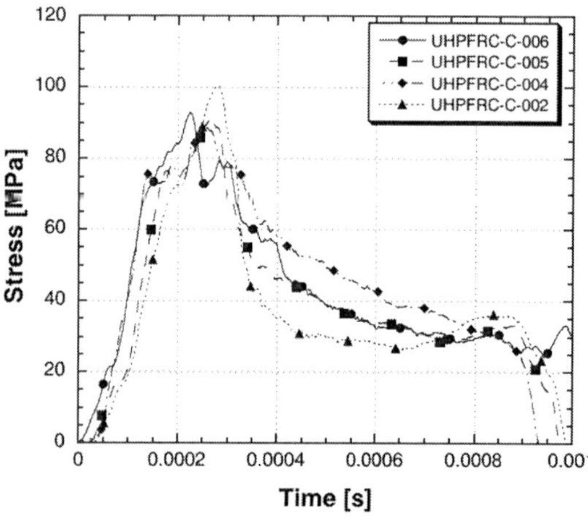

Figure 11. Results on UHPFRC at $110\,s^{-1}$.

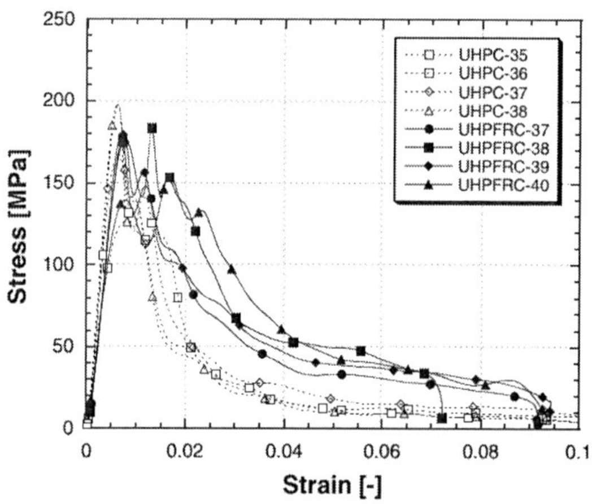

Figure 12. Comparison between UHPFRC and UHPC response at $466\,s^{-1}$.

influenced by the fibres presence. This can be reasonable because the principal role of the fibres is strictly related to the bridging effect.

The evolution of the strength in function of the stress rate (measured as the slope of the stress versus time curve) is shown in Fig. 14. The scattering of the results can be ascribed to several motivations. Firstly, the boundary

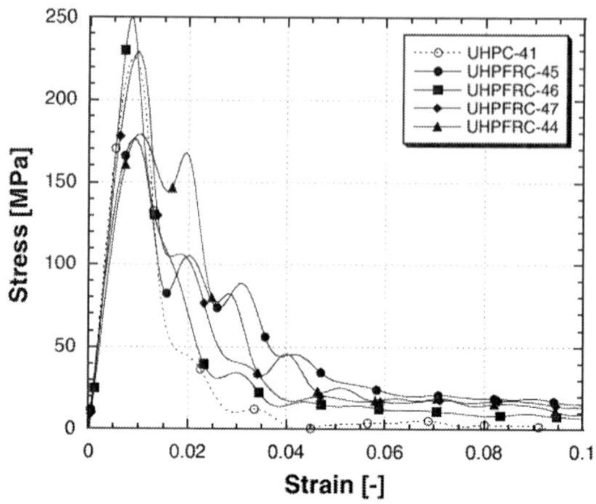

Figure 13. Comparison between UHPFRC and UHPC response at $688\,s^{-1}$.

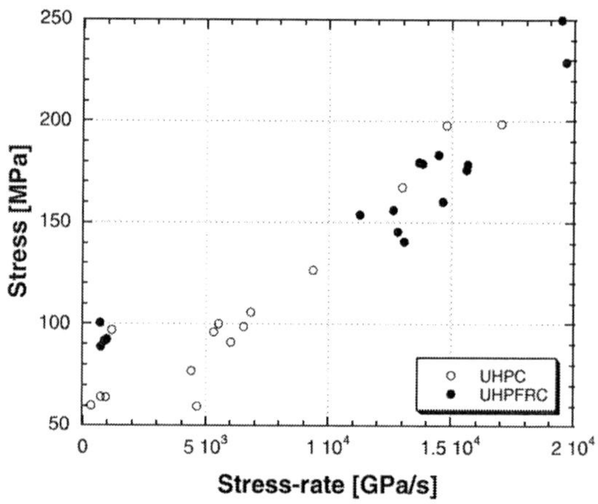

Figure 14. Strength versus stress-rate of UHPC and UHPFRC.

conditions of the specimens such as flatness, presence of grease or copper gasket. Secondly, the presence of the fibres can be considered a discontinuity in the materials causing the premature failure. Finally, the grade of dispersion of the fibres in the cross section is definitely different in each specimen.

6. Concluding remarks

The results of dynamic experimental investigations on ultra-high performance fibre reinforced concrete in compression have been described. The experiments were carried out using a Split Hopkinson Pressure Bar and a modified Hopkinson bar device.

UHPFRC shows an increase of the mechanical features as post-peak strength, failure time, and absorption energy respect to UHPC. The reason for this increase in the performance can be related to the contribution of the fibres distribution.

The results obtained at high strain rates are the basis to interpret the results acquired during blast experiments and to assess the goodness of the modelling.

The authors are grateful to Matteo Dotta and Nicoletta Tesio for their help with the test experiments. This research activity was partially financial supported by the Scientific & Technological Cooperation Programme Switzerland-Russia in the frame of Utilisation of Specific Infrastructure Project called "Construction materials under high strain rate". Swiss part of investigation was financed by *armasuisse*-Science and Technology of the Swiss Federal Department of Defence, Civil Protection and Sport (Grant 8003417679). Russian part of investigation was financed by the Russian Science Foundation (Grant 14-19-01096).

References

[1] A.E. Naaman, H.W. Reinhardt, Mat. Struct., **39**, 547 (2006).
[2] K. Habel, M. Viviani, E. Denarié, E. Brühwiler, Cem. Conc. Res. **36**, 1362 (2006).
[3] S.L. Yang, S.G. Millard, M.N. Soutsos, S.J. Barnett, T.T. Le, Constr. Build. Mater. **23**, 2291(2009).
[4] H. Katrin, G. Paul, Cem. Concr. Compos. **30**, 938 (2008).
[5] Z. Rong, W. Sun, Y. Zhang, Int. J. Impact Eng. **37**, 515 (2010).
[6] R. Yu, P. Spiesz, H.J.H. Brouwers, Const. Build. Mat. **68**, 158 (2014).
[7] S.H. Park, Kim D.J, Ryu G.S., Koh K.T., Cem. Concr. Compos. **34**, 172–84 (2012).
[8] E. Cadoni, A.M. Bragov, A. Caverzan, M. di Prisco, A. Konstantinov, A. Lomunov, Engng. Trans. **58**(3–4), 9 (2011).
[9] A. Caverzan, E. Cadoni, M. di Prisco, Mech. Mat. **59**, 87 (2013).
[10] A. Caverzan, E. Cadoni, M. di Prisco, Int. J. Impact Engng **45**, 28 (2012).
[11] E. Cadoni, D. Forni, EPJ-ST (to be published) (2015).
[12] A.M. Bragov, Y. Petrov, B.L. Karihaloo, A. Konstantinov, D.A. Lamzin, A.K. Lomunov, I.V. Smirnov, Eng. Fract. Mech. **110**, 477 (2013).

EPJ Web of Conferences **94**, 01021 (2015)
DOI: 10.1051/epjconf/20159401021

Experimental investigation of the behaviour of tungsten and molybdenum alloys at high strain-rate and temperature

Martina Scapin[1,a], Claudio Fichera[1], Federico Carra[1,2], and Lorenzo Peroni[1]

[1] Politecnico di Torino, Mechanical and Aerospace Engineering Department
[2] CERN, Engineering Department, Mechanical & Materials Engineering Group

Abstract. The introduction in recent years of new, extremely energetic particle accelerators such as the Large Hadron Collider (LHC) gives impulse to the development and testing of refractory metals and alloys based on molybdenum and tungsten to be used as structural materials. In this perspective, in this work the experimental results of a tests campaign on Inermet® IT180 and pure Molybdenum (sintered by two different producers) are presented. The investigation of the mechanical behaviour was performed in tension varying the strain-rates, the temperatures and both of them. Overall six orders of magnitude in strain-rate (between 10^{-3} and $10^{3}\,s^{-1}$) were covered, starting from quasi-static up to high dynamic loading conditions. The high strain-rate tests were performed using a direct Hopkinson Bar setup. Both in quasi-static and high strain-rate conditions, the heating of the specimens was obtained with an induction coil system, controlled in feedback loop, based on measurements from thermocouples directly welded on the specimen. The temperature range varied between 25 and 1000 °C. The experimental data were, finally, used to extract the parameters of the Zerilli-Armstrong model used to reproduce the mechanical behaviour of the investigated materials.

1. Introduction

The introduction in recent years of new, extremely energetic particle accelerators such as the Large Hadron Collider (LHC) required the development of advanced methods to predict the behaviour of components, such as the Beam Intercepting Devices (BID), in case of direct beam impact. For this reasons some components of the LHC machine are designed to operate in harsh radioactive environment highly solicited from thermo-structural point of view. As a matter of fact, they can directly interact with high energy beam and, in case of accident, the energy deposition on the bulk material could be potentially destructive. The impact produces a sudden energy deposition in the components generating outgoing of shock-waves and deforming the material in very high strain-rate and temperature conditions [1,2]. In order to study and predict the impact effects, the numerical simulations of the materials response via FE codes could be fundamental [3]. In the same way, they could represent the reference point for the design of complex experimental tests directly performed in the accelerator [4]. Obviously, reliable predictions need that the materials behaviour is described through confident constitutive relationships.

This context gives impulse to the development and testing of refractory metals and alloys based on molybdenum, tungsten, rhenium and iridium to be used as structural materials. In the scientific literature, few works can be found in which the mechanical characterization of these materials was performed in a wide range of temperature and strain-rate (e.g. [5–7]). Moreover, not always a strength model calibration was performed, which otherwise it is fundamental in FE simulations.

In this perspective, in this work the experimental results of a testing campaign on Inermet® IT180 and pure molybdenum are presented. Given the extreme loading conditions in which the materials could operate (e.g. thermo-mechanical shocks in accident situations), the investigation of the mechanical behaviour was performed in a wide range of strain-rates and temperatures. Finally, the experimental data are analyzed and used to extract the parameters of the Zerilli-Armstrong model [8].

2. Materials under investigation

The materials under investigation are a tungsten heavy-alloy and pure molybdenum.

The tungsten alloy is INERMET® IT180 (IT180 in the following), supplied by Plansee, which is actually used in the collimation system of the Large Hadron Collider at CERN. The material has a nominal composition of 95 wt% W, 3.5 wt% Ni and 1.5 wt% Cu. As it is possible to see from the microstructure of Fig. 1, the W grains are quite big, with an average dimension of about 100 μm, and are surrounded by the binder phase. The second phase is a W-Ni-Cu mixture, has a low melting point, represents the ductile connection between the brittle W grains and, finally, provides the necessary thermal and electrically continuity to the material. The material behaviour in terms of flow stress and damage is a consequence of the complex interaction between the behaviour of the two phases, depending on the values of strain-rate and temperature. The material could show different types of failure modes [7]: tungsten-tungsten grain boundary separation, tungsten

[a] Corresponding author: martina.scapin@polito.it

Article available at http://www.epj-conferences.org or http://dx.doi.org/10.1051/epjconf/20159401021

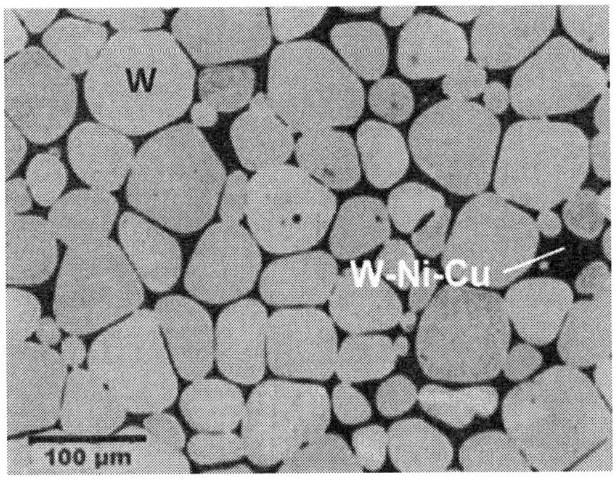

Figure 1. IT180 microstructure.

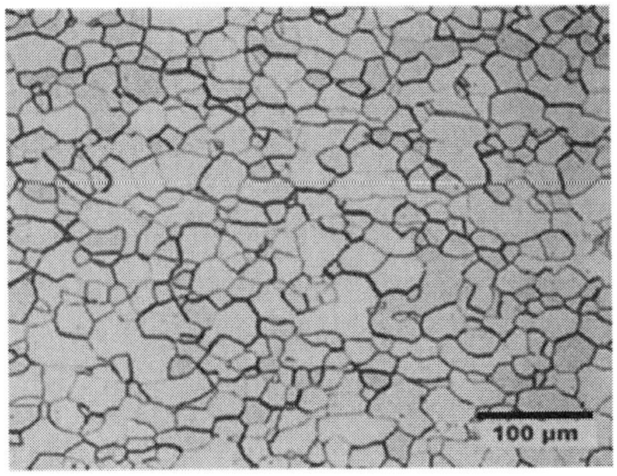

Figure 2. Pure molybdenum microstructure (Mo1).

grain cleavage, tungsten-binder interfacial separation and matrix fracture.

In Fig. 2, the microstructure of pure molybdenum is reported. The material is obtained starting from fine-grained metal powders of high purity which are pressed, sintered, hot worked and finally annealed to obtain a regular grain structure.

In more details, two different materials are analyzed, which nominally differ only for the producer: one is supplied by Plansee (in the following Mo1) and the second one by AT&M (in the following Mo2).

3. Experimental tests

The mechanical characterization was performed both in tension and compression, but in this work the attention is focused on the results obtained in tension. The specimen has a gage length of 5 mm and a diameter of 3 mm [9].

A series of tests at room temperature at different strain-rates is performed in order to obtain information about the strain and strain-rate sensitivity of the material. The strain-rate sensitivity is investigated starting from $10^{-3}\,s^{-1}$ up to $10^{1}\,s^{-1}$. For all the materials, the test at $10^{3}\,s^{-1}$ is not performed since in this condition the material is too brittle and no reliable results can be obtained. The medium-low strain-rate test are performed using electro-mechanical and servo-hydraulic testing machines. The high strain-rate condition is reached by means of a Hopkinson Bar setup in direct configuration.

In addition, a series of tests at different temperatures in both quasi-static and high strain-rate loading conditions is performed in order to obtain information about the thermal softening of the material. The atmosphere of the tests is air: this is not a problem for IT180, which is not subjected to heavy oxidation, while for molybdenum inert atmosphere should be used. The temperature range varies between 25 °C and 800 °C in quasi-static loading conditions, while temperatures up to 1000 °C are investigated at high strain-rate. The difference is mainly due to the fact that at low strain-rate the test duration is too high and the testing equipment could not withstand a high temperature for a long time; on the other hand, the dynamic test duration is about 400 μs.

The experimental setups and the testing methodology are widely described in [9].

For each testing condition at least three repetitions are performed in order to control the data scattering, then the average curves is reported in Figs. 3–5. The results are shown in terms of engineering stress vs. engineering strain curves at different loading conditions.

The evaluation of the data scatter is performed calculating the mean standard deviation for each loading condition, as reported in Table 1. The repeatability of the tests is good for the three materials, even if Mo2 is the material with the highest dispersion.

Looking at the results in terms of engineering stress vs. strain, it is possible to conclude that both IT180 alloy and pure molybdenum are temperature and strain-rate sensitive. Moreover the behaviour is such expected for BCC materials: the yield stress is a function of temperature and strain-rate and this justify the choice of the Z-A model.

Analyzing in more detail the IT180 results, it is possible to notice that the elongation at failure dramatically decreases increasing the strain-rate, while at high temperature (both in quasi-static and dynamic regimes), it grows increasing the temperature, reaches a maximum and then decreases again for very high temperature. Moreover, the tests are characterized by the absence of necking, which allows to analytically analyze the results. At high strain-rate, a temperature of 200 °C is needed for performing the test.

For what concerns the two grades of pure molybdenum, it is possible to notice that, in general, both the materials exhibit some yielding instabilities. The two materials have a similar behaviour, even if a higher stability at very high temperature is found for Mo2. Increasing the temperatures (both in quasi-static and dynamic cases), the materials strength decreases, showing a sort of plateau effect starting from 400 °C. In the same manner, also the elongation at failure increases. Increasing the strain-rate the materials harden and reduce the elongation at failure.

At high strain-rate, a temperature of 100 °C is sufficient for performing the test. Starting from medium-high strain-rate (above $10^{1}\,s^{-1}$), over a certain amount of deformation, the plastic work becomes relevant as well as the heat generated inside the specimen. In this condition, the

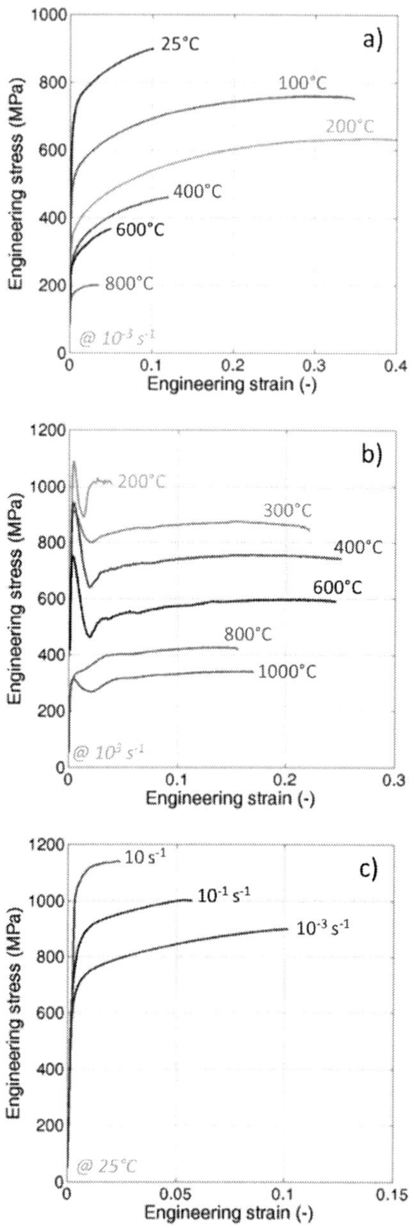

Figure 3. IT180 – Engineering stress vs. strain curves: a) quasi-static loading condition varying the temperature; b) dynamic loading condition varying the temperature and c) tests at room temperature varying the strain-rate.

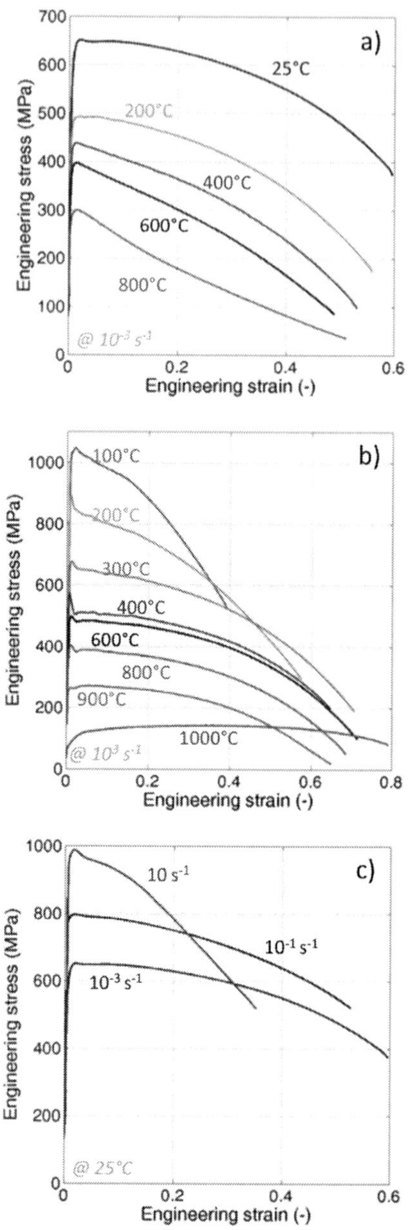

Figure 4. Mo1 – Engineering stress vs. strain curves: a) quasi-static loading condition varying the temperature; b) dynamic loading condition varying the temperature and c) tests at room temperature varying the strain-rate.

thermal softening balances and overcomes the effect of the strain-rate hardening: the deformation process starts to be adiabatic.

4. Data analysis

The experimental results showed that the materials are strongly dependent on strain, strain-rate and temperature. For these reasons, it is important to find an appropriate material model for the mechanical behaviour description and a suitable strategy for the parameters identification.

As it is well known, in the scientific literature, it is possible to find a great number of constitutive models for the viscoplastic flow description. One of the simplest models able to take into account a coupling effect between temperature and strain-rate and which is suitable for BCC materials is the Zerilli-Armstrong model. It expresses the flow stress as a function of the plastic strain, the strain-rate

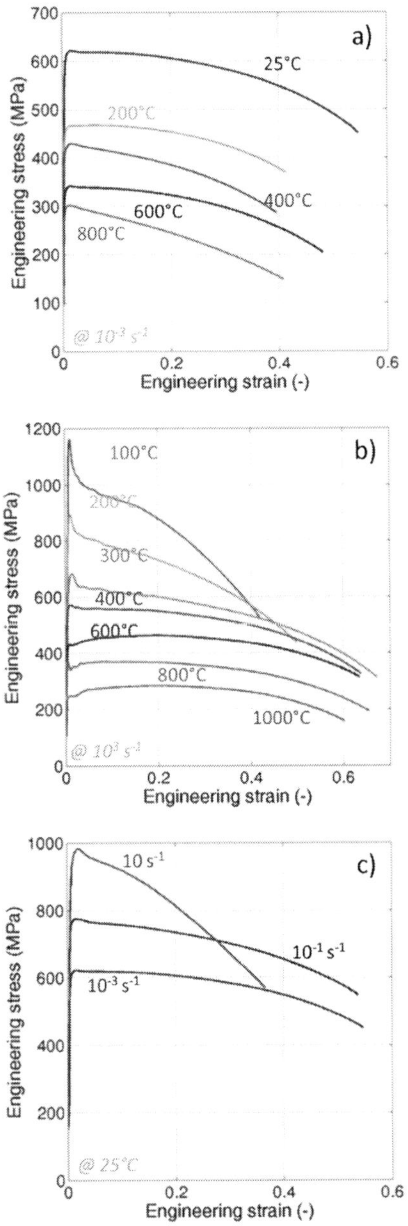

Figure 5. Mo2 – Engineering stress vs. strain curves: a) quasi-static loading condition varying the temperature; b) dynamic loading condition varying the temperature and c) tests at room temperature varying the strain-rate.

and the absolute temperature:

$$\sigma_y = C_1 + C_2 e^{(-C_3 + C_4 \ln \dot{\varepsilon})T} + C_5 \varepsilon^n \tag{1}$$

in which C_i are the material model parameters. In accordance with the original formulation, some of these parameters could be related to physical quantities of the material, but in this work all of them are considered as optimization variables. In more detail, they are obtained via a multi-objective optimization process, in which at the same time the data coming from different loading conditions are simultaneously used as targets. The optimization is based on the minimisation of the sum of the single normalized mean square errors.

Table 1. Data scatter for IT180, Mo1 and Mo2 in terms of mean standard deviation (MPa).

	IT180	Mo1	Mo2
10^{-3}s^{-1} @ 25 °C	11	7	25
10^{-3}s^{-1} @ 100 °C	2	-	-
10^{-3}s^{-1} @ 200 °C	3	8	26
10^{-3}s^{-1} @ 400 °C	12	10	33
10^{-3}s^{-1} @ 600 °C	10	9	24
10^{-3}s^{-1} @ 800 °C	11	27	8
10^{-1}s^{-1} @ 25 °C	15	10	28
10^{1}s^{-1} @ 25 °C	7	22	6
10^{3}s^{-1} @ 100 °C	-	11	7
10^{3}s^{-1} @ 200 °C	18	11	30
10^{3}s^{-1} @ 300 °C	14	14	20
10^{3}s^{-1} @ 400 °C	19	4	31
10^{3}s^{-1} @ 600 °C	2	7	14
10^{3}s^{-1} @ 800 °C	10	16	20
10^{3}s^{-1} @ 900 °C	-	36	-
10^{3}s^{-1} @ 1000 °C	3	25	9

Table 2. Optimized parameters of the Z-A model for the three materials.

	C_1 (MPa)	C_2 (MPa)	C_3 (10^{-3}/K)	C_4 (10^{-4}/K)	C_5 (MPa)	n (-)
IT180	21	2572	3.416	1.591	915	0.497
Mo1	0	4030	6.449	3.073	524	0.110
Mo2	0	2833	5.203	2.639	485	0.135

Due to the different plastic behaviour of IT180 and molybdenum, two different approaches are followed.

In case of IT180, an analytical approach is used, in which the data fitting is performed on true stress vs. true plastic strain curves calculated from the engineering ones (under the hypothesis of volume conservation). The range of applicability in strain is limited by the instability strain (maximum of the engineering stress vs. strain curves), which is sufficient in case of IT180. A series of different solutions are analyzed and good results, in terms of correlation between experimental and predicted responses, are obtained when the experimental data at temperature over 400 °C in quasi-static condition are omitted from the optimization process. In the optimization all the available repetitions are taken into account for the determination of 6 unknown (a more in-depth description of the procedure is reported in [9]). The optimized parameters are summarized in Table 2.

On the other hand, in case of Mo1 and Mo2, a numerical inverse approach is used. The parameters identification is reached varying the material strength parameters of the FE models (one for each experimental loading and constraint condition) and comparing the model results with the experimental data with the aim to obtain the best correlation. The great advantage of this procedure is that no hypothesis about the internal specimen stress-strain, temperature or strain-rate fields is made: in fact, the comparison is made in terms of macroscopic

Table 3. Percentage Normalized Mean Square Error for IT180, Mo1 and Mo2 obtained by the comparison between target and computed curves ("nc" means that data of the corresponding loading conditions is not considered in the optimization).

	IT180	Mo1	Mo2
$10^{-3}s^{-1}$ @ 25 °C	0.0342	0.0492	0.0582
$10^{-3}s^{-1}$ @ 100 °C	0.0339	-	-
$10^{-3}s^{-1}$ @ 200 °C	0.0424	nc	nc
$10^{-3}s^{-1}$ @ 400 °C	2.357	nc	nc
$10^{-3}s^{-1}$ @ 600 °C	nc	nc	nc
$10^{-3}s^{-1}$ @ 800 °C	nc	nc	nc
$10^{-1}s^{-1}$ @ 25 °C	0.00310	0.0404	0.129
$10^{1}s^{-1}$ @ 25 °C	0.0729	0.463	1.400
$10^{3}s^{-1}$ @ 100 °C	-	0.256	0.358
$10^{3}s^{-1}$ @ 200 °C	0.147	0.140	0.0868
$10^{3}s^{-1}$ @ 300 °C	0.0266	0.0728	0.321
$10^{3}s^{-1}$ @ 400 °C	0.0412	0.527	0.0540
$10^{3}s^{-1}$ @ 600 °C	0.0267	0.772	0.276
$10^{3}s^{-1}$ @ 800 °C	0.733	0.400	0.268
$10^{3}s^{-1}$ @ 900 °C	-	nc	-
$10^{3}s^{-1}$ @ 1000 °C	2.94	nc	nc

quantities which, in general, are force and displacement. The optimization of the parameters is performed with a dedicated algorithm included in the software LS-OPT.

Also for molybdenum, a series of different solutions are analyzed and the best results, in terms of correlation between experimental and predicted responses, are obtained when the experimental data at different temperatures in quasi-static condition are omitted from the optimization process, as well as the dynamic experimental data at temperature over than 800 °C. Each optimization process has 9 objectives and 5 variables to be determined (C_1 is set equal to 0). The target curves are obtained as the average behaviour of each loading condition. The optimized parameters are summarized in Table 2.

For the three materials, the qualitative comparison between experimental and computed quantities is reported in Fig. 6 for the dynamic tests at high strain-rate. As it is possible to notice, the optimized Z-A models are able to reproduce the materials behaviours with a good level of accuracy.

A more quantitative indication of the error is reported in Table 3 in terms of percentage Normalized Mean Square Error (NMSE), defined by the relationship:

$$NMSE = \frac{1}{n}\sum_{k=1}^{n}\left(\frac{x_k - \hat{x}_k}{\max(x)}\right)^2 \times 100 \qquad (2)$$

where the index k identifies one of the n points of the objective (target curve), x represents the experimental target and $\hat{x}$ represents the computed response.

In Fig. 7 the Z-A models obtained for the three materials are compared. In Fig. 7a the models are compared in terms of flow stress curves in quasi-static loading condition at room temperature. In Fig. 7b, the comparison is made in terms of equivalent stress as a function of temperature at 5% of strain. As expected, the models obtained for the two grades of molybdenum produce very similar results. For these materials, the quasi-static behaviour varying the temperature is extrapolated.

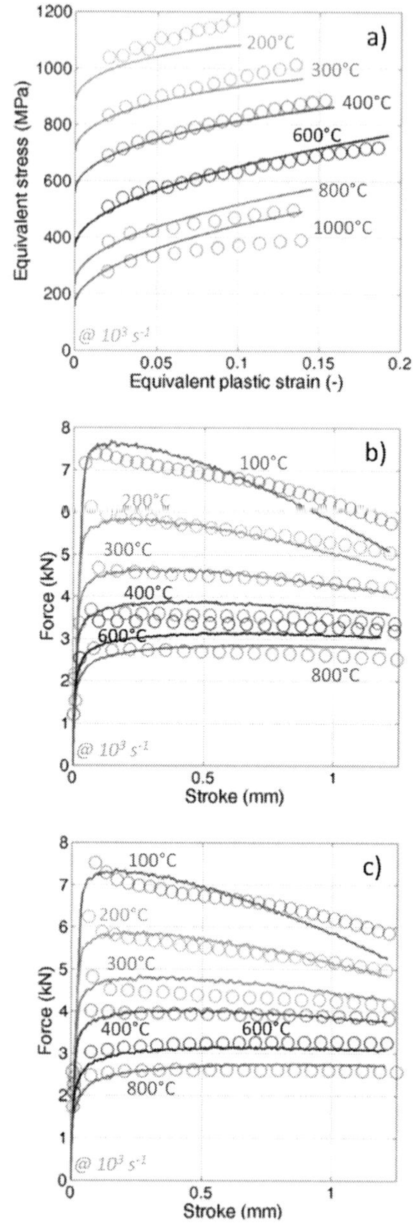

Figure 6. Comparison between experimental and computed quantities: a) for IT180 (analytical optimization); b) and c) for Mo1 and Mo2, respectively (numerical inverse optimization).

For what concerns the alloy IT180, the comparison reveals that the material shows a higher strength with respect to molybdenum at low temperatures, while the opposite is expected increasing the temperature, as a consequence of the presence of the second low-melting phase.

5. Conclusions

In this work an experimental test campaign was performed in order to study the strain-rate and temperature sensitivities of a tungsten alloy (IT180) and pure

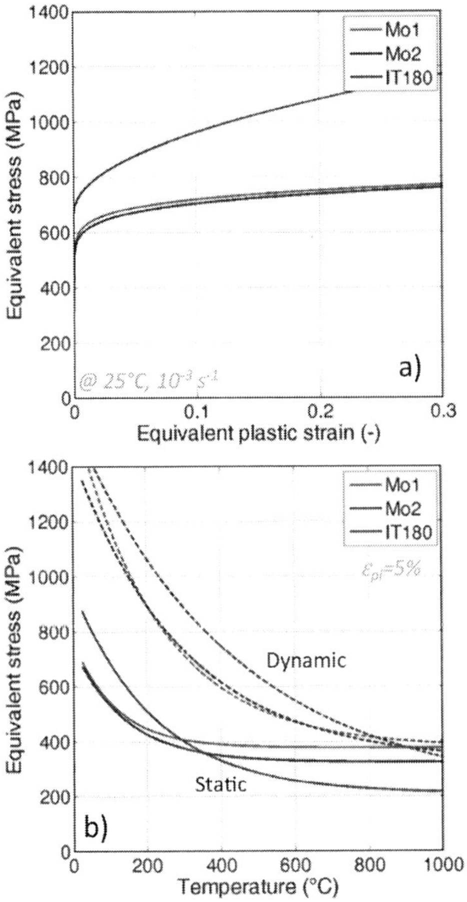

Figure 7. Comparison between optimized Z-A models obtained for the three materials: a) in terms of plastic flow stress in quasi-static loading condition at room temperature; b) in terms of equivalent stress as a function of temperature in quasi-static and dynamic cases.

molybdenum (in two different grades). The tests were performed in tension starting from quasi-static up to high strain-rate loading conditions. The dynamic tests were performed using a direct Hopkinson bar setup. The temperature sensitivity was investigated from room temperature up to 1000 °C both in quasi-static and dynamic regimes. The heating of the specimen was obtained by an induction coil system controlled in closed loop by means of thermocouples directly welded on the specimens. The results shown the materials were both strain-rate and temperature sensitive. The IT180 was characterized by the absence of necking with respect to molybdenum. All the materials exhibited a decrease of the elongation at failure increasing the strain-rate such that no reliable results were obtained from dynamic tests at room temperature. The experimental data were used to extract the coefficients for the Z-A model, used in the BCC formulation. The parameter identification was based on the definition of a multi-objective optimization. Two different approaches were used. In case of IT180 an analytical fitting was performed on the true stress vs. true plastic strain. On the other hand, due to instability and necking, in case of molybdenum, a numerical inverse approach was used. The optimized Z-A models were able to reproduce the materials behaviour with a sufficient level of accuracy. Comparing the results, it was possible to notice that the predicted results were in good agreement with the experimental one. The Z-A models obtained for the two grades of molybdenum produced very similar results, while the model obtained for IT180 predicted a higher strength at low temperatures associated to a more significant reduction in strength increasing the temperature.

References

[1] A. Bertarelli et al., "Limits for beam induced damage: reckless or too cautious?" in Proceeding of Workshop on LHC performance, Chamonix 2011.

[2] M. Scapin, L. Peroni and A. Dallocchio, J. Nucl. Mater. **420** (1–3) pp. 463–472 (2012).

[3] M. Scapin, L. Peroni, V. Boccone, F. Cerutti, Comput. Struct. **141** pp. 74–83 (2014).

[4] A. Bertarelli et al., Nucl. Instrum. Meth. B **308** pp-88–99 (2013).

[5] G. P. Skoro, J. R. J. Bennett, T. R. Edgecock, C. N. Booth, J. Nucl. Mater. **426** pp. 45–51 (2012).

[6] S. Nemat-Nasser, W. Guo, M. Liu, Scripta Mater. **40** (7) pp. 859–872 (1999).

[7] W-S. Lee, G-L. Xiea, C-F. Lin, Mater. Sci. Eng. A **257** pp. 256–267 (1998).

[8] R. W. Armstrong, F. J. Zerilli, J. Phys. **C3** (9) pp. 529–534 (1988).

[9] M. Scapin, Int. J. Refract. Met. H. **50** pp. 258–268 (2015).

EPJ Web of Conferences 94, 01022 (2015)
DOI: 10.1051/epjconf/20159401022

Influence of the temperature on the tension behaviour of EUROFER97 alloy at high strain rate

Ezio Cadoni[1,a], Matteo Dotta[1], Daniele Forni[1,2], and Philippe Spätig[3]

[1]DynaMat Laboratory, University of Applied Sciences of Southern Switzerland, 6952 Canobbio, Switzerland
[2]Department of Structural, Geotechnical and Building Engineering - Politecnico di Torino, 10129 Torino, Italy
[3]Laboratory for Nuclear Materials, Nuclear Energy and Safety Research Department,Paul Scherrer Institut, 5232 Villigen PSI, Switzerland

Abstract. This paper presents an experimental investigation on the influence of the temperature on the reduced activation steel Eurofer97 under uniaxial tensile loads at high strain rate. Round undamaged specimens of this material having gauge length 5 mm, diameter 3 mm, were tested in universal machine to obtain its stress-strain relation under quasi-static condition ($0.001\,s^{-1}$), and in modified Hopkinson bar to study its mechanical behaviour at high strain rates ($300\,s^{-1}$, $1000\,s^{-1}$) respectively. The tests at high strain rate were carried out at 450 °C and at nitrogen temperature. Finally, the parameters of the Zerilli-Armstrong constitutive material relationship were obtained.

1. Introduction

In a real fusion reactor, plasma disruptions are expected to occur that will yield disruption stress peaking in about 1 ms: that represents the typical loading rate of dynamical tests. Thus, up to now, not enough attention has been paid to characterize both the dynamic constitutive behaviour and dynamic fracture toughness behaviour of the tempered martensitic steels. As a first step to fill that gap, this study has been undertaken to investigate the tensile properties, yield stress and strain hardening, from static to highly dynamic regime of Eurofer97 steel. Those data are necessary to calculate the stress/strain field around the crack tip by finite element simulations to model the toughness-temperature behaviour in the transition region This paper presents an experimental investigation on the influence of the temperature of reduced activation steel Eurofer97 under uniaxial tensile loads at high strain rate. Round undamaged specimens of this material having gauge length 5 mm, diameter 3 mm, were tested in universal machine to obtain its stress-strain relation under quasi-static condition ($0.001\,s^{-1}$), and in modified Hopkinson bar to study its mechanical behaviour at high strain rates ($300\,s^{-1}$, $1000\,s^{-1}$) respectively. The test at high strain rate were carried out at 450 °C and at nitrogen temperature.

2. Material

The reduced activation steel Eurofer97 is a tempered-martensitic stainless steel of the 7–9wt% Cr class. The concentration of the main alloying elements is (in wt%) 0.1% C, 8.9% Cr, 1% W, 0.2% V and 0.15% Ta. In order to obtain the reduced-activation behaviour, several alloying elements commonly added to commercial martensitic stainless steels like Ni, Nb and Mo have been either removed (Ni), or replaced (Nb and Mo) by elements with shorter half lives (W, V, Ta). Eurofer97 has been selected by EU as reference structural material and will be used to fabricate the Test Blanket Modules of the International Thermonuclear Fusion Reactor (ITER) [1]. The material has been produced by Bölher AG as rolled plates of 8, 14 and 25 mm. In this work, we have studied the material coming from the 25 mm plate of the heat 9741. The final thermal treatment applied consisted of austenitization during 0.5h @ 980 °C + air cooling followed by tempering 1.5 h @ 760 °C + air cooling. A detailed description of the microstructure of the Eurofer97 can be found in [2] and [3]. Here, it suffices to say that the material features small prior austenitic grains, characterized by a mean intercept length of about 10 μm. The carbides, mainly M_23C_6 and TaC type, are not bigger than 400 μm. This steel is very clean and its inclusions level is extremely low.

3. Experimental set-up

The high strain rate tests were carried out by means of a Split Hopkinson Tensile Bar (SHTB) device shown in Fig. 1. It consists of two cylindrical high strength steel bars, having a diameter of 10 mm, with a length of 9 and 6 m for input and output bar, respectively and the thin sheet steel specimen is screwed to the two bars [4–8].

The test with the SHTB is performed as follows: 1) first a hydraulic actuator, of maximum loading capacity of 600 kN, is pulling part of the input bar (6 m) as pretension bar with a diameter of 10 mm; the pretension stored in this bar is resisted by the blocking device; 2) second operation is the rupture of the brittle bolt in the blocking device which gives rise to a tensile mechanical pulse of 2.4 ms duration with linear loading rate during the rise

[a] Corresponding author: ezio.cadoni@supsi.ch

Figure 1. SHTB device.

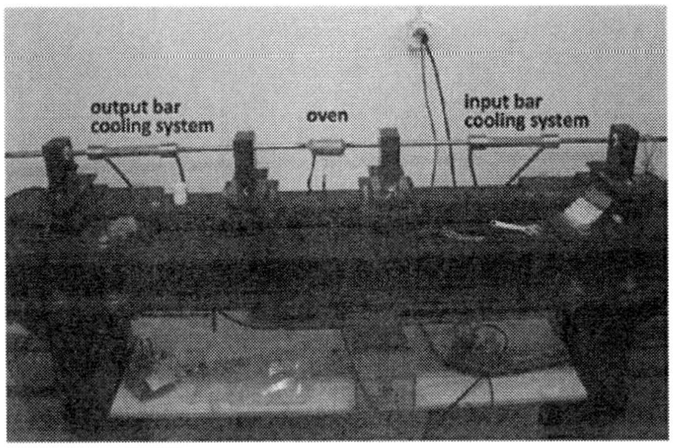

Figure 2. Set-up for high temperature high strain rate testing.

time (30 μs), propagating along the input and output bars bringing to fracture the specimen.

The input and output bars are instrumented with strain gauges, which measure the incident, reflected and transmitted pulses acting on the cross section of the specimen. As pre-tensioned bar is used part of the input bar. On the basis of the incident (ε_I), reflected (ε_R) and transmitted (ε_T) records, of the consideration of the basic constitutive equation of the input and output elastic bar material, of the one-dimensional wave propagation theory it is possible to calculate the stress, strain and strain-rate curves by the following equations [4–8]:

$$\sigma_E(t) = E_0 \frac{A_0}{A} \varepsilon_T(t) \tag{1}$$

$$\varepsilon_E(t) = -\frac{2C_0}{L} \int_0^t \varepsilon_R(t)\, dt \tag{2}$$

$$\dot{\varepsilon}(t) = -\frac{2C_0}{L} \varepsilon_R(t) \tag{3}$$

where: E_0 is the elastic modulus of the bars; A_0 their cross-sectional area; A is the specimen cross section area; L is the specimen gauge length; C_0 is the sound velocity of the bar material.

In order to test the EUROFER97 steel at high temperature and high strain rate a homemade oven was used. In Fig. 2 the high temperature set-up is shown. It consists of an oven, able to maintain constant the temperature (measured by a thermocouple in contact with the specimen), and cooling systems to avoid any influence on the strain gauges.

The low temperature tests were obtained by means of a system containing nitrogen liquid as shown in Fig. 3. The test starts when the temperature of the specimen is in equilibrium with the nitrogen liquid.

Figure 3. Experimental set-up for low temperature testing.

Table 1. Strain rate results at room temperature [6].

Strain-rate [s^{-1}]	$R_{0.2}$ [MPa]	R_m [MPa]	Uniform strain [%]	Reduction of area Z [%]
10^{-3}	560	652	4.8	80.7
5	665	753	8.6	77.2
30	680	783	6.4	78.1
300	716	814	10.0	77.7
600	720	837	11.5	77.3

4. Results

The results of the experiments carried out at room temperature (293°K) are summarized in Table 1 [6]. It can be noted as the flow stress increases with increasing strain rate, at room temperature.

The high strain rate results at high and low temperature are shown in Table 2 and Table 3, respectively.

It can be observed as the strength decrease increasing the test temperature. The reduction of area increase with increasing strain rate and temperature as also shown in Fig. 4, where the stress versus strain curves at high strain rate and different temperature are depicted.

The true stress and true strain were obtained by:

$$\sigma_{true} = \sigma_{eng}\left(1 + \varepsilon_{eng}\right) \tag{4}$$

$$\varepsilon_{true} = \ln\left(1 + \varepsilon_{eng}\right). \tag{5}$$

The true stress vs. strain curves must be regarded as significant until the point of ultimate tensile stress where the necking begins; after this point localization and fracture propagation governs the flow curve, which

Table 2. Strain rate results @723K.

Strain-rate $[s^{-1}]$	$R_{0.2}$ [MPa]	R_m [MPa]	Uniform strain [%]	Reduction of area Z [%]
568	321	473	5.07	86.53
548	383	517	5.57	86.28
1318	463	496	2.31	88.32
1955	404	412	4.00	89.66

Table 3. Strain rate results @77K.

Strain-rate $[s^{-1}]$	$R_{0.2}$ [MPa]	R_m [MPa]	Uniform strain [%]	Reduction of area Z [%]
228	1364		0.77	2.0
246	1203		0.65	0.0
982	858		0.39	0.0
1000	1288		0.61	0.0

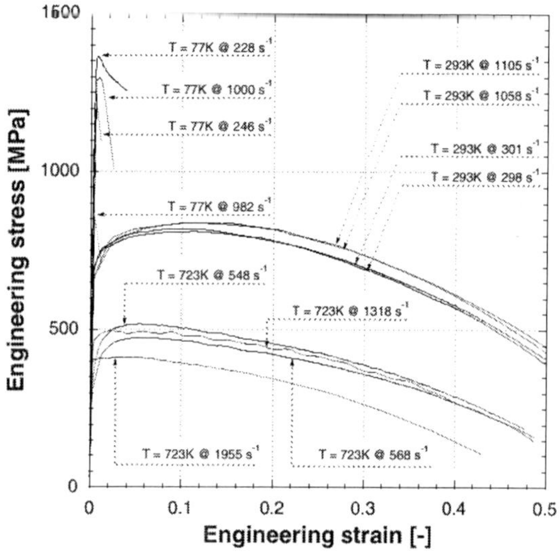

Figure 4. Stress vs. strain curves at high strain rate and different temperature.

is no more representative of homogeneous mechanical properties of the materials. In this case beyond the point of uniform straining of the engineering stress-strain curve the one-dimensional true stress-strain curve should be reconstructed by calculating the true stress and the true strain using the Bridgman formulae [9] which introduce the correction for the tri-axial stress state. At fracture the Bridgman formulae can be written as follows:

$$\sigma_{\text{true,fracture}} = \frac{\sigma_{\text{eng.,fracture}}}{(1 + 2R/a) \cdot \ln(1 + a/2R)} \quad (6)$$

where, a is minimum radius at fracture cross-section, R is the meridional profile radius at fracture neck (see Figs. 5–6), and $\sigma_{\text{true,fracture}} = P_{\text{fracture}}/\pi a^2 g$ the average stress at fracture and P_{fracture} the fracture force.

$$\varepsilon_{\text{true,fracture}} = 2 \cdot \ln\frac{a_0}{2 \cdot a} \quad (7)$$

where, a_0 is the initial diameter of the gauge length cross-section.

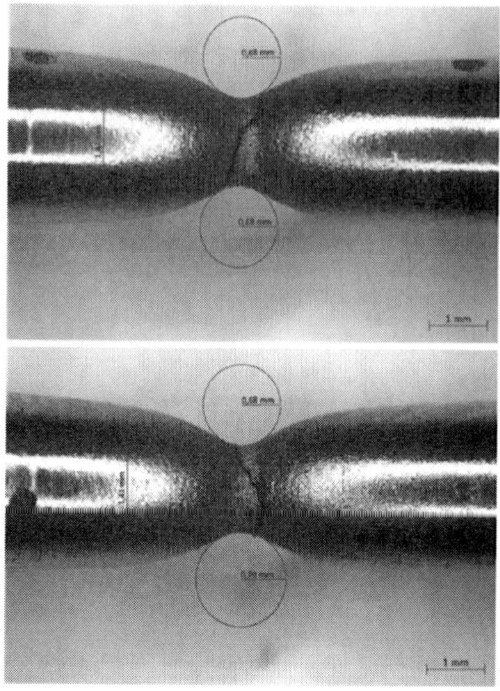

a)

b)

Figure 5. Dynamic test at 293K @: a) 1000 s^{-1} and b) 300 s^{-1}.

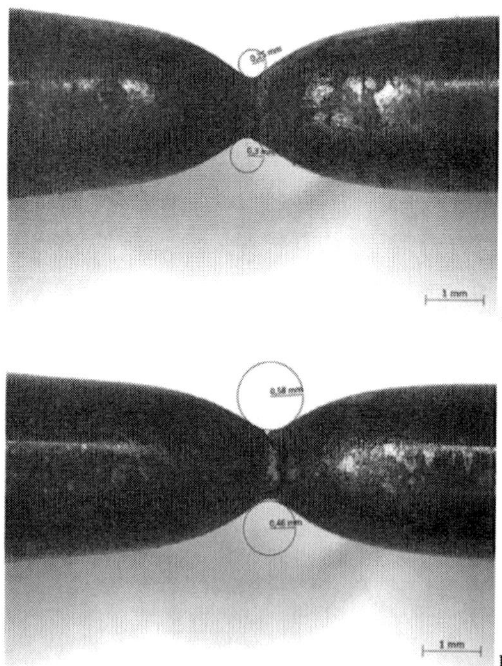

a)

b)

Figure 6. Dynamic test at 723K @: a) 1000 s^{-1} and b) 600 s^{-1}.

Having calculated the true stress and true strain at fracture with the Eqs. (4) and (5) for the complete construction of the true stress-strain curve during the necking deformation phase a straight line is drawn between

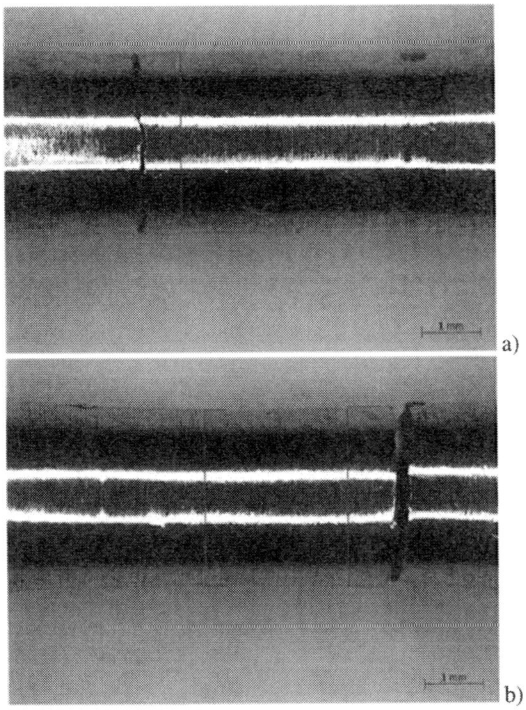

Figure 7. Dynamic test at 77K @: a) $1000\,s^{-1}$ and b) $300\,s^{-1}$.

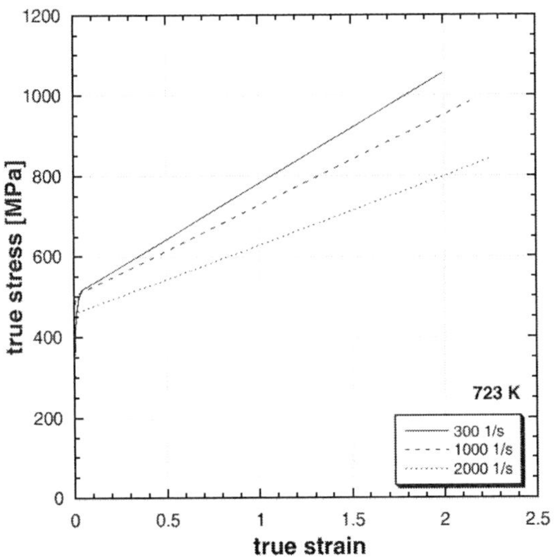

Figure 8. True stress vs. true strain curves @723K and different high strain rates.

the ultimate tensile strength/uniform strain point and the fracture point determined by application of the Eqs. (6) and (7).

In Fig. 8 the true stress versus true strain of the test at high strain rate and high temperature are shown. It can be noted as the strain hardening decreases with increasing the strain rate.

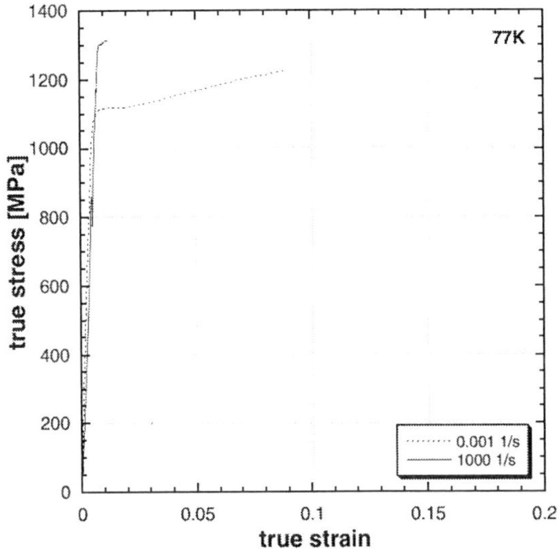

Figure 9. Comparison between quasi-static and dynamic true stress vs. true strain curves @77K.

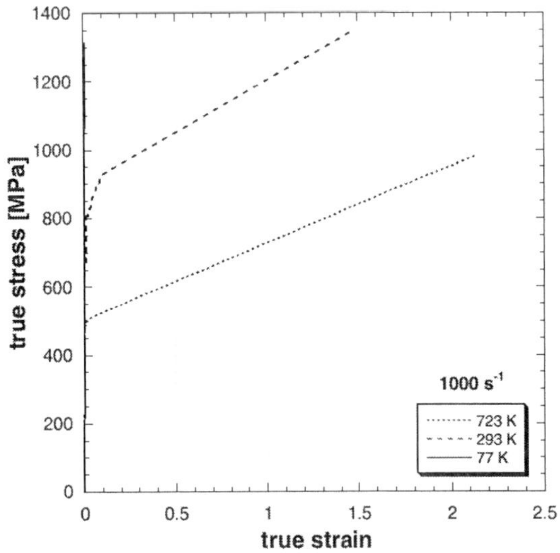

Figure 10. True stress vs. true strain curves @$1000\,s^{-1}$ and different temperatures.

Figure 9 shows the comparison between quasi-static and dynamic true stress versus strain curves at low temperature. The failure at high strain rate could be influenced by the presence of marks (see Fig. 7) produced manually by means of an electrical pencil device. At this temperature all failure start in correspondence of such surface discontinuity, and this could be the reason causing the premature failure of the test at $982\,s^{-1}$.

Comparing the true stress versus true strain curves (see Fig. 10) at the same strain rate but at different temperature is possible to observe the influence of the temperature. The brittleness of the material increases with the decrease of the temperature.

The influence of the temperature can be better understood analysing the yield stress in function of the

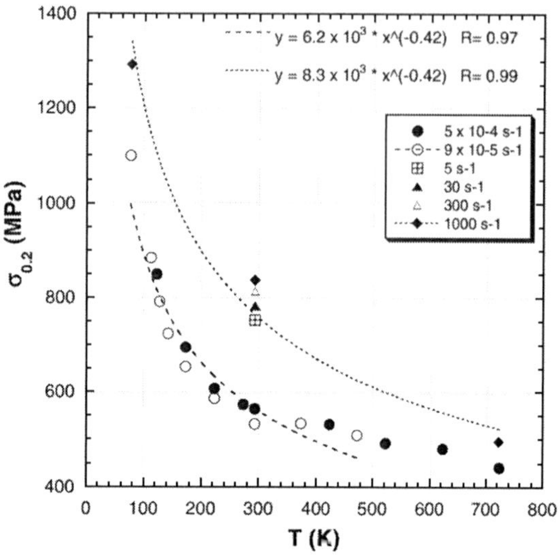

Figure 11. Yield stress vs. temperature for different strain rates.

temperature as illustrated in Fig. 11. Indeed, adding the results obtained at high strain rate to those obtained in quasi-static regime [2] at $9\cdot10^{-5}s^{-1}$ can be highlighted how the strain rate vertically moves the curve. The behaviour is well described by an exponential function.

5. Constitutive model

It is commonly known that many constitutive relations are simply a numerical fit to test data. At best these models include work hardening, strain-rate dependence as well as the thermal softening such as the well-known model proposed by Johnson and Cook in the eighties [10] and nowadays widely used. On the other hand, there are other models based upon physical concept such as Zerilli-Armstrong material constitutive model [11], based upon dislocation mechanics. Others authors based their constitutive model upon dislocation density [2,3,6].

To describe the influence of temperature effect on the dynamic behaviour of Eurofer97, that is a body centered metal [1], the Zerilli-Armstrong was used. The relationship proposed for bcc metals is:

$$\sigma = c_0 + c_1 \cdot \exp(-c_3 \cdot T + c_4 \cdot T \cdot ln(\dot{\epsilon})) + c_5\epsilon^n \quad (8)$$

where, c_1, c_3, c_4, c_5 and n are the five material constants for the bcc model, while c_0 take into account the influence of the dislocation density on the yield stress [2], evaluable as follow:

$$c_0 = \Delta_{\sigma'_G} + k \cdot l^{-1/2}. \quad (9)$$

This last parameter was chosen equal to 55 MPa thanks to a previous investigation of the plastic flow properties in tension of the Eurofer97 steel [2]. As a first approximation, the obtained five material constants are collected in the following Table 1.

In order to check the validity of the obtained constants, a comparison between experimental and Zerilli-Armstrong fit has been depicted in Fig. 12. Three

Table 4. Zerilli-Armstrong parameter for the Eurofer97.

c_1 [MPa]	c_3 [°K^{-1}]	c_4 [°K^{-1}]	c_5 [MPa]	n [-]
1075	$2{,}063\cdot10^{-3}$	$6{,}762\cdot10^{-5}$	673,1	0,5320

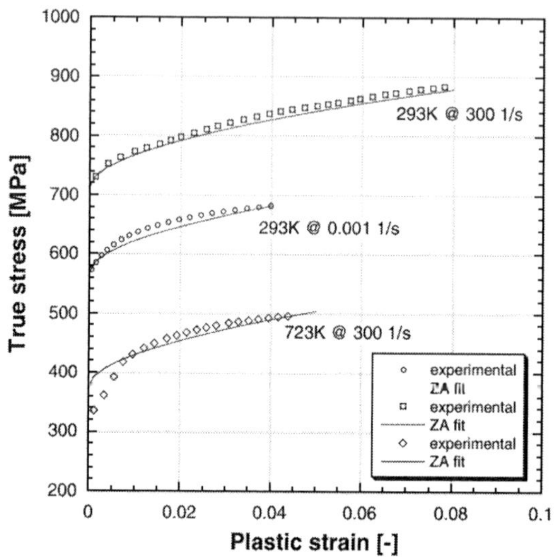

Figure 12. Comparison between Zerilli-Armstrong fits and experimental data.

different experimental tests and the constitutive model were compared: quasi-static at room temperature (293 K), high-strain-rate at room (293 K) and high temperature (723 K).

6. Conclusions

The effect of high and low temperature at high strain rates on the tensile properties of the tempered martensitic steel Eurofer97 was studied by means of a Split Hopkinson Tensile Bar device.

This tempered-martensitic stainless steel showed a quite high strain rate sensitivity. Furthermore, it shows a moderately high sensitivity to temperature at high strain rate.

Finally, the material parameters of the Zerilli-Armstrong constitutive equation were determined. The use of this calibration seems to fit relatively well with the experimental data.

References

[1] Standard Test Method for Determination of Reference Temperature, T_0, for Ferritic Steels in the Transition Range, E1921-08, Annual Book of ASTM Standards 2008, vol. 03.01, ASTM International, 2008.

[2] P. Spätig, R. Bonadé, G. R. Odette, J. W. Rensman, N. Campitelli, P. Mueller, J. Nucl. Mat. **367–370**, 527–538 (2007).

[3] E. Lucon, R. Chaouadi, M. Decréton, J. Nucl. Mater. **329–333** 1078 (2004).

[4] E. Cadoni, L. Fenu, D. Forni, Constr. Build. Mat. **35**, 399–407 (2012).
[5] D. Asprone, E. Cadoni, A. Prota, ACI Struct. J. **106**, 523–529 (2009).
[6] E. Cadoni, M. Dotta, D. Forni, P. Spätig, J. Nucl. Mat. **414**(3), 360–366 (2011)
[7] E. Cadoni, M. Dotta, D. Forni, N. Tesio, C. Albertini, Mat. Des. **49**, 657–666 (2013).
[8] C. Albertini, E. Cadoni, G. Solomos, *Phil. Trans. Roy. Soc. A:* **372**, 20130197, (2014).
[9] Bridgman PW (1952) *Studies in large plastic flow and fracture.* McGraw-Hill, New York
[10] G.R. Johnson, W.H. Cook, *Eng. Fract. Mech.* **21**, 31 (1985)
[11] F.J. Zerilli, R.W. Armstrong, *J. Appl. Phys.* **61**(5), (1987)

EPJ Web of Conferences **94**, 01023 (2015)
DOI: 10.1051/epjconf/20159401023

Experiments of cylindrical isentropic compression by ultrahigh magnetic field

Zhuowei Gu[a], Zhongyu Zhou, Chunbo Zhang, Xiaosong Tang, Yanjin Tong, Jianheng Zhao, and Chengwei Sun

Institute of Fluid Physics, Chinese Academy of Engineering Physics, Mianyang, Sichuan 621900, China

Abstract. The high Explosive Magnetic Flux Implosion Compression Generator (EMFICG) is a kind of unique high energy density dynamic technique with characters like ultrahigh pressure and low temperature rising and could be suitable as a tool of cylindrical isentropic compression. The Institute of Fluid Physics, Chinese Academy of Engineering Physics (IFP, CAEP) have developed EMFICG technique and realized cylindrical isentropic compression. In the experiments, a seed magnetic field of 5–6 Tesla were built first and compressed by a stainless steel liner which is driven by high explosive. The inner free surface velocity of sample was measured by PDV. The isentropic compression of a copper sample was verified and the isentropic pressure is over 100 GPa. The cylindrical isentropic compression process has been numerical simulated by 1D MHD code and the simulation results were compared with the experiments. Compared with the transitional X-ray flash radiograph measurement, this method will probably promote the data accuracy.

1. Introduction

The method of magnetic flux compression by cylindrical explosive implosion is a unique technique of ultrahigh magnetic field production [1–5]. The principle of this method is that of the conservation of magnetic flux enclosed within the boundary of a perfect conductor. Although the idea had been proposed more than half century, it is still advantaged comparing with other dynamic techniques. Besides, the magnetic pressure during the ultrahigh magnetic field production is a good tool of dynamic isentropic compression of samples. This technique will be used in many areas like high pressure physics, new material synthesizing and ultrahigh magnetic field physics [6–8].

2. Experimental set-ups

A schematic for the experiment set-up system is presented in Fig. 1 which includes condenser bank, the switch, a pair of coils, the liner, the explosive ring and a synchronous detonator. The initial fields within the conducting liners are produced and diffused by discharging a condenser bank through a pair of coil system at the two sides of liner. First, the condenser banks are triggered and discharged at the coils, when the discharging current is at the maximum and also the inner field is at the maximum, the cylindrical ring of explosive surrounding the liner is detonated. The subsequent implosion of the liner compresses the magnetic flux with a corresponding increasing of the magnetic field. The increasing magnetic field will exert continuous pressure on the surface of metal sample.

The capacitor bank consists of four 830-μf, 11-kv capacitors in all and is triggered by a semi-conductor switch system. A pair of coils was arranged at the two sides of the liner. Each coil was organized by 16 circles of solid copper wire which were surrounded and reinforced by insulator materials. The liner was made of stainless steel and its internal diameter is 97 mm, 1.5 mm in thickness. The explosive used in the experiment is RDX/TNT (60/40) and 55 mm in thickness. A 5–6 Tesla of initial magnetic field could be obtained in the center by our facilities.

3. Experimental results

Some shots of experiments of EMFICG and isentropic compression have been performed. The implosion movement of the stainless steel liner was recorded by an optical turn-mirror camera and shown in Fig. 2 and the average velocity of liner is about 4–6 km/s. From a detail frame photograph of Fig. 3 the stainless steel liner will happen a serous "buckling" phenomena during the implosion process. It was caused by thin liner elastic-plastic instability during the explosive driven process. In comparison a copper tube of 1 mm thickness was placed in the centre of the set-up and no clear "buckling" phenomena happen because of the isentropic compression.

To be emphasized, even the serious "buckling" phenomena of outside stainless steel liner will not influence the isentropic compression of inside sample because the magnetic flux inside the liner are even and the symmetrically isentropic compression of sample will also be realized.

We have carried out a 1-D hydrodynamic simulation of our experiments by a MHD code MC11D. The calculated result of liner movement and experiment result were shown in Fig. 4. From the results we could see the experiment and simulation of stainless steel liner movement are in good accord with each other.

[a] Corresponding author: zhuoweigu@caep.cn

Article available at http://www.epj-conferences.org or http://dx.doi.org/10.1051/epjconf/20159401023

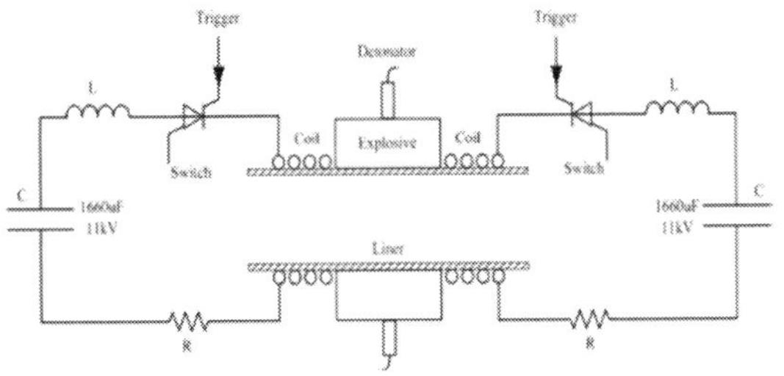

Figure 1. Schematic diagram of cylindrical implosion magnetic flux trapping system.

Figure 2. The liner movement recorded by optical turn-mirror camera.

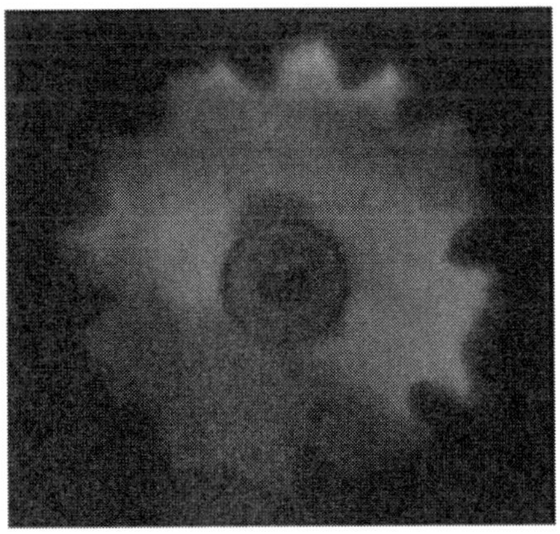

Figure 3. A copper sample was compressed by EMFICG (a PDV probe was placed inside the tube).

A typical magnetic field distribution in experiment is shown in Fig. 5 and a maximum field of about 1400 Tesla was obtained. The field measurements during the implosion were carried out by pickup coils. The $\phi 2$ mm probe includes 4 pickup coils of 5 mm interval each, so we could measure the histories of magnetic field at different positions on the central axis. From the Fig. 5 we know the maximum magnetic field was obtained at the centre of set-up where a maximum compression could be realized because of the edge attenuation during the explosive driven process.

A copper tube isentropic compression experiment was carried out. In the experiment, a $\phi 11$ mm copper tube with 2 mm thickness was compressed by ultrahigh magnetic field. The PDV technique was used to measure the inner free surface velocity of the tube and is shown in Fig. 6. From the experiment it is quite clear that the free surface velocity grows smoothly and the isentropic compression is verified.

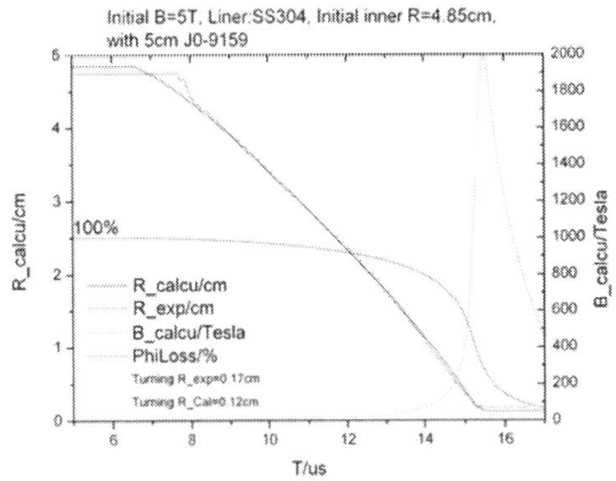

Figure 4. The simulation of liner displacement and maximum magnetic field.

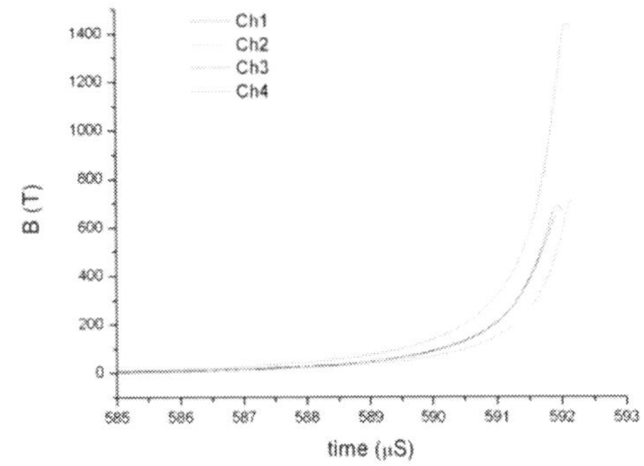

Figure 5. The magnetic field distribution in experiment (Ch2 is at the center of set-up).

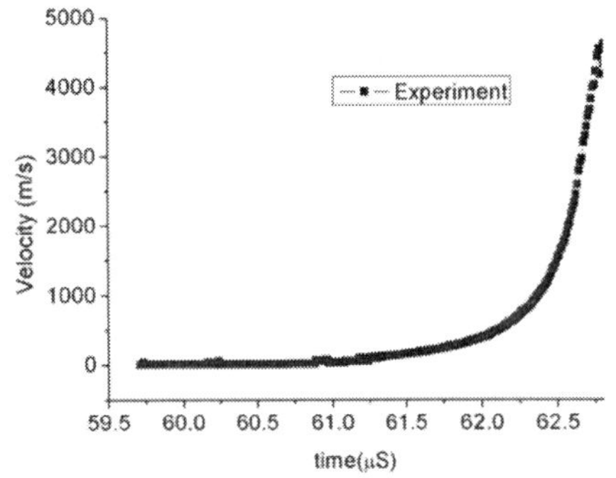

Figure 6. Inner free surface velocity of copper by isentropic compression.

4. Analysis and conclusions

In fact, the EMFICG technique will be very effective as a tool of isentropic compression method. Some numerical simulations have been done to show the effect of isentropic compression. In the simulation, the parameters of set-up were the same with experiment, the sample was solid deuterium, and its density is 0.1766g/cm^3. Two kinds of situation were considered, one is by magnetic compression, and the other is shock compression by explosive directly.

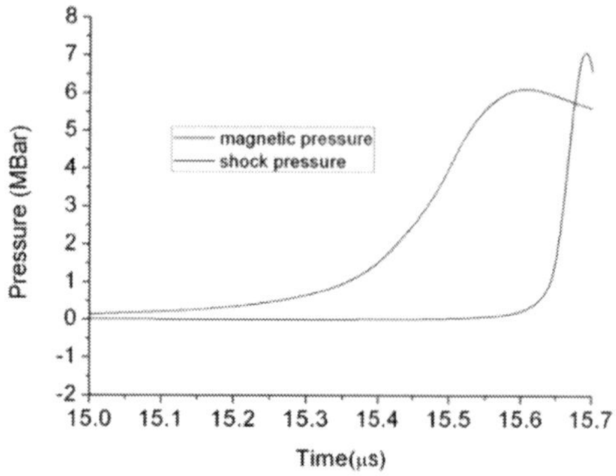

Figure 7. The pressure histories at the surface of sample.

The pressure histories on the surface of sample by magnetic compression and shock compression were shown in Fig. 7. Obviously, the magnetic compression was a kind of ramp wave loading method comparing with the shock loading. The primary calculations showed that the temperature in the sample by magnetic compression was about 10–20 times lower than that of shock situation.

Anyway, the EMFICG technique is verified by experiment and simulation as an effective isentropic compression method, and also as an ultrahigh magnetic field generator. This technique could be used widely in high pressure physics, new material synthesis and ultrahigh magnetic field physics. Some new isentropic compression experiments are under carried out and results will be published later.

References

[1] F. Herlach and H. Knoepfel, The review of scientific instruments, Vol. **36**, 8 (1965).

[2] C.M. Fowler, et al., Production of very high magnetic fields by implosion, J. Appl. Phys., Vol. **31**, 3 (1960).

[3] R.S. Hawke, et al., J. Appl. Phys., Vol. **43**, 6 (1972).

[4] A.I. Pavlovskii, M.I.Dolotenko, N.P. Kolokolchikov, et. al., Ultrahigh magnetic fields. Physics. Techniques. Eds. V.M. Titov and G.A. Shvetsov. Moscow: Nauka. 1984. P19-22.

[5] B.A. Boyko, et. al, proceeding of the VIIIth international conference on megagauss magnetic field generation and related topics, Edited by Hans J. Schneider-Muntau, 1998, 61.

[6] R.G. Clark, proceeding of the VIIIth international conference on megagauss magnetic field generation and related topics, Edited by Hans J. Schneider-Muntau, 1998, 12.

[7] Irvin R. Lindemuth, IEEE Transactions on Plasma Science, Vol. **25**, 6, 1997.

[8] G.V. Boriskov, et al. J. Low. Temp. Phys., vol. **159**, 307–310, 2010.

EPJ Web of Conferences **94**, 01024 (2015)
DOI: 10.1051/epjconf/20159401024

Failure stress criterion for adhesively bonded joint at different strain rates by using dynamic Arcan test device

Ludovic Dufour[1,a], Benjamin Bourel[2], Franck Lauro[1], Gregory Haugou[1], Nicolas Leconte[1], and Nicolas Carrere[2]

[1]University of Valenciennes and Hainaut Cambrésis, LAMIH, UMR CNRS 8201, 59313 Valenciennes, France
[2]ENSTA Bretagne, LBMS EA 4325, 29806 Brest Cedex 9, France

Abstract. The purpose of this work is to determine the failure criterion evolution of assembly bonded with a strain rate dependent adhesive. A new modified ARCAN device is then designed to obtain the average stress at failure under different loading angles and for strain up to $350\,s^{-1}$. Tests are performed on a hydraulic jack machine and a Digital Image Correlation measurement is used to control the opening and the sliding displacements of the two substrates.

1. Introduction

In order to achieve better fuel economy, the automotive industry is moving towards the use of lightweight materials, such as polymer/composites, in primary structural components. So, structural bonding often becomes an alternative solution to replace traditional assembly techniques. However, the successful integration of bonding materials in the automotive sector, and particularly for crash loading applications, requires a better understanding of their failure mechanisms at different strain rates. In finite element softwares, failure is mainly managed by three kinds of criteria based on a critical stress, a critical strain or a combination of the previous criteria by a critical energy. A brief summary of the major failure criteria is given below.

For the stress based criteria, the maximum principal stress is mainly used for very brittle materials [1,2] (since cracks initiate and propagate in a direction normal to these maximum principal stresses). Shear stress has also been extensively used to predict lap joint strength [3]. Other stress-based criteria, like the von Mises form has been used by Ikegami et al. [4] to study the strength of scarf joints between glass fibre composites and metals.

For ductile materials, strain criteria generally replace stress criteria, since joints can still carry large loads after adhesive yielding. Adams and Harris [5] used the maximum principal strain as the failure criterion for predicting joint strength. The maximum shear strain is also used in reference [6]. Da Silva et al. show that this criterion is very accurate for single lap joints assemblies [7].

Criteria based on strain energy are a combination of critical stress and strain criteria. They are generally more suitable since all the stress and strain components are taken into account. The energy approach derives from the Griffith's theory, where the failure mechanism is related to strain energy release rate G_c. Energy release rate of mode-I and mode-II can be obtained respectively from the Double Cantilever Beam (DCB) test and the End Notched Flexure (ENF) test. In the context of the automotive crash, the main difficulty comes from setting up these tests at high strain rates. In reference [8], Joudon et al. develop an experimental procedure to characterize the dynamic fracture toughness of advanced epoxy resins under dynamic loadings.

In the literature, still few studies deal with the strain rate dependence of the failure criterion. Some stress/strain based criteria are proposed, like the one extracted from Johnson and Cook works [9]. In reference [10], a mixed-mode failure criterion with strain rate dependency have been proposed for spot welds. In reference [11], Morin and al propose an equivalent failure strain which evolves with the triaxiality stress ratio and the strain rate.

Failure of adhesive is often characterized by assembly tests. Due to the simplicity to set it up, the most used tests are the simple lap shear and double lap shear test. But these tests provide a global response coming from both the behaviours of the adhesive and substrate. To obtain an *intrinsic* behaviour of the adhesive, it is then better to use a device with rigid substrate (Scarf, Butt joint, TAST, etc.). But for these tests, results in terms of failure depend on the edge effects. Otherwise, for crashworthiness, it is important to characterize adhesive for mixed mode loading and for a large range of loading velocities up to failure of the adhesive without edge effects. In this work, a new test device is proposed to characterize the failure initiation of a thick bonding joint under mixed mode loading, up to high strain rate.

The ultimate aim of this work is to propose a new strain based criterion with strain rate dependence. However, since the local strain being hardly measurable inside a very small joint thickness (0.3 mm in this study), the use of a numerical FE model is needed. Through this paper, the criterion will be presented to be a stress based criterion. By knowing the average stress in the adhesive, and consequently the load applied to make the assembly fail, it will be subsequently possible to recalculate the local

[a] Corresponding author: ludovic.dufour@univ-valenciennes.fr

Article available at http://www.epj-conferences.org or http://dx.doi.org/10.1051/epjconf/20159401024

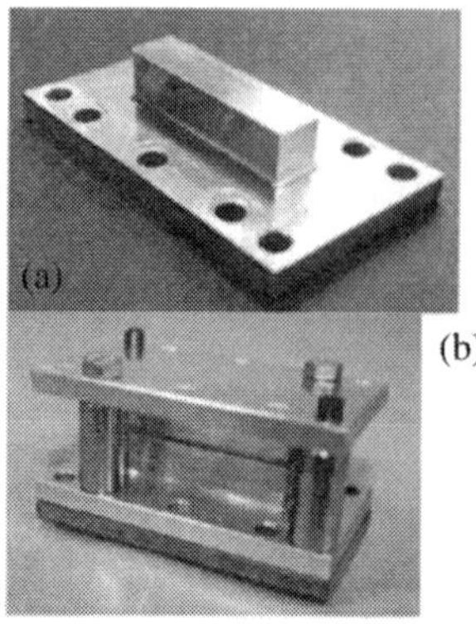

Figure 1. (a) Substrate, (b) positioning device for the bonding process.

failure strain thanks to a FE model, in which the adhesive behaviour is finely defined for a large strain rate range [12].

Then, this new failure criterion will be used to enrich the special cohesive element developed in reference [13].

2. Experimental device

To characterize the failure of the adhesive, an Arcan device is developed to obtain the mean stress state at failure with different loading angles (0°; 30°; 45°; 60°; 90°). This device is based on the modified Arcan fixture developed in [14] for quasi-static loads. The special feature of this new device falls within its low mass that required to obtain a noiseless dynamic response, i.e. an effort measurement not disturbed by the system resonance. It is composed by two substrates in aluminium (AU4G) bonded together with a bonding surface of 400 mm^2 (Fig. 1a).

For the bonded process, a specific device is designed to put one substrate relatively to the other one with a good thickness control. The substrates assembly is obtained with two centring pins and the thickness control with four spacers (Fig. 1b). After polymerisation, the thickness is checked with a calibrated binocular microscope. For all the specimens, a 0.3 mm (±0.05 mm) thickness is obtained.

The identification of the stress failure criteria depends on the accuracy at which the stress field is captured at the failure time. To obtain the best result, two conditions must be obtained. First, the stress state must be as much as possible homogeneous along the bonding joint and through the thickness. And secondly, the failure must occur inside the adhesive to be cohesive type. In order to obtain these conditions, the geometry of the substrates has to compensate the edge effects that makes a stress concentration and so in an early failure initiation in the assembly. Consequently, a beak all along the substrate is used according to previous studies performed by ENSTA laboratory [12] (Fig. 2).

In order to design the geometry of this beak, the simulation of a tensile test at 0° is carried out with Abaqus in plane strain with a fine mesh on a half model. The substrates are in Aluminium AU4G (E = 75Gpa, $\nu = 0.33$). The Epoxy Betamate1822 adhesive is modelled by an elasto-plastic behaviour law obtained by tensile test on bulk specimens. The optimised beak shape leads to a

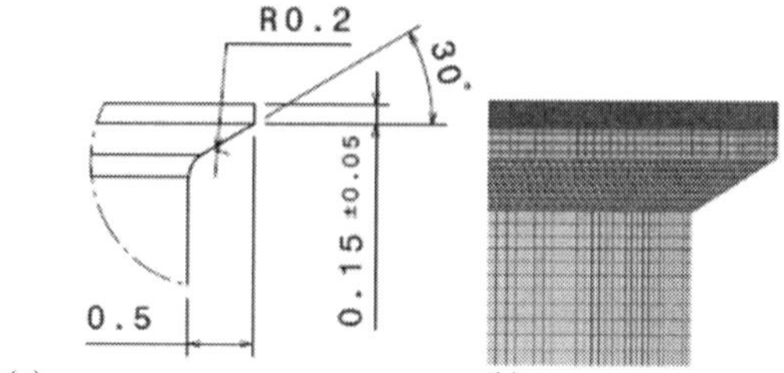

(a) (b)

Figure 2. (a) Beak geometry, (b) Mesh of the beak.

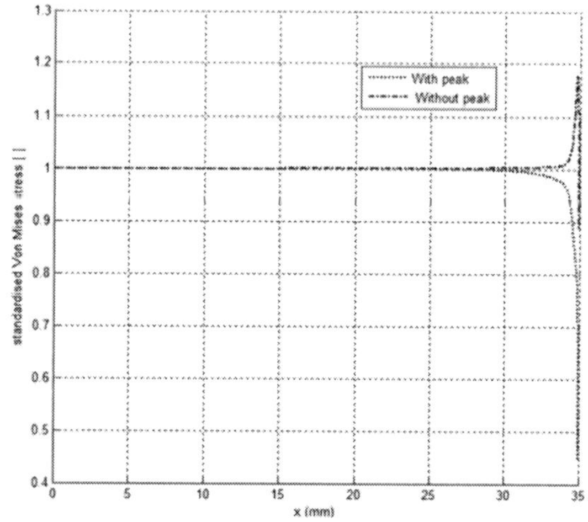

Figure 3. Normalized stress state in the mean plane of the adhesive: reduction of the edge effect by using a beak.

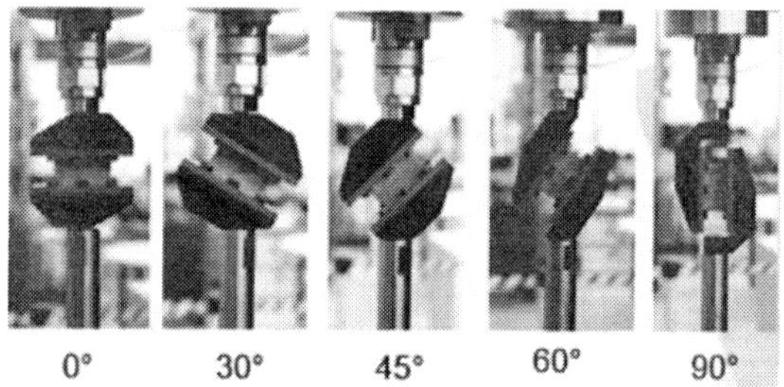

Figure 4. Dynamic Arcan devices.

deformed shape of the substrates which minimize the stress concentrations (Fig. 3) and allow a homogeneous stress state in the mean plane of the adhesive.

The substrates bonded together are then fixed on rigid supports. The supports are designed for different loading angles and to avoid resonance (Fig. 4).

3. Experimental study

3.1. Testing devices

The study is performed for an epoxy Betamate1822 from DOW Automotive. The thickness of the bonded joint is 0.3 mm and the polymerisation conditions are: 23 min from 70 °C to 200 °C and then 20 min at 200 °C.

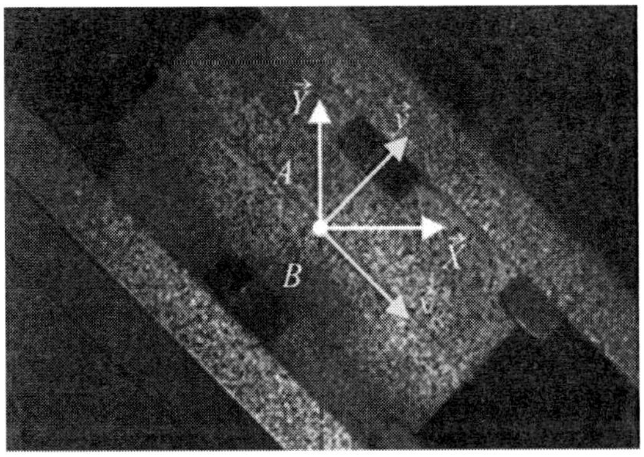

Figure 5. Post-processing by digital image correlation.

Some experiments are carried out with a tensile static machine SYNTECH 20D with an imposed load velocity of 0.5 mm/min for angles of 0°, 45° and 90° and with a dynamic tensile machine INSTRON VHS with imposed load velocity of 1 mm/s, 10 mm/s and 100 mm/s for angles of 0°, 45° and 90°. The force is measured by a KESTLER 30 KN uniaxial cell force, and the relative displacement of the substrates is measured with non- contact measurement technique by Digital Image Correlation.

3.2. Analysis

The pictures obtained by the high speed camera are post-processed with the software VIC2D in order to measure the displacement of the substrates. The direct measurement on the bonded joint is not possible due to its small thickness. The relative displacements are then measured on substrates according to Y-axis (loading direction). According to a rigid substrate assumption, the measurements of the displacement are done on the top substrate at point A and on the bottom substrate at point B (Fig. 5). DN and DT respectively are the normal relative displacement, and the tangential relative displacement between A and B according to the x and y frame. FN and FT are the corresponding normal and tangential strength to the adhesive bond.

The strain rates are then obtained from relative displacements DN and DT by the following expressions:

$$\dot{\varepsilon}_{yy} = \frac{d}{dt}\left(\ln\left(1 + \frac{DN}{e}\right)\right) \quad (1)$$

$$\dot{\varepsilon}_{xy} = \frac{d}{dt}\left(\frac{1}{2}\tan^{-1}\left(\frac{DT}{e}\right)\right) \quad (2)$$

with e the adhesive thickness.

3.3. Results

In all the experiments, the failure of the bonded joint occurs in the joint and corresponds to a cohesive failureas shown in Fig. 6.

The mixed mode failure surface obtained from different loading velocities is presented in Fig. 7. The results show a good repeatability of failure stress for the entire experimental campaign. The normal failure stress

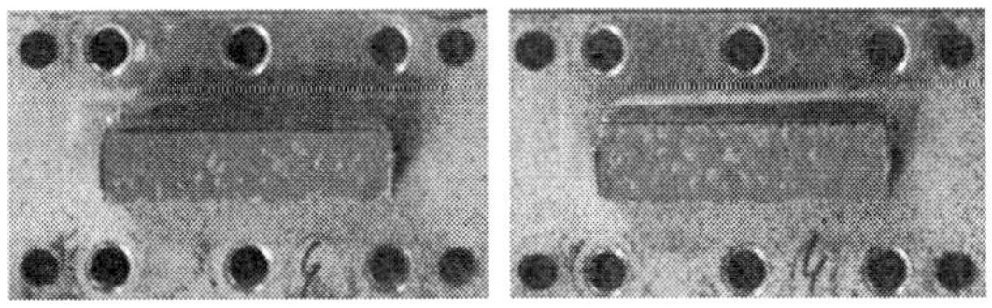

Figure 6. Cohesive failure.

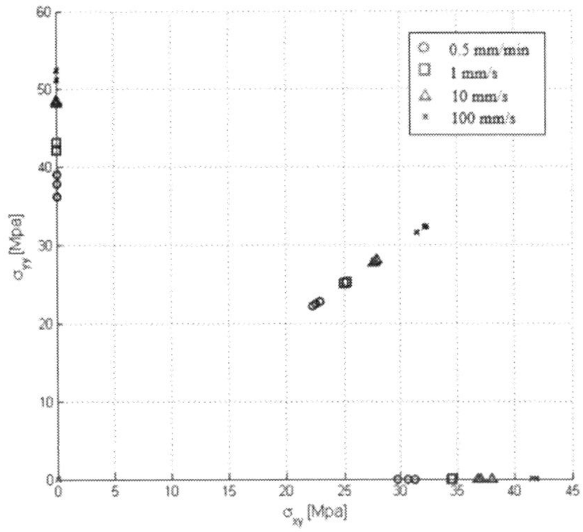

Figure 7. Failure surface.

is higher than the shear failure stress. The experimental results highlight the strain rate dependency of the failure criterion and also the loading mode dependency which could be related to the initial stress triaxiality ratio.

4. Failure criteria

According to these results, a new criterion based on a failure surface taking the mixed mode and strain rate dependencies into account is proposed:

$$\left(\frac{\sigma_{yy}}{\sigma_{yy}^{failure}(\dot{\varepsilon}_{yy})}\right)^a + \left(\frac{\sigma_{xy}}{\sigma_{xy}^{failure}(\dot{\varepsilon}_{xy})}\right)^b = 1 \quad (3)$$

where the parameters a and b give the failure surface shape. These parameters allow a better description of the criteria for mixed modes.

The strain rate effect is introduced in the tensile and shear failure stresses. Two evolution laws are used:

$$\sigma_{yy}^{failure}(\dot{\varepsilon}_{yy}) = \sigma_{yy}^{0} \cdot \left(1 + \left(\frac{\dot{\varepsilon}_{yy}}{\dot{\varepsilon}_0}\right)^{n_{yy}}\right) \quad (4)$$

$$\sigma_{xy}^{failure}(\dot{\varepsilon}_{xy}) = \sigma_{xy}^{0} \cdot \left(1 + \left(\frac{\dot{\varepsilon}_{xy}}{\dot{\varepsilon}_0}\right)^{n_{xy}}\right). \quad (5)$$

The strain rate value used in these equations is determined when the speed load is constant through the thickness.

Table 1 summarizes the set of parameters obtained from the model identification given by Eqs. (3–5). The fitting curves presented in Figs. 8 and 9, show the good agreement of the model chosen for the tensile and

Table 1. Parameter of the failure criteria.

a	b	σ^0_{yy}	σ^0_{xy}	n_{yy}	n_{xy}	$\dot{\varepsilon}$
[]	[]	[Mpa]	[Mpa]	[]	[]	[s-1]
2	1.5	16.56	12.71	0.04672	0.04839	1e-6

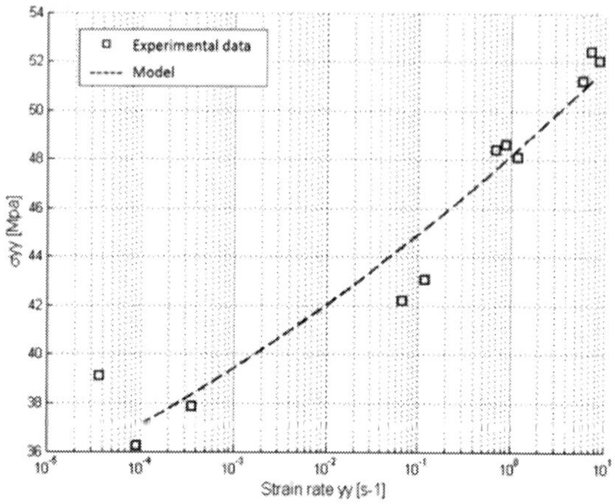

Figure 8. Strain rate dependency of the tensile failure stress.

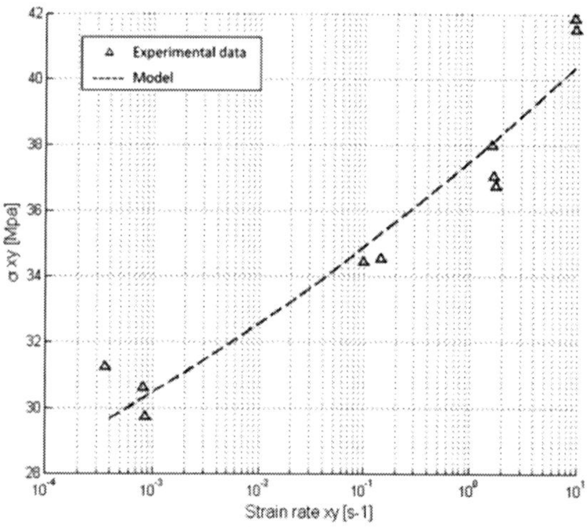

Figure 9. Strain rate dependency of the shear failure stress.

shear failure stresses evolution with the experiments. The scattering of the results is below 5% for tensile tests, and below 4% for shear tests.

Finally, Fig. 10 shows the failure stress criterion given by Eq. (3). The failure surface fits the experimental data with a maximum error of 6.6%.

5. Conclusion

A new Arcan device is proposed to characterize the failure of bonded joints. It is designed to minimize the edge effects and to ensure a good homogeneity of the stress field in the joint.

An experimental campaign is achieved for different modes of loading and for velocities until 100 mm/s. A low scattering on the failure stresses is obtained for all the experiments. Results also highlight the strain rate and mode of loading dependencies of the failure criterion.

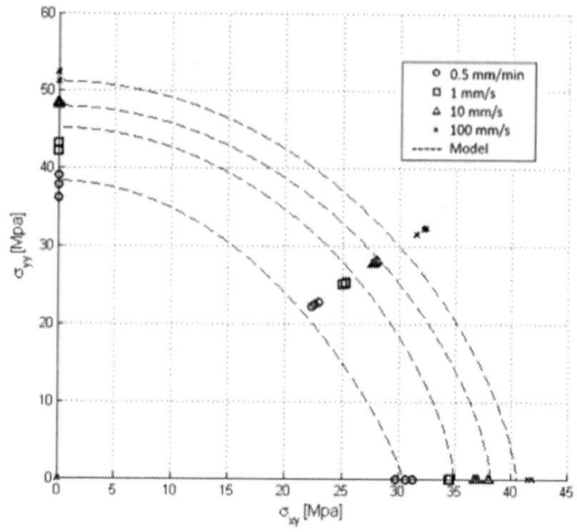

Figure 10. Failure criterion representation from 0.5 mm/min to 100 mm/s.

Consequently, a stress failure criterion with strain rate and mode of loading dependency is proposed, and identified with this new Arcan device. The identified failure surface fits well the experimental data with a maximum error of 6.6%.

In this paper, a failure criterion based on the maximum stress is proposed. This criterion will be compared in the future with other criteria based on strain or energy approaches.

The present research work call FASTLITE assembly has been supported by the "Programme véhicule du futur" from Projets d'Investissements d'Avenir coordinated by ADEME, the International Campus on Safety and Intermodality in Transportation, the Region Nord Pas de Calais, the European Community, the Delegation Regionale a la Recherche et a la Technologie, the Ministere de l'Enseignement Superieur et de la Recherche, and the Centre National de la Recherche Scientifique: the authors gratefully acknowledge the support of these institutions.

References

[1] M. Bocciarelli, P. Colombi, G. Fava, C. Poggi, Eng. Frac. Mech. **76**, 299–313 (2009).
[2] A. Carpinteri, P. Cornetti, N. Pugno, Eng. Struct. **31**, 2436–47 (2009).
[3] L. Greenwood, London: University of London Press, 1969.
[4] K. Ikegami, T. Takeshita, K. Matsuo, T. Sugibayashi, Int. J. Adhes. Adhes. **10**, 199–206 (1990).
[5] R.D. Adams, J.A. Harri, Int. J. Adhes. Adhes. **4**, 65–78 (1984).
[6] L.J. Hart-Smith, NASA contract report. NASA CR-112236 (1973).
[7] L.F.M. Da Silva, P.J.C. das Neves, R.D. Adams, A. Wang, J.K. Spelt, Int. J. Adhes. Adhes. **29**, 331–41 (2009).
[8] V. Joudon, G. Portemont, F. Lauro, B. Bennani, Eng. Fract. Mech. **126**, 166–177 (2014).

[9] G.R. Johnson, W.H. Cook, Eng. Fract. Mech. **21**, 31–48 (1985).
[10] B. Langrand, E. Markiewicz, Int. J. of Imp. Eng. **37**, 792–805 (2010)
[11] D. Morin, G. Haugou, B. Bennani, F. Lauro, Eng. Fract. Mech. **77**, 3481–3500 (2010).
[12] D. Morin, G. Haugou, B. Bennani, F. Lauro, Int. J. Adhes. sci. and Tech. **25**, 1581–1602 (2011).
[13] D. Morin, B. Bourel, B. Bennani, F. Lauro, D. Lesueur, Int. J. of Imp. Eng. **53**, 94–105 (2013).
[14] J.Y. Cognard, P. Davies, B. Gineste, L. Sohier, Comp. Sci. and Tech. **65**, 359–368 (2005).

EPJ Web of Conferences **94**, 01025 (2015)
DOI: 10.1051/epjconf/20159401025

Recording of particles velocity spectrum at the shock impact on different viscosity interface of liquids

A.V. Fedorov[a], A.L. Mikhailov, S.A. Finyushin, D.A. Kalashnikov, E.A. Chudakov, E.I. Butusov, and I.S. Gnutov

RFNC-VNIIEF, Nizhniy Novgorod reg., Sarov, Russia

Abstract. The results of experiments concerning the study of cavitational mechanism of liquid failure in a wide range of shock loading are presented in this paper. Free surface velocity of liquids and velocity spectrum of particles and jets were recorded using PDV method [1], their size was also determined. The value of spall strength of distilled water was defined.

1. Introduction

There are several mechanisms of shock ejecting: particle ejecting from the rough surface, spall fragmentation of material induced by internal tensile stresses, microcumulative jets appearing on gas bubbles, inhomogeneity of material structure etc. [2–5]. The case in study of shock ejecting in liquids and liquid metals with smooth and mirror-like surfaces is of special interest. In this case, there is no particle ejecting owing to the surface roughness.

The main goal of this paper is to study the dynamics of the drop cloud formation and to record the particle velocity spectrum as a result of cavitational damage of liquids which values of viscosity can differ in thousand times.

2. Experimental setup

Liquids with different values of viscosity (distilled water – 0.89 centipoise; ethanol – 1.07 centipoise; glycerin – 934 centipoise; oil VM1 – 300 centipoise) were chosen for experiments.

Liquids (ø80 mm) were loaded up to high (P = 7–11 GPa) and low (P = 0.1–0.9 GPa) pressures. In the first case, liquids were loaded by plane stationary shock wave (SW) from HE (ø120 × 40 mm) through an aluminum disk (ø120 × 1 mm), in the second case – by divergent wave with triangular section from explosion of detonator through an aluminum or steel disk (ø120 × 1 mm). The thickness of liquid layer was varied in a range of 5–30 mm. The velocities of the particles and the free surface (FS) were recorded using laser heterodyne-interferometer (PDV method). The PDV probe was placed at a distance of 35–50 mm from the surface.

3. Experimental results

Experimental results are shown in Table 1: W_{FS}, P_{SW}–the free surface velocity and shock pressure in liquids, V_{FF}^{max}–maximum velocity of the particle flow front at the moment of SW arrival to the free surface.

[a] Corresponding author: Fedorovsarov@mail.ru

3.1. Recording of particle velocity spectrum under loading of liquid up to high pressures

Experimental velocity spectrogram of ethanol drops (test #2) is shown in Fig. 1. The similar spectrograms are obtained for water, glycerin and oil. The PDV method allows to record the free surface velocity, the moments of shock impact and the flight time of liquid free surface and flow front (FF) to a collimator. Using the known base length, W_{FS} and the average flight velocity of drops, V_{FF} can be determined. The average velocity V_{FF} for three liquids was equal to 4.6–5.0 km/s. The average velocity V_{FF} for ethanol was equal to 5.6 km/s. High initial velocities of ejecting liquid particles V_{FF}^{max} (4.9–5.9 km/s) are recorded in spectrograms of the first four tests. According to particle braking law in the air at such high velocities, the particles must have high gradient of velocity decay what can be seen in the initial time. The process of particle size determination using their braking curves in gas is described in [6]. Spectrogram and deceleration pattern droplets for ethanol are shown in Fig. 1. Under hard braking of particles in gas, an air shock wave (ASW) runs rapidly and the particles get into the region of the air compressed by SW (P = 330 bar) where gas velocity must be equal to the piston velocity W_{FS} = 4.98 km/s.

But the free surface velocity is 6% lower than U_{air} = 5.27 km/s. The excess of the air velocity over SF velocity is connected, in our opinion, with the fact that the cloud of liquid drops has high values of density and velocity, and it additionally speeds up the air and increases its velocity in ≈6%.

According to the flight time of particle flow front and the free surface of liquid to collimator, the particle cloud thickness can be determined. The cloud thickness for liquids (ethanol, water, glycerin, oil) is 5 mm, 4 mm, 4.5 mm and 5.4 mm, correspondingly. The excess of maximum particles front velocity over free surface velocity in tests was equal to 25–35%. The free surface velocity W_{FS} cannot be recorded by PDV method in real experiments. It is connected with high density of ejecting particles, as laser beam is not able to illuminate.

Article available at http://www.epj-conferences.org or http://dx.doi.org/10.1051/epjconf/20159401025

Table 1. Experimental results.

#	Liquid	Liquid thick- ness Δ, mm	P_{SW}, GPa	W_{FS}, km/s	V_{FF}^{max}, km/s
1	Water	12	10.3	4.44	5.03
2	Ethanol	12	9.7	4.98	5.9
3	Glycerin	12	10.7	4.07	5.02
4	Oil	12	7.3	4.08	4.95
5	Water	20	0.18	0.195	0.25
6	Water	5	0.86	0.76	0.78
7	Water	10	0.093	0.11	0.3
8	Glycerin	30	0.22	0.17	–
9	Glycerin	20	0.35	0.26	0.25

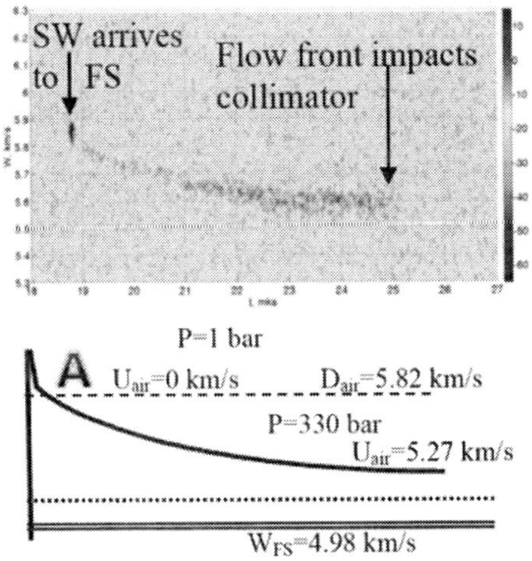

Figure 1. Spectrogram and deceleration pattern droplets for ethanol in the air.

Mechanism of dense drop cloud formation at the moment of SW arrival to FS of liquids is described in [5]. It is shown there that cavitational clusters grow in a tensile stress field and the liquid is gradually transiting into a singular foam phase in a form of thin films or a net of liquid bundles. Liquid net elements under further tension are divided into separate fragments and after that into jets which are failed into separate drops under effect of instability and surface tension.

3.2. Recording of particle velocity spectrum under loading up to low pressures

Three tests (#5, #6, #7) with water (see Fig. 2, spectrograms of the tests #5 and #7 are similar) and two tests (#8, #9) with glycerin (see Fig. 3) were conducted with loading up to low pressures ($P \approx 0.1$–0.9 GPa). A divergent wave with triangular section propagates from detonator, and it intensively decays when passing through a liquid. The free surface velocity is recorded in the tests #5 and #7. A dense layer of liquid particles blocks laser emission in the test #6. Velocity tracks were recorded in spectrograms of the tests #5 and #7 after SW arrival to the free surface.

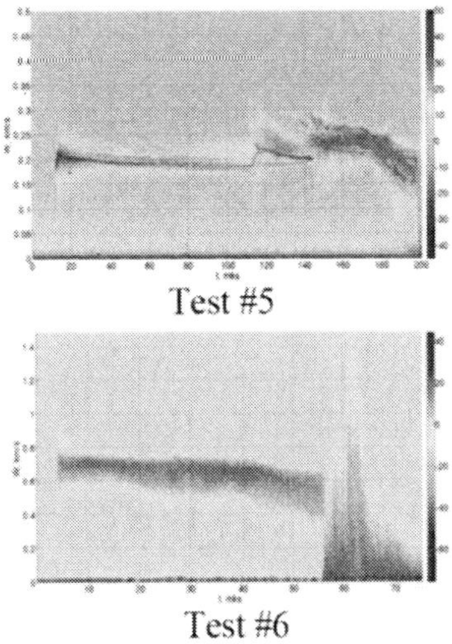

Figure 2. Spectrograms of the tests with water.

According to these tracks, the size of drops and jets was equal to 7–19 μm in the test #5 and 5–25 μm in the test #7 (see Fig. 2 and Fig. 4).

Drop size in these tests was determined taking into account gas compression after ASW passing, i.e., the drops move inside so called "air cushion". Cavitational failure of water is recorded in the test #6, and a cloud of drops moving with the maximum front velocity of 0.68 km/s and dispersion of the cloud velocity of $\approx$120 m/s is seen in a spectrogram. High velocity of the drop cloud if compared with the other tests is caused by higher decay gradient and shock parameters. It should be noticed that such failure pattern is observed in all the tests with water under the pressure of $P > 0.2$–0.3 GPa. The spall pulse is clearly seen in the tests with water at the moment of SW arrival to the free surface. The value of spall strength of water was equal to $\sigma = 30$ MPa, the thickness of the first spall layer was 0.1–0.25 mm in different tests.

The velocities of three spall layers are recorded by PDV in spectrogram of the test #5 ($W = 0.19$ km/s, $W = 0.199$ km/s and $W = 0.206$ km/s), the surfaces of these layers reflect laser emission. Moving spall layers during $\Delta t = 80\,\mu$s pass 15.2 mm, 16 mm and 16.8 mm, correspondingly. At $t = 80$–$100\,\mu$s it is seen that laser beam starts to be reflected not from the free surface of spall layers but from the cloud of particles. I.e., fragmentation of these spall layers into drops occurs under effect of the forces of surface tension and instabilities of boundaries.

The results of the tests with glycerin are shown in Fig. 3. Velocity tracks of the drops and jets of glycerin are clearly seen in spectrogram of the test #9. The absence of tracks in the test #8 is a subject of further research. It is seen in spectrogram of the test #9 that there are two regions: the region of front moving jets and the region of cavitating liquid. In this case, the front of cavitating liquid moves at 6.6 mm in the first 30 μs of motion and the front of the slowest jets moves at 7.8 mm. I.e., the jets tear away 1.2 mm from the main mass of liquid. The front of faster jets tears away further.

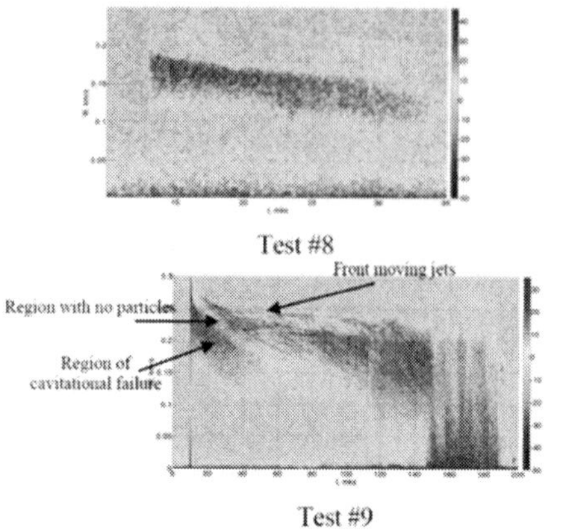

Figure 3. Spectrograms of the tests with glycerine.

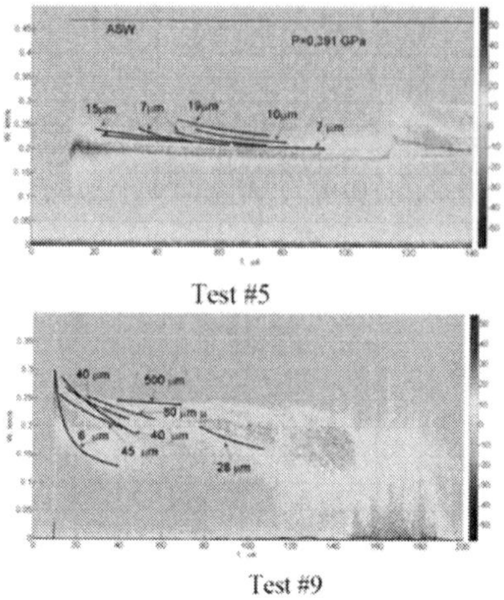

Figure 4. Spectrograms of the tests with water and glycerin.

Thus, the region with the velocity dispersion of $\approx$30 m/s is recorded by PDV (see "region with no particles" in Fig. 3). The size of fine drops in the test #9 was equal to 28–80 μm (see Fig. 4). The tracks corresponding to large-scale jets with the size of 200–550 μm are seen as well. The bigger (if compared with water) size of glycerin jets is probably connected with the higher value of liquid viscosity.

Spectrograms of the tests with water and glycerin are presented in Fig. 4. Particle tracks are shown by black lines. The proof of jet mechanism of liquid ejecting is an estimation of front displacement of the fastest jets. The maximum velocity of 0.25 mm/μs is recorded for jets with the characteristic size of 500 μm. Then these jets pass 25 mm at the moving interval of 100 μs. The average velocity of caviting liquid front does not exceed 0.18 km/s, i.e., it is equal to 18 mm at the moving interval of 100 μs. Thus, the large-scale jets will tear away 7 mm from the particle cloud. This effect was also recorded in [2] where the results of similar experiments conducted using the shadow optical technique are presented. The jet size in similar tests in [2] with glycerin was in a range from 100 μm to 1 mm, in tests with water it was equal to 7–19 μm. These values practically agree with our results. Thin jets of water moving in gas break up into drops, and glycerin jets do not break up.

4. Discussion

It is known that the average size of gas bubbles for water is of d $\approx$ 1.5 μm [5], and cumulative jets formed in bubbles at the moment of SW arrival to FS are not able to break through the water layer bordered on the surface (the ejecting process is not recorded in spectrogram at the beginning of the record). The bubble size of viscous glycerin is of d = 30–550 μm, and the ejecting is recorded just at the moment of SW arrival to FS. Thus, different ejecting mechanisms are realized in glycerin and water. The ejecting of microjets in glycerin occurs from the gas bubbles bordered on the liquid surface. At first, the free surface of water is mirror-like and there is no ejecting. After the wave reverberation in the first spall layer, the surface becomes rough, and the ejecting occurs when the following compression waves go out of it.

5. Conclusions

The experiments with PDV recoding of particle ejecting from the surfaces of liquids with different viscosity in a wide range of loading pressures were conducted. It is shown that cavitational failure of liquids occurs under effect of tensile stresses. The spectra of liquid particle velocity and the velocity tracks of jets and drops are recorded, their size is determined. The size of jets is equal to 5–25 μm for water and 30–550 μm for glycerin. The difference in a particle size can be caused by significant difference in the viscosity values of glycerin and water. The spall strength of water is determined to be $\sigma = 30$ MPa.

References

[1] O.T. Strand, D.R. Goosman, C. Martinez, and T.L. Whitworth. Compact system for high-speed Velocimetry using heterodyne techniques. Rev. Sci. Instrum. **77**. p. 083108-1–083108-8 (2006).

[2] N.V. Nevmerzhitskiy, E.A. Sotskov, E.D. Sen'kovskiy et al., Microscopic electron-optical recoding of particle ejecta from free surface of shock loaded metals and liquids. **XV** Khariton's topical scientific readings. Sarov (2013).

[3] V.M. Boiko and S.V. Poplavskii. Particle and Drop Dynamics in the Flow behind a Shock Wave. Fluid Dynamics. Vol. **42**, **#3**, pp.433–441 (2007).

[4] B.E. Gelfand. Droplet breakup phenomena in flows with velocity lag, Prog. Energy Combust. Sci. Vol. **22**, pp. 201–265 (1996).

[5] V.K. Kedrinskiy. Hydrodynamics of explosion. Experiment and simulations. Novosibirsk (2000).

[6] A.V. Fedorov, A.L. Mikhailov, S.A. Finyushin, D.V. Nazarov, E.A. Chudakov, D.A. Kalashnikov, E.I. Butusov. Study of lead characteristic feature under shock loading and the following unloading. **XV** Khariton's topical scientific readings. Sarov. p. 274 (2013).

EPJ Web of Conferences **94**, 01026 (2015)
DOI: 10.1051/epjconf/20159401026

The role of the modified taylor impact test in dynamic material research

Frank Bagusat[a] and Ingmar Rohr

Fraunhofer Institute for High-Speed Dynamics, Ernst-Mach-Institut, EMI, Eckerstrasse 4, 79104 Freiburg, Germany

Abstract. Dynamic material research with strain rates of more than 1000 1/s is experimentally very often done with a Split-Hopkinson Bar, Taylor impact tests or planar plate impact test investigations. At the Ernst-Mach-Institut (EMI), a variant of an inverted classical Taylor impact test is used by application of velocity interferometers of the VISAR type ("Modified Taylor Impact Test", MTT). The conduction of the experiments is similar to that of planar plate impact tests. The data reduction and derivation of dynamic material data can also be restricted to an analysis of the VISAR signal. Due to these properties, nearly each highly dynamic material characterization in our institute done by planar plate investigations is usually accompanied by MTT experiments. The extended possibilities and usefulness of a combined usage of these two highly dynamic characterization methods are explained. Recently, further developed MTT experiments with very small specimen sizes are presented. For the first time, Taylor impact and planar impact specimen can be used for which the load directions even in case of thin plate test material are identical and not perpendicular to each other. Consequences for testing construction elements are discussed.

1. Introduction

1.1. Taylor-test setup

Taylor-test investigations can be an important component in research examining material under dynamic loads. In terms of strain rate, Taylor-tests are a very useful addition to split Hopkinson and planar plate impact experiments. The classical method of execution [1–3] catapults the specimen perpendicularly onto a fixed flat target with a defined impact velocity. The impact process causes a deformation only on the impacted side of the specimen. Dynamical material data is calculated by using geometrical properties of the finally deformed specimen. In the past, changes of this setup have been introduced. One point was the inversion of the test, so that the specimen is staying at rest and serves as a target in an impact experiment. As impactor, a rod with identical shape ("symmetric test" [4]) or a large and heavy projectile is used ("anvil on rod test"). Furthermore, instantaneous data collection, i.e. during the experiment, has been made possible by adaption of strain gauges.

It is a very common idea already described by Taylor in his article introducing a new method for the determination of the dynamic yield stress [1] that due to the impact, elastic and plastic wave propagation processes are occurring in the specimen. Strain gauge technique can measure surface waves [4,5] on the specimen, from which dynamical material data can be calculated.

1.2. The modified Taylor-test (MTT)

Wave propagation processes can also be recorded by interferometric techniques. These have the advantages that they are contact free without any mechanical influence on the specimen and that they can measure continuously with high time resolution. At the EMI a free beam VISAR (velocity interferometer system for any reflector [6–10]) system is used [11–16], but in principle also fiber based VISAR setups are known to be used [4,17–19] in connection with inverted Taylor test experiments.

The inverted Taylor-Test with VISAR is denoted as modified Taylor-Test (MTT) in the following.

MTTs at the EMI are typically executed with a length-(L)-to-diameter-(D) ratio of the specimen of $L/D = 10{:}1$. The typical diameter of the specimen is 6 mm, but also much smaller specimens can be used [16], which will be demonstrated and discussed at the end of this article.

As a projectile, a sabot carrying a maximum hardened C45 steel cylinder with flat polished surface is used. This system has a mass of about 450 g.

Impacting a static specimen by a sabot driven projectile with sufficient high velocity will cause elasto-plastic wave processes in the specimen (a slower plastic and a faster elastic wave). If the elastic wave is reaching the rear side of the specimen, this side is accelerated; this acceleration can be observed by a VISAR [12–14]. It is assumed that the acceleration process causes a load of the specimen's material so that it reaches the yield point. Therefore, by measuring and evaluating the free surface velocity time signal of a specimen under impact, dynamic material properties can be derived. Typically, the acceleration is observed as a sequence of several velocity steps (Fig. 1).

1.3. Dynamic material properties derived from MTT investigations

It has been discussed elsewhere which kind of dynamic material properties can be derived from free surface velocity time curves [12–14]. Therefore, only the formulas are summarized here. The yield stress Y can be calculated by

$$Y = 0.5 * \rho * c_0 * \Delta u_{fs} \quad (1)$$

[a] Corresponding author: frank.bagusat@emi.fraunhofer.de

Article available at http://www.epj-conferences.org or http://dx.doi.org/10.1051/epjconf/20159401026

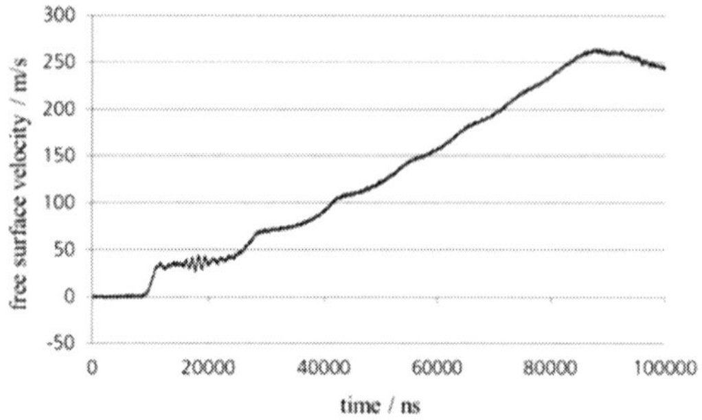

Figure 1. VISAR free surface velocity time curve, steel 11SMn30+C, rod size: 6 mm diameter, 60 mm length.

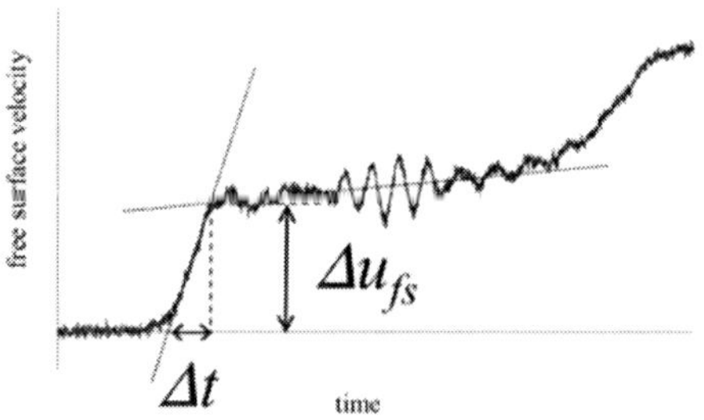

Figure 2. Principle of determination of the velocity jump Δu_{fs} and time interval Δt.

where ρ is the density of the specimen's material, c_0 denotes its longitudinal velocity of sound and Δu_{fs} is the height of the *first* velocity step visible in the free surface velocity time curve (Fig. 2). The strain εat the yield point can be estimated by

$$\varepsilon = \Delta u_{fs}/(2 * c_0). \tag{2}$$

The corresponding strain rate is therefore

$$d\varepsilon/dt = \varepsilon/\Delta t, \tag{3}$$

where Δt is the time interval needed for the velocity step Δu_{fs}. The determination of Δu_{fs} and Δt can be executed by a mathematical analysis or, much simpler, graphically depending on the needed accuracy (Fig. 2).

These material parameters are experimentally characterized by elastic wave processes in the specimen. Therefore, for a more accurate designation of strain and strain rate values, both are elastic values. Furthermore, due to their dependence to a velocity or time interval, these values are averaged properties.

2. Examples of MTT investigations

2.1. MTTs as one element for a strain rate dependent yield stress characterization

MTT specimen made of a high strength steel with a size of 6 mm diameter and 60 mm length were used in an extended investigation collecting yield stress data determined by various testing methods [12, 14]. One result of this measurement program was a yield stress-strain rate dependency depicted in Fig. 3. MTT results here delivered last points of a linear relationship of the yield stress strain rate behavior if the strain rate is scaled logarithmically. The yield stress is strongly increasing in the strain rate region of planar plate impact investigations. It is clear that the exact inflection point of the material behavior is only determined by two methods (MTT and planar plate impact measurements), but due to the strain rate dependency of the material tested by planar plate impact technique, the indicated linear fit seems not a bad choice. It is interesting that yield stress data of the second material presented in Fig. 3, a tungsten sinter alloy (WSA), show a linear relationship (with a logarithmic strain rate scaling) over all experimentally determined data points [15]. WSA specimens had a diameter of 5 mm and a length of 50 mm.

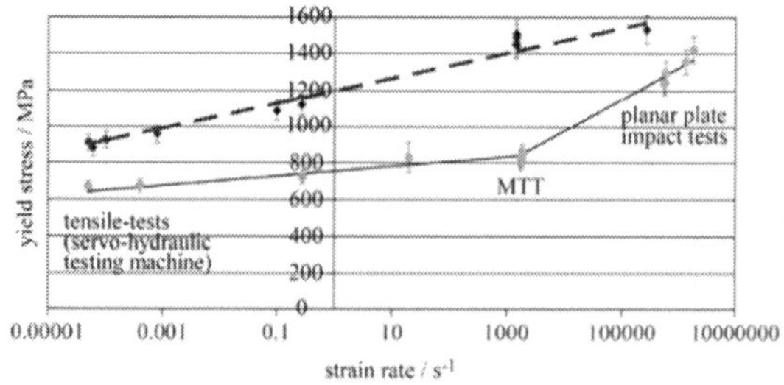

Figure 3. Yield stress as a function of strain rate for two materials: high strength steel (gray points; solid line [12, 14]) and tungsten sinter alloy (WSA, black points; dashed line [15]).

These results are a very impressive illustration for the need of methods which cover different strain rate regions for material characterization.

For the investigated materials, the yield stress dependence from increasing strain rate was not clear. Some high strength steels show almost no further increase of the yield stress with strongly increasing strain rate. One would assume that the yield stress of WSA is higher than that of steel because its physical or mechanical properties are mainly given by its content of tungsten. Tungsten is considerably superior over steel in respect to mechanical properties like hardness or elastic modulus. If one had not used (in comparison to quasistatic methods) laborious MTT and planar plate impact investigations, qualitative differences between the high strength steel and WSA would not have become visible. On the other hand, if the MTT investigation had been left out, the inflection point of the yield stress would probably shift to lower strain rates.

MTT and planar plate impact investigations can be seen as necessary and useful methods of material characterization. With a lack of one or two of these methods, a strain rate dependent modelling would miss essential material characteristics.

2.2. MTTs for modelling dynamic material properties

Classical Taylor tests deliver material related data only by an analysis of the deformed specimen after the test.

MTT tests of materials like tungsten sinter alloy (WSA) can show a damage of the rod on its impacted side. This can emerge to an almost complete failure, if the impact velocity is sufficiently high. A 'post impact' analysis is not possible in this case.

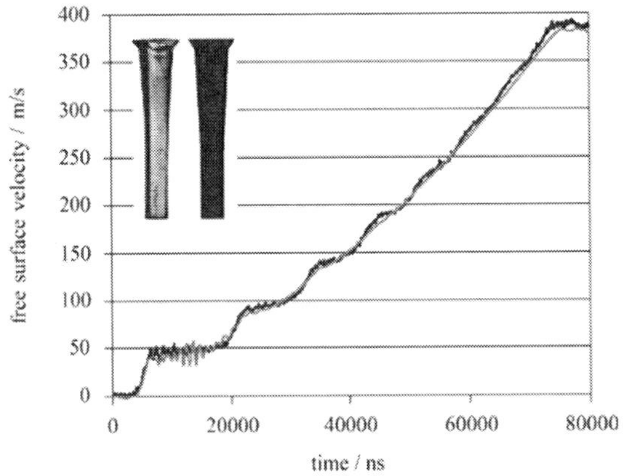

Figure 4. VISAR free surface velocity time curve [12–14] of a high strength steel and a simulation (gray curve) of the signal. Impact velocity 361 m/s. Material was the same as in experiments with results shown in Fig. 3. Experimental and simulated (black) shapes of the specimen are also shown.

A material data recording (based on wave propagation processes) during the deformation is only executable with strain gauges or with interferometric techniques as in case of the MTT. If the deformation itself is a main point of interest in a study, high speed photography might also be used in addition [17, 18].

For a validation of a chosen material model, the MTT free surface velocity time curve itself can be used. Modelling of a VISAR signal with several velocity steps (Fig. 4) representing elastic wave reflections in the material has been found to be much more sensitive than trying to simulate only the final diameter of the impacted side of the specimen correctly.

For the high strength steel used for experiments utilized in Fig. 3/Fig. 4, a phenomenological material model with a strength model and an equation of state has been developed [12, 14]. The agreement of simulated results including the VISAR curves and cross sections of deformed specimen with experimental findings has been found to be quite well, showing only differences of a few percent or less.

The Taylor rod deformation or failure with defined specimen geometry is always a highly dynamical process. It is essential that the modelling process incorporates implicitly the plastic deformation behavior which reaches (plastic) strain rates in the order of 10^5 1/s. The advantages of VISAR based MTT experiments are a continuous measurement (covering the whole acceleration process of the specimen), a very high time resolution (down to a few ns), a high absolute velocity measurement accuracy of the VISAR signals (typically ± 1–2% [9]) and an operation which is free of mechanical contact to the specimen. These properties are not given by camera or strain gauge measurements alone. Nevertheless, the experimenter is free to combine all these methods for data acquisition.

MTT investigations can therefore be a very interesting, powerful and accurate method by supplying Hopkinson and planar plate impact techniques in the intermediate strain rate regime especially for simulation purposes.

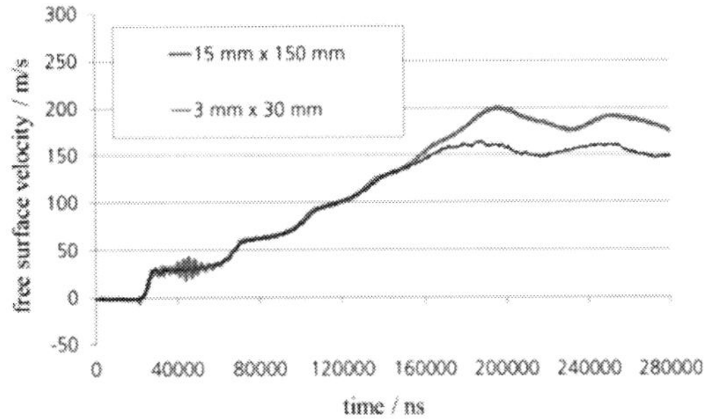

Figure 5. VISAR curves of MTT experiments with steel 11SMn30+C of different rod dimensions. The signal of the 15 mm × 150 mm specimen is shown with the original time scale, the time scale of the smaller rod was multiplied by a factor of 5 (in order to compensate a shorter travelling time of the elastic wave in the smaller rod, the scaling factor is given by the length ratio of the rods). Signal curves are smoothed.

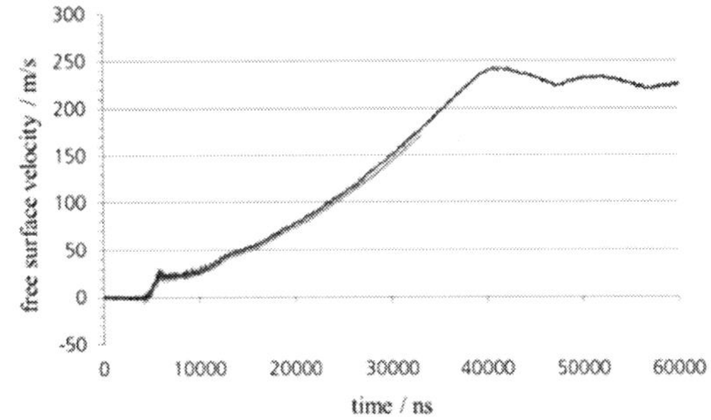

Figure 6. VISAR curves of MTT experiments with rods of pure, annealed Niobium (Nb), rod size 2 mm × 20 mm. Data of the experiment depicted in gray are cut due to signal overshoot.

3. MTT experiments with specimens of small dimensions

3.1. Considerations about the specimen size

EMI's modified Taylor tests are normally conducted with Taylor rods of 6 mm diameter and 60 mm length. Several reasons exist for reducing the geometric size of the specimens. Firstly, new materials which were very expensive and difficult to produce should be dynamically tested. Secondly, materials which are typically only available in the form of sheet material should be investigated. Thirdly, material properties of construction elements should be determined at locally different positions. This can be done much easier and with better local sensitivity by using smaller specimens.

It was an open question whether a size reduction would not have disturbing effects onto the VISAR signals, but tests with heavily varying rod sizes (Fig. 5) demonstrated a very good overlap of signal curves after adaption of the individual time scales according to the ratio of the specimen's lengths.

Rod sizes beyond the 3 mm × 30 mm dimension were tested with different materials and found to be quite easy to conduct. Figure 6 shows results of experiments with pure annealed Niobium rods of the dimension 2 mm × 20 mm.

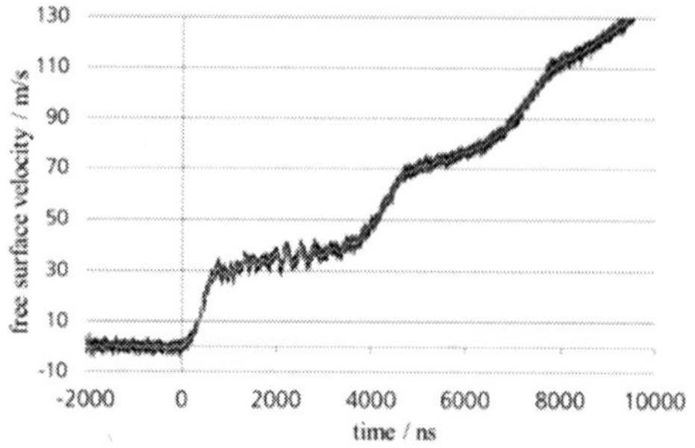

Figure 7. Effect of data smoothing on MTT VISAR curves. C45 steel rod of size 1.5 mm × 15 mm. Original data are shown in black, the smoothed curve is indicated by the fine white line. Data smoothing was applied before data reduction was executed.

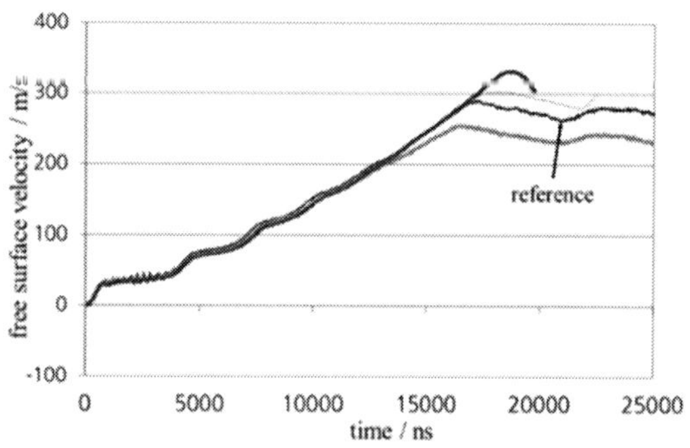

Figure 8. Signal curves (smoothed) of MTT experiments with C45 rods of dimension 1.5 mm × 15 mm. Data of an experiment with C45 steel (6 mm × 60 mm, time axis scaled) serves as reference and was not smoothed [16].

After the first elastic step, only tiny indications of more steps are observed in the acceleration process.

A further important point when working with smaller rod sizes is the signal quality. It is easy to imagine that a smaller sample could theoretically give less intense scattering signals due to the quality of adjustment of the VISAR focus on the backside of the rod or due to the obtainable quality of the polishing of smaller rods. This would result presumably in a reduced signal to noise (S/N) ratio of the measurements. In case of MTT investigations, there is a very simple way to compensate this by applying data smoothing as shown in Fig. 7. MTT signals do generally have no fast signal changes (as can be seen in planar plate impact experiments, where signal variations within a few nanoseconds occur). The typical time scale of MTT experiments falls in a range between several μs (Figs. 1, 6–9) and more than 160 μs in case of very large rods (Fig. 5). Data smoothing will therefore not eliminate essential data structures, but will improve the shape of the curve even in case of small rod sizes.

Figure 8 shows a series of experiments executed with C45 steel with rods of the dimension 1.5 mm × 15 mm. In addition, a curve of an experiment with a 6 mm × 60 mm rod is plotted. The time axis of this data set has been reduced according to the smaller samples. The figure illustrates a good reproducibility of the signal even with

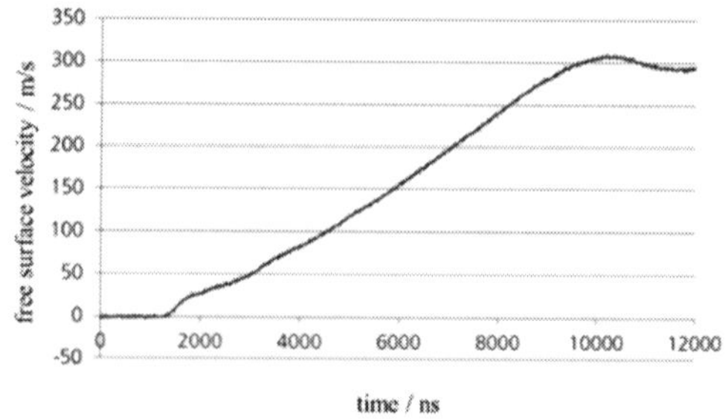

Figure 9. Free surface velocity time curve of a MTT experiment with specimen size of 0.7 mm diameter and 7 mm length (steel 115CrV3, material number 1.2210).

this rod size and also a good agreement of the signal with the 6 mm × 60 mm results.

Investigations with rod sizes of 1.5 mm × 15 mm showed in some cases a very poor S/N-ratio (2.1 to 1.1, normally 5:1 to 10:1 or even better is observed), but with data smoothing before data reduction, even these signals were usable at the end [16].

3.2. Effects limiting the miniaturization of MTT specimens

For a valid registration of a free surface velocity time curve, several conditions of the experimental setup should be fulfilled:

1) The illuminating laser focus should cover a sufficient amount of grains, so that averaging of the back scattered signal can take place. This is necessary in order to obtain reproducible free surface velocity time curves. That means that the grain size should be much smaller than the laser focus.

   The focus diameter is typically around 0.5 mm.
2) The specimen holder system has to be stable enough so that the illuminating focus will stay on the specimen for the duration of the whole experiment. Failing this, signal loss will occur.

   Otherwise, the holder system of the specimen should have no disturbing effect on the wave propagation within the specimen.
3) The orientation of the focus point relative to the specimen should be adjustable in very small steps in order to fulfill point 2).
4) In case of small specimen sizes, the mass of the specimens also is strongly reduced. This can lead to a heating of the specimens by the illuminating laser. Therefore, laser power and illumination time have to be reduced (or the specimens have to be cooled).

It was found that a free beam VISAR setup (the classical variant of a VISAR) could be used with our laboratory setup with a minimum specimen size of 1.5 mm × 15 mm [16]. In case of smaller specimens, loss of signal was predominantly observed, so that almost no signal or only a first step of the MTT signal was recorded.

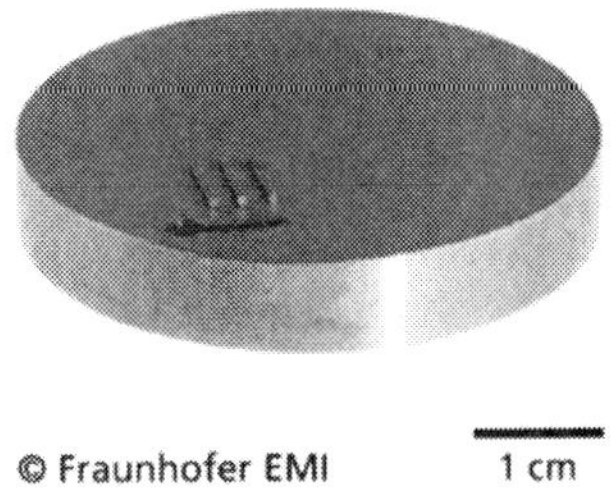

Figure 10. Specimen of C45 steel (50 mm diameter, 8 mm thickness) produced for a planar plate impact experiment and three Taylor rods with a size of 0.8 mm × 8 mm cut out of the outer border area by wire spark erosion technique. The load direction of both specimen types is identical.

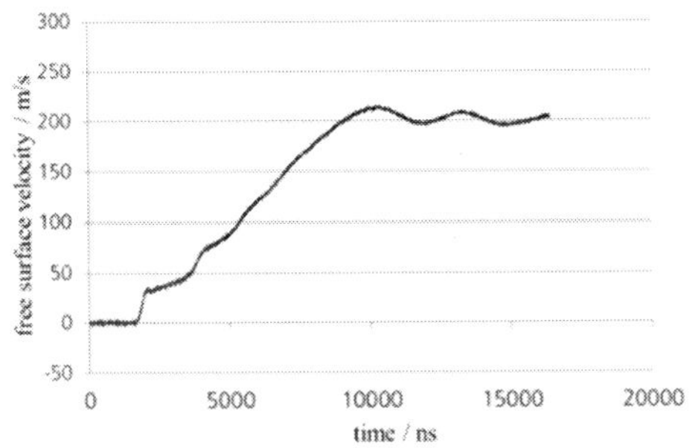

Figure 11. Free surface velocity time curve of a MTT experiment with specimen size of 0.8 mm diameter and 8 mm length (specimen taken from planar plate impact test geometry, Fig. 10).

As an alternative to a free beam VISAR, a fiber optic based VISAR (VALYN-VISAR [10]) with a focus length of 30 mm was tested. With this setup, even smaller specimens (1 mm × 10 mm and 0.7 mm × 7 mm) could be investigated successfully (Fig. 9).

These small specimen sizes enable the production of Taylor rods directly from thin plates like planar plate impact specimens (Fig. 10, experimental VISAR signal Fig. 11). The advantages of using Taylor rods manufactured from the planar plate impact test specimen can be summarized as follows. The load direction of test specimens is the same, the material is identical and the difficult polishing of small rods can be done during the preparation process of the planar plate impact test specimens so that they have perpendicular end surfaces. Furthermore, a series from MTT experiments can be produced from the outer region of each individual planar plate impact test specimen by wire spark erosion technique. This makes it possible to conduct a series of Taylor-tests of planar plate impact test specimens and can be used to control dynamic material data.

Of course, small specimen sizes offer a general benefit for the purpose of the determination of locally dynamic material properties of real construction elements.

All experimental examples shown in this article were tested with the impact facility at the EMI used for "normal size" MTT/planar plate impact test experiments.

4. Summary and conclusions

MTT experiments can deliver very accurate dynamic material properties like strain, strain rate and yield stress.

Regarding the strain rate, the method falls between split Hopkinson bar and planar plate test conditions and therefore gives additional information about the material behavior at high strain rates, which is especially helpful for purposes of material modelling and simulation of highly dynamic processes.

Results of MTT experiments with specimen sizes beyond the typical geometry of 6 mm by 60 mm show that measurements are possible with enhanced efforts concerning the specimen holder system. Minimal rod sizes of 0.7 mm × 7 mm can only be investigated with application of a VISAR system with a short observation range (fiber optic system with short focus length). That means all laboratories operating VISARS of the VALYN type can in principle execute MTT experiments with very small specimen sizes. With a VALYN-VISAR, MTT-experiments with small specimens (e.g. smaller than 2.5 mm × 25 mm) are easier to conduct than with larger specimens due to restrictions of the effective depth of field of the fiber optic probes.

Plate material with a thickness beyond 10 mm can now be investigated by planar plate impact test *and* MTT experiments executed with the same load direction due to the small possible size of the MTT specimen.

In case of specimens with sufficient small average grain size, experiments with even further reduced specimen size (e.g. 0.4 mm × 4 mm) seem to be possible, if a focus diameter optimized observation system is applied and great care is taken with laser power, so that heating up of the specimen can be neglected.

Working with small specimens offers the possibility to use also small accelerators, so that cost and time consuming handling of the specimens in larger impact facilities can be avoided.

The authors thank Siegfried Pfändler for his kind help in operating the impact facility. The contributions of Werner Stein, Reinhard Tham and Hartwig Nahme in many discussions regarding aspects of dynamical material characterization by MTT experiments are gratefully acknowledged. This work was partially conducted under an in-house research program.

References

[1] G. Taylor, Proc. Royal Soc. London A **194**, 289 (1948)
[2] A. C. Whiffin, Proc. Royal Soc. London A **194**, 300 (1948)
[3] W. E. Carrington, M. L. V. Gayler, Proc. Royal Soc. London A **194**, 323 (1948)
[4] L. C. Forde, W. G. Proud, S. M. Walley, Proc. Royal Soc. A **465**, 769 (2009)
[5] R. Tham, A. J. Stilp, Journal de Physique **49**, C3-85 (1988)
[6] L. M. Barker, R. E. Hollenbach, J. Appl. Phys. **43**, 4669 (1972)

[7] L. M. Barker, K. W. Schuler, J. Appl. Phys. **45**, 3692 (1974)
[8] W. F. Hemsing, Rev. Sci. Instrum. **50**, 73 (1979)
[9] L. M. Barker, *Shock Compression of Condensed Matter 1997, AIP Conference Proceedings CP 429, 833* (1998)
[10] `www.valynvisar.com`
[11] G. Kuscher, PhD-Thesis, RWTH Aachen, Aachen (1985)
[12] I. Rohr, PhD-Thesis, Universität der Bundeswehr München (2003)
[13] I. Rohr, I. H. Nahme; K. Thoma, J. Phys. IV France **110**, 513 (2003)
[14] I. Rohr, H. Nahme, K. Thoma, Int. J. Impact Eng. **31** 401 (2005)
[15] I. Rohr, H. Nahme, K. Thoma, C. E. Anderson Jr., Int. J. Impact Eng. **35**, 811 (2008)
[16] F. Bagusat, I. Rohr, *EPJ Web of Conferences, Volume 26, 2012, Article number 01007, 10th International Conference on the Mechanical and Physical Behaviour of Materials Under Dynamic Loading, DYMAT 2012; Freiburg; Germany; 2 September 2012 through 7 September 2012*
[17] D. E. Eakins, N. N. Thadhani, J. Appl. Physics **100**, 073503 (2006)
[18] M. Martin, T. Shen, N. N. Thadhani, Materials Science and Engineering A **494**, 416 (2008)
[19] A. Mishra, M. Martin, N. N. Thadhani, B. K. Kad, E. A. Kenik, M. A. Meyers, Acta Materialia **56**, 2770 (2008)

EPJ Web of Conferences **94**, 01027 (2015)
DOI: 10.1051/epjconf/20159401027

One testing method of dynamic linearity of an accelerometer

Jing-Yu Lei[1], Wei-Guo Guo[1], Xue-Ming Tan[1], and Yun-Bo Shi[2]

[1]School of Aeronautics, Northwestern Polytechnical University, Xi'an, Shaanxi 710072, China
[2]School of Instrument and Electronics, North University of China, Taiyuan 030051, China

Abstract. To effectively test dynamic linearity of an accelerometer over a wide rang of 10^4 g to about 20×10^4g, one published patent technology is first experimentally verified and analysed, and its deficient is presented, then based on stress wave propagation theory on the thin long bar, the relation between the strain signal and the corresponding acceleration signal is obtained, one special link of two coaxial projectile is developed. These two coaxial metal cylinders (inner cylinder and circular tube) are used as projectiles, to prevent their mutual slip inside the gun barrel during movement, the one end of two projectiles is always fastened by small screws. Ti6-AL4-V bar with diameter of 30 mm is used to propagate loading stress pulse. The resultant compression wave can be measured by the strain gauges on the bar, and a half –sine strain pulse is obtained. The measuring accelerometer is attached on the other end of the bar by a vacuum clamp. In this clamp, the accelerometer only bear compression wave, the reflected tension pulse make the accelerometer off the bar. Using this system, dynamic linearity measurement of accelerometer can be easily tested in wider range of acceleration values. And a really measuring results are presented.

1. Introduction

In the early 80s of last century, Robert D [1] in ENDEVCO corporation, Togami [2] and Ueda [3–5] had introduced the operation system for accelerometer calibration in the range of 10,000 g to greater than 100,000 g. This system works based on a Hopkinson bar, and calibration of an accelerometer is based on a strain gauge measurement of the wave. Li et al. [6] also presents the modified Hopkinson bar system, this system can calibrate the accelerometer in a range from 5,000 g to 200,000 g. In this system, a half-sine acceleration pulse is obtained by impacting a parabolic pointed projectile. To achieve a wider range acceleration calibration, different shapes of projectiles have to be used.

According to the measuring method of dynamic linearity for the accelerometer presented by references [7,8], in dynamic linearity measurement, except each projectile is alone fired, two projectiles should impinge against the end surface of the Hopkinson bar at the same time as when it is fired alone. In reality, it is hard to achieve such perfect results because two projectiles synchronization is not easy obtained by using independent double gun barrels.

In this paper, a more easy measuring method has been proposed, that is, we base on such a fact, the acceleration pulse is always located in first half of a loading strain pulse in the Hopkinson shock test [1–6], and combining with the relation between projectile shape and strain wave, two coaxial projectiles (outer circular tube and inner cylinder) are always fired by a single gun barrel during whole operation. This replace the coaxial double gun barrel, and the synchronization impact on the end surface of the bar is effectively ensured, so it is much more convenient for dynamic linearity measurement.

2. Measurement of accelerometer dynamic linearity

Reference [7] and [8] have introduced one disclosed method of dynamic linearity measurement, and a schematic diagram of measuring test is presented as Fig. 1. In the figure, a coaxial double gun barrel must be used, when an outer projectile (circular tube) impinges against the end surface of incident bar, the impact acceleration pulse $a_1(t)$ of the accelerometer is generated. In the same way, $a_2(t)$ is obtained when an inner projectile impacts to the bar. And then, when these two projectiles are lunched together to impinge against the surface of incident bar at the same time as when it is fired alone, third superimposition acceleration can be obtained, that is $a_{1+2}(t)$. And the following expression is held as long as the strain of the pulse waves keep the linear elastic deformation of the bar material.

$$\varepsilon_{1+2}(t) = \alpha\varepsilon_1(t) + \beta\varepsilon_2(t) \quad (1)$$

$$a_{1+2}(t) = \alpha a_1(t) + \beta a_2(t) \quad (2)$$

where α and β are arbitrary constants, formulas (1) and (2) are foundation of calibration of accelerometer dynamic linearity.

In Fig. 1, the experimental setup takes a double gun barrels (inner and outer tube) to fire coaxial projectiles. When inner projectile is first fired, the output signal from an accelerometer is $g_1(t)$ (here, $g_1(t) = a_1(t)/10$), then outer projectile is $g_2(t)$; finally inner and outer projectile are simultaneously fired, output pulse is $g_{1+2}(t)$. Based on the definition of accelerometer dynamic linearity, the

Article available at http://www.epj-conferences.org or http://dx.doi.org/10.1051/epjconf/20159401027

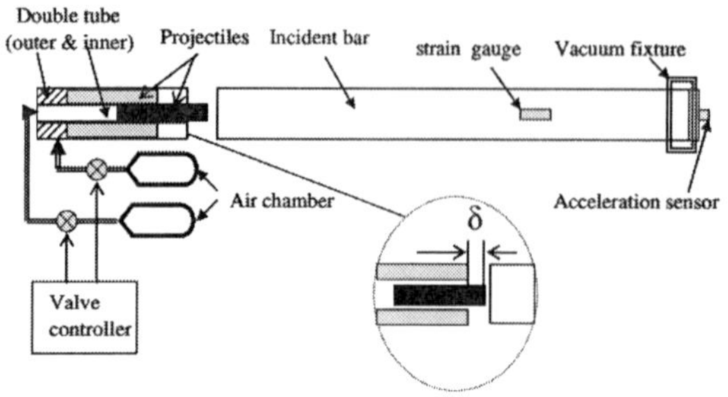

Figure 1. Accelerometer dynamic linearity by double gun barrels.

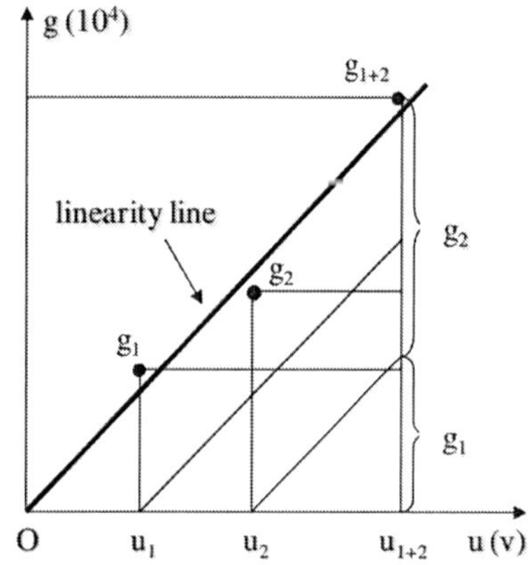

Figure 2. Diagrammatic sketch of dynamic linearity of accelerometer.

following formula can be established

$$g_{1+2}(t) = \alpha g_1(t) + \beta g_2(t). \tag{3}$$

Through computation, we can evaluate dynamic linearity of accelerometer. In testing procedure, $g_1(t)$ and $g_2(t)$ are obtained by independently firing oneself projectile (the inner or outer one alone); $g_{1+2}(t)$ is obtained by simultaneously firing inner and outer projectiles. If the max acceleration value only is used to evaluate an accelerometer, the dynamic linearity of accelerometer is obtained in Fig. 2.

The principal problem of this measuring method is, the synchronous impacting contact of two independent projectiles on the end of incident bar must be achieved in third testing step.

If there is a little synchronous difference, it is limited that this synchronous time difference is not allowed over $10\,\mu$s [7,8], but in practice we usually obtain the maximum displacement difference of the two strain pulse as shown in the Fig. 3. The δ_1 is usually more over $10\,\mu$s, which does not meet the requirement.

It is abstractly known that, in order to obtain $a_{1+2}(t)$, the outer and inner projectiles impact against the end surface of incident bar at the same time just as they are fired alone. In real tests, two projectiles are usually fired through valve controller, as shown in Fig. 1, and so through the independent air tubes, the gun barrels and air chambers etc., these will affect the moving synchronization of these

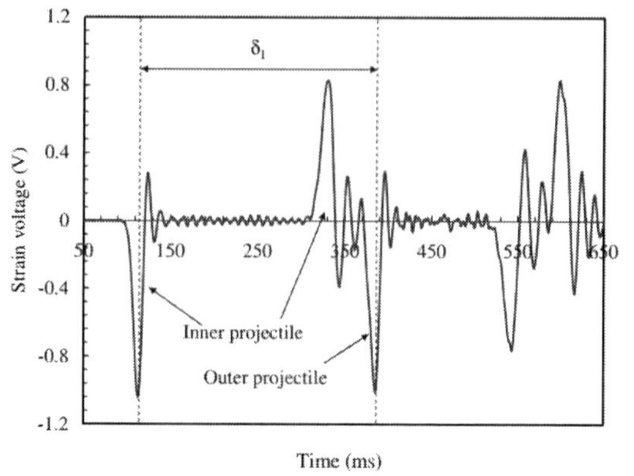

Figure 3. Strain pulse time difference under synchronous projectile firing.

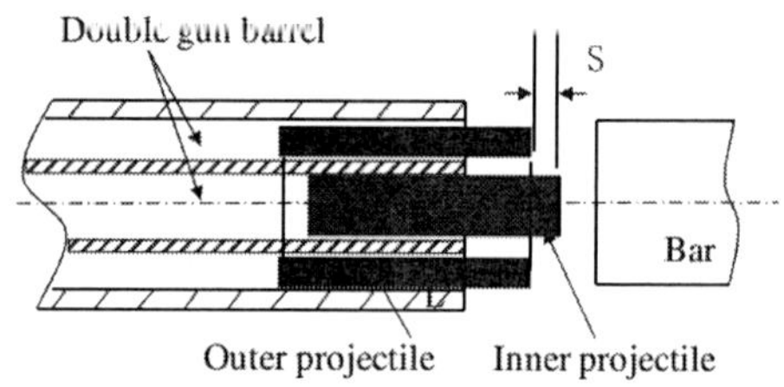

Figure 4. The schematic diagram of the displacement difference of two projectiles.

V (m/s)	△S(mm)
1	0.01
5	0.05
10	0.1
15	0.15
20	0.2
25	0.25
30	0.3
35	0.35

Figure 5. Max allowable displacement difference for two projectile impact.

two projectiles, and make it hard to synchronously impinge the end surface of incident bar under these uncertain factors.

Figures 4 and 5 further present a max allowable displacement difference if two projectiles synchronously impact in an allowable $10\,\mu$s time difference, e.g. a max

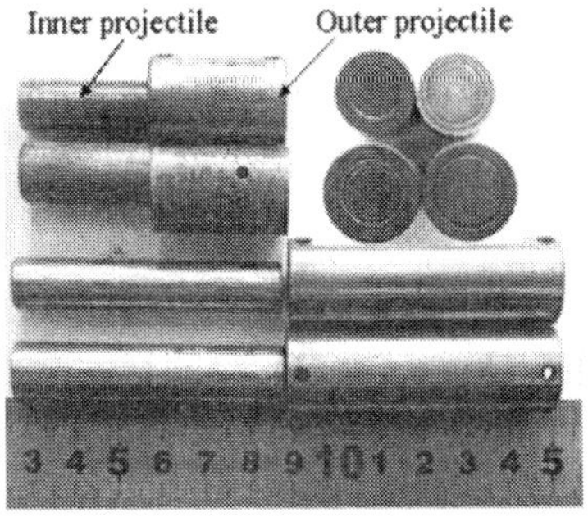

Figure 6. Pictures of the coaxial cylinder projectiles with different diameters and lengths.

allowable displacement difference is not over 0.1 mm ($\Delta S = 0.1$ mm) if the speed of two projectiles is about 10 m/s.

3. Modified measuring method

As known in the references 1 to 6, a wider range of acceleration values can be achieved by the single gun barrel device. However, in practice, the entire range of acceleration value up to about 200,000 g cannot be achieved by only one type projectile. Such two inevitable point are: (1) two types of projectiles have to be applied; (2) different types of projectiles can results in different acceleration pulse width, frequency components, etc.

Meanwhile, it is aware that, the acceleration pulse of the accelerometer is always located in the first half strain pulse on the bar. Based on this, we still take single barrel gun, but two projectiles (inner and outer projectiles) are simultaneously fired at any time, two coaxial cylinder projectiles in single barrel are illustrated in Fig. 6, and their contact faces are greased to reduce reaction friction force, while different length of projectiles can be used to obtain different wide acceleration pulse.

4. Real measuring demonstration

In this test, Ti-6Al-4V titanium rod was used as stress pulse propagating bar, it is 1200 mm long with 30 mm diameter. The projectiles are made into coaxial cylinders, the inner cylinder (inner projectile) and the circular tube (outer projectile, here its max diameter is 20 mm) can be the same cross section area, they can also be made into different section area. Two projectiles can be manufactured with different lengths, the length and cross section of projectiles depends on required g value and frequency width. Two sets real projectiles are shown in Fig. 6.

See Fig. 4, the overrun gap S of two coaxial cylinders projectiles is always set to about 4 mm, their rear-end is always fastened by small screws to prevent the relative slip of two projectiles when they moved inside the barrel. Whole experimental procedure is divided into three steps, in step 1, the inner projectile is fired at a fixed velocity of 10 m/s, the strain history $\varepsilon_1(t)$ on the bar and the accelerometer output pulse $g_1(t)$ are obtained; and then in step 2, the circular tube is fired at a same velocity of 10 m/s, $\varepsilon_2(t)$ and $g_2(t)$ are obtained, finally in step 3,

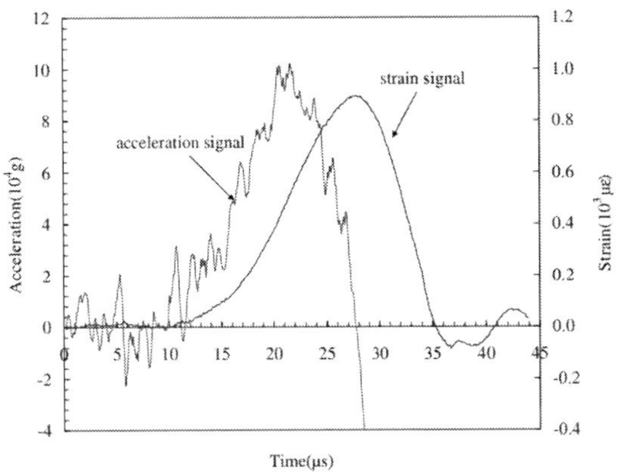

Figure 7. Acceleration and corresponding strain pulses on the bar by firing inner projectile.

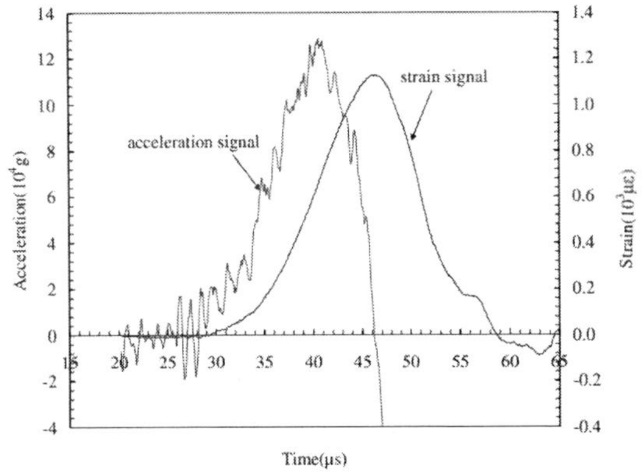

Figure 8. Acceleration and corresponding strain pulses on the bar by firing outer projectile.

two projectiles together without overrun gap are fired and impacted on the end of incident bar at the same time at a fixed velocity of 10 m/s, $\varepsilon_{1+2}(t)$ and $g_{1+2}(t)$ are obtained.

The actual operating procedure is as following:

First step, the inner projectile is ahead of outer projectile for 4 mm, and it first impacts on the bar at a fixed velocity of 10m/s, Acceleration and corresponding strain pulses obtained as shown in the Fig. 7. The corresponding max acceleration value is about 10.5×10^4 g.

Second step, the outer projectile is ahead of inner projectile for 4 mm, and it first impacts on the bar at a fixed velocity of 10 m/s, Acceleration and corresponding strain pulses obtained as shown in the Fig. 8. The corresponding max acceleration value is about 13×10^4 g.

Third step, the outer projectile is ahead of inner projectile for 0 mm, and they simultaneously impacts on the bar at a fixed velocity of 10 m/s, Acceleration and corresponding strain pulses obtained as shown in the Fig. 9. The corresponding max acceleration value is about 26×10^4 g.

In the Fig. 10, the acceleration pulses from first step and second step are added together, that is, an acceleration superimposition pulse, and corresponding strain superimposition are obtained. The corresponding superimposition acceleration value is about 23.5×10^4 g.

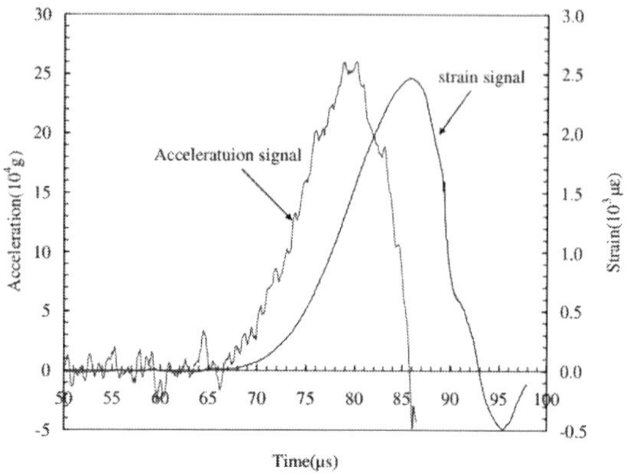

Figure 9. Acceleration and corresponding strain pulses on the bar by firing inner and outer projectiles at the same time.

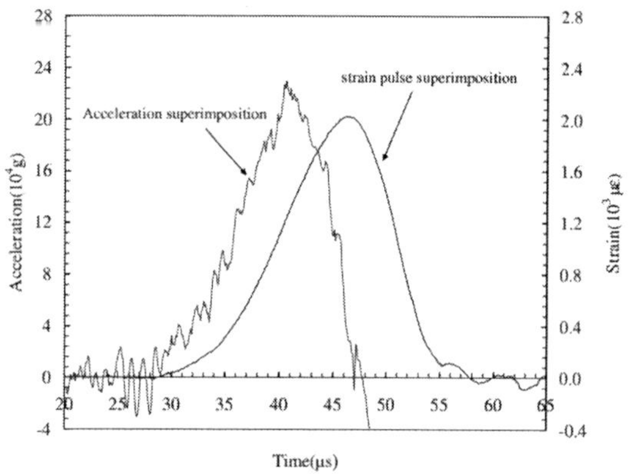

Figure 10. Acceleration and corresponding strain superimposition pulses from two projectile independent impacts on the bars.

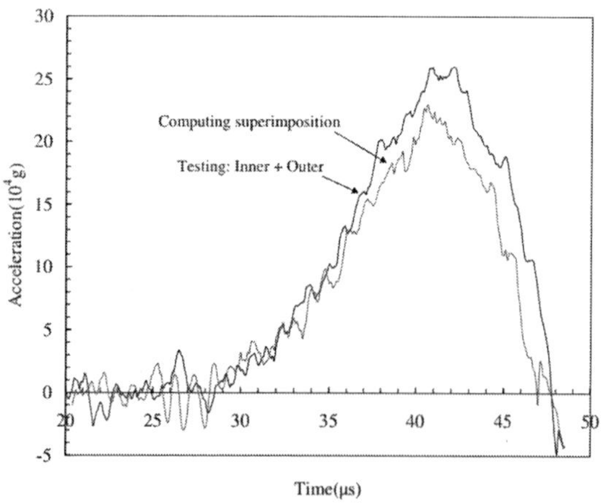

Figure 11. The comparison between testing acceleration pulse and computing superimposition pulses.

this is smaller 2.5×10^4 g than direct measurement value of Fig. 9.

In Fig. 11, the comparison between directly testing acceleration pulse and computing superimposition pulses

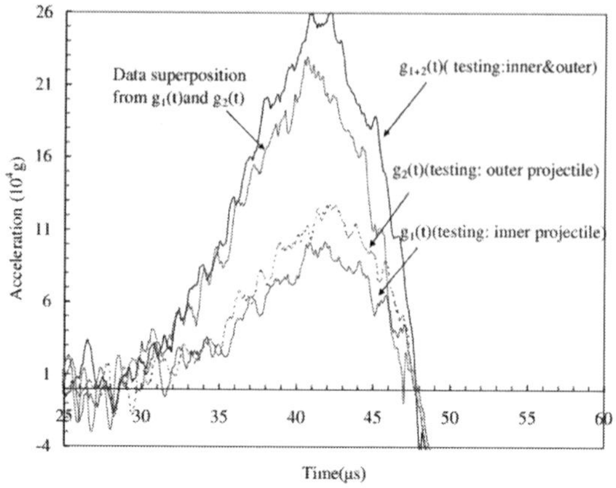

Figure 12. The comparison between testing acceleration pulses of two projectiles and computing superimposition pulses.

is presented, a good agreement of two pulses is obtained except their peak value difference.

In Fig. 12, all acceleration pulses are shown in the same figure, using the calculation method of dynamic linearity of an accelerometer as shown in Fig. 2, and using Eq. (3), the dynamic linearity differences of an accelerometer can be known at any acceleration value point.

5. Conclusions

In this work, to measure dynamic linearity of an accelerometer, a single barrel is taken to fire two coaxial cylinders at the any time and at a fixed velocity, the operation and control are easy. Using this method, it can correct a acceleration g value from order 10^4 g to about 20×10^4 g in accordance with standard procedure and also can measure dynamic linearity of the accelerometers by choosing different length and cross section area of the projectile. It should be remind that the dispersion on the bar, the front end shape of the coaxial projectiles (outer circular tube and inner cylinder) effect on the stain pulse, the measuring uncertainty etc.. Do not be further presented in the work due to the length limitation of the article.

This work is partly supported from the 111 project of China (B07050).

References

[1] Robert D. Still, Project Engineer, ENDEVCO. Albuquerque: International Instrumentation Symposium, Instrument Society of America (1983)

[2] T. C. Togami, W. E. Baker, M. J. Forrestal, Appl. Mech **63**, 353–356 (1996)

[3] K. Ueda, A. Umeda, in *Proceedings of the 9th International Conference on Experimental Mechanics, 1990*, p. 24

[4] K. Ueda, A. Umeda. Experimental Mechanics **33**, 228–233 (1993)
[5] Akira Umeda, Kazunaga Ueda. Method and Apparatus for Measuring Dynamic Response Characteristics of Shock Accelerometer [P]. United States Patent 5000030 (1991)
[6] Yulong Li, Weiguo Guo etc., Explosion and shock waves (in Chinese) **17**, 90–96 (1997)
[7] Akira Umeda, Method and Device for Measuring Dynamic Linearity of Acceleration Accelerometer [P]. United States Patent US2005/0160785 A1, (2005)
[8] CDV draft, Dynamic linearity measurement using an impact acceleration generator, IEC. Committee draft for vote, Project number IEC 60747-14-4 Ed. 1.0, 74–76 (2002)
[9] P. S. Follansbee, and C. Frantz, J. Eng. Mater. Technol **105**, 61–67 (1983)
[10] D. J. Frew, M. J. Forrestal, and W. Chen, Experimental

EPJ Web of Conferences **94**, 01028 (2015)
DOI: 10.1051/epjconf/20159401028

Crack initiation at high loading rates applying the four-point bending split Hopkinson pressure bar technique

Sebastian Henschel[a] and Lutz Krüger

Institute of Materials Engineering, TU Bergakademie Freiberg, Gustav-Zeuner-Str. 5, 09599 Freiberg, Germany

Abstract. Dynamic crack initiation with crack-tip loading rates of $\dot{K} \approx 2 \cdot 10^6\,\mathrm{MPa}\sqrt{\mathrm{m}}\mathrm{s}^{-1}$ in a high strength G42CrMoS4 steel was investigated. To this end, a previously developed split Hopkinson pressure bar with four-point bending was utilised. V-notched and pre-cracked Charpy specimens were tested. The detection of dynamic crack initiation was performed by analysing the dynamic force equilibrium between the incident and the transmission bar. Additionally, the signal of a near-field strain gauge and high-speed photography were used to determine the instant of crack initiation. To account for vibrations of the sample, a dynamic analysis of the stress intensity factor was performed. The dynamic and static analyses of the tests produced nearly the same results when a force equilibrium was achieved. Fracture-surface analysis revealed that elongated MnS inclusions strongly affected both the dynamic crack initiation and growth. Blunting of the precrack did not take place when a group of MnS inclusions was located directly at the precrack tip. Due to the direction of the elongated MnS inclusions perpendicular to the direction of crack growth, the crack could be deflected. The comparison with a 42CrMo4 steel without elongated MnS inclusions revealed the detrimental effect in terms of resistance to crack initiation. Taking the loading-rate dependency into consideration, it was shown that there was no pronounced embrittlement due to the high loading rates.

1. Introduction

For the investigation of dynamic fracture toughness properties at high loading rates, the split Hopkinson pressure bar principle is used, e.g. [1–5]. It enables the determination of forces and displacements on both the loading and support faces of the specimen [4,5].

Materials which exhibit linear-elastic behaviour under high loading rates are subject of intensive research [2,4–6]. However, investigation dealing with elastic-plastic behaviour are rare [7].

Dynamic crack initiation can be detected by a strain gauge in the near-field of the crack tip [8]. Furthermore, the strain gauge can be utilised to determine the stress intensity factor at the crack tip assuming a static calibration of the strain gauge [9].

In a study of a C45 low carbon steel, Klepaczko [10] observes a minimum of the plane strain fracture toughness at $10^4\,\mathrm{MPa}\sqrt{\mathrm{m}}\mathrm{s}^{-1}$. Furthermore, Klepaczko [11] points out that low strain hardening favours the formation of a minimum fracture toughness at high loading rates.

The aim of the present study is the determination of crack initiation at high rates of the stress intensity factor of $\dot{K} \approx 2 \cdot 10^6\,\mathrm{MPa}\sqrt{\mathrm{m}}\mathrm{s}^{-1}$ for a high-strength, tough and ductile G42CrMoS4 steel. To this end, a previously developed split Hopkinson pressure bar with four-point bending was utilised. V-notched and pre-cracked Charpy specimens were tested. Dynamic crack initiation was detected by a near-field strain gauge and by analysis of the force equilibrium. Furthermore, high-speed photography was utilised to evaluate the specimen loading.

[a] Corresponding author: sebastian.henschel@iwt.tu-freiberg.de

2. Materials and methods

2.1. Investigated steel

In this study, the resistance against crack initiation of the quenched and tempered cast steel G42CrMoS4 (1.7231) was evaluated. Heat treatment was performed by austenitising at 840 °C in a vacuum, followed by quenching in a stream of He. The samples were tempered at 560 °C in a N_2 atmosphere. After heat treatment, the microstructure of this steel consisted of tempered martensite. For comparison, a hot-rolled 42CrMo4 (1.7225) steel in the quenched and tempered state was tested.

The heat-treated samples were precracked up to $a_0/W \approx 0.5$ by utilisation of the crack length-dependent resonant frequency response. The precracking was started with a crack tip loading of $\Delta K = 18\,\mathrm{MPa}\sqrt{\mathrm{m}}$ at a stress ratio of $R = 0.1$. The size of the plastic zone in front of the crack tip was minimised by load reduction during the precracking procedure.

2.2. Apparatus

The set-up of the split Hopkinson pressure bar is shown in Fig. 1. Striker, incident bar, transmission bar and momentum trap were made of a high strength AA7075 aluminium wrought alloy. The incident bar and the transmission bar had the same cross sectional area (707 mm^2) due to the tubular shape of the transmission bar. Four-point bending of the specimen was achieved by two loading pins in the incident bar as well as in the transmission bar.

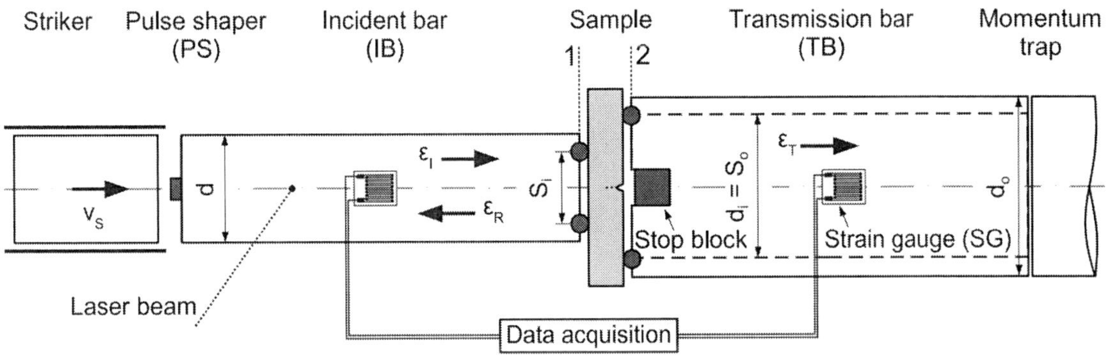

Figure 1. Schematic drawing of the four-point bending split Hopkinson pressure bar. Inner span $S_i = 20\,\mathrm{mm}$, outer span $S_o = 40\,\mathrm{mm}$.

The incident pulse was generated by the impact of the striker at the incident bar. Pulse shaping was applied to obtain a smooth incident pulse [12]. Due to the impedance mismatch at interface 1 (Fig. 1), the incident pulse (ε_I) was partially reflected (ε_R) and partially transmitted in the transmission bar (ε_T). Strain gauges (SG) at the mid-length of the bars were used to measure strain introduced by these pulses. Additionally, a laser system (Polytec OFV-525) was utilised to measure the surface velocity of the incident bar. The strain of the stress pulse was calculated by the measured surface velocity v_B and the sonic speed of the bar c_B [13]:

$$\varepsilon = v_B / c_B. \tag{1}$$

The small gauge length of the strain gauges (3.5 mm) in combination with a 1 MHz bandwidth amplifier (Tektronix AM502) and a high sampling frequency of 20 MHz enabled the measurement of small rise times.

Forces F_1, F_2 and displacements z_1, z_2 at the interfaces 1 and 2 were determined by

$$F_1(t) = F_I(t) + F_R(t) = \frac{\pi}{4} d^2 E_B (\varepsilon_I(t) + \varepsilon_R(t)) \tag{2}$$

$$F_2(t) = F_T(t) = \frac{\pi}{4}(d_o^2 - d_i^2) E_B \varepsilon_T(t) \tag{3}$$

$$z_1(t) = c_B \int_0^t \left(\varepsilon_I(t') + \varepsilon_R(t')\right) \mathrm{d}t' \tag{4}$$

$$z_2(t) = c_B \int_0^t \varepsilon_T(t') \mathrm{d}t'. \tag{5}$$

The modulus of elasticity of the bars $E_B = 72389\,\mathrm{MPa}$ was determined by ultrasonic methods. Force equilibrium was achieved when $F_1(t) = F_2(t)$, i.e. $\varepsilon_I(t) + \varepsilon_R(t) = \varepsilon_T(t)$. Hence, the relative displacement of the loading pins was then given by:

$$\delta(t) = z_1(t) - z_2(t) = -2c_B \int_0^t \varepsilon_R(t') \mathrm{d}t'. \tag{6}$$

Near-field strain gauges were utilised to detect the point of crack initiation, see Fig. 2. Furthermore, statically calibration of these strain gauges enabled the calculation of the force (F_{SG}) and the stress intensity factor ($K_{I,SG}$).

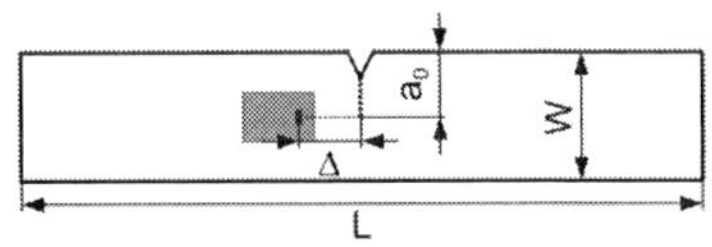

Figure 2. Near-field strain gauge used for the detection of crack initiation. Nominal sample dimensions: $W = 10\,\mathrm{mm}$, $a_0 = 5\,\mathrm{mm}$, $B = 10\,\mathrm{mm}$, $B_N = 8\,\mathrm{mm}$ (side-grooved sample), $L = 55\,\mathrm{mm}$.

The specimen loading was additionally analysed by means of high-speed photography. The high-speed camera Photron FASTCAM SA-Z enabled high frame rates of 330,000 fps in combination with a image size of 384× 96 pixels.

2.3. Calculation of the J integral

Under the assumption of quasi-static conditions, the J integral at the point of crack initiation was determined according to ISO 12135 [14]:

$$J = \frac{K^2(1-\nu^2)}{E} + \frac{2U_p}{B_N(W-a_0)}. \tag{7}$$

The quasi-static stress intensity factor K_I^{stat} was calculated for the four-point bend loading according to ASTM C 1421 [15]:

$$K_I^{stat} = \frac{F(S_o - S_i)}{BW^{3/2}} \cdot \frac{3(a_0/W)^{1/2}}{2(1-a_0/W)^{3/2}} \cdot f(a_0/W). \tag{8}$$

The plastic part of the energy U_p was determined from the force-displacement curve. During force equilibrium, F_1 and F_2 can be used to calculate K_I^{stat} and U_p.

Due to possible bending vibrations of the specimen, the dynamic stress intensity factor K_I^{dyn} was calculated according to Rokach [16–18]. The calculation of $K_I^{dyn}(t)$ involves both the force in the incident bar and in the transmission bar. An unsupported two-point loading by each bar was assumed. Finally, the stress intensity factor for four-point bending was obtained by superposition of the two separate loads.

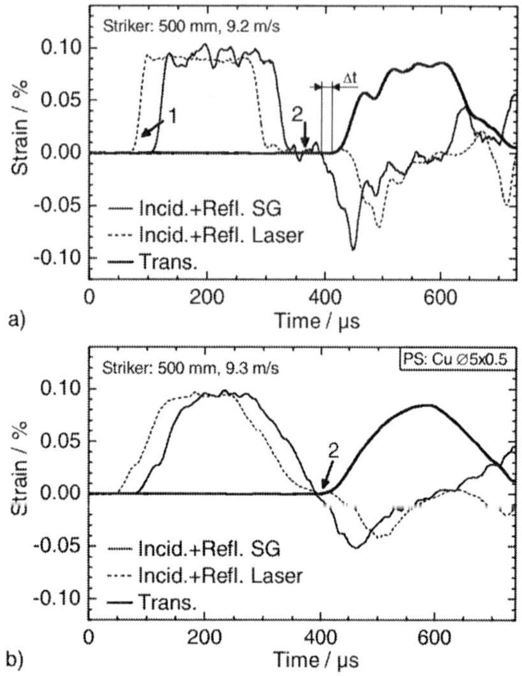

Figure 3. Strain history of the incident bar and the transmission bar measured by strain gauges at the mid-length of the bars (unnotched specimen). Additional application of laser measurement. a) unshaped pulse, b) shaped incident pulse.

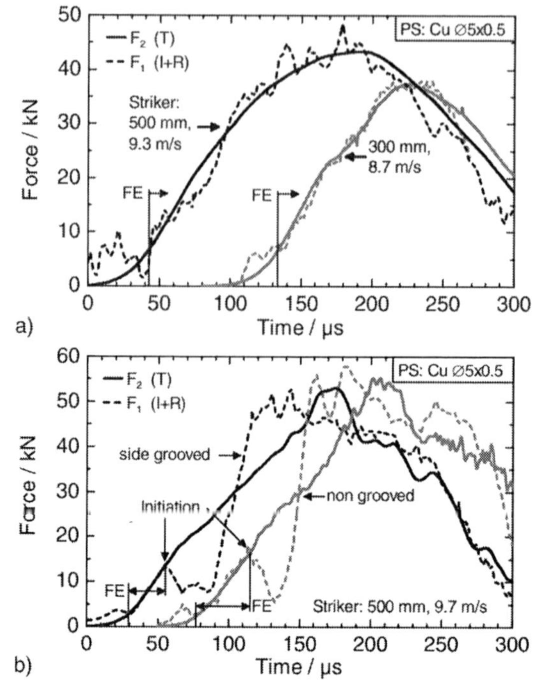

Figure 4. a) Force equilibrium (FE) is achieved after a short time. b) Loss of force equilibrium due to crack initiation. Signals are shifted for better visibility. Material: G42CrMoS4.

2.4. Fractography

Fracture surface analyses was performed by means of scanning electron microscopy (SEM). The SEM MIRA 3 XMU (TESCAN) enabled beam tilting and, therefore, the determination of height information of the fracture surface [19].

3. Results

3.1. Evaluation of laser measurement

Figure 3 shows stress pulses obtained from strain gauge and laser measurement. The offset in time (arrow 1) between the two measurements at the incident bar was defined by the distance between the laser measurement site and the position of the strain gauge. Using a pulse shaper and the strain gauge measurement, the end of the incident pulse interfered with the beginning of the reflected pulse (arrow 2). However, the beginning of the reflected pulse is of great importance for the calculation of F_1 and, therefore, the analysis of the force equilibrium at the early part of loading. The additional measurement site enabled the acquisition of separate pulses ε_I and ε_R also in the case of an intensively shaped pulse.

Both strain gauge measurement sites on the incident bar and the transmission bar have a distance of 750 mm to the loading pins. Hence, the offset in time between the start of the reflected pulse and the transmitted pulse was attributed to the time needed for the stress wave to propagate through the sample.

3.2. Force equlibrium and crack initiation

The applicability of the shaped incident pulse for achieving an early force equilibrium is shown in Fig. 4. After a short time of approximately 40 μs, force equilibrium was achieved in the case of unnotched samples (Fig. 4a). The force equilibrium was then maintained throughout the loading of the specimen. Due to a similar incident pulse generated by a different length of the striker (300 mm and 500 mm), the time to achieve the force equilibrium was independent of the striker length.

In the case of the cracked specimens (Fig. 4b), the force equilibrium was achieved at low forces. After further loading, the force/time history was characterised by a loss of force equilibrium. The force in the incident bar decreased rapidly. This was attributed to the increase in specimen compliance due to crack initiation. In order to compensate the force decrease due to compliance increase, an additional displacement of the incident bar would be necessary. However, the velocity of the incident bar was defined by the magnitude of the incident pulse. Hence, it was assumed that a considerably higher velocity of the incident bar results in the additional displacement needed to achieve further force increase. After the specimen came into contact with the stop block, there was an approximate force equilibrium.

The static calibration of the near-field strain gauge enabled the additional force measurement directly by the

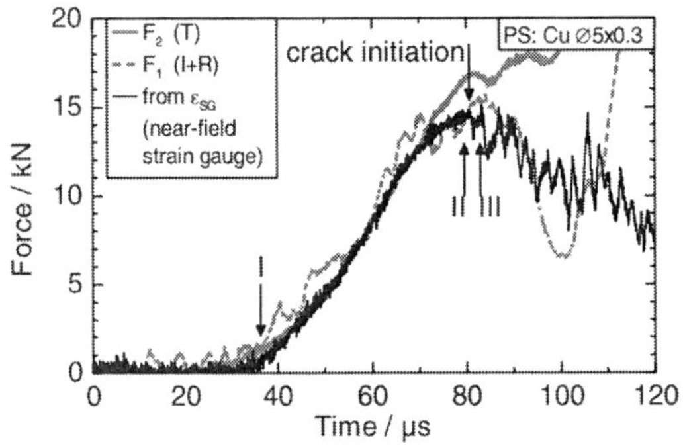

Figure 5. Force measurement by near-field strain gauge (ε_{SG}) equivalent to force measurement by the bars (Eqs. (2) and (3)). Material: 42CrMo4.

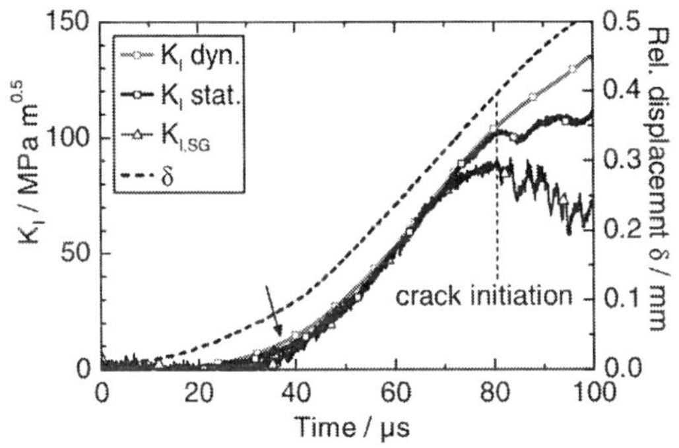

Figure 6. Almost no difference between K_I^{stat} (calculated with F_2) and $K_{I,SG}$ ($K_{I,SG} \sim F_{SG} \sim \varepsilon_{SG}$) during elastic sample behaviour. Good description of the crack-tip loading by K_I^{dyn} as calculated according to Rokach [16–18]. Material: 42CrMo4.

specimen. It was observed that the force measurement by the specimen essentially reflected the force measurement by the bars (Fig. 5). However, the beginning of the loading was characterised by an increase of forces measured by the bars whereas the force obtained from the specimen was approximately zero. This was attributed to the force disequilibrium introduced by stress wave propagation in the specimen. However, several reflections of the stress waves within the specimens were needed in order to load the sample and, therefore, the crack tip.

The calculation of the dynamic stress intensity factor revealed that static as well as dynamic evaluation of the specimen loading led to equal results if force equilibrium was achieved (Fig. 6). After crack initiation, the static and the dynamic calculation resulted in different stress intensity factors due to the force disequilibrium. Furthermore, the unknown crack extension rendered the calculation of K_I impossible.

The signal of the near-field strain gauge was analysed with respect to the point of crack initiation (Fig. 5). At the point of crack initiation, elastic strain energy was released due to the increase in specimen compliance. Hence, the measured strain was significantly reduced. Furthermore, the signal of the near-field strain gauge exhibited oscillations which were explained by reflections

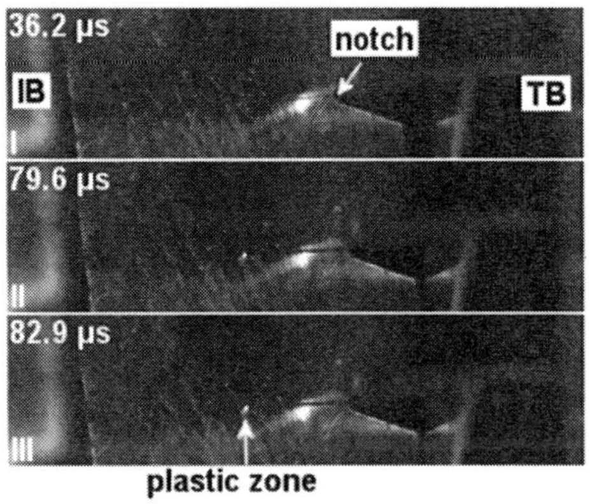

Figure 7. Evolution of plastic zone: I – beginning of loading, II – shortly before crack initiation, III – shortly after crack initiation. IB: incident bar, TB: transmission bar.

of the unloading wave within the specimen. One distinctive frequency component of the mixture of oscillations was, e.g. approximately 312 kHz. This frequency represents a shear wave which propagates along the length W and is reflected at the front and the back of the specimen.

The point of crack initiation was additionally investigated by means of high-speed photography. Figure 7 shows three frames which were shot at the times I, II and III marked in Fig. 5. It can be seen that the size of the plastic zone at the surface of the sample strongly increased after crack initiation. The crack initiation was not directly visible on the surface of the specimen. This was attributed to the low constraint at the free surface. In contrast, crack initiation was promoted in the centre of the specimen. Hence, the crack opening due to crack initiation in the centre then resulted in the strong increase in plastic zone size at the surface of the specimen.

The J-Integral was calculated at the point of crack initiation. The average crack resistance was 26 kJ/m^2 with a standard deviation of 6 kJ/m^2. The determination of the point of crack initiation was possible with an accuracy of approximately 1 μs. This time range results in a deviation of typically 4 kJ/m^2. Comparing this first estimation of dynamic crack initiation resistance at loading rates of about $2 \cdot 10^6$ MPa$\sqrt{m}$s^{-1} with results of tests performed at smaller loading rates no pronounced embrittlement was observed. Quasi-static tests with $\dot{K} = 2$ MPa$\sqrt{m}$s^{-1} and dynamic tests at $8 \cdot 10^4$ MPa$\sqrt{m}$s^{-1} resulted in values of the crack initiation toughness of approximately 45 and 40 kJ/m^2, respectively. According to Klepaczko [10], a minimum of toughness in the range of 10^3 to 10^5 MPa$\sqrt{m}$s^{-1} is observed for the steel C45. Hence, further investigations are necessary to characterise a possible minimum for the present steel.

3.3. Fractography

A fracture surface of the material is shown in Fig. 8. It was observed that the blunting of the precrack depended strongly on the local microstructure, i.e. the local inclusion distribution, in front of the crack tip. Regions which were characterised by an agglomeration of non-metallic

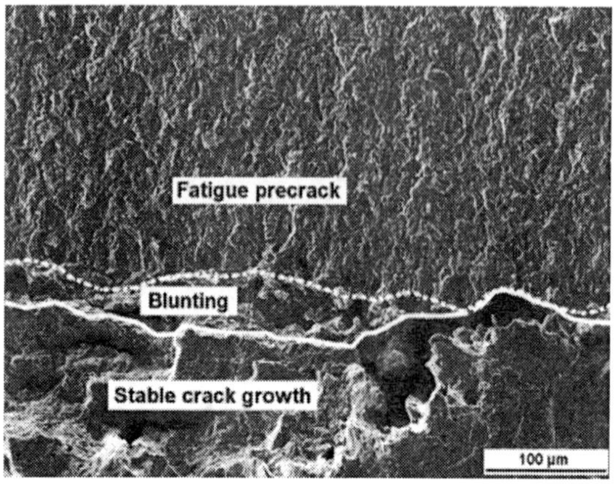

Figure 8. Crack initiation was strongly affected by the local distribution of non-metallic inclusions. Material: G42CrMoS4.

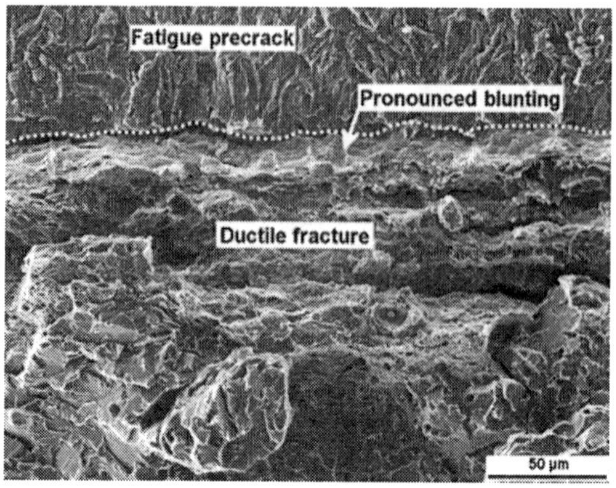

Figure 10. Blunting of precrack before crack initiation. Material: 42CrMo4.

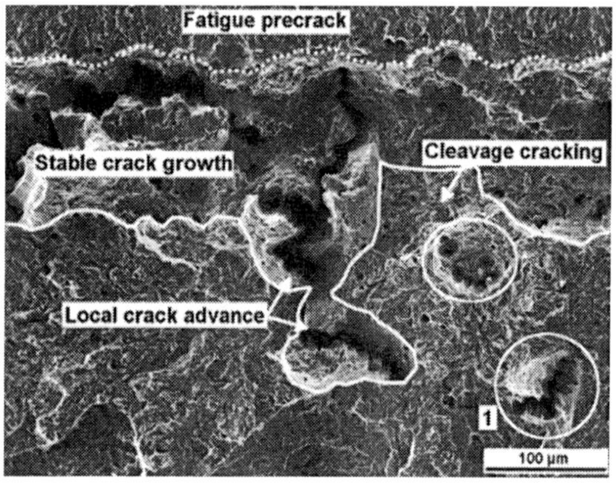

Figure 9. Local crack growth due to chains of non-metallic inclusions. Material: G42CrMoS4.

inclusions exhibited no crack tip blunting. Hence, the crack initiation resistance was not uniquely defined.

As shown in Fig. 9, local crack growth was another effect of the inhomogeneous distribution of the non-metallic inclusions. Chains of non-metallic inclusions resulted in locally enhanced crack growth rates. As a consequence, crack tip shielding decreased the crack driving force which eventually stopped the crack growth. Furthermore, ductile voids were observed which were surrounded by cleavage facets (label 1 in Fig. 9). The cleavage facets were attributed to the final cracking at low temperatures. However, the ductile voids were formed during the actual test which was confirmed by the oxidised surface due to heat tinting. Hence, there was a connection between the apparently separate ductile fracture surfaces. This connection was explained by secondary cracks.

The direction of the elongated MnS inclusion was mainly perpendicular to the plane of the fracture surface. As a result, crack path deflection occurred. However, this principally toughness-increasing mechanism was not considered to be effective due to the promotion of void initiation, growth and coalescence by the MnS inclusions.

In Fig. 10, the fracture surface of the 42CrMo4 steel used for comparison is shown. The fracture surface exhibited blunting of the precrack due to the relatively low inclusion content of this steel. Furthermore, crack path deflection was less distinctive than in the steel with the elongated MnS inclusions. No regions with local crack advance were observed.

4. Conclusions

In this paper, the resistance against crack initiation in a high-strength G42CrMoS4 cast steel was evaluated at very high loading rates of $2 \cdot 10^6\,\mathrm{MPa}\sqrt{\mathrm{m}}/\mathrm{s}$. The tests were performed by the application of a specially designed four-point bending split Hopkinson pressure bar. The main conclusions can be drawn as follows:

- The steel G42CrMoS4 exhibited crack-tip blunting (locally) followed by stable crack growth. Stable crack growth was achieved by ductile fracture. Crack path deflection was locally observed and was caused by elongated manganese sulphide inclusions.
- A significant embrittlement effect due to the elongated MnS inclusions was observed. Crack path deflection as a possible toughness-increasing mechanism was considered to be ineffective due to the damaging effect of the non-metallic inclusions.
- Due to irregularly developed crack-tip blunting, crack initiation was considered not to be a single step. Regions that exhibited non-metallic inclusions at the tip of the fatigue precrack showed no blunting. However, areas which were essentially characterised by the deformation of the metallic matrix did show blunting of the crack tip.
- The statistical distribution of the non-metallic inclusions along the crack front led to a scattering of the resistance against crack initiation. A slight decrease of toughness compared to J_i determined at smaller loading rates was observed. Further investigations of the loading rate dependence of the crack initiation toughness are necessary.

- When dynamic force equilibrium was achieved, the description of sample loading by the static and the dynamic stress intensity factors led to similar results.

The authors thank the German Research Foundation (DFG) for its financial support of the investigations at the Collaborative Research Center 920, subproject C05. The support of Markus Radajewski during the experimental work and the help of Thomas Meier (VKT Video Kommunikation GmbH) with the high-speed camera are greatly appreciated.

References

[1] K. Tanaka, T. Kagatsume, Bull. JSME **23**, 1736 (1980)
[2] L. Rubio, J. Fernández-Sáez, C. Navarro, Exp. Mech. **43**, 379 (2003)
[3] T. Weerasooriya, P. Moy, D. Casem, M. Cheng, W.W. Chen, J. Am. Ceram. Soc. **89**, 990 (2006)
[4] F. Jiang, K.S. Vecchio, Metall. Mater. Trans. A **38**, 2896 (2007)
[5] J.T. Foster, W.W. Chen, V.K. Luk, Eng. Fract. Mech. **78**, 1264 (2011)
[6] Z. Xu, Y. Li, Acta Mech. Sin. **27**, 559 (2011)
[7] S.F. Zhu, C. Yang, C.H. Guo, F. Jiang, Key Eng. Mat. **577–578**, 517 (2014)
[8] T. Demler, Ph.D. thesis, Universität Stuttgart (1990)
[9] J.F. Kalthoff, S. Winkler, W. Böhme, J. Phys. Colloques **46**, 179 (1985)
[10] J.R. Klepaczko, J. Eng. Mater. Techn. **104**, 29 (1982)
[11] J.R. Klepaczko, Int. J. Impact Eng. **3**, 191 (1985)
[12] S. Henschel, L. Krüger, K. Mandel, M. Radajewski, *Studie zur Impulsformung an Split-Hopkinson-Aufbauten*, in *15. Werkstofftechnisches Kolloquium*, edited by B. Wielage (2012), pp. 422–427
[13] G.T. Gray, in *ASM-Handbook*, edited by H. Kuhn, D. Medlin (ASM, Ohio, 2000), Vol. 8, pp. 462–476
[14] *ISO 12135:2002* (2002), Metallic materials – Unified method of test for the determination of quasistatic fracture toughness
[15] *ASTM C 1421-10* (2010), Standard test methods for determination of fracture toughness of advanced ceramics at ambient temperature
[16] I.V. Rokach, Fatigue Fract. Engng. Mater. Struct. **21**, 1007 (1998)
[17] I.V. Rokach, Fatigue Fract. Engng. Mater. Struct. **21**, 1015 (1998)
[18] I.V. Rokach, P. Łabędzki, Int. J. Fracture **160**, 93 (2009)
[19] A. Weidner, T. Mottitschka, H. Biermann, S. Henkel, Eng. Fract. Mech. **108**, 294 (2013)

EPJ Web of Conferences **94**, 01029 (2015)
DOI: 10.1051/epjconf/20159401029

A novel technique for performing symmetric Taylor impact

S.M. Walley[a], N.E. Taylor, D.M. Williamson, and A.P. Jardine

SMF Fracture and Shock Physics Group, Cavendish Laboratory, University of Cambridge, CB3 0HE, UK

Abstract. A novel spring tensioned wire cradle arrangement has been developed to hold the target rod lightly but firmly in place when performing rod-on-rod (symmetric Taylor) impact in a vacuum. In addition, a soft capture system has been designed and used to decelerate both rods while reducing the chance of them colliding a second time. High-speed photography was used to obtain the profile of a pure aluminium target rod as a function of time. Photon Doppler Velocimetry (PDV) was also deployed to record the velocity of the rear of the target rod as a function of time. Voiding in the interior of recovered rods was investigated non-destructively using X-ray tomography. The data provides comprehensive validation information for predictive constitutive models.

1. Introduction

The ideal symmetric Taylor impact experiment would be performed by firing two identical rods simultaneously at each other at the same velocity. Since there is a risk of wrecking both gun barrels, rod-on-rod impact is normally employed where one rod is fired against a stationary target rod [1,2]. In an ideal rod-on-rod impact, there would be no external constraint on the lateral deformation of the rods i.e. the impacting rod would be in free flight and the target rod would be levitated. It is also required for a valid experiment that the impact be planar and coaxial.

These requirements mean that (a) the impacting rod has to be as close to the end of the gun-barrel at impact as possible or even partially within the barrel (but protruding sufficiently that the section within the barrel does not plastically deform during impact) and (b) the target rod has to be mechanically supported so that the rods can be accurately aligned before the shot is performed. The mechanical support should put as little constraint upon the deformation as possible. In this paper, we describe an experimental setup which meets these criteria.

2. Experimental design

Previous researchers have used sabots to deliver the projectile rod to the target [3–8]. However, a sabot that only contains the rear section of a rod [9] will need to hold the rod firmly. This will change the boundary conditions for elastic waves reverberating up and down its length. So we decided to make the projectile rods good sliding fits within the gun barrel used. One aim was that the deformation and contact time should be determined as far as possible only by the mechanics of the impact between the two rods. Impacts were performed in a rough vacuum of about 40 mbar to minimise the effects of airblast ahead of the impacting rod due to residual gas in the barrel or from propellant gas leaking past the projectile rod.

[a] Corresponding author: smw14@cam.ac.uk

The main requirements for the design of the target rod holder were the following: (i) the mounting system should be sufficiently rigid to hold the weight of the rod and stabilise it against airblast from the gun; (ii) it should put as little mechanical constraint as possible on the dynamic expansion of the target rod; (iii) the mount should allow high-speed photography to be performed so that the development of the deformation can be followed; (iv) the mount should allow soft recovery of both rods so that their final lengths and profiles can be measured.

After some experimentation, a spring-tensioned wire cradle arrangement was found to work well (Fig. 1). The wires used had a PTFE coating giving them a low frictional interaction with the target rod. The cradle was mounted on top of a stack of three Thorlabs optical adjustment stages, namely x-, z-, and pitch-yaw. These allow sufficient adjustment of the position and orientation of the target rod so that it can be made both collinear and coaxial with a 25 cm long alignment rod of the same diameter partially projecting from the gun barrel.

The velocity of the impacting rod was measured in the space between the end of the gun barrel and the impact face of the target rod using a pair of light gates, each consisting of a diode laser and photodiode. The second laser beam was positioned close to the impact face. The distance between the end of the gun barrel and the impact face was set so that approximately 25% of the impacting rod (i.e. about 15 mm for a 63 mm long rod) was still in the barrel at the moment of impact. Having a short length of the moving rod still in the barrel at impact improves the planarity of impact without risking plastic deformation of the rod interacting with the end of the gun barrel. The output of the photodiode closest to the point of impact was used to trigger the camera and PDV system.

3. Results

Selected frames from a high-speed photographic sequence are presented in Fig. 2 for symmetric Taylor impact

Article available at http://www.epj-conferences.org or http://dx.doi.org/10.1051/epjconf/20159401029

Figure 1. Photograph of wire cradle for holding a target rod prior to mounting in the target chamber. The rod shown is 63.3 mm long and 12.66 mm diameter. In use, the projectile rod enters and strikes the target rod from the right hand side (see Fig. 2).

at 334 m s^{-1} of 99.999% pure aluminium supplied by Goodfellow Metals. The interframe time was set to 10 μs on the Vision Research Phantom 711 camera used. The time at which impact occurred (137.2 μs after trigger) was determined by digitising the coordinates of the ends of the rods in each frame and interpolating to the point of impact.

It can be seen that during the first stage of impact, the sides of the target rod are concave near the impact face. However, about 80 μs after impact (frame (d)) the rod begins to bulge outwards plastically some distance back from the impact face. This phenomenon has been seen before for copper [4] and is characteristic of fcc materials. In the full sequence, the rods can visibly be seen to start to separate about 200 μs after impact (340 μs after trigger) and continue to draw apart as the sequence continues (see Fig. 2(g)).

Photon Doppler Velocimetry (PDV) [10, 11] was also used to record the motion of the rear of the target rod (Fig. 3). The rear of the rod can be seen to accelerate smoothly but at an increasing rate. However, the PDV trace presented in Fig. 3 ends abruptly before the two rods separated. This is because the target rod struck the PDV's optical fibre (see Fig. 2(f)).

It is instructive to consider what velocity we would expect for two extremes of behaviour. If the system behaved as a 'Newton's cradle' [12, 13] (i.e. perfectly

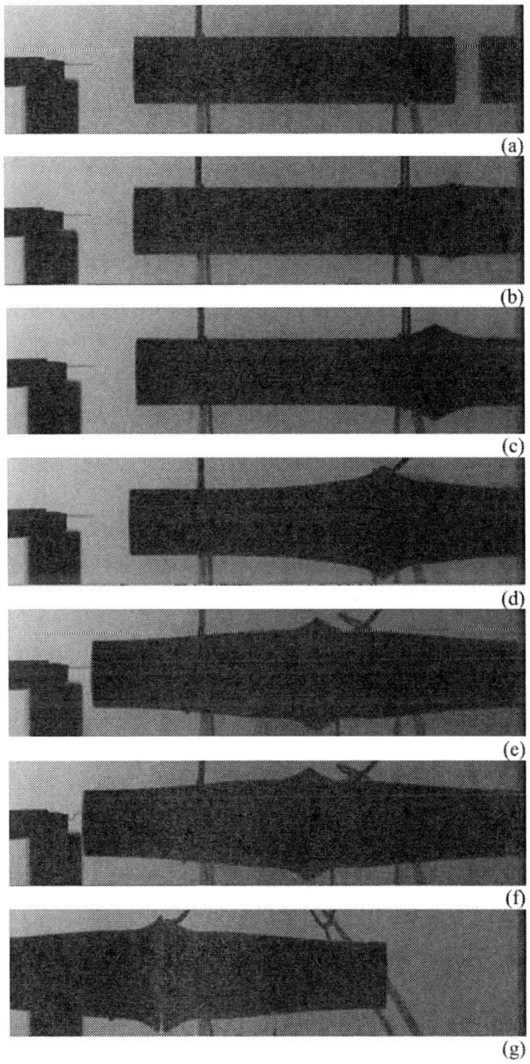

Figure 2. Selected frames from the high-speed photographic sequence of a 63.3 mm long, 12.66 mm diameter pure aluminium rod being impacted at 334 m s^{-1} by another identical pure aluminium rod. The camera used was a Vision Research, Phantom 711. The PDV probe can be seen to the left behind the target rod. Note that the block below the PDV optical fibre probe is not in the impact plane. Times of each frame from trigger: (a) 120 μs (b) 140 μs (c) 160 μs (d) 220 μs (e) 300 μs (f) 310 μs (g) 480 μs.

elastically), the projectile rod would come to a dead stop and the target rod would depart at the impact velocity of the projectile rod. On the other hand, if the two rods plastically deformed and joined together, they would continue on at half the impact velocity. Since the observed mechanical behaviour is intermediate between these two extremes, the projectile rod should continue moving at less than the impact velocity while the target rod exits the impact at slightly more than half the impact velocity. This was confirmed by another experiment on pure aluminium performed at 151 m s^{-1} where PDV showed that the final

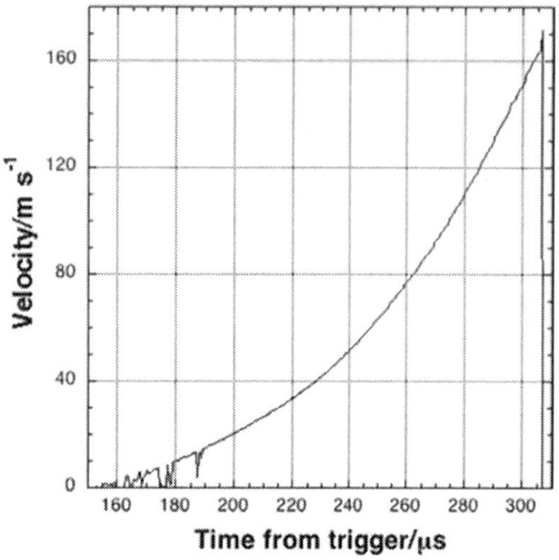

Figure 3. Velocity-time plot of the rear of target rod obtained using PDV. The sudden drop at 306.4 μs corresponds to when the target rod struck the optical fibre. This occurred before the two rods separated.

velocity of the target rod was 83.6 m s^{-1} as compared to 151/2 = 75.5 m s^{-1} if the two rods had stuck together.

It is clear from the argument just given and from Fig. 2(g) that the velocity difference between the two rods is sufficiently small that they only gradually separate from each other. Thus a soft capture method had to be developed to ensure they did not collide a second time during deceleration. After some experimentation, a tube containing approximately 40 cm of expanded polystyrene beads (diameter 2–5 mm) backed by rags to a similar depth was found to work in that both rods were recovered in almost the same condition that they left the impact zone (see Fig. 4). There is a small visible asymmetry in the deformation (see Fig. 4(b)) in that the impact surface of the target rod has a depression in the centre while the projectile rod has a corresponding bulge. One possible explanation is that the cylindrical surface of the projectile rod was frictionally heated during acceleration in the gun barrel producing a cylindrical annulus of softened metal in the projectile rod.

Rod profiles (Fig. 5) were obtained from the high-speed photographic sequence shown in Fig. 2. The profiles of both the upper and lower edges of the target rod are separately plotted in order to check for symmetry of deformation. These profiles are at present being compared with numerical predictions based on constitutive models.

It can be seen from Figs. 5(a) and 5(b) that the plastic deformation is initially close to symmetric, the profiles being concave near the impacted end. Figure 5(a) represents the deformation 2.8 μs and Fig. 5(b) 62.8 μs after impact. Figures 5(c) and 5(d) show that at a later stage the deformation becomes more complex with a second bulge forming between about 6 and 30 mm from the impact face. At the same time the difference between the upper and lower profiles became greater.

X-ray tomography was used to examine the recovered rods for voids (Fig. 6). Such voids had previously been

(a)

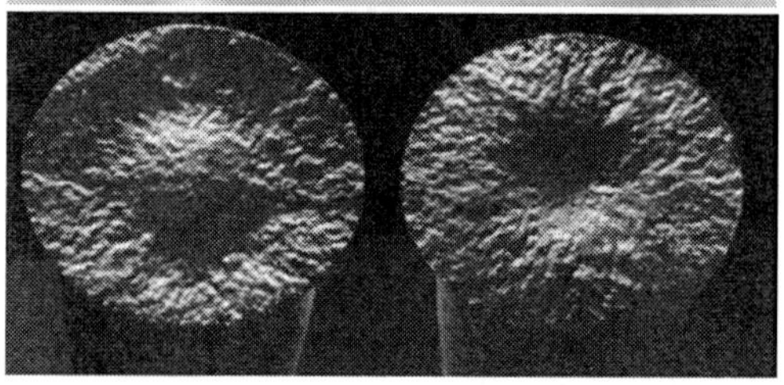
(b)

Figure 4. Photographs of pure aluminium rods soft-recovered after rod-on-rod impact at 334 m s^{-1} (same rods as shown in Fig. 2). (a) Side view: the right hand rod is the projectile rod; (b) Top view: the left hand rod is the projectile rod.

seen in symmetric Taylor impact [5]. It can be seen from Fig. 6 that voids were indeed formed. The voids show up as black dots close to the impact surface in Fig. 6(a). However, they are small and off-axis (see Fig. 6(b)). They might therefore easily have been missed if the rods had been physically sectioned.

4. Summary

(i) A novel wire-cradle arrangement has been designed and used to hold the target rod for symmetric Taylor impact experiments. The wire-cradle exerts a very small constraint on the deformation of the target rod.

(ii) Two diagnostics were applied to study the dynamic deformation of the target rod: high-speed photography and Photon Doppler Velocimetry (PDV).

(iii) A soft-recovery system was designed to ensure that the two rods did not collide a second time during deceleration. The system consists of a tube partially filled with expanded polystyrene foam beads backed with rags.

(iv) Recovered rods were examined using X-ray tomography to determine whether voiding had occurred. Aluminium rods subjected to an impact velocity of 334 m s^{-1} clearly contained voids close to the axis near the impact face.

(v) PDV showed that for the aluminium rods the rears of the target rods accelerated smoothly at an increasing rate

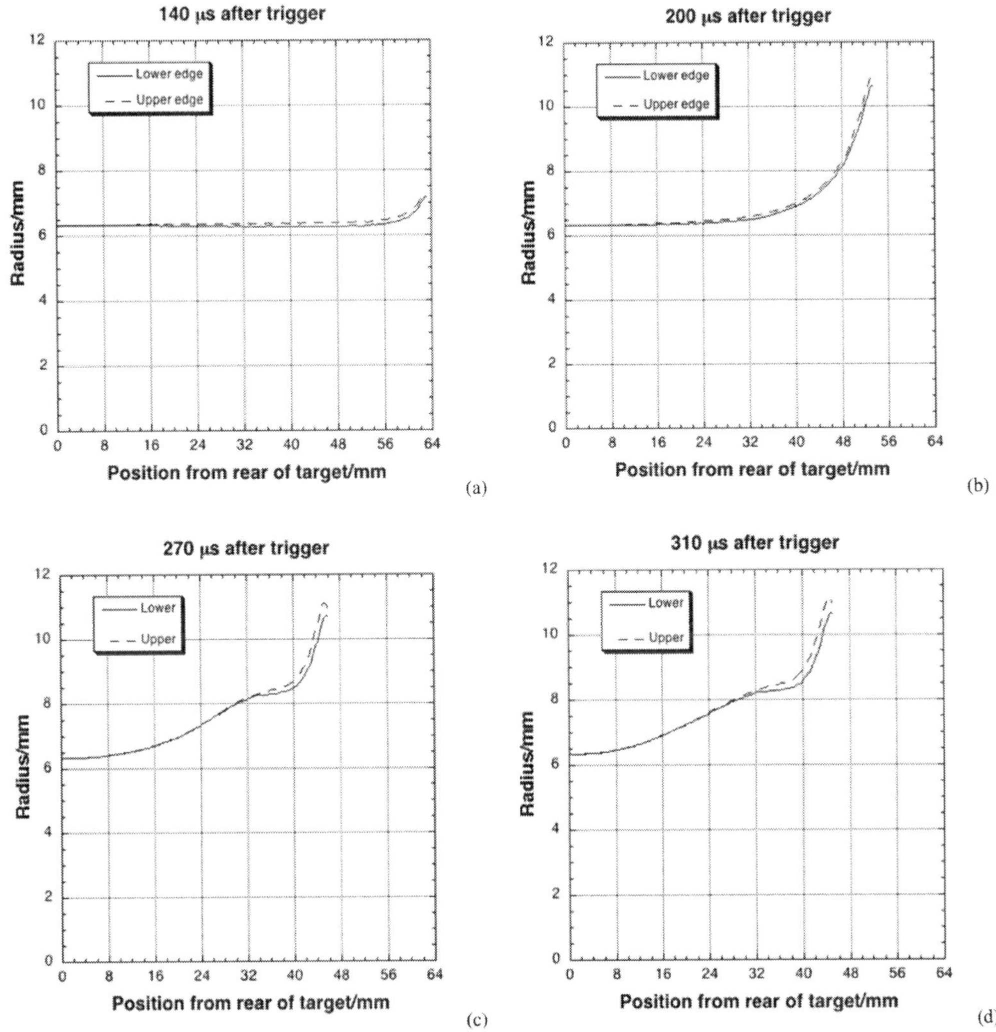

Figure 5. Profiles of selected frames from the high-speed photographic sequence of the impact shown in Fig. 2. Impact took place 137.2 μs after trigger. Distances (radii) to the upper and lower edges of the rod are measured from the centre line of the rod.

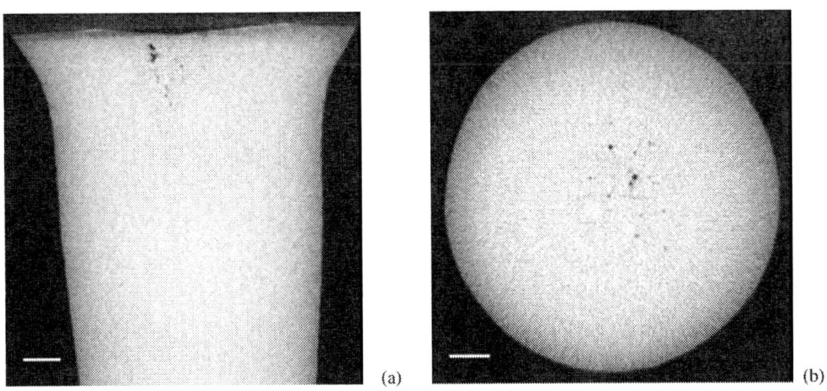

Figure 6. X-ray tomographs of aluminium projectile rod after impact with a stationary aluminium rod at 334 m s^{-1}. (a) Side view. (b) End view close to the impact surface. Scale bars represent 3 mm.

(characteristic of the deformation process) to a maximum velocity slightly greater than that for a perfectly plastic collision.

This research was supported by grants from the British Ministry of Defence through QinetiQ Fort Halstead. The authors would like to thank I.G. Cullis, P.D. Church and P.J. Gould for their interest.

References

[1] D.C. Erlich, D.A. Shockey and L. Seaman, *Shock Waves in Condensed Matter – 1981*, pp. 402–406 (American Institute of Physics, New York, 1982)

[2] D.C. Erlich, *Metals Handbook. Vol. 8 (9th edn.)*, pp. 203–207 (American Society of Metals, Metals Park, Ohio, 1985)

[3] H. Couque, J. Phys. IV France **10** (Pr. 9) 179–184 (2000)

[4] S.M. Walley, P.D. Church, R. Townsley and J.E. Field, J. Phys. IV France **10** (Pr. 9) 69–74 (2000)

[5] D.J. Chapman, D.D. Radford, M. Reynolds and P.D. Church, Int. J. Fract. **134**, 41–57 (2005)

[6] H. Couque, G. Nicolas and C. Altmayer, Int. J. Impact Engng **34**, 412–423 (2007)

[7] L.C. Forde, W.G. Proud and S.M. Walley, Proc. R. Soc. A **465**, 769–790 (2009)

[8] L.C. Forde, S.M. Walley, M. Peyton-Jones, W.G. Proud, I.G. Cullis and P.D. Church, *Proc. 9th Int. Conf. on the Mechanical and Physical Behaviour of Materials under Dynamic Loading (DYMAT 2009)*, pp. 1245–1250 (EDP Sciences, Les Ulis, France, 2009)

[9] W. Mocko, J. Janiszewski, J. Radziejewska and M. Grazka, Int. J. Impact Engng **75**, 203–213 (2015)

[10] M. Durand, P. Laharrague, P. Lalle, A. Le Bihan, J. Morvan and H. Pujols, Rev. Sci. Instrum. **48**, 275–278 (1977)

[11] O.T. Strand, D.R. Goosman, C. Martinez, T.L. Whitworth and W.W. Kuhlow, Rev. Sci. Instrum. **77**, 083108 (2006)

[12] D.R. Lovett, K.M. Moulding and S. Anketell-Jones, Eur. J. Phys. **9**, 323–328 (1988)

[13] V. Ceanga and Y. Hurmuzlu, J. Appl. Mech. **68**, 575–583 (2001)

EPJ Web of Conferences **94**, 01030 (2015)
DOI: 10.1051/epjconf/20159401030

Energy absorption at high strain rate of glass fiber reinforced mortars

Luigi Fenu[1,a], Daniele Forni[2,3], and Ezio Cadoni[3]

[1]Department of Civil Engg, Environment Engg, and Architecture, University of Cagliari, 09123 Cagliari, Italy
[2]Department of Structural, Geotechnical and Building Engineering – Politecnico di Torino, 10129 Torino, Italy
[3]DynaMat Laboratory, University of Applied Sciences of Southern Switzerland, 6952 Canobbio, Switzerland

Abstract. In this paper, the dynamic behaviour of cement mortars reinforced with glass fibers was studied. The influence of the addition of glass fibers on energy absorption and tensile strength at high strain-rate was investigated. Static tests in compression, in tension and in bending were first performed. Dynamic tests by means of a Modified Hopkinson Bar were then carried out in order to investigate how glass fibers affected energy absorption and tensile strength at high strain-rate of the fiber reinforced mortar. The Dynamic Increase Factor (DIF) was finally evaluated.

1. Introduction

Even if the use of glass fibers to reinforce concrete was first proposed in Russia before the 2-nd World War, the industrial use of Glass Fiber Reinforced Concrete (GRFC) dates back to the 1970's, after the development by Pilkington Corporation in 1967 of a suitable glass formulation to produce Alkali-Resistant glass fibers containing zirconia [1].

Nowadays glass fibers with improved alkali resistance allow a structural use of GFRC [1–3], otherwise limited by the embrittlement of glass fibers caused by the alkaline environment of the Portland cement paste.

Therefore, GFRC was extensively used in the industrial production of prefabricated elements, especially precast façade panels. Nowadays, an alternative solution to GFRC is basalt fiber reinforced concrete, that can similarly limit cracking and absorb impact energy [4].

While many articles describe the static behaviour of GFRC, its dynamic behaviour has been much less studied. A reference study on toughness and impact tests on GFRC panels was carried out by Mobasher and Shah [5].

Glinicki et al. [6] studied impact on GFRC plate specimens through drop weight instrumented test device that allowed to detect the maximum impact load, the energy absorbed up to the maximum load and the energy absorbed up to total failure, thus obtaining an impact-to-static energy absorption of 1.7–1.8 for the GFRC plate elements considered.

The impact behaviour of facade panels made of GFRC was investigated by Enfedaque et al. [7] by shooting steel spheres with high velocity on square samples of different GFRC panels, and by then calculating energy absorption as the kinetic energy difference of the projectile before and after impact.

Yldirim et al. [8] studied the impact behavior of different Fiber Reinforced Concretes (FRC) including GFRC. By using practically the same drop-weight method of ACI 544 2R-89 [9], impact test was performed time after time on cubical FRC samples with side 100 mm until failure occurred. Glass fibers were shown to be as effective as steel fibers and more effective than polypropylene fibers to prevent first cracking. While samples reinforced with steel fibers needed a much higher number of impact tests until failure occurred with respect to samples reinforced with both glass and polypropylene fibers, the latter ones needing almost the same number of impacts to failure. Adding glass fibers to steel fibers was very effective to delay failure, better than adding polypropylene fibers.

Sangeetha [10] studied the favourable effect of additives such as superplasticizer, air retaining agent and retarder on impact strength of GFRC plates.

In this paper the effect of adding glass fibers on the dynamic behaviour of cementitious mortars was investigated. Dynamic tensile tests at high strain rate were performed by means of a Modified Tensile Hopkinson Bar device. Also, reference static tests to evaluate flexural, tensile, and compression strength of the mortars were carried out.

2. Experimental procedure

Standard Portland cement (CEM I, 52.5 R in accordance with EN 197-1 [11]) and standard sand, as prescribed by the EN 196-1:2005, were used [12]. A reference cement mortar (water/cement ratio: 0.5) was prepared.

Fiber reinforced specimens were prepared through adding glass fibers to the cementitious mixture. Fiber content was 3 and 5%. Fibers with length 12 mm and diameter 14 μm were used. Immediately after mixing, the mixtures were introduced into the steel mould to manufacture the specimens, whose surface-were smoothed and covered by means of a polyethylene film. Water evaporation during the first hours was thus avoided, and the specimens were then cured at 20 °C and 95% R.H.

In accordance with EN 196-1:2005 [12], the flexural strength of prismatic specimens ($40 \times 40 \times 160\,mm^3$) was

[a] Corresponding author: lfenu@unica.it

Article available at http://www.epj-conferences.org or http://dx.doi.org/10.1051/epjconf/20159401030

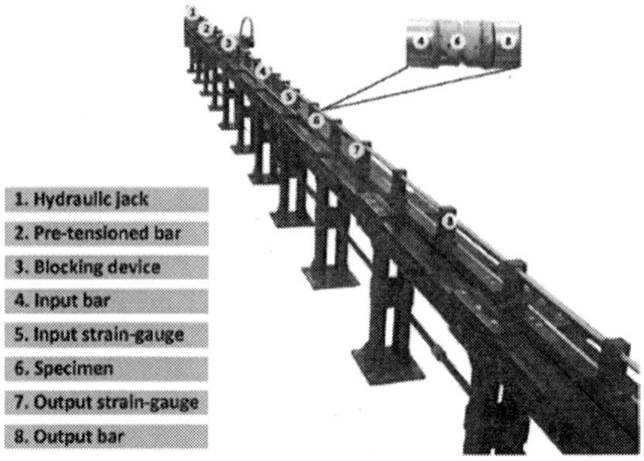

Figure 1. The Modified Hopkinson Tensile Bar used in this experimental research.

evaluated by means of the three point bending test. Then, compressive strength tests were carried out on the two prism halves obtained from the bending test.

After the same curing period, prismatic specimens were cored to obtain cylindrical specimens (h = d = 20 mm, h/d = 1) in order to perform dynamic test by means of a Modified Hopkinson Bar device of the DynaMat Laboratory (SUPSI) in Lugano.

The MHB device consisted of two circular aluminium bars, called input and output bars, both 20 mm in diameter but with different lengths of 3 m and 6 m, respectively. The sample was glued between these bars using a bi-component epoxy resin. The input bar was connected to a high strength steel pretension bar (6 m in length and 12 mm in diameter), that was used as pulse generator. The way of performing the test and a deep description of the MHB device used in this research is widely reported in [13]. The Hopkinson bar device is considered as a profitable technique in order to characterize materials behaviour at high strain rate, as proofed by a large amount of papers published in this field [14–16].

Thereafter, the stress, the strain and the strain-rate in the sample can be derived from Eqs. (1), (2) and (3):

$$\sigma_E(t) = E_0 \frac{A_0}{A} \varepsilon_T(t) \tag{1}$$

$$\varepsilon_E(t) = -\frac{2C_0}{L} \int_0^t \varepsilon_R(t) dt \tag{2}$$

$$\dot{\varepsilon}(t) = -\frac{2C_0}{L} \varepsilon_R(t) \tag{3}$$

where, L is the specimen length, A_0 is the cross-sectional area of output and input bars, A is the initial cross-sectional area of the specimen gauge length portion, E_0 is the elastic modulus of the bars, $C_0 = (E_0/\rho)^{1/2}$ is the bar elastic wave speed, ρ is the bar density and t is time.

In order to obtain reliable data for the evaluation of the energy absorption at high strain-rate, at least three tests for each material were performed.

Table 1. Flexural tensile and compressive strength of reference and fiber reinforced mortars from static tests.

	Flexural tensile strength MPa	Compressive strength MPa
Reference mortar	6,57 ± 0,13	70,20 ± 1,65
Glass fiber 3%	7,78 ± 0,31	57,58 ± 2,44
Glass fiber 5%	8,37 ± 0,69	50,46 ± 2,13

3. Results and discussion

The values of static flexural and compressive strength are summarized in Table 1, that shows that tensile strength was significantly increased by glass fibers, thus meaning that an effective bond between fibers and mortar was developed. Compressive strength of the reference mortar was instead significantly higher than that of the GFRC samples, meaning that the fiber reinforced samples behave as if fibers were inclusions scattered throughout the mortar.

Dynamic tests at high strain-rate were performed using the MHB device ranged between 500 and 600 GPa/s.

The study was conducted with the aim of investigating the capability of glass fibers of dissipating energy in the reinforced concrete of structures subjected to exceptional actions like blasts and collisions, where impact loading occurs at high strain-rate.

Through carrying out tensile tests at high strain-rate on both mortars reinforced with glass fibers and on the reference mortar, the influence of the addition of fibers on energy absorption and tensile strength of samples subjected to high impact loading was studied.

In order to achieve accurate stress-strain diagrams at high strain-rate, stresses and displacements were detected at high frequency (1 Msample/s), thus allowing to calculate energy absorption and get the maximum stress with good precision.

Tables 2–4 report the experimental results for each test carried out on samples made of glass fiber reinforced mortars and of the reference one. In particular, the Stress-Rate is the slope of the stress flow evaluated in function of time in the elastic regime, Max Tensile Stress is the highest value obtained for each test while Fracture strain, Fracture time and Displacement at fracture are the corresponding strain, time and displacement values respectively. Failure time is the time where the stress goes to zero. Eventually, Total energy is the energy represented by the area under the stress-crack opening displacement curve.

Figure 2 allows to compare the behaviour of the mortar reinforced with 3% and 5% glass fiber with that of the reference mortar.

It can be observed that the addition of 3% and even more of 5% of glass fibers (samples G3% and G5%) led to a better post- peak behaviour, with energy absorption much higher than in the reference mortar (Fig. 2a). The improved post-peak behaviour is also highlighted by the much higher failure time of both G3 and G5 mortars (Tables 3 and 4) with respect to that of the reference mortar. It can be also noted that tensile strength at high strain-rate resulted increased by the addition of 5% glass fibers, while was shown to be almost the same of that of the reference mortar

Table 2. Results for the reference mortar.

Sample	Stress Rate	Max Tensile Stress	Fract. strain	Fract. time	Failure time	Displ. at fract.	Failure displ.	Total energy
	GPa/s	MPa	‰	μs	μs	μm	μm	J/m^2
RM_01	556	12,2	0,30	32	44	6	25	177
RM_02	639	12,7	0,33	32	41	7	20	151
RM_03	476	12,4	0,32	35	46	6	25	177
RM_04	550	9,8	0,26	33	62	1	79	249
Average	555	11,8	0,30	33	48	5	37	188
STD	66	1,4	0,03	1	9	3	28	42

Table 3. Results for the mortar reinforced with glass fibers (3%).

Sample	Stress Rate	Max Tensile Stress	Fract. strain	Fract. time	Failure time	Displ. at fract.	Failure displ.	Total energy
	GPa/s	MPa	‰	μs	μs	μm	μm	J/m^2
G3%_01	873	15,7	0,18	30	58	4	48	195
G3%_03	496	14,1	0,33	36	272	6	867	1260
G3%_04	412	10,8	0,35	34	218	7	681	815
G3%_04	553	10,5	0,52	35	167	10	518	947
Average	584	12,8	0,34	33	179	7	528	804
STD	201	2,5	0,14	3	91	3	351	447

Table 4. Results for the mortar reinforced with glass fibers (5%).

Sample	Stress Rate	Max Tensile Stress	Fract. strain	Fract. time	Failure time	Displ. at fract.	Failure displ.	Total energy
	GPa/s	MPa	‰	μs	μs	μm	μm	J/m^2
G5%_01	470	14,2	0,58	35	222	12	686	1536
G5%_02	480	12,9	0,41	28	237	8	778	1214
G5%_03	566	11,6	0,34	32	241	7	760	909
Average	505	12,9	0,44	32	233	9	741	1220
STD	52	1,3	0,12	4	10	3	49	314

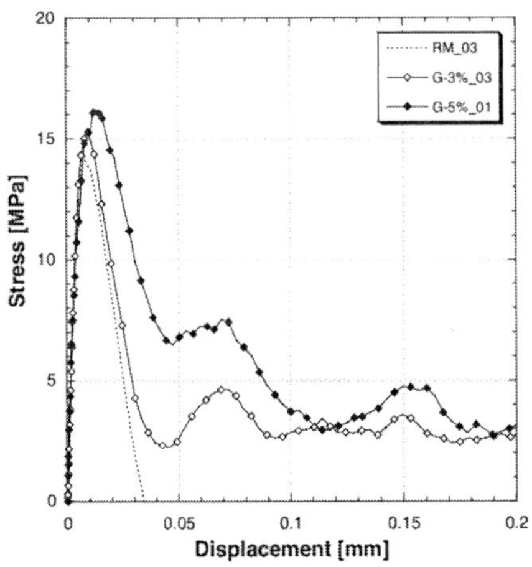

Figure 2. Stress as a function of displacement for reference and glass reinforced mortars.

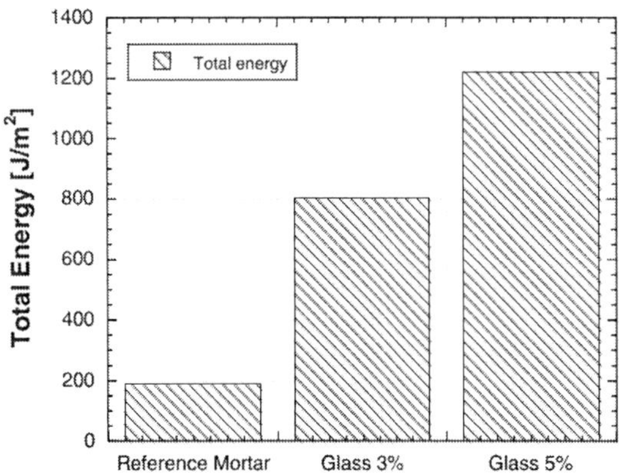

Figure 3. Comparison of total energy for different fiber content in glass.

(or even slightly lower) when the percentage of fibers was 3%.

Figure 3 allows to easier compare one another the values (reported in Tables 2–6) of total energy of the reinforced mortar and the reference one.

Figure 4 shows the images (obtained from the optical microscope) of the plan view and the lateral view of the failure surface of samples G3%_03. It can be noted that glass fibers of sample G3%_03 were cut and debonded by tension failure occurred under dynamic conditions.

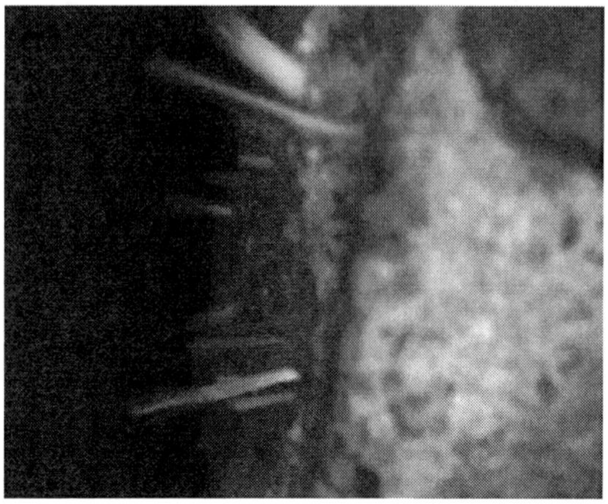

Figure 4. Images (obtained from the optical microscope) of lateral view of the failure surface of sample G3%_03.

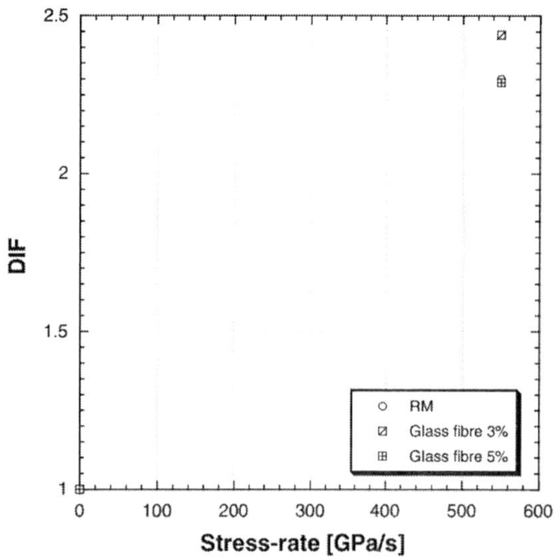

Figure 5. Dynamic Increase Factor vs. Stress-rate.

Finally, also the Dynamic Increase Factor (DIF) was evaluated (Fig. 5) as the ratio between the dynamic strength and the static strength of the material. For this aim, by using the same specimens geometry used for high strain-rate tensile tests, also direct static tensile tests (0.5 MPa/s) were performed.

Testing three specimens for each mortar RM, G3%, G5%, their tensile strength resulted 5.11 ± 0.0 MPa, 5.23 ± 0.5 MPa, 5.24 ± 1.2 MPa, respectively.

The DIF 2.30 for the reference mortar is a typical expected value for this material [17–19]. The lowest fiber addition (3%) led to a DIF slight increase for mortar G3% (2.44), while fiber addition until 5% brought on DIF values of 2.29 for mortars G5%, thus practically the same of the reference mortar. Therefore, DIF was shown to be little affected by fiber addition, and to vary little with fiber content.

It can be also noted a low DIF decrease with the fiber content increase from 3% to 5% of glass fibers. This means that, under dynamic conditions, the increase of fiber content was ineffective to improve micro-crack bridging [20,21].

Nevertheless, as seen before, comparison with plain mortars shows that the addition of glass fibers, while improving energy absorption under dynamic loading, only slightly increases tensile strength with respect to the reference mortar.

4. Conclusions

The dynamic behaviour in tension of mortars reinforced with glass fibers has been investigated. The experimental study has allowed to study the effect of the addition of glass fibers to a cementitious mortar on its dynamic behaviour.

From the dynamic tensile tests at high strain-rate carried out by means of a Modified Hopkinson Bar device, the following conclusions have been gathered:

- static flexural strength of the mortar was highly increased by the addition of glass fibers.
- dynamic tensile strength was significantly increased by the addition of 5% glass fibers, while was only slightly increased by the percentage of 3%.
- the addition of glass fibers to the cementitious mortar improved its post-peak behaviour under dynamic tension loading.
- fracture energy resulted highly increased by the addition of 5% glass fibers (7.2 times the fracture energy of the reference mortar), but was significantly increased also by a fiber content of 3% (4.6 times that of the reference mortar).
- DIF of mortars reinforced with glass fibers was practically the same as that of the reference mortar.

References

[1] J.P.J.G. Ferreira, F.A.B. Branco, Exp Techniques. **31**(3), 64–73 (2007).
[2] D. Cian, B. Della Bella, Structural Applications of GRC for Precast Floors. In: Proceedings of the 12th Int. GRCA Congress, Dublin Camberley (UK), October, 2001. p. 41–52.
[3] F. Branco, J. Ferreira, J. Brito, J. Santos. 'Building Structures with GRC. In: Proceedings of CIB World Building Congress, Wellington (NZ), April, 2001. p. 1–11.
[4] D. Asprone, E. Cadoni, F. Iucolano, A. Prota, Cem. Conc. Comp. **53**(10), 52–58 (2014).
[5] B. Mobasher, S.P. Shah, ACI Mat. J. **86**, 448–58 (1989).
[6] M.A. Glinicki, A. Vautrin, P. Soukatchoff, J. Francois-Brazier, Cem Concr Compos. **16**(4), 241–51 (1994).
[7] A. Enfedaque, D. Cendón, F. Gálvez, V. Sánchez-Gálvez, Eng Fail Anal. **18**(7), 1621–1908 (2011).
[8] S. Taner Yildirim, E. Ekinci Cevdet, F. Findik, Russ J Nondestruct+ **46**(7), 538–546, (2012).
[9] ACI 544 2R-89. American Concrete Institute (1989, Reapproved 1999).

[10] P. Sangeetha. J Eng Res & Stud. **2**(2), 36–40 (2011).
[11] EN 197-1. Cement – Part 1: Composition, specifications and conformity criteria for common cements. European Committee for Standardization 2011.
[12] EN 196-1. European Committee for Standardization (2005).
[13] E. Cadoni, Rock Mech Rock Eng. **43**, 667–676 (2010).
[14] E. Cadoni, L. Fenu, D. Forni, Constr. Build. Mat. **35**, 399–407 (2012).
[15] D. Asprone, E. Cadoni, A. Prota, G. Manfredi, ASCE J. Comp. Constr. **13**(6), 558–564 (2009).
[16] L. Coppola, E. Cadoni, D. Forni, A. Buoso, Appl. Mech. Mat. **82**, 190–195 (2011).
[17] D. Asprone, E. Cadoni, A. Prota, Struct. J. **106**, 106–113 (2009).
[18] D. Asprone, E. Cadoni, A. Prota, ACI Struct J **106**(4), 523–9 (2009).
[19] B. Erzar, P. Forquin, Exper Mech, **50**(7), 941–955, (2010).
[20] A. Caverzan, E. Cadoni, M. di Prisco, Int. J. Imp. Engng. **45**, 28–38 (2012).
[21] A. Caverzan, E. Cadoni, M. di Prisco, Mech Mat **59**, 87–109 (2013).

EPJ Web of Conferences **94**, 01031 (2015)
DOI: 10.1051/epjconf/20159401031

First application of the 3D-MHB on dynamic compressive behavior of UHPC

Ezio Cadoni[1,a], Matteo Dotta[1], Daniele Forni[1,2], Gianmario Riganti[1], and Carlo Albertini[3]

[1]DynaMat Laboratory, University of Applied Sciences of Southern Switzerland, 6952 Canobbio, Switzerland
[2]Department of Structural, Geotechnical and Building Engineering – Politecnico di Torino, 10129 Torino, Italy
[3]Dynalab-Impact Technologies, 21014 Laveno Mombello, Italy

Abstract. In order to study the dynamic behaviour of material in confined conditions a new machine was conceived and called 3D-Modified Hopkinson Bar (3D-MHB). It is a Modified Hopkinson Bar apparatus designed to apply dynamic loading in materials having a tri-axial stress state. It consists of a pulse generator system (with pre-tensioned bar and brittle joint), 1 input bar and 5 output bars. The first results obtained on Ultra High Performance Concrete in compression with three different mono-axial compression states are presented. The results show how the pre-stress states minimize the boundary condition and a more uniform response is obtained.

1. Introduction

Loading rate has effects on brittle material fracturing mechanism and mechanical properties including strength and absorbed energy.

The rate dependent behaviour of the brittle material in compression may be influenced by many factors including porosity, water content, boundary condition at the interface (specimen – bar) and confining pressure.

Understanding the dynamic response of Ultra High Performance Concrete (UHPC) has important applications on engineering, including protective structures and critical infrastructures [1,2].

The materials in a real structure [3,4] are usually subjected to a service load and often are also subjected to confinement. Dynamic loads (such as explosion, earthquake or impact) are loads applied to the structures together with the existing in situ stresses. Therefore, the test conditions should be as similar as possible to the actual conditions. This is the reason why a mono-, bi-, or tri-axial static stress state is applied before to the impact loading.

In order to study the dynamic behaviour of material in confined conditions a new machine was conceived and called 3D-Modified Hopkinson Bar (3D-MHB). It is a Modified Hopkinson Bar apparatus designed to apply dynamic loading in materials having a tri-axial stress state. It consists of a pulse generator (the pulse is generated by energy stored in a pre-tensioned bar and relaxed by the rupture of a fragile bolt), 1 input bar, and 5 output bars. The output bars are also providing static stresses to create a static tri-axial compression at required level.

Waves are measured by strain gauges on the input and all 5 output bars, to provide stress and strain.

The main innovation of the 3D-MHB apparatus is in the possibility to study the dynamic effects in confined materials (important for rocks, concrete, etc.). In fact, since 3 independent pairs of pressure bars will be used, the lateral confinement effect to the dynamic strength of materials can be investigated. The confinement effect is very crucial to brittle material failure. The confinement effect in dynamic conditions at high strain rates has not been experimentally examined largely due to the limitation of the existing equipment. This equipment was proposed in [5] and developed in the frame of the R'Equip program of the Swiss National Sciences Foundation [6]. This is innovative equipment and is the first piece of such kind of research equipment worldwide. The equipment was designed for a comprehensive use in future researches on dynamics of materials.

The theme tackled in this paper is the presentation of the first results obtained on UHPC in dynamic compression with three different mono-axial compression states.

2. Materials

The UHPC studied in this first application was the matrix of a commercial UHPFRC product [7]. It was a cementitious composite obtained mixing cement, water, quartz sand, micro-silica, and super plasticizers. The UHPFRC is usually obtained by reinforcing this composite with an amount of high strength steel fibres. The tests were carried out on cylindrical specimens having a diameter of 30 mm and height of 30 mm.

The quasi-static compressive strength was measured on three specimens by means of an electro-mechanical universal machine having 680 kN maximum load capacity obtaining a compressive strength of $f_c = 172 \pm 9$ MPa. The UHPFRC with 3% of fibre content had a compressive strength of 212 MPa.

3. The 3D-MHB apparatus

The 3D-MHB is a Hopkinson bar apparatus able to apply a quasi-static tri-axial stress before the dynamic test. The basic concept is shown in Fig. 1. It consists of one

[a] Corresponding author: ezio.cadoni@supsi.ch

Article available at http://www.epj-conferences.org or http://dx.doi.org/10.1051/epjconf/20159401031

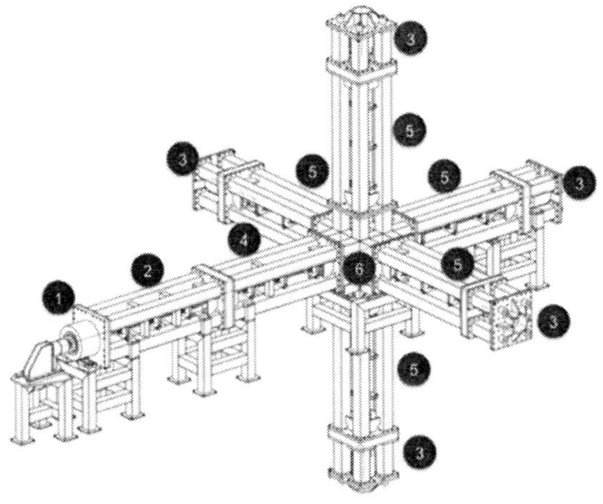

Figure 1. 3D-MHB apparatus scheme.

Table 1. Dimension of the 3D-MHB.

Element	dimension [mm]	area [mm^2]	length [mm]
Pre-tensioned bar (round)	$\varnothing = 56.5$	2507	1750
Input bar (square bar)	L=50	2500	2200
Output bar (square bar)	L = 50	2500	2100

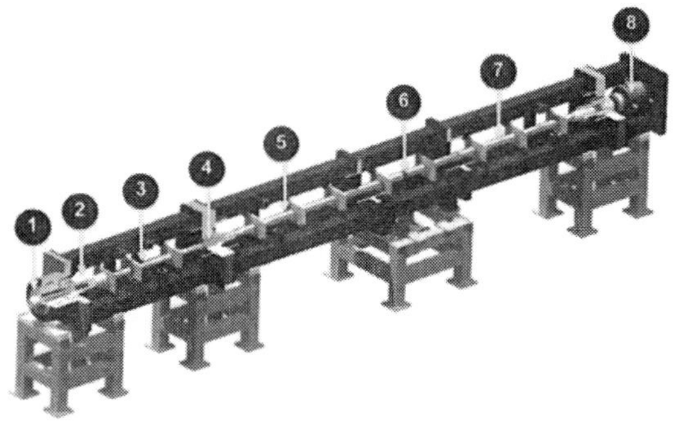

Figure 2. Experimental set-up for mono-axial testing.

Figure 3. 3D-MHB set-up view.

hydraulic actuator (1) connected to the pre-tensioned bar (2), and of five other hydraulic actuators (3) installed at the end of the five output bars (5). Finally the input bar (4) is connected directly to the pre-tensioned bar in one end and in the other end is in contact with the specimen (6).

A quasi-static tri-axial stress state is introduced in the specimen by the hydraulic actuators (3) of the output-confinement bars (5). The rupture of the brittle bolt (between (1) and (2)) gives rise to a rectangular square pulse propagating into the system and loading dynamically the specimen until fracture. The strain gauges on the input (4) and output-confinement bars (5) record the incident, reflected and transmitted pulses allowing the re-construction of the equivalent stress-strain curves.

It is important to realise that the measurements on all the output bars also provide information on wave/energy transmission and transformation at material failure. This allows to study energy/wave released during material fracturing.

3.1. The uniaxial set-up of 3D-MHB

The construction of the 3D-MHB apparatus started [6] with the uniaxial set-up. The new equipment was built and installed in the DynaMat Laboratory. In this configuration it represents a Modified Hopkinson Bar in compression with the addition of a hydraulic actuator at the end of the output bar which has the function of imposing the static pre-stress to the specimen. It consists of a pre-tensioned bar (cylindrical bar), input and output bars (with square cross-section). The dimensions of the three bars are reported in Table 1. The total length of the 3D-MHB apparatus along the impact axis is of 7.82 m. The length becomes 8.80 m when the bumper system is installed.

All bars are made of aged maraging steel. This material has several advantages compared to other high strength steels. First, it is possible to reach the required strength, which is 1′600 MPa in the measurement cross-section and locally higher at connections and other discontinuities. Than, the ageing treatment requires moderate heating (around 500 °C) and slow cooling minimizing thermal induced stress and thermal deformation of the pieces. It is possible to produce the bars with tiny details like threads without machining after ageing. It is also possible to treat the bars and the counter pieces to reach slightly different strength, minimizing the cost in case of damage. The drawbacks of the maraging steel are the cost and the limited choice of raw dimensions.

The 3D-MHB apparatus can generate a rectangular loading pulse of 3 MN amplitude and 0.800 ms duration which propagates through the input bar – specimen – output bar. It deforms the specimen at high strain rate up to fracture.

The operative sequence in order to carry out a test using the 3D-MHB is as follows (refers to Fig. 2):

(i) A quasi-static stress state is introduced in the UHPC specimen by the hydraulic actuator placed at the end of the output bar (8).

Table 2. Dynamic compressive strength of UHPC.

Preload [kN]	Dynamic compressive strength [MPa]
0	456 ± 65
40	504 ± 25
80	455 ± 12

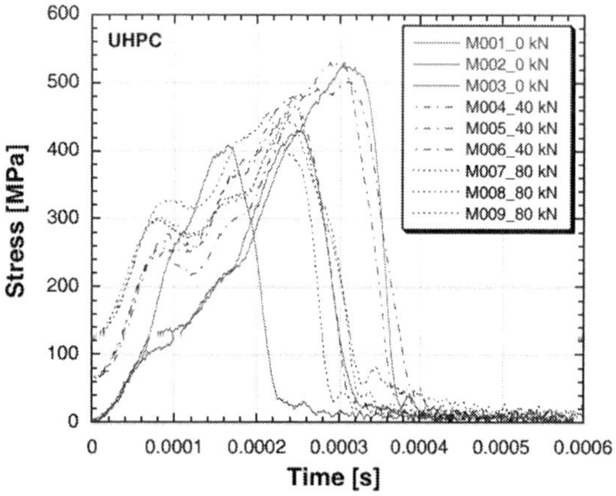

Figure 4. Stress versus time curves of UHPC with different pre-loading states.

(ii) By the principal hydraulic actuator (1) the pre-tensioned bar (3) is pulled till the established preload value. At this moment this load is withstand by the fragile bolt (2), the pre-tensioned bar and by the contrast ring (4). Because of the elastic movement of the entire system the pre-load on the specimen have to be checked and ruled by the actuator (8).

(iii) A mechanical system permits to continue to load the fragile bolt (2) without any influence on the rest of the measurement apparatus. In the 3D-MHB, the rupture of the fragile bolt gives rise to a rectangular square pulse propagating to the input bar (5) and output bar (7) and dynamically loading the specimen (6) until fracture.

(iv) The strain gages on the input and output-confinement bar record the incident, reflected and transmitted waves that allow the reconstruction of the equivalent stress-strain curves in the analytical way.

In Fig. 3 the experimental set-up is shown.

4. Results

The critical point of the dynamic compression test with the 2D-MHB is that common to all compression tests, that means it is affected by the variability of the boundary conditions at the interface loading member(bar)/specimen causing a remarkable dispersion of the test results. This high variability is due to the combined effects of contact between the specimen and the bars. Non-perfect shape of the specimen (such as imperfection of flatness and parallelism of the specimen faces) induces to a premature failure with anomalous crack paths. Also the contact

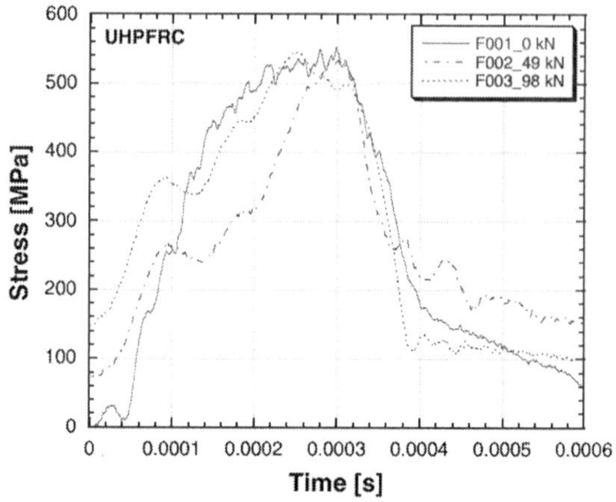

Figure 5. Stress versus time curves of UHPFRC with different pre-loading states.

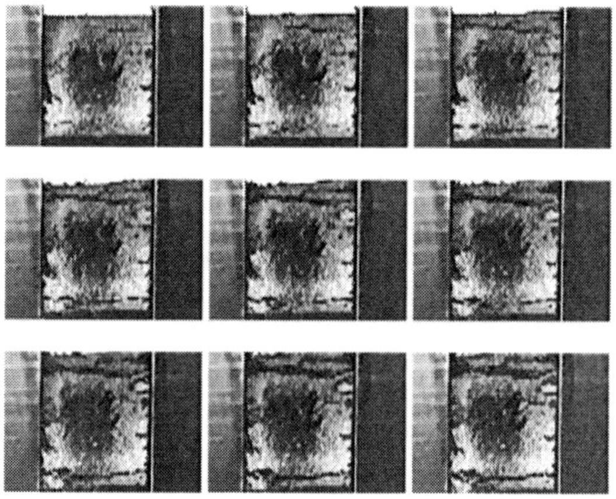

Figure 6. Photos of the UHPC specimen during the dynamic test with 0 kN pre-loading.

friction between bar surface and specimen surface has an influence on the transversal deformation of the specimen contributing to the above-mentioned variability of the test results. Eventual presence of interface layers (such as copper or grease) between specimen and the bars can reduce these problems, but this fact introduces an additional variable that must be taken into account in the analysis.

As first application with the new 3D-MHB apparatus has been chosen the study of the influence of the quasi-static preload on the dynamic compressive behaviour of the UHPC and the preliminary results are presented. The preliminary testing campaign consisted of a series of dynamic compressive tests performed varying the static pre-stress condition of the specimen. At least three tests for each pre-stress condition were carried out. Before the dynamic test quasi-static tests were performed. The average failure load in the quasi-static tests was 118.7 ± 6.2 kN. Two preload levels of 1/3 and 2/3 of the failure load were selected. The dynamic tests were carried out at the same velocity imposing 500 kN in the pre-tensioned bar giving a stress rate of 1600 GPa/s. The input and out bars

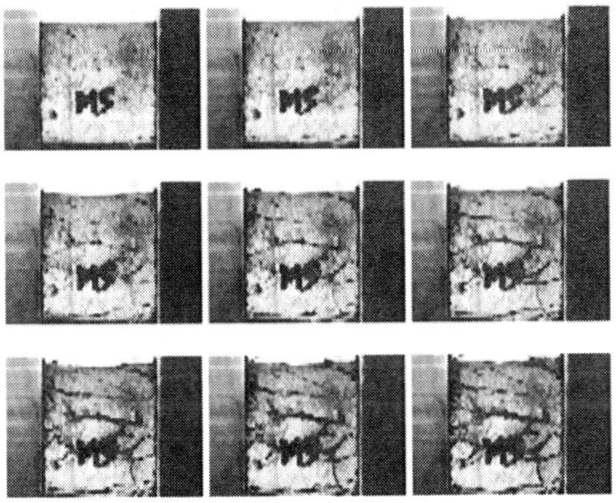

Figure 7. Photos of the UHPC specimen during the dynamic test with 40 kN pre-loading.

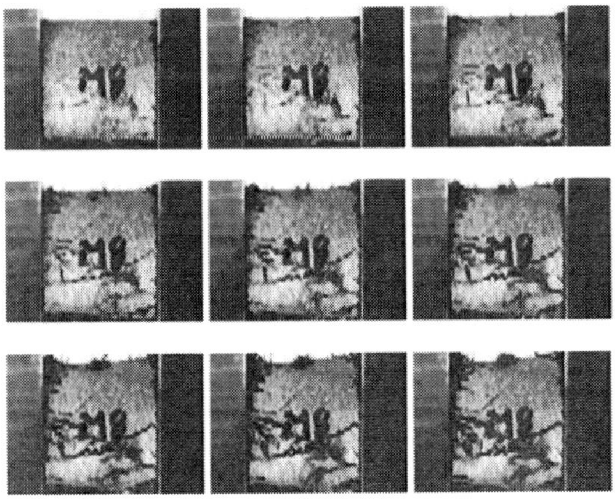

Figure 8. Photos of the UHPC specimen during the dynamic test with 80 kN pre-loading.

were instrumented by strain gauges in order to obtain strain history. These measurements were acquired by means of HBM-GEN2 data acquisition system with high sampling rates (1 Msample/s). The results in terms of maximum strength are collected in Table 2.

Figure 4 shows stress versus time curves of the performed tests. The standard deviation decreases with increasing the pre-load conditions. This means that the pre-load decreases the importance of the boundary conditions role.

Stress gradient is the same for all tests till the firsts 60 μs. After that time the slope of the stress versus time curve changes. This is due to the presence of the activated cracks during the pre-load [8]. In fact, stress rate as well as the number of activated cracks is a critical factor to enucleate the cracking process that induces specimen failure. The tests with higher static pre-loading value show a more uniform response of the material, but the strength was lower because of damage by the static pre-loading. This investigation was enlarged to few specimens of Ultra High Performance Fibre Reinforced Concrete (UHPFRC) to better understand this behaviour. Indeed the UHPFRC was pre-loaded with the same ratios of 1/3 and 2/3 of quasi-static failure load corresponding to 49 kN and 98 kN. In this case the UHPFRC presents not marked differences as the UHPC materials as shown in Fig. 5.

Figures 6, 7 and 8 shows some photos of the UHPC specimens during the failure process for 0 kN, 40 kN and 80 kN pre-loading, respectively. The photos were acquired by IDT Y4-S3 fast digital camera at 53′000 frame/s highlighting the different crack paths.

5. Conclusion

The influence of the preloading state on the compressive response of UHPC has been analysed by means of a new apparatus called 3D-MHB.

These preliminary results underline the critical influence of the boundary condition on the material response in compression; such critical influence shall be reduced by increasing planarity and parallelism of the specimen surfaces and the precision of the specimen/bar set-up.

The contact surfaces of the specimen shall be conditioned with special greases reducing friction effects. Increasing the pre-load state of the material a damage state can be induced especially if the pre-load is over the half of the failure quasi-static values. For pre-load lower, an effects homogenisation in the interface between specimen and experimental apparatus is obtained.

The new apparatus has been funded in the frame of the R'Equip program of the Swiss National Sciences Foundation with the Project SNSF No. 206021_128734 (2011).
These first tests have been supported by armasuisse- Science and Technology of the Swiss Federal Department of Defence, Civil Protection and Sport in the frame of the project *"Protection of Infrastructure Elements from the effects of IEDs and Blast Charges"*.

References

[1] A. Caverzan, E. Cadoni, M. di Prisco, Mech. Mat. **59**, 87–109 (2013).
[2] A. Caverzan, E. Cadoni, M. di Prisco, Int. J. Imp. Engng **45**, 28–38(2012).
[3] D. Asprone, E. Cadoni, A. Prota, ACI Struct. J. **106**, 523–529 (2009).
[4] D. Asprone, E. Cadoni, A. Prota, ACI Struct. J. **106**, 106–113 (2009).
[5] E. Cadoni, C. Albertini, Modified Hopkinson Bar Technologies Applied to the High Strain Rate of Rocks, in *Advances in Rock Dynamics and Applications*, Y. Zhou, J. Zhao eds. (Taylor & Francis, 2011), **4**, 79–104.
[6] J. Zhao, E. Cadoni, *Triaxially Compressed Hopkinson Bar (TriHB) for Geomaterial and Construction Material Testing*, Project SNSF No. 206021_128734 (2011).
[7] E. Cadoni, D. Forni, EPJ-ST (to be published) (2015).
[8] G. Riganti, E. Cadoni, Mat. Des. **57**, 156–167 (2014).

EPJ Web of Conferences **94**, 01032 (2015)
DOI: 10.1051/epjconf/20159401032

Experimental characterization and macro-modeling of mechanical strength of multi-sheets and multi-materials spot welds under pure and mixed modes I and II

Rim Chtourou[1,2,a], Gregory Haugou[1,b], Nicolas Leconte[1,c], Bassem Zouari[2,d], Fahmi Chaari[1,e], and Eric Markiewicz[1,f]

[1] LAMIH, UMR CNRS 8201, University of Valenciennes, 59313 Valenciennes Cedex 9, France
[2] LA2MP, National Engineering School of Sfax (ENIS), Route de Soukra, BP. W, 3038 Sfax, Tunisia

Abstract. Resistance Spot Welding (RSW) of multiple sheets with multiple materials are increasingly realized in the automotive industry. The mechanical strength of such new generation of spot welded assemblies is not that much dealt with. This is true in particular for experiments dedicated to investigate the mechanical strength of spot weld made by multi sheets of different grades, and their macro modeling in structural computations. Indeed, the most published studies are limited to two sheet assemblies.

Therefore, in the first part of this work an advanced experimental set-up with a reduced mass is proposed to characterize the quasi-static and dynamic mechanical behavior and rupture of spot weld made by several sheets of different grades. The proposed device is based on Arcan test, the plates contribution in the global response is, thus, reduced. Loading modes I/II are, therefore, combined and well controlled.

In the second part a simplified spot weld connector element (macroscopic modeling) is proposed to describe the nonlinear response and rupture of this new generation of spot welded assemblies. The weld connector model involves several parameters to be set. The remaining parameters are finally identified through a reverse engineering approach using mechanical responses of experimental tests presented in the first part of this work.

1. Introduction

Resistance spot welding (RSW) is the most common joining process for automotive body-in-white structures due to its production convenience and cost effectiveness. Around four thousands spot welds are indeed employed to assemble the body-in-white. With the demand of safety and lightweight vehicle structures, RSW of multiple sheets (more than two) and combining multiple materials (i.e. different steel grades) are increasingly realized. The literature survey shows that the mechanical strength of two sheet spot welded assemblies has been widely studied experimentally and numerically.

On the one hand, many tests are used, like pull-out, coach-peel and single-lap shear samples under quasi-static and dynamic loading conditions. Since 1958, mechanical tests have been discussed [1–3] in particular for pure tension, tension-shear and peel specimens. For this type of tests, a large part of the sheet sample plastically deforms, in particular far from the spot weld. Thus, it contributes significantly in the global behavior. Moreover, it is difficult to express the failure properties because the tensile/shear load ratio conditions varies during the test. More recently, many researchers [4–8] have proposed to adapt the Arcan tests to investigate the failure characteristics of spot welds under combined loading conditions. Indeed, Arcan tests have advantages over standard tests that the plates contribution around the weld nugget in the mechanical response is reduced. Moreover, it allows tensile (mode I) and shear (mode II) loads to be pure or combined and well controlled. However, these devices are characterized by a large mass that generates inertial forces during the dynamic loading. Thus, they can be used only under quasi-static loading conditions. To correct this limitation and to investigate the strain rate sensitivity of spot welds in dynamic conditions, Langrand and Markiewicz [7] have proposed an experimental device with a reduced mass to cope with these inertial forces.

On the other hand, researchers have developed various models to investigate the mechanical strength of spot welds. 3D FE detailed mesoscopic models allow a good representation of the nonlinear behavior of the assembly [9–11]. But such models are very expensive in terms of calculation time and are more dedicated to preliminary study, not for structural computations. Therefore, simplified models of the spot weld behavior and rupture are required. Some macro models have been developed and published in the literature [12–15]. One of the most interesting is that of Combescure et al. [6, 13], which is an elastic-plastic damageable Finite Element (FE) that models the global mechanical behavior of the whole "welded zone".

Almost all existing models have been evaluated for two-sheet assemblies. However, their performance to model multi-sheet multi-material spot welded assemblies remains challenging. Moreover, to the knowledge of the

[a] e-mail: Rim.Chtourou@etu.univ-valenciennes-.fr
[b] e-mail: gregory.haugou@univ-valenciennes.fr
[c] e-mail: Nicolas.Leconte@univ-valenciennes.fr
[d] e-mail: bzouari@yahoo.com
[e] e-mail: fahmi.chaari@univ-valenciennes.fr
[f] e-mail: eric.markiewicz@univ-valenciennes.fr

Article available at http://www.epj-conferences.org or http://dx.doi.org/10.1051/epjconf/20159401032

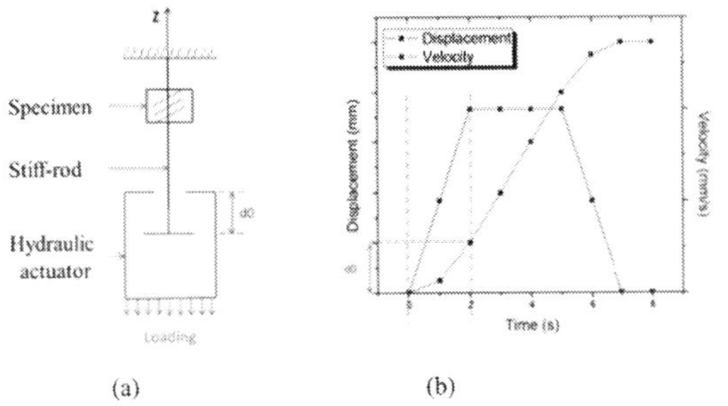

Figure 1. (a) Scheme of loading conditions (b) Typical curve of the displacement of the actuator.

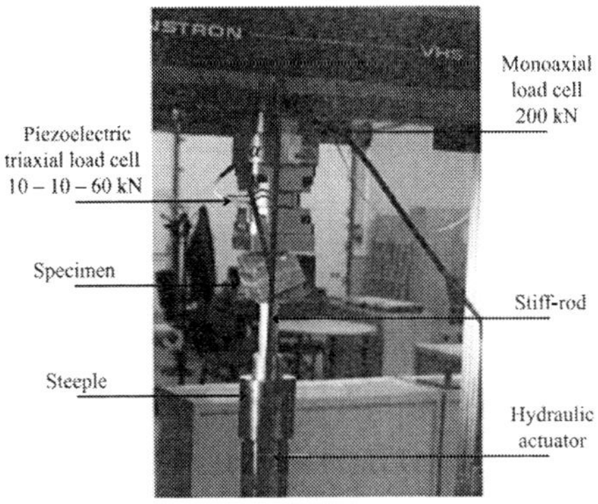

Figure 2. Experimental device for three-sheets assembly tensile testing.

authors, spot welds strength experiments that can be found in the literature deal only with two-sheet assemblies.

As a first part of this work an advanced experimental procedure is proposed to characterize the quasi-static and dynamic behavior of multi-sheet and multi-steel grades spot welds under pure and combined modes I/II loading conditions. An original experimental device based on the Arcan principle is designed to characterize specimens that consist of three-sheet spot welded assemblies. The second part of the work deals with the formulation of a connector model to simulate the global behavior and failure of spot welded assemblies. Following the modeling of this kind of assembly, the parameters of the joint equivalent model are identified based on the experimental results of the presented experiments.

Finally, some conclusions are presented.

2. New advanced experimental procedures

2.1. Experimental device

The proposed new device is designed to minimize inertia effects (optimized geometry and titanium alloy used as constitutive material). It allows to investigate the mechanical strength of these three-sheets assemblies under pure and combined modes I/II dynamic loading conditions. Note that an adequate test device was especially designed for each angular position.

The tests are carried out on a high speed hydraulic machine (INSTRON VHS 65/20). The hydraulic actuator moves along vertical axis (z) with an imposed speed to load the specimen. Under dynamic conditions, the displacement of the actuator increases progressively with the time until a linear regime. To ensure a constant speed loading during the tensile test, an experimental arrangement has been designed such as only the actuator moves without any loading of the sample along a distance a distance called here d_0 (Fig. 1(a)). Once the speed in established and the loading velocity becomes constant, the contact is effected, then the stiff-rod and the specimen are loaded as presented in Fig. 1(b).

This device allows tensile and shear loads to be combined and controlled. The angular position α between the loading direction (the vertical axis) and the normal to the surface of the specimen defines the tensile/shear ratio (Fig. 2).

The device was fixed to the machine with four screws. All device rotations are canceled. Forces along the three main directions of the machine (F_x, F_y and F_z) are measured during the tests, using a triaxial load cell (Kistler 9367C). The displacement is obtained by a LVDT sensor with a measurement range equal to 300 mm and error of 0.15 mm. Figure 2 shows the complete set up for the experiments.

Table 1. Mechanical properties of sheets metals.

	$\sigma_y(MPa)$	$\sigma_{max}(MPa)$	$A(\%)$
$DX54D$	171	301.6	57
$22MnB5$	1100	1500	6

2.2. Three-sheets spot weld specimen based on two constitutive materials

The proposed specimen consist of a three-sheets assembly: (P_1) 2 mm thick 22 MnB5, (P_2) 0.65 mm thick DX54D and (P_3) 1.6 mm thick 22 MnB5 (Fig. 3). The spot welding process was performed by our industrial partner Renault according to the know-how of the manufacturer. 22 MnB5 alloy is an ultra high strength steel, and it is the base material for the two external plates of the assembly. The middle plate is made of mild steel (DX54D). The mechanical properties of the two materials are presented in Table 1.

As it is shown in Fig. 3, two rigid heel joints have been designed similarly to the set up proposed by Langrand and Markiewicz [7] to ensure the connection between the specimen and the device. The assembly, rigid heel joint and specimen, is positioned on the device by two centering pins and fixed with six screws. The set-up is designed to limit the bending of sheets and ensure a good repeatability of the tests. The free part around the spot weld is circular for isotropy effect.

With three-sheets assemblies, three loading configurations (1–3), (1–2) and (2–3) are possible. For the first part of this project, only the configuration (1–3), which consist of loading the first and the third sheet of the assembly, has been studied and presented in this paper. Two other

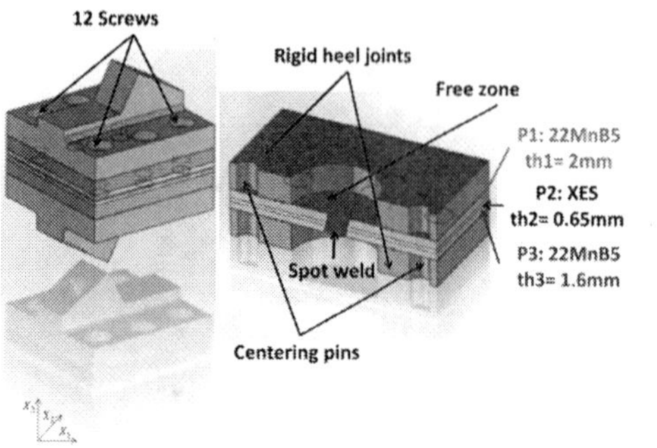

Figure 3. Three-sheets spot welds specimens based on two constitutive materials (22MnB5 and DX54D).

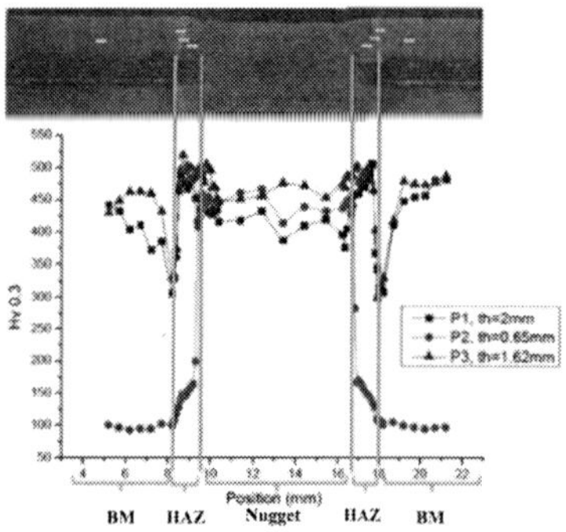

Figure 4. Hardness distribution along the diametrical direction of manufactured spot weld joint.

configurations (1–2) and (2–3) are currently in progress, and their results will be subject of future publications.

2.3. Hardness measurement

Vickers hardness testing was performed with Future Tech Hardness Tester (Model FM). The applied force level for hardness measurement was equal to 300 g. The hardest zone of the test samples (spot weld nugget) is indented with a pitch of 0.1 mm and the softest zone (base metal) parts indented with a pitch of 1 mm.

The averaged and linearised hardness distribution results revealed the nugget dimensions, and allows us to identify three zones of spot weld: Nugget, Heat Affected Zone (HAZ) and Base Material (BM) (Fig. 4). The shape of the nugget will be used for the numerical modeling which is presented in the later section.

2.4. Experimental results

To study the strain-rate sensitivity of spot welds, three velocities are employed in this work: v = [1 mm/s, 33 mm/s, 1 m/s]. So that, a jump of strain rate close to 30 is attended. For each loading speed four loading directions are chosen: pure tensile (mode I) with $\alpha = 0°$, pure shear (mode II) with $\alpha = 90°$, and combined loading with $\alpha = 25°$ and $\alpha = 45°$. A set of two or three tests has been performed for each angle/velocity combination.

Table 2. Experimental results of the different loading conditions.

		$1\ mm/s$	$33\ mm/s$	$1\ m/s$
$\alpha = 0°$	$F_{max}(kN)$	13.6	12.7	14.3
	$U_{rup}(mm)$	1.83	2.27	2.2
$\alpha = 25°$	$F_{max}(kN)$	13.8	13.7	14.2
	$U_{rup}(mm)$	1.52	1.85	2
$\alpha = 45°$	$F_{max}(kN)$	15.1	16.2	–
	$U_{rup}(mm)$	1.5	1.82	–
$\alpha = 90°$	$F_{max}(kN)$	38.8	–	–
	$U_{rup}(mm)$	2.18	–	–

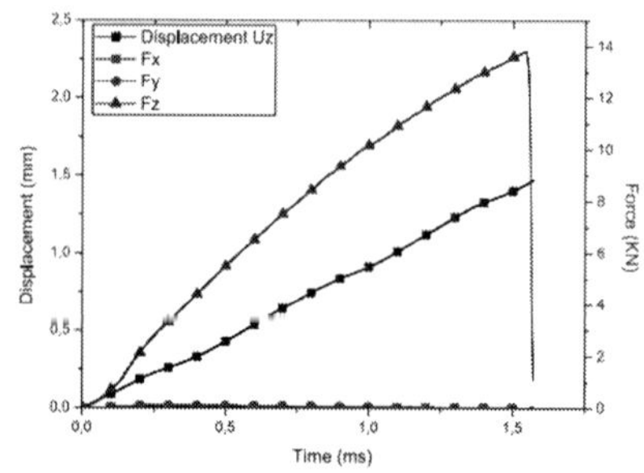

Figure 5. Typical forces and displacement versus time curves at 1 mm/s under pure tension loading.

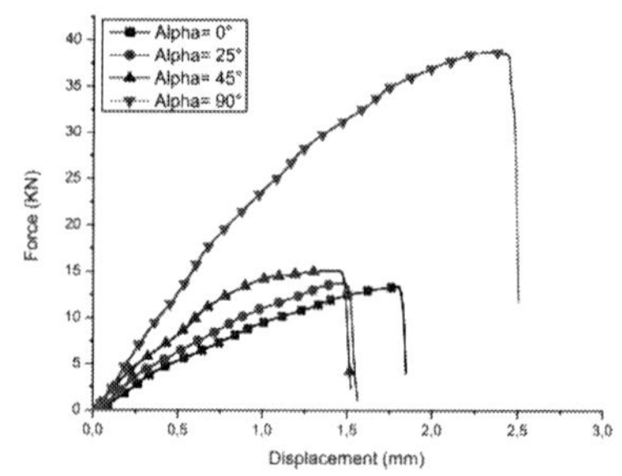

Figure 6. Typical force versus displacement diagrams measured under quasi-static ($V = 1$ mm/s) pure and combined tensile/shear modes.

2.4.1. *Influence of the load angle at a constant speed*

Figure 5 shows typical curves of forces and displacement versus time at 1 mm/s under pure tension loading. As it is expected, the displacements increase linearly during loading, thus, the speed remains constant during the test. Also, tangential forces (F_x and F_y) are negligible compared to the normal force F_z (loading direction).

Figure 6 shows typical experimental results of force versus displacement under different loading angles for a velocity at 1 mm/s. The different forces versus elongation responses have the same shape. Unlike the case of two-sheet assemblies [7], the failure occurs suddenly and there is no dissipated energy after the maximum load, for the shear test and the combined loading. That might be due to the fact that the tests are carried out in configuration (1–3). Indeed, the loading is applied to the plates made on 22 MnB5. Therefore, other failure modes can be detected for configurations (1–2) and (2–3).

(a) (b)

Figure 7. Failure mode (a) $\alpha = 45°$ (b) $\alpha = 90°$.

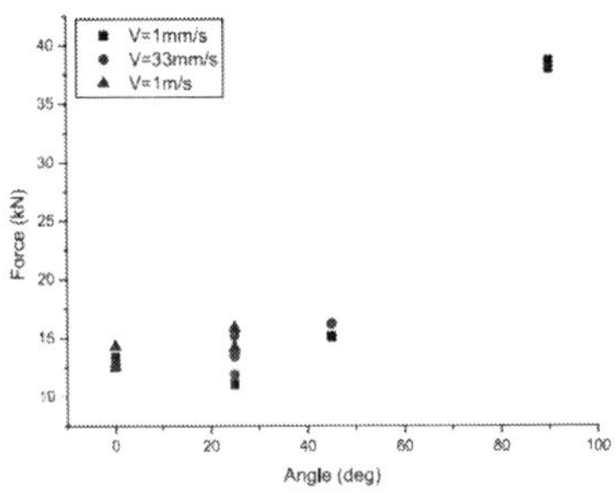

Figure 8. Evolution of the maximal load with the variation of the loading angle.

Note that, the maximum loading force is quasi-constant for $\alpha = 0°$ to $\alpha = 45°$ (mean value $F_{max} = 15$ kN), but brutally increases up to $F_{max} = 39$ kN for $\alpha = 90°$ (Fig. 8). This is because for loading angle between $\alpha = 0°$ and $\alpha = 45°$ the failure occurs suddenly when the metal tears at once all around the welded zone (Fig. 7(a)). However for shear test ($\alpha = 90°$) the core of the spot weld itself has been cutted off as it is shown in Fig. 7(b). Moreover, spot weld core is characterized by the higher hardness of the assembly because of the thermal treatment during the welding process.

Concerning ultimate elongation, its evolution is not perfectly established due to repeatability issues and centering defaults of the spot weld detected in the free zone (Fig. 3).

2.4.2. *Strain-rate effect*

To study the strain-rate effect, Table 2 summarize the evolution of the maximum loading force and the the motion at the rupture with variation of the loading speed. In fact, the maximum loading force increases slightly with velocity. For the case of loading angle $\alpha = 0°$, it increase by 5% between the lower and the higher velocity and by 2.8% for $\alpha = 25°$. For the case $\alpha = 45°$ we observe that the maximum load in the case of 33 mm/s velocity is lower of 7% than the case of 1 mm/s velocity. The displacement at the break increases, also, by 20% for the different loading angle Fig. 9 shows the evolution of the displacement at the break with the loading speed. Indeed, when the velocity increases from 1 mm/s to 33 mm/s.

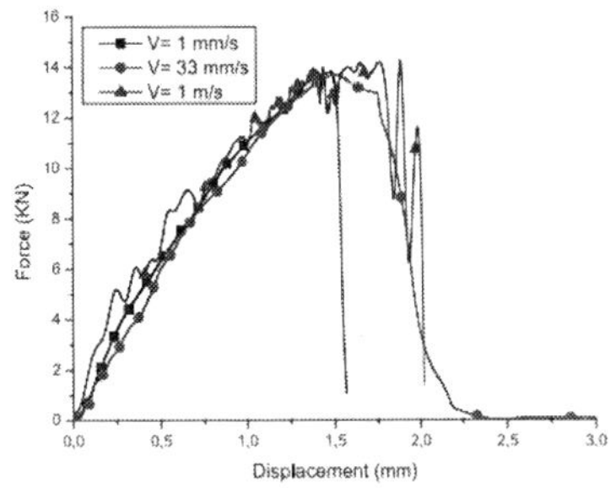

Figure 9. Influence of the loading speed in combining loading $\alpha = 25°$.

Table 3. Evolution of the dissipated energy.

	V = 1/ V = 33	V = 33/ V = 1000	V = 1/ V = 1000
$\alpha = 0°$	38%	1.5%	36%
$\alpha = 25°$	37%	8%	51%
$\alpha = 45°$	43%	–	–

The augmentation of the loading speed generates an increasing of the maximum loading force and the displacement at the break and consequently an increasing of the dissipated energy. Therefore, the energy dissipation is strain rate dependent. Table 3 represents the evolution of energy dissipation for each jump of strain, for different loading angles. The obtained results proves that the energy dissipation strain rate dependency is more important when the loading angle is larger.

Such results can't be yet confirmed because of the lack of experimental results. In the same way, any conclusion can be done for shear tests experiments.

Other experimental tests with other intermediate loading angle ($\alpha = 60°$ and $\alpha = 75°$) will be done to detect a representative trend for the evolution of the ultimate displacement and force with the loading angle. These complementary tests will also give information about the strain rate effect.

These results will be used to identify the parameters of the macro modeling approach presented in the following section.

3. Connector model formulation

The spot weld connector is modeled by an equivalent element featuring two nodes, similar to a generalized spring where the degrees of freedom are the relative generalized motion of referential to another.

As the local behavior of a spot weld is nonlinear, an elastic-plastic model has been introduced to define the mechanical behavior of the spot weld connector.

Stiffnesses of the spot weld connector model are assumed to be uncoupled, $\{f_i\} = [D_{ii}]\{u_i\}$.

The plasticity criterion is expressed by Eq. (1).

$$\phi(\bar{F}, \bar{u}^{pl}) = \bar{F} - F_H \leq 0 \qquad (1)$$

where $\phi(\bar{F}, \bar{u}^{pl})$ is the yield surface of the plastic model, $\bar{F}$ is the equivalent generalized force of spot weld connector and F_H is the hardening function. $\bar{F}$ is calculated using the

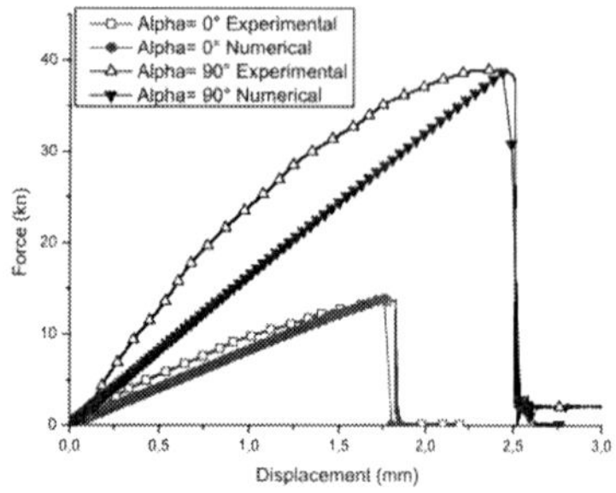

Figure 10. Load vs. displacement diagrams measured underpure tensile mode $\alpha = 0°$.

potential function P_1 (2).

$$\bar{F} = P_1(\{f\}) = \left[\left(\frac{F_N}{R_N}\right)^{\beta} + \left(\frac{F_S}{R_S}\right)^{\beta} + \left(\frac{M_T}{R_T}\right)^{\beta} + \left(\frac{M_B}{R_B}\right)^{\beta} \right]^{\frac{1}{\beta}} \quad (2)$$

where the F_N is normal force, F_S the shear force, M_B the bending moment, and M_T torsion moment.

Parameters β, R_N, R_S, R_B and R_T are chosen to fit the plastic behavior under different loading directions.

For this model an exponential Voce type law is chosen to model the isotropic hardening behavior of the spot weld connector model (Eq. (3)).

$$F_H = F_0 + Q(1 - \exp(-b\bar{u}^{pl})) \quad (3)$$

where F_0 is the yield value at zero plastic equivalent motion, Q is the saturation generalized force and b is a constant.

The failure of the connection is taken into account by a ductile damage model.

Damage is initiated when ($\bar{F} \geq F_i$), where F_i the critical force value. The damage evolution is defined by linear law that specifies how the damage variable d evolves. Then, the force response is expressed by Eq. (4):

$$\bar{F}_d = (1 - d)\,\bar{F},\ 0 \leqslant d \leqslant 1. \quad (4)$$

The set of parameters, which describe the elasto-plastic damage behavior of the spot weld colled, are called design parameters, have to be identified by reverse engineering approach.

4. Numerical modeling and results

The three-sheets two-materials spot weld assembly presents very big metallurgical and structural complexities. Therefore, different modeling methods can be proposed. In this work, the three-sheets spot welded assembly was modeled by two connector elements in series (Sect. 3). The same mechanical behavior was associated for the two connectors. Thus, just one set of parameters will be identified for the two connectors.

According to the description of the proposed experimental set-up, the spot-welded specimens are modeled by

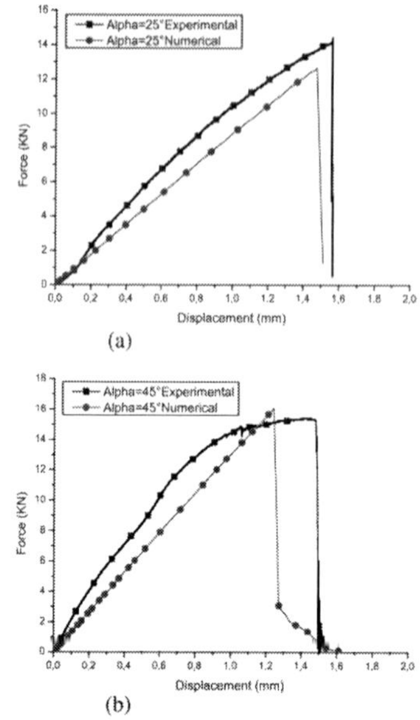

Figure 11. Load vs. displacement diagrams measured under combined tensile/shear mode(a) $\alpha = 25°$ (b) $\alpha = 45°$.

Table 4. Comparaison of total dissipated energy at rupture betwen numerical and experimental results.

$\alpha°$	0	25	45	90
Error in energy (%)	15.6	21.1	30	22.1

three parallel circular plates of 20 mm diameter each one corresponding to the free part sheet metal around the spot weld. The three plates are meshed with quadrangular shell elements of 8 mm size, type S4R with a reduce integration. The periphery of the top plate is defined as a rigid body. The extremity of the bottom plate is clamped and an imposed velocity $V_{\text{imp}} = 1$ mm/s is applied to the rigid body according to mode I and mode II directions (quasi-static loading).

The maximum force and the displacement at failure are very well described for the tensile test (mode I) and for shear test (mode II) (Fig. 10).

For the model validation, it is proposed a comparison of the experimental and numerical simulation results under mixed loads. Figure 1(a) and Fig. 1(b) present a comparison of the model results to experimental ones for angular positions $\alpha = 25°$ and $\alpha = 45°$, respectively.

Satisfactory results are obtained. The maximum loading force is underestimated by 10% for $\alpha = 25°$. However, for $\alpha = 45°$ it is overestimated by 7%.

In terms of energy, Table 4 summarizes the energy error for different loading angles.

5. Conclusion

An experimental procedure has been proposed in order to investigate the mechanical behavior of multi-materials multi-sheets spot welded assemblies under quasi-static and dynamic loading conditions and in pure and combined opening/shear modes. The experimental set-up is based on the Arcan test principle, with a reduced mass to investigate the strain rate of this type of assembly. A first part of the experimental tests has been done in four loading directions and for three loading velocities, and preliminary results have been obtained. Other experimental tests with other intermediate loading angle ($\alpha = 60°$ and $\alpha = 75°$) and two other configurations will be done to detect a trend for the evolution of the ultimate displacement and define a failure criteria.

Authors have also proposed to study numerically the nonlinear mechanical behavior of the three-sheets multi-materials spot welded assemblies. A simple spot weld connector model has been proposed. It is a macro modeling for applications to structural crashworthiness FEA. The parameters of this model have been identified using the experimental results of the proposed set-up in this work, and good results have been obtained.

This research is conducted through collaboration between the University of Valenciennes and the National Engineering School of Sfax. This collaboration is jointly financed in the frame of the Utique CMCU programme. The present research work has also been supported by the International Campus on Safety and Intermodality in Transportation, the Nord-Pas-de-Calais region, the European Community, the Regional Delegation for Research and Technology, the French National Research Agency, and by the Ministry of Higher Education and Research. The authors gratefully acknowledge the support of these institutions.

References

[1] Hartmann, E., Welding Journal (1958).

[2] VandenBossche, D. J., SAE paper 770214 (1977).

[3] Markiewicz, E., Drazetic, P., Mécanique et Industries **4**, 17–27 (2003).

[4] Lee, Y., Wehner, T., Lu, M., Morrissett, T., Pakalnins, E., Journal of Testing and Evaluation **26**, 213–219 (1998).

[5] Lin, SH., Pan, J., Tyan, J., Prasad, P., International Journal of Solids and Structures **40**, 5539–64 (2003).

[6] Langrand, B., Combescure, A., International Journal of Solids and Structures **41**, 6631–6646 (2004).

[7] Langrand, B., Markiewicz, E., International Journal of Impact Engineering **37**, 792–805 (2010).

[8] Song, J.H. Huh, H. International Journal of Mechanical Sciences **53**, 513–525 (2011).

[9] Langrand, B., Markiewicz, E., Drazetic, P., Geoffroy, P., Petitniot, J., Fabis, J., International Journal of Materials and Product Technology **12**, 428–446 (1997).

[10] Langrand, B., Geoffroy, P., Petitniot, J., Fabis, J., Markiewicz, E., Drazetic, P., Aerospace Science and Technology **3**, 215–227 (1999).

[11] Markiewicz, E., Ducrocq, P., Drazetic, P., Haugou, G., Fourmentraux, T., Berard, J.Y., International Journal of Materials and Product Technology **16**, 484 509 (2001).

[12] Langrand, B., Deletombe, E., Markiewicz, E., Drazetic, P., Finite Elements in Analysis and Design **38**, 21–44 (2001).

[13] Combescure, A., Delcroix, F., Caplain, L., Espanol, S., Eliot, P., International Journal of Impact Engineering **28**, 783–802 (2003).

[14] Yang, X., Xia, Y., Zhou, Q., Engineering Fracture Mechanics **77**, 1224–1239 (2010).

[15] Khandokera, N., Taklab, M., Materials and Design **54**, 323–3304 (2014).

EPJ Web of Conferences **94**, 01033 (2015)
DOI: 10.1051/epjconf/20159401033

Effect of strain rate on shear properties and fracture characteristics of DP600 and AA5182-O sheet metal alloys

Taamjeed Rahmaan[a], Cliff Butcher, Armin Abedini, and Michael Worswick

University of Waterloo, Mechanical and Mechatronics Engineering Department, Waterloo, Ontario, Canada

Abstract. Shear tests were performed at strain rates ranging from quasi-static (.01 s^{-1}) to 600 s^{-1} for DP600 steel and AA5182-O sheet metal alloys at room temperature. A miniature sized shear specimen was modified and validated in this work to perform high strain rate shear testing. Digital image correlation (DIC) techniques were employed to measure the strains in the experiments, and a criterion to detect the onset of fracture based on the hardening rate of the materials is proposed. At equivalent strains greater than 20%, the DP600 and AA5182 alloys demonstrated a reduced work hardening rate at elevated strain rates. At lower strains, the DP600 shows positive rate sensitivity while the AA5182 was not sensitive to strain rate. For both alloys, the equivalent fracture strain and elongation to failure decreased with strain rate. A conversion of the shear stress to an equivalent stress using the von Mises yield criterion provided excellent agreement with the results from tensile tests at elevated strain rates. Unlike the tensile test, the shear test is not limited by the onset of necking so the equivalent stress can be determined over a larger range of strain.

1. Introduction

Advanced High Strength Steels (AHSS) and aluminum alloys are currently of great interest in the automotive industry due to their superior strength-to-weight ratio and adequate formability. Dual phase (DP) steels can be found in many vehicle structural components while 5000-series aluminum alloys are commonly used in the fabrication of inner body panels. In automobile crash events or high speed metal forming, the local strain rates can vary from quasi-static conditions to over 1000 s^{-1}. Understanding the constitutive response and fracture mechanisms associated with dynamic loading is critical for the accurate design of sheet metal forming operations for structural components and their subsequent performance in vehicle crash simulations.

Sheet metal forming processes often involve large plastic strains. Optimization of the forming techniques requires an accurate knowledge of the elastoplastic behaviour of the metals until failure. Uniaxial tensile tests are used to characterize the mechanical properties of sheet metals until the onset of necking. However, diffuse or localized necking occurs in this test, limiting the accurate measurements of the behaviour of the materials to relatively small strains. Shear tests can investigate the material response until large plastic strains since localized necking does not occur and plane stress conditions are satisfied until fracture. In addition, the development of stress state-dependent fracture surfaces and damage models to predict failure in automotive forming and crash simulations has created great interest in developing experimental tests to characterize failure under shear stress states [1,2].

Unlike tensile tests, obtaining the shear stress of sheet metals is difficult as the shear stress distribution is generally inhomogeneous and there is no intrinsic gauge length that can define the local strain. Torsion tests are not suitable for sheet materials; therefore, various simple shear test geometries have been developed [3–8]. The most common shear test involves clamping a rectangular specimen and moving the clamps in opposite directions to impose a shear deformation. Alternatively, various experimental techniques and geometries have been proposed in the literature but most require a complex geometry that is inconvenient for dynamic testing [9–12].

Recently, Peirs et al. [12] developed a simple shear specimen geometry for both quasi-static and dynamic strain rates using numerical optimization and digital image correlation (DIC) to obtain the local shear strain. The design of this geometry is an evolution of the geometries proposed by Tarigopula et al. [13] and Bao and Wierzbicki [14]. The advantage of this geometry is that it is relatively simple to fabricate, does not require through-thickness machining and provides a uniform shear strain distribution that can be used to resolve the shear stress response of the material. Recent work by Abedini et al. [15] has demonstrated the advantages of the Peirs et al. [12] geometry for material characterization.

In the present work, the specimen geometry developed by Peirs et al. [12] has been modified in order to conduct experiments utilizing the testing apparatus with smaller grip size. The shear deformation behaviour and fracture characteristics of DP600 and AA5182-O sheet metal alloys have been investigated at quasi-static, 10, 100, and 600 s^{-1} strain rates. The dynamic shear tests served as a companion study to the recent high rate tensile characterization of these materials by the authors [16]. A method is proposed to determine the fracture strain based on hardening curve of the material.

[a] Corresponding author: trahmaan@uwaterloo.ca

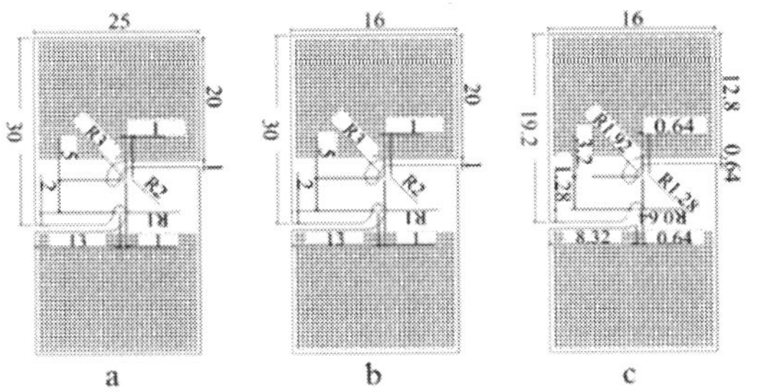

Figure 1. Shear specimen geometry: (a) regular size [12] (b) trimmed and (c) mini (scaled down) specimen. Dimensions are in mm. Shaded regions are the grip area.

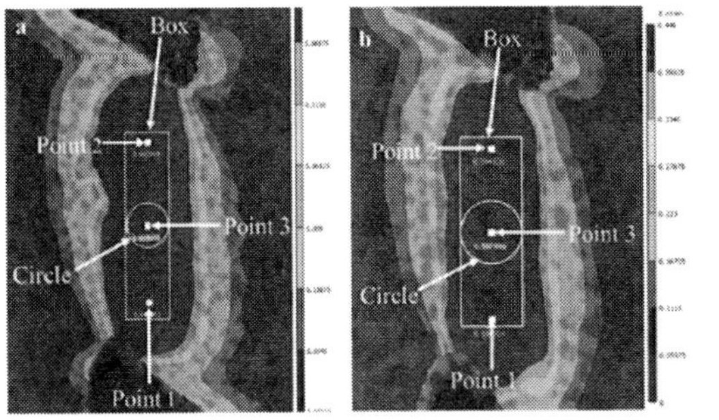

Figure 2. Different inspector tools of the DIC data processor software, Vic-2D positioned on the deformed DP600 shear specimens for a quasi-static strain rate test condition using (a) regular size [12] and (b) mini size specimen.

2. Experiments

2.1. Materials

Materials studied were DP600 and AA5182-O in nominal 1.5 mm thick sheets. They were selected due to their importance in lightweight automotive applications and their difference in observed rate sensitivity. Specimens were tested along the sheet rolling direction. The reader is referred to Rahmaan et al. [16] for the detailed mechanical properties and tensile behaviour for a wide range of strain rates.

2.2. Specimen geometry

The specimen geometry developed by Peirs et al. [12] was used in this work for the quasi-static shear testing. However, the width of the specimen was larger than the grip slot sizes of the dynamic strain rate experimental apparatus available at University of Waterloo. Thus, the shear specimen geometry developed by Peirs et al. [12] (referred to as regular shear specimen herein) was modified in two phases. In the first phase, the width of the regular shear specimen was trimmed down to match the size of the grip slots of the apparatus, while all other dimensions of the regular shear specimen were kept unchanged. In the second phase, all the dimensions of regular size specimen were scaled down by 36%, (referred to as mini shear specimen). The geometries of the three shear specimens are shown in Fig. 1.

2.3. Apparatus and procedure

An Instron model 1331 servo-hydraulic testing machine was used to conduct shear testing at nominal strain rates of 0.01 s^{-1} (or quasi-static). Cross-head velocities for the experiments were 0.03 mm/s for the regular and trimmed size shear specimen testing, and 0.019 mm/s was used for the mini size shear specimen testing.

A Hydraulic Intermediate Strain Rate (HISR) apparatus developed at the University of Waterloo was used to perform shear tests at strain rates of 10, 100, and 600 s^{-1}. The reader is referred to [17] for a detailed description of the HISR apparatus. Cross-head velocities of approximately 35 mm/s, 370 mm/s, and 2100 mm/s were used to achieve the nominal strain rates of 10, 100, and 600 s^{-1}, respectively.

The strains were measured using digital image correlation (DIC) techniques. A Point Grey Research GRAS-50S5M-C camera was employed for the quasi-static condition and a Photron SA5 high speed camera for the higher strain rates. In the present work, the frame rates used were 6 fps, 2,000 fps, 17,500 fps, and 40,000 fps for nominal strain rates of 1, 10, 100, and 600 s^{-1}, respectively. Full-field logarithmic (true) strain measurements were obtained using Vic-2D software from Correlated Solutions Inc. using the incremental correlation option to account for severe local strains. As discussed in Peirs et al. [12] and Abedini et al. [15], the average shear stress, τ_{avg}, is computed as:

$$\tau_{avg} = \frac{F}{A_{shear}} = \frac{F}{Lt} \tag{1}$$

where F is the force, t is the thickness, and L is the length of the shear region with nominal values of 1.9 mm for the mini-shear sample and 3.0 mm for the other geometries. The as-fabricated length and thickness of each sample was measured using an optical microscope. Three to six repeat tests were performed for each test condition and an average stress-strain curve was generated by interpolating the test data using Von Mises equivalent strain increments of 0.002.

3. Results and discussion

3.1. Shear stress-strain behaviour

Due to the small sample size and severe local strains within the gauge region it is instructive to investigate the procedure for extracting the local strain measurements. The local strain measurements can be extracted from single points (highly variable and noisy) or averaged over a user-defined area (rectangular or circular) which introduces a length scale into the analysis. To establish a consistent methodology for the DIC analysis of the shear samples, three different inspector tools were considered: a rectangular box (0.3 × 1.5 mm), a circle with a radius of 0.3 mm, and 3 points taken within the shear band as shown in Fig. 2 for a DP600 sample. The resulting shear stress vs. equivalent strain response obtained using the three measurement techniques is shown in Fig. 3 and

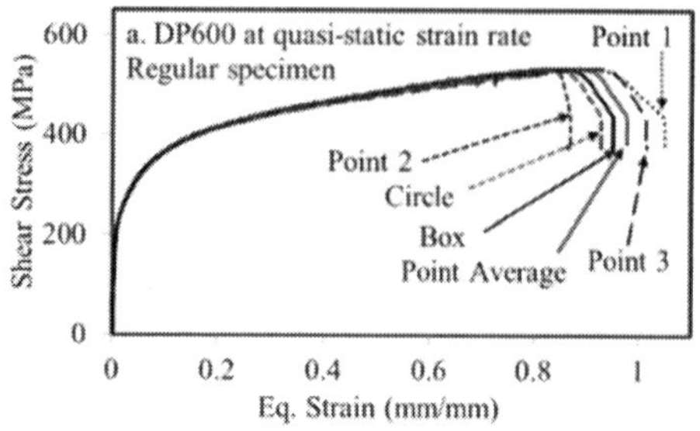

Figure 3. Quasi-static shear stress response of DP600 extracted using different inspector tools.

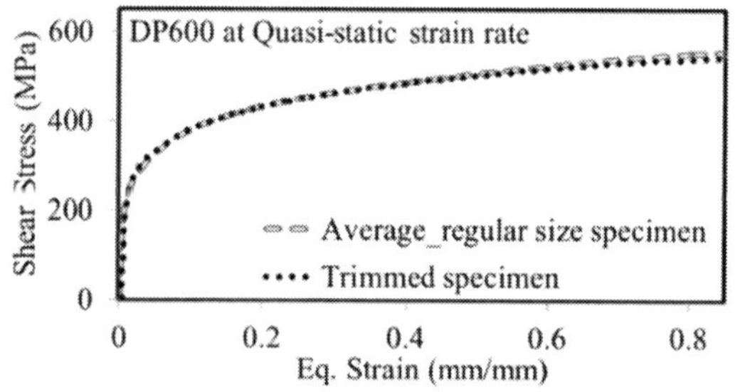

Figure 4. Comparison of the quasi-static shear stress responses of DP600 obtained using the regular and trimmed specimens.

show good agreement until very large strains on the order of 80% for DP600. The fracture strain computed using these different methods shows some dependency upon the measurement techniques with the point measurements showing some variation while the box and circle inspector tools give similar strains. The point measurements are inherently noisy so an average of the three points within the shear zone gives a very similar measurement as the area-averaged values. Thus, it would be safe to conclude that either the rectangular box or circular inspector tool provides accurate data as long as they are positioned within the shear region of the specimen.

Although the regular size shear specimen was successfully implemented for high strain rate testing by Peirs et al. [12], it was not possible to utilize this geometry for high strain rate testing due to narrow grip sizes of the current apparatus. In order to utilize the shear specimen in the existing dynamic strain rate testing apparatus, the width of the specimen was trimmed down from 25 mm to 16 mm, while all other dimensions remained the same. The shear stress vs. equivalent strain data for DP600 sheet material obtained using the trimmed specimen is presented in Fig. 4 and compared with the data from regular size shear specimen. The data obtained from both specimens are found be in excellent agreement. However, as shown in Fig. 5, the trimmed specimen does not have sufficient material in the grips to withstand an induced rotation due to severe shear deformation within the gauge section. This shear-induced rotation did not have an apparent effect on the stress-strain curves at quasi-static conditions but it could possibly introduce a misalignment and errors in a high rate shear test using a tensile split Hopkinson bar. As such, the trimmed sample geometry was discarded.

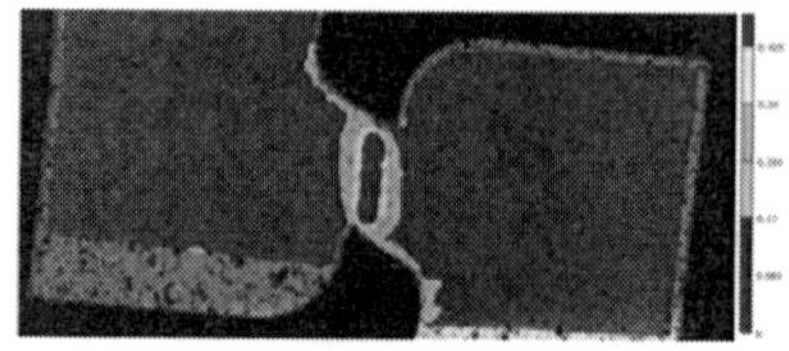

Figure 5. Equivalent strain distribution of the trimmed DP600 shear specimen prior to fracture.

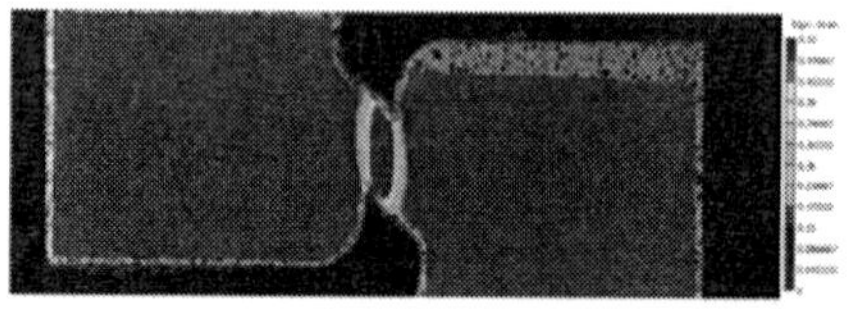

Figure 6. Strain contour plot of plastically deformed DP600 mini specimen at quasi-static strain rate test condition.

The regular size shear sample was then scaled down by 36% to obtain a sample width of 16 mm and is referred to as the mini-shear sample. The small dimensions of the new sample required fabrication using electric discharge machining. As shown in Fig. 6, the shear-induced rotation was removed and the same shear stress response was obtained for both the DP600 and AA5182 as shown in Figs. 7a and 7b, respectively. The experimental data obtained with the regular and mini-shear specimens is in excellent agreement at quasi-static conditions with very good repeatability. Thus, it is assumed that the mini-shear specimen geometry is suitable for dynamic shear tests as well since the regular sample was proven successful by Peirs et al. [12]. Different DIC strain inspector tools were also investigated for the miniature shear specimen with the same trends the regular specimen. A rectangular (0.2 × 0.7 mm) or small cicle with a radius of 0.2 mm placed within the shear zone gives consistent strain measurements.

The effect of strain rate on shear stress versus equivalent strain for DP600 and AA5182-O sheet materials is shown in Figs. 8a and 8b, respectively. For DP600, the yield stress scales upwards with increasing strain rate, while the shape of shear stress curve (or the hardening behavior) decreases with the increase of strain rate. As expected, the yield stress of AA5182-O is found to be strain rate insensitive; however, the hardening behavior of the material shows negative rate sensitivity.

The maximum temperature rise, ΔT using an adiabatic heating assumption was calculated using Eq. (2), where β is the ratio of plastic work converted into heat (taken as 0.9), ρ is the material density, C_p is the heat capacity of the material, and the remainder is the integral of the equivalent stress versus strain curve which is calculated as the area under the curve up to fracture strain. The adiabatic heat rise up to the fracture strain was calculated to be approximately 144 °C and 76 °C at 600 s^{-1} for DP600 and AA5182-O, respectively.

$$\Delta T = \frac{\beta}{\rho \cdot C_p} \int \sigma d\varepsilon. \qquad (2)$$

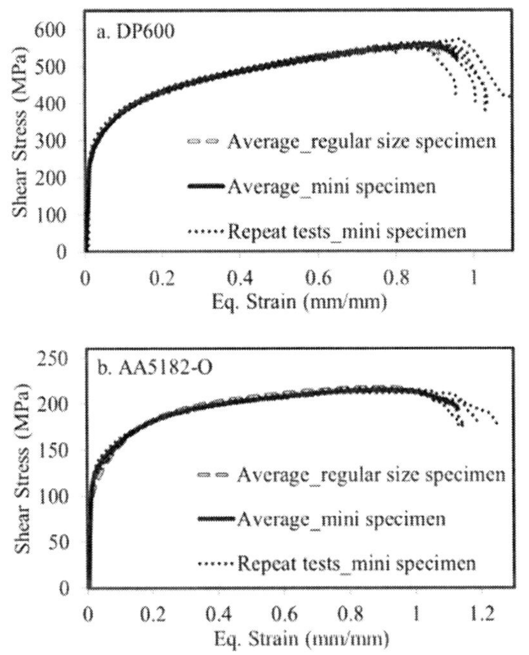

Figure 7. Comparison of the quasi-static shear stress response between the regular and mini-shear specimens for (a) DP600 and (b) AA5182-O.

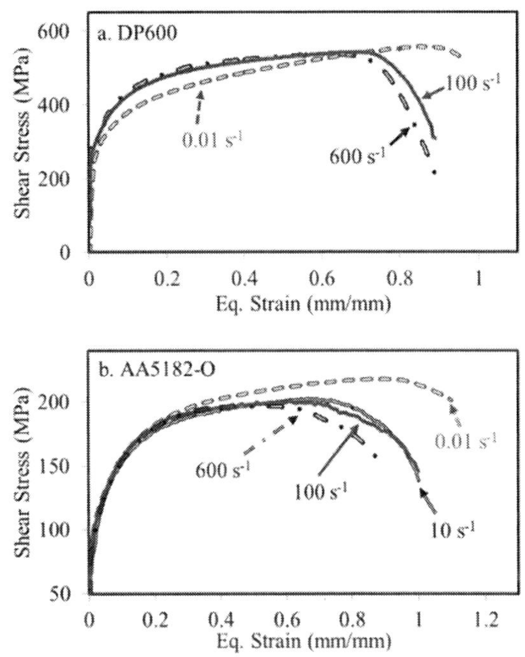

Figure 8. Averaged shear stress vs. equivalent strain curves for (a) DP600 and (b) AA5182-O sheet specimens at room temperature and strain rates from quasi-static to $600\,s^{-1}$.

The DP600 has positive rate sensitivity at strains up to approximately 30% and then the hardening rate becomes lower than quasi-static. AA5182-O is strain rate insensitive at strains up to approximately 25%, while the hardening rate decreases as the strain rate increases from quasi-static to $10\,s^{-1}$. The hardening rate of AA5182-O is rate insensitive for strain rates above $10\,s^{-1}$. The observed change in hardening behavior of the materials is thought to be due to either local temperature rise or strain localization within the microstructure leading to damage and failure. Understanding of such behavior requires further experimental analysis and therefore was left for future research.

3.2. Fracture criterion

The onset of fracture in a shear test is difficult to determine visually since the cracks do not appreciably open. The cracks propagate through the material with the two fracture surfaces maintaining contact. Visual means of detecting fracture are further complicated by the painted surface required for the DIC analysis. As shown in Fig. 9, shear cracks are evident on the unpainted AA5182-O sample while the cracks are difficult to discern in painted samples. In order to increase the accuracy of the fracture strain data, a fracture criterion for the experimental data set is proposed herein based on the hardening rate.

The logarithmic (true) finite strain tensor for simple shear deformation maintains a plane strain loading condition at all times and the thickness remains constant. The modified maximum force criterion (MMFC) of Hora et al. [18] is an extension of the Considere criterion and states that the strain path in tensile-dominated stress states will converge to plane strain loading before fracture. As a material begins to lose its ability to work harden, the stress state will transition to plane strain and then abruptly fracture. Since the simple shear test is already in a constant state of plane strain, the loss of work hardening should coincide with the onset of cracking and local failure. This criterion can be expressed as

$$\textit{Hardening rate} = d\tau/d\varepsilon_{eq} \leq 0 \textit{ for fracture onset} \quad (3)$$

where, τ is the average shear stress and ε_{eq} is the von Mises equivalent strain.

Figure 10a shows the stress response and hardening rate along with the onset of fracture using Eq. (3). The onset of fracture is taken at the strain level when Eq. (3) is satisfied and remains negative for the duration of the test. Shear fracture is not immediate for AA5182-O and DP600 as they have moderate ductility and sufficient crack resistance so that the load decreases during crack propagation. An inspection of the local equivalent strain rate in Fig. 10b also shows an increase in the local strain rate that is relatively in good agreement with onset of cracking by using Eq. (3). As such, Eq. (3) provides a very good estimate for the onset of fracture.

It is important to note that the fracture criterion in Eq. (3) is not required for materials with relatively low ductility that are susceptible to shear fracture. Abedini et al. [15] observed that shear fracture of ZEK100 magnesium and the higher strength DP780 exhibited abrupt shear fracture without a decrease in the load prior to failure.

The fracture strain of DP600 and AA5182-O were obtained with respect to the strain rates from the

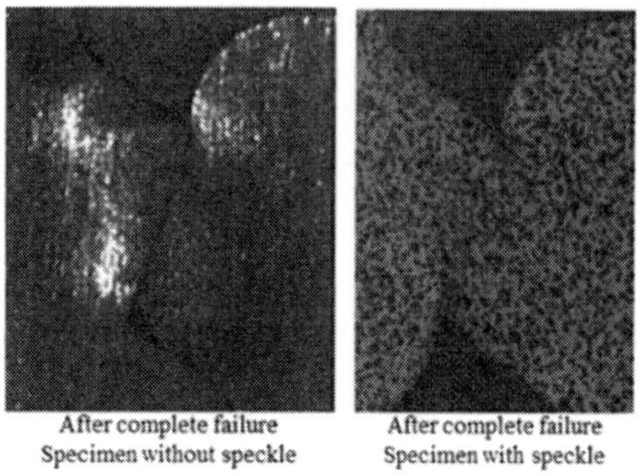

Figure 9. Fractured AA5182-O shear specimen with and without a painted speckle pattern for DIC strain analysis.

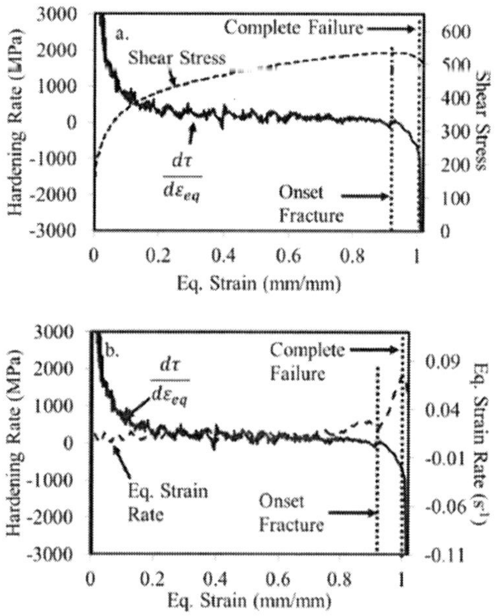

Figure 10. (a) Hardening rate and shear stress and (b) hardening rate and equivalent strain rate plotted against the equivalent strain indicating the onset fracture strain and a spike in strain rate at onset fracture for DP600 at a quasi-static strain rate.

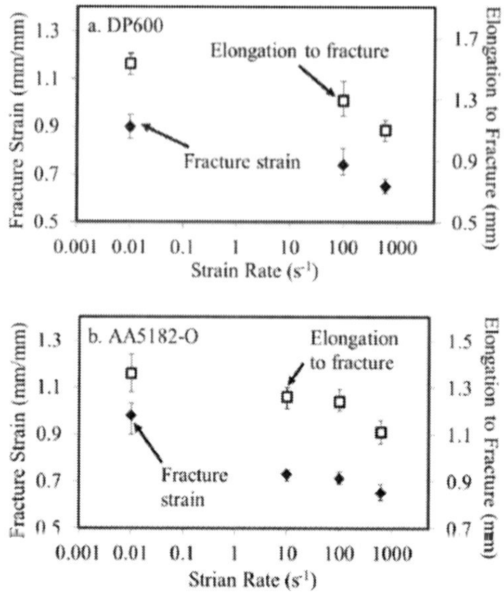

Figure 11. Equivalent fracture strain and elongation to fracture plotted against strain rates for (a) DP600 and (b) AA5182-O.

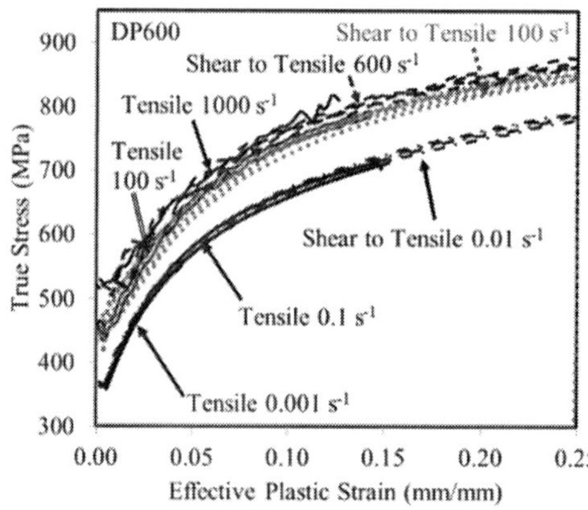

Figure 12. True stress vs. effective plastic strain comparison for experimental data from tensile [16] and shear testing at the strain rate ranging from 0.01 to 600 s^{-1}.

experimental data based on the proposed fracture criteria. The effect of strain rate on the fracture strain for DP600 and AA5182-O sheet material is shown in Fig. 11a and Fig. 11b, respectively. Total elongation to fracture is also plotted against the strain rate in the same figures. A negative rate sensitivity trend is observed in the fracture strain and elongation to fracture for both materials. For an increase in strain rate from 0.01 to 600 s^{-1}, the corresponding decrease in the equivalent fracture strain was found to be 0.25 and 0.33 for DP600 and AA5182-O sheet material, respectively.

3.3. Tensile and shear stress comparison

A conversion of shear data to tensile data can be useful to obtain flow stress curves for larger strains as well as to obtain flow stress curves for ultra-high ($>10^4$ s^{-1}) strain rate testing conditions that are not typically achievable in the uniaxial tensile testing. The shear stress data obtained in this work was converted to tensile stress using Von-Mises criteria assuming isotropic material behaviour. Shear stress up to 30% equivalent strain was converted to equivalent tensile stress in order to stay within the infinitesimal strain regime.

The converted shear to tensile data is compared with the experimental data obtained from uniaxial tensile testing for the same sheet materials along diagonal (45° to the rolling) direction [16] at comparable nominal strain rates (Figs. 12 and 13). Due to the relatively isotropic behaviour of DP600 sheet, the converted tensile data form shear testing shows an excellent agreement with the uniaxial tensile test data for strain rates ranging from low to high

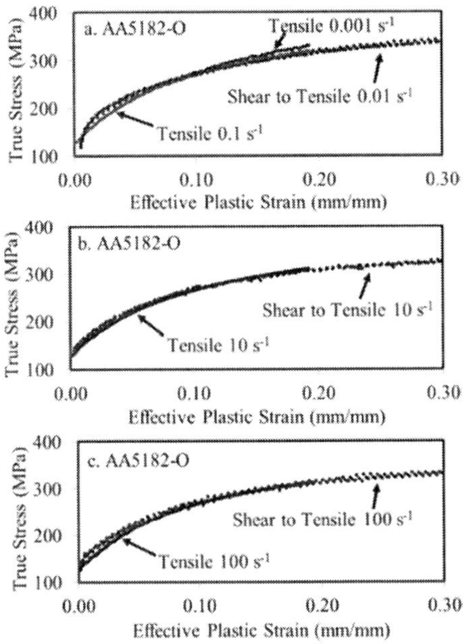

Figure 13. True stress vs. effective plastic strain comparison for experimental data from tensile [16] and shear testing at (a) quasi-static (b) 10 and (c) $100\,s^{-1}$ strain rate.

(as shown in Fig. 12). Figure 13(a-c) compares tensile data converted from shear test data with the tensile data obtained from uniaxial tensile tests for AA5182-O. The tensile data converted from the shear data using the von Mises yield criterion shows very good agreement with each other at the dynamic strain rate test conditions for AA5182-O, while only a satisfactory comparison is observed at the quasi-static strain rate condition. Further improvement in converting shear data to tensile data could be achieved using anisotropic yield criteria such as Barlat anisotropic material models [19]. Future work will focus on ultra-high strain rate experiments using the mini-shear specimen on a Hopkinson bar to obtain the tensile flow stress response of these materials.

4. Conclusions

I. The shear specimen developed by Peirs et al. [12] was successfully modified to a mini size shear specimen in order to perform high strain rate shear testing.

II. At equivalent strains lower than approximately 0.25, DP600 exhibited positive rate sensitivity, while AA5182-O is found to be strain rate insensitive. However, the hardening rate decreased at larger strains for both materials as the strain rate increased. This reduction in hardening behavior is thought to be due to thermal softening at higher strain rates.

III. For an increase in strain rate from 0.01 to $600\,s^{-1}$, the corresponding decrease in fracture strain was found to be approximately 0.25 and 0.33 for DP600 and AA5182-O sheet metals, respectively.

The authors greatly appreciate financial support from Automotive Partnerships Canada (APC), the Natural Sciences and Engineering Research Council (NSERC) of Canada, the Canada Research Chairs Secretariat, the Ontario Research Fund, Ford Research and Advanced Engineering, Amino Corporation, and ArcelorMittal.

References

[1] Y. Bai, T. Wierzbicki, Int. J. Impact Eng, **24**, 1071–1096 (2008)
[2] Y. Bai, T. Wierzbicki, Int. J. Fracture, **161**, 1–20 (2010)
[3] N. Iosipescu, J. Mater, **2**, 537–566 (1967)
[4] E.F. Rauch, J. Eng. Mater. Tech, **131**, 0111107-1-8 (2009)
[5] M. Gasperini, C. Pinna, W. Swiatnicki, Acta Mater, **40**, 4195–4208 (1996)
[6] S. Bouvier, H. Haddadi, P. Levee, C. Teodosiu, J. Mater. Tech, **172**, 196–103 (2006)
[7] J.R. Klepaczko, Int. J. Impact Eng, **15**, 25–39 (1994)
[8] J.R. Klepaczko, A. Rusinek, Int. J. Plasticity, **17**, 87–115 (2001) F. Gao, L. Gui, Z. Fan, Exp. Mech, **51**, 891–901 (2011)
[9] A. Reyes, M. Eriksson, O.G. Lademo, O.S. Hopperstad, M. Langseth, Mat. And Design, **30**, 596–608 (2009)
[10] J. Kang, D.S. Wilkinson, P.D. Wu, M. Bruhis, M. Jain, J.D. Embury, R.K. Mishra, J. Eng. Mater. Tech, **130**, 1–5 (2008)
[11] J. Peirs, P. Verleysen, J. Degrieck, Exp. Mech, **52**, 729–741 (2012)
[12] J. Peirs, P. Verleysen, W. Van Paepegem, J. Degrieck, Int. J. Impact Eng, **38**, 406–415 (2011)
[13] V. Tarigopula, O.S. Hopperstad, M. Langseth, A.H. Clausen, F. Hild, O.G. Lademo, M. Eriksson, Exp. Mech, **48(2)**, 181–196 (2008)
[14] Y.B. Bao, T. Wierzbicki, Int. J. Mech. Sci, **46(1)**, 81–98 (2004)
[15] A. Abedini, C. Butcher, D. Anderson, M.J. Worswick, T. Skszek, SAE World Congress, Paper # 2015-15H-0274 (2015)
[16] T. Rahmaan, A. Bardelcik. J. Imbert, C. Butcher, M. Worswick, submitted to Int. J. Impact Eng (2014)
[17] A. Bardelcik, Ph.D. thesis, Uni. of Waterloo (2012)
[18] P. Hora, L. Tong, J. Reissner, Proceedings of the Numisheet'96 Conference, 252–256 (1996)
[19] F. Barlat, Advanced Methods Mat. Forming, Springer Berlin Heidelberg, 1–18 (2007)

EPJ Web of Conferences **94**, 01034 (2015)
DOI: 10.1051/epjconf/20159401034

Tensile behaviour of geopolymer-based materials under medium and high strain rates

Costantino Menna[1,a], Domenico Asprone[1], Daniele Forni[2], Giuseppina Roviello[3], Laura Ricciotti[3], Claudio Ferone[3], Anna Bozza[1], Andrea Prota[1], and Ezio Cadoni[2]

[1]Department of Structures for Engineering and Architecture, University of Naples Federico II, 80125 Naples, Italy
[2]DynaMat Laboratory, University of Applied Sciences of Southern Switzerland, 6952 Canobbio, Switzerland
[3]Dipartimento di Ingegneria, University of Naples Parthenope, Centro Direzionale Napoli, 80143 Naples, Italy

Abstract. Geopolymers are a promising class of inorganic materials typically obtained from an alluminosilicate source and an alkaline solution, and characterized by an amorphous 3-D framework structure. These materials are particularly attractive for the construction industry due to mechanical and environmental advantages they exhibit compared to conventional systems. Indeed, geopolymer-based concretes represent a challenge for the large scale uses of such a binder material and many research studies currently focus on this topic. However, the behaviour of geopolymers under high dynamic loads is rarely investigated, even though it is of a fundamental concern for the integrity/vulnerability assessment under extreme dynamic events. The present study aims to investigate the effect of high dynamic loading conditions on the tensile behaviour of different geopolymer formulations. The dynamic tests were performed under different strain rates by using a Hydro-pneumatic machine and a modified Hopkinson bar at the DynaMat laboratory of the University of Applied Sciences of Southern Switzerland. The results are processed in terms of stress-strain relationships and strength dynamic increase factor at different strain-rate levels. The dynamic increase factor was also compared with CEB recommendations. The experimental outcomes can be used to assess the constitutive laws of geopolymers under dynamic load conditions and implemented into analytical models.

1. Introduction

Geopolymers, also known as "alkali-activated cements" or "alkali-activated aluminosilicates", represent a new class of high performance inorganic materials characterized by a three dimensional, CaO-free, silico-aluminate chemical structure [1]. They can be produced by mixing reactive aluminosilicate materials such as metakaolin [2,3], industrial [4] or natural waste products [5], and strongly alkaline solutions such as NaOH or KOH, resulting in a rigid geopolymeric network.

These materials are widely studied since they show interesting mechanical properties, thermal stability, freeze-thaw, chemical and fire resistance, long term durability and environmental advantages; for these reasons, their applications cover many fields.

First of all, geopolymer materials represent a "greener" alternative of conventional Ordinary Portland Cement (OPC) in new Sustainable Buildings due to several environmental advantages, such as a lower CO_2 footprint and low production temperature. Geopolymer technology also entails inexpensive and eco-friendly synthetic procedures, encompassing clays and other natural silico-aluminates as well as industrial wastes like coal fly ash and sediments [6]. Moreover, at elevated temperatures, geopolymer concretes exhibit less potential to suffer from spalling in comparison with OPC concrete.

Recently, geopolymer-based hybrid composite systems have been also investigated in order to overcome some issues linked to the brittle behaviour, the unknown long-term durability, and rheological behaviour of geopolymers that usually restrict the possible number of "in-situ" applications as structural materials. Among them, a great interest has been devoted to polymeric fillers due to their low density, chemical stability and easy processing. It was reported that hybrid composite materials based on geopolymers containing an organic epoxy resin content (up to 25%wt) can be effectively achieved by carrying out a direct polymerization in situ, simultaneously to the geopolymeric polycondensation reaction [7,8].

As far as geopolymer and geopolymer hybrid composite mechanical properties are concerned, most of the experimental works have been conducted within the quasi-static regime with the aim of determining the compressive strength, compressive Young's modulus, Poisson's ratio, splitting tensile strength, and fracture energy. Moreover, while strain rate effect on OPC concrete has been extensively studied, only a few studies have been conducted on the dynamic properties of geopolymer-based concretes, mainly focused on damping and compressive behaviour.

In particular, the damping behaviour of geopolymer pastes was investigated by Pan et al. [9], showing comparable performances with OPC pastes. Khandelwal et al. [10] investigated the strain rate effect on the Young's modulus, Poisson's ratio and compressive strength of low-calcium fly-ash-based geopolymer mortars. It was found that these properties substantially increase in a logarithmic manner as the strain rate increases. Recently, Feng et al. [11] investigated the effect of strain rate on

[a] Corresponding author: costantino.menna@unina.it

Article available at http://www.epj-conferences.org or http://dx.doi.org/10.1051/epjconf/20159401034

the compressive behaviour of geopolymer concrete and mortar by means of a split Hopkinson pressure bar in the range of 10^{-6}–$370\,s^{-1}$ of strain rate. They found that the dynamic increase factor for compressive strength and the corresponding critical strain increased with the increasing strain rates and CEB recommendations underestimated this factor in the case of geopolymer concrete.

Tensile behavior under high strain rates has not been investigated yet, as far as the authors know, in available literature. However dynamic tensile behavior is of large interest for understanding the fracture development and the energy absorbing capabilities of these kinds of materials, when subjected to impulsive loads and impacts.

In the present work, the tensile behavior of a metakaolin based geopolymer and two hybrid geopolymer composites is investigated. Direct tensile and splitting tests are performed using an Hydro-pneumatic machine and a modified Hopkinson bars, both installed at the DynaMat laboratory of the University of Applied Sciences of Southern Switzerland.

2. Materials and experimental methods

2.1. Geopolymer-based materials

Three different geopolymer-based materials are investigated and tested under dynamic conditions: a metakaolin based geopolymer (neat geopolymer, hereafter referred as GEO), an hybrid organic resin-geopolymer composite (GEO-Epoxy) and a hybrid inorganic resin-geopolymer composite (GEO-SIL). GEO samples were obtained starting form an alkaline activating solution prepared by dissolving solid sodium hydroxide into a sodium silicate solution with a composition expressed as $Na_2O\ 1.4\,SiO_2\ 10.5\,H_2O$ [12]. Then, metakaolin was incorporated into the activating solution with a liquid to solid ratio of 1.4:1 by weight, and mixed by a mechanical mixer for 10 min at 800 rpm. The composition of the whole geopolymeric system can be expressed as $Al_2O_3 \cdot 3.5\,SiO_2 \cdot 1.0\,Na_2O \cdot 10.5\,H_2O$, assuming that the geopolymerization occurred at 100%.

Hybrid-organic (Geo-Epoxy) and inorganic (Geo-SIL) samples were prepared by incorporating a commercial epoxy and siloxanic resin, according to the 10% of weight content, into the freshly prepared geopolymeric suspension under mechanical stirring, when the polycondensation reactions of the components were already started but far to be completed [8,13].

Afterwards, a set of samples was poured in cylindrical moulds and cured at 60 °C for 48 h and then kept in > 95% relative humidity conditions at room temperature for further 5 days. Finally, the specimens were kept further 21 days in air at room temperature. For all the samples, the evaporation of water was prevented by sealing the top of the moulds with a plastic film during the curing stage.

2.2. Experimental setup

The tensile dynamic behaviour of the investigated geopolymer-based samples was investigated by means of direct tensile and splitting tests. Both testing configurations were performed on 20 mm high cylindrical specimens with

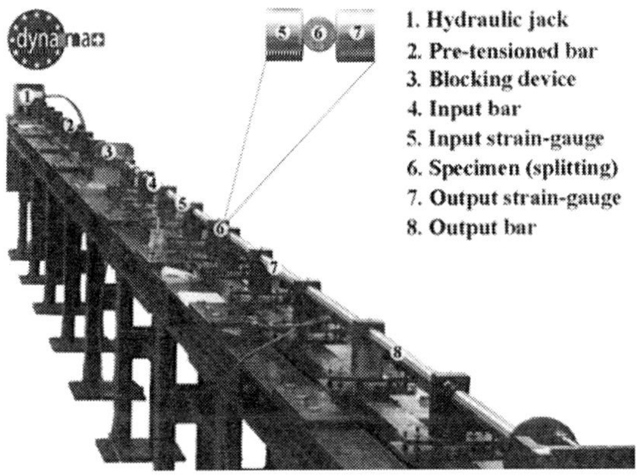

Figure 1. Modified Hopkinson bar setup for splitting tests.

a diameter of 20 mm. Each cylinder used for direct tensile test was also notched (with a notch depth of 1.5 mm). At least three specimens for each loading condition were considered.

A preliminary static characterization was initially conducted on the investigated specimens by means of a universal electro-mechanic machine with a constant strain-rate of $10^{-5}\,s^{-1}$.

Dynamic tensile failure tests were conducted, in the DynaMat laboratory of the University of Applied Sciences of Southern Switzerland, on all the specimens, investigating medium strain-rate and the high strain-rate behaviour. The former, commonly considered as ranging from 10^{-1} to $10\,s^{-1}$, can be induced by severe seismic actions, while the latter, varying between 10 and $10^2\,s^{-1}$, can be due to impact or blast loading. Two different experimental setups were used. For medium strain-rate tests, a hydro-pneumatic machine (HPM) was employed, whereas two modified Hopkinson bar apparatus, in tension (MHBt) and compression (MHBc), were used to conduct high strain-rate tensile and splitting tests, respectively.

The functioning of the hydro-pneumatic machine and the modified Hopkinson bar setup in tension have been widely reported [14,15], while the modified Hopkinson bar in compression, used for splitting tests (Fig. 1), consists of a hydraulic actuator (①) that put in tension a preloading bar (②) thanks to a blocking ring (③) placed at the extremity of this bar. The preloading bar, made of high strength steel, is 6 m long and 12 mm in diameter. This bar is connected to an aluminium bar with a diameter of 30 mm and a length of 3 m, working as a input bar (④). The specimen (⑤) is sandwiched between the input bar and another identical bar used as output bar (⑥). By pulling the pre-loading bar, it is possible to drive the test by the energy stored in it. The test starts when a fragile bolt positioned between the pre-stressed bar and the hydraulic actuator (①) by suddenly breaks. Consequently, a rectangular stress wave pulse is generated and propagates through the input bar, the specimen and the output bar. By the strain gauges placed on the input and output bars, the test signals are obtained.

Each dynamic test was filmed by means of a fast recording camera *MotionPro Y4-S3*.

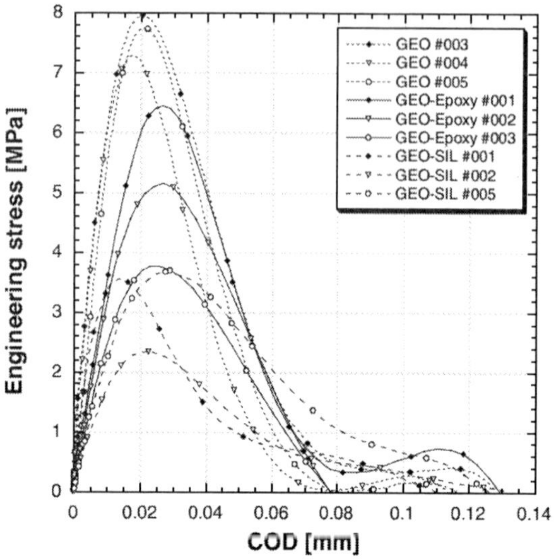

Figure 2. Tensile tests at $150\,s^{-1}$.

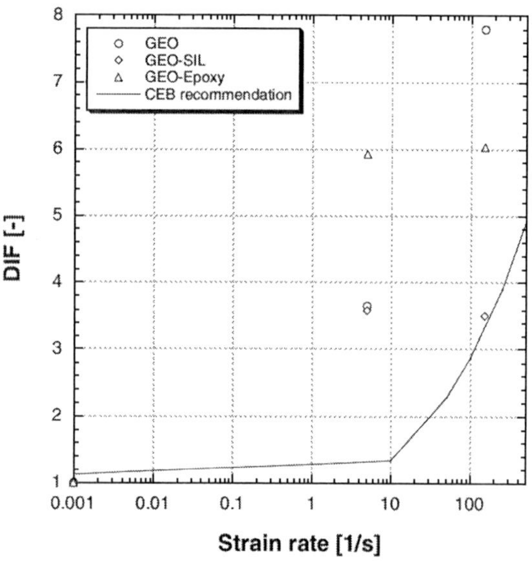

Figure 3. DIF versus strain rate.

The results are processed in terms engineering stress versus crack opening displacement (COD) and strength Dynamic Increase Factor (DIF) at different strain-rate levels.

3. Results and discussion

The strain rate behaviour of the three different geopolymer-based materials is hereafter discussed.

The results of the high strain rate direct tensile tests, in terms of engineering stress versus crack opening displacement, are depicted in Fig. 2. The obtained strain rate, measured in correspondence of the maximum tensile strength, is approximately $150\,s^{-1}$. Is it possible to observe that the addition of epoxy and siloxanic resins led to a decrease of the maximum tensile strength with respect to

Table 1. DIF experimental parameters.

	Tensile tests		Splitting tests
	$5\,s^{-1}$	$150\,s^{-1}$	$150\,s^{-1}$
GEO	3.66	7.80	3.99
GEO-SIL	3.59	3.51	4.26
GEO-Epoxy	5.93	6.04	3.35

Figure 4. Pictures of a splitting test (GEO) at 1/27100 fps.

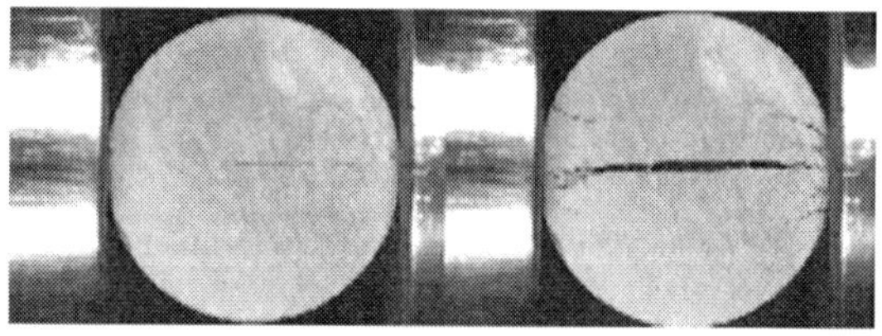

Figure 5. Pictures of a splitting test (GEO-SIL) at 1/28300 fps.

the metakaolin based geopolymer. This could be due to the different microstructures characterizing the different investigated materials [8, 13]. In particular, GEO samples have a rather irregular microstructure with diffused nano-pores due to the evaporation of the reaction water. On the contrary, the epoxy resin in the GEO-Epoxy samples consolidates within the geopolymer matrix as discrete micrometric spheres (i.e. a heterogeneous composite microstructure), affecting the overall dynamic behaviour due to their viscoelastic characteristics. GEO-SIL samples are characterized by a very compact and interconnected microstructure up to the nanometric scale that reduces the strain rate effect also for high levels of strain rate.

The strain rate behaviour can be also represented by means of the ratio between the dynamic and the static strength, commonly known as the Dynamic Increase Factor (DIF). The results in terms of DIF for both direct tensile and splitting tests have been collected in Table 1 and represented in Fig. 3. A comparison between the CEB [16] recommendation, widely used to predict the tensile behaviour of concrete under high strain rate, was also depicted in Fig. 3. It can be observed that GEO samples exhibited a remarkable change of DIF passing from the medium strain-rate and the high strain-rate level, whereas GEO-SIL and GEO-Epoxy seemed to not be affected by the two different ranges of strain rate.

The fracture development was filmed by means of a fast recording camera, highlighting different fracture evolutions: a single macro fracture was observed for both GEO (Fig. 4) and GEO-SIL (Fig. 5) materials, while a multi fracture was observed for GEO-Epoxy

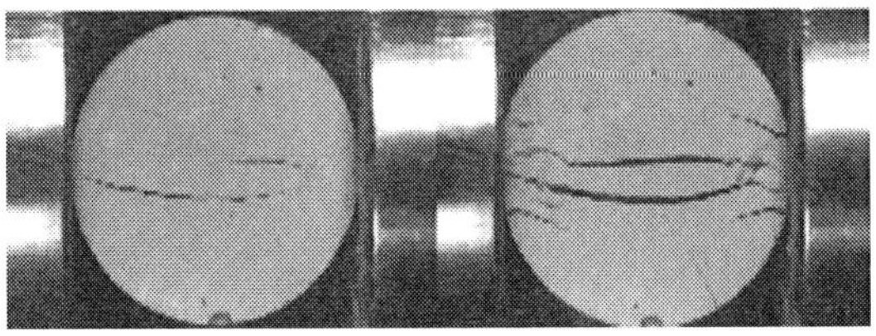

Figure 6. Pictures of a splitting test (GEO-Epoxy) at 1/28300 fps.

(Fig. 6) material. This could be due to the composite heterogeneous microstructure characterizing GEO-Epoxy samples. Indeed, different crack initiation sites could form under splitting tension loading, especially at the interface region between the epoxy micrometric spheres and the geopolymer surrounding matrix. The recorded DIF under splitting tensile loading was approximately 4 for all the three investigated materials.

4. Conclusions

The tensile behaviour of a metakaolin based geopolymer and two hybrid geopolymer composites was investigated at medium and high strain rates by means of dynamic tensile and splitting tests. The two types of tensile characterization tests pointed out the importance of the microstructure characterizing geopolymer-based materials with regards to strain rate effect.

Neat geopolymer material was strongly affected by strain rate (both medium and high levels), showing a remarkable shifting of the engineering stress versus crack opening displacement curves. The splitting dynamic test drew the attention also to the fracture development under high strain rate, resulting in multiple fracture sites only for GEO-Epoxy samples.

The dynamic behaviour of such geopolymer-based materials combined with aggregates for concrete products is currently under investigation by the authors.

References

[1] J. Davidovits, *Geopolymer Chemistry and Applications*, Institut Géopolymère 3rd ed., Saint Quentin, France (2014) `www.geopolymer.org`.

[2] J. Davidovits, J Therm Anal Calorim **35(2)**, 429–441 (1989)

[3] P. Duxson, A. Fernández-Jiménez, J. L. Provis, G. C. Lukey, A. Palomo, J. S. J. van Deventer, J Mater Sci **42(9)**, 2917–2933 (2007)

[4] S. Andini, R. Cioffi, F. Colangelo, C. Ferone, F. Montagnaro, L. Santoro, Adv Sci Technol **69**, 123–128 (2010)

[5] A. Buchwald, M. Hohmann, K. Posern, E. Brendler, Appl Clay Sci **3**, 300–304 (2009)

[6] B. Molino, A. De Vincenzo, C. Ferone, F. Messina, F. Colangelo, R. Cioffi, Materials **7**, 5603–5616 (2014)

[7] C. Ferone, G. Roviello, F. Colangelo, R. Cioffi, O. Tarallo, Appl Clay Sci **73**, 42–50 (2013)

[8] G. Roviello, L. Ricciotti, C. Ferone, F. Colangelo, R. Cioffi, O. Tarallo, Materials **6**, 3943–3962 (2013)

[9] Z. Pan, K.N. Feng, K. Gong, B. Zou, A.H. Korayem, J. Sanjayan, J Mater Sci **48(8)**, 3128–3137 (2013)

[10] M. Khandelwal, P.G. Ranjith, Z. Pan, J.G. Sanjayan, Arab J Geosci **6(7)**, 2383–2389 (2011)

[11] K.N. Feng, D. Ruan, Z. Pan, F. Collins, Y. Bai, C.M. Wang et al., Mater Struct **48**, 671–681 (2015)

[12] C. Ferone, F. Colangelo, G. Roviello, R. Cioffi, C. Menna, D. Asprone, A. Balsamo, A. Prota, G. Manfredi, Materials **6**, 1920–1939 (2013)

[13] G. Roviello, C. Menna, O. Tarallo, L. Ricciotti, C. Ferone, F. Colangelo, D. Asprone, R di Maggio, E. Cappelletto, A. Prota, R. Cioffi, sumbitted to Cement and Concrete Research (to be published)

[14] E. Cadoni, Rock Mech Rock Eng, **43**, 667–676 (2010)

[15] E. Cadoni, M. Dotta, D. Forni, P. Spätig, J. Nucl. Mat. **414(3)**, 360–366 (2011)

[16] FIB-International Federation for Structural Concrete, *Model Code 2010* (2012)

EPJ Web of Conferences **94**, 01035 (2015)
DOI: 10.1051/epjconf/20159401035

The effect of cutting speed on residual stresses when orthogonal cutting TC4

Chaofeng Liu[1,a], Zengqiang Wang[1], Guang Zhang[2], and Lei Liu[1]

[1]Key Laboratory of Ministry of Education for Contemporary Design and Integrated Manufacturing Technology, Northwestern Polytechnical University, Xi'an 710072, China
[2]Northwestern Polytechnical University, School of Aeronautics, Xi'an 710072, China

Abstract. As one of the most important parameters in mental cutting, cutting speed has a significant influence on residual stress. Finite element method and experiment method are used to study the relationship between cutting speed and residual stress when orthogonal cutting TC4 titanium alloy. The result of simulation and experiment shows that: when the cutting speed is low, the residual stress in axial direction is compressive stress and gradually converts to tensile stress with the increase of cutting speed, but it will convert to compressive stress again if the cutting speed continues to increase; the residual stress in tangential direction is constant to compressive stress and it will decrease with the increase of cutting speed.

1. Introduction

As aerospace materials, titanium alloy has a lot of excellent performances, such as low density, high strength-to-weight ratio, excellent temperature resistance, good corrosion resistance. Thus, titanium alloys is often used to manufacture aircraft engine bearing parts and rotating parts. However, the titanium alloys have low elastic modulus, seriously chilled property, high cutting temperature, large cutting force per unit area and serious tool wear, these may have a serious influence on the surface integrity of titanium components. Because of its poor intuitive and imperceptible, residual stress is often overlooked. However, residual stress has remarkable influence for the strength, fatigue life, corrosion resistance, dimensional stability of the structure, especially in the bearing components and the rotating components. The presence of residual stress often results in sudden destruction of the structure and the consequence is often very serious. Therefore, it is very important to deeply understand the relationship between cutting parameters and residual stress.

With the rapid development of computer graphics technology and finite element simulation technology, numerical simulation technology has been widely applied to the study of cutting force, cutting temperature and other metal cutting process. Valiorgue [1] developed a three-dimensional finite element model, by applying a thermal coupling load instead of the model material separation and chips forming to simulate the surface residual stress distribution of fine turning AISI304L; Sima [2] found that the constitutive equations of Ti-6Al-4V has a greatly effects on chip formation; Mohamed [3] studied the effects of four different nose radius on AISI316L orthogonal cutting surface residual stress high-speed machining process of titanium alloy; Outeiro [4] studied the effect of tool geometry, cutting, cutting thickness and feed on cutting surface residual stress by experimental and numerical methods; Shet [5] studied the effects of the rake angle and the friction coefficient on the machined surface residual stress and strain fields under the orthogonal metal cutting process under plane strain conditions; Lei [6] came up with a new material constitutive model for orthogonal cutting and studied the 1020 steel shear stress field, strain field and temperature field; Xie [7] used the finite element method to simulate the formation of the chip in metal cutting process and studied the effect of the shear banding, cutting force, rake angle on the shear angle, plastic strain field, and distribution of effective stress on tool rake face; Yu [8] using the finite element method to simulate pulling for cutting a workpiece after the stress changes in the surface residual stress; Kim [9] developed two-dimensional steady state orthogonal cutting finite element model based on metal cutting thermal coupling model; ALBERT [10] applied the finite element method to study contact mechanism and the workpiece deformation mechanism between the tool and the workpiece during metal cutting process; Sun [11] established a 3D finite element model according to the thermal-elastic-plastic finite element theory and studied the effect of cutting parameters and tool parameters on aerospace aluminum alloy surface residual stress. Huang [12] studied the effect of several special finite element techniques, such as the chip separation criteria and friction model, on the finite element simulation.

2. Numerical simulation procedures

2.1. Material constitutive model

Metal cutting process is a complex thermal elastic-plasticity of nonlinear problems, involving high strain,

[a] Corresponding author: chaofengliu@sina.com

Table 1. Johnson-cook material parameters of TC4 [13].

A	B	C	m	n
860	683	0.035	1.0	0.47

high strain rate and high temperature. Therefore, in order to reflect actual machining process, the constitutive model should be chosen under considering the influence of various factors on the flow stress of metallic materials. The Johnson-cook constitutive model is currently the most widely adopted constitutive model for fitting the mechanical properties of metal cutting, the constitutive equation is expressed as Eq. (1):

$$\bar{\sigma} = [A + B(\varepsilon^{pl})^n]\left[1 + C\ln\left(\frac{\bar{\varepsilon}}{\bar{\varepsilon}_0}\right)\right]\left[1 - \left(\frac{T - T_r}{T_m - T_r}\right)^m\right] \quad (1)$$

where $\bar{\sigma}$ is equivalent stress; A, B, C, m, n respectively represent the yield strength, strain hardening coefficient, strain rate hardening coefficient, temperature dependence coefficient and strain hardening exponent under quasi-static conditions, ε^{pl} is the equivalent plastic strain, $\bar{\varepsilon}$ is equivalent plastic strain rate, $\bar{\varepsilon}_0$ is quasi-static strain rate. Johnson-cook model material parameters forTC4 are listed in Table 1.

2.2. Chip separation criteria

Metal cutting process is a process, which chip continuously separated from the workpiece. In order to simulate the mechanical properties of metal cutting the chip separation criterion of simulation process must be determined. Separation rule of metal cutting finite element simulation is mainly divided into geometric criterion and physical standards. Geometric criterion defines the critical distance between tool tip and the element node in tool path, when the distance between the tool tip and node is less than the setting critical value, the node starts to separate. Physical standards define the critical value of physical quantity of simulation model of nodes, such as: stress, strain, strain energy density, when the physical quantity of the node of workpiece model reaches the critical value, the node begins to separate. In order to get better simulation, in this study the Johnson-cook shear failure model of metal cutting chip separation simulation was adopted, as Eq. (1):

$$\omega = \sum \frac{\Delta\varepsilon^{pl}}{\varepsilon_f^{pl}}. \quad (2)$$

2.3. Friction model

In the metal cutting process, there are two types of friction types: adhesive friction and sliding friction. These two different friction types based on Coulomb friction, we select the correction equation to simulate the friction behavior during the cutting process, as Eq. (3) and Eq. (4):

$$\begin{cases} \tau_f = \tau_s & \text{when} \quad \tau_f \geq \tau_s; \quad (3) \\ \tau_f = \mu\sigma_n & \text{when} \quad \tau_f < \tau_s; \quad (4) \end{cases}$$

Table 2. Mechanical properties and components of test specimen.

Mechanical properties	state	σ_b MPa	$\sigma_{0.2}$ MPa	σ_5 %	ψ %		
	M	895	825	10	25		
Components	Chemical components (%)		others (%)				
	Ti		single		sum		
	Allowance		≤0.1		≥0.4		
	Other ingredients (%)						
	Al	V	Fe	C	N	H	O
	5.7	4.0	0.14	0.02	0.04	0.001	0.17

Figure 1. The picture of cutting experiment.

where, τ_f is friction shear stress, τ_s is the critical shear yield stress of the workpiece, σ_n is the normal stress, μ is friction coefficient. In the simulation, if the tool and chip contact shear stress is greater than the critical shear yield stress of the workpiece, the point is in the adhesive friction area, otherwise the point is in the sliding friction.

3. Experiment and detection method of residual stress

Orthogonal cutting experiments were carried out on HK63 NC numerical control lathe. The test specimen employed in this work were TC4 bar with the dimension of 90 mm × 350 mm. The mechanical properties and components of test specimen were shown in Table 2.

The cutter type been used were LCMF160404-0400-FT CP500. The experiments were carried out with four different cutting speeds (60–90–120–150 m/min) and one cutting depth (0.1 mm). In this paper, the X-ray diffraction technology was used to detect the residual stress. Figure 1 shows the cutting process, and Fig. 2 shows the detection process of residual stress.

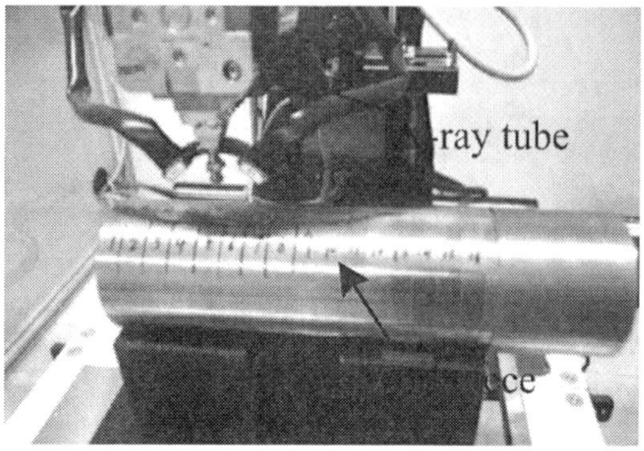

Figure 2. The picture of residual stress detection.

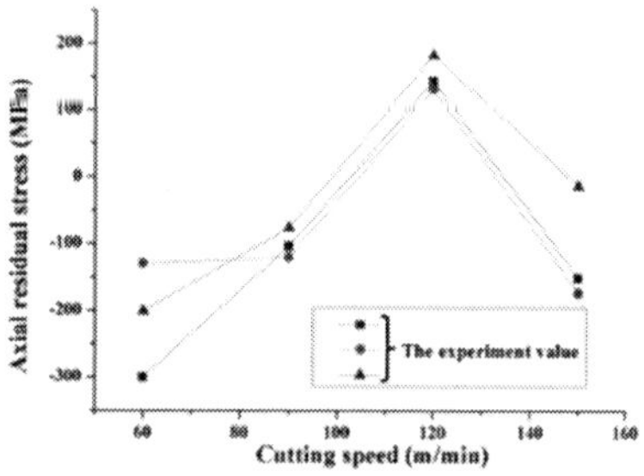

a. Axial direction

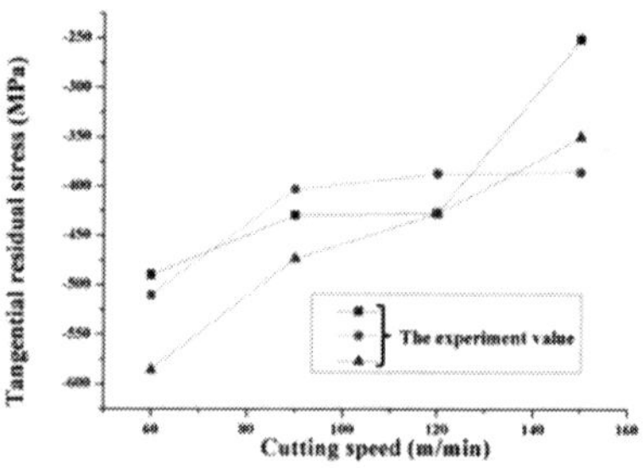

b. Tangential direction

Figure 3. The relationship between cutting speed and residual stress.

4. Results and discussion

In order to ensure the accuracy of detection results, we selected three points on the final surface after each set of experiments. As Fig. 3 shows, the three detection results are very close and have the same tendency. The deviations may be caused by the tiny internal structure difference in different detection- position.

When the cutting speed is lower, the residual stress in axial direction is compressive stress, and then gradually changes to tensile stress with the cutting speed increase, but finally it will convert to compressive stress again with the further increase in cutting speed (as it is shown in Fig. 3-a). As we all know, the residual stress are generated by the combination of thermal stress, mechanical stress and transformation factors. When the cutting speed is lower, the mechanical stress is dominant in the generation of residual stress which results in compressive stress. When the cutting speed goes higher, it seems that thermal stress and transformation will become more and more important.

The residual stress in tangential direction presents compressive stress and it will decrease with the cutting speed increase (as it is shown in Fig. 3-b), due to the temperature increase with cutting speed.

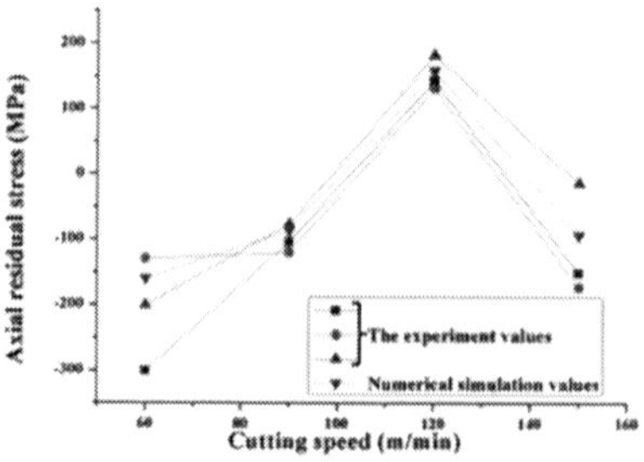

a. Axial direction

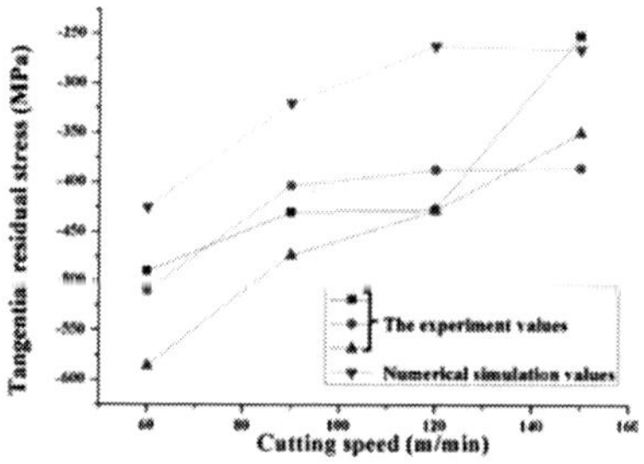

b. Tangential direction

Figure 4. The comparison between simulation and experiments.

Figure 4 shows the comparison between simulation and experiments.

As Fig. 4 show, the predicted results from numerical simulation are nearly of the same magnitude as the results obtained experimentally. Therefore, this numerical simulation method can be applied to study the relationship between cutting parameters and residual stress.

5. Conclusions

A numerical and experimental analysis of residual stress induced by orthogonal cutting of TC4 was performed in this investigation. The numerical results and experiments show good agreements. From the results obtained from this work, the following results can be derived.

- Cutting speed has a significant effect on residual stress.
- When the cutting speed changes from 60 m/min to 120 m/min, the axial residual stress tends to change from compressive stress to tensile stress gradually. When the cutting speed continues to increase, the tensile residual stress will decrease gradually.

– The tangential residual stress presents compressive stress and it will decrease with the cutting speed increase.

References

[1] F. Valiorgue, J. Rech, H. Hamdi et al., Int. J. Machine Tools and Manufacture, **53**(1), 77–90 (2012).

[2] M. Sima, T. Özel. Int. J. Machine Tools and Manufacture, **50**(11), 943–60 (2010).

[3] M.N.A. Nasr, E.G. Ng, M.A. Elbestawi, Int. J. Machine Tools and Manufacture, **47**(1), 401–11 (2007).

[4] J. Outeiro, D. Umbrello, R. M'Saoubi, Int. J. Machine Tools and Manufacture, **46**(14), 1786–94 (2006).

[5] C. Shet, X. Deng, Int. J. Machine Tools and Manufacture, **43**(6): 573–87 (2003).

[6] S. Lei, Y. Shin, F. Incropera, Int. J. Machine Tools and Manufacture, **39**(5), 731–50 (1999).

[7] J. Xie, A. Bayoumi, H. Zbib, Int. J. Machine Tools and Manufacture, **38**(9): 1067–87 (1998).

[8] X.X.Yu, W.S. Lau, T.C. Lee, Int. J. Machine Tools and Manufacture, **37**(10), 1525–1537 (1997).

[9] K.W. Kim, H-C. Sins, Int. J. Machine Tools and Manufacture, **36**(3): 379–97 (1996).

[10] A.J. Shih, Int. J. Machine Tools and Manufacture, **36**(1), 255–73 (1996).

[11] Y.Sun, H. Liu, Z. Lu, Chinese J Mech. Engng, **47**, 187–93 (2011).

[12] Z. Huang, Y. Ke, L. Wang, Acta Aer. Astr. sinica, **25**(3), 317–320 (2004).

[13] W-S Lee, C-F Lin, Mat. Sc. Engng: A, **241**(1), 48–59 (1998).

EPJ Web of Conferences **94**, 01036 (2015)
DOI: 10.1051/epjconf/20159401036

Mechanical behavior of ultrafine-grained materials under combined static and dynamic loadings

Y.Z. Guo[a], X.Y. Sun, J.G. Li, X. Yu, and Y.L. Li

School of Aeronautics, Northwestern Polytechnical University, Xi'an, China

Abstract. Ultrafine-grained (UFG) materials have extensive prospects for engineering application due to their excellent mechanical properties. However, the grain size decrease reduces their strain hardening ability and makes UFG materials more susceptible to deformation instability such as shear localization. In most cases, critical shear strain is taken as the criterion for formation of shear localization under impact loading or adiabatic shear band (ASB). Recently, some researchers found that the formation of ASB was determined only by the dynamic loading process and had nothing to do with its static loading history. They proposed for coarse-grained metals a dynamic stored energy-based criterion for ASB and verified it by some experiments. In this study, we will focus on the shear localization behavior of UFG metals such as UFG titanium and magnesium alloy AZ31. Quasi-static loading and dynamic loading will be applied on the same specimen alternately. The shear localization behavior will be analyzed and the criterion of its formation will be evaluated.

1. Introduction

Nanocrystalline/Ultrafine-grained (NC/UFG) materials have drawn extensive attention in the last few decades due to their superior mechanical properties, such as improved yield stress, high strength and good wear resistance (1). However, one drawback that hinders their engineering application is the decreased ductility. Many researchers have focused on this manner and made their contributions (2–4). The failure of UFG metals usually starts from necking under tension while from shear banding under compression. As in the case of dynamic loading, adiabatic shear (AS) localization is recognised as a trigger of catastrophic failure for most metals under compression (5–6). In fact, the phenomena of adiabatic shear banding (ASB) is widely encontered in industries such as metal forming, high-speed machining, balastic impact and so on (7). Whether ASB occurs or not in a material under impact depends on the competation between work hardening and thermal softening. Since the ability of strain hardening is greatly reduced or vanished for most NC/UFG materials, the propensity of ASB within these materials would be enhanced (8–9).

An important issue related to adiabatic shear localization is to build up the criterion of ASB formation. A few kinds of ctrterion have been established in the past few decades, which, however, were mainly based on the critical shear strain (6). Recently, Rittel and co-workers proposed a new point based on their experiments (10). They found that the shear localization of AM50 alloy was only determined by its dynamic loading history but had nothing to do with the fore stactic loading or thermal effect. They suggested that "dynamic mechanical energy" be a better criterion for ASB formation. In their following work, Rittel and co-workers (11) proposed a possible mechanism of ASB initiation, i.e., dynamic recrystalization. The microstructure of UFG materials are usually unstable compared to their coarse grain (CG) conterparts, thus it is reasonable to deduce that the ctriterion by Rittel et al. would be suitable for UFG materials. However, we did not find any published work on this aspect.

In this paper, combined static and dynamic compressive testing are carried out on two kinds of materials, ie. UFG titanium and fine-grained AZ31 magnesium alloy. The experimental results of CG AZ31 alloy are also given for comparison. The conditions of ASB formation are analyzed and the criterions are discussed. The results of this work would be instructive in understanding the nature of ASB.

2. Experiments

2.1. Materials

Two kinds of materials are employed in this work, ie. pure titanium and AZ31 magnesium alloy. The as-received Ti is a grade 2 commercial purity plate with an average equiaxed grain size of about 30 μm. The magnesium alloy is commercial Mg-Al-Zn alloy with chemical composition: Mg-3.28Al-1Zn-0.44Mn (all in wt.%, balance Mg) and grain size of about 16 μm. The as-received CG materials are then processed by equal channel anglar pressing (ECAP) method through Bc route to reduce their grain sizes (12–13). The details in materials fabrication are given elsewhere (14–15). Microstructures of the materials before and after processing are shown in Fig. 1 and Fig. 2. Clearly, the grain sizes are greatly reduced for both of the two kind of materials. Most of the grains of the ECAP-processed Ti are between 400 nm to 800 nm and the average grain size is about 500 nm, which falls in the range of UFG category.

[a] Corresponding author: guoyazhou@nwpu.edu.cn

Article available at http://www.epj-conferences.org or http://dx.doi.org/10.1051/epjconf/20159401036

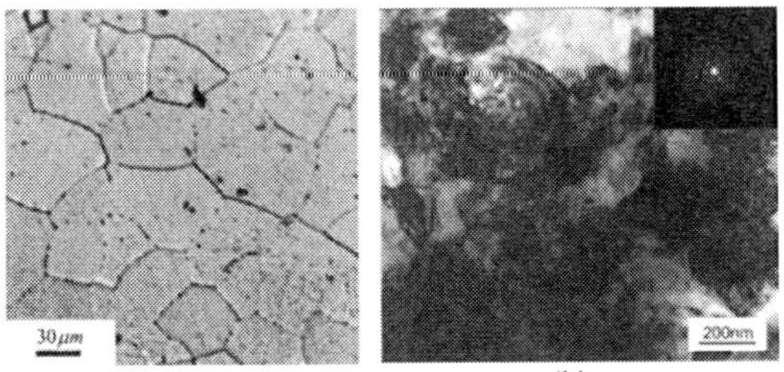

(a) (b)

Figure 1. The optical photographs of pure Ti. (a) as-received, (b) after ECAP. The average grain sizes were determined though lineal intercept procedure to be 30 and 0.5 μm, respectively.

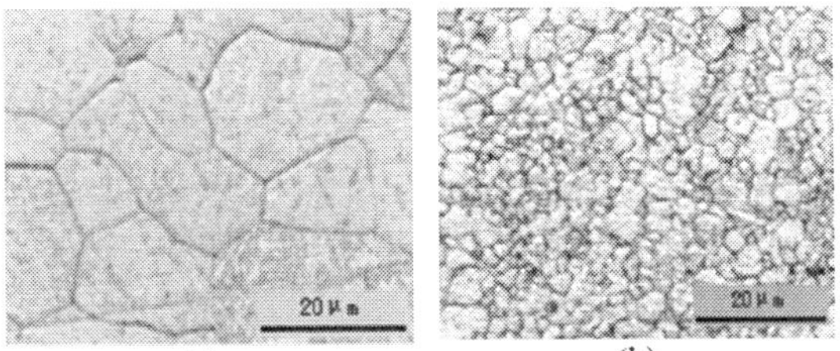

(a) (b)

Figure 2. The optical photographs of AZ31 Mg alloy. (a) as-received, (b) after ECAP. The average grain sizes were determined though lineal intercept procedure to be 16 and 1.7 μm, respectively.

The average grain size of ECAP-processed AZ31 is about 1.75 μm. The relatively larger grain is a result of the warm ECAP process.

2.2. Mechanical testing

Cylinderical specimens with dimension of Ø4 mm × 4 mm were prepared along the rolling/extrusion direction. The quasi-static uniaxial compressive experiments were carried out at room temperature on a servohydraulic testing machine and a extensometer was equipped to measure the deformation. The strain rate for all of the quasi-static tests are controlled to be $10^{-3}\,s^{-1}$. The dynamic testing were conducted via a split hopkinson pressure bar system at strain rates from $1500\,s^{-1}$ to $1800\,s^{-1}$. The strain-control during high strain rate testing is achieved by using stop-rings.

3. Results and discussion

Post-observation on the dynamically loaded specimens indicates that both the CG and UFG AZ31 are fractured due to adiabatic shear, shown in Fig. 3 as an example. The UFG Ti does not fracture under quasi-static compression in our tests (maximum strain 0.6), while adiabatic shear localization is observed in the specimen after high strain rate loading, shown in Fig. 4. The vein pattern on the fracture surface indicates material flow and adiabatic nature of the failure process. Since there is no apparent fracture in CG Ti under static or dynamic loading, these test results are not listed in this work.

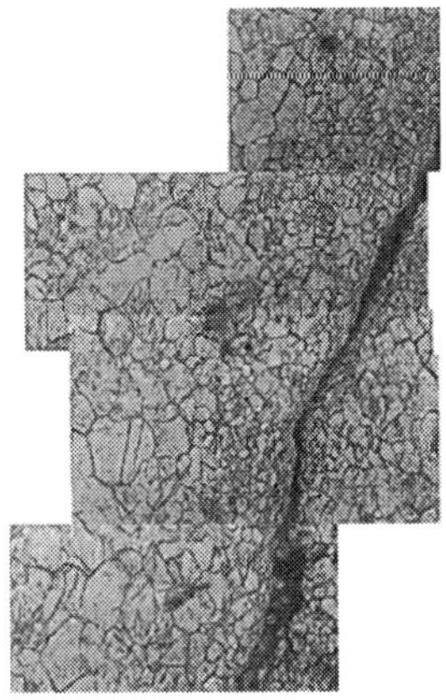

Figure 3. Representative dynamic compressive fracture mode of AZ31 alloy (CG).

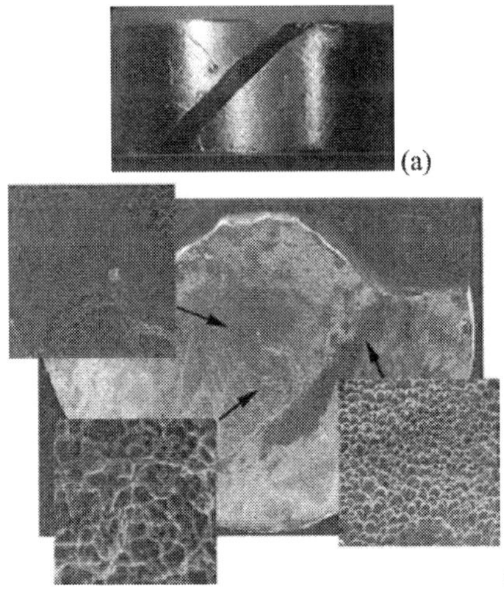

(a)

(b)

Figure 4. Representative compressive fractography of UFG Ti deformed at room temperature with strain rate $1500\,s^{-1}$. (a) Shear fracture, (b) fractography of the shear plane.

3.1. Mechanical properties of UFG titanium

The UFG Ti were loaded by combined static and dynamic loadings. Figure 5 displays the stress-strain curves of this material. Most of the specimens were loaded twice, either static loading followed by dynamic loading (Fig. 5 (a)) or dynamic loading followed by dynamic loading (Fig. 5 (b)), with curves of single loaded (static or dynamic) specimens for comparison. The charaters "S" and "D" within the icons of Fig. 5 indicate "Static" and "Dynamic", respectively. For example, "UFG6-S-D" denotes sample "UFG6" was firstly loaded statically to a certain strain (about 10% here) and then loaded dynamically until it was fractured. "UFG16-D-D-D" indicates sample "UFG16" was loaded three times at high strain rates ($1500\,s^{-1} \sim 1800\,s^{1}$). These two or three parts of the stress strain curves of one specimen are displaced with the same icon. The sharp increase of the curves in Fig. 5(b) is not the intrinsic behavior of the material but due to the contact of the stop-ring with the loading bars of SHPB. The arrows

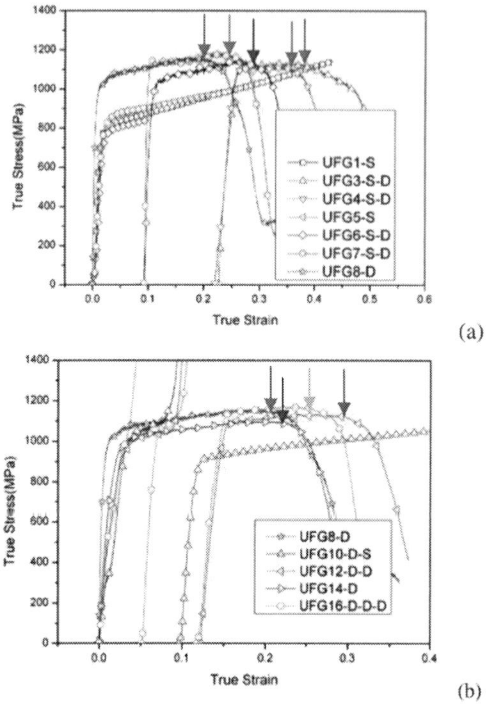

Figure 5. Mechanical behavior of UFG Ti under combined static and dynamic loading. (a) static loading followed by dynamic loading, (b) dynamic loading followed by dynamic loading. "S" and "D" indicate "Static" and "Dynamic", respectively. The abrupt rising up of the curves in (b) is not the intrinsic behavior of the material but due to the effect of stop-ring.

within the figure point out the strain where stress begin to fall.

Clearly, the fracture strain (or the critical strain for ASB initiation) of UFG Ti is not a constant but highly dependent on the loading history. Similar with the experimental results of Rittel et al. (10), the critical strain increases with increasing pre-strain caused by static loading, as shown in Fig. 5(a). The two-step or three-step dynamic loading also affects the failure strain, which is different from results of other researchers (10, 16–17). It should be noted that static loading itself did not lead to any catastrophic failure of UFG Ti. Moreover, we did not find any shear localization or crack within the specimen that experienced 10% dynamically pre-loaded stain and following static loading (specimen UFG10-D-S in Fig. 5(b)). In addition, the strain rate effect on mechanical properties of UFG Ti is significant. The yield stress of the material increases from 700 MPa at strain rate 0.001/s to 1000 MPa at 1500/s.

To investigate the fracture behavior of UFG Ti quantitatively, the total fracture strain and contributions from static and dynamic loadings are sumarried in Table 1. The initiation of ASB under single dynamic loading is around 0.2 uniaxial compressive strain. The static pre-loading before dynamic compression increases the total fracture strain. Moreover, as the static strain increases, the total strain increases accordingly. However,

Table 1. Fracture strain of UFG Ti.

No.	Loading history	Static strain	Dynamic strain	Total fracture strain
UFG1	S	0.43	0	No apparent fracture
UFG3	S~D	0.23	0.13	0.36
UFG4	S~D	0.22	0.12	0.34
UFG6	S~D	0.09	0.18	0.27
UFG7	S~D	0.09	0.14	0.23
UFG8	D	0	0.20	0.20
UFG14	D	0	0.21	0.21
UFG10	D~ S	0.57	0.10	No apparent fracture
UFG12	D~ D	0	0.12+0.14	0.26
UFG16	D~ D~ D	0	0.05+0.06+ 0.12	0.23

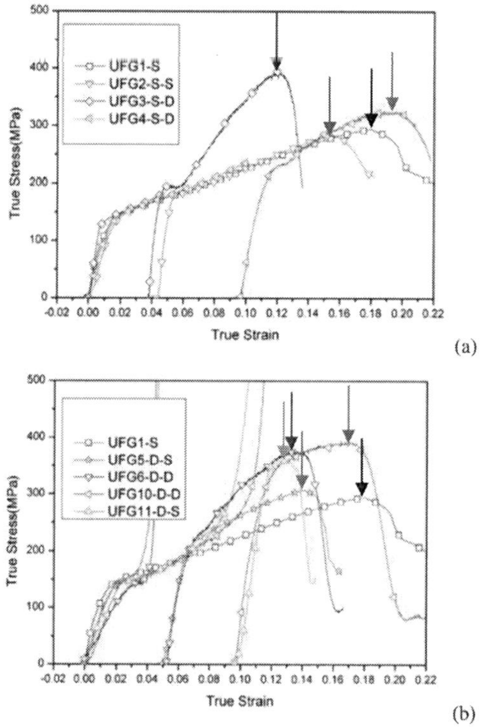

Figure 6. Mechanical behavior of UFG AZ31 alloy under combined static and dynamic loading. (a) static loading followed by dynamic loading, (b) dynamic loading followed by dynamic loading. "S" and "D" indicate "Static" and "Dynamic", respectively. The abrupt rising up of the curves in (b) is not the intrinsic behavior of the material but due to the effect of stop-ring.

the static loading process makes contribution to the final failure of the material. Larger static strain results in smaller dynamic strain in the deformation process of one specimen. Two-step or three-step dynamic loading affects the final ASB initiation strain too, although it is not so apparent as the static pre-loading.

3.2. Mechanical properties of UFG AZ31

Similar experiments were conducted on UFG AZ31 magnsium alloy. The stress strain curves are displayed in

Table 2. Fracture strain of UFG AZ31.

No	Loading history	Static Strain	Dynamic Strain	Total Fracture Strain
1	S	0.18	0	0.18
2	S~S	0.04+0.11	0	0.15
3	S~D	0.04	0.08	0.12
4	S~D	0.10	0.10	0.20
5	D~S	0.09	0.05	0.14
11	D~S	0.03	0.10	0.13
6	D~D	0	0.05+0.09	0.14
10	D~D	0	0.10+0.07	0.17

Fig. 6. Different from pure Ti, AZ31 Mg alloy fractures even under quasi-static loading. Obviously, the failure strains of specimens with different loading history differ a lot. The static pre-loading postpones the initiation of ASB and increases the total fracture strain. What's intresting is that the dynamic pre-loading has similar effect on the total fracture strain with that of static pre-loading. The contributions of static and dynamic loadings are summaried in Table 2. Although the experimental results seem to be irregular, one definite fact is that the contribution of dynamic strain to fracture of the sample is not a constant. The overall trend is that the dynamic part decreases with increasing static strain of pre-loading.

3.3. Mechanical properties of CG AZ31

Figure 7 presents the stress-strain curves of CGAZ31 under different loading process. The arrows within the figure point out the fracture strain, or initiation of shear localization. Different from Fig. 5 and Fig. 6, these arrows in Fig. 7 are rather concentrated, demonstrating the loading history has little effect on the final fracture strain. As shown in Fig. 7(a), a static pre-loading of 10% does not result in significant increase in total failure strain. The quantitative data of CG AZ31 are listed in Table 3. The fracture strain at static loading is identical with that at dynamic loading, indicating that the contribution of static loading is quantitatively similar with that of dynamic, even though they may lead to different deformation and failure mechanisms. The combined static~static, static~dynamic, dynamic~static or dynamic~dynamic loadings did not change this characteristic greatly. In this case, critical strain seems to be a good criterion for evaluating the formation of ASB.

3.4. Criterion on the deformation localization

Traditionally, most of the researchers take critical shear strain as a criterion of shear localization (or in dynamic case, ASB formation). Our experimental results indicate that this strain-based criterion could not describe precisely the fracture behavior of UFG pure Ti or AZ31. The fracture strain is highly related to the loading history.

For UFG pure Ti, which has comparatively good ductility under compression, the static deformation part plays a relatively small role in the initiation of ASB, yet it could not be ignored. The failure strain is about 0.21 under single dynamic loading. Take this value as a

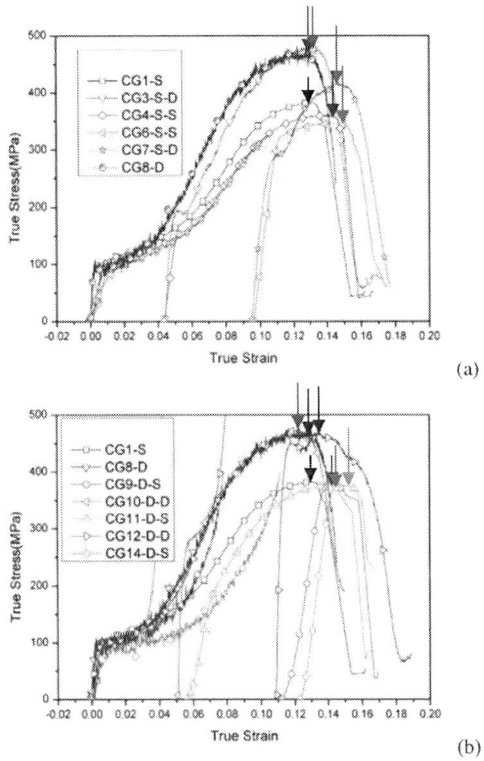

Figure 7. Mechanical behavior of CG AZ31 alloy under combined static and dynamic loading. (a) static loading followed by dynamic/static loading, (b) dynamic loading followed by dynamic/static loading. "S" and "D" indicate "Static" and "Dynamic", respectively. The abrupt rising up of the curves in (b) is not the intrinsic behavior of the material but due to the effect of stop-ring.

reference, the static strain could be converted to dynamic strain and the coefficient is about 0.4. The increase of total strain in the interrupted dynamic tests is probably due to the interruption of the heat accumulation within a specimen during dynamic loading. Dynamic stored energy is defined as the energy stored in the material by dynamic loading and could be calculated by integrating the stress-strain curve.The contribution of dynamic stored energy has identical trend with the dynamic strain in a failure process, thus it is not appropriate to use dynamic stored energy as a criterion for ASB formation for UFG Ti. The overall trend is similar for UFG AZ31 with UFG Ti, but the contribution of static loading to final failure is larger and comparable with that of dynamic loading. In that case, the strain-based criterion appears better than the criterion base on dyamic stored energy for UFG AZ31, but still the strain-based criterion could not describe the fractrue point accurately.

As to the CG AZ31, most of the failure strains are within 0.13 ~ 0.15 regardless of the loading process. The contribution of static strain in a failure process is identical with that of dynamic strain. Taking strain as the criterion for initiation of shear localization (ASB for dynamic case) is ideal for CG AZ31. The dynamic stored energy-based

Table 3. Fracture strain of CG AZ31.

No.	Loading history	Static Strain	Dynamic Strain	Total Fracture Strain
CG1	S	0.13	0	0.13
CG4	S~S	0.04+0.10	0	0.14
CG6	S~S	0.10+0.05	0	0.15
CG7	S~D	0.10	0.05	0.15
CG3	S~D	0.04	0.09	0.13
CG8	D	0	0.13	0.13
CG9	D~S	0.03	0.11	0.14
CG11	D~S	0.09	0.06	0.15
CG14	D~S	0.02	0.12	0.14
CG10	D~D	0	0.05+0.07	0.12
CG12	D~D	0	0.11+0.03	0.14

criterion is only related to the dynamic strain part, thus it is not applicable to this material.

4. Summary and conclusion

In this work, we evaluated two kind criterions for ASB formation, i.e. the strain-based criterion and dynamic stored energy-based criterion, on UFG materials as well as CG metals. The main conculsions are summaried in the following:

1. Neither strain-based nor dynamic stored energy-based criterion could describe accurately the failure behavior of UFG Ti or UFG AZ31 magnisium alloy.
2. Failure of the UFG materials is related to both static loading process and dynamic loading process. However, the contributions of the two processes are different and material dependent.
3. The strain-based criterion can describe the failure of CG AZ31 well regardless of loading history, indicating static loading has similar effect in ASB formation with dynamic loading.

Our experimental results demonstrate that the effects of loading history on ASB formation may be different for different materials. The dynamic stored energy is not a quantity for general application in evaluating shear failure, neither is the shear strain. Converting results of static loading to dynamic might be a feasible way to construct a general criterion, but the intrinsic relationship and microstructure-based mechanism need further investigating.

The authors would like to thank the financial support by the National Natural ScienceFoundation of China (Contract nos. 11472227 and 11102166), the 111 Project (Contract no. B07050), and the Fundamental Research Funds for the Central Universities (No.3102014O1JCQ01001).

References

[1] M. A. Meyers, A. Mishra, D. J. Benson, *Progress in Materials Science* **51**, 427 (May, 2006).
[2] Y. M. Wang, M. W. Chen, F. H. Zhou, E. Ma, *Nature* **419**, 912 (Oct. 2002).
[3] T. H. Fang, W. L. Li, N. R. Tao, K. Lu, *Science* **331**, 1587 (March 25, 2011, 2011).
[4] Y. T. Zhu, X. Z. Liao, *Nature Materials* **3**, 351 (2004).
[5] Y. L. Bai, *Journal of the Mechanics and Physics of Solids* **30**, 195 (1982).
[6] T. W. Wright, *The physics and mathematics of adiabatic shear bands*. (Cambridge University Press, Cambridge, New York, Port Melbourne, Madrid, Cape Town, 2002).
[7] Y. B. Xu, J. H. Zhang, Y. L. Bai, M. A. Meyers, *Metall. Mater. Trans. A-Phys. Metall. Mater. Sci.* **39A**, 811 (Apr. 2008).
[8] Q. Wei, *Journal of Material Science* **42**, 1709 (2007).
[9] Q. Wei *et al.*, *Acta Materialia* **52**, 1859 (Apr. 2004).
[10] D. Rittel, Z. G. Wang, M. Merzer, *Physical Review Letters* **96**, 075502 (2006).
[11] D. Rittel, P. Landau, A. Venkert, *Physical Review Letters* **101**, 165501 (2008).
[12] T. Suo, Y. Li, Y. Guo, Y. Liu, *Materials Science and Engineering A* **432**, 269 (2006).
[13] T. Suo *et al.*, *Mechanics of Materials* **43**, 111 (2011).
[14] X. Sun, Y. Guo, Y. Li, *Unpublished work* (2015).
[15] X. Yu, Y. Li, Y. Guo, *Unpublished work* (2015).
[16] D. Rittel, S. Osovski, *International Journal of Fracture* **162**, 177 (2010/03/01, 2010).
[17] J. A. Rodríguez-Martínez, G. Vadillo, D. Rittel, R. Zaera, J. Fernández-Sáez, *Mechanics of Materials* **81**, 41 (2015).

EPJ Web of Conferences **94**, 01037 (2015)
DOI: 10.1051/epjconf/20159401037

Effect of strain rate on bake hardening response of BH220 steel

Anindya Das[1,a], Soumitro Tarafder[1], S. Sivaprasad[1], and Debalay Chakrabarti[2]

[1]MST Division, CSIR-National Metallurgical Laboratory, Jamshedpur, India
[2]Metallurgical and Materials Engineering Department, IIT Kharagpur, Kharagpur, India

Abstract. This study aims at understanding the bake hardening ability of ultra low carbon BH220 steel at different strain rates. The as-received material has been pre-strained to four different levels and then deformed in tension under (a) as pre-strained state and (b) after baking at 170 °C for 20 minutes, at three different strain rates of 0.001, 0.1 and 100/s. In both the conditions, yield stress increased with pre-strain and strain rate, but bake hardening ability was found to decrease when strain rate was increased. The strain rate sensitivity of the material was also found to decrease with bake hardening. Generation of dislocation forests and their subsequent immobility during baking treatment enables them to act as long range obstacles during further deformation. At higher strain rates, less amount of dislocations are produced which can interact with themselves and produce hardening, because of which bake hardening ability and the strain rate drops. A dislocation based strengthening model, as proposed by Larour et al. 2011 [7], was used to predict the yield stress values obtained at different conditions. The equation produced excellent co-relation with the experimental data.

1. Introduction

Bake hardenable steels are popular grades in automobile industries due to their high strength to weight ratio and excellent formability. Generally, these steels are used in the outer body panels of the automobiles. The fabrication involves a cold forming process to get the required shape followed by an industrial paint baking cycle (170 °C for 20 minutes). The later process imparts an improved dent resistance of the component [1]. Cold deformation is essentially a pre-straining process during which fresh dislocations are produced. During baking process these dislocations are pinned down by the interstitial atoms by forming Cottrell atmosphere, thus providing an enhancement in strength during subsequent deformation through strain ageing phenomena [2].

This process of bake hardening is greatly influenced by the concentration of solute atoms, pre-strain levels, time and temperature of ageing, grain size of the material etc. The solute concentration helps in forming the Cottrell atmosphere over the generated dislocation cores, and the bake hardening ability increases if the solute concentration is increased up to a critical level where the Cottrell atmosphere tends to saturate [3]. Beyond the saturation limit, bake hardenability drops. Similar trend is also observed in case of pre-strain level, time and temperature of the ageing process [1,4,5]. On further increasing the solute concentration or ageing time and temperature, carbides starts precipitating which again increase the bake hardenability [4].

The bake hardening ability of these ultra low carbon steels is often quantified at quasi-static rates. But the nature of deformation is subjected to change when these components experience dynamic rate of loading, like in events of crash. This particular aspect has not been studied well and the available literature is scanty. This investigation aims at understanding the effects of strain rate on pre-strained as well as bake hardened ultra low carbon BH220 steel. The selected material, BH220 was obtained as sheets of 0.6 mm thickness and possesses 0.0027 wt.% carbon in the matrix. The steel showed a ferritic structure (Fig. 1), with an average grain size of 33.75 +2.3 μm.

2. Experimental design

Tensile specimens of 30 mm gauge length (ASTM E8M) were prepared from the as-received material along the rolling direction, to obtain the basic tensile properties as well as to decide the pre-strain levels. Tensile tests were performed at nominal strain rate of 0.001/s. Pre-strain levels of 2, 4, 6 and 8% were selected for the current study, since these levels fall within the hardening region of the plastic deformation portion, as shown in Fig. 2.

After pre-straining, one set of specimens (designated: PS) were directly pulled in tension at three different strain rates (0.001, 0.1 and 100/s).

The second pre-strained set of samples were aged at 170 °C for 20 minutes (designated: PS+BAK), and then pulled in tension at the same strain rates (0.001, 0.1 and 100/s). Tensile tests at 0.001 and 0.1/s strain rates were performed in a quasi-static servo-hydraulic test system, whereas the tests at 100/s were performed in a special high speed servo-hydraulic test system. The specimen design for all the tests are shown in Fig. 3.

3. Results and discussions

3.1. Tensile and bake hardening properties

The engineering stress-strain curves corresponding to PS and PS+BAK conditions, at three different strain rates

[a] Corresponding author: anindyadvc@gmail.com

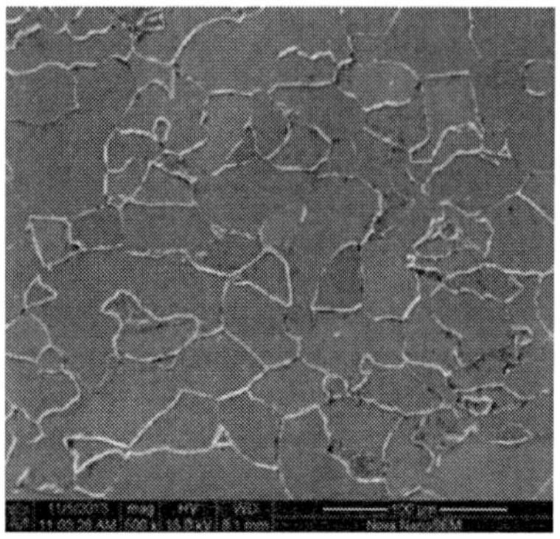

Figure 1. Microstructure of BH220 steel.

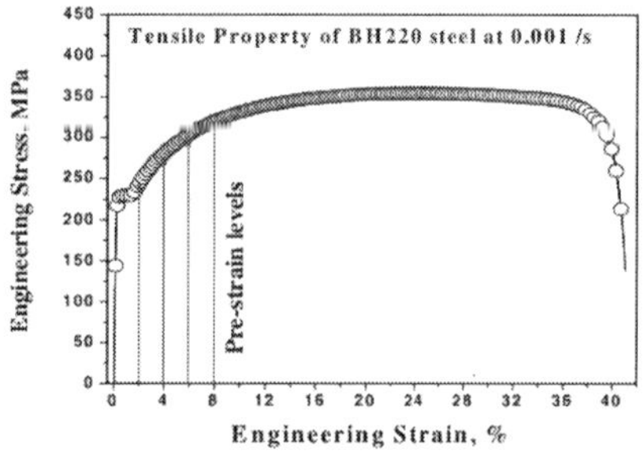

Figure 2. Tensile property of BH220 steel and the strain range selected for pre-straining deformation.

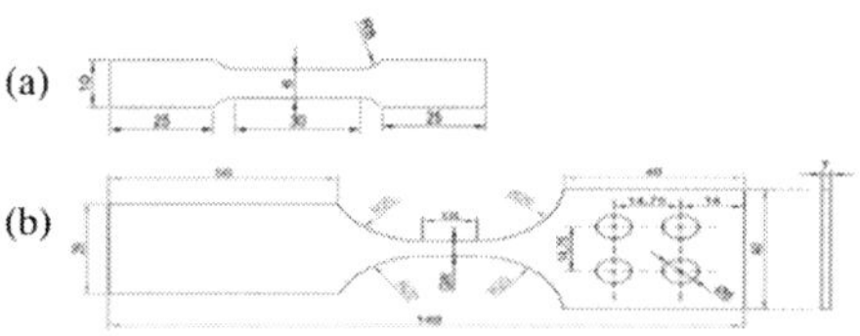

Figure 3. Specimen dimensions used for tensile deformation at (a) 0.001 and 0.1/s strain rate, and (b) 100/s *All dimensions are in mm.*

are shown in Fig. 4. The tensile curves for PS do not show any yield point effect unlike that of PS+BAK. This is because, in as pre-strained condition, the generated dislocations remain free and experience no obstacle as in the case after bake hardening. The solute atoms lock the dislocations during bake hardening of the material after pre-straining, which needs a critical stress to get unlocked while deforming. This process of unlocking produces the enhanced yield point phenomenon. This critical stress required to unlock the pinned dislocations in bake hardened condition is much larger than the stress required to move the mobile dislocations in as pre-strained material. Thus the yield stress for bake hardened condition is higher than that of as pre-strained material. In Fig. 5, the variation of yield stress is shown with pre-strain and strain rate for both PS and PS+BAK conditions. Yield stress increases with both pre-strain level and strain rate.

To understand the bake hardening ability of the material, the difference in yield stress has been calculated for bake hardened condition with as-received as well as

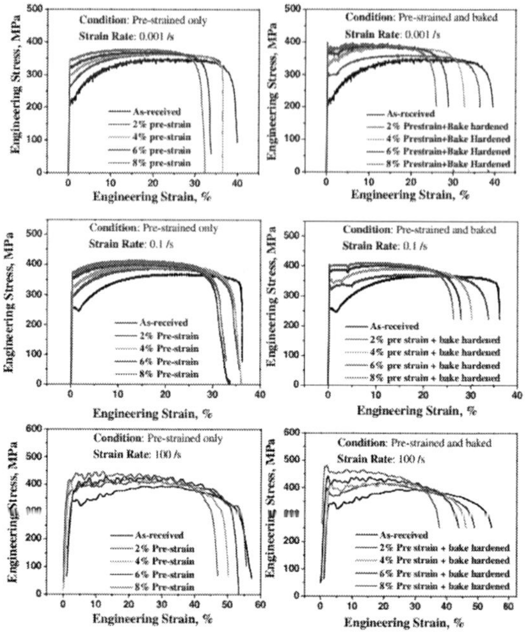

Figure 4. Engineering stress-strain curves obtained after pre-straining and pre-strain with baking at 3 different strain rates.

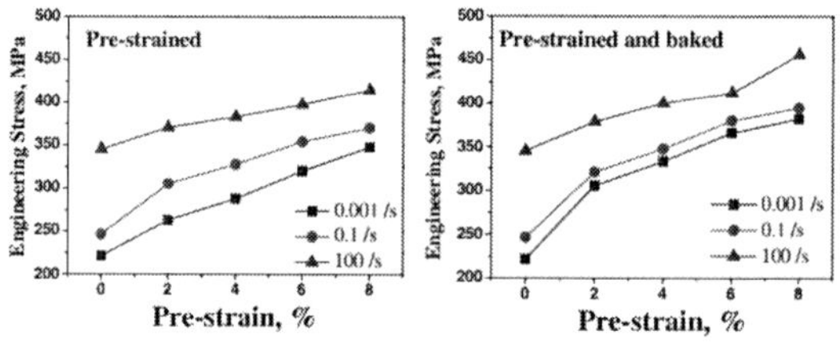

Figure 5. Yield stress values obtained after tensile deforming BH220 in as-received, PS and PS+BAK condition.

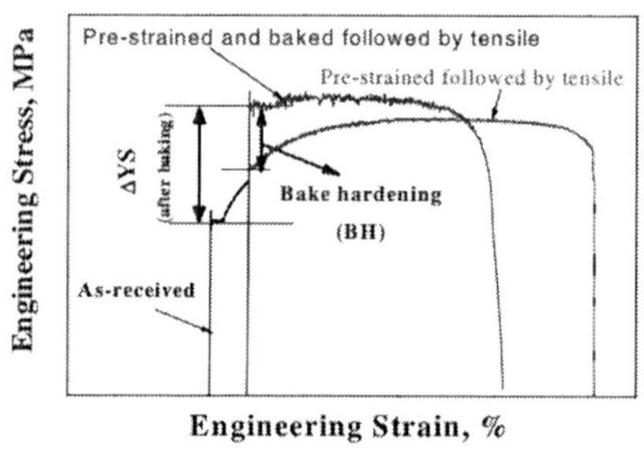

Figure 6. Schematic of different parameters.

with pre-strained material (as shown in Fig. 6), at each pre-strain level and strain rate.

The difference in yield stress between bake hardened and as-received state is designated as ΔYS and BH for the difference between bake hardened with its only pre-strained condition.

Figure 7 shows the variation of ΔYS with pre-strain and strain rate. The ΔYS values increase as the pre-strain range is increased at any particular strain rate but its value decreases as the strain rate increases at any

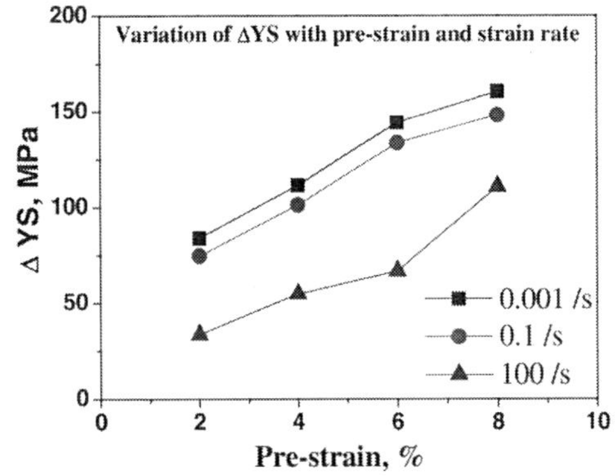

Figure 7. Variation of ΔYS with pre-strain and strain rate.

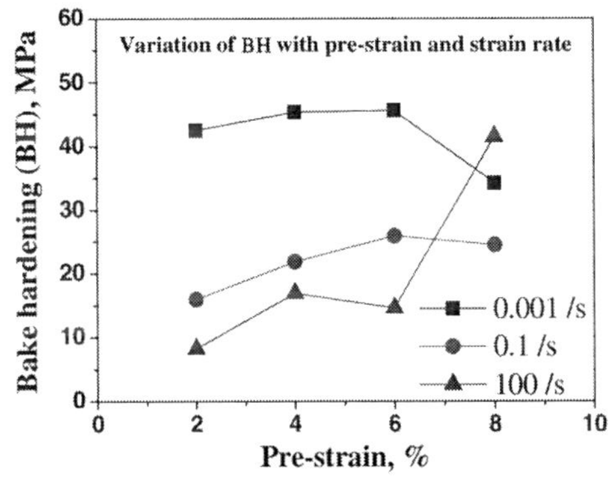

Figure 8. Variation of BH with pre strain and strain rate.

particular pre-strain level. The variation of BH with pre-strain (Fig. 8) is also similar to that of ΔYS except that the BH values tend to drop for pre-strain >6% for the quasi-static strain rates, but increase sharply after 6% pre-strain for high strain rate deformation. Pre-straining beyond 6%, tends to reduce the dislocation density since the dislocation annihilation mechanisms may start beyond that point. Due to the reduced dislocation density, fewer dislocations are getting locked at highest pre-strain levels. In other words, more free solute atoms remain in the matrix when bake hardening is done at the 8% pre-strain level. As strain rate increases, the fresh dislocations produced during deformation also gets locked by the free solute atoms, and the bake hardening ability or BH increases. But overall, the BH value decreases as the strain rate is increased. This is because, with pre-straining, dislocation forests are produced which will increase with the pre-strain magnitude. After bake hardening, the subsequent locking of the dislocations in the forest by solute atoms will restrict the mobility of these dislocation forests produced and ultimately these will act as long range barriers during further deformation. As strain rate increases during further deformation, less amount of fresh dislocations are produced due to less amount of available time for deformation, which can interact with themselves and can produce the hardening.

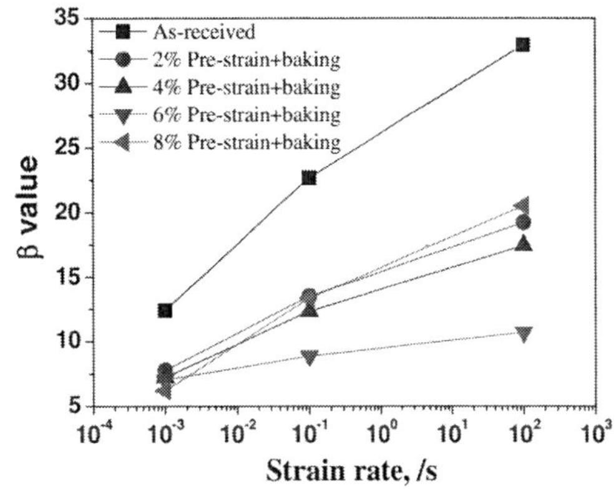

Figure 9. Variation of strain rate sensitivity with strain rate for as-received and bake hardened condition.

3.2. Strain rate sensitivity after bake hardening

Along with BH value, bake hardening also reduces the strain rate sensitivity of the material, as shown in Fig. 9. Semi-logarithmic strain rate sensitivity (β) was calculated on yield stress at all strain rates for as-received and PS+BAK condition using the following equation:

$$\beta = \frac{\partial \sigma}{\partial(\log \dot{\varepsilon})} \tag{1}$$

where, σ is the yield stress value and $\dot{\varepsilon}$ is the strain rate.

The strain rate sensitivity increases with strain rate at all conditions but, the magnitude of this increase is more for the as-received condition compared to the bake hardened condition. In case of as-received material, the strength of the material is purely derived due to dislocation generation and its interaction with the interstitials during deformation. When the strain rate is increased, dislocation density increases and more dislocations interact with solute atoms which in turn increase the strength. The higher strain rate sensitivity for as-received material suggests that there is an increase in the contribution of thermally activated dislocation processes to the overall stress required for dislocation movement [6] and thus at high strain rates the properties are dominated by the stress for individual dislocation movement. In case of bake hardening, the solute atoms are tied to the dislocations generated during pre-straining prior to the post baking deformation. The newly formed dislocations during the post baked deformation will have less chance of interacting with the solute atoms and would be available as free mobile dislocations. Thus even if the strain rate is increased, the hardening due to dislocation-solute or dislocation-dislocation interaction reduces leading to a lower strain rate sensitivity.

3.3. Experimental vs. mathematical form of yield stress during bake hardening

In general the flow stress during deformation can be decomposed into a thermally activated stress σ^* due to short range obstacles and an athermal stress component σ_i

Table 1. Model parameters obtained for different conditions.

Pre-strain	As-received	Only pre-strain 2%	4%	6%	8%	Bake hardened 2%	4%	6%	8%
σ_i(MPa)	213.8	303.78	328.95	366.58	378.42	269.584	295.084	323.32	347.12
σ_0^*(MPa)	612.86	336.52	321.55	336.33	336.45	463.54	406.74	273.54	244.73
$\dot{\varepsilon}_0(s^{-1})$	1.005E8	1.005E8	1.005E8	1.005E8	1.005E8	1.005E8	1.005E8	1.005E8	1.005E8
ΔG_0	1.21E-19	1.21E-19	1.21E-19	1.21E-19	1.21E-19	1.21E-19	1.21E-19	1.21E-19	1.21E-19
m'	2.21597	2.21597	2.21597	2.21597	2.21597	2.21597	2.21597	2.21597	2.21597

due to long range obstacles [7] as given below.

$$\sigma = \sigma_i + \sigma^* = \sigma_{i,0}.\frac{G(T)}{G(0K)} + \sigma^*(T, \dot{\varepsilon}) \quad (2)$$

where G(T): temperature dependent shear modulus, T: temperature, $\dot{\varepsilon}$: strain rate, $\sigma_{i,0}$: athermal stress component at zero thermal activation. Since, all the investigations have been made at room temperature (28 °C), the temperature dependency of the athermal flow stress component σ_i has been neglected in Eq. (2). The Eq. (2) reduces to,

$$\sigma = \sigma_{i,0} + \sigma^*(T, \dot{\varepsilon}). \quad (3)$$

According to Larour, [7],

$$\sigma^*(T, \dot{\varepsilon}) = \sigma_0^*.\left[1 - \frac{k.T.\ln\left(\frac{\dot{\varepsilon}_0}{\dot{\varepsilon}}\right)}{\Delta G_0}\right]^{m'} \quad (4)$$

where, σ_0^*: effective stress at zero thermal activation, $\dot{\varepsilon}$: critical strain rate or pre-exponential factor, ΔG_0: free energy required to overcome the obstacle without the help of thermal activation ($\Delta G = 0$), m' is strain rate exponent. Substituting Eq. (4) in Eq. (3), the expression for yield strength can be written as:

$$\sigma_{YS} = \sigma_i + \sigma_0^*.\left[1 - \frac{k.T.\ln\left(\frac{\dot{\varepsilon}_0}{\dot{\varepsilon}}\right)}{\Delta G_0}\right]^{m'}. \quad (5)$$

The equation has been used to model the YS variation observed in this investigation. Initial values of model parameters have been taken from the literature [8], and the parameters have been tuned for the convergence. The critical strain rate ($\dot{\varepsilon}_0$), ΔG_0 and strain rate exponent (m') have been kept constant in all the cases, since these thermal activation parameters are specific to the material, independent of their condition [9]. The fitting procedure produced a goodness of fit of more than 90% in all the cases. The experimentally obtained and modelled yield stress of PS and PS+BAK condition are shown in Figs. 10 and 11 respectively. The experimental and modelled yield strength values match reasonably well in all the conditions. Table 1 lists the model parameters obtained at different conditions. For as-received material the athermal flow stress component σ_i is very high for PS when compared to PS+BAK condition. But it increases with increase in pre-strain in both only pre-strained and pre-strain with bake hardened conditions. This stress due to long range obstacle can be attributed to the dislocation forest which is generating during pre-straining process, which also increase as the pre-strain range is increased.

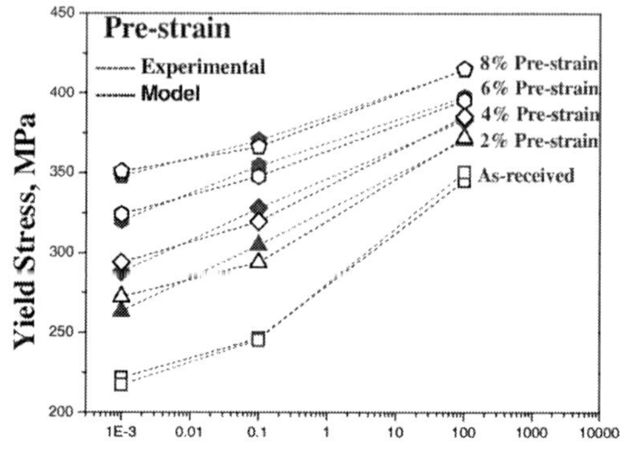

Figure 10. Experimental and predicted yield stress values of only pre-strained material at various strain rate.

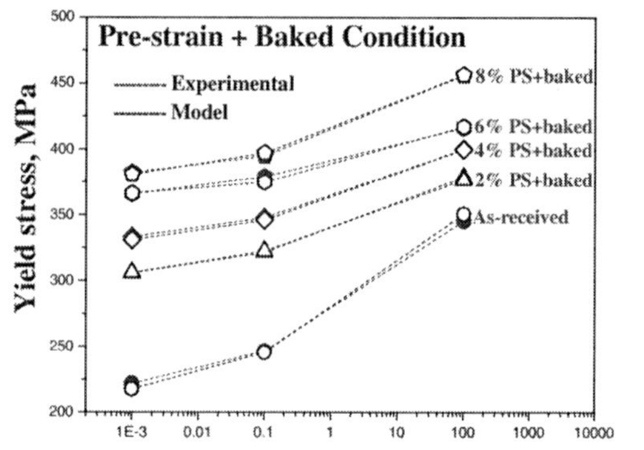

Figure 11. Experimental and predicted yield stress values of bake hardened condition at various pre-strain and strain rate.

4. Conclusion

Dislocation forest generated during pre-straining acts as long range obstacle after it has been arrested by solute atoms while bake hardening. These long range obstacles reduce the strain rate sensitivity, as well as bring down the bake hardening ability of the material.

References

[1] S. Berbenni, V. Favier, X. Lemoine, M. Berveiller, Scripta Materialia, **51** (2004)
[2] A. H. Cottrell, B. A. Bilby; Proc. of Phys Society A, 62 (1949)
[3] S. Hanai, N. Takemoto, Y. Tokunaga and Y. Mizuyama; Trans ISIJ, 24 (1984)
[4] P. Elsen and H. P. Hougardy; Steel Research (64) (1993)
[5] A. K. De, S. Vandeputte and B. C. De Cooman; Scripta Materialia, **41** (1999)
[6] R.S. Kircher, A. K. De, D. K. Matlock, J. G. Speer, MS&T conference proceedings (2004)
[7] P. Larour, K. Dahmen, W. Bleck, Steel research Int. **82** (2011)
[8] P. Larour, Ph.D. thesis, RWTH Aachen (2010)

EPJ Web of Conferences **94**, 01038 (2015)
DOI: 10.1051/epjconf/20159401038

Dynamic compressive behavior of foamed polyethylene film

Kohei Tateyama[1], Hiroyuki Yamada[2,a], Nagahisa Ogasawara[2], Ryo Okui[3], and Kinya Ogawa[4]

[1]Graduate Student, School of Systems Engineering, National Defense Academy, Kanagawa 239-8686, Japan
[2]School of Systems Engineering, National Defense Academy, Kanagawa 239-8686, Japan
[3]Foam Development Center, Sekisui Chemical Co., LTD., Saitama 616-8255, Japan
[4]Institute of Space Dynamics, Kyoto 616-8255, Japan

Abstract. The foamed film as the shock absorption material has attracted much attention because it is thin (100 μm ~ 400 μm) and has a closed cell structure. However, the dynamic mechanical properties have not been reported in the foamed film. The purpose of this study is to elucidate the compressive behavior of the foamed polyethylene film at the wide strain rate range. First, the new compressive test apparatus for the dynamic strain rate, the drop-weight testing machine with opposed load cell, was developed, which can be also evaluated the dynamic stress equilibrium of the specimen. It is confirmed that the compressive flow stress increased with increasing the strain rate, regardless of the film thickness. The foamed polyethylene film has the high strain rate sensitivity in the quasi-static deformation. On the other hand, there is almost no change of the strain rate sensitivity in the dynamic and the impact deformation. In order to investigate the mechanism of strain rate dependence, the foamed polyethylene film was observed by X-ray computed tomography scanner before and after compressive test. The fracture of the closed cell only occurred in the quasi-static deformation. It was clarified that the strain rate sensitivity of the foamed film depends strongly on that of the construction material, polyethylene.

1. Introduction

Smartphones and tablet computers have become more popular than cell phones and laptop computers in recent years. Most of these devices have a thin and large display, which show weakness in the impact such as falling or collision. The shock absorption ability is required in order to protect the display and substrate. Recently, a foamed polyethylene film as the shock absorption materials have attracted much attention, because it is thin (100 μm ~ 400 μm) and has a closed cell structure.

Mechanical properties of the general foamed material, such as polymeric and metal foams, have been intensively studied. It has been clarified that these materials have strain rate sensitivity of the stress-strain relationship [1–5]. In addition, the polymeric foams show remarkable strain rate sensitivity: an increased elastic modulus, an increased plateau stress and a decreased densification strain in comparison with metal materials, since the base material have viscoelastic properties [2–4]. However, the mechanical properties and measurement methods have not been reported in the foamed film, since this material is newly developed in recent years. Therefore, it is necessary to clarify the strain rate dependence of compressive deformation of the foamed film as the shock absorption materials.

The authors have reported the impact compressive properties of foamed polyethylene film using a split Hopkinson pressure bar (SHPB) method with the circular tube having a flange [5]. However, the SHPB method is a generally accepted test method for strain rates in the range 10^2–10^5 s^{-1}, which does not include the dynamic strain rate of 10^0–10^1 s^{-1}. The mechanical properties of the dynamic strain rate were required for the shock absorption materials.

In the present study, the effect of strain rate on the compressive behavior of the foamed polyethylene film was investigated. We attempt to use the drop-weight testing machine with opposed load cell for the dynamic compressive properties. Furthermore, the foamed polyethylene film of before and after compressive test was observed by X-ray computed tomography scanner to clarify the mechanism of strain rate dependence.

2. Experimental procedures

2.1. Material

A commercial closed cell foamed polyethylene film (hereinafter referred to as foamed PE film) with a density of 330 kg/m^3 was used (XLIM-WF01, Sekisui Chemical Co., Ltd.). The specimen with thickness of approximately 300 and 400 μm were prepared. The foamed PE film has features such as water resistance, chemical resistance and abrasion resistance. Figure 1 shows the surface and cross-sectional images of foamed PE film, which are observed using the digital microscope and the scanning electron microscope (SEM). It can be confirmed that the cell surfaces have no holes, and all cell is closed. In addition, the flattened cell was observed in the thickness direction, since the foamed PE film was produced by extrusion foaming. Compressive testing specimens with thickness of approximately 300 μm and 400 μm and diameter of 30 mm were prepared by using a hollow punch with circular razor.

[a] Corresponding author: ymda@nda.ac.jp

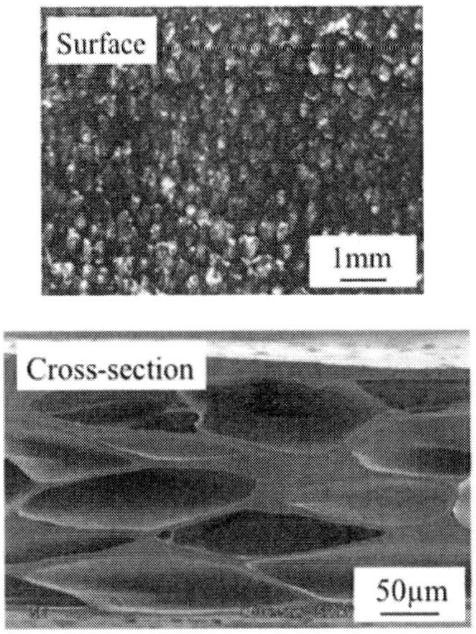

Figure 1. Surface and cross-section of Foamed film.

2.2. Quasi-static test

The quasi-static compression test was conducted using the universal testing machine.

The quasi-static compression test was carried out at the initial strain rates of approximately 1.7×10^{-3}, 1.7×10^{-2} and $1.7 \times 10^{-1}\,s^{-1}$. The test was conducted three times for each strain rate.

2.3. Dynamic test

The dynamic compression test was performed using the drop-weight testing machine with opposed load cell (hereinafter referred to as drop-weight testing machine). This testing machine was consisted of a drop-weight, guide rail and a load cell. A recording system is comprised of a Wheatstone bridge box, a differential amplifier (5305, NF Corporation), and a digital oscilloscope (DL750, Yokogawa Meters & Instruments Co.). Figure 2 shows the schematic view of a drop-weight testing machine and stress wave propagation in this testing machine. The load cell of drop-weight testing machine is composed of a stress detection part and a stress transmission part that was made of the T6-tempered 2024 aluminum alloy.

When the compressive stress wave reaches the interface between the detection part and the stress-transmitted bar, part of the stress wave reflects back into the detection part, while the remaining part of the wave propagates through the stress-transmitted bar. By considering the load equilibrium and the continuity of particle velocity at the interface, the following equations are obtained [6]:

$$A_1(\sigma_{11} + \sigma_{12}) = A_2\sigma_{21} \tag{1}$$

$$\frac{1}{\rho C}(\sigma_{11} - \sigma_{12}) = \frac{1}{\rho C}\sigma_{21} \tag{2}$$

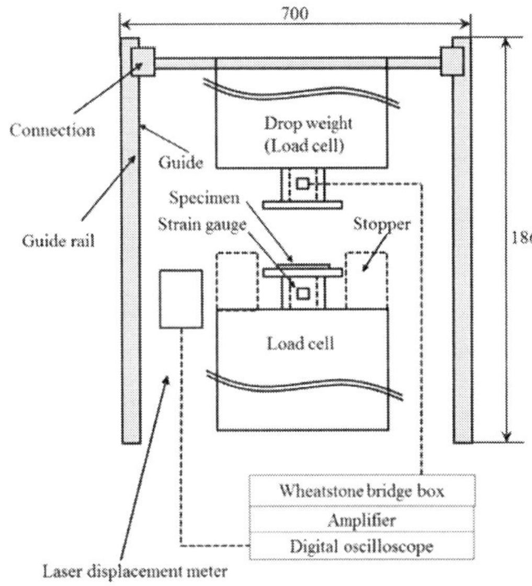

Figure 2. Setup of the drop-weight testing machine with opposed load cell for dynamic test.

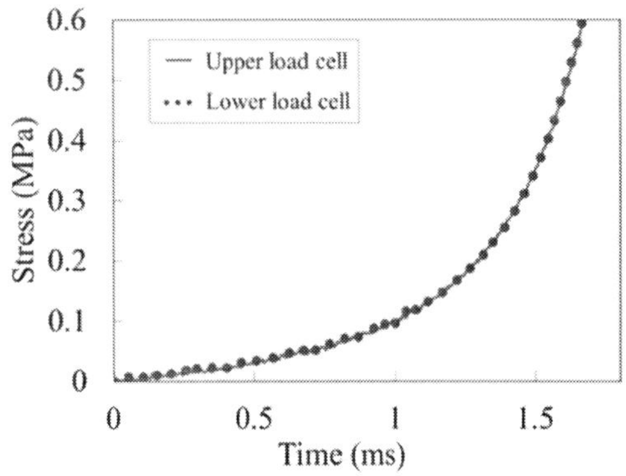

Figure 3. Time histories of applied stresses on each load cells at the dynamic test.

where σ_{11}, σ_{12} and σ_{21} are the incident, reflected and transmitted stress waves at the interface between the detection part and the stress-transmitted bar, respectively, A_1 is the cross-sectional area of the detection part, A_2 is the cross-sectional area of the stress-transmitted bar, and C and ρ are the velocity of the elastic wave and the density of the apparatus, respectively. From Eqs. (1) and (2), the following relationships are obtained:

$$\sigma_{21} = \frac{2}{1 + A_2/A_1}\sigma_{11}. \tag{3}$$

If $A_2 \gg A_1$, then $\sigma_{21} \ll \sigma_{11}$. Thus, the effect of the reflected stress wave from the bottom of the transmitted bar is very small when $\sigma_{21} \ll \sigma_{11}$. This implies that the compressive stress deformation can be detected for a long time without any disturbances since this apparatus ignores the reflected stress wave. In this study, the value of A_1 was $132\,mm^2$ and the value of A_2 was $13100\,mm^2$.

The displacement of the drop-weight is measured by a laser displacement meter (KEYENCE: LK-G5000V).

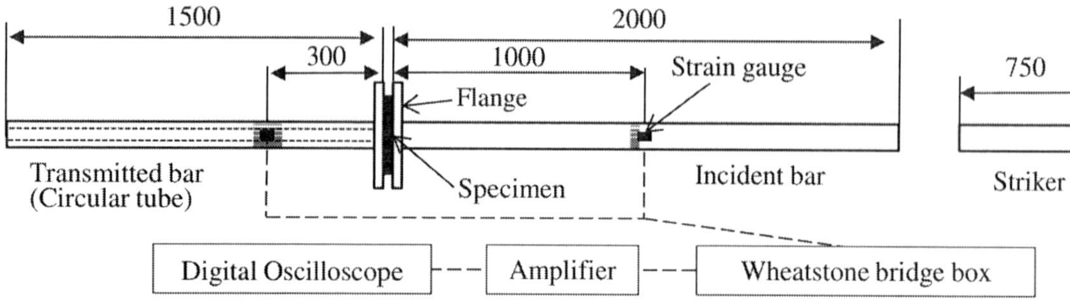

Figure 4. Setup of the SHPB apparatus for the impact compression test.

The impact compression test was carried out at approximately $5.0 \times 10^1\ s^{-1}$.

The drop-weight testing machine with opposed load cell have the two load cells at top and bottom of apparatus. The upper load cell used as drop-weight. It is possible to evaluate the dynamic stress equilibrium of the specimen by measuring the upper and lower load history in the same way as SHPB method. Figure 3 shows the stress-time relationship in dynamic test at both ends of the specimen. The compression test was carried out In the dynamic test, the dynamic stress equilibrium is achieved within the specimen, as shown in Fig. 3.

2.4. Impact test

The impact compression test was performed using the SHPB method established by Kolsky [7,8]. Generally, SHPB method is used solid bar adapted to the specimen diameter. If the impact test of the foamed film specimen with the diameter of 30 mm was performed by the solid bars, the poor measurement sensitivity was indicated because of the low strength. In the previous study [5], we proposed that the transmitted bar has a smaller cross-sectional area and a lower elastic modulus indicated the high sensitivity of the load measurements of specimen. The circular tube having a flange that was made of the 2024 aluminum alloy was used as the transmitted bar. In addition, an attempt was made to be high sensitivity of measurement of the incident stress wave using the incident bar having a flange.

Figure 4 shows the setup of the SHPB apparatus for the impact compression test, which consisted of a striker, an incident bar, a transmitted bar. The incident and striker bars made of stainless steel ISO 4301-304-00-I was 16 mm in diameter. The transmitted bar was consisted of the circular tube, which was made of the T6- tempered 2024 aluminum alloy with the outside diameter of 16 mm and the inside diameter of 12 mm. The incident bar and transmitted bar have a flange of diameter of 32 mm.

By applying elementary one-dimensional elastic wave propagation theory [9–11], the nominal stress ($\sigma(t)$), nominal strain ($\varepsilon(t)$), and strain rate ($\dot{\varepsilon}(t)$ in the foamed PE film specimen can be determined as follows:

$$\sigma(t) = \frac{A_t E_t \varepsilon_t(t)}{A_S} \quad (4)$$

$$\varepsilon(t) = \frac{1}{l_s} \int_0^t \left[2c_i \varepsilon_i(t) - \left(c_i \frac{A_t E_t}{A_i E_i} + c_t \right) \varepsilon_t(t) \right] dt \quad (5)$$

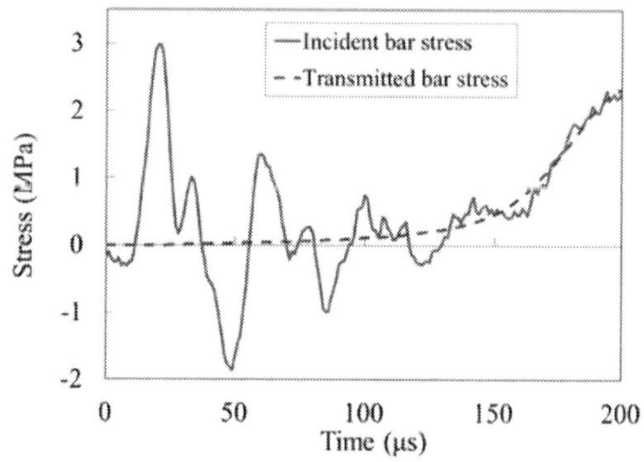

Figure 5. Time histories of applied stresses on each face of specimen at the impact test.

$$\dot{\varepsilon}(t) = \frac{1}{l_s} \left[2c_1 \varepsilon_i(t) - \left(c_1 \frac{A_2 E_2}{A_1 E_1} + c_2 \right) \varepsilon_t(t) \right] \quad (6)$$

where A_i and A_t are the cross-sectional area of the incident and transmitted bars, E_i andE_t are the Young's modulus of the incident and transmitted bars, respectively. $\varepsilon_i(t)$, $\varepsilon_r(t)$ and $\varepsilon_t(t)$ are the incident, reflected and transmitted waves, respectively. C_i (=5300 m/s) and C_t (=5100 m/s) are the velocity of the elastic wave in the incident and transmitted bars, respectively. In the present study, the nominal stress, nominal strain, and strain rate in the foamed PE film specimen were evaluated using Eqs. (4) to (6). The impact compression test was carried out at the typical initial strain rate of approximately $5.0 \times 10^3\ s^{-1}$.

Figure 5 shows the axial stress histories at both ends of the specimen in the impact test. The stress fluctuation was observed immediately after the loading, which is assumed due to the vibration of flange. In the SHPB method using the circular tube with flange, both ends of the specimen are achieved the stress equilibrium, which show the high sensitivity for measurement of the impact load.

3. Results and discussion

3.1. Stress-strain relationship

Figure 6 shows the typical stress-strain relationship for the foamed PE film with a thickness of 300 μm and 400 μm. The flow stress increased with the increase in strain rate and the plateau deformation was observed immediately after the compressive testing started in the foamed PE

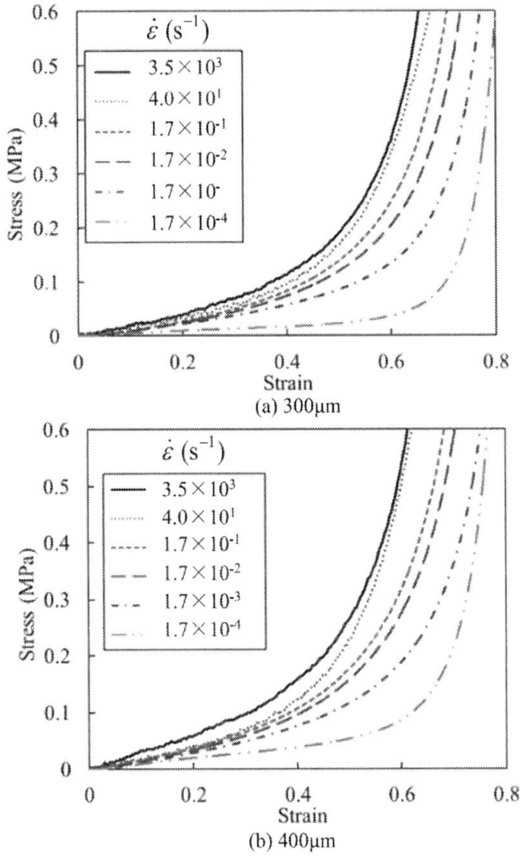

Figure 6. Typical stress-strain relationship of the PE film specimen with thickness of 300 μm (a) and 400 μm (b).

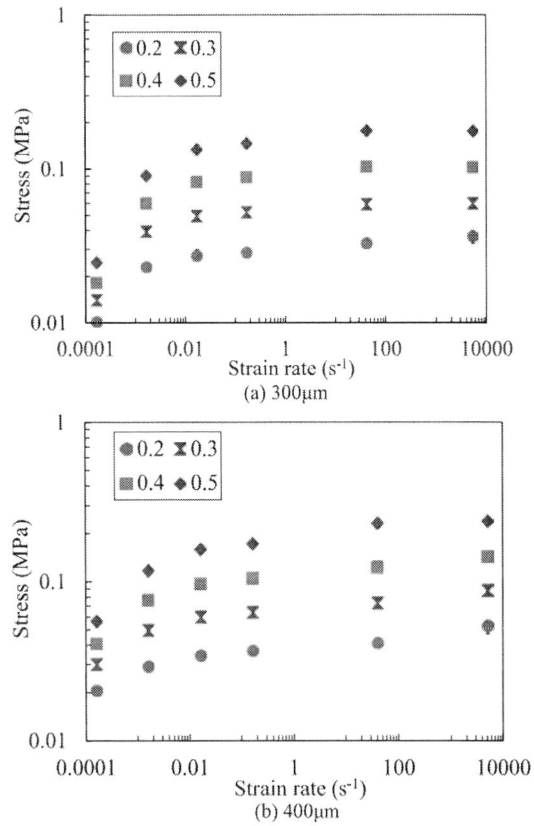

Figure 7. The variation in compressive stress plotted as a function of strain rate using a double logarithm at each strain with thickness of 300 μm (a) and 400 μm (b).

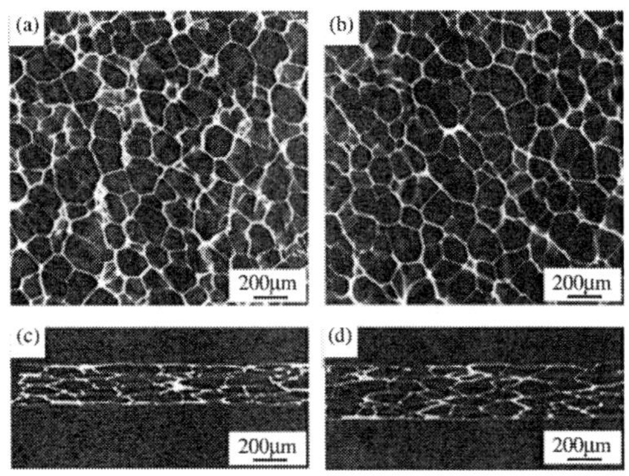

Figure 8. The cross-sectional CT images of the radial direction (a: 300μm, b: 400μm) and the thickness direction (c: 300μm, d: 400μm) in the foamed PE film.

film with a thickness of 300 μm and 400 μm. This trend of deformation process was the same as the thickness of 100 μm specimen [5]. Thus, it was revealed that the deformation mechanism of the foamed PE film was not affected by the thickness of specimen.

3.2. Effect of strain rate

In order to investigate the effect of the strain rate on the compressive stress of the foamed PE film in detail, the variation in compressive stress was plotted as a function of strain rate using a double logarithm at each strain (ε = 0.2–0.5), as shown in Fig. 8. Each plot is the average of 3–4 data. The rate of increase in the stress at each strain, which was caused by the increasing strain rate, was calculated as follows:

$$\sigma = C\dot{\varepsilon}^{\alpha} \quad (7)$$

where C is the constant value, which was unrelated to the strain rate. α is the increase rate in stress with increasing strain rate. In the present study, value of α was defined as the strain rate sensitivity at each strain. Table 1 shows the strain rate sensitivity at each strain obtained from the stress-strain relationship for the foamed PE film.

Both in the thickness of 300 μm and 400 μm specimens, the strain rate sensitivity at $\dot{\varepsilon} \leq 1.7 \times 10^{-2}\,\mathrm{s}^{-1}$ was increased with increasing the strain. On the other hand, the strain rate sensitivity at $\dot{\varepsilon} \geq 1.7 \times 10^{-2}\,\mathrm{s}^{-1}$, which indicates that the strain rate dependence of strength between the quasi-static and impact deformation, show the low value and done not change with increasing the strain. It has been reported [1,3] that the foamed polymer material such as the polystyrene foam had the high strain rate sensitivity above the dynamic strain rate when compared

Table 1. The strain rate sensitivity at each strain obtained from the stress-strain relationship for the foamed PE film.

Thickness (μm)	ε	α	
		$\leqq 1.7 \times 10^{-2}$ (Quasi-static)	$\geqq 1.7 \times 10^{-2}$ (Quasi-static →Impact)
300	0.2	0.080	0.020
	0.3	00.11	0.021
	0.4	00.14	0.023
	0.5	00.18	0.026
400	0.2	00.069	0.038
	0.3	00.079	0.030
	0.4	00.10	0.030
	0.5	00.13	0.032

Table 2. The detailed cell shape measured from the three-dimensional images in the foamed PE film.

Thickness (μm)	Max. length, L_{max} (μm)	Min. length, L_{min} (μm)	Aspect ratio	Cell size (μm)
300	368	60	6.1	234
400	398	70	5.6	214

to quasi-static strain rate. It was cleared that foamed PE film has the different strain rate sensitivity in comparison with general polymeric foam material.

The flow stress at the specimen with the thickness of 400 μm is higher than that of 300 μm, regardless of strain rate. Theoretically, the thickness of 300 μm and 400 μm specimens made by same material and the same density does not show the difference in flow stress. This difference was assumed to be due to the cell shape.

3.3. Effect of cell shape

In order to investigate the effects of cell shape on the strength, a three-dimensional measurement was performed using X-ray CT apparatus (Yamato Scientific Co., TDM1000H-II). The maximum image resolution of the X-ray CT is 2.4 μm. Figure 8 shows the cross-sectional CT images of the radial direction and the thickness direction in the thickness of 400 μm specimen. It can be confirmed that all cells are closed. Additionally, it is clarified that there is little anisotropy of the cell shape in the thickness direction.

Table 2 shows the detailed cell shape measured by the CT images in the thickness of 300 μm and 400 μm specimens. The aspect ratio was calculated as L_{max}/L_{min}, and the cell size was calculated as $(L_{max} + L_{min})/2$. Each parameter was an average of approximately 100 measurements. The cell size and aspect ratio at the thickness of 400 μm specimen is higher than that of 300 μm specimen. It is clear that thickness of 300 μm specimen is more flattened in the radial direction compared to that of 400 μm, which leads to the resistance due to the bending deformation of the cell wall.

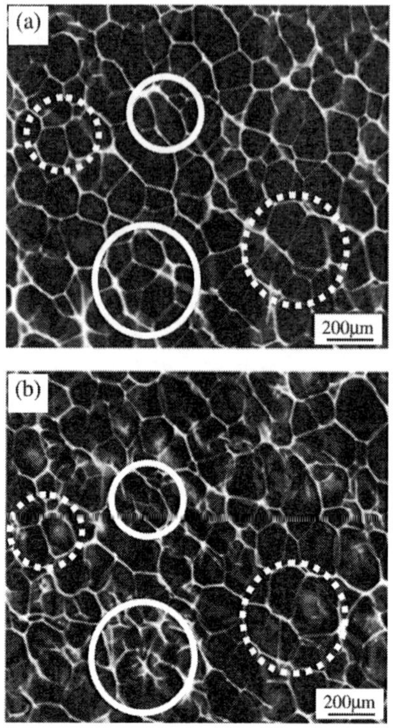

Figure 9. CT images at the radial direction of before or after quasi-static test at the foamed PE film with thickness of 400 μm (a) before, b) after).

3.4. Effect of the trapped gas in closed cell

It has been reported by Bouix *et al.* [2] that the trapped gas does not have enough time to escape from the foam until the cell is totally collapsed during the impact deformation using a fluid chamber, which is responsible for the strain rate dependence of strength in the polypropylene foam materials. However, it is necessary to confirm whether the inside air of cells flowed out due to the fracture of the closed cell, because the foamed PE film have very high ductility for the tensile deformation. We have to investigate the behavior of inside air of PE foamed film before and after the quasi-static compression test and the impact compression test, respectively. However, it is difficult to perform the direct observation of the outflow of air. Thus, the CT image was used to observe the deformation of the cell structure before and after compression test.

Figure 9 shows the CT images at the radial direction of before and after the quasi-static test in the thickness of 400 μm specimens. The test condition is the strain rate of 1.7×10^{-3} s^{-1} and the strain of 0.8. As shown in the white dotted circle, the cell structure keeps its shape before and after the quasi-static test. On the other hand, the cell wall was fractured by the compressive deformation in the area enclosed by the white solid line, which would lead to the outflow of air from the cell. It was reported [12] that the outflow from the pore was caused by the deformation resistance in the micro-cell structure. In the

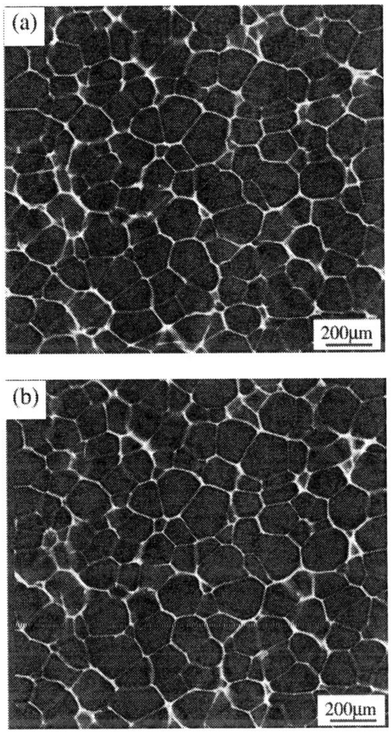

Figure 10. CT images at the radial direction of before or after impact test at the foamed PE film with thickness of 400 μm (a) before, b) after).

quasi-static compressive deformation of the foamed PE film, the increase in the strain rate sensitivity was shown due to the outflow resistance along with the fracture of the cell wall.

Figure 10 shows the CT image at the radial direction of before and after the impact test in the thickness of 400 μm specimens. The test condition was strain rate of $4.0 \times 10^2\,s^{-1}$ and the strain of 0.8. It was revealed that the cell collapse did not occur in the impact deformation. This is because the strength of the cell wall made by the polyethylene was increased due to the increase in strain rate. Thus, it is inferred that the inside air does not flow out during the impact compressive deformation.

4. Summary

To investigate the effect of the strain rate on the compressive behavior of the foamed PE film, a wide strain rate range test were performed. The results can be summarized as follows.

(1) The new compressive test apparatus for the dynamic strain rate, the drop-weight testing machine with opposed load cell, was developed, which can be evaluated the dynamic stress equilibrium of the specimen.
(2) The compressive flow stress increased with increasing the strain rate, regardless of the thickness.
(3) The plateau deformation was observed immediately after the compressive testing started and no linear-elastic response phase was observed in the foamed PE film specimen, regardless of the thickness.
(4) The foamed PE film has the high strain rate sensitivity in the quasi-static deformation, which is due to the outflow resistance along with the fracture of the cell wall.
(5) There is almost no change of the strain rate sensitivity in the dynamic and the impact deformation.

This study was supported in part by the Grant-in-Aid for Young Scientists (A) (Grant No. 25709004) from the Ministry of Education, Culture, Sports, Science and Technology, Japan. The authors are deeply grateful to Mr. Mizukami and Mr. Ishihara, undergraduate student at the National Defense Academy, for their assistance.

References

[1] J. Zhang, N. Kikuchi, V. Li, A. Yee, G. Nusholtz, Int. J. Impact Eng. **21** (1998), 369–386
[2] R. Bouix, P. Viot, J. L. Lataillade, Int. J. Impact Eng. **36** (2009), 329–342
[3] O. Ouelleta, D. Croninb and M. Worswick, Polymer Testing **25** (2006), 731–743
[4] P. Viot, Int. J. Imp. Eng. **36** (2009), 975–989
[5] H. Yamada, K. Tateyama, R. Okui, N. Ogasawara, K. Ogawa, J. Jpn. Soc. Exp. Mech. **14** (2014), 153–157
[6] K. Ogawa, J. Soc. Mat. Sci. Japan **54** (2005)
[7] H. Kobayashi, K. Horikawa, K. Ogawa, M. Hori, J. Sol. Mech. Mat. Eng. **3** (2009), 64–71
[8] B. Hopkinson, Phil. Trans. Roy. Soc., A **213** (1914), 437–452
[9] H. Kolsky, Proc. Phys. Soc. B **62** (1949), 676–700
[10] K. Ogawa, J. Jpn. Inst. Light Met. **52** (2001), 131–135
[11] K. Ogawa, H. Kobayashi, F. Sugiyama and K. Horikawa, JSME International Journal, A **48** (2005), 228–233
[12] L.J. Gibson, M.F. Ashby, (M. Otsuka translated), *Cellular Solids* (Uchida Roukakuho, 1993)

EPJ Web of Conferences **94**, 01039 (2015)
DOI: 10.1051/epjconf/20159401039

Dynamic tensile stress–strain characteristics of carbon/epoxy laminated composites in through-thickness direction

Kenji Nakai[a] and Takashi Yokoyama

Department of Mechanical Engineering, Okayama University of Science, Okayama, Japan

Abstract. The effect of strain rate up to approximately $\dot{\varepsilon} = 10^2$/s on the tensile stress–strain properties of unidirectional and cross-ply carbon/epoxy laminated composites in the through-thickness direction is investigated. Waisted cylindrical specimens machined out of the laminated composites in the through-thickness direction are used in both static and dynamic tests. The dynamic tensile stress–strain curves up to fracture are determined using the split Hopkinson bar (SHB). The low and intermediate strain-rate tensile stress–strain relations up to fracture are measured on an Instron 5500R testing machine. It is demonstrated that the ultimate tensile strength and absorbed energy up to fracture increase significantly, while the fracture strain decreases slightly with increasing strain rate. Macro- and micro-scopic examinations reveal a marked difference in the fracture surfaces between the static and dynamic tension specimens.

1. Introduction

In recent years, composite materials are increasingly replacing conventional metallic materials in aerospace, civil, marine and automotive industries, because of their higher specific strength and stiffness, higher fatigue properties and higher corrosion resistance. In many practical situations, composite structures are often subjected to dynamic loading. Therefore, it is required to characterize the dynamic behavior of composite materials. To date, the in-plane compressive [1–12], tensile [13–18] and interlaminar shear [19–24] properties of composite materials under dynamic loading have been determined with the conventional [25] or modified split Hopkinson pressure bar (SHPB). Nevertheless, except for Refs. [8, 11, 12, 26–28], their dynamic through-thickness stress–strain behavior has not been well investigated. This is because the through-thickness properties of composite materials are typically lower than their in-plane ones.

The primary objective of the present work is to determine the tensile stress–strain relations of unidirectional and cross-ply carbon/epoxy laminated composites in the through-thickness direction at strain rates up to nearly 10^2/s in the split Hopkinson bar (SHB). Waisted cylindrical specimens were used in both static and dynamic tests. The tensile stress–strain curves at low and intermediate strain rates were measured on an Instron 5500R testing machine. The influence of strain rate on the ultimate tensile strength, fracture strain and absorbed energy up to fracture was examined. Macro- and micro-scopic examinations were conducted to discuss the effect of strain rate on the fracture modes in the specimens.

[a] Corresponding author: `nakai@mech.ous.ac.jp`

2. Test composites and specimen preparation

42-ply unidirectional ($[0_{42}]$) and cross-ply ($[0/90]_{21s}$) carbon/epoxy composite laminates with a nominal thickness of 10.05 mm were used. They were chosen to assess the effect of the reinforcement geometry on their tensile behavior. The laminates were prepared from pre-impregnated tapes each of size $500 \times 500\,\text{mm}^2$ and processed in a heat press, i.e., the laminates were cured at a temperature of 125 °C for 1 hr under a pressure of 0.98 MPa and then cooled. The types of reinforcing fiber, matrix resin and fiber volume ratio are given in Table 1. The basic in-plane tensile characteristics of the unidirectional composite laminate in the longitudinal (L) and transverse (T) directions are listed in Table 2. The longitudinal (or fiber) and in-plane transverse directions are, respectively, represented by 1 and 2, and the through-thickness direction is represented by 3 (see, Fig. 1). Specimens were machined from the composite laminates in the through-thickness direction into short waisted cylinders. Figure 1 shows the shape and dimensions of the waisted tension specimen. The diameter ($d = 15$ mm) of the specimen ends was set equal to that of the Hopkinson bars. To ensure fracture in the gage (parallel) section, the specimen diameter was reduced to 5 mm. The gage length ($l_S = 4$ mm) to diameter ($d = 5$ mm) ratio of the specimen was taken as 0.8, approximately corresponding to the specimen geometry suggested by Medina and Harding [26], and Ferguson *et al.* [29]. To enable the specimens to be loaded in tension, magnesium end-caps were bonded to each end of the specimen using a high strength two-part epoxy adhesive (Sumitomo 3-M, Scotch weld® DP-460). A simple fixture (see, Fig. 2) was used to provide alignment of the end-caps with the specimen during bonding.

Article available at http://www.epj-conferences.org or http://dx.doi.org/10.1051/epjconf/20159401039

Table 1. Types of reinforcing fiber and matrix resin used in two kinds of carbon/epoxy laminated composites.

	Carbon/epoxy laminated composite	
	Unidirectional	Cross-ply
Fiber	T700 (Torayca)	
Matrix resin	Epoxy #2521	Epoxy #2500
Fiber volume ratio V_f	0.65	0.67

Table 2. In-plane tensile and physical properties of unidirectional carbon/epoxy laminated composite.

Tensile and Physical Properties	UD T700/2521	
	L (1-d)	T (2-d)
Young's modulus E (GPa)	135	8.6
Tensile strength σ_{TS} (MPa)	2016	37
Fracture strain ε_f (%)	1.8	0.5
Wave velocity c_o (m/s)	9273	2340
Mass density ρ (kg/m^3)	1570	

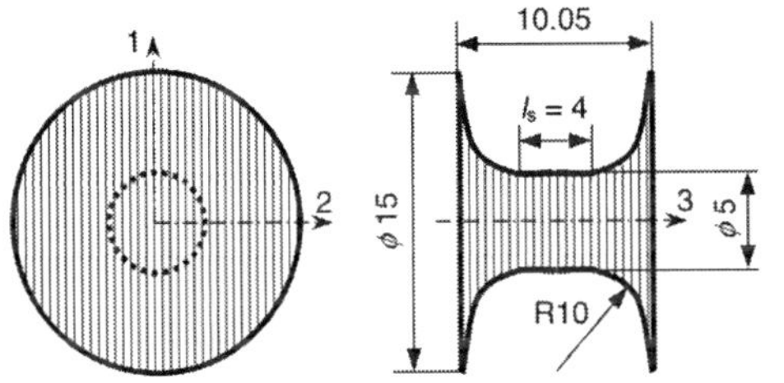

Figure 1. Geometry of static and dynamic waisted tension specimens (dimensions in mm).

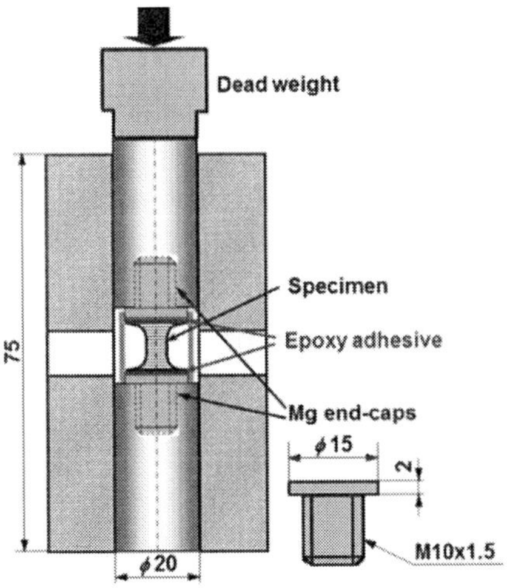

Figure 2. Alignment fixture for bonding end-caps to waisted specimen.

3. Experimental procedures

3.1. Low and intermediate strain-rate tension testing

The low and intermediate strain-rate tension tests were performed using the waisted specimens in the Instron 5500R testing machine at crosshead speeds V_c of 1 and 50 mm/min, respectively. The end-caps was screwed into two fixtures and loaded until fracture. The tensile stress in the specimen was obtained from the force recorded with a 100-kN capacity load cell (Instron: Model 2525-801). The tensile strain in the specimen was determined from strain gages (Kyowa: KFG-2-120-C1-5) attached directly to the specimen parallel section. Both stress and strain data were recorded at sampling rates of 5 Hz for $V_c = 1$ mm/min and 500 Hz for $V_c = 50$ mm/min.

3.2. Split Hopkinson bar testing

Figure 3 indicates a schematic of the tensile SHB system used, which consists principally of AZ80A–F Mg alloy input and output bars, an AZ80A–F Mg alloy striker tube, a gun barrel and an associated recording system (not shown here). The specifications for the SHB system are listed in Table 3. The Mg alloy bars with low mechanical impedance were used to reduce a drastic impedance mismatch between the composite specimen (through-thickness acoustic impedance $Z \fallingdotseq 4 \times 10^6$ kg/(m^2 s)) and conventional steel bars ($Z \fallingdotseq 40 \times 10^6$ kg/(m^2 s)), resulting in a transmitted strain signal with a very low signal-to-noise ratio. The end-caps are screwed into the Hopkinson bars (see the inset in Fig. 3). Two pairs of strain gages (Kyowa: KFG-2-120-C1-5) were attached directly to the specimen parallel section to accurately measure the strain in the specimen. A pulse shaping technique [30] is applied to generate well-defined tensile strain pulses without higher frequency components in the input bar. Namely, a 0.8 mm–thick 1050 Al ring of a 16 mm inside diameter and a 21 mm outside diameter is attached onto the impact (right) side of the loading block with a thin layer of petroleum jelly. Details of the test procedure are given elsewhere [31].

From the elementary one-dimensional theory of elastic wave propagation [32], we can determine the nominal strain $\varepsilon(t)$, strain rate $\dot{\varepsilon}(t)$ and stress $\sigma(t)$ in the specimen from the tensile SHB test records as

$$\varepsilon(t) = \frac{u_2(t) - u_1(t)}{l_S} = \frac{2c_o}{l_S}\int_0^t \left\{\varepsilon_i(t') - \varepsilon_t(t')\right\} dt' \quad (1)$$

$$\dot{\varepsilon}(t) = \frac{\dot{u}_2(t) - \dot{u}_1(t)}{l_S} = \frac{2c_o}{l_S}\{\varepsilon_i(t) - \varepsilon_t(t)\} \quad (2)$$

$$\sigma(t) = \frac{P_2(t)}{A_S} = \frac{AE}{A_S}\varepsilon_t(t). \quad (3)$$

Here u and P are the displacement and the axial force on both ends of the specimen, respectively, (where subscripts 1 and 2 denote the left and right interfaces, respectively; see the inset in Fig. 3); A, E and c_o are the cross-sectional area, Young's modulus and the longitudinal elastic wave velocity of the Hopkinson (Mg alloy) bars; A_S and l_S denote the cross-sectional area ($\fallingdotseq \pi\ 5^2/4 = 19.63$ mm^2) and original gage length (= 4 mm) of the specimen. Eqs. (1) to (3) are derived under the assumption of dynamic force equilibrium across the specimen, i.e.,

$$P_1(t) = P_2(t) \quad \text{or} \quad \varepsilon_i(t) + \varepsilon_r(t) = \varepsilon_t(t) \quad (4)$$

where

$$P_1(t) = AE\left[\varepsilon_i(t) + \varepsilon_r(t)\right], \quad P_2(t) = AE\varepsilon_t(t). \quad (5)$$

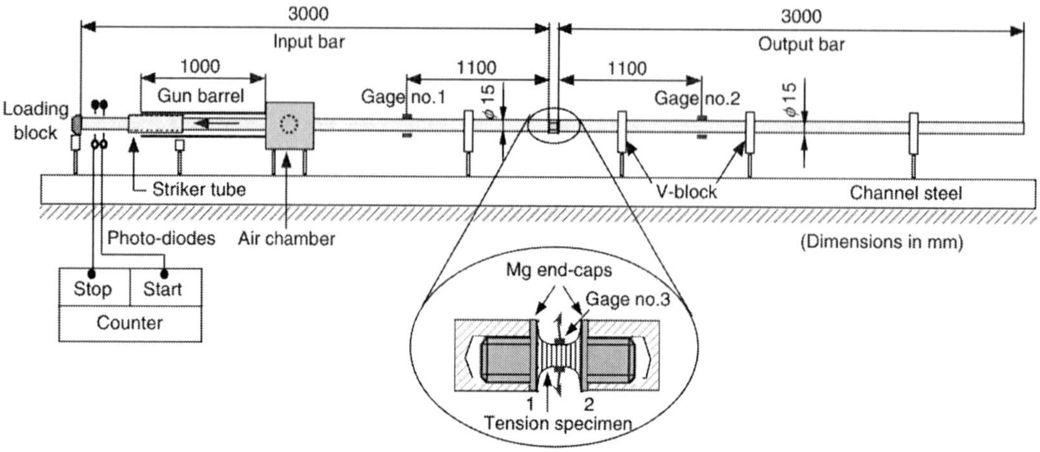

Figure 3. Schematic diagram of tensile split Hopkinson bar set-up (associated recording system not shown).

Table 3. Specifications for present SHB system.

	Striker tube	Input/output bar
Material	AZ80A–F Mg alloy	
Young's modulus E (GPa)	44.4	
Longitudinal wave velocity $c_o = \sqrt{E/\rho}$ (m/s)	4953	
Yield strength σ_Y (MPa)	275	
Mass density ρ (kg/m^3)	1810	
Acoustic impedance $Z = \rho c_o$ (kg/(m^2 s))	9×10^6	
Length L (mm)	200	3000
Diameter d (mm)	–	15
Inside diameter d_i (mm)	15.3	–
Outside diameter d_o (mm)	21.5	

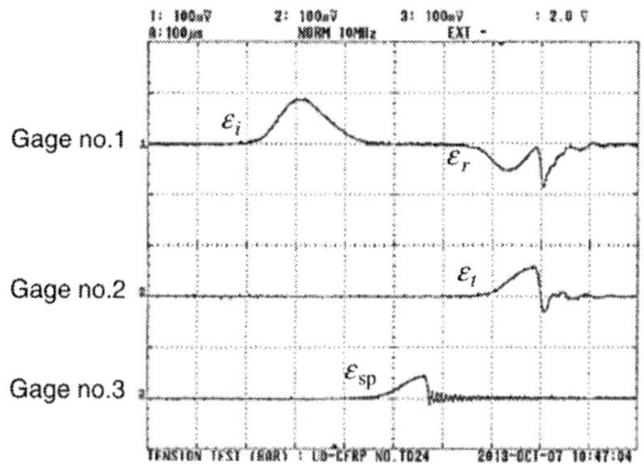

Sweep rate: 100μs/div
Vertical sensitivity:
Upper trace: 100mV/div (220με/div)
Middle trace: 100mV/div (228με/div)
Lower trace: 100mV/div (12977με/div)

Figure 4. Typical set of oscilloscope records from tensile SHB test on unidirectional carbon/epoxy laminated composite (striker tube velocity: $V_s = 3.9$ m/s).

In the above derivations, the incident and reflected strain pulses are time-shifted to the specimen-input bar interface, and the transmitted strain pulse is time-shifted to the specimen-output bar interface.

4. Results and discussion

4.1. Dynamic tensile behavior

A number of the tensile SHB tests were conducted using the composite specimens at room temperature. Figure 4 indicates a typical set of oscilloscope records from the SHB test on the unidirectional carbon/epoxy laminated composite. The upper trace gives the incident and reflected strain pulses (ε_i and ε_r), the middle trace gives the strain pulse (ε_t) transmitted through the specimen, and the lower trace gives the strain pulse (ε_{sp}) from the specimen strain gage. The output signals from the strain gages were recorded at a sampling rate of 1 MHz. The recorded signal data are neither smoothed nor averaged electronically. Note that the duration ($\fallingdotseq 80\,\mu$s) of the transmitted strain pulse is much shorter than that ($\fallingdotseq 250\,\mu$s) of the incident and reflected strain pulses. This implies that a tensile fracture took place in the specimen during stress wave loading, because if a fracture occurs, a tensile strain pulse can no longer be propagated to the output bar. Figure 5 gives the time variations of the applied tensile forces P_1 and P_2 obtained from Eq. (5). It is experimentally verified that the dynamic force equilibrium across the specimen is valid. Figure 6 presents the time histories for the dynamic tensile stress $\sigma(t)$ and strain $\varepsilon(t)$ determined from the SHB analysis, or Eqs. (1) and (3), and the strain $\varepsilon_{sp}(t)$ measured by the strain gage on the specimen. Note that the strain calculated from the SHB analysis is larger than that from the specimen strain gage. This is primarily due to the occurrence of deformation in the fillets at the end of the gage length of the waisted specimen. Eliminating time t through $\sigma(t)$ and $\varepsilon_{sp}(t)$ yields the dynamic tensile stress–strain relations. The crack initiation in the specimen took place at the peak stress (corresponding to the ultimate tensile strength σ_{UTS}) at 67 μs. The strain rate at fracture is expressed as $\dot{\varepsilon}_f = \varepsilon_f/t_f$ (t_f: fracture initiation time).

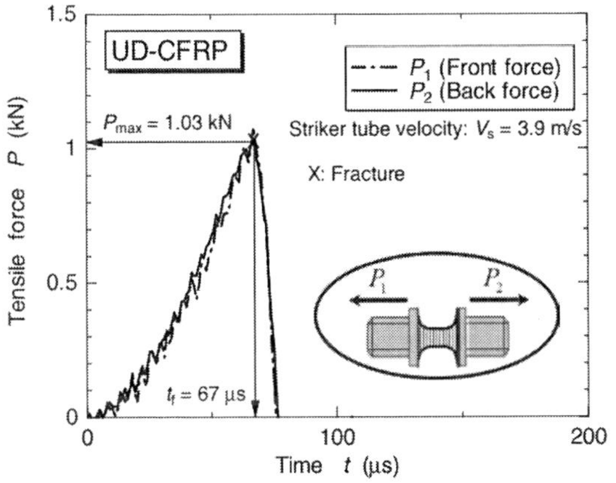

Figure 5. Axial tensile force histories at front and back ends of specimen.

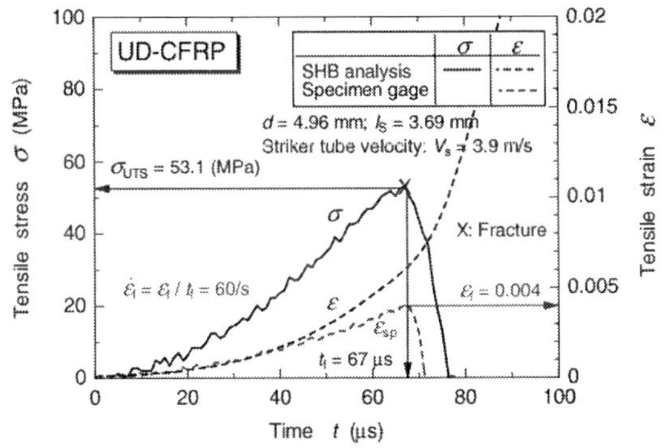

Figure 6. Dynamic tensile stress and strain histories for unidirectional carbon/epoxy laminated composite.

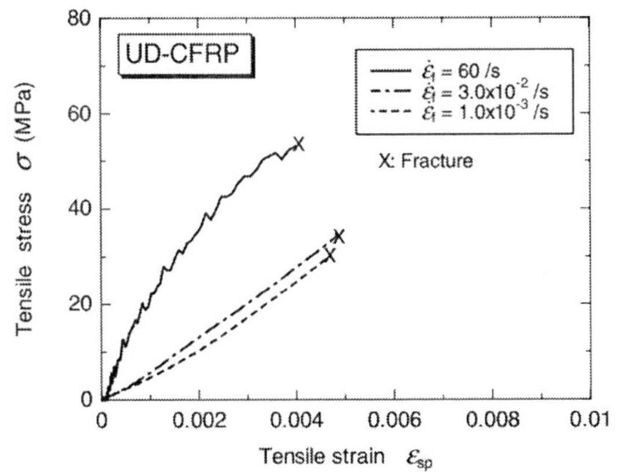

Figure 7. Through-thickness tensile stress–strain curves for unidirectional carbon/epoxy laminated composite at three different strain rates.

Figure 7 depicts the tensile stress–strain curves at three different strain rates. The initial slope (or initial modulus) and ultimate tensile strength σ_{UTS} increase greatly with increasing strain rate. This may be due to the viscoelastic nature of the epoxy resin matrix itself [4]. This result is consistent with a previous study [26] which reports that the initial modulus and tensile strength for a cross-ply

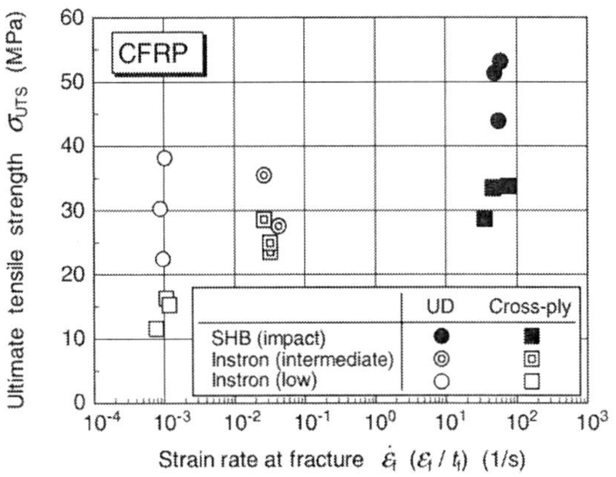

Figure 8. Effect of strain rate at fracture on ultimate tensile strength for two different carbon/epoxy laminated composites.

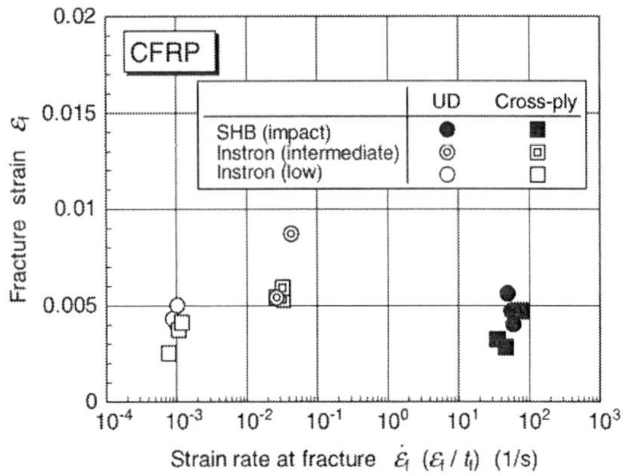

Figure 9. Effect of strain rate at fracture on fracture strain for two different carbon/epoxy laminated composites.

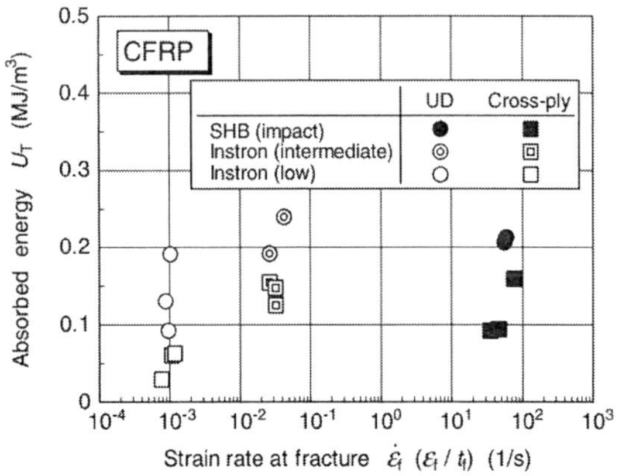

Figure 10. Effect of strain rate at fracture on absorbed energy for two different carbon/epoxy laminated composites.

carbon/epoxy (T300/924) laminated composite increase significantly with strain rate in the range of 10^{-4} to 10^3/s. In contrast, the fracture strain ε_f decreases slightly with increasing strain rate. The absorbed energy U_T is obtained by numerical integration of the tensile stress–strain curveup to fracture strain measured in the static

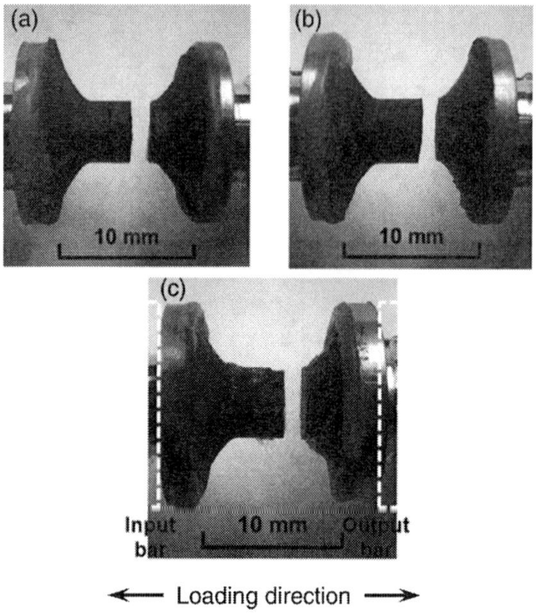

Figure 11. Photographs of tension specimens after static and dynamic testing for unidirectional carbon/epoxy laminated composite: (a) $\dot{\varepsilon}_f = 1.0 \times 10^{-3}$/s ($\sigma_{UTS} = 30.2$ MPa), (b) $\dot{\varepsilon}_f = 3.0 \times 10^{-2}$/s ($\sigma_{UTS} = 35.4$ MPa), (c) $\dot{\varepsilon}_f = 60$/s ($\sigma_{UTS} = 53.1$ MPa).

and dynamic tests. In an attempt to evaluate the effect of strain rate on the tensile behavior for the unidirectional and cross-ply carbon/epoxy laminated composites, σ_{UTS}, ε_f and U_T are plotted in Figs. 8 to 10 as functions of the strain rate at fracture. The test results display a significant amount of experimental scatter. This scatter is caused by inhomogeneity within the specimen. For both unidirectional and cross-ply carbon/epoxy laminated composites, the ultimate tensile strength and absorbed energy up to fracture increase dramatically, whereas the fracture strain decreases marginally with increasing strain rate at fracture. Furthermore, σ_{UTS}, ε_f and U_T for the unidirectional carbon/epoxy laminated composite are slightly higher than those for the cross-ply one at low and high rates of strain.

4.2. Effect of strain rate on fracture mode

In an effort to validate the test results, we observed the macro- and micro-scopic fracture modes in both the static and dynamic tension specimens. Figure 11 depicts fracture appearance of the unidirectional specimens after static and dynamic testing. Typical tensile fracture occurs in the plane perpendicular to the specimen axis at all strain rates. Little effect of strain rate on the fracture mode can be found from a macroscopic point of view. Figures 12 and 13 show SEM micrographs of the static and dynamic fracture surfaces. For both the unidirectional and cross-ply specimens, the dynamic fracture surface is much rougher and more uneven than the static one. At low strain rates, the fracture surfaces exhibit fiber-matrix interfacial fracture for both the specimens. In contrast, in the dynamic fracture surfaces, we can observe the

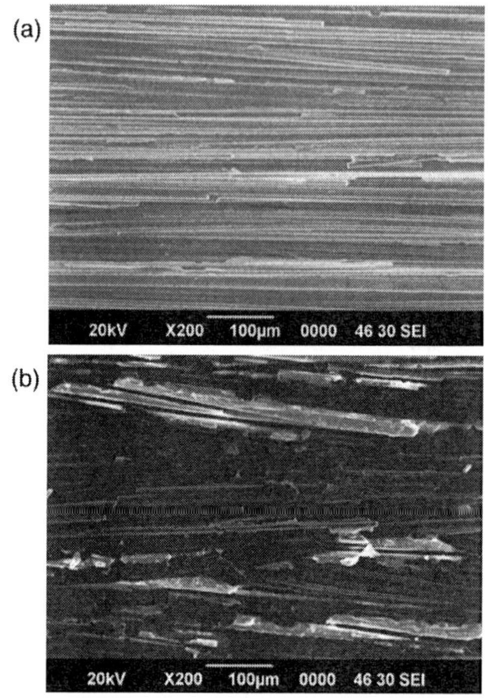

Figure 12. SEM micrographs of fracture surfaces in static and dynamic tension specimens of unidirectional carbon/epoxy laminated composite: (a) $\dot{\varepsilon}_f = 1.0 \times 10^{-3}$/s($\sigma_{UTS} = 30.2$ MPa), (b) $\dot{\varepsilon}_f = 60$/s ($\sigma_{UTS} = 53.1$ MPa).

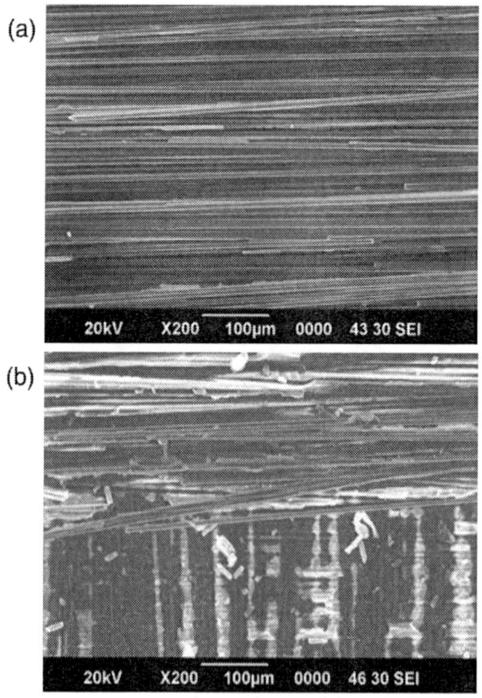

Figure 13. SEM micrographs of fracture surfaces in static and dynamic tension specimens of cross-ply carbon/epoxy laminated composite: (a) $\dot{\varepsilon}_f = 1.0 \times 10^{-3}$/s ($\sigma_{UTS} = 16.2$ MPa), (b) $\dot{\varepsilon}_f = 80$/s ($\sigma_{UTS} = 33.7$ MPa).

fiber-matrix interfacial cracking and matrix cracking for the unidirectional specimens, and interlaminar delamination for the cross-ply ones. The different tensile stress–strain behavior at low and high rates of strain is supported by the different fracture surfaces in the specimens. Furthermore, there is significant effect of reinforcement geometry on the fracture surface at high strain rates.

5. Conclusions

The effect of strain rate up to nearly $\dot{\varepsilon} = 10^2$/s on the tensile stress–strain characteristics of the unidirectional and cross-ply carbon/epoxy laminated composites in the through-thickness direction has been examined in detail. From the present experimental investigation, we can draw the following conclusions:

(1) The ultimate tensile strength and absorbed energy up to fracture exhibit a positive strain-rate sensitivity, whereas the fracture strain exhibits a negative one.
(2) The ultimate tensile strength, fracture strain and absorbed energy of the unidirectional carbon/epoxy laminated composite are slightly higher than those of the cross-ply one at any strain rate.
(3) The fracture surfaces are strongly affected by the strain rate and reinforcement geometry.

References

[1] L.J. Griffiths, D.J. Martin, J. Phys. D: Appl. Phys. **7**, 2329 (1974)
[2] A.M.A. El-Habak, Composites **22**, 129 (1991)
[3] J. Harding, Composites **24**, 323 (1993)
[4] T.E. Tay, H.G. Ang, V.P.W. Shim, Compos. Struct. **33**, 201 (1995)
[5] E. Woldesenbet, J.R. Vinson, AIAA J. **37**, 1102 (1999)
[6] H.M. Hsiao, I.M. Daniel, R.D. Cordes, J. Compos. Mater. **33**, 1620 (1999)
[7] B.A. Gama, J.W. Gillespie Jr., H. Mahfuz, R.P. Raines, A. Haque, S. Jeelani, T.A. Bogetti, B.K. Fink, J. Compos. Mater. **35**, 1201 (2001)
[8] M.V. Hosur, J. Alexander, U.K. Vaidya, S. Jeelani, Compos. Struct. **52**, 405 (2001)
[9] L. Ninan, J. Tsai, C.T. Sun, Int. J. Impact Eng. **25**, 291 (2001)
[10] R.O. Ochola, K. Marcus, G.N. Nurick, T. Franz, Compos. Struct. **63**, 455 (2004)
[11] T. Yokoyama, K. Nakai, *Proceedings of DYMAT 2009* **1/2**, 639 (2009)
[12] R. Gerlach, C.R. Siviour, J. Wiegand, N. Petrinic, Compos. Sci. Technol. **72**, 397 (2012)
[13] J. Harding, L.M. Welsh, J. Mater. Sci. **18**, 1810 (1983)
[14] I.S. Chocron Benloulo, J. Rodríguez, M.A. Martínez, V. Sánchez Gálvez, Int. J. Impact Eng. **19**, 135 (1997)
[15] A. Gilat, R.K. Goldberg, G.D. Roberts, Compos. Sci. Technol. **62**, 1469 (2002)
[16] T. Gómez-del Río, E. Barbero, R. Zaera, C. Navarro, Compos. Sci. Technol. **65**, 61 (2005)
[17] N.K. Naik, P. Yernamma, N.M. Thoram, R. Gadipatri, V.R. Kavala, Polym. Test. **29**, 14 (2010)
[18] R. Gerlach, C.R. Siviour, J. Wiegand, N. Petrinic, Mech. Adv. Mater. Struct. **20**, 505 (2013)
[19] J. Harding, Y.L. Li, Compos. Sci. Technol. **45**, 161 (1992)
[20] B. Bouette, C. Cazeneuve, C. Oytana, Compos. Sci. Technol. **45**, 313 (1992)
[21] J. Harding, L. Dong, Compos. Sci. Technol. **51**, 347 (1994)
[22] L. Dong, J. Harding, Composites **25**, 129 (1994)
[23] M.J. Hiley, L. Dong, J. Harding, Composites Part A: Appl. Sci. Manuf. **28**, 171 (1997)
[24] T. Yokoyama, K. Nakai, Appl. Mech. Mater. **3/4**, 191 (2005)
[25] H. Kolsky, Proc. Phys. Soc. B **62**, 676 (1949)
[26] J.L. Medina, J. Harding, *Journal de Physique IV*, 10-PR9, 275 (2000)
[27] T. Yokoyama, Appl. Mech. Mater. **1/2**, 11 (2004)
[28] D. Chen, F. Lu, B. Jiang, J. Compos. Mater. **46**, 3297 (2012)
[29] R.F. Ferguson, M.J. Hinton, M.J. Hiley, Compos. Sci. Technol. **58**, 1411 (1998)
[30] W. Chen, F. Lu, M. Cheng, Polym. Test. **21**, 113 (2002)
[31] T. Yokoyama, Strain **39**, 167 (2003)
[32] K.F. Graff, *Wave motion in elastic solids*, Chap. 2 (Clarendon Press, Oxford, 1975)

EPJ Web of Conferences **94**, 01040 (2015)
DOI: 10.1051/epjconf/20159401040

Experimental investigation of the failure envelope of unidirectional carbon-epoxy composite under high strain rate transverse and off-axis tensile loading

Peter Kuhn[a], Marina Ploeckl, and Hannes Koerber

Institute for Carbon Composites, Technische Universitaet Muenchen, Faculty of Mechanical Engineering, Boltzmannstrae 15, 85748 Garching bei Muenchen, Germany

Abstract. The mechanical response of the carbon-epoxy material system HexPly IM7-8552 was investigated under transverse tension and combined transverse tension / in-plane shear loading at quasi-static and dynamic strain rates. The dynamic tests of the transverse tension and off-axis tension specimens were carried out on a split-Hopkinson tension bar system, while the quasi-static reference tests were performed on a standard electro-mechanical testing machine. Digital image correlation was used for data reduction at both strain rate regimes. For the high rate tests, the strain rate in loading direction was adjusted to reach approximately the same strain rate value in the fracture plane for each specimen. The measured axial strengths were transformed from the global coordinate system into the combined transverse tension-shear stress space of the material coordinate system and compared with the Puck Mode A criterion for inter-fibre failure. A good correlation between the experimental data and the predicted failure envelopes was found for both investigated strain rate regimes.

1. Introduction

Over the past years, the number of applications in which fibre reinforced polymer matrix composites (FRPMCs) are used in primary automotive (e.g. BMW i-project) and aeronautical (e.g. Airbus A350 and Boeing 787) structures has significantly increased. In both areas of applications, components may be subjected to high speed loading events such as crash or foreign object impact. Therefore strain rate effects should be captured by advanced composite material models to predict initiation and evolution of damage accurately. The development of corresponding constitutive models and failure criteria requires an improved understanding of the mechanical behaviour of composites under high strain rate loading conditions, especially in the main material directions and under combined stress states. This critically needed data can be provided by high rate loading experiments such as the split-Hopkinson bar test.

In the present work the mechanical properties of a unidirectional (UD) carbon-epoxy composite are investigated under transverse tension and combined transverse tension / in-plane shear loading. Therefore tests at quasi-static and dynamic strain rates using transverse tension and off-axis tension specimens are performed. For the latter specimen type, the authors [1] have observed in a previous study that the strain rate acting in the fracture plane of the off-axis specimens is significantly higher than the applied axial strain rate and additionally depends on the off-axis angles of the specimens. Hence, in this work, the strain rate in the loading direction is adjusted to the reach same strain rate values in the fracture planes, allowing an accurate determination of the corresponding failure envelope.

The dynamic tensile properties of carbon fibre composites with thermoset and thermoplastic matrices in matrix-dominant directions have also been investigated in similar studies by Taniguchi et al. [2,3]. In addition, results from single off-axis and/or transverse tension tests under high strain rates have been published by several authors [4–8]. Dynamic compression tests with off-axis specimens have been performed in different works, as well [9–11]. In the study performed by Koerber et al. [9] the same material as in this work was used for the characterisation of the transverse compression, combined transverse compression and in-plane shear properties under dynamic loading. Together with the data of the present study, a comprehensive data set on the strain rate effect of the matrix dominated compressive and tensile response therefore now exists for this composite material system.

2. Material and experimental setup

2.1. Material and test specimens

The carbon-epoxy material system HexPly IM7-8552, which is commonly used in primary aerospace components, was selected for this work. In accordance with the specified heat cycle [12], a UD panel with a thickness of 1.5 mm was manufactured by curing a 12-ply prepreg layup in a hot press. A void content of less than 2% was determined for this panel by using digital image analysis of photomicrographs.

From the manufactured panel, off-axis tension specimens with fibre orientation angles $\theta = 15°$, 30°, 45° and

[a] Corresponding author: kuhn@lcc.mw.tum.de

Article available at http://www.epj-conferences.org or http://dx.doi.org/10.1051/epjconf/20159401040

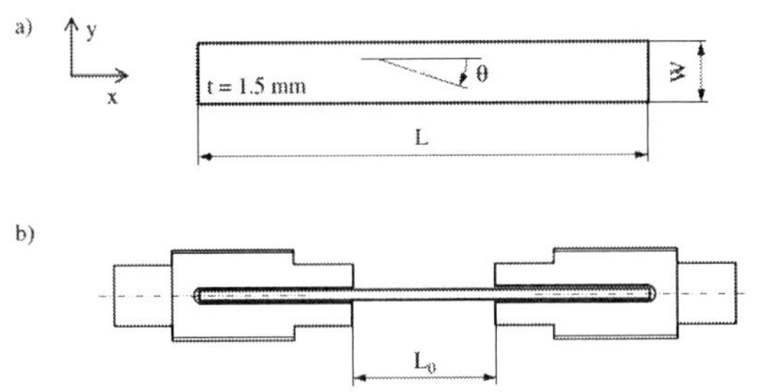

Figure 1. Tension specimen (a) and assembled specimen configuration (b).

transverse tension specimens ($\theta = 90°$) were cut on a water-cooled diamond saw. The nominal dimension of the 15° off-axis specimens were $72 \times 8 \times 1.5\,mm^3$ and $62 \times 8 \times 1.5\,mm^3$ for all other specimens (Fig. 1a). To attach the specimens to the split-Hopkinson tension bar system, slotted steel adapters with an M12 × 1.25 outside thread were developed. All specimens were glued into the adapters using the structural adhesive 3M Scotch-Weld DP 490, while proper alignment of the assembled configuration was ensured by a high accuracy bonding jig. The dimension of the free specimen length L_0 between the two adapters was chosen to obtain a section where no fibre is attached to either of the two adapters. Thus for the 15° off-axis specimens $L_0 = 30\,mm$ was used and for all other specimens $L_0 = 20\,mm$ was found to be sufficient.

The assembled specimen configuration is shown in Fig. 1b. The specimens were prepared by applying a random black-on-white speckle pattern to enable the use of digital image correlation (DIC) measurement. The fineness of the pattern was adapted to the available image resolutions of the recording camera systems. To ensure that the comparison between the quasi-static and dynamic test results is not affected by size or clamping effects, the same specimen type was used for both test configurations.

2.2. Quasi-static experimental setup

The quasi-static reference tests were peformed on a standard electro-mechanical testing machine (Hegewald & Peschke Inspect Table 100). The adapters of the assembled specimens were screwed into inserts, which in turn were connected to the test machine by a bolt construction (Fig. 2). A constant displacement rate of $0.5\,mm.min^{-1}$ was chosen. The GOM ARAMIS-4M system was used for DIC measurement of the three-dimensional in plane strain field. It consisted of two CCD cameras with a resolution of $1728 \times 2352\,pixel^2$, adjusted to capture a measuring volume of $35 \times 26\,mm^2$. The frame rate was 1 frame per second (fps) with a shutter speed of 60 ms.

2.3. Dynamic experimental setup

The high strain rate tests were carried out on a split-Hopkinson tension bar (SHTB) system. The setup, that is illustrated in Fig. 3, is based on a novel concept proposed by Gerlach et al. [13], using a U-shaped striker-bar.

Figure 2. Quasi-static tension test setup.

In this work, a configuration of titanium Ø20 mm loading-, Ø16 mm incident- and Ø16 mm transmission-bars with lengths of 2.15, 3 and 1.8 m, respectively, was chosen. The striker-bar had a length of 0.5 m. A ring of 2 mm thick silicon rubber, wrapped around the impact flange at the end of the loading bar, was used for pulse shaping. The resulting ramp-shaped pulse was ideally suited for the predominantly linear stress-strain behaviour [14], which was expected for the dynamic test due to the low plastic strains observed during the quasi-static material characterisation.

For the determination of the two-dimensional strain field using DIC, the deformation of the specimen was monitored by a single Photron SA5 high speed camera. Due to the fact that achievable frame rate and image resolution are contrary setup parameters of the camera, a suitable compromise had to be found for each specimen type. The chosen parameters are listed in Table 1.

3. Data reduction

3.1. Global coordinate system

Initially, the data reduction was done in the global coordinate system where xx coincides with the loading direction of the assembled specimen configuration. In the case of the quasi-static tests, the nominal axial stress σ_{xx} was therefore calculated by dividing the load F, measured from the load-cell of the testing machine, by the cross-section A_s of the specimen. For the high rate tests, σ_{xx} was determined according to the classic split-Hopkinson pressure bar analysis (SHPBA). According to Gray [15], the specimen stress can be calculated as

$$\sigma_{xx} = A_b E_b \varepsilon_T / A_s \quad (1)$$

where A_b, E_b and ε_b are the cross-section, Youngs modulus and the measured elastic strain wave of the transmission

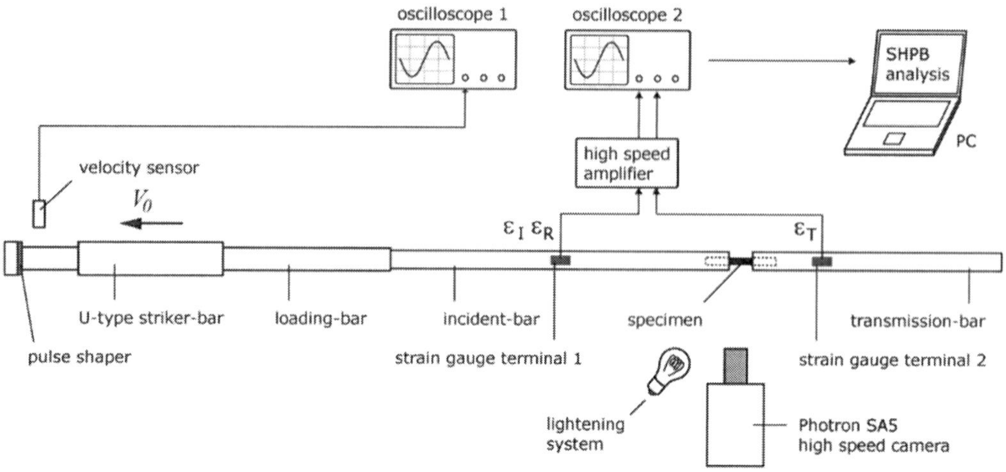

Figure 3. Split-Hopkinson tension bar setup for dynamic tests.

Table 1. Setup parameters of high speed camera.

Specimen type	Frames per second $[s^{-1}]$	Resolution $[pixel^2]$
15°	186.000	392 × 96
30°	186.000	256 × 112
45°	300.000	192 × 80
90°	300.000	192 × 80

Table 2. ARAMIS analysis parameters.

	Quasi-static	High rate
Facet size $[pixel^2]$	17 × 17	10 × 10
Facet step $[pixel^2]$	15 × 15	5 × 5
Computation size $[facets^2]$	5 × 5	5 × 5

bar. The in-plane strain field $\{\varepsilon_{xx}, \varepsilon_{yy}, \gamma_{xy}\}^T$ of the global coordinate system was obtained for all tests from DIC software GOM ARAMIS. To avoid calculation errors of the DIC at the specimen edges, an area of 5 × 5 mm^2 in the specimen centre was defined for the deformation analysis. The DIC analysis parameters were chosen accordingly to the resolutions of the camera images and are given in Table 2.

The strain component in loading direction was further used for the calculation of axial specimen strain rate $\dot{\epsilon}_{xx}$ by deriving ε_{xx} with respect to time:

$$\dot{\epsilon}_{xx} = d\varepsilon_{xx}/dt. \tag{2}$$

3.2. Material coordinate system

For the determination of material failure envelopes, it is necessary that the stress-strain response, calculated in the global coordinate system is transformed into the local, material system. For a transverse isotropic material, the orientation along the fibres is denoted as 1-direction and the 2-direction coincides with the in-plane transverse direction. According to Schuermann [16], the stress and strain components of the loading coordinate system can be transformed into the material coordinate system by

$$\sigma_{22} = \sigma_{xx}\sin^2\beta \tag{3}$$

$$\tau_{12} = -\sigma_{xx}\sin\beta\cos\beta \tag{4}$$

$$\gamma_{12} = -\varepsilon_{xx}\sin2\beta + \varepsilon_{yy}\sin2\beta + \gamma_{xy}\cos2\beta \tag{5}$$

where β denotes the transformation angle, consisting of the initial off-axis angle θ and the fibre rotation angle $d\theta$, that occurs during the test due to the extension shear coupling effect in off-axis specimens [17]. The fibre rotation angle can be obtained from DIC software GOM ARAMIS. In case of tensile tests, as performed in this study, the fibres in off-axis specimens tend to orientate in loading direction and therefore the actual transformation angle is

$$\beta = \theta_0 - d\theta. \tag{6}$$

The shear strain rate $\dot{\gamma}_{12}$ can be calculated analogous as for the axial strain rate by deriving the corresponding strain value with respect to time:

$$\dot{\gamma}_{12} = d\gamma_{12}/dt. \tag{7}$$

4. Results and discussion

For each specimen type and strain rate regime, three valid tests were performed still allowing a statistical treatment while keeping the experimental efforts within reasonable limits. A test was declared as valid when fracture occurred in the free specimen section.

As observed in a previous study [1], for an accurate comparison of the strain rate effects the strain rates acting in the failure plane should be the same for all tested specimens. To meet this requirement, the axial strain rate had to be increased with increasing off-axis angle of the specimens to capture the corresponding shift from an in-plane shear to a transverse tensile material response. Hence, for the 15°, 30° and 45° off-axis specimens, the

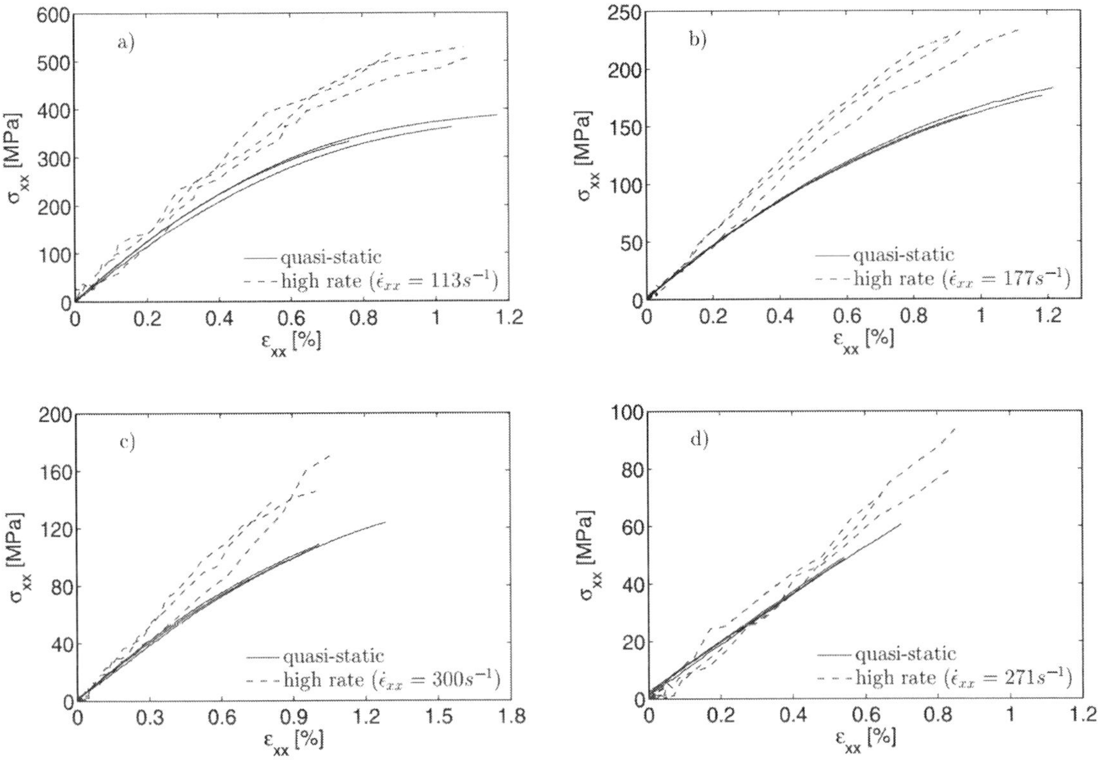

Figure 4. Quasi-static and high rate axial stress-strain curves for: (a) 15° off-axis tension, (b) 30° off-axis tension, (c) 45° off-axis tension and (d) transverse tension.

axial strain rate was adjusted to approximately $110\,s^{-1}$, $180\,s^{-1}$ and $300\,s^{-1}$, respectively, to reach a shear strain rate of $350\,s^{-1}$. In case of the transverse tension specimens, the axial strain rate is the governed value. Due to the striker-bar velocity limit of the used split-Hopkinson tension bar system, together with the chosen specimen geometry, the attainable strain rate for the 90° specimens was $271\,s^{-1}$ and therefore slightly lower than the target value.

The axial stress-strain curves of the off-axis tension and transverse tension specimens for both investigated strain rate regimes are plotted in Fig. 4. All off-axis loaded specimens show non-linear stress-strain behaviour at quasi-static test conditions and a tendency to more linear stress-strain curves under dynamic loading. For the transverse specimens, the stress-strain characteristic is linear at both strain rate regimes. A pronounced strain rate effect on the measured axial strength occurs for all tested specimen configurations. In detail, the strength increases by 45%, 34%, 35% and 32% for the 15°, 30°, 45° and 90° specimens, respectively, and therefore an approximately constant strain rate effect is found. The measured properties are summarized in Table 3.

The quasi-static and dynamic in-plane shear behaviour of the 15°, 30° and 45° off-axis specimens, which was calculated from Eqs. (4) and (5), can be seen for representative curves in Fig. 5. The initial part of the shear stress- shear strain curve is similar for both strain rate regimes and seems to be independent of the respective fiber orientation angle. However, the in-plane shear strength value is significantly affected by the superposed transverse stress ratio from Eq. (3).

Figure 6 shows the quasi-static and dynamic failure envelope for combined transverse tension/in-plane shear loading. The pure in-plane shear values were taken from Koerber et al. [9]. They found an increase of 42% for the in-plane shear strength under similar dynamic loading rates as realized in the present work. For the transverse tension specimens, pronounced scatter in the strength was found for both strain rate regimes. Nevertheless, the calculated mean value for quasi-static loading coincides well with the transverse strength listed in the HexPly IM7-8552 data sheet [12].

The Puck Mode A failure envelope for two-dimensional inter-fibre failure [18], that accounts for the combined transverse tension and in-plane shear stress state in the fracture plane, was constructed for the quasi-static results. For the construction of the dynamic failure envelope an average strength increase of 37% was used. It can be seen, that the experimental data and the predicted failure envelopes correlate well for both investigated strain rate regimes. For the 15° off-axis specimens, the predicted strength seems to be slightly over-predictive. A possible explanation for this could be that due to the extension-shear coupling effect, which is most pronounced for this specimen type, premature failure is induced by stress concentrations close to the adapters. It is assumed, however, that with the used specimen configuration, stress

Table 3. Quasi-static and high rate off-axis tension and transverse tension properties.

Loading rate	Fibre angle $\theta[°]$		Ultimate strength [MPa]	Shear angle $d\theta[°]$	Axial strain rate $\dot{\varepsilon}_{xx}[s^{-1}]$	Shear strain rate $\dot{\gamma}_{12}[s^{-1}]$
quasi-static	15°	mean	360	1.56	$2.1 \cdot 10^{-4}$	$7.8 \cdot 10^{-4}$
		STDV	22	0.32	$2.8 \cdot 10^{-5}$	$1.8 \cdot 10^{-4}$
		CV [%]	6.1	20.5	13.3	15.3
	30°	mean	173	0.82	$2.9 \cdot 10^{-4}$	$6.0 \cdot 10^{-4}$
		STDV	10	0.09	$7.7 \cdot 10^{-6}$	$4.2 \cdot 10^{-5}$
		CV [%]	5.8	11.0	2.7	7.0
	45°	mean	113	0.35	$2.6 \cdot 10^{-4}$	$3.8 \cdot 10^{-4}$
		STDV	8	0.03	$2.9 \cdot 10^{-5}$	$3.3 \cdot 10^{-5}$
		CV [%]	7.1	8.6	11.3	8.6
	90°	mean	62	-	$2.8 \cdot 10^{-4}$	-
		STDV	11	-	$1.7 \cdot 10^{-5}$	-
		CV [%]	17.7	-	5.9	-
high rate	15°	mean	521	1.45	113	358
		STDV	9	0.20	19	51
		CV [%]	1.7	13.8	16.8	14.2
	30°	mean	231	0.71	177	337
		STDV	3	0.05	10	10
		CV [%]	1.3	7.0	5.6	3.0
	45°	mean	153	0.34	300	410
		STDV	14	0.04	44	64
		CV [%]	9.2	11.8	14.7	15.6
	90°	mean	82	-	271	-
		STDV	10	-	13	-
		CV [%]	12.2	-	4.8	-

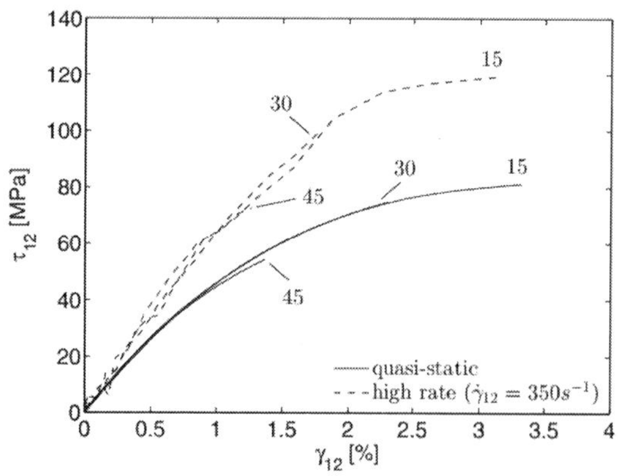

Figure 5. In-plane shear behaviour from off-axis tension tests.

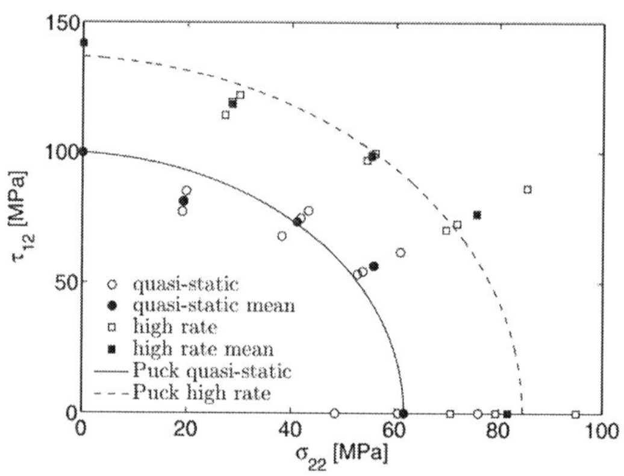

Figure 6. Experimental and predicted inter-fibre failure (IFF) envelope for quasi-static and high rate loading.

concentrations could be reduced due to the compliance of the adhesive layer, compared to rigid clamping boundary constraints at quasi-static standard tests.

5. Summary and conclusion

The mechanical response of the carbon-epoxy material system HexPly IM7-8552 was investigated under transverse tension and combined transverse tension / in-plane shear loading. Tests at quasi-static and dynamic strain rates using transverse tension and off-axis tension specimens were performed. A linear stress strain-strain response was found for the transverse specimens at both strain rates, whereas the off-axis loaded specimens showed non-linear stress-strain behaviour at quasi-static test conditions and a tendency to more linear stress-strain curves under dynamic loading. The measured axial strengths increased by 45%, 34%, 35% and 32% for the 15°, 30°, 45° and 90° specimens, respectively. An approximately constant strain rate effect was therefore found which can be attributed to same strain rate values in the fracture planes of all specimens. This was reached by adjusting the strain rate in the loading direction, accordingly.

In the combined transverse tension/in-plane shear stress space, good correlation between the experimental data and the Puck Mode A IFF failure envelope was observed for both strain rate regimes.

For all specimen configurations, an average increase of 37% for the ultimate strength under dynamic loading was found. This coincides well with the results for combined transverse compression/in-plane shear loading published by Koerber et al. [9]. They found an average increase of 40% under similar dynamic loading rates as realized in the present work. A comprehensive data set on the strain rate effect of the matrix dominated compressive and tensile response of HexPly IM7-8552 therefore now exists.

References

[1] H. Koerber, M. Vogler, P. Kuhn, P.P. Camanho, Proceedings of ECCM16 - 16th European Conference on Composite Materials, Seville, Spain (2014)

[2] N. Taniguchi, T. Nishiwaki, H. Kawada, Advanced Composite Materials **16**, 167–180 (2007)

[3] N. Taniguchi, T. Nishiwaki, N. Hirayama, H. Nishida, H. Kawada, Composites Science and Technology **69**, 207–213 (2009)

[4] A. Gilat, R.K. Goldberg, G.D. Roberts, Composites Science and Technology **62**, 1469–1476 (2002)

[5] S.V. Thiruppukuzhi, C.T. Sun, Composites Science and Technology **61**, 1–12 (2001)

[6] J. Harding, L.M. Welsh, Journal of Materials Science **18**, 181–1826 (1983)

[7] T. Gómez-del Río, E. Barbero, R. Zaera, C. Navarro, Composites Science and Technology **65**, 61–71 (2005)

[8] L.G. Melin, L.E. Asp, Composites: Part A **30**, 305–316 (1999)

[9] H. Koerber, J. Xavier, P.P. Camanho, Mechanics of Materials **42**, 1004–1019 (2010)

[10] I.M. Daniel, B.T. Werner, J.S. Fenner, Composites Science and Technology **71**, 357–364 (2011)

[11] J.L. Tsai, C.T. Sun, Composites Science and Technology **65**, 1941–1947 (2005)

[12] Material Data Sheet: HexPly 8852 Product Data (Feb. 2013)

[13] R. Gerlach, C. Kettenbeil, N. Petrinic, International Journal of Impact Engineering **50**, 63–67 (2012)

[14] H. Koerber, P.P. Camanho, Composites Part A **42**, 462–470 (2011)

[15] G.T. Gray III, *ASM Handbook Vol. 8 Mechanical Testing and Evaluation* (ASM International, Ohio, 2000) 462–476

[16] H. Schuermann *Konstruieren mit Faser-Kunststoff-Verbunden*, 2nd edn. (Springer Verlag, Berlin Heidelberg, 2007)

[17] C.T. Sun, S.P. Berreth, Journal of Composite Materials **22**, 766–779 (1988)

[18] A. Puck, H. Schuermann, Composites Science and Technology **58**, 1045–1067 (1998)

EPJ Web of Conferences **94**, 01041 (2015)
DOI: 10.1051/epjconf/20159401041

Characterization of unidirectional carbon fiber reinforced polyamide-6 thermoplastic composite under longitudinal compression loading at high strain rate

Marina Ploeckl[a], Peter Kuhn, and Hannes Koerber

Institute for Carbon Composites, Technische Universitaet Muenchen, Faculty of Mechanical Engineering, Boltzmannstraße 15, 85748 Garching bei Muenchen, Germany

Abstract. In the presented work, an experimental investigation has been performed to characterize the strain rate dependency of unidirectional carbon fiber reinforced polyamide-6 composite for longitudinal compression loading. An end-loaded compression specimen geometry, suitable for contactless optical strain measurement via digital image correlation and dynamic loading in a split-Hopkinson pressure bar, was developed. For the dynamic experiments at a constant strain rate of $100\,s^{-1}$ a modified version of the Dynamic Compression Fixture, developed by Koerber and Camanho [Koerber and Camanho, Composites Part A, 42, 462–470, 2011] was used. The results were compared with quasi-static test results at a strain rate of $3 \cdot 10^{-4}\,s^{-1}$ using the same specimen geometry. It was found that the longitudinal compressive strength increased by 61% compared to the strength value obtained from the quasi-static tests.

1. Introduction

Carbon fiber reinforced thermoplastic matrix composites are becoming increasingly attractive for automotive structural applications due to advantages in processing, damage tolerance and recyclability compared to conventional composites with thermoset resin systems. The design of composite parts for modern cars with corresponding safety precautions has to take into account high strain rate loading conditions especially for primary structures. In addition, finite-element simulations of impact and crash scenarios require dynamic material data to ensure a faithful reproduction of the mechanical response.

In this study the focus of the material characterization is on the unidirectional carbon fiber reinforced polyamide-6 (PA 6) composite material. This semi-crystalline thermoplastic polymer is one of the most commonly used polymers and currently receives much attention as a matrix material for future automotive composite structures. Common material properties of polyamide-6 and the quasi-static mechanical properties of short glass fiber reinforced polyamide-6 are well documented [1–3].

Walley and Field studied the strain rate sensitivity of the uniaxial compressive stress-strain response of moist and dry polyamide-6 using a direct impact Hopkinson bar (DIHB) [4]. The maximum stress increased with increasing loading rate. Similar results for the compressive [5] and tensile [5–8] behavior of the pure polymer under dynamic loading were investigated. Also the tensile strength of short and long glass fiber reinforced polyamide was found to increase with strain rate [7–11]. To the knowledge of the authors only few attempts were carried out to characterize the dynamic stress-strain response of carbon fiber reinforced polyamide-6 composites. Work done by Todo et al. [12] on plain weave fabric carbon and glass fiber reinforced PA 6 suggested a linear relationship between the tensile strength behavior of the composite and the logarithmic strain rate. A recent study of Jendli et al. [13] shows the characterization of twill woven carbon fiber reinforced polyamide-66 under 0°, 90° and ±45° tensile loading from quasi-static to high strain rates. They observed positive strain rate effects on the tensile strength for all loading directions. Both, Todo et al. [12] and Jendli et al. [13] used servo-hydraulic testing machines.

In the above mentioned studies the focus is on the characterization of the strain rate effect of the pure thermoplastic polymer, glass fiber reinforced or woven carbon fiber reinforced PA 6. The present study contributes to the description of the strain rate dependency of a unidirectional carbon fiber reinforced polyamide-6 composite. An experimental characterization method, composed of a specific end-loaded specimen geometry and a modified Dynamic Compression Fixture [14], has been developed. The composite material is loaded under longitudinal compression, using an electro-mechanical testing machine for the quasi-static tests and a split-Hopkinson pressure bar (SHPB) for the dynamic tests.

The developed specimen geometry offers the possibility to detect the strain-time behavior via digital image correlation. A valid longitudinal compressive failure mode, e.g. failure away from the loading surfaces, was obtained with the developed specimen geometry. This represents an improvement over the earlier study carried out by Koerber and Camanho [14], where the static and dynamic specimens failed at the loading faces. In [14] the measured

[a] Corresponding author: ploeckl@lcc.mw.tum.de

Article available at http://www.epj-conferences.org or http://dx.doi.org/10.1051/epjconf/20159401041

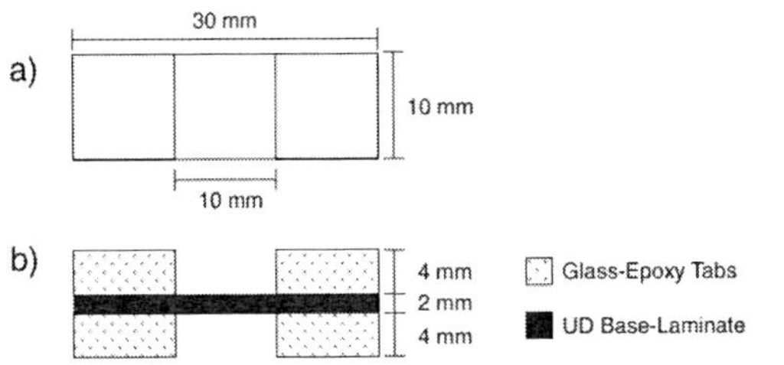

Figure 1. Top view a) and side view b) of end-loaded compression specimen geometry for quasi-static and high strain rate measurements.

strain rate effect on the longitudinal compressive strength therefore may have been somewhat under predicted. In the following text, the static tests performed with the end-loaded specimen geometry will be referred to as reference tests, which will be compared to the dynamic test series, using the same end-loaded specimen geometry.

2. Experimental procedure

2.1. Material

The specimens were prepared from Celstran® CFR-TP PA6 CF60-01, a 60% carbon fiber by weight polyamide-6 unidirectional reinforced thermoplastic composite tape [15]. According to the material data sheet [15] a fiber volume content of 48% is then obtained. In the present study the fiber volume content was determined to be 49%, using the test standard ASTM D3171, Method II, Procedure B (digestion of the matrix by a mixture of sulfuric acid and hydrogen peroxide) [16]. This material system is used in industrial, automotive and sporting goods applications and is commonly processed with automated fiber placement technologies. To manufacture 16-ply unidirectional plates, stripes of the tape were laid parallel and brick-like above each other such that the abutting edges of the stripes are at different locations in the individual layers. For a better handling of the stack the frame was welded at a few points with an ultrasonic welding apparatus (ProteUS, EM-Systeme GmbH) with a sonotrode frequency of 30 kHz. Using a RUCKS hot-press and a compression mold, the composite plates were consolidated at 260 °C with 0.5 MPa pressure for 15 minutes.

2.2. Specimen

In the presented work, an end-loaded compression specimen geometry suitable for contactless optical strain measurement via digital image correlation (DIC) and dynamic loading in a split-Hopkinson pressure bar was developed (Fig. 1). The specimen consists of a flat rectangular unidirectional (UD) base-laminate with dimensions of $30 \times 10 \times 2\,\text{mm}^3$ (length × width × thickness). To improve the stability of the specimen and to prevent premature failure at the loading ends, the cross-sectional area at the loading ends was increased by bonding two layers of glass-epoxy composite tabs with the adhesive Scotch-Weld™ DP 490 from 3M (for the quasi-static test specimens) and BETAMATE™ 1822 from Dow Automotive (for the dynamic test specimens) to both sides, leaving a free gauge section with a length of 10 mm at the middle of the specimen. The specimens were all cut on a water-cooled diamond saw and the surface parallelism was found to be within 0.02 mm. The loading surfaces were treated with a grinding machine to generate plane and parallel end-faces of good quality. A black-on-withe speckle pattern for the digital image correlation method was applied on the surface of each specimen, using aerosol spray painting. Based on the different image resolutions of the respective cameras the random speckles of the quasi-static specimens were smaller than the speckles of the dynamic specimens.

All specimens were dried in a drying oven under vacuum at a temperature of 80 °C for 72 hours. The dry specimens were stored in an aluminum multi-layer film bag until start of the experiment to avoid moisture absorption. Before each experiment a thin layer of molybdenum disulfide lubricant was applied on the end-surface of the specimen to reduce friction.

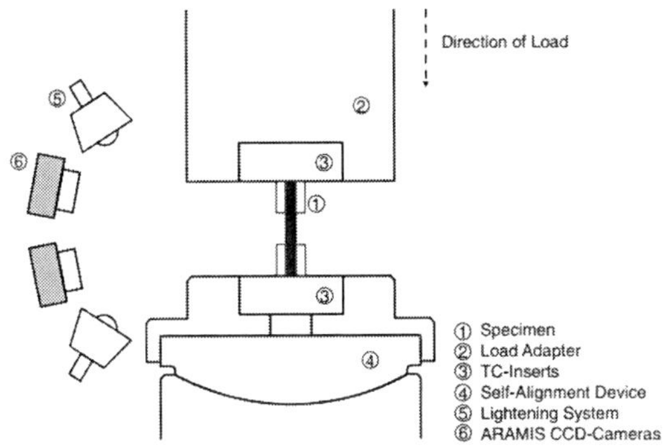

Figure 2. Quasi-static compression test setup with tungsten-carbide inserts.

2.3. Quasi-static test setup

Quasi-static reference tests were performed for five end-loaded compression specimens. An electro-mechanical testing machine of the type Inspekt Table 100 from Hegewald & Peschke was used with a displacement rate of 0.5 mm/min, which corresponds to a quasi-static axial strain rate of approximately $3 \cdot 10^{-4}\,\text{s}^{-1}$ by considering a nominal specimen length of 30 mm. The load was introduced by means of a self-aligning end-loaded compression setup with embedded polished tungsten-carbide inserts at the loading surfaces (Fig. 2). The latter prevent damage on the loading surfaces due to the high compressive strength of the specimen in fiber direction. A 100 kN load cell was used to measure the force. The GOM ARAMIS-4M system, consisting of two CCD cameras with a resolution of $1728 \times 2352\,\text{pixel}^2$ and two object lenses of type Titanar 50 mm, were used to obtain the three-dimensional quasi-static strain field of the specimen front surface. The optical strain measurement system was positioned at a distance of 225 mm away from the specimen surface. The cameras were mounted on a support bar in a distance of 140 mm to each other and oriented in such a way, that the light paths generate an angle of 25°.

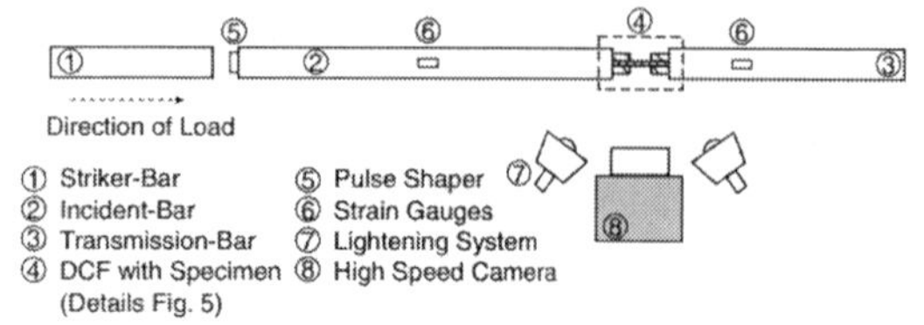

Figure 3. Split-Hopkinson pressure bar test setup with modified Dynamic Compression Fixture (DCF).

To ensure a uniform illumination of the specimen surface two LED lamps beside each camera were used. The acquisition rate of the cameras was set to 5 frames per second (fps) with a shutter speed of 75 ms. For post processing of the recorded pictures the digital image correlation software of the GOM ARAMIS-4M system with a facet size of $17 \times 17\,\text{pixel}^2$ and a facet step of $15 \times 15\,\text{pixel}^2$ was used. The computation size was set to 3×3 facets^2. For the determination of the axial specimen strain, used in the later shown stress-strain curves, the mean value of a $6 \times 6\,\text{mm}^2$ strain field at the center of the specimen surface was calculated. The quasi-static longitudinal compressive strength is defined as the quotient of the failure load and the specimen cross-section.

2.4. Dynamic test setup

High strain rate tests were performed for five end-loaded compression specimens. A split-Hopkinson pressure bar setup (Fig. 3) was used consisting of diameter 18 mm steel striker-, incident-, and transmission-bars with lengths of 0.6 m, 2.6 m and 1.6 m, respectively. The strain gauges for the detection of the elastic wave are located on the incident-bar at 1.3 m and on the transmission-bar at 0.3 m away from the bar-specimen interfaces and operate with a supply voltage of 7 V. The bar-gauge signals were amplified by using a high speed transducer amplifier of the type FYLDE FE-H379-TA with a gain setting of 100 and a frequency setting of 300 kHz. The amplified signals and the impact velocity of the striker-bar were each recorded with a TEKTRONIX TDS2004C oscilloscope. To generate an elastic wave with a triangular shape in the incident bar, a copper pulse shaper with a diameter of 4 mm and a thickness of 1 mm was used. The pulse shaper provides a triangular shape which is best suited for specimens exhibiting a linear stress-strain behavior up to failure. For the high strain rate tests an impact velocity of the striker-bar of approximately 7 m/s was reached by using an appropriate adjustment of the pressure at the gas gun. As a result of the nominal specimen length of 30 mm and the striker-bar velocity a strain rate of approximately $100\,\text{s}^{-1}$ was reached.

In the presented work, a modified version of the Dynamic Compression Fixture (DCF), developed by Koerber and Camanho [14] was used. The modified DCF (Fig. 4) consists of two tungsten-carbide (TC) inserts, surrounded by support-rings and placed between the bars and the specimen loading surfaces. Thus, indentation of the bar-ends and stress concentrations at the specimen end-surfaces could be avoided. To prevent an influence of the TC-inserts on the bar strain waves, the mechanical

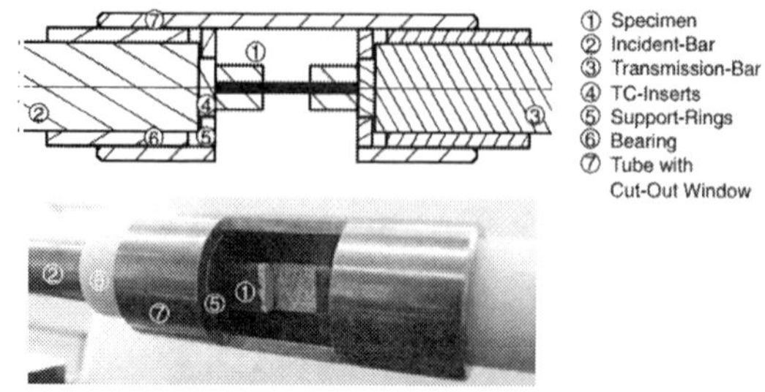

Figure 4. Modified Dynamic Compression Fixture used in the SHPB setup.

impedance between TC-inserts and the Hopkinson bars was matched by choosing an appropriate diameter ratio. The TC-inserts/support-rings are then aligned and held in place by a freely sliding tube, containing a cut-out window to monitor the specimen deformation and failure mode using a high speed camera. The specimen was clamped between the two inserts without any additional fixation.

A high speed camera of type FASTCAM SA 5 from Photron with a ZEISS Makro-Planar T* 2/100 mm ZF lens was positioned 950 mm away from the specimen surface to capture pictures with a frame rate of 372000 fps at a resolution of 128×80 pixel^2 during the experiment. The lens aperture was set to f/2.8. Two spotlights of type DEDOLIGHT 400D were used to provide a well illuminated specimen surface. For post processing of the recorded pictures the digital image correlation software of the GOM ARAMIS-4M system with a facet size of 10×10 pixel^2 and a facet step of 8×8 pixel^2 was used. The computation size was set to 5×5 facets^2. For the determination of the axial specimen strain, the mean value of a $5 \times 5\,\text{mm}^2$ strain field at the center of the specimen surface was calculated.

3. Results and discussion

3.1. Quasi-static experimental results

Five quasi-static reference specimens were loaded until failure at constant displacement rate. The longitudinal compressive stress-strain response for all tested specimens is shown in Fig. 5. The end point of each curve indicates the failure load and correlates with the quasi-static longitudinal compressive strength X_C^{qs} and the ultimate failure strain ϵ_C^u. The results of the reference tests, the mean values (Mean), the standard deviation (STDV) and the coefficient of variance (CV) are summarized in Table 1. The mean compressive strength of 604.86 MPa shows a relative high coefficient of variance with 16.30% but no correlation to varying failure modes could be observed. For all quasi-static reference tests, valid failure occurred in the form of fiber kinking away from the loading surfaces (Fig. 6). The kink-bands for every specimen are marked with yellow arrows respectively (Fig. 6). Three of five specimens lost one bonded tap with ultimate load. The taps fail adhesively with the black adhesive on the glass-epoxy tab (Fig. 6).

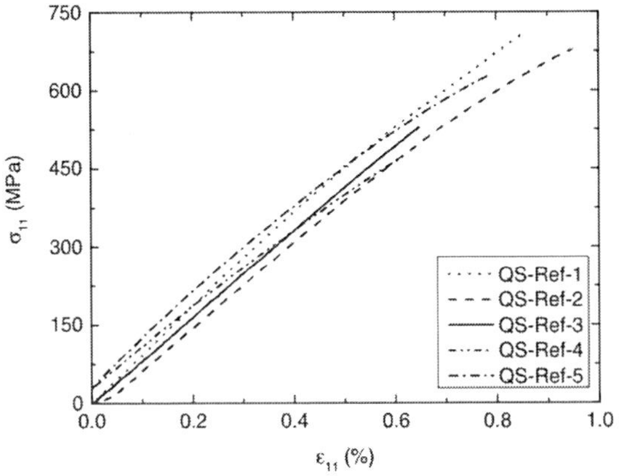

Figure 5. Quasi-static longitudinal compressive stress-strain curves from reference tests.

Table 1. Quasi-static reference test results.

Test	Strength X_C^{qs} (MPa)	Failure Strain ϵ_C^u (%)
QS-Ref-1	708.66	0.85
QS-Ref-2	678.49	0.95
QS-Ref-3	528.53	0.65
QS-Ref-4	477.63	0.62
QS-Ref-5	630.99	0.79
Mean	604.86	0.77
STDV	98.61	0.14
CV (%)	16.30	17.90

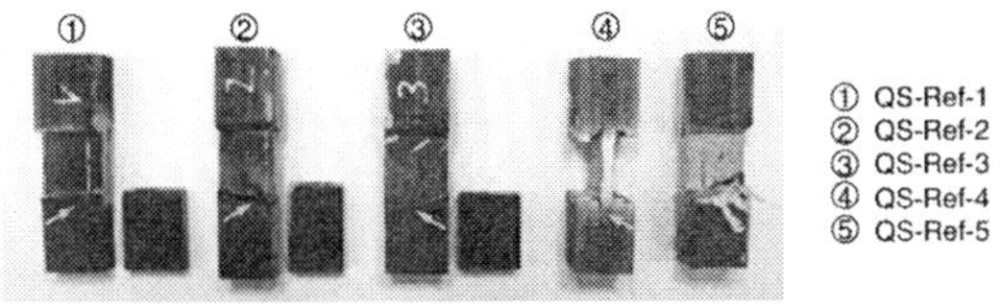

Figure 6. Quasi-static specimen failure mode of reference tests.

3.2. Dynamic experimental results

The dynamic test results were calculated by using the classic SHPB analysis, which is based on the one-dimensional wave propagation theory [17]. The forces acting at the incident-bar F_1 and at the transmission-bar F_2 side of the specimen respectively are defined as

$$F_1 = A_B E_B(\epsilon_I + \epsilon_R) \quad (1)$$

$$F_2 = A_B E_B \epsilon_T. \quad (2)$$

Here ϵ_I, ϵ_R and ϵ_T are the incident-, reflected and transmitted bar strains, A_B the bar cross-section and E_B the Young's modulus of the SHPB bars, respectively. It is known that the one-dimensional wave equation can be solved by using the D'Alembert's method and finally describes the specimen strain $\epsilon_S(t)$ and the specimen strain rate $\dot{\epsilon_S}(t)$ as follows:

$$\epsilon_S(t) = (c_B/l_S)\int(-\epsilon_I(t) + \epsilon_R(t) + \epsilon_T(t))dt \quad (3)$$

$$\dot{\epsilon_S}(t) = (c_B/l_S)(-\epsilon_I(t) + \epsilon_R(t) + \epsilon_T(t)) \quad (4)$$

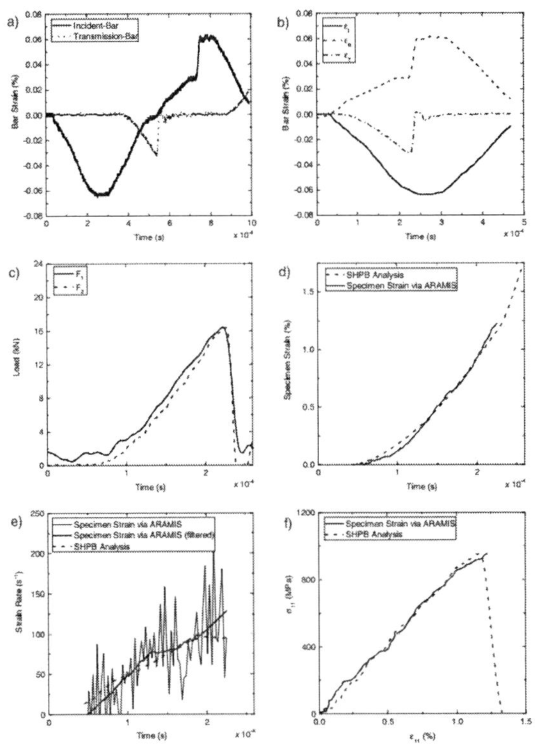

Figure 7. Incident and transmission bar signals of the specimen HR-1 a) and SHPB analysis results of the specimen HR-1 for the shifted bar strain waves b), the loads acting on the specimen c), the specimen strain behavior d), the specimen strain rate e) over time and the stress-strain response f).

where l_S represents the initial specimen length and c_B the elastic wave propagation velocity of the bar material. The wave velocity is calculated by using the equation $c_B = (E_B/\rho)^{0.5} = 5123$ m/s and the Young's modulus $E_B = 206000$ MPa and the density $\rho = 7850$ kg/m^3 of the steel bars used in the SHPB setup. It is noted that in the present work the Eqs. (3) and (4) are only used for comparison because the specimen strain and the specimen strain rate were directly obtained by contactless optical strain measurement via digital image correlation. The specimen stress σ_S is then calculated by the quotient of F_2 and the initial specimen cross-section A_S:

$$\sigma_S = F_2/A_S = (A_B E_B \epsilon_T)/A_S. \quad (5)$$

This parameter is also denoted as the 1-wave-stess or back-stress, because it represents the condition at the specimen/transmission-bar interface.

Figure 7 shows the post-processing results of the bar strain waves for a representative specimen (specimen number: HR-1). The graphs serve as illustration of the SHPB analysis. The unfiltered incident- and transmission-bar signals are shown in Fig. 7a). As a first step the bar strain waves detected at the two strain gauges on the incident- and transmission-bar are filtered with a Savitzky-Golay filter and then shifted to the time at which they occurred at the specimen (Fig. 7b)). The loads F_1 and F_2 were calculated with the Eqs. (1) and (2) and plotted over

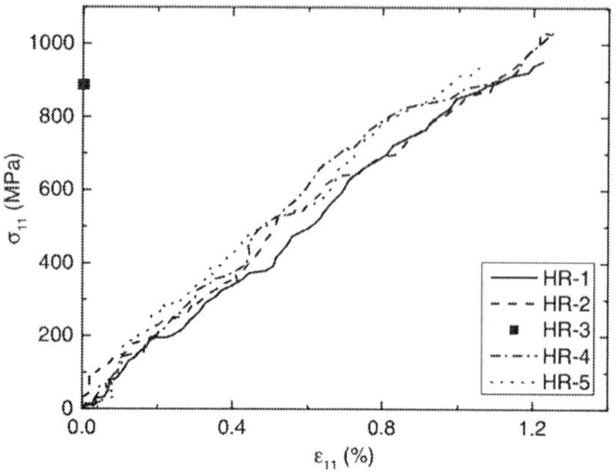

Figure 8. Dynamic longitudinal compressive stress-strain response of all tested specimens.

time (Fig. 7c)). The reason for the initial offset of the forces is the nonzero bar strain at the beginning of the reflected signal. Taking this into account, the load-time responses on both sides of the specimen are similar and thus dynamic equilibrium during the experiment could be presumed. A comparison of the specimen strain calculated with Eq. (3) and the specimen strain measured optically by using the ARAMIS system is shown in Fig. 7d). Both curves are very similar, but Koerber and Camanho [14] already showed that this is not always the case. Accompanied with the strain-time responses, the strain rate-time responses obtained from Eq. (4) and the slope of the ARAMIS strain signal, were compared (Fig. 7e)). Due to the small noise of the ARAMIS strain signal the strain rate signal oscillates significantly. The strain rate signal is filtered with a Savitzky-Golay filter. A constant strain rate could be achieved after half of the experiment. In Fig. 7f) the calculated specimen strain and the optically measured strain are plotted as a function of the specimen stress calculated with the Eq. (5).

The longitudinal compressive stress-strain response of all dynamically tested specimens (Fig. 8) is based on the optical measured strain and the calculated stress via the SHPB analysis. Due to the digitalization of the bar strain signals and the discrete number of strain values received from the digital image correlation the stress-strain curves are somewhat noisy. The failure load is indicated at the end point of each curve and corresponds to the longitudinal compressive strength X_C^{hr} and the ultimate failure strain ϵ_C^u. For specimen HR-3 only the failure load was recorded because of a non-functioning high-speed camera record. In Table 2 all dynamic experimental results are listed. A mean compressive strength of 974.93 MPa with a coefficient of variance of 6.47% for the unidirectional carbon fiber reinforced polyamide-6 composite was measured. All specimens fail similar to the quasi-static specimens away from the loading surface (Fig. 9) due to the formation of kink-band followed by through thickness fiber failure. The pictures of the high speed camera show, that with maximum stress the initiation of the fiber kinking starts and subsequently the development of the kink-band. The delamination of the base-laminate in the region of the glass-epoxy tabs occurred after the formation of the kink-band and additional longitudinal cracks.

Table 2. Dynamic test results.

Test	Strength X_C^{hr} (MPa)	Failure Strain ϵ_C^u (%)	Strain Rate $\dot{\epsilon}$ (s^{-1})
HR-1	952.88	1.22	96.05
HR-2	1034.52	1.22	85.93
HR-3	888.05	–	–
HR-4	1039.13	1.26	110.72
HR-5	960.05	1.06	92.56
Mean	974.93	1.19	96.32
STDV	63.10	0.09	10.48
CV (%)	6.47	7.45	10.88

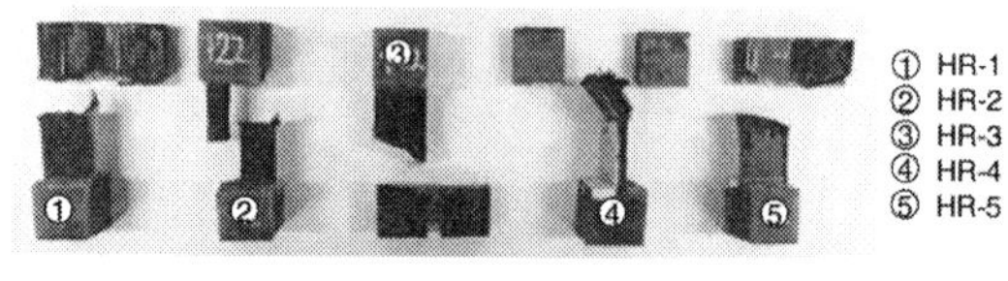

Figure 9. Dynamic specimen failure mode.

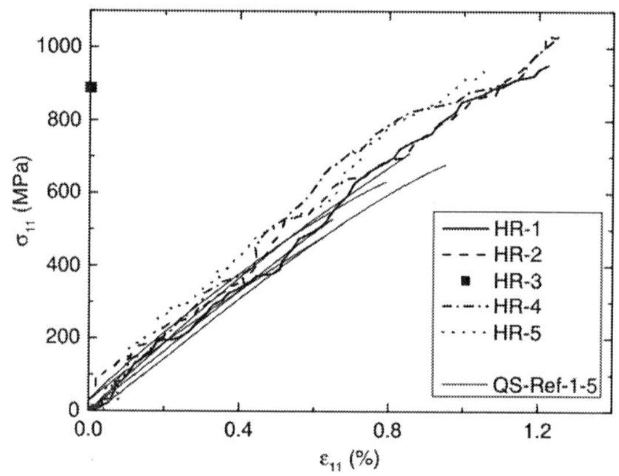

Figure 10. Comparison of the quasi-static longitudinal compressive stress-strain curves from reference and dynamic tests.

The comparison of all quasi-static reference test results and all dynamic test results indicates that the longitudinal compressive strength of the unidirectional carbon fiber reinforced polyamide-6 composite increased by 61% (Fig. 10). The Young's Modulus of the material at high strain rate shows a minor increase with increasing strain rate.

4. Conclusion

The strain rate dependency of the longitudinal compressive stress-strain response of a unidirectional carbon fiber reinforced polyamide-6 composite was investigated in the presented work. It was proven that the developed end-loaded compression specimen geometry, consisting of a flat rectangular unidirectional base-laminate and glass-epoxy tabs, is suitable for dynamic loading in a spit-Hopkinson pressure bar and additionally useful for contactless optical strain measurement via digital image correlation. Quasi-static reference tests were carried out at a strain rate of approximately $3 \cdot 10^{-4}\,s^{-1}$ using a self-aligning loading device with embedded polished tungsten-carbide inserts at the loading surfaces. For all quasi-static tests valid failure modes in form of fiber kinking away from the loading surfaces were observed. For the high strain rate tests the Dynamic Compression Fixture [14] was modified to provide a view field for the optical strain measurement via a high-speed camera to monitor the specimen deformation and failure mode. Similar to the quasi-static reference tests, all dynamic specimens failed due to fiber kinking away from the loading surfaces. It was observed that the longitudinal compressive strength of the unidirectional carbon fiber reinforced polyamide-6 composite increased by 61% compared to the strength value obtained from the quasi-static reference tests.

The authors would like to acknowledge the German Federal Ministry of Education and Research (BMBF) for providing funding within the project MAI Plast.

References

[1] M. Biron, *Thermoplastics and thermoplastic composites*, 1st edn. (Butterworth-Heinemann, Oxford, 2007)
[2] E. Baur, S. Brinkmann, T. Osswald, E. Schmachtenberg, *Saechtling Kunststoff Taschenbuch*, 30th edn. (Carl Hanser Verlag, München, 2007)
[3] L. Bottenbruch, R. Binsack, G. Becker, D. Braun, *Polyamide. Kunststoffhandbuch 3/4.* (Hanser, 1998)
[4] S.M. Walley, J.E. Field, DYMAT Journal **1**, 211–227 (1994)
[5] H. Pouriayevali, S. Arabnejad, Y.B. Guo, V.P.W. Shim, International Journal of Impact Engineering **62**, 35–47 (2013)
[6] G.-F. Shan, W. Yang, M. Yang, B. Xie, J. Feng, Q. Fu, Polymer **48**, 2958–2968 (2007)
[7] M. Hokka, V.-T. Kuokkala, S. Ihme, Proceedings of the SEM Annual Conference June 1-4, 2009 Albuquerque New Mexico USA, Society for Experimental Mechanics Inc. (2009)
[8] A. Vashchenko, I. Spiridonova, E. Sukhovaya, Metalurgija **39**, 89–92 (2000)
[9] Z. Wang, Y. Zhou, P.K. Mallick, Polymer Composites **23**, 858–871 (2002)
[10] B. Mouhmid, A. Imad, N. Benseddiq, S. Benmedakhène, A. Maazouz, Polymer Testing **25**, 544–552 (2006)
[11] S.J.H. Kuhlman, S.I. Hill, SAE Technical Paper 2014-01-1056 (2014)
[12] M. Todo, K. Takahashi, P. Béguelin, H.H. Kausch, Composites Science and Technology **60**, 763–771 (2000)
[13] Z. Jendli, J.-C. Walrick, M. Bocquet, J. Fitoussi, Proceedings of ECCM16 - 16th European Conference on Composite Materials, Seville, Spain (2014)
[14] H. Koerber, P.P. Camanho, Composites Part A **42**, 462–470 (2011)
[15] Material Data Sheet: Standard Product Values Celstran CFR-TP PA6 CF60-01 (7/29/2014)
[16] ASTM D3171: Standard test methods for constituent content of composite materials (1999)
[17] G.T. Gray III, *ASM Handbook Vol. 8 Mechanical Testing and Evaluation* (ASM International, Ohio, 2000) 462–476

EPJ Web of Conferences **94**, 01042 (2015)
DOI: 10.1051/epjconf/20159401042

Structural-temporal approach for dynamic strength characterization of gabbro-diabase

Y. Petrov[1,2], A. Bragov[3], A. Evstifeev[1], and E. Cadoni[4]

[1] Saint Petersburg State University, 199034 St. Petersburg, Russia
[2] IPME RAS, Extreme States Dynamics Department, 199178 St. Petersburg, Russia
[3] Lobachevski State University, 603950 Nizhny Novgorod, Russia
[4] DynaMat Laboratory, University of Applied Sciences of Southern Switzerland, 6952 Canobbio, Switzerland

Abstract. This paper presents some of the results of dynamic compression, splitting and direct-tensile tests of gabbro-diabase using two split Hopkinson bar apparatus. The results of laboratory experiments on specimens without and with notch for a wide range of loading parameters are given. Strength and deformation characteristics of gabbro-diabase were determined experimentally at high strain rates up to $10^3\,s^{-1}$. Mechanical characteristics were found to be strongly depended on the strain rate. A uniform interpretation of the rate effects of fracture of the tested gabbro-diabase is given on the basis of structural-temporal approach based on a set of fixed material constants. It is shown that temporal dependences of both the compressive and tensile strengths of the tested material can be effectively calculated using the incubation time fracture criterion.

1. Introduction

The dynamic characterization of rocks under high loading rates is fundamental to understand the material behaviour in case of dynamic events. The dynamic mechanical properties of rocks can be very different from those exhibited in quasi-static conditions. Specific investigations in such dynamic ranges appear necessary to correctly understand their behaviour under intermediate and high strain rate conditions. These data are necessary and require experimental techniques able to induce on the rock materials state of loading reproducing the actual dynamic condition. Also must be a simple criterion for a wide range of parameters of loadings.

The dynamic characterization has been carried out by means of traditional Split Hopkinson Pressure Bar (SHPB) and a Split Hopkinson Tensile Bar (SHTB) placed respectively at the Laboratory of Dynamic Investigation of Materials of Nizhny Novgorod and at the DynaMat Laboratory of the University of Applied Sciences of Southern Switzerland. These equipment are briefly described and the results of dynamic tension/compression tests at different high strain-rates on gabbro-diabase are discussed.

Thus, the objective of this work is the comparison between both tensile and compressive dynamic and quasi-static tests, and analysis obtained results with incubation time approach.

2. Materials

Gabbro is a dark, coarse-grained, intrusive igneous rock chemically equivalent to basalt. It is a plutonic rock, formed when molten magma is trapped beneath the Earth's surface and cools into a crystalline mass.

Gabbro is generally coarse grained, with crystals in the size range of 1 mm or greater. Finer grained equivalents of gabbro are called diabase, although the vernacular term "microgabbro" is often used when extra descriptiveness is desired. Gabbro may be extremely coarse grained to pegmatitic, and some pyroxene-plagioclase cumulates are essentially coarse grained gabbro, although these may exhibit acicular crystal habits.

Gabbro is usually equigranular in texture, although it may be porphyritic at times, especially when plagioclase oikocrysts have grown earlier than the groundmass minerals.

Gabbro – one of the most common intrusive rocks. This breed is composed of plagioclase and clinopyroxene in roughly equal proportions. Gabbro-diabase rock is the basic structure, with a massive texture, fine-medium grained, without cavities and irregularities uniformly painted in dark gray. Test specimens were taken from Drugoretskoe field (Republic of Karelia). Material properties are given in Table 1.

3. Experimental and analytical details

Static and dynamic loading tests have been conducted using different experimental set-ups on cylindrical rock specimens.

3.1. Dynamic compression and splitting tests

Dynamic compressive and splitting tests were performed at the Laboratory of Dynamic Investigation of Materials in Nizhny Novgorod by means of a SHPB shown in Fig. 1.

The experimental set-up consists of a compact gas gun (1), incident (2) and transmitter (5) steel (or duralumin) pressure bars with the specimen (3) sandwiched between

Table 1. Material properties.

Young's modulus, MPa	91550
Average Compressive strength, MPa	215
Average Tension strength, MPa	17,5

Figure 1. Experimental apparatus realizing SHPB for compressive and splitting tests.

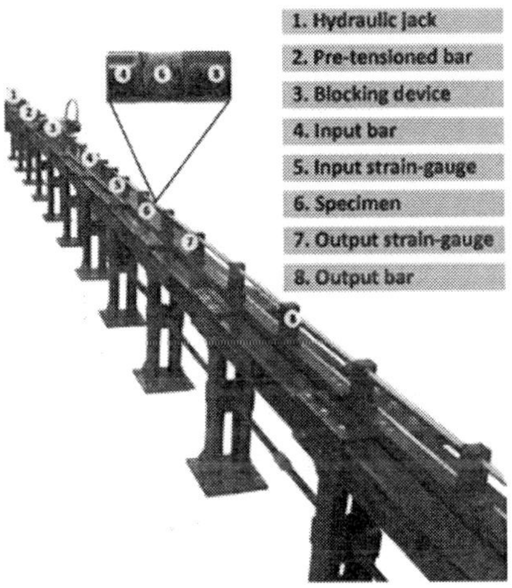

Figure 2. Split Hopkinson Tensile Bar used in this research.

them. Power supply and calibration of strain gauges was produced by an original scheme (4). In order to record the electrical signals from strain gauges, a multichannel digital oscilloscope (6) was used. Diameters of striker as well as pressure bar are 20 mm. The incident bar length was 1 m, whereas the length of transmitter bar was 3 m in order to provide correct registration of possible additional cycles of loading during the experiment [1].

Splitting test of disks, also called "Brazilian tests", is one of the available methods to measure the tensile strength of brittle materials [2]. Due to the stress-state of the disk, the failure is caused by tension when tensile stresses reach the tensile strength of the material on the diametric loading plane.

3.2. Dynamic tensile tests

The high strain-rate tests in tension were performed using a Split Hopkinson Tensile Bar (SHTB) [3], that consists of two circular aluminium bars, called input and output bars, having a length of 3 and 6 m respectively, with a diameter of 20 mm to which the specimen is glued on using a bi-component epoxy resin (see Fig. 2).

The input bar is connected with an high strength steel pre-tensioned bar 6 m in length, used as pulse generator with a diameter of 12 mm in order to obtain the same acoustical impedance of the input bar so that the pulse reffection, due to the interface, is avoided.

With reference of Fig. 2 a test with the SHTB is performed as follows [4]:

1. Firstly, a hydraulic actuator (1), of maximum loading capacity of 600 kN, is pulling the pretension high strength steel bar (2). The pretension stored in this bar is resisted by the blocking device (3).
2. Secondly, the rupture of a fragile bolt in the blocking device, gives rise to a tensile mechanical pulse of 2,4 ms with linear loading rate during the rise time, propagating along the input (4) and the output (7) bars bringing to fracture the specimen (6).

As well as in the classical Hopkinson bar apparatus, the pulse propagates along the input bar with the elastic wave velocity equal to C_0, and constant shape. When the incident pulse (ε_I) reaches the rock sample, it is partially reffected by it (ε_R). On the other hand, the portion of the pulse that is not reffected passes through the specimen (ε_T) and propagates into the output bar.

3.3. The concept of incubation time

Considering phenomena such as brittle failure of a defect free specimen, fracture of specimen with macro-defect (crack), we deal with strength properties of the materials. These very different processes exhibit some important common features. In the case of slowly applied load there exists the threshold value of load amplitude. In the case of very short durations the strength characteristics of materials are considerably different from those obtained in the case of quasi-static testing. This can be explained by the circumstance that the time of loading is of the same order as the typical time of certain processes on the micro-level. The above-mentioned time can be described by virtue of the material constant and will be referred to as "incubation time". An efficient criterion for the analysis of brittle fracture for defect-free material can be formulated in the following form [5–8]:

$$\frac{1}{\tau}\int_{t-\tau}^{t}\sigma(s)ds \leq \sigma_{st} \quad (1)$$

where τ is the incubation time, σ_{st} is a tensile strength for quasi-static loading and $\sigma(t)$ is the applied stress which for $t < 0$ is supposed to be zero. The instant of fracture t^* corresponds to the earliest realization of equality in Eq. (1) or, in the general case, violation of this condition.

For our tasks (uniaxial extension) take stress $\sigma(t)$ in this form

$$\sigma(i) = \dot{\sigma} III(l) \quad (2)$$

where $H(t)$ is a Heaviside function, $\dot{\sigma}$ is a Stress rate (supposed to be constant).

Here we will use a generalization of critical condition (1) based on taking into account two major factors: level (amplitude) and rate of local loading of the continuum. The corresponding criterion is given by following condition [7].

$$\frac{1}{\tau_\alpha}\int_{t-\tau_\alpha}^{t}\left[\frac{\sigma(s)}{\sigma_{st}}\right]^\alpha ds \leq 1 \quad (3)$$

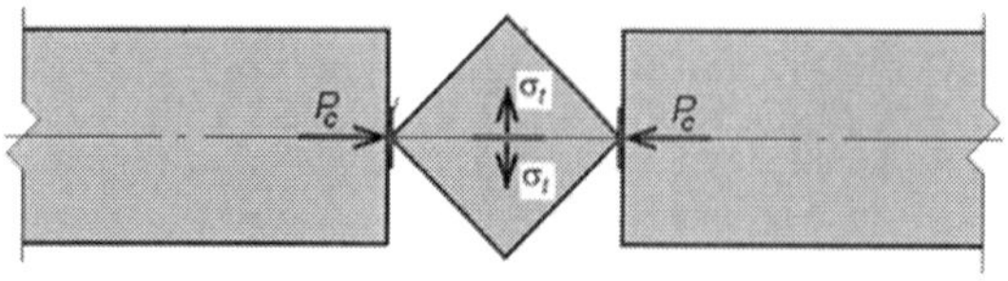

Figure 3. Experimental scheme for splitting tests.

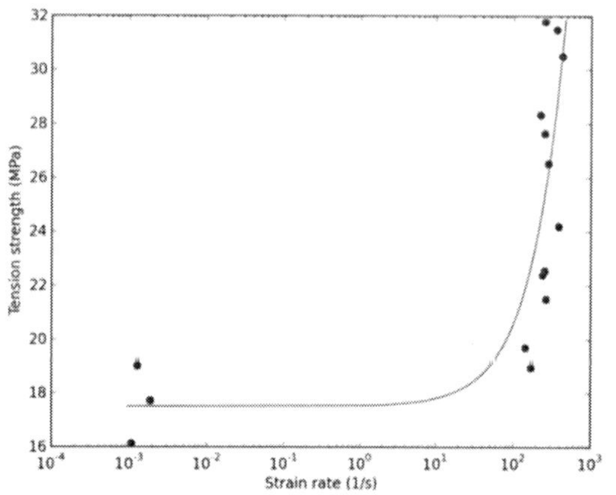

Figure 4. Tension strength.

where τ_α is the incubation time. Incubation time characterises sensitivity of material to rate of loading at the given point of continuum. The parameter α characterises sensitivity to the level (or amplitude) of local stress. For most of the solid continua $\alpha > 1$. The parameter σ_{st} is tensile strength for quasi-static loading and $\sigma(t)$ is the local stress which for $t<0$ is supposed to be zero like in the first case.

The general concept of incubation time works in both compressive and tensile cases, but the set material constants must be determined for compression and tension separately.

4. Results

The strain rate of the test is dependent on two variables: the behaviour of the material and the loading of the pre-stressed bar. A constancy of the preload has been chosen in order to verify the response of the rock specimen under the same loading condition.

4.1. Compression and splitting tests

Experiments for tension was completed like a model "Brazilian tests", but samples were in the form of rectangular parallelepipeds section 20 × 20 mm with length 10, 20, 30 mm. Experimental schemes on Fig. 3.

For tensile stress formula in the form

$$\sigma_t = 0.5187 \frac{P_c}{b\,h} \tag{4}$$

where b and h – dimensions of the sample, P_c – longitudinal force in the gauging rods, determined by indications on the support rod. The maximum value of the tensile stress is the required tensile strength of stone.

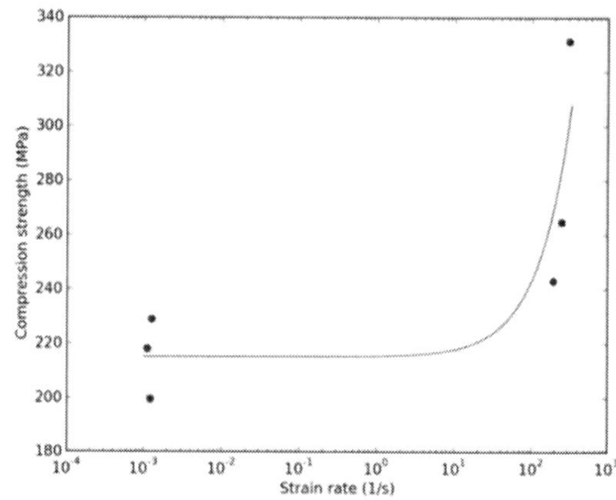

Figure 5. Compression strength.

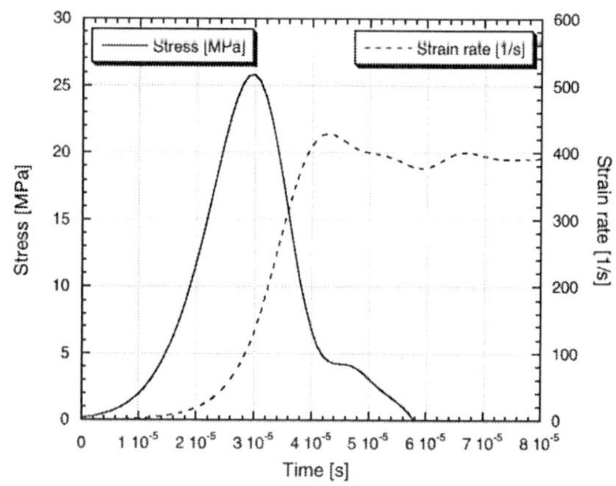

Figure 6. Tension strength and strain rate versus time.

Table 2. Dynamic tensile results ($\dot{\sigma}$ = 900 GPa/s).

Test	σ_{max} [MPa]	ε_u [‰]	Strain rate [s^{-1}]	W_{max} [J/m^2]	G_F [J/m^2]
004	28.36	1.09	54	936	531
005	29.05	0.77	71	921	621
006	29.05	0.82	73	935	634
Average (st. dev)	28.82 (0.4)	0.89 (0.17)	66 (10)	931 (8)	595 (56)

On Fig. 4 shows the effect of the dynamic tensile strength. Points show the experimental data. Curves show the theoretical line constructed by the formula (1) with parameters $\tau = 66\,\mu$s, $\sigma_{st} = 17.5$ MPa. We can see good compliance theoretical approach and experimental data. Parameter σ_{st} – static tensile stress. It is determines the horizontal section on theoretical curve. Parameter τ – can move theoretical curve to the right or to the left. Theoretical curve was built as a middle line of experimental data.

For compression tests were carried out standard experiments. On Fig. 5 shows the effect of the dynamic compression strength. Points show the experimental data. Curves show the theoretical line constructed by the formula (1) with parameters $\tau = 6\,\mu$s, $\sigma_{st} = 215$ MPa.

Table 3. Dynamic tensile results ($\dot{\sigma} = 1.5$ TPa/s).

test	σ_{max} [MPa]	ε_u [‰]	Strain rate [s^{-1}]	W_{max} [J/m²]	G_F [J/m²]
007	29.01	0.826	107	1472	1176
008	25.82	0.679	132	1602	1326
010	23.32	1.068	163	1175	911
Average (st. dev)	26.05 (0.3)	0.858 (0.19)	134 (28)	1416 (219)	1138 (210)

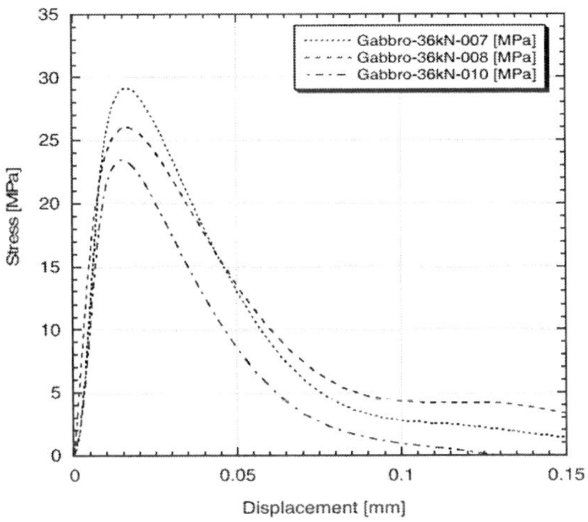

Figure 7. Stress vs. displacement curves in tension $\dot{\sigma} = 1.5$ TPa/s).

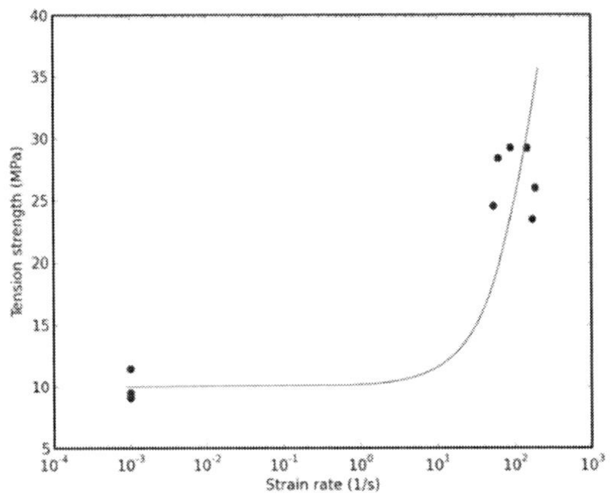

Figure 8. Tension strength.

We can see also good compliance theoretical approach and experimental data.

4.2. Tensile tests

Direct dynamic tensile tests on the gabbro-diabase were carried out at two strain-rates. The tests were obtained imposing 18 kN and 36 kN as pre-load in the pre-tensioned bar obtaining an average loading rate of 900 GPa/s and 1.59 TPa/s, respectively. As well known for brittle materials the strain rate changes during the test as shown in Fig. 6 where the stress and the strain rate in function of time are depicted. As proposed in [9] the conventional strain rate can be read in the figure such as 132 s^{-1}.

In Tables 2 and 3 are reported the results of the dynamic direct tensile tests.

In Fig. 7 the stress versus displacement curves are depicted for the higher stress rate.

On Fig. 8 shows the effect of the dynamic tensile strength. The specimens were notched. Points show the experimental data. Curves show the theoretical line constructed by the formula (1) with parameters $\tau = 3.4\,\mu s$, $\sigma_{st} = 10$ MPa. We can see good compliance theoretical approach and experimental data.

5. Conclusion

Dynamic tension, indirect and direct tensile tests of a gabbro-diabase were conducted using two set ups as Split Hopkinson Pressure Bar and Split Hopkinson Tensile Bar.

Rock tensile and compression strength is strain rate dependent. It has been veriffed that also for the gabbro-diabase this strength increases with the strain rate increasing within the wide strain rate range. The results show non-linear relationships between strength and strain rates.

It is shown that the mechanical characterization of tested rock can be effectively performed using two experimental setups.

A uniform interpretation of the rate effects of fracture of the tested gabbro-diabase is possible based on the structural-temporal approach and a set of fixed material constants. It is found that temporal dependences of both the compressive and tensile strengths of the tested material can be effectively predicted on the basis the incubation time fracture criterion.

Authors acknowledge Russian Foundation for Basic Research (14-01-00814), the program N25 of the Presidium of the Russian Academy of Sciences, FP7 EU MARIE CURIE TAMER (No.610547), and the Saint-Petersburg State University (6.38.243.2014) for the support.

A special acknowledgement goes to the Scientific & Technological Cooperation Programme Switzerland-Russia for the financial support of the Utilisation of Specific Infrastructure Project called "Dynamic fracture of rocks under high rate mechanical and electric breakdown induced loads" between DynaMat Laboratory of the University of Applied Sciences of Southern Switzerland and the St. Petersburg State University.

References

[1] J. Rodriguez, C. Navarro, V. Sanchez-Galvez, J. Phys. IV France, 04 (1994) 101.

[2] Bragov, A.K. Lomunov, I.V. Sergeichev, J. Appl. Mech. Tech. Phys. **42** (2001).

[3] E. Cadoni. Rock Mech. Rock Eng. **43** (2010) 667–676.

[4] A. Caverzan, E. Cadoni, M. di Prisco, Int. J. Impact Eng. **45** (2012), 28–38.
[5] Y. V. Petrov, and N. F. Morozov, J Appl. Mech. **61**, 710–712 (1994).
[6] Y. V. Petrov, N. F. Morozov, V. I. Smirnov, Fatigue Fract Engng Mater Struct. **26**, 363–372 (2003).
[7] A. Evstifeev, E. Cadoni, Y. Petrov, EPJ WoC **26**, 01041 (2012).
[8] Y. Petrov, I. Smirnov, A. Evstifeev, N. Selyutina, Frattura ed Integrità Strutturale **24**, 112–118 (2013).
[9] E. Cadoni, A. Meda, G. Plizzari. Mat. Struct. **42**(9), 1283–1294 (2009).

EPJ Web of Conferences **94**, 01043 (2015)
DOI: 10.1051/epjconf/20159401043

Mechanical response of a fibre reinforced earthen material under static and impact loadings

Francesco Aymerich[1], Luigi Fenu[2], Luca Francesconi[1,a], and Paola Meloni[1]

[1]Department of Mechanical, Chemical and Materials Engineering, University of Cagliari, Italy
[2]Department of Civil and Environmental Engineering and Architecture, University of Cagliari, Italy

Abstract. This study examines the improvements provided by the insertion of hemp fibres with different weight fractions and lengths in an earthen material. The structural response of the materials was investigated by means of static and impact bending tests carried out on notched samples. The main focus of the analyses was in the characterization of the structural properties of the materials in terms of fracture resistance, post-cracking performance and energy absorption capability. The results of the study show that hemp fibres improve significantly the mechanical and fracture properties of the earthen material under both static and dynamic bending. It was also found that the structural properties of unreinforced and reinforced earthen materials are highly sensitive to the stress-rate, with higher strength and fracture resistance under impact loading than under static loading.

1. Introduction

In the last decades earthen materials have attracted considerable interest and regained importance in many fields of the construction sector, especially in developing countries and in rural, protected and architectural heritage areas [1–3]. Earth is widely used as building material in many hot or temperate regions and recent surveys estimate that more than one third of the world's population live in houses built of mud or unfired earth [4].

The reasons for the widespread adoption of unbaked earth-based construction materials are found in the abundance, low cost and recyclability of soil, and in the characteristics of the typical production processes, which are technologically simple and highly energy-efficient, besides having a very low carbon footprint, with little or no impact on the ecosystem and environment [2,3]. In addition, these materials provide excellent sound and thermal insulation [5,6], together with good fire resistance and humidity control properties.

Despite these advantages, the wider use of earth as primary building material has been hampered by some inherent limitations of the material, such as low tensile and flexural strength, poor fracture resistance and ductility, high propensity to drying shrinkage cracking, strong sensitivity to water erosion. These issues are especially critical in seismic areas or in the presence of critical climate conditions.

The low strength and the brittle behaviour are probably the major issues for structural use of earth-based materials, particularly for buildings and constructions likely to be subjected to seismic events [7,8]. For this reason, vegetal or animal fibres, including straw, flax, sisal, jute and wool [9–12], have been often used to improve the mechanical properties of the base material for specific applications. However, only a very limited number of studies have been conducted to characterize the effect of fibrous reinforcement on fracture resistance, post-fracture performance and energy dissipation capacity of earthen materials [10–15]. As an example, Islam and Iwashita [13] showed that the insertion of straw or jute fibres effectively improved ductility and toughness of adobe materials, even if at the cost of a reduction in compressive strength. Increases in flexural strength and deformation capacity were also observed with the introduction of straw or wool fibres in plain [12] or notched earth-based samples [10, 14].

On the other hand, to the authors' knowledge, no study has been yet reported on the influence of fibrous reinforcement on the fracture resistance and energy absorption properties of earth-based materials under dynamic loading conditions.

This study examines the improvements in strength, crack resistance and post-cracking performance achievable under static and dynamic (impact) loading conditions through the introduction of hemp fibres in an earthen building material. Notched samples made with soil material and reinforced with different fibre percentages and fibre lengths were manufactured and tested in a three point bending configuration. The results obtained under quasi-static and low-velocity impact loading are illustrated and discussed in the paper.

2. Materials and method

2.1. Materials used

A soil from quaternary sediments alongside the Tirso River near Oristano (Sardinia, Italy), traditionally used in the past to manufacture earth bricks, was used in this study. The granulometric curve of the soil, which was obtained by sieving and hydrometer analyses and carried out in accordance with ASTM standard D422 [16] shows

[a] Corresponding author: lucafrancesconi@unica.it

Article available at http://www.epj-conferences.org or http://dx.doi.org/10.1051/epjconf/20159401043

Table 1. Mixture proportions.

Mixture designation	Water/Soil weight ratio	Fibre weight fraction	Fibre length (mm)
NFR (Unreinforced)	0.2	0%	N/A
FRC-21	0.3	2%	10
FRC-22	0.3	2%	20
FRC-23	0.3	2%	30
FRC-31	0.3	3%	10
FRC-32	0.3	3%	20
FRC-33	0.3	3%	30

that the soil may be classified as CL, i.e. inorganic clay with low plasticity and low liquid limit, in the Unified Soil Classification System (USCS) [17]. X-ray diffraction revealed the presence of dominant quartz, together with other constituents including illite and plagioclase. Grain size and values of Atterberg limits indicate that the soil is suitable for earth construction.

Hemp fibres with an average diameter of 0.2 mm were used to reinforce the earthen samples. Soil mixtures characterized by two fibre weight fractions (2% and 3%) and three different fibre lengths (10 mm, 20 mm and 30 mm) were produced to examine the effect of fibre length and weight ratio on the structural properties of the reinforced material.

2.2. Sample preparation

Soil, hemp fibres and demineralized water were used to manufacture the samples. The soil collected from the cave was first manually treated to remove inclusions such as roots, rocks and other unwanted organic materials. The resultant material was then crushed using a mortar and a pestle, sieved using a 4 mm mesh, and finally desiccated using a drying oven (40 Celsius degree for 2 hours) to eliminate the residual humidity. The hemp fibres were manually combed and cut in 3 different lengths (10 mm, 20 mm and 30 mm), and used to prepare six mixture compositions characterized by different weight percentages and fibre lengths. Table 1 shows the mixtures used for manufacturing of samples.

Previous experiences and analyses were considered to determine the water content of the various mixtures so as to obtain similar levels of workability for the different mixture compositions [14]. The manufacturing of the specimens consisted in two steps: the soil and the fibres were dry mixed and the water was gradually added, while continuously mixing, up to the desired water/soil weight ratio. The mixture was then casted into perspex molds and manually compacted to form blocks $70 \times 70 \times 160\,mm^3$ in size. All specimens were cured at room temperature for 30 days (for 48 h inside the moulds and for 28 days after demoulding) and at 40 °C in an oven for 15 days. A central notch 35 mm in depth, as visible in Fig. 1, was finally cut in the mid section of the samples by a diamond coated circular saw blade.

Figure 1. Fibre reinforced sample tested in the study.

2.3. Experimental testing

The notched specimens were tested to investigate the fracture performance of the unreinforced and reinforced materials under static and dynamic (impact) flexural loading. In particular, the experiments were focused on the analysis of the influence of different fibre contents and fibre lengths on the fracture resistance of the reinforced soil, and on the comparison of the energy absorption properties exhibited by the materials under different loading rates

During both static and impact testing, the bricks were loaded in a three-point bending configuration, using a distance between the supports (span length) of 140 mm. The centre load was imposed through a steel solid cylinder with a diameter of 10 mm.

Static tests were carried out using a closed-loop servo-electric 5 kN tension/compression testing machine. Tests were conducted under displacement control at a rate of 1 mm/min. The mid-span deflection values were recorded using both the built-in displacement transducer of the testing machine (which measures the deflection at the loaded side of the beam) and a linear variable differential transducer (LVDT) positioned at the bottom side of the beam in the vicinity of the notch [14]. Since no significant differences were observed between the two measurements, only the deflection data acquired on the loaded side of the samples were considered in the analyses.

Impact tests were carried out using an instrumented drop-weight impact testing machine, provided with a 4.83 kg falling mass (Fig. 2). A pneumatic braking system capable of capturing the falling mass after the first rebound was used in order to prevent multiple impacts on the samples. All samples were impacted with an energy of about 50 J, which corresponds to a drop height of approximately 1.05 m and to an impact velocity slightly higher than 4.5 m/s. The contact force between the impactor and the sample was measured by a piezoelectric force transducer mounted between the impactor tup and the drop weight. An infra-red sensor was used to measure the velocity of the impactor immediately before contact with the sample. The displacement and the kinetic energy of the impactor were calculated as a function of time by integration of the contact force signal.

Figure 2. Impact testing fixture.

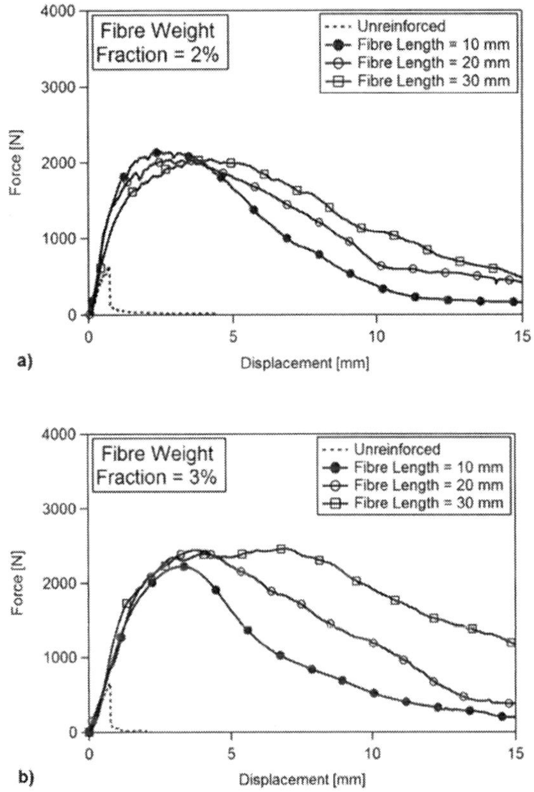

Figure 3. Force-displacement curves of unreinforced and reinforced samples subjected to quasi-static bending.

A minimum number of three samples were tested for any loading condition (static or dynamic) and for any of the mixtures investigated in the study.

3. Results and discussion

Typical force-deflection curves of unreinforced and reinforced samples subjected to quasi-static bending are presented in Fig. 3.

The plots of Fig. 3 clearly show that the introduction of hemp fibres greatly improves the mechanical properties of the soil material with regard to the residual strength and the energy absorption at large deflections.

As visible in the graphs, after an initial linear response, the force-deflection plot of the unreinforced material exhibits a sudden load drop, which corresponds to the fast propagation of a crack from the tip of the notch to the loaded side of the sample. Total loss of the flexural strength, with complete failure of the beam into two halves, was typically observed for maximum deflections of less than 1.5 mm. In reinforced samples, in contrast, after the onset of first cracking phenomena at the notch (which are signalled by the slight slope reduction visible in the force-deflection curves after the initial linear stage) the force keeps increasing under increasing deflection until it reaches a peak load; beyond this point the force starts to slowly decrease, even if samples are seen to retain some residual strength up to bending deflections of 15 mm and larger. The improvements in the load bearing capacity and in the fracture resistance produced by the fibrous reinforcement may be attributed to the bridging restraint provided by fibres across cracked surfaces (see Fig. 4), and to the energy dissipation associated to frictional sliding during pullout of debonded fibres at increasing deflections

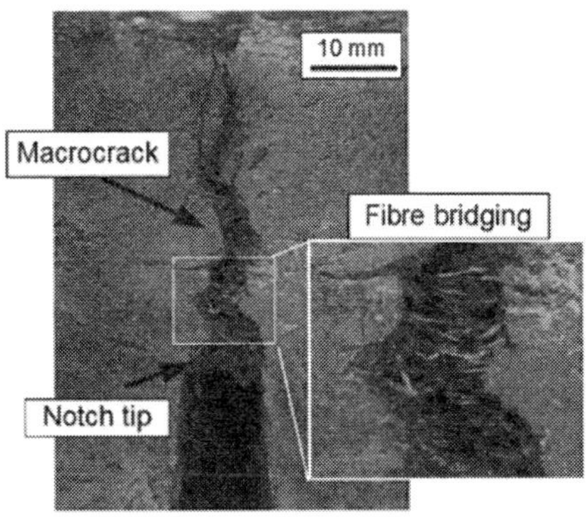

Figure 4. Fibre bridging across cracked surfaces in a sample reinforced with 2% of 10 mm fibres.

The experimental results summarized in the plots of Fig. 3 indicate that the post-cracking properties of the reinforced materials are controlled by the fibre weight ratio

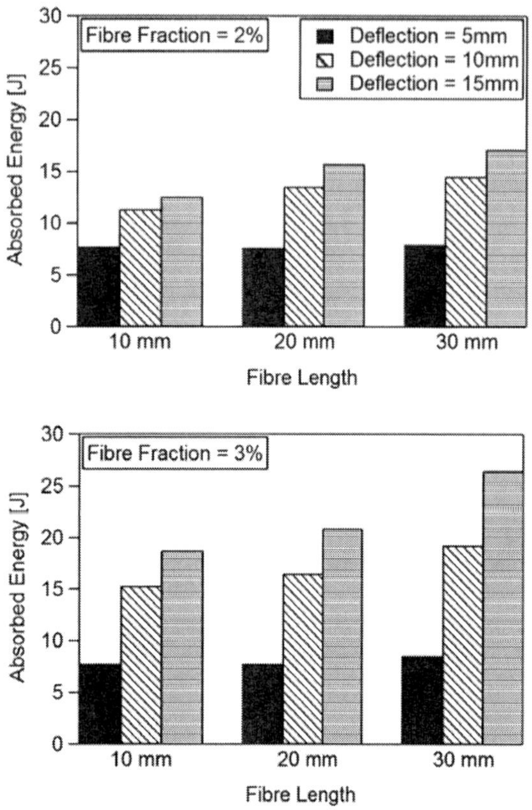

Figure 5. Average values of energy absorbed at different deflections by reinforced samples subjected to static bending.

and by the fibre length. It is clearly seen, in particular, that both the residual strength and the energy absorbed by the material at specific deflection levels increase with increasing fibre weight fraction and, for a constant fibre weight fraction, with increasing fibre length. As an example, the effect of fibre percentage and fibre length on the energy absorbed by the samples at different deflection levels is illustrated in the diagrams of Fig. 5. The energy absorbed by the samples was calculated as the area under the force–displacement curve up to the chosen deflection. We may observe that the energies absorbed at 10 mm and 15 mm deflections increase with increasing fibre content and fibre length. In contrast, the energy absorbed at 5 mm deflection does not appear to be significantly influenced by these parameters. These results indicate that large deflections (i.e. large crack openings) are required to fully activate the major mechanisms responsible for energy dissipation under bending loads, such as crack bridging and fibre debonding and pullout.

To examine the effect of the fibrous reinforcement on the fracture resistance of the materials under dynamic loading, three point bending tests were also carried out on the notched samples by an instrumented drop weight machine. The samples were simply supported on a support anvil using the same span length (140 mm) used during static tests and impacted with an energy of 50 J. Because of the sample inertial effect, the load measured by the force transducer mounted on the impactor does not represent the actual bending load acting on the sample [18].

Following [18], the true impact force for bending was obtained by subtracting the generalized inertial force from the force measured at the impactor. The generalized inertial force was calculated as a function of the acceleration at the mid-span, assuming a linear distribution of the acceleration along the sample. The history of the acceleration of the sample at mid-span during the impact event was measured by an accelerometer placed on the lower side of the sample in the vicinity of the notch.

Typical force-deflection curves obtained during impact tests for the different material configurations are reported in Fig. 6.

The plots of Fig. 6 show that the impact response of the unreinforced material exhibits an abrupt load drop at a deflection of about 4 mm, after which the load sustained by the sample quickly decreases to zero, as a result of the brittle collapse of the sample following the unstable propagation of a macrocrack from the notch tip. It should be observed however that the peak load measured during the impact is much larger than that recorded during quasi-static loading, thus indicating a strong sensitivity of the unreinforced material to the load rate.

A significant stress-rate sensitivity may also be observed in the structural response of the reinforced materials, as evident by comparison of the plots of Fig. 3 and Fig. 6. The analysis of static and dynamic force-deflections curves shows that all reinforced samples, irrespectively of fibre content and fibre length, exhibit higher peak loads, better residual strengths at large deflections, and larger energy dissipations under impact loading than under static conditions. As an example, a comparison of the static and impact responses of FRC-33 samples (fibre weight fraction = 3%; fibre length = 30 mm) reported in Figs. 3 and 6 indicate that the values of the peak load and of the residual strength at 15 mm deflection recorded during impact bending are, respectively, about 80% and 100% higher than those attained during static bending.

Average values of the energy absorbed by the samples at different deflection levels are shown in the diagrams of Fig. 7 as a function of fibre length.

The data reported in Figs. 5 and 7 confirm that the fracture resistance under high loading rates of this class of materials is significantly higher than that provided under static loading.

We may also notice that while fibre percentage and fibre length do not greatly affect the energy absorbed during impact for small deflections (see Fig. 7; 5 mm deflection), for larger deflections the energy absorbed during impact increases with increasing fibre weight fraction and, at a constant fibre weight fraction, with increasing fibre length (Fig. 7; 10 mm and 15 mm deflections).

The trends of these data are in accordance with those observed under static bending (Fig. 5), and this again indicates that the fracture response of the fibre-reinforced materials is governed by toughening mechanisms such as fibre bridging and fibre pullout only for sufficiently large crack opening displacements.

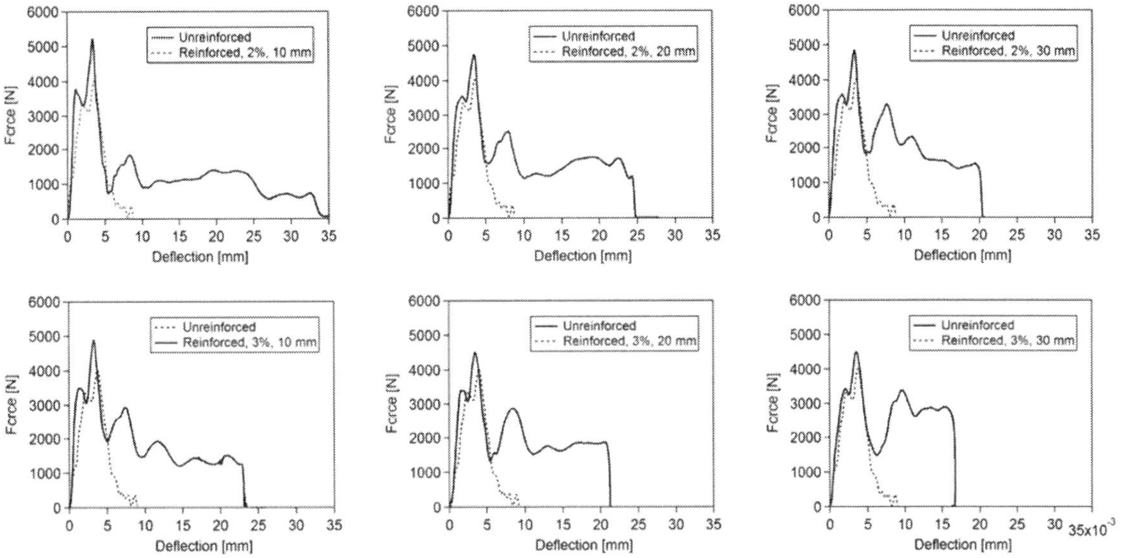

Figure 6. Force-displacement curves of reinforced and unreinforced samples subjected to impact bending.

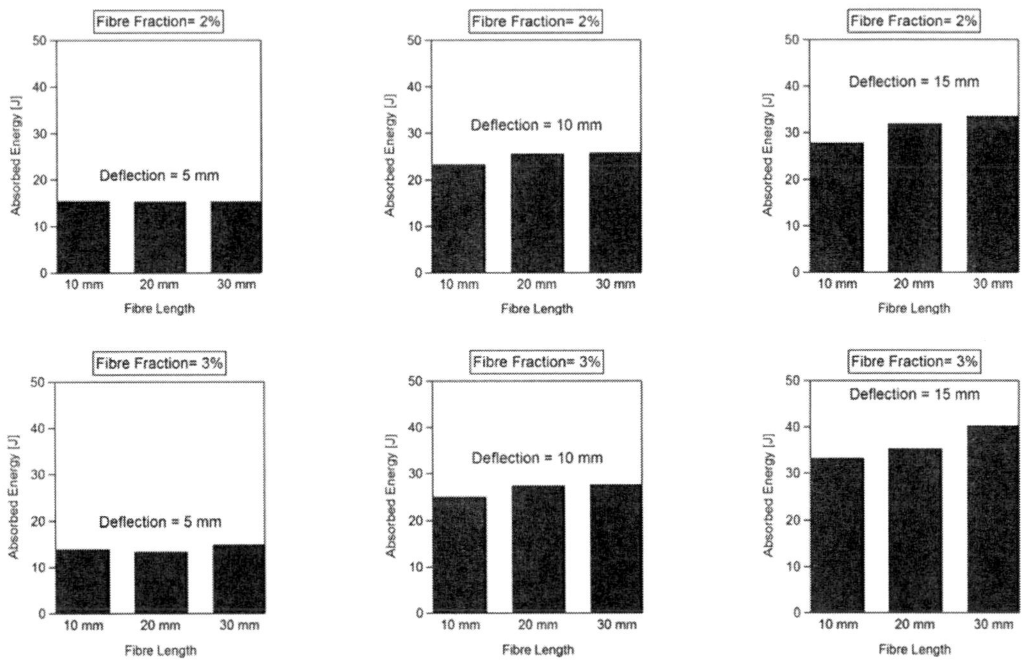

Figure 7. Average values of energy absorbed at different deflections by reinforced samples subjected to impact bending.

4. Conclusions

Notched earthen samples reinforced with hemp fibres were fabricated and tested under both static and dynamic bending. The main results of the analyses may be summarized as follows:

- The insertion of hemp fibres improves substantially the static and the dynamic mechanical properties of the earthen material.
- The strength and the fracture resistance of the earthen material increase with increasing fibre weight fraction and, for a constant fibre weight fraction, with increasing fibre length.
- Both unreinforced and reinforced earthen materials show a strong sensitivity to the loading rate, with significantly higher peak loads and energy absorptions under dynamic bending as compared to static bending.
- Bridging action, fibre debonding and fibre pullout are the key mechanisms responsible for the enhancement of strength and toughness of the base material.

The results indicate that reinforcement of earthen materials by hemp fibres represents an efficient way to enhance the structural performances of earth-based building materials under both static and dynamic loading conditions. Further investigations and more detailed analyses are however required to fully characterise the influence and relative importance of the different physical mechanisms by which fibres control strength and crack resistance properties of the material.

References

[1] H. Houben, H. Guillaud, *Traité de construction en terre* (Parenthèses, Marseille, 1989)

[2] G. Minke, *Earth construction handbook: the building material earth in modern architecture* (Wit Press, Southampton, 2000)

[3] G. Minke, *Building with earth. Design and technology of a sustainable architecture* (Birkhäuser, Basel, 2006)

[4] Avrami E., Guillaud H., Hardy M., *Terra literature review: an overview of research in earthen architecture conservation.* (The Getty Conservation Institute, Los Angeles, 2008)

[5] H. Binici,O. Aksogan, M. Nuri Bodur, E. Akca, S. Kapur, S. Thermal isolation and mechanical properties of fibre reinforced mud bricks as wall materials, Constr. Build. Mater. **21**, 5 (2007)

[6] M. Bouhicha, F. Aouissi, S. Kenai, Performance of composite soil reinforced with barley straw, Cem. Concr. Compos. **27**, 5 (2005)

[7] Q. Piattoni, E. Quagliarini, S. Lenci, Experimental analysis and modelling of the mechanical behaviour of earthen bricks, Constr. Build. Mater. **25**(4), 9 (2011)

[8] H. Binici, O. Aksogan, T. Shah, Investigation of fibre reinforced mud brick as a building material, Constr. Build. Mater. **19**, 8 (2005)

[9] K. Ghavami, R.D. Toledo Filho, N.P. Barbosa, Behaviour of composite soil reinforced with natural fibres, Cem. Concr. Compos. **21**, 10 (1999)

[10] F. Clementi, S. Lenci, T. Sadowski, Fracture characteristics of unfired earth, Int. J. Fract. **149**, 8 (2008)

[11] C. Galán-Marín, C. Rivera-Gómez, J. Petric, Clay-based composite stabilized with natural polymer and fibre, Constr. Build. Mater. **24**, 6 (2010)

[12] C. Galán-Marín, C. Rivera-Gómez, J. Petric, Effect of animal fibres reinforcement on stabilized earth mechanical properties, J Biobased Mater. Bioenergy **4**(2), 7 (2010)

[13] M.S. Islam, K. Iwashita, Seismic response of fiber-reinforced and stabilized adobe structures, *Proceedings of the Getty Seismic Adobe Project 2006 Colloquium* (2006)

[14] F. Aymerich, L. Fenu, P. Meloni, Effect of reinforcing wool fibres on fracture and energy absorption properties of an earthen material, Constr. Build. Mater. **27**, 6 (2012)

[15] F. Parisi, D. Asprone, L. Fenu, A. Prota, Experimental characterization of Italian composite adobe bricks reinforced with straw fibers, Compos. Struct. **122**, 300 (2015)

[16] ASTM Standard D422-63, Standard test method for particle-size analysis of soils (ASTM International, 2007)

[17] ASTM Standard D2487-11, Standard practice for classification of soils for engineering purposes (Unified Soil Classification System), (ASTM International, 2011)

[18] N. Banthia, S. Mindess, A. Bentur and M. Pigeon, M. (1989). Impact testing of concrete using a drop-weight impact machine, Exp. Mech. **29**(1), 7 (2010)

EPJ Web of Conferences **94**, 01044 (2015)
DOI: 10.1051/epjconf/20159401044

Experimental investigation of bond strength under high loading rates

Mathias Michal[1,a], Manfred Keuser[1], George Solomos[2], Marco Peroni[2], Martin Larcher[2], and Beatriz Esteban[1]

[1]Universität der Bundeswehr München, Werner-Heisenberg-Weg 39, 85577 Neubiberg, Germany
[2]European Commission – Joint Research Centre, Via E. Fermi 2749, 21027 Ispra (VA), Italy

Abstract. The structural behaviour of reinforced concrete is governed significantly by the transmission of forces between steel and concrete. The bond is of special importance for the overlapping joint and anchoring of the reinforcement, where rigid bond is required. It also plays an important role in the rotational capacity of plastic hinges, where a ductile bond behaviour is preferable. Similar to the mechanical properties of concrete and steel also the characteristics of their interaction changes with the velocity of the applied loading. For smooth steel bars with its main bond mechanisms of adhesion and friction, nearly no influence of loading rate is reported in literature. In contrast, a high rate dependence can be found for the nowadays mainly used deformed bars. For mechanical interlock, where ribs of the reinforcing steel are bracing concrete material surrounding the bar, one reason can be assumed to be in direct connection with the increase of concrete compressive strength. For splitting failure of bond, characterized by the concrete tensile strength, an even higher dynamic increase is observed. For the design of Structures exposed to blast or impact loading the knowledge of a rate dependent bond stress-slip relationship is required to consider safety and economical aspects at the same time. The bond behaviour of reinforced concrete has been investigated with different experimental methods at the University of the Bundeswehr Munich (UniBw) and the Joint Research Centre (JRC) in Ispra. Both static and dynamic tests have been carried out, where innovative experimental apparatuses have been used. The bond stress-slip relationship and maximum pull-out-forces for varying diameter of the bar, concrete compressive strength and loading rates have been obtained. It is expected that these experimental results will contribute to a better understanding of the rate dependent bond behaviour and will serve for calibration of numerical models.

1. Introduction

1.1. Bond between steel and concrete

For reinforced concrete structures under blast and impact not only the rate dependence of steel and concrete influences the structural behaviour. Their interaction determines significantly the structural response under static and dynamic loading. The transfer of forces between steel and concrete is always connected to a relative displacement. Already after a slight displacement the chemical adhesion is overcome and the main mechanism of bond, the mechanical interlock is activated. Thereby the ribs of the reinforcement are bracing concrete consoles in the concrete surrounding the steel. After shearing of the concrete consoles, only the frictional resistance remains for the transmission of forces. The bond stress is defined as the evenly over the circumference of the bar distributed force that is transferred between steel and concrete. For the description of bond the bond stress is related to the slip (Fig. 1). While for static loading with small deformation the assumption of a rigid bond mostly leads to satisfactory results, this can cause a wrong judgement of the structural behaviour under dynamic loading.

Only limited research has been done on bond under high loading rates in the past [1–6]. An overview is given in [7]. The same parameter influencing the static bond conditions are valid for bond behaviour under high loading rates. Some of them are of particular importance for the increase of dynamic bond strength as compared to static behaviour. With increasing *concrete compressive strength* a decreasing rate dependence of bond strength has been observed. Another parameter is given with the *geometry of the ribs*. In the majority of the reviewed research there seems to be no significant influence of loading rate for plain bars, the bond resistance of deformed bars increases with higher loading rates. The stress distribution under dynamic loading is more concentrated. With increasing *bond length*, the rate dependence decreases. Localized cracks due to concentrated transfer of bond stress for faster rate loading can lead to a brittle failure of structures showing a ductile behaviour under static loading. The spatial stress condition in the concrete leads to circular tensile forces, this can result in a change of the *failure type* from pull-out of the bar to splitting of the concrete cover. Tests showed increased ultimate loads under dynamic loading for specimens failing by pull-out but a higher dynamic increase for specimens failing by splitting. Only for a short *loading duration* the increase of bond strength can result in higher load bearing capacity. With increasing load duration the bond strength decreases.

1.2. Hopkinson Bar Technique

The Split-Hopkinson-Bar (SHB) is often used to determine the uniaxial material properties for medium strain rates. The theory of the method can be found in [8–11]. Thereby the specimen is sandwiched between an incident and an output bar (Fig. 2). With a loading device a stress wave

[a] Corresponding author: `mathias.michal@unibw.de`

Article available at http://www.epj-conferences.org or http://dx.doi.org/10.1051/epjconf/20159401044

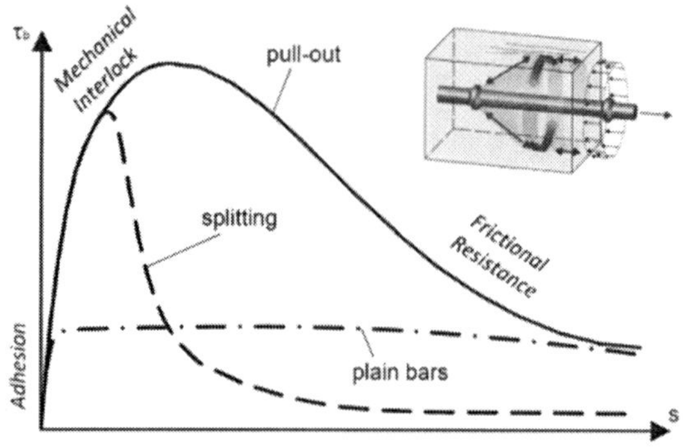

Figure 1. Idealized bond-slip-relationship [7].

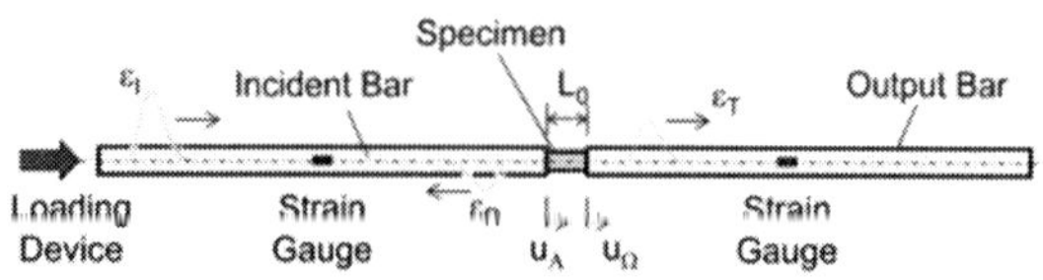

Figure 2. Schematic test configuration SHB.

is induced in the incident bar. At the interface between bar and specimen the wave is partly reflected and partly transmitted to the specimen. This process is repeated at the transition to the output-bar and leads to multiple reflection in the specimen. If the time the wave needs to pass through the specimen is essential short compared to its temporal extension, a uniform stress and strain distribution is assumed and the well known formulae for the analysis of the SHB-Test based on the propagation of a one-dimensional wave can be used.

The displacement of the initial (u_A) and end (u_Ω) cross-section of the specimen is obtained by integrating the particle velocity v_p (1)(2).

$$u_A(t) = \int_0^t v_{p,A}(t)dt = C_{0,Bar} \int_0^t [\varepsilon_I(t) - \varepsilon_R(t)]\, dt \quad (1)$$

$$u_\Omega(t) = \int_0^t v_{p,\Omega}(t)dt = C_{0,Bar} \int_0^t \varepsilon_T(t)dt. \quad (2)$$

Using the difference of the displacements and the original length of the specimen, the strain and thereby the strain rate and stress in the specimen can be calculated.

For the bond tests modified configurations of the Split-Hopkinson-Bar have been applied.

2. Rebar pull-out tests

Pull-out tests have been performed by employing Hopkinson bar techniques at the JRC in Ispra [6]. Confined and unconfined cylindrical specimens with a diameter of 100 mm and bars with $d_s = 20$ mm have been tested. The length of the specimen varied with the embedment length ($l_b = 100$ mm $= 5$ ds, 200 mm $= 10$ ds). A bond breaker of length 50 mm has been incorporated on both ends. These parts have been subdivided into three equal sectors.

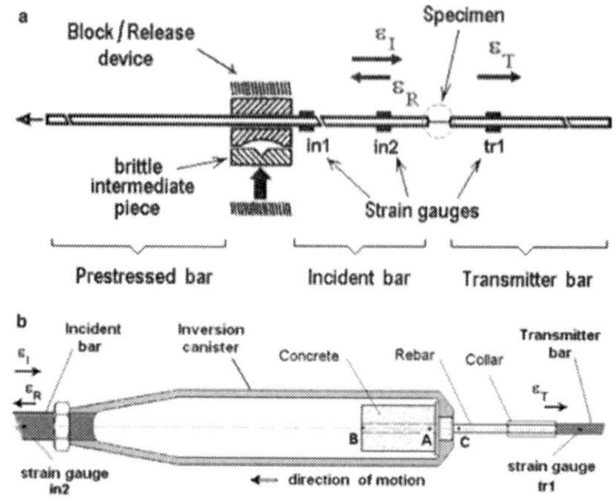

Figure 3. (a) Principle of functioning of the JRC modified Hopkinson bar; (b) details of mounting of the rebar specimen in the inversion canister between the incident and transmitter bar [6].

The specimens have been manufactured in two concrete compressive strengths (C25/30, C50/60).

An incident strain pulse was generated by pre-stressing and abruptly releasing a long bar, which was the continuation of the incident bar. The diameter of the incident bar was 72 mm and its length approximately 8 m. A 1.1 m long steel inversion canister, housing the specimen, was connected to its end (Fig. 3). The transmitter bar of diameter 25 mm and length 50 m was connected by a collar to the rebar of the specimen. The pull-out Force was measured by a strain gauge at the output bar and the displacement (slip) was referred to the interface point rebar/concrete at the front part of the specimen (point A).

The specimens confined by a steel tube with wall-thickness of 10 mm failed by pull-out. Due to the small concrete cover of $2.5 \cdot d_s$ the unconfined specimens failed by splitting of the concrete. While bond strength increased for C50/60 under high rate loading, the increase has been higher for the C25/30 specimens (Fig. 4). The increase of ultimate bond strength for doubling the embedment length under dynamic loading was lower compared to the increase under static loading. A detailed description of the tests and further results can be found in [6].

3. Rebar push-in tests

In a current research project, funded by the Federal Ministry of Economics and Technology (BMWi) the bond behaviour between steel and concrete is investigated. In collaboration between the University of the Bundeswehr and the Fraunhofer Ernst-Mach-Institute (EMI) push-in tests are performed using a Split-Hopkinson-Bar.

The push-in specimen is installed between the incident and the output bar (Fig. 5). The length of the incident bar is 5 m, that of the output bar 3.5 m and their diameter 75 mm. The bars and the transition cylinders are made of aluminium.

For the measurement of the displacement at the free end of the reinforcement bar a transition cylinder is placed

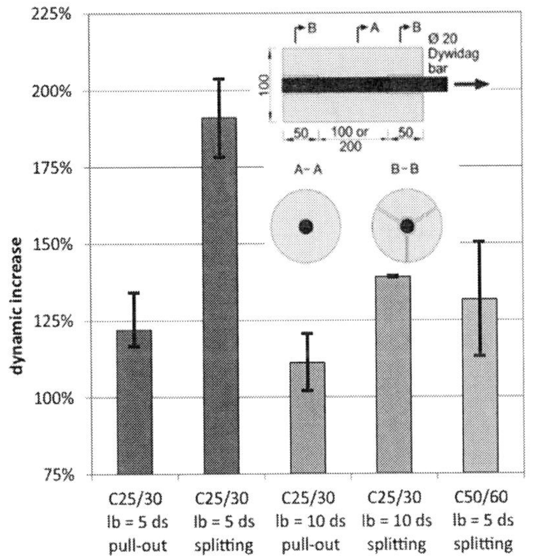

Figure 4. DIF for different concrete compressive strengths, bond lengths and failure mechanisms in [6].

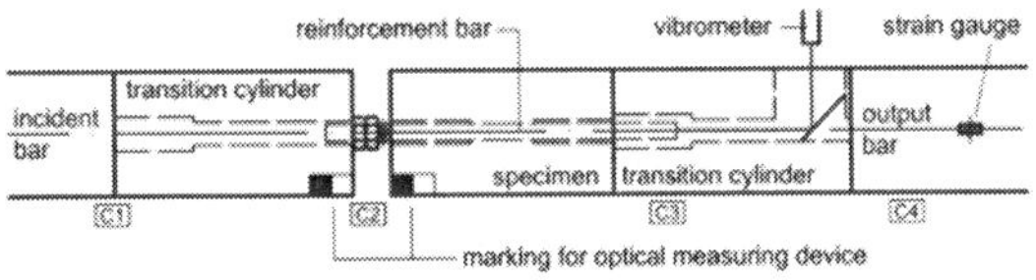

Figure 5. Schematic push-in test SHB configuration.

between specimen and output bar. With specially drilled holes and a mirror it is possible to measure the translation of the bar with a vibrometer. The slip is determined by the difference of the signal given by the vibrometer and the overall displacement of the specimen calculated with formula (2) from the time-shifted signal of the strain gauge mounted on the output bar. The dispersion in the bar has been considered with a VBA-Routine based on the method described in [8, 10–12].

The wave is separated in its frequency components using Fourier transformation (formulas 3–9).

$$f(t) = \frac{a_0}{2} + \sum_{n=1}^{\infty} d_n \cdot \cos(n\omega_0 t - \Phi_n) \tag{3}$$

$$\Phi_n = arctan\,(b_n/a_n) \quad phase\ angle \tag{4}$$

$$\omega_0 = 2\pi/T \quad angular\,frequency. \tag{5}$$

Fourier coefficients:

$$a_0 = \frac{2}{T}\int_{-T/2}^{T/2} f(t)dt \tag{6}$$

$$a_n = \frac{2}{T}\int_{-T/2}^{T/2} f(t)\cdot\cos(n\omega_0 t)dt \quad for\ n \geq 1 \tag{7}$$

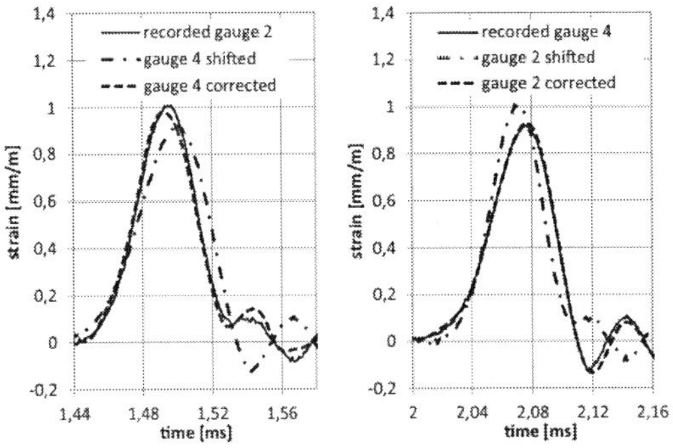

Figure 6. Backwards (left) and forwards (right) time-shifted and dispersion corrected signals $\Delta x = 3$ m, $C_0 = 5211$ m/s.

$$b_n = \frac{2}{T}\int_{-T/2}^{T/2} f(t)\cdot\sin(n\omega_0 t)dt \quad for\ n \geq 1 \tag{8}$$

$$d_n = \sqrt{a_n^2 + b_n^2}. \tag{9}$$

The phase velocity of the n-th frequency is determined by a numerical iteration. A wave length $\Lambda_{n,i}$ is assumed and the phase velocity $C_{n,i}$ is calculated with formula (10). With the interpolated phase velocity given by the table in [10] the wave length $\Lambda_{n,j}$ can be calculated again using formula (10). If $C_{n,i}$ and $C_{n,j}$ doesn't show sufficient agreement (e.g. $C_n/1000$), the calculation process is repeated with a mean wave length $\Lambda_{n,k} = 0.5 \cdot (\Lambda_{n,i} + \Lambda_{n,j})$. With the known wave velocity the phase angle difference to the phase angle without dispersion can be calculated with formula (11), where Δx is the distance over which the signal is shiftet.

$$C_{n,i} = \frac{n \cdot \omega_0 \cdot \Lambda_{n,i}}{2 \cdot \pi} = \frac{n \cdot \Lambda_{n,i}}{T} \tag{10}$$

$$\Phi_{dn} = n \cdot \omega_0 \cdot \left(\frac{\Delta x}{C_n} - \frac{\Delta x}{C_0}\right). \tag{11}$$

The new wave can be constructed with formula (12). In this study a number of 100 Fourier components has been chosen.

$$f(t) = \frac{a_0}{2} + \sum_{n=1}^{\infty} d_n \cdot \cos\left[n\omega_0 t - (\Phi_n + \Phi_{dn})\right]. \tag{12}$$

In the described procedure only the first mode vibration of the Pochhammer-Chree analysis is considered, but Fig. 6 shows the good agreement of the time-shifted, dispersion-corrected signal with the recorded signal.

In addition the displacement of the transition cylinder, used for load transfer to the reinforcement bar, and the specimen has been measured by an optical measuring device. Both methods showed good agreement. The transferred force is obtained from the strain gauge on the output bar. After a time-shift of the force, the bond-slip diagram could be drawn (Fig. 7).

From the performed experiments the maximum bond stress and a bond stress – slip relationship for loading rates of 100 to 400 MN/s could be obtained. Although no representative statements can still be made on the few experiments carried out, there is a clear tendency for the

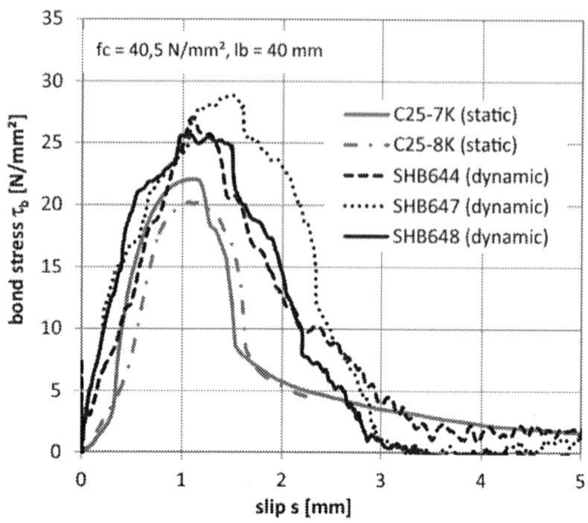

Figure 7. Schematic test configuration SHB.

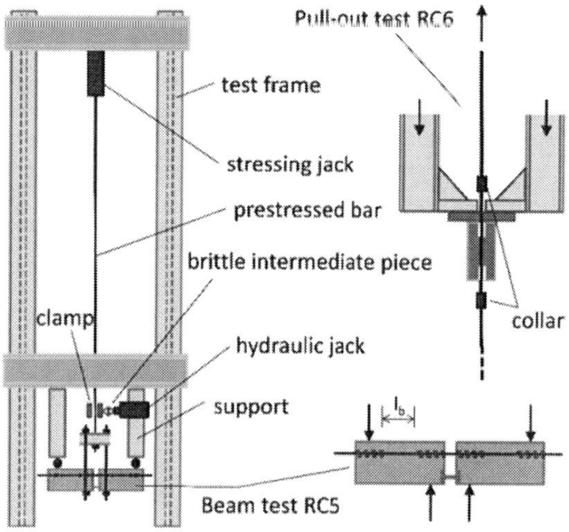

Figure 8. Schematic test configuration.

rate-dependent increase in bond strength. On this basis, an extensive experimental program will be conducted as a contribution to the research on bond stress under high loading rates.

4. Development of a new test setup

4.1. Test configuration

A new test setup (Fig. 8) based on the idea of Albertini [13] to induce a tensile wave by abruptly releasing the elastic mechanical energy stored inside a prestressed bar has been developed at the UniBw [15]. It is planned, to use specimens based on the RILEM pull-out test (RC6) and the RILEM beam test (RC5) [14].

The procedure of the tests is as follows: A prestressing bar is blocked by static friction between the bar and two bearing shells after a horizontal force is applied. Then the vertical force can be increased until the required prestressing is achieved. The prestressed bar is a DYWIDAG threadbar type 18WR with a diameter of 17.5 mm and a nominal cross-section area of 241 mm^2. The maximum pretension force is given by the manufacturer

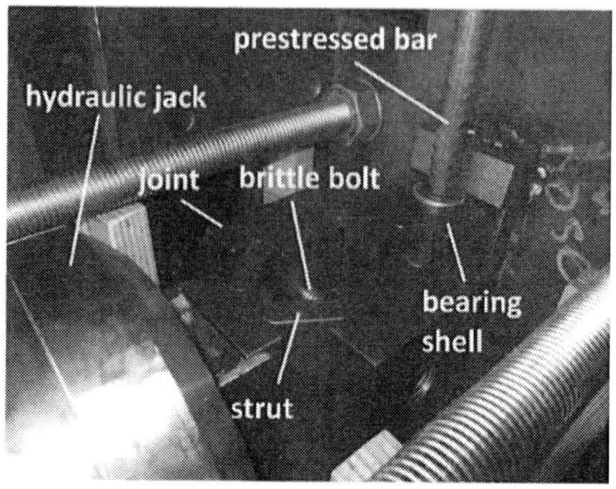

Figure 9. Blocking system.

with $0.9 \cdot F_{p0.1k} = 204$ kN and the nominal tensile strength with $R_m = 1050$ N/mm^2. For the tests the pretension force has to be within the elastic range of the material. After removing the blocking system a release wave is running through the prestressed part of the bar and a tensile wave propagates in the direction of the specimen. The length of the wave equals twice the length of the prestressed part of the bar and its amplitude is half of the prestressing.

While for the pull-out tests the induced wave runs directly through the specimen and principles of wave propagation will be used for the evaluation of the tests, there is no direct connection for the beam tests and inertia has to be considered.

A crucial part of the configuration is the brittle intermediate piece. To obtain high strain rates the preload force needs to be released abruptly. A blocking system similar to the theta-clamp, developed at the JRC [16,17] has been chosen (Fig. 9). The system consists of several elements. When the clamping force is applied, four struts are bracing against joints, held together by a brittle bolt. The force is increased up to a certain limit and the bar can be pre-tensioned. After a further increase of the force the bolt fails and the whole system collapses into pieces. This leads to a sudden release of the blocking and a steep rise of the wave. The bolts have been made of 42CrMo4 with a tensile strength of 1560 N/mm^2.

4.2. Pull-out Test (RC6)

First tests are based on the configuration for the RILEM Pull-out test [14]. The specimens have been modified as the focus in these tests lies on splitting failure. For the reinforcement a DYWIDAG prestressing steel 18 WR with a diameter of 17.5 mm has been chosen. It is thus possible to directly connect the specimen with the prestressed bar of the test setup. The rebar in the cylindrical specimens with a diameter of 85.6 mm has a bond length of 70 mm = 4 ds. The concrete cover is with 34.05 mm nearly 2 ds. On both ends of the concrete specimen there is a bond free length of 50 mm.

After releasing the blocking system, a tensile wave is transfered to the test bar. While the wave is running through the specimen a relative displacement between steel and concrete arises due to the translation of the bar.

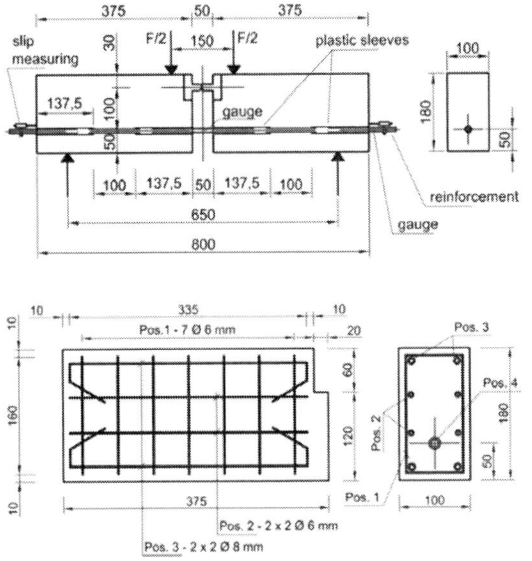

Figure 10. Specimen beam test.

Because of the small slip only a part of the wave can be transferred to the concrete. To prevent reflections, the bar is continued behind the specimen.

The static tests can be performed with the same configuration, which improves the comparability of the results. Tests with similar specimens and the method already described in chapter 2 will be performed at the JRC.

4.3. Beam Test (RC5)

With the RILEM Beam Test [14] the situation of a reinforcement bar in a bending beam should be simulated. The specimen consists of two identical reinforced halves of a beam connected by the reinforcement for the bond test and a steel hinge for the transfer of the compressive force (Fig. 10). The recommended dimensions of the specimen differ according to the diameter of the bar. For the dynamic beam test specimens with a bar diameter of 10 mm will be used.

For the static tests the bond stress can be calculated using the applied vertical load (13, 14).

$$F_s = 1,25 \cdot F \quad (10 \leq d_s < 16\,\text{mm}) \tag{13}$$

$$\tau_b = \frac{F_s}{\pi \cdot d_s \cdot l_b}. \tag{14}$$

In the dynamic test the pull-out force will be determined by a strain gauge mounted at mid span at the bar as due to inertia the formula used in the static test can't be applied.

5. Conclusions and outlook

The bond between steel and concrete is essential for reinforced concrete structures. It's based on adhesion, friction and mechanical interlock. Literature has been reviewed and static and dynamic tests on bond between steel and concrete have been performed with different methods by the University of the Bundeswehr Munich and the Joint Research Centre in Ispra. Due to the different test methods and boundary conditions it is difficult to compare results drawn from literature. For deformed bars an increased bond strength with increasing loading rate can be noted. The rate dependent increase of concrete strength seems to be of great importance for this phenomenon but there is no knowledge from the tests of the local strain rates in the concrete. The main parameter influencing the rate dependent bond behaviour can be identified as concrete compressive strength, diameter of the bar and geometry of the ribs, bond length, failure type and load duration. To quantify them there is still a considerable need for research.

For the design of reinforced concrete structures under blast and impact the knowledge of the rate dependent bond behaviour is important. Localized cracks or a decreased rotation capacity due to the increased bond can lead to undesirable effects on structures.

Further tests will be conducted with a new developed test configuration at the UniBw and a proven method at the JRC, using similar specimens. This can lead to a better understanding and interpretation of the results obtained by the different test methods.

References

[1] R.J. Hansen, A.A. Liepings, Proc. ACI J., Vol. **59**, No. 4, 563–583 (1962)
[2] H. Paschen, J. Steinert, O. Hjorth, Untersuchungen über das Verbundverhalten von Betonstählen bei Kurzzeitbeanspruchung, Tech. Rep., TU Braunschweig (1974)
[3] E. Vos, Reinhardt H.W., Rep. 5-80-6, Delft Univ. of Tech. (1980)
[4] C. Yan, PhD Thesis, Univ. of Brit. Col. (1992)
[5] J.H. Weathersby, PhD Thesis, Louisiana State Univ. (2003)
[6] G. Solomos, M. Berra, Mat. and Struct. **43**, 247–260 (2010).
[7] M. Michal, M. Keuser, in Proc. of 9th International Conference on Structural Dynamics, EURODYN 2014, Porto, Portugal (2014)
[8] R.M. Davis, Phil. Trans. of the Roy. Soc.of London, Series A, Vol. 240, No. **821**, 375–457 (1948)
[9] H. Kolsky, Proc. Phys. Soc. **B62**, 676–700 (1949)
[10] D. Bancroft, Phys. Rev., **59**, 588–593 (1941)
[11] J.C. Gong, L.E. Malvern, D.A. Jenkins, J. of Eng. Mat. and Tech., **112**, 309–314 (1990)
[12] S. Zheng, Dissertation, University Karlsruhe (1996)
[13] C. Albertini, M. Montagnani, in Proc. of the Conference on Mechanical Properties at High Rates of Strain, Oxford (1974)
[14] RILEM, Rec. for the Testing and Use of Constr. Materials, E & FN Spon, London (1994)
[15] M. Michal, Dissertation, University of the Bundeswehr Munich (to be published)
[16] C. Albertini, E. Cadoni, G. Solomos, Phil. Trans. R. Soc. A 372: 20130197 (2013)
[17] Y. Chen, A.H. Clausen, O.S. Hopperstad, M. Lagseth, Int. J. of Imp. Eng. **38**, 824–836 (2011)

EPJ Web of Conferences **94**, 01045 (2015)
DOI: 10.1051/epjconf/20159401045

Advances in experimental assessment of dynamic tensile strength of concrete by the spalling technique

... in tribute to Janusz Roman Klepaczko, Emeritus Professor at the Université Paul Verlaine in Metz, passed away on August 15, 2008.

Ahmed Brara[a]

National Center of Studies and Integrated Research on Building Engineering, Cité El Mokrani, Algiers, Algeria

Abstract. An experimental method to test concrete in dynamic tension by spalling with a Hopkinson bar as loading and measuring tool was developed in 1999. The dynamic strength of concrete specimen and strain rate were indirectly derived from an accurate data processing of the signals measured on the Hopkinson bar surface. This method suggested by late Prof. Klepaczko, allowed for reaching the highest strain rate reported in literature for which an intriguing tensile strength increase was highlighted. This simple and efficient technique has been adopted by many researchers around the world. Some significant improvements in terms of definition and reproducibility of the incident loading pulse travelling along the bar and direct and/or contactless measurements on concrete specimens have been introduced. The very high rate sensitivity of concrete tensile strength was corroborated by the additional experimental data obtained with this experimental technique during the last fifteen years.

1. Introduction

The most important phenomenon highlighted by the experimentation on concrete dynamics undertaken during the last decades is the rate sensitivity of the material, which means that its mechanical properties – notably the tensile strength, the Young's modulus, the Poisson ratio, the critical strain – depend on the applied loading rate. The phenomenon of increase of tensile strength with the loading rate, observed in the early sixties of the last century has been recognized as the "rate effect", and is typically represented as a dynamic increase factor (DIF), i.e., the ratio of dynamic to quasi-static strength versus strain rates. Globally, the rate dependency of concrete tensile strength can be subdivided into two regimes; a regime starting from the quasi-static range of strain rates up to about $1.0\,s^{-1}$ with moderate rate effects and a regime with extensive rate effects for higher strain rates for which a DIF increased from 2 to about 7 [1]. The highest strain rates were essentially obtained with techniques based on the spalling phenomenon. Spall is defined as the ejection of fragments of a material specimen from the opposite side from which the specimen was impacted and/or impulsively loaded.

The earliest experiment at very high strain rate reported by the literature was conducted on large plain concrete specimens (51 mm diameter and 838 mm length) impacted at one end by a steel sphere. The signals were recorded by means of strain gauges cemented on concrete specimen surface. In these tests, DIFs of 2.5 and 6 for $2.0\,s^{-1}$ and $23\,s^{-1}$ were respectively highlighted [2].

This considerable strength increase was also observed about two decades later for large concrete instrumented wall panels (4,57 m × 1,65 m), each which were mounted on the side of a rectangular box reaction structure, loaded by surface explosion of TNT charge [3]. Pullback velocity of the ejected fragments measured during the test by means of high speed photography (10^3 pictures/s) allowed estimating the spall strength whereas strain rates were assessed from the strain gauges measurements cemented in the vicinity of the explosion zone of the concrete panel. For an estimated strain rate of $157\,s^{-1}$ a DIF=7 was determined. A tenfold increase of tensile strength was also previously mentioned in 1978 for concrete structures impacted by missiles [4].

It should be noted that up to the mid 90ties, the experimental data concerning the dynamic tensile behavior of concrete were rather scarce or inexistent, notably at strain rates higher than $20\,s^{-1}$. This crucial need of additional experimental data on dynamic concrete tensile behaviour led to the development in 1999 of a new experimental arrangement in the Laboratory of Physics and Mechanics of Materials (LPMM) of Metz University (France). The idea inferred from an intensive study of the existing testing methods [5], was to combine the old spall technique for reaching high strain rate combined with a Hopkinson bar as reliable measuring tool by introducing the recent developments achieved in signals processing. This testing configuration was already adopted for characterizing the dynamic tensile behavior of ceramics in the 1D framework, amongst others [6].

One of the most difficult problems encountered for such kind of design, was to find an optimal compromise between the antagonistic effects induced by the dimensions (diameters ϕ and lengths L of striker, bar and specimen) of rods when used as mechanical waveguides. Indeed, the specimen size must be from one side sufficiently large and representative of the concrete material with avoiding scale effects due to its heterogeneity (aggregates essentially), from another side shortest striker lengths

[a] Corresponding author: ahmedbrara@hotmail.com

Article available at http://www.epj-conferences.org or http://dx.doi.org/10.1051/epjconf/20159401045

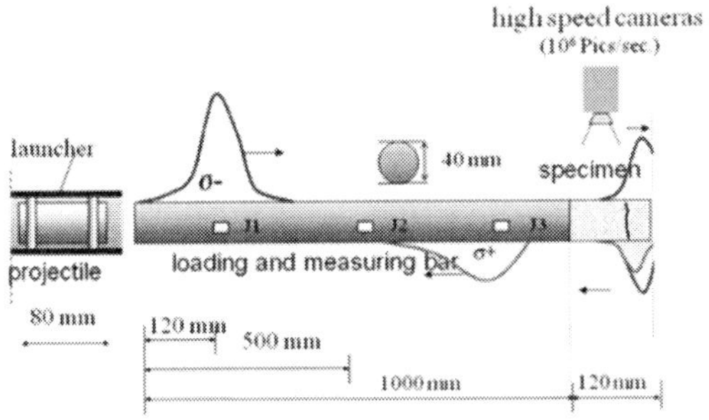

Figure 1. Scheme of experimental setup.

must be adopted in order to reach very high strain rates. However, the dynamic testing on rods (specimens and/or Hopkinson) with large diameters and short lengths (L_p) of striker (and then short wavelength λ) can introduce notable oscillations (radial and longitudinal dispersion of the propagating waves) which can spoil the measurements, and leads also to mechanical difficulties in controlling testing devices. It is known since the Davies work [7] that rod with finite lengths and short wavelengths induce in any case geometrical dispersion. From the cumulated experimental knowledge, it is generally admitted, for assuming reasonably an uniaxial longitudinal wave propagation in elastic bars, empirical conditions concerning their dimensions (length L and diameter ϕ) and wavelength λ must be satisfied, for instance [8]. Thus, the rod length-to-diameter and wavelength-to-rod ratios must respectively verify $L/\phi > 20$ for an uniaxial and homogeneous deformation along the cross section of the rod and $\lambda/\phi \sim 2L_p/\phi > 5$ for a propagation with a constant velocity $C_0 = (E/\rho)^{1/2}$ (with ρ being the mass density of the rod).

In this paper, the evolution of this method and some major modifications introduced during the last years are presented and discussed. Because of the numerous adaptations, only some representative configurations will be evoked in this paper.

2. Experimental methods based on spalling technique

2.1. The original setup developed in LPPM of Metz university

This experimental arrangement consisted of an air launcher of cylindrical projectile with a Hopkinson bar as a measuring tool and a cylindrical concrete specimen in contact with the bar, see the scheme on Fig. 1. The projectile, the Hopkinson bar made of the same aluminium alloy and the concrete specimens have the same diameter (40 mm) with respective length of 80 mm, 1000 mm and 120 mm. The incident compressive wave transmitted by the Hopkinson bar into the specimen is reflected as tensile wave causing spalling. The incident and reflected waves produced after striker impact are measured by three SR gage stations of high sensitivity cemented on the Hopkinson bar surface. The experimental technique is on the other hand combined with "up-to-date" instrumentation (adapted to dynamics strain gauges, high rate sampling oscilloscopes…) and equipment like high-speed video-cameras (10^6 pictures/sec.). The values of the chosen ratios Hopkinson bar diameter to its length $L_b/\phi_b = 25$ and wavelength to bar diameter $\lambda/\phi_b \sim 5$ correspond in fact to the lower limits for assumption of an unidimensional analysis of longitudinal wave propagation.

A special numerical procedure has been applied for analyses of wave process measured by the SR stations on the Hopkinson bar. The main task was the exact reconstitution of the transmitted wave σ_t into specimen via the interface bar-specimen by using the incident compression wave together with the reflected one in the Hopkinson bar. The signals are shifted by calculating phase velocity of their frequency components by the accurate resolution of the known Pochammer [9] – Chree [10] "frequency equation" and the correction of their geometrical dispersion of using Fourier analysis. The dynamic phase velocity in the Hopkinson bar was determined by the superposition method.

For same reasons (that is, avoiding of the effect of dispersive phenomena due to specimen dimensions and short wavelenght), indirect assessment of failure strength and loading rate by numerical simulation of the propagation of the transmitted pulse σ_t and deduction of spall strength σ_f and loading rate corresponding to the failure position Z_c, in an assumed framework of 1D analysis. The rate sensitivity of the Young's modulus E_{dyn} was calculated according to the relation already established by Toutlemonde [11],

$$E_{dyn} = E_0 + \alpha\phi(H)\log(\dot{\sigma}/\dot{\sigma}_0) \qquad (1)$$

where E_0 and $\dot{\sigma}_0$ are de quasi-static Young's modulus and loading rate respectively and α, ϕ (H) materials characteristics. More details can be found in [12] or in [13].

A large experimental program was conducted on wet and dry micro-concrete cylindrical specimens. The tested concrete was so called "micro-concrete" MB50, because of small size of the aggregates used (maximum size: 2 mm). Many experiments have been performed on this reference material within two successive research programs of the networks GEO and GEOMATERIAUX, carried out by several French laboratories [14, 15]. These important research programs conducted during the 90^{ties} were devoted to the general topic of "the behaviour of geomaterials and structures in their environment", including "behaviour of concrete in high rate dynamics" in which late Professor Janusz R. Klepaczko and his numerous co-workers in LPMM have taken an active part.

Several series of spall tests on micro-concrete specimens of diameter 40 mm and 120 mm length were performed with different impact velocities of striker. Depending on the impact velocity, one, two and even three flat failures were obtained. For the entire sample of specimens tested, no simultaneous fractures were observed. Moreover, the post mortem analysis of the broken specimens revealed only a single fracture plane, and no diffuse microfissuration was detected in its vicinity. For the data assessment, only the first fracture was taken into account, because the fracture process of specimen generates high frequency oscillations influencing strongly the residual tensile stress as experimentally observed by

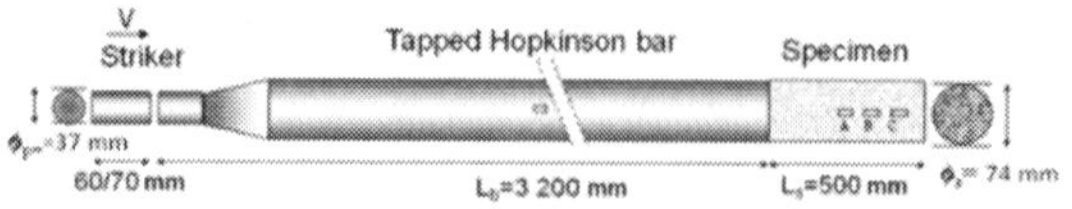

Figure 2. Experimental arrangement with tapped incident bar.

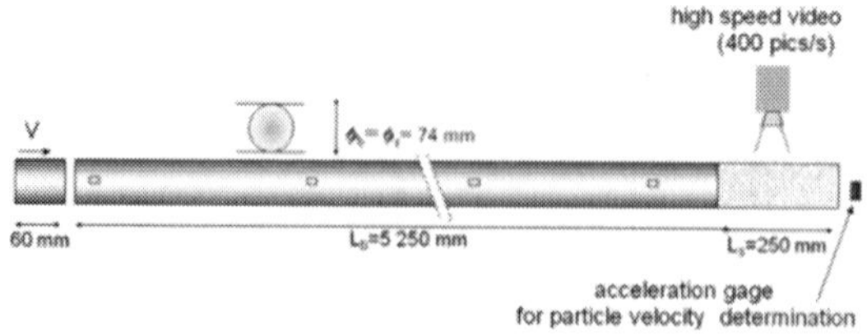

Figure 3. Experimental device with accelerometer for spall strength measurement.

Bierwirth [6] for ceramics and clearly established later by Galvez Diaz-Rubio [16], and Gomez del Rio [17]. The chronology of the fracture sequences were clearly established with the CCD captured pictures.

The experimental program by application of this technique at strain rates ranging between $20\,s^{-1}$ and $120\,s^{-1}$ has permitted to confirm very precisely the substantial increase of the tensile strength mentioned earlier by Birkimer [1] and Mac Vay [2]. Moreover, this increase was found without apparent limit, since at strain rates of about $120\,s^{-1}$, the dynamic tensile strength rises to 10 times the quasi-static strength! This strength enhancement was corroborated independently by the analysis of pictures obtained for every test by the fast video cameras. Series of captured pictures for each test during the failure process enabled to estimate the dynamic tensile strength through the fragment velocity (linearly related to the concrete impedance in 1D longitudinal uniaxial wave propagation approach) [12]. Numerical simulations of the tests with Discrete and Finite Elements Methods assuming an elastic brittle behaviour of concrete and a fracture plane were in accordance with the experimental results [13]. This confirmation via non contact measurements (video cameras) and the numerical simulations validates the reasonable assumption of an uniaxial wave propagation in the concrete specimen and the elastic-brittle behaviour of the material.

This efficient configuration to test concrete via spalling was adopted worldwide by numerous researchers with introducing some major modifications in the measurements and/or the compressive pulse generating methods, as chronologically (but non-exhaustively) reported below.

2.2. Advances in the experimental technique

The same arrangement was adopted by Wu et al. [18] consisting of a cylindrical projectile (70 mm length, ϕ 37 mm) impacting a partially taped incident pressure bar of 3200 mm length cemented to a 500 mm long cylindrical concrete specimen, Fig. 2.

The projectile length and dimensions of the taped section of the incident pressure bar were designed in order to ensure a transmitted stress wave in the concrete specimen in a repeatable form of an almost half periodic sine shaped wave of about $\sim 100\,\mu s$ duration. The concrete specimen and the incident pressure bar have the same diameter (74 mm). With the strain–time signal recorded by strain gauges glued on the concrete specimen, the dynamic tensile stress was calculated according the one-dimensional stress wave theory for viscoelastic material (however only the stress wave attenuation was taken into account in experimental data analyses, the geometrical dispersion was considered as not significant). The spall strength and strain rate at crack location were obtained by numerical simulation of the transmitted compressive pulse superposition process. For the current concrete (with a compressive strength of 34 MPa) tested, similar results as those obtained in LPMM were reported. At strain rates ranged from $20\,s^{-1}$ to about $90\,s^{-1}$, the DIF was found to vary from 2 to 9.

This experimental arrangement with partially tapped bar was also used to study the dynamic behavior of ultra-high performance cement based composites (UHPCCs) intended to defense works, with 200 MPa compressive strength [19]. The composites were prepared with natural sand and by replacing a large quantity of cement by industrial waste residues such as silica fume, fly ash and slag, and different fiber volume fractions. For these UHPCCs, DIF of 1.6 to 6.7 were found for strain rate increasing from $21\,s^{-1}$ to $66\,s^{-1}$.

To study tensile strength and fracture energy of current concrete (maximum aggregate size of 8 mm, compressive strength of 40 MPa), Schüler et al. (2006) [20] used a incident bar and notched and unnotched concrete specimens of 74.2 mm diameter with respective length of $L_b = 5500$ mm and $L_s = 250$ mm, Fig. 3. In order to measure the waves, four strain gage stations were applied at the Hopkinson bar surface, whereas the free surface velocity at the end of the specimen was determined via integration of the acceleration signal obtained with a piezoresistive accelerometer. The setup was also equipped with fast video cameras (400 pictures/sec.) in order to capture the fracture sequences.

For the data processing, almost the same procedure as in the LPMM experiment was adopted where the recorded compressive stress pulse is shifted at the bar end by taking into account the geometrical dispersion.

The wave velocity $C_0 = L_s/\Delta t$ and therefore the dynamic Young's modulus $E_{dyn} = \rho C_0^2$ of the tested concrete were calculated knowing the time Δt needed by the wave to propagate along the specimen. Having the measured free surface velocity ΔV_{pb} at the end of the specimen, the spall strength was derived as done for plate impact experiment for 1D strain conditions [21], that is

$$\sigma_f = 1/2\rho C_0 \Delta V_{pb}. \tag{2}$$

At strain rates ranging between $31\,s^{-1}$ to $68\,s^{-1}$ a DIF of 4 to 5.5 for current concrete (with 8 mm maximum aggregate size) was reported.

By means of this setup developed by Schuler et al. [20], dynamic spall experiments at strain rates varying between 140 and $180\,s^{-1}$ were conducted on notched and unnotched cylindrical specimens of strain hardening cement-based composites (SHCC) [22]. The SHCCs are characterized by

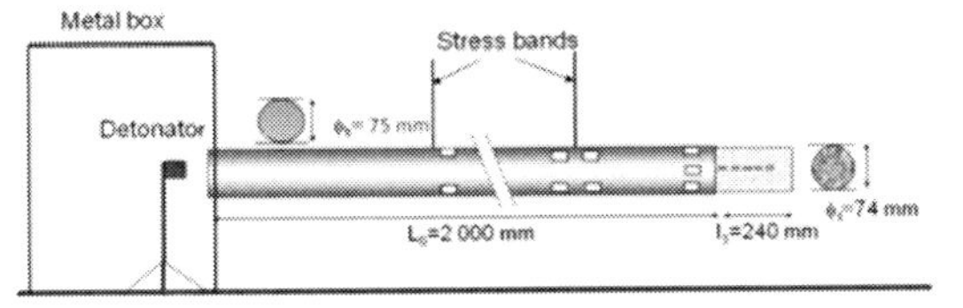

Figure 4. Setup with compressive wave generated by small explosive charge.

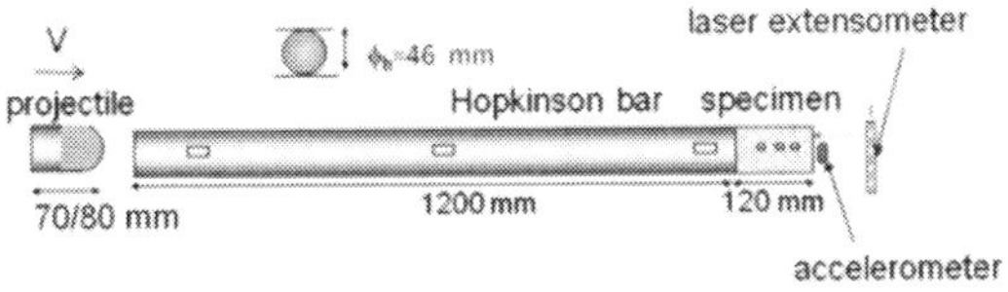

Figure 5. Spalling test with (and non-)contact measurements.

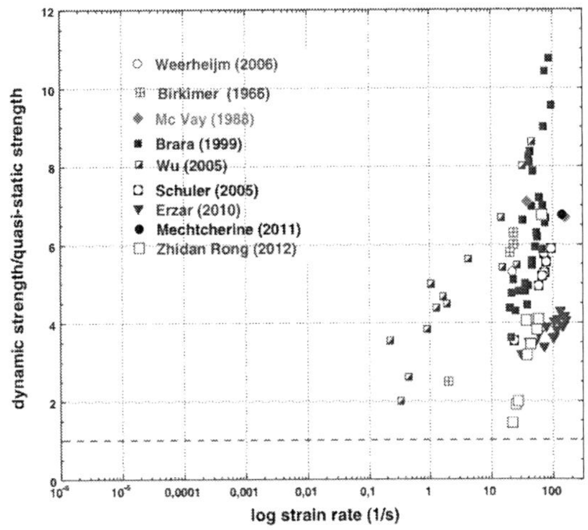

Figure 6. Dynamic increase factor as function of strain rate (log, 1/s).

very high binder and relatively high fiber volume contents in finely grained matrix composition (0.2 mm maximum aggregates size) with fly ash. At strain rate of about $\sim 140\,s^{-1}$, a DIF of 6.7 was highlighted for this material.

For the similar research aim as the precedent author, a setup based on the same principles was developed by Weerheijm et al. [23] consisting of an horizontal steel bar (length 2 000 mm, Ø 75 mm), supported by strings, at the end of which a concrete specimen of 240 mm length and 74 mm diameter was attached, Fig. 4.

The short compressive shock wave ($\sim 70\,\mu s$) was introduced into the rod by detonating a small explosive charge at other end of the bar. The loading rate and applied load were derived from the strain measurements on the steel bar and the specimen. The transmitted pulse was determined via the incident compressive pulse recorded with carbon gauges embedded in the concrete specimen and the dynamic Young modulus E_{dyn} (deduced from the wave velocity C_0 as done by Schuler et al. [20]), and calculated using the transmission time between the different strain gages locations distributed near the contact interface with the bar and along the rear half of the specimen intended to capture the failure zone. Assuming an uniaxial wave propagation, the transmitted tensile pulse was reconstructed theoretically, whereas the dynamic tensile strength was deduced from strain measurements just outside the analytically (and via exploring tests) predetermined failure zone. The geometrical dispersion was not taken into account in data processing.

A set of notched and un-notched concrete specimen were tested. The material studied was a current concrete with 8 mm maximum aggregate size and a compressive strength of 40 MPa. For the unnotched concrete specimen a DIF of 5.3 for strain rate of $20\,s^{-1}$ was found.

An experimental device that possesses almost the same features of the previous setup developed in the LPMM was rebuilt by Erzar and Forquin [24]. The dimensions of the Hopkinson and striking bars made of aluminum alloy and concrete specimens were nearly the same as in the original setup (projectiles with 46 mm diameter and 70/80 mm length, bar and specimens length of 1200 mm and 120/140 mm respectively with same diameter of 45 mm), Fig. 5. Some modifications were introduced like the use of hemispheric smooth-end projectiles in order to improve the homogeneity of tensile loading in the specimen and metrology instruments such as piezoresitive accelerometer and laser extensometers.

To measure particles velocities (and further the spall strength), laser extensometers were placed on the free end surface specimen. The wave speed C_0 was derived from the difference between the time corresponding to the minimum value of strain of the first gauge placed on the specimen near the interface and the time corresponding to the maximum value of velocity measured by the laser extensometer. Furthermore, the dynamic Young's modulus estimated from C_0 allowed for deducing the spall strength σ_f, assuming an uniaxial longitudinal wave propagation. The dynamic strength being known, the gauge signal was used to identify accurately the "failure time". The change of the strain-rate in the damaged area was time-derived from gauge signals and the strain at failure time was finally deduced. The position of failure plan responsible of velocity rebound recorded by the extensometer on the rear face of the specimen was precisely identified, starting from travel time of waves along the bar. This position permitted to locate the strain gauge for strain rate characterizing. In this investigation, the wave dispersion effects induced by the short loading pulse ($\sim 70\,\mu s$) were neglected.

For the several tests performed on a set of cylindrical MB50 micro-concrete (already studied in LPMM) specimens, relatively low values of DIF were found for strain rate varying from 3 to about 5 at strain rates ranged from $30\,s^{-1}$ to $150\,s^{-1}$.

Experimental results of the last decades are gathered on Fig. 6, where the tensile DIF versus the logarithm of strain rate is plotted. The studied material was generally dry and wet micro-, mini- or current and fiber reinforced concrete (from 2 mm to 25 mm maximum aggregate sizes), with different compressive strengths. Note that, normalizing data with respect to the quasi-static values may eliminate problems such as different maturities (relative to curing time), different mix strengths and eventual scale effects due to both specimen and aggregates sizes.

As illustrated by the Fig. 6, practically all the data obtained with the experimentation based on spalling technique show the same trend and corroborate the steep increase of the relative tensile strength at high strain rate.

2.3. Discussion

The retrospective of the development in recent years of experimentation on concrete based on the technique of spalling highlighted significant advances. Beside a better reproducibility of loading signals (via pulses shaping or loading with explosive charges), direct measurements on concrete specimens were introduced. Means of measurements with (or without) contact have been added such as accelerometers and laser extensometers next to those used previously, namely the strain gauges and fast video cameras but with largely improved performances. The accelerometer is a miniature device encapsulated in a small lightweight aluminum alloy box (less than $10\,mm^3$ in volume and weighing approximately 2.5 g) allowed to measure peak accelerations of $600\,000\,m/s^2$. Glued on the rear face of specimen, it allows, as in Schuler et al. [20] and Weerheijm et al. [23] tests, to measure the pullback velocity and next to calculate the dynamic resistance of the specimen knowing the wave velocity C_0 and the Young's modulus E_{dyn} of the concrete tested in the case of a 1-D analysis. The laser extensometer based on interferometry technique, combined with reflectors made of thin tiles of aluminium alloy glued on the rear face of the specimen, allow to directly obtaining the particles velocity. The signals recorded by the strain gauges and laser extensometers are sometime combined in order to obtain the history of wave propagation along the specimen and hence to determine their velocity C_0, and its linearly related Young's modulus E_{dyn} of the material tested, and also the failure time [24] .

It should be noted though that these new measuring instruments suffer from some limitations. Thus, the accelerometer glued to rear face of specimen, even of very small mass, may cause considerable local inertial forces when it reaches high accelerations that can locally disrupt the velocity fields or locally damage the specimen during the deceleration stage. As for the laser extensometer, the major disadvantage is that the velocity measurement is punctual (limited to a circular area of about $\sim 20\,\mu m$ diameter), which for an heterogeneous material such as concrete, composed of relatively large aggregates and sand in a cement matrix can represent an important source of uncertainty depending on where the laser is pointing. The same disadvantage with nonetheless less severity characterizes also measures with strain gauges cemented on concrete surface (or embedded in).

Concerning the data processing, spall strengths and the strain rates were deduced in some modified experimental arrangement, as in the original one, by numerical simulation of the waves superposition process in the specimen, knowing the fracture position experimentally determined [18,20]. In other experimental setups, strain rates were deduced from direct measurements obtained with strain gauges cemented on specimen surface. It is of course problematic to perform accurate strain measurement, because the sensor device should be located at the exact failure section. Even in this case, it will probably fail at the fracture instant without capturing the complete strain pulse. To overcome this difficulty, strain rates were in this case indirectly estimated from signals measured by the strain gauges located near the failure surface [23,24]. The question that arises naturally with this kind of processing is the level of precision of these data knowing that the value of net tensile pulse and the strain rate resulting from superposition process of the transmitted compressive wave can changes considerably along very short travelled distances. This approximation could explain the relative DIF's underprediction at very high strain rate found by Erzar and Forquin [24] when compared to those of other aforementioned authors, Fig. 7. The time deriving of strain-rate from gauge signals in an indirectly estimated position in the damaged area seems rather hazardous.

The need to deal with a representative concrete material to be studied in dynamics in the framework of 1D analysis of wave propagation explain to some extent the quasi-systematic choice of large dimensions of rods (respectively $240\,mm \div 1000\,mm$ and $2000\,mm \div 5250\,mm$ lengths for the specimen and the Hopkinson bar, for diameters of $45\,mm \div 75\,mm$). The dynamic compressive loading were generated by means of strikers (60–80 mm length) or detonation generating short lived pulses of $T = 70 \div 100\,\mu s$ duration, and with typical rise and fall times ranging from 20 to $50\,\mu s$. The ratio wavelength-to-specimen diameter takes then a value $\lambda/\phi \sim 1.2$, which let the assumption of a uniaxial propagation of compressive wave with a constant velocity C_0 rather questionable. Indeed, it was early established that impulsive loadings, with predominant high frequency components, associated with brittle failures may greatly increase dispersion (attenuation) effects on elastic (visco-elastic) wave propagation in the rods [7,12,16,17,25,26]. Because of the geometrical dispersion, the signals measured locally via strain gauges cemented on specimen surface and/or laser extensometer could be strongly distorted and attenuated even in a short traveled distance, and non-uniform along the specimen cross section.

Knowing that the dispersion phenomena have great importance in the data analysis of spalling tests of cylindrical bars, the correction procedures should applied be systematically and the approximation level of the local measurements and/or their extrapolations must also be carefully checked in order to get more precise and consistent results.

3. Conclusions

In this paper, a retrospective of the last fifteen years concerning the experimentation for assessing concrete dynamic tensile strength, based on the spalling technique developed in Metz University, highlighted to following main points

- Because its simplicity and efficiency this experimental technique becomes a reference in testing concrete at high strain rates in tension.
- During the last years, this experimentation has known remarkable advances by the introduction of

modifications like the pulse shaping and the use of detonation charge for a better reproducibility of loading signals, and also data processing based on recorded signals via strain gauges cemented on concrete specimens surfaces and/or contact (or non) measurements (accelerometer, laser extensometer, fast video).
- To deal with a representative concrete material, relatively large specimen diameter were tested, inducing necessarily the adoption of relatively long bars, for analyse on the 1D uniaxial wave framework. However, the geometrical dispersive phenomena induced by finite bars and short dynamic loading with high frequency content, are generally not taken into account in the data processing.
- The data gathered support clearly the considerable and puzzling strength increase of 600% observed six decades earlier for strain rate greater than $1.0\,s^{-1}$.

The experimentation is significantly improved by the introduction of direct measurements on concrete loaded specimens. However, the dispersive phenomena must be imperatively taken into account in the data processing. More research efforts are needed to improve the precision of measurements.

References

[1] L.J. Malvar, C. A. Ross, ACI Mat. J. **95** (6), 735–739 (1998)
[2] Birkimer, D.L., Ph.D. thesis, University of Cincinnati (1968)
[3] Mc Vay M.K., U. S. Army Corps of Engineers, Waterways Experiment Station, Vicksburg, Technical Report SL-88–22 (1998)
[4] Gupta Y. M., Seaman L., *Nuclear Eng. and Design* **45** (2), 507–514 (1978)
[5] Brara A., Klepaczko J. R.; Kruszka L., in: *Proceedings Brittle Matrix Composites 5 (BMC5)*, Brand, Li, Marshall Eds., Warsaw, Poland (1997), 281–290
[6] Bierwirth S., Najar J., VDI Fortschritt-Berichte n° 148, reihe **18**, 1–148 (1994)
[7] R. M. Davies, Phil. Trans. of the Roy. Soc. of London. Series A, Math. and Phys. Sci. (1948)
[8] Kolsky H. *Stress waves in solids*, (New York: Dover Publications; 1963)
[9] Pochhammer L., J. Reine Angew. Math., **81**, 324–36 (1876)
[10] Chree C., Cambridge Philos. Soc. Trans, **14**, 250–369 (1889)
[11] Toutlemonde F., PhD Thesis, Lab. of Public Works (LCPC), Paris, 1994 (in French)
[12] Brara A., Ph.D. thesis, Lab. of Phys. and Mech. of Mat. (LPMM), Metz 1999 (in French)
[13] Klepaczko J. R., Brara A., Int. J. Imp. Engng, **25**, 387–409 (2001)
[14] Bailly P., Lab. Netwok GEO, ENI , Bourges final report, Part 1 (1999)
[15] Darve F., Hicher P.Y., Reynouard J-M. *Mécanique des geomateriaux*, (Hermes Pub., Paris, 1995)
[16] F. Galvez Dıaz-Rubio, J. Rodrıguez Perez, V. Sanchez Galvez, Int. J. Imp. Engng, **27**, 161–177 (2002)
[17] Gomez del Rıo T. Rodrıguez Perez J., Int. J. Imp. Engng, **34** (2), 377–393 (2007)
[18] Wu H., Zhang O., Huang F., Jinb O., Int. J. Imp. Engng, **32**, 605–617 (2005)
[19] Zhidan Rong and Wei Sun, Construction and Building Materials, **31**, 168–173 (2012)
[20] Schüler H., Mayrhofer C., Thoma K., Int. J. Imp. Engng, **32**, 1635–1650 (2005)
[21] Novikov SA, Divnov II, Ivanov AG Fizika Metallov i Metallovedeniye **21**, 4 (1966)
[22] V. Mechtcherine, O. Millon, M. Butler, K. Thoma, Cem. & Conc.Comp. **33**, 1–11 (2011)
[23] Weerheijm J., Van Doormaal J.C.A.M., Int. J. Imp. Engng, **34** (3), 609–626 (2006)
[24] B. Erzar and P. Forquin, Experimental Mechanics **50**, 941–955 (2010)
[25] Franz, C., Follansbee, P.S., J. Eng. Mater. Technol. **105**, 61–66 (1983)

EPJ Web of Conferences **94**, 01046 (2015)
DOI: 10.1051/epjconf/20159401046

Effect of temperature and strain rate on the compressive behaviour of supramolecular polyurethane

Xuegang Tang[1,a], Clive R. Siviour[1], C. Paul Buckley[1], Antonio Feula[2], and Wayne Hayes[2]

[1] Department of Engineering Science, Oxford University, Parks Road, Oxford OX1 3PJ, UK
[2] Department of Chemistry, University of Reading, Whiteknights, Reading RG6 6AD, UK

Abstract. Supramolecular polyurethanes (SPUs) possess thermoresponsive and thermoreversible properties, and those characteristics are highly desirable in both bulk commodity and value-added applications such as adhesives, shape-memory materials, healable coatings and lightweight, impact-resistant structures (e.g. protection for mobile electronics). A better understanding of the mechanical properties, especially the rate and temperature sensitivity, of these materials are required to assess their suitability for different applications. In this paper, a newly developed SPU with tuneable thermal properties was studied, and the response of this SPU to compressive loading over strain rates from 10^{-3} to 10^{4} s^{-1} was presented. Furthermore, the effect of temperature on the mechanical response was also demonstrated. The sample was tested using an Instron mechanical testing machine for quasi-static loading, a home-made hydraulic system for moderate rates and a traditional split Hopkinson pressure bars (SHPBs) for high strain rates. Results showed that the compression stress-strain behaviour was affected significantly by the thermoresponsive nature of SPU, but that, as expected for polymeric materials, the general trends of the temperature and the rate dependence mirror each other. However, this behaviour is more complicated than observed for many other polymeric materials, as a result of the richer range of transitions that influence the behaviour over the range of temperatures and strain rates tested.

1. Introduction

The mechanical behaviour of polymeric materials has long been a topic of keen interest for scientists across many disciplines, and it is well-known that strain rate and temperature have significant influence on the mechanical properties [1,2]. Depending on their (processing) behaviour at elevated temperatures, normally, polymers are divided into three main classes: thermoplastics, thermosets and rubbers. Among them, rubbers, or alternatively elastomers, are soft, flexible and elastic materials at room temperature and are applied in a wide range of applications, such as tires, tubes, seals and gloves. Crosslinking is necessary to obtain the typical rubber properties, such as elasticity, high toughness and solvent resistance, since this severely restricts the motion of the chain molecules. Currently, the main industrial techniques for rubber cross-linking are sulfur vulcanization and peroxide curing, which results in a three-dimensional network via sulfur bridges in the first case and C-C bonds in the second. However, the formation of an irreversible network upon cross-linking prevents melt (re)processing of both thermosets and elastomers, and this gives them a distinct processing disadvantage compared to thermoplastics [3].

Supramolecular polymers, a newly-developing group of materials, can be loosely defined as those in which the monomers are held together by noncovalent interactions [4]. Therefore, a reversible network can be formed, which is sensitive to temperature, and can be destroyed and reconstructed easily during processing and post-processing. This provides those materials with unique thermoresponsive and thermoreversible properties, thereby delivering dramatic viscosity changes over well-defined and tunable temperature ranges: at room temperature they behave like elastomers, but at elevated temperature are low viscosity liquids. These characteristics offer huge advantages in processing and also provide unique properties, which are highly desirable in both bulk commodity and valued-added applications. By understanding the mechanical properties of these materials and the microscopic mechanisms controlling their mechanical behaviour, we will be better able to design supramolecular polymers and supramolecular polymer-based systems for specialized application. However, the mechanical properties, especially at high strain rate have been relatively uncharacterized to date.

In this study, a newly-developed supramolecular polyurethane (SPU) through hydrogen bond and pi-pi stacking was fabricated, and the effects of temperature and strain rate on the behaviour under compression were investigated. SPUs are sensitive to temperature, and the response at elevated temperatures is governed by the dissociation of non-covalent band such as hydrogen bonds and π-π stacking, as well as the time temperature superposition principle for traditional polymers [5]. These effects were considered when describing the rate and temperature dependence of the material, which was not expected to follow the usual form associated with 'traditional' polymeric materials.

[a] Corresponding author: xuegang.tang@eng.ox.ac.uk

Article available at http://www.epj-conferences.org or http://dx.doi.org/10.1051/epjconf/20159401046

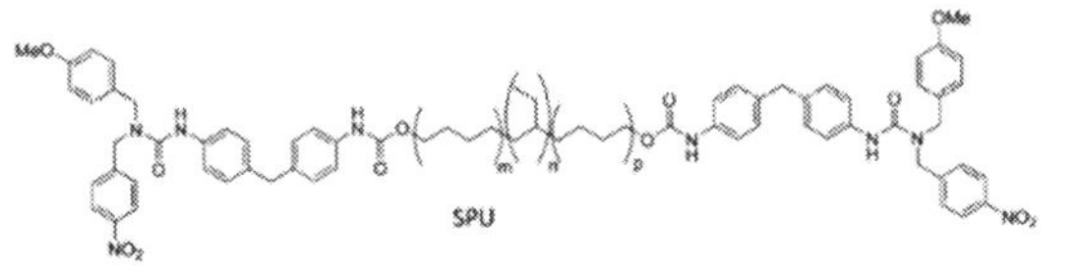

Figure 1. The chemical structure of SPU.

2. Experimental procedure

2.1. Chemical structure of SPU

Previous research suggests that hydrogen bond and π-π stacking can be introduced to polymer to form non-covalent interaction among polymer chains [6–9], and those non-covalent interactions present significant temperature sensitivity, which shows rich transitions at narrow temperature and frequency range. Therefore, those materials offer a unique chance to investigate the transitions of polymers, especially the effect of those transitions on stress-strain behaviour, at narrow temperature and/or frequency regime compared with traditional polymers such as polycarbonate and poly(methyl metacrylate) etc., which normally need large temperature and frequency range to observe those transitions [10–12].

In this study, a supramolecular polyurethane (SPU) was synthesized by an established approach using Krasol HLBH-P2000 and the diisocyanate MDI with the control of end group to achieve π-π stacking and hydrogen bond, at the University of Reading [8], and the specific chemical structure is shown in Fig. 1. Both Krasol HLBH-P2000 and the diisocyanate prepolymer are thick viscous liquids at room temperature whereas in contrast fabricated SPU is a solid material indicating the crucial role of the formation of supramolecular networks. Detail procedures of the synthesis and structure characterization will be reported in a separate paper.

2.2. Sample preparation

The fabricated SPU was moulded as cylindrical compression samples at the temperature of 50 °C for half hour with the diameter of 5mm and thickness of about 2.5 mm. The size of the test samples was based on previous work on compression testing, especially at high strain rate. This suggested that length-to-diameter ratios of 1:2 or less are desirable in the testing of low-impedance materials on high rate test systems, in order to provide adequate strain gage signals and improve the time taken to reach static stress equilibrium. Also, a length-to-diameter ratio of 1:2 has been shown to be optimal in negating the effects of radial and longitudinal inertia in the specimen [10, 13, 14].

2.3. Experimental set-up

The samples were tested across a wide range of strain rates from 10^{-3} to $10^{4}\,s^{-1}$, at room temperature (RT, 19.0 ± 0.5 °C) and a temperature range from −30 °C to 30 °C at the strain rate of $10^{-2}\,s^{-1}$. An Instron (model 5982) was used for quasi-static loading at the strain rates from $10^{-3}\,s^{-1}$ to $10^{-1}\,s^{-1}$. An environmental chamber with liquid nitrogen attachment was used during tests for temperature control. The specimens were loaded using hardened steel anvils. Standard Instron load cells were used to measure the forces applied to the specimen, whilst a clip extensometer was used to measure the displacement of the anvils, from which specimen strain was calculated, in addition a designed feedback loop between the extensometer and the actuator was used to ensure constant true strain rate control over the duration of the tests.

Medium rate experiments were performed using an in-house hydraulic system built at Oxford University to access strain rates between 1 and $50\,s^{-1}$. A strain gauge based load cell with a high natural frequency has been designed, and is against a rigid steel plate. A Linear Variable Differential Transformer (LVDT) was used to measure the displacement of the sample, from which strain can be calculated.

High strain rate experiments were performed using a split-Hopkinson bar system (SHPB). The SHPB consists of a 1000 mm long input bar and a 500 mm long output bar. The striker bar is 500 mm long, and all bars are 12.7 mm diameter. Both the input and output bars are instrumented with foil strain gauges, the input bar halfway along its length, and the transmitted bar 50 mm from the bar specimen interface. Magnesium alloy bars were used in this work for the test. This low stiffness, high strength material gives large gauge signals even when testing soft polymers, but are strong enough not to yield at the impact speeds required for the higher strain rate experiments. The tests and calculation was based on reference handbooks [15, 16].

All specimens were lubricated with a thin layer of paraffin wax on both faces to avoid barrelling effect.

Parallel plate oscillatory shear was performed with an Anton Paar Physica MCR 301 rheometer. For the temperature sweep at a single frequency of 5 Hz with the deformation of 0.1% to guarantee it's in the linear viscoelastic region, the samples were placed on the rheometer with well contact between parallel plates. The sample were then subjected to a cooling temperature ramp of 2 °C/min to −30 °C, and held at this temperature for 5 minutes. After this step they were heated up to 60 °C at 2 °C/min. For the isothermal frequency sweeps, the samples were again placed in the rheometer and maintained at the selected temperature for 5 minutes, and then the frequency sweeps were performed between 0.1 and 100 rad/s at a strain amplitude of 0.1 %. Three cycles of tests were measured, and the results show excellent reproducibility. During all of these steps the dynamic shear moduli (G', G") of the samples were recorded.

3. Results and discussion

3.1. Temperature dependence

Figure 2 shows the variation of complex shear modulus. It can be seen that the SPU presents the transition zone to flow, plateau zone (rubbery), and the transition zone to glassy behaviour from 60 to −30 °C, which are: lower than −20 °C, from −20 to 10 °C, and higher than 10 °C respectively. Especially, when the temperature is higher than 30 °C, a rapid drop of storage modulus can be observed owing to the dissociation of the supramolecular

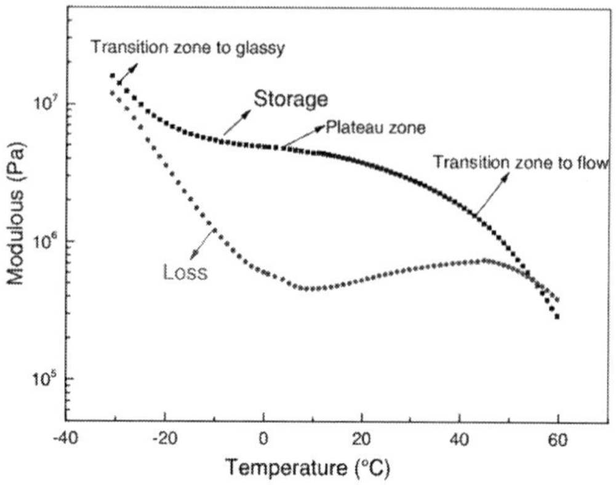

Figure 2. Variation of complex shear modulus with temperature at the frequency of 5 Hz.

polyurethane network, resulting in a large degree of softening. This shows that for SPU, rich transitions occur in a narrow temperature range, and may indicate different stress-strain behaviour with those transitions.

The compressive stress-strain curves of SPU at the strain rate of $10^{-2}\,s^{-1}$ with different temperature are shown in Fig. 3a. Generally, the compressive stress-strain response of SPU is similar to other elastomers, but it can also be observed that temperature affects the stress-strain behaviour of the SPU significantly. For example, at a temperature of 30 °C, after an initial deformation, the stress-strain curve is flat; there is a large flat region in the stress-strain curve following. However, with decreasing temperature, strain hardening at large deformations become more and more obvious. The compressive stress-strain behaviour can be divided into three categories: 1) a general flat regime at large deformation after initial deformation (about 25% strain) for the temperature from 30 to 10 °C; 2) almost linear behaviour during the whole deformation from 0 to −20 °C; and 3) a significant increase in stress at large deformations when the temperature is below −30 °C. In order to parameterise these curves for comparison to other data sets, the stress at 25% strain is plotted against temperature, Fig. 3b. These data show an almost linear increase of strength as the temperature drops from 30 to 0 °C, then a plateau down to −20 °C, with strength again increasing as the temperature drops further.

The observed behaviour agrees well with the rheometer test results, corresponding to the three distinguish physical status of SPU: transition zone to flow, plateau zone (rubbery), and transition zone to glassy behaviour at different temperature. Overall, the data strongly demonstrate the temperature sensitivity of the SPU.

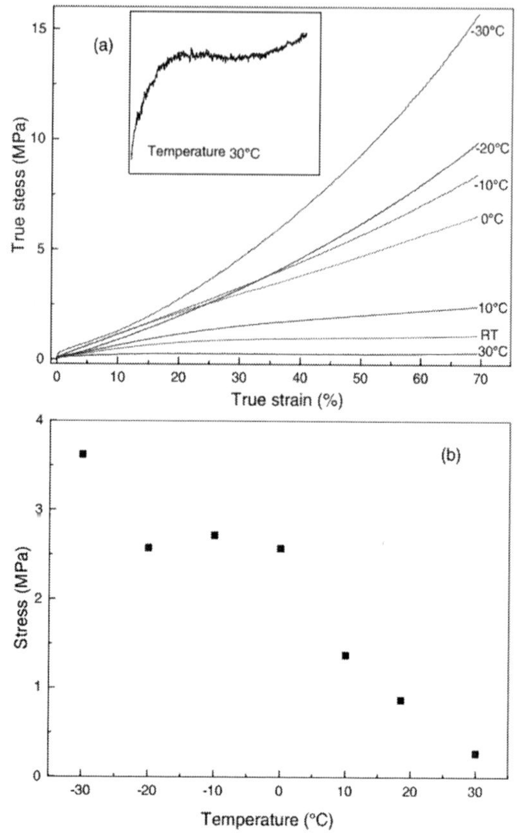

Figure 3. (a) Compressive stress-strain response of SPU and (b) stress (at 25% strain) at the strain rate of $10^{-2}\,s^{-1}$ with different temperature.

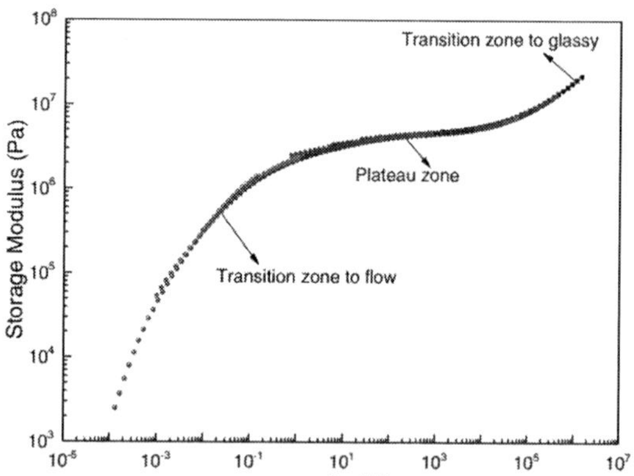

Figure 4. Variation of complex shear modulus with temperature frequency (master curve at the reference temperature of 20 °C). Colours indicate different temperatures.

3.2. Rate dependence

Polymers, due to their viscoelastic nature, present behaviour with both temperature and time (frequency) dependence during deformation. Time/temperature superposition, based on the equivalency between time (or frequency) and temperature, usually is applied to polymers, which suggests that the viscoelastic variations occur relatively quick at higher temperatures can be made to appear as if they occur at longer times or lower frequencies simply by shifting the data with respect to time (or frequency) [5]. In this work, frequency sweeps from temperatures −30 to 50 °C were used to generate a master curve exploiting the time/temperature superposition principle to predict the frequency dependence of SPU over a large frequency range, far beyond the reach of

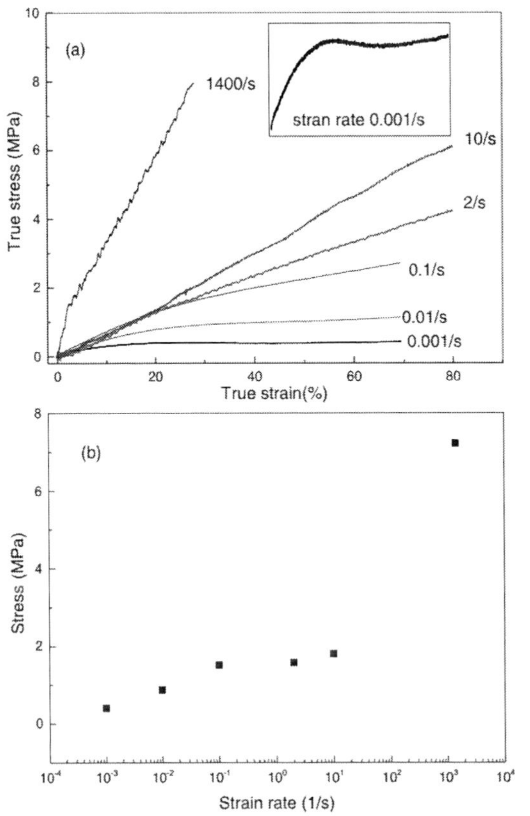

Figure 5. (a) Compressive stress-strain response of SPU and (b) stress (at 25% strain) at room temperature with different strain rate.

the test machine. Results show that time/temperature superposition works well for SPU, indicated by the satisfactory overlapping achieved during the shifting of the curves. The corresponding master curve is shown in Fig. 4. Similar to the temperature dependence, the master curve of frequency also shows three distinguish status with the variation of frequency: transition zone to flow, plateau zone (rubbery), and the transition zone to glassy behaviour at a narrow frequency range.

Figure 5a provides a comparison of representative high-rate true stress-true strain curve with representative low to moderate rate curves, which suggests significant changes of the deformation behaviour, indicating the high sensitivity of the SPU to strain rate. For example, at the lowest strain rate ($10^{-3}s^{-1}$), the stress-strain behaviour of SPU is similar to typical thermoplastic polymer: showing an initial elastic region, followed by a non-linear elastic region and yield (at a strain of about 25%), then strain softening followed by strain hardening [17]. However, as the strain rate increases, even still at low strain regime (10^{-2} and $10^{-1}s^{-1}$), yielding and strain softening disappear, and the behaviour is similar to traditional thermoplastic polyurethanes [18]. At moderate rate, the curves present linear behaviour during the whole deformation.

Figure 5b shows SPU stress (at the strain of 25%, which corresponds the yield point of SPU at the lowest strain rate) as a function of strain rate across all of the tested strain rates. Within the low rate regime, the stress is found to increase linearly with the logarithm of strain rate, indicating that the behaviour in this regime could be accurately explained and/or predicted in terms of a single activated process, which is in agreement with some previous works on polymers [10]. However, it is interesting to observe that there is a plateau regime at moderate rate, and followed by a sharp jump at high strain rates. Therefore, a bilinear behaviour as a function of strain rate is not enough to describe the rate dependence of the SPU. This is maybe due to the changes of response of SPU on strain rate: at low rate, it behaviours more like thermoplastic, and with the rising of strain rate, it behaviours more like rubber, and finally a transition to glassy. To some extent, these observations are entirely consistent with the observed temperature dependence, both of which exhibit an increase in the supported stress, then a plateau then a further increase. However, it is noted that the observed stresses and the shapes of the stress strain curves at the lowest rates are not the same as those at the highest temperatures (e.g. no strain softening at the highest temperature of 30 °C), and this will need further investigation.

4. Conclusions

A supramolecular polyurethane (SPU), which forms networks through hydrogen bonding and pi-pi stacking has been experimentally studied in compression at varying strain rates and temperatures, and in oscillatory shear rheometry. Time/temperature superposition was employed to construct the master curve for SPU to predict the properties at large frequency range. Dynamic rheometer results indicate that SPU presents a transition zone to flow, plateau zone (rubbery), and transition zone to glassy behaviour from 60 to −30 °C. Compression tests show that the SPU presents significant temperature and rate dependence, which is in agreement with the transitions observed in the rheometry. The results also suggest that a plateau regime has to be incorporated into the bilinear behaviour model to describe the rate dependence for SPU from low strain rate to high strain rate. However, additional testing and analysis is required to confirm these trends.

The authors would like to thank EPSRC for support (EP/J010715/1 and EP/J011436/1), in addition to the University of Reading for access to instrumentation in the Chemical Analysis Facility and Cray Valley for the supply of Krasol HLBH-P2000.

References

[1] E.M. Arruda, M.C. Boyce, R. Jayachandran, Mech. Mater. **19**, 193 (1995)

[2] L.C.A. van Breemen, T.A.P. Engels, E.T.J. Klompen, D.J.A. Senden, L.E. Govaert, J. Polym. Sci. Part B Polym. Phys. **50**, 1757 (2012)

[3] V.D. Mee, MAJ (Mark), Technische Universiteit Eindhoven (2007)

[4] A.W. Bosman, R.P. Sijbesma, E.W. Meijer, Mater. Today **7**, 34 (2004)
[5] J.D.V. Ferry, Viscoelastic properties of polymers (Wiley, New York, 1980)
[6] S. Burattini, B.W. Greenland, D.H. Merino, W. Weng, J. Seppala, H.M. Colquhoun, W. Hayes, M.E. Mackay, I.W. Hamley, S.J. Rowan, J. Am. Chem. Soc. **132**, 12051 (2010)
[7] S. Burattini, H.M. Colquhoun, J.D. Fox, D. Friedmann, B.W. Greenland, P.J.F. Harris, W. Hayes, M.E. Mackay, S.J. Rowan, Chem. Commun. **44**, 6717 (2009)
[8] P.J. Woodward, D.H. Merino, B.W. Greenland, I.W. Hamley, Z. Light, A.T. Slark, W. Hayes, Macromolecules **43**, 2512 (2010)
[9] P. Cordier, F. Tournilhac, C. Soulie-Ziakovic, L. Leibler, Nature **451**, 977 (2008)
[10] A.D. Mulliken, M.C. Boyce, M, Int. J. Solids Struct. **43**, 1331 (2006)
[11] C.R. Siviour, S.M. Walley, W.G. Proud, J.E. Field, Polymer **46**, 12546 (2005)
[12] A.D. Mulliken, in: Massachusetts Institute of Technology. Dept. of Mechanical Engineering (Massachusetts Institute of Technology, 2006)
[13] N.N. Dioh, P.S. Leevers, J.G. Williams, Polymer **34**, 4230 (1993)
[14] C.M. Roland, Rubber Chem. Technol. **79**, 429 (2006)
[15] G.T. Gray III, in *ASM Handbook. Vol. 8*, ASM International: Materials Park, Ohio, 2000, pp. 462–476
[16] G.T. Gray III, W.R. Blumenthal, in *ASM Handbook. Vol.8*, ASM International: Materials Park, Ohio, 2000, pp. 488–496
[17] J.L. Jordan, C.R. Siviour, B.T. Woodworth, Shock Compression of Condensed Matter – 2011, Pts 1 and 2, 1426 (2012) 4
[18] H.J. Qi, M.C. Boyce, Mech. Mater. **37**, 817 (2005)

EPJ Web of Conferences **94**, 01047 (2015)
DOI: 10.1051/epjconf/20159401047

The compressive behaviour and constitutive equation of polyimide foam in wide strain rate and temperature

Akifumi Yoshimoto[a], Hidetoshi Kobayashi, Keitaro Horikawa, and Kenichi Tanigaki

Grad. Sch. Of Eng. Sci., Osaka University, Toyonaka, Osaka 560-8531, Japan

Abstract. These days, polymer foams, such as polyurethane foam and polystyrene foam, are used in various situations as a thermal insulator or shock absorber. In general, however, their strength is insufficient in high temperature environments because of their low glass transition temperature. Polyimide is a polymer which has a higher glass transition temperature and high strength. Its mechanical properties do not vary greatly, even in low temperature environments. Therefore, polyimide foam is expected to be used in the aerospace industry. Thus, the constitutive equation of polyimide foam that can be applied across a wide range of strain rates and ambient temperature is very useful. In this study, a series of compression tests at various strain rates, from 10^{-3} to $10^3\,s^{-1}$ were carried out in order to examine the effect of strain rate on the compressive properties of polyimide foam. The flow stress of polyimide foam increased rapidly at dynamic strain rates. The effect of ambient temperature on the properties of polyimide foam was also investigated at temperature from −190 °C to 270 °C. The flow stress decreased with increasing temperature.

1. Introduction

Polymer foams, such as polyethylene foam and polystyrene foam, have excellent shock absorption and thermal insulation performance. Because of their properties they have been used as a shock absorber or thermal insulator in various daily situations [1,2]. Therefore, there are many studies focused on constitutive relationships of polymer foams [3–5]. Compressive properties at dynamic strain rates are very important for shock absorbers; therefore, many researchers have studied energy absorption performance of polymer foams at dynamic strain rates by using Split Hopkinson Pressure Bar (SHPB) equipment [6,7]. In general, these foams are suitable enough to absorb impact energy at room temperature. Over 100 °C, however, most polymer foams have insufficient strength as a shock absorber because their glass transition temperature (T_g) is lower than 100 °C.

Polymer foam made of polymer that has a higher T_g, such as polyimide, is suitable for use at high temperatures. Polyimide has a higher T_g(> 300 °C) and its mechanical properties do not vary greatly, even in low temperature environments. Therefore, polyimide foam is likely to be used in the aerospace industry and other fields [8]. The mechanical properties of polyimide foam filled honeycombs and properties of standalone polyimide foam at elevated temperatures were also examined [9,10]. Thus, determining the constitutive relationship of polyimide foam that can be applied across a wide range of strain rates and temperatures is very useful.

In this study, the effects of strain rate and temperature on the constitutive relationship of polyimide foam were experimentally examined by carrying out a series of compression tests. We performed compression tests at various strain rates ($10^{-3} \sim 10^3\,s^{-1}$) using three testing devices. The effect of temperature was investigated via high or low temperature tests using an electric furnace and liquid nitrogen.

2. Experiment

2.1. Materials and specimen

Polyimide foam specimen is made by being expanded a volume of polyimide to about hundred times firstly. Density of this foam is approximately 12 kg/m^3. By compressing this foam in one direction while applying heat, two types of polyimide foam specimens were made. In this molding process, open-cell structure with small pieces of thin sheets and short fibers, shown in Fig. 1, piled up in the compressive direction. Because of this structure, polyimide foam specimens in this study show pronounced anisotropy on the mechanical properties. Here, the direction in which small pieces (sheets and fibers) piled up was defined as the z-direction. The normal directions to the z-direction were defined as the x- and y-directions. Two kinds of polyimide foam specimen with different densities of 120 and 240 kg/m3 were provided by Industrial Summit Technology Co., Ltd. (I.S.T.). They have two different foaming magnifications of 10 and 5, denoted by Foam-A and Foam-B, respectively. We prepared cubic specimens with 20 mm sides. These specimens geometries are shown in Fig. 2(a).

To investigate the strain rate and temperature dependence of polyimide itself, the same base material used in the foam specimens, we prepared additional, bulk polyimide specimens. That specimen geometry is shown in Fig. 2(b). This second specimen was a cylindrical

[a] Corresponding author: akifumi.yoshimoto@impact.me.es.osaka-u.ac.jp

Article available at http://www.epj-conferences.org or http://dx.doi.org/10.1051/epjconf/20159401047

Figure 1. Microstructure of polyimide foam.

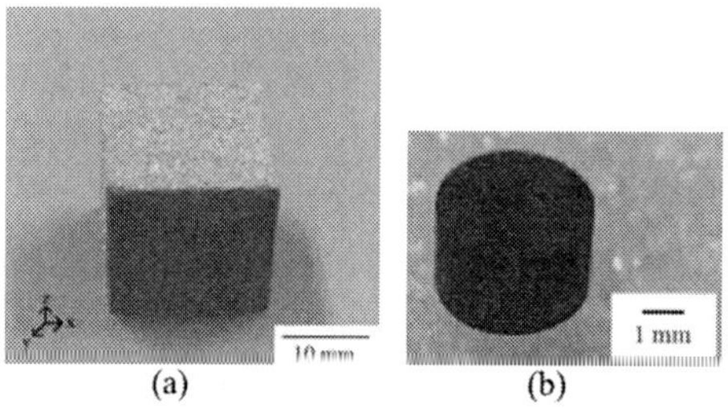

Figure 2. Specimen for compression tests. (a): Polyimide foam (b): Bulk-specimen.

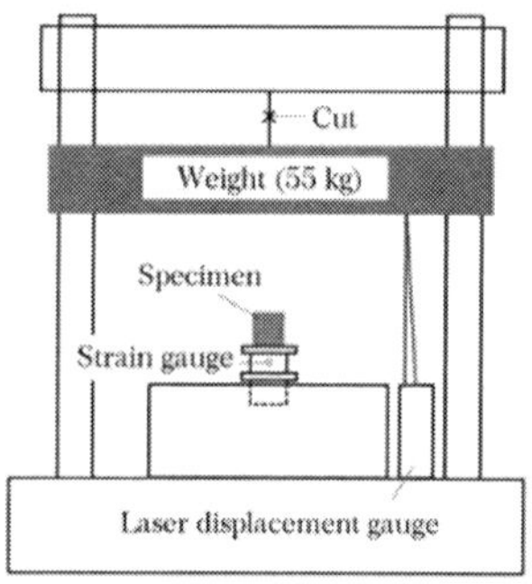

Figure 3. Dropping weight testing machine.

with a diameter of 5 mm and a height of 3 mm. This specimen is molded in the same way as the foam specimen. Compressive deformation is added until its density reached the same as polyimide.

2.2. Compression tests

By using a universal testing machine (Shimazu autograph), quasi-static compression tests were carried out. The strain rates evaluated were 1.7×10^{-3} and $1.7 \times 10^{-1}\,s^{-1}$. In high temperature quasi-static tests at 200 °C and 270 °C, specimens was heated to a desired temperature by an electric furnace. These temperature were determined within a range that does not exceed the glass-transition temperature T_g (= 310 °C) of polyimide foam. The heated specimen was compressed after being kept in the furnace at the desired temperature for fifteen minutes. In low temperature quasi-static tests, specimens were cooled to −190 °C by liquid nitrogen in a polystyrene foam box attached to the apparatus. Similarly, after maintaining the target temperature for fifteen minutes, the specimen was compressed. The same compression tests were also carried out on the Bulk-specimen.

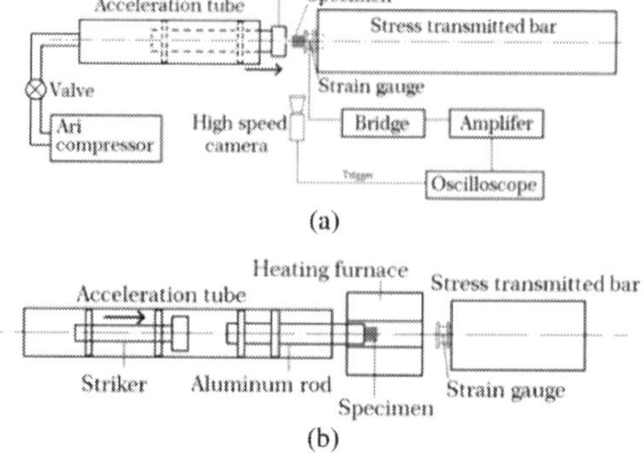

Figure 4. Dynamic test apparatus. (a): room temperature test (b): high temperature test.

A dropping weight testing machine, shown in Fig. 3, was used in intermediate speed tests. A dropping weight of 55 kg falls along a rigid frame and compresses the specimen. Its strain rate is approximately 50 s^{-1}. The deformation of the specimen was measured by laser displacement gauge. Load was measured by a special load cell, that contains a small detective section and stress transmitted block. This method when used in dynamic tests takes a relatively long time to apply the load to the specimen [11].

The dynamic compression test apparatus is shown in Fig. 4(a). A striker in an acceleration tube is fired by compressed air and impacts on the specimen which is attached to the special load cell. Its strain rate is about $1.0 \times 10^3\,s^{-1}$. The load was measured by the same method as that used in the dropping weight tests. The deformation of the specimen was measured by using a high-speed video camera (Phantom V4.2) with 25,000 frames per second.

Dynamic tests were carried out at high and low temperatures as well as at room temperature. Test temperatures were room temperature, 200 °C, 270 °C and −190 °C. The setup for dynamic tests at high temperature is shown in Fig. 4(b). Specimens were attached to the aluminum rod and heated to the desired temperature by an electric furnace and fired by impacting with a striker.

Its strain rate is about $250\,s^{-1}$. This method was adopted to avoid the elevated temperatures contacting the semiconductor strain gauge on the load cell. Low temperature dynamic tests were carried out by using the same apparatus. Before the test, specimens were cooled in a refrigerant box next to the testing machine. After a specimen was removed from the refrigerant box, it was attached to the aluminum rod immediately and compressed by striking the load cell.

This setup is used for tests that required a relatively long time to deformation and it is not suitable for small specimen like the Bulk-specimen because of the limited resolution of the high-speed video camera. Therefore, a Split Hopkinson Pressure Bar (SHPB) apparatus, as shown in Fig. 5, was used in dynamic tests for Bulk-specimens. The striker, input bar and output bar are made of SUS304 with a diameter of 30 mm. The nominal stress σ_n, nominal strain ε_n and strain rate $\dot{\varepsilon}$ relationship of the specimen in

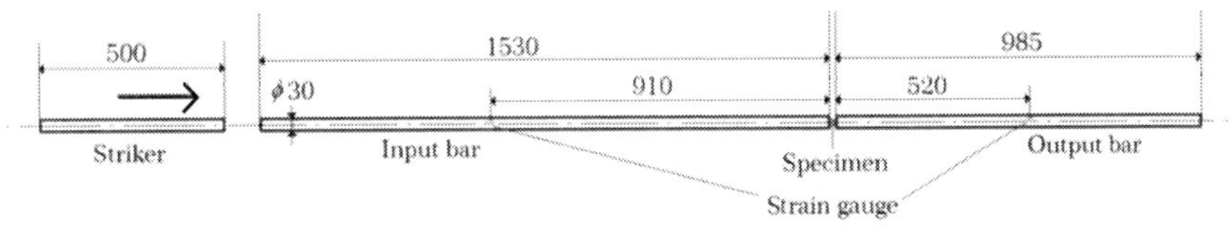

Figure 5. Scematic of SHPB apparatus.

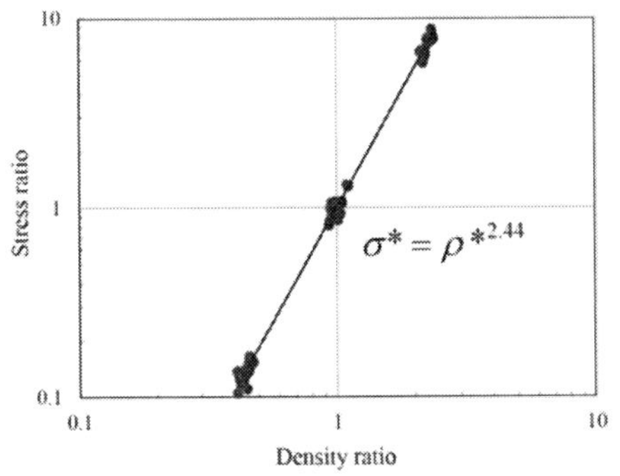

Figure 6. Relationship between strass ratio and density ratio.

SHPB is as follows

$$\sigma_m = \frac{A\sigma_t}{A_s} \tag{1}$$

$$\dot{\varepsilon} = \frac{2}{\rho C L_s}(\sigma_i - \sigma_t) \tag{2}$$

$$\varepsilon_n = \frac{2}{\rho C L_s}\int(\sigma_i - \sigma_t)dt. \tag{3}$$

Where, σ_i, σ_t and σ_r are the incident pulse, transmitted pulse and reflected pulse respectively. A, C and ρ are the cross-sectional area, the velocity of elastic wave and density of the input and output bar respectively. A_s and L_s are the cross-sectional area and height of specimen respectively.

3. Experimental results and discussions

3.1. Density normalization of specimen

Generally, compressive strength of foam material is sensitive to its relative density [12]. The polyimide foam specimens in this study have individual differences in the density. Therefore, there are differences in the constitutive relationship for the same specimen type. To evaluate the effects of strain rate and temperature it is necessary to control for the effect of density. We defined two ratios: one is the ratio of specimen density, $\rho^*(=\rho_1/\rho_2)$, and the other is the ratio of average stress, $\sigma^*(=\sigma_1/\sigma_2)$, where the subscripts of 1 and 2 indicate two arbitrary specimens compressed under the same conditions. Here, the average stress was defined as the mean stress in the range of 0.2 to 0.4 strain. Data plots of ρ^* and σ^* is shown in Fig. 6. From this graph we determined the equation expressing the relationship between ρ^* and σ^* as follows

$$\sigma* = \rho*^{2.44}. \tag{4}$$

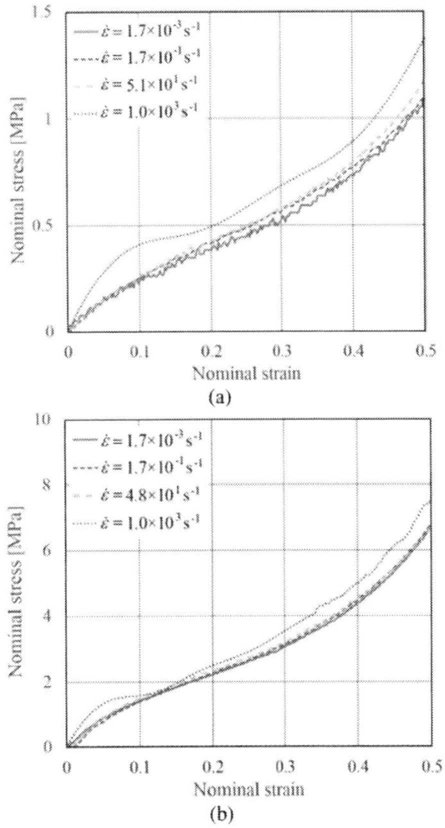

Figure 7. Stress-strain curves at various strain rates at room temperature. (a): Foam-A (b): Foam-B.

Experimental data of specimens that include the difference in density are normalized with the nominal density of Foam-A (120 kg/m^3) and Foam-B (240 kg/m^3). By doing so, it is possible to cancel the effect of density. In the following, we discuss based on the normalized data.

3.2. Strain rate dependence

Because of specimen anisotropy, its mechanical response indicates significant differences depending on the compression direction. If a specimen is compressed in the x- or y-directions, it cracks at about strain $\varepsilon = 0.3$ (Foam-A) or 0.1 (Foam-B). Specimens do not crack when compressed in the z- direction, and it is possible to be compressed into the densification region. Therefore, we focused on compressive strength in z-direction in this study.

The stress-strain curves of Foam-A and -B obtained from compression tests at various strain rates at room temperature are shown in Fig. 7.

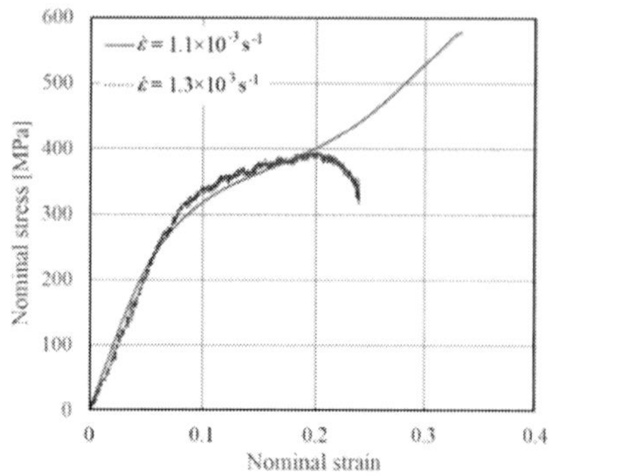

Figure 8. Stress-strain curves of Bulk-specimens at quasi-static and dynamic strain ratess at room temperature.

Generally, compressive behaviour of polymer foam consists of three stages: liner elastic, plateau and densification regions. Polyimide foam in this study also indicates roughly the same behaviour, but its liner elastic and plateau regions are unclear. In both Foam-A and -B, in the range of strain rate from quasi-static to intermediate, compressive flow stress has little difference. However compressive flow stress at dynamic strain rates is greater than others. This stress increment is sufficiently larger than the variation due to individual difference of specimen. Therefore, it can be said that the strain rate dependence of polyimide foam does not affect the compressive property up to the intermediate strain rate ($\dot{\varepsilon} \approx 50\,\mathrm{s}^{-1}$). And above intermediate strain rates the compressive strength of polyimide foam indicates salient strain rate dependence. It is considered that this dependence is related to the polyimide itself and the foam structure. Stress-strain curves of Bulk-specimens at quasi-static and dynamic strain rates are shown in Fig. 8. The differences due to strain rate are not observed in these graphs. This result indicates polyimide itself does not have strain rates dependence at strain rates from 10^{-3} to $10^{3}\,\mathrm{s}^{-1}$. Therefore, we can say that the effect of strain rate of compressive property of polyimide foam is related to the foam structure.

3.3. Temperature dependence

The stress-strain curves of Foam-A and -B obtained from compression tests at various temperatures at quasi-static strain rates are shown in Fig. 9. It is found that the compressive flow stress decreases with increasing temperature. These graphs indicate that compressive strength of polyimide foam has salient temperature dependence. Especially at 270 °C, the effect of temperature is appeared to be strong. The graph does not have an upward convex shape in the early region unlike others. At −190 °C, flow stress increases significantly and some Foam-A specimens cracked in the compression process. It is found that polyimide foam becomes brittle near the temperature of liquid nitrogen at quasi-static strain rates.

The stress-strain curves of Foam-A and -B obtained from compression tests at various temperatures at dynamic strain rates ($\dot{\varepsilon} \approx 250\,\mathrm{s}^{-1}$) are shown in Fig. 10.

At −190 °C, Foam-A specimen fractured, similar to quasi-static tests. Figure 11 shows the specimen after

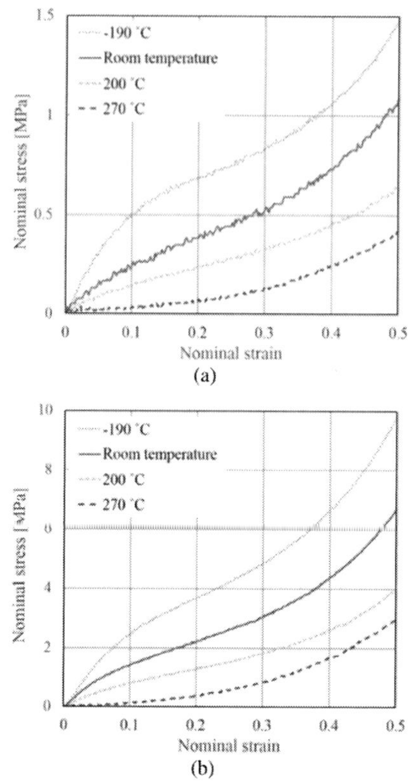

Figure 9. Stress-strain curves at various temperatures at quasi-static strain rates. (a): Foam-A (b): Foam-B.

compression tests. These results indicate that polyimide foam becomes brittle near the liquid nitrogen temperature at any strain rate. In dynamic tests, the effect of low temperature is less than that observed during quasi-static tests. The cause of this behaviour is that flow stress decreased due to temperature rise that accompanied strain rate. Flow stress of Foam-A at −190 °C became lower than the result at room temperature over strain 0.3. It is considered that this is caused by fracture and crushing during compression tests of Foam-A.

The stress-strain curves of Bulk-specimen at various temperatures are shown in Fig. 12. These graphs indicate that polyimide itself has significant temperature dependence. Therefore, it is considered that temperature dependence of polyimide foam is attributed to the polyimide material.

3.4. Constitutive equation

Based on these results, we devised constitutive equations of polyimide foam. However, the applicable range was limited to above room temperature. Results of low temperature tests may not be the correct since specimen cracked during compression tests at −190 °C. From the obtained stress-strain curves, the initial region of compressive deformation is expressed by the following equation.

$$\sigma = a(1 - \exp(-c\varepsilon)). \tag{5}$$

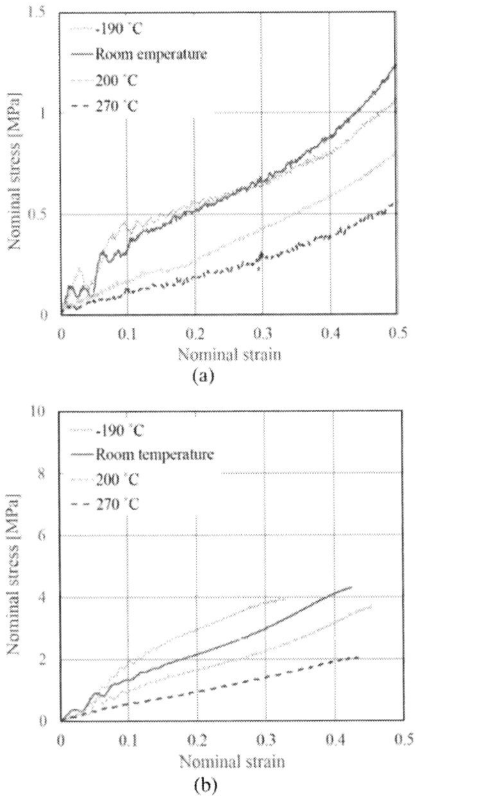

(a)

(b)

Figure 10. Stress-strain curves at various temperatures at dynamic strain rates. (a): Foam-A (b): Foam-B.

Figure 11. Specimens after compression tests.

In the study focused on the constitutive equation in initial region of PMMA foam, similar expression is used [13]. Where, a is constant. From the results of compression tests of polyimide foam, it is found that strain rate and temperature affect the initial gradient of stress-strain curves strongly. Therefore, in Eq. (5), we defined c as a variable of strain rate and temperature. Considering the test results, we expressed c in the variable which varies greatly in dynamic strain rate. The effect of temperature could be well expressed in the linear relationship. However, the effect of temperature on c varies by strain rate. Therefore, the function of strain rate described above is incorporated. To summarize these, variable c is expressed by the following equation.

$$c(\dot{\varepsilon}, T) = c_0 \left(1 + m\frac{\dot{\varepsilon}}{\dot{\varepsilon}_0}\right)\left(1 - p\frac{T}{1 + q\,(\dot{\varepsilon}/\dot{\varepsilon}_0)}\right). \quad (6)$$

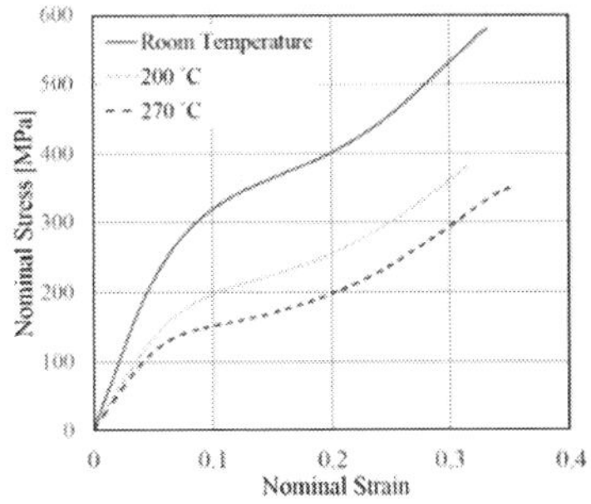

Figure 12. Stress-strain curves of Bulk-specimens at various temperatures at quasi-static strain rates.

Where, c_0, m, p and q are constant. $\dot{\varepsilon}_0$ is optional reference strain rate. In this case $\dot{\varepsilon}_0 = 0.001$. $\dot{\varepsilon}$ and T are strain rate and absolute temperature respectively.

Late region of compression is suitably expressed by an exponential function because the flow stress increases rapidly after densification. Therefore, the late region is expressed by the following equation.

$$\sigma = b\,\mathrm{cxp}(d\varepsilon). \quad (7)$$

Where, b is constant. From the test results, varying the strain rate and temperature affects the starting point of densification. Thus, we defined that d is a variable of strain rate and temperature. Since the effects of strain rate and temperature on late regions of stress-strain curves are almost the same as the effects on the initial gradient, d could be expressed by nearly the same equation as c. However, the effect of temperature on d does not vary by strain rate, d is expressed by a more simple equation than c. Therefore, variable d is expressed by the following equation.

$$d\,(\dot{\varepsilon}, T) = d_0\left(1 + r\frac{\dot{\varepsilon}}{\dot{\varepsilon}_0}\right)(1 - sT). \quad (8)$$

Where, d_0, r and s are constant.

As a result, we expressed the constitutive equation of polyimide foam by the following equation, it is sum of Eqs. (5) and (7).

$$\sigma = a(1 - \exp(-c\varepsilon)) + b\exp(d\varepsilon). \quad (9)$$

We calculated the variation of the flow stress at each strain rate and temperature. Based on those results the constants were determined by trial and error. The determined constants of Foam-A and -B are shown in Table 1. Results of calculating the constitutive relationship from Eq. (9) and experimental results at various strain rates at room temperature of Foam-A are shown in Fig. 13. Error between the experimental results and the calculated constitutive relationship at room temperature of Foam-A is shown in Fig. 14. In the region that indicates low stress, the error rate is larger, however, it is found that these results are consistent.

The same graphs of Foam-A at dynamic strain rates ($\dot{\varepsilon} \approx 250\,\mathrm{s}^{-1}$) at various temperatures are shown in Figs. 15 and 16. In these plots, experimental results

Table 1. Constants of constitutive equations of polyimide Foam-A and -B.

	Foam-A	Foam-B
a	0.73	2.9
b	5.8×10^{-3}	8.4×10^{-2}
c_0	7.7	12
d_0	10	9.5
m	7.8×10^{-7}	1.7×10^{-7}
p	1.76×10^{-3}	1.81×10^{-3}
q	3.0×10^{-7}	2.7×10^{-7}
r	7.1×10^{-8}	2.6×10^{-8}
s	4.1×10^{-4}	5.6×10^{-4}

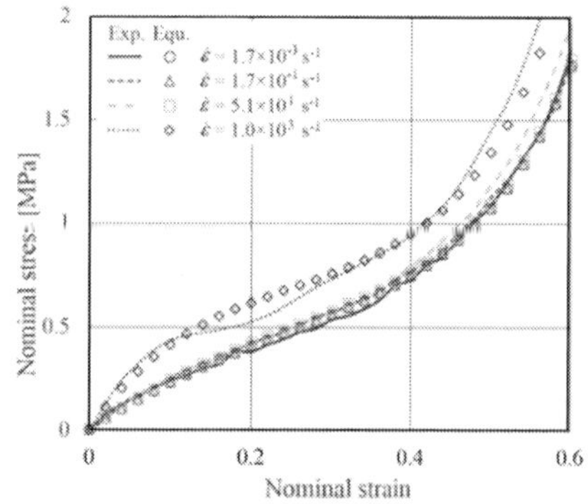

Figure 13. Experimental results and constitutive equations at room temperature of Foam-A.

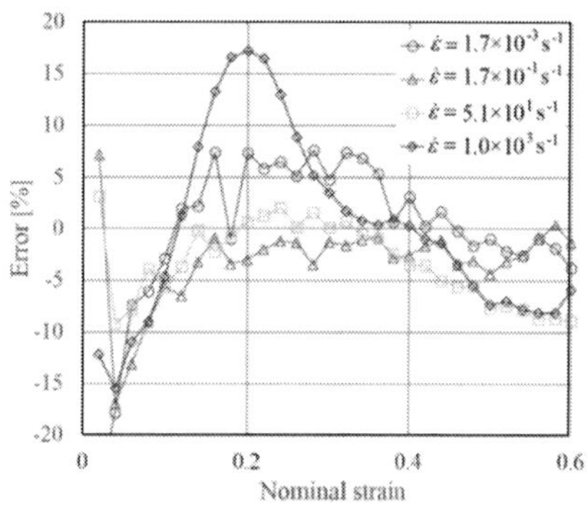

Figure 14. Error between the experimental results and the constitutive equations at room temperature of Foam-A.

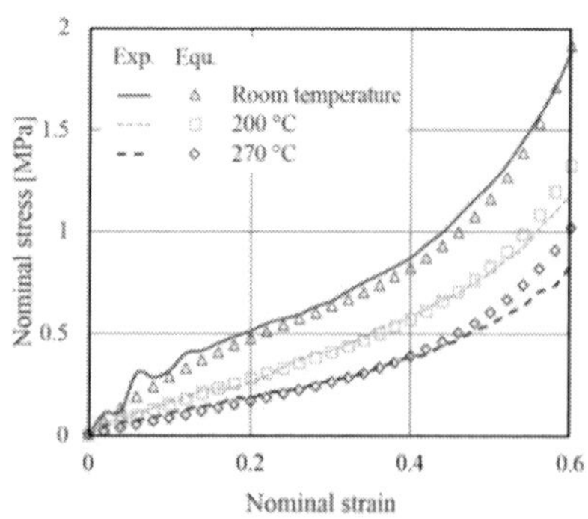

Figure 15. Experimental results and constitutive equations at dynamic strain rates of Foam-A.

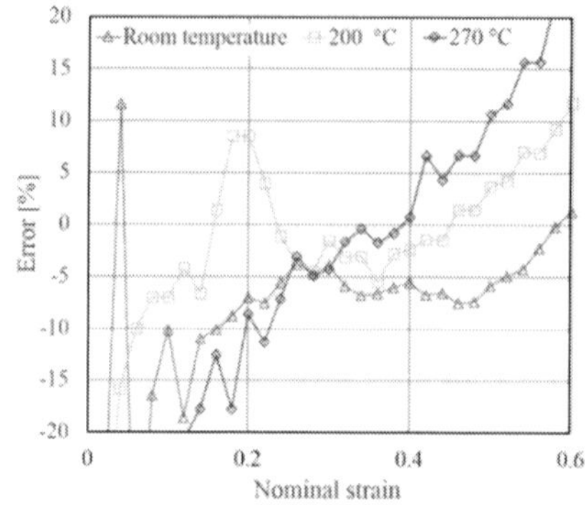

Figure 16. Error between the experimental results and the constitutive equations at dynamic strain rates of Foam-A.

and calculated values are found to be consistent. When evaluating all results, including results of quasi-static strain rates and Foam-B, this constitutive equation can express the constitutive relationship of polyimide foam with an error within ±15% at strain $\varepsilon < 0.6$.

4. Conclusion

The effects of strain rate and temperature on the compressive property of polyimide foam were investigated by a series of compression tests. The principal results obtained are as follows.

1) The flow stress during compressive deformation of polyimide foam indicates salient strain rate dependence above intermediate strain rates.

2) The flow stress during compressive deformation of polyimide foam indicates salient temperature dependence.

3) Over room temperature, the constitutive equation using an exponential function can express the constitutive relationship during compressive deformation of polyimide foam with an error of within ±15% at a strain $\varepsilon < 0.6$.

We would like to express our grateful appreciation to Mr. T. Sakata and I.S.T. Co. Ltd. for the provision of the polyimide foams.

References

[1] L. Sorrentino, M. Aurilia, S. Iannace, Polymer Testing, **26**, 878 (2007)
[2] P. Viot, F. Beani, J. Lataillade, J. Mat. Sci., **40**, 5829 (2005)
[3] Q. Liu, G. Subhash, Polymer Eng. Sci., **44**, 463 (2004)
[4] S. Ouellet, D. Cronin, M. Worswick: Polymer Testing, **25**, 731 (2006)
[5] R. Bouix, P. Viot, J.L. Lataillade, Int. J. Impact Eng., **36**, 329 (2009)
[6] B. Song, W. Chen, S. Dou, N. Winfree, J. Kang Int. J. Impact Eng., **31**, 509 (2005)
[7] J. Yi, M. Boyce, G. Lee, E. Balizer, Polymer, **47**, 319 (2006)

[8] Y. Takeda. T. Hakamada, N. Hama, K. Moriuchi: preprinted Coll. 3D10(JSASS-2009-5137)
[9] C. Resewski, W. Buchgraber, Materialwissen-schaft und Werkstofftechnik, **34**, 364 (2003)
[10] A. Kuwabara, M. Ozasa, T. Shimokawa, N. Watanabe and K. Nomoto, Adv. Compo. Mater., **14**, 343 (2005)
[11] S. Tanimura: Soc. of Mech. Eng. Coll. of papers A **68**, 1767 (2002)
[12] A. Gibson, M. Ashby, *CELLULAR SOLIDS: Structure & Properties* (1999)
[13] C. Jo, J. Fu, H. Naguib, Polymer, **46**, 11896 (2005)

EPJ Web of Conferences **94**, 01048 (2015)
DOI: 10.1051/epjconf/20159401048

Interrupted pulse electromagnetic expanding ring test for sheet metal

José Imbert[a], Taamjeed Rahmaan, and Michael Worswick

Department of Mechanical and Mechatronics Engineering, University of Waterloo, Waterloo, Ontario, Canada, N2L 3G1

Abstract. This paper describes the development of an interrupted pulse electromagnetic (EM) expanding ring experiment to study the high rate properties of AA5182 aluminum commercial sheet alloys at strain rates in excess of 5,000 s^{-1}. Experiments are performed to compare two commonly adopted methods of driving the expanding ring: EM expansion versus an exploding wire. After studying and testing both methods, it was determined that EM expansion had the greatest potential for being developed into a test that would result in free-flight of the samples. By interrupting the current pulse in the EM expanding ring test, the ring is allowed to achieve free-flight, thus eliminating the need to determine the induced EM forces and significantly reducing the uncertainty of the stress-strain behaviour determined from the test. Once the free-flight condition is established, the stress-strain behaviour of the material is determined from the free-flight deceleration of the sample, as calculated from the velocity measured using a Photon Doppler Velocimeter (PDV). Results are presented for AA5182 at strains rates between 1,000 to 5,500 s^{-1} and exhibit low strain rate sensitivity, are comparable to tensile split-Hopkinson bar results at strain rates of 1,000 s^{-1}.

1. Introduction

The current push for automotive vehicle light weighting has resulted in considerable interest in the application of aluminum alloys and advanced high strength steels in automotive mass production. Unfortunately, these material exhibit only moderate formability when compared to traditional low carbon steel alloys. The formability of some of these alloys has been shown to increase with the use of high speed forming techniques [1–4] such as electromagnetic (EM) and electrohydraulic forming. When these high speed techniques are used to form material into a die, strain rates in excess of 10,000 s^{-1} are predicted to occur during the interaction of the tool with the sheet [3,4]. The mechanisms that result in the observed increases in formability are not yet fully understood, in part due to the lack of understanding of the properties of sheet metals at such high rates of strain, which significantly limits the ability to model these processes. A practical test that can generate stress-strain data at these high rates is required to better understand high speed forming techniques and produce accurate material models. The expanding ring technique is one of a few with potential to satisfy this need and it is the subject of this work.

The expanding ring test generates uniaxial tensile stresses and high strain rates within a sample ring, by causing it to expand radially at high speeds (Fig. 1). If the ring is in free-flight or under the effect of a known driving force, the stress-strain behaviour of the material can be extracted by measuring its radial deceleration. Accurately determining the driving force can be very difficult in practice; thus, a test where no driving force is present, where the ring is allowed to decelerate in free-flight would present an ideal condition.

[a] Corresponding author: jmsimber@uwaterloo.ca

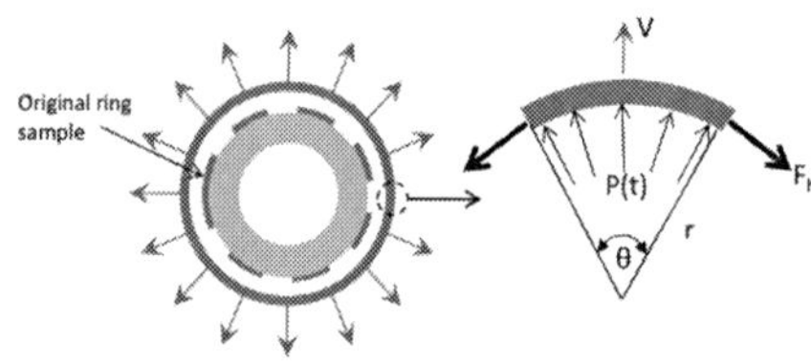

Figure 1. Schematic of an expanding ring test. V = radial velocity, F_h = internal force, P(t) = driving pressure.

In principle, obtaining the stress strain behaviour for a material from an expanding ring test is relatively easy, assuming the ring is in a uniaxial stress state and only deforming radially (Fig. 1). The governing equation is [5]:

$$F = ma. \quad (1)$$

Where is F = force, m = mass and a = acceleration. Applying equilibrium to a segment of the ring, as illustrated in Fig. 1;

$$-2F_h Sin\left(\frac{\theta}{2}\right) + F_{P(t)} = m\frac{dV}{dT}. \quad (2)$$

Where F_h = internal force, $F_P(t)$ = driving force generated by a driving pressure P(t) and v = velocity. The simplest and most accurate analysis is obtained when $F_P(t) = 0$, corresponding to free-flight, which results in;

$$-2F_h Sin\left(\frac{\theta}{2}\right) = m\frac{dV}{dT}. \quad (3)$$

After some basic manipulation we can obtain the stress, σ, in terms of the density, ρ, the radius, r, and the radial acceleration, $\ddot{r}$;

$$\sigma = \rho r \ddot{r}. \quad (4)$$

The strain and strain rate are respectively given by [5];

$$\varepsilon = \frac{\Delta r}{r} \tag{5}$$

$$\dot{\varepsilon} = \frac{\varepsilon}{\Delta t} = \frac{\dot{r}}{r}. \tag{6}$$

Obtaining the radial acceleration is crucial for the accurate determination of the stress. It is also difficult to measure during a ring expansion test, in which radial velocities can exceed 100 m/s and test durations are in the order of 50 μs. Early implementations of the tests used high speed photography or streak cameras to measure the displacement and determined the acceleration by differentiating the velocity twice. Determining acceleration from displacement is inherently noisy, which led to the adaptation of laser interferometry techniques to measure velocity directly [6,7].

Early ring expansion tests accelerated the ring by placing it on the outside of a hardened steel cylinder and using an explosive to create a stress wave in the cylinder that would expand the ring [5,8]. This method led to significant free-flight of the sample. Unfortunately, the health and safety constraints that result from using high explosives make this method impractical for most applications. Currently, the two preferred ways of performing the test are the EM expansion and the exploding wire tests. In EM expansion the ring is accelerated using the induced forces generated by current pulse flowing through a nearby spiral coil. This technique was introduced by [9,10]. EM expansion is effective for high conductivity, relatively low strength materials, such as copper and aluminium, but it is not as effective for lower conductivity, high strength materials, such as steels. High conductivity drivers can be used to test low conductivity materials [9], but they reduce the efficiency of the apparatus and introduce additional complications to the apparatus and analysis. In addition, spiral coils do not produce a uniformly distributed induced force, which will result in violations of the analysis assumptions. Recent implementation of EM expanding ring tests have begun to use Photon Doppler Velocimeters (PDV) to measure the velocity [12–17].

In the exploding wire test, a wire is placed within an elastomer and vaporized using a high frequency high current pulse [17,18]. The energy of the expanding gases is transmitted to the ring through the elastomer, which is referred to as a driver or actuator. Since the test does not rely on induced forces, it eliminates the effects of conductivity. However, the driving force produced by the driver is difficult to determine, as is the exact interaction between the driver and the specimen.

To account for the difficulties in determining the driving forces, both the EM and exploding wire expanding ring tests have been combined with finite element and optimization methods to achieve better constitutive data fits [17,19]. These methods have potential; however, significant errors can be introduced in the analysis.

To the authors' knowledge, there has been no significant effort to develop and fully understand the testing of sheet metal using the expanding ring technique.

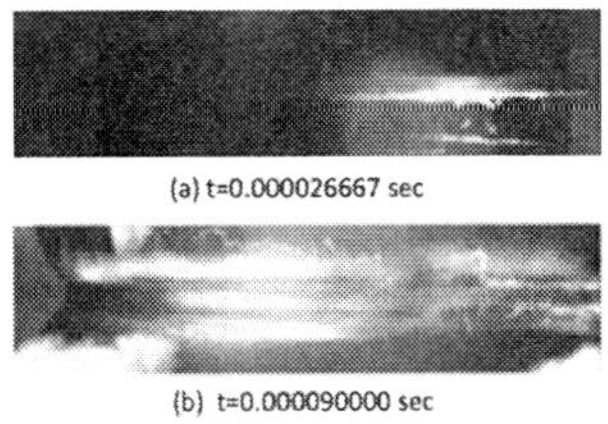

Figure 2. High speed images taken at 300,000 fps of an expanding ring test: a) as expansion begins and b) after the ring has expanded. Note how the polyurethane engulfs the ring in b).

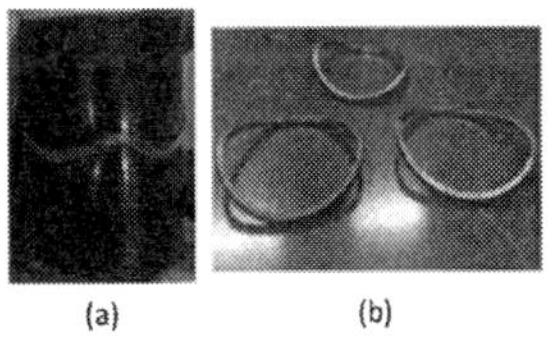

Figure 3. Out-of-plane ring deformation believed to be the result of material anisotropy and frictional forces.

Sheet metal poses challenges to the expanding ring technique, due to the anisotropy of the material and the difficulty of obtaining specimens that accurately represent the properties of the sheet metal. This work is part of an effort to develop such a test for AA 5182 and DP 600 sheet metal. Both the EM and exploding wire techniques were explored. The limitations of the EM expanding ring, as described above, were evident in the present work and exacerbated, since the AA 5182 and DP 600 are structural alloys with a significantly higher strength than the pure aluminum or copper used in many reported expanding ring experiments and in the case of the DP 600, have lower conductivity. The two main limitations of the EM expanding ring test for this work were the very low coil life and the presence of a driving force thought the expansion of the ring.

The evaluation of the exploding wire technique also revealed significant challenges. High speed video of the tests showed that the rings were in contact with the driver for the recorded expansion. In some cases the expanding driver can engulf the ring, as shown in Fig. 2, for a DP600 ring of 1 mm × 1 mm cross section, tested at 4.3 kJ, using 80A polyurethane. The tests also results in significant friction between the specimen and the driver that result in non-uniaxial stress states on the rings. These stress states can contribute to significant non-radial, out-of-plane deformation of the specimens, as can be seen in Fig. 3. A combination of a polyamide film and lubricant was used to reduce the friction; however, it was not eliminated. The result of the lubrication can be seen by comparing the results of a non-lubricated test (Fig. 3-a) to a test where the lubrication was used (Fig. 3-b).

After studying and testing both methods, it was determined that an interrupted current EM ring expansion test had the greatest potential for being developed into a test that would result in free-flight of the samples. By interrupting the current pulse in the EM expanding ring test, the ring is allowed to achieve free-flight, thus eliminating the need to determine the induced forces,

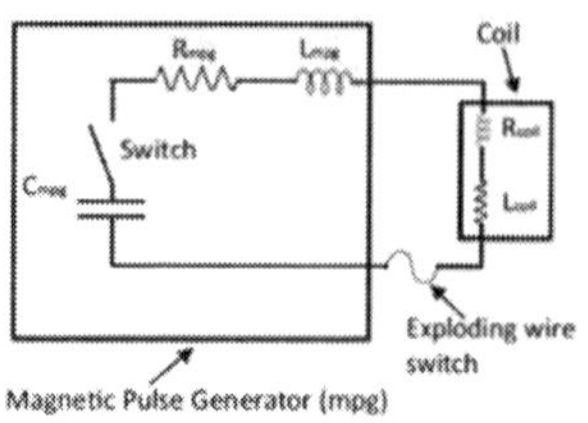

Figure 4. Schematic of the experimental apparatus.

significantly reducing the uncertainty of the stress-strain behaviour determined from the test. The following sections will describe the interrupted pulse test developed and present the results from tests on AA 5182, which yielded strain rates between 1,000 to 5,500 s^{-1}.

2. Experimental methods and procedures

2.1. Experimental apparatus

A schematic view of the expanding ring apparatus is shown in Fig. 4. It consists of a magnetic pulse generator (MPG), essentially a large capacitor bank, and a coil. When the charged capacitors are discharged through a coil, a current pulse is generated, which induces the driving forces on the sample ring. A typical EM expanding ring apparatus will have a similar schematic, but without the exploding wire switch used in the present work. For a more in depth overview on the basic principles the reader is referred to [1,20].

2.1.1. Magnetic pulse generator

The magnetic pulse generator used for this work was a Pulsar MPW 20 Research Edition Magnetic Pulse Generator (MPG). The MPG consists of a capacitor bank and a power supply to deliver the current to the capacitors at the required specifications. The MPG has a nominal maximum energy capacity of 20 kJ and a maximum charging voltage of 9,000 volts. The machine capacitance is 539.7 μF, inductance is 24.35 nH, and resistance is 2.98 mΩ. The nominal discharge frequency using a shorting bar across the output terminals of the capacitor bank was 24.51 kHz. For the present work the MPG was charged to 4.0 kV, which results in a stored energy of 4.3 kJ.

2.1.2. Pulse interruption

The current pulses used for an EM expanding ring test have peaks of in the order of 100 kA and frequencies above 10 kHz which makes interrupting the pulse very difficult since mechanical switches that can function at these speeds are not available. In principle, solid state devices could be used, but the authors were not able to find a source for one that could be used for this work. Another alternative is to use an explosive event to produce the switch. Gourdin [10] used a shaped explosive charge switch to interrupt the current by creating an alternate current path, with the intent of reducing the temperature rise on the samples caused by Joule heating. Switches based on exploding aluminum foil were developed by [21] to interrupt and shape pulses.

For this work exploding wires were used to open the circuit and interrupt the current pulse. When the current reached a certain value the wires ignited, opening the circuit. Aluminum wires of 0.8 mm diameter were used.

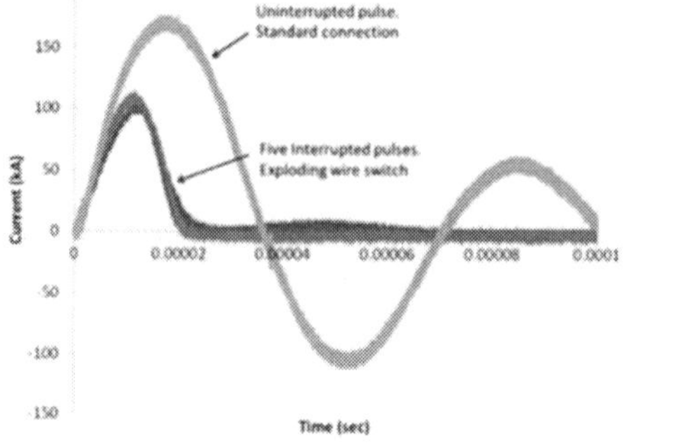

Figure 5. Comparison of uninterrupted and five interrupted current pulses.

Figure 5 shows a current pulse from a discharge with a solid connector (without the exploding wire switch), along with five recorded interrupted pulses achieved with the exploding wire switch. The interrupted current pulses have a lower peak and amplitude than the uninterrupted one.

This leads to less energy being transferred to the sample, when compared to the uninterrupted pulse.

2.1.3. Measurements of the sample velocity

The sample velocity was measured using an Ohio Manufacturing Institute Photon Doppler Velocimetry (PDV), which measures velocity based on the Doppler effect. A laser beam of 1550 nm in wave length is split into two; half is used to illuminate a test specimen and half as a reference. The beam that is used to illuminate the specimen is reflected and captured by a sensor. If the specimen is in motion the frequency of the reflected wave will be different than the original beam and the difference in the frequencies, the beat frequency, is then used to obtain the velocity of the specimen [22,23]. The PDV outputs a voltage versus time signal with a frequency equal to the beat frequency. The frequency spectrum of the signal was calculated using Matlab©and then the velocity was calculated using the formula [22,23];

$$V = beat\ frequency * wavelength/2. \qquad (7)$$

A representative velocity data set calculated from PDV readings is shown in Fig. 6. The ring velocity profile is clearly discernible and relatively smooth. Several gaps in the velocity profile are usually present due to gaps in the calculated spectrum of the signal. To determine the accelerations, a curve was fit to the data and the velocity obtained from the fitted curve was used to calculate the accelerations. Figure 6 also shows a fitted curve. Very good fits were obtained from the data.

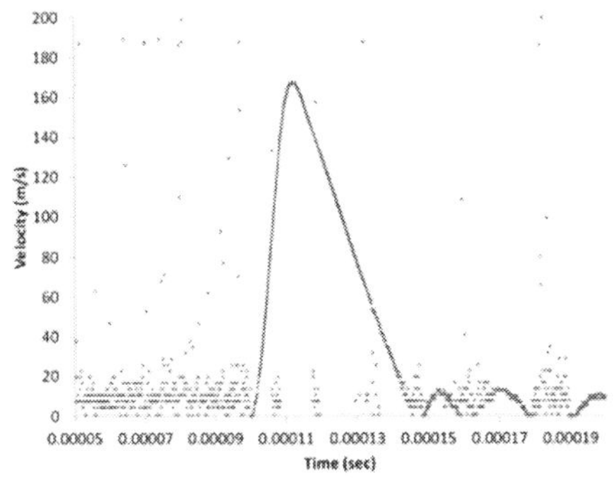

Figure 6. Velocity extracted from the experimental PDV readings (symbols) together with the data fit (solid line) used to calculate the accelerations.

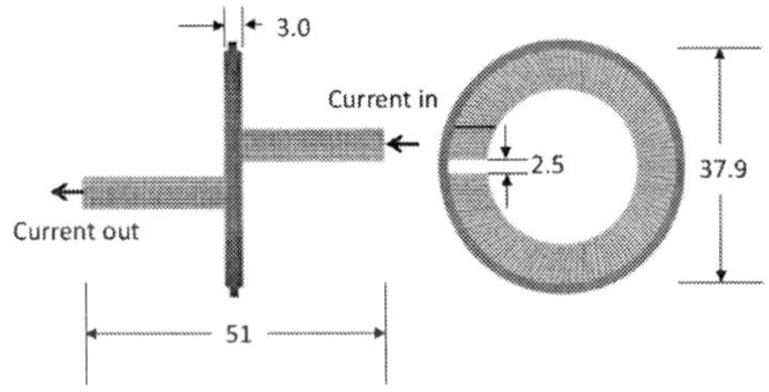

Figure 7. Single turn coil with ring in testing position. The figure is from the FEM mesh used for modelling the experiments. All dimensions in mm.

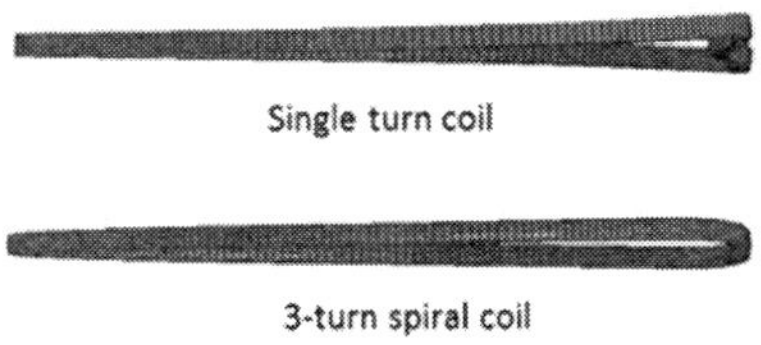

Figure 8. Predicted shapes of a sample ring formed with the single turn coil used in this work and a 3-turn spiral coil.

2.1.4. Coil

The coil is a critical part of the apparatus. The coil geometry and material will affect the amount of energy transmitted to the sample and play a significant role in determining the magnitude and direction of the forces induced on the sample ring. A single turn copper coil was designed and built for this work (Fig. 7). The goal of the design was to reduce the coil impedance, maximize strength and minimize non-radial (out-of-plane) movement of the sample. By reducing the windings to one, the inductance of the coil is reduced, allowing for higher peak currents and thus higher induced forces. In multiple winding coils the windings induce forces on each other, resulting in significant stress on the windings. With only one turn the winding only affects itself, reducing the forces on the coil and resulting in increased coil life. The non-radial movement of the ring was not eliminated, but simulations showed that it is reduced when compared to a 3-turn spiral coil, as can be seen in Fig. 8.

The 2.5 mm gap present in the single turn coil resulted in a reduction of the induced forces on the ring in the area

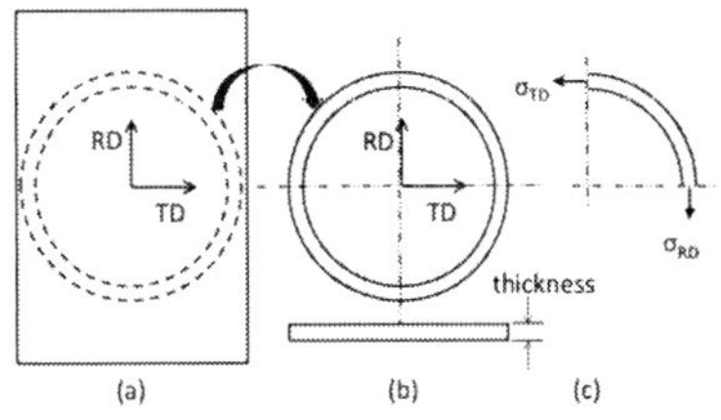

Figure 9. Rings are obtained from sheet metal (a) to (b). TD = transverse direction RD = rolling direction.

adjacent to it. The effect of this gap on the samples will be addressed in Section 4. The gap width was required due to machining limitations.

2.2. Material and sample geometry

The material tested in this work was 1 mm sheet of AA 5182-O, which has nominal yield strength of 130 MPa. Detailed constitutive characterization of this material at strain rates up to 1,000 s^{-1} has been performed by [24].

2.2.1. Sample geometry

Samples should have a cross section small enough to ensure that only hoop stresses are present [5] and should have a small aspect ratio to achieve plane stress [17]. For a sheet metal sample, these conditions are best satisfied by a square cross section, with the length of the sides equal to the thickness of the sheet. To minimize the effect of materials processing, the rings were cut from the stock sheet material (Fig. 9). This resulted in geometrically uniform and flat specimens.

It is well established that sheet materials have different strengths along the rolling direction (RD) and the transverse direction (TD) to rolling. Rings made as described above will have different strengths along the circumference, as the strength goes from the RD to the TD values (Fig. 9-c). Work on measuring the anisotropy is on-going and the results present herein do not include studies of anisotropy.

3. Numerical modelling

A numerical model was developed to gain insight into the general behaviour of a ring during the tests. An EM capable version of the commercial software LS-Dyna [25] was used. The material model used was a tested isotropic piece-wise-linear plasticity material model for a similar alloy (AA 5754) that was calibrated and tested for EM forming models [20]. No rate effects were introduced at this stage of model development. A resistivity of 55 nΩ.m was assumed for the samples. The mesh used for the model is shown in Fig. 7. Both the coil and ring were modelled with brick elements. The ring was modelled with 5 elements through both thickness and width, which resulted in 6,250 elements. The coil was modelled with 10 elements through thickness and width, for a total of 16,000 elements. An experimentally measured current profile was used as the current input. The model was not intended to exactly reproduce the test conditions, but to

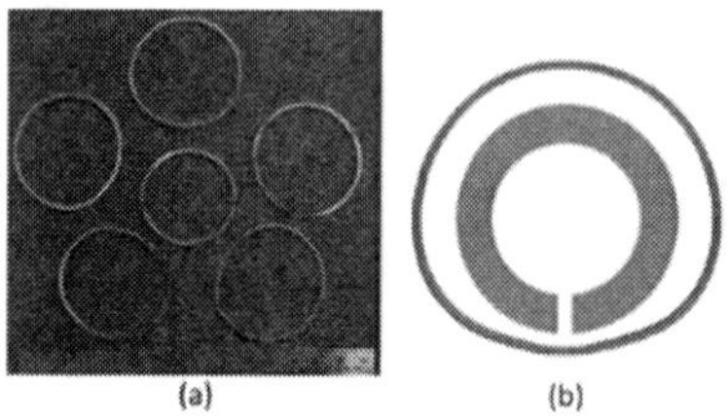

Figure 10. Tested samples surrounding an un-deformed ring (a) and final predicted shape of an expanded ring (b).

Table 1. Experimental peak currents and velocities, with final strains for the samples tested.

Sample	Peak Current (kA)	Peak Velocity (m/s)	Final strain
1	102	164	0.29
2	97	148	0.21
3	107	166	0.25
4	107	167	0.25
5	110	171	0.25

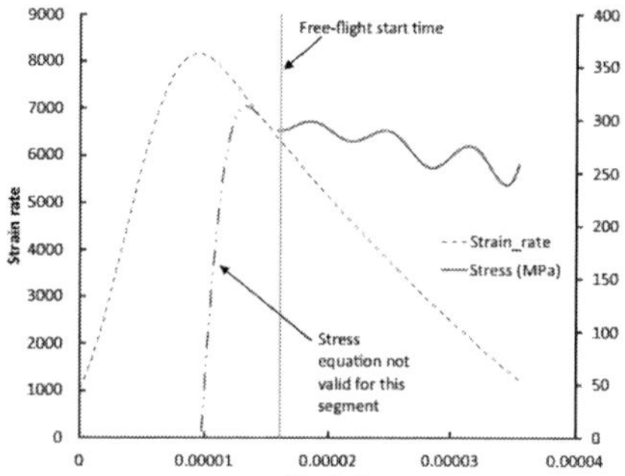

Figure 11. Strain rate and stress versus time for a single sample.

model the general behaviour of the test, thus the simplified material model used was deemed acceptable. A detailed description of the solution technique can be found in [20].

4. Results and discussion

The rings from five experiments are shown in Fig. 10-a. The final shape is not a perfect circle, with a flattened area visible on all samples. The numerical simulations indicate that the final shape is the result of the 2.5 mm gap discussed above (Fig. 10-b). An additional source of the final shape could be minor variations in the positioning of the sample on the coil.

The basic test parameters and final true strains of the sample are shown on Table 1. It can be seen the current pulse for sample 2 was lower than all the other ones, which resulted in the lowest measured peak velocity. This is reflected in the stress data shown below.

Experimental strain rate and stress data for sample 1 are shown plotted versus time in Fig. 11. The point where the current pulse stops and free-flight is reached is indicated with a line and shows that a significant period of free-flight was achieved. Strain rates peak at roughly

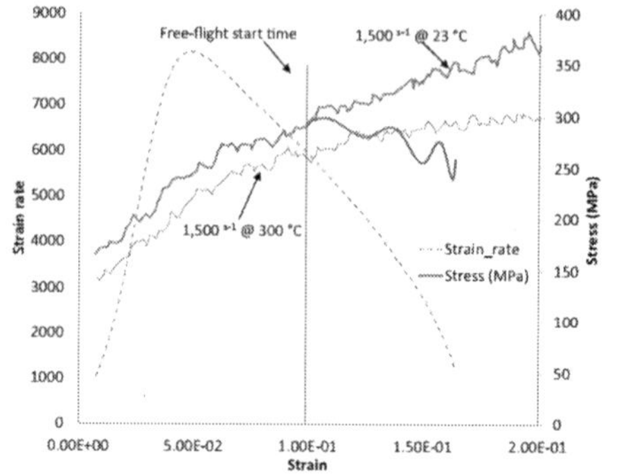

Figure 12. Strain rate and stress versus time for sample 1. Data for 1,500 s^{-1} from [26].

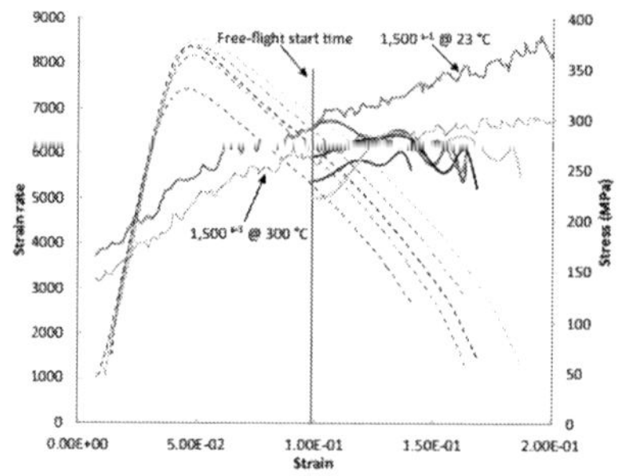

Figure 13. Strain rate and stress versus time for the five samples tested. Data for 1,500 s^{-1} from [26].

8,000 s^{-1} and decrease to approximately 5,500 s^{-1} at the start of free-flight, ending at 1,000 s^{-1}. The strain rates decay as the test progresses. This poses some challenges to data fitting that are being resolved as part of ongoing work. The part of the stress-strain curve shown in a dashed line represents the period were Eq. (4) is not valid due to the presence of induced forces. Subsequent graphs will not present this portion of the curve.

Figure 12 shows the strain rate and stress plotted versus the true strain calculated using Eq. (5). The figure also includes published data from Hopkinson bar tests carried out on AA 5182 at 1,500 s^{-1} at 23 and 300 °C [26].

Data for all five tests is shown in Fig. 13. There is relatively good repeatability, especially given the nature of this experiment. The tests are reasonably repeatable; with only one test (sample 2) deviating significantly from the rest. Of the published data presented in the figure, the data for the elevated temperature test is the closest to the current work. Although no firm conclusions can be drawn at this time, the data indicates that thermal softening could be affecting the dynamic strength of the material in the ring expansion experiment.

4.1. Joule heating

Significant temperature increases can be produced in the sample by the induced currents through Joule heating [10]. The temperature change on a ring sample is given by [10]:

$$\dot{T}j = \frac{RI^2}{C_p \Lambda \rho}. \tag{8}$$

Where $\dot{T}_j$ = temperature change, R = resistivity, I = sample current, C_p = heat capacity, Λ = sample volume, ρ = density.

Ideally, the current induced in the ring would be measured directly. Unfortunately, that capability was not available to the authors at the time of writing. In lieu of an actual measurement, a numerical model was used to obtain an approximate value of the temperature rise. Using the current in the sample predicted by the model and Eq. (8), it was determined that the sample temperature rise is on the order of 200 °C, which is in the range where thermal softening effects can be significant [27].

5. Conclusions

The interrupted pulse EM expanding ring test has significant potential for material characterization at high rates of strain. The combination of significant free-flight and velocity measurements using the PDV provide stress values from an analysis that is free of the additional assumptions that are common in expanding ring tests. The strain rates achieved were below the desired value of 10,000 s^{-1}, but the indications are that strain rates of 10,000 s^{-1} are possible. Future challenges will include fitting the data to a material model, due to the decaying strain rates experienced by the samples, and determining the sample temperature rise to account for the thermal softening effects produced by the Joule heating. The current coil design met the desired criteria and is a viable approach for ring expansion tests.

The authors greatly appreciate the financial support APC, NSERC, Canada Research Chairs Secretariat, ORF, Ford Research & Advanced Eng., Amino Corp., and ArcelorMittal.

References

[1] V. Psyk, D. Risch, B. Kinsey, A.E. Tekkaya and M. Kleiner, J. Mat. Proc. Tech., **211** (2011)
[2] V.S. Balanethiram, G.S. Daehn, 1994. Scripta Metal. Mater. **30** (1994)
[3] J.M. Imbert, M.J. Worswick, S.L. Winkler, S. Golovashchenko, V. Dmitriev, SAE Trans. J. Mater. Manuf. SAE paper number 2005-01-0082 (2005)
[4] J. Imbert, M. Worswick, J. Mat. Proc. Tech., **212-9** (2012)
[5] C. Hoggatt, R. Recht, Exp. Mechs., October (1969)
[6] R.H. Warnes, R.R. Karpp, P.S. Follansbee, J. Phys., C5, **46-8** (1985)
[7] R.H. Warnes, R.R. Karpp, P.S. Follansbee, J. Eng. Mat. Tech, **108** (1986)
[8] P.C. Johnson, B.A. Stein, R.S. Davis, U.S. Army Mat. Res. A. Rep. No. WAL TR 111.2/20-6 (1962)
[9] W.H. Gourdin, S.L. Weinland, R.M. Boiling, Rev. Sci. Instrum., **60-3** (1989)
[10] W.H.Gourdin, J. App. Phys., **65** (1989)
[11] D. Landen, S. Satapathy, D. Surls, 5th IEEE Int. Pulsed Power Conf. (2005)
[12] D. Landen, S. Satapathy, D. Surls, IEEE Trans. Mag., **43-1** (2007)
[13] D. Landen, D. Wetz, S. Satapathy, S. Levinson, IEEE Trans. Mag., **45-1** (2009)
[14] G. Daehn, Y. Zhang, S. Golowin, K. Banik, A. Vivek, J. Johnson, G. Taber, G. Fenton, I. Henchi, P. L'Eplattenier, 3rd Int. Conf. on High Speed Forming (2008)
[15] G. Taber, G, Daehn, A. Vivek, dissertation submitted to the Ohio State University (2008)
[16] J. Johnson, G. Taber, A. Vivek, Y. Zhang, S. Golowin, K. Banik, G. Fenton, G. Daehn, Steel Res. Int., **80-5** (2009)
[17] J. Johnson, G. Taber, G. Daehn, 4th Int. Conf. on High Speed Forming (2010)
[18] A. Rajendran, I. Fyfe, J. App. Mech., **49** (1982)
[19] I. Henchi, P. L'eplattenier, G. Daehn, Y. Zhang, A. Vivek, N. Stander, 10th Int. LS-DYNA Conf. (2008)
[20] J. Imbert, University of Waterloo Ph.D. thesis (2010)
[21] R. Bealing, P.G. Carpenter, J. Phys. E., **5** (1972)
[22] O. Strand, L. Berzins, D. Goosman, W. Kuhlow, P. Sargis, T. Whitworth, Vel. Using Heterodyne Techs. Conf., paper UCRL-CONF-206034 (2004)
[23] O. Strand, D. Goosman, C. Martinez, T. Whitworth, W. W. Kuhlow, Rev. Sci. Instrs., **77** (2006)
[24] T. Rahmaan, A. Bardelcik, J. Imbert, C. Butcher, M.J. Worswick, Submitted I. J. Impact. Eng. (2014)
[25] P. L'Eplattenier, G. Cook, C. Ashcraft, M. Burger, J. Imbert, M. Worswick, Steel Res. Int., **80-5** (2009)
[26] R. Smerd, S. Winkler, C. Salisbury, M. Worswick, D. Lloyd, M. Finn, I. J. Impact Eng., **32** (2005)

EPJ Web of Conferences **94**, 01049 (2015)
DOI: 10.1051/epjconf/20159401049

Investigations on specimen design and mounting for Split Hopkinson Tension Bar (SHTB) experiments

Noah Ledford[1,a], Hanna Paul[1,b], Georg Ganzenmüller[1,c], Michael May[1,d], Matthias Höfemann[2,e], Manuel Otto[2,f], and Nikica Petrinic[1,3,g]

[1] Fraunhofer Institute for High-Speed Dynamics, Ernst-Mach-Institut, EMI, Eckerstr. 4, 79104 Freiburg, Germany
[2] Salzgitter Mannesmann Forschung GmbH, Eisenhüttenstrasse 99, 38239 Salzgitter, Germany
[3] Department of Engineering Science, University of Oxford, Parks Road, Oxford OX1 3PJ, UK

Abstract. Split Hopkinson Tension Bar (SHTB) experiments can be used to test the material behavior with high strain rates in tension loading. The influence of the specimen mounting and the specimen design on the test results was investigated. Three mounting methods were tested. The best signal is achieve using a mounting based on form fit. The three tested specimen designs all lead to a valid fracture behavior, but result in a different local strain rate.

1. Introduction

For various applications, material characterization with strain rates of $1000\,s^{-1}$ and higher are of interest. Split Hopkinson Bar (SHB) testing methods are often used for this range of strain rates. In these experiments, a specimen is placed between two long elastic rods, the input and the output bar. The input bar is impacted with a projectile inducing a stress wave, which travels through the input bar into the specimen and into the output bar [1]. As opposed to the traditional Split Hopkinson Pressure Bar (SHPB) test, where the sample is simply placed between the input and output bars, a Split Hopkinson Tension Bar (SHTB) test requires a firm attachment of the sample to the bars, which is often difficult to realize. Another concern that arises with the SHTB is the tapering of the specimen from where it is attached to the input bar to the gauge section where within which the response of investigated materials is observed and quantified. Using a recent SHTB design derived from [2] which allows a longer loading pulse to be exerted along the loading bar with enhanced alignment by means of a lower pressure firing chamber, the effects of specimen design and mounting are investigated.

The challenge of attaching a specimen to the bars is addressed in different ways. One approach is to utilize an external thread on an axially symmetric specimen which engages an internal thread at the end of the bars. However, for materials without axial symmetry such as sheet metals or composites, other ways of mounting have to be employed. An alternative approach is to adhesively bond threaded metalic endcaps on both ends of the specimen and subsequently screw the specimen and the endcaps into the bars. If the adhesion between end caps and specimen is not sufficient, additional clamping mechanisms can be used [3]. Another way is to use specimens with holes and screws to fix them in the bar [4]. It also possible to use an interlocking geometry and directly insert the tapered specimen into the bar [5]. In the work presented here, various ways of mounting flat specimens into the bars are investigated. Special attention is given to the influence of different mounting methods upon the stress waves measured using strain gauges attached to the loading bars away from the specimen, in order to assess the impedance of proposed mounting methods.

In the design of specimen geometry several parameters play key roles. The most important are the specimen shape, the length and the specimen cross section areas as well as the way of the tapering of the specimen, i.e. the transition from the gauge section to the shoulders. The length of the specimen influences the strain rate. The cross section area of the specimen is directly related to the force required to deform and break the material. In addition a reduction of cross section area at the center of the specimen is necessary to realize failure of the specimen away from the gripping zone. However the reduction of width is limited and dependent on the type of material tested. With a poor taper on the specimen, failure initiates at the end of the transition due to the stress concentration rather than in the middle of the parallel section, thus complicating the analysis and possibly invalidating the test.

[a] e-mail: noah.ledford@emi.fraunhofer.de
[b] e-mail: hanna.paul@emi.fraunhofer.de
[c] e-mail: georg.ganzenmueller@emi.fraunhofer.de
[d] e-mail: michael.may@emi.fraunhofer.de
[e] e-mail: M.Hoefemann@sz.szmf.de
[f] e-mail: M.Otto@sz.szmf.de
[g] e-mail: nikica.petrinic@emi.fraunhofer.de

2. Experimental setup

2.1. The Split Hopkinson Tension bar (SHTB)

To investigate the effects of the geometry and mounting options on the test results a Split Hopkinson Tension bar based on the design proposed by Gerlach [2] is used.

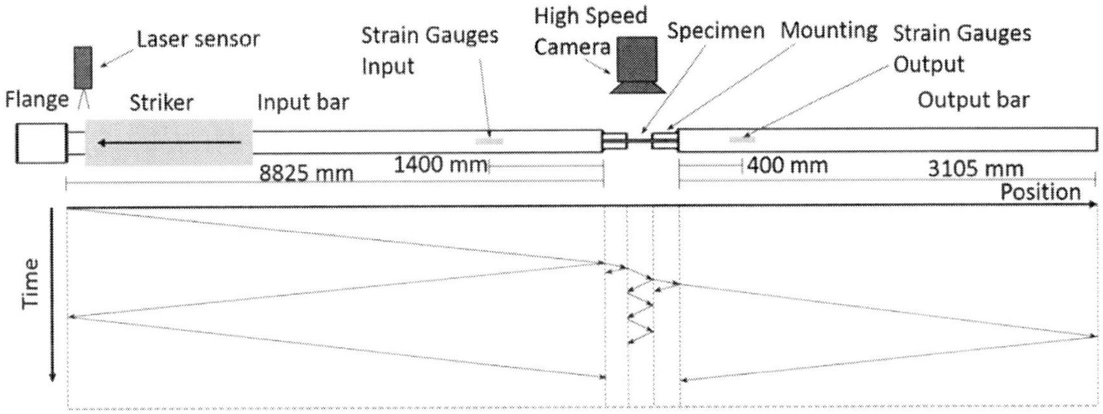

Figure 1. Schematic and Lagrange diagram of SHTB system.

Table 1. Mounting method qualitative comparison (++: better, +: good, -: not good, - -: worse).

Criteria	Adhesively bonded	Dowel-pin	Form-fit	Clamped-inertia	Clamped-friction
Stress Concentration	++	-	-	-	- -
Additional mass	++	++	+	-	- -
Impedance difference	+	+	-	-	- -
Wave symmetry	++	+	-	-	-
Thermal limit	-	+	+	+	++
Specimen alignment	-	+	+	-	-
Through put Speed	- -	+	++	+	-
Cost	-	+	+	+	+

Figure 1 show a schematic of the SHTB. The Striker is accelerated by a pressurized air system and impacts the transfer flange on the far left. This imparts a tension wave into the input bar which travels through the bar towards the specimen. The tension wave is transferred to the specimen after passing through the threaded interface, between the input bar and the mounting hardware. The stress of the specimen is determined using the strain gauges on the output bar. To get the force transferred to the output bar by the specimen a calibration for the bar was performed. The displacement of the specimen is directly observed using a high speed camera and then the strain computed using Digital Image Correlation DIC, since better results can be achieved if strain is measured locally [6]. To measure the strain in the specimen a Vision Research Phantom 1610 high speed camera was used. These photographs were then imported into Gom ARAMIS software and the displacements and strains computed.

2.2. Specimen mounting

Various mounting method have been used for flat specimens: adhesively bonded end caps [3], form-fit grips [5], bolted (dowel-pin) [4] specimen into bars and clamping the specimen. Clamping can be based on the inertia of the clamps at dynamic experiments or based on mechanical force and friction. To compare the general aspects of the different methods Table 1 has been included to give an overview of the trade-offs involved in the different designs. The trade-offs below are compared where a plus sign indicates a positive relative comparison with the other methods and a negative sign indicates a lower score. The categories are roughly ordered based on importance but that will naturally depend on the project at hand. The authors recognize that the Table 1 is a subjective evaluation but feel it helps identifying the strengths of the different methods. The research in this paper has been conducted in order to help quantify these distinctions.

A short description of the terms above will help clarify the thinking behind the table. Stress concentration denotes the departure from uniform stress distribution across any cross section affected by specimen mounting and specimen geometry; this will have an effect on the stress state of the sample and will often cause the failure outside of the gauge section. Additional mass is a negative factor for mounting specimen. It is ideal to minimize this term having to accelerate more mass will lower the loading rate of the specimen. Impedance difference is the change in impedance between the bar and the mounting hardware. With a large difference there will be more reflections in the hardware which will affect the measured stress signal in the output bar. By wave symmetry the loading path is considered and how symmetric the transition from the round bar to the flat specimen is. The thermal refers to the ability to use the mounting methods at temperatures significantly below or above the room temperature. Specimen alignment is how difficult it is to center the specimen in the mounting hardware. Throughput speed considers how quickly can the specimen be prepared and exchanged to conduct multiple tests. And cost considers the cost of the hardware to conduct a test using the mounting method.

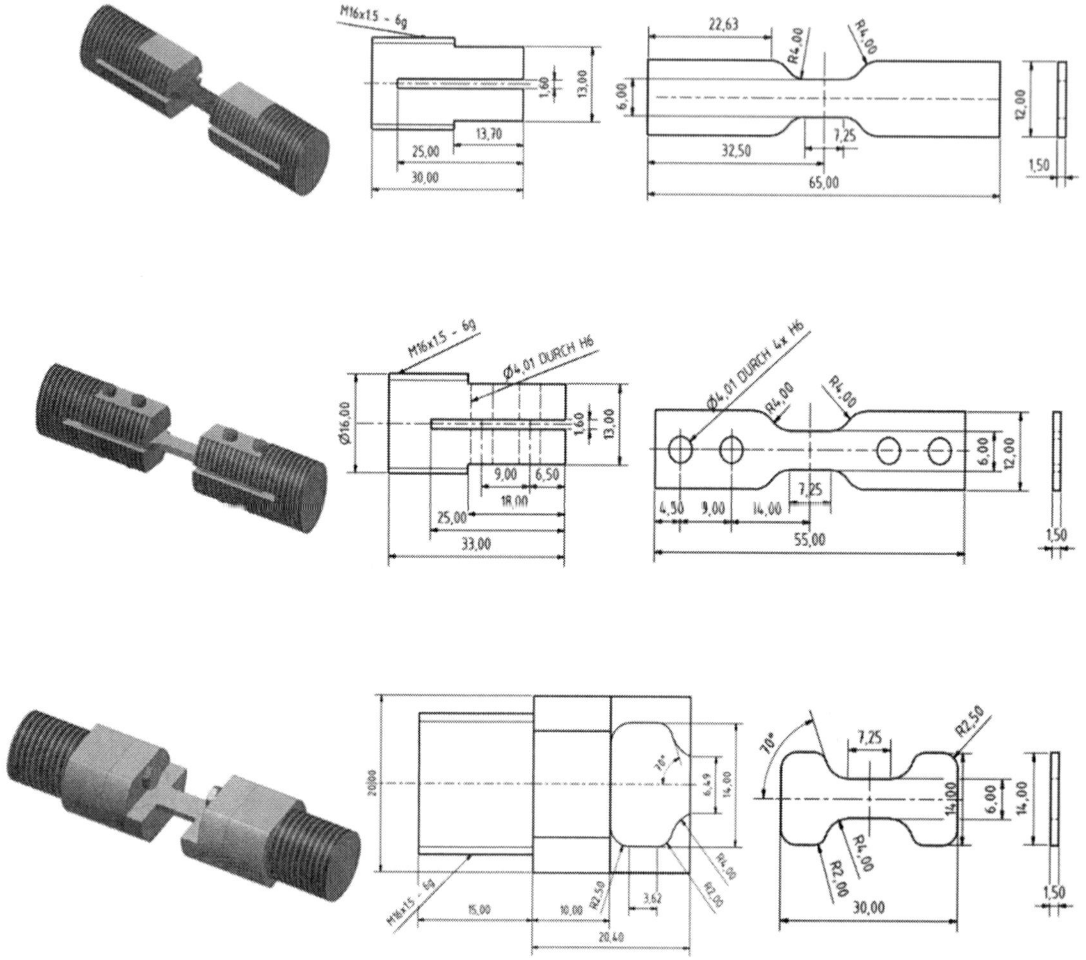

Figure 2. Different mounting methods: adhesively bonded mounting hardware (top), dowel-pin mounting hardware (middle), form-fit mounting hardware (bottom).

Since only room temperature tests were planned for the present project the adhesively bonded, dowel-pin and form-fit mounting were chosen for the investigations on the mounting methods.

For adhesively bonded specimens the mounting hardware consists of a threaded rod with a slit cut into the center where the specimen is adhesively bonded as shown in Fig. 2 (top). The adhesive used for the mounting of the specimen is LOCTITE EA 9514 which has a nominal shear strength of 50 MPa with a cure time of an hour at 150°C. The shear strength was used to determine the required surface area of the sample and consequently the length of the adhesively bonded mounting hardware. The amount of vertical space for the slit, 0.5 mm on both sides, was determined by the available tooling size and was found to be sufficient for a strong hold. The flat surfaces on the ends of the threaded rod are to allow a wrench to torque the mounting hardware into the end of the bars.

Dowel-pin mounting is using two dowels on each end of the specimen to attach it to the mounting hardware as shown in Fig. 2 (middle). Here the load is transferred through the dowel pins into the specimen. There is a stress concentration created on the pins. The design of the specimen needs to account for this and ensure that there is enough material at the connection to allow the specimen to break in the middle of the specimen. The tolerance on the dowel pin holes was selected allowing removal of the pins after a test so that the mounting hard ware could be reused.

The form-fit mounting option consists of a precisely cut specimen and mounting attachment that transfers the load on the tapered section of the specimen. This can be seen in Fig. 2 (bottom). In the form-fit mounting the specimen is placed into the mounting and the tension load is transferred through the tapered sides of the specimen and the mounting hardware. This will create a stress concentration and the geometry of the design is critical for successful use [5]. Finite element simulations were conducted to optimize the design to find an angle and end width that allows the stress to be spread in the ends of the specimen allowing for the failure to occur in the gauge section. A top is included on the hardware to ensure that

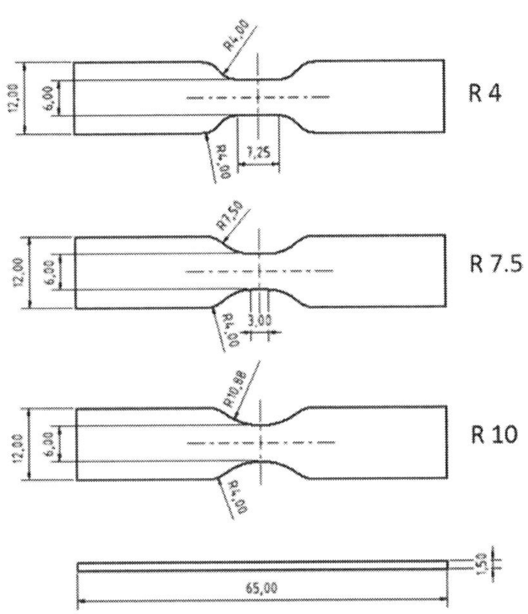

Figure 3. Dogbone specimen geometrics.

the specimen does not come out of the pocket during the test possibly causing injury or damaging the camera.

For all design options, specimens with the same gauge cross section area and length of specimens as well as the same initial radius for the tapering were used. They were all tested using the same method and striker speed.

2.3. Material and specimen geometry

To ensure that the specimen breaks in the middle of the specimen dogbone specimen are used where the sides of the center section are tapered to a narrower width than the initial section that is secured in the specimen mounting. This tapering leads to a stress concentration in the specimen. To observe the effects of this stress concentration three geometries were selected and can be seen in Fig. 3. These specimen held constant the open length between the mounting connections and varied the radius of the transition from the initial width of 12 mm to the gauge width of 6 mm. The radii chosen are a sharp transition of 4 mm radius and parallel gauge length of 7.25 mm, a middle value of 7.5 mm with a parallel length of 3 mm, and a large radius of 10.88 mm where there is no parallel gauge length. For the evaluation of the specimen geometry the adhesively bonded mounting was used. The material the specimen were made out of a high strength steel. The material is being developed by Salzgitter.

3. Results

3.1. Mounting method

Figure 4 shows the results of specimens tested with the different mounting methods. For all shown results the specimen geometry was R4 (see Fig. 3). There are ten results from tests with the adhesively bonded mounting and three from the form-fit and the dowel-pin mounting.

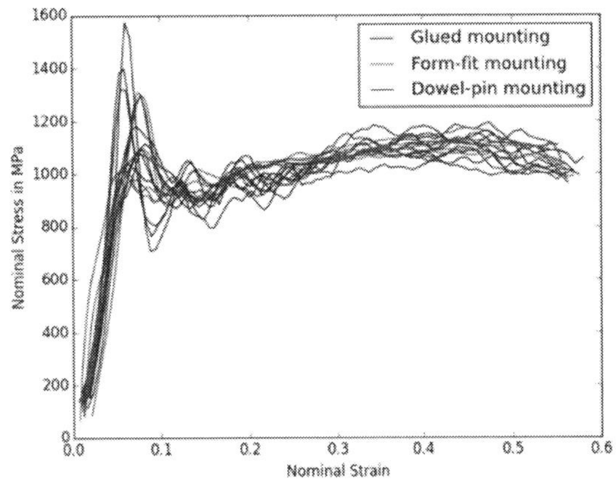

Figure 4. Nominal stress strain curves with different mounting hardware.

The results are shown as nominal stress versus nominal strain curves. This means that the stress is calculated using the initial cross section area of the specimen gauge section. The strains were measured using Digital Image Correlation. For the calculation of the strains displacements at two points with an initial distance of 7 mm were evaluated. With all mounting methods the same principal behavior is measured. The stresses and strains at the end of the tests are in the same range. There is a difference of the mounting methods at the beginning of the test. Each mounting method results in a different amount of oscillation. There is little oscillation for the form-fit mounting method. Both the dowel-pin and the adhesively bonded mounting produce results with significant oscillation.

It is obvious that the dowel-pin mounting results in oscillations, since the slit in the mounting is slightly bigger than the thickness of the specimen. In order to reduce the oscillations the slit needs to be reduced exactly to the specimen thickness. Since the specimen is firmly fixed for the adhesively bonded mounting the oscillations are not obvious in this case. One possibility is that the specimens are not positioned in the center of the endcaps (in hight and in line of the bars). This can result in lateral oscillations of the specimens which leads to the oscillations of the results. In the form fit mounting the correct positioning of the specimen is easily achieved.

The strain rate can be estimated with the striker velocity and the length of the parallel section of the specimen, since most of the elongation will happen in this area. The striker velocity was at 12 m/s. All specimens for the tests for the inverstigation of the mounting methods have a parallel section of 7.25 mm. This results in a expected strain rate of 1655 s^{-1}. Figure 5 shows the strain rate of the strains measured on the specimens. At all specimens the strain rate at the beginning and at the end of the test is at about 1200 s^{-1}. During the test there is a drop of strain rate. One reason for the lower actual strain rate is that the parts of the specimens beyond the parallel section are loaded as well. Another reason is that there is deformation in the mounting. For the form-fit mounting the drop is much higher than for the other two

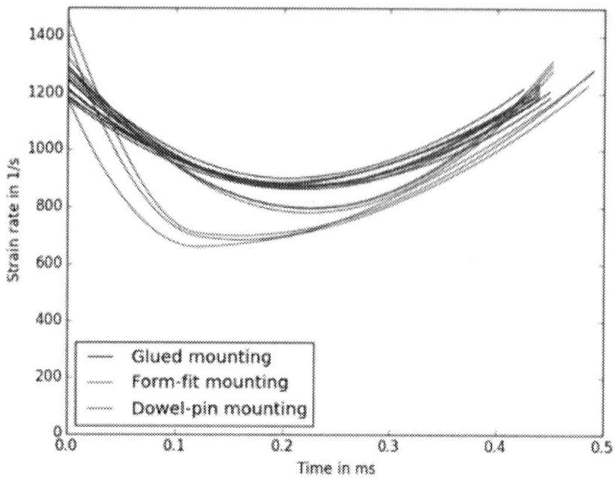

Figure 5. Strain rate with different mounting hardware.

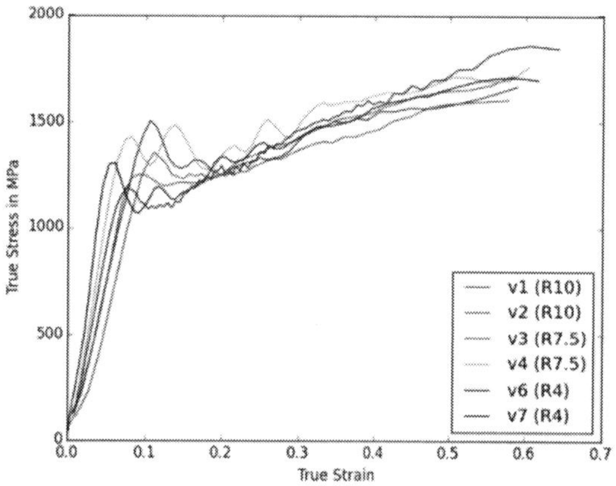

Figure 6. Local true stress strain curves for different specimen geometries.

mounting methods. There might be more deformation in the mounting than at the other mounting-methods.

3.2. Specimen geometry

Three different specimen geometries with different tapering were investigated (Fig. 3). Due to the chosen tapering, the length of the parallel section is different. For this reason the strains were evaluated as local true (logarithmic) strains in the fracture area. Additionally the strains perpendicular to the loading were used to calculate actual cross section area of the specimen and with that the true stresses. Figure 6 shows local true stress strain curves for the three tested specimen geometries. There is no significant influence of the specimen geometry on the results.

Figure 7 shows the local strain rates calculated from the local strains in the fracture area of the specimens. In contrast to the strain rates displayed in Fig. 5 which were calculated at the ends of the parallel section the local strain rates have a significant rise before failure. This can be explained with the localization of the material. There is a significant difference in strain rates for the different

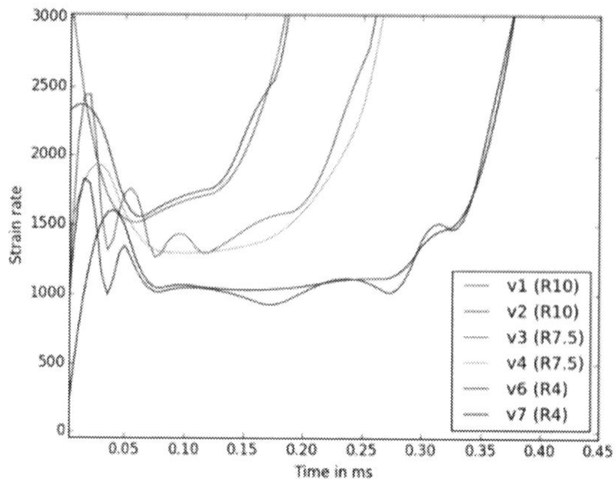

Figure 7. Local strain rates for different specimen geometries.

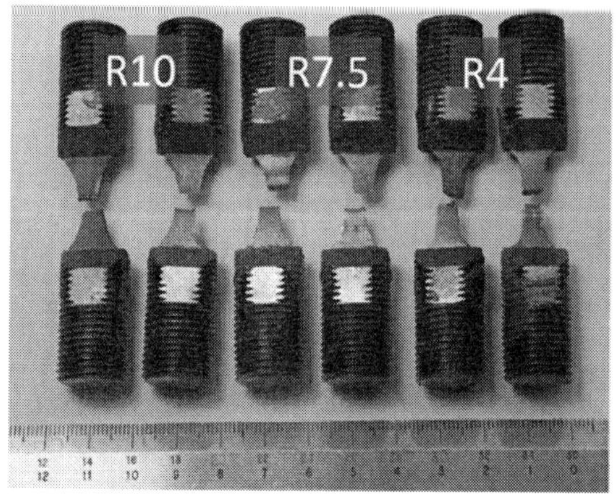

Figure 8. Investigation of influence of specimen geometry: Failed specimen.

specimen geometries. The specimen with the bigger radius (R10) has a higher localization of strain and with this a higher strain rate. Since the failure for all shown specimens was in the middle (see Fig. 8) the geometry with the smaller radius (R4) is favored.

4. Conclusions

In this work different mouting methods for Split Hopkinson Tension Bar Experiments (SHTB) were investigated. For the investigated sheet metal all three mounting methods (adhesively bonded, with dowel pin and with form-fit) give reproducible results. The adhesively bonded and the dowel-pinned hardware lead to oscillations at the beginning of the tests. The best results were gained with the form-fit mounting. The strain rate has a minimum in the middle of the test. The drop is higher for the form-fit mounting. Summing up all mounting methods are suitable for SHTB experiments. But since the form fit mounting is cost effective and leads to the best signal this mounting should be used for this kind of materials.

In the second part three different specimen geometries were investigated. For this specific material all specimen geometries lead to a failure in the center and are therefore possible to use. For the specimen with the smallest radius and the longest parallel section the loading is less localized and therefore this geometry is favored.

It has to be highlighted that the results are valid only for this kind of material with a fairly high fracture strain. For materials such as carbon fiber composites with a linear elastic material behavior and projects with other constraints other mounting methods and specimen geometries might be appropriate. A possibility to gain better results with the adhesively bonded mounting method would be pulse shaping.

In order to get a better understanding on the influence of positioning and the mass of the mounting on the results it is intended to preform finite element simulations of the Split Hopkinson Tension Bar experiments.

References

[1] Chen W., Song B., *Split Hopkinson (Kolsky) Bar – Desing, Testing and Applications* (Springer, New York Dordrecht Heidelberg London, 2011)

[2] Gerlach R., Kettenbeil C., Petrinic N., International Journal of Impact Egnineering **50**, 63–67 (2012)

[3] Gerlach R., Siviour C.R., Wiegand J., Petrinic N., Composites Science and Technology **72**, 397–411 (2012)

[4] Huh H., Kang W.J., Han S.S., Experimental Mechanics **42**, 8–17 (2002)

[5] Owens A.T., Tippur H.V., Experimental Mechanics **49**, 799–811 (2009)

[6] Schmidt T., Gilat A., Walker A., Seidt J., Tyson J., Proceedings of the Xth International Congress and Exposition, Society of Experimental Mechanics, Orlando (2008)

EPJ Web of Conferences **94**, 01050 (2015)
DOI: 10.1051/epjconf/20159401050

Specimen Inertia in high strain rate tensile testing

Meyrick Hockly and Clive R. Siviour[a]

Department of Engineering Science, University of Oxford, Parks Road, Oxford OX1 3PJ, UK

Abstract. It is well established that measurements of the dynamic mechanical response of materials are susceptible to errors due to the inertia of the specimen, which causes stresses in addition to the intrinsic material strength. A number of authors have derived equations for these stresses in compression experiments; these equations can be used as a guideline for good specimen design. However, no such equations have been presented in the literature for the equivalent effects in tensile experiments. This paper begins by considering and rationalising the equations available for compression, before producing a set of equations which can be used in design of specimens for, e.g., tensile Hopkinson bar experiments.

1. Introduction

It has long been known in the high rate testing community [1,2] that the apparent strength of any test specimen is a combination of the intrinsic material strength, and any effects due to specimen inertia. For loading at low rates of strain, the effect of inertia is minimal; however, it can be shown [1–5] that for large specimen sizes, dense materials, or materials with low intrinsic strength, the additional stresses required to overcome the axial and radial inertia of the specimen can represent a significant fraction of the measured loads in a high rate characterisation experiment. Indeed, as the inertial loads scale with the square of strain rate ($\dot{\varepsilon}^2$) and rate of change of strain rate ($\ddot{\varepsilon}^2$), their effects can rapidly become significant in dynamic testing.

A number of authors have derived equations to quantify the effects of specimen inertia in compressive Hopkinson bar experiments. The calculations can be divided into two categories: those which use an energy based approach, and those which use calculations of the stress-field in the specimen. Both approaches require some approximations to the behaviour of a real specimen.

This paper will begin by examining the different equations derived for compressive inertia, reconciling the different expressions obtained by the energy and stress based approaches. Then, extending the energy based analysis, expressions will be derived for the effects of inertia in a tensile experiment. Firstly, tension will simply be considered as a reverse compression on a long cylinder. Then, point masses will be added to mimic the ends of a dog-bone tensile specimen, before moving to an analysis which considers a more realistic specimen geometry.

2. Inertia in compression specimens

2.1. Definition of variables

This paper will adopt a similar nomenclature to Gorham, which in the first instance is designed to simplify the equations in compression. Later some of the definitions will be altered for the tension work. The symbols are defined as in Fig. 1. Strain is positive in tension.

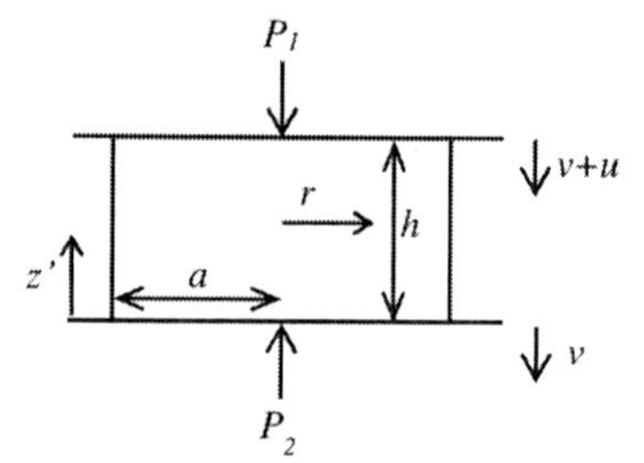

Figure 1. Definition of symbols used in this paper. P_1 and P_2 are the mean pressures between the specimen and input and output bars respectively. The output bar moves at speed v. The (θ, r, z') coordinate system is shown and is defined in the moving frame such that $z' = 0$ (always) at the bottom of the specimen and h (instantaneously) at the top; $r = 0$ at the axis and a at the outer edge. There is also an inertial (θ, r, z') system in which the bottom of the specimen has axial velocity $-v$, the rest of the specimen $-v - u_z(z) = -v - uz/h$. The radial velocity is $u_r(r)$.

2.2. Energy based approaches

A number of authors have considered the effects of specimen inertia by comparing the work done by the forces applied to the two ends of the specimen by the Hopkinson bars to the combination of kinetic and deformation energy in the specimen. The first approaches by Davies and Hunter [2] and Samanta [3] made a number of assumptions which ignored potentially significant terms: in particular assuming that the output end of the specimen was stationary. Gorham [4] considered a more general case, but still assuming that the strain and stress fields in the specimen are everywhere the same at any time. He derived the following equations for the stresses at the interface

[a] Corresponding author: clive.siviour@eng.ox.ac.uk

Article available at http://www.epj-conferences.org or http://dx.doi.org/10.1051/epjconf/20159401050

between the input bar and the specimen (P_1) and the interface between the specimen and the output bar (P_2)[1]

$$P_1 = \sigma_y - \rho\ddot{\varepsilon}\left(\frac{a^2}{8} + \frac{h^2}{3}\right) + \rho\dot{\varepsilon}^2\left(\frac{a^2}{16} - \frac{h^2}{3}\right) + \frac{\rho h\dot{v}}{2}; \quad (1)$$

$$P_2 = \sigma_y - \rho\ddot{\varepsilon}\left(\frac{a^2}{8} - \frac{h^2}{6}\right) + \rho\dot{\varepsilon}^2\left(\frac{a^2}{16} - \frac{h^2}{6}\right) + \frac{\rho h\dot{v}}{2}; \quad (2)$$

where a and h are the specimen radius and height, ρ the density, ε the strain rate and v the velocity of the interface between the specimen and the output bar. It is noted that in this case, the volume of the specimen is assumed to be constant during the experiment and that the strain and strain rate are negative in compression. Further, following Gorham, σ_y is the compressive yield stress, i.e. it is positive in compression, as are P_1 and P_2, whilst $\dot{\varepsilon}$ and $\ddot{\varepsilon}$ will typically have negative numerical values.

Earlier papers [2,3], by ignoring the uniform velocity field v, derived equations which only contained the first two expressions, e.g.

$$P_2 = \sigma_y - \rho\ddot{\varepsilon}\left(\frac{a^2}{8} - \frac{h^2}{6}\right). \quad (3)$$

A useful outcome of this result is that if $h = a\sqrt{3}/2$, the inertia term is cancelled. Since, in a normal Hopkinson bar experiment, $\ddot{\varepsilon}$ is only significant early on in the test, where the specimen dimensions have not changed significantly, this is often used as a useful guide for specimen design. However, one must also confirm that, for the specimen design chosen, the other inertia terms in Eq. (2) are not significant compared to σ_y.

2.3. Stress based analysis

Forrestal produced equations which explicitly consider the radial dependence of the stress in the specimen. Assuming again that the material is incompressible, and using an elastic analysis, he produced the following equation for the hydrostatic pressure, p, in the specimen due to radial inertia:

$$p(r) = \frac{\rho(a^2 - r^2)}{4}\ddot{\varepsilon}_z; \quad (4)$$

here, r is the radial position measured from the axis of the specimen and ε_z the longitudinal strain. The obvious implication of this equation is that the hydrostatic pressure is largest at the centre of the specimen, as might be expected, and that the use of annular specimens might reduce the effect of inertia when performing measurements on weak or dense materials. This has been confirmed by experimental studies [6].

2.4. Comparison of approaches

The total additional stress predicted by the stress analysis may be calculated by integrating over the specimen surface, to give

$$P_{additional} = \frac{1}{\pi a^2}\int_0^a \frac{\rho\left(a^2 - r^2\right)}{4}\ddot{\varepsilon}\, 2\pi r\, dr = \frac{\rho\ddot{\varepsilon}a^2}{8}, \quad (5)$$

such that, in the terminology of the energy based approach

$$P_1 = P_2 = \sigma_y - \frac{\rho\ddot{\varepsilon}a^2}{8}. \quad (6)$$

Because the analysis considers only a thin slice, dz, in the axial direction of the specimen, and because it assumes that the stress is constant in z, it does not give any difference between the stresses at the two bar-specimen interfaces, nor does it take into account the effects of the specimen height.

We will now modify the stress-based analysis, and show that it gives results consistent with the energy based approach. Firstly, it will be assumed that the material is behaving plastically according to a Mises yield criterion, and that the uniaxial yield stress is σ_y. It will also be assumed that $\sigma_r = \sigma_\theta$ and $\varepsilon_r = \varepsilon_\theta$. These assumptions are proved in Forrestal's elastic analysis, but will instead be assumed in this plastic treatment. We define the true strains in the axial and radial directions as $\varepsilon_z = ln[h/h_0]$ and $\varepsilon_r = ln[r/r_0]$, and assume that the specimen is incompressible, so that $\varepsilon_r = -\varepsilon_z/2$. The relationships between velocity and acceleration and true strain rate are given by

$$u_z = -u = h\dot{\varepsilon}_z; \dot{u}_z = -\dot{u} = h[\ddot{\varepsilon}_z + \dot{\varepsilon}_z^2] \quad (7)$$

and

$$u_r = r\dot{\varepsilon}_r; \quad \ddot{u}_r = r[\ddot{\varepsilon}_r + \dot{\varepsilon}_r^2] = -\left[\frac{\ddot{\varepsilon}_z}{2} - \frac{\dot{\varepsilon}_z^2}{4}\right]. \quad (8)$$

The calculations will proceed assuming that it is possible to treat the radial and axial motions separately, and a brief discussion of this assumption will be given later on.

Firstly, we write the equilibrium equation in the radial direction

$$\frac{d\sigma_r}{dr} + \frac{\sigma_r - \sigma_\theta}{r} = -\rho\frac{d^2u_r}{dt^2}, \quad (9)$$

and, given that $\sigma_r = \sigma_\theta$ we note that

$$\frac{d\sigma_r}{dr} = -\rho r\left[\ddot{\varepsilon}_r + \dot{\varepsilon}_r^2\right] \quad (10)$$

given that $\sigma_r = 0$ at $r = a$,

$$\sigma_r = -\frac{\rho}{2}\left[\ddot{\varepsilon}_r + \dot{\varepsilon}_r^2\right](a^2 - r^2) = \frac{\rho}{2}\left[\frac{\ddot{\varepsilon}_z}{2} - \frac{\dot{\varepsilon}_z^2}{4}\right](a^2 - r^2). \quad (11)$$

Assuming a Mises yield criterion, and noting that whilst the σ_r above is negative in compression the convention in this work is for compression to be positive, we can write that $\sigma_z + \sigma_r = \sigma_y$, such that

$$\sigma_z = \sigma_y - \sigma_r = \sigma_y - \frac{\rho}{2}\left[\frac{\ddot{\varepsilon}_z}{2} - \frac{\dot{\varepsilon}_z^2}{4}\right](a^2 - r^2). \quad (12)$$

Compared to the analysis of Forrestal, this equation contains an extra term in $\dot{\varepsilon}_z^2$; this term is essentially due to

[1] A mistake in Gorham's Eq. (9a) has been corrected here, the original reads $P_1 = \sigma_y + \rho\ddot{\varepsilon}\ldots$

the change in axial velocity required to deliver a constant true strain, and so appears in the relationship between axial acceleration and true strain rate.

This stress can now be integrated over the specimen surface to give

$$P_1' = P_2' = \sigma_y - \frac{\rho}{8}a^2\left[\ddot{\varepsilon}_z - \frac{\dot{\varepsilon}_z^2}{2}\right]$$
$$= \sigma_y - \frac{\rho}{8}a^2\ddot{\varepsilon}_z + \frac{\rho}{16}a^2\dot{\varepsilon}_z^2, \qquad (13)$$

which agrees with the terms in a^2 from the energy based analysis.

In order to introduce axial variation, we note that

$$(P_1 - P_2)\pi a^2 = m\left(\frac{\ddot{u}}{2} + \dot{v}\right)$$
$$\rightarrow (P_1 - P_2) = \rho h\left(\frac{h[\ddot{\varepsilon}_z + \dot{\varepsilon}_z^2]}{2} + \dot{v}\right), \qquad (14)$$

and more generally that

$$(P(z) - P_2)\pi a^2 = m\left(\frac{\dot{u}z}{2h} + \dot{v}\right)$$
$$\rightarrow (P(z) - P_2) = \rho h\left(\frac{-[\ddot{\varepsilon}_z + \dot{\varepsilon}_z^2]z}{2h} + \dot{v}\right), \qquad (15)$$

where $P(z)$ is introduced as the mean axial pressure at any axial position in the specimen.

Overall, then, we can say that

$$P_2 = \sigma_y - \frac{\rho}{8} + \frac{\rho}{16}\dot{\varepsilon}_z^2 a^2 + constant \qquad (16)$$

$$P_1 = \sigma_y - \frac{\rho}{8}a^2\ddot{\varepsilon}_z + \frac{\rho}{16}a^2\dot{\varepsilon}_z^2 + constant \qquad (17)$$
$$- \frac{\rho h^2\ddot{\varepsilon}_z}{2} - \frac{\rho h^2\dot{\varepsilon}_z^2}{2} + \rho h\dot{v}.$$

These equations are therefore entirely consistent with the results from the energy-based analysis *except* that they give no indication of how to derive the constant, which we can see from the other analysis is in fact

$$\frac{\rho h^2\ddot{\varepsilon}_z}{6} + \frac{\rho h^2\dot{\varepsilon}_z^2}{6} - \frac{\rho h\dot{v}}{2}.$$

3. Inertia in tension specimens

The description of inertia in tension specimens will be split into three parts: use of the compressive equations with the directions reversed; addition of lumped masses; consideration of the shape of the specimen to give lower and upper bounds.

A typical tensile specimen design is a flat or cylindrical dogbone with a gauge section, shoulders and endcaps, Fig. 2(a). The simplest assumption is that the shoulders and endcaps do not exist, and that only the gauge length of the specimen is required, Fig. 2(b). This would be the case, for example, if the forces were measured at the end of the gauge section. The second assumption is equivalent to saying that the shoulders and endcaps do not deform, which might be a good approximation to the real behaviour, and instead just act as point masses on the end of the gauge length, Fig. 2(c). More realistically, some undefined portion of the shoulders deforms significantly, Fig. 2(d). In a perfect specimen of an elastic perfectly plastic material, plastic deformation would be confined to the gauge length, but because the radial inertia increases the axial stress required to yield the gauge length, plastic deformation would in fact enter the shoulders in a real tensile scimen deformed at high rates of strain.

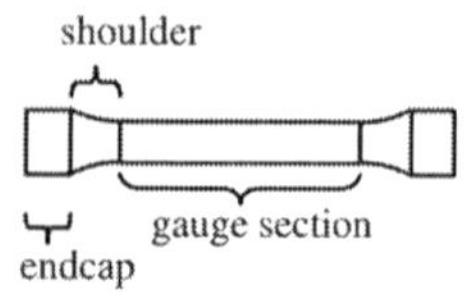

Figure 2(a). Schematic diagram of a typical tensile test specimen.

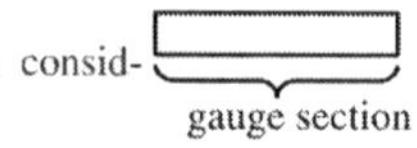

Figure 2(b). Specimen considered with no endcaps.

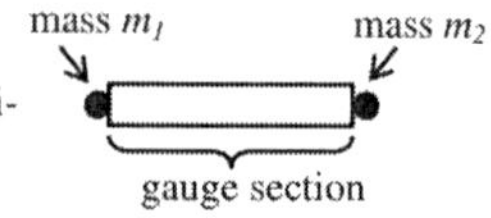

Figure 2(c). Specimen approximated with non-deforming endcaps.

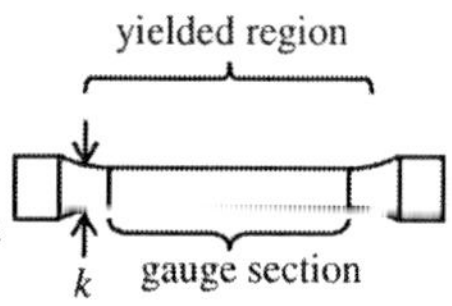

Figure 2(d). Specimen in which shoulders are allowed to yield.

3.1. Reverse compression

A detailed analysis is not presented here, but following the arguments of Samanta and Gorham shows that the mean end stresses in a cylindrical specimen loaded in tension are the reverse of those found in compression

$$P_1 = \sigma_y + \rho\ddot{\varepsilon}\left(\frac{a^2}{8} + \frac{h^2}{3}\right) - \rho\dot{\varepsilon}^2\left(\frac{a^2}{16} + \frac{h^2}{3}\right) + \frac{\rho h\dot{v}}{2} \qquad (18)$$

$$P_2 = \sigma_y + \rho\ddot{\varepsilon}\left(\frac{a^2}{8} - \frac{h^2}{6}\right) - \rho\dot{\varepsilon}^2\left(\frac{a^2}{16} + \frac{h^2}{6}\right) - \frac{\rho h\dot{v}}{2}, \qquad (19)$$

where the directions of P_1, P_2 and v are reversed and σ_y is positive in tension. We note now that that $\dot{\varepsilon}$ and $\ddot{\varepsilon}$ will typically have positive numerical values.

3.2. "Lumped Mass"

If the endcaps do not deform, they may simply be considered as masses on the end of the cylindrical specimen. They must be accelerated, so work must be done to provide their kinetic energy; however, there is

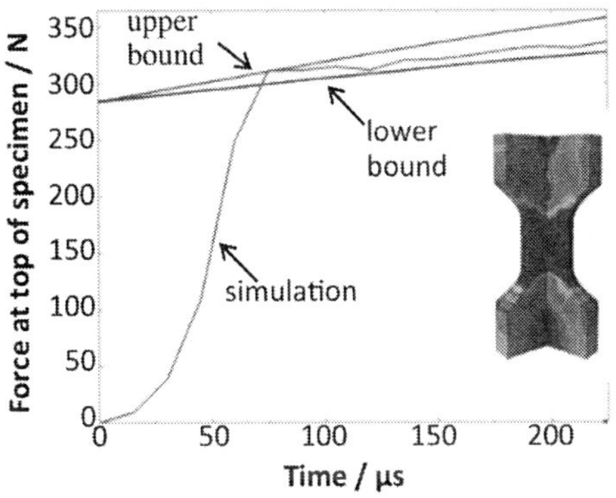

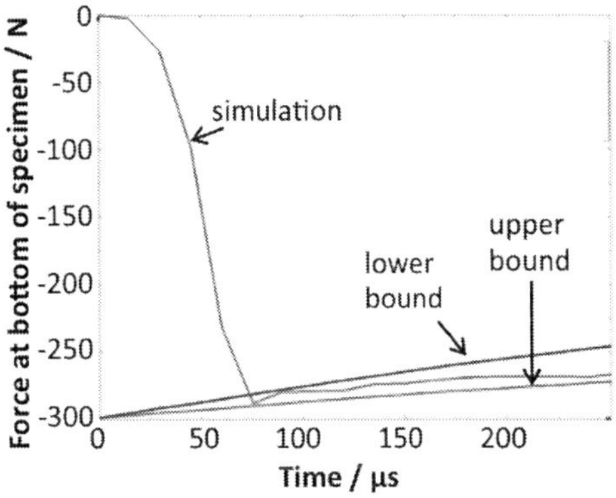

Figure 3. Comparison of specimen end forces taken from an FE simulation with upper and lower bound solutions. The specimen is shown.

no dormation energy. The two modifications made to the analysis of Samanta or Gorham are to state that

$$(P_1 - P_2)\pi b^2 = \rho\pi a^2 h\left(\frac{\dot{u}}{2} + \dot{v}\right) + m_1(\dot{u} + \dot{v}) + m_2\dot{v} \tag{20}$$

where b is the radius of the endcaps, and m_1, m_2 their masses (usually $m_1 = m_2$), and also to express the kinetic energy as

$$\begin{aligned}\dot{K} &= \dot{K}_{m1} + \dot{K}_{m2} + \dot{K}_{cylinder} \\ &= m_1(u + v)(\dot{u} + \dot{v}) + m_2 v\dot{v} + \dot{K}_{cylinder}\end{aligned} \tag{21}$$

where $\dot{K}_{cylinder}$ is the change in kinetic energy of the gauge section as calculated by Samanta and Gorham. It is noted that P_1 and P_2 are now defined at the end of the endcaps. The result of the analysis is

$$\begin{aligned}P_1 = {} & \frac{a^2}{b^2}\sigma_y + \frac{m_1}{\pi b^2}(\dot{u} + \dot{v}) \\ & + \rho\frac{a^2}{b^2}\left[\ddot{\varepsilon}\left(\frac{a^2}{8} + \frac{h^2}{3}\right) - \dot{\varepsilon}^2\left(\frac{a^2}{16} + \frac{h^2}{3}\right) + \frac{\dot{v}h}{2}\right]\end{aligned} \tag{22}$$

$$\begin{aligned}P_2 = {} & \frac{a^2}{b^2}\sigma_y - \frac{m_2}{\pi b^2}\dot{v} \\ & + \rho\frac{a^2}{b^2}\left[\ddot{\varepsilon}\left(\frac{a^2}{8} - \frac{h^2}{6}\right) - \dot{\varepsilon}^2\left(\frac{a^2}{16} + \frac{h^2}{6}\right) - \frac{\dot{v}h}{2}\right]\end{aligned} \tag{23}$$

so that the force on each end of the specimen is further modified by the mass of the endcap on that end of the specimen.

3.3. Deformable shoulders

In an ideal tensile specimen, the axial stress at the point where the gauge length meets the shoulders would be σ_y (and all other stress components would be zero). In fact, this is not the case, and the axial stress is $\sigma_y + \sigma_i$ (where σ_i is the inertial stress). Because the shoulders deform less than the specimen, it is likely that the inertial stress in the shoulders is less than σ_i, and therefore there is potential for the shoulders themselves to yield if the reduction in σ_i is not sufficiently countered by an increase in radius. This leads to an increase in the work done to deform the material, and therefore an increase in the measured pressures at the ends of the bars.

The actual situation is complex, but may be simplified by thinking of upper and lower bounds. The lower bound on the additional stress is provided in the previous section (Eqs. (22) and (23)), because no additional deformation energy is required to be provided by P_1 and P_2. The upper bound will be calculated by assuming that the specimen yields to a new section whose radius is k, such that $\pi k^2\sigma_y = \pi a^2(\sigma_y + \sigma_i)$, where σ_i is the axial stress owing to inertia in the gauge length. An overestimate of the additional work required is provided by $\dot{W} = \pi k^2 u\sigma_y = \pi a^2 u(\sigma_y + \sigma_i)$; this is an overestimate because it assumes that the radius of the *whole* deforming region is k.

In these calculations, the assumption is made that we may use u as the representative velocity for the work calculations. Until now, u has been the velocity of the gauge section, now it is the velocity of the interface between yielded and yielded material in the shoulder. In practice, we may reconcile this by stating that u is the relative velocity of the ends of the Hopkinson bars, which, during plastic deformation of the gauge section, will indeed be a good representation of the velocity of this interface (whether or not it coincides with the end of the gauge section). In any case, as the gauge section now moves more slowly than the shoulders, the real value of σ_i will be less than we assume in this calculation. We will continue to use the equations described above for the relationship between u, $\dot{\varepsilon}_z^2$ and $\ddot{\varepsilon}$, even though we know that they are no longer precisely correct for the strains in the gauge length. It is possible to adopt an iterative approach to correct this, but this is likely to be too much effort in practice, and the current equations should be suitable as a design guideline. Despite these issues, the approximation is used because it gives a neat result which will be particularly suitable as a design guideline: the contribution of the cylinder inertia to the measured stress simply doubles:

$$\sigma = \sigma_y + 2\sigma_i \tag{24}$$

so that

$$P_1 = \frac{a^2}{b^2}\sigma_y + \frac{m_1}{\pi b^2}(\dot{u} + \dot{v}) + 2\rho\frac{a^2}{b^2}\left[\ddot{\varepsilon}\left(\frac{a^2}{8} + \frac{h^2}{3}\right) - \dot{\varepsilon}^2\left(\frac{a^2}{16} - \frac{h^2}{3}\right) - \frac{\dot{v}h}{2}\right] \tag{25}$$

$$P_2 = \frac{a^2}{b^2}\sigma_y - \frac{m_2}{\pi b^2}\dot{v} + 2\rho\frac{a^2}{b^2}\left[\ddot{\varepsilon}\left(\frac{a^2}{8} - \frac{h^2}{6}\right) - \dot{\varepsilon}^2\left(\frac{a^2}{16} + \frac{h^2}{6}\right) - \frac{\dot{v}h}{2}\right]. \tag{26}$$

A number of Finite Element simulations have been performed with a range of different specimen geometries. The results from one of these simulations are compared to the upper and lower bound solutions in Fig. 3.

4. Conclusions

Equations have been derived for the effects of specimen inertia on the observed mechanical response of dog-bone specimens under dynamic tensile loading. In particular, by making assumptions about the yielding behaviour of the specimen outside the gauge length, it is possible to calculate quantitative lower and upper bounds on the extra stress observed owing to this inertia. The equations produced can be used as design guidelines for high strain rate tensile specimens.

References

[1] H. Kolsky, Proc. Phys. Soc. B**62** (1949)

[2] E.D.H. Davies and S.C. Hunter, J. Mech. Phys. Solids **11** (1963)

[3] S.K. Samanta, J. Mech. Phys. Solids **19** (1971)

[4] D.A. Gorham, J. Phys. D: Appl. Phys. **22** (1989)

[5] M.J. Forrestal, T.W. Wright, W. Chen, International Journal of Impact Engineering **34** (2007)

[6] B. Song, Y. Ge, W.W. Chen T. Weerasooriya, Experimental Mechanics **47** (2007)

EPJ Web of Conferences **94**, 01051 (2015)
DOI: 10.1051/epjconf/20159401051

Experimental and numerical study of a bird strike against a windshield

Fabien Plassard[a], Pierre-Louis Hereil, Pierric Joseph, and Jérôme Mespoulet

THIOT INGENIERIE, Route Nationale, 46130 Puybrun, France

Abstract. This paper describes a bird strike study performed at THIOT-INGENIERIE laboratory. For aeronautic requirements, the 220 mm version of the gas gun TITAN is used to perform bird strike on instrumented structures. This paper shows a bird strike on a polycarbonate windshield instrumented with high speed cameras, velocity and force sensors. A crossed work with numerical simulation has been performed to design target support and diagnostic tools. It permits also to demonstrate reliability of the numerical tools.

1. Introduction

Since the beginning of the aeronautic age, around 250 deaths and 80 destroyed civil aircrafts have been caused by bird strikes [1]. Moreover, with the increase of the traffic, the cost of bird strike damages and delays is higher than 1 billion dollars in the world. Civil Aircraft in the United States reported 30 bird strikes per day in 2013 [2,3]. The vulnerable parts of planes or helicopters are windshield, nose, fuselage panels, wing and empennage leading edges, rotor blades, fan blades and engines inlets (Fig. 1). Bird strike resistance has to be proved for the dynamic certification tests. During the design of aeronautic parts, numerical simulations allow studying the best technical solution to get the best structural behaviors.

A simultaneous work between experiments and numerical simulations is necessary to well understand the behavior of a bird during an impact and also the reaction of the impacted material and structure.

This paper deals with the study of a bird strike against a windshield performed at THIOT-INGENIERIE shock physics laboratory.

The first part presents the experimental configuration and the metrology of the test and the second part shows the associated numerical simulation performed to anticipate and fit the results.

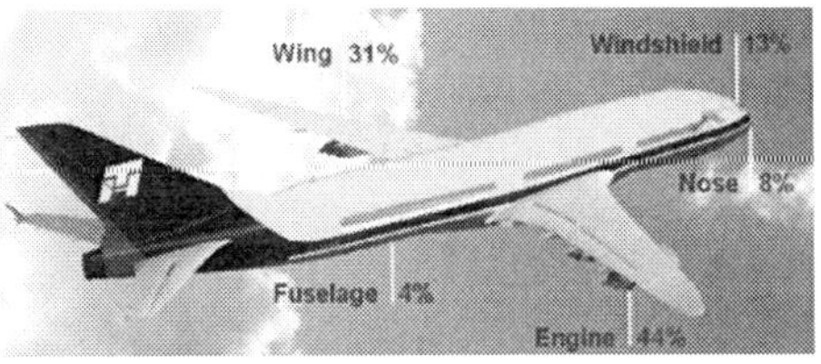

Figure 1. Aircraft component exposed to bird strike.

Figure 2. TITAN light gas gun with 220 mm calibre.

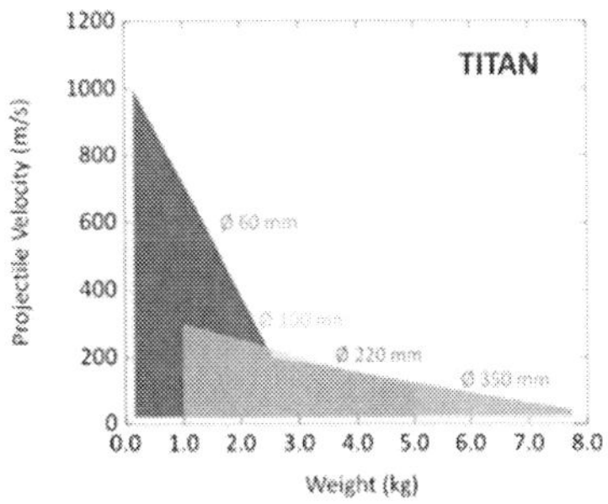

Figure 3. Projectile velocities achievable for each configuration of the TITAN LGG.

2. Experimental setup

The TITAN Light Gas Gun (LGG) is a single stage launcher (Fig. 2) which can be configured using different barrel diameters: Ø60 mm, Ø100 mm, Ø 220 mm and Ø 350 mm [4–6]. Thanks to this particularity, this LGG allows to obtain several impact velocities with numerous types of projectiles or targets. For each TITAN LGG configuration, the projectile velocity achievable with a particular projectile weight is summarized on Fig. 3.

The objective of the experimental configuration is to launch a 622 g bird at a velocity of 125 m/s on a polycarbonate plate which is representative of a windshield. A schematic description of the experimental set-up is presented on Fig. 4. The windshield was positioned on a dedicated tool which provides both angular positioning and suitable impact location. Target positioning was realized using a laser alignment.

[a] Corresponding author: `technique@thiot-ingenierie.com`

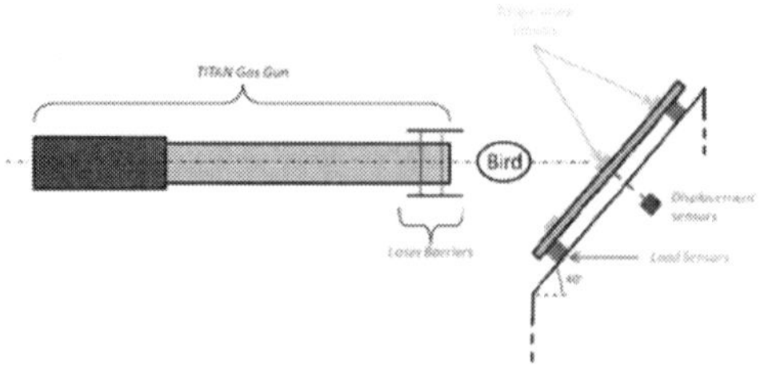

Figure 4. Schematic description of the experimental setup.

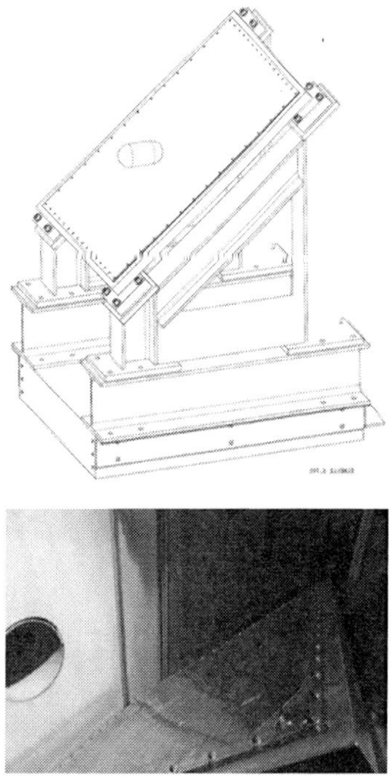

Figure 5. 3D view of the dedicated tool used to position the test parts (top) – Gas gun exit and windshield target (bottom).

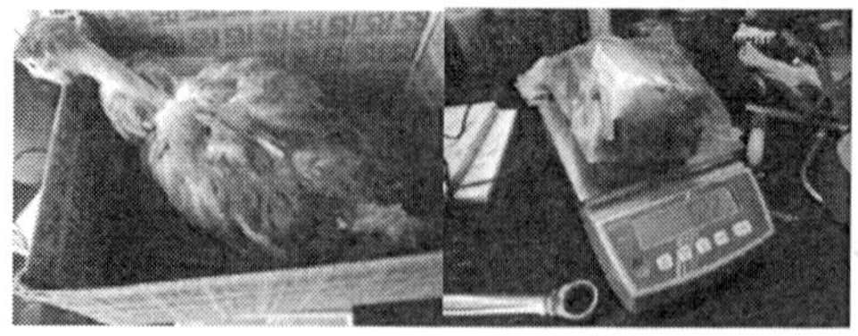

Figure 6. Alive bird (on the left) - Wrapped dead bird before test (right).

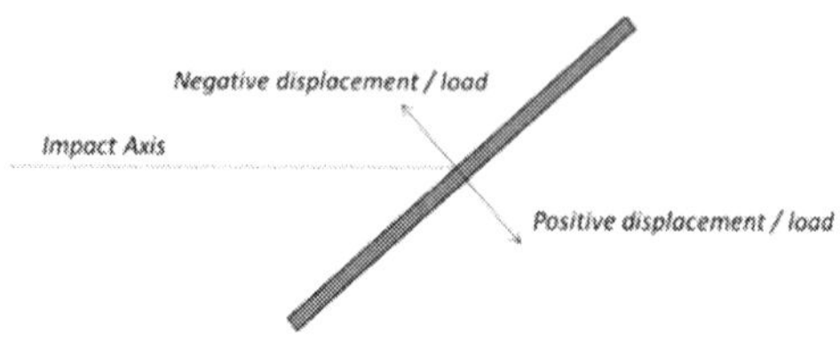

Figure 7. Sign convention for displacement and load measurements.

Table 1. Precision of CFW/330kN sensors with the full measuring system.

Sensor number	Precision (%)
F1	0.04
F2	0.01
F3	0.08
F4	0.02

Table 2. High speed video cameras settings.

Camera ID	Optics	Frames per sec	Resolution
Photon Ultima APX	Lens 17–35 mm, diaphragm 2.8	6000	512 × 512
Phantom v710	Lens 24–70 mm, diaphragm 2.8	6000	1280 × 800
Phantom M320S	Lens 17–35 mm, diaphragm 2.8	6000	832 × 544

Laser barriers are used for impact velocity measurement and displacement sensors, load sensors and high speed cameras are used to analyse the behaviour of the projectile and the target during impact.

The impactor is a whole dead chicken wrapped with adhesive tape. It is enclosed in a sabot that is designed to carry it at the right location on the windshield target. The total weight of the impactor is 622 g.

The target is a 1046 mm × 826 mm polycarbonate plate of 8 mm in thickness. It is screwed on the tool in order to avoid the lateral displacement and to have a well-known condition at the periphery of the plate.

Photographs of the windshield target and the bird projectile are presented on Fig. 5 and Fig. 6.

2.1. Dynamic load sensors

Four dynamic load sensors were positioned at the four corners of the windshield (green arrows on Fig. 4). Table 1 summarizes actual precision observed on each load sensor with the full measuring system. Figure 7 indicates sign convention for load measurements.

2.2. Velocity sensor

The displacement of the windshield was measured using a heterodyne velocimetry system (PDV). This kind of system measures the local velocity of the windshield wall at the impact point through Doppler Effect. Displacement is then calculated using this measure of velocity. Typical error is 1% in the $-2000\,\mathrm{m.s^{-1}}$ / $+2000\,\mathrm{m.s^{-1}}$ range. Bandwidth of this system is estimated to 31 MHz. Sign convention for displacement measurement is indicated on Fig. 7.

2.3. High speed cameras

Three high speed video cameras were used during tests:

- Photron ULTIMA APX,
- Phantom v710,
- Phantom M320S

The positions of the cameras around the target are schematized on Fig. 8 and cameras settings are reported in the following table.

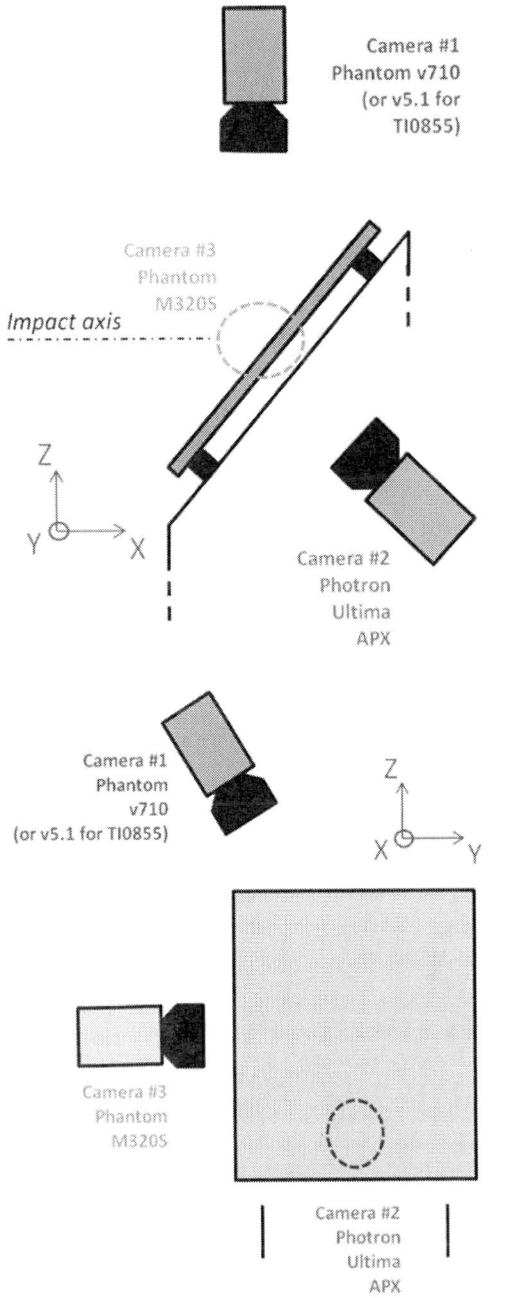

Figure 8. Schematic description of the cameras locations during tests (top – side view; bottom – front view).

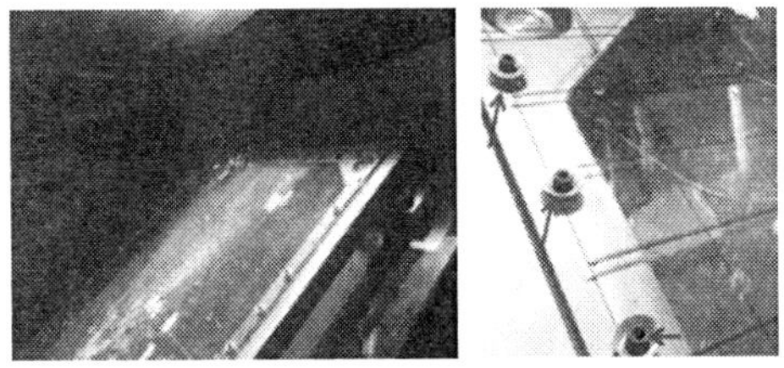

Figure 9. Windshield after test.

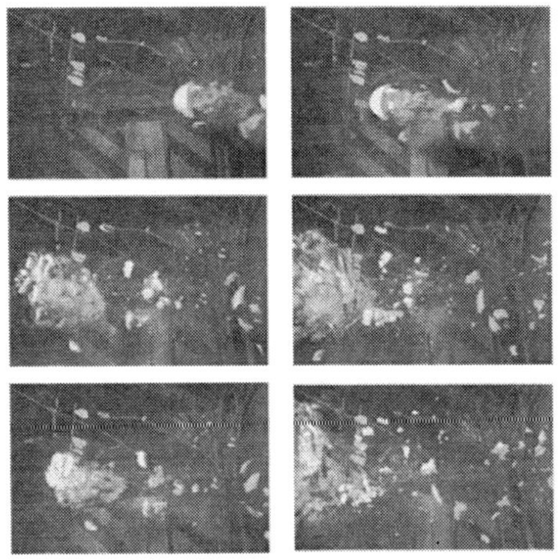

Figure 10. Pictures taken from high speed camera #1 (top view).

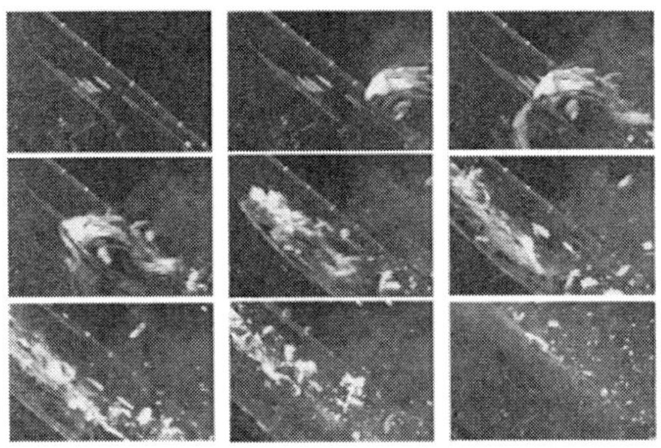

Figure 11. Pictures taken from high speed camera #3 (lateral view).

3. Experimental results

Experimental results obtained during this bird strike test #TI0849 are indicated on Fig. 9 to Fig. 14. Measured projectile velocity is 128 ± 2 m/s.

Photographs of windshield after the test (Fig. 9) indicates:

- no break of the polycarbonate plate and no visible residual strains: the material remained elastic during the impact,
- strong residual strains on the washers of the screws (Red arrows on Fig. 9).

Pictures taken from the three high speed cameras (Fig. 10, Fig. 11 and Fig. 12) indicate some white parts corresponding to the sabot debris but the mass of this pieces made of light foam has no effect.

The bird impacts the center of windshield. It is totally smashed, but it is more consistent than water.

A strong deflection of the polycarbonate plate and an elastic spring back can be observed: the polycarbonate plate is very flexible. Shock wave in polycarbonate plate is generated at the impact as can be observed on frame #4 and #5 of Fig. 11.

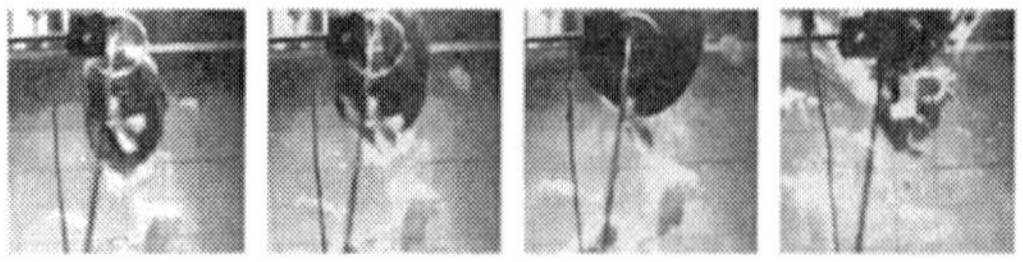

Figure 12. Pictures taken from high speed camera #2 (rear view).

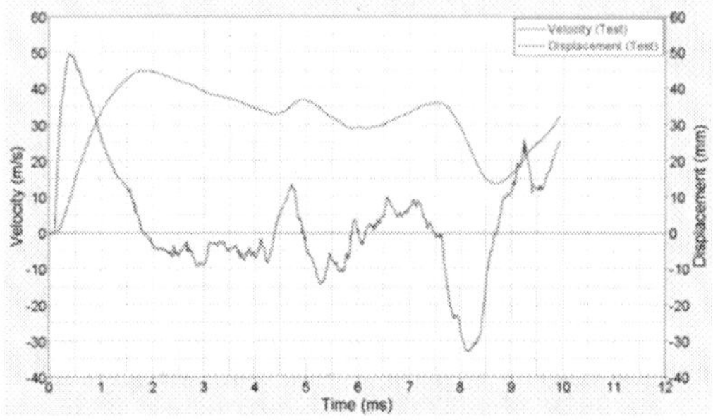

Figure 13. Velocity measurement.

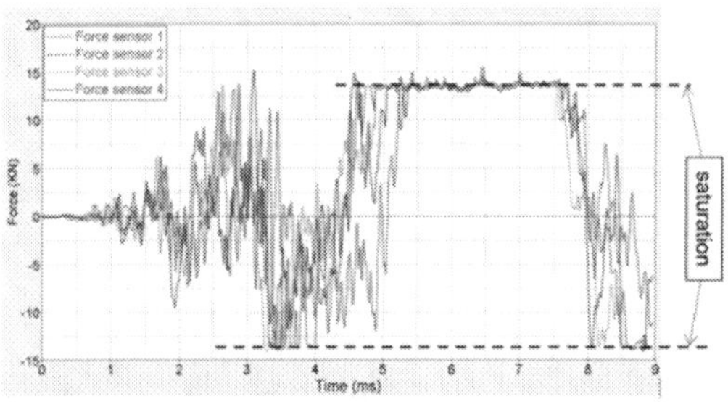

Figure 14. Force measurement (only values between ±13 kN are recorded).

PDV measurement behind the windshield shows a maximum velocity of 50 m/s and by time integration, a maximum deflection of 45 mm (Fig. 13).

For this test, a mistake has been done on the parameters of the load sensors which results in a saturation of this sensors at a value of ±13 KN (Fig. 14).

4. Numerical simulations

4.1. Model

Simulations of test #TI0849 have been done using the nonlinear explicit code LS-DYNA. Experimental results are compared to numerical results given by 3D lagrangian elements and 3D-SPH. The SPH solver was firstly introduced in the late 70's for astrophysics. It is a Lagrangian method with a variable nodal connectivity. It is gridless and thus can handle severe distortions without grid tangling and so does not need the use of unphysical erosion algorithms. This technique is then well adapted for bird strike simulations.

Some studies have been performed to determine what is the right shape to represent a bird in simulations [7]. The choice of using SPH formulation to model the bird allows us to use a more complex shape closer than the shape of a real bird (Fig. 15).

Figure 15. SPH model for the bird.

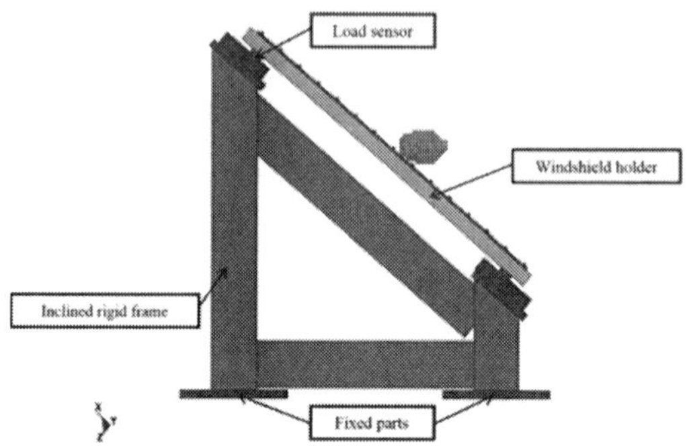

Figure 16. Numerical model side view.

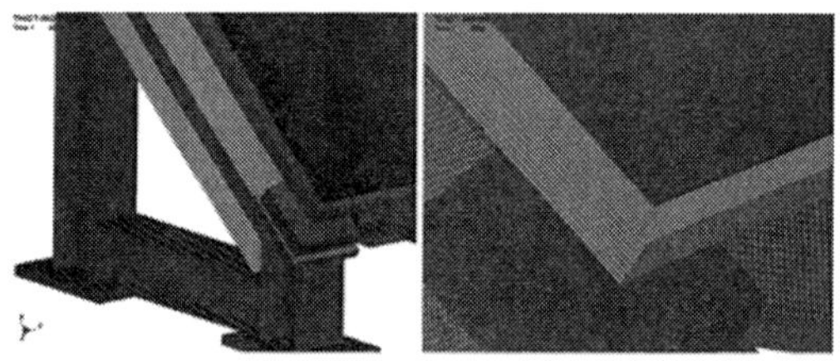

Figure 17. Isometric-closed views of the mesh.

An homogeneous material model with a low yield strength has been used to simulate the bird. This allows keeping the consistency during the impact. In our case, local behaviors are not preponderant because no local break of the windshield appears and no cutting of the bird occurs: it is why this material model is sophisticated enough for our case study [8].

The entire structure holding the windshield has been modeled with a millimeter sized mesh (Fig. 16, Fig. 17). The force sensors are set between the windshield holder and the inclined rigid frame. During the dynamic impact, mass and rigidity of each part are essential to reproduce load measurements.

The 46 screws are modeled without prestresses. This unstressed assumption makes possible the observation of the minimal needed tension in the screw during the impact.

4.2. Results and perspectives

Comparisons between simulation and the three high speed video cameras pictures are indicated on Fig. 18 and Fig. 19. These comparisons shows that the global dynamic responses of the bird impactor and the polycarbonate target are correlated. In particular, visible waves of the polycarbonate plate on the middle and between screws are well represented.

Figure 20 shows correlation between velocity measurement and simulation. The intensity and the shape of the first wave (up to 2 ms) are well simulated and the following waves are correctly reproduced. Comparison

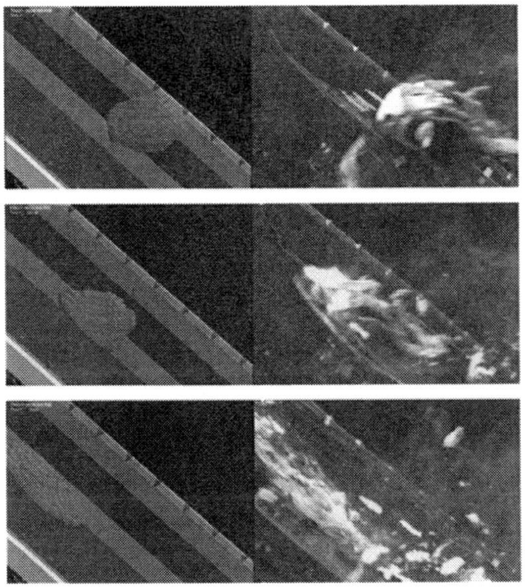

Figure 18. Simulation compared with pictures from camera #3.

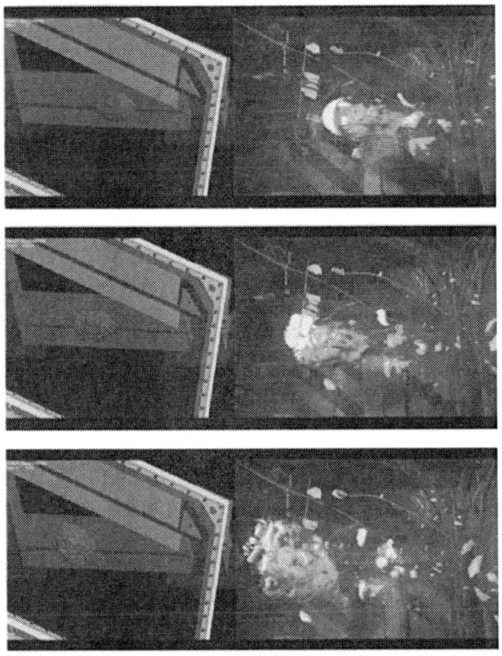

Figure 19. Simulation compared with pictures from camera #1.

between displacement profiles (obtained by integration of velocity profiles) is very good.

Figure 21 shows correlation for the load measurement on one corner. For values below 13 kN, level and global shape of the curve are well reproduced. Simulation shows the maximal load expected on each corner is 40 kN and that one screw can have a maximal tension of 14 KN. This value is high enough to explain the distorsion of the washers observed after the test.

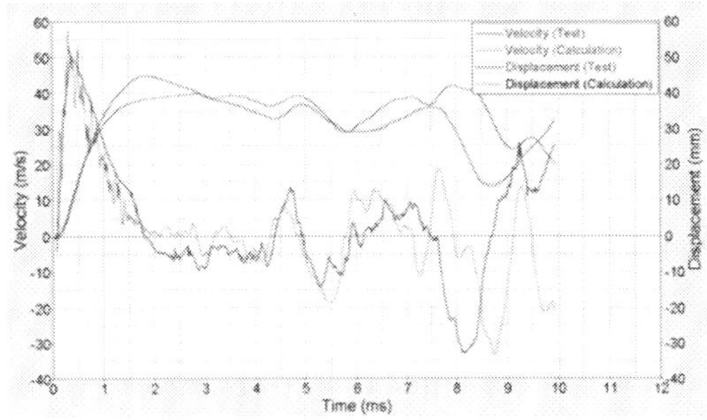

Figure 20. Correlation between test and calculation of the velocity measurement and displacement.

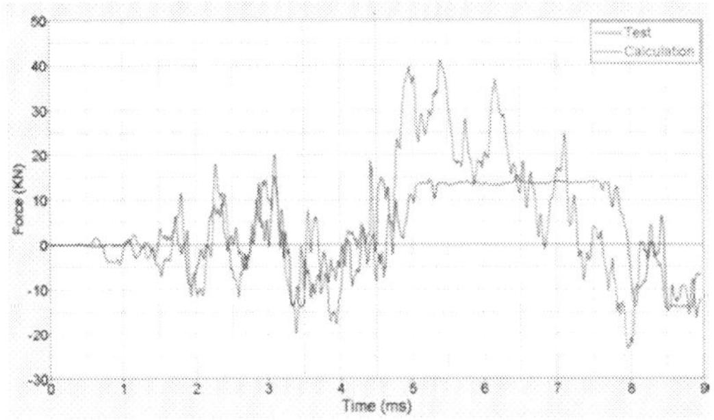

Figure 21. Correlation between test and calculation of load measurements (one corner).

All these comparisons show that there is a good fitting between experiment and calculation.

This study has shown that the simulation is refined enough (mesh size, material models…) to represent a bird strike on a polycarbonate windshield. It is possible to foresee the behavior of the windshield and design the fastenings.

The limits of this model are:

- A material subjected to local brittle breaks due to reaction to Hugoniot pressure cannot be represented with this material model [9].
- The validity of bird material model is not proved for situation where the bird is cut. Especially for local breaks on turbo fans during motor ingestion or on leading edge. A heterogeneous model in this case may be needed and should be a mandatory [10].

5. Conclusions and future works

TITAN gas gun using the 220 mm caliber has shown its ability to perform bird strike impacts in the big cubic target chamber to work with large instrumented aeronautics structures.

A crossed work between experiments and calculations has been performed to developed test facilities and numerical tools for bird strike studies.

Correlations between experiment and calculations show a good fitting of the polycarbonate target velocity and load measurements at the corners of the plate. Our numerical tools are able to give expertise assistance to aeronautic designer for parts development subjected to bird strikes. Calculations can be also reliable to prepare complex specific experiments (tools and metrology).

It is planned to go forward bird strike studies to improve models when local phenomena occur due to a shock especially for local breaks on glass windshield or composite parts. A homogeneous model of bird cannot be sufficient for every bird strike cases especially for turbo fan ingestion.

References

[1] J. Thorpe: *International bird strike committee: "Fatalities and destroyed civil aircraft due to bird strike*, 1912-2002 (2003)

[2] Report of Associate Administrator of Airports, Office of Airport Safety and Standards, Airport Safety & Certification "Wildlife strikes to civil aircraft in the United States 1990-2013", Washington (2014)

[3] J.R. Allan, *The costs of bird strikes and bird strike prevention* (2000)

[4] P. Thiot, P.L. Héreil, J. Mespoulet, G. Darnis, V. Gasquet, W. Chan and K. Taylor, *New's member ARA presentation*, Proceeding of the 59th ARA meeting, Cap Town (2008)

[5] P. Thiot, *Technical description and capabilities of THIOT INGENIRIE's Two-Stage Light-Gas-Gun*, Proceeding of the 60th ARA meeting, Baltimore (2009)

[6] P. Thiot, *Description of a new 220 mm gas launcher at THIOT INGENIERIE's plant and first results*, Proceeding of the 61st ARA meeting, Thun (2010)

[7] S.C. McCallum, C. Constantinou, *The Influence of Bird Strike Analysis* (2005)

[8] L. Jun, L. Yulong, G. Xiaosheng, *The Influence of bird-strike on sidewall structure on an aircraft nose*, Chin. Soc. Of Aero. & Astro., Beihang University

[9] L.S. Nizampatnam, *Models and methods for bird strike load prediction*, Aero. Eng. Wikita State University (2009)

[10] J.P. Barber, H.R. Taylor, J.S. Wilbeck, *Characterization of bird Impacts on a rigid Plate* (1975)

EPJ Web of Conferences **94**, 01052 (2015)
DOI: 10.1051/epjconf/20159401052

Single rod impact tests: Analytical, experimental, and modelling investigation

Yongchao Huang and Clive R Siviour[a]

Department of Engineering Science, University of Oxford, UK

Abstract. This work investigates the single rod impact test technique, from which wave propagation characteristics, i.e. the propagation coefficient and wave speed, can be obtained and used to identify the material modulus as a function of frequency. To accomplish this, an elastic Titanium rod and a viscoelastic Polymethyl Methacrylate (PMMA) rod have been tested. With a known modulus, the Titanium rod modulus was successfully recovered from the wave propagation coefficient in a frequency range of 5 kHz ~ 30 kHz, which verifies the validity of this technique. The PMMA modulus identified by this technique agrees reasonably with the modulus derived from DMA tests in the range 2 kHz ~ 20 kHz. The good agreement between the theoretical and simulation-based modulus further confirms the effectiveness of the single rod impact test in this use. It is also demonstrated that the single rod impact technique is capable of detecting the minor mechanical changes induced by temperature variations as small as 2 °C.

1. Introduction

A number of mechanical testing techniques have become available to obtain intrinsic material properties from low to high strain rates, these include classic tension, compression and torsion tests, drop weight test, SHPB test, Taylor and plate impact tests, etc. By looking at material behaviour at various loading speeds, some of these techniques directly yield material properties such as modulus, whist others present the stress-strain curves. However, classic testing methods have been developed for most materials but are confined to low strain rate testing and may pose specific requirements on the material shape being tested due to clamping arrangements. High rate testing methods such as SHPB have very limited capacity in testing soft materials such as rubber-like materials due to some un-resolved critical issues such as stress equilibrium in specimens. Impact tests on a single long rod, as performed earlier by Korsky [1], Bacon [2], and more recently Hillstrom [3] and Benatar [4], have been validated to be effective in deriving material properties such as modulus and Poisson's ratio. The experimental setup for a single rod impact test is much more simple and convenient compared with other material testing techniques, normally a long specimen rod is needed (otherwise wave separation techniques may be required), and any excitations, such as pendulum, air gun, or even hand, that can generate tiny disturbance in the bar are theoretically applicable, the monitored quantities can be either displacement, strain, or particle velocity, etc. By applying frequency domain analysis to the measured waves, wave propagation characteristics such as damping and wave speed can be easily obtained, and based on the propagation characteristics the frequency-dependent material properties, in particular the complex modulus, can be derived. Instead of directly measuring the material stress-strain behaviours under various loading conditions, this approach provides a different way to look at the frequency-dependent material behaviours. Although the frequency, another concept comparable to strain rate, that can be achieved from a single rod impact test is limited by a number of factors such as the experimental setup itself and the numerical analysis method, this technique has the potential to be improved for testing a wide range of materials such as metals and polymers.

[a] Corresponding author: clive.siviour@eng.ox.ac.uk

2. Theoretical basis

The work is based on the wave propagation theories in a slender rod. Many theories have been established for characterizing wave propagation in a rod, either elastic [5] or viscoelastic [6]. These theories vary in whether it is 1D or 3D, what cross-section it can be applied to, etc. The complexity of these theories is out of this paper's scope, and only basic wave propagation theories will be introduced here.

The time domain displacement and frequency domain strain expressions are normally employed as solutions to the wave equation of a propagating wave:

$$u(x,t) = Ae^{-\gamma x}e^{i\omega t} + Be^{+\gamma x}e^{i\omega t} \tag{1}$$

$$\tilde{\varepsilon}(x,\omega) = \tilde{P}(\omega)e^{-\gamma x} + \tilde{N}(\omega)e^{\gamma x}. \tag{2}$$

In which

$$\gamma(\omega) = \alpha(\omega) + i\kappa(\omega), \quad (\alpha > 0). \tag{3}$$

The functions A, B, $\tilde{P}$, $\tilde{N}$ are determined by initial conditions. γ is termed the propagation coefficient, the attenuation coefficient α is a positive even function while

Article available at http://www.epj-conferences.org or http://dx.doi.org/10.1051/epjconf/20159401052

the wavenumber κ is an odd function (positive for $\omega > 0$). The wavenumber κ is associated with the wave speed by

$$\kappa(\omega) = \frac{\omega}{c(\omega)}. \tag{4}$$

Furthermore, the propagation coefficient γ can be related to the complex modulus E^* by

$$\gamma^2(\omega) = \frac{-\rho\omega^2}{E^*(\omega)}. \tag{5}$$

Which further yields

$$E'(\omega) = \frac{\rho\omega^2(\kappa^2 - \alpha^2)}{(\kappa^2 + \alpha^2)^2}, \quad E''(\omega) = \frac{\rho\omega^2(2\kappa\alpha)}{(\kappa^2 + \alpha^2)^2},$$

$$\tan\delta = \frac{2}{\frac{\kappa}{\alpha} - \frac{\alpha}{\kappa}} \tag{6}$$

or

$$\alpha^2 = \frac{\rho\omega^2}{2} * \frac{\sqrt{E'^2 + E''^2} - E'}{E'^2 + E''^2},$$
$$\kappa^2 = \frac{\rho\omega^2}{2} * \frac{\sqrt{E'^2 + E''^2} + E'}{E'^2 + E''^2}. \tag{7}$$

Therefore, if either the modulus and/or the wave propagation coefficient is known, the other one can be calculated by using Eq. (6) or Eq. (8). In this paper, Eq. (6) is used in single rod impact test analysis while Eq. (8) is adopted in the DMA analysis. Note that Eq. (5), Eq. (6) and Eq. (8) are all 1D-based, thus appropriate revision may be required if 3D effects, such as geometric dispersion, are to be included.

Imagine a wave is propagating in an infinite rod, and 2 strain pulses are measured at 2 sections x_1 and x_2 without overlapping, then the 2 sub-coefficients can be obtained by

$$\alpha = -\frac{\ln\left|\frac{\tilde{\varepsilon}(x_1,\omega)}{\tilde{\varepsilon}(x_2,\omega)}\right|}{x_1 - x_2}, \quad \kappa = -\frac{angle\left[\frac{\tilde{\varepsilon}(x_1,\omega)}{\tilde{\varepsilon}(x_2,\omega)}\right]}{x_1 - x_2}. \tag{8}$$

Besides, based on the general strain solution Eq. (2), it is convenient to derive the general expressions for displacement, particle velocity, acceleration, and force in frequency domain:

$$\tilde{u}(x,\omega) = \frac{1}{\gamma}\tilde{P}(\omega)e^{-\gamma x} + \frac{1}{\gamma}\tilde{N}(\omega)e^{\gamma x} \tag{9a}$$

$$\tilde{\dot{u}}(x,\omega) = \frac{i\omega}{\gamma}\left[\tilde{P}(\omega)e^{-\gamma x} - \tilde{N}(\omega)e^{\gamma x}\right] \tag{9b}$$

$$\tilde{a}(x,\omega) = \frac{\omega^2}{\gamma}\left[\tilde{P}(\omega)e^{-\gamma x} - \tilde{N}(\omega)e^{\gamma x}\right] \tag{9c}$$

$$\tilde{F}(x,\omega) = \frac{\rho A\omega^2}{\gamma^2}\left[\tilde{P}(\omega)e^{-\gamma x} + \tilde{N}(\omega)e^{\gamma x}\right]\cdot \tag{9d}$$

Thus the strain in Eq. (8) can be replaced by displacement, particle velocity, acceleration, or force if these quantities are to be measured instead.

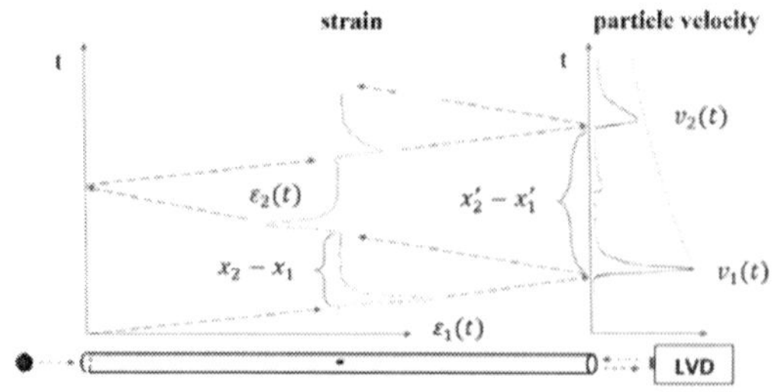

Figure 1. Wave propagation in a single rod impact test.

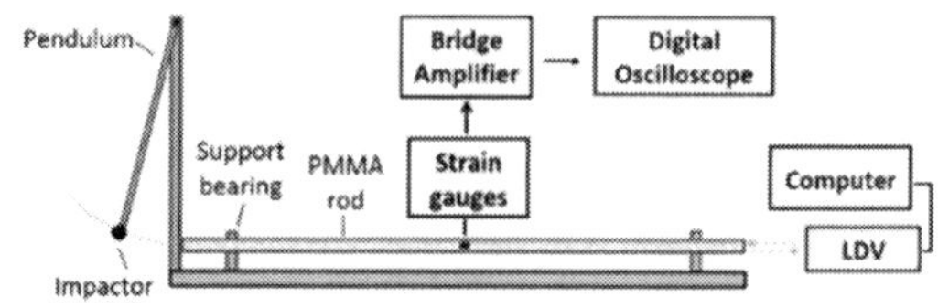

Figure 2. Experimental setup of the single rod impact test.

One assumption that was made in deriving the propagation coefficient γ Eq. (8) is that the 2 pulses can be measured separately by using an "infinite" rod. In a real test, an "infinite" rod is impossible but approximations can be made. A relatively long linear rod (either elastic or viscoelastic), with both ends free after impact, can serve as an ideal approximation of an "infinite" rod because waves reflected at a linear free end either remain the same (e.g. particle velocity) or with sign inversed (e.g. strain). Figure 1 graphically shows the wave traveling path during an impact test, pulse duration should be controlled to avoid wave superposition at the gauge station. Also, the distance between 2 pulses sets a lower limit on the useful frequency that can be measured without overlapping.

Another assumption associated with modulus calculation (Eq. (6)) is 1D wave propagation in the rod. This excludes 3D effects such as the radial inertia-induced dispersion, and poses requirements on rod geometry. A "slender" rod is normally required for minimizing the 3D dispersion effect, in fact it has been reported that dispersion effects can be ignored as long as the rod diameter is at least 10 times less than the wavelength [7]. For a rod with a fixed diameter, this requirement in turn poses an upper limit on the useful frequency without appreciable dispersion. Though 3D solutions have been reported [6], it is our intension here that the dispersion effect be cancelled at the initial stage of data processing, this was achieved by applying Love's 3D theory [5] and adopting the phase-only dispersion correction methodology [8].

3. Experimental implementation

3.1. Experimental setup

The experimental setup of a single rod impact test is shown in Fig. 2. A portable pendulum setup, with the geometry of about 1.5 × 0.1 × 0.5 m (L × W × H), was designed. The base is made of aluminium and the pendulum arm has a length of about 30 cm. 2 rods were used in the

Table 1. Impactors and corresponding drop heights.

Impactors	Drop heights (cm)
Ø12.50 mm PMMA ball	30, 28, 26, 24, 22, 20, 18
Ø12.34 mm Aluminium ball	27, 25, 23, 21, 19

pendulum impact tests: 1 PMMA rod with geometry of Ø13 × 1000 mm and 1 Titanium rod of Ø12.7 × 1000 mm. They were supported by 2 ring bearings and are axially movable during the tests. The commercial PMMA rod has a mass density of 1180 kg/m^3, while the Titanium rod has a density of 4500 kg/m^3. Grease was used to reduce the friction at the ring bearings to avoid unexpected energy dissipation.

Two pairs of axially mounted strain gauges, each consists of 2 diametrically opposite strain gauges, were attached on the middle surface. A Laser Doppler Vibrometer (LDV) was made use of in the 23.4 °C PMMA rod impact tests to measure particle velocities at the non-impact end. The collected particle velocity signals were recorded separately by a computer, while strain signals were first amplified and then recorded by a digital oscilloscope.

Spherical balls are generally preferred in this impact tests as they can generate higher frequency components than cylinder impactors [2]. Also, as during wave propagation significant extension of pulse duration may appear as a result of the viscoelastic nature, the ball diameter is expected to be as small as possible. By taking these considerations into account, an impactor selection process was performed and after examining wave superposition in the recordings produced by different impactor types, finally 2 impactors made of different materials, i.e. the Ø12.50 mm PMMA ball and Ø12.34 mm Aluminium ball, were employed in our tests. And various drop heights (Table 1) have been adopted to generate different strain amplitudes within the linear viscoelastic range (LVR).

PMMA modulus can also be reported by Dynamic Mechanical Analysis (DMA). PMMA films, with the geometry of 1 × 5 × 30 mm, were tested in tension mode; a frequency scan was performed at 5 frequencies (1, 5, 10, 25, 40 Hz) while temperature sweep was done at a ramping rate of 2 °C/min.

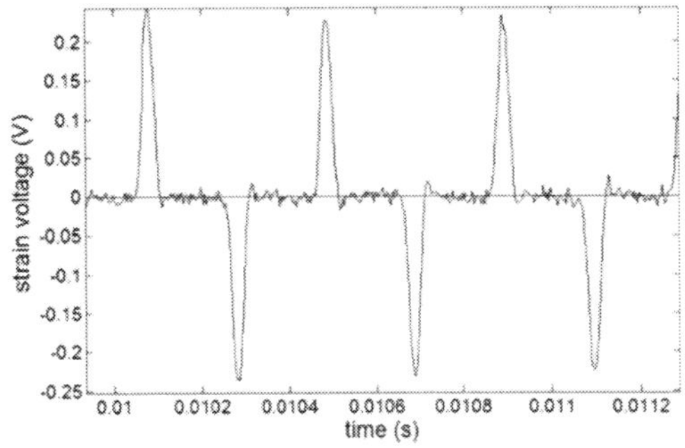

Figure 3. Strain waves in a Titanium rod impact test.

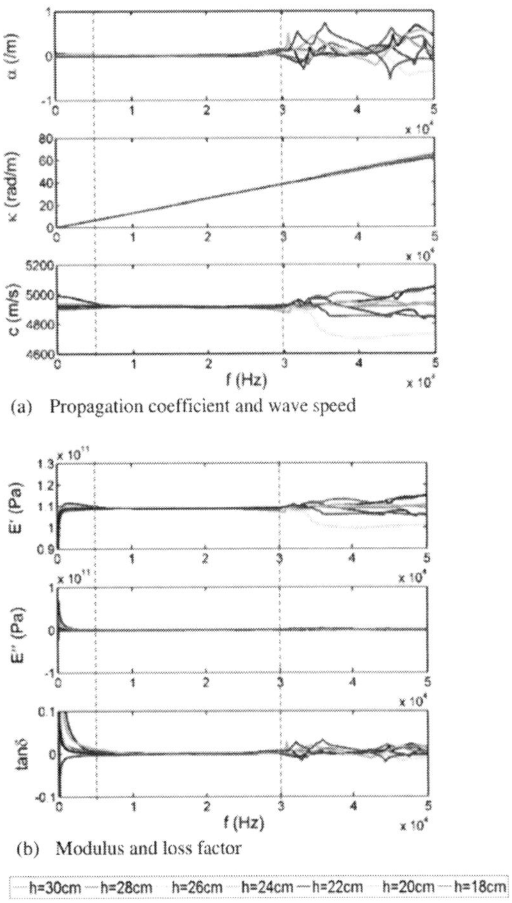

(a) Propagation coefficient and wave speed

(b) Modulus and loss factor

Figure 4. Titanium rod impact results (T = 22.6 °C).

3.2. Experimental results and discussions

The Ø12.7 × 1000 mm Titanium rod, with a known modulus of about 110 GPa, was tested at $T = 22.6$ °C to verify the effectiveness of the single rod impact technique and the validity of the frequency domain analysis method. A typical Titanium rod impact strain recording is shown in Fig. 3, the nearly constant amplitudes indicate that no energy dissipation appeared during the strain wave propagation in the elastic Titanium rod.

The resulting wave propagation coefficient, wave speed, modulus and loss factor of the Titanium rod are presented in Fig. 4. It is observed that different drop heights yield almost the same results, with the attenuation coefficient, loss modulus and loss factor being zero, while the storage modulus being about 110 GPa, which yields a wave speed of about 4944 m/s. The experiment-derived modulus agree well with the theoretical one in a frequency range of 5 kHz ∼ 30 kHz, which effectively verifies the validity of this technique. The lower useful frequency limit is affected by the distance between 2 adjacent pulses, while the highest useful frequency might be limited by dispersion effects and the excitation.

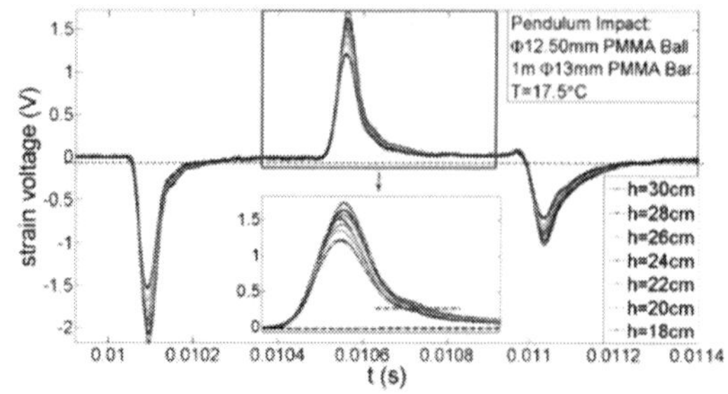

Figure 5. Strain recordings from PMMA rod impact tests.

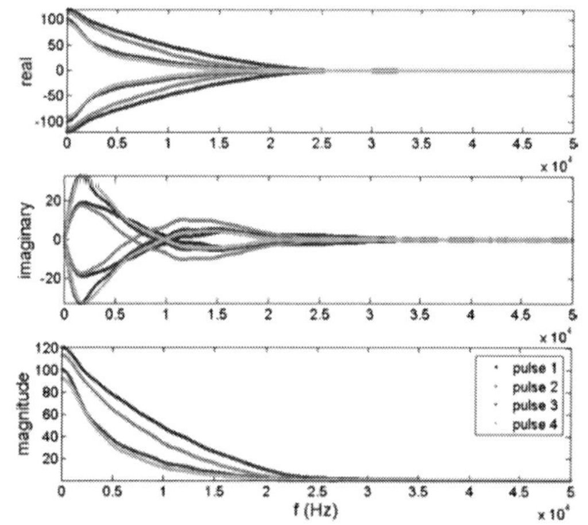

Figure 6. Fourier transform of the first 4 pulses of a PMMA rod impact signal.

PMMA rod impact tests were performed at 4 temperatures: 17.5 °C, 19.5 °C, 23.4 °C, 27 °C. Figure 5 shows an example of some typical signals generated by different drop heights, and shown in Fig. 6 is an example of the Fourier transform of the collected signal, attenuation can be clearly seen in the magnitudes of the propagating pulses after Fourier transform.

Figure 7 presents the results obtained at these 4 temperatures. It can be seen that, the damping characteristics such as the attenuation coefficient, loss modulus and loss factor are observed to increase with temperature, while an inverse trend can be seen in the storage modulus and wave speed. This complies with our common sense that most materials tend to be more solid-like and become stiffer as temperature decreases, and stress waves propagates faster in solids than in soft medium. Evidence from molecular analysis also help support this phenomenon: molecular chains become more active and more space become available for mobility when temperature arises. It is thus concluded that this technique is sensitive enough for capturing the small variations in mechanical behaviours.

A comparison of the results derived from different drop heights confirms the independence of the wave propagation coefficient and modulus on strain amplitudes within the LVR ($\varepsilon \leq 0.1\%$). Also consistent

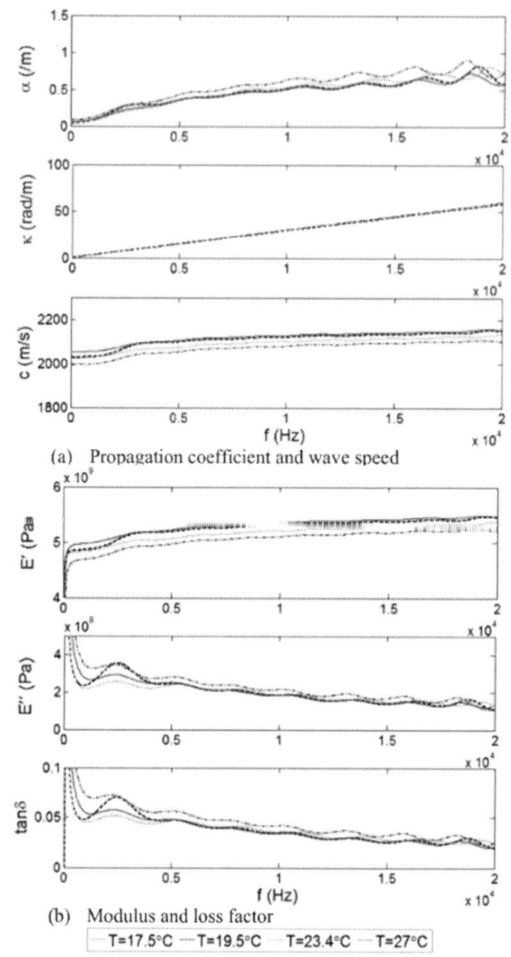

Figure 7. PMMA rod impact results (Ø12.34 mm Aluminium ball).

are the results produced by the 2 impactors as no major difference has been found between these results. Small deviations occur when comparing the strain-based and velocity-based attenuation coefficients (Fig. 8): the velocity measurements yield larger damping than strain measurements. The deviation in attenuation coefficient leads to a visible gap in loss modulus and loss factor but not in storage modulus, as storage modulus is less sensitive to the variation in attenuation coefficient than the other two. Noted that the useful frequency range for identifying this PMMA rod properties by using the single rod technique is 2 kHz ~ 20 kHz.

The DMA-reported modulus are both temperature-dependent and frequency-dependent. By using both manual shifting and the formula-based Arrhenius shifting methods, 2 master curves were constructed at the reference temperature $T = 23.4$°C. Comparison between the single rod impact results and DMA test results was made in Fig. 9. Deviations can be seen in wave speed and storage modulus while the damping characteristics (α, E'' and $tan\delta$) show less variations.

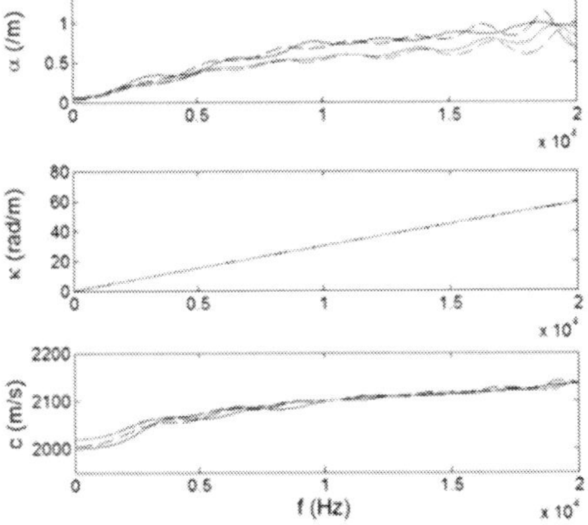

(a) Propagation coefficient and wave speed

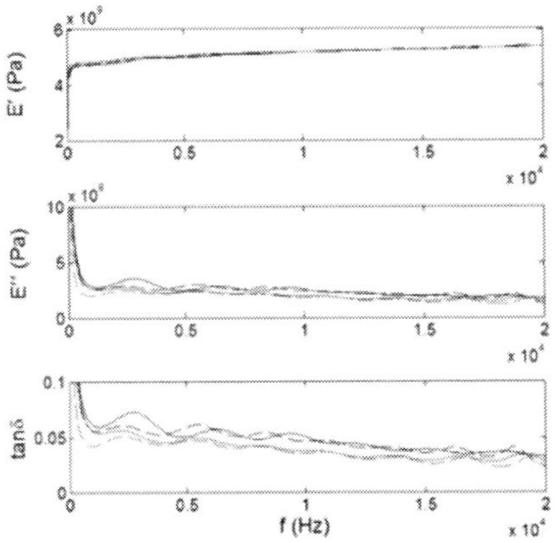

(b) Modulus and loss factor

Black: laser, red: strain gauge;
Solid: PMMA ball, dashed: aluminium ball

Figure 8. Comparison of results by 2 measurement methods ($T = 23.4\,°C$).

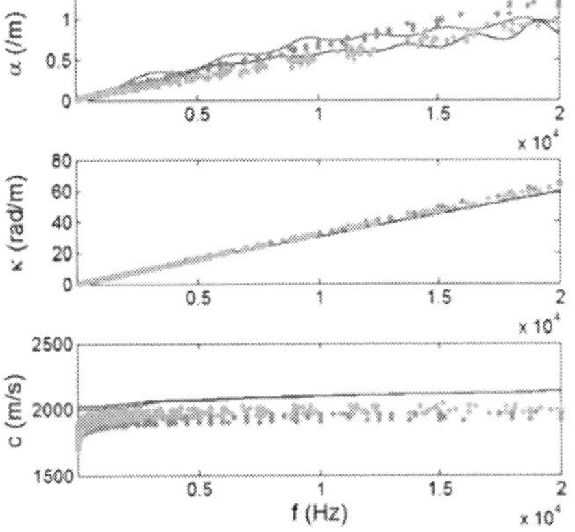

(a) Propagation coefficient and wave speed

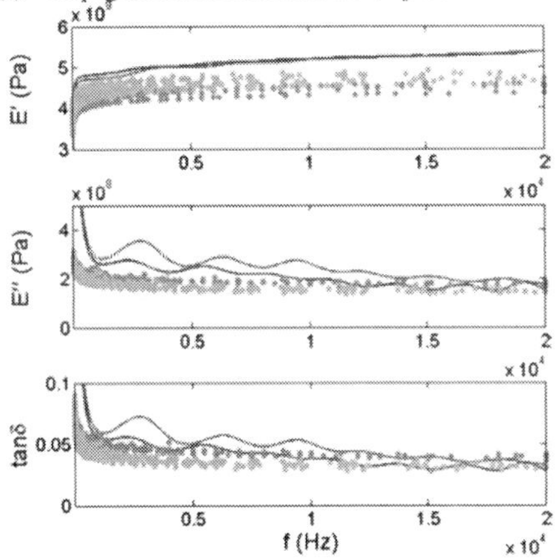

(b) Modulus and loss factor

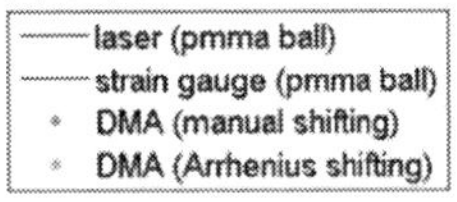

Figure 9. Comparison of DMA and single rod impact results ($T = 23.4\,°C$).

4. Simulation investigation

4.1. Model definition

Viscoelastic impact behaviours were simulated using the time domain PRONY model in ABAQUS/Explicit. The PRONY model is a parallel network model consisting of a group of springs and dashpots. It was defined by directly specifying the PRONY constants $\bar{g}_i^P$, $\bar{k}_i^P$ and τ_i. These dominating constants, as shown in Table 2, were obtained by fitting the complex modulus from the $T = 23.4\,°C$ PMMA rod impact tests in the frequency range $2\,kHz \sim 30\,kHz$. Note the shear and bulk behaviours were defined separately in ABAQUS.

For the 2-term PRONY model, the frequency-dependent theoretical modulus of are formulated as:

$$E'(\omega) = E_0\left(1 - \sum_{i=12}^{2} \bar{g}_i^P\right) + E_0 \sum_{i=1}^{2} \frac{\bar{g}_i^P \tau_i^2 \omega^2}{1 + \tau_i^2 \omega^2}$$
$$E'(\omega) = E_0 \sum_{i=1}^{2} \frac{\bar{g}_i^P \tau_i \omega}{1 + \tau_i^2 \omega^2}. \tag{10}$$

Table 2. PRONY model constants.

$\bar{g}_i^P$	$\bar{k}_i^P$	τ_i(s)
0.031379	0	5.88e-006
0.092126	0	5.292e-005
0	0.031379	5.88e-006
0	0.092126	5.292e-005

Where E_0 is the instantaneous modulus, $\bar{g}_i^P$ and τ_i are the PRONY model constants specified in Table 2.

The 1st pulse in the $T = 23.4\,°C$ recordings was employed as the simulation input. In real case the deformation becomes uniform across the across-section in a distance longer than the largest cross-sectional dimension according to the Saint-Venant's principle, it is thus reasonable to distribute the input loading uniformly on the impact end cross-section.

4.2. Simulation results and discussions

The strain and particle velocity outputs are shown in Fig. 10 and the resulting wave propagation coefficient, wave speed, modulus and loss factor presented in Fig. 11,

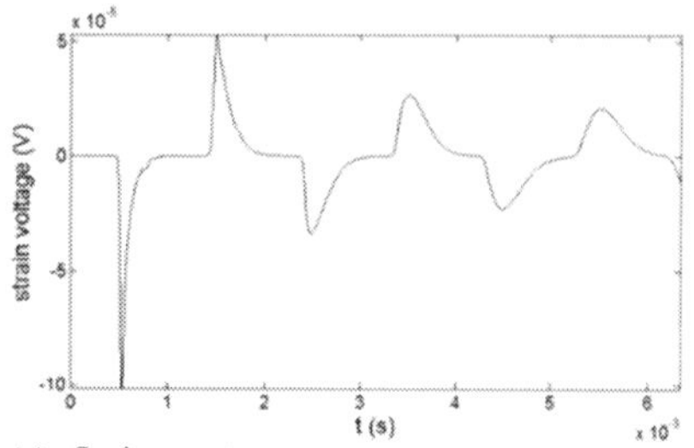
(a) Strain output

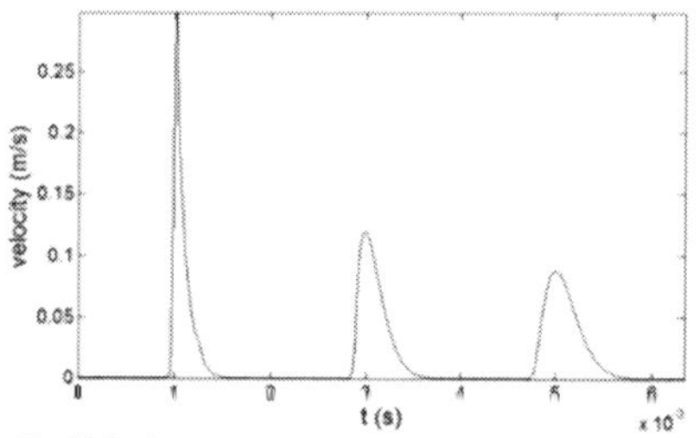
(b) Velocity output

Figure 10. Simulation signals.

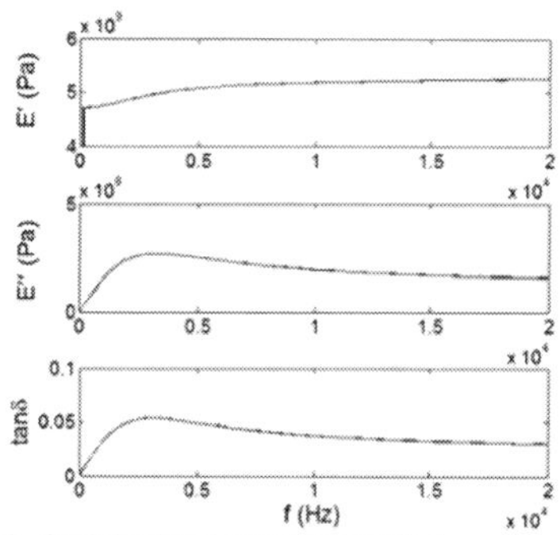
Black: simulation-based, magenta: theoretical

Figure 11. Comparison of the simulation-based and theoretical modulus.

accompanied by the theoretical modulus derived from Eq. (9). It can be seen that the simulation-derived moduli agree well with the theoretically predicted ones.

5. Conclusions

The ability of the single rod impact technique to derive the wave propagation characteristics and identify the material modulus was demonstrated in this paper. The Titanium rod moduli was recovered in a frequency range of 5 kHz ~ 30 kHz while the PMMA rod moduli were identified in the range 2 kHz ~ 20 kHz. Temperature-control tests also show that the single rod impact test technique is capable of capturing small changes in material mechanical properties induced by temperature variations as small as 2 °C. The good agreement between the theoretical and simulation results further confirms the effectiveness of this technique and the validity of the frequency domain analysis method adopted.

Effort sponsored by the Air Force Office of Scientific Research, Air Force Material Command, USAF, under grant number FA8655-12-1-2015. The U.S Government is authorized to reproduce and distribute reprints for Governmental purpose notwithstanding any copyright notation thereon. The authors thank S Fuller of AFOSR and M Snyder and R Pollak of EOARD for their support.

References

[1] H. Korsky, Phil. Mag 8, **1** (1956)
[2] C. Bacon, Exp Mech **38** (1998)
[3] L. Hillstrom, M. Mossberg, B. Lundberg, J. Sound Vib 3, **230** (2000)
[4] A. Benatar, D. Rittel, A.L Yarin, J. Mech Phys Solids **51** (2003)
[5] A.E.H. Love, *A treatise on the mathematical theory of elasticity* (4th Edition, Cambridge University Press, 1927)
[6] H. Zhao, G. Gary, J. Mech Phys Solids **43** (1995)
[7] R.M. Davies, Phil. Trans. A, 240, **375** (1948)
[8] D.A. Gorham, J. Phys. E: Sci. Instrum., **16** (1983)

EPJ Web of Conferences **94**, 01053 (2015)
DOI: 10.1051/epjconf/20159401053

Strain rate sensitivity of autoclaved aerated concrete from quasi-static regime to shock loading

Jérôme Mespoulet[a], Fabien Plassard, and Pierre Louis Hereil

THIOT INGENIERIE, Route Nationale, 46130 Puybrun, France

Abstract. The quasi-static mechanical behavior of autoclaved aerated concrete is well-known and can be expressed as a function of its density. There are however not much studies dealing with its dynamic behavior and its damping ability when subjected to a mechanical shock or a blast. This study presents experimental results obtained at the Shock Physics Laboratory of THIOT INGENIERIE company. The test specimens are made of YTONG$^{(TM)}$ cellular concrete with porosity in the range of 75 to 80%. Experimental tests cover a large strain rate amplitude (higher than $10^4\,s^{-1}$) for specimens up to 250 mm. They were carried out with a small compression press and with two facilities dedicated to dynamic material characterization: JUPITER dynamic large press (2 MN, 3 ms rising time) and TITAN multi-caliber single-stage gas gun. Results in un-confined conditions show an increase of the compressive strength when strain rate increases (45% increase at $5.10^2\,s^{-1}$) but dynamic tests induce damage early in the experiment. This competition between dynamic strength raise and specimen fracture makes the complete compaction curve determination not to be done in unconfined dynamic condition. A 25% increase of the compressive strength has been observed between unconfined and confined condition in Q.S. regime.

1. Introduction

Autoclaved Aerated Concrete (AAC) has been produced industrially for more 90 years since the actual process was patented by a Swedish architect in 1923. About 15 million m^3 of AAC are now produced in Europe per year on 100 production site [1]. This precast product is strong, lightweight, inorganic and non-toxic building material. The increasing needs of thermal efficiency to achieve zero energy consuming housing make the AAC one of the favourite actors in new building construction. A simple confirmation of this fact is the increasing volume production that now attain the equivalent of 300,000 homes built per year in Europe [1].

Numerous quasi-static characterization of this material can be found in the literature for rough or fibber reinforced AAC [2–4] but only few studies have been done in dynamic regime. In this field works have been by Belouettar [5,6] and Pontiroli [7] who investigated its dynamic behaviour.

Utilization of this building material as a damping candidate against shock and penetration events is still explored to mitigate explosive and ammunition threats (blast and fragments impacts). The AAC is referenced as a limiting sympathetic reaction material in ammunition storage in the UK Ministry of Defence (MOD) explosives regulations JSP482 [8] for the safe storage and processing of ordnance, munitions and explosives regulations. A complete and detailed analysis has also been done by Pontiroli et al. in this subject [7]. They presented a work on three densities of AAC with oedometric and triaxial quasistatic characterisation campaign combined with penetrations experiments and associated simulations (Fig. 1).

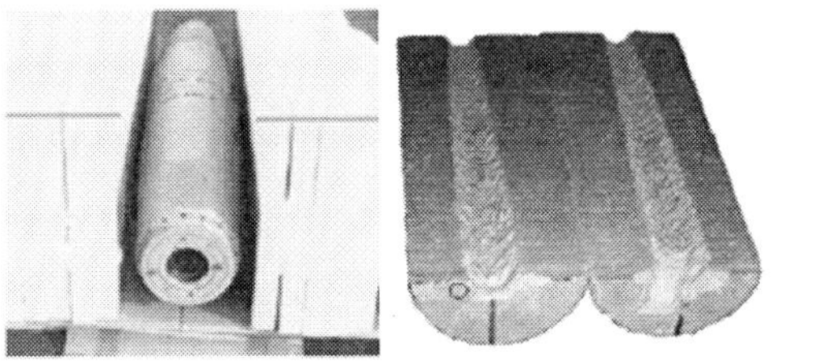

Figure 1. Picture of ammunition individual cell storage (left) and fragments penetration test made in A.A.C. (right) [3].

Works proposed in this paper are not directly related to those explosive applications but the shock absorption and fragments catching ability of this porous low-cost material makes it a potential actor in the various test configurations that are performed in our laboratory. The collected data obtained during this study will be essential to choose material laws and adjust their parameters for explicit simulation using dedicated software code such as LSDYNA$^{(TM)}$.

The other benefit of this work is to check the ability of our testing machines to be able to test and characterize high area density material in a wide strain rate range.

2. Material description

AAC is composed of sand, limestone, cement and water. A small amount of aluminum powder is then added which induces the generation of hydrogen and by expansion makes the characteristic open cells of aerated concrete.

[a] Corresponding author: mespoulet@thiot-ingenierie.com

Figure 2. Pictures of a studied material (left: initial block; right: close view of the cells on a recovered fragment).

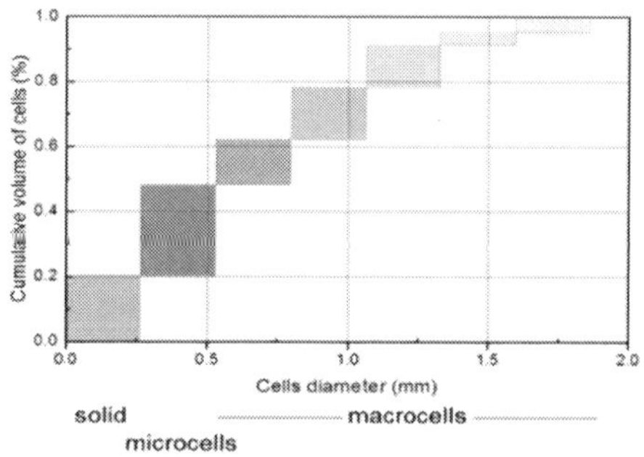

Figure 3. Cumulative cells volume repartition versus solid and cells diameter (reproduced from [10]).

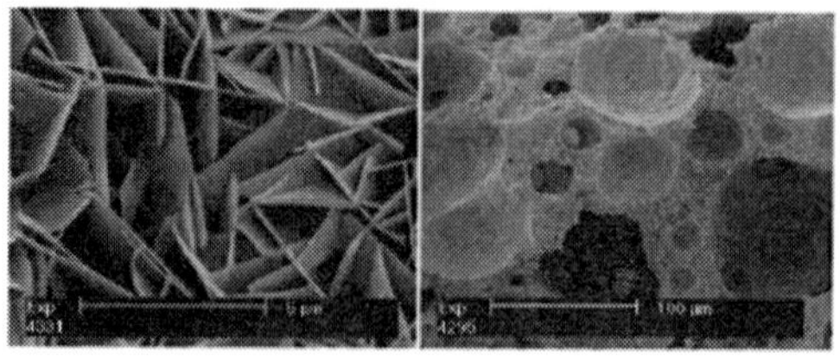

Figure 4. SEM pictures of a AAC with microcells (left) and macrocells (right) (from [10]).

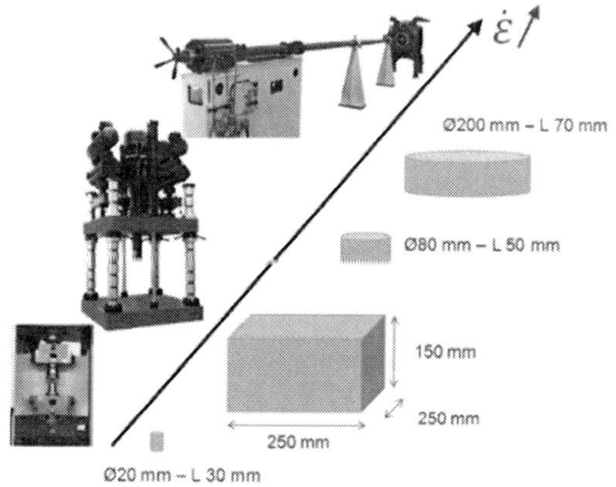

Figure 5. Investigated area: description of testing facilities and associated specimen dimensions versus increasing strain rate.

Hydrogen is then rapidly evacuated by ambient air. Final product is then sliced to appropriate dimensions and inserted in an autoclave that cures it at 180 °C and 10 bar. This process fixes sands and lime and creates Tobermolite crystals or Hydrated Calcium Silicate ($C_5S_6H_5$).

The material investigated in this study is an AAC manufactured by Xella thermopierre [9] with a nominal density of 550 kg/m^3. Pictures of an initial block and a postmortem fragment of the tested material are proposed on Fig. 2.

The repartition of the voids is not uniform in terms of dimension and space position. Two kinds of cells are identified [10]: macro-cells with diameters between 0.5 and 2 mm and micro-cells (below 0.5 mm) (Fig. 3). Pictures on Fig. 3 show two SEM views at mesoscopic and microscopic scales [10].

The compression strength given by the provider technical sheet is 4 MPa for this density of AAC.

3. Experimental works

3.1. Investigated area

The investigated area covers from quasi-static regime (3 $10^{-3}\,s^{-1}$) up to shock loading (>6.0 $10^3\,s^{-1}$) and different sizes of specimen have been tested to collect maximum data taking into account the capabilities of each facility and associated metrology (peak force for load cell for example).

The quasi-static oedometric tests have been done at CRITT at Toulouse on small sample (Ø20–L30 mm). Two facilities in Thiot Ingenierie Shock Physics laboratory have been involved to cover the large strain rate amplitude (Fig. 5):

- Dynamic compression tests on un-confined 250 ∗ 250 ∗ 150 mm cubic specimen have been done using the large dynamic press JUPITER. This unique equipment is a large dynamic press that is able to generate a stabilized load up to 200 kdaN in a few ms on 1 m^3 samples,
- Direct impact on Ø80 mm–L50 mm specimen and reverse plate impact experiment with Ø200 mm–L70 mm have been performed with TITAN single stage light gas gun with the 100 and the 220 mm caliber versions.

3.2. Quasistatic oedometric tests

The purpose of the oedometric compression test is to realize a uniaxial compression by keeping the external diameter of specimen constant. For porous material such as AAC, a dedicated confining steel cell is placed below the press stroke and the diameter of the cylindrical specimen is finally adjusted with a very small gap so as to make the lateral expansion of the studied material restricted. The test is performed by applying a force on the top of specimen with a cylindrical steel piston that is well-adjusted to the cell's inside diameter.

The compression press at CRITT is a ZWICK Z150 with 150 kN maximum allowable load (Fig. 6 – left). This peak value fixed the maximum allowable specimen diameter that can be characterized by this facility. A 20 mm

Figure 6. Quasistatic oedometric compression test pictures: initial configuration with oedometric cell (left); example of initial and post-mortem recovered sample (right).

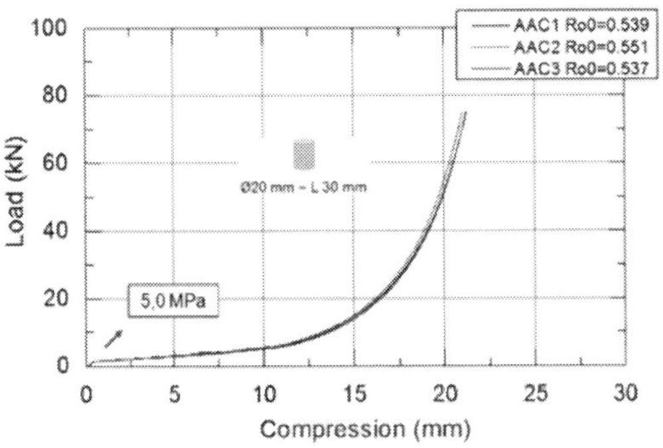

Figure 7. Comparison of the 3 QS oedometric compression tests performed in CRITT on Ø20 mm –L 30 mm specimen.

diameter by 30 mm length specimen has been selected to get the full compaction of AAC with the beginning of the solid slope with only half of the allowable press power.

Three specimens have been characterized and all the specimens have been recovered after test (Fig. 6 right). The compaction curves recorded during those tests are presented on Fig. 7. Tests were really reproducible and an average crushing pressure at the beginning of the "crushing plateau" has been estimated at 5 MPa in this quasistatic oedometric configuration at 3 10^3 s^{-1} strain rate.

3.3. Quasi-uniaxial dynamic compression test

Dynamic large un-confined compression tests have been done on our large dynamic compression press but dynamic load could reach 5 MN during experiment. Shape and dimensions of the specimens have been updated to take benefit of the extended capacity given by this facility in comparison of what was done at CRITT.

The square shape of the base of specimen was imposed by the compression tools that was available during the tests campaign. A 250 * 250 mm cross section of the AAC sample was selected to obtain the crush pressure and ensure enough stroke power to get the initial portion of the dense response of the AAC once compacted. Pictures of initial setup and recovered specimen post-test are presented on Fig. 8.

A high speed camera has been setup to record the compression event at 5,900 frames/s. Frames with 2.04 ms time step for one test are shown on Fig. 9 between 0 and 18.56 ms. The various frames show that damage occurs very soon in the compaction process and integrity of the main specimen is conserved up to 8 ms after the beginning of the compression load. At this time, the displacement of the stroke is 20 mm.

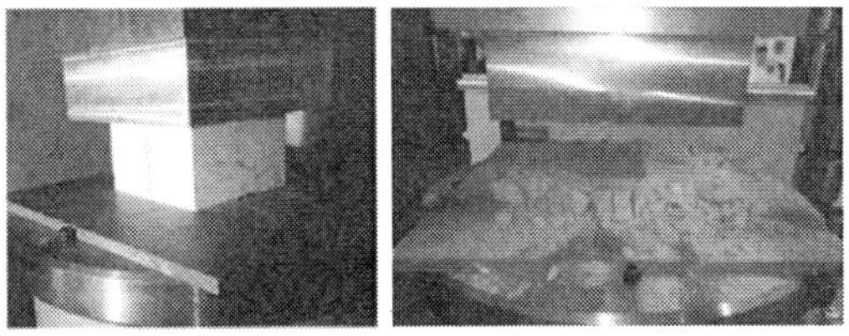

Figure 8. Top: large view of the test configuration. Bottom: example of dynamic large unconfined compression test: picture of initial setup (left) and recovered specimen (right).

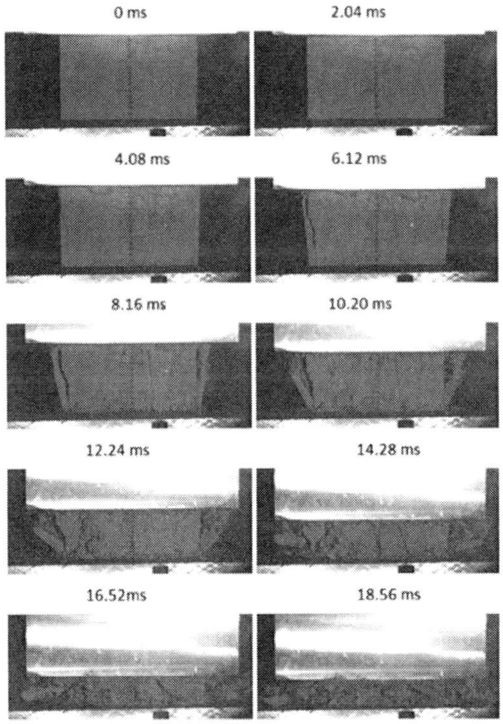

Figure 9. Examples of high speed camera pictures between 0 and 18.56 ms (2.04 ms interframe) obtained during a dynamic compression test with the 250 * 250 * 150 mm cubic specimen.

Two tests have been performed with two acting pressure. On this facility, the higher the pressure, the faster the compression is and so the higher the strain rate will be. Load versus compaction obtained for those two tests are given on Fig. 10. Profiles of those curves are quite similar to the shapes obtained during quasistatic oedometric experiments even if oscillations induced by the dynamic loading appear in this new configuration. The dashed line on this graph show an increase of the transmitted load due to the AAC compaction but it includes also the dynamic damage that amplifies all along the experiment. This competition between strength and damage makes this slope less steep than what can be observe in QS condition when material has enough time to

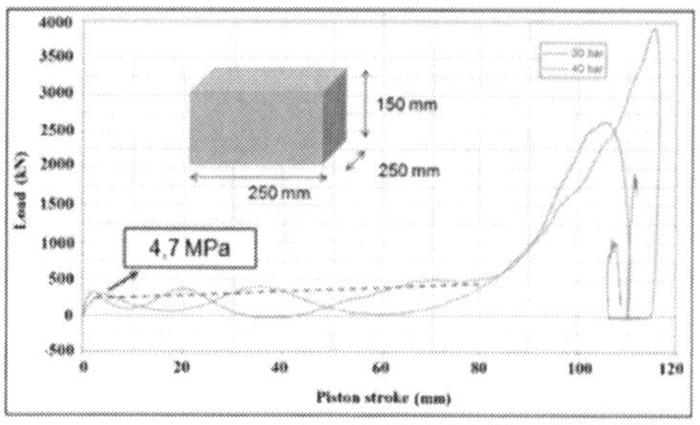

Figure 10. Comparison of the two dynamic compression tests performed with JUPITER large dynamic press.

Figure 11. Pictures of initial set-up (left) and after test (right) configuration of direct impact loading configuration.

ensure stress and strain equilibrium. However, the crushing load and by consequence the compressive strength can be evaluated at a value of 4.7 MPa in this unconfined condition.

3.4. Direct impact loading

Direct impact loading on AAC specimen has been explored by Belouettar using a direct impact technique and an instrumented split Hopkinson pressure bar [5]. Tests have been performed with 40 mm cubic specimens made in two different AAC and in dry and wet conditions. The dried tests show a positive strain rate sensitivity of AAC until damage occurs and pulverized the specimen.

For this study, direct impact loading on Ø80 * 50 mm specimen has been done at 25.7 m/s with a 10 mm aluminum thick impactor that gives a 600 g total weight rigid hammer. The nominal strain rate for this experiment $\varepsilon/dt = V_{imp}/l_s$ with V_{imp} the impact velocity and l_s the specimen length is evaluated at $515\,s^{-1}$.

Pictures before and after experiment are shown on Fig. 11. The specimen is mounted on a special device that is instrumented with a dynamic load cell. The compression is recorded with an independent laser based velocity system that gives specimen compression after time integration of the hammer deceleration velocity. A high speed camera has been setup to record the impact event at 20,000 frames/s. All the instrumentation devices are time synchronized using dedicated oscilloscopes.

High speed cameras pictures of the experiment are given on Fig. 12. The various pictures with 50 μs time step show that the specimen is compacted all along the experiment but cracks appear and damage occurs and expands 150 μs after impact (fourth image on Fig. 12). At this time, the specimen is subjected to the same competition between increasing transmitted load by porous compaction and weakness generated by failure

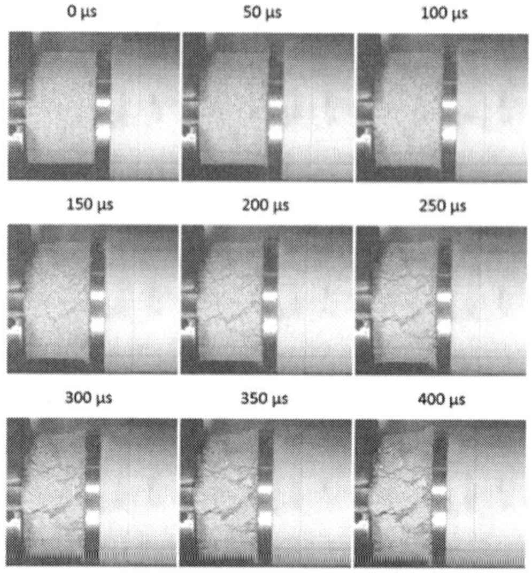

Figure 12. High speed camera snapshots of direct impact test between 0 and 400 μs (50 μs interframe).

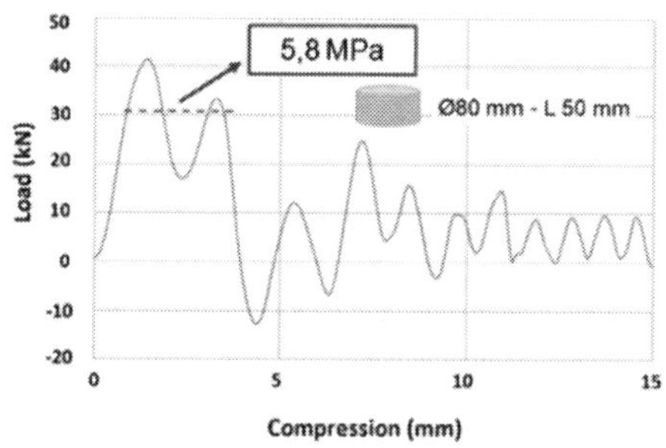

Figure 13. Load-compression curve obtained during direct impact loading at 25.7 m/s on the Ø80 mm – L 50 mm specimen.

and damage expansion than what was observe with the dynamic press n large samples.

Analysis and comparison of those high speed camera pictures with load given by the force sensor (Fig. 13) confirm the shape of recorded load. On this graph it can be observed that more than 30 kN are transmitted by the specimen up to 4 mm compaction before damage occurs and after the transmitted load decreases down to 10 kN.

The equivalent crushing pressure in unconfined condition is evaluated to be 5.8 MPa based on the 80 mm initial diameter of the specimen.

3.5. Plate impact loading

High strain rate behavior of YTONG material has been analysed with plate impact loading. A reverse plate impact configuration has been selected because instrumentation of porous material with stress gages or particle velocity with this nature of material is quite difficult. In this technique, a disk of the studied material is thrown on a target that is made in a well-known material (aluminum, steel, copper…). This witness material stays in elastic regime during all the experiment and the free surface velocity at the rear face is measured by using Photonic

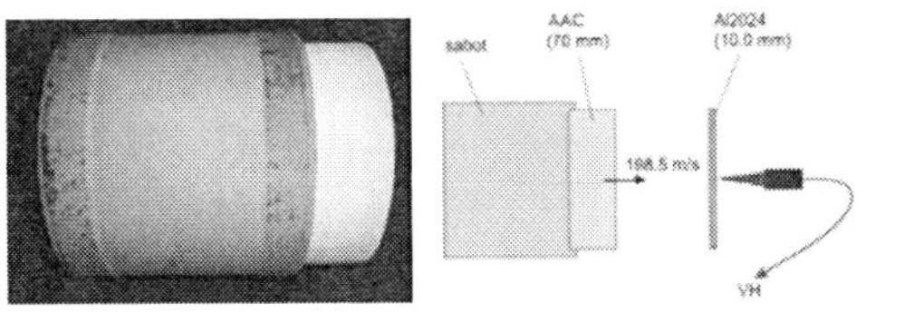

Figure 14. Side view of the AAC sample inside sabot (left) and schematic of the reverse plate impact configuration (right).

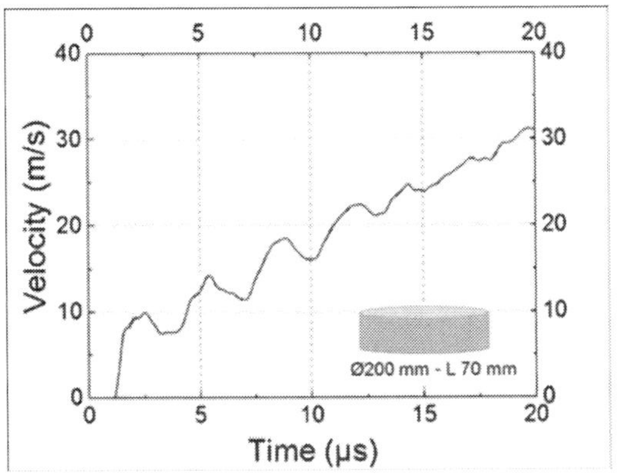

Figure 15. Free surface velocity- versus time obtained for the reverse plate impact experiment at 198.5 m/s on Ø200 mm – L70 mm specimen.

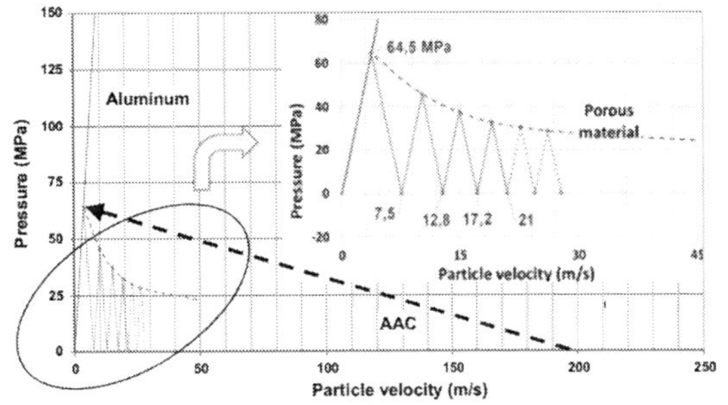

Figure 16. Analytical determination of pressure versus particle velocity generated during the reverse plate impact experiment at 198.5 m/s with the Ø200 mm – L 70 mm specimen on 10 mm aluminum plate.

Table 1. Synthesis of crushing stress obtained during experimental works on 550 kg/m^3 AAC.

Strain rate (s^{-1}) Configuration	3 10^{-3}	0.3–1.0 10^2	5.15 10^2	>6 10^3
Un-confined compression	4.0	4.7	5.8	X
Confined compression	5.0	To be done	To be done	Done

Doppler Velocimetry (PDV, or Heterodyne Velocimetry, VH in French).

A side view of the AAC specimen inside the sabot before impact and schematic of the reverse plate impact experiment on the 220 mm launcher are presented here below (Fig. 14).

It was decided to do the reverse plate impact experiment with the 220 mm caliber instead of the 60 mm to collect a maximum number of data. This change of diameter allows a longer time record before release lateral waves reach the axis of the aluminum witness plate.

Figure 15 shows the free surface velocity recorded at of the rear of the aluminum target during the experiment. The fourth first velocity steps observed during the experiment illustrate the reflection of the waves generated by the AAC inside the aluminum target. The velocity steps after 13 μs are less pronounced and then show the effect of the release wave that come from the outer diameter of the aluminum witness plate: the 1D condition of wave propagation in the target is then not ensure and so signal is no more representative of the AAC response.

Those reflections are highlighted on Fig. 16 with the analytical analysis of pressure versus particle velocity of the reverse plate impact configuration. The pressure at the interface between AAC and aluminum built up following a straight line (dashed arrow on the figure from [198.5 m/s-0MPa] to [3.75 m/s-64.5 MPa]. This curve interacts the aluminum curve and then multi-reflections of shock waves occurs in the Aluminum plate with the aluminum pressure slope. Each reflection at the rear of the Al target is the theoretical level of velocity that will be measured by the PDV with values of 7.5, 12.8, 17.2 and 21 m/s (given in the top right closed view).

To conclude with this graph, the plate impact test gives only one characteristic point for the AAC in this configuration. This point is defined by using the 3.75 m/s particle velocity (half value of the first free surface velocity record by the PDV) and the 64.5 MPa pressure generated in the aluminum plate. By assuming an error of $\pm$2 m/s on the AAC impact velocity and a $\pm$0.5 m/s error on the PDV record, the deduced point is the stress versus specific volume curve of the ACC is 64.5 $\pm$ 2.3 MPa for a specific volume of 1.22 $\pm$ 0.3 cm^3/g).

4. Analysis and future works

A synthesis of crushing stress obtained during this study is presented in Table 1. The results show an increase of compressive stress with the strain rate rising (+ 45% at 515 s^{-1}). Confined compression versus unconfined property reveal a 25% increase in quasistatic condition. Those tendencies have to be confirmed by future experiments like described here after.

The curves proposed on Fig. 17 gives all the compaction curves results obtained during this study. Only the oedometric quasi-static tests have allowed the complete compaction curves to be obtained. The dashed or dotted lines on unconfined tests show the pressure obtained after damage appears in the experiment. Those phenomenon appear really soon in the compaction process and that means that only the compression strength is achievable with unconfined configurations because complete dynamic response is tricky to perfom with brittle materials.

Future works on AAC are numerous but confirmation of quasi-static property on large samples is the first step.

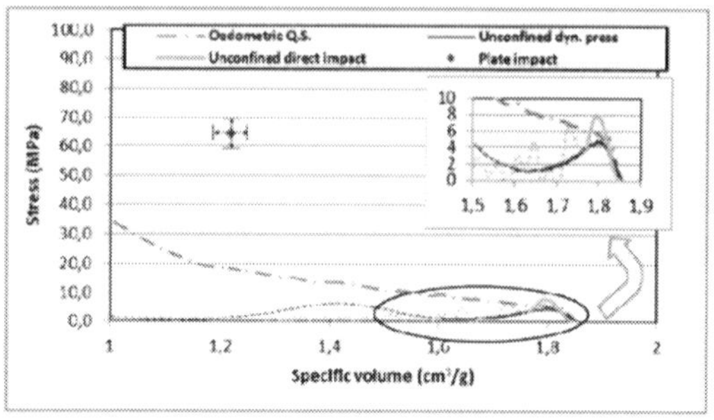

Figure 17. Stress (MPa) versus specific results obtained during all the experiments performed in this study.

Reduced scale unconfined and confined configuration using our Ø20 mm Split Hopkinson bars could be an option and can directly be compared to oedometric quasistatic tests performed at CRITT.

A dedicated oedometric cell for the large dynamic press is planned to be done to get AAC response at $10^2\ s^{-1}$ regime in confined condition and on large samples.

Complementary tests with direct impact loading could be performed at different impact velocities but also in oedometric conditions to collect more data in this low velocity direct impact configuration.

Additional plate impact tests down to 50 m/s and up to 600 m/s need to been done on this launcher. Those tests will give essential data of the reaction of this porous material in high velocity shock loading configuration.

Ballistic tests on AAC blocks will be performed between 10's m/s and 2 km/s to evaluate the real performance of AAC blocks against penetration events.

Finally, all the collected data have to be finely analysed to select the best material models that will be used in simulation.

5. Conclusion

Wide range of strain rate tests have been done on a dried 550 kg/m3 Autoclaved Aerated Concrete using various facilities. The investigated strain rate range from $10^{-4}\ s^{-1}$ in quasi-static condition up to shock loading higher than $6.10^3\ s^{-1}$ using reverse plate impact configuration.

Results in un-confined conditions show an increase of the compressive strength when strain rate increases but dynamic tests induce damage early in the compressive process (45% increase at $515\ s^{-1}$). This competition between dynamic strength raise and specimen fracture makes the complete compaction curve determination difficult to be done in unconfined dynamic conditions especially because the material show a real brittle and not ductile behaviour once damaged. This study reveal that oedometric experiment are a mandatory with such kind of porous and breakable material.

Comparison in Q.S. regime between unconfined (provider specification) and confined (experiments performed in this study) properties reveals a 25% increase of the compressive stress induces the confinement.

Future works will enrich this study with more dynamic tests and especially in confined/oedometric conditions to obtain the compaction curve in dynamic regime.

Ballistics tests have planned to be done between 10's m/s and 2 km/s on AAC blocks to validate the capacity of this low-cost and easy to provide material for shock absorption and fragments catching.

Finally, this work confirms the ability of our testing facilities to be able to test and characterize high area density materials in a wide range of strain rates.

The authors thank Thiot Ingenierie for funding and express gratitude to Thiot Ingenierie Facility technical staff for doing the experiments.

References

[1] European Autoclaved Aerated Concrete Association, www.eaaca.org
[2] J.F. Argudo thesis, Master of Science Engineering thesis, The university of Texas at Austin (2003)
[3] U.M. Cancino thesis, Master of Science Engineering thesis, The university of Texas at Austin (2003)
[4] H. Bagheri, Licentiate thesis, Royal Institute of Technology, School of Architecture and the Built Environment, Stockholm (2006)
[5] R. Belouettar, J.R. Klepaczko, M.T. Abadlia, Revue Française de génie civil, **5** (2001), 667–675
[6] R. Belouettar, J.R. Klepaczko, Proceeding of the 9th DYMAT conference, Bruxelles, **1** (2009), 503–509
[7] C. Pontiroli, R. Huet, X.H. Vu, Y. Malécot, L. Daudeville, Proceeding of the 22nd MABS symposium, Bourges (2013)
[8] JSP 482: Ministry of Defense explosives regulations for the safe storage and processing of ordnance, munitions and explosives, UK (2013)
[9] www.xella.fr
[10] www.febecel.be

EPJ Web of Conferences **94**, 01054 (2015)
DOI: 10.1051/epjconf/20159401054

Vulnerability analysis of a pressurized aluminum composite vessel against hypervelocity impacts

Pierre-Louis Hereil[a], Fabien Plassard, and Jérôme Mespoulet

THIOT INGENIERIE, Route Nationale, 46130 Puybrun, France

Abstract. Vulnerability of high pressure vessels subjected to high velocity impact of space debris is analyzed with the response of pressurized vessels to hypervelocity impact of aluminum sphere. Investigated tanks are CFRP (carbon fiber reinforced plastics) overwrapped Al vessels. Explored internal pressure of nitrogen ranges from 1 bar to 300 bar and impact velocity are around 4400 m/s. Data obtained from Xrays radiographies and particle velocity measurements show the evolution of debris cloud and shock wave propagation in pressurized nitrogen. Observation of recovered vessels leads to the damage pattern and to its evolution as a function of the internal pressure. It is shown that the rupture mode is not a bursting mode but rather a catastrophic damage of the external carbon composite part of the vessel.

1. Introduction

The context of this paper is the vulnerability of high pressure vessels subjected to high velocity impact of space debris. The increase number of these debris leads to a new regard on these phenomena. From an operational point of view, it is essential to know the limit between the perforation regime and the bursting regime of the vessel as a function of the internal pressure and of the impact characteristics (velocity, projectile weight, impact angle).

One of the first study on this topic has been performed by NASA in 1963 [1]. The context of this paper is the vulnerability of high pressure vessels subjected to high velocity impact of space debris. The increase number of these debris leads to a new regard on these phenomena. From an operational point of view, it is essential to know the limit between the perforation regime and the bursting regime of the vessel as a function of the internal pressure and of the impact characteristics (velocity, projectile weight, impact angle, ...).

One of the first study on this topic has been performed by NASA in 1963 [2–5].

Since 1995, EMI and ESA group have presented a lot of experimental, numerical and analytic works on the vulnerability of high pressure vessels [6–10]. Interesting experimental results are the visualization of the debris cloud in high pressure gas and the deceleration of this debris by this gas. Most of these tests have been realized at normal incidence, at a velocity around 7 km/s with aluminum spheres as projectiles. The impact tests were performed on cylindrical pressure vessel which were unshielded or shielded and made of AL5754, Al2219, unalloyed Ti or Ti6Al4V. Most of the cylindrical vessels had a wall thickness of 1 mm, a diameter of 150 mm and a length of 350 mm.

In 2001, a French group (EADS, INSA, CNRS and CNES) has presented some numerical simulations with SPH algorithm on the behavior of high pressure composite tank under an hypervelocity impact. This numerical study has been used to prepare a following experimental campaign on composite tanks [11,12].

The aim of the work presented in this paper, is to analyze the different damage process of a pressurized composite/metal vessel subjected to an hypervelocity impact. These regimes are the penetration mode, the critical rupture and fragmentation mode, and finally the bursting mode. Driving parameters are impact velocity, debris characteristics and internal pressure level.

As for the previous paper, the effects of internal nitrogen (N2) pressure vessel on the response of the vessel is analyzed for an 8 mm aluminum sphere at an impact velocity of 4400 m/s. The first difference with previous published studies is the tested vessel which is an overwrapped carbon fiber on an aluminum liner. The second difference concerns the internal pressure, which ranges from 1 to 300 bar.

The results presented in this paper are a first part of a more important program supported by CNES and which includes numerical simulations with adapted hydrocodes and also characterization of material behavior and equation of state. In this idea, hypervelocity impact experiments are performed with specific measurements in order to have enough data to discriminate the calculations.

During the test, two types of metrology are used in order to analyze the physical phenomena: flash X rays to observe the postimpact behavior of the projectile through the vessel and PVD interferomer system to monitor particle velocity profile at the back side of the vessel.

2. Experimental study

2.1. Test setup

The hypervelocity impact tests were performed with the HERMES two-stage light gas gun at Thiot Ingenierie Laboratory [13]. This gun differs from others standard

[a] Corresponding author: hereil@thiot-ingenierie.com

Article available at http://www.epj-conferences.org or http://dx.doi.org/10.1051/epjconf/20159401054

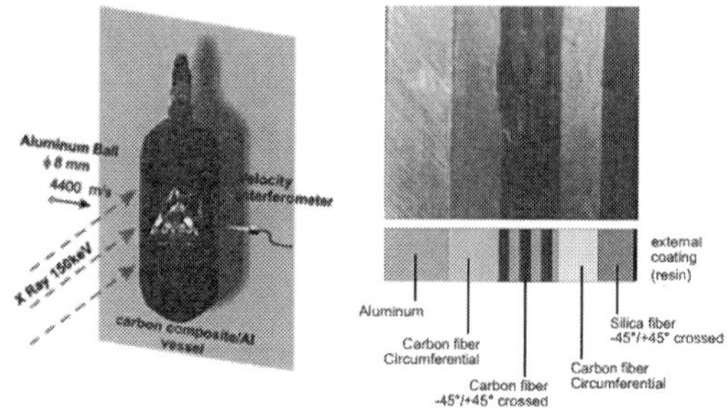

Figure 1. Cut view with description of the Aluminum-Composite vessel (left) and comparison of cut-view with model (right).

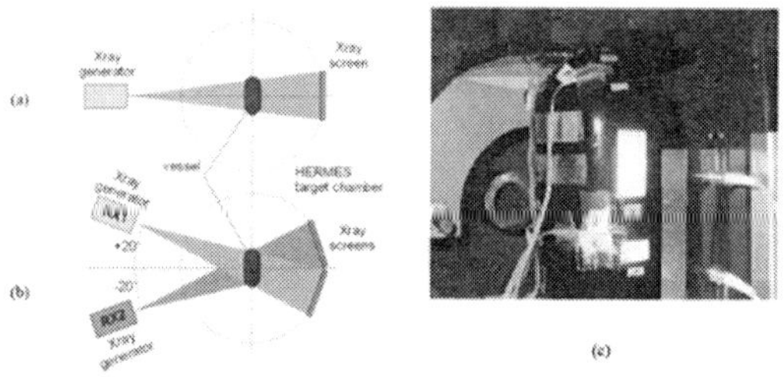

Figure 2. Schematic view of Xray configuration: (a) single beam (b) double beam (c) Composite-Aluminum pressure vessel inside the target chamber.

two-stage light gas guns by its "whole gas version" which avoids the use of powder and therefore the problems due to pyrotechnic safety.

A schematic view of the test apparatus as positioned in the target chamber is illustrated in Fig. 1 – left. It consists in the impact of a 8 mm diameter Al ball, on a CFRP/aluminium vessel at a velocity around 4400 m/s. Maximum allowed internal pressure in the vessel is 300 bar. Pressure was monitored before the test by using a calibrated pressure gauge.

Target vessel is manufactured as "paintball tank". The outer diameter of the vessels is 100 mm and its length 310 mm. The thickness of the vessel is higher near the top and the bottom of this cylindrical shape and lower in the central part. In this central part, the aluminium liner thickness is around 2.3 mm and composite thickness around 4.8 mm (Fig. 1-right). The composite layer is composed of four layers of wrapped composite fibers. From external to internal, the first and third layer are 45 °Cross wrapping while the second one and the fourth one are circumferential wrappings. The thickness of these different layers is about 1.2 mm (Fig. 1-right).

Triggering of the metrology was done by optical barriers and by self-shorting contacts. Impact velocity of the aluminum ball was performed by calculation of the chronometry between all these time information.

Xray 150 keV system (EUROPULSE EP8005 [14]) was used to visualize the state of the vessel and the projectile after the impact. It is triggered at the time of impact and its time duration is 30 ns. Two configurations have been used:

- one Xray channel perpendicular to the shot axis for the first shots (Fig. 2a),

Table 1. Parameters of the hypervelocity impact tests campaign.

Test n°	Projectile velocity (m/s)	Vessel pressure (bar)	Impact Angle (°)
#HE0183	4334 ± 60	1	90
#HE0184	4325 ± 60	200	90
#HE0187	4322 ± 60	250	90
#HE0188	4310 ± 60	300	90

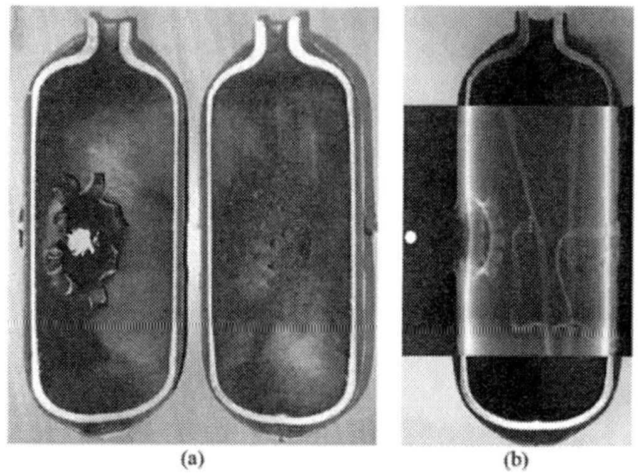

Figure 3. Test #HE0183: (a) Recovered vessel cut in two parts – Left: impact side – Right: rear side (b) Xray radiography at 41 μs after impact.

- two Xrays channels (named RX1 and RX2) tilted at $+20°$ and $-20°$ from the horizontal for the last shots (Fig. 2b).

The second metrology is a velocity interferometer system characterized by a response time of 1 ns and a error of 1% on the measured velocity. This system is referred as PDV (Photonic Doppler Velocity) or VH (Velocimeter Heterodyne). It leads to the profile of the free surface velocity at the rear face of the vessel.

A photograph of the Composite-Aluminum pressure vessel in the target chamber is displayed in Fig. 2c. Strain gauges measurements, not included in this paper, have also been used during this experimental campaign.

2.2. Experimental results

Experimental parameters of the four experiments performed in this study are presented in Table 1. For all experiments, impact have been performed at normal incidence.

2.2.1. HE0183

The first test #HE0183 was performed at an impact velocity of 4334 m/s and an internal pressure of 1 bar in the vessel. The measurement location of VH optical head is not in the axis of the impact but has been moved downward by 20 mm. The recovered vessel was cut in two parts in order to observe the impact zone and the back side of the internal aluminum liner (Fig. 3a).

The corresponding photograph shows the petaling shape of the impact area and the highly damaged

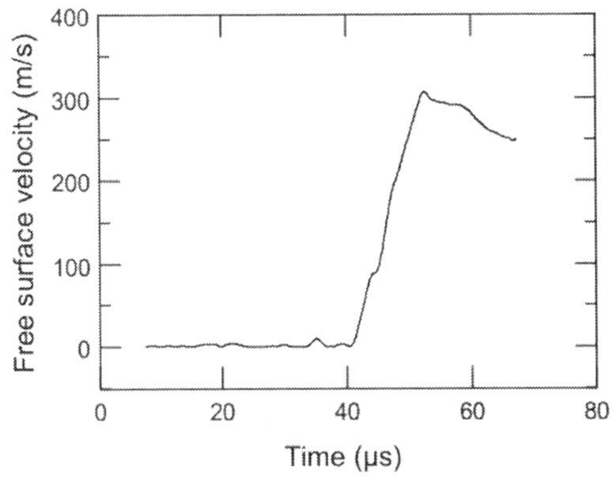

Figure 4. #HE0183: free-surface particle velocity diagram obtained at the rear face of the vessel.

composite around the impact hole. The back side is characterized by many craters done by the debris cloud.

The radiography obtained at 41 μs after the impact, is depicted in Fig. 3b with superposition of a half portion of the initial vessel. This picture shows the shapes of the retrojet, of the impact crater and of the debris cloud. The white line corresponds to others measurements (strains gauges) not included in this paper. The shape of the projectile has been added on this picture (white circle on the left of the radiography).

The particle velocity profile obtained with VH metrology at the rear face of the vessel (Fig. 4) shows an increase at a velocity of 300 m/s at time 41 μs after the impact of the projectile on the rear face of the vessel. This value of velocity corresponds to the impact of at least one large fragment on the rear face of the debris cloud generated after the impact. The fragment velocity is estimated at 2300 m/s with this chronometry.

2.2.2. HE0184

The second test #HE0184 was performed at an impact velocity of 4425 m/s and an internal N2 pressure of 200 bar in the vessel. The recovered vessel (Fig. 5) is not fully failed but a large portion has been removed. The dimension of this portion corresponds to the impact hole: its width is the same as the diameter of the impact hole and its length is a half diameter of the vessel. The composite is fully damaged in this portion. Observations of the internal liner of this recovered vessel show no impact or crater specifically at the rear side.

The radiography obtained at 15 μs after the impact (Fig. 6) shows the shapes of the retrojet, of the impact crater and of the debris cloud. The front wave which is visible forward to the debris cloud, is interpreted as the shock wave transmitted in the pressurized gas.

The particle velocity profile obtained with VH metrology at the rear face of the vessel (Fig. 7) has an oscillating shape. The maximum value corresponds to a velocity of 40 m/s at time 95 μs after the impact. This value of velocity is about eight times lower than the equivalent value obtained during the previous test. This leads to the conclusion that the debris are slowed down by the pressurized gas.

Figure 5. Test #HE0184 (N_2 pressure of 200 bar): recovered vessel.

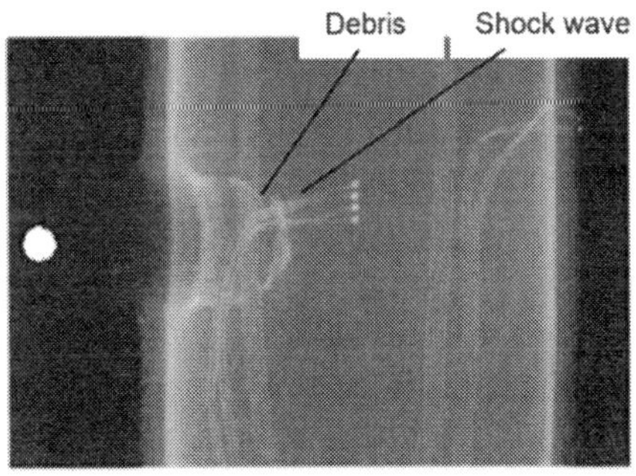

Figure 6. Test #HE0184 (N_2 pressure of 200 bar): Xray radiography at 15 μs after impact.

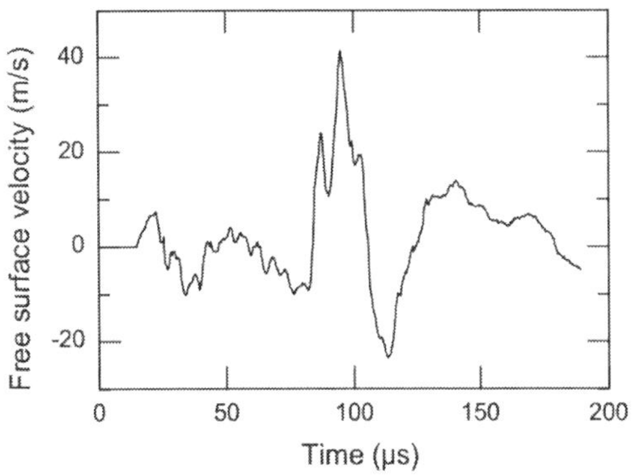

Figure 7. Test #HE0184 (N_2 pressure of 200 bar): Free surface velocity diagram obtained at the rear face of the pressurized vessel.

2.2.3. HE0187 and HE0188

For the two last tests #HE0187 and #HE0188, the experimental set-up has been improved by using two Xray channels. Test #HE0187 was performed at an impact velocity of 4322 m/s and an internal pressure of 250 bar in the vessel and test #HE0188 at 4310 m/s and 300 bar.

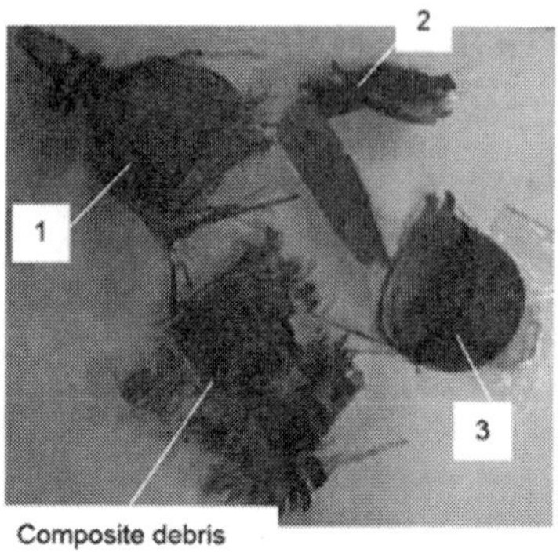

Figure 8. Test #HE0187 (N_2 pressure of 250 bar): recovered vessel.

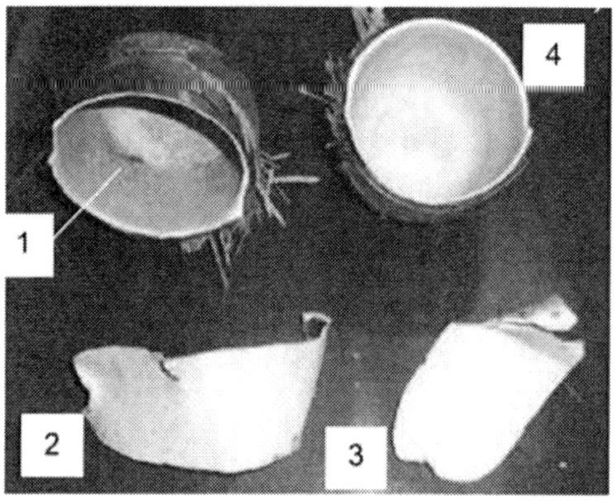

Figure 9. Test #HE0188 (N_2 pressure of 300 bar): recovered vessel.

The tested vessels (Fig. 8 and Fig. 9) are recovered in three and four parts:

- a bottom part,
- a top part,
- a central part (one piece for #HE0187, two pieces for #HE0188).

The external composite layers are highly damaged especially in the central area where they are fully broken. Inspection of the internal face of the aluminium liner reveals no craters and no trace of impact specifically on the rear side.

Radiographies obtained at 13 μs and 28 μs after the impact (Fig. 10 and Fig. 11) shows (from the left to the right) the retrojet near the impact zone, the shape of the impact crater, the debris cloud and the shock wave in the pressurized nitrogen. Due to the tilted configuration of the Xray stations, the shape of these different points seems more asymmetric than in the two previous shots.

The particle velocity profiles obtained with VH metrology at the rear face of the vessel (Fig. 12) have an oscillating shape similar to the previous test. The maximum value corresponds to a velocity of 18 m/s at time 95 μs for test #HE0187 and 16 m/s at time 94 μs for test #HE0188.

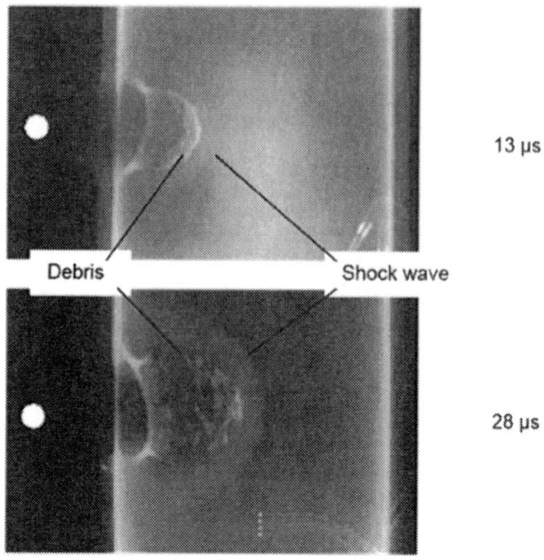

Figure 10. Test #HE0187 (N_2 pressure of 250 bar): Xrays radiographies at 13 μs and 28 μs after impact.

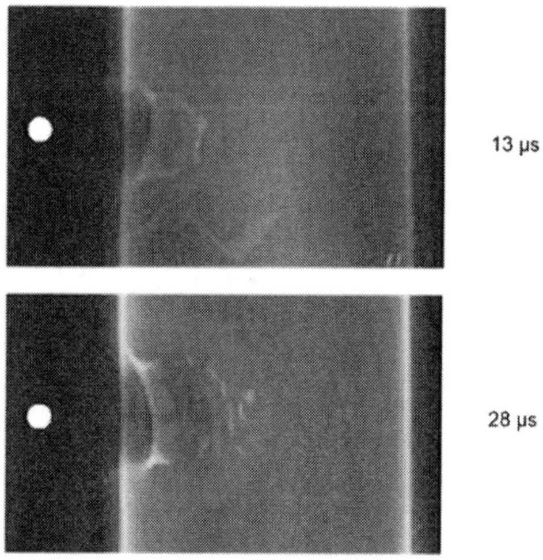

Figure 11. Test #HE0188 (N_2 pressure of 300 bar): Xrays radiographies at 13 μs and 28 μs after impact.

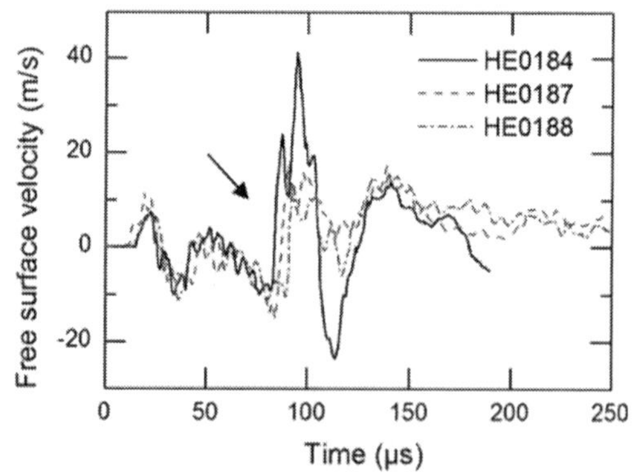

Figure 12. Free surface velocity diagrams obtained at the rear face of the pressurized tank for the three tests with pressurized vessels.

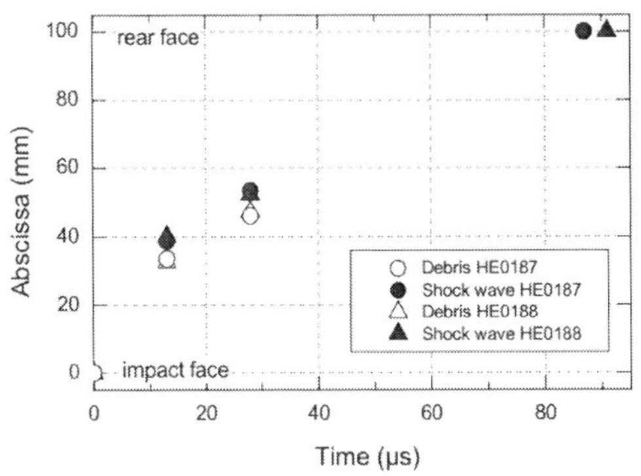

Figure 13. Evolution of the shock wave velocity and the debris in the N2 pressurized vessels.

2.3. Analysis

Analysis of the data and the results obtained during the four hypervelocity impacts on composite-Aluminum vessel confirms the results already presented on similar studies: the velocity of the debris cloud generated by the impact is reduced by the effect of the gas. Two observations confirm this analysis:

- the internal face of the aluminium liner shows no trace of impact when pressure vessel is higher than 200 bar,
- the velocity of the debris decreases between two radiographies.

Measured points for test #HE0187 and test #HE0188 are plotted in a XT diagram in Fig. 13. Abscissa corresponds to the distance in the vessel from the impact face up to the rear face. Open points correspond to debris and black points to the shock wave in pressurized nitrogen. Three points have been selected: first radiography, second radiography and free surface velocity profile. For this last point, the time of arrival has been identified with the third wave at a time between 90 μs and 100 μs (arrow in Fig. 12). This last point is an hypothesis which has to be confirmed by complementary measurements. Nevertheless, this diagram shows the consistency of the measured points.

A mean value of 900 ± 50 m/s is calculated for the debris velocity and for the shock wave velocity between RX1 and RX2 and a mean value of 780 ± 20 m/s for the shock velocity between RX2 and rear face.

Failure of the pressurized vessel increases with the inflated pressure, but for all the pressure investigated (200–300 bar) failure mode is the same: it leads to 2, 3 or 4 main debris of the aluminium layer of the vessel and a highly damaged state for the composite layers.

Failure mode is directly dependent of the composite stacking sequence. The penetration of the aluminum sphere leads to the rupture of the composite fibers. The confining stress induced by these overwrapped fibers is then released in this circumferential portion which results in the rupture of the aluminum liner at this same location.

3. Conclusion

The response of gas-filled pressure composite-aluminum vessels subjected to hypervelocity impacts has been analyzed from experimental results.

Investigated tanks are manufactured CFRP (carbon fiber reinforced plastics) overwrapped Al vessels. Explored internal pressure of nitrogen ranges from 1 bar to 300 bar and impact velocity are around 4400 m/s at normal incidence. Projectiles consists in a 8 mm diameter Al sphere. Data obtained from Xrays radiographies and particle velocity measurements show the evolution of debris cloud and shock wave propagation in pressurized nitrogen. Observation of recovered vessels leads to the damage pattern and to its evolution as a function of the internal pressure.

Analysis of the data and the results obtained during the hypervelocity impacts confirms the results already presented on similar studies: the velocity of the debris cloud generated by the impact is reduced by the effect of the gas. At an internal pressure level of 250 bar, debris cloud is fully stopped by the gas and no debris reaches the rear face of the vessel. It has been found that failure of the pressurized vessel increases with the inflated pressure and that failure mode is directly dependent of the composite stacking sequence. It is shown that the rupture mode is not a bursting mode but rather a catastrophic damage of the external carbon composite part of the vessel.

This first analysis has to be confirmed by future calculations and complementary experiments.

The authors thank CNES and Thiot Ingenierie for funding. In addition, the authors appreciate the support provided by Thiot-Ingenierie Facility and personnel including Valérie Berger, Ghislain Darnis and Fabrice Vigié.

References

[1] F.S. Stepka, R.P. Dengler, C. R. Morse, *Preliminary Investigation of Catastrophic Fracture of Liquid-Filled Tanks Impacted by High-Velocity Particles*, NASA Technical Note D-1537, LeRC, OH (1963)

[2] J.P. Whitney and P.D. White, *Designing hypervelocity impact testing for thin-walled pressure vessels*, Proceeding of the 43rd ARA meeting, Columbus, OH (1992)

[3] J.P. Whitney, *Hypervelocity impact tests of shielded and unshielded pressure vessels*, NASA Johnson Space center,Report JSC-32294 (1993)

[4] E. Christiansen, J.H. Kerr and J.P. Whitney, *Debris cloud ablation in gas-filled pressure vessels*, Int. J. of Impact Engng **20**, 7 (1997)

[5] S.A. Mullin, H. Couque, B.G. Cous-Palais, B.G. Grosch and J.D. Walker, *Bursting of shielded pressure vessels subject to HVI*, Int. J. of Impact Engng **20**, 7 (1997)

[6] M. Lambert and E. Schneider, *Hypervelocity impacts on gas filled pressure vessels*, Int. J. of Impact Engng **20**, 7 (1997)
[7] F.E. Schäfer, E. Schneider and M. Lambert, *An experimental study to investigate HVI on pressure vessels*, Second European Conference on Space Debris, Darmstadt, Germany (1997)
[8] F.E. Schäfer, E. Schneider and M. Lambert, *Impact fragment cloud propagating in a pressure vessel*, Act. Astr. **39**, 1 (1997)
[9] I.Y. Telitchev, F.E. Schäfer and E. Schneider, *Hypervelocity impacts on gas filled pressure vessels*, Int. J. of Impact Engng **23**, 7 (1999)
[10] M. Palmieri, F.E. Schäfer, S. Hiermaier and M. Lambert, *Numerical simulation of non-perforating impacts on shielded gas-filled pressure vessels*, Int. J. of Impact Engng **26**, 7 (2001)
[11] C. Maveyraud, J.P. Vila, D. Sornette, C. Le Floch, J.M. Dupilier and R. Salomé, *Numerical modelling of the behavior of high pressure vessel tank under an hypervelocity impact*, Proceeding of the 3rd Int. Conf. on Spacecraft Propulsion, ESA (2000)
[12] R. Salomé V. Albouys, C. Le Floch, D. Sornette and J.P. Vila, *High pressure composite tank behavior under an hypervelocity impact*, ESA (2001)
[13] P. Thiot, P.L. Héreil, J. Mespoulet, G. Darnis, V. Gasquet, W. Chan and K. Taylor, *Technical description and capabilities of THIOT INGENIERIE's Two-Stage Light-Gas-Gun*, Proceeding of the 60th ARA meeting, Baltimore (2009)
[14] `www.euro-pulse.com`, EUROPULSE, route de Gignac, 46500 Cressensac, France

EPJ Web of Conferences **94**, 01055 (2015)
DOI: 10.1051/epjconf/20159401055

Research of dynamic properties of alloys of AMg6BM and AMg6M in shock-wave experiment on a gas gun

S.S. Mokrushin, E.I. Karnaukhov, S.N. Malugina, D.N. Kazakov, O.E. Kozelkov, and A.V. Pavlenko[a]

FSUE Zababakhin RFNC-VNIITF, 456770, Snezhinsk

Abstract. Spall strength and elastoplastic parameters of aluminum alloys AMg6BM and AMg6M were studied in shock wave experiments with light gas guns LGG-1200 and LGG-2500 and results of these experiments are provided. Strength wave profiles were registered simultaneously by VISAR and interferometer PDV. Consideration was given to dependences of spall strength and strength parameters of aluminum alloys AMg6BM and AMg6M on the amplitude of the shock-wave loading in the range of 1.1 to 14.2 GPa and on deformation rate in the range of 0.2 to $3.6 \cdot 10^5\,s^{-1}$. Threshold damage levels for alloys AMg6M and AMg6BM are determined. It was shown that alloy AMg6BM softens under the pressure more than 5 GPa.

1. Introduction

Aluminum alloys of the system Al-Mg (magnalia) combine good strength and high plasticity, have high processibility, are well distorted, have satisfactory corrosion resistance and good fusability. This determines wide application of these alloys in atomic industry, building industry, aviation, rocket building, in constructions subjected to high thermal and force loadings [1,2]. Russian industry proves semifinished alloyAMg6 in the form of sheets, rods, foils, profiles in thermally unprocessed (AMg6), annealed (AMg6M), with technological cladding of sheets (AMg6BM) or without cladding (AMg6M), in cold-worked (AMg6N) and other states of products. Prediction of the results of pulse impact on such structures is an important practical task. In order to solve this task, it is necessary to know dynamic, kinematic and thermophysical properties of such structures [3,4].

Authors of [5–8] got the first experimental data on strength characteristics of alloys AMg6 and AMg6M in the conditions of explosive impact with microsecond duration of loading pulse at deformation rate in the range of 10^3 to $10^6\,s^{-1}$ and the temperature in the range of ($-13\,°C$) to $+550\,°C$. However, specialists performing experiments described in [5–8] used various approaches in measuring techniques. In activities [5–7], the fact of presence or absence of a spall was determined by visual examination or metallographic study of the samples that were recovered after experiments, and determination of the values of destructing stresses was made by assessment methods (using acoustic spall theory or EOS of the sample material and impactor). In [8], for determination of values of destructing stresses, the authors used profiles of rate $W(t)$ that were obtained with the help of capacitive pick-up of pressure. The use of various measuring methods led to sufficient disagreement of experimental results [5–8], especially while determining threshold levels of samples destruction ($\sim$ by a factor of 2).

While experimental equipment was developed in [9, 10], the study of dynamic properties of AMg6M was continued. Registration of profiles of free surface of samples was made with the help of laser-interferometric measuring methods. As the result, new experimental data on dynamic properties of AMg6M at deformation rate in the range of 10^3 to $10^6\,s^{-1}$, pressure from 2 to 56 GPa, temperature from $+20$ to $+600\,°C$ were obtained at qualitatively new level. Dynamic properties of AMg6 at deformation rate $190 - 1450\,s^{-1}$ using split Hopkinson bar were studied by the authors of [11].

However, high deformation rates and loading levels realized under using explosive devices did not permit to study in details the regions with transition from elastic to plastic deformation. That is why it is more expedient to use such experimental data for building wide-range mathematical models [3,4]. The development of up-to-date models of elasto-visco-plastic deformation of structural materials [12,13] that are used at our Institute is performed by comparison of experimental profiles of free surface velocity for materials under study and calculation data. In order to accurately determine numerical codes [12], select parameters of material model [13], one needs detailed investigation into dynamic properties of materials with smooth change of deformation rate and amplitude of shock-wave impact. It is necessary to note that comparison of strength properties of AMg6 samples fabricated out of rod and sheet material was performed only in [5,7] in the field of pressures up to 3 GPa while investigating threshold levels of the sample destruction. That is why the question about the influence of technological processing of materials on dynamic properties of AMg6M and AMg6BM samples remains unclear. Authors of [14,15] paid much attention to this topic.

This presentation gives the results of investigation into dynamic properties of aluminum alloy AMg6M and alloy AMg6BM. Contrary to [6,8–10], in order to clarify the

[a] Corresponding author: avpavlenko@vniitf.ru

Article available at http://www.epj-conferences.org or http://dx.doi.org/10.1051/epjconf/20159401055

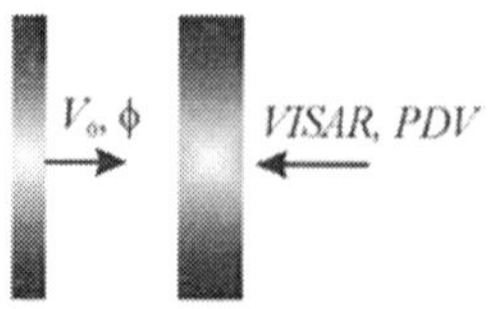

Figure 1. Scheme of loading and measuring methods.

influence of technological processing on the properties of materials under study, the samples were fabricated out of AMg6M KR40 rod (GOST 21488-97) and AMg6BM 5 sheet (GOST 21631-76).

Investigation was performed on two one-stage light gas guns [16] according to method of impact with simultaneous use of two-channel interferometer VISAR [17] and interferometer PDV [18]. The pressure in the samples was varied in the range from 1.1 to 14.2 GPa, deformation rate – in the range from 0.2 to $3.6 \cdot 10^5\,s^{-1}$.

Experimental data were obtained in the form of profiles of free surface velocity $W(t)$ for each type of samples of AMg6M and AMg6BM. Experimental data were processed and analyzed, the values of spall strength, dynamic yield limit, elastic limit and shear stress were determined. Threshold values of pressure were also determined, under which, spall destruction begins. For each material, dependences of spall strength on amplitude of loading pulse and deformation rate were built. Experimental profiles were used for determination of parameters of Johnson-Cook material model with Mie-Gruneisen.

2. Experimental equipment

Experimental studies of dynamic properties of aluminum alloy samplesAMg6M and AMg6BM were performed on light gas guns (LGG) with caliber 44 mm. Collision velocity was varied in the range of 0.2 to 2 km/s. Figure 1 shows the scheme of alloys samples loading and applied measuring methods.

In all experiments, evolution of shock waves in the samples of AMg6M and AMg6BM was studied. In experiments with aluminum alloy AMg6M, the technology of forming high deformation rates was probed at the cost of using impactors out of tantalic foils. Data on experiment set-up are given in Tables 1 and 2.

The velocity of impactor collision with the sample V_0 was determined with the help of electro-contact gauges. In addition, they determined the value of a slope, which characterized disalignment (distortion φ) of collided surfaces of impactor with the sample. Registration of profiles of stress waves was made simultaneously with the help of two-channel laser-interferometric complex VISAR and interferometer PDV. That permitted to make registration of velocity profiles $W(t)$ with an error less than 9 m/s and time resolution at least 2,2 ns.

We are investigated aluminum alloy made out of a sheet AMg6BM 5 (GOST 21631-76) and rod AMg6M KR40 (GOST 21488-97). Metallographic analysis of the check test piece for alloy AMg6M confirmed the presence

Table 1. Data on experiment set-up with the samples out of aluminum alloy AMg6BM.

No	Impactor AMg6M, mm	V_0, m/s	φ, mrad	Sample AMg6BM, mm
025	2.02	599	3.00	3.80
026	1.99	869	2.34	3.80
027	2.01	383	1.42	3.80
028	2.01	227	1.54	3.80
029	2.02	164	0.92	3.81
031	2.00	231	0.78	3.80
032	2.01	223	0.25	3.80
036	2.01	158	0.93	3.80
040	2.04	196	0.30	3.81
220	2.06	1688	10.37	3.81
221	2.02	1449	3.69	3.80

Table 2. Data on experiment set-up with the samples out of aluminum alloy AMg6M

No	Impactor material	Thickness, mm	V_0, m/s	φ, mrad	Sample AMg6M, mm
044	AMg6M	2.05	855	1.10	3.99
047	AMg6M	2.00	570	0.75	3.99
048	AMg6M	2.04	369	0.25	4.00
050	AMg6M	2.04	232	0.9	3.99
051	AMg6M	1.98	184	1.32	4.00
053	AMg6M	2.01	212	0.94	4.00
054	Tantalum	0.05	568	1.02	4.00
058	Tantalum	0.05	569	1.92	2.03
229	AMg6M	2.02	1252	1.74	4.02
230	AMg6M	2.03	1844	21.4	4.01
231	Tantalum	0.05	1285	11.0	3.99
234	AMg6M	2.04	2000	–	4.00

Table 3. Chemical composition (in %) of alloy AMg6 (GOST 4784-97).

Al	Cu	Mg	Mn	Fe
Base	<0.1	5.8–6.8	0.5–0.8	<0.4
Si	Zn	Ti	Be	Sum of impurities
<0.4	<0.2	<0.10	<0.005	0.1

Table 4. Chemical composition (in %) of cladding layer (GOST 21631-76).

Alloying components	Impurities, not more than				
Al	Fe	Sr	Cu	Mn	Sum of permissible impurities
Not less than 99.30	0.30	0.30	0.02	0.025	
	Zn	Ti	Mg	Other	
	0.1	0.15	0.05	impurities 0.02	0.70

of alternating big and small grains, and this fact speaks about thermal processing of material. The size of big grains was $(76 \pm 23)\,\mu m$, the size of small ones was $(15 \pm 4)\,\mu m$. Table 3 shows the chemical composition of alloy AMg6 (GOST 4784-97).

For cladding of sheets, aluminum with chemical composition shown in the Table 4 is used.

Before experiments, mass and geometrical sizes of each sample we are measured. According to them, the

Table 5. Data on the sound velocity in the samples of alloysAMg6BM and AMg6M.

Material	C_l, m/s	C_s, m/s	C_0, m/s
AMg6BM	6460 ± 15	3167 ± 15	5325
AMg6M	6487 ± 15	3197 ± 15	5334

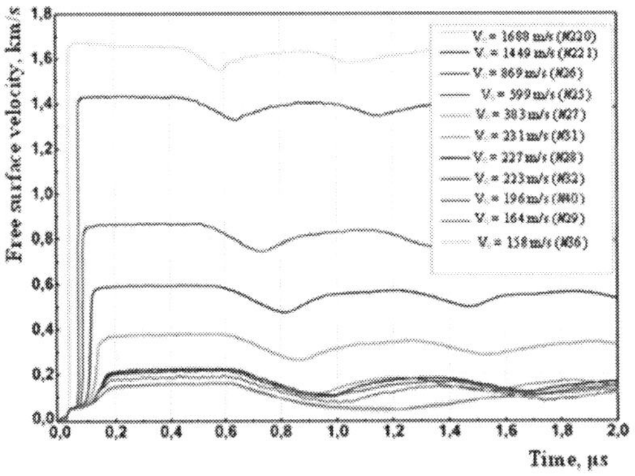

Figure 2. Profiles of free surface velocity in experiments with AMg6BM samples.

density of samples was calculated. The mass of samples was measured with the help of laboratory electronic scales of HR-I series. The diameter and thickness of all samples was measured by a linear displacement transducer with 10 μm absolute accuracy. Measurement of each parameter of the sample was made 10 times. The error of measurements was calculated at reliability $P = 0.95$ (Student's coefficient $t_\alpha = 2.26$, at $n = 10$ measurements of a parameter). Measured density of AMg6M samples was $\rho = 2.62 \pm 0.02$ g/cm^3, and density of AMg6BM samples was $\rho = 2.62 \pm 0.01$ g/cm^3.

Preliminary, the sound velocity was determined for the samples of AMg6M and AMg6BM. The velocity of longitudinal wave C_l was determined using ultra sound thickness gauge T-MIKE E, and measurement of transverse wave velocity C_s – using electromagnetic acoustic method of excitation and pick-up of ultra sound waves. Table 5 shows the data about sound velocity for mentioned samples.

3. The results of studies

The studies of dynamic properties of alloys AMg6B and AMg6M we are made in two series of experiments (Tables 1 and 2). The presence of spall destruction was fixed according to appearance so called "spall" signals on amplitude-time dependences of free surface velocity $W(t)$ that speak about relaxation of stress waves under destruction in the samples [3]. The amplitude of shock-wave impact was varied in the range of 1.1 to 14.2 GPa (the range of collision velocity was from 0.15 to 2 km/s), deformation rate was from 0.2 to $3.6 \cdot 10^5$ s^{-1}. Figures 2 and 3 show the profiles of free surface velocity $W(t)$ obtained in experiments with samples of AMg6BM and AMg6M, respectively.

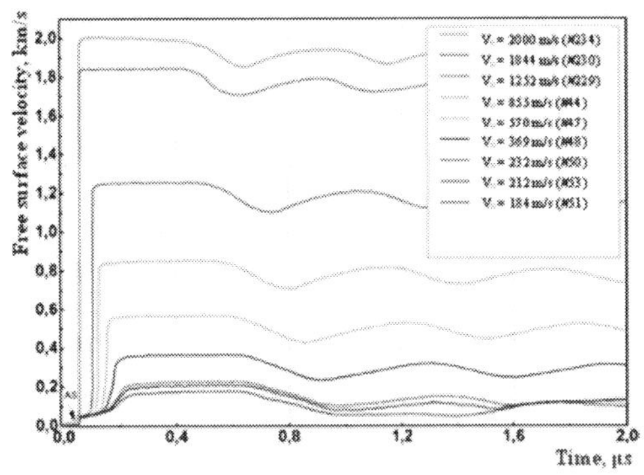

Figure 3. Free surface velocity profiles in the experiments with AMg6M alloy samples.

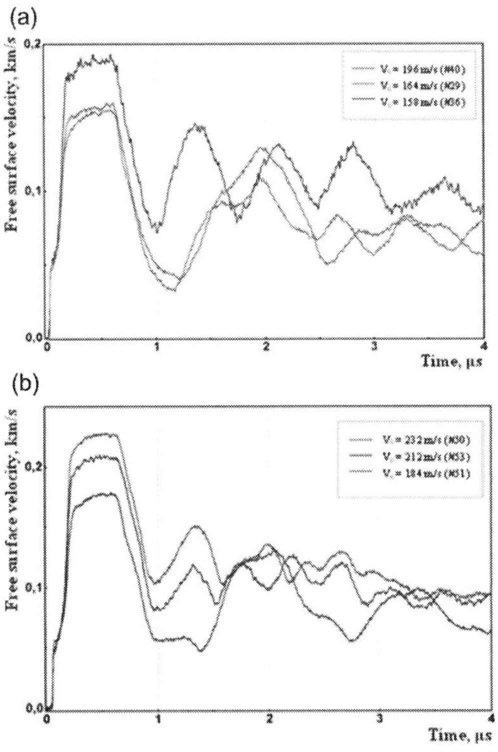

Figure 4. Velocity profiles $W(t)$ in the experiments with AMg6BM (a) and AMg6M (b) alloy samples.

Threshold damage levels for tested materials were determined during the work. Figure 4(a,b) contains velocity profiles $W(t)$ for AMg6BM and AMg6M alloys.

In Fig. 4(a,b) we can see, that spall pulse can be registered at a collision velocity higher than 196 m/s (for AMg6BM alloy) and 212 m/s (AMg6M alloy). No damages, able to form spall signal, are caused at a collision velocity lower than 164 m/s (for AMg6BM alloy) and 184 m/s (for AMg6M alloy).

Deformation rate change in the relief part of the compression pulse was within the range from $0.2 \cdot 10^5$ to $1.6 \cdot 10^5$ s^{-1} at collision velocity range from 183 to 2000 m/s. In our work, we tried to extend the deformation

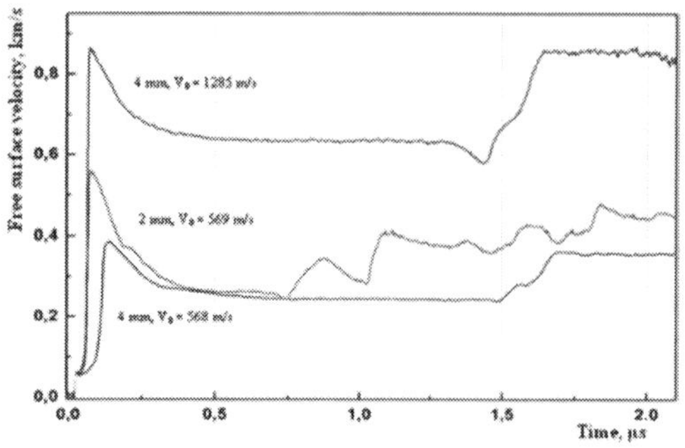

Figure 5. $W(t)$ velocity profiles in the experiments with AMg6M alloy samples.

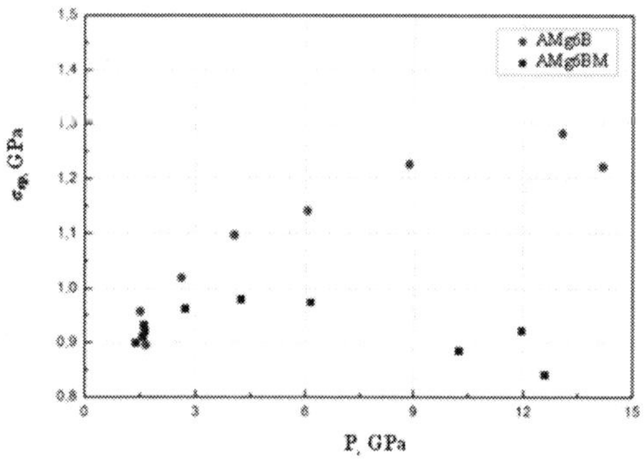

Figure 6. Spall strength of AMg6M and AMg6BM alloys samples depending on the amplitude of shock compression.

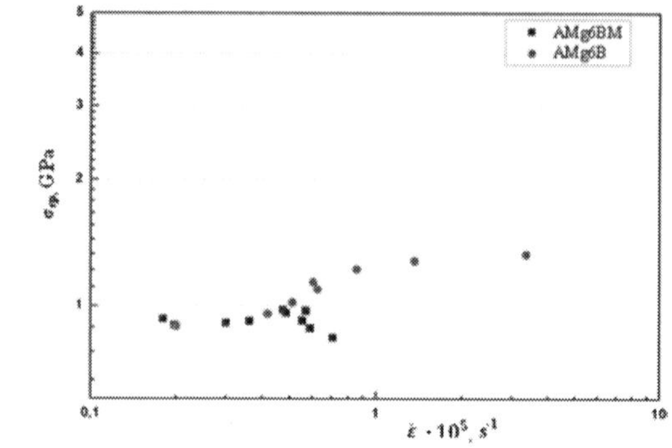

Figure 7. Spall strength of AMg6M and AMg6BM aluminum alloys samples depending on deformation rate.

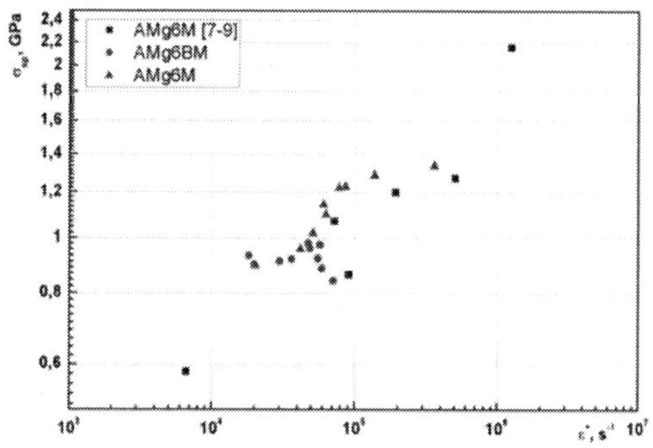

Figure 8. Spall strength of AMg6M and AMg6BM aluminum alloys samples depending on deformation rate.

rate range using loading technology of AMg6M alloy sample loading with tantalum foils 50 μm thick. In Fig. 5, we can see free surface velocity profiles for AMg6M alloy; these profiles are formed in the samples with these strikers.

Technology used to form load pulses in the samples allowed to extend the deformation rate range to $3.6 \cdot 10^5\ s^{-1}$. But spall signal under conditions shown in Table 2 was registered only in one experiment 48. AMg6M alloy spall strength for this experiment was $\sigma_{sp} = 1.33$ GPa.

Large number of experiments allowed to determine statistical straggling for the Hugoniot limit values (σ_{hel}), shear stress values (τ_T) and dynamic yield limit values (σ_τ). These parameters for AMg6M alloy were the following: $\sigma_{hel} = 0.42 \pm 0.03$ GPa, $\tau_T = 0.10 \pm 0.01$ GPa, $\sigma_\tau = 0.20 \pm 0.01$ GPa. For AMg6BM alloy: $\sigma_{hel} = 0.40 \pm 0.02$ GPa, $\tau_T = 0.10 \pm 0.01$ GPa, $\sigma_\tau = 0.19 \pm 0.01$ GPa.

In Figs. 6 and 7, we can see dependencies of spall strength for alloys AMg6M and AMg6BM on shock compression amplitude and on deformation rate, respectively.

In Figs. 6–7, we can see that spall strength of aluminum alloy AMg6M monotonously increases with pressure increase in the sample in the range from 1.4 to 14.2 GPa. Nevertheless, changes in spall strength of AMg6BM aluminum alloy bear clear-cut softening character at shock compression amplitudes higher than 5 GPa, and deformation rate higher than $0.5 \cdot 10^5\ s^{-1}$. It is hardly probable that cladding of AMg6BM sheets will change the material strength. The difference between AMg6M and AMg6BM lies only in the technologies of mechanical processing, which influences the material texture. All samples of AMg6BM alloy were manufactured out of the same sheet, thus intercrystalline corrosion of the material couldn't cause noticed alloy softening. It seems that material rolling changed its interior structure, which led to softening of the material AMg6BM.

Comparison of the experimental results, obtained using LGG, with the results, published in literature [8–10] (see Fig. 8), has shown that spall strength of bar material AMg6M is significantly higher than the one of the sheet material AMg6M. The same destruction character was observed in earlier works [5,7] while studying threshold values for destruction of AMg6 alloy samples, made of bar and of sheet. But some new previously unnoticed peculiar features in material response to shock-wave loading were found in sheet material AMg6M and AMg6BM. These features concerned deformation rate 10^4–$10^5\ s^{-1}$.

4. Conclusion

This work covers the results of the study and comparison of the dynamic properties of the aluminum alloys AMg6M and AMg6BM. Aluminum alloys loading was performed by shock waves with submicrosecond length, amplitude from 1.1 to 14.2 GPa and deformation rate from $0.2 \cdot 10^5$ to $3.6 \cdot 10^5\ s^{-1}$. As opposed to the works [6,8–10], for this study, we used samples of alloys AMg6M and AMg6BM made of bars of AMg6M KR40 (GOST 21488-97) and sheet AMg6BM 5 (GOST 21631-76).

Experimental data is given in the form of sample free surface velocity profiles. Experimental results are analysed. Dependencies of spall strength on amplitude of shock-wave action $\sigma(P)$ and deformation rate $\sigma(\dot{\varepsilon})$ are constructed.

The work results reveal that:

- the dependencies σ(P) and $\sigma(\dot{\varepsilon})$ for both sample materials monotonously increase up to the pressure of 5 GPa and $\dot{\varepsilon} = 0.5 \cdot 10^5\ s^{-1}$;
- when pressure is higher than 5 GPa and deformation rate is more than $0.5 \cdot 10^5\ s^{-1}$, softening of AMg6BM alloy takes place;
- threshold values of spall strength were $\sigma = 1.4$ GPa for AMg6BM alloy, and $\sigma = 1.5$ GPa for AMg6M alloy, respectively.

Experimental velocity $W(t)$ profiles obtained were used to determine the parameters of Jonson-Cook model of the material with Mie-Grüneisen equation of state. Experimental data satisfactorily agreed with numerical data.

References

[1] *Physical material science*, 6. Construction materials of nuclear power industry/Executive editor B. Kalin (2012)

[2] V. Belezkii, G. Krivov, *Reference book: Aluminum alloys (composition, properties, technology, application)* (2005)

[3] G. Kangel, V. Fortov, S. Razorenov, A. Utkin, *Shock-wave phenomena in condensed media* (1996)

[4] E. Avrorin, B. Vodolaga, V. Simonenko, V. Fortov, Progress in Physical Science Magazine, **163**, 5, 1–34 (1993)

[5] B. Tarasov, Strength problems, 3, 121–122 (1974)

[6] Yu. Batkov, S. Novikov, V. Sinitsin et al., PMTF, 3, 133–137 (1979)

[7] V. Golubev, S. Novikov, Yu. Sobolev, N. Yukina, Strength problems, 2, 53–59 (1983)

[8] G. Kanel, S. Razorenov, V. Fortov, PMTF, 5, 60–64 (1984)

[9] G. Kanel, S. Razorenov, V. Fortov, DAS USSR, **294**, 2, 350–352 (1987)

[10] S. Razorenov, G. Kanel, V. Fortov, FMM, **95**, 1, 91–96 (2003)

[11] B. Glushak, O. Ignatova, V. Pushkov et al., PMTF, **41**, 6, 139–142 (2000)

[12] A. Petrovtsev, G. Kovalenko, G. Zadorozhniy, *Zababakhin Scientific Talks: Sourcebook of VII International conference – Snezhinsk; RFNC-VNIITV publishers* (2003)

[13] Johnson G.R., Cook W.H., *Proc. Of 7th symposium of ballistics*, Den Hague, Niderlands (1983)

[14] V. Chuveldeev, M. Gryaznov, V. Kopilov, A. Sisoev, Vestnik Nizhegorodskogo Universiteta, **4**, 35–42 (2008)

[15] B. Kolachev, V. Elagin, V. Livanov. *Physical metallurgy and thermal treatment of nonferrous metals and alloys* (2005)

[16] A. Pavlenko, S. Balabin, O. Kozelkov, D. Kazakov, Instr. and Exp. Tech., **4**, 122–124 (2013)

[17] A. Pavlenko, S. Malyugina, V. Pereshitov, I. Lisitsina, Instr. and Exp. Tech., **2**, 127–129 (2013)

[18] S. Mokrushin, N. Anikin, S. Malyugina, A. Pavlenko, A. Tyaktev, Instr. and Exp. Tech., **4**, 107–109 (2013)

EPJ Web of Conferences **94**, 01056 (2015)
DOI: 10.1051/epjconf/20159401056

Explosive fragmentations of alumina (Al_2O_3) under quasistatic compressive loading

Qingyan Zhang[1], Xiaoqing Jin[1], and Fenghua Zhou[1,2,a]

[1] MOE Key Laboratory of Impact and Safety Engineering, Ningbo University, Ningbo 315211, Zhejiang, China
[2] The Faculty of Mechanical Engineering and Mechanics, Ningbo University, Ningbo 315211, Zhejiang, China

Abstract. Quasistatic compression tests for alumina (Al2O3) cylinders were conducted for the investigations of the compressive strengths and the dynamic fragmentation properties of the material. We focused on the post-failure dynamic fragmentation phenomenon. Most of the fragments were collected after tests, the shapes and sizes of these fragments were measured and statistically analyzed. The fragments were divided into three types on basis of their shapes and sizes, namely: the flaky medium sized fragments, the tiny debris, and the remaining large blocks, each type of the fragments were formed at different stages of the compressive failure-fragmentation process. The tiny debris were mainly generated from the "explosion" of the cylindrical specimen, in this stage the stored elastic energy within the specimen was released rapidly. The tiny fragments accounted the most part of the fragments in numbers. The average fragment size calculated by the proposed formula agree well with the experimental data.

1. Introduction

Due to their excellent mechanical properties of high hardness, high modulus and strength, low density, and good heat resistance, modern ceramics find broad applications in the area of protective engineering. Nevertheless, the materials are essentially brittle: having little fracture toughness and containing strength-limiting internal defects; Moreover, the tensile and compressive behaviors of the materials are asymmetric, with the compressive strength typically one order of magnitude higher than the tensile strength. The tension-compression asymmetry, in combinations with the material's brittleness, make ceramics easy to break into many pieces under different loading conditions. It is well known that under impact loading, the failure of a ceramic material is accompanied by a fragmentation process, when the material breaks (fragmentizes) into pieces millimeter or less in sizes [1–3]. Recent studies show that the compressive fragmentation are not limited to the impact loading cases. Under quasi-static compressive loading a ceramic specimen also fail in an "explosive" way [4,5].

The compressive fragmentation phenomenon can be understood from the energy point of view. Ceramics are strong to compression, with the compressive strength at the gigapascal (GPa) level. This means that at the failure point, the strain energy storage of the specimen is $1 \sim 10\,\mathrm{MJ/m^3}$. On the other hand, the high elastic modulus ($\sim$370 GPa) and the relatively low density ($4000\,\mathrm{kg/m^3}$) result in a very high fast crack speed (> 5000 m/s). Once failure starts, many cracks initiate and propagate within the specimen in a flash of time, fragmentizing the specimen and opening a way for the strain energy to release. This energy release process may occur in a time scale of $10^0 \sim 10^2\mu$s order, making the failure more like an "explosive".

Understanding the failure behavior of the ceramics under quasi-static or impact loadings are crucial to the design and performance evaluation of a protective structure. One important issue in this study is to understand the dynamic fragmentation and the energy release process of ceramics under compressive loading. In the previous study [5], we experimentally investigated the compressive failure and fragmentation of brittle alumina under quasistatic compression. High speed videos showed that the failure of the specimen started with specimen splitting, and ended with dynamic fragmentations. An energy based fragmentation model was proposed to calculate the average post-failure fragment size.

In this paper, we extend the previous research, focusing on the post-failure dynamic fragmentation phenomena. The experimental set-up was improved so that most of the fragments can be recovered during a test. After each test, the sizes of these fragments were measured and statistically analyzed. Different types of fragments were found to be related to the different stages of compressive fragmentation. Theoretical model was used for the estimation of the average fragment size.

2. Quasistatic compression test

2.1. Specimen and test arrangement

Three kind of Al_2O_3 cylindrical specimens were made by hot-sintering 99.5% purity alumina powder. The dimensions of the specimen were 10mm diameter (Φ10) by 10 mm height, 10 by 20, and 20 by 40. The error of the upper and lower surfaces parallelism was less than 5 μm. Quasi-static compressive tests were conducted

[a] Corresponding author: zhoufenghua@nbu.edu.cn

Article available at http://www.epj-conferences.org or http://dx.doi.org/10.1051/epjconf/20159401056

Figure 1. Quasi-static compression tests of Al_2O_3 cylinders: (a) the specimen and the WC anvils; (b) typical failure process.

on a servo-controlled hydraulic test machine with the maximum loading capacity 2000 KN. Tungsten carbide (WC) anvils were used between the specimen and loading platens to avoid indentations. Lubricant was applied on the interfaces between the anvils and the specimen to reduce friction. During the test process a high-definition video camera and a high-speed video camera were used to record the specimen failure process. The high speed video camera was triggered manually with 1 second pre-trigger time. To ensure that most fragments will be collected for further analysis, a rectangular cover made of transparent plexiglass was used to encapsulate the specimen, the WC anvils, and part of the loading platens, as shown in Fig. 1(a).

Quasi-static compression tests were conducted at an equivalent strainrate of 10^{-4} 1/s. The loading-strain curves show that the compressive deformation of the specimen is dominantly linear, with no observable signs of damage nucleation within the specimen. The compressive failure of the specimen occur as a sudden event, with immediate drop of loading stress, and always accompanied by explosive fragmentation. Figure 1(b) show a typical compressive failure/fragmentation process. The complete deform and failure process of a specimen include the stages of: 1) uniform deforming without observable damage initiation; 2) chipping when small debris fly out of the specimen surface; and sudden explosive failure.

2.2. Strength and fragmentation process

A total of 14 tests were conducted on different types of specimen. Table 1 collects the strengths of these specimen. Specimen with polished surface are apparently stronger than those unfinished. Most finished cylindrical specimen have strengths above 1GPa, except one specimen ($\Phi 20 \times 40$, No. 6) which had an observable surface damage near corner, and exhibited significantly low strength (0.22GPa). The strength scatters is shown in Fig. 2. Excluding the anomaly, smaller specimen tend to have higher strength than the larger ones

The failure of cylindrical specimens always accompany with fragmentations. The process when a specimen

Table 1. Strengths of Al_2O_3 cylinder specimen (GPa).

Specimen \ Test Number		1	2	3	4	5	6 (*)
Φ20x40	Polished	1.38	1.03	1.08	1.33	1.01	0.22
Φ10x20	Polished	1.58	0.8	1.14	-	-	-
	Unpolished	0.83	0.9	0.86	-	-	-
Φ10x10	Polished	1.71	1.03	-	-	-	-

* There existed an apparent surface damage area in this specimen.

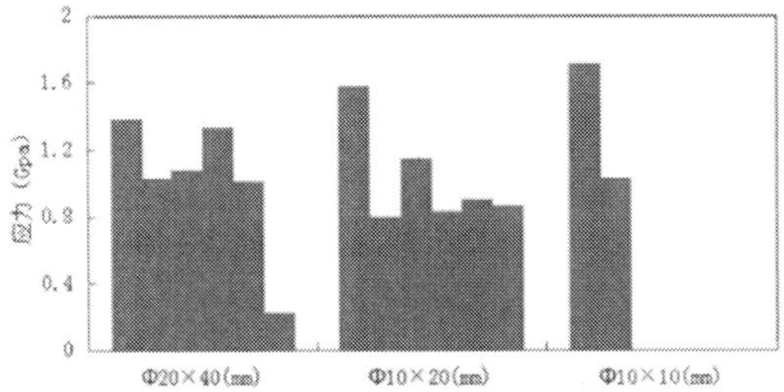

Figure 2. Scatters of compression strength.

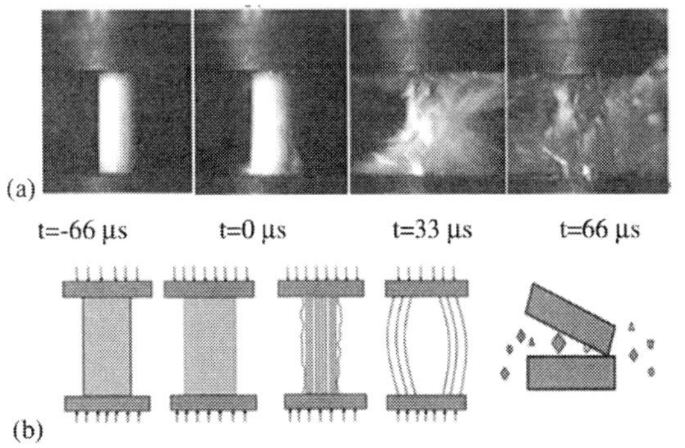

Figure 3. (a) High speed videos showing fragmentation process; (b) a schematic diagrams.

experience "explosive" fragmentation is shown in details by high speed videos in Fig. 3(a), and schematically explained in Fig. 3(b). In this process, the material fail at first with axial splitting, when many longitudinal cracks occur inside the specimen, splitting it into many columns. Under very large compression, the columns buckled and break into small pieces to release the stored strain energy.

The explosive fragmentation process were identical in all tests, whether the specimen was large or small. Typical fragmentation process occur within several dozens of μs. Always the bursting of specimen gave out a loud explosion sound and flashing light. There were discussions about whether the strain energy stored within the test machine would contribute to the fragmentation process. It appears that this energy flux is unlikely to happen. As Fig. 4(a) shows, once the failure occurs, the sudden loss of specimen loading capacity created an unloading wave in the elastic structures. This unloading wave would reflect at the specimen/anvil/platen interfaces as a tensile stress if they keep in contact, or cause separation if the interface

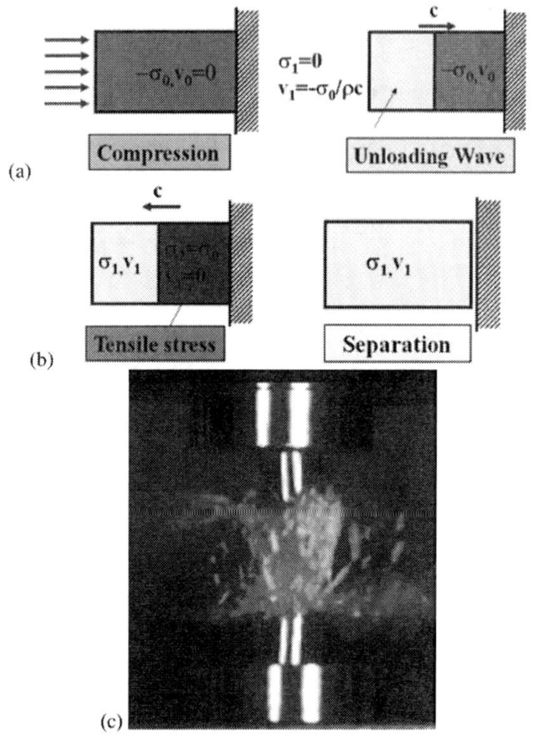

Figure 4. (a) Unloading waves propagation in elastic parts; (b) The reflections of the unloading wave on the interface, causing tensile stress of separations; (c) Separations of specimen-WC anvils.

cannot sustain tensile force (Fig. 4(b)). Eventually the specimen-anvil-loading platens interfaces separate, cutting off the energy flow from outside (Fig. 4(c)).

3. Fragment analysis

3.1. Fragment statistics

In each test, the fragments were contained in the plexiglass box and carefully collected. By comparing the total weight of the collected debris with the specimen weight, it can be concluded that most of the fragments were recovered. For each specimen, the debris statistical characteristics are similar, reflecting the fragmentation and energy release process. In this paper, we chose the three tests, each exhibiting the maximum compressive strength of same geometrical specimen group, namely, Sample **A** for the $\Phi 20 \times 40$ group specimen with compressive strength of 1.38GPa, Sample **B** for the $\Phi 10 \times 20$ group specimen with compressive strength of 1.58 GPa, and Sample **C** for the $\Phi 10 \times 10$ group specimen with compressive strength of 1.71 GPa.

For each sample, the sizes of the collected debris vary significantly. In current study we uses the debris size characterization method similar to that used for sediment grain size analysis in hydrological study [7]. Firstly different sieves were used to screen the fragments larger than 1.18 mm, and create the mass distributions of these "large" fragments. For the particles smaller than

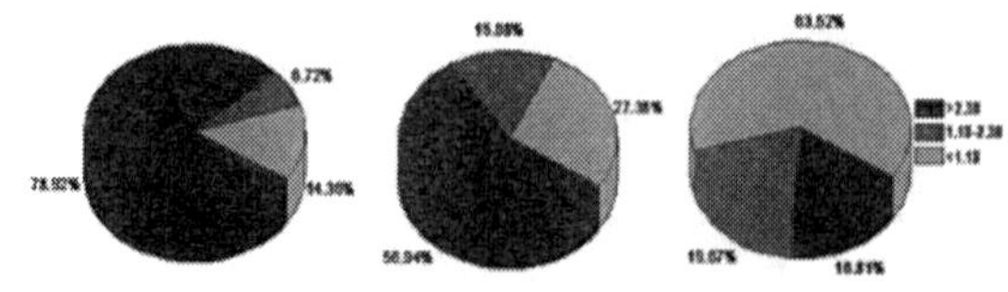

Figure 5. Mass distribution of large, medium, and small fragments in different samples: Sample A ($\Phi 20 \times 40$), B ($\Phi 10 \times 20$), and C ($\Phi 10 \times 10$).

1.18 mm size, a micro laser particle size analyzer was used to produce detailed quantitative statistical data.

In Fig. 5, the mass percentages of the large (sizes > 2.36 mm), medium (sizes between 1.18 mm and 2.36 mm), and small (sizes < 1.18 mm) fragments are shown for different samples. It is seen that the large/medium/small fragments ration are different for different types of specimen. For big specimen (Sample A), the large fragments (> 2.36 mm) account for the major part of the total mass (~79%), while for small specimen (Sample C) the major mass part were the small fragments (< 1.18 mm, mass ratio ~64%). This phenomenon can be explained as following: On one side, the small specimen (Sample C, failed at 1.71 GPa) had less internal imperfections and failure at higher compressive stress, so that its dynamic crushing process was more "explosive", producing large quantities of small pieces. On the other side, the large specimen (Sample A) failed at relatively lower stress (1.38 GPa) resulting in larger fragments. Besides, after fragmentation some large chunks may remain at the specimen inner parts, accounting for the major mass percentage.

3.2. Fragment categorizations

The compressive fragmentation of a ceramic cylinder is a quite complex process, with different stages resulting fragments of different shapes and sizes. As Fig. 6(a) illustrates, the failure of the specimen starts at surface peeling in which some flake-shape (size in one direction significant smaller the sizes in the other two directions). At the ultimate "explosion" process, the split ceramic columns buckle-fractured into small, cubical shaped (the sizes in three directions similar) fragments. Most tiny fragments were created in this stage. If the stored strain energy is not so high, after the specimen broke, some inner parts of the specimen may remain, results in large blocks of rectangular and cubic shapes. The larger blocks of the specimen remnants are easy to identify. The middle sized flake debris and cubical debris can be identified by their shapes. Figure 6(b) shows the medium sized fragments collected in Sample C.

3.3. Medium and small fragment size statistics

As discussed above, the large chunks of the fragments, being the remnants of the explosive fragmentations, were not directly related to the dynamic fragmentation process. The middle and small size fragments are statistically analyzed. For the small fragments < 1.18 mm, a laser particle size analyzer was used to measure the equivalent size of fragments. Here the equivalent fragment size was

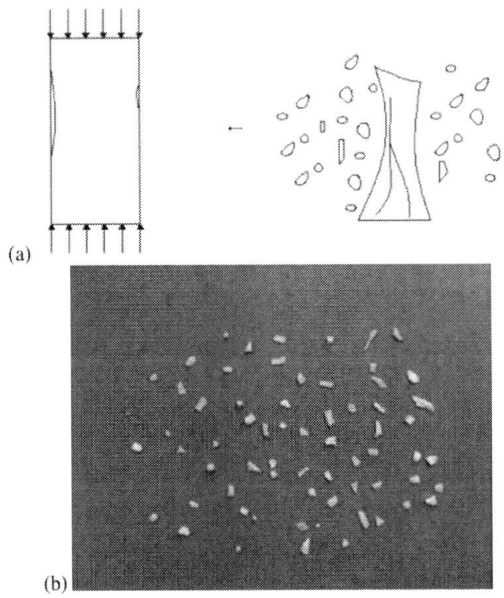

Figure 6. The different size fragments resulted from different fragmentation process: (a) surface fragment spallation; (b) fragmentation processes; c) The medium sized fragments (1.18 ~ 2.36 mm) from sample C.

taken as the diameter of the sphere having the same volume of the fragment. Figure 7(a) gives the statistical distribution of the small fragments (< 1.18 mm) with their equivalent size. The fragments were in Sample B. It is seen that there exist two peaks for the small fragments, at 250 μm and 800 μm. Though detailed analysis of these distributions might be interesting, here we only focus on the average size of a total 3067 small fragments. Figure 7(b) shows the morphological appearances of some small fragments under scanning electron microscope. The geometrical shapes of these fragments were more or less like cubes or spheres, justifying the use of the equivalent size as the fragment measure.

For the group of medium size (1.18 ~ 2.36 mm) fragments, the average fragment size were obtained by taking the picture of them (similar to Fig. 6(C)), then measuring the sizes using a image process software. Table 2 lists the average fragment sizes of all group, in each samples. The final average fragment size are taken as the number-weight average of the medium size group and the small size group.

4. Fragment size estimations

There exists many formulae for the estimations of fragment sizes [7–10]. We have proposed a model that linked the non-dimensional fragment size to the non-dimensional strain rate [11]:

$$\frac{s_{ave}}{s_0} = 4.5\left[1 + 4.5\left(\frac{\dot{\varepsilon}}{\dot{\varepsilon}_0}\right)^{2/3}\right]^{-1} \qquad (1)$$

where $s_0 = EG_c/\sigma_t^2$, $\dot{\varepsilon}_0 = c\sigma_t^3/E^2G_c$ are the characteristic fragment size and the characteristic strain rate of

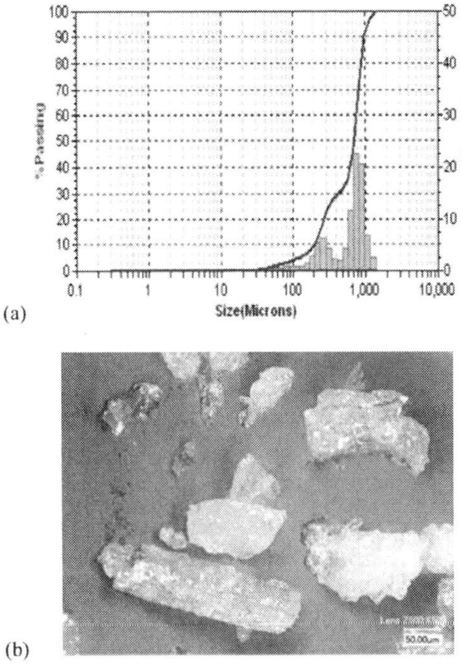

Figure 7. (a) Particle diameter analysis diagram of tiny debris (from Sample B, small size fragments); (b) Tiny fragments observed under SEM.

Table 2. Statistics of average fragments size.

Group \ Sample		A	B	C
1.18-2.36 (mm)	Mass (g)	3.295	0.954	0.660
	Numbers	320	96	63
	Ave. Size (mm)	1.7061	1.6985	1.7114
<1.18(mm)	Mass (g)	7.038	1.660	1.957
	Numbers	17696	3067	6335
	Ave. Size (mm)	0.5796	0.6437	0.5286
Experimental Average Size (mm)		**0.6000**	**0.6757**	**0.5402**
Theoretical Fragment Size (mm)		**1.31**	**0.922**	**0.63**

Table 3. Material parameters and characteristic quantities of Al_2O_3.

ρ g/cm^3	E GPa	c m/s	G_c N/m	σ_t MPa	$\dot{\varepsilon}_0$	s_0 mm
3.96	370	9700	40	150	5978	0.66

the material; E is the material's Young's modulus, G_c the fracture energy, σ_t the tensile strength, $c = \sqrt{E/\rho}$ the elastic wave velocity, and ρ the density. Table 2 listed the material data for the material. The s_0 and $\dot{\varepsilon}_0$ values thus calculated are 6000 s^{-1} and 0.66 mm, respectively.

It is notable that although the ceramic specimen was loaded quasi-statically, the sudden failure and fragmentation occur in a very short time and is undoubtedly dynamical. In the explosive fragmentation

Table 4. Estimation of fragments average size according to the theoretical formula.

Test	σ_c GPa	h mm	$\dot{\varepsilon}$ s^{-1}	$\dot{\varepsilon}/\dot{\varepsilon}_0$	s_{ave}/s_0	s_{ave} mm
A	1.36	40	*891.35*	0.1491	1.9860	*1.31*
B	1.58	20	*2071.08*	0.3464	1.3977	*0.92*
C	1.71	10	*4482.97*	0.97499	0.9546	*0.63*

process, the compressed columns of ceramic specimen buckle-bend rapidly, causing a tensile strainrate well above $10^3\ s^{-1}$. Following the method used in [5], The equivalent column expanding strainrate $\dot{\varepsilon}_0$ can be estimated by the energy balance theory: at the failure point the column has been pre-compressed to the compressive stress σ_c. Assuming that the total compressed energy has been transformed into the outward (radial) expanding kinetic energy, the expanding velocity of the fragmentizing part of the specimen is: $v = c\sigma_c/E$. Also assume that the column curvature radius is comparable to the specimen height h, the equivalent tensile strainrate is estimated by:

$$\dot{\varepsilon} = v/h = c\sigma_c/Eh. \tag{2}$$

Using the material parameters listed in Table 3, and the test conditions, the theoretical fragmentation parameters (equivalent tensile strainrate and average fragment size) can be estimated, as collected in Table 4.

The theoretically evaluated fragment sizes were also listed as the last row in Table 3 for comparisons. It is seen that experimental measured fragment size were generally smaller than, but pretty close to the theoretical estimations. This difference could be attributed to the experimental measurements that excluded the large fragment chunks without accounting for the stored strain energy in them. As the specimen size decrease, the theoretical estimations are closer to the experimental data, because the small specimen fragmentizes more significantly, namely, the small specimen (sample B and C) fragmentized completely. In whole, the theoretical model provided good fragment size estimations, better than simple order of magnitude. This model can be further extended to other ceramics and to the dynamical loading cases.

5. Summary

In this research we conducted quasi-static compression tests on three sizes Al_2O_3 cylinders. The compressive strength of the cylinders were measured. These strength show slight size dependence. The specimen explosively fragmentize after compression failure. The physics of these fragmentation phenomenon were explained from energy viewpoint. The fragments in each test were collected and analyzed. The fragments can be roughly classified into three types, each corresponding to different stage of the fragmentation process: the flaky medium sized fragments originated from surface spallation, the tiny debris generated from the "explosion" of the cylindrical specimen, and the large chunks remaining in the centre of the specimen. The tiny fragments accounted the most part of the fragments in numbers. The average fragment size calculated by using the standard formula agree reasonably well with the experimental measurement.

This research was supported by the National Science Foundation of China (NSFC) under Grants No 11272103, and by K.C. Wong Magna Fund of Ningbo University. The work has been presented in the 11th National Conference of Impact Dynamics in China.

References

[1] S. Sarva S. and S. Nemat-Nasser, Mat. Sci. Engng, **A317**(1-2): 140–144 (2001)

[2] W. Chen and G. Ravichandran, J. Mech. Phy. Solids, **45**(8): 1303–1328 (1997)

[3] H. Wang H. and K.T. Ramesh, Acta Materialia, **52**(2): 355–367 (2004)

[4] M.Y. Lee, R.M. Brannon and D.R. Bronowski, Sandia Report, SAND2004-6005 (2005)

[5] F. Zhou and Y. Wang, Proceeding DYMAT 2009: 1311–1316 (2009)

[6] Z. Niu and R. He, J. Sediment Res. 2002(5): 6–14 (in Chinese)

[7] D.E. Grady D.E., J. Appl. Phy. **53**: 322–325 (1982)

[8] L.A. Glenn and A. Chudnovsky, J. Appl. Phy. **59**: 1379–1380 (1986)

[9] W.J. Drugan, J. Mech. Phy. Solids, **49**: 1181–1208 (2001)

[10] V.B. Shenoy and K-S. Kim, J. Mech. Phy. Solids, **51**: 2023–2035 (2003)

[11] F. Zhou F., J-F. Molinari and K.T. Ramesh, Int. J. Frac. **139**:169–196 (2006)

EPJ Web of Conferences **94**, 01057 (2015)
DOI: 10.1051/epjconf/20159401057

Impact performance of FRC slabs under various strain rates

Alena Horska[a], Pavel Jiricek, and Marek Foglar

CTU in Prague, Fac. Of Civil Engineering, Thákurova 7, 166 29 Praha 6, Czech Republic

Abstract. Impact resistance of plain concrete and FRC composites gains high importance in the present days. This paper presents outcomes of the experiments focused on performance of FRC slabs subjected to impact loading of various strain rates. The FRC slabs, thickness 30, 60 and 120 mm were subjected to various drop-weight impacts. The different drop-hammer weights provided different loading speeds and strain rates. The performance of the slabs was recorded with the use of high speed cameras. The cameras were used for impact speed control and specimen deflection measurement. With the use of the recordings, the force-time response of the slabs could be plotted. The plots correspond to the mechanical behaviour obtained by static load tests (loading speed 0.2 mm/min) but several differences can be found. Several types of FRC (varying fiber material, fiber content, etc.) were tested and evaluated. The experimental program is supplemented by numerical modelling which provides good agreement with the experimental results.

1. Introduction

Tensile resistance of concrete is significantly dependent on the speed of loading and type of reinforcement. The increase of strength of concrete subjected to loadings of higher strain rates is a well-known phenomenon and is described in the design standards [1,2] and literature, e.g. [3,4]. The dynamic increase factor (DIF) corresponds to the strain rate the concrete is subjected to and is used for quantification of the increase of tensile and/or compressive strength when describing the behaviour of concrete under extreme loadings such as blasts and impact. To obtain real values of impact resistance of the variously reinforced concrete samples, drop – weight experiments were performed.

A drop-weight frame was built for investigation of impact resistance of concrete and fiber reinforced cementitious composites. The device is as well used for investigation of fatigue loading on impact resistance.

a)

b)

Figure 1. a) Steel frame with the hoist and the burden; b) Specimen support.

2. Experimental set up

2.1. Dropping device

For the impact loading, a steel frame (Fig. 2a) carrying drop weight was built in the Experimental Centre of the CTU. The construction consists of a braced steel frame to ensure maximal stability of the dropping device. Due to laboratory build-in crane, height of the frame is limited to 4.1 m. Horizontal base of the construction is made of a steel frame anchored to the floor of the laboratory. Vertical stability is provided by bracing diagonals. The frame was ordered, delivered and built in 2013, the drop weight and the hoist and burden in 2014. The burden is considered as a set of different weights to enable maximal variety of the drop-weight.

The tested concrete specimen is located in the biaxial centre of the horizontal frame. The sample is supported at two parallel sides by reinforced steel girders which were anchored to the floor (Fig. 1b). Lower surface of the specimen is 160 mm above the laboratory floor and is supported by flanges of the girders 15 mm on each side. The exact centre of the concrete sample under the burden was set up using a plummet.

The steel burden used during the experiment consisted of a central pole and a main cylinder/hammer, which supports steel discs of different weights (Fig. 2).

The required weight was obtained by combination of the steel discs. On the lower surface of the steel cylinder, a hemispherical tip (R = 20 mm) is placed, which concentrates the applied impact loading into one point (Fig. 3).

[a] Corresponding author: alena.horska@fsv.cvut.cz

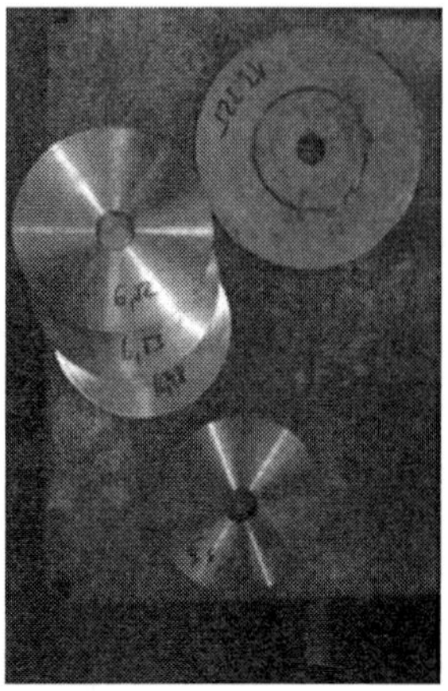

Figure 2. Steel discs of different weights.

Figure 3. Hemispherical tip of the burden.

Figure 4. Concrete mixture in the forms.

2.2. Tested specimens

The specimens, loaded with the force impulse, were concrete tiles. Supplementary cubes used to determine compressive strength of evaluated concrete mixtures were cast. Tiles and cubes were cast in laboratories of Faculty of Civil Engineering. Dimensions of the tiles were 300 × 300 mm with variable height (30/60/120 mm) (Fig. 4). The fibers used as reinforcement of several specimens were made of polypropylene (54 mm long with tensile strength 570–660 MPa and steel (60 mm long with tensile strength 1,115 MPa for NSC and 13 mm long with tensile strength 2,750 MPa for HPC); a basalt fabric was used to reinforce some of the concrete tiles. The fabric was placed in two layers, lower surface and/or both surfaces. For each mixture, 3 identical specimens of all the used thicknesses were prepared (Table 1). Two of the specimens were loaded by different impact loading and one sample was used for three-point bending test.

In the first quarter of 2014 production of the burden and dropping device was ordered and production of

Table 1. Tested concrete specimens.

Concrete mixtures	Reinforcement	Volume of fibers	Tested (Feb 2015)
NSC	-	-	YES
NSC	PP	9 kg/m^3	YES
NSC	FE	80 kg/m^3	YES
NSC	PP	4,5 kg/m^3	NO
NSC	FE	40 kg/m^3	NO
NSC	TEX	-	NO
NSC	TEX II	-	NO
HPC1	FE	100 kg/m^3	NO
HPC1	TEX	100 kg/m^3	NO
HPC1	TEX II	100 kg/m^3	NO
HPC2	FE	100 kg/m^3	NO
HPC2	TEX	100 kg/m^3	NO
HPC2	TEX II	100 kg/m^3	NO

TEX – basalt fabric – lower surface.
TEX II – basalt fabric – both surfaces.

Figure 5. Impact process; 46.4 kg burden, 60 mm thick tile made of steel fiber reinforced concrete.

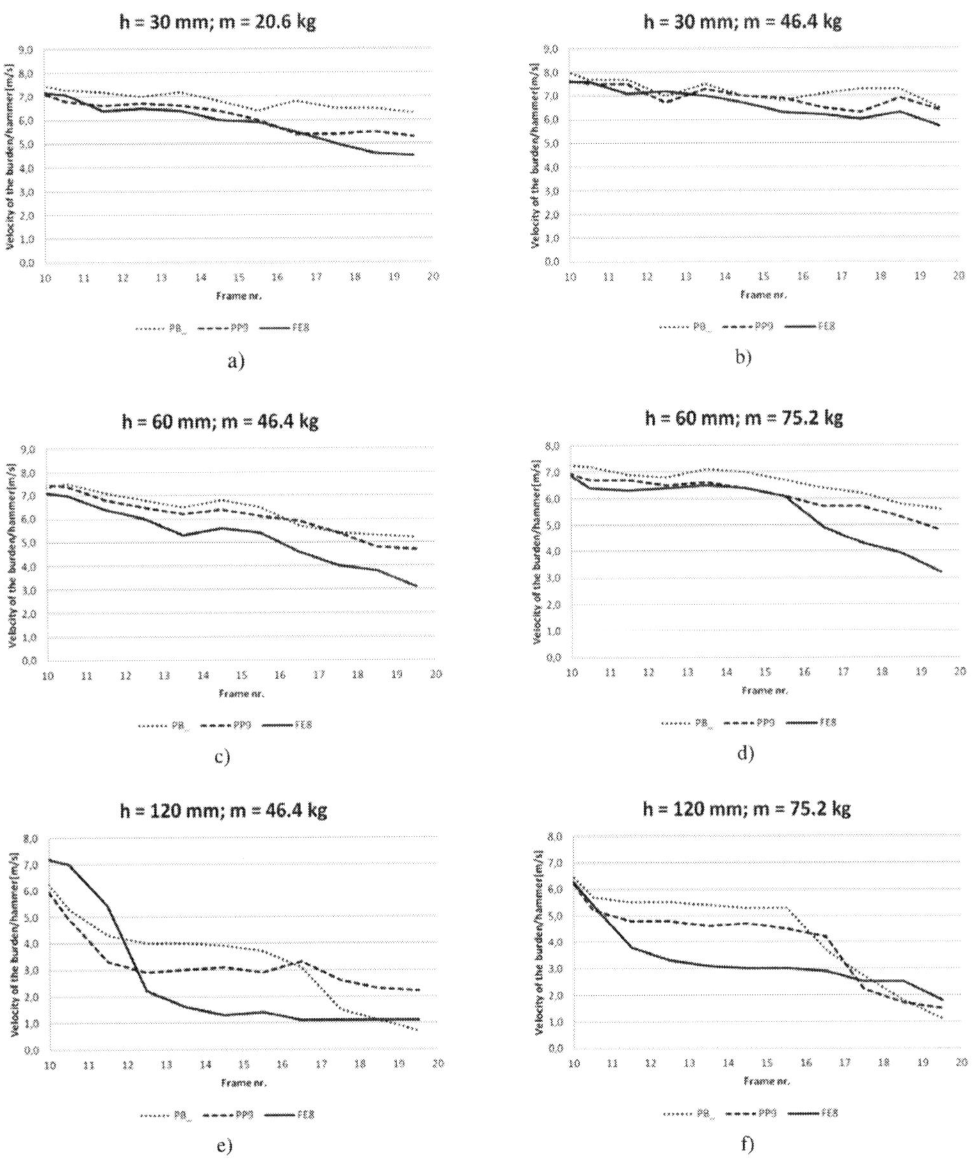

Figure 6. Comparison of different material type with the same burden weight and the same thickness; a) Thickness 30 mm, 20.6 kg burden; b) Thickness 30 mm, 46.4 kg burden; c) Thickness 60 mm, 46,4 kg burden; d) Thickness 60 mm, 75.2 kg burden; e) Thickness 120 mm, 46.4 kg burden; f) Thickness 120 mm, 75.2 kg burden.

the concrete samples has begun. The production of the concrete tiles took almost half of the year; in total 117 pcs of concrete tiles were cast. Each one of the specimens had to reach maximal compressive strength (min. 28 days).

2.3. Recording of the experiments

The release of the burden/hammer was recorded by a high speed camera. The camera was located 2–3 m from the specimen. To eliminate the effect of perspective, the horizontal level of the camera was set to the level of the tested specimen. The scale of the shots made by the camera was determined from the known height of the samples and from the lath with black and white stripes (50 mm wide). The camera was recording 1000–10 000 fps.

The impact loading was recorded (Fig. 5), the captured decomposed frames were graphically processed. The velocity of the burden during impact and deflection of the sample were obtained and plotted in graphs.

3. Evaluation

To determine the impact resistance of specimens, the deceleration of the burden was observed (Fig. 6). According to known weight of the burden and height of the dropping device, the momentum was evaluated. From the change of the momentum (deceleration of the burden),

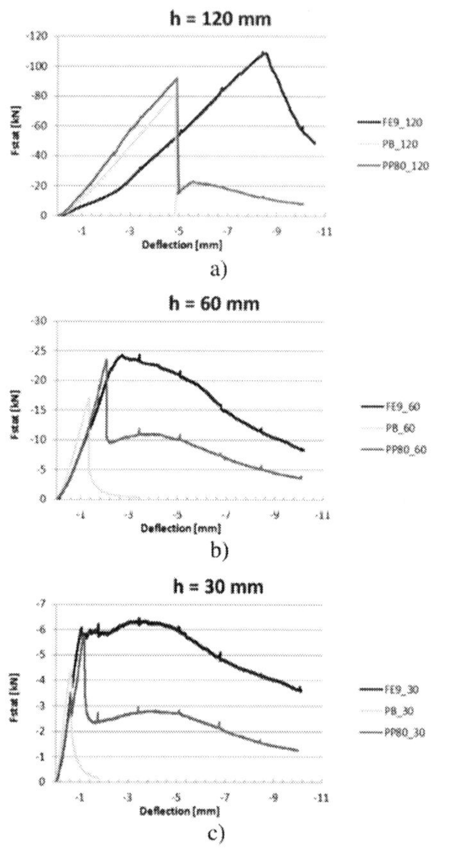

Figure 7. Results of three-point bending test (speed of loading 0.2 mm/min); a) Specimens 120 mm thick; b) Specimens 60 mm thick; c) Specimens 30 mm thick.

the impact force leading to collapse of the specimen was determined.

In the graphs PB_ refers to NSC mixture without reinforcement, PP9 to NSC with 9 kg/m^3 polypropylene fibers and FE8 to NSC with 80 kg/m^3 steel fibers.

The dynamic force required to break the specimen is simply obtained from the change of the velocity of the burden/hammer during impact.

$$F = ma \tag{1}$$

$$a = \Delta v / \Delta t. \tag{2}$$

The results of three-point bending test are shown in Fig. 7 in graph plotting static force to deflection (0.2 mm/min). Comparison of static and dynamic forces obtained from the experiment is presented in Table 2.

Fiber reinforced concrete (FRC) is a inhomogeneous material which mechanical properties can vary from specimen to specimen according to fiber content and distribution in the decisive cross-sections. In the presented experimental program, 3 specimens were cast of every concrete mix and specimen dimension. During probe experiments, the results showed only negligible variations, therefore the three specimens were used to study the effect of varying load on the performance of FRC subjected to impact loading.

Table 2. Comparison of static and dynamic strength.

Concrete mixtures	h [mm]	Burden [kg]	F_{dyn} [kN]	F_{stat} [kN]	F_{dyn}/F_{stat}
NSC	30	20.6	5.2	4.4	**1.18**
NSC	30	46.4	6.2	4.4	**1.41**
NSC	60	46.4	15.5	17.2	**0.90**
NSC	60	75.2	20.7	17.2	**1.20**
NSC	120	46.4	74.2	81.0	**0.92**
NSC	120	75.2	75.2	81.0	**0.93**
NSC-PP-9	30	20.6	6.9	5.7	**1.21**
NSC-PP-9	30	46.4	7.7	5.7	**1.35**
NSC-PP-9	60	46.4	23.2	23.5	**0.99**
NSC-PP-9	60	75.2	30.1	23.5	**1.28**
NSC-PP-9	120	46.4	116.0	92.2	**1.26**
NSC-PP-9	120	75.2	120.2	92.2	**1.30**
NSC-FE-80	30	20.6	8.2	6.5	**1.26**
NSC-FE-80	30	46.4	7.7	6.5	**1.18**
NSC-FE-80	60	46.4	23.2	24.5	**0.95**
NSC-FE-80	60	75.2	33.8	24.5	**1.38**
NSC-FE-80	120	46.4	102.1	109.0	**0.94**
NSC-FE-80	120	75.2	137.8	109.0	**1.26**

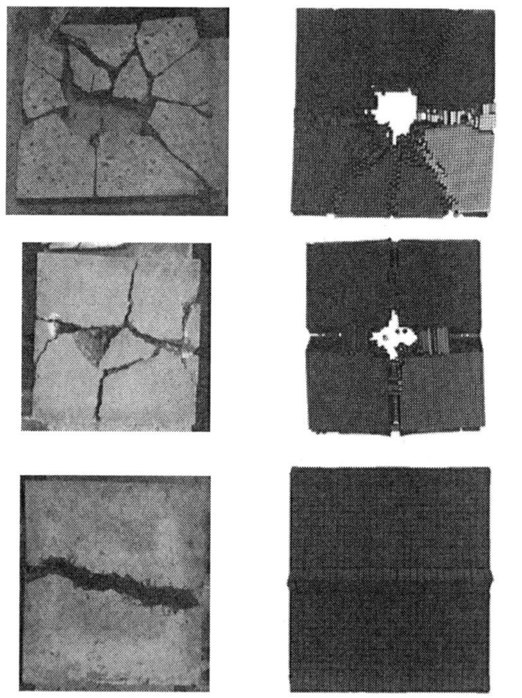

Figure 8. Illustrative comparison of the experiment and the numerical simulation.

4. Numerical simulations

Numerical simulations were prepared together with the experiment (Fig. 8), material models are calibrated according to the results of drop weight test. These results, together with results obtained from specimens not tested yet, will allow detailed description of behaviour of concrete under impact loading. Impact resistance of concrete reinforced with different types of fibers and basalt fabric will enable full-scale numerical simulations of extreme loading cases in civil engineering and increasing its impact resistance.

5. Conclusions

Concrete specimens reinforced with steel fibers slow the burden the most and plain concrete the least. The 30 mm thick tiles are the least impact resistant; on the other hand, the samples with height 120 mm almost stop the burden. The ratio of dynamic and static force needed to penetrate the element is 1.16 in average.

As a part of the experiment ongoing, numerical simulations were made to calibrate material model of normal strength concrete and different types of fiber reinforced concrete. Results of the simulation lead to good agreement with the experimental data.

The financial support of the Czech Science Foundation grant project 13-30441S is gratefully acknowledged.

References

[1] CEB-FIP Model Code 1990, Comité Euro-International du Béton (1993).
[2] CEB-FIP Model Code 2010, Comité Euro-International du Béton (2010).
[3] L. J. Malvar, C. A. Ross, ACI Material Journal, **95(6)**, 735–739 (1998).
[4] J. W. Tedesco, C. A. Ross, Journal of Pressure Vessel Technology, **120(4)**, 398–405 (1998).
[5] P. Jiříček, M. Foglar, Bridge Maintenance, Safety, manegement and Life Extension 2014, 762–768 (2014).
[6] P. Jiříček, M. Foglar, Proceeding of 23rd Military Aspects of Blast and Shock 2014 (2014).

EPJ Web of Conferences **94**, 01058 (2015)
DOI: 10.1051/epjconf/20159401058

Impact performance of specimens subjected to fatigue loading – experimental investigation

Jakub Göringer[a], Marek Foglar, and Pavel Jiříček

Czech Technical University in Prague, Faculty of Civil Engineering, 16629 Thakurova 7, Prague 6, Czech Republic

Abstract. The impact performance of reinforced concrete specimens subjected to fatigue loading has not been quantified properly yet. This topic is significant in the field of vehicle impact or similar applications. The paper aims to fill this gap by presenting the on-going experimental program. The paper presents outcomes of the experiments focused on the performance of RC beams subjected to drop-weight impact loading. The behaviour of the beams which were prior to the impact testing subjected to cyclic loading was compared to the behaviour of the beams which were not subjected to cyclic loading.

1. Introduction

Fatigue is the process of permanent progressive changes in the structure of a material subjected to cyclic loading. The influence of fatigue on changes in material structure and their descriptions have been studied by many authors, e.g. [1,2], but the effect of the matrix damage due to cyclic loading on the behaviour of specimens under impact loading has not been properly quantified yet. The combination of these two effects can be expected especially at the structures of transport infrastructure.

The assessment of the structures for impact loading (vehicles, trains) was already studied e.g. in [3] and is already incorporated into the design standards e.g. EN 1991-1-7 [4], but the assessment approaches do not take into account the material damage due to the loads that occur within the lifetime of the structures. In addition, many of these structures are often located in aggressive environment (e.g. de-icing salts for road maintenance). The phenomenon of chemical deterioration has been extensively investigated with regard to the effect of acid rain on concrete structures in [5,6].

The paper presents an on-going experiment dealing with the combined damage due to cyclic loading and aggressive environment together with the impact loading.

2. Experimental program

The experimental program is focused on three deteriorative effects reinforced concrete specimens can be exposed to – deterioration due to aggressive environment, cyclic loading and impact loading. The paper presents results of the impact behaviour of the specimens stored in different environments.

2.1. The reinforced concrete specimens

Reinforced concrete specimens were designed for the experimental program. The strength class of concrete C25/30-X0 (with low resistance against aggressive environment) was chosen. The specimens were designed as beams with dimensions of 300 × 150 × 1300 mm (width × height × length). The mix proportions of the concrete mixture are provided in Table 1.

The specimens were designed as over-reinforced (6 ⌀ 16 grade B500 reinforcing steel), thus a failure by compressive-zone crushing should occur and the fatigue failure of concrete (bending/shear) can be assumed during the cyclic loading. The scheme of specimen reinforcement can be seen in Fig. 1.

The specimens were stored in dry or aggressive environment to allow the determination of the influence of the aggressive environment on the impact behaviour. The specimens were subsequently exposed to cyclic loading.

In total 17 specimens were cast (see Table 2 for the list of the specimens) and three were already tested for impact loading; specimen #1 – stored in dry condition, not exposed to cyclic loading (R1); specimen #2 – stored in dry condition, exposed to cyclic loading (C3) and specimen #3 – stored in acidic environment, exposed to cyclic loading (A7).

2.2. Aggressive environment

For the purpose of simulation of the aggressive environment on structures, several specimens were stored in a solution of hydrochloric acid. The acidity of the aggressive environment was chosen according to the exposure classes specified in the European standards, namely EN 206-.

For the acceleration of the chemical deterioration, the solution of hydrochloric acid was maintained with pH = 2.0. The methodology for the evaluation of the influence of the aggressive environment on the concrete specimens is described in [7].

2.3. The fatigue testing

The arrangement of the cyclic loading is the four-point bending test with span length 1000 mm and overhangs with

[a] Corresponding author: jakub.goringer@fsv.cvut.cz

Article available at http://www.epj-conferences.org or http://dx.doi.org/10.1051/epjconf/20159401058

Table 1. Mix proportions of the concrete mixture.

CEM II 32.5 [kg/m3]	320
Sand [kg/m3]	836
Fine aggregate [kg/m3]	495
Coarse aggregate [kg/m3]	443
Water [kg/m3]	185
w/c ratio	0.58

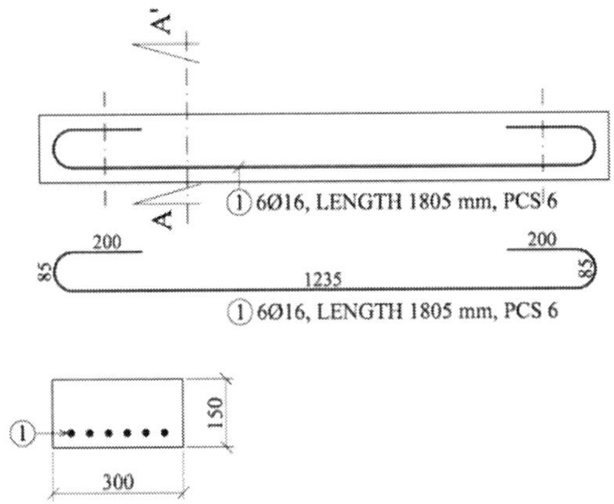

Figure 1. Reinforcement scheme of the specimens.

Table 2. List of all specimen sets.

Dry environment			Acidic environment			
Spec	Cyclic testing age [day]	Impact testing [Y/N]	Spec	Cyclic testing age [day]	Acid exp. [day]	Impact testing [Y/N]
C1	28	N	**A1**	133	7	N
C2	35	N	**A2**	182	14	N
C3	42	Y	**A3**	224	30	N
C4	63	N	**A4**	364	60	N
C5	77	N	**A5**	406	90	N
C6	105	N	**A6**	728	180	N
R1	-	Y	**A7**	777	360	Y
R2	-	N	**A8**	1099	720	N
			A9	1141	1080	N

Figure 2. The arrangement of the fatigue testing.

Table 3. Specification of the fatigue testing settings.

Loading Frequency	Cyclic force		Eccentricity
	Min	max	
Hz	kN		m
5	5	100	0.3

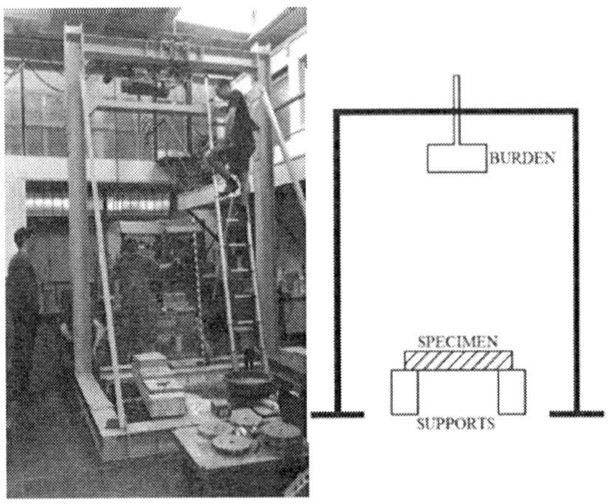

Figure 3. Steel frame and the scheme or the experiment.

length of 150 mm. This testing layout has been chosen for its several advantages described in [7].

Experiments are conducted in the Experimental Centre of Faculty of Civil Engineering, Czech Technical University in Prague. The arrangement of the fatigue testing can be seen in Fig. 2, for its parameters see Table 3.

Due to the high time demands of the experiments the maximum number of applied cycles for each specimen is around 400 thousand cycles.

These phenomena mentioned above – cyclic loading and aggressive environment are reflected in the discussion of the behaviour during the impact loading.

2.4. The impact loading

For the impact loading, a braced steel frame (Fig. 3) carrying drop weight was built in the Experimental Centre of the CTU. Due to laboratory build-in crane, height of the frame is limited to 4.1 m. Horizontal base of the construction is made of steel frame anchored to the floor of the laboratory. Vertical stability is provided by bracing diagonals. The burden is considered as a set of different weights to enable maximal variety of the drop-weight.

The tested concrete specimen is located in the biaxial centre of the horizontal frame. The sample is supported on two parallel sides by reinforced steel girders which were anchored to the floor. To ensure similar layout as in case of the cyclic loading two reinforcing bars were attached to the steel girders. This ensured the span length of 1000 mm and simple supported beam behaviour. The difference between the cyclic loading and impact loading is in the force position (four-point bending for cyclic loading and three-point bending for impact loading).

The steel burden used during the experiment consisted of a central pole and a main cylinder/hammer, which supports steel discs of different weights. The required weight was obtained by combination of the steel discs.

Figure 4. Steel burden, discs and hemispherical tip.

On the lower surface of the steel cylinder, a hemispherical tip (R = 20 mm) is placed, which concentrates the applied impact loading into one exact point. The burden used for the testing weights 90 kg. The burden, discs and tip can be seen on Fig. 4.

3. Interaction of the impact loading, fatigue and chemical deterioration

The rebound hammer (as described in Sect. 2.4) was used for the testing of the influence of fatigue and chemical deterioration on the impact of RC specimens.

As mentioned before, as far in total 3 specimens were tested; specimen #1 – stored in dry condition, not exposed to cyclic loading; specimen #2 – stored in dry condition, exposed to cyclic loading and specimen #3 – stored in acidic environment, exposed to cyclic loading. The specimens were made from the same concrete mix.

The fatigue loading resulted into an increase of deflections. The increase due to the deterioration caused by the aggressive environment was observed as well, but in smaller scale. It can be assumed that with the increase of strain rates (reflected by the dynamic increase factor (DIF)), can cause that the increase in deflection or damage will be higher for the deteriorated specimens. This assumption is based on the fact that the value of DIF increases along with the increase in strain rates. Thus the decrease in mechanical properties due to chemical deterioration of concrete can become more significant. Values of the DIF for the cyclic and impact testing were determined according to the standards for concrete, Eq. (1) – Model Code 1990 [8] and Eq. (2) – Model Code 2010 [9].

$$f_{c,imp,k} \big/ f_{cm} = \left(\dot{\varepsilon}_c / \dot{\varepsilon}_{c0}\right)^{0.014} \text{for } \dot{\varepsilon}_c \leq 30\, s^{-1} \quad (1)$$

$$f_{ct,imp,k} \big/ f_{ctm} = \left(\dot{\varepsilon}_{ct} / \dot{\varepsilon}_{ct0}\right)^{0.018} \text{ for } \dot{\varepsilon}_{ct} \leq 10\, s^{-1}. \quad (2)$$

Table 4. Specification of the fatigue testing settings.

Load type	Strain rate	DIF compression		DIF tension	
		MC90	MC10	MC90	MC10
cyclic	1.17e-02	1.180	1.093	1.292	1.184
impact	2.89e+00	1.350	1.174	1.532	1.307

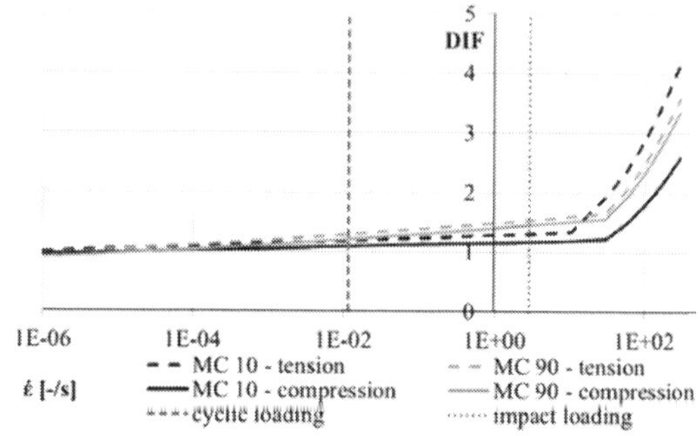

Figure 5. DIF according to MC90 [8] and MC10 [9].

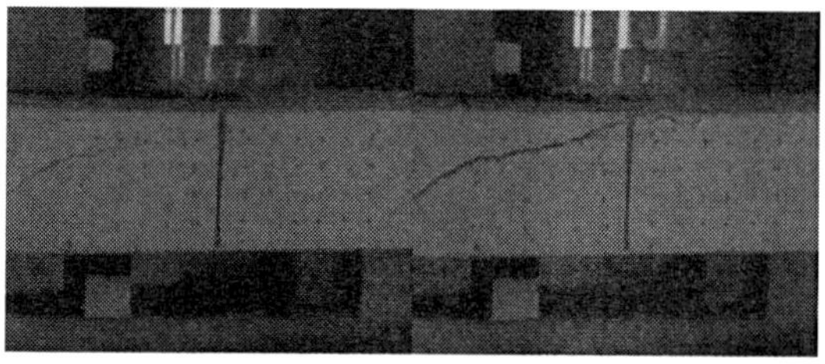

Figure 6. Two sample frames from the high-speed camera.

For both types of the testing, the DIF was evaluated on the base of the strain rates. The strain rate was calculated from the change of the specimen deflections in time; the scheme of the calculation is represented by Eq. (3).

$$\Delta w(t) \rightarrow \Delta 1/r(t) \rightarrow \dot{\varepsilon}. \quad (3)$$

The evaluated strain rates with corresponding values of DIF are listed in Table 4. The development of DIF as a function of different strain rates can be seen in Fig. 5. The results of the first run of tests are described further.

3.1. First set of impact testing

The decrease in instantaneous velocity within the impact together with the increase in deflections were chosen as a main factors for the evaluation of the specimens behaviour and resistance to impact loading. The experiment was recorded with high-speed camera with 10.000 fps. The video sequence was used for the evaluation of the behaviour. Two sample frames with time interval 1 ms can be seen in Fig. 6.

Results for the specimen #1, not exposed to cyclic loading, are in Fig. 7.

After the impact (frame 12) the steep decrease in velocity can be observed. The absorbed energy is reflected in deflection increase. The impact lasts about 8 ms (each frame corresponds to 1 ms). The shear cracks appeared similarly to the fatigue testing as can be seen on the Fig. 8.

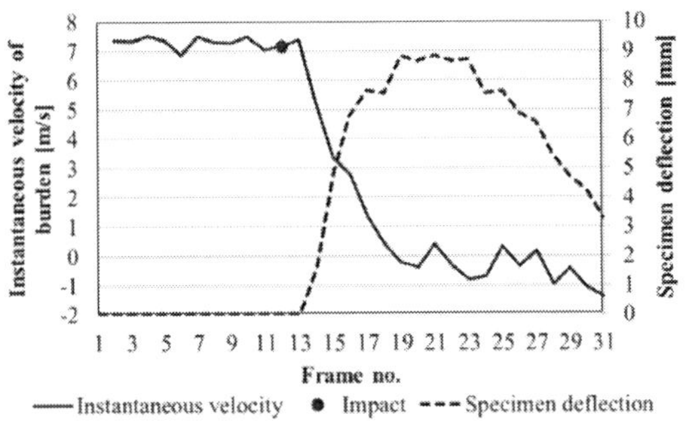

Figure 7. Development of burden velocity and specimen deflections during impact (specimen #1).

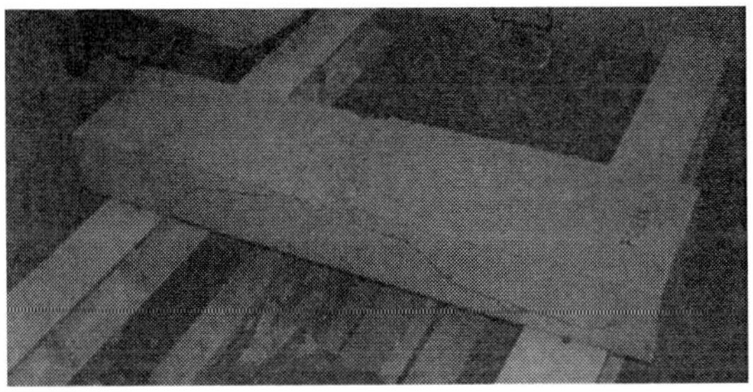

Figure 8. Condition of the specimen after impact (spec. #1).

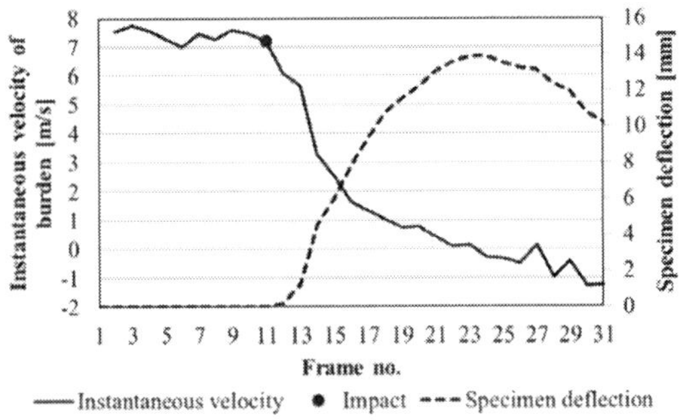

Figure 9. Development of burden velocity and specimen deflections during impact (specimen #2).

Figure 10. Condition of the specimen after impact (spec. #2).

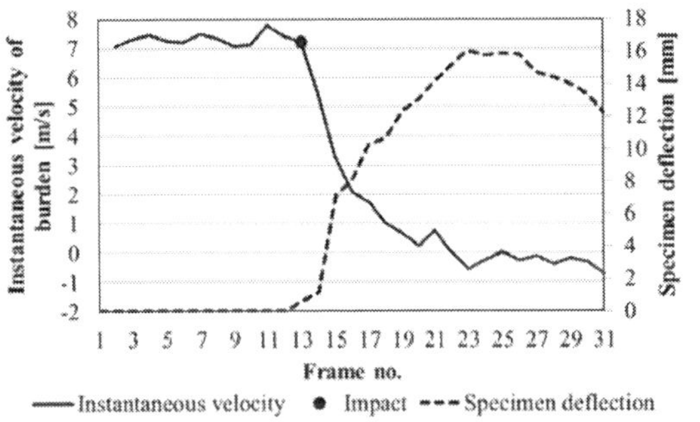

Figure 11. Development of burden velocity and specimen deflections during impact (specimen #3).

Figure 12. Condition of the specimen after impact (spec. #3).

After the rebound of the burden (velocities lower than 0) the specimen deflection partially decreased.

Results for the specimen #2, exposed to cyclic loading, are in Fig. 9. The decrease in velocity after the impact is slower than for the specimen #1. Shear cracks from the fatigue testing were opening, thus the energy was absorbed at slower rate, but with result of higher deflections. The impact lasts about 12 ms. In comparison of the specimen #1 and specimen #2 the increase in deflections in impact loading due to the damage caused by fatigue testing is up to 40%. Condition of the specimen #2 after the impact loading can be seen in Fig. 10.

Results for the specimen #3, chemically deteriorated specimen exposed to cyclic loading, can be found in Fig. 11. The decrease in velocity after the impact is slower than for the specimen #1 and #2. Shear cracks from the fatigue testing were opening and the deteriorated layers of concrete crushed in contact with the burden. Duration of the impact was about 14 ms. In comparison of the specimen #2 and specimen #3 the increase in deflections in impact loading due to the damage caused by chemical deterioration is up to 15%. When comparing the specimen #1 and #3 the increase in deflections due to the combined deterioration due to fatigue and aggressive environment is up to 60%. The specimen #3 after the impact loading is in Fig. 12.

From the results depicted in Figs. 7, 9 and 11 preliminary conclusions can be drawn. The comparison between the behaviour of all three tested specimen from the viewpoint of the velocity or deflections can be seen on the Figs. 13, 14 respectively. The Figures depict only the time interval after the impact. The impacts last cca 8, 12 and 14 ms respectively for the specimen #1, #2 and #3. The assumed ends of the impacts are marked with points.

The differences in the decrease in the velocity of the specimens can be observed. The impact in case of

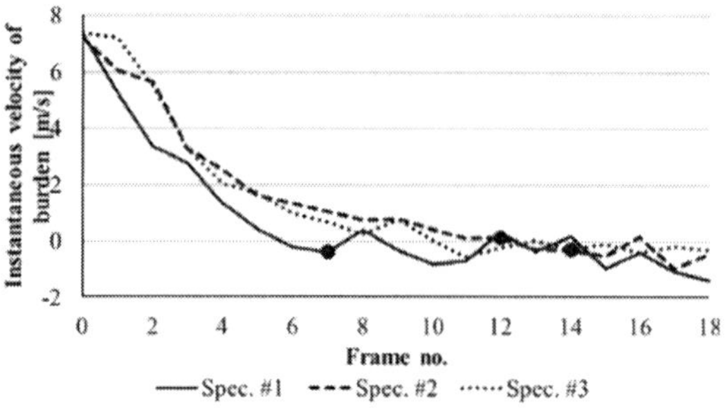

Figure 13. Comparison of the specimens in view of the velocity.

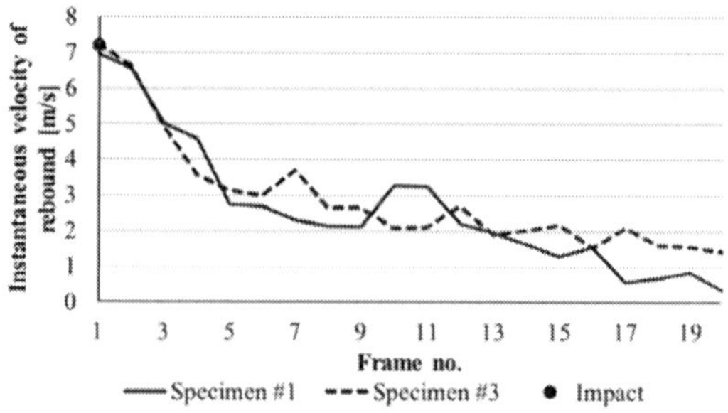

Figure 15. Results of the 2nd impact loading – inst. velocity.

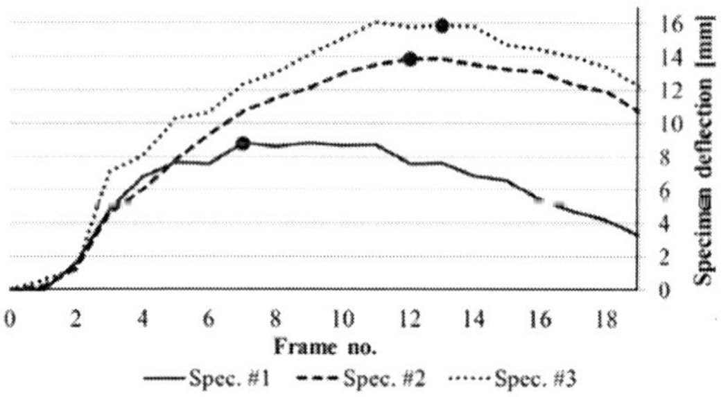

Figure 14. Comparison of the specimens in view of the velocity.

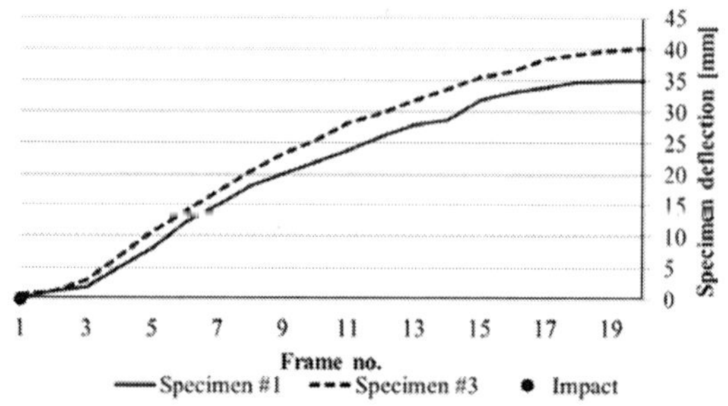

Figure 16. Results of the 2nd impact loading – spec. deflections.

the specimen #1 last only 8 ms. After that the hammer rebounded and the velocity on the graph acquires negative values (movement against the direction of gravity). The impact interval of the specimens #2 and #3 last up to 75% longer (14 ms in case of specimen #3).

The comparison with respect to deflections corresponds to the velocity comparison. The deflection of the specimen #1 is significantly lower than in case of the other specimens.

The cracks development transforms the problem from the linear elasticity to the problem of the fracture mechanics and plasticity. It can be assumed that the behaviour of the specimens is affected by the fracture toughness. Furthermore it can be assumed that the fracture toughness of the specimens with partially damaged material matrix due to chemical deterioration or cyclic loading is lower, thus the energy required for the crack developing is lower as well.

The phenomenon of lower decrease in velocities of the specimens #2 and #3 can be explained due to the opening of the cracks in structure and microstructure of the specimens, which formed due to cyclic loading and the chemical deterioration. In the case of the specimen #1 there can be assumed only micro cracks formed due to shrinkage and chemical processes during the hydration of the concrete. It can be assumed that these cracks occur in considerably lesser extent. Thus the propagation of the new cracks significantly increases the required energy.

The phenomenon of the higher deflection of the deteriorated specimens can be explained with an initial damage as well. The necessity for the formation of the new cracks highly absorbs the kinetic energy obtained during the impact. The initial increase in deflection (until 5th frame) corresponds with the formation of visible shear crack and its opening. For the specimen #1 it is almost the only damage visible and from the graph it is obvious that this crack is dominant in the deflection value. In the case of the specimens #2 and #3 the shear crack opened and other cracks formed due to the cyclic loading or the chemical deterioration were further opening and developing. This results in the higher deflections and lower rate of the energy absorption.

3.2. Residual resistance to the impact loading

For the purposes of the residual resistance to the impact loading two additional impact loads were applied to the above mentioned specimens.

Due to technical difficulties, the video sequencing used for the evaluation was not triggered at each time, thus the results for the 2nd and 3rd impact loading of the specimen #2 and 3rd impact loading for the specimen #1 are not evaluated. Comparison for the behaviour of the specimens #1 and #3 during the second impact loading can be seen on Figs. 15 (instantaneous burden velocity) and 16 (specimen deflections). The graphs take into account only the interval of the impact.

As can be observed on the Figs. 13 and 14 the behaviour after the first impact loading is similar for the both specimens. The decrease in the velocities is slightly lower in the case of the specimen #3 which corresponds with the higher cracks distribution in the specimen, thus lower rate of the energy absorption.

In case of the deflections the pattern is the same for both specimens. This is mainly due to the fact, that the specimens are already severely damaged especially in the tensile zone, thus the magnitude of deflections is primarily affected by the steel reinforcement, which is the same in both specimens (6 × ⊘16 B500B). The higher deflection of the specimen #3 can be attributed to the coupled chemical and mechanical damage (due to the aggressive environment and cyclic loading) of the concrete matrix in compression.

4. Conclusions

The paper presented preliminary results of the on-going experimental program focused on the combined effect of the cyclic loading, deterioration due to aggressive environment and impact loading.

The exposure to the cyclic loading resulted in higher deflection under the impact loading (increase up to 40%) and lower rate of the energy absorption. The specimen #3 which was exposed to cyclic loading and the aggressive environment (360 days in solution of hydrochloric acid) exhibited even higher deflection (increase up to 15% compared to specimen #2 and up to 60% compared to specimen #1). The rate of energy absorption was lower as well. These phenomena can be explained by cracks in specimens after cyclic loading (shear cracks and micro cracks) and by the material matrix damage caused by the deteriorative chemical reactions due to aggressive environment.

The behaviour after the first impact loading has similar pattern for all the evaluated specimens. This can be attributed to the significant influence of the reinforcement, which was not affected by the cyclic loading or chemical deterioration.

For the proper quantification of the mentioned phenomena all the remaining specimens will be tested by rebound hammer impact loading in 2015.

The financial support of the Czech Science Foundation grant project 13-30441S is gratefully acknowledged.

References

[1] A. Alliche, Int. J. Fat., **26**, 915–921 (2004)
[2] M. Foglar, *Bridge Maintenance, Safety, Management and Life Extension*, 3185–3189 (2010)
[3] Y. Chen, I.M. May, Struct. Build., **162**, 45–46 (2009)
[4] CEN, *Eurocode 1: Part 1-7* (Brussels, 2006)
[5] Y.F. Fan, Z.Q. Hu, Y.Z. Zhang, Const. Build. Mat., **24**, 1975–1983 (2010)
[6] M.C. Chen, K. Wang, L. Xie, Eng. Fail. Anal., **27**, 272–285 (2013)
[7] J. Göringer, M. Foglar, Adv. Mat. Res., **891**, 494–499 (2014)
[8] fib, *Model Code 1990* (Lausanne, 1993)
[9] fib, *Model Code 2010* (Lausanne, 2010)

EPJ Web of Conferences **94**, 01059 (2015)
DOI: 10.1051/epjconf/20159401059

A study on ship impacting a flexible crashworthy device for protecting bridge pier

Liming Yang[a] and Jun Liu

Mechanics and Material Science Research Center, Ningbo University, Ningbo, Zhejiang Province 315211, China

Abstract. As the accident of a vessel impacting a bridge pier will cause serious disaster, such as destroyed bridge, sinking ship and polluting environment, the technology and method to protect bridge pier from ship collision have been widely investigated recently. Due to the huge kinetic energy of large-tonnage ship and the short time duration in the collision, the studies involve impact mechanics. A developed flexible crashworthy device has been developed to protect bridges, which consists of an outer steel-periphery, an inner steel-periphery and the rubber coating SWRCs(soft elements) installed between them. When the SWRC crashworthy device is installed, the collision duration under low impact force is prolonged due to its high compliance, which results in the ship having enough time to turn its navigation direction and most of the remainder kinetic energy being carried off by the turned away ship. Consequently, both impact forces on the ship and on the bridge pier decrease markedly. This is the key reason as to why the SWRC crashworthy device can avoid the destruction of both the bridge and the ship. Based on our results of theoretical studies and numerical simulations, the present paper will propose an experiment-adopted a real ship to impact a flexible crashworthy device. The collision test has been performed 12 times with different speed, carrying capacity, and impact angle of the ship. After the experiments, the ship, flexible crashworthy device and the pier are not damaged. The experiments show that the flexible crashworthy device can turn away the impact ship, so that the ship moves along the outer part of the device, which reduces the ship impact force on the bridge pier obviously. It not only protects bridges but also avoids the damage to ships.

1. Introduction

As the accident of a vessel impacting a bridge pier will cause serious disaster, such as destroyed bridge, sinking ship and polluting environment, the technology and method to protect bridge pier from ship collision have been widely investigated recently [1]. Due to the huge kinetic energy of large-tonnage ship and the short time duration in the collision, the studies involve impact mechanics. The model experiments, the theoretical analyse and numerical simulations have attracted tremendous research interest in the last few decades [2–5]. However, the results obtained could not be accepted by bridge managements, due to the devices have not been undergone a impact test with a real ship.

Based on our results of theoretical studies and numerical simulations [2], a flexible crashworthy device to protect bridge pier against ship collision has been developed. The present paper will introduce an experiment: adopted a real ship to impact the flexible crashworthy device to validate its capability of protecting the bridge pier and the impacting ship.

2. Flexible crashworthy device for protecting bridge pier

2.1. Structure and principle of the flexible crashworthy device

Resent years, a flexible crashproof device has been created developed by authors [2], which essentially consists of many steel-wire-rope coil (SWRC) connected in parallel and series. A developed flexible crashworthy device is shown in Fig. 1, which consists of an outer steel-periphery, an inner steel-periphery and the rubber coating SWRCs installed between them. The outer steel-periphery will be collided with a coming ship. Impacting force which bridge pier withstood will be buffered by SWRC.

Figure 1. Structure of a flexible crashworthy device.

When the flexible crashworthy device is installed, the collision duration under low impact force is prolonged due to its high compliance, which results in the ship having enough time to turn its navigation direction and most of the remainder kinetic energy (60–95% depending on the actual incidence collision angle) being carried off by the turned away ship. Consequently, both impact forces on

[a] Corresponding author: yangliming@nbu.edu.cn

Article available at http://www.epj-conferences.org or http://dx.doi.org/10.1051/epjconf/20159401059

Figure 2. Flexible crashworthy device for test.

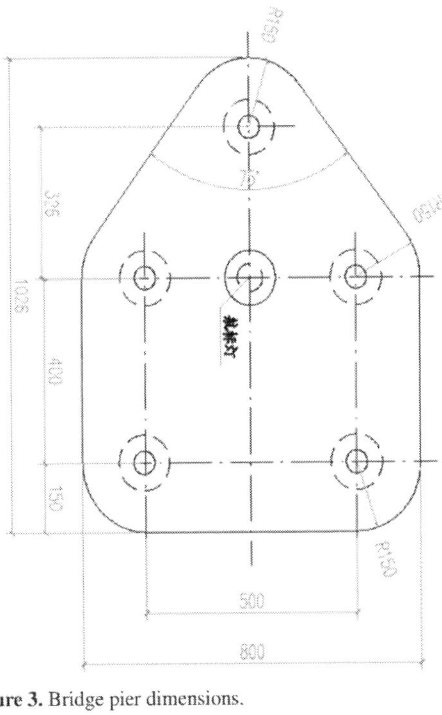

Figure 3. Bridge pier dimensions.

the ship and on the bridge pier decrease markedly. This is the key reason as to why the present SWRC crashworthy device can avoid the destruction of both the bridge and the ship. On the other hand, the stiffness of the outer steel-periphery is strong so that the ship collision cannot produce a deformation location and a localized instability.

2.2. Flexible crashworthy device for test

In order to test the validity and reliability of the developed bridge anti-ship collision flexible protection technology, in Ningbo Xiangshan White Pier Port the tests have been performed to use a real ship to impact the flexible crashworthy device which is shown in Fig. 2. The force sensors are installed between SWRC and inner-periphery, as well as the acceleration sensors are arranged in the experimental ship. The ship weight is 250 ton and its deadweight (dwt) is 400 ton. The track of the ship after impact has also been measured by high speed camera.

Figure 4. The force sensors are installed in the device.

(a)

(b)

Figure 5. The collision experiments.

The detail dimension of the bridge pier is shown in Fig. 3. The device width is about 13.7 m, length is about 14 m (Fig. 2). The structures of outer steel-periphery and inner steel-periphery are box shape. The device is floatable, so that its capability to protect the bridge pier cannot be affected by the tide. Figure 4 shows the force sensors are located in the device.

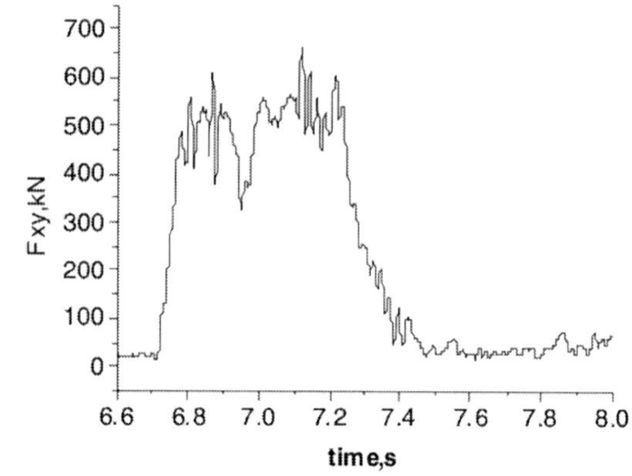

(a) Measurement

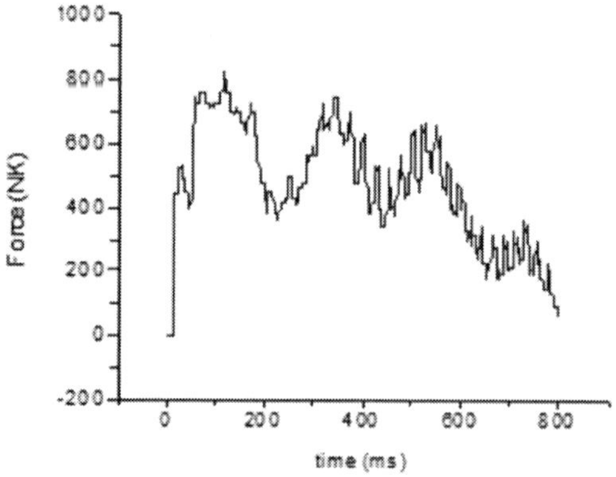

(b) Simulation

Figure 6. The curves of time-impact force in case 1.

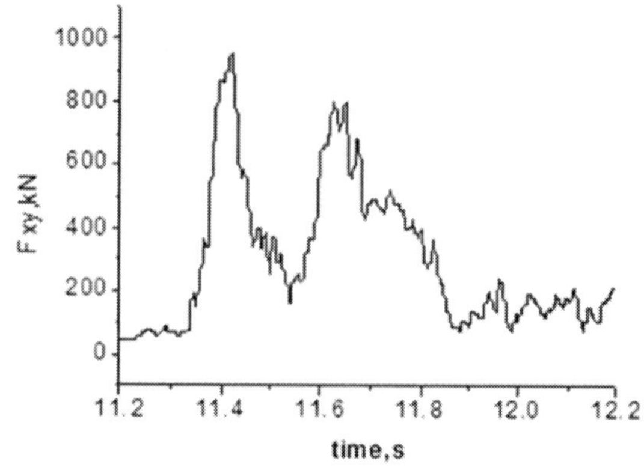

(a) Measurement

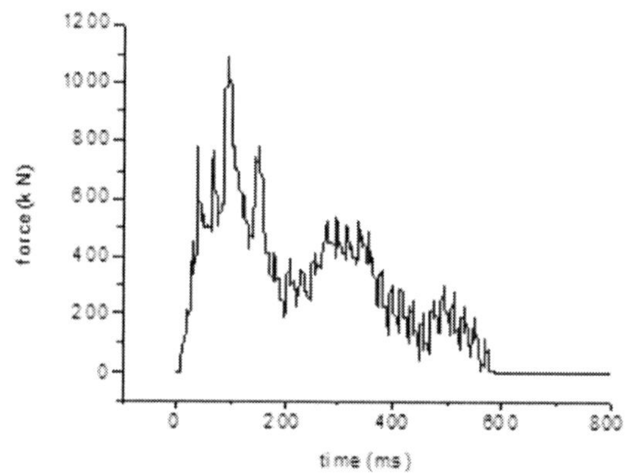

(b) Simulation

Figure 7. The curves of time-impact force in case 2.

2.3. Collision experiments

The collision test has been performed 12 times with different speed, carrying capacity, and impact angle of the ship. The ship speed is increased from 1 m/s to 4 m/s. The range of ship weight is 250,000 kg ~ 400,000 kg. The impact angle changed from 0° to 26°. In the tests, the impact forces, the ship acceleration and the ship track are measured. Tests show that:

(1) The device can let ship turn its navigation direction and move along the outside of the outer steel-periphery (Fig. 5), even if the ship impact angle reached 26°.
(2) During impact, the ship almost keeps its constant speed. For the tests in which the ship impact angle is less than 5°, the speed only decreased 5% after impact. For the tests of ship impact angle is 26°, the speed decreased 20% after impact. The results means that the dissipation of ship kinetic energy is small during impacted.
(3) After the 12 experiments, the ship, flexible crashworthy device and the pier are all not damaged. It is found that the developed crashworthy device has the capability to protect the bridge pier and the impact ship.

Table 1. Conditions of calculation results compared to test data.

Ship weight (ton)	Ship speed (m/s)	Ship impact angle	Impact force without protecting device (ton)	Impact force with protecting device (ton)
250	2.3	26°	270	65
250	3.0	0°	160	65
400	3.5	0°	230	100

2.4. Comparison with simulation

The finite element software Ls-DYNA has been adapted to simulate the dynamic responses of the ship-bridge collision. In the finite element model, the bridge pier is assumed to be rigid, steel mechanical property is elastic-plasticity, and the SWRC is modelled by a nonlinear visco-elastic body. The material parameters used are same as those shown in the reference [2]. Some numerical simulations and experimental results are shown below.

Case 1: the ship weight is 250,000 kg, its speed is 2.3 m/s, and the ship impact angle is 26°. Figure 6(a) shows the measurement data, and Fig. 6(b) is numerical simulation results.

Case 2: the ship weight is 400,000 kg, its speed is 3.5 m/s, and the ship impact angle is 0°. Figure 7(a) shows the measurement data, and Fig. 7(b) is numerical simulation results.

Compared above figures, it is found that the numerical simulation results are in reasonable agreement with the measurement results. The numerical simulation can be considered to be reliable and valid. Thus, the numerical simulation is used to calculate the impact force during a ship crashes a bridge pier without any protecting device. The results are shown in Table 1.

Table 1 shows that the flexible crashworthy device can decrease markedly contact force.

3. Conclusion

The experiments show that the flexible crashworthy device can turn away the impact ship, the ship moves along the outer side of the device, which reduces the ship impact force on the bridge pier obviously. It not only protects bridges but also avoids the damage to ships.

The support from the National Natural Science Foundation of China under its Key Project No. 11032001 is gratefully aknowledged.

References

[1] G. Y. Chen, L. L. Wang, L. M. Yang, M. D. Chen, *The Protection Theory for Ship-Bridge Collisions and the Design of Crashworthy Devices* (China Communications Press, 2014)

[2] L. L. Wang, L. M. Yang, D. J. Huang, Z. W. Zhang and G. Y. Chen, I. J. Impact Engn. **35**, 895 (2008)

[3] D. Larry, P. E. Olson, *Dynamic Bridge Substructure Evaluation and Monitoring*. Report No. FHWA-RD-03-089, U.S. Federal Highway Administration, (2005)

[4] P. T. Pedersen, Marine Struct. **23**, 241 (2010)

[5] F. Yang, L. M. Yang, Chinese J Solid Mech. **32**, 399 (2011)

EPJ Web of Conferences **94**, 01060 (2015)
DOI: 10.1051/epjconf/20159401060

Effects of aspect ratio and specimen size on uniaxial failure stress of iron green bodies at high strain rates

Yuki Kuroyanagi[1,a], Masahiro Nishida[1], Takashi Ogura[1], H.-Å. Häggblad[2], P. Jonsén[2], and G. Gustafsson[2]

[1]Nagoya Institute of Technology, Gokiso-cho, Showa-ku, Aichi, 466-8555, Japan
[2]Luleå University of Technology, 971 87 Luleå, Sweden

Abstract. Powder metallurgy is used for the production of a number of mechanical parts and is an essential production method. These are great advantages such as product cost effectiveness and product uniqueness. In general, however parts created by powder metallurgy have low strength because of low density. In order to increase strength as well as density, new techniques such as high-velocity-compaction (HVC) was developed and further investigation has been conducted on improvement of techniques and optimum condition using computer simulation. In this study, the effects of aspect ratio and specimen size of iron green bodies on failure strength of uniaxial compression and failure behavior were examined using a split Hopkinson pressure Bar. The diameters of specimens were 12.5 mm and 25 mm the aspect ratios (thickness/diameter) were 0.8 and 1.2.

1. Introduction

Powder metallurgy is the serial process of blending fine powder (mixture), compressive them into a desired shape (compacting), and then heating the compressed material (sintering). This is used for the production of a number of mechanical parts and is an essential production method. There are great advantages such as product cost effectiveness and product uniqueness. In general, parts created by powder metallurgy have low strength because of low density. The production of high strength parts having complex shape is required [1–4]. In order to increase strength following the increment of density, new techniques such as high velocity compaction (HVC) were developed and further investigation have been conducted on improvement of techniques and optimum condition using computer simulation [5,6]. In general, the typical material model of iron green bodies used in the simulation is modified Drucker-Prager cap model (modified DPC model) is often used. In order to estimate the modified DPC model parameter is necessary tested about uniaxial compression, shear, diametral compression, uniaxial tension test. The authors' group has been examining the strain rate sensitivity of strength and determination of material parameter of iron green bodies (iron based powder compacts) at high strain rate [7].

In this study, the effects of aspect ratio and specimen size of iron green bodies on failure strength of uniaxial compression were examined using a split Hopkinson Pressure Bar (SHPB). During dynamic compression, the failure behavior of specimens was also observed using a high speed video camera. The failure stress and failure behavior of dynamic compressive tests were compared with those of static tests.

[a] Corresponding author: k_yuki_0227@yahoo.co.jp

2. Experimental methods

2.1. Specimens

The powder material used for the experiments was a press-ready premix containing Distaloy AE, 0.5% graphite (uf-4) and 0.6% Kenolube. Distaloy AE is a pre-alloyed water atomized iron powder from Höganäs AB, Sweden. Kenolube is a lubricant from the same supplier. The theoretical pore free density of this mix is 7.52 g/cm^3. The specimens were manufactured by conventional compaction (Dartec RK 250). Cylindrical specimens were manufactured with a thickness of 10 mm and 15 mm, a diameter of 12.5 mm with 6.9 g/cm^3. Table 1 listed the size of specimen. Figure 1 shows a photograph of specimens. To confirm the effect of only the aspect ratio, aluminum alloy (A2024-T4) showing less strain rate dependence was used for examining the effects of aspect ratio and the diameter of specimens by SHPB.

2.2. Experimental setup

The dynamic compressive tests were carried out using a SHPB method. Figure 2 is shown experimental set up for dynamic compressive tests. The input and output bars for the SHPB experiments were made of steel, and their diameter and length were 32 mm and 2300 mm/1700 mm respectively. Strain gages were put on both sides of input bar and output bar 1150 mm and 300 mm away from the specimen. The stress and strain histories of specimens were calculated from the strain of bars measured by the strain gages based on following equations [8–11].

$$\sigma(t) = \frac{AE}{A_s}\varepsilon_t(t) \tag{1}$$

$$\varepsilon = \frac{c_0}{l_s}\int_0^t [\varepsilon_i(t') - \varepsilon_r(t') - \varepsilon_t(t')] \tag{2}$$

Article available at http://www.epj-conferences.org or http://dx.doi.org/10.1051/epjconf/20159401060

Table 1. Size of specimens.

Density [g/cm^3]	Diameter [mm]	Thickness [mm]	Aspect ratio
6.9	12.5	10	0.8
		15	1.2
	25.0	20	0.8

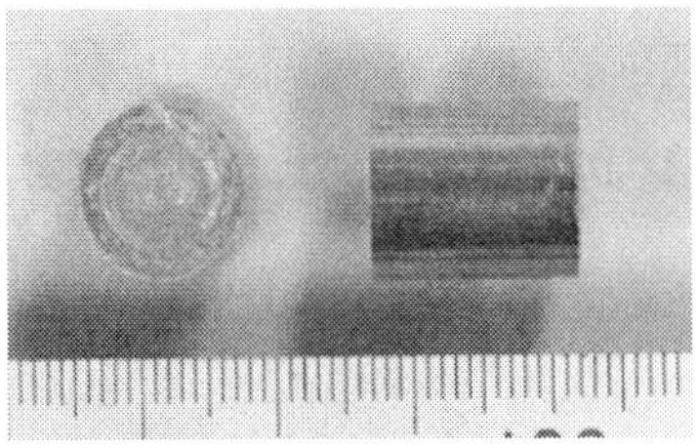

Figure 1. Photograph of specimen.

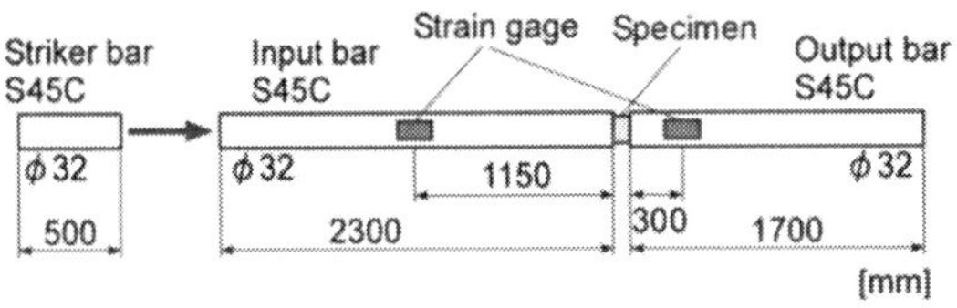

Figure 2. Experimental set up for dynamic compressive tests.

Table 2. Material constants of input and output bars.

Density	Elastic wave velocity of the input/output bar, c_0	Young's modulus, E
7.85 g/cm^3	5110 m/s	208 GPa

where ε_i, ε_r and ε_t are the axial strains in the input bar induced wave and reflected wave from the specimen and the axial strain in the output bar induced by the transmitted wave. E and c_0 are Young's modulus and elastic wave velocity of the input and output bars, respectively. l_s is the specimen thickness. A and A_s are the areas of cross sections of the input/output bars and the specimens. Material constants of the steel bars used in the calculation are shown in Table 2. We used steel strikers with a diameter of 32 mm and length of 500 mm. During dynamic compression, we observed failure behaviour of specimens using a high speed video camera (Photron FASTCAM-1024PCI 100K, frame rate of 18000 frames /sec).

3. Result and discussion

Figures 3–5 are shown a stress-strain curve of a dynamic compressive test. Figure 3 shows the results of aspect ratio 0.8. Figure 4 shows the results of aspect ratio 1.2. Figure 5 shows the results of specimen diameter 12.5 mm. After the strain history and stress history of the 6.9 g/cm^3 specimen were calculated from the strain gages using Eqs. (1) and (2), the stress-strain curve was obtained. Figures 3–5 show first, the stress increased with strain. Second, the nominal stress-strain curve showed a peak. Finally, after the peak, the stress decreased gradually with strain. An image at strain 0.009 (56 μs after impact)

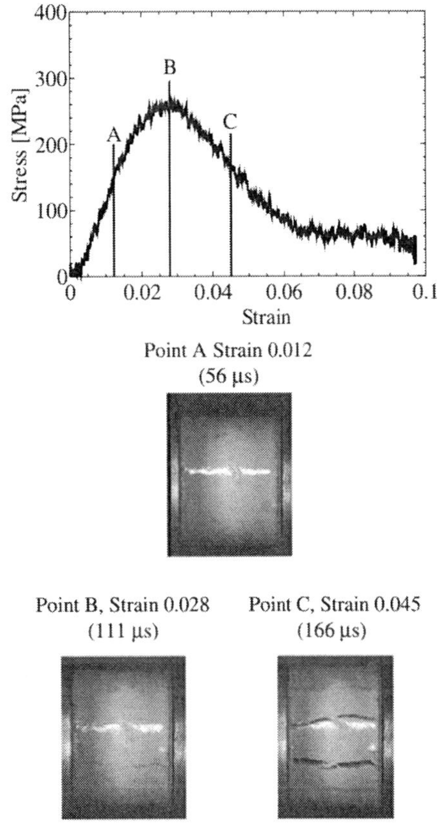

Figure 3. Stress-strain curve of dynamic compressive test and observe specimen (aspect ratio 0.8).

showed that any cracks on the surface of specimen were not observed. At 111 μs, small cracks in the compression direction on the surface of specimen started to be observed. At 166 μs, the cracks in the compression direction was progressed, finally specimens broke into small pieces. There was no difference of aspect ratio and the diameter of specimens in the failure behaviour. Table 4 listed the maximum stresses of iron green bodies.

It is common knowledge that aluminium alloys show small strain rate dependence. At first, the effects of aspect ratio and specimen size on yield stress were examined using aluminium alloy A2024-T4. Table 3 shows the yield stresses of the aluminum alloy A2024-T4 by SHPB. The yield stress of aspect ratio 0.8 was about 1.09 times higher than that of the aspect ratio 1.2. The yield stress of the diameter of specimen 12.5 was about 1.15 times higher than that of the diameter of specimen 25 mm. The maximum stresses of iron green bodies were corrected using the values obtained by the ratio of yield stress in Table 3. Table 4 shows that the corrected maximum stress of aspect ratio 1.2 was larger than the maximum stress of aspect ratio 0.8 when the diameter was 12.5 mm. The corrected maximum stress of specimen diameter 25 mm is larger than that of 12.5 mm.

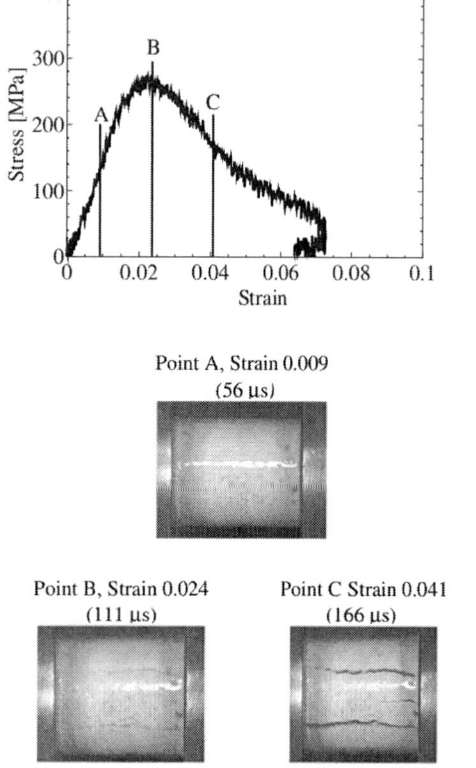

Figure 4. Stress-strain curve of dynamic compressive test and observe specimen (aspect ratio 1.2).

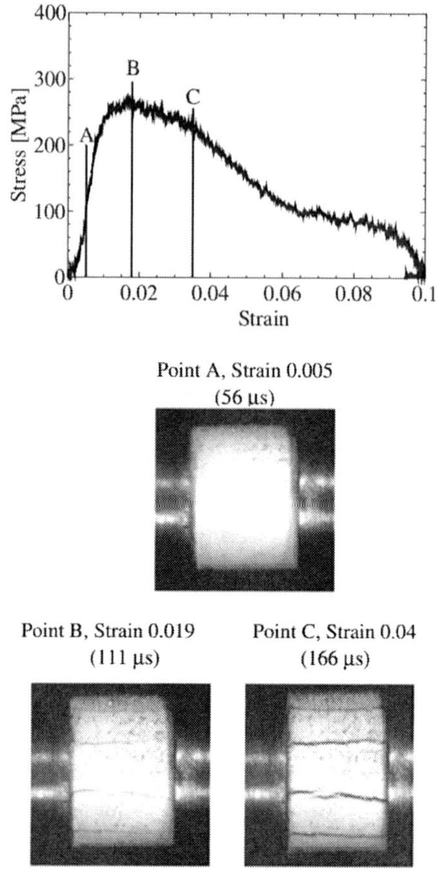

Figure 5. Stress-strain curve of dynamic compressive test and observe specimen (the diameter of specimen 25 mm).

Table 3. Effects of aspect ratio and specimen size (Aluminum alloy A2024-T4).

Diameter [mm]	Thickness [mm]	Aspect ratio	Yield stress [MPa]	Ratio of yield stress	
12.5	10	0.8	543	1.09	1.15
	15	1.2	500	1.00	
25.0	20	0.8	473		1.00

Table 4. Effects of aspect ratio and specimen size (Iron green body).

Diameter [mm]	Thickness [mm]	Aspect ratio	Maximum stress [MPa]	Corrected value of maximum stress [MPa]
12.5	10	0.8	267	
	15	1.2	275	300
25.0	20	0.8	261	298

4. Conclusions

The effects of aspect ratio and specimen size on the maximum stress and failure behaviour were examined at high strain rate. The aspect ratio and the diameter of specimens clearly affected the maximum stress. However, there is no difference in failure behaviour. It was found that the maximum stress increased with aspect ratio.

The authors are also greatly indebted to Associate Professor Fumihiro Itoigawa at Nagoya Institute of Technology for his help with taking images using a high speed camera.

References

[1] G. Gustafsson, M. Nishida, H-Å. Häggblad, H. Kato, P. Jonsén, T. Ogura, *Powder Technology* **268**, 293 (2014)
[2] Höganäs handbook for sintered components, Höganäs AB (1997)
[3] MPIF Standard 35, Materials Standards for P/M Structural Parts (2000) edition
[4] MPIF Powder Metallurgy Design Solutions (1999)
[5] P. Jonsén, H-Å. Häggblad, L. Troive, J. Furuberg, S. Allroth and P. Skoglund, Materials Science Forum, **534–536**, 289–292 (2007)
[6] M. Nishida, H. Kato, H.-Å. Häggblad, P. Jonsén, G. Gustafsson, Proc. PM2012, Yokohama (2012)
[7] G. Gustafsson, M. Nishida, H.-Å. Häggblad, H. Kato, P. Jonsén, T. Ogura, Powder Technology **239**, 231–238 (2013)
[8] B. Hopkinson, Philosophical transactions of the Royal Society of London **213**, 437–456 (1914)

[9] H. Kolsky, Proceedings of the Physical Society of London **62**, 676–700 (1949)
[10] G.T. Gray III, *Classic split-Hopkinsonbar testing, AMS Handbook*, Vol.8, Mechanical Testing and Evaluation, H. Khunand D. Medlin, eds, Material Park, Ohio, ASM International, pp. 462–476 (2000)
[11] W. Chen and B. Song, *Split Hopkinson(Kolsky) bar design, testing and applications* (Springer 2010)

EPJ Web of Conferences **94**, 01061 (2015)
DOI: 10.1051/epjconf/20159401061

Experimental and calculated approach to the study of deformation and strength characteristics of elastoviscoplastic materials by direct impact method

V.G. Bazhenov[1], M.S. Baranova[1,a], E.V. Nagornykh[2], and D.L. Osetrov[2]

[1] Research Institute of Mechanics Lobachevsky State University of Nizhni Novgorod, Pr. Gagarina 23, 603950 Nizhni Novgorod, Russia
[2] Lobachevsky State University of Nizhni Novgorod, Pr. Gagarina 23, 603950 Nizhni Novgorod, Russia

Abstract. It is proposed to develop experimental and calculated approach to the study of the strength characteristics of elastoviscoplastic materials in a non uniform strain stress state. Integral characteristics (forces, displacements and displacement speed) of the deformation process of hat-shaped specimens in tension are determined by a direct impact method, and their strain – stress states are determined by numerical solution of the axisymmetric problem. The results of experimental and theoretical study of the deformation and failure of hat-shaped specimens in the presence of stress concentrators are obtained.

1. Introduction

Originally vertical gas-gun testing stands solved the problem of metal forming technologies. The results of these tests are not very suitable for the mathematical simulations of deformation due to roughness measurements and methods of processing the results of the experiment. Refinement methodologies occurred after implementation to measure the resistance forces Hopkinson bar in 1914. Kolsky proposed split bar and data processing in 1924. This methodology is most popular in the last 50 years in dynamic tests of materials for compression and tension [1–3]. The methodology allows to implement a range of strain rates $5 * 10^2$–$5 * 10^3$ 1/s at moderate degrees of deformation. Range of strain rates further narrowed due to the lack of impact energy, implemented through the transmission rod on the specimen for fracture in tests of metals and alloys in tension to failure. Modification of the split Hopkinson bar by removing the transmission bar and the direct impact of striker of the same diameter as the measuring rod for the test specimen was proposed for increasing realized experimentally strain rates above $5 * 10^3$ 1/s and strains. This methodology combines the advantages of a split Hopkinson bar method and a direct impact on the accuracy of measurements of displacement rod ends. But it requires a fairly large length striker, which imposes restrictions on the technical possibilities of its application. To achieve greater impact energy in a direct impact method striker is selected significantly larger diameter than the measuring rod. The direct impact method which realized on the vertical gas-gun testing stand allows to increase the strain rates and strains, but also to cover the range of low strain rates up to 10^1 1 / s at the impact energy sufficient to fracture of the hat-shaped specimens in tension testing. Klepaczko suggested to use optical gauges for measuring the velocity of the striker end which contacts with the specimen in 2001 [4]. Methodologies of experimental results (build diagrams of deformation by a direct impact) are investigated and underdeveloped compared to the procedure for the Kolsky methodology for split Hopkinson bar. Processing of the results of tests of the hat-shaped specimens in tensile conducted under the assumption of homogeneity of stress-strain state in the test portion of the specimens. What is impracticable because of the difficulties in preparing the specimens with slit-like ports and not widely used and further develop.

In Russia there is no practical realization of direct impact on the vertical gas-gun testing stand for construction of dynamic diagrams of deformation and strength characteristics of the study materials in tension. Those who have previously been established, do not meet the accuracy requirements for use in the mathematical simulation of deformation and strength assessment of modern constructions.

2. Influence of the form striker in the process of impact tension of hat-shaped specimens

In order to study the possibilities of increasing the pulse width while maintaining the length of the striker numerical simulations of various forms of the impactor (cylindrical, with slotted opening and balls) is done.

The problem is solved in an axisymmetric formulation. Z – axis of symmetry. Various types of striker (transfer ring and cylinder, transfer ring and cylindrical striker with slotted opening) are simulated. Striker and transfer ring move with an initial velocity $V_0 = 3$ m/s.

The gas-gun testing stand for tensile testing by direct impact method (Fig. 10) contains the following elements:

[a] Corresponding author: mari.baranova18@gmail.com

Article available at http://www.epj-conferences.org or http://dx.doi.org/10.1051/epjconf/20159401061

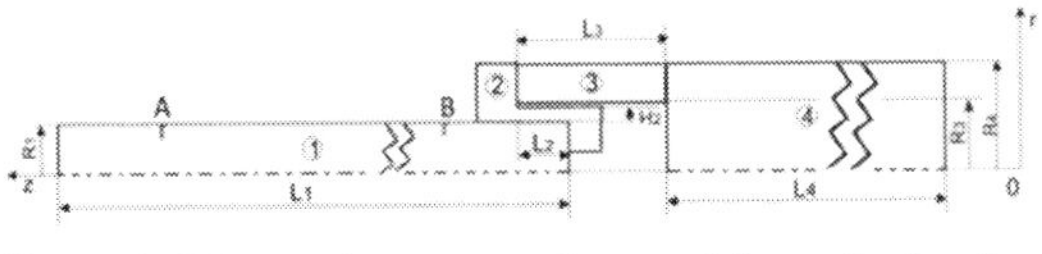

Figure 1. Scheme of gas-gun testing stand for testing by direct impact method.

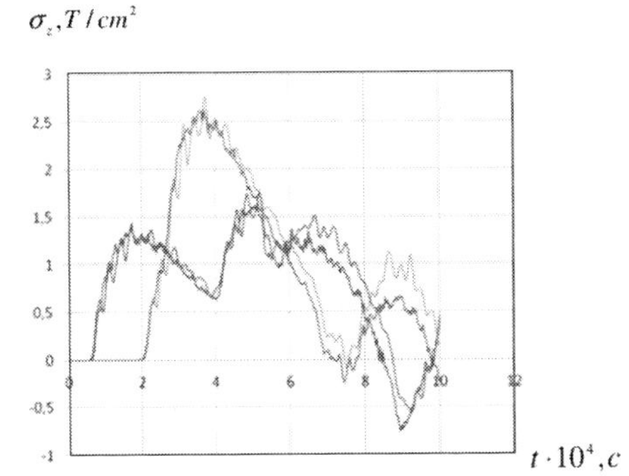

Figure 2. Stresses from gauges in sections A and B of the measuring rod.

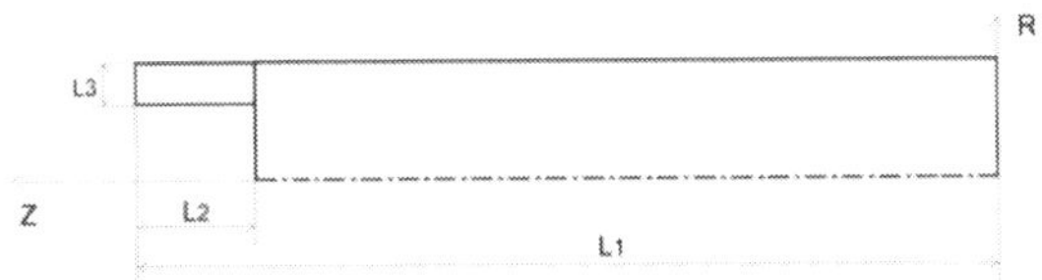

Figure 3. Cylindrical striker.

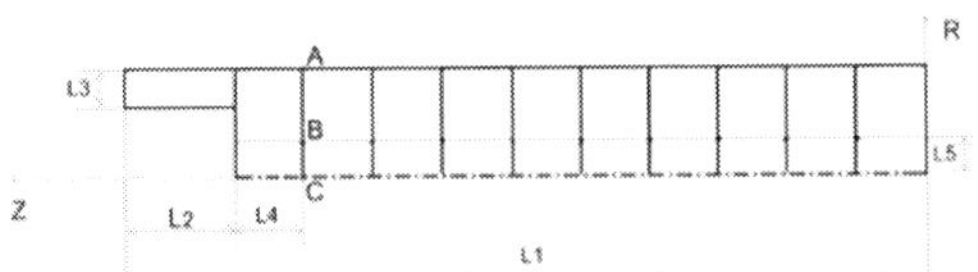

Figure 4. Cylindrical striker with slotted opening.

supported measuring rod (marked 1), hat-shaped specimen (2), transfer ring (3), striker (4).

Geometric parameters of the elements of gas-gun testing stand are described in [5].

Comparing the results of calculations using "Dynamics-2" [6] and LS-DYNA is done. Figure 2 shows stresses from gauges in sections A and B of the measuring rod (Fig. 1), located at a distance of 105.8 cm and 30.8 cm from the impacted end of measuring rod on the same finite element mesh in the initial impact velocity of 4.52 m / s. Red and blue marked stresses in the gauges in the section B and A (Fig. 2) using "Dynamics-2." Purple and green – using "LS-DYNA".

The calculation results are in good agreement, which confirms the accuracy of the "Dynamics-2."

Two forms (cylindrical and cylindrical with slotted opening) of strikers are modelled. Figures 3–4 show forms of strikers.

$L_1 = 0,3585$ м $L_2 = 0,05$ м $L_3 = 0,0148$ м $L_4 = 0,0303$ м, $L_5 = 0,023$ м.

BC is a contact surface.

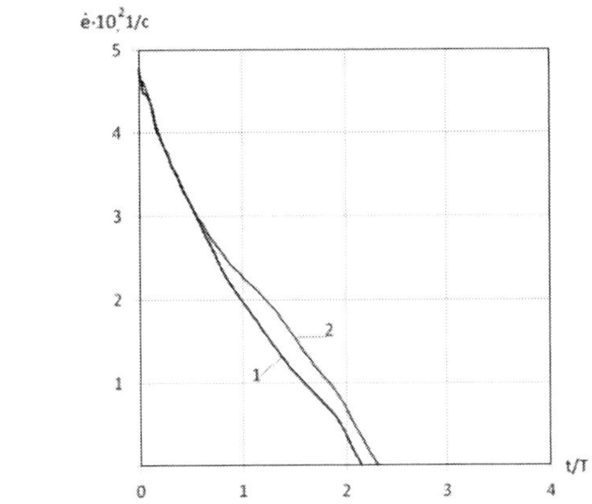

Figure 5. Strain rate.

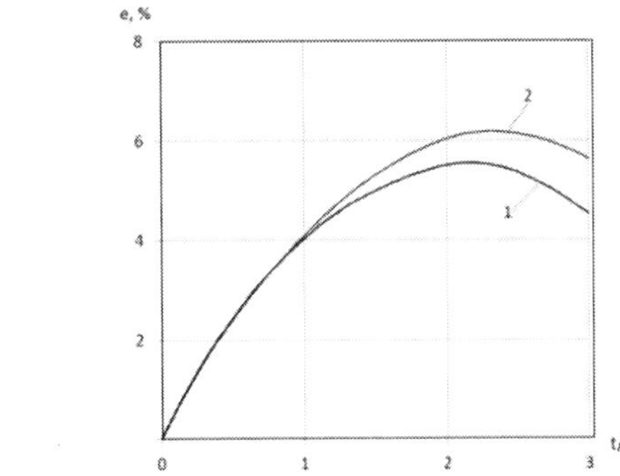

Figure 6. Strain.

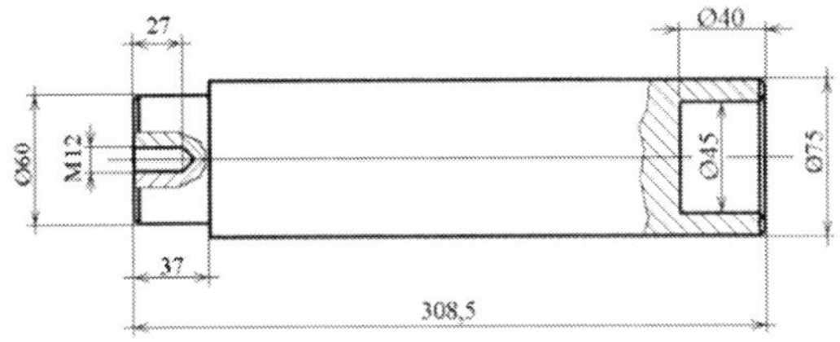

Figure 7. Striker.

Z – axis of symmetry.

Figures 5–6 show strain rate and strain for two types of strikers.

1 – cylindrical striker with a tube, 2 – striker with slotted opening.

Striker with pits gives some effect of increasing strain and strain rate due to stretching of the pulse due to the dispersion of waves.

In [7] the numerical simulation of cylindrical workpieces using standard hammer and hammer with a filler in the form of steel balls. Based on the above studies, it is concluded that the complexity of form striker does not have much effect. Striker was made from practical considerations, ease of manufacture and use in the experiments. It is shown in Fig. 7.

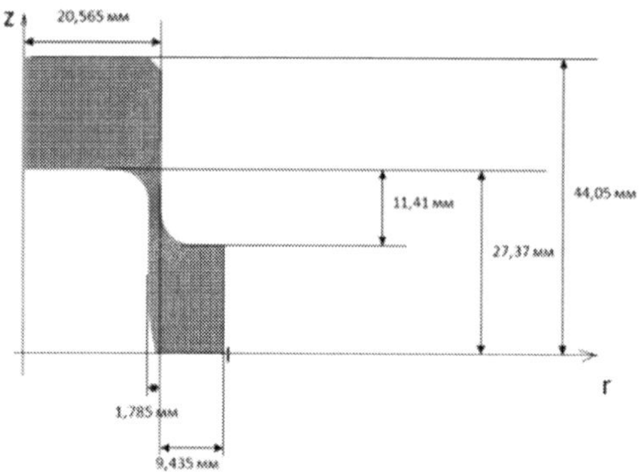

Z - axis of symmetry.

Figure 8. Hat-shaped specimen.

Figure 9. Hat-shaped specimens after testing.

The application of one-dimensional model of the wave process for the recovery of the experimental parameters of loading and deformation of the hat-specimen are given in [5].

3. Investigation of the stress-strain state in the hat-shaped specimens

Experiment on tensile of hat-shaped specimens is performed by gas-gun testing stand by a direct method. The scheme of gas-gun testing stand is shown in Fig. 8. The scheme and dimensions of hat-shaped specimen is shown in Fig. 8.

Geometric parameters of the elements of gas-gun testing stand were described in [5].

Figure 9 shows photographs of specimens after tensile testing by gas-gun testing stand. Failure of specimens is a conical surface of the test portion at an angle from the concentrator is 45° on the inner surface of the stress to the stress on the concentrator outer surface.

Figures 10–12 show the change of forces in the strain gauge at a distance of 21.66 cm from the impacted end of the measuring rod (Fig. 10), on the contact surface of the measuring rod - hat-shaped specimen (Fig. 11), in the middle of the test portion (Fig. 12) versus time in the calculation without failure (marked 1) and with failure (2).

Figures 13–14 show effective plastic strains and effective stresses in the times $t = 6,7494 * 10^{-5}$ s (Figs. 13a, 14a), $6.8992 * 10^{-5}$ s (Figs. 13b, 14b), $7.1998 * 10^{-5}$ s (Figs. 13c, 14c).

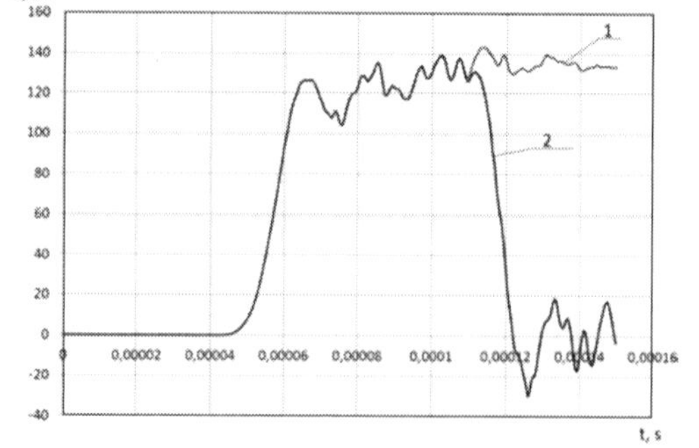

Figure 10. Force in the strain gauge.

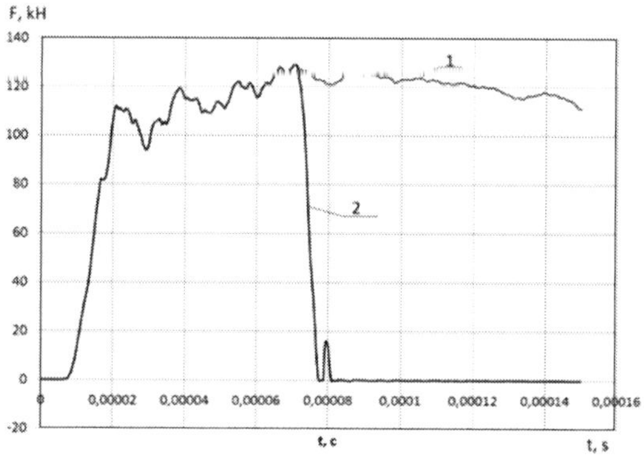

Figure 11. Force on the contact surface of the measuring rod – hat-shaped specimen.

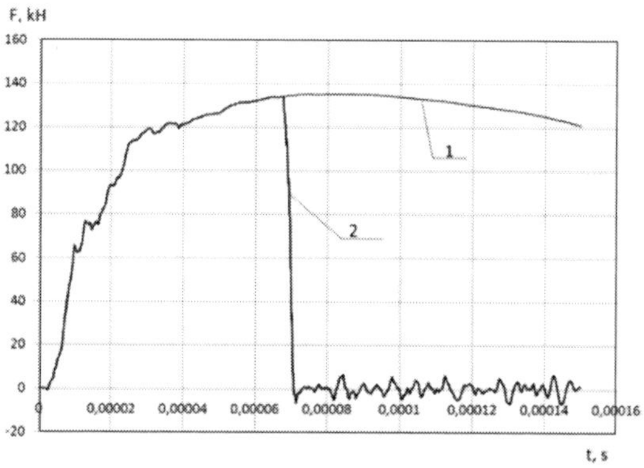

Figure 12. Force in the middle of the test portion.

The maximum intensity of plastic strains of 16% is used as criterion.

When this criterion is achieved the specimen is destroyed and normal to the failure surface stresses in the element are zero.

Studying of strength parameters of the material using of the hat-shaped specimens have some advantages as compared to solid axisymmetric corset specimens because it allows to implement a wider range of variation of the

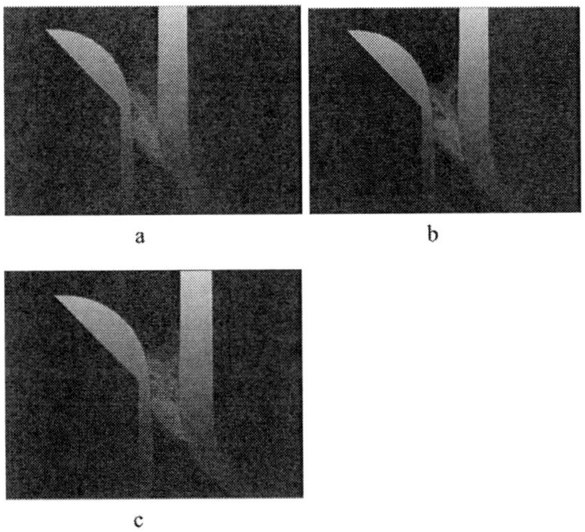

Figure 13. Effective plastic strains.

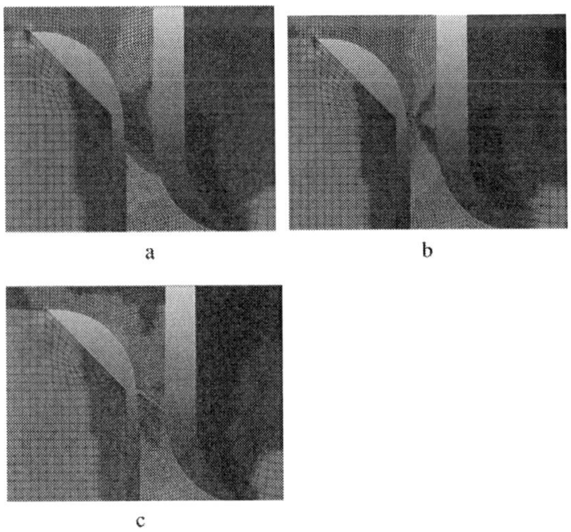

Figure 14. Effective stresses.

stress state by varying the thickness and length of the test portion. At low wall thickness of specimen compared to the length of the test portion the stress-strain state is closed to uniaxial. And if there are stress concentrators the stress-strain state volumetric. Volumetric stress state is characterized for corsetry specimens. A viscous and brittle failure can be realized in hat-shaped specimens.

4. Conclusion

Numerical and physical simulations of the tensile deformation process of hat-shaped specimens on the vertical gas-dynamic testing stand are done. The influence of the form striker on the tensile deformation process and choose the best form for the realization of experiments on tensile specimens on the gas-gun testing stand by a direct method. A wider range of varying types of the stress-strain state in the study of strength parameters of materials of hat-shaped specimens compared with corset axisymmetric solid specimens is described. Sufficiently high sensitivity of defined experimental parameters is shown in the experimental and theoretical analysis of the stress-strain state of the specimens at the time of destruction by numerical simulation of elastic-plastic deformation and failure of hat-shaped specimens.

The research was financially supported by the Federal Targeted Programme for Research and Development in Priority Areas of Development of the Russian Scientific and Technological Complex for 2014–2020 under the contract 14.578.21.0036 (unique identifier RFMEFI57814X0036).

References

[1] U.S. Lindholm, L.M. Yeakley Exp. Mech. **8**, 1 (1968)
[2] D. Mohr, G. Gary J. Phys. IV **134** (2006)
[3] T. Nicholas Exp. Mech. **21**, 5 (1981)
[4] J. R. Klepaczko Trends in Mechanics in Materials / eds. W.K. Novacki, J.R. Klepaczko. Warsaw (2001)
[5] V.G. Bazhenov, M.S. Baranova, E.V. Nagornykh Hopkinson Centenary Conference, Cambridge, UK, September 9–11, Fraunhofer EMI, Freiburg, Germany (2014)
[6] V.G. Bazhenov, S.V. Zefirov, A.V. Kochetkov, Problems of strength and plasticity. Research and optimization of constructions. Gorkovsky university, 4–13 (1987)
[7] V.Y. Lavrinenko, V.G. Bazhenov, E.V. Pavlenkova High Tech in mechanical engineering, No 5 (23) (2013)

EPJ Web of Conferences **94**, 01062 (2015)
DOI: 10.1051/epjconf/20159401062

Behavior of RC-slabs under impact-loading

Tino Kühn[a] and Manfred Curbach

TU Dresden, Institute of concrete structures, 01069 Dresden, Germany

Abstract. In this paper the current state of a research project on the behavior of reinforced concrete slabs subjected to impact loads are presented. As one part of the project several slabs were tested under varying conditions. As a result there failure behavior is analyzed and parameter dependences are studied. Variations like the drop height and drop mass and there substituted drop velocity and energy, the influence of the concrete strength and the level of reinforcement are part of the full project. Additional parameters like impactor size and shapes are evaluated in peripheral. Due to the large number of slabs and parameters combination a consistent database will be created for later benchmark tests and calibration of prediction models. The presented paper gives some insight in the methodology, the problems which need to be solved and first principal example results for public discussion.

1. Principles

The behavior of concrete slabs subjected to dynamic loading are experimentally and numerically investigated at the Institute of Concrete Structures, TU Dresden. The projects are focused on the local material behavior at different loading conditions as well as on their influence on the global structural behavior caused by impact loads such as rock fall or vehicle impact at moderate strain velocities. The higher velocity in dynamic experiments leads to assumed strain rate effects of the concrete material which is verified by many authors with different significance within the compressive [1] and tensile [2] domain. Nevertheless of the structural influences this effect can be described by numerical models for example as a viscous effect at low rates or a crack retardation effect at higher velocities [3]. With numerical benchmarks of standard experiments it could be shown that the structural and material response could not really separated and therefore the results are difficult to transfer to other structures [4]. As shown by experimental drop tests in [5] the knowledge of the detailed behavior of the sample interaction with the test facility is crucial to achieve best and publishable results. The influence of the material is somehow recognizable but not really to separate. To understand the dynamic behavior numerical simulation even with rough material assumptions are indispensable which gives more local insight and a feeling for the global reaction without disturbance [6]. In this work sample experiments on reinforced concrete slabs using a newly build high energy drop tower are presented as a subset of the underlying projects aims. This paper gives short introductions to the experimental facility with reflecting basic experimental results.

[a] Corresponding author: `tino.kuehn@tu-dresden.de`

1.1. The experimental facility

Concerning those requirements, a high energy drop tower was newly designed and implemented at the Otto Mohr Laboratory in Dresden. With a drop height up to 11 meters and a designed drop mass of up to 2.5 tons, a free fall energy of approximately 200 kJ can be used testing sample structures with a geometry of approx. 2×2 meters. The moderate impact velocity of up to 15 m/s in combination with a variable impactor system spans a wide dynamic load range useful for experiments with pure compressive to pure spall failure modes.

In addition a pneumatic accelerometer pipe can be vertically included in the 4 stories building. Two parallel pneumatic reservoirs with a capacity of 1000 l can be loaded up to 16 bar of pressure which represents the nearly same energy storage of 200 kJ. The pipe diameter is fixed to 100 mm. A variety of projectiles up to designed 100 kg can be included and accelerated up to the velocity of 150 m/s. With this facility high dynamic impact problems can be investigated. Material modelling and model validation as well as specific optimization or design changes for protective wall layers or other such structures might be performed. In order to have a wide range of experimental setup a 3×3 m flexible span was designed. The underlying shock absorbing mass of 30 tons is mounted on four air springs with a maximum amplitude of 500 mm and adjustable height and stiffness.

Figure 1 gives an impression of the setup area with a destroyed 1.5×1.5 m slab sample, the drop carriage with a 100 mm diameter impactor and the vertical guiding girders to the left and right side. Beside the impactor in the middle, the booth rescue shock dampers can be seen which are activated after punching through the specimen to adsorb excessive energy and prevent further destruction.

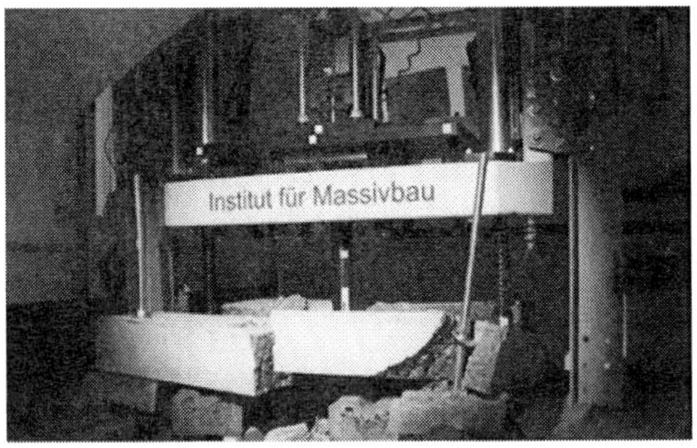

Figure 1. view to the setup field with drop carriage and damaged sample.

1.2. Energy and impulse relations

Static experiments are force or displacement controlled in most cases. Free fall drop tower tests are energy driven and can be controlled by the drop height and the falling mass. Energy E is simply related to the potential energy at specified high h as $E = m \cdot g \cdot h$ with Mass m and g as natural acceleration. Ignoring frictional effects the impact velocity v is given by:

$$v = \sqrt{2 \cdot g \cdot h} \tag{1}$$

whereas the free fall time t is given by:

$$t = \sqrt{\frac{2 \cdot h}{g}}. \tag{2}$$

The impulse induced by the impactor while hitting the slab is partly transferred to the specimen. Ignoring surface effects leads to a resultant impactor velocity v_{IR} of:

$$v_{IR} = k \cdot \frac{m_I \cdot v_I + m_S \cdot v_S - m_S \cdot (v_I - v_S)}{m_I + m_S}. \tag{3}$$

The resultant specimen velocity v_{SR} is given by:

$$v_{SR} = k \cdot \frac{m_I \cdot v_I + m_S \cdot v_S - m_I \cdot (v_S - v_I)}{m_I + m_S} \tag{4}$$

with k indicating the impulse transfer between 1 (elastic) and 0 (plastic), the masses m and the initial velocity v of the specimen and impactor in each case. The experimentally determined coefficient of restitution $k = 0.1$ for example, qualifies an experimental standard setup of 2 m drop height with a 590 kg drop mass hitting a C40 concrete slab of $1.5 \times 1.5 \times 0.3$ meters. This leads to a small backflip of the impactor with the velocity $v_{IR} = -0.48\, m/s$ and a resultant foundation velocity of $v_{IS} = 0.14\, m/s$. Thus the measurement of the foundations seismic mass velocity gives a first integral value for a global damage behavior description of the specimen.

1.3. Foundation and springs

The foundation is mounted on 4 double gaiter air springs with variable pressures and several stabilizers. In this example the mass displaces 24 mm within a time of approx. 200 ms. The spring behavior is highly nonlinear and depends on the geometrical change at different strokes and is pressure dependent as well. This leads to a static pressure load of $p_S = 3.1$ bar at 235 mm stroke length to carry the foundation, payload and specimens mass with a force of 80 kN for each spring. Due to the dynamic load the additional amplitude of 5.5 kN is calculated from the mass acceleration for each spring. The air system is closed during the testing so that force changes lead to air compression within the system. In this case the isentropic pressure-volume relation is given by:

$$p_D = \frac{p_S(h_S) \cdot V_S(h_S)^\kappa}{V_D^\kappa} \tag{5}$$

with the pressure $p_S(h_S)$, the spring volume $V_S(h_S)$ at static conditions and p_D and V_D the dynamic response. The parameter κ is the isentropic exponent with a value of $\kappa = 1.4$ for dry air. In the given example the pressure increases to $p_D = 3.7$ bar with a stroke length of 210 mm. The temperature within the spring increases during compression. Assuming adiabatic condition the relation is given as follows:

$$T_D = \frac{T_S}{\left(\frac{V_D}{V_S}\right)^\kappa} - T_S \tag{6}$$

with T_S and T_D the temperature at static and dynamic load conditions. For the sample experiment the temperature increase is given by 41 K assuming 20 degree ambient temperature. The volumetric work during compression is given by:

$$dW = \frac{-p_S \cdot V_S}{\kappa - 1} \cdot \left(1 - \left(\frac{V_S}{V_D}\right)^{\kappa - 1}\right). \tag{7}$$

For the sample experiment it is given as 1.8 kJ for each spring which means, that approx. 65% of the drop energy is transferred to the springs. Figure 2 illustrates the measured foundation data from the sample experiment which fit well to the analytical assumption. The knowledge about the reaction, and the adjustable foundation stiffness could be a first indicator for a global experimental judgment.

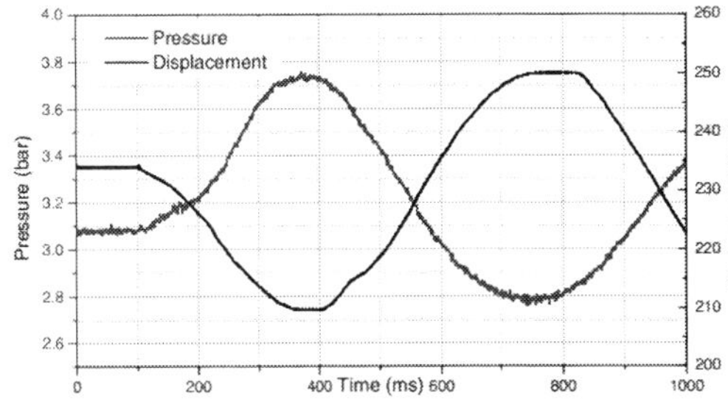

Figure 2. Pressure and displacement distribution of the accelerated foundation.

1.4. Second hit prevention

Since the foundation follows Eq. (4), the impactor reaction velocity can be calculated with Eq. (3) where an increasing

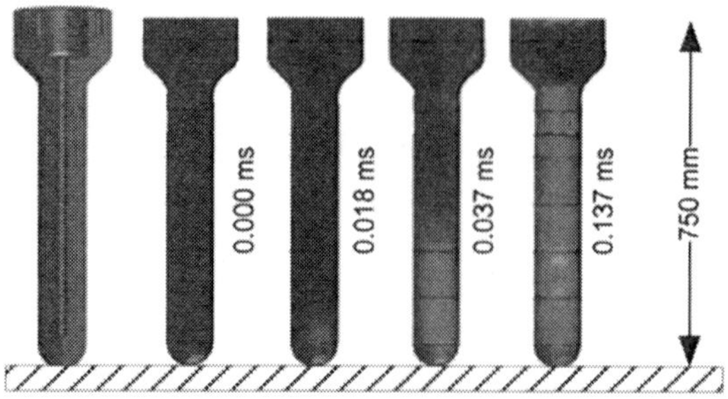

Figure 3. Segmented Impactor, stress state at different timesteps.

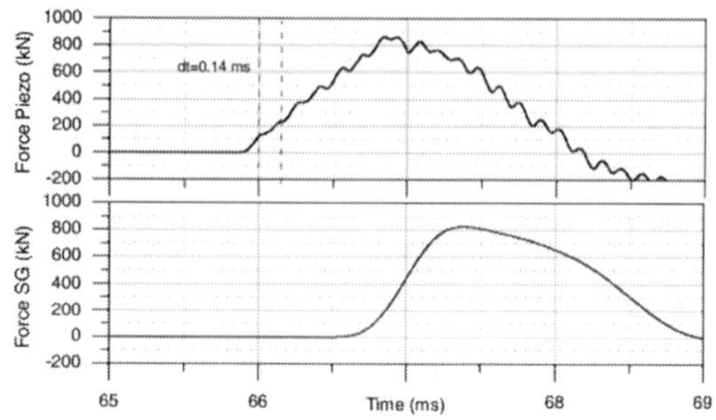

Figure 4. Sample impactor forces for piezoelectric and resistance based sensor types.

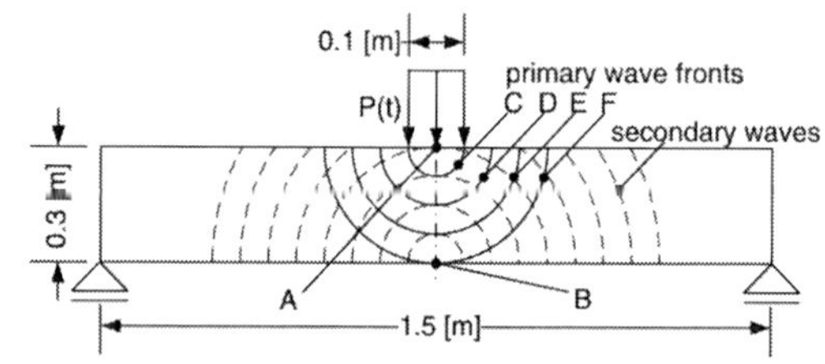

Figure 5. Principal sketch of the wave propagation in the slab during the first few milliseconds.

coefficient of restitution leads to a backflip velocity increase in negative direction. In the sample experiment the negative velocity v_{IR} leads to impactor reversal followed by a second (or more) hit onto the sample. To prevent this from the study the impactor carrier can be automatically stopped with self-locking breaks exactly at the upper death point of the movement ad zero forces.

1.5. Impactors

Following this approach just the first hit should be relevant for the stiffest experiment. The length of this hit is given by the length of the impactor assuming a linear elastic compression and expansion. The time is given by:

$$t_I = 2\frac{l_I}{c_I}, \quad (8)$$

with the impactor length l_I and the impactor sound speed c_I given by:

$$c_I = \sqrt{\frac{E \cdot (1 - \mu)}{\rho \cdot \left(1 - \mu - 2 \cdot \mu^2\right)}}, \quad (9)$$

the elastic modulus E of steel, its poison ratio μ and its density ρ. With a sound speed of approx. $c_I =$ 5500 mm/ms, the impulse duration for this example is given with t_I =0.274 ms. Further assuming purely elastic undamaged response, the maximum stress can be calculated independently of the mass by:

$$S_I = \frac{1}{2} \cdot \rho \cdot c_I \cdot v_I. \quad (10)$$

The impactor stress for this sample experiment is $S_I =$ 134 MPa which is simply correlated to the impactor force of $F_I = 1.1$ MN on its relevant area.

Figure 4 compares the measured impactor forces during the first 4 ms given by piezoelectric strain sensors at the top and semiconductor based strain gages at the bottom. The maximum force of approx. 0.8 MN is less than the theoretical assumed value. The impactor frequency can be found at the top curve as overlaid oscillation of the nearly 1 ms mean impulse. The stiffness of the impactor system can be modified to extent the load duration and impulse shape. This is done by additional layers between impactor and drop masses and adjustable springs.

2. Principle slab behavior

Until this point the outer system was considered with abstracting the sample slab behavior into the coefficient of restitution $k = 0.1$ as the root point of investigations. The slab is mounted on the foundation mass with four solid bearings which are also functioning as load cells. This gives a stiff linkage between sample and basement. Comparing Fig. 2 and Fig. 4, the load impulse finished at approx. 70 ms while the foundation movement starts 100 ms later. Hence the bearings can assumed to be fixed in vertical direction during the interesting inspection time.

Figure 5 sketches the load conditions during the first load impulse from the upper center point A. With equation (9) the longitudinal wave speed for the concrete sample is given by $c_{CL} = 4025$ mm/ms. With a slab thickness of $H_C = 300$ mm, the time for the first wave reflection is given by $t_{CL} = 0.075$ ms. At this time the pressure wave reaches the lower point B and reflects back as a tensile wave.

The numerically computed vertical and horizontal stress components in the top position (A) and bottom position (B) are given as a function of time at Fig. 6. As the wave front spreads spherically within the slab the stress values decrease with increasing distance r from the upper side roughly according to r^{-1}. After the initial phase a reflection of stresses arises at the bottom side with a reversal of compression into tension. The reflected wave front interferes with the incoming wave and the stress state within the structure becomes more complex. Thus, the local strain rate values will span from a high level at the impacted zone to a low level distributed over the whole structure.

The vertical displacements of position (A) and (B) are shown in Fig. 7. Damage or plasticity was not allowed in

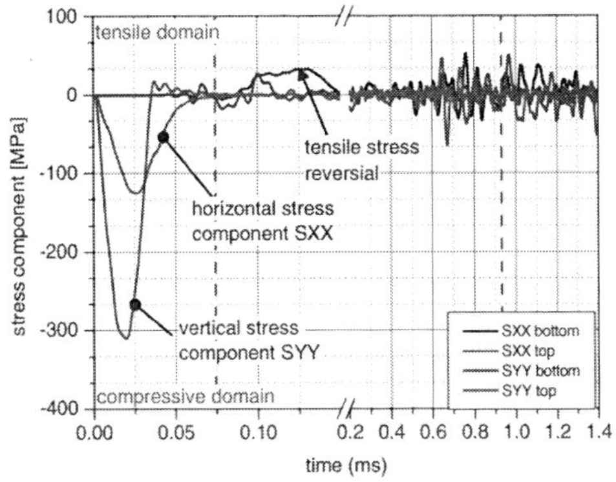

Figure 6. Computed stress distribution during the first 1.4 ms.

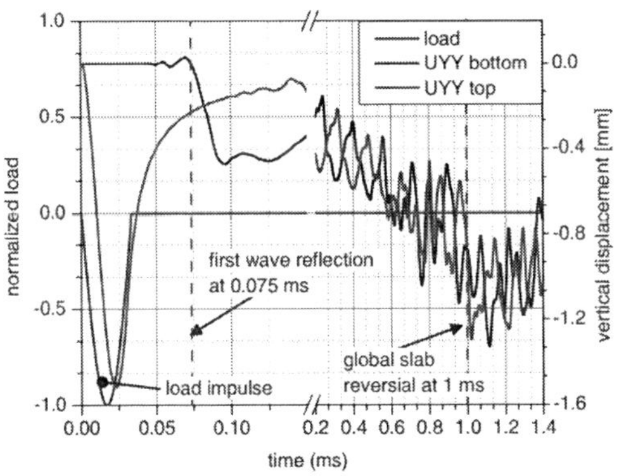

Figure 7. Computed displacement distribution during the first 1.4 ms.

this principal simulation. An indentation effect can be seen during the loading time with much larger displacement of (A) compared to (B). The indentation relieves after the end of the loading time until (A) and (B) have the same displacement values superposed by oscillations. A global deflection goes on with a maximum value at approximately t = 1 ms, which corresponds to the natural period of this eigenmode.

Resuming Fig. 4 the impulse load duration fits well with the global slab displacement time which could lead to more static like behavior and tensile bending cracking.

3. Experimental results

The investigation aims to deliver robust experimental reference datasets of impacting slabs under different conditions according to the following questions:

- How is the influence of the impact velocity,
- the influence of the mass and its related energy,
- the scale influence of the sample and impactor geometry,
- the influence of the rebar placement and design
- and the strength influence of different concrete grades?

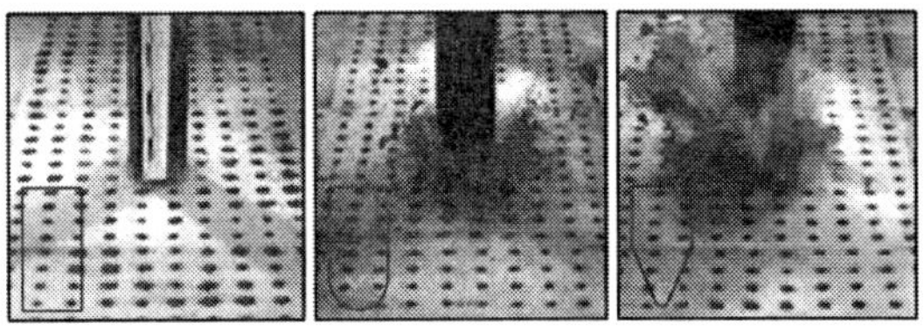

Figure 8. Impactor shape influence on the crushing behavior at the same time step.

3.1. Experimental setup

Beginning with a reference state several parameters are varied with respect to a statistically significant accuracy. Reference data is given:

- Sample
  - Concrete grade: C40
  - Geometry: 1.5 × 1.5 × 0.3 m
  - Reinforcement Type BST500, D 10 mm
  - 2 × 100 × 100 mm with 25 mm wall distance
- Impactor
  - Drop height 2 m
  - Drop weight 800 kg
  - Impactor length 750 mm
  - Impactor diameter 100 mm
  - Flat impactor shape
- Foundation
  - Weight plus payload 32.000 kg
  - Position 235 mm at 3.1 bar pressure

Standard data records are available for the impactor forces and its tip acceleration where forces are simultaneously measured with different sensor types. Foundation behavior is recordable with its displacement, acceleration measure and spring pressure distribution. The sample reaction forces to the foundation are measured at four bearing points with three types of sensors to observe low static changes as well as fast dynamic changes at the same position.

The data is usually recorded with a sampling rate of 2 MHz for the relevant time of approx. 1 second. In addition two high speed cameras are triggered to directly measure displacements as needed. In the following a subset of principal experimental results is shown without the claim to completeness. The underlying project and data acquisition process is still running and experimental improvements and data analysis are continuous tasks.

3.2. The Impactor shape influence

The shape of the impactors tip plays a major role for the damage behavior of the slab sample. In the following it was varied at reference conditions with impactor diameter 100 mm and slab thickness 200 mm. Figure 8 compares the crushing behavior by snapshots at 100 ms post triggering. The impactors already intruded approx. 100 mm in each case and one can distinguish between two principal failure modes. The flat tip leads to a punching like failure while the round and sharp shape tip causes crushing at the top surface.

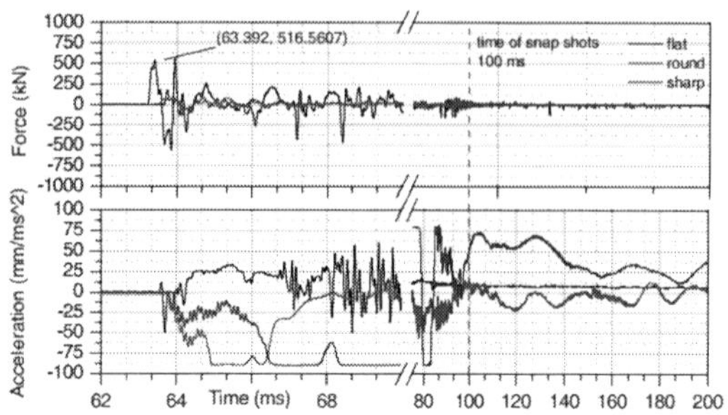

Figure 9. Impactor force and acceleration distribution for different tip shapes.

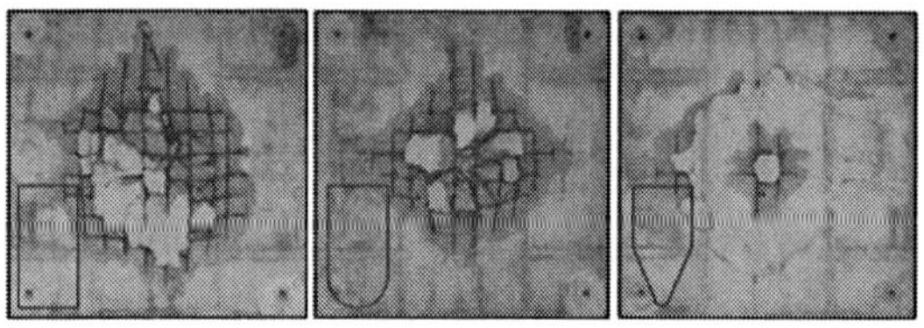

Figure 10. Impactor shape influence to the damage behavior on the slabs rear side.

Taking a closer look to the impactor forces at Fig. 9 shows a high value of its compressive maximum at 534 kN for the flat, 187 kN for the round and 152 kN for the sharp shaped tip. The acceleration distributions within the tip are low for the flat tip with 4000 g and exceed the sensors ultimate value of 5.000 g in the other cases. More interesting than the maximum values is their further distribution. While the flat tip values are nearly zeroed after approx. 10 ms, the round and sharp values are still higher, indicating the still ongoing crushing process.

The process finishes with the impactor fully punched through the slab for the flat, and sticking for the other cases. Figure 10 compares the slabs rear sides where the punched cone can be recognized to the left and totally different shapes are found at the right side.

3.3. Drop mass variation

The drop mass was extended to 1800 kg with respect to the reference configuration with flat impactor tip. The slab thickness was again reduced to 200 mm. The drop energy increases from 15 kJ to 35 kJ. The resultant impactor forces and the acceleration distributions are shown in Fig. 11.

The force maximum of the first relevant impulse increases from 534 kN to 671 kN which is moderate according to the energy increase but without any statistical consideration. This is due to the fact that the slab sample is already totally damaged by the lighter drop mass. Totally damaged means at this point, that the impactor punches through the whole thickness of the slab and only the underlying rebar layer is carrying the load. With higher energy this additional amount of energy leads to deeper penetration until the rescue dampers adsorb this to prevent further demolition. The energy absorption by the foundation increases from approx. 11 kJ to 17 kJ but the ratio of the introduced drop energy reduces from approx.

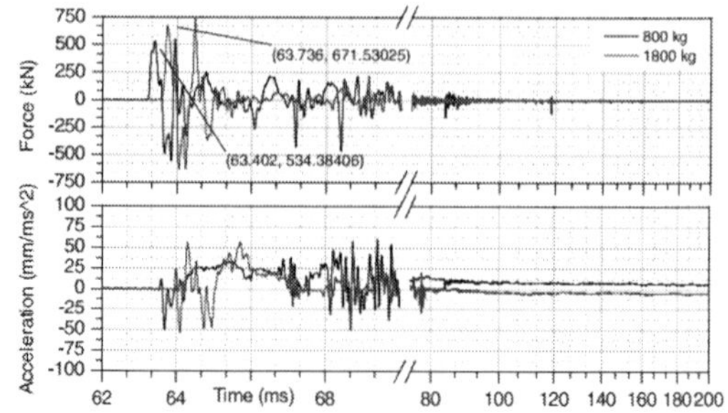

Figure 11. Impactor force and acceleration distributions for different drop masses.

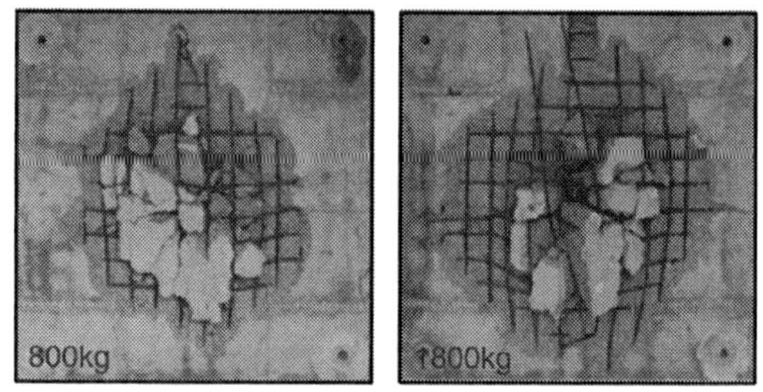

Figure 12. Drop mass influence on the rear side.

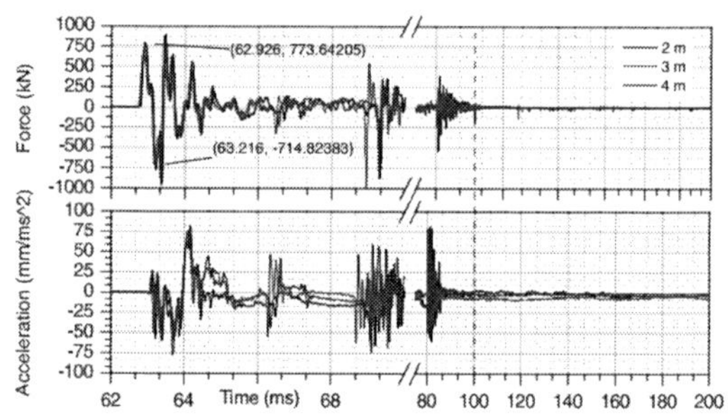

Figure 13. Drop high influence on the damage behavior on the rear side.

70% to 50%. This difference should be carried by activated dampers and needs to be considered globally.

Figure 12 again compares the rear sides of the samples for high and low masses. The impactor punched through the whole slab at high mass. This leads to the massive damage with destroyed reinforcement partially reaching its tensile strength. At the left side the impactor is sticking within the slab. The rear side reinforcement is moderately deformed but carries the resultant energy.

3.4. Drop high variation

By increasing the drop height the energy and velocity also increases. At this sample configuration three steps at 2, 3 and 4 m were tested which led to energies of 15; 24 and 32 kJ at impactor speeds of 6.3, 7.7 and 8.6 m/s. The slab thickness is 300 mm in all cases. The starting times in Fig. 13 were adjusted to fit the impulse start of the 2 m data. It is shown that the velocity increase does

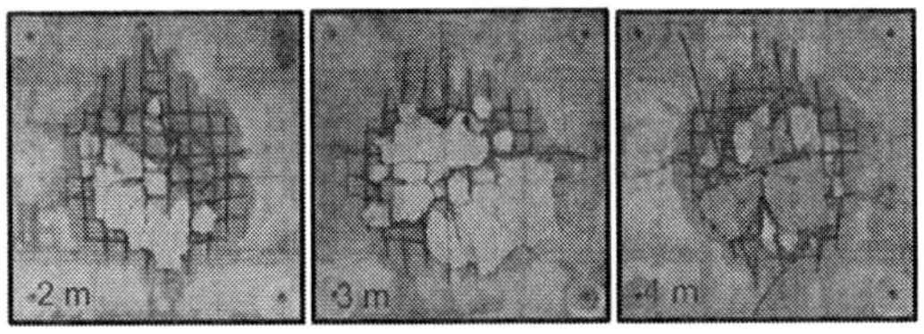

Figure 14. Drop high influence to the damage behavior on the rear side.

not really affect the mass increase or the impulse shape. The amplitudes vary about 5% around a mean value of 750 kN. The impactor tip accelerations vary from 57, 49 to 34 mm/ms^2 for heights of 2, 3, and 4 m which is significant but not fully analyzed yet.

Comparing the rear sides for variable drop heights shows the large indentation at the right side. The high energy at 4 meters again leads to a total punching with activating the rescue dampers. Nevertheless is the data valid during the most relevant first few milliseconds. The energy is close to the energy at the higher drop mass of 1800 kg and with ignoring the thinner slab the damaged shape rear sides are similar to each other. Comparing with 2 and 3 m drop height the lower rebar layer is cracked exciting the tensile strength of the steel.

4. Conclusions

The remarks given show the relevance of knowing in detail what is going on while measuring data in high speed processes. The theoretical deduction of fundamental times and frequencies are the most important factor establishing a valid measuring concept for such experiments. The knowledge of the sample behavior and its interaction with the test facility leads to much better understanding and interpreting of the sometimes as noise devalued data.

A sample interpretation of the data was done by only two measured values of the impactor since some kind of static spare force are most relevant for civil engineers. In real this restricted view are not valid since much more parameters are affected by the sample-facility interaction. For example the force measurement between impactor and drop masses are affected by cross forces since bending was not considered until now. To compensate this, additional sensors need to be integrated.

The three examples show the difficulties to distinguish between the three main parameters energy, mass and velocity and there influence to the damage behavior. Even to define the term "damage" is not really easy in this type of experiment and needs more consideration. Keeping this in mind, experiments without extensive demolition could give more information about the load capacity of the sample than experiments with energy overloads. To understand this more in detail, extensive numerical simulation of the experiment and the facility are in process leading to improved measure and evaluation concepts for the next series.

References

[1] P. Bischoff, S. Perry, *Materials and Structures* **24** (1991)

[2] L. J. Malvar, C. A. Ross, *ACI Materials Journal* **95** (1998)

[3] U. Häussler-Combe, T. Kühn, *International Journal of Impact Engineering* (2012)

[4] T. Kühn, U. Häußler-Combe, *WCCM proceedings* **10** (2012)

[5] B. Beckmann, A. Hummeltenberg, T. Weber, M. Curbach, *Structural Engineering International*, **22** (2012)

[6] T. Kühn, B. Beckmann, A. Hummeltenberg, M. Curbach, *fib proceedings Tel Aviv* (2013)

EPJ Web of Conferences **94**, 01063 (2015)
DOI: 10.1051/epjconf/20159401063

Two-wave photon Doppler velocimetry measurements in direct impact Hopkinson pressure bar experiments

Lewis J. Lea[a] and Andrew P. Jardine

SMF Fracture and Shock Physics Group, Cavendish Laboratory, JJ Thomson Avenue, Cambridge CB3 0HE, UK

Abstract. Direct impact Hopkinson pressure bar systems offer many potential advantages over split Hopkinson pressure bars, including access to higher strain rates, higher strains for equivalent striker velocity and system length, lower dispersion and faster achievement of force equilibrium. Currently advantages are gained at a significant cost: the fact that input bar data is unavailable removes all information about the striker impacted specimen face, preventing the determination of force equilibrium, and requiring approximations to be made on the sample deformation history. Recently photon Doppler velocimetry methods have been developed, which can replace strain gauges on Hopkinson bars. In this paper we discuss an experimental method and complementary data analysis for using Doppler velocimetry to measure surface velocities of the striker and output bars in a direct impact bar experiment, allowing similar data to be recorded as in a split bar system, with the same level of convenience. We discuss extracting velocity and force measurements, and improving the accuracy and convenience of Doppler velocimetry on Hopkinson bars. Results obtained using the technique are compared to equivalent split bar tests, showing improved stress measurements for the lowest and highest strains.

1. Introduction

Two general types of Hopkinson Bar systems exist for compression experiments: Split Hopkinson Pressure Bars (SHPBs, also known as Kolsky Bars [1]) and Direct Impact systems (DIHBs) [2]. In a typical SHPB test, a sample is placed between an input and output bar as depicted in Fig. 1 (a). The input bar, B, is struck by a striker bar, A, passing a pressure wave through the input bar, into the sample and out through the output bar, C. A fourth bar, D, traps the momentum transferred by the striker, protecting the experiment. The local strain at some point on the input and output bars is typically measured using strain gauges. The bars remain elastic during the experiment, allowing the force and velocity histories of the end faces of the bars, and therefore the sample, to be found from contact measurements at a distance.

In direct impact bars, the input bar is removed and the sample is struck directly by the striker bar as shown in Fig. 1 (b). Striking the sample directly has many benefits over using an input bar. Higher strains be achieved with an equivalent system as the striker bar unloads over multiple passes of the wave, rather than one in an SHPB. Higher strain rates can be achieved as striking the sample directly bypasses the collision between the striker and input bar, which limits the firing velocity based on the properties of the bars,

$$V_{max} = \frac{Y_{bar}}{Z_{bar}} \tag{1}$$

where Y_{bar} is the yield stress of the bars and Z_{bar} is their acoustic impedance. In a DIHB experiment the striker decelerates more slowly as it continues moving with the

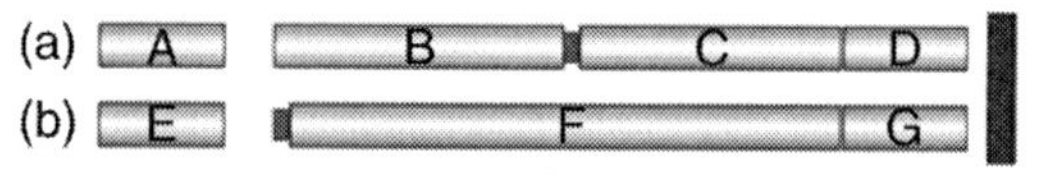

Figure 1. The co-axial bar arrangements in typical SHPB (a) and DIHB (b) systems. The experiment is initiated by the striker bars, A and E respectively, colliding with the system.

deforming sample, and does so for multiple passes of the wave, meaning the peak and average stresses experienced by the striker bar for a given velocity are lower, and higher shot velocities can be used given the same bar material.

The removal of the input bar substantially reduces the length of the system, making direct impact systems more convenient to use and the geometry also removes the Pochhammer-Chree dispersion that occurs as the pulse travels though the input bar [3]. Some specimen types may reach force equilibrium faster in a DIHB setup as removing dispersion makes the rising edge of the stress pulse steeper, equilibrium is still however dominated by the elastic properties of the specimen.

Currently, the advantages gained in a direct impact setup come at a significant cost. Removing the input bar removes all gauge information about the input face, meaning approximations have to be made about the striker face velocity and therefore strain and strain rate. Similarly the time at which sample equilibrium is reached cannot be experimentally verified. High speed photography has been used since the earliest direct impact experiments to measure the velocity of the striker bar [2] however this provides no information about the force at the striker face, has limited resolution and is time consuming. Progress has recently been made using strain gauges to instrument striker bars [4]. However these measure the particle

[a] Corresponding author: 11379@cam.ac.uk

Article available at http://www.epj-conferences.org or http://dx.doi.org/10.1051/epjconf/20159401063

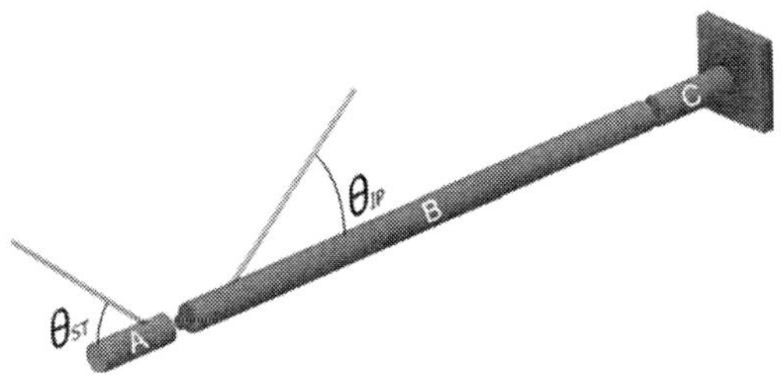

Figure 2. The incidence angles between the probing lasers and the bar axis in a DIHB system. The spot probing the striker bar, A, is positioned based on the location of the bar at the moment of impact.

velocity in the rest frame of the striker bar, providing no direct information on the bulk velocity of the bar, and strain gauges remain susceptible to damage even in specialised barrels. Both methods make DIHB experiments much less convenient to perform, both during the experiment and in post-processing, sacrificing one of the main attractions of this technique.

Recently, Photon Doppler Velocimetry (PDV) methods have been developed by Avinadav et al. [5] which provide equivalent sets of measurements to those made by strain gauges on an SHPB. This is achieved by making a non-contact measurement of the surface velocity of the bar through reflecting a laser beam off the surface. The source and returning light are combined, creating a time varying interference pattern from which the velocity history of a bar can be deduced. These methods measure local velocity, not strain, making them well suited to obtaining data from a striker bar during a direct impact experiment where there are velocity components due to bulk motion and elastic deformation.

In this paper we will demonstrate the use of PDV to measure both the velocity and force on the striker face of a direct impact bar specimen, allowing data to be collected that is similar to that obtained using an SHPB, with similar convenience. These results are compared with traditional direct impact analysis of the same data and with SHPB tests under the same conditions. Finally we will discuss general methods for improving the convenience and accuracy of all PDV-based bar systems.

2. Experimental techniques

2.1. The bar systems

The DIHB system was used in a standard arrangement [2], depicted in Fig. 2. The bars were a Dural alloy, 25 mm in diameter, the output bar, B, was 1 m long and the striker, A, was 25 cm long. The sample of interest was attached to the face of the output bar using Vaseline lubricant.

The SHPB system consisted of 12.7 diameter mm bars, with 50 cm input and output bars, and a 25 cm striker bar. Vaseline was similarly used to provide an equivalent test. In both cases the bars were mounted coaxially on V-blocks and the striker was propelled using a gas gun. The striker velocity before impact was measured using a pair of light gates, the second of which was placed such that it was

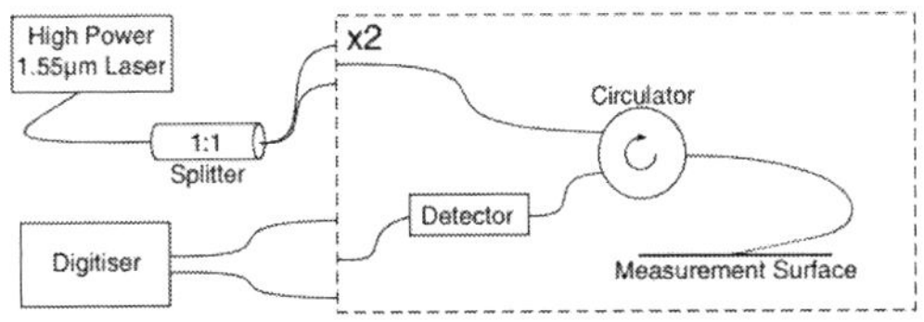

Figure 3. The photon Doppler interforometer: Light is passed by the circulator to the measurement surface, and then from the surface to GHz response photodiodes, which output an intensity proportional voltage. Interferometer probes were orientated as shown in Fig. 2.

partially obscured by the test sample and was and used to trigger the optical diagnostics.

2.2. Optical velocity measurements

A standard [6] $1.55\,\mu$m optical fibre interferometer, depicted in Fig. 3, was used to probe the bar surfaces. Light reflecting off of the bar surface is classically Doppler shifted by

$$\Delta f = \frac{2V}{\lambda}. \qquad (2)$$

Unshifted light is partially reflected back at the cleaved fibre end, and added to surface reflected light collected by the same fibre. The combination creates an interference pattern with beat frequency dependant on the shift. The entire pattern is modulated by a faster component averaged out during measurement. The resultant measured intensity profile is

$$I(t) = I_0 + I_1 \cos(2\pi \Delta f t), \qquad (3)$$

where I_0 and I_1 are dependant on many variables including the incident intensity of the laser and the reflectivity of the surface. However they are unimportant, as all required information is encoded in the frequency of the pattern. Frequency measurement makes the technique resilient to typical noise.

Velocity histories of the faces of the striker bar and output bar in contact with the sample were deduced from surface velocity measurements made further down the bars as depicted in Fig. 2. Surface velocities were extracted from the PDV data using a standard procedure [6]. Each interference pattern, exemplified in Fig. 4 (a) was split into a series of overlapping time windows, each of which were fast fast Fourier transformed, creating a series of spectra, depicted in Fig. 4 (b), where each vertical slice is a windowed spectrum. The final velocity measurements are made by treating the spectra as probability density functions and fitting a Gaussian curve to each window's spectrum. Background spectra are removed and regions of interest are defined to reduce the effects of noise on the measurement. A strong signal near zero velocity, visible in Fig. 4 (b), caused by imperfections in the optical components is both subtracted and avoided during the selection of regions of interest. Interpolation is used to fit a curve between the mean points of the Gaussian curves fit to each window. Window lengths of $8.2\,\mu$s were used with a $7.2\,\mu$s overlap, giving a $2.1\,\mu$s step size.

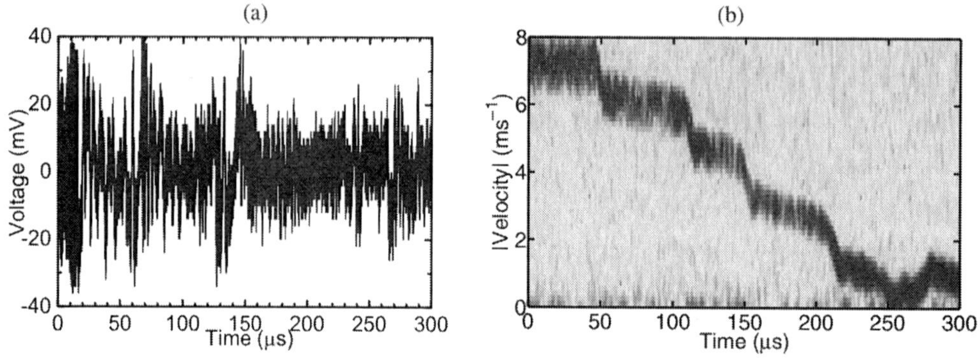

Figure 4. (a) The raw interferance pattern measured at the photodiode in the photon Doppler interferometer. (b) The probability map of the velocities obtained from Fourier transforming segments of the interferance pattern as described in the body text. The velocities depicted relate to the frequencies measured as defined by Eqs. (2) and (3).

True velocities are found by accounting for the the projection of the bar axis onto the laser axis

$$V_{true} = \frac{V_{meas}}{\cos\theta}, \tag{4}$$

where θ is incidence angle of the probing laser depicted in Fig. 2. Light gate measurements were used to confirm the measurement of striker velocity immediately before impact, which is required for correct analysis as shown in Sect. 3.

2.3. Input & output bar measurements

The strain gauges usually placed on the input and output bars were replaced with retro-reflective paint, which consists of a layer of small glass spheres that cause incident laser light to reflect back on itself. Retro reflection allowed the fibres to be placed at glancing angles (typically 10°) with reference to the bar surface without the severe loss of light that normally occurs in diffuse reflection. Glancing angle measurements lead to a reduction in errors caused by the precision of the laser incidence angle and bending velocities during the experiment.

Bending errors are quantified by pointing fibres at opposite sides of the same point on the bar, allowing the measurement projections to be separated into axial and radial (bending) velocities [7]. Bending modes cause a velocity towards one probe, and way from the other, resulting in the measured velocities

$$V_1 = V_{axial}\cos\theta + V_{bend}\sin\theta, \tag{5}$$

and

$$V_2 = V_{axial}\cos\theta - V_{bend}\sin\theta, \tag{6}$$

allowing for the separation of the axial and bending velocities. The trace from a single probe can then be compared with the pure axial velocity extracted from a double probe measurement. At an incident angle, θ, of 10°, and careful alignment of the bar system, the $\sin\theta$ projection of bending velocities reduced the related errors below 1% in 12.7 mm bars (tested using input bar in the SHPB setup), and less for 25 mm bars. The reduction in the projection of bending modes allowed each bar to be probed with a single laser, reducing the number of velocimeter channels required from four to two.

Measurements are carried out at points along the bar surface, as depicted in Fig. 2, and are time shifted back to the bar faces. Probe positions are chosen to be sufficiently far for the stress wave to satisfy Saint-Venant's principle [8], making the stress uniform across the entire cross section of the bar, and thus making surface measurements representative of the bulk.

Applying the laser probe at glancing incidence causes the spot size of the laser on the bar to increase, and therefore the time resolution of the probe to decrease. Infra-red visualising card allowed the spot size at the surface to be measured to be of the order of a millimetre in diameter, which given a typical bar sound speed of over 1 km s^{-1} still allows for better than 1 μs resolution.

2.4. Striker bar measurements

As the striker bar is repeatedly fired from the barrel, it is not coated in any kind of reflective or retro reflective paint. Instead the incident laser was shone at a steeper angle (typically 30°) to increase the diffuse return. Despite the absence of any paint, a signal was still collectable with the fibre placed 5 mm or closer to the bar surface. The laser probe was placed at the mid point of the bar, 12.5 cm from the sample firstly to satisfy Saint-Venant's principle as discussed above, and to aid wave separation, discussed in Sect. 3.1, as internal reflection in the striker requires additional wave separation after time shifting the trace to correspond to the face in contact with the sample.

3. Method of analysis

3.1. Internal reflection in the striker bar

Due to the nature and timescale of the unloading in the striker bar during direct impact tests, measurements of the sample face velocity will usually be superimposed with the tensile wave reflections of compressional waves returning from the free end of the bar. Separation of these waves becomes further complicated by bulk velocity, the velocity of the wave medium with reference to the probe. The velocity measured by the probe is the combination of three

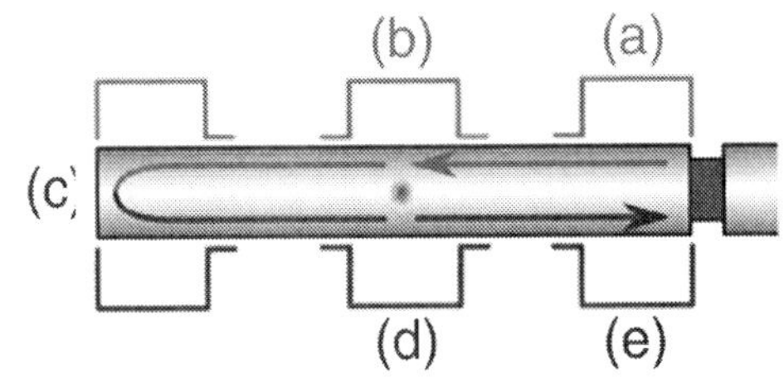

Figure 5. The complete path of a compression wave from the sample to the free face and back as a tensile wave, passing the measurement probe each way.

separate components,

$$V_{probe} = V_{bulk} + V_{compressive} + V_{tensile}. \quad (7)$$

In order to correctly restore the force and velocity history of the sample face, these profiles must be correctly separated and time shifted. Separation is possible as the initial bulk velocity is measurable, and the initial tensile velocity is zero.

Figure 5 depicts the complete path of one wave in the striker. The compression wave is formed at the sample, (a), at t_1 and passes by the measurement probe (b) at t_2. It then reaches the back surface (c) where the free end boundary condition forces it to be perfectly reflected into a tensile wave. The tensile wave passes back past the measurement probe at (d) at a time t_3 and contributes a further change in the velocity at that point, equal to that the compression wave created at (b). After passing as a reflected wave, the tensile wave returns to the sample at (e) and any effects of the wave at that point can then be considered as part of the motion at the sample face, and do not require separating and correcting.

The initial velocity at the struck face of the specimen is the signal in the absence of the reflected trace, as that is the only velocity not to originate from the sample. The corrected trace is then given by

$$V_{st}(t_1) = V_{probe}(t_2) - V_{tensile}(t_2), \quad (8)$$

where V_{st} is the velocity of the struck face of the sample, and timeshifting from t_2 to t_1 is applied as discussed above. Given perfect reflections, the inversion of both wave type and travel direction at the striker free end means the velocity change due to the tensile component is the same as that of the equivalent earlier compressive component,

$$V_{tensile}(t_3) = V_{compressive}(t_2). \quad (9)$$

For the purpose of discussing the correction procedure simply, we will define a dimensionless time such that one unit of dimensionless time is the time taken for a sound wave to travel one length of the striker bar,

$$\tau_i = \frac{c_{str}}{l_{str}} t_i, \quad (10)$$

where c_{str} is the sound speed and l_{str} is the length of the striker bar. For a probe at the midpoint of the striker bar, the dimensionless time between τ_1 and τ_2 is 0.5 and between τ_2 and τ_3 is 1.

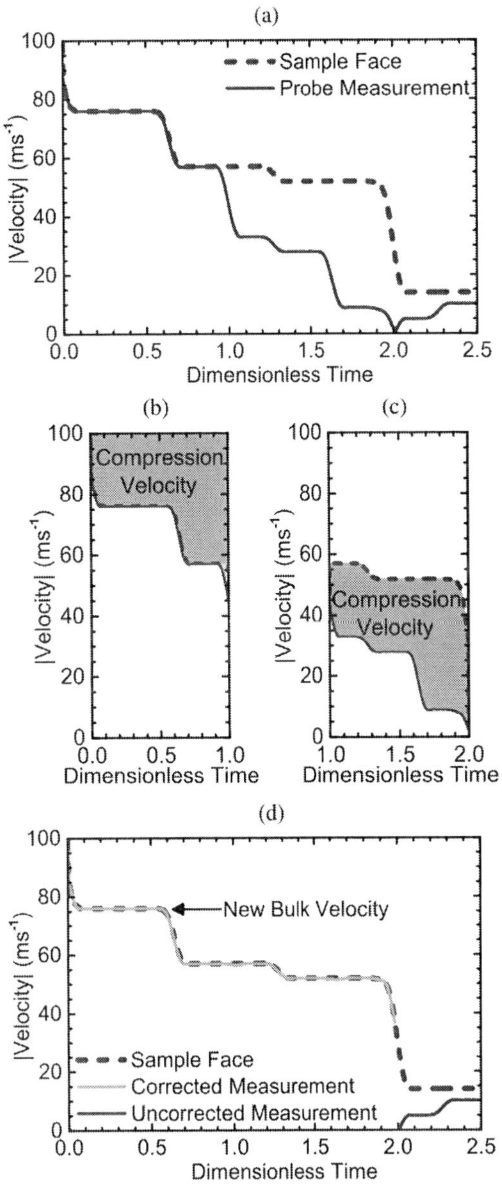

Figure 6. A series of "cartoon" velocity traces explaining how the correction procedure is performed. (a) depicts the uncorrected trace, (b) and (c) show the correction, and (d) shows the trace for the next unit of dimensionless time. The initial striker velocity is $100\,\mathrm{m\,s^{-1}}$.

Figure 6 depicts the correction procedure. In (a), both the probe measurement (solid line) and the true face velocity (dashed line) are shown. The probe trace has been time shifted back by τ=0.5 to overlap it onto the corresponding sample face velocity. During the first unit of dimensionless time, the trace does not need any correction, as no tensile waves have returned. Using this correct measurement of the face velocity, and knowledge of the bulk velocity, the compressional velocity can be found for the first unit of dimensionless time, shown by the shaded area in Fig. 6 (b). The compressional velocity is in the opposite direction to the bulk velocity, making it negative if the bulk velocity is considered positive.

In Fig. 6 (c), the next segment of dimensionless time is corrected by subtracting the previously measured compressional velocity, assuming perfect reflection, to account for it returning as a tensile velocity. As the measured compressional velocity was negative, subtracting it from the later trace increases the velocity above the measured velocity. Practically this results in adding the shaded area from (b) on to the top of the probe measurement in (c). This gives the corrected trace as depicted in (d), from which the new bulk velocity of the bar can be determined.

The bulk velocity of a bar for any unit of dimensionless time is equal to last completed velocity step of the sample face in the previous unit of dimensionless time, as shown in Fig. 6 (d). This is due to completed steps in velocity being the ones which have traversed the entire bar, from (a) to (e) in Fig. 5.

Time shifting can be done more reliably in the case of striker reflections than reflections in the input bar in an SHPB, as the shifting takes place between parts of the same continuous wave; any incorrect shifting produces obvious spikes in the recovered velocity.

3.2. Sample deformation calculation

Once measurements have been made of the front, or striker impact face, V_{st}, and the back, or output bar face, V_{op}, the sample length as a function of time is determined by the difference between these two velocities

$$\frac{\partial L(t)}{\partial t} = V_{op} - V_{st}, \tag{11}$$

and

$$\varepsilon(t) = ln\left(\frac{L_0}{L(t)}\right), \tag{12}$$

where compressive strains are considered positive.

The stress history of the sample is measured as per a classic DIPB test, using the elastic relation

$$\sigma(t) = V_{op}(t)\, Z_{op} \frac{A_{op}}{A_{sa}(t)}, \tag{13}$$

where Z_{op} is the acoustic impedance of the output bar and A_{op} and A_{sa} are the cross sectional areas of the output bar and sample respectively.

3.3. Force measurements and equilibrium

The force at the output face can be found directly from the measured velocity

$$F_{op}(t) = V_{op}(t)\, Z_{op}\, A_{op}, \tag{14}$$

the same relationship holds true for the striker bar. However, the only velocity component due to the force from the sample is the compressive wave

$$F_{st}(t) = Z_{str}\, A_{str}\, V_{compressive}(t). \tag{15}$$

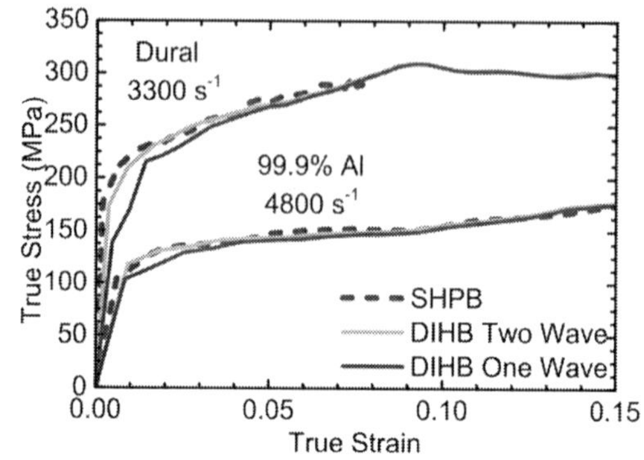

Figure 7. The stress-strain relationships for equal striker velocitiy tests in SHPB and DIHB setups. Two-wave analysis of the data appears to be a better match to SHPB data early on. However, one-wave measurements reach similar accuracy after a true strain of approximately 0.05. The largest SHPB strain reached by the 99.9% Al is 0.13.

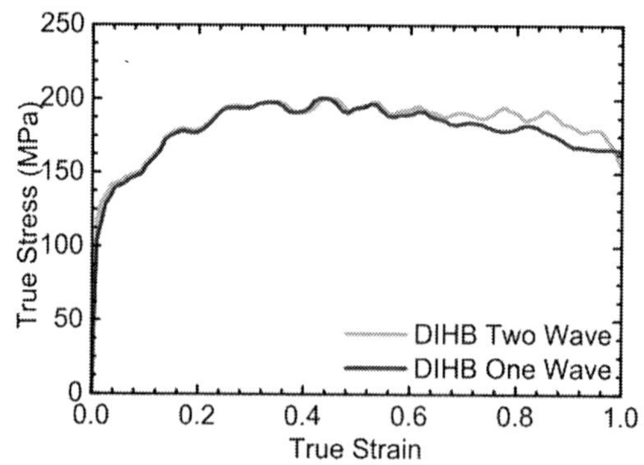

Figure 8. The stress strain relationships derived from one and two-wave measurements of the same DIHB test on 99.9% Al at $4800\,s^{-1}$. One-wave measurements show a notable drop in measured stress at high strain due to an overestimate of the striker face velocity and thus an overestimate of the sample area when conserving volume.

These measurements allow force equilibrium analysis to be performed in exactly the same way as classic split bar tests. It should be noted that bars of equal acoustic impedance and area, a compressional velocity comparison is equivalent to a force comparison.

4. Experimental results & discussion

Sample disks of 1 mm thickness and 3 mm diameter made from a 99.9% pure aluminium and a commercial Dural alloy and were tested at equal striker velocities in both SHPB and DIHB arrangements. SHPB velocity traces from the surface were analysed using standard three-wave methods [5]. DIHB data was analysed using the two-wave analysis described in Sect. 3, and traditional one-wave analysis [2] where the striker face velocity is approximated by

$$V_{st}(t) = V_0 - V_{op}(t) \tag{16}$$

where V_0 is the firing velocity of the striker.

The stress-strain relationships obtained using these methods are depicted in Figs. 7 and 8. Figure 7 is cropped

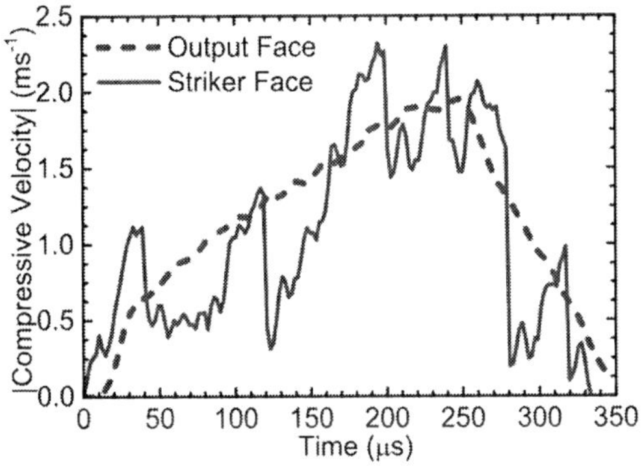

Figure 9. The compressional wave velocies leaving the sample through the striker bar and output bar as a function of time. These are directly proportional to the forces on the sample faces as shown in Eqs. (14) and (15). The spikes every $80\,\mu s$ in the striker face correspond to the wave in the striker bar making new passes of the sample after traveling two bar lengths (Striker Length 0.25 m, sound speed $6.3\,\mathrm{km\,s^{-1}}$).

to a true strain of 0.2 to allow clearer comparison between the DIHB methods and the SHPB results. Complete curves for both DIHB methods are depicted in Fig. 8.

At true strains lower than 0.05, two-wave analysis of the DIHB data appears to be a closer match to SHPB data. This is likely due to the one-wave DIHB measurement incorrectly assuming force equilibrium during ring up in the sample. Figure 9 depicts the forces on the sample faces in the 99.9% aluminium DIHB test, approximately $10\,\mu s$ is taken for the back face of the sample to reach equilibrium with the front face. This time corresponds to a true strain of approximately 0.05 at a strain rate of $4800\,\mathrm{s^{-1}}$, making it similar to the time-scale over which the two-wave and one-wave traces differ.

In Fig. 8 for strains greater than 0.5 the one-wave trace is seen to consistently decrease below the two-wave trace. By this point the bulk velocity of the striker bar has been reduced by multiple passes of decelerating waves. In one-wave analysis the striker's bulk velocity is assumed constant, as in Eq. (16). This leads to an over-estimation of strain and therefore sample area due to volume conservation, leading to an underestimate of the flow stress in Eq. (13).

5. Conclusions

We have presented a method for directly measuring the velocities and forces at the impacted face of a specimen in a DIHB experiment. Analysis of the data was demonstrated including the recovery of the sample face trace in the presence of free end striker reflections. Experimental data comparing the method to current approximations was presented with reference to equivalent SHPB data. Direct measurements were observed to more closely correspond to SHPB measurements for low strains and to not suffer from the same high strain inaccuracies as assuming constant striker inertia. Our method makes DIHB measurements viable for samples which do not quickly reach force equilibrium. These include polymers, foams and cellular materials, potentially allowing access to previously untested strain regimes.

This research was supported by grants from QinetiQ and the Engineering and Physical Sciences Research Council. The authors would like to thank P.J. Gould and P.D. Church for their interest.

References

[1] H. Kolsky, Proceedings of the Physical Society B **62**, 676–700 (1949)
[2] D. Gorham, P. Pope, J. Field, Proceedings of the Royal Society A **438**, 153–170 (1992)
[3] D. Gorham, Institute of Physics Conference Series **47**, 16–24 (1980)
[4] R. Govender, Unpublished Presentation at DYMAT Hopkinson Centenary Conference (2014)
[5] C. Avinadav, Y. Ashuach, R. Kreif, Review of Scientific Instruments **82**, 073908 (2011)
[6] O. Strand, D. Goosman, C. Martinez, T. Whitworth, W. Kuhlow, Review of Scientific Instruments **77**, 083108 (2006)
[7] D. Casem, M. Zellner, Experimental Mechanics **53**, 1467–1473 (2013)
[8] A. Love, *A treatise on the mathematical theory of elasticity* (Cambridge University Press, 1927) 257–258

EPJ Web of Conferences **94**, 01064 (2015)
DOI: 10.1051/epjconf/20159401064

Dynamic behaviour of HPFRCC: The influence of fibres dispersion

Alessio Caverzan[1,2,a], Marco di Prisco[1], and Ezio Cadoni[3]

[1] Department of Civil and Environmental Engineering, Politecnico di Milano, 20133 Milano, Italy
[2] Present Address: European Commission, Joint Research Centre (JRC), Institute for the Protection and Security of the Citizen (IPSC), European Laboratory for Structural Assessment (ELSA), 21027 Ispra, Italy
[3] DynaMat Laboratory, University of Applied Sciences of Southern Switzerland, 6952 Canobbio, Switzerland

Abstract. The promise of fibre-reinforced cementitious composites for dynamic loading application stems from their observed good response under static loading mainly due to fibre contribution. An experimental research aimed at contributing to the understanding of the behaviour of advanced fibre-reinforced cementitious composites subjected to low and high strain rates was carried out underlining the influence of fibres. The material behaviour was investigated at three strain rates (0.1, 1, and 150 s^{-1}) and the tests results were compared with their static behaviour. Tests at intermediate strain rates (0.1–1 s^{-1}) were carried out by means of a hydro-pneumatic machine (HPM), while high strain rates (150 s^{-1}) were investigated by exploiting a modified Hopkinson bar (MHB). Particular attention has been placed on the influence of fibre and fibre dispersion on the dynamic behaviour of the materials: matrix, HPFRCC with random fibre distribution and aligned fibres were compared. The comparison between static and dynamic tests highlighted several relevant aspects regarding the influence of fibres on the peak strength and post-peak behaviour at high strain rates.

1. Introduction

The mechanical behaviour of fibre-reinforced cementitious composites when subjected to impact or blast still has many aspects requiring further studies [1], with specific reference to large and socially-sensitive structures, as sheltering structures, high-rise buildings, bridges, off-shore platforms, pipelines, gasification reactors, secondary containment shells for nuclear power plants, and tunnels. In the last decade a particular class of cementitious materials defined High Performance Fibre Reinforced Cementitious Composites (HPFRCC) has been taken into consideration as a promising solution by many researchers due to the materials characteristics [2]. HPFRCCs are characterized by high ductility and tension hardening behaviour in statics. The high energy absorbed during the fracture process is associated to multiple cracking and pull-out phenomena. The good response highlighted in the static field by many researchers suggests using these materials also for dynamic loading applications. However, very few investigations have been carried out in order to guarantee the improvement of the dynamic response. Current understanding of the impact resistance of cementitious composites, and especially of high-strength concrete, is very limited [3,4]. Moreover, the results available in the literature are also often contradictory. For example, while some researchers [5–7] reported a progressively reduced sensitivity to strain rate with the growing of the static compressive strength of concrete, other studies contrast this opinion [8].

Fibers enhance the ductility of brittle materials like concrete, and this improvement is strictly related to the process by which load is transferred from matrix to fibers and the bridging effect of fibers across the cracks. Hence fiber pull-out is the principal mechanism contributing to the high toughness of the material, thus preventing fiber failure mechanism. Since, the behavior of the fiber, the cement matrix and the bond interface (or the pull-out mechanism) are likely dependent on the strain rate, it is expected that the strain rate sensitivity moves to the composite level [9].

However, the researches quoted were not focused on the influence of fiber reinforcement in the strength increase at high strain rates, if any. Caverzan et al. [10] have pointed out that fibers increase the peak strength in static, favoring the stable crack propagation, but when the strain rate is increased, fiber influence on the stable crack propagation phenomena reduces: at high strain rates the peak strength ratio between HPFRCC and HPCC decreases. Starting from this observation, in this work the fiber role is investigated at different strain rates comparing the results obtained for a plain and fiber reinforced HPCC. In particular, two different kinds of HPFRCCs have been considered in order to underline the influence of fibres distribution in the matrix. The two reinforced materials are characterized by different casting procedures that have led to different fibre dispersion and orientation, HPFRCC-A was characterized by a homogeneous dispersion of fibres which were all aligned to the load direction. On the contrary, HPFRCC-R had a random fibre distribution and no preference fibre direction can be highlighted.

2. Material and casting procedure

This research studies a high performance cementitious composite optimised with steel fibres. The mix design of

[a] Corresponding author: alessio.caverzan@jrc.ec.europa.eu

Table 1. Mix design.

Constituent	dosage [kg/m^3]
Cement type I 52.5	600
Slag	500
Water	200
Sand 0–2 mm	983
Fibres (l_f = 13 mm; d_f = 0.16 mm)	100
Super plasticiser	33 l/m^3

the HPFRCC material is specified in Table 1. The steel fibres were high carbon straight fibres, 13 mm long with a 0.16 mm diameter (aspect ratio l_f/d_f equal to 80); their proportion in the mix was equal to 100 kg/m^3.

The sand used in the material was sieved up to 2 mm and it can be classified as mixed quartz sand. A 1.6 m × 0.60 m in plane slab, 30 mm thick, was manufactured. The slab was cast by applying a unidirectional flow as shown. In order to guarantee a certain fibre orientation, the properties of the self compacting material were used, taking advantage of the flow direction. Three prismatic beam samples, 40 mm wide and 600 mm long, were sawed from the slab and tested in bending to perform a proper mechanical characterisation of the material according to Italian Guidelines [11]. The high fibre content and the favourable orientation imposed by the casting flow control allow us to guarantee a small dispersion of the response before and after single-crack localisation and a hardening behaviour in uniaxial tension. From the bent specimens, several small cylinders 20 mm long, labeled HPFRCC-A, were cored in the direction of tensile stresses. Their diameter was nominally equal to 20 mm. Each cylinder cored from each prismatic specimen was notched (notch depth = 1.5 mm), to be tested in uniaxial tension at different loading rates. HPFRCC-R samples were cored from different prismatic specimens used to perform the mechanical characterization of the material according to National Recommendations [12], three prismatic specimens were cast. Specimens were 150 mm × 150 mm cross-section and 600 mm long. Due to the high viscosity of the material at fresh state, it was not possible to cast them using a trowel with a minimum opening equal to 200 mm, as suggested by UNI 11039 [12], but they were cast in steel formworks by applying a flow at the right angle with the longer formwork side. After 28 days, the beams were notched in the middle of the face close to the casting surface by means of a diamond saw. The notched specimens were tested in bending. After performing the bending tests, the specimens were divided into two halfs and more cores were taken from each half. The cores object of the present work have a dimeter equal to 20 mm, then these cores where cut and notched in order to obtain samples identical to the samples HPFRCC-A previously described. The matrix of the HPFRCCs materials previously described has been used as a reference plain high performance composite (HPCC). The casting procedure adopted, and described above, for the HPFRCC-A sample has been used to produce the nothced cylindrical samples of HPCC. The cylindrical samples were tested in static and dynamic range.

3. Static tests

Uniaxial tensile tests on notched cylindrical specimens have been performed with an electro-mechanical testing machine INSTRON 5867 at the laboratory of the Politecnico di Milano – Polo Regionale di Lecco. The press has a maximum loading capacity of 30 kN. The samples have been glued to the press platens by means of an epoxy resin. Two aluminum cylinders, connected to the press by a knuckled joint, have been used as press platens. In both cylinders, a 5 mm deep cylindrical cavity with a 22 mm diameter was made in order to increase the glued sample surface. Stroke was considered as feedback parameter during the tests. The displacement rate imposed during the tests was equal to 5×10^{-5} mm/s up to 1.5 mm and then it was progressively increased to 10^{-3} mm/s. Test results are shown in Fig. 1 in terms of nominal stress (σ_N) versus crack opening displacement (COD) for the HPFRCC-A and HPFRCC-R materials, while in Table 2 are reported the peak strength values, the effective cross section area and the number of fibers counted in the failed sections for the HPFRCC aterials and the HPCC as well. Due to a defect identified in the sample 3 only two specimens were tested for the HPCC. Considering the cavity depth of the platens (5 mm for each side, where samples are glued) as a rigid zone, a clear span equal to 10 mm can be used to compute the strain, starting from the COD values.

On the basis of this assumption, it is interesting to observe that aligned steel fibers are able to favor stable propagation up to a strain of about 2% in the HPFRCC-A material (Fig. 1(a)). The pre-peak behavior is well described by a parabola rectangular model. This would mean that, even if a notch depth ratio equal to 0.15 was used, the material is hardening and it distributes over the whole clear length the cracking process. Of course, the geometrical defect introduced by the notch and the only stroke measure prevents a correct evaluation of the effective ductility measured in terms of strain of the peak plateau. New tests without any notch are in progress. A peak strength increase of about 40% can be pointed out comparing the results obtained for the HPCC. The fibers favor a more stable crack propagation leading to the increase of peak strength in the static field. Moreover, it is interesting to highlight the low peak strength measured and the softening post-peak behaviour in the HPFRCC-R which is characterised by a random fibre distribution. Figure 1(b) shows a drop to a stress plateau of about 1.5 MPa after the peak. In these specimens, fibres were not able to promote the stable crack propagation mainly due to their random distribution. The number of fibres and their orientation was completely different from that obtained by applying an unidirectional flow during the cast procedure. Also the decay in the peak strength could be partly related to the random fibre distribution, now fibres can be assimilated to an internal defect, which could cause the observed decrease in the peak strength. These considerations underline that the cast procedures play a relevant role in this kind of materials. Performing the same tests on specimens with the same geometry a decrease of about 50% in the peak strength is observed.

On the basis of this assumption, it is interesting to observe that aligned steel fibers are able to favor stable

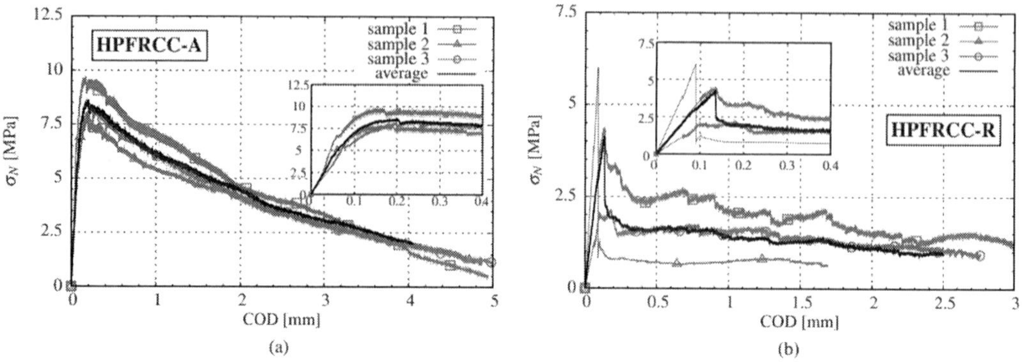

Figure 1. Test results: material with alligned fibres (HPFRCC-A) (a); material with random fibre distribution (HPFRCC-R) (b).

Table 2. Peak strengths and number of fibres.

Material type	sample n°	fibre n°	σ_{Peak} [MPa]	$\sigma_{Peak,av}$ [MPa] (std)
HPFRC-A	1	75	9.71	
HPFRC-A	2	73	7.83	8.62
HPFRC-A	3	77	8.33	(0.97)
HPFRC-R	1	20	4.32	
HPFRC-R	2	31	6.03	4.15
HPFRC-R	3	27	2.10	(1.97)
HPC	1	–	6.81	
HPC	2	–	5.50	6.15
HPC	3	–	–	(–)

propagation up to a strain of about 2% in the HPFRCC-A material (Fig. 1(a)). The pre-peak behavior is well described by a parabola rectangular model. This would mean that, even if a notch depth ratio equal to 0.15 was used, the material is hardening and it distributes over the whole clear length the cracking process. Of course, the geometrical defect introduced by the notch and the only stroke measure prevents a correct evaluation of the effective ductility measured in terms of strain of the peak plateau. New tests without any notch are in progress. A peak strength increase of about 40% can be pointed out comparing the results obtained for the HPCC. The fibers favor a more stable crack propagation leading to the increase of peak strength in the static field. It is interesting to highlight the low peak strength measured and the softening post-peak behaviour. Figure 1(b) shows a drop to a stress plateau of about 1.5 MPa after the peak. In these specimens, fibres were not able to promote the stable crack propagation mainly due to their random distribution. The number of fibres and their orientation was completely different from that obtained by applying an unidirectional flow during the cast procedure. Also the decay in the peak strength could be partly related to the random fibre distribution, now fibres can be assimilated to an internal defect, which could cause the observed decrease in the peak strength. These considerations underline that the cast procedures play a relevant role in this kind of materials. Performing the same tests on specimens with the same geometry a decrease of about 50% in the peak strength is observed.

4. Dynamic tests

In order to identify the influence of fibers and fibres dispertion on the material response under dynamic conditions an experimental program has been carried out. Two different mechanical testing machines have been used to investigate the dynamic field: a hydro-pneumatic machine (HPM) has been employed to investigate the strain rates equal to 0.1 and 1 s^{-1} while a modified Hopkinson bar (MHB) [13] has been exploited to carry out high strain rate tests (150 s^{-1}). The HPM functioning is widely described in [14], here only a briefly description of the device is reported. At the beginning of the test, a sealed piston divides the cylindrical tank into two chambers, one being filled with gas at high pressure (viz. 150 bars), and the other with water. An equal pressure is initially established in the water and gas chambers so that the forces acting on the two faces of the piston are in equilibrium. The test starts when the second chamber discharges the water through a calibrated orifice that is activated by a fast electro-valve. The piston starts then to move, expelling the water. The specimen is connected on one side to the piston shaft and, on the other side, to the end of an elastic bar which is rigidly fixed to a supporting structure. The elastic bar is instrumented with a strain-gauge that provides, through the elastic properties of the bar, the force acting on the specimen during the test. Two targets are attached at both ends of the specimen and their movement is measured by two contact-less displacement transducers. The MHB consists of two circular aluminum bars, called input and output bars (with a diameter of 20 mm and having length of 3 and 6 m, respectively) between which the specimen is glued using a bi-component epoxy resin. The input bar is connected to a high strength steel pretension bar (having 6 m length and 12 mm diameter), used as pulse generator. A test with the MHB starts pulling the high strength steel bar by means of a hydraulic actuator (of maximum loading capacity of 600 kN); the pretension stored in this bar is assured by the blocking device. In the second operation the fragile bolt in the blocking device is ruptured, giving rise to a tensile mechanical pulse of duration 2.4 ms and with a linear loading rate during the rise time. The pulse

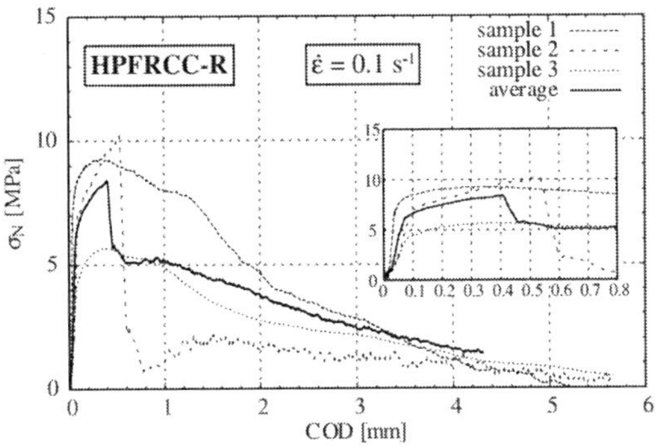

Figure 2. Stress vs. strain curves for the dynamic tests with a strain rate $\dot{\varepsilon}$ equal to 0.1 s^{-1}.

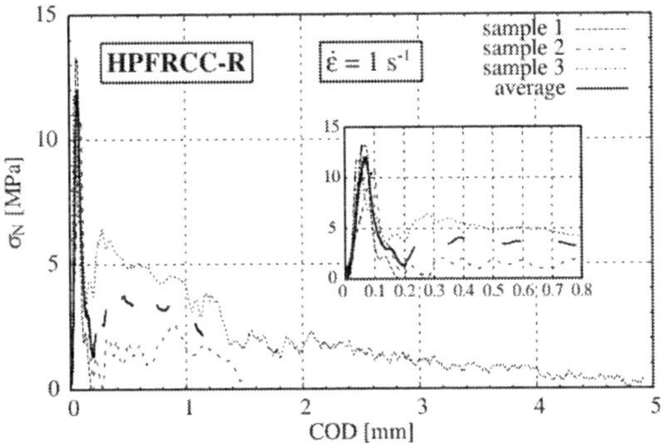

Figure 3. Stress vs. strain curves for the dynamic tests with a strain rate $\dot{\varepsilon}$ equal to 1 s^{-1}.

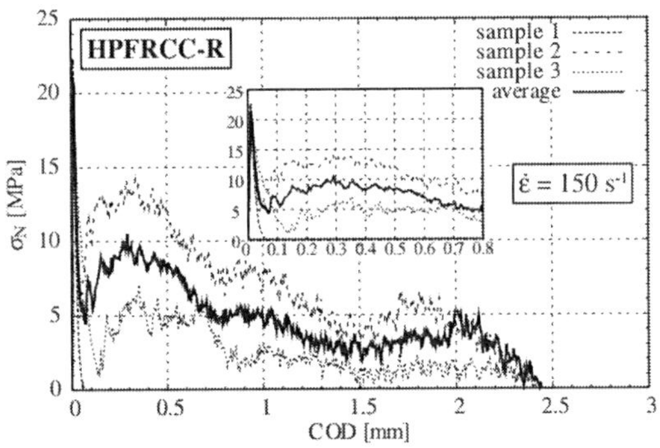

Figure 4. Stress vs. strain curves for the dynamic tests with a strain rate $\dot{\varepsilon}$ equal to 150 s^{-1}.

then propagates along the input and output bars, leading the specimen to failure. In Figs. 2, 3 and 4 the stress versus COD curves of the HPFRCC-R specimens tested at 0.1, 1 and 150 s^{-1} respectively are reported. In Fig. 5 only the average curves for the HPFRCC specimens are presented [3] while the complete trends for the dynamic behaviour of HPCC are reported in [10].

The peak strength values and the number of fibers counted in the failed cross sections are listed in Table 3. Analyzing the test results, some interesting considerations can be highlighted. An increasing of peak strength can

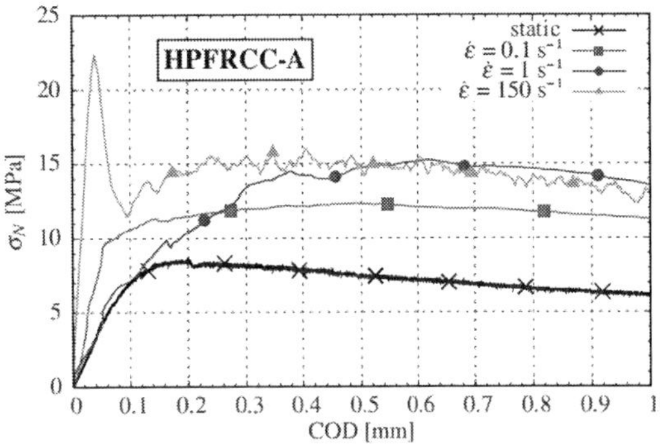

Figure 5. Stress vs. strain curves for the dynamic tests with a strain rate $\dot{\varepsilon}$ equal to 150 s^{-1}.

Table 3. Peak strengths and number of fibres.

Strain rate [s^{-1}]	Material type	fibre n°	σ_{Peak} [MPa]	$\sigma_{Peak,av}$ [MPa] (std)
0.1	HPFRC-A	75	9.71	8.62 (0.97)
		73	7.83	
		77	8.33	
	HPFRC-R	50	9.28	8.40 (2.40)
		22	10.25	
		33	5.68	
	HPC	–	10.54	9.12 (1.34)
		–	7.87	
		–	8.95	
1	HPFRC-A	76	14.29	15.25 (1.65)
		71	14.09	
		73	17.67	
		77	14.96	
	HPFRC-R	26	13.21	12.00 (1.19)
		20	10.84	
		27	11.95	
	HPC	–	10.22	11.31 (1.07)
		–	11.33	
		–	12.36	
150	HPFRC-A	79	20.46	22.36 (1.89)
		90	22.28	
		79	24.96	
		81	21.74	
	HPFRC-R	20	22.40	21.04 (2.50)
		61	22.57	
		26	18.16	
	HPC	–	15.89	17.05 (1.32)
		–	18.48	
		–	16.77	

be observed with respect to the static one, nevertheless specimen HPFRCC-R showed a peak strength lower than that highlighted for HPFRCC-A samples as it can be seen in Table 3, where the peak strength values and the number of fibers counted in the failed cross sections are listed. Moreover a peak strength increase has been observed increasing the strain rate from 0.1 to 1 s^{-1} for the same material type. The comparison between dynamic response of specimens HPFRCC-A and HPFRCC-R showed to what

Table 4. Dynamic Increase Factors of peak strengths.

Strain rate	$DIF_{HPFRCC-A}$	$DIF_{HPFRCC-R}$	DIF_{HPC}
0.1	1.43	2.03	1.48
1	1.77	2.90	1.84
150	2.59	5.08	2.77

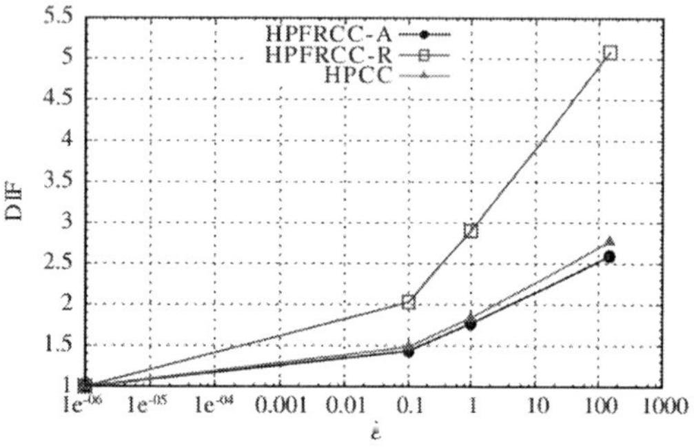

Figure 6. Dynamic increase factors (DIFs) versus strain rates.

extent the cast procedure, and the resulting random fibre distribution, can influence not only the static material behaviour but also the dynamic one. Considering high strain rate (150 s^{-1}) carried out on specimens HPFRCC-A (see Fig. 5) stresses drop down to a stress plateau after the peak strength. Because of fibre random distribution, in the HPFRCC-R samples the stress plateau has a value of about 10 MPa (30% less than the HPFRCC-A plateau) up to a crack opening displacement of about 4 mm.

In order to evaluate the peak strength trend in the dynamic field the dynamic increase factor (DIF) can be considered. The dynamic increase factors are listed in Table 4 and they have been computed as the ratio between the dynamic strength of the material ($\sigma_{Peak,d}$) and the static strength of the material ($\sigma_{Peak,s}$). In Fig. 6 DIFs versus strain rate are plotted for the HPFRCCs and the HPCC as well. The DIF parameter can be used to quantify the rate sensitivity of these materials, HPFRCC-A and HPC have showed a comparable trend in the strain rate range investigated. However the HPCC material denoted a higher values of DIF for all the strain rates. Moreover, HPFRCC-R exhibites high values of DIF comperate to what observed for the material characterised by aligned fibres (HPFRCC-A). The main reason is associated with the low peak strength recorded in the static regime which have led to an increase of the DIF values.

To clarify contribution of fibre and their dispetion on the peak strength increase, the ratio between HPFRCCs and the HPCC peak strengths ($\sigma_{Peak}^{HPFRCC}/\sigma_{Peak}^{HPCC}$) for all the strain rates investigated are plotted in Fig. 7. Several interesting considerations can be highlighted. In the static field aligned fibers (HPFRCC-A) strongly influence the stable crack propagation leading to an increase of the peak strength of about 40%. However, when random fibres dispersion and orientation (HPFRCC-R) are considered the stable crack propagation has not been favored, fibres acted as a defect in the matrix leading to a decay in the peak strength of about 30%. When the strain rate increases a reduction of the fibers influence can be observed. Passing

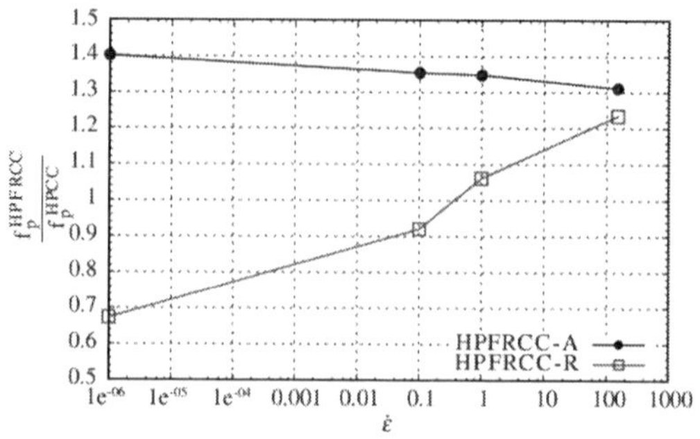

Figure 7. Peak strength ratio versus strain rate.

from 0.1 to 150 s^{-1} peak strength of HPFRCC-A material grows less (35–30%) with respect to the static one. On the contrary, HPFRCC-R exhibits a sensible increase of the peak strength ratio when strain rates increases. Hence, in the dynamic regime the influence of fibres on the peak strength is reduced as the reduction of gap between HPFRCC-A and HPFRCC-R peak ratio suggests (Fig. 7). When high strain rate (150 s^{-1}) is considered the peak ratio values are almost coincided. Fibres, specially inf randomly distributed and oriented, could be considered as defects in the matrix which become less important when strain rates are increased. A plausible physical explanation of this behaviour is that at higher strain-rates all particles of the material are accelerated under the action of the very fast growing loading pulse, resulting in a more uniform distribution of strain and strain energy. Consequently the influence of existing single defects and voids becomes less important, as the whole material volume works together to resist the applied load which is similar to the theory of multi-activation of the fracture process at impact regime [15].

5. Conclusions

The behaviour of HPFRCC and HPCC materials when subjected to different strain rates was investigated. From the experimental results presented in this research work some important conclusions can be drawn:

- the comparison between static and variable strain rate tests, carried out by means of three different mechanical devices, exhibits high values of the dynamic increase factor for the HPCC and the HPFRCCs characterised by aligned and random distributed fibres;
- the rate sensitivity of HPCC seems to be higher than what observed for HPFRCC-A with aligned fibres. In fact, HPCC material has denoted a weakly higher values of DIF for all the strain rates investigated. Moreover, HPFRCC-R showed the highest trend of DIF characterised by increments between the 37% and the 103%;
- aligned fibers increase the peak strength in static of about 40%, favoring the stable crack propagation. When the strain rate is increased, fiber influence on the stable crack propagation phenomena reduces: at high strain rates the peak strength ratio between HPFRCC-A and HPCC decreases to 30%;

- random fibres dispersion and orientation (HPFRCC-R) have not favored the stable crack propagation, fibres acted as defects in the matrix leading to a decay in the peak strength of about 30% with respect to what exhibited by HPCC;
- in the dynamic regime the influence of fibres on the peak strength is reduced as the reduction of gap between HPFRCC-A and HPFRCC-R peak ratio has suggested; when high strain rate (150 s^{-1}) is considered the peak ratio values are almost coincided.

References

[1] V. Bindiganavile, N. Banthia, Tech. rep., ACI 544 Technical Commission (2007)
[2] D.j. Kim, S. El-Tawil, A. Naaman, Materials and Structures **42**, 399 (2009)
[3] A. Caverzan, E. Cadoni, M. di Prisco, International Journal of Impact Engineering **45**, 28 (2012)
[4] A. Caverzan, E. Cadoni, M. di Prisco, Mechanics of Materials **59**, 87 (2013)
[5] N. Banthia, S. Mindess, A. Bentur, M. Pigeon, Experimental Mechanics **29**, 63 (1989)
[6] A. Bentur, S. Mindess, N. Banthia, Society for Experimental Mechanics pp. 449–458 (1987)
[7] C. Ross, *Review of Strain Rate Effects in Materials*, in *Structures under Extreme Loading Conditions* (1997), ASME Pressure Vessels and Piping Conference, pp. 255–262
[8] P.H. Bischoff, S.H. Perry, J. Engrg. Mech. **121**, 685 (1995)
[9] D. Kim, S. El-Tawil, A. Naaman, ACI Materials Journal **105**, 576 (2008)
[10] A. Caverzan, E. Cadoni, M. di Prisco, in *RILEM State of the Art Reports*, edited by G. Parra-Montesinos, H. Reinhardt, A. Naaman (Springer Netherlands, 2012), Vol. 2, pp. 339–346–, `http://dx.doi.org/10.1007/978-94-007-2436-5_41`
[11] A. Caverzan, M. Colombo, M. di Prisco, B. Rivolta, Materials and Structures **On line first**, 1 (2015)
[12] UNI 11039, *Concrete Reinforced with Steel Fibres. Part II: Test Method for the Determination of First Cracking Strength and Ductility Indexes* (2003)
[13] E. Cadoni, C. Albertini, G. Solomos, J. Phys. IV France **134**, 647 (2006)
[14] D. Asprone, E. Cadoni, A. Prota, G. Manfredi, International Journal of Rock Mechanics and Mining Sciences **46**, 514 (2009)
[15] E. Cadoni, G. Solomos, C. Albertini, Magazine of Concrete Research **61**, 221 (2009)

EPJ Web of Conferences **94**, 01065 (2015)
DOI: 10.1051/epjconf/20159401065

Tailored ramp wave generation in gas gun experiments

Matthew Cotton[1,a], David Chapman[2], Ron Winter[1], Ernie Harris[1], and Daniel Eakins[2]

[1] AWE Plc, Aldermaston, Reading RG7 4PR, UK
[2] Institute of Shock Physics, Imperial College London, Prince Consort Road, London SW7 2AZ, UK

Abstract. Gas guns are traditionally used as platforms to introduce a planar shock wave to a material using plate impact methods, generating states on the Hugoniot. The ability to deliver a ramp wave to a target during a gas gun experiment enables access to different regions of the equation-of-state surface, making it a valuable technique for characterising material behaviour. Previous techniques have relied on the use of multi-material impactors to generate a density gradient, which can be complex to manufacture. In this paper we describe the use of an additively manufactured steel component consisting of an array of tapered spikes which can deliver a ramp wave over 2 μs. The ability to tailor the input wave by varying the component design is discussed, an approach which makes use of the design freedom offered by additive manufacturing techniques to rapidly iterate the spike profile. Results from gas gun experiments are presented to evaluate the technique, and compared with 3D hydrodynamic simulations.

1. Introduction

It is well known that experiments to measure the changing profiles of ramp compression waves can provide useful data on the material through which the wave propagates (for example see [1,2]). By introducing a shockless compression to a material the temperature change is minimised, and the pressure states generated in the target are described more closely by the isentrope. Quasi-isentropic loading has traditionally been achieved through the use of magnetic or laser drive systems, which operate over very short (ns) timescales. Gas gun generated plate impacts can provide longer duration loading of materials of interest, and also benefit from the relative availability and cost of the experimental platforms. However, gas guns are better suited to introducing shock waves to a target due to the discontinuous change in pressure generated by high speed impact between two plates. A feasible method of generating a ramp wave under impact conditions is through the use of a flyer whose density is low at the leading face but increases with distance from that face. Previously this has been achieved through the use of a layered impactor consisting of multiple materials of varying density [3–5], using powder metallurgical or sedimentation techniques. These have produced good results but can be complex to manufacture, with layer thickness being a significant factor in the quality of the ramp profile. In this paper we explore the ability of a single material graded density flyer produced by an additive manufacturing technique known as selective laser melting (SLM) to generate ramp waves in a target, while minimising the initial shock rise.

A stainless steel flyer was designed which consisted of an array of tapered spikes attached to a solid base plate, referred to as the Bed-of-Nails (BoN) design. The tapered outer profile produces the continuous density gradation which is required to deliver a ramp wave to the target upon impact. Using a 3D CAD model of the BoN component the Sandia National Laboratories code CTH was used to model the wave delivered to a series of solid targets at a range of impact velocities. Based on this computational study experiments were designed with the objective of demonstrating the bed-of-nails concept for generating ramp waves and of determining the validity of CTH to compute the configuration.

2. BoN design

The choice of SLM as the manufacturing route for the BoN components was based on the ability of the process to produce complex geometries at small scales. The process involves depositing 40 μm layers of stainless steel powder onto a base plate, which is then selectively melted by application of a 200 W CW laser. The regions to be melted are specified by a 3D model of the desired component. When one layer is complete another powder layer is deposited until the structure is complete. In this case the BoN design consists of an array of tapered spikes of length 5.5 mm with a 1 mm^2 base, lying on a 2.5 mm thick base plate. A 3D model of the design is shown in Fig. 1.

For the initial design a parabolic outer profile was chosen for the spikes to deliver a linear variation in areal mass along the impact axis. The parabola which defines the outer surface can be expressed as:

$$z = 25 - 6x^2. \tag{1}$$

Subsequent investigations have explored the effects of varying the profile to change the mass distribution and thus the shape of the wave delivered to the target upon impact, the results of which are described later in this paper.

Upon receipt of the completed BoN specimens, a sample was sectioned and a series of micrographs taken to

[a] Corresponding author: matthew.cotton@awe.co.uk

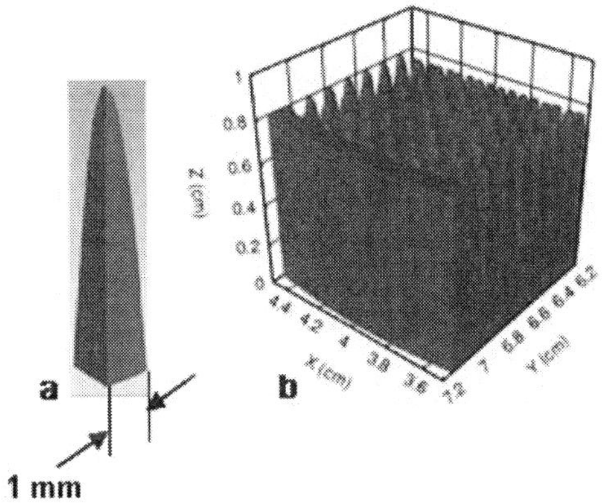

Figure 1. (a) Individual spike with parabolic outer profile. (b) a section of the BoN array with outer retaining ring.

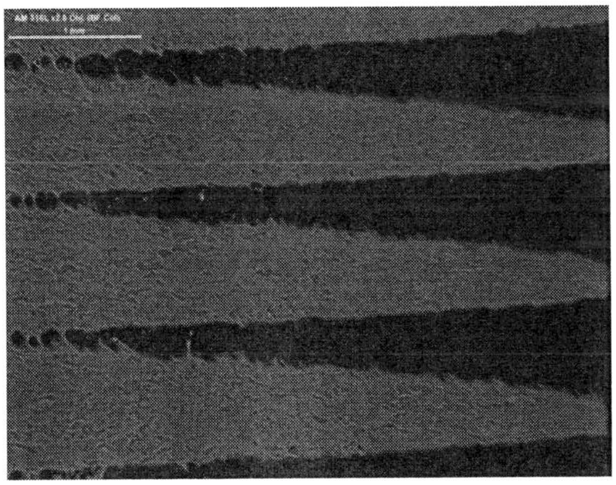

Figure 2. Cross-section of the spikes. The spikes join together above the nominal contact point due to the additional layer of powder particles.

establish how accurately the SLM process had reproduced the features in the CAD model. These images displayed some additional deposition of partially melted powder particles on the outer surface of the spikes which led to an increase in the overall mass of the BoN compared to nominal values. An example of this phenomenon is shown in Fig. 2.

Similar observations regarding surface finish have been made by several authors who have utilised SLM [6–8]. From these investigations it can be concluded that a number of factors including laser power, movement speed and scanning direction all contribute to characteristics of the completed part. Options exist for post-processing of the SLM part to minimise the effects of surface roughness, but in this case the BoN components were used in the as-received condition.

3. CTH simulations

A series of simulations of the experimental geometry have been conducted using CTH, a 3D Eulerian hydrocode [9]. This code was selected due to its ability to import the same 3D CAD models which were used to manufacture the

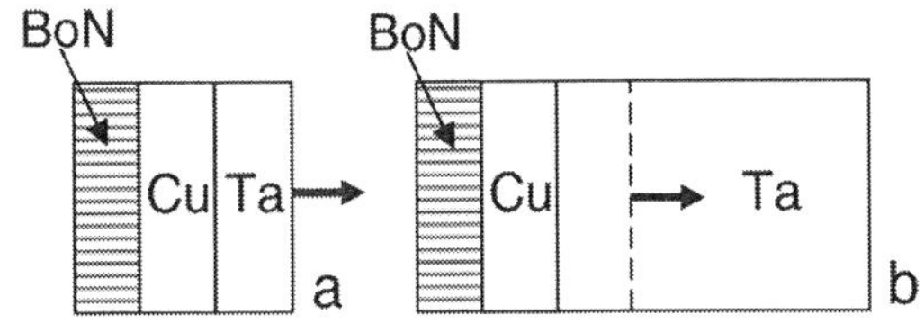

Figure 3. Calculational setup for (a) free surface and (b) in-situ wave profiles.

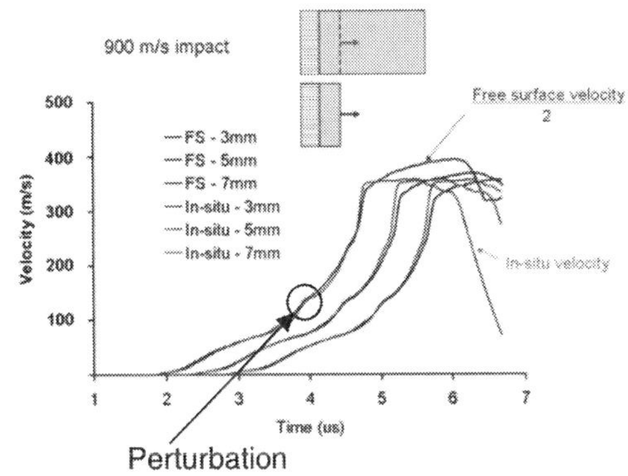

Figure 4. Calculated velocity profiles for both in-situ particle velocity and half the free surface velocity.

BoN components during the SLM process, as well as its suitability for modelling high strain rate impact processes.

In order to reduce the computational workload a reduced version of the BoN CAD model, consisting of a 3x3 array of spikes, was used for most simulations. It was established beforehand that the use of the reduced model did not introduce any significant variation in the results. For all simulations a 50 μm mesh was used, and a Johnson Cook strength model was assigned to the materials. As discussed above, some discrepancies were noted between the CAD model and the manufactured components. An updated model with an increased total mass was generated by adjusting the equation of hte parabola and used to make comparisons with the experimental data.

The calculational geometry is illustrated in Fig. 3. The BoN component impacts a 6 mm thick copper transmitter backed by a tantalum plate of varying thickness, at an impact velocity of 900 m/s. The data was output in the form of velocity-time profiles from the rear surface of the target (a) to allow it to be compared to the experimental results. In-situ velocities (b) were also output to estimate the error introduced by using a free surface approximation (in-situ velocity, u_p, is half the free surface velocity, u_{fs}) in our analysis.

A copper plate was placed between the BoN impactor and the tantalum target to smooth out any discontinuities in the input wave prior to it entering the target.

The results of these simulations are shown in Fig. 4.

The two profiles are near identical during the ramp phase, which gives confidence in the use of free surface velocity data in our subsequent analysis. An evaluation of the velocity discontinuities was based on the velocity plots from the 3 mm target, shown in Fig. 5.

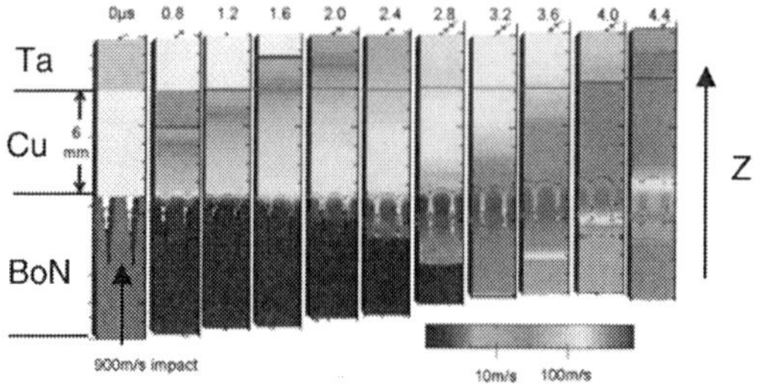

Figure 5. Velocity plots for a 900 m/s impact on to a copper/tantalum target.

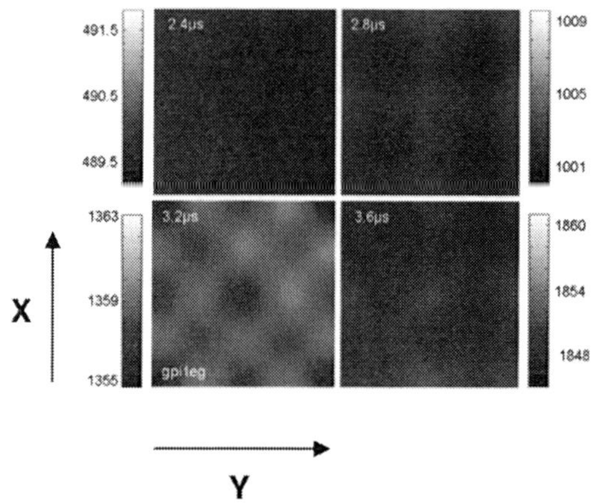

Figure 6. Velocity distribution across a 2 mm^2 region in the x-y plane at the rear of a 3 mm thick tantalum disc.

The results demonstrate that velocity perturbations along the z-axis are rapidly smoothed out as the wave propagates into the copper buffer plate. A small discontinuity appears to be introduced to the copper around 2.8 μs after impact, which may be responsible for the small discontinuity visible in the trace approximately half way along the ramp.

The effect of perturbations in the x-y plane (where z is the impact axis) were also assessed to determine if the non-uniform pressure distribution at the impact face propagated though the target to the proposed diagnostic recording location. Figure 6 shows the calculated velocity distribution at the rear surface of the tantalum disc as the wave is reflecting from the free surface.

These results indicate that the in-plane velocity varies by only ~0.5% from the mean velocity, giving confidence that any perturbations introduced by the spikes upon impact will have been damped as they pass through the target.

The results from these simulations provided confidence in the ability of the design to deliver a ramp wave, and experimental trials were conducted using this configuration.

4. Experimental data

The initial trial series consisted of two experiments conducted at nominal impact velocities of 900 and 1100 m/s. The target consisted of a copper transmitter plate with three tantalum discs of varying thickness mounted on the rear surface. The principal diagnostic used was heterodyne velocimetry (HetV) which recorded velocity-time profiles from the rear surface of the tantalum discs. Piezoelectric pins were also used to provide time of arrival information at the copper/tantalum interface. The principal features of the experimental assembly are illustrated in Fig. 7. The BoN impactor is shown mounted on a polycarbonate sabot in Fig. 8.

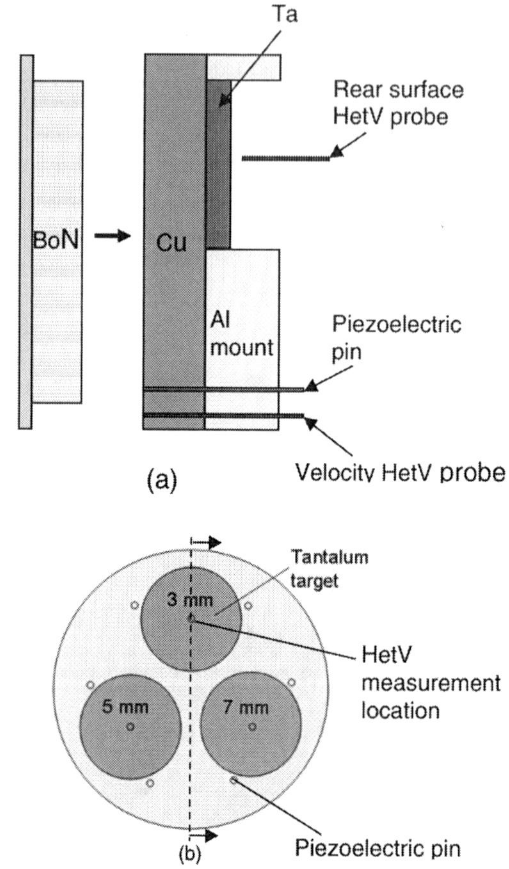

Figure 7. (a) Schematic section and (b) rear surface views of the experimental assembly.

The two experiments were fired at impact velocities of 900 and 1100 m/s. The higher velocity impact experiment supplied poor timing data, so the subsequent analysis shown here is confined to the lower velocity shot.

The measured velocity profiles from the three tantalum targets from the lower velocity shot are displayed in Fig. 9, alongside the CTH predictions.

The experimental data confirms that the BoN impactor has delivered a ramp wave to the target over a period of ~2.5 μs. An initial elastic wave can also be distinguished up to ~50 m/s, representing the isentropic elastic limit. The ramp wave transitions to a shock near to the peak velocity, apparent as a notable steepening of the gradient. For the purposes of ramp wave analysis, this transition marks the end point of the experiment. The arrival of

Figure 8. BoN impactor mounted on a polycarbonate sabot – note machined edge to provide a reflective surface for velocity measurement using HetV.

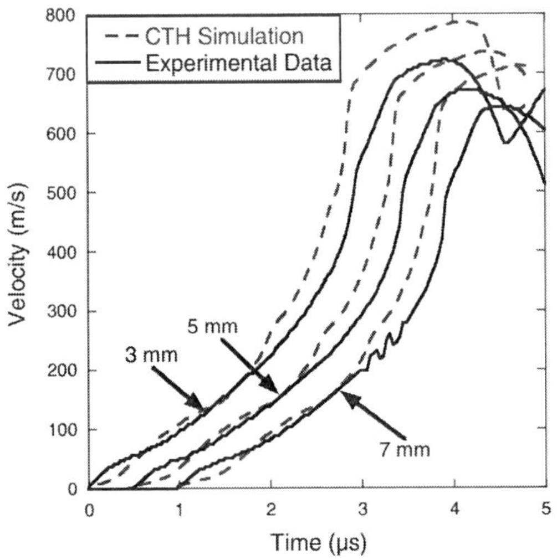

Figure 9. Experimental rear surface velocity data for each Ta target, and CTH simulations for comparison. 0 μs corresponds to wave arrival at the rear of the 3 mm Ta target.

a release wave from the rear of the flyer causes a further change in gradient corresponding to a reduction in acceleration. Peak velocity is observed to decrease as target thickness increases, an effect caused by the faster moving release wave overtaking the compression wave as it propagates further into the target. Similar features were observed in ramp loading data gathered using magnetic drive techniques by Asay [10].

The simulations agree with the experimental data reasonably well over the course of the ramp, providing increased confidence in our predictive capability using CTH. The main discrepancy is an increase in velocity at $\sim$2 μs after shock arrival at the rear surface which does not occur in the experimental data. This leads to a vertical

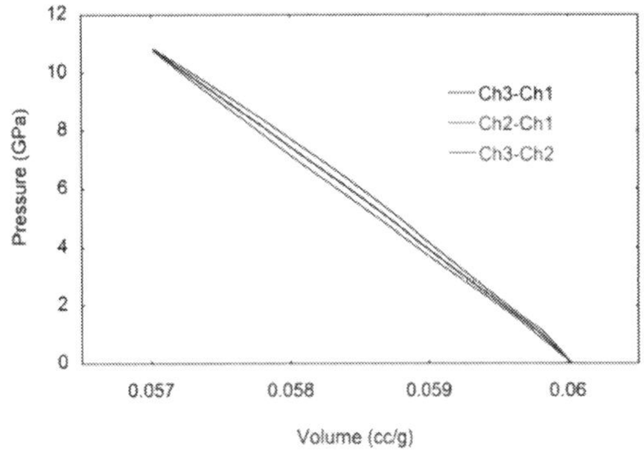

Figure 10. P-V plots of the data from the three pairs of HetV channels.

offset between the simulation and experiment for the remainder of the ramp. The time of arrival suggests that the velocity increase originates from a wave reflection at the BoN/copper interface caused by the density discontinuity. The fact that this feature does not appear in reality may indicate that the density change at the BoN/copper interface smoothes out post impact in a manner which is not captured in CTH.

Using the free surface approximation, discussed above, the HetV data can be analysed to extract stress-volume points. For the region of ramp wave compression these data points should lie on the material isentrope. The following equations are used [11]:

$$C_L = (z_2 - z_1)/(t_2(u_p) - t_1(u_p)) \tag{2}$$

$$dP = \rho_0 C_L du_p \tag{3}$$

$$dV = -V_0(du_p/C_L). \tag{4}$$

Where:

C_L – Lagrangian sound speed
z – target thickness
u_p – Particle velocity
$t(u_p)$ – time at which velocity u_p is reached
P – Pressure
V – Volume

Subscripts 1 and 2 refer to the two Ta targets which are being used in the analysis.

The isentrope can be calculated using velocity profiles at two different points on the z axis, therefore the three targets fielded in this experiment provide three routes to accomplish this. By comparing the isentropes produced by these routes the consistency of the data can be evaluated. This data is shown in Fig. 10.

Encouragingly the data does appear to be fairly consistent, and the small deviations that do occur are likely to result from timing errors which contribute to the majority of the $\sim$2–3% uncertainty in C_L.

5. Revised design

Following the initial series of experiments the design of the BoN was adjusted with the aim of improving the linearity

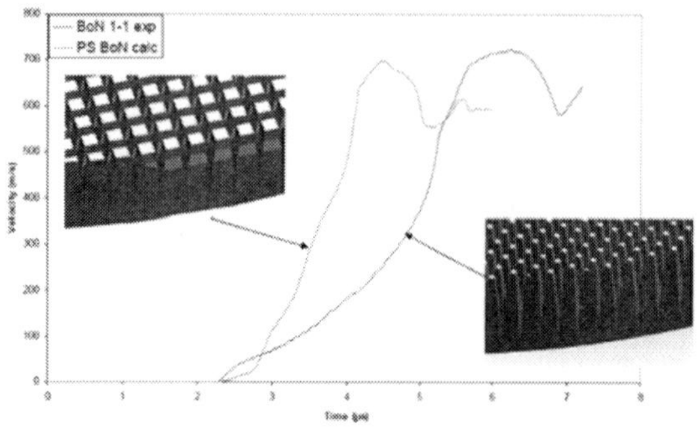

Figure 11. Comparison of BoN (right) and PSBoN (left) wave profiles.

of the ramp wave supplied by the technique. The objective was to introduce a wave which delivers a shorter ramp with a more linear increase in velocity. This adjustment is intended to reduce the effects of wave reflections from the rear of the flyer which can complicate the analysis. To achieve this the original BoN design went through a number of iterations which were simulated in CTH to assess the changes in the wave profile. The changes in mass distribution along the nail led to a more complex stepped profile being generated, referred to as the Parabolic Step Bed-of-Nails (PSBoN). A number of modifications were also made to the target design with the aim of minimising the errors in timing. Principal among these was the introduction of additional HetV probes to provide timing information at the copper/tantalum interface. A comparison of the original BoN wave profile with the predicted PSBoN profile is shown in Fig. 11.

This method enables the desired input wave to be specified based on the experimental requirements. As the technique matures it is expected to deliver a capability to generate more complex wave profiles based on a combination of shock, ramp and release stages. These options will be investigated further following the conclusion of the PSBoN experiments.

6. Conclusions

A novel technique for delivering ramp waves to a target in a gas gun plate impact experiment has been demonstrated. The BoN design delivered a smoothly ramped wave to the tantalum targets with minimal initial shock upon impact, and data quality is comparable to that produced by other gas gun graded density impactors. Subsequent analysis of free surface wave profiles has yielded data on the tantalum isentrope. A series of simulations of the experiments has also demonstrated that the CTH hydrocode can reproduce the principal features of the wave profiles, and can therefore be applied as a predictive tool in the design phase.

The ramp wave generation relies on the use of an additively manufactured component which utilises the design freedom offered by this processing route to make the BoN impactors. This technique also offers the opportunity to tailor the input wave profile by altering the design of the impactor, a method which has been demonstrated through the development of the PSBoN variant. Future work will explore the possibility of creating more complex wave profiles using the same design methodology.

The authors would like to thanks Neil Bond (AWE) for supplying the optical microscopy, Nick Gibbs (AWE) for assisting with the design of the BoN component, and Steve Johnson (Imperial College London) for experimental support. AWE and Imperial College London are acknowledged for their continued support of the Institute of Shock Physics.

References

[1] J-P. Davis, J. Appl. Phys. **99**, 103512 (2006).
[2] J.R. Asay, T.J. Vogler, T. Ao, J.L. Ding, J. Appl. Phys. **109**, 073507 (2011).
[3] J.H. Nguyen, D. Orlikowski, F.H. Streitz, J.A. Moriarty, N.C. Holmes, J. Appl. Phys. **100**, 023508 (2006).
[4] H. Xiong, L. Zhang, L. Chen, R. Yuan, T. Hirai, Metall. Mater. Trans. A **31**, 2369 (2000).
[5] L.P. Martin, J.R. Patterson, D. Orlikowski, J.H. Nguyen, J. Appl. Phys. **102**, 023507 (2007).
[6] G. Strano, L. Hao, R.M. Everson, K.E. Evans, J. Mat. Proc. Tech. **213**, 589 (2013).
[7] M. Smith, W.J. Cantwell, Z. Guan, S. Tsopanos, M.D. Theobald, G.N. Nurick, G.S. Langdon, J. Sand. Struct. Mat. **13**, 479 (2010).
[8] S. Van Bael, G. Kerckhofs, M. Moesen, G. Pyka, J. Schrooten, J.P. Kruth, Mat. Sci. Eng. A **528**, 7423 (2011).
[9] J.M. McGlaum, S.L. Thompson, M.G. Elrick, Int. J. Imp. Eng. **10**, 351 (1990).
[10] J.R. Asay, T. Ao, T.J. Vogler, J-P. Davis, G.T. Gray, J. Appl. Phys. **106**, 073515 (2009).
[11] S.D. Rothman J.R. Maw J. De Physique IV **134**, 745 (2006).

EPJ Web of Conferences **94**, 01066 (2015)
DOI: 10.1051/epjconf/20159401066

Dynamic high-temperature Kolsky tension bar techniques

Bo Song[1,a], Kevin Nelson[2,b], Ronald Lipinski[1,c], John Bignell[1,d], G.B. Ulrich[3,e], and E.P. George[3,f]

[1] Sandia National Laboratories, Albuquerque, NM 87185, USA
[2] Sandia National Laboratories, Livermore, CA 94550, USA
[3] Oak Ridge National Laboratory, Oak Ridge, TN 37831, USA

Abstract. Kolsky tension bar techniques were modified for dynamic high-temperature tensile characterization of thin-sheet alloys. An induction coil heater was used to heat the specimen while a cooling system was applied to keep the bars at room temperature during heating. A preload system was developed to generate a small pretension load in the bar system during heating in order to compensate for the effect of thermal expansion generated in the high-temperature tensile specimen. A laser system was applied to directly measure the displacements at both ends of the tensile specimen in order to calculate the strain in the specimen. A pair of high-sensitivity semiconductor strain gages was used to measure the weak transmitted force due to the low flow stress in the thin specimen at elevated temperatures. As an example, the high-temperature Kolsky tension bar was used to characterize a DOP-26 iridium alloy in high-strain-rate tension at $860\,s^{-1}/1030\,°C$.

1. Introduction

Kolsky bar, also called split Hopkinson bar, was originally developed for compression testing [1] and subsequently modified for use in tension and torsion testing [2,3]. A variety of Kolsky tensile bar techniques have been developed since the 1960s [4]. However, the Kolsky tensile bars have been mostly utilized for room-temperature testing. It has been very challenging to conduct high-temperature Kolsky tension bar experiments. In a Kolsky tension bar test, the specimen has to be firmly attached to the bar ends before dynamic loading. This makes it nearly impossible to heat the specimen individually, which has been executed for high-temperature Kolsky compression bar tests [5]. Su et al. [6] applied thermal-protective coating with a very low heat transfer coefficient to the bar surface to mitigate the heat transfer to the bars. Using the same method, the temperature in the bars was reduced to below 300 °C when the specimen was heated to 527 °C [7]. This means the heat was still transferred from the high-temperature specimen to the bars through the threads, which generated a thermal gradient in the bars. Such a thermal gradient may become more significant when the testing temperature increases. A significant thermal gradient will result in erroneous stress and strain measurements in the specimen, particularly when the test temperature is over 600 °C where the Young's modulus of steel, a typical bar material, decreases significantly. These challenges limit current Kolsky tension bar tests to temperatures below 600 °C [6–10], where the effects of temperature gradients on the steel bars can be neglected [9]. However, special experimental design considerations are required for Kolsky tension bar experiments at higher temperatures.

In this study, the conventional Kolsky tension bar was modified for dynamic high-temperature tensile characterization of thin-sheet alloys. As an example, a thin-sheet DOP-26 iridium alloy was characterized in tension at $\sim 860\,s^{-1}/1030\,°C$.

2. Modified high-temperature Kolsky tension bar system

In this study, the Kolsky tension bar described in [11] was modified for dynamic high-temperature characterization of thin-sheet alloys. Figure 1(a) shows a schematic of the modified high-temperature Kolsky tension bar system. Since it is not possible to directly thread a thin-sheet specimen into the bar ends, a pair of specimen fixtures was designed and shown in Fig. 1(b). This design used for a flat dog-bone shaped tensile specimen is similar to the one used in [12]. The fixture was machined with a slot with the same dimensions as the non-gage section of the specimen such that the whole non-gage section of the specimen was placed into the slot of the fixture. In this design, the specimen shoulder takes the load to pull the specimen. Since the area of the shoulder is much larger than that of the specimen gage section, this design minimizes the deformation in the non-gage section during dynamic loading. The specimen was then covered with a semicircular cap. The depth of the fixture was made the same as the specimen thickness such that the semicircular cap did not provide additional perpendicular force on the specimen but retained the specimen during dynamic

[a] e-mail: bsong@sandia.gov
[b] e-mail: knelso@sandia.gov
[c] e-mail: rjlipin@sandia.gov
[d] e-mail: jbignel@sandia.gov
[e] e-mail: ulrichgb@ornl.gov
[f] e-mail: easo.george@rub.de

Article available at http://www.epj-conferences.org or http://dx.doi.org/10.1051/epjconf/20159401066

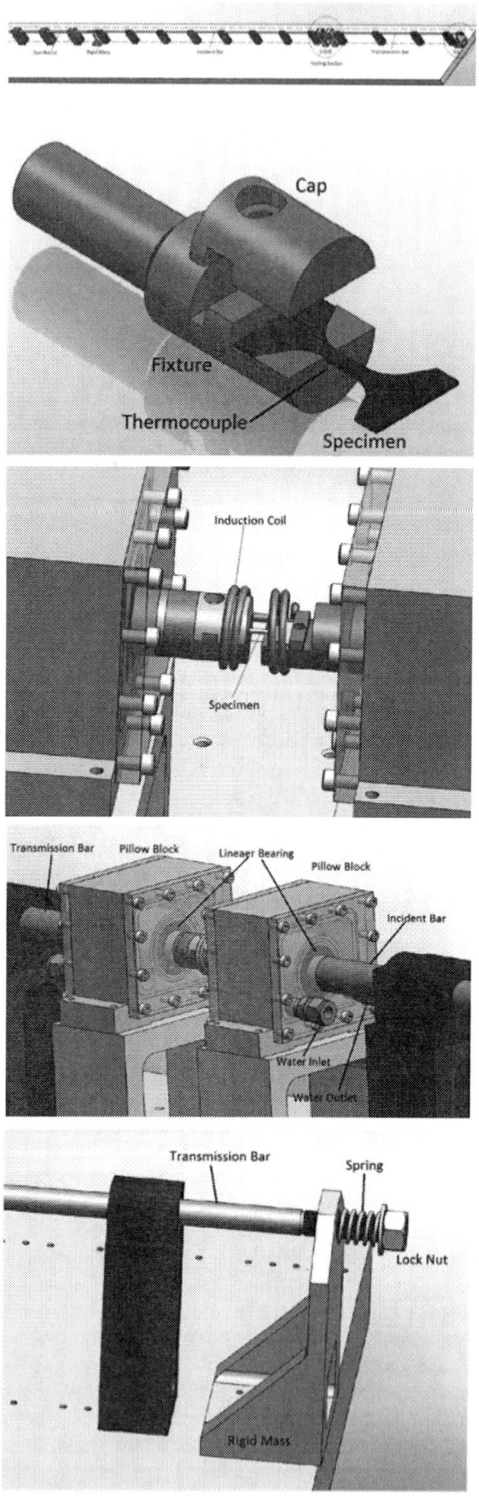

Figure 1. High-temperature Kolsky tension bar setup.

loading. Both the fixtures and the semicircular caps were made of Inconel 718 steel, which has relatively high strength at elevated temperatures.

An induction coil heater was installed on the testing section. Due to the small size of the specimen under investigation, the induction coil was set to heat the relatively large fixtures and then let the heat transfer to the specimen (Fig. 1(c)). This experimental design enables direct measurement of displacements at the specimen ends using a laser. When the fixtures are heated with the induction coil, the heat is transferred to both the specimen and the bars simultaneously. In order to prevent heating of the bars, a pair of hollow water-cooled pillow blocks was installed on the bar ends, as shown in Fig. 1(d). This design is similar in principle to that developed by Scapin et al. [13]. The difference is that Scapin et al. [13] applied a Cortex-tube-based air cooling system to cool down the bars for testing up to 400 °C. The water cooling system used in this study was shown to be capable of cooling the bars below room temperature when the testing temperature was as high as 1030 °C.

As shown in Fig. 1(c), the high-temperature fixtures attached to the bar ends still generate a thermal gradient between the bars and the specimen even though the bars are kept at room temperature, which may modify the wave propagation. In addition, the complicated design of the fixtures themselves exhibits many interfaces that may also modify the wave propagation. Both result in an unreliable reflected pulse for specimen strain calculation. A laser system that has been recently used for direct displacement measurement in Kolsky bar experiments [12,14] was implemented to the high-temperature Kolsky tension bar to directly measure the displacement histories at the specimen ends of the fixtures. The laser system used in this study was the same as that presented in [14] and demonstrated a bandwidth of 100 kHz or even higher, depending on the resolutions, which is sufficiently high for Kolsky bar experiments. The specimen strain can be calculated as

$$\epsilon = \frac{L_1 - L_2}{L_s} \tag{1}$$

where L_1 and L_2 are displacements of the specimen ends attached to the incident and transmission bars, respectively; L_s is the gage length of the specimen.

When the thin tensile specimen is heated to high temperatures, the specimen may become longer due to thermal expansion. However, the force generated by such a thermal expansion may not be sufficiently high to overcome the friction between the bars and the bar supports and push the incident and transmission bars back, which can consequently buckle the thin tensile specimen. The buckling generated in the specimen will produce an erroneous stress-strain response and is difficult to correct. A spring-loaded pre-tension system, as shown in Fig. 1(e), was developed to prevent the specimen from buckling during heating. The spring was placed between a rigid mass and a flange screwed onto the free end of the transmission bar. Screwing the flange toward the rigid mass compresses the spring and in turn generates a tension load in the tension bar system. Another rigid mass was placed against the gun barrel (Fig. 1(a)) to prevent the bar system from moving backwards when the whole bar system is pre-loaded in tension [15]. In this

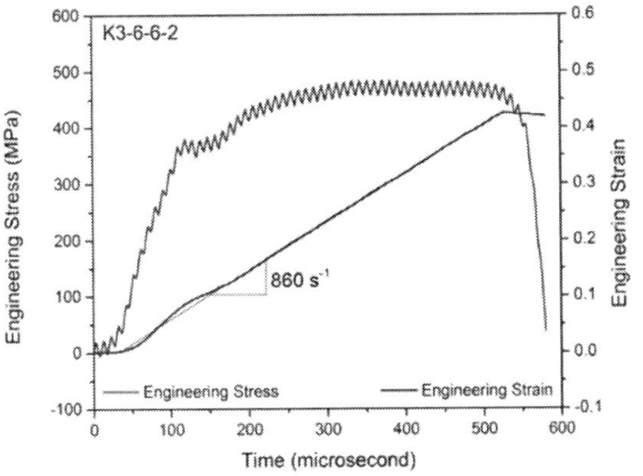

Figure 2. Engineering stress and strain histories at 860 s^{-1} and 1030 °C.

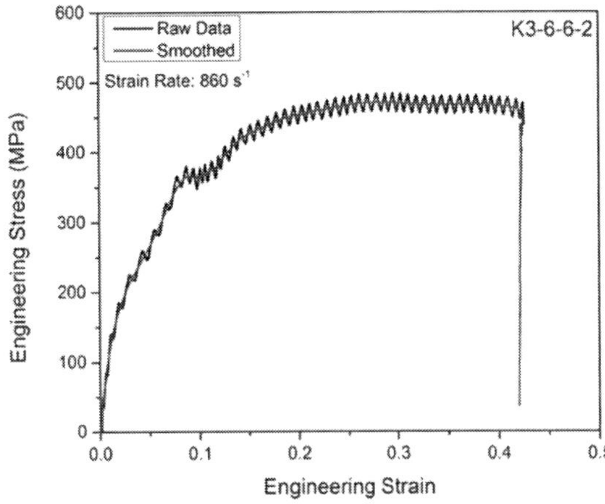

Figure 3. Engineering stress and strain curve at 860 s^{-1} and 1030 °C.

study, the spring was set to generate a pre-tension load of approximately 18 N which is sufficient to straighten the iridium specimen during heating but insufficient to produce further stretch on the iridium specimen.

Another high temperature testing issue is thermal softening of the specimen. In general, the flow stress in metallic materials decreases significantly at elevated temperatures, which results in a very weak transmitted signal in Kolsky bar experiments. Particularly when the cross-sectional area of the thin tensile specimen is much smaller than that of the pressure bars, the amplitude of the transmitted signal becomes further lower. In order to measure the weak transmitted signal with relatively high resolution, a pair of semiconductor strain gages was used to replace the regular resistor strain gages on the transmission bar. The semiconductor strain gages have a gage factor of 139, which is approximately 70 times more sensitive than the regular resistor strain gages. The specimen stress is calculated as

$$\sigma = \frac{E_0 A_0 \epsilon_t}{A_s} \quad (2)$$

where E_0 is Young's modulus of the bar material; A_0 is the cross-sectional area of the transmission bar; ϵ_t is the transmitted strain; and A_s is the cross-sectional area of the specimen. Combining the measurements of the semiconductor strain gages (Eq. (2)) and the laser system (Eq. (1)) for specimen stress and strain histories, respectively, yields the stress-strain curve of the specimen under investigation.

3. Dynamic high-temperature tensile characterization of iridium alloy

As an example, we employed the high-temperature Kolsky tensile bar to characterize a DOP-26 iridum alloy at 1030 °C. The tensile specimen that is shown in Fig. 1(b) had a thickness of 0.66 mm, a width of 2.54 mm in gage section, and a gage length of 6.35 mm.

We performed preliminary tests to attach three thermocouples to different locations in the gage section of the iridium specimen to check the temperature uniformity in the specimen during heating. The results showed that the temperatures were very consistent in these three locations during the heating process. Therefore, it is reasonable to consider the uniform temperature across the whole specimen gage section. In actual iridium alloy testing, only one thermocouple was attached to the fixture surface under the specimen, as shown in Fig. 1(b), to avoid spot welding the thermocouple to the surface and causing specimen microstructure changes.

The engineering stress and strain histories in the specimen, which were calculated with the transmitted and laser signals by using Eqs. (2) and (1), respectively, are shown in Fig. 2. The strain rate was then calculated with the slope of the strain history as a nearly constant 860 s^{-1}. It is noted that when the reflected pulse becomes unreliable, it is difficult to compare the force histories at both ends of the specimen for force/stress equilibrium check. In this study, we used high-rate high-temperature digital image correlation (DIC) approaches to qualitatively check the deformation uniformity across the whole specimen gage section. The DIC results showed that, under the same loading condition, the specimen with a gage length of 6.35 mm was deformed uniformly until necking occurred. However, due to the interaction between the high-intensity light for the DIC measurement and the laser system used for the direct strain measurement in the specimen, we were not able to apply DIC in each test. Based on the stress and strain histories shown in Fig. 2, the tensile stress-strain curve at 860 s^{-1} and 1030 °C was obtained and shown in Fig. 3. The stress-strain curves exhibit oscillations because of the effect of electromagnetic field generated by the induction coil on the highly-sensitive semiconductor strain gage signals. The raw stress-strain data was filtered to remove the oscillations, the result of which is also shown in Fig. 3.

The dynamic high-temperature stress-strain curve of the iridium alloy show different profiles than quasi-static curves [16]: there is an initial elasticity followed by significant work hardening behavior when the strain is below 10%. This phenomenon may be related to a change in microstructure or deformation mechanism at high strain rates and elevated temperatures, which is still under investigation. When the strain increases, the stress-strain response exhibits a typical plastic flow behavior.

4. Conclusion

The conventional direct-tension Kolsky bar was modified for dynamic high-temperature tensile characterization of thin sheet alloys. An induction coil was applied to heat the iridium specimen to elevated temperatures while the specimen ends of the incident and transmission bars were cooled to reduce the thermal gradient in the bars. A spring-loaded pretension system was installed on the free end of the transmission bar to prevent the high-temperature specimen from buckling during heating. A pair of semiconductor strain gages on the transmission bar was used to directly measure the force/stress in the specimen while a laser system was employed to calculate the specimen strain. Dynamic tensile stress-strain curves of the iridium alloy were obtained at $\sim 860\,\mathrm{s}^{-1}/1030\,^{\circ}\mathrm{C}$. The iridium alloy shows high ductility at elevated temperature and strain rate.

The authors would like to acknowledge the support of Dr. Helena Jin for the preliminary DIC work and Kevin Connelly for his initial specimen and fixture design support.

This work was sponsored by the United States Department of Energy (DOE) Office of Space and Defense Power Systems (NE-75). The authors gratefully acknowledge the support and guidance of Ryan D. Bechtel of the US Department of Energy.

Sandia National Laboratories is a multi-program laboratory managed and operated by Sandia Corporation, a wholly owned subsidiary of Lockheed Martin Corporation, for the U.S. Department of Energy's National Nuclear Security Administration under contract DE-AC04-94AL85000.

Oak Ridge National Laboratory is a multi-program research laboratory managed by UT-Battelle, LLC, for the US DOE under contract DE-AC05-00OR22725.

References

[1] Kolsky H, Proceeding of Royal Society of London **B62** (1949) 676–700

[2] Harding J, Wood EO, Campbell JD, Journal of Mechanical Engineering Science **2** (1960) 88–96

[3] Baker WE, Yew CH, Journal of Applied Mechanics **33** (1966) 917–923

[4] Chen W, Song B, *Split Hopkinson (Kolsky)Bar, Design, Testing and Applications*, (Springer, New York, 2011)

[5] Song B, Nelson K, Lipinski R, Bignell J, Ulrich G., George EP, Strain, (2014) 539–546

[6] Su J, Guo W, Meng W, Wang J, Mechanics of Materials **65** (2013) 76–87

[7] Guo W-G, Gao X, Materials Science and Engineering A **561** (2013) 468–476

[8] Feng, F, Huang S, Meng Z, Hu J, Lei Y, Zhou M, Wu D, Yang Z, Materials and Design **57** (2014) 10–20

[9] Whittington WR, Oppedal AL, Turnage S, Hammi Y, Rhee H, Allison PG, Crane CK, Horstemeyer MF, Materials Science and Engineering A **594** (2014) 82–88

[10] Vilamosa V, Clausen AH, Fagerholt E, Hopperstad OS, Bϕrvik T, Strain **50** (2014) 223–235

[11] Song B, Antoun BR, Connelly K, Korellis J, Lu W-Y, Measurement Science and Technology **22** (2011) 045704

[12] Guzman O, Frew DJ, Chen W, Measurement Science and Technology **22** (2011) 045703

[13] Scapin M, Peroni L, Fichera C, Materials at High Temperatures **31** (2014) 131–140

[14] Nie X, Song B, Loeffler CM, Journal of Dynamic Behavior of Materials, (2015), DOI: 10.1007/s40870-015-0005-7

[15] Song B, Lu W-Y, in *Proceedings of SEM XII International Congress and Exposition on Experimental and Applied Mechanics, Costa Mesa, CA 2012*

[16] Schneibel JH, Carmichael CA, George, EP, *Oak Ridge National Laboratory Report*, (2007) ORNL/TM-2007/81

EPJ Web of Conferences **94**, 01067 (2015)
DOI: 10.1051/epjconf/20159401067

Separation of the elastic waves' system in pre-stretched bar devices

Grégory Haugou[a], Nicolas Leconte, and Hervé Morvan

Laboratory of Industrial and Human Automation Control, Mechanical Engineering and Computer Sciences, UMR CNRS 8201, University of Valenciennes, Le Mont-Houy, France

Abstract. The loading mode generated by Kolsky bars apparatus is limited due to physical reasons to a duration time of around 0.4 ms. This is not suited in the characterization of high ductility metallic alloys up to fracture at moderate strain rates, since the duration time is too short. The pre-stretched bar device is a good candidate to characterize these materials up to the moderate strain rates. Unfortunately, the elastic waves are generally not separated in time at measurements points in the input bar, and post treatment leads to assumptions. A configuration of the pre-stretching conditions that ensures no superposition of the elastic waves' system developed along the input bar is thus proposed so as to fulfill standard data analysis requirements, i.e. without any assumptions. However, a drawback of the proposed device is that one of its hydraulic devices generates a loosen tail. This statement comes from a FE numerical model of the whole set-up including the generation of the loosen tail. Finally, an analytical correction to cancel the loosen tail is proposed and validated with respect to a numerical model featuring ideal pre-loading conditions.

1. Introduction

The loading generated by Kolsky bars apparatus – developed in the 50s' either for tension or compression loadings – is transient by nature since it is generated by the impact of a striker on a incident bar [1–3]. Since the length of the striker is 1 m at most due to technical reasons, the duration of the pulse generated by the metallic striker is generally limited to 0.4 ms, excepted for tension tests using a long U-shape projectile [4]. Indeed, the duration time depends on the ratio between the length of the striker and the wave speed of the elastic pulse generated just after impact. For instance, for a strain rate aimed at 100/s and for a duration time of 0.4 ms, the total strain cannot exceed 0.04. Therefore, it is generally admitted that Kolsky bars cannot be employed or adapted in order to characterize high ductility metals up to fracture at moderate strain rates.

The pre-stretched bar technique allows to load specimens with upper duration times than those generated by strikers. To the knowledge of the authors', the first device of this kind has been developed by Albertini [5]. Note that the ratio between the pre-stretched part and the input bar is 0.75 in this version, so as to maximize the duration time. This ratio became more or less a standard ratio to be used with this kind of set-up. The duration times of the pre-stretched technique are commonly found to be between 2 and 4 ms [2,5–9]. For instance, with a duration time of 1.2 ms, it is possible to reach a strain rate of 160/s for a total strain of 0.2. Thanks to these durations, it is possible to reach moderate rates of strain that can be divided at least by 3 when compared to Kolsky bars. However, a major drawback in the use of pre-stretched bars is that the incident pulse is cancelled by the arrival of the reflected one and cannot be obtained without approximations in the post processing [8]. In fact, it is due to a superposition of the elastic waves' system in the input bar. By the way, the signals can still be exploited by rebuilding the incident signal [2,5–9]. The rebuilt incident pulse is considered to be valid and representative of the expected one by some authors. The shape of the incident pulse is still subjected to discussions (Fig. 1).

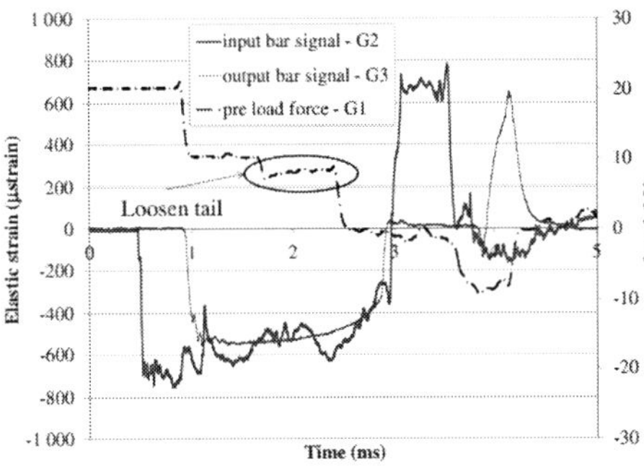

Figure 1. Typical recording of a tensile test using a pre-stretched bar with superposition of the elastic waves' system (20 MnCr5 rounded specimen at 100/s – pre-stretched length of 5556 mm).

To the authors' knowledge, none of the presented devices [2,4–9] allows avoiding both the superposition of the elastic waves' system and the existence of an intrinsic tail. The aim of the article is thus to propose a pre-stretched bar set up for the fast dynamics loading of high ductility metals, which does not feature a superposition of the elastic waves' system. In fact, the aim is to be able to benefit from the advantages of such loadings modes (long duration time) without the corresponding drawbacks (superposition of the elastic waves' system), i.e., being

[a] Corresponding author: `gregory.haugou@univ-valenciennes.fr`

Article available at http://www.epj-conferences.org or http://dx.doi.org/10.1051/epjconf/20159401067

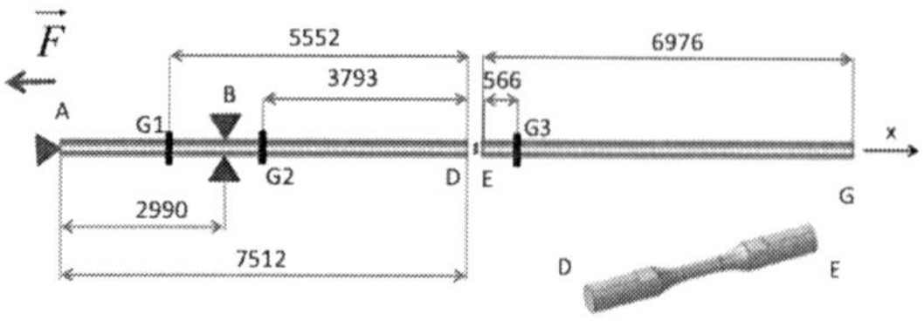

Figure 2. Pre-stretched tensile bar and sample – in mm.

Table 1. Properties of the input and output bars.

	C (m/s)	E (GPa)	ν
Input/output bars	4612	163.5	0.3

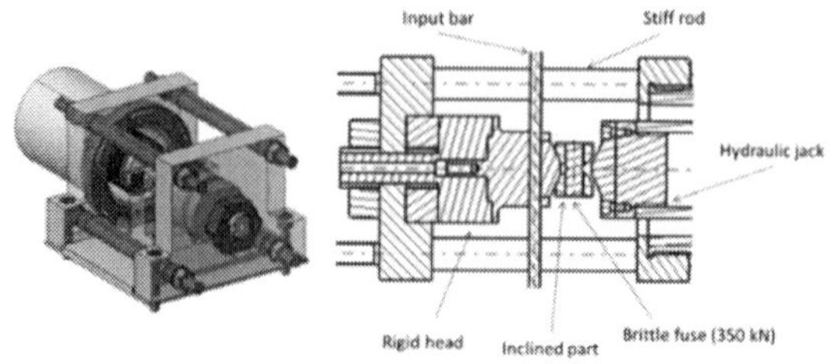

Figure 3. Locking mechanism placed in B – Details of the fuse system.

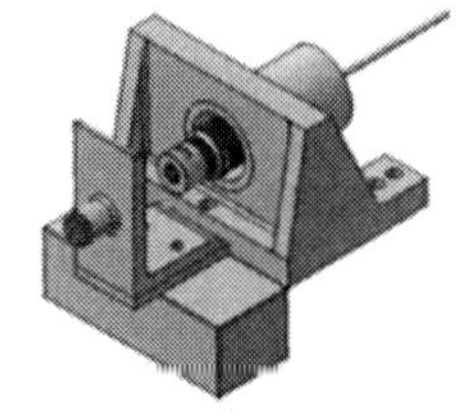

Figure 4. Pre-stretched bar mechanism placed in A.

able to post-process the waves in a standard way. It is foreseen that it can be achieved by suitably setting the ratio between the pre-stretched part and the length of the input bar. However, since the development of this kind device is not straightforward [10], the authors propose to couple the experiments together with numerical simulations in order to validate some assumptions, conclusions, and technical choices. Europlexus software [11] is used to perform the numerical simulations. It's based on a finite-element computer program for the analysis of fluid-structure systems subjected to transient dynamic loadings. Europlexus software is being jointly developed by the French Commissariat à l'Energie Atomique (CEA – Saclay) and by the Joint Research Centre of the European Commission (JRC – Ispra).

The article is organized as follows. The pre-stretched bar device proposed and its corresponding numerical model are presented in Sect. 2. In Sect. 3, the results obtained from the experiments are compared to the numerical model. Finally, a correction in the raw signals analysis provided by the set-up is proposed and validated with respect to the numerical model given in Sect. 4.

2. Pre-stretched device and its modeling

2.1. Loading principle

A sketch of the Split Hopkinson Tension Bar installed in 2006 at L.A.M.I.H. laboratory, France, is given in Fig. 2.

Here, the tensile testing scenario consists of partly pre-stretching (A to B, here 2990 mm in length) the input bar (A to D), which length is 7512 mm. The length of the output bar (E to G) is 6976 mm. Note that both bars were obtained from the same cast process (Table 1).

Both bars have an average diameter equal to 10.98 mm and are precisely aligned and guided on a rigid I-frame. The specimen is inserted between positions D and E using M7 threads. The pre-loading force F – stabilized between A and B – is generated by means of two quasi-static loadings applied on the input bar:

- A locking mechanism based on an hydraulic jack (Max. capacity: 600 kN) ensures the fastening of the input bar thanks to a brittle fuse, which geometric details are given in Fig. 3.
- After being fixed using the locking mechanism, the input bar is pre-stretched using another hydraulic device composed by a hydraulic hollow cylinder (Max. capacity: 300 kN), as illustrated in Fig. 4. The pre-load force is recorded on a full strain bridge bonded at 5552 mm from the specimen (position G1). The force value is limited to 45 kN so as to keep the pre-stretched part of the input bar in its elastic domain and to avoid any risk of sliding inside the locking system.

The combination of these two operations ensures the storage of an elastic energy that can be brutally released when the brittle fracture of the fuse occurs at the position B. Then, an incident elastic wave in tension develops with a rise-time of about 26 μs thanks to a specific design of the connectors mounted on the fuse (Fig. 3). The elastic wave propagates along the input bar with a constant duration time depending on the length of the pre-stretched part, and loads the specimen up to fracture. A set of two additional full strain bridges (HBM LY 41 350 Ω–length: 1.5 mm) are strategically positioned along the experimental set-up for the post-treatment of the waves (see Sect. 2.3):

- The position of gauge G2 (Fig. 2) is defined to detect both incident and reflected waves separately.
- The gauge G3 (Fig. 2) is placed to measure the transmitted signal.

Here, the position in B is chosen so that no superposition of the reflected wave with the incident one can occur at the position of the strain gauge bonded in G2. The strain rate value can be changed by varying the pre-loading force F of the pre-stretched device. Note that to evaluate the proposed configuration; rounded specimens made of 20MnCr5 material are tested in the next experiments.

2.2. Finite element modeling

This section details the modeling of an experiment. The whole set-up is modeled together with the specimen. Note that a specific care is used in the modeling of the set-up,

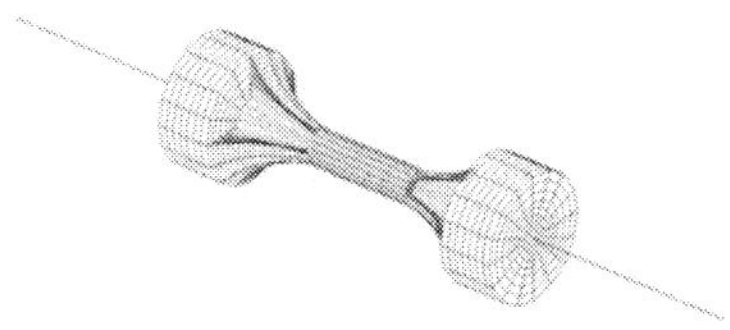

Figure 5. Mesh of the rounded specimen fixed between the cylindrical bars.

while a simple approach is employed for the modeling of the specimen, since the focus of this work is based on the development of a configuration where no superposition of the elastic waves occurs.

The experimental device is respectively modeled by 7512 (input bar) and 6976 (output bar) 1-D bar elements. The material behaviour of the bars is considered to be elastic (see Table 1). Note that the connections between 1-D bar elements (at points D and E, Fig. 2) and the faces of the specimen consisting of 3-D hexahedral elements are exactly imposed on the corresponding node velocities by means of Lagrange multipliers (Fig. 5) [12].

The boundary conditions are imposed on the set up as follows:

- an initial stress condition is imposed in the bar elements corresponding to the stretched part of the input bar,
- an unilateral constraint is imposed at the beginning of the input bar (point A) by means of pinballs (one attached to the ground and one massless attached to the input bar) in order to prevent the motion of the bar in the positive x-direction (Fig. 2).

Finally, the bar elements of the numerical model corresponding to the strain gauges bonded on the experimental set up at positions G1 to G3 are chosen as "numerical gauges" in order to ensure a similar post-treatment for both experiments and numerical simulations. The specimen consists of 9408 fully-integrated 3-D hexahedral finite elements (Fig. 5). Its material behaviour is modeled by a Ludwik law (where A = 750 MPa, B = 820 MPa and n = 0.525 at a strain rate of 300/s). The failure is simply predicted by means of element erosion when a threshold von Mises stress is reached.

2.3. Elastic waves' analysis

The strain gauges are commonly used to measure the elastic strain waves propagating along the bars (commonly called incident, reflected and transmitted pulses). Gauge G2 is used to detect incident and reflected pulses, while gauge G3 is used to measure the transmitted pulse. These measurements are required to calculate the current velocities and forces at interfaces, and stress/strain engineering relations based on the one-dimensional elastic wave theory [13, 14]. The incident elastic wave propagates along the input bar with a constant velocity and square shape. Its amplitude is function of the pre-stretched force value applied on the input bar. When the incident pulse reaches the specimen, it is partly reflected towards the input bar, and partly transmitted along the output bar after crossing throughout the specimen. The amplitudes of the incident, reflected, and transmitted pulses depend on the mechanical and geometrical properties of the specimen, so that the gauge section is calibrated to balance reflected and transmitted pulses. It is assumed that the specimen, which is short enough in comparison with the input and output bars, reaches a state of forces' equilibrium after several reflections of elasto-plastic waves inside its gauge length. The following equations are used to compute the global values as a function of time, and are based on the elastic pulses named $\varepsilon_{inc(t)}$, $\varepsilon_{ref(t)}$, $\varepsilon_{ref(t)}$ assuming that a separation of the elastic waves in time is ensured:

$$V_{OB}(t) = C_{OB} \cdot [\varepsilon_{tra}(t)] \tag{1}$$

$$V_{IB}(t) = C_{IB} \cdot [\varepsilon_{inc}(t) + \varepsilon_{ref}(t)] \tag{2}$$

where V_{OB}(t) and V_{IB}(t) are the output and input bars velocities associated to the wave's speeds C_{OB} and C_{IB}, respectively.

$$F_{OB}(t) = S_{OB} \cdot E_{OB} \cdot [\varepsilon_{tra}(t)] \tag{3}$$

$$F_{IB}(t) = S_{IB} \cdot E_{IB} \cdot [\varepsilon_{inc}(t) + [\varepsilon_{ref}(t)] \tag{4}$$

where F_{OB}(t) and F_{IB}(t) are output and input bars forces associated to the elastic moduli E_{OB} and E_{IB} and cross sections S_{OB} and S_{IB}, respectively. The average engineering stress and the engineering strain rate are presented in relations (5) and (6), respectively:

$$\sigma_{eng}(t) = \frac{E_{OB} \cdot S_{OB}}{S_0} \cdot [\varepsilon_{tra}(t)] \tag{5}$$

$$\dot{\varepsilon}_{eng}(t) = \frac{V_{IB}(t) - V_{OB}(t)}{L_0} \tag{6}$$

where S_0 and L_0 are respectively the initial cross section and length in the gauge zone. Plastic strains are determined as a function of time using a first order integration of strain rate relation (6).

The validity of the data analysis presented above is based on two hypotheses. First, the stresses and velocities are assumed to propagate through the bars without dispersion. If this first hypothesis is true, a linear relation between the diameter of the bar and the pulse amplitude is fulfilled. Secondly, the time required by the elastic wave to propagate through the specimen must be short compared to the total time of the test; in these conditions, many reflections inside the specimen are produced. Note that these equations are regardless used in the data analysis process of the experiments (Sect. 2.1) and of the finite element approach (Sect. 2.2).

3. Typical responses

3.1. Experiments

A typical recording of a tensile test performed with a preload value of 25 kN on a 20 MnCr5 rounded specimen (diameter: 3.5 mm – gauge length: 10 mm) is presented in Fig. 6. In these conditions, a rate of strain of $300/s$ is reached. In that case, the elastic waves' superposition is

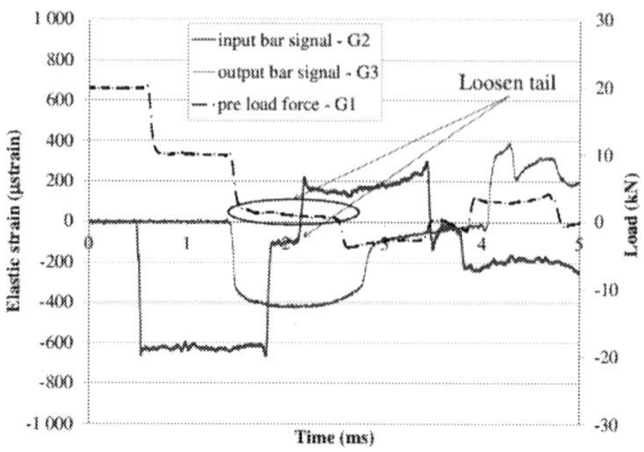

Figure 6. Typical recording of a tensile test using the pre-stretched bar without superposition of the elastic waves' system (20 MnCr5 rounded specimen tested at $300/s$ – pre-stretched length of 2990 mm).

avoided since the incident wave reveals a quasi-constant amplitude when compared to Fig. 1. These conditions are in accordance with the classical data analysis employed for Kolsky bars apparatus. The post-treatment presented in Sect. 2.3 can thus be used without any assumption. The main goal of this work has thus been achieved.

However, a drawback of the proposed set up is that it generates a loosen tail. It starts in the current experiment at around 1.5 ms and it is highlighted in the force signal recorded at gauge G1 by red dotted lines (Fig. 6). In fact, some observations using high speed imaging have revealed that this phenomenon may be generated when releasing the hydraulic system used to pre-stretch the input bar (Point A, Fig. 2). The loosen tail is directly detected in the gauge G1 used for the measurement of the pre load force applied on the input bar. As observed in the input signal of Fig. 6, the loosen tail is also detected at gauge G2 once the incident pulse decreases to zero, and the consequences on the amplitude of the reflected wave are not negligible. In the space-time diagram (not shown here for the sake of conciseness), the loosen tail and its effects on the elastic waves' system can be observed easily. In fact, it affects the shape of the reflected single and therefore the strain, strain rate and input force expressions given by relations 2, 4 and 6. Note that for a longer pre-stretched part, this problem is delayed due to the position of the point of measurement of the incident and reflected waves.

In classical configurations of Kolsky bars, a separation in time of the incident pulse with the reflected pulse is intrinsic. However, a tail can be generated between the incident and reflected pulses due to imperfections in the loading conditions or/and in the material properties of the bars [15]. Here, the authors have highlighted the existence of a typical tail for pre-stretched bar devices obtained under tensile testing conditions (Fig. 6).

3.2. FEM results

First, it is important to highlight that the model presented in Sect. 2.2 is ideal, i.e., it does not take into account any spurious effect that could be produced by the hydraulic device placed in A (Fig. 2). An additional boundary condition is thus introduced at this location in the numerical model to this end. A force is thus added in the

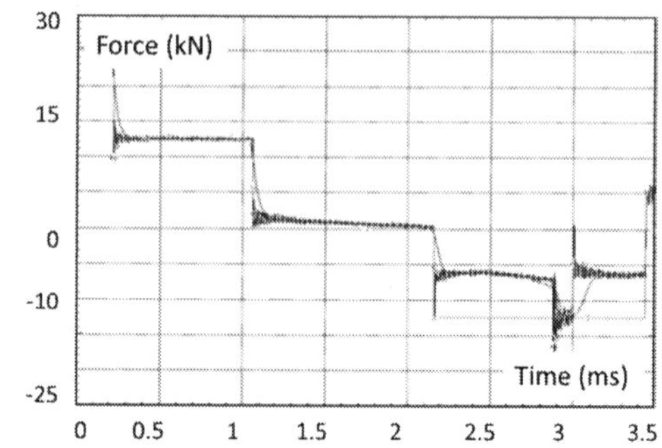

Figure 7a. Typical responses recorded at position of strain gauge G1 considering tail effects (elastic prediction in red, experiment in black, and computation in blue).

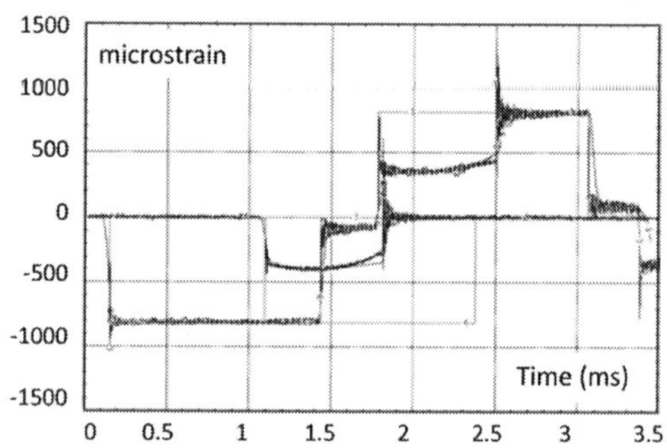

Figure 7b. Typical responses recorded at position of strain gauges G2 and G3 considering tail effects (elastic prediction in red, experiment in black, and computation in blue).

FE model at the hydraulic device location (in A). Note that the amplitude of the tail depends on the pre-load force value. Its time-expression for a pre-load force of 25 kN (see Sect. 3.1) needs to be defined in the numerical model. To do so, it is proposed to identify the loosen tail based on the force signal recorded at gauge G1 (Fig. 6). It is assumed that a second order polynomial is suited to model the loosen tail [16]. A least square approach is employed to identify its coefficients. On the one hand, the force-time expression of the loosen tail is fully identified at gauge G1, and on the other hand, the phenomenon generating it is supposed to be located at the beginning of the input bar (point A). It is thus necessary to shift the identified expression in time. The time shift is simply obtained from the wave speed in the input bar and the distance between the strain gauge G1 and point A. The results – summarized in Figs. 7a,b – compare theoretical elastic predictions, the experimental results, and the computations. A good agreement is observed between the computations and the signals recorded experimentally. It can thus be concluded that:

- Europlexus software provides accurate predictions of elastic waves' propagation principle.
- The birth place of the loosen tail is located at point A,
- The loosen tail is entirely generated by the hydraulic device located at point A.

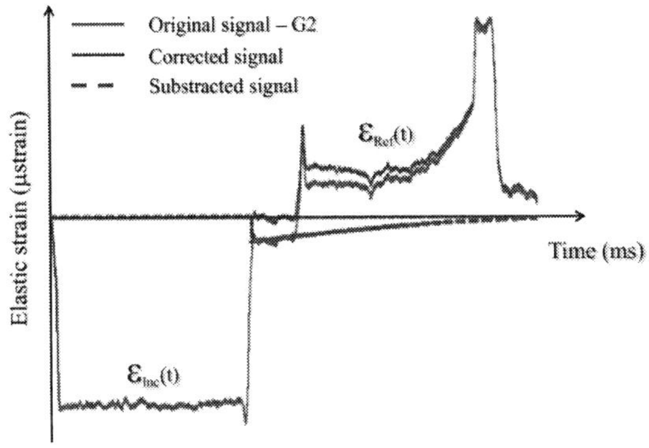

Figure 8. Tail correction effect on reflected pulse.

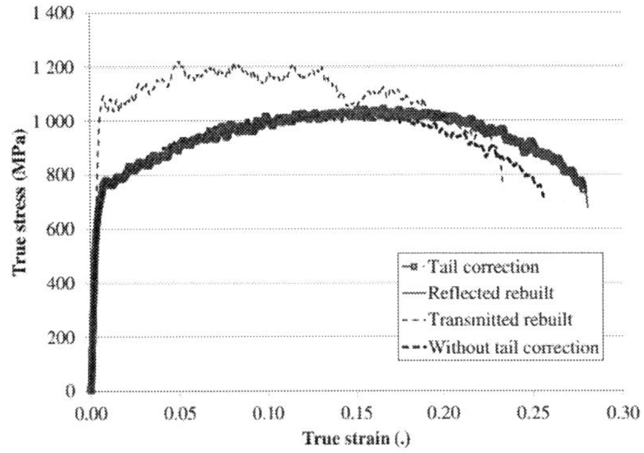

Figure 9. Validation of the tail correction process for the stress-strain curves calculation.

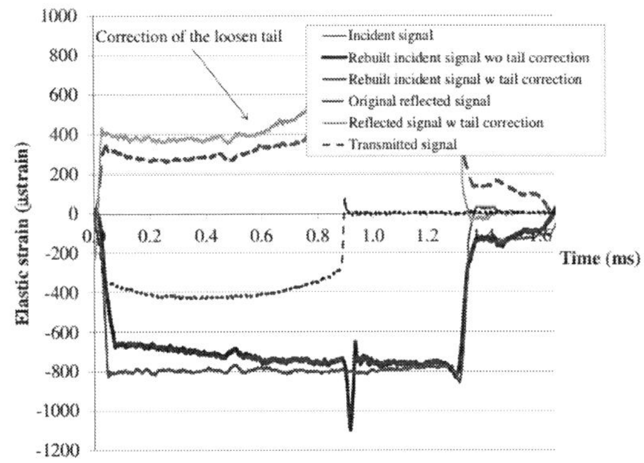

Figure 10. Comparison of the (un)corrected incident signals (see relation. 7).

– The methodology employed to generate the loosen tail at its birth place in the numerical model is correct.

4. Analytical correction of the loosen tail

4.1. Correction of the typical responses

A methodology to correct the tail identified previously (Sect. 3.2) that is similar to the one dedicated to the correction of the tail of Kolsky bars is proposed. In the typical loading conditions of the pre-stretched input bar, the reflected signal (and thus the material laws identified when post-processing these signals, see Sect. 2.3) is affected by the loosen tail. Thus, it must be corrected so as to fulfill the relation (7):

In Sect. 3.2 the profile of the loosen tail has been accurately identified at gauge G1. The tail recorded at gauge G2 is thus similar to the one identified at gauge G1 according to 1D elastic wave propagation theory, but is cut by the reflected wave. Consequently, the reflected wave is underestimated and the tail is used to correct the reflected signal according to the developments of Gary [16]. It is then simply subtracted on the raw signals (red curve in Fig. 8). Finally, the Eq. (7) must be checked after applying the correction.

Note that this correction is based on empirical observations and must be confirmed with respect to the achievement of forces' equilibrium assuming that the amplitude of the transmitted pulse is not modified by the tail. To do so, that, the authors propose to determine a typical strain-stress curve based on:

- the determination of the reflected wave from incident and transmitted signals,
- the determination of the transmitted signal from the incident and reflected signals,
- no tail correction is applied on the reflected signal as described in Fig. 8.

It is then clearly put into evidence the consequences of the loosen tail on a typical material response. In Fig. 10 are presented the results of the different means to calculate the strain-stress relation according relation given in (7).

In details, the loosen tail affects directly the reflected signal thus the strain rate and strain values (see Eq. (2) and (6)) and given in Fig. 10. Indeed, a correction of the reflected signal reveals being unavoidable. It is confirmed by the fact that the determination of the reflected signal based on the incident and transmitted signals (see relation. 7) matches well with the tail correction process proposed here. Finally, the transmitted signal is correct because any modification will conduce to important errors in the strain-stress relations.

4.2. FEM verification and final results

The tail correction process has been validated in the previous Section and it is now applied on the reflected signal of the experiments. The aim is to verify if the correction applied allows recovering an ideal configuration, i.e., a configuration in which no tail is generated. Note that the numerical model of the ideal configuration simply corresponds to a case in which the force F(t) defined in Sect. 3.2 is removed. The results – given in Figs. 10a,b confirm a good agreement between the computations and the signals recorded experimentally. It is thus verified that the ideal configuration is recovered thanks to the correction proposed. Thus, the waves' analysis can be performed in a standard way (see Sect. 2.3).

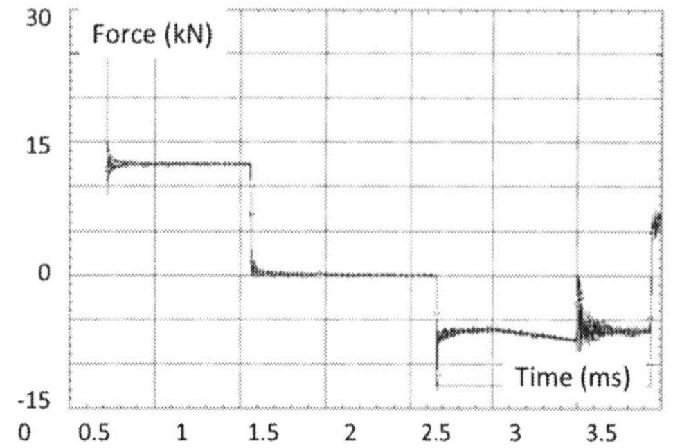

Figure 10a. Typical responses recorded at position of strain gauge G1 without tail effects (elastic prediction in red, experiment in black, and computation in blue).

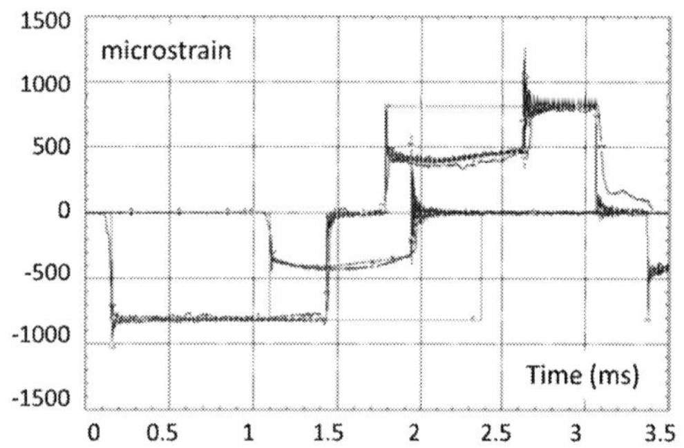

Figure 10b. Typical responses recorded at position of strain gauges G2 and G3. Experimental tail correction (elastic prediction in red, experiments in black, and computations in blue).

5. Conclusions

A principle based on a pre-stretched bar that ensures no superposition of the elastic waves' system has been proposed together with a particular technical solution. This sort of experimental set-up is interesting in particular for the characterization of high ductility metallic alloys up to fracture at moderate strain rates. However, the selected technical solution generates a loosen tail initiated at the end of the incident pulse which induce a disturbance of the reflected pulse. A correction that allows recovering an ideal configuration has thus been proposed and validated with respect to a numerical model developed using Europlexus software. Future works will focus on technological solutions avoiding the emergence of loosen tails. The numerical modeling approach of the device will be also used to investigate on other kinds of loadings.

References

[1] H. Kolsky, "Stress waves in solids", Dover Publications Inc., New York, 1963.

[2] U.S. Lindholm, L.M. Yeakley, "Some experiments with the Split Hopkinson Pressure Bar", Experimental Mechanics, Vol. **12**, pp 317–355, 1964.

[3] W. Chen, B. Song, "Split Hopkinson (Kolsky) bar. Design, testing and Applications." New York (USA): Springer; 2011.

[4] R. Gerlach, C. Kettenbeil, N. Petrinic, International Journal of Impact Engineering, Vol. **50**, pp 63–67, 2012.

[5] C. Albertini, M. Montagnani, Mechanical Properties at High Strain Rates, Vol. **21**, pp 22–32, 1974.

[6] Tore Børvik, Arild H. Clausen, Odd Sture Hopperstad, Magnus Langseth, International Journal of Impact Engineering, Vol. **30**, Issue 4, pp 367–384, 2004.

[7] E. Cadoni, M. Dotta, D. Forni, N. Tesio, C. Albertini, Materials and Design, Vol. **49**, pp 657–666, 2013.

[8] Y. Chen, A.H. Clausen, O.S. Hopperstad, M. Langseth, International Journal of Impact Engineering, Vol. **38**, Issue 10, pp 824–836, 2011.

[9] D. Saletti, S. Pattofatto, H. Zhao, Mechanics of Materials, Vol. **65**, pp 1–11, 2013.

[10] G. Riganti, E. Cadoni, Materials and Design, Vol. **57**, pp 156–167, 2014.

[11] `http://www-epx.cea.fr/`

[12] F. Casadei, N. Leconte. International Journal for Numerical Methods in Engineering, **86** (1), pp 1–17, 2011.

[13] J. Achenbach. (1973), "Wave propagation in elastic solids". North-Holland.

[14] K. Graff. (1975), "Wave motion in elastic solids". Ohio State University Press.

[15] H. Zhao, G. Gary, J.R. Klepaczko, International Journal of Impact Engineering, Vol. **19**, n° 4, pp 319–330, 1997.

[16] G. Gary, V. Degreef, DAVID, Users' manual version, Labview version, LMS Polytechnique, Palaiseau, France, version 12341.

EPJ Web of Conferences **94**, 01068 (2015)
DOI: 10.1051/epjconf/20159401068

Investigating strength of materials at very high strain rates using magnetically driven expanding cylinders

Zev Lovinger[1,2,a], Ron Nemirovsky[3], Eyal Avriel[1,2], Avraham Dorogoy[2], Yehezkel Ashuach[1], and Daniel Rittel[2]

[1] Rafael- Advanced defence systems, Haifa, Israel
[2] Faculty of Mechanical Engineering, Technion – Israel Institute of Technology
[3] Solid State Institute and Physics Department, Technion – Israel Institute of Technology

Abstract. Dynamic characterization of strength properties is done, in common practice by the means of a Split-Hopkinson Pressure Bar (also named Kolsky-Bar) apparatus. In such systems, strain rates are limited up to $\sim 5 \cdot 10^3\,\mathrm{sec}^{-1}$. For higher strain rates, the strain rate hardening is assumed to be the same as that measured at lower rates, with no direct measurement to validate the assumptions used for this extrapolation. In this work we are using a pulsed current generator (PCG) to create electro-magnetic (EM) driving forces on expanding cylinders. Most standard techniques for creating EM driving forces on cylinders or rings, as reported in the literature, reach strain rates of 1e3-1e4. Using our PCG, characterized by a fast rise time, we reach strain rates of ~1e5, thus paving the way to a standard technique to measure strength at very high strain rates. To establish the experimental technique, we conducted a numerical study of the expanding cylinder set up using 2D hydrodynamic simulations to reach the desired high strain rates.

1. Introduction

It is well known that strain rate is an important factor influencing the strength and failure properties of materials. For most materials, the dynamic behavior is significantly different from their behavior in the quasi-static regime. Furthermore, at high strain rates, new phenomena which were not relevant or with a weak influence at lower strain-rates can arise and change significantly the behavior of the material. In the work of Follansbee & Kocks [1], examining the flow stress dependence on strain rate in copper, it was shown that the flow stress increases significantly when strain rates rise above $\sim 5 \cdot 10^3\,\mathrm{sec}^{-1}$. Such behavior, evident in many materials, is related to a change in the dislocation mobility mechanism. In the low rate regime, dislocation glide is controlled by thermal activation, while at the high rate regime the dynamics are controlled by viscous effects such as the mechanism of phonon drag. Experimental techniques today reach strain rates of up to $10^9\,\mathrm{sec}^{-1}$ (e.g. loading by thin layer ablation using high energy lasers [2]), extending the behavior shown in [1] to higher strain rates with a general two slope behavior which meet a transition at some rate between 10^3–$10^4\,\mathrm{sec}^{-1}$. Dynamic characterization of strength properties is done, in common practice by the means of a Split-Hopkinson Pressure Bar (SHPB, also known as Kolsky-Bar) apparatus. In such systems, strain rates are of the order of $\sim 10^3\,\mathrm{sec}^{-1}$. Higher rates could be reached in plate impact experiments and in laser driven experiments with characteristic strain rates of $\sim 10^6$ and $\sim 10^9$, respectively. While plate impact tests are common, no agreeable method for measuring strength exists for those tests. Furthermore, in such experiments very high pressures evolve, affecting dramatically the strength of the material aside the high strain rates. The pressure dependence is usually the main parameter one seeks to investigate in such experiments [3,4]. In the laser experiments, strength is measured from the evolution of instabilities ([2]). Such experiments require a high-end facility making them to be very complicated and expensive to conduct. In addition, these experiments again, incorporate combined conditions of very high pressures and ultra-high strain rates. Without any experimental data taken at very high strain rates, modeling strength at this regime can only take into account the rate sensitivity measured at lower rates using SHPB and make the needed extrapolation to higher rates. In order to validate the model at high rates, usually Taylor tests are conducted. In a Taylor test, a short rod is accelerated into a rigid wall which causes the rod to deform plastically, flowing in a mushroom-like geometry. The impacted end is characterized by high strains and by strain-rates which reach values of about $10^5\,\mathrm{sec}^{-1}$. For model validation, the profile of the projectile is traced and compared with that obtained from the simulations (e.g. Ref. [5]).

For different applications there is a true need for a strength model which could represent properly the strain rate hardening of materials in hydrodynamic simulations. Such models must take into account a large span of strain rates since different parts of the problem can reach either low or very high strain rates, at least for a certain period of time. For example, simulations of an EFP (Explosively formed Projectile) formation show that different areas of the deforming liner are characterized by very large strains and strain rates which span between 10^3–$10^5\,\mathrm{sec}^{-1}$ [6].

[a] Corresponding author: cvloving@gmail.com

Article available at http://www.epj-conferences.org or http://dx.doi.org/10.1051/epjconf/20159401068

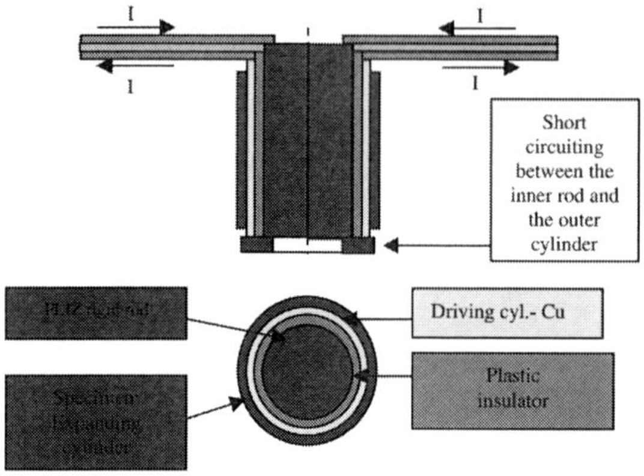

Figure 1. Schematic of the expanding cylinder test for strength measurements.

In this work, we seek to develop a simple and reliable experimental technique which will provide direct measurements of strength at strain rates above 10^4 in expanding cylinder experiments. The experiments are conducted with a pulse current generator (PCG) providing magnetic pressures as the driving force. Expanding cylinders were used in various works in the literature to measure strength [7] or fragmentation [8,9] at high rates, yet these experiments reach at most values of $10^4\,\text{sec}^{-1}$. Using our PCG, characterized by a fast rise time of about $1\,\mu$s [10], we manage to reach strain rates of ~1e5, with no significant temperature rise, thus paving the way to a standard technique to measure strength at very high strain rates.

2. Experimental set-up

The pulsed current generator uses a capacitor bank system charging 20–30 kVolts which are released using high voltage dischargers allowing fast and simultaneous discharge into the conductor system, achieving current flows of about 1 MA. The PCG was further detailed in [10]. The experimental set-up for expanding cylinders using the PCG is shown in Fig. 1.

The conductor system is made of two plates, each attached to an opposite pole, separated by an isolating layer. The specimen is an assembly of coaxial cylinders: an inner rigid rod is attached to the upper plate, the outer cylinder is attached to the bottom one and an isolating plastic cylinder is put in between them. The tested specimen is driven by the expanding outer cylinder. On the bottom side of the specimen, the cylinders, each attached to an opposite pole, are short-circuited. When the voltage is discharged, an electrical current flows from one pole, through the upper plate and the inner rod, to the outer cylinder and to the bottom plate. As a result, a high current flows through the rod and outer cylinder, but in *opposite* directions, creating repulsive magnetic forces between them, causing the expansion of the outer cylinders. The main diagnostics is velocity interferometry measurements of the expanding specimen. The strength is calibrated and detrmined as an *inverse problem*, through hydrodynamic simulations, using different strength models. A main

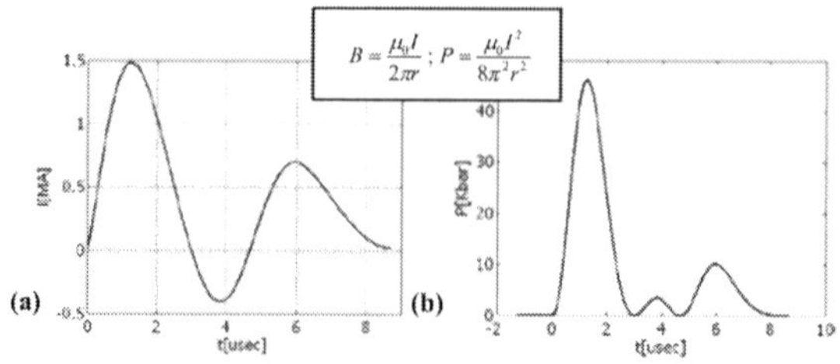

Figure 2. Typical histories derived from B-dot measurements on the PCG (from [10]): (a) measured current-flow history, (b) calculated boundary pressure.

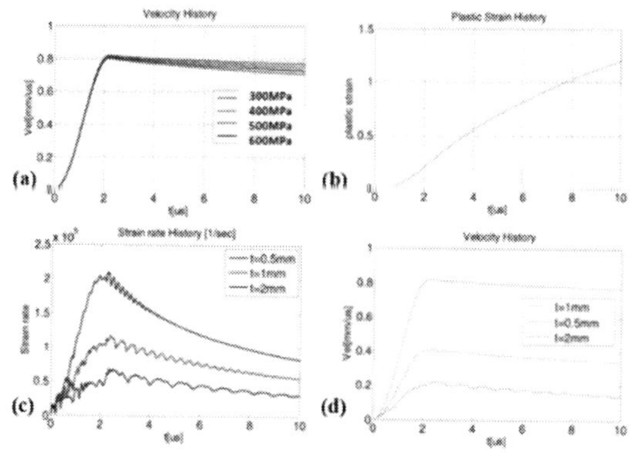

Figure 3. Simulation results of an expanding cylinder: (a) velocity history with different values of strength (b) plastic strain history, (c) vtrain rate and (d) velocity histories as function of the cylinder wall thickness.

demand for such calibration is to provide accurate loading conditions in the experiment. The boundary magnetic pressure is extracted from current flow measurements. The current flow is calculated from Bdot measurements and calibrated under the assumption of an ideal RLC circuit.

A typical Bdot signal and calculated pressure boundary condition in such a test are shown in Fig. 2. Typical pressures are 20–40 kbars with a rise time of $\sim 1\,\mu$s.

3. Initial results

To design the experimental set-up, we conducted numerical simulations, specifically, we calculated the particle velocity sensitivity to changes in the strength of the material and the strain rates that are expected to evolve.

We carried out 2D simulations of the expanding cylinder with axial symmetry. Cylinder dimensions and pressure boundary conditions were chosen to be the same as those of the collapsing cylinder experiments we conducted on the PCG [10]. The cylinder is made from copper and has an inner diameter of 7 mm with a thickness of 0.5 mm. We used a simple EPP strength model with strengths of 300-400-500 and 600 MPa. The pressure boundary condition is a half sine function with a duration of $2.2\,\mu$s and a peak pressure of 2.5 GPa. The results are shown in Figs. 3a–3c. The strain rates achieved are $(1–2)\cdot 10^5\,\text{sec}^{-1}$. The particle velocity sensitivity to strength changes is $\sim 10\,\text{m/s}$ per 100 MPa. From former measurements done with our interferometry system, 10 m/s is estimated to be the limit resolution, thus determining that *our PDV measurements is sensitive to changes of 100 MPa in strength*.

The expected velocities using the above mentioned boundary conditions are $\sim 800\,\text{m/s}$. Furthermore, we examined the strain rates and velocities as function of the wall thickness as shown in Figs. 3c,d, showing that we reach strain rates above 1e5 sec^{-1} with cylinder thicknesses under 1 mm.

4. Summary

In this work we present an experimental set up of an expanding cylinder, using a pulsed current generator to produce the magnetic driving forces. This new set-up enables to measure strength at very high strain rates, above 1e5 sec^{-1}, where a significant rise in the material strength is expected. Presenting the geometry of our initial specimen, we demonstrated by numerical simulations, that we manage to reach strain rates of ~1e5, with no significant temperature rise, thus paving the way to a standard technique to measure strength at very high strain rates. Initial tests are currently conducted on copper specimens.

The financial support of the Pazi Foundation is kindly acknowledged.

References

[1] P. S. Follansbee, U. F. Kocks, Acta Metall, Vol **36**, pp. 81–93 (1988).
[2] H. S. Park et al., Physics of Plasmas. Vol **17**, No. 5 (2010).
[3] J. Lipkin, J. Asay, J. Appl. Phys., Vol **48**, No. 1 (1977).
[4] J. C. F. Millet et al., J. Appl. Phys, Vol **101**, 033520 (2007).
[5] F. J. Zerilli, R. W. Armstrong, J. Appl, Phys. Vol **61**, pp. 1816 (1987).
[6] H. Couque, R. Boulanger, Int. Sym. Ballistics, Spain (2007).
[7] J. Petit et al. J. Phys IV, France (1997).
[8] D. E. Grady, D. A. Benson, Exp. Mech., Vol **23**, pp. 393 (1983).
[9] H. Zhang, K. Ravi-Chandar, Int. J. Fracture, Vol **163**, pp. 41 (2010).
[10] Z. Lovinger et al., IJIE, Vol **38**, 918–929 (2011).

EPJ Web of Conferences **94**, 01069 (2015)
DOI: 10.1051/epjconf/20159401069

Strain rate effects on tensile strength of iron green bodies

Masahiro Nishida[1,a], Yuki Kuroyanagi[1], Hans-Åke Häggblad[2], Pär Jonsén[2], and Gustaf Gustafsson[2]

[1] Nagoya Institute of Technology, Gokiso-cho, Showa-ku, Aichi 466-8555, Japan
[2] Luleå University of Technology, Division of Mechanics of Solid Materials, SE-97187 Luleå, Sweden

Abstract. Impact tensile strength of iron green bodies with densities of 7.2 and 7.4 g/cm^3 was examined by Brazilian test using the split-Hopkinson pressure bar (Kolsky bar) method. The powder material used for the experiments was a press-ready premix containing Distaloy AE, graphite, and lubricant. During dynamic compression, the failure behavior of specimens was observed using a high-speed video camera. The failure stress and failure behavior of dynamic compressive tests were compared with those of static compressive tests.

1. Introduction

Powder metallurgy is the serial process of blending fine powders (mixture), compressing them into a desired shape (compacting), and then heating the compressed material (sintering). This results in great advantages such as product cost effectiveness and product uniqueness. In general, parts created by powder metallurgy have low strength because they have relatively low density and high porosity. The production of high-strength parts having complex shapes is required. In order to increase both strength and density, new techniques such as high velocity compaction (HVC) were developed and further investigation has been conducted on the improvement of techniques and optimum conditions using computer simulation. The authors' group has been examining the strain rate sensitivity of uniaxial failure strength and determination of material parameter of unsintered compacted powder objects made of iron (iron-based powder compacts), so-called "green bodies" in the field of powder metallurgy, at high strain rates [1].

In this study, impact tensile strength of iron green bodies was measured by Brazilian test using the split-Hopkinson pressure bar (Kolsky bar) method, so-called impact splitting test. The powder material used for the experiments was a press-ready premix containing Distaloy AE, graphite (uf-4), and lubricant (Kenolube). Iron green bodies with densities of 7.2 and 7.4 g/cm^3 were employed. During dynamic compression, the failure behavior of specimens was also observed using a high-speed video camera. The failure stress and failure behavior of dynamic compressive tests were compared with those of static compressive tests.

2. Experimental method

In order to measure tensile strength of brittle materials at high strain rates, the spalling tests of long bar specimens (for example, ceramic [2–4], glass [5], concrete [6–8], and so on) and the Brazilian tests using the split-Hopkinson pressure bar (for example, ceramic [9] and concrete [10–15] before or during 2001) are often used. Because it is difficult to make long green bodies (long specimens) with a homogeneous density for specimens of the spalling tests, we employed the Brazilian tests using the split-Hopkinson pressure bar for measuring tensile strength of iron green bodies, even though specimens during experiments are under the biaxial stress state.

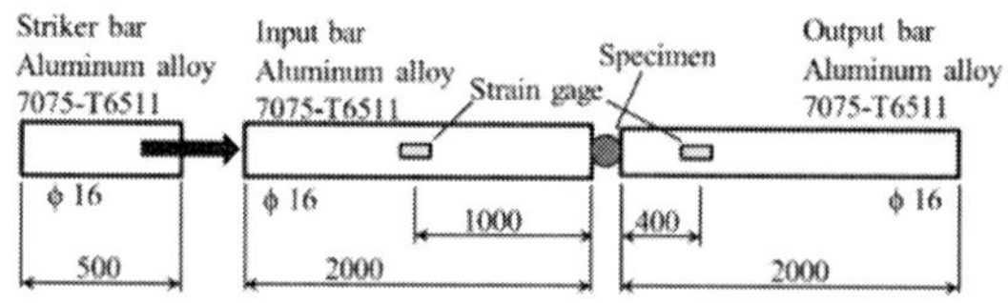

(a) Size of striker, input bar, and output bar

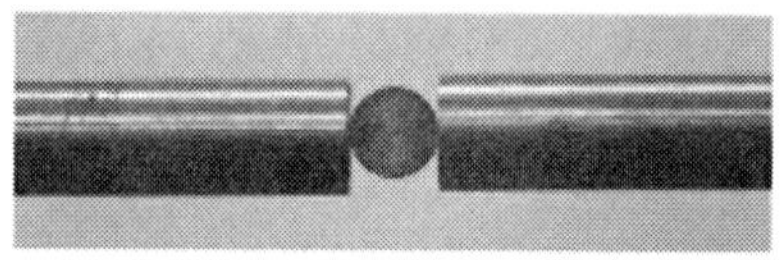

(b) Photograph near specimen

Figure 1. Experimental set up for Brazilian tests using split-Hopkinson pressure bar.

Figure 1 shows a split-Hopkinson bar test as the Brazilian test. The diameters and lengths of input and output bars were 16 mm and 2000 mm, respectively. Strain gages were applied to both sides of the input and output bars at distances of 1000 mm and 400 mm from the disc specimen, respectively. The input and output bars were made of aluminum alloy (7075-T6511). The applied force F and displacement x of the specimens were calculated from the strains of the bars measured by strain gauges, based on the following equations [16,17]:

$$F_{av}(t) = A\ E\ \varepsilon_T(t) \qquad (1)$$

$$x_{av}(t) = c_3 \int_0^t [\varepsilon_I(t) - \varepsilon_R(t) - \varepsilon_T(t)]dt. \qquad (2)$$

[a] Corresponding author: nishida.masahiro@nitech.ac.jp

Article available at http://www.epj-conferences.org or http://dx.doi.org/10.1051/epjconf/20159401069

Table 1. Material constants of input and output bars used in calculation.

Density	Elastic wave velocity in bar	Young's modulus E
$2.77 \times 10^3\,kg/m^3$	5150 m/s	73.6 GPa

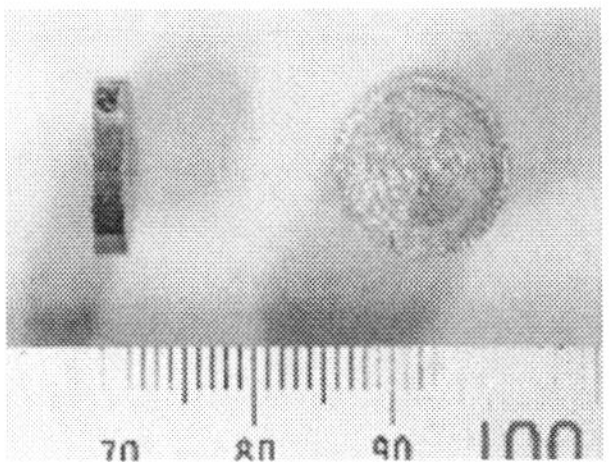

Figure 2. Photograph of disc specimen.

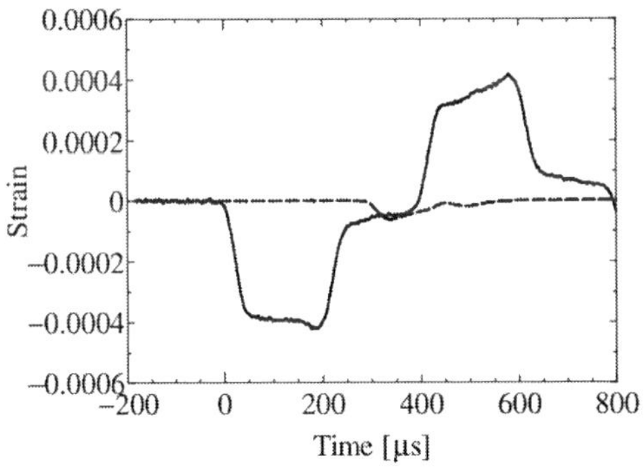

Figure 3. Strain histories of strain gauges for specimens with $7.2\,g/cm^3$ (Solid line: input bar; dashed line: output bar).

Here, ε_I, ε_R, and ε_T denote the axial strains induced in the input bar by the incident wave and reflection wave, and in the output bar by the transmitted wave, respectively. E and c_3 respectively denote the Young's modulus and elastic wave velocity of the input and output bars. A denotes the cross-sectional areas of the input/output bars. The material constants of the aluminum alloy (7075-T6511) bars used in the calculations are listed in Table 1. Aluminum alloy (7075-T6511) strikers with a diameter of 16 mm and lengths of 500 mm were used. During dynamic compression, we observed failure behavior of specimens using a high-speed video camera (Photron FASTCAM-1024PCI 100 K, frame rate of 18000 frames/sec).

The powder materials used in the experiments were a press-ready premix consisting of Distaloy AE, 0.5% graphite (uf-4), and 0.6% Kenolube. Distaloy AE is a pre-alloyed water atomized iron powder from Höganäs AB, Sweden. Kenolube is a lubricant from the same supplier. The theoretical pore-free density of this mix is $7.52\,g/cm^3$. The behavior in the compression state of powder mixes was characterized by static and dynamic tests. The static tests were performed in a universal testing machine.

Cylindrical disc specimens were manufactured with a height of 2.5 mm and a diameter of 12.5 mm with two different densities, 7.2 and $7.4\,g/cm^3$. A photograph of specimens is shown in Fig. 2. The specimens were manufactured using a universal testing machine (DARTEC RK 250).

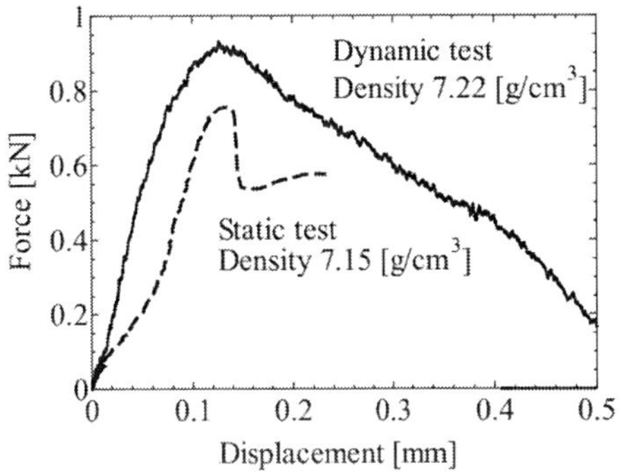

Figure 4. Force-displacement curves of specimen with $7.2\,g/cm^3$ (Compression speed of dynamic test 3.6 m/s, static test 0.002 mm/s).

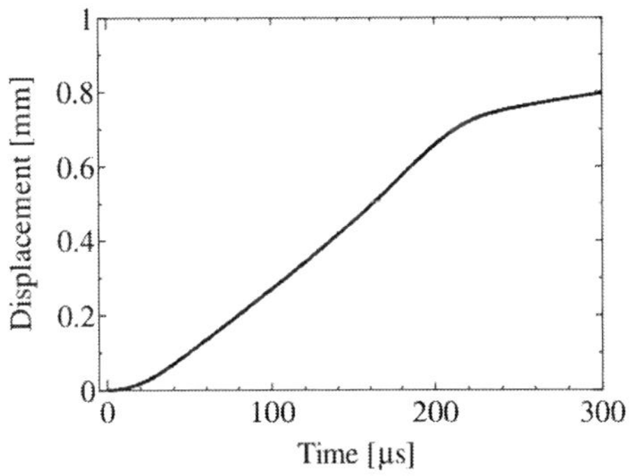

Figure 5. Displacement history of specimen with $7.2\,g/cm^3$ at high strain rate.

3. Results and discussion

Figure 3 shows strain histories of strain gauges using a specimen with a density of $7.2\,g/cm^3$. It was found from the reflection wave that displacement rate is almost constant. Figure 4 shows force-displacement curve for the specimen with a density of $7.2\,g/cm^3$ at a compression speed of 3.6 m/s. Compared with a static test result at a compression speed of 0.002 mm/s, the maximum stress (failure stress) increased with compression speed. Here, the compression speed was determined by a slope of displacement history in Fig. 5. The slope was almost constant. Figure 6 shows a force-displacement curve of the specimen with a density of $7.4\,g/cm^3$. In the case of density of $7.4\,g/cm^3$, the maximum stress (failure stress) also increased with compression speed. The measured maximum stresses are shown in Table 2. In the case of density of $7.2\,g/cm^3$, the maximum stress at high compression speed was 1.2 times larger than that at low compression speed. In the case of density of $7.4\,g/cm^3$, the maximum stress at high compression speed was 1.6 times larger than that at low compression speed. Regardless of compression speed, the maximum stress increased with density.

Table 2. Maximum stress of impact splitting tests.

	Low speed compression	High speed compression
Lower density	0.76 kN (7.15 g/cm^3)	0.96 kN (7.22 g/cm^3) 1.27 kN (7.14 g/cm^3)
Higher density	0.90 kN (7.39 g/cm^3)	1.48 kN (7.34 g/cm^3)

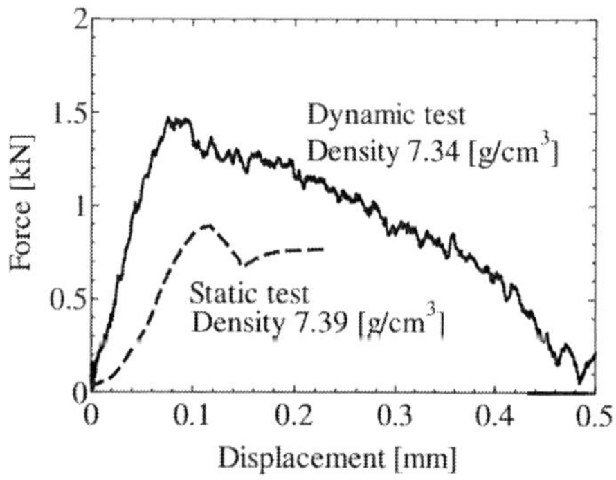

Figure 6. Force-displacement curves with specimen of 7.4 g/cm^3 (Compression speed of dynamic test 3.6 m/s, static test 0.002 mm/s).

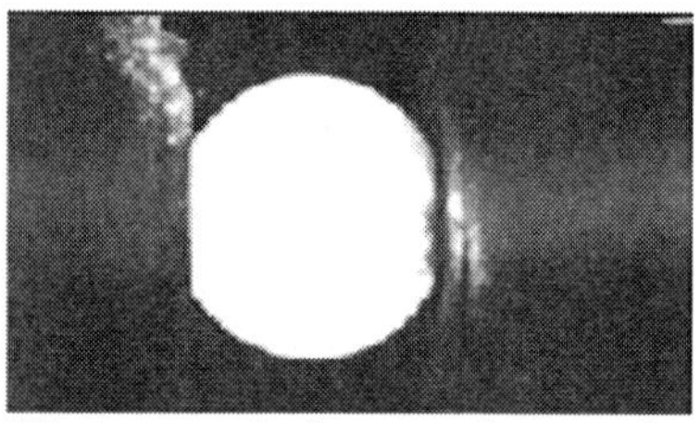

111 microseconds after onset of compression

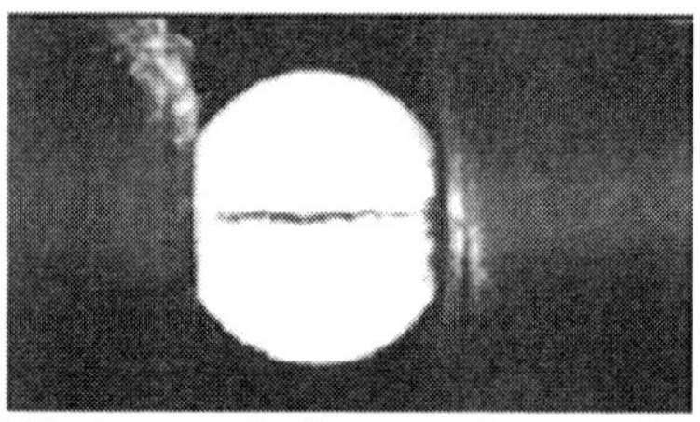

166 microseconds after onset of compression

Figure 7. Photographs of high-speed video camera with specimen of 7.14 g/cm^3, compression speed 3.1 m/s, lateral direction is compression direction.

Finally, Fig. 7 shows photographs (18,000 frames/second) of specimens with a density of 7.14 g/cm^3 and maximum stress at 1.27 kN in Table 2, using a high-speed video camera. A crack began to form near the center of the specimen and moved in both directions along the central axis of the specimen. Failure behavior at high compression speed was almost the same as that at low compression speed in Fig. 8, with force-displacement shown in Fig. 4 and maximum stress is 0.76 kN, as shown in Table 2.

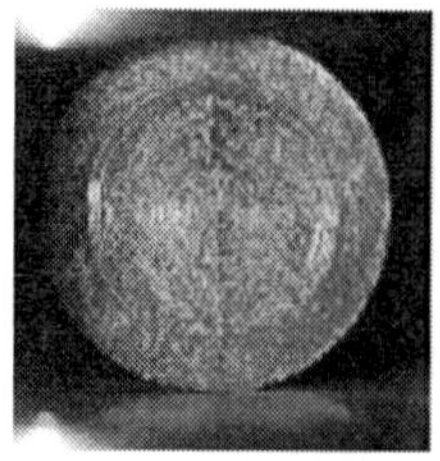

(a) Displacement 0.079 mm

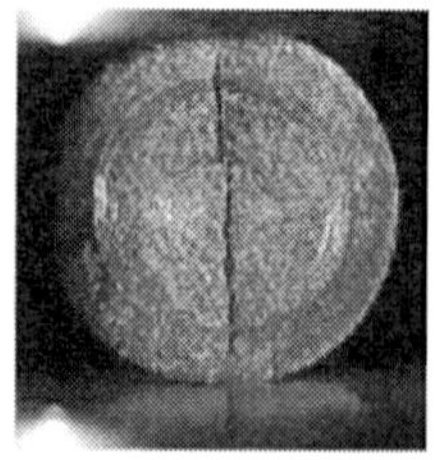

(b) Displacement 0.132 mm

Figure 8. Photographs of static splitting test with specimen of 7.15 g/cm^3, vertical direction is compression direction.

4. Conclusions

The impact tensile strength of iron green bodies was measured by the impact splitting test and was compared with results of normal Brazilian test (static splitting test). The maximum stress (failure stress) also increased with compression speed. Even at high compression speed, the maximum stress (failure stress) increased with density. Failure behavior at high compression speed was almost the same as that at low compression speed.

The authors are greatly indebted to Professor Masashi Daimaruya of Muroran Institute of Technology for his valuable advice with respect to the split Hopkinson bar. The authors are also greatly indebted to Associate Professor Fumihiro Itoigawa of Nagoya Institute of Technology for his help with taking images using a high speed camera.

References

[1] G. Gustafsson, M. Nishida, H.-Å. Häggblad, H. Kato, P. Jonsén, T. Ogura, *Powder Technology* **268**, 293 (2014)

[2] J. Najar, *J. Phys. IV*, Colloque C8, Supplément au Journal de Physique III, **4**, 647 (1994)

[3] F. Díaz-Rubio Gálvez, J. Rodríguez, V. Sánchez-Gálvez, *J. Phys. IV*, Colloque C3 Supplément au Journal de Physique III, **7**, d'Août, 151 (1997)

[4] F. Díaz-Rubio Gálvez, J. Rodríguez, Pérez, V. Sánchez-Gálvez, *Int. J. Impact Engineering*, **27**, 161 (2002)

[5] C. Johnstone, C. Ruiz, *Int. J. Solids Struct.*, **32**, 17/18 2647 (1995)

[6] M. Diamaruya, H. Kobayashi, T. Nonaka, *J. Phys. IV*, Colloque C3 Supplément au Journal de Physique III, 7, d'Août, 253 (1997)

[7] M. Daimaruya, H. Kobayashi, *J. Phys. IV*, **10** Pr9 173 (2000)

[8] A. Brara, J.R. Klepaczko, *Mechanics of Materials*, **38**, 3 253 (2006)

[9] T. Nojima, K. Ogawa, *Proc. 4th Conference on Mechanical Properties of Materials at High Rates of Strain*, J. Harding, ed.. Oxford, John Harding, 371 (1989)

[10] J.W. Tedesco, C.A. Ross, R.M. Brunair, Computers and Structures, **32**, 3-4 609 (1989)

[11] C.A. Ross, S.T. Kuennen and J.W. Tedesco, *Micro-mechanics of Failure of Quasi-Brittle Materials*, (eds. S.P. Shah, S.E. Swartz and M.L. Wang), Elsevier Applied Science 353 (1990)
[12] M.L. Hughes J.W. Tedesco C.A. Ross, *Computers and structures*, **47**, 653 (1993)
[13] K. Jitsu, K. Shirai, C. Ito, H. Ohnuma, *Proc. 9th symposium on impact problems in civil engineering*, **4**, 119 (1998) (in Japanese).
[14] D.E. Lambert and C.A. Ross, *Int. J. Impact Engineering*, **24**, 985 (2000)
[15] M. Daimaruya, H. Kobayashi, H. Shizawa, R.A. Siriegar and Y. Ishihata, *Proc. 10th International Congress on Fractures*, (2001), CD-ROM
[16] G.T. Gray, III, ASM Handbook Vol. 8, *Mechanical Testing and Evaluation*, ASM 462 (2000)
[17] W. Chen, B. Song, *Split Hopkinson (Kolsky) Bar: Design, Testing and Applications*, Springer (2010)

EPJ Web of Conferences **94**, 01070 (2015)
DOI: 10.1051/epjconf/20159401070

Investigation of strength properties of freshwater ice

A. Bragov[1], L. Igumnov[1], A. Konstantinov[1], A. Lomunov[1,2], A. Filippov[1], Yu. Shmotin[3], R. Didenko[3], and A. Krundaeva[3]

[1] Research Institute of Mechanics, Lobachevsky State University of Nizhny Novgorod, Russia
[2] Nizhny Novgorod State University of Architecture and Civil Engineering, Nizhny Novgorod, Russia
[3] NPO Saturn JSC, Rybinsk, Russia

Abstract. A study of the strength and deformation properties of freshwater ice under compression, tension and shear in a wide range of strain rates ($10^{-4} - 3 \cdot 10^3\,s^{-1}$) and temperatures of $-5°$ C, $-20°$ C, $-40°$ C and $-60°$ C was performed. Static stress-strain curves of ice under compression were obtained on which the identified strength properties of ice as well as compressive modulus. To determine the mechanical properties of ice at high-speed loading the Kolsky method was used with various embodiments of split Hopkinson bar. The deformation curves were obtained at various loading conditions. Thereon breaking points were defined as well as their dependence on the strain rate and temperature. Also static and dynamic strength properties of ice at splitting and circular shear were defined. Increase in the dynamic strength properties upon the static ones for all loading conditions was marked.

1. Introduction

When designing objects of aviation technology a lot of attention paid to emergency situations, accompanied by impact influence. Such situation may arise, for example, at ingress of ice pellets (hail) into the gas-turbine engine and hit the fan blade, which can lead to destruction of the turbine, the motor housing and the fuselage of the aircraft, and as a result, to casualties. For reliable strength calculation of dynamically loaded constructions aircraft engine experimental data are required on the mechanical properties (strain diagram, yield strength, tensile strength, modulus of hardening, extreme characteristics of plasticity, etc.) of metallic components of the engine as well as the mechanical properties of the ice fragments.

Mechanical properties of metallic structural elements of the engine have already been studied in detail, and to obtain the missing dynamic properties can be used known dynamic methods (the Kolsky method and split Hopkinson bar). Investigation of mechanical properties of ice is much more difficult, especially in the field of dynamic loading [1–3].

Ice is one of well-known materials, but, nevertheless, its properties are still not sufficiently studied. This is due to a large variety of structure, composition and physico-mechanical state of ice. Timing processes play significant role in the behavior of the ice [4]. Analysis of the known publications on the mechanical properties of ice indicates that there is a large variation of the strength, since on their quantitative value, beyond the main characteristics of ice (temperature, density and structure), are significantly influenced by testing procedure and dimensions of tested specimens [5].

There are two methods to determine the elastic moduli and shear: dynamic and static. The dynamic method is based on calculating of modules on the propagation velocity of the longitudinal and transverse elastic waves in the ice. The essence of the static method is to measure the magnitude of deformation of ice samples under static load.

The first experiments to determine the elastic modulus of ice under static loads were carried out at the beginning of the XIX century by Thomas Young. The obtained values of the elastic modulus of ice were varied in a very wide range – from 10^3 to 10^4 MPa. For many years a spread of values of the elastic modulus of ice was surprising and sometimes it was distrust to research results of different authors. More stable modulus values obtained by the dynamic test methods based on measurement of the velocity of propagation of elastic waves. The propagation velocity of sound waves in the ice is practically independent of the load and is determined only by the density of the ice, its structure, as well as by the temperature. The speed of longitudinal waves in pure ice is about 3800 m/s at a temperature of 0° C and rises to 4200 m/s at $-20°$ C.

The mechanical properties of the ice, i.e. ability to resist to external forces, are varying greatly depending on the temperature. The closer the temperature of the ice to melting point, the more pronounced its plastic properties and reduced strength. Lowering the temperature increases the strength of the ice [4].

In the work it was conducted a comprehensive study of the strength and deformation properties of freshwater ice in compression, tension (splitting) and circular shear in a wide range of deformation rates ($10^{-4} \div 3 \cdot 10^3\,s^{-1}$) and at temperatures of $-5°$ C, $-20°$ C, $-40°$ C and $-60°$ C.

2. Materials

For static and dynamic tests of ice was made four batches of ice samples of different configurations:

Article available at http://www.epj-conferences.org or http://dx.doi.org/10.1051/epjconf/20159401070

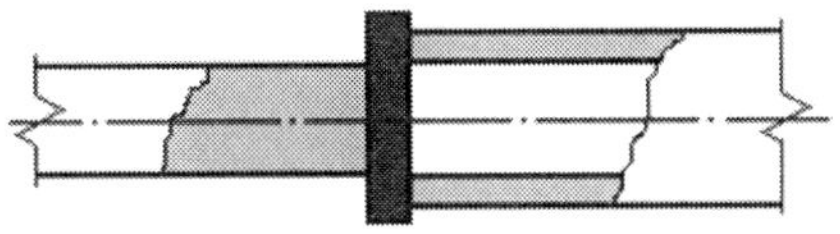

Figure 1. The test circuit for circular shear.

Type of test	Diameter, mm	Length, mm
Static compression	20	40
Dynamic compression	20	10
Static and dynamic splitting	20	20
Static and dynamic circular shear	30	10

For the manufacture of ice samples distilled water was used. Freezing water was done in a freezer at a temperature of −18° C. For each type of test the kits were made of thin-walled metal collars desired configuration. The rate of freezing of water in the collar was reduced by placing them in a heat insulating containers. In this case, the resulting ice samples have a uniform solid structure without cracks. The resulting density of the ice samples was 0.91–0.93 g/cm^3.

3. Test methods

Static properties were investigated using a universal testing machine LR5KPlus with temperature chamber. Maintaining a negative temperature in the chamber is provided by liquid nitrogen vapor.

To determine the mechanical properties of ice at high strain rates the Kolsky method was used with various realizations of the split Hopkinson bar. The tests were performed in compression by the traditional way [6], at splitting – by using Brazilian test [7]. In conventional tests at compression of cylindrical sample a compressive load is applied along the longitudinal axis of the sample. In experiments on splitting the cylindrical sample is rotated on 90° about a transverse axis, and the load applies along the diametral plane of the sample.

To determine the shear properties of ice based on a modification of the Kolsky method, wherein the supporting bar is replaced to a measuring tube (Fig. 1). The incident bar and supporting tube are working as a matrix and a punch, and in the sample thus realized circular shear (cut) by an annular surface having a diameter equal to the diameter of the bar.

Shear stress is the ratio of force $F(t)$, acting on the sample, to the square S_{sp} of circular surface corresponding to the medial surface of the sample:

$$\tau(t) = \frac{F(t)}{\pi \cdot D_{sp} \cdot L},$$

where D_{sp}– diameter of the middle surface.

The time dependence of the shear deformation is given by:

$$\gamma = \frac{U_1(t) - U_2(t)}{h},$$

here h – the height of the working part of the sample.

For comparison of diagrams obtained in this kind of tests with diagrams obtained at tension and compression, you must go to the equivalent of strain and stress through Mises criterion using the following formulas:

$$\varepsilon_{eqv} = \frac{\gamma}{\sqrt{3}}, \quad \sigma_{eqv} = \sqrt{3} \cdot \tau.$$

Sets of pressure bars for testing at compression and splitting each consist of two bars of high-strength aluminum alloy with a diameter of 20 mm and a length of 1 m. In the case of test materials in a circular shear instead supporting bar the measuring tube is used made of aluminum alloy with an inner diameter of 20.5 mm. For accurate registration of pulse passing through the ice sample and having very small amplitude (fractions of a millivolt) on a supporting bar (when tested in compression and splitting) or on the supporting tube (with shear tests) instead of foil strain gages there are glued semiconductor gages having a high gauge factor.

For the tests at low temperatures the original heat insulating chamber is used, enveloping the ends of measuring bars and the test sample placed between them. The chamber was purged with liquid nitrogen vapors for a few minutes in order to the ends of the bars before testing had a required negative temperature. The thermocouple for the sample temperature measuring was placed directly inside the chamber near the contact zone of one of the bar with the sample but without touching of the metal bar.

4. Results of investigation

After series of static tests the strength properties of ice under compression, splitting and shear were obtained. In the compressive experiments the elasticity modulus of ice was defined (by the displacement of test machine grips), which amounted to about 200 MPa at a temperature of −5°C and ~310 MPa at a temperature −60°C.

Dynamic tests of ice samples were conducted at compression, tension and splitting at three values of the strain rate (~ $0.5 \cdot 10^3\,s^{-1}$, ~ $1 \cdot 10^3\,s^{-1}$ and ~ $2 \cdot 10^3\,s^{-1}$ in compression) and at four negative temperatures (−5°C, −20°C, −40°C and −60°C). Required strain rates of the sample were provided by varying the impactor speed accelerated in the gas gun barrel and the required negative temperature – by controlling the rate of evaporation of liquid nitrogen.

Since during the splitting determine accurately the deformation of the sample (and, accordingly, the rate of deformation) is not possible, the stress-strain curves for this test mode were not built. Instead of the strain rate, the compression rate by the diameter was determined as the quotient of the impactor velocity (mass flow rate of end of the loading bar) to the length of the splitting plane (the diameter of the sample).

For each mode of loading (strain rate, temperature) were conducted a several (at least five) retests. Then, based on a series of curves for each mode of the test the average curve was determined with the characteristics of scatter of the experimental data (confidence intervals).

As an example in Fig. 2 the average compressive stress-strain curves at strain rate of $2 \cdot 10^3\,s^{-1}$ and at various test temperatures are shown. The Fig. 2a shows the time processes of stress in the sample and the strain rate (dotted

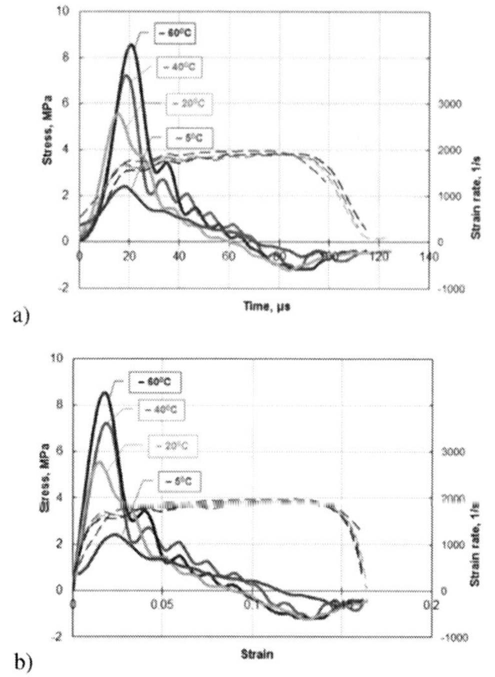

Figure 2. Effect of temperature on the dynamic stress-strain curves of ice under compression at strain rate $2000\,\mathrm{s}^{-1}$.

line) as well as in the Fig. 2b – there are the same curves as a function of sample deformation.

It's clearly that temperature has small influence on the start time of destruction of the sample (~ 20 mks), whereas the strength at low temperatures increases significantly.

As a result of the cycle of the static and dynamic tests of samples of ice in different types of stress-strain state and at the four values of the negative temperature the respective stress-strain curves were obtained. According to the curve obtained for each mode of loading strength of ice was determined at compression (the maximum stress on the diagram $\sigma \sim \varepsilon$), during splitting (the maximum stress on the curve $\sigma \sim$ t) and at the annular shear (maximum stress on the curve $\tau \sim$ t).

It is interesting to compare the dynamic and static strength properties of ice. The following shows the effect of strain rate and temperature on the mechanical properties of ice under compression (Fig. 3), splitting (Fig. 4) and shear (Fig. 5).

Obtained properties of ice will be used to create a model of ice in modern computational systems for modeling situation of flying of ice fragment into a turbine of aircraft engine.

5. Summary

Mechanical properties of ice both at compression and at splitting and at shear increases with increasing of strain rate, and at lower temperatures.

Observable quite a large scatter of strength properties of ice under different loading conditions, apparently

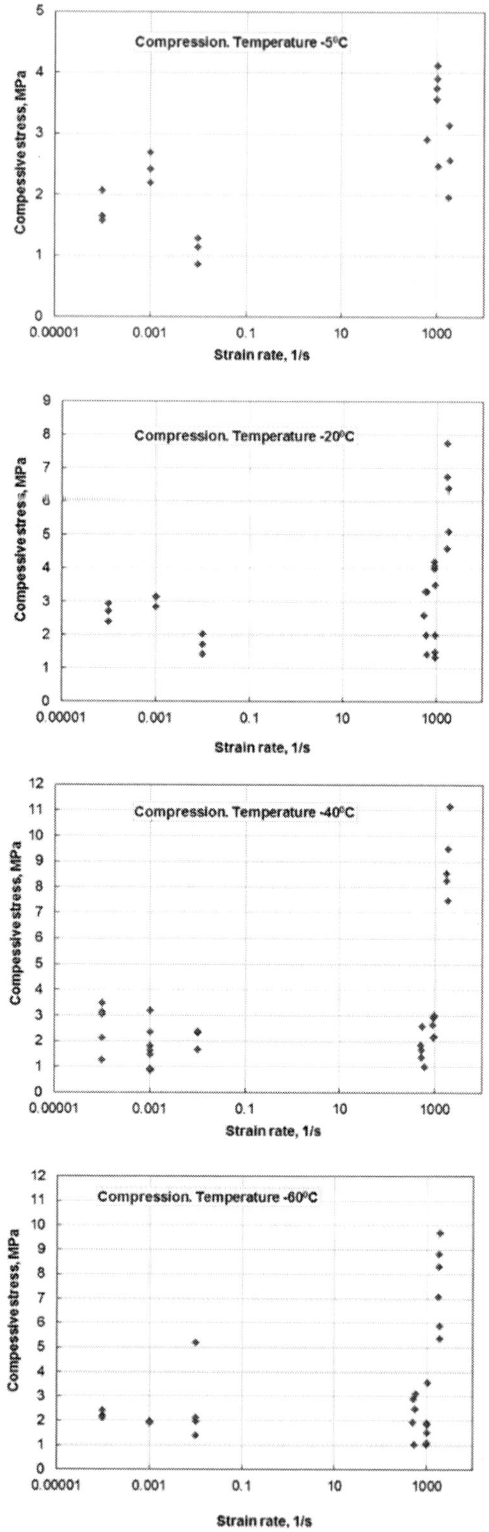

Figure 3. Effect of strain rate and temperature on strength of ice at compression.

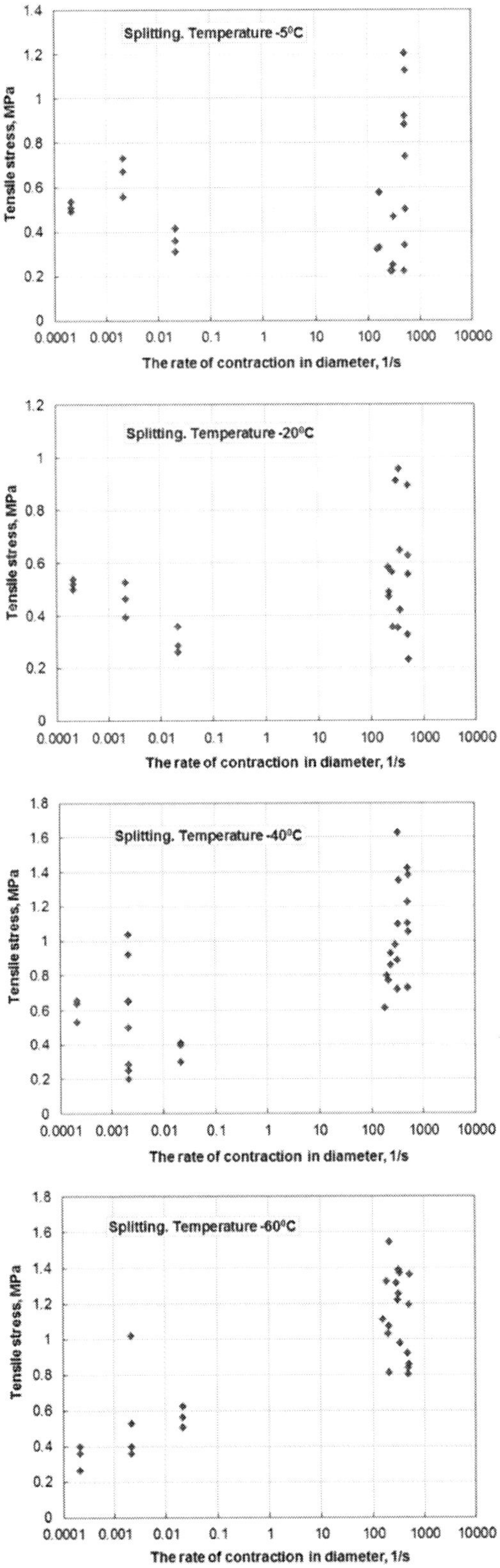

Figure 4. Effect of strain rate and temperature on strength of ice at splitting.

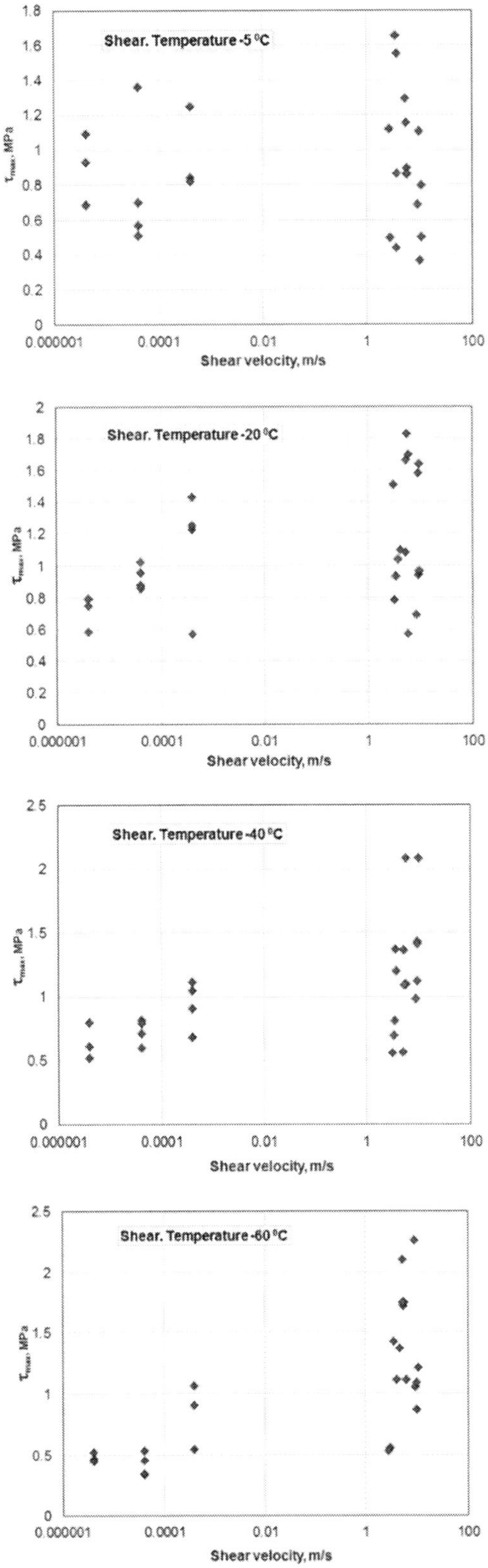

Figure 5. Effect of strain rate and temperature on strength of ice at shear.

associated with the possible heterogeneity of the structure of the frozen samples.

Static tests of ice were made with partial financial support of the Ministry of education and science of Russia (application 7.846.2014 K).

Dynamic experiments and results analysis were performed under support of the Russian Science Fund (grant 14-19-01096).

References

[1] J.H. Currier, E.M. Schulson, Acta Metall **30**, 1511 (1982)
[2] E.M. Schulson, J. Miner Metals Mater Soc **51**, 21 (1999)
[3] J.J. Petrovic, J. Mater. Sci. **38**, 1 (2003)
[4] N.A. Gorbushin, Yu.V. Petrov, Doklady Physics **59**(6), 283 (2014)
[5] X. Wu, V. Prakash, Int. J. Imp. Engng **76**, 155 (2015)
[6] A.M. Bragov, A.K. Lomunov, Int. J. Imp. Engng **16**(2), 321 (1995)
[7] T. Rodriguez, C. Navarro, V. Sanchez-Galvez, J. Phys. IV **4**, 101 (1994)

EPJ Web of Conferences **94**, 01071 (2015)
DOI: 10.1051/epjconf/20159401071

Dynamic testing of concrete under high confined pressure. Influence of saturation ratio and aggregate size

P. Forquin[1,a], E. Piotrowska[1], and G. Gary[2]

[1] Laboratoire Sols, Solides, Structures et Risques, Université Grenoble Alpes, BP. 53, 38041 Grenoble Cedex 9, France
[2] Laboratoire de Mécanique des Solides, Ecole Polytechnique, 91128 Palaiseau, France

Abstract. Concrete structures can be exposed to intense pressure loadings such as projectile-impact or detonation near a concrete structural element. To investigate the mechanical behaviour of concrete under high confining pressure, dynamic quasi-oedometric compression tests have been performed with a large diameter (80 mm) Split Hopkinson Pressure Bar apparatus. The concrete sample is placed within a steel confining ring and compressed along its axial direction. Hydrostatic pressures as high as 800 MPa and axial strain of about −10% are reached during the tests. In the present work, experiments have been conducted on two types of concrete: MB50 microconcrete with a maximum grain size of 2 mm and R30A7 ordinary concrete of maximum grain size about 8 mm. Both concretes are tested in dry or saturated conditions. According to these dynamic experiments it is noted that grain size has a small influence whereas water content has a strong effect on the confined behaviour of concrete.

1. Introduction

In many military and civil applications related to the safety of buildings or structures (power plants, head-quarters, tanks of dangerous materials, dams...), regarding accidental or intentional hazards such as ballistic impact, blasting, rock fall, plane crash or also earthquake, concrete can be exposed to extreme loading involving high hydrostatic pressure up to 1 GPa and high strain-rate levels (several hundreds of 1/s) [1]. Under such loading conditions shear mechanisms and collapse of pores are observed. So the deviatoric strength (the deviatoric stress being defined as the difference between the maximal and the minimal principle stresses) and the compaction law of concrete need to be investigated taking into account the free water saturation ratio within the concrete. Quasi-oedometric compression test allows testing the mechanical response of the material under a quasi-uniaxial deformation loading path [2]. During a QOC (Quasi-Oedometric Compression) test, a cylindrical specimen tightly enclosed in a confinement vessel is axially compressed by means of high strength compression plugs (Fig. 1). As it tends to expand, both radial and axial stresses increase in the specimen. Thus, the test provides a reading of the strength of the material at different levels of pressure.

Several experimental set-ups have been proposed in the literature to perform quasi-oedometric compression tests. One may cite the experimental work performed by Burlion et al. [3] with steel confining ring. Dynamic tests have been conducted by Gatuingt [4] with confining cells made of steel or brass. A new analysis of experimental data was proposed by Forquin et al. [2,5], taking into account the shortening of the specimen and a possible plastic deformation of the confining cell. This methodology was applied to dynamic QOC tests performed with a Split-Hopkinson Pressure Bar device [6]. The accuracy of the method and the influence of friction were specially discussed.

In the present paper, the dynamic testing results obtained on an ordinary concrete (R30A7 concrete, max. grain size equal to 8 mm) are compared to experimental data obtained with a MB50 microconcrete [7]. The concrete composition, the specimen size, the experimental procedure, the data processing and the results are detailed in the next parts.

2. Experimental method and processing of data

2.1. Concrete composition

Two types of concrete with very different mesostructures have been tested. On the one hand, the MB50 micro-concrete, is characterized by a water to cement ratio of 0.5 and a maximum grain size of 2 mm. The aggregate distribution is similar to a standard concrete but at a lower scale. On the other hand, the R30A7 concrete is closer to a standard concrete with siliceous aggregates up to 8 mm diameter and a water to cement ratio of 0.64 resulting in a compressive strength of about 30 MPa [8,9]. Four levels of free-water saturation ratio are considered in the present study in the case of R30A7 concrete: dry, 60%, 80%, 100%. The composition of each concrete is gathered in Table 1.

2.2. Specimen size

Different sample sizes are used for both concretes. The cylindrical samples of MB50 microconcrete were 29 mm

[a] Corresponding author: pascal.forquin@3sr-grenoble.fr

Article available at http://www.epj-conferences.org or http://dx.doi.org/10.1051/epjconf/20159401071

Table 1. Composition of MB50 microconcrete and R30A7 concrete.

Concrete	MB50 [7]	R30A7 [8,9]
Aggregates (kg/m^3)	0	1008
Sand (kg/m^3)	1783	838
Cement (kg/m^3)	400	263
Water (kg/m^3)	200	169
Admixture (kg/m^3)	12	0
Water/Cement	0.5	0.64
Maximum grain size (mm)	2	8

Figure 1. Steel confining ring, compression plugs and cylindrical sample used in a quasi-oedometric compression test.

in diameter and 40 mm in length and the steel confining ring was 60 mm in outer diameter and 45 mm long (Table 2) [7].

In the present work, the confining ring and the sample used for testing R30A7 concrete are larger ones. The cylindrical samples, 40 mm in diameter and 50 mm in length, are used. The confining ring made of high-strength steel (elastic limit of about 1800 MPa) is 40 mm in inner diameter, 80 mm in outer diameter and 60 mm long (Fig. 1, Table 2). The outer diameter of the ring is large enough to prevent any plastic strain within the ring for levels of internal pressure applied on the internal surface as high as 800 MPa. The length of the ring (50 mm), higher than that of the specimen, prevents the expulsion of the specimen from the ring during the test (especially in dynamic tests). The specimen is compressed by means of 2 cylindrical plugs (Fig. 1) made of high-strength steel (elastic limit of about 1800 MPa) allowing maximum axial stress level during the tests of about 1600 MPa. The steel plugs have the same diameter than the concrete specimens (40 mm) and are 30 mm long.

2.3. Setting of the sample

The R30A7 concrete specimens were machined in large blocks 40 days after the mix was poured into watertight plywood moulds. A special procedure was set to align the ring, the sample and the two plugs. First, the concrete sample is scotch taped to the upper plug. The concrete specimen is slowly introduced within the ring previously partially filled by a bi-components epoxy resin named "Chrysor® C6120" so this interface product is slowly extruded during this stage and the internal gap between the specimen and the ring is totally filled. 24 hours later, the Chrysor® product is hardened so the ring, the specimen and the two plugs can be safely handled.

2.4. Dynamic testing facility

The dynamic quasi-oedometric compression experiments have been conducted in the LMS laboratory with a large SHPB (Split Hopkinson Pressure Bar) facility. The SHPB system employed in this work and the processing method are detailed in [Forquin et al., 2008b]. It consists in a striker, an input bar and an output (transmitter) bar all made of steel bars (elastic limit 1200 MPa) with a diameter of 80 mm. The striker, the input bar and the output bar are 2.2 m, 6 m and 4 m long, respectively. A so-called pulse-shaper technique is used to increase the rising time of the loading force and to ensure correct stress equilibrium in the interesting part of the experiment. For this purpose a thin disk of lead is placed on the impact end of the incident bar. As long as the lead is not totally flattened, the maximum stress induced by the striker is not reached.

2.5. Processing method

Since our specimen is located between the two plugs, and not directly sandwiched between the input and transmitter bars, forces and velocities have to be calculated at the faces of the specimen. Their calculation is based on the analysis of the wave propagations in a bar with non-constant acoustic impedance [10–12]. Knowing the basic waves, the waves at specimen faces are computed, allowing for the calculations of forces and displacements at compression plugs-specimen interfaces. Input and output forces provide the axial stress $\overline{\sigma}_{zz}^S$ in the sample whereas the axial strain ε_{zz}^S is deduced from the axial displacements at both ends of the sample.

Furthermore, strain-gauges glued on the ring provide the circumferential (hoop) strain used to estimate the level of internal radial stress $\overline{\sigma}_{rr}^S$ and strain ε_{rr}^S within the specimen according to the data processing proposed by [2,6,7]. In particular, the calculation of the mean radial stress $\overline{\sigma}_{rr}^S$ within the specimen is calculated as function of the external hoop strain $\varepsilon_{\theta\theta}^{(z=0)}$, the shortening of the specimen (of length h) deduced by the axial strain, and the defect of symmetry (δ_z) that could result from an initial imperfect positioning or a slip of the specimen inside the ring during the test.

$$\bar{\sigma}_{rr}^S = \left(P_\sigma(h) + Q_\sigma(h) \times (\delta_z)^2\right) \varepsilon_{\theta\theta}^{(z=0)} \quad (1)$$

where P_σ and Q_σ are two polynomial functions of degree 2, which coefficients have been identified for the considered cell from numerical simulations considering a range of h (45 to 50 mm) and δ_z (0 to 3 mm). The axial stress and radial stress being calculated, it is possible to evaluate the average deviatoric stress:

$$\bar{\sigma}_{deviatoric}^S = \left|\bar{\sigma}_{zz}^S - \bar{\sigma}_{rr}^S\right| \quad (2)$$

Table 2. Dimensions of samples and steel ring used in previous work and present work.

Experiments	[7]	**Present study**
Concrete, max. grain size	MB50, 2 mm	R30A7, 8 mm
Sample: Diameter × length (mm)	29 × 40	40 × 50
Steel ring: Outer diameter × length (mm)	60 × 45	80 × 60

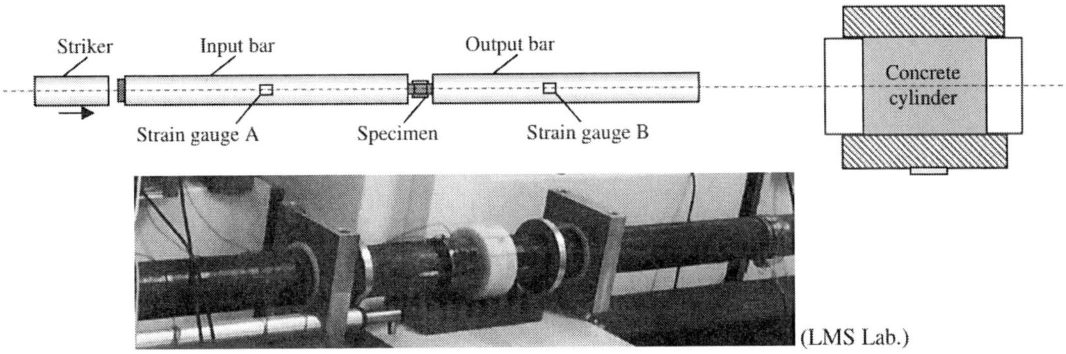

Figure 2. Diagram of SHPB device and dynamic testing facility used in LMS laboratory.

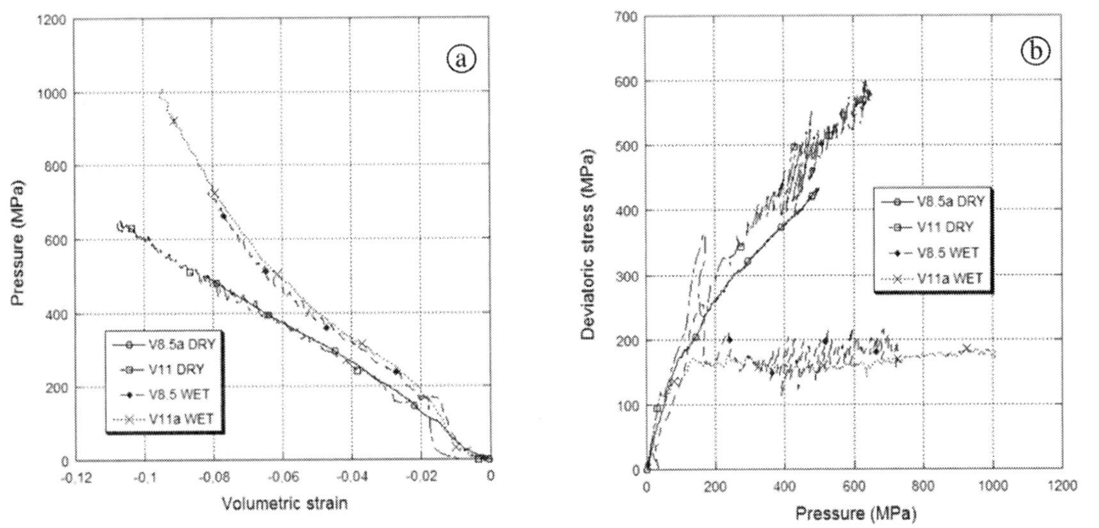

Figure 3. Dynamic tests performed on dried and wet microconcrete (Tests V8.5: strain rate = 90/s, Test V11: strain rate = 150/s) [Forquin et al., 2010]: (a) Compaction curve and (b) Deviatoric behaviour.

as well as the average hydrostatic pressure and the volumetric strain according to the formulas:

$$\bar{P}^S_{hydrostatic} = -\frac{1}{3}\left(\bar{\sigma}^S_{zz} - 2\bar{\sigma}^S_{rr}\right) \tag{3}$$

$$\bar{\varepsilon}^S_{volumetric} = \left(1 + \varepsilon^S_{zz}\right)\left(1 + \varepsilon^S_{rr}\right)^2 - 1. \tag{4}$$

Then the deviatoric stress can be plotted as function of the hydrostatic pressure (deviatoric behaviour) and the hydrostatic pressure can be plotted as function of the volumetric strain (hydrostatic behaviour).

3. Dynamic experiments performed on MB50 microconcrete [7]

The confined behaviour of dry and wet (water-saturated) microconcrete was investigated by Forquin et al., [7] with the testing facility and experimental procedure described in the part 2. The change of deviatoric stress is plotted as function of the hydrostatic pressure (Fig. 3b) and the hydrostatic pressure is plotted as function of the volumetric strain (Fig. 3a). First, the hydrostatic response of wet microconcrete is stiffer than the one of dry microconcrete (Fig. 3a). Moreover, the deviatoric strength of wet microconcrete is not exceeding about 200 MPa even under hydrostatic pressure as high as 1000 MPa

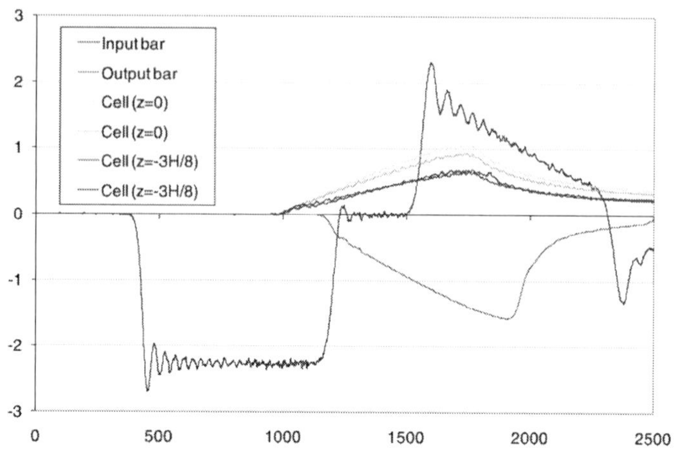

Figure 4. Dynamic experiments performed on R30A7 concrete with the SHPB facility. Blue curve: incident and reflected pulse, pink curve: transmitted pulse (strain-rate of about 100/s)

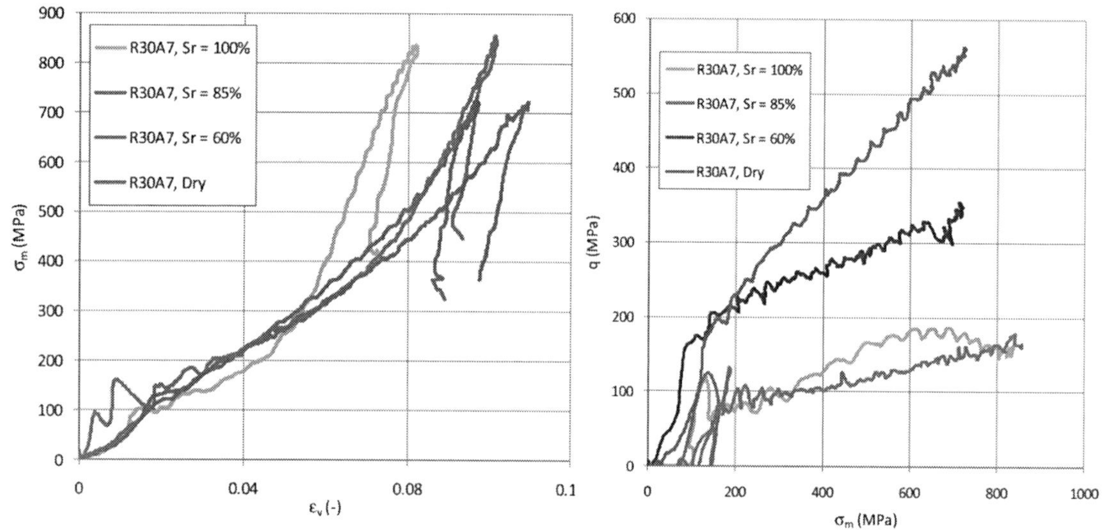

Figure 5. Dynamic quasi-oedometric experiments performed on wet R30A7 concrete: (a) Compaction curve and (b) Deviatoric behaviour.

(Fig. 3b). On the other hand, dry microconcrete exhibits a deviatoric strength continuously increasing with the increase of pressure (Fig. 3b). As explained by [7,9], the limited strength in saturated concrete is due to an increase of pore-pressure within the microstructure.

4. Results of dynamic experiments performed on wet R30A7 concrete

The confined behaviour of R30A7 concrete was investigated with the testing facility and experimental procedure described in the part 2 considering four different saturation ratios (dry concrete, 60%, 80%, 100%). As mentioned in the part 2, larger cylindrical samples (40 mm in diameter and 50 mm in length) and confining ring (80 mm in outer diameter and 60 mm long) are used than with microconcrete. An example of experimental data (strain gage on the input bar, strain gage on the output bar and strain gages on the confining cell) obtained with a dry sample is given in Fig. 4.

Figure 5 illustrates the dynamic response of R30A7 concrete samples considering four saturation ratios (dry concrete, 60%, 80%, 100%). Similarly to what has been observed with microconcrete, both hydrostatic and deviatoric responses are strongly influenced by the water content. In particular, the deviatoric strength of wet concrete is not exceeding 200 MPa even under the maximum hydrostatic pressure reached during the test: 800 MPa. On the other hand, dry concrete shows a continuous increase of strength with hydrostatic pressure, up to 550 MPa at 700 MPa of hydrostatic pressure (Fig. 5b). In addition, the hydrostatic response of saturated concrete is much stiffer than that of concrete with lower saturation degrees (Fig. 5a).

Finally, it is observed that the both dynamic hydrostatic and deviatoric responses of R30A7 concrete are very similar to the ones of MB50 microconcrete whatever the water content (dry or saturated). It confirms that the dynamic behaviour of concrete at high level of confinement is primarily a question of free water content rather than an effect of aggregate size.

5. Conclusion

Dynamic quasi-oedometric compression tests have been conducted on R30A7 ordinary concrete (maximum aggregate size: 8 mm) and MB50 microconcrete (maximum aggregate size: 2 mm) to investigate their mechanical response under high levels of pressure at high strain-rates. The experimental procedure allows evaluating the hydrostatic behaviour (compaction curve) and deviatoric strength under pressure levels up to 800 MPa. The experimental results have shown a major influence of free water content on the confined behaviour of both concretes. The wet or saturated concretes exhibit stiffer hydrostatic response than dry concrete. Moreover their deviatoric strength is weaker than in dry concrete. Finally, the present study confirms the importance to take into account the free-water content in concrete to correctly investigate the mechanical response of concrete structures exposed to intense pressure loadings such as projectile-impact or detonation near a concrete structural element.

This research has been performed with the financial support of the CEA: French Alternative Energies and Atomic Energy Commission (Commissariat à l'énergie atomique et aux énergies alternatives). We would like to thank Dr. Eric Buzaud and Dr. Christophe Pontiroli (CEA- Centre de Gramat) for their sound technical and scientific advices. SHPB dynamic experiments have been performed in the LMS (Laboratoire de Mécanique des Solides, Palaiseau, France). The authors want to deeply thank Philippe Chevallier for its technical support, and Prof. Patrick Le Tallec, director of the LMS, for the warm welcome.

References

[1] P. Forquin, A. Arias, R. Zaera., R., 2008a. *Int. J. Impact Eng.* **35** (3), 133–146.

[2] P. Forquin, A. Arias, R. Zaera., 2007. *Int. J. Solids Struct.* **44**, 4291–317.

[3] N. Burlion, G.Pijaudier-Cabot, N. Dahan, 2001. Int. J. Numer. Anal. Meth. Geomec. **25**, 467–1486.

[4] F. Gatuingt, Ph.D. thesis, ENS Cachan, 1999.

[5] P. Forquin, A. Arias, R. Zaera. Experimental Mechanics **49**, pp. 613–625, 2009.

[6] P. Forquin, G. Gary, F. Gatuingt, 2008b. Int J Imp Eng. **35**(6), 425–446.

[7] P. Forquin, K. Safa, G. Gary, 2010. Cement Concrete Res. **40**(2), 321–333.

[8] T. Gabet, Y. Malécot, L. Daudeville, 2008. Cem. Conc. Res. **38**, 403–412.

[9] X.H. Vu, Y. Malécot, L. Daudeville, E. Buzaud, 2009. Int J Solids Struct. **46**, 1105–1120.

[10] B. Lundberg, J. Carlsson, K. G. Sundin, 1990. J Sound and Vibration. **137**(3), 483–493.

[11] G. Bacon, J. Carlsson, J.L. Lataillade, 1991. J of Phys IV. **1**, 395–402.

[12] G. Gary, 2005. DAVID, instructions manual, Palaiseau, France. (`http://www.lms.polytechnique.fr/EQUIPE/dynamique/index.html`)

EPJ Web of Conferences 94, 01072 (2015)
DOI: 10.1051/epjconf/20159401072

On the characterisation of the dynamic compressive behaviour of silicon carbides subjected to isentropic compression experiments

Jean-Luc Zinszner[1,a], Benjamin Erzar[2], Pascal Forquin[1], and François Barthélémy[3]

[1] Laboratoire Sols Solides Structures – Risques (L3SR), Université Grenoble Alpes, BP. 53, 38041 Grenoble Cedex 9, France
[2] CEA, DAM, GRAMAT, BP. 80200, 46500 Gramat, France
[3] DGA Techniques Terrestres, 18021 Bourges Cedex, France

Abstract. Ceramic materials are commonly used as protective materials particularly due to their very high hardness and compressive strength. However, the microstructure of a ceramic has a great influence on its compressive strength and on its ballistic efficiency. To study the influence of microstructural parameters on the dynamic compressive behaviour of silicon carbides, isentropic compression experiments have been performed on two silicon carbide grades using a high pulsed power generator called GEPI. Contrary to plate impact experiments, the use of the GEPI device and of the lagrangian analysis allows determining the whole loading path. The two SiC grades studied present different Hugoniot elastic limit (HEL) due to their different microstructures. For these materials, the experimental technique allowed evaluating the evolution of the equivalent stress during the dynamic compression. It has been observed that these two grades present a work hardening more or less pronounced after the HEL. The densification of the material seems to have more influence on the HEL than the grain size.

1. Introduction

Ceramic materials are commonly used as protective materials since the sixties and the Vietnam War [1]. Indeed, thanks to their outstanding hardness and compressive strength [2] combined to their low density, their use in bilayered armour configurations allows, for a same level of protection, significant weight benefit in comparison to monolithic steel plate armours [3]. However, the microstructure of the ceramic may have a strong influence on its penetration resistance [4] and the links between the microstructure and the ballistic efficiency are not fully understood. In order to characterise the dynamic compressive strength of brittle materials, plate impact experiments are generally performed. This kind of experiments allows determining the Hugoniot elastic limit (HEL), i.e. the elastic limit of the material under uniaxial strain state, and the Hugoniot curve which is the loci of all possible equilibrium states under shock loading. Moreover, the determination of yield strength of ceramics with the flyer plate technique requires intrusive techniques like lateral stress gauges embedded in the material. Several tests are necessary to determine accurately shock response of ceramics. In this work, dynamic compression experiments have been performed on two silicon carbide grades using a high-pulsed power generator called GEPI. The GEPI machine is a low inductance electric generator, based on the strip-line concept and composed of two electrodes separated by dielectric foils [5]. The current, reaching progressively an intensity up to 3.3 MA in nearly 0.5 μs, generates on the internal faces of the electrodes a high-pressure loading through Laplace effect. Using this generator, the same compressive pulse is applied on the two electrodes. Thus, by performing a compression experiment on two samples of a same material with different thicknesses, one can use the lagrangian analysis method [6]. This method integrates the conservation equations (mass, momentum, energy) between two velocity signals measured on ceramic samples with different thicknesses and allows accessing to the whole loading path without any assumption. Therefore, a single GEPI experiment with 2 samples is sufficient to determine the dynamic compressive behaviour of ceramics up to more than 20 GPa. Two silicon carbide grades have been characterised in order to highlight the links between the microstructure of ceramics and their dynamic compressive behaviour. These grades present a high density, but they have been made using pressureless or spark plasma sintering (SPS). These two processes lead to different grain sizes and hardnesses. The differences of dynamic compressive behaviour observed between the grades provide data on the role of the microstructure at very high loading rates. Moreover, all the experimental data have been compared to numerical results allowing validating the constitutive models. First, a brief description of the microstructural characteristics of the two silicon carbide grades is realised. The principles of the testing machine, the experimental method as well as a description of the lagrangian analysis method are then summarized. In a third part, the experimental results obtained on the two silicon carbides grades are presented before comparing it with numerical results.

[a] Corresponding author: jeanluc.zinszner@3sr-grenoble.fr

2. Presentation of the two silicon carbide grades

The two silicon carbide grades are sintered by Saint-Gobain using pressureless sintering process (PS) or spark plasma sintering process (SPS). They are both sintered in

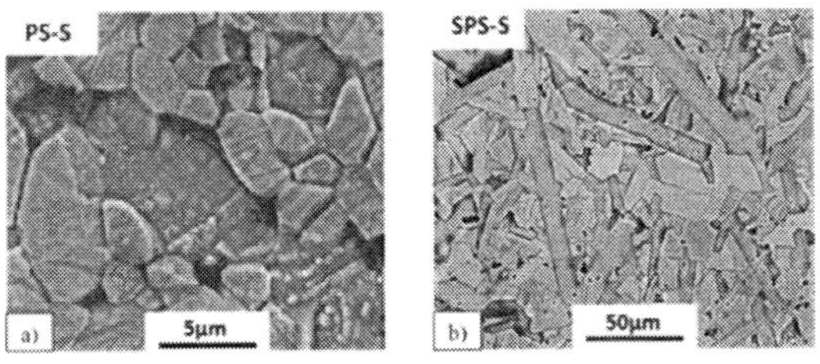

Figure 1. SEM pictures of microstructures from grades (a) PS-S and (b) SPS-S [7].

solid state (SSS) and are called PS-S for the pressureless sintered ceramic and SPS-S for the spark plasma sintered ceramic. These two ceramic grades present different microstructures. The pressureless sintered silicon carbide presents a density greater than 98% of the theoretical density of the SiC. The microstructure is fine and the grain size is in the range 4–10 μm (cf. Fig. 1a). The SPS-S grade presents a density about 99% the theoretical density. Elongated grains, with size beyond 100 μm on the long axis, can be observed in its microstructure (cf. Fig. 1b) [7].

3. Principle of isentropic compression experiments

3.1. Principle of GEPI device

The GEPI machine is based on High Pulsed Power (HPP) technologies. This generator applies the strip line concept [5]: two plane electrodes made of aluminium, separated by dielectric foils and locally connected by a short circuit (cf. Fig. 2) are subjected to an intense current focused on the load region. In those experiments, the current reaches a maximum intensity of 3.3 MA in nearly 0.5 μs. The association of current and magnetic field produces Lorentz forces applied on the internal face of the electrodes. The pressure pulse applied to the electrode can be directly computed (cf. Eq. (1)) knowing the evolution of the current with time $I(t)$, the width of the terminal part of the electrode W, the vacuum permeability $\mu 0$ and an edge effect factor kp (near 1).

$$P_{mag}(t) = \frac{1}{2} k_p \mu_0 \left[\frac{I(t)}{w} \right]^2 . \tag{1}$$

This machine is generally devoted to isentropic compression experiments thanks to the ramp loading. During experiments, the compressive pulse propagates through the electrodes and the ceramic specimens before reaching the LiF windows. The LiF window is necessary if one want to perform a dynamic compressive test. When no window is placed after the ceramic, the reflection of the compressive loading on the free surface leads to a tensile fracture of the ceramic. This technique has been used by Erzar and Buzaud [8] to perform shockless spalling tests on alumina.

3.2. Experimental configuration

Due to the symmetry of the GEPI configuration (Fig. 2), two specimens can be tested during one GEPI compression

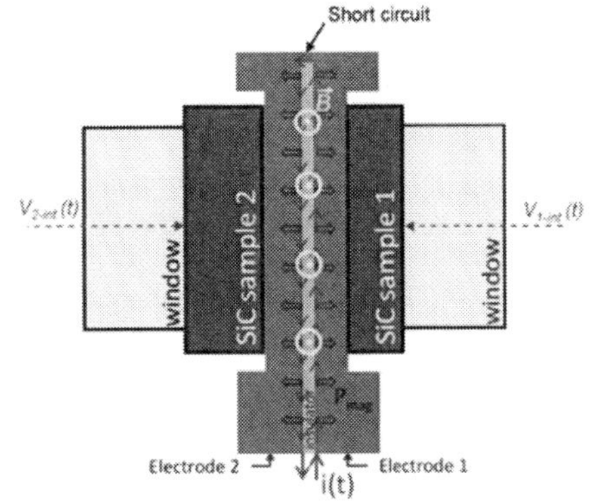

Figure 2. Experimental configuration used to perform GEPI compression tests and allowing using lagrangian analysis method.

test. This configuration also allows using the lagrangian analysis method when a GEPI compression test is performed on two specimens of a same material but having different thicknesses. In this work, one GEPI compression test is performed on each grade. The thicknesses of the specimens are 3 mm and 4 mm.

The electrode width and the specimen's widths are set to 15 mm. This width allows ensuring a compressive pulse sufficiently high in order to exceed the HEL of the material but also allows ensuring that the interface velocities measured on the two specimens are not affected by radial release waves during the compressive loading.

3.3. Lagrangian analysis method

The lagrangian analysis method realised after one GEPI compression test allows determining the whole loading path of the material whereas only one point on this loading path can be obtained after one plate impact experiment. This method is based on the integration of the equations of conservation of mass, momentum and energy. Moreover, this method does not need any assumption on the behaviour of the material. The principle is based on a comparison of velocity signals obtained on two different thicknesses of a material.

For each particular velocity u_p, an associated wave velocity $C_L(u_p)$ can be calculated from times Δt (cf. Fig. 3). This wave velocity then allows us to compute increments of longitudinal stress σ_x, longitudinal strain ε_x (stress and strain being negative in compression) or specific volume associated to each increment of particular velocity following:

$$d\sigma_x = -\rho_0 C_L(u_p) du_p \tag{2}$$

$$dv = -\frac{1}{\rho_0 C_L(u_p)} du_p \tag{3}$$

$$d\varepsilon_x = \frac{du_p}{C_L(u_p)} . \tag{4}$$

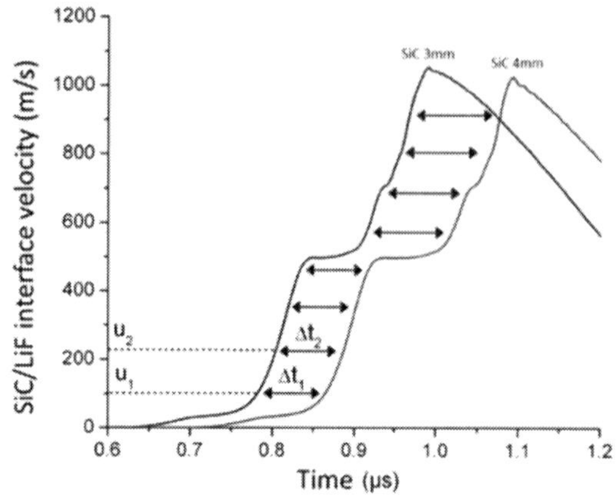

Figure 3. Principle of lagrangian analysis method: determination of the wave velocity $C_L(u_p)$ associated to each particular velocity u_p (arbitrary origin of time).

It can be noted that a correction based on the mechanical impedances of the ceramic and of the window material have to be realised to determine the particular velocity using the interface velocity measured.

4. Experimental results

4.1. GEPI compression on PS-S grade

After the GEPI compression test performed on the PS-S grade, one can observe that the velocity signal obtained on the 4 mm thick specimen presents a shock for velocities comprises between 0 and 460 m/s (cf. Fig. 4). Moreover, the rest of this velocity signal presents an uneven profile. It has nevertheless been possible to filter it and to perform a lagrangian analysis using the two velocity signals for velocities comprises between 500 and 850 m/s. The release signal cannot be used due to the arrival of radial release waves.

In a first time, two methods can be used to determine the HEL of the material. The first one consists in determining the velocity at which a slope change is observed. The HEL is then determined using Eq. (5):

$$HEL = \rho_0 C_L u_p^{LEH}, \tag{5}$$

where ρ_0 is the initial density, C_L is the longitudinal wave velocity in the elastic regime and u_p^{LEH} is the particular velocity at the slope change (obtained after correction).

In the case of the PS-S grade, a slope change is observed at about 569 m/s. That corresponds to a HEL of 14.9 GPa.

The second method for determining the HEL consists in exploiting the lagrangian analysis data. By plotting the stress-strain curve, the elastic/inelastic transition is observed starting from 15.1 GPa. These two HEL values are similar but a better confidence is given to results obtained with the lagrangian analysis. In all cases, the uncertainty is evaluated at ±0.2 GPa.

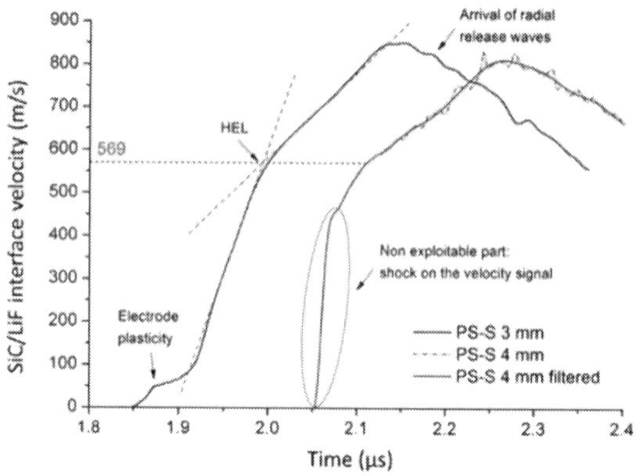

Figure 4. SiC/LiF interface velocity signals obtained on the two PS-S specimens after GEPI compression test (arbitrary origin of time).

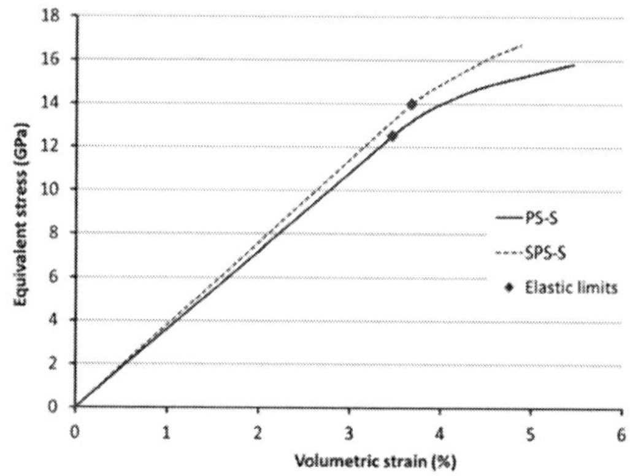

Figure 5. Equivalent stress/strain curves obtained with lagrangian analysis for the two SiC grades.

The longitudinal stress obtained with the lagrangian analysis method is not sufficient to determine the equivalent stress. However, during a uniaxial deformation or for all triaxial loading such as $\sigma_y = \sigma_z$, one can write:

$$\sigma_{eq} = \frac{3}{2}(-\sigma_x - P), \tag{6}$$

where P denotes the hydrostatic pressure.

By setting an equation of state for the SiC material, one can thus evaluate the evolution of the equivalent stress. In the case of the PS-S grade, a low variation of the K +(4/3)G modulus is observed during the elastic compression. Thus, a constant bulk modulus K is assumed, meaning the pressure is known during the whole experiment. Using Eq. (6), one can deduce the equivalent stress in the ceramic specimen during the compressive loading. The equivalent stress/strain curve obtained is presented in Fig. 5.

One can see that the transition between the elastic and the inelastic behaviour is smooth and that a work hardening of the material is observed after this transition.

A numerical validation has been performed using the Abaqus/Explicit finite element code by imposing as

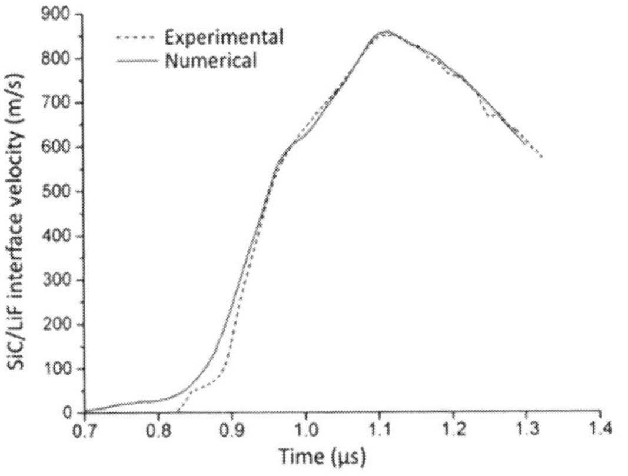

Figure 6. Comparison between experimental and numerical interface velocity signal in the case of the PS-S grade.

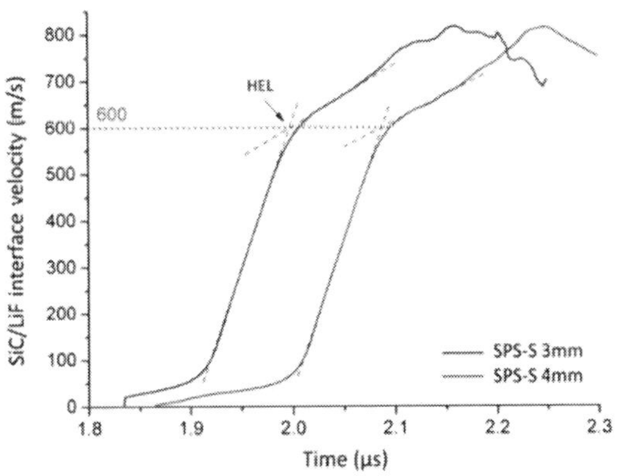

Figure 7. SiC/LiF interface velocity signals obtained on the two SPS-S specimens after GEPI compression test (arbitrary origin of time).

plasticity law the stress/strain curve determined above. One can observe that the interface velocity signals obtained experimentally and numerically are similar, validating the plasticity law identified using the lagrangian analysis (cf. Fig. 6).

4.2. GEPI compression on SPS-S grade

After the GEPI compression test performed on the SPS-S grade, no shock is observed (cf. Fig. 7). Thus, a lagrangian analysis can be realised using the whole velocity signals. A HEL of 16.8 GPa is obtained with the lagrangian analysis; HEL which is greater than the one of the PS-S grade. This result tends to show that the grain size is not a dominating factor on the HEL of silicon carbides. Indeed, despite the greater grain size of the SPS-S grade and the presence of elongated grains, this grade presents the higher HEL. The better densification of the SPS-S grade is probably the reason why the HEL is higher.

Using Eq. (6), the equivalent stress/strain curve of the SPS-S has been identified. As in the case of the PS-S, a constant bulk modulus was used. Due to the higher HEL,

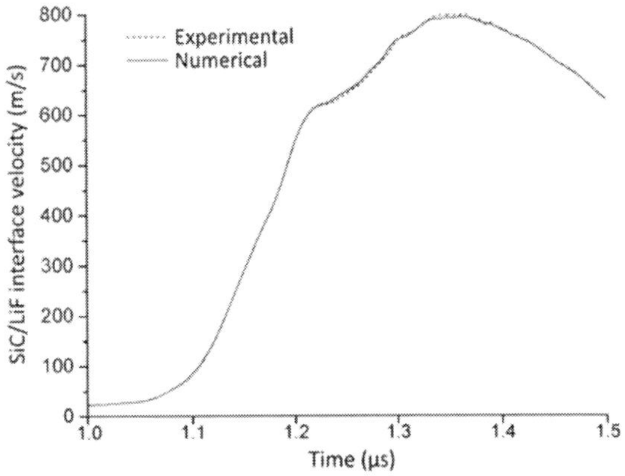

Figure 8. Comparison between experimental and numerical interface velocity signal in the case of the SPS-S grade.

the inelastic part of the loading path obtained with the SPS-S is shorter than the one obtained with the PS-S grade. However, one can anyway see a better work hardening of SPS-S comparing to the PS-S.

Moreover, one can see a very good correlation between the velocity signals obtained experimentally and by simulating the GEPI compression test on SPS-S using the identified plasticity law (cf. Fig. 8).

5. Conclusions

In this work, isentropic compression experiments have been performed on two silicon carbide grades using a high pulsed power generator, called GEPI. The use of the GEPI device allows realising a lagrangian analysis with the velocity signals obtained and thus determining the whole loading path in the material. In a first time, the Hugoniot elastic limit (HEL) of the two SiC grades has been determined. The SPS-S grade, having the denser microstructure but composed of big and elongated grains, presents the highest HEL. One can first conclude that the grain size is not the dominating factor on the HEL of silicon carbides and that the densification is probably one more important parameter. In a second time, the lagrangian analysis allows calculating the equivalent stress and identifying a plasticity law. These plasticity laws have then been validated by performing numerical simulations of the compression experiments. It has been observed that the two materials present different work hardening after the elastic limit; work hardening which is more important in the case of the SPS-S grade. However, it is not possible to conclude on the influence of microplasticity and microcracking on the transition between the elastic and the inelastic behaviour and on the work hardening. Shock recovery experiments and microscopic analyses are needed.

This work was supported by DGA (French General Delegation for Armament, French Ministry of Defense). The authors are also grateful to P.-Y. Chanal, J.-J. Royer and P. Rey (CEA GRAMAT) for their technical contribution to this project.

References

[1] E.R. Barron, A.L. Alesi, Tech. Report 69-43-CE, U.S. Army Natick Laboratories (1969)

[2] R. Feng, G.R. Raiser, Y.M. Gupta, J. Appl. Phys. **83**, 79 (1998)

[3] C.J. Roberson, Ceramic materials and their use in lightweight armour systems, *Lightweight armour system symposium* (1995)

[4] P.J. Hazell, C.J. Roberson, M. Moutinho, Mat. Design **29**, 1497 (2008)

[5] C. Mangeant, F. Lassalle, P. L'Eplattenier, P.L. Héreil, D. Bergues, G. Avrillaud, *Shock compression of condensed matter*, 1173 (2001)

[6] J. Cagnoux, P. Chartagnac, P.L. Héreil, M. Perez, Ann. Phys. Fr. **12**, 451 (1987)

[7] G. Rossiquet, P. Forquin, J.L. Zinszner, F. Barthélémy, *Personal Armour System Symposium* (2014)

[8] B. Erzar, E. Buzaud, Eur. Phys. J. Special Topics **206**, 71 (2012)

EPJ Web of Conferences **94**, 01073 (2015)
DOI: 10.1051/epjconf/20159401073

Influence of loading-rate and steel fibers on the shear strength of ultra high performance concrete

Lukic Bratislav and Forquin Pascal[a]

Laboratoire 3SR, Grenoble Alpes University, BP. 53, 38041 Grenoble Cedex 9, France

Abstract. The paper describes quasi-static and dynamic experimental methods used to examine the confined shear strength of an Ultra High Performance Concrete, with and without the presence of steel fibers in the concrete composition. An experimental setup was created to investigate the concrete shear strength under quasi-static loading regime using a hydraulic press Schenk while dynamic shear strength was characterized by subjecting concrete samples to dynamic loading through a modified Split Hopkinson Pressure Bar. Both methods are based on a Punch Through Shear (PTS) test with a well-instrumented aluminum passive confinement ring that allows measuring the change of radial stress in the shear ligament throughout the test. Firstly, four equally distributed radial notches have been performed in order to deduce the radial stress by suppressing a self-confinement of the sample peripheral part. However, by analyzing the strain gauge data from the confinement ring, it has been noticed that these were apparently insufficient, especially for fiber-reinforced samples, resulting in subsequently practicing eight radial notches through the sample peripheral part. The results obtained from both procedures are reported and discussed.

1. Introduction

Concrete structures such as nuclear power plants, dams, bridges or bunkers can be exposed to intensive loadings such as earthquakes, blasting or projectile-impact. During these dynamic loading conditions, in addition to tensile failure, shear damage modes can be observed as cratering on the front face and shear strain close to the crater [1]. Mode II shear fracturing has also been observed in ultra-high performance concrete during edge-on impact experiments [2]. However, the shear strength of ultra-high performance concrete remains unknown and the influence of steel fibers as well as strain-rate on the shear strength need to be investigated. The Punch Through Shear (PTS) experimental technique was first proposed by Watkins [3]. The cylindrical specimen contains two cylindrical notches on the top and the bottom face such that the inner diameter of the upper notch coincides with the outer diameter of the lower notch consequently forming a cylindrical shear surface. In further works, the specimen was first subjected to a pure hydrostatic pressure whereas an axial load was added in the second step to create a shear fracturing of the ligament. This technique was more recently used by Montenegro [4] to evaluate the fracture energy of conventional concrete (maximum aggregate size 6 mm) as function of the hydrostatic pressure. In the present work, quasi-static and dynamic Punch Through Shear (PTS) tests have been conducted with an ultra-high strength concrete, with and without short steel fiber reinforcement.

A well-instrumented passive confining ring has been applied to the PTS samples in a similar way than in quasi-oedometric compression tests [5] allowing to deduce radial stress in the fractured zone. An overview of the two experimental procedures is given with presenting and discussing the results obtained.

[a] Corresponding author: pascal.forquin@3sr-grenoble.fr

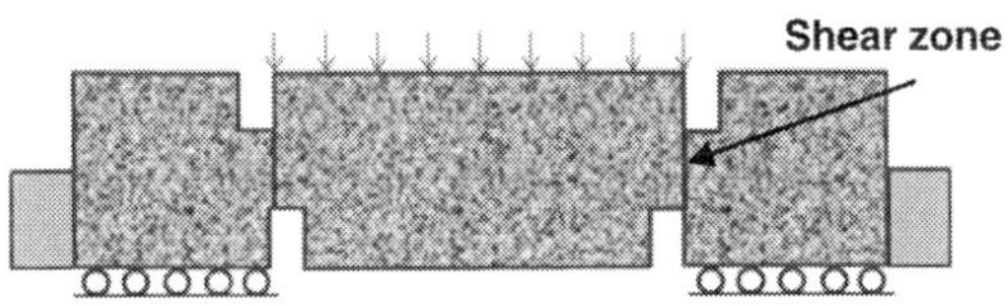

Figure 1. Punch Through Shear (PTS) specimen: Geometry of the sample used.

2. Concrete sample

2.1. The PTS sample

The specimen geometry is shown in the Fig. 1, inner diameter of the upper notch and outer diameter of the bottom notch coincide to form a straight cylindrical fracture surface. The length of the ligament ensures an almost perfectly homogeneous shear stress field in the shear zone. Moreover, the confining ring is decentered with respect to the specimen symmetry plane as to obtain uniform pressure field between the confining ring and the concrete sample and allows deducing the radial stress in the shear zone. The aluminum confining ring used in both experimental procedures is 10 mm in thickness and 15 mm in height and is instrumented with two strain gauges. The sample is carefully introduced into the confinement ring and the gap between the sample and the confining ring is filled by a bi-component resin (Chrysor®). Furthermore, equally distributed radial notches are performed through the peripheral part of the specimen in order to avoid self-confinement.

2.2. Tested material

The material tested is an Ultra High Performance Concrete (UHPC) whose mechanical properties and composition are gathered in Table 1. In contrast to conventional concrete,

Article available at http://www.epj-conferences.org or http://dx.doi.org/10.1051/epjconf/20159401073

Table 1. Material Composition and Mechanical Properties of an Ultra High Performance Concrete [6].

Composition		Ductal®
Cement	[kg/m³]	730
Silica Fume	[kg/m³]	235
Crushed quartz grains	[kg/m³]	220
Sand	[kg/m³]	885
Superplasticizer	[kg/m³]	10
Water	[l/m³]	160
Steel fibers	[vol. %]	2%
W/(C+SF)		0.17
Mechanical Properties		
Density	[kg/m³]	2396
Young's Modulus	[GPa]	55
Compressive strength	[MPa]	200

the mixture contains finely crushed quartz grains and silica fume. The optimized granular mixture of UHPC, where the maximal grain size is approximately 0.6° mm (sand), and extremely low water to cementious material (cement and silica fumes) ratio allow casting a finely graded and highly homogeneous concrete matrix with outstanding mechanical properties. The use of superplasticizer in order to increase the workability is of great importance. In the present work, two specimen categories have been considered for shear testing: one which contains 2% volumetric ratio of short steel fibers (0.2° mm diameter and 7° mm in length) and the other without additional fiber reinforcement.

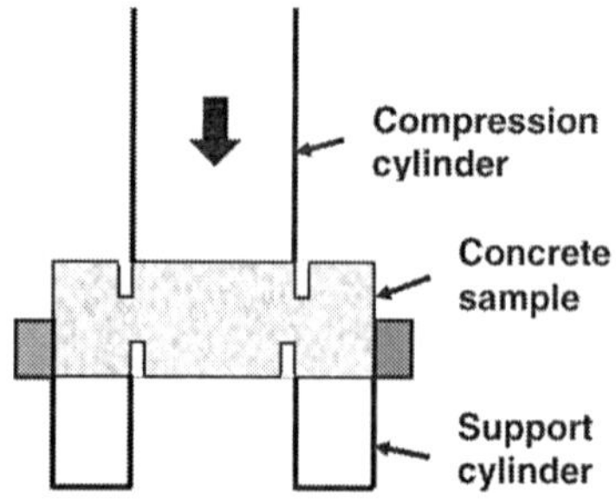

(a)

(b)

Figure 2. Quasi-static experimental set-up used in 3SR Laboratory (a) sketch of the PTS sample configuration, (b) the Schenk press.

3. Quasi-static shear testing

3.1. Principle of the experimental procedure

The quasi-static experiments have been conducted in 3SR Laboratory. The PTS sample was placed on a steel support cylinder, ensuring a perfectly flat contact surface such that the driving force is equally distributed throughout the peripheral part of the sample. The inner diameter of the support cylinder corresponds to the outer diameter of specimen bottom notch, allowing the axial displacement of sample middle part with respect to its peripheral part. Additionally, a compression cylinder is attached to the sample upper so that its diameter coincides with the inner diameter of the upper notch. A sketch of the sample configuration cross section is shown in (b), Fig. 2a. The sample is then placed in The Schenk press, (b) Fig. 2b, whose capacity of 100 t is more than sufficient for the purpose of the quasi-static testing. Two perfectly synchronized linear variable differential transformers (LVDTs) were used for measuring the displacement between press pressure caps. In order to achieve a desirably low strain-rate that corresponds to quasi-static loading regime, the load speed was set to 0.1 μm/s, which corresponds approximately to rate of strain less than $10^{-4}\,s^{-1}$.

The test was driven until the value of displacement was equal to 10 mm, for which it is considered that the total failure of the shear ligament is reached. The mean shear stress in the shear ligament is deduced directly from the axial load that is transmitted onto the shear surface:

$$\sigma_{shear} = \frac{F_{axial}}{S_{ligement}}. \tag{1}$$

Furthermore, as reported in [7], the mean radial stress in the ligament is deduced from the average contact pressure between the sample and the confinement cell, knowing the contact surface and the area of the shear zone:

$$\sigma_{radial} = \frac{S_{contact}}{S_{ligament}} P_{contact}. \tag{2}$$

The contact pressure between the sample and the cell was obtained in a similar way as in the quasi-oedometric tests presented in [5], by post processing the data from the strain gauges glued on the aluminum confinement ring.

3.2. Experimental results

A series of PTS experiments have been carried out on Ultra High Performance Concrete (UHPC) with and without addition of short steel fibers. In order to skirt around

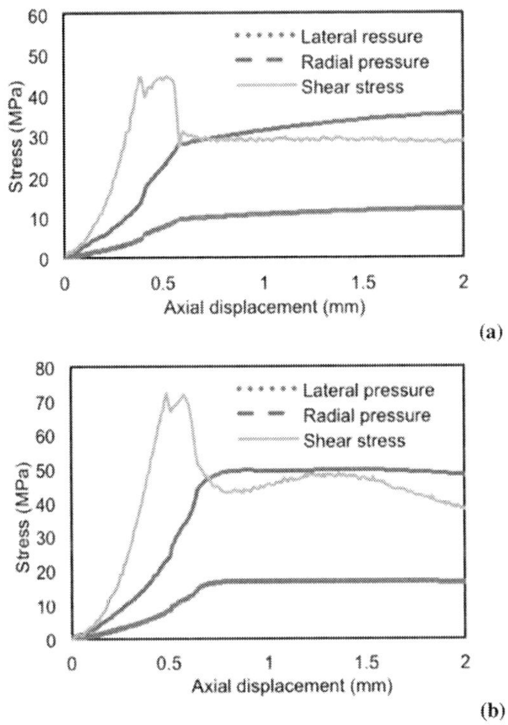

Figure 3. Results of quasi-static shear tests performed on *four-notch* specimen: (a) UHPC without additional steel fibers and (b) FUHPC with addition of short steel fibers (2% volumetric content).

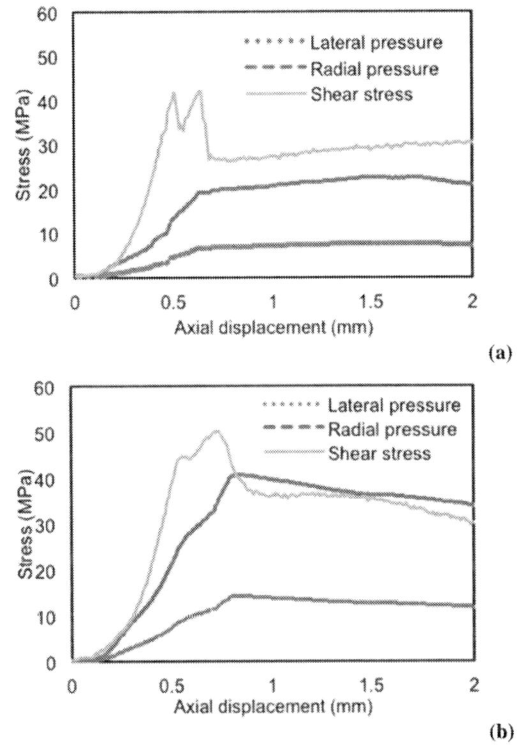

Figure 4. Results of quasi-static shear tests performed on *eight-notch* specimen: (a) UHPC without additional fibers and (b) UHPC with addition of 2% volumetric addition of short steel fibers.

the self-confinement of the peripheral part, four equally distributed radial notches have been performed. The results of the quasi-static shear tests are reported in Fig. 3.

The experimental data indicates a maximum shear stress in specimen without addition of steel fibers of 45 MPa while on the other hand the sample with 2% volumetric fiber reinforcement exhibited a markedly higher shear stress of about 70 MPa. A similar radial stress at the peak shear stress of about 25 MPa is observed in both tests.

As the question raised from analyzing the four-notched quasi-static results was weather the increase of shear strength is induced solely by the presence of steel fibers in the shear zone or by the influence of an additional self-confinement (both series of samples exhibited tensile radial cracking pattern), further experiments have been carried out on samples with eight peripheral radial notches and the results are reported in Fig. 4. The maximum shear stresses reached, in this case, were remarkably similar for both concrete mixtures, non-fibered sample showed shear strength of 42 MPa, while the reinforced sample exhibited shear stress of 49 MPa.

3.3. Discussion on quasi-static results

A double peak stress response is observed in all tested samples. At the first peak, a Mode I fracturing is expected, which is then followed by pure Mode II cracking at the second stress peak. A noticeable softening behaviour after

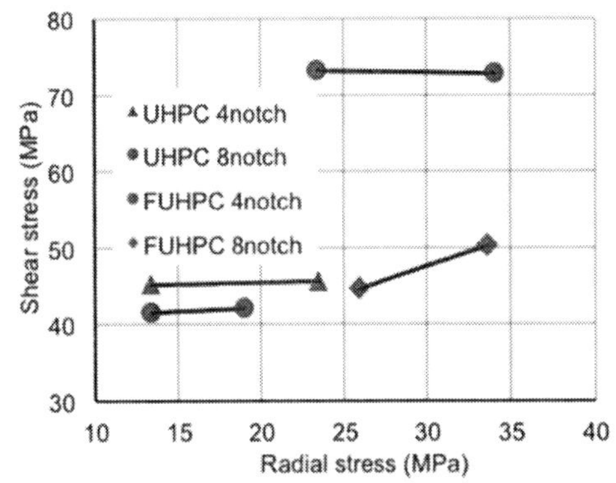

Figure 5. Results from quasi-static shear tests performed on UHPC without additional fibers and UHPC with 2% volumetric addition of short steel fibers for both four-notch and eight-notch samples representing double shear stress peak behavior with their corresponding radial stress.

the second stress peak has been observed in both series of samples. However, samples with steel fiber reinforcement have shown a larger post peak shear stress and a more gradual decrease towards the same residual stress than in samples with no addition of steel fibers. A possible explanation is that after the second peak, the pull-out strength of fibers is still active.

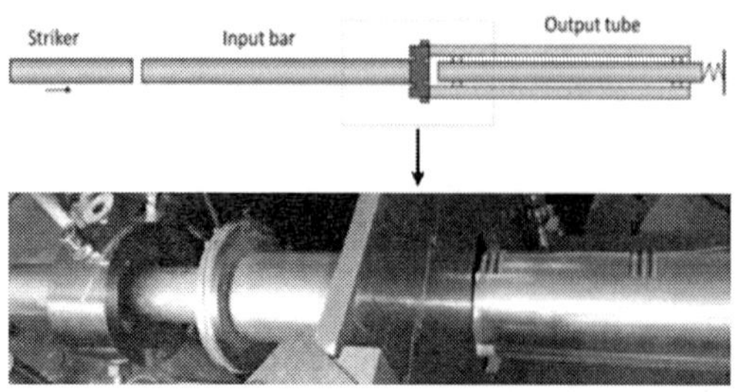

Figure 6. Experimental device used in 3SR Laboratory to perform dynamic PTS tests.

Points representing both first and second shear stress peaks with their corresponding radial stress, summarizing the results from quasi-static experiments are reported in Fig. 5.

A lower overall shear stress is noted for samples with eight-notches, apparently due to the presence of self-confinement of the peripheral part in four-notch samples causing higher shear strength. This is especially observed in fibered four-notch samples. The possible explanation is that the crack-bridging effect resulting from the presence of fibers may probably increase the self-confinement, which cannot be explicitly measured, resulting in higher shear strength in fibered samples with four-notch compared to eight-notch. It can be concluded that it is necessary to practice eight radial notches on ultra-high performance concrete samples due to the presence of an additional apparent self-confinement.

4. Dynamic shear testing

4.1. Principle of experimental procedure

The dynamic experimental setup built in 3SR Laboratory is presented on Fig. 6. The sample is subjected to dynamic shear load using a modified Split Hopkinson Pressure Bar (SHPB). It is composed of a projectile, input bar and output tube all made out of the same aluminum alloy. The cylindrical Hopkinson bar and the tube are both instrumented with strain gauges located in their half-lengths which purpose is to register wave propagation in the bars due to projectile impact.

A pressure tank is used to launch the striker at impact-speeds up to 20 m/s that generates an incident compression pulse in the input bar. The strain gauge measurements on the input bar provide the incident and reflected waves, so the input and output forces between the specimen and the Hopkinson bars are obtained as following:

$$F_{input} = S_{input} E_{bar}(\varepsilon_{incident} - \varepsilon_{reflected})$$
$$F_{output} = S_{output} E_{bar}(\varepsilon_{transmited}). \quad (3)$$

Finally, the shear stress in the ligament is deduced by averaging the input and output forces:

$$\sigma_{shear} = \frac{F_{input} + F_{output}}{2.S_{ligament}}. \quad (4)$$

The axial displacement of the sample middle part is obtained by integrating the particle velocity measured by the laser interferometer pointed out toward the middle of the output tube.

$$V_{axial} = V_{reflected} + V_{incident} - V_{transmited}. \quad (5)$$

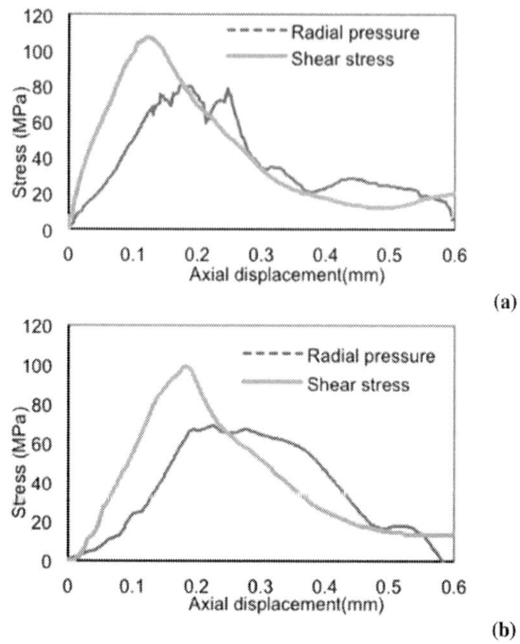

Figure 7. Results of dynamic shear tests performed on UHPC without additional fibers: (a) four-notch sample (b) eight-notch sample.

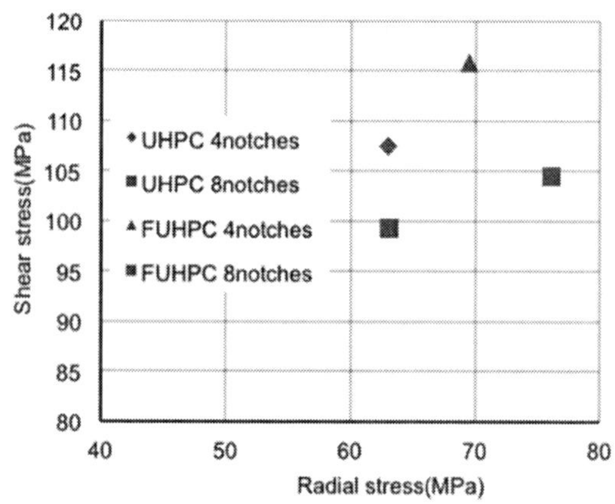

Figure 8. Results of dynamic shear tests performed on UHPC without additional fibers and UHPC with 2% volumetric addition of short steel fibers for both four-notch and eight-notch samples representing both shear stress peaks with their corresponding radial stress.

4.2. Experimental results

A series of experiments have been carried out on ultra high performance concrete with and without addition of short steel fibers. As in quasi-static experimental procedure, first the samples with four equally distributed radial notches have been tested and then, eight notches have been performed in order to avoid self-confinement as explained in previous part. The results of the dynamic shear tests on non-reinforced samples are reported in Fig. 7.

The maximum shear stress reached in four-notch sample without addition of steel fibers was 105° MPa while on the other hand, the sample with fibers exhibited higher shear strength of about 116° MPa. The maximum shear stress reached in non-fiber reinforced eight-notch sample was around 100 MPa while the fibered sample exhibited a shear stress of 108° MPa which again is fairly similar.

4.3. Discussion of the dynamic experiments

Points representing shear stress peaks with their corresponding radial stress summarizing the results from dynamic tests are reported in Fig. 8. Higher shear strength is observed in comparison to quasi-static results. This is due to a much higher radial stress reached in the ligament. Similar to quasi-static experiments, a decrease of shear strength is observed by increasing the number of radial notches, indicating a presence of apparent self-confinement especially in fiber reinforced samples. However, in this case only one-stress peak behaviour is observed. A possible explanation may be that the presence of tensile cracking prior to pure Mode II is being delayed, therefore affecting the radial stress in the ligament. This, furthermore, could be explained by a higher tensile strength observed in concrete like materials when subjected to high strain rate tensile loading, that has been reported in literature by several authors [8–10].

5. Conclusion

A series of PTS experiments have been conducted on an ultra-high performance concrete in quasi-static and dynamic regimes in order to investigate the influence of strain rate on the confined shear strength. Furthermore, half of the samples contained additional short steel fibers in order to study their influence on material behaviour. At the beginning, four radial notches have been performed to reduce the influence of self-confinement. The obtained results in quasi-static testing showed a major difference in shear strength depending on the fiber content. Practicing additional notches, eight in total, and testing the samples in the same regime, resulted in significantly lower shear strength, especially in fibered concrete. A clear positive influence of fibers until the first stress peak has been reported, that reflects in the closure of tensile cracks (crack-bridging effect) prior to mode II cracking, increasing the radial stress and consequently influencing the overall shear strength. Moreover, the samples with addition of steel fibers showed a higher post-peak shear stress. Dynamic test results showed much higher shear strength compared to quasi-static results. This difference is thought to be the consequence of a higher radial confinement that could be induced by adjournment of tensile cracking prior to Mode II cracking occurring in the ligament, as tensile strength is substantially increased at high strain rates.

This work has been partially supported by the CEA (French Alternative Energies and Atomic Energy Commission) – Centre de Gramat and by the LabEx Tec 21 (Investissements d'Avenir – grant agreement n° ANR-11-LABX-0030).

References

[1] Q. M. Li, S. R. Reid, H. M. Wen, and A. R. Telford, Int. J. Impact Eng. **32**, 224–284, (2005)
[2] P. Forquin and F. Hild, J. Eng. Mech. **4**, 302–315, (2008)
[3] J. Watkins, Int. J. Fract. **23**, 135–138, (1983)
[4] O. Montenegro, D. Sfer, C. M. López, and I. Carol, *Proceedings FraMCoS-8*, (2013)
[5] P. Forquin, G. Gary, and F. Gatuingt, Int. J. Impact Eng. **35**, 425–446, (2008)
[6] P. Forquin, These Doctorat, LMT Cachan, (2003)
[7] P. Forquin, Appl. Mech. Mater. **82**, 148–153, (2011)
[8] F. Toutlemonde, C. Boulay, and P. Rossi, *Proceedings of FRAMCOS-2*, (1995)
[9] J. Weerheijm and J. C. a. M. Van Doormaal, Int. J. Impact Eng. **34**, 609–626, (2007)
[10] B. Erzar and P. Forquin, Exp. Mech. **50**, 941–955, (2010)

EPJ Web of Conferences **94**, 01074 (2015)
DOI: 10.1051/epjconf/20159401074

Dynamic response of reverse Taylor impact based on DIC technology

Jiancheng Liu, Aiguo Pi[a], Haijun Wu, and Fenglei Huang

State Key laboratory of Explosion Science and Technology, Beijing Institute of Technology, Beijing 10081, PR China

Abstract. Reverse ballistic impact test, which can obtain the response data of rod/projectile more comprehensive and quantitative than forward impact test, was widely used for the measurement of material dynamic and structure response. Based on the DIC technology and traditional optical measurement (high-speed camera measurement), the Taylor experiment of reverse ballistic with different length-diameter ratio and different impact velocities were carried out by 57 mm compression-shear type light-gas gun, which provides the instantaneous response data of the Taylor rod in microsecond level. Then, the transient structural deformation of the specimen and the characteristics of plastic wave propagation were analysed by DIC technology and compared with traditional optical measurement. Applying the theory of reverse Taylor impact deformation and combining with the simulation results by LS-DYNA, the rules of structure deformation and plastic wave propagation were obtained. The method above can be applied for the structure response of penetrator under the condition of reverse ballistic penetration.

1. Introduction

The structure response of projectile during the impact process includes the deformation, the propagation and interaction of stress wave, the destruction mechanism and so on. For hard target (such as concrete target, reinforced concrete plate, metal plate, etc.) penetration, there are only a few researches on the structure response during the penetration process. For the existing researches using forward impact to analyze the structure response rules of kinetic projectile with a large slender ratio, the testing device such as stress sensor can't be installed on the projectile because of the characteristic of high-speed penetration, which leads to the result that the response parameters can't be measured. Therefore, it is necessary to develop the reverse ballistic impact tests to analyze the structure response quantificationally. It is possible to discuss the structure response of large slender kinetic projectile by reverse impact experiment.

Reverse Taylor impact test is becoming an important way to analyze the dynamic response and structure response. For the structure response of Taylor impact tests, the specimen is placed statically and the testing devices can be set to test the response parameters. Moreover, high-speed digital image correlation (DIC) is a kind of non-contact strain testing system, which break through the limitation of traditional electrical tests and improve the optical deformation measurement. It is fast, simple and accurate to process the data, and also improves the efficiency.

Dynamic response of material was researched in 1948 by Taylor [1], the test was conducted by using a rod to impact a rigid plate and plastic deformation occurred at the end of the rod after impact. In recent years, high speed camera was used for improving the reverse Taylor impact test [2,3]. In the first improvement, Erilich et al. [4] applied high speed camera to capture the outline of specimen deformation and provided the theory model associated with time. Besides, House et al. [2] proved that the transient image in the different stage of impact test can be used for determining the strain and stress, and derived the stress as a function of time at the fixed sample length. The OFHC copper is used by Jones et al. [5,6] in the Taylor impact test which obtained the three stages of plastic wave propagation. The early time behavior, denoted by Phase I, was shock dominated nonlinear motion of the plastic wave front. This was followed by steady motion of the plastic wave front, denoted by Phase II. The stages of deformation of the Taylor specimen are characterized by the addition of a third regime, Phase III, in which deceleration of the plastic wave front occurs and the deformation stopped. But these analyzes only apply to the first wave, and the following plastic strain wave does not apply. Different from the traditional Taylor impact tests, the specimen and rigid plate are unconstrained in reverse ballistic tests, therefore, the images shoot by high-speed camera don't contain any projection and fixed boundary. The original Taylor theory is no longer applicable and a new theory must be put forward for the reverse Taylor test. For the reverse test, Eakins et al. [7,8] used the specimen outline data caught by high-speed camera to calculate the radius strain and plastic wave position on the direction of length. By developing the original Taylor theory, the flow stress along the rod was calculated by radius strain. Reverse ballistic technology was performed by Anderson et al. [9] to analysis the influence of constraints of target, extent of target pre-damage and radial size of rod on penetration velocity during the process of silicon carbide impact rod. Lidén [10] studied the geometry and motion of long rod projectiles after penetrating thin obliquely oriented and moving armour-plates, which indicated that the velocity, direction and thickness of target have great influence on the failure or fracture of projectile.

Digital image correlation, namely DIC, which mainly used for measure the areal deformation under the action of external force, is a rapidly developed modern optical measurement mechanical method. Compared with traditional methods, the characteristics of DIC includes

[a] Corresponding author: aiguo_pi@bit.edu.cn

Article available at http://www.epj-conferences.org or http://dx.doi.org/10.1051/epjconf/20159401074

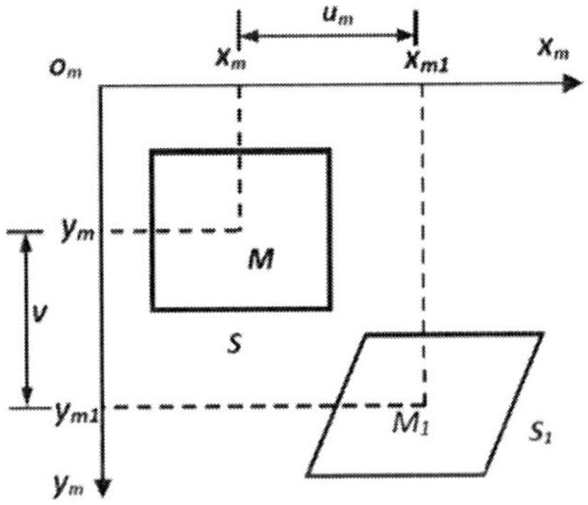

Figure 1. sub-region before and after deformation.

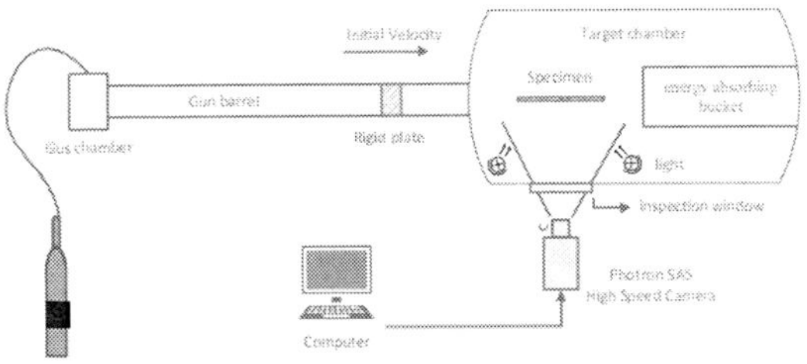

Figure 2. Reverse Taylor impact experiment layout.

full-field measurements, non-contact, nondestructive to specimen, simple optical path, measurement fields which can be independently adjusted as needed, no interference fringe processing, wide application range, low requirement to environment, simple operation and so on. The pixel was employed by Asundi [11] to measure the displacement of object. Cofaru et al. [12] elaborated on the calculation principles of DIC method, using the method to analyze the test data under low strain rate, and then, obtained the accurate results. High-speed DIC method was applied to three-point bending test which was loaded by Hopkinson bar, and the mechanical properties of aluminum under the action of bending is obtained by Pierron et al. [13]. Kumar et al. [14] applied the DIC method to high strain rate experiment with explosion loading, the bending response of aluminum was measured and agreed well with the simulation results.

Through Taylor impact test, which is a typical plastic dynamic fundamental problem, reverse impact experiments were established. The deformation rules in the Taylor experiments was obtained and simulation was employed to the impact. The reverse ballistic experiment design principles were obtained to act better on the structure response of penetrator. In order to research the plastic wave propagation of specimen in reverse Taylor impact test, DIC was applied to measure the deformation rules of specimens with different velocities and structure, in addition, the plastic wave propagation was analyzed. Then, compared with the simulation results, the deformation of Taylor rod influenced by wave propagation characteristics were further studied, which proves the feasibility of DIC method in reverse ballistic test.

2. Digital image correlation matching principle

The basic principle of DIC method is matching the geometric point on the digital speckle image of object surface under different status, tracking the movements of point and analyzing the deformation information. Speckles are random distributed in the speckle field, each scattered spots is different with the surrounding point, and the small area surrounding the point is called the sub-region. In the speckle field, a point at the center of sub-region can be used as the displacement carrier, by searching and analyzing the movement and the change of the sub-region to get the displacement of the point. As shown in Fig. 1, M is the center point of sub-region S which size is $m \times m$ pixels. When the surface is moving, sub-region S moves to the location of sub-region S_1, thus the location of M_1 which comes from the feature point M after moving can be obtained and the deformation of M can be measured.

Usually a correlation coefficient C will be defined (as shown in formula (1)) to represent the characteristic speckle pattern matching degree of M and M1, its extreme value point shows the best match. DIC matching search algorithm can reach the accuracy of pixel level, but the precision also can't satisfy the precision of the conventional experimental requirements. Thus, after the pixel level matching, in order to obtain a higher measurement accuracy, special mathematical methods are used to get the position of sub-pixel level matching point. For high quality speckle figure, the measurement precision of DIC displacement can reach 0.01 pixels. The $f(x, y)$ and $g(x^*, y^*)$ are the pixel matrix of M and M_1, respectively.

$$C = 1 - \frac{\sum f(x, y) \cdot g(x^*, y^*)}{\left[\sum f^2(x, y) \cdot \sum g^2(x^*, y^*)\right]^{1/2}}. \quad (1)$$

3. Reverse Taylor impact experiment

$\phi 57$ mm Compression-shear type light-gas gun was adopted in the test system as a loading platform, as shown in Fig. 2, nitrogen was used as the emission gas in the reverse impact test. Nitrogen bottle was connected to the gas chamber, and nitrogen was compressed to the gas chamber first, then open the valve between chamber and the gun barrel. Gas drives plate to move and makes it accelerate to the predetermined speed. In the target chamber, the specimen aligned with the center of the gun barrel, which means the axes of both centers must be on one line, then, made it in a state of relative freedom. In order to offset the specimen's gravity, only a thin stick was used to hold it up at the bottom. At the back-end of the specimen a recycling basket filled with soft energy absorption material was placed to prevent the specimen from a secondary injury after impact. An inspection window was located at one side of the target and high speed camera can shoot through it during the impact process. $\phi 57$ mm Compression-shear type light-gas gun is shown in Fig. 3(a), the rated highest air pressure of gas chamber is 15 MPa, rated launch mass is $1 \sim 8$ kg, valve open time ≤ 2 ms. Protron SA5 high-speed camera

(a) Launcher system

(b) Target chamber and high-speed camera

(c) Reverse impact specimen in target chamber

Figure 3. Test equipment.

was used to record the experiment process, as shown in Fig. 3(b), shooting frame was 75000 fps, exposure time was 1/424000 s. Because of the high filming frames, two strong illuminants made of tungsten iodine lamps were needed in the target chamber to fill light. For the reverse impact test, the rigid plate was made of 45# steel with a diameter of 48 mm and thickness of 10 mm, which was put into the nylon sabot to launch and polished at the impact end. The specimen was located about 50 cm to the muzzle as shown in Fig. 3(c), the gas pressure of the test is $0.6 \sim 1.8$ Mpa, impact velocity is $104\,m/s \sim 201\,m/s$. High- speed camera was used to read the launch velocity, a lot of images were selected before impact to calculate the velocity ranging at pixel level and take average.

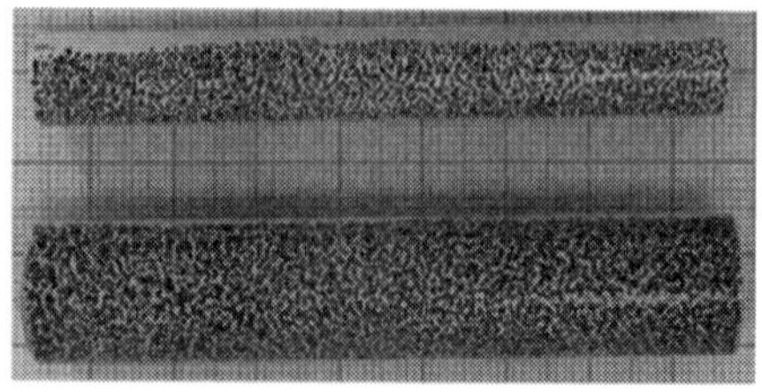

Figure 4. Specimens with speckle in reverse impact test.

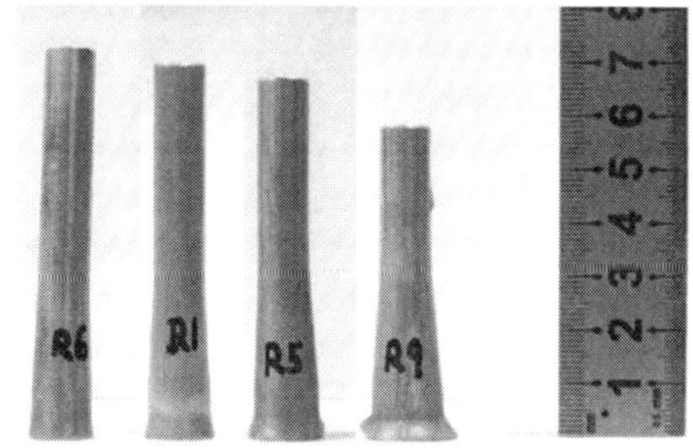

Figure 5. D = 10 mm specimens' deformation after impact.

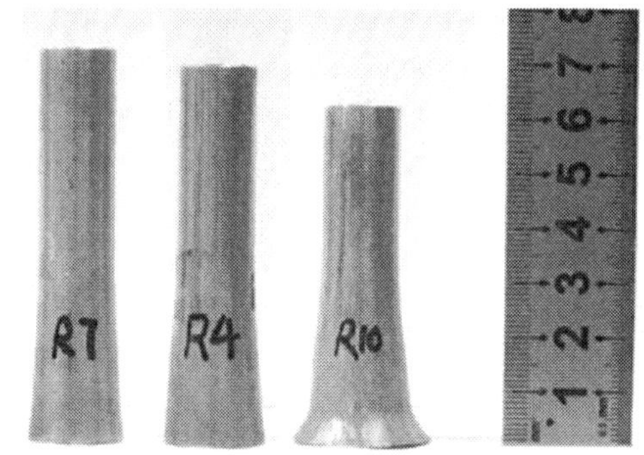

Figure 6. D = 15 mm specimens' deformation after impact.

Because of the use of DIC test system, speckles should be sprayed on the side near to the high-speed camera of the specimens, as shown in Fig. 4, speckles whose size was about $1 \sim 5$ pixels need to be sprayed evenly and irregularly. High-speed camera and gas gun were triggered synchronously by trigger device, and the shot pictures can be used for analyzing the structure response of specimens in microsecond level. The propagation speed and mode of plastic wave front can be obtained through the pictures. Then, the strain and displacement distribution diagram can be calculated by DIC method, and the deformation rules of different structure and plastic wave propagation characteristics of specimens were explored.

Figure 5 and Fig. 6 show the comparison of the specimens' deformation after reverse impact. Along with the increase of impact velocity, obvious upsetting phenomenon is produced in the impact end, and a barreled region was generated at the upper side of the end, such as

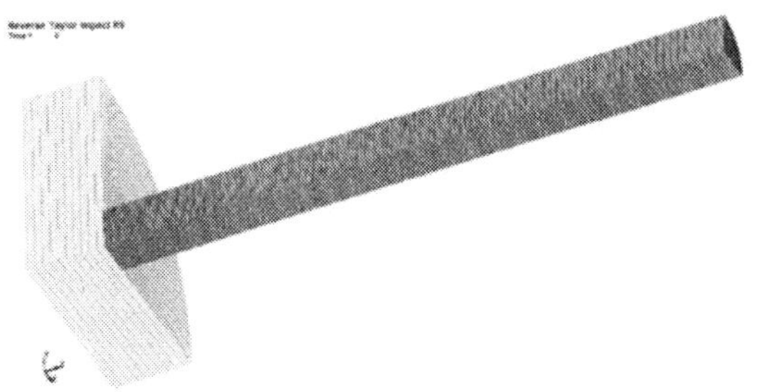

Figure 7. Finite element models of reverse Taylor impact.

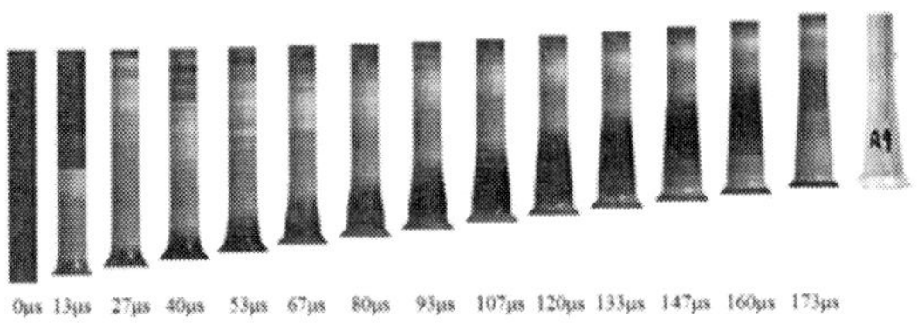

Figure 8. Von Mises stress of No.R9 reverse impact by simulation and specimen after experiment.

No.R10 specimen. The comparison between the thin rods and the thick rods in Fig. 5 and Fig. 6 show that the plastic deformation was more easily occurred in thin rods. For the thin rod, when the impact velocity is close to 150 m/s, the barreled region has already appeared. But for the thick rod, the barreled region appeared only when the velocity is above 190 m/s. The appearance of the barreled region explains the plastic wave propagation characteristics in a certain extent.

4. Simulation of reverse Taylor impact test

In order to further research the reverse Taylor impact, the numerical simulation of Taylor impact was conducted by LS-DANY3D under different working condition and mass ratio.

4.1. Finite element model

The finite element models are shown in Fig. 7, because of the symmetry, a quarter model was adopted. For the reverse impact, a solid element and rigid material model was selected for the target which was treated as rigid without deformation. Non-proportional meshing was chosen for the copper specimens, with a dense grid close to the impact end and sparse grid far away. Eroding surface to surface contact was taken between the rod and target during the impact process.

4.2. The contrast between simulation and experiment result

Figure 8 shows the von mises stress given by simulation and specimens after experiment of No.R9, and the velocity of No.R9 is 201 m/s. At the beginning of impact, when $t = 13\,\mu s$, the mushroom head was generated. Both the elastic and plastic wave propagated to the far end, but plastic wave velocity is much lower than elastic wave.

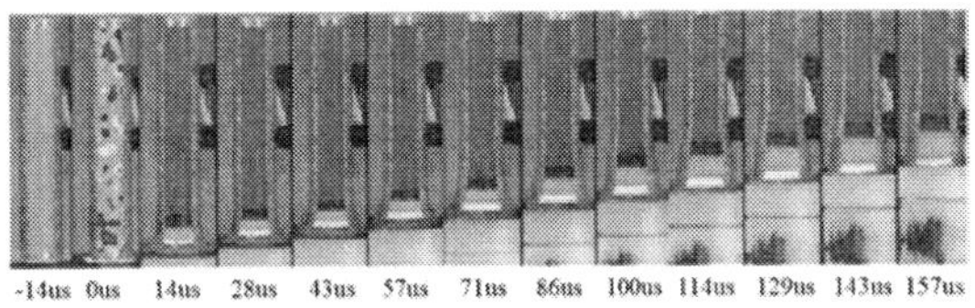

Figure 9. Radial strain of No.R10 in the impact process.

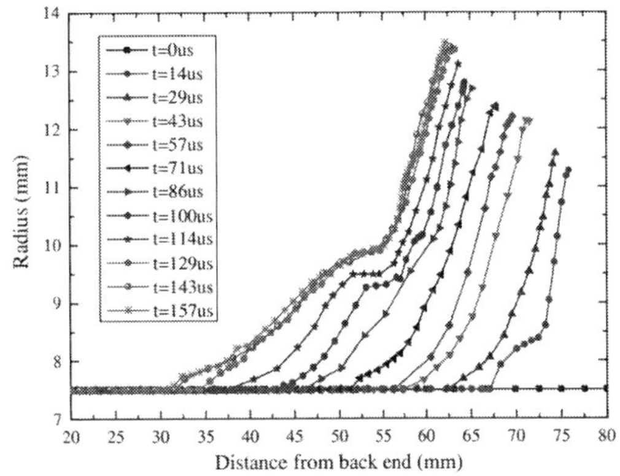

Figure 10. The instantaneous outline of No.10.

At $t = 40\,\mu s$, the elastic wave arrived at the far end and reflected, the plastic wave still propagated to the far end. Whereafter, the elastic wave bounced and overlapped at back and forth inside the specimen, but the plastic wave propagated in a relatively lower velocity to the far end as before. After $t = 133\,\mu s$, obvious movement occurred on reverse impact specimen, the total energy gradually transformed to kinetic energy. At $t = 173\,\mu s$, a very weak plastic wave still propagated in reverse impact. As shown in Fig. 8, the deformation in simulation and experiment have a good agreement, which confirms the feasibility of simulation.

5. Results and discussion

Eakins [8] adopted the analysis method similar to Taylor [1], an expression for the stress and strain was derived using the conservation equations of the elastic and elastic/plastic transition fronts. It should be noted that in all cases when conservation laws of mass and momentum are used, the equalities are performed between adjacent time-frames, rather than at a particular instance in time. At the elastic boundary, the rate of mass flow entering the elastic front must be equal to the rate of mass flow leaving the front; hence

$$\frac{d}{dt}(M^{in}) = \frac{d}{dt}(M^{out}) \tag{2}$$

$$\frac{d}{dt}(\rho_0 A_0 C dt) = \frac{d}{dt}[\rho A_0 (C - U_{p1}) dt] \tag{3}$$

where M^{in}, M^{out}, A_0, ρ_o and ρ are, respectively, the mass entering the front, mass leaving the front, original cross-sectional area, initial density, and elastic strain density. This expression contains the unknown residual particle velocity behind the elastic wave front U_{p1}. Applying the

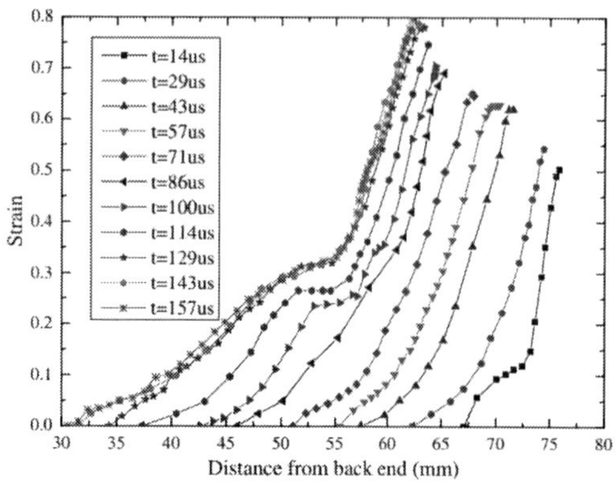

Figure 11. The instantaneous strain of No.10.

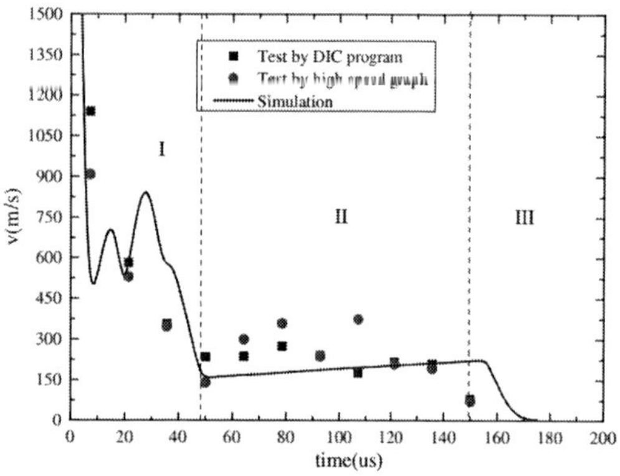

Figure 12. The elastic-plastic interface velocity of No.R10.

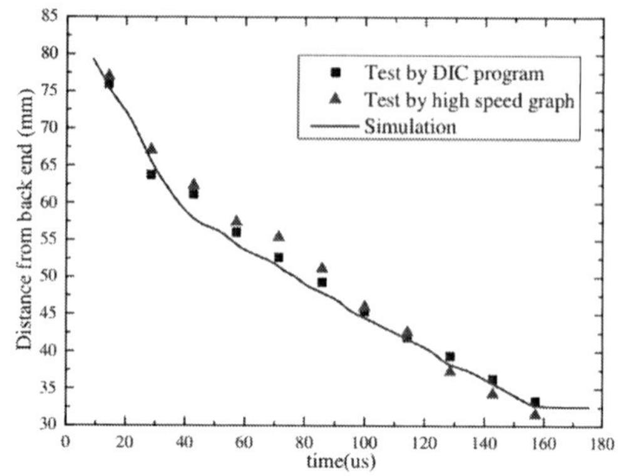

Figure 13. The elastic-plastic interface position of No.R10.

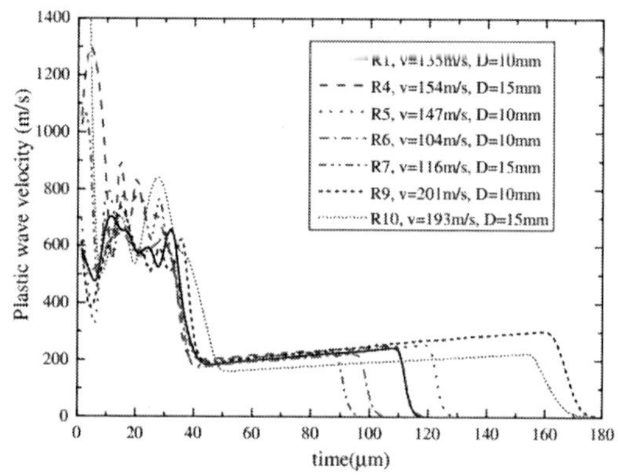

Figure 14. Time history curve of plastic wave velocity with different impact velocity and structure.

conservation of mass to the elastic/plastic transition front, the Eq. (7) gives an equivalent expression for the rate of mass flow when entering and exiting the front

$$\frac{d}{dt}[\rho A_0(v-U_{p1})dt] = \frac{d}{dt}[\rho A(v-U_{p2})dt] \qquad (4)$$

where, v represents the speed when the elastic/plastic transition front (leading plastic wave) moves into elastically strained material, and A is the increased cross-sectional area just within the plastic zone. Again, the initial area A_0 is taken at a position ahead of the elastic/plastic transition front at t_1, while the strained area A is taken at that same position after the front has passed at a later time t_2.

Considering the conservation of momentum across the elastic/plastic transition front, by equating the rate of change in momentum to the net impulse. The impulse can be expressed as

$$\mathrm{F}dt = \sigma(A_0 - A)dt \qquad (5)$$

where, F is force, σ is stress. Then, the rate of momentum change across the elastic/plastic transition front can be written as

$$\frac{d}{dt}(p^{in} - p^{out}) = \frac{d}{dt}(M^{in}v^{in} - M^{out}v^{out}) \qquad (6)$$

where, p is momentum, M and v are the mass and corresponding velocity of material entering or exiting the transition front. Considering the density change across the elastic front is negligible, the stress expression can be given by the Eqs. (2)–(6) as

$$\sigma = -\frac{\rho_0 A_0 v^2}{A} = -\rho_0 v^2(1-\varepsilon) \qquad (7)$$

then,

$$\varepsilon = \frac{A - A_0}{A}. \qquad (8)$$

Calculation and analysis were conducted on the results of DIC measurement, and the radius strain pictures were obtained. Figure 9 reflects the process of plastic wave propagation in the rod. It can be seen that a tiny vibration was occurred at the beginning of the impact, then the expansion happened at the impact end, and the strain showed a trend of layered. The nearer to the impact side, the greater the strain gradient was. The experiment results are similar to the simulation, which proves the correctness of the above simulation work on the other side.

Through the measurement of experiment, the curves of instantaneous outline and strain were obtained, as shown in Fig. 10 and Fig. 11. At $t = 14\,\mu s$, a barreled region caused by elastic wave appeared at the position of 70 mm away from the non-impact end. Along with the impact process, because of the serious deformation at the impact end, the elastic barreled region rebounded during $t = 29 \sim 86\,\mu s$. After $t = 86\,\mu s$, the barreled region appeared as a plastic deformation. Then, the deformation gradually slowed down and the impact process ended.

The elastic-plastic interface velocity of No. R10 is shown in Fig. 12, considering the theory of Jones et al. [6], the deformation of a Taylor specimen was characterized by three patterns of behavior. The early time behavior, denoted by Phase I, was shock dominated nonlinear motion of the plastic wave front, as shown in Fig. 12 and Fig. 13 at $0 \sim 50\,\mu s$. This was followed by steady motion of the plastic wave front, denoted by Phase II during 50–150 μs. This simple description is mostly correct, but it is evident that the velocity of the plastic wave front must come to zero at the end of the event. Hence, a description that includes nonlinear behavior of the wave front at the end of the event is appropriate. This description is represented as the third phase, namely Phase III, in which deceleration of the plastic wave front occurs in 150–180 μs. Notice that the velocity comes to zero at the end of the event. According to the distance from elastic-plastic wave front to impact end in Fig. 13, the trend of wave propagation agree well with the theory of Jones et al. [6], which proves that the DIC method is effective.

Comparing the rods with different impact velocities and different structures during the Taylor impact process, the length of the stable phase of plastic wave propagation is different, but the values of the duration are slightly different. As shown in Fig. 14, the stable phase of plastic wave propagation is long when it is at a high impact velocity. For 15 mm diameter rods, the length of stable phase is shorter than that for 10 mm diameter rods, and the propagation velocities are similar. It illustrates that the deformation of 15 mm diameter rods is smaller than that for 10 mm diameter rods.

6. Conclusions

Through reverse Taylor impact tests, the dynamic response data of Taylor rod in microsecond level time resolution was obtained, and the deformation and outlines of the rod at different times were analyzed.

It is better to use the high speed DIC method than traditional optical method to measure the reverse impact tests. According to the contrast of experiments and simulation, which proved the three stages of plastic wave propagation are basically identical, the propagation characteristic of plastic wave was analyzed. The rules of plastic wave propagation in different rod structures and impact velocities were obtained.

This work was supported by the National Natural Science Foundation of China under Grant (No.11221202) and the National Natural Science Foundation of China (No.11202029, No.11390362).

References

[1] G. Taylor, Proc. R. Soc. London, Ser. A **194** (1038), 289 (1948).
[2] J. W. House, B. Aref, J. C. Foster, and P. P. Gillis, J. Strain Anal. Eng. Des **34** (5), 337 (1999).
[3] Davila, J. M. Huntley, G. H. Kaufmann, and D. Kerr, Appl. Opt. **44** (19), 3954 (2005).
[4] D. C. Erlich and D. A. Shockey, presented at the Shock Waves in Condensed Matter – 1983. *American Physical Society Topical Conference*, 18–21 July 1983, Amsterdam, Netherlands, 1984 (unpublished).
[5] S. E. Jones, P. P. Gillis, J. J. C. Foster, and L. L. Wilson, J. Eng. Mater. Technol **113** (2), 228 (1991).
[6] S. E. Jones, P. J. Maudlin, and J. C. Foster, Int. J. Imp. Eng **19** (2), 95 (1997).
[7] D. Eakins and N. N. Thadhani, J. Appl. Phys. **100** (7), 073503 (2006).
[8] D. Eakins and N. N. Thadhani, Int. J. Imp. Eng **34** (11), 1821 (2007).
[9] E. Anderson Jr, T. Behner, T. J. Holmquist, and D. L. Orphal, Int. J. Imp. Eng **38** (11), 892 (2011).
[10] E. Lidén, B. Johansson, and B. Lundberg, Int. J. Imp. Eng **32** (10), 1696 (2006).
[11] Asundi, Opt. Lett. **25** (4), 218 (2000).
[12] Cofaru, W. Philips, and W. Van Paepegem, Meas. Sci. Technol. **23** (10), 105406 (2012).
[13] F. Pierron, M. A. Sutton, and V. Tiwari, Exp. Mech. **51** (4), 537 (2010).
[14] P. Kumar, J. LeBlanc, D. S. Stargel, and A. Shukla, Int. J. Imp. Eng **46**, 74 (2012).

EPJ Web of Conferences **94**, 01075 (2015)
DOI: 10.1051/epjconf/20159401075

A comparative study on the restrictions of dynamic test methods

GH. Majzoobi and S. Lahmi

Mechanical Engineering Department, Bu-Ali Sina University, Hamedan, Iran

Abstract. Dynamic behavior of materials is investigated using different devices. Each of the devices has some restrictions. For instance, the stress-strain curve of the materials can be captured at high strain rates only with Hopkinson bar. However, by using a new approach some of the other techniques could be used to obtain the constants of material models such as Johnson-Cook model too. In this work, the restrictions of some devices such as drop hammer, Taylor test, Flying wedge, Shot impact test, dynamic tensile extrusion and Hopkinson bars which are used to characterize the material properties at high strain rates are described. The level of strain and strain rate and their restrictions are very important in examining the efficiency of each of the devices. For instance, necking or bulging in tensile and compressive Hopkinson bars, fragmentation in dynamic tensile extrusion and petaling in Taylor test are restricting issues in the level of strain rate attainable in the devices

1. Introduction

A wide variety of dynamic materials testing machines with different capabilities have been developed across the world. Generally, dynamic testing of materials can be categorized as intermediate (10^{-1}–$10^{2}\,s^{-1}$) or high strain rate (10^{2}–$10^{4}\,s^{-1}$). Strain rates higher than $10^{4}\,s^{-1}$ usually correspond to the phenomena involved in shock wave propagation. The Intermediate strain rate testing techniques utilise air, hydraulic or mechanical testing machines. The piston type machine designed by Maiden and Green [1] is a typical example of the first category. The second category comprises hydraulic testing machines such as the advanced hydraulic torsion testing machine developed by Lindholm [2], which was capable of creating a continuous range of strain rates from quasi-static to $500\,s^{-1}$. In mechanical testing machines, the required energy is usually stored in a large moving mass such as a fly wheel. One of the earliest types of dynamic mechanical testing machines was built by Nadia and Manjoine [3]. In this machine, the flywheel rotates at a constant angular velocity. At the time of impact, a hammer which is supported by a spring is suddenly released and is engaged with the base of the specimen. Another testing machine which is capable of producing a uniform strain rate is the cam plastometer [4]. In Much of the experimental work on the high strain rate response of materials involves the use of the most universally accepted high rate testing apparatus, the split Hopkinson bar (SHB), also known as the Kolsky bar, with different loading modes, including tension [5], compression [6], and shear [7]. Strain rates up to $10^{4}\,s^{-1}$ can be attained by the SHB. An alternative dynamic tensile testing facility, known as the Flying Wedge, which is capable of generating strain rates from $10^{2}\,s^{-1}$ up to in excess of $10^{4}\,s^{-1}$, has been developed and used at Leeds University for several decades [8].

2. The high strain rate features

Some of the objectives in dynamic tests which look to be of more importance are described in following sections.

2.1. Strain and strain rate distribution

Strain rate is often obtained approximately from V/H where V is the deformation velocity and H is the gauge length of the specimen. But this relation is too rough and in some cases not true at all. In Hopkinson bar strain rate is calculated from:

$$\frac{d\varepsilon_s(t)}{dt} = -\frac{2C_0}{L}\varepsilon_R(t).$$

Where, $\varepsilon_R(t)$is the stress wave reflected into the input bar. The strain rate is obviously constant before necking in tension and bulging in compression. The distribution of strain rate obtained in Taylor test is illustrated in Fig. 1. As it is seen strain rate is not uniform within the specimens. Therefore, talking of strain rate in a test is basically nonsense. As a matter of fact, strain rate is a local quantity that is nonuniformly distributed within the specimen and varies from point to point in time and space.

2.2. Stress-strain curve and material models

Material models are widely used in finite element codes for analysis of material deformations particularly at high strain rates and elevated temperatures. Each material model has some constants which are determined using stress-strain curves obtained by experiment. The problems such as necking and bulging limit the conventional test techniques to measure the stress-strain curves only up to small strains. This is while, in some deformation processes, the strain can be greater than 1. Over the past recent years, a combined experimental, numerical and optimization approach has been used for determination of the constants. The method is explained in Sect. 5. The flow stress is defined in terms of strain, strain rate and temperature through some material models a number of which are given below:

Johnson-Cook [9]:

$$\sigma = \left(A + B\varepsilon^{n}\right)\left(1 + C Ln(\dot{\varepsilon}/\dot{\varepsilon}_0)\right)\left(1 - T^{*m}\right)$$

Article available at http://www.epj-conferences.org or http://dx.doi.org/10.1051/epjconf/20159401075

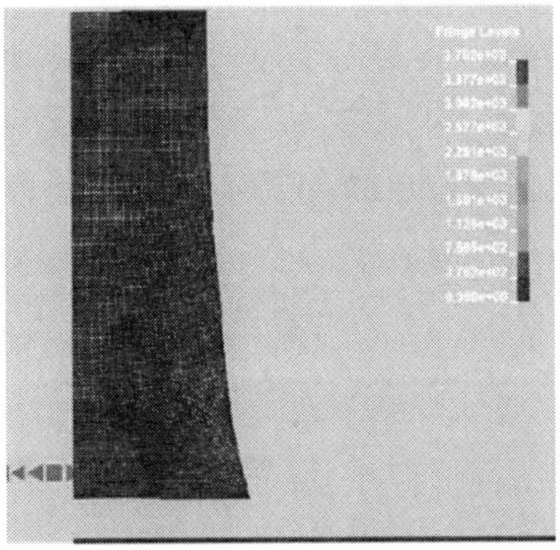

Figure 1. Strain rate distribution in a Taylor test specimen.

Zerilli-Armstrong [10]:

$$\sigma = C_1 + C_2 \exp(-C_3 T + C_4 T \mathrm{Ln}(\dot{\varepsilon})) + C_5 \varepsilon^n \quad \text{for} \quad \text{fcc}$$

$$\sigma = C_1 + C_2 \sqrt{\varepsilon} \exp(-C_3 T + C_4 T \mathrm{Ln}(\dot{\varepsilon})) \quad \text{for} \quad \text{bcc}.$$

2.3. Damage models

Determination of the constants of damage models are also one of the main objectives in dynamic tests. Two of the most important models are:

Analytical Gurson, Tvergaard and Needleman model [11]:

$$\phi = \left(\frac{\sigma_e}{\sigma_0(\bar{\varepsilon})}\right)^2 + 2q_1 f \cosh\left[-\frac{3}{2}\frac{\sigma_m}{\sigma}\right] - \left(1 + q_1^2 f^2\right) = 0.$$

Empirical Johnson-Cook damage model [12]:

$$\varepsilon_f = \left(D_1 + D_2 EXP\left(D_3\left(P/\sigma_y\right)\right)\right) \left(1 + D_4 Ln\dot{\varepsilon}\right)\left(1 + D_5 T^*\right).$$

2.4. Fracture mechanism

Fracture mechanism of a material may change from a pure ductile under quasi-static conditions to pure brittle under dynamic loading. Typical fracture surfaces are shown in Fig. 2. Therefore, fracture mechanism is one of the major concerns in high strain rate deformations.

2.5. Fragmentation

Fragmentation occurs in high rate impacts or deformations. Typical example is shown in Fig. 3. Fragmentation is a restricting event in high rate testing techniques. For instance, fragmentation in Taylor test limits the plasticity in the specimens. However, some testing techniques such dynamic tensile extrusion have been innovated to study the fragmentation behavior of materials at high strain rates.

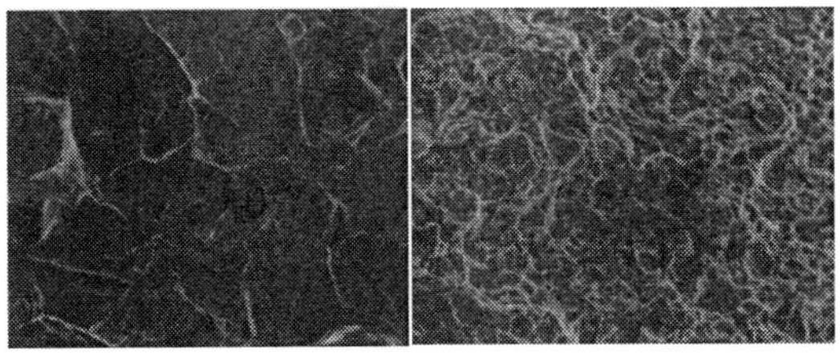

Figure 2. A pure brittle and (b) a pure ductile fracture mechanism.

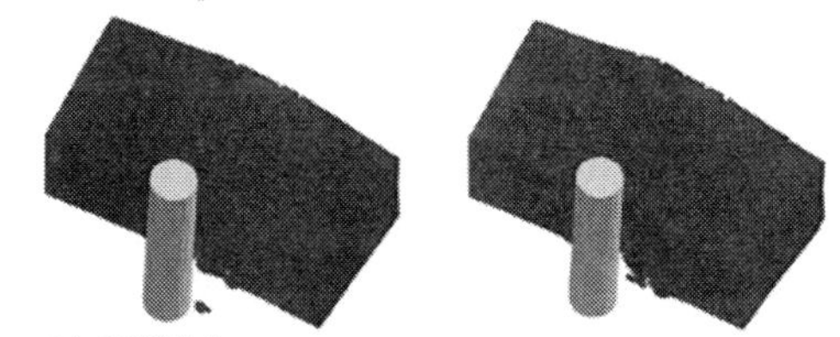

Figure 3. Fragmentation in a ballistic test [13].

Figure 4. A general view of the "Flying wedge".

3. Experimental methods

The experimental methods as explained in Sect. 1 are various. However, some of the most important ones which have wider application and the newest ones and their features are explained in the following sections.

3.1. Flying wedge

A general view of this apparatus is shown in Fig. 4. The slider mechanism, shown in Fig. 5, consists of two sliders with chamfered faces, the angle of which match up precisely with the wedge angle, together with two rails which allow the sliders to move freely, and two shock absorbers. Specimens are fitted between two specimen holders that are located within the sliders. When the gas gun is fired, the wedge plate connected to the piston rod accelerates quickly and hits the sliders. As a consequence of the impact, the sliders move rapidly away from each other resulting in tensile loading of the specimen.

By changing the gas gun pressure and using notched specimens strain rates up to $10^3\ s^{-1}$ can be achieved with this testing rig. This device is very useful for the study of fracture mechanism and also damage evolution [14 & 15] at high strain rates. However, the device is not suitable for obtaining stress-strain curve.

Figure 5. The slider mechanism and the specimen.

V=328 m/s V=487.2 m/s

Figure 6. Typical specimens after Taylor test.

3.2. Taylor test

Taylor test was initially proposed by Taylor [16] in 1948 aiming at the study the ballistic behavior of metals. It relates the dynamic properties of the materials to the quasi-static one. His analysis is based on stress wave propagation in the specimens due to the impact of the projectile against a rigid target. The analysis takes account of the mushroomed profile of the specimen after impact. Today, Taylor test is still widely used for the study of dynamic behavior of materials at high strain rates along with numerical simulation. It is also used in inverse engineering such as the method explained in Sect. 5 for determining the parameters of material models.

Typical specimens deformed in Tayor test are illustrated in Fig. 6. As the figure indicates, when the impact velocity increases, the specimens deformation changes from a quite mushroomed shape to a shape which consists of two quite distinct regions; an undeformed region and a petal zone. The petaling limits the strain and strain rates achievable in Taylor test on one hand but provides means to study the petaling phenomenon itself on the other hand.

3.3. Inverse Taylor test

In this technique, the target (specimen) and the projectile (striker) swap their positions as usually considered in the conventional Taylor test (see Fig. 7) [17]. This type of test configuration eliminates the difficulties which are normally associated with *in situ* heating up the specimen. However, strain rates in inverse Taylor test is considerably lower than that in Taylor test. The reason is that the projectile (Target) in inverse Taylor test is much heavier than that (specimen) in Taylor test and consequently, the impact velocity in Taylor test would be much higher.

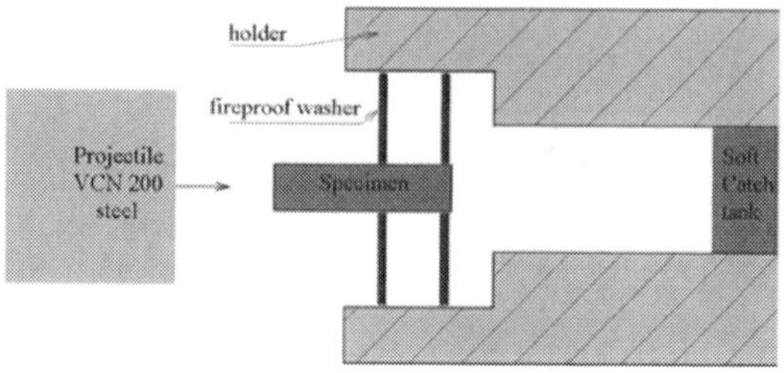

Figure 7. The schematic view of the inverse Taylor test [17].

Figure 8. A tensile Hopkinson bar apparatus.

3.4. Hopkinson bar

Hopkinson bar consists mainly of three bars, striker, incident and transmission bars [5–7]. The general view of a Hopkinson bar is shown in Fig. 8. The striker bar is driven by a gas gun and hits the incident bar. A stress wave is produced and travels down the incident bar toward the specimen. In Hopkinson test method, the signals including the incident wave, ε_I, reflected wave, ε_R, and transmitted wave, ε_T are used to calculate the torque, strain, stress and strain rate in the specimen as follows [5].

$$\frac{d\varepsilon_s(t)}{dt} = -\frac{2C_0}{L}\varepsilon_R(t)$$

$$\varepsilon_s(t) = -\frac{2C_0}{L}\int_0^t \varepsilon_R(t).dt$$

$$\sigma_s(t) = E\frac{A_b}{A_s}\varepsilon_T(t).$$

Where, E, C_0, A_b, A_S, L are the Young modulus, elastic wave speed, cross sectional area of the bars, initial cross sectional area and initial length of the specimen, respectively.

3.5. Shot impact test

The problems such as necking and bulging limit the conventional test techniques to measure the stress-strain curves only up to small strains. This is while, in some deformation processes, the strain can be greater than 1. In shot impact method, high strength shots are impacted on specimens at high impact velocities and at elevated temperatures using a gas gun [18]. The section of the shot inside a crater is illustrated in Fig. 9. Strain rates up to about $10^5\,s^{-1}$ are achieved in this work. The geometry of the crater created by the shot impact on the specimen is used for determination of the constants of the material model. This method will eliminate the need to extrapolate

Figure 9. The section of the shot inside the crater [19].

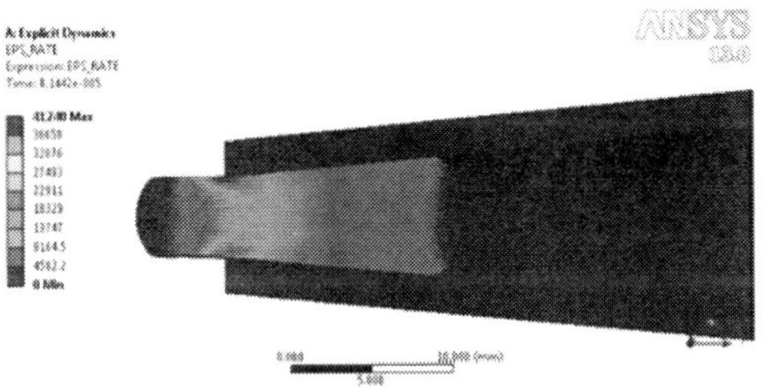

Figure 10. The process of DTE [18].

the experimental stress strain curve to large strains which may not be true in reality.

The constants of the material model correspond to the case when the experimental and numerical geometries of the craters created by the shot impact on the specimen coincide. The coincidence can be achieved by an optimization process such as genetic algorithm. The shot impact test is confined to plasticity without damage. Therefore, it appears to remain dominated solely by plasticity even at the upper velocities.

3.6. Dynamic tensile extrusion

In dynamic tensile extrusion (DTE) proposed by Gray *et al.* [19] several years ago, a spherical projectile (specimen) is shot through an open conical die in which the exit diameter of the die is smaller than the sphere diameter, the projectile undergoes severe dynamic tension at high strain rates. A simulation of the process is shown in Fig. 10. The order of strain rate is 10^4/s which is less than those seen in shot-impact tests and even Taylor tests. Dynamic Tensile Extrusion (DTE) is mostly used for studying the fracture mechanism particularly for determining the ductile to fracture transition in materials especially in polymers, although it has been used by Bonora et al and a few others for adjusting some material models. DTE is quite an interesting method by which the effect of material's microstructures on macro-structural response of materials such as fragmentation and damage can be studied.

3.7. Drop hammer

In this device a dropping weight as the hammer hits the specimen which is mounted on an anvil. This device is

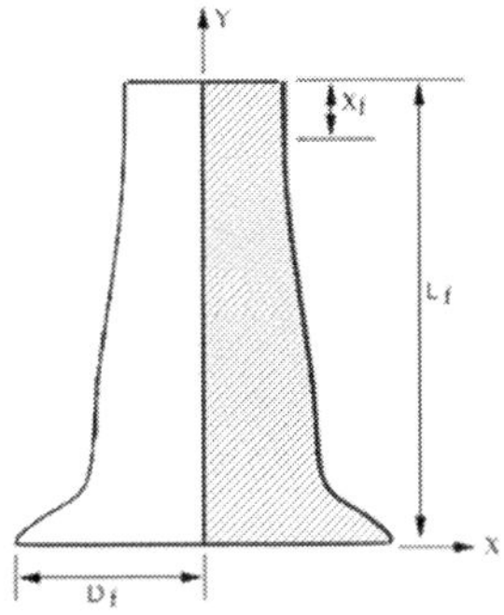

Figure 11. The dimensions used for optimization of final profile.

mainly used to study the behavior of energy absorbers. However, it can be used for compressive test and even tensile test too. This test provides medium strain rates. It lacks most of the capabilities of the dynamic test devices and even universal testing rigs.

3.8. Universal devices

Universal machines are the commercial testing apparatuses which are mainly used mechanical characterization of materials under quasi-static and low rate loadings. The universal machines such as Instron, Zwick, MTS, Avery, etc are among the most frequently used machines in the world. These machines lacks most of the capabilities of high rate testing machine such as the capabilities to study fragmentation, fracture mechanism transformation, erosion and other features which occur only in high rate deformations. However, these machines enjoy capabilities such as providing stress-strain curve, cyclic test for studying the damage evolution or isotropic/kinematic hardening of material.

4. Inverse engineering and material model

As explained above, dynamic stress-strain curves are rigorously converted into material models which are highly important elements in the simulation of material deformation. In a new strategy called as "inverse engineering", the dynamic stress-strain curve is eliminated from the test procedure and the constants of the material model are determined in a way that the difference between the experimental and the numerical load-displacement curves or the shape of dynamically deformed specimens is minimized [20,21]. This is performed by a combined experimental, numerical and optimization technique. The difference is taken as the objective function of an optimization problem. which can be defined by a function such as a polynomial as follows:

$$OBJ(x) \cong a_0 + \sum_{i=1}^{n} a_i x_i + \sum_{i,j=1}^{n} b_{ij} x_i x_j. \quad (1)$$

In which x_i (i = 1 to n) is the model parameters vector, the coefficients a_i and b_{ij} are identified from the numerical simulations and n is the number of parameters of the material model under investigation. The number of

Table 1. The evaluation of various parameters in dynamic testing devices.

Device Name	Loading mode	Strain Rate (s^{-1})	Elevated temperature test	Fracture mechanism	$\sigma-\varepsilon$ curve	Hysteresis curve	Inverse engineering	Damage evolution	Fragmen Tation	Phase Transformation
Flying Wedge	Tension	10^3	√√	√√√	√	–	√	√√√	–	√√√
Taylor test	Compression	10^4	√	–	–	–	√√√	–	√	√√√
Inverse Taylor test	Compression	10^3	√√ √	–	–	–	√√√	–	√	√√√
Hopkinson Bar	Tension Compression Shear Torsion	10^3	√√√	√√√	√√√	– –	√√	√√√ (tensile)	– –	√√√
Universal devices	Tension Compression Shear Torsion	10^1	√√√	–	√√√	√√√	√	√√	–	√√√
Shot impact test	–	10^5	√√√	√	–	–	√√√	–	–	√√√
Dynamic tensile extrusion	–	10^4	√	√√√	–	–	√√√	–	√√√	√√√
Drop Hammer	Tension Compression Shear	10^2	√√√	√	–	–	√√	–	√	√√
√√√ **Excellent**					√√ **Good**			√ **Weak**		

coefficients a_i which required to be computed from the simultaneous solution of a system of n algebraic equations determines the number of the numerical simulations needed for each tensile or compressive test. In Taylor test, the optimization is performed by designating three dimensions on the final profile of the specimen as shown in Fig. 11.

In this figure, $a_1 = L_f$ and $a_3 = D_f$ are the final length and the final end diameter of the specimen, respectively. The variable $a_2 = X_f$ is the length of the specimen which has not experienced any deformation during the impact. The objective function is defined as follows:

$$OBJ = \sqrt{OBJ^{I^2} + OBJ^{II^2} + OBJ^{III^2}} \quad (2)$$

$$\begin{aligned} OBJ^{I} &= a_{1\text{experimental}} - a_{1\text{numerical}} \\ &= L_{f\text{experimental}} - L_{f numerical} \end{aligned} \quad (3)$$

$$\begin{aligned} OBJ^{II} &= a_{2\text{experimental}} - a_{2\text{numerical}} \\ &= X_{f\text{experimental}} - X_{f\text{numerical}} \end{aligned} \quad (4)$$

$$\begin{aligned} OBJ^{III} &= a_{3\text{experimental}} - a_{3\text{numerical}} \\ &= D_{f\text{experimental}} - D_{f numerical}. \end{aligned} \quad (5)$$

The terms $a_{i\text{Experimental}} - a_{i\text{Numerical}}$ (i = 1,2,3) represent the differences between the numerical and experimental profiles of the specimen for the three designated dimensions shown in Fig. 11.

5. Summary and Conclusions

The results of this study are summarized in Table 1. Each method clearly covers a range of strain rate. Therefore, the methods may not be comparable with each other in some respects. However, it is extremely important to know the capabilities of each method and its restrictions. As Table 1 indicates, some test methods are excellent for the study of some quantities or phenomenon but are useless for others. For instance, although a few techniques can be used for investigating fragmentation but dynamic tensile extrusion method is the best choice for this purpose. This is while this method is inefficient in other respects. It can be observed that the best equipments for capturing stress-strain curve are Hopkinson bars. Nevertheless, all methods can be employed for determining the constants of material models by inverse engineering as explained in Sect. 4. Flying wedge and tensile Hopkinson bar are the most appropriate devices for the study of fracture mechanism at high strain rates. These devices are also the best choices for the study of damage evolution which is a subject of damage mechanics. The specimens should be deformed to a controlled level and voids initiation, growth and coalescence is examined to obtain the damage evolution. This can be done using Flying Wedge of Hopkinson bar. Of course, damage evolution is also evaluated by cyclic test and hysteresis curve which can solely be obtained by universal testing machines. Phase transformation such as martensite formation of deformation mechanism such as twinning can be studied by nearly all test methods.

References

[1] C.J. Maiden and S.J.Green, J. app. mech., **33**, 4961–4970 (1966).
[2] U.S. Lindholm Techniques of metal research, 199–253 (1967).
[3] Nadia A. and Manjoine M.J., Trans. ASME, **63**, A77 (1941).
[4] E. Orowan, B.I.S.R.A. Report MW/F/22/50, (1950).
[5] J. Harding, E.D. Wood and J.D.Campbell, J. of Mech. Eng. Sci., **2**, 89, 88–96 (1960).
[6] T.V.T Yeung-Wye-Kong, Ph.D. Thesis, Mech. Eng. Dep. Leeds Univ. (1972).
[7] J. Duffy, J.D. Campbell and R.H. Hawley, Trans. ASME: J. Appli. Mech., **38**, 83–91 (1971).
[8] B.N. Cole and J.L. Sturges, Int. J. Imp. Eng., **28**, 891–908 (2003).
[9] G.R. Johnson and W.H. Cook, 7th int. symp. Ballistics, 541–547 (1983).
[10] F.J. Zerilli, R.W. Armstrong, J. Appl. Phys. **61** 1816–1825 (1987).
[11] A.L. Gurson, Ph.D. Thesis, Brown University, USA (1975).
[12] G.R. Johnson, W.H. Cook, Engng Fract Mech, **21**, 31–48 (1985).
[13] `http://www.cusatis.us/?p=820`
[14] G.H. Majzoobi, A Hosseini, A Shahvarpour, Part I: experiments: Steel research int. **79**, 685–692 (2008).
[15] G.H. Majzoobi, A Hosseini, A Shahvarpour, Part II: Simulation, Steel research int. **79**, 712–718 (2008).
[16] G.I. Taylor, Proc. R. Soc. Lond. A, **194**, 289-299 (1948).
[17] G.H. Majzoobi, P. Kazemi, M.K. Pipelzadeh, J. Exp. tech., In press, (2014).
[18] G.H. Majzoobi, A. Hosseinkhani, M.K. Pipelzadeh, S.j. Hardy, J Strain Analysis, **49**(5) 342–351 (2014).
[19] Gray, G.T., Cerreta, E., Yablinsky, C.A., Addessio, L.B., Henrie, B.L., Sencer, B.H., Burkett, M., Maudlin, P.J., Maloy, S.A., Trujillo, C.P., Lopez, M.F., AIP Conf. Proc., **845**, 725–728 (2005).
[20] G.H. Majzoobi, F. Freshteh Saniee, S. Khosroshahi, H. Beikmohammadloo, Part I: Experiments and simulations, Computational Material science, **49**, 192–200 (2010).
[21] G.H. Majzoobi, S. Khosroshahi, H. Beikmohammadloo, Part II, Optimization, Computational Material science, **49** 201–208 (2010).

EPJ Web of Conferences **94**, 01076 (2015)
DOI: 10.1051/epjconf/20159401076

Study of phase transitions in cerium in shock-wave experiments

M.V. Zhernokletov[a], V.A. Borisenok, V.G. Simakov, V.A. Bragunets, E.E. Shestakov, A.M. Podurets, and M.I. Tkachenko

Russian Federal Nuclear Center – All-Russia Scientific and Research Institute of Experimental Physics, Sarov, Russia

Abstract. Cerium has a complex phase diagram that is explained by the presence of structure phase transitions. Planar gauges were used in various combinations in experiments for determination of sound velocity dependence on pressure in cerium by the technique of PVDF gauge. The data of time dependence on pressure profiles with use of $x(t)$ diagrams and the $D(u)$ relation for cerium allowed the definition of the Lagrangian velocity of the unloading wave C^{Lagr} and the Eulerian velocity C^{Eul} by taking into account the compression σ. These results accords with data obtained by using the technique of VISAR and a manganin-based gauge, and calculated pressure dependence of isentropic sound velocity according to the VNIITF EOS. Metallography analysis of post experimental samples did not find any changes in a phase composition.

1. Introduction

Cerium is a very convenient material to carry out shock wave studies aimed at deriving a multiphase equation of state and strength characteristics [1].

Analysis of earlier theoretical and experimental researches testifies that cerium properties have been rather completely investigated in the pressure area above 5 GPa. There was no actually experimental data in the pressure area lower 5 GPa, where, by the theory, extra strong anomaly of thermodynamic characteristics of cerium should be observed. This fact determined the objectives of this work. They are investigation of isomorphic phase γ-α transformation in cerium and its abnormal behavior under shock-wave loading in the pressure range up to 3 GPa; determination of sound velocity in compressed material; determination of phase transition pressure; metallography of post-experimental samples.

2. Registration of pressure profiles in cerium

Cerium samples were loaded under pressures lower than 3.8 GPa with the use of a ballistic facility based on a light-gas gun. Helium was used as a working gas in the gun. The facility is equipped with a laser device intended for shell velocity measurement and an equipment complex intended for the technique of a PVDF gauge.

Samples (targets) with diameter of 52 mm and with thickness from 2 mm to 6 mm were used in the tests. The samples were manufactured using electrolytic cerium having purity of 99.83%. A content of impurities (mass %) was the following: La – 0.07%, Nd –0.05%, Fe–0.02%. Density of the material amounted to 6.75 g/cm^3.

Tests were performed with various experimental setups, namely, symmetric impact, asymmetric impact (Figs. 1, 2).

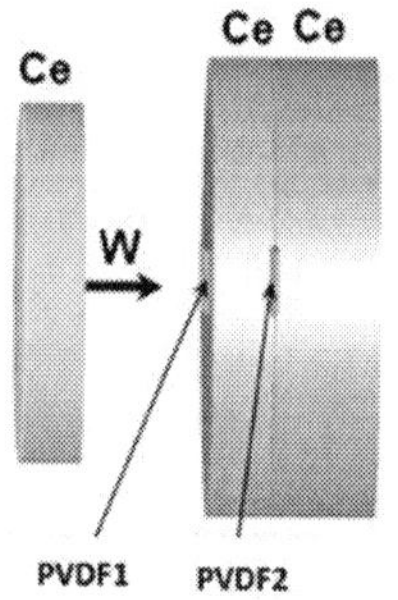

Figure 1. Experimental setup with symmetric impact.

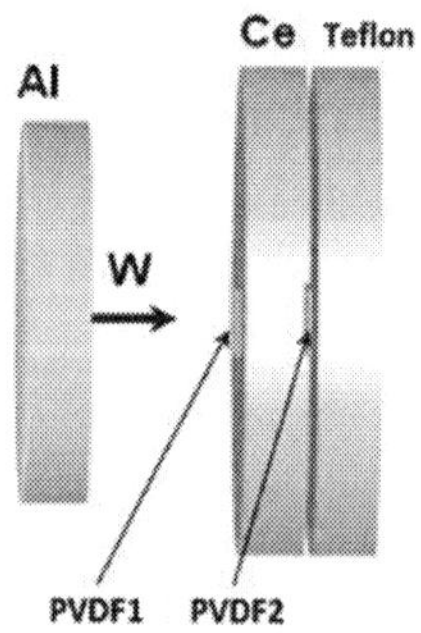

Figure 2. Experimental setup with asymmetric impact.

Velocities of aluminum and cerium impactors were changed from 312 m/s to 654 m/s. It corresponds to loading pressure range in cerium samples 0.81 GPa to 2.28 GPa.

Pressure profiles obtained in cerium samples in "gun" and "explosive", test are present in Fig. 3.

Analysis of the $P(t)$ dependence structures shows that a two-wave structure is recorded in cerium (Fig. 3). It consists of the head wave of isentropic compression and the shock wave following after it. Formation of this structure is caused by isomorphic (γ–α) phase transition.

[a] Corresponding author: MVZhernokletov@vniief.ru

Article available at http://www.epj-conferences.org or http://dx.doi.org/10.1051/epjconf/20159401076

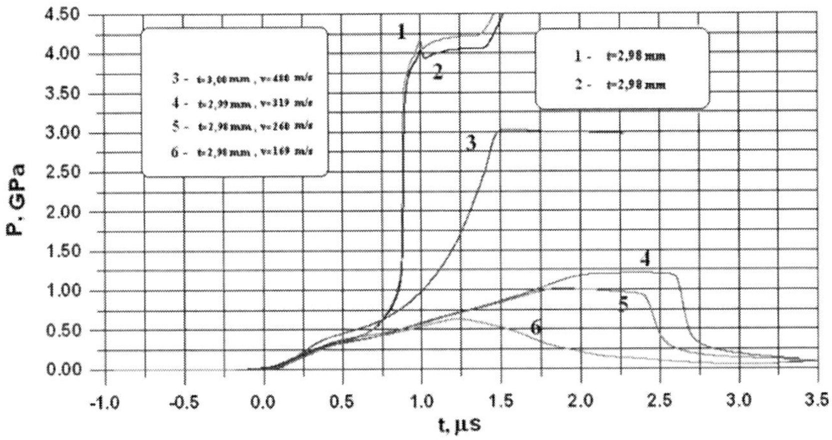

Figure 3. Pressure profiles in cerium under varios loading conditions ranging from (0.8–4.1) GPa.

Table 1. Experimental results. Symmetric impact.

W, m/s	$X\mathrm{Ce}_{\mathrm{imp}}$, mm	X_{Ce}, mm	U_{Ce}, km/s	D_{Ce}, km/s	P_{Ce}, GPa	P_{Ce}, GPa phase	C, km/s
312	3.0	2.97	0.156	1.16	1.01	0.89	1.25
484	1.93	1.95	0.242	0.95	1.18	0.86	1.07
564	1.78	1.93	0.282	0.99	1.33	0.77	1.04

Table 2. Experimental results. Asymmetric impact.

W, m/s	$\mathrm{XAl}_{\mathrm{imp}}$, mm	X_{Ce}, mm	D_{Ce}, km/s	P_{Ce},GPa	C, km/s
466	3.0	3.97	0.96	1.47	0.97
550	3.0	3.94	1.17	1.63	0.99
654	3.0	3.99	1.32	1.93	1.04
519	4.0	1.85	1.59	-	1.26 (1.3) 1.93 (4.8)
438	4.0	1.71	1.55	-	1.35 (1.32)

It has incomplete character at impactor velocities up to W=654 m/s, which take place in the tests. As a result, the material is a mixture of γ and α phases, which determines its anomalous properties. As the impactor velocity grows, shape of the pressure profile in the loading phase approaches its "explosive" analog [2]. Basing on results of all tests, value of the phase transition pressure is 0.83 ± 0.06 GPa.

Decrease in pressure occurs at all $P(t)$ dependences in the zone of constant flow. To all appearances, it can be explained by the occurred isomorphic phase transition. Rarefaction shock wave is formed in the unloading phase.

3. Sound velocity in compressed cerium

The information about the sound velocity C is required along with the basic parameters of the equation of state of a substance such as pressure P, volume V or density ρ, temperature T and entropy S. The sound velocity is one of the quantities characterizing a substance behavior under shock compression and subsequent unloading.

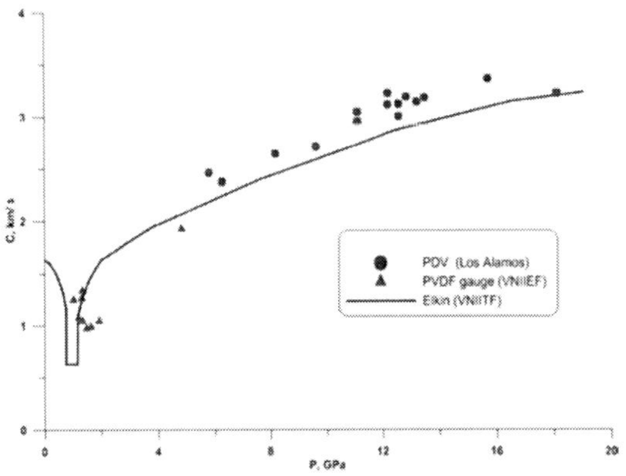

Figure 4. Analytical dependence of bulk sound velocity on pressure at shock-wave loading of cerium [1], data from LANL tests [8] and data from this work.

Planar PVDF gauges were used in various combinations in experiments for determination of pressure dependence on a sound velocity in cerium by the use of the technique of the PVDF gauge.

The data of time dependence on pressure profiles with the use of $x(t)$ diagrams and the $D(u)$ relation for cerium [5] allow the definition of the Lagrangian velocity of the unloading wave C^{Lagr} and the Eulerian velocity C^{Eul} by taking into account the compressibility σ. Obtained experimental data are presented in Table 1 and Table 2.

One of the demonstrations of anomalous properties of cerium in the loading phase is the inverse dependence of rarefaction wave velocity on pressure. In [1], results of calculation of bulk sound velocity C_{B} are presented versus pressure along the curve of isentropic compression of cerium. Experimental confirmation of this phenomenon was obtained in the performed tests (Fig. 4).

Wave velocities D were determined basing on results of the tests, which had been performed in the regime of symmetric impact. The obtained values were plotted at shock adiabat [3] (Fig. 5). Calculation was accomplished with use of the relation $P = \rho_0 \cdot D \cdot u$ (ρ_0 is the initial density of material; u· is the particle velocity).

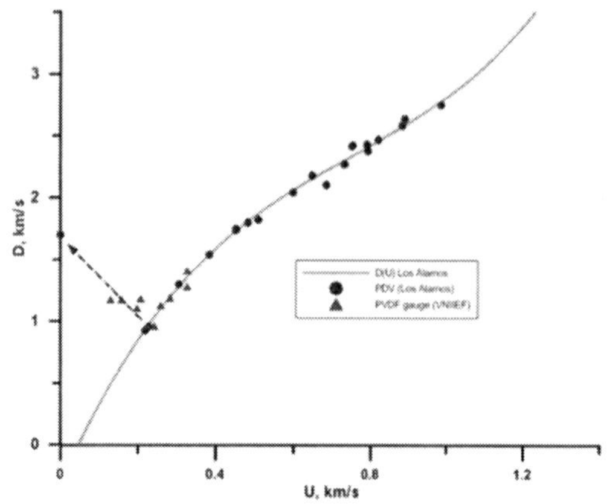

Figure 5. Shock adiabat in D-U coordinates [3] and data from this work.

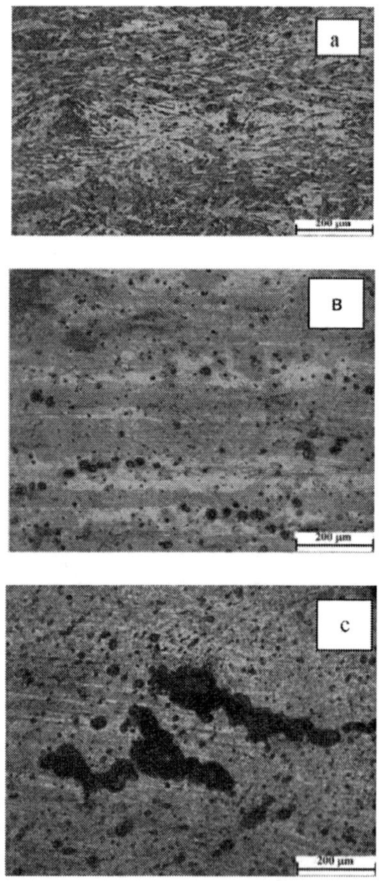

Figure 6. Microstructures of samples in initial state (a); after tests with impactor velocities W = 550 m/s (b) and W = 654 m/s (c).

It follows from the figure that, at $u > 0.2$ km/s, the results of this work are well described by the cubic relation, which was obtained in [3]. Significant difference from dependence [3] is observed at lower values of particle velocity. It corresponds to the pressures of ~1 GPa.

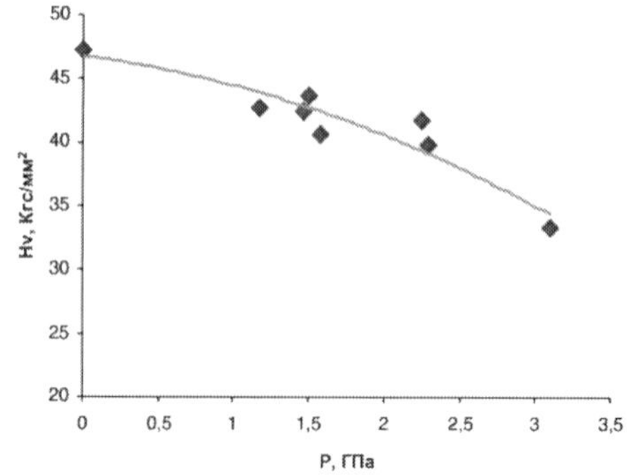

Figure 7. Dependence of microhardness on pressure.

The possible reason is anomalous behavior of cerium at pressures of ~1GPa, which was predicted by the theoretical model [1].

4. Metallography of post-experimental samples

In some cases, cerium samples were recovered after shock loading. These samples were investigated by optical metallographic microscope Axiovert 40 MAT at the magnification of 50× – 500× in light background and polarized light. Smoothing and polishing of microsection surfaces were mechanically provided by the grinding-and-polishing machine "PHOENIX BETA". Microsection etching was chemically accomplished with use of the mixture of nitric acid and glycerin followed by rinsing in ethyl alcohol (Fig. 6).

At the initial stage of formation, damage is accumulation of pores having spherical shapes with the characteristic pore diameters of 20–40 μm. Microcracks are joined spherical pores.

Measurements of microhardness of a source cerium sample and all samples recovered following tests showed (Fig. 7) its monotone decrease with a rise of a pressure.

5. Conclusions

It is demonstrated in the work that incomplete phase (γ–α) transition occurs at the mentioned loading pressures. As a result, the substance is a mixture of γ and α cerium phases, which determines it anomalous properties. Basing on results of all tests, pressure of the phase transition is 0.83 ± 0.06GPa. Rarefaction shock wave is formed in the unloading phase. Lagrangian velocities of rarefaction wave were determined basing on the experimental results. Wave velocities D were determined in the tests in the regime of symmetric impact.

An X-ray structural analysis did not find any changes in a phase composition of the tested samples. The γ-phase is a main phase is Ce in all the samples, namely in initial samples and samples after tests. The γ-phase Ce has a face-centered cubic structure.

References

[1] M.V. Zhernokletov, A.E. Kovalev, V.V. Komissarov et al. JETPh, **112**(2), 212–219 (2011)

[2] V.M. El'kin, E.A. Kozlov, E.V. Kakshina, Yu.S. Moreva, Physics of metals and physical metallurgy. **101**(3), 1–10 (2006)

[3] A.R. Kutsar, FMM. **33**(5), 1104–1108 (1972)

[4] I.L. Aptekar', E.G. Ponyatovsky, FMM. **25**(5), 777–786 (1968). II. FMM. **25**(6), 1049–1057 (1968)

[5] V.A. Borisenok, V.G. Simakov, V.A. Volgin et al. FGV. **43**(4), 121–126 (2007)

[6] B.J. Jensen, F.J. Cherne, J.C. Cooley, M.V. Zhernokletov, A.E. Kovalev, Phys. Rev. B. **81**, 214109 (2010)

[7] M.V. Zhernokletov, A.E. Kovalev, V.V. Komissarov, et al. *Shock compression of condensed matter* (M. Elert, M.D. Furnish, editor, 117–120, 2007)

EPJ Web of Conferences **94**, 01077 (2015)
DOI: 10.1051/epjconf/20159401077

Determination of sound velocities of "overcompressed" detonation in HMX-based explosive

Mikhail V. Zhernokletov, Alexey E. Kovalev[a], Vladimir M. Bel'sky, and Evgeny N. Bogdanov

Russian Federal Nuclear Center – VNIIEF, Russia

Abstract. The authors present results of determination of sound velocities in explosion products (EP) of HMX-based explosive overcompressed up to the pressures of 50–85 GPa by overtaking unloading method. The radiowave and optical methods are used to record the time when a front of overcompressed detonation wave in investigated sample of high explosive (HE) is overtaken by expansion wave, which propagates from the back surface of impactor with sound velocity. The data on sound velocities, which were independently obtained by two different methods, were in agreement. The methods with use of radiointerferometer and indicator liquid are rather effective for determination of sound velocities in overcompresed EP and for investigation of parameters at the Jouget point of various HEs, which are required for calibration of their equations of state (EOS).

1. Introduction

When investigating explosives, to understand physics of detonation and formulate equations of state (EOS) of explosion products, plane-wave experiments are most often fulfilled. However, equilibrium thermodynamics is closely associated with rates of a chemical reaction in such experiments. In other words, to simulate and analyze experiments, there is a need to separate the data relative to rates of a reaction, from the data typical for equations of state. The values of sound velocities, measured with the same accuracy, as the values of velocities of detonation in plane-wave experiments with HE provide the determination of derivatives at a surface of an equation of state with the same accuracy, as the surfaces of EOS. The data on a sound velocity together with the Hugoniot adiabat for overcompressed detonation make it possible to calculate the Grueneisen coefficient and a pressure at the Chapman-Jouguet point (Ch-J). The Ch-J pressure can be deduced with a good accuracy through thermodynamic state parameters, which are not connected with the effects of a reaction zone [1].

The overtaking unloading method with the use of indicator liquids for optical recording of thermal radiation caused by substance compression in a shock wave is the most informative method for measurement of sound velocities in metals compressed up to pressures higher 30 GPa. The current method is also used for measurements of sound velocities in explosive compositions (EC). In addition, it is possible to define a sound velocity in explosion products by using a radiointerferometer in the millimeter wavelength range.

The objective of the present work is to measure sound velocities in overcompressed explosion products of the HMX-based explosion composition containing a plastic bond made of fluoroplastic by using two independent methods. A regime of overcompressed detonation is performed at the expense of loading of a "cold" composition with metal plates up to pressures of 100 GPa in a shock adiabat.

[a] Corresponding author: aekovalev@vniief.ru

2. The rarefaction overtake technique for determination of sound velocity in explosion products

The rarefaction overtake technique for determination of a sound velocity in substances is used to record the point of time when a front of an overcompressed detonation wave in the sample under study is overtaken by an rarefaction wave travelling from the back surface of an impactor with a sound velocity. Figure 1 presents the X–t – diagram of flows occurring when a thin metal plate strikes the explosive charge with the thickness δ.

Let us assume that a plate moves at the speed W and "comes across" the explosive charge [2] at the point of time $t = 0$ (point O). While a plate impacts on HE, an overcompressed detonation wave is excited in HE, which propagates at the speed D_C. The shock wave OA is propagated through a plate at the speed relative to the fixed coordinate system $D_l - W$, which goes towards a free surface of a plate at the point A. A centered rarefaction wave arises at the point A, whose head characteristic arrives at the interface between a plate and explosion products at the point B. An expansion wave begins to propagate in explosion products. The expansion wave overtakes a detonation front at the point B, following which a front starts to slow down (see the insertion in Fig. 1).

In this case, there exist two ways for determination of an overtake moment (or a coordinate) of the detonation front by the help of a rarefaction wave in an explosive. The first way consists in the direct recording of detonation front movement in an explosive; a radio-wave technique

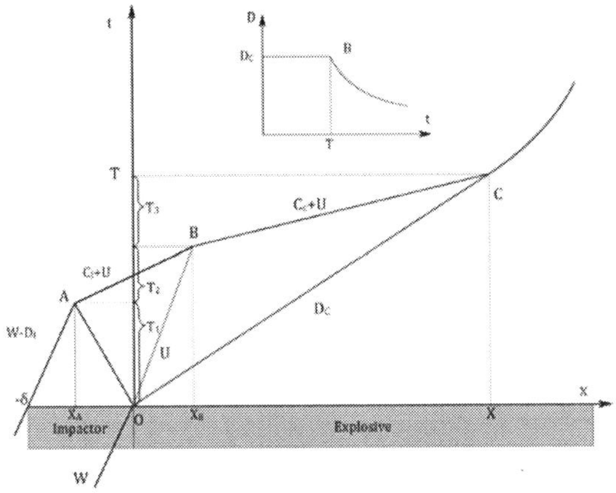

Figure 1. *X–t* – diagram of impact of thin plate on explosive.

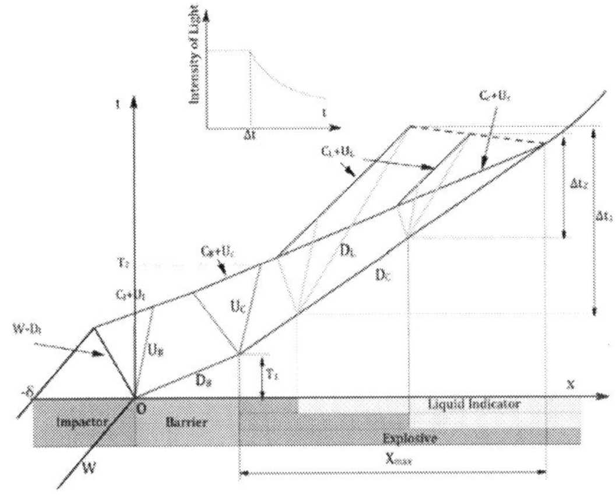

Figure 2. *X–t* diagram of indicator technique.

was used for this purpose in the present paper [3–5]. A thickness of an explosive sample decreases in the second way, and a layer of an indicator liquid is placed above the sample. The time when a rarefaction wave overtakes a shock wave front [6,7] is recorded via the optical technique.

2.1. Radio-wave technique

To determine the arrival time of a longitudinal rarefaction wave T into a detonation front in HE, we use the X–T – diagram of a process in Fig. 1:

$$T = T_1 + T_2 + T_3, \tag{1}$$

where T_1 is the time of a shock wave entering the back surface of a plate from a moment of impact; T_2 is the traveling time of a longitudinal rarefaction wave over a plate prior to its exit to the interface; T_3 is the traveling time of a longitudinal rarefaction wave in explosion products to a front of an overcompressed detonation wave.

Then:

$$T_1 = \frac{X_A}{W - D_l} = \frac{-\delta + W\,T_1}{W - D_l}.$$

From which

$$T_1 = \frac{\delta}{D_l}. \tag{2}$$

The time T_2 can be determined using the relation:

$$T_2 = \frac{x_B - x_A}{C_l + U}, \tag{3}$$

where C_l is the velocity of a longitudinal rarefaction wave in impactor's material; is the velocity of the interface; $X_A = -\delta + W\,T_1$, $X_B = U\,(T_1 + T_2)$ are the coordinates of the points A and B respectively.

Therefore

$$T_2 = \frac{\delta}{D_l} \cdot \frac{D_l - W + U}{C_l}. \tag{4}$$

Evidently the value T_3 is defined as

$$T_3 = \frac{X_B - X_B}{U + C_C}, \tag{5}$$

where $X_B = D_C\,(T_1 + T_2 + T_3)$, C_C is the sound velocity in explosion products behind a detonation front in HE.

After a series of elementary transformations using relations (2)–(5) the formula can be obtained:

$$T = \frac{\delta}{D_l} \cdot \frac{C_C}{C_l} \cdot \frac{C_l + D_l - W + U}{C_C - D_C + U}. \tag{6}$$

Hence it follows the expression for a sound velocity in explosion products:

$$C_C = \frac{D_C - U}{1 + \dfrac{\delta}{T} \cdot \dfrac{C_l + D_l - W + U}{D_l\,C_l}}. \tag{7}$$

To determine experimentally the sound velocity C_C behind a front of an over compressed detonation wave in HE according to relation (7), it is necessary to record the instant of time T, after which a detonation wave velocity begins to decrease in HE (see the insertion in Fig. 1). The longitudinal sound velocity C_l in impactor's material may be deduced as per a reference book (for example [8]) by using known state parameters at shock break out. In the case if a plate strikes a screen and the HE sample is placed behind a screen, the expression for the sound velocity C_C is complicated a little.

2.2. Optical technique

The gist of the optical technique of measurement of sound velocities by using an indicator is presented in papers in detail [6,7]. Figure 2 demonstrates the X–t – diagram of flows existing at the instant of time when a thin metal impactor strikes the system of screen-HE-indicator. Despite the more complicated character of processes in a system, the instant of the overtake of a detonation front is determined not in HE in the present technique, but in indicator liquid, which covers a sample made of HE under study. Constant or increasing radiation intensity is being recorded in the indicator prior to the moment of arrival of the first characteristic of a rarefaction wave at a shock wave front. Then radiation intensity begins to fall further in an indicator.

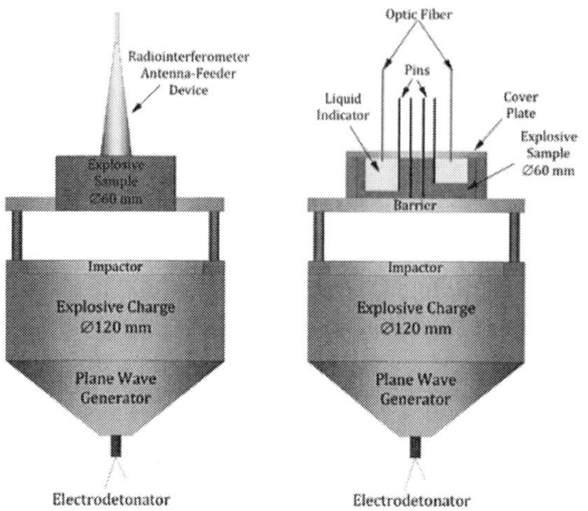

Figure 3. Sketches of experimental assemblies when using radiowave technique and optical technique of recording.

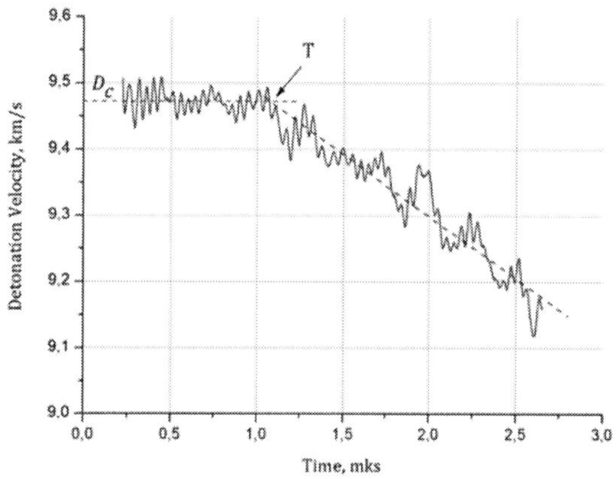

Figure 4. Dependence of overcompressed detonation wave velocity on time in sample.

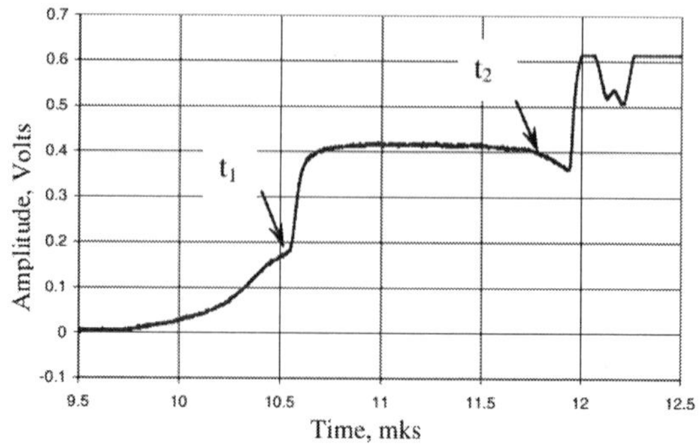

t_1 - time of shock wave entering the indicator; t_2 - time when the rarefaction wave overtakes the shock wave front.

Figure 5. Recording of luminescence of carbogal behind a step made of HE sample.

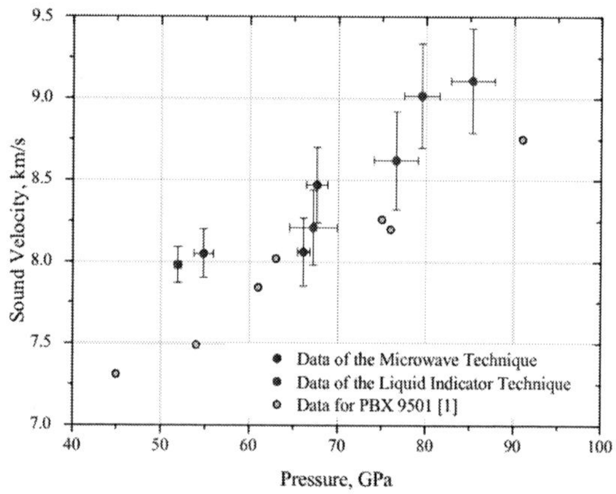

•- data obtained by using a radio interferometer; •- data obtained by indicator liquid technique; •- data of the USA for the composition PBX 9501 [1].

Figure 6. Dependence of sound velocities on pressure in overcompressed explosion products.

The signal duration of indicator luminescence Δt_i is determined for each first step of the sample (i = 1, 2, ... n, where n is the number of steps) in the experiment from the beginning of luminescence to the beginning of luminescence fall. The values Δt_i decrease linearly with increase in thicknesses Δx_i of steps. This takes place always, independently of the relation of dynamic impedances of an indicator and a sample. A task consists in the determination of the sample thickness X_{MAX} from the experiment; a rarefaction wave will overtake a shock wave front that moves ahead. A thickness of sample steps should be smaller than X_{MAX}. The n of independently measured points (Δx_i, Δt_i) is recorded at the pressure P in the sample in the experiment. By using these points the a and b coefficients of linear dependence are found through a method of linear approximation: $\Delta t = a - b\Delta x$. Then we obtain $X_{MAX} = a/b$ at $\Delta t \to 0$ that is the sample's thickness, at whose interface an unloading wave will overtake a shock wave front (SWF) in a sample. The value of the sound velocity C_C is determined in explosion products in this case by the formula:

$$C_C = \frac{X_{MAX}(D_C - U_C)}{D_C \cdot \left(T_1 - T_2 + \dfrac{X_{MAX}}{D_C}\right)}, \qquad (8)$$

where T_1 is the time of a shock wave entering the sample; T_2 is the time of rarefaction wave entering the interface between a screen and HE; D_C is the velocity of a detonation wave in a sample; U_C is the mass velocity behind a detonation front; X_{MAX} is the thickness of a step, when overtaking occurs at the sample/indicator interface.

The values T_1 and T_2 are defined by the help of the $X - t$ diagram of Fig. 2. The calculations were made with the use of reference data from [8] in the gasdynamic code "UP" [9], in which equations of state of materials are given in the form of Mie-Grueneisen.

Table 1. Sound velocity in overcompressed explosion products (radiowave technique).

Test num	ρ_0, g/cm^3	W, km/s	D_c, km/s	U, km/s	P, GPa	T, μs	C_c, km/s
1	1,885	5,22 ± 0,03	9,35 ± 0,04	2,95 ± 0,03	51,99 ± 0,40	2,02 ± 0,02	7,98 ± 0,11
2	1,886	5,45 ± 0,09	9,42 ± 0,03	3,08 ± 0,07	54,72 ± 1,10	2,81 ± 0,02	8,05 ± 0,15
3	1,885	4,96 ± 0,04	9,80 ± 0,07	3,58 ± 0,04	66,13 ± 0,69	1,89 ± 0,06	8,06 ± 0,21
4	1,885	4,96 ± 0,04	10,02 ± 0,13	3,58 ± 0,04	67,62 ± 1,19	2,53 ± 0,13	8,47 ± 0,23

Where ρ_0 is the initial density of HE samples; W is the impactor velocity; D_c is the velocity of an overcompressed detonation wave in HE; U is the mass velocity behind a front of an overcompressed detonation wave in HE; P is the pressure of an overcompressed detonation wave in HE; T is the time of the arrival of a lateral rarefaction wave at a detonation front; C_c is the sound velocity in overcompressed explosion products of the composition.

Table 2. Sound velocity in overcompressed explosion products (indicator liquid technique).

Test num	ρ_0, g/cm^3	W, km/s	T_1, μ s	T_2, μ s	X_{MAX},mm	D_c, km/s	U, km/s	P, GPa	C_c, km/s
1	1,882	4,94 ± 0,04	0,289	0,644	10,7	9,43	3,78	67,2	8,21
2	1,887	4,94 ± 0,04	0,278	0,635	11,1	9,95	4,09	76,6	8,62
3	1,880	5,00 ± 0,05	0,286	0,648	10,5	10,07	4,18	79,5	9,02
4	1,887	5,58 ± 0,06	0,162	0,410	7,68	10,40	4,35	85,3	9,11

Where T_1 is the time of a shock wave entering the sample; T_2 is the time of rarefaction wave entering the interface between a screen and HE; X_{MAX} is the thickness of a step, when overtaking occurs at the sample/indicator interface.

3. Experimental setup

To generate over-compressed detonation waves in samples of HMX-based explosives, the experts used plane shock wave generators and cylindrical active HE charges of diameter 120 mm based on RDX and HMX with different power; their explosion products were accelerated by aluminum (AD-1) impactors or steel (St 3) impactors. Sketches of experimental assemblies are given in Fig. 3.

The values of the flying up impactor velocity W were determined in experiments as per a scheme of Fig. 3 (left) with the use of a radiowave technique. The authors determined also the velocities of the overcompressed detonation wave D_C and the time T of overtaking of a detonation front by a rarefaction wave using a radiowave technique. Figure 4 presents a typical experimental dependence of a detonation wave velocity on time.

The HE samples under study were placed into a capsule, whose construction consists of: a screen, a technological composite cartridge and a cap with outputs for optical fiber and electrocontact pins in experiments with the use of the optical method (Fig. 3, right). Four blind holes ("wells") of diameter 16 mm were milled in the HE sample 60 mm in diameter. Each "well" had its depth, namely, a thickness of a bottom – a step of a sample. "Wells" were filled with indicator liquid. Carbogal (perfluor-1.3-dimethylcyclohexane, the initial density ρ_0 = 1.856 g/cm^3) was used as indicator liquid in experiments. Photodiode sensors recorded luminescence of an indicator via optical outputs and quartz-polymer fibers located above each well. In this case, walls of the "wells" excluded lateral flare from adjoining sample steps. Electrocontact pins and optical sensors diagnosed state parameters in an explosive in the experiment. The accuracy of time measurements of motion of shock wave front in liquid indicators is not worse than 2 ns. Figure 5 presents one of the oscillograms of a test, recorded by the help of a photodiode sensor.

By using indicator liquid, a relative error of measurements of sound velocities by the help of the optical method does not exceed 7%.

4. Experimental results

The data obtained in tests for measurements of sound velocities in overcompressed explosion products of the explosive composition being studied are tabulated (Tables 1 and 2).

Figure 6 presents the dependence of the obtained sound velocities in explosion products of the composition under study on a pressure in overcompressed detonation front. Besides, the graph contains the data for sound velocities in the composition PBX 9501 from [1].

As is possible to see in the graph of Fig. 6, the data on sound velocities in explosion products of the composition under study, which were obtained by two different techniques independently from each other, are in agreement with one another and they do not contradict the data for PBX 9501 with the similar HE composition from work [1].

5. Conclusion

By using the overtaking unloading method, sound velocities were determined through radiowave and optical techniques in explosion products of the HMX-based composition, which were overcompressed up to 50–85 GPa. The data on sound velocities in explosion products of the composition under study, which were obtained by two different methods independently from each other, are in agreement with one another and with the data for a foreign analog PBX 9501. The achieved results testify to the fact that techniques with the use of a radiointerferometer and indicator liquid, are usable for determination of sound velocities in overcompressed explosion products and for research of parameters at the Jouget point of various HEs.

References

[1] J.N. Fritz, R.S. Hixson, M.S. Shaw, C.E. Morris, and R.G. McQueen, J. Appl. Phys. **80**, 11 (1996)

[2] V.P. Chelyshev, B.I. Shekhter, FGV, **4**, p.68–74 (1966)

[3] A.L. Mikhaylov, V.E. Kostyukov, Yu.I. Orekhov et al., *Proceedings of International Conference "VII Khariton's Topical Scientific Readings. Extreme states of substance. Detonation. Shock waves"*, Sarov, RFNC-VNIIEF, p.649–654 (2005)

[4] A.V. Rodionov, V.M. Bel'skiy, E.N. Bogdanov et al., *Proceedings of International Conference "XI Khariton's Topical Scientific Readings. Extreme states of substance. Detonation. Shock waves"*, Sarov, RFNC-VNIIEF, p.69–73 (2009)

[5] V.M. Bel'skiy, A.L. Mikhaylov et al., *Proceedings of International Conference "XIII Khariton's Topical Scientific Readings. Extreme states of substance. Detonation. Shock waves"*, Sarov, RFNC-VNIIEF, p.785–794 (2011)

[6] R.G. McQueen, J.W. Hopson, J.N. Fritz, Rev. Sci. Instrum, **53**, 2, p.245–250 (1982)

[7] M.V. Zhernokletov, A.E. Kovalev, V.V. Komissarov et al., SCCM-2007, ed. by M.L. Elert, M.D. Furnish, R. Chau, N. Holmes, J. Nguyen. – AIP, p.117–120 (2007)

[8] R.F. Trunin, *Experimental data on shock-wave compression and adiabatic expansion of condensed materials,* RFNC- VNIIEF, Sarov, p.531 (2006)

[9] N.F. Gavrilov, G.G. Ivanova, V.I. Selin, V.N. Sofronov, Problems of atomic science and engineering. Series: Methods and codes for numerical solution of problems of mathematical physics, **3**, 11, p.11–14 (1982)

EPJ Web of Conferences **94**, 01078 (2015)
DOI: 10.1051/epjconf/20159401078

Experimental investigation on shear fracture at high strain rates

Christian C. Roth[1,2] and Dirk Mohr[1,2,3]

[1] Solid Mechanics Laboratory (CNRS-UMR 7649), Department of Mechanics, École Polytechnique, Palaiseau, France
[2] Impact and Crashworthiness Lab, Department of Mechanical Engineering, Massachusetts Institute of Technology, Cambridge MA, USA
[3] Department of Mechanical and Process Engineering, ETH Zurich, Switzerland

Abstract. Adiabatic shear banding is a well-understood failure mechanism of metals at high strain rates. In addition, recent research on the ductile fracture of metals has demonstrated that shear localization at the microscale is also an important precursor of fracture initiation at low strain rates. This talk presents a new shear fracture specimen which is used to conduct fracture experiments on advanced high strength steel sheets at strain rates of up to 1/s in a hydraulic testing machine and for strain rates of up to 2500/s in a Split Hopkinson Bar system. The experimental result for a 22 MnB5 steel show a significant increase in ductility as a function of strain rate. Results from scanning electron microscopy are also shown to provide insight into the effect of the strain rate on the shear localization at the microscale.

1. Introduction

A new shear specimen is presented that is suitable for characterizing shear fracture in sheet metals at low, intermediate and high strain rates. Due to the good observability of the flat specimen surface, the strains to fracture can be directly measured through digital image correlation (DIC). Experiments at three different strain rates are performed and analyzed. The strain rates pursued include slow quasistatic, intermediate (~1/s) and fast (~2500/s) loading rates. To shed more light on the shear band formation, the specimens are examined post mortem by electron backscatter microscopy. The material examined is a 1.02 mm thick quench-hardenable manganese boron steel (22 MnB5).

2. Shear specimen

For in-plane shear experiments on flat specimens, fracture is very likely to commence from the free boundary edge of the gage section. Despite the overall shear loading, plane strain tension stress states often prevail near the free boundaries, which normally have lower strains to fracture than pure shear. As a result, fracture initiates prematurely from the boundaries and the shear fracture strains obtained from such experiments can often only be interpreted as a lower bound. To address this well-known challenge, many authors make use of a reduced gage section thickness, so that the strain will be much higher in the specimen center, increasing the likeliness of fracture initiation under shear.

Inspired by the work of Till and Hackl (2013), Roth and Mohr (2015a) presented an in-plane shear fracture specimen of uniform thickness which features two parallel gage sections. This was achieved through an FEA-based optimization of the outer contour of the shear gage sections, making use of the Hosford-Coulomb fracture initiation model. The final shape of the specimen is shown in Figs. 1 and 2.

3. Experiments

Figure 2 shows the "smiley" specimens as extracted from a 1 mm thick 22 MnB5 sheet. To obtain good dimensional accuracy, wire electric-discharge machining (EDM) is used. To perform digital image correlation, a random pattern that can be tracked by the DIC software (Vic2D, Correlated Solutions) is applied to the specimen surface with an average speckle size of 15 μm.

3.1. Setup for low and intermediate strain rates

All low and intermediate strain rate experiments are carried out on a MTS hydraulic testing machine with a 10 kN load cell. A digital camera AVT Pike F-505B/C with a 90 mm 1:1 macro lens is used to observe the overall displacement of the specimen at the positions highlighted through solid dots in Fig. 2. It is positioned at a distance of approximately 500 mm from the specimen, so that a spatial resolution of approximately 19 μm/pix is obtained. A second camera AVT Pike F-505B/C with a 105 mm 1:1 macro lens is utilized to closely observe the two gage sections of the specimen. It is positioned perpendicularly to the specimen surface at a distance of approximately 20 cm, allowing a high spatial resolution of 3.5 μm/pix.

All slow speed experiments are carried out at a crosshead displacement of 0.4 mm/min. Images are taken at a frequency of 0.5 Hz in conjunction with the commercial software VicSnap (Correlated Solutions). For the tests at intermediate strain rates, corresponding to a crosshead displacement of 8.3 mm/s, a high speed camera (Vision Research, Phantom 7.3) with a 105 mm 1:1 macro lens is utilized. With a resolution of 800 × 600 pixels, the images for DIC are acquired at a frequency of 1,000 Hz

Article available at http://www.epj-conferences.org or http://dx.doi.org/10.1051/epjconf/20159401078

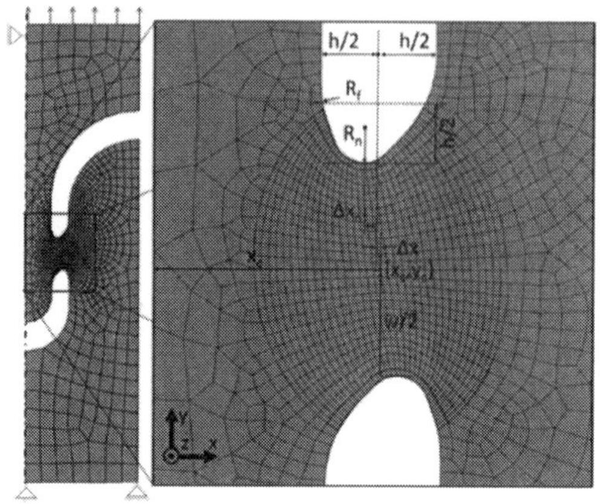

Figure 1. Finite element model of the optimized shear specimen geometry, including the parameterized variables.

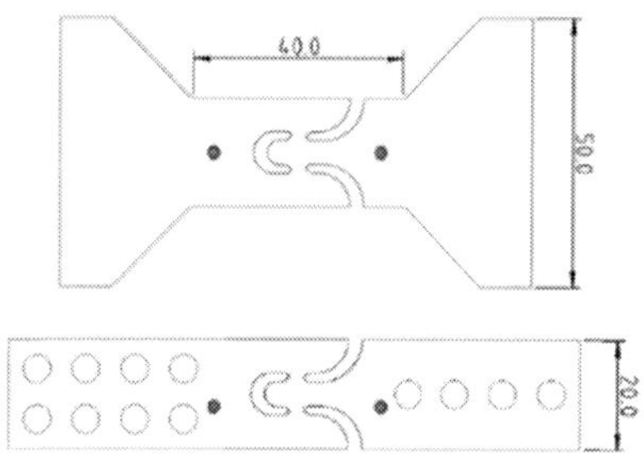

Figure 2. Specimens used for the experiments at low and intermediate strain rates (on the top), and high strain rates (on the bottom). Blue solid dots highlight the position of the virtual extensometer for relative displacement and speed measurements.

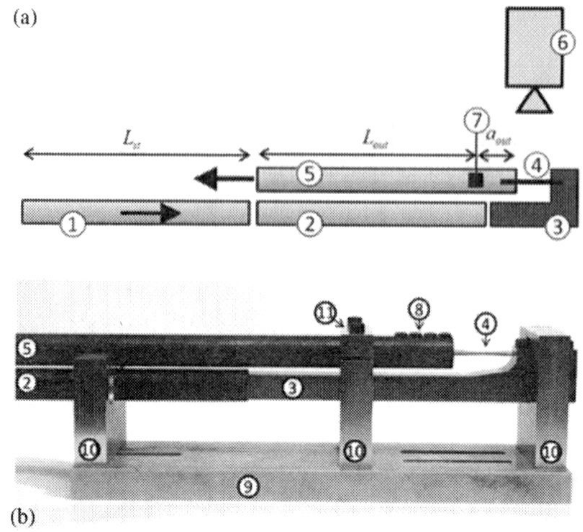

Figure 3. (a) Illustration of the experimental setup (b) Photograph of the Load Inversion Device – ① striker bar, ② input bar, ③ pusher of the LID, ④ specimen, ⑤ output bar, ⑥ high speed camera, ⑦ position of strain gauge, ⑧ grip section with slit, ⑨ base plate, ⑩ bearings, ⑪ removable clamp.

and a square pixel edge length of approximately 77 μm. The image acquisition is triggered by the rise in the force signal recorded by the load-cell.

3.2. Setup for high strain rates

A Split-Hopkinson Pressure Bar (SHPB) setup is used for the experiments at high strain rates in conjunction with a custom-made load-inversion device (Roth and Mohr, 2015b). The system comprises of a striker bar (3.80 m long, 20 mm diameter), an input bar (4.49 m long, 20 mm diameter) a load-inversion device and an output bar (4.35 m long, 20 mm diameter), with the latter positioned on top of the input bar. All bars are made of steel. In this configuration the setup allows for a valid experiment duration of 1.53 ms. The geometry of the specimen used is shown in Fig. 2. As shown in Fig. 3, the load-inversion device consists of a pusher, which converts the compressive load from the input bar into a tensile load on the mounted specimen.

The opposite grip section of the specimen fits into a machined slit in the output bar, which is closed by four counter-sunk M5-12.9 screws, thus maintaining a symmetric mass distribution with respect to the output bar's center axis. The assembly is guided through bearings with lubricated contact surfaces. Strain gages positioned at a distance of 403 mm from the specimen/output bar interface are used to record the output bar strain histories ε_{tra} [t]. The history of the axial force F[t] acting on the specimen is then calculated as

$$F[t] = A_{out} E_{out} \varepsilon_{tra} [t] \qquad (1)$$

with E_{out} denoting the Young's modulus, and A_{out} the cross section area of the output bar. For the displacement measurements, the same high speed camera (Vision Research, Phantom 7.3) as for the experiments at intermediate speeds is employed. Two different setups are used: (i) To measure the global displacement history of the specimen, the camera is positioned at a distance of approximately 510 mm from the specimen, resulting in a spatial resolution of approximately 60 μm/pixel. Images from an area of 576 × 32 pixels are obtained, which corresponds to approximately 35.1 × 2.0 mm of the specimen surface. It is slightly offset from the center of the specimen to observe the onset of fracture. The camera is set to the highest possible acquisition frequency of 133 kHz, what corresponds to an image obtained every 7.5 μs. (ii) To measure the local displacement fields in the gage section of the specimen, the camera is positioned at a distance of approximately 250 mm. Close-up images with a spatial resolution of 23 μm/pixel are obtained from an area of 224 × 72 pixels at an acquisition speed of 133 kHz.

3.3. Results

It is observed that with increase in strain rate, the yield stress of the material increases as well – for the intermediate case the force-level is approximately 8% higher than for the slow one, while in the fast case it is approximately 25% higher at a displacement of 0.1 mm. The shapes of the force-displacement curves for the slow and intermediate experiments are similar up to the force maximum; thereafter, the intermediate one

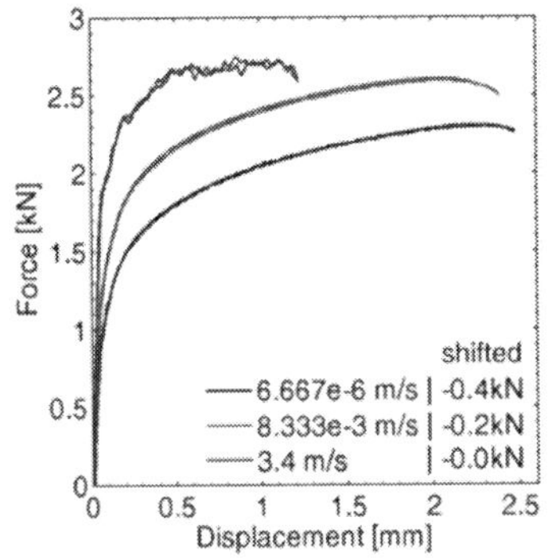

Figure 4. Plot of the force-displacement curves of the experiments (solid lines) and simulations (dots) at different strain rates and room temperature.

decreases slower than the slow ones towards fracture. The force-displacement curves of the experiments at high speed exhibit a different behavior. They show a plateau in the force level, before it decreases towards the onset of fracture. It is noteworthy that the displacement to fracture decreases with an increase in the loading speed.

A slow as well as one high strain rate experiment are examined with digital image correlation using the commercial software VIC2D (Correlated Solutions). Herein, the surface displacement is calculated by tracking the gray intensity distribution for a subset over subsequent images in conjunction with subset shape functions.

The error between the function and the image is minimized using a normalized squared differences approach algorithm. To obtain sub-pixel accuracy, the discrete greyscale distribution within a subset is interpolated by means of an 8-tap filter function. To assure comparability a step size of 0.1 mm is chosen. A resolution of approximately 28 pixel/0.1 mm ($\sim$3.5 μm/pixel) is achieved for the slow experiments, while the high speed camera allowed obtaining only 4 pixel/0.1mm ($\sim$22.6 μm/pixel) for the fast experiment. To improve the continuity of the displacement field and the correlation, the subset size is chosen to overlap by 1/8th. It is noteworthy that due to the limited acquisition frequency as well as the limited resolution of the high speed camera system, a minimum subset size of 9 × 9 pixel ($\sim$0.2 × 0.2 mm) had to be chosen for the fast case, while in the slow case a subset of approximately 0.1 × 0.1 mm was utilized. No additional threshold options are chosen, while for the strain calculation with the Green-Lagrangian strain tensor a center weighted Gaussian filter smoothing over 5 data points is used

Figure 5 shows the evolution of the displacement field before the onset of fracture for the slow and the fast experiment. Clearly visible is the diagonal crack propagation which is seen as an indicator for shear fracture. The evolution of the shear strain along the white line in the displacement field is shown in the second row, while the third row depicts the evolution of the shear strain rate at the point where the shear strain has its maximum. It is observable that for both experiments the strain rate remains constant (approximately 0.005 for the slow case and 2500 for the fast case), up to the point of localization. Here the strain rate begins to increase, reaching a value of an order of magnitude more at fracture. It has to be noted that the limited available resolution of the images for the high strain rate experiments, have a direct impact on the obtainable detail in the displacement measurement. Nevertheless, the location and width of the localization can be measured.

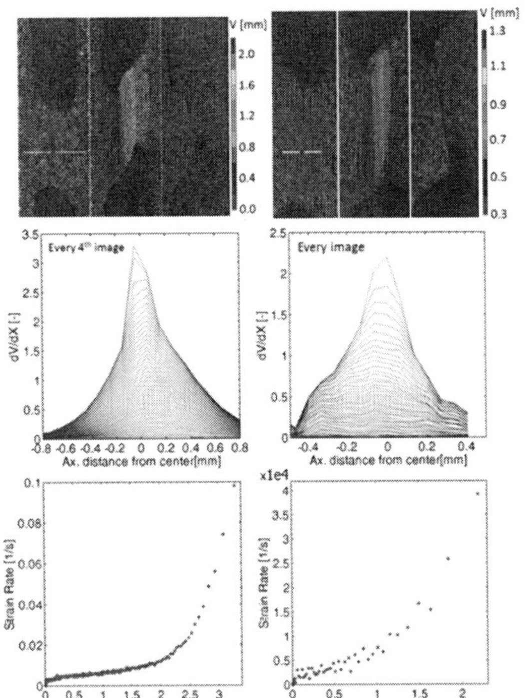

Figure 5. Displacement field and the extracted evolution of the strain along the white line and the strain rate at the location of the red dot (slow case left column, fast case right column).

4. Microscopic analysis

One specimen each from the slow, intermediate and fast tests is examined closely using electron backscatter diffraction (EBSD) microscopy. The EBSD technique can be used to determine the crystallographic orientation of a material, thus effectively making orientation mapping and grain boundary identification possible. After extracting the zone of interest around the fractured gage sections with a diamond wire saw, the specimens are polished with SiC paper to grit #4000 and a 3 μm diamond paste. Final polishing is performed by means of ion beams on a Gatan Pecs II ion polishing device. The specimens are exposed to the beam with an energy of 5 keV stepwise from 6° for 20 min, 4° for 20 min and finally at a 2° angle for 1 hour.

The observations are carried out on a scanning electron microscope FEI Quanta 600 F at 20 kV and a spot size of 4.5. Using the HKL Channel software package, a 700 μm × 200 μm wide area perpendicular to the rolling direction and close to the fracture plane is analyzed. Every data point is scanned three times and then averaged.

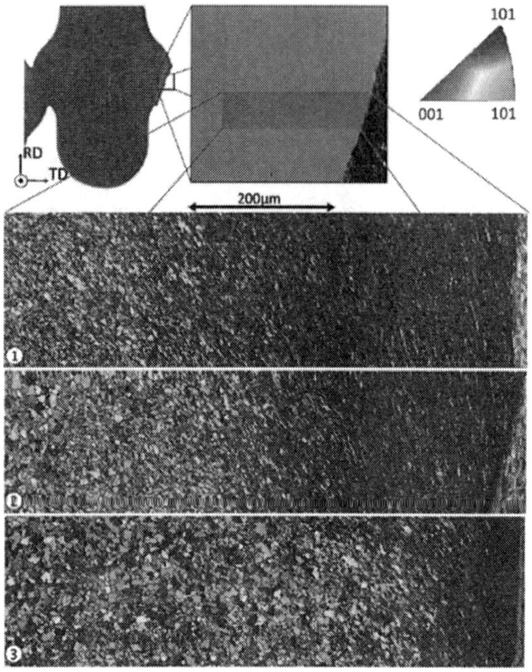

Figure 6. IPF-Y plots of the ① slow, ② intermediate and ③ fast experiments with added grain boundaries. Colors represent the grain orientation normal to the sample surface as defined by the stereographic triangle inset in the top right corner of the image.

With a step size of $0.43\,\mu$m each image is obtained in around 16 hours and a resolution of approximately 1628×465 points per image is achieved. The data is post-processed using very slight noise-reduction of the HKL Tango software (iteratively up to level 2 of 8).

5. Conclusions

A new flat shear fracture specimen is presented to perform experiments at low, intermediate and high strain rates. Its design is derived using an FEA based shape optimization methodology and allows for the direct observation of the surface strain field by means of DIC.

The experimental results show a positive strain rate hardening behavior of the material, while the fracture displacement decreases. Utilizing electron backscatter diffraction microscopy, it is shown that all experiments exhibit several deformed zones. It can be seen that with increasing strain rate the localization of deformation becomes more pronounced. In particular, the zone of localized deformation at low strain rates is about three times wider than that observed for high strain rates.

References

[1] D. Mohr, M. Dunand, and K.-H. Kim, *Evaluation of associated and non-associated quadratic plasticity models for advanced high strength steel sheets under multi-axial loading*. International Journal of Plasticity, (2010) **26**(7): p. 939–956.

[2] C.C. Roth and D. Mohr, *Effect of Strain Rate on Ductile Fracture Initiation in Advanced High Strength Steel Sheets: Experiments and Modeling*. International Journal of Plasticity **56** (2014) 19–44.

[3] C.C. Roth, G. Gary and D. Mohr, *Compact SHPB System for Intermediate and High Strain Rate Plasticity and Fracture*, Experimental Mechanics (2015b) under review.

[4] C.C. Roth and D. Mohr, *Basic Experiments on Sheet Metal with Propor-tional Loading Paths Until Fracture Initiation* (2015a) to be submitted.

[5] R. Hill, *A theory of the yielding and plastic flow of anisotropic metals*. roceedings of the Royal Society of London. Series A, Mathematical and Physical Sciences, (1948) **193**(1033): p. 281–297.

[6] T.B. Stoughton, *A non-associated flow rule for sheet metal forming*. International Journal of Plasticity, (2002) **18**(5–6): p. 687–714.

[7] Till E., B. Hackl (2013). *Calibration of plasticity and failure models for AHSS sheets*, Proceedings of the International Deep Drawing Research Conference IDDRG 2013.

[8] G.R. Johnson and W.H. Cook. *A Constitutive Model and Data for Metals Subjected to Large Strains, High Strain Rates and High Temperatures*. in *7th International Symposium on Ballistics* (1983). The Hague.

[9] G.R. Johnson and W.H. Cook, *Fracture characteristics of three metals subjected to various strains, strain rates, temperatures and pressures*. Engineering Fracture Mechanics (1985). **21**: p. 31–48.

EPJ Web of Conferences **94**, 01079 (2015)
DOI: 10.1051/epjconf/20159401079

On loading velocity oscillations during dynamic tensile testing with flying wheel systems

Borja Erice[1,a], Christian Roth[1], Gerard Gary[1], and Dirk Mohr[1,2,3]

[1] Solid Mechanics Laboratory (CNRS-UMR 7649), Department of Mechanics, École Polytechnique, Palaiseau, France
[2] Impact and Crashworthiness Lab, Department of Mechanical Engineering, Massachusetts Institute of Technology, Cambridge MA, USA
[3] Department of Mechanical and Process Engineering, ETH Zurich, Switzerland

Abstract. Flying Wheels (FW) provide a space-saving alternative to Split Hopkinson Bar (SHB) systems for generating the loading pulse for intermediate and high strain rate material testing. This is particularly attractive in view of performing ductile fracture experiments at intermediate strain rates that require a several milliseconds long loading pulse. More than 50 m long Hopkinson bars are required in that case, whereas the same kinetic energy (for a given loading velocity) can be stored in rather compact flying wheels (e.g. diameter of less than 1.5 m). To gain more insight into the loading capabilities of FW tensile testing systems, a simple analytical model is presented to analyze the loading history applied by a FW system. It is found that due to the presence of a puller bar that transmits the tensile load from the rotating wheel to the specimen, the loading velocity applied onto the specimen oscillates between about zero and twice the tangential loading speed applied by the FW. The theoretical and numerical evaluation for a specific 1.1 m diameter FW system revealed that these oscillations occur at a frequency in the kHz range, thereby questioning the approximate engineering assumption of a constant strain rate in FW tensile experiments at strain rates of the order of 100/s.

1. Introduction

Flying Wheels (FW) or flywheels [1] provide a space saving alternative to conventional Hopkinson bar systems [2]. FW systems have been successfully used to characterize the mechanical response of different materials at intermediate and high strain rates either compressively [3,4] or in tension [5–7]. The main advantage of such systems is that testing at the same average strain rate, the FW saves a significant amount of space. Where more than 50 m long Hopkinson bars are needed, a 1.5 m diameter FW system can often achieve the same loading speed and experimental duration.

However, when FW systems in their tensile configuration, a short input bar with an anvil needs to be introduced to grip and apply the loading onto the specimen. It is shown in this conference paper that this deviation from the direct impact configuration (which is the preferred choice for compression testing) creates spurious oscillations in the specimen loading velocity history. This problem is first analyzed in an approximate manner using a simple one-dimensional theoretical model before presenting the results from large scale finite element analysis of a FW system.

The set-up of conventional Split Hopkinson Tensile Bar (SHTB) systems is similar, i.e. the function of their input bar corresponds to that of the puller bar in a FW system. However, the main conceptual difference is that the length of the input bar in a SHTB system is chosen such that the experiment is terminated after completion of a single wave round-trip in the bar. In a FW system, on the other hand, the puller bar is very short and subject to multiple round-trips during an experiment. It is also worth noting that the use of short puller bars is also common practice on so-called fast hydraulic testing machines where the piston first needs to accelerate before hitting a puller bar that applies the loading onto the specimen.

[a] Corresponding author: borja@lms.polytechnique.fr

2. Flying wheel system configuration

The FW analyzed here is a solid disk of diameter D_w and width W_w with an angular velocity $\omega = 2\pi f$ which has a small extremity called hammer that hits the anvil of a puller bar attached to the specimen (Figure 1). We note that the flange of the puller bar shall be much shorter than its overall length. Throughout the experiment, the hammer drags the puller, which at the same time stretches the specimen. It is tempting to assume that the velocity of the puller-specimen interface is constant and equal to the tangential velocity $v = \omega D_w/2$ of the hammer. It will be shown in the sequel that this is not completely true.

The specimen geometry used is depicted in Fig. 2. With these initial geometries one can make some rough energy estimations that can be used to estimate the strain rates at which the use of the FW is effective for a particular specimen geometry and material stress-strain response. The kinetic energy that is stored in the FW is uniquely dependent on its angular velocity:

$$E_k = \frac{1}{2} I\omega^2 = \frac{1}{16} m_w D_w^2 \omega^2 = \frac{\pi}{64} \rho W_W D_w^4 \omega^2. \quad (1)$$

Article available at http://www.epj-conferences.org or http://dx.doi.org/10.1051/epjconf/20159401079

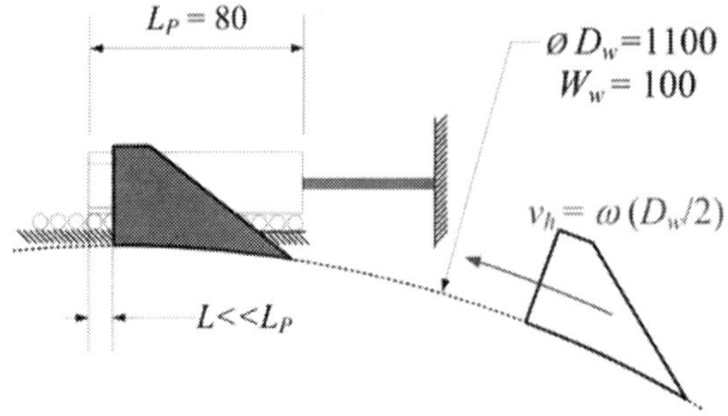

Figure 1. Schematic design of the flying wheel.

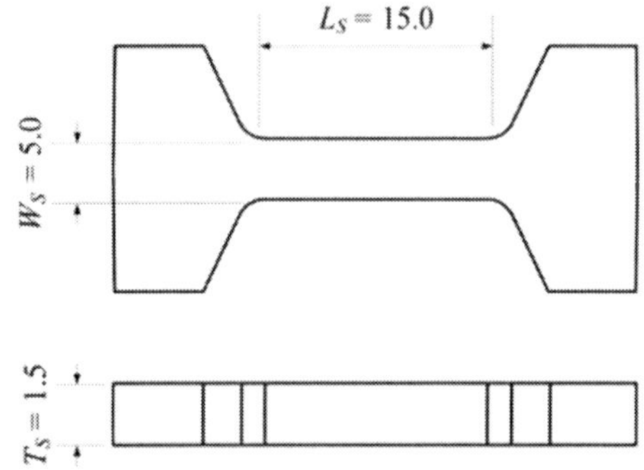

Figure 2. Geometry and dimensions of the specimen.

For the sake of simplicity all the elastic constants ($E = 210\,\text{GPa}$ and $\nu = 0.3$) as well as the mass density ($\rho = 7850\,\text{kg/m}^3$) were considered to be the same for all the materials. The specimen was considered as a generic rate-independent steel with power-law hardening,

$$Y\left(\bar{\varepsilon}_p\right) = A + B\bar{\varepsilon}_p^n \quad (2)$$

where $A = 250\,\text{MPa}$, $B = 800$, $n = 0.1$ and $\bar{\varepsilon}_p$ is the equivalent plastic strain. Let us assume, using von Mises plasticity, that the strain energy that needs to be consumed to fully fracture the specimen is:

$$E_s = A_s L_s \int_0^{\bar{\varepsilon}_p^f} \bar{\sigma}(\varepsilon)\, d\bar{\varepsilon}_p \quad (3)$$

being $\bar{\sigma}$ the von Mises equivalent stress. Assuming an engineering failure strain of $u/L_S = 50\%$ an approximate energy of 41.6 Joules is calculated to be necessary to fracture the specimen. Therefore the energy of the wheel needs to be significantly larger than that.

Given the dimensions of the wheel, for rotation frequencies of 0.1, 0.5, 1.0 and 1.5 Hz the wheel has a kinetic energy of 22.27, 556.81, 2227.27 and 5011.32 J respectively. In principle, the last three frequencies would allow the wheel to store enough energy to fracture the specimen. The first frequency however would not be enough to do it and therefore the wheel would stop rotating.

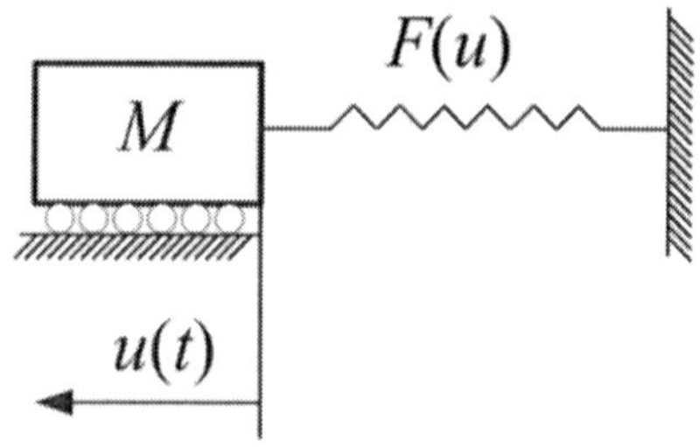

Figure 3. SDOF system representing the puller-specimen system.

3. Theoretical analysis

3.1. Problem set-up

We consider the Flying Wheel as an infinite bar with a diameter much bigger than that of the puller that travels at a velocity v and that hits the stationary puller. The flange where the impact occurs has a way smaller dimension than the rest of the components involved in the calculation so it will be neglected. The wave generated inside the puller reaches the specimen-puller interface end with velocity v at a time L_P/c and it initially is fully reflected with a velocity twice the initial one. The time that the wave takes to do so is $2L_P/c$ Note that $c = \sqrt{E/\rho}$ is the elastic wave propagation velocity in the material. When the reflected wave reaches the hammer-puller contact interface the leftmost puller end detaches from the hammer as it moves with a velocity of $2v$. The force applied by the specimen slows the puller down from $2v$ to some speed αv until the puller is caught by the hammer again at time t_i, where i is the number of contact interactions between the hammer and the puller. It is important to note that the duration of the free-flying periods is much longer than the duration of a single wave round-trip within the puller bar. Consequently, the wave propagation within the puller will not be described and quasi-static conditions assumed instead. The set of velocity conditions of the specimen-puller interface can be written as:

$$u'(t) = \begin{cases} 0 & 0 \le t < L_P/c \\ v \to 2v & t = L_P/c \\ 2v & L_P/c < t \le 3L_P/c \\ u'(t) & 3L_P/c < t < t_1 + L_P/c \\ \alpha v \to 2v - \alpha v & t = t_1 + L_P/c \\ 2v - \alpha v & t_1 + L_P/c < t \le t_1 + 3L_P/c \\ u'(t) & t_1 + 3L_P/c < t < t_2 + L_P/c. \end{cases} \quad (4)$$

Note that the same set of conditions applies again after time $t_2 + L_p/c$. In order to determine the unknown velocity $u'(t)$ of the puller bar in the periods when it is not in contact with the hammer, a simplified single degree of freedom system representing the specimen-puller system is considered (see Fig. 3). More specifically, the puller bar is considered as a lumped mass that is attached to a non-linear spring representing the specimen. The equation of motion of such system reads:

$$F_s\{u(t)\} + Mu''(t) = 0 \quad (5)$$

with the function $F(u)$ characterizing the specimen response. The logarithmic axial strain in the specimen is

$$\varepsilon = \ln\left(1 + \frac{u(t)}{L_s}\right) \tag{6}$$

where L_s was the initial gage length of the specimen. Additionally, the elastic strain was considered negligible compared to the plastic strain ($\varepsilon \simeq \varepsilon_p$). Therefore, making use of the plastic incompressibility,

$$A_S L_S = a_S l_S \tag{7}$$

where A_S and L_S are the area and gage length of the specimen respectively, while the lower case a_S and l_S are their deformed counterparts, the force in the specimen F_s could be written as a function of the applied displacement as:

$$\begin{aligned} F_s\{u(t)\} &= \sigma(\varepsilon)\, a_s \\ &= A + B\left(\ln\left(1 + \frac{u(t)}{L_s}\right)\right)^n \frac{a_s L_s}{L_s + u(t)}. \end{aligned} \tag{8}$$

The ODE was solved numerically considering the following initial conditions:

$$\begin{aligned} &3L_P/c < t < t_1 + L_P/c \left| \begin{array}{l} u'(3L_P/c) = 2v \\ u(3L_P/c) = 0 \end{array} \right. \\ &t_1 + 3L_P/c < t < t_2 + L_P/c \\ &\quad \times \left| \begin{array}{l} u'(t_1 + 3L_P/c) = 2v - \alpha v \\ u(t_1 + 3L_P/c) = u(t_1 + 3L_P/c). \end{array} \right. \end{aligned} \tag{9}$$

3.2. Results

Equation (4) along with the non-linear boundary conditions (9) was solved to obtain the puller velocity history for FW rotation frequencies of 1 and 1.5 Hz. The calculations were stopped at a displacement of around 5 mm which would correspond to an engineering strain of 33%. The displacement (red) and the velocities (blue) at the specimen-puller interface obtained from solving the ODE are plotted in Fig. 4. The black curves represent the constant linear velocity of the hammer and its displacement calculated as its integral with respect time. It should be noted that such black curves would represent the puller ideal behavior as stated in the previous sub-section. In other words, it would be the behavior we sought in the design of the FW system. The solid curves correspond to the ODE solution, while the dashed curves correspond to the velocity conditions summarized Eq. (4).

The plots of Fig. 4 give a clear idea of what is happening during the specimen loading in the FW system. Regardless of the rotation frequencies, the phenomenon is basically the same. When the wave generated in the hammer-puller impact reaches the specimen-puller interface the velocity jumps from 0 to $2v$ almost instantly. It can also be observed that the displacement is increasing linearly with time. After that period the velocity remains constant until the reflected wave reaches the puller free end (starts the free-fly of the puller) and comes back. The displacement increases due to the velocity jump, whereas the velocity is reduced progressively. The puller is caught again by the hammer after about 0.3ms at the point where the displacements of specimen-puller interface and hammer intersect (see Fig. 4). Then the velocity jumps

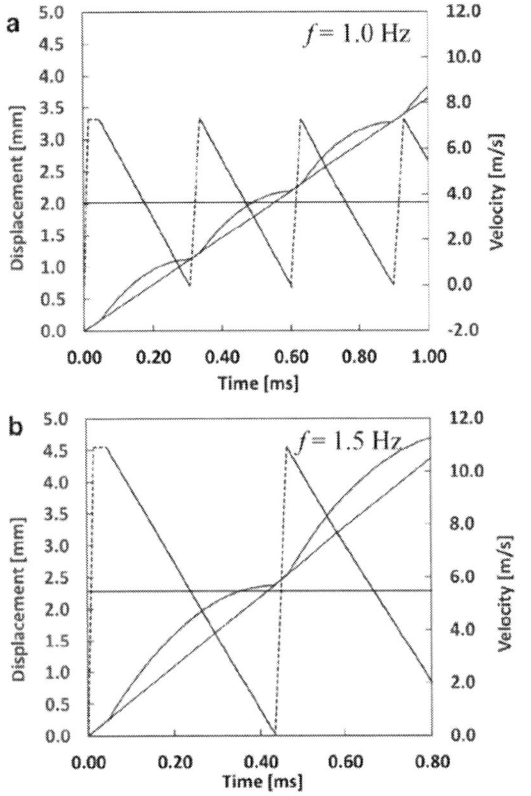

Figure 4. Displacement and velocity of the puller as a function of time as a result of using the simplified model for rotational frequencies of 1.0 (a) and 1.5 Hz (b).

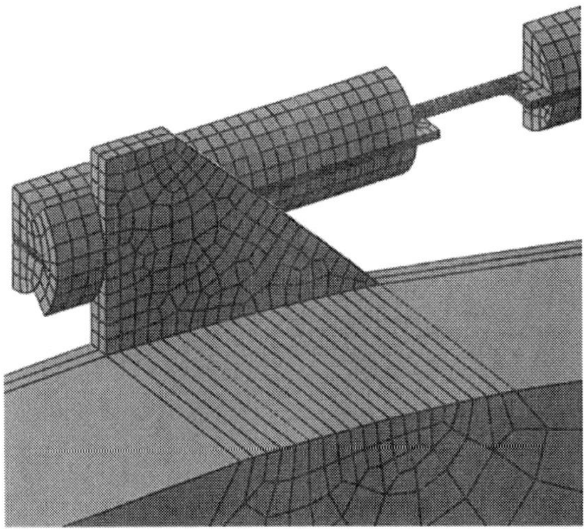

Figure 5. Finite element model of the FW system depicted in Fig. 1.

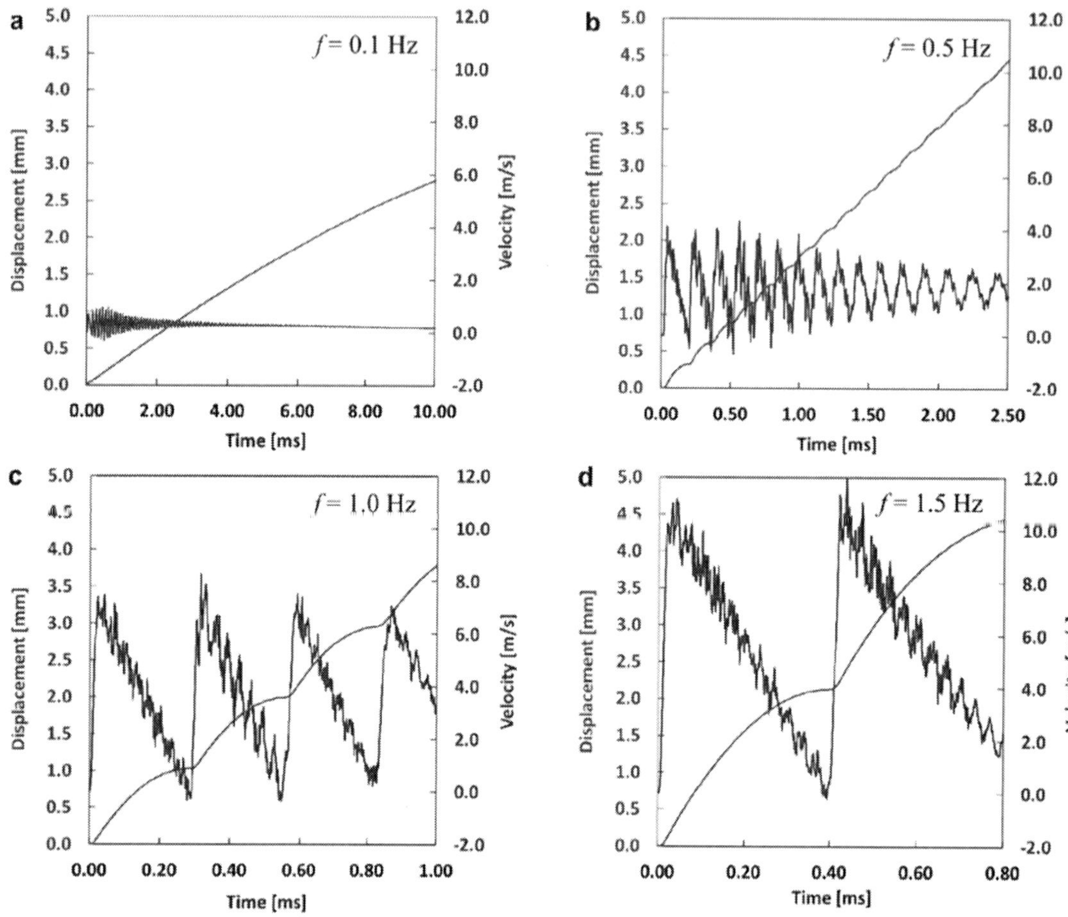

Figure 6. Velocity and displacement of the puller as a function of time obtained from the numerical simulations of the FW system of rotation frequencies of 0.1 (a), 0.5 (b), 1.0 (c) and 1.5 Hz (d).

again and the process starts all over again. The puller free-fly period increases with increasing angular velocity as can be seen by comparing Figs. 4(a) and (b). Those longer "free-fly periods" increase the deviation of the puller displacement history from the ideal linear path, which may be seen as a loss of displacement control during the test.

4. Numerical simulations

4.1. Numerical set-up

Four numerical simulations of the FW system shown in Fig. 1 are performed. Each one of them corresponds to a specific rotational frequency: 0.1, 0.5, 1.0 and 1.5 Hz. The geometry was discretized with eight-node reduced integration elements with stiffness-based hourglass control. The specimen gage length was discretized with 0.2 mm elements, while the rest was meshed with an approximate element size of 3 mm (see Fig. 5). The specimen behavior was modelled according to what is exposed in the subsequent sub-section. The rest was considered as elastic. Due to symmetry conditions only one half of the entire system was modelled. The specimen clamping to the puller and load cell was done sharing nodes. The displacement of the non-clamped load cell end nodes was constrained. The angular velocity was initially imposed through a coupling between the wheel and an additional rigid body that rotates at the same velocity. The rail on which the puller is mounted was simulated as a non-frictional analytical rigid surface. The contacts between surfaces were modelled employing a surface-to-surface algorithm [8]. A J2 plasticity model with associated flow rule and isotropic power-law hardening is employed to describe the specimen material.

4.2. Results

First, let us compare the results obtained from simulations with those from the theoretical analysis. The results of both simulations and theory are remarkably close. The Fig. 6(a) would correspond with Fig. 6(c) where the frequency of the wheel is 1.0 Hz, while the Fig. 6(b) should be compared to Fig. 6(d), both with a wheel frequency of 1.5 Hz. The color of the curves shown in Fig. 6 has the same significance as those plotted in Fig. 6. The solutions calculated for the specimen-puller interface displacement and velocity in the theoretical analysis capture the trends of those obtained in the numerical simulations. The values of displacement and velocity are also close. It is thus concluded that the basic analytical model provides a reasonable approximation of the specimen loading process in a FW tensile system.

The results obtained from the simulations run with frequencies of 0.5 (see Fig. 6 (a)) and 0.1 Hz (see Fig. 6(b)) extrapolate the results obtained for higher velocities. The lower the angular velocity of the wheel, the lower the amplitude and frequency of the specimen-puller interface velocity. For very slow wheel speeds (e.g. 0.1 Hz) the displacement of the specimen-puller interface converges towards the ideal one. However, different from loading at higher speeds, a noticeable drop of the velocity is present due to the lower wheel energy. Recall that the energy of the wheel was not enough to fracture the specimen according to the energy calculation made in the previous section. Therefore, the wheel stops spinning and the displacement of the puller remains constant.

5. Conclusions

The basic working principle of the specimen loading in a tensile FW experiment has been analyzed. An analytical model is developed to predict the displacement history that is imposed onto the specimen boundaries for a FW rotating at a constant speed. Additionally, numerical simulations are performed of the same FW system for wheel rotational frequencies of 0.1, 0.5, 1.0 and 1.5 Hz to confirm the validity of the analytical model.

The analysis clearly demonstrates that the loading speed applied onto the specimen boundary by a FW system is not constant even if the wheel rotates at constant angular velocity. This is due to the necessary insertion of a short puller bar that transmits the load from the rotating wheel to the specimen boundary. For a 1.1m diameter FW system with an 80mm long puller bar and a small 15mm long steel specimen, the velocity oscillations are negligibly small in experiments with an average strain rate of 20/s. However, already at 120/s, the loading speed oscillations are very significant oscillating between 0 and 240/s at a frequency of about 6.4kHz. At very high speeds, i.e. if the duration of the experiment is shorter than the round-trip travel within the puller bar, the set-up of the FW system is conceptually the same as an Split Hopkinson Tensile Bar system and the observed phenomenon is no longer relevant.

The first author is grateful for the partial financial support through the French National Research Agency (Grant ANR-11-BS09-0008, LOTERIE).

References

[1] M.A. Meyers, *Dynamic Behavior of Materials*. (1994): John Wiley & Sons.

[2] B. Song and W. Chen, W., *Split Hopkinson (Kolsky) Bar*. (2011): Springer.

[3] P. Viot, F. Beani, and J.L. Lataillade, *Polymeric foam behavior under dynamic compressive loading*. J Mater Sci, (2005). **40**(22): p. 5829–5837.

[4] R. Bouix, P. Viot, and J.-L. Lataillade, *Phenomenological study of a cellular material behaviour under dynamic loadings*. J. Phys. IV France, (2006). **134**: p. 109–116.

[5] R.R. Ambriz, C. Froustey, and G. Mesmacque, *Determination of the tensile behavior at middle strain rate of AA6061-T6 aluminum alloy welds*. International Journal of Impact Engineering, (2013). **60**(0): p. 107–119.

[6] C. Froustey, M. Lambert, J.L. Charles, and J.L. Lataillade, *Design of an Impact Loading Machine Based on a Flywheel Device: Application to the Fatigue Resistance of the High Rate Pre-straining Sensitivity of Aluminium Alloys*. Exp Mech, (2007). **47**(6): p. 709–721.

[7] L.W. Meyer and S. Abdel-Malek. *Deformation and Ductile Fracture of a Low Alloy Steel under High Strain Rate Loading*. in *2nd International Conference on High Speed Forming*. (2006).

[8] ABAQUS, *ABAQUS theory manual v6.12*. (2012).

EPJ Web of Conferences **94**, 01080 (2015)
DOI: 10.1051/epjconf/20159401080

A rate-dependent Hosford-Coulomb model for predicting ductile fracture at high strain rates

Stephane J. Marcadet[2], Christian C. Roth[1,2], Borja Erice[1], and Dirk Mohr[1,2,3]

[1] Solid Mechanics Laboratory (CNRS-UMR 7649), Department of Mechanics, École Polytechnique, Palaiseau, France
[2] Impact and Crashworthiness Lab, Department of Mechanical Engineering, Massachusetts Institute of Technology, Cambridge MA, USA
[3] Department of Mechanical and Process Engineering, ETH Zurich, Switzerland

Abstract. The Hosford-Coulomb model incorporates the important effect of the Lode angle parameter in addition to the stress triaxiality to predict the initiation of ductile fracture. A strain-rate dependent extension of the Hosford-Coulomb model is presented to describe the results from low, intermediate and high strain rate fracture experiments on advanced high strength steels (DP590 and TRIP780). The model predictions agree well with the experimental observation of an increase in ductility as function of strain rate for stress states ranging from uniaxial to equi-biaxial tension.

1. Introduction

Recent research on ductile fracture has clearly demonstrated that the effect of the Lode parameter is as important as that of the stress triaxiality. For example, the drastic loss in ductility of metals when changing the stress state from equi-biaxial tension (triaxiality of 2/3) to plane strain tension (triaxiality of 0.58) cannot be explained using conventional stress triaxiality dependent models such as the Johnson-Cook failure model. However, Lode angle dependent theories such as the Hosford-Coulomb fracture initiation model (Mohr and Marcadet, 2015) can fully explain this phenomenon through the increased stability of plastic flow under axisymmetric stress states such as equi-biaxial tension as compared to generalized shear stress states such as plane strain tension.

It is the objective of the present work to make use of these important recent findings on the stress-state dependency of ductile fracture to improve the predictions of ductile fracture initiation under dynamic loading conditions. For this, we present a rate-dependent version of the Hosford-Coulomb model. The proposed model may be seen as a Lode-angle dependent extension of the rate- and temperature dependent Johnson-Cook (1985) failure model. In most parts, this conference paper is compiled from extracts from the journal papers of Roth and Mohr (2014), Mohr and Marcadet (2015) and Gu and Mohr (2015). We first present the model formulation before demonstrating the model's ability to describe the experimental data from fracture experiments on advanced high strength steels that have been performed on a tension Hopkinson bar system.

2. Theory

2.1. Definition of the Lode parameter and stress triaxiality dependency

The stress state is described by the stress triaxiality and the Lode angle parameter. The stress triaxiality is defined as the ratio of the mean stress σ_m and the von Mises stress $\bar{\sigma}$,

$$\eta = \frac{\sigma_m}{\bar{\sigma}}. \tag{1}$$

It may be interpreted as a measure of the ratio of the first and second stress tensor invariants. The Lode angle parameter on the other hand measures the ratio of the third and second stress tensor invariants,

$$\bar{\theta} = 1 - \frac{2}{\pi}\arccos\left[\frac{3\sqrt{3}}{2}\frac{J_3}{(J_2)^{3/2}}\right]. \tag{2}$$

According to the above definition, the Lode angle parameter varies between -1 (axisymmetric compression) and 1 (axisymmetric tension). Based on the *modified* Haigh-Westergaard coordinates, the ordered principal stresses ($\sigma_i \geq \sigma_{ii} \geq \sigma_{iii}$) may be constructed as

$$\sigma_i = \bar{\sigma}\,(\eta + f_i) \tag{3}$$

with the Lode angle parameter dependent trigonometric functions

$$f_1[\bar{\theta}] = \frac{2}{3}\cos\left[\frac{\pi}{6}(1-\bar{\theta})\right],\quad f_2[\bar{\theta}] = \frac{2}{3}\cos\left[\frac{\pi}{6}(3+\bar{\theta})\right],$$
$$f_3[\bar{\theta}] = -\frac{2}{3}\cos\left[\frac{\pi}{6}(1+\bar{\theta})\right]. \tag{4}$$

2.2. Stress-state dependent fracture initiation model

Dunand and Mohr (2014) subjected a unit cell of a Levy-von Mises material with a central void of 1.2% porosity to combinations of shear and normal loading to determine the macroscopic equivalent plastic strain at the onset of localization as a function of the stress state. They

Article available at http://www.epj-conferences.org or http://dx.doi.org/10.1051/epjconf/20159401080

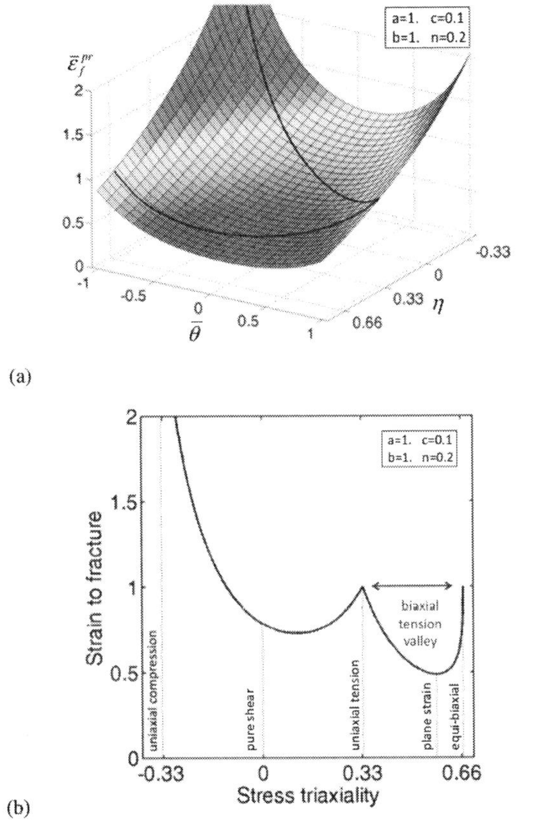

Figure 1. Hosford-Coulomb model for proportional loading: (a) strain to fracture as a function of the stress triaxiality and the Lode angle parameter; (b) fracture envelope for plane stress conditions.

performed this type of analysis for more than 160 different stress states for stress triaxialities ranging from 0 to 1 and Lode angle parameters ranging from −1 to 1. Their results demonstrate that a Hosford-Coulomb criterion of the type

$$\bar{\sigma}_{HF} + c(\sigma_I + \sigma_{III}) = \beta \tag{5}$$

provides a good approximation of their simulation data, with the Hosford equivalent stress

$$\bar{\sigma}_{HF} = \left\{\frac{1}{2}\left((\sigma_I - \sigma_{II})^a + (\sigma_{II} - \sigma_{III})^a + (\sigma_I - \sigma_{III})^a\right)\right\}^{\frac{1}{a}} \tag{6}$$

and $1 \leq a \leq 2$ denoting the Hosford exponent. The onset of ductile fracture is considered to be imminent with the onset of localization. The above localization criterion (5) is therefore utilized to predict the onset of fracture under proportional loading. The results from ductile fracture experiments are typically presented in terms of the equivalent plastic strain, the stress triaxiality and the Lode angle parameter. We transform the localization criterion from the principal stress space to the mixed strain-stress

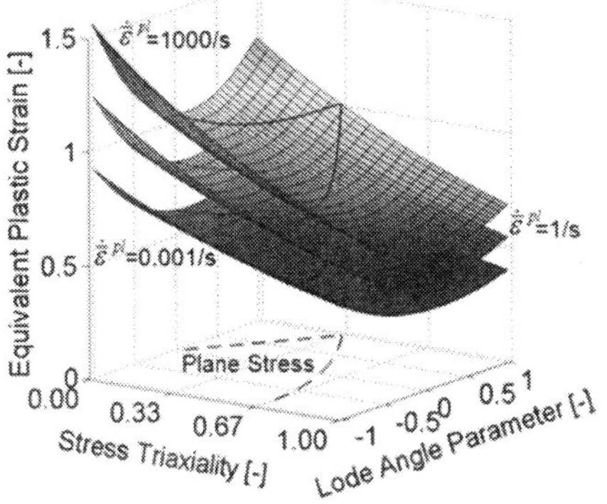

Figure 2. Illustration of the fracture surfaces (for proportional loading) for different strain rates.

space (for details, see Roth and Mohr, 2014),

$$\begin{aligned}\bar{\varepsilon}_f^{pr}[\eta, \bar{\theta}] &= b(1+c)^{\frac{1}{n}} \ldots \\ &\ldots \left(\left\{\frac{1}{2}\left((f_1 - f_2)^a + (f_2 - f_3)^a + (f_1 - f_3)^a\right)\right\}^{\frac{1}{a}} \right. \\ &\left. + c\,(2\eta + f_1 + f_3)\right)^{-\frac{1}{n}}\end{aligned} \tag{7}$$

Figure 1a shows a plot of the isotropic Hosford-Coulomb fracture initiation model for proportional loading. The strain to fracture is a monotonically decreasing function of the stress triaxiality and a convex non-symmetric function of the Lode angle parameter. The corresponding strain to fracture for plane stress conditions (see black curve in Fig. 1a) is shown as a function of the stress triaxiality only in Fig. 1b. Note that the apparent non-monotonic dependence on the stress triaxiality is deceiving because of the underlying Lode angle effect that is due to the functional relationship $\bar{\theta} = \bar{\theta}[\eta]$ for plane stress. The main model parameters are $\{a,b,c\}$. The Hosford exponent a controls the effect of the Lode angle parameter, while the friction coefficient c primarily controls the effect of the stress triaxiality on the strain to fracture. The model parameter $b$is a multiplier controlling the overall magnitude of the strain to fracture; it is defined such that it is equal to the strain to fracture for uniaxial tension (which is the same as that for equi-biaxial tension). The parameter $n = 0.1$ is a fixed transformation parameter. Note that we introduced the superscript "pr" to indicate that the above expressions are applied to predict the onset of fracture under proportional loading only. For non-proportional loading, the integral form

$$\int_0^{\bar{\varepsilon}_f} \frac{d\bar{\varepsilon}_p}{\bar{\varepsilon}_f^{pr}[\eta, \bar{\theta}]} = 1 \tag{8}$$

is used, where the expression for the strain to fracture for proportional loading serves as weighting function.

2.3. Rate-dependent Hosford-Coulomb model

In loose analogy with the Johnson and Cook (1985) fracture model, the effect of strain rate is incorporated into

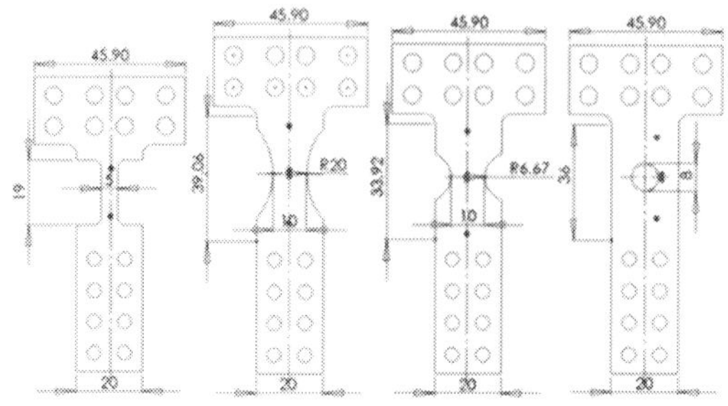

Figure 3. Specimen geometries for high strain rate testing.

the Hosford-Coulomb model through the coefficient b,

$$b = \begin{cases} b_0 & \text{for } \dot{\bar{\varepsilon}}_p < \dot{\varepsilon}_0 \\ b_0 \left(1 + \gamma \ln \left[\frac{\dot{\bar{\varepsilon}}_p}{\dot{\varepsilon}_0}\right]\right) & \text{for } \dot{\bar{\varepsilon}}_p \geq \dot{\varepsilon}_0 \end{cases} \tag{9}$$

$b_0 > 0$ is the strain to fracture for uniaxial tension at low strain rates, while the model parameter $\gamma \geq 0$ controls the strain rate sensitivity of the fracture initiation model. Figure 2 illustrates the effect of the strain rate on the resulting "fracture surface" in the space of equivalent plastic strain, stress triaxiality and Lode angle parameter.

3. Experimental results

3.1. Material, specimens and experimental procedures

The tested materials are 1.4 mm thick DP590 steel and 1.4 mm thick TRIP780 steel. Uniaxial tension (UT), notched tension (NT) and central hole (CH) tension specimens (Fig. 3) are extracted from these sheets. Prior to testing, a random speckle pattern is applied to the specimens. The relative displacement of two points on the respective upper and lower specimen shoulders (see blue dots in Fig. 3) is measured in all experiments by means of digital image correlation (VIC2D, Correlated Solutions). The distance between these points is equal to $l_0 = 17.5$ mm for the UT-specimens, and $l_0 = 30$ mm for the NT and CH specimens. The experiments at low and intermediate strain rates are carried out on a MTS hydraulic testing machine with a 250 kN load cell and custom-made high pressure clamps.

3.2. Experimental results

The effect of the strain rate on the engineering stress-strain curve for DP590 steel for uniaxial tension along the rolling direction is illustrated by Fig. 4. Observe that the material's yield stress is about 17% higher in the intermediate strain rate experiment than in the low strain rate experiment. It is also worth noting that the elongation to fracture (engineering strain at the onset of fracture as evaluated for an extensometer length of 17.5 mm) is significantly higher in the low strain rate experiment than in the intermediate and high strain rate experiments.

Figure 5 shows the force-displacement curves experiments for three different speeds of loading which are also qualitatively representative for the NT6 and CH experiments; All experiments show a force maximum prior to the onset of fracture, which increases as a function of the loading speed. The shapes of the force-displacement curves for low and intermediate strain rates are similar up to the force maximum, i.e. a dynamic increase factor could be used to describe the effect of the loading speed.

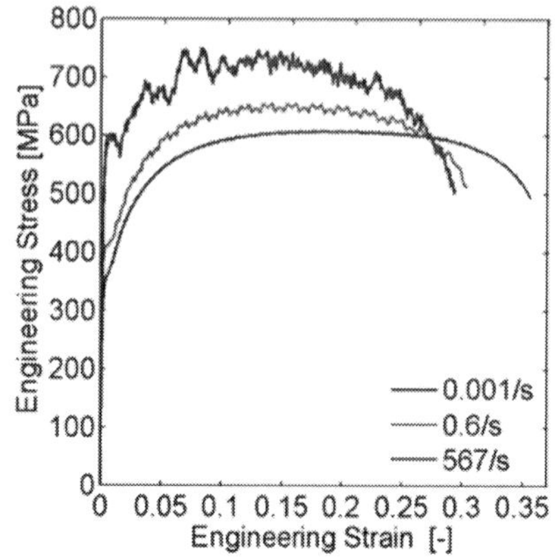

Figure 4. Stress-strain response for DP590.

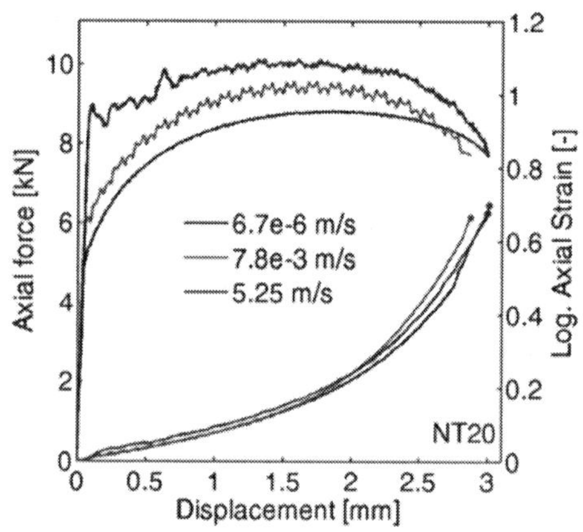

Figure 5. Force-displacement curve for notched tension (NT20).

The shape of the curve for the fast loading speed is different, i.e. it exhibits a significantly higher force level at the beginning of the experiment followed by a less pronounced hardening regime. Note that after the force maximum, the decrease in force is higher the faster the loading speed. However, it is remarkable to see that for all three specimen types, the displacement to fracture appears to be almost independent of the loading speed. To gain further insight into the local deformation fields, we also report the strain determined from a 1 mm long virtual extensometer on the specimen surface (see secondary axis in Fig. 4). The evolution of the local strain as a function of the axial displacement is very similar irrespective of the speed of loading. In particular, only small variations are observed at the onset of fracture.

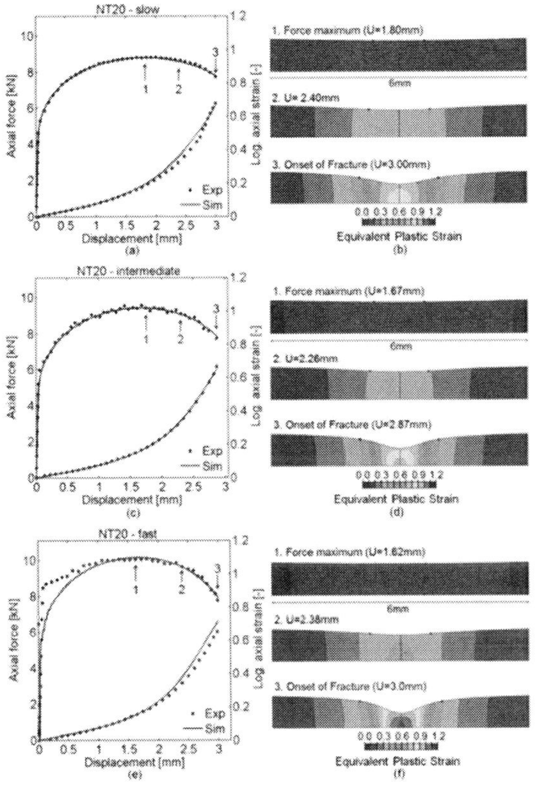

Figure 6. Force-displacement curve for notched tension NT20 (left column) and longitudinal cross-sectional views (right column).

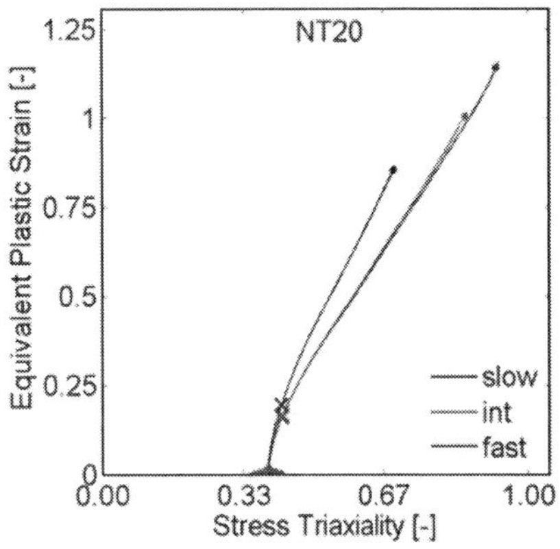

Figure 7. Loading paths to fracture for notched tension (DP590 steel).

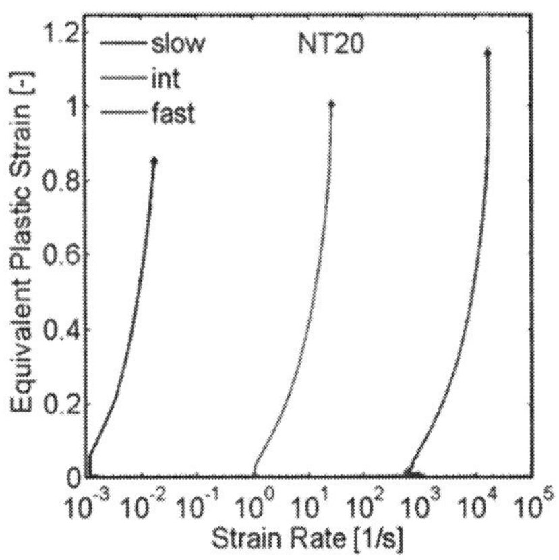

Figure 8. Force-displacement curve for notched tension (NT20).

4. Model calibration and validation

Numerical simulations are performed of all experiments using the finite element software Abaqus/explicit along with a strain rate and temperature-dependent plasticity model (Roth and Mohr, 2014). An example for the comparison of measured (solid dots) and computed (solid lines) force-displacement curves is shown in Fig. 6. The final result from each simulation is the so-called *loading path to fracture*, i.e. the evolution of the equivalent plastic strain $\bar{\varepsilon}_p$ as a function of the strain rate $\dot{\bar{\varepsilon}}_p$, the temperature T, the stress triaxiality η and the Lode angle parameter $\bar{\theta}$. It is assumed that fracture initiates in the simulation at the integration point with the highest equivalent plastic strain at the instant of specimen fracture in the experiment. For the NT20 and NT6 specimens, this point is positioned at the very center of the specimen, i.e. at the intersection point of all planes of specimen symmetry. In the CH specimens, this point is positioned on the specimen mid-plane at a distance of about 0.5 mm from the central hole boundary.

Figure 7 shows the loading paths to fracture in terms of the equivalent plastic strain and the stress triaxiality for NT20 specimens. The cross symbols on the loading paths indicate the point at which the force maximum is reached. Irrespective of the specimen type (NT20, NT6, CH), the stress state is independent of the strain rate and remains approximately constant up to the force maximum. Thereafter, the stress triaxiality increases while the Lode angle parameter decreases. This is attributed to the built-up of an out-of-plane stress during through-thickness necking. Due to the absence of thermal softening for slow loading, necking is less pronounced for this case than for intermediate and fast loading. The longitudinal cross-sectional views of the specimens (e.g. right columns in Fig. 6) show the strain distribution throughout the necking phase. Up to the force maximum, plane stress conditions prevail and the strain distribution is approximately uniform along the thickness direction. Throughout necking, the deformation localizes within a small zone along the longitudinal axis and through-thickness gradients built up.

The comparison of the cross-sectional views at the onset of fracture for different loading velocities clearly shows that the localization severity increases the higher the speed of loading. The corresponding plots of the equivalent plastic strain versus the strain rate (Fig. 8) show that the strain rate increases by at least one order of magnitude after the onset of necking. For example, the pre-necking strain rate in the fast NT20 experiment is about $10^3/s$, while a strain rate of about $10^4/s$ is reached at the onset of fracture.

The analysis of the loading paths to fracture for different stress states (NT20, NT6, CH) and different materials (DP590, TRIP780) supports the assumption made in the rate-dependent Hosford-Coulomb model: the strain to fracture increases as a function of the loading speed. The solid dots in Fig. 8 are the predictions of the instant of fracture initiation by the Hosford-Coulomb model; these dots almost coincide with the end of the solid curves which corresponds to the instant of fracture initiation observed in the experiments.

5. Conclusions

A rate-dependent Hosford-Coulomb fracture initiation model is presented and used to predict the onset of fracture in static and dynamic experiments on advanced high strength steel (DP590, TRIP780) for different stress states and different loading velocities. The model may be seen as an extension of the Johnson-Cook failure model; however, it accounts not only for the effect of strain rate and stress triaxiality, but it also incorporates the important effect of the Lode parameter on fracture initiation. The experimental results demonstrate that for biaxial tension, the strain to fracture increases as a function of the loading velocity. Due to the adiabatic heating during the dynamic experiments, the effects of strain rate and temperature could not be separated in the present analysis.

CR and BE are grateful for the partial financial support through the French National Research Agency (Grant ANR-11-BS09-0008, LOTERIE). The authors are also grateful for the support of the MIT Industrial Fracture Consortium.

References

[1] C.C. Roth and D. Mohr, *Effect of Strain Rate on Ductile Fracture Initiation in Advanced High Strength Steel Sheets: Experiments and Modeling*. International Journal of Plasticity **56** (2014) 19–44.

[2] D. Mohr and S. J. Marcadet, *Micromechanically-motivated Phenomenological Hosford-Coulomb Model for Predicting Ductile Fracture Initiation at Low Stress Triaxialites*, Int. J. Solids Struct (2015), in press.

[3] G. Gu and D. Mohr, *Anisotropic Hosford-Coulomb Fracture Initiation Model: Theory and Application* (2015), under review.

[4] Dunand M, Mohr D (2014). *Effect of Lode parameter on plastic flow localization after proportional loading at low stress triaxialities*, J. Mech. Phys. Solids **66**, 133–153.

[5] Dunand M, Gary G, Mohr D (2013), *Load-Inversion Device for the High Strain Rate Tensile Testing of Sheet Materials with Hopkinson Pressure Bars*, Exp. Mech. **53**, 1177–1188

[6] G.R. Johnson and W.H. Cook, *Fracture characteristics of three metals subjected to various strains, strain rates, temperatures and pressures*. Engineering Fracture Mechanics (1985), **21**, p. 31–48.

EPJ Web of Conferences **94**, 02001 (2015)
DOI: 10.1051/epjconf/20159402001
© Owned by the authors, published by EDP Sciences, 2015

Mechanism of grain refinement and its effect on Adiabatic Shear Bands in 4340 steel and pure copper during impact

Nabil Bassim[a] and Solomon Boakye-Yiadom

Department of Mechanical Engineering, University of Manitoba, Winnipeg, MB, Canada, R3T5V6

Abstract. Pre-deformation and post-deformation microstructure characterization was conducted on tempered 4340 steel and commercial pure copper specimens under impact to determine the microstructural changes and the mechanism of grain refinement that occur during the evolution of ASBs. It was observed that the movement and multiplication of dislocations, elongation of grains, breaking of elongated grains, rotation, carbide fragmentation and boundary refinement of broken grains occur simultaneously during the evolution of ASBs in the impacted 4340 steel specimens. The extent of these mechanisms depends on the imposed local strain and strain rate. Extensive grain refinement coupled with high density of dislocations results in the shear band structures being more susceptible to crack nucleation and propagation. In copper, it was observed that sequential occurrence of emergence of dislocations, dislocation cell formations with varying cell boundaries and cell interiors, dynamic recovery and extensive micro-twinning results in the formation of the shear bands. The structure within the evolved shear bands becomes less brittle after the onset of dynamic recovery and micro-twinning. The differences in the mechanism of grain refinement and evolution of the shear bands in both materials is attributed to the differences in the mobility of dislocations, the rate of strain hardening and strain hardening exponents in both materials studied.

1. Introduction

Metals and other materials exhibit the formation of narrow bands of extreme strains known as Adiabatic Shear Bands (ASBs) when deformed at high strain rates and large strains [1]. Mechanical and material factors have been reported to influence the appearance and susceptibility of materials to the formation of ASBs [1]. It has been reported that BCC metals exhibit an increased propensity to ASB when their grain sizes are reduced into the ultrafine-grain and nano-crystalline regime [2]. This is attributed to the associated changes in the strength, strain hardening behaviour and strain rate sensitivity when the grain size changes [2]. Also, it has been observed that the ASBs associated with FCC metals, such as aluminium and copper, are diffused and exhibit a plastic flow phenomenon [3]. The influence of these factors on the formation of ASBs during dynamic loading makes the phenomenon of ASBs very complex.

In this study, pre-impact and post-impact microstructure characterization was conducted on tempered 4340 steel (BCC) and pure copper (FCC) specimens to determine the mechanism of evolution of ASBs. This was done in order to determine the effect crystal structure and microstructure on the mechanisms and appearance of ASBs in both materials.

2. Experimental procedure

Heat treatable AISI 4340 steel (C: 0.38-0.43%, Mn 0.6-0.8%, Si 0.15-0.35%, Cr 0.7-0.9%, Mo 0.2-0.3, Ni 1.65-2.00%) and copper of commercial purity (99.9% purity) were used in this study. As received rod forms of both specimens were machined into a diameter of 9.5 mm and a length of 10.5 mm. The machined steel specimens were heat treated using an austenitize-quench-and-temper procedure prior to impact. The machined specimens were impacted at high strain rates and large strains using the Direct Impact Hopkinson Pressure Bar (DIHPB) at different projectile velocities resulting in several impact momentums. During the impact, an AISI 4340 steel projectile (18.67 N, 38 mm diameter by 127 mm length) with a hardness of 47HRC is fired by a gas gun, and strikes the specimen attached to a transmitter bar at a high impact velocity [4,5].

Selected specimens of both materials were studied prior to and after impact using metallographic and electron microscopy techniques. The microstructural analyses were conducted using a ZIESS Axiovert 25 inverted-reflected light microscope, JEOL 5900 Scanning Electron Microscope (SEM), a JEOL 2100F TEM/STEM and a JEOL 2010F TEM to document the structure and appearance of the ASBs after impact.

3. Results

3.1. Microstructural evolution of shear Bands in steel during impact

The tempered steel specimens prior to impact consisted of precipitated platelet and spherical carbides within ferrite matrix (tempered martensite structure) as shown in Fig. 1. The predominant platelet carbides were aligned on the lath boundaries (interlath carbides) and within the laths (intralath carbides). Details of the initial structure of

[a] Corresponding author: author@email.org

Article available at http://www.epj-conferences.org or http://dx.doi.org/10.1051/epjconf/20159402001

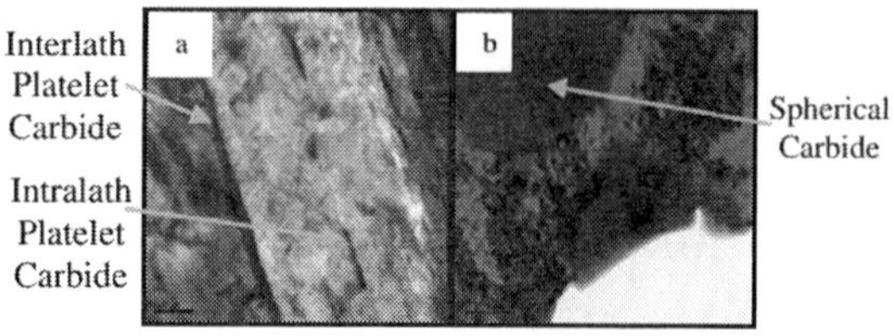

Figure 1. TEM bright field micrographs of the tempered steel specimens prior to impact.

Figure 2. Optical micrographs of adiabatic shear band in a single steel specimen impacted at 51 kg · m/s.

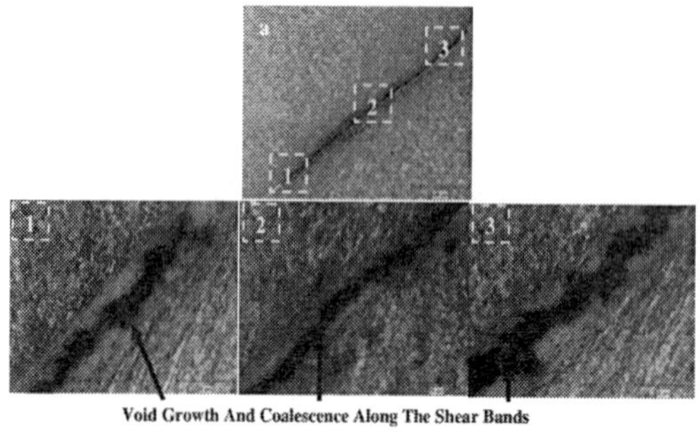

Figure 3. Optical micrographs of microvoids coalescing to form cracks along a shear band as shown in regions 1, 2 and 3.

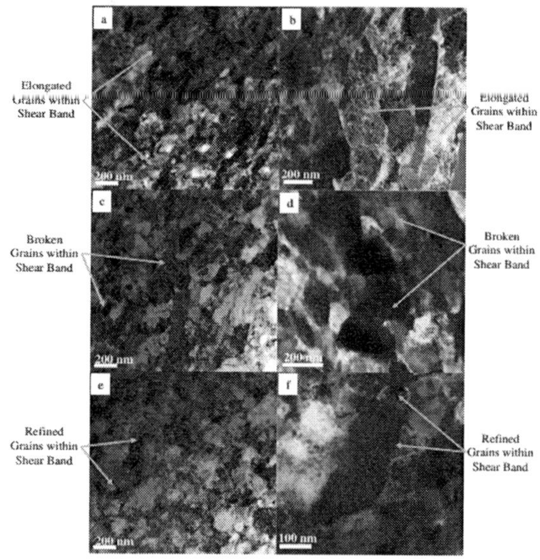

Figure 4. TEM micrographs showing (a), (b) elongated grains within shear band in the steel specimen impacted at 43 kg · m/s. (c), (d) broken grains within shear band in the steel specimen impacted at 47 kg · m/s. (e) (f) refined grains within shear bands in the steel specimen impacted at 50 kg · m/s.

the tempered steel specimens have been published in references [4,5].

Figure 2 shows the trajectory of an evolved ASB in a single steel specimen impacted at 51 kg · m/s. The evolved shear bands were very hard and brittle compared to the regions outside the shear bands. This creates a hardness gradient across the microstructure of the impacted material. The structure of the shear bands become metallographically distinct as the impact momentum (strain rate) is increased.

At higher impact momentums (strain rates), cracks propagated along the lengths of the shear bands. Figure 3 shows cracks propagating along a shear band that evolved in a steel specimen after impact. The wavy nature of the propagating crack demonstrates void nucleation, growth and coalescence which develop into cracks along the length of the band. It is inferred that the connection between adjacent voids give the cracks their wavy nature prior to fracture and fragmentation as shown in Fig. 3.

Extensive microstructural analysis on the impacted steel specimens revealed that the evolved shear band microstructures were not uniform at any of the stages of the impact because several types of grains and structures were observed coexisting within the ASBs. At relatively lower impact momentum/strain rates, after the initiation of strain localization, the structure is comprised of mainly elongated grains and few broken grains within the ASB. After this stage, the structure that evolved after increasing the impact momentum/strain rate was made of predominantly broken grains; a few refined grains and nanograins, while yet another structure after increasing the strain rate was made of dominant refined grains, nanograins and subgrains. Detail description has been published in references 11. and 12. Figure 4 shows TEM micrographs of the microstructures that evolved within the shear bands in the steel specimens after impact at 43, 47 and 50 kg · m/s.

Figure 5 is a schematic of the stages of grain evolution that occurs within the evolving shear bands during impact of the steel specimens. It is based on the observations of the structure within ASBs as a function of strain rate/impact momentum. It is emphasized here that each of the stages represents an already failed material because of the presence of either a diffused or distinct ASB after impact. The occurrence of strain localization starts with texture development and the emergence of dislocations as shown in Fig. 5(a) and (b) which depicts the elongated grains with initiated transverse and random dislocation boundaries. At this stage, the elongated grains are decorated with high density of dislocations including clusters of dislocations along the predominant transverse dislocation

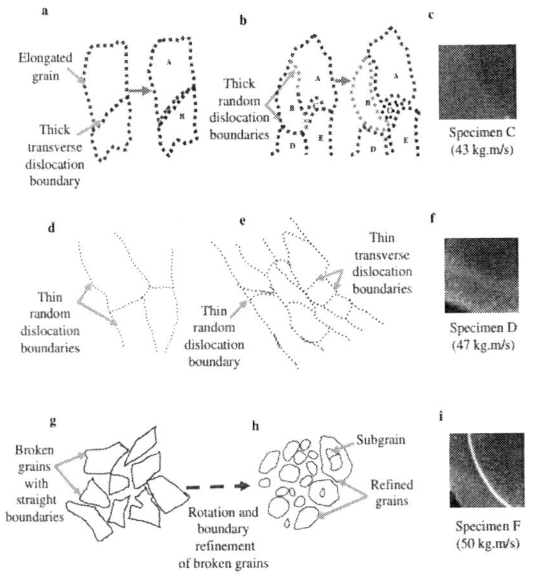

Figure 5. Schematic of the stages of grain refinement during the evolution of ASBs. (a), (b) Occurs after the initiation of strain localization which corresponds to the diffused ASB in (c). (d), (e) Occur prior to breaking of the elongated grains which corresponds to the ASB in (f). (g) Represents broken grains which undergo rotation and boundary refinement to produce refined grains (h) Refined grains and subgrains which corresponds to the ASB in (i).

boundaries. This represents a unique structure of ASB which corresponds to the observed metallographically diffused ASB shown in Fig. 5(c). Figure 5(d) and (e) illustrates a later stage after grain elongation prior to the breaking of the elongated grains along the initiated dislocation boundaries. The observed boundaries at this stage have thin clusters of dislocations along the initiated dislocation boundaries and correspond to the shear band in Fig. 5(f). The broken grains with smaller sizes and straight boundaries are shown in Fig. 5(g). At this stage, the observed dislocations were within the broken grains and not concentrated on the boundaries. Rotation and boundary refinement of the broken grains results in refined grains with round boundaries as shown in Fig. 5(h) which corresponds to the distinct ASB in Fig. 5(i).

3.2. Microstructural evolution of shear bands in pure copper during impact

Pre-impact microstructure of the pure copper specimens revealed that the structure is made of large grains and deformation twins as shown in Fig. 6. Dislocations were also observed both within the large grains and on the twins.

Figure 7(a) shows an optical micrograph of a copper specimen after impact at 45 kg · m/s. The shear bands were wide and appeared metallographically diffused from the surrounding regions. The thickness and appearance of the shear bands remained relatively the same (wide-diffused bands) regardless of the impact momentum/strain rate. This is in contrast to the shear bands observed in the impacted steel specimens. The shear bands in the steel specimens were wide and metallographically diffused at lower impact momentum/strain rates while they become thin and distinct from the surrounding regions as the impact momentum/strain rate increase. Figure 7(b) and (c) show Scanning Electron Images (SEI) of the shear band in Fig. 7(a). The effect of shear strain localization on the structure within the shear band compared to the structure outside the shear band becomes apparent (micrographs taken under the same conditions). The regions within the shear bands are highly strained compared to the regions outside the bands.

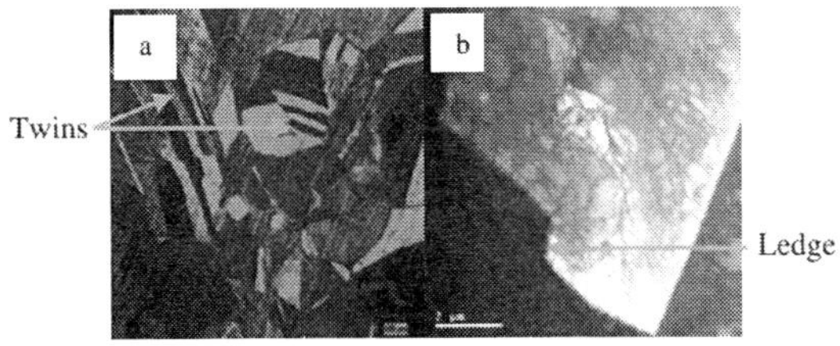

Figure 6. (a) Optical micrograph (b) TEM bright field micrograph of copper specimen prior to impact.

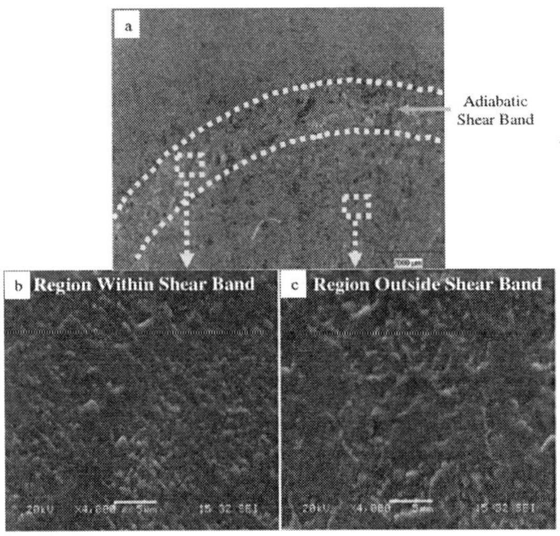

Figure 7. (a) Optical micrograph of ASB in a pure copper after impact at 45 kg · m/s (b), (c) SEI of regions within and outside ASB respectively.

A comprehensive Transmission/Scanning Transmission Electron Microscopy (TEM/STEM) analysis on the impacted copper specimens revealed that there were extensive differences between the regions within the shear bands in the copper specimens impacted at relatively lower impact momentum/strain rates compared to the regions within the shear bands in the copper specimens impacted at high impact momentum/strain rates.

Figure 8(a) to (f) show the microstructure of the regions within the adiabatic shear bands that developed in the copper specimens impacted at 45 kg · m/s. Dislocation cell structures were the predominant features within the shear bands in this region. The dislocation cells had

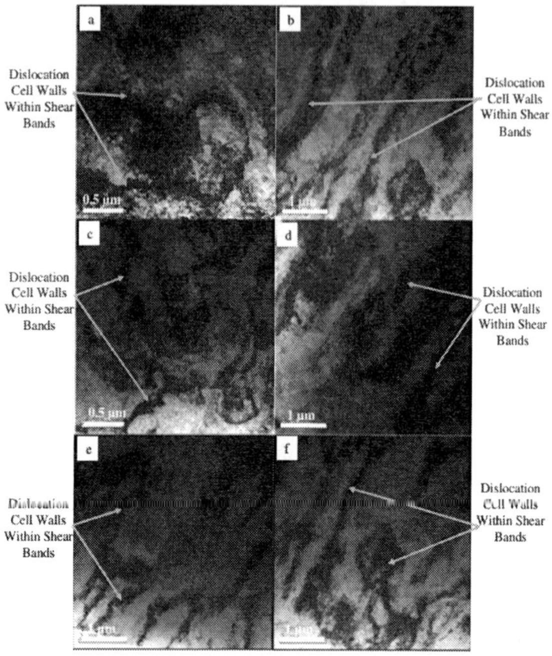

Figure 8. TEM micrographs of regions within an ASB that evolved in a copper specimen after impact at 45 kg · m/s showing dislocation cell formation.

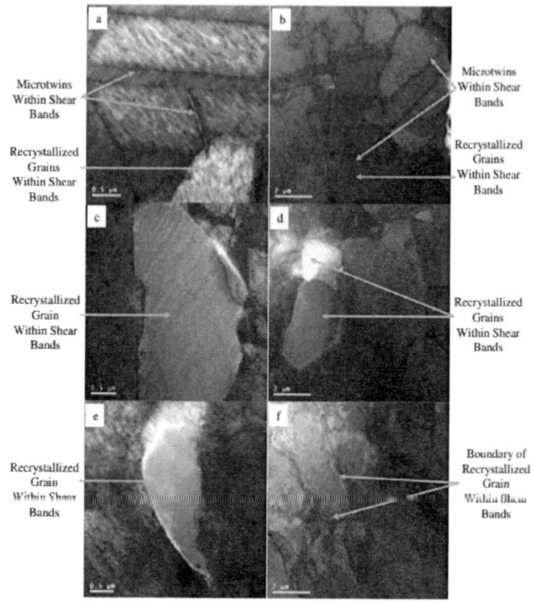

Figure 9. TEM micrographs of regions within an ASB that evolved in a copper specimen after impact at 57 kg · m/s showing evolved recrystallized grains and annealing twins.

thick boundaries with clean cell interiors. Some of the dislocation cells had thicker cell boundaries with more dislocations within the cell interiors. It is inferred that the emergence of dislocations and the formation of dislocation cell boundaries are the predominant structures within adiabatic shear bands that evolve in copper at relatively lower strain rates as shown in Fig. 8.

At relatively higher strain rates, the adiabatic shear bands that evolved within the copper specimens were characterized by recrystallized grains and microtwins.

Figure 9 shows the microstructure of regions within evolved adiabatic shear bands in a copper specimen that was impacted at 57 kg · m/s. These recrystallized grains were well developed with distinct grain boundaries. The dislocation cells with predominantly thicker boundaries disappeared from the regions within the shear bands due to associated localized recovery and dynamic recrystallization mechanisms.

Annealing twins were also predominant within the shear bands. The annealing twins did not have dislocation structures within the grains compared to the observed dislocation structures within the deformation twins that were observed in the pre-impact copper specimens.

4. Discussions

The current study shows that the mechanism of evolution of adiabatic shear bands in tempered steel specimens are different from those observed in pure copper specimens. The onset of strain localization in the tempered steel specimens was characterized by the emergence of dislocations and alignment of the grains in the shear direction (texture development). Elongated grains, broken grains, refined grains, subgrains and nanograins decorated with high density of dislocations were observed within the shear band structures that evolved after the onset of strain localization. However, at every other impact momentum/strain rate, after the onset of strain localization, there were predominant grain structures and microstructural features that characterized the shear band structure. This demonstrate that different mechanisms occur simultaneously including the emergence of dislocations, breaking of grains, and grain refinement during the evolution of the shear bands in the tempered steel specimens depending on the imposed local strain and strain rate.

On the other hand, dislocation multiplication and cell formation, and dynamic recovery and recrystallization characterized the structure within the shear bands that evolved after impact in the copper specimens. There was only a single mechanism that was dominant within the microstructure of the evolved shear band after the onset of strain localization. At relatively lower strain rates, dislocation multiplication and dislocation cell formation characterized the microstructure of the evolved shear bands within the copper specimens. As the impact momentum/strain rate increased, the dislocations and dislocation cells were replaced by recrystallized grains and microtwins. Fine equiaxed grains with low dislocation density and well-defined grain boundaries are the microstructural characteristics of the occurrence of Dynamic Recrystallization (DRX) [1,6]. DRX has been defined as a structural rearrangement as a result of refinement of initial grain structures of materials to a point where new nanograins form and grow [1]. It is inferred that the high impact momentum/strain rate of deformation results in the onset of dynamic recrystallization. Prior to the onset of dynamic recovery and recrystallization, the microstructure within the shear band reaches the saturation dislocation density due to dislocation multiplication and dislocation cell formation processes. The excess deformation energy coupled with adiabatic heating raises the temperature of the copper specimens which triggers dynamic recovery and recrystallization processes [7].

5. Conclusion

In this study, pre-impact and post-impact microstructure characterization was conducted on tempered 4340 steel (BCC) and pure copper (FCC) specimens to determine the mechanism of evolution of ASBs. This was done in order to determine the effect of crystal structure and microstructure on the mechanisms and appearance of ASBs in both materials. Following the tests, the specimens were examined using optical and electron microscopy. Specimens from both materials exhibited failure by the development of adiabatic shear bands.

The evolution of the shear band structure in the tempered steel specimens can be considered as a simultaneous layering of microstructures initially driven by dislocations which produce the final structures observed in the shear bands at the end of passage of the stress wave. On the other hand, the shear band in the copper specimens evolved by the occurrence of sequential mechanisms including dislocation multiplication and cell formation, and dynamic recovery and recrystallization. The differences in the mechanism of grain refinement and evolution of the shear bands in both materials is attributed to the differences in the mobility of dislocations, the rate of strain hardening and strain hardening exponents in both materials studied.

The financial support of Defense Research and Development Canada (DRDC DND) for this study is greatly appreciated.

References

[1] B. Dodd, and Y. Bai, Adiabatic Shear Localization: Frontiers and Advances, Elsevier (2012).

[2] Q. Wei, Z. L. Pan, X. L. Wu, B. E. Schuster, L. J. Kecskes And R. Z. Valiev, Microstructure And Mechanical Properties At Different Length Scales And Strain Rates Of Nanocrystalline Tantalum Produced By High-Pressure Torsion, Acta Materialia **59** (2011) 2423–2436.

[3] P. W. Leech, Observations Of Adiabatic Shear Band Formation In 7039 Aluminum Alloy, Metallurgical Transactions A **16** (1985) 1900–1903.

[4] S. Boakye-Yiadom, A. K. Khan and N. Bassim, (2014) A systematic study of grain refinement during impact of 4340 steel. Materials Science and Engineering A **605** (2014) 270–285.

[5] S. Boakye-Yiadom, A. K. Khan and N. Bassim, Effect of pre-impact microstructure on the nucleation and initiation of Adiabatic Shear Bands (ASBs) during impact, Materials Science and Engineering A **615** (2014) 373–394.

[6] Y. Xu, J. Zhang, Y. Bai And M. A. Meyers, Shear Localization In Dynamic Deformation: Microstructural Evolution, Metallurgical And Materials Transactions A **39** (2008) 811–841

[7] A. Mishra, M. Martin, N. N. Thadhani, B. K. Kad, E.A. Kenik, M.A. Meyers, High-strain-rate response of ultra-fine-grained copper, Acta Materialia **56** (2008) 2770–2783.

EPJ Web of Conferences **94**, 02002 (2015)
DOI: 10.1051/epjconf/20159402002

Beryllium strain under dynamic loading

Victor Pushkov[1,2,a], Alexey Yurlov[1,2], Andrew Okinchits, and Tatiana Naydanova[1]

[1] Russian Federal Nuclear Center - VNIIEF, 607190, Sarov, Nizhniy Novgorod region, Russia
[2] Sarov Physico-Technical Institute NIYaU MEPhI, 607186, Sarov, Nizhniy Novgorod region, Russia

Abstract. There are some data (not much) on dynamic characteristics of beryllium that are important, for example, when estimating construction performance at NPP emergencies. A number of data on stress-strain curves, spall strength, shear strength, fracture and structure responses of shock loaded beryllium have obtained in US and Russian laboratories. For today the model description of this complex metal behavior does not have a reasonable agreement with the experimental data, thus a wider spectrum of experimental data is required. This work presents data on dynamic compression-test diagrams of Russian beryllium. Experiments are performed using Hopkinson bar method (SHPB). Strain rates were $\dot{\varepsilon} \sim 10^3\,s^{-1}$.

1. Introduction

Beryllium has an asymmetric hexagonal close-packed (hcp) lattice and has a number of unique properties: the highest (among all other metals) specific strength and heat capacity [1]. The combination of low density, high modulus of elasticity, strength and heat conductivity makes beryllium needed in aeronautical and space engineering [2]. Due to small beryllium atomic mass, small capture cross section and radiation resistance, it is one of the best materials for reflectors and moderators in nuclear engineering [2]. However, data on its mechanical properties are mainly obtained at static loading. There are some data (not much) on dynamic characteristics of beryllium that are important, for example, when estimating construction performance at NPP emergencies. A number of data on deformation curves for different types of beryllium, in particular S200F (USA) at strain rates $\dot{\varepsilon} = 1500 - 8000\,s^{-1}$ are given in [3,4]. For the same type of beryllium, mechanical characteristics at higher strain rates $\dot{\varepsilon} = 10^4 - 10^5\,s^{-1}$ (spall strength) [4,5] as well as fracture and structure responses of shock loaded beryllium [6] are known. In [7–9] the data are given on investigation of spall and shear strength of Russian beryllium at strain rates $\dot{\varepsilon} \cdot \sim 10^4 - 10^5\,s^{-1}$ using various methods.

Researchers perform a model description of beryllium behavior under dynamic and shock-wave loading, e.g. [4–6]. For today the model description of this complex metal behavior does not have a reasonable agreement with the experimental data [4], [6], thus a wider spectrum of experimental data is required.

This investigation provides data on dynamic compression-test diagrams of Russian beryllium. Experiments are performed using Hopkinson bar method (SHPB). Strain rates were $\dot{\varepsilon} \sim 10^3\,s^{-1}$.

[a] Corresponding author: postmaster@ifv.vniief.ru

2. Experimental

Beryllium is prepared using the method of hot vacuum pressing [1], [7] with addition of preprepared beryllium powder. Beryllium density is $1,85\,g/sm^3$, Be content is > 98 weight %, $O_2 \sim 1,5$ weight%, other major impurities are Fe and C, and grain size is $\sim 50\,\mu m$.

Experiments were performed using the SHPB method at strain rates $1000–1600\,s^{-1}$. The experimental setup is given in Fig. 1.

The bars were made of titanium BT-20 (Ø20 × 1500 mm), samples had the size Ø10 × 7 mm.

3. Results

Figure 2 presents dynamic "σ-ε" compression-test diagrams of Be based on experimental results. It is obvious from Figure 2 that Be has considerable strain hardening. The degree of hardening is 7,5–12,0 GPa/rel. units, which is close to data in [3]. In tests 1–5 samples did not fail, the residual strain measured after tests was $\varepsilon_{res} = 4{,}3–9$ %.

Dynamic "σ-ε" compression-test diagrams of Be were used to determine yield strength values presented in Table I. The yield strength dependence on strain rate was not revealed within the range of $\dot{\varepsilon} = 1000–1600\,s^{-1}$.

In terms of the behavior of diagrams and the values of strength yield $\sigma_{0.2}$, beryllium is very similar to preloaded (P = 59 GPa) uranium [10]. However, the values $\sigma_{0.2}$ are considerably higher than in [3].

In test No. 6 at $\dot{\varepsilon} = 1280\,s^{-1}$, the sample failed; Be strength yield was $\sigma_в = 1490$ MPa, the residual strain was 9%. Failure behavior was quasi-brittle.

Figure 3 presents a photo of the damaged sample in test No. 6.

In test No. 4 the sample deformed by ~ 7,5%. The sample from test No. 4 was subjected to the secondary compression at the strain rate $\dot{\varepsilon} = 1150\,s^{-1}$. Diagrams of Be double dynamic compression are provided in Fig. 4. In the second test the sample failed. In that case Be strength was $\sigma_в = 1450$ MPa.

Article available at http://www.epj-conferences.org or http://dx.doi.org/10.1051/epjconf/20159402002

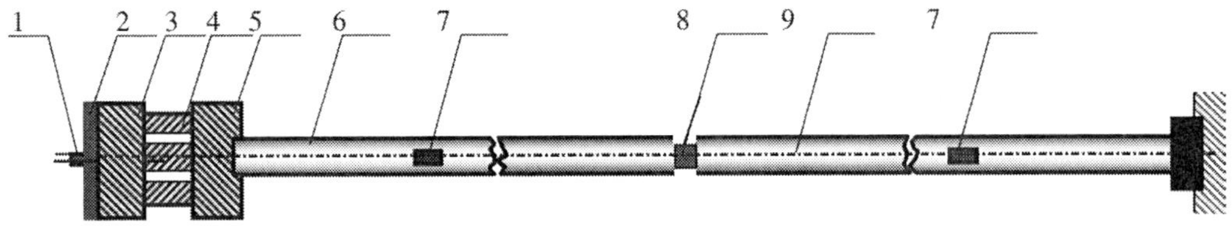

Figure 1. Experimental setup. 1,2 – HE; 3 – impactor; 4 – aluminum damper; 5 – connecter; 6 – loading bar; 7 – strain gauges; 8 – sample; 9 – support bar.

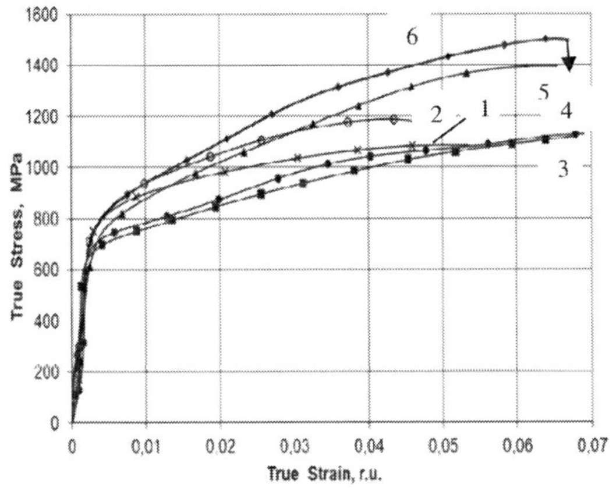

Figure 2. Be dynamic compression-test diagrams at various strain rates. $1 - \dot{\varepsilon} = 1600\,s^{-1}$; $2 - \dot{\varepsilon} = 1000\,s^{-1}$; $3 - \dot{\varepsilon} = 1550\,s^{-1}$; $4 - \dot{\varepsilon} = 1200\,s^{-1}$; $5 - \dot{\varepsilon} = 1250\,s^{-1}$; $6 - \dot{\varepsilon} = 1280\,s^{-1}$.

Table 1. Be yield strength values.

Test No.	Strain rate $\dot{\varepsilon}$, s^{-1}	Yield strength $\sigma_{0.2}$, MPa	Yield strength average value $\sigma_{0.2}$, MPa	Yield Strength σ_{B}, MPa
1	1600	790	772 ± 43	
2	1000	810		
3	1550	720		
4	1200	730		
5	1250	765		
6	1280	815		1490

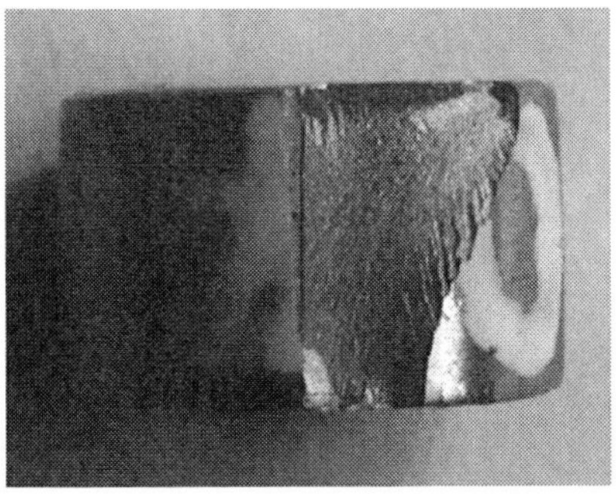

Figure 3. Fracture pattern of the sample in test No. 6.

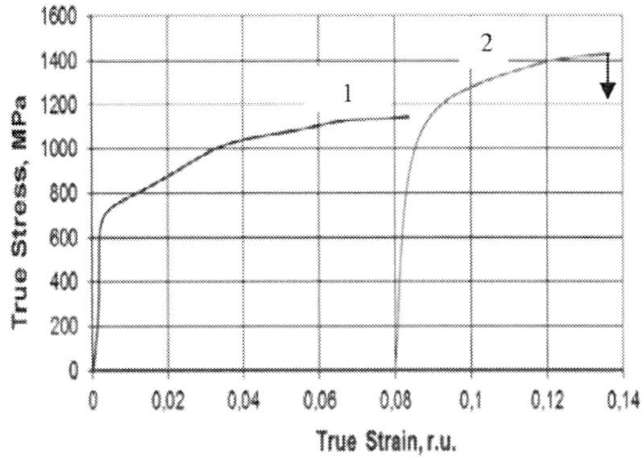

Figure 4. Diagrams of sequential double compression of beryllium at $\dot{\varepsilon} = 1200\,s^{-1}$ (1) and $\dot{\varepsilon} = 1150\,s^{-1}$ (2).

4. Conclusion

Using SPHB method, dynamic compression "stress-strain" diagrams of Russian beryllium were examined. Beryllium was prepared using the method of hot vacuum pressing with admixture of preprepared beryllium powder. Experiments were made at strain rates of $1000–1600\,s^{-1}$. Obtained data will be useful for testing the available models of Be behavior and developing new ones which describe Be behavior under various loading, including those typical for accidents in nuclear engineering, more adequately.

References

[1] *Great Soviet Encyclopedia (Rus)*, **3**, 232–233 (Izd-vo Sovetskaya Entsiklopediya, Moscow, 1970)

[2] *Great Russian Encyclopedia (Rus)*, **3**, 365–366 (Nauchnoye Izd-vo "Great Russian Encyclopedia", Moscow 2005)

[3] W.R. Blumenthal, S.P. Abeln, D.D. Cannon, G.T. Gray III, R.W. Carpenter, *Proc. SCCM 1997*, **429**, 411–414, American Physical Society (1997)

[4] M.B. Prime, S.R. Chen, C.D. Adams, *Proc. SCCM 2011*,**1426**, Part Two, 1035–1038, American Physical Society (2011)

[5] C.D. Adams, W.W. Anderson, G.T. Gray III, W.R. Blumenthal, C.T. Owens, F.J. Freibert, J.M. Montoya, P.J. Contreras, *Proc. SCCM 2009*, **1195**, 509-512, American Physical Society (2009)

[6] C.M. Cady, C.D. Adams, L.M. Hull, G.T. Gray III, M.B. Prime, F.L. Addessio, T.A. Wynn, P.A. Papin, E.N. Brown, *Proc. DYMAT 2012*, Freiburg, Germany, September 2-7, pp. 01009/p.1–01009/p.6, EDP Science, "The European Physical Journal", (2012)

[7] V.A. Arinin, D.V. Kryuchkov, V.A. Ogorodnikov, V.A. Raevsky, K.N. Panov, V.V. Peshkov, V.I. Skokov, O.A.Tyupanova, *Proc. SCCM 2011*, **1426**, Part Two, 1073–1076, American Physical Society (2011)

[8] V.A. Arinin, Yu.V. Batykov, D.V. Kryuchkov, A.N. Malyshev, K.N. Panov, V.I. Skokov, *Proc.* XIII *International Workshop "Fundamental properties of Plutonium"*, Sarov, Russia, September 9–14, 2013, p. 174, RFNC-VNIIEF (2013)

[9] E.A. Kozlov, V.I. Tarzhanov, I.V. Telichko, D.G. Pankratov, S.A. Brichikov, D.P. Kuchko, M.A. Ralnikov, A.Yu. Nikolaev, A.E. Shirobokov, *Proc.* XIII*International Workshop "Fundamental properties of Plutonium"*, Sarov, Russia, September 9-14, 2013, pp. 178–180, RFNC-VNIIEF (2013)

[10] V.A. Pushkov, M.L. Andreeva, A.V. Yurlov, A.V. Kalmanov, I.V. Shiberin, *Proc.* XIII *International Workshop "Fundamental properties of Plutonium"*, Sarov, Russia, September 9–14, 2013, pp. 197–199, RFNC-VNIIEF (2013)

EPJ Web of Conferences **94**, 02003 (2015)
DOI: 10.1051/epjconf/20159402003

Effects of grain size on the spallation behavior of pure copper under plate-impact loading

ZhaoXiu Jiang[1], Hui Peng[2], MingZhi Xing[1], HaiTing Shen[1], HongLiang He[2], and YongGang Wang[1,a]

[1] Key Laboratory of Impact and Safety Engineering, Ministry of Education of China, Ningbo University, Zhejiang 315211, People's Republic of China
[2] Key Laboratory for Shock Wave and Detonation Physics Research, Institute of Fluid Physics, China Academy of Engineering Physics, Mianyang 621900, People's Republic of China

Abstract. The effects of grain size on the dynamic tensile fracture (or spall) response were investigated for high purity copper materials by plate-impact experiments. The spall strength estimated from the free surface velocity profile is nearly constant with no significant effect from the grain size. However, differences are observed in the acceleration rate of velocity rebound beyond the minima. This may be attributed to the effect of grain size on the growth rate of damage. Metallographic analyses of the fracture surface show that the characteristic feature of the fracture surface clearly depends on the grain size. In the smaller samples, the fracture surfaces are decorated with large, high-density ductile dimples suggesting that the preferential failure mode is ductile intergranular fracture. In the larger samples, the fracture surfaces have a rock candy appearance with small, brittle, high density dimples as well as large ductile dimples suggesting that the fracture mode is a mix of both brittle intergranular fracture and ductile transgranular fracture.

1. Introduction

The dynamic tensile fracture (or spall) behaviors of metallic materials produced by the interaction of release waves after shock loading is a complicated physical process that can be strongly influenced by many variables related to the applied loading, temperature, and the material itself [1–4]. This extensive work has shown that the grain size plays an important role in spall failure. Because many parameters and mechanisms are intertwined, previous studies regarding grain size effects on spall behavior have reported different and sometimes contradictory results [5,6]. Recently, a comprehensive set of incipient spall experiments on copper was performed by Escobedo et al. [7]. The free surface velocity measurements show no significant grain size dependence on the spall strength of 30-, 60-, 100- and 200-μm copper samples. However, there were obvious differences in the free surface velocity behavior after the pull-back velocity minima. This is attributed to the different active mechanisms of damage, some of which depend on the grain size and grain boundary types. To ensure a good correlation between a particular microstructural feature and the spall response, it is important that only one parameter be varied at a time. In this paper, plate impact experiments were performed on high purity copper samples to examine the relationship between grain size and spall response. In order to eliminate the effect of loading on spall behavior, the peak compressive stress and tensile strain rate in plate impact experiments are carefully controlled and held constant. The results are presented and discussed in terms of free surface velocity measurements as well as metallographic examinations of the spall surface with scanning electron microscopy (SEM).

2. Experimental procedures

2.1. Materials

Polycrystalline copper plate material with 99.99% purity was subjected to an additional cold-rolling procedure followed by recrystallization to produce a wide range of grain sizes. Samples were annealed in vacuum with furnace cooling at 500 °C for 3.5 hours to produce an average grain size of 78 μm, 700 °C for 3.0 hours to produce a grain size of 273 μm, and 850 °C for 3.0 hours to produce a grain size of 400 m. Microstructural characterization of different heat-treated copper samples is shown in Fig. 1. The grain size was measured using the line intercept method.

2.2. Plate-impact experiments

A 57-mm light gas gun was used for the planar symmetrical impact spall experiments. The diameters of the flyer and the target are 50 mm and 35 mm with thicknesses of 3 mm and 6 mm, respectively. Time-resolved free surface velocity profiles of the shocked sample were measured with a compact all-fiber single point displacement interferometer system for any reflector (DISAR) to monitor spall fracture [8]. A 1550-nm laser beam was used as the light source for the DISAR system. A single-mode 1-mm optic fiber pigtail was used to collect light reflected from the moving surface. The output DISAR signals were recorded using a high-frequency digital Tektronix with a time resolution of 0.5 ns. To examine

[a] Corresponding author: wangyonggang@nbu.edu.cn

Article available at http://www.epj-conferences.org or http://dx.doi.org/10.1051/epjconf/20159402003

(a) 78μm (b) 273μm (c) 400μm

Figure 1. Optical micrographs of high purity copper with the average grain sizes: (a) 78 μm, (b) 247 μm, and (c) 400 μm.

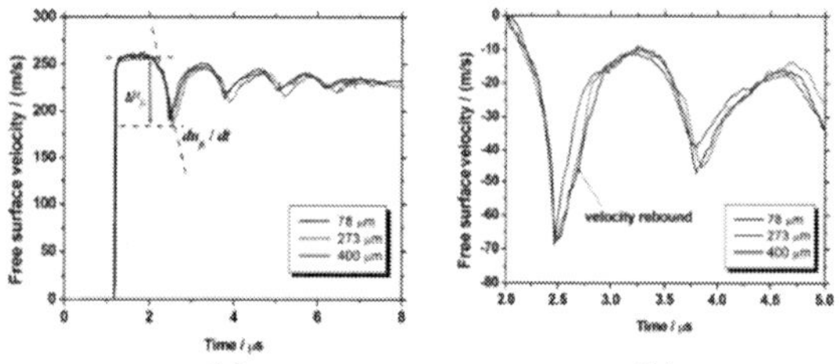

(a) (b)

Figure 2. (a) Free surface velocity profiles for copper samples with different grain sizes. (b) Pullback wave measured at different grain sizes.

the metallographic characterization of the fracture surface, a soft recovery technique for the shocked specimen was adopted. All experimental conditions are listed in Table 1.

3. Results and discussions

3.1. Free surface velocity profiles

Figure 2(a) shows the measured free surface velocity profiles for the different grain size copper samples. The peak free surface velocities range from 254 to 258 m/s, corresponding to peak compressive stresses of 4.68–4.76 GPa. Figure 2(b) shows the free surface velocity profiles once significant yielding begins causing a drop in the free surface velocity. The times and velocities are shifted so that the drop from the peak state starts at t = 0. From the slope of the drop in free surface velocity profiles (Fig. 2(a)) the tensile strain rate of $2.7 \times 10^4\,s^{-1}$ is estimated in all cases via [9]:

$$\dot{\varepsilon} = \frac{1}{2c_0}\frac{du_{fs}}{dt} \tag{1}$$

where $c_0 = 3940\,m/s$ is copper's bulk sound speed. The velocity difference between the first peak plateau and the first valley, commonly referred to as the pullback velocity (Δu_{fs}), estimates the spall strength or stress threshold for fracture (Fig. 2(a)). For materials undergoing elastic-plastic deformation either during compression or tension, the spall strength is calculated from the observed pullback velocity using [9]:

$$\sigma_s = \rho_0 c_l \Delta u_{fs}\left(1 + \frac{c_l}{c_0}\right)^{-1} \tag{2}$$

where $c_l = 4770\,m/s$ is copper's longitudinal elastic wave speed. The measured and calculated key parameters are listed in Table 1. A spall strength of 1.29 ~ 1.33 GPa was calculated for different grain size samples by holding the loading profile constant, i.e., peak stress and tensile strain rate are the same for each experiment. Figure 3 shows the spall strength as a function of the average grain size for pure copper samples along with previous results from Escobedo et al., which indicates no significant effect of grain size on the spall strength.

Table 1. Summary of the experimental parameters and results.

Shot No.	Grain size (μm)	Impact velocity (m/s)	σ_e/GPa	σ_s/GPa
1	78	258	4.76	1.32
2	273	254	4.68	1.29
3	400	256	4.72	1.33

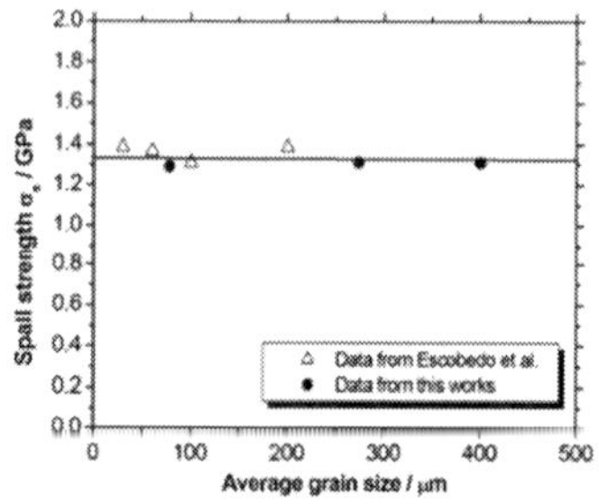

Figure 3. Spall strength as a function of the average grain size for pure copper materials.

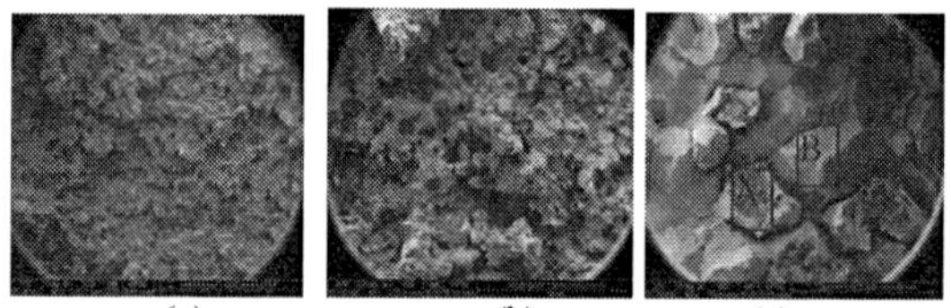

(a) (b) (c)

Figure 4. Scanning electron micrographs of the fracture surface at low magnification (40×): (a) 78 μm, (b) 273 μm and (c) 400 μm.

3.2. Post-impact metallurgical characterization

Careful fractography of the recovered samples was conducted by SEM to illustrate the damage characteristics and fracture mechanism. Although grain size does not affect spall strength of the high purity copper, grain size does markedly influence the type of fracture surface produced during spall. Figure 4(a)-(c) shows SEM images of the fracture surface at low magnification (40×) with different grain sizes. From Fig. 5(a) and (b), we see that the fracture surfaces of the 78-μm and 273-μm samples are completely decorated with high-density dimples indicative of microvoid coalescence. The fracture surface of the 400-μm sample shown in Fig. 4(c) is mainly filled with rock candy-like grains and exhibits a mixture of brittle inter-granular fracture facets and ductile transgranular fracture regions. The higher magnification image reveals a dimpled

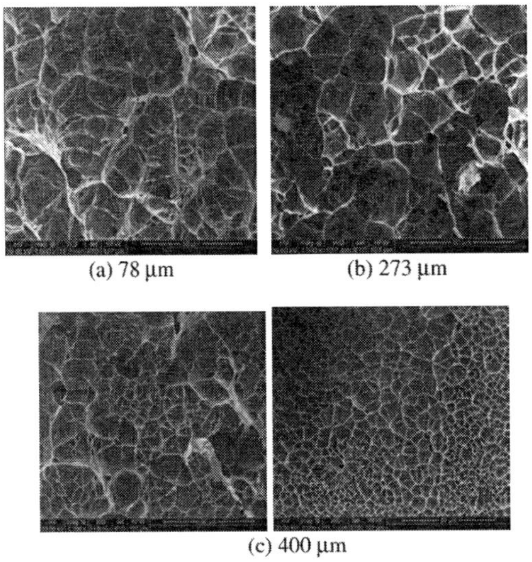

Figure 5. Scanning electron micrographs of the fracture surface at high magnification (1000×): (a) 78 μm, (b) 273 μm and (c) 400 μm.

topology on the grain facets. Figure 5(a)-(c) shows SEM images of the fracture surface at high magnification (1000×) with different grain sizes. This illustrates the change of the dimple structure as a function of the grain sizes. For the samples with smaller grain size (78 μm and 273 μm), the fracture surfaces have larger ductile dimples. These are produced by microvoid coalescence along the grain boundaries, which indicates that the fracture is a ductile intergranular fracture. The dimple size (average diameter and average depth) of the 78-μm sample is smaller than the 273-μm sample. Void coalescence will be enhanced in a sample with relatively small grain sizes (78 μm in this study) in which voids are nucleated on different grain boundaries and are close together. On the contrary, in samples with intermediate grain sizes (273 μm), the voids can grow larger but remain isolated before void coalescence that results in larger dimples on the fracture surfaces. For the 400-μm grain sample (Fig. 5(c)), the fracture surface has larger ductile dimples that suggest a ductile transgranular fracture mode (region A in Fig. 4(c)). In contrast, in region B of the same figure the fracture surface exhibits small, high density brittle dimples suggesting a brittle intergranular fracture mode.

Usually, voids preferentially nucleate along the grain boundaries, and grain boundaries are the preferred path for spall fracture. The reason for ductile transgranular fracture in coarse grain size samples is that the grain boundary area available for fracture near the spall plane decreases with increasing grain size. This forces the fracture to occur at large grain sizes. In other words, if the width of the tensile peak near the spall plane is smaller than the grain size, then a transgranular fracture is induced rather than an intergranular fracture. In turn, brittle intergranular fractures correspond to smaller dimples in coarse grain size samples because many voids are nucleated along the same grain boundary very close together. Thus, the void coalescence is strongly enhanced to restrain void growth.

4. Conclusion

Plate impact experiments and microscopic observations on post-impact recovered samples were used to study the influence of grain size on the dynamic tensile/spall behavior of high purity copper samples. By holding the loading conditions constant, the free surface velocity profiles were studied as a function of grain size. The spall strength estimated from the pullback velocity has no significant effect of the grain size. The post-impact metallographic analyses show that samples with smaller grain sizes (78 μm and 273 μm) have fracture behavior dominated by ductile intragranular fractures with many large ductile dimples on the fracture surface. In the 400-μm samples, the fracture mode is dominated by a mixture of both brittle intergranular fractures and ductile transgranular fractures that combine smaller dimples with larger ductile dimples on the fracture surface.

This research was supported by the National Science Foundation of China under Grant No. 11272164, 11172221 and 11472142, and sponsored by K. C. Wong Magna Fund and K. C. Wong Eduation Fund in Ningbo University, China.

References

[1] D.D. Koller, R.S. Hixson, G.T. Gray III, P.A. Rigg, L.B. Addession, E.K. Cerreta, J. Appl. Phys. **98**, 103518 (2005)
[2] Y.G. Wang, Z.X. Jiang, M.Z. Xing, L.L. Wang, Mater. Sci. Eng. A **596**, 222 (2014)
[3] X.Chen, J.R. Asay, and S.K. Dwivedi, J. Appl. Phys. **99**, 023528 (2006)
[4] C.L. Williams, K.T. Ramesh, and D.P. Dandekar, J.Appl.Phys. **111**, 123528 (2012)
[5] S. Christy, H.R. Pak, and M.A. Meyers, in Metallurgical Applications of Shock-Wave and High-Strain-Rate Phenomena, edited by L.E. Murr, K.P. Staudhammer, and M.A. Meyers (Dekker, New York, NY, 1986), p. 835
[6] R.W. Minich, J.U. Cazamias, M. Kumar, and A.J. Schwartz, Metall. Mater. Trans. A **35**, 2663 (2004)
[7] J.P. Escobedo, D. Dennis-Koller, E.K. Cerreta, B.M. Patterson, C.A. Bronkhorst, B.L. Hansen, D. Tonks, and R.A. Lebensohn, J. Appl. Phys. **110**, 033513 (2011)
[8] J.D. Weng, X. Wang, Y. Ma, H. Tan, L.C. Cai, J.F. Li, C.L. Liu, Rev. Sci. Instru. **79**, 113101 (2008)
[9] T. Antoun, L. Seaman, D.R. Curran, G.I. Kanel, S.V. Razorenov, A.V. Utkin, *Spall Fracture* (Springer, Berlin, 2003)

EPJ Web of Conferences **94**, 02004 (2015)
DOI: 10.1051/epjconf/20159402004

An experimental investigation of energy absorption in TRIP steel under impact three-point bending deformation

Hang Pham[1,3,a] and Takeshi Iwamoto[2]

[1] Graduate School of Engineering, Hiroshima University, 1-4-1 Kagamiyama, Higashi-Hiroshima, Hiroshima 739-8527, Japan
[2] Institute of Engineering, Hiroshima University, 1-4-1 Kagamiyama, Higashi-Hiroshima, Hiroshima 739-8527, Japan
[3] Faculty of Engineering, Vietnam National University of Agriculture, Trauquy, Gialam, Hanoi 131004, Vietnam

Abstract. TRIP (Transformation-induced Plasticity) steel is nowadays in widespread use in the automobile industry because of their favorable mechanical properties such as high strength, excellent formability and toughness because of strain-induced martensitic transformation. Moreover, when TRIP steel is applied to the components of the vehicles, it is expected that huge amount of kinetic energy will be absorbed into both plastic deformation and martensitic transformation during the collision. Basically, bending deformation due to buckling is one of the major crash deformation modes of automobile structures. Thus, an investigation of energy absorption during bending deformation at high impact velocity for TRIP steel is indispensable. Although TRIP steel have particularly attracted the recent interest of the scientific community, just few studies can be found on the energy absorption characteristic of TRIP steel, especially at impact loading condition. In present study, experimental investigations of bending deformation behaviors of TRIP steel are conducted in the three-point bending tests for both smooth and pre-cracked specimen. Then, energy absorption characteristic during plastic deformation and fracture process at high impact velocity in TRIP steel will be discussed.

1. Introduction

Over the past century, the most important and challenging task for design considerations in the automotive industry is vehicle crashworthiness and an occupant safety with a decrease in weight of automobiles. Thus, energy-absorption characteristic of structural materials has become a critical issue to protect the passengers from a high-speed crash of automobiles. During a crash of automobiles, a dominant mechanism to keep the safety in automobiles by structures upon the collision is buckling because kinetic energy with higher speed of the vehicles can be consumed by inelastic deformation of components for protection of the passengers [1]. Bending deformation due to buckling is one of the major collapse modes of the automotive structures [2]. Therefore, the performance of structural materials under bending deformation needs to be investigated carefully, especially at high rate of deformation.

TRIP (Transformation-induced Plasticity) steel is nowadays in widespread use in the automobile industry [3] because of favorable mechanical properties and high energy-absorption characteristic. Although, TRIP steel has particularly attracted the considerable attention of researchers [4–6], just few studies can be found on the bending deformation behavior and energy absorption of TRIP steel, especially under impact loading condition. Thus, an investigation of energy-absorption characteristic at high impact velocity for TRIP steel is indispensable expected from viewpoint of a safety design. On the other hand, in general, energy-absorption characteristic of materials can be evaluated from a stress-strain curve obtained by a tensile test. However, an unstable deformation at high strain rate induces difficulties to determine onset of necking in a specimen, which is used to evaluate energy absorption. At the same time, three-point bending test is considered to be relatively simple for investigation of performance of smooth specimen as well as becomes a useful tool in the determination of fracture parameters [7,8]. As a result, energy absorption in TRIP steel can be calculated from the three-point bending tests.

Previously, few studies have investigated experimentally on the energy absorption in TRIP steel. According to Rong et al. [9] and Curtze et al. [10], a rate sensitivity of energy absorption in TRIP steel can be observed from results of the tensile tests. Based on the observations of perforation tests, Rodríguez-Martínez et al. [11] reported that a decreasing testing temperature leads to larger values of the impact energy-absorption for TRIP steel. On the other hand, energy absorption characteristic during fracture process of materials is considered to be associated with fracture toughness [12,13]. Recently, Zhu and Joyce [14] evaluated energy absorption of materials by J-integral defined by Rice [15] from the area under a force-displacement as a result of the presence of a crack initiation.

In this study, the performance of both smooth and pre-cracked specimens made of type 304 austenitic stainless steel, a kind of TRIP steel, is investigated experimentally at various deflection rate. The three-point bending tests are conducted by an INSTRON-type conventional material

[a] Corresponding author: author@email.org

Article available at http://www.epj-conferences.org or http://dx.doi.org/10.1051/epjconf/20159402004

testing machines for quasi-static loading, a drop-weight testing machine, and a testing apparatus based on the split Hopkinson pressure bar (SHPB) method for smooth specimen. Experimental procedures for the three kinds of bending tests are described in detail. Then, the three-point bending tests are conducted under quasi-static and impact conditions for pre-cracked specimen based on the ASTM standard. Energy absorption for TRIP steel during plastic deformation and fracture process will be discussed by examination of characteristic of force-deflection curves and value of J-integral. Additionally, stretch zone width is also investigated at different deflection rate to check the validity for the result of J-integral. Finally, conclusions on energy-absorption characteristic in TRIP steel are introduced.

2. Methodology

2.1. Methodology to evaluate energy absorption characteristic

In present study, energy absorption of smooth specimen is evaluated in terms of the increase in the external force because the area surrounding the normalized force–normalized deflection curve increases until a certain level of normalized deflection is reached because of a rate-sensitive hardening effect. Additionally, value of energy-absorption characteristic in fracture process is calculated by J-integral defined by Rice [10] from experimental results for pre-cracked specimen at the onset point of crack extension. Thus, it is necessary to determine an onset point of crack extension during deformation of the pre-cracked specimen. Here, a direct current potential drop technique is used for a detection of the onset of crack extension. A constant current is supplied through the specimen. The potential difference across the crack will change as the result of crack growth since a resistance increases because of a decrease in the cross-section of the specimen. Consequently, the output voltage may be related directly to area of the crack extension, and the onset point of crack growth is defined by the point that the tendency of a change in voltage.

Then, J-integral is evaluated by obeying ASTM E1820 [7] as following equation.

$$J = \frac{K^2(1-\nu^2)}{E} + \frac{2A_{pl}}{B(W-a_0)} \quad (1)$$

where E is Young's modulus, ν is Poisson's ratio and A_{pl} is area under force-deflection curve as shown in Fig. 1. K is calculated as,

$$K = \frac{PS}{BW^2} \times 3\left(\frac{a_o}{W}\right)^{1/2} \times \frac{\left[1.99-\left(\frac{a_0}{W}\right)\left(1-\frac{a_0}{W}\right)\left(2.15-3.93\left(\frac{a_0}{W}\right)+2.7\left(\frac{a_0}{W}\right)^2\right)\right]}{2\left(1+2\frac{a_0}{W}\right)\left(1-\frac{a_0}{W}\right)^{3/2}} \quad (2)$$

where P is external force, S is span length, B and W are the initial thickness and width of the specimen, respectively, a_o is the initial crack length including both the fatigue pre-cracked and the notch length.

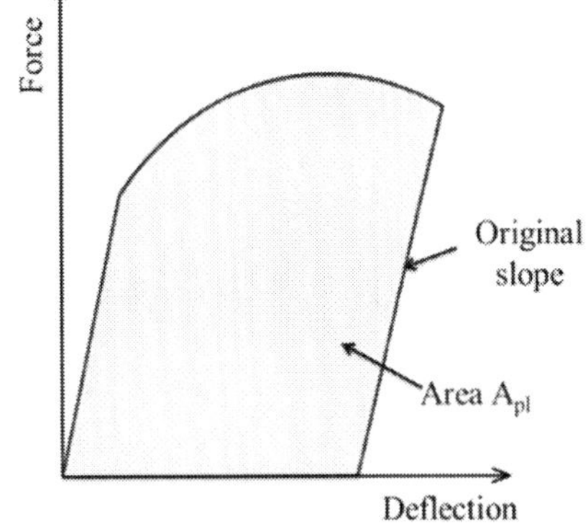

Figure 1. The definition of area under force-deflection curve for the calculation of J-integral [7].

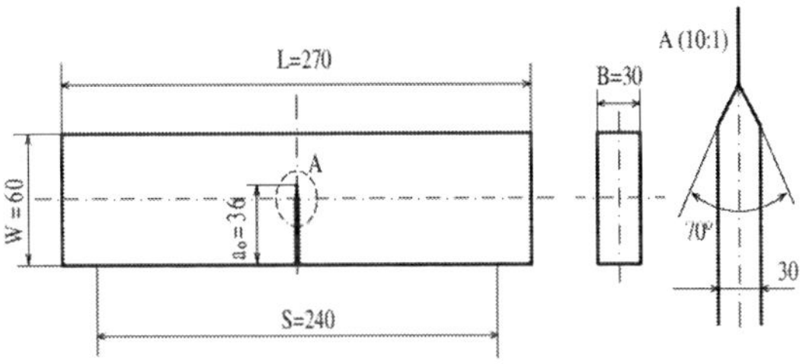

Figure 2. The dimensions of the pre-cracked specimen.

2.2. Specimens

Smooth plate specimens (length: 100 mm, height: 15 mm, width: 7.5 mm) and pre-cracked specimens with the dimensions obeyed the ASTM standard as shown in Fig. 2 are used in the present study. These specimens are made of type 304 austenitic stainless steel, a kind of TRIP steel. To obtain a uniform austenitic structure, after machining, the specimens are subjected to a solution-heat treatment at 1323 K for 30 min in an electric furnace (Full-tech FT100) and then quenched in cold water.

2.3. Experimental apparatus

2.3.1. Quasi-static test

Figure 3 shows a photograph of an Instron-type conventional material testing machine (Shimadzu AG-X250KN) for quasi-static three-point bending test. The deflection rate is controlled by using constant crosshead speeds of 0.008, 0.08, 0.8, and 8 mm/s for smooth specimen and 36, 360, 3600, 36000 mm/s for pre-cracked specimen. Deflection is measured as the displacement of the crosshead of the testing machine. In addition, experimental apparatus to determine the onset of crack extension of pre-cracked specimen are shown in this figure. The location of probes for supplying current and measuring output voltage can be seen here. The probes for measuring output voltage are fixed around the crack-tip on the opposite side in the thickness direction of the specimen to make the output voltage signals as sensitive as possible. They are spot-welded on the specimen. The supplying current in experiment is set to 12A by a parallel

Figure 3. A photograph of testing machine for quasi-static test.

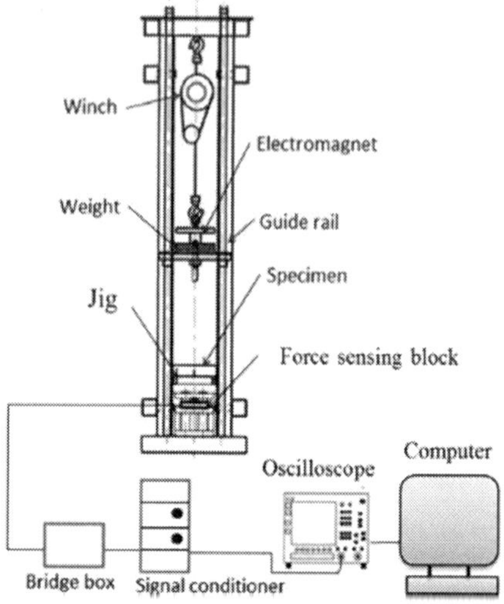

Figure 4. A schematic illustration of the drop-weight test.

connection of two DC power supplies (KENWOOD PA18-6A).

2.3.2. *Drop-weight impact test*

Figure 4 shows a schematic illustration of the drop-weight testing setup.

A force-sensing block introduced by Chuman et al. [16] is placed under the center of the jig to capture the impact force for long time periods by protecting the reflected wave that propagates back into the force-sensing part. A low-pass filter (Nippon Avionics 9B02) with a cut-off frequency of 400 Hz is used to filter oscillations of output signals. A weight with a mass of 140 kg is allowed to freely fall from a height of 45 mm for smooth specimen and 400, 450 and 550 mm for pre-cracked specimen. The optical fiber sensors (Keyence FU-77V) and amplifiers (Keyence FS-V31 and 32) are used to measure the velocity of the weight just before it hits the specimen. Deflection is recorded using a high-speed camera (Photron Fastcam-512PCI32K) during impact deformation with respect to time. A linear magnetic scale system including a magnetic scale (MACOME Laboratory SIS-310S) and a sensor head (MACOME Laboratory SIH-410) is used to measure the displacement of the weight. The deflection rate is calculated from the obtained deflection with respect to time by the finite difference method.

2.3.3. *Impact test based on split Hopkinson pressure bar method*

The impact test based on SHPB technique is used to investigate deformation behavior of smooth specimen at relatively high defection rate. The schematic diagram of the modified apparatus for the three-point bending test based on the SHPB method proposed by Yokoyama and Kishida [17] for smooth specimen is shown in Fig. 5. A striker bar (length: 500 mm), an input bar (length: 4000 mm), and two output bars (length: 4000 mm) are used. All the pressure bars are 16 mm in diameter. The axial strain in the three pressure bars obtained from the gages is recorded using the above mentioned system. The velocity of the striker bar just before it hits the input bar is measured using the same system mentioned above. The impact velocity of the striker is controlled by controlling the pressure in the air tank. Air pressures of 2 and 5 MPa correspond to impact velocities of 4000 and 8000 m/s, respectively. To suppress the initial oscillation of the transmitted and reflected stress waves, pulse shapers made of zinc are used. By using the incident, reflected, and transmitted stress pulses, the deflection at the center of the specimen is measured by a conventional method [17] and the deflection rate is calculated with respect to time by a finite difference method.

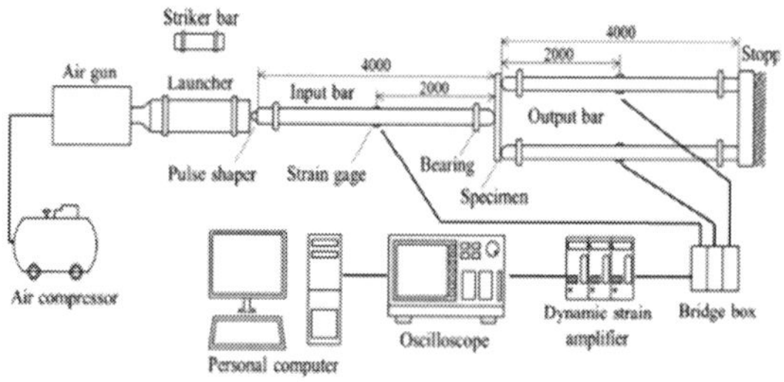

Figure 5. A schematic of the three-point bending test based on the SHPB method.

2.4. Measurement of stretch zone width

The plastic deformation at the crack tip leads to blunting the crack tip and finally the formation of the stretch zone. It is considered that the stretch zone width is strongly related to the resistance of the crack initiation [13]. Therefore, the stretch zone width is also investigated in present study to confirm validity of the value of J-integral. After bending deformation, the fracture surface of the pre-cracked specimen is observed by an analysis of fractography taken by a scanning electron microscope (SEM, JEOL, JSM-6510). If an image of fracture surface is at relatively-higher magnifications, then the crack stretch zone exhibits a typical featureless appearance easily distinguishable from the surface of the fatigue pre-crack.

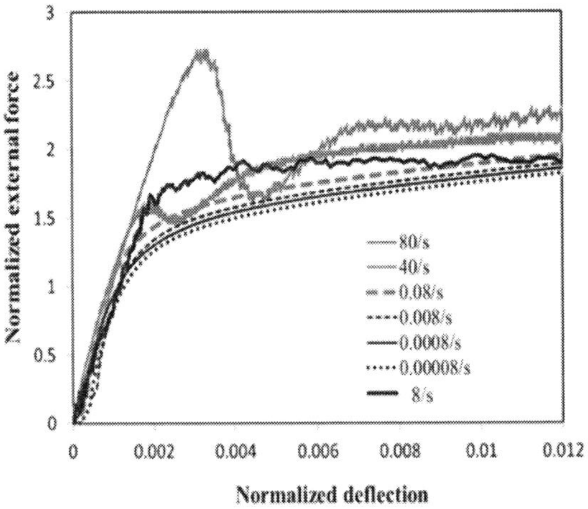

Figure 6. Normalized external force–normalized deflection curves obtained from the experiment for various normalized deflection rate for smooth specimen.

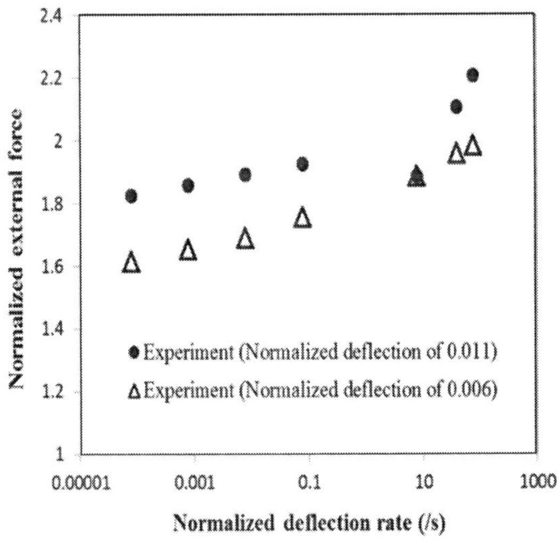

Figure 7. A plot of the normalized external force for a normalized deflection of 0.006 and 0.011 with respect to the normalized deflection rate.

Finally, the value of the stretch zone width is obtained by counting the pixels on the images.

3. Results and discussion

The measured external force and the deflection are normalized. An external force P is normalized as the ratio of the maximum bending stress at the middle of the specimen to the initial yield stress at room temperature σ_y. The normalized force P_{nor}, the normalized deflection δ_{nor} and the normalized deflection rate $\dot{\delta}_{nor}$ are expressed as following equation.

$$P_{nor} = \frac{PL}{4Z\sigma_y},\ \delta_{nor} = \frac{4h\delta}{L^2},\ \dot{\delta}_{nor} = \frac{4h\dot{\delta}}{L^2}, \qquad (3)$$

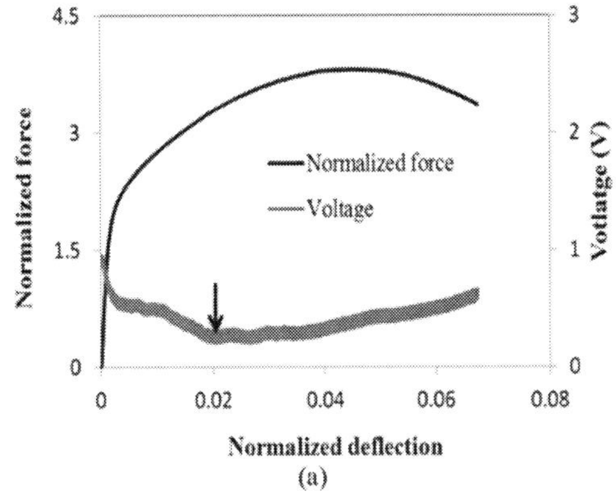

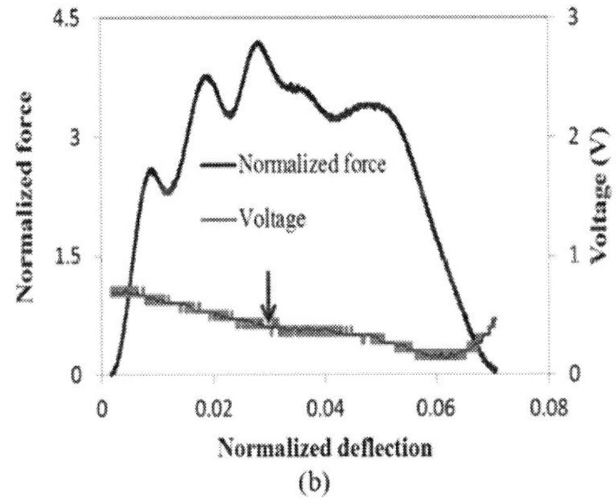

Figure 8. Force-deflection curve and determination of onset point of crack extension at normalized deflection rate of (a) 0.04/s and (b) 10.44/s.

where L is the span length and Z is the section modulus, and h is the height of the specimen. δ is deflection and $\dot{\delta}$ is the deflection rate obtained from experiment. In the case of smooth specimen, the corresponding normalized deflection rate is approximately 0.00008, 0.0008, 0.008, and 0.08/s for the quasi-static deformation, and the maximum normalized deflection rate is approximately 8, 40, and 80/s for the impact deformation. These values of the normalized deflection rate are used as representative values for smooth specimen.

Figure 6 shows the normalized external force–normalized deflection curves at various normalized deflection rate for smooth specimen. In cases of impact deformation, oscillations of external force are insufficiently suppressed although output signals are filtered by a low-pass filter in drop-weight impact test and the pulse shaper is used to generate incident wave in impact test based on SHPB method. Moreover, to obtain a certain level of normalized deflection, the normalized external force needs to be higher at higher normalized deflection rate. Next, value of the normalized external force for the normalized deflection of 0.006 and 0.011 at different normalized deflection rate is discussed.

Figure 7 shows the relationship between the normalized external force and normalized deflection rate in a semi-logarithmic plot for two cases of normalized

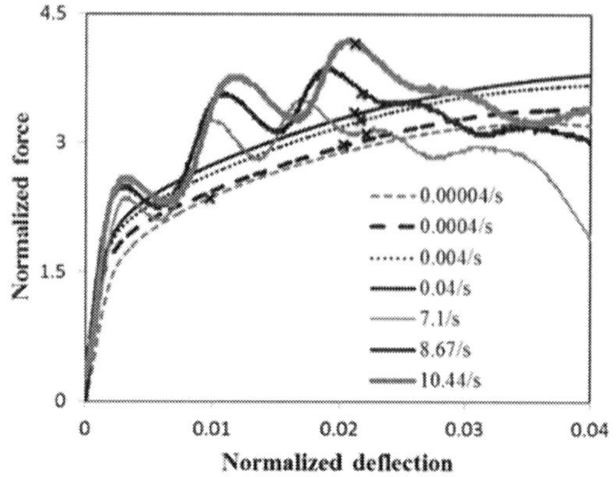

Figure 9. The relationship between normalized force and normalized deflection at various normalized deflection rate for pre-cracked specimen.

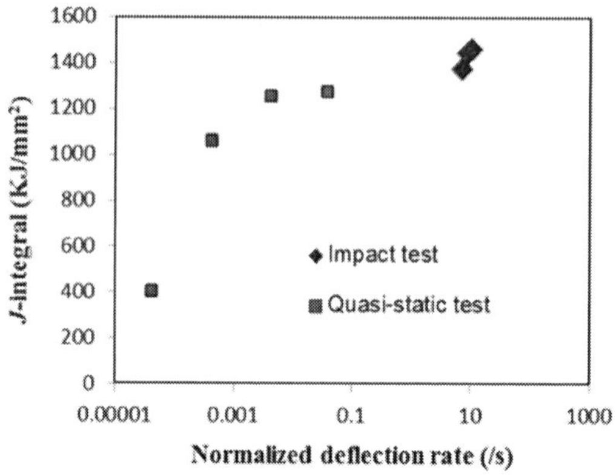

Figure 10. Value of J-integral at different normalized deflection rate.

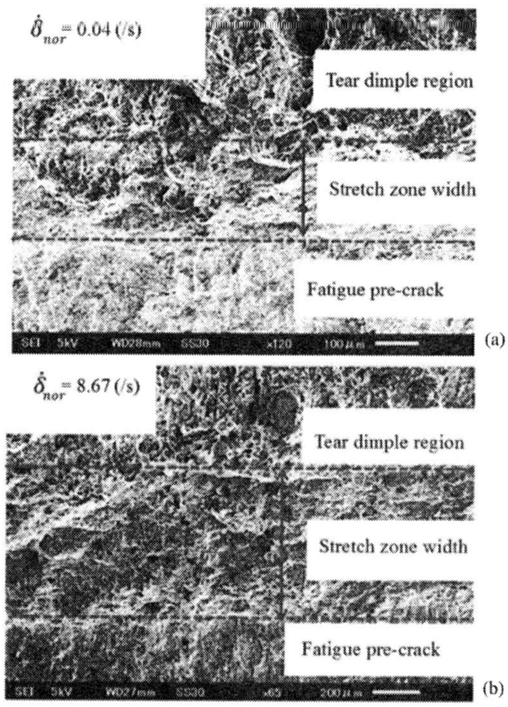

Figure 11. Photograph of microstructure near the crack tip including stretch zone in the case of normalized deflection rate of (a) 0.004/s and (b) 8.67/s.

deflection. From this figure, TRIP steel might have higher energy absorption at a higher deflection rate. Moreover, a positive rate sensitivity of energy absorption in TRIP steel is clearly observed from quasi-static to impact deformation. An almost linear relationship between the normalized external force and normalized deflection rate can be seen at the lower level of normalized deflection. Thus, TRIP steel might show excellent characteristic of energy-absorption under impact loading condition. This can be explained that more energy might be consumed by inelastic deformation processes for a higher defection rate because of the hardening effect. The linear relationship between the normalized external force and normalized deflection rate in the cases of the quasi-static condition can be seen at higher normalized deflection. However, the slope of this relationship is slightly larger in the impact condition, and a sudden drop in the force can be seen for a normalized deflection rate of 8/s. A higher temperature rise due to adiabatic heating by inelastic irreversible work at high deflection rate might induce thermal softening of materials in general. Furthermore, an increase in the temperature under impact condition at higher normalized deflection might suppress martensitic transformation in TRIP steel, which is expected to improve performance of this steel. Consequently, the relationship between normalized external force and normalized deflection rate becomes nonlinear. Therefore, it is predicted that the non-monotonic positive rate sensitivity of the energy absorption in TRIP steel might not be observed with an increase in the normalized deflection.

In the case of pre-cracked specimen, the obtained normalized deflection rate is approximately 0.00004, 0.0004, 0.004, and 0.04/s for the quasi-static deformation, and the maximum normalized deflection rate is approximately 7.1, 8.67, and 10.44/s for the impact condition. They are also used as representative normalized deflection rate for pre-cracked specimen.

Figure 8 shows the normalized force-normalized deflection curve and determination of the onset point of crack extension at normalized deflection rate of (a) 0.04/s and (b) 10.44/s. The change in the tendency of output voltage signals can be seen and is denoted by arrow in the figure. The onset point of crack extension can be detected by the point of the change in the tendency.

Figure 9 shows the normalized force-normalized deflection curve for pre-cracked specimen at different normalized deflection rate. The onset point of crack extension for each case of normalized deflection rate is denoted by " **x** " on each curve. Oscillations of normalized force-normalized deflection curves in impact deformation are observed in this figure. Clearly, a similar tendency with values for smooth specimen can be seen here. To obtain the same value of normalized deflection, value of normalized force required higher at higher normalized deflection rate.

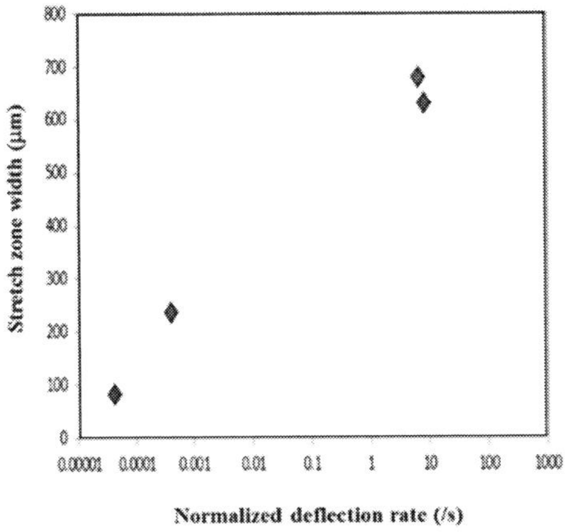

Figure 12. Relationship between stretch zone width and normalized deflection rate in a semi-logarithmic plot.

Additionally, value of the normalized deflection at the onset point of crack extension under impact loading is higher than that under quasi-static deformation in general. From this figure, energy absorption is calculated by J-integral. Then, the value of J-integral with respect to normalized deflection rate in a semi-logarithmic plot is shown in Fig. 10.

From Fig. 10, a higher value of J-integral can be seen at a higher normalized deflection rate with nonlinear relationship. Furthermore, this figure shows a positive rate sensitivity of J-integral from quasi-static to impact deformation. This result is suitable with observations from Fig. 7 for smooth specimen. As a result, TRIP steel might have better performance of energy absorption under impact loading condition.

Figure 11 shows photograph of the microstructure and definition of stretch zone width at normalized deflection rate of (a) 0.004/s under quasi-static deformation and (b) 8.67/s under impact loading deformation. From this figure, it can be seen clearly that the value of stretch zone width in impact case is relatively higher than that in quasi-static condition.

Figure 12 shows the value of stretch zone width at different normalized deflection rate. A considerably higher value of stretch zone width can be seen under impact condition than that under quasi-static deformation. Thus, a positive rate sensitivity of J-integral from quasi-static to impact deformation as above discussion can be confirmed.

4. Concluding remarks

In this study, the bending deformation behavior of both smooth and pre-cracked specimens made of TRIP steel is investigated experimentally at various deflection rate to evaluate energy-absorption characteristic. Characteristic of normalized force-normalized deflection curve and value of J-integral, which may be equivalent to the energy absorption, are discussed in quasi-static as well as impact condition. A positive rate sensitivity of the external force in TRIP steel can be observed from quasi-static to impact deformation. Moreover, TRIP steel indicates higher value of J-integral at higher normalized deflection rate. The positive rate sensitivity of J-integral is confirmed by an investigation of stretch zone width of fracture surface after bending deformation. As a result, it can be concluded that TRIP steel might have excellent energy-absorption characteristic under impact loading deformation.

We gratefully acknowledge financial supports from the 23rd ISIJ Research Promotion Grant provided by the Iron and Steel Institute of Japan.

References

[1] Z. Kazanci, K. Bathe, Int. J. Imp. Eng. **42**, 80 (2012)
[2] Z. Shen, X. Qiao, and H. Chen, SAE-china and FISITA (Eds), *Proc. FISITA 2012 World Automotive Congress* **197**, 13 (2013)
[3] L. Durrenberger, D. Even, A. Molinari, A. Rusinek, J. Phys. IV. **134**, 1287 (2006)
[4] T. Iwamoto, M. Cherkaoui, T. Sawa, Int. J. Modern Phys. B **22**, 5985 (2008)
[5] F. D. Fischer, G. Reisner, E. Werner, K. Tanaka, G. Cailletaud, T. Antretter, Int. J. Plast. **16**, 7231 (2000)
[6] Y. Tomita, T. Iwamoto, Int. J. Mech. Sci. **37**, 1295 (1995)
[7] *Standard Test Method for Measurement of Fracture Toughness*, `http://www.astm.org/Standards/E1820`
[8] J. F. Kalthoff, Int. J. Fract. **27**, 277 (1985)
[9] T. Rong, L. Lin, B. C. De Cooman, W. Xi-chen, S. Peng, ISIJ Int. **13**, 51 (2006)
[10] S. Curtze, V. T. Kuokkala, M. Hokka, P. Peura, Mater. Sci. Eng. A **507**, 124 (2009)
[11] J.A. Rodríguez-Martínez, R. Pesci, A. Rusinek, A. Arias, R. Zaera, D. A. Pedroche, Int. J. Solids and Struct. **47**, 1268 (2010)
[12] J. Nakayama, Japanese J. Appl. Phys. **3**, 422 (1964)
[13] T. Iwamoto, T. Tsuta, Int. J. Plast. **18**, 1583 (2002)
[14] X. K. Zhu, J. A. Joyce, U.S. Navy Res. Paper **49** (2012)
[15] J. R. Rice, J. Appl. Mech. **35**, 379 (1968)
[16] Y. Chuman, K. Minuras, K. Kaizu, S. Tanimura, Int. J. Impact Eng. **19**, 165 (1997)
[17] T. Yokoyama, K. Kishida, Exp. Mech. **29**, 188 (1989)

EPJ Web of Conferences **94**, 02005 (2015)
DOI: 10.1051/epjconf/20159402005

Characterization of the dynamic behaviour of ALGOTUF armour steel during impact and in torsion

Nabil Bassim[1,a], Solomon Boakye-Yiadom[1], Genevieve Toussaint[2], and Manon Bolduc[2]

[1] Department of Mechanical Engineering, University of Manitoba, Winnipeg, Canada
[2] DRDC Valcatier, Department of National Defense, Quebec, Canada

Abstract. Algotuf is a new steel which is proposed as a candidate for armour material. To assess this application, a study of the impact properties of this steel was conducted at the University of Manitoba using two types of Hopkinson Bar systems, namely a torsional bar equipment and a direct impact system capable of producing high strain rates and large strains. Stress strain curves for the steels were obtained in pure shear and in compression. Temperatures of 25 °C, 200 °C and 500 °C were used in the testing. Following the testing, a microstructural examination of the specimens tested was carried out to investigate the effect of microstructure on the mechanism of failure of this material. It was found that, above a value of impact momentum corresponding to a high strain rate, adiabatic shear bands are formed. The microscopic examination showed that the initiation of these shear bands corresponded at locations where martensitic laths were present and around regions of maximum shear stresses. Generally, the shear bands act as precursors to the formation of microcracks that may lead to failure. On the other hand, the high strength and formability of the steel makes it suitable for use as an armour material.

1. Introduction

The behavior of materials and their mechanical properties at high strain rates of deformation differ considerably from that observed at quasi-static or intermediate strain rates. Deformation of materials at quasi-static or intermediate strain rates is characterized by slip and twinning mechanisms [1,2]. However, at high strain rates such as impact, explosive forming, and ballistics, plastic deformation is characterized by strain localization along narrow bands [3,4]. This results in crack nucleation and propagation and subsequent fracture and fragmentation.

In this study, two types of Hopkinson Bar systems, namely a torsional bar equipment and a direct impact system were used to test Algotuf steel specimens at room temperature (25 °C), and at high temperatures of 200 °C and 500 °C. Following the testing, a microstructural examination of the specimens tested was carried out to investigate the effect of dynamic deformation and testing temperature on the mechanism of failure of this material.

2. Experimental procedures

2.1. Dynamic Torsion Tests of Algotuf

High strain rate torsion tests were conducted on Algotuf specimens using a Torsional Split Hopkinson Bar (TSHB) under room and high temperature conditions. Specimens were cut from received plates and machined to specifications. The specimens were grouped and tested under room temperature, at 200 °C and 500 °C using the TSHB. The test specimens were thin-walled tubes with a thickness of 0.4 and a length of 3.8 mm. Details of the testing procedure are provided in [5,6]. The torque needed to rapidly load the specimen is determined by the angle of twist and stored in the loading end of the input bar [6]. When the loading arm is released, an elastic wave is generated which travels rapidly through the input bar and is recorded as incident wave by attached strain gages. Part of this incident wave is used in deforming the specimen and part is transmitted through the specimen on to the output bar as transmitted wave. The transmitted wave is captured by the strain gage attached to the output bar. The fraction of the elastic wave that was used in deforming the specimen is reflected back and captured by the strain gage attached to the input bar as reflected wave. These three waves provide the strain signals which are conditioned, amplified and recorded by the data acquisition system connected to the bars. Dynamic stress-strain curves are generated from the captured strain wave signals.

2.2. Dynamic impact tests of Algotuf

High strain rate compression tests were conducted on the Algotuf steel specimens using a Direct Impact Hopkinson Pressure Bar (DIHPB). Cylindrical test specimens (10.5 × 9.5 mm) were impacted using a 1.905 kg quench-hardened AISI 4340 steel bar projectile. The projectile travels through a light gun barrel and strikes the test specimen attached to the output bar of the Hopkinson bar system. The output bar is also made from quench-hardened AISI 4340 steel with the same cross-sectional area as the striking projectile. The elastic waves generated when the specimen is impacted travels through the output bar and is captured by strain gage attached to it. This provides the strain signals which are conditioned, amplified and captured by the data acquisition system connected to the bar. The data acquisition system consists of strain pulse conditioner/amplifier and a mixed signal digital

[a] Corresponding author: bassimnabil@hotmail.com

Article available at http://www.epj-conferences.org or http://dx.doi.org/10.1051/epjconf/20159402005

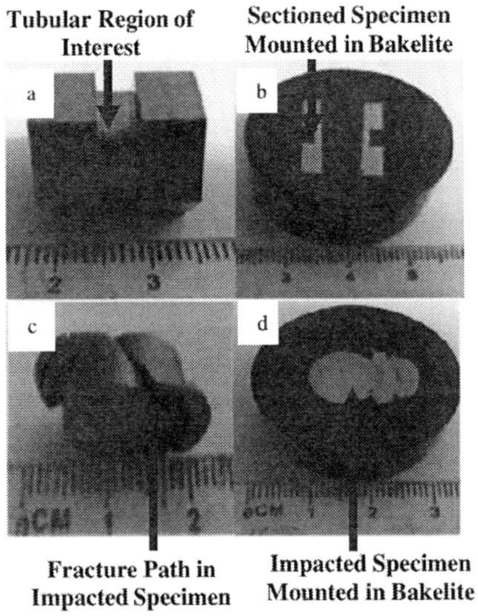

Figure 1. Images of the deformed (a), (b) torsional (c), (d) impacted Algotuf specimens showing the regions where microstructural analyses were carried out.

oscilloscope. Dynamic stress-strain curves are generated from the captured strain signals.

2.3. Metallographic analysis

Metallographic and electron microscopy (SEM) analysis of the deformed specimens were carried out to determine the effect of dynamic deformation and testing temperature on the mechanism of failure of this material. Selected deformed specimens were cut, grinded, polished, and etched with a 2% Nital solution. Micro- hardness measurements were performed with a Buehler micromet 5100 series under a test load of 300 gf. Microstructural investigations were carried out with an inverted reflected light optical microscope (OM) equipped with a CLEMEX vision image analyzer and a JSM 5900 scanning electron microscope (SEM) with an Oxford EDS system. Figure 1 shows the steel specimens after torsional and impact tests as well as the mounted specimens where the microstructural analyses were done.

3. Results

3.1. Results of dynamic torsion tests

Figure 2 shows the true shear stress vrs shear strain graphs for specimens all deformed at 6° angle of twist at Room Temperature (RT), and at High Temperstures (HT) of 200 °C and 500 °C. In general, shear stress increases initially with increasing shear strain, reaches a maximum and decreases with subsequent increase in shear strain irrespective of the testing temperature. The average strain rates increased as the testing temperature increase. The average strain rate was $580\,S^{-1}$ for the steel specimens tested at 6 ° angle of twist at room temperature. At the same angle of twist, the average strain rates increased to $638\,S^{-1}$ for the specimens tested at 200 °C and $702\,S^{-1}$ at 500 °C.

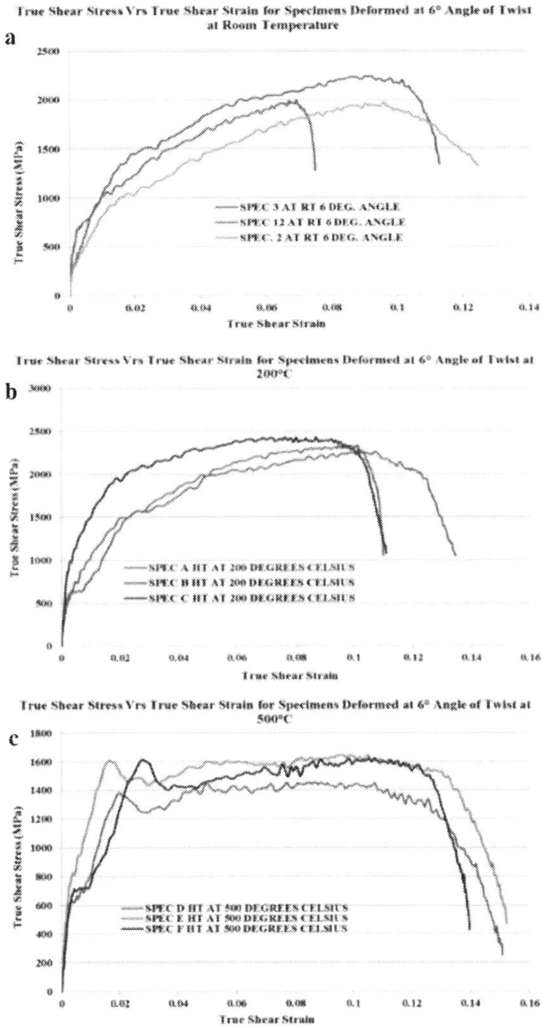

Figure 2. True shear stress – shear strain graphs for specimens all deformed at 6° angle of twist at (a) Room Temperature (RT), (b) 200 °C and (c) 500 °C.

Table 1 shows the strain rates and maximum shear stresses of the specimens after deformation. At room temperature (RT), the average maximum shear stress was 2050 MPa. This increased to 2304 MPa after deformation at 6 ° angle of twist at 200 °C. However, there was a decreased in the average maximum shear stress to 1564 MPa at 500 °C. These changes in the average maximum shear stresses are attributed to the effect of temperature on the microstructure and properties of the steel during deformation.

3.2. Results of dynamic impact tests

Table 2 shows the impact momentum and maximum strain rates of the room temperature and high temperature impacted Algotuf steel specimens. The specimens were all impacted at the same impact momentum of 32.35 kg · m/s.

Table 1. Strain rates and maximum shear stresses of the steel specimens after deformation.

Specimen	Angle Of Twist	Strain Rate (S^{-1})	Temperature of Test	Maximum Stress (MPa)
2	6°	603	RT	1970
3	6°	507	RT	1946
12	6°	631	RT	2235
A	6°	628	200 °C	2288
B	6°	703	200 °C	2228
C	6°	583	200 °C	2396
D	6°	709	200 °C	1446
E	6°	720	200 °C	1636
F	6°	677	200 °C	1611

Table 2. Impact momentum and maximum strain rates of impacted specimens at room and high temperatures.

Specimen	Impact Momentum (kg · m/s)	L_O- L_F (mm)	Nominal Strain	Strain Rate (S^{-1})
00-R1 RT	32.35	1.2667	0.11965	1132
01-R2 RT	32.35	1.2650	0.11973	1133
02-R3 RT	32.35	1.2233	0.11537	1092
00-HT1-200 °C	32.35	1.1200	0.10576	1001
01-HT2-200 °C	32.35	1.0175	0.09777	925
02-HT3-200 °C	32.35	1.0325	0.09831	930
02-HT1-500 °C	32.35	1.4050	0.13311	1260
03-HT2-500 °C	32.35	1.4833	0.14029	1328
04-HT3-500 °C	32.35	1.5850	0.14932	1413

The average strain rate at room temperature was 1119 S^{-1} which decreased to 952 S^{-1} after impact at the same momentum at 200 °C. However, at 500 °C the average strain rate increased to 1334 S^{-1}.

The impact resistance graphs were similar to the true shear stress vrs shear strain graphs in the dynamic torsion tests. In response to the impact, flow stress increased initially with strain, reached a maximum and decreased with subsequent increase in strain. This trend for the flow stress was prevalent in all the impacted specimens regardless of the tests temperature as shown in Fig. 3. The amount of deformation can be approximated to be the area under the impact resistance graphs.

Table 3 shows the average yield stresses and maximum shear stresses of impacted specimens at room and high temperatures. Increasing the impact temperature decreased the average yield stress while the strain at yield increased.

3.3. Metallographic analysis on the specimens prior to and after dynamic torsion tests

The morphology of the steel specimens prior to testing consisted of large martensite laths and packets. The average hardness of the steel prior to deformation was 425 ± 14 (HV).

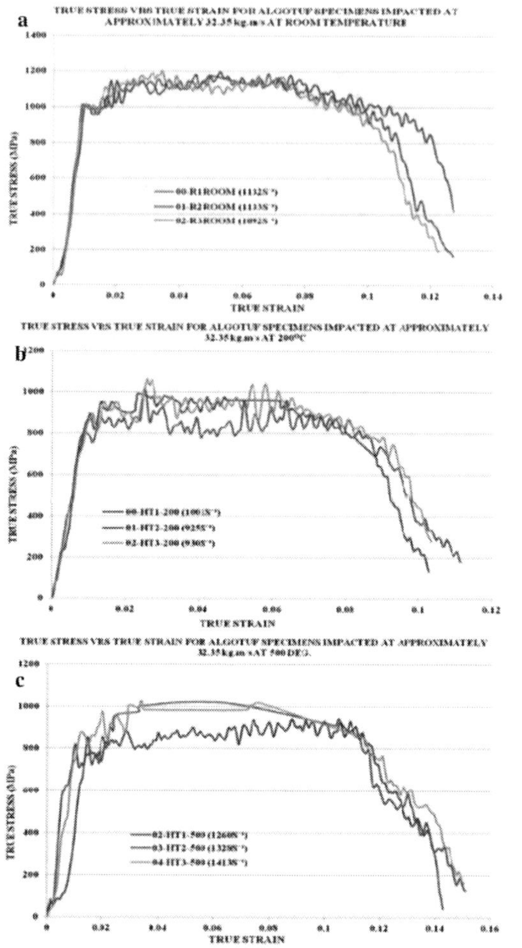

Figure 3. Impact resistance graphs of specimens impacted at 32.35 kg · m/s at (a) Room Temperature (RT), (b) 200 °C and (c) 500 °C.

Table 3. Average stresses and strains of impacted specimens.

Specimen	Yield Stress (MPa)	Yield Strain (%)	Maximum Stress (MPa)	Strain Rate (S^{-1})
00-R1 RT	1013	0.997	1162	1132
01-R2 RT	1012	0.999	1176	1133
02-R3 RT	1003	0.960	1128	1092
00-HT1-200	814	0.931	920	1001
01-HT2-200	899	1.098	960	925
02-HT3-200	880	1.065	959	930
02-HT1-500	807	1.105	1002	1260
03-HT2-500	754	1.367	916	1328
04-HT3-500	867	1.179	1011	1413

Figure 4 is a Secondary Electron Image (SEI) of the steel speceimen prior to dynamic torsion tests showing the morphology of the martensite laths. This was used as a baseline for comparison to the microstructure of the

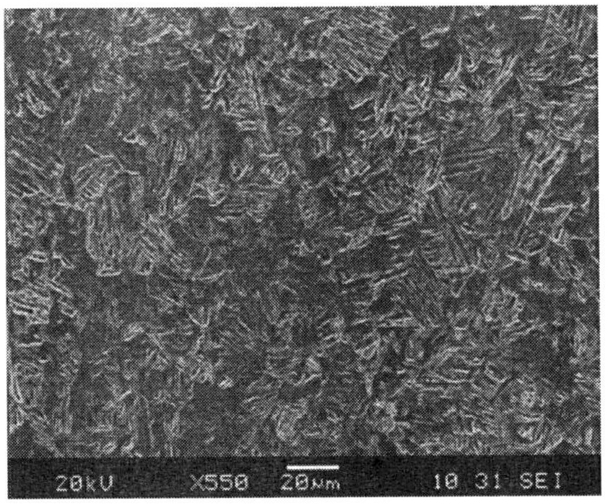

Figure 4. Scanning Electron Micrograph of Algotuf specimen prior to testing showing the morphology of the martensite laths.

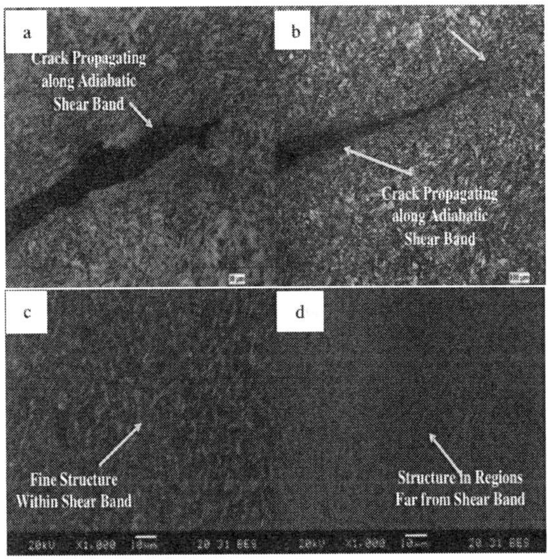

Figure 5. (a), (b) Optical micrographs of steel specimen 10 showing cracks propagating along adiabatic shear bands after deformation at 10° angle of twist at room temperature. (c) BSE micrograph of regions within the shear band (d) BSE micrograph of regions outside the shear bands.

deformed specimens to determine the relative degree of deformation and the changes in the microstructure that occurs after dynamic torsional testing.

Metallographic analysis on the specimen 12 after deformation at 6° angle of twist at room temperature revealed that there was no adiabatic shear band in the specimen but there was a relative increase in hardness due to the amount of deformation. There was little or no observable straining effects within the microstructure when compared to the microstructure of the untested specimens. This was predominant in all the steel specimens tested at room temperature at 6° angle of twist.

On the other hand, steel specimens that were tested at relatively higher angles of twist developed adiabatic shear bands. Figure 5 shows a crack propagating along an adiabatic shear band observed in specimen 10 which was deformed at 10° angle of twist at room temperature. The shear band leads the cracks and extensive plastic flow was observed at the tips of the cracks. The specimen fractured completely after testing. Figure 5(c) and (d) are Backscattered Electron (BSE) images that compare the structure of the shear band region to the regions surrounding the shear bands. The fine structure of the regions within the shear bands are apparent on Fig. 5(c) and (d). This is attributed to the effect of shear strain localization which results in a change from global distribution of deformation to localized distribution of deformation within the adiabatic shear bands.

No adiabatic shear band was observed in the steel specimen ddeformed at 6° angle of twist at 200 °C. However, there was a relative increase in hardness compared to the untested specimen and the room temperature tested specimens due to carbide precipitation at the relatively higher deformation temperature. Figure 6 shows the Secondary Electron Image (SEI) of the steel specimen C after deformation at 6° angle of twist at 200 °C showing the presence of precipitated carbides which were absent from the microstructure of the pre-deformation specimens.

At 500 °C, even though the specimen did not fracture completely, there were cracks and microcracks that

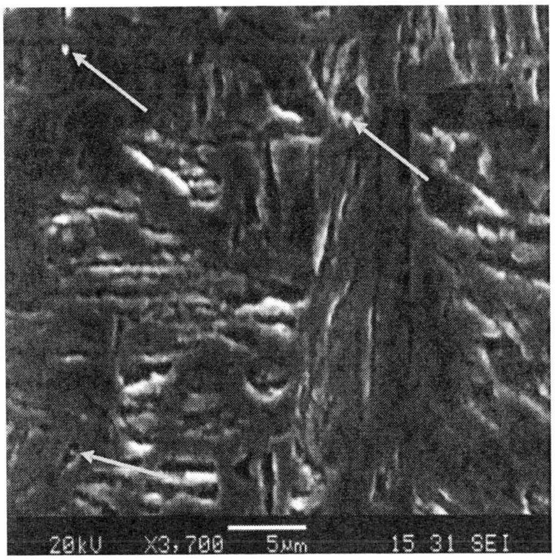

Figure 6. Scanning electron micrograph of steel specimen C showing precipitated carbides, orientation and shape of the martensite laths after deformation at 6° angle of twist at 200 °C.

propagated along the length of evolved adiabatic shear bands. The testing temperature of 500 °C is above the static recrystallization temperature of the steel. Testing the steel specimens at 500 °C led to a significant reduction in hardness and strength due to recovery and recrystallization processes. Plastic flow in the shear bands was easily observable as shown on Fig. 7.

Average hardness measurements (HV) on the specimens prior to testing and after testing are shown on

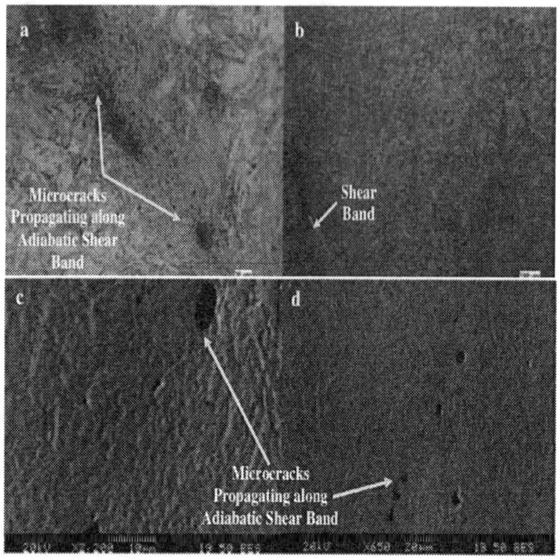

Figure 7. (a), (b) Optical micrographs (c), (d) BSE micrographs of steel specimen F showing cracks and microcracks propagating along adiabatic shear bands after deformation at 6 ° angle of twist at 500 °C.

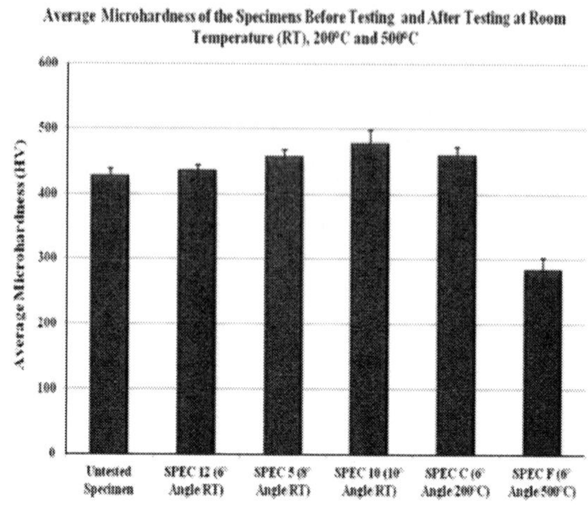

Figure 8. Average hardness (HV) for the steel specimens prior to testing and after testing at room and high temperatures.

Table 4 and Fig. 8. It was observed that hardness increased as the deformation energy increase. However, it was observed that deformation at 6 ° angle of twist at 200 °C, resulted in higher hardness and higher average maximum shear stress as was shown on Table 1. The increase in hardness and maximum shear stress is attributed to the observed precipitated carbides when the specimens were deformaed at 200 °C as shown in Fig. 8 and Table 4.

Table 4. Average hardness distribution of the steel specimens prior to and after deformation at room and high temperatures.

Specimen	Average Microhardness (HV)
Untested Specimen	425 ± 14
SPEC 12 (6 ° Angle/RT)	433 ± 11
SPEC 5 (8 ° Angle/RT)	455 ± 13
SPEC 10 (10 ° Angle/RT)	474 ± 24
SPEC C (6 ° Angle/200 °C)	457 ± 15
SPEC F (6 ° Angle/500 °C)	281 ± 21

4. Discussions

Understanding material behaviour under large plastic strains requires measuring the strains and stresses above the tensile necking limit. A compression test and a torsion test are alternative approaches that overcome necking instability that occurs in tension tests [6].

In this study, testing the steel specimens at the same angle of twist at room temperature and high temperature revealed that prior to recrystallization, there is a relative increase in the strength of the steel specimens as well as their resistance to shear strain localization. This is evident in the steel specimen tested at 6 ° angle of twist at room temperature and at 200 °C. The strength of the steel specimens increased during the test at 200 °C which was confirmed by the increase in hardness. This is attributed to the size and distribution of the precipitated carbides. There were no carbides observed within the pre- deformation microstructure. However, at 200 °C, precipitated carbides were observed within the deformed steel specimens. It is inferred that a prior heat treatment on the Algotuf steel could increase its strength and impact resistance because of carbide precipitation. It has been reported that the prior deformation microstructure has a lot of effect on the dynamic properties and susceptibility of structural steels to shear strain localization during deformation at high strain rates and large strains [3,4]. Thus, structural steels can be heat treated to produce a strong microstructure that is resistant to strain localization during deformation at high strain rates [4].

Strain localization has been identified as a major failure mechanism in materials under high strain rates and large strains. In the current study, the dynamic impact and torsional testing on Algotuf steel specimens revealed that deformation at high strain rates and large strains causes the material to fail by the formation of narrow adiabatic shear bands (ASBs). It was observed that adiabatic shear bands preceded fracture and fragmentation leading to catastrophic failure. The higher the strain rate, the more severe the damage and the failure that occurs. Multiple shear bands with multiple cracks were prevalent in the impacted steel specimens compared to the specimens tested in torsion due to the very high strain rates that occur during impact. It is concluded that the evolution of adiabatic shear bands are the cause of damage that occurs in this steel leading to fracture and fragmentation.

5. Conclusion

In this study, two types of Hopkinson Bar systems, namely a torsional bar equipment and a direct impact system were used to test Algotuf steel specimens at room temperature (25 °C), and at high temperatures of 200 °C and 500 °C. Following the tests, the specimens were examined using optical and electron microscopy. Most of the tested specimens exhibited failure by the development of adiabatic shear bands. Some of the bands contained microcracks which definitely confirm that the material undergoes intense plastic deformation in the shear bands before fracture occurs. This demonstrates that ASBs can serve as precursors and preferential sites for crack nucleation and propagation. Excessive grain refinement which occurs within the shear bands during impact can create potential weak points. These weak points can form microvoids which can grow and form cracks.

The financial support of Defense Research and Development Canada (DRDC DND) for this study is greatly appreciated.

References

[1] B. Dodd, and Y. Bai, Adiabatic Shear Localization: Frontiers and Advances, Elsevier (2012).

[2] Y. Xu, J. Zhang, Y. Bai, and M. A. Meyers, Shear Localization in Dynamic Deformation: Microstructural Evolution, Metallurgical And Materials Transactions A **39** (2008) 811–841.

[3] S. Boakye-Yiadom, A. K. Khan and N. Bassim, Effect of pre-impact microstructure on the nucleation and initiation of Adiabatic Shear Bands (ASBs) during impact, Materials Science and Engineering A **615** (2014) 373–394.

[4] S. Boakye-Yiadom and M. N. Bassim, Effect of prior heat treatment on the dynamic impact behavior of 4340 steel and formation of adiabatic shear bands. Materials Science and Engineering A **528** (2011) 8700–8708.

[5] M. Nabil Bassim and N. Panic, High strain rate effects on the strain of alloy steels, Journal of Materials Processing Technology **92** ± **93** (1999) 481–485.

[6] E. Cepus, M.Sc. Thesis, University of Manitoba, Manitoba, Canada, 1995.

EPJ Web of Conferences **94**, 02006 (2015)
DOI: 10.1051/epjconf/20159402006

Structure/property (constitutive and dynamic strength/damage) characterization of additively manufactured 316L SS

G.T. Gray III[1], V. Livescu[1], P.A. Rigg[2], C.P. Trujillo[1], C.M. Cady[1], S.R. Chen[1], J.S. Carpenter[1], T.J. Lienert[1], and S. Fensin[1]

[1] Los Alamos National Laboratory, Los Alamos, New Mexico 87545, USA
[2] Institute for Shock Physics, Washington State University, Pullman, Washington 99164, USA

Abstract. For additive manufacturing (AM), the certification and qualification paradigm needs to evolve as there exists no "ASTM-type" additive manufacturing certified process or AM-material produced specifications. Accordingly, utilization of AM materials to meet engineering applications requires quantification of the constitutive properties of these evolving materials in comparison to conventionally-manufactured metals and alloys. Cylinders of 316L SS were produced using a LENS MR-7 laser additive manufacturing system from Optomec (Albuquerque, NM) equipped with a 1kW Yb-fiber laser. The microstructure of the AM-316L SS is detailed in both the as-built condition and following heat-treatments designed to obtain full recrystallization. The constitutive behavior as a function of strain rate and temperature is presented and compared to that of nominal annealed wrought 316L SS plate. The dynamic damage evolution and failure response of all three materials was probed using flyer-plate impact driven spallation experiments at a peak stress of 4.5 GPa to examine incipient spallation response. The spall strength of AM-produced 316L SS was found to be very similar for the peak shock stress studied to that of annealed wrought or AM-316L SS following recrystallization. The damage evolution as a function of microstructure was characterized using optical metallography.

1. Introduction

Additive manufacturing (AM) technology is a key enabling technology fueling the revolutionary transformations occurring in rapid prototyping, freeform and net-shape manufacturing, and local production at a global scale. AM offers increased efficiencies in energy, cost, and material savings in manufacturing [1–5]. However, for AM materials or components to supplant conventional manufactured materials and processes, the certification and qualification paradigm needs to evolve as there exists no "ASTM-type" additive manufacturing certified process or AM-material produced specifications. Even for small changes in starting feed material (powder or wire), component geometry, build process variables, and post-build thermo-mechanical processing, the qualification cycle can be complicated leading to long implementation times for even minor changes. This is due in large part to the fact that we are not able to predict and control processing-structure-property-performance (PSPP) relationships [6–8]. Metallic-component certification requirements have been documented elsewhere in the case of some manufacturing processes but generally involve meeting engineering and physics requirements tied to the functional requirements of the engineering component and finally process and product qualification. For AM materials and components to meet qualification and certification requirements for critical engineering applications, key microstructural parameters and defects must be quantified and quantitatively linked to processing and equipment parameters to establish minimum performance properties [9].

316 stainless steel is a ubiquitous austenitic SS that is often used in marine and medical environments because of its combination of strength and corrosion resistance. The pervasiveness of use and wide availability of inexpensive feedstock for additive manufacturing, has lead to several recent publications that focus on this particular alloy system [10–13]. In several of these studies connections are made between defects, such as porosity, and performance in standard mechanical behavior tests [10,12]. Although this property data is important in qualifying AM products, a major concern when comparing AM parts with those made in traditional methods is the failure and damage mechanisms that arise from the unique microstructures. Some studies exist in the area of high cycle fatigue that begin to make these connections. [14,15] but none for the 316 stainless steel series. Further, no studies of the dynamic damage evolution due to shock spallation loading of an AM-produced 316L SS material has been conducted to date. The purpose of this study is to report initial constitutive and dynamic fracture (spallation) properties of 316L SS produced by Laser Engineered Net Shaping (LENS) additive manufacturing in comparison to wrought 316L SS as a result of their microstructures.

2. Experimental procedure

2.1. Material – processing

This study involved production, characterization and testing of solid cylinders of 316L austenitic stainless steel(hereafter 316L SS) produced by laser additive manufacturing and contrasted to wrought 316L SS plate.

Article available at http://www.epj-conferences.org or http://dx.doi.org/10.1051/epjconf/20159402006

Table 1. Chemical Analysis of Wrought 316L SS plate and 316L SS Powder.

Wt.%	**C**	**Cr**	**Cu**	**Mn**	**Mo**
Wrought	0.022	16.16	0.39	1.70	2.08
Powder	0.02	20.7	0.19	1.32	2.45
Wt.%	**N**	**Ni**	**P**	**S**	**Si**
Wrought	0.063	10.03	0.029	0.0004	0.40
Powder	0.09	11.4	0.02	0.01	0.50

The wrought 316L SS was procured in 12.5 mm-thick plate form. The analyzed chemical composition of the wrought plate 316L SS (wt. pct.) is shown in Table 1. The crystallographic texture of the 316L SS investigated was investigated using X-ray diffraction to be almost random in nature. The 316L powder was procured from Carpenter Powder Products (Bridgewater, PA) and was produced using atomization techniques. The analyzed chemical composition of the 316L SS powder (wt. pct.) is also provided in Table 1. An inert gas fusion method (ASTM E 1019-11) was used to determine the wt.% oxygen for the powder (0.013%) and the deposit (0.018%). The powder particle size distribution was −80/ +270 mesh (53 um to 180 um) and was determined using standard sieve analysis of metal powders.

Cylinders of 316L SS were produced using a LENS MR-7 laser AM system from Optomec (Albuquerque, NM) equipped with a 1kW Yb-fiber laser. The laser beam was delivered to the collimator with a 200 um stepped index fiber optic and was focused via a lens to a focus length of 160.4 mm. The laser focus condition was arranged so that the minimum waist of the focused beam was positioned ~3.8 mm below the substrate surface. The laser power employed throughout the trials was 380 W.

The laser beam was directed normal to the substrate and passed through a nozzle ~20 mm in diameter accompanied by anargon shield gas with a flow rate of 20 lpm to prevent damage to the optics. Powder was delivered to the deposition region through four nozzles with exit orifices positioned concentrically around the larger nozzle. The exit orifices of the nozzles were positioned approximately 9.5 mm above the substrate surface. The four nozzles were arranged symmetrically about the laser beam axis and were angled downward toward the substrate at ~45 degrees. A mass flow rate of the powder of $\sim 6.3\,\text{gm}\cdot\text{min}^{-1}$ was determined by averaging three measurements conducted by capturing powder for one minute from the nozzles in a plastic bag.

The AM components produced were solid cylinders 2.5 cm in diameter and 3.3 cm tall. The cylinder was deposited onto a substrate comprised of a 0.95 cm thick plate of 304L SS. A z step height of 0.3 mm was used with a hatch spacing of 0.46 mm. Each layer was produced using linear "hatching" passes at $1.12\,\text{cm s}^{-1}$. The direction of the hatching passes was alternated between 0° and 90° directions for every other layer. Oxygen levels were monitored throughout building and did not exceed 6 ppm at any point. The powder feed rate was set at 3.5 RPM and was held constant during the build. After AM fabrication, the 304L substrates were machined away. One cylinder was sectioned to provide AM-as-built samples while the other was heat-treated. The heat treatment was 1060 °C for one hour under vacuum followed by cooling to room temperature in $2\frac{1}{2}$ minutes by rapid Argon gas quenching. The heat treatment leads to an AM-recrystallized microstructure (hereafter referred to as AM-Rx).

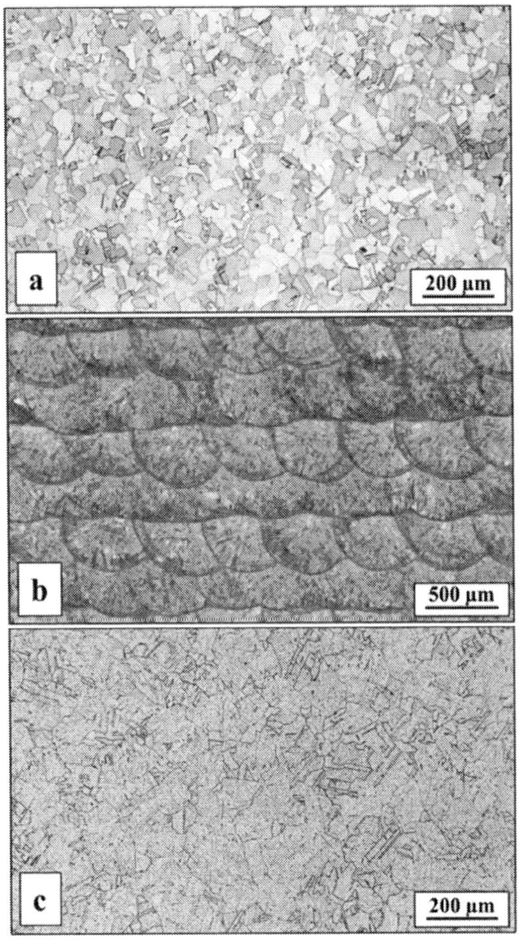

Figure 1. Optical microscopy of the 316L SS materials studied: a) annealed wrought plate, b) AM-(As-Built), and c) AM-following recrystallization heat-treatment at 1060C for 1 hour (AM-Rx).

Samples from the three starting 316L SS materials were sectioned and prepared following standard metallographic procedures including polishing to a mirror finish and etching. Optical microscopy images were acquired at various magnifications using a Zeiss Axio Imager M2m optical microscope. The microstructures of the three 316L SS materials investigated are shown in Fig. 1. The annealed wrought plate material is seen to exhibit a fine-grained equiaxed microstructure. The AM-As-Built displays a macroscopic squamous or fish-scale type morphology indicative of the deposit interfaces of the alternate layers of the build. Each "fish scale" exhibits a fine-grained microstructure with elongated grains along the periphery and equiaxed grains in the center. The as-deposited

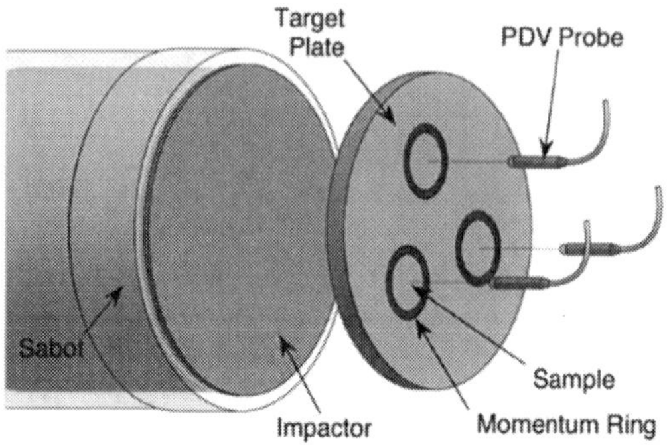

Figure 2. Experimental configuration used to produce incipient damage or complete spall in three samples simultaneously. Wrought 316L SS impactors were accelerated in a 80 mm gas gun and impacted onto the target. PDV probes were placed over the center of each sample to provide free-surface velocimetry data from each sample in each experiment.

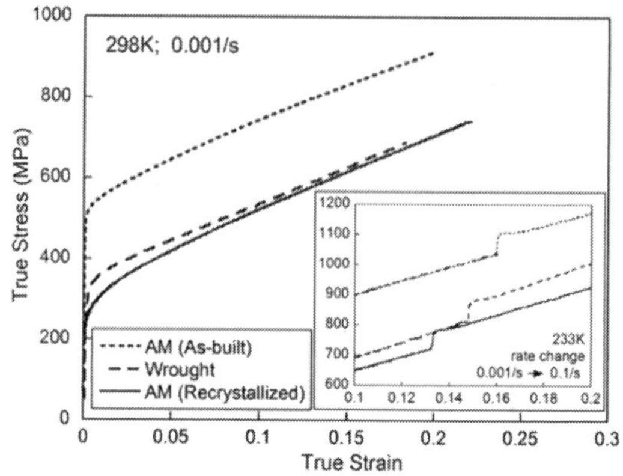

Figure 3. Compressive True Stress-True Strain of 316L SS in the AM-as-built condition, AM + recrystallization heat-treatment, and annealed wrought 316L SS.

microstructure was characterized by a network of lacy ferrite (BCC) phase in a matrix of austenite (FCC) due to the primary solidification to ferrite [16]. The AM-Rx displayed a nominally polycrystalline microstructure albeit coarser and not as faceted as the equiaxed microstructure of the wrought 316L SS.

2.2. Constitutive/spallation characterization

Cylindrical samples, with the compression axis parallel to the AM cylinder axis and thru-thickness to the wrought plate, were electro-discharge machined to dimensions of 5 mm in height and 5 mm in diameter. Quasi-static compression tests were conducted using an Instron screw-driven test system at a strain rate of 0.001 s^{-1} and temperature of 298 K, with $MoSi_2$ lubrication to minimize barreling. Strain-rate jump tests from 0.001 to 0.1 s^{-1} were conducted at 233K to assess the strain-rate sensitivity of the three materials.

Spallation experiments were performed using the configuration shown in Fig. 2. Impactors of wrought 316L SS were accelerated using an 80 mm bore gas gun at a velocity of nominally 250 m/s and impacted onto the 316L SS targets. All impactors were roughly 2.5 mm thick. Each target consisted of three separate components: a target plate, one, two, or three samples, and a momentum ring for each sample. Targets were 5mm thick and designed specifically such that recovery of the shocked samples was facilitated for post-mortem examination. The samples were press-fit into the momentum rings and the rings press-fit into the target plate to ensure intimate contact of all components initially. Single sample spallation experiments were done as described elsewhere [17]. For the two- and three-sample targets, samples were machined as right circular cylinders 12.7 mm in diameter from wrought, as-built additively manufactured, and the AM-Rx 316L material. In the experiment, the impactor thickness was one-half the sample thickness to cause tensile damage to occur in the center of the samples. Samples were surrounded by momentum trapping rings made from wrought material with an inner diameter (ID) 0.1 mm less than the outer diameter (OD) of the sample. The OD of the rings was 16.5 mm with the holes in the target plate made to 16.4 mm diameter to accommodate press fitting of all parts. The assembled targets were lapped flat and parallel to within 5 micrometers.

Samples were placed in the target ring such that the shortest distance between sample edges to each other and to the edge of the target plate were equal. Free surface velocities were measured in each experiment using Photon Doppler Velocimetry (PDV) [18,19]. Collimated probes provided by AC Photonics (Part #: 1CL15P020LC-C01) were positioned to collect velocimetry data from the center of each sample and connected to a LANL-built 4-channel PDV system. One additional PDV probe (not shown in figure) was placed next to each target plate to measure the projectile velocity directly to an accuracy of 0.1% [19]. A single PZT trigger pin (Dynasen, Inc.) was also positioned next to each target plate with a typical standoff distance of a few millimeters to provide a trigger to the digitizer used to collect the PDV data.

3. Results and discussion

3.1. Constitutive response

The quasi-static compressive constitutive response of the three 316L SS materials is shown in Fig. 3. The as-built AM material exhibits twice the yield strength of the AM-Rx material and ~60% higher yield than the annealed wrought plate. Following yielding, all three materials display similar rates of work hardening (work-hardening slopes). To evaluate if the higher flow stress levels in the AM-as-built material was indicative of a higher starting dislocation density or level of residual stress, strain rate jump tests were conducted at 233 and 298 K.

The magnitude of flow stress increase with the strain rate jumps was seen to be invariant for all three materials; see the rate-jump insert in Fig. 3 conducted at 233 K. The similarity in rate jump response suggests the higher starting yield and flow stress displayed by the AM-as-built material is due to an intrinsic barrier component to the strength [20] consistent with the extremely fine

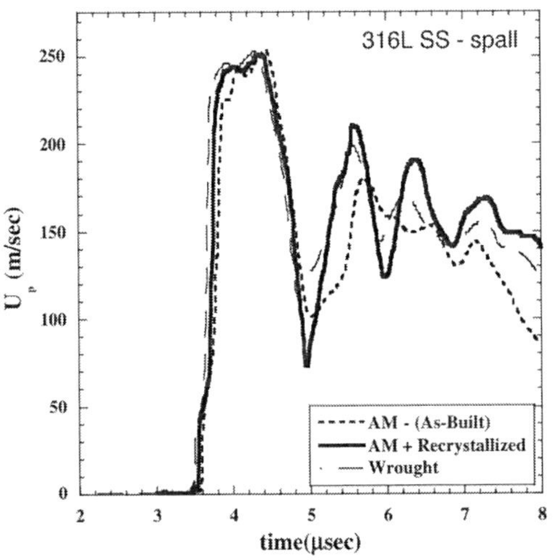

Figure 4. PDV wave profiles (particle velocity versus time plot) for the three 316L SS materials impacted at 257 m/sec.

dendritic substructure, within the coarser "fish scale" macrostructure, formed during the additive manufacturing solidification and not related to either a high starting defect population or residual stresses.

3.2. Velocimetry

The PDV data for the three 316L SS materials spallation experiments conducted at a peak impact velocity of 257 m/s is presented in Fig. 4. The wave profiles for the three materials are seen to display: 1) nominally similar elastic-plastic transitions, or Hugoniot Elastic Limits (HEL's), as seen in the loading portion of the profile, 2) non-constant responses upon the peak stress on the Hugoniot suggestive of non-uniform plastic deformation with the samples, most evident in the AM-as-built material, 3) slightly different magnitudes(depth) of "pull-back" signals suggesting different damage nucleation and growth responses with the wrought and AM-as-built displaying bi-linear pull-back slopes perhaps indicative of differing energy partitioning during nucleation and growth of damage as compared to the AM+Rx displaying a linear "pull-back" signal, and 4) the AM-as-built displaying a shifted time interval of the "pull-back" signal reload peak consistent with the evolved damage in the sample not located solely near the center line of the sample, thereby altering the "ringing" interval in the pull-back signal, in contrast to the wrought and AM+Rx samples where the incipient damage is centered near the mid-plane of the sample thickness. The difference in the magnitude of the "pull-back" signals for the three materials and the calculated spall strengths is presented for the three 316L SS samples in Table 2.

The spall strength (σ_{spall}) was calculated using the relationship for a material that exhibits an elasto-plastic behavior [21]:

$$\sigma_{\text{spall}} \equiv \rho_0 C_L \Delta FSV \left(1 + \frac{C_L}{C_o}\right)^{-1} \qquad (1)$$

Table 2. 316L SS Spall Data.

316L SS Material	C_0 (m/sec)	C_L (m/sec)	Peak Stress (GPa)	ΔFSV (m/sec)	Spall (GPa) (GPa)
AM	4474	5734	4.54	143.3	2.87
AM-Rx	4464	5722	4.54	172.1	3.43
Wrought	4358	5701	4.44	131.0	2.57

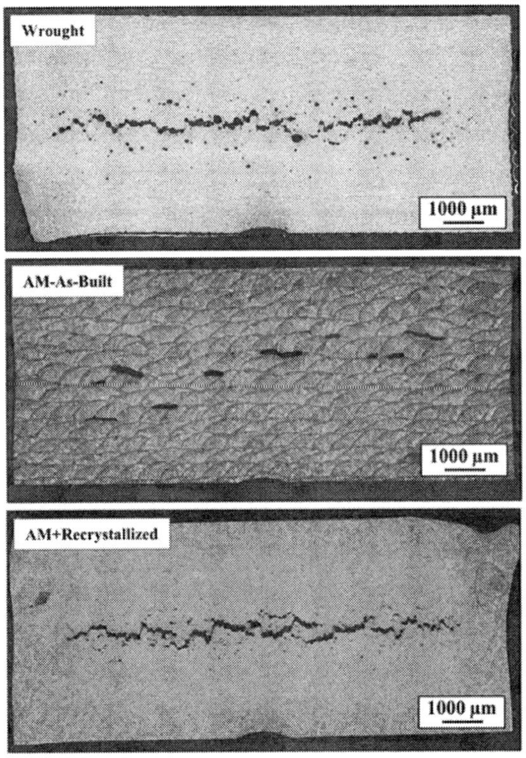

Figure 5. Macroscopic optical metallography of incipiently spalled 316L SS in the AM-as-built condition, AM + Rx, and annealed wrought 316L SS.

where $C_L = 5730$ (AM), 5720 (AM-Rx) and 5700 (wrought) m/s is the longitudinal sound speed and ΔFSV is the difference in the free surface velocity from the peak state to the minima.

3.3. Post-mortem metallurgical analysis

The three samples from the spallation experiment were cross-sectioned along the diameter, then prepared and inspected as described in Sect. 2.1. The damage evolution in the incipiently spalled 316L SS materials was seen to vary between the three materials more significantly than suggested by the three spall strengths. Figure 5 presents the macroscopic optical metallographic cross-sections, following polishing and etching, of the incipiently spalled three 316L SS spall samples. On a coarse scale the first apparent difference is in the position of the incipient damage. In the wrought and AM+Rx samples the damage field is centered in the thickness direction. This is consistent with the selection of the impactor thickness

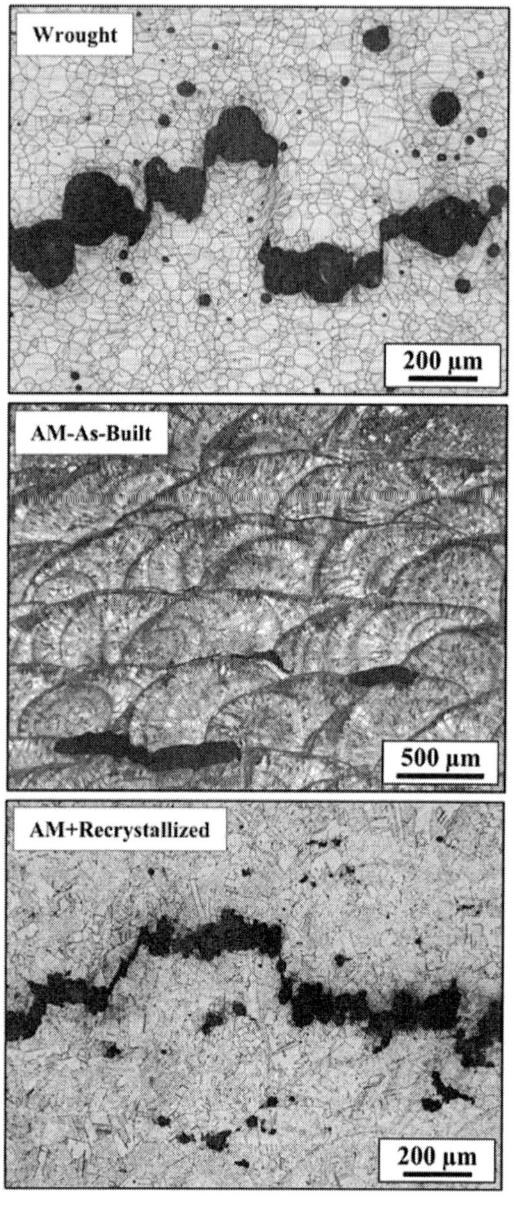

Figure 6. Higher magnification optical micrographs of damage evolution in incipiently spalled 316L SS in the AM-as-built condition, AM+Rx, and annealed wrought 316L SS.

being one-half the sample thickness and thereby placing the maximum tensile stress field in the middle of the sample. Conversely, the AM-as-built displays incipient crack formation following layer solidification boundaries located along the sample mid-plane as well as further from the center of the sample thickness.

The macroscopic damage evolution of the three 316L SS samples is observed to vary between the samples. In Fig. 6, higher magnification optical images of the damage evolution mechanisms and correlation to the microstructure are presented. In the wrought 316L SS the damage field is comprised of disperse round voids, with void coalescence evident along the spall plane. Regions of coalesced voids linked by a network of shear localized plastic flow regions were observed. This mode of damage evolution in 316L SS is similar to that observed previously on wrought 316L SS [22, 23]. Higher magnification imaging of the ductile voids show that most of the voids are associated with grain boundaries suggesting that boundaries are the primary void nucleation sites similar to past observations in pure copper [24].

The damage evolution in the AM-as-built, seen in Fig. 6, displays void formation and cracks along layer solidification boundaries, the "fish-scale structure" and as well as cracks following some other layer solidification boundaries as seen in Fig. 6. The cracks and voided areas are seen to lie along boundaries nominally directly orthogonal to the shock direction. This suggests selection of regions experiencing the maximum tensile loading. Future spallation testing of samples loaded perpendicular to the AM-build direction will examine how the damage evolution is affected by the orientation of loading relative to the layer deposition direction. The microstructure within the individual solidified layer regions reveals a very fine dendritic microstructure comprised of primary and tertiary dendrites with no further evidence of evolved damage nor localized plasticity.

The damage evolution in the AM+Rx displayed a nominally equiaxed microstructure albeit the grains exhibited a more irregular morphology. Evidence of spherical void formation at grain boundaries is seen. Some void coalescence is observed as well as jagged cracks and localized shear and cracks connecting voids was evident. In contrast to the wrought specimen, these cracks are the result of many small voids simultaneously nucleated and coalesced before significant void growth. Higher magnification of the AM-Rx incipiently spalled sample following longer chemical etching also revealed residual memory of the AM chemical segregation formed during solidification in the AM process which the recrystallization at 1060 °C for 1 hour clearly did not homogenize. Future research on 316L SS produced by AM+Rx will examine the effect of alternate heat-treatments on microstructure development.

4. Summary

The constitutive and spallation response of 316L SS fabricated via LENS additive manufacturing is compared to that of annealed wrought 316L SS and the AM-as-built material following recrystallization, termed AM-Rx. Preliminary results indicate that the as-built additive 316L SS material exhibits 10% higher spall strength while displaying a 60% higher yield strength, pronounced macroscopic solidification boundary structure, and chemical segregation evident in the as-built microstructure compared to the equiaxed annealed wrought 316L SS. The recrystallized additive 316L SS material was seen to display similar quasi-static constitutive stress-strain response to the annealed wrought plate but displayed differences in terms of damage evolution methodology following spallation loading.

Through post-build heat treatments, design of 316L SS material for specific performance objectives appears a fruitful line of research.

Los Alamos National Laboratory is operated by Los Alamos National Security, LLC, for the National Nuclear Security Administration of the U.S. Department of Energy under contract DE-AC52-06NA25396. This work was partially sponsored by the Joint DoD/DOE Munitions Technology Development Program.

References

[1] Y. Zhai, D.A. Lados, J.L. Lagoy, Journal of Metals, **66**, 808 (2014)

[2] A.N. Hopkinson, P. Dickens, J. Mechanical Engineering Science, **217**, 31 (2003)

[3] B.R. Berger, *Additive Manufacturing: A Game Changer for the Manufacturing Industry?* (Berger STrategy Consultants, Munich, Germany, 2013)

[4] T.J. Horn, O.A. Harrysson, Science Progress, **95**, 255 (2012)

[5] I. Gibson, D.W. Rosen, B. Stucker, *Additive Manufacturing Technologies: Rapid Prototyping to Direct Manufacturing* (Springer, New York, 2010)

[6] D.L. Bourell, M.C. Leu, D.W. Rosen, *Roadmap for Additive Manufacturing: Identifying the Future of Freeform Processing* (University of Texas at Austin, Austin, TX, 2009)

[7] T. Foresight Project, *The Future of Manufacturing: A New Era of Opportunity and Challenges for the UK* (Science TGOf, London, England, 2013)

[8] T. Energetics Incorporated, *Measurement Science Roadmap for Metal-Based Additive Manufacturing* (Technology NIoS, Washington, DC, 2013)

[9] N.R. Council, *3D Printing in Space* (The National Academy Press, Washington, DC 20001, 2014)

[10] E. Yasa, J. Deckers, J.-P. Kruth, M. Rombouts, J. Luyten, VIrtual and Physical Prototyping, **5**, 89 (2010)

[11] B. Verlee, T. Dormal, J. Lecomte-BEckers, Powder Metallurgy, **55**, 260 (2012)

[12] J.A. Cherry, H.M. Davies, S. Mehmood, N.P. Lavery, S.G.R. Brown, J. Sienz, International journal of Advanced Manufacturing Technology, **76**, 869 (2015)

[13] A.S. Wu, D.W. Brown, M. Kumar, G.F. Gallegos, W.E. King, Metallurgical and Materials Transactions A, **45A**, 6260 (2014)

[14] E. Brandl, U. Heckenberger, V. Holzinger, D. Buchbinder, Materials and Design, **34**, 159 (2012)

[15] B. Baufeld, E. Brandl, O. van der Biest, Journal of Materials Process Technology, **211**, 1146 (2011)

[16] J.A. Brooks, J.C. Williams, A.W. Thompson, *Trends in Welding Research* (ASM, 1982)

[17] G.T. Gray III, In: *ASM Handbook. Vol. 8: Mechanical Testing and Evaluation*, edited by Kuhn H, Medlin D, ASM International, Materials Park, Ohio(2000), p. 530

[18] O.T. Strand, D.R. Goosman, C. Martinez, T.L. Whitworth, W. W. Kuhlow, Rev. Sci. Instrum., **77**, 083108 (2006)

[19] B.J. Jensen, D.B. Holtkamp, P.A. Rigg, D.H. Dolan, J. Appl. Phys., **101**, 013523 (2007)

[20] P.S. Follansbee, *Fundamentals of Strength* (John Wiley & Sons, Inc., Hoboken, N.J., 2104)

[21] S.A. Novikov, I.I. Divnov, A.G. Ivanov, Phys. Metals Metallog., **21(4)**, 122 (1966)

[22] G.T. Gray III, N.K. Bourne, B.L. Henrie, J.C.F. Millett, J. Phys. IV France, **110**, 773 (2003)

[23] G.T. Gray III, N.K. Bourne, B.L. Henrie, J. Appl. Phys., **101**, 093507 (2007)

[24] J.P. Escobedo, D. Dennis-Koller, E.K. Cerreta, B.M. Patterson, C.A. Bronkhorst, B.L. Hansen, D. Tonks, R.A. Lebensohn, J. Appl. Phys., **110**, 033513 (2011)

EPJ Web of Conferences **94**, 02007 (2015)
DOI: 10.1051/epjconf/20159402007

Effects of surface cracks and strain rate on the tensile behavior of Balmoral Red granite

Ahmad Mardoukhi[a], Mikko Hokka, and Veli-Tapani Kuokkala

Tampere University of Technology, Department of Material Science, POB 589, 33101 Tampere, Finland

Abstract. This paper presents an experimental procedure for studying the effects of surface cracks on the mechanical behavior of Balmoral Red granite under dynamic and quasi-static loading. Three different thermal shocks were applied on the surface of the Brazilian Disc test samples by keeping a flame torch at a fixed distance from the sample surface for 10, 30, and 60 seconds. Microscopy clearly shows that the number of the surface cracks increases with the duration of the thermal shock. After the thermal shock, the Brazilian Disc tests were performed using a servohydraulic materials testing machine and a compression Split Hopkinson Pressure Bar (SHPB) device. The results show that the tensile strength of the rock decreases and the rate sensitivity of the rock increases as more cracks are introduced to the structure. The DIC analysis of the Brazilian disc tests shows that the fracture of the sample initiates at the center of the samples or slightly closer to the incident bar contact point. This is followed by crushing of the samples at both contact points with the stress bars.

1. Introduction

Geothermal energy has recently attracted a lot of attention among engineers and scientists. However, there are still significant technological and scientific challenges to deal with before the lucrative energy resources can be easily reached. One of the biggest challenges is the drilling of the wells as deep as five kilometers to reach the required temperature to produce electricity. This challenge becomes bolder in the areas far away from the tectonic plate boundaries and in regions where the bedrock consists of hard rocks such as granite typically found in the Nordic countries. The current drilling technologies are based on rotary and percussive drilling. To improve the current drilling technology, the engineers need to know the rock strength and other properties at the drilling conditions, which include high hydrostatic pressures and elevated temperatures. However, since the straightforward experimental testing at these conditions is scientifically and practically very difficult, various numerical simulations and constitutive modeling approaches are currently being used for predicting the fracture behavior and the rock-tool interactions such as tool wear. Saksala et al. [6] studied the fracture behavior of Kuru granite by analyzing the Brazilian Disc tests with numerical and experimental approaches. In a separate study, Saksala et al. [7] extended the modeling approach to 3D and studied the bit-rock interactions and rock fracture in percussive drilling. Tang et al. [8,9] carried out numerical simulations to study the influence of microstructure on rock failure in uniaxial compression. They analyzed the effect of the structure heterogeneity of the rock on the fracture behavior and reported that the macro-fracture occurs abruptly as soon as the specimen reaches the peak stress. They also reported observations of micro-cracks over the whole volume of the specimens prior to the macro-fracture, and that the splitting failure is the dominant mechanism in uniaxial loading. However, due to the heterogeneous nature of rocks, other kinds of failure mechanisms were also observed.

The construction of material models and validation of the simulation results, however, relies on good scientific understanding of the material behavior as well as experimental data for calibrating the material models. The dynamic testing of rocks, however, is challenging, especially with the bar methods, since the material fails at very low strains. Therefore, it is very difficult to achieve the stress equilibrium, and cracks tend to initiate on the incident loading side of the sample before the other parts of the sample are fully loaded. Luckily, non-contact full-field deformation analysis techniques, such as photoelasticity, Moire, coherent gradient sensor, and digital image correlation (DIC) can be used to analyze the fracture behavior during a dynamic impact [1–5].

From the scientific point of view, there are still various problems in the understanding of rock behavior. For instance, the crack initiation, propagation, and coalescence are not completely understood, and many observations cannot be explained by any solid theory. Therefore, a material model that takes into account these factors and predicts the material behavior is needed for high quality drilling simulations. For example, the simulations of several repeated impacts of the drill on the same location of the rock surface require understanding of the effects of microcracks formed during the first impact. However, producing reliable experimental data that accurately describes the rock behavior at the impact speeds is not an easy task. Percussive drilling occurs at impact speeds of around 5–20 m/s, and each impact causes dynamic loading on the rock. Therefore, understanding of the

[a] Corresponding author: ahmad.mardoukhi@tut.fi

Article available at http://www.epj-conferences.org or http://dx.doi.org/10.1051/epjconf/20159402007

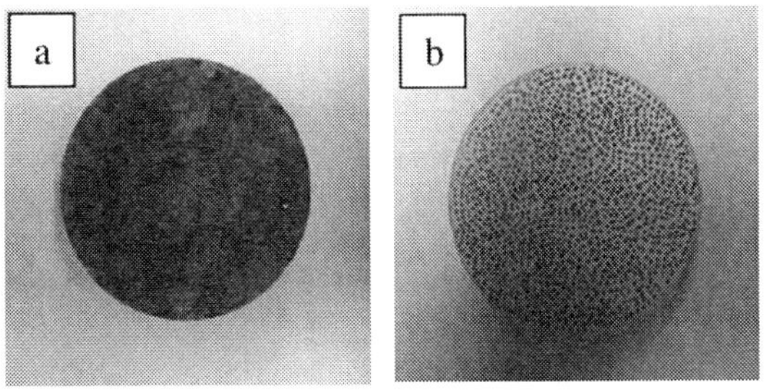

Figure 1. a) Natural surface of a Balmoral Red granite sample and b) DIC spackle pattern on the sample.

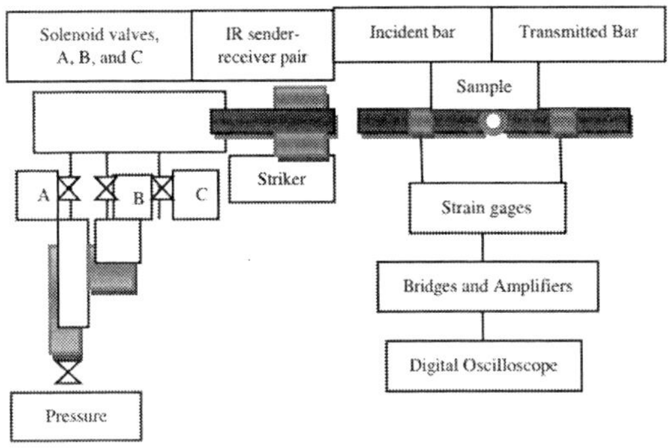

Figure 2. Schematic picture of the Split Hopkinson Pressure Bar device at Tampere University of Technology.

dynamic material behavior and obtaining numerical data are crucially important for the simulation purposes and material modeling, but also for the designing of new and more efficient drills and methods for faster and easier drilling. In this paper, the effects of microstructure, and especially the influence of surface cracks and the crack patterns on the mechanical behavior of the rock material were studied. The mechanical behavior of Balmoral Red granite was studied using Brazilian Disc (BD) tests at a wide range of strain rates. High speed photography and DIC were used to analyze the fracture behavior of the rock samples. The experimental setup, data acquisition, and interpretation of the results are presented in details.

2. Experimental

The material studied in this work was coarse grained Balmoral Red granite. This granite comprises Potash Feldspar, Quartz, Plagioclase, Biotite, and Hornblende. The mean value of the open porosity is 0.38% and the quasi-static compression strength is close to 180 MPa. The rock microstructure has no texture, the minerals are distributed homogeneously, and the mechanical properties are essentially isotropic. The Brazilian Disc samples with a diameter of 40.5 mm were cut from granite plates with a thickness of 21 mm. Figure 1a shows the natural surface of the BD sample, and Fig. 1b the sample with the DIC spackle pattern. The samples were tested at different strain rates, and the effect of surface cracks was studied by applying different heat shocks to the samples. Before the thermal shocks, liquid penetrant and optical microscopy were used to analyze the pattern of the surface cracks. Thermal shocks of 10, 30, and 60 seconds were applied on the BD samples by keeping the samples at a fixed distance from an oxygen-acetylene flame torch. After the heat shock, the samples were allowed to cool down to room temperature in air. After the cooling the liquid penetrant was re-applied on the surface to see the changes in the crack patterns caused by the thermal shock. The images were obtained using a stereomicroscope using UV light. Fractal dimensions of the surface crack patterns were calculated from the obtained images using a box counting method in Matlab.

The low strain tests were performed using an Instron 8800 servohydraulic materials testing machine with a displacement rate of 1 mm/s, and the high strain rate tests with a compression Split Hopkinson Pressure Bar (SHPB). The pressure bars were of Maraging steel with a diameter of 22 mm and a length of 1200 mm. The tests were performed using a 300 mm striker bar with an impact speed of 10 m/s. Each test was repeated five times. Two Photron SA-X2 high speed cameras were used to record the deformation and fracture of the samples during the high strain rate tests. As the rock surface itself does not provide a very strong contrast for the correlation algorithm, the surfaces of the samples were painted with white base coat and the black speckles were applied on it using a permanent ink marker, as shown in Fig. 1b. The images were recorded at 160 kfps, the size of the image at this rate being 256 *176 pixels. Two Decocool lights were used for illuminating the sample. The recorded images were analyzed using the LaVision StrainMaster (DaVis) 3D-DIC software. The displacements were calculated by comparing the deformed images to the first (reference) image using a step size of 9 pixels and a subset size of 25 pixels. The analyzed area was masked from the left reference image using a circular mask to cut out the outermost edges of the sample. The seed points were manually selected from areas that did not move much during the test. Figure 2 shows a schematic picture of the high strain rate testing setup.

The tensile stress in the center of the sample can be calculated when the force acting on the specimen is known. In this work, the forces on the sample surfaces were calculated from the incident and reflected pulses, and from the transmitted pulse. The forces on the incident and transmitted sides of the sample were compared with each other to verify stress equilibrium. However, the transmitted pulse typically contains fewer oscillations, and therefore it was used for calculating the tensile stress in the sample. The mathematical description for calculating the tensile stress from the contact forces is presented for example in ref. [6].

3. Results and discussion

Figure 3 shows an example of the optical analysis used to characterize the crack patterns on the sample surface. Figures 3a and 3c are the original grayscale optical images obtained from the sample before and after a 30 second heat shock, and Figs. 3b and 3d show the black and

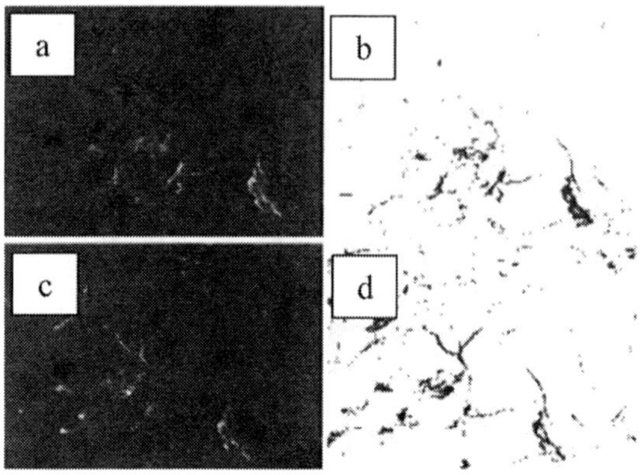

Figure 3. Optical images a) before and c) after the thermal shock, and identified cracks b) before and d) after the thermal shock.

Table 1. Fractal dimensions for the heat shocked samples.

	10 Seconds		**30 Seconds**		**60 Seconds**	
	Before	**After**	**Before**	**After**	**Before**	**After**
	1.3097	1.2583	1.1796	1.3631	1.4464	1.5132
	1.2779	1.349	1.2498	1.2783	0.9976	1.4426
	1.3254	1.3313	1.3819	1.4733	1.3375	1.4001
	1.3192	1.3094	1.3567	1.4296	1.4912	1.5821
	1.3239	1.324	1.4424	1.6094	1.518	1.5241
Average	**1.31122**	**1.3144**	**1.3406**	**1.406**	**1.390**	**1.405**
STD	**0.245**	**0.0345**	**0.106**	**0.124**	**0.213**	**0.0938**
Change		**0.3%**		**8%**		**8%**

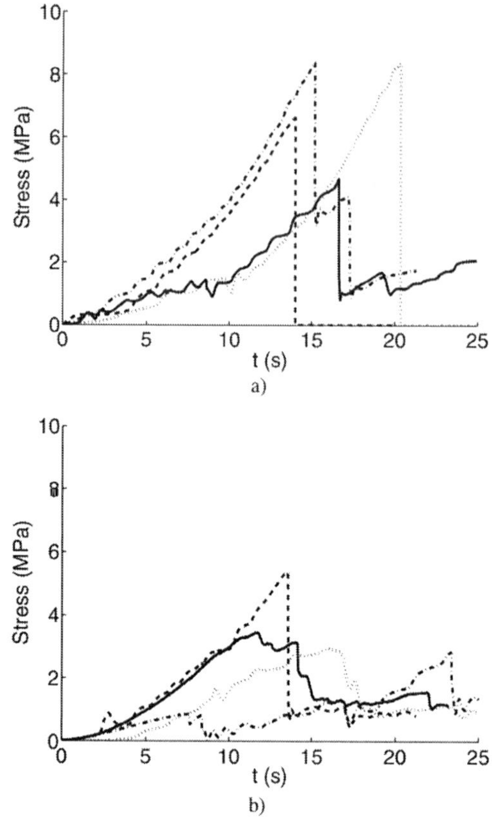

Figure 4. Low strain rate test results showing the tensile stress as a function of time for a) samples without a thermal shock and b) samples with a 60 second thermal shock.

white images after thresholding of the images shown in Figs. 3a and c. The fractal dimensions of the crack patterns were calculated from the black and white images. The fractal dimensions of the crack patterns increase during the thermal shock, indicating that the pattern of the cracks after the thermal shock is more complex and the network contains more interconnected surface cracks.

The fractal dimensions were obtained for a total of 15 samples, i.e., five samples for each heat shock of 10, 30, and 60 seconds. The average of the fractal dimensions prior to the heat shock for all 15 samples was 1.334 ± 0.13. The effect of the heat shock on the crack pattern can be estimated from the change in the fractal dimensions. After the 10 second heat shock, the fractal dimensions are essentially the same as before the heat shock. However, after the 30 and 60 second heat shocks the fractal dimensions increase by 8%. Unfortunately, there was a significant scatter in the fractal dimensions obtained prior to the heat shocks, and therefore the comparison of the effect of different heat shocks on the mechanical properties is somewhat limited. The relative change in the fractal dimensions, however, seems to correlate quite well with the change in the mechanical properties. Table 1 summarizes the fractal dimensions for the heat shocked samples before and after the heat shock.

The results of the low strain rate tests are shown in Fig. 4. The stress as a function of time for the granite specimens without any heat shocks are shown in Fig. 4a, while Fig. 4b shows the corresponding data for the samples with a 60 second heat shock. The tensile strength of the rock without a heat shock varies between 8 MPa and 4 MPa, and the fracture of the sample always occurs as a very clear and rapid drop of the stress to essentially zero. However, after the 60 second heat shock, siginificant scatter appears in the fracture behavior of the rock, as can be seen in Fig. 4b. The heat shocked samples do not fail by only one rapidly propagating fracture, as observed for the rock samples without a heat shock (Fig. 4a), but the failure occurs by several smaller consecutive drops in the tensile stress. The strength of the rock also drops by about 50%, with only one sample showing a significantly higher strenght. This behavior is evidently due to the increase in the number of cracks generated during the heat shocks, which facilitates the fracture and realease of the energy absorbed by the sample during the impact.

During the dynamic loading, the fracture behavior of the studied material is different and the scatter is significantly reduced. Figure 5 shows the tensile stress as a function of time for the samples without a heat shock (5a) and for the samples with a 60 second shock (5b). The average tensile strength obtained from the five dynamic tests on the non-shocked specimens is 28 MPa. The heat shock, however, decreases the dynamic strength significantly, as the average tensile strengths for the heat

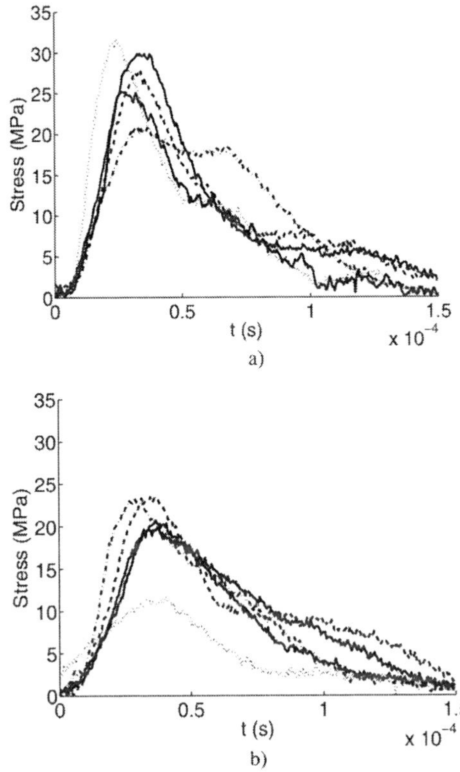

Figure 5. High strain rate test results showing the tensile stress as a function of time for samples a) without a heat shock and b) samples with a 60 second thermal shock.

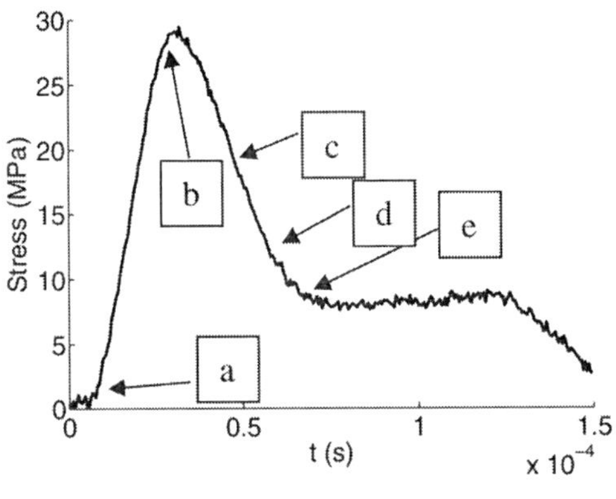

Figure 6. Tensile stress as a function of time for a sample with a 10 second thermal shock.

shocked samples are 23 MPa, 22 MPa, 19 MPa for the heat shocks of 10, 30, and 60 seconds. As described before, the fractal dimensions of the surface cracks increase with the duration of the heat shock. This can be interpreted as an increase in the number of cracks on the surface, and more importantly as an increase of the complexity and interconnectivity of the crack network. Based on the mechanical testing, the increase in the fractal dimensions correlates with the duration of the heat shock and also with the decrease in the strength of the tested samples.

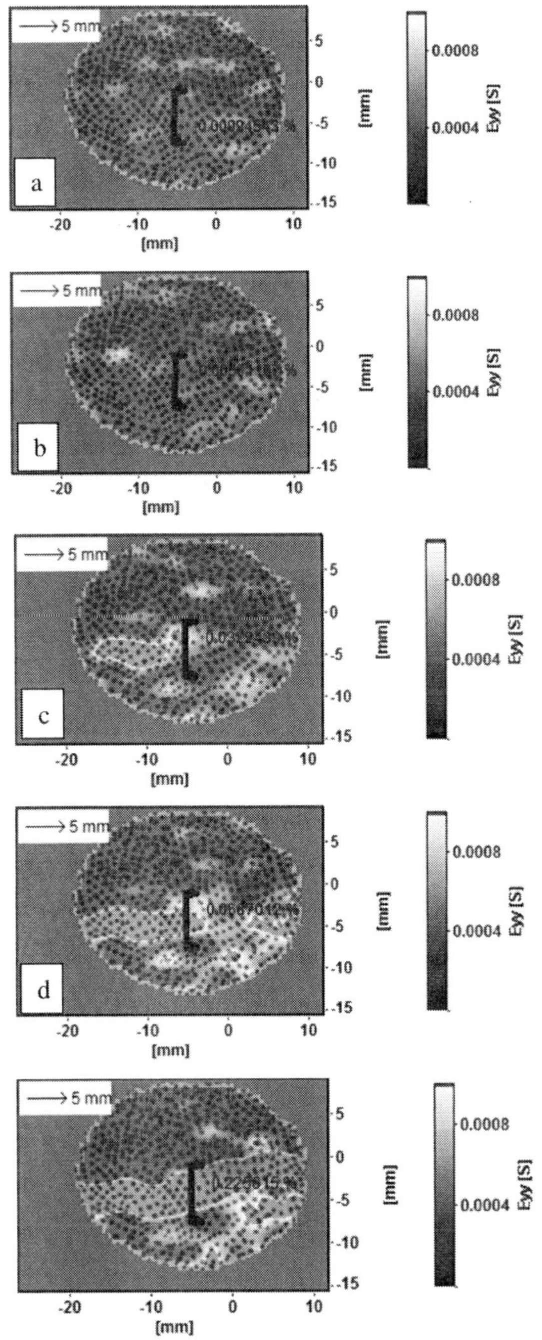

Figure 7. DIC images obtained during the dynamic loading of the sample for which the force-time plot is shown in Fig. 6. a) t = 0 μs, b) t = 31.25 μs, c) t = 50 μs, d) t = 56.25 μs, and e) t = 62.5 μs.

The 60 second heat shock decreases the low strain rate strength of the rock from 7 ± 0.43 MPa for the samples without a heat shock to 3.5 ± 2.5 MPa for the samples tested after the heat shock. The decrease of strength is therefore about 48%. When the loading rate is increased, the same 60 second heat shock decreases the rock strength from 28 ± 3.35 MPa to 19 ± 0.53 MPa, i.e., by 32%. The increased loading rate therefore increases the rock strength by about 400% for the samples tested without a heat shock and by about 550% for the samples tested after a 60 second heat shock. Therefore, the relative increase in strength due to the rate effect seems to be stronger for the samples with a higher content of surface cracks.

The DIC analysis of the BD samples shows that the fracture of the specimens initiates at the center or slightly closer to the incident bar contact point. This observation and comparison of the forces on the incident and transmitted sides of the sample indicate adequate stress equilibrium. The engineering strain in the direction perpendicular to the loading direction prior to fracture was measured by placing a virtual strain gage at the center of the specimen using LaVision StrainMaster (DaVis) 3D-DIC software. The results reveal a decreasing trend of failure strain as the duration of the heat shock increases. For the non-heat treated samples the maximum strain prior to failure is 0.8%, while for the heat shocked samples the corresponding numbers are 0.6%, 0.6%, and 0.5% for the 10, 30, and 60 second heat shocks. The relative decrease in the maximum strain before fracture seems to be quite similar to that observed for the decrease in strength.

Figure 6 shows the tensile stress as a function of time for a sample tested after a 10 second heat shock. Point (a) in the Figure corresponds to the start of the loading and (b) indicates the moment just before the maximum stress. The high speed images corresponding to the time stamps indicated in Fig. 6 are shown in Fig. 7. Also the strain maps in the Y-direction (vertical direction, perpendicular to loading) are overlaid with the original high speed images. It can be seen from the images that the strains in the two first images (7a and b) are very low and essentially constant over the sample surface. The first strain localization does not yet show in the image obtained just microseconds before the strength of the rock starts to decrease. This indicates that the fracture initiates inside the sample and does not appear on the surface at this point. As the fracture propagates, it becomes visible also in the strain maps (Fig. 7c) as a strong concentration of deformation. The crack then propagates through the sample very quickly in Figs. 7d and e. This is followed by crushing of the contact points of the sample with the stress bars.

4. Conclusions

The mechanical behavior of Balmoral Red granite was studied using the Brazilian disc tests at quasi-static and dynamic strain rates. The effect of surface cracks on the mechanical properties was studied by generating Brazilian disc samples with different surface crack patterns by modifying the rock surface structure using thermal shock treatments. The optical images and fractal dimensions obtained from the surface of the samples before and after the heat shock clearly show an increase in the amount of surface cracks and in the complexity of the crack network.

At quasi-static strain rates the fracture of the rock samples without a heat shock occurs as a sharp drop of strength at maximum stress. The samples tested after the heat shock show significant scatter in the strength and the final fracture can occur in several short steps instead of one clear drop. However, at high strain rates the rock shows more consistent behavior and the scatter is significantly reduced. Based on the results, the strain rate effect is stronger for the samples with a 60 second heat shock than for the samples without any heat shock.

Digital image correlation was used to observe the crack initiation and propagation on the surface of the samples during the high rate tests. Time stamps from the test were recorded on the images, and the data from the HSPB was synchronized with the obtained images. The cracks initiate typically inside the sample and propagate to the surface only after the strength of the sample has already decreased significantly. The DIC can also be used to estimate the fracture strain of the material during a high strain rate test. This can be done, for example, with a virtual extensometer placed on the image. The obtained maximum strains prior to the fracture also decrease with the heat shock duration.

This work was supported by the Finnish Research Foundation for Natural Resources under the grant no. 1768/14.

References

[1] W.H. Peters, W.F. Ranson, Opt. Eng. **21**, 427 (1982)

[2] J.E. Field, S.M. Walley, W.G. Proud, H.T. 2. Goldrein, C.R. Siviour, Impact. Eng. **30**, 725 (2004)

[3] F. Hild, S. Roux, Exp. Mech. **42**, 69 (2006)

[4] V. Tiwari, M.A. Sutton, S.R. McNeill, Exp. Mech. **47**, 561 (2007)

[5] P.L. Reu, T.J. Miller, Strain Anal. Eng. Des. **43**, 673 (2008)

[6] T. Saksala, M. Hokka, V.T. Kuokkala, J. Mäkinen, Rock Mech. Mining Sci. **59**, 128 (2013)

[7] T. Saksala, Numer. Anal. Met. **37**, 309 (2012)

[8] C.A. Tang, H. Liu, P.K.K. Lee, Y. Tsui, L.G. 8. 8. Tham, Rock Mech. Mining Sci. **37**, 555 (2000)

[9] C.A. Tang, L.G. Tham, P.K.K. Lee, Y. Tsui, H. Liu, Rock Mech. Mining Sci. **37**, 571 (2000)

EPJ Web of Conferences **94**, 02008 (2015)
DOI: 10.1051/epjconf/20159402008

High-power laser shock-induced dynamic fracture of aluminum and microscopic observation of samples

Zhang Fan[1], Huang Xiuguang[1,a], Shu Hua[1], Xiao Dawu[2], He Lifeng[2], Xie Zhiyong[1], Ye Junjian[1], Dong Jiaqin[1], Jia guo[1], Fang Zhiheng[1], and Zhou Huazhen[1]

[1]Shanghai Institute of Laser Plasma(SILP), ShangHai 201800, China
[2] Materials Research Institute, China Academy of Engineering Physics, MianYang 621900, China

Abstract. High-power laser induced shocks generated by "ShenGuang II" laser facility has been used to study spall fracture of polycrystalline aluminum at strain rates more than 106/s. The free surface velocity histories of shock-loaded samples, 150 μm thick and with initial temperature from 293 K to 873 K, have been recorded using velocity interferometer system for any reflector (VISAR). From the free surface velocity profile, spall strength and yield stress are calculated, it demonstrates that spall strength will decline and yield strength increase with initial temperature. The loaded samples are recovered to obtain samples' section and free surface metallographic pictures through Laser Scanning Confocal Microscopy. It is found that there are more micro-voids and more opportunity to appear bigger voids near the spall plane and the grain size increases with temperature slowly but smoothly except the sharply change at 893 K (near melting point). Besides, the fracture mechanisms change from mainly intergranular fracture to transgranular fracture with the increase of initial temperature.

1. Introduction

The research of materials dynamic damage characteristics has great significance in many fields such as high speed collision of automobile, aerospace, new material synthesis, satellite protection, weapons design etc. Spall is a typical dynamic fracture that takes place inside a solid body as a result of tensile stresses that develop due to the interaction of rarefaction waves. Brittle fracture occurs through the generation and propagation of micro cracks while ductile fracture proceeds through the nucleation, growth, and coalescence of micro voids [1]. In 1914, Hopkinson firstly observed the spall phenomenon [2] that there were planar separation of material parallel to the wave front when materials under dynamic loads. After the Second World War, driven by engineering and military applications, spall research gets widespread concern and has a great progress [3–5]. Since the 1970s, the concepts of damage accumulation and evolution have been proposed as being responsible for the dynamic fracture of ductile metals. Many scientists have endeavored to develop a theoretical equation for damage evolution on the basis of statistical analysis of damage of recovered samples, and more detailed information based on experimental observation is needed in order for one to better understand or modify these theoretical results. Different methods of dynamic loading have been tried to find materials' damage characteristics, mainly of these studies use SHPB(split Hopkinson pressure bar) with $10^4\,s^{-1}$ strain rates, and gas gun and explosives with strain rates up to $10^4\,s^{-1}$–$10^6\,s^{-1}$. Recent years, high-power lasers also have been used to generate spall in materials at much higher strain rate, sometimes up to $108\,s^{-1}$, the laser is focused onto and absorbed at the front surface of the target, heating, and ablating the surface. The backwards-expelled material acts as a rocket engine pushing the front surface with a pressure strong enough to launch a shock wave into the material. This paper attempts to study aluminum dynamic damage characteristics through spall experiments at "ShenGuang II" laser facility. Meanwhile, microscopic observations of recovered samples have been done to get damage information and verify the results of macroscopic measurements such as spall strength and yield strength.

2. Experimental technique

The present study was performed using the "Shenguang-II" Nd: Glass laser (converted to $\lambda = 0.351\,\mu$m) of the National Laboratory on High Power Laser and Physics. The"Shenguang-II" laser facility provides one-dimensional compression by smoothed laser beams with short wavelength and high intensity. The temporal profile of the laser is nearly square with a rise and fall time of ~300 ps and a full width at half maximum (FWHM) of ~3 ns. Lens-array (LA) [6–8] was used to eliminate the large scale spatial modulation and to obtain a flat-topped profile in the focal plane. Characteristics of the optical system (Lens+LA) were such that the focal spot had a ~1 mm diameter flat region. The absorbed laser intensity in the focal spot was $\sim 10^{11}$ W/cm^2.

Foils of polycrystalline aluminum of 99.99% purity with 150 μm thickness and different initial temperature were used in the experiment. Samples were measured to be fully dense (2.71 g/cm^3) to within an accuracy of 1%. The longitudinal sound velocity $c_l = 6450 \pm 30$ m/s and the bulk sound velocity $c_b = 5360 \pm 30$ m/s were measured using a dynamic experiment. The metallography examination of the foils shows that the foils contain

[a] Corresponding author: innocentman001@163.com

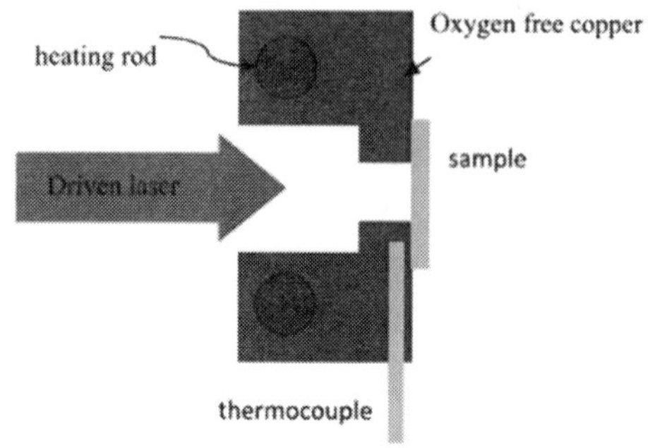

Figure 1. Schematic of the target.

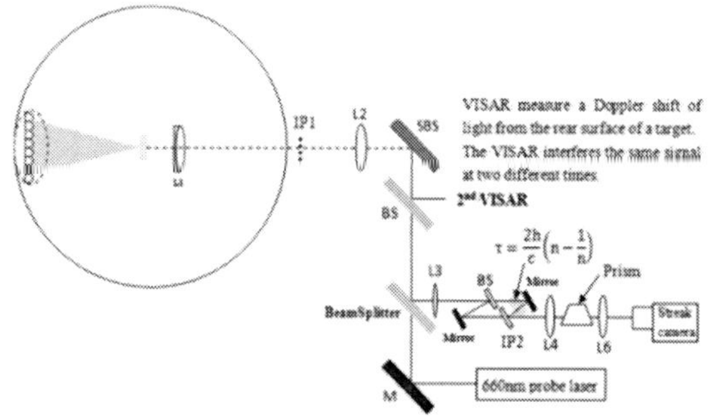

Figure 2. Schematic of the VISAR. L1-L6:Image Lens; IP1-IP2:Image Position; LA: Lens Array; SBS: Special Beam Splitter.

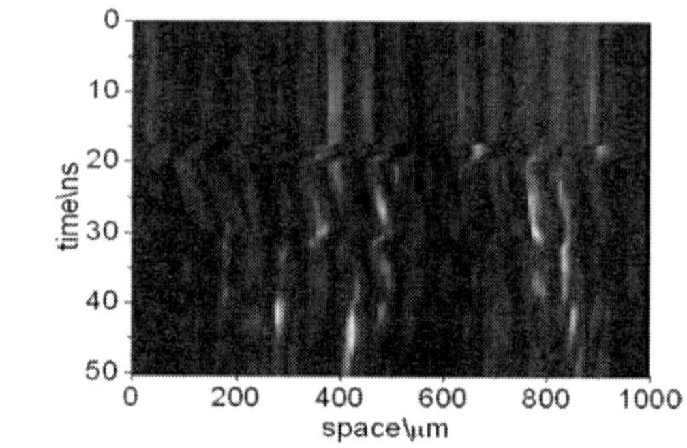

Figure 3. VISAR image of 873 K Aluminum sample.

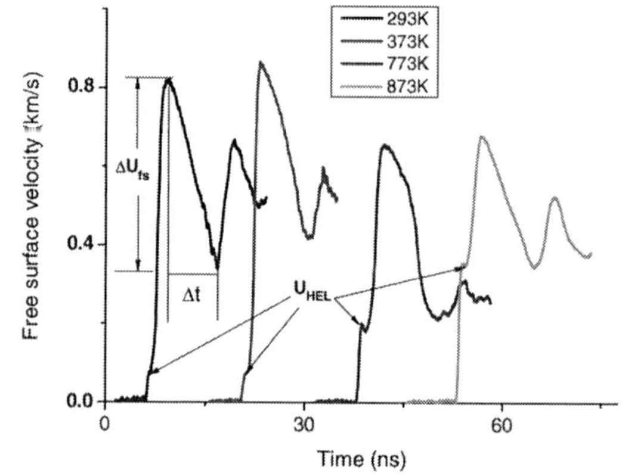

Figure 4. Free surface velocity profile of different initial temperature samples.

globular grains, whose average diameter is approximately 20 μm. The shock-loaded samples were 10 mm × 10 mm squares of different thicknesses. Each sample was glued to an oxygen-free copper target holder. The target holder had two heating rods. The heater power was sufficient to reach the melting temperature of aluminum after 15 min of heating. The temperature was controlled by a thermocouple placed near the sample. The experimental setup is shown in Fig. 1. The heating system could heat up the sample to 973 K with temperature controlling accuracy of 2 K.

The experiment diagnostic system is a line VISAR [9, 10]. The schematic optical setup of the system is shown in Fig. 2. The total magnification of the image system is 11×. The minimum scale of the resolution pattern is 7 μm and the time resolution of the VISAR system was ~100 ps in this experiment.

After laser loading, recover the samples from target frame. Then obtain the metallographic samples through sampling, mosaic, grinding, polishing and etching. Finally make metallographic analysis using fracture surface and section pictures got by LSCM(laser scanning confocal microscope).

3. Experiment results and discussion

3.1. Measurements of spall strength and yield stress

Typical VISAR image of the experiment is shown in Fig. 3. The horizontal direction of the image is the space. The vertical direction is the time. The temporal scale of the image is 50 ns.

According to the measurement principle of VISAR, we can get the free surface velocity profile of different temperature initial spall samples shown in Fig. 4.

From the free surface velocity profile, we can get information like HEL (Hugoniot Elastic Limit) velocity, pullback velocity and so on listed in the Table 1.

Table 1. Parameters of free surface velocity profile.

Initial Temperature (K)	Strain Rate (10^6/s)	Spall strength σ_{sp} (GPa)	HEL strength σ_{hel} (GPa)	Yield Strength Y(GPa)
293	480	78	7.59	4.85
373	450	73	7.25	4.85
773	450	200	8.25	4.16
873	340	360	8.12	3.16

Using the acoustic approximation formula (1), (2), (3), spall strength and yield stress of different initial temperature aluminum samples can be calculated through the HEL velocity and pullback velocity. The results are shown in Fig. 5.

$$\sigma_{HEL} = \frac{1}{2}\rho_0 c_l u_{HEL} \quad (1)$$

$$Y = \frac{3}{2}\sigma_{HEL}\left(1 - \frac{c_b^2}{c_l^2}\right) = \frac{3}{4}\rho_0\left(c_l - \frac{c_b^2}{c_l}\right)\mu_{HEL} \quad (2)$$

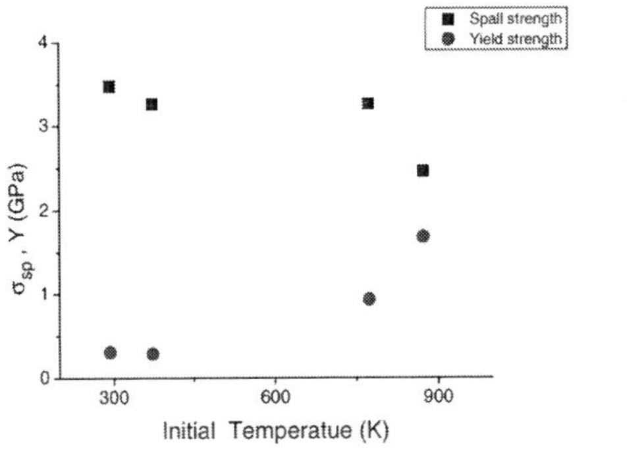

Figure 5. Yield stress and spall strength as a function of the initial temperature.

$$\sigma_{sp} = \frac{1}{2}\rho_0 c_b u_{fs}. \tag{3}$$

We can find that the amplitude of elastic precursor increased with the increasing of the temperature and the spall strength decreased with temperature.

At low strain rates, overcoming obstacles by dislocations is due to the joint action of applied stress and thermal fluctuations. The increase of the fluctuations with temperature facilitates the motion of dislocations and leads to the decrease of the yield stress. At sufficiently high strain rates, phonon friction is the dominant mechanism that controls dislocation motion. The phonon-related dislocation drag is proportional to the temperature, so there is an increase of the yield stress with temperature. pure cobalt, silver, titanium, and some other crystalline solids also have similar characteristics [11–13].

On the other hand, Due to the theoretical models, the incipient stage of a high-rate spall fracture is the nucleation and growth of numerous voids or cracks. The resistance to growth of voids is dominated by the flow stress of the material around the void. So, the decrease of the yield strength with heating should lead to a decrease of spall strength. Since the spall strength data of Fig. 5 show a reverse trend. we can conclude that the spall strength of preheated aluminum is controlled more by the rate of void nucleation rather than its growth [14].

3.2. Metallographic observation and analysis

The recovered samples are first used to take pictures of target surface like Fig. 6. It is obvious that the front surface has a characteristic of melting and solidification reveal that the target has been ablated by high-power laser. The back surface (or free surface) has convex character, and incomplete spall samples have relatively complete morphology while complete spall samples have pits cause a part of materials parallel to surface have flown out.

Cut the samples along the impact direction symmetrically, take one piece and make cold mosaic with epoxy resin. Then, set the sample of epoxy resin column in the polishing machine fixture, using different particle size of sand paper step by step from coarse to fine polishing. When the surface polishing scratches is very small, polish with diamond polishing liquid; Finally, reveal

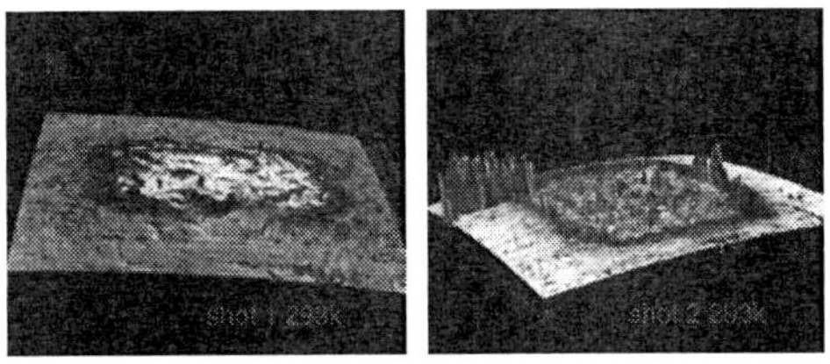

Figure 6. The back surface of incomplete and complete spall samples.

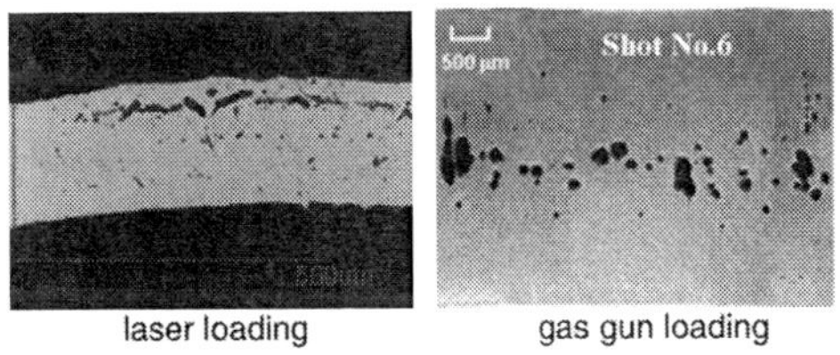

Figure 7. Recovered samples of laser loading and gas gun loading.

metallographic structure by the erosion corrosion liquid, and further metallographic analysis can be done.

3.2.1. Influence of strain rate on material strength

Figure 7 is the metallographic section pictures of laser loading and gas gun driven flyer impact samples [15] respectively.

It is obvious that the voids are smaller and more irregular in samples of high-power laser loading than the gas gun loading, demonstrate that the growth time of voids of laser loading is much shorter which is closely related to strain rate, the strain rate of high-power laser loading is higher than that of gas gun loading. Considering the dynamic fracture of material damage cumulative time effect, to achieve the same material damage threshold, the laser loading spall need greater tensile stress than gas gun loading because of its shorter duration. Therefore under the same initial temperature, materials loading by high-power laser should have higher yield stress and spall strength, and this have been proved by lots of researches. Figure 8 is the comparison of this work and E. B. Zaretsky's work [16] which agree our prediction well.

3.2.2. Effect of temperature on grain size

The grain size of the sample is an important factor that affect the strength of materials properties, so we observed the grain size of the sample with the temperature changes, as shown in Fig. 9. It can be found that with the rise of temperature, grain size increases slowly and smoothly, but dramatically when close to the melting point. P. B. Trivedi et al. [17] found that with the increase of grain size, the proportion of grain boundary in materials decreases, and the nucleation point in high purity metal is mainly on the grain boundary, therefore with reduced nucleation point number, spall strength increases. In our work, with the increase of grain size, the spall strength decreases. Shows that other factors effect such as flow stress caused by temperature exceeds the grain size effect on the strength

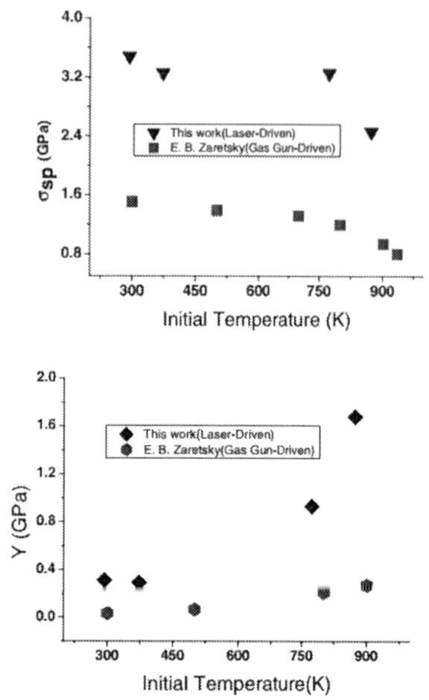

Figure 8. Spall strength and yield stress comparison of different loading.

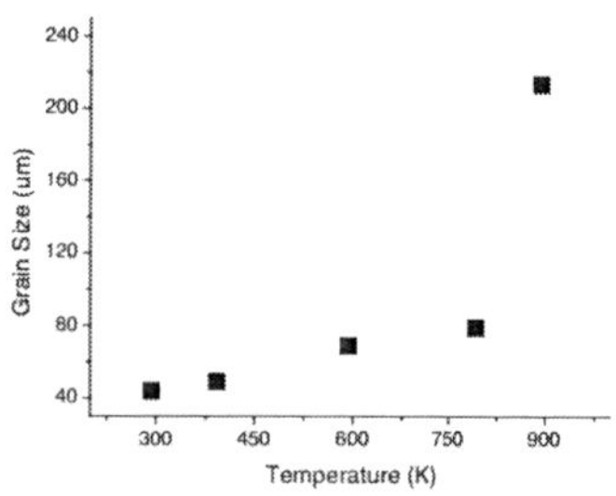

Figure 9. Grain size as a function of initial temperature.

of materials, shows that material damage characteristic is a combination of many kinds of factors, it also provides a useful reference information for damage evolution model.

3.2.3. *Effect of temperature on the material fracture mode*

Strain rate effects on material failure has been studied a lot [18, 19]. With the increase of strain rate, the dynamic fracture mode will change from the intergranular fracture primarily to transgranular fracture. Dynamic fracture is a time related to evolution process, time effect or rate effect mainly influence the grow process of the microvoids; However when the strain rate is in the same order of magnitude, the influence of thermal effects on the material fracture mode needs more consideration. Figure 10 is section picture of samples with the initial temperature of 373 K, 573 K and 873 K respectively. It can be seen that the material fracture mode changes from mainly the intergranular fracture (shot 3 and shot 4) to mainly the transgranular fracture (shot 7). With the increase of temperature, "fluid" properties of the material appear more, and the difference of damage threshold of grain boundary and inside grain will be smaller. On

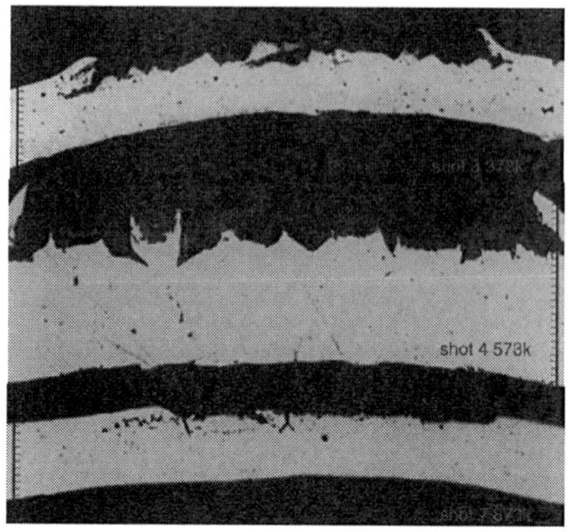

Figure 10. Effect of initial temperature on the material fracture mode

the other hand, thermal fluctuation becomes the main factor of nucleation in high purity metal, making the void distribution more dense and the distance of the adjacent voids shorter, so the coalescence of void in different position like inside grain and on grain boundary will be much easier as temperature increasing. Therefore, the material will show the transgranular fracture mode more. It is important to note that the original sample thickness is about 150 μm, and in 873 K material's grain size is about 200 μm, thus the change of fracture mode may be a scale effect. whether thicker samples have the same nature need follow-up experiments to verify.

4. Conclusion

Spall experiments are conducted on "ShenGuang II" laser facility to study the effect of initial temperature on aluminum at a approximately same strain rate with same thickness samples. It can be concluded that with the increase of initial temperature, spall strength decrease and the yield stress increases; Material grain size increases with temperature slowly, but wheb near the melting point it increases dramatically; Materials fracture mode changes from mainly the intergranular fracture to mainly the transgranular fracture with the increase of initial temperature.

References

[1] T. Antoun, L. Seaman, D. Curran, G. Kanel, S. Razorenov, A. Utkin, *Spall Fracture*, p 3–4 (2002)
[2] HoPkinson, B. Philosophical Transactions of the Royal Society of London, **213(A)**, 437 (1914)
[3] Lee Davison, A. L. Stevens, J. Appl. Phys. **43(3)**, 988 (1972)
[4] Lynn Seaman, Donald R. Curran, Donald A. Shockey, J. Appl. Phys. **47(11)**, 4814 (1976)
[5] Grady, D.E. J. Mech. Phys. Solids **36(3)**, 353 (1988)
[6] Deng Ximing, Liang Xiangchun, Chen Zunze, et al., Chinese Journal of Lasers **12(5)**, 257 (1984)
[7] Fu Sizu, Sun Yuqin, Huang Xiuguang, et al., Chinese Journal of Lasers **30(2)**, 129 (2003)

[8] Huang Xiuguang, Fu Sizu, Wu Jiang, et al., High Power Laser and Particle Beams **18(5)**, 811 (2006)
[9] Shu Hua, Fu Sizu, Huang Xiuguang, et al., Chinese Journal of Lasers **37(1)**, 176 (2010)
[10] Shu Hua, Fu Sizu, Huang Xiuguang, et al., Acta physica sinica **61(11)**, 114102 (2012)
[11] G.V. Garkushin, G.I. Kanel, S.V. Razorenov, Phys. Solid State. **52(11)**, 2369 (2010)
[12] E.B. Zaretsky, J. Appl. Phys. **108**, 083525(2010)
[13] E.B. Zaretsky, G.I. Kanel, J. Appl. Phys. **110**, 073502 (2011)
[14] Hua Shu, Sizu Fu, Xiuguang Huang et al., J. Appl. Phys. **116**, 033506 (2014)
[15] Meilan Qi, Chao Luo, Hongliang He, et al., J. Appl. Phys. **111**, 043506 (2012)
[16] E.B. Zaretsky, G.I. Kanel, J. Appl. Phys. **112**, 073504 (2012)
[17] P.B. Trivedi, J.R. Asay, Y.M. Gupta, et al., J. Appl. Phys. **102**, 083513 (2007)
[18] E. Moshe, S. Eliezer, E. Dekel, A. Ludmirsky, et al., J. Appl. Phys. **83(8)**, 4004 (1998)
[19] S.V. Razorenov, G.I. Kanel, K. Baumung, et al., *AIP Conference Proceedings*, **620**, 503 (2002)

EPJ Web of Conferences **94**, 02009 (2015)
DOI: 10.1051/epjconf/20159402009

Investigation on mechanical behaviour of ECAPed 2A12 aluminium alloy

Cun-xian Wang[1], Chao Zhang[1], Lei Wang[1], Zhong-bin Tang[1,2], and Tao Suo[1,2,a]

[1] School of Aeronautics, Northwestern Polytechnical University, 710072 Xi'an, China
[2] Fundamental Science on Aircraft Structural Mechanics and Strength Laboratory, Northwestern Polytechnical University, 710072 Xi'an, China

Abstract. In the present work, the uniaxial compressive behavior of aluminum alloy processed by equal channel angular pressing (ECAP) for 1–8 passes are investigated experimentally under both quasi-static and dynamic loading conditions via an electronic universal testing machine with a maximum load capacity of 10KN and the split Hopkinson pressure bar (SHPB). The strain hardening rate as well as strain rate sensitivity the ECAPed with different passes have been determined and compared with annealed coarse grained counterpart. The experimental results show a continuously increase of both flow stress and strain rate sensitivity for the aluminum alloy subjected to ECAP process as the pressing pass number increasing. It is proposed that the reduction in grain size plays an important role in the enhancement of flow stress and strain rate sensitivity. However, the strain hardening rate of the ECAPed materials decreases remarkably. Meanwhile, compressive experiments at elevated temperatures indicate the temperature sensitivity of the material increases as the grain size is refined into fine grain regime. Based on thermal activation theory, it is proposed that the enhanced temperature and strain rate sensitivity of ECAPed aluminum alloy can be related to the reduction in activation volume due to grain refinement.

1. Introduction

In the past few years, ultra fine-grained/nanocrystalline materials show many unusual physical and mechanical properties due to grain refinement and the larger proportion of the grain boundary, which attracted the attentions of many domestic and foreign scholars. A lot of research results indicated that grain refinement can lead to the increase of both strength and the change of deformation mechanism [1–4]. For example, it is observed that the tensile strength of ECAPed 7075 aluminium alloy can reach as high as 677 MPA while the ECAPed 2024 aluminium alloy can also reach 618 MPa; (Zheng et al. [5]). There are also many literatures pay attention on the strain rate sensitivity of the ultra fine-grained/nanocrystalline materials. Experimental studies show that the strain rate sensitivity of the face-centred cubic crystal metals decreases with the grain size increases, while the body-centered cubic on the contrary, with the reduction in grain size [6].

Strain hardening is also an important mechanical property of materials, which reflects the relations between the flow stress and the flow strain. From the microscopic mechanism, the material microstructure hinder the movement of the movable dislocations, then the strain hardening is generated. The temperature, strain rate will affect the strain hardening properties of aluminium alloys. For ultrafine-grain aluminium fabricated by ECAP, the biggest factor of the strain hardening properties is the high dislocation density [7]. But the investigations of the effect on strain hardening by temperature at the high strain rate are still relatively rare.

[a] Corresponding author: suotao@nwpu.edu.cn

Previous studies are mainly concentrated on the mechanical properties of aluminum alloys in the quasi-static condition, few works pay attention to the dynamic mechanical behavior of aluminum alloy. The limited works indicates that the microstructure changes will not only cause a significant improvement of the mechanical properties in the quasi-static condition, dynamic mechanical properties are also observed to change accordingly [8].

Although the research on the mechanical behaviour of fine-grained aluminium alloy has been made in the past few years, there still remain many issues especially the coupling effects of the strain rate and temperature on the mechanical behaviour of the fine grained materials. The primary aim of this work is to investigate the coupling effects of the strain rate and temperature on compressive behaviour of ECAPed-2A12 aluminium alloy. The strain rate and temperature sensitivities as well as the temperature sensitivity of the material are analyzed based on the experimental results.

2. Material preparation and experimental procedure

In the present work, the commercial 2A12 aluminium alloy was selected as the experimental materials. The raw rods with the dimension of 9.8 mm × 9.8 mm × 60 mm were annealed for two hours at the temperature of 573 K. After that, the ECAP method was employed using a hydraulic machine at a speed of ∼15 mm/min at 473 K to produce fine grained material. The ECAP die used in this work has the geometry of an inner corner angle of $\Phi = 90°$ and an outer angle of $\psi = 0°$ (see in Fig. 1). To achieve ultrafine

Article available at http://www.epj-conferences.org or http://dx.doi.org/10.1051/epjconf/20159402009

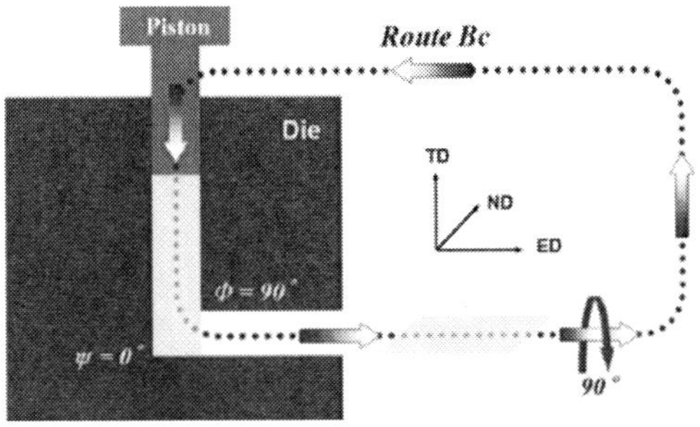

Figure 1. Schematic sketch of ECAP.

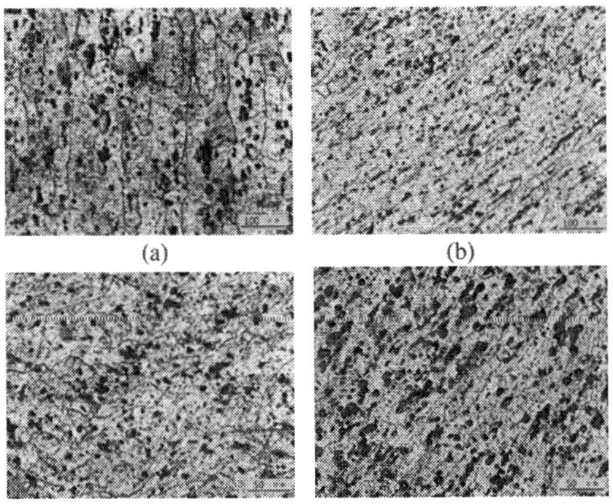

Figure 2. Micrographs of ECAPed-2A12 with different passes (a) 0 pass, (b) 1 pass, (c) 4 passes, (d) 8 passes.

grained microstructure, each specimen was pressed 8 passes using route Bc.

Quasi-static compressive experiments were performed using an electronic universal testing machine with a maximum load capacity of 10KN. The split Hopkinson pressure bar (SHPB) was employed for high strain rate experiments. According to the one dimensional elastic stress wave theory, the dynamic strain, stress and strain rate of tested specimen can be calculated respectively [9]. The specimens for both quasi-static and dynamic compressive tests were taken from the center part of the ECAPed rods and machined to cylinders with dimensions of $\Phi 4 \times 4$mm. The loading direction was parallel to the ECAPed rod axis. During the tests, the strain rates were controlled from 10^{-3} to 10^3/s.

3. Results and discussion

3.1. Microstructure of ECAPed-2A12

Figure 2 gives the metallographic micrographs of 1 pass, 4 pass, 8 pass ECAP aluminium alloy as well as the annealed alloy. It can be seen from the pictures that the grain size of annealed 2A12 aluminium alloy is about 100 μm. However, most of grains are elongated after the first ECAP process and the large grains transform gradually to sub-grains accompanied by small angle grain boundaries (see in Fig. 2(b)). After the further passes (see in Fig. 2(c)), those elongated grains were broken into fine grains with the average grain size of 3 μm. Comparison of the microstructure of 4 pass ECAPed-2A12 with that of 8 pass alloy indicates that the additional 4 passes give almost no further grain refinement to the aluminium alloy. However, more homogenous distribution of equaxial grains is believed to be achieved by the further plastic deformation [8,10]. From the figures, many Al_2CuMg compounds with the length 0.8 μm and width 0.4 μm can also be found in α-Al the substrate. The further plastic deformation does not change the size of these Al_2CuMg compounds but the distribution of them seems more uniform in the 8pass ECAPed-2A12 if compared with that in annealed alloy.

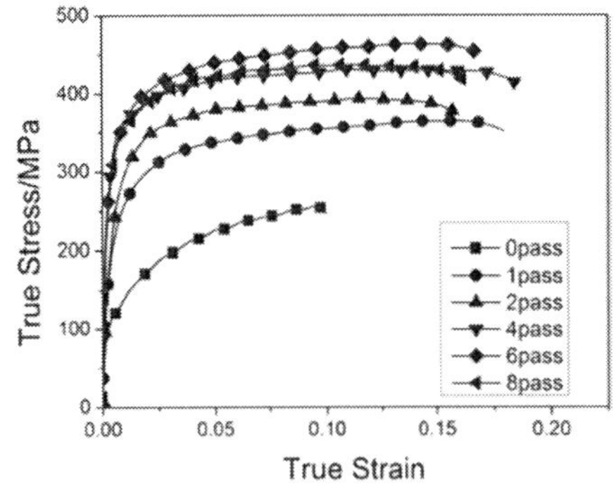

Figure 3. The true stress versus true strain curves of ECAP-2A12 aluminum alloy at the strain rate of 0.001/s

3.2. Strength enhancement after ECAP process

Figure 3 shows the true stress versus true strain curves of ECAPed-2A12 aluminium alloys with different passes under quasi-static loading. It can be seen that the yield stress and flow stress of the material increases with pass number until reach the maximum values after 6 passes. If comparing the flow stress of 8 pass ECAPed-2A12 alloy with that of 6 pass, there is no significant increment. This is in good agreement with the microstructural observation that the grain size of ECAPed-2A12 reaches the minimum value after 4 passes. The fact that the ECAP-2A12 aluminium alloy displays higher strength as the grain size is reduced into ultrafine grain regime can be explained by the remarkable increment of dislocation density after sever plastic deformation as well as the classical Hall-Petch relationship from which the flow stress of metals is inversely proportional to square root of grain size.

3.3. Strain rate effects

Figure 4 shows true stress-strain curves of ECAPed-2A12 aluminium alloy deformed at 293 K under both quasi-static and dynamic loading conditions. For comparison, the true stress-strain curves of the annealed coarse grained aluminum alloy at different strain rates are also shown in Fig. 4(a). It can be easily found that the ECAPed-2A12 aluminium alloy show almost elastic-ideal plastic behaviour at quasi-static strain rate while for the annealed coarse-grained aluminium alloy remarkable

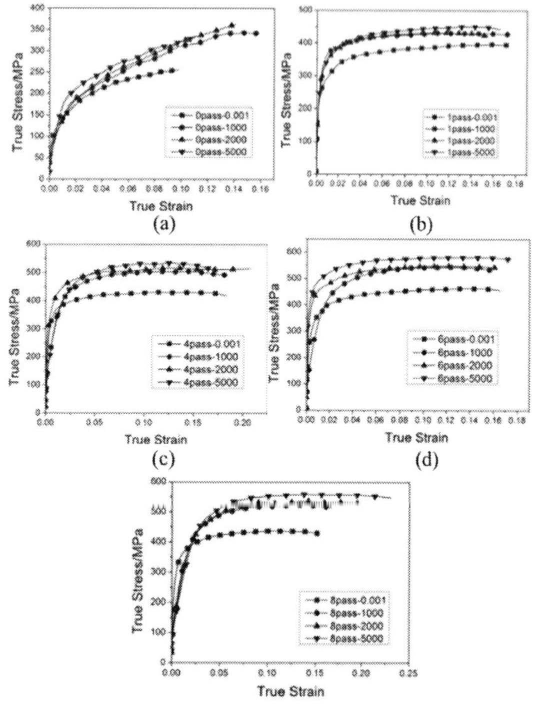

Figure 4. True stress-strain curves of ECAPed-2A12 aluminum alloy at different strain rates (a) 0 pass, (b) 1 pass, (c) 4 passes, (d) 6 passes, (e) 8 passes.

strain-hardening can be observed. Wang et al. [7] also carried out systemic study on the strain hardening behaviour of the material after grain refinement and the results showed that the strain hardening behaviour of the material decreased rapidly as the grains is refined. This is in good consistent with the results of the present work. The reason for the poor strain hardening capability is believed to be that the internal dislocation density of the material is very high after severe plastic deformation which leads to higher dislocation dynamic recovery rate than coarse-grained materials. During the initial stage of plastic deformation, a large number of dislocations generate from dislocation sources which subsequently interact with grain or sub-grain boundaries, causing high strain hardening rate. However, the enhanced dynamic recovery process makes the balance between generation and annihilation of dislocations quickly, leading to a steady-state dislocation density and hence the material undergoes a long deformation stage with slight variation in strain hardening rate [3,7,11]. It should be pointed out that with increasing strain rate, the ECAPed-2A12 aluminium alloys with different passes show strain hardening behaviour more or less, and the strain hardening rate increases with strain rate.

As can be seen from the true stress-strain curves of ECAP-2A12 aluminium alloy, the change of strain rate affects the strength of the material. Some studies have shown that the grain refinement will lead to the changes of strain rate sensitivity (SRS) [6,7,12]. In general, the strain

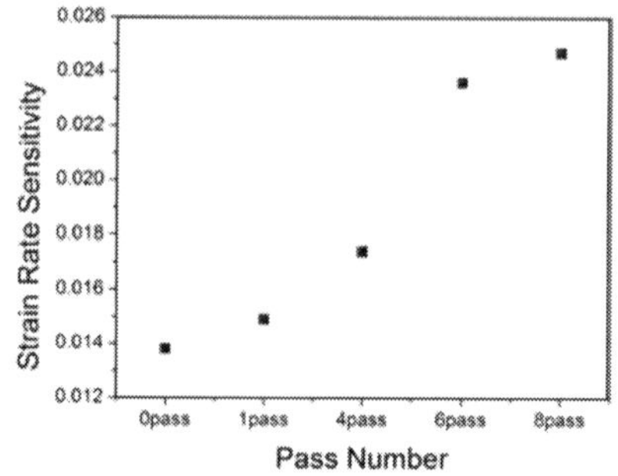

Figure 5. Strain rate sensitivity of ECAPed-2A12 with different passes.

rate sensitivity generally increases with grain refinement on the face-centred cubic (FCC) metals while strain rate sensitivity decreases with grain refinement on the body-centred cubic (BCC) metals [6]. It is believed that higher strain rate is not conducive to the slip spread continuity between the grains and may cause large stress concentration in the vicinity of grain boundaries easily which will increase the deformation resistance [13]. The dependence of flow stress on the applied strain rate is usually described quantitatively by a coefficient named strain rate sensitivity (SRS) factor m, which can be expressed as

$$m = \left.\frac{\partial \ln \sigma}{\partial \ln \dot{\varepsilon}}\right|_{\varepsilon} \tag{1}$$

where σ and ε are flow stress and strain respectively. According to the formula, the values of SRS can be determined approximately by using log-log flow stress versus strain rate plots at a certain strain from which m was derived as the slope of a linear regression fit [6,14]. It should be point out that the value of m is a function of applied strain according to the definition. In this work, the flow stresses corresponding to the fixed true strain of 0.07 are employed to calculate the strain rate sensitivity factor.

Figure 5 shows the SRS factor for ECAPed-2A12 aluminium alloy after different passes. It can be seen from the figure that the strain rate sensitivity of ECAPed-2A12 aluminium alloy increases remarkably with the pass number until to 6 passes. After that, slight increment of SRS is found as the pass number increases from 6 to 8. As mentioned above, although the latter passes may not change the subgrain size and perhaps the overall dislocation density, the misorientation among the subgrains is expected to be changed and the dislocations may be rearranged. Consequently, the dislocation density in the vicinities of grain boundaries can be rather high, which may lead to a larger contribution to the thermally activated deformation processes (higher SRS value) [8,10].

Since plastic deformation of metals is thermal activated process, for studying the physical mechanism of plastic deformation of crystals, an effective method is thermal-activation analysis. Employing the concept of thermal-activation theory, the strain rate sensitivity factor can also be given by

$$m = \frac{\sqrt{3}kT}{\sigma V^*} \tag{2}$$

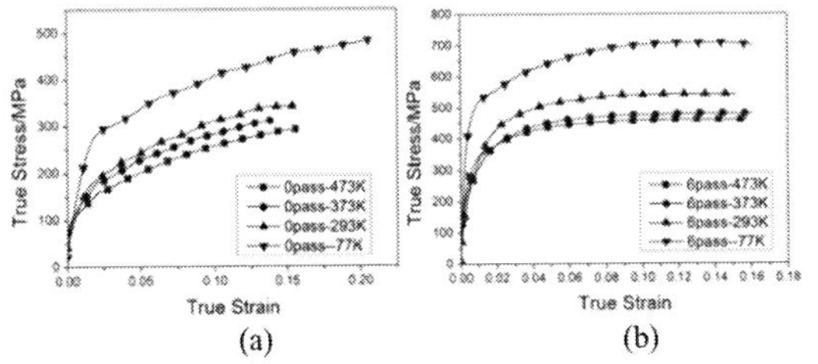

Figure 6. Stress-strain curves of annealed and 6 pass ECAPed-2A12 alloy at strain rate of 1000/s and at different temperatures.

where $\dot{\varepsilon}$ is applied strain rate, T is the absolute temperature and V^* is thermal activation volume. It can be seen from the equation that the strain rate sensitivity can be affected by the combined effects of stress and activated volume at fixed temperature. We all know that the fully annealed metals with FCC structure usually have a large activation volume, i.e. $\sim 10^2$–10^3 b3, and the dominant rate-controlling process is the intersection of moving dislocations cutting through forest dislocations [13]. While for fine grained aluminium, the activation volume is believed to be at least one order lower than that of its coarse grained counterpart although the thermally activated forest dislocation intersections tend to occur in the vicinity of grain boundaries or/at the grain boundaries where the density of dislocations is supposed to be higher [8]. Therefore, it's easy to understand that the strain rate sensitivity of ECAPed-2a12 alloy can be enhanced since the enhancement in flow stress is limited if compared with the annealed aluminium alloy.

3.4. Temperature effects

Figure 6 shows the true stress-strain curves of annealed coarse-grained aluminium alloy and 6 pass ECAPed-2A12 aluminium alloy at strain rate of 1000/s and at different temperatures. It can be seen that with the decreasing of test temperature, flow stresses of the two materials significantly increase. For annealed coarse-grained aluminium alloy, strain hardening is more obvious. However, the flow stress of 6pass ECAPed-2A12 aluminium alloy increases rapidly only at the beginning of plastic deformation. With the rising temperature, the strain hardening rate of annealed alloy keeps almost unchanged. While for the ECAPed alloy, it shows remarkably strain hardening behaviour at the cryogenic temperature (77 K). It is believed that the low temperature can suppress the dynamic recovery of dislocations which cause obvious strain hardening behaviour [15]. The fact that materials can regain the ability of strain hardening at cryogenic temperature has been proved by Wang [7] as a strategy to achieve larger uniform tensile deformation.

Figure 7 shows the flow stress variation with temperature of the ECAPed-2A12 aluminium alloy and the annealed coarse-grained aluminum alloy at the strain rate of 10^3/s at the fixed strain of 7%. It can be seen from the Fig. 7 that the flow stress of the material tends to increase with the decreasing temperature. Meanwhile, we can clearly see that as the temperature rises the flow stress of the material with finer grains changes more significantly, indicating the flow stress is more sensitive to temperature for the ECAPed-2A12 aluminium alloy.

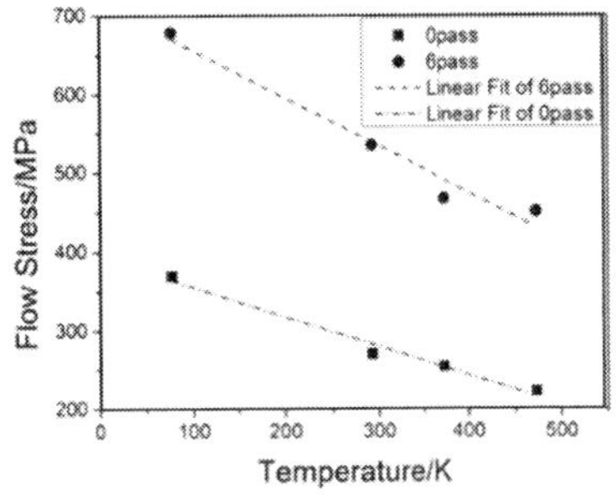

Figure 7. Variation of flow stress of the ECAPed and the annealed coarse-grained aluminium alloy at fixed strain of 7% with temperature.

In order to analyze the influence on the temperature sensitivity of flow stress by the grain size changes, the temperature sensitivity factor can be defined as [15]:

$$S_T = \left(\frac{\partial \sigma}{\partial T}\right)_{\varepsilon}. \tag{3}$$

It can be seen from Eq. (3) that the temperature sensitivity changes with strain. Therefore, the corresponding strain must be indicated when using the formula. The slope of the linear fit line in Fig. 7 is the estimation of the temperature sensitivity factor of the material. The temperature sensitivity factor of the ECAPed-2A12 aluminium alloy and the annealed coarse-grained aluminum alloy can be calculated to be −0.51 and −0.83 respectively where "-" means that the flow stress decreases with the increasing temperature. This indicates that the temperature sensitivity increases obviously with the grain refinement. It is believed that the unusually small activation volume in NC or UFG FCC metals implies a stronger temperature dependence of the yield and flow stress. Also based on thermal-activation theory, the temperature sensitivity factor can be expressed as [16]

$$\frac{\partial \sigma}{\partial T} = -\sqrt{3}\,K \ln \frac{\dot{\varepsilon}_0}{\dot{\varepsilon}} / V^*. \tag{4}$$

The above equation does indicate that at fixed strain rate the S_T depends on the evolution of activation volume. For UFG materials with FCC crystal structure, the activation volume has been proven to decrease with the reduction in grain size. Therefore, the smaller the grain size, the higher the temperature sensitivity of flow stresses is expected. This conclusion is in good agreement with the experimental data. Since the pre-exponential strain rate, $\dot{\varepsilon}_0$, is of the order of 10^6–10^8/s [17] which is several orders of magnitude higher than aare expectedin rates in the present tests, values of temperature sensitivity is expected to be negative according to Eq. (4). This indicates the reduction in flow stress with increasing temperature.

4. Conclusions

In the present work, we have carried out uniaxial compressive experiments on ECAPed-2A12 aluminium alloy fabricated by the ECAP method over a wide range of temperatures and strain rates. It can be seen from the experimental results that the yield stress of the 2A12 aluminium alloy increases with ECAP pass number. However, the yield stress of ECAPed-Al peaks after six passes and changes slightly in further passes. Due to fine grain aluminum alloys own great dynamic recovery ability, their strain hardening capacity decreases remarkably. The strain rate sensitive factor of ECAPed-2A12 aluminium alloy increases with pass number. It is believed that the reduction in grain size as well as increase of dislocation density after SPD process contribute to the enhanced strain rate sensitivity. If comparing with the annealed coarse grained aluminium, the temperature sensitivity of ECAPed-2A12 aluminium alloys is relatively high. It is believed that the enhanced temperature sensitivity of fine-grained aluminium alloy can be related to reduction in activated volume.

This work was performed under the financial support from the National Science Foundation of China (grant Nos. 11272267, 11372206 and 11102168) and 111 Project (grant No. B07050) of Northwestern Polytechnical University.

References

[1] R.Z. Valiev, T.G. Langdon, Prog. Mater Sci. **51**, 881–981 (2006)

[2] J.C.M. Li. *Mechanical Properties of Nanocrystalline Materials* (Pan Stanford Publishing, New York, 2011)

[3] M.A. Meyers, A. Mishra, D.J. Benson, Prog. Mater. Sci. **51**, 427–556 (2006)

[4] R.Z. Valiev, R.K. Islamgaliev, I.V. Alexandrov, Prog. Mater. Sci. **45**, 103–189 (2000)

[5] L.J. Zheng, C.Q. Chen, T.T. Zhou, Mater. Characterization **49**, 455–461 (2002)

[6] Q. Wei, J. Mat. Sci. **42**, 1709–1727 (2007)

[7] Y.M. Wang, E. Ma, Mat. Sci. Eng. A **375**, 46–52 (2004)

[8] T. Suo, Y.Z. Chen, Y.L. Li, C.X. Wang, X.L. Fan, Mat. Sci. Eng. A **560**, 545–551 (2013)

[9] H. Kolsky, Proc. Phys. Soc. B **62**, 676–700 (1949)

[10] F.H. Dalla Torre, E.V. Pereloma, C.H.J. Davies,. Acta Mater. **54**, 1135–1146 (2006)

[11] L. Zhu, J. Lu, Inter. J. Plasticity **30-31**, 166–184 (2012)

[12] C.C. Koch. Scripta Mater. **49**, 657–662 (2003)

[13] D. Jia, K.T. Ramesh, E. Ma, Acta Mater. **51**, 3495–3509 (2003)

[14] G.T. Gray III, T.C. Lowe, C.M. Cady, R.Z. Valiev, I.V. Aleksandrov, Nanostructured Mater. **9**, 477–480 (1997)

[15] T. Suo, Y. Li, F. Zhao, X. Fan, W. Guo, Mech Mater, **61**, 1–10 (2013)

[16] G. Malygin, Physics Solid State **49**, 1013–1033 (2007)

[17] H. Conrad, D. Yang, J Electron Mater. **31**, 304 (2002)

EPJ Web of Conferences **94**, 02010 (2015)
DOI: 10.1051/epjconf/20159402010

When do interfaces become important for failure?

S.J. Fensin[1,a], E.K. Walker[1,2], E.K. Cerreta[1], and G.T. Gray III[1]

[1] Los Alamos National Laboratory, Los Alamos, NM 87544, USA
[2] School of Materials Engineering, Purdue University, West Lafayette, IN 47907, USA

Abstract. Previous experimental research has shown that microstructural features such as interfaces, inclusions, vacancies and heterogeneities can all act as void nucleation sites. However, it is not well understood how important these interfaces are to the damage evolution as a function of the surrounding parent materials. In this work, we present results on three different materials: 1) Cu, 2) Cu-10 wt%Ag, and 3) Cu-15 wt%Nb examined to probe the influence of bi-metal interfaces on void nucleation and evolution. These materials were chosen due to the differences in the stacking fault energy between the two phases. The initial results suggest that when there are significant differences between the bulk properties (for example: stacking fault energy and melting temperature etc) the type of interface between the two parent materials does not influence the damage process. Rather, it is the "weaker" material that dictates the dynamic spall strength of the material.

1. Introduction

Multi-phase alloys are one class of materials that are widely utilized in modern engineering applications. The two-phase nature of such engineering alloys is balanced to enhance the best properties of each phase. For example, strength from one and ductility from the other. In this work, we consider two Cu matrices with a small amount of a softer (Ag) and a harder second phase (Nb), respectively. The aim of this study is to not only understand the role of a second phase in void nucleation and growth but also to investigate conditions (soft/soft vs. soft/hard) under which interfaces between the two phases are not important to the damage evolution and failure process. In addition, we are also interested in studying the mechanisms and how they change when the bi-metal interface changes from FCC/FCC to FCC/BCC.

In the past, extensive research has focused on understanding the dynamic fracture response of both metals and alloys, especially in cases where the second phase additions are stiffer than the primary metallic phase. In the case of single-phase ductile materials, it is widely accepted that microstructure affects the response of a material to dynamic loading [1–7]. For example, experiments by Kanel et al. [8] on high-purity copper indicate that the onset stress for dynamic fracture (i.e. spall strength) of single-crystal copper is two to three times higher than that of polycrystalline copper. This observation suggests that the threshold stress for void nucleation at a grain boundary (GB) might be significantly lower than for nucleation at other defects like dislocation cell walls [8]. Recent experiments by Escobedo et al. [9] on polycrystalline Cu additionally demonstrate that certain special ($\Sigma 3$) boundaries are more resistant to void nucleation during dynamic loading. This work also corroborates the experimental results of Wayne et al. [10] on Cu, which shows that boundaries with certain misorientations are preferred locations for intergranular damage. These results collectively suggest that the types of grain boundaries in a sample can strongly influence the amount and location of damage in a material during dynamic loading. However, less is understood about this effect under conditions of two-phase boundaries or at bimetallic interfaces.

In contrast, work by Minich et al. [5] on two phase materials like single crystal copper with SiO_2 inclusions shows that the presence of small, hard SiO_2 precipitates lowers the stress required to nucleate voids in this material as compared to that for pure Cu. Similarly, work by Christy et al. [11] that studies inclusions present at a grain boundary in Cu, reveals that voids prefer to nucleate at these inclusions along the grain boundaries. Gray et al. [12] show an interface dependence of this phenomena based on the response of aluminum with alumina inclusions and within a Cu/Nb material. The spall strength of the Al/Al_2O_3 system is found to be lower than that of pure aluminum and is highly dependent on the volume fraction of inclusions and their specific particle morphology. On the other hand, the spall strength of a Cu/Nb system remains unchanged by the presence of 15 vol % Nb particles embedded in the Cu matrix in comparison to high purity copper. Although these and other experimental observations show that the presence of a second phase can impact the spall strength of a material, the specific mechanisms behind damage nucleation and its evolution in these two-phase materials remain poorly understood.

To thoroughly understand the role of bi-metal or metal/inclusion interfaces on damage behavior, we have conducted a series of systematic plate impact experiments on Cu-10 wt%Ag and Cu-15 wt%Nb. These results have been compared with similar experiments performed on high-purity Cu. The remainder of this paper is organized as

[a] Corresponding author: saryuj@lanl.gov

Article available at http://www.epj-conferences.org or http://dx.doi.org/10.1051/epjconf/20159402010

Figure 1. Optical image showing the equiaxed grain structure of the annealed high purity Cu.

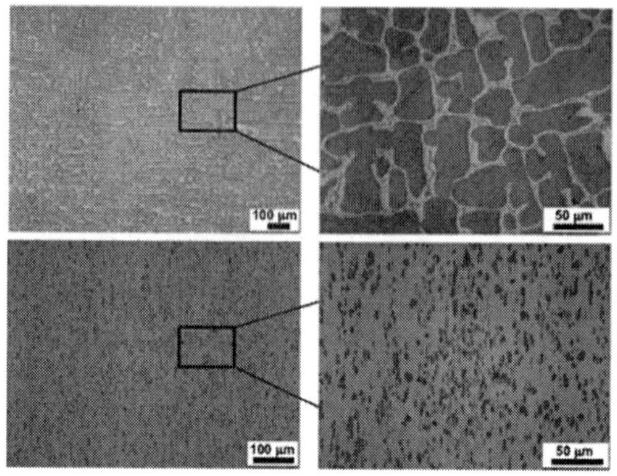

Figure 2. Optical images showing the initial microstructure of Cu-10%Ag (top) and Cu-15%Nb (bottom), respectively. In the top image the white regions show silver and the grey particles in the bottom image depict Nb.

follows. The next section discusses the details of the plate-impact experiments and resulting analysis undertaken in this study. The results and conclusions are presented in Sects. III and IV, respectively.

2. Experimental methodology

In this section, we outline the materials, as well as experimental and analytical approaches utilized in this investigation.

2.1. Initial material characterization

The target materials were prepared from 1) high purity annealed Cu, 2) annealed Cu-Ag alloy with 10 wt%Ag and 3) Cu-Nb alloy with 15 wt%Nb. It is important to note that the CuNb was not heat treated in this case.

The microstructure of the annealed specimens was examined with optical microscopy using a Zeiss microscope. It was difficult to obtain EBSD data for the CuAg and CuNb alloys due to their two-phase nature. Standard metallographic techniques were used to prepare samples for inspection. For all samples, a final polish of 1 μm Al_2O_3 was followed by vibratory polishing using a solution of 0.04 μm colloidal silica. The Cu samples were lightly etched using a solution of $FeCl_3$ and HCl whereas the CuAg and CuNb samples were not etched.

Figure 1 shows an optical image for the high-purity Cu. This micrograph shows that Cu has equiaxed grains with an average grain size of 60 μm, whereas Fig. 2 shows the optical micrographs for CuAg and CuNb. In contrast to pure Cu, CuAg has a eutectic microstructure with a small effective grain size (typical for eutectic materials) and Nb is present in precipitate form in the Cu matrix in the CuNb material. Even though the initial microstructures for these materials are different, phenomenological comparison of the locations for void nucleation is made to determine the role of a second phase in void nucleation and growth processes.

2.2. Plate impact experiments

A plate impact experiment was conducted using a 80-mm bore gas launcher and soft-recovery techniques as shown using the configuration in Fig. 3 [13]. An impactor of Cu was accelerated using an 80 mm bore gas gun at a velocity of 190 m/s and impacted onto the Cu, CuAg and

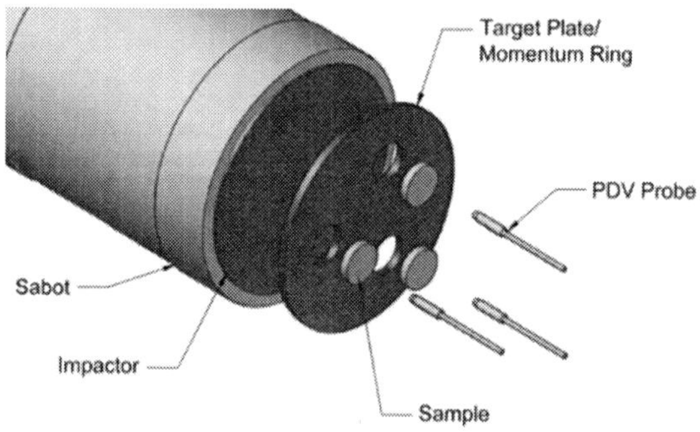

Figure 3. Experimental configuration used to produce incipient damage or complete spall in three samples simultaneously. Copper impactors were accelerated in an 80 mm gas gun and impacted onto the target. PDV probes were placed over the center of each sample to provide free-surface velocimetry data from each sample in each experiment.

CuNb targets. The impactor was roughly 1.29 mm thick. The target consisted: a target plate with three samples, and a momentum ring for the whole target. Targets were 2.59 mm thick and designed specifically such that recovery of the shocked samples was facilitated for post-mortem examination. The momentum rings were press-fit into the target plate to ensure intimate contact of all the components. For the three-sample targets, samples were machined as right, regular cylinders 12.7 mm in diameter. In the experiment, the impactor thickness was one-half the sample thickness to cause tensile damage to occur in the center of the sample thickness. A single momentum trapping ring made out of Cu surrounded the whole assembly. This was done because the sample diameters were large such that the sample had ample time to release before the edge waves came back to the sample. Placement of the samples in the target ring were set such that the shortest distance between sample edges to each other and to the edge of the target plate were equal.

Free surface velocities were measured in each experiment using Photon Doppler Velocimetry (PDV)

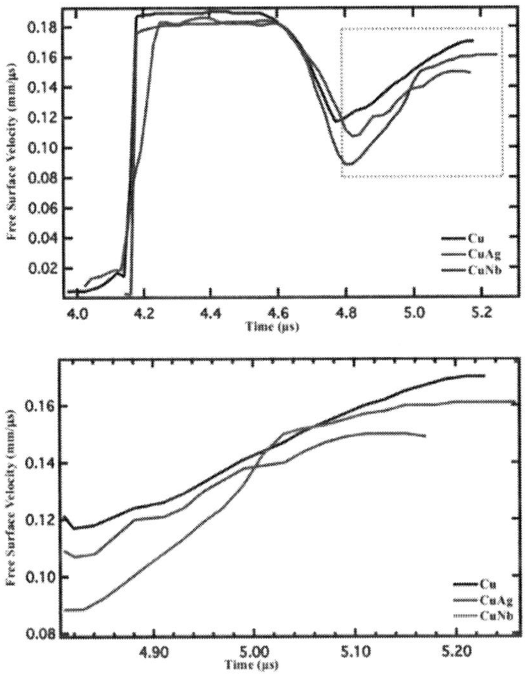

Figure 4. (Top) traces of the free surface velocity showing shock loading for Cu, CuAg and CuNb experiments. (Bottom) free surface velocity rise after the minima.

[14,15]. Collimated probes provided by AC Photonics (Part #: 1CL15P020LC-C01) were positioned to collect velocimetry data from the center of each sample and connected to a LANL-built 4-channel PDV system. One additional PDV probe (not shown in figure) was placed next to each target plate to measure the projectile velocity directly to an accuracy of 0.1% [15]. A single PZT trigger pin (Dynasen, Inc.) was also positioned next to each target plate with a typical standoff distance of a few millimeters to provide a trigger to the digitizer used to collect the PDV data. Following impact, all samples were "soft recovered" by decelerating them into low-density foam.

2.3. Postimpact metallurgical characterization

Subsequent characterization of the damage in the spalled samples included both optical microscopy. In preparation for the optical analyses, each recovered specimen was diametrically sectioned. They were subsequently mounted in an epoxy resin and prepared following standard metallographic techniques as listed above in Sect. 2.1.

3. Results

3.1. Free surface velocity analysis

Free surface velocity (FSV) profiles provide in-situ information regarding the motion of shock waves in a sample and can be used to infer the stresses and kinetics associated with void nucleation and growth. Figure 4 shows the FSV histories for the incipient spall experiments on Cu, CuAg and CuNb.

Table 1. Parameters used to calculate peak stress and spall strength. Parameters for CuAg and CuNb were calculated through using a weighted average of the individual parameters for each component.

Experiment	ρ (g/cm^3)	C_B (mm/μs)	C_L (mm/μs)	s	γ	h (mm)
Cu	8.924	3.94	4.72	1.489	1.96	1.08
CuAg	9.081	3.87	4.72	1.495	1.96	0.43
CuNb	8.873	4.02	4.77	1.445	1.96	1.06

The wave profiles for the three materials are seen to display a number of features: 1) Lack of elastic-plastic transitions, or Hugoniot Elastic Limits (HEL's), as seen in the loading portion of the profile, 2) CuAg has a different slope in the loading part of the velocity profile. This could be related to non-uniform dissipation of energy through plasticity which is further related to the higher propensity of Ag to undergo extensive deformation twinning, 3) Constant responses upon the peak stress on the Hugoniot suggestive of uniform plastic deformation with all the samples, 4) Upon release all profiles exhibit a classic pull-back signal and ringing, which generally is associated with spall or dynamic fracture in the sample. However, there are different magnitude of the "pull-back" signal quantified by ΔFSV suggesting different damage nucleation responses in the materials and 5) Similar slopes for the free surface velocity rise for Cu and CuAg are observed. These slopes are different from the one observed in CuNb which shows two different slopes in this rise region.

Due to the different composition of the materials, the peak free surface velocities are 0.187, 0.182, 0.182 mm/μs in Cu, CuAg and CuNb, respectively. The corresponding average peak compressive stresses are 3.40, 3.31 and 3.35 GPa in Cu, CuAg and CuNb, respectively. These values are calculated using the Mie Grüneisen equation of state with parameters listed in Table 1. Parameters for CuAg and CuNb were calculated by using a weighted average of the individual parameters for each component obtained from the LASL Hugoniot EOS database [16].

The spall strength (σ_{spall}), which is one measure of dynamic failure stress, is calculated using the relationship proposed by Novikov [17] and the correction proposed by Kanel [18]:

$$\sigma_{spall} \equiv \frac{1}{2}\rho_o C_B(\Delta FSV + \delta), \qquad (1)$$

where ρ_o is the ambient density, C_B is the bulk sound speed and ΔFSV is the difference in the free surface velocity from the peak state to the pull-back minima (Fig. 4). The accuracy of the spall strength calculation is improved by correcting for the ΔFSV in Eq. 1 [18]:

$$\delta = h\left(\frac{1}{C_B} - \frac{1}{C_L}\right) * \frac{|\dot{u}_1 * \dot{u}_2|}{|\dot{u}_1| + \dot{u}_2}, \qquad (2)$$

where h is the thickness of the spalled region (measured from optical images), C_L is the longitudinal sound speed, and $\dot{u}_1$ and $\dot{u}_2$ are the unloading and re-compression rates calculated as

$$|\dot{u}_1| = -1/2\frac{dFSV}{dt} \text{ and } \dot{u}_2 = 1/2\frac{dFSV}{dt}, \qquad (3)$$

Table 2. Calculated and measured parameters from the FSV data.

	Pull back characteristics					
Shot	ΔFSV (mm/μs)	$\dot{u}_1$ (mm/μs^2)	$\dot{u}_2$ (mm/μs^2)	δ (mm/μs)	Average Spall Strength (GPa)	Corrected Spall Strength (GPa)
Cu	69	0.39	0.148	0.0052	1.328	1.333
CuAg	74	0.37	0.155	0.0020	1.421	1.423
CuNb	94	0.49	0.284	0.0069	1.806	1.813

Here, $\frac{dFSV}{dt}$ is measured from the pullback signal as described in Ref. [19] and these values for the current experiments are listed in Table 2. The unloading rates $\dot{u}_1$ are indicators of the kinetics of the tensile pulse imposed on the target. These results (Table 2) indicate that the CuNb experiences a faster tensile stress rate as compared to the Cu and CuAg.

The corrected spall strength values calculated using Eqs. (1), (2) and (3) are listed in Table. 2. The data reveals that the spall strength of Cu increases with the addition of 10% Ag and 15 %Nb. Specifically, the spall strengths for CuAg and CuNb are 1.42 and 1.81 GPa, respectively. These values are 6% and 26% higher than the average spall strength for Cu. The slight increase in the spall strength of CuAg could be due to second phase hardening (not precipitate hardening) especially because in general silver has a slightly higher tensile strength in comparison to Cu [20]. In addition, the effective grain size in CuAg is much lower than Cu due to its eutectic microstructure and as such can increase the spall strength of the material. In contrast, increase in the spall strength of CuNb is related to precipitation hardening since Nb is dispersed through the matrix in the form of fine precipitate particles (Refer Fig. 2). Also, Nb is much harder than both Cu and Ag and is an efficient barrier to dislocation motion. In general, the microstructures of these three materials differ to the extent that it would not be fair to do a one-to-one quantitative comparison. Nonetheless, the trends in spall strengths are reasonable.

In addition, differences are also observed in the rate at which the free surface velocity rises beyond the pull-back signal minima between the three samples. Specifically, the rate of increase is similar for Cu and CuAg (0.148 and 0.155 mm/μs^2) but is 23% higher in CuNb (0.194 mm/μs^2). In fact, the rise rate of CuNb can be divided into two parts; the first part rises faster (0.284 mm/μs^2) and the second part is much flatter (0.054 mm/μs^2). This bi-linear pull-back slope in CuNb is perhaps suggestive of differing energy partitioning during nucleation and growth of damage as compared to Cu and CuAg which display a linear "pull-back" signal.

3.2. Post-impact metallurgical examination

Even though the in-situ, FSV measurements provide insight into the stress history of the dynamically loaded materials, these continuum level measurements can only provide limited information regarding the specific details of the operative, internal damage evolution within a material. For this reason, metallographic characterization of the incipiently spalled samples was performed to gain insight into the relationship of damage evolution and microstructure as a function of material and the imposed drive condition.

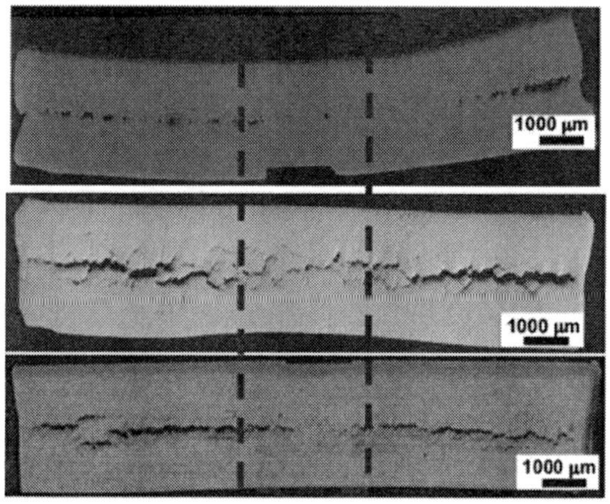

Figure 5. Optical micrographs of spalled high purity Cu, Cu-10%Ag and Cu-15%Nb respectively. The spall plane is approximately located in the center of the samples except in the case of CuPb.

Figure 5 shows optical micrographs of the cross sections of the recovered, incipiently, spalled samples. It is important to note that since there was only one momentum trapping ring on the entire target assembly, the regions on the outside of the samples, which are all showing higher coalescence of voids can be influenced by to late time, radial release tensile loading affects which are known for increase void linkage. Hence, we will focus on the damage in the center of the samples also marked by the blue box in Fig. 5. On a coarse scale the first apparent difference is in the thickness of the spall plane. Even though the spall planes lie in the center of the sample for all materials, the area occupied by the spall plane decreases in the order: CuAg, CuNb and Cu, respectively. This area directly reflects the extent of plastic deformation in each material. For example, in CuNb, where the spall strength is the highest, most of the damage is localized to the area where the material experiences maximum tension. From Fig. 5 it is also evident that Cu has the least number of coalesced voids. This is consistent with a low spall strength and lower void growth rate (inferred from the rate at which the free surface velocity rises beyond the minima). In this case, void nucleation is dominant over void growth. Whereas in the case of CuAg, whose spall strength is 6% higher than Cu, larger areas of coalesced voids are observed. This could be due to the higher rate of free surface velocity rise and also due to the fact that the effective grain size is much smaller in this material (eutectic microstructure) which

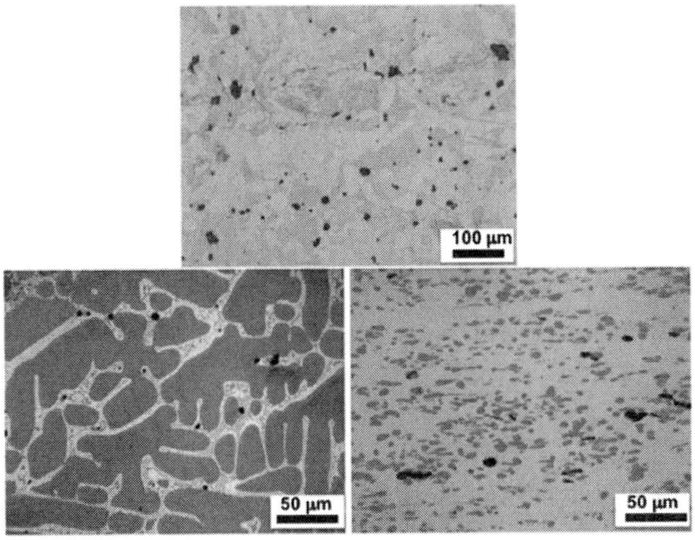

Figure 6. (Top) optical image for spalled Cu with black regions representing void. (Bottom right) optical image of spalled CuAg where white regions represent silver and black regions are voids. (Bottom left) optical image of spalled CuNb where the grey particles are Nb and the black areas are voids.

makes it easier for the voids to coalesce. In CuNb, there is a larger number of coalesced voids in comparison to Cu but a lower number of coalesced voids in comparison to CuAg. This could be because even though it is much harder to nucleate voids in this material (highest spall strength) the rate of growth of these voids is the highest (highest slope for FSV rise) once damage is formed. However, since the microstructures are not exactly the same, other factors like texture and grain size could also be influencing these results.

Figure 6 shows higher magnification optical images of the spalled Cu, CuAg and CuNb. These images show that in Cu voids tend to nucleate at grain boundaries as previously observed [9] whereas in CuAg voids nucleate in Ag and in CuNb voids tend to be located on the Cu side of the Cu/Nb interface. Although, the majority of the voids in CuNb were too well developed to unambiguously determine their exact nucleation location. Despite the different microstructures, the phenomenological observations regarding the location for void nucleation appear reasonable.

4. Conclusions

In this research, we present the spallation response of three different materials: 1) Cu, 2) Cu-10%Ag, and 3) Cu-15%Nb to investigate the influence of bi-metal interfaces on void nucleation and evolution. These materials were chosen due to the differences in crystallographic structure (FCC/FCC vs. FCC/BCC) and strength of the individual parent phases. Preliminary results indicate that the spall strength of Cu increases with the addition of Ag and Nb by 6% and 26%, respectively. This increase was attributed to the smaller effective grain size and precipitate hardening in CuAg and CuNb, respectively. The spall strength along with the void growth rates (inferred from the rise in the FSV) were also used to determine that void nucleation was the dominant process for stress dissipation in Cu (lowest spall strength) whereas void growth was more dominant in CuNb (highest spall strength). In contrast, in CuAg, which has a slightly, higher spall strength and void growth rate than Cu, the small effective grain size of the microstructure dominated its spallation response. Most importantly the location for the void nucleation was observed to vary with material. In Cu, voids were observed to nucleate at grain boundaries, whereas in CuAg and CuNb voids nucleated in Ag (the softer phase) and on the Cu size of the Cu/Nb interface. As such, this work demonstrates that there are cases where bi-metal interfaces (Like in CuAg) do not control void nucleation in contrast to cases where interfaces can be important as shown in CuNb. In our future work, we plan to quantify the mechanisms that control void nucleation and growth in these materials using molecular dynamic simulations.

Los Alamos National Laboratory is operated by Los Alamos National Security, LLC, for the National Nuclear Security Administration of the U.S. Department of Energy under contract DE-AC52-06NA25396. This work was partially sponsored by the Joint DoD/DOE Munitions Technology Development Program.

References

[1] L. Seaman, D.R. Curran and D.A. Shockley, Phys. Today **30**, 46–55 (1977)
[2] L. Seaman, D.R. Curran and D.A. Shockley, J. Appl. Phys. **47**, 1414–26 (1976)
[3] M.A. Meyers and C. T. Aimone, Prog. Mater. Sci. **28**, 1–96 (1983)

[4] D.R. Curran, L. Seaman and D.A. Shockley, Phys. Reports **147**, 253–388 (1987)

[5] R.W. Minich, J. Cazamias, M. Kumar and A. Schwartz, Metall. Mater. Trans. A **35**, 2663–73 (2004)

[6] R.W. Minich, M. Kumar, A. Schwartz and J. Cazamias, AIP Conf. Proc. **845**, 642–45 (2006)

[7] G.T. Gray III, Ann. Rev. Mater. Res. **42**, 285–303 (2012)

[8] G.I. Kanel, S.V. Rasorenov and V.E. Fortov, *Shock Wave and High-Strain-Rate Phenomena in Materials*(Marcel Dekker, New York, NY 1992)

[9] J.P. Escobedo, D. Dennis-Kolle, E.K. Cerreta, B.M. Patterson, C.A. Bronkhorst, B.L. Hansen, D. Tonks and R.A. Lebensohn, J. Appl. Phys. **110**, 033513 (2011)

[10] L. Wayne, K. Krishnan, S. DiGiacomo, N. Kovvali, P. Peralta, S.N. Luo, S. Greenfield , D. Byler, D. Paisley, K.J. McClellan, A. Koskelo and R. Dickerson, Scripta Mater. **63**, 1065–68 (2010)

[11] S. Christy, H.R. Pak and M.A. Meyers, *Metallurgical Applications of Shock-Wave and High-Strain-Rate Phenomena* (Marcel Dekker, New York, NY, 1986) 835

[12] R.S. Hixson, J.N. Johnson, G.T, Gray III and J.D. Price, AIP Conf. Proc. **370**, 555–8 (1996)

[13] G.T. Gray III, ASM International **8**, 530 (2000)

[14] O.T. Strand, D.R. Goosman, C. Martinez, T.L. Whitworth, W.W. Kuhlow, Rev. Sci. Instrum. **77**, 083108 (2006)

[15] B.J. Jensen, D.B. Holtkamp, P.A. Rigg, D.H. Dolan, J. Appl. Phys. **101**, 013523 (2007)

[16] S.P. Marsh, *LASL Shock Hugoniot Data* (University of California Press, 1980)

[17] S.A. Novikov, I.I. Divnov. and A.G. Ivanov, Phys. Met. Metallogr. **21**, 122 (1966)

[18] G.I. Kanel, J. Appl. Mech. Tech. Phys. **42**, 358 (2001)

[19] J.P. Escobedo, E.N. Brown, C.P. Trujillo, E.K. Cerreta and G.T. Gray III, J. Appl. Phys. **113** (2013)

[20] E.B. Zaretsky and G.I. Kanel, J. Appl. Phys. **110**, 073502 (2011)

EPJ Web of Conferences **94**, 02011 (2015)
DOI: 10.1051/epjconf/20159402011

The effect of grain size on dynamic tensile extrusion behaviour

Leeju Park[a], Hack Jun Kim, and Seok Bong Kim

Agency for Defense Development, PO Box 35-42, Taejon 305-600, Korea

Abstract. Dynamic tensile extrusion (DTE) tests were conducted on coarse grained and ultrafine grained (UFG) OFHC Cu, Interstitial free (IF) Steel, and pure Ta. Equal channel angular pressing (ECAP) of 16passes with Bc for Cu, IF Steel and 4 passes for Ta was employed to fabricated UFG materials. DTE tests were carried out by launching the sphere samples (Dia. 7.62 mm) to the conical extrusion die at a speed of ~500 m/sec. The fragmentation behavior of the soft-recovered fragments were examined and compared with each other. The DTE fragmentation behavior of CG and UFG was numerically simulated by the LS-DYNA FEM code.

1. Introduction

DTE technique is a newly developed mechanical test [1]. In the ordinary DTE test, a spherical sample launched at high velocity passes through an open conical die. Due to the smaller die exit diameter than the sample diameter, the sample experiences severe tensile deformation. Therefore, the DTE test can characterize the mechanical response of materials under both high strain rate and high strain circumstances. The DTE technique has been applied to coarse grained (CG) pure metals such as Cu [1], Ta [2], Zr [3], etc..

Extensive and intensive researches during past two decades clearly reveal that ultrafine grained (UFG) materials exhibit very different mechanical and thermal responses from CG materials. There are several studies on mechanical behavior of UFG materials at high strain rates [4]. However, the strain rate employed in those studies (typically $10^3\,s^{-1}$order) was quite lower than that being attainable in the dynamic tensile extrusion (DTE) test.

Ultrafine grained (UFG) materials usually exhibit higher strength due to the Hall-Petch strengthening but lower ductility due to shear localization than coarse grained (CG) counterparts at room temperature [5]. Meanwhile, some UFG materials show high strain rate super plasticity (HSRS) at high temperatures [6]. The shear localization is beneficial for the self-sharpening of the kinetic energy penetrator. HSRS is possible to operate on the metal jet formation of the metal liner in the chemical energy penetrator [7]. Accordingly, UFG materials are promising as the high performance penetrator materials. In this study, the metal jet formability of UFG OFHC Cu, IF steel and pure Ta were compared to those of CG counterparts by means of DTE test.

2. Experimental

Commercial OFHC annealed Cu and IF steel bars (18 mm diameter) were subjected to 16 passes equal channel angular pressing (ECAP) with route B_c in order to fabricate equiaxed UFG samples. Pure Ta bars were conducted with 4 passes ECAP with route B_c for fabricating UFG samples. The sphere samples of 7.62 mm diameter were machined from the central part of unECAPed and ECAPed bars for DTE tests. DTE tests were carried out by using an all-vacuumed gas gun system which consists of the gas gun, the sample flying barrel, the DTE die chamber, and the sample recovery station; the details of the DTE equipment are described elsewhere [8]. The velocity of sample in this experiment was ~500 m/sec upon reaching the DTE die. After DTE tests, the sample fragments were soft recovered. The numbers and the order of fragments exiting the die were confirmed by the high speed photography. Besides, the complete fragment recovery was ensured by comparing the weight of all fragments with that of the initial sample. A routine microstructural observation were made on CG and UFG materials before the DTE tests with optical microscopy and electron backscattered diffraction (EBSD). The DTE behavior of CG and UFG Cu was numerically analyzed by using a commercial finite element code (LS-DYNA 2-dimensional axis-symmetric model [9]). The Johnson–Cook model was employed in the numerical analysis. Five unknown parameters in the Johnson–Cook model were obtained by conducting tensile tests at $10^3\,s^{-1}$ and $1\,s^{-1}$ and compression tests at $2000\sim4000\,s^{-1}$ on CG and UFG materials; tensile tests and compression tests were carried out on a hydraulic universal testing machine and a split Hopkinson pressure bar tester, respectively. In the numerial simulation process, the 2D R-adaptive remeshing was done in order to prevent severe distortion of the mesh. That is, a completely new mesh was created every $1\,\mu$sec in order for the elements to keep a regular shape and a characteristic dimension. The new mesh is initialized from the old mesh by a least square approximation. The simulation results were compared with the experimetal ones in terms of total DTE elongation (sum of the axial elongation of individual fragment) and the number and the dimension of fragments).

[a] Corresponding author: ljparkr@add.re.kr

Article available at http://www.epj-conferences.org or http://dx.doi.org/10.1051/epjconf/20159402011

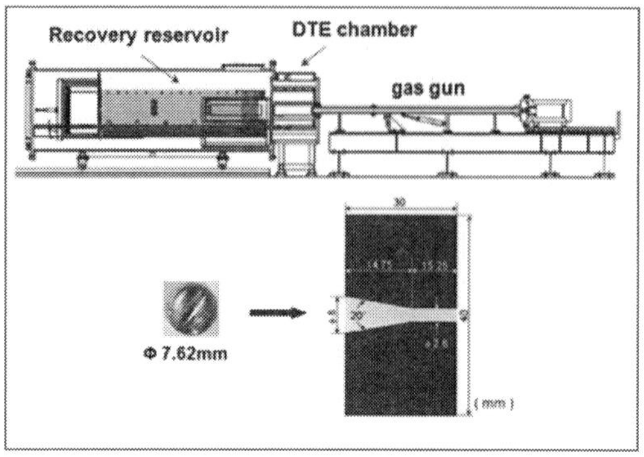

Figure 1. (a) The schematic illustration of the DTE facility consists of gas gun system, (b) Configuration of the DTE die (dimension in mm).

Figure 2. IPF maps (a) CG OFHC Cu, (b) UFG Cu, (c) CG IF Steel, (d) UFG IF Steel, (e) CG Ta, and (f) UFG Ta.

3. Results and discussion

3.1. Examine the grain size of CG and UFG materials

The Inverse pole figure (IPF) maps of the plane parallel to the extrusion axis of the CG and UFG samples are shown in Fig. 2. The average grain size of CG OFHC Cu was ~1mm and that of UFG Cu was ~0.35 μm. The average grain size of IF steel was ~163 μm and that of UFG IF was ~0.61 μm. The average grain size of CG Ta was ~53 μm and that of UFG Ta was ~3.94 μm. IF steel and Ta has a weak texture after ECAP. The grain size difference of CG and UFG samples were range from a few times to 10^6 times.

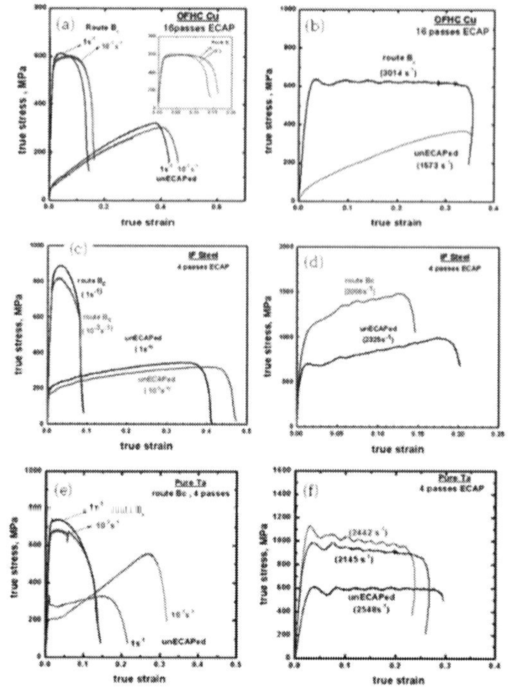

Figure 3. True stress-strain curves of (a) OFHC Cu at static, (b) OFHC Cu at dynamic, (c) IF steel at static, (d) IF Steel at dynamic, (e) Ta at static and (f) Ta at dynamic.

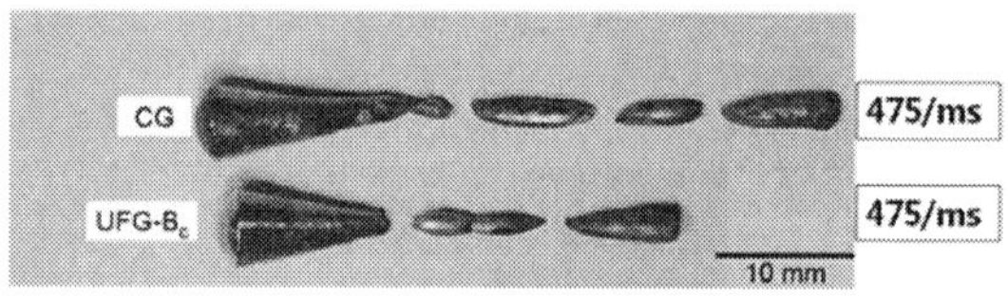

Figure 4. DTE fragments of CG Cu and UFG OFHC Cu in sequence exiting the DTE die. The DTE direction is from left to right.

3.2. Mechanical properties of CG and UFG materials

The true stress-strain curves of CG and UFG sample are shown in Fig. 3. The yield stress of UFG materials were increased a few ten times that of CG materials. Meanwhile, the elongation of UFG materials were decreased a few times that of CG materials. As usual, regardless of the strain rate, CG Cu and IF steel exhibited extensive strain hardening after low stress yielding while near-perfect plasticity without strain hardening after high stress yielding occurred in UFG Cu and Ta. While UFG IF steel showed strain hardening at high strain rate test (~2000/s) with SHPB.

3.3. DTE behaviour: Fragmentation, ductility

The representative soft-recovered fragments of the CG Cu and UFG Cu after DTE are shown in Fig. 4: the conical fragment is the remnants remained in the DTE die. For all three runs of each sample, the CG Cu was fragmented

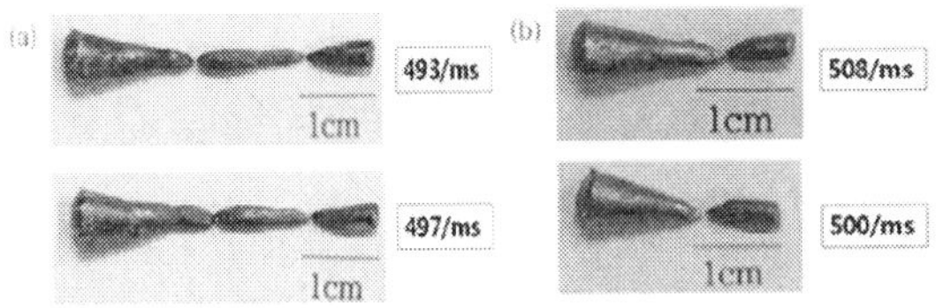

Figure 5. DTE fragments of (a) CG IF steel and (b) UFG IF steel in sequence exiting the DTE die.

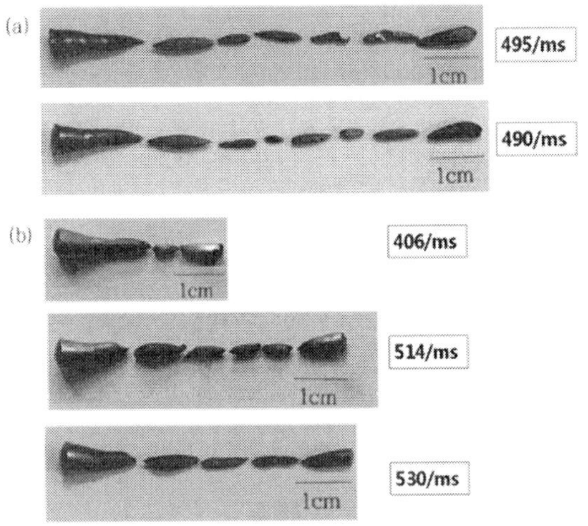

Figure 6. DTE fragments of CG and UFG pure Ta in sequence exiting the DTE die.

into 4 pieces while the UFG Cu was fragmented into 3 pieces. All fragments except the conical remnants were in the lenticular shape indicating that fragmentation occurred by plastic instability (i.e. necking) rather than void coalescence.

The average DTE ductility of each sample were 45.09 mm (~592%) and 33.331 mm (~437%) for the CG and UFG Cu, respectively: the DTE ductility is the sum of the axial ductility of each fragment with respect to the initial sample diameter, i.e. DTE ductility = $(Sd_i - d_o)/d_o$ where d_i is the longitudinal length of the ith fragment and d_o is the initial sample diameter. The soft-recovered fragments of the CG and UFG IF Steel after DTE are shown in Fig. 5: the conical fragment is the remnants remained in the DTE die. For all two runs of each sample, the CG IF steel was fragmented into 3 pieces while the UFG IF Steel was fragmented into 1 piece. All fragments except the conical remnants were in the lenticular shape similar to OFHC Cu. The DTE ductility of each sample was 41.31 mm (542%), 41.31mm (518.8%) and 23.22 mm (~305%), 23.16 mm (~304%) for the CG and UFG IF steel, respectively.

The recovered fragments of the CG and UFG pure Ta after DTE are shown in Fig. 6. The CG tantalums were fragmented into 6 or 7 pieces while the UFG tantalums were fragmented into 4 or 5 piece. All fragments except the conical remnants were in the lenticular shape similar to the other above samples. The DTE ductility of each sample was 70.71 mm (~930%), 72.2 mm (~947.5%) for the CG

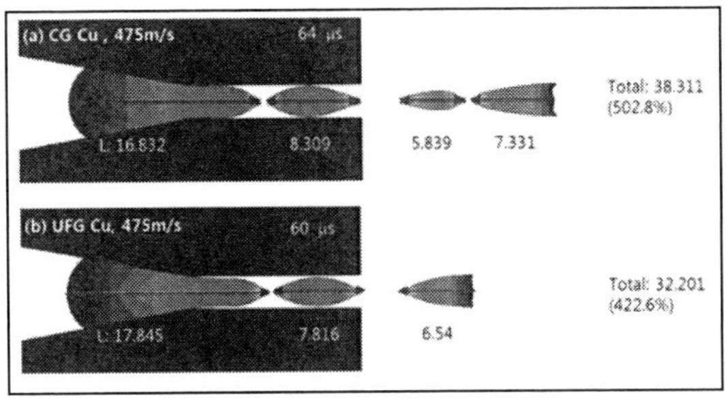

Figure 7. FEM simulation result showing the fragmentation of the DTE fragment of (a) CG Cu, and (b) UFG Cu.

and 33.51 mm (~440%), 51.82 mm (680.05%), 49.35 mm (~647.6%) % for the UFG tantalums, respectively.

3.4. Numerical simulation results

The present DTE behavior of CG and UFG OFHC Cu was numerically simlulated by the LS-DYNA FEM code with the 2D R-adaptivity remeshing. A simulation example for strain after complete fragmentation at ~60 μsec is presented in Fig. 7. The number of fragment is correctly prediced by simulation, i.e. 4 fragments for CG Cu and 3 fragments for UFG Cu. The simulated total length of CG Cu and UFG Cu was 38.3 mm (DTE elongation ~503%) and 32.2 mm (DTE elongation ~423%), respectively. The simulated DTE elonagtion of UFG Cu is in reasonable agreement with the experimental one (~437%). In contrast, for CG Cu, the experimental DTE elongation (~592%) was larger than the simulated one (503%). The maximum strain was developed at the necked region in both sample with the similar value. The strain at the necked region upon fragmentation reached ~5.5. Simulation revealed more localized necking in UFG Cu, possibly causing smaller DTE elongation than CG Cu. In both sample, the strain rate was also maximum at the necked region with $10^5\,s^{-1}$ order which is at least one order or more higher than that achievable by the ordinary Hopkinson test. The maximum strain rate of UFG Cu was slightly higher than that of CG Cu, corresponding to more diffused strain distribution in the latter. The stress imposed by impacting the die was higher in UFG Cu due to its higher yield and flow stresses. The samle velocity was maximum at the exiting tips by the inertia effect. The tip (i.e. maximum) velocity of CG Cu was faster than that of UFG Cu. As expected considerable temperature rise occurred by adiabatic heating. Temperature at the stretched portion in the straight channel was close to or even higher than 700 °K which is about 0.5 T_m. It was locally over 800 °K (~0.6 T_m of Cu) upon fragmentation.

A simulation example DTE behavior of CG and UFG IF steeels were shown in Fig. 8. The number of fragment is correctly prediced by simulation, i.e. 3 fragments for CG steel and 2 fragments for UFG steel. The simulated total length of CG steel and UFG steel was 34.92 mm (DTE elongation ~458%) and 27.17 mm (DTE elongation ~357% respectively. The simulated DTE elonagtion of CG and UFG steel is in reasonably well agreement with the experimental one ~519% for CG and ~304% for UFG.

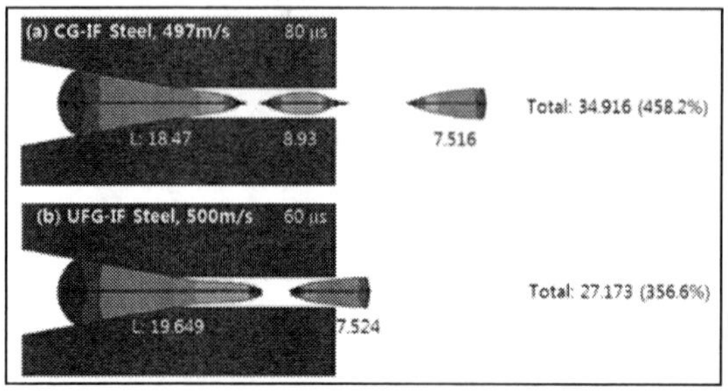

Figure 8. FEM simulation result showing the fragmentation of the DTE fragment of (a) CG IF steel, and (b) UFG IF steel.

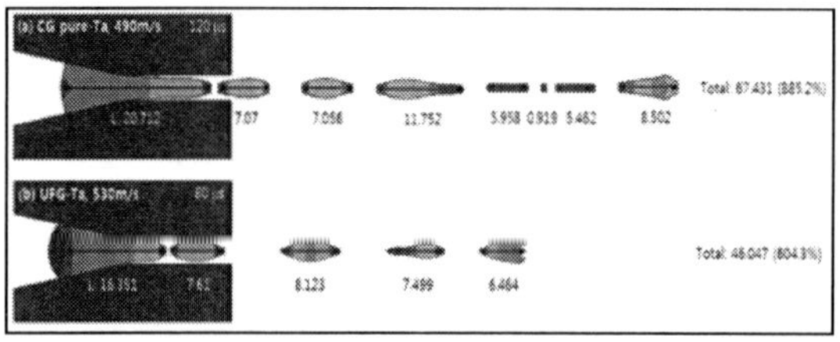

Figure 9. FEM simulation result showing the fragmentation of the DTE fragment of (a) CG Ta, and (b) UFG Ta.

A simulation example DTE behavior of CG and UFG pure tantalums were shown in Fig. 9. The number of fragment is correctly prediced by simulation, 7 fragments for CG tantalum and 5 fragments for UFG steel. The simulated total length of CG tantalum and UFG tantalums was 67.43 mm (DTE elongation ∼885%) and 46.05 mm (DTE elongation ∼604%), respectively. The simulated DTE elonagtion of CG and UFG tantalums are in reasonably good agreement with the experimental one (∼948%) for CG and (∼648%) for UFG tantalum repectively.

4. Summary

1. A series of dynamic tensile extrusion (DTE) tests, the newly developed mechanical test at high strain rate, was conducted on coarse grained (CG) and ultrafine grained (UFG) OFHC Cu, IF steel, and pure Ta. CG materials exhibited higher DTE elongation than UFG materials.
2. The inferior metal jet stability of UFG materials are caused by its initial higher strength and lack of strain hardenability compared to CG materials.
3. Numerical simulation empoying the mesh adaptivity predicted the fragmentation behavior of CG and UFG materials in terms of DTE elonagtion and the number of fragments. The results of numerical simulation are in reasonsably well agreement with the those of experimental.

References

[1] Gray III G.T. et al. 2006 *Shock Compression of Condense Matter* ed by Furnish M.D. et al. (American Institute of Physics) p. 725

[2] Cao F. et al. 2008 *Acta Mater.* **56**, 5804

[3] Escobedo J.P. et al. 2012 *Acta Mater.* **60**, 4379

[4] Farrokh B. and Kahn A.S. 2009 *Int. J. Plasticity* **25**, 715

[5] Q. Wei, T. Jiao, K.T. Ramesh and E. Ma, Scripta Mater. **50**, 359 (2004)

[6] D.H. Shin, D.Y. Hwang, Y.J. Oh, K.-T. Park, Metall. Mater. Trans. A **35A**, 825 (2004)

[7] A.H. Chokshi, M.A. Meyers, Scripta Metall. Mater. **24**, 605 (1990)

[8] Park K.T. et al. 2013 Mater SciEng A **569**, 61

[9] LS-DYNA® *Keyword User's Manual*, 2007 Vol. 1. Version 97 (Livermore Software Technology Corp)

EPJ Web of Conferences **94**, 02012 (2015)
DOI: 10.1051/epjconf/20159402012

Static and dynamic tensile behaviour of aluminium processed by high pressure torsion

Patricia Verleysen[1,a], Wouter Oelbrandt[1], Soroosh Naghdy[1], and Leo Kestens[1,2]

[1]Ghent University, Faculty of Engineering and Architecture, Department of Materials Science and Engineering, Technologiepark 903, 9052 Zwijnaarde
[2]Delft University of Technology, Faculty of 3mE, PO Box 5025, 2600 GA Delft, The Netherlands

Abstract. High pressure torsion (HPT) is a severe plastic deformation technique in which a small, disk-like sample is subjected to a torsional deformation under a high hydrostatic pressure. In present study, the static and dynamic tensile behaviour of commercially pure aluminium (99.6 wt%) processed by HPT is studied. The high strain rate tensile behaviour is characterized using a purpose-developed miniature split Hopkinson tensile bar setup by which strain rates up to $5 \times 10^3\,s^{-1}$ can be reached. During the tests, the deformation of a speckle pattern applied to the samples is recorded, by which local information on the strain is obtained using a digital image correlation technique. Electron back scatter diffraction images are used to investigate the microstructural evolution, more specifically the grain refinement obtained by HPT. The fracture surfaces of the tensile samples are studied by scanning electron microscopy. Results show that the imposed severe plastic deformation significantly increases the tensile strength, however, at the expense of ductility. The strain rate only has a minor influence on the materials tensile behaviour.

1. Introduction

Processing of metals using severe plastic deformation (SPD) is now a well-established procedure for refining the grain size of bulk samples to the submicrometer or nanometer level. Among different SPD approaches, equal channel angular pressing (ECAP) [1], high pressure torsion [2,3] and accumulated roll bonding (ARB) [4] have been mostly applied. In high pressure torsion (HPT) processes a small, thin disk-like sample is subjected to a torsional deformation under high hydrostatic pressure. Neglecting thinning of the sample, a shear strain is obtained which linearly evolves as a function of the distance to the centre of the sample.

The strain-induced grain refinement gives rise to materials with exceptional properties. Strength increases typically with a factor of 5 have been reported in literature [5]. Grain refinement is the main strengthening mechanism for the thus-obtained nanocrystalline (NC, grain size smaller than 100 nm), ultrafine grained (UFG, d $\sim$ 200 nm) or microcrystalline (MC, d $>$ 1 micron) metals, which follow in general the well-known Hall-Petch relation which relates the strength of polycrystalline materials with the grain size [6].

Next to a higher strength, for some metals, significant changes in strain rate sensitivity (SRS) are reported. Independent of the processing route used to refine the grain size, the SRS of face-centered cubic metals is found to be higher whereas for body-centered cubic metals a reduced SRS is observed compared to their coarse-grained counterparts [7]. An increased SRS contributes to the tensile ductility of materials which is most often low for fine-grained materials. Indeed, although reasonable fracture elongations are commonly obtained, major drawback of fine-grained materials is that they suffer from low uniform strains in tensile tests. This is mainly attributed to the fact that the size of SPD evolved grains is too small to allow for efficient storage of dislocations and further strain hardening [8]. An increased strain rate sensitivity is proposed as one of the strategies to achieve higher uniform tensile deformation in a nanostructured metal. Strain rate sensitivity is dependent on the deformation mechanisms and, as such, dependent on the strain rate [7].

The m-value is determined using strain rate jump experiments or experiments carried out at different strain rates [7,9]. The reported strain rates during the tests are all in the low, quasi-static strain rate regime ($\leq$10-1/s). To the authors knowledge no information can be found on the high strain rate tensile behaviour of fine-grained materials. Characterisation of the mechanical behaviour of HPT processed samples is not obvious, because of the intrinsic heterogeneity and small size of the samples. Both necessitate a local measurement approach for the strain. At high strain rates, the short test duration and the generally reduced deformation capacity of the sample, make accurate characterization of the tensile response even more difficult.

In present work, static and dynamic tensile experiments are performed on commercially pure aluminium (99.6 wt%) processed by HPT. From small disks subjected to several levels of deformation by HPT, miniature dogbone tensile specimens are extracted. The equivalent strain achieved by HPT in the gage section varies from 0 to 32. The resulting grain refinement is assessed by EBSD measurements. Static tensile tests are carried out

[a] Corresponding author: Patricia.Verleysen@UGent.be

Article available at http://www.epj-conferences.org or http://dx.doi.org/10.1051/epjconf/20159402012

Table 1. Chemical composition of the pure aluminium.

Fe	Si	Cu	Mg	Mn	Al
0.280	0.050	0.049	0.018	0.005	99.6

at room temperature at a strain rate of $2.8 \times 10^{-3}\,s^{-1}$. To obtain local information on the strain, a speckle pattern applied on the samples is recorded during deformation and processed by a Digital Image Correlation (DIC) technique. To study the high strain rate tensile behaviour, a miniature split Hopkinson tensile bar setup is developed by which strain rates up to $4.4 \times 10^{3}\,s^{-1}$ can be reached. Fracture surfaces of the static and dynamic tensile samples are observed using Scanning Electron Microscopy (SEM).

2. Materials and experiments

2.1. Material

The tested material is a commercial purity aluminium with a chemical composition given in Table 1. The material was rolled to a plate with an initial thickness of 1.1 mm. From the rolled plate, disks with a diameter of 15 mm were punched. The disks were annealed for 1 h in air at 823 K to reach an initial grain size of approximately 300 μm.

The disks were subjected to a rotational deformation of 1/8, 1/4, 1/2, 1, 2 and 3 rotations in a HPT device under a nominal pressure of 2.5 GPa. The anvils of the device are designed to ensure semi-confined deformation conditions [2]. To estimate the equivalent strain during torsion the von Mises strain is widely used [2]:

$$\varepsilon_{vM} = \gamma/\sqrt{3}. \qquad (1)$$

In this equation γ is the engineering shear strain. Assuming that the thickness h of the disk is independent of the rotation angle θ and does not change significantly during rotation, γ can be calculated at a radial distance r from the centre by $\lambda = r\,\theta/h$.

2.2. Static and dynamic tensile test techniques

The experiments at low strain rates are carried out on a 5 kN Deben™ microtensile test device at TUDelft. Miniature dogbone shaped samples with a gage length of 3 mm and width of 1 mm were extracted from the HPT processed aluminium disks by electro discharge machining. The centre of the gage length is at a distance of 3 mm from the centre of the disk, see Fig. 1. In Table 2 the equivalent von Mises strain in the centre of the samples is given calculated using Eq. (1). The samples are mounted horizontally. The tests are performed at room temperature with a crosshead velocity of $8.3 \cdot 10^{-3}$ mm/s.

The clamp displacement and force are recorded during testing, from which respectively the average strain and stress in the gage section of the sample are calculated. To obtain reliable local strain data a speckle pattern is applied to the sample surface prior to testing. During a test, the deforming speckle pattern is recorded using an Olympus SZX9 light optical microscope equipped with a camera, see Fig. 2. After testing the images are processed using a Digital Image Correlation (DIC) technique to calculate the local strain field.

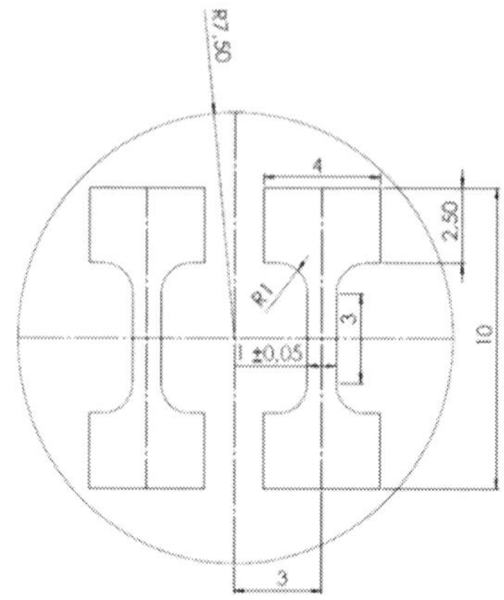

Figure 1. Miniature tensile specimens extracted from HPT processed aluminium disks with a diameter of 15 mm.

Table 2. Equivalent strain in the centre of the gage section of the samples calculated using Eq. (2), grain size measured by EBSD, number of grains a cross section of gage section of the tensile samples and dimple size observed in SEM images of the fracture surface of the statically loaded samples.

Applied rotations	Equivalent strain	Measured grain size	Dimple size fracture
[rot]	[–]	[micron]	[micron]
0	0	300.12	12.04
1/8	1.36	184.76	13.42
1/4	2.72	103.39	12.93
1/2	5.44	3.37	9.87
1	10.88	2.42	8.47
2	21.77	1.04	9.67
3	32.65	1.05	8.37

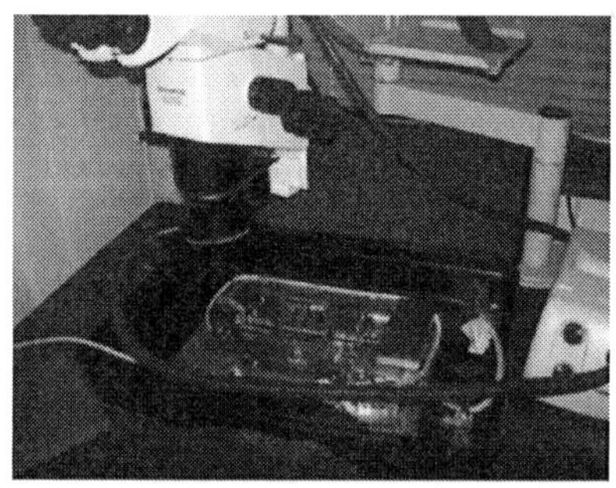

Figure 2. Microtensile test device equipped with a light optical microscope used for the static tensile tests.

High strain rate tensile test were carried out using a purpose-developed Split Hopkinson Tensile Bar (SHTB) setup. The setup consists of two Hopkinson bars – an input and output bar – between which the sample is fixed, see Fig. 3. A tube-like impactor is put around the input bar. To perform a test, the impactor is accelerated towards an anvil at the free end of the input bar, thus generating an

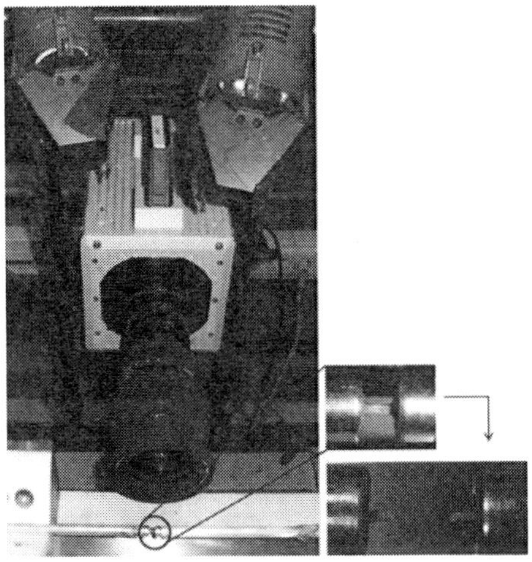

Figure 3. Tensile specimen glued between the Hopkinson bars together with high speed camera used to record the specimen deformation (left), before and after testing (right).

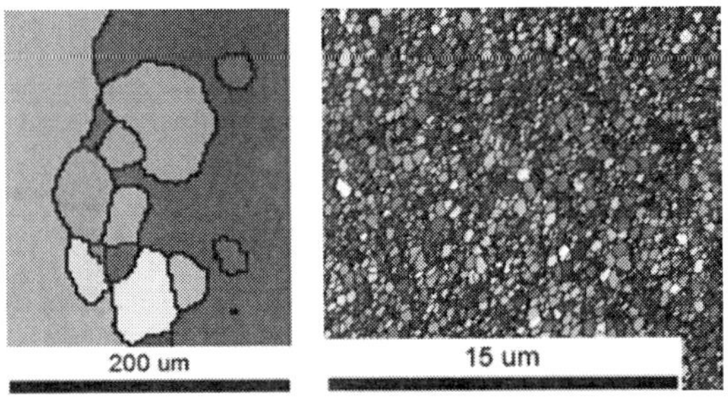

Figure 4. EBSD images of the undeformed material with a mean grain size of 300 micron (left) and the material deformed up to an equivalent strain of 50 with a mean grain size of 1.01 micron (right).

incident tensile wave with a length of 1 m and an amplitude proportional with the impactor velocity. In order to limit the impedance mismatch between the Hopkinson bars and the sample, and to capture the low force needed to deform the sample with a sufficiently high accuracy, the input and output bars consist of aluminium tubes with lengths of 2.1 m and 1.2 m respectively, and outer diameter of 8 mm and wall thickness of 1 mm.

3. Results

3.1. Grain refinement obtained by HPT

The grain size of the HPT processed samples is determined by electron back scatter diffraction (EBSD). In Fig. 4 EBSD images are represented of the annealed aluminium and a material sample in which the saturation grain size is reached (approximately after an equivalent strain of 15). Comparison of the two images clearly shows the obvious grain refinement obtained by the SPD process. In Table 2 values of the grain size are given at a distance of 3 mm away from the centre the HPT disk samples. This location corresponds with the centre of the gage section of the tensile specimens.

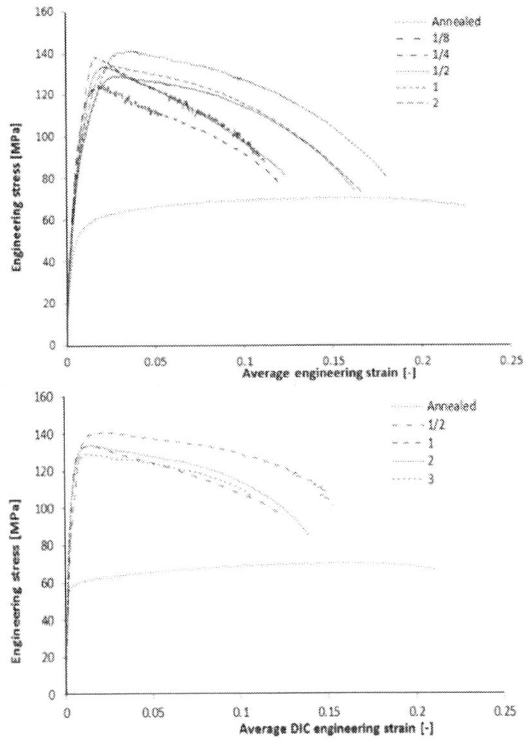

Figure 5. Static stress-strain curves of annealed and HPT processed material. The strain is calculated from the clamp displacement (top) or by DIC (bottom).

3.2. Static and dynamic tensile curves

In Fig. 5 the static tensile curves for the annealed and HPT samples are represented. In the top figure the strain values are averaged values obtained by dividing the clamp displacements by the initial gage length.

In Fig. 6 images of a deforming sample are given together with the results of DIC by which local strain values are obtained. The strain distribution is clearly non-uniform: in the top picture strains vary from 1 to 15%, in the bottom picture from 1 to 47%. Unfortunately, at high deformations, paint flaked off in the neck region and DIC strain values become unreliable.

In the bottom Fig. 5, curves are presented for the tests for which reliable DIC strain values are obtained. Before necking, the average strains obtained from the clamp displacements (top Fig. 5) overestimate the average strains in the gage section (bottom Fig. 5). For the dimensions of the tensile samples used here, deformation of the transitions zones of the sample, i.e. the shoulders, is limited [10]; the differences observed are mainly attributed to the slip in the clamps and a limited stiffness of the tensile device.

HPT processing has a pronounced influence on both the stress and elongation values. In the annealed sample the deformation is uniform up to a strain of 15%. In the HPT samples on the other hand, very soon after the onset of plastic deformation the curves reach a maximum and necking sets in. For the tensile tests represented in

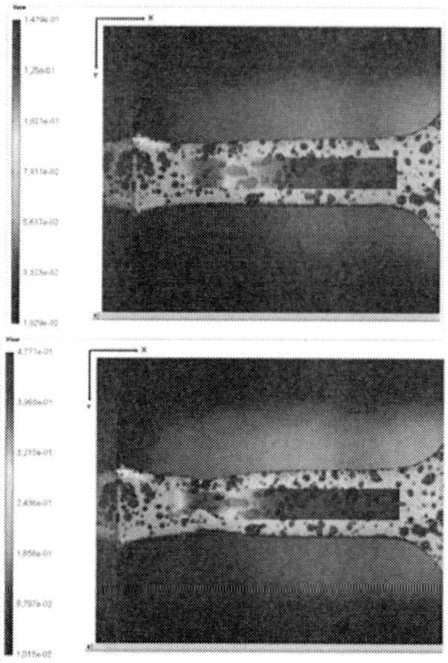

Figure 6. Sample during deformation with axial strain obtained by DIC processing (using MatchID^{TM}) of the speckle pattern applied to the sample.

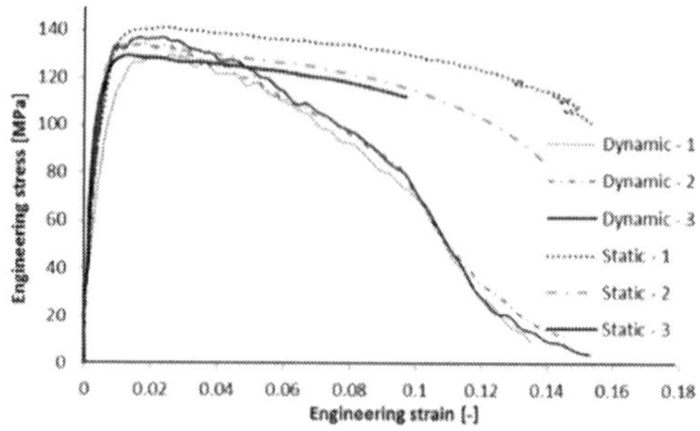

Figure 7. Static and dynamic stress-strain curves for the HPT samples subjected to 1, 2 and 3 rotations.

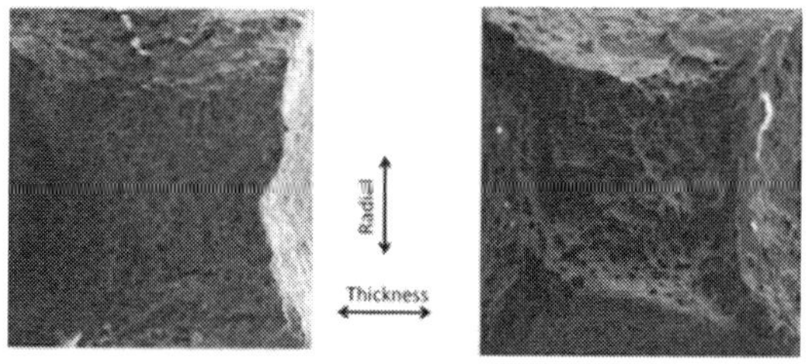

Figure 8. Fracture surface of a sample subjected to $^1/_2$ rotations in HPT, corresponding with an equivalent strain of 5.4, after a static (left) and dynamic (right) tensile test.

the bottom of Fig. 5, uniform elongations between 1.2% and 2.2% are found. A higher HPT deformation indeed results in lower uniform strains. From the tensile results a considerable increase in the ultimate tensile strength (UTS) of the material, even at lowest HPT deformation, is obtained compared with annealed samples.

The annealed sample has a tensile strength of 70 MPa. The strength increases significantly to a value of 125 MPa for the $^1/_8$ rotated sample, to 138 MPa for the $^1/_4$ sample. For higher HPT deformations, the changes in maximum stress level cannot directly be related with the applied HPT deformation and/or obtained grain refinement. Indeed, although no differences in grain size exist between the 2 and 3 rotations samples, the maximum stress of the sample of 3 rotations is lower than that of the 2 rotations sample. This might be attributed to the value of the cross section used to calculate the strain from the measured force. Here, the dimensions of the central section are used. However, in the HPT processed samples, a higher reduction of the sample thickness is obtained away from the centre, so this will result in an underestimation of the actual stress. The differences in thickness over the gage length are more pronounced for high levels of HPT deformation. This is in line with the observation that for higher HPT deformations, the strain localisation preceding fracture, occurs further away from the centre of the gage section.

In Fig. 7 next to static tensile curves, dynamic curves are given obtained by the split Hopkinson bar setup. The average strain rate during plastic yielding is 4400/s. When the static curves are compared with the dynamic curves, some trends can be observed. The effect of the strain rate on the onset of yielding and uniform deformation lies within the experimental scatter. HPT processing does not seem the affect the materials strain rate sensitivity. However, once necking sets in, the stress decreases more rapidly at higher strain rates. This might be attributed to thermal softening of the material due to adiabatic heating of the samples in the dynamic tests.

3.3. Post mortem analysis

In all samples a cup and cone fracture is observed with clear dimples typical for ductile fracture preceded by void nucleation, growth and coalescence. Fracture surfaces of the static and dynamic tensile samples are observed using scanning electron microscopy (SEM). In SEM images of the fracture surfaces equiaxed dimples around the centre and parabolic shear dimples in the narrow shear lips close to the borders are present, as can be seen in Fig. 8 for a sample subjected to $^1/_2$ rotations in HPT after a static (left) and dynamic (right) tensile test.

In Fig. 9 the engineering radial and thickness strain as measured in the symmetry planes in the fracture surface are represented as a function of applied strain in the HPT process. The radial strain is in all static (Fig. 9, top) and dynamic tests (Fig. 9, bottom) higher than the thickness strain, although there is some evolution as a function of applied deformation. The anisotropy as observed in the fracture surface, reaches its highest value at a HPT von Mises strain of 2.7. However, it should be mentioned that the strain values at higher deformations, are less accurate because of the thickness variation over the gage length of the sample.

From Fig. 9 it is also clear that, although the uniform deformation of the sample is limited, at fracture high deformations are obtained. The lowest values of the radial and thickness strain are obtained in the dynamic experiments; however, their value is well above 50%. Consequently, although the tensile curves suggest a low

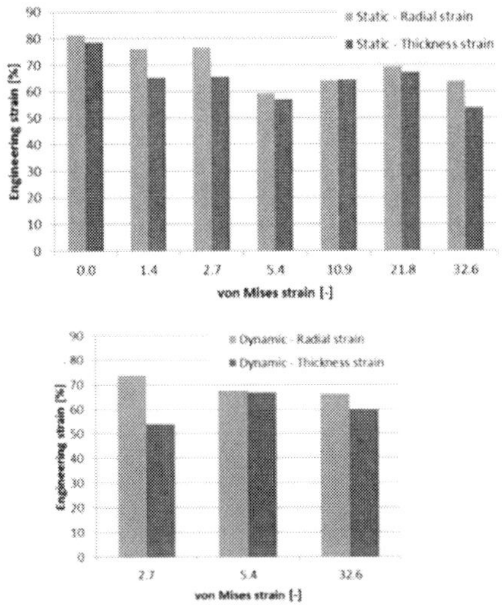

Figure 9. Fracture strain as a function of strain applied by HPT in the radial and thickness direction measured from SEM images of the fracture surfaces for the static (top) and dynamic (bottom) samples.

tensile ductility of the HPT processed aluminium, the material is intrinsically ductile. The low uniform tensile deformation is the result of a geometrical instability.

For low HPT deformations, the dimples in the fracture surface are approximately 12 microns, which is much smaller than the grain size. Independent of the strain rate, the dimple size decreases with increasing HPT strain, until it saturates at $\approx$9 micron, approximately nine times the grain size for samples subjected to an equivalent strain higher than 10.

4. Conclusions

In present study, the static and dynamic tensile behaviour of commercially pure aluminium processed by HPT is studied. The deformation applied to the samples by HPT processing prior to tensile testing results in a significant grain refinement: from a grain size of 300 micron in the annealed state to a saturation size of $\sim$1 μm for equivalent strains higher than 15. The grain size reduction results in a strength increase of the order of 2, in both the static and dynamic tests. For the commercially pure aluminium considered here, the influence of the strain rate on the onset of plastic deformation in the samples and the uniform elongation is found to be negligible. At higher strains, beyond necking, the stress drops more rapidly in the dynamically loaded samples, which might be attributed to thermal softening. Observation of the fracture surfaces shows that, independent of deformation level or strain rate, a ductile cup and cone fracture is obtained. The dimple size decreases with increasing HPT deformation to reach a saturation level when also the grain size saturates. Measurement of the strains in the fracture surface shows a slight difference in strain along the thickness and radial direction of the samples. The anisotropy seems to be more pronounced for low levels of deformation. The fracture strains are lower in the dynamic tests, although also here values well above 50% are found. Consequently, although the tensile curves suggest a low tensile ductility of the HPT processed aluminium, the HPT processed pure aluminium is intrinsically ductile.

The authors would like to acknowledge the Interuniversity Attraction Poles Program (IUAP) of the Federal Science Policy of Belgium and the partners of IUAP-VII-project P7/21 "Multiscale mechanics of interface dominated materials".

Also the help of Dr. Farideh Hajy Akbary of the Department of Materials Science and Engineering at TUDelft (The Netherlands) for carrying out the static tensile experiments is gratefully acknowledged.

References

[1] V. Segal, Mater. Sci. Eng. A **197** (1995)
[2] A. Zhilyaev, T. Langdon, Progr. Mat. Sci. **53** (2008)
[3] T. Lowe, R. Valiev, Nanostruct. Mater. **10** (1998)
[4] Y. Sait, H. Utsunomia, N. Tsuji, T. Sakai, Acta Mater. **47** (1999)
[5] E. Ma, J. Mat. Eng. Perf. **14** (2005)
[6] G. Dieter, Mechanical Metallurgy. 8th ed. New York, McGrw-Hill (1998)
[7] Q. Wei, S. Cheng, K.T. Ramesh, E. Ma, MSEA **381** (2004)
[8] Y. Wang E. Ma, Acta Mat. **52** (2004)
[9] Y.M. Wang, E. Ma, MSEA **375–377** (2004)
[10] P. Verleysen, J. Degrieck, T. Verstraete, J. Van Slycken, Exp. Mech. **48** (2008)

EPJ Web of Conferences **94**, 02013 (2015)
DOI: 10.1051/epjconf/20159402013

Impact behaviour of an innovative plasticized poly(vinyl chloride) for the automotive industry

C.A. Bernard[1], N. Bahlouli[1,a], C. Wagner-Kocher[2], S. Ahzi[1,3], and Y. Rémond[1]

[1] University of Strasbourg/CNRS, ICUBE Laboratory, 2 rue Boussingault, 67000 Strasbourg, France
[2] University of Strasbourg,UHA Mulhouse, LPMT, 11 rue Alfred Werner, 68093 Mulhouse Cedex, France
[3] Qatar Environment and Energy Research Institute, Qatar Foundation, PO Box 5825, Doha, Qatar

Abstract. Plasticized poly(vinyl chloride) (PPVC) is widely used in the automotive industry in the design of structural parts for crashworthiness applications. Thus, it is necessary to study and understand the influence of the mechanical response and mechanical properties of PPVC over a wide range of strain rate, from quasi-static to dynamic loadings. The process is also investigated using different sample thicknesses. In this work, the strain rate effect of a new PPVC is investigated over a wide range of strain rates at three temperatures and for three thicknesses. A modelling of the yield stress is also proposed. The numerical prediction is in good agreement with the experimental results.

1. Introduction

Polymeric materials are widely used in automotive industry in order to reduce the mass of vehicles and, by consequence, reduce the fuel consumption. Several polymers are used in the manufacturing of a vehicle. Among the polymers used in the automotive industry, poly(vinyl chloride) (PVC) is largely used for crashworthiness applications. Several authors [1–3] measured the glass transition temperature of PVC around 82 °C. Thus, at room temperature, the PVC exhibits a glassy behaviour.

A large part of the PVC used in automotive industry is modified by plasticizing in order to obtain flexible PVC at room temperature. This new material, so-called plasticized PVC (PPVC), is obtained by mixing powdered resin with stabilizer, plasticizer, lubricant and dioctyl phthalate due to lack of thermal stability of PVC at processing temperatures [4,5]. Then, this compound is melt mixed in a bench kneader at a rotor speed of 100 r.p.m. and at a temperature of 100 °C for 15 min.

The influence of the level of plasticizer on the mechanical response of the PPVC has been of particular interest for academic researchers and industrials because of the interest of PPVC applications in automotive crashworthiness. Kendall and Siviour [6,7] investigated the mechanical response of a PPVC under uniaxial compression, with a high level of plasticizer, from quasi-static to dynamic loadings at room temperature. At high strain rates, the PPVC exhibits a viscoelastic-viscoplastic response. However, no strain softening is observed unlike PVC. At low strain rates, the mechanical behaviour of the PPVC is viscohyperelastic. Thus, a change of behaviour is observed between quasi-static and dynamic loadings. Kendall and Siviour [6,7] have also investigated the temperature sensitivity if PPVC at $10^{-2}\,s^{-1}$. At high temperature, the PPVC has viscohyperelastic behaviour. For temperatures lower than −40 °C, the PPVC exhibits a glassy behaviour with a yield stress clearly identified and followed by a strain softening. They also show that the yield stress of PPVC is bilinearly dependent on the log of strain rate. Mulliken et al. [8] investigated on the mechanical behaviour of a 20% PPVC. They found that the addition of 20% of plasticizers in PVC blends decreased the glass transition temperature. Moreover, contrary to Kendall and Siviour [6,7], they found that the PPVC exhibited a glassy behaviour with the yield point followed by the strain softening at room temperature and high strain rate. Thus, the percentage of plasticizers in the PVC blend is of great importance for the mechanical response of material. Moreover, polymeric materials are well known to be strain rate and temperature dependent [9,10]. Thus, plasticized PVC exhibits a triple dependence on strain rate, temperature and plasticizers content.

At high stain rate, other phenomenon could affect the mechanical response of polymers. Dioh et al. [11] investigated on the thickness effects in Split Hopkinson Pressure Bars tests. Using samples with two different thicknesses, they performed quasi-static and dynamic compressive tests, at room temperature, from $10^{-4}\,s^{-1}$ to $10^{4}\,s^{-1}$. For several polymers (polycarbonate, high density polyethylene, medium density polyethylene and poly(ether ether ketone)), they observed, for strain rates greater than $10^{3}\,s^{-1}$, a discrepancy in the yield stress sensitivity to the log of strain rate between the two thickness.

In this study, the strain rate sensitivity of new plasticized PVC, for the automotive industry, is investigated over a wide range of temperatures and strain rates. This material was manufactured by Faurecia. It is well known that the mechanical properties of a polymer are dependent

[a] Corresponding author: nadia.bahlouli@unistra.fr

Article available at http://www.epj-conferences.org or http://dx.doi.org/10.1051/epjconf/20159402013

of the process, particularly if the final structure can have different thicknesses. A gradient in temperature through the thickness can lead to different microstructure through the thickness, and thus, a gradient of the mechanical properties. In order to verify that the PPVC processing does not affect the mechanical properties of the material, the manufacturer has prepared three different thicknesses.

2. Experimental procedures

2.1. Material

The plasticized poly(vinyl chloride) (PPVC) is composed of 50% of PVC, 40% of plasticizers and 10% of additives. The PPVC sheets are provided by Faurecia with three different thicknesses: 1, 1.25 and 1.5 in normalized values, respectively called TH1, TH2 and TH3.

A previous investigation has been performed by Bernard et al. [12] on the PPVC used in this study. They investigated the temperature sensitivity of PPVC materials over a wide range of strain rate and temperature for the three thicknesses. They performed thermogravimetry analysis (TGA), differential scanning calorimetry (DSC), quasi-static uniaxial tensile tests, dynamic uniaxial compressive tests and observed the morphology of this material. For each thickness, they measured the glass transition temperature around $-32\,°C$ using DSC and two degradation peaks, around $330\,°C$ and $450\,°C$ using TGA. Thus, Bernard et al. [12] concluded that, for the three thicknesses, thermal properties are similar. Thus, the process used to manufacture the new PPVC does not affect the thermal properties. However, quasi-static uniaxial tensile tests and dynamic uniaxial compressive tests have shown a non-negligible discrepancy in the mechanical response between the three thicknesses. Moreover, a non-linear relationship has been found between the temperature and the elastic modulus, yield stress and yield strain.

In the present paper, we investigate the strain rate dependence by considering a much larger range of loading rates compared to our previous study [12] and considering three temperatures ($-30\,°C$, $23\,°C$, $85\,°C$). We also use the cooperative model to predict the yield stress and compare predicted results with experimental ones for the considered wide range of temperatures and strain rates.

2.2. Mechanical tests

Quasi-static tensile tests and dynamic compressive tests have been conducted in order to investigate on the strain rate sensitivity of the material for each thickness. The mechanical characterization of the three thicknesses of PPVC has been performed at three temperatures: $-30\,°C$, close to the glass transition temperature, $23\,°C$ and $85\,°C$ in the rubbery region.

The low strain rate tests were carried out using an INSTRON 3384 universal testing machine equipped with a temperature chamber. For each temperature, tests were performed at two constant displacement rates: 50 mm/min and 500 mm/min. At the yield point, these displacement rates lead to true strain rates from $0.01\,s^{-1}$ and $0.3\,s^{-1}$. The tensile test samples are cut in PPVC sheets by Faurecia. The specimen geometry follows the ASTM D412 Standard.

The high strain rate tests were carried out on a homemade Split Hopkinson Pressure Bars (SHPB) for a wide range of strain rate from $1000\,s^{-1}$ to $7000\,s^{-1}$. The test fixture is composed of three 22 mm steel bars. The strikers are 0.5 m and 1 m long and the input and output bars are 3 m long. A description of the experimental setup can be found in [9,10]. In order to perform dynamic compressive tests at low and high temperatures, two devices were adopted on our SHPB. The device for high temperatures consists in positioning the sides of the incident bar and transmitted bar in contact with the specimen in a furnace containing two electrical resistances. For low temperatures, the material sample was sandwiched between the two bars inside a temperature chamber. Then, the specimen and a short part of the compressive bars were cooled.

To reach the thermal equilibrium, the sample were heated or cooled in the appropriate chamber during 15 min before testing. The temperature inside the chamber is measured using four thermocouples. The compressive samples, cut from the PPVC sheets, are cylindrical specimens with a diameter of 10 mm and a thickness depending of the material (TH1, TH2 or TH3).

3. Results and discussion

3.1. Strain rate sensitivity of the mechanical properties

The stress-strain behaviour of PPVC, with thickness TH2, under uniaxial tension at low strain rate and uniaxial compression at high strain rate is presented in Fig. 1 for different temperatures. A change of behaviour could be observed between quasi-static loadings and dynamic loadings for all temperatures. At $-30\,°C$ and at low strain rate, in Fig. 1(a), the PPVC exhibits viscoelastic-viscoplastic behaviour. For this temperature and this range of strain rate, the PPVC is tested in its glass transition region. At high strain rate and at $-30\,°C$, PPVC exhibits glassy polymer behaviour with the yield point clearly identified before the strain softening. For strain rates greater than $3800\,s^{-1}$, the orientational hardening of the material is also observed.

At $23\,°C$ and low strain rate, in Fig. 1(b), PPVC exhibits viscohyperelastic behaviour. At this temperature and this strain rate range, the material is tested at the beginning of its rubbery region. At high strain rates, PPVC exhibits viscoelastic-viscoplastic behaviour similar to the one obtained at $-30\,°C$ and at $0.03-1$. Thus, at high strain rates and at $23\,°C$, PPVC is tested in its glass transition region. An important strain hardening is noticed in Fig. 1(b).

At $85\,°C$, at low and high strain rates, PPVC exhibits a rubbery behaviour. Moreover, comparing Figs. 1(a), 1(b) and 1(c), we can observed that at low strain rates and at $85\,°C$, due to limitation of the machine, no failure was reached for any specimens.

Comparing Figs. 1(a), 1(b) and 1(c), we can clearly identify the temperature sensitivity of PPVC from glassy to rubbery regions through the glass transition region.

In Fig. 2, the strain rate sensitivity on the elastic modulus, yield stress and yield strain are plotted for the

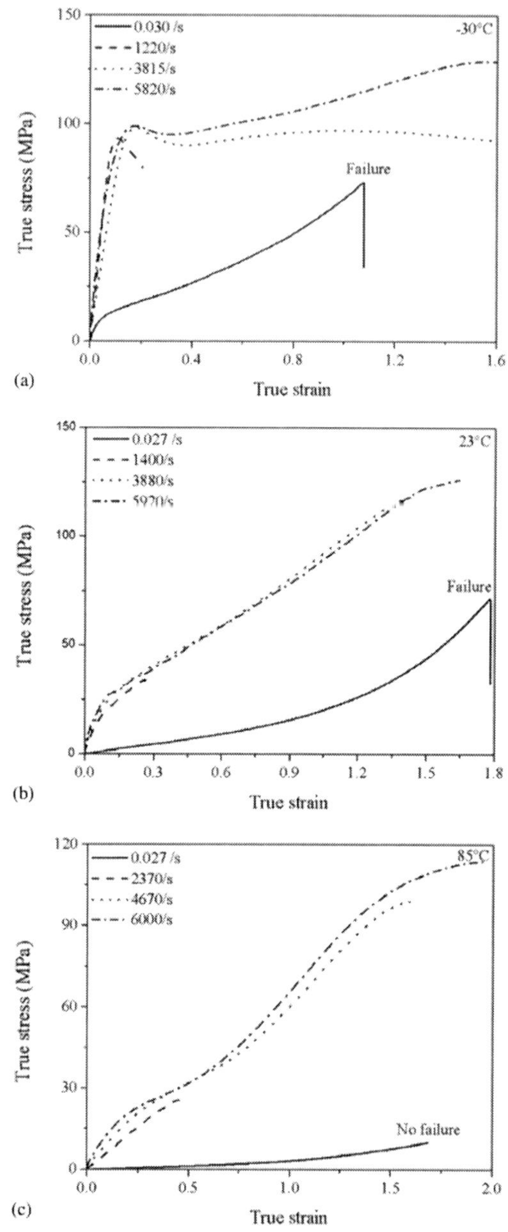

Figure 1. Stress-strain behaviour of the PPVC over a wide range of strain rate at (a) −30 °C, (b) 23 °C and (c) 85 °C for the thickness TH2.

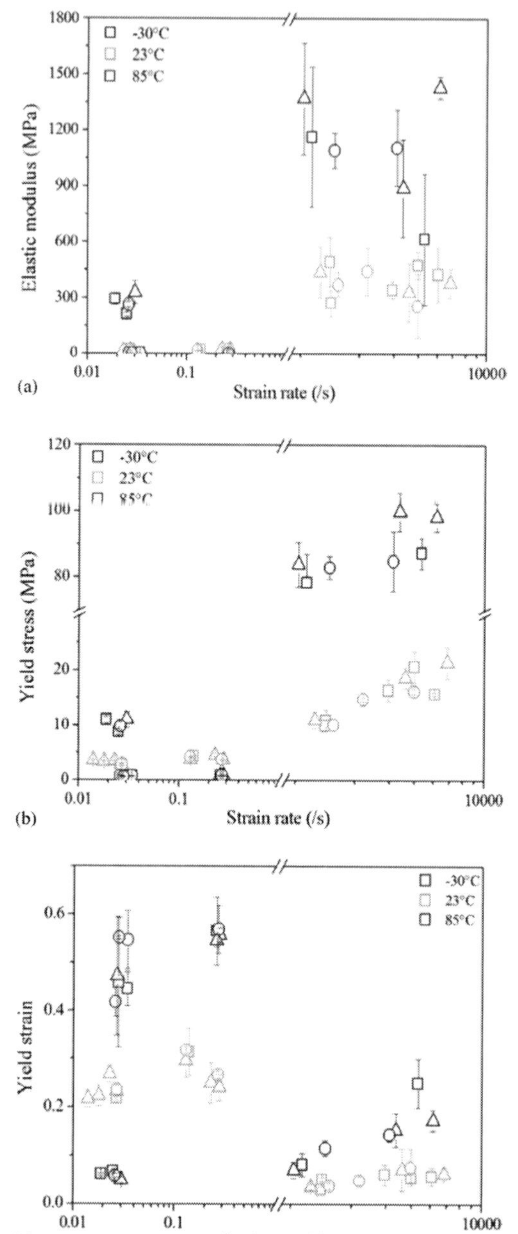

Figure 2. Strain rate sensitivity on (a) elastic modulus, (b) yield stress and (c) yield strain of PPVC for the three temperatures and the three thicknesses with ⊡ TH1, △ TH2, ⊙ TH3.

three thicknesses. For the elastic modulus, in Fig. 2(a), large discrepancy is observed, especially at −30 °C which can be attributed to the proximity of the glass transition region and at non-optimal dynamic equilibrium. However, we can still note an increase of the elastic modulus when the temperature decreases and the strain rate increases. The thickness TH2 presents a higher elastic modulus than the two other thicknesses.

In Fig. 2(b), the discrepancies are less important than for the elastic modulus (Fig. 2(a)). A bilinear dependence of the yield stress in function of log of strain rate is observed, in Fig. 2(b), particularly at 23 °C. At −30 °C and at 23 °C, we measured higher yield stress for the thickness TH2. Moreover, we can note that the yield stress increases when the strain rate increases and the temperature decreases.

In Fig. 2(c), the yield strain is plotted as function of log of strain rate from $0.01\,s^{-1}$ to $10000\,s^{-1}$. For quasi-static tests, the yield strain increases with the temperature and the strain rate. However, at high strain rates, a modification of the mechanical behaviour is observed. The yield strain still increases with the stain rate. However, the yield strain is greater at −30 °C than at 23 °C. The modification of behaviour could be explained by the transition from quasi-static to dynamic loadings, from rubbery to glass transition region at 23 °C and from glass transition to glassy region at −30 °C. In order to validate this assumption, experimental tests need to be performed at temperature lower than −30 °C.

3.2. Yield stress modelling

At 23 °C, the bilinear dependence of the yield stress to the log of the strain rate is clearly identified in Fig. 2(b). Numerous theories have been developed in the literature [13–16] in order to model the strain rate and temperature dependence of the yield stress of amorphous polymers over a wide range of strain rate and temperature. However, few models allow reproducing the yield stress of polymers in their glass transition or rubbery regions. The cooperative model, proposed by Richeton et al. [16], is adopted here. In this model, it is proposed to consider individual segment motions as separated events and built a model based on n cooperative motions. In their formulation of the cooperative model, Richeton et al. [16] considered the time/temperature superposition principle in order to build a master curve at a chosen temperature. The cooperative model assumes the existence of an internal stress σ_i such that the effective stress σ^{eff} is given by (Fotheringham and Cherry [17], Richeton et al. [16]):

$$\sigma^{eff} = \sigma_y - \sigma_i \tag{1}$$

with σ_y the yield stress. The internal stress is a structural parameter which depicts the arrangement of defects inherited from past thermal history. The expression of the cooperative model below and above the glass transition temperature is given by (Richeton et al. [16]):

$$\frac{\sigma_y}{T} = \sigma_i(0) + \frac{2k}{V}\sinh^{-1}\left(\frac{\dot{\varepsilon}}{\dot{\varepsilon}^*(T)}\right) \tag{2}$$

where

$$\sigma_i(T) = \begin{cases} \sigma_i(0) - mT & \text{if } T < T_g \\ 0 & \text{if } T \geq T_g \end{cases} \tag{3}$$

and

$$\dot{\varepsilon}^*(T) = \begin{cases} \dot{\varepsilon}_0 \exp\left(-\frac{\Delta H_\beta}{RT}\right) & \text{if} \quad T < T_g \\ \dot{\varepsilon}_0 \exp\left(-\frac{\Delta H_\beta}{RT_g}\right) \exp\left(\frac{\ln 10 c_1^g (T - T_g)}{c_2^g + T - T_g}\right) \\ \text{if} \quad T \geq T_g \end{cases} \tag{4}$$

with $\sigma_i(0)$ the internal stress at 0 K, m a material parameter, T the absolute temperature, T_g the glass transition temperature, ΔH_β the glass transition temperature, k the Bolzmann constant c_1^g and c_2^g the WLF parameters and $\dot{\varepsilon}_0$ a constant pre-exponential strain rate.

In order to build the master curve of PPVC, we first chose a reference temperature and we performed the time/temperature superposition principle. The material parameters are presented in Table 1. Due to a lack of data in the glassy region the internal stress and the parameter m are assumed null.

In Fig. 3, the master curve is built for the PPVC at −30 °C. All thicknesses are fused to obtain more data. A good agreement is found between the experimental data and the numerical model in spite of some discrepancies observed at intermediate and high strain rates.

Table 1. Parameters for the cooperative model. The WLF parameters are from Richeton et al. [16] for PVC.

Parameters	PPVC
ΔH_β (kJmol^{-1})	92
$\sigma_i(0)$ (Mpa)	0
m (MPaK^{-1})	0
V (m^3)	2×10^{-28}
$\dot{\varepsilon}_0$ (s^{-1})	2.3×10^{20}
n	3.6
T_g (K)	242
c_1^g	17.44
c_2^g (°C)	51.6

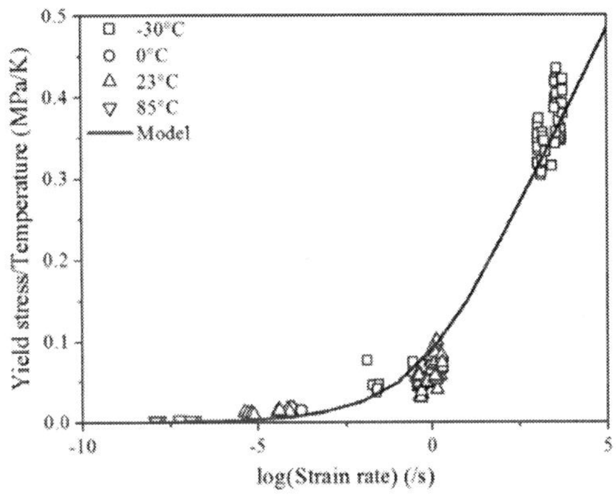

Figure 3. Master curve built at −30 °C for the PPVC tested in uniaxial compression at high strain rate and uniaxial tension at low strain rate. All thicknesses are fused.

4. Conclusion

Strain rate investigation has been performed on a new PPVC dedicated to the automotive industry. The experimental results show that the mechanical properties especially at high strain rates are different for the different thicknesses. For all the investigations, it seems that the elastic properties (elastic modulus and yield stress) are greater for the thickness TH2. The temperature gradient through the thickness reached during the process for TH2 leads to a microstructure which improves the elastic properties. The yield stress of PPVC exhibits a bilinear dependence to the log of strain rate and a master curve at −30 °C has been built and modelled using the cooperative model. A good correlation is obtained between the experimental data and the numerical prediction. For the yield strain, a change of the behaviour, in the strain rate dependence, is observed between low and high strain rate. This phenomenon could be explained by the change of behaviour of the material at the considered strain rates from rubbery to glass transition regions and from glass transition to glassy regions.

The authors would like to acknowledge Faurecia for providing the materials.

References

[1] Y. Wang, E.M. Arruda, P.A. Przybylo, Rubber Chem. Technol. **74**, 560 (2001)

[2] M. Aubin, R.E. Prud'homme, Macromolecules **21**, 2945 (1988)

[3] X. Lu, R.A. Weiss, Macromolecules **25**, 3242 (1992)

[4] K.T. Varughese, G.B. Nando, S.K. De, S. K. Sanyal, J. Mater. Sci. **24**, 3491 (1989)

[5] C.-S. Ha, Y. Kim, W.-K. Lee, W.-J. Cho, Y. Kim, Polymer **39**, 4765 (1998)

[6] M.J. Kendall, C.R. Siviour, EPJ Web of Conferences **26**, 02009 (2012)

[7] M.J. Kendall, C.R. Siviour, P. Roy. Soc. Lon. A: Mat. **470** (2014)

[8] A.D. Mulliken, S.Y. Soong, M.C. Boyce, R.E. Cohen, J. Phys. IV **134**, 217 (2006)

[9] M.R. Boumbimba, E. Hablot, K. Wang, N. Bahlouli, S. Ahzi, L. Avérous, Compos. Sci. Technol. **71**, 674 (2011)

[10] K. Wang, F. Addiego, N. Bahlouli, S. Ahzi, Y. Rémond, V. Toniazzo, Compos. Sci. Technol. **95**, 89 (2014)

[11] N.N. Dioh, P.S. Leevers, J.G. Williams, Polymer **34**, 4230 (1993)

[12] C.A. Bernard, N. Bahlouli, C. Wagner-Kocher, J. Lian, S. Ahzi, Mater. Design (under review) (2015)

[13] H. Eyring, J. Chem. Phys. **4**, 283 (1936)

[14] R.E. Robertson, J. Chem. Phys. **44**, 3950 (1966)

[15] A.S. Argon, Philos. Mag. **28**, 839 (1973)

[16] J. Richeton, S. Ahzi, L. Daridon, Y. Rémond, Polymer **46**, 6035 (2005)

[17] D.G. Fotheringham, B.W. Cherry, J. Mater. Sci. **13**, 951 (1978)

EPJ Web of Conferences **94**, 02014 (2015)
DOI: 10.1051/epjconf/20159402014

Behavior of metals Induced by magnetic pulse loading

Atroshenko Svetlana[1,a], Morozov Viktor[2], Gribanov Denis[1], Lukin Anton[2], and Petrov Yuriy[1]

[1]IPME RAS, V.O., Bolshoy, 61, St. Petersburg 199178, Russia
[2]Saint Petersburg state university, 7-9, Universitetskaya nab., St. Petersburg 199034, Russia

Abstract. The investigation of copper and aluminum ring samples was carried out using magnetic pulse loading. Two modifications of the magnetic pulse technique were used. They were based on a GKVI-300 high-voltage narrow-pulse generator Morozov et al. (2011) [1]. It is possible using these two approaches to decrease the period of the harmonic load up to 100 ns. The study of fracture surfaces of aluminum and copper samples after the test was carried out on an optical microscope Axio-Observer-Z1-M in a dark field, and study of the cross sections structure – in the bright field or C-DIC. The structure has been studied in cross sections after appropriate etching. Grain size and the number of pores on the surface of cross sections were determined after etching. Microhardness was measured on a PMT-3 device with a load of 20 g. The optical micrographs of aluminum demonstrate that the long pulse causes almost fully ductile fracture. In the case of the short pulse, the number of fibers decreases: the fracture surface exhibits the signs of both ductile cup fracture and brittle crystalline fracture with cracks, which are sometimes rather deep. In addition, the short pulse results in twinning, which seems surprising for aluminum featuring a high stacking fault energy. It is seen that under short loading dynamic recrystallization occurs. As for copper samples before loading they were in the form of single crystal and after loading their structure due to dynamic recrystallization consists of small grain. The specimen with notch has more developed dynamic recrystallization shear bands.

1. Introduction

Zhang and Ravi-Chandar [2–5] reported recent experimental data for high-speed tangential extension to rupture of thin metallic rings by a magnetic pulse technique. Electro-magnetic loading provided radial expansion rates of the ring in the range 80–200 m/s or strain rates on the order of $10^4\,s^{-1}$. Necking and fragmentation were recorded with a high-resolution streak camera. It was shown that the ring fails for a time much shorter than the first period of current oscillations in it. Fragmentation lasted less than 20 μs. The radius of the ring as a function of time was determined photographically; the velocity, by differentiating this function. The ring rapidly picked up a speed of 100–250 m/s and then slowed down when the current decreased and the energy dissipated as a result of plastic deformation.

Strain rate $\varepsilon' = r'/r$ was roughly estimated as falling into the interval $(0.5\text{–}1.3)\times10^4\,s^{-1}$. In uniformly deforming regions of the ring, the strain was found to never exceed a necking threshold. An increase in the cross-sectional area of the sample considerably increased the strain preceding the onset of strain localization. The strain before localization did not depend on the strain rate. In samples with a cross-sectional area ratio of more than 5 the strain localized in the form of shear bands. As the sample grew in size, the strain distribution became more uniform. It was shown that polymeric coatings slightly influence the onset of necking at high-rate deformation. Tubes exhibited uniform expansion to a critical strain with subsequent formation of shear bands and failure at the sites of their intersection. Nomenclature is given in Table 1.

In this work, the approach suggested in Morozov et al. (2011) [1] is elaborated.

2. Experimental technique

Two modifications of the magnetic pulse technique were used. They were based on a GKVI-300 high-voltage narrow-pulse generator providing voltage amplitudes of 30–300 kV [1]. They could realized a sinusoidal load with a period of 7.5 μs and with a period of 1 μs.

The loading scheme is shown in Fig. 1. A current passing through a solenoid on which a coaxial ring sample is placed induces a current in the sample, and interaction between these currents gives rise to a repulsion force between the ring and solenoid. The solenoid was made of a copper wire 1 mm in diameter. It had five turns and a diameter of 25 mm. The current through the solenoid was measured by Rogowski coil and displayed on digital oscilloscope, and information from the oscilloscope was written in electronic format. When the sample coaxially placed on solenoid broke, photodiode recorded a flash indicating the instant of rupture.

The study of fracture surfaces of aluminum and copper samples after the test was carried out on an optical microscope Axio-Observer-Z1-M in a dark field, and study the structure of cross sections – in the bright field or C-DIC. The structure has been studied in cross sections after appropriate etching. Grain size and the number of pores on the surface of cross sections were determined after etching. Microhardness was measured on a PMT-3

[a] Corresponding author: author@email.org

Article available at http://www.epj-conferences.org or http://dx.doi.org/10.1051/epjconf/20159402014

Table 1. Nomenclature.

Symbol	Title
T	the period of the harmonic load, μs
S	is the shear area at the fracture surface
d	sample diameter
D	grain size
n	number of pores in the area of 400 μm^2

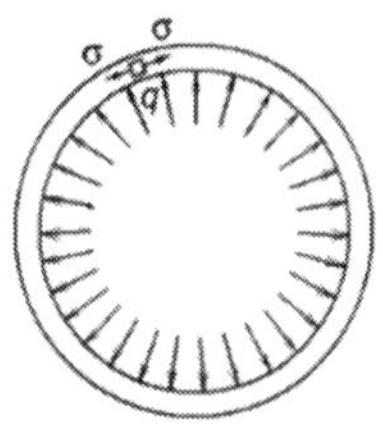

Figure 1. Sample loading scheme: q, load; σ, tensile stress [1].

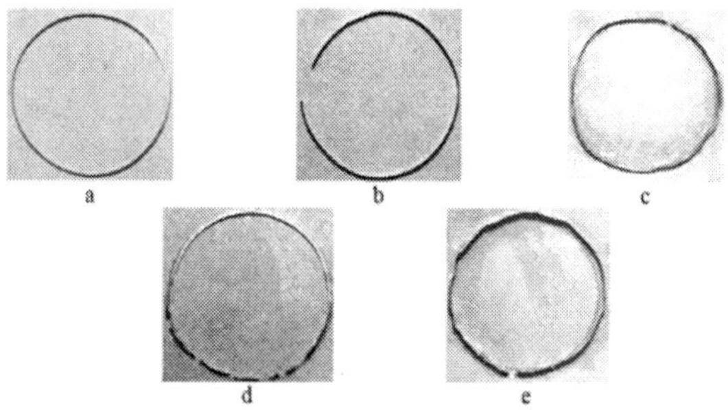

Figure 2. View of the sample: (a) before loading; (b) fracture at a threshold energy, $T = 7.4\,\mu s$; (c) fracture at a threshold energy, $T = 1\,\mu s$; (d) fracture at an energy far exceeding the threshold, $T = 7.4\,\mu s$; and (e) fracture of the sample made of aluminum foil 0.03 mm thick, $T = 1\,\mu s$ [1].

device with a load of 20 g. The ductile fracture surface was matt gray with characteristic "fibers." The brittle fracture surface is seemingly crystalline without noticeable signs of plastic deformation.

Amount B (%) of the ductile component on the fracture surface was determined by the formula (1)

$$B = 100 - X \tag{1}$$

given in the State Standard (X is the percentage of the brittle component on the fracture surface). The surface area of the brittle component was measured on the photograph.

Test rings were made of aluminum alloy and had an inner diameter of 28 mm. Rings to be tested with long-period voltages were 0.11–0.40 mm thick and 0.5–1.0 mm wide, and those subjected to short-period voltages were 0.11 mm in thickness and 1.0 mm in width. It was investigated also rings made of foil 0.03 mm thick and 1.5–2.0 mm wide.

3. Results and discussion

Figure 2 shows the samples in initial state and those having been subjected to loading under different conditions. Both for the 7.4-μs-long pulse (Fig. 2b) and for the 1.0-μs-long

Table 2. Percentage of fibers on the fracture surface of aluminum.

Number	Loading technique	T, μs	S, %
1	Magnetic pulse loading	7.5	98.2
2	The same	1	91.5
3	Light gas gun	1.3	94.03

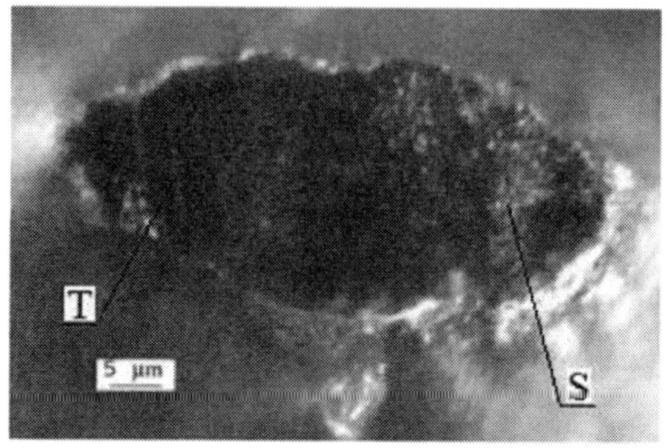

Figure 3. Fracture surface of the sample subjected to 1-μm-wide loading pulses: S – shear area and T – twins.

Table 3. Font styles for a reference to a journal article.

Material	T, μs	d, mm	h x b, μm^2	HV, MPa	D, μm	n, 1/400 μm^2
Al	7.5	28	193.5×10^3	1098	1.2	121
Al	1	28	68.4×10^3	1202	3.0	70
Al	0.08	28		303.3	1.05	36
Al initial		28	120×10^3	135.76	4.6	53
Cu	1	28	29.9×10^3	405.4	2.5	96
Cu notch	1	28	75.8×10^3	705.8	0.8	380
Cu notch	0.08	28		1070.5	0.84	111
Cu initial		28	15×10^3	142.46		282

pulse (Fig. 2c), rupture is seen to occur at the same site; the only difference is that many necks are observed in the latter case. When the energy of the 7.4-μs-long pulse exceeds the threshold value, fragmentation becomes more intense. When the foil sample is loaded by the 1.0-μs-long pulse (Fig. 2e), fragmentation is accompanied by necking.

The measured data for the ductile component of fracture are given in Table 2. The notation in Table 2 is the following: T is the pulse width, S is the shear area at the fracture surface. As the pulse gets shorter, the fraction of fibers in fracture decreases monotonically; that is, the samples embrittle.

For comparison, Table 2 gives experimental values of T and S for aluminum loaded by an air–gas impact machine with an intermediate pulse width of 1.3 μm. The results of fracture surface examination are given in Table 2. In the case of the long pulse, fracture is seen to be mostly ductile; that is, the samples are more ductile than for the short pulse. The shortening of the current period the share area on the fracture surface decreases monotonically, i.e. samples become more brittle.

The optical micrographs demonstrate that the long pulse causes almost fully ductile fracture. In the case

a

b

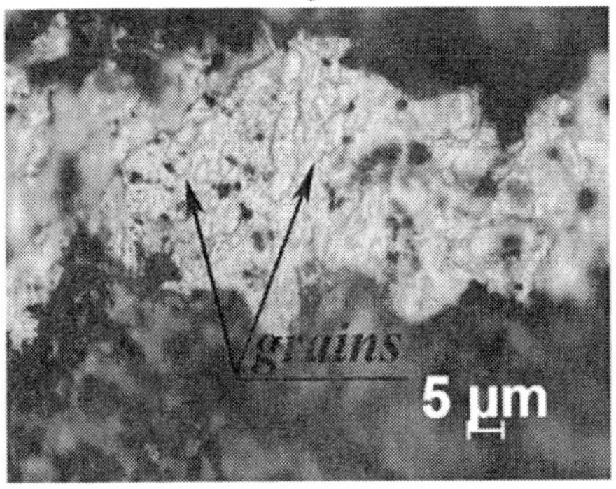

c

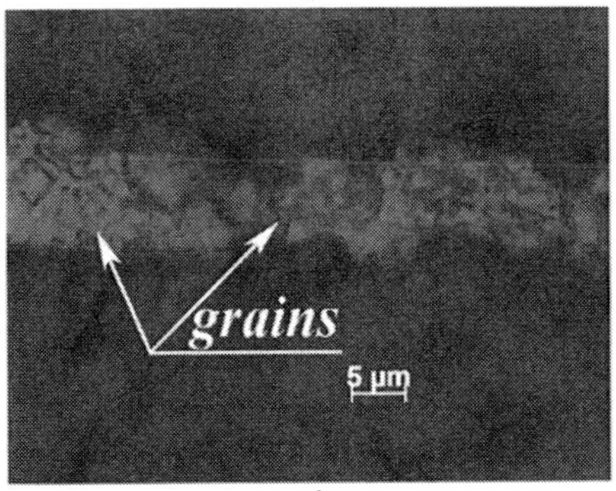

d

Figure 4. The structure of aluminum: (a) – in the initial state, (b) – after loading with the period T = 7.5 μs (S = 0.19 mm^2); (c) – after loading with the period T = 1 μs (S = 0.07 mm^2); (d) – after loading with the period T = 0, 08 μs (S = 0.02 mm^2).

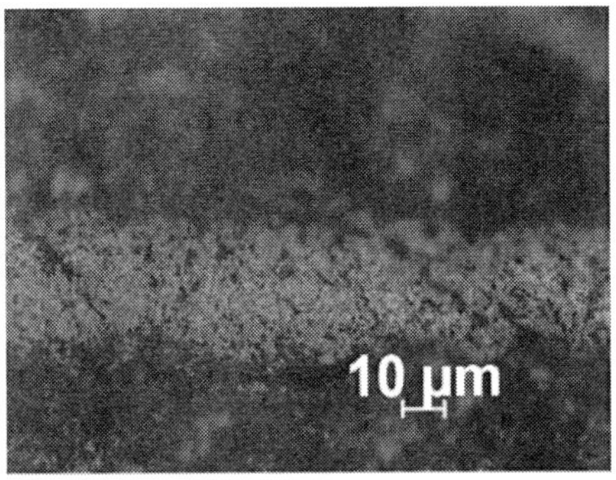

a

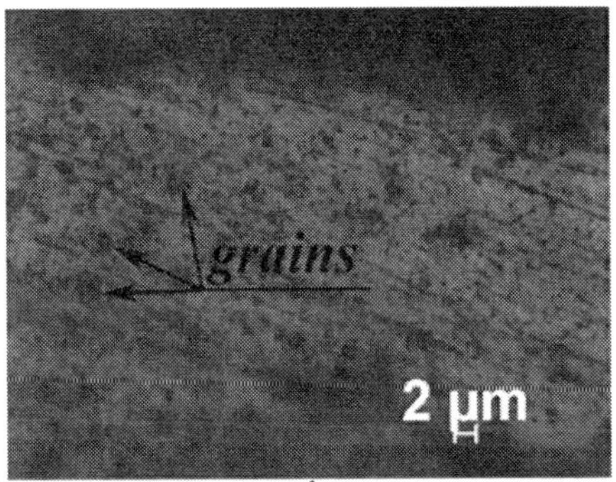

b

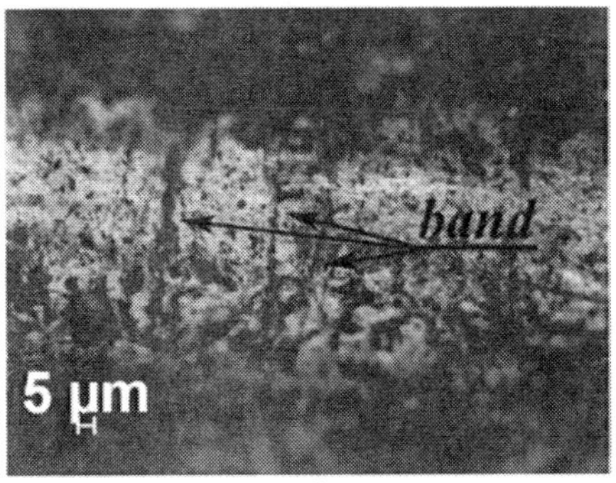

c

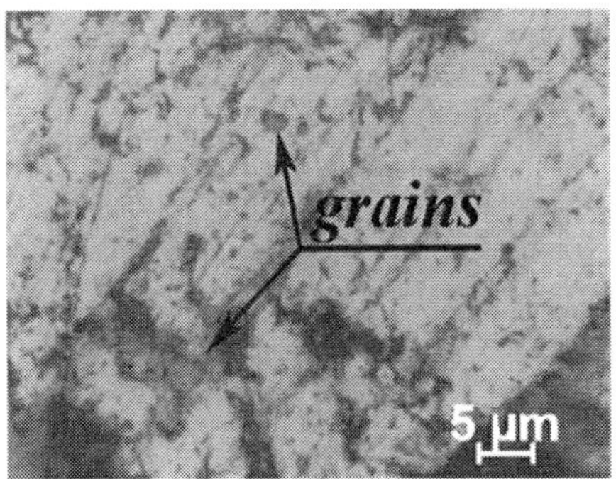

d

Figure 5. Structure of copper: a – in the initial state, b – nucleation of new grains during dynamic recrystallization after loading with the period T = 1 μs (S = 0.03 mm^2); c – shear bands in a sample with a notch after loading with period T = 1 μs (S = 0.01 mm^2), d – nucleation of new grains during dynamic recrystallization after loading with the period T = 0.08 μs (S = 0.02 mm^2).

of the short pulse, the shear area decreases: the fracture surface exhibits the signs of both ductile cup fracture and brittle crystalline fracture with cracks, which are sometimes rather deep. In addition, the short pulse results in twinning (Fig. 3), which seems surprising for aluminum alloys featuring a high stacking fault energy. Twinning is a typical deformation mechanism at fast loading of alloys with a low or medium stacking fault energy. As for materials with a high stacking fault energy, they twin only under a high load (above the twinning threshold) or at a very high rate of loading.

Characteristics of the samples and results are summarized in Table 3, where T – the period of oscillation of the current in the coil; d – ring diameter; S = h × b – cross-section of the sample; D – grain size; n – number of pores in the area of 400 μm^2; HV – microhardness.

Figure 4 shows the structure of aluminum samples at initial state and after loading with different periods and different cross-section (scale factor).

It is seen the dynamic recrystallization – the formation of new small grains for short-time loading. The highest degree of dynamic recrystallization is in the samples of aluminum with longer duration of loading and a maximum cross-section (scale factor). Aluminum samples after loading showed a greater tendency to porosity with increasing loading period compared with baseline.

Porosity in copper after magnetic pulse loading was below than in the initial state. In addition, with increasing loading duration the samples had the generation of multiple cracks.

Figure 5 shows the structure of copper in the initial state and after the magnetic pulse loading.

The copper sample before loading was a single crystal, and after loading as a result of dynamic recrystallization small grains appear with the size – 0.7–2.5 μm. And the more advanced dynamic recrystallization is in a sample with a higher period of the current – smaller grains are here. Furthermore, in the sample with a notch there are large amount of shear bands.

4. Conclusions

1. It was developed and tested a magnetic pulse method with simultaneous photographic registration of the moment of fracture metallic ring samples with short (about 1 μs) pulse tensile influences.

2. It was found that shortening the exposure time leads to the fracture of the samples. Samples become more brittle. Dynamic recrystallization processes and the formation of cracks observed in the aluminum samples, whereas twinning and shear bands formation occur in copper.

Work was financially supported by RFBR grants 14-01-00814, 12-01-91154-NSFC, RFBR 13-01-00335, No.25 RAS Presidium program and FP7 EU MARIE CURIE Project TAMER No. 610, 547.

References

[1] V.A. Morozov, Yu.V. Petrov, A.A. Lukin, V.M. Kats, A.G. Udovik, S.A. Atroshenko, D.A. Gribanov, G.D. Fedorovsky, Technical Physics **56**, (2011) 797–802

[2] O.H. Zhang and K. Ravi Chandar, Int. J. Fract. **142**, (2006) 183–217

[3] O.H. Zhang and K. Ravi Chandar, Int. J. Fract. **150**, (2008) 3–36

[4] O.H. Zhang and K. Ravi Chandar, Int. J. Fract. **155**, (2009) 101–118

[5] O.H. Zhang and K. Ravi Chandar, Int. J. Fract. **163**, (2010) 41–65

EPJ Web of Conferences **94**, 02015 (2015)
DOI: 10.1051/epjconf/20159402015

Strength and failure of a damaged material

Ellen K. Cerreta[1,a], George T. Gray III[1], Carl P. Trujillo[1], Mark L. Potocki[2], Shraddha Vachhani[1], Daniel T. Martinez[1], and Manual L. Lovato[1]

[1]MST-8, Los Alamos National Laboratory, Los Alamos 87545, New Mexico
[2]XTD-PRI, Los Alamos National Laboratory, Los Alamos 87545, New Mexico

Abstract. Under complex, dynamic loading conditions, damage can occur within a material. Should this damage not lead to catastrophic failure, the material can continue to sustain further loading. However, little is understood about how to represent the mechanical response of a material that has experienced dynamic loading leading to incipient damage. Here, this effect is examined in copper. Copper is shock loaded to impart an incipient state of damage to the material. Thereafter compression and tensile specimens were sectioned from the dynamically damaged specimen to quantify the subsequent properties of the material in the region of intense incipient damage and in regions far from the damage. It is observed that enhanced yield stresses result from the damaged material even over material, which has simply been shock loaded and not damaged. These results are rationalized in terms of stored plastic work due to the damage process.

1. Introduction

There are a number of applications that expose structural materials to the extreme environment of dynamic loading. These applications range from aerospace, automotive, civil, and even manufacturing needs. Often times under dynamic loading conditions, which typically impart high stresses (multiple GPa) at high rates, failure of these structural materials can be catastrophic [1,2]. This is even true for ductile, structural metals. For this reason, it is desirable to not only predict the catastrophic failure of these materials but also move toward developing such a robust understanding that will enable next generation materials to be designed to mitigate failure.

Currently, many predictive failure models do not accurately represent failure under dynamic loading conditions [3–5]. This is, in part, attributed to a lack of understanding about the developing microstructure under dynamic loading. One aspect of this microstructure that is of particular importance is the strength of the dynamically loaded or shocked microstructure prior to failure.

For a number of years, multiple investigations have examined the influence of shock loading on the subsequent properties of a material [6–8]. These studies have focused primarily on examining the quasi-static, uni-axial stress response of a metal that had been loaded to peak shock stresses in excess of 5 GPa and then soft recovered for post mortem examination. It has been shown, for high purity metals, that the way in which a material stores defects, determines the role of shock loading. For example, FCC metals that tend to deform through dislocation glide and cross slip, can exhibit enhanced hardening after shock loading [9]. Prestraining FCC metal, by shock loading has been shown to develop a dislocation cell structure and debris within the microstructure. This reduces available glide distances within the material upon any subsequent reloading and thus leads to an observed hardening response over as-annealed material. BCC metals that tend to deform through planar slip, like Ta, display little enhanced hardening after shock loading [10]. Due to the high Peierls stresses inherent in many BCC metals, like Ta and Nb, shock prestraining stores arrays of planar dislocations. These arrays do not substantially work harden the material and therefor do not lead to an enhanced hardening response.

While significant work has been performed to investigate the role of shock loading on subsequent properties of a metal, much less is understood about the role of shock-induced damage on subsequent properties of a material [11]. Specifically, the strength of an incipiently spalled material has received less attention. To this end, here a study is presented that is aimed at understanding the quasi-static, uni-axial stress-strain response of incipiently spalled, high purity, copper. Copper has been selected for this study as its dynamic response and corresponding microstructural evolution have been well characterized, previously.

2. Anisotropic loaded damage and failure models

Simple, maximum, tensile failure criteria through the Johnson-Cook failure models are isotropic, treating all components of the stress tensor equally. The result of failure by these models is the material within a computational element becomes effectively a gas that can not resist further tension. However, when materials fail, due to anisotropic loading, the failure has a directional component although the material may be modified in the directions perpendicular to the failure. By incorporating directional failure in 2D codes, a more 3D material behaviour is achieved. That is by incorporating anisotropic failure models into 2D codes, out-of-plane loads such

[a] Corresponding author: ecerreta@lanl.gov

as hoop stress can be computed correctly, and failure of material within a zone does not result in the material becoming a fluid, but only having failed in directions that exceed failure criteria and the element remains able to resist loads in non-failed directions.

For most materials a non-linear material behaviour model may be adopted with two distinct material mechanical processes: plasticity and damage mechanics. These two processes are influenced by microcrack, microcavity nucleation and coalescence, inclusions, grain boundaries and other discontinuities. With anisotropic loading, the development of degradation becomes complex, and frequently is based on anisotropic damage evolution. The Cauchy stress tensor $\underline{\underline{\sigma}}$ is for a material consisting of metal matrix and voids. This composite is related to the matrix stress through the operation of the damage effect tensor **M:**

$$\underline{\underline{\sigma}} = \boldsymbol{M} : \underline{\underline{\tilde{\sigma}}} \tag{1}$$

Where the overscore tilde designates a pure material variable defined over the same area or volume as the analogue composite variable, tensor order is denoted by the number of underbars and bolded symbols are fourth-order tensors. The simplest selection for $\boldsymbol{M}$ is the isotropic damage form:

$$\boldsymbol{M} = (1 - D)\boldsymbol{\delta} \tag{2}$$

where D is the amount of isotropic damage, assumed to be the porosity (a ratio of the void volume to the current volume ignoring elastic volume changes) and $\boldsymbol{\delta}$ is the fourth order identity tensor ($\boldsymbol{\delta}_{ijkl} = \boldsymbol{\delta}_{ij}\ \boldsymbol{\delta}_{kl}, \boldsymbol{\delta}_{ij}$ designating the Kronecker delta function).

To obtain failure directions, the material is first rotated by polar decomposition to the material frame with principal stress directions. For maximum stress failure to occur the n_1 component of the traction in the n_1 direction $t_1 = \sigma \cdot n_1$ reaches a certain failure strength τ_1:

$$\sigma^* = \sigma - (n_1 \cdot \sigma \cdot n_1) n_1 \otimes n_1 \tag{3}$$

where n_1 is the normal to the failure surface in material coordinates. When:

$$t_1 \cdot n_1 = (n_1 \cdot \sigma \cdot n_1) \geq \tau_1 \tag{4}$$

the modified stress tensor satisfies

$$\left(n_1 \cdot \sigma^* \cdot n_1\right) \geq 0. \tag{5}$$

Damage evolution often accompanies the relaxation of stress in the direction of failure, due the formation of damage. The modified tensor σ^* is symmetric, but the trace is no longer pressure. To calculate plastic deformation, new traceless deviators are computed:

$$s^* = \sigma^* - \left(\frac{1}{3}\right) Tr\sigma^* I \tag{6}$$

where I is the identity matrix. After a failure in one direction, failure in a second direction occurs when:

$$\sigma^{**} = \sigma^* - (n_1 \cdot \sigma \cdot n_2) n_2 \otimes n_2 \tag{7}$$

$$t_1 \cdot n_2 = (n_1 \cdot \sigma \cdot n_2) \geq \tau_2. \tag{8}$$

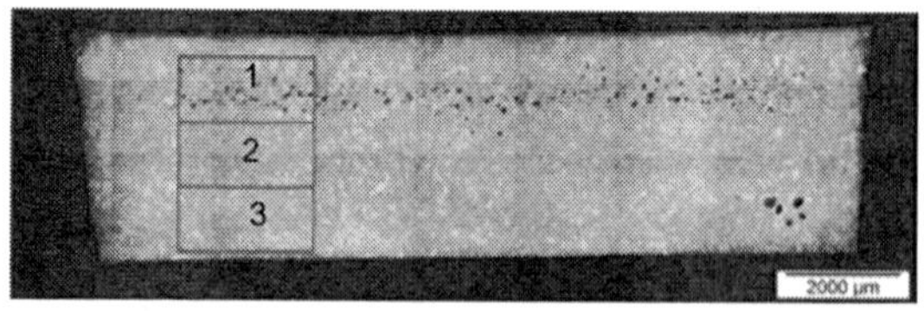

Figure 1. Optical image of the incipiently damaged, shock loaded Cu specimen.

The values τ_1 and τ_2 are material properties that could be functionals that are updated as the damage effect tensor, and are typically dependent on temperature, strain rate, total strain, initial conditions, and more. These functionals have to be determined by experiment, be non-negative, and are then entered into the model as parameters.

3. Experiments

Oxygen free, high conductivity (OFHC) copper was utilized in the as-annealed condition (600 °C, 1 hour, in vacuum) for all experiments. This as-annealed Cu had a nearly random texture and a 60 μm, equiaxed grain structure. Details about this material, such as processing, chemistry and microstructure, can be found elsewhere [12]. From this as-annealed material, 4 mm in height by 40 mm in diameter right cylinders were sectioned. These cylinders were assembled into a momentum trapping fixture made of a low purity Cu material and schematically described elsewhere [13]. This material was shocked loaded using a plate impact configuration, in a 100 mm gas gun, and soft recovered. The impactor was a 2 mm thick quartz flyer that was launched at 131 m/s. This plate impact velocity and plate material were chosen as they had been previously shown to impart an early stage of incipient damage to copper [12]. This damage was known to be in the form of uncoalesced, small voids, indicative of the early stages of ductile damage.

The soft-recovered specimen was cross-sectioned and prepared using standard metallographic techniques. The final polishing step was performed with a 0.5 μm colloidial silica solution. The specimen was then viewed optically to examine location and extent of the incipient damage within the material. An image of the entire cross section of the specimen is given in Fig. 1.

To further analyse deformation and damage imparted to the Cu specimen, electron back-scattered diffraction (EBSD) was performed. Data was collected from the three, boxed regions labelled in Fig. 1. Data was utilized to generate both inverse pole figure (IPF) as well as kernel average misorientation (KAM) maps to assess meso-scale deformation and stored work in the form of misorientation, respectively. KAM is the average mis-oreintation between a point and its neighbors [14].

Finally, quasi-static, uni-axial test specimens were machined from the soft recovered Cu sample. Tension and compression specimens were machined. Figure 2 shows the locations of each tensile and compression specimen with respect to the primary damage field. Compression specimens were right, regular cylinders that were 2.4 mm in height and 2.6 mm in diameter. Dimensions for the tensile specimens are given in Fig. 3. All specimens were

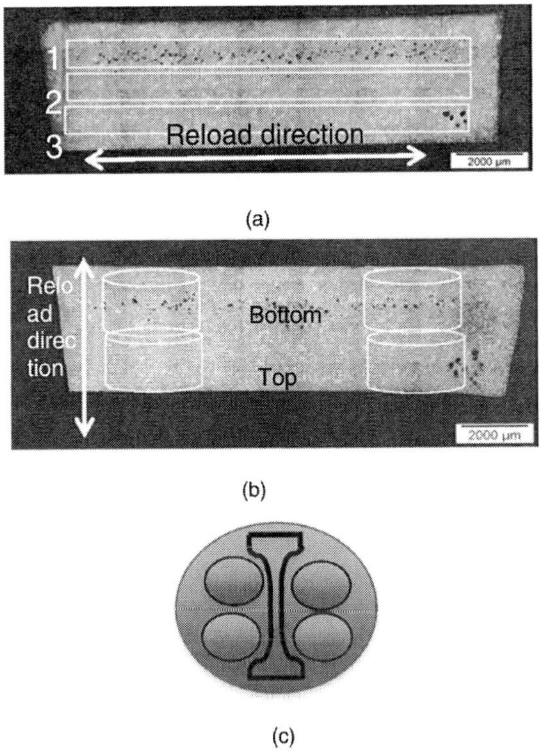

Figure 2. Schematics for the location in which (a) tensile and (b) compression specimens were cut, (c) gives a plan view of all specimen types sectioned from the shock loaded sample.

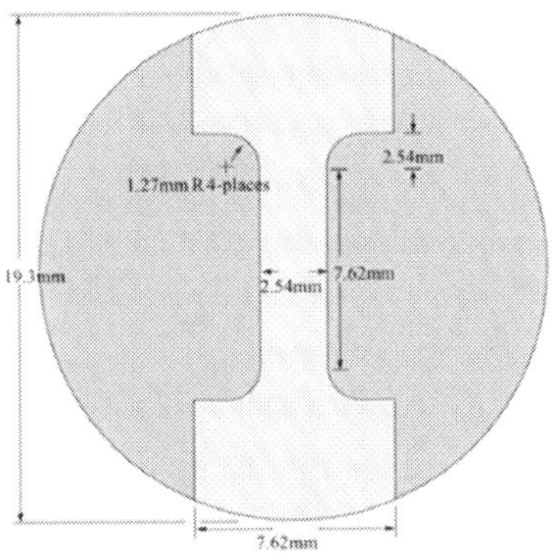

Figure 3. Dimensions for the tensile specimens sectioned from the shock loaded Cu sample.

tested at room temperature and 0.001/s. The three tensile specimens were sectioned from regions 1(most damage) and then 2 and 3 (least damage) as shown in Fig. 2a. Eight compression specimens were sectioned as shown in Figs. 2b and c. Specimens labelled "top" had less damage than specimens labelled bottom.

4. Results

4.1. Microstructure analysis

To quantify the incipient damage imparted to the shock loaded Cu, optical images such as Fig. 1, were used.

Table 1. Damage Statistics for Regions Labelled in Fig. 1.

Region	Avg. size (μm^2)	% Area of void
1	263.6	0.33
2	33.4	0.01
3	19.3	0.02

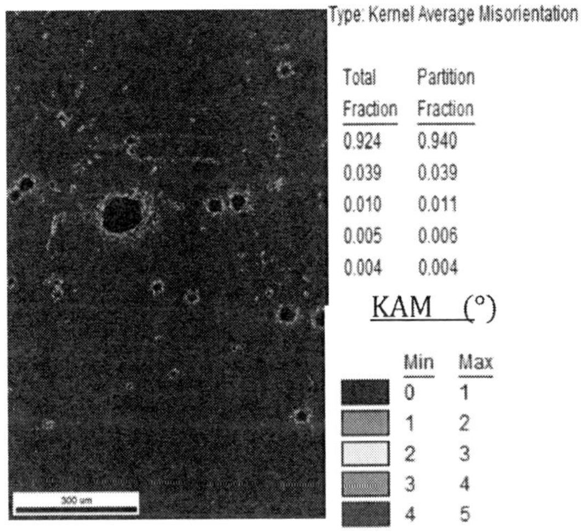

Figure 4. KAM map of region 1 from Fig. 1 showing mis-oreintation around voids in the incipiently damaged Cu sample.

Images revealed a ductile damage field of uncoalesced voids expected for Cu shocked to the conditions described above. Damage was quantified in terms of void count, total area of voids, average void size, and percent area of void. These data were collected from regions 1–3, given in Fig. 1. J-image software tool was utilized to perform the quantitative analysis of optical images that had been thresholded to black and white contrast. Average void size and percent area of voids data are given in Table 1. Region 1 was observed to have the most damage, with less damage in regions 2 and 3.

After the optical analysis of the damage had been completed, the cross-sectioned Cu specimen was examined via EBSD. Initial analysis revealed that there was little evidence of damage beyond void formation. No shear bands were present. Additionally, KAM maps were generated for regions 1–3. These maps showed little evidence of misoreintation developed during shock loading, except in regions near the voids. An example of one of the KAM maps is given in Fig. 4 for region 1. This was the region that displayed the most significant amount of misorientation as it had the most damage. This is important, because misorientation mapping is a tool for examining stored defects within a microstructure. It would be expected in a dynamically deformed microstructure, that higher degrees of misorientation would correlate with increased density of stored defects, such as dislocations [15,16].

4.2. Quasi-static mechanical testing

Quasi-static mechanical tests were conducted on the eight compression specimens and three tension specimens

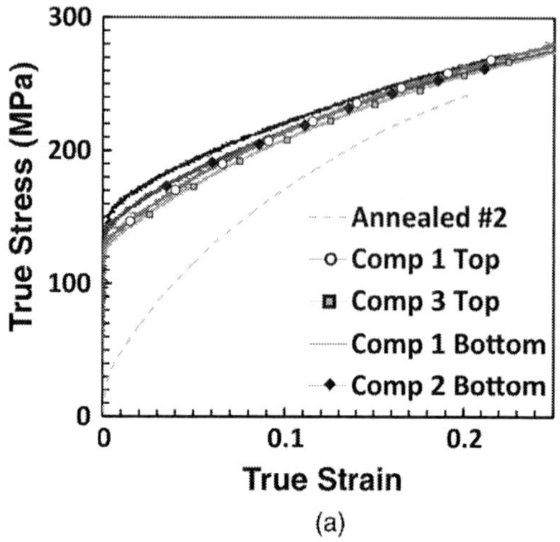

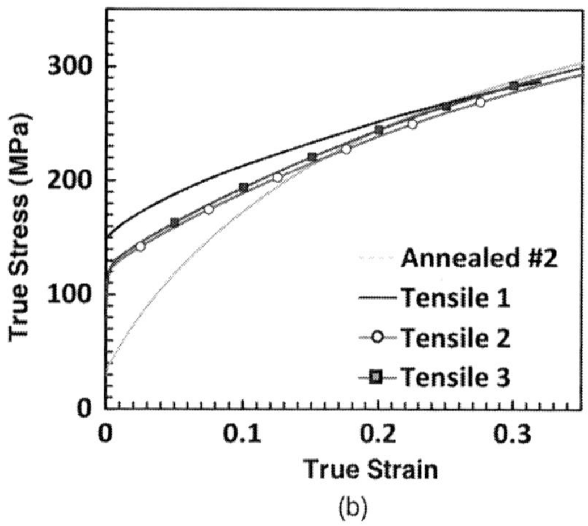

Figure 5. Quasi-static stress-strain data of incipiently damaged Cu: (a) compressive and (b) tensile response.

sectioned from the incipiently damaged Cu sample. For the tension specimens, Tensile 1 corresponds to the tensile specimen sectioned from region 1 shown in Fig. 2a. Tensile 2 and 3 were sectioned from regions 2 and 3, respectively. Compression specimens were tracked with regard to the top and bottom of the shock loaded specimen. For labelling purposes, compression specimens that came from the top of the shock loaded sample were the specimens that contained less incipient damage. The converse was true for compression specimens with a "bottom" designation. The mechanical test data in compression and tension is given in Fig. 5a and b. Only four representative compression tests are shown in Fig. 5a for the sake of clarity. For the purpose of this comparison, tensile and compressive stress-strain responses of as-annealed Cu are also provided.

It can be seen from Fig. 5 that all the quasi-static reload specimens displayed an enhanced hardening response over as-annealed Cu. Interestingly, this is most pronounced in the specimens with the highest volume fraction of incipient damage due to shock loading: "bottom" quasi-static, reload compression specimens and tensile 1. This enhanced hardening is observed upon initial yield. The work hardening response appears to be similar for all of the quasi-static, reload specimens and significantly lower as compared to the work hardening response observed in the as-annealed Cu specimens.

5. Discussion

The quasi-static reload data in tension and compression shows that all of the shock loaded specimens display an enhanced hardening as compared to the as-annealed, high-purity Cu. This indicates that during the shock process, significant work hardening of the Cu sample occurs. This work hardening is directly tied to the isotropic damage term D in Sect. 2 above. This yields material that contains a significant degree of stored plastic work. In turn, this stored work leads to enhanced strength of the material. This response was expected in tensile specimens 2 and 3 and from compression specimens labelled "top". These specimens did not display a high degree of damage imparted to them during the shock loading process, and work by Gray et al. [8, 9, 17] has already shown that stored work during shock loading leads to enhanced hardening in FCC metals like Cu.

Prior to this study, the response of material that had sustained dynamic damage was less well understood than material that had only been shock loaded but had not experienced the early stages of failure, such as void nucleation and growth [11]. In this case, the behaviour of "bottom" compression specimens and tensile 1 was less predictable. From Fig. 5, it is clear that these specimens, with a modest amount of damage, display enhanced hardening in excess of the shock loaded specimens from other regions in the Cu sample. This is rationalized in terms of the response of the quasi-static reload specimens with relatively less damage and the KAM map, Fig. 4. KAM maps from region 1, showing enhanced misoreintation as compared to regions 2 and 3. This enhanced misoreintation does not exist in the bulk of the shock loaded microstructure, but is measureable around the voids. This is postulated to be caused by the significant slip activity required for void growth. All of the regions experienced significant work hardening due to shock loading and therefore display enhanced hardening as compared to the as-annealed Cu material. In the case of specimens that encompassed material from region 1, that enhanced hardening would be expected to be higher based on the measureable additional plastic work in the form of higher dislocation density around the voids. The small enhancement in hardening observed for tensile 1 and "bottom" compression specimens correlate with the expected contribution to hardening from the small volume of significantly misoriented material around the voids.

The stored work due to shock loading does not only influence yield, but also influences the work hardening response of the Cu material. Here, it is observed to lead to lower work hardening rates than in the as-annealed copper. This lower rate of work hardening can be associated with a less stable dislocation substructure that can promote catastrophic failure [16, 18]. It follows that there is a critical amount of stored work in a microstructure after which the mechanical response of the material is unstable and failure readily results. This is observed in material that has received significant rolling reductions [18]. If shock loaded material behaves similarly, then there should be a critical amount of damage after which enhanced hardening over shock loaded material with no damage and material may not be observed. Future studies will probe the effect of increasing volume fraction of damage on subsequent mechanical properties.

6. Conclusions

This study examined the role of ductile, incipient, dynamic damage imparted during a spallation loading experiment on subsequent mechanical properties of copper. It was found that the mechanical response of Cu is significantly altered due to shock loading. This correlates well with what has been observed previously for shock loaded FCC metals that had not incipiently spalled. Interestingly, material that had experienced a modest degree of dynamic damage displayed an enhanced hardening even in excess of the shock-loaded material without damage. This is attributed to the contribution of the additional stored plastic work contained in the volumes surrounding voids.

Los Alamos National Laboratory is operated by LANS for the NNSA of the US Department of Energy under Contract No. DE-AC52-06NA25396.

References

[1] M. Meyers, Y. Xu, Q. Xue, M. Perez-Prado, T. McNelley. Acta Mater. **51**, (2033).

[2] N. Bourne, *Materials in Mechanical Extremes*. (Cambridge University Press, Canbridge, 2013).

[3] L. Murr, K. Staudhammer, M. Meyers, *Metallurgical Applications of Shock Wave and High Strain Rate Phenomena*. (Marcel Dekker, Inc., New York, 1986).

[4] C. Bronkhorst, B. Hansen, E. Cerreta, J. Bingert. J. Mech. Phys. Sol. **55**, 2351–2383 (2007).

[5] B. Boyce, S. Kramer, H. Fang. Int. J. Fracture **186**, 5–68 (2104).

[6] E. Cerreta, G.T. Gray, R. Hixson, P.A. Rigg, D.W. Brown. Acta Mater. **53**, 1751–1758 (2005).

[7] G.T. Gray. (American Institue of Physics, 1994), pp. 1103–1106.

[8] G.T. Gray. Mat. Res. Symp. Proc. **499**, 87–98 (1998).

[9] G.T. Gray, K. Vecchio. Metal. Mater. Trans. A **26A**, 2555–2563 (0ctober 1995, 1995).

[10] S.R. Chen, G.T. Gray III. Metal. Trans. A **27**, 2994–3006 (1996).

[11] G. Gray, N. Bourne, paper presented at the Dislocations, Plasticity, Damage, and Metal Forming: Material Response and Multiscale Modeling, 2005.

[12] J. Escobedo, D. Dennis-Koller, E. Cerreta, B. Patterson, C. Bronkhorst, B. Hansen, D. Tonks, R. Lebensohn. J. Appl. Phys. **110**, 033513 (2011).

[13] A.K. Zurek, W.R. Thissell, D.L. Tonks, R. Hixson, F. Addessio. J. Phys. IV **7**, 903–908 (1997).

[14] S.I. Wright, M.M. Nowell, D.P. Field. Microscopy and Microanalysis **17**, 316–329 (2011).

[15] J. Bingert, V. Livescu, E. Cerreta, in *Electron Backscatter Diffraction in Materials Science*, A. Schwartz, M. Kumar, B. Adams, D. Filed, Eds. (Springer, New York, 2009), pp. 301–316.

[16] E. Cerreta, J. Bingert, G. Gray, C. Trujillo, M. Lopez, C. Bronkhorst, B. Hansen. Inter. J. Plasticity **40**, 23–38 (2013).

[17] G.T. Gray, N. Bourne, J. Millett. J. Appl. Phys. **94**, 6430–6436 (2003).

[18] H. Fujiwara, T. Sato, A. Kamio. J. Japan Inst. Metals **64**, 641–650 (2000).

EPJ Web of Conferences **94**, 02016 (2015)
DOI: 10.1051/epjconf/20159402016

Penny-shaped crack propagation in spallation of Zr-BMGs

Z. Ling[1,a], X. Huang[2], and L.H. Dai[1]

[1] LNM, Institute of Mechanics, Chinese Academy of Sciences, Beijing 100190, PR China
[2] Institute of Systems Engineering, China Academy of Engineering Physics, Mianyang 621900, PR China

Abstract. Typical penny-shaped microcracks at their propagating in spallation of Zr-based bulk metallic glass (Zr-BMG) samples were captured by a specially designed plate impact technique. Based on the morphology and stress environment of the microcrack, a damaged zone or propagation zone around the crack tips, similar to the cohesive zone in classical fracture theories, is applied. Especially the scale of such a damaged zone represents a scale of the crack propagation. Its fast propagation would quickly bring a longer crack or cause cracks coalesce to form another longer one. The estimated propagation scales of microcracks are reasonable compared with what occurred in the Zr-BMG samples.

1. Introduction

Spallation in bulk metallic glasses (BMGs) is an important dynamic fracture under tensile pulse in uniaxial strain condition [1]. In contrast to the shear-sensitive fractures in BMGs under tensile stress in uniaxial stress condition [2,3], the spallation is controlled by hydro-static tension stresses. Therefore, spallation fractures take place in a plane of the maximum tensile stress, perpendicular to the tensile direction [1]. Spallation in BMGs is finally attributed to microdamage evolution occurred in a narrow layer where tensile stress amplitude reaches to its maximum. The narrow layer, alike a shear band in BMGs [4,5] supplies a hydro-tensile stress environment for microdamage nucleation, propagation and coalescence.

The spallation behavior of BMGs has been investigated according to spallation strength, spallation setup as well as microdamage evolution [6–9]. In the literatures on the spallation behavior in BMGs, little information exists concerning the microdamage nucleation, growth and coalescence of spallation [8]. Studies on microdamage evolution of Zr-based bulk metallic glasses' (Zr-BMG) spallation exhibited that, as having a unique amorphous structure or disorder atomic structure, Zr-BMG's microdamage evolution of spallation seems to be unlike the evolution of traditional crystalline metals. In particular, in the Zr-BMG samples the whole microdamage evolution, including nucleation, growth and coalescence, would take 10^2ns [8]. The microdamages in the material are microvoids in the size of 10^{0-1}nm [8], $10^{1\sim2}$nm [10] and 10^3 nm, in their respective to nucleation, growth and coalescence. Obviously, the microvoids' scale grows about 10^{1-2}nm during the growth process of 100 ns. This scale and duration are several orders smaller than those of traditional crystalline materials' microdamage evolution [1]. In this case, for further studies of microvoid evolution of spallation in the BMGs, it is extremely important to obtain accurate data for the voids and cracks during the evolution of 10^2 ns in the materials [1].

Plate-impact technique is an effective method to capture the evolution of microdamage of spallation in laboratories. This technique provides an experimental configuration of uniaxial tensile strain and a tensile hydro-static stress pulse. Especially, by controlling tensile stress amplitude and its duration in the technique, one can make microdamage at different evolution stage to be "frozen" in the samples [1] for further study. Recently, a plate-impact technique with specially designed flyers has been developed [11] to efficiently capture information of microdamage evolution of a 10^2ns duration in Zr-BMG samples. By this technique, mcrocracks of 10^{2-3}nm were "frozen" in the samples. In this paper, an investigation to the microcrack propagation occurred in Zr-BMGs under the plate impact is presented based on classical fracture theories.

2. Microdamage evolution experiments

2.1. Material and plate impact experiment

Spallation microdamage evolution tests were performed with a 101 mm single-stage light gas gun. To study microdamage evolution of a 10^2ns duration in Zr-BMGs, the short-multi-stress pulse technique was applied. In the plate-impact experiments, specially designed double-flyers were employed [11]. Figure 1 is a schematic of the specially designed apparatus, where three BMG samples are targeted by the three sets of double-flyers with gaps of varied sizes, so that microdamage in different stress durations, under the same stress amplitude, could be "frozen in" the BMG samples. The details of experimental implementation and the double-flyers technique can be referred to the paper [11].

The material applied in current experiments was $Z_{r41.2}Ti_{13.8}Cu_{12.5}Ni_{1}0Be_{22.5}$ [Vit.1]. Table 1 presents mechanical properties of Zr-BMG, where ρ is density; E elastic modulus and E_L lateral elastic modulus; υ Poisson's

[a] Corresponding author: lingz@lnm.imech.ac.cn

Article available at http://www.epj-conferences.org or http://dx.doi.org/10.1051/epjconf/20159402016

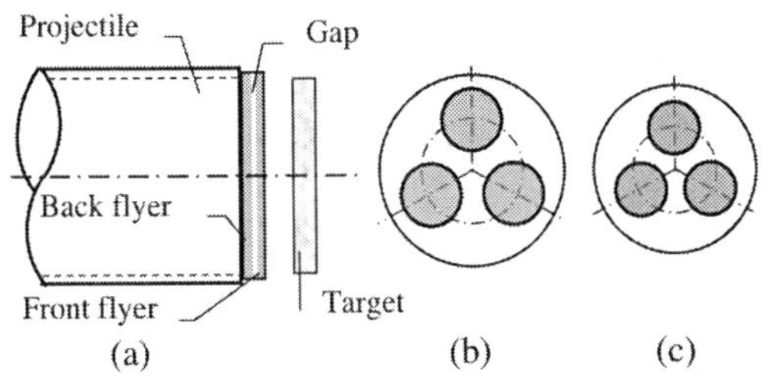

Figure 1. A schematic of the special designed plate impact apparatus, (a) Plate impact apparatus with the specially designed double flyers; (b) Impactor and (c) Target.

Table 1. Mechanical properties of the Zr-BMG samples.

Materials	ρ (kg/m)^3	E(GPa)	E_L(GPa)	υ	c_L(m/s)
Zr-BMG	6125	98.6	160	0.36	5182

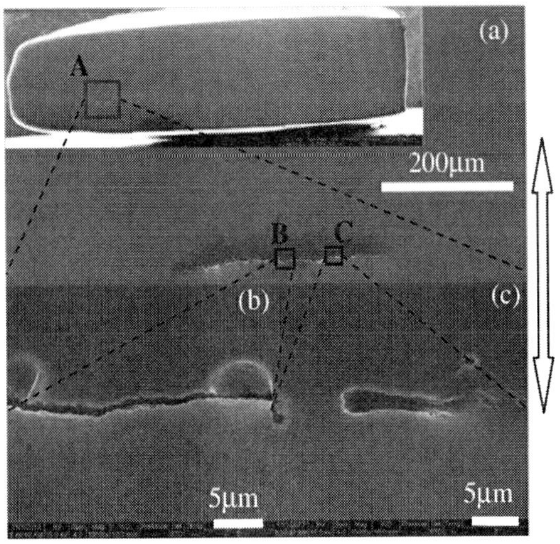

Figure 2. Close-up observations of a part of the cross-section of a sample (σ3.2 GPa, $\Delta t = 250$ ns), where (a) details of zone-A in the cross-section; (b) an enlarged show of zone-B and (c) zone-C. The arrow is along the loading direction.

ratio; c_L longitudinal wave speed [12]. After the impact experiments were performed, samples were recovered and sectioned to observe the induced microdamage. Each of the tested samples was cut along its diameter and then the cut-section was polished carefully. A scratch-free cross-section of the samples was obtained after being etched [3] for microscopic examination under higher magnification.

2.2. Micro-observations

Figure 2 presents typical micrographs of a partial cross-section of one microdamaged sample under a stress duration of 250 ns with stress amplitude of 3.2 GPa. As shown in the insert, under the given stresses with duration of 250 ns, strip-like concave regions have appeared in the sample. The strip-like concave regions have been verified as microdamage regions, $10^2\,\mu$m in length and parallel to each other. In particular they are perpendicular to the loading direction and located at where maximum tensile stress is [11]. Meanwhile, no any other strip-like region can be seen on the section, either near by or far from the microdamage regions along the loading direction.

Figure 2a is an enlarged image of zone-A in the inserted square. The strip-like microdamage region consists of several disconnected micro-flaws. The longest flaw is about $150\,\mu$m with the others' length varied about 10–30 μm (Fig. 2a). Such micro-flaws are distributed along a strip located in the layer or plane of the maximum tensile stress, perpendicular to the impact direction.

Figure 2b is an enlarged show of zone-B marked in Fig. 2a, a part of a $150\,\mu$m crack, almost perpendicular to the loading direction. Part of the crack is slightly curvilinear, inferring the long crack may be formed by coalescence of several short cracks of $10^{0-1}\,\mu$m.

Figure 2c is a detail of zone-C in the square in Fig. 2a. This is a crack with a length of $10\,\mu$m. Dimples of 10^{2-3}nm, without any tortuous, distributed along the internal edge of the crack. A few dimples of 10^2nm can be seen on the upper and lower surfaces of the crack. The crack is obviously a typical penny-shaped crack located at where the maximum tensile stress is. Especially, the crack's tips are blunt and dimples of $2\,\mu$m are visible at the rim of one tip (Fig. 2c).

The penny-shaped crack shown in Fig. 2c is a typical microcrack in samples under plate impact loading [1], resulted from development of micro-damages. Previous researches on spallation of Zr-BMGs illustrates that microdamamges of the materials under hydro-static tension are microvoids with initial nucleation size of 10^0 nm [8]. Then tiny microvoids would grow up or coalesce to form larger voids, say 10^{1-3} nm and so on [10] and finally to leave dimples of $10^0\mu$m in the spallation morphology [8,9]. The penny-shaped crack of $10^1\,\mu$m in Fig. 2c seems not to be formed by microvoids' growth and coalesce only. The dimples of 10^2 nm along the crack's internal edge and those of dimples of $10^0\,\mu$m on the crack's blunt tips suggest that the penny-shaped crack would propagate after the crack's initial formation. To understand what observed in experiments, an investigation is presented in the following section to show how such a penny-shaped crack, like what in Fig. 2c, propagates under a given stress environment.

3. Penny-shaped crack's propagation

The given stress environment of a Zr-BMG sample under plate impact is uniaxial strain, that is, assuming loading direction is along z-axis,

$$\varepsilon_z \neq 0,\ \varepsilon_r = \varepsilon_\theta = 0 \tag{1}$$

where z, r and θ denote directions along thickness, radius and tangential direction of the sample, respectively. Stresses corresponding to the uniaxial strain are

$$\begin{aligned} \sigma_z &= (K + \tfrac{4}{3}G)\varepsilon_z = E_L\varepsilon_z \\ \sigma_r &= \sigma_\theta = \tfrac{\nu}{1-\nu}\sigma_z \end{aligned} \tag{2}$$

where K, G and υ are bulk modulus, shear modulus and Poisson's ratio; E_L is lateral elastic modulus, which is always higher than the elastic modulus. In the current case,

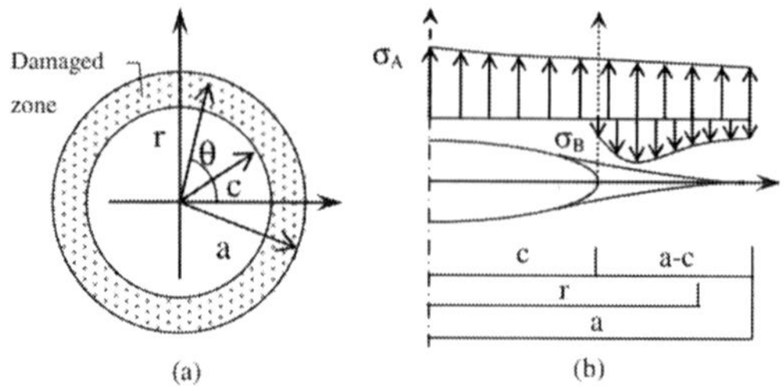

Figure 3. Schematic of a penny-shaped crack, (a) an overlook of the crack, where the shadow is damaged zone at rim of the crack; (b) stresses in the rim and surfaces of the crack.

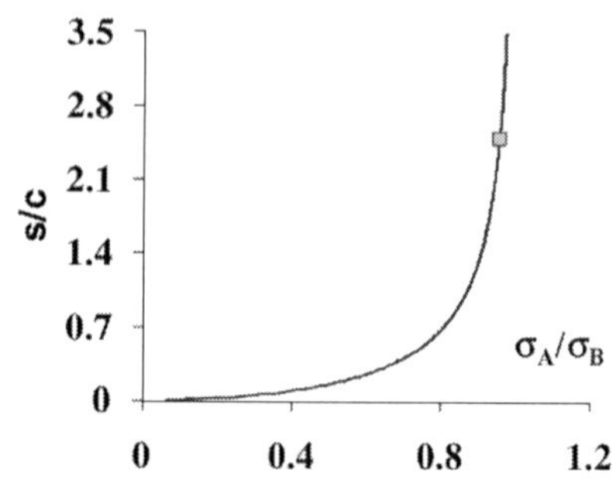

Figure 4. Plots of $\frac{s}{c} \sim \frac{\sigma_A}{\sigma_B}$ shown as Eq. (7).

samples are under hydrostatic tension with a maximum tensile stress of 3.2 GPa, less than Zr-BMGs' Ugoniot Elastic Limit 6.7 GPa [12]. Therefore, the following investigations are quasi-static elastic.

3.1. Damaged zone

As shown in Fig. 2c, the penny-shaped crack is isolated and no any other visible microdamage clutters around its upper and lower surfaces, except for those at its bilateral with distances greater than 5 μm. This suggests that the crack's forming and propagating are in the layer or plane of the maximum tensile stress without being disturbed by other micro-damages or -cracks. It is worth noticing that, the penny-shaped crack is propagating under the tensile stress of 3.2 GPa within 250 ns, a stress environment of fixed stress amplitude within a given duration. Hence the current crack propagation in the short stress duration is approximately quasi-static. In such a case, an isolated crack's propagation can be described by classical fracture theories.

In these fracture theories, a crack with blunt tips illustrates that the stresses at the tip of the crack are limited, inferring at front of the crack's tip a small cohesive zone, compared with the crack's size [13,14]. The morphology and stress environment of the penny-shaped crack shown in Fig. 2c, suggest that a damaged zone or propagation zone, analogous to the cohesive zone above, is at the front of the crack's tip. In contrast to the case of the small cohesive zone [13,14], the current case is of a large-scale involving in micro-damaging [1], since microvoids have been observed further away around the microcrack tip (Fig. 2c).

Figure 3 is a schematic of a penny-shaped crack and its rim [15]. Figure 3a is an overlook of the crack, where the shadow represents a latent damaged zone at the rim of the crack. In the figure, c is radius of the crack, s size of the damaged zone, and a total length of the crack and its damaged zone, $a = c + s$. Figure 3b shows stress distributions around the crack, in which σ_A is stress acting on the surface $a = c + s$ to cause the crack's opening along the tension direction; σ_B is the cohesive stress acting partially over the crack's surface in the damaged zone attempting to close up the crack, in the direction opposite the tension. Stress σ_B also represents material resistance to microstructure evolution, intrinsic defects motion or microvoids nucleation in the region over the damaged zone [14,15]. Both stresses σ_A and σ_B are supposed to be uniform in the affected region. In addition, as being a hydro-static tension, the crack is subjected to other two stresses σ_r and σ_θ. Apparently, these two stresses are acting in planes perpendicular to the crack's opening direction, so do not contribute to the opening.

The stress intensity factor of the total stress on the tip of the penny-shaped crack is[14]

$$K_T = K_A - K_B \tag{3}$$

where K_A is contributed by the stress σ_A applied on $a = c + s$, K_B is by the stress σ_B acted on $s = a - c$. For a 3-D penny-shaped crack, one has

$$K_A = \frac{\sigma_A \sqrt{a}}{\pi} \tag{4}$$

$$K_B = \frac{\sigma_B \sqrt{a}}{\pi}\left(1 - \frac{c^2}{a^2}\right)^{1/2}. \tag{5}$$

A crack's blunt tip infers stresses on the tip are limited, only valid when $K_A = K_B$. Thus,

$$\frac{\sigma_A \sqrt{a}}{\pi} = \frac{\sigma_B \sqrt{a}}{\pi}\left(1 - \frac{c^2}{a^2}\right)^{1/2} \tag{6}$$

Noticing $a = c + s$, this leads to

$$\frac{s}{c} = \left[1 - \left(\frac{\sigma_A}{\sigma_B}\right)^2\right]^{-1/2} - 1 \tag{7}$$

As shown in Eq. (7), a penny-shaped crack's damaged zone is dependent on the crack's length and the loading on the crack as well as the cohesive stresses. Moreover in current case, once $K_A > K_B$, the crack propagates along where the damaged zone is under a given stress environment. The scale of the damage zones also represents a propagation scale of the crack. In other words, a current crack of c would grow up to become a new one in the scale of a, thereupon $s = a - c$ corresponds to the propagation scale of the current crack.

Figure 4 is a plot of Eq. (7), between s/c and $\frac{\sigma_A}{\sigma_B}$. Clearly, stresses in the rim of the crack are always higher than that on the crack surface, i.e. $\sigma_A < \sigma_B$. However, with increasing $\frac{\sigma_A}{\sigma_B}$, s/c varies. While $\sigma_A \leq \frac{\sqrt{3}}{2}\sigma_B$, $s/c < 1$, which illustrates that the scale of damaged zone increases but is still less than that of the crack. Obviously this process corresponds to a slow propagation of the crack.

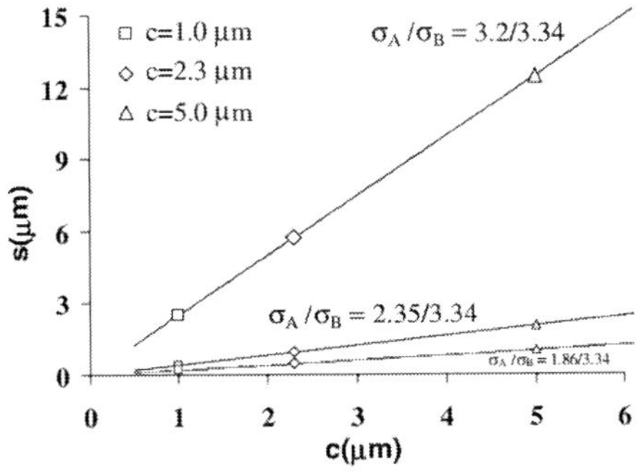

Figure 5. Plots of $s \sim c$ for the current Zr-BMGs.

After $\sigma_A > \frac{\sqrt{3}}{2}\sigma_B$, $s/c > 1$, the scale of the damaged zone increases, always greater than that of the crack. In particular, with$\sigma_A \to \sigma_B$, a little incremental in $\frac{\sigma_A}{\sigma_B}$ would lead to the damaged zone to develop fast. Thereby this of course corresponds to the fast propagation of the crack.

Further understand to this procedure would require more details about the current microcrack and its stress environment.

3.2. Propagation scale

As stated above, the scale of the damaged zone s also represents a scale of the crack propagated, verified by the current experimental data. The maximum tensile stress in current samples is 3.2 GPa. Under this tensile stress, resultant penny-shaped crack shown in Fig. 2c is about 5 μm radius and at the crack tip (right side) a dimple pattern about 1 μm radius (Fig. 2c) appears. The dimple pattern at the crack tip hints that the stress amplitude in the zone is 3.34 GPa, the critical stress of a microvoid in its critical size [10]. So $\frac{\sigma_A}{\sigma_B} = \frac{3.2}{3.34}$ is chosen for the current case of the crack's fast propagation, as shown in the square in Fig. 4.

In addition, there is no any other visible microdamage around the microcrack on both upper and lower surfaces in this sample, even if under high magnifications (Fig. 2). Nevertheless, as being demonstrated in previous studies, the microvoids nucleated size is about 10^{0-1}nm, too tiny to be visible with currently available microscopic instruments [8,9]. The loading stress of 3.2 GPa is obviously higher than the nucleation stress of 2.35 GPa [8] and static tensile strength of Zr-BMGs, 1.86 GPa [16]. Therefore in this case, probably some tiny microvoids have been nucleated, especially around the crack's upper or lower surfaces. Therefore two stresses of 2.35 GPa and 1.86 GPa are also supposed on the crack surfaces.

Figure 5 exhibits plots of $s \sim c$ of current Zr-BMG samples corresponding to the above stresses. As exhibiting, $\frac{\sigma_A}{\sigma_B} = \frac{3.2}{3.34}$ represents the current case and the other two,$\frac{\sigma_A}{\sigma_B} = \frac{2.35}{3.34}$ and $\frac{\sigma_A}{\sigma_B} = \frac{1.86}{3.34}$ are cases of microvoid nucleation and static tensile strength of this Zr-BMG, respectively. Three symbols represent the current crack size (radius) $c = 5\,\mu$m (Fig. 2c); nucleated microvoid's critical size (radius) $c = 2.3\,\mu$m [12] as well as the maximum void size at the crack tip in Fig. 2c $c = 1\,\mu$m.

The line with a tangent of 3.2/3.34 in Fig. 5 corresponds to the current case. In this case, $s > c$, that is, the damaged zone develops faster than the crack, or the scale of a crack propagation is greater than that of the crack. While a crack is very short, say c = 1–2 μm, the damaged zone is about 2–5 μm and a resultant crack would be of 3–7 μm in radius. This is almost as the same as of the crack in Fig. 2c, that is, a 5 μm radius crack would be developed from propagation of a 1–2 μm radius crack. While a crack becomes longer, say 5 μm in radius, the length of its damaged zone would be 2.5 times of the current crack. This indicates that the crack propagates exceeding 17 μm radius and a resultant crack is to be 30 μm in its length. This seems to be beyond what was observed in Fig. 2c, whereas microcracks of 20–30 μm appeared in the same sample (see Fig. 2a). This also illustrates that a crack of 30 μm in length is probably resulted from the propagation of another crack of 5 μm in radius, like that in Fig. 2c.

In Fig. 5, crack propagation along the line with a tangent of 2.35/3.34 is the case of microvoids nucleated around the upper and lower surfaces of the crack. It is a linear dependence of the damaged zone and the crack. Compared with the fast propagating in the case of 3.2/3.34, the crack's damaged zone develops much slower than the crack, i.e. $s/c < 1$. Especially, the estimated damaged zone of the current crack of 5 μm radius is about 2.0 μm, less than the distance between the crack and its adjacent crack shown in Fig. 2c. A reasonable explanation is probably that, the nucleated microvoids around the crack caused stress relaxation, resisting or slowing the development of the damaged zone.

The bottom line with a tangent of 1.86/3.34 in Fig. 5 is the case of that the material around the crack is at its static tensile strength. Clearly, along this line, one possible damaged zone develops slower than those of 2.35/3.34 and 3.2/3.34. Hence, for a given crack under the given stress environment, the crack propagating may be always limited on local, unless increasing the stress amplitude or extending the stress duration.

3.3. Discussions

Based on the morphology and stress environment of a penny-shaped crack in Zr-BMG samples, a simplifying approximation to the damaged zone around the crack tips is applied. Actually, the latent damaged zone is resulted from nucleation, growth, and coalescence of millions of tiny voids in nano- or micro-scales [1]. Therefore the damaged zone's scale represents the scale of crack propagation scale. Following this way, further understanding of the damaged zone's developing is presented as following:

As being obvious in Fig. 5, there are two areas, fast crack (3.2/3.34) propagation and slow propagation (2.35/3.34, 1.86/3.34). In the fast propagation, along the line with a tangent of 3.2/3.34, the crack's propagation zone is fast increased so that it will meet with another crack's propagation zone. In this case, linking or coalescing between two adjacent cracks develops readily, or more adjacent cracks link each other simultaneously. This has been verified by the crack of 150 μm exhibited in Fig. 2a, formed as coalescence of several cracks of $10^1\,\mu$m distributed in the same layer. Furthermore, in view of microdamage evolution, the crack in Fig. 2c probably was formed later compared to other cracks [1]. While this crack just grows to $5\,\mu$m radius from 1–$2\,\mu$m, the other cracks have been coalesced to form a longer one, say $150\,\mu$m (Fig. 2a).

Meanwhile, in the slow propagation (the line of 2.35/3.34), the crack propagation is always less than the current crack, implying that cracks' linking may not take place between two adjacent cracks. Compared with those in the fast propagation (the line of 3.2/3.34), a crack in the same size in slow propagation probably is always isolated, like that in Fig. 2c. Evidently, cracks under or below the stress of 2.35 GPa would not develop to be linked and form longer cracks.

This understanding of the fast propagation (3.2/3.34) and slow propagation (2.35/3.34, 1.86/3.34) of a current crack is in fact consistent with the understanding of critical criteria of spallation [17]. It is well known that spallation is resulted from microcracks nucleation, growth and coalescence under a given stress amplitude within a stress duration. If the applied stress amplitude is high enough, the resultant internal microcracks would propagate and link in sudden so that spallation occurs. However, in case of a low stress amplitude, possible nucleated microcracks would propagate too slowly to coalesce, even if the stress duration maintains longer. Hence, in case of these isolated microcracks, spallation could not be completed due to lack of cracks' coalescing or linking.

4. Conclusions

Propagation of a penny-shaped microcrack in Zr-BMG samples under uniaxial strain or hydrostatic tension has been described. The followings are deduced:

1. Typical penny-shaped microcrack propagation was "frozen" in Zr-BMG sample by a specially designed plate impact technique; blunt tips of the isolated microcrack indicated limited stresses at the crack's tips;
2. According to the morphology and stress environment of the penny-shaped crack in Zr-BMG samples, a damaged zone similar to the cohesive zone at the rim of a penny-shaped crack is applied to characterize the crack propagation;
3. The scale of the damaged zone in this work represents the propagation scale of a penny-shaped crack. Under a given stress amplitude, the damaged zone's fast propagation would quickly bring a longer crack or cause cracks coalesce to form another longer one. The estimated scales are reasonable compared to the micro-observations in the experiments;
4. The fact that damaged zone's fast propagation can lead to adjacent microcracks' coalescing or linking each other under a given stress amplitude is in accordance with what the critical criteria of spallation stipulated.

This work is financial supported from the NSFC (Grants Nos.: 11272328, 10872206, 11472287 and 11402245) and the National Natural Science Foundation of China-NSAF. Grant No: 10976100.

References

[1] D.R. Curran, L. Seaman, D.A. Shockey, Physics Reports, **147**, 5–6 (1987) 253–388
[2] F. Spaepen, Acta Metallurgica, **25**, 4 (1977) 407–415
[3] C.A. Pampillo, J. Mat. Sci. **10** (1975) 1194–1227
[4] J. Li, F. Spaepen and T.C. Hufnagel, Phil. Mag. A, **82** (2002) 2623–2630
[5] K.M. Flores, R.H. Dauskardt, Acta Mater. **49** (2001) 2527–2537.
[6] S.T. Turneaure, S.K. Dwivedi, Y.M. Gupta, J. Appl. Phys., **10**, 4 (2007) 043514
[7] J.P. Escobedo, Y.M. Gupta, J. Appl. Phys. **107**, 12 (2010) 123502
[8] X. Huang, Z. Ling, H.S. Zhang, J. Ma, L.H. Dai, J. Appl. Phys. **110**, 10 (2011) 103519
[9] Z. Ling, X. Huang, L.T. Shen, L.H. Dai, *Proceedings of DYMAT2012*, EPJ Web of Conferences **26** (2012) 02003
[10] X. Huang, Z. Ling, L.H. Dai, Int. J. Solids and Structures, **50** (2013) 1364-1372
[11] Z. Ling, X. Huang, L.H. Dai, *Hopkinson Centenary Conference Cambridge, UK* S.Hiermaier (eds.) (2014) 211–224
[12] X. Huang, Doctoral dissertation, University of CAS, (2013)
[13] G.I. Barenblatt, *Advances in Applied Mechanics*, H.L. Dryden, T. Von Karmen and G. Kuerti (eds.), **7** (1962)
[14] I.N. Sneddon, M. Lowengrub, *Crack problems in the Classical Theory of Elasticity* (John Wiley & Sons. Inc.1969)
[15] G.G. Chell, Engineering Fracture Mechanics, **9** (1977) 55–63
[16] J. Lu, G. Ravichandran, W.L. Johnson, Acta Mater., **51** (2003) 3429–3443
[17] F.R. Tuler, B.M. Butcher, Int. J. Fracture Mechanics, **4**, 4 (1968) 431–437

EPJ Web of Conferences **94**, 02017 (2015)
DOI: 10.1051/epjconf/20159402017

Effect of composition on the high rate dynamic behaviour of tungsten heavy alloys

Kesemen Latif[1,a], Çalışkan N. Kaan[1], Konokman H. Emrah[1], and Durlu Nuri[2]

[1] TÜBİTAK SAGE, P.K. 16 06261 Mamak Ankara, Turkey
[2] TOBB University of Economics and Technology, Department of Mechanical Engineering, 06560 Ankara, Turkey

Abstract. Tungsten heavy alloys are currently used as kinetic energy penetrators in military applications due to their high density and superior mechanical properties. In the literature, quasi-static properties of different tungsten heavy alloys based on W-Ni-Cu and W-Ni-Fe ternary systems are well documented and presented. However, comparison of the dynamic behaviour of these alloys in terms of the correlation between quasi-static mechanical characterization and dynamical properties is lacking. In the present study, dynamic properties of tungsten heavy alloys having different binder phase compositions (90W-7Ni-3Cu and 90W-8Ni-2Fe) at different projectile velocities were investigated. The examined and tested alloys were produced through the conventional powder metallurgy route of mixing, cold compaction and sintering. Mechanical characterization of these alloys was performed. In the ballistic tests, cylindrical tungsten heavy alloys with L/D ratio of 3 were impacted to hardened steel target at different projectile velocities. After the ballistic tests, deformation characteristics of test specimens during dynamic loading were evaluated by comparing the change of length and diameter of the specimens versus kinetic energy densities. The study concluded that 90W-8Ni-2Fe alloy has better perforation characteristics than 90W-7Ni-3Cu alloy.

1. Introduction

Perforation is a phenomena related to the characteristics of materials under dynamic conditions. Perforation characteristics of a material can be correlated to material properties such as density and strength as well as thickness of the projectile and impact velocity. There are several theories in the literature which determine and predict perforation characteristics of materials. Cylinder impact on an almost rigid surface is a very simple and useful experimental method for the material characterization under dynamic conditions. It is mostly called as Taylor impact [1] due to name of first applicant of this test. Although Hopkinson [2,3] method is widely preferred in similar studies, it is limited by the strain rate range of 10^2–$10^4\,s^{-1}$. By using Taylor method, higher strain rates can be achieved and the material behaviour in high strain rate levels can be observed.

It is known that during the impact of a Taylor cylinder, firstly elastic and plastic wave propagation occurs. With plastic wave propagation, specimen begins to deform plastically from the impact end. Its deformation depends on the impact velocity and material's strength dependency to such as strain hardening, strain rate and temperature. Hence, the shape of the specimen after the impact is governed by these material characteristics. Taylor impact projectiles are impacted to a hard wall with high speeds and very high strain rates occurs during impact process. Strain rate may reach values up to 10^3 to 10^6 depending on the impact velocity. Thus, high strain rate characteristic of the material is very important for the deformation of the projectile.

[a] Corresponding author: `latif.kesemen@tubitak.gov.tr`

Tungsten heavy alloys are two phase metal matrix composite materials which are generally manufactured by liquid phase sintering methods. The main phase is high density and high strength tungsten particles; the second phase, i.e, binder phase, is tough and ductile nickel based alloys which usually contain some iron, copper or cobalt. Tungsten based alloys characterized by high density, high strength and ductility which make these composite materials suitable for applications such as penetrators, radiation shields and balance weights [4–7]. In the manufacturing of tungsten heavy alloys, the use of Ni-Cu alloys as binder phase were first developed by Price et al. in 1938 [8]. Liquid phase sintered W-Ni-Cu alloys, however, showed low strength and ductility, and in the 1950's, W-Ni-Fe alloys having better mechanical properties were developed due to the higher tungsten solubility in binder phase. For tungsten heavy alloy, higher tungsten solubility in binder phase leads to increases in room-temperature yield strength, tensile strength, and ductility [9]. In this study, perforation performance of two different tungsten alloys, 90W-8Ni-2Fe and 90W-7Ni-3Cu, was investigated under dynamic conditions by using Taylor impact test technique.

2. Experimental procedure

In this study, two tungsten based alloys namely given as 90W-7Ni-3Cu and 90W-8Ni-4Fe were prepared for the experiments. Proper amounts of W, Ni, Cu and Fe powders were weighted by a precision balance that has $+\ 0.01$ g accuracy (CP3200S, Sartorius, Germany) in order to give the alloy compositions. Mixing was performed in a turbula mixer (T2F, Glenn Mills, Switzerland) at a speed of 67 rpm

Article available at http://www.epj-conferences.org or http://dx.doi.org/10.1051/epjconf/20159402017

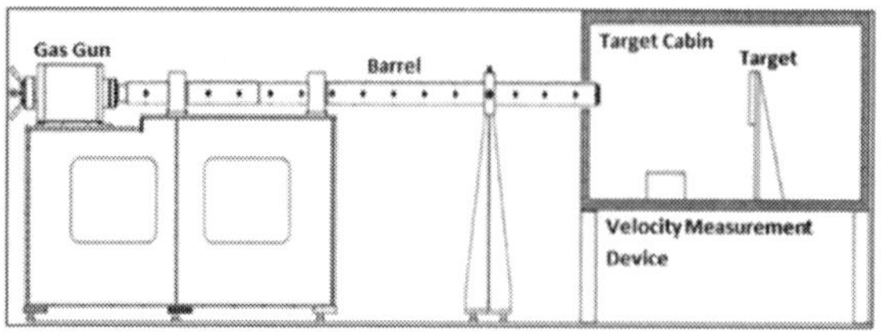

Figure 1. Set-up used in gas gun experiments.

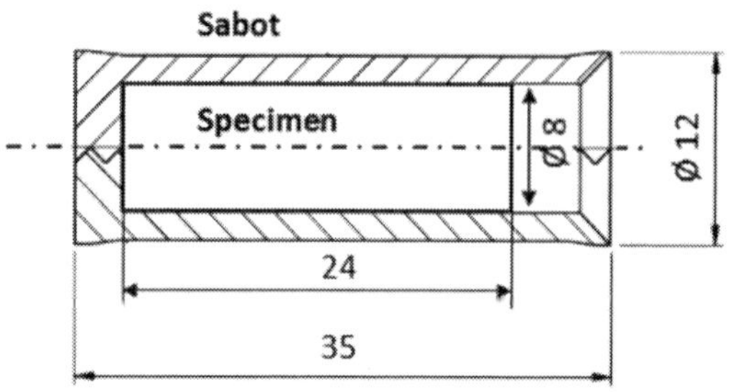

Figure 2. Sabot and specimen dimensions used in gas gun experiments.

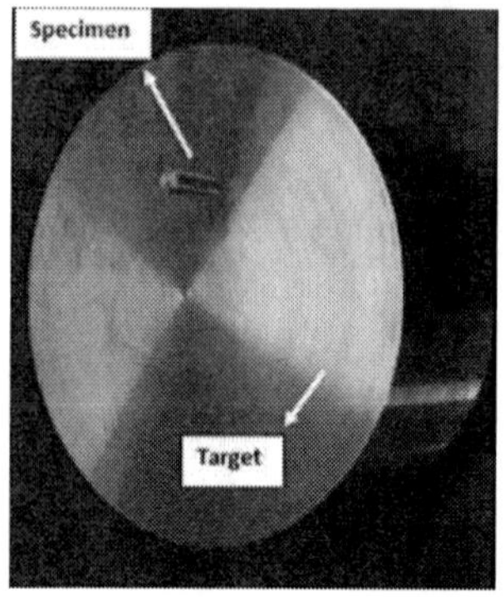

Figure 3. The target and the specimen.

Table 1. Ductility and tensile strength of liquid phase sintered tungsten alloys.

Alloy Composition Specimen Code	**Ductility (% Elongation)**	**Ultimate Tensile Strength (MPa)**
90W-8Ni-2Fe	29 ± 8	952 ± 3
90W-7Ni-3Cu	4 ± 2	729 ± 50

for 45 minutes. Mixed powders were compacted by cold isostatic pressing (CIP 42260, Flow Autoclave Engineers, USA) at 310 MPa for 15 seconds. The compacted samples of 90W-7Ni-3Cu and 90W-8Ni-2Fe were sintered for 30 minutes at 1430 and 1490 °C (Linn HT-1800, Germany), respectively. After sintering, samples were heat treated at around 1100 °C under nitrogen atmosphere in order to provide intermetallic free and chemical homogeneous binder metallic phase.

Sintered alloys were characterized by evaluating their mechanical properties such as tensile strength and % elongation. Density of the alloys was measured by Archimedes water immersion technique by using xylene with a density of 0.86 g/cm3. The tensile test specimens were prepared as per ASTM E8 and were tested using a tensile testing machine (Zwick, Germany) with a crosshead speed of 0.5 mm/min at room temperature. At least 3 specimens were tested and average value of the 3 specimens was reported.

The experiments were conducted by using a gas gun, a target and a velocity measurement device shown as Fig. 1. The specimens of tungsten alloys were accelerated by pressurized gas and impacted to a target in various velocities.

The gas gun used in the experiments has 12 mm barrel diameter. Moreover, 8 mm diameter specimens and 12 mm diameter sabots were manufactured to be accelerated along this barrel. The sabot and specimen sizes and their assembly are given in Fig. 2. The sabots are designed as two pieces to separate from the specimen before it reaches to the target.

At the exit of the gun barrel, the specimens used in the experiments reached to the final velocity measured by a velocity measurement device given in Fig. 1. Then, the specimens were impacted to a target having a close distance to the barrel exit. Target used for the specimen impact was produced from SAE 4340 steel with a hardness of 45–50 HRc. The target was prepared large enough in order to prevent deformation due to kinetic energy of the specimens and to behave as an almost rigid body.

3. Experimental results

The ductility and ultimate tensile strength of the alloys are given in Table 1. The ductility of the 90W-8Ni-2Fe and 90W-7Ni-3Cu alloys is approximately 29% and 4%. The difference in the ductility of the alloys can be attributed to the tungsten solubility in binder matrix phase. The presence of higher tungsten solubility in the binder matrix phase led to higher tensile strength as observed in the study. Necking was observed during tensile testing for 90W-8Ni-4Fe alloys.

Taylor impact tests were conducted by two types of tungsten alloys (90W-8Ni-2Fe and 90W-7Ni-3Cu) under different velocities. After the specimens were impacted to the target, their diameters and lengths have been changed as shown in Fig. 4. The impact surfaces were enlarged and the total lengths were shortened due to deformation under dynamic conditions.

The final values of the diameter and length of the tested materials with different velocities were represented in Fig. 5. Three velocity levels 200 m/s, 225 m/s, 250 m/s were defined to compare the results in the same conditions. Taylor tests in higher velocities cannot be achieved since both materials broke around 266 m/s as shown in Fig. 6. This result also showed that these two alloys, 90W-8Ni-3Fe and 90W-7Ni-3Cu, were broken into several pieces at the same limit of impact velocities.

As shown in Fig. 5, W-Ni-Cu alloy specimens had larger impact surface area than W-Ni-Fe alloys after the impact. In other words, W-Ni-Cu alloys were exposed to more deformation than the W-Ni-Fe specimens. It is observed that W-Ni-Cu alloy tungsten specimens

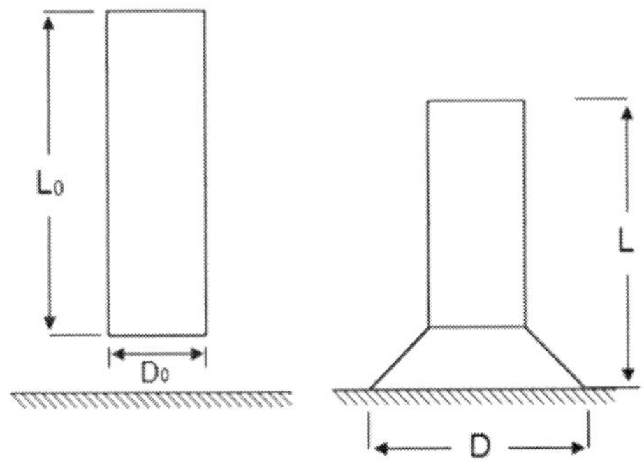

Figure 4. Taylor test scheme.

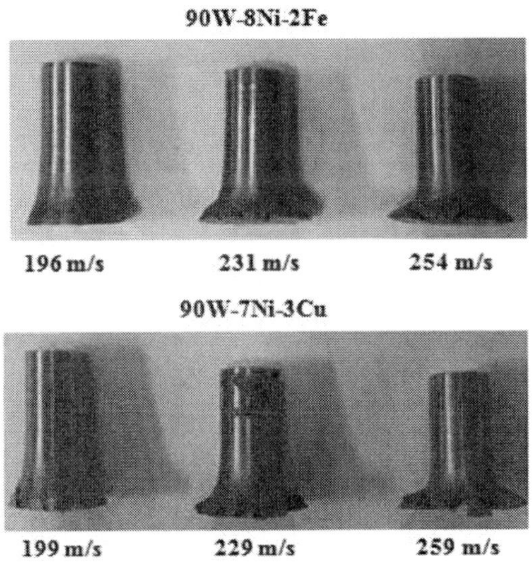

Figure 5. Final shapes of the specimens at different impact velocities.

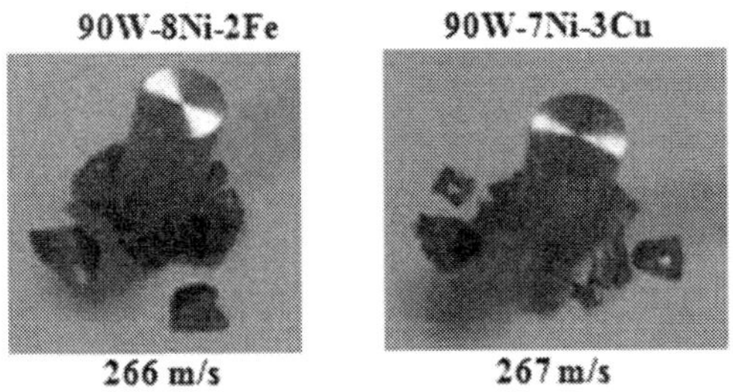

Figure 6. Fracture of tungsten alloy specimens above a certain velocity.

shortens 4–7% more and its impact end diameter becomes 15–24% larger compared to W-Ni-Fe alloy tungsten for the same initial kinetic energies. Furthermore, it is known that deformation and the ductility inversely effects perforation and penetration of the materials. Deformation causes the larger contact face and the target applies higher resistance to penetrative material. For example, according to one of the most famous perforation equation Thor [11], larger presented area decreases in final velocity. This means that the tungsten alloy with the content of W-Ni-Fe is more effective in the perforation characteristics compared to the alloy consisting of W-Ni-Cu.

Table 2. Experiment results data.

Specimen	Velocity, V [m/s]	Diameter, D [mm]	Length, L [mm]
90W-8Ni-2Fe	196	11.8	20.7
90W-8Ni-2Fe	231	14.1	19.7
90W-8Ni-2Fe	254	15.6	18.8
90W-7Ni-3Cu	199	12.7	20.6
90W-7Ni-3Cu	229	14.9	19.5
90W-7Ni-3Cu	259	16.8	18.4

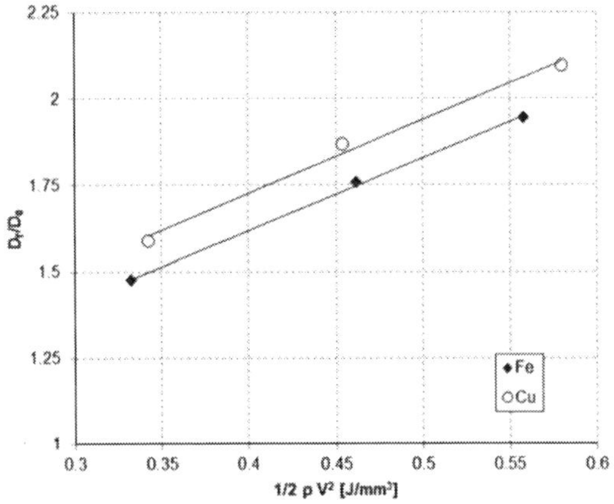

Figure 7. Diameter ratio change by initial kinetic energy.

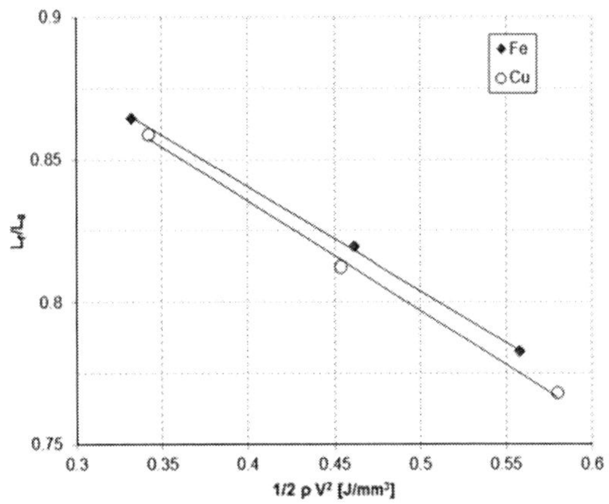

Figure 8. Length ratio change by initial kinetic energy.

In Table 2, change in the diameter and length of specimens were given with change in the velocity of the projectile. According to these results, it can be stated that W-Ni-Cu alloy is more severely deformed than W-Ni-Fe alloy.

In order to compare the results given in Table 2, diameter and length ratios change by initial kinetic energy and initial energy were drawn graphically. All the specimens had the same volume and then the following kinetic energy formula can be used;

$$K.E. = {}^{1}/_{2}(\rho V^2). \quad (1)$$

Since the experiments were performed at the same temperature, internal energies are assumed to be equal for all the tests. The changes in kinetic energy with diameter and length ratio variations upon impact were given in Fig. 7 and Fig. 8 respectively. As mentioned by Banarjee [10], the relation between the energy and the diameter and the relation between the energy and length ratios are linear. Our results verified the theory by Banarjee.

4. Conclusions

In this study, Taylor method has been utilized in order to understand the dynamic behaviour of two different types of tungsten alloys. The study showed that, 90W-8Ni-2Fe alloy having higher strength and ductility at quasi-static conditions has also higher strength than 90W-7Ni-3Cu alloy at dynamic conditions. Furthermore, another significant result obtained by the experiments is that at the same impact velocity levels (266 m/s), both alloys were fractured into several pieces. This indicates taht both of them have similar type of fracture behaviours under severe dynamic conditions. As a summary, by using Taylor impact test technique, 90W-8Ni-2Fe alloy having a higher perforation characteristics than 90W-7Ni-3Cu alloy was determined.

References

[1] Taylor G. The Use of Flat-Ended Projectiles for Determining Dynamic Yield Stress. I. Theoretical Considerations. Proceeding of the Royal Society of London. Series A, Mathematical and Physical Sciences, **194**, 288–299 (1948).

[2] Hopkinson B. A Method of Measuring the Pressure Produced in the Detonation of High Explosives or by the Impact of Bullets. Phil. Trans. R. Soc. London A, **213**, 437–456 (1914).

[3] Kolsky H. An Investigation of the Mechanical Properties of Materials at Very High Rates of Loading, Proc. Phys. Soc. London, B62, 676 (1949).

[4] German RM, Bourguignon LL, Rabin BH. Microstructure limitations of high tungsten content heavy alloys. J Metals; **37**(8):36–9 (1985).

[5] N. K. Çalışkan, N. Durlu, Ş. Bor. Swaging of Liquid Phase Sintered 90W–7Ni–3Fe Tungsten Heavy Alloy. Int. Journal of Refractory Metals and Hard Materials, **36**, 260–264 (2013).

[6] Upadhyaya A., Processing strategy for consolidating tungsten heavy alloys for ordnance applications. Mater. Chem. Phys.; **67**, 101–10 (2001).

[7] N. Durlu, N. K. Çalışkan, Ş. Bor,. Effect of Swaging on Microstructure and Tensile Properties of W-Ni-Fe Alloys, Journal of Refractory Metals and Hard Materials, **42**, 126–131 (2014).

[8] G. H. S. Price, C. J. Smithells, S. V. Williams. Copper-Nickel-Tungsten Alloys: Sintered with a Liquid Phase Present. Journal of the Institute of Metals, **62**, 239–264 (1938).

[9] K. E. Knipling, G. Zeman, J. S. Marte, S. M. Kelly, S. L. Kampe. Effect of Dissolved Tungsten on the Deformation of 70Ni-30Fe Alloys, Metallurgical and Materials Transactions: A. **35**, 2821–2828 (2004).

[10] Project THOR, The Resistance of Various Metallic Materials to Perforation by Steel Fragments; Emprical Relationships for Fragment Residual Weight, Ballistic Analysis Lab. The Johns Hopkins Univ., Proj. THOR Technical Rept. #47 Baltimore, MD, (1961).

[11] Biswajit Banerjee. An Evaluation of Plastic Flow Stress Models for the Simulation of High-Temperature and High-Strain-Rate Deformation of Metals. Department of Mechanical Engineering, University of Utah, 50 S Central Campus Dr., MEB 2110, Salt Lake City, UT 84112, USA (2005).

EPJ Web of Conferences 94, 02018 (2015)
DOI: 10.1051/epjconf/20159402018

Twinning in magnesium under dynamic loading

Neha Dixit[1], Kavan Hazeli[2], and Kaliat T. Ramesh[1,2,a]

[1] Department of Mechanical Engineering, Johns Hopkins University, Baltimore, MD, USA
[2] Hopkins Extreme Materials Institute, Johns Hopkins University, Baltimore, MD, USA

Abstract. Twinning is an important mode of deformation in magnesium (Mg) and its alloys at high strain rates. Twinning in this material leads to important effects such as mechanical anisotropy, texture evolution, tension-compression asymmetry, and sometimes non-Schmid effects. Extension twins in Mg can accommodate significant plastic deformation as they grow, and thus twinning affects the overall rate of plastic deformation. We use an experimental approach to study the deformation twinning mechanism under dynamic loading. We perform normal plate impact recovery experiments (with microsecond pulse durations) on pure polycrystalline Mg specimens. Estimates of average TB velocity under the known impact stress are obtained by characterization of twin sizes and aspect ratios developed within the target during the loading pulse. The measured average TB velocities in our experiments are of the order of several m s^{-1}. These velocities are several orders of magnitude higher than those so far measured in Mg under quasi-static loading conditions. Electron back-scattered diffraction (EBSD) is then used to characterize the nature of the twins and the microstructural evolution. Detailed crystallographic analysis of the twins enables us to understand twin nucleation and growth of twin variants under dynamic loading.

1. Introduction

The low density, high specific strength (strength/weight ratio) [1], and good damping properties [2,3] of Mg and its alloys makes these materials attractive for applications such as automobile and aerospace components. For successful use of these materials in applications in which dynamic loading conditions (e.g. crash) are expected, an understanding of the deformation mechanisms under dynamic loading is important. Dynamic loading generally involves varying strain rates, short time durations and complex stress states. It is known that the stress–strain response of Mg alloys is highly anisotropic as well as sensitive to strain rate [4]. Tucker et al. [5] reported strong strain rate dependences of the compressive yield, strain hardening rate and ductility for loading along the normal direction of a rolled Mg sheet. In the rolling and transverse directions, no strain rate dependence of the yield strength was observed. Instead a marked increase in strain hardening was observed beginning at a characteristic strain level. Agnew et al. [6] investigated the effect of rare-earth-elements on the ballistic properties of Mg alloys using a Kolsky bar technique. This study demonstrated that the effect of strain rate on the yield strength of a textured Mg WE43B-T5 alloy is small and the strain-hardening behavior is only mildly rate sensitive. The review article by Prasad et al. [7] suggested that the flow stress at yield is relatively insensitive to strain rate in the cases where deformation twinning was the dominant mechanism, but the flow stress exhibits some sensitivity to the strain rate when slip was the dominant deformation mechanism. Despite the great interest and importance of deformation mechanisms associated with dynamic behavior, there are limited studies on the dynamic response of pure Mg and very little information on the relative contributions to plastic deformation of twinning versus slip as a function of strain rate. Our objective is to build a fundamental understanding of the dynamic deformation of Mg and its alloys. We begin by focusing on pure Mg as a model material in this study. The insights gained from this work can then perhaps be applied to the analysis of the dynamic behavior of various Mg alloys.

2. Normal plate impact experiments

A hot-extruded Mg rod of commercial purity (99.9%) was used in this study. The microstructure of the material consists of equiaxed grains with an average grain size of about 20 μm. Normal plate impact experiments were performed, with impact in the extrusion direction at impact velocities of 60–70 m/s. At this velocity, there is elastic-plastic wave propagation in the material but no shocks are generated. In the normal plate impact experiment, a cylindrical plate of material (called the flyer) is carried on a projectile that is launched at high velocities, producing an impact onto a stationary plate (called the target) that is supported in a target holder within an evacuated target chamber. The two plates are parallel to each other and perpendicular to the direction of projectile motion, and thus the impact causes longitudinal wave propagation in both plates.

A schematic of the wave propagation in the plates is shown in the Lagrangian diagram of Fig. 1. Upon impact, longitudinal elastic and plastic waves are generated in both flyer and target plates. This happens when the impact velocity generates stresses above the yield strength of the material in this uniaxial strain configuration. The

[a] Corresponding author: ramesh@jhu.edu

Article available at http://www.epj-conferences.org or http://dx.doi.org/10.1051/epjconf/20159402018

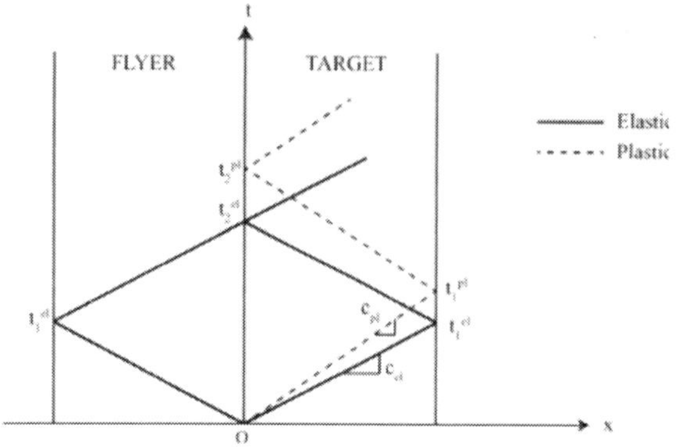

Figure 1. Elastic-plastic wave propagation in the symmetric normal plate impact configuration (the plastic wave is not shown in the flyer for simplicity).

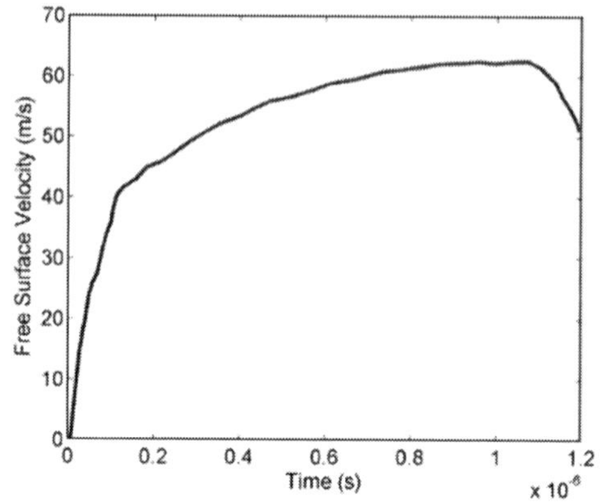

Figure 2. Experimentally measured particle velocity history at the free surface of a 3.5 mm target after an initial impact at 70 m s^{-1}.

elastic wave represented by a solid line travels at a larger speed (c_{el}) than the plastic wave (c_{pl}) represented by the dashed line. Both these waves travel through the thickness of the plate to the rear free surface. The rear surface has a stress-free boundary condition, and therefore the reflected waves from the free surface are of opposite sign. Similar wave propagation occurs in the flyer plate as well, but only the elastic wave in the flyer is shown in the figure for simplicity. The reader interested in the plate impact technique is referred to the following article in the Handbook of Experimental Solid Mechanics [8].

The flyer and the target plates in these experiments are both made of pure Mg and are of the same dimensions (25.4 mm diameter, 3.5 mm thickness) for this symmetric impact. Therefore, the reflected waves from the rear surfaces of the flyer and target plates will arrive at the impact face at approximately the same time and therefore unload each other. Note that the normal plate impact experiment imposes a condition of uniaxial strain at the centre of the plate before unloading (release) waves from the edges arrive at the point being interrogated on the rear surface of the target. The experimental measurements consist of the projectile velocity before impact (measured using a Laser Line Velocity Sensor), the tilt/misalignment between the impacting plates (measured using tilt pins) and the normal particle displacement at the free surface of the target (measured using a laser-based Normal Displacement Interferometer).

The density of Mg is 1740 kg/m^3, the elastic Young's modulus is 44 GPa and the Poisson's ratio is 0.35 [9]. The elastic longitudinal wave speed in Mg is $c_{el} = 6370$ m/s, and so the round trip travel time for the 3.5 mm thick plate is 1.1 μs for the elastic wave (t_2^{el}). The effective pulse duration for this experiment is the time needed for the plastic wave to arrive again at the impact face after traveling through the target thickness (t_2^{pl}), and is about 1.3 μs. The plates are recovered after impact and examined using optical and electron microscopy to determine the deformation mechanisms active within these loading times.

3. Results and discussion

3.1. Free surface velocity

The particle displacement at the free surface of the target is measured by the normal displacement interferometer (NDI), and is then numerically differentiated to obtain the rear surface particle velocity. An example of the free surface velocity history for a target plate of 3.5 mm thickness is shown in Fig. 2. There is a rapid increase in the free surface velocity (Fig. 2) at first in response to the arrival of the elastic wave, and then the rate of increase of the velocity slows down as plasticity sets in. Due to the wave propagation associated with the plate impact experiment, the stress, strain and strain rates are inhomogeneous throughout the thickness of the plates and evolve with time at each location. Computational estimates of the dynamic plastic deformations [REF Neha's thesis] suggest that the local plastic strain rates are of the order of 10^4–10^5 s^{-1} in these experiments, but these high strain rates are present in specific regions of the plates for only small fractions of a microsecond.

We now examine the deformation mechanisms active within the target plate during this dynamic loading.

3.2. Microstructural evolution

The deformed target plates were recovered after the plate impact experiment and sectioned by electrical discharge machining (EDM) to make samples for microstructural analysis. The microstructure of these targets was analysed using electron back-scattered diffraction (EBSD), with the focus on the characterization of the observed twins (dislocations cannot be effectively observed using this method, and TEM results on the dislocation motion will be reported in subsequent work.

Note that Mg has two commonly observed modes of twinning, leading to extension and contraction twins [10, 11]. In addition to primary twinning, secondary twinning can occur within the reoriented primary twins, known as double twinning [12]. The material being tested has an initial texture as a consequence of the extrusion. As a consequence, compressive loading along this direction

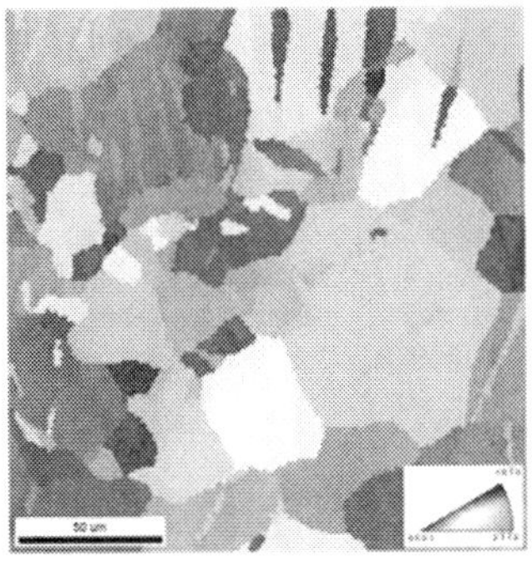

(a)

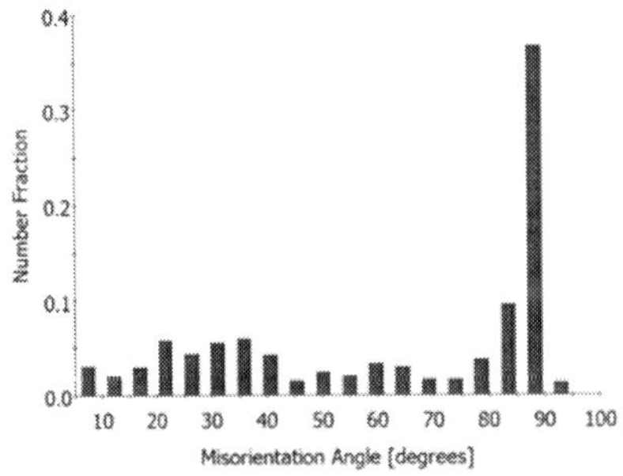

(b)

Figure 3.) Inverse pole figure in the through thickness direction of the specimen deformed under normal plate impact loading. (b) Calculated misorientation angle distribution for the section in part (a).

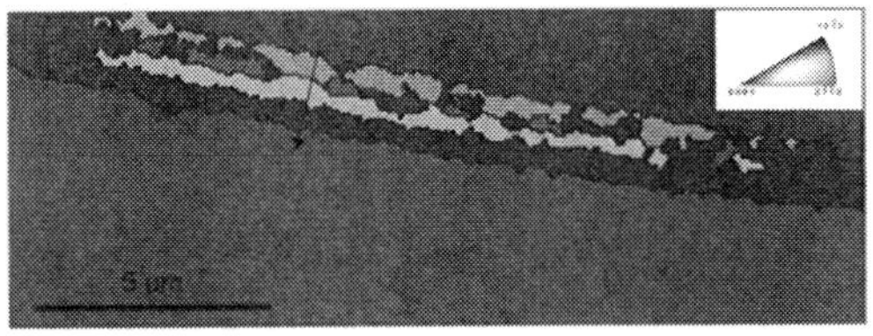

(a)

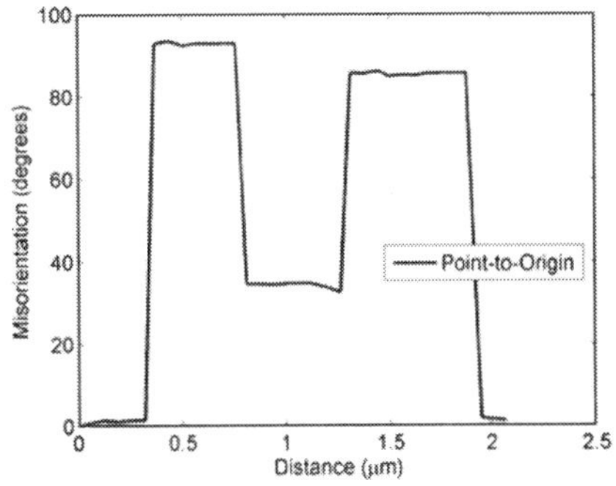

(b)

Figure 4. (a) EBSD inverse pole figure map of a region in a grain within the deformed target. (b) Calculated misorientation angle profile along the black line across the twinned region in the inverse pole figure map shown in (a).

should result in the development of some extension twinning. However, the microstructure before impact was devoid of twins (as observed on other samples cut from the same rod).

The inverse pole figure of one region of the deformed sample as measured by EBSD is shown in Fig. 3a. The image shows the through thickness region of the sample such that the impact direction is horizontal in the plane of the figure. Note the grain size of the material, and the presence of many lenticular twins in the deformed microstructure. Note that some grains have multiple twins. Since the microstructure before impact was devoid of twins, the observed twins must have been induced by the impact loading. It is found that most of the twins are extension twins. The morphology of these twins appears to be different than those formed under the high strain rate loading imposed by Kolsky bars and reported earlier by the authors [13]. We now discuss the orientation and size of these impact-induced twins in this pure Mg material.

Misorientation angle distribution analysis was conducted to characterize the observed twins, and the results are shown in Fig. 3b. The strong peak in the in the 85°–90° range in the misorientation angle distribution chart in Figure 3b corresponds to the extension twin boundaries present in the microstructure.

The average length and width of the twins were measured by analyzing the EBSD and optical images using the commercial image analysis software ImageJ. The average twin length L_{avg} was found to be $27.5 \pm 2.4\,\mu$m and the average twin width W_{avg} was $4.4 \pm 0.7\,\mu$m. From these length measurements and the pulse duration, we can obtain estimates for the twin tip velocity and twin boundary velocity. The maximum amount of time for the twins to grow to the measured sizes is the pulse duration corresponding to the plastic wave propagation. Therefore, the lower bound for the twin tip velocity is given by ($v_{tip} = L_{avg}/t_2^{pl}$) and for the twin boundary velocity is given by ($v_{boundary} = W_{avg}/t_2^{pl}$). In this case, the estimated lower bounds are $v_{tip} = 20.8$ m/s and $v_{boundary} = 3.3$ m/s for this Mg alloy under these impact conditions.

To obtain detailed characterizations of the twins developed in plat impact, we performed high resolutions EBSD scans (step size $\leq 0.3\,\mu$m) in the regions of interest. The inverse pole figure map of one of the regions studied is presented in Fig. 4a. The different colors represent different crystallographic orientations as represented by the map key in the inset of Fig. 4. It is seen that the twin is in fact composed of different regions that correspond to different orientations. The orientation of the parent grain is represented by the red color and is close to the (0001) orientation. The point-to-origin misorientation profile across the black line in Fig. 4a is illustrated in Fig. 4b.

The point-to-origin profile represents the misorientation between each point on the black line with respect to the parent grain orientation. In the line profile, an ~86° change in the misorientation is observed upon crossing the first extension twin boundary to enter the green twin in Fig. 4a. Upon crossing the boundary between the green and the yellow twins, a smaller misorientation of ~38° (with respect to the parent orientation) is observed. The yellow twin is actually a contraction twin most likely formed through a double twinning mechanism [14]. The blue and green regions are both extension twins that represent the first and second order prismatic planes respectively, and therefore they have a 30^0 in-plane misorientation between them.

This morphology of extension twins is significantly different from that observed in the same material under uniaxial stress Kolsky bar loading [13]. Typically only one twin variant is observed to grow significantly and then coalesce with other twins under the Kolsky bar loading. The uniaxial strain condition in the plate impact experiment imposes a constraint on the lateral strains. Therefore, stresses build up in the lateral directions. Within the uniaxial strain configuration an overall higher stress amplitude is necessary to induce significant plastic deformation in this highly plastically anisotropic metal. The uniaxial strain configuration coupled with the extremely short times available for twin nucleation and propagation are likely the major reasons for the observed twin morphology.

4. Summary

The dynamic behavior of pure Mg under normal plate impact loading of microsecond duration was studied. Extension twinning is observed in many grains. The twins are long and thin, and twin coalescence is generally not observed. Estimates of twin tip and twin boundary velocities provide a lower bound on the velocities: $v_{tip} = 21$ m/s and $v_{boundary} = 3.3$ m/s for this Mg alloy under these impact conditions.

The observed twins often consist of multiple variants of extension twins. Double twinning is also observed. We speculate that due to the high strain rates (on the order of 10^4–10^5 s^{-1}), the high amplitude of stresses due to the uniaxial strain condition and the short times available for deformation ($\sim 1\ \mu s$), nucleation of multiple twin variants is preferred under the plate impact loading instead of the growth and coalescence of the same variant. Therefore, the dynamic loading and nature of the stress state is likely to have a significant effect on the specific deformation mechanisms (especially twinning) activated within magnesium under dynamic loading.

The authors gratefully acknowledge financial support from the NSF DMR Materials World Network program and support from the MEDE CRA, US Army Research Laboratory, to KTR and KH.

This research was sponsored in part by the Army Research Laboratory and was accomplished under Cooperative Agreement Number W911NF-12-2-0022. The views and conclusions contained in this document are those of the authors and should not be interpreted as representing the official policies, either expressed or implied, of the Army Research Laboratory or the U.S. Government. The U.S. Government is authorized to reproduce and distribute reprints for Government purposes notwithstanding any copyright notation herein.

References

[1] B.L. Mordike and T. Ebert, Mat. Sci. & Eng. A, **302** (1): 37–45 (2001).

[2] K. Kainer, B.L. Mordike, Wiley-VCH, (2000).

[3] K. Hazeli, A. Sadeghi, M.O. Pekguleryuz, and A. Kontsos, Mat. Sci. & Eng. A, **589**: 275–279 (2014).

[4] G. Wan, B.L. Wu, Y.D. Zhang, G.Y. Sha, and C. Esling, Mat. Sci. & Eng. A, **527**(12): 2915–2924 (2010).

[5] M.T. Tucker, M.F. Horstemeyer, P.M. Gullett, H. El Kadiri, and W.R. Whittington, Scripta Materialia, **60**(3): 182–185 (2009).

[6] S. Agnew, W. Whittington, A. Oppedal, H. El Kadiri, M. Shaeffer, K.T. Ramesh, J. Bhattacharyya, R. Delorme, and B. Davis, JOM, **66**(2): 277–290 (2014).

[7] K. Eswar Prasad, B. Li, N. Dixit, M. Shaeffer, S.N. Mathaudhu, and K.T. Ramesh, JOM **66**(2), pp. 291–304 (2014).

[8] K.T. Ramesh (2008), High rates and impact experiments, Handbook of Experimental Solid Mechanics, Springer US (pp. 929–960).

[9] M.M. Advesian and H. Baker, ASM international, **244** (199).

[10] M.R. Barnett, Mat. Sci. & Eng. A, Part I, **464**, pp. 1–7 (2007).

[11] M.R. Barnett, Mat. Sci. & Eng.: A, Part II, **464**: 8–16 (2007).

[12] L. Jiang, J.J. Jonas, R.K. Mishra, A.A. Luo, A.K. Sachdev, S. Godet, Acta Materialia **55** (11), 3899–3910 (2007).

[13] N. Dixit, K.Y. Xie, K.J. Hemker, K.T. Ramesh, Acta Materialia, **87**: 56–67 (2015).

[14] M.D. Nave and M.R. Barnett, Scripta Materialia, **51.9**: 881–885 (2004).

EPJ Web of Conferences **94**, 02019 (2015)
DOI: 10.1051/epjconf/20159402019

Strain rate dependency of laser sintered polyamide 12

J.E.T. Cook[1,a], R.D. Goodridge[2], and C.R. Siviour[1]

[1]Department of Engineering Science, University of Oxford, Parks Road, Oxford, UK
[2]Additive Manufacturing & 3D-Printing Research Group, School of Engineering, University of Nottingham, Nottingham, UK

Abstract. Parts processed by Additive Manufacturing can now be found across a wide range of applications, such as those in the aerospace and automotive industry in which the mechanical response must be optimised. Many of these applications are subjected to high rate or impact loading, yet it is believed that there is no prior research on the strain rate dependence in these materials. This research investigates the effect of strain rate and laser energy density on laser sintered polyamide 12. In the study presented here, parts produced using four different laser sintered energy densities were exposed to uniaxial compression tests at strain rates ranging from 10^{-3} to 10^{+3} s^{-1} at room temperature, and the dependence on these parameters is presented.

1. Introduction

The principle advantage of Additive Manufacturing (AM) is the ability to form complex geometries that would be impossible or highly expensive compared to conventional subtractive or formative methods. In Additive Manufacturing objects are directly built from CAD models by depositing material in a layer-by-layer process. Laser sintering (LS) is a form of AM whereby a thin layer of powder material is deposited onto a pre-heated build area (normally pre-heated), a CO2 laser is then used to selectively consolidate the powder. The consolidated regions form the final product, whilst the remaining powder is removed.

LS is increasingly being used in demanding applications, with the aviation and F1 industry adopting LS for processing parts [1]. Detailed analysis of the stiffness, strength and surface finish must be further understood for LS parts to meet future in-service loading and operational requirements. The performance of components under high rate and impact loading is a particularly important design criterion for crash worthiness.

The mechanical properties of LS can in part be attributed to the degree of particle melt (DPM) [2], which defines the quantitative variations in the consolidation of sintering. Powder properties (e.g. morphology) and build parameters (e.g. laser energy, density, temperature) can affect the DPM of a given material in laser sintering [3].

Differential scanning calorimetry (DSC) is frequently used to predict the DPM of a material [3]. However, it is believed that there is no prior research on the dependency of strain rate on a part's mechanical properties. By understanding the effect of a wide range of strain rates it is hoped that a more versatile constitutive model for LS components can be developed. A range of energy densities are investigated to help determine the particular effect of strain rate, where it is hoped that this will provide further information on the role of the microstructure.

1.1. Calculating the energy density

Also known as the Andrew number, it has been shown that the energy per unit area supplied by the laser to the powder bed is dependent on the laser power (W), scan speed (mm/s) and scan spacing (mm) [4].

$$ED(J/mm^2) = \times Laser\ power/(scan\ speed * scan\ spacing). \quad (1)$$

1.2. Effect of the energy density

The energy applied to the powder causes coalescence by tipping the material into its molten state. The typical T_M for polyamide 12 is 175°C. The powder bed is heated to approximately 152°C to prevent rapid cooling of the part. As polyamide 12 is a semi-crystalline polymer, rapidly cooling through its glass transition will cause significant shrinkage and inherent thermal stress concentrations.

In reality, powder cannot be uniformly consolidated due to limitations of the scan spacing, precision of levelled powder and variation in powder morphology. In partial melting only the shell of the powder melts allowing the molten polymer to form necks between adjacent powders. However, the core can remain solid which is characterised by an un-molten core surrounded by spherulites.

By imparting a different energy density one can alter the inter-crystal microstructure through (a) powder coalescence (necking of particles) (b) and crystallised particles (spherulites) [2]. By investigated various energy densities, it is possible to better understand the role of a) and b) on mechanical properties.

2. Experimental approach

Four different energy densities were investigated in this study, listed in Table 1. Only the laser power and scan

[a] Corresponding author: james.cook@eng.ox.ac.uk

Article available at http://www.epj-conferences.org or http://dx.doi.org/10.1051/epjconf/20159402019

Table 1. Energy densities investigated.

	ED1	ED2	ED3	ED4
Laser Power (W)	19	21	23	21
Scan spacing (mm)	0.25	0.25	0.25	0.25
Scan speed (mm/s)	2500	2500	2500	1500
Energy Density (J/mm²)	**0.03**	**0.034**	**0.037**	**0.056**
Physical density (kg/m³)	**889**	**876**	**873**	**884**

speed was varied. Each energy density was used to form a 6 × 6 mm right cylinder, where a range of strain rates were applied to investigate the relationship between strain rate and relative density in compression. All samples were processed on an EOS P100 machine. Each sample was orientated in the z-axis, which will be the subsequent loading axis, during part build.

The mass and volume of each energy density was measured five times to calculate the respective mean physical density, also listed in Table 1. The mass was found using a XS105 dualrange analytical balance with a readability of 0.01 mg. A micrometer of readability 0.001 mm was used to gauge the length, height and width of each cube specimen.

The *quasi-static compression* testing was undertaken using an Instron testing machine. For these low strength materials machine compliance is not an issue, and true strain control from the cross head was used however, an extensometer was also attached to the loading anvils close to the specimen to verify the specimen extension. The total resisting force on the specimen as a function of time was obtained from a 100 kN load cell with a stated precision of ±0.5 N.

Medium strain rate was obtained through a custom built hydraulic load frame that is used to access strain rates between 1 and 50 s^{-1}. A linear Variable Differential Transformer (LVDT) measured the displacement of the sample; the signal suffered no significant distortion from load cell ringing and other machine noise.

High strain rate compression experiments were performed using a Split Hopkinson Pressure Bar (SHPB). For the SHPB system, the input and output bars were made of silver steel. The input bar is 1 m long, and gauged halfway along its length; the output bar is 500 mm long and gauged 50 mm from the bar-specimen interface. The reflected and transmitted gauge signals are used to derive the stress strain relationship using the standard analysis [6].

To obtain reliable data, good lubrication of the anvils is essential. A control test of lubricants for the semi-crystalline polymer – Polypropylene was investigated by Okereke [7]. In the post-yield region, petroleum jelly gave the lower yield/flow stress indicating that least friction was generated and so this lubricant was applied for all testing in the current study.

Barrelling of the sample was not apparent post compression. However, occasionally slanting was seen, in which one end of the recovered specimen was observed to have a larger diameter than the other. This was connected to defects of the cell wall during the LS process.

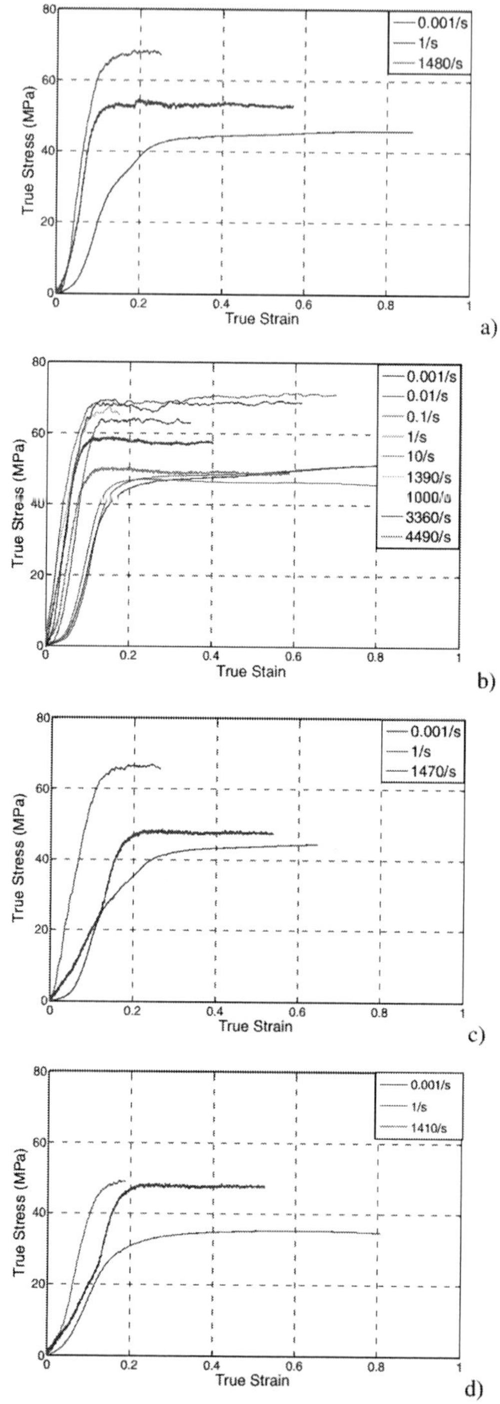

Figure 1. Compressive stress-strain response of ED, from top: a) 0.030 J/mm² b) 0.034 J/mm² c) 0.037 J/mm² d) 0.056 J/mm².

3. Results and discussion

The stress-strain response of four different energy densities can be seen in Fig. 1, where a wide range of strain rates are applied in uniaxial compression. Results are plotted astrue

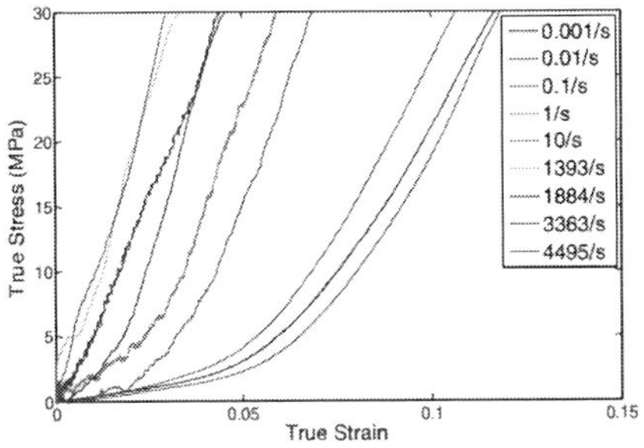

Figure 2. Effect of surface roughness for 0.034 J/mm^2 over a range of strain rate.

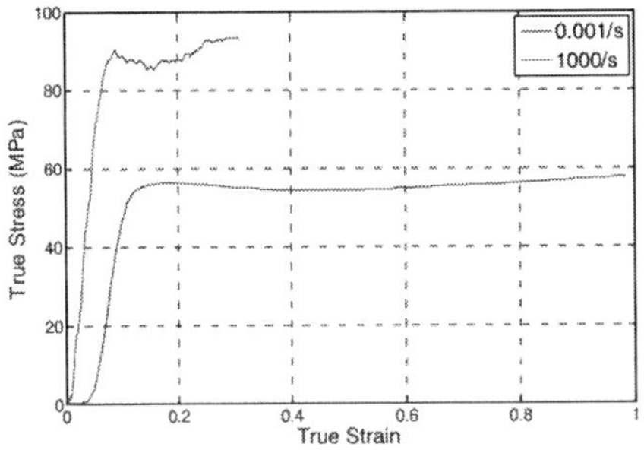

Figure 3. Extruded polyamide 12 at low and high strain rates.

strain against true stress. The true strain is determined by assuming constant volume. Whilst three of the energy densities were tested at rates c.a. 0.001, 1 and 1500 s^{-1}, one energy density, 0.0034 J/mm^2, which is currently the most commonly used, was tested over a wide range of rates, Fig. 1(b).

The point at which yield stress is initiated is seen to shift to a lower true strain as the strain rate increases for ED1, 2 and 3, particularly between low and medium strain rates. The shift is unusual compared with other semi-crystalline polymers such as polypropylene, where yield stress is observed to start at 10% true strain across all strain rates (see e.g. [7]). Looking at Fig. 2, it was observed that this is due to a low strength 'bedding in' before the main deformation. This is attributed to surface roughess in the specimen, which is either more easily observed, or has a greater effect, in the quasi-static experiments.

The LS compression tests can be compared to the mechanical response of extruded polyamide 12 at low and high rate, which is shown in Fig. 3.

Compared to the data from the SL material, the extruded polyamide shows a small amount of strain softening after yield. This is common for glassy polymers. A mechanism developed by Boyce and Haward showed that plastic strain-induced structural evolution of the glass can create a less compact but higher energy structure that produces lower flow stresses. [8].

This phenomenon is characteristic of a glassy polymer deformed at a temperature below its glass transition, and causes post yield softening in the material.

In Fig. 3, only slight softening can be seen. However, the degree of strain softening appears to increase from strain rate 10^{-3} to 10^{+3} s^{-1}. According to Wang and

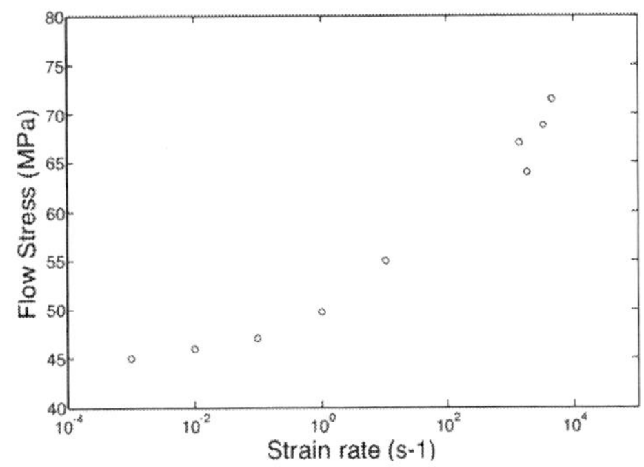

Figure 4. Eyring plot showing stress at 25% strain vs log $\dot{\varepsilon}$ for 0.034 J/mm^2.

Arrruda [9], high strain rate deformation can cause strain softening due to adiabatic heating within the sample.

Interestingly, no strain softening or hardening was seen in the LS polymer. The consistent flow stress could be due to the larger degree of porosity in the specimen, making it more akin to a foam, which will require further investigation. LS is known to produce parts that have significant porosity due to partial coalescence of powder, also known as necking. This will be discussed further below.

With a wide range of strain rate testing on the LS sample, it is possible to conduct an Eyring plot, showing the yield stress versus the logarithm of strain rate, as seen in Fig. 4.

Here the flow stress was chosen to be the stress at 25% strain. The Eyring plot is usually linear for most polymers at room temperature when tested over a modest range of strain rate. However, when constructed over a wide range, many polymers show a significant increase in strain-rate dependence at higher rates and a similar effect is seen here.

This response, often described as bi-linear, has been shown in other polymers due to the transition between two different relaxation times governing the response. These can be assumed to be the α-(glass) and β-processes with $\tau_\alpha > \tau_\beta$, since the glass temperature is measured in the literature to be 155 [°C]. The increased strain rate dependence of the yield strength of LS polyamide 12 at high strain rates is therefore expected to be a material property and is due to the movement of the β transition to room temperature at these rates. An understanding of this transition is crucial to predicting the material's response to high strain rate deformation [10].

It has been shown that the degree of a) crystallisation and b) coalescence of powder are related to the rate of cooling. A change of energy density will vary the thermal gradient between powder temperature and the temperature of the molten region, the steeper this gradient, the more likely it is that higher rates of cooling will occur [3]. The high rate of cooling will inhibit coalescence of powder causing high porosity in the part. It would be expected that ED4 = 0.056 J/mm2 would have the highest thermal gradient and thus the highest level of porosity. However, on measuring the physical densities there was no trend between the physical density and energy density, as shown in Fig. 5 suggesting that the energy density between 0.03 and 0.56 J/mm^2 has little effect on the degree of porosity.

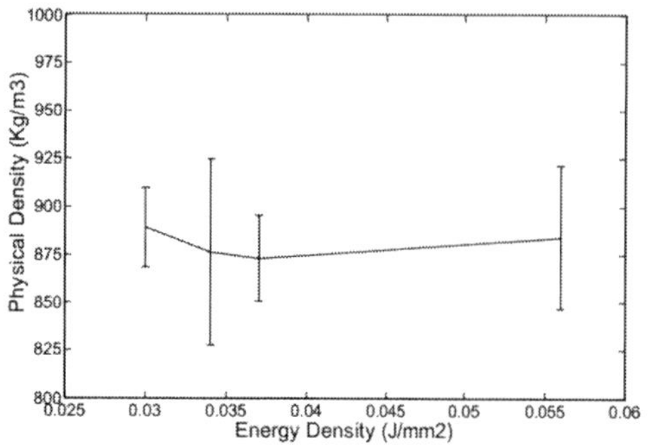

Figure 5. The physical density for four different energy densities. Error bars represent standard deviation from 5 measurements.

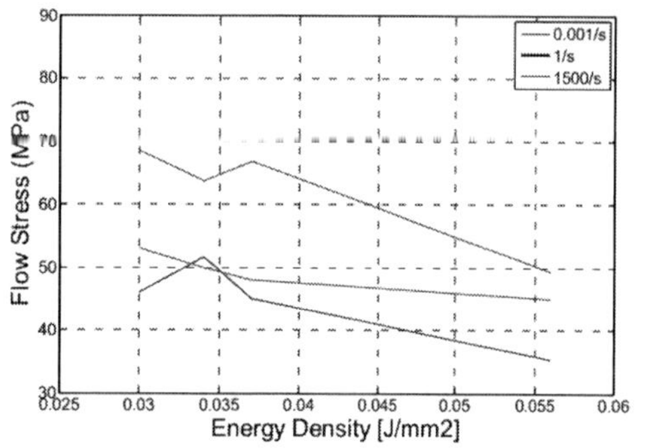

Figure 6. Relationship between energy density and flow stress. Flow stress is obtained at the initial point of plateau.

According to Ashby and Gibson [11] the yield strength for open macro cellular foams can be predicted from the degree of porosity as shown in equation 2.

$$\sigma^* = C_1 \sigma_s (p^*/p_s) \quad (2)$$

C_1 is dependent on the material, σ_s is the yield strength for fully dense material and (p^*/p_s) is the relative density of the foam. However, in this case, there was no significant change in physical density between the different specimens, and it is therefore likely that changes in properties between the energy densities are a result of smaller scale microstructural changes in the polymer.

Figure 6 shows the dependence of flow stress on energy density at three strain rates. Because some of the samples were not tested to 25% strain, the flow stress was obtained at the initial point of plateau. There appears to be a general, but weak, trend of decreasing strength with increasing energy density. However, there is no observed correlation between ED1, ED2, ED3 and their respective flow stresses. One possible cause of energy density dependence could be the change in activation volume at the molecular level due to the degree of crystallinity in the microstructure, which has been shown to vary according to the energy density [2], and may affect the activation volume and hence the rate dependence.

4. Conclusions

AM allows the construction of more complex geometries than traditional manufacturing techniques, giving designers and engineers the freedom to further optimise parts. AM lattices are used in the automotive industry as they provide an enhanced mechanical response in impact absorption. This paper has two objectives; firstly, to investigate the strain rate dependency of polyamide 12, and secondly to further understand the relationship between energy density and strain rate.

The flow stress was found to be consistent after yield, showing no strain softening or strain hardening. The plateau of flow stress can be related to the porosity as observed in foams and under SEM of polyamide 12 [12]. The apparent true strain at which yield stress occurred varied at different strain rates, this observation is believed to be due to the surface roughness from processing rather than a material property.

Processing at an energy density of 0.056 J/mm^2 significantly reduced the flow stress; this energy density was achieved by reducing the scanning speed from 2500 mm/s to 1500 mm/s. Energy densities, 0.03, 0.034 and 0.037 J/mm^2 were achieved by altering the laser power from 19 to 21 W and had a limited effect on the mechanical response.

Since there was no relationship between energy density and physical density, it seems that porosity is not the driving factor in determining the mechanical response within the aforementioned range of ED.

Other factors such as DPM that defines the completeness of melting within the microstructure may have a more significant influence on mechanical properties.

A deeper understanding of the relationship between energy density and rate dependency could provide an alternative insight into the role of the microstructure, whilst also allowing optimisation of process parameters for a particular strain rate environment.

References

[1] N. Hopkinson, R. Hague, P. Dickins, Rapid Manufacturing, Wiley, (2005)

[2] Kamperman, J.J. de Vlieger, Materials Science and Engineering, **435**, 172 (2006)

[3] H. Zarringhalam, C. Majewski, N. Hopkinson, Rapid Prototyping J, **15**, 126 (2009)

[4] R.D. Goodridge, C.J. Tuck, R.J.M. Hague, Progress in Materials Science, **57**, 229, (2012)

[5] J.C. Nelson, phD Dissertation, Uni of Texas at Austin, Texas, (1993)

[6] C. Yan, Y. Shi, L. Hao, Polymer processing J, **4**, 416 (2011)

[7] G.T. Gray III, ASM Handbook, Mechanical Testing and Evaluation' **8**, 560, (2000)

[8] M.I. Okereke, C.P. Buckley, C.R. Siviour, Mechanics of Time-Dependent Materials, **16**, 361, (2012)

[9] M. Boyce, R. Haward, The physics of glassy polymers, 508 (1997)

[10] Y. Wang, E.M. Arruda, Eng. Mater. Technol. J, **128**, 551(2006)

[11] C.R. Siviour, S.M. Walley, W.G. Proud, N.F. Kamperman, J.J. de Vlieger, Materials Science and Engineering, **435**, 172 (2006)

[12] E.W. Andrews, L.J. Gibson, M.F. Ashby, Acta Mater, **47**, 2853 (1999)

[13] B. Caufield, P.E. McHugh, S. Lohfeld, Mater. Processing Technol. J, **182**, 477 (2006)

EPJ Web of Conferences **94**, 02020 (2015)
DOI: 10.1051/epjconf/20159402020

Deformation of zirconium – niobium alloy E635 in sub-microsecond shock waves

D.N. Kazakov, O.E. Kozelkov, A.S. Mayorova, S.N.Malyugina[a], S.S. Mokrushin, and A.V. Pavlenko

RFNC-VNIITF, 456770, Snezhinsk, Chelyabinsk region, Russia

Abstract. Strength characteristics of zirconium - niobium alloy E635 were measured under shock - wave loading conditions at normal and elevated temperatures and results of these measurements are presented. Measurements were taken in conditions when samples were impacted by plane shock waves with the pressure up to 13 GPa and duration from ~0.05 μs up to 1 μs. Free-surface velocity profiles were recorded with the help of VISAR and PDV laser Doppler velocimeters having nanosecond time resolution. Evolution of elastic precursors with samples thickness varying from 0.5 up to 8 mm is also considered. Measured attenuation of the elastic precursor was used to determine plastic strain rate behind the precursor front. Temperature effect on the value of dynamic elastic limit and spall strength at normal and elevated temperatures is studied.

This work is implemented with the support of the State Atomic Energy Corporation "Rosatom" under State Contract H.4x.44.90.13.1111.

1. Introduction

Zirconium and alloys on its basis are main structural materials in nuclear power engineering. Small thermal neutron capture cross-section, high-temperature strength, high resistance to creep and irradiation-induced swelling, and also high corrosion resistance together with good compatibility with nuclear fuel make these materials to be most suitable for shells and other elements of fuel assemblies [1,2].

Use of products made from zirconium and its alloys under high temperatures and also the necessity to predict behavior of these materials both in severe operational conditions and probable emergency situations highlights importance of the issue how temperature effects elastic-plastic and mechanical strength characteristics of these materials under high-strain-rate deformation. Investigation into processes of elastic-viscous-plastic deformation of metals and alloys under shock-wave loading [3] allows us to measure velocity-temperature relationships how these materials resist to deformation and fracture. The structure of shock waves in solid bodies depends on processes of their elastic-viscous-plastic deformation, possible phase transformations, as well as on kinetics of fracture incipience and development [4–6].

Available experimental data on dynamic characteristics of zirconium alloys in the range of high strain rates are insufficient to construct wide-range models of elastic-viscous-plastic deformation. At the same time when strain rates are higher than $10^4\,s^{-1}$, dependence of the flow and fracture stresses sharply increases in many metals due to a changed mechanism of dislocation movement [3,4].

Zirconium alloy E635 (Zr–1.0% Nb; ~ 1% Sn; ~ 0.5% Fe) as well as E110 (Zr - 1.0% Nb) – are the main commercial alloys for nuclear power engineering. This paper presents results given by investigations into strength characteristics of zirconium alloy E635 in sub-microsecond shock-waves at strain rates of 10^4–$10^6\,s^{-1}$ and also compares its characteristics with those of zirconium alloy E110.

Measurements were taken in the experimental setup when plane shock waves impacted samples in well-controlled loading conditions within 4–12 GPa. Shock load duration varied from 0.05 up to 1 μs. The material was studied in the α – phase under normal and elevated temperatures and in the β – phase (under higher than 900 °C temperature).

2. Material and experimental setup

Zirconium – is a polymorphous metal that has a hexagonal close-packed lattice under normal conditions and belongs to transition metals. Zirconium alloy E635 contains niobium (0,9–1,1 mass%), tin (1.1..1.4 mass%), and iron (0.3..0.47 mass%) as main alloying components. The rest admixtures (hafnium, nickel, chromium, carbon, silicon, etc.) in the aggregate do not exceed 0.25%. The alloy has 6.49 g/cm^3 density, 500 MPa yield strength, and 590 MPa ultimate strength [7]. The X-ray diffractometry demonstrated that the test material is a two-phase material. Insignificant presence of the β – phase (bcc lattice) is observed in addition to the α – phase (hcp lattice).

The material was studied in the as-received condition. The rod with the 35-mm diameter was sliced into disks with the thickness of 0.3, 0.5, 1.0, 2, 4, and 8 mm to serve as zirconium alloy E635 samples for tests. The ultrasonic technique and samples with the 8-mm thickness were used to determine longitudinal c_l and shear c_s sound velocities: $c_l = 4711 \pm 12$ m/s and $c_s = 2274 \pm 2$ m/s. Volume sound velocity was estimated to be $c_0 =$ 3.91 km/s. In order to estimate stresses arising in the

[a] Corresponding author: `svetlana_malyugina@mail.ru`

Table 1. Setting Data on experimental setups and measured strength characteristics of zirconium alloy E635.

V_0, m/s	h_i, mm	h_s, mm	W_h, m/s	σ_h, GPa	σ_τ, GPa	σ_{sp}, GPa	SR_{rw}, s^{-1}
630	1,99	3,977	64	0,98	0,46	2,88	0,13
303	1,972	3,972	66	1,01	0,47	2,79	0,1
644	0,116	0,476	85	1,30	0,57	3,38	1,16
652	0,980	0,981	80	1,22	0,57	-	-
655	0,997	1,981	62	0,95	0,44	2,81	0,24
646	0,454	0,983	71	1,09	0,49	3,06	0,58
945	2,013	3,985	58	0,89	0,41	2,67	0,14
930	0,993	1,993	69	1,05	0,49	2,89	0,31
616	0,116	0,426	89	1,36	0,67	3,82	1,50
616	0,116	0,315	89	1,36	0,63	3,83	1,30
595	3,984	7,963	57	0,87	0,41	3,06	0,067
606	3,976	7,963	56	0,86	0,40	2,83	0,73
638	1,992	3,962*	71	1,09	0,46	2,72	0,127
555	2,00	4,00**				2,23	0,07

* Temperature of sample 331°C.

** Temperature of sample 960°C.

samples, we used Hugoniot adiabat of zirconium $D = 3.91 + 0.91\,u$ (km/s), where D – is shock wave velocity, u – is mass velocity [8] with a correction for sound velocity data obtained for alloy E635.

Shock-wave experiments to record the sample deformation dynamics used a light-gas gun of the 44-mm caliber [9] in the range of impact velocities from 300 to 950 m/s. The plane shock wave in samples having 1, 2, 4, and 8-mm thickness was generated by an impactor of zirconium alloy E635 with the thickness of 0.5, 1, 2, and 4 mm, respectively. With samples having 0.5 and 0.3-mm thickness, an impactor made from the zirconium foil with the 0.1-mm thickness was used to generate a shock wave.

Impact velocity was measured with the error at least 0.5%. In experiments out-of-parallelism of impacting surfaces was ~1..2 mrad. Experiments were performed in conditions of vacuum.

Resistance heaters were used for tests under elevated temperatures in order to heat up samples to the required temperatures. Temperature was monitored with the accuracy of ±5°C by thermocouples fixed at the lateral surface of samples.

In experiments, velocity of the sample free-surface was continuously recorded with the help of VISAR [10] and PDV [11] techniques. Time resolution of VISAR-interferometer is at least 2 ns, velocity resolution of PDV-interferometer is at most 1% within the velocity range under consideration. Amplitude resolution of the interferometer is at most 1%, time resolution – is at least 4 ns. The Hilbert method was used to process PDV data.

3. Measurement results

Table 1 gives basic data on the setup of shock-wave experiments, as well as certain measured strength characteristics of zirconium alloy E635.

We also used the following designation: V_0 – impact velocity, h_i – impactor thickness, h_s – sample thickness, W_h – velocity at the front of the elastic precursor, σ_h = σ_{HEL} – dynamic elastic limit, σ_τ – dynamic yield stress, σ_{sp} – spall strength, SR_{rw} – strain rate in a rarefaction wave (V'/V_0).

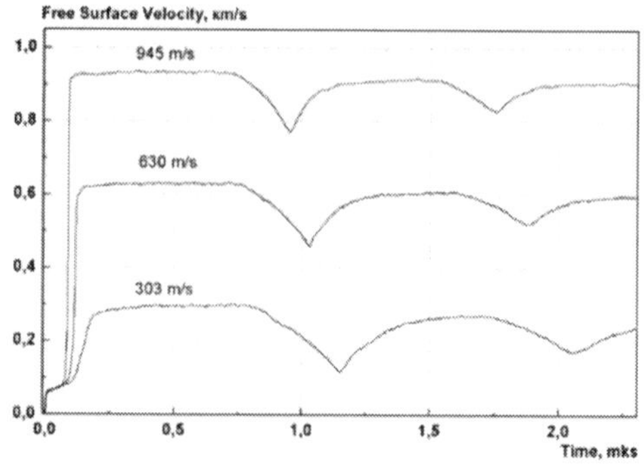

Figure 1. Free-surface velocity profiles for zirconium alloy E635 samples.

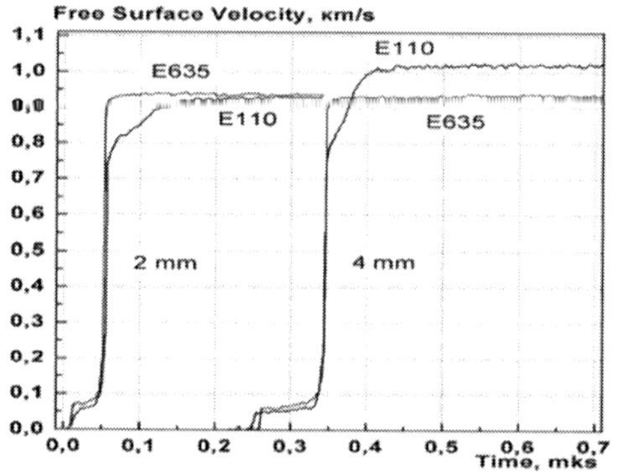

Figure 2. Free-surface velocity profiles for samples from zirconium alloys E635 and E110. Thickness of samples: 2 and 4 mm.

The value of the dynamic elastic limit

$$\sigma_{HEL} = \rho_0 c_l W_h / 2 \quad (1)$$

agrees with conditions of uniaxial deformation; this value is converted to the yield strength in standard conditions of uniaxial stressed state σ_τ using a simple relationship

$$\sigma_\tau = 3/2\sigma_{HEL}(1 - (c_b/c_l)^2. \quad (2)$$

Figure 1 shows characteristic free-surface velocity profiles $W(t)$ for samples from zirconium alloy E635 at different-intensity impacts. An impactor velocity is given above each profile.

Free-surface velocity profiles for samples made from zirconium alloy E635 are observed to have a two-wave configuration of the shock wave with its splitting into the elastic precursor and the plastic compression wave propagating behind the precursor. For comparison in samples from zirconium alloy E110 under loading amplitudes above 10.6 GPa, the elastic and phase precursors are recorded in front of the principal plastic wave due to the polymorphous $\alpha \rightarrow \omega$ transition with the decrease of the specific volume (see Fig. 2) [12, 13].

The portion of steady-state flow (Fig. 1) is followed by the decrease of rarefaction wave parameters. Interaction

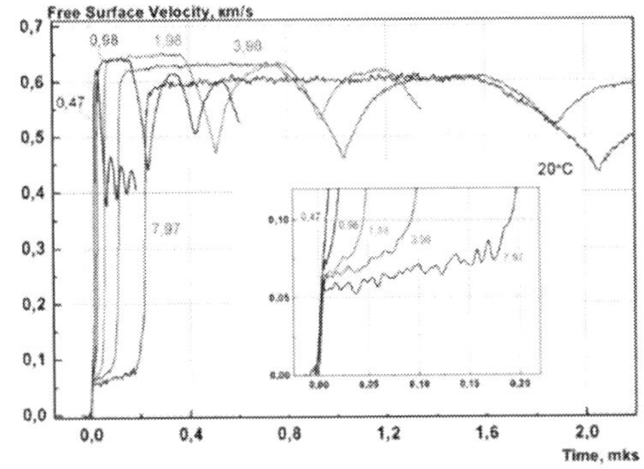

Figure 3. Free-surface velocity profiles for Zirconium alloy E635 samples. Profiles have nominal thickness of samples (T = 20°C).

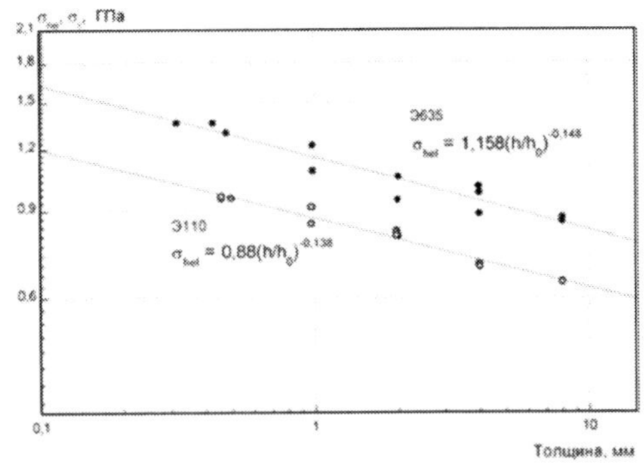

Figure 4. Elastic precursor attenuation in zirconium alloys E635 and E110 [12].

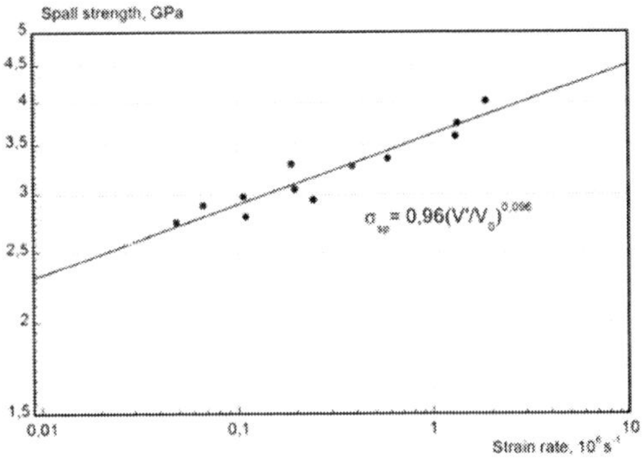

Figure 5. Measured spall strength of zirconium alloy samples versus deformation rate under expansion.

of the incident and reflected rarefaction waves causes generation of tensile stresses inside the sample and thus its spall fracturing. За областью стационарного течения (рисунок 1) следует спад параметров в волне разрежения. Relaxation of tensile stresses due to fracturing causes initiation of the compression wave in the stretched material and arrival of this wave at the surface is accompanied by the repeated increase of its velocity, which is called spall pulse. The subsequent periodic fluctuations of the free-surface velocity are conditioned by waves reverberation, i.e. reflection of rarefaction and compression waves from the surface on the separated layer of the material.

Spall strength was estimated based on measured free-surface velocity decrease under unloading from the shock-compressed state taking into account correction for the velocity profile distortion caused by the difference in the velocity of spall pulse and plastic wave propagation in the unloading area [5].

3.1. Elastic precursor and plastic flow behind its front

Figure 3 shows free-surface velocity profiles for different-thickness samples from zirconium alloy E635 (V_0–610..655 m/s). Each profile has designated thickness of samples.

Our experimental data demonstrate decrease in the elastic precursor amplitude with the compression wave propagation. This is due to stress relaxation during plastic deformation.

Measurement results for the elastic precursor attenuation are given in Fig. 4.

The dynamic elastic limit versus thickness relationship for alloy E110 is given for comparison [13]. Measurement results demonstrate that zirconium alloy E635 has 25..35% higher dynamic elastic limit compared to E110.

Similar to previous studies [6–9,14–16], attenuation measurement results are approximated by the following empirical relation:

$$\sigma_h = S(h/h_0)^{-\alpha}, \tag{3}$$

where $S = 1.158$ GPa, $\alpha = 0.148$, and $h_0 = 1$ mm. The similar data for zirconium alloy E110 - $S = 0.87$ GPa, $\alpha \approx 0.14$.

Analysis of elastic precursor attenuation provides information on the initial plastic deformation rate directly after the elastic limit is attained. Measurements of compression velocity in the plastic shock wave provide data on the material behavior at the follow-on stages and usually at higher deformation rates [17,18].

$$\gamma'_p = {}^3\!/_4(\tau E'/(SG))^{(\alpha+1)/\alpha} S\alpha c_l/(h_0 G). \tag{4}$$

The plastic strain rate behind the elastic precursor front in alloy E635 ($\alpha = 0.148$) is comparable with the plastic stain rate in alloy E110 [12.13] with the determining parameter $\alpha \approx 0.14$.

Higher dynamic strength of alloy E635 can be conditioned by the effect of such alloying admixtures as tin and iron.

3.2. Spall strength

Figure 5 shows results of processing the stress wave profiles as the "spall strength versus the material expansion rate in the unloading part of the incident compression pulse" relationship.

Spall strength is calculated using approaches described in [5]. Experimental data are acceptably approximated by the following empirical relation:

$$\sigma_{sp} \sim 0.96(V'/V_0)^{0.096}\,\text{GPa}. \tag{5}$$

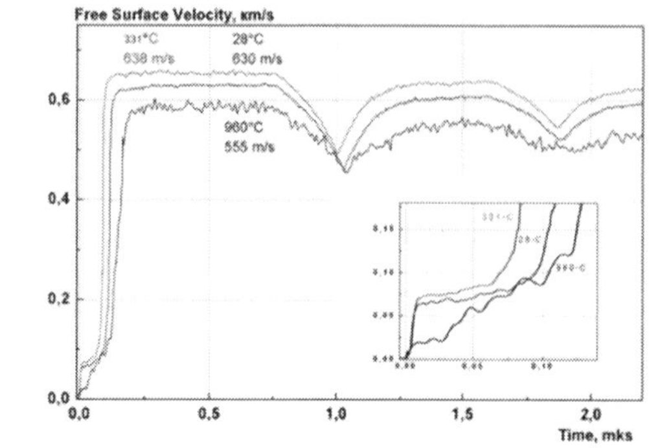

Figure 6. Free-surface velocity profiles $W(t)$ for zirconium alloy E635 samples having 4-mm thickness loaded by the impact of the 2-mm plate from the same material at a rate of 555.. 630 m/s under normal and elevated up to 315 °C and 960°C temperatures.

Relationship (5) also determines how the initial rate of discontinuities growth depends on stresses and can be used to construct a macrokinetic models of fracturing.

Experimental data given in Fig. 4 cover the impact load duration range from $\sim 0.05\mu s$ up to 1 μs.

Compared to E110, spall strength of zirconium alloy E635 is less dependent of the strain rate.

3.3. Temperature effects

Figure 6 illustrates how the dynamic elastic limit and spall strength of zirconium alloy E635 depend on temperature. With the temperature increase up to 331°C, the dynamic elastic limit is observed to increase whereas spall strength remains the same.

Wave profiles were processed taking into account temperature relationships of elastic moduli given in [20]. Particularly in this temperature range, shear modulus of zirconium decreases by 20% and the longitudinal sound velocity – by 3.2%.

Comparison of wave profiles in figure 6 demonstrates 10% increase in the dynamic elastic limit of zirconium at 331°C. Spall strength measured at normal and elevated temperatures in the very similar loading conditions underwent no changes and are equal to 2.7 GPa with the regard for measurement error.

At 960°C, the material shall be in the β – phase (bcc lattice). Measurement results demonstrate the two-wave profile of the shock wave with its splitting into the elastic precursor and the plastic wave. Spall strength of the material at 960°C decreases by 15..20%.

4. Conclusion

The paper presents measurement results for dynamic characteristics of zirconium alloy E635 in the range of loading duration from $\sim 0.05\,\mu s$ up to $1\,\mu s$ when the pressure amplitude in the shock wave varies from 4 up to 12 GPa. Measured relationships of the elastic precursor attenuation were used to estimate the plastic strain rate with the sample thickness from 0.3 mm to 8 mm. The obtained values are comparable with the plastic strain rate in alloy E110 with samples having similar thickness.

Experiments at temperatures elevated up to 960°C are conducted for the first time.

Compared to E110, zirconium alloy E635 has the 25..35% higher dynamic elastic limit. A basic distinctive feature of these alloys is availability in alloy E635 of tin and iron as alloying admixtures. Influence of alloying admixtures is essential. Values of spall strength of zirconium alloys E635 and E110 are comparable.

Spall strength of the zirconium alloy grows with the strain rate as $\sigma_{sp} \sim 0.96\,(V'/V_0)^{0.096}$. Spall strength values for zirconium alloys E635 and E110 are comparable in the considered range of strain rates.

Plastic wave splitting with phase precursor separation registered for alloy E110 under shock-wave amplitudes above 10.6 GPa ($\alpha \rightarrow \omega$ trnsition) is not observed on the stress-wave profiles for alloy E635 under the shock-compression pressure of 11-12 GPa.

Spall strength changes insignificantly with the temperature increase up to 331°C. At 960°C, spall strength decreased by 15..20%.

Authors express their sincere gratitude to G.I. Kanel for problem statement and useful discussions during this work implementation.

Very helpful in experiments were E.I. Karnaukhov, M.I. Serkov, M.F. Geraschenko, S.Yu. Filatov.

References

[1] G.P. Kobylyansky, A.E. Novoselov. Radiation strength of zirconium and zirconium-based alloys. Reference materials on reactor material science. Dimitrovgrad: State Science Center NIIAR. (1996) 176 p.

[2] A.S. Zaimovsky, A.B. Nikulina, N.G. Reshetnikov. Zirconium alloys in nuclear power engineering.- M.: Energoatomizdat. (1994).

[3] G.I. Kanel, V.E. Fortov, S.V. Razorenov. Rus.J. UFN ("Uspekhi Fizicheskikh Nauk"- Advances in Physical Sciences), v. **177**, № **8**, pp. 809–830 (2007).

[4] G.I. Kanel, S.V. Razorenov, A.V. Utkin, V.E. Fortov. Shock-wave phenomena in condensed media. M.: "Yanus –K", p. 408 (1996).

[5] G.I. Kanel. R.J. Rus.J. PMTF (Applied Mechanics and Technical Physics), v. **42**, № **2**, c. 1–5 (2001).

[6] G.I. Kanel, S.V. Razorenov, A.V. Utkin, V.E. Fortov. Experimental profiles of shock waves in condensed matter. M.: FIZMATLIT (2008).

[7] Material physics: Coursebook for higher educational institutions. Volume 6, part 1. Structural materials for nuclear power engineering./ B.A. Kalin, P.A. Platonov, I.I. Chernov, Ya. I. Shtrombakh. – M.: MIFI. (2008). 672 p.

[8] M.V. Zhernokletov, V.N. Zubarev, R.F. Trunin, V.E. Fortov. Experimental data on shock compressibility and adiabatic expansion of condenced matter under high energy densities. Chernogolovka Publisher (1996).

[9] A.V. Pavlenko, S.I. Balabin, O.E. Kozelkov, D.N. Kazakov. Rus.J. PTE, (Instruments and experimental technique), № **4**. pp. 122–124 (2013).

[10] A.V. Pavlenko, S.N. Malyugina, V.V. Pereshitov, I.N. Lisitsyna. Rus.J. PTE, (Instruments and experimental technique), № **2**. pp. 127–129 (2013).
[11] S.S. Mokrushin, N.A. Anikin, S.N. Malyugina, A.A. Tyaktev, A.V. Pavlenko. Rus.J. PTE, (Instruments and experimental technique), № **4**, pp.107–110 (2014).
[12] D.N. Kazakov, O.E. Kozelkov, A.S. Mayorova, S.N.Malyugina, S.S. Mokrushin and A.V. Pavlenko, Rus. J. MTT, **6**, 77–86 (2014)
[13] S.N.Malyugina, D.N. Kazakov, O.E. Kozelkov, A.S. Mayorova, S.S. Mokrushin and A.V. Pavlenko, In Proc this Conf (2015).
[14] E.B. Zaretsky, G.I. Kanel. Plastic flow in shock-loaded silver at strain rates from 104 s-1 to 107 s-1 and temperatures from 296 K to 1233 K. J. Appl. Phys. 110 (7), 073502 (2011).
[15] E.B. Zaretsky and G.I. Kanel. Effect of temperature, strain, and strain rate on the flow stress of aluminum under shock-wave compression. J. Appl. Phys. 112, 073504 (2012).
[16] E.B. Zaretsky and G.I. Kanel. Response of copper to shock-wave loading at temperatures up to the melting point. J. Appl. Phys. 114, 083511 (2013).
[17] E.B. Zaretsky and G.I. Kanel. Tantalum and vanadium response to shock-wave loading at normal and elevated temperatures. Non-monotonous decay of the elastic wave in vanadium. Journal of Applied Physics 115, 243502 (2014).
[18] S.V. Razorenov, G.I. Kanel, G.V. Garkushin, O.N. Ignatova. Rus.J. FTT (Solid State Physics), v. 54, issue 4, pp. 742–749 (2012).
[19] Duvall G.E. In: Stress Waves in Anelastic Solids, edited by H. Kolsky and W. Prager, Berlin: Springer-Verlag. p. 20 (1964).
[20] Guinan M.W. and Steinberg D.J. Pressure and temperature derivatives of the isotropic polycrystalline shear modulus for 65 elements. J. Phys. Chem. Solids, v. **35**. pp. 1501–1512 (1974).

EPJ Web of Conferences **94**, 02021 (2015)
DOI: 10.1051/epjconf/20159402021

Dynamic behavior of zirconium alloy E110 under submicrosecond shock-wave loading

D.N. Kazakov, O.E. Kozelkov, A.S. Mayorova, S.N. Malyugina, S.S. Mokrushin, and A.V. Pavlenko[a]

RFNC-VNIITF, 456770, Snezhinsk, Chelyabinsk region, Russia

Abstract. Stress waves have been measured under shock wave loading of zirconium alloy E110 samples with the 0.5 – 8 mm thickness at normal and elevated temperatures. Duration of shock loading pulses varied from ~0.05 up to 1 μs with the amplitude varying from 3.4 up to 23 GPa. Free-surface velocity profiles have been registered using VISAR and PDV interferometers with nanosecond resolution. Attenuation of the elastic precursor has been measured to determine plastic strain rate behind the elastic precursor front. The plastic strain rate was observed to decrease with propagation from $10^6\ s^{-1}$ at the 0.46-mm distance down to $2 \cdot 10^4\ s^{-1}$ at the 8-mm distance. Spall strength has been measured under normal and elevated temperatures. Spall strength versus strain rate relationships have been constructed in the $10^5\ s^{-1}$ – $10^6\ s^{-1}$ range. Under shock compression higher than 10.6 GPa, the three-wave configuration of the shock wave has been registered and the polymorphous $\alpha \rightarrow \omega$ transition is considered to be the reason of this phenomenon.

This work was supported by State Atomic Energy Corporation "Rosatom" within State Contract # H.4x.44.90.13.1111

1. Introduction

Investigation into processes of elastic-viscous-plastic deformation of metals and alloys under shock-wave loading [1] allows us to measure velocity-temperature relationships how these materials resist to deformation and fracture. In shock waves, a wide range of strain rates of solid bodies can be covered and thus we can accumulate experimental data that can serve as the basis for construction of defining relationships needed to predict materials behavior under high-intensity shock loads [2]. The structure of shock waves in solid bodies depends on processes of their elastic-viscous-plastic deformation, possible phase transformations, as well as on kinetics of fracture incipience and development [3–5]. In recent years, shock-wave methods were used for [6–9] systematic investigations into the velocity-temperature relationships for yield stresses and fracture of pure metals with face-centered and body-centered cubic structures. These data are practically absent for metals with hexagonal close-packed structure.

Interest to zirconium and alloys on its basis is inspired by a wide application of this material in nuclear power engineering due to happy combination of its nuclear physical (small thermal-neutron capture cross-section), physical-and-mechanical characteristics, processing behavior and also resistance to corrosion. Zirconium – is a polymorphous metal that has a hexagonal close-packed lattice under normal conditions and belong to transition metals. Zirconium alloy E110 (Zr – 1.0% Nb) – is one of the main commercial alloys for nuclear power engineering. Experimental data on dynamic properties of zirconium alloys in the area of high-strain rate deformation are practically unavailable. The considered experimental investigation results for zirconium alloy E110 samples loaded in well-controlled conditions is an attempt to fill this gap.

[a] Corresponding author: avpavlenko@vniitf.ru

2. Material and experimental setup

Zirconium alloy E110 contains niobium (Nb 1%) as an alloying component; principal impurities – are iron and oxygen (O and Fe less than 0.1%), the rest admixtures (tin, hafnium, nickel, chromium, carbon, silicon, etc.) in the aggregate do not exceed 0.25%. The alloy has 6.5 g/cm^3 density, 160 MPa yield strength, and 270–310 MPa ultimate strength.

The material was studied in the as-received condition. The rod with the 35-mm diameter was sliced into disks with the thickness of 0.5, 1.0, 2, 4, and 8 mm to serve as zirconium alloy E110 samples for tests. Unevenness of the sample surface after processing was within 2.5–6 μm. The ultrasonic technique and samples with the 8-mm thickness were used to determine longitudinal c_l and shear c_s sound velocities: $c_l = 4700 \pm 15$ m/s and $c_s = 2262 \pm 5$ m/s. Volume sound velocity was estimated to be $c_0 = 3.91$ km/s. In order to estimate stresses arising in the samples, we used Hugoniot adiabat of zirconium $D = 3.91 + 0.91u$ (km/s), where D – is shock wave velocity, u– is mass velocity [10] with a correction for sound velocity data obtained for alloy E110.

Our experiments used a 44-mm bore one-stage gun [11] and the "symmetric" impact approach when strikers and samples are made of the same material apart from experiments with 0.5-mm thick samples, wherein strikers were cut our from the zirconium foil having 0.1-mm thickness. Shock waves in samples with 1-; 2-, and 4-mm

Article available at http://www.epj-conferences.org or http://dx.doi.org/10.1051/epjconf/20159402021

Table 1. Setting Data on experimental setups and measured strength characteristics of zirconium alloy E110.

$V_0, m/s$	h_i, mm	h_s, mm	$W_h, m/s$	σ_h, GPa	σ_τ, GPa	σ_{sp}, GPa	SR_{rw}, s^{-1}
254	1.979	3.995	46	0.709	0.32	2.56	0.13
302	1.985	7.947	38	0.58	0.27	2.75	0.08
307	1.974	7.977	42	0.65	0.30	-	0.08
548	2.0	1.965	54	0.82	0.38	-	
577	2.004	3.988	46	0.70	0.32	2.58	0.13
591	0.468	0.971	55	0.85	0.39	3.31	0.57
607	0.972	1.972	44	0.68	0.31	2.93	-
618	1.993	7.943	38	0.58	0.27	2.98	0.09
619	0.114	0.495	63	0.96	0.44	4.07	1.48
~620	1.975	3.917 (315 °C)	51	0.78	0.36	2.60	0.13
627	1.994	7.970	40	0.62	0.29	-	0.09
918	0.11	0.457	63	0.95	0.44	4.09	1.64
950	0.980	1.983	53	0.80	0.37	3.28	0.25
955	0.462	0.972	60	0.92	0.42	3.00	0.71
1050	2.00	3.996	46	0.69	0.32	3.20	0.18
1181	1.993	7.991	42	0.65	0.30	3.23	0.08
1225	1.043	1.963	53	0.80	0.37	3.35	0.26
1252	1.998	3.962 (368 °C)	46	0.70	0.32	3.01	0.18
1363	0.115	0.457	63	0.96	0.44	4.43	1.45
1524	2.00	3.99	46	0.71	0.33	3.01	0.13
1575	0.991	1.962	53	0.82	0.38	3.03	0.27

thickness were formed by the striker having 0.5-; 1-, and 2-mm thickness and experiments with 8-mm thick samples used 2-mm thick strikers. Impact velocity varied from 0.25 to 1.6 km/s. Striker velocity was measured with the error at least 0.5%. In experiments with the up to 1 km/s velocity, out-of-parallelism of impacting surfaces was ~1 mrad with the velocity higher than 1 km/s–2..4 mrad. Experiments were performed in conditions of vacuum.

Resistance heaters were used for tests under elevated temperatures in order to heat up samples to the required temperatures and this heating took not more than 15 minutes. Temperature was monitored with the accuracy of ±5 °C by T-type thermocouples fixed at the lateral surface of samples.

Laser Doppler interferometers VISAR [12] and PDV [13] recorded velocity profiles $W(t)$ for the free surface of samples. Surface sounding area for VISAR-interferometer was in the center of the sample and diameter of this area was not more than 0.8 mm. Sounding area for PDV-interferometer was shifted from the sample center by 3 mm and was not more than 0.4 mm. Time resolution of VISAR-interferometer is at least 2 ns, its velocity resolution is not more than 1% when velocities are measured within 0.3–1.6 km/s. The Hilbert method was used to process PDV interferometric data.

3. Measurement results

Table 1 gives (new and previous [14]) basic data on the setup of shock-wave experiments, as well as certain measured strength characteristics of zirconium alloy E110.

We also used the following designation: V_0–impact velocity, h_i–impactor thickness, h_s–sample thickness, W_h–velocity at the front of the elastic precursor,

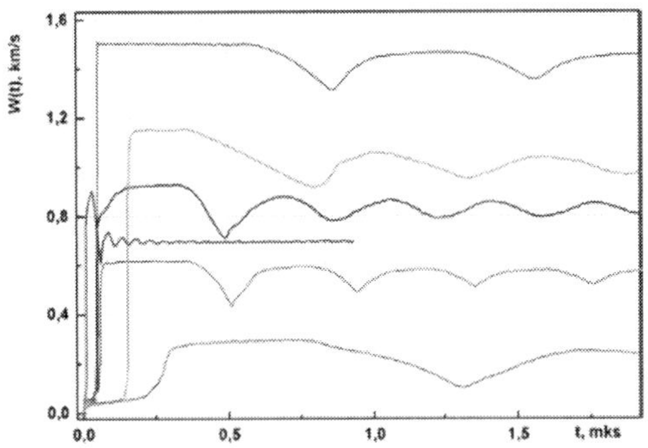

Figure 1. Free-surface velocity profiles for Zirconium alloy E110 samples.

$\sigma_h = \sigma_{HEL}$–dynamic elastic limit, σ_τ–dynamic yield stress, σ_{sp} - spall strength, SR_{rw}–strain rate in a rarefaction wave (V'/V_0).

The value of the dynamic elastic limit $\sigma_{HEL} = \rho_0 c_l \; W_h/2$ agrees with conditions of uniaxial deformation; this value is converted to the yield strength in standard conditions of uniaxial stressed state σ_τ using a simple relationship $\sigma_\tau = 3/2\sigma_{HEL}(1-(c_b/c_l)^2)$.

Figure 1 shows free-surface velocity profiles $W(t)$for Zirconium alloy E110 samples having 0.5-; 2-; 4- and 8-mm thickness under different-intensity impacts.

Each profile has designated velocity of the striker. The two-stage elastic-plastic compression wave is registered when the impact velocity is up to 0.6 km/s. When shock compression pressure goes higher than 10.6 GPa, the plastic shock wave looses its steadiness and splits due to the known polymorphous transformation $\alpha \rightarrow \omega$ with the decrease of the specific volume. When shock compression

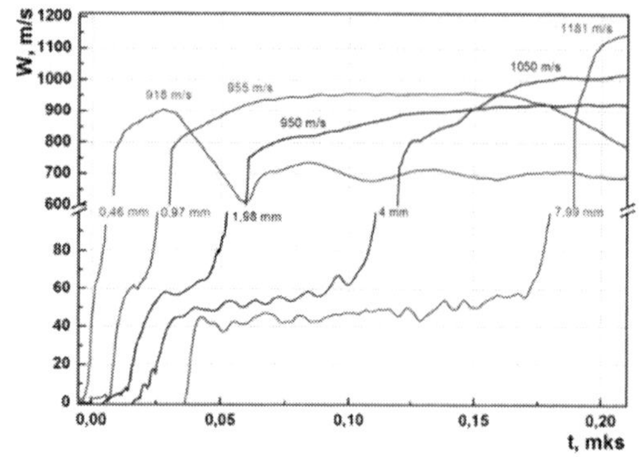

Figure 2. Free-surface velocity profiles for Zirconium alloy E110 samples. Profiles have designated velocity of the striker and nominal thickness of samples; exact values of thickness for samples and strikers are given in Table 1.

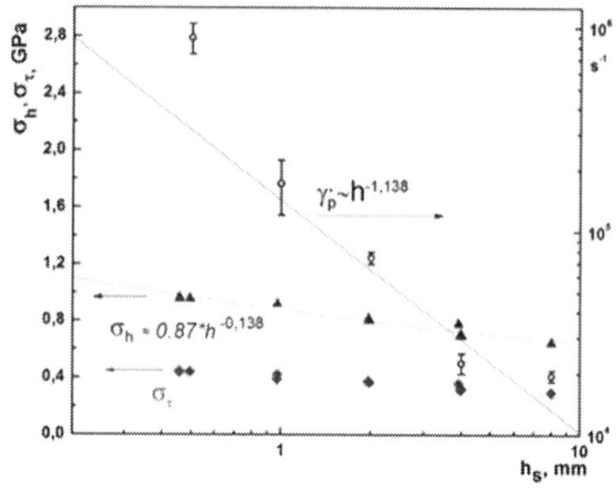

Figure 3. Relationships of measured dynamic elastic limit σ_h, yield strength σ_τ, calculated plastic train rate γ_p', and measured strain rate behind the elastic precursor front

pressure is higher than 16 GPa, the two-wave configuration "closes" due to the fact that the velocity of the wave with phase transformation becomes equal to or higher than the plastic shock wave velocity. When impact velocity varies, the compression stress on the elastic precursor remains practically unchanged. It follows that in this range of parameters, attenuation of the elastic precursor is practically independent of the final shock compression pressure.

All sample free-surface velocity profiles demonstrate decrease of parameters in the rarefaction wave behind the area of steady-state flow. Interaction of the incident and reflected rarefaction waves leads to generation of tensile stresses inside the sample and to its spall fracturing. Relaxation of tensile stresses due to fracturing causes initiation of the compression wave in the stretched material and arrival of this wave at the surface is accompanied by the repeated increase of its velocity, which is called spall pulse. The subsequent periodic fluctuations of the free-surface velocity are conditioned by waves reverberation, i.e. reflection of rarefaction and compression waves from the surface on the separated layer of the material.

Spall strength was estimated based on measured free-surface velocity decrease under unloading from the shock-compressed state taking into account correction for the velocity profile distortion caused by the difference in the velocity of spall pulse and plastic wave propagation in the unloading area [4].

Elastic precursor and plastic flow behind its front. Figure 2 gives measured profiles for the free surface velocity of zirconium alloy E110 samples having different thickness. Each profile has designated thickness of samples and velocity of the striker that has generated the compression wave in the sample. Amplitude of the elastic precursor decreases with its propagation. Elastic precursor attenuation is conditioned by relaxation of tresses due to plastic deformation development.

Measurement results for the elastic precursor attenuation are given in Fig. 3. Similar to previous studies [6–9, 14–16], attenuation measurement results are approximated by the following empirical relation:

$$\sigma_h = S(h/h_0)^{-\alpha}, \tag{1}$$

where $S = 0.87$ GPa, $\alpha = 0.138$, and $h_0 = 1$ mm. The elastic precursor attenuation is related to the plastic deformation rate γ_p' just behind its front by the relationship [15, 17]:

$$d\sigma_x/dh|_{HEL} = -4/3G\gamma_p'/c_l, \tag{2}$$

where G–is shear modulus, σ_x – is compression stress in the direction of wave propagation, h–is distance covered by the wave. Taking into account relationship (1), the plastic deformation rate can be expressed as:

$$\gamma_p' = -3/4(S\alpha c_l(h/h_0))^{-(\alpha+1)}/(h_0 G). \tag{3}$$

After the wave-covered distance h from relationship (1) is substituted into (3), we have the initial velocity of the "plastic shear deformation versus shear stress behind the precursor front" relationship:

$$\gamma_p' = 3/4(\tau E'/(SG))^{(\alpha+1)/\alpha} S\alpha c_l/(h_0 G). \tag{4}$$

Figure 3 shows data on the dynamic elastic limit attenuation, empirical relationship (1), and appropriate plastic deformation rate.

Plastic deformation rate behind the elastic precursor front changes from 10^6 s^{-1} at the 0.46-mm thickness of samples up to $2 \cdot 10^4$ s^{-1} at the 8-mm thickness of samples and this is almost one order of magnitude higher than the plastic deformation rate in magnesium alloy Ma-2 [15].

Analysis of elastic precursor attenuation provides information on the initial plastic deformation rate directly after the elastic limit is attained. Measurements of compression velocity in the plastic shock wave provide data on the material behavior at the follow-on stages and usually at higher deformation rates. The maximum plastic deformation rate of alloy E110 in the plastic shock wave is $5 \cdot 10^5$ s^1 at the impact velocity of 254 m/s (compression stress behind the plastic shock wave is 3.4 GPa) and $1.3 \cdot 10^7$ s^{-1} at the impact velocity of 577 m/s (compression stress – is 7.8 GPa). The maximum shear stresses in the plastic shock wave were 173 GPa and 284 GPa, respectively.

Phase transformation. Splitting of the plastic shock wave (Fig. 2) indicates polymorphous transformation,

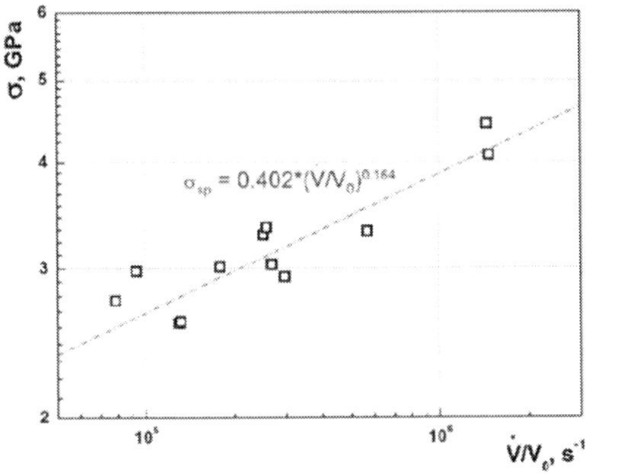

Figure 4. Measured spall strength of zirconium alloy samples versus deformation rate under expansion.

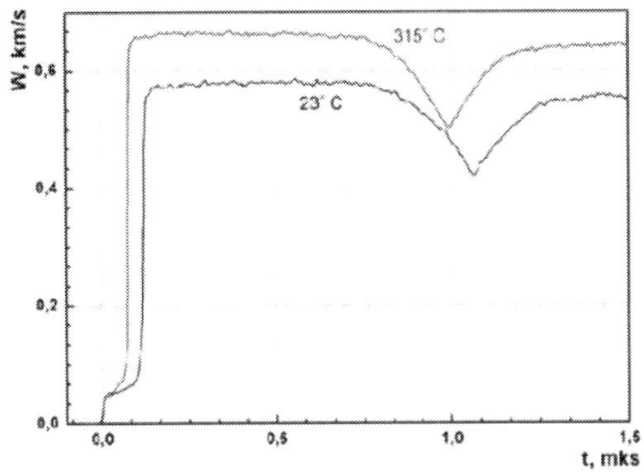

Figure 5. Free-surface velocity profiles $W(t)$ for Zirconium alloy E110 samples having 4-mm thickness loaded by the impact of the 2-mm plate from the same material at a rate of about 600 m/s under normal and elevated up to 315°C temperatures.

probably a well-known transformation $\alpha \rightarrow \omega$ [18]. As the first plastic shock wave propagates, pressure at its front, which corresponds to transformation onset, decreases from 11.2 GPa for the sample having 0.46-mm thickness down to 10.6 GPa when this wave leaves the sample having 4-mm thickness. Pressure decrease in the first plastic shock wave as it is propagating is indicative of the final velocity of phase transformation. When striker velocity is 1181 m/s, loading amplitude becomes sufficient for the principal wave to begin to "close" the first plastic shock wave.

Spall strength. Figure 4 shows results of processing the stress wave profiles as the "spall strength versus the material expansion rate in the unloading part of the incident compression pulse" relationship.

Measurement results for the elastic precursor attenuation are given in Fig. 3. Similar to previous studies [6–9, 14–16], attenuation measurement results are approximated by the following empirical relation:

$$\sigma_{sp} = 0.402\ (V'/V_0)^{0.164}\ \text{GPa}. \qquad (5)$$

Relationship (5) also determines how the initial rate of discontinuities growth depends on stresses and can be used to construct a macrokinetic model of fracturing.

Experimental data given in Fig. 4 cover the impact load duration range from ~0.05 μs up to 1 μs and, respectively, spall thicknesses – from 0.08 mm up to 2 mm.

Figure 5 illustrates how the dynamic elastic limit and spall strength of zirconium alloy E110 depend on temperature. We were forced to limit experiments to the temperature of 370°C as zirconium looses its reflectivity even if its surface is protected by deposited gold.

In the meantime, zirconium alloy members are usually in service just in this temperature range.

Wave profiles were processed taking into account temperature relationships of elastic moduli given in [19]. Particularly in this temperature range, shear modulus of zirconium decreases by 20% and the longitudinal sound velocity–by 3.2%.

Comparison of wave profiles in Fig. 5 shows 10% growth in the dynamic elastic limit of zirconium, i.e. from 640 MPa at room temperature up to 710 MPa at 315°C. Calculation of the appropriate yield strength $\sigma_\tau = 1.5\sigma_{HEL}(1 - (c_b/c_l)^2)$ taking into account the time-dependent longitudinal sound velocity demonstrated that yield strength decreases from 300 MPa at room temperature down to 280 MPa at 315°C. Spall strength was measured at normal and elevated temperatures in very similar loading conditions taking into account measurement errors and it proved to remain unchanged and to be equal to 2.7 GPa.

4. Conclusion

The paper presents measurement results for dynamic characteristics of zirconium alloy E110 in the range of loading duration from ~0.05 μs up to 1 μs when the pressure amplitude in the shock wave varies from 3.5 up to 23 GPa. Measured dependences of elastic precursor attenuation allowed us to estimate the plastic deformation rate that changes from 10^6 s^1 at the 0.46-mm thickness of samples to $2 \cdot 10^4\ s^{-1}$ at the 8-mm thickness and this practically one order of magnitude higher than the plastic deformation rate in zirconium alloy E635, VT1-0 titanium and magnesium alloy Ma-2.

The reason why the compression wave is split into the three-wave configuration under pressure higher than 10.6 GPa (striker velocity is higher than 780 m/s) is the polymorphous $\alpha \rightarrow \omega$ transformation. Attenuation of the first plastic shock wave indicates the final rate of phase transformation. Splitting of the plastic shock wave "closes" when the impact velocity is more than 1.2 km/s.

Spall strength of zirconium alloy increases with the deformation rate according to $\sigma_{sp} = 0.402\ (V'/V_0)^{0.164}$. Spall strength of the material is practically independent of heating up to 315°C.

Authors express their sincere gratitude to G.I. Kanel for problem statement and useful discussions during this work implementation.

Very helpful in experiments were E.I. Karnaukhov, M.I. Serkov, M.F. Geraschenko, S.Yu. Filatov.

References

[1] G.I. Kanel, V.E. Fortov, S.V. Razorenov. Rus.J. UFN ("Uspekhi Fizicheskikh Nauk"- Advances in Physical Sciences), v. **177**, **8**, pp. 809–830 (2007).

[2] E.N. Avrorin, B.K. Vodolaga, V.A. Simonenko, V.E. Fortov. Rus.J. UFN, v. **163**, **5**, pp. 1–34 (1993).

[3] G.I. Kanel, S.V. Razorenov, A.V. Utkin, V.E. Fortov. Shock-wave phenomena in condensed media. M.: "Yanus -K", p. 408 (1996).

[4] G.I. Kanel. R.J. Rus.J. PMTF (Applied Mechanics and Technical Physics), v. **42**, **2**, c. 1–5 (2001).

[5] G.I. Kanel, S.V. Razorenov, A.V. Utkin, V.E. Fortov. Experimental profiles of shock waves in condensed matter. M.: FIZMATLIT (2008).

[6] E.B. Zaretsky, G.I. Kanel. J. Appl. Phys. **110** (7), 073502 (2011).

[7] E.B. Zaretsky and G.I. Kanel. J. Appl. Phys. **112**, 073504 (2012).

[8] E.B. Zaretsky and G.I. Kanel. J. Appl. Phys. **114**, 083511 (2013).

[9] E.B. Zaretsky and G.I. Kanel. Journal of Applied Physics 115, 243502 (2014).

[10] M.V. Zhernokletov, V.N. Zubarev, R.F. Trunin, V.E. Fortov. Experimental data on shock compressibility and adiabatic expansion of condenced matter under high energy densities. Chernogolovka Publisher (1996).

[11] A.V. Pavlenko, S.I. Balabin, O.E. Kozelkov, D.N. Kazakov. Rus.J. PTE, (Instruments and experimental technique), **4**. pp. 122–124 (2013).

[12] A.V. Pavlenko, S.N. Malyugina, V.V. Pereshitov, I.N. Lisitsyna. Rus.J. PTE, (Instruments and experimental technique), **2**. pp. 127–129 (2013).

[13] S.S. Mokrushin, N.A. Anikin, S.N. Malyugina, A.A. Tyaktev, A.V. Pavlenko. Rus.J. PTE, (Instruments and experimental technique), **4**, pp.107–110 (2014).

[14] D.N. Kazakov, O.E. Kozelkov, A.S. Mayorova, S.N.Malyugina, S.S. Mokrushin and A.V. Pavlenko, Rus. J. MTT, **6**, 77–86 (2014).

[15] G.V. Garkushin, G.I. Kanel, S.V. Razorenov. Rus.J. FTT (Solid State Physics), v. **54**, **5**, pp. 1012–1018 (2012).

[16] S.V. Razorenov, G.I. Kanel, G.V. Garkushin, O.N. Ignatova. Rus.J. FTT (Solid State Physics), v. **54**, issue **4**, pp. 742–749 (2012).

[17] Duvall G.E. In: Stress Waves in Anelastic Solids, edited by H. Kolsky and W. Prager, Berlin: Springer-Verlag. p. 20 (1964).

[18] P.A. Rigg, C.W. Greeff, M.D. Knudson and G.T. Gray III. In Proc. 16th Int. Conf. of American Institute of Physics. Shock Compression of Condensed Matter–p. 1171–1174 (2009).

[19] Guinan M.W. and Steinberg D.J. Pressure and temperature derivatives of the isotropic polycrystalline shear modulus for 65 elements. J. Phys. Chem. Solids, v. **35**. pp. 1501–1512 (1974).

EPJ Web of Conferences **94**, 02022 (2015)
DOI: 10.1051/epjconf/20159402022

Electronic phases of substances. Phase transitions with change of electron and crystalline structure

Boris Nadykto[a]

Russian Federal Nuclear Center – VNIIEF, Sarov, Nizhni Novgorod region, 607190, Russia

Abstract. There is plenty of experimental data on high-pressure phase transformations in various materials. Variations in materials characteristics (for example, equilibrium density and bulk modulus), while the crystalline structure remains unchanged, are indicative of energy variations in outer-shell electrons of solid atoms. In experiments with crystalline structure variations, the dependence of pressure on density in some cases can be described by the same curve, the parameters of which are independent of the crystalline structure. Examples of such transformations in some materials at static compression and in shock-wave experiments are given.

1. Introduction

In the last decade, numerous publications have appeared describing complex variations in the crystalline structure of "simple" alkaline and alkaline-earth elements under compression [1–4]. These variations drive changes in many characteristics, in particular, electrical conductivity. This paper will focus on the variations in the P(ρ) relationship, and the behavior of such elements in this respect is similar to the behavior of many other elements. Shock and static compression experiments reveal high-pressure variations in the structure of materials that manifest themselves as kinks or even discontinuities in the P(ρ) curve. In some cases, when the crystalline structure of the material is unchanged, such kinks can be attributed only to a change in the electronic structure of solid atoms (i.e. redistribution of electrons between inner and outer atomic shells).

The notion of solid electronic phases appeared in the late 1940s, when it was detected from X-ray diffraction measurements of samples under pressure [5] that different cerium phases had the same type of crystalline lattice, but different density. Fermi suggested that different phase characteristics were due to the different electronic structure of solid atoms in different phases. During a phase transition, electrons are redistributed between atomic shells, while the crystalline structure is preserved. Based on Fermi's concept, Steinheimer in [6] calculated characteristics of two isostructural phases of cesium.

2. Computational analysis of the structure of solids

Nadykto [7–9] suggested a method for calculating the energy of a compressed atom. For the case of confined volume per atom (i.e. for an atom within a solid), a system of energy levels distinct from those of a free atom was obtained. The model explicitly includes the quantum nature of electron motion in solid atoms. Application of this model to a unit cell of a solid atomic lattice leads to a semiempirical equation of state that relates material compressibility to compression of the outer electron shell of an atom within a solid. Solids display a peculiar effect of both quantum properties of the system (discrete angular momentum of electrons and related energy discontinuity) and continuous energy and pressure variation with a change in external conditions (for example, during compression), where the quantum state remains unchanged.

Elastic energy of a solid is a sum of the kinetic energy of electrons provided that their angular momentum is quantized and potential energy of their interaction with the nucleus charge and each other. Different materials and their different phases are described by the same analytical expression for the equation of state. Only parameters of the equation of state differ: equilibrium density and bulk modulus (at P = 0, T = 0) or energy of outer-shell electrons of the atomic cell of a solid.

Energy and pressure are written as a sum of cold (elastic, potential) and thermal components:

$$E = E_X + E_T \tag{1}$$

$$P = P_X + P_T. \tag{2}$$

The elastic components of energy and pressure are defined as

$$E(\sigma) = \frac{9B_0}{\rho_n}\left(\frac{\sigma^{2/3}}{2} - \sigma^{1/3}\right) \tag{3}$$

$$P(\sigma) = 3B_0\left(\sigma^{5/3} - \sigma^{4/3}\right) \tag{4}$$

where $\sigma = \rho/\rho_n$ is the compression ratio, and ρ_n is the equilibrium material density at $P = 0$, $T = 0$.

[a] Corresponding author: nadykto@vniief.ru

Table 1. Equation-of-state parameters for some elements.

Element	ρ_n, g/cm^3	B_0, GPa	E_n, eV
Li	0.548	13.7	8.07
Li	0.415	6.5	3.84
Na	1.00	7.8	8.4
Na	1.18	15.7	14.3
K	0.892	4,0	8.11
Rb	1.616	3.1	7.7
Rb	2.02	4	7.9
Zn	7.21	62.7	26.5
Zn	7.5	114	46.5
Zn	9.18	337	112
Th	11.78	50	50.8
Th	12.31	76	66.8
Th	18.63	360	209
Pa	15.37	115	80.6
Pa	19.83	400	217
Am	13.67	28.5	23.4
Am	11.89	13	12.3
Am	18.59	80	48.4
Am	24.46	350	161

The Mie-Grueneisen form of the thermal component of pressure is $P_T = \Gamma \rho E_T$. The Grueneisen coefficient Γ is computationally estimated using relations (3), (4) in the Dougdale-Macdonald approximation.

With a known bulk modulus B_0, it is possible to determine the effective energy of state of outer-shell electrons in the atomic cell of a solid:

$$E_n = \frac{9AB_0}{2N_A\rho_n}. \quad (5)$$

Varying the number and quantum state of outer-shell electrons results in a non-monotone behavior of compressibility (electronic phase transition), which is observed when comparing simulations with experimental data.

Table 1 shows the equation-of-state parameters thus obtained for the elements discussed below.

2.1. Alkaline metals

2.1.1. Lithium

At low temperatures (200 K), lithium undergoes a series of transformations at pressures up to 130 GPa: bcc – fcc – cI16 – oC88 – oC40 – oC24 [4]. However, at pressures up to 80-100 GPa, the P(ρ) diagram of statically compressed lithium can be described by common parameters of equation of state (4) that do not depend on the crystalline structure. The complex behavior of lithium is observed under shock compression [10], [11]. The experiments were performed in a limited range of pressures. The results of measurements in [10] taken at pressures up to 25 GPa closely match the equation of state describing the experimental isotherm from [4]. However, at higher pressures (30–70 GPa) the data from [11] on the Hugoniot deviate significantly from the computational curve obtained using the initial elastic compression curve. The description at once of the Hugoniot and the isotherm (or cold curve) is indicative of a potential change in the electronic structure of compressed Li. The data in [11] point at a change in the slope of the D-u curve. However, in the range, where the data from [10] and [11] overlap, the data from [11] lie 300 m below [10] along D.

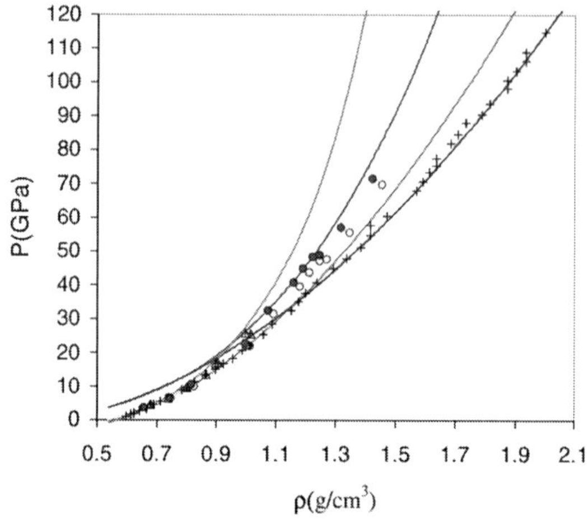

Figure 1. Isotherm at 190 K [4] and Hugoniot [10, 11] of Li. The curves represent calculations, and the points show experimental values.

If this is associated with a systematic error in [11], then at high pressure the values of P(ρ) from [11] shift a little bit up. Fig. 1 shows our computational curves along with the experimental data of [10], [11] (Hugoniot) and [4] (Diamond anvil sell), and well as corrected data from [11]. Above the 1.2 g/cm^3 density, the data on the isotherm and the corrected data from [11] are described by the same equation-of-state parameters, which differ from the parent phase parameters. The new parameters point at a transition of electrons from the $1s^2 2s$ state to $1s^2 2p$ in compressed lithium. To render conclusions more definite, new shock-wave experiments on Li need to be conducted at pressures of 30–100 GPa. The calculated parameters of different phases of Li are given in Table 1.

2.1.2. Sodium

The experimental data on static compression of Na [4, 12] up to 160 GPa (compression ratio 5) can be described by two curves. The parent phase parameters describe the experimental data up to P=7.5 GPa. Note that at 7.5 GPa the crystalline structure of Na remains unchanged. At P=7.5–160 GPa, experimental points are accurately described by the high-pressure phase parameters (see Table 1). Figure 2 shows a comparison of calculated and experimental data.

The bcc-to-fcc transition in Na observed experimentally at 65 GPa does not essentially affect the P(ρ) curve. The presented parameters of the electronic phases of Na accurately describe the data of shock experiments [10], [11], which closely match each other. The following series of transformations is observed in [4] up to 160 GPa: bcc – fcc – cI16 – oP8 – tI19. All these phases at 7.5 – 180 GPa are described by the same equation-of-state parameters, which differ from those of the parent phase.

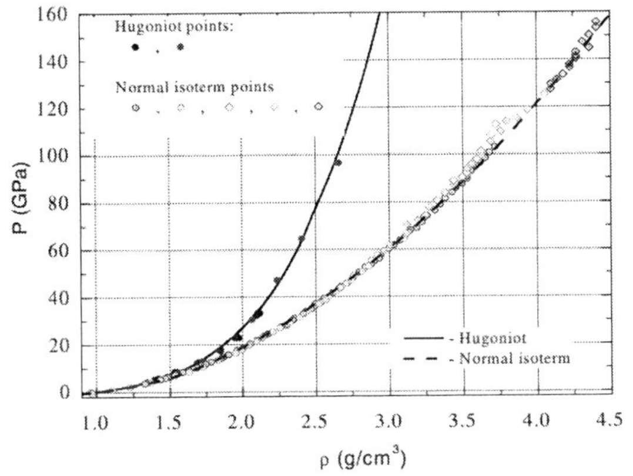

Figure 2. Isotherm at 300 K [4, 12] and Hugoniot [10, 11] of Na. The curves represent calculations, and the points show experimental values.

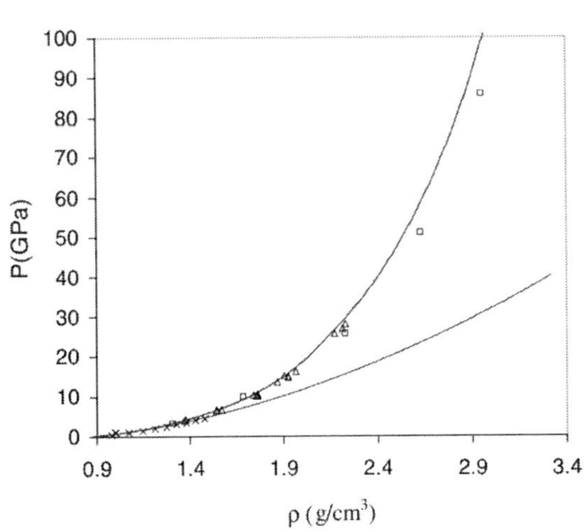

Figure 3. Isotherm at 300 K [4] and Hugoniot [10, 11] of potassium. The curves represent calculations, and the points show experimental values.

2.1.3. Potassium

The experimental data on compression of potassium are rather scarce. Reference [13] shows the results of processing the data of static-compression experiments conducted by P. Bridgeman at pressures up to 4.5 GPa. Simulations with parent-phase equation-of-state parameters provide a satisfactory description of these data and results of shock experiments [10, 11]. Figure 3 shows a comparison of calculated and experimental data. No phase transformation discontinuities are observed in the P(ρ) curve up 85 GPa.

2.1.4. Rubidium

For rubidium, in addition to the Hugoniot from [10] and static-compression data [13], there are diamond-anvil measurement results [14], which clearly point at a phase transition at 13–18 GPa. Figure 4 shows calculated and experimental data. As one can see in Table 1, the energy of atoms in these phases of Rb is virtually the same. Therefore, one can suppose that the transition

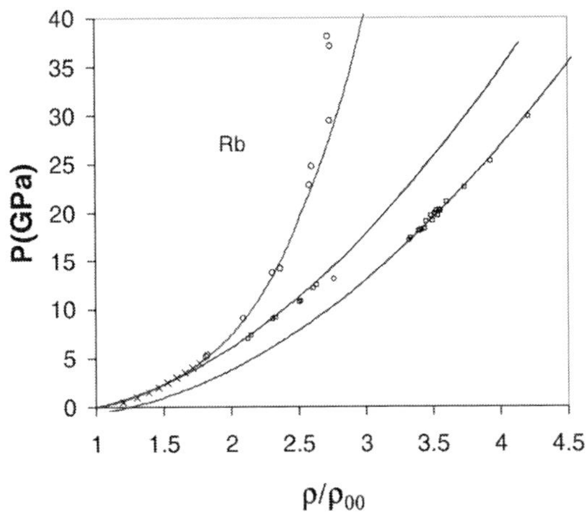

Figure 4. Isotherm at 300 K [13, 14] and Hugoniot [10, 11] of rubidium. The curves represent calculations, and the points show experimental values.

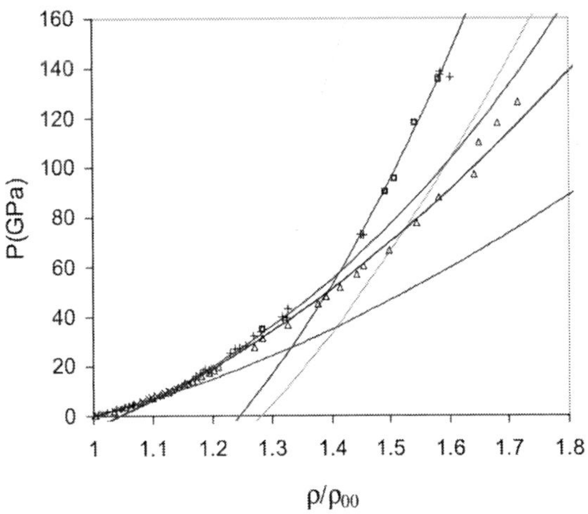

Figure 5. Pressure $P(\rho)$ on the normal isotherm and Hugoniot in zinc. The points represent experimental data. The curves represent calculations for different electronic phases of zinc.

occurs, while the electronic structure of Rb atoms remains unchanged.

2.2. Zinc

The equation-of-state parameters for zinc obtained by processing the data of shock-wave experiments [10], [15] differ from the parent-phase parameters of zinc. Description of the experimental points on the Hugoniot suggests that there are phase transitions at 7.5 GPa and 57 GPa. The parent- and the high-pressure-phase parameters of zinc are given in Table 1. Calculated Hugoniots and experimental points are shown in Fig. 5. These parameters refer to the equilibrium state at $P = 0$, $T = 0$. Considering thermal expansion, the density of zinc under normal conditions is $\rho_{00} = 7.14$ g/cm^3 and the bulk modulus is $B_0 = 61.0$ GPa.

Figure 5 also shows experimental points on the normal isotherm of zinc from [13,16]. Calculations for the parent phase are in perfect agreement with [13] up to 4.5 GPa. Data on static compression of zinc in [15] coincide with the computational curve for the parent phase at low pressures and with the computational curve for the high-pressure phase at $P > 7.5$ GPa.

The data in [17] suggest that the hcp lattice of zinc is stable up to 126 GPa. Hence, one can say that the observed phase transitions in zinc are electronic phase transitions with unchanged crystalline structure. The parent phase atomic cell energy 26.5 eV in zinc is close to the energy of two outer-shell electrons of the free atom of zinc in the $3d^{10}4\,s^2$ state. The energy of three outer-shell electrons of the free atom of zinc is 67 eV and that of four electrons is 129 eV.

The electronic phase transition with material crystalline structure remaining unchanged was experimentally detected in zinc [16] from the measured ratio of axes c/a of the hexagonal close-packed structure of zinc as a function of V/V_0, which showed a non-monotonicity at 9 GPa. Existence of the electronic phase transition in zinc at $P = 6.6$ GPa and $T = 4.2$ K also follows from Moessbauer spectroscopy [18], which suggests a drastic change in the crystalline lattice dynamics in this transition. In [16], [18], this transition is treated as a Lifshitz topological phase transition (TPT) [19]. Since 1995, there have been published a lot of experimental and theoretical investigations to prove the existence of TPT in zinc, which yielded contradictory results (see, e.g., [20]). Our analysis of zinc compressibility clearly points to this electronic phase transition in zinc at $P = 7.5$ GPa, as there is a kink in the slope of the $P(\rho)$ curve (jump in the bulk compression modulus). In addition, the analysis infers the existence of another electronic phase transition at $P = 57$ GPa, with the crystalline structure remaining unchanged, too. The electronic phase transitions similar to those in zinc can be observed in Cd, Tl, Pd, Ag, Sb, and other elements.

2.3. Actinides

2.3.1. *Thorium*

In [21], the $P(\rho)$ diagram on the normal isotherm in thorium was measured up to a pressure of 300 GPa. At pressures above 25 GPa, thorium displays high-pressure phases, first with 69.5 eV energy corresponding to four electrons in the outer shell, and then there is a transition to a phase of 19.1 g/cm^3 equilibrium density, 400 GPa bulk modulus, and 227 eV outer-shell electron energy per atom. This energy corresponds to six electrons in the outer shell, and this means that the electron rearrangement involves completed shells.

In the presence of phase transitions, different phase states of the same density can be observed in static and shock-compression experiments. Thus, the kink in the $P(\rho)$ curve on the normal isotherm of thorium [21] that can be attributed to a change in the quantum state of outer-shell electrons occurs at $P = 150$ GPa and compression ratio $\sigma = 2.2$. The kink in the Hugoniot of thorium at $P = 80$ GPa is adequately described by parameters of the high-pressure phase appearing on the normal isotherm at $P = 150$ GPa. This means that in the compression range of $\sigma = 1.7$–2.2 the elastic component of pressure on the Hugoniot in thorium differs from the values on the normal isotherm (see Fig. 6).

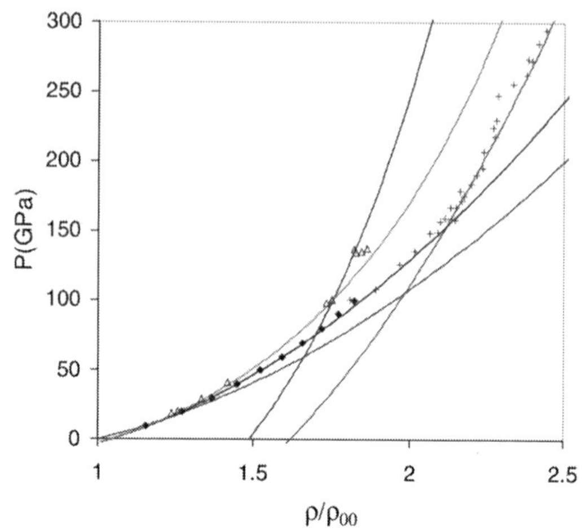

Figure 6. Isotherm at 300 K [21] and Hugoniot [10] of thorium. The curves represent calculations, and the points show experimental values.

2.3.2. *Protoactinium*

In [22], compressibility of protoactinium was studied up to a pressure of 129 GPa by the diamond-anvil-cell method and X-ray diffraction with a synchrotron source of radiation. The computational analysis indicates that the experimental points lie on the calculated curve with $\rho_0 = 15.37$ g/cm^3 and $B_0 = 115$ GPa up to $P \approx 95$ GPa. At higher pressures, the experimental points deviate significantly from the calculated curve and can be described as states of a different electronic phase of protactinium with parameters from Table 1. Figure 7 shows calculated curves and experimental points on the normal isotherm of protactinium from [22].

The authors of [22] note that the tetragonal structure of protactinium transfers to the low-symmetry orthorhombic structure of α-uranium at $P = 77$ GPa. As one can see in Fig. 7, there is no noticeable change in the slope of the $P(\rho)$ curve at 77 Gpa, which is indicative of the unchanged initial electron structure of protactinium. An abrupt change in the slope and electronic structure occurs at 94 GPa, apparently, with the unchanged orthorhombic crystalline structure of α-uranium.

2.3.3. *Americium*

In [23], Lindbaum et al. obtained high-pressure experimental data for americium up to $P = 100$ GPa. The transition from the AmI phase to the AmII and AmIII phases leads to a change in the slope of the $P(\rho)$ curve that corresponds to a drop in the bulk modulus (more than by a factor

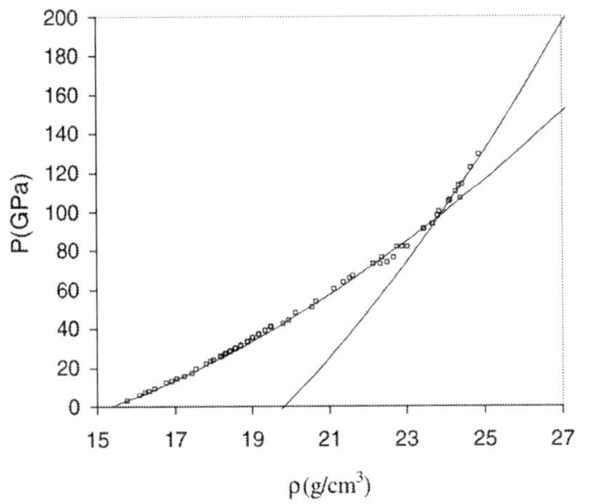

Figure 7. Isotherm at 300 K [22] for protactinium. The curves represent calculations, and the points show experimental values.

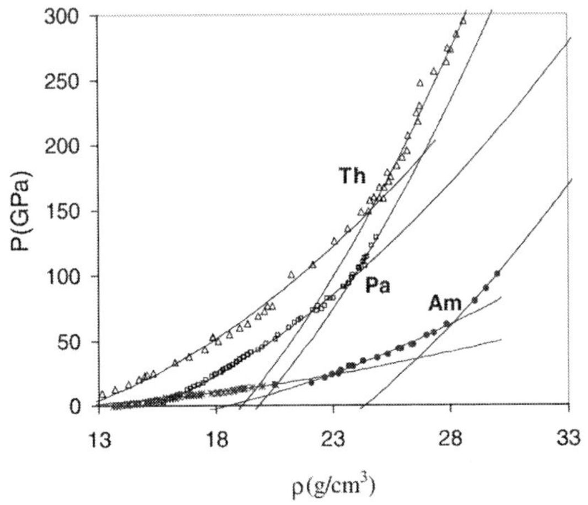

Figure 8. Isotherm at 300 K [23] for Th, Pa, U, Am. The curves represent calculations, and the points show experimental values.

of 2) and increase in the equilibrium volume for these phases. Figure 8 shows calculated curves and experimental points on the normal isotherm of americium from [23]. Despite the different crystalline structure of the AmII and AmIII phases, their $P(\rho)$ diagram can be described by a single curve with the same values of bulk modulus, $B_0 = 13.0$ GPa, and equilibrium density, $\rho_n = 11.9$ g/cm^3.

In americium, an abrupt increase in the bulk modulus occurs with a transition to the AmIV phase. Most of experimental points for the phase lie on the calculated curve with $B_0 = 80$ GPa and $\rho_n = 18.6$ g/cm^3. However, three upper experimental points dramatically deviate from the calculated curve and are described by $B_0 = 350$ GPa and $\rho_n = 24.5$ g/cm^3. This change suggests a significant rearrangement of the electronic structure of the outer electron shell under pressure.

3. Conclusion

The analysis shows that many materials exhibit peculiar properties under compression, which is due to different states of atomic cell electrons in solids. The variety of electronic phases in solids is a reflection of the diversity of energies of excited electronic states observed in atomic spectroscopy.

Simple alkaline metals, just as transition metals and actinides, are characterized by a complex behavior under compression, with changes both in the crystalline and in the electronic structure.

In the experiments with crystalline structure variation, the pressure-density relationship can be described by the same curve, the parameters of which are independent of the crystalline structure.

References

[1] E.G. Maksimov, M.V. Magnitskaya, V.E. Fortov, Uspekhi Fiz. Nauk **175**, 793 (2005)
[2] E. Gregoryanz, O. Degtyareva, M. Somayazulu et al. Physical Review Letters **94**, 185502 (2005)
[3] E. Gregoryanz, L.F. Lundegaard, M.I. McMahon et al. Science **320**(5879), 1054 (2008)
[4] C.L. Guillaume, E. Gregoryanz, O. Degtyareva et al. Nature Physics Letters, **7**(3), 211 (2011)
[5] A.W. Lawson, T.Y. Tang, Phys. Rev. **76**, 301 (1949)
[6] R. Steinheimer, Phys. Rev. **78**, 235 (1950)
[7] B.A. Nadykto, Uspekhi Fiz. Nauk **163**, 37 (1993)
[8] B.A. Nadykto, Vopr. Atomnoi Nauki i Tekhniki, Sr. Teor. I Prikl. Fizika **3**, 58 (1996)
[9]. B.A. Nadykto, Rusian Chemical Journal **XLV** (5-6), 7 (2001)
[10] LASL Shock Hugoniot Data. Ed. Marsh S.P. Berkeley - Los Angeles - London: University of California Press, 1980. 658.
[11] A.A. Bakanova, I.P. Dudoladov, R.F. Trunin, Fizika Tverdogo Tela **7**, 1615 (1965)
[12] M. Hanfland, I. Loa, K. Syassen, Phys. Rev. B **65**, 184109 (2002)
[13] S.N. Vaidya, G.C. Kennedy, Phys. Chem. Solids **31**, 2329 (1970)
[14] U. Schwarz, A. Grechnik, K. Syassen et al., Physical Review Letters **83**, 4085 (1999)
[15] L.V. Altshuler, A.A. Bakanova, R.F. Trunin, ZheTF **42**, 91 (1962)
[16] K. Takemura, Phys. Rev. Lett. **75**, 1807 (1995)
[17] K. Takemura, Phys. Rev., B **56**, 5170 (1997)
[18] Potzel W., Steiner M., Karzel H et al., Phys. Rev. Lett. **74**, 1139 (1995)
[19] I.M. Lifshitz, Zh. Eksp. Theor. Fiz. **38**, 1569 (1960) [Sov. Phys. JETP, 11, 1130 (1960)]
[20] B.K. Godval, P. Modak, R.S. Rao, Bull. Mater. Sci. **26**, 175 (2003)
[21] Y.K. Vohra, J. Akella, Phys. Rev. Lett. **67**, 3563 (1991)
[22] R.G. Haire, S. Heathman, M. Idiri et al. Phys. Rev., 2003, B **67**, 134101
[23] A. Lindbaum, S. Heathman, K. Litfin. et al., Phys. Rev. B **63**, 214101 (2001)

EPJ Web of Conferences **94**, 02023 (2015)
DOI: 10.1051/epjconf/20159402023

Microstructure and local mechanical property evolution during high strain-rate deformation of tantalum

Shraddha J. Vachhani[1,a], Nathan Mara[1,2], Veronica Livescu[1], and Ellen Cerreta[1]

[1] Materials Science and Technology, Los Alamos National Laboratory, Los Alamos, NM 87544, USA
[2] Center for Integrated Nanotechnologies, Los Alamos National Laboratory, Los Alamos, NM 87544, USA

Abstract. Shear localization is often a failure mechanism in materials subjected to high strain rate deformation. It is generally accepted that the microstructure evolution during deformation and the resulting heterogeneities strongly influence the development of these shear bands. Information regarding the development of local mechanical heterogeneities during deformation is difficult to characterize and as such, constitute is a critical missing piece in current crystal plasticity models. With the recent advances in spherical nanoindentation data analysis, there is now an unprecedented opportunity to obtain insights into the change in local mechanical properties during deformation in materials at sub-micron length scales. In this work, we quantify the evolution of microstructure and local mechanical properties in tantalum under dynamic loading conditions (split Hopkinson pressure bar), to capture the structure- property correlations at the sub-micron length scale. Relevant information is obtained by combining local mechanical property information captured using spherical nanoindentation with complimentary structure information at the indentation site measured using EBSD. The aim is to gain insight into the role of these microstructural features during macroscopic deformation, particularly their influence on the development of mechanical heterogeneities that lead to failure.

1. Introduction

Shear localization is often a failure mechanism in materials subjected to high strain rate deformation. It is generally accepted that the microstructure evolution during deformation and the resulting heterogeneities strongly influence the development of these shear bands. However, current crystal plasticity models [1–3] fail to effectively capture the heterogeneous plastic deformation. One critical missing piece in these models is the information on the development of local structural heterogeneities during deformation. Consideration of local microstructure evolution through electron back scattered diffraction (EBSD) measurements [4,5] constitutes only one aspect of the characterization of a deformed material. Another equally important attribute, which is somewhat difficult to capture, is change in local mechanical properties at the sub-micron length scale that this microstructural evolution causes.

With the recent advances in spherical nanoindentation data analysis [6,7], there is now an unprecedented opportunity to obtain insights into the change in local mechanical properties during deformation in materials at sub-micron length scales. Recent advances in instrumentation (e.g., the availability of the continuous stiffness measurement (CSM)) [8,9] have now made it possible to convert the measured load-displacement data from spherical nanoindentation into highly reproducible and consistent indentation stress-strain (ISS) curves [6]. This ISS data allows one to follow the local elastic, yield and post elastic behavior of materials during the indentation process. With this new characterization tool at our disposal, we are now in a position to gain insights into the grain-scale evolution of microstructure during the deformation of polycrystalline metals. In recent work, the use of ISS curves for nanoindentation data analysis has demonstrated tremendous promise in providing new insights into material behavior, including the role of crystal orientation [10] and grain boundaries during macroscale deformation in metals [11], buckling behavior of carbon nanotube forests [12], and lamellar level properties in bone [13].

In this work, evolution of microstructure and local mechanical properties in dynamically deformed (strain rates $\sim 10^3$); high-purity tantalum was investigated. More specifically, a novel combination of spherical nanoindentation and EBSD was used to characterize changes in the local slip resistance in deformed, polycrystalline samples of tantalum. The indentation yield strength (Y_{ind}) extracted from ISS curves is influenced by both local crystal orientation and the local dislocation density. To decouple the two contributions, a two-step procedure presented in [14] is followed. In the first step, the effect of local crystal orientation on Y_{ind} is accounted for by focusing on annealed samples (negligible dislocation density) of the material system of interest. The attention is shifted to deformed samples in the second step. However, the knowledge gained in step-1 is extremely important to accurately relate the Y_{ind} measured in the deformed samples to the changes in the slip resistance caused by the increased dislocation density.

The paper is organized as follows: Sect. 2 provides a brief overview of the indentation data analysis protocols

[a] Corresponding author: Shraddha@lanl.gov

Article available at http://www.epj-conferences.org or http://dx.doi.org/10.1051/epjconf/20159402023

used in this work. The details of the materials and experimentation methodologies used in this study are provided in Sect. 3. In Sect. 4, we describe the experimental observations and discuss the results. The concluding remarks are presented in Sect. 5.

2. Indentation stress-strain (ISS) curves

The raw nanoindentation load-displacement data was converted into ISS curves following the procedures developed by Kalidindi and Pathak [6]. A brief summary of the same is provided here. This procedure follows a two-step approach consisting of (i) the determination of an effective zero point followed by (ii) the calculation of the contact radius used for computing the indentation stress and strain values. The aim of the first step is to determine the effective point of initial contact, i.e., a clear identification of a point that makes the measurements in the initial elastic loading segment consistent with the predictions of Hertz's theory [15,16]. Typically, the default protocols in the vendor-provided analyses software do not perform adequately in this respect, mostly resulting in the overestimation of the sample surface [4]. The problem is often made worse by surface roughness issues, non-ideal geometry of the indenter and the effects of the surface preparation techniques [17]. The load and displacement corrections (P^* and h^* respectively) are calculated by performing a linear regression analysis of the expected relationship between the measured load signal, $\tilde{P}$, the measured displacement signal, $\tilde{h}$, and the continuous stiffness measurement, S. For the initial elastic loading segment in the spherical indentation of a flat isotropic sample surface this relationship is given by Hertz theory as:

$$S = \frac{3P}{2h_e} = \frac{3(\tilde{P} - P^*)}{2(\tilde{h}_e - h^*)}. \quad (1)$$

In the second step, Hertz's theory is recast in the following set of equations to calculate the contact radius, a, and indentation stress, σ_{ind}, and indentation strain, ε_{ind}, values:

$$a = \frac{S}{2E_{eff}} \qquad \sigma_{ind} = E_{eff}\varepsilon_{ind},$$
$$\sigma_{ind} = \frac{P}{\pi a^2}, \qquad \varepsilon_{ind} = \frac{4}{3\pi}\frac{h_e}{a} \approx \frac{h_e}{2.4a}. \quad (2)$$

Here, h_e is the elastic indentation depth, S is the elastic stiffness described earlier, a, the radius of the contact boundary at the indentation load P, the effective radius and the effective stiffness of the indenter and the specimen system, R_{eff} and E_{eff} respectively are given by the following equations:

$$\frac{1}{E_{eff}} = \frac{1 - \nu_s^2}{E_s} + \frac{1 - \nu_i^2}{E_i}$$
$$\frac{1}{R_{eff}} = \frac{1}{R_i} + \frac{1}{R_s} \quad (3)$$

where ν and E are the Poisson's ratio and the Young's modulus, R refers to the radius and the subscripts s and i refer to the specimen and the indenter, respectively.

Note that the protocols described above rely on the CSM to obtain a continuous estimate of the radius of contact, a, at every point on the load-displacement curve. It has been shown [18] that the most reliable (ISS) curves can be obtained using the CSM at 45 Hz oscillation frequency and a 2 nm oscillation amplitude. These test conditions were utilized in all of the measurements reported here.

3. Materials and methodology

High-purity tantalum was used in this investigation to reduce the number of factors that can influence the evolution of microstructure during deformation such as second phases, twinning, and inclusions. The details of this tantalum plate material (chemistry, texture, grain size) can be found elsewhere [19]. Right, cylindrical compression samples were cut from the as-received plate of tantalum (5mm in diameter by 5mm in height). A split Hopkinson pressure bar (SHPB) was used to dynamically (strain rates ~1800/s) impart predetermined levels of strain into the tantalum specimens. A typical stress-strain response for tantalum is shown in Fig. 1. Also marked in Fig. 1 are the strains to which interrupted compression tests were performed. Characterization, presented here focused on the as-annealed state and then again of the compression cylinder deformed to 20% strain. The other samples (intermediate strain levels) will be the focus of future work and are included here only to illustrate the evolution in the microstructure. All specimens tested in the SHPB were sectioned along the compression axis and mounted in epoxy. Since a high quality surface finish is critical for both nanoindentation [17] and EBSD, mechanical grinding was followed by polished using alumina and subsequently silica suspensions. The samples were then lightly etched using a mixture of water (10 ml), nitric acid (10 ml), hydrochloric acid (10 ml) and hydrofluoric acid (5 ml).

Grain orientation maps for the samples were obtained using orientation imaging microscopy (OIM), which is based on the automatic indexing of EBSD patterns. Orientation at each point was denoted using the Bunge-Euler angles [20], an ordered set of three angles (φ_1, Φ, φ_2), which describe the rotations for the transformation from the sample reference frame to the crystal reference frame. Inverse pole figure (IPF) maps as well as grain orientation spreads maps for the tantalum samples are also shown in Fig. 1.

Nanoindentation tests were then performed at pre-selected locations on these samples using the MTS nanoindenter XP®equipped with the CSM attachment. A 100 μm spherical indenter tip was used to perform the indentation tests. Over 160 indentations in the fully annealed samples and about 50 indentations in the deformed samples (20% strain) were done on multiple days, in multiple batches to ensure repeatability. After the indentation tests were performed, the regions of interest were imaged again in the SEM using OIM, to confirm the location of the indents and determine the local crystal orientation at the indentation site. Any indent that landed on the grain boundary was excluded from the analysis. From the ISS curves, Y_{ind} was determined as the point of intersection between the modulus line and the best-fit line for the stress-strain curve within the strain range of 0.01

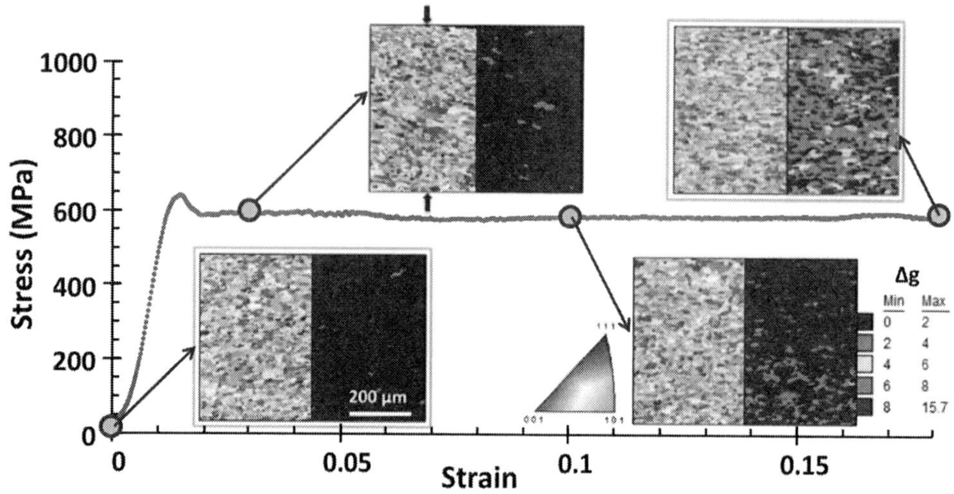

Figure 1. Typical stress-strain response for high purity polycrystalline tantalum deformed in a split Hopkinsin bar at a strain rate of ~1800/s. Inverse pole figure maps and grain orientation spreads for samples after a strain of 0, 3, 10 and 20% strain are also shown. The samples with 0 and 20% strain were used for mechanical characterization using spherical nanoindentation.

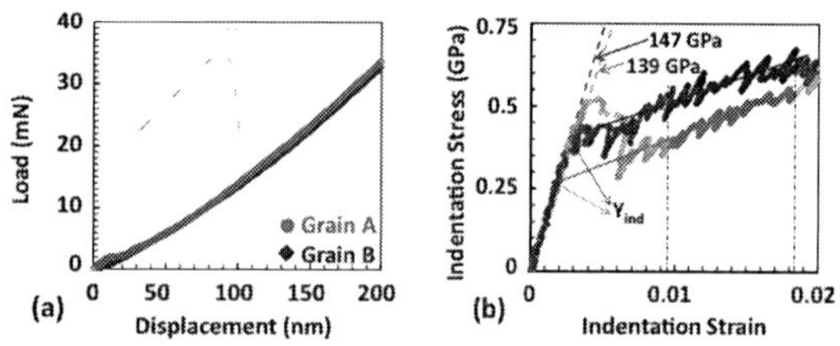

Figure 2. Representative (a) load-displacement and (b) corresponding indentation stress-strain response for two grains in the as received tantalum sample show that the nanoindentation response is strongly influenced by the local crystal lattice orientation. The inset in (a) shows the IPF map which indicates the orientation of the grains with respect to the indentation axis. The dashed vertical lines in (b) show the segment of the stress-strain curve used for the back extrapolation.

and 0.017. This protocol was standardized for all the tests reported in this paper.

4. Results and discussion

Representative load displacement curves and corresponding ISS curves for two orientations tested in the as-received condition are presented in Fig. 2. The inset in Fig. 2(a) shows the orientations of grains A and B plotted on the standard [001] inverse pole figure (IPF) map. Specifically, the position of grains A and B on the IPF map represent the crystal direction in the grain that is parallel to the indentation direction. Note that since we use an axisymmetric indenter tip, in plane rotations of the crystal orientation do not alter our indentation response. Consequently, only two (Φ, φ_2) of the three Bunge-Euler angles used to represent the crystal orientation have an effect on the indentation response.

In Fig. 2, grain A is close to the [001] corner of the IPF map, meaning that the [001] crystal direction is parallel to the indentation axis and grain B is close to the [111] corner of the IPF map and has the [111] crystal direction almost parallel to the indentation axis. The misorientation between these two orientations is close to the maximum possible value for cubic materials and as expected, the two orientations show a very large difference in their nanoindentation response.

While the nanoindentation response for only two orientations is shown in Fig. 2, a total of 160 orientations were tested. The local crystal anisotropy affects both the elastic and plastic response. The values of E_s were found to be in the range of 139 to 150 GPa and the Y_{ind} values range from 240 to 400 MPa. Values for both E_s and Y_{ind} are the lowest for orientations close to [001] and highest close to the [111] crystal direction.

Figures 3 and 4 are contour plots showing the effect of crystal orientation on E_s and Y_{ind}, respectively. The contour plots were created by interpolating between the measured E_s and Y_{ind} using generalized spherical harmonics (GSH) [20]. The contour plot in Fig. 4, henceforth referred to as the indentation yield surface, is of tremendous value in the studies on deformed samples as it provides a baseline and helps decouple the effects of orientation and local dislocation density on the measured Y_{ind} in deformed materials. Using this map, Y_{ind} for any given orientation in high-purity tantalum can now be estimated. Having established a quantitative relationship between local crystal orientation and Y_{ind}, we focus our attention now on the deformed tantalum sample.

As described in the previous section, a high purity tantalum sample was deformed to 20% strain in a SHPB (strain rate ~1800/s). The plastic deformation leads to an increased dislocation density as well as a higher degree of misorientation within the grains as compared to the as-received state. While the latter is evident in the EBSD maps (Fig. 1), the increased dislocation density manifests as hardening in the sample, leading to an increase in the Y_{ind}. Once again, a series of nanoindentation tests

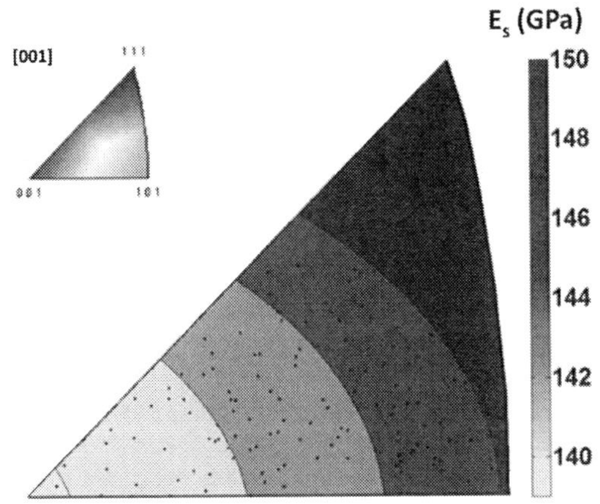

Figure 3. Contour plot generated by interpolating between the Es values extracted for about 160 orientations in the as received condition. The black circles represent the positions of the tested orientations on the standard IPF plot.

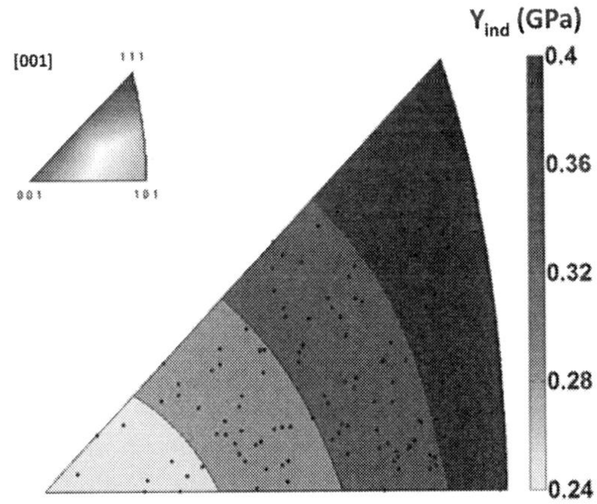

Figure 4. Contour plot generated by interpolating between the Yind values extracted for about 160 orientations in the as received condition. The black circles represent the positions of the tested orientations on the standard IPF plot.

were performed and EBSD was used to determine the local crystal orientation at the indentation site. Y_{ind} values were extracted using the back-extrapolation method, as in the previous case. Figure 5 shows the difference in the nanoindentation response for four grains, two tested in the as received condition (same data as Fig. 2) and two in the 20% deformed sample. As seen from the inset in Fig. 5(a), grains A2 and B2, tested in the deformed condition are almost the same orientation as grain A and B respectively, which were tested in the as received condition. As seen here, the increment in the Y_{ind} due to the imposed deformation can be as high as 60% depending on the location of the indent.

For each of the grains tested in the deformed condition, the Y_{ind} in the as received condition was predicted using the indentation yield surface presented in Fig. 4. This is the Y_{ind} that would have been measured for the orientations in question, if they were tested in the as received condition.

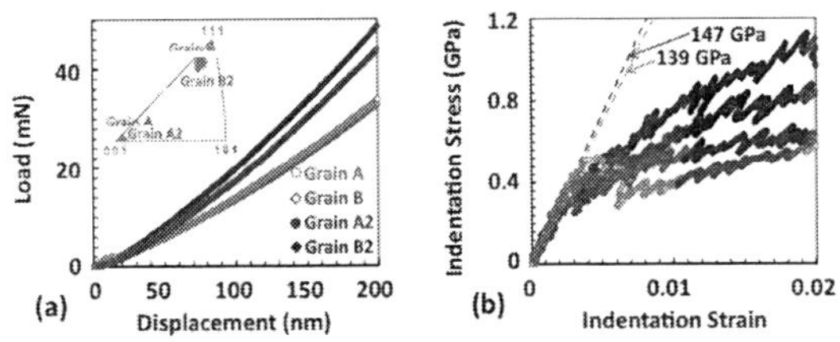

Figure 5. (a) The load-displacement data, and (b) corresponding indentation stress-strain curves for two pairs of almost identically oriented grains showing the effect of deformation on the ISS response. For each pair, one grain was tested in the as received condition (Grain A and B) and the other grain was tested after the sample was deformed by 20% (Grain A2 and Grain B2).

It is important to accurately establish this value since the Y_{ind} can vary by as much as 40%, even in the as received condition, depending on the local crystal lattice orientation at the indentation site. A reliable estimate of the increment in the indentation yield strength (ΔY_{ind}) for any given orientation can then be determined as the difference between the measured indentation yield point in the deformed condition ($Y_{ind}(g, d)$) and the estimated indentation yield point in the fully annealed condition Y_{ind} (g,0) and is given as

$$\Delta Y_{ind}(g, \varepsilon) = Y_{ind}(g, d) - Y_{ind}(g, 0) \tag{4}$$

where ε refers to the amount of deformation- 20% strain under dynamic loading conditions and g refers to the local crystal lattice orientation at the indentation site.

The simplest relationship that one can establish between the increment in the indentation yield point (ΔY_{ind}) and the local dislocation content (ρ) is through the increment in the critical resolved shear strength ($\Delta \tau_{crss}$) of the slip system. In a highly simplified manner, this relationship can be expressed as

$$Y_{ind} = M(\Phi, \phi_2)\, \tau_{crss} \tag{5}$$

$$\Delta \tau_{crss} \propto \sqrt{\rho} \tag{6}$$

and

$$M = \frac{\sum \Delta \gamma}{\Delta \varepsilon} \tag{7}$$

where M is a Taylor factor that depends only on the grain orientation with respect to the indentation direction (in this case, only two of the three Bunge-Euler angles describing local crystal orientation), τ_{crss} is the average critical resolved shear stress in the crystal, $\Delta \tau_{crss}$ is the increment in the local averaged critical resolved shear strength between the annealed and deformed conditions, and ρ is the local dislocation density. $\Sigma \Delta \gamma$ is the sum of the shears on all the slip planes and $\Delta \varepsilon$ is the imposed macroscopic strain.

Combining Eqs. (4) and (5), the percentage increase in the critical resolved shear stress is given as

$$\frac{\Delta Y_{ind}}{Y_{ind}(g, 0)} = \frac{\Delta \tau_{crss}}{\tau_{crss}(g, 0)}. \tag{8}$$

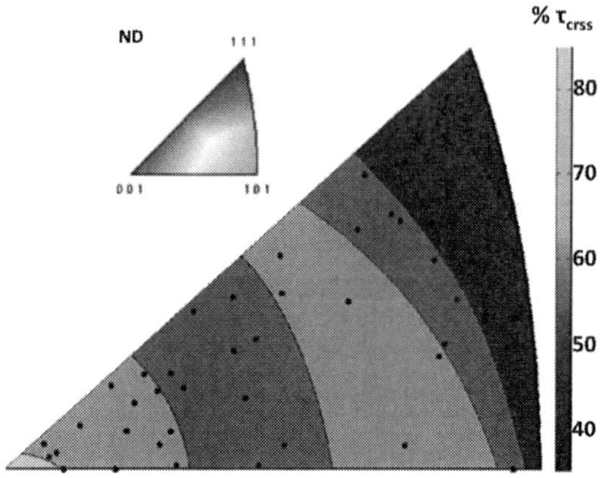

Figure 6. Contour plot generated by interpolating between the change in slip resistance values extracted for about 70 orientations in the 20% deformed sample. The black circles represent the positions of the tested orientations on the standard IPF plot.

Since the change in the average critical resolved shear stress in the crystal can be related to the local dislocation density (Eq. (5)), the percentage increase in Y_{ind} provides an indirect measure of the local dislocation content at the indentation site.

A total of about 50 orientations were tested in the 20% deformed sample and the $\%Y_{ind}$ (or $\%\tau_{crss}$) was computed in each case. It is seen that the computed $\%\tau_{crss}$ is highly non-uniform across all the tests and varies significantly depending on the grain orientation. The variation of increment in the local critical resolved shear strength as a function of crystal orientation is plotted in Fig. 6. It is seen from this figure that the highest values for $\%\tau_{crss}$ occur for orientations close to the [001] orientation.

It has been hypothesized that orientations with a lower Taylor factor (soft grains) develop lower dislocation densities compared to grains with a higher Taylor factor (hard grains) and that for a given increment of imposed plastic strain, the increment in local dislocation density is directly proportional to the Taylor factor. The relevant Taylor factor, in this case is that for uniaxial compression, and is not to be confused with the Taylor factor for nanoindentation, referred to in Eq. (4). Taylor factor is a parameter frequently used for indicating slip activity for a particular orientation, during a given deformation mode and is defined in Eq. (7). Thus, an orientation with a lower Taylor factor is expected to undergo a lower amount of slip activity in order to accommodate the imposed strain, as compared to an orientation with the higher Taylor factor. Consequently, an orientation with a lower Taylor factor is expected to have a lower amount of stored dislocations. This hypothesis [14] has been verified using the approach described here on large grained (2–4 mm), high-purity, aluminum samples tested under quasi-static conditions, when nanoindentation tests were performed in the middle of the grains and well away from grain boundaries. At grain boundaries however, it has been shown that the soft grains, when present next to hard grains, show a hardened layer close to the boundary, as a result of grain interactions [21]. This hardened layer was seen to be up to 20 μm thick.

It the present study however, the highest values for $\%\tau_{crss}$ are seen for orientations close to the [001], which is a soft orientation and the highest values are seen close to the [111] orientation, which is a hard orientation. The trends observed are opposite to those seen for grain interiors in the case of aluminum. In other words, they are the trends that were seen close to grain boundaries for aluminum. Note that the hardened layer in the case of aluminum extended up to 20 μm from the grain boundary. And, in the case of the tantalum samples used in the present study, the grain size is ~40 μm. It appears that smaller grain size causes the grain interactions to dominate the local mechanical response within entire grains in the deformed sample, unlike the case of aluminum, where the effect of grain interactions was limited to a relatively thin (as compared to the grain size) layer and the grain interiors deformed as per Taylor predictions.

5. Concluding remarks

A novel approach to characterize the percentage increase in the slip resistance during dynamic deformation of tantalum is presented in this work. This two-step process combines local mechanical property information obtained using spherical nanoindentation with corresponding structure information gathered using EBSD at increasing levels of strain. The increase in the local slip resistance (due to the dynamic deformation) can then be linked to the change in Yind. It was seen that for the tantalum studied here, with grain sizes of ~40 μm, grain interactions dominated the change of local slip resistance. While only one sample, deformed to 20% strain was characterized in this work, characterization of samples strained to intermediate levels (5% and 10%) will be done in the future and is likely to shed more light into the evolution of slip resistance and its dependence on crystal orientation. This work will also serve to establish an initial, comparative basis for future studies focused on quantifying the strength of dynamically damaged tantalum.

References

[1] Kalidindi, S.R., A. Bhattacharya, and R. Doherty, *Detailed Analysis of Plastic Deformation in Columnar Polycrystalline Aluminum Using Orientation Image Mapping and Crystal Plasticity Models.* Proceedings of the Royal Society of London: Mathematical, Physical and Engineering Sciences., 2004. **460**(2047): p. 1935–1956.

[2] Asaro, R.J., *Crystal Plasticity.* Journal of Applied Mechanics, 1983. **50**(4b): p. 921–934.

[3] Kalidindi, S.R. and L. Anand, *An Approximate Procedure for Predicting the Evolution of Crystallographic Texture in Bulk Deformation Processing of Fcc Metals.* International Journal of Mechanical Sciences, 1992. **34**(3): p. 309–329.

[4] Adams, B.L., *Orientation imaging microscopy: Emerging and future applications.* Ultramicroscopy

Proceedings of the 1996 6th Conference on Frontiers in Electron Microscopy in Materials Science, Jun 4-7 1996, 1997. **67**(1-4): p. 11–17.

[5] Adams, B.L. *Measurement and representation of polycrystalline microstructure*. in *Winter Annual Meeting of the American Society of Mechanical Engineers, Nov 8–13 1992*. 1992. Anaheim, CA, USA: Publ by ASME, New York, NY, USA.

[6] Kalidindi, S.R. and S. Pathak, *Determination of the effective zero-point and the extraction of spherical nanoindentation stress-strain curves*. Acta Materialia, 2008. **56**: p. 3523–32.

[7] Pathak, S., J. Shaffer, and S.R. Kalidindi, *Determination of an effective zero-point and extraction of indentation stress-strain curves without the continuous stiffness measurement signal*. Scripta Materialia, 2009. **60**: p. 439–42.

[8] Li, X. and B. Bhushan, *A review of nanoindentation continuous stiffness measurement technique and its applications*. Materials Characterization, 2002. **48**(1): p. 11–36.

[9] Fischer-Cripps, A.C., *Study of analysis methods of depth-sensing indentation test data for spherical indenters*. Journal of Materials Research, 2001. **16**: p. 1579–84.

[10] Pathak, S., D. Stojakovic, and S.R. Kalidindi, *Measurement of the local mechanical properties in polycrystalline samples using spherical nanoindentation and orientation imaging microscopy*. Acta Materialia, 2009. **57**: p. 3020–8.

[11] Pathak, S., et al., *Studying grain boundary regions in polycrystalline materials using spherical nanoindentation and orientation imaging microscopy*. Journal of Materials Science, 2012. **47**(2): p. 815–823.

[12] Pathak, S., et al., *Viscoelasticity and high buckling stress of dense carbon nanotube brushes*. Carbon, 2009. **47**: p. 1969–1976.

[13] Pathak, S., et al., *Assessment of lamellar level properties in mouse bone utilizing a novel spherical nanoindentation data analysis method*. Journal of the Mechanical Behavior of Biomedical Materials, 2012. **13**(0): p. 102–117.

[14] Vachhani, S.J. and S.R. Kalidindi, *Grain-Scale Measurement of Slip Resistances in Aluminum Polycrystals using Spherical Nanoindentation* Acta Materialia, 2015. accepted.

[15] Hertz, H., *Miscellaneous Papers*. New York: MacMillan and Co., Ltd, 1896.

[16] Johnson, K.L., *Indentation Contact Mechanics*. 1985: Cambridge University Press, Cambridge.

[17] Pathak, S., et al., *Importance of surface preparation on the nano-indentation stress-strain curves measured in metals*. Journal of Materials Research, 2009. **24**: p. 1142–55.

[18] Vachhani, S.J., R.D. Doherty, and S.R. Kalidindi, *Effect of the continuous stiffness measurement on the mechanical properties extracted using spherical nanoindentation*. Acta Materialia, 2013. **61**(10): p. 3744–3751.

[19] T. Bucchiet E.K. Cerreta, L.D., S.R. Chen, J. Michael, *Characterization of Tri-Lab Tantalum*. Sand2014-17645.

[20] Bunge, H.-J., *Texture analysis in materials science. Mathematical Methods*. 1993, Göttingen: Cuvillier Verlag.

[21] Kalidindi, S.R. and S.J. Vachhani, *Mechanical characterization of grain boundaries using nanoindentation*. Current Opinion in Solid State and Materials Science, 2014. **18**(3): p. 196–204.

EPJ Web of Conferences **94**, 02024 (2015)
DOI: 10.1051/epjconf/20159402024

On the importance of recrystallization to reproduce the Taylor impact specimen shape of a pure nickel

Hervé Couque[a]

Nexter Munitions, 7 route de Guerry, 18023 Bourges, France

Abstract. Taylor tests are a mean to investigate the dynamic plastic and failure behaviour of metals under compression. By taking in account the strengthening occurring at high strain rates, the Taylor final diameter of a pure nickel impacted at 453 m/s have been numerically reproduced by 13%. Through post-mortem observations of the specimen impacted at 453 m/s, a recrystallization process has been found to occur resulting in a softening of the pure nickel. Subsequent numerical simulations taking in account this softening have been found to reduce the difference between experimental and numerical diameter by 10%.

1. Introduction

There are known challenges with the numerical simulation of large plastic deformation of metals generated under high strain rates [1]. One of the challenges is the stabilization of plastic deformation when the viscous behavior of the dislocations prevailed. Such challenge has been answered by taking in account the strengthening occurring in the viscous regime with the use of adequate constitutive models such as the Zerrilli Amstrong or modified Johnson-Cook models [2]. However, a challenge remains with regard to the accuracy of the simulations. This challenge has been quantified with the numerical simulations of symmetric Taylor tests conducted with a pure nickel. At an impact speed of 453 m/s, the numerical simulation was found to underestimate the final diameter of the Taylor specimen by 13% [2].

Recent post-mortem analyses of pure nickel compression specimens tested in the viscous regime reveal the occurrence of recovery/ recrystallization [3]. In this work the recrystallized nickel was found to have a similar dislocation density and strength that the initial state of the nickel.

These results have motivated to reconsider the Taylor test conducted at an impact speed of 453 m/s.

In a first step, post-mortem analyses of the Taylor specimen have been conducted to quantify ductile failure occurring in the center of the specimen and to provide recrystallization evidences based on hardness measurements. In a second step, numerical simulations have been conducted with a tensile failure criterion and a modified Johnson-Cook model taking in account the softening when recrystallization occurred.

2. The symmetric taylor test

The Taylor test is a mean to validate constitutive models at strain rate ranging from 10^3 to $10^4\,s^{-1}$ and to evaluate dynamic compression failure properties of materials. Taylor testing has been intensively used in the original configuration consisting of one cylindrical specimen impacting an anvil made of hard steel as to reproduce a rigid wall, referred as the single Taylor test [4,5]. To simulate the single Taylor test it is required to make an estimation of the friction coefficient between the specimen and the anvil, as well as, to assume that the anvil is rigid.

With the symmetric loading procedure introduced by Erlich et al. [6], such hypotheses are no longer needed. When compared to single Taylor testing, symmetric Taylor testing involves one cylindrical specimen sent with a gas gun against another identical specimen located at the gun muzzle. For the symmetric Taylor test, it is equivalent either to send two specimens against a second specimen at rest. Evidences of the importance of friction have been revealed in a comparison between the two Taylor test configurations [7].

One inconvenient of Erlich's approach is the use of ceramic pins to locate the specimen sitting at the gun muzzle making difficult coaxial and planar impact of the two specimens. To facilitate planar and coaxial impacts, two testing procedures involving Teflon sabots have been developed by the author [8]. The first procedure addresses materials of strength greater than 800 MPa in which the Teflon sabots are deformed with the Taylor specimens. The second procedure was developed for low strength materials, 100–800 MPa, in which a Teflon sabot and Teflon supports are designed to not interfere with the specimens. It is the last procedure that was employed for the pure nickel impacted at 453 m/s.

2.1. Pure nickel

A nickel elaborated by Eramet in plate 12 mm thick with a 99,99% content of nickel was employed. The nickel was used in a heat treated condition providing a quasi-static yield stress of 240 MPa [2].

[a] Corresponding author: h.couque@nexter-group.fr

Article available at http://www.epj-conferences.org or http://dx.doi.org/10.1051/epjconf/20159402024

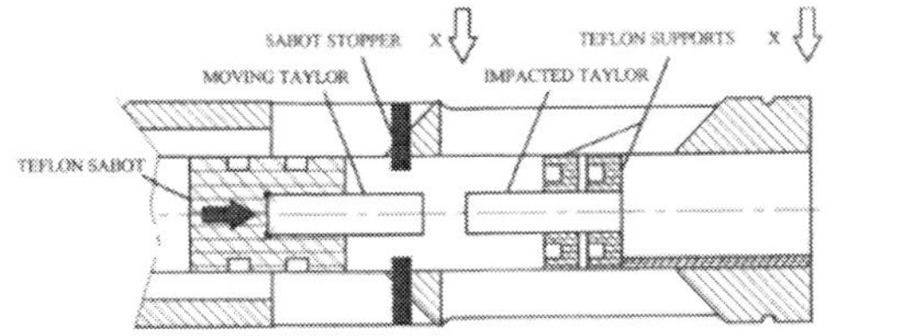

Figure 1. Schematic of the symmetric Taylor test for low strength materials.

2.2. Experimental procedure

The experimental set-up is shown in Fig. 1 with two Taylor specimens 9 mm in diameter and 36 mm in length. The loading system consist of a 25 mm caliber gas gun launching one Taylor specimen guided with a Teflon sabot which is arrested prior impact using copper inserts. The velocity of the launched specimen was measured using two laser beams separated by 18 mm located 30 mm behind the impact plane. The Taylor specimen located at the gun muzzle is positioned with two crushable Teflon supports. X ray observations were conducted with a Teflon witness bar in contact with the second Teflon support as to verify that the Teflon supports of the impacted Taylor specimen did not move prior impact [8].

A numerical investigation has been conducted to establish that the Teflon supports did not interfere during the plastic deformation of the Taylor specimens made of aluminium for a test conducted at an impact speed of 335 m/s [8]. The equivalent plastic strain profiles of the Taylor specimens are provided when passing the Teflon supports and after passing the Teflon supports, see Figs. 2a and 2b. The strain profiles remain unchanged confirming that the stress applied by the Teflon supports on the Taylor specimens remains lower than the yield stress. This is verified with the von Mises and equivalent plastic strain records of the two material points I and II located in the plastic deformed region of the impacted Taylor specimens, see Figs. 2c and 2d.

3. Results

Previous numerical results obtained with modified Johnson-Cook models are first reported. Then, results from post–mortem analyses of the Taylor specimen impacted at 453 m/s are provided revealing damage in the impact region composed of a tensile failure zone and a recrystallization zone. Finally, numerical results taking in account these two damage phenomena are reported.

3.1. Numerical results with a modified Johnson-Cook model

Numerical simulations of the Taylor test were conducted with a modified Johnson-Cook model (JCM) taking in account the strengthening occurring in the viscous regime [2]. The simulations were conducted with AUTODYN with an optimized numerical model of the Taylor specimen composed of cells 100 μm in size in the plastically loaded region and 300 μm in size for the elastically loaded region. The modified Johnson-Cook model includes a power strain

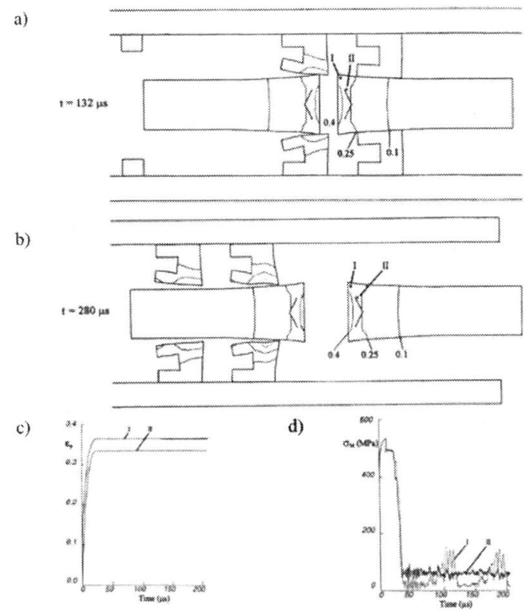

Figure 2. Numerical results of Taylor specimens made of aluminium tested at an impact speed of 335 m/s: a) and b) equivalent strain profiles at 132 and 280 μs, respectively; c) and d) von Mises stress history and equivalent plastic strain history of material points I and II, respectively.

rate component added to the logarithm strain rate term of the original Johnson-Cook (JC) formulation, $D(\dot{\varepsilon})/\dot{\varepsilon}_j)^k$, with D and k two constants. This component is normalized by a reference strain rate, $\dot{\varepsilon}_1$, equal to 10^3 s^{-1} delimitating the thermally activated regime and the viscous regime. The equivalent stress function of plastic strain, ε_p, strain rate, $\dot{\varepsilon}$ and temperature, T, is expressed as:

$$\sigma = (A + B\varepsilon_p^n)(1 + C\ ln(\dot{\varepsilon}/\dot{\varepsilon}_o) + D(\dot{\varepsilon}/\dot{\varepsilon}_1)^k)$$
$$\times(1 - [(T - T_r)/(T_m - T_r)]^m)$$

with the usual constants of the classical Johnson-Cook formulation: T_r,T_m the room and melting temperatures, $\dot{\varepsilon}_o$, a reference strain rate equal to 1 s^{-1} and A, B, C, n, m five material constants describing the mechanical response. This formulation enables to go back to the classical Johnson-Cook formulation when strain rates are lower than 10^3 s^{-1}.

At low impact speeds, numerical simulations of the final diameter of the Taylor specimens were improved with the use of the JCM model when compared to the JC model with difference less than 10%, see Fig. 3. In an another hand, numerical simulations at high impact speeds were found to provide a significant difference, with an underestimation of the final diameter by 17% for the test conducted at 453 m/s.

To improve the numerical simulation at high impact speeds, a saturation stress in the 10^4 s^{-1} strain rate regime was introduced based on Steinberg's argument which stipulates that at a given strain rate all effects of strain rate have saturated and the material strength becomes independent of strain rate [9]. Strength levelling observed

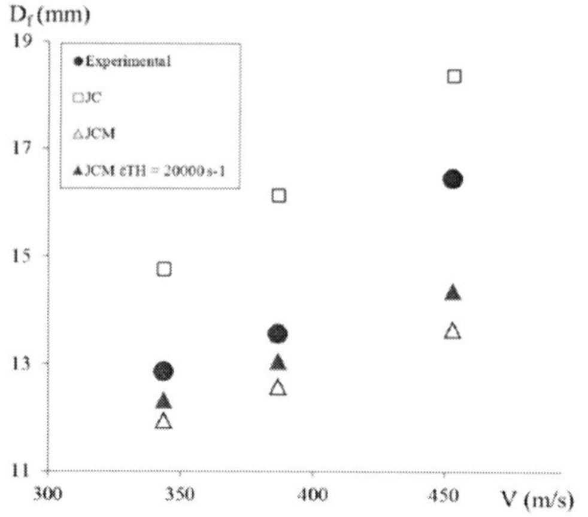

Figure 3. Experimental and numerical final diameter of the Taylor specimens versus impact speed. Numerical diameters were generated using three materials models: Johnson-Cook classical formulation (JC), Modified Johnson-Cook model (JCM), Modified Johnson-Cook model with a strain rate threshold, $\dot{\varepsilon}_{TH}$, of $20000\,s^{-1}$.

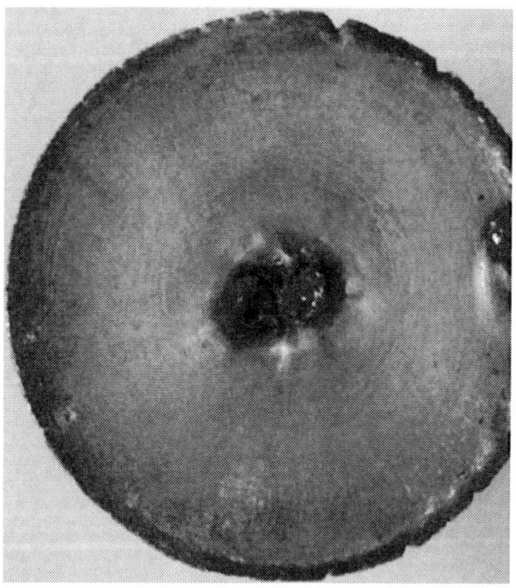

Figure 4. Tensile failure zone of the Taylor specimen tested at 453 m/s. The final diameter of the specimen is 16,45 mm.

at high impact speeds with the compression test has motivated the choice of a strain rate threshold of $20000\,s^{-1}$ [2]. The use of the JCM model with the strain rate threshold provides a better fit at high impact speeds with an underestimation of the Taylor final diameter reduced to 13% for the test conducted at 453 m/s, see Fig. 3.

3.2. Post-mortem analyses

Figure 4 reveals a tensile failure zone in the center of the impact faces of the Taylor specimens, 2,5 mm in size. This

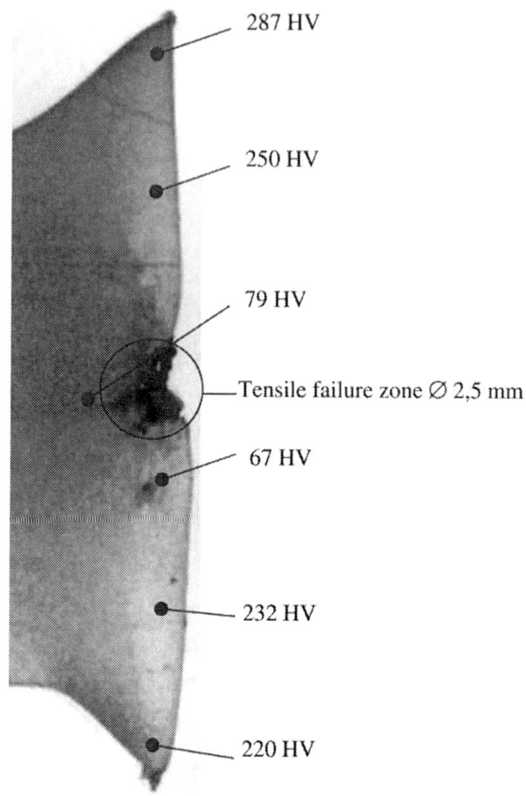

Figure 5. Sectioned Taylor specimen revealing the tensile failure zone and the variation of the hardness.

zone extends inside the specimen to about 2,5 mm in depth, see Fig. 5.

Vickers hardness measurements were generated in the impact zone with an applied force of 0,005 kgf, see Fig. 5. The hardness of the initial nickel was measured to be 179. A decrease of the hardness to 67 was measured near the tensile failure zone, see Fig. 6. Such low hardness confirms that recrystallization occurred. Based on a yield stress of 240 MPa for the initial nickel, the yield stress associated with the recrystallized nickel was estimated to be of the order of 50 MPa.

3.3. Numerical results with a tensile failure criterion

Numerical simulations were conducted with a failure strain criterion applied to the principal strain ε_{11}, ε_{22} and ε_{33}, see Fig. 6a. A failure criterion of 0,63 has to be applied to reach the planar size of tensile failure zone, D_{tf}, of 2,5 mm, see Fig. 5. For this failure criterion, no significant increase of the Taylor final diameter has been noticed, see Fig. 6b.

The failure zone was not quite reproduce as it was limitated to a tensile failure of 1 mm in depth. Based on the work of Chapman and all, the extend of tensile failure inside the specimen comes from the coalescence of radial release waves at the specimen axis [10]. As suggested in this work, simulations conducted with a critical treshold

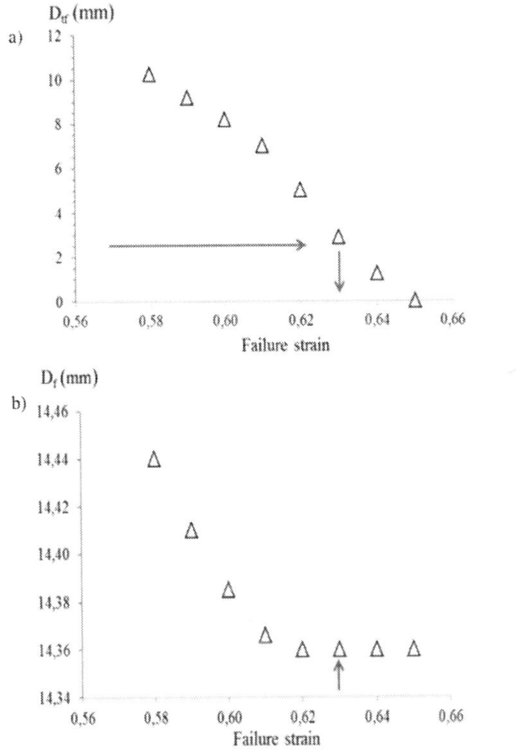

Figure 6. Tensile failure: a) planar size function of the failure criterion, b) Taylor final diameter function of the failure criterion.

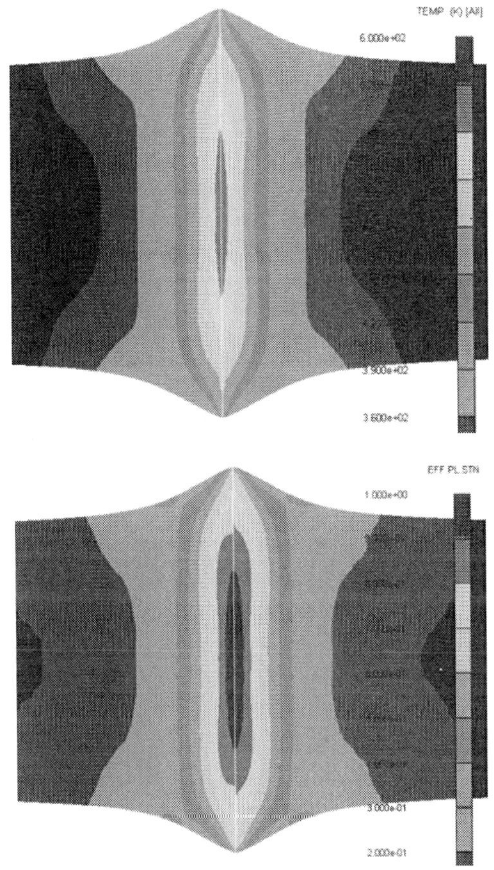

Figure 7. Temperature and effective plastic strain at 12 μs after impact.

pressure appear to be a mean to generate this bulk tensile failure.

However, the ability to reproduce the radial tensile failure demonstrates that the difference between the experimental and numerical diameters does not originate from the failure process occuring in the Taylor specimen.

3.4. Numerical results with a recrystallization criterion

The decrease in strength associated with the recrystallization phenomena was taken in account in the numerical simulation by introducing a recrystallization temperature criterion. The work of Lichtenberger provides recrystallized hardness temperature curves for different level of initial plastic deformation [11]. From these curves, quasi-static recrystallization temperatures of 290 and 230°C are deduced for a plastic deformation of 30 and 70%, respectively. Based on this work, a temperature of 230°C was selected as a recrystallization temperature criterion. Such temperature is reached 12 μs after impact, see Fig. 7. At this instant the plastic deformation is about 80%.

Numerical simulations were consequently generated incorporated two type of JCM models, with the first model representative of the original material, see Sect. 3.1, and the second model representative of the recrystallized material. The yield and hardening coefficients for the recrystallized material were identified as to provide a rigid plastic behaviour with a quasi-static yield stress of 50 MPa. Precisely, the A and B coefficients of the JCM models were 200 and 800 MPa for the first model, and 100 and 400 MPa for the second model, respectively. The simulations were conducted by monitoring the temperature. At the recrystallization temperature criterion of 230°C, the model representative of the recrystallized material was applied.

Results conducted with these models are provided in Fig. 8. To describe the difference in final diameter, a simulation was conducted with one specimen without recrystallization criterion (a2) and the other with a recrystallization criterion (a1). A larger final diameter was generated with the specimen incorporating the recrystallization criterion which is less than 10% the experimental diameter, see Fig. 9. The same diameter was obtained when the simulation of the two Taylor specimens involving the recrystallization temperature criterion, see Fig. 8b.

These results provide a framework to measure the dynamic recrystallization temperature of metals. Further investigations are required to improve the accuracy of the numerical data. These investigations will have to provide

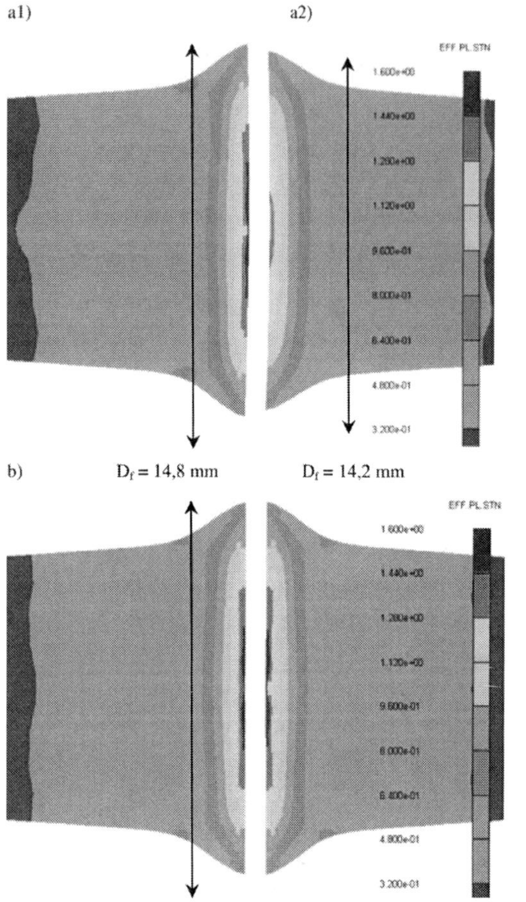

Figure 8. Final effective plastic strain maps with JCM models without a recrystallization criterion (a2) and with a recrystallization criterion (a1) and (b). The numerical diameters are indicated. The experimental diameter is 16,4 mm.

more precise data on the size of the recrystallized zone and on the mechanical responses of the recrystallized material.

4. Conclusion

Taylor tests have been found to be a mean to investigate the dynamic deformation of a pure nickel in the 10^3–$10^4 s^{-1}$ strain rate regime. Numerical experimental investigations taking in account the strengthening occurring in this regime have been a mean to reproduce the Taylor final diameter by 7% when considering an impact speed of 350 m/s. However at higher impact speeds, significant difference between numerical and experimental diameters remained, up to 13% for an impact speed of 453 m/s. Through post-mortem observations of the specimen impacted at 453 m/s, a recrystallization process has been found to occur. Such recrystallization process results in a softening of the pure nickel. Subsequent

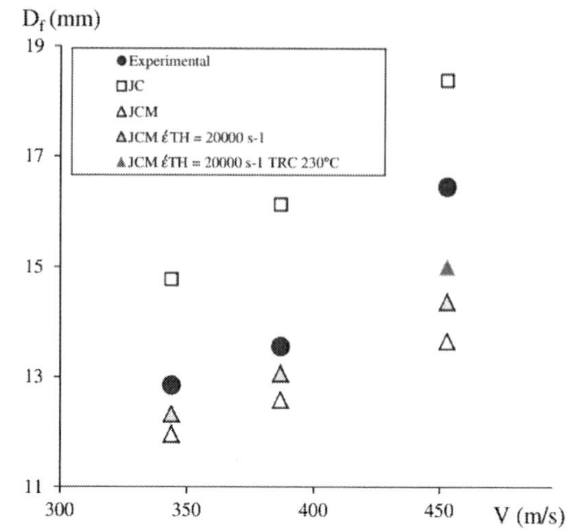

Figure 9. Experimental and numerical final diameter of the Taylor specimens versus impact speed incorporating the data generated with the recrystallization temperature criterion of 230°C.

numerical simulations taking in account this softening have been found to be a mean to reduce the difference between experimental and numerical final diameters from 13% to 10%.

The author thanks M. Jean-Charles Saint Supery from Nexter Systems for generating the hardness measurements.

References

[1] H. Couque, Phil. Trans. R. Soc. A **372**, 218 (2014).
[2] H. Couque, R. Boulanger, 23rd Int. Symp. Ballistics, Eds F. Galvez & V. Sanchez-Galvez, 255 (2007).
[3] G. Dirras, H. Couque, J. Gubicza, A. Ouarem, T. Chauveau, P. Jenei, Mat. Sc. & Eng. 1 **527**, 4128 (2010).
[4] G. Taylor, Proc. Roy. Soc. London A **194**, 289 (1948).
[5] S. Jones, J. Drinkard, W. Rule, L. Wilson, Int. J. Imp. Engng **21**, 1 (1998).
[6] D. Erlich, D. Shockey, L. Seaman, Conf. Shock Waves Condensed Matter, Eds Amer. Inst. Phys. **402** (1982).
[7] H. Couque, A. Lichtenberger, E. Lach, B. Salesse J. Phys; IV France **10**, 353 (2000).
[8] H. Couque J. Phys; IV France **10**, 179 (2000).
[9] D. Steinberg, S. Cochran S. G., M. Guinan, *J. Appl. Phys.*, **561**, 1498 (1980).
[10] D. Chapman, D. Radford, P. Church, Int. Jr. Fr. **134**, 41 (2005).
[11] A. Lichtenbeger, ISL Technical Report "Etude du cuivre et du nickel pour revêtements de charges creuses mise en forme par matriçage", S-R 902/90 (1990).

EPJ Web of Conferences **94**, 02025 (2015)
DOI: 10.1051/epjconf/20159402025

Dynamic mechanical characterization and modelling of polypropylene based organoclay nanocomposite

Kui Wang[1], Rodrigue Matadi Boumbimba[2], Nadia Bahlouli[3,a], Said Ahzi[1], and René Muller[4]

[1] Qatar Environment and Energy Research Institute, Qatar Foundation, PO Box 5825, Doha, Qatar
[2] National Engineering School of Metz, Laboratory of Mechanics, Biomechanics, Polymers and Structures, 1 route d'Ars Laquenexy, 57078 Metz, France
[3] ICUBE Laboratory-CNRS, University of Strasbourg, 2 rue Boussingault, 67000 Strasbourg, France
[4] Laboratoire d'Ingénierie des Polymères pour les Hautes Technologies, ECPM-LIPHT, Université de Strasbourg, 25 rue Becquerel, 67087 Strasbourg Cedex 2, France

Abstract. In order to investigate the dynamic behaviour of polypropylene based organoclay nanocomposite, the polypropylene matrix and a master batch of polypropylene modified anhydrid maleic were mixed by means of melt mixing technique. The experimental characterization was performed by using split Hopkinson pressure bars (SHPB), at different strain rates and temperatures. A significant increase of the yield stress of nanocomposite was shown with the present of organoclay, comparing to neat PP. A three-phase approach based on the micromechanical formulation of the cooperative model is proposed to model the yield behaviour of the polymer nanocomposite. Our proposed approach accounts for strain rate and temperature effects as well as the organoclay exfoliation effect. The predictions of models for the nanocomposite yield behaviour showed a good agreement with experimental data.

1. Introduction

Polymer based nanocomposite is one of the hottest worldwide research field from the last decades. This is mainly because the introduction of a low amount of nanoparticles can effectively improve the properties of polymer. Among different nanoparticles, montmorillonite clays are frequently used. Although few works were realized on polypropylene (PP) based organoclay nanocomposite, most of them were focused on the processing, thermal and physical characterization. There is no research on the dynamic mechanical behaviour of PP based organoclay nanocomposites. In this paper, the dynamic mechanical behaviour of PP based organoclay nanocomposites was characterized by means of split Hopkinson pressure bars (SHPB). A three-phase model was used to predict the yield behaviour of the nanocomposites.

2. Material and sample preparation

Polypropylene (PP Moplen HP500N) matrix was supplied by Basell. The modified montmorillonite masterbatch composed by 50 wt% of organoclay, 25 wt% of polypropylene and 25 wt% of polypropylene-graft-maleic anhydride (pp-g-MA) was also supplied by Basell. The neat PP reinforced with 0.5, 1, 3, 6 wt% of organoclay was prepared using a mixer (BUSS Knider) at 50 rpm and a mixing temperature of 200 °C. The samples were injection-moulded at 200 °C. In the following, the composites will be denoted by PP Nanocor X wt%, where X is the weight fraction of organoclay.

[a] Corresponding author: nadia.bahlouli@unistra.fr

The dispersion of the organoclay in PP matrix was characterized by X-ray diffraction (XRD) and transmission electron microscopy (TEM) in our previous studies [1]. The results show an exfoliated structure for PP Nanocor with 0.5 wt% organoclay and a partially exfoliated morphology for the organoclay content above 0.5 wt%.

3. Experimental investigations

The high rate tests of PP based organoclay nanocomposites were carried up at various strain rates from $360\,s^{-1}$ to $2400\,s^{-1}$ as following : Strain rate $1 = 360 \pm 6.7\%\,s^{-1}$, Strain rate $2 = 816 \pm 4.5\%\,s^{-1}$, Strain rate $3 = 1450 \pm 4.8\%\,s^{-1}$ and Strain rate $4 = 2400 \pm 4.2\%\,s^{-1}$. It should be noted that all the tests were achieved at variation temperatures from 20 °C to 80 °C. For neat PP and PP organoclay nanocomposites, the yield stresses were measured from the maximum stress of stress-strain curves.

4. Modelling

Gueguen et al. [2] developed an extended formulation based on Richeton model [3] to describe the yield behaviour of semi crystalline polymers. In this formulation, the semi crystalline polymer is composed by amorphous phase and crystalline phase:

$$\frac{\sigma_y}{T} = \frac{\sigma_i(0) - m \cdot T}{T} + \frac{2k}{V_M} \sinh^{-1}\left(\frac{\dot{\varepsilon}}{\dot{\varepsilon}_0 \exp\left(-\frac{\Delta H_M}{RT}\right)}\right)^{1/n} \quad (1)$$

$$\Delta H_M = \frac{\varphi \cdot \Delta H_c \cdot \Delta H_a}{\Omega \cdot \Delta H_a + (1-\Omega) \cdot \Delta H_c} + (1-\varphi) \cdot \Delta H_a$$
$$V_M = \frac{\varphi \cdot V_c \cdot V_a}{\Omega \cdot V_a + (1-\Omega) \cdot V_c} + (1-\varphi) \cdot V_a \tag{2}$$

where σ_y is the compressive yield stress, T refers to the absolute temperature, k is the Boltzmann's constant, $\dot{\varepsilon}$ is the test strain rate, $\dot{\varepsilon}_0$ is a constant pre-exponential factor. ΔH_M and V_M are the effective activation energy and effective activation volume, respectively. These two parameters were obtained from the activation parameters of amorphous phase ΔH_a, V_a, and of the crystalline phase, ΔH_c and V_c, following the Takayanagi micromechanical model (Eq. (2)) [4].

Considering that the microstructure of polypropylene based organoclay nanocomposite is composed by three phases: a homogenized semi-crystalline matrix, the nanoclay fillers and the interphase between the matrix and the fillers. The Pukanszky formulation, Eq. (3) is incorporated into Eq. (1) where the interfacial interaction can be quantitatively characterized [5] as described by Eq. (4).

$$\frac{\sigma_{y,c}}{\sigma_{y,M}} = \frac{1-\varphi_f}{1+2.5\,\varphi_f} Exp\,(B\varphi_f) \tag{3}$$

$$\frac{\sigma_{y,c}}{T} = \left[\frac{1-\varphi_f}{1+2.5\varphi_f}\exp(B\varphi_f)\right] \times \left[\frac{\sigma_i(0) - m \cdot T}{T} + \frac{2k}{V_M}\sinh^{-1}\left(\frac{\dot{\varepsilon}}{\dot{\varepsilon}_0 \exp\left(-\frac{\Delta H_M}{RT}\right)}\right)^{1/n}\right] \tag{4}$$

where $\sigma_{y,c}$, $\sigma_{y,M}$ are the yield stresses of the nanocomposite and the PP matrix, respectively, φ_f is the fillers' volume fraction and B is a parameter characterizing the interfacial interaction, including the interlayer thickness and interfacial strength. For poor interfacial bonding, the particles do not carry any load, so that B = 0. B increases for the low organoclay concentration, however B decreases for high organoclay clay concentration, showing that B can be related with the exfoliation degree. In fact the low organoclay filled polymer generally leads the exfoliation of the organoclay and a great interfacial surface adhesion between the clay and the polymer matrix.

5. Results and discussion

Firstly, PP matrix was considered as a two phase material, composed by an amorphous and a crystalline phase. The yield behaviour of neat PP is then described by Gueguen model [2]. For identification the parameters, the experimental data have to be superposed horizontally and vertically in an Eyring plot at a chosen reference temperature to build a mater curve. Once the master curve is built, the parameters of Gueguen model for neat PP can easily be chose. The parameters are listed in Table 1. The detail have been reported elsewhere [6]. In Fig. 1, the predicted results of yield behaviour of neat PP are compared to our experimental results. The cooperative model predictions of the yield stress are in good agreement with the experimental data.

Table 1. Gueguen models parameters for neat PP.

Parameters	Value
n	2.3
φ	0.81
$\dot{\varepsilon}_0$	1.55E+16
$\sigma_i(0)$ (MPa)	80
m	0.165
ρm (g/cm^3)	0.90
Tref (K)	313
Cristallinity degree (%)	35
ΔH_a (kJ/mol)	69.15
V_a	1.17E-28
ΔH_c (kJ/mol)	119.55
V_c	6.57E-28
V_M (m^3)	2.34E-28
ΔH_M (kJ/mol)	85.50

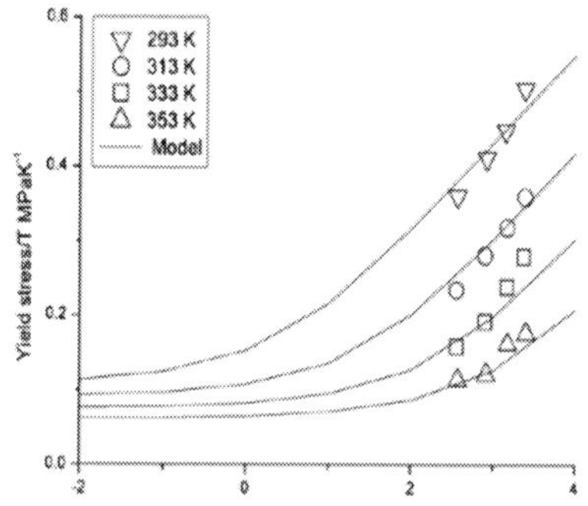

Figure 1. Yield stress/temperature versus strain rate of neat PP.

Knowing the parameters for the effective pure PP matrix estimated previously, the yield stress of nanocomposite can be calculated for a given organoclay content, according to Eq. (4), for different values of the parameter B.

According to Aït Hocine et al. [7], the parameter B could be determine by using the following approximation:

$$B = \frac{1}{\varphi}\ln\left(\frac{\sigma_{y,c}}{\sigma_{y,M}}\frac{1+2.5\varphi}{1-\varphi}\right). \tag{5}$$

Where $\sigma_{y,M}$ and $\sigma_{y,c}$ are the experimentally measured matrix and composite yield stresses, respectively. Using our experimental results for pure PP and PP/organoclay nanocomposites, we can therefore estimate B. Because the parameter B just depends on the nature of matrix and filler, and process condition. The test temperature and strain rate have no significant effect on the parameter B. For this raison we chose the average values of parameter B at each organoclay concentration. In order to use parameter B to estimate the extent of exfoliation of PP organoclay nanocomposite, according to the DRX results in our previous studies [1], the highest average B value 71.4 is chosen for fully exfoliation (PP Nanocor 0.5%),

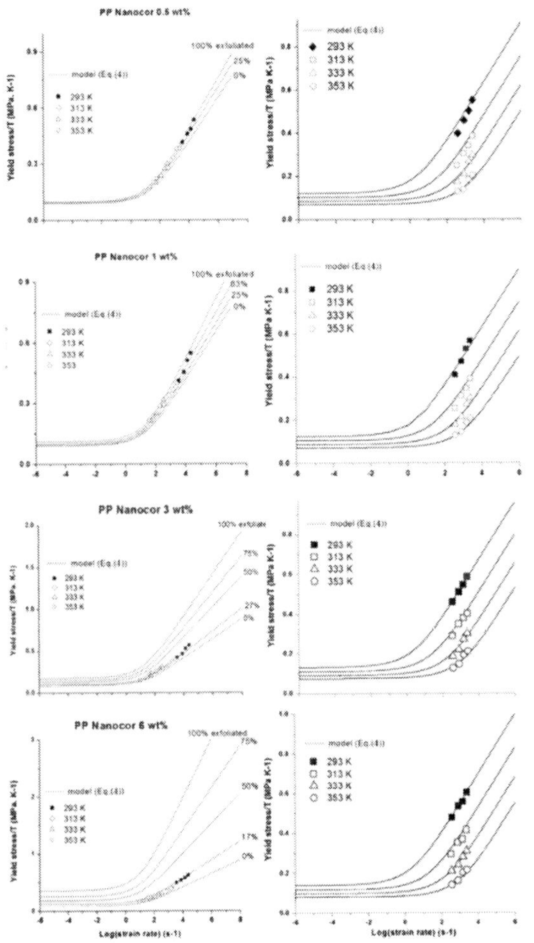

Figure 2. Master curve of PP organoclay nanocomposites build at a reference temperature of 313K (Left), Strain rate dependence for the compressive yield stress of PP organoclay nanocomposites models, predictions and experimental data (Right).

other lower average B values are chosen for partially exfoliation with the organoclay content above 0.5%. Then knowing the value of B, the yield stress of the organoclay nanocomposite can be predicted for a given organoclay concentration.

Figure 2 represents the master curves of PP organoclay nanocomposites built at a reference temperature of 313 K with our three phase model described by Eq. (4) and the comparison between the modelling prediction and experimental investigation of strain rate dependence for the compressive yield stress of PP organoclay nanocomposites. As shown in master curves, the experimental data lies between the two extremes of the extent of exfoliation for all the concentration. In Fig. 2, the three phase model described by Eq. (4), which takes in account the interphase contribution though the parameter B, predicts results that spans from low to high exfoliation degree. Thus, the three phase model yields good agreement with the experimental results for 100%, 63%, 27% and 17% exfoliation for 0.5, 1, 3 and 6 wt% nanocomposites, respectively. These correspond to B values of 71.4, 46.6, 20.4, and 13.1 respectively. In Fig. 2, it can also be noticed that the range of the yield stress between the minimum and the maximum degree of exfoliation increases with increasing organoclay concentration. This observation is caused by the fact that increasing both particle concentration and extent of exfoliation, increases the specific surface area and lead to enhancement of the yield stress. The strain rate dependence for the compressive yield stress of PP organoclay nanocomposites were plotted on the right side of Fig. 2. The three phase model provides an excellent agreement with experimental data for different strain rates and temperatures. This very good agreement obtained with the three phase model allows us to conclude that the yield behaviour of PP organoclay nanocomposite depends heavily on the yield stress of the virgin matrix PP and on the extent of exfoliation. The fact that the three phase model accounts for the degree of exfoliation making a more robust model for the yield behaviour of polymer nanocomposites.

6. Conclusion

Dynamic compressive yield behaviour of a melt mixing polypropylene based organoclay nanocomposites has been studied using split Hopkinson pressure bars (SHPB). Experimental results have shown that the yield stress increases with increasing strain rate, organoclay concentration and the extent of exfoliation. However, as expected, the yield stress decreases with increasing temperature. The modified three phase cooperative model was used to predict the yield stress of PP/organoclay nanocomposites. In this three phase model, a parameter B allows for the effect of the extent of exfoliation. It is shown that parameter B increases with the increase of extent of exfoliation and decreases with organoclay concentration. The three phase model gives a fairly good prediction of experimental data within the investigated range, which account for the degree of exfoliation. This is an important factor in the determination of the properties of polymer organoclay nanocomposites. Therefore, we believe that this modified three phase cooperative model is more appropriate for the modelling of the yield behaviour of polymer nanocomposites.

The authors would like to thank the Alsace Region for their financial support. The authors also acknowledge the Fond National de la Recherche (FNR) of Luxembourg for the Ph.D. financial support for K. Wang.

References

[1] K. Wang, R. M. Boumbimba, N. Bahlouli, S. Ahzi, R. Muller, and M. Bouquey, Journal of Engineering Materials and Technology **134** (1, 010905), 11 (2012).

[2] O. Gueguen, J. Richeton, S. Ahzi, and A. Makradi, Acta Materialia **56**, 1650 (2008).

[3] J. Richeton, S. Ahzi, L. Daridon, and Y. Remond, Polymer **46** (16), 6035 (2005).

[4] M. Takayanagi, Kyushu Univ **23**, 1 (1963).
[5] B. Pukánszky Jr, K. Bagdi, Z. Tóvölgyi, J. Varga, L. Botz, S. Hudak, T. Dóczi, and B. Pukánszky, European Polymer Journal **44** (8), 2431 (2008).
[6] R. M. Boumbimba, K. Wang, N. Bahlouli, S. Ahzi, Y. Rémond, and F. Addiego, Mechanics of Materials, submitted (2012).
[7] N. Aït Hocine, P. Médéric, and T. Aubry, Polymer Testing **27** (3), 330 (2008).

EPJ Web of Conferences **94**, 02026 (2015)
DOI: 10.1051/epjconf/20159402026

Dynamic behaviour and shock-induced martensite transformation in near-beta Ti-5553 alloy under high strain rate loading

Lin Wang[1,2,a], Yangwei Wang[1,2], Xin Xu[1], and Chengze Liu[1]

[1] School of Materials Science and Engineering, Beijing Institute of Technology, 100081 Beijing, China
[2] National Key Lab of Science and Technology on Materials under Shock and Impact, Beijing Institute of Technology, 100081 Beijing, China

Abstract. Ti-5553 alloy is a near-beta titanium alloy with high strength and high fracture toughness. In this paper, the dynamic behaviour and shock-induced martensite phase transformation of Ti-5553 alloy with alpha/beta phases were investigated. Split Hopkinson Pressure Bar was employed to investigate the dynamic properties. Microstructure evolutions were characterized by Scanning Electronic Microscopy and Transmission Electron Microscope. The experimental results have demonstrated that Ti-5553 alloy with alpha/beta phases exhibits various strain rate hardening effects, both failure through adiabatic shear band. Ti-5553 alloy with Widmannstatten microstructure exhibit more obvious strain rate hardening effect, lower critical strain rate for ASB nucleation, compared with the alloy with Bimodal microstructures. Under dynamic compression, shock-induced beta to alpha″ martensite transformation occurs.

1. Introduction

Ti-5553 (Ti-5Al-5Mo-5V-3Cr) alloy is a near-beta titanium alloy with wide applications in aerospace industry due to its high strength and high fracture toughness [1,2]. The alloy is metastable and near-beta type as its molybdenum equivalent is about 10.5%, which indicates beta matrix phase is unstable and easy to transform to various alpha phases under different heat treatments or loading conditions. Till now, the effects of heat treatment procedures on the static mechanical properties and fatigue properties of Ti-5553 alloy have been widely investigated [3–5]. The researches on the titanium alloy have shown that alpha/beta microstructures and their contents induce various mechanical properties, similar with other titanium alloys. Such phase transformations as $\beta \rightarrow \alpha$, $\beta \rightarrow \alpha'$ or $\beta \rightarrow \alpha''$ usually occur under the mentioned loading conditions [6–8].

The dynamic properties of Ti-5553 alloy and other near-beta alloys have also been studied in recent years because of their potential applications in impact loading conditions [9–12]. The researches have demonstrated that adiabatic shear band is easy to occur in Ti-5553 alloy, similar with other titanium alloys. Also, in other metastable near-beta Ti alloy, such as TB10, when in pure beta phase condition, the occurrence of shock-induced martensite (SIM) phase transformation of $\beta \rightarrow \alpha''$ prohibits the occurrence of ASB [9]. However, due to different compositions and microstructures of metastable near-beta titanium alloys, the relationships among dynamic properties, ASB and SIM transformation have not been well understood yet. Also, for Ti-5553 alloy, the research about SIM transformation has not been reported. These situations limit the further applications of Ti-5553 alloy and the like near-beta titanium alloy in high strain rate loading environment. Therefore, in this paper, the dynamic behaviours of Ti-5553 alloy with Bimodal and Widmannstatten microstructures were investigated. Adiabatic shear band and SIM transformation and their tentative correlation were discussed.

2. Experimental procedures

Ti-55531 (Ti-5Al-5Mo-5V-3Cr-1Fe) alloy is chosen as the experimental material. The transformation temperature of beta phase to alpha phase is 814 °C. The heat treatment is as follows: (1) 770 °C/3h/AC+ 600 °C/3h/AC, Bimodal structure obtained and labelled as TL. (2) 850 °C/3h/AC+ 600 °C/3h/AC, Widmannstatten structure obtained and labelled as TA.

The static mechanical properties of Ti-5553 alloy were measured with MTS. Split Hopkinson Pressure Bar (SHPB) was used to carry out dynamic compression and to obtain dynamic compression properties. The sample is with the dimension of ϕ4 mm × 4 mm. Soft recovered cylinders were prepared with standard metallurgical method for SEM observation, with etch solution of $HF:HNO_3:H_2O$=2:8:90, in volume ratio. For TEM samples, cylindrical samples were cut to 0.3 mm thickness, followed by mechanical thinning and Ion thinning to the thickness about 30 μm and diameter of 3 mm. The acceleration voltage of TEM microscopy is 200 KV.

3. Experimental results and discussions

3.1. Static mechanical properties of Ti-5553 alloy

Static mechanical properties of Ti-5553 alloy through different heat treatment procedures are shown in Table 1.

[a] Corresponding author: `linwang@bit.edu.cn`

Article available at http://www.epj-conferences.org or http://dx.doi.org/10.1051/epjconf/20159402026

Table 1. Static mechanical properties of Ti-5553 alloy with different microstructures.

Microstructure	Tensile strength (MPa)	Compressive strength (MPa)	Reduction of area section (%)	Elongation (%)
Bimodal structure	1200	1670	23	7.0
Widmannstatten structure	1265	1670	6	2.5

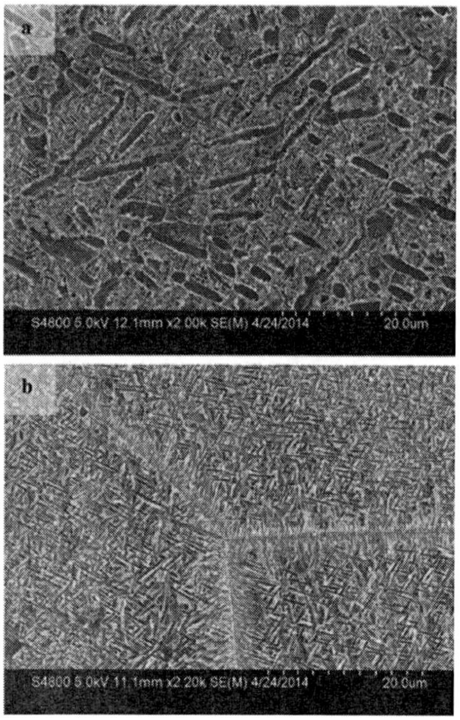

Figure 1. SEM observation of Ti-5553 alloy (a) TL Bimodal microstructure (b) TA Widmannstatten microstructure.

It can be seen that the alloy with Bimodal $\alpha + \beta$ phase exhibit relatively good combination of high tensile strength, compression strength and good ductility. The alloy with Widmannstatten microstructure has relatively higher tensile strength, same compression strength, very low ductility.

The microstructures of Ti-5553 alloy are shown in Fig. 1. Different solid solution temperatures induce various $\alpha + \beta$ phase microstructure contents. For TL Bimodal microstructure, it includes $\alpha_p + \alpha_s + \beta$, i.e. primary alpha, secondary alpha and beta phases, with primary alpha located discontinue in the beta matrix. For TA Widmannstatten microstructure, it has clear grain boundary and includes $\alpha_s + \beta$, i.e. secondary alpha and beta phases. There are more and finer needle second alpha phases and no primary alpha phase in TA microstructure. Therefore, with increasing solid solution temperature, matrix beta phase transforms more to the primary alpha and then to secondary alpha phase, the content of primary alpha phase decreases.

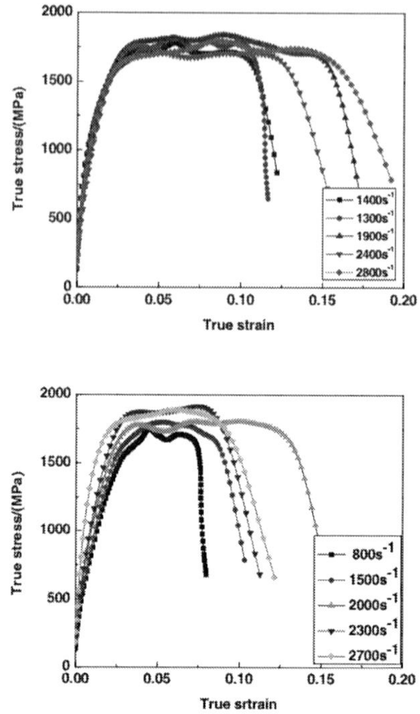

Figure 2. True strain-true stress curves of Ti-5553 alloy under different high strain rates (a) TL Bimodal microstructure (b) TA Widmannstatten microstructure.

3.2. Dynamic mechanical properties with SHPB

The true strain-true stress curves of Ti-5553 alloy are shown in Fig. 2. For the alloy with both microstructures, the stresses developing with strains exhibit similar trend. The stresses increase quickly with the increasing strain, with no obvious yield point, keep steady till stress collapse. Also, the strain rate hardening and strain hardening effects balance with thermal softening effect, therefore the plastic deformation stages keep steady, with the maximum compression strength is about 1800 MPa for TL and about 1900 MPa for TA. In the strain rates ranging from $800\,s^{-1}$ to $2700\,s^{-1}$, the stress increase more obviously for TA Widmannstatten microstructure alloy, compared with that TL Bimodal microstructure, which indicate the more obvious strain rate hardening effect for TA alloy. This is possibly because TA microstructure has similar high strength, but lower ductility, compared with TL alloy. Meanwhile the secondary alpha phase in TA Widmannstatten microstructure is more and finer than that in TL Bimodal microstructure.

3.3. Microstructure evolutions

3.3.1. Adiabatic shear band under SHPB compression

The microstructure evolutions are shown in Fig. 3 and Fig. 4. For TL Bimodal alloy, the critical strain rate for

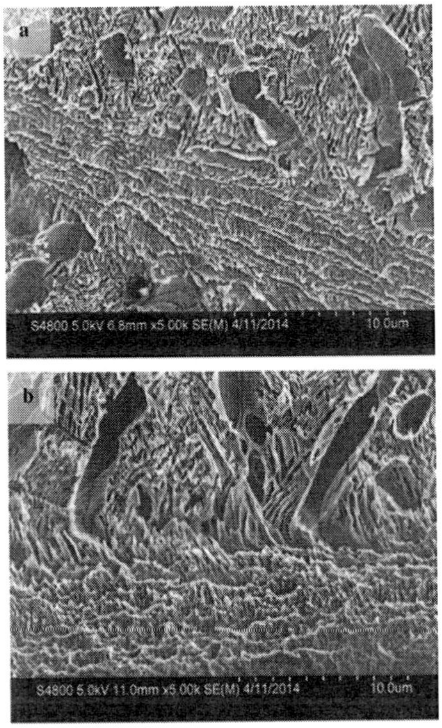

Figure 3. Adiabatic shear band of TL Bimodal alloy (a) ASB under strain rates of $1500\,s^{-1}$; (b) ASB under strain rates of $2600\,s^{-1}$.

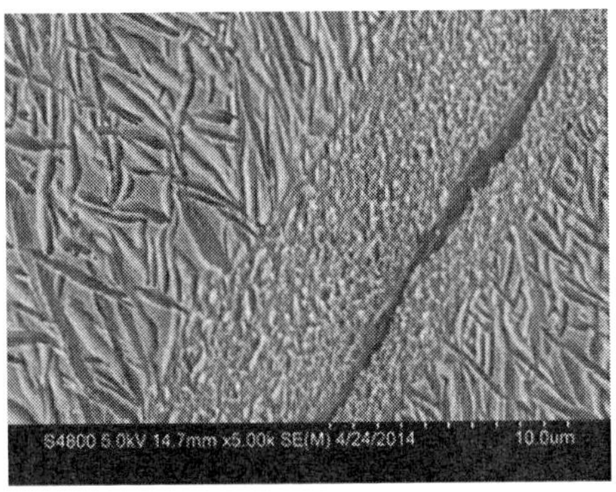

Figure 4. Adiabatic shear band of TA Widmannstatten microstructure alloy under strain rate of $1500\,s^{-1}$.

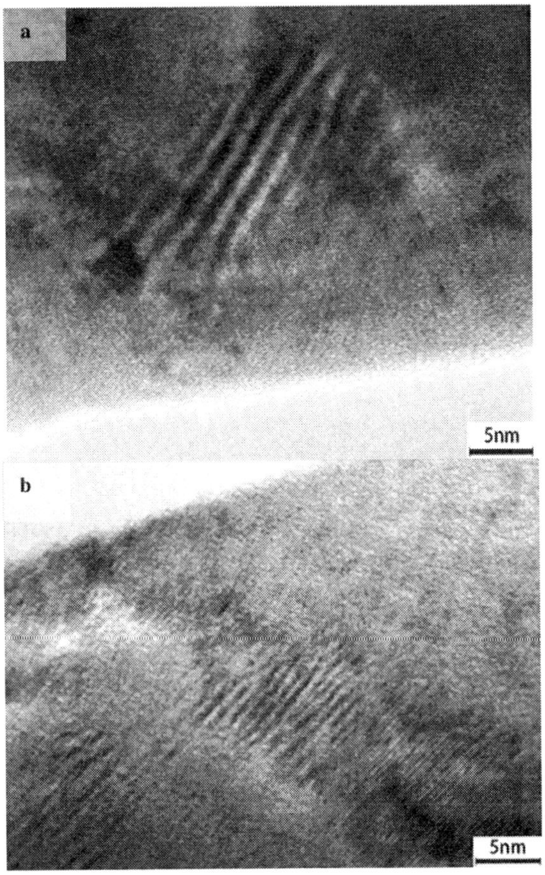

Figure 5. TEM observation of Ti-5553 alloy under strain rate of $1500\,s^{-1}$ (a) TL Bimodal microstructure; (b) TA Widmannstatten microstructure.

adiabatic shear band nucleation is about $1500\,s^{-1}$. TL cylinder fractures under strain rate of $2600\,s^{-1}$, along with transformed shear band which develop from deformed to transformed shear band with the increasing strain rate. For TA Widmannstatten alloy, the critical strain rate for ASB nucleation is $1000\,s^{-1}$, and the sample fracture occurs under $1500\,s^{-1}$, with wider transformed shear band occur, comparing with TL alloy. This indicates that for Ti-5553 alloy, even with both alpha and beta phases, TA alloy which is solid solution treated at temperature higher than phase transformation exhibits more sensitivity to adiabatic shear localization, lower critical strain rate for ASB nucleation.

The possible reason for the different critical strain rate for ASB nucleation is the content of primary and second alpha phases. With the increasing solid solution temperature from 770 °C for TL alloy to 850 °C for TA alloy, primary alpha phase decrease or even disappear, more secondary alpha phase produce. The needle-shape secondary alpha phase is relatively difficult to deform, compared with ellipse-shape primary alpha phase, which induce easily uneven plastic deformation or localized deformation under high strain rate loading. The uneven plastic deformation corresponds to the adiabatic shear banding nucleation and development.

3.3.2. Shock-induced phase transformation under SHPB compression

In SEM observation, only microstructure evolution of ASBs could be observed. However, although no obvious transformed phase could be observed by means of SEM, shock-induced martensite (SIM) transformation could be found in TEM, as shown in Fig. 5. The transformation product of plate-shape α'' phase exists [9], which is different from the secondary α phase observed in SEM. The plate width of α'' phase in TA Widmannstatten alloy is finer than that in TL Bimodal alloy. Also, the density number of dislocation lines served as the nuclei sites for α'' phase is much more in TA alloy than in TL alloy, which means more α'' SIM phase possible occur in TA alloy. These two microstructure characteristics contribute to the obvious strain rate hardening effect in TA Widmannstatten alloy, compared with TL Bimodal alloy.

The occurrence of SIM phase in near-beta titanium alloy may also influences the nucleation of ASB to some extents. However, in this paper, more SIM phase occur TA Widmannstatten alloy, which is also more sensitive to ASB nucleation. This result is opposite to SIM transformation happened in pure beta phase in TB10 alloy, where SIM phase transformation suppress the nucleation of ASB [9]. The possible reason is that in this study, the microstructure combine both α and β phases. Although β phase is meta-stable and easy to transform to α'' phase under shock loading, but the primary α phase restrain the SIM transformation. In TA alloy, as almost no primary α phase exist, thus has no restrain effect for occurrence of SIM phase transformation. Therefore, more SIM phase was observed in TA alloy. At the meantime, the alloy has lower critical strain rate for ASB nucleation. So, it is the alpha/beta phases microstructure that concerns the main difference from the pure beta phase in metastable titanium alloy.

4. Conclusions

In summary, we have found that Ti-5553 alloy with Widmannstatten microstructure exhibit more obvious strain rate hardening effect, lower critical strain rate for ASB nucleation, compared with the alloy with Bimodal microstructures. Under dynamic compression, more shock-induced martensite (SIM) beta to alpha'' transformation occurs in the alloy with Widmannstatten microstructure compared with the alloy with Bimodal microstructures, which is related with the original alpha/beta microstructures.

References

[1] R Panza-Giosa. Sc. thesis. McMaster University & University of Toronto (2009)
[2] B A Welk. Sc. Thesis, Ohio State University, (2010).
[3] J Huang, Z R Wang, K M Xue. Mater. Sci. Eng. A, **528** (29–30) (2011)
[4] N G Jones, et al. Acta Materialia, **57**(13) (2009)
[5] S K Kar, et al. Materials Characterization, **81** (2013)
[6] F Xue, et al. Chinese J Nonferrous metals, **20**(1) (2010)
[7] Y Ren, et al. Mater. Sci. Eng A, **562** (2013)
[8] E Aeby-Gautier, et al. J. Alloy and Compounds, **577** (2013)
[9] Z P You, et al. Chinese J. Rare Metal, **35**(2) 2011
[10] Duvall et al. Shock-induced phase transitions in solids: propagation of shock waves in solids, ASME, **17** (1976)
[11] Q Y Sun, et al. J. Mater. Sci., **37** (12) (2002)
[12] X Xu, L Wang, et al. SEM Fall Conference & International Symposium on Intensive Loading and Its Effects (2014)

EPJ Web of Conferences **94**, 03001 (2015)
DOI: 10.1051/epjconf/20159403001

Influence of strain rate on the mechanical behaviour in tension of bovine cortical bone

C. Latella[1], M. Dotta[2], D. Forni[2,3,a], N. Tesio[2], and E. Cadoni[2]

[1] University of Genova, Italy
[2] DynaMat Lab, University of Applied Sciences of Southern Switzerland, 6952 Canobbio, Switzerland
[3] Department of Structural, Geotechnical and Building Engineering, Politecnico di Torino, Torino, Italy

Abstract. The mechanical behaviour of bones when subjected to tension loading in a wide range of strain-rates is fundamental to develop protection systems. The paper presents the preliminary tests on the tensile behaviour of bovine cortical bone at medium and high strain rates. Two special apparatus, both installed at the DynaMat Laboratory of the University of Applied Sciences of Southern Switzerland, a Hydro-Pneumatic Machine and a Modified Hopkinson Bar respectively for medium and high strain-rate tests have been used. Flat shape specimens (having 10 mm of gauge length, 5 mm width and 3 mm thickness) have been obtained from 15 bovine femurs with the same age. The paper describes the preparation techniques of the samples and the experimental results obtained. The bovine cortical bone shown a quite important strain rate dependency.

1. Introduction

In human and other mammal bodies, bone is generally classified into two different types, trabecular and cortical bone, known also as cancellous or spongy bone and compact bone respectively.

On the basis of the porosity and the unit microstructure classification, the cortical bone, as is known, is much denser, harder and stiffer than the cancellous bone. This type of bone, that is the major component of the human body, makes up nearly 80% of skeletal mass and is of vital importance to weight bearing and as a protective structure because of its high bending and torsional resistance.

Even if the mechanical properties measurement is a very difficult task, their quantification can be useful to study and develop theories concerning fracture mechanism that can predict failures.

For these reasons this material has been the subject of mechanical investigation for many years. In numerous studies the rate dependency nature of its behaviour has been acknowledged.

For example Pitioux [1] compared the bone failure under two different loading rates, developing a statistical model. Ferreira [2], by using a traditional SHPB, investigated the compressive mechanical properties in static and dynamic conditions of both trabecular and cortical bones. Other authors, [3], conducted a preliminary and fracture toughness test on dry and wet bovine bones under quasi-static and dynamic loading conditions. A viscoelastic, viscoplastic model of the cortical bone valid in a wide range of strain rates was proposed by Johnson [4] by fitting different sets of stress-strain measurement from the available literature data. Hansen [5], investigated the tensile and compressive properties of human cortical bone at different strain rates.

Another consideration concerns the fracture behaviour. It is widely reported [6] that the bone is strain rate sensitive while increasing the loading rate. This behaviour is relevant to cortical bone response in both tension [7] and compression [8–10].

The purpose of this paper is to present the preliminary tests on the tensile behaviour of bovine cortical bone in a wide range of strain rates ($10^{-3} \div 100\,s^{-1}$). This could be helpful for the assessment of finite element analysis in order to improve the development of advantageous body protection systems.

2. Experimental program

Tensile tests at three different strain rates [10^{-3}, 3, $100\,s^{-1}$] were performed at room temperature on 15 flat dog-bone samples (Fig. 1) obtained from bovine cortical bone femurs. All the specimens are stored in a frozen conditions, except during the preparation and the test. In order to reduce conditioning problems connected to the post-mortem ageing (PMA) all dynamic tests were performed during the same day. Only static tests were performed few days later.

2.1. Preparation techniques

The cortical femur bones were acquired in the Genova abattoir from adult bovines of nearly the same age. The specimens were extracted so that their axis was longitudinal-oriented with respect to the whole femur bone. The articulate ends (*epiphysis*) were removed in order to obtain only the central part (*diaphysis*). The derived pieces were split in a half for a more useful sample realisation and secondly, these half-pieces were

[a] Corresponding author: daniele.forni@supsi.ch

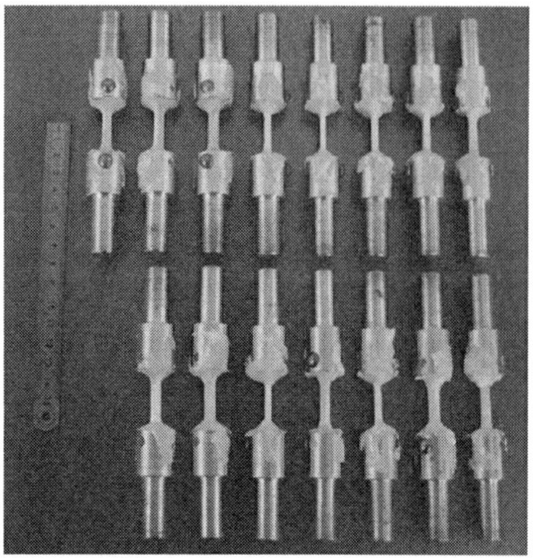

Figure 1. Samples ready for tensile tests.

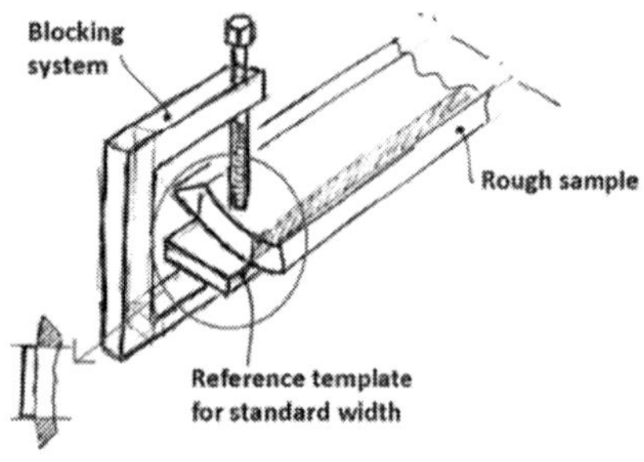

Figure 2. Sample's preparation sketch.

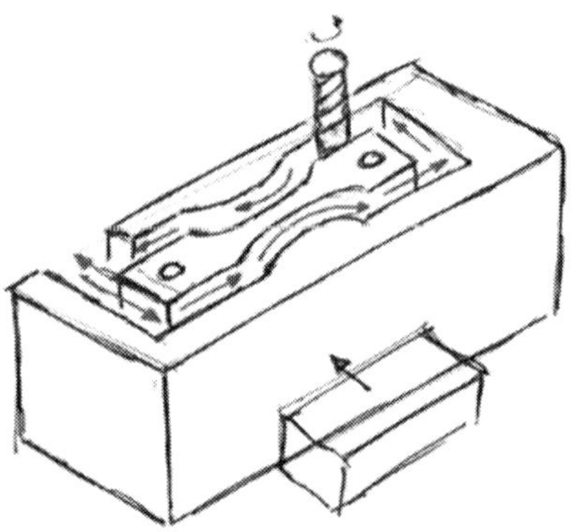

Figure 3. Sample's preparation sketch.

machined (Fig. 2) with the purpose of obtaining equal thin specimens. Then, by means of a numerical controlled machine the requested sample geometry was obtained (Fig. 3).

Lastly, with a sterile radiopaque acrylic cement for orthopaedic surgery (*Biocemium I*) the samples were glued and then immediately joined to the supporting holders, previously designed for the connection with the testing machines.

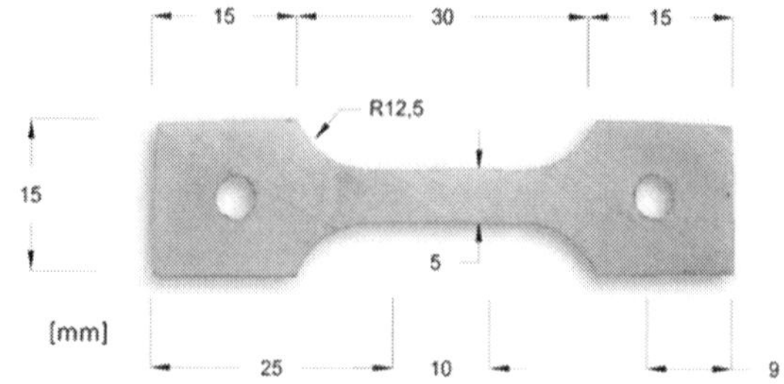

Figure 4. Sample's geometry.

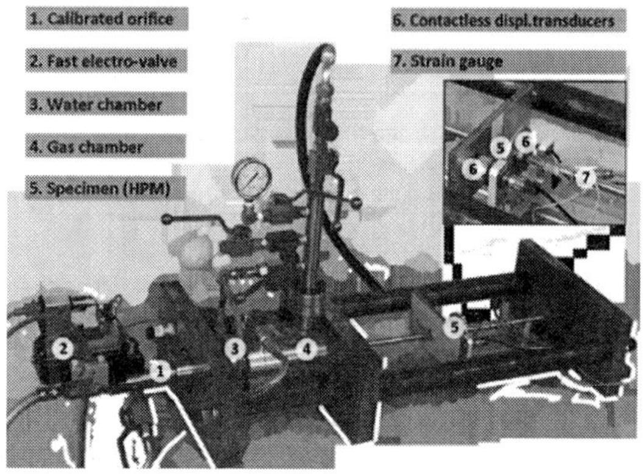

Figure 5. Hydro-pneumatic machine used in this research.

2.2. Experimental set-ups

Different experimental set-ups for tensile testing at several strain-rates have been used. All tests were performed at room temperature, and the target strain-rates were set at the following three levels: 10^{-3}, 3 and $100\,s^{-1}$.

2.2.1. *Quasi-static tests*

Quasi-static tests have been performed by means of a universal electromechanical testing machine *Schenck FC-400* installed at the University of Genova.

2.2.2. *Intermediate strain rate tests*

Intermediate strain-rate behaviour in tension has been investigated by means of a hydro-pneumatic machine (HPM), whose scheme is reported in Fig. 5. The functioning of the machine is well described here [11, 12] and is briefly reported as follow.

Firstly, the water is filled in the upper tank-chamber (3) and the gas is filled in the lower tank-chamber (4). At this moment, equal pressure is established in both chambers so that the forces acting on the two piston faces are in equilibrium. Secondly, the specimen (5) is fixed to both the piston shaft and the elastic bar instrumented with a strain gauge (7). Lastly, by activating a fast electro-valve (2) closing the water chamber, the force exerted by the gas pressure on one face of the piston prevails, accelerating the piston, which simultaneously loads the specimen and pushes water to flow out through the calibrated orifice (1) at a constant speed, with the result of imposing a strain to the specimen with a constant strain-rate. The stress

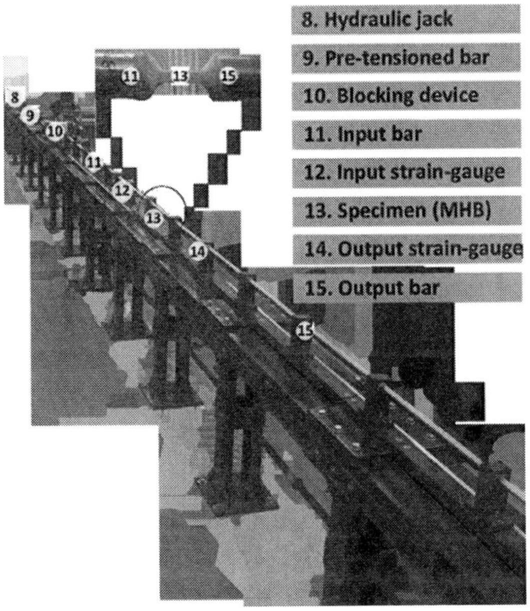

Figure 6. Split Hopkinson Tensile bar used in this research.

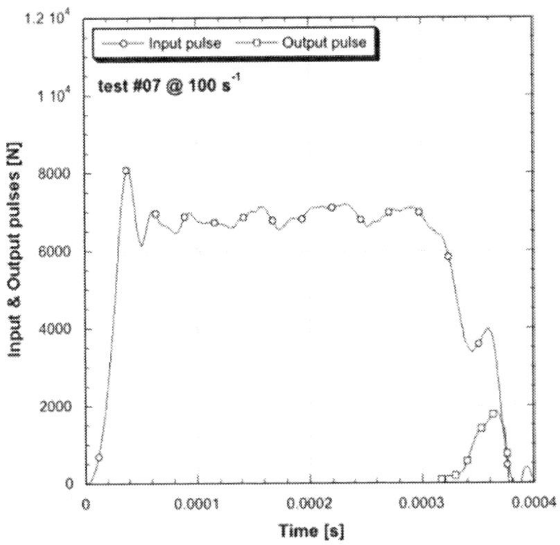

Figure 7. Input and output pulses during an high strain rate test.

measurements were obtained by the strain gauge of the elastic bar connected to the specimen while the strain measurements were obtained by means of two contact-less displacement transducers (6) placed at the ends of the specimen.

2.2.3. *High strain rate tests*

The high strain-rate tests in tension were performed using a Split Hopkinson Tensile Bar (SHTB) [13], that consists of two circular aluminium bars, called input and output bars and having a length of 3 m and 6 m respectively, with a diameter of 20 mm. The specimen is screwed between these bars (Fig. 6). The input bar is connected with an high strength steel pre-tensioned bar, 6 m in length, used as pulse generator with a diameter of 12 mm in order to obtain the same acoustical impedance of the input bar so that the pulse reflection, due to the interface, is avoided.

With reference of Fig. 6, a test with the SHTB is performed as follows: firstly, the hydraulic actuator (8) of maximum loading capacity of 600 kN is pulling the pretension high strength steel bar (9). The pretension stored in this bar is resisted by the blocking device (10). Secondly, the rupture of a fragile bolt in the blocking device, gives rise to a tensile mechanical pulse of 2,4 ms with linear loading rate during the rise time, propagating along the input (11) and the output (15) bars bringing to fracture the specimen (13).

As well as in the classical Hopkinson bar apparatus, the pulse propagates along the input bar with the elastic wave velocity equal to C_0 and constant shape. When the incident pulse (ϵ_I) reaches the bone sample, it is partially reflected by it (ϵ_R). On the other hand, the portion of the pulse that is not reflected, passes through the specimen (ϵ_T) and propagates into the output bar.

The clean resolution (Fig. 7) of the incident, reflected and transmitted pulses, the sharp rise time of the incident pulse, as well as the almost constant amplitude of the incident pulse have been proved in past researches [11–13].

Thereafter, the stress, the strain as well as the strain rate in the sample can be derived from the following equations:

$$\sigma(t) = E_0 \cdot \frac{A_0}{A_s} \cdot \epsilon_T(t) \tag{1}$$

$$\epsilon(t) = -\frac{2 \cdot C_0}{L_s} \int_0^t \epsilon_R(t)\, dt \tag{2}$$

$$\dot{\epsilon}(t) = -\frac{2 \cdot C_0}{L_s} \epsilon_R(t). \tag{3}$$

Where, L_S is the specimen length, A_S is the specimen cross-section area, A_0 is the output bar cross-section area, E_0 is the bar elastic modulus and t is the time.

Using the above mentioned equations, the stress and the strain-rate versus time are plotted in Fig. 8. Due to the brittle behaviour of the bone, the conventional strain rate has been adopted as the value in correspondence of the maximum tensile stress.

3. Results and discussion

Tensile tests on flat dog-bone samples obtained from bovine cortical bone femurs were performed at room temperature and at three different strain rates: 10^{-3}, 3, $100\,s^{-1}$.

Literature data report that usually high loading rates lead to an increase of yield and pre-yield properties as well as the elastic modulus [5]. On the other hand, post-yield properties are less consistent, and both increases and decreases in strength are reported [1].

Our experimental data are depicted in Fig. 9 and Fig. 10 for tests performed at medium and at high strain rate respectively. With the purpose of pointing out

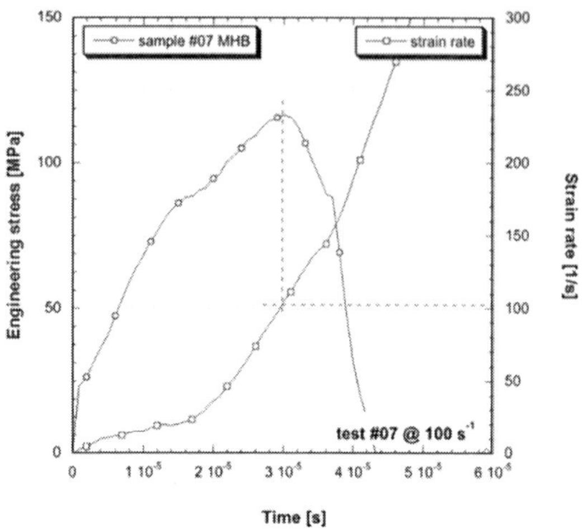

Figure 8. Stress and strain-rate versus time.

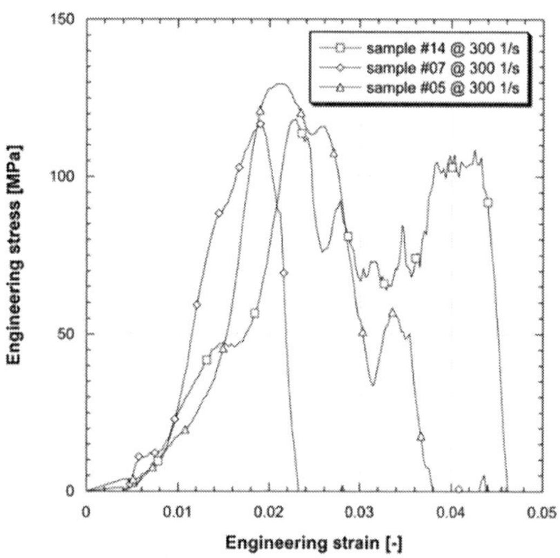

Figure 9. Comparison plot of high strain rate tests.

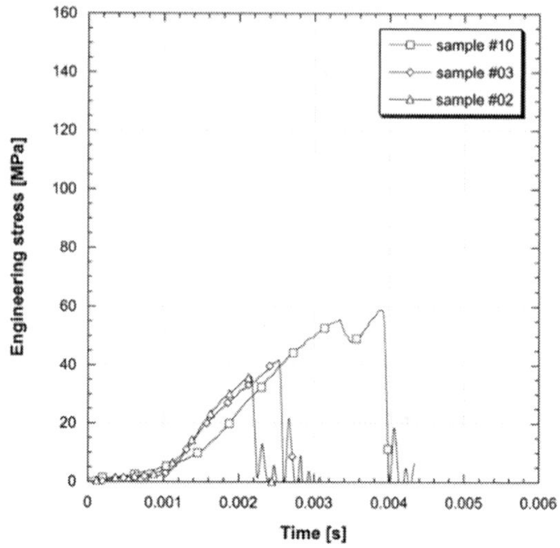

Figure 10. Comparison plot of intermediate strain rate tests.

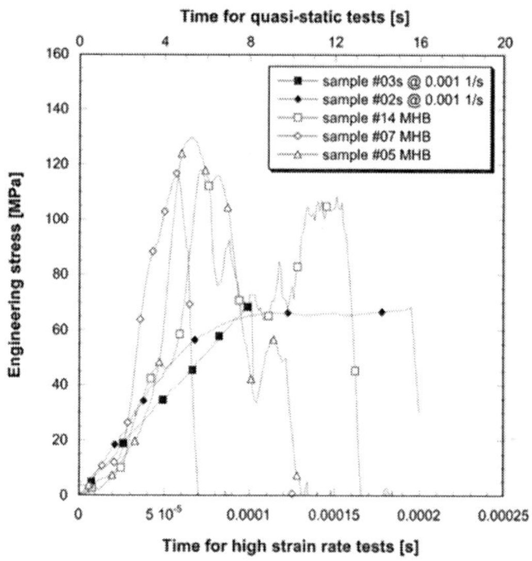

Figure 11. Different behaviour of quasi-static and high strain rate tests.

the strain rate sensitivity of the bovine cortical bone, a comparison between static and dynamic experimental data has been performed (Fig. 11).

In static condition the maximum engineering stress is 69.33 ± 1.46 MPa, while at high strain rate this value become 121.50 ± 7.04 MPa and at intermediate strain rate is 46.65 ± 11.81 MPa. The elastic modulus obtained in quasi-static condition was about 8.8 GPa, while at high strain rate was about 14.9 GPa. It has not been possible to estimate the elastic modulus at medium strain rate because of problems connected to the displacement measurement during these tests.

With the exception of intermediate strain rate data, whose results show a marked sensitivity to strain rate. Tests performed at $100\,s^{-1}$ highlight an increase of elastic modulus, maximum stress, as well as the corresponding deformation.

As far as we know, not many researches are focused on the direct tensile behaviour of bone at high strain rate. Some authors [1] performed direct tensile tests at different strain rates, finding an increase of elastic modulus $11.3 \div 17.5$ GPa to $10 \div 21.7$ GPa and a decrease of both maximum stress $105 \div 193$ MPa to $33 \div 50$ MPa and strain at failure $0.93 \div 1.8\%$ to $0.23 \div 0.44\%$. These results are only partially in accordance with our high strain rate tests, but seems to be in conformity with the intermediate strain rate results.

Wood [14], employing an hydraulic testing machine performed tensile tests at different strain rates ($0.005 \div 150\,s^{-1}$) on more than 120 specimens obtained from human cranial bone. A regression line for elastic modulus data is given in function of the strain rate. In this research is reported that statistical analysis of the data reveals no important differences in the modulus of elasticity, the breaking stress, the breaking strain, or the

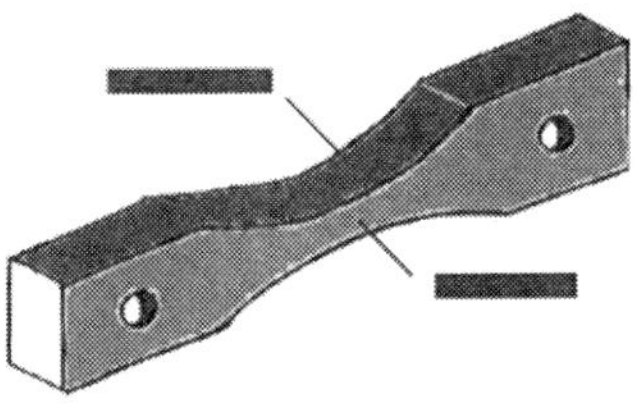

Figure 12. Sample's geometry.

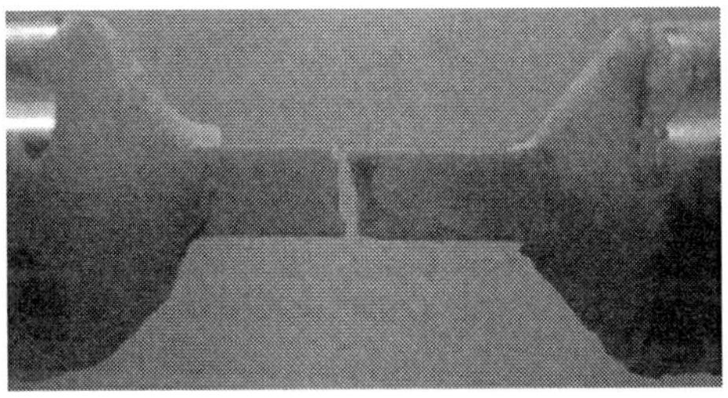

Figure 13. Sample's failure after an high strain rate test.

Figure 14. Dark streak lines.

energy absorbed to failure when compared according to type of bone, side of body or age of the individual. For that reason this relationship has been applied to our data. For strain rates ranging from $10^{-3}\,s^{-1}$ to $100\,s^{-1}$ gives 10.2 GPa and 19.8 GPa respectively, in good agreement with our results.

Crowninshield [7] conducted a study on bovine cortical bone in tension ($10^{-3} \div 250\,s^{-1}$) finding an increase of elastic modulus with strain rate from 10 to 12.5 GPa, that slightly underestimate our results.

With the purpose of both a better interpretation of our results and an evaluation of possible influence of samples preparation, a post-mortem analysis with an optical microscope has been performed. This analysis highlighted that the front side of the sample presents more flaws with respect to the lateral side. This could be due to the preparation process. Even though these imperfections, the most interesting aspect has been observed in the lateral side, where the direction of the fracture seems to be correlated with dark streak lines (Fig. 14). The nature of these bands should be connected with the damage morphology development during the bone fracture. As a confirmation of our hypothesis, this diffuse damage was pointed out and discussed in previous researches [15, 16].

4. Conclusions

Although intermediate strain rate data shows a negative sensitivity to strain rate, tests at $100\,s^{-1}$ highlight an increase of elastic modulus, maximum stress as well as the corresponding deformation. With reference of comparable researches, it has been possible to point out that our results are in a satisfactory agreement with previously published data.

With the purpose of obtaining samples with a strict geometry useful for the mechanical characterisation, the most difficult task of this research has been the preparation of the samples. For that reason only few samples were satisfactory well prepared and ready for mechanical tests. Even if no more than 15 samples were prepared (Fig. 1), only 3 tests at each strain rate have been considered in the analysis. Future development will be focused on the improvement of samples preparation process, as well as a in-depth understanding with regards to the damage morphology development during the bone fracture.

References

[1] M. Pithioux, D. Subit and P. Chabrand, Med. Eng. Phys., **26**, 647–567 (2004)

[2] F. Ferreira, M.A. Vaz and J.A. Simoes, Mater. Charact., **57**, 71–79 (2006)

[3] R.R. Adharapurapu, F. Jiang and K.S. Vecchio, Mat. Sci. Eng. C, **26**, 1325–1332 (2006)

[4] T.P.M. Johnson, S. Socrate and M.C. Boyce, Acta Biomater., **6** (2010)

[5] U. Hansen, P. Zioupos, R. Simpson, J.D. Currey and D. Hynd, J. Biomech. Eng. **130** (2008)

[6] A. van der Westhuizen, Ph.D. Thesis (2008)

[7] R.D. Crowninshield and M.H. Pope, Ann. Biomed. Eng., **2**, 217–225 (1974)

[8] J.H. McElhaney, J. Appl. Physiol, **21**, 1231–1236 (1966)

[9] A. Bekker, T.J. Cloete, A. Chinsamy-Turan, G.N. Nurick and S. Kok, Mat. Sci. Eng. C, **46**, 443–449 (2015)

[10] M. Prot, D. Saletti, S. Pattofatto, V. Bousson and S. Laporte, J. Biomech., **48**, 498–503 (2015)

[11] E. Cadoni, M. Dotta, D. Forni, and P. SpŁtig, J. Nucl. Mater., **414**, 360–366 (2011)

[12] E. Cadoni, M. Dotta, D. Forni, N. Tesio and C. Albertini, Mater. Design, **49**, 657–666 (2013)

[13] E. Cadoni, L. Fenu and D. Forni, Constr. Build. Mat., **35**, 399–407 (2012)

[14] J.L. Wood, J. Biomech., **4**, 1–12 (1971)

[15] T. Diab and D. Vashishth, Bone, **37**, 69–102 (2005)

[16] D.B. Burr, C.H. Turner, P. Naick, M.R. Forwood, W. Ambrosius, M.S. Hasan and R. Pidaparti, J. Biomech., **31**, 337–345 (1998)

EPJ Web of Conferences **94**, 03002 (2015)
DOI: 10.1051/epjconf/20159403002

Evaluation of the performance of three elastomers for non-lethal projectile applications

N. Thota[1,2,a], J. Epaarachchi[2], and K.T. Lau[2]

[1] CAD-CAE Consultants, 45 Horley Tce, Kilburn, South Australia, 5084, Australia
[2] Centre of Excellence in Engineered Fibre Composites, University of Southern Queensland, Toowoomba, 4350, Queensland, Australia

Abstract. Less lethal kinetic ammunitions with soft noses such as eXact iMpact 1006, National Sports Spartan and B&T have been commonly used by military and law enforcement officers in the situations where lethal force is not warranted. In order to explore new materials to be used as nose in such ammunitions, a scholastic study using finite element simulations has been carried out to evaluate the effectiveness of two rubber like elastomers and a polyolefinic foam (low density, highly compressible, stiff and closed cell type of thermos plastic elastomer). State-of-the art thorax surrogate MTHOTA has been employed for the evaluation of blunt thoracic trauma. Force-rigid wall method was employed for the evaluation of head damage curves for each material. XM 1006 has been used as the benchmark projectile for the purpose of comparison. Both blunt thoracic trauma and head damage criterion point of view, both rubbers (R1 and R2) have yielded high values of VC_{max} and peak impact force. Polyolefinic foam (F1) considered in the study has yielded very promising VC_{max} values and very less peak impact force when compared with those of bench mark projectile XM 1006.

1. Introduction

In the past few decades, law enforcement agencies, military and defense forces have started using less-lethal weapons which were designed to temporarily incapacitate the subject in the situations where lethal force is not warranted. The very basic requirement of non-lethal weapons are that the projectile impact should give short duration pain, be sufficient to deter the subject and should not cause any serious injuries, which require hospitalization and medical treatment [1–3].

Non-lethal munitions, though made up of foam and PVC, are very expensive (cost of each round of XM 1006 is approximately 30 – 55 USD depending upon the order quantity [4,5], and the cost of TASER-XREP round is more than 99 USD [6]). Therefore, a scholastic study was carried out to produce cheaper alternatives to XM 1006 which can be fired using the same M79 grenade launcher.

The projectile of interest XM 1006 (benchmark projectile in the present study), in the laboratory tests using 3-RCS, has produced a VC_{max} of 0.21 at an impact velocity of 91.5 m/s. Though the design (dimensions) of the projectile is available in the technical documents published by Defense Tech, as there is no mention of the foam nose material data, the development of an XM 1006 equivalent foam nosed projectile becomes an arduous task and entails extensive physical tryouts with grenade launchers and thorax surrogates (Cadavers, surrogates, anesthetized animals) and experimental setup with costly equipment. Therefore, in the present study, authors have used a state of the art thorax surrogate MTHOTA [7,8] and experimentally obtained and validated material data of few elastomers (from the own database [9] and published articles by other researchers [10]) and validated 3 materials (Rubbers R1, R2 and Polyolefin F1) in a virtual testing environment whether or not they suitable as nose material in manufacturing less lethal ammunition.

Though XM 1006 is widely used and aerodynamically stable (spin stabilized), impacts on the head (due to dispersion effects) can be lethal as the peak impact force on the head at the speed of 60 m/s is > 20 kN, which indicates very high probability for skull fracture and coma [11]. Therefore, in the present study, authors have employed force wall method and evaluated head damage characteristics of the projectile with R1, R2 and F1 noses.

In this paper, authors have presented blunt thoracic trauma and head damage characteristics of the projectile with R1, R2 and F1 noses and compared with those of the benchmark projectile.

2. Methodology

Non-linear finite element analyses were carried out for two purposes. Firstly, to evaluate the blunt thoracic trauma in terms of the peak viscous criterion and secondly, to evaluate head damage criterion in terms of peak impact force caused by the impacts of the projectiles with R1, R2 and F1 noses so that they can be compared with those of the benchmark projectiles XM 1006.

2.1. Evaluation of the blunt thoracic trauma

In order to evaluate the suitability of the materials under consideration as energy absorbing mechanisms

[a] Corresponding author: tnmurthy@cae-consultants.com.au

Article available at http://www.epj-conferences.org or http://dx.doi.org/10.1051/epjconf/20159403002

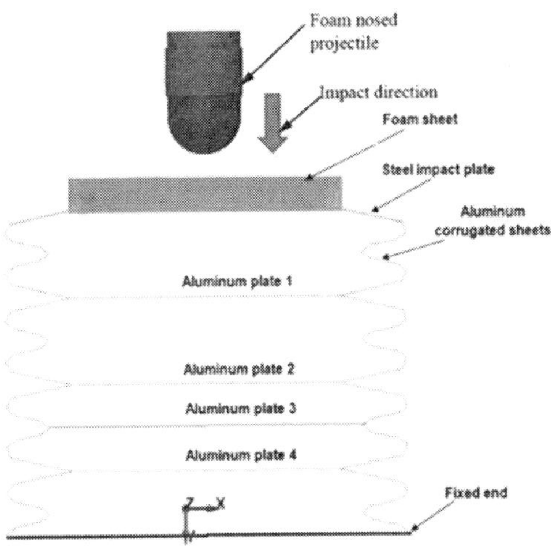

Figure 1. Details of MTHOTA thorax surrogate.

(compressible nose) in manufacture of non-lethal impact munitions, Mechanical THOrax for Trauma Assessment (MTHOTA), a state-of-the art surrogate of the thorax was subjected to the following cases impacts with projectiles whose design is similar to that of XM1006, in a virtual testing environment. Design details of the XM1006 were obtained from the published research papers and technical reports [12,13].

1. An FEA non-linear simulation run with the projectile (nose made up of Rubber-1 material) impacting MTHOTA at 99 m/s speed.
2. Repeat the above simulation run with Rubber-2 and Poly-olefin Foam as nose materials.
3. For better comparison, all of the above three cases were repeated with equivalent impact velocities. Equivalent velocities were evaluated such a way that initial kinetic energy of the projectiles were equal.

Construction details of the MTHOTA surrogate are illustrated in the Fig. 1.

MTHOTA surrogate is validated with human biomechanical response corridors developed by researchers at Wayne State University [14,15] such a way that the peak viscous response (VC_{max}) is given by the Eq. (1).

$$VC_{max} = S(Y_{max}/D)(Y_{max}/T). \tag{1}$$

Where,

VC_{max} = Peak viscous injury in m/s

S = Scaling factor = 0.366 for MTHOTA

Ymax = maximum deflection of the impact plate in mm

D = Dummy constant = 110 mm for MTHOTA

T = time at which deflection is maximum in ms.

Complete details of MTHOTA, its development, validation and its applications can be found in the research articles published by the present authors [7,8,16–18].

Table 1. Human head tolerance limits when subjected to blunt ballistic impacts [11].

Type of head damage	Peak impact force (kN)
Insignificant	≤ 2.5
Unconsciousness	$2.5 \leq$ peak impact force ≤ 5.0
Meningeal	$5.0 \leq$ peak impact force ≤ 7.5
Bone fracture and coma	≥ 7.5

2.2. Evaluation of head damage criterion by force wall method

Due to dispersion effects, though most of the projectiles' designs are aerodynamically stable, there is a possibility of "subjects of interest" getting hit on the head. Such incidents can be fatal, if intracranial pressure caused by the impact is ≥150 kPa which corresponds to ≥7.5 kN peak impact force on the head. Table 1 delieneates relation between Peak impact force and the type of head damage, which was experimentally obtained by researchers of Direction Generale de l'Aremement (DGA) of the French Ministry of Defense.

The researchers at DGA have developed head damage curves for non-lethal ammunitions using both head and animal head biological models. The data was very well correlated with that obtained from the force wall method (impacting a perfectly rigid wall with the projectile). Oukura et al. [11] expressed the relation between peak impact force and impact velocity as given in the Eq. (1).

$$(F)_{Head} = K * (V/10)^N. \tag{2}$$

Where,

K = experimetnally obtained coefficient

N = experimentallyobtained coefficient

$(F)_{Head}$ = Peak impact force on the head in kN

V = impact velocity of the projectile in m/s.

The following non-linear transient dynamic simulations were carried out for the evaluation of head damage criterion by force wall method.

1. Projectile with R1 nose impacting a rigid fixed plate at 20–100 m/s, with an increment of 10 m/s.
2. The above analyses were repeated with the projectile with R2 and F1 as nose materilas.

2.3. Material data and material models used in the scholastic study

Material properties and material models used for the FEA modelling of Rubbers (R1 and R2) and Polyolefin (F1) are given in the Table 2 and Table 3 respectively.

Engineering stress versus Engineering strain curve (as shown in the Fig. 2) obtained from the quasi static compression tests used in the modelling of MAT_057. In case of MAT_057 of LS-Dyna, a simple one term relaxation function (optional relaxation young's modulus) defines the rate effects of the material. Rate effects were ignored in the present study. In order to make the material model robust, some variables (HY, BETA, DAMP, etc., given in the Table 3) were obtaned by correlating the CAE simulations with rigourous experiments [10,19].

Table 2. Material properties and material data of Rubber-1 and Rubber-2.

Material parameter	Details	
Material model	MAT_MOONEY_RIVLIN_RUBBER (MAT_027 of LS-DYNA 9.71 software)	
Strain energy density function	$W = A(I-3) + B(II-3) + C(III^{-2}-1) + D(III-1)^2$ I, II and III are invariants of right Cauchy-Green Tensor $C = 0.5\,A + B$ $D = [A(5\,\mu-2) + B(11\,\mu-5)]/(2-4\,\mu)$	
Shear modulus of linear elasticity	$2\,(A+B)$	
μ (Poisson's ratio)	0.495	
	Rubber-1	Rubber-2
ρ (mass density)	1.07×10^{-6} kg/mm^3	0.97×10^{-6}
A	0.0001243	0.000334
B	6.14×10^{-4}	0.00167

Table 3. Material data and material model of the polyolefin (Foam-1) used in the study.

Polyolefin	Foam-1
Mass Density (kg/mm^3)	3.7×10^{-8}
Modulus of Elasticity (GPa)	0.0105
Shape factor for unloading (SHAPE)	15
Hysteretic unloading factor (HU)	1.0
Decay constant (BETA)	0.101
Viscous coefficient (DAMP)	0.225
Material model (MAT_057)	MAT_LOW_DENSITY_FOAM

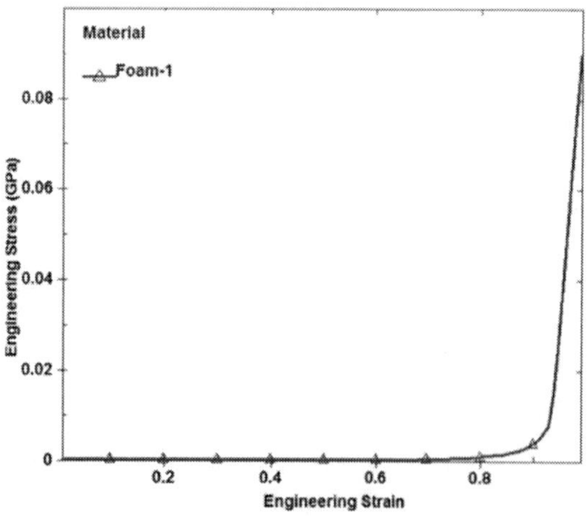

Figure 2. Stress-Strain curve for the Polyolefin Foam-1 material (obtained from the quasi-static compression test at 500 mm/min strain rate.

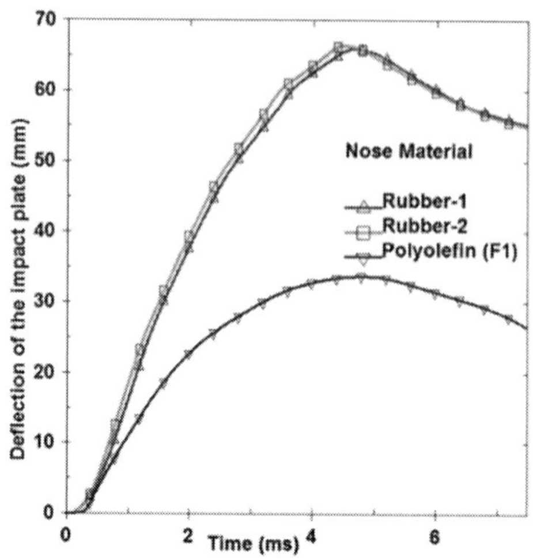

Figure 3. Dynamic deflection response of the MTHOTA when impacted at 99 m/s impact speed.

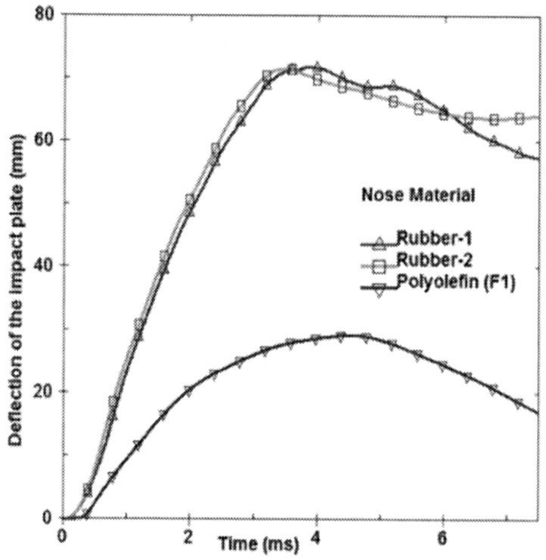

Figure 4. Dynamic deflection response of the MTHOTA when impacted with projectile with 147.03 J of initial Kinetic Energy.

3. Results and discussion

3.1. Evaluation of the blunt thoracic trauma

Blunt thoracic trauma in terms of VC_{max} was evaluated for all impact cases (3 cases of projectile nose material and 2 cases of impact velocity) using the displacement time histories of the impact plate of the MTHOTA surrogate. Deflection-time responses of impact cases for are illustrated in the Fig. 3 and Fig. 4 Stages of MTHOTA and the projectile with polyolefin foam nose during the impact are shown in the Fig. 5.

Using the equation-1 and maximum deflection of the impact plate and impact duration at which deflection is maximum, VC_{max} values were elicited for all projectile cases for impact speeds of 99 m/s and equivalent speed and provided in the Table 4 and Table 5 respectively.

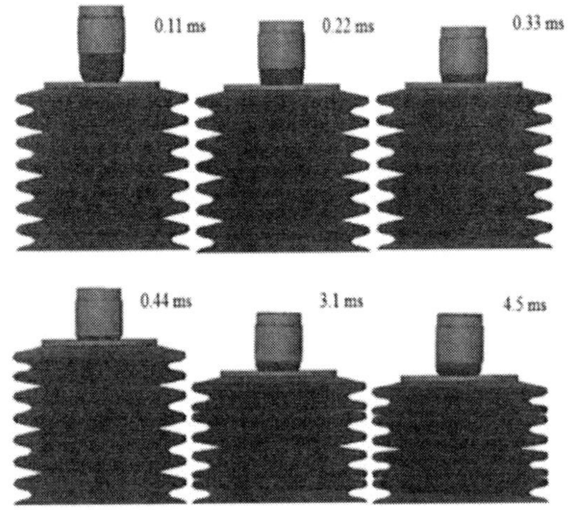

Figure 5. Stages of the MTHOTA when impacted with projectile with Polyolefinic nose at 99 m/s impact speed.

Table 4. VC_{max} values evaluated for 99 m/s impact speed.

Nose material	Maximum deflection (mm)	Time (ms)	VC_{max} (m/s)
Rubber-1	45.6	3.94	3.81
Rubber-2	44.8	3.29	4.41
Polyolefinic foam	17.6	4.64	0.51

Table 5. VC_{max} values evaluated for equivalent impact speed.

Nose material	Maximum deflection (mm)	Time (ms)	VC_{max} (m/s)
Rubber-1	39.5	4.38	2.57
Rubber-2	40.1	3.19	3.64
Polyolefinic foam	19.1	4.74	0.71

$VC_{max} = 1.0$ m/s indicate that blunt thoracic trauma point of view the 25% probability for AIS3+ injuries. Both elastomeric rubbers have yielded very high VC_{max} values and therefore, they are not suiable for making noses of the projectiles. Polyolefinic foam considered for the scholastic study has shown promising results as VC_{max} values are very much less than the 1.0 m/s. It is important to note that at 99 m/s, benchmark projectile's VC_{max} was evaluted as 0.21 m/s using 3-RCS thorax surrogate. In order to evaluate at what impact speed, projectile with the polyolefinic foam nose would provide the similar viscous response, further simulations were carried out using the projectile with the polyolefinic foam nose with impact speeds of 70 and 75 m/s. VC_{max} values evaluated for these two impact cases were 0.25 and 0.31 m/s respectively.

3.2. Head damage criterion

In the present study, non-linear finite element simulations were carried out by impacting a rigid fixed plate with the projectile with the softnose made up of Rubber-1. Impact speeds considered for this analysis were 20–100 m/s (with an increments of 10 m/s). From the deceleration time histories of the projectile, peak impact forces were evaluated and illustrated in the Fig. 6.

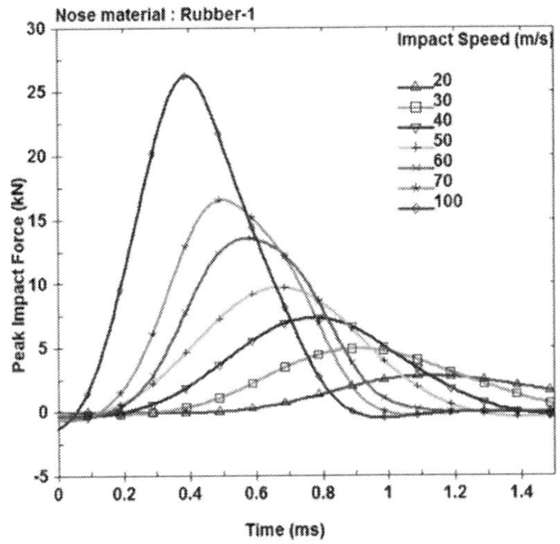

Figure 6. Force-time response of MTHOTA when impacted with the projectile with nose of Rubber-1.

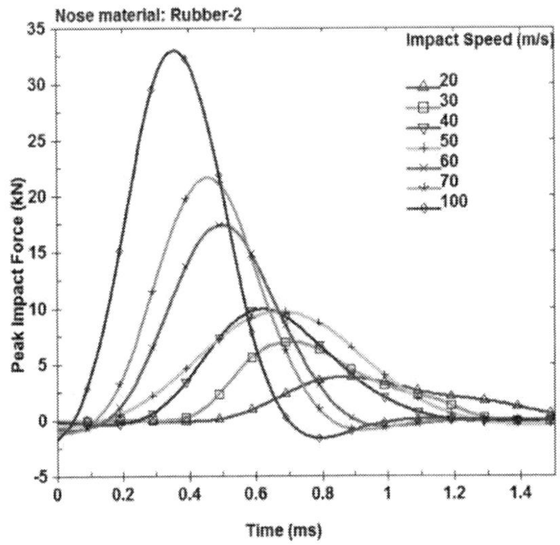

Figure 7. Force-time response of MTHOTA when impacted with the projectile with nose of Rubber-2.

Similar analyses were carried out for nose with materials Rubber-2 and Poly-olefinic foam. Force-time plots obtained from the analyses for these projectile cases are shown in the Fig. 7 and Fig. 8 respectively.

Head damage curves (peak impact force versus impact velocity) for all three cases of projectile (R1, R2 and F1 as nose materials), along with the bench projectile (XM 1006 as evaluated by Oukura et al. [11]) are illustrated in the Fig. 9.

From the Fig. 9, it is evident that the polyolefin Foam (F1) produced very promising results as at 60 m/s impact has yielded 7.35 kN of peak impact force, when compared to the 20.12 kN that is produced by the benchmark projectile. Projectiles with R1 and R2 noses also yielded

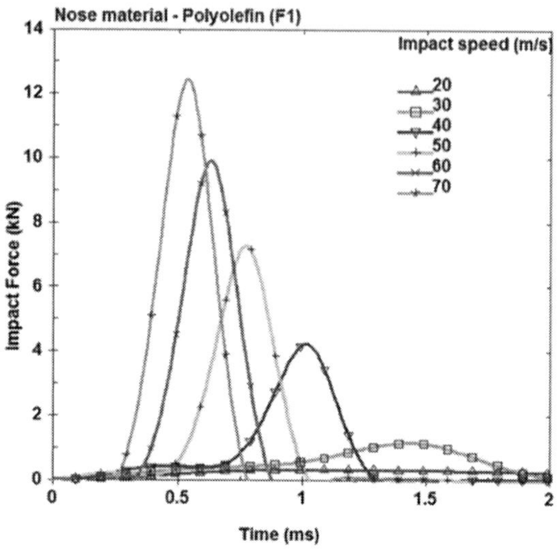

Figure 8. Force-time response of MTHOTA when impacted with the projectile with nose of Polyolefin foam (Foam-1).

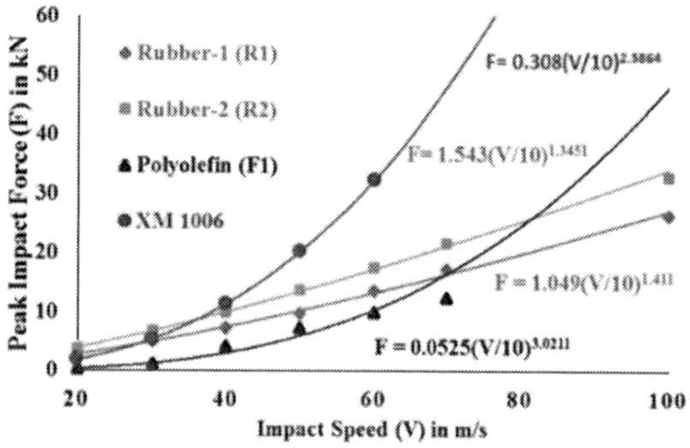

Figure 9. Head damage curves produced using force-wall method.

lesser peak impact forces than that of the benchmark projectile.

4. Conclusions

From the synthesis of the outcome of the FEA simulations, the following conclusions were drawn.

1. Projectiles with R1 and R2 noses have produced lesser peak impact forces on the head when compared to that produced by benchmark projectile. But, as far as the blunt thoracic trauama is concerned these projectiles have produced comparatively very high VCmax values, which disqualifies them as nose materials.
2. Projectile with polyolefin (F1) foam nose, as far as the head damage point of view, exibited better characteristics. Though the VCmax values produced by this projectile at 99 m/s is 0.51 m/s which is more than that of the benchmark projectile, the results are very promising.
3. Projectile with polyolefin foam (F1) has produced VCmax value of 0.25 m/s and peak impact force 12.2 kN. Blunt thoracic trauma point of view, at this impact speed, the projectile with F1 nose produced almost an equal effect and head trauma point view, the peak impact force was reduced by 75% when compared with those of the benchmark projectile.

The principal author gratefully acknowledges Mr Velizar Sedev and Mr Jason Tamas, Directors of CAECON of Adelaide of South Australia for complete sponsorship to attend the DYMAT-2015 conference. He also thanks Mr Srinivasa Rao Koya, Chairman of Amerigo Structural Engineers Private Limited, Bangalore, India' for providing the necessary funds and resources for the successful completion of the research project.

References

[1] J. Widder, C. Perhala, and J. Rascoe, "Caseless, Variable Velocity Systems for Precision Non-Lethal Fires to beyond 100 Meters. Part-II: Terminal effects Study, Selection of Projectile Parameters," *AMMTIAC-WSTIAC Journal* vol. 1, pp. 13–16 (2003).

[2] J.M. Widder, D.J. Butz, and J.M. Milosh, "Assessing the Blunt Trauma Potential of Free Flying Projectiles for Development and Safety Certification of Non-Lethal Impactors," Battelle Columbus Operations, Columbus, OH 43201, USA (1997).

[3] B. Koene, F. Id-Boufker, and A. Papy, "Kinetic non-lethal weapons," *Netherlands annual review of military studies*, pp. 9–24 (2008).

[4] (2011, 15/02/2015). *X1006 Sponge Round (Point)*, http://www.globalsecurity.org/military/systems/munitions/40-nl.htm

[5] (2015, 15/02/2015). *Defnse Technology eXact iMpact Sponge Round, 40 mm* [Electronic]. Available: http://www.chiefsupply.com/defense-technology-exact-impact-sponge-round-40mm.html

[6] S. Johnson. (2010, February 1, 2015). Taser XREP Up Close (and pricing!). *TFB – The Firearms Blog*. Available: http://www.thefirearmblog.com/blog/2010/02/10/taser-xrep-up-close-and-pricing/

[7] N.M. Thota, J.A. Eepaarachchi, and K.T. Lau, "Develop and validate a biomechanical surrogate of the human thorax using corrugated sheets: a feasibility study," in *Proceedings of the 7th Australasian Congress on Applied Mechanics (ACAM 7)*, (2012).

[8] N. Thota, J. Epaarachchi, and K.T. Lau, "Development and validation of a thorax surrogate FE model for assessment of trauma due to high speed blunt impacts," Journal of Biomechanical Science and Engineering, vol. 9, pp. JBSE0008-JBSE0008 (2014).

[9] N. Thota, "Development and Validation of a Mechanical Thorax Surrogate for the Evaluation of the Blunt Trauma Due to Ballistic Impacts" Doctor of Engineering, Centre of Excellence in Engineered

Fibre Composites, University of Southern Queensland, Australia (To be published).
[10] G. Slik, G. Vogel, V. Chawda, and D. Automotive, "Material model validation of a high efficient energy absorbing foam," in *Proceedings of the 5th LS-DYNA Forum* (2006).
[11] A. Oukura, C. Robbe, N. Nsiampa, and A. Papy, "Comparison of the lethality of different kinetic energy non-lethal projectiles using a new French assessment approach for head impacts," presented at the *27th International symposium on ballistics Freiburg*, Germany (2013).
[12] D.H. Lyon, "Development of a 40mm nonlethal catridge," *Defense Technology Federal Laboratories Research Journal*, pp. 64–78 (1997).
[13] D.H. Lyon, C.A. Bir, and B.J. Patton, "Injury Evaluation Techniques for Non-Lethal Kinetic Energy Munitions," *DTIC Document* (1999).
[14] C. Bir, D. Viano, and A. King, "Development of biomechanical response corridors of the thorax to blunt ballistic impacts," J Biomech, vol. 37, pp. 73–9 Jan. (2004).
[15] C.A. Bir, "The evaluation of blunt ballistic impacts of the thorax," Doctor of Philosophy, Biomedical Engineering, Wayne State University, Detroit, Michigan, USA (2000).
[16] N.M. Thota, J.A. Eepaarachchi, and K.T. Lau, "Effect of energy absorbing mechanisms on the blunt thoracic trauma caused by ballistic impacts," in *Proceedings of the 8th Australasian Congress on Applied Mechanics (ACAM 8)*, Melbourne Convention Centre, Melbourne, Victoria, Australia (2014).
[17] N.M. Thota, J.A. Eepaarachchi, and K.T. Lau, "Evaluation of the blunt thoracic trauma due to baseball impacts - Review of the blunt criterion," in *Proceedings of the 8th Australasian Congress on Applied Mechanics (ACAM 8)*, Melbourne Convention Centre, Melbourne, Victoria, Australia (2014).
[18] N.M. Thota, J.A. Eepaarachchi, and K.T. Lau, "Viscous criterion and its relation with the projectile-thorax energy interactions," in *Proceedings of the 8th Australasian Congress on Applied Mechanics (ACAM 8)*, Melbourne Convention Centre, Melbourne, Victoria, Australia (2014).
[19] F. Pan, S. Quander, and Z. Ma, "Use of material models for elastomer foams in ABAQUS and LS-DYNA," in *Constitutive Models for Rubber IV – Austrell & Kari (eds)*. vol. IV, A. Kari, Ed., ed London: Taylor & Francis Group, 2005, pp. 593–597.

EPJ Web of Conferences **94**, 03003 (2015)
DOI: 10.1051/epjconf/20159403003

Dynamic compressive properties of bovine knee layered tissue

Masahiro Nishida[1,a], Yuki Hino[1], and Mitsugu Todo[2]

[1] Nagoya Institute of Technology, Gokiso-cho, Showa-ku, Aichi 466-8555, Japan
[2] Kyushu University, 6-1 Kasuga-koen, Kasuga, Fukuoka 816-8580, Japan

Abstract. In Japan, the most common articular disease is knee osteoarthritis. Among many treatment methodologies, tissue engineering and regenerative medicine have recently received a lot of attention. In this field, cells and scaffolds are important, both ex vivo and in vivo. From the viewpoint of effective treatment, in addition to histological features, the compatibility of mechanical properties is also important. In this study, the dynamic and static compressive properties of bovine articular cartilage-cancellous bone layered tissue were measured using a universal testing machine and a split Hopkinson pressure bar method. The compressive behaviors of bovine articular cartilage-cancellous bone layered tissue were examined. The effects of strain rate on the maximum stress and the slope of stress-strain curves of the bovine articular cartilage-cancellous bone layered tissue were discussed.

1. Introduction

In Japan, because of articular diseases, more than one million people go to hospital and 80% of them are elderly persons. On the other hand, the main reason of elderly persons for needing long-term care is articular diseases in Japan. The most common articular disease is knee osteoarthritis (degenerative arthritis). The knee osteoarthritis is defined as a degenerative change of bones based on the progressive degenerative lesion of articular cartilage [1]. The articular cartilage consists of hyaline cartilage, surrounds articular surfaces and is translucent white. The main functions of articular cartilage are energy absorption, stress redistribution and lubrication. In general, articular cartilage is worn down by advancing age, obesity and muscle deterioration. In the early stage of the knee osteoarthritis, conservative therapy is effective; heat-retention and knee supporter in the case of mild symptoms, and antiphlogistic analgetic and injection of hyaluronan drug under a painful condition. In the case of joint degeneration after symptom progression, operative therapy such as high tibial osteotomy (HTO) is provided. For patients of further joint degeneration, total knee arthroplasty (artificial joint replacement surgery) is usually performed [2].

Artificial joints have been widely used and their parts are made of metals such as titanium alloys and cobalt chrome alloys, ultrahigh molecular weight polyethylene (UHMWPE) and ceramics such as alumina, zirconia and hydroxyapatite. Therefore, when artificial joints are used for long time, contact surfaces are always worn down and revision surgery are usually needed. Recently, the tissue engineering and regenerative medicine have received a lot of attention. In the field of tissue engineering and regenerative medicine, cells and scaffolds which are artificial structures supporting tissue formation are important both ex vivo as well as in vivo. From the view point of effective treatment, in addition to histological feature, compatibility of mechanical properties is also important.

Several studies have been conducted on static and dynamic mechanical properties of body tissues near knee joints. Kobayashi et al. measured static and dynamic compressive properties of bovine cancellous bones using universal testing machine and Hopkinson pressure bar method and reported that elastic modulus and yield stress increased with density and strain rate [3]. Shimizu et al. measured static and dynamic compressive properties of porcine patellar cartilages using universal testing machine and Hopkinson pressure bar method and the mechanical properties were compared with those of double network gels for artificial articular cartilage [4]. Parish et al. studied dynamic tensile behavior of porcine cortical bones [5]. Laporte et al. [6] and Prot et al. [7] examined dynamic behavior and microstructure of bovine cancellous bone. They have focused on mechanical properties of each tissue.

However, because body tissues are always layered structures, mechanical properties of layered structures are essential. Hwang and Todo [8] studied static compressive properties of layered structures. In this study, dynamic and static compressive properties of bovine articular cartilage-cancellous bone layered tissue were measured.

2. Experimental methods

A part of knee joint of bovine thigh bone was used as specimens. Figure 1 shows a photograph of the cross-sectional surface of the distal end of femoral bone (external and internal condyles) which is upper side of knee joint. The surface of the distal end of femoral bone is surrounded by the white cartilage tissue and red cancellous bone is observed on the inner side of femoral bone.

[a] Corresponding author: nishida.masahiro@nitech.ac.jp

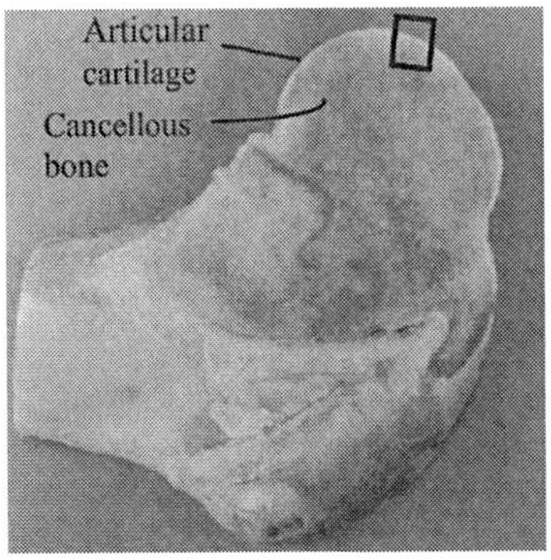

Figure 1. Photograph of distal end of femoral bone (Upper side of knee joint).

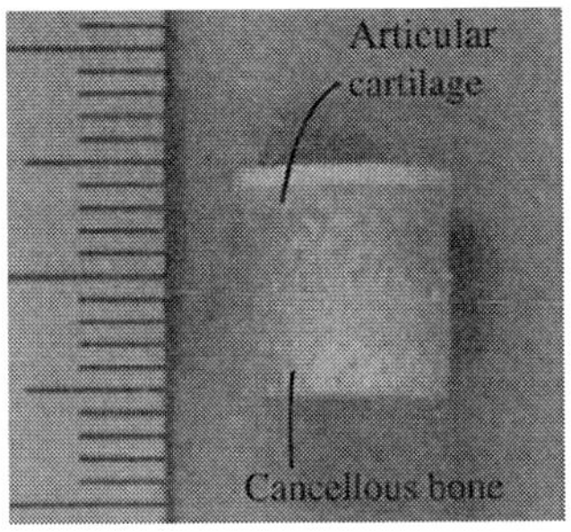

Figure 2. Compressive test specimen.

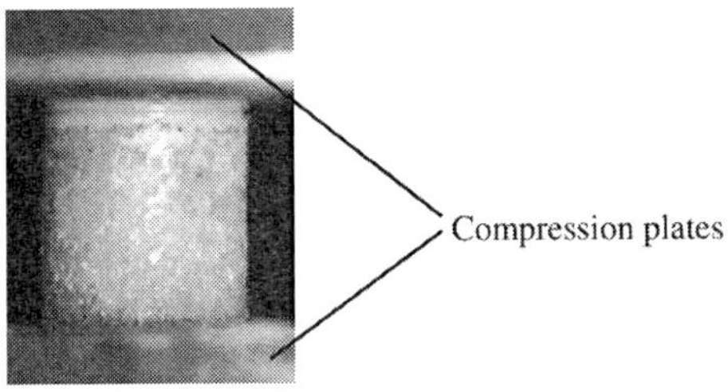

Figure 3. Photograph of specimen during compression test.

Articular cartilage-cancellous bone layered tissue indicated by the black quadrangle part in Fig. 1 was cut out by machining. Compressive test specimens with a diameter of 10 mm and a height of 10 mm were made as shown in Fig. 2. The averaged thickness of four points for articular cartilage was 0.87 mm.

Quasi-static compressive tests were conducted at a crosshead speed of 0.05 mm/min (strain rate of $8.3 \times 10^{-5}\,s^{-1}$) using a universal testing machine. The displacement of compression plates was measured by displacement transducer (Kyowa Electronic Instruments Co., Ltd., DTH-A-30). Each strain of articular cartilage and cancellous bone was measured by image analysis of photographs during quasi-static compressive tests, using ImageJ. Figure 3 shows a photograph of specimen during quasi-static compressive tests.

At high strain rates over $100\,s^{-1}$, the stress-strain curves were measured by the split Hopkinson pressure bar (Kolsky bar) method, as shown in Fig. 4. The

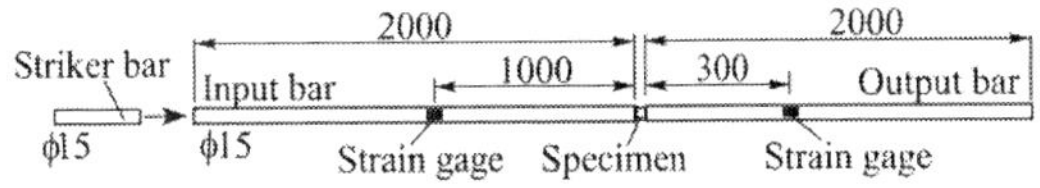

Figure 4. Split Hopkinson pressure bar (Kolsky bar) method.

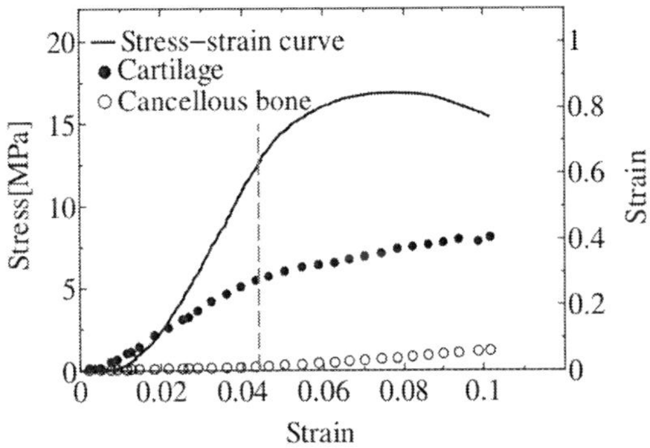

Figure 5. Stress-strain curve of specimen and each strain of cartilage and cancellous bone.

Table 1. Material constants of input and output bars used in calculation.

Density	Elastic wave velocity in bar	Young's modulus E
2.81×10^3 kg/m^3	5150 m/s	71.7 GPa

input and output bars were made of an aluminum alloy (7075-T6511), and they had a diameter of 15 mm and respective lengths of 2000 mm and 1300 mm. Strain gages were placed on the input and output bars at distances of 1000 mm and 300 mm away from the specimen, respectively. When the stress histories were almost equal on both sides of the specimens, the stress and strain of the specimens can be calculated from the strain on the bars using Eqs. (1) and (2) [9, 10].

$$\sigma(t) = \frac{AE}{A_S}\varepsilon_T(t) \quad (1)$$

$$\varepsilon(t) = \frac{2c_3}{L}\int_0^t [\varepsilon_I(t) - \varepsilon_T(t)]\,dt. \quad (2)$$

Here, ε_I and ε_T denote the axial strain in the input bar induced by the incident wave and the axial strain in the output bar induced by the transmitted wave, respectively. E and c_3 denote the Young's modulus and elastic wave velocity of the input and output bars, respectively. L denotes the specimen thickness. A and A_s denote the cross-sectional areas of the input/output bars and the specimens. The material constants of the aluminum alloy (7075-T6511) bars used in the calculations are shown in Table 1. The strikers with a diameter of 15 mm and lengths of 490 mm made of aluminum alloy (7075-T6511) were used.

3. Results and discussion

Figure 5 shows static stress-strain curves. The right axis shows each strain of articular cartilage, black circles, and cancellous bone, white circles, as a function of specimen

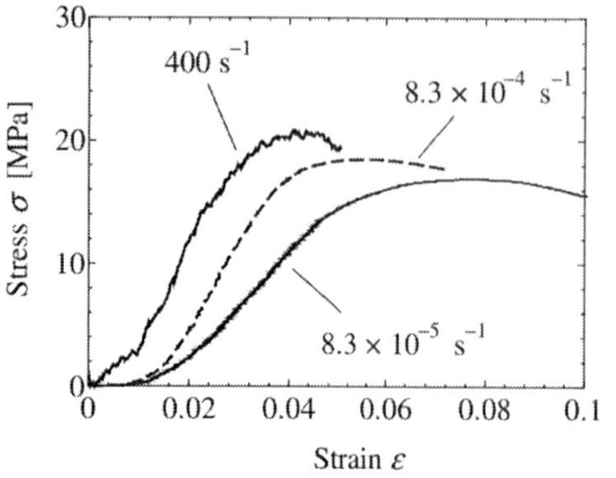

Figure 6. Stress-strain curve of specimens at high and low strain rate.

strain on the horizontal axis. Each strain of articular cartilage and cancellous bone was measured by image analysis. The stress-strain curves were divided into two stages. In the early stage until strain of 0.04, the stress of specimen increased with strain on the horizontal axis and the strain of articular cartilage also increased as shown by black circles. However the strain of cancellous bone was very small as shown by white circles. At the strain of 0.04, the increase rate of the strain of articular cartilage lessened as shown by black circles and the strain of cancellous bone started to increase as shown by white circles.

Until the strain of 0.04, the slope of stress-strain curve continuously increased and, when cancellous bone began to increase, the slope of stress-strain curve started to decrease. At the specimen strain of 0.075, the stress-strain curves showed a peak and the stress at the peak (the maximum stress) was approximately 17 MPa.

Laporte et al. reported that the maximum stress of bovine cancellous bones was approximately 8 MPa and the strain showing the maximum stress was 0.05 in the static compression tests [5]. Kobayashi et al. reported that the maximum stress of bovine cancellous bones was approximately 25 MPa and the strain showing the maximum stress was 0.04. Our results of articular cartilage-cancellous bone layered tissues showed that the maximum stress was 17 MPa and the strain of only cancellous bones (right axis) at the maximum stress was 0.032. Those values were similar to the results of Laporte et al. and Kobayashi et al. It is assumed that the main reason for the different is an individual difference in bones.

As a result, the stress-strain curves of articular cartilage-cancellous bone layered tissues were roughly divided into two regions. In the first part, the articular cartilage mainly deformed and the cancellous bone seldom deformed. In the next part that the deformation of specimens progressed and the stress of specimens approached to the failure stress of cancellous bone, the deformation of articular cartilage and the fracture of cancellous bone occurred simultaneously. Because the slope of stress-strain curves decreased when the cancellous bone started to deform, the fracture of cancellous bones progressed just after the onset of deformation.

Compressive tests were also conducted at a crosshead speed of 0.5 mm/min (strain rate of $8.3 \times 10^{-4}\,s^{-1}$) using a universal testing machine. Results using a universal testing machine were compared with results of SPPB. Figure 6 shows the effects of strain rate on stress-strain curves. The strain rate slightly increased the maximum stress and clearly increased the initial slope of stress-strain curves. It seems that the viscoelastic behavior of articular cartilage and/or cancellous bone increased the slope.

4. Conclusions

The dynamic and static compressive properties of bovine articular cartilage-cancellous bone layered tissue were measured using a universal testing machine and a split Hopkinson pressure bar method. The stress-strain curves of articular cartilage-cancellous bone layered tissue were roughly divided into two stages at low strain rate. The strain rate slightly increased the maximum stress and clearly increased the initial slope of stress-strain curves.

The authors are greatly indebted to Professor Tadashi Inaba and Professor Yuichi Kasai of Mie University for their valuable advice with respect to the split Hopkinson bar. The authors are also greatly indebted to Mr. Masayuki Hayashi, an assistant researcher of Nagoya Institute of Technology, for his help with the experimental work.

References

[1] Y. Tachibana, Physical Therapy Science, **20**, 235 (2005) (In Japanese)

[2] T. Suzuki, Y. Hayashi, Bone Encyclopedia, Asakura Publishing, 409 (2003) (In Japanese)

[3] K. Kobayashi, Y. Tanabe and T. Hara, Transactions of the Japan Society of Mechanical Engineers, Series A, **63**, 160 (1997) (In Japanese)

[4] D. Shimizu, H. Fujiki, M. Daimaruya, O. Shiozaki, Proc. 47th JSME Hokkaido branch conference, 109 (2008)

[5] A. Parish, W. Chen and T. Weerasooriya, Proc. DYMAT 2009, **2**, 917 (2009)

[6] S. Laporte, F. David, V. Bousson and S. Pattofatto, Proc. DYMAT 2009, **2**, 895 (2009)

[7] M. Prot, T.J. Cloete and S. Pattofatto, EPJ Web of Conference, **26**, 03003 (2012)

[8] S. Hwang and M. Todo, Journal of Solid Mechanics and Materials Engineering, **7**, 293 (2013)

[9] G.T. Gray, III, ASM Handbook Vol. 8, *Mechanical Testing and Evaluation*, ASM 462 (2000)

[10] W. Chen, B. Song, *Split Hopkinson (Kolsky) Bar: Design, Testing and Applications*, Springer, (2010)

EPJ Web of Conferences **94**, 03004 (2015)
DOI: 10.1051/epjconf/20159403004

A preliminary investigation of the dynamic viscoelastic relaxation of bovine cortical bone

T.J.C. Loete[1,3,a], G. Paul[1,3,b], and E.B. Ismail[2,3,c]

[1] Blast Impact and Survivability Research Unit, University of Cape Town, Private Bag X3, Rondebosch 7701, South Africa
[2] Centre for Computational and Applied Mechanics, University of Cape Town, Private Bag X3, Rondebosch 7701, South Africa
[3] Department of Mechanical Engineering, University of Cape Town, Private Bag X3, Rondebosch 7701, South Africa

Abstract. A new experimental approach is proposed to characterize the dynamic viscoelastic relaxation behaviour of cortical bone. Theoretical models are presented to show that a linear viscoelastic material, when allowed to relax between two long elastic bars, will produce stress, strain and strain rate histories that contain characteristic features. Furthermore, typical experimental results are presented to show that these characteristic features are observed during split Hopkinson bar tests on bovine cortical bone using a Cone-in-Tube striker. The interpretation of this behaviour in the context of a standard linear viscoelastic model is discussed.

1. Introduction

The design of protective structures to prevent injury during impact loading events, such as sporting accidents and vehicle collisions, requires a detailed understanding of the mechanisms of dynamic bone deformation and fracture. Traditional approaches to quantify responses of the human body to impacts have typically included full scale experimental testing involving the use of a post mortem human subject (PMHS) [14]. However, such tests are expensive and time consuming which has lead to the increasing use of numerical modelling [6], for which detailed material models of bone are required.

At the macroscopic level, bone is classified into cortical (hard/compact) bone and cancellous/trabecular (soft/spongy) bone, of which the former is the primary skeletal load-bearing component. At a microscopic level, bone is a complex composite material composed of mineral crystals (hydroxyapatite) embedded in an organic matrix (primarily collagen fibres). This combination of organic and mineral components gives rise to the viscoelastic quasi-brittle behaviour of bone, *i.e.* bone displays highly rate sensitive viscoelastic behaviour but fails via progressive brittle fracture. Additional viscous contributions may arise during dynamic loading events due to the fluid component contained within the bone cavities.

Given its viscoelastic nature, the development of a material model for cortical bone requires knowledge of its mechanical properties over a wide range of strain rates, from quasi-static (QS) to high strain rate (HSR), *i.e.* $10^{-3}\,s^{-1} - 10^{3}\,s^{-1}$. Johnson *et al.* [8] reported an extensive survey of the rate dependence of the apparent Young's modulus of cortical bone. They note that most of the work has focused on strain rates in the range of $10^{-3}\,s^{-1} - 0.1\,s^{-1}$, which are relevant to activities such as walking and running [2]. By contrast, high strain rate bone characterization tests, using drop testing [10] or split Hopkinson bar (SHB) [1,3] techniques, are typically restricted to strain rates above $300\,s^{-1}$, which correspond to severe impact loading conditions. Consequently, the data showed that the intermediate strain rate (ISR) range (*i.e.* $1\,s^{-1} - 100\,s^{-1}$) is sparsely populated. Furthermore, despite the relatively large quantity of HSR data available, there are some concerns regarding the quality of the older data since it is only within the last decade that techniques for conducting dynamic compression tests on bone at near-constant strain rates have been developed [1,3].

The issue of the lack of data in the ISR regime has recently been addressed by Cloete *et al.* [4], who presented results for the dynamic compression of bovine cortical bone. A novel Cone-in-Tube (CiT) striker design was introduced that allows constant strain rate compression tests to be conducted in the upper ISR regime (*i.e.* $50\,s^{-1} - 100\,s^{-1}$) using a conventional split Hopkinson bar arrangement, as shown in Fig. 1, while a Wedge-Bar apparatus was used in the lower ISR regime (*i.e.* $10\,s^{-1} - 50\,s^{-1}$). The results show a rapid transition in the viscoelastic strain rate sensitivity of bovine cortical bone under ISR conditions, as shown in Fig. 2.

Several viscoelastic models for cortical bone have been proposed based on compression data such as that shown in Fig. 2 [3,8]. However, for a more complete description of bone behaviour, knowledge of the viscoelastic relaxation response is also required. This can be accomplished using a stress relaxation tests or a cyclic test using a dynamic mechanical analyser (DMA) in which a small

[a] e-mail: trevor.cloete@uct.ac.za
[b] e-mail: graeme.paul@angloamerican.com
[c] e-mail: ernesto.ismail@uct.ac.za

Article available at http://www.epj-conferences.org or http://dx.doi.org/10.1051/epjconf/20159403004

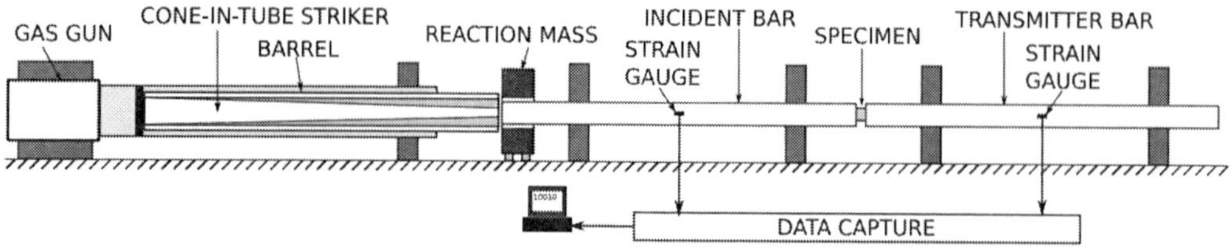

Figure 1. Layout of the split Hopkinson bar setup using a CiT striker.

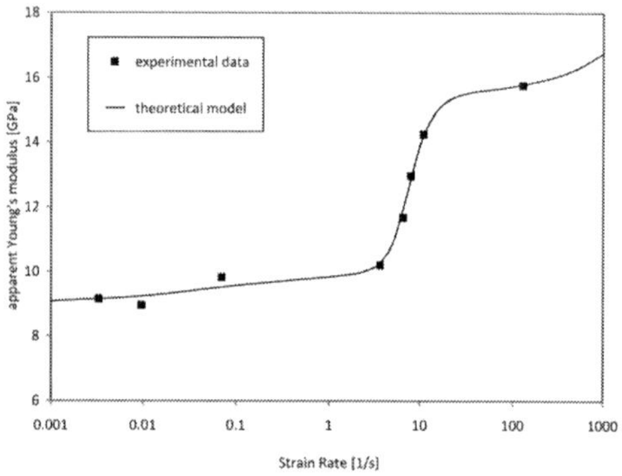

Figure 2. The strain rate sensitivity of the apparent Young's modulus of bovine cortical bone specimen as reported by Cloete *et al.* [4].

specimen is placed under a cyclic load [13]. An analysis of the displacement response provides values for the storage and loss moduli, which are viewed as indicative of the elastic and viscous components of the material. However, these techniques are typically limited to low strain rates and are therefore not suitable for the HSR or ISR regimes [12]. Furthermore, Saraf *et al.* [11] note that techniques involving oscillatory loading may examine different mechanisms than those developed by the single HSR loading pulse typically generated during an impact.

In this paper, a new experimental approach is proposed to investigate the dynamic viscoelastic relaxation of cortical bone in the upper ISR regime. This work is a continuation of that reported by Cloete *et al.* [4] and makes use of the same split Hopkinson setup with a CiT striker. It will be shown that a dynamically loaded bone specimen, when allowed to relax between two Hopkinson bars, produces stress, strain and strain rate histories that contain characteristic features which can be interpreted in the context of a standard linear viscoelastic model. Finally, a few preliminary experimental results are presented to illustrate the proposed approach.

2. Specimen preparation

Whole fresh femurs of slaughter-age cattle were sourced from a local butcher and stored in a frozen condition. During specimen preparation, the bones were thawed and cortical specimens were extracted from the mid diaphyses of two femurs. The specimens were machined to nominal dimensions of 4 mm diameter and 5 mm length and individually inspected, weighed, measured and refrozen. The results reported in this paper are for specimens originally orientated in the longitudinal direction relative to the bone axis. As mentioned above, this paper is a continuation of previous work reported by Cloete *et al.* [4] and will focus on the subset of the data in the upper ISR regime. In total, five specimens were tested at a strain rate of approximately $130\,s^{-1}$ of which two did not fail and can be used to investigate dynamic viscoelastic relaxation.

3. Experimental method

Dynamic bone compression and relaxation tests were performed using the classic SHB technique [7] with a CiT [4], as shown in Fig. 1. As the name suggests, a CiT striker consists of a slender cone made from 6063-T6 aluminium coaxially mounted inside a tube of the same material, where the large end of the cone is threaded to the tube. The CiT striker had a length of 870 mm, with inner cone end diameters of 9 mm and 25 mm. The outer tube has a standard section and was chosen to have approximately the same cross-sectional area as the large end of the inner cone. The Hopkinson bars had a length of 3 m and a diameter of 19.05 mm and were machined from maraging steel with a density of $7950\,kg\,m^{-3}$ and a longitudinal wave propagation speed of $4575\,ms^{-1}$. Diametrically opposed foil strain gauges (2 mm, 120 Ω) were bonded to the midpoint of the Hopkinson bars. The signals were amplified using custom-built amplifiers (gain of 1000, bandwidth of 100 kHz) based on the Burr Brown INA110 chip and digitized using an ADLINK PCI-9812 card (10 MHz, 12 bit).

The CiT striker design is intended to produce an incidence pulse that allows the apparent Young's modulus of a viscoelastic specimen to be measured at small strains and near-constant strain rates in the upper ISR range. This requires a shaped incidence pulse that has a steeply rising slope with the same gradient as that of the specimen response. Furthermore, the end of the pulse is sharply truncated to allow rapid load removal, which has several benefits. Firstly, it allows the test duration to be maximized for a given SHB length, *i.e.* the strain rate can be minimized for a given final strain. Secondly, it would allow for the recovery of an intact specimen for microstructural analysis. Thirdly, it allows for the possibility of conducting dynamic viscoelastic relaxation tests. In other words, the cone-in-tube striker design provides a means to control both the shape and duration of the incidence wave such that a bone specimen can be loaded at a constant strain rate up to a point just before fracture occurs and then fully relax without any further induced loading. This has not been possible with previously published pulse shaping

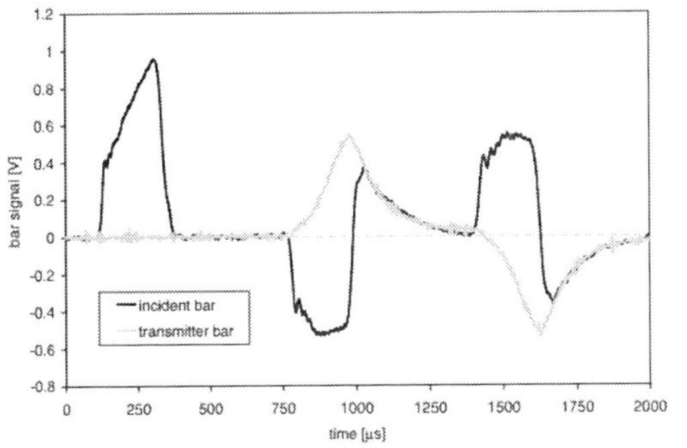

Figure 3. Raw SHB data for a dynamic relaxation test on a bovine bone specimen.

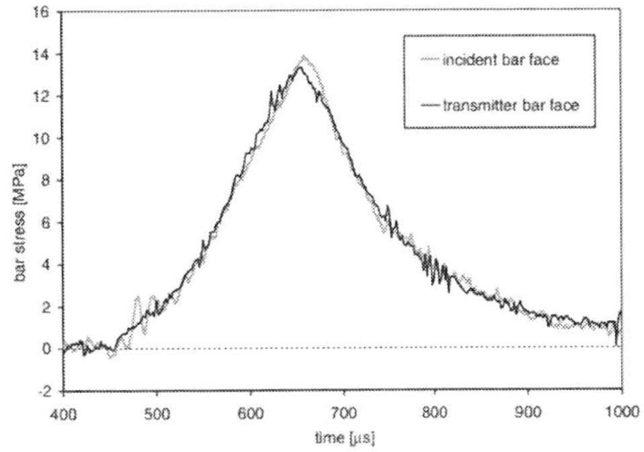

Figure 4. Specimen equilibrium during a dynamic relaxation test on a bovine bone specimen.

concepts which have only been able to meet the steep slope requirement [1,3].

The CiT striker is launched in a conventional manner using a gas gun. However, only the inner cone impacts upon the incident bar, whereas the outer tube impacts upon a large reaction mass mounted coaxially around the incident bar, as shown in Fig. 1. The resulting stress waves in the cone and tube simultaneously reach the free end of the striker where the cone and tube are connected. In isolation, the stress wave produced in a slender cone with a large ratio between its end diameters is insufficient to reverse the velocity of the large end, resulting in further loading of the incident bar. However, in the CiT striker, the reflected stress waves of the cone and tube interact and, if the tube is large enough, result in the velocity of the large end of the conical striker being reversed. This velocity reversal induces a tensile stress wave in the cone that, upon arrival at the impact face, causes the conical striker to separate from the incident bar, thus abruptly terminating the incidence pulse. In this way, a steeply rising and sharply truncated incidence pulse, with a large ratio between the initial and final stress wave magnitudes, can be created, as shown in Fig. 3.

A minor limitation of the CiT striker technique is that the required cone angle is not known *a priori*, and, in general, a set of inner cones with a range of cone angles would be required to cover possible variation in specimen responses. The cone angle and diameter ratios used for the CiT striker was selected based on previous experimental results and a simple spreadsheet-based one-dimensional mass-spring code and did not require any iterations [3].

4. Experimental results

In addition to the shaped incidence pulse, Fig. 3 shows typical reflected and transmitted signals for the case where a cortical bone specimen does not fracture. Note that after the passage of the reflected wave the stress signals in both Hopkinson bars show the same decay behaviour. This is due to the specimen exerting equal but opposite forces on the bar faces during the unloading phase of the test and shows that the specimen is in quasi-equilibrium. A similar degree of specimen equilibrium was obtained during the loading phase as shown in Fig. 4, where the stress histories at the specimen - bar interfaces have been obtained using standard SHB data processing formulae, such as those given by Gray [7].

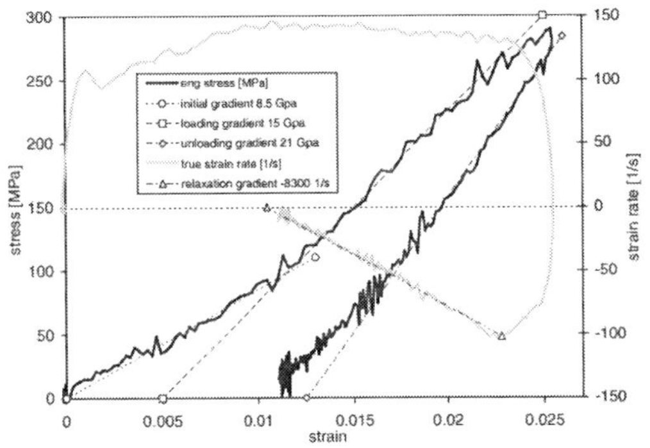

Figure 5. Processed data for a dynamic relaxation test on a bovine bone specimen.

The processed results corresponding to the raw data in Fig. 3 are given in Fig. 5, which shows the resulting stress and strain rate curves as a function of strain. The stress curve has an initial gradient of 8.5 GPa up to a strain of 0.01, followed by a loading gradient of 15 GPa. Similar initial gradients of varying duration were observed in three of the other four tests but was absent in the fourth. By contrast, there was less variation in the main portion of the loading curves where all five experiments resulted in gradients of 15.2 ± 0.7 GPa. This implies that the latter part of the loading curve represents the actual mechanical properties of the bone specimen, while the initial gradient represent transient "settling" behaviour as the specimen experiences slight adjustments until it is in full contact with the Hopkinson bar faces. This transient behaviour is also evident in the strain rate curve which rises slightly during the initial loading and then remains near-constant at approximately $140\,s^{-1}$.

During unloading the bone specimen displays significant hysteresis. After a peak stress of 290 MPa, the strain rate changes rapidly to a value of $-100\,s^{-1}$, while the stress initially falls with a gradient of 21 GPa. There is a noticeable linear relationship between the strain and strain rate during the unloading phase, which will be referred to as the relaxation gradient and discussed in Sect. 5.

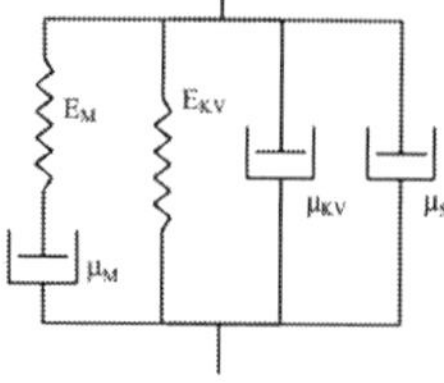

Figure 6. Schematic diagram of the standard linear viscoelastic model used in this work. The damper element μ_{SHB} represents the role of the Hopkinson bars during unloading and is deactivated during the loading phase.

Due to the limited test duration of a SHB apparatus, the complete specimen unloading behaviour could not be captured. In lieu of this, a simple extrapolation of the relaxation gradient would appear to indicate a residual strain of approximately 0.01, although the "settling" behaviour during the loading phase suggests a value closer to 0.005. Furthermore, preliminary measurement of the recovered specimen immediately after the test indicated that there was some residual deformation, though not as much as indicated by the SHB data. However, further measurement of the specimen several days later showed that the specimen had relaxed completely to its original length. The implication is that the deformation of cortical bone is a complex combination of several competing deformation mechanisms with some acting within less than a millisecond while others act over hours or even days.

5. Theoretical model

While there is clear stress relaxation behaviour shown in Figs. 3 to 5, it does not conform to the strict definition of a viscoelastic relaxation test for which, per definition, the strain is kept constant [5]. Hence, none of the standard equations for assessing viscoelastic relaxation can be applied, which possibly explains the scarcity of data in the literature regarding the dynamic viscoelastic relaxation of bone. In lieu of this, it is proposed that the SHB test results can be evaluated using a standard linear viscoelastic model, which consists of a Kelvin-Voigt element (E_{KV} and μ_{KV}) in parallel with a Maxwell element (E_M and μ_M), as shown in Fig. 6.

During the unloading phase, the specimen applies equal but opposite forces to the Hopkinson bars, which generate stress waves, as shown in Fig. 3. Hence, based on the well known stress wave formulae [9], the particle velocities in the Hopkinson bars will be proportional to the stress applied by the specimen. In other words, the Hopkinson bars will behave like perfect linear dampers and their effect can be modelled by adding an equivalent damper μ_{SHB}, as shown in Fig. 6. Note, however, that the equivalent SHB damper does not represents a component of the material model for the specimen and is, therefore, not included when calculating the specimen stress.

The magnitude of the equivalent damper coefficient can be obtained by expressing the relationship between stress and particle velocity in a Hopkinson bar in terms of the specimen stress and strain, which gives,

$$\sigma_b = \frac{A_s}{A_b}\sigma_s = (\rho c)_b v_b = (\rho c)_b \dot{\varepsilon}_s \tfrac{1}{2} L_s \qquad (1)$$

where the subscripts s and b denote respectively the specimen and Hopkinson bar stress σ, cross-sectional area A, density ρ, wave speed c, strain rate $\dot{\varepsilon}$ and length L. Note that only half the specimen length is considered since both of the Hopkinson bars will be moving with the same speed v_b but in opposite directions.

Hence the equivalent damper coefficient can be expressed as,

$$\mu_{SHB} = \tfrac{1}{2} L_s \frac{A_b}{A_s}(\rho c)_b. \qquad (2)$$

From the data given in Sects. 2 and 3, the value of the equivalent damper coefficient is 2.062 MPa s^{-1}, which is used throughout the work presented in this paper.

Given the above, an explanation for the linear relationship between the strain and the strain rate during the unloading phase is proposed. Consider an unloading scenario where the strain rate is sufficiently high to cause the Maxwell damper μ_M to "lock-up", *i.e.* its deformation relative to the Kelvin-Voigt element is negligible such that the Maxwell spring E_M is effectively operating in parallel to the Kelvin-Voigt spring E_{KV}. Since there is no external force, equilibrium dictates that the spring and damper forces must be equal and opposite,

$$(E_{KV} + E_M)\varepsilon_s = -(\mu_{KV} + \mu_{SHB})\dot{\varepsilon}_s. \qquad (3)$$

Hence, the ratio between the specimen strain and the strain rate is constant and has the form,

$$\frac{\dot{\varepsilon}_s}{\varepsilon_s} = -\frac{E_{KV} + E_M}{\mu_{KV} + \mu_{SHB}}. \qquad (4)$$

Assuming that the loading gradient of 15 GPa in Fig. 5 is approximately equal to a combination of the Kelvin-Voigt and Maxwell springs, while neglecting the Kelvin-Voigt damper as small in comparison to equivalent SHB damper, gives an estimate for the relaxation gradient of -7274 s^{-1}. Given the simplicity of the model, this result compares well with the measured value of -8300 s^{-1}.

6. Theoretical model results

A discrete version of the standard linear viscoelastic model described above was implemented in a spreadsheet and numerically integrated to replicate the observed behaviour of bovine cortical bone in dynamic relaxation tests. During the loading phase the strain rate was specified and the equivalent SHB damper was not considered, whereas during the unloading phase the equivalent SHB damper was included and the strain rate was determined via the interaction of the spring and damper components.

Initial hypothetical test cases were simulated to illustrate the effect that the various parameters have on the overall response. Figure 7 shows the result for a simulation with parameter values E_{KV} = 10500 MPa, E_M = 0 MPa, μ_{KV} = 0.5 MPa s^{-1} and μ_M = 0.01 MPa s^{-1}. In other

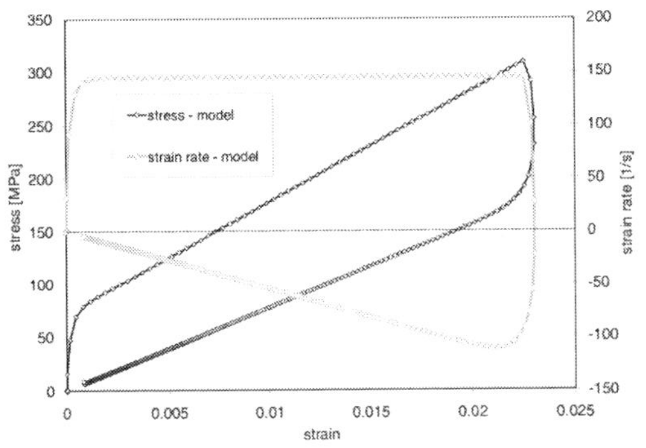

Figure 7. A simulated test case where the Kelvin-Voigt element dominates.

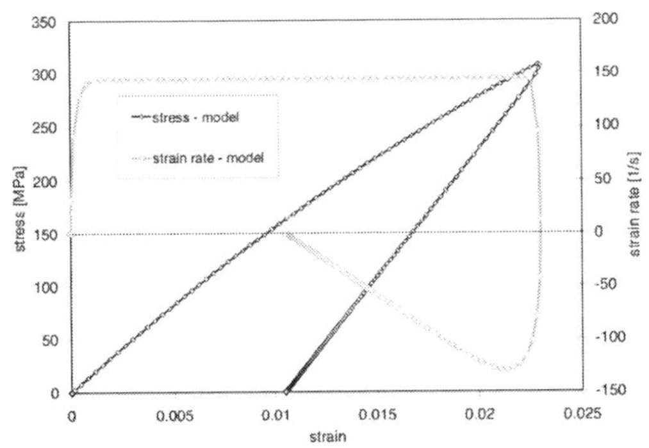

Figure 8. A simulated test case where the Maxwell element dominates.

words, the effect of the Maxwell element is removed while the effect of the Kelvin-Voigt damper is emphasized. The Kelvin-Voigt damper causes positive and negative stress jumps to occur at the beginning of the loading and unloading phases respectively. The apparent unloading gradient is slightly smaller than that of the loading phase, which is due to the decreasing strain rate during the unloading phase. The relaxation gradient is linear, as expected following the argument presented in Sect. 5, and the residual strain tends to zero at the end of the simulation.

Figure 8 shows the result for a simulation with parameter values $E_{KV} = 0\,\text{MPa}$, $E_M = 17500\,\text{MPa}$, $\mu_{KV} = 0\,\text{MPa}\,\text{s}^{-1}$ and $\mu_M = 5\,\text{MPa}\,\text{s}^{-1}$. In contrast to the previous test case, the effect of the Maxwell element has been emphasized while the effect of the Kelvin-Voigt element has been removed. Due to the absence of the Kelvin-Voigt damper, no stress jumps occur at the beginning of the loading or unloading phases. Furthermore, unlike the previous case, the apparent unloading gradient is significantly larger than that of the loading phase, due to the residual strain remaining at the end of the simulation. Note that the relaxation gradient is again linear, and, during the loading phase, where the strain rate is constant, the gradient varies slightly whereas during the unloading phase, where the strain decreases, the gradient appears to be constant.

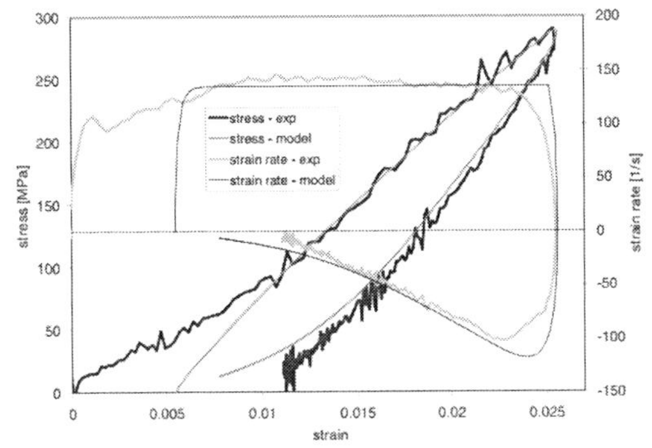

Figure 9. A simulation replicating the experimental result discussed in Sect. 4.

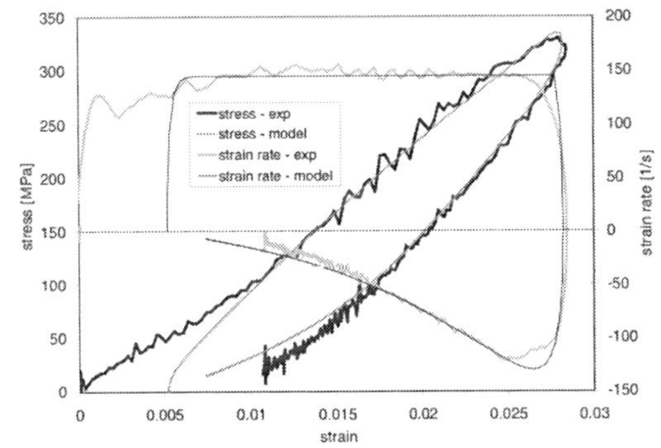

Figure 10. A simulation replicating an experimental result for a specimen that was nominally identical to that used to obtain the results presented in Sect. 4.

Figure 9 shows the result of a simulation replicating the experimental result discussed in Sect. 4 with parameter values, $E_{KV} = 9800\,\text{MPa}$, $E_M = 7500\,\text{MPa}$, $\mu_{KV} = 0.03\,\text{MPa}\,\text{s}^{-1}$ and $\mu_M = 1\,\text{MPa}\,\text{s}^{-1}$.

Figure 10 shows a similar simulation replicating the experimental result of a different specimen from the test series reported by Cloete *et al.* [4]. The specimen was nominally identical to that used to obtain the results in Fig. 9 with parameter values $E_{KV} = 10500\,\text{MPa}$, $E_M = 7000\,\text{MPa}$, $\mu_{KV} = 0.1\,\text{MPa}\,\text{s}^{-1}$ and $\mu_M = 0.75\,\text{MPa}\,\text{s}^{-1}$.

In both Figs. 9 and 10 the origin of the numerical results has been shifted to compensate for the "settling" behaviour in the experimental curve.

7. Discussion of results

The numerical results in both Figs. 9 and 10 show good correlation with the experimental results and have consistent parameter values.

In both tests there is a relatively small drop in the stress after the point of maximum stress, which indicates that the contribution of Kelvin-Voigt μ_{KV} damper is small at strain rates up to $100\,\text{s}^{-1}$. This result is consistent with the data in Fig. 2 and similar results reported in the literature [3,8] where the apparent modulus begins to increase exponentially above the ISR regime.

Significant hysteresis is observed in both tests, which is reflected by the large contribution made by the Maxwell element in both simulations. This result is also consistent with the data in Fig. 2 where there is a sudden transition in the apparent modulus in the ISR regime.

The numerical results begin to deviate appreciably from experimental results toward to end of the unloading phase. In particular, the simulated relaxation curves do not remain linear below a strain rate of $50\,s^{-1}$. Furthermore, the numerical simulations do not predict a significant residual strain such as that observed in the bone specimens immediately after the SHB tests. Based on the correlation in Fig. 2, which required the use of a nonlinear Maxwell element [4], it is likely that the model presented in this paper requires further development including additional elements with nonlinear properties.

8. Conclusions and recommendations

An experimental approach to characterize the dynamic viscoelastic relaxation behaviour of cortical bone was proposed. A standard SHB incorporating a CiT striker was used to perform dynamic relaxation tests on bovine cortical bone samples. The use of a CiT striker allowed the bone sample to be compressed at a near constant strain rate during the loading phase. During the unloading phase, the specimen relaxed between the Hopkinson bars without any further interaction with stress waves resulting from the loading phase. The technique displayed good specimen equilibrium and provided consistent results.

The Hopkinson bars were shown to act like linear viscous dampers during the unloading phase, which provided an explanation for the observed near-linear relaxation curves. A standard linear viscoelastic material model was presented and used to replicate the experimental results. The model was able to capture several important features of the stress and strain rate curves, although detailed features toward the end of the unloading phase were not captured and further development of the model is required. Nevertheless, the preliminary results are encouraging and further use of CiT striker concept, or similar techniques, to conduct dynamic relaxation test on bovine cortical bone is recommended.

The authors wish to thank Mr S. Sacks of Sacks Butcheries for supplying the bovine femurs, Mr C. Harris, of the University of Cape Town (UCT) Medical School, for the machining of the bone specimens and Mr P. Smith, of the Department of Mechanical Engineering UCT, for machining the CiT striker and reaction mass. The financial assistance of the National Research Foundation and the University of Cape Town towards this research is hereby acknowledged. Opinions expressed and conclusions arrived at are those of the authors and are not necessarily to be attributed to the National Research Foundation.

References

[1] Adharapurapu R.R., Jiang F. and Vecchio K.S., Mater. Sci. Eng. C **26**, 1325–1332 (2006)

[2] Carter D. and Hayes W., J. Bone Joint Surg. Am. **59**, 954–962 (1977)

[3] Cloete T.J., van der Westhuizen A., Kok S. and Nurick GN, 9th Int. Conf. on the Mechanical and Physical Behaviour of Materials under Dynamic Loading, Brussels, Belgium (DYMAT2009) **1**, 901–907 (2009)

[4] Cloete T.J., Paul G. and Ismail E.B., Phil. Trans. R. Soc. A **372**, 20130210 (2014)

[5] Fung Y.C., *A First Course in Continuum Mechanics, 3rd Edition* (Prentice-Hall, New Jersey, 1994) 193–201

[6] Golman A.J., Danelson K.A., Miller L.E. and Stitzel J.D., Accident Analysis and Prevention **64**, 1–8 (2014)

[7] Gray G.T. III, *ASM handbook, vol. 8, Mechanical testing and evaluation* (ASM International, Materials Park, O.H., 2000) 462–476

[8] Johnson T., Socrate S. and Boyce M., Acta Biomater. **6**, 4073–4080 (2010)

[9] Marais S.T., Tait R.B., Cloete T.J. and Nurick G.N., Lat. Am. J. Solids Struc. **1**, 319–339 (2004)

[10] McElhaney J.H., J. Appl. Physiol. **21**, 1231–1236 (1966)

[11] Saraf H., Ramesh K.T., Lennon A.M., Merkle A.C. and Roberts J.C., Exp. Mech. **47**, 439–449 (2007)

[12] Siviour C.R., Walley S.M., Proud W.G. and Field J.E., Polymer **46**, 12546–12555 (2005)

[13] Yamashita J., Furman B.R., Rawls H.R., Wang X. and Agrawal C.M., J. Biomed. Mater. Res. **58**, 47–53 (2001)

[14] Yoganandan N., Pintar F.A., Stemper B.D., Gennarelli T.A. and Weigelt J.A., Journal of Biomechanics **40**, 227–243 (2007)

EPJ Web of Conferences **94**, 03005 (2015)
DOI: 10.1051/epjconf/20159403005

Effect of micromorphology of cortical bone tissue on crack propagation under dynamic loading

Mayao Wang, Xing Gao[1], Adel Abdel-Wahab[1], Simin Li[1], Elizabeth A. Zimmermann, Christoph Riedel[2], Björn Busse[2], and Vadim V. Silberschmidt[1,a]

[1] Loughborough University, Leicestershire LE11 3TU, UK
[2] University Medical Center Hamburg-Eppendorf, Lottestr. 59, 22529 Hamburg, Germany

Abstract. Structural integrity of bone tissue plays an important role in daily activities of humans. However, traumatic incidents such as sports injuries, collisions and falls can cause bone fracture, servere pain and mobility loss. In addition, ageing and degenerative bone diseases such as osteoporosis can increase the risk of fracture [1]. As a composite-like material, a cortical bone tissue is capable of tolerating moderate fracture/cracks without complete failure. The key to this is its heterogeneously distributed microstructural constituents providing both intrinsic and extrinsic toughening mechanisms. At micro-scale level, cortical bone can be considered as a four-phase composite material consisting of osteons, Haversian canals, cement lines and interstitial matrix. These microstructural constituents can directly affect local distributions of stresses and strains, and, hence, crack initiation and propagation. Therefore, understanding the effect of micromorphology of cortical bone on crack initiation and propagation, especially under dynamic loading regimes is of great importance for fracture risk evaluation. In this study, random microstructures of a cortical bone tissue were modelled with finite elements for four groups: healthy (control), young age, osteoporosis and bisphosphonate-treated, based on osteonal morphometric parameters measured from microscopic images for these groups. The developed models were loaded under the same dynamic loading conditions, representing a direct impact incident, resulting in progressive crack propagation. An extended finite-element method (X-FEM) was implemented to realize solution-dependent crack propagation within the microstructured cortical bone tissues. The obtained simulation results demonstrate significant differences due to micromorphology of cortical bone, in terms of crack propagation characteristics for different groups, with the young group showing highest fracture resistance and the senior group the lowest.

1. Introduction

Bone has a highly heterogeneous structure at a micro-scale level, consisting of Haversian canals, osteons, cement lines and an interstitial matrix. This unique microstructure is essential not only for the daily functionality such as the exchange of nutrition and substance, but also for the physical strength and rigidity of bones. However, aging and degenerative bone diseases can increase the risk of bone-related injuries, which can influence mobility and the quality of life. These two conditions can directly affect the balance of a remodelling process leading to a change of micro-morphology and quality of microstructural constituents. In previous research, bone mineral density (BMD) method [15] was mainly used to evaluate the toughness of cortical bone. It is well known that factors such as disease and aging could reduce bone mineral density [1]. Wachter [2] further pointed out that the BMD has a positive correlation with a yield stress and an overall osteonal area. Therefore, bone specimens from senior and diseased groups behave more fragile than those from a health group. Still, though measurements using only BMD give an imperial relationship between the overall bone's mineral density and its fracture risk, it could not predict a fracture process accurately at the micro-level. Therefore, in recent papers, the distribution of micro-structural constituents of cortical bone became a main target to understand its fracture behaviour related to heterogeneity properties of bone [9,14,16,17]. Experiments employed in these studies included both scanning electron microscopy (SEM) [16] and nanoindentation [17] to analyse microstructural constituents of human cortical bone and their mechanical properties, in order to understand theirs functions and mechanisms in the fracture process. Watanabe [3] tested with X-ray and light microscopy cortical-bone specimens of 72 male and 26 female cadavers' donors, whose age range was from 9 to over 90. The morphological size of osteons and Haversian canals was measured, using collection data to estimate the age of bone. Their result showed that the number of osteons increased with age, while their volume fraction decreased. According to Busse's [4] investigation, the area of Haversian canals in cortical bone grew persistently with the increase of age, resulting in a higher porosity ratio and a lower osteonal ratio in the senior group. In addition, Busse and his group [5] pointed out that different remodelling processes between the senior and young groups could further widen the difference in the porosity ratio between these groups. In the Bernhard's study [6], the dimensions of osteons from four groups, including young, senior, diseased and medically treated ones, were measured and compared with each other. The result showed that the young group had

[a] Corresponding author: V.Silberschmidt@lboro.ac.uk

Article available at http://www.epj-conferences.org or http://dx.doi.org/10.1051/epjconf/20159403005

a smallest size of Haversian canals and the largest size of osteons, while the diseased group had the opposite result.

A better understanding of the effect of micromorphology on bone's behaviour could be achieved with advanced numerical simulations. A numerical study of the crack propagation in heterogeneous materials is a rather challenging task that requires adequate modelling tools. One of the most suitable approaches is an extended finite-element method (X-FEM) that was introduced by Belyschko and Black in 1999 [7]. Since then, it has been used increasingly in simulation of discontinuity problems such as crack propagation. It employs a locally enriched element area to handle a numerical singularity associated with crack opening thus eliminating a need for remeshing. The authors [8,9] employed X-FEM to study the function of cement lines in bone fracture. Budyn *et al.* [12] developed a multiple scale statistical X-FEM simulation method to evaluate crack propagation of cortical bone under uniaxial tension. In another study, the authors [9, 18] utilized X-FEM to evaluate the facture behaviour of cortical bone under three-point bending. However, our previously developed models were based on animal (bovine) bones; there is yet a validated human bone model to have. Additionally, quantitative correlations between the type and distribution of microstructural constituents and the fracture behaviour of human cortical bones in different groups are still unknown. Therefore, in this paper, images from four groups originated from University Medical Center Hamburg-Eppendorf in Germany were measured using Image-Pro and Photoshop software. Then, the data was analysed to establish statistical bone models. Finally, numerical simulations with X-FEM were used to simulate the facture behaviour of cortical bone under dynamic loading condition to understand the relationships between microstructural constituents and crack propagation.

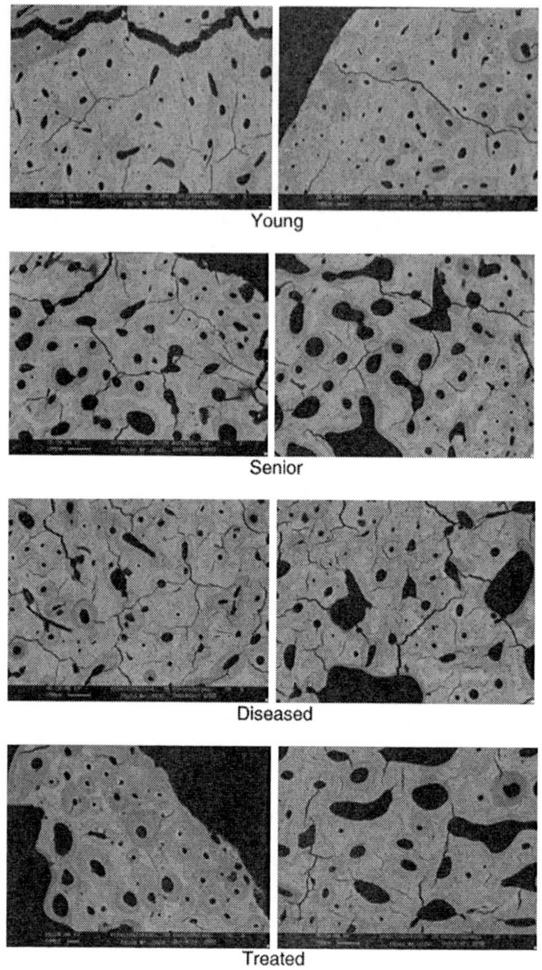

Figure 1. Micro-images of human cortical bones in transverse section of four groups.

2. Materials and methods

2.1. Materials

Cortical-bone specimens acquired from proximal femoral diaphysis (female donors, age between 24 and 88) was scanned using SEM through the horizontal cross-section after the grinding and polishing process [6]. The images can be divided into four groups with two images of each group: young group, senior group, diseased group and medically treated group, as shown in Fig. 1.

2.2. Analysis method

The first step of analysis was to evaluate the volume fractions of microstructural constituents of cortical bones. In order to calculate them accurately, dark edges around the captured images were removed with software prior to measurements to eliminate the effect on porosity. Then, the contrast of images was enhanced to facilitate the differentiation between different micro-constituents using a combination of image process, image automatic recognition and manual operations. In Fig. 2, the original and enhanced images demonstrate clearly that red, green, blue and purple represent porosity (resorption cavities, Haversian canals and osteocytes), osteonal areas, interstitial matrix and microcracks, respectively. Then, the overall number of pixels of various colours could be obtained to estimate the volume fractions of microstructural constituents. In the next step, the size of individual osteons (approximated with elliptical shapes), including dimensions of their long/short axes and the radius of Haversian canals, were measured accurately using Image-Pro software (Fig. 3). Then, the evaluation process utilises mathematical functions to approximate (with best-fit functions) the distribution of each measured parameters statistically (Fig. 4). Finally, these parameters were used as inputs to developed models of human cortical bones with randomly distributed microstructural elements for four studied groups.

3. Model data

This study adopted the X-FEM simulation technique to analyse crack propagation in human cortical bone under dynamic loading condition. The multi-body dynamic

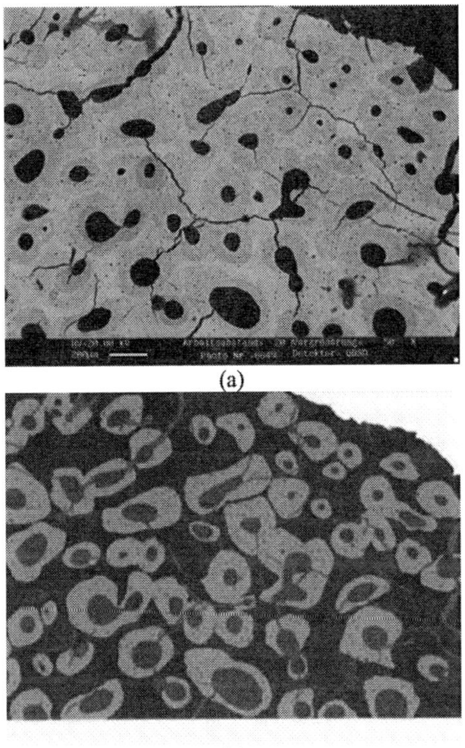

Figure 2. Calculation method of volume fraction using Image-Pro and Photoshop software: initial (a) and enhanced (b) images.

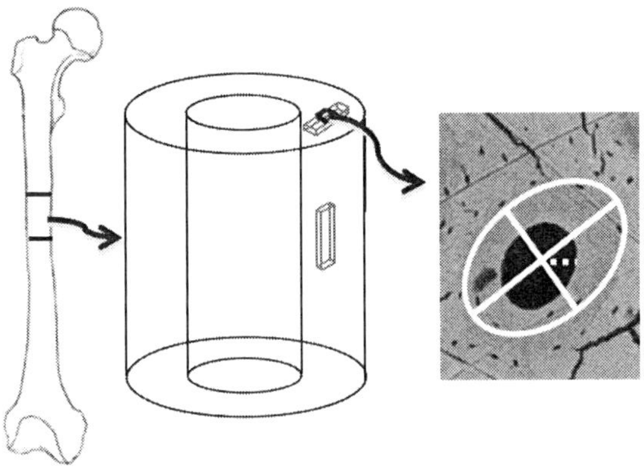

Figure 3. Method of measuring morphological parameters of individual osteons.

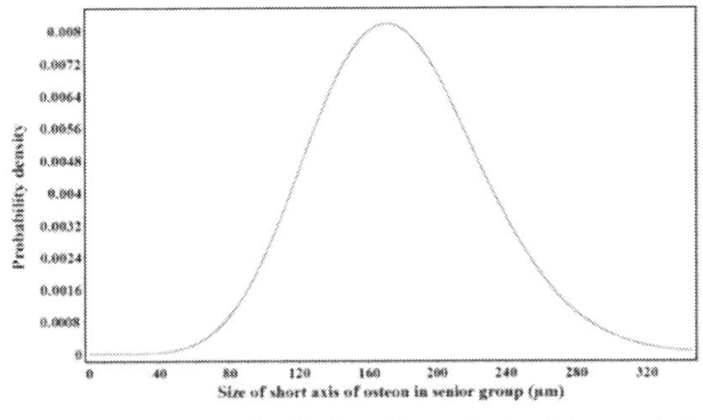

Figure 4. Statistical distribution of magnitude of long axis for osteons in aged group.

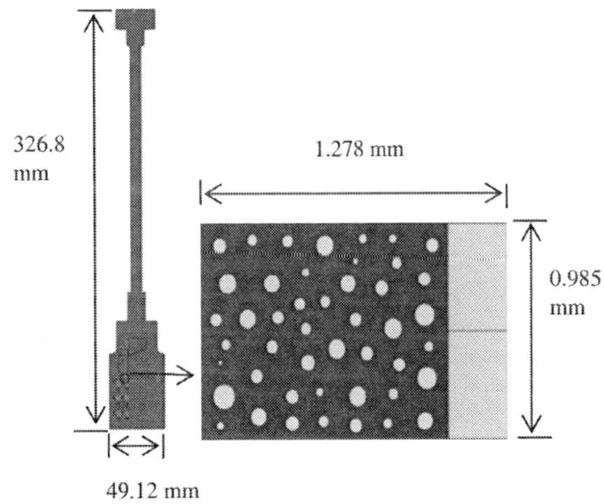

Figure 5. Models of hammer size and microstructured domain in front of initial notch.

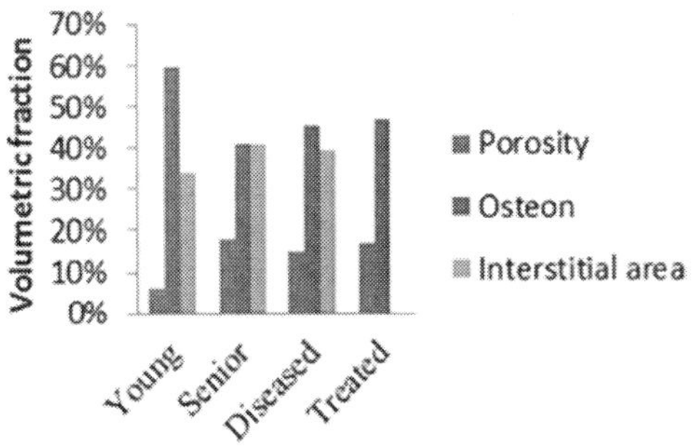

Figure 6. Volume fractions of microstructure constituents of human cortical bone for four groups.

model was established according to the Izod test configuration shown in Fig. 5 [11].

The cortical-a bone specimen was modelled as rectangular area with a pre-notch. The specimen was divided into two domains: a microstructured area of 1.278 mm in length and 0.958 mm in width as an area of interest was embedded in front of the pre-notch into a homogenized bone material domain (50 mm in length and 8 mm in width) to reduce the overall computational cost. Following the setup of the Izod test, the loading conditions were those of a rotational impact with 5.33 rad/s immediately before the moment of contact between a hammer and the specimen. The hammer was modelled as made of carbon steel with isotropic material properties. Its elastic modulus, Poisson's ratio and density were 210 GPa, 0.3 and 7850 kg/m^3, respectively. A master-slave contact interaction between the hammer and the specimen was defined during the impact process. The bottom half of the specimen was fixed using an encastre type of boundary

Table 1. Material properties of microstructural constituents of cortical-bone tissue and homogenous material [12,19–21].

Model	Elastic modulus (GPa)	Poisson's ratio	Strain energy-release rate (N/mm)
Homogeneous material	10.46	0.167	0.422
Osteon	9.13	0.17	0.86
Interstitial matrix	14.122	0.153	0.238
Cement line	6.85	0.49	0.146

Table 2. Average sizes of individual osteons.

	Long axis (μm)	Short axis (μm)	Haversian canal (μm)
Young	219.48	159.8	20.77
Senior	222.5	170.7	43.57
Diseased	205.52	172.28	34.24
Treated	212.52	185.08	30.36

conditions. The element type used in this simulation was 4-node bilinear plain-strain quadrilateral (CPE4R) element. The mechanical properties of different microstructural constituents, such as osteons, interstitial area, and cement line were based on the authors' previous research [9]. The elastic modulus of cement line was 25 percent lower than that of the osteon following the suggestion in Budyn's paper [12].

These simulations were based on the damage initiation and evaluation criteria of X-FEM using a cohesive traction- separation law. The initial horizontal notch with the length of 300 μm was embedded into the homogenized area next to the microstructural area (Fig. 5). The damage in this model was determined based on the elastic-plastic fracture mechanics with the critical maximum principal strain of 0.6% [9] and strain energy release rate in Table 1.

4. Results and discussion

In this study, the results from the analysis showed that the young group had the highest fraction of osteonal area (60.03%), but with the lowest porosity (5.96%) and interstitial area (34.01%). However, the fraction of osteonal area in the senior group was 40.79%, lowest of all groups, while the volume fraction of interstitial area was the highest (41.11%).

The volume fraction of porosity in the diseased group was the second-lowest (after the young group), only 15.22%, lower than that for the senior (18.10%) and the treated groups (17.13%). The results demonstrated that the senior group had the higher porosity area and lower osteon area than the young group [4].

For individual osteons, the radius of Haversian canals in the young group was the lowest (20.77 μm) of the four groups leading to the lowest fraction of porosity. The senior group had the highest size of the long axis – about 222.5 μm, while the highest size of the short axis was in the treated group: 185.08 μm. The results for the volume fractions and the size of individual osteons reflected the

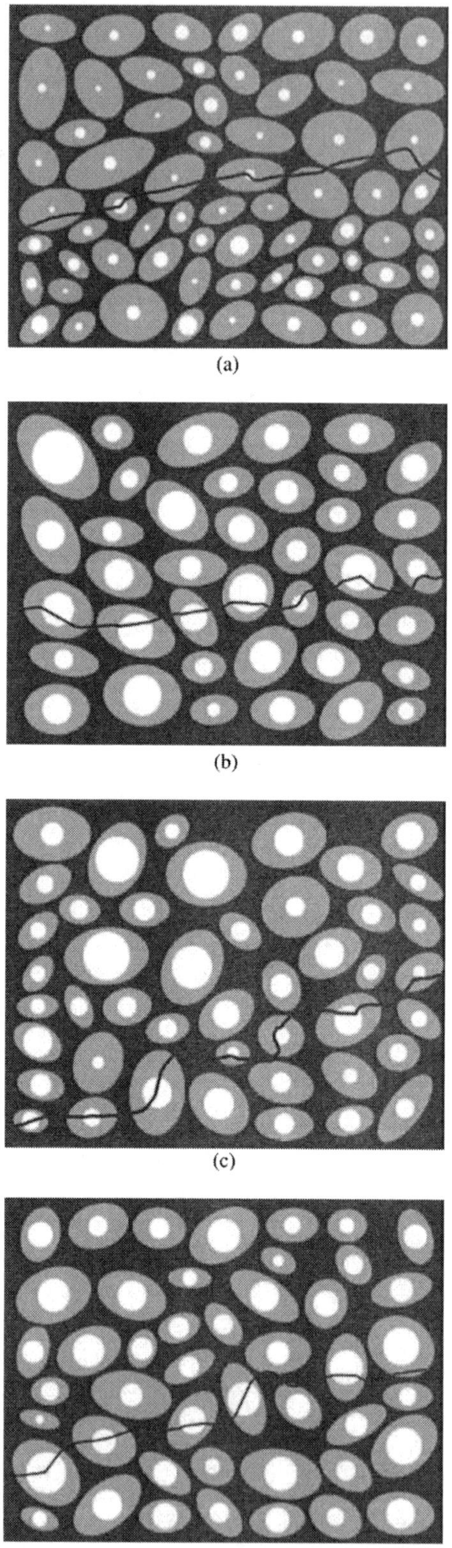

(a)

(b)

(c)

(d)

Figure 7. Crack propagation for X-FEM models of four groups: (a) young; (b) senior; (c) diseased; (d) treated.

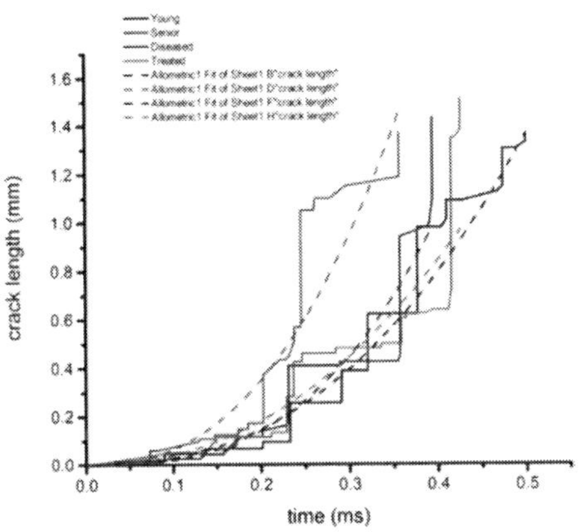

Figure 8. Evolution of crack length during impact.

fact that the aging process and diseases could influence the balance of the remodelling process and directly affect the morphological parameters of osteons.

Four models with different statistical realizations based on the experimental measurements were developed using random distributions of microstructural constituents. The numerical results (Fig. 7) indicated that the crack propagation paths were different in bones of different groups, due to variations in distributions of micro-structural constituents. Crack paths in the four groups demonstrated different crack deflection characteristics [13], with the young group having the least crack deflection path of all other groups. For the diseased and treated groups, the crack paths demonstrated more deflections kinks comparing to the other two groups.

The crack propagation process was additionally influenced by cement lines, and cracks tended to grow through Haversian canals. Previously it was indicated that the presence of cement lines may prevent the crack from destroying other Haversian system during fracture process [14].

Comparing the evolution of crack length with time during the impact for all the groups, the difference was not large, with a standard deviation of 6.75% for the four groups, with an average crack length of 1.419 mm. Among the groups, the young group had the lowest crack length (1.37 mm), while it took the longest time for the crack to propagate through the microstructured area. This means that the structure and parameters of the microstructural constituents in this group resisted the crack propagation through the cortical bone. According to the trend lines for the four groups, the senior group had the lowest toughness, while the young group the highest.

During the first 0.1 ms, the four groups had a similar crack growth rate, while later the senior group demonstrated a higher crack growth rate. Apart from the senior group, the similar trends of crack propagation were noticeable in other three groups up to a crack length of some 0.4 mm; then, the crack in the diseased group grew quicker while the young and treated groups had comparable rates of crack propagation, demonstrating a considerable improvement over the diseased group. The result shows that micromorphology of individual osteons plays a key role in influencing the crack propagation during the facture process of the cortical bone tissue.

5. Conclusion

The relationships between the micromorphology of microstructural constituents of cortical bone and crack propagation under dynamic loading conditions were investigated using the X-FEM simulation method. The statistical characteristics of volume fractions and sizes of microstructural constituents changes due to aging, disease and medication. These morphological changes could affect features of the crack propagation under dynamic loading conditions, such as the deflection crack path and – as in case of cement line – prevent the crack from damaging the bone tissue. Analysis of crack growth with time reflects that the young and treated groups have a higher resistance capability to the crack growth in the cortical bone than the senior and diseased groups. The treated group showed a considerable improvement in this regard compared to the diseased group.

References

[1] M. Grynpas, Calcif. Tiss. Intern. **53**(1), S57–S64 (1993).
[2] N.J. Wachter, G.D. Krischak, M. Mentzel, M.R. Sarkar, T. Ebinger, L. Kinzl, L. Claes, P. Augat, Bone **31**(1), 90–95 (2002).
[3] Y. Watanabe, M. Konishi, M. Shimada, H. Ohara, S. Iwamoto, Forens. Sci. Intern. **98**(1), 55–65 (1998).
[4] B. Busse, M. Hahn, T. Schinke, K. Püschel, G. N. Duda, M. Amling, J. Biomed. Mater. Res. Pt A, **92**(4), 1440–1451 (2010).
[5] A. Bernhard, P. Milovanovic, E. A. Zimmermann, M. Hahn, D. Djonic, M. Krause, S. Breer, K. Püschel, M. Djuric, M. Amling, B. Busse. Osteop. Intern. **24**(10), 2671–2680 (2013).
[6] D. Mulhern, M. Rib. Amer. J. Phys. Anthrop. **111**(4), 519–530 (2000).
[7] T. Belytschko, T. Black, Int. J. Numer. Meth. Engng. **45**(5), 601–620 (1999).
[8] A.A. Abdel-Wahab, A.R. Maligno, V.V. Silberschmidt, Comput. Mater. Sci. **52**(1), 128–135 (2012).
[9] S. Li, A. Abdel-Wahab, E. Demirci, V. V. Silberschmidt, Int. J. Fract. **184**(1–2), 43–55 (2013).
[10] E. Budyn, T. Hoc, J. Jonvaux, Comput. Mech. **42**(4), 579–591 (2008).
[11] A.A. Abdel-Wahab, V.V. Silberschmidt, Eur. Phys. J. Special Topics **206**(1), 41–50 (2012).
[12] E. Budyn, T. Hoc, Eur. J. Comput. Mech. /Rev. Eur. Numér. **16**(2), 213–236 (2007).
[13] M.E. Launey, M.J. Buehler, R.O. Ritchie, Ann. Rev. Mater. Res. **40**, 25–53 (2010).
[14] S. Mohsin, F.J. O'Brien, T.C. Lee, J. Anat. **208**(1), 81–89 (2006).
[15] J.Y. Rho, M.C. Hobatho, R.B. Ashman, Med. Engng Phys. **17**(5), 347–355 (1995).

[16] S. Mischinski, A. Ural, Computer Meth. Biomech. Biomed. Engng **16**(1), 81–94 (2013).
[17] J.Y. Rho, M.E. Roy, T.Y. Tsui, G.M. Pharr, J. Biomed. Mater. Res. **45**(1), 48–54 (1999).
[18] S. Li, A. Abdel-Wahab, V.V. Silberschmidt, Engng Fract. Mech. **110**, 448–458 (2013).
[19] J.L. Katz, H.S. Yoon, S. Lipson, R. Maharidge, A. Meunier, P. Christel, Calcif. Tiss. Intern. **36**(1), S31–S36 (1984).
[20] R.O. Ritchie, J.H. Kinney, J.J. Kruzic, R.K. Nalla, Fatigue Fract. Engng Mater. **28**(4), 345–371 (2005).
[21] A.A. Abdel−Wahab, A.R. Maligno, V.V. Silberschmidt, In: Proc. ASME 2010 10th Biennial Conference on Engineering Systems Design and Analysis (ESDA2010), July 12–14, 2010, Istanbul, Turkey. Paper no. ESDA2010-25329, **1**, 821–829 (2010).

EPJ Web of Conferences 94, 03006 (2015)
DOI: 10.1051/epjconf/20159403006

Intermediate strain rate behaviour of cancellous bone: Links between microstructural and mechanical properties

Marianne Prot[1,a], Trevor Cloete[2], Dominique Saletti[1,3], and Sebastien Laporte[1]

[1] LBM/Institut de Biomécanique Humaine Georges Charpak, Arts et Métiers ParisTech, 151 Boulevard de l'Hôpital, 75013 Paris, France
[2] Blast Impact and Survivability Research Unit (BISRU), Department of Mechanical Engineering, University of Cape Town (UCT), Private bag X3, Rondebosch 7701, South Africa
[3] Univ. Grenoble Alpes, 3SR, 38000 Grenoble, France

Abstract. Relationships between the micro-architecture description of cancellous bone, obtained from medical imaging, and its mechanical properties can be used to assess the compression fracture risk at high and low strain rate. This study extends the rupture prediction to the intermediate strain rate regime. The micro-architecture description was obtained with a CT-scan, for which geometry, topology, connectivity and anisotropy parameters were computed and compared to mechanical identified parameters in order to confirm their usefulness. Three strain rates were investigated: 1/s, 10/s and 100/s using two different devices: a Wedge-Bar apparatus and a conventional split Hopkinson pressure bar implemented with a Cone-in-Tube striker and a tandem momentum trap. This setup provides a constant strain rate loading with routine specimen recovery allowing the fracture zone to be investigated. This study reveals that a transition in the response behaviour occurred in the intermediate regime and confirms the significant porous organization influence through the regimes.

1. Introduction

Human musculoskeletal modeling requires accurate knowledge of bone failure response including cancellous bone fracture. Literature studies have focused on extreme loadings (low [1] and recently high [2,3] strain rates). The intermediate strain rate (ISR) regime (1 and 10/s) has been characterized [4,5] but the observed scatter was not investigated or linked to the upper ISR limit (100/s) and extreme strain rates. Detailed features of the micro-architecture of porous cancellous bone [6] have been linked to the quasi-static behaviour, to the dynamic and confined dynamic loading [7] to predict the fracture localization [8], the micro-damage [9] and the global mechanical response [10].

In this study, bovine bone specimens were compressed up to the point of trabeculae collapse, at 3 strain rates in the ISR regime (1/s to 100/s) using two different devices: a Wedge-Bar apparatus [11,12] and a conventional split Hopkinson pressure bar (SHPB) implemented with a Cone-in-Tube (CiT) striker [12] and a tandem momentum trap [13]. Both testing techniques allow for specimen recovery after a single loading event. All the specimens were micro-CT scanned before and after testing to investigate the pre- and post-compression architecture.

The aim of this study is to quantify the links between cancellous architectural descriptors and macroscopic mechanical behaviour including fracture initiation under ISR compression loading.

[a] Corresponding author: marianne.prot@ensam.eu

2. Methods

2.1. Samples

42 cylindrical bovine cancellous bone specimen (length: 7.5 mm; diameter: 10.5 mm) were slowly cut from 6 different femoral bones in order not to damage the peripheral trabeculae. The anatomical position and orientation of each specimen was recorded (i.e.: femoral head versus great trochanter and perpendicularity with respect to one of 3 anatomic planes: sagittal, transversal, frontal). The non-defatted specimens were stored in a frozen condition in a saline solution. Samples were thawed for 24 hours at + 5 °C before being brought to room temperature prior to experiments.

2.2. Pre-compression microstructural properties

A General Electric phoenix v|tome|x L240/NF180 micro-CT scanner was used to identify the specimen architecture with a low dose of radiation. X-ray settings were 70 kV and 350 microA, acquisition time of 500 ms per image, with no averaging and no skipping of images, for a resolution of approximately 80 μm. The Otsu [14] multi-threshold segmentation technique was used to automatically separate bone from marrow and background. In accordance with recommendations in the literature [1,15], architectural parameters were extracted from 3D and skeletonized images using the codes CTan and BoneJ. The eleven architectural parameters summarized in Table 1, are divided into descriptors of geometry, topology, connectivity and anisotropy.

Table 1. Summary of architectural parameters.

Parameter	Description	Unit
BV/TV	Bone Volume/ Total Volume	%
Tb.Th	Mean thickness of trabeculae	mm
Tb.Sp	Trabecular Separation	mm
Conn.D	Connectivity Density or number of trabeculae per unit volume	/mm3
FD	Fractal Dimension	
SMI	Structure Model Index	
N.Nd	Number of Junctions	/mm3
N.Tp	Number of Triple Point, i.e. Number of Junctions with 3 branches	/mm3
N.Qp	Number of Quadruple Point, i.e. Number of Junctions 4 branches	/mm3
DA	Normalized Degree of Anisotropy	
MIL	Mean Intercept Length (3 parameters)	mm

Table 2. Specimen test repartition.

Regime	Intermediate		
Strain Rate	1/s	10/s	100/s
Apparatus	Wedge	Wedge	SHPB
Number of Specimens	11	13	18

2.3. Experimental technique

Experiments were performed as described by the authors in a previous study [16]. Two different devices were used. A conventional SHPB system implemented with a CiT striker and a momentum trap system was used to compress specimens at 100/s while a Wedge-Bar apparatus was used to load specimens at 1/s and 10/s. The recovery method was set in order to stop the loading after 10% strain. ISR compression tests using the Wedge-Bar were filmed using a Photron Fastcam -APX RS250k at 50 000 frames per seconds for a resolution of 128 per 192 pixels^2.

The samples with observed architectural abnormalities or with too much noise in the signal were discarded from the study. The number of specimen tested at the three strain rates is given in Table 2.

2.4. Mechanical properties

A robust automated procedure was developed to determine the mechanical properties through all the three different strain rates. Inspired by a logistic risk curve [17], the stress strain behaviour was model with a 5 parameters law. A robust non-linear least square method was used to set the apparent Young's Modulus, E_{app}, as the inflexion point slope, the yield Stress and Strain, ε_y & σ_y, the ultimate Stress and Strain, ε_u & σ_u. In addition, error estimates were obtained using the bootstrap method [18].

2.5. Data analysis

A Kruskal-Wallis test was used to assess that there is no difference between the micro-architectural values in the 3 sets of strain rates. A p-value of 0.05 was chosen as the upper threshold of significance. Statistical tests were performed using XLSTAT© software.

The limited number of specimens dedicated to each strain rate could adversely affects the accuracy of the Spearman tests usually used to highlight correlation between architectural and mechanical parameters. The Mann-Whitney statistical test was previously performed to appraise the strain rate grouping into one intermediate regime more trustworthy for cross-correlations.

Table 3. Mean values and standard deviation (italic) of architectural parameters. Non-exhaustive list.

	BV/TV	Tb.Th
1/s	54.8 (*7*)	0,32 (*0.003*)
10/s	51.5 (*9*)	0,32 (*0.003*)
100/s	51.2 (*14*)	0,33 (*0.003*)

The mechanical parameters sensitivity to strain rate needs to be carefully considered as 6 different bones and 3 anatomical orientations were used for sampling. A variance analysis was used to qualitatively investigate the relative influence of the bone provenance, the anatomic direction, the location in the femur and the strain rate. The influence quantification was appraised using the Kruskal-Wallis test.

2.6. Post- compression investigation

A similar procedure to the pre-compression architecture acquisition was completed after testing. Specimen scanning thresholds were set to the same level as the pre-compression procedure. Pre- and post-compression specimen image registrations were done on Avizo©. Specimen scans were superimposed using a rigid transformation and used to compute a histogram-based correlation between the model and the reference.

3. Results and discussion

3.1. Microstructure properties

Mean values and standard deviations of microstructure parameters are given in Table 3.

The magnitude of the observed variations is consistent with previously published data on bovine cancellous bone [15].

The difference between architectural parameters at the three strain rates was found to be insignificant by the Krusal-Wallis test. Then the data can be treated as a single dataset.

3.2. Mechanical properties

A typical example of stress-strain behaviour with its mechanical parameters identification is presented in Fig. 1.

The method, providing good results and was not biased by noise while effectively dealing with the toe region of the curve. A maximum estimation error was observed at 10/s due to load cell noise.

The values of mechanical parameters are broadly in agreement with the literature dedicated to bovine femur properties under quasi-static loading [19]. Cancellous bone is known to be strain rate dependent [20]. The results are lower than those for the dynamic characterization of bovine tibia [21] but higher than human tibia results [3]. Results were in accordance with the previous statement

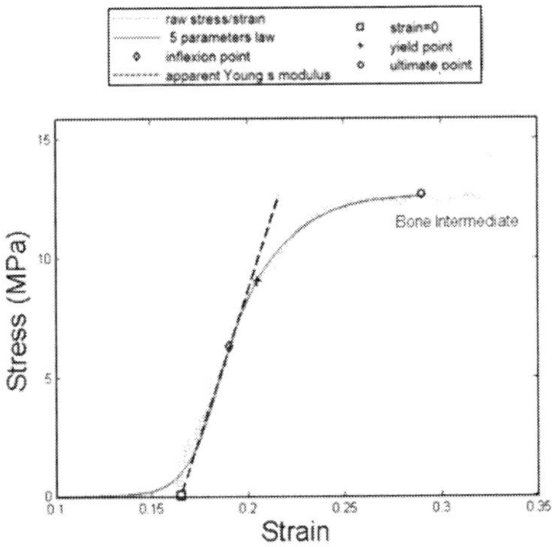

Figure 1. Mechanical parameters determination at 1/s.

regarding E_{app}, σ_y & σ_u while the yield and ultimate strains appear higher at 1/s. For this specific behaviour, stick and slip mechanism and time effect were discarded thanks to footage and the raw data. Value variability is also higher at 1/s. Further investigations on the marrow effect at intermediate strain rate loadings will be developed in order to explain this difference.

The Mann Whitney test was used to appraise the strain rate grouping. The results for ε_y and ε_u confirm that a distinct transition occurs between the three strain rates. The value of E_{app}, appears to be similar for 10/s and 100/s while results at 1/s are close to the quasi-static case. Finally the yield and ultimate stress confirmed that all the three strain rates can be considered as one intermediate regime.

3.3. Sensitivity of the behaviour

The Fig. 2 presents a non-exhaustive illustration of the ultimate stress sensitivity to the anatomic planes (Fig. 2(a) and the sampling location in the proximal femur (Fig. 2(b).

The variance results confirms that the plane supporting the body weight is presenting a higher σ_u. The higher ultimate stress value from the femoral head location versus the values from the great trochanter location also supports the anatomic hypothesis regarding the real loading [22]. Mechanical parameters were consistent between the six bones with a maximum Kruskal-Wallis p-value of 0.54 for σ_y.

Due to the limited number of specimen tested in regard to all the configurations (bone provenance, femoral head/great trochanter, anatomic plan, strain rates), no cross-interaction could be studied.

Samples extracted from the same bone and adjacent anatomical positions are compared in Fig. 3.

As suggested by the literature focused on quasi-static [23,24] and dynamic strain rates [7], an architecture effect (geometry, connectivity, topology, anisotropy) contributes to the experimental scatter observed in Fig. 3.

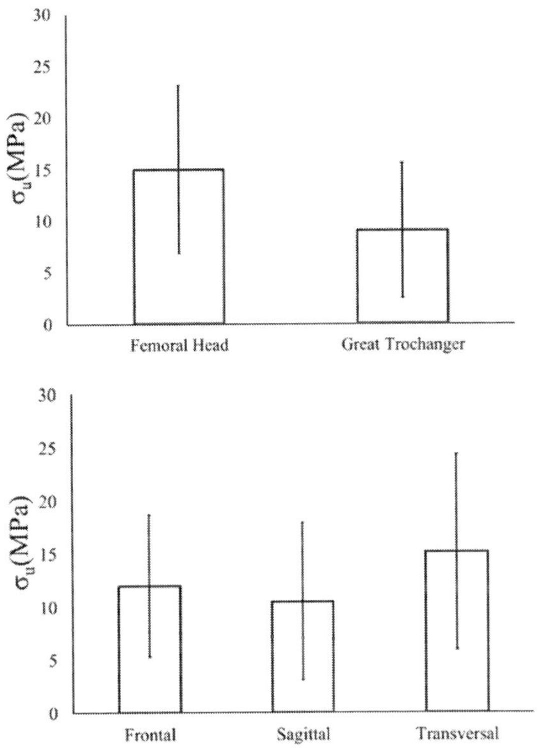

Figure 2. Ultimate stress sensitivity: a) to the anatomic planes, b) to sampling location in the proximal femur.

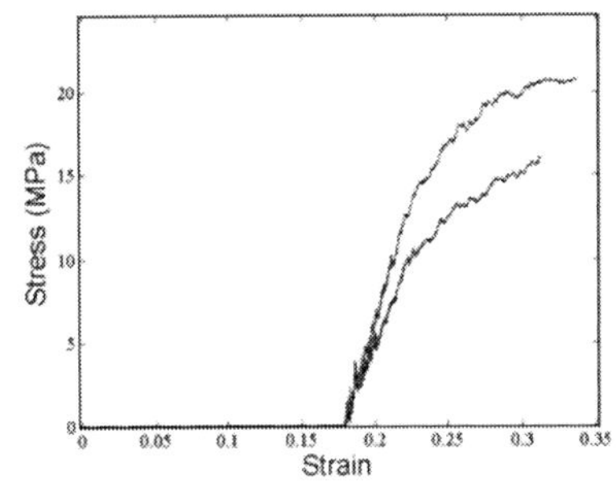

Figure 3. Comparison of two closely extracted specimens at strain rate of 1/s (raw curve).

Table 3. Mean values and standard deviation (italic) of identified mechanical parameters.

	E_{app}(MPa)	ε_y(%)	σ_y(MPa)	ε_u(%)	σ_u(MPa)
1/s	286 (7)	4 (2)	8 (5)	11 (5)	12 (8)
10/s	543 (9)	2 (1)	10 (6)	5 (2)	14 (8)
100/s	629 (14)	2 (0)	11 (7)	3 (1)	13 (9)

3.4. Relationships between parameters

The micro-architecture description obtained with a CT-scan was compared to mechanical identified parameters in order to confirm their usefulness. Spearman test results for combined strain rates are presented in Table 4.

The mechanical parameters presented are linked with geometry and topology architectural parameters such as

Table 4. Spearman test results. Non-exhaustive list. *: Correlation between the parameters (p-value <0.05).

Variables	BV/TV	Tb.Th	Tb.Sp	Tb.N	Conn.D	SMI	FD
E_{app}	*	*			*	*	*
σ_y	*	*	*	*	*	*	*
σ_u	*	*	*	*	*	*	*

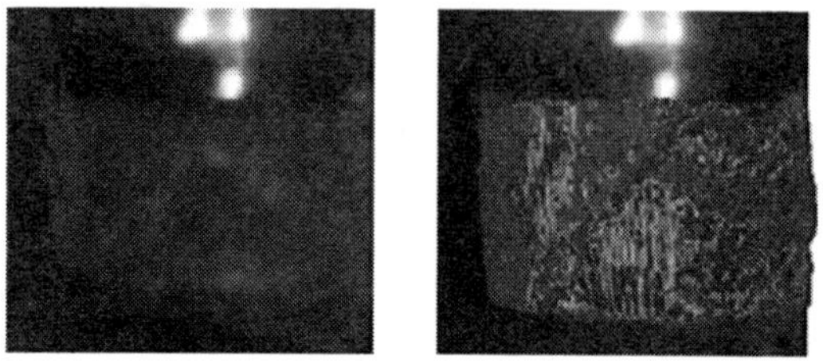

Figure 4. a) Footage at 10/s. b) Distance illustration between the surface of the post and pre- registered compressed reconstructed 3D CT-scans.

the connectivity parameters (see Table 1) not presented in this Table 4. However, the anisotropy parameters (3 MILs and DA) are not linked with the behaviour. A projection along the compression axis will overcome the bias inherent in the 3 anatomical plane considered in the study.

Architectural parameter outliers were included in the data set as their effect, while noticeable, was not found to significantly affect the correlations.

The resulting correlations generally concur with the published quasi-static [23,24] and dynamic [7] published study. Trabecular separation, Tb.Sp, was not found to correlate with the Young's Modulus. This result is in contrast with result reported in the literature where such correlation was obtained. The resolution of this issue will require further test using the robust data processing methodology presented in this paper.

3.5. Post- compression investigation

Specimens were successfully recovered after a single loading event of a predetermined duration at a strain of approximately 10%. This allows the fracture localization to be investigated as depicted in Fig. 4.

The displacement evaluation of the pre- and post-registered surface extracted from the micro-scans appears to provide a good correlation with the observed fracture zone. Ongoing work to quantifying this fracture behaviour will allow a more precise description that could be implemented in cancellous bone architecture-based model.

4. Conclusion

This study presents the relationship between cancellous bone architectural parameters and mechanical response parameters at low and high strain rates in the intermediate strain rate regime (ISR: 1 to 100/s). This work presents a new approach to characterizing the mechanical strength of cancellous bone subjected to intermediate loading rates via architectural parameters descriptions. These results can be implemented in material models for cancellous bone for the prediction of typical dynamic fractures in both the lower to the higher strain rate regimes.

The authors wish to thank Mr. S. Sacks of Sack Butcheries for supplying the bovine femurs, Mr. C. Harris (UCT, Medical School) for the machining of the specimens, Mr. P. Smith and Mr. G. Newins (UCT, Department of Mechanical Engineering) for machining the setup components and the UCT Research Committee for the financial support to conduct this work.

References

[1] H. Follet, S. Viguet-Carrin, B. Burt-Pichat, et al., J. Orthop. Res., **29(4)**, 481–488 (2011)
[2] A. Pilcher, W. Wang, Z. Kaltz, et al., J. Biomech. Eng., **132(8)** (2010)
[3] C.K. Teja, A. Chawla, S. Mukherjee, Int J Crashworthiness, **18**(1), 11–18 (2013).
[4] D.R. Carter, W.C. Hayes, J. Bone Joint Surg. Am, **59**, 954–962 (1977)
[5] F. Linde, P. Norgaard, A. Odgaard, et al., J. Biomech., **24(9)**, 803–809 (1991)
[6] M. Stauber, A. Nazarian, R. Muller, J. Bone Min Res, **29(1)**, 134–141 (2006)
[7] M. Prot, D. Saletti, S. Pattofatto, V. Bousson, S. Laporte, J. Biomech, **48 (3)**, 498–503 (2015)
[8] C.J. Hernandez, F.M. Lamber, J. Widjaja, C. Chapa, C.M. Rimnac, Bone, **66**, 205–213 (2014)
[9] S. Tassani, G.K. Matsopoulos, J. Biomech., **60**, 78–86 (2014)
[10] B. Zhou, X. Sherry Liu, J. Wang, et al., J. Biomech., **47(3)**, 702–708 (2014)
[11] T.J. Cloete, M. Stander, *EPJ Web of Conf*, **26**, 01025 (2012)
[12] T.J. Cloete, G. Paul, E.B. Ismail, Philos Trans A Math Phys Eng Sci, **372** (2015)
[13] M. Prot, T.J. Cloete, S. Pattofatto, *EPJ Web of Conf*, **26**, 03003 (2012)
[14] N. Otsu, Automatica, **11(285–296)**, 23–27 (1975)
[15] A. Syahrom, MRA. Kadir, J. Abdullah & A. Ochsner, **49(12)**, Med. Biol. Eng. & Comp. 1393–1403 (2011)
[16] M. Prot, T.J. Cloete, D. Saletti, S. Laporte, Comput. Meth. Biomec., **17(S1)**, 50–51 (2014)
[17] J.H. Ricketts, G.A. Head, Am. J. Phy. **277(2)**, R441-R454 (1999)
[18] B. Efron, R.J. Tibshirani, *An introduction to the bootstrap, CRC press*, **57** (1994)
[19] V. Kefalas, D.A. Eftaxiopoulos, J. Mech. Behav. Biomed., **6**, 41–52 (2012)
[20] R.M. Guedes, J.A. Simoes & JL. Morais, J. Biomech., **39(1)**, 49–60 (2006)
[21] A. Pilcher, X. Wang, Z. Kaltz, J.G., Garrison, G.L., Niebur, J., Mason, W. Chen, J. Biomed. Eng. **132(8)**, 081012 (2010)
[22] P.A. Hulme, S.K. Boyd, S.J. Ferguson, Bone **41(6)**, 946–957 (2007)
[23] E. Mittra, C. Rubin, Y.X. Qin, J. Biomech. **38(6)**, 1229–1237 (2005)
[24] J. Teo, K.M. Si-Hoe, J.E.L. Keh, S.H. Teoh, Clin. Biomech., **21(3)**, 235–244 (2006)

EPJ Web of Conferences **94**, 04001 (2015)
DOI: 10.1051/epjconf/20159404001

Evaluation of a strain based failure criterion for the multi-constituent composite model under shock loading

Christopher T. Key[1,a], Shane C. Schumacher[2], and C. Scott Alexander[2]

[1]HI-TEST Laboratories Inc., Applied Technologies Group, 32 Murphys Drive, Groton, CT 06340, USA
[2]Sandia National Laboratories, PO Box 5800, Albuquerque, NM 87185, USA

Abstract. This study details and demonstrates a strain-based criterion for the prediction of polymer matrix composite material damage and failure under shock loading conditions. Shock loading conditions are characterized by high-speed impacts or explosive events that result in very high pressures in the materials involved. These material pressures can reach hundreds of kbar and often exceed the material strengths by several orders of magnitude. Researchers have shown that under these high pressures, composites exhibit significant increases in stiffness and strength. In this work we summarize modifications to a previous stress based interactive failure criterion based on the model initially proposed by Hashin, to include strain dependence. The failure criterion is combined with the multi-constituent composite constitutive model (MCM) within a shock physics hydrocode. The constitutive model allows for decomposition of the composite stress and strain fields into the individual phase averaged constituent level stress and strain fields, which are then applied to the failure criterion. Numerical simulations of a metallic sphere impacting carbon/epoxy composite plates at velocities up to 1000 m/s are performed using both the stress and strain based criterion. These simulation results are compared to experimental tests to illustrate the advantages of a strain-based criterion in the shock environment.

1. Introduction

Composite materials are often used in shock loading applications caused by high-velocity impacts, where the strain rates can reach on the order of 10^7 and the pressures are often in the hundreds of kbar. Hydrocodes are commonly used to simulate material response under such conditions. Hydrocodes typically separate a material's response into volumetric and deviatoric components as given in Eq. (1). The deviatoric (s_{ij}) or distortional response of the material is modeled with a constitutive model, while the volumetric (P) response is modeled using an equation of state (EOS).

$$\sigma_{ij} = s_{ij} + P\delta_{ij}. \tag{1}$$

The work presented in this paper utilizes the shock physics hydrocode CTH [1] to model unidirectional composite materials under high velocity impacts where the pressures can reach very high levels. As a result of these high pressures, the pressure term on the right hand side of Eq. (1) can dominate the material's stress response. As a consequence, damage and failure predictions using traditional stress based failure criterion can often times be misleading. Factors which influence these misleading results include; stiffness and strength increases at elevated pressures [2], the retention of shearing resistance under compression and the fact that these materials often times cannot react on the time scales of the pressure wave propagations.

Figure 1 shows an example of the pressure dependent strength behavior for a graphite/epoxy composite material. This data shows a nearly 2 times increase in the matrix dominated failure modes of the material as a result of the applied pressure being increased from standard atmospheric conditions to 6 kbar. Figure 1 also shows that fiber dominated failure modes are less influenced by the effect of pressure than are matrix modes. These observations lead to the suggestion that a deformation or strain-based criterion may be more suitable for shock loading problems.

In what follows, we briefly outline the composite material constitutive model and the equation of state model used for these initial studies. We then detail both the stress and strain-based failure criterion used along with the experimental test details used to support that analytical work. Finally, we compare the results of the two failure criterion against the experimental test results of a carbon-epoxy panel subjected to spherical projectile impacts.

2. Composite material models

2.1. Constitutive model

To simulate the anisotropic response of a unidirectional composite material, the multi-constituent Composite Model (MCM) within the CTH constitutive material library was utilized. The MCM model is based on the work of Hill [3] and adaptation of Garnich and Hansen [4], where a detailed derivation of the approach and corresponding decomposition relationships can be found.

This MCM model is based upon a multicontinuum concept, which provides constituent (fiber and matrix) averaged stress and strain fields at the continuum level. To do this, the MCM model assumes that at each point

[a] Corresponding author: chris.key@hitestlabs.com

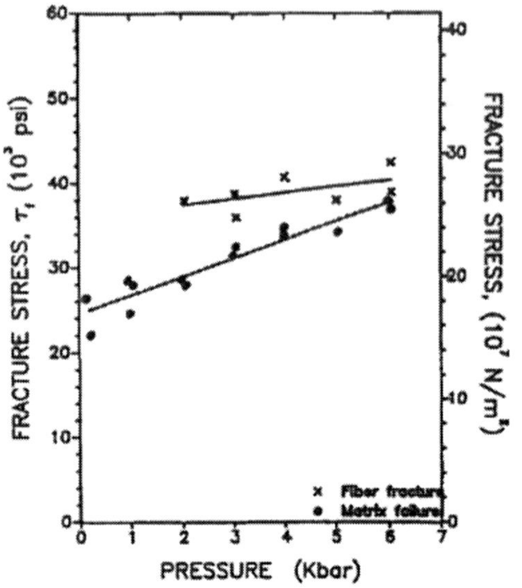

Figure 1. Pressure dependent strength data for fiber and matrix fracture modes in composite materials [2].

within a structure, there exists a representative volume element (RVE). For the unidirectional composite material considered in this work, Fig. 2 illustrates the hexagonally packed basic repeating unit microstructure used, where the red regions represent the carbon fibers and the blue regions are the epoxy matrix.

The volume averaged stress fields for each of the constituents, which comprise the RVE, can be expressed as:

$$\bar{\sigma}_i = \frac{1}{V_i}\int_{D_i} \bar{\sigma}(X)dV. \quad (2)$$

For the case of a unidirectional composite as shown in Fig. 2, i = 2, representing the fiber and matrix constituent.

Volume averaging leads to the relationship in Eq. (3) between the homogenized composite and constituent stress fields.

$$\bar{\sigma} = \phi_{matrix}\bar{\sigma}_{matrix} + \phi_{fiber}\bar{\sigma}_{fiber}. \quad (3)$$

Where in this equation ϕ_{matrix} and ϕ_{fiber} are the volume fractions of the matrix and fiber, respectively.

Combining Eqs. (2) and (3) with the linear elastic constitutive relationships for the composite and constituents, the following strain decomposition equation can be derived.

$$\{\varepsilon_{matrix}\} = \left(\phi_{matrix}[I] + \phi_{fiber}[A]\right)^{-1}\{\varepsilon\} \quad (4)$$

where,

$$[A] = -\frac{\phi_{matrix}}{\phi_{fiber}}\left([C] - \left[C_{fiber}\right]\right)\left([C] - [C_{matrix}]\right). \quad (5)$$

Equation (4) provides the volume averaged continuum strain field in the matrix constituent as a function of the composite strain field (ε) and the fiber (C_{fiber}), matrix (C_{matrix}) and composite (C) stiffness matrices. Next, using the matrix averaged strain field, the fiber averaged strain

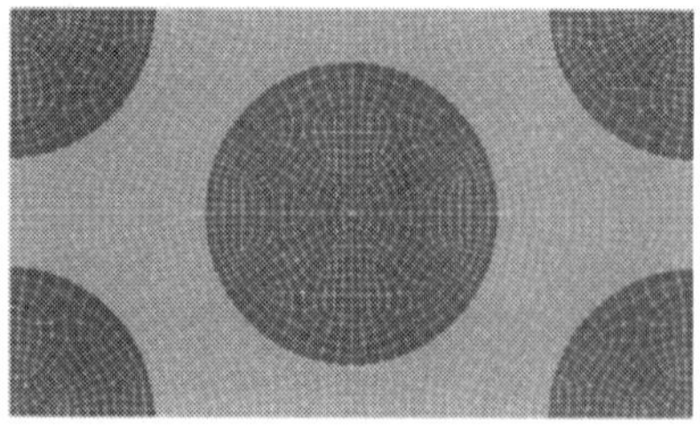

Figure 2. Hexagonally packed unidirectional composite microstructure.

field can be found by applying a similar volume averaging scheme as used in Eq. (3) to yield:

$$\left\{\epsilon_{fiber}\right\} = \frac{1}{\phi_{fiber}}\left(\{\varepsilon\} - \phi_{matrix}\{\varepsilon_{matrix}\}\right). \quad (6)$$

Finally, the average stress fields for the fiber and matrix can be calculated using their respective linear elastic constitutive relationship.

2.2. Equation of state coupling

One implication of the anisotropy of the unidirectional composite material considered in this work, is the inherent coupling between the deviatoric and volumetric responses of the material. This coupling causes a problem in a hydrocode, where the deviatoric and volumetric responses are separated. Therefore, the MCM model must account for the fact that hydrostatic pressure applied to a finite volume causes not only a volume change, but also a shape or distortional change.

The coupling is accounted for within the MCM model using the relationships developed by Lukyanov [5]. Lukyanov's relationships modify the pressure term, which results from the EOS, such that an applied pressure only causes a volume change without a distortional change. This approach therefore retains the traditional definition of pressure.

Lukyanov expressed the pressure in term of a tensor rather than the traditional scalar according to Eq. (7).

$$\tilde{P} = -\left(P^*\right)\alpha_{ij}. \quad (7)$$

Where P^* is the scalar pressure defined as:

$$P^* = P^{EOS} + \frac{\beta_{ij}s_{ij}}{\alpha_{ij}\beta_{ij}}. \quad (8)$$

In Eq. (8), the scalar pressure is defined as the summation of the Mie-Grüneisen [6] EOS pressure (P^{EOS}) and a coupling term which is a function of the deviatoric stress tensor (s_{ij}). This coupling term is a function of the α_{ij} and β_{ij} tensors, which are composed of the material stiffness and compliance, respectively. These coupling terms are formulated such that when combined with the deviatoric stress as shown in Eq. (8), the deviatoric strains are zero. The reader is referenced to Lukyanov [5] for a complete derivation and description of these terms, as the details are not included here for brevity.

Finally, combining the relationship of Eq. (8) with the stress decomposition relationship of Eq. (1), Lukyanov arrives at the following relationship for the deviatoric stress:

$$s_{ij} = \sigma_{ij} - \alpha_{ij} \cdot \frac{1}{||\alpha||}\sigma_{kl}\alpha_{kl}. \tag{9}$$

Equation (9) accounts for the coupling effects of the pressure on the deviatoric stresses just as Eq. (8) accounted for the coupling effects of the deviatoric stresses on the pressure.

The reader is referenced to Schumacher and Key [7] for a detailed description of the anisotropic equation of state coupling model and its implementation within CTH.

2.3. Failure criteria

The failure criteria used within the MCM model for this work is based on the unidirectional composite material failure criterion of Hashin [8]. Hashin's criterion predicted constituent level failure modes based on the stress fields of the homogenized composite. Mayes and Hansen [9] modified Hashin's criterion such that the fiber and matrix failure modes were predicted as a function of their individual volume averaged continuum stress fields, which are a direct result of the MCM model. Here we provide a brief review of the constituent level failure criteria of Mayes and Hansen. Further details of the development of these criterions can be found in Mayes [10] and Mayes and Hansen [9].

2.3.1. Stress-based failure criterion

The MCM criteria is based on the assumption that unidirectional fiber reinforced composite materials are transversely isotropic. Therefore, the criteria can be expressed in terms of transversely isotropic stress invariants [11] given in Eq. (10).

$$\begin{aligned} I_1 &= \sigma_{11}, \\ I_2 &= \sigma_{22} + \sigma_{33}, \\ I_3 &= \sigma_{22}^2 + \sigma_{33}^2 + 2\sigma_{23}^2, \\ I_4 &= \sigma_{12}^2 + \sigma_{13}^2, \\ I_5 &= \sigma_{22}\sigma_{12}^2 + \sigma_{33}\sigma_{13}^2 + 2\sigma_{12}\sigma_{13}\sigma_{23}. \end{aligned} \tag{10}$$

Using the quadratic terms from Hashin's original criterion, the transversely isotropic stress invariants from Eq. (10) and simplifications detailed by Mayes and Hansen, the following general form of the MCM failure criterion is arrived at:

$$A_1 I_1^2 + A_2 I_2^2 + A_3 I_3 + A_4 I_4 = 1. \tag{11}$$

Where the A terms are strength coefficients which are functions of the tensile, compressive and shear strength of the material.

Equation (11) can be simplified to predict matrix failure using the observation that conventional composite materials are composed of fibers which have significantly higher strengths than the matrix which bonds them together. Therefore, transverse failure of a unidirectional composite material is controlled by the strength of the matrix. If one assumes that the fibers of a unidirectional fiber reinforced composite run in the 1-direction, the A_1 coefficient in Eq. (11) can be set to zero, as it is only a function of the longitudinal stress (σ_{11}) and the resulting matrix failure criterion used in the MCM model is then:

$$^{\pm}A_2^m \left(I_2^m\right)^2 + A_3^m I_3^m + A_4^m I_4^m = 1. \tag{12}$$

The matrix strength coefficients in Eq. (12) are determined through application of in-plane shear, transverse shear, transverse tension and transverse compression uniaxial load cases resulting in the matrix failure strength coefficients give in Eq. (13).

$$\begin{aligned} ^{\pm}A_2^m &= \frac{1}{\left(^{\pm}S_{22}^{22m} + {}^{\pm}S_{33}^{22m}\right)^2} * \left(1 - \frac{\left(^{\pm}S_{22}^{22m}\right)^2 + \left(^{\pm}S_{33}^{22m}\right)^2}{2\left(S_{23}^m\right)^2}\right) \\ A_3^m &= \frac{1}{2\left(S_{23}^m\right)^2} \\ A_4^m &= \frac{1}{\left(S_{12}^m\right)^2}. \end{aligned} \tag{13}$$

In Eqs. (12) and (13) the $\pm$ symbol indicates a dependence on the parameters for tensile vs. compressive stress.

Similar to the matrix failure criterion approach, one can assume that the higher strength of the fibers means the longitudinal failure of the composite material is controlled by the strength of the fibers. Hence, the A_2 and A_3 terms in Eq. (11) are set equal to zero as they are related to the transverse stresses (σ_{22} and σ_{33}). The resulting fiber failure criterion used in the MCM model is then:

$$^{\pm}A_1^f \left(I_1^f\right)^2 + A_4^f I_4^f = 1. \tag{14}$$

The coefficients for the fiber failure criterion of Eq. (13) are determined by applied uniaxial loading conditions of in-plane shear, longitudinal compression and longitudinal tension, resulting in the fiber failure strength coefficients given in Eq. (15).

$$\begin{aligned} ^{\pm}A_1^f &= \frac{1}{\left(^{\pm}S_{11}^f\right)^2} \\ A_4^f &= \frac{1}{\left(S_{12}^f\right)^2}. \end{aligned} \tag{15}$$

2.3.2. Strain-based failure criterion

The strain-based failure criterion is identical in form to the stress-based failure criterion presented previously. The main difference being that the invariants are now the transversely isotropic strain invariants and the resulting strength variables (S) in Eqs. (13) and (15) are the ultimate strain values of the composite material rather than the ultimate stress values.

3. Experimental testing details

The high velocity impact experiments for this work were performed at Sandia National Laboratories using single and two stage light gas guns. The gas guns were used to fire 1/4 inch diameter steel spheres at velocities from 250 to 1000 m/s. The test setup included flash x-ray or high-speed cameras to capture the penetration event, panel motion and projectile residual velocity if applicable. Following each impact test, the composite panel was inspected using ultrasound to measure the extent of damage that occurred during the impact event.

The composite material system used to fabricate the panel was IM7 carbon fibers pre-impregnated with a toughened epoxy (8552) from Hexcel. Each composite target panel was fabricated to be 7.5 inches by 7.5 inches square with a fiber volume fraction of 68% and thicknesses ranging from 1/4 to 1 inch. The lamination schedule of each panel was quasi-isotropic $[0/90/+45/-45/-45/+45/90/0]_a$, with the repetitions of this base layup (a) being repeated as necessary to achieve the overall desired panel thickness.

In this paper we focus only on two of the tests conducted which demonstrate the extremes of the composite damage. These two tests, impacting 1/2 inch thick panels, include an impact velocity of 506 m/s where the projectile did not penetrate the composite target panel and a 1073 m/s impact where the projectile perforated the composite panel.

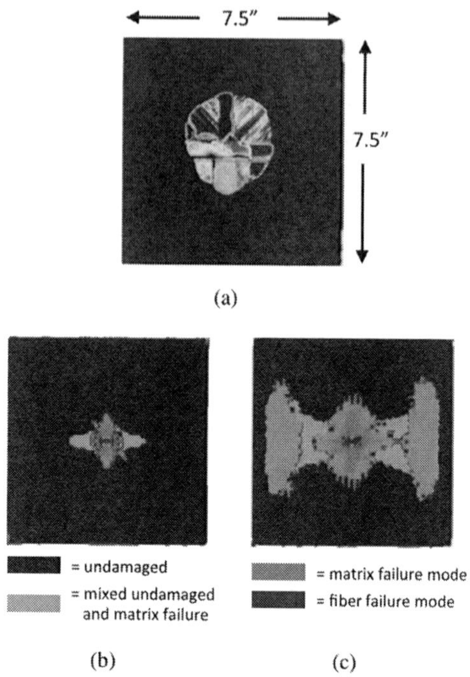

Figure 3. Damage patterns in (a) ultrasound data, (b) strain-based and, (c) stress-based predictions (Mesh 1) for 506 m/s impact.

4. Numerical model details

The CTH numerical simulations used to evaluate the failure criteria where performed using a full 3-dimensional model of both the composite target panel and spherical projectile. The projectile was modeled using a simple rate-dependent plasticity constitutive model and a Mie-Grüneisen EOS with an initial density of 8.13 g/cm^3 to represent high-density steel.

The composite target panel was modeled using the previously described MCM composite model with a quasi-isotropic layup and an initial density of 1.56 g/cm^3. The stiffness parameters for the fiber and matrix constituents were taken from Key et al. [12], while the composite stiffness properties were generated using the constituent properties and the finite element micromechanics model shown previously in Fig. 2. The initial constituent material strength properties were also determined from the micromechanics model and the composite material strength properties of Key et al. [12].

As a simple attempt to account for the strength increases that occur due to the elevated pressures and strains rates, a strength increase factor estimate of 1.75 was applied to each of the ultimate stress and strain properties. This factor was estimated from the data of Shin et al. of Fig. 1, where although the peak pressures in our problem are at or above 100 kbar, there was no experimental data found to support further strengthening above the 6 kbar pressure of Shin.

Finally, each of the two impact velocity conditions was modeled using various different Eularian mesh densities as part of a larger effort. For this paper we will use the following naming convention when referring to each of the different densities:

Mesh 1: 0.273 cm × 0.273 cm × 0.1524 cm
Mesh 2: 0.136 cm × 0.136 cm × 0.0762 cm
Mesh 3: 0.0683 cm × 0.0653 cm × 0.0635 cm
Mesh 4: 0.0453 cm × 0.453 cm × 0.0423 cm.

As part of a larger effort, the different mesh densities were chosen to study their effect on both the predicted damage patterns and the residual velocity predictions. However, the main focus of this paper is on the differences in the numerical predictions as related to the stress and strain-based failure criterion. Therefore, the mesh density differences are only provided in select results.

5. Results

In order to evaluate the performance of the MCM model using both the stress and strain-based failure criteria, two evaluation metrics were chosen. First, the damage patterns predicted by the analytical simulations were compared to the observed ultrasonic post-test patterns. Second, the measured projectile residual velocity from the experimental test was compared directly against the analytically predicted residual velocity.

It is noted here that in all subsequent numerical simulation figures, green indicates matrix failure modes such as matrix microcracking or delamination, while red indicates fiber failure modes. The colors in the ultrasound data indicate time of flight which is related to the depth of the damage region.

Table 1. Residual velocity comparisons for 506 m/s impact.

	Mesh 1	Mesh 2	Mesh 3
Experimental	0 m/s		
Strain-based criterion	0 m/s	0 m/s	0 m/s
Stress-based criterion	0 m/s	20 m/s	100 m/s

5.1. 1/2 inch panel; 506 m/s impact velocity

Figure 3(a) shows the damage patterns measured from the ultrasonic inspection of the 1/2 inch panel after the spherical projectile impacted the composite target panel at 506 m/s. From this image one can see that the damage envelope through the thickness of the target panel is nearly circular in shape and consists of striations, which align with the quasi-isotropic lamination schedule (0°, 90°, + 45° and −45°) of the panel. The damage extent is also evident in this figure, as the entire 7.5 inch by 7.5 inch domain of the panel is shown.

Figure 3(b) shows the predicted damage envelope from the CTH-MCM simulation when using the strain-based failure criterion over the entire 7.5 inch by 7.5 inch panel domain. The damage envelope in this figure and all subsequent numerical images was generated by superimposing 2D damage slices of the laminate at the center of each ply. This approach was necessary as there is currently no capability available within the CTH post-processing package to generate the through thickness envelopes.

From Fig. 3(b) one can see that the strain-based criterion captures the basic characteristics of the damage envelope that are seen in the ultrasound results. Specifically, the predicted striations and extent of the damage in the 0° direction (horizontal) is very similar to the extent in the ultrasound. The predicted damage extents or striation in the 90°, + 45° and −45° directions is also evident. However, the extent of these directions appears to be approximately 1/2 the extent seen in the experimental data for the 90°, + 45° and −45° orientations.

Figure 3(c) shows the predicted results of the same simulation using the stress-based criterion. From this figure it is evident that the stress-based criterion significantly over-predicts the damage extent in the composite panel in the 0° direction (horizontal) as compared to the ultrasound results. In this direction the damage is observed to extend nearly to the plate free-edge, with large regions of damage far from the central impact zone. Also, there is very little evidence of the directional striations seen in the experimental results and the strain-based results.

For this experimental configuration, the projectile was defeated by the composite panel such that the residual velocity of the projectile was 0 m/s. The residual velocities for the spherical projectile predicted by the CTH-MCM simulations are given in Table 1 for each of the failure criteria. The results in Table 1 show that the strain-based criterion predicts that the composite panel will defeat the spherical projectile, as was observed in the experimental testing, for each of the mesh densities. However, for the stress-based criterion, it is evident that as the mesh density increases, the model predicts that the projectile defeats the composite panel and the residual velocity of the penetrator is approximately 100 m/s for the highest mesh density.

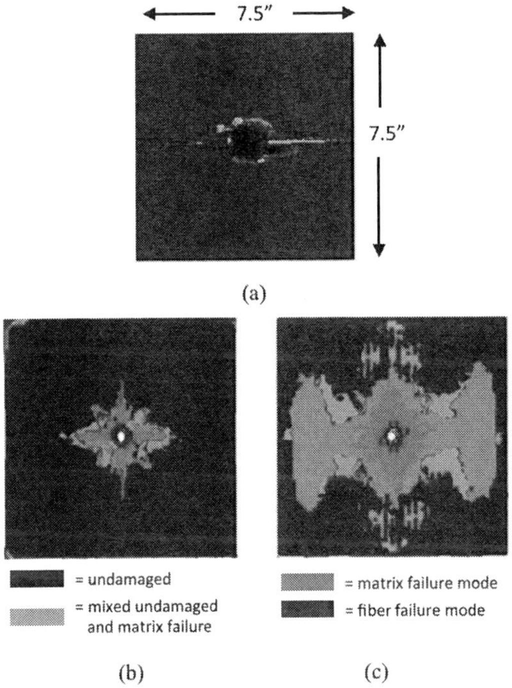

Figure 4. Damage patterns in (a) ultrasound data, (b) strain-based and, (c) stress-based predictions (Mesh 1) for 1076 m/s impact.

Table 2. Residual velocity comparisons for 1076 m/s impact.

	Mesh 1	Mesh 2	Mesh 3	Mesh 4
Experimental	562 m/s			
Strain-based criterion	25 m/s	300 m/s	525 m/s	560 m/s
Stress-based criterion	150 m/s	325 m/s	450 m/s	500 m/s

5.2. 1/2 inch panel; 1076 m/s impact velocity

Figure 4(a) shows the post-test ultrasound evaluation of the 1/2 inch panel after being impacted at 1076 m/s. Unlike the slower velocity condition discussed in the previous section, under this increased velocity the projectile easily defeats the panel. As a result of the projectile penetrating the panel, the ultrasound image of Fig. 4(a) can be misleading. First, looking at the figure one might interpret the black regions as holes or perforations in the panel; however, this is not the case. The black regions in the ultrasound are regions on the panel where the ultrasound was unable to make a conclusive measurement due to the large and varying amounts of interference present in the ultrasound signal. Visual examination of the panel confirms that the perforation in the panel was almost exactly the same diameter as the spherical projectile (1/4 inch).

Similar to the previous slower impact condition, the damage envelope predicted by the strain-based criterion

(Fig. 4(b)) shows damage extents and patterns that are similar to those observed in the tested panel. The stress-based criterion results shown in Fig. 4(c) were again much greater in extent than what was observed in the experimental results. These prediction results were similar to the previous condition in that they extended nearly to the free-edges of the panel in the 0° and 90° directions.

Table 2 shows the residual velocity prediction for this impact velocity along with the experimentally measured residual velocity. These results show that both criterions are converging towards the experimental results. However, the strain-based criterion appears to be converging on the experimental result quicker than the stress-based criterion.

Tables 2 and 3 together also show that at the less dense mesh configurations (Mesh 1, 2) a similar trend is observed where the stress-based failure criterion predicts a higher residual velocity. However, between Mesh 2 and Mesh 3, the trend reverses for the high velocity shot and the residual velocity for the strain-based criterion is higher than the stress-based residual velocity. Efforts are currently ongoing to better understand the mesh dependent behaviors seen in for each of these criterion results.

6. Conclusions

The preliminary results of this work show that a stress-based failure criterion tends to over predict the damage extent of the composite panels when subjected to shock loading. The results also show that this increased damage leads to an over-prediction of the projectile residual velocity. In contrast, the strain-based criterion appears to capture the basic characteristics of the damage patterns observed in the experimental testing and also does a good job of converging on the projectile residual velocity.

These results lead to the conclusion that when modeling composite materials under high velocity impact conditions, a strain based failure criterion may be more appropriate than a stress-based, pressure independent criterion. Future work will include the evaluation of a pressure-dependent, stress or strain-based criteria [13] to eliminate the artificial strengthening parameter applied during these early studies.

References

[1] J.M. McGlaun, S.L. Thompson, L.N. Kmetyk, M.G. Elrick, Int. J. Impact Eng. **10**, 351 (1990)

[2] E.S. Shin, K.D. Pae, J. Compos. Mater. **26**, 828 (1992)

[3] R. Hill, J. Mechan. Phys. Solids **12**, 199–212 (1964)

[4] M.R. Garnich, A.C. Hansen, J. Compos. Mater. **31**, 71 (1997)

[5] A.A. Lukyanov, Int. J. Plasticity **24**, 140 (2008)

[6] E.S. Hertel, G.I. Kerley, *Sandia National Laboratories Report* SAND98-0947 (1998)

[7] S.C. Schumacher, C.T. Key, *Sandia National Laboratories Report* SAND2012-7714 (2012)

[8] Z. Hashin, J. Appl. Mech. **47**, 329 (1980)

[9] J.S. Mayes, A.C. Hansen, Comp. Sci. Tech. **64**, 379 (2004)

[10] J.S. Mayes, *University of Wyoming Doctoral Dissertation* (1999)

[11] A.C. Hansen, D.M. Blackketter, D.E. Walrath, J. Appl. Mech. **58**, 881 (1991)

[12] C.T. Key, M.R. Garnich, A.C. Hansen, Compos. Struct. **65**, 357 (2004)

[13] S.T. Pinho, P. Robinson, C Schuecker, P.P. Camanho, ICCM 17 Proceedings (2009)

EPJ Web of Conferences **94**, 04002 (2015)
DOI: 10.1051/epjconf/20159404002

Perforation of aluminium foam core sandwich panels under impact loading: A numerical and analytical study

Ibrahim Elnasri[1,2,a] and Han Zhao[3]

[1]Laboratoire de Génie Mécanique, Ecole Nationale d'Ingénieurs de Monastir, Av. Ibn ElJazzar, Monastir 5019, Tunisia
[2]Institut Supérieur des Systèmes Industriels de Gabès, 214 rue Slaheddine Ayoubi, 6011 Gabès, Tunisia
[3]Laboratoire de Mécanique et Technologie, ENS-Cachan/CNRS-UMR8535/Université Paris 6, 61, avenue du président Wilson, 94235 Cachan Cedex, France

Abstract. This paper reports the numerical results of the inversed perforation test instrumented with Split Hopkinson Pressure Bar SHPB with an instrumented pressure bar on the AlSi7Mg0.5 aluminium foam core sandwich panels with 0.8 mm thick 2024 T3 aluminium top and bottom skin. The numerical models are developed in order to understand the origin of the enhancement of the top skin loads found under impact loading (paper published by [1]). Numerical predicted piercing force vs displacement curves are compared with experimental measurements (tests at impact velocities at 27 and 44 m/s). The simulation catches all process of the perforation of the sandwich panels (top skin, foam core, and bottom skin). Within experimental scatter, there is a good agreement between numerical predictions and experimental measurements. Virtual tests with different impact velocities up 200 m/s are presented and showed a significant enhancement of the piercing force under impact loading (top skin peak and foam core plateau loads). In order to understand the origin of these force enhancements, any difference of detailed local information between static and dynamic loading is studied and showed that a shock front effect is responsible for the enhancement piercing force. An analytical model using an improved RPPL shock model based a power law densification assumption is proposed to calculate the top skin piercing force. The improved RPPL shock model agrees with the FE results for small velocities and gives better prediction of the piercing force than the RPPL shock model for large velocities (>100 m/s).

1. Introduction

Sandwich plates and shells are important elements in modern lightweight construction. The main advantage of the sandwich principle is the fact that structures with high bending stiffness and rather low specific weight are obtained. The impact behaviours (energy absorption, perforation limits) are the important design features for sandwich panels used in aeronautics structures. The core materials of such panels (honeycombs, foams, hollow sphere agglomerates, etc.) play often dominant roles in the whole energy absorption as well as their penetration behaviour. [2] Conducted experimental and finite element studies on aluminium foam sandwich panels subjected to both quasi-static and impact loadings. Four deformation modes were summarised as global bending, localised indentation, localised indentation with global bending, and localised indentation with bending along clamping edge. However, key features of force–displacement curves, detailed deformation in face-sheets and foam core were not highlighted. [3] Carried out experimental tests and numerical simulations of the bird strike on sandwich panels made from aluminium foam core with aluminium face sheets. There is no perforation observed from the tests. [4] investigated the impact response of sandwich panels with Alporas foam cores and fibre-reinforced thermoplastic, and fibre-metal laminate (FML) face-sheets. The indentation resistance of these sandwich panels was found to be rate sensitive over the full range of conditions examined. Zhao et al. [1] tested the perforation behaviour of AlSi7Mg0.5 aluminium foam core sandwich panels with 0.8 mm thick 2024 T3 aluminium top and bottom skin using a split Hopkinson pressure bar, and recorded the piercing force displacement history. Quasi-static, impact tests (around 20 m/s and around 45 m/s) were performed. A significant enhancement of the top skin peak loads under impact loading was observed, although the skin sheet as well as foam cores are nearly rate insensitive. They suggested that a possible reason is the difference of compressive strain level of foam core reached before the perforation of top skin under static and impact loading because of different face-foam core interaction mechanism. Such localised foam core strength enhancement leads to the increase of the top skin peak loads. However, no detailed quantitative analysis has been presented in this work and there is no detailed local information due to experimental difficulties to better understand the original of the enhancement of the top skin peak loads observed under impact loading. Additionally, the explanation of the puzzling results have not convinced the scientists, meaning that they still need an in depth study.

For this purpose, a numerical model is built and it is calibrated by testing the results to be sure that the main features of the perforation of the aluminium foam core sandwich panels are well represented. Afterwards, numerical testing at various impact velocities is performed and the local information, which are not experimentally available, (stress, strain, etc.) are analysed in order to

[a] Corresponding author: ibrahim.nasri@issig.rnu.tn

Article available at http://www.epj-conferences.org or http://dx.doi.org/10.1051/epjconf/20159404002

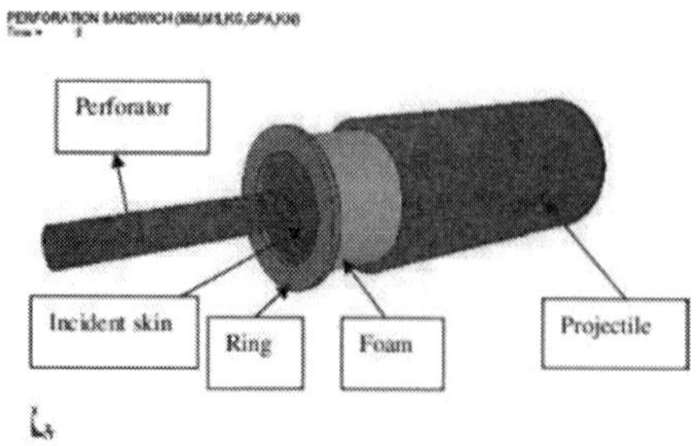

Figure 1. Numerical model for the perforation of the sandwich panels.

understand the piercing process. Finally, an analytical model using an improved RPPL shock model based on a power law densification assumption allows for an easy determination of its parameters from experimental data that are applied to calculate the resistance force during the perforation process.

2. Numerical studies

2.1. Finite element modelling

The numerical model is defined on the explicit code LS-DYNA [5]. Figure 1 shows the numerical model of the perforation of the sandwich panel which consists of 6 parts. The sandwich panel is integrated between two rigid parts. The aluminium hollow tube-like projectile (part 6) and aluminium clamping ring (part 2) are modelling as the rigid body by the material type 20 * (MAT_RIGID) in LS-DYNA codes. The rigid parts are merged for one Part 6. The perforator is represented as the elastic by the material type 1 * (MAT_ELASTIC, density = 8050 Kg/m^3, Young moduli: 188 GPa, Poisson ratio = 0.3). The aluminium 2024 T3 sheet are represented by shell elements with five integration points in the thickness. This model consists of 4800 2-D elements for the two sheets. The foam is modelled by using brick elements. This model consists of 90195 nodes, 76536-3D elements divided in 28416 for the part of the perforator, 120 for the two rigid parts and 48000 brick elements for the foam core. As the most important zone is the penetration area just under the perforator, the sandwich panels is modelled with a fine mesh along with the length. The number of the elements on the thick is 21 elements for the foam core. The number of the elements on the transverse direction is 51 elements for the sheet and for the foam core. Tiebreak constraints were applied between the faceplate and the foam core. Here, the nodal tiebreak failure in the interface was given by the following quadratic failure criterion;

$$\left(\frac{\sigma_{tb}}{N_{tb}}\right)^2 + \left(\frac{\tau_{tb}}{S_{tb}}\right)^2 \geq 1 \quad (1)$$

where σ_{tb} and τ_{tb} are the nodal normal and shear stress components and N_{tb} and S_{tb} are the strength of the adhesive in pure tension or pure shear, respectively. Here, $N_{tb} = S_{tb} = 5$ MPa was used. Contact between SHPB perforator bar and sandwich plate was established using eroding surface to surface contact. To model

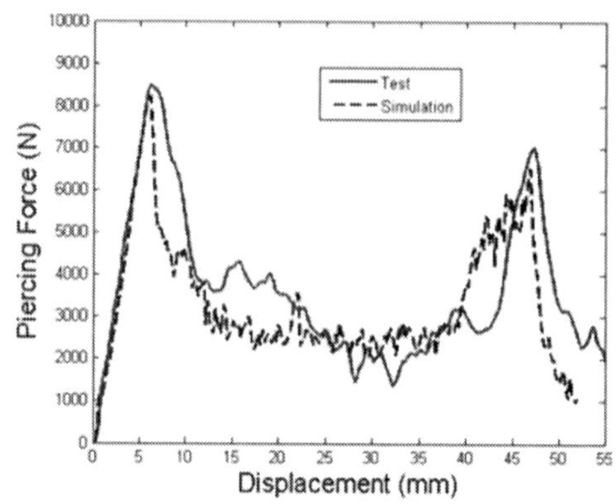

Figure 2. Comparison between experimental and simulated piercing force vs. displacement curves of the sandwich panel perforated at impact velocity 44 m/s.

contact between faceplate and the foam core after failure, eroding single surface contact was defined. Segment based contact option (SOFT = 2) was used for simulating the sandwich plate impact. Frictionless contact was used in all simulations. All the parts in present work were modelled with single integration point (constant stress solid element, ElFORM = 1) because the reduced integration elements are robust choices for nonlinear analysis to overcome negative volumes, which are prone to happen in full integration elements. In spite of the afore mentioned benefits, the reduced integration elements suffer from non-physical spurious/hourglass modes. Control hourglass type 1 control cart is used to overcome this non-physical behaviour.

The boundary conditions in the impact simulation are the following: the perforator is fixed in z-translation and z-rotation in degrees of freedom in the no-impact end. The contact between the sandwich panel and the two parts is defined as tied. The velocity of the sandwich panel and the projectile are modelled and put as initial velocity nodes for the sheets and for the foam and via the command INATIAL_RIGIG_BODY_VELOCITY for the projectile.

2.2. Numerical models for the aluminum skin sheet and for cymat foam core

The behaviour of the skin sheet is modelled by a damageable isotropic hardening elastic-plastic constitutive law proposed by [6], available in Ls-Dyna code [5] as the model of type 104 (MAT_DAMAGE_1). The parameters used for 2024 T3 aluminium sheet were identified from normalised tension tests by [7]. The foam core is modelled by a honeycomb constitutive model (MAT_26) available on the Ls-Dyna code with appropriate tensile failure strain. The normal stress vs. volumetric strain is obtained from quasi-static compression test (density = 235 kg/m^3, velocity = 0.01 mm/s). Due to the fact that they do not have a specific device to perform a shear test, the shear stress vs volumetric strain curve is deduced from the quasi-static compressive curve which is equal to the half. The fracture of the cell of the foam is achieved by using available tensile failure strain. Iteratively different values

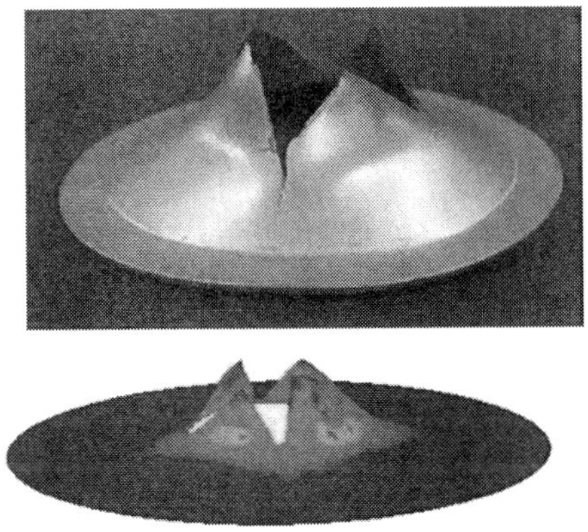

Figure 3. Breaking "petal" shape by simulation and by post mortem observation (Bottom skin, v = 44 m/s).

Figure 4. Breaking "petal" shape by simulation and by post mortem observation (top skin, v = 44 m/s).

were used to achieve the appropriate failure strain value. The constitutive properties of the foam are:
Density = 235 kg/m^3, Young Modulus = 70 GPa, Poisson ratio = 0.285, SIGY = 240 MPa, Vf = 0.23, TSEL = 5%.

2.3. Results and discussions

With appropriate selection of constitutive model of the foam core, Fig. 2 shows the simulated piercing force-displacement curves at 44 m/s compared with the experimental result. The piercing force is obtained by using the so-called command: *DATABSE_ NODAL_ FORCE_ GROUP. All nodes on the perforated end were integrated in the group nodes which can be compared to the piercing force recorded by the strain gauge instrumented on the Split Hopkinson Pressure Bar. It was observed that the simulated piercing curve was in good agreement with experiments. One can note that the simulated peak piercing force for the incident sheet is small lower than the experiment case but the difference was not significant to affect the final result and is apriority due to errors for the modelling. It can be seen the simulation catch ell the petalling failure mode observed on the experimental test (Fig. 3 for the bottom skin and Fig. 4 for the top skin).

3. Analysis of virtual numerical tests

The present numerical model provides the virtual tests where all the detailed information are not available in a real test (strain, stress, etc.) is known. Five numerical tests are conducted in dynamic loading up to the impact velocity 200 m/s. Piercing impact force vs. displacement curves for impact velocities 44, 100, 140 and 180 m/s are given in Fig. 5. It is observed that:

- The piercing top skin peak force and level of the plateau of the perforation of the foam core increases regularly with the impact speed. It indicates that this model contains probably the hints for the understanding of this force enhancement, which is probably due to the propagation of the shock front because the constitutive relation used for the foam cores as well as aluminium sheets are all rate insensitive.

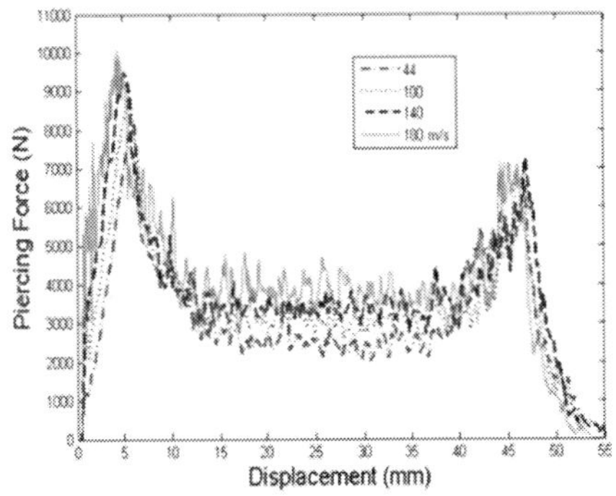

Figure 5. Comparison of simulated piercing force vs displacement curves of the sandwich panel perforated at 44, 100, 140 and 180 m/s.

- The maximum displacement of the top skin before breaking is unchanged with the impact velocity increase and the displacement of the bottom skin breaking is the same of all curves, these can prove that the foam core isn't sustained for a compression.
- The peak load of the piercing force of the bottom skin is nearly the same value for all curves; this is confirmed that the aluminium face sheet is rate insensitive.

In order to understand the origin of these force enhancements, any difference of detailed local Information between static and dynamic loading is interesting. A noticeable difference is to study the stress level in the of the foam core in the contact with the perforator. Figure 6 shows the equivalent Von Mises Stress vs displacement for one element located in the middle of the foam core. The coordinates of the element along the loading are (0, 2 mm). A significant increase in stress amplitude is found for impact loading case and is in increase with impact velocity, the huge enhancement on the strength is prior due to the propagation of the shock front in the foam core, which is the possible reason, because the uncompressing of the foam core is confirmed by simulation, which leads to a higher local foam core strength under the perforator and can increase the top skin peak loads.

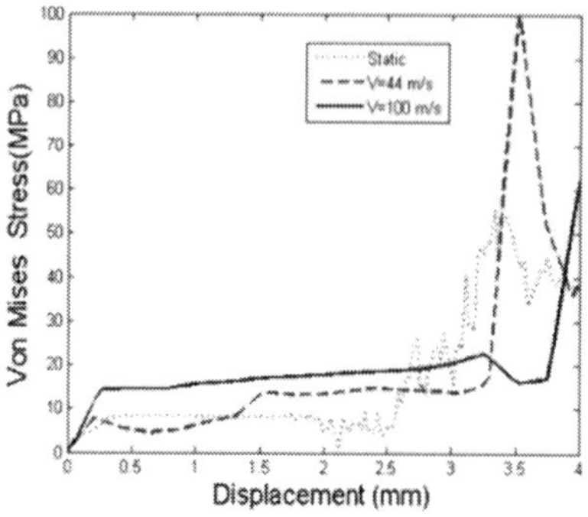

Figure 6. Comparison of Von Mises Stress vs. displacement curves for the element for three impact velocities (static, v = 44 m/s, v = 100 m/s). The coordinates of the element along the loading direction are (0,2 mm).

4. Basic model of the impact piercing force enhancement

In order to determine the model of the impact piercing force enhancement, analytical model for failure load predictions from [8] are considered for comparison with numerical predictions and experimental measurements. To take account the effect of the front shock, an improved shock model with a hardening locking proposed by [9] is also considered on the model. When the impact is normal to the surface of the sandwich panel, the motion of the projectile is:

$$M\frac{dv}{dt} = -F_p. \tag{2}$$

Where F_p is the dynamic perforations force and M is is the mass of the projectile (sandwich panel+ hollow tube-like projectile+ ring). The dynamic perforation force can be expressed as the sum of the force needed to stretch the aluminium face sheet plate and the force needed to crush the foam beneath the indender F_{cruch}, the force required to tear the cells around the edge of the indenter F_{tear} and the force enhancement caused by crushing the cellular material in the shock front F_{shock}.

$$F_p = F_{strech} + F_{cruch} + F_{tear} + F_{shock}. \tag{3}$$

Where;

$$F_{strech} = \sqrt{2}A\ l_{df}(\varepsilon_{crdf})^{\frac{3}{2}}. \tag{4}$$

Where:

$$l_{df} = \pi\ D_C. \tag{5}$$

$$A = \frac{E h}{1-\upsilon^2}. \tag{6}$$

Where l_{df}, A, E, h, υ and ε_{crdf} are the length of the damage, stretching stiffness, Young's modulus, thickness, Poisson's ratio and dynamic fracture strain of the faceplate material, respectively, D_C is the diameter of the volume crashed zone.

$$F_{cruch} = \pi\ K\ D_C^2\ \sigma_{pl}/4. \tag{7}$$

Where:
K the constraint factor for core crushing. [8] Use an average value of 2.0 (from a range between 1.7 and 2.5) for the constraint factor, K in their work, which is based on the compression and indentation tests [10].

σ_{pl} is the compression plateau stress.

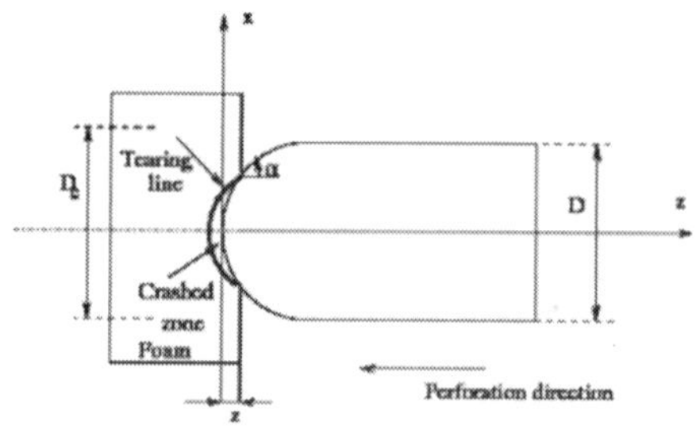

Figure 7. Cross-section of an axisymmetric perforator with a semi-spherical nose.

$$F_{tear} = \pi\ D_C\ \gamma. \tag{8}$$

Where:
γ is the tearing energy per unit newly created area.

The tearing energy can be approximated by the following empirical formula:

$$\gamma = 550.8\left(\frac{\rho_0}{\rho_s}\right)^{1.8}\ (kJ\ m^{-2}). \tag{9}$$

Where ρ_s is the density of base material of aluminium Cymat foam. For aluminium Cymat foam with a nominal relative density of 8.7%, the tear energy is 6.8 KJ/m^2.

For the contribution of the force due to propagation of the front shock:

$$F_{shock} = \frac{\pi\ D_C^2\ \rho_0\ v^2}{4\ \varepsilon_{shock}}. \tag{10}$$

Where:
ρ_0 is the density of the aluminium foam before crushing, ε_{shock} is the shock strain, v is the impact velocity of the sandwich panel projectile.

D_C can be expressed by the following formula [10]:

$$D_C = D + 2[1.39\,z\ \tan\alpha]. \tag{11}$$

Where D is the diameter of the perforator, z is the indentation depth and α is the angle between the tearing line and the vertical axis. Here, we can use the value of the angle used in the indentation of the foam with semi-spherical nose ($\alpha = 10.31°$). Figure 6 showed the simplified scheme of the perforation processes and the parameters presented in Eq. (11).

Substituting Eqs. (4)–(11) into Eq. (3) leads to:

$$F_p = \pi\ D_C^2\left(\frac{4\sqrt{2}}{D_C}A\ (\varepsilon_{crdf})^{\frac{3}{2}} + K\ \sigma_{pl} + \frac{4\gamma}{D_C} + \frac{\rho_0\ v^2}{\varepsilon_{shock}}\right)/4. \tag{12}$$

As it is mentioned by [9], the ε_{shock} can be expressed en function of the basic continuity equation of the shock front:

$$\sigma_{shock} - \sigma_y = \frac{\rho_0 V_{impact}^2}{\varepsilon_{shock}}. \tag{13}$$

Table 1. Identified parameters for power law locking.

Material	Plateau stress (MPa)	m	K (MPa)
Cymat foam core	4	8.99	48.3

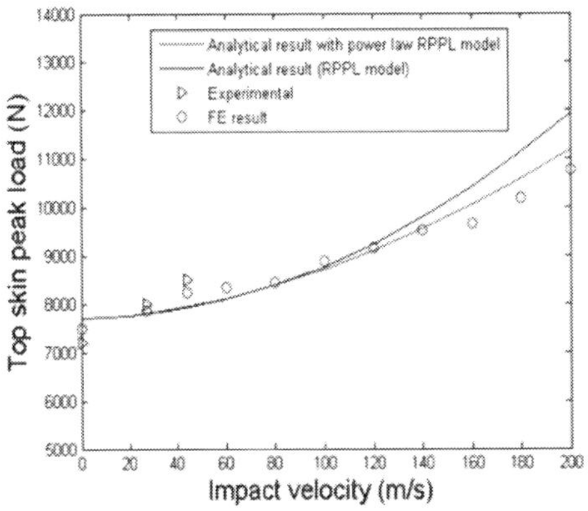

Figure 8. Comparison between FE results and the analytical models with power law RPPL model and with RPPL model the piercing force with the impact velocity with experimental results.

And the solution is defined with stress-strain of the materiel:

$$\sigma_{shock} = f(\varepsilon_{shock}). \quad (14)$$

And we can use a model using power law without locking strain to replace the RPPL model as used by [9]. The densification curve is defined by the initial yield stressσ_{pl}, the power m, and the coefficient K:

$$\sigma = \sigma_{pl} + k\ \varepsilon^{m}. \quad (15)$$

Equations (13) and (14) then lead to:

$$K\ \varepsilon_{shock}^{m} = \frac{\rho_0 v^2}{\varepsilon_{shock}}. \quad (16)$$

The derivation of the shock strain ε_{shock} is straightforward:

$$\varepsilon_{shock} = (\frac{\rho_0\ v^2}{k})^{\frac{1}{m+1}}. \quad (17)$$

From experimental stress–strain data, it is easy to identify such model. Table 1 provides parameters (m, k) for the studied Cymat foam. Substituting Eq. (17) into Eq. (12) leads to:

$$F_p = \pi\ D_C^2 \left((4\frac{\sqrt{2}}{D_C} A\,(\varepsilon_{crdf})^{\frac{3}{2}} + k\,\sigma_{pl} + \frac{4\,\gamma}{D_C} + \ldots \right.$$

$$\left. + (\rho_0\ v^2)^{\frac{m}{m+1}}\ k^{\frac{1}{m+1}} \right) /4. \quad (18)$$

The dynamic perforation force from Eq. (12) with RPPL and the modify shock theory model is compared with the corresponding results from FE model and the experimental results (v = 27, 44 m/s) as shown in Fig. 7. The structure modifies shock theory and the RPPL models are agrees with FE results up impact velocities close to 100 m/s. It can observed that the RPPL overestimates the dynamic perforation load for large velocities (< 100 m/s) but the structure modify shock theory can ameliorate the prediction on the perforation force than the RPPL shock theory for a large impact velocities.

5. Conclusions

The numerical models of the inversed perforation test instrumented with SHPB with an instrumented pressure bar on the AlSi7Mg0.5 aluminum foam core sandwich panels with 0.8 mm thick 2024 T3 aluminum top and bottom skin are developed in order to understand the origin of the enhancement of the top skin loads found under impact loading (paper published by [1]).The numerical simulation is done using 3D finite element models in Ls-Dyna. The behaviour of the skin sheet is modeled by a damageable isotropic hardening elastic-plastic constitutive law as the model of type 104 (MAT_DAMAGE_1), where the foam core is modeled by a honeycomb constitutive model (MAT_26) with appropriate tensile failure strain.

Numerical predicted piercing force vs displacement curves are compared with experimental measurements (tests at impact velocities at 27 and 44 m/s). The simulation catch all processes of the perforation of the sandwich panels (top skin, foam, and bottom skin). Within experimental scatter, there is a good agreement between numerical predictions and experimental measurements. Virtual tests with different impact velocities up 200 m/s are presented and showed:

- The piercing top skin peak force and level of the plateau of the perforation of the foam core increases regularly with the impact speed. It indicates that this model contains probably the hints for the understanding of this force enhancement, which is probably due to component interaction because the constitutive relation used for the foam cores as well as aluminum sheets are all rate insensitive.
- The maximum displacement of the top skin before breaking is unchanged with the impact velocity increase and the displacement of the bottom skin at breaking is the same of all simulated tests, these can prove that the foam core isn't sustained to a compression.
- The peak load of the piercing force of the bottom skin is nearly the same value for all curves; this is confirmed that the aluminum face sheet is rate insensitive.

In order to understand the origin of these force enhancements, a stress level in the element of the foam core on the contact with the perforator for three impact velocities (static, 40 m/s and 100 m/s) for three impact velocities (static, 40 m/s and 100 m/s) is analysed. A significant increase in the stress amplitude is found for impact loading case and is in increase with impact velocity, the huge enhancement on the strength is prior due to propagation of shock front in the foam core and is the one possible reason, because the uncompressing of the foam core is confirmed by simulation, which leads to a higher local foam core strength under the perforator and can increase the top skin peak loads and the foam core loads.

An analytical model using an improved RPPL shock model based on a power law densification assumption is proposed to calculate the top skin piercing and foam core piercing forces. The improved RPPL shock model agrees with the FE results for small velocities and gives better prediction of the piercing force than the RPPL shock model for large velocities (> 100 m/s).

References

[1] H. Zhao, I. Elnasri, Y. Girard, I. J. of Impact Eng, **34** (7): 1246–57 (2007)

[2] W. Hou, H. Zhang, G. Lu, X. Huang X. In: Proceedings of the 6thAsia-Pacific conference on shock and impact load on structures, Perth, 7–9 December, p. 275–82 (2005)

[3] A.G. Hanssen, Y. Girard, L. Olovsson, T. Berstad, M. Langseth M., I. J. of Impact Eng. **32**(7):1127–47 (2006)

[4] W.J. Cantwell, H. Kiratisaeve. J., Reinf Plast Comp; **24**(10):1057–72 (2005)

[5] LS-DYNA keyword user's manual. Version 971(2007)

[6] Lemaitre, J. 2nd ed. Berlin: Springer, ISBN 3-540-60980 (2006)

[7] S. Pattofatto, H.B, Zeng, H. Zhao. J. of Sand. Struct and Materiels (2011)

[8] M.S. Fat, K.S Park, Comp Struct; **52**(3–4):353–64 (2001)

[9] S. Pattofatto, I. Elnasri, H. Zhao, H.Tsitsiris, F. Hild, Y. Girard, J. Mech. Phys. Solids, **55**; 2672–2686 (2007)

[10] Reddy T.Y., Soden P.D., Reid S.R., Sadighi M. Collaborative research program on the costeffective use of fibre-reinforced composite offshore, marintech research, phase1, report no. CP04 (1991)

EPJ Web of Conferences **94**, 04003 (2015)
DOI: 10.1051/epjconf/20159404003

Overstress and flowstress approaches to dynamic viscoplasticity

Yehuda Partom[a]

Retired from RAFAEL, Israel

Abstract. Viscoplasticity is mostly modelled by the flowstress approach, where the flowstress (Y) is a function of pressure, temperature, plastic strain and strain rate Y(P,T, ε^p, $\dot{\varepsilon}$). For dynamic Viscoplasticity the flowstress approach is used in hydrocodes together with the radial return algorithm, to determine deviatoric stress components in each computational cell and for each time step. The flowstress approach assumes that during plastic loading, the flowstress in stress space follows the current stress point (current Y). Unloading of a computational cell is therefore always elastic. The overstress approach to dynamic viscoplasticity was used in various versions in the 1950s and early 1960s, before the advent of hydrocodes. By the overstress approach a state point may move out of the quasistatic flow surface upon loading, and hence the term overstress. When this happens, the state point tends to fall back (or relax) onto the quasistatic flow surface through plastic flow, and the rate of this relaxation is an increasing function of the amount of overstress. In the paper we first outline in detail how these two approaches to dynamic viscoplasticity work, and then show an example for which the overstress approach has an advantage over the flowstress approach. The example has to do with elastic precursor decay in planar impact, and with the phenomenon of anomalous thermal strengthening, revealed recently in planar impact tests. The overstress approach has an advantage whenever plastic flow during unloading is of importance.

1. Flowstress and overstress approaches

Dynamic viscoplastic response of ductile materials is usually modelled by assuming a constitutive law for the flowstress (Y):

$$Y = Y\left(P, T, \varepsilon^p, \dot{\varepsilon}\right) \qquad Y^2 = \tfrac{3}{2} s_{ij} s_{ij} \tag{1}$$

where P = pressure, T = temperature, ε^p = effective plastic strain, $\dot{\varepsilon}$ = effective strain rate, s_{ij} = stress deviator. The initial value of Y denoted by Y_0 is called yield stress. The second equation of (1) is a closed surface in deviatoric stress space, known as the flow surface. With the flowstress approach the current stress point is always inside or on the flow surface.

Most (if not all) hydrocodes use the flowstress approach. For any computational cell, and for each time cycle, the new stress deviator is determined by a two-step algorithm introduced by Wilkins [1], and known as radial return. Step 1 calculates an intermediate stress deviator s^*_{ij} and flowstress Y^*, assuming that the current deformation rate is fully elastic. For $Y^* \leq Y$, the deformation is indeed elastic, and $s_{ij} = s^*_{ij}$. For $Y^* > Y$, the new stress deviator is on the intersection of the flow surface Y and the ray from s^*_{ij} to the origin:

$$\beta = Y/Y^* \ ; \ s_{ij} = \beta s^*_{ij}. \tag{2}$$

To evaluate Y we need the new values of ε^p and $\dot{\varepsilon}$. These are not available at this stage, and old values are used instead. This often introduces noise into the solution. It can be shown that for an unchanging flow surface, radial return is equivalent to the flow rule:

$$d^p_{ij} = \lambda s_{ij} \tag{3}$$

where d^p_{ij} is the plastic deformation rate, and λ is a positive scalar. But when the flowstress changes by Eq. (1), radial return is not equivalent to Eq. (3). Equation (1) expresses isotropic hardening/softening. But it can be generalized to express also kinematic hardening and/or shape change. This has no effect on the radial return algorithm, except that in some extreme cases the ray to the origin may not cross the flow surface. Once the new s_{ij} are obtained, plastic deformation rate follows by:

$$\delta^{pe}_{ij} = d^p_{ij} = \delta_{ij} - \delta^e_{ij} \qquad \delta^e_{ij} = s_{ij}/2G\left(P, T\right) \qquad \left(\dot{\varepsilon}^p_{eff}\right)^2 = \tfrac{2}{3} d^p_{ij} d^p_{ij} \tag{4}$$

where G is the shear modulus, and δ_{ij} is the deformation rate deviator. Finally, the stress deviator is corrected for rigid body rotation by:

$$\dot{s}_{ij} = s_{i\alpha}\omega_{\alpha j} + s_{j\alpha}\omega_{\alpha i} \tag{5}$$

where ω_{ij} is the spin tensor.

We see that the flowstress approach is simple and easy to apply, and is therefore used in most if not all hydrocodes. Accordingly, most viscoplastic constitutive models have the form of Eq. (1). Also, because upon loading (deviatoric stress increase) from the flow surface, the current state becomes the new flow surface, unloading is always elastic by the flowstress approach. In

[a] Corresponding author: ypartom@netvision.net.il

Article available at http://www.epj-conferences.org or http://dx.doi.org/10.1051/epjconf/20159404003

this respect the overstress approach is quite different from the flowstress approach.

The overstress approach is motivated by Orowan's equation, and it distinguishes the quasistatic yield stress (that sets stationary dislocations into motion) from the current stress (that maintains the motion of already moving dislocations). The quasistatic yield stress is:

$$Y_{qs} = Y_{qs}(P, T) \tag{6}$$

and the effective plastic deformation rate is given by:

$$\begin{aligned} d^p_{eff} &= d^p_{eff}\left(\sigma_{eq} - Y_{qs}\right) \\ \sigma_{eq} &= \tfrac{3}{2} s_{ij} s_{ij} \end{aligned} \tag{7}$$

where $\sigma_{eq}-Y_{qs}$ is the overstress. The plastic deformation rate components are then given by Eq. (3) where λ is:

$$\lambda = \tfrac{3}{2} d^p_{eff} / \sigma_{eq}. \tag{8}$$

The rates of the deviatoric stress components are then:

$$\begin{aligned} \delta^e_{ij} &= \delta_{ij} - d^p_{ij} \\ \dot{s}_{ij} &= 2G(P,T)\,\delta^e_{ij} + s_{i\alpha}\omega_{\alpha j} + s_{j\alpha}\omega_{\alpha i} \end{aligned} \tag{9}$$

where the first right hand side term in the second equation of (9) is an approximation good for low deviatoric elastic strains.

We see that the overstress approach is expressed in terms of rate equations. These need to be integrated for each computational cell and for each time step separately, which can be done with a standard ODE solver like Runge-Kutta. Many times a single integration step over a computational time step is sufficient.

The deviatoric rate equations are coupled to the equation of state (EOS) both ways. First, the deviatoric part needs P and T, determined from the EOS. Second, plastic work is added to the volumetric internal energy, thereby influencing P, T and V (specific volume). The rate equation for P is:

$$E = E(P, V); \quad \dot{E} = -P\dot{V} + \dot{W}^p$$

$$\dot{W}^p = V\, s_{ij} d^p_{ij}$$

$$\therefore \dot{P} = -\frac{P + \partial E/\partial V}{\partial E/\partial P}\dot{V} + \frac{\dot{W}^p}{\partial E/\partial P} \tag{10}$$

where $\dot{V}$ during a time step is known from mass conservation, and T can be determined from the temperature EOS T(P,V).

Equation (7), which we call the flow function, is the main equation of the overstress approach. Its meaning is:

- In a dynamic situation the stress point is allowed to move out of the quasistatic yield surface, hence the term overstress.
- When this happens, the stress point tends to fall back (or relax) onto the quasistatic yield surface at a rate that increases with the amount of overstress.
- The rate of relaxation determines the strain rate sensitivity of the deformation.
- During plastic flow the stress point is mostly outside the quasistatic yield surface.

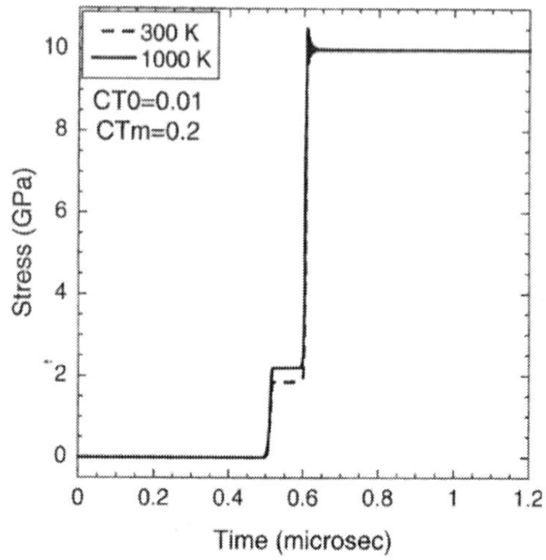

Figure 1. Longitudinal stress history at 1 mm into the target, calculated with the flowstress approach and a temperature dependent strain rate coefficient.

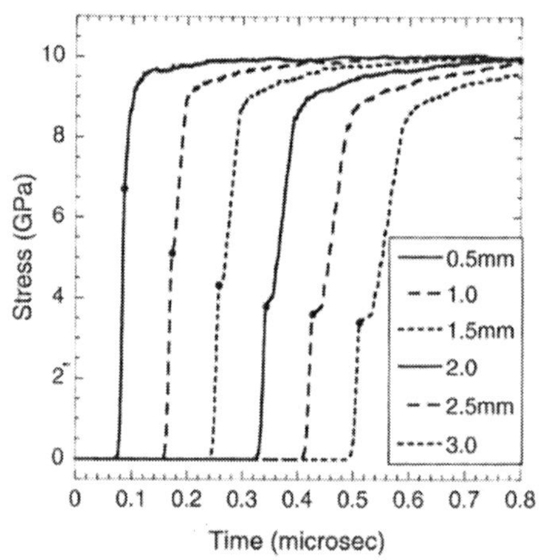

Figure 2. Longitudinal stress histories, every 0.5 mm into the target, calculated with the overstress approach.

- Also, plastic flow may continue upon unloading, and there is no noise from abrupt changes between elastic unloading and plastic reloading, as often happens with the flowstress approach.

We use for the flow function the simplest forms possible:

$$\begin{aligned} d^p_{eff} &= A\left(\sigma_{eq} - Y_{qs}\right) \\ d^p_{eff} &= A\left(\sigma_{eq} - Y_{qs}\right)^{\alpha}. \end{aligned} \tag{11}$$

The flow function, which reflects dislocation motion through the crystal lattice, determines both strain and strain rate hardening/softening. For a given A, when the total deformation rate is higher, there is more elastic deformation rate, and the stress increases (strain rate hardening). Also, for a given total deformation rate, when A decreases (with $\varepsilon^p_{\text{eff}}$), elastic strain rate increases (strain hardening).

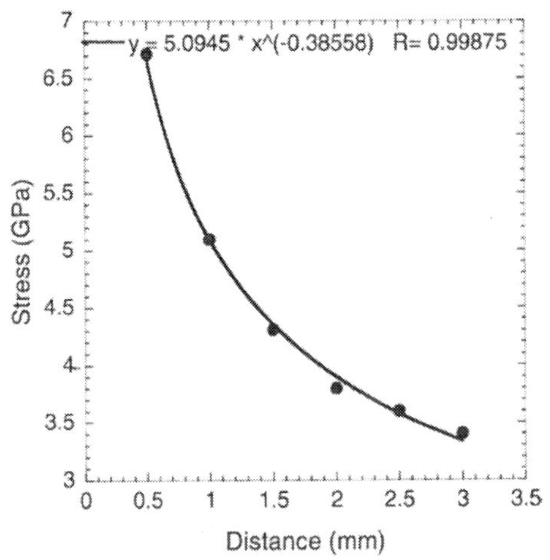

Figure 3. Precursor decay as function of distance from the calculation of Fig. 2.

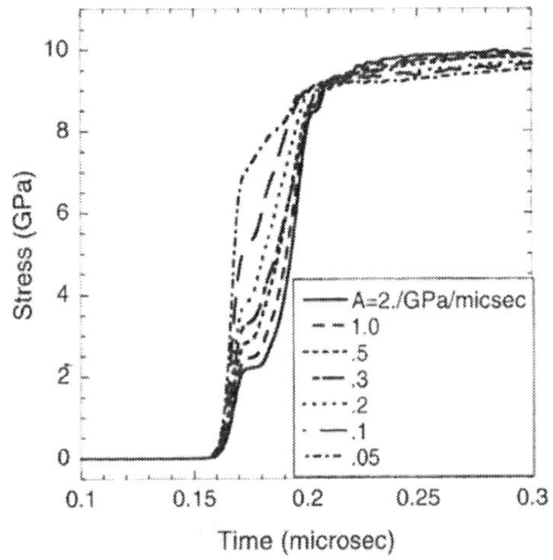

Figure 4. Longitudinal stress histories at 0.5 mm into the target for different values of the coefficient A.

In what follows we give an example for which use of the overstress approach has an advantage over the flowstress approach. The example is about elastic precursor wave decay in planar impact and the phenomenon of anomalous thermal strengthening.

2. Anomalous thermal strengthening

Planar impact tests on thin metal targets (1 mm or less), and at different temperatures have been recently reported in [2]. In these tests the authors focused on the elastic precursor wave amplitudes, and observed that for higher temperatures the rate of elastic precursor decay is slower. Their own interpretation was that dynamic strength increases with temperature, and accordingly they referred to the phenomenon as "Anomalous Thermal Strengthening" (ATS).

Planar impact tests have been used for over 70 years to study the dynamic response of materials to high pressure shocks. In such tests a thin plate disc (the projectile) is accelerated against a thicker plate disc (the target). A shock, or a structured shock, then propagates through the target. By monitoring the velocity history of the back surface of the target on the axis, it is possible to acquire information on the pressure (EOS) and on the shear stress (strength) response of the target material. The response of viscoplastic materials to planar impact has been described and understood since the 1960s [3,4].

Dynamic response to impact (planar or non-planar) is usually simulated with a hydrocode (or wavecode), and as stated above, hydrocodes usually use the flowstress approach to calculate the shear response. As a result, hydrocodes are not able to reproduce the phenomenon of elastic precursor decay in planar impact. We show this by running a planar impact problem using the flowstress approach. At the same time we also show that ATS can be reproduced with the flowstress approach by making the strain rate term of Eq. (1) depend explicitly on temperature (in addition to the usual temperature dependence factor):

$$Y = Y_{qs} Y_P(P)\, Y_T(T)\, Y_\varepsilon\left(\varepsilon^p_{eff}\right)\left[1 + C_T(T)\, \ell n\, \dot{\varepsilon}\right]$$
$$C_T(T) = C_{Tm} + (C_{T0} - C_{Tm})\left(T_m - T/T_m - T_0\right) \quad (12)$$

where the index m signifies melting. In Fig. 1 we show stress histories at 1 mm into the target from two such runs. The target is stainless steel with 100 cells per mm. The entering shock is 10 GPa. We use Yqs = 0.9 GPa, G = 77 GPa, $C_{T0} = 0.01$ and $C_{Tm} = 0.2$. The two runs are with T = 300 K and 1000 K, respectively.

We see from Fig. 1 that the precursor level from the run with 1000 K is indeed higher. We don't show it here, but stress history curves at longer distances into the target look the same, except that the precursor wave duration is longer, and precursor decay with distance is not reproduced.

We adapted our hydrocode to use the overstress approach for the same planar impact problem. We use Eq. (11) with $\alpha = 2.4$ and different values of A. In Fig. 2 we show stress histories every 0.5 mm into the target and with $A = 0.1/\text{GPa}/\mu s$. Elastic precursor decay is clearly seen.

In Fig. 3 we show the precursor decay curve as function of distance from the run of Fig. 2. In Fig. 4 we show stress history curves at 0.5 mm into the target for different values of the coefficient A. For A=A(T) this would be equivalent to curves of different temperatures. We see that for lower values of A (higher temperatures) precursor decay is slower, and precursor stress at a given distance is higher.

3. Summary

The flowstress and overstress approaches to dynamic viscoplasticity have an entirely different structure and logic, but produce equivalent results in many cases, as in monotonic loading. Flowstress is the standard approach in hydrocodes, and it is always used in conjunction with the radial return algorithm. By the flowstress approach, the flow surface in stress space is dynamic. It follows the equivalent stress upon loading, and unloading is always elastic. The overstress approach, on the other hand, recognizes only a quasistatic flow surface, which the stress point may move out of upon loading, and hence the term overstress. When it does, the stress point tends to fall back onto the quasistatic flow surface at a rate that increases with overstress. By the overstress approach the rate of plastic flow may be nonzero upon unloading.

The flowstress and overstress approaches may predict entirely different results whenever plastic flow during unloading is of importance. One such situation is elastic precursor wave decay in planar impact. In chapter 2 we demonstrate this by simulating a planar impact example with both approaches. We show that by using the overstress approach we can explain the anomalous thermal strengthening phenomenon recently obtained in planar impact tests of thin metal discs heated to high temperatures.

References

[1] M. Wilkins, *Calculation of Elastic-Plastic Flow*, LLNL Report UCRL-7322 Rev. (1969)

[2] G. I. Kanel, in Shock Compression of Condensed Matter, 937 (2011)

[3] W. Herrmann, D. L. Hicks, E. G. Young, *Sock Waves and the Mechanical Properties of Solids*, Syracuse Univ. Press, **23** (1971)

[4] Y. Partom, J. Appl. Phys., **59**, 2716 (1986)

EPJ Web of Conferences **94**, 04004 (2015)
DOI: 10.1051/epjconf/20159404004

Field tests and computational simulations of the explosion of buried charges

Eve Roger[1,a], Benjamin Loret[2], and Jean Paul Calvel[1]

[1] DGA Techniques terrestres, Rocade Est, Echangeur de Guerry, 18000 Bourges Cedex, France
[2] Laboratoire 3SR, Domaine Universitaire, BP. 53, 38041 Grenoble Cedex 9, France

Abstract. Modelling buried explosion is a matter of concern for vehicle protection. Indeed, in the battlefield, Improvised Explosive Devices (IEDs) are one of the major threats for land vehicles and, more specifically, for their underbelly. Two series of field tests using several masses of explosives have been performed, varying certain geometrical parameters, the nature and the physical properties of the soil. These controlled tests have shown that the impulse transmitted to the vehicle is a function of the saturation of the soil as well as of depth of burial of the explosive. In an effort to simulate the phenomena that take place during the explosions, these tests have been used to feed the data requested in computational simulations in a finite element context. Soil modelling presents its own difficulties, especially because soil is a porous medium and the three phases (solid grains, water and air) must be considered. A non linear viscoplastic cap model has been developed where the degree of saturation is variable. The yield surface includes a failure part, a cap and a tension cutoff. Soil stiffening associated with the air expulsion has been observed to be an important aspect of the model.

1. Introduction

In the last few years, Improvised Explosive Devices (IED) became one of the major threats on theatres of operations. Some of them can be buried and impact the belly of vehicles. However the type of soil, whether it is cohesive or not, its water content and other parameters like the depth of burial of the explosive device largely influence the energy released by the detonation and thus the impulse transmitted to a vehicle and the deformations of its hull.

Several experimental researches have addressed the issue: Hlady [1] and Peles et al. [2] performed buried experiments in fine and coarse grained soils in dry and saturated conditions. The type of soil influences the energy transmitted by the detonation. Fine grained soils form chunks that contribute significantly to the impulse and kinetic energy (Hlady [1]). Sand projections provide much less energy than fine grained soil ejecta.

Moreover a soil with a higher density and/or a moisture content transfers more energy. Zakrisson et al. [4] highlighted that the water content of the soil influences not only the impulse transmitted by the detonation but also the deformation of a plate standing above the buried explosion. For a sandy soil, Assaf et al. [5] showed that when the degree of saturation is close to 1, the difference is really significant on the impulse. Besides a shot in water provides the largest value of impulse and deformation (Fourney et al. [3]).

The depth of burial has also an influence on impulse and deformation. As the depth of burial of the explosive increases, the aerial pressures decrease. However, the kinetic energy transmitted upwards increases with the thickness of the soil above the explosive (Hlady [1], Zakrisson et al. [4]). Still, there exists a depth of burial at which the impulse is the largest. Beyond this value, increasing the depth of burial results in a decrease of impulse. This depth is directly linked to the quantity of explosive. As for the deformation, Zakrisson et al. [4] showed that it decreases as the depth of burial increases, because it is mostly associated with the shock wave and less with the impact of projections.

Experiments involving vehicles are expensive. Modelling can be a good compromise if the numerical simulations are validated by experiments. However soil modelling presents its own difficulties because of the composition of the soil which is a porous medium composed of three phases (solid particles, water and air). The three phases are important to be considered to model a buried explosion: while the air is expelled during the detonation, the water is trapped in the soil because the detonation is a so quick phenomenon that water does not have the time to escape.

A soil model has been developed taking into account the influence of the water content. Two experimental campaigns have been carried out to provide data to validate the simulations. The first campaign involved small scale detonations at three depths in dry and saturated sandy gravel. These experiments are close to the work of Bergeron et al. [6]. Pressures have been recorded and the propagation of the shock wave has been tracked by shadowgraphy. The second experimental campaign addressed the impulse transmitted by buried explosions to a test rig and the deformations of plates. Attention has been laid on the influences of the depth of burial of the explosive, of the stand-off distance and of the water content of the soil on the impulse and deformation of plates.

[a] Corresponding author: eve.roger@intradef.gouv.fr

The same STANAG soil has been used in the two experimental campaigns. It is defined in the NATO procedures AEP 55 [7] by a granulometry, a density and a water content related to the Proctor optimum:

- 100% of the mass of the soil particles must pass the 40 mm sieve;
- between 60 and 40% must pass the 5 mm sieve;
- a maximum of 10% must pass the 80 μm sieve;
- a density of 2200 $\pm$ 100 kg $\cdot$ m^{-3} and a water content less than 1.5% of the optimum Proctor are required.

While the requirements on the granulometry were respected during the experiments, various densities and water contents were tested.

2. A non linear elastic viscoplastic cap model

A viscoplastic cap model with an isotropic hardening has been developed from the existing model of Tong and Tuan [8]. It is associated with a stiffening curve depending on water content to model non linear elasticity. The initial state of the soil is taken into account with the initial void ratio and the bulk modulus at zero stress. The model was written in a user-defined material subroutine and incorporated in LS Dyna [9].

2.1. Overview of the model

The constitutive yield surface is an extension of the Drucker-Prager model. Time dependence (viscoplasticity) is obtained by Perzyna's overstress idea.

The yield surface is divided into three regions, Fig. 1: a cap surface f_1 (Eq. (1)) for $I_1 > L$ which can harden when stresses increase and has the form of an ellipse, a failure surface f_2 (Eq. (2)) for $0 < I_1 < L$ which limits shear, and a cutoff surface f_3 (Eq. (3)) for $I_1 < 0$ of elliptical shape to control tensile stresses. Attention has been laid to ensure continuity of the outward normals at the transition between two surfaces.

$$f_1(I_1, \sqrt{J_2}) = \sqrt{J_2 + \frac{1}{R^2}(I_1 - M)^2} - \frac{1}{R}(X - M) \quad (1)$$

$$f_2(I_1, \sqrt{J_2}) = \sqrt{J_2} + \alpha - \gamma \exp(-\beta I_1) + \theta I_1 \quad (2)$$

$$f_3(I_1, \sqrt{J_2}) = \sqrt{J_2 + \frac{B^2}{A^2}(I_1 - (A - T))^2} - B \quad (3)$$

$$A = T \frac{T\, df_2/dI_1(0) - f_2(0)}{2T df_2/dI_1(0) - f_2(0)} \quad (4)$$

$$B = T \frac{A^2}{A - T} \sqrt{1 - \frac{(A - T)^2}{A^2} \frac{df_2}{dI_1}(0)}. \quad (5)$$

In the above equations, I_1 is the first stress invariant, J_2 is the second deviatoric stress invariant, X corresponds to the intersection between the cap surface and the I_1 axis, M is the point of the cap with a horizontal tangent, L is the value of I_1 such as $f_1(L) = f_2(L)$ and α, β, γ and θ are soil parameters. T is the limit tensile stress.

The hardening law for the cap surface uses the viscoplastic volume change as a hardening variable:

$$X(k) - X(0) = \frac{1}{D}\left(\left|1 - \frac{k}{W}\right|^{-1} - 1\right), \quad k = -\mathrm{tr}\,\boldsymbol{\varepsilon}_{vp}. \quad (6)$$

The parameter W is the volumetric fraction of void in the soil and D a material parameter. Altogether, the model requires 12 parameters.

2.2. Stiffening during air expulsion

The detonation of explosives involves very large stresses in the soil of the order of the gigapascal. Elastic moduli highly depend on stresses: De [10] and Zakrisson et al. [4] showed that elastic moduli could be multiplied by more than 1000 when the soil undergoes such large stresses. A constitutive stiffening of the elastic properties was therefore developed.

During a detonation, while the air is expelled, the water is trapped in the soil because the detonation is so fast that it does not have the time to escape. Therefore, as the stresses increase in the soil, the three phase medium progressively turns to a two phase medium composed of soil particles and water. In the fully compacted state, the elastic properties of the soil-water mixture are obtained by averaging the elastic properties of soil and water by their volume fractions.

The bulk modulus of the crushed soil matter K_{wg} therefore depends on the bulk modulus of water K_w (around 2 GPa) and solid particles K_g (around 36 GPa):

$$K_{wg}(w) = K_g(1 - \sqrt{n_w^\infty})^2 + K_w n_w^\infty \quad (7)$$

with $n_w^\infty = V_w/V^\infty$ the final volume fraction of water. Here V_w is the current volume of water, V^∞ is the final volume of the soil mixture. The volume fraction n_w^∞ can also be written

$$n_w^\infty = \frac{w\, \rho_g/\rho_w}{1 + w\, \rho_g/\rho_w} \quad (8)$$

where w is the water content of the soil, ρ_g and ρ_w are the mass densities of the solid particles and water respectively.

Figure 2 presents the stiffening curve relating the pressure (equal to the mean soil stress) to its density. The bulk modulus K which corresponds to the slope of this curve progressively increases to finally be equal to the bulk modulus $K_{wg}(w)$ of a mixture of water and solid particles. From this point onwards, the bulk modulus of the crushed soil is constant. The expression includes two parameters, which are fixed by the bulk modulus at zero pressure and the knowledge of a particular point.

3. Pressure and shock wave propagation in small scale buried charge experiments

The detonation of a small quantity of explosive (100 g of C4) buried in a soil tank with either dry or saturated STANAG soil can now be addressed. The explosive

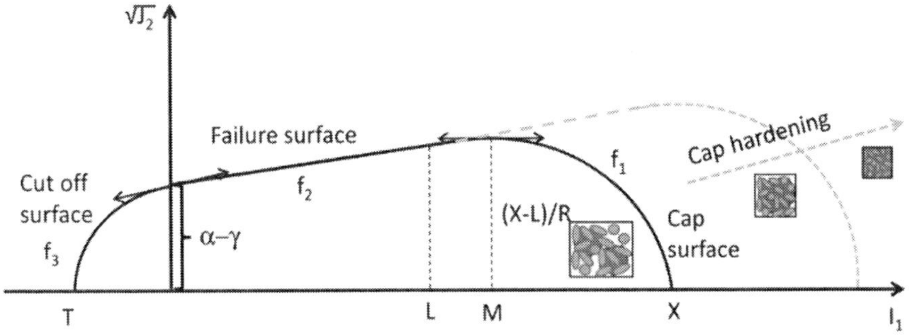

Figure 1. The yield surface of the soil model. The region of the cap is the only one which can harden.

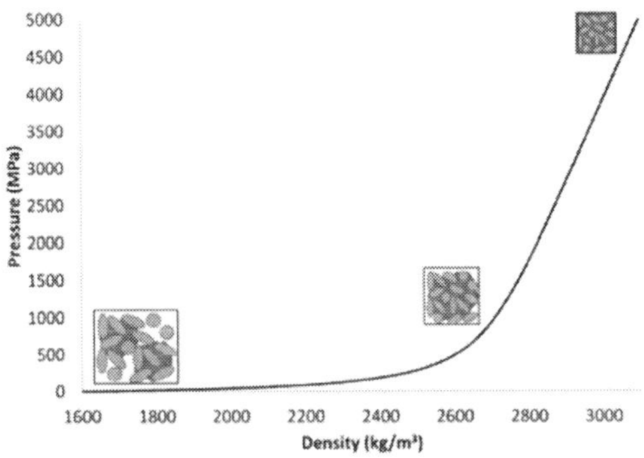

Figure 2. Curve relating the mean soil stress p to its density ρ underlining the increase of the bulk modulus K of the soil.

was 20 mm high and had a diameter of 64 mm. It was surrounded by polystyrene to maintain the detonator in place.

Three depths of burial were tested: a flush charge (its top surface is at the same level as the free surface of the soil), and charges buried at 3 and 8 cm. The depth of burial was measured between the top surface of the explosive and the free surface of the soil.

Two states of the soil were considered: a STANAG soil with its natural water content (the soil was stored in the field) and a saturated STANAG soil.

A soil tank of 1 m × 1 m × 0.5 m was designed so as to mimimize the quantity of soil to be prepared while limitating the influences of boundaries. Particular attention was paid to the soil preparation. To obtain acceptable reproducibility, the same methodology was followed for all the shots. The soil was prepared in two layers. Each of them was moistened (for a saturated soil) and tamped. Tamping is used to increase the mass density.

The density and the water content were controlled in two ways. A sample was extracted from the soil tank. Its weight was measured to obtain its density and it was oven dried to get its water content. A gamma densitometer was also used. This nuclear device can measure both density and water content. A probe is placed in the soil and two different sources are emitted: Cesium 137 and Americium 241 are sent from the probe. The first one provides the density of the soil: gamma rays are scattered when colliding with soil particles. The dispersion of gamma rays is directly linked to the density. As for the water content,

Figure 3. Experimental setting of small charge experiments.

the fast neutrons from Americium 241 are naturally slowed down by hydrogen atoms. Since water contains the largest quantity of hydrogen atoms in a soil, the number of slow neutrons can be related to the water content of the soil.

After each shot, the resulting crater was scanned to get its profile.

Figure 3 presents an overview of the experiments. A shaft was mounted above the soil tank to place five pressure gauges. They were located at 30, 70, 110, 150 and 190 cm above the free surface.

A reflective panel was placed in the background. The shock wave drives a difference of density. Its reflection on the reflective panel can be followed by a high speed camera. A rate of 20 000 pictures per second was chosen. Figure 4 shows the resulting shock wave propagation observed for a detonation of a flush explosive in a dry sandy gravel, and for a detonation of an explosive buried

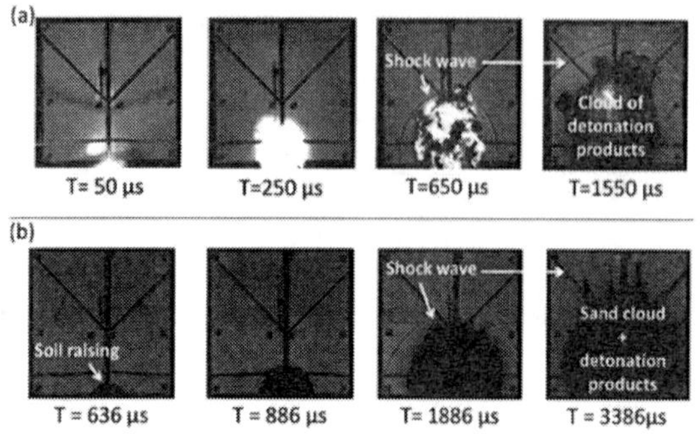

Figure 4. Pictures taken at 4 instants during the detonation (a) in dry sandy gravel for a flush explosive and (b) in a saturated soil for an explosive buried at 3 cm.

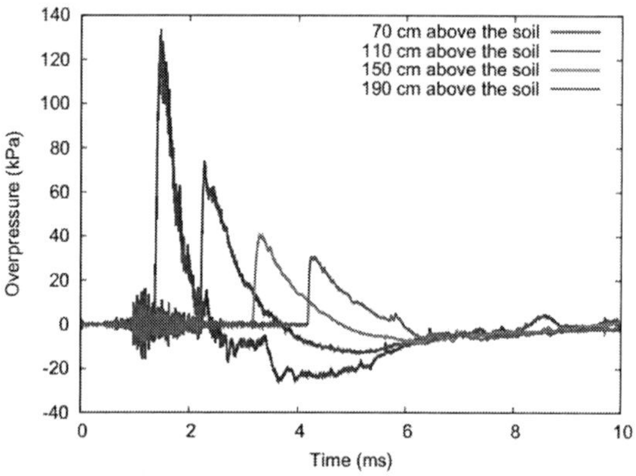

Figure 5. Pressures obtained at the four highest gauges above the soil surface for an explosive buried at 8 cm.

at 3 cm in a saturated soil. For the detonation of a flush explosive, the light of the detonation is first visible, then the shock wave followed by the detonation products appears. In the case of a buried explosion, the light cannot be observed anymore since it occurs in the soil. At variance the raise of the soil is first clearly visible. Then the shock wave propagates ahead of a cloud composed of detonation products and soil projections.

Figure 5 presents the pressures measured at the four highest gauges of the shaft for the detonation of a charge buried at 8 cm. Apart from a vibration of the structure which appears at the beginning of each signal, the rest of the measures is not disrupted by soil projections. On the other hand, the first pressure can be very noisy due to the impacts of soil particles on the gauge.

The spatial decrease of the maximum pressures is exponential. The maximum pressures plot almost linearly in a logarithm scale in Fig. 6. The pressures for a buried charge at 3 cm are more scattered than for a depth of 8 cm. The STANAG soil contains grains with a size up to 4 cm. For a depth of burial smaller than this maximum size, the aerial pressures may depend on the way the soil is replaced on the explosive and how the grains are arranged.

Except on crater sizes where differences could be noticed between dry and saturated soils, the pressures show negligible difference between unsaturated and saturated.

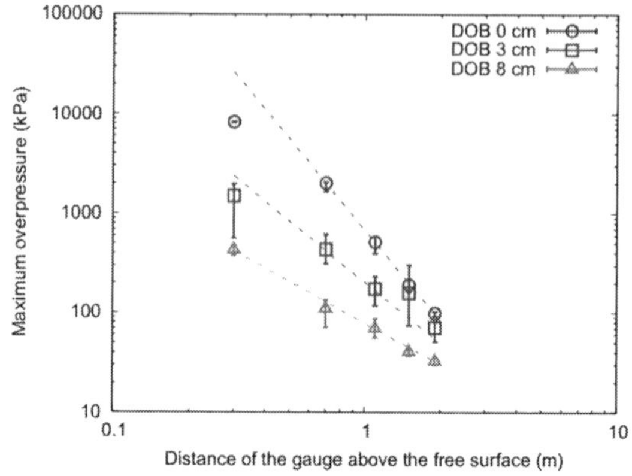

Figure 6. Maximum overpressures in logarithmic scale obtained at 5 heights for 3 depths of burial (DOB): a flush charge and charges buried at 3 and 8 cm.

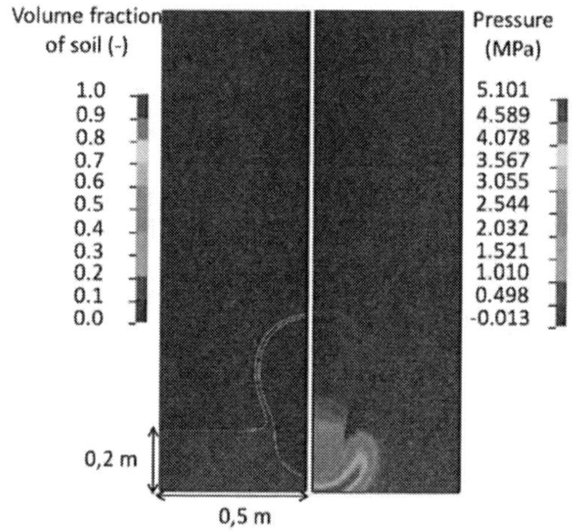

Figure 7. Contours of volume fraction of soil (left) and pressure (right) at 0.5 ms after the detonation of 100 g of C4 buried at 3 cm. In the left plot, the red (resp. blue) color denotes the soil (resp. air).

4. Small scale experiments versus simulations

Axisymmetric Eulerian simulations using the viscoplastic cap model have been run with LS Dyna to reproduce the experiments. The simulation outputs were the pressures at the five heights above the soil. While the explosive was modelled with a JWL equation, a polynomial equation was chosen for the air part. The polystyrene surrounding the explosive was taken into account with the same dimensions as in the experiments considering a space of air around the explosive. The explosive was initiated at its bottom center. As for the boundary conditions, a condition of zero displacement was chosen at the border of the soil to take into account the reflections on the soil tank while the atmospheric pressure was applied on the free surface.

Figures 7 and 8 show the contours of pressure and of volume fraction of soil at 0.5 ms and 1.5 ms after the detonation of 100 g of C4 buried at 3 cm. The

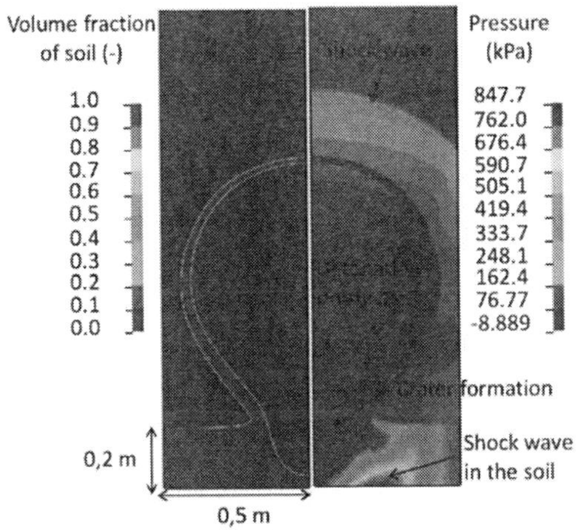

Figure 8. Same as Fig. 7 but at 1.5 ms after the detonation.

simulations reproduce qualitatively and quantitatively well the phenomena that are obversed during the experiments: since the explosive is buried, a soil cap moves upward and is visible in pressure contours (darker blue) and in the volume fraction creating a kind of soil bubble containing the detonation products. As time goes by, the bubble grows up until it has reached a maximum height and it subsequently drops. Since no fracture model has been introduced, the bubble remains closed during the whole simulation, and the detonation products cannot escape.

As in experiments, once the shock wave has reached the free surface, it propagates in the air in front of the soil projections. In the simulations, the shock wave is clearly noticeable on the pressure contours, while the motion of the soil, in particularly above the explosive can be deduced by the volume fraction of soil.

As indicated by the pressure contours, the soil undergoes large stresses. The elastic stiffening of the soil makes it able to resist the detonation. If the elastic moduli had kept a standard and fixed value throughout the simulation, the soil would have not resisted and would have been pushed away by the detonation.

The mesh size was determined with a sensitivity study. The simulation of the detonation of a flush explosive requires a finer mesh than for a buried charge. In the soil, the energy dissipation is more important than in the air, resulting in a faster widening of the shock wave. Since several elements must be contained in the shock wave, a coarser mesh could be selected to model buried detonations. Therefore the mesh sizes were equal to 1 mm and 5 mm for flush charges and buried charges respectively.

The influence of each parameter of the model on the pressures was studied. Two parameters were found to be very important, namely the initial density and the size X of the yield surface.

The relative size of the explosive is small compared to the thickness of the soil placed back above the charge. This is particularly the case when the 20 mm thick charge is buried at 8 cm. The initial density of the soil plays an important role: a denser soil requires a larger part

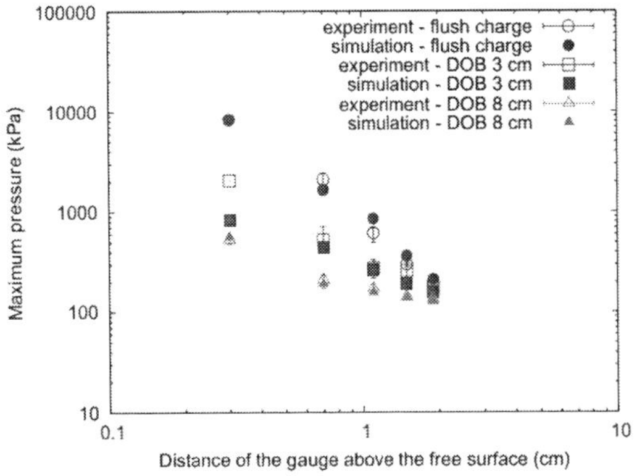

Figure 9. Maximum pressures for simulations and experiments at five heights above the soil surface for three depths of burial (0, 3 cm and 8 cm).

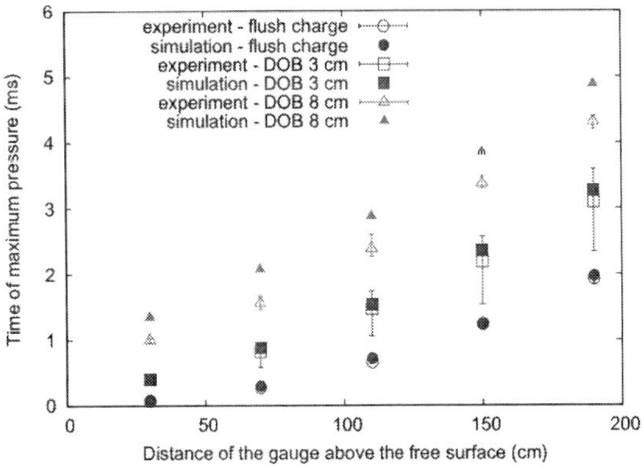

Figure 10. Times of arrival of maximum pressures between simulations and experiments at five heights above the soil surface for three depths of burial (0, 3 cm and 8 cm).

of the explosive energy to be spent in its deformation. Correspondingly, the resulting aerial pressures are smaller than those emanating from a lighter soil.

As the size X of the yield surface increases, the soil is more elastic. The speed of propagation of the shock wave is increased. Larger pressures result and the arrival of the shock wave is faster.

The maximum pressures and their corresponding times of arrival are compared to the small scale experiments for the three depths of burial in Figs. 9 and 10.

For a flush charge and a charge buried at 3 cm, the simulations reproduce well the experiments. For a charge buried at 8 cm, while the pressure outputs are similar to the experiments, the simulation outputs are slightly delayed with respect to the experiments. Since it remains constant for the five gauges, this delay may be due to soil preparation.

5. Experimental study of impulse and plate deformation

The above small scale experiments provide essentially aerial pressures. A second experimental campaign aimed at evaluating the impulse transmitted by larger masses of explosive to a test rig and at measuring the deformation of plates which stand for vehicle bellies.

Various parameters linked to the soil and to the plate were tested: two depths of burial (100 and 250 mm), two stand off distances measured between the plate and the soil (600 and 800 mm), various thicknesses of armored steel plates (15, 20, 25, 30 and 50 mm) and various masses of explosives (2, 4, 6 and 8 kg) were studied.

In the first experimental campaign the aerial pressures emanating from a soil stored in the field and a saturated soil did not show much differences. At variance in this campaign, the water contents were clearly distinct, which had noticeable consequences on deformations and on impulses.

6. Conclusion

Small scale experiments have been carried out to provide data for simulations of the explosions of buried landmines. Pressures were measured at various heights above the free surface of the soil where the explosive was buried at three different depths in a dry or saturated soil. However, the dry soil was stored in the field. So the difference on water content between what was called a dry soil, and a saturated soil is not large enough to influence the pressures significantly.

A model of soil has been developed. It takes into account the stiffening of the soil due to the shock wave and the water content. The outputs of pressure in the simulations were compared to the measurements. They correlate well the experiments. For the charge buried at 8 cm, the time of arrival of the shock wave predicted by the simulations is delayed with respect to the experiments. This delay may be due to the way the soil is placed on the top of the explosive.

A second experimental campaign with larger explosive masses has been launched with emphasis on impulse and plate deformation. A dedicated test rig has been developed. Further finite element analyses will be run to simulate the measurements obtained during these large scale experiments.

References

[1] Hlady S. L., 18th Military Aspects of Blast and Shock (2004)

[2] Peles S., Touati D., Azulay I., Neuberger A., 24th International Symposium on Ballistics (2008)

[3] Fourney W. L., Leiste H. U., Hauch A., Jung D., IMPLAST Conference (2010)

[4] Zakrisson B., Haggblad H. A., Jonsen P., Central European Journal of Engineering **2**, 532–550 (2012)

[5] Assaf Z., Ran E., Golan G., Drori O., Katalan I. K., 28th International Symposium on Ballistics (2014)

[6] Bergeron D., Walker R., Coffey C., *Detonation of 100-Gram Anti-Personnel Mine Surrogate Charges in Sand* (DRES, Canada, 1998)

[7] AEP-55, *Procedures for evaluating the protection level of logistic and light armoured vehicles, Volume 2, Edition C, Version 1* (NATO, 2014)

[8] Tong X., Tuan C., Journal of Geotechnical and Geoenvironmental Engineering **133**, 206–214 (2007)

[9] Livermore Software Technology Corporation (LSTC), *LS-DYNA theoretical manual* (Livermore, California, 2014)

[10] De A., Computers and Geotechnics **43**, 72–79 (2012)

EPJ Web of Conferences 94, 04005 (2015)
DOI: 10.1051/epjconf/20159404005

Scale and size effects in dynamic fracture of concretes and rocks

Y. Petrov[1,2,a] and N. Selyutina[1,b]

[1]Saint Petersburg State University, 7/9, Universitetskaya nab., St. Petersburg 199034, Russia
[2] IPME RAS, Extreme States Dynamics Department, V.O., Bolshoj pr., 61, St. Petersburg 199178, Russia

Abstract. Structural-temporal approach based on the notion of incubation time is used for interpretation of strain-rate effects in the fracture process of concretes and rocks. It is established that temporal dependences of concretes and rocks are calculated by the incubation time criterion. Experimentally observed different relations between ultimate stresses of concrete and mortar in static and dynamic conditions are explained. It is obtained that compressive strength of mortar at a low strain rate is greater than that of concrete, but at a high strain rate the opposite is true. Influence of confinement pressure on the mechanism of dynamic strength for concretes and rocks is discussed. Both size effect and scale effect for concrete and rocks samples subjected to impact loading are analyzed. Statistical nature of a size effect contrasts to a scale effect that is related to the definition of a spatio-temporal representative volume determining the fracture event on the given scale level.

1. Introduction

The behaviour of the material strength dependent on the dimensions of the specimen is problem of fracture mechanics, construction, mining engineering and other fields engineering. To conduct of tests at the laboratory scale economically sound. The understandings of the translation data on material strength from microanalysis to macroexamination determine new concepts of material properties. These concepts allow to description of the material behaviour at any geometrical dimension.

At the equilibrium compression material is characterized by static strength; however, in the case of dynamic loading the failure stress is different depending on the strain rate. According to in [1–3], inertial confinement is the main response of increase of the fracture stress subject to dynamic loading relative quasi-static strength concrete. In this paper, it is proposed to use a notion of the incubation time, independent of inertial effect. As referred to in [4], the incubation time is a result of relaxation processes that precede the development of micro-structural defects in the material. It is a characteristic time of the fracture process invariably for any strain rate. On the basis of the incubation time criterion, it is possible to predict the behaviour of fracture stress over a wide range of rates.

The comparison of the fracture stress at quasi-static and dynamic conditions between two materials is conducted. It should be noted that material with low strength of material in case quasi-static tests involved higher strength on dynamic loading. Grote et al. [5] observed between specimens aggregated concrete and mortar. In [5], this phenomenon is explained by the acute effect on the plate impact constraints of compressive loads that inhibit crack growth, despite the quasi-static uniaxial compression increases the probability of cracking. It is interesting to note that in the static tests constraint loads also occur, but their influence on the impact plate experiment is so great there, as described in [5]. In this paper is shown that the improvement of the concrete properties at high strain rates relative to mortar can be explained without the influence of the inertia effect (inertia at the micro-level of the material).

In [6] detailed the studying of the size effect is presented. Chasiotis and Knauss [6] observed the sensitivity of polysilicon films tensile strength to a change of specimen dimensions and identified by size and scale effects as the translation of data from micro-mechanical to macroscopic material properties and associated with the enormous influence of molecular or atomistic phenomena on the failure process of the material. It should be noted that notions of "size effect" and "scale effect" in [6] were nonselective. In this paper, "size effect" and "scale effect" concrete and rocks is considered based on structural-temporal approach, described in [7–14]. The static compressive strength and the fracture stress decreases at the growth of specimen's length (size effect) and rises at the increases of specimen's width (scale effect). Petrov et al. [13] proposed, that the main characteristics of the material (incubation time and static strength) of incubation time criterion establish at a given scale level.

In the present research it is regarded results tests on Split Hopkinson bar of different rocks and concrete, obtained in [5,15–17]. According to structural-temporal approach, we evaluate the incubation time concrete and mortar plot dependences (high-speed curve) of the fracture stress on strain rate [5]. The different behaviour of concrete and mortar under static and dynamic loading is exhibited. Using fracture criterion we calculate incubation time and high-speed curve for each specimen of concrete [15] and rocks [16,17]. Comparisons of material strength of fixed length or width specimen are presented.

2. Incubation time criterion

The fracture criterion based on the concept of incubation time, proposed in [7–14], makes it possible to predict the

[a] e-mail: yp@YP1004.spb.edu
[b] e-mail: nina.selutina@gmail.com

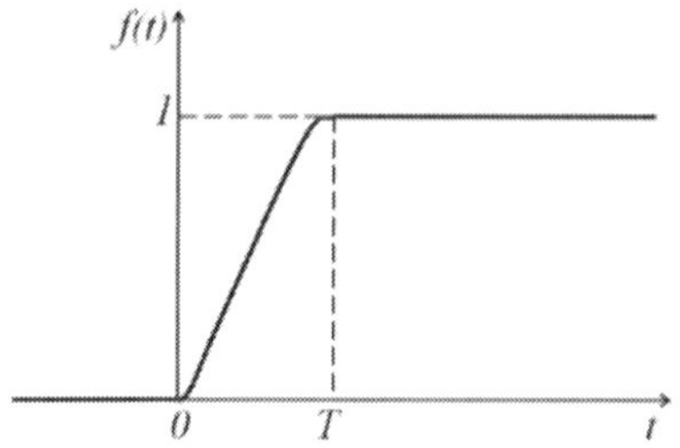

Figure 1. Schematic of fracture kinetics at the place of rupture.

unstable behaviour of the dynamic-strength characteristics observed in experiments on the dynamic fracture of solids. The fracture criterion can be written in the following general form:

$$\frac{1}{\tau}\int_{t-\tau}^{t}\left(\frac{F(t')}{F_c}\right)^{\alpha}dt' \leq 1. \qquad (1)$$

Here, $F(t)$ is the intensity of the local force field causing the fracture (or structural transformation) of the medium, F_c is the static limit of the local force field, and τ is the incubation time associated with the dynamics of the relaxation processes preceding the fracture. It actually characterizes the strain (stress) rate sensitivity of the material. The fracture time t^* is defined as the time at which the equality sign is reached in Eq. (1). The parameter α characterizes the sensitivity of the material to the intensity (amplitude) of the force field causing the fracture (or structural transformation). Often, $\alpha = 1$ gives a good agreement with test data.

One of the possible means of interpreting and determining the parameter τ is proposed here in the example of the mechanical rupture of a material. Let us assume that a standard test specimen made of the material in question is subjected to tension and is broken into two parts under a stress P arising at a certain time $t = 0$: $F(t) = PH(t)$, where $H(t)$ is the Heaviside step function. In the case of quasi-brittle fracture, the material would unload, and the local stress at the break point would decrease rapidly (but not instantaneously) from P to 0. In this case, a corresponding unloading wave is generated that propagates over the sample and can be detected by standard (e.g., interferometry) methods. The stress variation at the break point can be conditionally represented by the relation $\sigma(t) = P - Pf(t)$, where $f(t)$ varies from 0 to 1 (Fig. 1) within a certain time interval T.

The case $f(t) = H(t)$ corresponds to the classical strength theory. In other words, according to the classical approach, rupture occurs instantaneously ($T = 0$). In practice, the rupture of a material (sample) is a process in time, and the function $f(t)$ describes the micro-scale level kinetics of the transition from a conditionally defect-free state ($f(0) = 0$) to a completely broken state at the given point $f(t^*) = 1$ that can be associated with the macro-fracture event. On the other hand, application of the fracture criterion (1) to the macro-level situation ($F(t) = PH(t)$), gives the time to fracture $t^* = T = \tau$ at $P = F_c$. The physical meaning of incubation time is related to the relaxation process of growth of microdefects in the structure of material, which provides its non-reversible deformation. In this case characteristic time of relaxation can be considered as the incubation time [14]. This duration can be measured experimentally by statically fracturing the samples and controlling the rupture process by different possible methods, e.g., by measuring the time of the increase in pressure at the unloading wave front based on the recorded velocity profile of points (by interferometry) on the sample boundary.

2.1. Definition fracture stress on incubation time criterion

Let us present the application of the incubation time approach to a definition of strength under compression in a wide range of strain rate. As a rule, the concept of strength defines the growth limit value of the local stresses (maximum stress), at which the material does not fall into decay. In the incubation time approach, dynamic strength is characterized by the incubation time, but not maximum stress.

According to experimental data [5, 15] the criterion (1) rewritten for case the defect-free sample in the following form:

$$\frac{1}{\tau}\int_{t-\tau}^{t}\left(\frac{\sigma(t')}{\sigma_c^{compr}}\right)^{\alpha}dt' \leq 1, \qquad (2)$$

where $\sigma(s)$ is the time dependence of the average compressive stress in the specimen, σ_c^{compr} is the static compressive strength of the material, and τ is the incubation time of material, α – parameter of sensitive to the level (or amplitude) of the intensity of the local stress and $\alpha \geq 1$.

According to the typical dependence of loading stress on time at the moment of fracture, described in [5, 15], an increase the compressive stress with strain rate can be assumed to be linear until it reaches the maximum value, so that

$$\sigma(t) = E \cdot \dot{\varepsilon} \cdot t \cdot H(t) = \dot{\sigma} \cdot t \cdot H(t), \qquad (3)$$

where $H(t)$ is the Heaviside function, E is Young's modulus, $\dot{\varepsilon}$ is the strain rate, $\dot{\sigma}$ is the stress rate.

The material fracture occurs if equality is attained in criterion (3). Defining an ultimate stress σ_d as a stress at the moment of fracture $\sigma_d = \sigma(t_*)$, we obtain the function of limiting compressive stress on strain rate:

$$\sigma_d(\dot{\varepsilon}) = \begin{cases} \left((\alpha+1)\left(\sigma_c^{compr}\right)^{\alpha} E\dot{\varepsilon}\tau\right)^{\frac{1}{\alpha+1}}, \dot{\varepsilon} \geq \frac{\sigma_c^{compr}}{E\tau}(\alpha+1)^{\frac{1}{\alpha}} \\ \sigma_c^{compr} + \left(1 - \frac{1}{(\alpha+1)^{\frac{1}{\alpha}}}\right)E\dot{\varepsilon}\tau, \dot{\varepsilon} < \frac{\sigma_c^{compr}}{E\tau}(\alpha+1)^{\frac{1}{\alpha}}. \end{cases} \qquad (4)$$

Grote et al. [5] were carried out plate impact experiments at strain rates of the order $10^4 - 10^5\,s^{-1}$ of concrete and mortar, and tests on a SHPB at strain rates of the order $10^2 - 10^4\,s^{-1}$ of mortar samples. It was determined during the quasi-static test the static strength of mortar specimens in compression 46 MPa and Young's modulus 20 GPa.

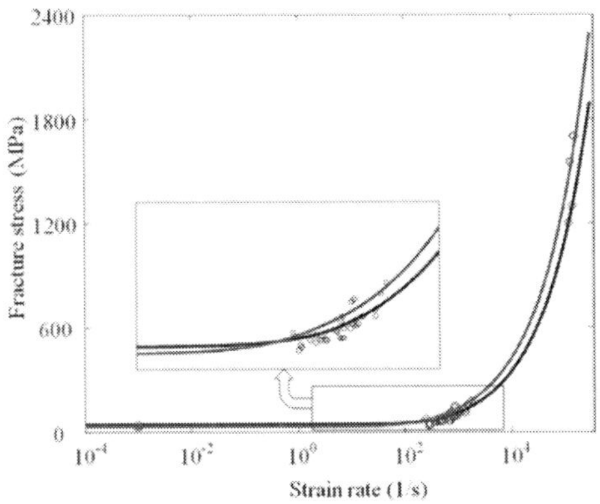

Figure 2. Compressive strength of concrete (red color) and mortar (black color) [5] as function of strain rate.

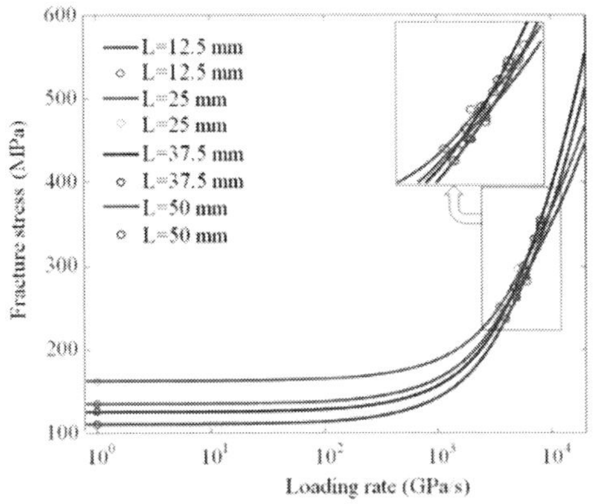

Figure 3. Compressive strength of Laurentian granite [16] as function of strain rate.

Table 1. Comparison of the behaviour of concrete and mortar in quasi-static tests and plate impact experiments [5].

Material	σ_d, MPa	E, GPa	σ_c^{compr}, MPa	τ, μs
concrete	1700	45	30	6.5
mortar	1300	20	46	6.5

Using the function (4) and the test data [5] is obtained the incubation time for the mortar and concrete specimens $\tau = 6.5\,\mu s$ as an arithmetical middling from all samples. The dependence of the fracture stress on the strain rate (4) for concrete and mortar is plotted in Fig. 2. The comparison of mechanical properties two materials for test data [5] in quasi-static and dynamic experiments are presented in Table 1.

3. Size and scale effect

Let us consider the concept of the size effect is defined as a decrease of the material static strength in the specimen length growth and the concept of the scale effect is understood as an increase of the material static strength in the specimen width growth. Observe that notions of size and scale effects in [6] were nonselective and associated with the enormous influence of molecular or atomistic phenomena in the failure process of material and identified as the translation of data from micro-mechanical properties of material to macroscopic. In current paper is purposed that a change of the static strength dependent on material dimensions described of two different effects.

It is recalled that according to the "weakest link hypothesis", reported in [18], the static strength depends on the most damageable defect in the material under the application of a specific load. According to the statistical approach of strength (Weibull statistics), the massive body (or fiber) consists in set of structural elements and each component has own a local strength. In addition the fracture of a massive body (or fiber) is defined as elements group failure (or single element) under the application of a specific load.

The size effect can be explained using "weakest link hypothesis" [19]. The increase of sample length can be provided for the largest possibility of the propagation of the material defect. The growth of the defects is weakened the strength properties of the material. It can be expected that response of the size effect result in growth of fracture probability in the same structure of the material.

Petrov et al. [20,21] and Morozov et al. [14] proposed the structural-temporal discretisation of the failure process and introduced the micro-structural fracture time (incubation time of the dynamic fracture process) and the spatial scale (characteristic size of the fracture zone) determined at a given scale level. According to the supposed concept in [13], fixed temporal and spatial characteristics of a scale level define lower and upper bounds of the characteristic size of the test specimen. Taking into account the change of material dimensions the failure characteristics of the incubation time approach can be modified within the limits only one or several failure levels. In the present paper the fracture level is characterized by incubation time calculated based on criterion (2).

Dai et al. [16] were conducted dynamic tests on compression for Laurentian granite cylindrical samples at loading rates 2000–10000 GPa/s. Samples of length 12.5 mm, 25 mm, 37.5 mm, 50 mm and diameter 25 mm were employed. Static strength 220 MPa and Young's modulus 92 GPa were determined in [16,22].

Static and SHPB tests on cylindrical specimens of granite, sandstone and limestone different dimensions ($D = 22$ mm and $L = 11$ mm, $D = 36$ mm and $L = 18$ mm, $D = 75$ mm and $L = 37.5$ mm) were undertaken in [17] at strain rate 36–360 s^{-1}. Young's modulus of granite, sandstone and limestone is equal 63 GPa, 5.4 GPa, 24 GPa, respectively.

On the basis of function (4), the dependence of the compressive strength on strain rate and estimated incubation time and loading parameter each specimen for rocks [16,17] are presented in Fig. 3 – Fig. 6 and Table 2–Table 3.

Figure 3 shows that the quasi-static strength and average fracture stress decrease in the growth specimen's length. The increase of the failure stress in the growth specimen's width is presented in Fig. 4 – Fig. 6.

Hao et al. [15] were performed quasi-static and SHPB compression tests in the range of strain rates 50–500 s^{-1} on cylindrical concrete samples with

Table 2. Mechanical properties of Laurentian granite [16] dependent on specimen dimension.

L, mm	D, mm	τ, μs	α
12.5	25	69.01	1.812
25	25	58.96	1.836
37.5	25	62.07	1.014
50	25	64.3	1.07

Table 3. Mechanical properties of rocks [16] dependent on specimen dimension.

L mm	D mm	τ, μs granite	α	τ, μs sandstone	α	τ, μs limestone	α
11	22	23.2	1.04	393.7	1.08	146.8	1.06
18	36	34.6	1.22	345.2	1.08	96.18	1.06
37.5	75	94.9	1.26	1000	1.21	18.89	1

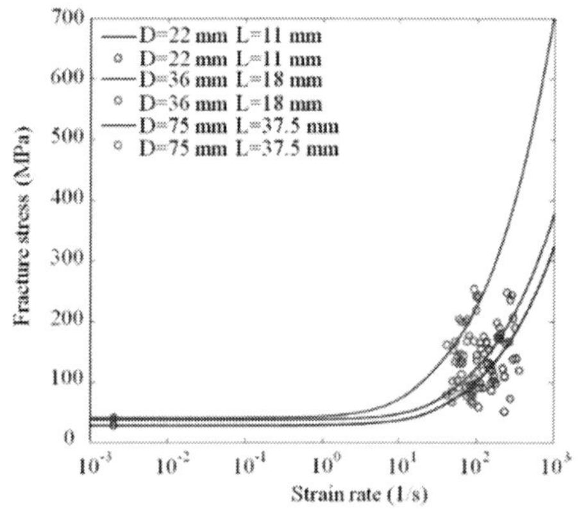

Figure 4. Compressive strength of granite [17] as function of strain rate.

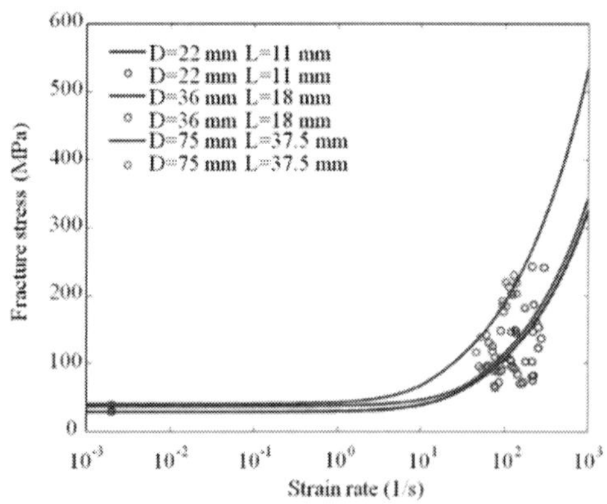

Figure 5. Compressive strength of sandstone [17] as function of strain rate.

different dimensions: $D = 40$ mm and $L = 20$ mm, $D = 22$ mm and $L = 11$ mm, $D = 32$ mm and $L = 20$ mm, $D = 32$ mm and $L = 10$ mm. According to the bulk modulus and the shear modulus of concrete [15] is obtained Young's modulus 45 GPa. Table 4 shows static strength for each specimen, described in [15].

The results represented in the Table 4 show that an essential increase in sample diameter from 22 mm

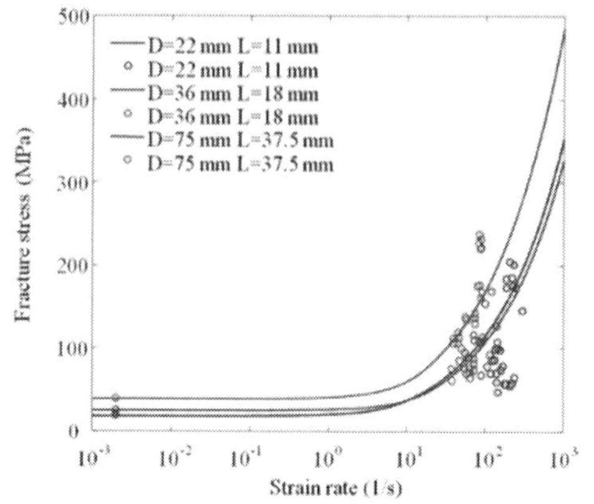

Figure 6. Compressive strength of limestone [17] as function of strain rate.

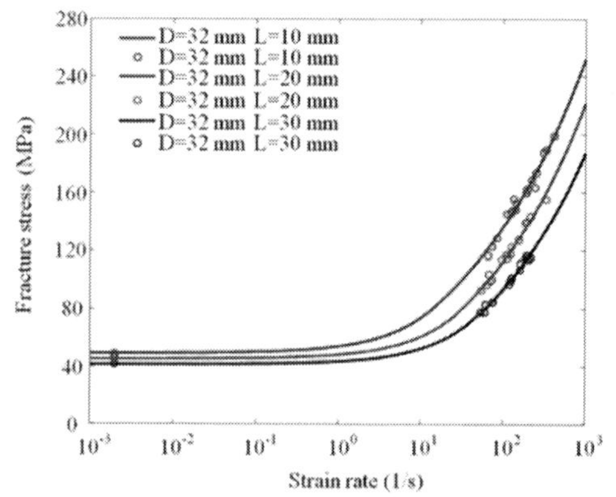

Figure 7. Experimental data [15] for concrete and dependences of compressive strength on strain rate at fixed width 32 mm and different lengths of the concrete specimen.

Table 4. Mechanical properties of concrete [15] dependent on specimen dimension.

L, mm	D, mm	σ_c^{compr}, MPa	τ, μs	α
11	22	45.5	50.07	3.709
10	32	48.8	128	2.703
20	32	45.3	63.02	2.377
30	32	41.2	40.54	2.301

to 32 mm (at nearly the same length 10 mm–11 mm) leads to a noticeable increase of measured static strength and incubation time. On the other hand the greater is the sample length for giving diameter, the less is the compressive strength as well as the material incubation time.

4. Discussion

Figure 2 – Fig. 6 show that average fracture stress is increased with strain rate. It should be noted that the critical comprehensive strength of material begins to increase drastically with a strain rate of the order 10^2–10^3 s^{-1}. According to high-speed curves, it can get an average value of the dynamic strength of the material at different loading rate or strain rates. This is phenomena

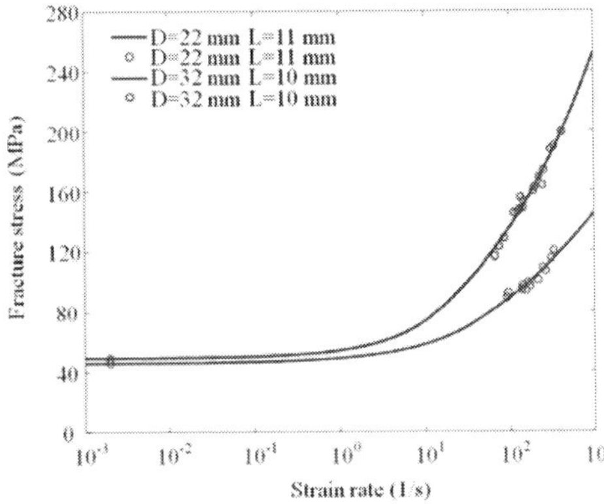

Figure 8. Experimental data [15] for concrete and dependences of compressive strength on strain rate at fixed length 10 (11) mm and different widths of the concrete specimen.

is fully explained by incubation time criterion at different values α.

As can be seen from Table 1 the impact experiment [5] showed that concrete was stronger than mortar by 30%.

Grote et al. [5] explain the effect of the existence of high confining pressure in the specimens. These stresses are growth drivers of micro-cracks in the static compression that critical compression stress reduces; also, these stresses are retard growth factors of microdefects in tests at high strain rates whereby material strength increases. Concrete is more inhomogeneous than in the case of high strain rates this influence of modulus becomes determinant mortar, so the fracture stress of mortar is less than that of concrete in plate impact experiments.

The results [5] in Table 1 demonstrate that the aggregated concrete has smaller static strength and larger elastic modulus compared to mortar. It is seen that the static compressive strength does not influence much of the critical stress at high strain rates of the order 10^4–10^5 s^{-1}. Here it is considered that the local stress history follows the linear law (3) where Young's modulus defines the growth rate of compressive stress. Thus, it turns out that the competitive "strength inversion" between concrete and mortar is defined by the existence of incubation period and by the difference in elastic modulus. Taking in mind similar values of τ for mortar and aggregated concrete we get that the greater is Young's modulus the bigger is ultimate critical stress, which is attained during the incubation period. It has little effect at low impact velocities, but in the case of high strain rates this influence of modulus becomes determinant.

Figure 4 – Fig. 6 and Table 3 show that the behaviour of the incubation time in case increase in diameter and length of specimen is difficult. According to Table 2, incubation time of granite at a fixed diameter of the specimen is contained in the short range of 59–69 μs^{-1}. The decrease of static strength in Fig. 3 can be associated with one scale level.

The different behaviour of static strength (known experimentally for each specimen concrete) at a fixed length and width of specimen is observed. Note that decrease and growth of the static strength of specimen are presented in Fig. 7 and Fig. 8, respectively. Table 4 shows that the growth of the incubation time and decrease of loading parameter are characterized by scale effect (in case of size effect, there will be a different (an opposite) relation between material strength and sample's sizes). Hence, variation of static strength can be exhibited using parameter material τ. The set material parameters containing static strength and incubation time define spatial-temporal representative volume determining the fracture event on the given scale level.

5. Conclusion

The average critical compressive stress concrete and rocks at different loading rate and strain rate based on the structural-temporal approach was predicated using the incubation time of the fracture as a material parameter and coefficient of sensitive to amplitude of loading impulse.

The behaviour of both materials concrete and mortar over the course of quasi-static tests and plate impact experiments was investigated and compared. It is obtained that limiting compressive stress of mortar at a low strain rate was smaller than that of concrete, but at a high strain rate the opposite is true. It should be noted that the reason for this "strength inversion" effect can be the combined influence of incubation time and material elastic modulus along with the growth of strain rate.

Using experimental data from a split Hopkinson bar test was established incubation time for concrete, mortar samples of the different specimen dimension. On the basis of incubation time criterion characteristic curves in a wide range of strain rates were calculated. The static strength dependent on a sample length based on criterion (1) was evaluated. The size effect based on weakest link hypothesis was explained.

Size and scale effect as different phenomena was explained by structural-temporal approach. According to dynamic test data, definition of incubation time for different specimens can be predicated behaviour static strength of material. It is noted that concept of the size effect of static strength concerns a consideration of material failure in limit of one scale level. The physical meaning of the static strength of scale effect is defined as a transition between fracture levels.

This research was supported by the RFBR (14-01-00814) and Saint-Petersburg State University (grants 6.38.243.2014; 6.39.319.2014).

References

[1] H. Le Nard, P. Bailly. Mech Cohes-Frict Mater **5**, 491–510 (2000)
[2] T. Tang, L.E. Malvern, D.A. Jenkins. ASCE J Eng Mech **118**(1), 108–24 (1992)
[3] C. Huang, G. Subhash. J Mech Phys Solids **51**, 1089–105 (2003)
[4] A.M. Bragov, Yu.V. Petrov, B.L. Karihaloo, A.Yu. Konstantinov, D.A. Lamzin, A.K. Lomunov, I.V.

Smirnov. Engineering Fracture Mechanics **110**, 477–88 (2013)

[5] D.L. Grote, S.W. Park, M. Zhou. Int J of Impact Eng **25**, 869–86 (2001)

[6] I. Chasiotis, W.G. Knauss. J Mech Phys Solids **51**, 1551–72 (2003)

[7] Yu.V. Petrov, A.A. Utkin. Mater Sci **25**(2), 153–6 (1989)

[8] Yu.V. Petrov. Doklady Physics **49**(4), 246–9 (2004)

[9] Yu.V. Petrov, E.V. Sitnikova. Technical Physics **49**(1), 57–60 (2004)

[10] V. Bratov, Y. Petrov. International Journal of Fracture **146**(1), 53–60 (2007)

[11] A.N. Berezkin, S.I. Krivosheev, Yu.V. Petrov, A.A. Utkin. Doklady Physics **45**(11), 617–9 (2000)

[12] Yu.V. Petrov, P.A. Glebovski. Technical Physics **49**(11), 1447–51 (2004)

[13] Y.V. Petrov, B.L. Karihaloo, V.V. Bratov, A.M. Bragov. International Journal of Engineering Science **61**, 3–9 (2012)

[14] N. Morozov, Y. Petrov. Berlin-London-New York: Springer Verlag (2000)

[15] Y. Hao, H. Hao, G.P. Jiang, Y. Zhou. Cement and concrete research **52**, 63–7 (2013)

[16] F. Dai, S. Huang, K. Xia, Z. Tan. Rock Mech Rock Eng **43**, 657–66 (2010)

[17] L. Hong, X.B. Li, C.D. Ma, T. Yin, Z. Ye, G. Liao. Chin J Rock Mech Eng **27(3)**, 526–33 (2008)

[18] Yu.N. Rabotnov, *Mechanics of solids* (Moscow: Nauka, 1988) 712

[19] D.A. LaVan, T. Tsuchiya, G. Coles, W.G. Knauss, I. Chasiotis, D. Read. Cross Comparison of Direct Strength Testing Techniques on Polysilicon Films. In: Muhlstein, C., Brown, S.B. (Eds.), Mechanical Properties of Structural Films, ASTM STP 1413. American Society for Testing and Materials,West Conshohocken, PA, 1–12 (2001)

[20] Y. Petrov, N. Morozov. ASME Journal of Applied Mechanics **61**, 710–2 (1994)

[21] Y.V. Petrov. Doklady Akademii Nauk USSR **321**(1), 66–8 (1991)

[22] M.H.B. Nasseri , B. Mohanty, P.-Y.F. Robin. Rock Mech Min Sci **42**, 450–60 (2005)

EPJ Web of Conferences **94**, 04006 (2015)
DOI: 10.1051/epjconf/20159404006

Strain rate effects for spallation of concrete

Ulrich Häussler-Combe[a], Evmorfia Panteki[b], and Tino Kühn[c]

Institut für Massivbau, Technische Universität Dresden, 01062 Dresden, Germany

Abstract. Appropriate triaxial constitutive laws are the key for a realistic simulation of high speed dynamics of concrete. The strain rate effect is still an open issue within this context. In particular the question whether it is a material property – which can be covered by rate dependent stress strain relations – or mainly an effect of inertia is still under discussion. Experimental and theoretical investigations of spallation of concrete specimen in a Hopkinson Bar setup may bring some evidence into this question. For this purpose the paper describes the VERD model, a newly developed constitutive law for concrete based on a damage approach with included strain rate effects [1]. In contrast to other approaches the dynamic strength increase is not directly coupled to strain rate values but related to physical mechanisms like the retarded movement of water in capillary systems and delayed microcracking. The constitutive law is fully triaxial and implemented into explicit finite element codes for the investigation of a wide range of concrete structures exposed to impact and explosions.

The current setup models spallation experiments with concrete specimen [2]. The results of such experiments are mainly related to the dynamic tensile strength and the crack energy of concrete which may be derived from, e.g., the velocity of spalled concrete fragments. The experimental results are compared to the VERD model and two further constitutive laws implemented in LS-Dyna. The results indicate that both viscosity and retarded damage are required for a realistic description of the material behaviour of concrete exposed to high strain effects [3].

1. Introduction

Experimental observations demonstrate an increase in concrete strength with rising strain rates. This phenomenon is referred to as strain rate effect and is particularly noticeable in tension. The strength increase is normally characterized by the Dynamic Increase Factor (DIF), which expresses the ratio of dynamic strength to the corresponding quasistatic value. In practise this aspect is often relevant for concrete structures exposed to traffic accidents, earthquakes, rockfalls and similar cases of impact and explosions.

A summary of experimental work on the material behavior of concrete under high strain rates is given in Fig. 1 for compression [4] and tension [6]. Layed on logarithmic axes determined DIF values can be approximated by a bilinear curve. DIF-values of up to 2 for strain rates up to $\approx 10\,\mathrm{s}^{-1}$ can be approximated by the first flat branch. Strain rates above $10\,\mathrm{s}^{-1}$ can be approximated by the second steeper branch with DIF-values up to 10 for stain rates of $\approx 100\,\mathrm{s}^{-1}$.

Corresponding on the assumption the two branches of the DIF-diagram are in connection with different physical mechanisms, the VERD model was developed. Under the terms of this concept, the flat branch results from pore water bound in the capillary systems of cement. Viscous effects take place during the deformation of pores and allow the material to withstand higher stresses under high strain rate conditions [7]. In this regard the second steeper branch is contributed to retarded damage [8,9]. Cracks cannot spread arbitrarily fast, which implies that crack speeds below the materials wave speed permit the lowering of stresses even before the formation of cracks.

The paper examines the VERD model and evaluates their suitability for the modelling of concrete behavior under high strain rates in contrast to two further constitutive models implemented in LS-Dyna. The VERD model was added to the program with a user-defined subroutine. The KCC (Karagozian & Case Concrete model) and the RHT (Riedel, Hiermaier and Thoma) model were chosen for comparative reasons, as they are among the most commonly used concrete constitutive models in LS-Dyna. Both are phenomenological and provide a wide range of parameters which ideally can be adjusted to reproduce the desired concrete behavior. In terms of simplification the option of automatic parameter generation is given. This approach has been practiced here and its suitability as a first estimate was tested. The basic required parameter was the unconfined compressive strength, which was set to $30\,\mathrm{N/mm}^2$.

2. Homogenous behavior

Main comparative analysis of the material behavior is done on the basis of the Split Hopkinson Bar. Prior to that homogeneous material behavior is modelled and evaluated with a one element setup using a 8-node cube. Structural inertia effects are minimized by a relatively short edge length of 1 mm. The load is applied with a prescribed displacement of the four upper nodes, as shown in Fig. 2. This leads to a uniaxial stress state and controllable strain rates. Quasi static strengths arise for a strain rate of

[a] e-mail: `ulrich.haeussler-combe@tu-dresden.de`
[b] e-mail: `evmorfia.panteki@tu-dresden.de`
[c] e-mail: `tino.kuehn@tu-dresden.de`

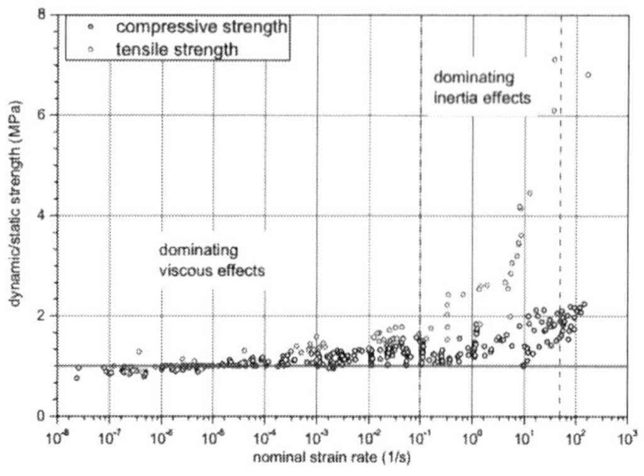

Figure 1. DIF-diagram.

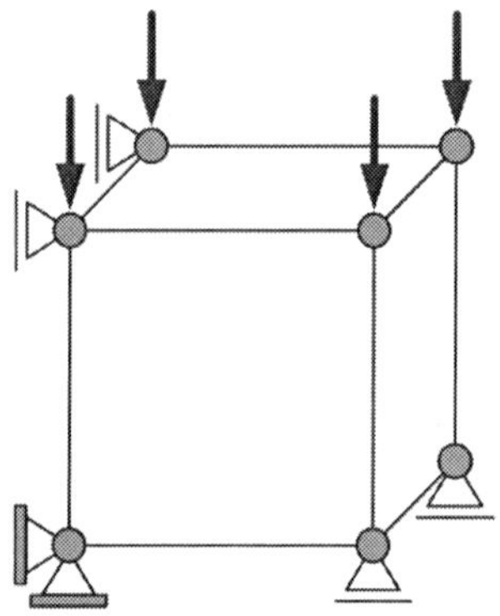

Figure 2. One element setup.

about 1 10^{-5} 1/s, whereas increasing the strain rates comes with an increasing maximum stress. The resulting uniaxial stress-strain relations in tension and compression are given for sample strain rates in Fig. 3 and Fig. 4, distinguished by material model.

A critical point concerning the VERD model are the high computed strength increases in particular observed in case of compressive loading. This can however be considered of minor importance in the mainly tensile testing configuration of the Split Hopkinson Bar. Also noticeable is the constant stress level the RHT element maintains after having reached its compressive stress peak. The formulations of the KCC and the VERD model on the other hand describe pre-peak nonlinearities as well as post-peak softening. For the given element size and with use of the automatically generated parameters no noteworthy strain rate dependency could be noted for the KCC material model. Computed strength increases with bigger element sizes in case of compressive loading were attributed to inertia effects.

3. Split Hopkinson Bar modelling

3.1. Experimental setup

Dynamic material response evaluation is continued with the set up of the Split Hopkinson Bar (SHB) [5]. This configuration has widely been used to study the material behavior of concrete under high strain rate conditions.

The classical version of the setup consists of three parts in line. These are the aluminium incident bar, the concrete specimen and the aluminium output bar, which here is omitted for the considered spallation configuration. The loading occurs while a steel impactor hits the incident bar and subjects it to a pulse load. This results in a compressive wave traveling in longitudinal direction through the configuration. With the intension to study the materials tensile response prescribed velocities are used to generate pulse intensities below the concretes compressive strength but above its corresponding tensile value. Considered are pulse intensities after the partial reflection at the contact area between aluminium and concrete. This approach leads to concrete failure shortly after the wave fully reflects at the specimens right free end. Figure 5 schematically shows the course of events.

The SHB analysis is mostly evaluated under the assumption of uniaxial wave propagation, whereas in reality triaxial stress states will takes place. This is encouraged by relatively large specimen diameters, which are necessary in order to counteract the material heterogeneity. This in turn leads to considerable wave form modifications, which is covered by the triaxiality of the material models.

3.2. Simulation model

The Split Hopkinson Bar dimensions used for the study correspond to the setup used by Schuler, Mayrhofer & Thoma (2006) [2]. The diameter of all three parts was thus set to 75 mm. The length was 60 mm for the steel projectile, 5500 mm for the incident bar and 250 mm for the specimen.

Experimental results are given for impactor velocities of 4.1, 7.6 and 11.1 m/s respectively. For comparison purposes these initial conditions were also applied in the computations with LS-DYNA.

The material behavior of the steel impactor and the aluminum impactor bar was assumed linear elastic, whereas for the concrete specimen VERD, KCC and RHT material models were taken into account. For the discretization of all three components 8-node single integration point elements (ELFORM = 1) were used. Preliminary numerical results revealed no significant changes below an element size of approximately 6 mm. Modelling was therefore performed with 1050 elements for the projectile, 94500 elements for the impact bar and 4305 elements for the specimen.

3.3. Results for stresses

The numerical spallation results are presented by material model. First considered is the KCC model. In the quasistatic case the use of automatically generated material parameters resulted in a quite ductile uniaxial material behavior, as shown in Fig. 3. Regarding the largest compressive stress the first load stage (4.1 m/s) resulted in $\approx 28\,\mathrm{kN/m^2}$ whereas the largest tensile strength was $\approx 3.4\,\mathrm{kN/m^2}$, as shown in Fig. 6.

The spallation process could potentially be problematically influenced by the ductile material behavior. Cracking is being artificially delayed, whilst further transversal

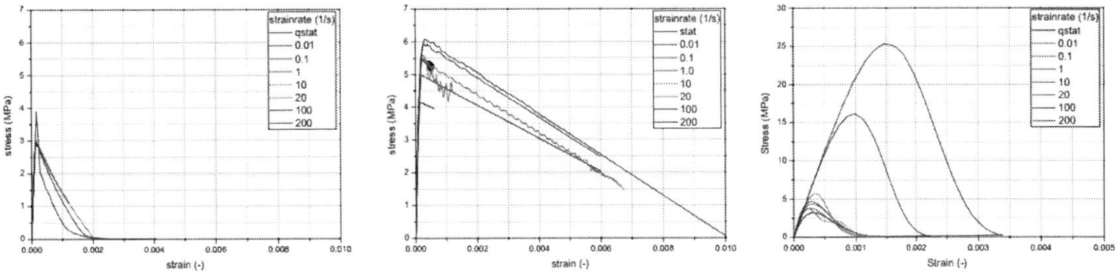

(a) KCC (b) RHT (c) VERD

Figure 3. Uniaxial stress-strain relations for sample strain rates in tension.

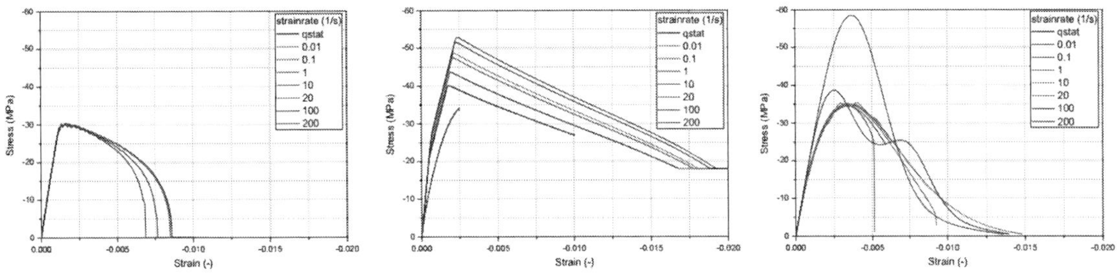

(a) KCC (b) RHT (c) VERD

Figure 4. Uniaxial stress-strain relations for sample strain rates in compression.

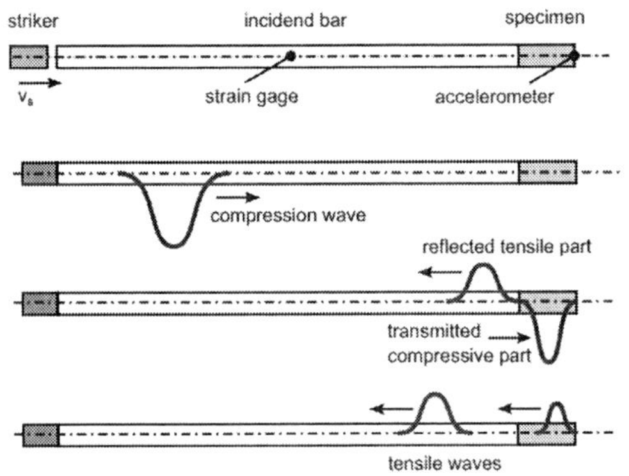

Figure 5. Course of events in the spallation test.

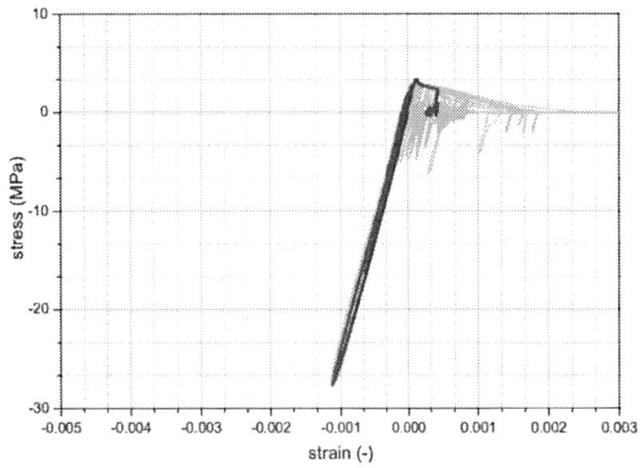

Figure 6. KCC material model: Stress-strain diagram of the specimens centre elements in longitudinal direction for loading stage one.

element strain arises from the assumed constant Poissons ratio. This in turn favors multi axial stress states, which affect the elements failure strength.

Loading stage two (7.6 m/s) showed similar results in tension and maximum compressive stresses of $\approx 51\,\mathrm{kN/m^2}$. The simulation performed with impactor velocity three (11.1 m/s) lead to compressive failure of the material. In this cases large permanent deformations are caused by the entry of the compressive pulse, which thus reaches the materials compressive strength. At this point the numerical results do not correspond to the experimental results, where no compressive failure was observed.

Next considered is the RHT model. The study of its response revealed an even more ductile behavior than the KCC material model. The largest compressive stresses increased with increasing projectile velocity and were computed as ≈ 30, 53 and $72\,\mathrm{kN/m^2}$. Validation of these computations is possible to a limited extent only, since assumed stress states in experimental investigations rely on the adoption of assumed stress strain relations. According to such assumptions maximum compressive stresses were given by Schuler [10] as ≈ 25, 49 and $64\,\mathrm{kN/m^2}$ for loading stages one, two and three.

At a closer consideration of loading stage three it becomes apparent that the longitudinal stress strain relations differ strongly from one element to another. This is due primary to the compaction of the porous assumed RHT material and the deviations from the purely longitudinal wave propagation scenario.

Last considered is the VERD material model. Here (for all three loading stages) the maximum tensile strength significantly exceeds the earlier computed stresses,

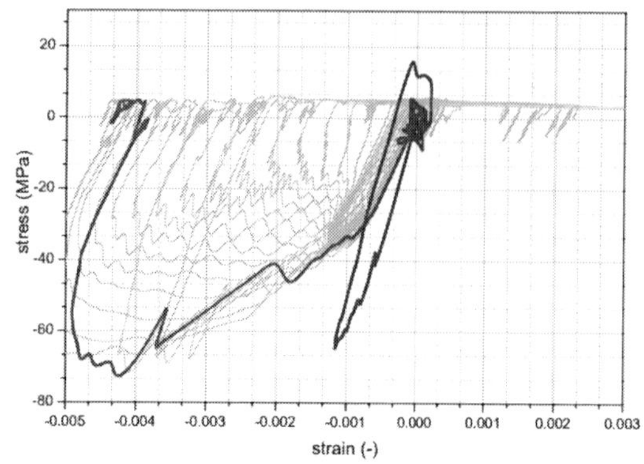

Figure 7. RHT material model: Stress-strain diagram of the specimens centre elements in longitudinal direction for loading stage three.

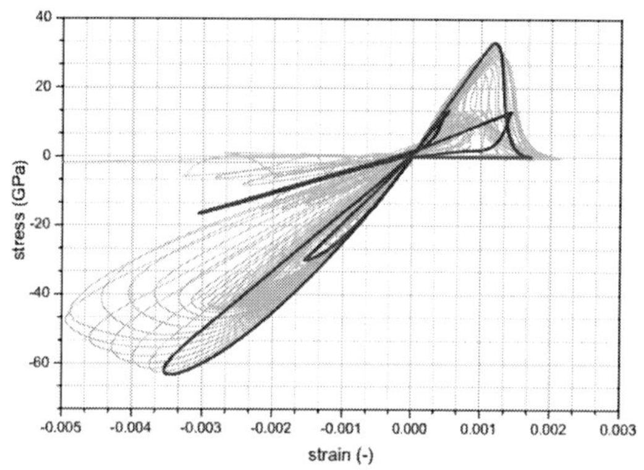

Figure 8. VERD material model: Stress-strain diagram of the specimens centre elements in longitudinal direction for loading stage three.

whereas the maximum compressive stresses come close to the values given by Schuler. Computed were ≈ 25.8, 45.6 and $63.3\,\text{kN/m}^2$. Load stage one computes a maximum tensile failure strength of $\approx 25.8\,\text{kN/m}^2$, which implies an dynamic increase factor of ≈ 7 based on the quasistatic value.

3.4. Results for velocities

A major point of the Split Hopkinson experiment evaluation considers the determination of the dynamic tensile strength. Since this is experimentally not directly possible an indirect method has been developed based on the velocity data of the right free specimen end.

Particle displacements and velocities arise from the wave entering the specimen as a compression pulse. Provided that its pulse intensity is kept bellow the materials compressive strength tensile failure occurs after the reflection at the specimens free end. It takes place at some distance from the free end and leads to spallation i.e. cracking and fragmentation.

The critical evaluation value is the particle velocity reduction or pull pack velocity Δv_{pb}, measured after the first velocity maximum. This value is an indicator for the maximum reached tensile strength inside the specimen. Basis of the estimation is the uniaxial wave propagation in linear elastic material with brittle tensile fracture. Subject to this requirements the relation

$$f_{ct,dynamic} = \frac{1}{2}\,\rho\, c\, \Delta v_{pb} \tag{1}$$

accurately reflects the failure tensile strength, where ρ is the density of concrete and c the wave speed. In contrast to this physical nonlinearities and multiaxial stress states restrict the validity of the relation, turning it into an estimation.

Focal point of the material model comparison is the evaluation based on the right free end velocity. At this point available experimental data and relatively high measurement accuracy provide a suitable foundation for data valuation. Figure 9 shows the velocities computed for all three material models together with the experimentally recorded ones. It should be noted that velocities of the KCC model are not relevant after compressive failure and are therefore plotted for loading stage one and two only. Following observations can be made:

- Plotting of the right end velocities for the RHT material model results in non-smooth data, which in this case illustrates the deviations from the uniaxial dispersionless wave propagation.
- The VERD model computes the highest free end velocities and the largest pull-back-velocities compared to the other material models. Both measurands are in the same order of magnitude as the corresponding experimental values.
- In this way evaluation of Eq. (1) suggests the highest dynamic tensile stresses for the VERD model. They were calculated as ≈ 11.9, 15.4 and $14.5\,\text{kN/m}^2$ in the numerics, whereas Schuler [10] determined the corresponding experimental values to ≈ 12.9, 16.2 and $18\,\text{kN/m}^2$.
- Considerations must be given to the fact that while dynamic tensile strengths of the VERD model determinded over equation 1 show good agreement with experimental values, detailed insight into the specimens stress states reveals higher longitudinal stresses (see e.g. Fig. 6).

3.5. Local insights on the VERD formulation

Figure 10 is taking a closer look at the specimens behaviour according to the VERD material formulation. The red curve indicates the stress time distribution for the first element carrying the maximum compressive load with underlying greyed average results. The back flip of the stress into the tensile branch shows the relaxing of the free incident surface while the outer pressure load releases. The later oscillating indicates the intactness of the specimens left side. In contrast to this, the blue curve describes the stress distribution at the position of the later failure. Undergoing a compressive load, it reflects on the free right end and induces the tensile cracking at a high stress level.

Figure 11 further extracts this position and illustrates the stress strain relation with its highlighted strain rates distribution. Starting at (0) the point caries compression (1)

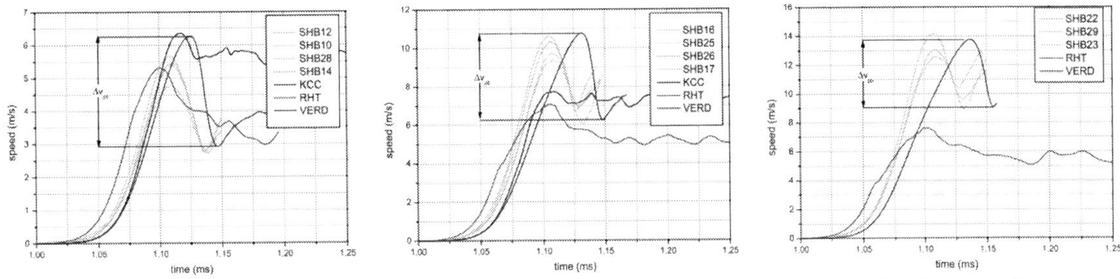

(a) loading stage one (b) loading stage two (c) loading stage three

Figure 9. Right free end velocities from computations and experiments.

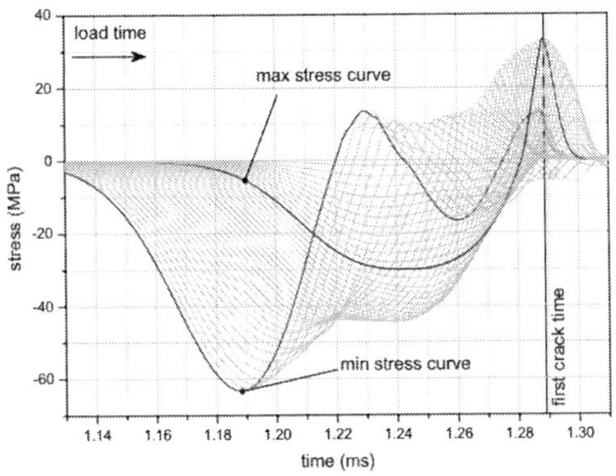

Figure 10. Specimens time based stress distribution.

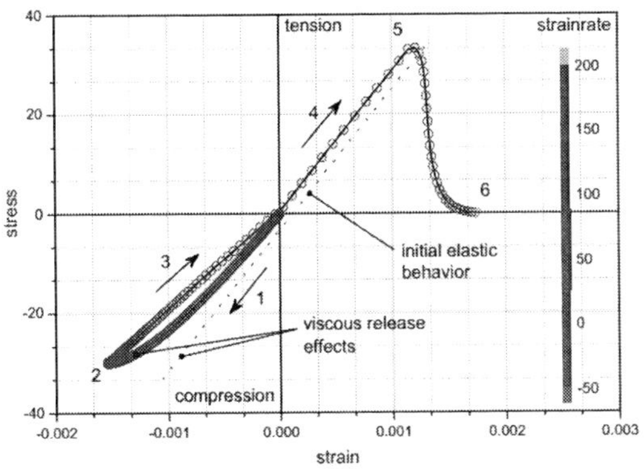

Figure 11. Specimens stress strain relation.

at a moderate strain rate of $\approx 50\,s^{-1}$. This low rate leads to relaxing elasticity as a result of the viscous formulation and is indicated by the opening gab towards the initial elastic slope. At point (2) the maximum pressure releases and a fast unloading occurs at (3) back to (0). The slope (3) is different to (1) and describes elastic hysteresis rather than damage. This can be shown at the following tensile load. The slope increase in (4) due to the high strain rate of $\approx 200\,s^{-1}$ and leads to a stiffening with respect to (3). In (5) the tensile stress leads to the damage and stresses relaxes toward zero at (6).

4. Conclusions

Performance in simulationing high speed dynamics of concrete was carried out with three different constitutive laws. Examined were the LS-Dyna supported models KCC and RHT and the newly developed material model VERD. Validation was in all three cases carried out by means of relevant experimental data derived from SHB tests in the spallation configuration. Considerable importance is attached to the particle velocities of the right free end of the specimen. The resulting characteristic curves indicate the closest match to the critical velocity decline for the VERD material model. Though better results with proper adjustment of the KCC and RHT material parameters are assumed, the VERD material model achieved superior results regarding the derived failure strength.

Further investigations require variating viscosity and crack retardation related parameters in order to improve the order of magnitude to which computed absolute velocity and strength values match to experimental data.

References

[1] Häussler-Combe, U.; Kühn, T.: Modelling of Strain Rate Effects for Concrete with Viscoelasticity and Retarded Damage. International Journal of Impact Engineering **50** (2012), 17–28

[2] Schuler, H.; Mayrhofer, C.; Thoma, K.: Spall experiments for the measurement of the tensile strength and fractureenergy of concrete at high strain rates, International Journal of Impact Engineering **32**, (2006), 1635–1650

[3] Hussler-Combe, U.: A novel strain-rate model for concrete and its influence on wave. In: Bicanic, N.; Meschke, G.; Mang, H., de Borst, R. (Hrsg.): Proceedings of EURO-C 2010, 24–27 March 2014, St. Anton am Arlberg, Austria., 2014, 97–108

[4] Kolsky, H. (1949). An investigation of the mechanical properties of materials at very high rates of loading. *Proc. Physical Society, Section B* **62**, 676–700

[5] Bischoff, P. & S. Perry (1991). Compressive behavior of concrete at high strain rates. *Materials and Structures* **24**, 425–450

[6] Malvar, L.J. & C.A. Ross (1998). Review of strain rate effects for concrete in tension. *ACI Materials Journal* **95**, 735–739

[7] Rossi, P. (1991) A physical phenomenon which can explain the mechanical behavior of concrete under high strain rates. *Materials and Structures* **24**, 422–424

[8] Reinhardt, H.W. & J. Weerheijm (1991). Tensile fracture of concrete at high loading rates taking account of inertia and crack velocity effects. *International Journal of Fracture* **51**, 31–42

[9] Eibl, J. & Schmidt-Hurtienne (1999). Strain-rate-sensitive constitutive law for concrete. *Journal of Engineering Mechanics* **125**, 1411–1420

[10] Schuler, H. (2004). *Experimentelle und numerische Untersuchungen zur Schdigung von stoßbeanspruchtem Beton* (Doctoral dissertation, Universität der Bundeswehr München, Universitätsbibliothek)

EPJ Web of Conferences **94**, 04007 (2015)
DOI: 10.1051/epjconf/20159404007

Atomistic modeling of the dislocation dynamics and evaluation of static yield stress

A.V. Karavaev[a], V.V. Dremov, and G.V. Ionov

Russian Federal Nuclear Center – Zababakhin Institute of Technical Physics (RFNC-VNIITF), 13 Vasiliev st., Snezhinsk, Chelyabinsk reg. 456770, Russia

Abstract. Static strength characteristics of structural materials are of great importance for the analysis of the materials behaviour under mechanical loadings. Mechanical characteristics of structural materials such as elastic limit, strength limit, ultimate tensile strength, plasticity are, unlike elastic moduli, very sensitive to the presence of impurities and defects of crystal structure. Direct atomistic modeling of the static mechanical strength characteristics of real materials is an extremely difficult task since the typical time scales available for the direct modeling in the frames of classical molecular dynamics do not exceed a hundred of nanoseconds. This means that the direct atomistic modeling of the material deformation can be done for the regimes with rather high strain rates at which the yield stress and other mechanical strength characteristics are controlled by microscopic mechanisms different from those at low (quasi-static) strain rates. In essence, the plastic properties of structural materials are determined by the dynamics of the extended defects of crystal structure (edge and screw dislocations) and by interactions between them and with the other defects in the crystal. In the present work we propose a method that is capable to model the dynamics of edge dislocations in the fcc and hcp materials at dynamic deformations and to estimate the material static yield stress in the states of interest in the frames of the atomistic approach. The method is based on the numerical characterization of the stress relaxation processes in specially generated samples containing solitary edge dislocations.

1. Introduction

Plastic deformations of crystals are controlled by the presence of extended defects of crystal structure, primarily, dislocations. Dynamics of the dislocations due to external stresses determines the kinetics of plastic deformations and thus plays an important role in the controlling of the mechanical characteristics of structural materials. It is well known that the yield stress depends substantially on the deformation rate. This becomes more evident at the deformation rates exceeding $\sim 10^3 - 10^4\,\mathrm{s}^{-1}$. At low plastic deformation rates (that is, at low dislocation sliding velocities) the dislocations overcome potential barriers impedimental their sliding as a result of combined action of the applied external stress and thermal fluctuations. While for the high rate plastic deformations one needs to apply much higher stress. At the high plastic deformation rates (exceeding $10^4\,\mathrm{s}^{-1}$ for the most of elemental hcp and fcc metals) acting stresses are sufficient to provide for dynamic potential barriers overcoming without additional help of the thermal fluctuations. In the high plastic deformation rate regimes the dominant mechanism of the dislocation drag is the dislocation energy transfer to the lattice vibrations (phonon excitations). Here we present a method that allows not only to study the dynamics of the edge dislocations in the hcp and fcc materials under dynamic loading, but also estimate static yield stress of the materials in the frames of atomistic modeling. The method is based on the numerical characterization of the stress relaxation processes in specially constructed samples containing solitary edge dislocations.

2. Traditional methods of edge dislocation dynamics simulations

The dynamics of edge dislocations in close-packed materials under dynamic loading has been studied earlier in the frames of classical molecular dynamics (CMD) by other researchers [1–7] as following. Typical scheme of the simulation model is presented in Fig. 1. The lab-frame axes are oriented in congruence to the dominant sliding system of edge dislocations in the fcc crystals. The full edge dislocation Burrgers vector is aligned along the x-axis ($[1\bar{1}0]$-direction). The dislocation sliding occurs in the xz – (111) plane. In order to generate the sample containing solitary edge dislocation the following procedure is widely used. In ideal non-defective crystal structure one removes two mono-atomic half-planes (110). Then the sample is slightly compressed, and atomic layers nearest to the cutaway draw together and cohere. Then the thermalization of the sample at the conditions of interest takes place. During the thermalization a constitutive process occurs, namely, a splitting of the ideal edge dislocations to a pair of partial dislocations $\frac{a}{2}[110] \rightarrow \frac{a}{6}[121] + \frac{a}{6}[21\bar{1}]$ with the formation of stacking fault area between them. In order to reproduce in the frames of CMD the processes related to the plastic deformations of fcc crystals one needs to be sure that the model of the interatomic interactions used is capable to describe basic properties of the edge dislocations. One of the constitutive

[a] Corresponding author: a.v.karavayev@vniitf.ru

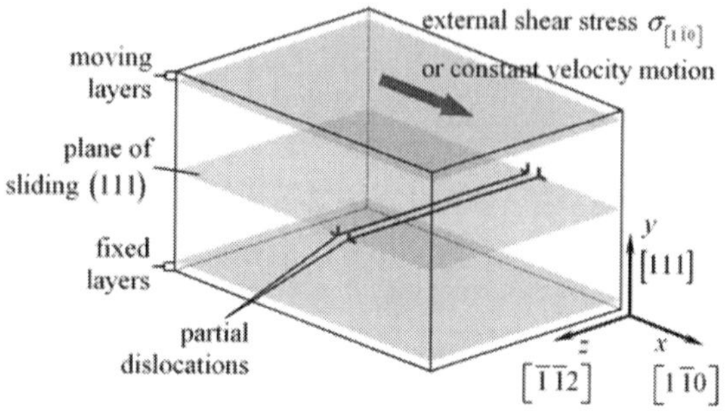

Figure 1. Typical schematic of the model for the edge dislocation dynamics in fcc crystal simulations. Along the x and z directions periodic boundary conditions are used. The sample part in-between two red planes consists of freely moving atoms. Atoms laying bellow the lower red plane are fixed, while the atoms laying above the upper red plane are allowed to move along the x-axis only either with a constant velocity v_x or due to a constant external force F_x. The applied external force or the constant velocity motion of the upper part of the sample cause plastic deformation of the sample (the edge dislocation sliding along x-axis). In the first case the shear stress acting on the sample is controlled, while in the second – the plastic deformation rate.

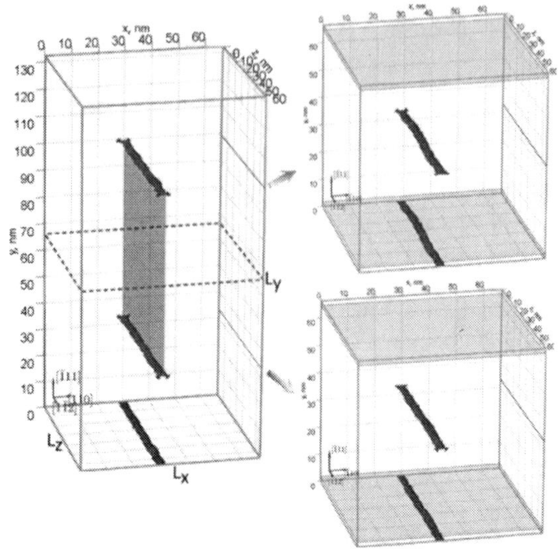

Figure 2. Initial samples generation for the shear stress relaxation method by example simulations of copper with the EAM interatomic potential [8]. From the ideal non-defective samples with the sizes $L_x \times L_y \times L_z$ and the orientation shown on the left in periodic boundary conditions in all directions we removed two mono-atomic layers of (110) type with $\frac{1}{4}L_y < y < \frac{3}{4}L_y$. As a result two edge dislocation of opposite signs are formed in the sample. Then the sample was thermalized at the conditions of interest for at least 2 ns. During the thermalization the dislocations obtain their equilibrium width and all the undesirable elastic waves caused by the artificial removing of the atomic layers decay. After that we divide the sample as it is shown, and change the boundary conditions in the new smaller samples: in the x and z directions periodic boundary conditions are used, while in the upper and bottom parts of the samples we fix all the atoms separated by the red planes. So we get two samples containing solitary relaxed equilibrated edge dislocations at the conditions of interest.

characteristics of the edge dislocations in the close-packed materials is its energetics in other words the energy stored in the defect itself and energy of elastic deformation of the crystal around it. The energy balance between the dislocation core and elastic shear deformation energy caused by the presence of the dislocation determines the dislocation core structure.

The described above approach to the CMD modeling of the edge dislocation dynamics in close-packed lattices was demonstrated to be quite an effective method. However, despite of its effectiveness the method has significant drawbacks. All the simulations had been performed so far with the rather small samples, that corresponds to the gigantic deformation rates and rather high dislocation densities considerably exceeding dislocation densities in real materials. Moreover the CMD simulations with the fixed deformation rate (dislocation velocity) or fixed shear stress provides for only one point of $v(\sigma)$ dependence in a stationary regime. In the stationary simulations it is impossible to obtain an estimate for the static yield stress (Peierls-Nabarro stress). Here we propose instead of the fixed deformation rate or fixed shear stress calculations to perform CMD simulations of the dislocation motion during the relaxation of shear stress. In such approach one can obtain in a single simulation an entire dependence of the dislocation velocity on the actual shear stress and get Peierls-Nabarro stress as the limit when the dislocation stops.

3. Stress relaxation approach

Detailed description of the initial sample generation procedure for the proposed method is given in the caption of Fig. 2 by example simulations of copper with the Embedded Atom Model (EAM) interatomic potential [8]. All the CMD simulations presented here were carried out using CMD code MOLOCH [9]. The samples containing solitary relaxed equilibrated edge dislocations at the conditions of interest are subjects for the instantaneous shear deformation ϵ_{xy} of various values. As the result of the instantaneous shear deformation ϵ_{xy} there are elastic shear stresses σ_{xy} in the samples, that cause the edge dislocation sliding along the x-direction. As a result of the dislocation motion (plastic deformation) the relaxation of the elastic shear stresses σ_{xy} takes place in the samples.

In Fig. 3(a) a typical time dependence of the shear stress σ_{xy} in the sample at the ambient conditions is presented. Initial shear deformation in the simulation was set $\epsilon_{xy} = 0.004$. At the moment $t = 0$ the shear stress has its maximum. Then the dislocation starts to move, and the shear stress decreases because of the plastic deformation. It is important that the shears stress decrease does not reach zero value, but stops somewhere around 5 MPa. This value corresponds to the shear stress when the dislocation motion stops (Peierls-Nabarro stress), while the Peierls-Nabarro stress corresponds to the minimal estimate of the engineering yield stress. In Fig. 3(b) the corresponding time dependence of the dislocation position in the same simulation is presented. The position of

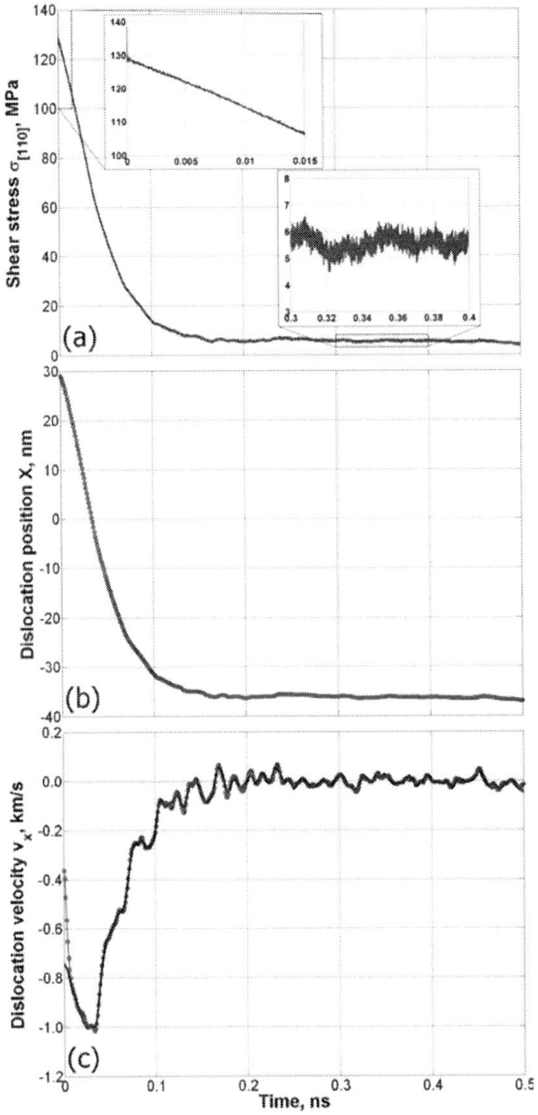

Figure 3. Time series of the shear stress $\sigma_{[110]}$ – (a), dislocation position x – (b), and dislocation sliding velocity v_x – (c) in the shear stress relaxation simulation of copper with the deformation $\epsilon_{xy} = 0.004$ at the ambient conditions.

the dislocation was determined using Adaptive Template Analysis (ATA) [10] which allows to analyze precisely crystal structure of virtual samples at finite temperatures even approaching melting. In particular, the ATA method determines with high confidence all the stacking fault atoms forming the edge dislocation. The dislocation position x is determined as average of x coordinates of all atoms in the stacking fault positions. If one takes time derivative of the dependence presented in Fig. 3(b) one gets the velocity of the edge dislocation sliding as a function of time shown in Fig. 3(c) by dots. Combining the time series from Fig. 3(a) and (c) we get the dependence of the dislocation velocity v_x on the applied external shear stress σ_{xy} presented in Fig. 4. In Fig. 4 one can see non-stationary

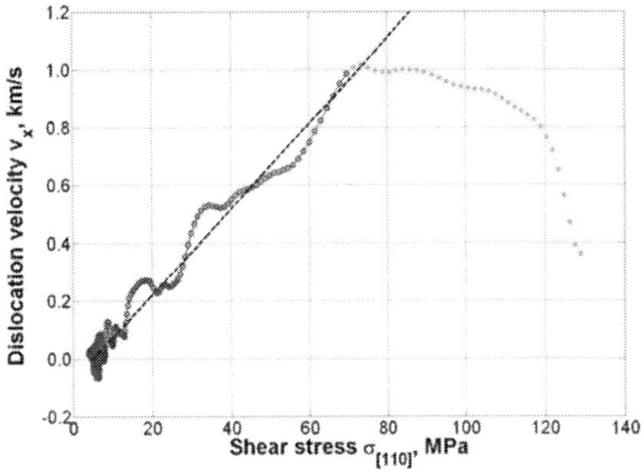

Figure 4. Dependence of the dislocation sliding velocity on the shear stress in the copper sample after the shear deformation $\epsilon_{xy} = 0.004$ at the ambient conditions.

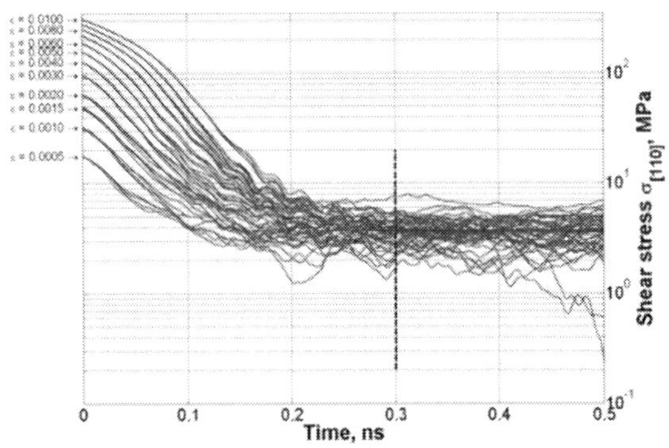

Figure 5. Time dependence of the shear stress in the copper samples containing solitary edge dislocations after various shear deformations ϵ at the ambient conditions. Blue solid line represents overall time average for all the curves calculated for the last 0.2 ns. The Peierls-Nabarro stress estimate is $\sigma_0 = 3.65$ MPa with the standard deviation $\Delta\sigma_0 = 1.08$ MPa.

initial stage when the dislocation velocity increases while the shear stress decreases. One can see also oscillations of the dependence due to circulations of elastic waves caused by the motion of the dislocation. Notice one more time that the dislocation stops not at zero shear stress but at some other one corresponding to the Peierls-Nabarro stress.

Besides the direct observation of the dislocation motion we can determine its velocity indirectly using the rate of the shear stress relaxation. At low deformations the total deformation of the sample ϵ can be expressed as the sum of plastic ϵ_{pl} and elastic deformation ϵ_{el} components. Taking time derivative of the total deformation ϵ and retaining that the total deformation of the samples does not change during the shear stress relaxation simulations we get $\dot{\epsilon}_{pl} = -\dot{\epsilon}_{el}$. In the case of low deformation the elastic shear stress is determined by the Hooke's law. Thus the rate of the plastic deformation in our simulations can be calculated as

$$\dot{\epsilon}_{pl} = \frac{\dot{\sigma}_{xy}}{G_{xy}}, \tag{1}$$

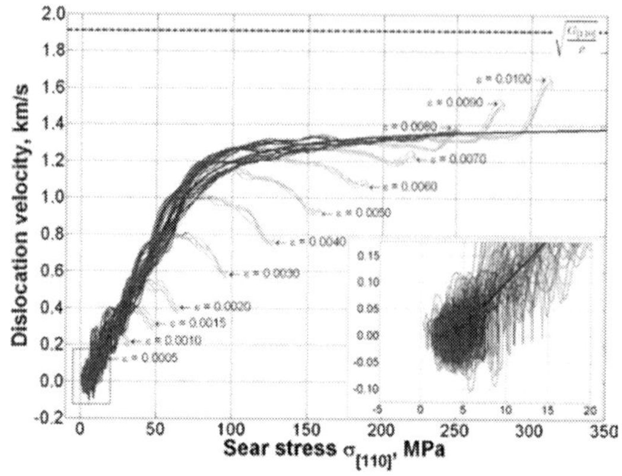

Figure 6. Dislocation sliding velocity v_x *vs.* shear stress σ_{xy} in the copper samples containing solitary edge dislocations after various shear deformations ϵ at the ambient conditions. Black dashed line represents shear sound velocity, blue solid line is approximation of the red brunches of the v_x on σ_{xy} curves. The light gray sections of the v_x on σ_{xy} curves were not used for the approximation construction.

where G_{xy} is corresponding to the deformation elastic shear modulus. In the case of the pure shear deformation of the sample with the solitary edge dislocation the plastic deformation rate $\dot{\epsilon}_{pl}$ directly proportional to the linear velocity the dislocation sliding. Thus, knowing the sample size and the shear modulus G_{xy} we can reconstruct the time dependence of the dislocation sliding velocity from the shear stress relaxation curve. In Fig. 3(c) one can see a comparison of the time dependence of the dislocation velocity obtained as the time derivative of the dislocation positions from ATA-analysis (line with dots) and one calculated from the shear stress relaxation curve (black solid line).

In Fig. 5 the time series of the shear stress in the copper samples obtained for various initial shear deformations are presented. All the curves consist of the branch where the shear stress decreases with time and stationary brunch with nearly constant non-zero value of the shear stress. That value corresponds to the stop of the dislocation sliding, *i.e.* the estimate of the Peierls-Nabarro stress σ_0. Blue solid line in Fig. 5 represents overall time average for all the curves calculated for the last 0.2 ns. The Peierls-Nabarro stress estimate is $\sigma_0 = 3.65$ MPa with the standard deviation $\Delta\sigma_0 = 1.08$ MPa. The yield stress is scaled as $\sigma_y = 2\sigma_0$. Thus, according to the EAM model with the parameterization [8] the estimate of the static yield stress of copper at the ambient conditions is $\sigma_y = (7.3 \pm 2.2)$ MPa. It should be noted that the choice of copper as a material for current investigation was made to check the sensitivity of the proposed relaxation method, because copper possesses extremely low yield stress. Indeed, the experimentally measured yield stress in high purity well annealed single crystal copper at the ambient conditions is in the range from 1.0 MPa to 4.4 MPa [11–15]. Thus, the obtained here value is only slightly higher than the experimental ones.

From the time series of the shear stress presented in Fig. 5 one can reconstruct the dependence of the dislocation sliding velocity v_x on the shear stress σ_{xy} presented in Fig. 6. Black dashed line represents shear sound velocity, blue solid line is the approximation of the red brunches of the v_x on σ_{xy} curves. The light gray sections of the $v_x(\sigma_{xy})$ curves were not used for the approximation construction. Notice that the approximation crosses zero velocity line not at zero shear stress point, and the shear stress corresponding to the zero dislocation velocity is the Peierls-Nabarro stress.

4. Conclusion

In the present work we introduce a novel approach to the simulations of the dislocation dynamics in close-packed materials. The method is based on the numerical characterization of the shear stress relaxation processes in specially generated samples containing solitary edge dislocations of particular orientations. The method allows obtaining in a single numerical experiment the entire dependence of the dislocation sliding velocity on the applied shear stress. Besides, the method provides for the capability to estimate the shear stress when the dislocation sliding stops, *i.e.* to estimate the Peierls-Nabarro stress at the condition of interest.

References

[1] Yu.N. Osetsky and D.J. Bacon, Modell. Simul. Mater. Sci. Eng. **11**, 427 (2003)
[2] Yu.N. Osetsky and D.J. Bacon, J. Nucl. Mater. **323**, 268 (2003)
[3] Yu.N. Osetsky, D.J. Bacon, and V. Mohles, Philos. Mag. **83**, 3623 (2003)
[4] D.J. Bacon and Yu.N. Osetsky, J. Nucl. Mater. **329–333**, 1233 (2004)
[5] T. Hatano and H. Matsui, Phys. Rev. B **72**, 094105 (2005)
[6] T. Hatano, Phys. Rev. B **74**, 020102(R) (2006)
[7] A.Yu. Kuksin, V.V. Stegailov, and A.V. Yanilkin, Doklady Phys. **53**, 287 (2008)
[8] Y. Mishin *et al.*, Phys. Rev. B **63**, 224106 (2001)
[9] F.A. Sapozhnikov *et al.*, EPJ WoC **10**, 00017 (2010)
[10] F.A. Sapozhnikov, G.V. Ionov, and V.V. Dremov, Rus. J. Phys. Chem. B **2**, 238 (2008)
[11] F.W. Young, J. Appl. Phys. **33**, 963 (1962)
[12] J.D. Livingston, Acta Metall. **10**, 229 (1962)
[13] Z.S. Basinski and S.J. Basinski, Philos. Mag. **9**, 51 (1964)
[14] J.W. Steeds, Proc. Roy. Soc. A **292**, 343 (1966)
[15] Y. Kawasaki, J. Phys. Soc. Japan **36**, 142 (1974)

EPJ Web of Conferences **94**, 04008 (2015)
DOI: 10.1051/epjconf/20159404008

A viscoelastic-viscoplastic model for short-fibre reinforced polymers with complex fibre orientations

M. Nciri[1,2,a], D. Notta-Cuvier[1], F. Lauro[1], F. Chaari[1], B. Zouari[2], and Y. Maalej[3]

[1]University of Valenciennes (UVHC), LAMIH UMR CNRS 8201, 59313 Valenciennes, France
[2] National Engineering School of Sfax (ENIS), L2MP, BP. W3038, Sfax, Tunisia
[3] University of Tunis El Manar, ENIT, MAI (LR11ES19), 1002 Tunis, Tunisia

Abstract. This paper presents an innovative approach for the modelling of viscous behaviour of short-fibre reinforced composites (SFRC) with complex distributions of fibre orientations and for a wide range of strain rates. As an alternative to more complex homogenisation methods, the model is based on an additive decomposition of the state potential for the computation of composite's macroscopic behaviour. Thus, the composite material is seen as the assembly of a matrix medium and several linear elastic fibre media. The division of short fibres into several families means that complex distributions of orientation or random orientation can be easily modelled. The matrix behaviour is strain-rate sensitive, i.e. viscoelastic and/or viscoplastic. Viscoelastic constitutive laws are based on a generalised linear Maxwell model and the modelling of the viscoplasticity is based on an overstress approach. The model is tested for the case of a polypropylene reinforced with short-glass fibres with distributed orientations and subjected to uniaxial tensile tests, in different loading directions and under different strain rates. Results demonstrate the efficiency of the model over a wide range of strain rates.

1. Introduction

Thermoplastics reinforced with short fibres are more and more appealing for a wide-range of industrial applications, in particular in automotive industry in parts possibly subjected to sever loading conditions (e.g. crash…). As a consequence of complex reinforcement configurations in terms of orientation, geometrical and mechanical properties and complex polymeric matrix behaviour, SFRC behaviour is governed by numerous interdependent phenomena, such as plastic flow, ductile damage and strain-rate dependency of the matrix, fibre breakage, debonding or pull-out... Beyond the limits of linear elasticity, these phenomena can obviously not be modelled by a simple rule of mixture and more elaborated models, like the extended homogenization methods to components' non-linear behaviour have been developed. However, when all or some of phenomena listed above are interfering, the aforementioned methods become too complex to implement to allow to take all these relatively sophisticated features into account. To overcome these limitations, an original approach was recently proposed by Notta-Cuvier et al. [1,2], where complex fibre orientations can be easily dealt with thanks to the consideration of an assembly of the matrix medium with several fibre media, each of them being characterized by a volume fraction, v_f^α, and a unit orientation vector. Stress transfer from the matrix to fibres is governed by a modified shear-lag model. Note that the model was then extended to strongly anisotropic ductile damage in the matrix material [3]. In this work, the modelling is now extended to take the rate-dependency of the matrix behaviour into account.

2. Constitutive model

2.1. Stress state of matrix material

Rate dependency of the matrix behaviour is predicted with coupled viscoelastic-viscoplastic schemes. The viscoelastic part of the response is taken into account using the linear Wiechert model (i.e. Maxwell Generalized) consisting of a finite number of Maxwell elements arranged in parallel with a linear elastic Hooke element. This response is updated, when necessary by a viscoplastic scheme. The hereditary integral over the time, t, of the viscoelastic stress response is formulated as:

$$\sigma(t) = \int_{-\infty}^{t} R^{ve}(t-\Gamma) : \frac{d\varepsilon(\zeta)}{d\zeta} d\zeta \qquad (1)$$

where σ and ε are the matrix stress and strain tensors, respectively. R^{ve} is the fourth order relaxation tensor expressed by:

$$R^{ve}(t) = L_\infty^{ve} + \sum_{i=1}^{N} L_i^{ve} exp\left(-\frac{t}{\tau_i}\right) \qquad (2)$$

where N is the number of Maxwell elements. L_∞^{ve} is the fourth order long-term elasticity tensor of the Hooke's element defined by:

$$L_\infty^{ve} = 2G_\infty I_d + K_\infty I \otimes I \qquad (3)$$

[a] Corresponding author: mariem.nciri@etu.univ-valenciennes.fr

τ_i and L_i^{ve} are the relaxation time and the fourth order elasticity tensor of the ith Maxwell element, respectively, with:

$$L_i^{ve} = 2G_i I_d + K_i I \otimes I \tag{4}$$

I_d is the deviatoric projection tensor such as:

$$I_d = I_s - \frac{1}{3} I \otimes I. \tag{5}$$

The matrix viscoplasticity is taken into account following Perzyna's viscoplastic model [4]. The pressure dependent and nonsymmetric (visco) plastic flow, i.e. sensitivity of the polymeric matrix behaviour to the nature of loading (e.g., tension, compression..), are modelled by Raghava's yield surface [5]. Plastic flow occurs as soon as the first invariant of the stress tensor, $I_1(\sigma)$, and the second invariant of the deviatoric stress tensor, $I_2(\sigma)$, reach a critical combination. Expression of the yield surface, f, is given by:

$$f(\sigma, R) = \frac{(\eta - 1) I_1(\sigma) + \sqrt{(\eta-1)^2 I_1^2(\sigma) + 12\eta I_2(\sigma)}}{2\eta} - \sigma_t - R(\kappa). \tag{6}$$

with

$$I_1(\sigma) = tr(\sigma) \tag{7a}$$

and

$$I_2(\sigma) = \frac{1}{2} S : S \tag{7b}$$

where S is the deviatoric stress tensor. The hydrostatic pressure dependency parameter, η, is defined using the ratio between the quasi-static initial yield stress in compression and tension, σ_c and σ_t, respectively, and:

$$\eta = \frac{\sigma_c}{\sigma_t}. \tag{8}$$

In the expression of the yield surface, $R(\kappa)$, is a nonlinear isotropic hardening function, defined by Balieu et al. [6]:

$$R(\kappa) = Q_1 \kappa exp(-b_1\kappa) + Q_2(1 - exp(-b_2\kappa)) + b_3\kappa^3 + b_4\kappa^2 + b_5\kappa \tag{9}$$

where Q_1, Q_2, b_1, b_2, b_3, b_4 and b_5 are material parameters. The hardening variable κ is the equivalent viscoplastic strain defined by:

$$\kappa = \sqrt{\frac{3}{2}\varepsilon^{vp} : \varepsilon^{vp}} \tag{10}$$

with ε^{vp} the viscoplastic strain tensor. Since the viscoplastic deformation of polymer materials is not an isochoric phenomenon, a non-associated viscoplastic flow rule is used. The asymmetry of this volume change (i.e., not similar for positive and negative hydrostatic pressures) is modelled by a hyperbolic viscoplastic dissipation potential [6], defined by:

$$F(\sigma) = \sqrt{3I_2(\sigma) + \frac{1}{3}\left(a^+ < I_1 >^2 + a^- < -I_1 >^2\right)} \tag{11}$$

where a^+ and a^- are volume variation parameters for positive and negative hydrostatic pressures, respectively. The viscoplastic strain rate tensor can therefore evolve differently for positive and negative hydrostatic pressures. The symbol $\langle . \rangle$ is the Macauley braket, that is, for any scalar x, given by $\frac{(x+|x|)}{2}$. According to this non-associated flow rule the viscoplastic strain rate tensor is given by:

$$\dot{\varepsilon}^{vp} = \dot{\lambda} n. \tag{12}$$

Where n is the viscoplastic flow direction tensor expressed by:

$$n = \frac{\partial F}{\partial \sigma}. \tag{13}$$

Consequently, with the given dissipation potential, the viscoplastic strain rate tensor becomes:

$$\dot{\varepsilon}^{vp} = \dot{\lambda} \frac{\frac{3}{2}S + \frac{1}{9}\left(a^+ < I_1 > + a^- < -I_1 >\right) I}{\sqrt{3I_2(\sigma) + \frac{1}{27}\left(a^+ < I_1 >^2 + a^- < -I_1 >^2\right)}}. \tag{14}$$

Matrix material's strain rate dependency is introduced by a viscoplastic formulation. The static yield surface, f, is extended according to an overstress based viscoplastic theory and a dynamic yield surface is then defined, as follows [4]:

$$F^{vp}(\sigma, R, \dot{\kappa}) = \frac{(\eta - 1) I_1(\sigma) + \sqrt{(\eta-1)^2 I_1^2(\sigma) + 12\eta I_2(\sigma)}}{2\eta} - (\sigma_t + R(\kappa)) - \tilde{\sigma}^{vp} \equiv 0. \tag{15}$$

With $\tilde{\sigma}^{vp}$ the overstress: As postulated in Perzyna's model, the viscoplastic multiplier $\dot{\lambda}$ is expressed in terms of overstress as follows:

$$\dot{\lambda} = \frac{\dot{\kappa}_0}{\sqrt{\frac{2}{3}n : n}} \left(\frac{\tilde{\sigma}^{vp}}{2\eta(\sigma_t + R(\kappa))}\right)^{\frac{1}{n}} \tag{16}$$

and equivalently:

$$\tilde{\sigma}^{vp} = (\sigma_t + R(\kappa)) \left(\frac{\dot{\kappa}}{\dot{\kappa}_0}\right)^n \tag{17}$$

where n and $\dot{\kappa}_0$ are the strain rate sensitivity and viscosity parameters, respectively, and $\dot{\kappa}$ is the equivalent viscoplastic strain rate, defined by:

$$\dot{\kappa} = \sqrt{\frac{2}{3}\dot{\varepsilon}^{vp} : \dot{\varepsilon}^{vp}} = \dot{\lambda}\sqrt{\frac{2}{3}n : n}. \tag{18}$$

The expression of the dynamic yield surface can therefore be rewritten as follows:

$$F^{vp}(\sigma, R, \dot{\kappa}) = \frac{(\eta - 1) I_1(\sigma) + \sqrt{(\eta-1)^2 I_1^2(\sigma) + 12\eta I_2(\sigma)}}{2\eta} - (\sigma_t + R(\kappa)) \left(1 + \frac{\dot{\kappa}}{\dot{\kappa}_0}\right)^n \equiv 0. \tag{19}$$

The standard Kuhn-Tucker loading/unloading conditions (i.e., $F^{vp} \leq 0$, $\dot{\lambda} \geq 0$, $\dot{\lambda} F^{vp} = 0$) are applied to the dynamic yield surface. It can be noted that both static and dynamic yield surfaces are updated all along the implementation. As a consequence, during unloading overstress does not systematically vanish and stress state can remain in the viscoplastic domain (i.e. stress state above the static yield surface, on updated dynamic surface). The rate form of the viscoplastic multiplier is obtained by substituting its expression into the viscous yield surface and is as follows:

$$\dot{\lambda} = \begin{cases} 0, & \text{if } f < 0, \\ \frac{\dot{\kappa}_0}{\sqrt{\frac{2}{3} n:n}} \left(\frac{(\eta-1) I_1(\sigma) + \sqrt{(\eta-1)^2 I_1^2(\sigma) + 12\eta I_2(\sigma)}}{2\eta(\sigma_t + R(\kappa))} \right)^{\frac{1}{n}}, & \text{if } f \geq 0. \end{cases} \tag{20}$$

2.2. Stress state of the composite material

Once matrix behaviour is solved, contributions of fibres media to SFRC behaviour are introduced according to the original method developed by Notta-Cuvier et al. [1]. Each fibre family is subjected to the projection of the total deformation gradient tensor along its orientation and behaves in a one−dimensional linear elastic way. The average axial stress in the fibres is computed based on Bowyer-Bader theory [7], where a progressive degradation of the fibre-matrix interface is taken into account. As previously stated, the stress state of the composite material is computed as a combination of the different constituent contributions, starting from a state potential, the Helmholtz free energy (Eq. (21)), that is divided into as many terms as there are media constituting the composite.

$$\rho\phi = v_M \rho_M \phi_M + \sum_{\alpha=1}^{n_{fam}} v_F^\alpha \rho_F^\alpha \phi_F^\alpha \tag{21}$$

where ρ, ρ_M and ρ_f^α are respectively the densities of the composite material, the matrix material and the fibres family α. Using a thermodynamic formulation a particular solution of the Clausius-Duhem inequality can be expressed by Eq. (21) thus giving the expression of composite macroscopic stress tensor.

$$\sigma = v_M \sigma_M + \sum_{\alpha=1}^{n_{fam}} v_F^\alpha A^\alpha \sigma_F^\alpha A^\alpha. \tag{22}$$

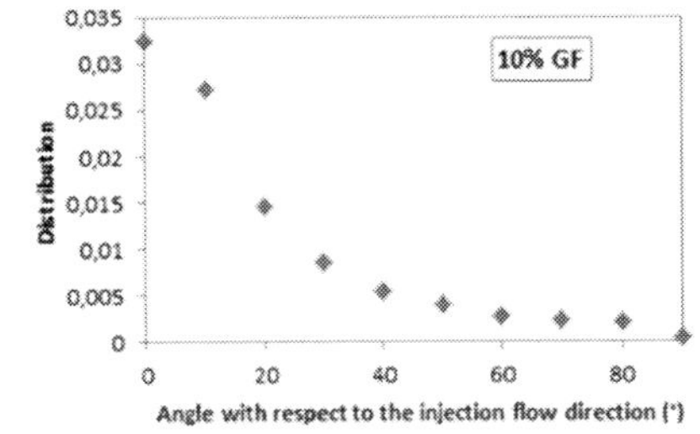

Figure 1. Distribution of orientations [8].

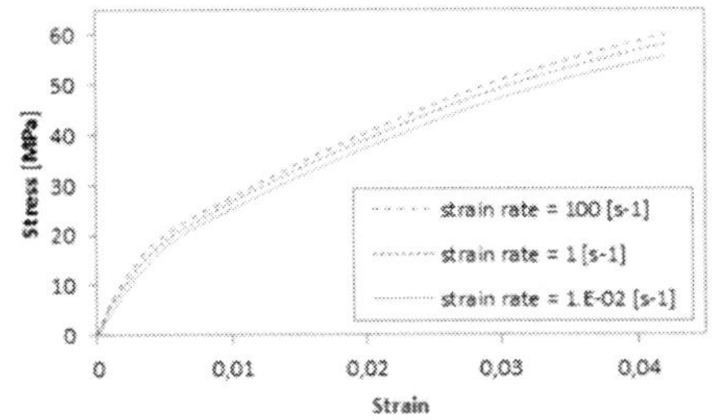

Figure 2. $\theta = 0°$.

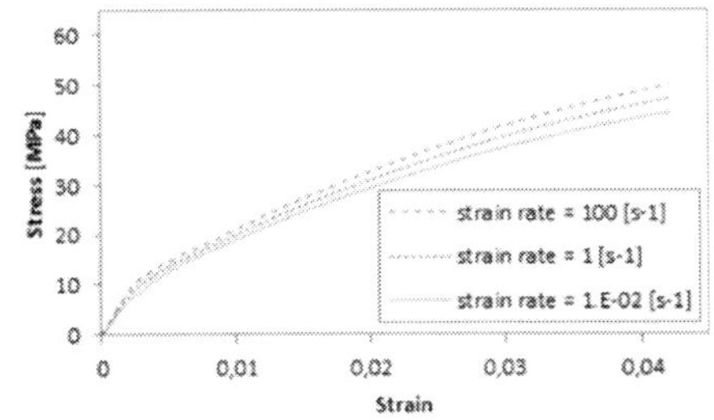

Figure 3. $\theta = 30°$.

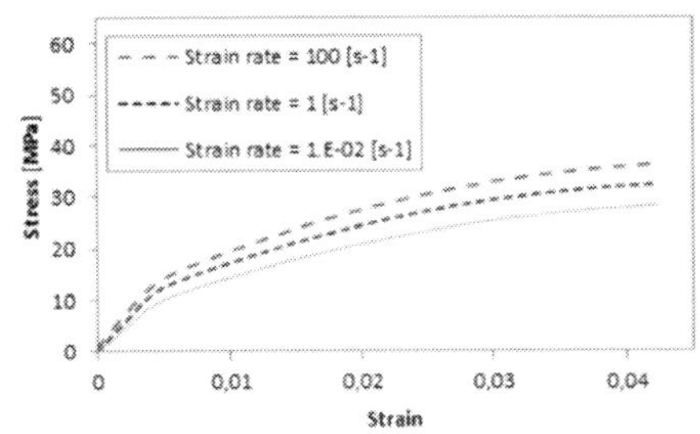

Figure 4. $\theta = 45°$.

3. Example of application

The model is implemented as a VUMAT (Abaqus Explicit) and is tested for a fictitious injection moulded SFRC, namely a polypropylene reinforced with 10 vol.% glass fibres. Fibre configuration is extracted from work by Kammoun et al. [8] ($E_F = 76$ GPa, distributed orientations with preferential orientation in the injection flow direction). In the present implementation, the fibre distribution is modelled using 10 fibre media with angles of orientation varying from 0 to 90°, by step of 10°, with respect to the injection flow direction and a suitable volume fraction is associated to each fibre media, as shown in Fig. 1. The matrix viscoelastic−viscoplastic material parameters are chosen in accordance with the model proposed by Balieu et al. [6], for a 20% mineral filled polypropylene co−polymer.

In order to assess the model accuracy, uniaxial tensile tests are performed on one element for different strain rates and loading directions (angle of loading, θ, successively equals to 0°, 30°, 45° and 90° with respect to injection flow direction) as shown in Figs. 2 to 5.

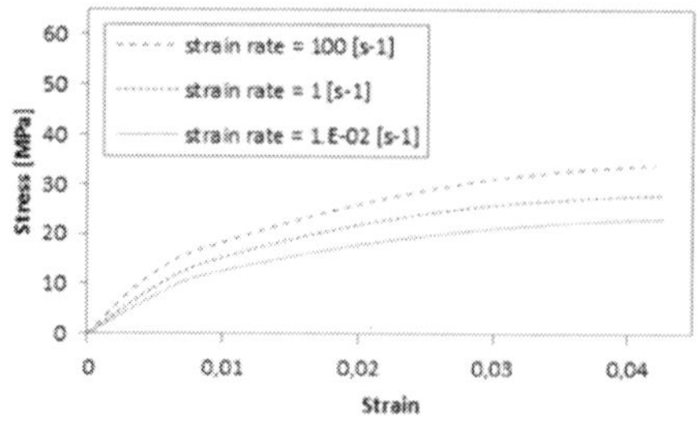

Figure 5. $\theta = 90°$.

Responses' dependency on the strain rate is highlighted in the linear and non-linear phases according to the hereinabove presented implementation. This evolution is obviously dependent on the loading direction regarding the principal fibre direction (i.e. the injection flow direction). It can be seen that, as the angle of loading direction with respect to fibres' one increases, the strain rate dependency of the whole material's modelled response is dimmed down, which is in accordance with the fact that the rate independent linear elastic response of the fibres is of greater influence in its principal direction, while at increased angles of orientation the response is ratherdominated by that of the matrix.

4. Conclusions

A behaviour model for strain-rate dependent short fibre reinforced composites is presented, based on an original approach that aims to be an efficient alternative to more complex procedures of homogenisation. The matrix has viscoelastic-viscoplastic and pressure dependent behaviour and complex reinforcement configurations are considered. Current development concerns a physical modelling of fibre/matrix interface degradation during loading, with possible strain-rate effects. The determination of the material constants is the subject of an already started static and dynamic experimental investigation.

This research is conducted through collaboration between the University of Valenciennes and the National Engineering School of Sfax. This collaboration is jointly financed in the frame of the Utique CMCU programme. The present research work has been supported by International Campus on Safety and Intermodality in Transportation, the Région Nord Pas de Calais, the European Community, the Délégation Régionale á la Recherche et á la Technologie, the Ministére de l'Enseignement Supérieur et de la Recherche, and the Centre National de la Recherche Scientifique: the authors gratefully acknowledge the support of these institutions.

References

[1] D. Notta-Cuvier, F. Lauro, B. Bennani, R. Balieu. International Journal of Solids and Structures **50**, 2857–2871 (2013)

[2] D. Notta-Cuvier, F. Lauro, B. Bennani. Composites: Part A **62**, 60–66 (2014)

[3] D. Notta-Cuvier, F. Lauro, B. Bennani, R. Balieu. Mechanics of materials **68**, 193–206 (2014)

[4] P. Perzyna. Advances in Applied Mechanics **9**, 243–377 (1966)

[5] R. Raghava, R.M. Caddell, G.S.Y. Yeh. Journal of Materials Sciences **8**, 225–232 (1973)

[6] R. Balieu, F.Lauro, B. Bennani, R. Delille, T. Matsumoto, E. Mottola. International Journal of plasticity **51**, 241–271 (2013)

[7] W.H. Bowyer, M.G. Bader. Journal of Materials Science **7**, 1315–1321 (1972)

[8] S. Kammoun, I. Doghri, L. Adam, G. Robert, L. Delannay. Composites Part A **42**, 2892–2902 (2011)

EPJ Web of Conferences **94**, 04009 (2015)
DOI: 10.1051/epjconf/20159404009

Statistics of dynamic fragmentation for a necking instability

Jean-Lin Dequiedt

CEA, DAM, DIF, 91297 Arpajon, France

Abstract. In the aim of predicting the fragmentation of expanding structures, a Mott type approach is applied to the potential failure points driven from an instability analysis. Substantially, fragmentation is preceded by the localization of plastic deformation in multiple necks forming potential sites for failure. This pattern is predicted by a combination of periodic instable eigenmodes of the structure, the necks being the maxima with varying spacing and amplitudes of the instable perurbation. Obscuration conditions, linked to the propagation of unloading waves, are next written for this set of potential failure points and yield a distribution of fragments. The approach is developed for an expanding ring experiment justifying the scatter in fragment sizes. It shows incidentally that the unloading waves must propagate at a celerity much lower than the elastic wave speed.

1. Introduction

The fragmentation of structures under dynamic expansion is governed by the competition between the activation of failure points when deformation increases and the inhibition of failure in areas obscured by the unloading waves emerging from these points. The phenomenon has been modeled with various assumptions since the early work of Mott [1], usually postulating random positions in the structure for the potential failure sites [2]. It yielded statistic distributions for the fragment sizes at the end of the fragmentation process. However, for ductile materials, fragmentation is initiated by the localization of plastic deformation and these sites are clearly related with the necking pattern emerging in this first phase. The onset of localization is ruled by the existence of instable eigenmodes of the structure and it is treated in a linear stability analysis framework. The aim of the present work is to apply the approach of Mott to a discrete set of potential failure points identified as the maxima of an instable perturbation (forming necks in thickness of the structure): a combination of instable modes is retained in order to get maxima of various amplitudes and spacing as shown in [3]. This justifies the scatter in failure times which is at the origin of the obscuration process. The one dimensional case of an expanding ring, assimilated to a stretching rod, is developed and, as an illustration, the predictions are compared to the sizes of fragments recovered after an explosive expansion experiment.

2. Potential failure points generated by a plastic necking instability

The linear stability analysis of a stretching cylindrical rod representing an expanding ring (Fig. 1) is conducted in a one-dimensional framework as in Zhou et al. [4] in order to identify the instable eigenmodes of the structure and next the positions and failure times of the potential failure points associated with a multimode perturbation. The rod

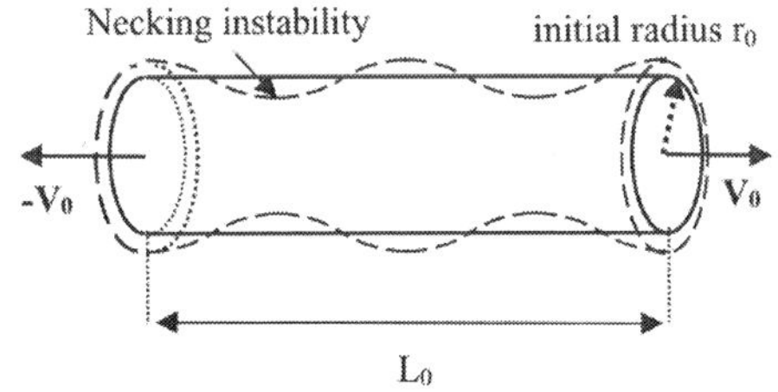

Figure 1. Stretching rod and necking instability.

is submitted to a constant tensile velocity at the extremities and the initial strain rate is $\dot{\varepsilon}_0 = V_0/L_0$. The material is supposed to be incompressible with mass density ρ and strain hardening. Elasticity and thermal effects are neglected in the localization phase. All mechanical variables are functions of the Lagrangian coordinate X along the rod and of time t.

The constitutive relations satisfied by the homogeneous and perturbed response of the rod are obtained in the following way. The yield stress is supposed to be a function of the logarithmic strain ($\sigma_Y(\varepsilon)$) and the axial stress is derived by taking into account the triaxiality of stress in necked regions with the formula of Bridgman [5]:

$$\sigma = \left(1 + \frac{1}{\theta}\right) \ln(1+\theta)\sigma_Y \qquad (1)$$

in which θ measures the local curvature of the rod external profile. It expresses as a function of the strain derivatives and the initial radius r_0 of the rod cross section:

$$\theta = \frac{r_0^2}{8}\left(-2\left(\frac{d^2\varepsilon}{dX^2}\right) + \left(\frac{d\varepsilon}{dX}\right)^2\right)\exp(-\varepsilon). \qquad (2)$$

The strain and material velocity satisfy:

$$\frac{dv}{dX} = \exp(\varepsilon)\frac{d\varepsilon}{dt} \qquad (3)$$

Article available at http://www.epj-conferences.org or http://dx.doi.org/10.1051/epjconf/20159404009

and the momentum balance writes:

$$\frac{d}{dX}(\sigma \exp(-\varepsilon)) = \rho \frac{dv}{dt}. \quad (4)$$

Instable perturbations $(\delta\varepsilon, \delta v)$ of the uniform stretching of the rod are searched in the form of periodic eigenmodes growing exponentially with time:

$$\delta\varepsilon(X,t) = \delta\hat{\varepsilon}\exp(ikX + \eta t)$$
$$\delta v(X,t) = \delta\hat{v}\exp(ikX + \eta t). \quad (5)$$

Equations (1) to (4) yields to the following relation between the growth rates and wavenumbers of perturbations at any time t_0:

$$\eta^2 - \frac{1}{\rho}k^2 \exp(-2\varepsilon(t_0))\left(\sigma_Y(t_0) - \frac{d\sigma_Y}{d\varepsilon}(t_0)\right) + \frac{r_0^2}{8\rho}k^4 \exp(-3\varepsilon(t_0))\sigma_Y(t_0) = 0. \quad (6)$$

The curve $\eta(k)$ evolves with time t_0 but all curves are similar with a maximum $\eta_{\max}(t_0)$ for a wavenumber $k_{\max}(t_0)$ (or wavelength $\lambda_{\max}(t_0)$):

$$\frac{\eta(k)}{\eta_{\max}} = \sqrt{2\left(\frac{k}{k_{\max}}\right)^2 - \left(\frac{k}{k_{\max}}\right)^4} \quad (7)$$

with:

$$k_{\max}(t_0) = \frac{2}{r_0}\exp\left(\frac{\varepsilon(t_0)}{2}\right)\sqrt{1 - \frac{1}{\sigma_Y(t_0)}\frac{d\sigma_Y}{d\varepsilon}(t_0)} \quad (8)$$

$$\eta_{\max}(t_0) = \frac{\sqrt{2}}{r_0}\sqrt{\frac{\sigma_Y(t_0)}{\rho}}\exp\left(-\frac{\varepsilon(t_0)}{2}\right) \times \left(1 - \frac{1}{\sigma_Y(t_0)}\frac{d\sigma_Y}{d\varepsilon}(t_0)\right). \quad (9)$$

The instable perturbations belong to the domain of positive $\eta(k)$, namely of wavenumbers $k < \sqrt{2}k_{\max}$. Let us assume that the whole localization process is governed by the spectrum of instable modes identified at time t_0 and that these modes develop much faster than the homogeneous solution, in such a way that the time evolution of the growth rates can be ignored. The fastest growing mode $(\eta_{\max})$ is usually considered as dominating the localization process and $\lambda_{\max}$ gives an estimation of the average neck spacing.

The circumference of the ring is supposed to be $N_L\lambda_{\max}$ and the instable perturbation to be initiated by a "white noise", i.e. a combination of all instable modes with wavenumbers $k_j = (j/N_L)\,k_{\max}$, equal amplitudes and arbitrary phases φ_j. The perturbation is thus a multimode one with amplitudes growing in accordance with the respective growth rates of the modes (represented on Fig. 2). The time evolution of this perturbation then writes:

$$\delta\varepsilon(X,t) = \frac{\delta\varepsilon_0}{\sqrt{E\left(\sqrt{2}N_L\right)}} \times \sum_{j=1}^{E\left(\sqrt{2}N_L\right)} \cos(k_j X + \varphi_j)^* \exp\left(\eta\left(k_j\right)t\right). \quad (10)$$

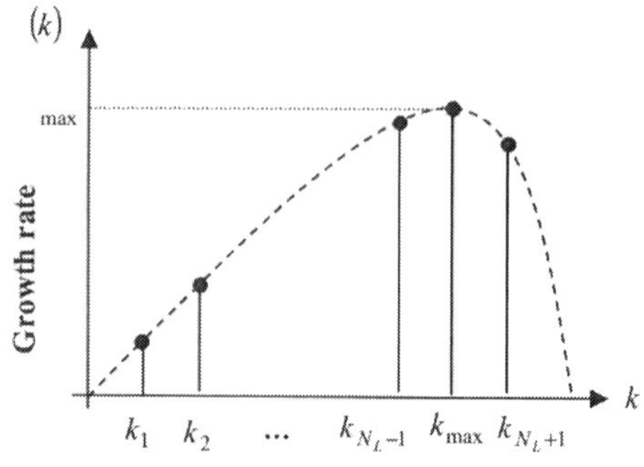

Figure 2. Spectrum of the instable perturbation and growth rates.

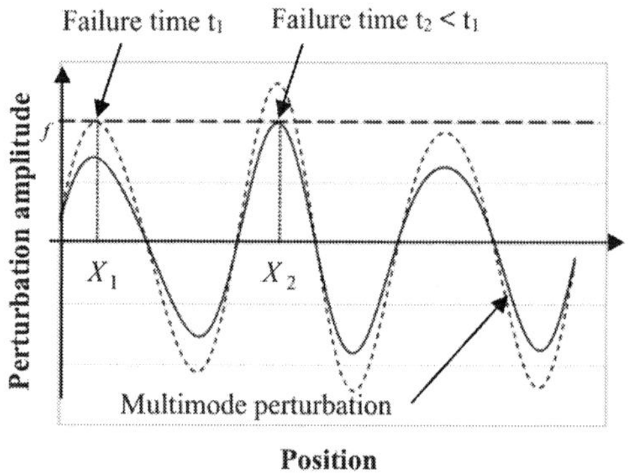

Figure 3. Determination of failure points for a multimode perturbation.

The perturbation has maxima with amplitudes and spacing varying with time.

The failure of the rod is ruled by a critical value $\delta\varepsilon_f$ for the perturbation in strain. One potential failure point is identified each time that a maximum of the perturbation reaches $\delta\varepsilon_f$. The failure points X_l and failure times t_l thus satisfy (as schematized on Fig. 3):

$$\delta\varepsilon(X_l, t_l) = \delta\varepsilon_f \quad \text{and} \quad \frac{d\delta\varepsilon}{dX}(X_l, t_l) = 0$$
$$\left(\text{such that: } \frac{d^2\delta\varepsilon}{dX^2}(X_l, t_l) < 0\right). \quad (11)$$

These points depend, among others, on the ratio between the failure strain $\delta\varepsilon_f$ and the initial amplitude of the perturbation $\delta\varepsilon_0$.

3. Fragmentation of a discrete set of failure points

Let us consider a discrete set of potential failure points with different failure times. When one of these points breaks, two unloading waves are emitted on both sides. They are supposed to propagate at a constant celerity c. They cover a so-called "obscuration cone" in the (X,t) plane in which failure is inhibited [6]. It can be easily seen that the obscuration cone of a point contains the

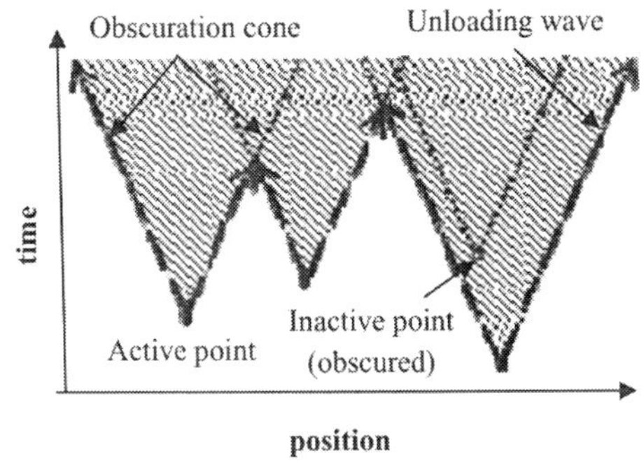

Figure 4. Area obscured by a family of failure points.

obscuration cones of all the points it obscures. Therefore, the area obscured by a set of points is the union of all obscuration cones no matter whether the points are active or not (Fig. 4).

One point l actually breaks if (X_l, t_l) is outside the area obscured by all the preceding and following points, namely if:

$$t_{l-p} + \frac{1}{c}\left|X_l - X_{l-p}\right| > t_l \quad \text{and}$$
$$t_{l+p} + \frac{1}{c}\left|X_l - X_{l+p}\right| > t_l$$
$$\text{for any } p > 0. \qquad (12)$$

4. Fragment size distribution for an expanding copper ring

An expanding copper ring experiment designed by CEA/Valduc is considered [7]. The potential failure points are identified from the analysis of the instable modes of the ring. The obscuration conditions are written in order to estimate which of these points actually break and comparison is made with the recovered fragments.

More precisely, the experimental set-up is given in Fig. 5. The ring deforms under free expansion after separating from the transmitter and multiple necks progressively appear. Some of these necks concentrate the deformation up to failure. The fragments were recovered after the test and allowed to reconstitute the entire ring with 7 failing points and 7 arrested necks (which could be interpreted as obscured failure points). The final elongation was about 46% giving an estimate of the ring deformation at the onset of necking.

The experiment is simulated and the evolution of the instable eigenmodes with time is investigated (namely of $k_{\max}(t_0)$ and $\eta_{\max}(t_0)$ given by (8) and (9)).

The localization process is supposed to be governed by the modes growing around time t_0 such that circumference of the ring is 14 times the dominant wavelength $\lambda_{\max}(t_0)$ (which would give an average of 14 necks in accordance with experiment). Then, the failure points associated with the multimode perturbation given by relation (10) and a random set of phases are displayed (a ratio $\delta\varepsilon_f/\delta\varepsilon_0 = 1000$ is assumed). Their positions and failure times are reported on Fig. 6.

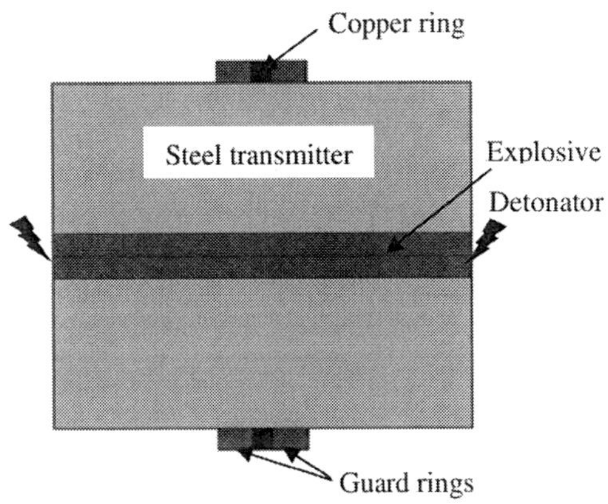

Figure 5. Expanding ring set-up.

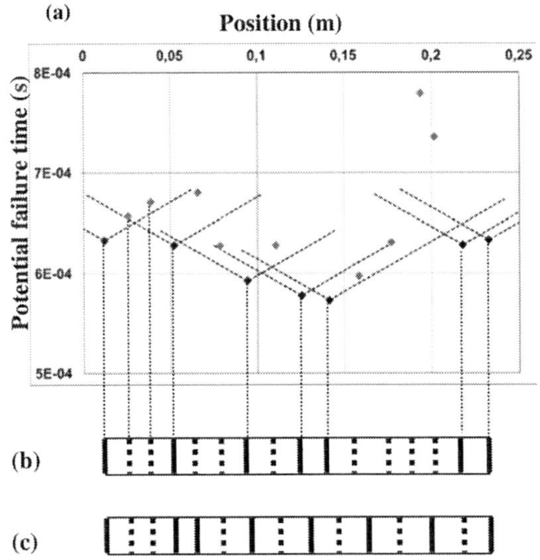

Figure 6. (a) Potential failure points and unloading waves (dashed lines) for the copper ring (active points in black and obscured points in gray). Distribution of broken and arrested (dotted) necks in (b) the prediction and (c) the experiment.

The obscuration process is next investigated by considering different values for the unloading wave celerity.

The most intuitive is to consider that they are elastic waves as in [8]; then the celerity in uniaxial stress is given by $c = \sqrt{E/\rho} \approx 3700\,\mathrm{m/s}$ where E is the Young modulus. However, this leads to unloading waves propagating very fast in such a way that the first breaking neck would obscure all the others, which does not match with experimental results. Thus, the celerity is progressively reduced down to $c = 1000\,\mathrm{m/s}$ when about a half of the necks are inhibited.

In this case, the release waves are plotted and the active and obscured points are identified on Fig. 6. Below, the predicted distribution of broken and arrested necks is drawn and compared with the experimental one.

It comes that both experiment and prediction exhibit heterogeneity of fragment sizes but less pronounced in the experiment. More generally, the influence of many parameters such as the ratio $\delta\varepsilon_f/\delta\varepsilon_0$ or the spectrum of the initial perturbation should be addressed more precisely before rigorous comparison with experiment really makes sense. Incidentally, the sensitivity to the set of phases, not illustrated here, may explain a part of the differences.

5. Conclusion

The prediction of the fragment size distribution resulting from the fragmentation of a ductile expanding structure may be addressed by coupling an instability analysis to predict the onset of localization and a fragmentation approach to investigate the obscuration process governing the inhibition of some weak points.

Some elements are given by showing that a multimode linear analysis yields to the formation of necks of varying amplitudes which may break at different times. It gives rise to the formation of unloading waves obscuring the necks with the latest failure times.

Comparison with experiment, although depending on many assumptions, seems to show, however, that the unloading waves cannot propagate as fast as the elastic wave speed, which is consistent with the progressive unloading observed in the simulation of necking [9].

At last, the stochastic character of failure, usually assumed in fragmentation analyses, has been disregarded here. It would introduce an additive scatter of failure times and favor obscuration.

References

[1] N.F. Mott, In: *Proceedings of the Royal Society*. **A**, 300–308, 1947

[2] D.E. Grady, J. Geophys. Res., **86**, B2, 1047–1054, 1981

[3] S. El Maï, S. Mercier, J. Petit, A. Molinari, Int. J. Sol. Struct., **51**, 3491–3507, 2014

[4] F. Zhou, J.F. Molinari, K.T. Ramesh, Int. J. Imp. Eng., **33**, 880–891, 2006

[5] P.W. Bridgman, *Studies in large plastic flow and fracture, with special emphasis on the effect of hydrostatic pressure* (McGrawHill Book Company Inc., New York, 1952)

[6] C. Denoual, F. Hild, Eur. J. Mech. A/Solids, **21**, 105–120, 2002

[7] F. Llorca, A. Juanicotena, In: *J. Phys IV France 7*, **C3**, 235–240, 1997

[8] E.H. Lee, *Energetics in Metallurgical Phenomena III* (Ed. W.M. Mueller, M. Shaw, Gordon and Breach, New York, 1967)

[9] F. Dalle, In: *J. Phys IV France*, **134**, 429–434, 2006

EPJ Web of Conferences 94, 04010 (2015)
DOI: 10.1051/epjconf/20159404010

The terminal effects of chisel-shape projectile penetrating into metallic target plates

Xu Tao[1,2], Xiaohu Yao[1], and Wei Ma[2,a]

[1] School of Civil Engineering and Transportation, South China University and Technology, Guangzhou 510640, China
[2] Institute of Mechanics, Chinese Academy of Sciences, Beijing 100190, China

Abstract. This work performs the analysis and simulation investigations of penetration behaviors of chisel-shape projectile. In analysis, the projectile is assumed to be a rigid body and the target plate is elastic-plastic material. By introducing the velocity potential function, the velocity field in target is first obtained. Then, the momentum equation is solved for determining the pressure and stress fields in the elastic and plastic regions in target. The variation of the resultant force subjected by the projectiles with the penetration depth is studied. The approximate expressions of penetration depth and the residual velocity with the initial impacting velocity are obtained for the exploration of the penetration mechanisms of the chisel-shape projectile. In numerical simulation, the main attention focuses on the dissipation mechanisms of the kinetic energy of the chisel-shape projectile in penetration process.

1. Introduction

The geometric effect of projectile on the failure behaviors of the target material has drawn great attention in academia and engineering since it determines, to a great extent, the terminal effects of penetration. The previous studies mainly deal with the penetration behaviors of projectiles with conical, hemispherical, flat and ogival configurations. Bishop et al. [1] proposed the cavity expansion model and used the model to estimate the forces of conical-shape projectile of punching slowly into metal targets. Goodier [2] used the cavity-expansion theory to predict the penetration depth of rigid spheres launched into metal targets and to study the penetration behaviors of non-compressible strain hardening material. Forrestal et al. [3,4] studied by penetration experiments the influences of target material constitutive behaviors on the penetration mechanism of different projectile configurations. Chen et al. [5] studied the terminal effects of flat projectile penetrating into metallic target and demonstrated the shear failure mechanism of target board material in penetration. Alekseevskii [6] and Tate [7,8] developed an inviscid flow model for projectile and target material which depicts the velocity field of target material in high-speed collision between projectile and target plate. Ravid et al. [9,10] used this model to research the failure mechanism of metallic plates and the evolution of velocity field at high impacting velocity. Yarin and Rubin [11, 12] established the penetration model by introducing a velocity potential function and analyzed two-dimensional transient stress fields in target for several projectile nose shapes and dynamic phenomena in highly penetration process. Recently, Jordan and Naito [13] carried out an experimental investigation of the effects of nose shape on fragments penetrating E-Glass/Phenolic glass fiber reinforced plastic targets. Initial velocity, residual velocity, and ballistic limit velocities were determined for each nose shape and target thickness. However, it is still a lack of the documents on the studies of the penetration mechanism of chisel-shape projectile into metal target plate.

In this work, the penetration mechanisms of chisel-shape projectile into metal target plate are studied. The penetration model of Yarin et al. [9] is developed to obtain an approximate solution of normal penetration problem of a chisel-shape projectile into elastic-plastic target plates. Through the analysis of stress state in the elastic and plastic regions of target material, the equation of motion of the projectile in the axial direction is solved. The approximate analysis reveals the variations the resultant pressure, penetration depth, kinetic energy dissipation with the initial impact velocity and time. The penetration behaviors of chisel-shape projectile to a finite target plate are investigated.

2. Model and analysis

Consider the problem of normal penetration of a rigid chisel-shape projectile into an elastoplastic target plate. Presently, we assume that the material in plastic deformation region of target is an incompressible fluid and rate-insensitive. The velocity and stress fields in plane strain state are assumed for simplicity. Figure 1 shows the analytic model and the coordinate system (o; x, y, z) with the unit vectors ($\boldsymbol{i}$, $\boldsymbol{j}$, $\boldsymbol{k}$) which has a fixed origin o at the front surface of the target. The tip of the projectile is located by $z = w(t)$. The initial impacting velocity is U_0. Furthermore, the projectile body is assumed to have a rectangle cross-section such that to obtain the approximate solution of problem is possible.

[a] Corresponding author: watwm@imech.ac.cn

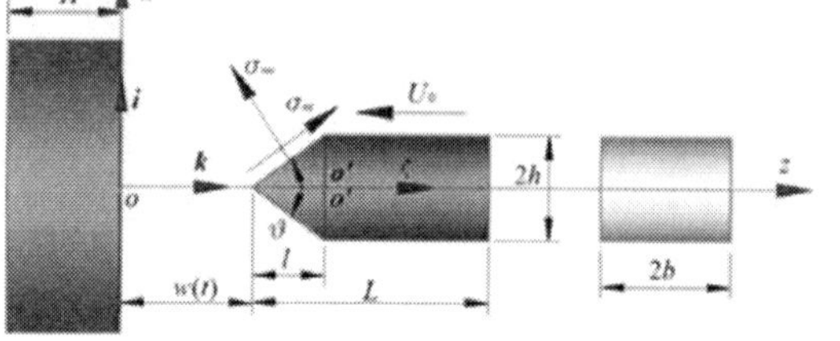

Figure 1. The model for the analysis of the chisel-shape projectile penetration behaviors.

In the following analysis of the penetration behaviors of chisel-shape projectile into metallic target, we adopt the analytic method of Yarin et al. [9] to the mathematic derivation and calculus. The approximate velocity field is first determined by introducing a velocity potential $\phi(x, z, t)$ and then the stress field in target is calculated. Thus, we have the velocity vector expression as follows

$$\mathbf{v} = \nabla\phi, \qquad \nabla^2\phi = 0 \tag{1}$$

the momentum balance equations:

$$\nabla\left[\rho\left\{\frac{\partial\phi}{\partial t} + \frac{1}{2}\nabla\phi\cdot\nabla\phi\right\} + p\right] = \text{divs} \tag{2}$$

the conservation condition of mass:

$$\text{div } \boldsymbol{\nu} = 0. \tag{3}$$

The Cauchy stress tensor σ has been separated into a hydrostatic pressure p and the stress deviatoric part $\mathbf{s}$:

$$\boldsymbol{\sigma} = -p\mathbf{I} + \mathbf{s}, \qquad \mathbf{s}\cdot\mathbf{I} = 0 \tag{4}$$

where $\mathbf{I}$ denotes the unit tensor and the deviatoric stress $\mathbf{s}$ is determined as

$$\mathbf{s} = 2\mu\boldsymbol{\varepsilon}^e + \left(\frac{2}{3}\right)^{1/2}\frac{\tau_S}{(\dot{\boldsymbol{\varepsilon}}^p\cdot\dot{\boldsymbol{\varepsilon}}^p)^{1/2}}\dot{\boldsymbol{\varepsilon}}^p \tag{5}$$

where ε^e is the elastic strain and $\dot{\varepsilon}^p$ is the rate of plastic deformation tensor. τ_s is the shear yield stress. The chisel-shape projectile is assumed to be a two-dimensional body for the simplicity in below analysis and the chisel-shape nose lateral profile is defined by

$$x = \begin{cases} \tan\theta\xi + h & (-l \le \xi \le 0, 0 \le y \le b) \\ h & (0 \le \xi \le L - l, 0 \le y \le b) \end{cases} \tag{6a}$$

$$\xi = z - w(t) - \frac{h}{\tan\theta} \tag{6b}$$

ξ is an axial coordinate measured from the point o' in projectile (Fig. 1) and 2θ denotes the chisel angle.

The velocity field is found as

$$v_x = -\dot{w}(t)\frac{h}{\tan\theta}\frac{x}{\xi^2 + x^2}, \; v_z = -\dot{w}(t)\frac{h}{\tan\theta}\frac{\xi}{\xi^2 + x^2} \tag{7}$$

where $\dot{w}(t)$ is the moving velocity of the projectile tip. The divergence of deviatoric stress is

$$div\,\mathbf{s} = \nabla\left[-\tau_s \ln\left(\frac{\xi^2 + x^2}{h^2}\right)\right] \tag{8}$$

where τ_s is the shear yield stress. The pressures in the elastic and plastic regions of target and the stress components are respectively determined as follows

$$p^p = f(t) + \Omega(\zeta, x, t) - \tau_s \ln\left(\frac{\xi^2 + x^2}{h^2}\right)$$
$$p^e = g(t) + \Omega(\zeta, x, t)$$
$$\sigma_{nn} = -p + \frac{\tau_s\Psi_1}{\Psi_0}, \sigma_{tt} = -p - \frac{\tau_s\Psi_1}{\Psi_0}, \sigma_{nt} = \frac{\tau_s\Psi_2}{\Psi_0} \tag{9}$$

where

$$\Omega(\zeta, x, t) = -\frac{\rho h\dot{w}^2}{2\tan^2\theta}\frac{h + 2\xi\tan\theta}{\xi^2 + x^2} - \frac{\rho h\ddot{w}}{2\tan\theta}\ln\frac{h^2}{\xi^2 + x^2}$$
$$\Psi_0 = \left(\bar{X} - 1\right)^2 + \tan^2\theta\bar{X}^2$$
$$\Psi_1 = \left(1 + \tan^2\theta\right)\bar{X}^2 - 2\bar{X} + \frac{1 - \tan^2\theta}{1 + \tan^2\theta}$$
$$\Psi_2 = \tan\theta\cos 2\theta\bar{X}^2 - \tan\theta\,(2 + \cos 2\theta)\,\bar{X} + \sin 2\theta \tag{10}$$

in (9), the functions of time $f(t)$ and $g(t)$ are determined by boundary conditions and in (10), $\bar{X} = X/h$ an $\bar{Y} = Y/h$ are the normalized coordinates.

The moving equation of projectile in z direction is given by

$$[M + A(w, \dot{w})]\,\ddot{w} = F(w, \dot{w}) \tag{11}$$

where $M = 2\rho_P hb(2L - h/\tan\theta)$ and ρ_P are the projectile mass and density. $F(w, \dot{w})$ is the resultant force subjected by projectiles in the positive z axial direction. According to the method of Yarin et al. (9), the motion of chisel-shape projectile with angle 2θ penetrating into the metallic target plate with thickness H can be solved. Thus, we obtain the resultant force $F(w, \dot{w})$ is

$$F(w, \dot{w}) = \begin{cases} 2bh\tau_s C_1(\eta_1) & (\zeta_0 \le w \le 0) \\ 2bh\tau_s C_2(\eta_2) & (-H \le w < \zeta_0) \\ 0 & (w < -H). \end{cases} \tag{12}$$

Furthermore, we may determine the relationships of the residual velocity and penetration depth varying with the initial impacting velocity as follows

$$U_r = \frac{1}{C_0}\left[C_0U_0^2 + C_3(\eta_2)\right]^{1/2} \quad (w \le -H)\cdot \tag{13}$$

and

$$P = \frac{hC_0^2U_0^2}{\eta_0\left[1 + 2\ln\left(2\mu/\tau_s\right)\right]}\cdot \tag{14}$$

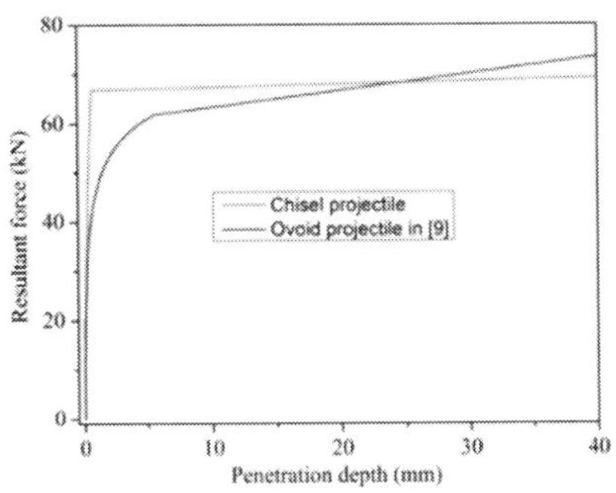

Figure 2. The variation of the resultant forces of the projectiles with different nose shapes with the penetration depth. The target thickness is 40 mm.

In (12) to (14),

$$C_0 = \left(\frac{M + A}{4bh^2\tau_s}\eta_0\right)^{1/2},$$

$$C_1(\eta_1) = 1 + \ln\left[\eta_0^2\left(\frac{2\eta_1^2}{1+\eta_1-2\eta_1^2}\right)^{2/3}\right]$$

$$C_2(\eta_2) = 1 + \ln\left[\eta_0^2\left(\frac{2\eta_2^2}{\tan\theta(\eta_2+1)+2\eta_2^2}\right)^{2/3}\right]$$

$$C_3(\eta_2) = 1 - \frac{\eta_0}{h}\left[2\eta_2\ln\eta_0 - (H+\zeta_0)C_2(\eta_2)\right]$$

$$\zeta_0 = -\frac{h\eta_2^2}{\eta_0\tan\theta(\eta_2+1)},\ \eta_0 = \frac{\tau_s}{2\mu}$$

$$\eta_1 = \eta_0\left(1 - \frac{\zeta_0 - w}{h}\tan\theta\right),\ \eta_2 = \eta_0\left(1 + \frac{H}{h}\tan\theta\right) \quad (15)$$

where ζ_0 denotes the boundary between the elastic and plastic regions in target ahead the projectile nose.

3. Results and discussion

Figure 2 illustrates the variation of the resultant forces respectively caused by the chisel- and ovoid-shape projectiles with the penetration depth as striking the target of 40 mm thickness. The lengths of the projectile noses and bodies are respectively 5 mm and 25 mm. Since the contact area between the chisel nose and target is sufficient large, the force amplitude of the chisel-shape projectile reaches its peak value as the penetration depth is only 2 mm, and then it remains constant. The force of ovoid-shape projectile non-linearly increases in amplitude in initial penetration distance of the nose length 5 mm. Then, as the ovoid nose completely gets into the target which is denoted by the inflexion on the cover, the force increases linearly with the penetration depth. When the whole body of projectile gets into the target, the force amplitude exceeds that of the chisel-shape projectile, which implies that the chisel-shape projectile is favorable to the penetration of the target plate with large thickness comparing to the ovoid-shape projectile.

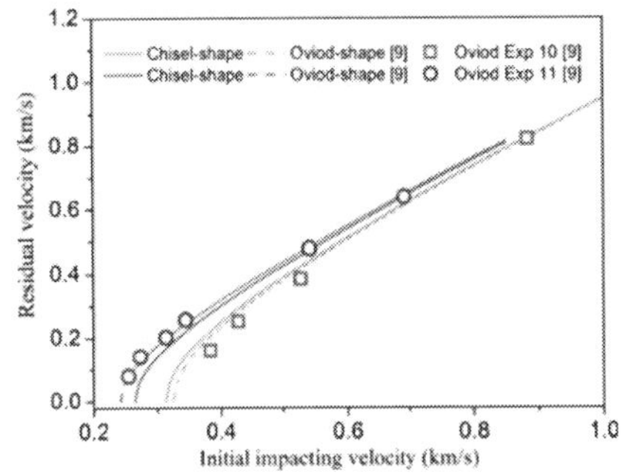

Figure 3. The effects of projectiles with different nose shapes on the residual velocities.

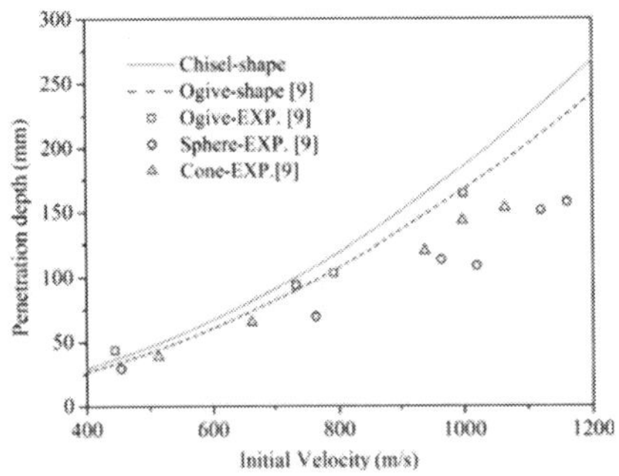

Figure 4. The relationships between the penetration depth and the initial impacting velocity.

Figure 3 shows the effect of the chisel- and ovoid-shape projectiles on the residual velocity. For the comparison, these projectiles are assumed to have same masses but different lengths, and the target thickness is taken as 25.4 mm. Clearly, in the penetration process at high impacting velocities, the influence of projectile nose shapes on the residual velocities is not evident. However, at lower impacting velocities, the residual velocities of the chisel-shape projectile with larger length (blue line) are lower evidently then that of the ovoid-shape projectile. This is because that the work done by the drag force caused by target acting on the projectile is larger, whereas for the shorter ovoid-shape projectile in length (red line), the work done by drag force is less so that the influence is slight.

Figure 4 compares the theoretical result in this study and the studying results in [9] for the penetration mechanisms of different nose shape projectiles into semi-infinite target plates. It predicts that the chisel-shape projectile may have better penetration efficiency than that with other nose shapes at high impacting velocities. Particularly, the chisel-shape projectile has evident increasing tendency for the material dynamic failure of target plates with large thickness in the penetration process at high impacting velocity.

4. Numerical simulation

In order to further study the energy dissipation mechanisms of different projectiles striking a metallic target plates, numerical simulations were performed using the LS-DYNA finite element code [1]. Figure 5 shows the

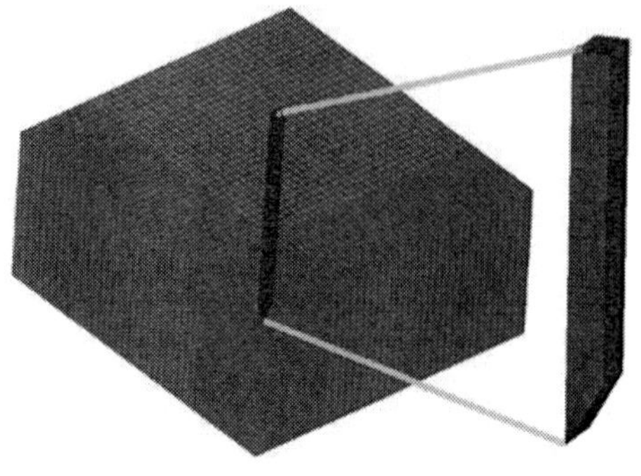

Figure 5. The finite element model of the chisel-shape projectiles penetrating into the target plate.

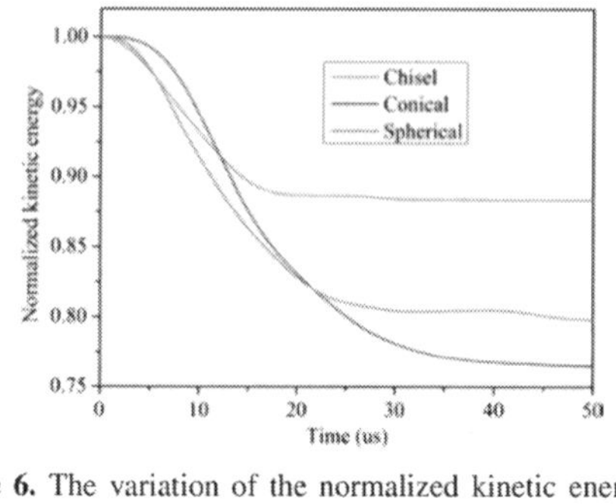

Figure 6. The variation of the normalized kinetic energy with time in the penetration process of the target plate in 6 mm thickness.

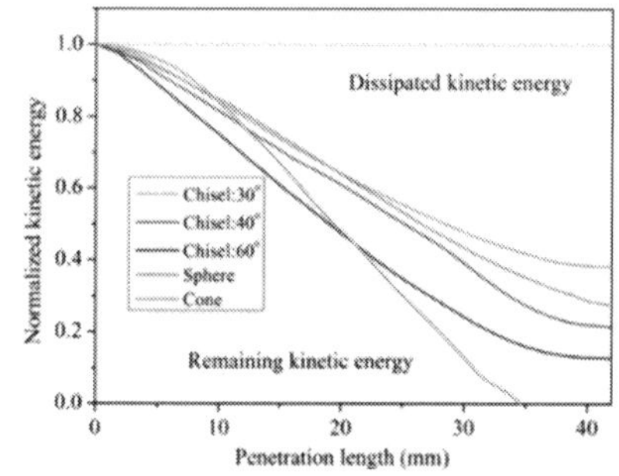

Figure 7. The variation of the normalized kinetic energy with the depth of penetration in the penetration process of the target plate in 40 mm thickness. The chisel angles are respectively 30°, 45° and 60°.

finite element model used in this simulation. Consider the symmetric property of the problem; a quarter part of the model is calculated for conserving computational time. Similarly, the projectiles with different nose shapes are assumed as rigid body and the nose shapes of projectiles simulated are chosen as spherical, conical and chisel-shapes. The mass of the projectile is about 63 g, diameter is 7.2 mm and the length is 48 mm. The target material is the 4340 steel and two kinds of target plate sizes are $240 \times 240 \times 6\,mm^3$ and $240 \times 240 \times 40\,mm^3$ respectively. The mesh comprises 8-node brick elements with reduced integration and stiffness-based hourglass control with exact volume integration. In the simulation for penetrating 6 mm target plate, the smallest element in the impact region has the initial size of $1 \times 1 \times 0.6\,mm^3$ and a total of 10 layer elements are distributed over the target thickness. In the case of the 40 mm target plate, the element size is $1 \times 1 \times 1\,mm^3$ and 40 layer elements are distributed in thickness.

Figure 6 shows the variation of the normalized kinetic energy with time in the penetration process of the target plate in 6 mm thickness. The kinetic energy of chisel- and spherical-shape projects decreases quickly when the penetration depth is less than the thickness of target plate and it begins to decrease for the conical-shape projectile when the penetration depth is near the thickness of target plate. For the chisel-and spherical-shape projectiles, when the penetrable lengths equal to 18.6 mm and 27.4 mm, the dissipation of the kinetic energy finish and, for the conical-shape projectile it is stopped as the penetrable length is larger than 38.2 mm. Thus, we can say that the chisel-shape projectile remains larger residual velocity, which means that it is in favor of the penetration of thick target plate. The simulation results in Fig. 7 shows that the sphere- and chisel-shape projectiles with the chisel angles 30°, 45° and 60° can penetrate through the target plate of 40 mm thickness but the penetration depth of the cone-shape projectile is 35 mm. The normalized kinetic energies of all kinds of projectiles decrease linearly with the penetration distance. The dissipations of kinetic energy for the sphere-shape and the chisel-shape 30° projectiles are almost identical. The loss of kinetic energy of chisel-shape projectile increases with the chisel angle increasing. The cone-shape projectile dissipates the less kinetic energy at the beginning stage of the penetration, however, most of kinetic energy quickly dissipates in the later penetrating stage so that it cannot penetrate the target plate of 40 mm thickness. Therefore, the chisel-shape projectile has the ability for the penetration of thicker target plates and the cone-shape projectile is favorable to the penetration of thin target plate.

5. Conclusions

Through the analysis and numerical simulation study of the chisel-shape projectile penetrating a target plates with finite thickness, we have shown that, comparing to different nose shape projectiles, the chisel-shape projectile is favorable to penetrate the thick target plates at high impacting velocities; contrarily, the cone-shape projectile is to that of thin target plates at low impacting velocities. The simulation result shows that the sphere-shape and the chisel-shape 30° projectiles are almost identical of the energy dissipation behaviors in the penetration process of thick target plates.

References

[1] R.F. Bishop, R. Hill, N.F. Mott, J. Proceedings of the Physical Society, **57**(3): 147 (1945)
[2] J.N. Goodier, M. Stanford Research Institute (1964)
[3] M.J. Forrestal, D.Y. Tzou, J. International Journal of Solids and Structures, **34**,(31): 4127–4146 (1997)
[4] M.J. Forrestal, K. Okajima, V.K. Luk, J. Journal of applied mechanics, **55**: 755 (1988)

[5] Xiaowei, Chen, Guanjun, Liang, Yong Yao, et al, J. Chinese journal of theoretical and applied mechanics, **41**(1): 84-90 (2009)

[6] V.P. Alekseevskii, J. Explosion, and Shock Waves. 15 July 63 (1965)

[7] A. Tate, J. Me &. Phys. Solids, **17**, pp. 141 to 150 (1969)

[8] A. Tate, J. Me &. Phys. Solids, **15**, pp. 387 to 399 (1967)

[9] M. Ravid, S.R. Bodner, I. Holcman, J. International journal of engineering science, **25**(3): 473–482 (1987)

[10] M. Ravid, S.R. Bodner, I. Holcman, J. International journal of impact engineering, **15**(4): 587–603 (1994)

[11] A.L. Yarin, M.B. Rubin, Roisman I.V., J. International journal of impact engineering, **16**(4): 801–831 (1995)

[12] G. Yossifon, M.B. Rubin, A.L. Yarin, J. International journal of impact engineering, **25**(3): 265–290 (2001)

[13] M.B. Rubin, J. International Journal of Impact Engineering, **40**: 1–9 (2012)

[14] J.B. Jordan, C.J. Naito. J. International Journal of Impact Engineering, **63**: 63–71 (2014)

EPJ Web of Conferences **94**, 04011 (2015)
DOI: 10.1051/epjconf/20159404011

Finite strain formulation of viscoelastic damage model for simulation of fabric reinforced polymers under dynamic loading

S. Treutenaere[1,a], F. Lauro[1], B. Bennani[1], T. Matsumoto[2], and E. Mottola[2]

[1] University of Valenciennes and Hainaut Cambrésis, LAMIH, UMR CNRS 8201, 59313 Valenciennes, France
[2] TOYOTA MOTOR EUROPE, 1140 Brussels, Belgium

Abstract. The use of fabric reinforced polymers in the automotive industry is growing significantly. The high specific stiffness and strength, the ease of shaping as well as the great impact performance of these materials widely encourage their diffusion. The present model increases the predictability of explicit finite element analysis and push the boundaries of the ongoing phenomenological model.

Carbon fibre composites made up various preforms were tested by applying different mechanical load up to dynamic loading. This experimental campaign highlighted the physical mechanisms affecting the initial mechanical properties, namely intra- and interlaminar matrix damage, viscoelasticty and fibre failure.

The intralaminar behaviour model is based on the explicit formulation of the matrix damage model developed by the ONERA as the given damage formulation correlates with the experimental observation. Coupling with a Maxwell-Wiechert model, the viscoelasticity is included without losing the direct explicit formulation.

Additionally, the model is formulated under a total Lagrangian scheme in order to maintain consistency for finite strain. Thus, the material frame-indifference as well as anisotropy are ensured. This allows reorientation of fibres to be taken into account particularly for in-plane shear loading. Moreover, fall within the framework of the total Lagrangian scheme greatly makes the parameter identification easier, as based on the initial configuration.

This intralaminar model thus relies upon a physical description of the behaviour of fabric composites and the numerical simulations show a good correlation with the experimental results.

1. Introduction

In these last decades, the use of carbon fabric reinforced polymers (CFRP) in the automotive industry increased very significantly. Their high specific stiffness and strength, their great energy absorption as well as the reduced manufacturing cost widely encourage their diffusion.

Previously limited to small runs (premium vehicles, racing), last advances in highly productive manufacturing process lead to the use of CFRP for high volume automotive production. Therefore, the behaviour understanding and modelling of these materials become essential for their implementation into the design loop, needed for the deployment on mass-produced vehicles. In order to ensure the protection of pedestrians and drivers/passengers in case of collision with a CFRP panel, a model dedicated to the finite element analysis (FEA) of impacts is needed.

The impact properties of layered composite materials are well reviewed by Richardson [1] and Cantwell [2]. The nonlinear material behaviour which leads to differences in the impact response of composites is attributed to fibre failure, intra- and interlaminar matrix cracking, fibre-matrix debonding and strain rate sensitivity of the matrix. Additionally, the textile composite materials are capable of large shearing prior to the ultimate failure due to the sliding and reorientation of yarns. The modelling of all these phenomena is essential to describe the impact behaviour of layered fabric composites.

To develop a material model at the ply scale achieves a balance between CPU-time and physical meaning needed for structural analysis. Indeed, the properties of the above mentioned phenomena (except for the interlaminar matrix cracking) are common to a ply but can vary from one ply to another. Among these, this study is focused on the finite strain modelling of the intralaminar matrix damage and the strain-rate dependency of fabric plies.

The use of the Continuum Damage Mechanic (CDM) for anisotropic and composite materials was introduced by Chaboche [3] and Ladevèze [4]. Further works of Chaboche [5,6] introduced the damage deactivation after the closing of cracks. Marcin [7] adapted and extended the micromechanics based CDM model proposed by Chaboche [8] to the fabric reinforced materials for implicit simulations.

In addition, Marcin extends the matrix damage model by taking into account the resin-induced viscoelasticity. To do this, a bi-spectral viscoelastic model is added. However, this formulation is not adapted to an explicit analysis because of the Newton-Raphson iterative loop to achieve the equilibrium.

Accordingly, the spectral viscoelastic model is replaced by a generalized Maxwell model. It has been successfully extended and used in finite strain for anisotropic [9,10] materials.

[a] Corresponding author: sylvain.treutenaere@univ-valenciennes.fr

In this paper, the matrix damage model proposed by Marcin is coupled to a generalised Maxwell model. This present damageable viscoelastic model is then expressed in the finite strain framework and implemented in the commercial finite element software LS-DYNA. A comparison between experimental and numerical results is done in the final part.

2. Continuum matrix damage model

Experimentally the matrix cracks are observed to be oriented in the directions of reinforcement or transverse to them. As the *Onera Damage Model MicroStructure* (ODM_MS) [7] is based on this assumption, the present model relies on a close version of the ODM_MS.

2.1. Constitutive relation

The model is formulated in strain-space to ensure a good efficiency for FEA. The constitutive relation, by using the Voigt notation, is defined as follows

$$\sigma = \tilde{\boldsymbol{C}}^m : (\varepsilon) - \boldsymbol{C}^0 : (\varepsilon^r + \varepsilon^s) \tag{1}$$

where σ and ε are respectively the infinitesimal stress and strain tensor. $\boldsymbol{C}^0$ is the initial stiffness tensor and $\tilde{\boldsymbol{C}}^m$ is the effective stiffness tensor.

The stored strains ε^s are needed in order to avoid the discontinuity of the (σ, ε) response for bi-axial loading, and to ensure the recovery of the initial stiffness after the damage deactivation.

The evolution of the residual strains ε^r is linearly dependent on the damage evolution.

2.2. Effective stiffness tensor

The damage is introduced by adding additional compliance to the initial compliance tensor $\boldsymbol{S}^0$. The effective stiffness tensor is thus defined with $(\tilde{\boldsymbol{C}}^m)^{-1} = \boldsymbol{S}^0 + \Delta \boldsymbol{S}^m$. The additional compliance tensor due to the matrix damage is given by

$$\Delta \boldsymbol{S}^m = \sum_i \eta_i^m d_i^m \boldsymbol{H}^m{}_i \tag{2}$$

where $\boldsymbol{H}^m{}_i$ is the compliance tensor associated with the damage variable d_i^m. η_i^m represents the crack closure index which varies from 0 (closed crack) to 1 (opened crack).

2.3. Damage evolution

In order to describe separately the different fracture modes, two driving forces affect the evolution of each damage variable: the normal y_i^{nm} and the tangential y_i^{tm} to the damage directions. The thermodynamic forces so defined are dependent on the effective strain tensor, which corresponds to the positive part of the spectral decomposition of the strain tensor.

Therefore, the damage criterion is defined as follows

$$F_i^m = f_i^{nm}(y_i^{nm}) + f_i^{tm}(y_i^{tm}) - di^m \leq 0 \tag{3}$$

with f the cumulative distribution function of Weibull.

This formulation – compared to a standard frame where the thermodynamic forces are based on the derivative of the Helmholtz free energy – makes easier the parameter identification, retains the explicitness of the formulation and ensures the thermodynamic coherence.

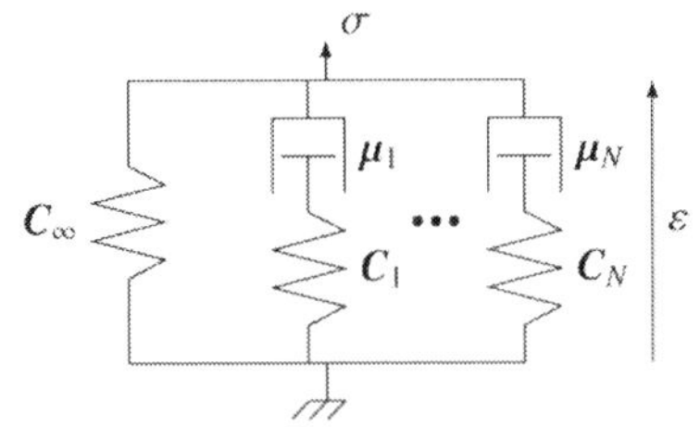

Figure 1. Generalised Maxwell model.

3. Viscoelastic damageable continuum model

The generalised Maxwell model is well adapted to the explicit finite element method because of the direct dependence of the total stress with the strain. It is symbolically represented by a spring of stiffness $\boldsymbol{C}_\infty$ paralleled with N Maxwell elements (Fig. 1).

Hence, the total stress, provided by the coupling between the above mentioned matrix damage and the generalised Maxwell viscoelastic models, is given by

$$\sigma = \sigma_\infty + \sum_{j=1}^{N} \tilde{h}_j \tag{4}$$

where σ_∞ is the long time stress, defined by means of the ODM_MS with the equation (1), and $\tilde{h}_j$ is the effective viscoelastic stress of the j-th branch. This tensor is determined thanks to the Boltzman superposition principle and is expressed in an iterative form [9, 10] by

$$\tilde{h}_j(t^{n+1}) = \exp\left(-\frac{\Delta t}{\tau_j}\right)\tilde{h}_j(t^n) + \tilde{\boldsymbol{C}}_j^m \frac{1 - \exp\left(-\frac{\Delta t}{\tau_j}\right)}{\frac{\Delta t}{\tau_j}} \Delta\varepsilon \tag{5}$$

with $\Delta t = t^{n+1} - t^n$, $\Delta\varepsilon = \varepsilon(t^{n+1}) - \varepsilon(t^n)$ and $\tilde{\boldsymbol{C}}^m{}_j$ the effective viscoelastic stiffness tensor. This tensor is defined in a same manner as the effective stiffness tensor presented in the subsection 2.2. Consequently,

$$\tilde{\boldsymbol{C}}_j^m = \left(\tilde{\boldsymbol{S}}_j^m\right)^{-1} \quad \text{and} \quad \tilde{\boldsymbol{S}}_j^m = \boldsymbol{S}_j^0 + \sum_{i=1}^{3} d_i^m \boldsymbol{H}_{i,j}^{ve,m} \tag{6}$$

where d_i^m is the same damage variable as that used during the computation of the long time stress in the Eq. (1), $\boldsymbol{S}_j^0$ is the initial viscoelastic compliance tensor and $\boldsymbol{H}_{i,j}^{ve,m}$ is the viscoelastic compliance tensor associated with the damage variable d_i^m.

To use the same damage variables is not unreasonable as the matrix damage and the viscoelasticity are both inherent in the resin.

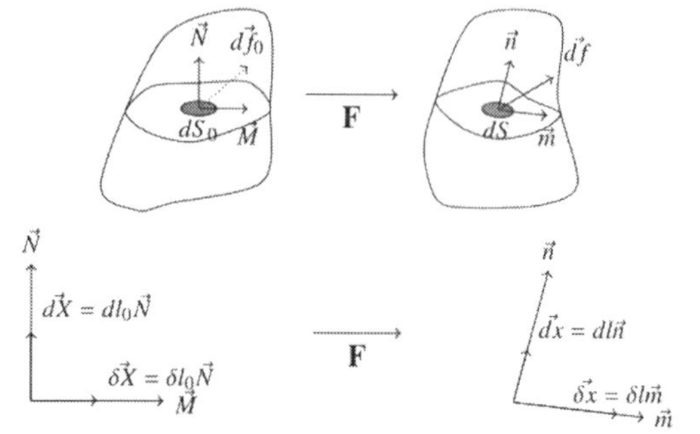

Figure 2. Deformation of a body with representation of undeformed and current configurations.

4. Material model formulation for finite strain

Given the shear behaviour of the fabric composite ply with the possibility of large rotations of the yarns, the base (or undeformed) and the current (deformed) configurations cannot be assumed identical. It is therefore essential to extend the model in finite strain.

With few exceptions, Lagrangian (or Material) coordinates are used to describe the deformations of a solid. This approach facilitates the formulation of material constitutive models since the position and the physical properties of solid particles are described according to a reference position of these material particles and time.

4.1. Expression of tensors

As the objective is the formulation of a constitutive model in finite strain for structural analysis, it is appropriate to recall the definition of strain and stress tensors.

Let $\vec{dX}$ be a position vector which describes material points in the undeformed configuration. The material points in the deformed configurations are now described by $\vec{dx}$ (Fig. 2). The change of the material points is defined by the function $x_i = m_i(X_I, t)$. Hence, the differential is

$$dx_i = \frac{\partial x_i}{\partial X_J} dX_J$$

which leads to the definition of the deformation gradient tensor $\mathbf{F}$ by

$$\vec{dx} = \mathbf{F}\vec{dX} \qquad \text{with} \qquad \mathbf{F}_{iJ} = \frac{\partial x_i}{\partial X_J}. \tag{7}$$

The transformation of a volume element dv_0 to a volume element dv is given by the relation

$$dv = J dv_0 \qquad \text{with} \quad J = \det \mathbf{F}.$$

The elongation of a position vector in the $\vec{N}$ direction is defined as

$$\varepsilon(\vec{N}) = \frac{dl - dl_0}{dl_0}.$$

Lastly, the shear angle is determined by

$$\gamma(\vec{M}, \vec{N}) = \widehat{(\vec{M}, \vec{N})} - \widehat{(\vec{m}, \vec{n})}.$$

From now, a distinction should be made between the tensors expressed according to the base configuration or the current configuration.

First of all, suppose that the reference is the base configuration. The scalar product of the material vectors in the current configuration, evaluated according to these material vectors in the base configuration, is given by

$$\begin{aligned} \vec{dx}.\vec{\delta x} &= dx_i.\delta x_i = F_{iJ} dX_J F_{iK} \delta X_K \\ &= dX_J F^{\mathrm{T}}_{Ji} F_{iK} \delta X_K \\ &= \vec{dX} \mathbf{F}^{\mathrm{T}} \mathbf{F} \vec{\delta X} = \vec{dX} \cdot \mathbf{C} \cdot \vec{\delta X} \end{aligned} \tag{8}$$

where $\mathbf{C} = \mathbf{F}^{\mathrm{T}}\mathbf{F}$ is the right Cauchy-Green deformation tensor. Similarly, the variation of the scalar product according to the material vectors in the base configuration is defined with

$$\begin{aligned} \vec{dx}.\vec{\delta x} - \vec{dX}.\vec{\delta X} &= d\vec{X}_J(\mathbf{C}_{JK} - \delta_{JK})\delta\vec{X}_K \\ &= 2\vec{dX} \cdot \mathbf{E} \cdot \vec{\delta X} \end{aligned} \tag{9}$$

with $\delta_{JK} = 1$ when $J = K$, otherwise zero, and where $\mathbf{E} = \frac{1}{2}(\mathbf{C} - \mathbf{1})$ is called the Green-Lagrange strain tensor with $\mathbf{1}$ the identity tensor.

In a consistent manner, but by considering the current configuration as reference, the strain can be described by the Almansi tensor $\mathbf{A} = \frac{1}{2}(\mathbf{1} - \mathbf{B}^{-1})$ where $\mathbf{B} = \mathbf{F}\mathbf{F}^{\mathrm{T}}$ is the left Cauchy-Green deformation tensor.

Now that the different strain tensors are defined, the stresses have to be evaluated according to both configurations too.

$\vec{df}$ is a force which acts on the current body and is the only one measurable from the experiments. By expressing the force vector according to the base or the current configuration, this leads to

$$\begin{cases} \vec{df} = \mathbf{T}\vec{n} dS \\ \vec{df} = \mathbf{\Pi}\vec{N} dS_0 \end{cases} \tag{10}$$

with $\mathbf{T}$ the Cauchy stress tensor and $\mathbf{\Pi}$ the first Piola-Kirchoff stress tensor (PK1). Note that whereas $\mathbf{T}$ is symmetric, $\mathbf{\Pi}$ because of its mixed nature is not.

Let $\vec{df_0}$ be a virtual force, seen as the equivalent of $\vec{df}$ which may act on the reference configuration. $\vec{df_0}$ has no physical existence and is the transposition of $\vec{df}$ in the base configuration: $\vec{df_0} = \mathbf{F}^{-1}\vec{df}$.

Hence, a new stress tensor integrally based on the reference configuration can be defined through the relation $\vec{df_0} = \mathbf{S}\vec{N} dS_0$ with

$$\mathbf{S} = \mathbf{F}^{-1}\mathbf{\Pi} \tag{11}$$

where $\mathbf{S}$ is called the second Piola-Kirchoff stress tensor (PK2) and $\mathbf{S}$ is symmetric.

As a result, the behaviour law in finite strain can be expressed either according to the current configuration by finding the relation $\mathbf{T} = t(\mathbf{A})$ or according to the reference configuration by finding the relation $\mathbf{S} = s(\mathbf{E})$.

4.2. Iterative formulations

By means of the Finite Element Method (FEM), the final solution is obtained from the base configuration

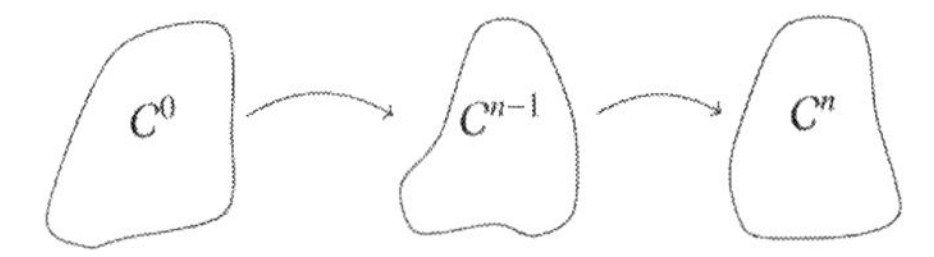

Figure 3. The major configurations used in Updated Lagrangian framework.

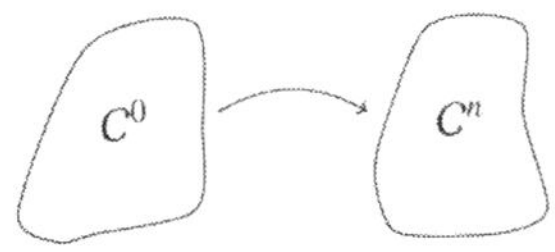

Figure 4. The major configurations used in Total Lagrangian framework.

C^0 by stepping computations in an incremental solution process. At step n, a current configuration C^n is computed based on the reference configuration $C^{\mathcal{R}}$. The choice of this reference configuration leads to different Lagrangian formulations.

4.2.1. Updated Lagrangian formulation

In the updated Lagrangian formulation, the reference is assumed to be the previous current configuration C^{n-1} and is usually updated after each incremental step. Therefore, the new current configuration C^n is directly getting from the previous current configuration C^{n-1} (Fig. 3).

The reference configuration being at a deformed state, the behaviour law is formulated in terms of the conjugate stress-strain pair $\mathbf{T} : \mathbf{A}$.

But this scheme leads to additional computations to achieve the incremental objectivity. In addition, objective stress rate as the commonly employed Green-Naghdi or Jaumann, do not correctly follow the directions of anisotropy of the composite material [11,12].

LS-DYNA, for example, uses the updated Lagrangian framework as stress-return algorithm.

4.2.2. Total Lagrangian formulation

Otherwise, in the total Lagrangian formulation, the reference is assumed to be the base configuration at each step. Therefore, the current configuration C^n is always getting from the base configuration C^0, regardless the current configuration computed at the previous step (Fig. 4).

The reference configuration being at the undeformed state, the behaviour law is formulated in terms of the conjugate stress-strain pair $(\mathbf{S}, \mathbf{E})$.

With this formulation the objectivity is ensured by the use of total Lagrangian tensors and the direction of fibres is well followed. This formulation is used to formulate the present model in finite strain.

4.2.3. Coupling

The formulation used by the finite element program differs from the chosen scheme for the material model. The links between the total and the updated Lagrangian frameworks need to be established.

This is done by converting the second Piola-Kirchoff stress tensor to the Cauchy stress tensor through the push-forward operation by using the formula

$$\mathbf{T} = J^{-1}\mathbf{F}\mathbf{S}\mathbf{F}^{\mathrm{T}}. \tag{12}$$

4.3. Constitutive relation

The matrix damage and the viscoelastic models share the same damage variables. Therefore, it is essential that both models are expressed by using the same formulation. Due to the anisotropy and as explained in the Sect. 4.2.2, the model is formulated in the total Lagrangian framework.

The second Piola Kirchoff stress tensor $\mathbf{S}$ and the Green-Lagrange strain tensor $\mathbf{E}$ are symmetric and can be represented through the Voigt notation, respectively s and ε_G. By following the methods given by Kaliske [9,10] and according to the damageable vicoelastic model formulated in the Sect. 3, the constitutive relation becomes

$$s = s_\infty + \sum_{j=1}^{N} \tilde{g}_j \tag{13}$$

with

$$s_\infty = \tilde{\mathbf{C}}^m : (\varepsilon_G) - \mathbf{C}^0 : (\varepsilon_G^r + \varepsilon_G^s) \tag{14}$$

and

$$\tilde{g}_j(t^{n+1}) = \exp\left(-\frac{\Delta t}{\tau_j}\right)\tilde{g}_j(t^n) + \tilde{\mathbf{C}}^m{}_j \frac{1 - \exp\left(-\frac{\Delta t}{\tau_j}\right)}{\frac{\Delta t}{\tau_j}} \Delta\varepsilon_G. \tag{15}$$

The parameter identification have to be adapted to this finite strain framework.

4.3.1. Shear locking

In the particular case of textile fabric, which is when the yarns are interlaced, a locking phenomenon appears after the shear reaches a critical angle. Accordingly, a significant rise of the elastic shear stiffness is observed. To take into account this phenomenon the constitutive relation becomes for the in-plane shear stress component

$$s_{12} = s_{\infty,12} + \sum_{j=1}^{N} \tilde{g}_{j,12} + \alpha^{sl} \left\langle \varepsilon_{G,12} - \varepsilon_G^{sl} \right\rangle^2 \tag{16}$$

with $\langle x \rangle = x$ if $x > 0$ and 0 if $x < 0$ and where α^{sl} and ε_G^{sl} are material parameters.

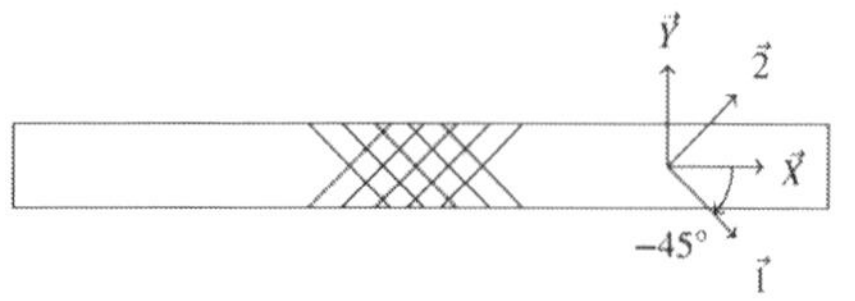

Figure 5. Global ($\vec{X}$, $\vec{Y}$) and material ($\vec{1}$, $\vec{2}$) coordinate systems.

4.4. Identification

The large rotations of the yarns appear under in-plane shearing and other loading do not lead to large deformations. Therefore, the identification procedure presented in this article only concerns the parameters determined from cyclic in-plane shear test which is derived from the standardized NF EN ISO 14129 monotonic in-plane shear test.

All measurements are carried out in the global coordinate system and provide:

- $\varepsilon_X = \varepsilon(\vec{X})$ the elongation in the centre of the coupon according to the $\vec{X}$ direction,
- $\varepsilon_Y = \varepsilon(\vec{Y})$ the elongation in the centre of the coupon according to the $\vec{Y}$ direction,
- $f = \vec{f} \cdot \vec{X}$ the measured force applied to the coupon.

As a result, the deformation gradient tensor **F** and the first Piola-Kirchoff stress tensor **Π** are in the global frame

$$\mathbf{F} = \begin{bmatrix} 1+\varepsilon_X & 0 \\ 0 & 1+\varepsilon_Y \end{bmatrix}, \quad \mathbf{\Pi} = \begin{bmatrix} \frac{f}{S_0} & 0 \\ 0 & 0 \end{bmatrix}. \tag{17}$$

After a change of basis to be placed in the material frame and by using the Eqs. (9) and (11), the Lagrangian tensors are determined and given by

$$\mathbf{E}_m = \frac{1}{2}\begin{bmatrix} \varepsilon_X + \varepsilon_Y + \dfrac{\varepsilon_X^2 + \varepsilon_Y^2}{2} & \varepsilon_X - \varepsilon_Y + \dfrac{\varepsilon_X^2 - \varepsilon_Y^2}{2} \\ \varepsilon_X - \varepsilon_Y + \dfrac{\varepsilon_X^2 - \varepsilon_Y^2}{2} & \varepsilon_X + \varepsilon_Y + \dfrac{\varepsilon_X^2 + \varepsilon_Y^2}{2} \end{bmatrix} \tag{18}$$

$$\mathbf{S}_m = \frac{f}{(1+\varepsilon_X)S_0}\begin{bmatrix} 1 & 1 \\ 1 & 1 \end{bmatrix}. \tag{19}$$

Hence, it becomes possible to identify the parameters of the total Lagrangian behaviour law $\mathbf{S} = s(\mathbf{E})$.

The damage evolution law is determined through the evolution of the shear stiffness on quasi-static tests and without considering any viscoelastic effects. However in the case of textile preforms and when the angle of locking is reached, it becomes impossible to distinguish the effects of the damage and of the shear locking. So after this critical shear angle, the damage evolution law is approximated to best fit the evolution of the residual strains. Indeed, the model is formulated in such a way that the residual strains are directly dependent on the damage variables. The gap in the shear stiffness is eventually considered as be due to the shear locking.

(a) Non-Crimp Fabric (b) 3K woven

Figure 6. View of the upper face of the preforms.

The initial viscoelastic parameters are identifies through Dynamic Mechanical Analysis (DMA) tests [13] and the viscoelastic compliance tensors associated to the damage variables are determined by an iterating process to best fit the shear behaviour.

5. Numerical examples and discussions

The formulation proposed is illustrated by the simulation of in-plane shear tests carried out on two different fabric preforms.

Both are made up carbon fibres but the process to get the fabric preforms differ. The first one is a Non-Crimp Fabric (NCF) where the carbon fibre are arranged in two uni-directional orthogonal plies and then stitched together by a glass fibre tow (Fig. 6a). The second one is a plain-weave preform made up interlaced yarn including 3000 fibres each (Fig. 6b).

The experimental tests are carried out by means of a high-speed jack facility on coupons defined by the standard NF EN ISO 14129. Three speed loadings were set: 1.7 mm/s, 41 mm/s and 1000 mm/s. The shear strain is measured by an optical extensometer. The reference is taken as the mean value of the experimental results of a given configuration.

The Fig. 7 shows the comparison between the experimental results and the finite element analysis results of in-plane shear tests on both fabric materials at various strain-rate. The values are not stated because of confidentiality reasons. The model correlates well the material behaviours, even for completely different fabric preforms.

6. Concluding remarks

In this paper, a finite strain damageable viscoelastic model for fabric reinforced polymers has been presented.

The model is a combination between the Onera Damage MicroStructure model and a generalised Maxwell viscoelastic model. Because of the large rotation of the yarns for shear loading, the model is extended in finite strain. The total Lagrangian formulation is used for both matrix damage and viscoelastic models in order to well track the fibre orientation and ensure the objectivity. Moreover, the damage variables can be shared between the both element of the finite strain model. The phenomenon of shear locking in the particular case of textile preforms is also taking into account by a adding a non-linearity to the elastic shear stiffness.

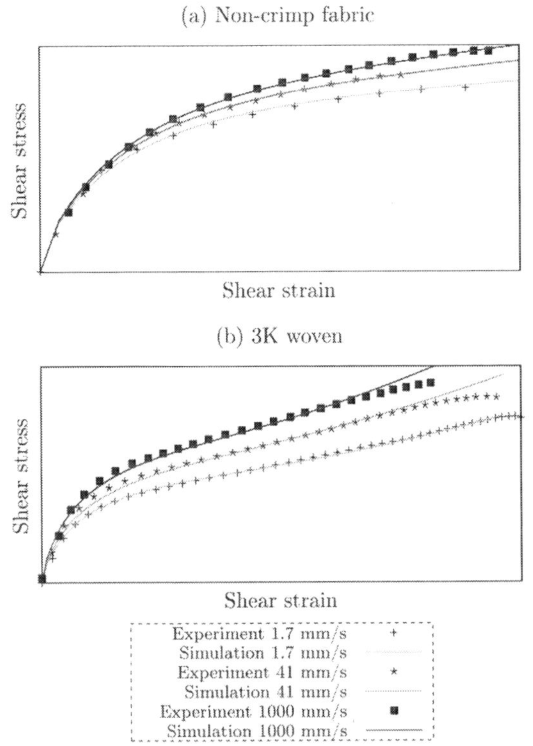

Figure 7. Resultant stress-strain curves of in-plane shear test simulation compared to experimental results at various strain-rates.

From the numerical point of view, the model is implemented in the commercial finite element software LS-DYNA. It is validated through experimental tests carried out at various strain-rate and reaching the loading rate of 1 m/s. The simulation results show the good efficiency of the proposed model.

However, in order to fully simulate the behaviour of layered fabric composites, additional physical phenomenon have to be taking into account. Such is the case, for instance, of the fibre failure or the intralaminar damage.

The present research work has been supported by International Campus on Safety and Intermodality in Transportation, the Region Nord Pas de Calais, the European Community, the Delegation Regionale a la Recherche et a la Technologie, the Ministere de l'Enseignement Superieur et de la Recherche, the Centre National de la Recherche Scientifique and TOYOTA MOTOR EUROPE: the authors gratefully acknowledge the support of these institutions.

References

[1] M.O.W. Richardson, M.J. Wisheart, Composites Part A: Applied Science and Manufacturing **27**, 1123 (1996)

[2] W.J. Cantwell, J. Morton, Composites **22**, 347 (1991)

[3] J.L. Chaboche, Nuclear Engineering and Design **79**, 309 (1984)

[4] P. Ladeveze, E. LeDantec, Composites Science and Technology **43**, 257 (1992)

[5] J.L. Chaboche, International Journal of Damage Mechanics **1**, 148 (1992)

[6] J.L. Chaboche, S. Kruch, J.F. Maire, T. Pottier, International Journal of Plasticity **17**, 411 (2001)

[7] L. Marcin, Ph.D. thesis, Université Bordeaux 1 (2010), `http://tel.archives-ouvertes.fr/tel-00481601/`

[8] J.L. Chaboche, J.F. Maire, Aerospace Science and Technology **6**, 131 (2002)

[9] M. Kaliske, H. Rothert, Computational Mechanics **19**, 228 (1997)

[10] M. Kaliske, Computer Methods in Applied Mechanics and Engineering **185**, 225 (2000)

[11] P. Boisse, Journal of Materials Science **41**, 6591 (2006)

[12] R.H.W. ten Thije, R. Akkerman, J. Huétink, Computer Methods in Applied Mechanics and Engineering **196**, 3141 (2007)

[13] R. Balieu, Ph.D. thesis, Université de Valenciennes et du Hainaut-Cambresis (2012)

EPJ Web of Conferences **94**, 04012 (2015)
DOI: 10.1051/epjconf/20159404012

3D finite element simulation of effects of deflection rate on energy absorption for TRIP steel

Asuka Hayashi[1], Hang Pham[1,3], and Takeshi Iwamoto[2]

[1] Graduate School of Engineering, Hiroshima University, 1-4-1 Kagamiyama, Higashi-Hiroshima, Hiroshima 739-8527, Japan
[2] Institute of Engineering, Hiroshima University, 1-4-1 Kagamiyama, Higashi-Hiroshima, Hiroshima 739-8527, Japan
[3] Faculty of Engineering, Vietnam National University of Agriculture, Trauquy, Gialam, Hanoi 131004, Vietnam

Abstract. Recently, with the requirement of lighter weight and more safety for a design of automobile, energy absorption capability of structural materials has become important. TRIP (Transformation-induced Plasticity) steel is expected to apply to safety members because of excellent energy absorption capability and ductility. Past studies proved that such excellent characteristics in TRIP steel are dominated by strain-induced martensitic transformation (SIMT) during plastic deformation. Because SIMT strongly depends on deformation rate and temperature, an investigation of the effects of deformation rate and temperature on energy absorption in TRIP is essential. Although energy absorption capability of material can be estimated by J-integral experimentally by using pre-cracked specimen, it is difficult to determine volume fraction of martensite and temperature rise during the crack extension. In addition, their effects on J-integral, especially at high deformation rate in experiment might be quite hard. Thus, a computational prediction needs to be performed. In this study, bending deformation behavior of pre-cracked specimen until the onset point of crack extension are predicted by 3D finite element simulation based on the transformation kinetics model proposed by Iwamoto et al. (1998). It is challenged to take effects of temperature, volume fraction of martensite and deformation rate into account. Then, the mechanism for higher energy absorption characteristic will be discussed.

1. Introduction

Recently, a decrease in weight and improvement of safety for automobiles are challenging tasks for automobile industry [1]. Thus, energy-absorption capacity of structural materials has become important to protect the passengers from a high-speed crash of automobiles [2]. From the perspective of safety, nowadays, TRIP (transformation induced plasticity) steel is commonly applied for such structures and components [2] because of the high strength and excellent ductility due to strain-induced martensitic transformation (SIMT) [3]. In addition, TRIP steel might have high energy-absorption capacity because huge amount of kinetic energy might be dissipated during phase transformation process [4]. On the other hand, deformation mode of safety members during crash includes not only axial tension but also bending deformation. Therefore, an investigation of effect of deflection rate on energy-absorption characteristic during bending deformation process for TRIP steel is necessary.

Although many past research works have focused on TRIP steel [5–7], the effect of SIMT on energy-absorption has still insufficiently investigated. Energy-absorption capacity of TRIP steel can be estimated by J-integral derived by Rice [8] experimentally by using pre-cracked specimen [9]. However, SIMT process itself is considered quite complicate and it is difficult to determine volume fraction of martensite and temperature rise during the crack extension of tested specimens. Additionally, their effects on energy-absorption characteristic, especially at a high deformation rate might be quite hard to observe in experiment. Thus, a computational prediction needs to be performed.

With the purpose of predicting and controlling the mechanical properties of TRIP steel, Olson and Cohen [10] established a model for SIMT kinetics which can express the temperature dependent of SIMT. After that, Stringfellow et al. [11] generalized the model of Olson and Cohen by including the effect of stress state into the driving force of the martensitic transformation. Then, Tomita and Iwamoto [12] incorporated the effects of the strain rate sensitivity into the model proposed by Stringfellow et al. Iwamoto et al. [13] incorporate the stress state dependence based on the model proposed by Olson and Cohen [10], and Stringfellow [11]. The validity of the model is confirmed by some furthermore researches [12, 13].

In this study, bending deformation behaviour of pre-cracked specimen is investigated by 3D finite element simulation based on the transformation kinetics model proposed by Iwamoto et al. [14]. After confirming the validity of computation, effects of deflection rate, temperature, volume fraction of martensite on energy absorption of TRIP steel until the onset point of crack extension of specimen are examined. Then, the mechanism for energy absorption characteristic in TRIP steel will be discussed.

2. Methodology

2.1. The constitutive model for TRIP steel

The rate of the volume fraction of martensite, $\dot{f}^{\alpha'}$ is calculated as shown in the following equation [14].

$$\dot{f}^{\alpha'} = \left(1 - f^{\alpha'}\right)\left(A\dot{\bar{\varepsilon}}_{(\gamma)}^{pslip} + B\dot{g}\right) \tag{1}$$

$$A = \alpha\beta(f^{sb})^{n-1}(1 - f^{sb})$$
$$B = \eta n_v(g)(f^{sb})^n H(\dot{g})$$
$$\alpha = (\alpha_1 T^2 + \alpha_2 T + \alpha_3 - \alpha_4 \Sigma)\left[\frac{\dot{\bar{\varepsilon}}_{(\gamma)}^{pslip}}{\dot{\varepsilon}_y}\right]^M$$
$$\beta = \frac{\eta}{\sqrt{2\pi}\sigma_g}\int_{-\infty}^{g}\exp\left\{-\frac{(\acute{g} - g_0)^2}{2\sigma_g^2}\right\}dg'$$
$$\eta = \frac{K\bar{v}^{\alpha'}}{\bar{v}^{sb}},\ g = -T + g_1\Sigma,\ \Sigma = \sigma_{ii}/3\bar{\sigma}$$

where $\dot{\bar{\varepsilon}}_{(\gamma)}^{pslip}$ is equivalent plastic strain rate by slip deformation in the austenite, f^{sb} is the volume fraction of the shear band, g is driving force for SIMT and T is the absolute temperature. Σ is the stress triaxiality parameter, M is the strain-rate sensitivity exponent, $\dot{\varepsilon}_y$ is the reference strain rate and $\alpha_1 - \alpha_4$ are material parameters.

To express the deformation mode dependence of the deformation behaviour of TRIP steels, the thermo-elasto-viscoplastic constitutive equation for large deformation including transformation strain rate can be formulated as [15]

$$\breve{S}_{ij} = D_{ijkl}^{v} d_{kl} - B_{ij}^{e}\dot{T} - P_{ij}\Theta_1 - \left(Q_{ij} - \sigma_{ij}\right)\Delta v \dot{f}^{\alpha'} \tag{2}$$

$$D_{ijkl}^{v} = D_{ijkl}^{e} - \Theta_2 P_{ij} P_{kl},\ B_{ij}^{e} = \frac{E}{1-2v}\alpha_T\delta_{ij},$$
$$P_{ij} = \frac{3E}{2\bar{\sigma}(1+v)}\frac{\partial f}{\partial\sigma_{ij}},\ Q_{ij} = -P_{ij}\Sigma + \frac{1}{3}\delta_{ij}\frac{E}{1-2v} \tag{3}$$

where $\breve{S}_{ij}$ is Jaumann rate of Kirchhoff stress, D_{ijkl}^{e} is the elastic stiffness tensor, α_T is the thermal expansion coefficient, E is the Young's modulus and v is the Poisson's ratio. According to Tomita and Iwamoto [12], the heat conduction equation accounting for the latent heat due to SIMT $l_{\alpha''}$ is can be delivered as

$$\rho C_v\dot{T} = \zeta\sigma_{ij}\dot{\varepsilon}_{ij}^{p} + \kappa'\nabla^2 T - \rho l_{\alpha'}\dot{f}^{\alpha'} \tag{4}$$

where C_v is the specific heat at constant volume, ρ is the mess density and ζ is Taylor-Quinney coefficient.

2.2. Finite element equations

In impact condition, inertia force is considered and added into finite element equation as following,

$$M\ddot{d}^e + K\dot{d}^e = \dot{f}_1 + \dot{f}_2 + \dot{f}_3 + \dot{f}_t + \dot{f}_g \tag{5}$$

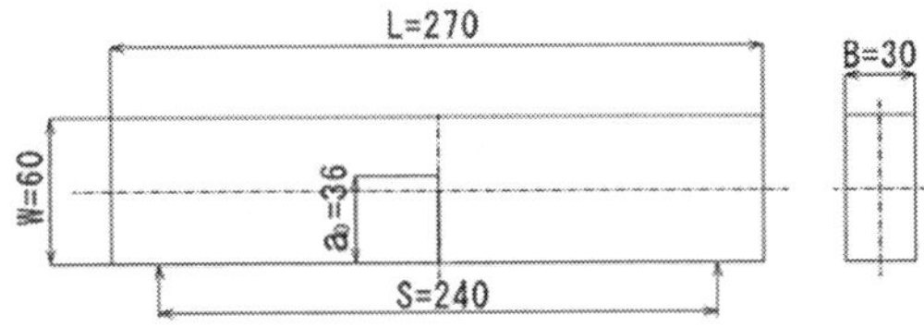

Figure 1. The dimensions of a pre-cracked specimen using for the finite element simulation [16].

$$M = \int_V^{\rho} \rho\Phi^T\Phi dV$$
$$K = \int_V \left\{B^T\left(D^v - F'\right)B + E^T Q' E\right\}dV,$$
$$\dot{f}_1 = \int_V \dot{T}B^T B^e dV, \qquad \dot{f}_2 = \int_V \Theta_1 B^T P dV,$$
$$\dot{f}_3 = \int_V \Delta v\dot{f}^{\alpha'}B^T(Q - \sigma)dV,$$
$$\dot{f}_t = \int_{S_t}\Phi^T\dot{F}dS, \qquad \dot{f}_g = \int_V \Phi^T\dot{G}dV$$

where D^v, D^e, F', Q', P, Q, B^e are the matrices corresponding to D_{ijkl}^{v}, D_{ijkl}^{e}, F_{ijkl}', σ_{mj}, P_{ij}, Q_{ij}, B_{ij}^{e} respectively. $\dot{d}^e$ is the nodal displacement rate and Φ is the shape function of an element for the displacement rate. E and B are corresponding transformation matrices. $\rho\dot{G}$ and $\dot{F}$ are the rates of the body and surface forces, respectively.

Additionally, the FE heat conduction equation can be derived as

$$\bar{K}_1\dot{t} + \bar{K}_2 t = \bar{R} + \bar{Q} - \bar{L}, \tag{6}$$
$$\bar{K}_1 = \int_V \rho C\bar{\Phi}^T\bar{\Phi}dV,$$
$$\bar{K}_2 = \int_V \kappa'\bar{B}^T\bar{B}dV, \qquad \bar{R} = \int_V \bar{\Phi}^T\zeta\bar{\sigma}\dot{\bar{\varepsilon}}^p dV,$$
$$\bar{Q} = \int_{S_q}\bar{\Phi}^T\bar{Q}dS, \qquad \bar{L} = \int_V \bar{\Phi}^T\rho l_{\alpha'}\dot{f}^{\alpha'}dV$$

where t is the nodal temperature vector, $\dot{t}$ is the rate of the nodal temperature vector, and $\bar{\Phi}$ is the shape function for a temperature field. By using the Houlbolt Method, the second derivative of nodal displacement rate in time at moment $t + \Delta t$, $\dddot{d}_{t+\Delta t}^{e}$ can be rewritten.

3. Three-dimension finite element model

Figure 1 shows the dimension of a pre-cracked specimen using for the finite element simulation. Because of symmetric deformation, a quarter of the whole of the specimen is simulated. Figure 2 shows the 3D FE model established for the pre-cracked specimen. A hexahedral quadratic element with twenty nodes and reduced integration are used. The size of the mesh becomes smaller exponentially with approaching to crack-tip. The nodal displacement rate at the symmetrical boundary condition and the supporting point are set to be zero. The

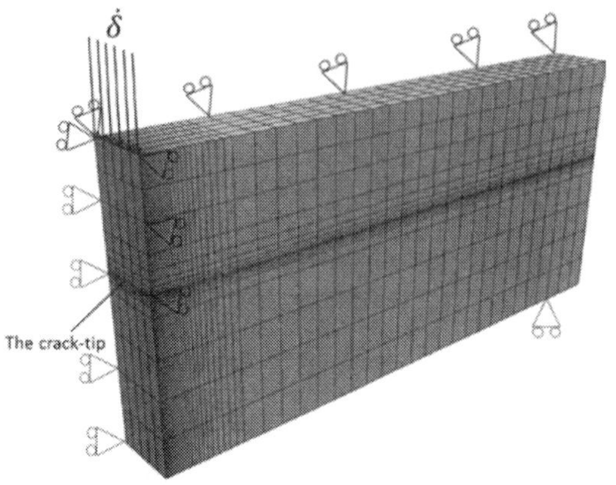

Figure 2. The 3D FE model for computational simulation.

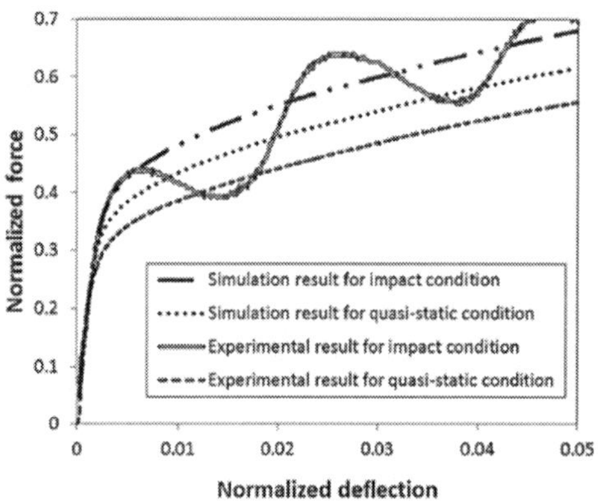

Figure 3. Normalized force vs normalized deflection under two cases of loading condition in experiment and simulation.

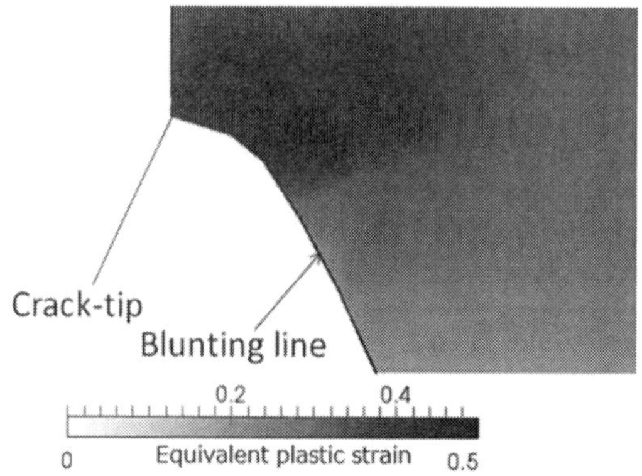

(a) Under quasi-static deformation

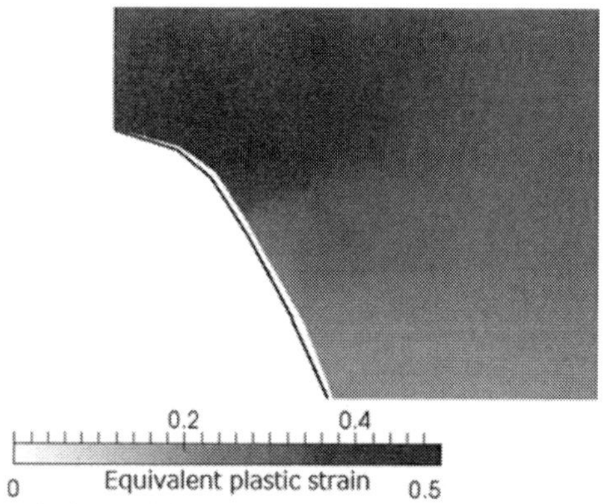

(b) Under impact deformation

Figure 4. The distribution of equivalent plastic strain under quasi-static and impact deformation at normalized deflection of 0.047 near crack-tip.

normalized deflection rate of 3.74×10^{-2} and 10.44 /s are set for simulation in the case of quasi-static and impact deformation, respectively. The environmental temperature is 298 K.

4. Results and discussion

The value of force P and deflection δ obtained from computational simulation is normalized as shown in the following equations.

$$P_n = \frac{S}{4Z\sigma_y}P, \quad \delta_n = \frac{4W}{S^2}\delta \tag{7}$$

where σ_y is the initial yield stress at room temperature, Z is the modulus of section, S is the span lengths and W is the width of the specimen.

In this study, energy-absorption is evaluated from normalized force-normalized deflection curve until the onset point of crack extension of specimen. Thus, the characteristic of normalized force-normalized deflection curve is examined. Figure 3 shows the relationship between normalized force and normalized deflection at normalized deflection rate of 3.74×10^{-2} under quasi-static deformation and 10.44 /s under impact loading deformation in experiment and simulation. Here, experimental results are obtained from Pham et al. [16]. A quite agreement between experimental and simulation results can be seen. Therefore, the validity of computation is confirmed. From this figure, it is clear that to obtain the same value of normalized deflection, external normalized force is required higher under impact condition than under quasi-static condition.

Next, the onset point of crack extension is determined through the blunting of crack tip. Figure 4 shows the distribution of equivalent plastic strain in the case of (a) quasi-static deformation, (b) impact deformation near crack-tip at normalized deflection of 0.047. It is clear that the crack has not extended yet. Here, the plastic deformation around the crack-tip leads to blunting the crack-tip. The blunting of crack-tip is larger under impact condition compare with under quasi-static condition at the same normalized deflection. As a result, the onset point of crack extension in case of impact deformation increases because the blunting of crack-tip induces resistance against crack initiation. From observations of Figs. 3 and 4, it can be said that energy-absorption in TRIP steel increases at a higher deflection rate.

Figure 5 shows the volume fraction of martensite with respect to normalized deflection. It is clear that the volume fraction of martensite under impact deformation

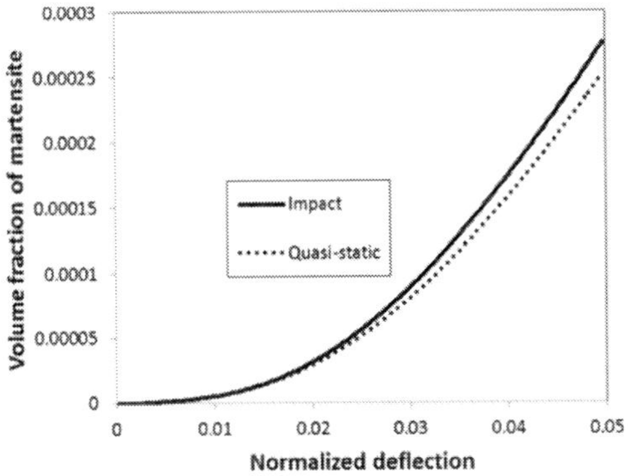

Figure 5. The relationship between volume fraction of martensite and normalized deflection under quasi-static and impact condition.

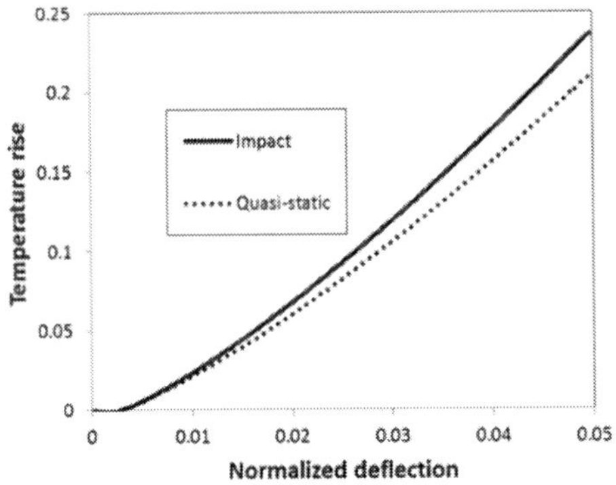

Figure 6. The relationship between temperature rise and normalized deflection under quasi-static and impact condition.

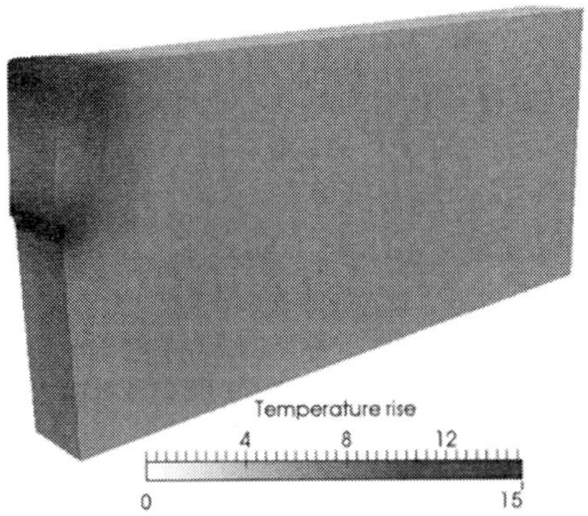

(a) Under quasi-static deformation

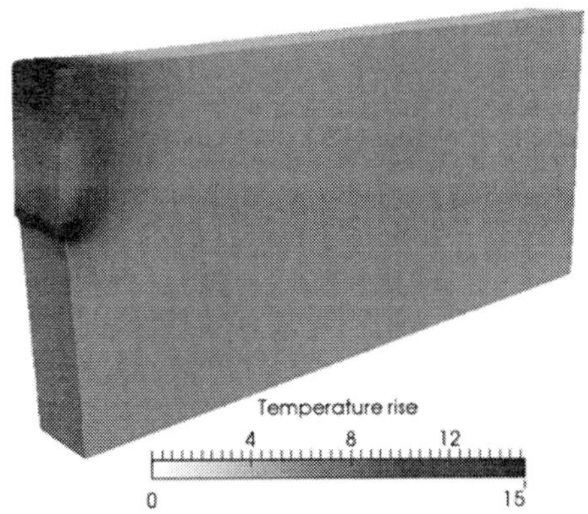

b) Under impact deformation

Figure 7. The distribution of temperature rise under quasi-static and impact deformation at normalized deflection of 0.047.

is higher than that under quasi-static deformation. A considerable value of volume fraction of martensite might be responsible for excellent energy-absorption characteristic in TRIP steel under impact loading as discussed above.

Figure 6 shows temperature rise with respect to normalized deflection for two cases of deformation. Here, temperature rise under impact deformation is higher than that under quasi-static deformation. This can be explained that at higher normalized deflection rate, adiabatic heating because of an inelastic irreversible work induces temperature rise. The distribution of temperature rise in specimen is shown in Fig. 7 in the case of (a) quasi-static deformation and (b) impact deformation at the normalized deflection of 0.047. The distribution of temperature can be seen around the crack tip and loading point. Moreover, temperature rise is significantly higher under impact condition compare with under quasi-static condition. Although temperature increases considerably during impact deformation process, a high value of the volume fraction of martensite can be seen in Fig. 5. Thus, effect of other factors, except for temperature, needs to be considered.

Figure 8 shows the distribution of triaxiality for case of (a) quasi-static deformation and (b) impact deformation at the normalized deflection of 0.047. In this figure, the triaxiality is distributed within three regions with value of -0.33, 0 and 0.33 corresponding with compression, shear and tension, respectively. Compare with case of quasi-static condition, the compressive region in impact condition is slightly larger.

Next, Fig. 9 shows the distribution of volume fraction of martensite at the normalized deflection of 0.047. Obviously, the volume fraction of martensite is distributed within compressive region, which is near the crack-tip and around the loading-point region. It can be said that compressive deformation creates more favorable situations for SIMT than tensile deformation because triaxiality is effective for shear band formation in the former case. Additionally, even though temperature rise is significantly high, the volume fraction of martensite is larger at a higher normalized deflection rate under impact loading. A larger compressive deformation as shown in Fig. 8 can be responsible for this result. Here, the stress triaxiality parameter Σ strongly effects on SIMT rather than temperature. Moreover, at higher rate of deformation, the number of shear-band intersection is induced more as expressed in Eq. (1). As a result, volume fraction of martensite in impact condition is higher in quasi-static condition as observation in Fig. 5.

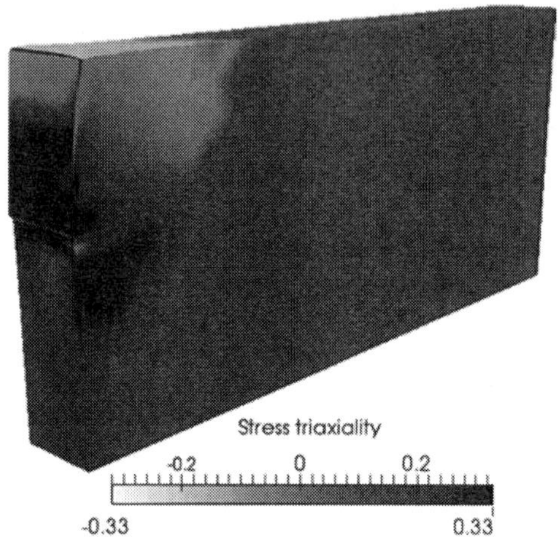

(a) Under quasi-static deformation

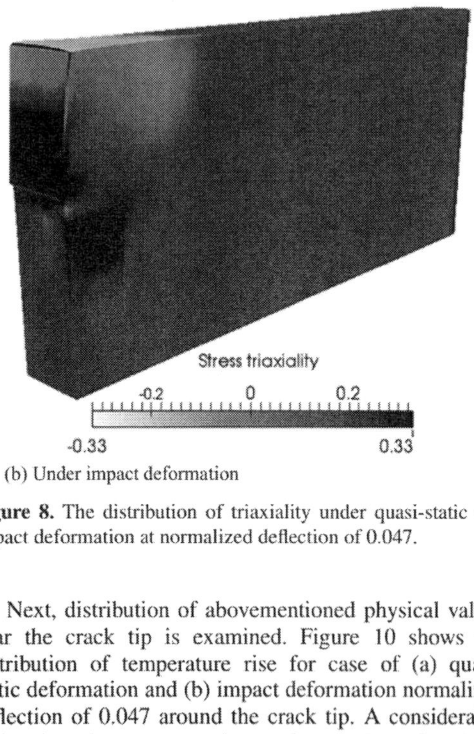

(b) Under impact deformation

Figure 8. The distribution of triaxiality under quasi-static and impact deformation at normalized deflection of 0.047.

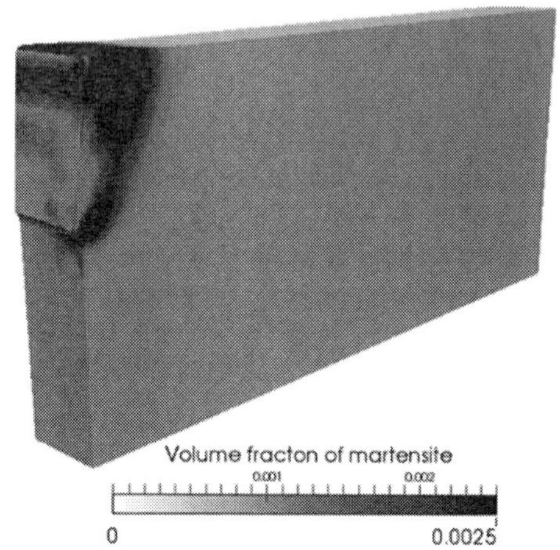

(a) Under quasi-static deformation

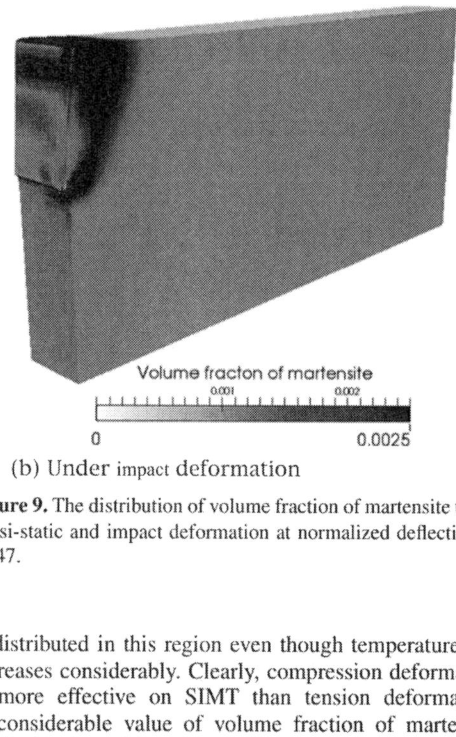

(b) Under impact deformation

Figure 9. The distribution of volume fraction of martensite under quasi-static and impact deformation at normalized deflection of 0.047.

Next, distribution of abovementioned physical values near the crack tip is examined. Figure 10 shows the distribution of temperature rise for case of (a) quasi-static deformation and (b) impact deformation normalized deflection of 0.047 around the crack tip. A considerably high value of temperature rise can be seen here. Specially, under impact deformation, temperature increases higher than M_d temperature. This might induce a softening thermal effect which leads a higher equivalent plastic strain near crack tip as shown in Fig. 4(b). Furthermore, it is expected that SIMT is suppressed under impact deformation around the crack tip because of high temperature rise.

Figure 11 shows the distribution of (a) triaxiality and (b) volume fraction of martensite near crack tip in case of impact condition at normalized deflection of 0.047. A compression region can be seen in the stretch zone of the crack tip. At the same time, volume fraction of martensite is distributed in this region even though temperature rise increases considerably. Clearly, compression deformation is more effective on SIMT than tension deformation. A considerable value of volume fraction of martensite might lead to delay the crack initiation, thus, energy-absorption of TRIP steel can be improved.

5. Concluding remarks

In this study, bending deformation behaviour of pre-cracked specimen is investigated by 3D finite element simulation based on the transformation kinetics model. The concluding remarks are introduced as follows:

(1) To obtain same normalized deflection, external force is required higher under impact condition than

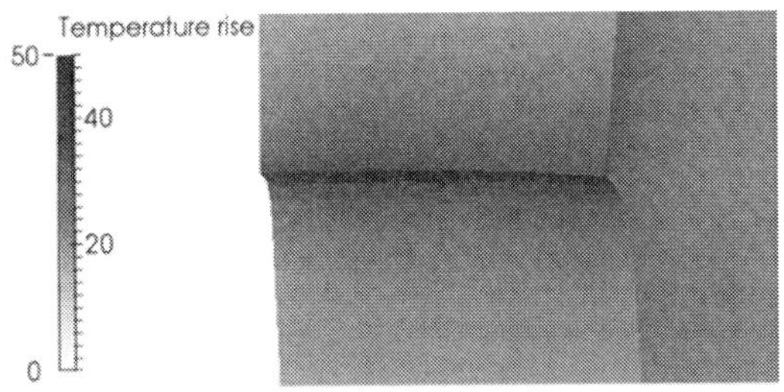

(a) Under quasi-static deformation

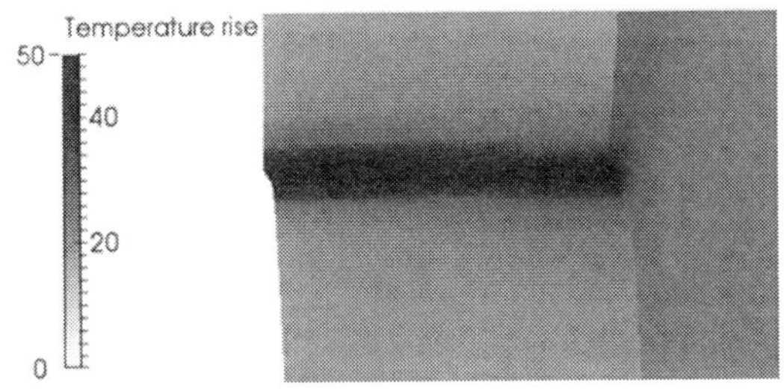

(b) Under impact deformation

Figure 10. The distribution of temperature rise under quasi-static and impact deformation at normalized deflection of 0.047.

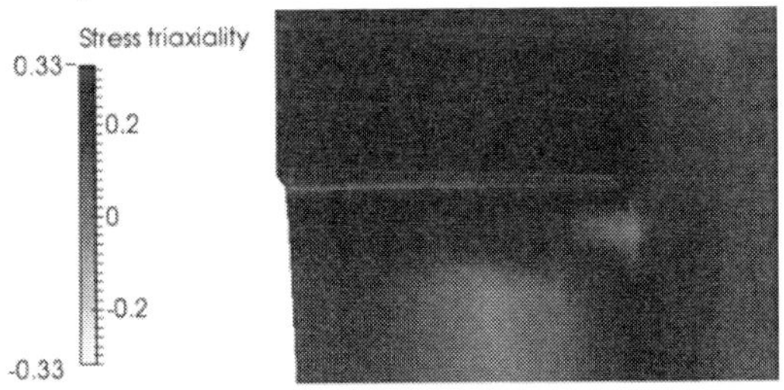

(a) The distribution of triaxiality

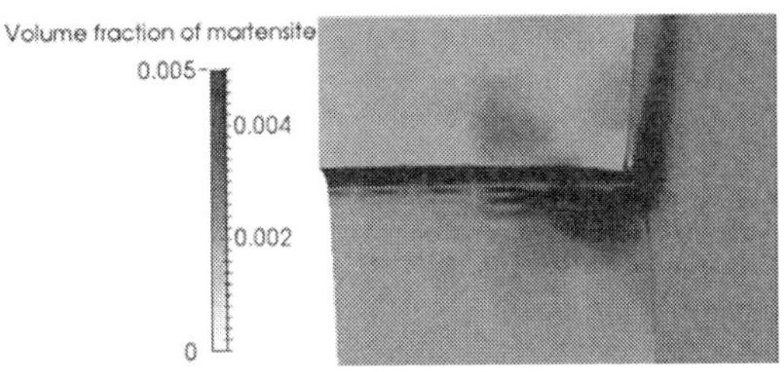

(b) The distribution of volume fraction of martensite

Figure 11. The distribution of triaxiality and volume fraction of martensite under impact deformation at normalized deflection of 0.047.

under quasi-static condition. Moreover, blunting near crack tip under impact deformation is larger than that under quasi-static deformation. Therefore, it is considered that TRIP steel indicates better energy-absorption under impact condition than under quasi-static condition.

(2) The total volume fraction of martensite under impact deformation is higher than that under quasi-static deformation. This can be responsible for excellent energy-absorption capacity in TRIP steel at a higher deflection rate.

(3) Although temperature rise increases over M_d around the crack tip under impact loading, volume fraction of martensite is distributed in compressive region. Here, triaxiality factor strongly effects on SIMT rather than temperature. Martensitic phase formed around the crack tip might lead to delay the crack initiation, thus, energy absorption of TRIP steel is improved.

We gratefully acknowledge financial supports from the 23rd ISIJ Research Promotion Grant provided by the Iron and Steel Institute of Japan.

References

[1] M. Hori, JSAE Review **25**, 3 (2004)
[2] IISI AUTOCO Report : www.worldautosteel.org
[3] I. Tamura, Met. Sci. **16**, 245 (1982)
[4] T. Iwamoto, T. Sawa, M. Cherkaoui, Int. J. Mod. Phys. B **22**, 5985 (2008)
[5] M.R. Berrahmoune, S. Berveiller, K. Inal, A. Moulin, E. Patoor, Mater. Sci. Eng. A **378**, 304 (2004)
[6] F.D. Fischer, G. Reisner, E. Werner, K. Tanaka, G. Cailletaud, T. Antretter, Int. J. Plasticity **16**, 723 (2000)
[7] S. Curtze, V.T. Kuokkala, M. Hokka, P. Peura, Mater. Sci. Eng. A **507**, 124 (2009)
[8] J.R. Rice, J. Appl. Mech. **35**, 379 (1968)
[9] ASTM E813-81, *Annual Book of ASTM Standards*, 810 (1981)
[10] G.B. Olson, M. Cohen, Metall. Trans. A **6**, 791 (1975)
[11] R.G. Stringfellow, D.M. Parks, G.B. Olson, Acta Metall. Mater. **40**, 1703 (1992)
[12] Y. Tomita, T. Iwamoto, Int. J. Mech. Sci. **37**, 1295 (1995)
[13] T. Iwamoto, T. Tsuta, Int. J. Plasticity **16**, 791 (2000)
[14] T. Iwamoto, T. Tsuta, Y. Tomita, Int. J. Mech. Sci. **40**, 173 (1998)
[15] T. Iwamoto, T. Tsuta, Int. J. Plasticity **18**, 1583 (2002)
[16] H.T. Pham, L. Shi, T. Iwamoto, Key. Mat. Eng. **626**, 340 (2014)

EPJ Web of Conferences **94**, 04013 (2015)
DOI: 10.1051/epjconf/20159404013

Cavitation instability in bulk metallic glasses

L.H. Dai[1,a], X. Huang[2], and Z. Ling[1]

[1] State Key Laboratory of Nonlinear Mechanics, Institute of Mechanics, Chinese Academy of Sciences, Beijing 100190, China

[2] Institute of Systems Engineering, China Academy of Engineering Physics, Mianyang, Sichuan 621999, China

Abstract. Recent experiments have shown that fracture surfaces of bulk metallic glasses (BMGs) usually exhibit an intriguing nanoscale corrugation like fractographic feature mediated by nanoscale void formation. We attribute the onset of this nanoscale corrugation to TTZs (tension transformation zones) mediated cavitation. In our recent study, the spall experiments of Zr-based BMG using a single-stage light gas gun were performed. To uncover the mechanisms of the spallation damage nucleation and evolution, the samples were designed to be subjected to dynamic tensile loadings of identical amplitude but with different durations by making use of the multi-stress pulse and the double-flyer techniques. It is clearly revealed that the macroscopic spall fracture in BMGs originates from the nucleation, growth and coalescence of micro-voids. Then, a microvoid nucleation model of BMGs based on free volume theory is proposed, which indicates that the nucleation of microvoids at the early stage of spallation in BMGs is resulted from diffusion and coalescence of free volume. Furthermore, a theoretical model of void growth in BMGs undergoing remote dynamic hydrostatic tension is developed. The critical condition of cavitation instability is obtained. It is found that dynamic void growth in BMGs can be well controlled by a dimensionless inertial number characterizing the competition between intrinsic and extrinsic time scales. To unveil the atomic-level mechanism of cavitation, a systematic molecular dynamics (MD) simulation of spallation behaviour of a binary metallic glass with different impact velocities was performed. It is found that micro-void nucleation is determined TTZs while the growth is controlled by shear transformation zones (STZs) at atomic scale.

1. Introduction

Bulk metallic glasses (BMGs) are a relatively young class of alloy materials, which have amorphous atomic structure induced by fast cooling from the liquid to solid states. For the excellent physical and mechanical properties such as high strength and hardness, BMGs have potential of wide applications in the engineering areas, and has raised much attention in recent years. Especially, their fracture mechanism is widely investigated. As the atomic structures are disordered, plastic deformation of this kind of materials is prone to be localized into thin shear bands. Hence the fracture process of BMGs is usually undergoing highly localized deformation via formation and rapid propagation of shear bands [1].

However, recent studies have shown that metallic glasses can also fail by cavitation instability, which is the principle ductile fracture mechanism in conventional polycrystalline materials. Bouchaud et al. [2] conducted a quasistatic tension experiment on a Zr-based metallic glass (Vit. 1) and typical dimple patterns were observed on the fracture surfaces of tested samples. Through the study of the mismatch between two opposite fracture surfaces, it is shown that the fracture is controlled by growth and coalescence of damage cavities. To reveal the fracture mechanism of bulk metallic glasses, Jiang et al. [3] performed a systematic study on fracture patterns of Vit. 1 via compression, tensile and plate impact experiments. Dimple and honeycomb structures were observed on dynamic mode I fracture surfaces, which indicated that the fracture is a void dominated process. Particularly, fine 100 nm size dimples and nanoscale periodic corrugations were found as well. Based on the fracture patterns, a new atomic motion unit called as tension transformation zone (TTZ), which may be interpreted as nucleation of nanoscale voids ahead of crack tip, was proposed. Similar phenomenon has been observed in spallation experiments studied the spallation behaviour of Zr-based metallic glass (Vit. 1) via plate-impact experiments. Equiaxed cellular patterns were found on the spall surface of recovered samples, which indicated that the spallation in the materials is induced by nucleation, growth and coalescence of micro-voids. Moreover, this micro-voids dominated fracture process has also been observed in atomic scale during molecular dynamics simulations. Arman et al. [4] investigated the dynamic response of a CuZr metallic glass with MD simulations and found that spallation process of this material is accompanied by nucleation and growth of micro-voids. In order to reveal the intrinsic mechanism that governs the fracture process in different metallic glasses, Murali et al. [5] performed MD simulations on two typical metallic glasses which were brittle FeP and ductile CuZr. Their work showed that even the brittle FeP glass can fracture through nucleation and coalescence of multiple voids for the high degree of atomic scale spatial fluctuations.

It is clear that void dominated fracture process is a common failure mode in metallic glasses. To understand the fracture mechanism, our recent experimental results

[a] Corresponding author: lhdai@lnm.imech.ac.cn

Article available at http://www.epj-conferences.org or http://dx.doi.org/10.1051/epjconf/20159404013

are reported in this paper. Some theoretical models are proposed to reveal the void nucleation and growth mechanism. Furthermore, MD simulations of spallation behaviour of a binary metallic glass are perform, which show the cavitation mechanism at atomic level.

2. Spallation experiments

To obtain the physical picture of cavitation instability, spallation experiments are conducted on a single-stage light gas gun. The sample material we used was a Zr-based amorphous alloy with the nominal composition $Zr_{41.2}Ti_{13.8}Cu_{12.5}Ni_{1}0Be_{22.5}$ (Vit. 1). The flyer material was H62 brass, which is similar to C28000 brass (e.g., 62% Cu and 38% Zn).

To obtain stress pulses with the identical amplitude and different durations, a double-flyer technique is used. With this technique, the samples are impacted by two flyers, and the stress durations are controlled accurately by the gap between them. In our experiments, the impact velocity was 200 m/s, corresponding to a stress amplitude of ~3.2 GPa. And the stress durations are controlled to be 125 ns, 250 ns and 372 ns.

Close-up observations with a high resolution scanning electronic microscopy (HRSEM) (FEI-Sirion NC HRSEM with 1.5 nm resolution) were performed to recovered samples. The spallation damage evolution process was observe in the recovered BMG samples. While the stress duration increases, there are more micro-cracks observed on the cross-section. Details of the cracks shows that they are formed due to nucleation, growth and coalescence of micro-voids. Now, the question is what is mechanism of void nucleation and evolution in BMGs.

3. Void nucleation mechanism

To understand the void nucleation mechanism, we proposed a we propose a void nucleation model of amorphous alloys developed from traditional stress temperature-driven models widely used in polycrystalline materials [6]. We also consider free-volume theory to understand and characterize microvoid nucleation in BMGs.

In our model, the process of void nucleation can be divided into three stages: First, the frozen free volume is activated by applied tensile stress; second, free volume diffuses in the material and causes a net diffusional coalescence; and third, a void that is larger than a critical size forms and can grows under tensile loading as shown in Fig. 1. At this stage, the void is nucleated.

Based on the model, we derived the nucleation rate theoretically written as [6]:

$$\dot{N} = \frac{A}{\sigma_T^2}\exp\left(B\sigma_T - \frac{C}{\sigma_T^3}\right) \tag{1}$$

where σ_T the applied tensile stress under uniaxial strain condition, A is a material constant at a certain temperature, B describes the influence of tensile loading on the activation barrier, C is a coefficient associated with the influence of tensile loading on the number of subcritical nuclei.

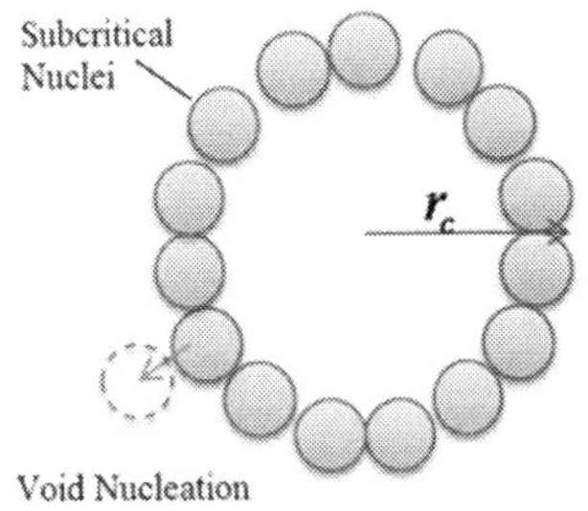

Figure 1. Schematic of void nucleation [6].

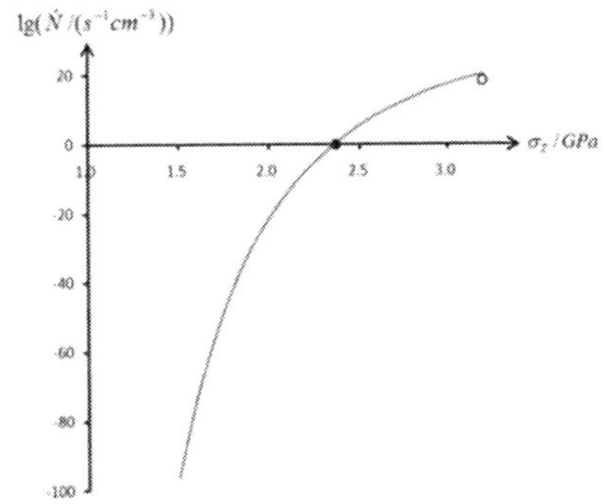

Figure 2. Schematic of void nucleation [6].

To test the reliability of our void nucleation model, we choose the BMG Vit. 1 as a model material. Using typical values of the material parameters, we obtain the curve of nucleation rate as a function of the applied tensile stress, as shown in Fig. 2.

If the value zero on the vertical scale in Fig. 2 is designated as the critical value below which void nucleation is so slow that it can be neglected, the threshold stress of void nucleation for Vit. 1 is about 2.35 GPa, a little higher than the quasistatic tensile strength of Vit.1, 1.86 GPa. As the spall strength is usually thought to be higher than the quasistatic tensile strength because of the loading rate effect, the threshold stress obtained by Eq. (3) is reasonable. We also compared the calculated result with our experimental data. From the observed cellular pattern, we estimated the nucleation rate (marked as a white point in Fig. 2) in our spallation experiments. It is obvious that the theoretical result obtained by Eq. (1) agrees quite well with the experimental data.

4. Void growth mechanism

Next, void growth in BMGs under remote tensile loading was investigated [7].

4.1. Basic model

We consider a spherical void of radius a in an infinite metallic glass under the remote hydrostatic tensile pressure p^{app}, as shown in Fig. 3. The void grows as the hydrostatic tensile pressure is applied. It is assumed that the void remains spherical throughout the growth process. Once the void surface yields, the elastic-viscoplastic boundary is considered as a spherical surface propagating outwards.

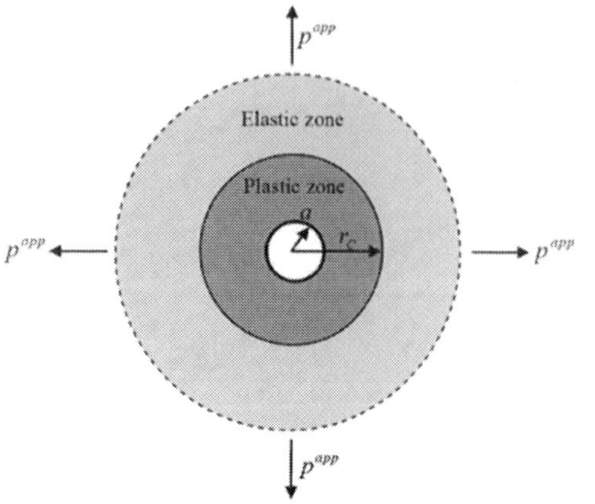

Figure 3. Schematic of void growth under tensile pressure [7].

Then, according to continuum mechanics and void growth model in conventional polycrystalline materials, we can write the governing equations [7]. Here we just give some key equations.

To describe the matrix material (BMG), an elastic-viscoplastic constitutive law is adopted as:

$$\varepsilon_{ij} = \frac{s_{ij}}{2\mu} \text{ if } \tau + Qp \leq \hat{\tau} - (C_1 T/T_g)^{1/2} \tag{2}$$

$$\dot{\varepsilon}_{ij} = \frac{\dot{s}_{ij}}{2\mu} + \frac{s_{ij}}{2\eta} \text{ if } \tau + Qp \geq \hat{\tau} - (C_1 T/T_g)^{1/2} \tag{3}$$

where ε_{ij} is the strain tensor, s_{ij} is the stress deviator, μ is the shear modulus, and η is the viscosity. To characterize the pressure sensitivity in plastic flow, the yield criterion of BMGs suggested by Sun et al. [8] is used. In this criterion, $\tau = (\sigma_\theta - \sigma_r)/2$ is the maximum shear stress (σ_θ and σ_r are the principle stresses), $p = (\sigma_r + 2\sigma_\theta)/3$ is the hydrostatic pressure, Q is the pressure sensitivity coefficient and the term $\hat{\tau} - (C_1 T/T_g)^{1/2}$ represents the yield strength ($\hat{\tau}$ is the barrier shear resistance of a STZ, C_1 is a coefficient that reflects the temperature dependence of strength, T is the temperature, and T_g is the glass transition temperature).

And the stress dependent viscosity is defined as:

$$\eta \equiv \frac{\tau_e}{\dot{\gamma}^p} = \frac{\tau_e}{2f \exp\left(-\frac{\Delta G^m}{k_B T}\right) \sinh\left(\frac{\tau^e \Omega}{2k_B T}\right) \exp\left(-\frac{1}{\xi}\right)} \tag{4}$$

where $\tau_e = \sqrt{J_2} = \sqrt{s_{ij}s_{ij}/2}$ is the effective shear stress, $\dot{\gamma}^p$ is the plastic strain rate, f is the frequency of atomic vibration (~Debye frequency), ΔG^m is the activation energy, k_B is the Boltzmann constant, Ω is the atomic volume, and ξ is the concentration of free volume ($\xi = v_f/\chi v^*$, here v_f, χ and v^* are, respectively, the free volume, a geometric factor and the effective hard-sphere size of an atom).

The free volume concentration is adopted as an order parameter, which is a function of the radius r in the current configuration and the time t. The free volume evolution equation is written as:

$$\frac{\partial \xi}{\partial t} = D\nabla^2 \xi + G(\xi, T, \tau, p) \tag{5}$$

where D is the diffusion coefficient of free volume concentration and $G(\xi, T, \tau, p)$ is the net generation rate of free volume.

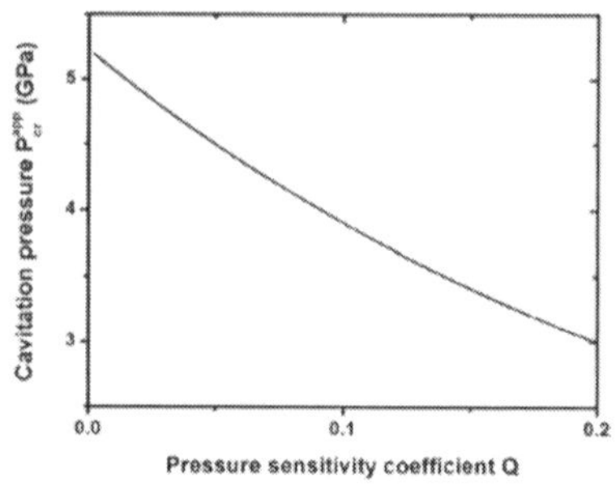

Figure 4. The cavitation pressure as a function of pressure sensitivity coefficient [7].

4.2. Cavitation instabilities

Cavitation instability can be regarded as the unbounded growth of a pre-existing void. Usually a critical pressure determines whether cavitation instability occurs. To get out this criterion, we consider the quasistatic case, and derived the critical pressure in BMGs as follows [7]:

$$p_{cr}^{app} = \frac{\hat{\tau} - (C_1 T/T_g)^{1/2}}{Q} - \frac{1}{Q\mu^{4Q/3}} \left(\frac{3\hat{\tau} - 3(C_1 T/T_g)^{1/2}}{3 + 4Q}\right)^{(3+4Q)/3}. \tag{6}$$

Using the material parameters of a typical metallic glass Vit. 1, the critical pressure for cavitation instabilities versus the pressure sensitivity coefficient is shown in Fig. 4. We found that the cavitation pressure decreases as the pressure sensitivity coefficient increases.

4.3. Dynamic growth of voids

To reveal the dominant factor that controls the void growth, we proposed an dimensionless number by scaling the governing equations [7]:

$$I_{inertia} = \frac{\sqrt{\rho A^2/p_S}}{t_a} + \frac{\sqrt{\rho A^2/p_S}}{\eta/\mu}. \tag{7}$$

Here, $I_{inertia}$ is a dimensionless number similar to the Deborah number and consists of three characteristic time scales: (1) the inertial time scale $t_{inertia} = A/\sqrt{p_S/\rho}$, a characteristic time scale of the outward flux of matter; (2) the loading time scale t_a, which is just the rise time of the loading history; and (3) the relaxation time scale $t_r = \eta/\mu$, a characteristic time for viscous flow. The three time scales represent inertial, loading rate and viscous effects respectively.

To quantify the effects of dominant factors, finite difference method simulations on dynamic void growth were carried out. The loading which is applied on the outer boundary is divided to two stages: the rise stage in which the applied pressure increases linearly until the desired loading p_S is achieved after a rise time t_a, and the steady stage in which the loading is held constant at p_S during the hold time t_S. It is found that at the rise stage of loading history, the competition between the inertial effects and the loading rate effects controls the growth process. While the inertial effects become more dominant, the vibration of growth rate is observed. As the viscous effects begin to work, the loading rate effects are gradually replaced, and the viscous effects will induce higher growth rate and disappearance of vibrating growth. At the steady stage, the competition between the inertial effects and the viscous effects controls the growth process. While the inertial effects become dominant, the growth of void is obviously impeded.

5. Atomic-level mechanism of cavitation

To unveil the atomic-level mechanism of cavitation, a systematic molecular dynamics (MD) simulation of spallation behavior of a binary metallic glass was performed [9].

5.1. MD simulation of spallation

We choose a simple binary MG $Zr_{50}Cu_{50}$ as the model material. To describe the atomic interactions, the Finnis-Sinclair type interatomic potential with parameters given by Mendelev et al. [10] is used. Calculations are carried out with the open source code LAMMPS. For simulations of spallation, we construct the traditional flyer-target configurations. The size of flyer plate is about $100 \times 20 \times 20\,\mathrm{nm}^3$, and the target has the same cross-section area ($20\,\mathrm{nm} \times 20\,\mathrm{nm}$) but its thickness is twice as that of flyer. To achieve shock loading with different amplitudes, we choose different impact velocities V of 600, 900, 1200, 1500, and 1800 m/s respectively.

Figure 5 illustrates the free surface velocity histories on the target side. Typical "pull-back" waves, which are signatures for spallation, are observed in all cases, except the case of V = 600 m/s. It indicates that spallation has occurred in the cases of V = 900, 1200, 1500, and 1800 m/s. To reveal the spallation mechanism at atomic level, the case of V = 900m/s is further characterized.

5.2. Atomic-level mechanism

Figure 6 shows the damage evolution process at V = 900 m/s, and the colours represent value of $D^2_{\min}$ calculated from the same reference (t = 81 ps). It is clear that the spallation occurs via nucleation and growth of voids in the MG.

Here the critical size for void nucleation is determined to be ~1 nm in diameter. Thus, according to the void size, Figs. 6(a–c) illustrate the nucleation process, but Figs. 6(d–e) exhibit the growth process. As shown in Figs. 6(a–c), during the void nucleation process, the atoms with a larger nonaffine displacement are randomly

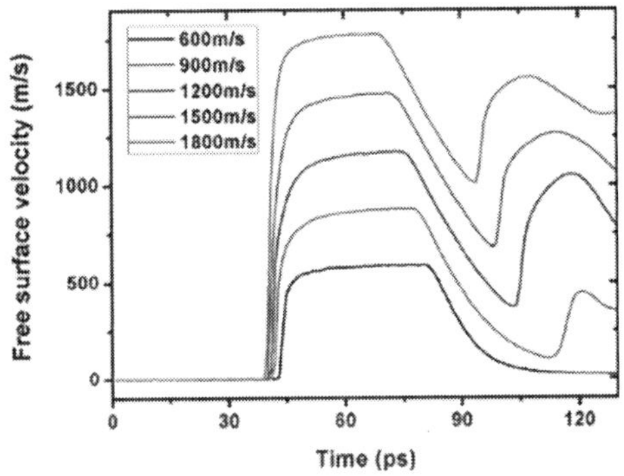

Figure 5. Damage evolution process at V = 900 m/s [9].

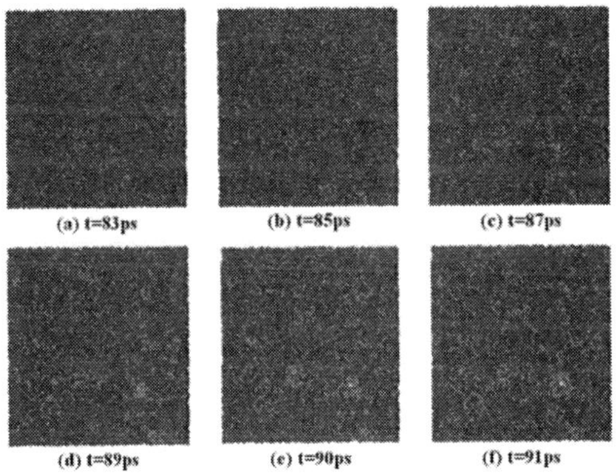

Figure 6. Damage evolution process at V = 900 m/s [9].

distributed in the material. With increasing the time interval (the reference configuration is the same), the number of atoms with larger nonaffine displacement increases. No apparent difference is observed between the location where void nucleation occurs and other region. It implies that the nonaffine displacement is induced by temperature (or structural relaxation) instead of stress. However, during the void growth process, nonaffine displacement of atoms in the region around the void is much larger than that away from the void. It indicates that plastic deformation of the material is mainly induced by void growth, there is nearly no contribution from void nucleation.

Further investigation reveals that the nucleation and growth of voids is closely related to the fundamental unit-processes of collective atomic motion in MGs. Figure 7 shows some close-up views of the atomic cluster motion around the void. As shown in Fig. 7(a), during the nucleation process, transformation of the atomic structure at the centre of the void is corresponding to the picture of tension transformation zone (TTZ) proposed by Jiang et al. [3] But during the growth process as shown in Fig. 8(b), the motion of the atomic cluster at the edge of the void is close to the picture of shear transformation zone (STZ) proposed by Argon et al. [11] Thus, we argue that cavitation instability originates from TTZs and develops from STZs at atomic scale.

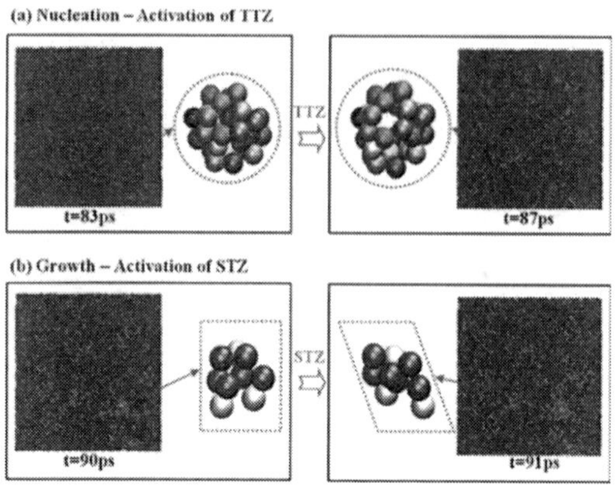

Figure 7. Motion of atomic clusters during the damage evolution process at V = 900 m/s: (a) nucleation; and (b) growth [9].

6. Conclusion

In this paper, we have investigated cavitation instability in MGs. Our recent experimental results shows that spallation in BMGs is due to nucleation, growth and coalescence of micro-voids. Based on free volume theory, a void nucleation model is proposed, and the nucleation can be interpreted as a diffusion and coalescence process of free volume. Theoretical and numerical study of void growth reveals that the void growth rate is controlled by inertial, loading rate and viscous effects. Via MD simulations, we found that micro-void nucleation is determined TTZs while the growth is controlled by shear transformation zones (STZs) at atomic scale.

Financial support was from the National Key Basic Research Program of China (2012CB937500), the NSFC (Grants Nos.: 11272328, 11472287, and 11402245), and the CAS/SAFEA International Partnership Program for Creative Research Teams.

References

[1] L.H. Dai, M. Yan, L.F. Liu and Y.L. Bai, Appl. Phys. Lett. **87** (14), 141916 (2005).
[2] E. Bouchaud, D. Boivin, J.L. Pouchou, D. Bonamy, B. Poon and G. Ravichandran, Europhys. Lett. **83** (6), 66006 (2008).
[3] M.Q. Jiang, Z. Ling, J.X. Meng and L.H. Dai, Philos. Mag. **88** (3), 407–426 (2008).
[4] B. Arman, S.-N. Luo, T.C. Germann and T. Çağın, Phys. Rev. B **81** (14), 144201 (2010).
[5] P. Murali, T. Guo, Y. Zhang, R. Narasimhan, Y. Li and H. Gao, Phys. Rev. Lett. **107** (21), 215501 (2011).
[6] X. Huang, Z. Ling, H.S. Zhang, J. Ma and L.H. Dai, J. Appl. Phys. **110** (10), 103519 (2011).
[7] X. Huang, Z. Ling and L.H. Dai, Int. J. Solids Struct. **50** (9), 1364–1372 (2013).
[8] L. Sun, M.Q. Jiang and L.H. Dai, Scripta Mater. **63** (9), 945–948 (2010).
[9] X. Huang, Z. Ling and L.H. Dai, J. Appl. Phys. **116**, 143503 (2014).
[10] M.I. Mendelev, D.J. Sordelet and M.J. Kramer, J. Appl. Phys. **102**(2), 043501 (2007).
[11] A.S. Argon, Acta Metall. **27**(1), 47–58 (1979).

EPJ Web of Conferences **94**, 04014 (2015)
DOI: 10.1051/epjconf/20159404014

A constitutive model for the compressive response of metallic closed-cell foams including micro-inertia effects

Romain Barthélémy[1,2,a], Nicolas Jacques[2], François Vermeersch[1], and Steven Kerampran[2]

[1] MBDA France, 1 avenue Réumur, 92350 Le Plessis-Robinson, France
[2] ENSTA Bretagne, LBMS, 2 rue François Verny, 29800 Brest, France

Abstract. Metallic foams have known a keen interest in the last decades. Their ability to undergo very large deformations while transmitting low stress levels make them capable of performing functions of protective layers against intense loadings and of energy absorbers, for instance. The behaviour of metal foams varies considerably between quasi-static and dynamic regimes. Those differences can be linked to the strain-rate sensitivity of the skeleton material and to micro-inertial effects (induced by the crushing of the foam cells).

In the present work, a micromechanical model has been developed to take into account micro-inertia effects on the macroscopic behaviour of closed-cell foams under dynamic loading conditions. The proposed modelling is based on the dynamic homogenisation procedure introduced by Molinari and Mercier (J. Mech. Phys. Solids 49 (2001) 1497–1516). Within this framework, the macrostress is the sum of two terms. The first one is a static stress, that can be described with any existing model of metal foam. The second contribution is a dynamic stress related to micro-inertia effects. Considering an initially spherical shell as a Representative Volume Element (RVE) of the foam material, a closed-form expression of the dynamic stress was obtained. The proposed modelling was applied to shock propagation in aluminium foams (it should however be noted that the present theory is not restricted to uniaxial deformation but can be applied to arbitrary loadings). From experimental data of the literature, it is observed that incorporating micro-inertia effects allows one to achieve a better description of the foam shock response. This indicates that micro-inertia may have a significant influence on the dynamic behaviour of metallic foams.

1. Introduction

Among recent materials, cellular ones have known large developments in the last decades. They are found at natural state (wood, cancellous bones, sponges, cork...) or man-made (bread, foams...). Cellular solids are often classified according to their relative density (the ratio of the density of the foam sample to the one of its constitutive material). Cellular materials with high relative density are called "porous media" while "foam" corresponds to materials with a relative density lower than 0.3 [1]. The latter is the focus of this study.

Foam skeletons can be made of different materials such as polymer, ceramic, glass or metal and present two types of structure (open-cell or closed-cell), see Fig. 1.

Metallic foams have many attractive features which make them reliable for applications such as vibration suppression, heat transfer, electromagnetic shielding or blast protection [2]. The last application, which is of our concern, involves high strain-rate deformation.

The quasi-static behaviour of metal foams is well-known and documented [3–9]. Some studies deal with the high strain-rate response of cellular materials [10–12]. A common conclusion of these studies is that metal foams exhibit some strain-rate sensitivity. One can also find, in literature, studies regarding the origins of this phenomenon [13–15]. Zhao & al. [16] give four elements susceptible to explain the dependency of metallic foams to strain-rates:

- the rise in pressure of the internal fluid (particularly for closed-cell foams);
- the strain-rate sensitivity of the constitutive material of the skeleton;
- the shock-enhancement mechanism [14];
- micro-inertia effects appearing when foam cells are rapidly crushed.

The influence of the entrapped gas is negligible in the case of metal foam (because of the high strength of the skeleton material). Material strain-rate sensitivity (viscoplasticity) and the shock enhancement mechanism can be easily accounted for by numerical simulations using standard explicit codes. This is not the same for micro-inertia effects. Indeed, the role of this mechanism is not fully understood and, to our knowledge, constitutive models accounting for microscale inertia have been developed only for open-cell foams [17].

The aim of the present work is to investigate the influence of micro-inertia on the shock response of closed-cell metal foams. For this purpose, a new model based on micromechanics has been developed. It will be shown that taking micro-inertia effects into account when modelling aluminium foams leads to a better agreement with published experimental data.

[a] Corresponding author: romain.barthelemy@mbda-systems.com

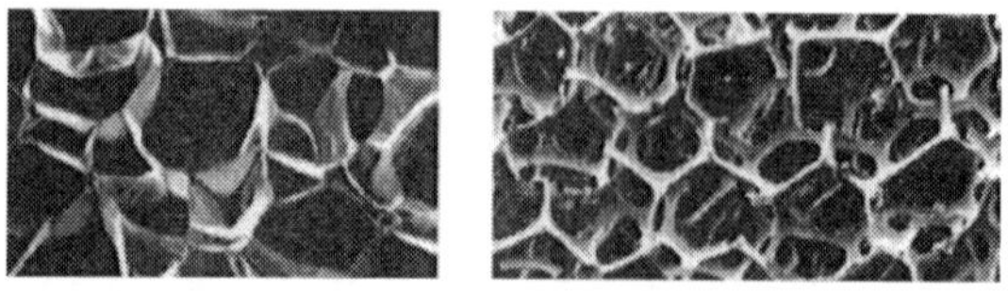

Figure 1. Microstructures of a closed-cell (on the left) and an open-cell (on the right) metallic foam [1].

2. Modelling of metal foam behaviour under uniaxial compression

2.1. Micro-inertia effects

To account for micro-inertia effects, an original model is proposed. This model has been developed using a micromechanical analysis and the dynamic homogenization procedure introduced by Molinari and Mercier [18] (see also Sartori & al. [19]). The aim of this procedure is to link microscopic quantities (written in lower case letters hereafter) to macroscopic ones (written in capital letters) through averaging over a Representative Volume Element (RVE). In the present case, the RVE is chosen to be an initially spherical shell.

Within the framework proposed by Molinari and Mercier [18], the macroscopic (nominal) stress tensor P is defined as:

$$\underline{\underline{P}} = < \underline{\underline{p}} >_0 + < \rho_0 \underline{\ddot{u}} \otimes \underline{x_0} >_0, \tag{1}$$

where ρ_0 denotes the initial density of the foam skeleton material, x_0 is the initial position of a material point in the RVE and $\underline{\ddot{u}}$ its acceleration. $< . >_0$ stands for the averaging operator:

$$< X >_0 = \frac{1}{\Omega_0} \int_{\Omega_0} X . d\Omega_0. \tag{2}$$

As discussed in [19], Eq. (1) is not appropriate to obtain a closed-form expression of the macrostress as the microscopic acceleration and stress field are in general unknown. A more convenient way to characterize the macroscopic response of the material is to use the virtual work principle, combined with an approximate trial velocity field. Indeed, the macroscopic stress (as defined by Eq. (1)) is conjugated with the macroscopic velocity gradient $\underline{\underline{F}}$ through the virtual work principle:

$$\delta \underline{\underline{F}}^T : \underline{\underline{P}} = < \delta \underline{\underline{f}}^T : \underline{\underline{p}} >_0 + < \rho_0 \delta \underline{u} . \underline{\ddot{u}} >_0. \tag{3}$$

From this relation, the macrostress can be written as the sum of two components, a static stress $\underline{\underline{P^{sta}}}$ related to the constitutive response of the skeleton material and a dynamic (or inertial) stress $\underline{\underline{P^{dyn}}}$ induced by the local acceleration of the material within the RVE:

$$\underline{\underline{P}} = \underline{\underline{P^{sta}}} + \underline{\underline{P^{dyn}}}. \tag{4}$$

We consider as a Representative Volume Element of the foam material an initially spherical shell of radius Rc and the following acceleration and virtual displacement fields are considered:

$$\underline{\ddot{u}} = \underline{\underline{\ddot{F}}} . \underline{x_0} \quad \text{and} \quad \delta \underline{u} = \delta \underline{\underline{F}} . \underline{x_0}. \tag{5}$$

Finally, the following expression for the dynamic stress is obtained:

$$\underline{\underline{P^{dyn}}} = \frac{\overline{\rho_0}}{3} . R_c^2 . \underline{\underline{\ddot{F}}}^T, \tag{6}$$

where $\overline{\rho_0}$ is the initial average mass density of the foam material. We observe that the dynamic stress also depends on the initial radius R_c of the foam cells. This quantity can be identified from micrographic observations of foam microstructure. More details about the derivation of the micro-inertia model will be given in a future paper.

Because the present paper is focused on the propagation of planar shocks, we consider in the following a uniaxial deformation state of the form:

$$\underline{\underline{F}} = \begin{bmatrix} F & & \\ & 1 & \\ & & 1 \end{bmatrix}. \tag{7}$$

In this case, the component of the (nominal or Cauchy) stress in the shock propagation direction is given by:

$$\Sigma^{dyn} = P^{dyn} = \frac{\overline{\rho_0}}{3} . R_c^2 . \ddot{F}. \tag{8}$$

2.2. Static response (uniaxial deformation)

In the present study, we do not intend to develop a micromechanical model to describe the quasi-static stress. Indeed, the static behaviour of foam materials can be easily characterized using standard experimental techniques. For this reason, it is believed that it is better to employ a phenomenological constitutive model.

The quasi-static compressive behaviour of metal foams generally exhibits three phases: an elastic one observed in the early deformation, a "plateau" regime in which the stress is almost constant and a densification (or hardening) phase with a strong increase in stress. Several simplified models exist in the literature (see the Rigid Perfectly Plastic Locking model [20] and the Elastic Perfectly Plastic Hardening model [21] for example). In the present work, an elastic-plastic model with non-linear isotropic hardening has been used. This model is based on the following hypothesis:

- the logarithmic strain $\epsilon = \ln(F)$ is used;
- the strain is described as the sum of an elastic (ϵ_e) and a plastic (ϵ_p) part;
- the elastic strain is calculated by using the following linear relation $\epsilon_e = \Sigma / E^*$;
- the plastic part $\dot{\epsilon}_p$ evolves as $\dot{\epsilon}_p = \dot{p} . sgn(\Sigma)$;
- the effective plastic strain $\dot{p}$ is defined by the Kuhn-Tucker complementary conditions [22]:

$$\dot{p} \geq 0, \quad |\Sigma| - \Sigma_y \leq 0, \quad (|\Sigma| - \Sigma_y) . \dot{p} = 0; \tag{9}$$

- the evolution of the yield stress Σ_y of the foam as a function of the effective plastic strain p is given by the following relationships:

$$\Sigma_y = \Sigma_{pl} \ \text{if} \ p \leq \epsilon_d \tag{10}$$

$$\Sigma_y = \Sigma_{pl} + K_p . (p - \epsilon_d)^n . \frac{\epsilon_l - \epsilon_d}{\epsilon_l - p} \ \text{if} \ p > \epsilon_d. \tag{11}$$

Table 1. Parameters of the elastic plastic model with non linear hardening corresponding to an Alporas® foam of initial relative density 0.11

Parameter's name	Description	Value
E^*	Elastic modulus	3.120 GPa
Σ_{pl}	Initial yield stress	3 MPa
ϵ_d	Densification strain	0.6328
K_p	Hardening modulus	24.6 MPa
n	Hardening exponent	2.11
ϵ_l	Locking strain	4.16

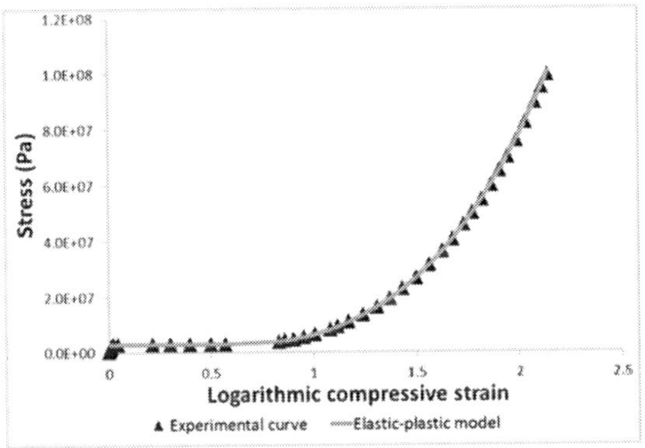

Figure 2. Experimental quasi-static compressive stress-strain curve of an Alporas® foam of initial relative density 0.11 [23] along with the prediction of the elastic-plastic model with non-linear hardening.

The considered model involves six parameters which are the elastic modulus E^*, the initial yield stress Σ_{pl}, the densification strain ϵ_d, the hardening modulus K_p, the hardening exponent n and the locking strain ϵ_l. We have identified these parameters for a closed-cell aluminium foam of initial density 0.11 (Alporas® foam produced by Shinko Wire Co. Ltd.) using the experimental data of Radford et al. [23].

These authors conducted uniaxial compressive tests on cylindrical foam samples at a nominal strain-rate of $10^{-3}\,s^{-1}$. The identification of the model parameters was carried out using a least-square fitting technique. The obtained parameters are presented in Table 1. Figure 2 shows a comparison between the experimental stress-strain curve and the prediction based on the elastic plastic model with non linear hardening.

3. Results

3.1. Presentation of the considered configuration

In order to assess the accuracy of the proposed model, we have used the results of the impact tests carried out by Radford et al. [23]. The experimental set-up is presented in Fig. 3: a cylindrical foam projectile is launched by a gas gun and impacts a steel bar initially at rest. The foam samples had a diameter of 28 mm and a length of 50 or 100 mm. Impact velocities ranging from $50\,m\,s^{-1}$ to $500\,m\,s^{-1}$ were considered. At the impact against the bar, a shock wave is generated and propagates across the

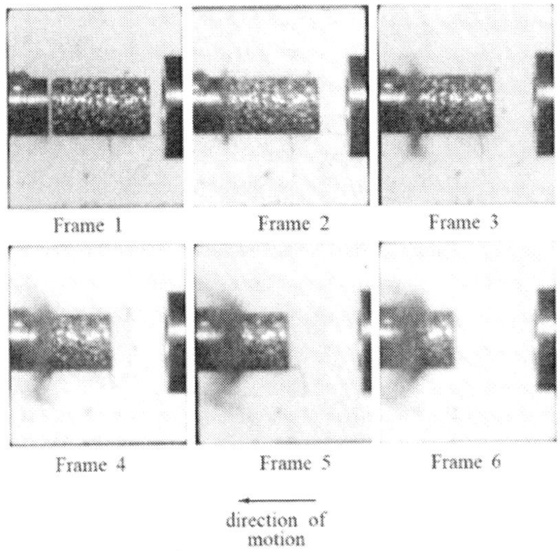

Figure 3. Photographic sequence of the impact of an Alporas® sample against a steel bar at about $380\,m\,s^{-1}$. Inter-frame time: $20\,\mu s$. From [23].

foam sample. Strain gauges were mounted on the bar at about 285 mm of the impact surface. The authors deduced the time-evolution of the contact pressure from the strain gauge recordings.

From Fig. 3, no lateral expansion of the specimen is observed during the shock compaction of the sample. This behaviour can be related to the results presented by Deshpande and Fleck [24], who observed that the plastic Poisson ratio of low-density metal foams is close to zero.

3.2. One-dimensional finite element simulations

Standard finite element dynamic codes do not account for micro-inertia effects. For this reason, we have developed a one-dimensional finite element code in Fortran 90 to simulate the impact tests described in the previous section. Standard linear finite elements are used to approximate the displacement and velocity fields. The time integration is based on the central difference scheme.

It is worth emphasizing that the present simulations do not involve any adjustable parameters. Indeed, the parameters related to the static response of the foam material were identified from a single quasi-static compression curve (see Sect. 2.2). The dynamic stress (micro-inertia effects) depends on the radius R_c of the cells of the foam microstructure (in the modelling, all cells are assumed to have the same size). From micrographic observations, Radford et al. [23] indicated that the average cell size of the Alporas® foam is about 4 mm. Thus, we have taken $R_c = 2$ mm.

In the computations, the output bar is not meshed, but is modelled using an impedance boundary condition (assuming a linear elastic response of its constitutive material). Some precautions should be taken in order to make consistent comparisons between numerical and experimental results. Indeed, as explained previously, in experiments the contact pressure history is deduced from

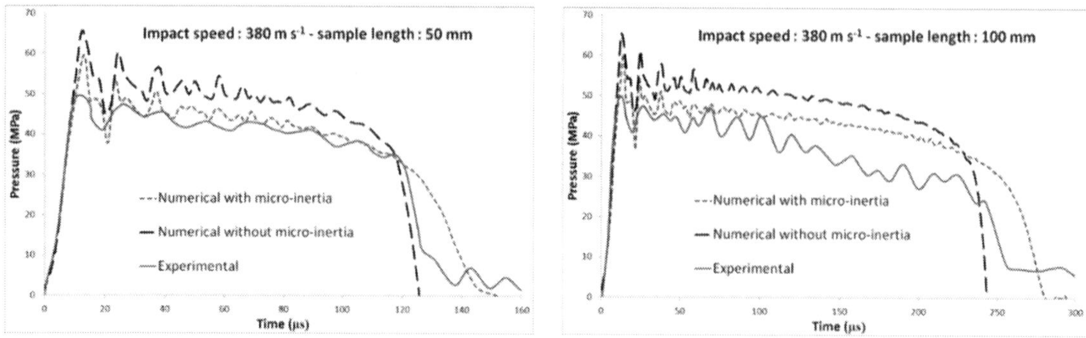

Figure 4. Influence of wave dispersion in the steel bar. This figure presents a comparison between the contact pressure history applied at the end of the bar and the stress history at 285 mm of this surface, as predicted by an axisymmetric Autodyn® simulation.

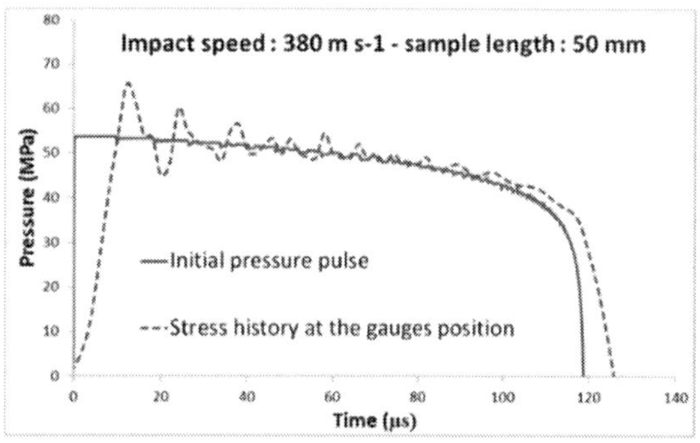

Figure 5. Impact of aluminium foam samples of lengths 50 and 100 mm at 380 m s^{-1}. Contact pressure histories obtained with one-dimensional finite element computations, with and without taking micro-inertia into account, are compared to the experimental data of Radford et al. [23].

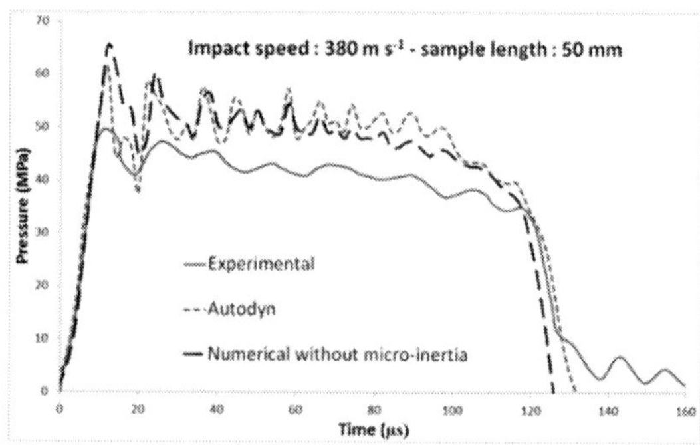

Figure 6. Comparison of experimental measurements and numerical predictions (Autodyn® and Fortran without micro-inertia) for the impact of an Alporas® foam sample of length 50 mm at a velocity of 380 m s^{-1}.

signals of strain gauges placed at about 285 mm of the impact surface. It is well known that the propagation of elastic waves in cylindrical bar is dispersive, meaning that some distortion of the stress pulse occurs during propagation.

In order to analyse the influence of wave dispersion in the experiments of Radford et al. [23], we have carried out axisymmetric simulations of wave propagation in the steel bar using the Autodyn® code. The contact pressure history derived from the Fortran code is applied on the end of the bar and we analyse the evolution of the stress pulse along the bar. Figure 4 shows a comparison between the pressure pulse applied on the end of the bar and the axial stress history observed at the position of the strain gauges (285 mm of the end). It appears that wave dispersion induces significant fluctuations of the stress. All the results presented hereafter take wave dispersion in the steel bar into account (using axisymmetric Autodyn® simulations). Let us now discuss the influence of micro-inertia. Figure 5 shows the experimental stress history and the corresponding numerical results, obtained with ($R_c = 2$ mm) and without ($R_c = 0$) micro-inertia effects, for the impact of Alporas® foam samples of lengths 50 and 100 mm at 380 m s^{-1}. Firstly, one can observe the influence of micro-inertia on the pressure pulse. Accounting for micro-inertia in the simulation leads to:

- a 5% decrease in the peak and average values of the contact pressure;
- an increase of about 15% of the pressure pulse duration.

Secondly, it can be seen that numerical predictions (both with and without micro-inertia) are in a quite well agreement with experimental measurements. Nevertheless, a better correlation is achieved when micro-inertia effects are included in the modelling.

3.3. Autodyn simulations

For comparison purpose, we also carried some simulations of the impact tests using the dynamic code ANSYS Autodyn®. In these simulations, both the foam projectile and the Kolsky bar are modelled. The foam behaviour is described by the "crushable foam" model (the version of the "crushable foam model" available in Autodyn® is the same than the one in LS-Dyna, that has been discussed by Pattofatto & al. [15]). This model requires the axial stress-volumetric strain curve and the initial bulk and shear moduli. The Autodyn® computations do not take microscale inertia into account.

The numerical prediction obtained with Autodyn® for the impact of an Alporas® foam sample (length:

50 mm – impact velocity: 380 m s^{-1}) is presented in Fig. 6, together with the experimental stress history and the result of a simulation performed with the 1D fortran code, neglecting micro-inertia. On the one hand, it appears that the Autodyn® simulation underpredicts the pressure level. On the other hand, the Autodyn® results are very close to those obtained with the 1D Fortran code. This suggests that taking micro-inertia into account in the Autodyn® simulations may improve the agreement with the experimental results.

4. Concluding remarks

An original model was developed to describe the response of foams under highly dynamic solicitations. Through dynamic homogenization techniques, micro-inertia effects induced by the rapid crushing of foam cells are taken into account. The macroscopic stress in the foam is the sum of a quasi-static and a dynamic (inertial) term. The dynamic stress involves only physical parameters, as it depends on the initial foam density and the mean radius of the cells of the foam microstructure.

This constitutive model was implemented using a one-dimensional finite element code written in Fortran 90. Numerical simulations of the impact tests conducted by Radford & al. [23] for aluminium closed-cell foams have been carried out. Despite some discrepancies, the results obtained with the Fortran code are in a good agreement with the experimental results when micro-inertia effects are included in the modelling.

Simulations were also conducted using ANSYS AUTODYN®. The foam is modelled by using the "crushable foam" model. Micro-inertia effects are not taken into account in these simulations (it might be possible to implement the inertia the proposed modelling in Autodyn® with the use of user-subroutines). The Autodyn® results are close to those obtained by the Fortran code when no micro-inertia is included.

Future works will include the implementation of micro-inertia effects into ANSYS AUTODYN®. To check and extend the validity range of the model presented in this paper, impact experiments will be conducted at velocities ranging from 500 to 2000 m s^{-1}. Results of these tests will be the scope of another paper.

The work reported was supported by the "Association Nationale de la Recherche et de la Technologie" (ANRT) through grant CIFRE no 0843/2013.

References

[1] L. Gibson, M. Ashby, *Cellular solids: Structure and properties - second edition* (Cambridge University Press, 2005)

[2] M. Ashby, A. Evans, N. Fleck, L. Gibson, J. Hutchinson, H. Wadley, *Metal Foams : A Design Guide* (Butterworth-Heinemann, 2000)

[3] E. Andrews, W. Sanders, L. Gibson, Materials Science and Engineering A **270**, 113 (1999)

[4] T. Nieh, K. Higashi, J. Wadsworth, Materials Science and Engineering A **283**, 105 (2000)

[5] D. Ruan, G. Lu, F. Chen, E. Siores, Composite structures **57**, 331 (2002)

[6] J. Blazy, Ph.D. thesis, École Nationale supérieure des Mines de Paris (2003)

[7] L. Gong, S. Kyriakides, International Journal of Solids and Structures **42**, 1381 (2005)

[8] L. Gong, S. Kyriakides, W. Jang, International Journal of Solids and Structures **42**, 1355 (2005)

[9] Z. Zou, S. Reid, P. Tan, S. Li, J. Harrigan, International Journal of Impact Engineering **36**, 165 (2009)

[10] K. Dannemann, J. Lankford, Materials Science and Engineering A **293**, 157 (2000)

[11] T. Mukai, T. Miyoshi, S. Nakano, H. Somekawa, K. Higashi, Scripta Materialia **54** (2006)

[12] F. Yi, Z. Zhu, F. Zu, S. Hu, P. Yi, Materials Characterization **47**, 417 (2001)

[13] P. Tan, S. Reid, J. Harrigan, International Journal of Solids and Structures **49**, 2744 (2012)

[14] I. Elnasri, S. Pattofatto, H. Zhao, H. Tsitsiris, F. Hild, Y. Girard, Journal of the Mechanics and Physics of Solids **55**, 2652 (2007)

[15] S. Pattofatto, I. Elnasri, H. Zhao, H. Tsitsiris, F. Hild, Y. Girard, Journal of the Mechanics and Physics of Solids **55**, 2672 (2007)

[16] H. Zhao, I. Elnasri, S. Abdennadher, International Journal of Mechanical Sciences **47**, 757 (2005)

[17] P.A. Romero, S.F. Zheng, A.M.C. no, Journal of the Mechanics and Physics of Solids **56**, 1916 (2008)

[18] A. Molinari, S. Mercier, Journal of the Mechanics and Physics of solids **49**, 1497 (2001)

[19] C. Sartori, S. Mercier, N. Jacques, A. Molinari, Mechanics of Materials **80**, 324 (2015)

[20] S. Reid, C. Peng, International Journal of Impact Engineering **19**, 531 (1997)

[21] J. Harrigan, S. Reid, A.S. Yaghoubi, International Journal of Impact Engineering **37**, 918 (2010)

[22] J. Simo, T. Hugues, *Computational Inelasticity* (Springer, 1998)

[23] D. Radford, V. Deshpande, N. Fleck, International Journal of Impact Engineering **31**, 1152 (2005)

[24] V. Deshpande, N. Fleck, Journal of the Mechanics and Physics of Solids **48**, 1253 (2000)

EPJ Web of Conferences **94**, 04015 (2015)
DOI: 10.1051/epjconf/20159404015

Perforation of welded aluminum extrusions

Numerical prediction and experimental validation

Jens Kristian Holmen[1,a], Tore Børvik[1], Ole Runar Myhr[1,2], and Odd Sture Hopperstad[1]

[1] Structural Impact Laboratory (SIMLab), Department of Structural Engineering, Norwegian University of Science and Technology, NO-7491 Trondheim, Norway
[2] Hydro Aluminium, Research and Technology Development (RTD), NO-6601 Sunndalsora, Norway

Abstract. A purely numerical procedure for predicting the perforation resistance of welded extruded AA6082-T6 aluminum profiles is presented in this study. The numerical work was conducted completely independent of the experimental tests that were only used for validation purposes. The outline of the three-step numerical procedure is as follows: (1) the temperature development due to a specified welding process is predicted by a thermal solver, (2) based on the chemical composition of the alloy, the temperature-time history during aging and the welding analysis, the yield strength and flow stress of the material were determined; and (3) the ballistic limit velocity is found using explicit finite element simulations. The experimental validation program is described and it shows that the ballistic limit velocities found from the impact experiments correlate closely with the numerically predicted values obtained without any physical material or component tests. Further, welding of the 10 mm thick extrusions gives a 10% degradation of the capacity in terms of ballistic limit velocity compared to the unaffected material.

1. Introduction

Johnsen et al. [1] showed that from knowing only the chemical composition of an AA6xxx aluminum alloy and the thermal history during aging we can predict how rolled plates behave under impact loading without conducting any experiments whatsoever. However, real structural parts are commonly connected by bolts, nuts, or weldments, and connections are in many cases weaker than the rest of the structure and require special attention. Welding generates a relatively weak zone compared to the base material, known as the heat affected zone (HAZ). Under normal circumstances ballistic performance is governed by material strength [2,3], so welding may significantly decrease the capabilities of aluminum extrusions to withstand impacts by small-arms bullets.

This study describes the numerical procedure employed to predict the ballistic behavior of welded, flat extruded aluminum profiles of 10 mm thickness. Figure 1 illustrates the three-step procedure where the chemical composition of the material, the artificial aging history of the extrusion, and the welding procedure of the final specimen are used to determine the spatial distribution of yield strength and the flow stress which in turn serves as input to solve the impact problem. Experimental data has only been used to validate the numerical results in this study.

[a] Corresponding author: jens.k.holmen@ntnu.no

2. Numerical procedure

2.1. Welding simulation

How the temperature evolves through the welding process is predicted by the thermal solver WELDSIM [4]. Welding of the 10 mm thick plates in this study requires three weld-passes. The result is a complex thermal situation in the heat affected zone which is accounted for in the simulations. Symmetry along the welding line was assumed; an arc efficiency of 0.8 was applied and the same values for current, voltage and welding speed as in the experimental program in Sect. 3 were used. The temperature-time history found from the WELDSIM simulation was used as input in the subsequent work.

2.2. Strength evaluation

Evaluation of the strength evolution was done in the nano-scale material model NaMo [6]. This tool predicts the microstructural changes taking place during artificial aging, welding, and room temperature storing. NaMo consists of three separate modules. The first module determines the particle size distribution and the solute content. Using these data and relations from dislocation mechanics, the two last modules calculate the yield strength and the work hardening of the alloy, respectively. A brief description of the model can be found in Ref. [7].

NaMo-analyses of the base material and the material at locations $x = 4.1$ mm, 6.1 mm, 8.1 mm, 10.1 mm and 14.1 mm away from the weld center line were conducted. The output from the analyses were the equivalent stress-equivalent plastic strain curves shown in Fig. 2. NaMo predicts that the base material is the strongest, while the

Article available at http://www.epj-conferences.org or http://dx.doi.org/10.1051/epjconf/20159404015

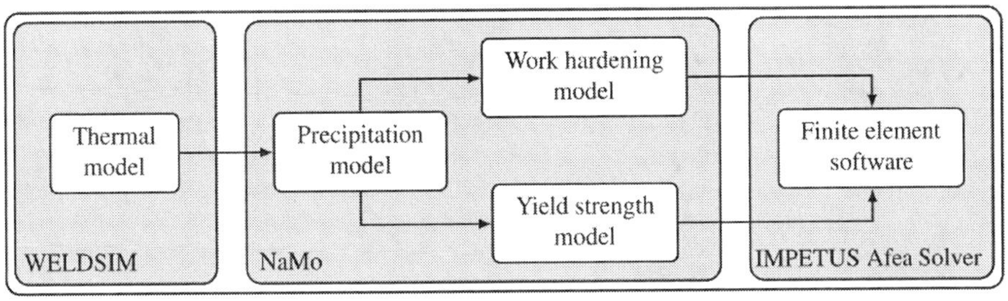

Figure 1. Systematic illustration of the various tools used to predict the ballistic behavior of the aluminum extrusions in this study. Adapted from Ref. [5].

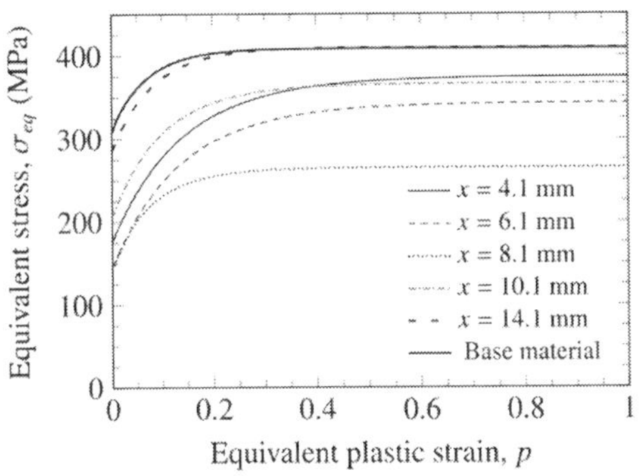

Figure 2. The resulting flow-stress curves obtained from NaMo. These curves determine the hardening behavior used in IMPETUS Afea Solver. Here x is defined as the distance from the weld center line.

weakest part of the HAZ is predicted to be 8.1 mm away from the weld center line.

2.3. Finite element simulations

IMPETUS Afea Solver [8], an explicit finite element solver, was used to simulate the impact process. A thermoelastic-thermoviscoplastic constitutive model [9] based on the well-known Johnson-Cook flow-stress relation [10] was implemented:

$$\sigma_{eq} = \Big(\sigma_0 + R(p)\Big)\left(1 + \frac{\dot{p}}{\dot{p}_0}\right)^c \Big(1 - T^{*m}\Big)$$
$$R(p) = \sum_{i=1}^{2} Q_i \Big(1 - \exp(-C_i p)\Big) \qquad (1)$$

where p is the equivalent plastic strain, σ_0 is the quasi-static yield strength and Q_1, C_1, Q_2, C_2 are hardening parameters in the extended Voce work hardening rule. The increased strength due to elevated strain rate $\dot{p}$ is controlled by the strain rate parameter c and $\dot{p}_0$ where the latter is a user defined reference strain rate. $T^* = (T - T_r)/(T_m - T_r)$ is the homologous temperature which is a function of the temperature T, the ambient temperature T_r, and the melting temperature of the material T_m. The superscript m is a material parameter controlling strength degradation due to rising temperature. No failure criterion can be calibrated based on the available numerically obtained data, so a tiny pinhole was introduced through the plate at the point of impact as seen in Fig. 3. Only the hard steel-core part of the entire AP bullet was modeled, partly because the core is rigid relative to the target plate, thus we do not have to rely on additional material data, and partly because it is numerically much easier. It has been shown experimentally that the effect of omitting the brass jacket is limited [11]. Symmetry was employed to save computational time, i.e., a 30° slice of the extrusion was modeled with appropriate symmetry conditions. High-velocity impact is known to be a localized process, so we did not model the whole target plate. This is not expected to affect the simulation results.

The extended Voce hardening rule in Eq. (1) was adapted to the results in Fig. 2, and simulations such as the one shown in Fig. 3 were run with the obtained material parameters found from curve fitting to Fig. 2. The parameters are given in Ref. [7]. Note that each simulation included only one set of material parameters, i.e., no variation of strength was present within a single simulation in this study. The initial-residual velocity pairs are plotted as markers in Fig. 4 where the solid lines follow the Recht-Ipson model for rigid sharp projectiles [12]

$$v_r = \left(v_i^2 - v_{bl}^2\right)^{1/2}. \qquad (2)$$

We see that the ballistic limit velocity v_{bl} is the only unknown parameter in Eq. (2). Figure 2 and Fig. 4b suggest that the ballistic limit velocity increases with increasing material strength, as expected.

3. Experimental validation

3.1. Material processing

The AA6082-T6 specimens used in this study were provided by Hydro Aluminium as 0.2 m wide and 2.0 m long extrusions. The main alloying elements are shown in Table 1. To obtain the desired strength of the material the specimens were heated to 175°C for 5 h and 30 min after extrusion.

Automatic MIG-welding was carried out by Marin Aluminium AS. A Safra 5182 welding wire was used as the welding consumable. Measures were taken to ensure that the temperature in the HAZ was lower than 100°C before

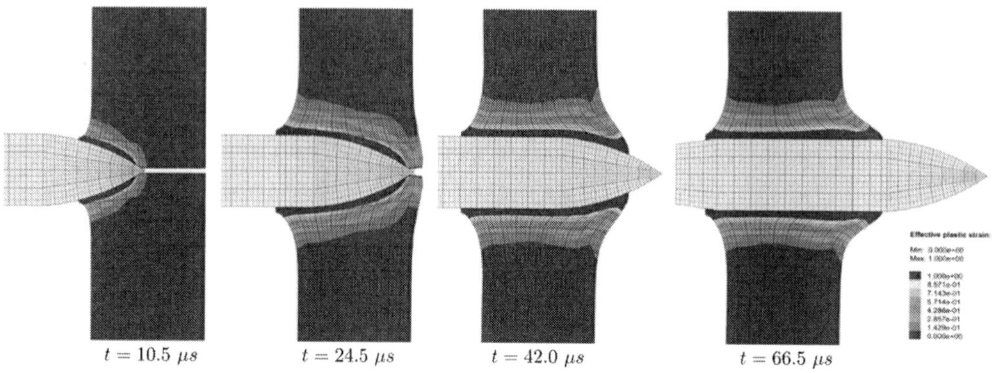

Figure 3. Image series from a typical simulation. The colors show accumulated plastic strain in the range $p \in [0; 1]$, and the mesh is reflected about the centerline to illustrate the pinhole ($v_i = 450.0\,\text{m/s}$, $v_r = 290.0\,\text{m/s}$).

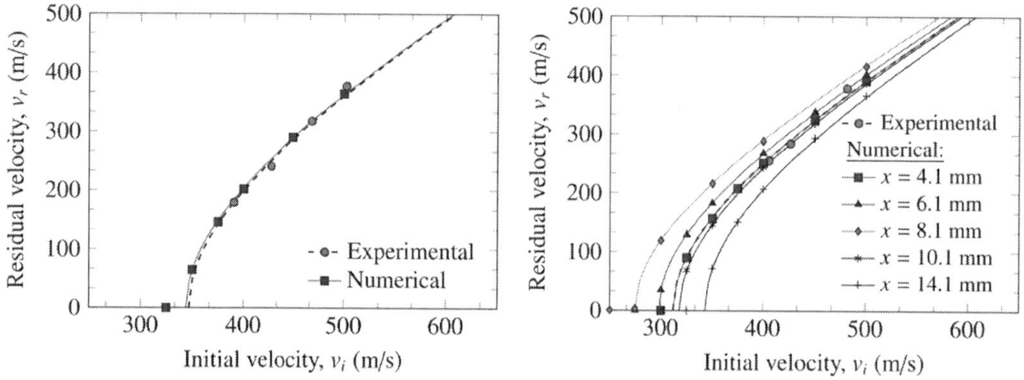

(a) Base material

(b) HAZ

Figure 4. Experimentally determined ballistic limit curves compared to their numerical counterparts for (a) the base material and (b) the HAZ. Here x is defined as the distance from the weld center line.

Table 1. Chemical compostition of the 10 mm thick extrusions of AA6082 aluminum alloy.

Si	Mg	Mn	Fe	Ti	Zn	Cu	Cr	Al
0.93	0.60	0.55	0.18	0.011	0.002	0.008	0.011	Balance

the next weld seam was initiated, this was checked by thermocouples that measured the temperature throughout the welding process.

3.2. Material testing

Three uniaxial tension tests were performed in two directions at 0° and 90° with respect to the extrusion direction. A laser-scan micrometer measured the diameter reduction in two perpendicular directions all the way to specimen failure while the force was measured by a calibrated load cell in a Zwick/Roell 30 kN testing machine. No significant difference in yield strength, flow stress or failure strain was seen between the tests in the 0° and 90°-directions. A typical result is shown in Fig. 5 where the force F is plotted against the averaged diameter reduction of a specimen taken from the extrusion direction.

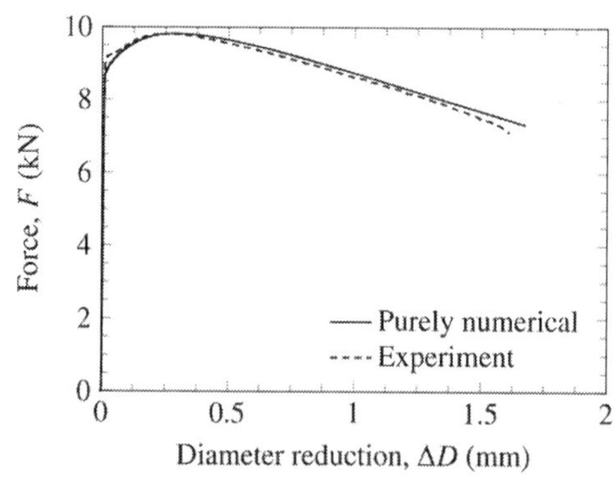

Figure 5. Force-diameter reduction curve for the 0°-direction from the experiment compared to a simulation based on NaMo results.

Figure 5 also shows that a simulation of the tension test using only numerically obtained data behaves similarly to the experimental result, suggesting that NaMo predicts accurate material behavior for the base material.

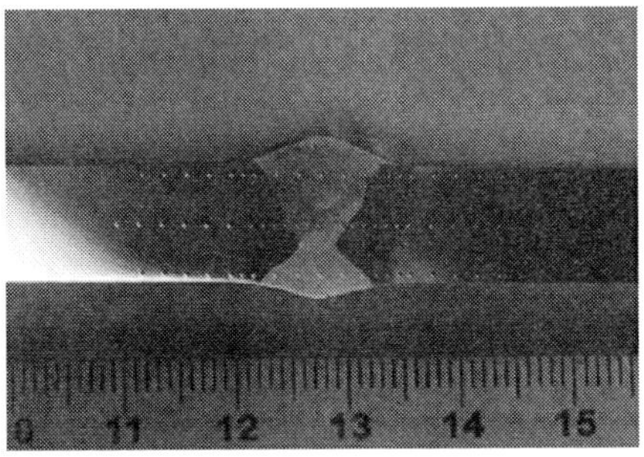

Figure 6. Cross section of the weld. The dots are indentations made by the hardness-test machine.

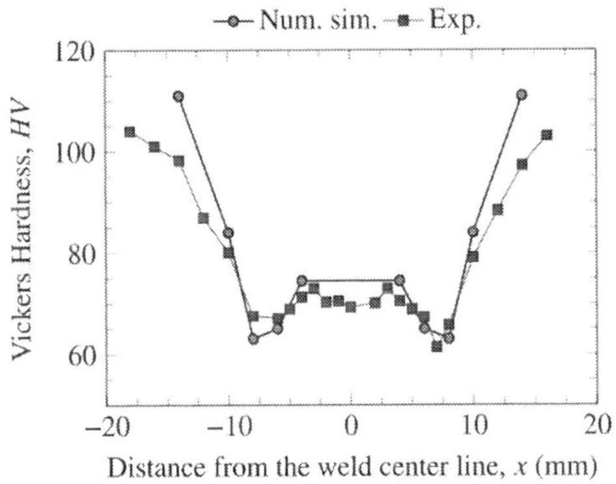

Figure 7. Vickers hardness values obtained from NaMo compared to the hardness measurements.

Hardness tests were performed, and the indented surface is shown in Fig. 6. This was done to get some form of validation of the material behavior in the HAZ. To this end, we applied the following conversion from yield strength (in MPa) to hardness (in VPN) [13]:

$$HV = 0.33\sigma_0 + 16.0 \quad (3)$$

where HV is the Vickers hardness and σ_0 is the initial yield strength found from Fig. 2. Figure 7 compares the hardness profile through the HAZ at the center of the extrusion (as seen in Fig. 6) to the numerically obtained hardness profile. The numerical results provide a good fit in the HAZ.

3.3. Ballistic testing

The ballistic testing was carried out at the ballistic laboratory at SIMLab. A specially designed smooth-bored Mauser rifle fired 7.62 mm armor piercing (AP) bullets towards the target plates that were mounted in a rigid rack inside a bullet proof impact chamber. The point of impact was adjusted by moving the target, not the rifle. The initial velocity could be predefined with an accuracy of ±20 m/s by varying the amount of powder in the cartridge. A Phantom v1610 high-speed video camera was, however, used to measure the exact initial (striking) and residual velocities. Figure 8 shows pictures of the perforation process.

A total of nine shots were fired into the 10 mm thick flat aluminum extrusion: four in the base material, four in the HAZ and one in the weld. However, no simulations were done of the weld consumable, so this result is omitted in the following[1]. We did not have enough material to experimentally determine the ballistic limit velocities for the cases mentioned above so they were approximated by curve-fitting Eq. (2) in the same way as for the numerical results. Figure 4 compares the numerically obtained results to the experiments, and the correspondence is in some cases excellent, but generally good.

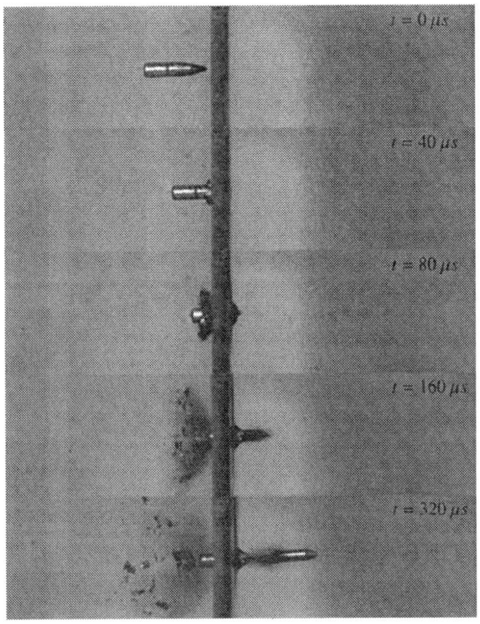

Figure 8. Image sequence showing the perforation process of the 10 mm thick aluminum extrusion.

The material is affected by heat, and is thus weaker than the base material, up to approximately 20 mm from the weld center. This means that potentially a 40 mm wide area exists where the ballistic capacity of the welded component is weakened. Of course the material has a strength gradient in the HAZ which makes is difficult to determine exactly what the strength is at the point of impact in the experiments. Figure 4b shows that the experimental ballistic tests lie within the spread from the numerical simulations which is sensible considering that the experiments were in fact affected by the entire extent of the HAZ.

The numerically predicted ballistic limit velocity for the base material is only 3 m/s higher than the experimentally determined one (347 m/s compared to 344 m/s). In fact, this gives a relative difference of less than 1% which is a remarkable result.

As mentioned, only one ballistic limit velocity was determined experimentally in the HAZ and the value is 312.0 m/s, i.e., 9.3% lower than in the base material. The numerical simulations provided enough data to predict ballistic limit velocities for a range of material behaviors at different distances from the weld center line. The weakest

[1] The approximated v_{bl} from the weld is slightly lower than the base material, but significantly higher than the HAZ [7].

material was found 8.1 mm from the weld center line and the strongest was found closest to the base material, i.e., at $x = 14.1$ mm. The averaged simulated ballistic limit velocity from the HAZ was 309.0 m/s which is approximately 1% lower than that of the experiments.

4. Concluding remarks

In this study the perforation resistance of welded aluminum extrusions was successfully predicted based only on numerical simulations.

The base material, as expected, exhibited the highest capacity in both the numerical simulations and the experiments. The material behavior was validated and a tension test was simulated giving excellent correspondence, as shown in Fig. 5. Figure 4a shows that the numerical simulations predicted the ballistic behavior accurately with almost no discrepancy from the experimental result. Of course, many uncertainties exist in the experimental work, and several assumptions have been made in the numerical procedure.

The predominant failure mode in all the experiments was ductile hole-growth, suggesting that the modeling approach as described in Sect. 2 with an initial pin hole in the material is an acceptable approximation.

Figure 4b shows the experimental results at the center of the spread from the numerical simulations. Despite our best aiming-efforts, we can not change that the diameter of the bullet is nearly the same size as the entire extent of the HAZ and only one ballistic limit velocity can be found from the experiments. This result should, in theory, reflect the average behavior of the HAZ – which it does.

To conclude: The numerical results are in good agreement with the experimental data, and the deviation is never more than 10%. We showed that without carrying out a single experiment, we can predict the ballistic behavior of welded Al-Mg-Si aluminum alloy extruded profiles subjected to small-arms bullets with good accuracy. It is believed that the proposed modeling approach can become very useful in preliminary design of aluminum protective structures.

The financial support for this work from the Structural Impact Laboratory (SIMLab), Centre for Research-based Innovation at the Norwegian University of Science and Technology and the Norwegian Defence Estates Agency, is gratefully acknowledged. So is the contribution from the MSc students Steffen Breivik and Espen Thomsen.

References

[1] J. Johnsen, J.K. Holmen, O.R. Myhr, O.S. Hopperstad, T. Børvik, Computational Materials Science **79**, 724 (2013)

[2] T. Børvik, S. Dey, A.H. Clausen, International Journal of Impact Engineering **36**, 948 (2009)

[3] J.K. Holmen, J. Johnsen, O.R. Myhr, O.S. Hopperstad, T. Børvik, International Journal of Impact Engineering **57**, 119 (2013)

[4] H.G. Fjær, O.R. Myhr, S. Klokkehaug, S. Holm, *Advances in aluminum weld simulations applying WeldSim*, in *Proceedings of the 11th International Conference on Computer Technology in Welding*, edited by T.A. Siewert, C. Pollock (NIST Special Publication, Washington, 2001), pp. 251–263

[5] S.M.W. Breivik, E.F. Thomsen, Master's thesis, SIMLab - Norwegian university of Science and Technology, NTNU (2014)

[6] O.R. Myhr, Ø. Grong, K.P. Pedersen, Metallurgical and Materials Transactions **41A**, 2276 (2010)

[7] J.K. Holmen, T. Børvik, H.G. Fjær, O.R. Myhr, O.S. Hopperstad, Submitted for possible journal publication (2015)

[8] *IMPETUS Afea Solver* (Cited 2014-20-10), `http://www.impetus-afea.com`

[9] T. Børvik, O.S. Hopperstad, T. Berstad, M. Langseth, European Journal of Mechanics - A/Solids **20**, 685 (2001)

[10] G.R. Johnson, W.H. Cook, *A constitutive model and data for metals subjected to large strains, high strain rates and high temperatures*, in *Proceedings of the 7th International Symposium on Ballistics* (The Hague, Netherlands, 1983), pp. 541–547

[11] M.J. Forrestal, T. Børvik, T.L. Warren, Experimental Mechanics **50**, 1245 (2010)

[12] R.F. Ipson, T.W. Ipson, Journal of Applied Mechanics **30**, 384 (1963)

[13] O.R. Myhr, Ø. Grong, S.J. Andersen, Acta Materialia **49**, 65 (2001)

EPJ Web of Conferences **94**, 04016 (2015)
DOI: 10.1051/epjconf/20159404016

Experiments and modeling of ballistic penetration using an energy failure criterion

M. Dolinski and D. Rittel[a]

Faculty of Mechanical Engineering, Technion – Israel Institute of Technology, Israel

Abstract. One of the most intricate problems in terminal ballistics is the physics underlying penetration and perforation. Several penetration modes are well identified, such as petalling, plugging, spall failure and fragmentation (Sedgwick, 1968). In most cases, the final target failure will combine those modes. Some of the failure modes can be due to brittle material behavior, but penetration of ductile targets by blunt projectiles, involving plugging in particular, is caused by excessive localized plasticity, with emphasis on adiabatic shear banding (ASB). Among the theories regarding the onset of ASB, new evidence was recently brought by Rittel et al. (2006), according to whom shear bands initiate as a result of dynamic recrystallization (DRX), a local softening mechanism driven by the stored energy of cold work. As such, ASB formation results from microstructural transformations, rather than from thermal softening. In our previous work (Dolinski et al., 2010), a failure criterion based on plastic strain energy density was presented and applied to model four different classical examples of dynamic failure involving ASB formation. According to this criterion, a material point starts to fail when the total plastic strain energy density reaches a critical value. Thereafter, the strength of the element decreases gradually to zero to mimic the actual material mechanical behavior.

The goal of this paper is to present a new combined experimental-numerical study of ballistic penetration and perforation, using the above-mentioned failure criterion. Careful experiments are carried out using a single combination of AISI 4340 FSP projectiles and 25[mm] thick RHA steel plates, while the impact velocity, and hence the imparted damage, are systematically varied. We show that our failure model, which includes only one adjustable parameter in this present work, can faithfully reproduce each of the experiments without any further adjustment.

Moreover, it is shown that the most common failure criterion based on a critical strain is simply inadequate to reproduce the results, due to the linear nature of the damage evolution. The advantages of the energy-based failure criterion are discussed in detail.

1. Introduction

One of failure modes in high strain rates applications consists of localization of the plastic deformation into a narrow band or surface. This failure mode is referred to as adiabatic shear banding (ASB), and was first documented by Tresca (1879), who observed it while forging platinum billets. ASB occurs in various applications such as machining, high velocity plastic processing, and plays a major role in the context of terminal ballistics. ASB's are frequently encountered in materials that have thermal softening and quite low strain hardening. Zener and Hollomon (1944) suggested that ASB initiates when the adiabatic thermal softening exceeds the strain hardening. Consequently, the traditional way to model ASB formation in a dynamically deformed solid consists of introducing a critical failure strain, the latter being calculated based on the competing mechanisms of strain hardening and thermal softening.

In the recent years, Rittel et al. (2006) challenged this assumption and proposed that the dynamic stored energy (density) of cold work can be considered as an alternative physical criterion. Later work (Rittel et al., 2008) identified a microstructural evolution, called dynamic recrystallization (DRX) as a local softening factor leading to the dynamic shear localization. The relative contribution of microstructural and local thermal evolutions was assessed by means of a numerical model by Osovski et al. (2013).

From a practical point of view, the concept of a critical strain energy density as a criterion for the initiation of the ASB, as well as a damage evolution criterion were illustrated by Dolinski et al. (2010), who demonstrated its feasibility by modeling 4 different laboratory experiments involving dynamic shear failure. Noam et al. (2014) extended this proof of feasibility by investigating the potential of the criterion to undergo scaling, and showed a simple penetration example based on the experimental work of Børvik et al. (1999).

One commonly used material that undergoes ASB is Rolled Homogeneous Armor (RHA), as defined in the military standard: MIL-DTL-12560J (MR). This steel is widely used in the assessment of the ballistic performance of various ammunitions. Shear bands in RHA have been reported in various works from microstructural and mechanical standpoints (Bassim et al. (2013); Fellows and Harding (2001); Gray et al. (1994)).

As of today, the emerging picture is that reliable prediction, through numerical modeling, of ballistic failure in general is still in its infancy. Putting aside extremely complex numerical models which are not implemented by the engineering community, one can find several (over)

[a] Corresponding author: rittel@me.technion.ac.il

simplified simulations, generally of thin plates, which are not always supported by experimental work, not to mention purely numerical works whose validity has not been verified.

What would be desirable is a simple (as much as can be) approach to model ballistic failure in a realistic way on the one hand, to an extent where the simulations can also have a predicting power. On the other hand, it is the authors' opinion that such an approach should be based on iterative numerical-experimental work, consisting of the following phases. The first phase is the calibration of the failure model, criteria and parameters based on one or more experiments, once a satisfactory convergence of the solution has been defined. The next phase, validation of the above-mentioned choices, consists of successfully replicating additional experimental tests results without changing any numerical parameter. Once this validation phase has been established, it can be concluded that the numerical tool has a predictive capability for the selected set of materials (target-projectile). Moreover, once validated, such an approach can be extended in a straightforward manner to additional materials systems.

The above paragraph is precisely the goal of the present paper. We investigate by means of carefully controlled ballistic impact experiments, the response of thick RHA steel. The chosen projectile is Ø20[mm] fragment simulating projectile (FSP), a standard and commonly used projectile. Furthermore, the classical failure criterion based on a critical strain is evaluated vs. the energy density criterion, and shown to be inadequate for the purposes of this research.

2. Energy failure criterion and its implementation

The failure criterion consists of a critical level of the plastic strain energy density W_{crit} reached in each element, given by:

$$W_{crit} = \int_0^\alpha S_{ij} d\varepsilon_{ij}^P = \int_0^\alpha \sigma_M d\varepsilon_{eq}^P. \quad (1)$$

Where S_{ij} and ε_{eq}^P are the deviatoric stress and plastic strain tensor components, σ_M and ε_{eq}^P are the Mises stress and equivalent plastic strain. The upper integration limit α is selected as the level of strain at which the structural strength starts to deteriorate, $\bar{\varepsilon}_{eq}^{crit}$. From this point on, the element starts to fail gradually, rather than abruptly, according to:

$$\tilde{\sigma}_{eq} = \tilde{\sigma}_{eq}^*(1 - D^b). \quad (2)$$

Where $\tilde{\sigma}_{eq}^*$ is the current equivalent stress level at a given equivalent strain $\bar{\varepsilon}_{eq} \geq \bar{\varepsilon}_{eq}^{crit}$, D is the level of damage in the element and b is a constant. In the present work the parameter b=1, therefore only one adjustable parameter is needed for calibration, i.e. W_{frac} as mentioned in the formulation below. W_{frac} is the plastic strain energy density when the stress drops to zero.

The damage evolution is given by:

$$D = \left\{ \begin{array}{ll} 0 & W \leq W_{crit} \\ \frac{W - W_{crit}}{W_{rac} - W_{crit}} & W \geq W_{crit} \end{array} \right\} \quad (3)$$

This criterion was implemented via a user-subroutine (VUSDFLD) in the commercial finite element code Abaqus explicit V6.12 (2012), as presented in the appendix of Dolinski et al. (2010).

Another important point is the strength decay, damage evolution, after the critical energy density is met. The decay is exponential due to the formulation of the damage. Note that $W = \int_0^\alpha \sigma_M d\varepsilon_{eq}^P$, and although the material's strength might be constant in some cases like the steel of the current work, when $D > 0$ $D > 0$, the strength decreases causing a smaller dW for each increment $d\varepsilon_{eq}^P$.

3. Experimental setup

The selected experimental setup was dictated by controllability, reproducibility, and selection of commonly used armor steel that fails by adiabatic shear. Moreover, we performed the experiments in a way that their simulation would be as simple as possible, meaning, axisymmetric simulation with simple homogeneous and isotropic materials.

Projectile: A standard 20[mm] FSP (Fragment Simulating Projectile) was used. In order to simplify the simulation, the "nose" of the projectile was trimmed prior the experiment, such as to obtain an axisymmetric problem. The FSP is made of 4340 steel, with a hardness of $R_C = 30$. The constitutive equation for this material was chosen as Johnson Cook equation, and the parameters were taken from their original article (Johnson and Cook, 1983). Note that this material is very ductile, so that damage and fracture of the projectile are not taken into account in the simulations.

Target: The experimental target was a 25 ± 0.1[mm] thick RHA (Rolled Homogeneous Armor) plate, with a hardness of $R_C = 39$.

4. Experimental results

A total of 9 shots were fired towards two different target plates. Three shots were fired at the first plate (Fig. 1a), and the remaining 6 shots stroke the second plate, as shown in Fig. 1b. All the relevant experimental parameters are listed in Table 1. The range of velocities spanned from 1150[m/s] to 1920[m/s].

At velocities of 1322[m/s] and below, the plate was penetrated but not perforated. The depth of penetration (DOP) depends of course on the impact velocity. It is interesting to note that, even at quite low impact velocity (1248[m/s]), the rear of the plate contains a crack, which may remind of brittle fracture. This fracture is due to the tensile waves that are reflected from the free surface of the plate's back.

Another interesting result occurred when the plate was hit at a velocity of 1394[m/s] (plate #2). This is the *minimal velocity* at which the plate was totally perforated in the experiments. The plate's failure mechanism was undoubtedly by adiabatic shear banding. First of all, the fracture surface is observed to undergo extensive plastic deformation, for which a plug was formed and recovered.

Table 1. Summary of the experimental results. "P" indicates full perforation.

Shot number	Measured impact velocity ±20[m/s]	Depth of penetration or (P) perforation ±0.05[mm]	FSP weight ±0.01[gr]	FSP length ±0.02[mm]	Inner crater diameter ±0.05[mm]	Outer crater diameter ±0.05[mm]
1	1920	P	46.74	20.68	49.20–54.05	28.95–30.45
2	1715	P	47.68	21.06	48.35–49.60	27.15–28.90
3	1460	P	46.60	20.95	41.70–42.50	23.60–26.70
4	1466	P	46.36	20.5	41.05–43.90	22.65–24.75
5	1256	14.09	46.89	20.73	39.45–41.15	–
6	1248	13.74	46.75	20.66	38.45–39.10	–
7	1322	16.8	45.94	20.0	39.45–40.70	–
8	1394	P	46.51	20.58	39.85–41.70	21.35–26.90
9	1151	11.05	46.79	20.32	36.30–37.40	–

Figure 1. Plate #1 (upper) and plate #2 (lower). The numbers near the holes indicate the impact velocity in [m/s].

Table 1 summarizes the experimental results, including the impact velocity, FSP parameters, perforation or DOP and crater diameters.

It must be noted that the results are not "perfect", meaning, the craters are not circular but rather elliptic, with an axes ratio of up to 1.1. Therefore the crater dimensions are given as a range of values in Table 1. Another noticeable scatter lies in the projectile dimensions and weight. The scatter in its length is about ±2.5%. A representative length of 20[mm] was chosen for all the simulated projectiles, as shown in the next section.

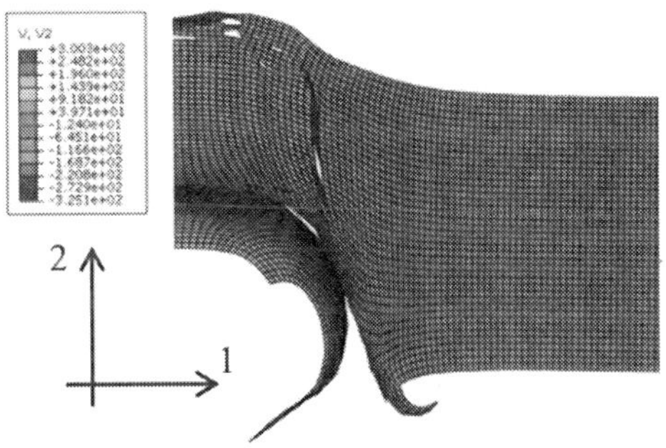

Figure 2. Velocity (V2) contour for impact velocity of 1400[m/s] at t = 42[μs].

5. Numerical results and comparison to experiments

Impact velocity of 1400 [m/s]:

At this impact velocity, the plate was fully perforated by plugging, i.e. adiabatic shearing. Figure 2 shows the velocity contour along the impact axis. This picture is taken 42[−] after the impact. A fully developed plug is visible. The calculated plug dimensions and velocity are very similar to the experimental values, as summarized in Table 2. A fracture inside the plug causes elements to overlap, so that actual dimensions of the plug cannot be assessed. Moreover, the plug is not perfectly cylindrical but rather conical, and the simulation does not include frictional contact between the plug and the crater, as mentioned earlier regarding interaction between bodies.

Yet, the same energetic failure criterion is able to reproduce the plugging process with a high resemblance (qualitative and quantitative) to the experimental results. Very sensitive experimental results, such as the plug exit velocity, are captured with a relative error of less than 10%.

Table 2. Comparison of experimental and numerical results for impact velocity of 1400[m/s].

	Experiment	Simulation
Impact velocity	1394 ± 20[m/s]	1400[m/s]
Crater maximal diameter	39.45–40.7 $\pm$ 0.1[mm]	39.53[mm]
Plug exit velocity	196 ± 20[m/s]	180 ± 20[m/s]
Plug diameter	20.9–26.2 $\pm$ 0.1[mm]	26.05[mm]
Plug height	22.2 ± 0.1[mm]	20.1[mm]

6. Discussion and conclusions

This paper has presented an alternative approach to the treatment of dynamic shear localization to the well accepted procedures based on a critical strain to failure.

We use here a simple, yet physical, criterion based on the plastic strain energy density for the onset of shear localization. A preliminary validation of the criterion can be found in Dolinski et al. (2010) who analyzed relatively simple examples of experiments involving dynamic shear localization.

The case studied in this paper is far more complex and therefore rather challenging. In this work, we have presented a careful series of experiments in which, all parameters being identical, the impact velocity of a standard projectile is carefully controlled.

The physical failure mechanisms are well defined and consist of adiabatic shear and spalling (tensile). For each experiment, a different final state of the impacted plate differs, ranging from partial penetration, through penetration and full penetration with fragmentation, according to the impact velocity.

The goal of this study was to demonstrate that those three kinds of final states can be successfully modeled using the same approach and a fixed set of parameters.

The first issue to be addressed is that of the pathological mesh sensitivity of dynamic shear localization problems. While several sophisticated methods have been devised, they are often only partly supported by experimental evidence, and almost always quite complex to include in a finite element code. Consequently, an engineering approach was adopted here, namely an iterative determination of the mesh parameters and material properties that would yield an accurate replication of all the measurable experimental parameters, mostly of a geometrical nature. This was achieved based on a preliminary set of experiments. Once a satisfactory prediction was reached, all the subsequent experiments were modeled with this exact same set of parameters without further adjustment. The outcome was quite satisfactory, and validated the selected approach.

It is therefore suggested that a similar approach be adopted to model additional penetration/perforation problems for other sets of target materials and projectiles.

Another outcome of this work was a re-assessment of a critical strain to failure as the dynamic shear failure criterion. Indeed, when the above-mentioned approach was applied using this calibrated parameter on one set of experiments, it was found that the other experiments could not be faithfully replicated unless the failure strain is readjusted, which is not physical.

A comparison of the dynamic strain energy density and the damage evolution model with the critical strain criterion revealed that the underlying reason for the inadequacy of the latter and the adequacy of the former lies in the evolution of the damage. Namely, it was found that the stress decrease is linear in strain for the strain criterion, whereas with the energy criterion, strength decreases exponentially, as shown in Fig. 1. One should note that this exponential decay not only leads to satisfactory numerical results, but it also matches the experimental observations of .()Fellows and Harding (2001) who investigated RHA steel in the same procedure of Marchand and Duffy (1988).

Finally, to highlight the specific nature of the present work, one can note that:

- The present work is not an experimental study with a large statistical database. It nevertheless carefully emphasizes specific impact velocities for which specific failure mechanisms operate.
- The same criterion predicts the initiation and dictates the propagation of the ASB.
- The overall approach exposed here is relatively simple and physical, with a minimum number of adjustable parameters.
- This ASB failure criterion is not rate dependent, as shown experimentally by Osovski et al. (2012).

The following conclusions can be now drawn from this work:

- Ballistic penetration/perforation can be successfully modeled using a simple plastic strain energy density and damage model with one adjustable parameter.
- This study extends and validates further the proposed criterion.
- A detailed combined experimental-numerical approach is proposed to calibrate the parameters of the numerical model.
- The concept of a critical strain to failure is shown to be inadequate for the purpose of numerically simulating and predicting ballistic failure.
- All in all, the proposed approach is relatively simple and applicable, and should therefore be of practical interest to the engineering community.

EPJ Web of Conferences **94**, 04017 (2015)
DOI: 10.1051/epjconf/20159404017

Deformation and failure of bulk metallic glasses under different initial temperatures

J.C. Li[1], X.W. Chen[1,a], and F.L. Huang[2]

[1] Institute of Structural Mechanics, China Academy of Engineering Physics, Mianyang, Sichuan 621900, China
[2] State Key Laboratory of Explosion Science and Technology, Beijing Institute of Technology, Beijing 100081, China

Abstract. Based on the coupled thermo-mechanical model, a constitutive model for bulk metallic glasses (BMGs), which is generalized to the multi-axial stress state and considers the effects of free volume, heat and hydrostatic stress, has been modified in the present paper. Besides, a failure criterion of critical free volume concentration is introduced based on the coalescence mechanism of free volume. The constitutive model as well as the failure criterion is implemented into the LS-DYNA commercial software by user material subroutine (UMAT). Then FEM simulations for different initial material temperatures are conducted and the evolutions of material parameter as well as corresponding macroscopic mechanical behaviour of material are analyzed. Relative analysis shows that the initial material temperature significantly affects the deformation and failure of material.

1. Introduction

Bulk metallic glasses (BMGs) have many excellent mechanical and physical performances. It usually show good plasticity even superplasticity while under the condition of high temperature (within the supercooled liquid region) and low strain rate [1–3]. However, in the case of low temperature (below the glass transition temperature T_g) and high stress or high strain rate, seriously localized shear bands are easily to be produced during the deformation process. Fracture occurs immediately along the propagation of shear bands and thus results in a catastrophic failure phenomenon. Moreover, the material usually shows an asymmetry between tension and compression [4,5]. All these particular properties promote that BMGs may have many engineering applications, e.g., the good plasticity is beneficial for the precise microforming process [6], and the localized shear bands may be employed in the material optimization of kinetic energy penetrator (KEP) [7,8].

Related analyses demonstrated that the microstructure of material may have an important effect on the shear banding process in BMGs [9–11]. Besides, it is usually found a significant temperature rise in the fracture surface [5]. Presently, a coupled thermo-mechanical shear model [5,11], deriving from the free volume model and the adiabatic deformation theory, could predict the shear banding behaviour of BMGs at room temperature relatively well. Comparatively, for the homogeneous flow of BMGs at high temperature ($T \geq T_g$), the related analyses usually follow the models used for the crystalline alloys [1,3]. Regarding the yielding and failure of BMGs, some macroscopic failure criteria have been suggested and successfully describe the deviation of fracture surface from the maximum shear stress plane [12,13]. However, the micro-mechanism of deformation and failure of BMGs is still under the exploration process [14–17]. Recently, the simulation based on finite element method (FEM) is gradually employed for investigating the mechanical behavior of BMGs. Related work includes the theoretical constitutive modeling and the construction technique of the geometric model, etc. [18,19].

Integrating with related investigations, the present manuscript further generalizes the coupled thermo-mechanical shear model into the multi-axial stress status, and takes all the contributions of free volume, temperature and hydrostatic stress into account. Besides, based on the coalescence mechanism of free volume in BMGs, a failure criterion of critical free volume concentration, ξ_c, is introduced. Furthermore, the generalized multi-axial-stress-status model is extended to describe the mechanical behavior of BMGs at high temperature ($T \geq T_g$) by estimating related material parameters. Then the constitutive model and failure criterion are implemented into LS-DYNA commercial software through the user material subroutine (UMAT), and FEM simulations for different initial material temperatures are conducted integrated with the related test data. The evolutions of parameters as well as their effects on the mechanical behavior of material are analyzed, and corresponding mechanism of the macroscopic behavior is discussed. Related analyses show that the initial temperature has a significant effect on the deformation and failure of BMGs. Especially, the microstructure as well as corresponding mechanical behavior of material is extremely different while the initial temperature is below or within the supercooled liquid region.

[a] Corresponding author: chenxiaoweintu@yahoo.com

2. Constitutive model

2.1. Constitutive equations and failure criterion

Basing on the free volume model, Dai et al. considered the effect of temperature additionally and suggested a coupled thermo-mechanical model for BMGs [5, 11]. However, the model is still limited within the one-dimensional simple shear mode. Gao [18] generalized the one-dimensional free volume model into the multi-axial stress status but ignoring the effect of temperature. Actually, during the deformation process of BMGs, the hydrostatic stress (or the normal stress acting on the shear plane), which derived from the shear-induced dilatation of material, also contributes to the deformation and failure [4, 13, 17]. Thus, the effect of hydrostatic stress should be considered in the constitutive model. By further considering the effect of hydrostatic stress on the deformation, the evolution of free volume and temperature, a modified thermo-mechanical model for BMGs which is generalized to multi-axial stress status could be obtained:

$$\sigma_e^{'} = \Lambda\sigma_m + \sqrt{J_2} \tag{1}$$

$$\dot{\varepsilon}_{ij} = \dot{\varepsilon}_{ij}^{e} + \dot{\varepsilon}_{ij}^{p} \tag{1a}$$

$$\dot{\varepsilon}_{ij}^{e} = \frac{1+\nu}{E}\left(\dot{\sigma}_{ij} - \frac{\nu}{1+\nu}\dot{\sigma}_{kk}\delta_{ij}\right) \tag{1b}$$

$$\dot{\varepsilon}_{ij}^{p} = f\exp\left(-\frac{\Delta G^{m}}{K_B T}\right)\exp\left(-\frac{1}{\xi}\right)\sinh\left(\frac{\sigma_e^{'}\Omega}{2K_B T}\right)\frac{S_{ij}}{\sigma_e} \tag{1c}$$

$$\sigma_e = \sqrt{3J_2} \tag{1d}$$

$$\sigma_e^{'} = \Lambda\sigma_m + \sqrt{J_2} \tag{1e}$$

$$\dot{\xi} = D\nabla^2\xi + \frac{1}{\alpha}f\exp\left(-\frac{\Delta G^{m}}{K_B T}\right)\exp\left(-\frac{1}{\xi}\right) \times\left\{\frac{2K_B T}{\xi v^{*} s}\left[\cosh\left(\frac{\sigma_e^{'}\Omega}{2K_B T}\right)-1\right]-\frac{1}{n_D}\right\} \tag{2}$$

$$\dot{T} = \kappa\nabla^2 T + \frac{\beta_{TQ}}{\rho C_v}\sigma_e^{'}\dot{\varepsilon}_e^{p} \tag{3}$$

In Eqs. (1)–(3), a dot over a quantity () denotes differentiation with time.

Equation (1) implies the material is assumed to be elastically isotropic, and the strain is decoupled into the elastic and plastic parts. E and ν are the Young's modulus and Poisson's ratio, respectively; f is the frequency of atomic vibration and usually adopts the Debye frequency; ΔG^{m} is the activation energy; K_B is the Boltzmann constant and T the absolute temperature; $\xi = \bar{v}_f/(\alpha v^{*})$ is the free volume concentration in BMGs (here $\bar{v}_f$, $v^{*}$and α are, respectively, the average free volume per atom, critical volume (hard-sphere volume of an atom) and a geometrical factor) [10]; Ω is the average atomic volume; σ_e is the von-Mises effective stress, in which $J_2 = \sqrt{S_{ij}S_{ij}/2}$ is the J_2 invariant of stress, $S_{ij} = \sigma_{ij} - \delta_{ij}\sigma_m$ is the deviatoric stress tensor and $\sigma_m = \sigma_{kk}/3$ the mean stress; $\sigma_e^{'}$ is an effective stress which considers the contribution of hydrostatic stress, in which Λ is the hydrostatic stress sensitivity factor.

Equation (2) is the free volume evolution equation, in which D is the diffusion coefficient of free volume; $s = E/[3(1-\nu)]$ is the Eshelby modulus and n_D the number of diffusive jumps necessary to annihilate a free volume as v^{*}. Equation (2) indicates there are three different physical processes that alter the local free volume concentration: diffusion (the first term), generation induced by stresses (the first term in bracket) and annihilation by the atomic rearrangement (the second term in bracket).

Equation (3) is the temperature evolution equation, where $\kappa = K/(\rho C_v)$ is the thermal diffusivity (here K, ρ and C_v are the thermal conductivity, mass density and specific heat at constant volume, respectively); $\varepsilon_e^{p} = \sqrt{2\varepsilon_{ij}^{p}\varepsilon_{ij}^{p}/3}$ is the effective plastic strain. β_{TQ} is the Taylor-Quinney coefficient which is usually taken as 0.9 or considered to be related to the strain rate. Since the adiabatic deformation usually occurs in the case of high strain rate, the Taylor-Quinney coefficient β_{TQ} may be defined as a function of irreversible part of the effective plastic strain rate [11]:

$$\beta_{TQ} = \frac{1}{\pi}\arctan\left(\frac{1}{3}\tan\frac{2\pi}{5}\lg\frac{\dot{\varepsilon}_e^{p}}{\dot{\varepsilon}_0}\right) + \frac{1}{2} \tag{4}$$

in which $\dot{\varepsilon}_0$ is the referenced strain rate and takes a value of 1 s-1. Note from Eq. (4) that it satisfies $\lim_{\dot{\varepsilon}_e^{p}\to 0}\beta_{TQ} = 0$ and $\lim_{\dot{\varepsilon}_e^{p}\to\infty}\beta_{TQ} = 1$. Moreover, Zhou et al. [20]'s analysis demonstrated that when the material yields, the stress located within the shear band begins to collapse and unloads its neighboring material. Then the unloading wave propagates towards the neighboring zone of the shear band. When failure occurs in the shear band, the deformation energy in the neighboring zone where the unloading wave has arrived will release on the fracture surface and induces extreme temperature rise. Thus, in the simulation we could further modify the coefficient β_{TQ} when the material is going to fail.

Furthermore, different from crystalline alloys, BMGs display a few special failure properties. The related macroscopic failure criteria are difficult to describe the failure mechanism completely. Researchers analyzed the failure process based on the change of microstructure of BMGs [15, 16], and demonstrated that the free volume will be created during the deformation process and then coalesce together. Micro-voids will be formed due to the coalescence of free volumes and further leads the damage and failure eventually. Thus, the failure criterion in the present model will be based on the evolution of microstructure. It is assumed that failure occurs when the net free volumes exceed a critical value, i.e. corresponding failure criterion is that the free volume concentration ξ achieves the critical value ξ_c:

$$\xi \geq \xi_c \tag{5}$$

Table 1. Mechanical properties and parameters of $Zr_{41.25}Ti_{13.75}Cu_{12.5}Ni_{1}0Be_{22.5}$ (Vitreloy 1) BMG.

Parameter	Notation	Unit	Value
Young's modulus	E	GPa	96 for $T_0 < T_g$ $E(T_0, \dot{\varepsilon})$ for $T_0 \geq T_g$
Poisson ratio	ν	—	0.36
Glass transition temperature	T_g	K	625
Melting temperature	T_m	K	993
Density	ρ	kg/m^3	6125
Specific heat at constant volume	C_v	$J/(kg \cdot K)$	400
Frequency of atomic vibration	f	s^{-1}	1×10^{13}
Average atomic volume	Ω	A^3	25
Critical volume	v^*	A^3	20
Initial temperature	T_0	K	295-683
Initial free volume concentration	ξ_0	—	0.05
Critical free volume concentration	ξ_c	—	0.065
Geometrical factor	α	—	0.05 for $T_0 < T_g$ $\alpha(T_0)$ for $T_0 \geq T_g$
Activation energy	ΔG^m	eV	ΔG^m ($\dot{\varepsilon}$)
Number of diffusive jumps	n_D	—	$n_D(T_0)$
Hydrostatic stress sensitivity factor	Λ	—	$\Lambda_C = 0.05$

2.2. Finite element implementation of constitutive model

Through the UMAT subroutine, the above constitutive model and the failure criterion are implemented into the LS-DYNA commercial software. As mentioned in [18], if the thermal conduction and the diffusion of free volume are both considered, it will be difficult to match up the two different length scales in the FEM simulation. As the influences of the diffusion of free volume and the thermal conduction are small [10,11], to avoid the trouble of matching up the different length scales, the two diffusion issues are ignored in the FEM implementation. After such a predigesting treatment, the corresponding FEM simulation is independent on the length scale.

2.3. Estimation of material parameters

For convenience of comparative analysis with the corresponding test data, the typical $Zr_{41.2}Ti_{13.8}Cu_{12.5}Ni_{10.0}Be_{22.5}$ (Vitreloy 1) BMG is chosen as the model material. Integrated with the related literatures [10,11,15,18,19], corresponding material parameters are listed in Table 1. The authors had conducted a detailed analysis on the material parameters, and the related discussions could be referred in [21].

3. Deformation and failure of BMGs under different initial temperatures

In order to validate the applicability of the constitutive model, FEM simulations will be conducted and also the effect of various parameters on the mechanical behavior of BMGs will be analyzed. For the computational convenience, a one-element geometrical model is adopted. As the characteristic size of the test sample is usually in millimeter scale, we choose a hexahedral element with side length of 1 mm.

The mechanical properties of BMGs under different strain rates and temperatures varies significantly. In the following he evolutions of various material parameters at different initial temperatures will be analyzed in detail, especially the inner variables. Then corresponding mechanism of the macroscopic mechanical behavior is discussed based on the parameter analysis. In the simulations the failure of material will be ignored for observing the evolutions of parameters thoroughly and discussing their influences. Besides, the analysis mainly focuses on the case of uni-axial compression, and deformation and failure characteristics of BMGs under tension or shear are similar to that under compression.

For convenience to compare with corresponding test data, two typical strain rates, $\dot{\varepsilon} = 10^{-1}\,s^{-1}$ and $10^3\,s^{-1}$, are selected. The evolutions of various parameters under different initial temperatures are shown in Figs. 1–2, respectively, in which the abscissa represents the engineering strain. Besides, the test data are symbolized as "test" and indicate the results in [3]. The data with no special symbol represent the simulation results. Same symbols will be used in the latter.

It can be seen in Fig. 1(a) the simulation results agree with the test data well. The yielding stress σ_Y decreases gradually with the increase of initial temperature, and the reduction is much more significant after $T_0 \geq T_g$. Besides, within the supercooled liquid region, the softening matter becomes slower and the collapse of stress after softening decreases gradually with increasing the initial temperature. Almost no softening matter occurs at 683 K. It further confirms that BMGs within the supercooled liquid region are suitable for precise microforming, and it can be inferred that a slight increase of material temperature will lead a significant reduction of difficulty in microforming process.

The evolution of temperature T is similar to that of ε_e^p in Fig. (1c). However, due to the strain rate dependence of the Taylor-Quinney coefficient β_{TQ} in Eq. (4), the temperature rises ΔT under various initial temperatures are all relatively low. It deviates from the experimental results as the fracture surface of BMGs usually shows a heat phenomenon. The possible reason is that, the

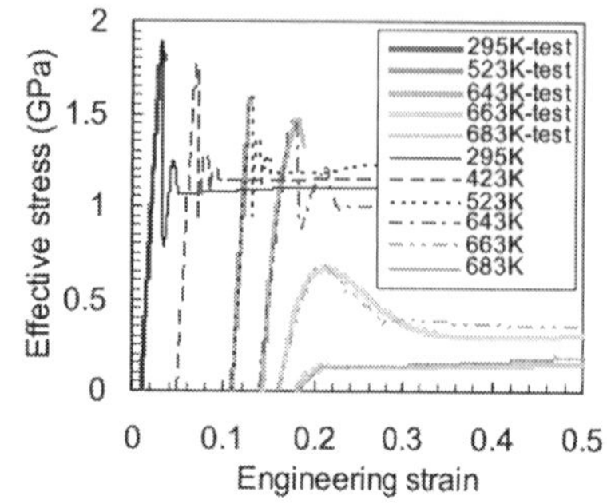

(a) Effective stress

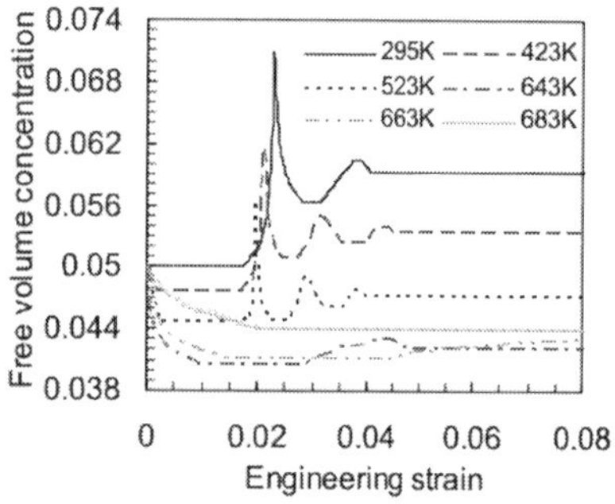

(b) Free volume concentration

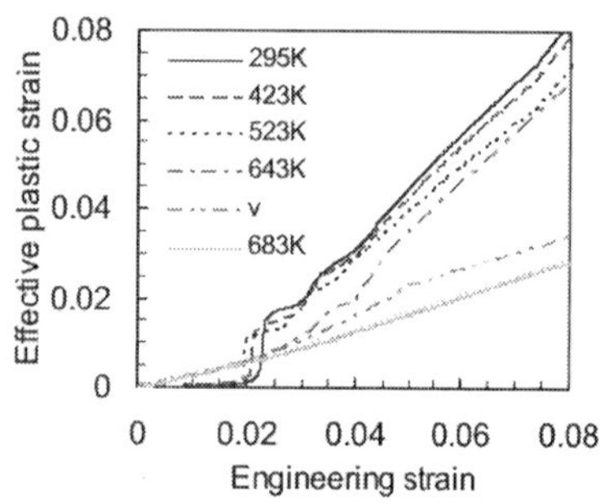

(c) Effective plastic strain

Figure 1. Evolutions of parameters for uniaxial compression at $\dot{\varepsilon} = 10^{-1}\ s^{-1}$ and various initial temperatures.

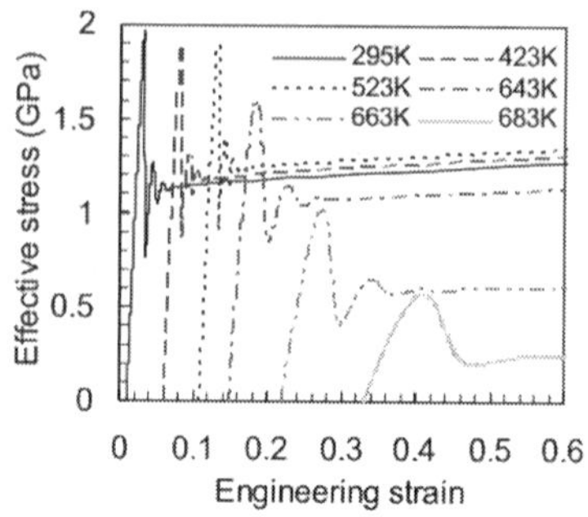

(a) Effective stress

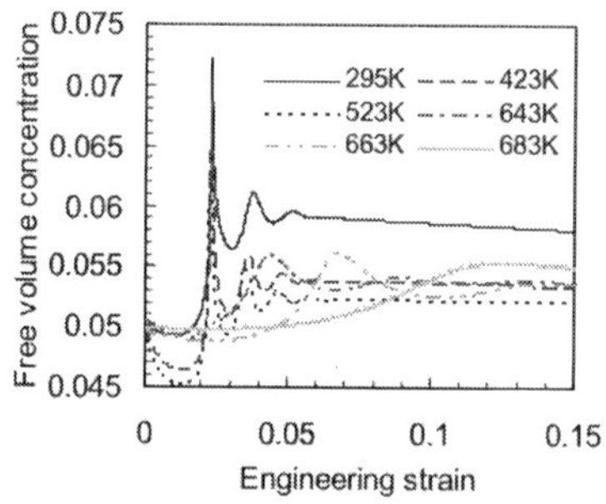

(b) Free volume concentration

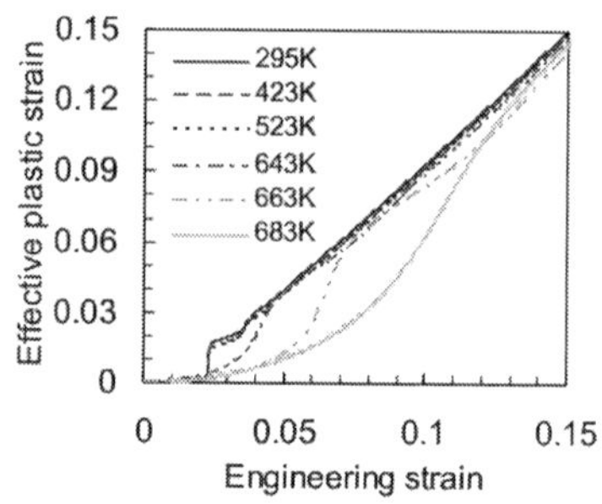

(c) Effective plastic strain

Figure 2. Evolutions of parameters for uniaxial compression at $\dot{\varepsilon} = 10^{3}\ s^{-1}$ and various initial temperatures.

simulations only consider the deformation energy of one element, whilst the neighboring deformation energy also releases on the fracture surface and induce significant temperature rise for the actual material [20]. When the simulation considers the failure of BMG, we can change the value of β_{TQ} and correspondingly ΔT is modified. Related analysis is seen in [21].

By comparing Fig. 2 with Fig. 1 it can be found that while $T_0 < T_g$, the evolutions of parameters are similar to each other for the two strain rates. The yielding stress and strain increase slightly for the higher strain rate, and the values of ξ_{peak} as well as ε_e^p after the precipitous increase also have a slight increase, i.e. deformation localization becomes more significant. Comparatively, at the temperatures of $T_0 \geq T_g$, the differences between evolutions under the two strain rates are distinct. While under high strain rate, the yielding stress and strain increase distinctly, and the softening of stress becomes much more significant. Moreover, the value of ξ at the latter stage exceeds the initial value ξ_0, and ε_e^p exhibits a certain precipitous increase characteristic. These properties indicate that though within the supercooled liquid region, the inner structure of material will become more disordered after deformation under high strain rates, and deformation localization occurs, i.e. the homogeneous flow begin to translate into inhomogeneous deformation. It accords with the actual mechanical behavior in the tests [3].

To further investigate the deformation and failure characteristics, the variations of peak free volume

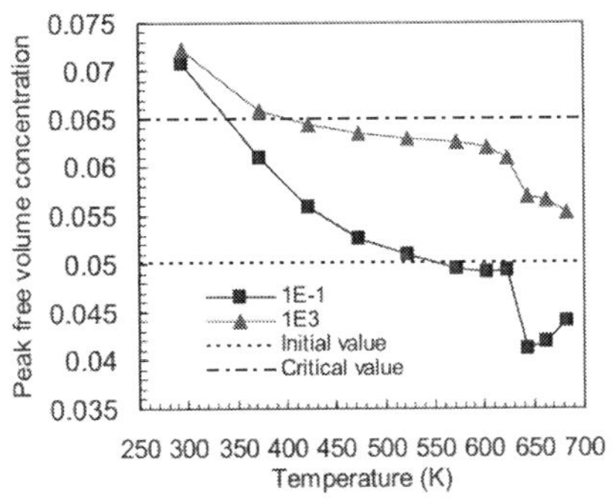

Figure 3. Variations of peak free volume concentration ξ_{peak} via the initial temperature at different strain rates.

concentration ξ_{peak} via the initial temperatures at the two strain rates are listed in Fig. 3. As ξ no longer changes after decreasing to a certain value at $\dot{\varepsilon} = 10^{-1}\,s^{-1}$ and $T_0 \geq T_g$ (see Fig. 1(b)), the stable value is adopted as the value of ξ_{peak}.

As shown in Fig. 3, ξ_{peak} decreases with increasing the initial temperature, and the decrease is more speedy for a lower strain rate. Besides, in the case of $\dot{\varepsilon} = 10^{-1}\,s^{-1}$, ξ_{peak} is smaller than ξ_0 when $T_0 \geq 550\,K$, implying that the deformation of material may display as a homogeneous flow even if it hasn't been within the supercooled liquid region. Being as the test evidence, Nieh et al. [2] demonstrated when T_0 increases up to a value of $0.7\,T_g$, the BMG deformation indeed displayed as a homogeneous flow. In contrast, in the case of $\dot{\varepsilon} = 10^3\,s^{-1}$, ξ_{peak} is significantly larger than ξ_0, indicating the inner structure of material becomes more disordered after deformation, and deformation localization occurs, i.e., the homogeneous flow translates into inhomogeneous deformation. However, the value of ξ_{peak} is still less than that of ξ_c when $T_0 \geq 423\,K$, which indicates the onset of failure is relatively more difficult for a higher initial temperature.

4. Conclusions

Based on the coupled thermo-mechanical model, a multiaxial-stress-status constitutive model which considers the effects of the free volume, temperature and hydrostatic stress on the deformation and failure of BMGs is further suggested in the present manuscript. Besides, a failure criterion of critical free volume concentration is introduced. The constitutive model and the failure criterion are implemented into the LS-DYNA commercial software through UMAT subroutine. The deformation and failure characteristics of BMGs under different initial temperatures are analyzed based on the FEM simulations.

Related analysis shows that the initial temperature significantly affects the mechanical behavior of BMGs. While under the condition of $T_0 < T_g$, the material displays a precipitous softening of stress after yielding, and the deformation and failure process is very speedy and highly localized. In contrast, within the supercooled liquid region ($T_0 \geq T_g$), the softening character of material is weakened and the material prefers to flow homogeneously. However, the deformation will translate into inhomogeneous status if distinctly increasing the strain rate. Generally, when the initial temperature increases, the deformation localization character is weakened and failure is more difficult to occur in BMGs.

The project is supported by the National Natural Science Foundation of China-NSAF (10976100), the National Outstanding Young Scientists Foundation of China (11225213) and the National Natural Science Foundation of China (11221202).

References

[1] Inoue A., Kawamura Y., Shibata T., et al. Viscous flow deformation in supercooled liquid state of bulk amorphous $Zr_55Al_10Ni_5Cu_30$ alloy. Materials Transactions JIM, 1996, **39** (6): 1337–1341.

[2] Nieh T.G., Wadsworth J., Liu C.T., et al. Plasticity and structural instability in a bulk metallic glass deformed in the supercooled liquid region. Acta Materialia, 2001, **49**; 15: 2887–2896.

[3] Lu J., Ravichandran G., Johnson W.L. Deformation behavior of the $Zr_{41.2}Ti_{13.8}Cu_{12.5}Ni_{10}Be_{22.5}$ bulk metallic glass over a wide range of strain-rates and temperatures. Acta Materialia, 2003, **51**: 3429–3443.

[4] Zhang Z.F., Eckert J., Schultz L. Difference in compressive and tensile fracture mechanisms of $Zr_{59}Cu_{20}Al_{10}Ni_8Ti_3$ bulk metallic glass. Acta Materialia, 2003, **51**:1167–1179.

[5] Dai L.H., Bai Y.L. Basic mechanical behaviors and mechanics of shear banding in BMGs. International Journal of Impact Engineering, 2008, **35**: 704–716.

[6] Nishiyama N., Inoue A. Glass transition behavior and viscous flow working of $Pd_{40}Cu_{30}Ni_{10}P_{20}$ amoephous alloy. Materials Transactions JIM, 1999; **40** (1): 64–70.

[7] Conner R.D., Dandliker R.B., Scruggs V., et al. Dynamic deformation behavior of tungsten-fiber / metallic–glass matrix composites. International Journal of Impact Engineering. 2000; **24**: 435–444.

[8] Choi-Yim H., Conner R.D., Szuecs F., et al. Quasistatic and dynamic deformation of tungsten reinforced $Zr_57Nb_5Al_10Cu_{15.4}Ni_{12.6}$ bulk metallic glass matrix. Scripta Materialia, 2001; **45**: 1039–1045.

[9] Spaepen F. A microscopic mechanism for steady state inhomogeneous flow in metallic glasses. Acta Metallurgica, 1977, **25** (9): 407–415.

[10] Huang R., Suo Z., Prevost J.H. Inhomogeneous deformation in metallic glasses. Journal of the Mechanics and Physics of Solids, 2002, **50**: 1011–1027.

[11] Jiang M.Q., Dai L.H. On the origin of shear banding instability in metallic glasses. Journal of the Mechanics and Physics of Solids, 2009, **57**: 1267–1292.

[12] Zhang Z.F., Eckert J. Unified Tensile Fracture Criterion. Physical Review Letters, 2005, PRL **94**: 094301-1-4.

[13] Chen Y., Jiang M.Q., Wei Y.J., et al. Failure criterion for metallic glasses. Philosophical Magazine, 2011, **91**(36): 4536–4554.

[14] Zhang Z.F., Wu F.F., Gao W., et al. Wavy cleavage fracture of bulk metallic glass. Applied Physics Letters, 2006, **89**: 251917-1-3.

[15] Wright W.J., Hufnagel T.C., Nix W.D. Free volume coalescence and void formation in shear bands in metallic glass. Journal of Applied Physics, 2003; **93** (8):1432–1437.

[16] Xue Y.F., Cai H.N., Wang L., et al. Effect of loading rate on failure in Zr-based bulk metallic glass. Materials Science and Engineering A, 2008, **473**: 105–110.

[17] Jiang M.Q., Ling Z., Meng J.X., et al. Energy dissipation in fracture of bulk metallic glasses via inherent competition between local softening and quasi-cleavage. Philosophical Magazine, 2008, **21**(8): 407–426.

[18] Gao Y.F. An implicit finite element method for simulating inhomogeneous deformation and shear bands of amorphous alloys based on the free-volume model. Modelling and Simulation in Materials Science and Engineering, 2006, **14**: 1329–1345.

[19] Chen Y., Jiang M.Q., Dai L.H. How does the initial free volume distribution affect shear band formation in metallic glass?. Science China: Physics, Mechanics & Astronomy. 2011, **54** (8): 1488–1494.

[20] Zhou F., Wright T.W., Ramesh K.T. A numerical methodology for investigating the formation of adiabatic shear bands. Journal of the Mechanics and Physics of Solids, 2006, **54**: 904–926.

[21] Li J.C., Wei Q., Chen X.W., et al. On the mechanism of deformation and failure in bulk metallic glasses. Materials Science and Engineering A, 2014, **610**: 91–105.

EPJ Web of Conferences **94**, 04018 (2015)
DOI: 10.1051/epjconf/20159404018

Behaviour and modelling of aluminium alloy AA6060 subjected to a wide range of strain rates and temperatures

Vincent Vilamosa, Tore Børvik, Odd Sture Hopperstad, and Arild Holm Clausen[a]

Structural Impact Laboratory (SIMLab), Department of Structural Engineering, Norwegian University of Science and Technology (NTNU), Trondheim, Norway

Abstract. The thermo-mechanical behaviour in tension of an as-cast and homogenized AA6060 alloy was investigated at a wide range of strains (the entire deformation process up to fracture), strain rates (0.01–750 s^{-1}) and temperatures (20–350 °C). The tests at strain rates up to 1 s^{-1} were performed in a universal testing machine, while a split-Hopkinson tension bar (SHTB) system was used for strain rates from 350 to 750 s^{-1}. The samples were heated with an induction-based heating system. A typical feature of aluminium alloys at high temperatures is that necking occurs at a rather early stage of the deformation process. In order to determine the true stress-strain curve also after the onset of necking, all tests were instrumented with a digital camera. The experimental tests reveal that the AA6060 material has negligible strain-rate sensitivity (SRS) for temperatures lower than 200 °C, while both yielding and work hardening exhibit a strong positive SRS at higher temperatures. The coupled strain-rate and temperature sensitivity is challenging to capture with most existing constitutive models. The paper presents an outline of a new semi-physical model that expresses the flow stress in terms of plastic strain, plastic strain rate and temperature. The parameters of the model were determined from the tests, and the stress-strain curves from the tests were compared with the predictions of the model. Good agreement was obtained over the entire strain rate and temperature range.

1. Introduction

Aluminium alloys in the AA6xxx series are often used for extruded profiles and rolled sheets or plates. Such components occur frequently in fields of engineering where lightweight designs are required, for instance safety parts in vehicles and different protective structures. Such parts have to resist rapid loading. High strain rate occurs often in combination with elevated temperatures in impact situations, but also in metal forming operations like extrusion and rolling. Common for all these fields of application is that numerical tools and the finite element method are important in the design process. Accurate predictions require material models that represent the physical response in an adequate way.

There are very few systematic experimental studies of coupled effects between strain rate and temperature in the literature. A particular feature associated with tension tests at high temperatures is that necking occurs at a comparatively small deformation, calling for local measurements of the strains in the neck in order to determine the true stress-strain curve.

This paper presents results from uniaxial tension tests on an AA6060 alloy at a wide range of strain rates and temperatures. The test data are fitted to a recently proposed constitutive model that provides a close representation of the observed behaviour. A more comprehensive presentation of the dynamic test rig allowing for different temperatures, the uniaxial test results and the material model is provided in three articles by Vilamosa et al. [1–3].

[a] Corresponding author: `arild.clausen@ntnu.no`

2. Materials and methods

This investigation involves the two aluminium alloys Al-0.5Mg-0.45Si and Al-0.45Mg-0.4Si. Both are within the window of the AA6060 alloy. Their mechanical response is similar [2], and they are therefore treated in common in the numerical part of this paper.

The material was delivered as cast and homo-genised billets by Hydro Aluminium, and was naturally aged for about 18 months prior to testing. The sample shown in Fig. 1 was applied in all tests.

The thermo-mechanical test series consisted of tension tests at different temperatures (20 °C, 200 °C, 250 °C, 300 °C and 350 °C) and nominal strain rates (0.01 s^{-1}, 1 s^{-1}, 350 s^{-1} and 750 s^{-1}). The tests at low to moderate strain rates were carried out in a universal testing machine, while a SHTB system was employed for the dynamic tests. In advance, it was checked that the materials are isotropic [2].

2.1. Thermo-mechanical tension tests

2.1.1. Quasi-static tests

The tests at nominal strain rates $\dot{e}$ of 0.01 s^{-1} and 1 s^{-1} were carried out under displacement control in a Zwick-Roell testing machine, applying a cross-head velocity of 0.05 mm/s and 5 mm/s. A water-cooled induction heating system delivered by MSI Automation was employed to heat the samples. The heating rate was about 10 °C/s. The temperature was kept stable during the test with a feed-back loop provided by the temperature measurement system (a laser-based pyrometer delivered

Article available at http://www.epj-conferences.org or http://dx.doi.org/10.1051/epjconf/20159404018

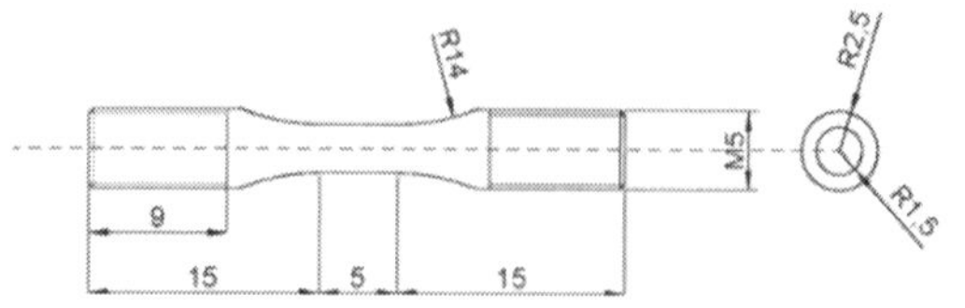

Figure 1. Tensile test specimen. Measures in mm.

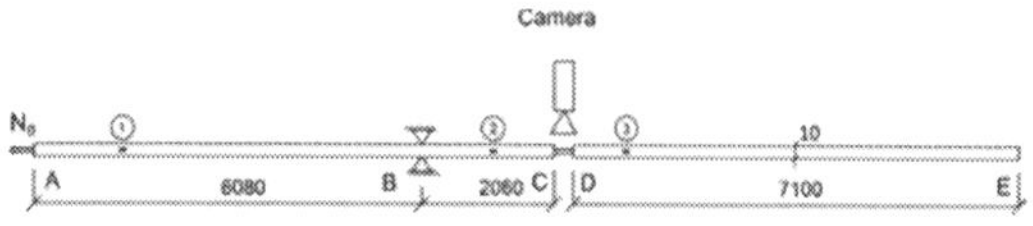

Figure 2. Sketch of SHTB system. Measures in mm.

by LumaSense Technologies). Vilamosa et al. [2] provide more information on this experimental set-up.

It was pointed out in Sect. 1 that necking occurs at an early stage of deformation at elevated temperatures. Therefore, it was required to determine the local strain inside the neck The deformation of the samples was captured by a digital camera (Prosilica GC2450) having a 5 megapixel Sony ICX625 CCD sensor. Frames were recorded until fracture with a sampling rate of 2 Hz ($\dot{e} = 0.01\,s^{-1}$) and 15 Hz ($\dot{e} = 1\,s^{-1}$). The pixel size was determined before each test by measuring the initial diameter D_0 of the samples both with the camera and a digital calliper. After the tests, an edge detection script was applied to determine the minimum cross section diameter D_s of the sample [1]. Assuming constant volume during plastic deformation, the local logarithmic strain ε was determined from

$$\varepsilon = \ln\left(\frac{A_0}{A_s}\right) = 2\ln\left(\frac{D_0}{D_s}\right) \quad (1)$$

where A_0 is the initial area of the sample's cross-section and $A_s = (\pi/4)\,D_s^2$ is the minimum cross-section area during testing. The true stress σ was found by dividing the force F measured by the load cell with the current area, viz.

$$\sigma = \frac{F}{A_s}. \quad (2)$$

Finally, the plastic strain ε^p was determined from the following relation

$$\varepsilon^p = \varepsilon - \frac{\sigma}{E} \quad (3)$$

where E is Young's modulus at room temperature.

2.1.2. Dynamic tests

A split-Hopkinson tension bar (SHTB) system, see Fig. 2, was used in the dynamic tests at approx. $\dot{e} = 350\,s^{-1}$ and $\dot{e} = 750\,s^{-1}$. It consists of an 8140 mm long input bar (ABC) and a 7100 mm long transmission bar (DE). The sample is located between points C and D. Both bars have diameter 10 mm, and are made of high-strength steel with Young's modulus 210 GPa at room temperature.

The bars are instrumented with three strain gauge stations at the locations ①, ② and ③. Strain gauge ① is used to monitor the level of the pre-tension force N_0. The stress wave propagation during the test is measured with the gauges at ② and ③. Applying the classical Kolsky equations, the data from these gauges can be used to determine the nominal stress, strain and strain rate in the specimen. Provided that the deformation of the sample is homogeneous, i.e., before the onset of necking, it is straightforward to calculate the true stress and logarithmic strain from the nominal data.

Also the dynamic tests had to be instrumented with a camera in order to find σ and ε in the comparatively large phase of the test after necking. A SA1.1 Photron high-speed camera with frame rate 100 kHz was applied for this purpose. The minimum diameter of the sample during the test was found from the digital pictures in the same way as in the quasi-static tests. Equation (1) was thereafter applied to find the logarithmic strain. The force in the sample was determined by use of the conventional method for analysis of SHTB tests. It has been shown by Chen et al. [4] that there is no dispersion in the SHTB system shown in Fig. 2. The force F in the specimen is therefore proportional with the strain ε_T measured at position ③ of the transmission bar, i.e.

$$F = A_b E_b \varepsilon_T \quad (4)$$

where A_b and E_b respectively are the cross-section area and Young's modulus of the bar. Subsequently, the true stress was found with Eq. (2), and Eq. (3) gives the plastic strain.

The induction heating equipment presented in Sect. 2.1.1 was used also in the dynamic tests. It was demonstrated by Vilamosa et al. [1] that the heated sample did not disturb the wave propagation in the bars. Further, the camera-based measurement technique was validated against the conventional approach (Kolsky equations) for strain determination before necking. A more comprehensive presentation of the experimental set-up for the thermo-mechanical tests is provided by Vilamosa et al. [1,2].

2.2. Data processing

All samples exhibited pronounced necking that increased with increasing temperature. This results in a three-dimensional stress state in the neck. This is handled with the Bridgman relation, which takes the shape of the neck and the diameter of the smallest cross-section into account. Such information is available from the digital pictures. The equivalent stress σ_{eq} is determined from

$$\sigma_{eq} = \frac{\sigma}{\left(1 + \frac{4R}{D_s}\right)\ln\left(1 + \frac{D_s}{4R}\right)} \quad (5)$$

where R is the mean radius of the curvature of the neck. This radius was estimated from each frame with a least squares method [1].

Small oscillations were observed in the equivalent stress – plastic strain curves from all tests, in particular the dynamic ones from SHTB. As the subsequent analysis requires values of σ_{eq} at certain levels of plastic strain ε^p, see Sect. 4.2, these oscillations should be removed. For this

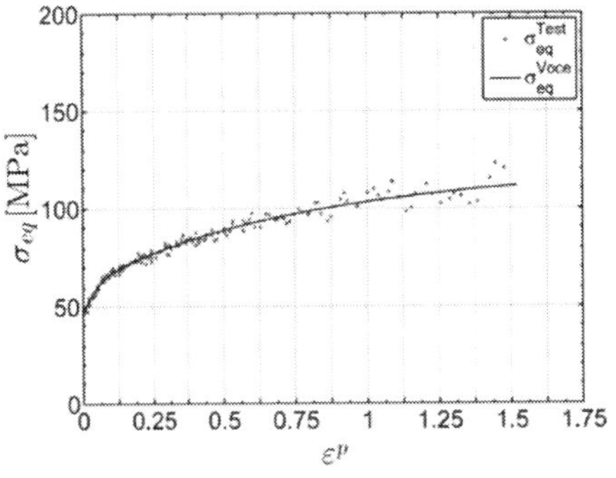

Figure 3. Equivalent stress – plastic strain curves determined from a test (discrete data points) and fitted to Eq. (6) for a representative test at a nominal strain rate at approx. $350\,s^{-1}$ and temperature of 350 °C.

purpose, all stress-strain curves were fitted to a two-term Voce relation extended with a linear hardening term

$$\sigma_{eq} = \sigma_0 + \sum_{i=1}^{2} Q_i \left(1 - \exp\left(-\frac{\theta_i}{Q_i}\varepsilon^p\right)\right) + H\varepsilon^p \tag{6}$$

The five parameters Q_i, θ_i and H were fitted to the part of the $\sigma_{eq} - \varepsilon^p$ curve with plastic strain larger than 0.01, thereby omitting the initial part of the curve where there is a lack of equilibrium in the SHTB tests. Considering a representative dynamic test, Fig. 3 shows the $\sigma_{eq} - \varepsilon^p$ curves as obtained in the test and as fitted with Eq. (6).

In all tests, the local plastic strain rate $\dot{\varepsilon}^p$ in the necked section was obtained by numerical differentiation of the plastic strain vs. time curve.

3. Experimental results

Equivalent stress – plastic strain curves from all tests are shown in Fig. 4. The solid and dotted lines refer to the two slightly different AA6060 alloys, respectively Al-0.5Mg-0.45Si and Al-0.45Mg-0.4Si.

Figure 4 reveals that 1 or 2 samples of each alloy were tested at each combination of strain rate and temperature. The scatter between replicate tests was small. It appears that the response of the two alloys is almost identical. They are therefore treated in common in the subsequent data analysis. Some curves are clipped at the onset of necking. In these cases, the neck was not visible in the digital photos because it was hidden behind the coil of the induction heater.

Figure 4 shows that the material exhibits a decrease in both yield strength and work hardening with increasing temperature for the four levels of nominal strain rate. It is interesting to notice that the curves at different rates are rather similar at room temperature, while the work hardening increases considerably with $\dot{e}$ at 350 °C. This interaction effect between strain rate and temperature is in general not well captured by existing constitutive models for use in finite element (FE) simulations [2].

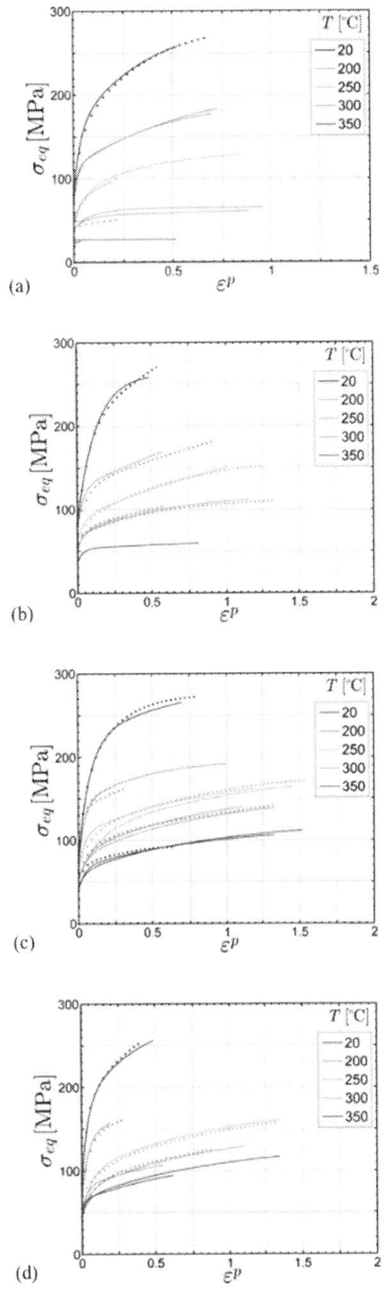

Figure 4. Equivalent stress – plastic strain curves for Al-0.5Mg-0.45Si (solid lines) and Al-0.45Mg-0.4Si (dotted lines) at all temperatures and nominal strain rate levels of (a) $0.01\,s^{-1}$, (b) $1\,s^{-1}$, (c) $350\,s^{-1}$ and (d) $750\,s^{-1}$.

4. Constitutive model

The most employed constitutive models in FE codes are of phenomenological nature. Such models are empirical, i.e., based on experimental observations, they have few parameters, and are often expressed as algebraic equations that are easy to implement in the code. The physically based models provide a different approach. They seek to relate the flow stress to the evolution of dislocation density and the thermal activation energies that are required for dislocations to overcome obstacles.

As an attempt to capture the response of AA6060, see Fig. 4, in FE simulations, a thermoelastic-thermo-viscoplastic model has been developed. It assumes co-rotational hypoelastic plasticity, implying that the elastic strains should be small. Since the considered material is isotropic, the model incorporates isotropic tensors for the thermo-elastic moduli and also the isotropic Hershey yield function. Vilamosa et al. [3] provides a more comprehensive presentation of the general framework of the model. The next section will, however, outline the main ideas of how the flow stress is modelled.

4.1. Flow stress and work hardening

Bergström [5] suggested to split the flow stress σ_f into three terms

$$\sigma_f\left(R, \dot{\varepsilon}, T\right) = \sigma_a(T) + \sigma_v\left(\dot{\varepsilon}, T\right) + R. \qquad (7)$$

The athermal yield stress $\sigma_a(T)$ is dependent on temperature only; not strain rate. It is assumed that the temperature sensitivity of σ_a is the same as for the shear modulus (and Young's modulus) [3]. This stress captures the strengthening effect from particles.

The instantaneous viscous stress $\sigma_v\left(\dot{\varepsilon}, T\right)$ depends on both strain rate and temperature. Thus, a change of either of these variables, or both, will cause a change of σ_v. As outlined by Vilamosa et al. [3], the expression for σ_v combines an Arrhenius type of expression with an activation energy profile and also a regularization to avoid instabilities at low strain rates. The viscous stress represents the presence of obstacles.

The work hardening R is related to the dislocation density ρ, i.e., R increases with ρ. It depends indirect-ly on strain rate and temperature through the evolution rule for ρ, and in addition on temperature through the shear modulus [3]. The final expression for the evolution of work hardening as function of plastic strain reads

$$\frac{dR}{d\varepsilon^p} = \theta_0 \exp\left(-\frac{R}{R_s}\right) \qquad (8)$$

where θ_0 is a temperature sensitive term representing the initial slope of the work hardening and R_s is a saturation value which depends both on temperature and strain rate. At the strain rates covered in this study, the dislocation density is believed to depend mainly on dynamic recovery which is a thermally activated process. The significant strengthening observed in the dynamic tests, see Fig. 4(c) and (d), is attributed to a change of the strain rate sensitivity of the material resulting in a reduced dynamic recovery [3,6]. This is implemented in the model. The new constitutive model might therefore be applied for this class of materials up to relatively high strain rates of order $10^4\,\mathrm{s}^{-1}$.

4.2. Evaluation of the model

As the two AA6060 alloys exhibit similar response at $350\,\mathrm{s}^{-1}$ and $750\,\mathrm{s}^{-1}$, the constitutive model is calibrated with the test data obtained at nominal strain rates of

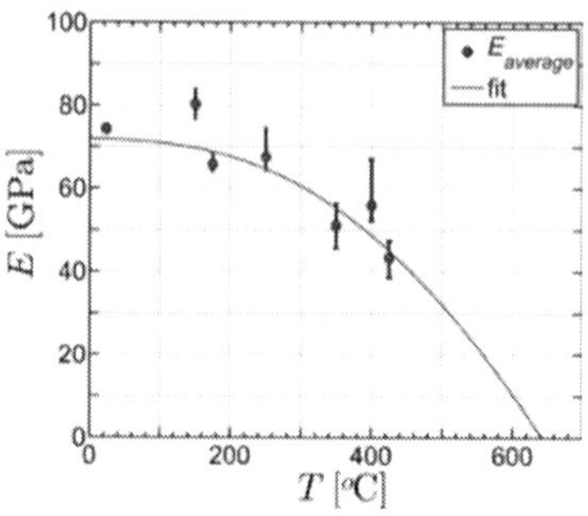

Figure 5. Temperature sensitivity of Young's modulus. The bars denote the range of the measured values.

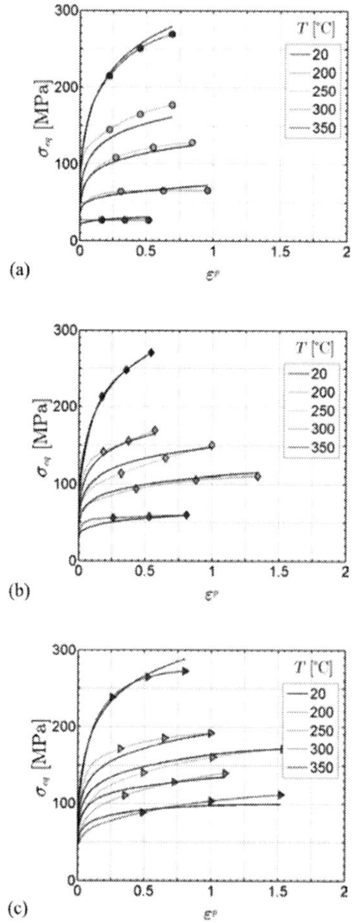

Figure 6. Comparison of equivalent stress – plastic strain curves from representative tests (lines with symbols) and the constitutive model (solid lines without symbols) at all temperatures and nominal strain rate levels of (a) $0.01\,\mathrm{s}^{-1}$, (b) $1\,\mathrm{s}^{-1}$ and (c) $350\,\mathrm{s}^{-1}$.

$0.01\,\mathrm{s}^{-1}$, $1\,\mathrm{s}^{-1}$ and $350\,\mathrm{s}^{-1}$. In addition, information about the temperature sensitivity of Young's modulus is required, see Fig. 5. Vilamosa et al. [3] provide a more comprehensive outline of the parameter identification procedure.

Figure 6 compares results from representative tests (taken from Fig. 4) with equivalent stress – plastic strain

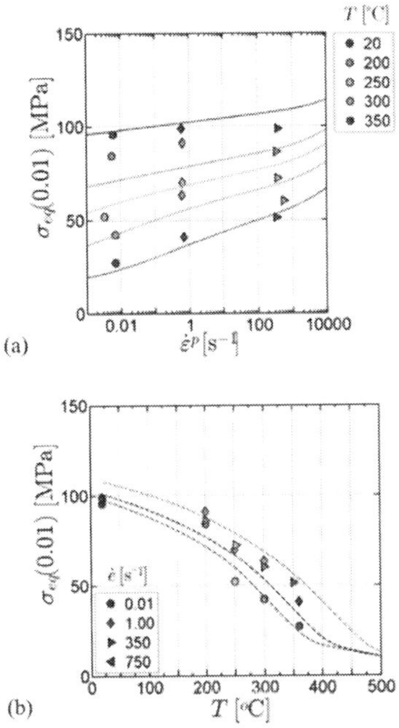

Figure 7. Comparison of equivalent stress at a plastic strain of 0.01 in the tests (discrete points) and the model (lines) at (a) different strain rates, and (b) different temperatures.

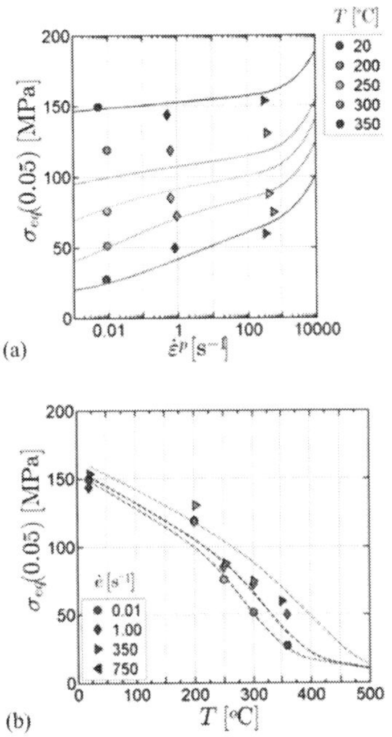

Figure 8. Comparison of equivalent stress at a plastic strain of 0.05 in the tests (discrete points) and the model (lines) at (a) different strain rates, and (b) different temperatures.

curves predicted by the model. The entire spectre of strain rates and temperatures is covered in the figure. The agreement between the test data and the model is in general good at all levels of plastic strain. In particular, the model captures that the work hardening at the highest temperatures almost vanishes at $\dot{e} = 0.01\,\mathrm{s}^{-1}$, while it is significantly larger at $\dot{e} = 350\,\mathrm{s}^{-1}$.

A closer look reveals, however, that the model underestimates the yield strength in some cases, in particular at temperatures around 200 °C. Such a mismatch has also consequences for the prediction of the flow stress. On the other hand, the flow stress is overestimated at high temperatures under dynamic condi-tions at low plastic strains.

An alternative way to compare the results is to extract the equivalent stress at certain levels of plastic strain. The values of $\varepsilon^p = 0.01$, $\varepsilon^p = 0.05$ and $\varepsilon^p = 0.6$ were chosen for this purpose because these strains represent different stages of work hardening. These three levels are addressed in Figs. 7, 8 and 9, respectively, where sub-figures (a) address the response as function of plastic strain rate, while sub-figures (b) have temperature at the abscissa axis. The three dashed lines in sub-figures (b) refer to strain rates of $0.01\,\mathrm{s}^{-1}$, $1\,\mathrm{s}^{-1}$ and $350\,\mathrm{s}^{-1}$.

At high strain rates, there is a significant adiabatic heating during a tension test. The associated increase of temperature ΔT is estimated by

$$\Delta T = \int_0^{\varepsilon_p} \chi \frac{\sigma_{eq} d\varepsilon_p}{\rho C_p} \tag{9}$$

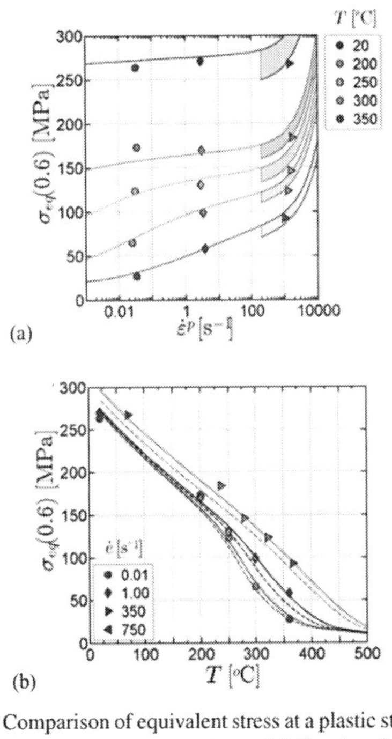

Figure 9. Comparison of equivalent stress at a plastic strain of 0.6 in the tests (discrete points) and the model (lines) at (a) different strain rates, and (b) different temperatures. The shaded areas in (a) address the difference between adiabatic and iso-thermal conditions. The solid lines in (b) represent the increase of strain rate at the neck.

where ρ is the density, C_p is the specific heat capacity, and χ is the Taylor-Quinney coefficient. Both C_p and χ are assumed to be temperature independent. The temperature increase gives a softening effect on σ_{eq}, in particular at $\varepsilon^p = 0.6$. The effect of ΔT is addressed through the shaded area in Fig. 9(a).

Sub-figures (a) in Figs. 7–9 show clearly that the material is close to rate insensitive at room temperature, while σ_{eq} increases significantly with rate at the high temperatures. This tendency is present around the onset of yielding (Fig. 7) as well as after considerable work hardening (Fig. 9). The model captures these observations. Sub-figures (b) show that the softening with increasing temperature is more prominent at quasi-static loading conditions than at high strain rates. The model represents also this feature rather accurately.

In addition to a more comprehensive derivation of the model, Vilamosa et al. [3] also compares the predict-tions of the model with experimental results for an AA6082 alloy. This material has a higher content of the alloying elements Mg and Si than AA6060. Also here, the model gives a close reconstruction of the stress-strain curves. Moreover, finite element simulations of both the quasi-static and dynamic tension tests, in the latter case including the entire SHTB set-up, show that the model is capable of describing the behaviour at different combinations of strain rate and temperature.

5. Conclusion

This paper presented a series of thermo-mechanical tension tests on two slightly different AA6060 alloys. The tests were performed at strain rates between $0.01\,s^{-1}$ and $750\,s^{-1}$ and at temperatures between 20 °C and 350 °C. A universal testing machine was used at low to medium strain rates, while a split-Hopkinson tension bar system was employed at the dynamic strain rates. The samples were heated with an induction apparatus. A camera-based system was used to determine the local strains also after the onset of necking.

A strong coupling between strain rate and temperature sensitivity was found in the tests. The material is close to insensitive to strain rate at room temperature. At the high temperatures, however, there is a significant increase of the yield stress and work hardening with strain rate.

The paper also outlined the main features of a new constitutive model. The model gave in general a faithful representation of the experimental observations.

The authors would like to express gratitude to M.T. Auestad at SIMLab, NTNU, for his assistance with the experimental work. The contributions from PhD E. Fagerholt at SIMLab, NTNU, PhD S.R. Skjervold at SAPA and Professor B. Holmedal at Department of Materials Science and Engineering, NTNU, are also acknowledged.

References

[1] V. Vilamosa, A.H. Clausen, E. Fagerholt, O.S. Hopperstad, T. Børvik, Strain **50**, 223–235 (2014)
[2] V. Vilamosa, A.H. Clausen, T. Børvik, S.R. Skjervold, O.S. Hopperstad, Submitted for journal publication (2015)
[3] V. Vilamosa, A.H. Clausen, T. Børvik, B. Holmedal, O.S. Hopperstad, Submitted for journal publication (2015)
[4] Y. Chen, A.H. Clausen, O.S. Hopperstad, M. Langseth, Int. J. Impact. Eng. **38**, 824–836 (2011)
[5] Y. Bergström, Mat Sci Eng. **5**, 193–200 (1970)
[6] P.S. Follansbee, U.F. Kocks, Acta Metall. **36**, 81–93 (1988)

EPJ Web of Conferences **94**, 04019 (2015)
DOI: 10.1051/epjconf/20159404019

Craters in concrete slabs due to detonation – drawbacks of material models with a Mohr-Coulomb yield surface

Markus Conrad[a]

SAAB Bofors Dynamics Switzerland Ltd., 3600 Thun, Switzerland

Abstract. Numerical simulations have been performed with a commercial distributed explicit FE-solver and the results have been compared with experiments. High explosive was placed in front of different concrete slabs with the dimension 100 × 100 × 16 cm. Some of the results of the simulations, in particular the profile of the craters, are not in agreement with the test results. Therefore the key characteristics of the constitutive equation based on Mohr-Coulomb yield surfaces and a damage evolution linked to the plastic strain has been reviewed.

1. Introduction

This paper shows drawbacks of finite element simulations of concrete walls: Five different composite concrete walls following to Fig. 1 have been investigated and compared to experiments.

The amounts of explosive have been chosen in order to create a hole through the wall thickness (called "big" quantity hereafter) or to have undamaged rear surface (called "small" quantity hereafter). Since the corresponding amount was not known in advance, a huge number of tests have been performed and compared to numerical results.

The constitutive relation with parameters from in the literature and already implemented in the material libraries of commercially distributed finite element programs yield in general good results – even with isotropic material behavior! The amount of explosive or the kinetic energy of a penetrator [1] needed for perforation of concrete is predicted well. However, very simple empirical formulas based on experiments yield the same results. Unfortunately, the standard models built in commercial finite element programs did not correctly predict the cross section of a crater inside concrete. The diameters of the crater on the front and rear surface of the concrete can be calculated precisely, but the diameter of the hole inside the concrete wall was always much bigger than in reality.

This work gives a short introduction into constitutive relations which are used in commercial available finite element codes. The material models are based on Mohr-Coulomb yield surfaces and an isotropic damage parameter which is linked to the effective plastic strain. Emphasis was put on the key characteristic, in particular on the additive decomposition of the stress tensor into pressure and stress deviators as well as the dependence of the model on the fineness of the finite element. It will be shown, that the sensitivity of material parameters onto the dimension of the crater is completely different for each type of composite concrete slab shown in Fig. 1. The introduction of an orthotropic damage parameter which is able to reproduce the orthotropic behaviour of a crack only improved the qualitative issues of a crater. Since the orthotropic damage is based on the principal strain, the result is still dependent on the fineness of the finite element. There is no convergence of the results by mesh refinement. On the contrary the material parameters have been optimized for element size of 5–10 mm.

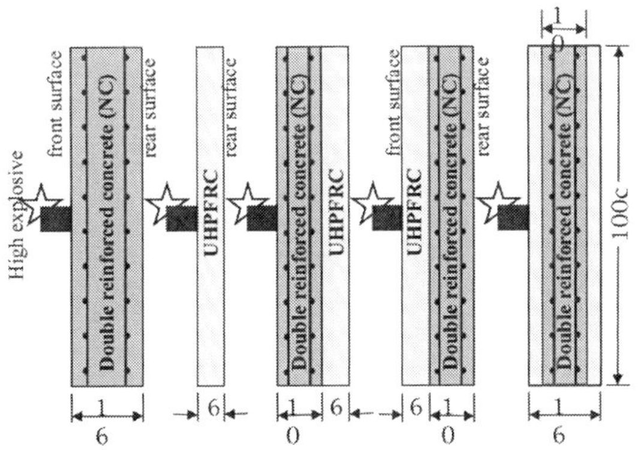

Figure 1. Five different concrete walls with high explosive attached on the front surface. All walls are 16 cm thick except the ultra high performance fibre reinforced (UHPFRC) wall which is only 6 cm thick. The normal concrete is double reinforced, i.e. a mesh of steel reinforcement bars are placed approximately 3 cm below each the surfaces with dimension 100 cm × 100 cm.

2. Constitutive relations

Our FE-Programs uses an additive decomposition of the stress and strain tensor into dilation and deviation. The stress deviator is linked to the strain deviator by the shear modulus. Beyond the yield surface the total strain tensor is decomposed into an elastic strain and plastic strain. The compression is linked to the pressure by the bulk modulus.

[a] Corresponding author: markus.conrad@saabgroup.com

Article available at http://www.epj-conferences.org or http://dx.doi.org/10.1051/epjconf/20159404019

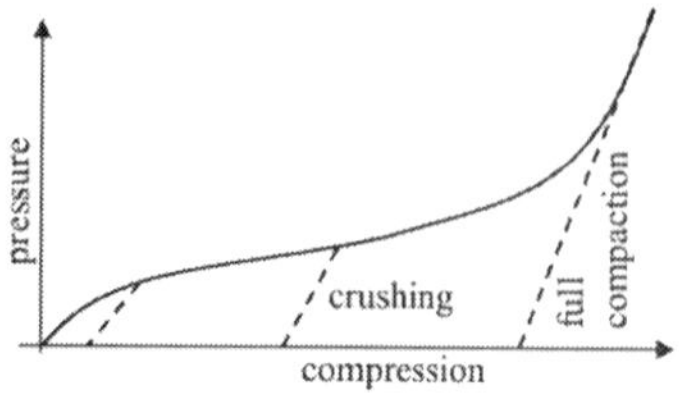

Figure 2. The pressure is a function of compression. The dashed lines – which slopes are the bulk moduli – define the volumetric crushing.

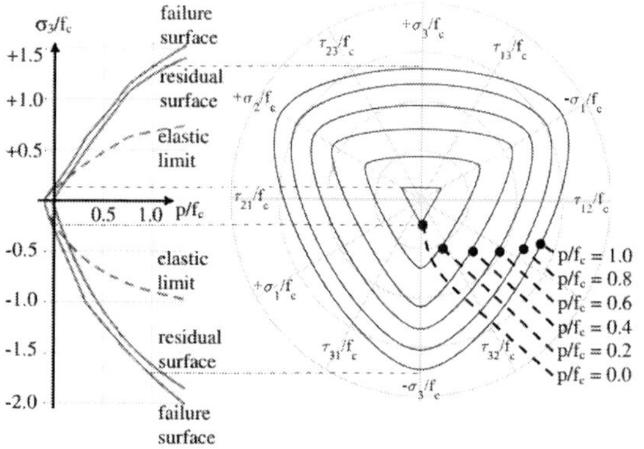

Figure 3. Section plane of the elastic limit, failure surface, and residual surface characterised as Mohr-Coulomb surfaces (on the left hand side). Contours of the failure surface (on the right and side). f_c is the compressive strength of the concrete, p the applied pressure and σ_1, σ_2, and σ_3 the principal stresses.

The equation of states allows for crushing. However, there is no cross link between deviator and dilation in the constitutive relation. There is no volumetric expansion or compression linked to pure shear loads.

2.1. Equation of State

The equation of state links the pressure to the compression. After a tiny elastic zone with reversible compression the concrete starts to crush (irreversible compaction) up to full compaction. During crushing, the bulk modulus varies from the initial to the fully compacted value. Autodyn's RHT-Concrete model uses the p-a equation of state [1], some concretes in LS-Dyna [2] (e.g. *mat_concrete_damage_rel3) use look up table. In other application the bulk modulus is the first constant of a cubic equation of state.

2.2. Strength

The strength of the Material is described by three surfaces: Elastic Limit, failure surface and residual surface. Beyond the elastic limit, there are plastic deformations with strain hardening effect up to the failure surface. Hereafter the material strength softens up to fully damage where the residual surface is applied. The three yield surfaces are shown in Fig. 3 as a function to the pressure and the principal stresses. The residual surface is relatively close to the failure surface except in region with low pressure. However, exactly this low pressure region and the expanding regime is in our focus, i.e. $-0.1 < p/f_c < 0.2$.

The use of yield surfaces is established for ductile materials such as metals. The surfaces in Fig. 3 are based on Mohr-Coulomb's, Johnson-Cook's, or a similar law and take the brittleness of concrete in a threefold manner into account: (1) The residual surface is zero for negative pressure, i.e. broken concrete is expanded without any forces. (2) The elastic and failure surfaces allow for reaction forces in expanding finite elements. (3) The contours of the surfaces are not circles but take a triaxial loading into account using a function of the principal stresses following to Willam and Warnke [1,2]. The absolute value of the yielding is reduced for tensile stresses.

Malvar et al. [2] shows how the surfaces have to be adapted to static tests with compressive strength f_c and tensile strength f_t. For example, the absolute value $\Delta\sigma_m$ of the compressive failure surface is given by

$$\Delta\sigma_m = a_0 r + \frac{p}{a_1 + p \cdot a_2/r} \quad (1)$$

with r the relation of the compressive strength of a "new" concrete to the compressive strength of a concrete for which the material parameters a_0, a_1, and a_2 are validated. Finally, the strength increases with high strain rate.

2.3. Damage

The damage is a parameter in volumetric finite elements which characterises the influence of cracks in brittle materials. The stresses σ are calculated on the bases of the nodal forces and the cross section A_0 of the element. However, an expanding crack reduces this cross section by its surface A_R. As a consequence the local stresses in the element $\tilde{\sigma}$ are increased as

$$\tilde{\sigma}(A_0 - A_R) = \sigma \cdot A_0. \quad (2)$$

The damage is linked to the relation $(A_0 - A_R)/A_0$. As a consequence, a fully damaged material does not anymore resist to expansion or shear. There is *either* a continuing softening between the failure and residual surfaces *or* the strength follows to the failure surface until a critical value of the accumulated damage (typically $D_c = 0.7$) is reached. Beyond this value the material strength follows the residual surface. The first technique is called continuously softening and the latter is called spontaneous failure.

2.3.1. *Damage evolution*

The damage growth follows to an empirical formula. Each of the authors in the reference [1–3] proposes its own formula. They agree that the damage is linked to the plastic strain and total expansion (or compression). Figure 4 shows the damage as a look up table between effective plastic strain and pressure. The damage is given by the user as a function of a modified plastic strain λ with

$$\lambda = \begin{cases} \int_0^{\bar{\varepsilon}_p} \frac{d\bar{\varepsilon}_p}{r_f[1+p/(r_f f_t)]^{b_1}} & \text{if} \quad p > 0 \\ \int_0^{\bar{\varepsilon}_p} \frac{d\bar{\varepsilon}_p}{r_f[1+p/(r_f f_t)]^{b_2}} & \text{if} \quad p \leq 0 \end{cases} \quad (3)$$

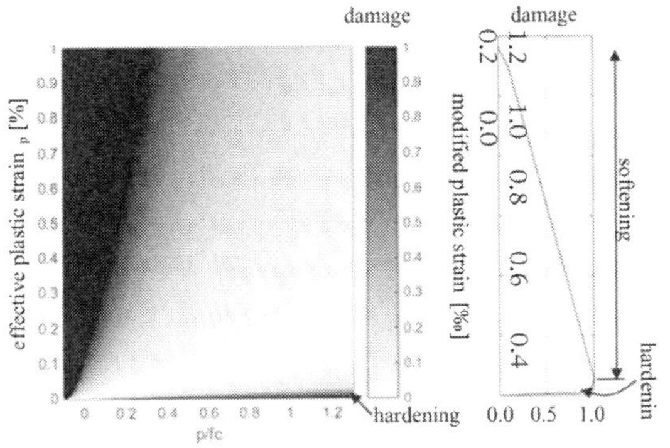

Figure 4. The damage is a function and plastic strain ε_p. The modified plastic strain λ is a function of the plastic stain and pressure. Depending on λ, the strain hardening (transition from elastic limit to failure surface) and softening (transition from failure surface to residual surface) effect is evaluated.

With constants b_1, b_2, and r_f which takes the compressive strength of a "new" concrete into account. Malvar et al. [2] put a lot of emphasis on the damage evolution:" *If the analysis yields a different localization width than anticipated, this should be corrected and the calculation restarted. These parameters will be of importance when the structure analyzed is lightly reinforced or is tension- or shear- critical.*"

Malvar et al. [2] describe the strain hardening also by the damage parameter: For very small plastic strain, there is a continuous transition from the elastic limit to the failure surface. Since the strain hardening effect is accomplished at really small effective plastic strains a simplified strength algorithm without separate elastic limit and strain hardening gives nearly the same results: The energy accumulated during strain hardening is small compared to the total energy needed for perforation of concrete walls.

2.3.2. Localisation

The energy consumed by the crack growth should be equal to the energy dissipated by the damage and the plastic deformation. Lemaître et al. [3] showed that this has an impact to the correct mesh size: The energy Φ^c released by crack growth is proportional to the increase of the crack surface and *equal* to the energies released by plastic deformation Φ^p and damage growth Φ^D which are proportional to the volume involved. A characteristic length l_0 results by equating these two energies, i.e.

$$\left.\begin{array}{l} \Phi^c = \int_0^{A_c} G dA \approx l_0^2 G_c \\ \Phi^p = l_0^3 \int_0^{fracture} \sigma_{ij} d\varepsilon_{ij}^p \approx l_0^3 f_t \varepsilon_u \\ \Phi^D = l_0^3 \int_0^{D_c} \frac{f_t^2}{2E} dD \approx l_0^3 \frac{f_t^2}{2E} D_c \end{array}\right\}$$

$$\Rightarrow G_c = l_0 \left(f_t \varepsilon_u + \frac{f_t^2}{2E} D_c \right) \tag{4}$$

With ε_u the ultimate strain at rupture, G_c the fracture energy release rate, f_t the tensile strength and D_c the critical Damage. Lemaître et al. [3] showed that the characteristic length in concrete is in the order of 10–100 mm, depending on f_t, ε_u and Young's modulus E. In case the mesh size is refined, the parameters for the damage evolution law have to be adapted. The simulation results are mesh dependent, unless some average value of the damage over a region of the characteristic length is used. For a fine mesh, the damage is concentrated in few heavily distorted elements. The volume involved is small and thus the total energy dissipated by plastic deformation and damage growth is much less than in reality. Several techniques to overcome this drawback have been developed and published in the literature. Peerlings et al. [4] compare the different averaging techniques with different weight functions with the gradient enhance method. In the present work, the simplest method has been chosen: The parameters for damage evolution have simple been optimized for mesh sizes between 5–10 mm.

2.4. Anisotropy

A fracture reduces the strength of a structure only for tensional and shear loads which have components along the normal to the fracture surface. Compressive loading with closed cracks are equal to loadings without any cracks. In addition tensional loads parallel to the crack surface are also not altered by the presence of crack.

2.4.1. Damage evolution

The anisotropic influence of cracks in a structure is taken into account with a second order damage tensor $\mathbf{D}$. Following to Lemaître et al. [3], the damage grows proportional to the strain rate only in the direction with positive principal strain.

$$\dot{\mathbf{D}} = A \left[1 + \left(\frac{\varepsilon^*}{a} \right)^2 \right]^{-1} \frac{\langle \boldsymbol{\varepsilon} \rangle_+^2}{\varepsilon^{*2}} \dot{\varepsilon}^* \tag{5}$$

$$\text{with } \varepsilon^* = \sqrt{\langle \boldsymbol{\varepsilon} \rangle_+ : \langle \boldsymbol{\varepsilon} \rangle_+} \text{ and } \langle \varepsilon \rangle_+ = \begin{cases} \varepsilon \text{ if } \varepsilon > 0 \\ 0 \text{ if } \varepsilon \leq 0 \end{cases} \tag{6}$$

and the material constants A and a. The operator ":" is the scalar product of two tensors: $\mathbf{x}{:}\mathbf{x} = x_{ij} x_{ij}$ with Einstein's convention of summation over repeated indices. The symbol $\langle \cdot \rangle$ is the MacCauley bracket.

Finally the threshold of damage growth does not simply depend on a ignition limit strain ε_0^{lim}, but also on the progress of the fracture (damage). Damage growth starts if the strain exceeds a critical value ε^{crit}.

$$\varepsilon^{crit} = a \cdot \tan \left[\frac{\text{trace}(\mathbf{D})}{a \cdot A} + \arctan \left(\frac{\varepsilon_0^{lim}}{a} \right) \right]. \tag{7}$$

2.4.2. Local stress tensor

The local stress $\tilde{\boldsymbol{\sigma}}$ is increased due to that fact, that the crack reduces the original cross section. With A_0 the cross section normal to the direction $\mathbf{t}$ of the traction and A_R the surface of the crack projected onto A_0 the local stress tensor is derived from the global stress tensor $\boldsymbol{\sigma}$ as

$$\begin{array}{c} (A_0 - A_R)\ \langle \tilde{\boldsymbol{\sigma}} \rangle_+ \mathbf{t} = (A_0)\ \langle \boldsymbol{\sigma} \rangle_+ \mathbf{t} \\ \langle \tilde{\boldsymbol{\sigma}} \rangle_- \mathbf{t} = \langle \boldsymbol{\sigma} \rangle_- \mathbf{t} \\ \langle \tilde{\boldsymbol{\sigma}} \rangle_+ + \langle \tilde{\boldsymbol{\sigma}} \rangle_- = \boldsymbol{\sigma}. \end{array} \tag{8}$$

The direction of the normal vector of the crack surface is generally not parallel to the direction **t** of the traction. The term $A_0/(A_0 - A_R)$ is consequently a second order tensor and in principle the inverse of the damage tensor **D**. A formal relationship taking also the direction of the traction and the normal vector of the crack surface into account is given by Voyiadjis et al. [5]:

$$\langle \tilde{\sigma} \rangle_+ = \mathbf{M} \langle \sigma \rangle_+ . \quad (9)$$

The components of the fourth order tensor **M** consists on terms of the components of the damage tensor only. Details are given in the original reference [5]: Using the Voigt's notation for the stress tensor $\sigma = (\sigma_x, \sigma_y, \sigma_z, \tau_{xy}, \tau_{xz}, \tau_{zx})$ the fourth order tensor **M** is transformed to a second order tensor as:

$$\mathbf{M} = \frac{1}{\nabla} \begin{bmatrix} \mathbf{M}_{11} & 2\mathbf{M}_{12}^T \\ \mathbf{M}_{12} & \mathbf{M}_{22} \end{bmatrix} \quad (10)$$

with $\nabla = P_{xx}P_{yy}P_{zz} - D_{yz}^2 P_{xx} - D_{xz}^2 P_{yy} - D_{xy}^2 P_{zz} - 2D_{xx}D_{yy}D_{zz}$ and D_{ij} the components of the orthotropic damage tensor and P_{ij} is chosen such that $D_{ij} + P_{ij} = \delta_{ij}$ (with Kronecker symbol δ_{ij}). The sub-matrices $\mathbf{M}_{11}$, $\mathbf{M}_{12}$, $\mathbf{M}_{22}$ are:

$$\mathbf{M}_{11} = \begin{bmatrix} 2\left(P_{yy}P_{zz} - D_{yz}^2\right) & 0 & 0 \\ 0 & 2\left(P_{xx}P_{zz} - D_{xz}^2\right) & 0 \\ 0 & 0 & 2\left(P_{xx}P_{yy} - D_{xy}^2\right) \end{bmatrix}$$

$$\mathbf{M}_{12} = \begin{bmatrix} D_{xz}D_{yz} + D_{xy}P_{zz} & D_{xz}D_{yz} + D_{xy}P_{zz} & 0 \\ D_{xy}D_{yz} + D_{xz}P_{yy} & 0 & D_{xy}D_{yz} + D_{xz}P_{yy} \\ 0 & D_{xy}D_{xz} + D_{yz}P_{xx} & D_{xy}D_{xz} + D_{yz}P_{xx} \end{bmatrix}$$

$$\mathbf{M}_{22} = \begin{bmatrix} \left(P_{yy}P_{zz} - D_{yz}^2\right) + \left(P_{xx}P_{zz} - D_{xz}^2\right) & D_{xy}D_{xz} + D_{yz}P_{xx} & D_{xy}D_{yz} + D_{xz}P_{yy} \\ D_{xy}D_{xz} + D_{yz}P_{xx} & \left(P_{xx}P_{yy} - D_{xy}^2\right) + \left(P_{yy}P_{zz} - D_{yz}^2\right) & D_{xz}D_{yz} + D_{xy}P_{zz} \\ D_{xy}D_{yz} + D_{xz}P_{yy} & D_{xz}D_{yz} + D_{xy}P_{zz} & \left(P_{xx}P_{zz} - D_{xz}^2\right) + \left(P_{xx}P_{yy} - D_{xy}^2\right) \end{bmatrix}$$

3. Results

3.1. Results using models with isotropic damage

The results of the finite element analysis were only roughly precise: The amount of explosive for creation of a hole is predicted quite well even with isotropic material models. Even the biggest diameter of the craters measured on the front and rear surfaces are close to the reality.

However the cross section of the crater in the model was wrong not only in quantity but also in quality. The hole inside the concrete wall is much smaller than the diameter of the crater on the front surface. In case the damage is taken as a measure for the crater, the calculated diameter of the hole is close to biggest diameter of the craters! This is a consequence of the localization and its characteristic length l_0 which is in the order of 10–100 mm. The material parameters and mesh size are linked such that the damaged is smeared over a rather big region. Let us discuss on behalf

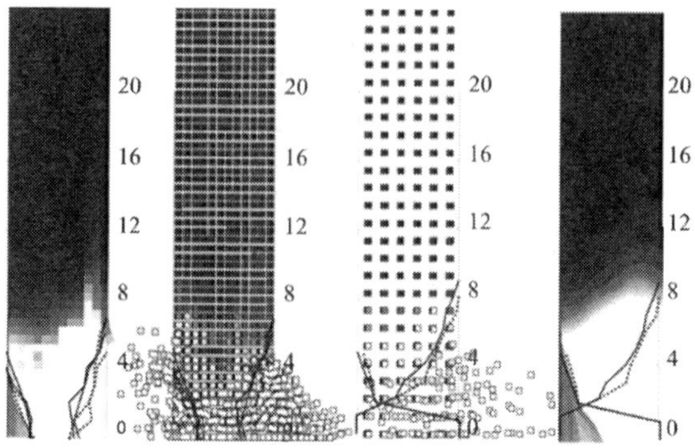

Figure 5. Crater in UHPFRC following to the second configuration of Fig. 1 in the scale 1:5. On the left hand side are shown a Lagrangian model and an SPH-Model with 5 mm element size and a small quantity of explosive which did not perforate the wall. The dotted and solid lines are the crater profile of the field tests. On the right hand side are shown the SPH and Lagrangian model with 10 mm element size and the double amount of explosive needed for perforation. The models shows the damage between 0 (black) and 0.7 (white) which corresponds to the critical damage D_c.

of Fig. 5 the nonlinear synergetic effect of the mesh size and the solver used for model set-up:

The craters of the two figures in the middle have almost the same dimension (Ø 12–16 mm) although the quantity of the explosive is different by factor two. Indeed the model with the fine mesh is "soft". However, the damage zone is expected to increase not proportional to the quantity of the explosive. On the contrary, following facts are evident: The detonation shock pressure inside the explosive is constant and a characteristic of the explosive. The mass of the explosive is proportional to the volume and the momentum applied to the concrete is roughly proportional to the contact surface between explosive and concrete.

The explosive and its detonation have been modelled with an Eulerian mesh fine enough to resolve the over pressure of the shock pressure which is reflected by the concrete all. The concrete of the models on the outer left and outer right in Fig. 5 is modelled with Lagrangian mesh, of course including the reinforcement bars. Autodyn has a built-in algorithm which transfers the momentum from the Eulerian mesh to the Lagrangian mesh and viz. versa. Unfortunately, there is no such implementation between an Eulerian mesh and a SPH part. However there is a work around: The momentum is first transferred from the Eulerian zone to a thin foil modelled with Lagrangian elements. This foil is attached by a contact algorithm to the SPH-part. This twice fold transfer from the detonation shock in the Eulerian mesh to the SPH-solver dissipates numerically more energy than the direct transfer to the Lagrangian mesh. As a consequence, the momentum transferred to a concrete wall modelled with smooth hydrodynamic particles is less than the one transferred to concrete wall modelled with Lagrangian elements.

3.1.1. Parameter study

The influence of three settings in the models has been investigated fully factorial as follows:

- Element length: 5 mm or 10 mm
- Solver: Lagrangian or Smooth Particle Hydro-dynamics (SPH)
- Damage growth rate: "quick" or "slow"; "quick" means, that the element is fully damaged at small plastic strain such that the residual strength is applied already in an early state of the simulation.

The pairs of the variation above have a weaker (5 mm element length, Lagrangian Solver, quick damage growth rate) or a stronger (10 mm element length, SPH Solver, slow damage growth rate) resistance to perforation. Simulations with weak parameters are expected to yield bigger dimensions in craters as simulations with stronger parameters. Unfortunately some simulations yielded the opposite behavior. The fully factorial evaluation showed the influence of the parameters onto the characteristic of the craters was different or even contrary for each configuration shown in Fig. 1. The evaluation of the performance is based on quantities such as diameter of the crater and zone of delamination of the different simulation are within $\pm 15\%$ of the measurements.

One example on the basis of the configuration in the middle in Fig. 1: A quick damage growth rate should yield an increased diameter of the crater compared to the one of a simulation with a slow damage rate. The same was anticipated for the mesh refinement. This was true for small quantities of the explosive (called "small" hereafter) which were not enough for the perforation of the wall. In contrast, the anticipation did not correspond for quantities of explosive (called "big" hereafter) needed for perforation. To make matters worse, a simultaneous refinement of the mesh and quick damage growth rate yielded an opposite result, i.e. a smaller diameters of craters for the "small" quantity of explosive. Finally a combination of coarse mesh and slow damage growth rate yielded a smaller diameter in crater for the "big" quantity of explosive, as expected by assuming positive synergetic effects.

3.1.2. Composite normal concrete with UHPFRC

A delamination of a composite concrete slab occurred not exactly between the interface of the normal concrete and the UHPFRC but inside the normal concrete. The failure occurred in typically 1–2 cm below the interface such that the rear plane of reinforcement bars is still covered by normal concrete. The delamination is not planar but there are concentric rings detected as shown in Fig. 6.

The simulation gives some hints how the concentric rings emerge. Even the dimension and location of the delamination of *some* simulations was in quite good correspondence with the experimental results.

3.2. Influence of anisotropy in normal concrete

In the beginning of Sect. 3.1 has been shown that profile of the crater of the simulation are even qualitatively wrong. This drawback is mainly observed for the first two types of concrete walls shown in Fig. 1. The main reason is assumed to be due to the characteristic length l_0 lined to the damage parameter. In this section the improvements due to introduction of an orthotropic damage parameter will be show on the basis of Fig. 8. There is some qualitative improvement of the profile. However a good set of parameters (damage growth rate, critical damage D_c for spontaneous failure, mesh size, Mohr-Coulomb yield surfaces) have not been found yet. Since the list of parameters is so long, a set of parameters optimized of the basis of some tests may yield wrong results for another test setup. In addition, a result of the fully factorised parameter

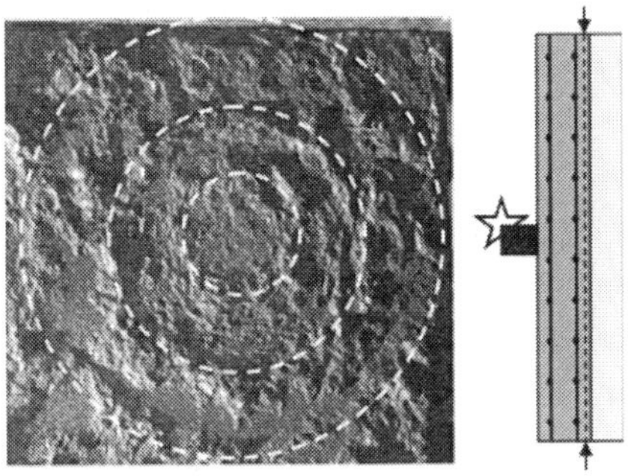

Figure 6. The delamination is approximately 1–2 cm below the interface between the double reinforced normal concrete and the UHPFRC as shown with the dashed line marked by two arrows on the right hand side. The photograph on the right hand side shows the delamination surface with three concentric rings as hills and valleys. The white dashed circles mark the valleys.

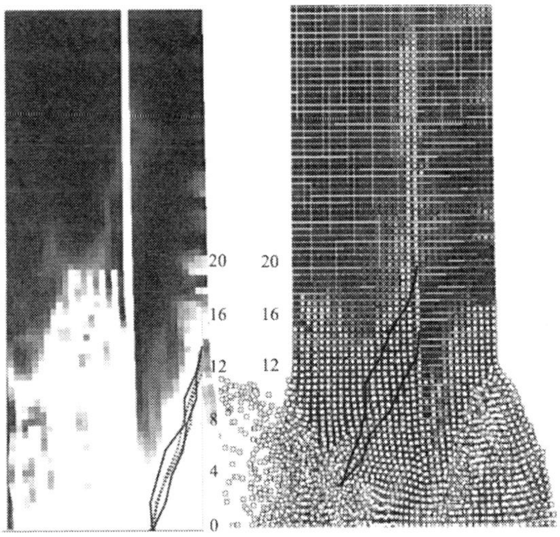

Figure 7. Crater and delamination in composite concrete slabs. This is the configuration in the middle of Fig. 1 in the scale 1:5. On the left hand side is shown an example with a Lagrangian mesh and a small quantity of explosive which did not create a hole thru the complete wall. On the right hand side is shown an example with SPH model and a big quantity of explosive far above the quantity needed for perforation. The solid and dotted lines show the profile of the craters of corresponding field test. The simulation shows the damage between 0 (black) and 0.7 (white) which corresponds to the critical damage D_c.

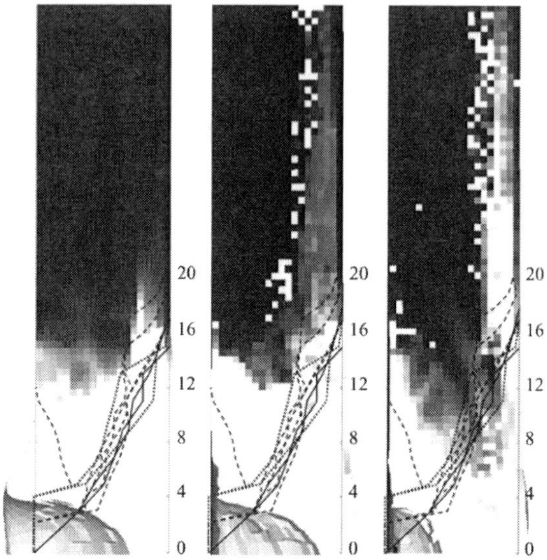

Figure 8. Crater in double reinforced normal concrete corresponding to the first configuration in Fig. 1 in the scale 1:5. On the left hand side is a Lagrangian model with isotropic damage parameter. The models in the middle and on the right hand side are examples with orthotropic damage parameter. Shown is the first principal damage from 0 (black) to 0.7 (white) which corresponds to the critical damage D_c. The curves (solid, dotted, and dashed lines) are the profiles of the craters measured after corresponding field tests.

study is the dependency of the numerical results on the combination of parameters in a way which is not at all intuitively predicable.

The example above shows the significant the difference of a simulation with continuous decrease of the strength from failure to limit yield surface (with increase of the damage from zero to one) compared to a simulation with spontaneous failure with only two distinct state of the strength: The accumulated damage within the "strong" sate is below a critical value and the strength is defined by the residual surface. After the accumulated damage exceeds a critical value the material response is "weak" and the strength follows the residual surface. As a consequence, the first principal stresses, which result from the reflection of the shock wave and are detected close to the rear surface and in a distance above 16 cm from the centre, are higher, because the strength material in the centre is hold on the failure surface for a "quite long" duration. Therefore, the last model in Fig. 8 predicts on the one hand a damaged zone just under the rear surface in a distance beyond 16 cm from the centre. On the other hand, the diameter of the hole in the concrete is closer to the reality. However the shape of the crater in the simulation is "distorted".

4. Conclusions

It has been shown that the results of the simulations with commercial available programs with standard parameters for the constitutive relation of concrete yields generally good results. However, the carter profile of the simulation does not correspond to the reality. The introduction of orthotropic damage improved the results only slightly. The finite element models are dependent on the meshing because the damage evolution is linked to the plastic strain. Moreover, it is quite cumbersome to define a right set of parameters not only because of numbers of parameters but also due to their synergetic effects such that the relationship with the simulation results is highly nonlinear or even chaotic.

E. Cadoni and G. Riganti [6] are developing a constitutive equation for concrete under highly dynamic loading which is not based on a damage parameter. They reconsider to the definition of the damage parameter and calculate the fracture surface using a crack propagation speed, a defined density of nucleation, and a limit load at which a crack starts to evolve from a nucleation. Spontaneous failure starts if the local stress tensor based on formula (8) exceeds a limit. In principle, such a model does not depend on the fineness of the mesh. Therefore, the results shall converge with mesh refinements.

This study has been financed by the arma suisse and is a part of a long term research project ARAMIS 042-39. The author was always invited to participate to meetings and field tests. There was free access to internal reports, in particular to the evaluation of modified Split Hopkinson Bar tests (6) and field tests (6) sketched in Fig. 1.

References

[1] W. Riedel, *EMI-Bericht 6/00* (2000)
[2] L. Malvar, J. Crawford, J. Wesewich, D. Simons, Int. J. Impact Eng., **19**, pp 847–873 (1997)
[3] J. Lemaître, R. Desmorat, *Engineering Damage Mechanics*, Springer (2005)
[4] R. Peerlings., M. Geers M.G.D., R. de Borst R., W. Brekelmans, int. J. of solid and structures (2001)
[5] G. Z. Voyiadjis, P. Kattan, *Advances in Damage Mechanics: Metals and Metal Matrix Composites*, Elsevier (1999)
[6] E. Cadoni, G. Riganti, *report rap 7222 04* (2013)
[7] B. Cavelti, *FoPro 3 – TP3* (2012)

EPJ Web of Conferences **94**, 04020 (2015)
DOI: 10.1051/epjconf/20159404020

Mesoscale numerical modeling of plastic bonded explosives under shock loading

Hailin Shang, Feng Zhao[a], Guangfu Ji, and Hua Fu

National Key Laboratory of Shock Wave and Detonation Physics, Institute of Fluid Physics, China Academy of Engineering Physics, Mianyang 621999, Sichuan, China

Abstract. Mesoscale responses of plastic bonded explosives under shock loading are investigated using material point method as implemented in the Uintah Computational Framework. The two-dimensional geometrical model which can approximately reflect the mesoscopic structure of plastic bonded explosives was created based on the Voronoi tessellation. Shock loading for the explosive was performed by a piston moving at a constant velocity. For the purpose of investigating the influence of shock strength on the responses of explosives, two different velocities for the piston were used, 200 m/s and 400 m/s, respectively. The simulation results indicate that under shock loading there forms some stress localizations on the grain boundary of explosive. These stress localizations lead to large plastic deformations, and the plastic strain energy transforms to thermal energy immediately, causing temperature to rise rapidly and form some hot spots on grain boundary areas. The comparison between two different piston velocities shows that with increasing shock strength, the distribution of plastic strain and temperature does not have significant change, but their values increase obviously. Namely, the higher the shock strength is, the higher the hot spot temperature will be.

1. Introduction

Plastic bonded explosives are heterogeneous and often consist of a mixture of polycrystalline explosives and binder materials. Thus, these heterogeneous explosives exhibit obviously different mechanical, thermal and chemical behaviour compared to pure explosives because of the discontinuity at mesoscale which associated with the granular nature of the explosive constituents. Under shock loading, significant temperature and pressure rise at the defect areas of these heterogeneous explosives, forming a lot of hot spots. Thereafter, the chemical reaction and combustion come up, finally the detonation occurs. Because of the difficulties in experimental study at mesoscale, numerical simulation is an available way to investigate the mesoscale response of explosives under shock loading.

Menikoff [1] studied the mechanism for hot-spot formation occurs as a shock wave passes over a high-density impurity using a two-dimensional hydrodynamic simulation performed with the xRage code. The results pointed out that interactions generated by reflected waves from neighboring beads can significantly increase the peak hot-spot temperature when the beads are suitably spaced. Baer [2] studied the mesoscopic processes of consolidation, deformation, and reaction of shocked porous energetic materials using the shock physics code, CTH. Numerical simulations indicate that "hot-spots" are strongly influenced by multiple crystal interactions. Conley et al. [3] investigated the shock compaction in granular porous and cast HMX using a multi-material, two dimensional Eulerian hydrocode, RAVEN. In their simulation results, no significant particle size effect upon HMX temperature was observed in either of the two compacts. But the presence of the binder significantly altered the response of intragranular voids.

In this paper, we studied the shock responses of plastic bonded explosives using material point method as implemented in the Uintah Computational Framework [4].

2. Material point method

The material point method (MPM) was described by Sulsky et al. [5,6] as an extension to the FLIP (Fluid-Implicit Particle) method of Brackbill [7], which itself is an extension of the particle-in-cell (PIC) method of Harlow [8].

The basic idea of MPM is: objects are discretized into particles, or material points, each of which carries the history-dependent state variables such as stresses and strains, as well as mass and kinematic variables such as position, velocity and acceleration [5,9]. A regular background mesh covers the computational domain. The cells of the grid behave like finite elements, and the grid points behave like finite element nodes. During each time step, the mass and kinematic variables at the material points are mapped to the grid, where momentum equations are solved. The acceleration and velocity at the grid are then mapped back to the material points. State variables are then calculated at the material points using the updated kinematic variables.

The use of a regular background grid in MPM has a lot of computational advantages. Computational of spatial gradients is simplified. For heterogeneous materials like explosives, MPM is more suitable for solving problems involving contact between colliding objects, cracks, large

[a] Corresponding author: ifpzf@163.com

deformations and other high velocity problems. It has an advantage over FEM in that the use of the regular grid eliminates the need for doing costly searches for contact surfaces [10].

The discrete momentum equation for MPM is given as:

$$ma = F^{ext} - F^{int} \tag{1}$$

where m is the mass matrix, $\mathbf{a}$ is the acceleration vector, F^{ext} is the external force vector, and $\boldsymbol{F}^{int}$ is the internal force vector resulting from the divergence of the material stresses

The solution begins by accumulating the particle state on the nodes of the computational grid, to form the mass matrix $\boldsymbol{m}$ and to find the nodal external forces $\boldsymbol{F}^{ext}$, and velocities, $\boldsymbol{v}$. These quantities are calculated at individual nodes by the following equations, where $\sum$ represents a summation over all particles:

$$m_i = \sum_p S_{ip} m_p \tag{2}$$

$$v_i = \frac{\sum_p S_{ip} m_p v_p}{m_i} \tag{3}$$

$$F_i^{ext} = \sum_p S_{ip} F_p^{ext} \tag{4}$$

and i refers to individual nodes of the grid. m_p is the particle mass, $\mathbf{v}_p$ is the particle velocity, and $\boldsymbol{F}_p^{ext}$ is the external force on the particle. S_{ip} is the shape function of the ith node evaluated at x_p.

F^{int} is computed at the nodes as a volume integral of the divergence of the stress on the particles, specifically:

$$F_i^{int} = \sum_p G_{ip} \sigma_p v_p \tag{5}$$

where $\mathbf{G}_{ip}$ is the gradient of the shape function of the ith node evaluated at $\boldsymbol{x}_p$, and σ_p and v_p are the time n values of particle stress and volume respectively.

Using Eq. (1) the momentum equations for grid nodes are solved, and then mapped back to the material points. State variables are finally calculated at the material points using the updated kinematic variables.

3. Results and discussion

Two-dimensional simulation model for plastic bonded explosives is shown in Fig. 1. The model was created based on the Voronoi tessellation. It has the same width and length of 10 mm. There are 1024 (32 × 32) HMX grains in total, which contain 939,999 material points. And there are also 60,001 binder points. Considering the density of HMX is 1.9×10^3 kg/m^3 and for binder is 1.1×10^3 kg/m^3, we can get the mass fraction for both constituents is 96.44% and 3.56%, respectively.

An elastic-plastic constitutive-model with a Mie-Gruneisen equation-of-state is used for HMX, and a hyper-elastic constitutive-model is used for binder. The properties for HMX and binder are given in Table 1. Where S_1–S_3 are the coefficients of the slope of the u_s–u_p curve, u_s is shock velocity, and u_p is particle velocity.

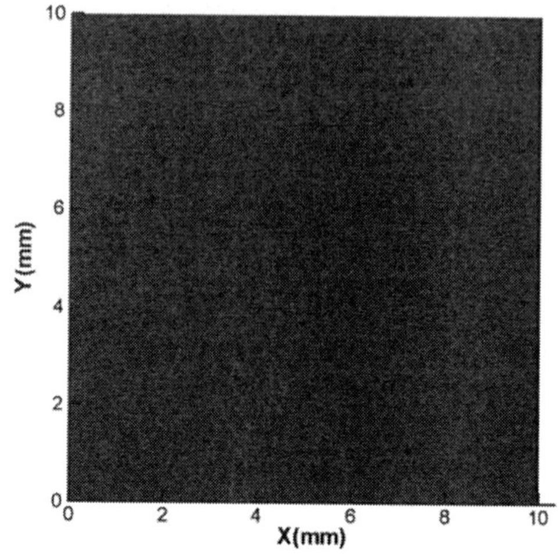

Figure 1. Simulation model for plastic bonded explosive.

Table 1. Parameters for HMX and binder.

	HMX	binder
Density (kg/m^3)	1.90E3	1.10E3
Shear modulus (Pa)	7.20E9	2.70E8
Bulk modulus (Pa)	10.2E9	3.65E9
Yield strength (Pa)	3.70E8	—
Sound speed (m/s)	2901.0	—
Gruneisen gamma	1.1	—
S_1	2.058	—
S_2	0	—
S_3	0	—

3.1. Low shock strength results

Shock loading is performed by a piston moving at a constant velocity from the left boundary to right. The top and bottom boundaries are free slip boundaries. The right boundary is a free surface. In this section, the piston velocity is set to 200 m/s. As shock wave reaches the free surface, the effective stress, plastic strain and temperature distribution of the explosive are shown in Figs. 2–4 (only show the result for HMX grains, not contain the binder).

The results shown in Figs. 2–4 indicate that under shock loading there forms a lot of stress localizations at the interfaces between HMX grains and binder layers. These stress localizations lead to large plastic deformations. Simultaneously, the plastic strain energy transforms to thermal energy, causing the temperature to rise rapidly and form hot spots on grain boundary areas.

3.2. High shock strength results

To increase the shock strength, a higher velocity of 400 m/s for the piston is used. The model properties and other simulation conditions are the same as in Sect. 3.1. The effective stress, plastic strain and temperature distribution

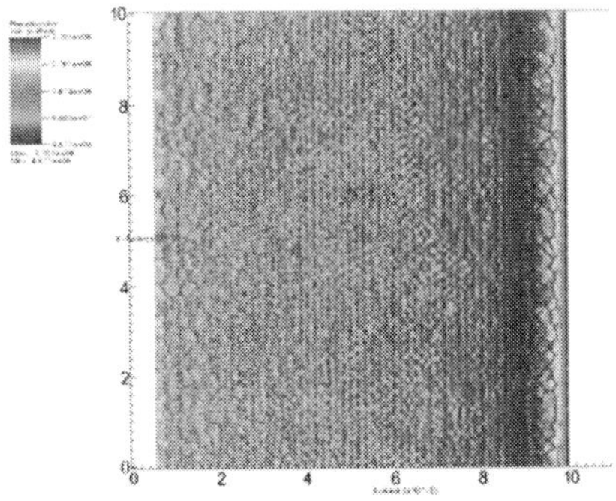

Figure 2. Effective stress distribution for the plastic bonded explosive as shock wave reaches the free surface (piston velocity is 200 m/s).

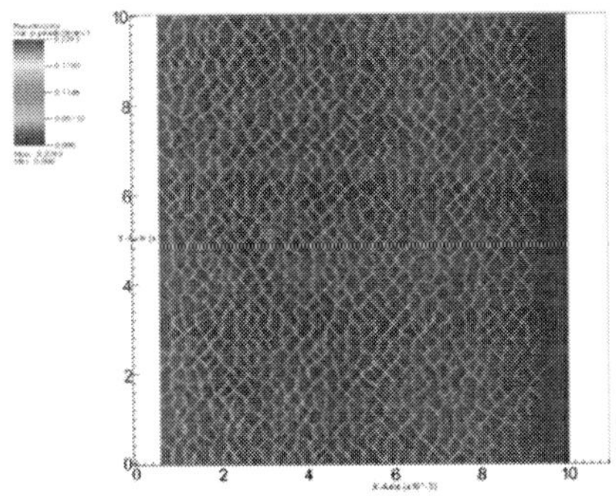

Figure 3. Plastic strain distribution for the plastic bonded explosive as shock wave reaches the free surface (piston velocity is 200 m/s).

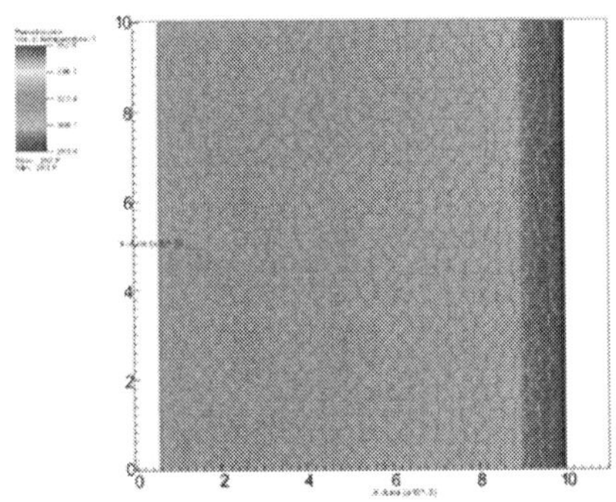

Figure 4. Temperature distribution for the plastic bonded explosive as shock wave reaches the free surface (piston velocity is 200 m/s).

of the explosive as shock wave reaches the free surface are shown in Figs. 5–7 (also only show the result for HMX grains, not contain the binder).

Results shown in Figs. 5–7 indicate that with increasing shock strength, the distributions of stress, plastic strain and temperature do not have significant changes, but their values do have large increases. Thus, a conclusion can be summarized as, the higher the shock strength is, the higher the hot spot temperature will be.

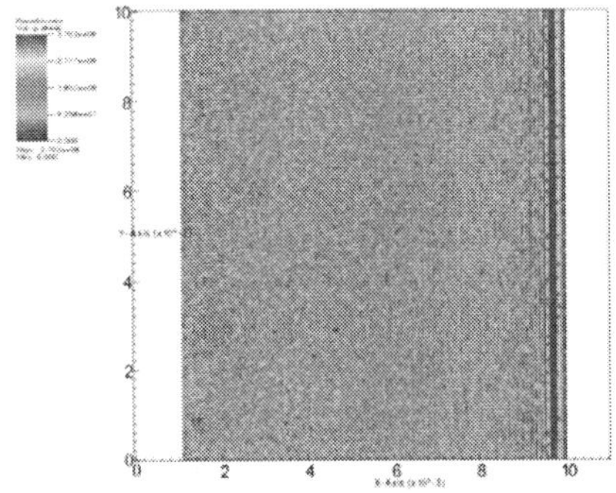

Figure 5. Effective stress distribution for the plastic bonded explosive as shock wave reaches the free surface (piston velocity is 400 m/s).

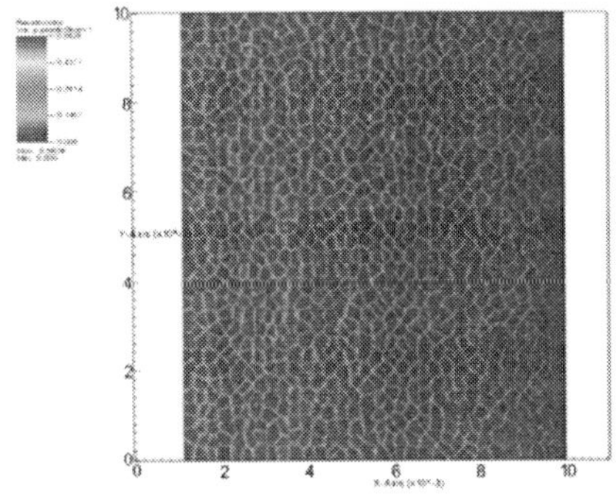

Figure 6. Plastic strain distribution for the plastic bonded explosive as shock wave reaches the free surface (piston velocity is 400 m/s).

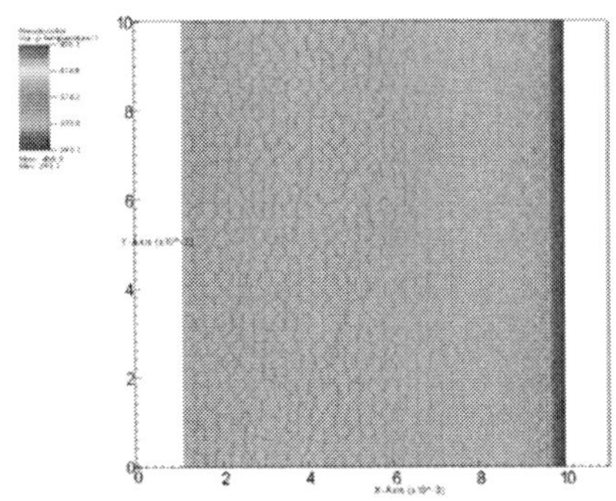

Figure 7. Temperature distribution for the plastic bonded explosive as shock wave reaches the free surface (piston velocity is 400 m/s).

4. Conclusion

In this paper, mesoscale responses of plastic bonded explosives under shock loading are investigated using material point method. Simulation results show that under shock loading there forms a lot of stress localizations at the interfaces between HMX grains and binder layers. And these stress localizations lead to large plastic deformations. Simultaneously, the plastic strain energy transforms to thermal energy, causing the temperature to rise rapidly and form hot spots on grain boundary areas. With increasing shock strength, the distributions of stress, plastic strain and temperature do not have significant changes, but their values do have large increases. We can conclude that, the higher the shock strength is, the higher the hot spot temperature will be.

This work was supported by National Natural Science Foundation of China (11272296, 11272294), Science and Technology Foundation of CAEP (2014A0201008, 2012A0201007), National Key Laboratory Foundation (2012-Z-05), and Province Programming Foundation of Institute of Fluid Physics (LYGH201402).

References

[1] R. Menikoff, Shock Waves **21**, 141 (2011).
[2] M.R. Baer, Thermochimica Acta **384**, 351 (2002).
[3] P.A. Conley, D.J. Benson, Proceedings of the 11th International Detonation Symposium, 768 (1998).
[4] P. Dickson, B. Asay, B. Henson, L. Smilowitz, Proc. Roy. Soc. Lond. A **460**, 3447 (2004).
[5] D. Sulsky, Z. Chen, H.L. Schreyer, Comput. Methods Appl. Mech. Eng. **118**, 179 (1994).
[6] D. Sulsky, S. Zhou, H.L. Schreyer, Comput. Phys. Commun. **87**, 236 (1995).
[7] F.H. Harlow, Methods Comput. Phys. **3**, 319 (1963).
[8] J.U. Brackbil, H.M. Ruppel, J. Comp. Phys. **65**, 314 (1986).
[9] H.L. Jonah, H. Daisy, IOP Conf. Series: Mater. Sci. Eng. **10**, 012093 (2010).
[10] S.G. Bardenhagen, J.U. Brackbill, D. Sulsky, Comput. Methods Appl. Mech. Eng. **187**, 529 (2000).

EPJ Web of Conferences **94**, 04021 (2015)
DOI: 10.1051/epjconf/20159404021

Using Laplace transform to solve the viscoelastic wave problems in the dynamic material property tests

Yuxuan Zheng and Fenghua Zhou[a]

MOE Key Laboratory of Impact and Safety Engineering, Ningbo University, Ningbo, Zhejiang, China

Abstract. In relation to the dynamic tests of materials, the approach to solve the viscoelastic wave propagations in a one dimensional viscoelastic rod was summarized. By conducting Laplace transform, the governing partial differential equations were transformed to ordinary differential equations for the image functions, which were solved analytically with suitable boundary equations. Inversely transforming these image functions gives the results of the stress, velocity, and strain in the bar. Two wave problems occurred in split Hopkinson pressure bar (SHPB) tests are analyzed: 1) the problem of evaluating the internal stress distributions in a viscoelastic specimen; and 2) the problem of stress wave propagations in a viscoelastic bar. Both problems were solved numerically by way of numerical inverse Laplace transform. For the first problem, the special case when the specimen is pure elastic was solved analytically, giving the exact solution to the problem of elastic wave propagation in a sandwich elastic media.

1. Introduction

Split Hopkinson Pressure Bar (SHPB, also known as Kolsky bar) test technique is widely used in the dynamic material properties tests under strainrate 10^2–10^4 s^{-1}. The exact analysis of SHPB experimental results is based on the two basic assumptions: one-dimensional stress wave propagation in the bar, the uniform stress/strain of the specimen along the length direction. While the specimen is viscoelastic material, the key of the problem is summed up in solving the viscoelastic wave propagation. Usually, the approach to solve the viscoelastic wave propagations is divided into two categories: the method of characteristic and the method of propagation coefficient. The first method is a finite difference method, solving the wave problem by the compatibility relationships along the characteristic lines [1–3]. The latter method is a analytical method by way of Fourier series. The stress wave is decomposed into different frequency harmonic wave. Then, considering the attenuation and dispersion of each harmonic wave in the process of propagating through media, we can restructure the harmonic wave [4,5]. Lili Wang has expounded 1-D viscoelastic wave theory [1]. He analyzes the linear and nonlinear viscoelastic constitutive relation.

Wave equations as a set of partial differential equation, can be solved solved numerically by way of some methods such as integral transform [6]. Laplace transform as a kind of common integral transform, can be used for the solving the wave propagation problems of SHPB or specimen. By conducting Laplace transform, we analyzed the viscoelastic wave propagation rule of short specimens in the SHPB experiment.

[a] Corresponding author: zhoufenghua@nbu.edu.cn

2. Laplace transform in the 1D viscoelastic wave propagation

2.1. Control equations and initial conditions

Lagrange coordinate system is employed in the analysis where any mass point is identified by its original location X. At any time t (t≥0), the control equations of the 1D viscoelastic wave propagation are

$$\text{kinematic equation: } \frac{\partial v}{\partial t} = \frac{1}{\rho}\frac{\partial \sigma}{\partial x} \tag{1}$$

$$\text{continuity equation: } \frac{\partial v}{\partial x} = \dot{\varepsilon} = \frac{\partial \varepsilon}{\partial t} \tag{2}$$

and constitutive equation:

$$\sigma(t) = \int_{-\infty}^{t} G(t-\tau)\,\dot{\varepsilon}(\tau)\,d\tau \tag{3}$$

where σ, v, ε, $\dot{\varepsilon}$ denote stress, mass velocity, strain and strain-rate. ρ and $G(t)$ denote density of the material and the stress relaxation function. For generalized Maxwell viscoelastic solid, the stress relaxation function is

$$G(t) = E_e + \sum_{i=1}^{n} E_i e^{-\frac{t}{\theta_i}}. \tag{4}$$

Where $\theta_i = \eta_i/E_i$ is the relaxation time of the each Maxwell element, and the parameter E_e of the model means that the solid materials cannot reduce the relaxation stress to zero.

As a simplified model, the stress relaxation function for a standard 3-element viscoelastic solid is

$$G(t) = \frac{E_2}{E_1+E_2}\left(E_1 + E_2 e^{-\frac{t}{\theta}}\right), \quad \theta = \frac{\eta}{E_1+E_2}. \tag{5}$$

Article available at http://www.epj-conferences.org or http://dx.doi.org/10.1051/epjconf/20159404021

At time zero (t = 0), the viscoelastic solid is static. The initial conditions are

$$\sigma(x,0)=0, \varepsilon(x,0)=0, v(x,0)=0, \dot{\varepsilon}(x,0)=0. \quad (6)$$

The boundary conditions will be depending on the different cases.

2.2. Solution of the image functions

For the time variable, we can transform the partial differential equation of the initial unknown function into ordinary differential equation of the image function by Laplace transforms method. The image function is denoted by the function with overline, whose transform variable is denoted by s. Combined with the initial conditions Eq. (6), the partial differential Eqs. (1)–(3) can be transformed to the follow equations:

$$\bar{v}=\frac{1}{\rho s}\frac{d\bar{\sigma}}{dx} \quad (7)$$

$$\bar{\dot{\varepsilon}}(s)=\frac{d\bar{v}}{dx}=\frac{1}{\rho s}\frac{d^2\bar{\sigma}}{dx^2} \quad (8)$$

$$\bar{\varepsilon}=\frac{1}{s}\frac{d\bar{v}}{dx}=\frac{1}{\rho s^2}\frac{d^2\bar{\sigma}}{dx^2} \quad (9)$$

$$\bar{\sigma}(s)=s\bar{G}(s)\bar{\varepsilon}(s). \quad (10)$$

Where $\bar{G}(s)$ denotes the stress function transformed by Laplace transform method. For the generalized Maxwell viscoelastic solid or the standard 3-element viscoelastic solid,

$$\bar{G}(s)=\frac{E_e}{s}+\sum_{i=1}^{n}\frac{E_i\eta_i}{E_i+\eta_i s}. \quad (11)$$

$$or, \bar{G}(s)=\frac{E_2(E_1+\eta s)}{s(E_1+E_2+\eta s)}. \quad (12)$$

The differential equation of the image function $\bar{\sigma}(x,s)$ gives the result of the stress based on the Eqs. (7)–(10):

$$\frac{d^2\bar{\sigma}}{dx^2}=\frac{\rho s}{\bar{G}(s)}\bar{\sigma}. \quad (13)$$

In particular, for a standard 3-element viscoelastic solid,

$$\frac{d^2\bar{\sigma}}{dx^2}=\rho s^2\frac{E_1+E_2+\eta s}{E_2(E_1+\eta s)}\bar{\sigma}. \quad (14)$$

The general solution of Eq. (13) is

$$\bar{\sigma}(x,s)=Ae^{\Lambda(s)x}+Be^{-\Lambda(s)x} \quad (15)$$

where $\Lambda(s)=\sqrt{\rho}\sqrt{\frac{s}{G(s)}}$ is a root of the equation

$\Lambda^2=\frac{\rho s}{G(s)}$. For a standard 3-element viscoelastic solid,

$$\Lambda(s)=s\sqrt{\rho}\left[\frac{E_2+(E_1+\eta s)}{E_2(E_1+\eta s)}\right]^{\frac{1}{2}}. \quad (16)$$

Equation (15) gives the result of the stress for the image function $\bar{\sigma}(x,s)$, where the coefficients $A=A(s)$, $B=B(s)$ would be solved with suitable boundary equations. More, we can confirm the image functions $\bar{v}(x,s)$, $\bar{\dot{\varepsilon}}(x,s)$ and $\bar{\varepsilon}(x,s)$ giving the results of the mass velocity, strainrate and strain. Then, the problems of the wave propagation would be solved numerically by way of numerical inverse Laplace transform.

3. 1D wave propagation in the viscoelatic specimen occurred in SHPB experiment

3.1. Derivation of the image function based on coupling boundary conditions

At time zero (t = 0), the incident wave propagates to the left of the specimen along the right-bound characteristic line. According to the laws of the wave propagation along the right-bound characteristic line, we can confirm

$$\sigma_L(t)-\sigma_i(t)=\rho_0c_0[v_L(t)-v_i(t)]=$$
$$\rho_0c_0\left[v_L(t)+\frac{\sigma_i(t)}{\rho_0c_0}\right].$$

Accordingly, the left boundary condition is

$$\sigma_L(t)=2\sigma_i(t)+\rho_0c_0v_L(t). \quad (17.1)$$

Where ρ_0c_0 denotes wave impedance of the elastic bar. At the right of the specimen, the right boundary condition is $\sigma_R(t)-0=-\rho_0c_0[v_R(t)-0]$, which means

$$\sigma_R(t)=-\rho_0c_0v_R(t). \quad (17.2)$$

Combining with Eq. (7), Eq. (17) are transformed by Laplace transform to the follow image functions:

$$\begin{cases}\bar{\sigma}(s,0)=2\bar{\sigma}_i(s)+\frac{\rho_0c_0}{\rho s}\bar{\sigma}_x(s,0)\\ \bar{\sigma}(s,L)=-\frac{\rho_0c_0}{\rho s}\bar{\sigma}_x(s,L)\end{cases}. \quad (18)$$

The general solution of image function for the stress Eq. (15) is substituted into the boundary condition Eq. (18). The coefficients $A=A(s)$, $B=B(s)$ would be solved:

$$A=\frac{2\bar{\sigma}_i(s)[1-\Lambda_1(s)]e^{-\Lambda(s)L}}{[1-\Lambda_1(s)]^2e^{-\Lambda(s)L}-[1+\Lambda_1(s)]^2e^{\Lambda(s)L}}$$
$$B=\frac{-2\bar{\sigma}_i(s)[1+\Lambda_1(s)]e^{\Lambda(s)L}}{[1-\Lambda_1(s)]^2e^{-\Lambda(s)L}-[1+\Lambda_1(s)]^2e^{\Lambda(s)L}}. \quad (19)$$

Where $\Lambda_1(s)=\frac{\rho_0c_0}{\rho s}\Lambda(s)=\frac{\rho_0c_0\sqrt{s}}{\sqrt{\rho}}\frac{1}{\sqrt{G(s)}}$. Especially, for the standard 3-element elastic solid, the result is:

$$\Lambda(s)=s\sqrt{\rho}\left[\frac{E_2+(E_1+\eta s)}{E_2(E_1+\eta s)}\right]^{\frac{1}{2}}$$
$$\Lambda_1(s)=\frac{\rho_0c_0}{\sqrt{\rho}}\left[\frac{E_2+(E_1+\eta s)}{E_2(E_1+\eta s)}\right]^{\frac{1}{2}}. \quad (20)$$

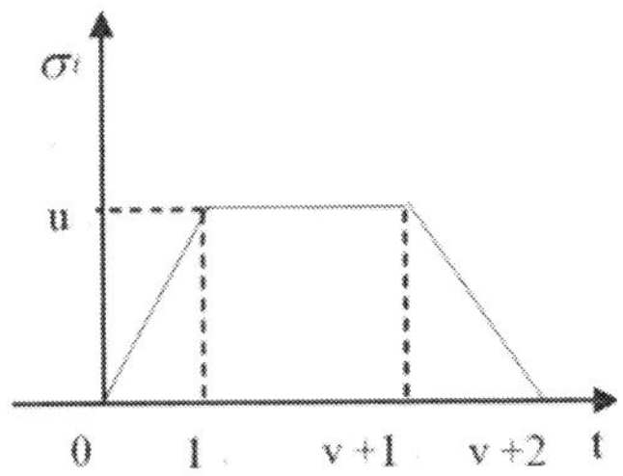

Figure 1. A trapezoidal incident stress wave.

3.2. Exact solution to the problem of wave propagation in the elastic specimen

Due to space limitation, a special case when the specimen is pure elastic was solved analytically, giving the exact solution to the problem of elastic wave propagation in a sandwich elastic media. Here, the standard 3-element solid degrades into the pure elastic solid whose Young's modulus is E_2 and elastic wave is

$$c = \sqrt{E_2/\rho}. \text{ So, } \Lambda(s) = s\sqrt{\frac{\rho}{E_2}} = \frac{s}{c},$$
$$\Lambda_1(s) = \frac{\rho_0 c_0}{\sqrt{\rho E_2}} = \frac{\rho_0 c_0}{\rho c}.$$

The wave impedance ratio between the elastic bar and the specimen is denoted by $\lambda = \frac{\rho_0 c_0}{\rho c}$, and the elastic wave propagation time in the specimen is denoted by $T = \frac{L}{c}$. Equation (20) is simplied:

$$A(s) = \frac{2}{1-\lambda}\frac{\bar{\sigma}_i(s)}{1-\mu^2 e^{2Ts}}$$
$$B(s) = \frac{-2}{1-\lambda}\frac{\bar{\sigma}_i(s)\mu e^{2Ts}}{1-\mu^2 e^{2Ts}}. \tag{21}$$

The image functions give the result of the stress of the two ends of the specimen:

$$\bar{\sigma}_{Left} = A(s) + B(s) = \frac{2\bar{\sigma}_i(s)}{1-\lambda}\frac{1-\mu e^{2Ts}}{1-\mu^2 e^{2Ts}}$$

$$\bar{\sigma}_{Right} = Ae^{\Lambda(s)L} + Be^{-\Lambda(s)L} = \frac{-4\lambda\bar{\sigma}_i(s)}{(\lambda-1)^2}\frac{e^{Ts}}{1-\mu^2 e^{2Ts}}. \tag{22}$$

In SHPB experiment, the incident wave is generally an approximate trapezoid. A incident wave is defined as Fig.1, and u, v denote the amplitude and width of the trapezoid wave. The rising and falling time is defined as 1 (non-dimensional value). The original function and image function of the incident wave are

$$\sigma_i(t) = u\begin{bmatrix} th(t) - (t-1)\,h(t-1) - \\ (t-1-v)h(t-1-v) + \\ (t-2-v)h(t-2-v) \end{bmatrix}$$
$$\bar{\sigma}_i(s) = \int_0^{+\infty} \sigma_i(t)e^{-st}dt$$
$$= u\frac{1-e^{-s}}{s^2} + u\frac{e^{-s(2+v)} - e^{-s(1+v)}}{s^2}$$
$$= \frac{u}{s^2}\left[1 - e^{-s} - e^{-s(1+v)} + e^{-s(2+v)}\right].$$

From the above equations, we can confirm the stress image functions of the two ends of the specimen. Using series expansion method, the problems was solved by way of inverse Laplace transform. The stress of the two ends of the elastic specimen is

$$\sigma_{Left}(t) = \frac{2u}{1-\lambda}\sum_{n=0}^{\infty}\mu^{-2n-1}\{[(t-2nT)h(t-2nT)$$
$$-(t-2nT-1)h(t-2nT-1)$$
$$+(t-2nT-v-1)h(t-2nT-v-1)$$
$$+(t-2nT-v-2)h(t-2nT-v-2)]$$
$$-\mu^{-1}[(t-2nT-2T)h(t-2nT-2T)$$
$$-(t-2nT-2T-1)h(t-2nT-2T-1)$$
$$-(t-2n-2T-v-1)h(t-2nT-2T-v-1)$$
$$+(t-2nT-2T-v-2)h(t-2nT-2T-v-2)]\}$$
$$\sigma_{Right}(t) = \frac{4u\lambda}{(\lambda-1)^2}\sum_{n=0}^{\infty}\mu^{-2n-2}$$
$$[(t-2nT-T)h(t-2nT-T)$$
$$-(t-2nT-T-1)h(t-2nT-T-1)$$
$$-(t-2nT-T-v-1)h(t-2nT-T-v-1)$$
$$+(t-2nT-T-v-2)h(t-2nT-T-v-2)]. \tag{23}$$

The analytical solution is compared with the numerical solution for the stress history at the left end of the specimen in Fig. 2a. Figure 2b shows the differences of the stress history between the two ends of the specimen.

If we define SHPB system as a wave propagation bar, the specimen between incident bar and transmission bar can be viewed as a impurity in the wave propagation bar which has imparity wave impedance. Figure 2b shows that the incident wave may change into the dispersive wave because of the transmission wave disturbed by the specimen when the incident wave pass through the elastic specimen. The phenomenon shows that the pure elastic may cause the dispersion.

3.3. Stress wave propagation in the viscoelastic specimen

Laplace transform of the governing equations turns the partial differential equations into ordinary differential equations, which can be solved analytically together with the transformed boundary conditions. The inverse Laplace transform of the solution gives the time histories of stress at arbitrary site. A modified Fixed Talbot (FT) algorithm is proposed to realize the inverse transform numerically. Results from this algorithm agree well with those obtained from the finite difference computation. Using this technique, parametric investigations are conducted to study the influences of specimen length and material parameters on the transmitting stress waveforms.

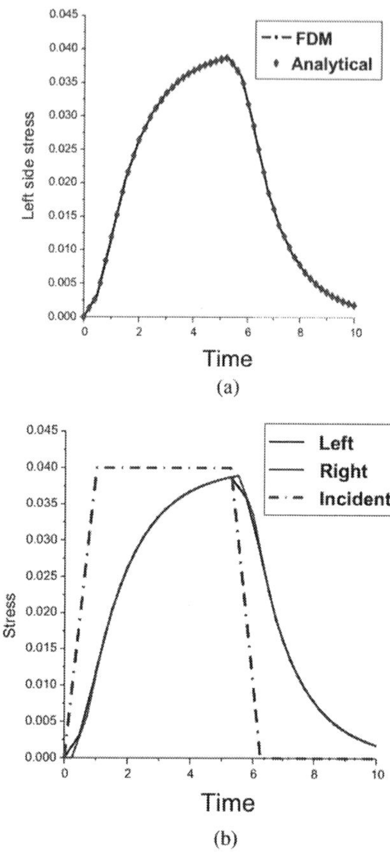

Figure 2. (a) Stress profile on the left end of the specimen, obtained by using series expansion method and finite difference simulation; (b) Stress profiles on both ends of the specimen.

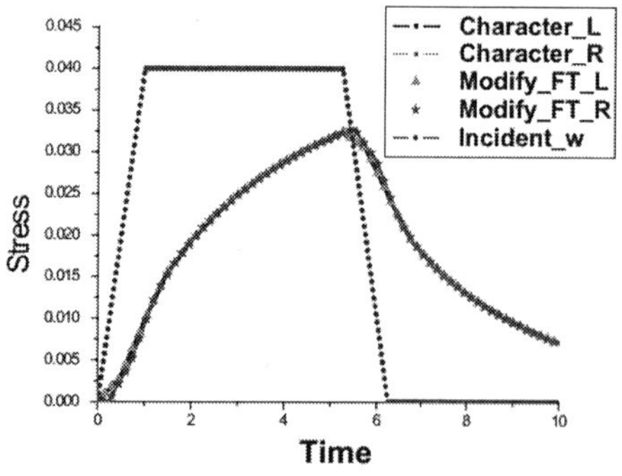

Figure 3. Stress profile on both ends of the specimen, obtained by using modified FT algorithm and finite difference simulation.

4. 1D stress wave propagation in the semi-infinite viscoelatic bar

In the Laplace transform domain, the image functions at different positions in the bar are formulated by (15) and (16). Assuming that the viscoelastic bar is semi-infinited, we can ignore the effect of the right-end. It means $A = 0$. The boundary conditions of the left-end is $\sigma(0, t) = \sigma_L(t)$, meaning $B = \bar{\sigma}(0, s) = \bar{\sigma}_L(s)$. The image function of stress in the bar is

$$\bar{\sigma}(x, s) = \frac{\sigma_0 e^{-\Lambda(s)x}}{s}. \tag{24}$$

Where $\Lambda(s) = \sqrt{\rho}\sqrt{\frac{s}{G(s)}}$, specially, for a standard 3-element viscoelastic solid,

$$\Lambda(s) = s\sqrt{\rho}\left[\frac{E_2 + (E_1 + \eta s)}{E_2(E_1 + \eta s)}\right]^{\frac{1}{2}}.$$

In the case of strong discontinuity viscoelastic wave propagation in the Semi-Infinite bar, the image function of stress in the bar is

$$\bar{\sigma}(x, s) = \frac{\sigma_0 e^{-\Lambda(s)x}}{s}. \tag{25}$$

For a standard 3-element viscoelastic solid, Eq. (25) is equivalent to

$$\bar{\sigma}(s) = \frac{\sigma_0 e^{-\Lambda(s)x}}{s} = \frac{\sigma_0}{s}\exp\left[-Ts\sqrt{\frac{E_2 + (E_1 + \eta s)}{E_1 + \eta s}}\right] = \frac{\sigma_0}{s}e^{-Ts}e^{Ts\left(1-\sqrt{1+\frac{E_2}{E_1 + \eta s}}\right)} \tag{26}$$

where, $c = \sqrt{E_2/\rho}$, $T = x/c$ denote the velocity of viscoelastic wave propagation and the time of the wave arriving at the location x. According to the lag theorem, the stress wave at the location x is

$$\sigma(x, t) = \sigma_1(t - T)\mathrm{h}(t - T). \tag{27}$$

The Laplace transform for the $\sigma_1(t)$ is

$$\bar{\sigma}_1(s) = \frac{\sigma_0}{s}\exp\left[Ts\left(1 - \sqrt{1 + \frac{E_2}{E_1 + \eta s}}\right)\right]. \tag{28}$$

Assuming the time is infinite, we can get

$$\sigma_1(\infty) = \lim_{s\to 0}[s\bar{\sigma}_1(s)] = \sigma_0 \exp\left\{\lim_{s\to 0} Ts\left[1 - \sqrt{1 + \frac{E_2}{E_1+\eta s}}\right]\right\} = \sigma_0. \tag{29}$$

At the time of the stress wave arriving the location x, the stress of the wavefront is

$$\sigma_1(0^+) = \lim_{s\to\infty}[s\bar{\sigma}_1(s)] = \sigma_0 \exp\left\{\lim_{s\to\infty} Ts\left[1 - \sqrt{1 + \frac{E_2}{E_1+\eta s}}\right]\right\}. \tag{30}$$

Using modified Fixed Talbot (FT) algorithm, we can give some cases showed in Fig. 4.

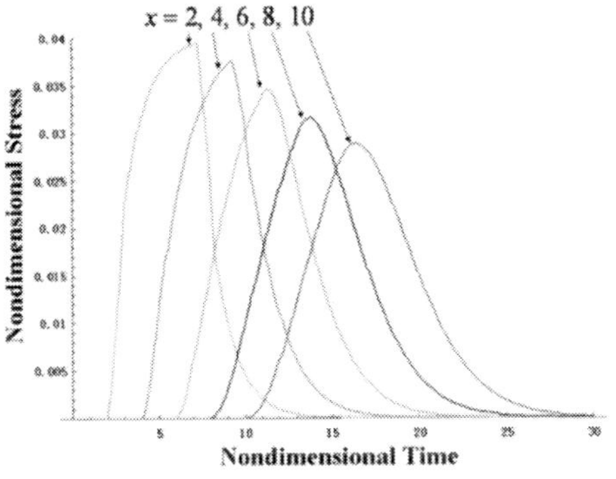

(a)

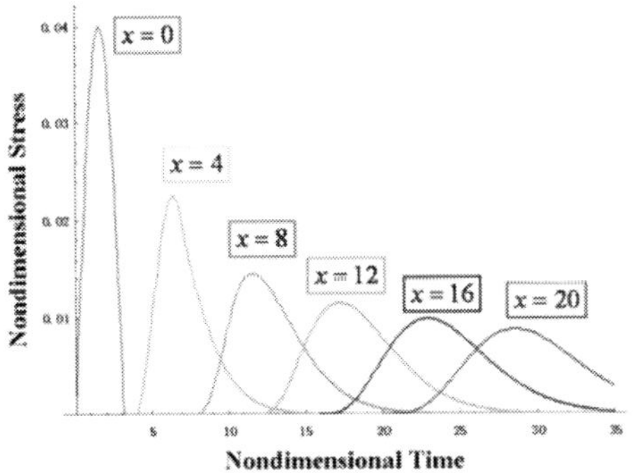

(b)

Figure 4. Stress histories at different sites of a viscoelastic bar impacted at left end by (a) a trapezoidal stress pulse showed in Fig. 1; (b) a half-sine stress pulse.

5. Summary

The viscoelastic wave propagation in the specimen of SHPB experiment was analyzed by Laplace transform method. By conducting Laplace transform, the governing partial differential equations were transformed to ordinary differential equations for the image functions, which were solved analytically with the boundary equations of the two ends of the specimen. For viscoelastic specimen ,the problems was solved numerically by way of numerical inverse Laplace transform. For pure elastic specimen, it was solved analytically by Laplace transform.

The research was supported by National Science Foundation of China (11272163, 11390361). The part of the paper has been published in Chinese Journal of Theoretical and Applied Mechanics.

References

[1] Wang Lili. Stress Wave Foundation. Beijing: National Defense Industry Press, 2005 (in Chinese)

[2] Zhou Fenghua, Wang Lili, Hu Shisheng. On the effect of stress non-uniformness in polymer specimen of SHPB tests. Experimental Mechanics, 1992, **7**(1):23–29 (in Chinese)

[3] Wang L, Labibes K, Azari Z. Generalization of split Hopkinson bar technique to use viscoelastic bars. International Journal of Impact Engineering, 1994, **15**: 669–686

[4] Zhao H, Gary G, Klepaczko JR. On the use of a viscoelastic splitHopkinson pressure bar. International Journal of Impact Engineering, 1997, **19**: 319–330

[5] Subhash G, Liu Q, Gao XL. Quasistatic and high strain rate uniaxial compressive response of polymeric structural foams. International Journal of Impact Engineering, 2006, **32**: 1113–1126

[6] Du_y DG. Transform Methods for Solving Partial Di_erential Equations.2nd Edition. Boca Raton, Florida: Chapman and Hall/CRC, 2004

EPJ Web of Conferences **94**, 04022 (2015)
DOI: 10.1051/epjconf/20159404022

Strain-rate sensitivity of foam materials: A numerical study using 3D image-based finite element model

Yongle Sun[1], Q.M. Li[1,a], and P.J. Withers[2]

[1] School of Mechanical, Aerospace and Civil Engineering, The University of Manchester, Sackville Street, Manchester M13 9PL, UK
[2] Henry Moseley X-ray Imaging Facility, School of Materials, The University of Manchester, Grosvenor Street, Manchester M13 9PL, UK

Abstract. Realistic simulations are increasingly demanded to clarify the dynamic behaviour of foam materials, because, on one hand, the significant variability (e.g. 20% scatter band) of foam properties and the lack of reliable dynamic test methods for foams bring particular difficulty to accurately evaluate the strain-rate sensitivity in experiments; while on the other hand numerical models based on idealised cell structures (e.g. Kelvin and Voronoi) may not be sufficiently representative to capture the actual structural effect. To overcome these limitations, the strain-rate sensitivity of the compressive and tensile properties of closed-cell aluminium Alporas foam is investigated in this study by means of meso-scale realistic finite element (FE) simulations. The FE modelling method based on X-ray computed tomography (CT) image is introduced first, as well as its applications to foam materials. Then the compression and tension of Alporas foam at a wide variety of applied nominal strain-rates are simulated using FE model constructed from the actual cell geometry obtained from the CT image. The stain-rate sensitivity of compressive strength (collapse stress) and tensile strength (0.2% offset yield point) are evaluated when considering different cell-wall material properties. The numerical results show that the rate dependence of cell-wall material is the main cause of the strain-rate hardening of the compressive and tensile strengths at low and intermediate strain-rates. When the strain-rate is sufficiently high, shock compression is initiated, which significantly enhances the stress at the loading end and has complicated effect on the stress at the supporting end. The plastic tensile wave effect is evident at high strain-rates, but shock tension cannot develop in Alporas foam due to the softening associated with single fracture process zone occurring in tensile response. In all cases the micro inertia of individual cell walls subjected to localised deformation is found to have negligible effect on the macro strain-rate sensitivity of Alporas foam.

1. Introduction

Aluminium foams possess unique properties, which are attractive for various engineering applications such as energy absorption and sandwich construction. Their compressive properties are of particular importance when they are used to absorb energy and attenuate impact/blast loads. The understanding of their tensile strength is also important for the design of load-bearing structures (e.g. sandwich structures). Under dynamic loading, their cell deformation may be distinct from the one under quasi-static loading, contributing to macroscopic strain-rate effect. Strain-rate effect on the mechanical behaviour of cellular solids is one of the major concerns for their applications associated with dynamic loads, which has been studied extensively in recent years. However, the actual causes of strain-rate effect have not been completely recognised.

Despite the lack of a general conclusion on the strain-rate sensitivity of aluminium foams, the strain-rate hardening of compressive strength of closed-cell aluminium Alporas foam has been widely observed [1–6]. Meanwhile, a number of explanations have been proposed. For instance, Paul and Ramamurty [3] suggested that the strain-rate hardening is related to both the strain-rate sensitivity of the aluminium matrix and the micro inertia of the individual cell walls. Cady et al. [5] argued that many factors, e.g. the cell-wall interaction, pore architecture and the rate dependence of the cell-wall material, are involved. Dannemann and Lankford [2] attributed the strain-rate effect to the flow of gas through ruptured cell walls. On the other hand, Elnasri et al. [7] examined the structural response during high speed impact and proposed that the strong strain-rate hardening is mainly a shock wave related phenomenon.

From these previous studies, the rate dependence of cell-wall material, the micro inertia, the shock wave, and the trapped gas are identified as the most likely causes responsible for the macro strain-rate effect. However, the individual contributions of these factors cannot be easily distinguished in experiments. Therefore, theoretical and numerical analyses are needed, which usually use simplified or idealised cell structures, e.g. cubic cell and Voronoi structure, but they may not be able to sufficiently capture the actual structural effect of foam materials, since some factors such as cell-wall rate dependence and micro inertia depend on the actual structural characteristics [8]. This motivates the geometrically

[a] Corresponding author: qingming.li@manchester.ac.uk

realistic simulations based on cell structures obtained from computed tomography (CT) images (so-called image-based modelling [9]), which has been employed to investigate the strain-rate sensitivity of open-cell foams [10, 11] and the quasi-static compressive behaviour of closed-cell foams [12, 13]. The dynamic behaviour of Alporas foam has yet to be examined in this way.

The objective of this work is to investigate the strain-rate sensitivity of closed-cell aluminium foam (Alporas) and the associated mechanisms. 3D finite element (FE) model is developed using the actual cell structure obtained from CT image. Then compression and tension are simulated at the applied nominal strain-rates ranging from $1 \times 10^{-3}\,s^{-1}$ to $3 \times 10^{3}\,s^{-1}$. The numerical prediction of the strain-rate sensitivity of the compressive strength is also compared with test data reported in the literature.

2. Image-based 3D FE model

2.1. Geometrical characterisation and meshing

A cylindrical Alporas foam sample with dimensions of Ø15.0 × 14.4 mm was scanned using a Nikon Metris CT system housed in a customised bay at the Henry Moseley X-ray Imaging Facility (HMXIF, Manchester, UK). The accelerating voltage and current of the X-rays were set as 75 kV and 125 μA, respectively. The effective voxel size was 10.9 μm and the reconstruction of the X-ray radiographs (2000 projections) was accomplished using Nikon Metris CT-Pro software. To eliminate the surface damage incurred by machining during sample preparation, the central portion of the foam sample with dimensions of Ø11.0 × 13.5 mm was selected as the region of interest (ROI). Greyscale-based segmentation was used to extract the solid phase in ROI. The tiny pores having diameters smaller than 400 μm are neglected, which greatly simplifies the meshing and reduces element number. In addition, the cell walls were slightly dilated (by 44 μm in this case) to facilitate image down-sampling (i.e. increasing voxel size) without losing thin walls. The down-sampled CT image has a voxel size of 50.0 μm. The number of voxels is significantly reduced by the down-sampling, such that the number of elements in the mesh can be greatly reduced.

Figure 1 shows the transverse CT slices at the mid-height of the virtual foam sample after reconstruction and image processing, which manifests the extremely complex cell morphology and topology. In Fig. 1b the colours indicate the local thickness of cell walls measured in 3D [14]. The cell structure constructed here maximally preserves the actual cell morphology and topology, although some fine features (e.g. extremely tiny pores and thin walls) are neglected to restrict complexity.

Table 1 lists the statistic data of the cell size and local cell-wall thickness, for which the maximum value is about one order larger than the minimum value. The high scatter in cell size and cell-wall thickness is also clearly seen in Fig. 1. The thickness of the walls at the junctions is much larger than that at other locations, and many small cells are located in the thick walls.

The FE discretisation of the foam sample was carried out by using ScanIP (Simpleware Ltd, UK), which is a commercial software for directly converting CT images into high quality 3D meshes [15]. The 3D profile of the foam sample after meshing is shown in Fig. 3. Quadratic elements were used for meshing since they have high numerical precision which is necessary to capture the bending and buckling of thin walls which are unlikely to be very finely meshed due to the geometrical complexity and the limited mesh density that can be adopted. The relative density (RD, i.e. the ratio of the cell wall volume to the foam volume) of the meshed sample is 0.109, around 11% larger than that of the scanned sample. This mainly results from errors introduced by the image processing and meshing steps. A similar volume increase of closed-cell foam model also occurred when other image processing and meshing techniques were used (e.g. around 21% volume increase occurring in the model by Jeon et al. [12]).

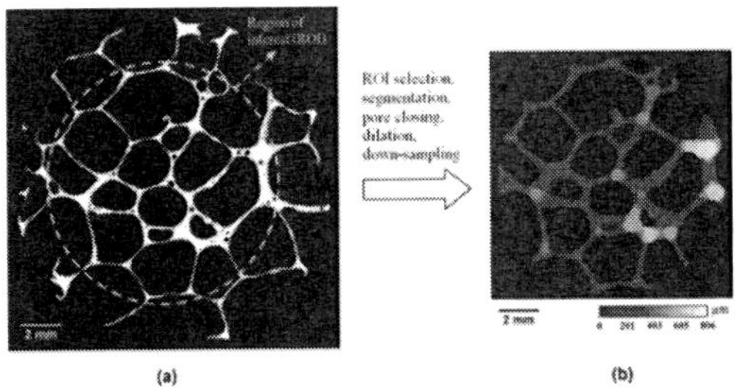

Figure 1. (a) Transverse slice CT image of the scanned foam (Ø15.0 × 14.4 mm, voxel size is 10.9 μm); (b) corresponding slice CT image of the ROI (Ø11.0 × 13.5 mm, voxel size is 50.0 μm) after the image processing with the local cell-wall thickness (measured in 3D) represented as a colour-scale.

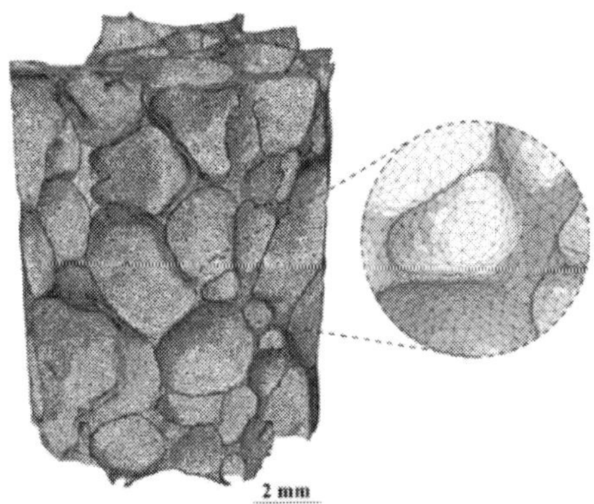

Figure 2. 3D FE mesh (493150 quadratic tetrahedral elements and 957697 nodes) of the foam sample created from the CT image and enlarged view of the surface cells to show the mesh density and quality.

Table 1. Geometrical data of the foam sample (unit: mm).

	Mean ± SD	Max.	Min.
Cell size	2.01 ± 1.25	5.01	0.43
Cell-wall thickness	0.28 ± 0.30	0.81	0.10

2.2. FE model setup

The material properties of the cell walls were assumed isotropic and homogeneous. A power-law was adopted to describe the strain hardening of the yield stress, and von Mises plasticity was used. The elastic modulus, yield strength and hardening exponent are 68 GPa, 35.5 MPa and 8.5, respectively [16]. The rate dependence of the yield stress was determined by the Cowper-Symonds equation with parameters of $\dot{\varepsilon}^* = 6500\,\mathrm{s}^{-1}$ and $m = 4$ [11] for typical aluminium. The case where the cell-wall material is rate-independent was also considered for comparison. The loading processes were treated as dynamic in order to incorporate inertia effects at various strain-rates, and the cell-wall density was taken to be $2710\,\mathrm{kg/m^3}$.

The foam sample was loaded on the top surface by applying a constant velocity to the associated nodes, while for the bottom surface the displacement was fixed in the vertical direction. In-plane motion was allowed on the end surfaces and no constraints were exerted on the lateral sides of the sample. The nominal strain ε and stress σ were defined as the displacement divided by the original height and the end-surface reaction force divided by the original cross-sectional area, respectively. The nominal strain-rate $\dot{\varepsilon}$ was defined as the ratio of the loading speed to the original height and varied from $1 \times 10^{-3}\,\mathrm{s}^{-1}$ to $3 \times 10^{3}\,\mathrm{s}^{-1}$ in the simulations. To compare with previous experimental results [1–4], the compressive strength was defined as the first peak stress (i.e. collapse stress), while the 0.2% offset yield point was used to characterise tensile strength.

The general purpose FE code, Abaqus, was employed to perform the simulations considering large deformation effect. The Abaqus/Implicit solver capable of highly accurate stress solutions was adopted for the dynamic simulations so as to take account of inertia effect.

3. Results and discussion

3.1. Stress-strain relationship

The high strain-rate can lead to force imbalance between the two end surfaces. For instance, at a strain-rate of $3 \times 10^{3}\,\mathrm{s}^{-1}$, as shown in Fig. 3, the stress at the loading end is significantly larger than that at the supporting end for both compression and tension, whether the cell-wall material is rate-dependent or not. This phenomenon is known as shock enhancement for cellular solids subjected to compression [7]. For tension, the inertial force associated with the propagation of plastic stress wave also produces higher stress level at the loading end. However, it is not a shock phenomenon since the tangential modulus of the tensile stress-strain curve decreases continuously with the strain, and eventually, the deformation is dominated by the softening mechanism due to the formation of single fracture process zone [17]. The shock phenomenon in dynamic compression is caused by the localised dynamic densification of the collapsed cells leading to the overall increase of compressive stress and shock discontinuity [18]. Another observation from Fig. 3 is that both the loading and supporting stresses for the foam with rate-dependent cell walls are larger than those for the foam with rate-independent cell walls.

To exclude the inertial force occurring at the loading end (i.e. behind shock or plastic wave front), the stress measured at the supporting end is used to characterise

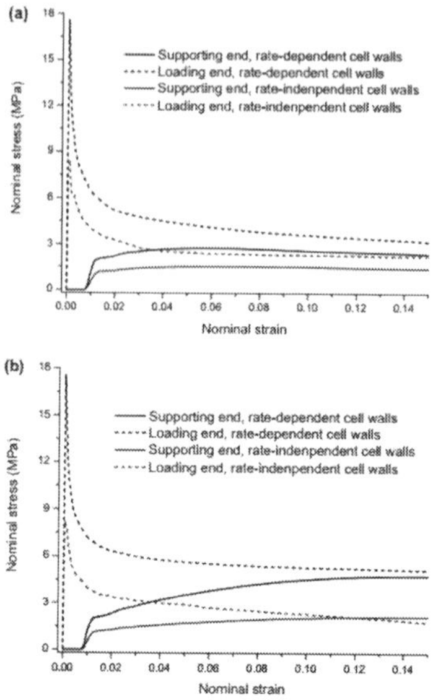

Figure 3. Stress-strain curves at a strain-rate of $3 \times 10^{3}\,\mathrm{s}^{-1}$: (a) compression; (b) tension.

the foam strength at high strain-rates. Nevertheless, it should be noted that at low and intermediate strain-rates ($1 \times 10^{-3} - 1 \times 10^{2}\,\mathrm{s}^{-1}$), the loading and supporting forces (and stresses) are equal. This balancing of forces at the macro-scale, however, can co-exist with the micro inertia of individual cell walls. The micro inertia arises from the nonuniform deformation of the heterogeneous cell structure and was often surmised to affect the macro strength. Figure 4 shows the compressive stress-strain curves at different strain-rates, which are typical foam responses under uniaxial compression [2,3,19,20], i.e. an essentially linear relationship at low stress before attaining a peak stress after which the stress drops to a plateau due to cell crushing. The quasi-static compressive strengths at a strain-rate of $1 \times 10^{-3}\,\mathrm{s}^{-1}$ obtained for foams with cell-wall materials being rate-dependent and rate-independent are 1.61 MPa and 1.57 MPa, respectively, which lie within the scatter band of the compressive strength (1.3–2.3 MPa) reported for Alporas foam of similar density [21,22]. Figure 5 shows the tensile stress-strain curves, which capture the stress-strain trend observed in tension experiments [17]. However, as cell-wall cracking is not considered in the present FE model, the peak stress and the stresses afterward, which are controlled by single fracture process zone [17], are expected to be overestimated in the tension simulations, as shown in Fig. 6. Nevertheless, the fracture area is small when strain is smaller than 0.01, and local plastic deformation dominates at small strain [17]. Thus the 0.2% offset yield point (see Fig. 5) is acceptable to evaluate the tensile strength using present FE model. For compression, the plastic buckling of cell walls controls the cell collapse and thus fracture effect is secondary to determine the collapse stress.

For the foam having rate-independent cell walls the stress-strain curves are identical at the strain-rates between $1 \times 10^{-3}\,\mathrm{s}^{-1}$ and $1 \times 10^{2}\,\mathrm{s}^{-1}$ (see Figs. 4a and

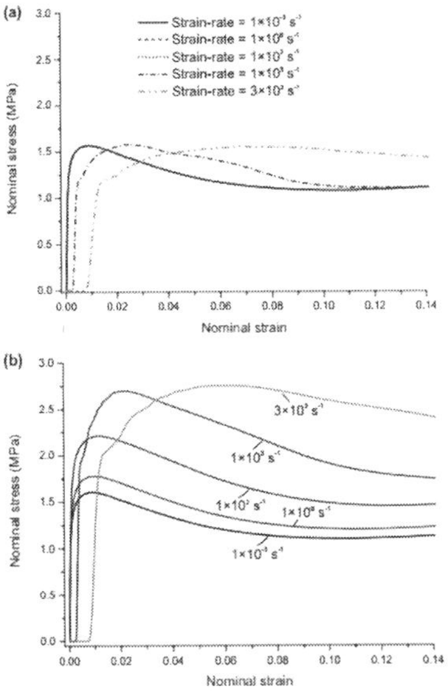

Figure 4. Compressive stress-strain curves measured at the supporting end at different strain-rates with the yield stress of the cell-wall material being rate-independent (a) and rate-dependent (b). Note that the stress-strain curves are identical at the strain-rates of $1 \times 10^{-3} - 1 \times 10^{2}\,s^{-1}$ in (a).

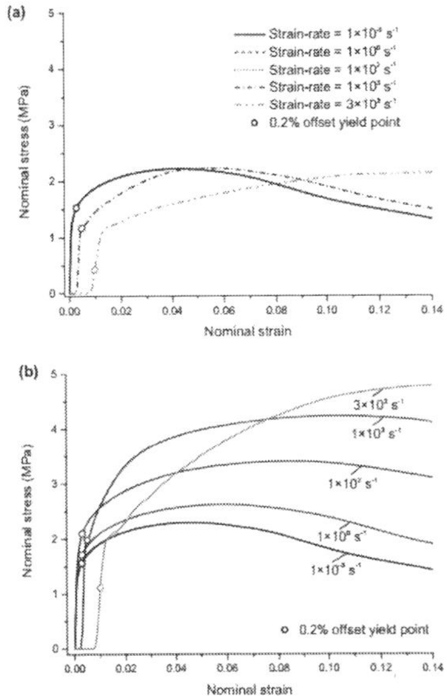

Figure 5. Tensile stress-strain curves measured at the supporting end at different strain-rates with the yield stress of the cell-wall material being rate-independent (a) and rate-dependent (b). Note that the stress-strain curves are identical at the strain-rates of $1 \times 10^{-3} - 1 \times 10^{2}\,s^{-1}$ in (a).

5a), indicating the negligible influence of micro inertia for Alporas foam in this strain-rate range. By contrast, differences arise at strain-rates above $1 \times 10^{3}\,s^{-1}$ due to the inertia effect. For instance at a strain-rate of $3 \times 10^{3}\,s^{-1}$, the elastic precursor wave does not arrive at the supporting end until the nominal strain reaches a critical value, after which the stress becomes non-zero and increases rapidly as a result of the transmission of load through the deforming cell walls. The result is different if the cell-wall material is taken to be rate-dependent. In this case, the foam strength generally increases with strain-rate (Figs. 4b and 5b), but the increase of collapse stress becomes subtle from $1 \times 10^{3}\,s^{-1}$ to $3 \times 10^{3}\,s^{-1}$ for compression and the 0.2% offset yield point decreases for tension. This is because when shock compression or plastic tensile wave is initiated at high strain-rates, the deformation is concentrated at the loading end and the cells near the support end are much less deformed.

3.2. Strain-rate sensitivities of the compressive and tensile strengths

To identify the strain-rate effect, the foam strength measured at the supporting end is normalised by that at a strain-rate of $1 \times 10^{-3}\,s^{-1}$ (regarded as quasi-static). The normalised strength is plotted against strain-rate in Fig. 6, alongside compression test data from the literature. Reasonable agreement is achieved between the compression simulations (for rate-dependent cell walls) and experimental tests although the experimental strain-rate sensitivity appears higher. Furthermore, the foam has slightly higher strain-rate sensitivity than the base material itself (Cowper-Symonds curve in Fig. 7) until the shock compression or plastic tensile wave occurs. When the

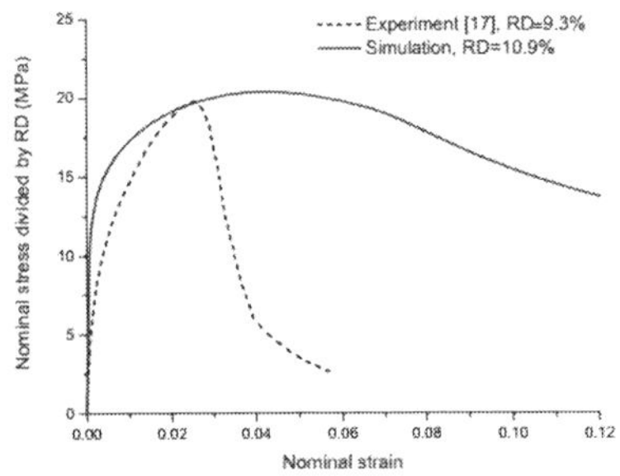

Figure 6. Comparison of the quasi-static tensile stress-strain curves obtained from experiments and simulations. The stress is normalised to eliminate the density effect (the tensile stress is approximately proportional to relative density [23]).

cell-wall material is rate-independent, the foam strength is insensitive to strain-rate although a drop in strength occurs at high strain-rates.

The contributions of three factors, i.e. micro inertia, stress wave (i.e. shock compression and plastic tensile wave) and rate dependence of cell-wall material, to the macro strain-rate effect are discussed as follows. The micro inertia, which is related to the local cell-wall motion, cannot be the cause of the strain-rate hardening, since if it had any effect the foam strength should depend on the strain-rate even when the cell-wall material is rate- independent, which obviously does not occur in the simulations (see Figs. 4–7). Unsurprisingly, the shock compression or plastic tensile wave enhances the stress measured at the loading end, but it results from inertial force, irrespective of the strain-rate sensitivity of the cell wall material. However, the stress wave effect on

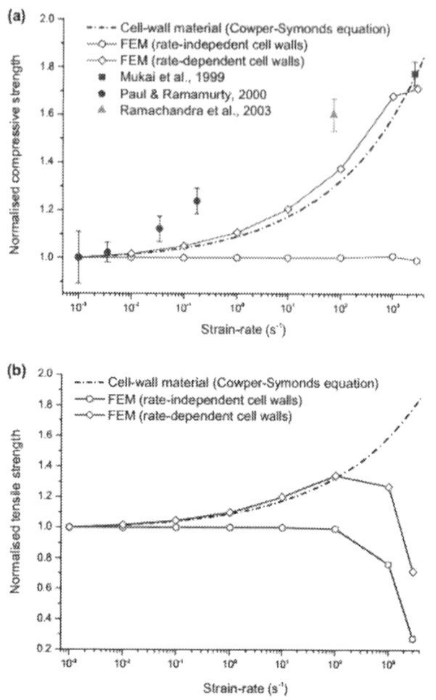

Figure 7. Predicted strain-rate sensitivity of the compressive (a) and tensile (b) strengths of Alporas foam for rate dependent and independent material models alongside available compression test data [1,3,4]. The dashed curve corresponds to the rate dependence of the cell-wall material. The dynamic strength is normalised by the quasi-static one at a strain-rate of $1 \times 10^{-3}\,s^{-1}$.

the foam strength measured at the supporting end is complicated. When the cell-wall material properties are rate-independent, it leads to a discernible reduction in compressive collapse stress and a significant drop of 0.2% offset yield point for tension. The cause of this effect is the concentrated large cell deformation at the loading end and the constrained small deformation at the supporting end, which arise from the inertia-controlled load transmission in the cell structure. This effect competes with the hardening effect of the rate dependence of cell-wall material when both are operative, which complicates the observed strain-rate sensitivity of foam strength.

It is evident from Fig. 7 that it is the rate dependence of cell-wall material that dominates the strain-rate effect of the foam strength. Furthermore, the foam strength can exhibit slightly higher strain-rate sensitivity than that of the cell-wall material at low and intermediate strain-rates (e.g. $< 1000\,s^{-1}$). This is because the local strain-rate can be higher than the nominal one. When the related cell walls act as the critical load-bearing elements, more significant strain-rate hardening of foam strengths will be observed. In addition, it is found that the compressive strain-rate sensitivity is underestimated by the FE model, which is mainly due to the assumption of the rate dependence of the cell walls described by the Cowper-Symonds equation with typical parameters for aluminium. The cell-wall material may exhibit higher rate dependence than that assumed in the simulations, since the components and microstructure of the cell-wall material (with aluminium matrix) of Alporas foams is complicated due to the foaming process during which additional elements are introduced and special processing is applied [24].

Although the effect of internal gas is neglected in the present FE model, it has been confirmed that the trapped gas hardly affects the foam strength of aluminium foams according to the previous analytical estimate [25] and numerical simulation [26]. However, the gas effect could be significant at the compression densification stage [26] or when the initial gas pressure is much higher than atmosphere pressure [27].

4. Conclusions

Three-dimensional geometrically realistic FE model based on CT image has been developed to investigate the strain-rate effect on the dynamic properties of closed-cell aluminium foam (Alporas). Uniaxial compression and tension have been simulated at different nominal strain-rates ($1 \times 10^{-3} - 3 \times 10^{3}\,s^{-1}$). The simulations show that the rate dependence of the cell-wall material dominates the macro strain-rate effect on the foam strengths measured at the supporting end, whereas micro inertia has negligible contribution for Alporas foam. At sufficiently high strain-rates, shock compression and plastic tensile wave occur, which lead to higher stresses measured at the loading end but has a complex effect on the stress at the supporting end, depending on the strain-rate sensitivity of the cell-wall material. In general, the cell deformation close to the supporting end is restrained at high strain-rates, thereby limiting the stress level and the strain-rate sensitivity there. The numerical results also demonstrate that foams can be more strain-rate sensitive than their cell-wall materials.

References

[1] T. Mukai, H. Kanahashi, T. Miyoshi, M. Mabuchi, T. Nieh, K. Higashi, Scripta Mater. **40**, 921–928 (1999).
[2] K.A. Dannemann, J. Lankford Jr, Mat. Sci. Eng. A **293**, 157–164 (2000).
[3] A. Paul, U. Ramamurty, Mat. Sci. Eng. A **281**, 1–7 (2000).
[4] S. Ramachandra, P. Sudheer Kumar, U. Ramamurty, Scripta Mater. **49**, 741–745 (2003).
[5] C.M. Cady, G.T. Gray Iii, C. Liu, M.L. Lovato, T. Mukai, Mat. Sci. Eng. A **525**, 1–6 (2009).
[6] T. Mukai, T. Miyoshi, S. Nakano, H. Somekawa, K. Higashi, Scripta Mater. **54**, 533–537 (2006).
[7] I. Elnasri, S. Pattofatto, H. Zhao, H. Tsitsiris, F. Hild, Y. Girard, J. Mech. Phys. Solids **55**, 2652–2671 (2007).
[8] C.R. Calladine, R.W. English, Int. J. Mech. Sci. **26**, 689–701 (1984).
[9] E. Maire, P.J. Withers, Int. Mater. Rev. **59**, 1–43 (2014).
[10] A.D. Brydon, S.G. Bardenhagen, E.A. Miller, G.T. Seidler, J. Mech. Phys. Solids. **53**, 2638–2660 (2005).
[11] M. Vesenjak, C. Veyhl, T. Fiedler, Mat. Sci. Eng. A **541**, 105–109 (2012).
[12] I. Jeon, T. Asahina, K.-J. Kang, S. Im, T.J. Lu, Mech. Mater. **42**, 227–236 (2010).

[13] C. Veyhl, I.V. Belova, G.E. Murch, T. Fiedler, Mat. Sci. Eng. A **528**, 4550–4555 (2011).
[14] T. Hildebrand, P. Rüegsegger, J. Microsc. **185**, 67–75 (1997).
[15] P.G. Young, T.B.H. Beresford-West, S.R.L. Coward, B. Notarberardino, B. Walker, A. Abdul-Aziz, Philos. T. Roy. Soc. A **366**, 3155–3173 (2008).
[16] I. Jeon, K. Katou, T. Sonoda, T. Asahina, K.-J. Kang, Mech. Mater. **41**, 60–73 (2009).
[17] C. Motz, R. Pippan, Acta Mater. **49**, 2463–2470 (2001).
[18] L. Wang, Foundations of stress waves (Elsevier Science Ltd, Amsterdam, 2007).
[19] I. Jeon, T. Asahina, Acta Mater. **53**, (2005) 3415–3423.
[20] J. Shen, G. Lu, D. Ruan, Compos. Part. B-eng. **41**, 678–685 (2010).
[21] U. Ramamurty, A. Paul, Acta Mater. **52**, 869–876 (2004).
[22] M.F. Ashby, A.G. Evans, N.A. Fleck, L.J. Gibson, J.W. Hutchinson, H.N.G. Wadley, Metal foams: a design guide (Elsevier, 2000).
[23] L.J. Gibson, M.F. Ashby, Cellular solids: structure and properties, 2 ed. (Cambridge University Press, 1997).
[24] A.E. Simone, L.J. Gibson, Acta Mater. **46**, 3109–3123 (1998).
[25] Y.D. Liu, J.L. Yu, Z.J. Zheng, J.R. Li, Int. J. Solids. Struct. **46**, 3988–3998 (2009).
[26] Q. Fang, J. Zhang, Y. Zhang, Z. Gong, H. Wu, Int. J. Impact. Eng., doi: 10.1016/j.ijimpeng.2014.10.009, (2014).
[27] W. Zhang, Z. Xu, T.J. Wang, X. Chen, Int. J. Plasticity. **25**, 1231–1252 (2009).

EPJ Web of Conferences **94**, 04023 (2015)
DOI: 10.1051/epjconf/20159404023

MD modeling of screw dislocation influence upon initiation and mechanism of BCC-HCP polymorphous transition in iron

V.V. Dremov[1,a], G.V. Ionov[1], F.A. Sapozhnikov[1], N.A. Smirnov[1], A.V. Karavaev[1], M.A. Vorobyova[1], and M.V. Ryzhkov[2]

[1] Russian Federal Nuclear Center – Zababakhin Institute of Technical Physics (RFNC-VNIITF), Snezhinsk, Chelyabinsk reg., Russia
[2] Institute of Solid State Chemistry, Ural Branch of RAS, Yekaterinburg

Abstract. The present work is devoted to classical molecular dynamics investigation into microscopic mechanisms of the bcc-hcp transition in iron. The interatomic potential of EAM type used in the calculations was tested for the capability to reproduce *ab initio* data on energy evolution along the bcc-hcp transformation path (Burgers deformation + shuffle) and then used in the large-scale MD simulations. The large-scale simulations included constant volume deformation along the Burgers path to study the origin and nature of the plasticity, hydrostatic volume compression of defect free samples above the bcc to hcp transition threshold to observe the formation of new phase embryos, and the volume compression of samples containing screw dislocations to study the effect of the dislocations on the probability of the new phase critical embryo formation. The volume compression demonstrated high level of metastability. The transition starts at pressure much higher than the equilibrium one. Dislocations strongly affect the probability of the critical embryo formation and significantly reduce the onset pressure of transition. The dislocations affect also the resulting structure of the samples upon the transition. The formation of layered structure is typical for the samples containing the dislocations. The results of the simulations were compared with the *in-situ* experimental data on the mechanism of the bcc-hcp transition in iron.

1. Introduction

A simulation of polymorphous transitions with Classical Molecular Dynamics (CMD) is complicated by the absence of an interatomic potential which could describe at least two phases with the same accuracy. Indeed, the transition from one crystal symmetry to another one is accompanied by changes in the electronic structure. Thus, it is rather difficult to use the same functional form and parametrization of an interatomic potential to describe interatomic interactions provided by different electronic structures.

Nevertheless classical MD has proved to be useful tool for the study of mechanisms of polymorphous transitions. Although interatomic potentials used cannot describe with high accuracy all the scope of properties of phases involved in the transition they are capable to describe/explain qualitatively and sometimes quantitatively the nature of processes that take place on the atomic level.

First MD simulations of polymorphous transition in iron under shock compression conditions were reported in [1–3]. The EAM potential by Voter & Chen [4] was used in these simulations. In these papers the peculiarities of bcc-hcp polymorphous transition in monocrystalline and polycrystalline samples were investigated. As for the monocrystalline samples the effect of loading in different crystallographic directions was studied. The main result of that work is that within the CMD approach it is possible to reproduce basic characteristics of shock-wave loading accompanied by the polymorphous transition. The kinetics of the transformation proved to be rather fast to allow the considerable amount of the material to be transformed into the new phase within the timescale of about 10–20 ps. The degree of the metastability at that (*i.e.* the distance from the equilibrium line or, in other words, the degree of overheating and overcompressing) proved to be reasonable.

The kinetics of the transformation is extremely sensitive to the difference of Gibbs potentials of the phases and when a Lagrangian particle crosses equilibrium boundary the rate of the martensitic transformation grows exponentially while the particle moves away from the boundary. When the calculations were carried out with polycrystalline samples the degree of metastability was even lower because grain boundaries serve as generators of plastic deformation favoring the formation of the new phase embryos of overcritical size.

In more recent paper [5] the polymorphous transition in iron was investigated under uniform uniaxial compression conditions at the strain rate $10^9\,s^{-1}$. Polycrystalline samples were used in the simulations. The investigation was focused on a comparison of a number of interatomic potentials from the point of view of reproducing basic properties of $\alpha-$, $\gamma-$, and ϵ-phases, as well as the transformation of initial bcc structure into a close-packed ones provided. The following potentials were tested in [5]: Mendelev [6], MEAM-p [7], Ackland [5] and Voter [4].

[a] Corresponding author: V.V.Dryomov@vniitf.ru

Article available at http://www.epj-conferences.org or http://dx.doi.org/10.1051/epjconf/20159404023

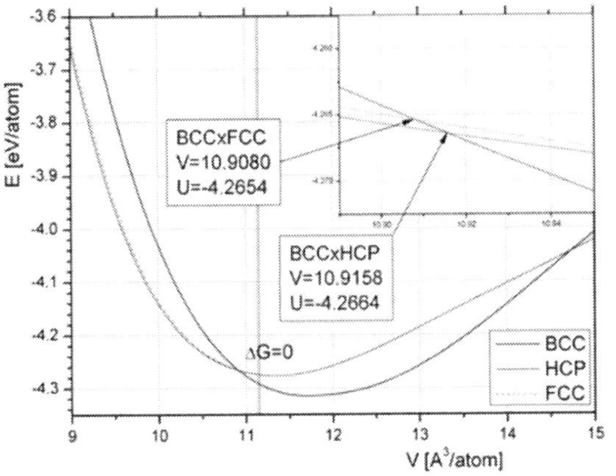

Figure 1. Cold (T = 0 K) energy curves for bcc, fcc, and hcp crystal structures for the Ackland potential. Volume and energy at the intersection of bcc-fcc and bcc-hcp curves are shown in boxes.

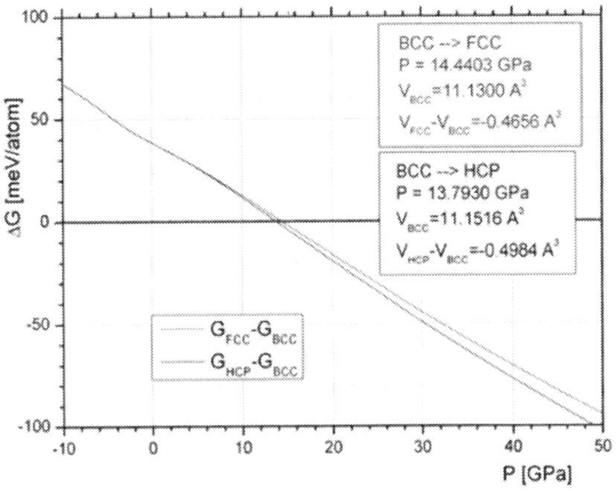

Figure 2. Cold (T = 0 K) enthalpy $E + PV$ curves for bcc, fcc, and hcp crystal structures for the Ackland potential. Volume of bcc phase and volume change at the transitions bcc-fcc and bcc-hcp are shown in boxes.

The goal of present work is an investigation into microscopic mechanisms of the bcc-hcp polymorphous transformation in iron, namely, the role of Burgers mechanism accompanied by the shuffle of atomic layers (see [11]) and preexisting dislocations. In particular, the generation of dislocations and the plasticity mechanisms taking place during pure Burgers deformation of α-iron samples and the effect of dislocations on the time of ϵ-phase critical embryos generation are investigated.

2. Interatomic potential and basic properties of hcp and bcc iron

In the following simulations we used interatomic potential of the EAM type that was modified in [5] as an improvement of Finnis-Sinclair potential [8]. The high pressure behavior of the potential [8] was improved in [5] to provide for $\alpha - \epsilon$ transition at 13.75 GPa. In [5] this potential was characterized as quite well reproducing main features of the bcc-hcp transition in iron as well as dislocation properties. Hereafter this potential is referenced as Ackland potential. In Fig. 1 energy-volume curves $E(V)$ calculated at T = 0 K are presented. In Fig. 2 enthalpy-pressure $G(P)$ curves at T = 0 K for iron phases are shown relative to bcc phase.

Next, we calculated energy dependence on the deformation ϵ along the Burgers path and compared the obtained results with the results of calculations performed with the VASP code [9,10]. In VASP calculations 8 atom supercell analogous to that from [11] was used. The bcc-phase was taken as having ferromagnetic structure and hcp-phase as having antiferromagnetic AFM I structure. Pseudopotential PAW GGA taking into account 8 outer electrons as valent was used. The cut-off energy was chosen in a standard way. A $18 \times 4 \times 26$ k-point grid was used.

The calculations of the deformation along the Burgers path were performed for the bcc structure undergoing the deformation from $\epsilon = 0$ to $\epsilon = 1$ (Shuffle coordinate remains zero) and for the hcp structure undergoing deformation from $\epsilon = 1$ to $\epsilon = 0$ (Shuffle coordinate remains equal 1, i.e. all four atomic layers are fully shuffled). For the details of the Burgers deformation see [11].

The CMD calculations were carried out with the sample composed of $10 \times 10 \times 10$ supercells (containing 8 atoms) with periodic boundary conditions. To provide for a direct comparison with the results of the VASP code the energy was calculated for a number of fixed deformations ϵ and the MD relaxation was turned off. For a given deformation ϵ the corresponding deformed 8 atom supercell was constructed and then the whole sample was constructed of these deformed supercells. In such a way we excluded the generation of any elastic waves in the sample. The volume of the supercell at which the calculations were carried out corresponds to the intersection of bcc and hcp energy curves, *i.e.* for the VASP and the EAM [5] calculations the volumes slightly differ and are equal 10.53 and 10.9158 A^3/atom, respectively. The comparison of the calculations obtained with the potential [5] and with the VASP code is presented in Fig. 3.

One can see that the potential [5] considerably underestimates energy growth along the Burgers path for the bcc-phase. For the hcp-phase the agreement is much better but the underestimation is also takes place. The considerable underestimation of the energy change for α-phase along the Burgers path is due to magnetic energy contribution that is not taken into account in the EAM potential. Note here, that it is quite difficult to account for the magnetic contribution to the total energy in the frame of semiempirical approach.

In the next Fig. 4 we plotted energy of 8 atom supercell as dependent on shuffle coordinate at a number of fixed deformations ϵ. The calculations were carried out for the volume V = 10.9158 A^3/atom corresponding to the intersection of bcc and hcp energy curves $E(V)$. This figure is analogous to Fig. 4 from [11] and provide an information about transition thresholds along the Burgers+Shuffle pathway. The general conclusion is that

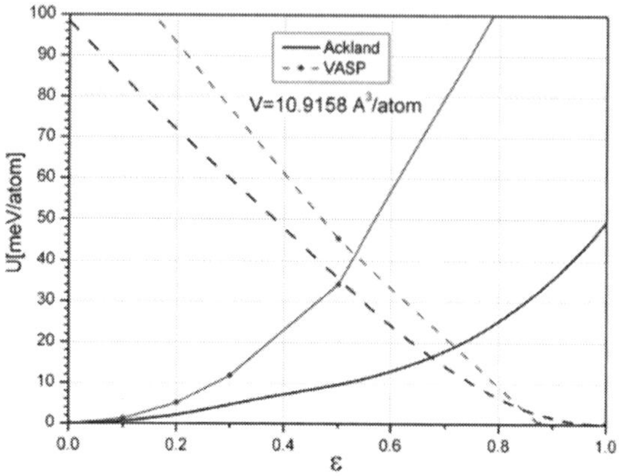

Figure 3. Energy dependence on deformation ϵ along the Burgers path for the Ackland potential. Red is for VASP results, black is for Ackland potential. Solid lines are for bcc initial state ($\epsilon = 0$, Shuffle = 0 and dashed lines are for hcp initial state ($\epsilon = 1$, Shuffle = 1).

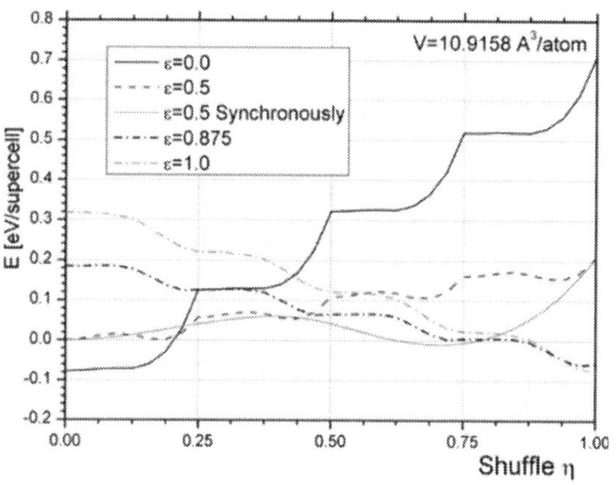

Figure 4. Energy of 8 atom supercell as dependent on shuffle coordinate at fixed deformation ϵ for the Ackland potential. The curve designated as "Synchronous" is for simultaneous shuffling of all 4 atoms in the unit cell. Other curves are for the case when layers shuffle takes place in a consecutive way: coordinate 0.25 corresponds to the configuration where one layer is fully shuffled, 0.5– two layers etc..

the thresholds are smaller for the EAM potential used if compared with *ab initio* calculations [11].

3. Large scale CMD simulations

The first series of the simulations was carried out to investigate into the mechanism of plasticity proper to the Burgers deformation path. In the simulations the samples constructed of $50 \times 25 \times 100$ supercells (8 atom per supercell) were used and periodic boundary conditions were applied. To provide for a quasistatic regime of loading the deformation from $\epsilon = 0$ to $\epsilon = 1$ takes one million MD step (MD step = 1.5 fs). Prior to the deformation modeling the samples were relaxed at zero stress tensor and T = 300 K during 100,000 MD steps.

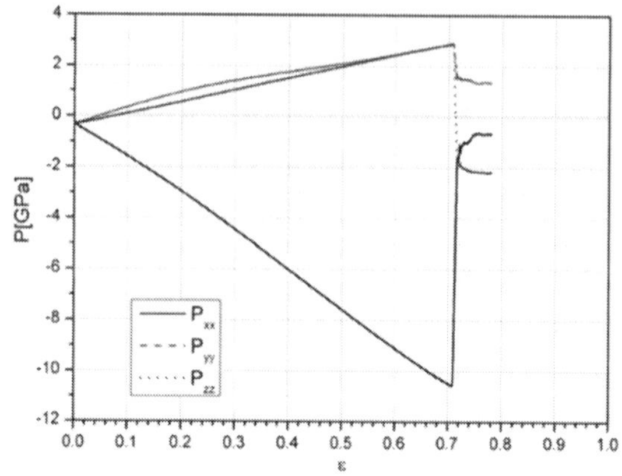

Figure 5. Stresses vs. deformation along the Burgers path.

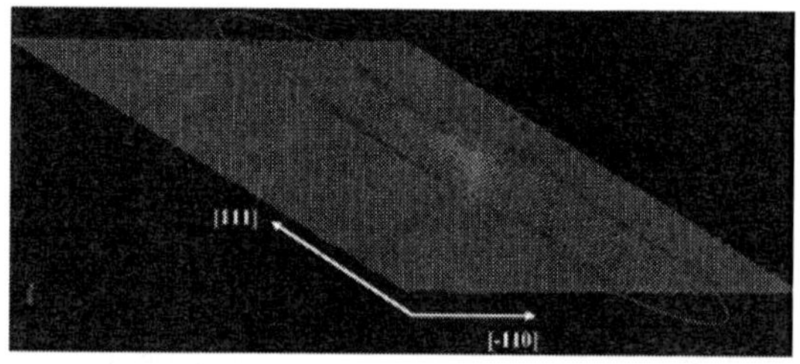

Figure 6. Thin cut of the samples after Burgers deformation. Red oval designates the twin formed in the sample as a result of the Burgers deformation. Yellow is for the bcc structure. Structures were recognized by the ATA.

The dependence of normal stresses during the Burgers deformation is presented in Fig. 5. It is seen that at ϵ 0.72 the sharp release of the stresses takes place due to the commencement and development of plastic deformation. The character of the plastic deformation is shown in Fig. 6. Analysis of the structural changes showed that the release of the stresses took place due to formation of twins. Results of structure recognition with the Adaptive Template Analysis (ATA) [16] show that the twins form tube-like structure oriented along [111] direction (see Fig. 7). The orientation of crystal lattice with respect to the twin boundary is presented in Fig. 8.

The MD simulations of bulk compression were carried out with big samples (dimensions of the samples were $60 \times 30 \times 30$ nm, ~ 5 mln particles) in periodic boundary conditions. The simulations were performed for three deformation rates $\dot{\epsilon}$: $666.6\,\mu s^{-1}$ (Fast), $66.66\,\mu s^{-1}$ (Medium), $6.666\,\mu s^{-1}$ (Slow). Temperature was kept constant T = 300 K. The deformation rates provide for uniform compression since the duration of the simulation is much longer than the time of acoustic waves propagation across the sample. Moreover, we completely excluded the possibility of the formation of elastic waves due to simultaneous compression of computing box and recalculation of atomic coordinates inside the box. Before the simulations of the compression the samples were relaxed at zero stress tensor. Two types of samples were used: defect free monocrystallines and monocrystallines containing a pair of screw dislocations oriented along

Figure 7. Tube-like structure of twins in the bulk of the sample. Only boundaries of twins are shown. Structures were recognized by the ATA.

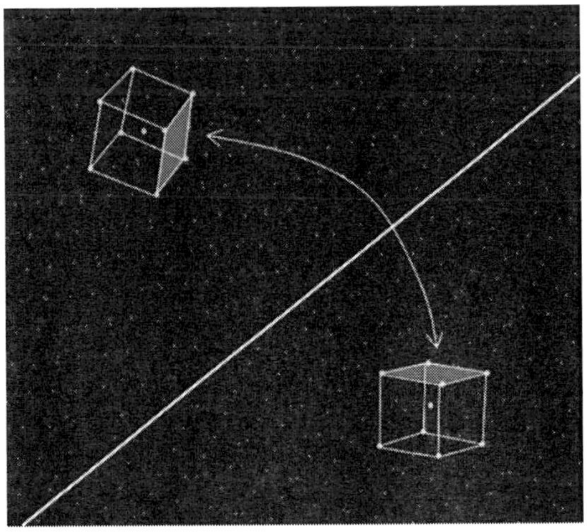

Figure 8. Orientations of crystal lattice (cubic unit cells) above and below the boundary between twins.

[110] axis. To generate the screw dislocations we used the method described in [15].

The pressure-density plot for the bulk compression of the sample free of defects is presented in Fig. 9. As it is seen in the figure in the absence of defects there is a noticeable degree of overcompressing before the transition starts. The equilibrium transition pressure for this potential is ~14 GPa. But because of the small system size and low probability of formation of critical size embryo of the new phase at pressures slightly above the equilibrium pressure the transition does not start (in the timescale of the MD simulation) until some degree of the metastability is achieved. In other words the probability of critical size embryo formation rapidly grows as the current state gets further away from the equilibrium line, *i.e.* with the difference increase of Gibbs potentials between the $\alpha-$ and the $\epsilon-$ phases. At some degree of overcompression the

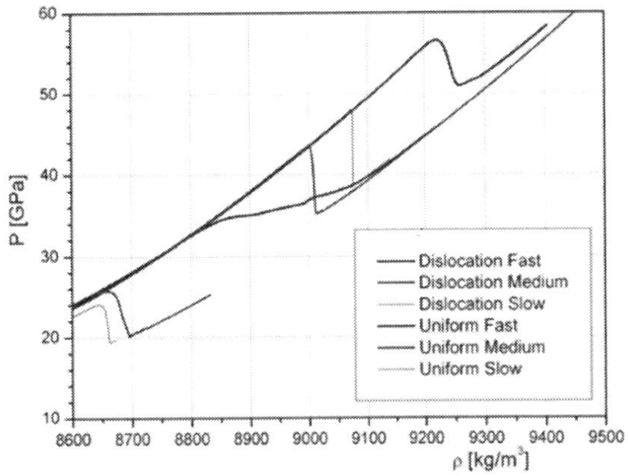

Figure 9. Pressure vs. density for the bulk compression.

Figure 10. Fragment of the sample initially free of defects upon completion of the bcc-hcp transition. Medium loading rate, MD step = 360000: P = 35.261 GPa, $\rho = 9.02\,g/cm^3$. ATA colors: Blue – hcp structure, green and red – boundaries between the hcp grains.

probability becomes high enough for the transition occurs at the given size of the sample and the timescale of the simulation.

At the fast compression rate the transition starts only at ~56 GPa. There is a noticeable dependence on the rate of loading. Indeed, at the medium rate the characteristic "delay time" at each pressure is effectively an order of magnitude greater, and the transition occurs at smaller pressure ~47 GPa. At the slowest rate of loading the transition starts at ~43 GPa.

Let's now look at the structure of the samples during the transition. A fragment of the sample initially free of defects is shown in Fig. 10. The picture presented here is for the medium rate of loading. But for all the rates the pictures look qualitatively the same. The transition starts from the embryos of the new phase randomly produced in the bulk of the sample. The rapid growth of the embryos leads to the formation of polycrystalline structure composed mostly of the hcp grains chaotically oriented relative to each other.

In the case of the sample containing the dislocations the transition starts at 33, 26, and 24 GPa for the fast, medium, and slow loading respectively (see Fig. 9).

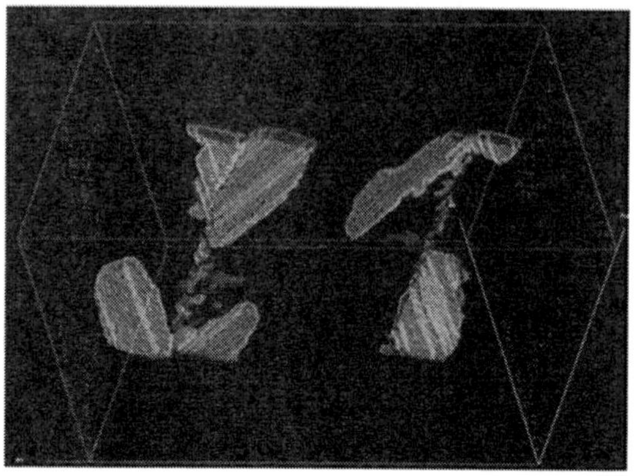

Figure 11. Fragment of the sample containing dislocations. Medium loading rate. ATA colors: blue – hcp structure, green and red – boundaries between the hcp and the bcc structures and stacking faults in growing hcp grains. Dislocations are shown by arrows. Atoms belonging to the bcc structure are not shown.

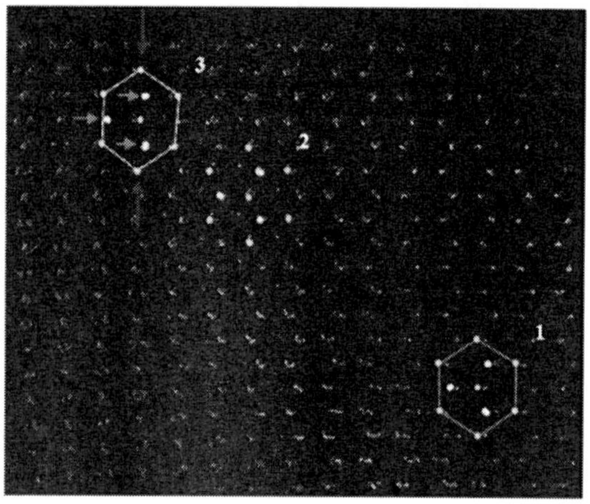

Figure 12. Boundary between growing hcp plate (small blue points) and bcc matrix (small yellow points). The atomic arrangements are viewed from the [110] direction. Elemental structures (central atom and its 12 neighbors) of hcp, transition, and bcc type are designated as 1,2,3. Red arrows show Burgers deformation, green arrows show shuffle direction.

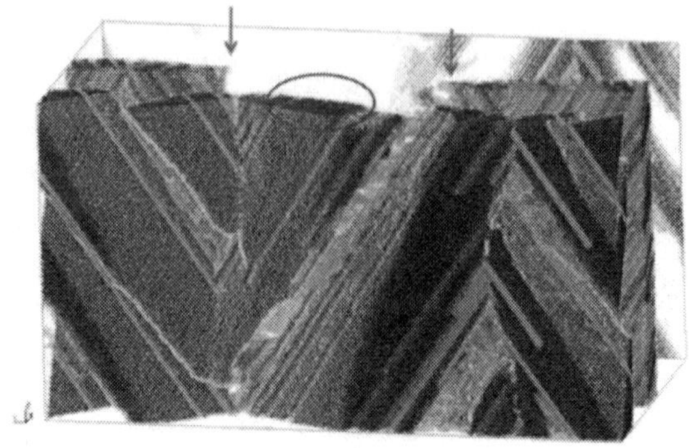

Figure 13. Fragment of sample containing dislocations. MD Step 280,000. Transition is in progress. ATA colors: Blue – hcp-structure, green and red – boundaries between the hcp and the bcc structures and between hcp grains. The bcc structure is not shown. Dislocations have vertical orientation; their positions are shown by arrows.

Figure 14. Fragment of sample containing dislocations MD. Step 290,000. Transition is completed. ATA colors: Blue – hcp-structure, green and red – boundaries between the hcp grains and stacking faults in the hcp grains.

Thus, the presence of the dislocations strongly affects the probability of critical embryo formation and significantly decreases the degree of the metastability achieved during the loading. In Fig. 11 the initial stage of the transition is shown. It is well seen in the figure that thin plates of the new phase starts to grow at the dislocations. The dislocations are 1D sources of the transition. As it was concluded in [17] the shear stress is important to initiate the bcc-hcp transition. In the vicinity of the screw dislocations the shear component of deformation is non-zero. The mechanism of the transition from the bcc to hcp structure is realized as the boundary separating the newly formed hcp plate from the bcc matrix propagates through the sample. It is shown in Fig. 12. Here one can see three elemental structures (central atom and its 12 nearest neighbors): 1) belonging to the bcc, 2) located at the boundary, and 3) belonging to the just born hcp structure. It is interesting to note that the hcp structure obtained is slightly distorted due to unfinished shuffle. The similar behaviour was registered in quiasistatic experiments [17] whereas the shear motion finished at the beginning of the transition. In our simulations the newly born hcp structure encapsulated in the thin plates is not in hydrostatic conditions (see below) and we relate the unfinished shuffle with the conditions of unrelaxed shear stress. At later time of the simulation in the bulk of the hcp grains the structure is close to the perfect hcp distorted only by thermal vibrations.

The rapid growth of thin hcp plates (originated at the dislocations) in specific crystallographic directions leads to a creation of layered structure shown in Figs. 13, 14. The process of the transition results in polycrystalline sample most of the material in which is in the hcp-phase. The hcp grains are not perfect monocrystallines i.e. the regions with the hcp structure are separated by 2D defects produced due to plastic deformation of newly formed hcp grains. This plastic deformation takes place because of the non-hydrostatic conditions the hcp grains are being immersed just upon the formation. Comparing regions designated in Figs. 13, 14 one can see that the plasticity inside the hcp grain takes place at later times than the grain formation.

The final layered structure produced due to the growth of thin plates originated at dislocations forms the polycrystalline samples the grains of which are not chaotically oriented relative to each other. This orientation is dictated by the orientation of dislocations (and of course in some sense by the periodic boundary conditions) and is produced by the overlapping of the growing hcp plates.

4. Conclusion

The CMD simulations were carried out to investigate into the mechanism of polymorphous transition bcc-hcp in iron. The CMD simulations of the Burgers deformation showed that the bcc crystal undergoes plastic transformations during the deformation. We observed twinning as the mechanism of plasticity.

The simulations of volume compression demonstrated high level of metastability achieved when compressing the samples. The transition starts at the pressure much higher than the equilibrium pressure. Screw dislocations strongly affect the probability of the formation of new phase critical embryos and significantly reduce the pressure at which the transition begins. The dislocations affect also the resulting structure of the samples upon the transition. The formation of layered structure is typical for the samples containing the dislocations due to the origin of the transition at the dislocations in the form of the thin plates of the hcp phase growing outward the dislocation line.

References

[1] Kadau K., Germann N.C., Lomdahl P.S., Holian B.L., Science, **296**, (2002) 1681.

[2] Kadau K., Germann N.C., Lomdahl P.S., Holian B.L., Cherne F.J., Atomistic simulation of shock induced phase transitions, AIP Conf. Proc., **706** (2003), 229–234.

[3] Hawreliak J., Rosolankova K., Belak J.F., Collins G., et al., Shock induced alpha-epsilon transiton in iron: analysis of MD simulations and experiment. AIP Conf. Proc., **845** (2005) 220–223.

[4] Harrison R.J., Voter A.F., Chen S.-P., in Atomistic Simulations of Materials: Beyon pair potentials, edited by Vitek V. and Srolovitz D.J., Plenum Press, New York (1989), 219.

[5] Gunkelmann N., Bringa E.M., Kang K., Ackland G.J., Phys. Rev. B, **86** (2012), 144111.

[6] M.I. Mendelev, S.Han, D.J. Srolovitz, G.J. Ackland, D.Y. Sun, and M. Asta, Philos. Mag. **83**, 3977 (2003).

[7] T. Lee, M.I. Baskes, S.M. Valone, D.J. Doll, J. Phys.: Condens. Matter. **24**, 225404 (2012).

[8] G. J. Ackland, D. J. Bacon, A. F. Calder, and T. Harry, Philos. Mag., *A75*, (1997) 713.

[9] G. Kresse and J. Furthmuller, Comput. Mater. Sci., **6**, 15 (1996).

[10] G. Kresse and J. Furthmuller, Phys. Rev. B, **54**, 11169 (1996).

[11] B. Dupé, B. Amadon, Y.-P. Pellegrini, and C. Denoual, Phys. Rev. **B 87**, 024103 (2013).

[12] D.F. Johnson, E.A. Carter, J. Chem. Phys., **128**, 104703 (2008).

[13] J.C. Boettger, D.C. Wallace, Phys. Rev. B, **55**, 2840-2849 (1997).

[14] M.S. Duesbery, V. Vitek, Acta mater., **46**, 1481 (1998).

[15] W. Cai, V.V. Bulatov, J. Chang, J. Li, and S. Yip, Philos. Mag., **83**, 539 (2003).

[16] F.A. Sapozhnikov, G.V. Ionov and V.V. Dremov, Russian Journal of Physical Chemistry B, **2**, 238-245 (2008).

[17] N. Ishimatsu, Y. Sata, H. Maruyama, T. Watanuki, N. Kawamura, M. Mizumaki, T. Irifune, H. Sumiya, Phys. Rev. B, **90**, 014422 (2014).

EPJ Web of Conferences **94**, 04024 (2015)
DOI: 10.1051/epjconf/20159404024

Development and validation of model for sand

P. Church[1,a], V. Ingamells[1], A. Wood[1], P. Gould[2], J. Perry[3], A. Jardine[3], and A. Tyas[4]

[1] QinetiQ, Fort Halstead, Sevenoaks, Kent TN14 7BP, UK
[2] QinetiQ, Bristol Business Park, Coldharbour Lane, Bristol BS16 1FJ, UK
[3] Surface, Microstructure & Fracture Group Cavendish Laboratory, JJ Thomson Ave, Cambridge CB3 0HE, UK
[4] Department of Civil and Structural Engineering, University of Sheffield, Sheffield S1 3JD, UK

Abstract. There is a growing requirement within QinetiQ to develop models for assessments when there is very little experimental data. A theoretical approach to developing equations of state for geological materials has been developed using Quantitative Structure Property Modelling based on the Porter-Gould model approach. This has been applied to well-controlled sand with different moisture contents and particle shapes. The Porter-Gould model describes an elastic response and gives good agreement at high impact pressures with experiment indicating that the response under these conditions is dominated by the molecular response. However at lower pressures the compaction behaviour is dominated by a micro-mechanical response which drives the need for additional theoretical tools and experiments to separate the volumetric and shear compaction behaviour. The constitutive response is fitted to existing triaxial cell data and Quasi-Static (QS) compaction data. This data is then used to construct a model in the hydrocode. The model shows great promise in predicting plate impact, Hopkinson bar, fragment penetration and residual velocity of fragments through a finite thickness of sand.

1. Introduction

QinetiQ has a growing requirement for developing assessments of granular materials when there is little experimental data to fit a model. This has necessitated the development of theoretical tools which have universal application to a wide range of materials.

This paper describes the use of Quantitative Structure-Property Modelling (QSPM) in deriving physically based equations of state directly, from knowledge of the constituents of the material. For geological materials this is based on an implicit assumption that many geological materials are a derivative of different crystal forms of silica (i.e. crystobalite, coesite, α-quartz, stishovite). The constitutive model is then fitted to existing Quasi-Static (QS) triaxial cell data. The ensuing model is then tested against a range of experiments covering different regimes of behaviour. The main issue with this approach is the compaction behaviour which is very variable and difficult to quantify.

2. Quantitative structure property modelling of silica

Silica is a molecule which exhibits 4 main crystal forms. These are in ascending order of density: crystobalite, quartz, coesite and stishovite and their molecular structure is illustrated in Fig. 1. CASTEP [1] reproduced the minimum energy structures of the different polymorphs of silica with the specified space group of the crystal forms and predicted their correct density values. However, CASTEP was unable to simulate changes in the crystal structure of silica with different boundary conditions such as pressure or volume imposed. The practical consequence of this problem was that the predicted bulk elastic modulus of all four polymorphs was identical, with a value of about 100 GPa.

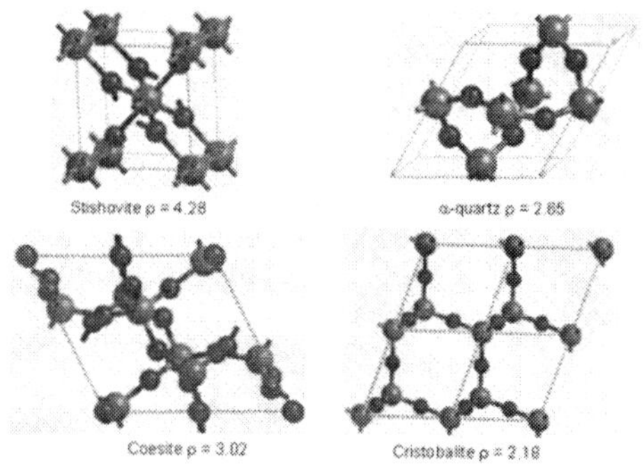

Figure 1. Different Crystal Forms of Silica.

Most quantum mechanics programs try to change the length or angle of a chemical bond to simulate a change in structure, since the energy associated with these changes is large. However, the Si-O bond is very mobile and can rotate easily around relatively rigid chemical bonds with very small changes in energy. Thus, changes in crystal structure or dimensional changes in silica are dominated by rotation of torsional bond angles that are very difficult to reproduce with programs such as CASTEP, unless a crystal structure is specified in advance to direct the calculations.

The failure of molecular mechanics or quantum mechanics to predict the volumetric properties of silica polymorphs requires a different modelling technique that

[a] Corresponding author: pdchurch@qinetiq.com

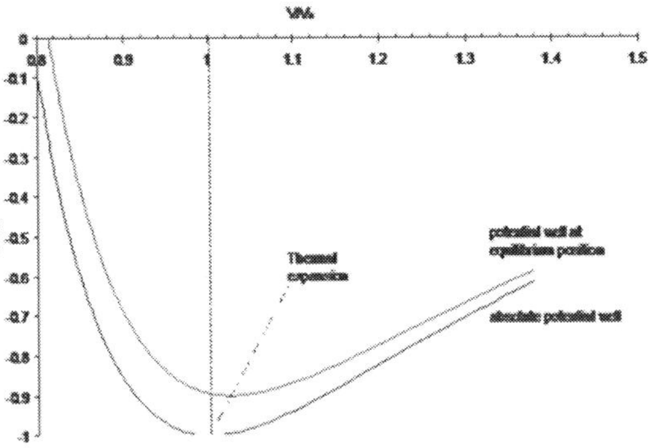

Figure 2. A Potential Well for Intermolecular Energy.

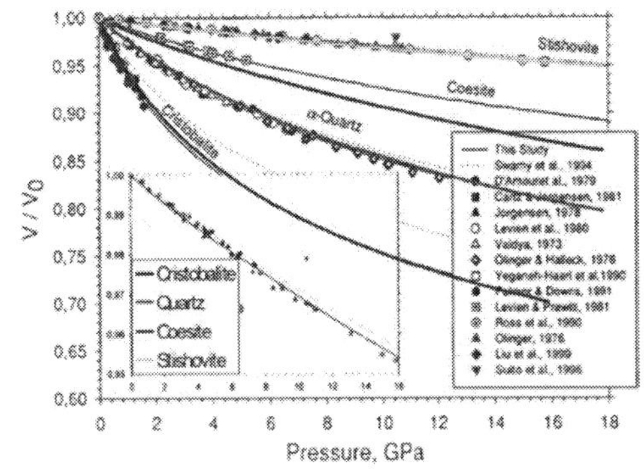

Figure 3. Comparison of Model with Experiment for Different Crystal Forms of Silica.

can predict properties associated with the relatively small changes in energy associated with physical (Van der Waals) bonding between atoms that are not chemically bonded in a structure. This type of bonding dominates the properties of polymers, since the preferred mode of deformation is through the weak van der Waals bonding normal to the chemical bonding in the chain axis of the polymer macromolecules. A direct analogy is made between a polymeric chain and linked chains of silica tetrahedra.

2.1. Potential function method

A modelling technique has been developed to predict the structural properties of polymers, called Group Interaction Modelling, GIM [2]. This technique has been used to predict equations of state for polymer-based materials [3]. The technique uses an empirical equation to describe the relationship between energy and separation distance between adjacent (but not chemically bonded) groups of atoms in a molecular structure. Fortunately, a very simple power-law function called the Lennard-Jones function works very well for most polymers that are based upon carbon or silicon lattices. To a first approximation, the method assumes that chemical bonds do not deform significantly, relative to the weak physical bonds, due to electronic interactions between atoms or molecules. Note that, although the potential function here is an empirical relation for simplicity, all the fundamental contributions to bond energy are embodied in that function, and a potential function can be constructed from "first principles" by a series of quantum mechanics simulations.

Figure 2 shows the general form of a potential energy well for Van der Waal's bonding. Energy, E, is expressed relative to the depth of the potential energy well, Eo, and dimensions are given as volume, V, relative to the volume at the absolute depth of the potential energy well, Vo. It should be noted that the energies for bonding are negative.

As positive energy (such as thermal energy, HT) is increased, the interaction dimensions move away from the minimum energy position to two possible values on either side of the minimum. Generally, the potential well is asymmetric, such that the equilibrium position of the new well minimum moves to higher volumes to cause thermal expansion in a material. The full method is described in [4] and the pressure can be expressed as a function of volume through the highly non-linear relation where E_T and V_T are energy and volume at constant temperature T.

$$P = \frac{d\varepsilon}{dV} = \frac{4E_T}{V_T}\left\{\left(\frac{V_T}{V}\right)^3 - \left(\frac{V_T}{V}\right)^5\right\}. \quad (1)$$

2.2. Application of model to geological materials

On close inspection, the parameters show a qualitative trend that the reference parameter of volume for the different silica forms scales with the inverse of the density of the polymorphs and the energy parameter stays remarkably constant. The conclusion from this observation is that the many polymorphs of silica can be modelled as a single material in terms of a single adjustable parameter of volume at zero pressure, V_T, and two absolute reference parameters of energy and molar volume of an underlying polymeric form of silica, Er and Vr respectively. The key point is that each polymorph adopts a crystal structure as if it were constrained by an energy or pressure, Pr, of configuration associated with the non-zero energy of each particular set of metastable chemical bond angles in the polymorph structure.

The remarkably simple conclusion from the pressure-volume relations for silica can be translated into a single predictive equation for any given silica polymorph of the following form

$$P(GPa) = 0.34\left\{\left(\frac{V_r}{V}\right)^3 - \left(\frac{V_r}{V}\right)^5 - \left(\frac{V_r}{V_T}\right)^3 + \left(\frac{V_r}{V_T}\right)^5\right\}. \quad (2)$$

This implies that the only variable required to predict the equation of state (EOS) is the initial density of the material. The prediction using this approach for the pressure/volume relationship for each of the crystal forms compared to experimental data is shown in Fig. 3. The results are highly encouraging which gives great confidence in the approach.

The model has also been used to predict shock Hugoniot data for a range of geological materials as shown in Fig. 4, again with very encouraging results.

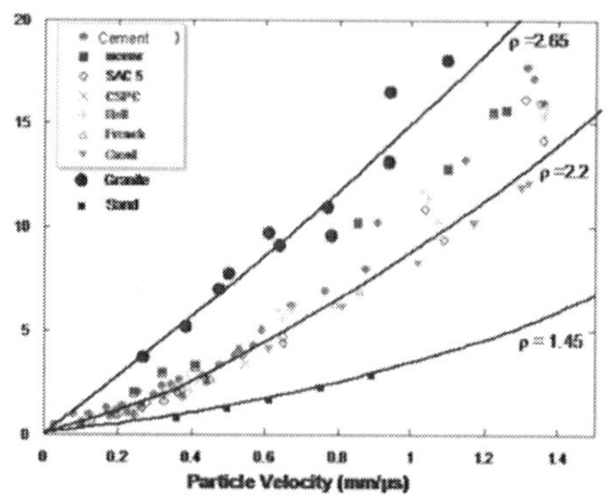

Figure 4. Comparison of Model with Shock Hugoniot Data.

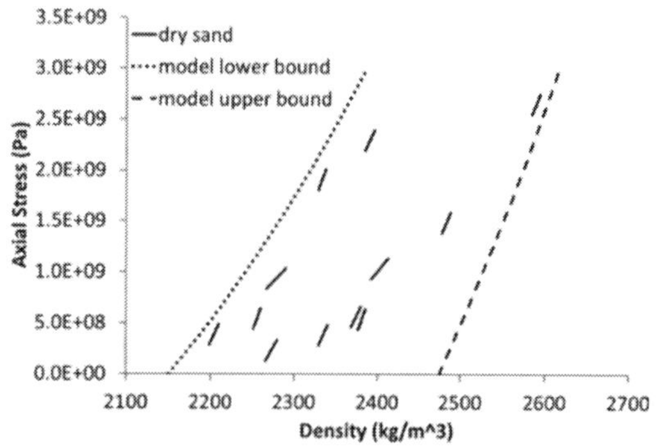

Figure 5. Comparison of Model with Elastic Unloading Paths.

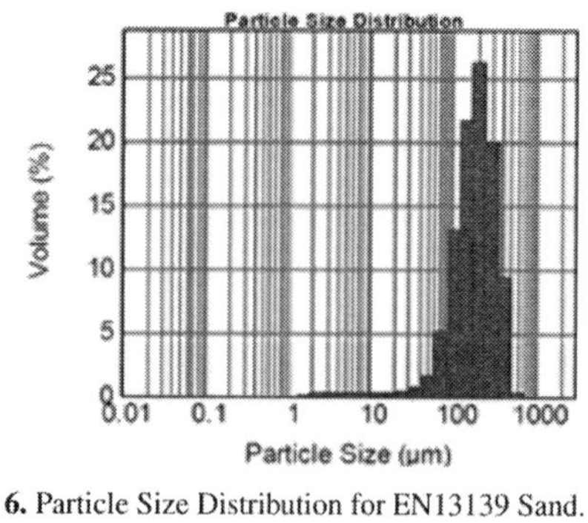

Figure 6. Particle Size Distribution for EN13139 Sand.

To emphasise the success of the elastic model in comparison with data Fig. 5 compares the predictions of the Porter-Gould density dependent model with unloading paths both in plate impact and in quasi-static 1-D deformation. The unloading paths are presumed to be elastic. The lines on the figure for the data are not the full unloading but indicate the direction of unloading in stress-density space with the top of the line being the point from which the sand unloaded. Comparing the gradients of these lines with the upper and lower bounds shows that most of the unloading paths lie between those bounds.

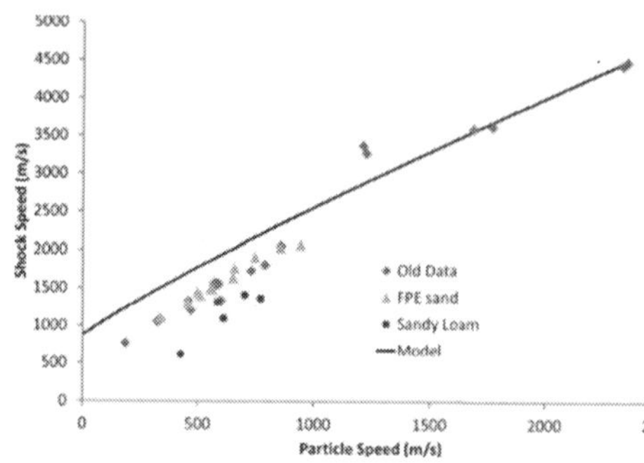

Figure 7. Shock Response Data for Various Dry Sands.

3. Development of model for sand

The sand used in this study is based on sand to a specification of EN13139 with a particle size distribution (PSD) as shown in Fig. 6 and angular shaped particles.

The Porter-Gould model is capable of predicting the shock EOS response of sand which is dominated by an elastic molecular response at higher impact velocities, as shown in Fig. 7. However, at lower impact velocities the model over predicts the behaviour.

The reason for this is that at the lower impact velocities the behaviour of the sand is dominated by compaction, which is a micromechanical response based around re-organisation of sand grains, fracture and potentially grain plasticity. The compaction response is different for different sands due to particle shape and PSD. At present there is no reliable theoretical method for predicting the compaction response given an initial PSD and shape. Therefore some simplifying assumptions are required.

As a first approximation an analytic approach was taken where it was assumed that all of the complexity of compaction for a specific sand could be represented by a single parameter. The idea being to have a high-rate, high pressure parameter that was characteristic of the sand in question and to avoid having to produce a model that required significant number of characterisation experiments. Equation (3) shows how the parameter was calculated. The compaction parameter, C, determines how much of the density increment is inelastic and how much is elastic: that is how much the reference density changes in the transition to the shock state and therefore what the pressure in the elastic equation of state should be.

$$\frac{1}{V_T} \rightarrow \frac{1}{V_T} + C\left(\frac{1}{V} - \frac{1}{V_T}\right). \tag{3}$$

It was decided that a single plate impact experiment could be used to measure C and then use that value to predict all of the other shock experiments. This is shown in Fig. 8. It can be seen that this analytic approach gives good agreement with the whole data set.

Whilst this approach succeeded analytically and showed that there was a potential way forward should C be able to be calculated knowing the microstructure, plastic behaviour and friction coefficients for the sand there was a problem with implementing it into the hydrocodes for which it was meant. The available implementation

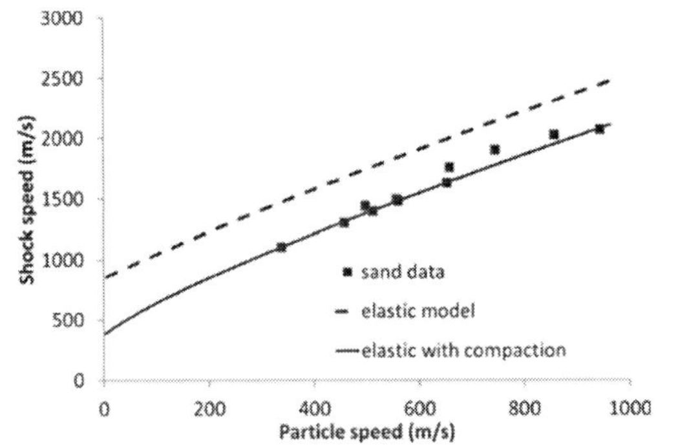

Figure 8. Comparison of Compaction Model with Experiment.

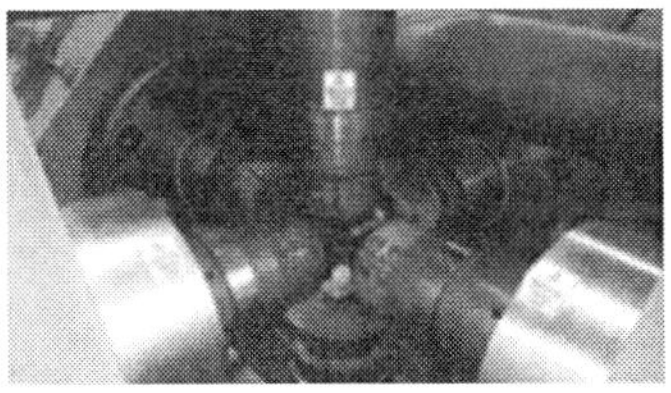

Figure 9. MACET Rig Showing Plattens.

schemes could not be structured to use this finite deformation approach and so another way to develop the model for numerical simulations was needed.

The approach adopted for the hydrocode model is to fit the compaction behaviour directly to Quasi-Static data using the Multi-Axial Compression of Concrete at Elevated Temperature (MACET) triaxial rig at Sheffield University. The MACET rig, Fig. 9, comprises six platens which can be controlled independently in both load and displacement control which gives access to many triaxial load paths.

This approach has the advantage of using the current models with updated physical understanding whilst still only needing low rate data to access the high-rate high pressure regime. The EOS QS compaction curve fitted to the MACET data is shown in Fig. 10 for an initial density of 1.54 g/cc. The curve from the MACET data is extended to lock-up on the quartz Hugoniot which has been predicted using the Porter-Gould model. The two points chosen in the region not covered by data are based upon a monotonically increasing bulk modulus. The initial point after yield is chosen so that any realistic starting sand density can be used.

The constitutive response is based on a Johnson-Holmquist model [5] and fitted to QS data as shown in Fig. 11 for dry sand.

4. Validation of model

The model was implemented in the Eulerian hydrocode GRIM and the Lagrangian hydrocode DYNA. The validation of the model was conducted against experiments in different stress states and regimes. The validation experiments were the plate impact, Split Hopkinson Pressure Bar (SHPB) and fragment penetration.

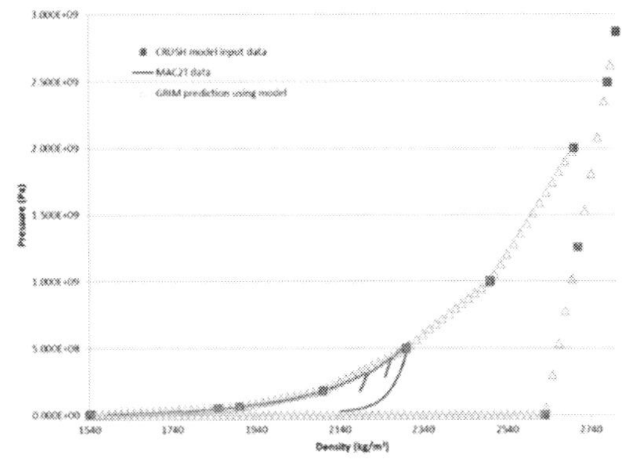

Figure 10. EOS Curve in Hydrocode Compared to MACET Data.

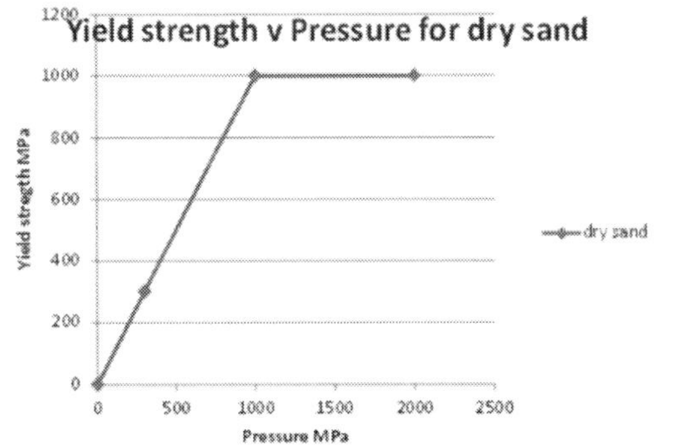

Figure 11. Constitutive Model for Dry Sand.

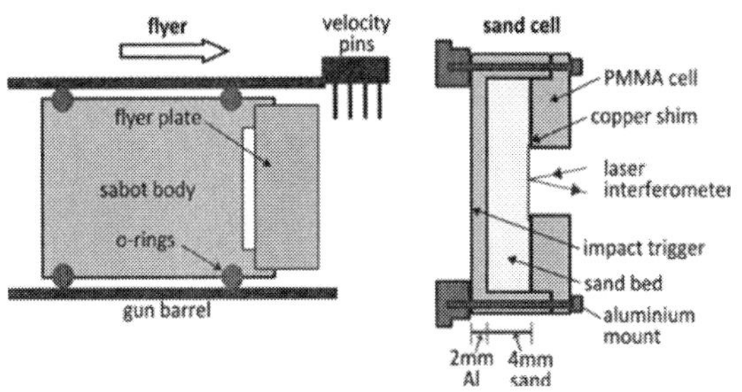

Figure 12. Plate Impact Experimental Arrangement

4.1. Plate impact

This is a standard test for measuring the shock response of the material and the experimental arrangement is shown in Fig. 12 and the experiments were performed on the single stage gas gun at the Cavendish Laboratory at Cambridge University.

This particular configuration is also used to measure the release behaviour of the sand under shock loading. The response for dry sand is shown in Fig. 13 and the release path follows the slope of the solid quartz Hugoniot, which is also similar to the assumptions in the model.

Given that there are various assumptions in generating these curves it was felt more realistic to compare the model with the raw heterodyne velocimetry (het-v) laser interferometer data. This comparison is shown in Fig. 14.

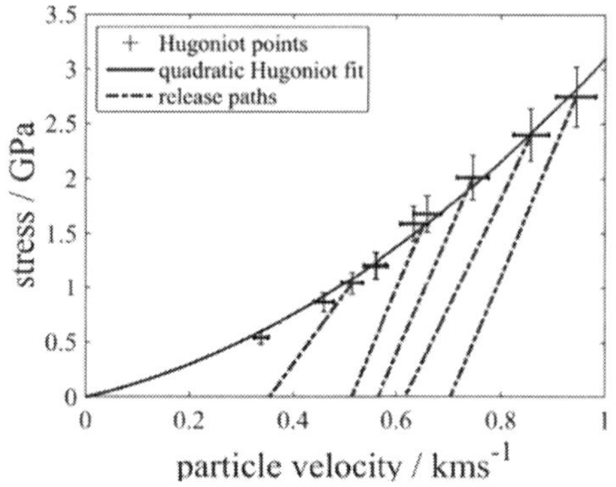

Figure 13. Shock and Release Paths in Dry Sand.

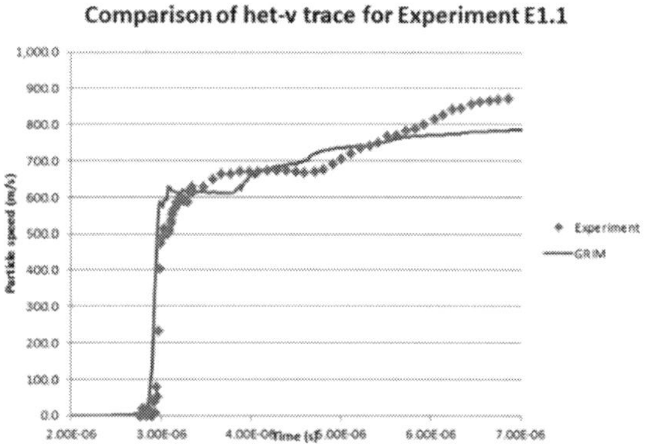

Figure 14. Comparison of Model and Experiment for Het-v Trace.

The level of agreement is encouraging. The arrival time is very close indicating that the wavespeed is correct and also the general structure of the trace is reproduced, particularly the first peak. This suggests that the assumptions in the model concerning the release behaviour are reasonable.

4.2. Fragment penetration

The experiments comprise various fragments such as sphere, cubes and Fragment Simulating Projectiles (FSP) impacting finite thickness sand targets (i.e. 50 to 200 mm) at a range of velocities (i.e. up to 1500 m/s). The penetration process is illustrated for two impact velocities for a steel cube as shown in Fig. 15.

At higher velocities the fragment is eroded which affects the presented area during penetration and hence the residual velocity. The comparison of the model with experiment for a range of impacts is shown in Fig. 16, which shows excellent agreement.

For semi-infinite penetration the model is within the experimental spread for a range of impact conditions and sand moisture contents.

4.3. Split Hopkinson Pressure Bar (SHPB)

The SHPB tests were performed at Sheffield University and the sample was confined in a steel jacket such that the deformation was essentially uniaxial strain as opposed to uniaxial stress. The validation is a direct comparison with

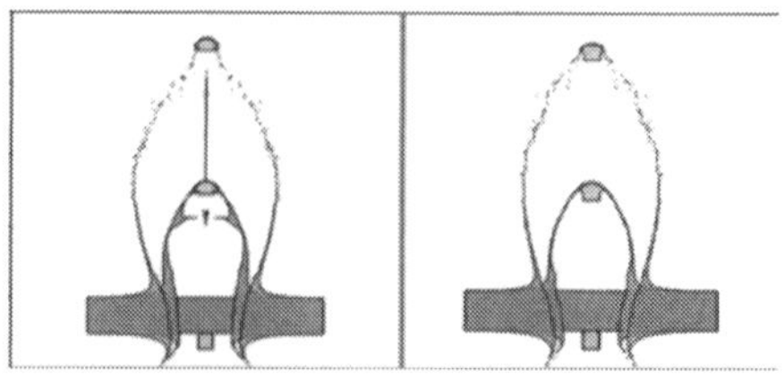

Figure 15. Illustration of Fragment Penetration Process.

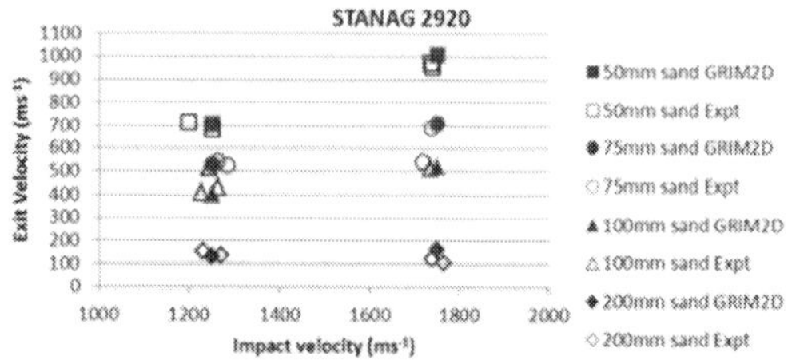

Figure 16. Comparison of Residual Velocity for Fragment.

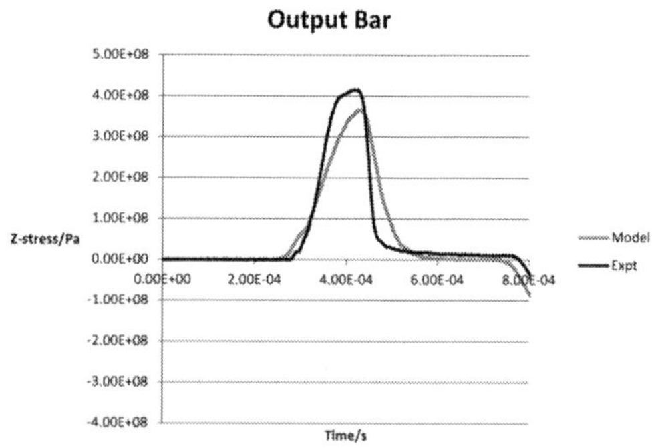

Figure 17. Comparison of Model with Output Trace in SHPB.

the output strain gauge trace which negates any need for assumptions on equilibrium in the sample. The comparison is shown in Fig. 17.

The results are encouraging in terms of general stress levels and also suggests that the wavespeed in the material is correct.

5. Discussion

The theme of this paper was to develop a model for a new soil when there is a lack of experimental data to fit.

As discussed the major obstacle to developing a model is the compaction behaviour, given that the elastic response is well predicted by the Porter-Gould density-dependent equation of state. This compaction is a function of particle size and particle shape and there are significant barriers to predicting this behaviour including physical properties of the sand, understanding the evolution of sand grain re-arrangement under hydrostatic and other loads, and implementation of relevant schemes into hydrocodes.

The previous sections have demonstrated that the existing models fitted to a QS compaction curve work really well over a range of regimes including shock response and high strain rate behaviour.

The idea therefore is to generate a family of compaction curves for a range of sands and soil incorporating different particle sizes and shapes. In addition this would need to include moisture content.

Thus for the new material one would use this data to provide the closest fit for the compaction data. This would allow a rapid assessment to be performed for the new material. One issue is that it is not known how many different compaction curves are required to capture the spectrum of response for granular materials.

In terms of developing better theoretical tools the Porter-Gould model is being used to assess compaction behaviour. It has been shown that the elastic response is well predicted by the density-dependent equation of state and so, given experimental data on the response of sand along various load paths it should be possible to deconvolute the inelastic components of the response and thereby quantify both the elastic (i.e. reversible) and inelastic (i.e. irreversible) components. Primary amongst this needed set of data is true hydrostatic compaction of sand coupled to X-ray to understand the trade-off between grain rearrangement, grain fracture and grain plasticity in this highly constrained loading mode. This can then be extended to other interesting loading modes such as isochoric shear as another way in to understanding how these models may be used in current hydrocodes that explicity split hydrostats from deviators. If this, then, allows 1-D deformation to be predicted all the better.

Despite the generally small strains found in geological materials it is considered best to implement this model using a finite deformation scheme which would allow tracking of the inelastic components and leave the stress response to be a hyperelastic outcome.

6. Conclusions

A sand model has been created based on a theoretical Porter-Gould approach for the EOS combined with a Johnson-Holmquist constitutive model.

The biggest issue is compaction which is a strong function of particle size distribution and shape.

The compaction is fitted to a QS triaxial test (i.e. MACET).

The model has then been shown to predict the behaviour of controlled sand over a wide range of regimes ranging from shock to penetration behaviour giving confidence that this approach works.

It is recommended to generate a family of compaction curves together with additional theoretical developments on separating the reversible and irreversible compaction effects.

This work has been supported by UK MoD and Dstl.

References

[1] `www.Accelrys.com`

[2] Porter D., *Group Interaction Modelling of Polymer Properties* Marcel Dekker Inc, New York (1995)

[3] Porter D., Gould P., *Jnl de Phys IV, DYMAT* (2003)

[4] Church P. et al., *Cratering in the Solar System,* Noordvjck, The Netherlands, (May 2006)

[5] Johnson G., Holmquist T., *14th Int Symp Ballistics*, (1983), pp. 591–600

EPJ Web of Conferences **94**, 04025 (2015)
DOI: 10.1051/epjconf/20159404025

Three-dimensional BEM for transient dynamic analysis of piezoelectric and anisotropic elastic solids

Leonid Igumnov[a], Ivan Markov, Igor Vorobtsov, Svetlana Litvinchuk, and Anatoly Bragov

Research Institute for Mechanics, Lobachevsky State University of Nizhni Novgorod, Russia

Abstract. A boundary element approach based on the Green's functions in integral representation and the convolution quadrature method is presented. Proposed approach is designed for analyzing 3D initial boundary-value problems of the dynamics of general anisotropic elastic and piezoelectric linear homogeneous solids with mixed boundary conditions. Numerical modelling of transient dynamics of elastic anisotropic and piezoelectric three-dimensional solids is carried out to demonstrate the potential of the developed boundary element software. Obtained solutions are compared with the corresponding FEM results and results of the dynamic experiment. A numerical technique based on the exact Laplace-domain boundary integral equations for the direct approach of 3D linear theories of anisotropic elasticity and piezoelectricity is employed. The BEM scheme is constructed using the collocation method and the convolution quadrature method in the form of a stepping method for numerical inversion of integral Laplace transform. Results of the stepped BE-modelling of the problems when a transient force is acting on 3D piezoelectric and anisotropic elastic homogeneous solids are presented.

1. Introduction

Three-dimensional transient problems of linear anisotropic elasticity and piezoelectricity can be successfully treated by the direct Boundary Element Method (BEM). Yet such BEM formulations hardly reported in the literature due to the lack of the exact closed-form expressions of dynamic fundamental solutions (often referred to as Green's functions) and high complexity of their integral representations [1,2]. The Dual Reciprocity BEM (DRBEM) formulations which require only static fundamental solutions were proposed [3–5] to avoid this complication. In this study we present a Laplace domain direct BEM formulation based on exact singular boundary integral equation (BIE) and on integral expressions of corresponding dynamic fundamental solutions.

2. Basic equations

2.1. Linear anisotropic elasticity

First we consider a three-dimensional linear and homogeneous anisotropic elastic solid of volume Ω and boundary Γ. In the absence of body forces, equations of motion are given by

$$C^E_{ijkl}u_{k,lj} = \rho\ddot{u}_i \quad \text{in} \quad \Omega \subset R^3, \quad i,j,k,l = \overline{1,3}, \quad (1)$$

where C^E_{ijkl}, u_i and ρ represent the elasticity tensor, displacement vector and mass density, respectively.

The initial and boundary condition are given as follows

$$u_i(t=0) = \dot{u}_i(t=0) = 0 \text{ in } \Omega, \quad (2)$$

$$u_i = u_i^* \text{ on } \Gamma^u, \quad (3)$$

$$t_i = t_i^* \text{ on } \Gamma^t, \quad (4)$$

where $\Gamma = \Gamma^u \cup \Gamma^t$ and **t** is the traction vector, defined as

$$t_i = \sigma_{ij}n_j, \quad (5)$$

where **n** is the unit outward normal vector to the boundary Γ and σ_{ij} is the Cauchy stress tensor.

2.2. Linear piezoelectricity

Under quasi-electrostatic assumption and in the absence of body forces, the coupled equations of motion for a homogeneous linear piezoelectric solid take the following form [5,6]:

$$C^E_{ijkl}u_{k,li} + e_{lij}\phi_{,li} = \rho\ddot{u}_j, \quad (6)$$

$$e_{ikl}u_{k,li} - \kappa_{il}\phi_{,li} = 0, \quad (7)$$

where e_{ijk} and κ_{ik} are piezoelectric and dielectric permittivity tensors, respectively. In addition, ϕ represents the electric potential and it has been taken into account that piezoelectric materials possess no free electric charges.

In order to combine elastic and electric variables into corresponding generalized vectors and tensors we employ the following conventional contracted notations [5]

$$U_k = \begin{cases} u_k, & k = \overline{1,3}, \\ \phi, & k = 4, \end{cases} \quad (8)$$

$$T_j = \begin{cases} t_j, & j = \overline{1,3}, \\ q, & j = 4, \end{cases} \quad (9)$$

[a] Corresponding author: igumnov@mech.unn.ru

Article available at http://www.epj-conferences.org or http://dx.doi.org/10.1051/epjconf/20159404025

$$C_{ijkl} = \begin{cases} C^E_{ijkl}, & i,j,k,l = \overline{1,3}, \\ e_{lij}, & i,l,j = \overline{1,3}, k = 4, \\ e_{ikl}, & i,l,k = \overline{1,3}, j = 4, \\ -\kappa_{il}, & i,l = \overline{1,3}, k = j = 4, \end{cases} \tag{10}$$

where q denotes the surface charge density, defined as

$$q = D_j n_j \tag{11}$$

and D_j is the electric displacements.

We now can rewrite the equations of motion (6)-(7) in compacted form as follows

$$C_{ijkl} U_{k,il} = \rho \delta^*_{jk} \ddot{U}_k \text{ in } \Omega \subset R^3, i,l = \overline{1,3},\ j,k = \overline{1,4}, \tag{12}$$

$$\delta^*_{jk} = \begin{cases} \delta_{jk}, & j = k = \overline{1,3}, \\ 0, & \text{otherwise}. \end{cases} \tag{13}$$

Initial conditions for the elastic quantities have to be prescribed

$$u_i(t=0) = \dot{u}_i(t=0) = 0 \text{ in } \Omega. \tag{14}$$

The following boundary conditions are assumed for the elastic field

$$u_i = u^*_i \text{ on } \Gamma^u, \tag{15}$$

$$t_i = t^*_i \text{ on } \Gamma^t, \tag{16}$$

for the electrical field

$$\phi = \phi^* \text{ on } \Gamma^\phi, \tag{17}$$

$$q = q^* \text{ on } \Gamma^q, \tag{18}$$

where Γ^ϕ and Γ^q are the parts of the boundary with the prescribed electric potential and surface charge density, respectively.

2.3. Laplace transform formulation

As is can be observed, Eq. (12) has essentially the same form as Eq. (1) and for the sake of simplicity we now formulate the generalized initial-boundary value problem for three-dimensional linear and homogeneous anisotropic elastic or piezoelectric solid of volume Ω and boundary Γ, assuming zero body forces and zero initial conditions:

$$C_{ijkl} U_{k,il} = \rho \delta^*_{jk} \ddot{U}_k \text{ in } \Omega \subset R^3, i,l = \overline{1,3},\ j,k = \overline{1,N}, \tag{19}$$

where N denotes the dimension of the problem; $N = 3$ for elasticity, $N = 4$ for piezoelectricity and we assume that the proper boundary conditions are prescribed.

Applying the Laplace transform in time variable to Eq. (19) we get

$$C_{ijkl} \bar{U}_{k,il} = \rho s^2 \delta^*_{jk} \bar{U}_k \text{ in } \Omega \subset R^3, \tag{20}$$

where a bar over a variable hereinafter represents its Laplace transform and s is the transform variable.

Equation (20) along with corresponding boundary conditions represents the generalized boundary value problem in *Laplace domain*.

3. Boundary integral equation and BEM formulation

Besides the case of elasticity, the generalized Betti-Rayleigh reciprocal theorem also holds for piezoelectricity. Thus, assuming zero body forces and zero initial conditions a boundary integral representation of the solution of Eq. (20) can be obtained in the form

$$c_{jk}\bar{U}_k(\mathbf{y},s) + \text{p.v.} \int_\Gamma \bar{h}_{jk}(\mathbf{x},\mathbf{y},s)\bar{U}_k(\mathbf{x},s)d\Gamma(\mathbf{x}) = \int_\Gamma \bar{g}_{jk}(\mathbf{x},\mathbf{y},s)\bar{T}_k(\mathbf{x},s)d\Gamma(\mathbf{x}), \quad \mathbf{y} \in \Gamma, \tag{21}$$

where c_{jk} are the free term coefficients, p.v. indicates that the integral is treated in the sense of its Cauchy principal value, $\bar{g}_{ij}(\mathbf{x},\mathbf{y},s)$ and $\bar{h}_{ij}(\mathbf{x},\mathbf{y},s)$ represent the corresponding (elastic or piezoelectric) displacement and traction fundamental solutions in Laplace domain.

In this study we employ matched boundary element approach [7]: the boundary is approximated by eight-node quadratic quadrilateral elements, the boundary variables $\bar{U}_k$ and $\bar{T}_k$ are approximated by linear and constant elements, respectively.

Once Eq. (21) is discretized and boundary conditions are taken into account, the resulting system of linear equations is solved for unknown boundary generalized displacements and tractions *for the given value of Laplace transform variable s*. In order to obtain solution in time-domain we use convolution quadrature method in the form of stepping method for numerical inversion of Laplace transforms [8–10].

4. Dynamic anisotropic elastic and piezoelectric fundamental solutions

Laplace domain displacement fundamental solutions for anisotropic elastic or piezoelectric solids are obtained by applying the Laplace transform to the corresponding time-domain expressions [1,2]:

$$\bar{g}_{jk}(\mathbf{x},\mathbf{y},s) = \bar{g}_{jk}(\mathbf{r},s) = \int_0^\infty g(\mathbf{x},\mathbf{y},t)e^{-st}dt = \bar{g}^S_{jk}(\mathbf{r}) + \bar{g}^R_{jk}(\mathbf{r},s), \tag{22}$$

$$\mathbf{r} = \mathbf{x} - \mathbf{y},\ r = |\mathbf{r}|, \tag{23}$$

where $\bar{g}^S_{jk}$ and $\bar{g}^R_{jk}$ denote the static and dynamic parts of the fundamental solution, respectively.

Decomposition of dynamic fundamental solution into two parts Eq. (22) allows using the interpolation scheme for calculation of static part first proposed by Wilson and Cruse [11]. In this study we employ linear Lagrange interpolation on the 800 × 800 grid.

For 3D anisotropic elastic problems, $\bar{g}^S_{jk}$ and $\bar{g}^R_{jk}$ are expressed as follows [12]:

$$\bar{g}^S_{jk}(\mathbf{r}) = \frac{1}{8\pi^2 r} \int_{|\mathbf{d}|=1} \Gamma^{-1}_{jk}(\mathbf{d})dL(\mathbf{d}), \quad j,k = \overline{1,3}, \tag{24}$$

$$\bar{g}^R_{jk}(\mathbf{r},s) = -\frac{1}{8\pi^2}\int\limits_{|\mathbf{n}|=1\,\mathbf{n}\cdot\mathbf{r}>0}\sum_{m=1}^{3}\frac{k_m E_{jm}E_{km}}{\rho c_m^2}e^{-k_m|\mathbf{n}\cdot\mathbf{r}|}dS(\mathbf{n}), \tag{25}$$

$$c_m = \sqrt{\lambda_m/\rho}, \quad k_m = s/c_m, \tag{26}$$

$$\mathbf{n} = \sqrt{1-b^2}\,\mathbf{d} + b\mathbf{e}, \quad \mathbf{e} = \mathbf{r}/|\mathbf{r}|, \tag{27}$$

$$\mathbf{e}\cdot\mathbf{d} = 0, \mathbf{d} = \begin{bmatrix} e_2\cos\varphi + e_1e_3\sin\varphi \\ -e_1\cos\varphi + e_2e_3\sin\varphi \\ -\left(1-e_3^2\right)\sin\varphi \end{bmatrix} \Big/ \sqrt{1-e_3^2}, \tag{28}$$

$$\Gamma_{jk}(\mathbf{d}) = C_{ijkl}d_id_l, \tag{29}$$

$$dS(\mathbf{n}) \in D^R = \{0 \le b \le 1;\ 0 \le \varphi \le 2\pi\}, \tag{30}$$

$$dL(\mathbf{d}) \in D^S = \{0 \le \varphi \le 2\pi\}, \tag{31}$$

where λ_m, E_{jm} are eigenvalues and corresponding eigenvectors of Christoffel tensor $\Gamma_{jk}(\mathbf{d})$; c_m and k_m are the phase velocities and wave numbers, which are functions of the propagation vector $\mathbf{n}$.

And for 3D piezoelectric problems $\bar{g}^S_{jk}$ and $\bar{g}^R_{jp}$ are expressed as follows [13]:

$$\bar{g}^S_{jp}(\mathbf{r}) = \frac{1}{8\pi^2 r}\int\limits_{|\mathbf{d}|=1}\Gamma^{-1}_{jp}(\mathbf{d})dL(\mathbf{d}), \quad j,p=\overline{1,4}, \tag{32}$$

$$\bar{g}^R_{jp}(\mathbf{r},s) = -\frac{1}{8\pi^2}\int\limits_{\substack{|\mathbf{n}|=1\\ \mathbf{n}\cdot\mathbf{r}>0}}\sum_{m=1}^{Q}\frac{k_m\tilde{P}^m_{jp}}{\rho c_m^2}e^{-k_m|\mathbf{n}\cdot\mathbf{r}|}dS(\mathbf{n}), \tag{33}$$

$$\tilde{P}^m_{jp} = \begin{cases} P^m_{jp}, & j,p=\overline{1,3}, \\ -\Gamma_{k4}P^m_{jk}/\Gamma_{44}, & j=\overline{1,3},\ p=4, \\ \Gamma_{4k}P^m_{kl}\Gamma_{l4}/\Gamma^2_{44}, & j=p=4, \end{cases} \tag{34}$$

$$P^m_{jp} = \frac{A^m_{jp}}{A^m_{ii}}, \quad L_{jp} = \Gamma_{jp} - \frac{\Gamma_{j4}\Gamma_{4p}}{\Gamma_{44}}, \tag{35}$$

where A^m_{jp} is the adjugate of matrix $\left[L_{jp} - \rho c_m^2\delta_{jp}\right]$ and it was taken into consideration that the eigenvectors need not to be evaluated explicitly, hence λ_m are the eigenvalues of matrix L_{jp} and Q is number of distinct eigenvalues.

In addition, the traction fundamental solution $\bar{h}_{jk}$ is obtained using the following relation

$$\bar{h}_{jm}(\mathbf{x},\mathbf{y},s) = C_{ijkl}\bar{g}_{km,l}n_i. \tag{36}$$

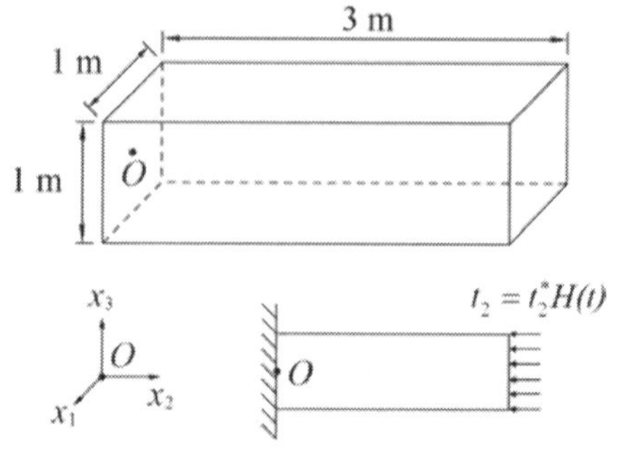

Figure 1. Example A: geometry and boundary conditions.

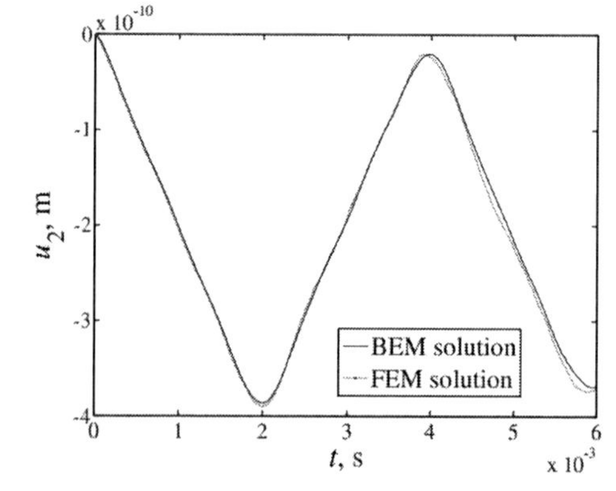

Figure 2. Transient response of displacements u_2 (t).

5. Numerical examples

5.1. Example A

First we consider 3D anisotropic elastic rod which is clamped at one end, and subjected to Heaviside-type traction $t_2 = t_2^*H(t)$, $t_2^* = -1\text{N/m}^2$ at the opposite free end, as illustrated in Fig. 1. The remaining surfaces are traction free. The material considered is an orthotropic 65% graphite-35% epoxy [14] with density $\rho = 1600$ kg/m^3 and whose material principal axes are rotated sequentially by 30° around x_2 and then by the same angle around x_3'. This yields the following fully populated elasticity tensor, which is given in the Voigt notation as:

$$\mathbf{C} = \begin{bmatrix} 60.23 & 18.67 & 18.96 & -7.69 & 15.60 & -25.28 \\ & 21.26 & 9.36 & -3.74 & 4.21 & -8.47 \\ & & 47.04 & -8.82 & 15.28 & -8.31 \\ & & & 10.18 & -9.54 & 5.69 \\ & \text{sym.} & & & 21.19 & -8.54 \\ & & & & & 20.75 \end{bmatrix}\text{GPa}. \tag{37}$$

The employed BEM mesh consists of 504 boundary elements and 506 nodes. The transient response of displacements $u_2(t)$ at boundary point (0, 3, 0) m is shown in Fig. 2 where BEM solution are given in comparison with the corresponding FEM results. It can be observed from

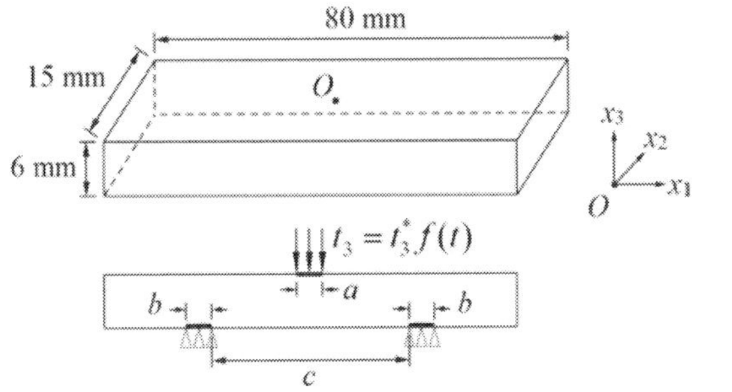

Figure 3. Example B: geometry and boundary conditions.

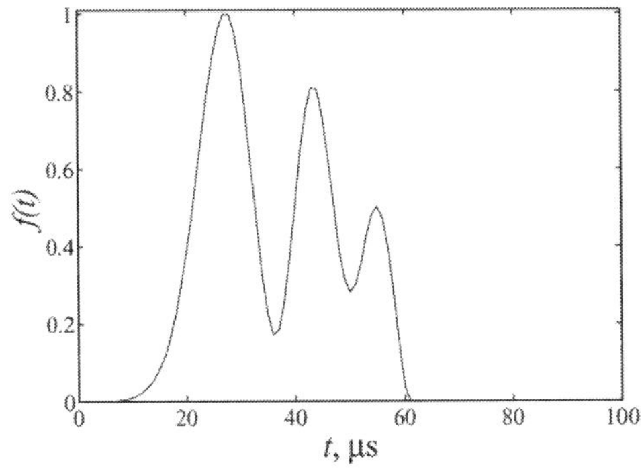

Figure 4. Function $f(t)$.

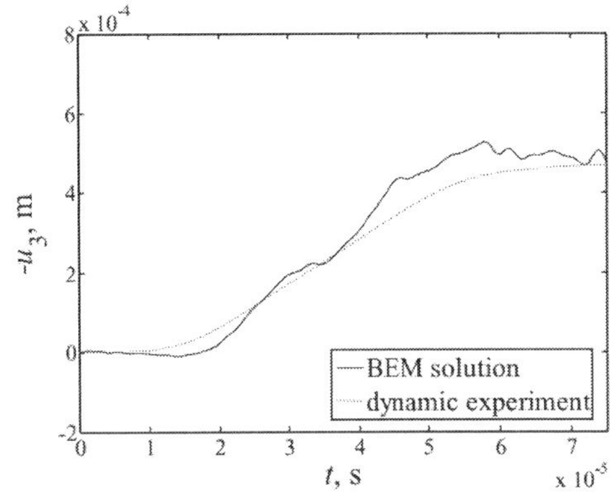

Figure 5. Transient response of displacements u_3 (t).

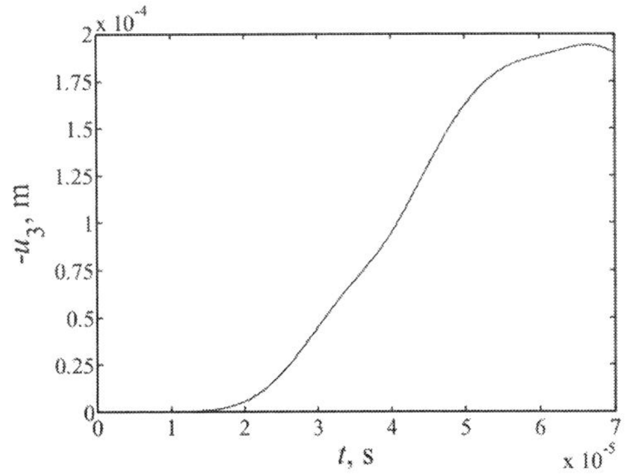

Figure 6. Transient response of displacements u_3(t).

this figure that the agreement between the BEM and FEM results are very good.

5.2. Example B

As another example, consider the problem depicted in Fig. 3: an anisotropic elastic prismatic bar is simply supported at one side on two areas with width b = 1.5 mm each, that are located symmetric with respect to the x_3 axis at a distance c = 45 mm from each other. Bar is subjected to a uniformly distributed normal load $t_3 = t_3^* f(t)$, $t_3^* = -9.724 \cdot 10^7/\text{m}^2$ acting on the area with width a = 1.4 mm located exactly in the middle on the opposite side. Function $f(t)$ is given in Fig. 4.

In the first case we consider an orthotropic material for which we also conducted dynamic experiment. The employed BEM mesh consists of 540 boundary elements and 542 nodes. The results from the BEM analysis and dynamic experiment for the vertical displacements $-u_3(t)$ at boundary point O (0, 0, 0) m are in the good agreement as it shown in Fig. 5.

In the second case we consider the same triclinic material as in Example A. Obtained BEM results for the vertical displacements $-u_3(t)$ at boundary point O (0, 0, 0) m are given in Fig. 6.

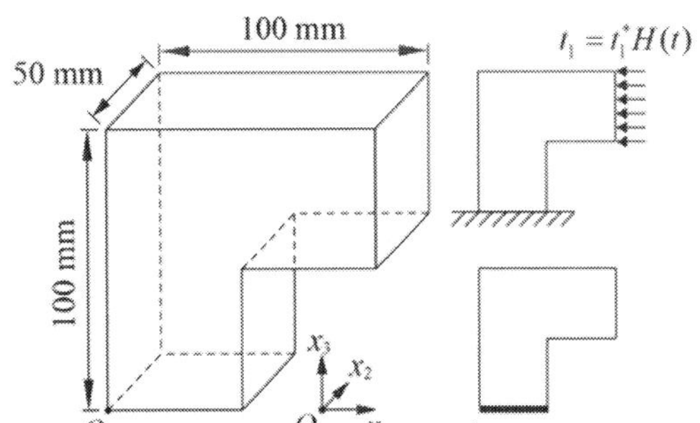

Figure 7. Example C: geometry and boundary conditions.

5.3. Example C

For the final example we consider a Γ-shaped piezoelectric solid [5] under Heaviside-type mechanical load $t_1(t) = t_1^* H(t)$, $t_1^* = -10^8\ \text{N}/\text{m}^2$ at one face and applied voltage $\phi(t) = \phi^* = 0$ V at another face, which is clamped (Fig. 7). Remaining surfaces are free of generalized tractions.

The piezoelectric material is PZT-4 with mass density $\rho = 7500$ kg/m^3 whose material principal axes are rotated sequentially by 36° around x_1, 60° around x_2' and finally by 18° around x_3''. Thus, the material tensors

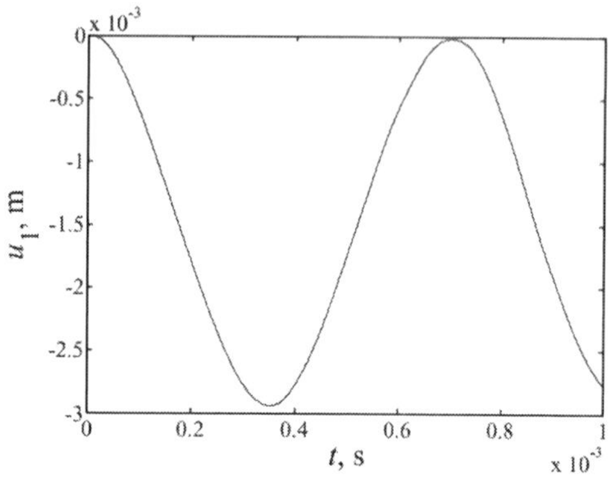

Figure 8. Transient response of displacements u_1 (t).

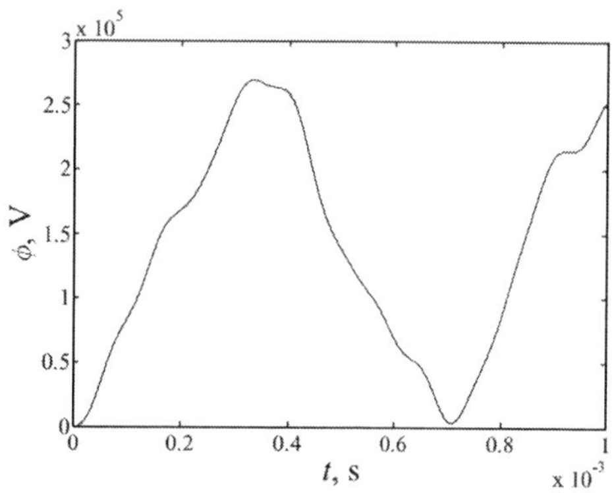

Figure 9. Transient response of electric potential $\phi(t)$.

are expressed as follows

$$\mathbf{C} = \begin{bmatrix} 132.8 & 75.3 & 76.52 & -0.88 & 2.51 & 4.81 \\ & 123.8 & 75.42 & -3.67 & 0.33 & 4.4 \\ & & 134.7 & -4.1 & 2.55 & 1.13 \\ & & & 27.1 & 1.69 & 0.63 \\ & \text{sym.} & & & 28.72 & -1.35 \\ & & & & & 26.84 \end{bmatrix} \text{GPa}, \quad (38)$$

$$\mathbf{e} = \begin{bmatrix} 9.21 & -4.962 & -2.588 \\ 4.01 & 13.287 & -3.344 \\ 2.925 & -4.68 & 7.834 \\ 0.776 & 3.897 & 9.202 \\ 4.653 & 0.776 & -5.751 \\ 8.92 & -4.67 & 0.776 \end{bmatrix}^{\mathrm{T}} \mathrm{C}/\mathrm{m}^2, \quad (39)$$

$$\kappa = \begin{bmatrix} 6.263 & 0.316 & 0.165 \\ & 5.955 & -0.264 \\ \text{sym.} & & 6.323 \end{bmatrix} \cdot 10^{-9}\mathrm{C}/\mathrm{Vm}. \quad (40)$$

Mesh used in computations is regular and consists of 350 boundary elements and 352 nodes. Obtained BEM solutions for $u_1(t)$ and $\phi(t)$ at boundary point (0, 0, 0.1) m are given in Fig. 8 and Fig. 9, respectively.

6. Concluding remarks

The proposed Laplace domain direct BEM formulation for transient piezoelectric and anisotropic elastic analyses produce highly accurate results, which was demonstrated comparing the BEM solutions with the corresponding FEM results and with the results of dynamic experiment.

The work is financially supported by the by the Russian Foundation for Basic Research (RFBR) under grants No. 14-08-31410, No. 15-08-02814, No. 15-08-02817 and by the Ministry of Education and Science of the Russian Federation in the frame of the main part of the governmental task under project No. 2014/134 2226.

References

[1] C.Y. Wang, J.D. Achenbach, Proc. R. Soc. Lond. A **449**, 441 (1995)
[2] A.N. Norris, Proc. R. Soc. Lond. A **447**, 175 (1994)
[3] D. Nardini, C.A. Brebbia, Appl. Math. Modelling **7**, 157 (1983)
[4] M. Kogl, L. Gaul, Eng. Anal. Bound. Elem. **24**, 591 (2000)
[5] L. Gaul, M. Kogl, M. Wagner, *Boundary element methods for engineers and scientists* (Springer Verlag, 2003)
[6] V.Z. Parton, B.A. Kudryavtsev, *Electrocmagnetoelasticity* (Gordon and Breach Science Publishers, 1988)
[7] V.G. Bazhenov, L.A. Igumnov, *Boundary Integral Equations & Boundary Element Methods in treating the problems of 3D elastodynamics with coupled fields* (PhysMathLit, 2008) (in Russian)
[8] C. Lubich, Numerische Mathematik. **52**, 129 (1988)
[9] C. Lubich, Numerische Mathematik. **52**, 413 (1988)
[10] L. Banjai, M. Messner, M. Schanz, Comput. Methods Appl. Mech. Engrg. **245-246**, 90 (2012)
[11] R.B. Wilson, T.A. Cruze, Int. J. Numer. Meth. Engng. **12**, 1383 (1979)
[12] M. Dravinski, Y. Niu, Int. J. Numer. Meth. Engng. **53**, 455 (2002)
[13] C.Y. Wang, Ch. Zhang, Eng. Anal. Bound. Elem. **29**, 454 (2005)
[14] A.H. Nayfeh, *Wave Propagation in Layered Anisotropic Media* (Elsevier, 1995)

EPJ Web of Conferences **94**, 04026 (2015)
DOI: 10.1051/epjconf/20159404026

A combined application of boundary-element and Runge-Kutta methods in three-dimensional elasticity and poroelasticity

Leonid Igumnov[a], Aleksandr Ipatov, Aleksandr Belov, and Andrey Petrov

Research Institute for Mechanics, Lobachevsky State University of Nizhni Novgorod, 23 Prospekt Gagarina, Bld. 6, 603950 Nizhni Novgorod, Russia

Abstract. The report presents the development of the time-boundary element methodology and a description of the related software based on a stepped method of numerical inversion of the integral Laplace transform in combination with a family of Runge-Kutta methods for analyzing 3-D mixed initial boundary-value problems of the dynamics of inhomogeneous elastic and poro-elastic bodies. The results of the numerical investigation are presented. The investigation methodology is based on direct-approach boundary integral equations of 3-D isotropic linear theories of elasticity and poroelasticity in Laplace transforms. Poroelastic media are described using Biot models with four and five base functions. With the help of the boundary-element method, solutions in time are obtained, using the stepped method of numerically inverting Laplace transform on the nodes of Runge-Kutta methods. The boundary-element method is used in combination with the collocation method, local element-by-element approximation based on the matched interpolation model. The results of analyzing wave problems of the effect of a non-stationary force on elastic and poroelastic finite bodies, a poroelastic half-space (also with a fictitious boundary) and a layered half-space weakened by a cavity, and a half-space with a trench are presented. Excitation of a slow wave in a poroelastic medium is studied, using the stepped BEM-scheme on the nodes of Runge-Kutta methods.

1. Problem formulation

The governing equations for the elasticity problem in Laplace domain are as follows

$$L^0_{ik}\bar{u}_k(x,s) + \bar{F}_i = \rho s^2 \bar{u}_i(x,s), \quad (1)$$

$$\bar{u}(x,s) = \tilde{u}, \quad x \in \Gamma^u, \quad \bar{t}_n(x,s) = \tilde{t}_n, \quad x \in \Gamma^\sigma, \quad (2)$$

where $L^0_{ik} = G\Delta + (K + G/3)\text{grad div}$, Γ^u denotes Dirichlet boundary and Γ^σ – Neumann boundary, G, K are elastic moduli, $\bar{F}_i$ is bulk body force, ρ is material density, s is the Laplace transform parameter.

The boundary-value problem for full Biot's model of linear saturated poroelastic continuum in Laplace domain concerning 4 basic functions – skeleton displacements $\bar{u}_i$ and pore pressure $\bar{p}$ – takes the following form [1]:

$$\begin{aligned} &G\bar{u}_{i,jj} + \left(K + \tfrac{G}{3}\right)\bar{u}_{j,ij} - (\alpha - \beta)\bar{p}_{,i} \\ &= s^2(\rho - \beta\rho_f)\bar{u}_i - \bar{F} \end{aligned} \quad (3)$$

$$\frac{\beta}{s\rho_f}\bar{p}_{,ii} - \frac{\phi^2 s}{R}\bar{p} - (\alpha - \beta)s\bar{u}_{i,i} = -\bar{a}, \quad x \in \Omega, \quad (4)$$

$$\bar{u}'(x,s) = \tilde{u}', \quad x \in \Gamma^u, \quad \bar{u}' = (\bar{u}_1, \bar{u}_2, \bar{u}_3, \bar{p}), \quad (5)$$

$$\bar{t}'_n(x,s) = \tilde{t}'_n, \quad x \in \Gamma^\sigma, \quad \bar{t}' = (\bar{t}_1, \bar{t}_2, \bar{t}_3, \bar{q}) \quad (6)$$

where ϕ is porosity, $\bar{F}_i$, $\bar{a}$ are bulk body forces,

$$\beta = \frac{\kappa\rho_f\phi^2 s}{\phi^2 + s\kappa(\rho_e + \phi\rho_f)}, \quad \alpha = 1 - \frac{K}{K_s}, \quad (7)$$

$$R = \frac{\phi^2 K_f K_s^2}{K_f(K_s - K) + \phi K_s(K_s - K_f)} \quad (8)$$

– constants describe the interaction between the skeleton and filler, κ is permeability, ρ, ρ_e, ρ_f are material density, apparent mass density and filler density respectively, K_s, K_f are elastic bulk moduli of the skeleton and filler respectively.

The governing equations of partially saturated poroelasticity in the Laplace domain with five unknowns – solid displacements u_i, the pore wetting fluid pressure p^w, and the pore non-wetting fluid pressure p^a – given by [2]

$$\begin{aligned} &G\bar{u}_{i,jj} + (K + \tfrac{G}{3})\bar{u}_{j,ij} - (\rho - \beta S_w\rho_w - \gamma S_a\rho_a)s^2\bar{u}_i = \\ &= (\alpha - \beta)S_w\bar{p}^w_{,i} - (\alpha - \gamma)S_a\bar{p}^a_{,i} - \bar{F}_i, \end{aligned} \quad (9)$$

$$\begin{aligned} &-(\alpha - \beta)S_w s\bar{u}_{i,i} - (\zeta - S_{aa}S_w + S_u)s\bar{p}^a + \frac{\beta S_w}{\rho_w s}\bar{p}^w_{,ii} \\ &= \left(\zeta S_{ww}S_w + \frac{\phi}{K_w}S_w - S_u\right)s\bar{p}^w - \bar{I}^w, \end{aligned} \quad (10)$$

$$\begin{aligned} &-(\alpha - \gamma)S_a s\bar{u}_{i,i} - (\zeta S_{ww} + S_u)s\bar{p}^w + \tfrac{\gamma S_a}{\rho_a s}\bar{p}^a_{,ii} = \\ &= \left(\zeta S_{aa} + \frac{\phi}{K_a}S_a - S_u\right)s\bar{p}^a - \bar{I}^a, \ x \in \Omega, \end{aligned} \quad (11)$$

$$\begin{aligned} &\bar{u}'(x,s) = \tilde{u}', \quad x \in \Gamma^u, \quad \bar{u}' = (\bar{u}_1, \bar{u}_2, \bar{u}_3, \bar{p}^a, \bar{p}^w), \\ &\bar{t}'_n(x,s) = \tilde{t}'_n, \quad x \in \Gamma^\sigma, \quad \bar{t}' = (\bar{t}_1, \bar{t}_2, \bar{t}_3, \bar{q}^a, \bar{q}^w), \end{aligned} \quad (12)$$

[a] Corresponding author: Igumnov@mech.unn.ru

Where K_w and K_a are bulk moduli of the fluid, ϕ is porosity, $\bar{F}_i$, $\bar{I}^w$, $\bar{I}^a$ are bulk body forces. The bulk density is denoted by

$$\rho = (1-\varphi)\rho_s + \varphi S_w \rho_w + \varphi S_a \rho_a, \quad (13)$$

where ρ_s is the density of the solid, ρ_w is wetting fluid density, ρ_a is non-wetting fluid density. The saturation degrees are defined as the ratios of the volume occupied by the fluid V_w or V_a to the void volume, i.e. it holds

$$S_w = \frac{V_w}{V_{void}}, \quad S_a = \frac{V_a}{V_{void}}, \quad S_w + S_a = 1. \quad (14)$$

The following abbreviations

$$\zeta = \frac{\alpha - \phi}{K_s}, \qquad S_{ww} = S_w - \vartheta(S_w - S_{rw}), \quad (15)$$

$$S_{aa} = S_a + \vartheta(S_w - S_{rw}) \quad (16)$$

$$S_u = -\frac{\vartheta(S_{ra} - S_{rw})}{p^d}\left(\frac{S_w - S_{rw}}{S_{ra} - S_{rw}}\right)^{\frac{\vartheta+1}{\vartheta}} \quad (17)$$

are introduced, where S_{rw} is the residual wetting fluid saturation and S_{ra} is the non-wetting fluid entry saturation. The symbol p^d is non-wetting fluid entry pressure, ϑ is the pore distribution index while the value of ϑ lies between 0.2 and 3, normally. The symbols β and γ are Laplace parameter dependent variables and expressed as

$$\beta = \frac{\kappa_w \phi \rho_w s}{\phi S_w + \kappa_w \rho_w s}, \qquad \gamma = \frac{\kappa_a \varphi \rho_a s}{\phi S_a + \kappa_a \rho_a s}, \quad (18)$$

where κ_w and κ_a the phase permeability of the wetting and the non-wetting fluid are given by $\kappa_w = K_{rw}k/\eta_w$ and $\kappa_a = K_{ra}k/\eta_a$ respectively. K_{rw} and K_{ra} denotes the relative fluid phase permeability, k denotes the intrinsic fluid permeability, η_w and η_a are viscosity of the fluid. To evaluate relative phase permeability following equations are used

$$K_{rw} = S_e^{(2+3\vartheta)/\vartheta}, \quad K_{ra} = (1-S_e)^2\left[1 - S_e^{(2+\vartheta)/\vartheta}\right]. \quad (19)$$

S_e denotes the effective wetting fluid saturation degree given by

$$S_e = \begin{cases} 0 & S_w \le S_{rw} \\ \dfrac{S_w - S_{rw}}{S_{ra} - S_{rw}} & S_{rw} < S_w < S_{ra} \\ 1 & S_w \ge S_{ra} \end{cases} \quad (20)$$

For the extreme case $S_w = 0$, the equations turn out to be simple a elasticity problem, for the extreme case $S_w = 1$ – saturated poroelacticity problem.

2. Boundary integral equations and discretization

The boundary-value problem can be reduced to the following boundary integral equation [1,3,4]:

$$\int_\Gamma \left(\tilde{T}_{ik}(x,y,s)\bar{u}_i(y,s) - \tilde{T}^0_{ik}(x,y,s)\bar{u}_i(x,s)\right.$$
$$\left. - \tilde{U}_{ik}(x,y,s)\bar{t}_i(y,s)\right) d\Gamma = -\alpha_\Omega \bar{u}_k(x,s), \quad (21)$$

where $\tilde{U}(x,s)$, $\tilde{T}(x,s)$ are fundamental and singular solutions for boundary integral equation (BIE), $\tilde{T}^0(x,s)$ contains the singularities isolation, $(x \in \Gamma)$, $\bar{t} = (\bar{t}_1, \bar{t}_2, \bar{t}_3, \bar{q})^T$, $\bar{u} = (\bar{u}_1, \bar{u}_2, \bar{u}_3, \bar{p})^T$ for linear saturated poroelastic model, and $\bar{t} = (\bar{t}_1, \bar{t}_2, \bar{t}_3, \bar{q}^a, \bar{q}^w)^T$, $\bar{u} = (\bar{u}_1, \bar{u}_2, \bar{u}_3, \bar{p}^a, \bar{p}^w)^T$ for partially saturated poroelasticity.

To approximate the boundary consider its decomposition to a set of quadrangular and triangular 8-node biquadratic elements, where triangular elements are treated as singular quadrangular. Every element is mapped to a reference one (square $\xi = (\xi_1, \xi_2) \in [-1,1]^2$ or triangle $0 \le \xi_1 + \xi_2 \le 1$, $\xi_1 \ge 0$, $\xi_2 \ge 0$). Interpolation nodes for boundary unknowns are a subset of geometrical boundary-element grid nodes. Local approximation follows the Goldshteyn's generalized displacement-stress matched model [4]: generalized boundary displacements are approximated by bilinear elements whereas generalized tractions are approximated by constant. For BIE discretization the collocation method is used; the set of collocation nodes coincides with the set of approximation nodes for the boundary functions. Integrals in discretized BIE's are calculated using Gaussian quadrature in combination with singularity decreasing and eliminating algorithms [4].

3. Laplace transform inversion

CQM and time-step method for numerical Laplace transform inversion are quite close to each other but whereas CQM is based on the convolution theorem and intended for convolution integral calculation, time-step method for numerical Laplace transform inversion is based on the integration theorem and dedicated to the calculation of the original function integral.

In order to employ time-step method for obtaining the original function $f(t)$ such as $f(0) = 0$ from known image $\bar{f}(s)$ we need to apply the integration theorem to its derivative:

$$f(t) = \int_0^t f'(\tau)d\tau$$
$$= \frac{1}{2\pi i}\lim_{R\to\infty}\int_{c-iR}^{c+iR} \bar{f}(s)s\int_0^t e^{s(\tau)}d\tau ds = \ldots \quad (22)$$

Following the theorem, the integral can be found as (basing on [2–4]):

$$f(0) = 0, \quad f(n\Delta t) = \sum_{k=1}^{n} \omega_k(\Delta t), \quad n = 1, \ldots N, \quad (23)$$

$$\omega_n(\Delta t) = \frac{R^{-n}}{L}\sum_{l=0}^{L-1} \bar{f}(s)se^{-in\varphi}, \quad s = \frac{\gamma(z)}{\Delta t}, \quad (24)$$

$$z = Re^{i\varphi}, \quad \varphi = 2\pi\frac{l}{L}. \quad (25)$$

Here R is the radius of the analyticity region for $\bar{f}(\gamma(z)/\Delta t)$ and $\gamma(z)$ is the characteristic function for the linear multistep method applied to the Cauchy problem arising within the integral evaluation. Backward

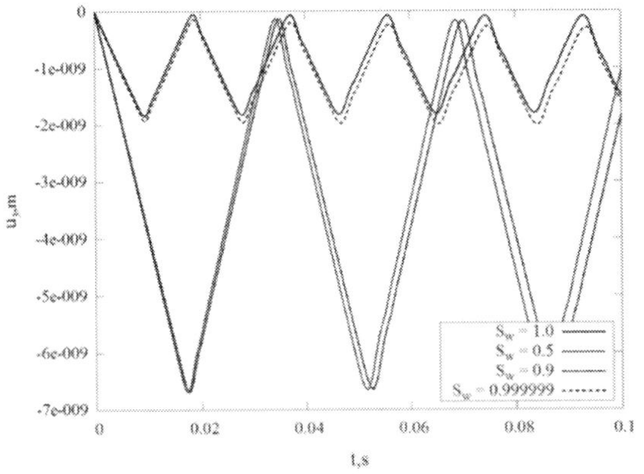

Figure 1. Displacement u_3.

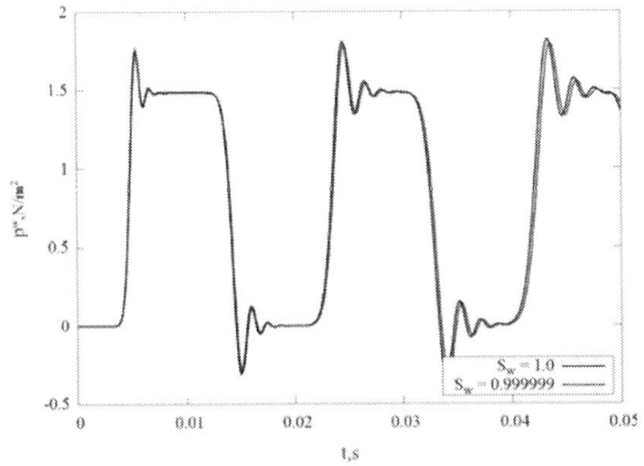

Figure 2. Pore pressure p^w.

differentiation (BDF) based methods of order ≤ 6 are applicable in scope of this solution scheme. For BDF-2 we have $\gamma(z) = 3/2 - 2z + z^2/2$. For Runge-Kutta method applied instead of linear multistep method we obtain (based on [5–7]):

$$f(0) = 0,\ f(n\Delta t) = b^T A^{-1} \sum_{k=1}^{n} \omega_k(\Delta t),\ n = 1, ...N. \tag{26}$$

Runge-Kutta method is referred to as Butcher tableau

$$\frac{c|\ A^T}{b^T},\quad A \in R^{m\times m},\quad b, c \in R^m. \tag{27}$$

Formula for coefficients $\omega_n(\Delta t)$ retains its written form with the only difference that now it's a matrix formula. The characteristic function writes as

$$\gamma(z) = A^{-1} - zA^{-1}[1]b^T A^{-1},\quad [1] = (1, \ldots, 1)^T. \tag{28}$$

4. Poroelastic column

Consider a problem of a poroelastic column of length l with one clamped end and a Heaviside-shaped in time load $f(t) = S_0 H(t)$ to another end [8]. Assuming the values $S_0 = 1\,\mathrm{N/m^2}$ and $l = 10\,\mathrm{m}$, the pore pressures p^w, p^a at the clamped end and displacement u_3 at the another

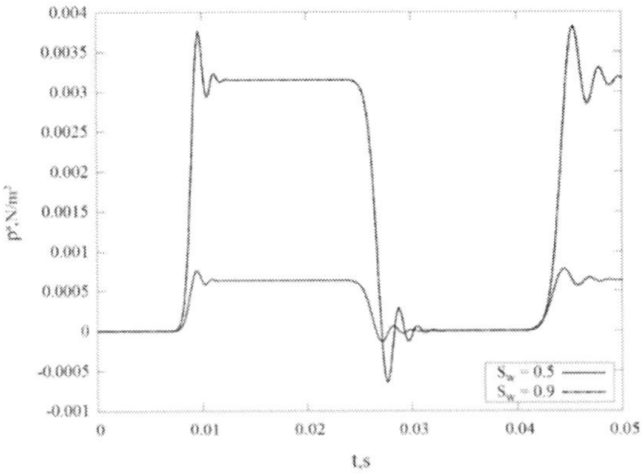

Figure 3. Pore pressure p^a.

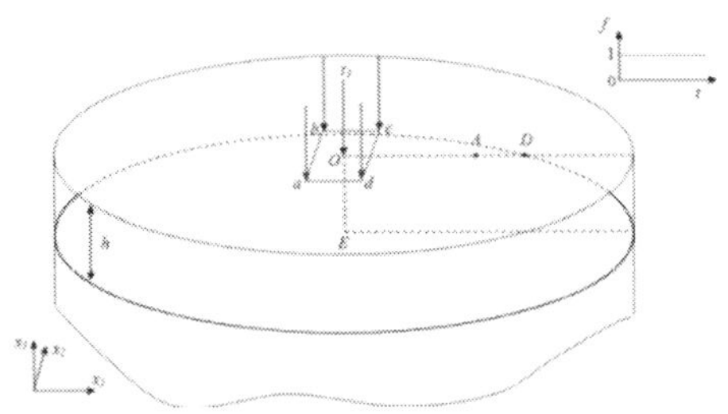

Figure 4. 2-layer half-space.

end are calculated with varying saturation degree. The pore size distribution index ϑ is set to 1.5, the residual water saturation S_{rw} is set to 0, and the air entry saturation S_{ra} is set to 1. The column is made of Massilon sandstone with the following properties: $K = 1.02 \cdot 10^9\,\mathrm{N/m^2}$, $G = 1.44 \cdot 10^9\,\mathrm{N/m^2}$, $\phi = 0.23$, $\rho_s = 2650\,\mathrm{kg/m^3}$, $\rho_w = 997\,\mathrm{kg/m^3}$, $\rho_a = 1.10\,\mathrm{kg/m^3}$, $K_s = 3.55 \cdot 10^{10}\,\mathrm{N/m^2}$, $K_w = 2.25 \cdot 10^9\,\mathrm{N/m^2}$, $K_a = 1.10 \cdot 10^5\,\mathrm{N/m^2}$, $\kappa = 2.5 \cdot 10^{-12}\,\mathrm{m^2}$, $\eta_w = 1.0 \cdot 10^{-3}\,\mathrm{Ns/m^2}$, $\eta_a = 1.8 \cdot 10^{-5}\,\mathrm{Ns/m^2}$.

The problem is solved with the help of BDF-2 scheme with Ł=4N=1000; a mesh of 896 boundary elements is chosen. Figures 1–3 show the displacements and pore pressures.

For the nearly saturated case $S_w = 0.99999$, the results are in good agreement with those of the saturated case.

5. Homogeneous half-space

A problem of vertical load $t_3 = t^0 f(t)$, $t^0 = -1000\,\mathrm{N/m^2}$ to a surface element *abcd* of homogeneous poroelastic half-space is considered, where $f(t)$ is Heaviside function (Fig. 4). The area of *abcd* is $1\,\mathrm{m^2}$. The rest of the surface is free and permeable: $p = 0$. The poroelastic material is rock with the following properties: $K = 8 \cdot 10^9\,\mathrm{N/m^2}$, $G = 6 \cdot 10^9\,\mathrm{N/m^2}$, $\rho = 2458\,\mathrm{kg/m^3}$, $\phi = 0.19$, $K_s = 3.6 \cdot 10^{10}\,\mathrm{N/m^2}$, $\rho_f = 1000\,\mathrm{kg/m^3}$, $K_f = 3.3 \cdot 10^9\,\mathrm{N/m^2}$, $\kappa = 1.9 \cdot 10^{-10}\,\mathrm{m^4/(N \cdot s)}$. The problem is treated on several boundary element meshes.

One of the research aspects was modeling slow compression wave in terms of the half-space. For this purpose point E at a depth 3 m under the excitation area is considered. To calculate generalized displacement and

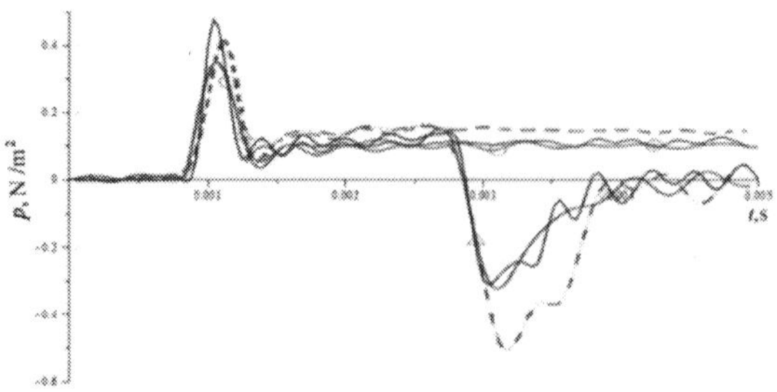

Figure 5. Pore pressure at point E. $\kappa = 1.9 \cdot 10^{-10}\,\mathrm{m}^4/(\mathrm{N} \cdot \mathrm{s})$ ......Radau, -o- BDF2, - - - Solution from [9]; $\kappa = 1.9 \cdot 10^{-6}\,\mathrm{m}^4/(\mathrm{N} \cdot \mathrm{s})$ — Radau, -Δ- BDF2, — · · · — · · · —Solution from [9].

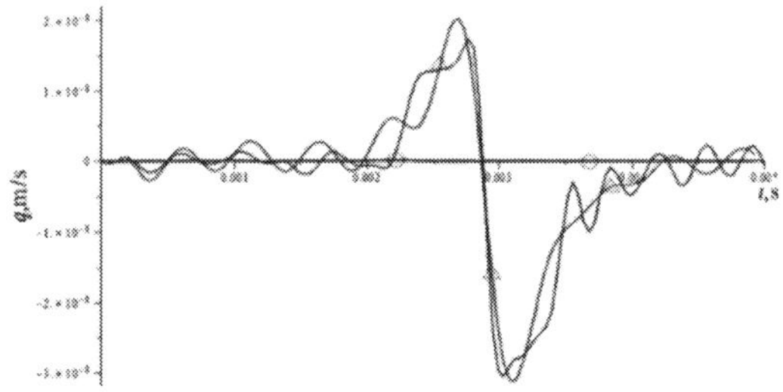

Figure 6. Pore flux at point E. $\kappa = 1.9 \cdot 10^{-10}\,\mathrm{m}^4/(\mathrm{N} \cdot \mathrm{s})$ — Radau, -o- BDF2; $\kappa = 1.9 \cdot 10^{-6}\mathrm{m}^4/(\mathrm{N} \cdot \mathrm{s})$ ......Radau, -Δ- BDF2.

traction responses at this point a fictitious boundary at a depth of 3 m is introduced in parallel with the surface. In this formulation the problem is quite close to the problem of layered half-space. Radau scheme is used with N = 125 and Ł = 1000 on $[0, \pi]$, BDF2 demands of N = 1000. The results are shown in Figs 5–6 in comparison with other solutions where available.

The study validates and improves the results of [9]. As compared to [9], here all basic functions including displacements, tractions, pore pressure and flux are drawn; a mesh of 1176 boundary elements is used whereas [9] uses 342 triangular elements.

6. Layered half-space

A problem of a 2-layer poroelastic half-space subjected to a Heaviside-type load is considered (Fig. 4). All the parameters are the same as for the problem of homogeneous half-space except that here the layers are of different material properties: a layer of sandstone is located over the rock half-space. The sandstone properties are: $K = 2.1 \cdot 10^8\,\mathrm{N/m}^2$, $G = 9.8 \cdot 10^7\,\mathrm{N/m}^2$, $\rho = 1884\,\mathrm{kg/m}^3$, $\phi = 0.48$, $K_s = 1.1 \cdot 10^{10}\,\mathrm{N/m}^2$, $\rho_f = 1000\,\mathrm{kg/m}^3$, $K_f = 3.3 \cdot 10^9\,\mathrm{N/m}^2$, $\kappa = 3.55 \cdot 10^{-9}\,\mathrm{m}^4/(\mathrm{N} \cdot \mathrm{s})$. Figures 7, 8 represent vertical displacements at point A 10 m from the excitation area for sandstone layer of thickness h = 5 m and h = 10 m respectively.

The Rayleigh wave appearance (t ≈ 0.043 sec) can be observed after the fast compression wave arrival (t ≈ 0.01 sec). Then (t > 0.05 sec), some perturbation can as well be seen, that argues for the arrival of compression waves, reflected by the sandstone/rock boundary.

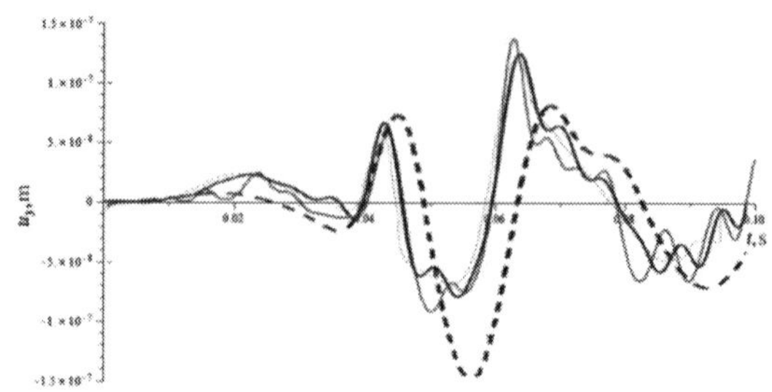

Figure 7. Vertical displacements at point A, h = 5 m. ......Radau, Ł=1000, N = 270 ······· BDF2, L = N = 1000; — Durbin's method ($\alpha = 5$) - - - - solution from [9].

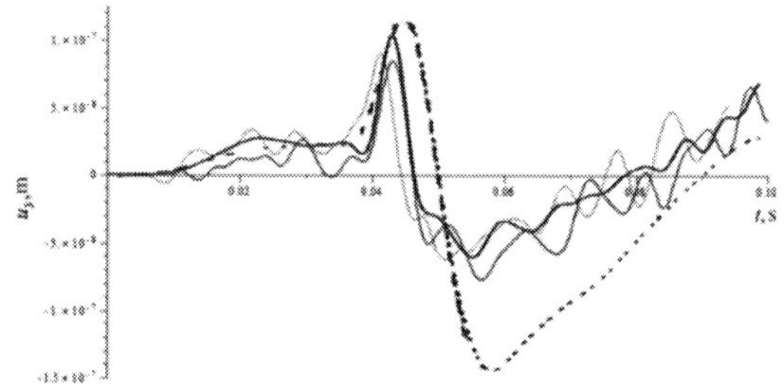

Figure 8. Vertical displacements at point A, h = 5 m. ......Radau, Ł= 1000, N = 270 ······· BDF2, L = N = 1000; — Durbin's method ($\alpha = 5$) – – – – solution from [9].

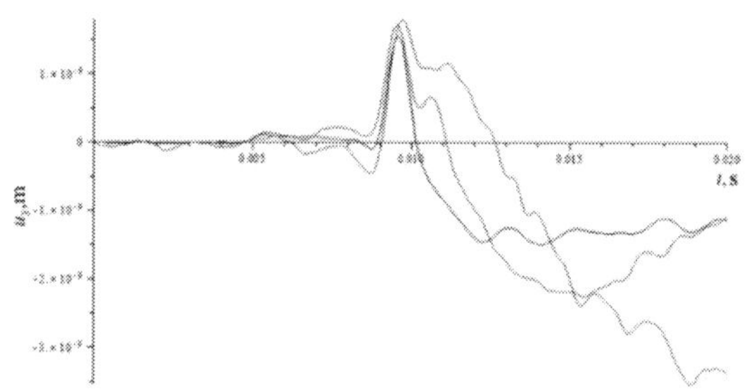

Figure 9. Vertical displacements. — Spherical cavity — Cubic cavity — No cavity.

The chosen sandstone layer thickness still allows distinguishing the effect of Rayleigh wave thanks to the fact that reflected compression waves arrive later. The results here as compared to ones in [9] rely on the matched displacement-stress model whose usage is better motivated than of isoparametric one.

7. Half-space with a cavity

A problem of vertical load to a surface element of poroelastic rock half-space with spherical or cubic cavity is considered. The load $P(t) = P_0 f(t)$, $P_0 = 1000\,\mathrm{N/m}^2$ acts on the area $1\,\mathrm{m}^2$; the cube side/sphere diameter is 10 m; the center of cavity is located at a depth of h = 7.5 m under the excitation area. The half-space surface is impermeable: $q = 0$. The problem is treated with the help of Radau-based scheme with N = 250 and Ł= 1000 on $[0, \pi]$. The boundary element mesh takes into account 2 symmetry planes; a quarter of the mesh contains

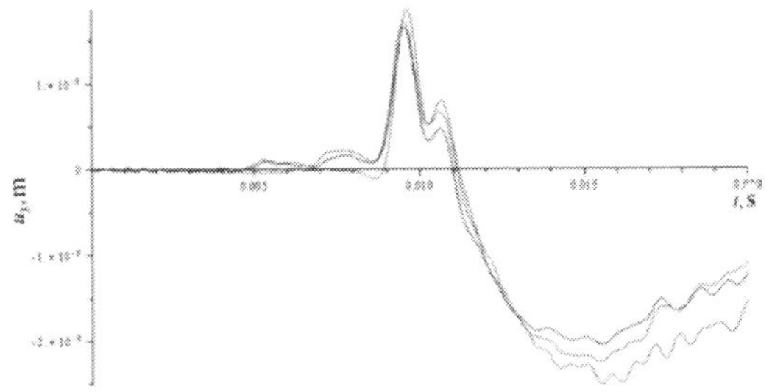

Figure 10. Vertical displacements for case spherical cavity. — poroelastic solution, — drained model, — undrained model.

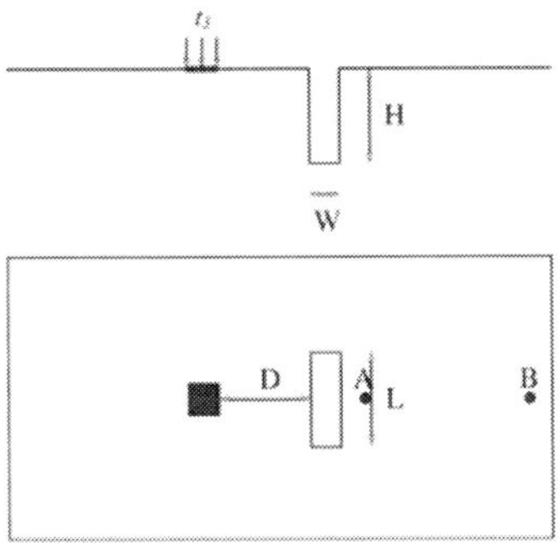

Figure 11. Problem layout.

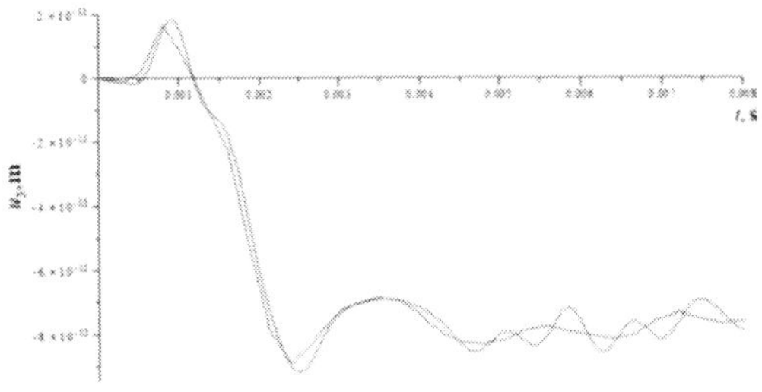

Figure 12. Vertical displacements at point A. — BDF2 — Radau scheme.

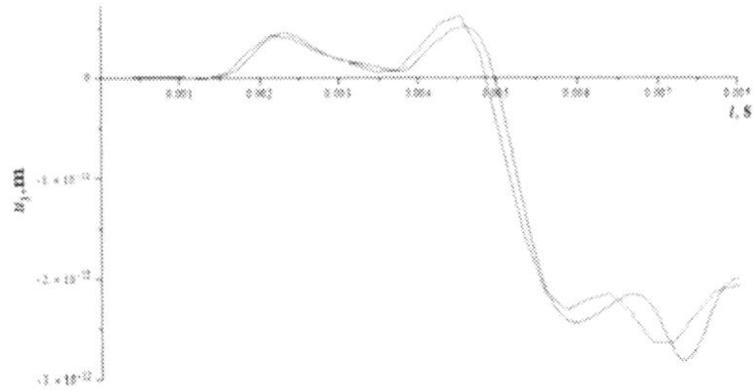

Figure 13. Vertical displacements at point B. — BDF2 — Radau scheme.

420 elements on the surface and 324 elements on the cavity boundary. Responses of displacements and pore pressure at point 15 m from the excitation area are shown in Fig. 9 in comparison with the case of poroelastic half-space without cavity. Figure 10 presents the results obtained using the drained and undrained models. It's seen from the diagrams that the shape of the cavity affects the responses form qualitatively. Displacements for the case without cavity are closer to the spherical cavity-free case than to the cubic cavity case. However there's still certain difference. For the cavity case vertical displacements gain one more peak ($t \approx 0.011$ sec) following the arrival of Rayleigh wave in the central part of the diagram ($t \approx 0.0095$ sec); in the horizontal displacements diagram the peak right before the arrival of Rayleigh wave becomes much stronger ($t \approx 0.009$ sec).

8. Half-space with a trench

A problem of vertical load applied to a surface element of poroelastic half-space with open trench is considered [8]. The load $P(t) = P_0 f(t)$, where $P_0 = 1000\,\mathrm{N/m^2}$ and $f(t)$ is Heaviside function, acts on the area $0.25\,\mathrm{m^2}$ and remaining surface is traction free. The pore pressure are assumed to be zero all over the surface. The trench with parameters $\mathrm{L} = 1\,\mathrm{m}$, $\mathrm{W} = 0.5\,\mathrm{m}$, $\mathrm{H} = 1\,\mathrm{m}$ is located at a distance $\mathrm{D} = 1\,\mathrm{m^2}$ from the loading area (Fig. 11). The observation points A and B are 1.75 m and 6.83 m behind the trench along the center line of the trench. The sandstone properties are: $\mathrm{K} = 1.02 \cdot 10^9\,\mathrm{N/m^2}$, $\mathrm{G} = 1.44 \cdot 10^9\,\mathrm{N/m^2}$, $\rho_s = 2650\,\mathrm{kg/m^3}$, $\phi = 0.23$, $\mathrm{K}_s = 3.5 \cdot 10^{10}\,\mathrm{N/m^2}$, $\rho_f = 997\,\mathrm{kg/m^3}$, $\mathrm{K_f} = 2.25 \cdot 10^9\,\mathrm{N/m^2}$, $\kappa = 2.5 \cdot 10^{-12}\,\mathrm{m^4/(N \cdot s)}$. The mesh contains 580 elements on the surface. The problem is treated with the help of Radau-based scheme with with $\mathrm{N} = 30$, Ł$= 242$ on $[0, 2\pi]$ and BDF2 with $\mathrm{N} = 100$, Ł$= 415$ on $[0, 2\pi]$.

The Radau-based time-step scheme of the BEM gives results that are close to those based on the traditional stepped scheme but is much more economical: it requires only 30 time-steps as compared with 100 for the Euler-based scheme, and 121 double nodes for argument φ as compared with 415 single ones for the Euler-based scheme.

9. Concluding remarks

A boundary element approach utilizing Runge-Kutta based time-step method for numerical Laplace transform inversion is developed and applied to 3d poroelastodynamic boundary value problems. In combination with varied integration step and highly oscillatory quadrature employed here it helps to reduce computational costs for solution. It's seen from figures that the scheme gives results close to ones from the analogous scheme relying on traditional Euler, and better. As compared to BDF2, Runge-Kutta gives no delay of the final solution and better keeps its form in time.

The research is supported by the Russian Foundation for Basic Research (RFBR) under grants 14-08-31415, 14-08-00811, 15-08-02814, 15-08-02817.

References

[1] M. Schanz, *Wave Propagation in Viscoelastic and Poroelastic Continua* (Springer Berlin, 2001).

[2] P. Li, M.Schanz, Wave propogation in 1-D partially saturated poroelastic column, Geophys. J. Int. **53** (2011).

[3] V.G. Bazhenov, L.A. Igumnov, *Metody granichnyh integralnyh uravneniy i granichnyh elementov v reshenii zadach tryohmernoy dinamicheskoy teorii uprugosti s sopryazhyonnymi polyami* (Boundary Integral Equations & Boundary Element Methods in treating the problems of 3D elastodynamics with coupled fields) (PhysMathLit, Moscow, 2008) (in Russian).

[4] R.V Goldshteyn, K voprosu o primenenii metoda granichnyh integral'nyh uravneniy dlya resheniya zadach mehaniki sploshnyh sred (To the issue of BIE method application in solving problems of continuum mechanics), *Metod granichnyh integralnyh uravneniy: Vychislitelniye aspekty i prilozheniya v mehanike (Mehanika: novoye v zarubezhnoy nauke), sbornik statey*, pp. 183–209, (Mir: Moscow, 1978), (in Russian).

[5] C. Lubich. A. Ostermann, Runge-Kutta methods for parabolic equations and convolution quadrature, Math. Comput., **60**, 105–131 (1993).

[6] M.P. Calvo, E. Cuesta, C. Palencia, Runge-Kutta convolution quadrature methods for well-posed equations with memory, Numer. Math., **107**, 589–614 (2007).

[7] L. Banjai, M. Messner, M. Schanz, Runge-Kutta convolution quadrature for the boundary element method, Comput. Methods Appl. Mech. Engrg., **245–246**, 90–101 (2012).

[8] P. Li, M. Schanz, Time domain boundary element formulation for partually saturated poroelasticity, Eng. Anal. Bound. Elem. **37**, 1483–1498 (2013).

[9] M. Schanz, H. Antes, *Porous Media: Theory, Experiments and Numerical Applications* (Ehlers W., Bluhm J. eds.), (Springer, Berlin, 383, 2002).

EPJ Web of Conferences **94**, 04027 (2015)
DOI: 10.1051/epjconf/20159404027

The interaction of the projectile with moving plates and rods

A. Gerasimov and S. Pashkov

Tomsk State University, Research Institute of Applied Mathematics and Mechanics, 634050 Tomsk, Russia

Abstract. The study of the problem of protecting the elements of constructions from impact loadings is very important due to the need of constant perfection of the means of shock-wave impact on the objects of modern technology. The problem of creating a reliable protective system dictates necessity of studying different ways to counteract to high-velocity elongated projectiles. The interaction of projectiles with plates and rods which are thrown towards by HE is investigated. The strain and fracture of the projectile sharply reduce its penetrating ability.

1. The equations describing movement of compressed elastic-plastic body taking into account probabilistic fracture behavior

The problem is solved in 3-D statement in view of natural heterogeneity of real materials structure affecting distribution of physical-mechanical characteristics along the volume of construction elements and being one of the factors, defining destruction character of the latter.

The necessity to account the given factor for equations of deformable solid mechanics dictates the application of probabilistic laws of physical-mechanical characteristics distribution along the volume of the construction under consideration.

The equations describing spatial adiabatic movement of the strong compressed solid are differential result of fundamental laws of conservation of mass, pulse and energy. In general they have the following form [1–3].

The equation of continuity

$$\frac{1}{\rho}\frac{d\rho}{dt} + \frac{\partial v_i}{\partial x_i} = 0. \tag{1}$$

The equations of movement

$$\rho\frac{dv_i}{dt} = \rho F_i - \frac{\partial P}{\partial x_i} + \frac{\partial S_{ij}}{\partial x_j}. \tag{2}$$

The equation of energy

$$\rho\frac{dE}{dt} = S_{ij}\varepsilon_{ij} + \frac{P}{\rho}\frac{d\rho}{dt}, \tag{3}$$

where x_i – coordinates; t – time; ρ_0 – initial density; ρ – current density; v_i – velocity vector components; F_i – components of the vector of mass forces; S_{ij} – stress deviator components; E – specific internal energy; ε_{ij} – components of the strain rate tensor; P – pressure.

Equations (1)–(3) must be added by equations that take into account the relevant thermodynamic effects related to the adiabatic compression of the medium and the strength of the medium. In general at action of forces on a solid deformable body both the volume (density) and shape of the body change, and at different dependences. Therefore the tensor of stress is represented as the sum of a pressure and the stress deviator components.

$$\sigma_{ij} = S_{ij} - P\delta_{ij}, \qquad i, j = 1, 2, 3,$$
$$\delta_{ij} = 1, \quad i = j,$$
$$\delta_{ij} = 0, \quad i \neq j,$$

where δ_{ij} – Kronecker symbol.

To describe resistance of a solid to shear, we use the following ratios:

$$2\mu(e_{ij} - \frac{1}{3}e_{kk}\delta_{ij}) = \frac{DS_{ij}}{Dt} + \lambda S_{ij}; \tag{4}$$

$$\frac{DS_{ij}}{Dt} = \frac{dS_{ij}}{dt} - S_{ik}\omega_{jk} - S_{jk}\omega_{ik}; \tag{5}$$

$$2\omega_{ij} = \frac{\partial v_i}{\partial x_j} - \frac{\partial v_j}{\partial x_i}; \tag{6}$$

$$2e_{ij} = \frac{\partial v_i}{\partial x_j} + \frac{\partial v_j}{\partial x_i}, \tag{7}$$

and also plasticity condition

$$J_2 = \frac{1}{2}S_{ij}S_{ij} = \frac{1}{3}\sigma^2, \tag{8}$$

where e_{ij} – strain rate tensor components; μ – shear modulus; σ – dynamic yield stress; D/Dt – Jaumann derivative.

The equation of state was chosen in the form of Mi-Gruneisen

$$P = \frac{K(1 - G_0\xi/2)}{(1 - c\xi)^2}\xi + \rho_0 G_0 E, \tag{9}$$

where G_0 – Gruneisen coefficient; c, K – material constants; ρ_0 – initial density; $\xi = 1 - \rho_0/\rho$.

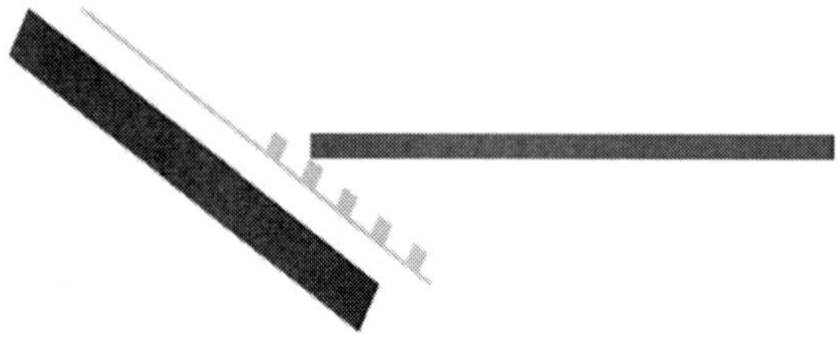

Figure 1. Initial system configuration "target – ribbed plate – projectile".

Figure 2. The three-dimensional picture interaction of projectile with ribbed plate and target.

Figure 3. 2-D cross-section of three-dimensional calculated area interaction of projectile with a throwing plate and target.

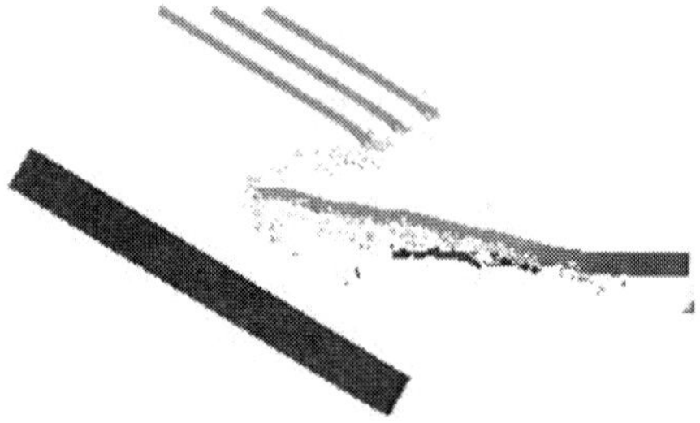

Figure 4. 2D cross section of a three- dimensional calculated area interaction of projectile with three plates.

The limiting equivalent of plastic deformation was used as a criterion of shear fracture [4] $\varepsilon^p = \varepsilon^p_*$. The system of Eqs. (1)–(9) is written in general form for the spatial motion of a deformable body.

The process of destruction of real materials is largely determined by internal structure of the solid, presence of heterogeneity, usually caused by different orientation of the grains in a polycrystalline material or irregularities in the composition of composite materials, the difference in micro-strength inside the grain and on inter-grain or inter-phase boundary. Therefore to the adequacy of the process numerically modeled with the experimental data it is necessary to bring in perturbations to physical-mechanical characteristics of the destroyed environment, i.e. to set casual distribution of the factors determining the strength properties of the material. The natural fragmentation of projectiles and target is calculated by introducing probability distribution mechanism of initial defects in the material structure to describe separated and shear cracks. The reaching its limiting value by equivalent plastic deformation is used as a criterion of fracture under intensive shear deformations.

The initial heterogeneities were modeled by that limiting equivalent plastic deformation was distributed across cells using a modified random number generator issuing a random variable obeying the chosen law of distribution.

The system of the basic equations is supplemented with appropriate initial and boundary conditions. At the initial moment all projectile points have an axial velocity V_0 with regard to its sign, the plate or rods had normal to the surface of the target velocity V_1, the target state is assumed unperturbed. Boundary conditions are as follows: on borders, free from pressure, conditions are satisfied $\sigma_n = \tau_n = 0$. On the contact site between bodies we condition ideal sliding of one material relative to another along the tangent: $\sigma_{n1} = \sigma_{n2}$, $\upsilon_{n1} = \upsilon_{n2}$, $\tau_{n1} = \tau_{n2} = 0$, where σ_n, τ_n-normal and tangent components of stress vector; υ_n-normal component of velocity vector in a point of contact; indexes 1 and 2 refer to contacting bodies.

To calculate elastic-plastic flows one used the technique implemented on tetrahedral cells and based on joint use of Wilkins method for calculating interior points of the body and Johnson method for calculating contact interactions [2,5,6]. Three-dimensional area was sequentially split into tetrahedrons by means of automatic grid construction subroutines.

The most complete ideology and methodology of the probabilistic approach to the problem of solids fracture is given in [7,8].

2. The results of calculations

Three possible impacts on high-speed elongated projectiles of tungsten alloy, approaching the target at a velocity 1500 km/sec, have been calculated. In the first case steel ribbed plate, the second case three plates of tungsten alloy, in the third case three-layer plate (middle layer – ceramics) were thrown towards the projectile.

In these cases velocity of the throwing elements was equal to 1000 m/sec and was directed along the normal to the target. The thickness of the target was 5 cm and the angle of deviation from a vertical was 60° (Fig. 1).

Ribbed plate is an combination of two systems – plates and rods (Figs. 1, 2).

The use the ribbed plate for protection changes the picture of its interaction with a projectile. The ribs deform the projectile to form a system of waves that can be clearly seen in Fig. 3, with their further destruction and

Figure 5. The three-dimensional picture of projectile interactions with three plates.

Figure 6. Interaction of projectile with the three-layer plate (middle layer – ceramics) at time t = 159: initial system configuration "target – two plates + ceramics – projectile".

ricocheting of the obtained fragments from the surface of the target. The damages observed in the calculations are insignificant and capture only the surface layer of the material.

In this paper we have calculated two more possible scenarios of the action on approaching the target high-velocity elongated projectile made of tungsten alloy. In the first case three tungsten alloy plates were thrown towards the projectile (Fig. 4), in the second case – a three-layer plate was thrown towards the projectile (Fig. 6). A ceramic layer, equal in weight to a middle layer of the three-layer system, was placed between the tungsten layers. Physical-mechanical properties of ceramic materials are given in [7,8].

In all scenarios the velocity of the thrown elements was equal to 1000 m/s and it was directed along the normal to the target. The thickness of the target was 5 cm and the angle of deviation of target from vertical was 60°. The length of the projectile was 65.4 cm, diameter – 2.4 cm and velocity – 1800 m/sec. In all cases plates of tungsten alloy had a size of 25 cm and 50 cm in width and length and 1 cm thickness.

When interacting with the three plates are amplified (Figs. 4,5) the destruction of the contact surface of the projectile and the degree of deviation from the direction of impact, leading eventually to the ricocheting projectile from the surface target and the preservation of its integrity.

Such character of deformation projectile and fracture can be explained by a sequence of shocks on the plates. Each subsequent plates to act already on the pre-deformed and partially destroyed part of the rod, reinforce the effect of the preceding plates.

In contrast to the above results for the three layers of tungsten alloy, the replacement of a middle layer by

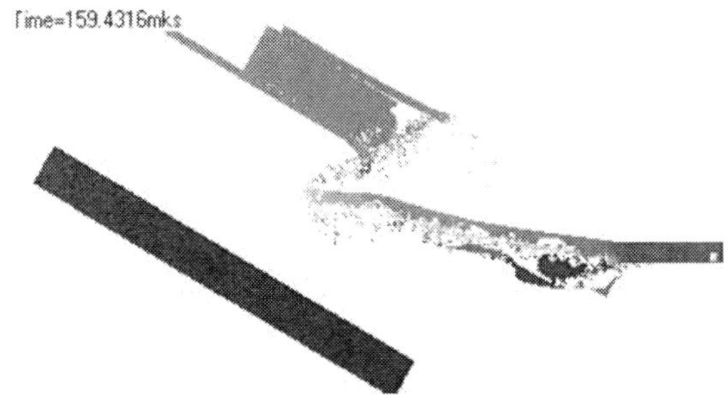

Figure 7. 2-D cross-section of three-dimensional calculated area interaction of projectile with the three-layer plate.

a thicker ceramic layer (Figs. 6, 7) prolonged the time of interaction of projectile with the system "two plates + ceramics". Impact of the ceramic layer caused more intensive destruction of the projectile in the zone of their contact and significant deviation of the projectile from the initial trajectory toward the target.

3. Conclusions

The aim of the present work is the numerical modeling of the processes of deformation and fracture of high-speed projectiles by elements of multilayer composite systems and different combinations of plates, as well as investigation of the impact of high-velocity projectile fragments on the main protected target. The problems were solved in three-dimensional formulation as such approach enabled us to study the processes of dynamic loading by the most adequate way to reality. In the multilayer plates included the layers of metal alloys and ceramic materials.

Calculation results showed the possibility of the proposed approaches to reduce the penetrating power of the projectile. At vigorous dynamic interaction one observes deformation and destruction of the projectiles, as well as their deviation from the line of collision with a target. Consequently the projectiles either ricochet from the target surface or deviate from the protected object and do not interact with the target. All these factors decrease the penetrating power of the projectiles and reduce probability of perforation of the protected object.

The calculations also showed that the developed 3-D numerical method enables to simulate in complex the processes of high-velocity interaction of long projectiles with layered plates containing ceramic insets, spaced plates, rods and ribbed constructions, as well as interaction of remnants of the projectiles with protected objects in the collision along a normal and at an angle to the surface of target in a wide range of velocities and angles of impact and also to investigate the processes of fragmentation of projectile and target and a character of the fields of fragmentation.

The method takes into account probabilistic nature of the fragmentation of the colliding bodies, which leads to three dimensionality of the problem even at the impact along the normal to the target surface. Theoretical and experimental data were found to be in good agreement, which enables to use the method developed to calculate engineering structures subjected to impact loads and to select the most effective combinations of materials and geometrical parameters of the thrown elements.

This work was partially supported by RFBR grant No 13 -08-00296-a.

References

[1] K.P. Stanjukovich (Ed), *Physics of explosion* (Nauka, Moscow, 1975) [in Russian]

[2] M.L. Wilkins, *Fundamental Method in Hydrodynamics* (Mir, Moscow, 1967) [Russian translation]

[3] M.L. Wilkins, *Structural impact and crashworthiness* (New York, 1984)

[4] K.N. Kreyenhagen, M.H. Wagner, J.J. Piechocki, R.L. Bjork, AIAA J. **8**, 2147 (1970)

[5] G.R. Johnson, D.D. Colby, D.J. Vavrick, Int. J. Numer. Methods Engng., **14**, 1865 (1979)

[6] G.R. Johnson, Trans. ASME. J. of Appl. Mech., **48**, 30 (1981)

[7] A.V. Gerasimov (Ed), *Theoretical and experimental researches of high-speed body interactions* (Tomsk State University, Tomsk, 2007) [in Russian]

[8] A.V. Gerasimov, S.V. Pashkov, Composites: Mechanics, Computations and Applications. An International Journal, **4**, 97(2013)

EPJ Web of Conferences 94, 04028 (2015)
DOI: 10.1051/epjconf/20159404028

Adiabatic shearing failure of explosively driven metallic cylinder shell, from experiments to simulation

Xinlong Dong[1,a], Xinlu Yu[1], Yingqian Fu[1], Fenghua Zhou[1], Tiegang Tang[2], and Haibo Hu[2]

[1] MOE Key Laboratory of Impact and Safety Engineering, Ningbo University, Ningbo, Zhejiang 315211, China
[2] Institute of Fluid Physics, China Academy of Engineering Physics, Mianyang 621900 Sichuan, China

Abstract. The adiabatic shear bands (ASB) of the tubular metal specimens expanded explosively have been studied by many researchers in the recent years. The onset and evolutions of the multiple shear failure of metal cylinder under explosive loadings are affected by many factors such as the characteristics of the impulsive loadings, the dynamic behavior of the materials, etc. In this work, we investigate the failure and fragmentation of 45# steel cylinder shell driven by the JOB9003 explosive. Experimental and FEM numerical simulation investigations are made for cylinder modeled shell. The results show that for the perfect homogeneous FEM model, the failure mechanism of cylinder shell is differs from that of the experiments, in which the spalling originates is oriented by high intensity of rarefaction wave. Through numerical experiments, it was found that distributed geometrical defects of cylinder shell affect the fragmentation process and mechanism, in which the strain localization controlled by the defects and shear bands initiate on the inter-surface of the cylinder shell.

1. Introduction

The subject of fragmentation of explosively driven cylinder has long been of interest in the military field, which relate to the applications including the design of fragment and blast resistant structures and protective facilities [1,2]. It is a highly complex phenomenon in which the fragmenting material is plastically deformed by the passage of an intense shock (~ 30 ms of GPa) followed by high-rate expansion deformation that ultimately leads to fracture. The previous works shown that several failure modes in the fragment cross-section could be occurs, including radial fracture, internal micro cracks, shear-lip fracture and spall fracture, which are related to the cylindrical geometry, materials behavior and loading characteristic of the experiment [2–5]. As one of the major failure mechanism in explosively driven metallic expansion cylinder, adiabatic shear bands (ASB) have receved increasing attention in recent years [6–8]. This phenomenon is characterized by a localization of the plastic deformation in narrow band. The onset of thermo-plastic instability has been examined, either through a maximum stress criterion (e.g., Zener and Hollomon [9]), through linear stability analyses (e.g., Bai [10]). However, the questions of how an ASB formation and develops after instability and how it interacts with the boundaries or neighboring localizations still require further understanding. Nesterenko, Xue et al. [11–14] proposed a controlled way to create multiple shear bands in collapsing thick-walled cylinder (TWC). Another approach of study of ASBs is to model it by numerical simulating via numerical computations and constitutive modeling [15–18], because a numerical simulation can easily include nonlinear effects or check the limits of a theoretical solution. One of the major issues or difficulties in the simulation of ASB propagation is the onset criterion for ASB growth and propagation [15,18].

In this work, we investigate the failure and fragmentation of medium carbon 45 steel cylinder shell driven by the JOB9003 explosive. The high-speed photography has been applied to observer the expanding fracture of the steel cylinder shell. The metallurgical analysis reveals adiabatic shear fracture mechanism. The FEM simulation was conducted to investigate the evolution process of strain localization. It was found that distributed geometrical defects of cylinder shell affect the fragmentation process and mechanism, in which the strain localization controlled by the defects and shear bands initiate on the inter-surface of the cylinder shell. The effects of the defect on the ductile fragmentation process were investigated.

2. Experiment

2.1. Dynamic expansion of cylinder shell

In order to study the initiation of failure and fragmentation of metal cylinder shell during dynamic loading, an axisymmetric cylindrical geometry driven by inner explosive is used in the current experiments. Figure 1 presents a schematic view of the experimental device. The cylinder has an internal diameter of 40 mm and wall-thickness of 5, all made from medium carbon 45# steel. The microstructure of the 45# steel contains dual phases of ferrite grains and pearlite colonies.The high explosive is bonded into the cylinder. A nominal cylinder length of 140 mm is used to achieve a plane-strain load path. A frustum shaped booster comprising a conventional high explosive, RHT-901, initiates the JOB9003 main charge at

[a] Corresponding author: dongxinlong@nbu.edu.cn

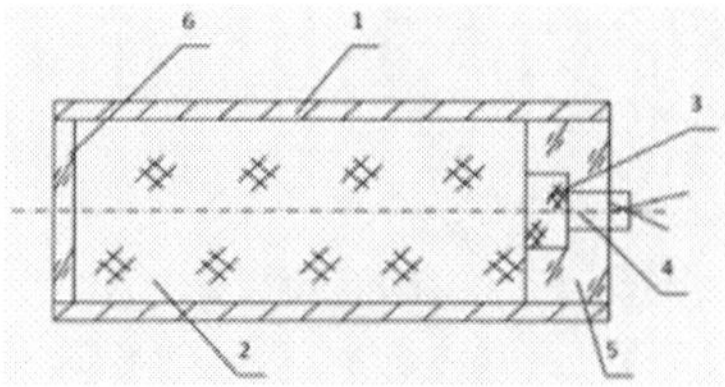

Figure 1. Schematic view of the experimental device.

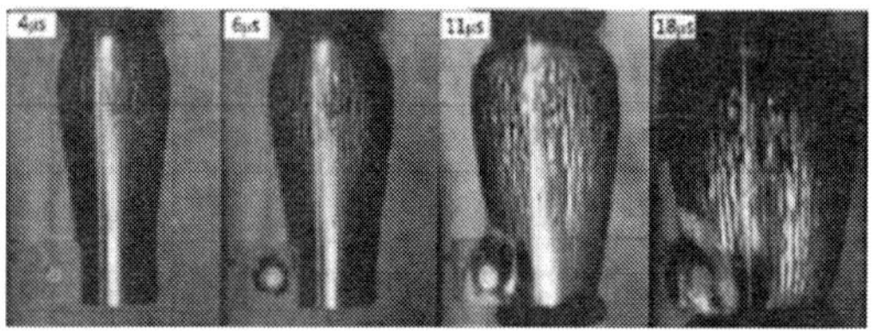

Figure 2. Typical framing records for an exploding 45# steel cylinder with fully charged JOB9003.

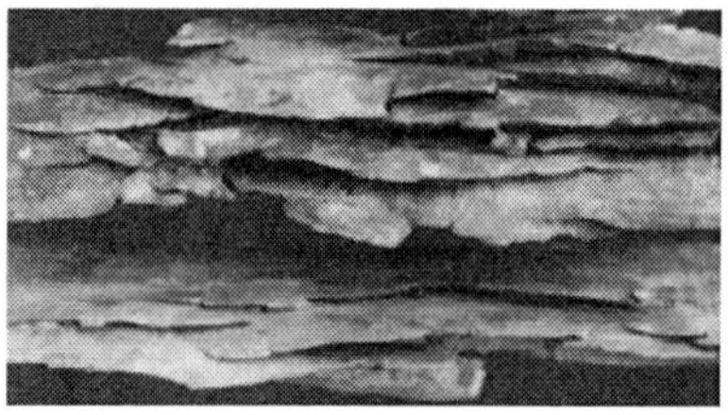

Figure 3. The recover samples show shear failure characteristics.

one end. The purpose of this booster is to provide a near-planar detonation to the main charge.

The high-speed photography (sampling time 0.67 ms) has been applied to observer the expanding fracture of the steel cylinder shell. The fiber Doppler velocimeter was used to measure the expanding velocity on outer surface of the metal cylinder.

2.2. Experimental results

Figure 2 shows typical high-speed photography records for a fully charged cylinder of 45# steel expanding symmetrically. A number of crack initiation and axial propagation are seen on the surface at 4 and 18 ms, these features were also seen in other cylinders of TC4 titanium alloy[7]. Typical fragments from an 45 steel cylinder areshown in Fig. 3. The fragment widths are typically a few mm. The processes of the appearance of initial shear fracture and their developments on inner surfaces were recorded. and the fractured surfaces of recovery specimen are analyzed, showing the longitude shear fracture cell characteristics typical for 45 steel, as shown in Fig. 3.

Shear failure running parallel to the fragment length define the fragment width. The fragment cross-sectioned perpendicular to its length, shows fracture surfaces that are inclined at about 45 or 135 degrees with the original inner and outer surfaces. Shear bands are also found within the fragment width, as shown in Fig. 4. Within the fragment interior many adiabatic shear bands are observed in parallel (see Fig. 4), which their separations are on

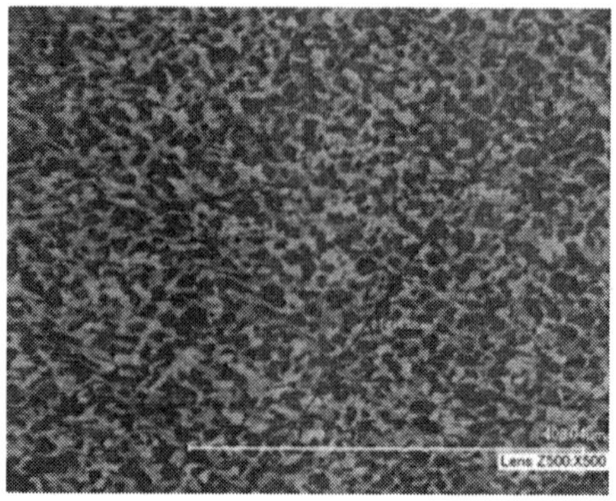

Figure 4. Cross-sectional micrograph of the fragment show shear bands patterns.

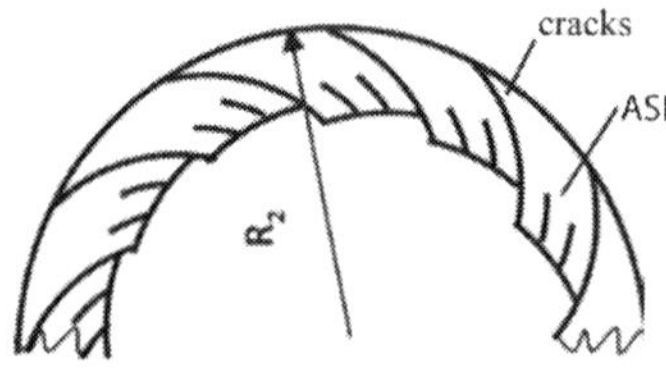

Figure 5. Schematic diagram of shear cracks and shear bands spacing.

the order of 110–210 mm. These ASBs do not associated with cracks and propagate the entire way through the shell thickness.

The fracture and metallurgical analysis reveals adiabatic shear fracture mechanism, i.e. the self-organized shear bands deform in inter-surface of cylinder and develop outwards either in the clockwise or anticlockwise direction, at an angle of about 45° or 135° with the radius, the crack propagate along the shear bands, as shown in Fig. 5.

3. Numerical simulation and discussion

3.1. FEM model

To understand the adiabatic shear fracture phenomena and fragmentation size of the cylinder shell, computer simulation for the initial stage of explosion was made using Abqus software, Fig. 6, the FEM element mesh size is $40\,\mu m \times 40\,\mu m$. The 2D numerical modeling of shock dynamic processes were carry out. Explosive behavior was described by the radial shock loading $P(t)$ applied on inner surface of the shell,

$$p = p_0 e^{-(t-t_d)/t_d} \tag{1}$$

here, P_0 is Chapman-Jouguet pressure 35 Gpa, t_d is decay time $1.25\,\mu s$.

The GRUNEISEN state equation and Johnson–Cook constitutive equation are used to describe for the homogeneous behavior of 45# steel:

$$\sigma = (A + B\varepsilon^n)\left[1 + C\ln\frac{\dot{\varepsilon}}{\dot{\varepsilon}_0}\right]\left[1 - \left(\frac{T - T_t}{T_t - T_m}\right)^m\right] \tag{2}$$

where A, B, n, m and C are the material parameters, $\dot{\varepsilon}_0$ is a reference strain rate, T_0 is the initial temperature

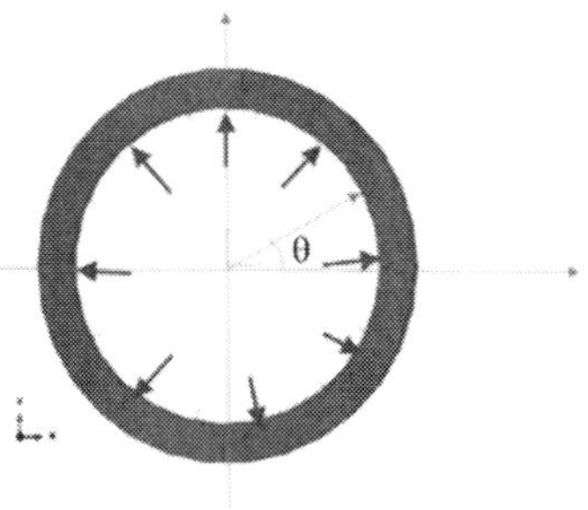

Figure 6. FEM model.

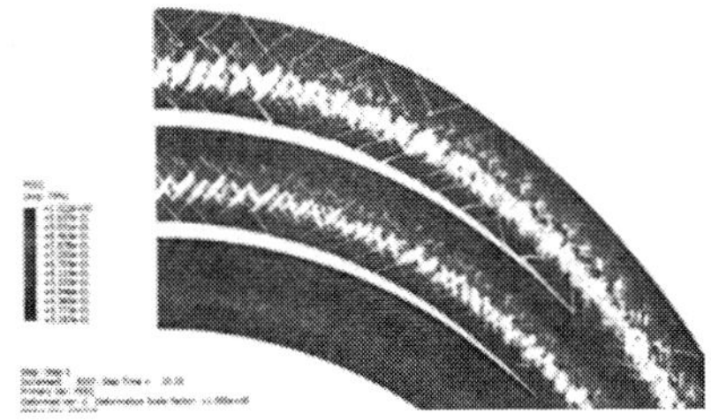

Figure 7. Configuration of the equivalentic plastic strain and damages evolution.

Table 1. Parameters of the 45# steel.

A MPa	B MPa	n	c	m	Melting Temp [K]	Transition Temp [K]	Reference Strain Rate
507	320	0.28	0.064	1.06	1793	293	1
state equation	C_0/m/s	S	$g\gamma_0$				
$u_s - u_p$	4600	1.49	2.17				

(room temperature in this case) and T_m the melting temperature. The parameters of the J-C model we used in the simulations are listed in Table 1.

Shear band formation is complicated and can involve failure of the material in the developing band. The thermal softening, obtained using the J-C model, is sufficient to correspond to the heating of the bulk material. However, in order to obtain localization of the ASB's, a more enhanced softening is essential–accomplished by the use of the shear-soften model. Briefly stated, shear band formation and failure can be seen as an cohesion instability failure of material which is characterized by the progressive deggradation of the material stiffness [],

$$D = \frac{\varepsilon - \varepsilon_{cr}}{\varepsilon_f - \varepsilon_{cr}} \qquad \varepsilon_{cr} < \varepsilon < \varepsilon_f. \tag{3}$$

Here, $\varepsilon_{cr} = 0.3$ adiabatic shear initiation plastic strain, $\varepsilon_f = 0.7$ as the adiabatic shear failure strain of the material. Once the initial of damage, the flow stress in the shear band is,

$$\sigma = \sigma_0 (1 - D) \tag{4}$$

which provides the positive feedback needed to cause adiabatic shear localization. Once $D = 0.9$, the material is no longer allowed to carry hydrostatic tension or deviatoric stresses.

3.2. Simulation results and discussion

3.2.1. Results of the perfect homogeneous shell model

Considering a perfect homogeneous shell model, i.e. the cylinder shell is without any geometric imperfections and strain loclization cause from the random difference of the elememt meshes. Figure 7 shows the results, where we can observe that the damages of micro cracks and voids were produced firstly in the middle area of metal shell. The failure mode of cylinder shell is differs from that of the experiments, in which the self-organized shear bands deform in inter-surface of cylinder and develop outwards either in the clockwise or anticlockwise direction, at an angle of about 45° or 135° with the radius, the crack propagate along the shear bands. The spall fracture originates on this case is oriented by high intensity of rarefaction wave, and propagates in the material inward and outward to surface of shell.

3.2.2. Results of the shell with geometric imperfections

Actually during the fabrication and the machining process, metallic components are inevitably brought with defects or inhomogeneities. Generally such defects or inhomogeneities have an geometric distributions. In this paper, the cylinder shell is prescribed with random geometrical defects, i.e. inner radius of cylinder shell $r = r_0 + P\ rand(1)$, which *rand(1)* is random function and $P = \Delta r / r_o$ are 0.1% and 0.5% respectively.

Through numerical experiments, it was found that failure mechanism of the shell with the initial defects usually distingish from the shell without defects (see Fig. 8 and Fig. 7). The simulation results show that the equivalent plastic strain localization occure in the inner surface of metal tuber. All ASBs randomly propagate outwards either in the clockwise or anticlockwise direction, at an angle of about 45° or 135° with the radius. The lengths of those shear bands vary greatly, and a shielding effect is noticed that areas adjacent to the well developed shear bands are characterized by smaller shear bands as they enter stress-relieved regions. The FEM numerical results are consistent trend with the expermental phenomenon.

Futhermore, the size of the defect also affect the failure process. It was found that the shell with larger the initial defects usually broke into pieces earlier than that of relatively small defects. It show that numbers of shear band decrease with increasing defect size (see Fig. 8(a) and (b)).

4. Summary

In this work, we investigate the failure and fragmentation of metal cylinder shell driven by the JOB9003 explosive. Experimental and FEM numerical simulation investigations are made for cylinder modeled shell with external diameter of 50 mm and internal diameter 40 mm, all made from medium carbon 45 steel. The high-speed photography (sampling time 0.67 ms) has been applied to observer the expanding fracture of the steel cylinder

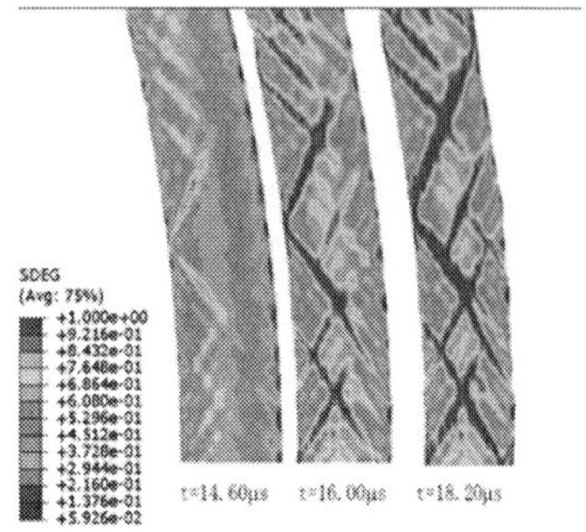

(a) the equivalentic plastic strain evloution with p=0.1%

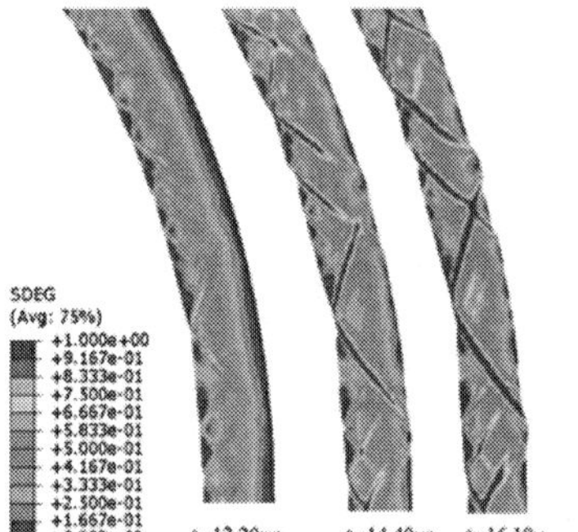

(b) the equivalentic plastic strain evloution with p =0.5%

Figure 8. Configuration of calculation area at different times.

shell. The metallurgical analysis reveals adiabatic shear fracture mechanism, i.e. the self-organized shear bands deform in inter-surface of cylinder and develop outwards either in the clockwise or anticlockwise direction, at an angle of about 45° or 135° with the radius, the crack propagate along the shear bands. The FEM simulation was conducted to investigate the evolution process of strain localization. For the perfect homogeneous model, the failure mode of cylinder shell is differs from that of the experiments, in which the spalling originates is oriented by detonation wave, fracture initiates on the case and propagates in the material inward and outward to surface of shell. Through numerical experiments, it was found that distributed geometrical defects of cylinder shell affect the fragmentation process and mechanism, in which the strain localization controlled by the defects and shear bands initiate on the inter-surface of the cylinder shell. The effects of the defect on the ductile fragmentation process were investigated.

The authors greatly appreciate the financial support from the National Science Foundation of China (No. 11172144, U1230122), the Doctoral Foundation of Ministry of education of china (No.20113305110004).

References

[1] Gurney R.W. The initial velocities of fragments from bombs, shell and grenades[R]. Army Ballistic Research Lab Aberdeen Proving Ground MD, 1943

[2] Mott N.F. Fragmentation of shell cases. Proc Royal Soc Lond A (1947);189:300–8

[3] Taylor G.I. The fragmentation of tubular bombs. In: Scientific papers of G.I. Taylor, vol. III. Cambridge: Cambridge University Press; 1963. No. 44, pp. 387—390

[4] Hoggatt R.H., Recht R.F. Fracture behavior of tubular bombs. J. Appl. Phys. (1968); **39**(2):1856–62

[5] Singh M., Suneja H.R., Bola M.S., et al. Dynamic tensile deformation and fracture of metal cylinders at high strain rates. International journal of impact engineering (2002), **27**(9): 939–954

[6] Tang T., Li Q., Sun X., Sun Z., Jin S. and Gu Y., Strain-rate effect of expanding fracture of 45 steel cylinder shells driven by detonation, Explsion and shock Waves (2006), **26**(1):129–133 (in Chinese)

[7] Tang T., Gu Y., Li Q., et al. Expanding fracture of steel cylinder shell by detonation driving, Explsion and shock Waves (2003), **23**(6): 529–533 (in Chinese)

[8] D.M. Goto*, R. Becker, T.J. Orzechowski, H.K. Springer, A.J. Sunwoo, C.K. Syn, Investigation of the fracture and fragmentation of explosively driven rings and cylinders, International Journal of Impact Engineering **35** (2008) 1547–11556

[9] C. Zener, J.H. Hollomon, Effects of strain rate upon plastic flow of steel. J. Appl. Phys. (1944) **15**, 22–32

[10] Y. Bai, B. Dodd. Adiabatic shear localization. Oxford: Pergamon press (1992) p. 24

[11] V. F. Nestrenko. Shear localization and shear bands pattern in heterogeneous materials. In: Dynamics of Heterogeneous Materials, New York, (2001) p. 307–384

[12] Q. Xue, M.A. Meyes, V.F. Nestrenko. Self-organization of shear bands in stainless steel. Material Science and Engineening A (2004) **384**, 35–46

[13] V.F. Nestrenko, M.A. Meyes, T.W. Wright. Self-organization in the initiation of adiabatic shear bands. Acta materiallia (1997) **46**, 327–340

[14] Hoggatt C.R., Recht R.F. Fracture behavior of tubular bombs[J]. Journal of Applied Physics (2003), **39**(3): 1856-1862

[15] R.C. Batra, C.H. Kim. Analysis of shear banding in twelve materials. Int. J. Plasticity (1992) **8**, 425–452

[16] M. Zhou, A.J. Rosakis. G. Ravichandran. Dynamically propagating shear bands in impact-loaded pre-notched plates–1. Experimental investigations of temperature signaturesand propagation speed. J. Mech. Phys. Solids (1996) **44**, 981–1006

[17] F. Zhou, T.W. Wright and K.T. Ramesh. A numerical methodology for investigating the formation of adiabatic shear bands, Journal of Mechanics and Physics of Solids (2006) **54**, 904–926

[18] N. Medyanika, W.K. Liua and S. Li. On criteria for dynamic adiabatic shear band propagation, Journal of the Mechanics and Physics of Solids (2007) **55**, 1439–1461

EPJ Web of Conferences **94**, 04029 (2015)
DOI: 10.1051/epjconf/20159404029

3D finite element simulations of high velocity projectile impact

Joško Ožbolt[a], Barış İrhan, and Daniela Ruta

Institute of Construction Materials, University of Stuttgart, 70569 Stuttgart, Germany

Abstract. An explicit three-dimensional (3D) finite element (FE) code is developed for the simulation of high velocity impact and fragmentation events. The rate sensitive microplane material model, which accounts for large deformations and rate effects, is used as a constitutive law. In the code large deformation frictional contact is treated by forward incremental Lagrange multiplier method. To handle highly distorted and damaged elements the approach based on the element deletion is employed. The code is then used in 3D FE simulations of high velocity projectile impact. The results of the numerical simulations are evaluated and compared with experimental results. It is shown that it realistically predicts failure mode and exit velocities for different geometries of plain concrete slab. Moreover, the importance of some relevant parameters, such as contact friction, rate sensitivity, bulk viscosity and deletion criteria are addressed.

1. Introduction

It is well known that in concrete structures the resistance, failure mode, crack pattern and crack velocity are strongly influenced by loading rate [1–13]. The rate dependent response of concrete is controlled through three different effects: (i) through the rate dependency of the growing micro-cracks (influence of inertia at the micro-crack level), (ii) through the viscous behaviour of the bulk material between the cracks (viscosity due to the water content) and (iii) through the influence of inertia of different kind, e.g. structural inertia, inertia due to the softening or hardening of the material or inertia related to the crack propagation. From the numerical point of view, assuming macro or meso scale analysis, the first two effects can be accounted for by the rate dependent constitutive law. Assuming that the resolution of underlying spatial discretization is fine enough, the third effect should be automatically accounted for through dynamic analysis where the constitutive law interacts with inertia [12].

Principally, with increase of loading rate failure mode tends to change from mode-I to mixed mode. As discussed in [12] and [13], responsible for this is inertia, which homogenizes the material of the impact zone and forces damage (crack) to move away from the zone of high inertia. Furthermore, when crack starts to propagate relatively fast, inertia at the crack tip tends to prevent crack propagation. Consequently, crack branching occurs and single crack splits into two inclined cracks. Moreover, significant inertia can be generated as a consequence of material softening and hardening. Elastic-brittle materials, such as glass, exhibit almost no influence of the cracking induced inertia on the rate dependent resistance. However, in quasi-brittle materials (e.g. concrete) and especially in ductile materials (e.g. steel) the inertia related to the propagation of macro-cracks significantly influence resistance and ductility of structures. For instance, it is well known that although steel exhibits almost no strain rate sensitivity, compared to quasi-static loading, its resistance increases with increasing loading rate and the response becomes more brittle. The reason for such behaviour is due to the fact that with increasing loading rate the size of the plastic zone at the crack tip reduces in size due to the action of inertia related to highly non-linear material response.

Recently, a 3D explicit finite element code has been developed for simulation of high velocity impact and fragmentation events [14]. In the code the rate sensitive thermo-mechanical microplane material model for concrete is used. In order to investigate the behaviour of concrete under high loading rates and verify the computational procedures developed, different problems were studied in the past, e.g. Compact Tension Specimen, Split Hopkinson Bar and Hammer Drop Tests [12], [15–17]. The comparisons between experimental and numerical results have shown that the code is able to correctly predict the phenomena related to high loading rate such as rate dependent resistance, crack branching, crack velocity and rate dependent failure mode. In the present study range of applicability of the code and validity of the underlying theories is tested by performing finite element simulations for some of the projectile perforation experiments documented in [18]. Note that exactly the same experiments have been studied numerically also by [19,20].

2. 3D multi-body dynamic rate dependent FE analysis

2.1. Constitutive law-rate sensitive microplane model for concrete

In the microplane model the material response is computed based on the monitoring of stresses and strains in different predefined directions. Integrating microplane stresses in a thermodynamically consistent way it is possible to

[a] Corresponding author: ozbolt@iwb.uni-stuttgart.de

Article available at http://www.epj-conferences.org or http://dx.doi.org/10.1051/epjconf/20159404029

calculate macroscopic stress tensor. In the model, material is characterized by the uni-axial relations between stress and strain components on planes of various orientations. The microplane model used in this work is based on so-called relaxed kinematic constraint concept [21], which is a modification of the M2 microplane model proposed by Bažant and Prat [22]. More details on the model can be obtained from [21].

The rate dependency in the here used version of the microplane model for concrete accounts for two effects: (1) the rate dependency related to the formation (propagation) of the micro-cracks, which is the effect of inertia forces at the level of the micro-crack tip, and (2) the rate dependency due to the viscosity of concrete (bulk material) between the micro-cracks. The rate dependency for each microplane component is modeled based on the rate process theory [23–26]. It should be kept in mind that the rate sensitivity defines the influence of loading rate on the true material strength only, i.e. without any consideration of inertia at the macro scale. The inertia at the macro scale is automatically considered through dynamic analysis. The model parameters are calibrated based on the uniaxial compressive tests performed by [3], where the rate of applied loading was kept within the range so that inertia does not play a major role (less than 1/s). The model was verified against experimental results for compressive and tensile behavior under different strain rates [9, 15, 16]. For example, in the case of compact tension specimen [16] it was confirmed that the progressive increase of resistance is due to the inertia generated at the tip of the macro crack, which causes crack branching and leads to progressive increase of resistance. Moreover, it was shown that the true tensile strength and fracture energy increase approximately linear in semi-log scale with increasing strain rate. The linear increase is controlled by the constitutive law (inertia at micro scale and influence of viscosity) whereas the progressive increase comes automatically from dynamic finite element analysis.

2.2. Finite element formulation

The dynamic nodal equilibrium equations are obtained by discretizing the weak (integral) form of balance equation for linear momentum in space. By constructing equilibrium equations at a known state and employing mass lumping, it is possible to update position of the nodes directly, without performing any iteration, using explicit time integration. In the presence of contact interactions, unfortunately, position update cannot be performed in single step because contact forces are not known beforehand. One can remedy such a situation by setting up a predictor-corrector type algorithm [14, 27].

Artificial bulk viscosity should be applied to the numerical solution for two reasons: (i) to prevent elements from collapsing down due to high velocity gradients and (ii) to quiet truncation frequency "ringing" [28]. Bulk viscosity pressure is not a part of the constitutive response and can be used conveniently for all material models. It has been found in numerical simulations performed for concrete perforation experiments that the strengthening due to rate dependent micro-crack growth (material rate sensitivity) is not sufficient to produce necessary resistance against penetration. In addition bulk viscosity effects should also be present.

During projectile penetration motion function starts immediately losing its regularity around the contact region. In such a situation one should firstly consider if it is possible to heal motion function such that Lagrangian framework can be retained throughout the rest of deformation process. Among various possible alternatives, method based on adaptive element deletion has been employed for this purpose. In this technique underlying finite element discretization is kept in place. The problematic part of the body, where motion function started to lose its regularity, is simply removed from the system. Element deletion criteria play an important role and should be set with extreme care in order to avoid non-physical solutions.

In all penetration simulations performed in this study, maximum principal strain has been used as a deletion criteria and its value has been set to 1.0. From the numerical results, it has been observed that stress state in some concrete elements around the contact region is almost hydrostatic compression. Such elements normally undergo very large deformations and still carry very high level of compressive stresses in the order of 1.0 GPa when the maximum principal strain reaches the failure value. Since such elements can still transmit force and resist projectile, deletion of them might numerically reduce the resistance of concrete slab. Consequently, failure mode, exit velocity and all other relevant features might drastically change and simulation results turn out to be failure criteria value dependent.

Based on above discussion, deletion of elements with very high stress is acceptable if and only if it does not change the results of numerical simulation significantly. Elements with very high stress can be conveniently deleted for penetration simulations if the following two conditions are satisfied. (i) Contact algorithm should not be influenced from the deletion of elements with very high stress too much. As a concrete element gets into contact with projectile, it deforms and also its kinetic energy is increased. If this element is continuously in contact, after a while nodes of this element, which contribute to contact, start to move together with projectile and, therefore, contact forces transmitted over such an element gradually lose their intensity. Only after then deletion of such an element will not disturb contact behavior and therefore is allowed even if it still carries very high stress. (ii) Response of concrete slab should not be altered too much as the concrete elements with very high stress are deleted. This is only possible if there is an internal mechanism which brings back approximately the nodal forces removed due to element deletion. This effect, of course, should be produced by the undeleted elements connected to those nodes. The nodal internal forces in those elements should be adjusted such that dynamic equilibrium prior to element deletion is nearly recovered. Such an artificial strengthening is possible due to existence of material rate dependency and bulk viscosity pressure. However such a recovery has not been observed at projectile penetration simulations because at the onset of element deletion strain rates are already very high and as a consequence artificial strengthening is negligibly small.

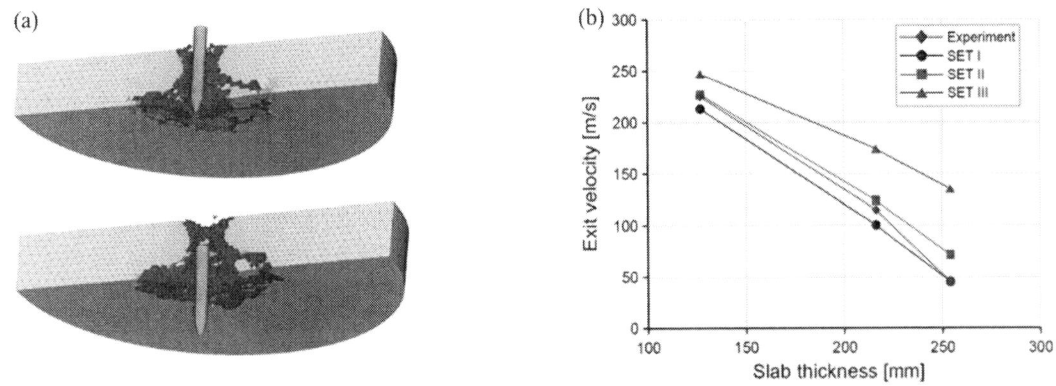

Figure 1. (a) 3D finite element discretization with two typical sequences of penetration events for SET-I/CS216 (section cut); (b) Relation between exit velocity and slab thickness for all three sets of analyses.

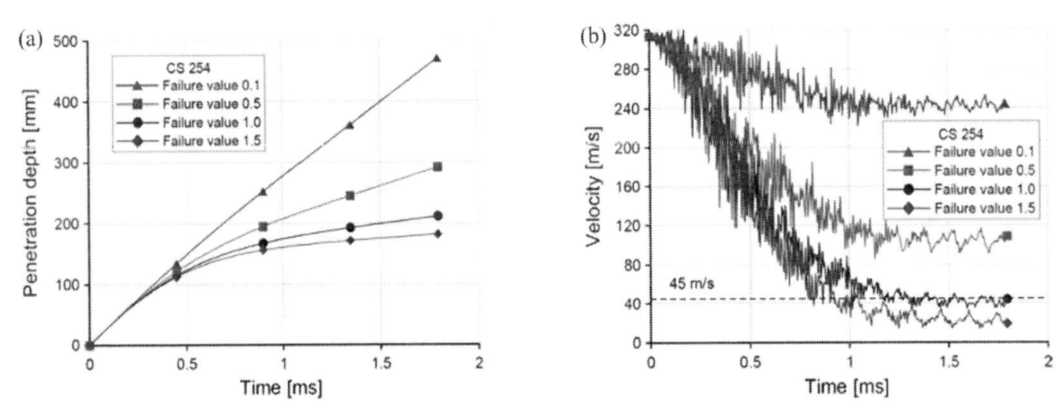

Figure 2. Time history graphs for different failure values: (a) Penetration depth vs. time and (b) Projectile velocity vs. time.

3. Projectile perforation of concrete slab

3.1. Experimental investigations

To test the performance of the developed 3D explicit finite element research code at very high loading rates series of numerical simulations have been performed for some of the projectile perforation experiments documented in [18]. In the experiments plain concrete circular slabs of different thickness were tested against penetration by using ogive-nose shaped projectiles with an initial impact velocity of approximately 313 m/s. Further details about experiments can be found in [18].

3.2. Numerical investigations

The numerical simulations have been carried out for three slab thicknesses (t = 127, 216 and 254 mm, Fig. 1a) with entire geometry, i.e. no symmetry was utilized. The objectivity of the results with respect to the mesh size is assured by the use of the crack band method [29]. In the analysis maximum principal strain has been used as failure criteria for the deletion of elements.

Failure value was set as equal to 1.0. To simplify notation we will denote 127 mm circular slab as CS127 and so on. Rate dependent microplane material model [21] has been used for concrete.

For projectile Saint-Venant Kirchhoff hyper-elastic material model is adopted assuming that projectile would undergo little or no deformation during perforation. First, a parametric study has been conducted in order to investigate the significance of concrete rate sensitivity and friction between circular slab and projectile on the results. For each slab thickness three sets of simulations have been performed: (I) rate-on/friction-on, (II) rate-on/friction-off and (III) rate-off/friction-off. Friction coefficient is set to 0.5.

The typical failure mode SET-I/CS216 that is in very good agreement with experimental observation is shown in Fig. 1a. As can be seen from Fig. 1b, for SET-I (rate-on/friction-on) exit velocities are in very good agreement with experiments for all slab thicknesses. The results for SET-II (rate-on/friction-off) also exhibit relatively good agreement with test results. It can be seen that friction does not significantly influence the exit velocity. On the other hand, results obtained from SET-III (rate-off/friction-off) are not objective. The exit velocity becomes very sensitive to rate effect as the thickness of the slab increases. This suggests that in case of high velocity impact problems not only inertia but also rate sensitivity influences the

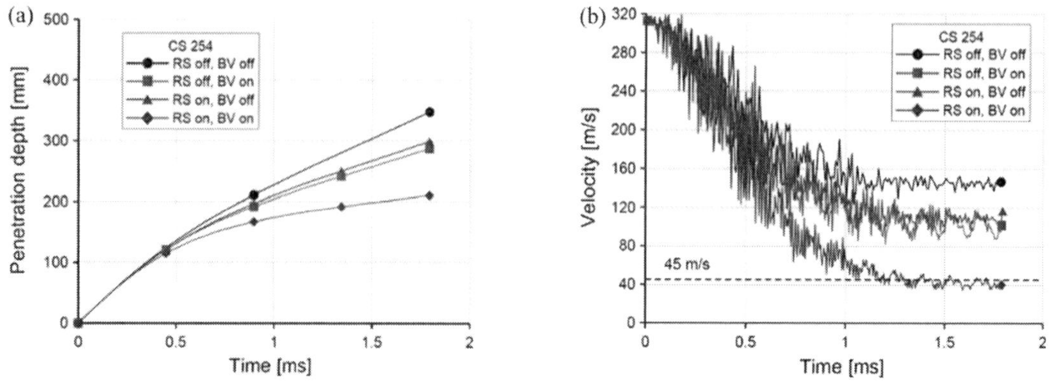

Figure 3. Influence of rate sensitivity and bulk viscosity on time history response: (a) Penetration depth vs. time and (b) Projectile velocity vs. time.

response. It is reasonable that with increase of structural size (thickness of the slab) rate sensitivity becomes more important. Namely, due to the large volume of activated material the structural resistance increases.

As has been discussed in previous section, it is very difficult to predict an optimum failure value beforehand for deletion criterion based on maximum principal strain, especially from contact algorithm point of view, if there are some elements with very high stress at the onset of element deletion. For this reason, set of numerical simulations have been performed to study the sensitivity of numerical results with respect to failure value. Simulations have been performed for failure values (maximum principal strain) 0.1, 0.5, 1.0 and 1.5 using SET-I/CS254.

As can be seen from Fig. 2, numerical simulations are very sensitive to the failure value especially for smaller ones. On the other hand, there is a convergence of results as the failure value increases. Based on these observations, it can be concluded that optimum value should be somewhere between 1.0 and 1.5.

To avoid numerical stability problems and maintain the robustness of underlying contact algorithm, failure value has been set as equal to 1.0 in all the simulations. It should be noted that even if the failure value is set to 1.5, no significant change in the overall response is observed.

To investigate the importance of material rate sensitivity (RS) and bulk viscosity (BV) three additional simulations, first one with RS-on/BV-off, the second one with RS-off/BV-on and the third one with RS-off/BV-off have been performed using SET-I/CS254.

The important results of these simulations are shown in Fig. 3 together with those for the case RS-on/BV-on. As can be seen from Fig. 3, the resistance of the slab is reduced dramatically if both contributions are set to off. Moreover, the resistance is reduced almost for the same amount when one parameter, either RS or BV, is set to off. Based on this observation it can be concluded that both RS and BV are very important and should be included in the numerical simulations for correct prediction of concrete resistance against penetration. Moreover, this suggests that the volumetric strain rate of concrete in front of the projectile tip significantly contributes to the resistance against penetration.

4. Conclusions

In the present article recently developed 3D explicit finite element code for simulation of high velocity impact and fragmentation events is briefly discussed. The code is then used in the simulation of the projectile perforation of the plain concrete slab. Based on the results of the numerical simulations the following conclusions can be drawn out. (1) The comparison between experimental and numerical results shows that the employed 3D FE code based on the rate sensitive microplane model for concrete is able to realistically predict exit velocities and crater profiles for all three slab thicknesses. (2) Based on the results of the parametric study it can be concluded that in the case of high velocity, impact rate sensitivity of concrete cannot be neglected. It significantly contributes to the resistance, especially if the size (slab thickness) of concrete specimen is large. (3) In contrast to rate sensitivity, friction between projectile and concrete does not significantly contribute to the resistance. (4) The choice of deletion criteria of finite elements significantly influences the results of the analysis. It is shown that maximum principal strain of 1.0 is a reasonably good choice for the failure value of concrete elements. It is shown that failure values larger than 1.0 do not significantly influence exit velocities and failure mode. (5) It is found that both material rate sensitivity and bulk viscosity influence the response. For the investigated case the influence of bulk viscosity of concrete is approximately the same as the influence of concrete rate sensitivity. For realistic analysis both contributions should be accounted for. This suggests that the volumetric strain rate of concrete in front of the projectile tip significantly contributes to the resistance against penetration.

References

[1] L. Freund, Crack propagation in an elastic solid subjected to general loading – I. Constant rate of extension, *Journal of the Mechanics and Physics of Solids*, **20**(3), 129–140 (1972)

[2] L. Freund, Crack propagation in an elastic solid subjected to general loading – II. Non-uniform rate of extension, *Journal of the Mechanics and Physics of Solids*, **20**(3), 141–152 (1972)

[3] W. Dilger, R. Koch, and R. Kowalczyk, Ductility of plain and confined concrete under different strain rates, *ACI Journal Proceedings*, **81**(1), 73–81 (1984)
[4] N. Banthia, S. Mindess, and A. Bentur, Impact behaviour of concrete beams, *Materials and Structures*, **20**(4), 293–302 (1987)
[5] H.-W. Reinhardt, Concrete under Impact Loading, Tensile Strength and Bond, *HERON*, **27**(3) (1982)
[6] P. Bischoff and S. Perry, Compressive behaviour of concrete at high strain rates, *Materials and Structures*, **24**(6), 425–450 (1991)
[7] J. Weerheijm, *Concrete under impact tensile loading and lateral compression*, TU Delft, the Netherlands (1992)
[8] J. Ožbolt and H.-W. Reinhardt, Rate dependent fracture of notched plain concrete beams, *CONCREEP 7*, 57–62 (2005)
[9] J. Ožbolt, K. K. Rah, and D. Meštrović, Influence of loading rate on concrete cone failure, *International Journal of Fracture*, **139**(2), 239–252 (2006)
[10] M. Larcher, Development of discrete cracks in concrete loaded by shock waves, *International Journal of Impact Engineering*, **36**(5), 700–710 (2009)
[11] R. R. Pedersen, *Computational modelling of dynamic failure of cementitious materials*, Delft University of Technology (2010)
[12] J. Ožbolt, A. Sharma, and H.-W. Reinhardt, Dynamic fracture of concrete-compact tension specimen, *International Journal of Solids and Structures*, **48**(10), 1534–1543 (2011)
[13] J. Ožbolt and A. Sharma, Numerical simulation of dynamic fracture of concrete through uniaxial tension and L-specimen, *Engineering Fracture Mechanics*, **85**, 88–102 (2012)
[14] B. İrhan, High velocity impact and fragmentation of concrete: Numerical simulation, Institut für Werkstoffe im Bauwesen der Universität Stuttgart (2014)
[15] J. Ožbolt, A. Sharma, B. İrhan, and E. Sola, Tensile Behavior of Concrete under High Loading Rates, *International Journal of Impact Engineering*, **69**, 55–68 (2014)
[16] J. Ožbolt, J. Bošnjak, and E. Sola, Dynamic Fracture of Concrete Compact Tension Specimen: Experimental and Numerical Study, *International Journal of Solids and Structures*, **50**, 4270–4278, (2013)
[17] N. Bede, J. Ožbolt, A. Sharma, and B. İrhan, Dynamic fracture of notched plain concrete beams: 3D finite element analysis, *International Journal of Impact Engineering*, **77**, 176–188 (2015)
[18] J. D. Cargile, *Development of a constitutive model for numerical simulation of projectile penetration into brittle geomaterials*, Technical Report SL-99-11, U.S. Army Engineer Research and Development Center, Vicksburg, MS (1999)
[19] F. C. Caner and Z. P. Bažant, Impact comminution of solids due to local kinetic energy of high shear strain rate: II-Microplane model and verification, *Journal of the Mechanics and Physics of Solids*, **64**, 236–248 (2014)
[20] A. O. Frank, M. D. Adley, K. T. Danielson, and H. S. McDevitt Jr, The high-rate brittle microplane concrete model: Part II: application to projectile perforation of concrete slabs, *Computers and Concrete*, **9**(4), 311–325 (2012)
[21] J. Ožbolt, Y. Li, and I. Kožar, Microplane model for concrete with relaxed kinematic constraint, *International Journal of Solids and Structures*, **38**(16), 2683–2711 (2001)
[22] Z. P. Bažant and P. C. Prat, Microplane model for brittle-plastic material: I. Theory, *Journal of Engineering Mechanics*, **114**(10), 1672–1688 (1988)
[23] H. Mihashi and F. Wittmann, Stochastic Approach to Study the Influence of Rate of Loading on Strength of Concrete, *HERON*, **25**(3) (1980)
[24] A. S. Krausz and K. Krausz, Fracture kinetics of crack growth, *Springer*, **1** (1988)
[25] Z. P. Bažant et al., Large-strain generalization of microplane model for concrete and application, *Journal of Engineering Mechanics*, **126**(9), 971–980 (2000)
[26] Z. P. Bažant, F. C. Caner, M. D. Adley, and S. A. Akers, Fracturing rate effect and creep in microplane model for dynamics, *Journal of Engineering Mechanics*, **126**(9), 962–970 (2000)
[27] N. J. Carpenter, R. L. Taylor, and M. G. Katona, Lagrange constraints for transient finite element surface contact, *International Journal for Numerical Methods in Engineering*, **32**(1), 103–128 (1991)
[28] L. M. Taylor and D. P. Flanagan, PRONTO 3D: A three-dimensional transient solid dynamics program (1989)
[29] Z. P. Bažant and B. Oh, Crack band theory for fracture of concrete, *Materials and Structures:* 155–177 (1983)

EPJ Web of Conferences **94**, 04030 (2015)
DOI: 10.1051/epjconf/20159404030

Strain-rate dependence for Ni/Al hybrid foams

Anne Jung[1,a], Martin Larcher[2,b], Ondrej Jirousek[3,c], Petr Koudelka[3,d], and George Solomos[2,e]

[1] Saarland University, Institute of Applied Mechanics, Campus A4.2, 66123 Saarbrcken, Germany
[2] European Commission, Joint Research Centre (JRC), Institute for the Protection and Security of the Citizen (IPSC), European Laboratory for Structural Assessment, via E. Fermi 2749, 21027 Ispra (VA), Italy
[3] Academy of Sciences of the Czech Republic, v.v.i., Department of Biomechanics, Institute of Theoretical and Applied Mechanics, Prosecka 76, 190 00 Prague 9, Czech Republic

Abstract. Shock absorption often needs stiff but lightweight materials that exhibit a large kinetic energy absorption capability. Open-cell metal foams are artificial structures, which due to their plateau stress, including a strong hysteresis, can in principle absorb large amounts of energy. However, their plateau stress is too low for many applications. In this study, we use highly novel and promising Ni/Al hybrid foams which consist of standard, open-cell aluminium foams, where nanocrystalline nickel is deposited by electrodeposition as coating on the strut surface. The mechanical behaviour of cellular materials, including their behaviour under higher strain-rates, is governed by their microstructure due to the properties of the strut material, pore/strut geometry and mass distribution over the struts. Micro-inertia effects are strongly related to the microstructure. For a conclusive model, the exact real microstructure is needed. In this study a micro-focus computer tomography (μCT) system has been used for the analysis of the microstructure of the foam samples and for the development of a microstructural Finite Element (micro-FE) mesh. The microstructural FE models have been used to model the mechanical behaviour of the Ni/Al hybrid foams under dynamic loading conditions. The simulations are validated by quasi-static compression tests and dynamic split Hopkinson pressure bar tests.

1. Introduction

Cellular materials like metal foams are a very interesting class of bionic lightweight materials. Open-cell foams mimic the construction elements of human femur bones or the star fish skeletons. Nature uses these cellular structures as advanced, evolution-optimised lightweight materials. Based on their special microstructure, consisting of a network of interconnected pores, metal foams are able to undergo large deformations at a nearly constant stress, forming a stress plateau [1,2]. The stress plateau evolves by the successive collapse of single pore layers. Due to the stress plateau, the foams are able to absorb considerable amounts of kinetic energy by plastic dissipation, whereas metal foams can be used as crash absorbers.

Unfortunately, at the moment, foams have to compete with established materials that were characterized better, up to now and the commercially available foams do not provide the required mechanical properties. In the last few years, there were efforts to improve the mechanical properties of metal foams by coating light substrate foams with hard materials [3–8]. Ni/Al hybrid foams are a highly novel and innovative material. They consist of standard, open-cell aluminium foams with a coating of nanocrystalline nickel. The innovative Ni/Al hybrid foams skillfully combine three design strategies of materials design: cellular lightweight construction, nanotechnology and composites. When these strategies are coupled cleverly and the synergies are used, nanocrystalline coated hybrid foams become an innovative and multifunctional nanomaterial for future applications in the fields of lightweight construction and energy absorption.

Cellular materials are called microheterogeneous materials, because their mechanical behaviour can be described on different scales. For metal foams, three different hierarchical scales can be defined. The macro scale deals with the complete sample or component, the meso scale with several pores and the micro scale describes the mechanical behaviour of the struts. Hence, the global mechanical properties of metal foams strongly depend on the microstructure, especially under higher strain-rates. Under high strain rates, the strut geometry and mass distribution over the struts plays a significant role, since micro-inertia effects are strongly related to these microstructural features. Hence, for conclusive models, especially to describe the global mechanical behaviour of the foams under high strain rates, the real microstructure is needed.

In this study, Ni/Al hybrid foams were experimentally and numerically investigated for their strain-rate dependency. Quasi-static and dynamic compression tests were performed. To determine the microstructure, a micro-focus computer tomography system has been used. Based on the μCT measurements, a microstructural Finite Element (micro-FE) mesh was developed. The microstructural FE models have been used to model the mechanical behaviour of the Ni/Al hybrid foams under dynamic

[a] e-mail: anne.jung@mx.uni-saarland.de
[b] e-mail: martin.larcher@jrc.ec.europa.eu
[c] e-mail: jirousek@itam.cas.cz
[d] e-mail: koudelkap@itam.cas.cz
[e] e-mail: george.solomos@jrc.ec.europa.eu

Article available at http://www.epj-conferences.org or http://dx.doi.org/10.1051/epjconf/20159404030

loading conditions and were validated by the macro-mechanical experiments.

2. Materials and methods

2.1. Synthesis of the Ni/Al hybrid foams

Aluminium alloy foams ($AlSi_7Mg_{0.3}$, Celltec Materials, Dresden, Germany) with a pore size of 30 ppi (pores per inch) were investigated. To produce Ni/Al hybrid foams, the aluminium foams were coated with different theoretical overall coating thicknesses ranging from 25 to 75 μm nano nickel with a crystallite size of about 50 nm, using direct current electrodeposition. A commercial nickel sulfamate electrolyte (Enthone GmbH, Langenfeld, Germany) with a nickel content of 110 g/L nickel was used at a pH of 3.8 and a temperature of 40 °C. For a qualitatively good coating on the aluminium, the pretreatment steps of pickling and electroless plating preventing the dissolution of the aluminium in the acid nickel electrolyte, have been performed [5]. To guarantee a quite homogeneous coating, a cage-like anode filled with nickel pellets from Ampere GmbH, Dietzenbach, Germany, was used as described by Jung et al. [6–8].

For the μCT measurements, cubic samples with an edge length of 15 mm were coated with 75 μm nickel. According to a bar diameter of 20 mm for the split Hopkinson pressure bar apparatus, cylindrical samples with a diameter of 20 mm and a height of 5 mm were used for the dynamic as well as for the quasi-static compression tests. The foams were coated with 25, 50 and 75 μm Ni, respectively.

2.2. Micro-focus computer tomography

High resolution microCT measurements of the samples were performed using a state-of-the-art DSCT/DECT X-ray imaging device developed in the Centre of Excellence Telč (patent pending, European Patent Office No. 14002662.6-1559 and Industrial Property Office of Czech republic No. PV2013-607). The device (depicted in Fig. 1) was designed to achieve maximum flexibility of the measurement in terms of specimen dimensions, magnification and beam characteristics. For this reason, a flexible modular design based on an orthogonal assembly of two tube-detector imaging pairs with independent geometry setting and a shared rotational stage was mounted on a complex 16-axis CNC positioning system. This enables unprecedented variability of measurement for highly detailed tomographical measurements, which is particularly advantageous in the case of heterogeneous materials with complex microstructures (i.e. composites, foams, stones, etc.).

In this work, samples of the metal foam were irradiated using a reflection type X-ray tube XWT-240-SE (X-ray Worx, Germany; voltage range 20–240 kV, minimum focal spot size 5 mm, maximum target power 280 W) with voltage and target current set to 80 kV and 300 μA resulting in a target power of 24 W. Acquisition was performed using large area GOS scintillator flat panel detectors XRD1622 (PerkinElmer, USA) with an active area 410 × 410 mm, resolution 2048 × 2048 px (pixel size 200 mm) at 2 × 1000 ms acquisition time. Setup geometry was set to obtain the maximum possible magnification leading to a source-object distance of 79.7 mm and a source-detector distance of 1345.4 mm, which provided a 16.87× magnification. Tomographic scanning was performed in 1200 projections with a 0.3° angular step to improve the accuracy and reliability of the tomographic reconstruction due to the very complex microstructure of the investigated material.

Figure 1. Laboratory setup of DSCT/DECT X-ray imaging device used for the tomography measurements. The sample was fixed on the rotational table in the middle (the sample is not depicted in the figure).

The linearization of the attenuation range (beam hardening correction) [9] was applied to all acquired radiographic projections. During the correction procedure, a set of aluminium calibrators (thicknesses 0.4–2 mm) covering the whole attenuation range of the specimens in their widest part was irradiated. The image data were corrected using the attenuation curve fitted through the measured Al calibrators' thicknesses.

2.3. Quasi-static and dynamic compression tests

Quasi-static compression tests were performed on 30 ppi aluminium and Ni/Al hybrid foams using an INSTRON 4204 universal testing machine at a strain-rate of about $5 \times 10^{-3}\,s^{-1}$.

The dynamic compression tests were conducted using a classical split Hopkinson pressure bar (SHPB) apparatus. The cylindrical bars are made of Zicral (AlZn 7075) to guarantee a comparable mechanical impedance between the bars and the sample. Using the SHPB setup, strain-rates of up to $5000\,s^{-1}$ were applied to cylindrical samples (diameter = 20 mm; height = 5 mm) of pure aluminium foams and Ni/Al hybrid foams with a pore size of 30 ppi and varying coating thicknesses.

2.4. Dynamic material modelling

The strain-rate dependency of the foam has also be investigated by using numerical methods with the explicit finite element code EUROPLEXUS (`http://www-epx.cea.fr/`). This is done using a micro model, which means that the microstructure of the material is dissolved in order to obtain the material behaviour on a macro level. This kind of investigations are very powerful as they can be used as a kind of virtual laboratory testing in order to investigate the material behaviour. They need a strong computational power and adapted material laws for the microstructure.

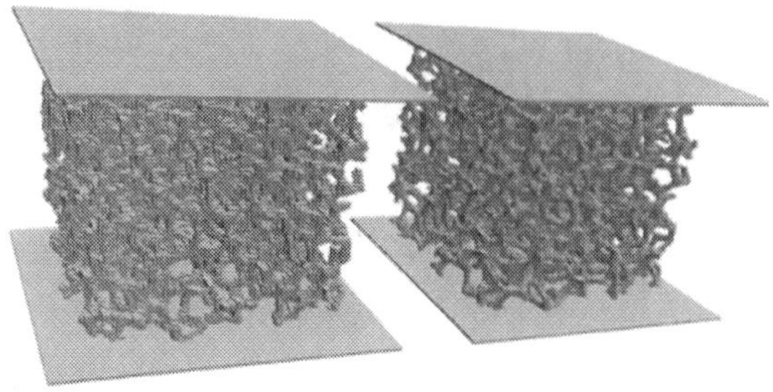

Figure 2. FEM meshes with element sizes of 170 μm (left) and 85 μm (right).

Table 1. Used material parameters for aluminium.

	Unit	Value
Density	kg/m^3	2660
Young's modulus	N/m^2	7e10
Poisson ratio	–	0.35
Elastic limit	N/m^2	230e6
Erosion (principal strain)	–	0.1

Similar investigations are done also for concrete [11], sand [12] and laminated glass [13].

The microstructure of the foam gained by micro-focus computer tomography (μCT) is used in order to create a solid 3D mesh from the μCT hexahedron cell mesh. These meshes use quite a large number of elements. The meshes contain about 130,000 elements for a mesh size of 170 μm and 5.4 million elements for a mesh size of 85 μm (Fig. 2). While the coarser meshes are manageable, the calculation time for the finest mesh is very high. The coating thickness for the investigated foams is in the range between 25 and 75 μm, while the scanned one has a coating thickness of about 75 μm. Therefore, it is not possible to build up the coating in the raw resolution. Further approaches could use beam elements with much lower calculation times, for which the coating is defined via the cross section of the beam, similar to layered shell elements. Nevertheless, it will be quite complex to build beam element meshes basen on the scan results.

A quite generic elastic-plastic material law is used to represent the behaviour of the aluminium. No hardening is taken into account. In addition, an erosion criterion based on principal strain is introduced (Material parameters, see Table 1). The numerical virtual experiment is conducted displacement controlled. The displacements are introduced via rigid load plates similar to real experiments. Between all aluminium elements and these plates, a contact is defined by using a pinball method [14]. This very powerful method attaches spheres around each concerned element. The contact between these spheres is checked each step. In addition to the contact between the sample and the plates, a self contact is introduced in order to allow contact between the struts after larger displacements or failure of them.

The loading is applied in two different loading speeds: a moderate loading rate of 1 m/s and a faster one of 10 m/s are used. For a numerical specimen size of 14 mm, these loading rates correspond to strain rates of 71 and 714 1/s resp. The objective of the virtual experiment is to obtain the macro stress-strain relation of a particular experiment. The strain can be calculated by dividing the actual displacement of the plates by the initial distance of the plates. The contact forces between the plate and the sample (aluminium hexahedrons) divided by the initial cross section of the sample provide the current stress.

Figure 3. Voxel model of the foam. Each voxel was directly converted to a linear hexahedral element.

3. Results and discussions

3.1. Evaluation of the microstructure by means of micro-focus computer tomography

For the development of the microstructural FE models, the radiograms were reconstructed with a cone-beam backprojection algorithm using Volex software (Fraunhofer-Allianz Vision, Germany). The resolution of the reconstructed three dimensional images was approximately 11.85 mm^3. The reconstructed data were then used to develop the FE model for the simulation of the dynamic compression tests. The resolution of the original images 1445 × 1459 × 1330 px was progressively reduced to 401 × 401 × 343 px, 200 × 200 × 171 px and 101 × 101 × 85 px to lower the complexity of the FE model for the simulations. The resampled image data were used to develop microstructural voxel models using an in-house open source software tool [10] by directly converting every voxel (spatial pixel) into a linear hexahedral element (see Fig. 3).

3.2. Experimental investigation of strain-rate dependency

Figure 4 shows a comparison of the stress-strain diagrams obtained from the quasi-static and dynamic compression tests. For both strain-rates, there is a clearly visible increase in the plastic collapse stress (PCS), corresponding to the strength of the foams, and of the plateau stress with increasing coating thickness. Comparing one sample type/coating thickness at different strain-rates outlines that there is a pronounced strain-rate sensitivity for the coated foams, which becomes visible by increasing PCS and plateau stress and even by a change in the deformation mechanism. The change in the deformation mechanism

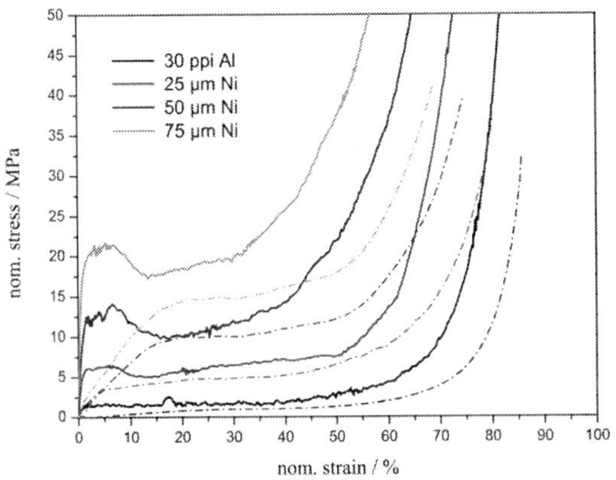

Figure 4. Comparison of the stress-strain diagrams for quasi-static and dynamic compression tests (dashed lines: quasi-static $5 \times 10^{-3}\,s^{-1}$, compact lines: dynamic $5000\,s^{-1}$).

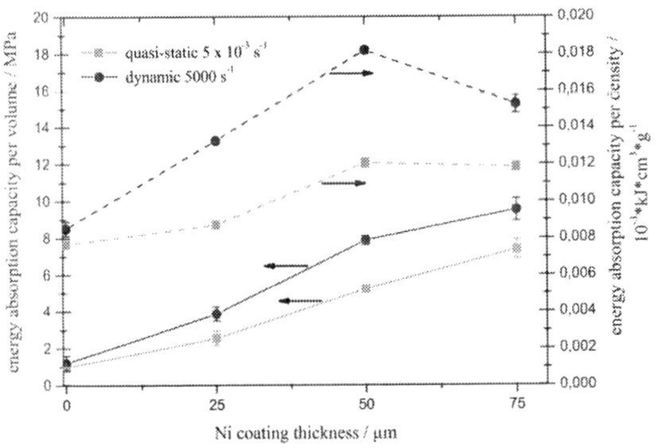

Figure 5. Specific energy absorption capacities up to the densification point as function of the Ni coating thickness, expressed per volume (lower curves) and per density (upper curves).

from quasi-static to dynamic mode is expressed by the pronounced plastic collapse peak in the dynamic curves. This comes from a time-delayed collapse of the pores, because, for high strain rates, the struts will first be stretched before they fail by buckling. The plateau stress is less strain-rate sensitive for small coating thicknesses. At higher coating thicknesses, rate effects occur. This rate dependency is also reflected in the energy absorption capacity, outlined in Fig. 5. For both strain rates, a linear increase of the energy absorption capacity per volume with increasing coating thickness can be observed. The rate dependency is pronounced by the higher energy absorption capacity under higher strain-rates. The pure aluminium foams show no strain-rate sensitivity, whereas the strain-rate sensitivity of the Ni/Al hybrid foams increases with increasing coating thickness. For open-cell metal foams, strain rate dependency can only arise by micro-inertia effects and strain-rate sensitive strut material. In contrast to aluminium, nickel is strain-rate sensitive and hence, increasing the amount of nickel leads to a pronounced strain-rate dependency of the Ni/Al hybrid foams. The change in the deformation mechanism for the hybrid foams, expressed by the higher stiffness and the

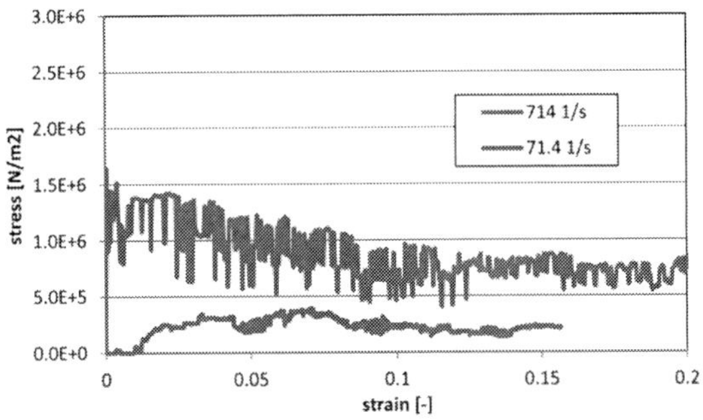

Figure 6. Influence of strain rate on stress-strain curve for coarse mesh (170 μm) – detail for small strains.

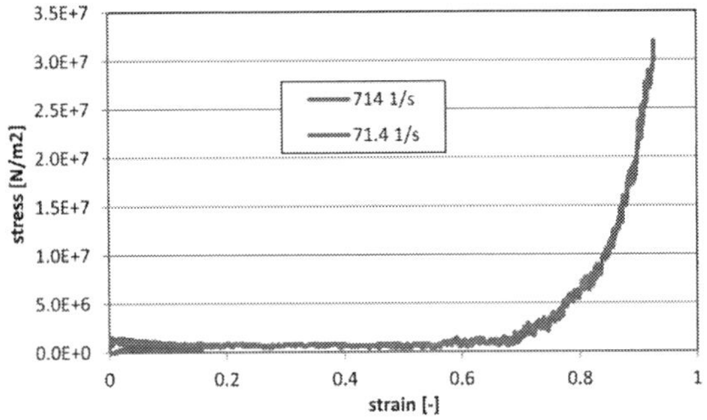

Figure 7. Influence of strain rate on stress-strain curve for coarse mesh (170 μm).

appearance of a plastic collapse stress peak under dynamic load conditions, indicates that micro-inertia effects become important for the Ni/Al hybrid foams, but not for the pure aluminium foams.

3.3. Numerical investigation of strain-rate dependency

The first calculation with the smaller strain rate is run up to a displacement of about 2 mm corresponding to a strain of about 0.15 is reached. The classical foam behaviour can be observed (Fig. 6). After a linear increase of the stress, a plateau in the stress-strain curve can be observed. The calculation is not done long enough to also see the steepening up of the curve. The calculation was performed on 6 processors of a very powerful desktop PC and took about 12 days.

The calculation with a strain rate of about 714 1/s was done until a much higher strain was reached. It can be observed (see Figs. 6 and 7) that the stress increases much faster at the beginning of the stress-strain relation. In addition, the curve is much more scattered due to the oscillations of the loading wave. Figure 7 shows also that the model can represent quite well the steepening up of the stress-strain curve after a certain point.

Figure 8 shows the displacement history of the virtual experiment. It can be observed that, after the first 0.25 ms, some of the struts have failed and the first strut layers are compacted. This mechanism proceeds unit about 1 ms. At that time, the sample has somehow reached the point at which the stiffness is increasing again. At 1.25 ms, the

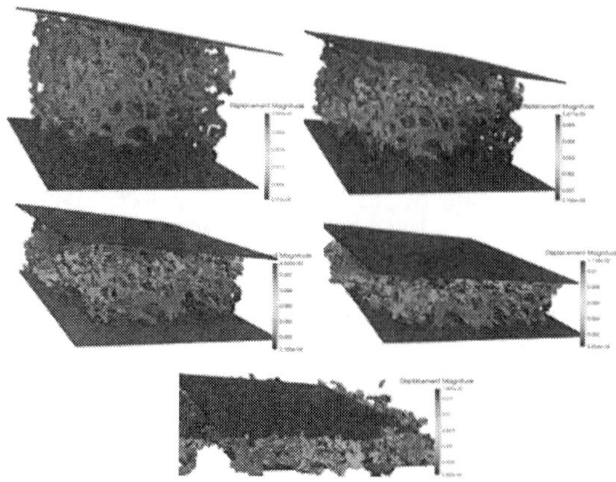

Figure 8. Displacement history for strain-rate of about 714 1/s: 0.25 ms, 0.5 ms, 0.75 ms, 1.0 ms, 1.25 ms.

sample is strongly compacted. The stiffness of the material should not be higher than the one from pure aluminium, which is still the case. Further calculation is very time consuming due to the large number of contacts between all elements.

4. Conclusion

In this contribution, the strain-rate dependence of pure aluminium and Ni/Al hybrid metal foams was studied by experimental und numerical dynamic compression tests. The SHPB tests outline a significant strain-rate sensitivity for the hybrid foams, which increases with the coating thickness, but not for the aluminium foams. For the evaluation of a micro model as virtual testing laboratory, μCT measurements were performed to determine the microstructure of the foams, since micro-inertia effects comming from the microstructure are crucial for strain-rate dependence. The numerical simulations show a strain-rate sensitivity for the alumminium foams, too. Hence, the strain-rate dependence of the Ni/Al hybrid foams results on the one hand from micro-inertia effects and mainly on the other hand from the strain-rate sensitivity of the coating metal nickel.

References

[1] M.F. Ashby, Metall. Trans. A **14**, 1755–1769 (1983)

[2] M.F. Ashby, A. Evans, N.A. Fleck, L.J. Gibson, J.W. Hutchinson, H.N.G. Wadley, *Metal foams: design guide* (Butterwoth-Heinemann, Oxford, 2000) 1–40

[3] Y. Boonyongmaneerat, C.A. Schuh, D.C. Dunand, Scripta Mater. **59**, 336–339 (2008)

[4] B.A. Bouwhuis, J.L. McCrea, G. Palumbo, G.D. Hibbard, Acta Mater. **57**, 4046–4053 (2009)

[5] A. Jung and M.R. Koblischka and E. Lach and S. Diebels and H. Natter, Int. J. Mater. Sci. **2**, 97–107 (2012)

[6] A. Jung, H. Natter, R. Hempelmann, S. Diebels, M. R. Koblischka, U. Hartmann E. Lach, ECS Transactions **25**, 165–172 (2010)

[7] A. Jung, H. Natter, S. Diebels, E. Lach R. Hempelmann, Adv. Eng. Mater. **13**, 23–28 (2011)

[8] A. Jung and H. Natter and R. Hempelmann and E. Lach, Metal foams, EP 2261398 A1

[9] D. Vavrik and J. Jakubek, Nucl. Instrum. Meth. A. **607**, 212–214 (2009)

[10] O. Jirousek, T. Doktor, D. Kytyr, P. Zlamal, T. Fila, P. Koudelka, I. Jandejsek and D. Vavrik, J. Instrum. **8**(2), C02012 (2013)

[11] W. Riedel, M. Wicklein, K. Thoma, Int. J. Impact Eng. **35**(3), 155–171 (2008)

[12] M. Larcher, N. Gebbeken, EPJ Web of Conferences **26** (2012)

[13] Martin Larcher, George Solomos, Folco Casadei, Norbert Gebbeken, Int. J. Impact Eng. **39**, 42–50 (2012)

[14] F. Casadei. A Hierarchic Pinball Method for Contact-Impact in Fast Transient Dynamics. VI Congresso Nazionale della Societ'a Italiana di Matematica Applicata e Industriale (SIMAI 2002), Chia (Cagliari), Italy, 27–31 May 2002

EPJ Web of Conferences **94**, 04031 (2015)
DOI: 10.1051/epjconf/20159404031

Effect of structurally-induced lateral confinement on split Hopkinson pressure bar test specimens of concrete-like materials

E.A. Flores-Johnson[1] and Q.M. Li[2,a]

[1]School of Civil Engineering, The University of Sydney Building J05, Sydney NSW 2006, Australia
[2]School of Mechanical, Aerospace and Civil Engineering, The University of Manchester, Manchester M13 9PL, UK

Abstract. In dynamic testing of concrete-like materials, there is a need in distinguishing structural effects from genuine strain-rate effects. In this paper, this generic problem is studied by numerical simulations based on a phenomenological material model available in the commercial finite element (FE) code Abaqus. The numerical results show that the increase of the dynamic increase factor (DIF) with the increase of strain-rate in concrete-like materials in a Split Hopkinson Pressure Bar (SHPB) test is a phenomenon related not only to material strain-rate effects but also to structural effects. It was found that dilation, surface friction and lateral inertia cause lateral confinement, which enhances DIF when the strain-rate is greater than a transition strain-rate in the order of $10^2\,s^{-1}$. Although, genuine strain-rate effect may exist as suggested by meso-scale simulations in previous investigations, the findings in this study show that structural effects have a significant contribution to the increase of DIF, and therefore, it is necessary to correctly calibrate existing phenomenological models and interpret the results obtained from split Hopkinson pressure bar (SHPB) tests.

1. Introduction

It is well-known that compressive properties of concrete and similar brittle materials, e.g., mortar or rock, exhibit strain-rate dependence [1–3]. The increase of the compressive strength in dynamic testing is commonly described by using the dynamic increase factor (DIF), which is defined by the ratio of the dynamic strength to the quasi-static strength in unconfined uniaxial compression [4]. The DIF of concrete-like materials in the high strain-rate range of 10^1–$10^3\,s^{-1}$ is often obtained by performing dynamic test using a split Hopkinson pressure bar (SHPB); however, it has long been debated whether the increase of compressive strength observed in SHPB tests of unconfined specimens is an intrinsic material property or is related to structural or inertial effects [4–6]. It has been shown that the induced lateral confinement in SHPB test, which causes increase of DIF, may be caused by friction between the bars and the specimen [5,7], large specimen diameters [8,9], and radial inertia [10,11]. The enhanced DIF observed when the induced lateral confinement is present should not be related to material property and the SHPB data should be correctly interpreted.

Some recent investigations using meso-scale modelling have shown that the enhancement of mechanical properties in dynamic testing is also related to intrinsic material properties. A recent study by Snozzi et al. [12] shows than the increase of strength in concrete with the increase of strain-rate is associated with the fast extension of micro-cracks, which is related to the intrinsic opening time in the cohesive law used to define the interface strength between mortar and aggregates. Qin and Zhang [13] modelled dynamic compression of concrete using the particle element method. They observed the formation of reticular and dispersed cracks at high strain-rate and related the increase of strength to higher force chain intensity and oblique bifurcations of the force chain distribution. In their study, the increase of strength was also attributed to higher frictional forces due to the increase of contact bond breaks [13]. Both studies agree with experiments reported by Ross [14] where crack pattern, density and distribution increase with the increase of strain-rate. Song and Lu [15] reported that the increase of DIF with strain-rate is related to the heterogeneity of multi-phase materials.

Although meso-scale models bring a better understanding of the mechanisms responsible for the increase of DIF with strain-rate in concrete-like materials, they are impractical for the macroscopic modelling of large engineering structures subjected to dynamic actions, e.g. blast and impact. In this case, a computationally-effective continuum model is needed for the macroscopic response. However, the phenomenological models should be able to capture material's mesoscopic and microscopic behaviours using parameters that have clear physical meanings.

Dilatancy of concrete-like materials refers to the volume increase with shear stress, which may result from the formation and growth of micro-cracks in compression [1,16]. This suggests that dilation may be related to the increase of DIF in concrete-like materials in a SHPB test because of the lateral confinement due to the volume increase. A recent numerical study by Elmer VII et al. [17] showed that the enhancement of concrete strength in SHPB tests is strongly related to shear dilation. They investigated the influence of the dilation parameter of the K&C model

[a] Corresponding author: qingming.li@manchester.ac.uk

Article available at http://www.epj-conferences.org or http://dx.doi.org/10.1051/epjconf/20159404031

Table 1. Geometrical details, material properties and material model parameters.

Material properties	Specimen [3]	Input/output bar [4]
Density ρ (kg/m^3)	2000	7800
Young's modulus E (GPa)	20	210
Poisson's ratio ν	0.2	0.3
Friction angle β(°)	0, 30, 50	–
Dilation angle ψ(°)	0, 30, 50	–
Geometrical details		
Diameter D (mm)	12, 20	20
Length L (mm)	6	1000

available in LS-DYNA and found that the DIF of concrete increases when the dilation parameter is increased.

As aforementioned, there is a need to improve our understanding of structural effect contributions to the observed strain-rate effects using phenomenological material models, but nonetheless with meaningful parameters that can be easily determined for engineering numerical simulations. In this paper, this problem is addressed by presenting numerical simulations dealing with the effect of induced lateral confinement using a phenomenological material model available in the FE code Abaqus.

This work focuses on a numerical study of a material model that takes into account dilation for continuum finite element modelling. A parametric study is performed to further investigate the effect of dilation, friction angle, friction coefficient and specimen dimensions on the increase of DIF of concrete-like materials in SHPB tests. FE simulations of SHPB test using Abaqus/Explicit employs the Extended Drucker-Prager material model. In all numerical simulations, strain-rate enhancement factor is not employed in order to study only structural effects. FE models are described in Sect. 2. Numerical results and discussion are presented in Sect. 3, which are followed by conclusions in Sect. 4.

2. Numerical simulations of SHPB tests

The FE simulations of SHPB tests in this work were performed with a similar setup as that presented by Grote et al. [3]. The geometrical details of the specimen, input and output bars are depicted in Table 1. The numerical simulations were performed using the commercial (FE) software Abaqus/Explicit [18]. To reduce computational costs, a representative quarter-model was constructed using 8-node brick elements with reduced integration (Fig. 1). The elements of the specimen and the elements in the impact areas of the bars (Fig. 1) have an average size of $0.2 \times 0.2 \times 0.2\,\text{mm}^3$. The total number of elements in a 12-mm specimen is 17250, while it is 43365 for each bar. A mesh sensitivity analysis showed that smaller element size did not provide an increase in accuracy.

The input and output bars of the SHPB were modelled using an isotropic elastic material model with input parameters showed in Table 1. The specimen was modelled in Abaqus/Explicit using the extended Drucker-Prager material model with a linear yield criterion with parameters showed in Table 1 and hardening input curve

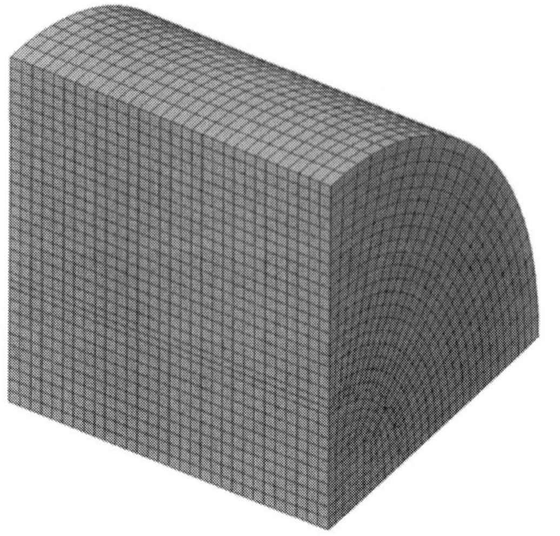

Figure 1. Finite element mesh of the split Hopkinson pressure bar test specimen.

showed in Fig. 2. An overview of the model is described in the next section.

2.1. Extended Drucker-Prager constitutive model

The extended Drucker-Prager material with linear yield criterion is given by [18],

$$F = t - p\tan\beta - d = 0 \tag{1}$$

where β is friction angle of the material, which is the slope of the linear yield surface in the p–t stress plane. The cohesion of the material d is given by $d = \sigma_c(1 - (1/3)\tan\beta)$, where σ_c is the compressive yield stress. The deviatoric stress measure t is defined as,

$$t = \frac{1}{2}q\left[\frac{1}{K} - \left(1 - \frac{1}{K}\right)\left(\frac{r}{q}\right)^3\right] \tag{2}$$

where q is the Mises equivalent stress, r is the third invariant of the deviatoric stress and K is the ratio of the yield stress in triaxial tension to the yield stress in triaxial compression [18]. The parameter K has to be within the range $0.788 \leq K \leq 1$ to ensure the convexity of the yield surface. The flow rule of the extended Drucker-Prager model is given by,

$$d\boldsymbol{\varepsilon}^p = \frac{d\varepsilon^p}{c}\frac{\partial G}{\partial \boldsymbol{\sigma}} \tag{3}$$

where ε^p is the equivalent plastic strain, c is a constant given by $c = 1 - (1/3)\tan\psi$ if hardening is defined in uniaxial compression and $\boldsymbol{\sigma}$ is the stress tensor. The flow potential G is defined as,

$$G = t - p\tan\psi \tag{4}$$

where p is the equivalent pressure stress and ψ is the dilation angle. If $\psi = 0$, the plastic deformation is incompressible and when $\psi \geq 0$ the material dilates.

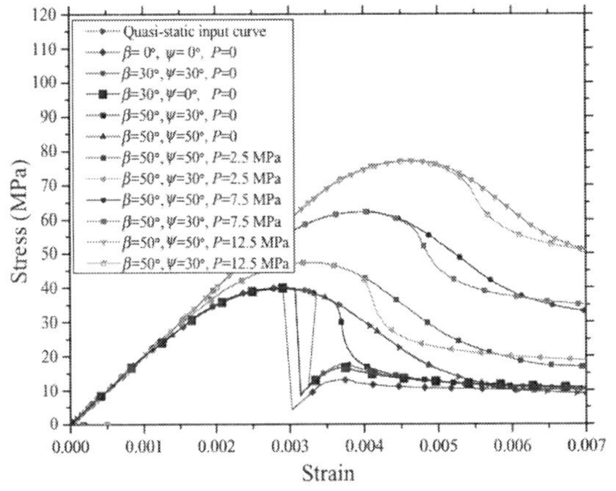

Figure 2. Stress-strain curves for different input parameters and lateral confinements (P) at $\dot{\varepsilon} = 8 \times 10^{-2}\,\mathrm{s}^{-1}$ for the Drucker-Prager model.

When $\beta = \psi$ and $K = 1$, a fully associated flow is obtained and the original Drucker-Prager model is reinstated [18]. For concrete-like materials, it has been found that β and ψ are in the ranges of 20–60° and 20–50°, respectively [19–21].

2.2. Numerical analyses of SHPB test

A parametric study comprising the effects of the friction angle, dilatancy, friction coefficient and specimen diameter on the DIF of concrete subjected to strain-rates in the range of 10^1–$10^3\,\mathrm{s}^{-1}$ is performed. For the numerical simulations, three friction angle values β (0°, 30°, 50°), three dilation angles values ψ(0°, 30°, 50°) and three friction coefficient values μ (0, 0.1, 0.2) are used.

To achieve different strain-rates, trapezoidal stress pulses with different rising time and intensity are applied to the end of the incident bar. The rising time varies from 60 to 200 μs while the duration and unloading times are fixed to 20 μs as they do not have significant effect on the strain-rate and only strain at strength is of interest in this study. The intensity of the pulse varies from 50 to 900 MPa.

No strain-rate enhancement factor is employed in both material model and software in order to study only structural effects. To measure the axial stress and strain rate in the specimen in the numerical simulations, the formulae described by Lu and Li [22] are used. These formulae use the nodal forces and velocities at the specimen interfaces to obtain the time histories of the average engineering stress $\sigma(t)$ and average strain-rate $\dot{\varepsilon}(t)$over the specimen, i.e.,

$$\sigma(t) = \frac{P_I(t) + P_O(t)}{2A_S} \qquad (5)$$

$$\dot{\varepsilon}(t) = \frac{V_I(t) - V_O(t)}{L_S} \qquad (6)$$

where $P_I(t)$ and $V_I(t)$ are the force and velocity time histories at the interface between the input bar and the specimen, respectively; and $P_O(t)$ and $V_O(t)$ are the force and velocity time histories at the interface between the output bar and the specimen, respectively. The constant A_S is the cross-sectional area of the specimen and L_S is the specimen length.

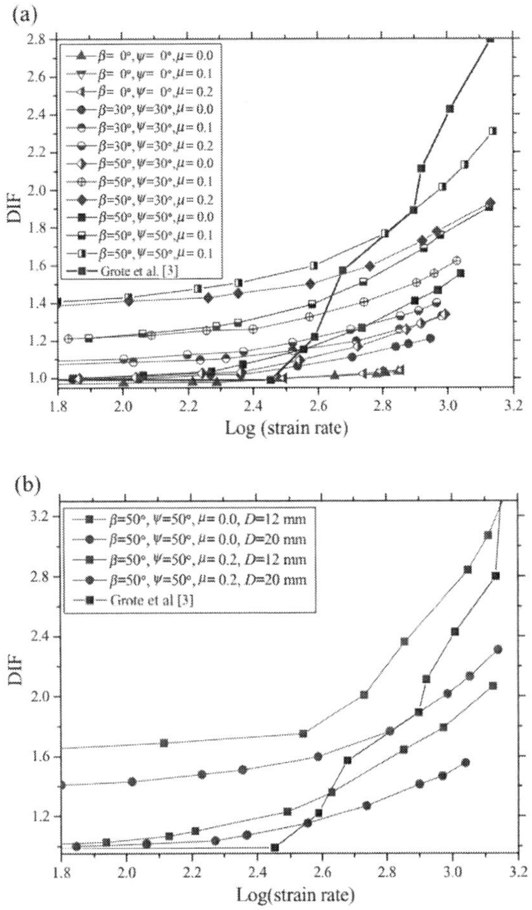

Figure 3. Numerical results of SHPB test using different parameters: (a) Effect of dilation parameter β, dilation angle ψ and friction coefficient μ; (b) Effect of specimen diameter D.

It is known that if the peak stress in the specimen is reached before the stress equilibrium is reached, the test may not be valid. Therefore, for all the simulations, the dynamic stress equilibrium was verified using the parameter $R(t)$ defined as [22,23],

$$R(t) = 2\left|\frac{P_I(t) - P_O(t)}{P_I(t) + P_O(t)}\right|. \qquad (7)$$

It is generally accepted that when there is difference of 5% or less in stress across the SHPB specimen with respect to the average stress, i.e., $R(t) \leq 0.05$, the equilibrium has been reached [24].

3. Numerical results and discussion

Figure 2 shows quasi-static compression stress-strain curves performed at low strain-rate ($\dot{\varepsilon} = 8 \times 10^{-2}\,\mathrm{s}^{-1}$). It can be seen that although parameters β and ψ define the shape of the softening part of the curves as expected, these parameters have negligible influence on the

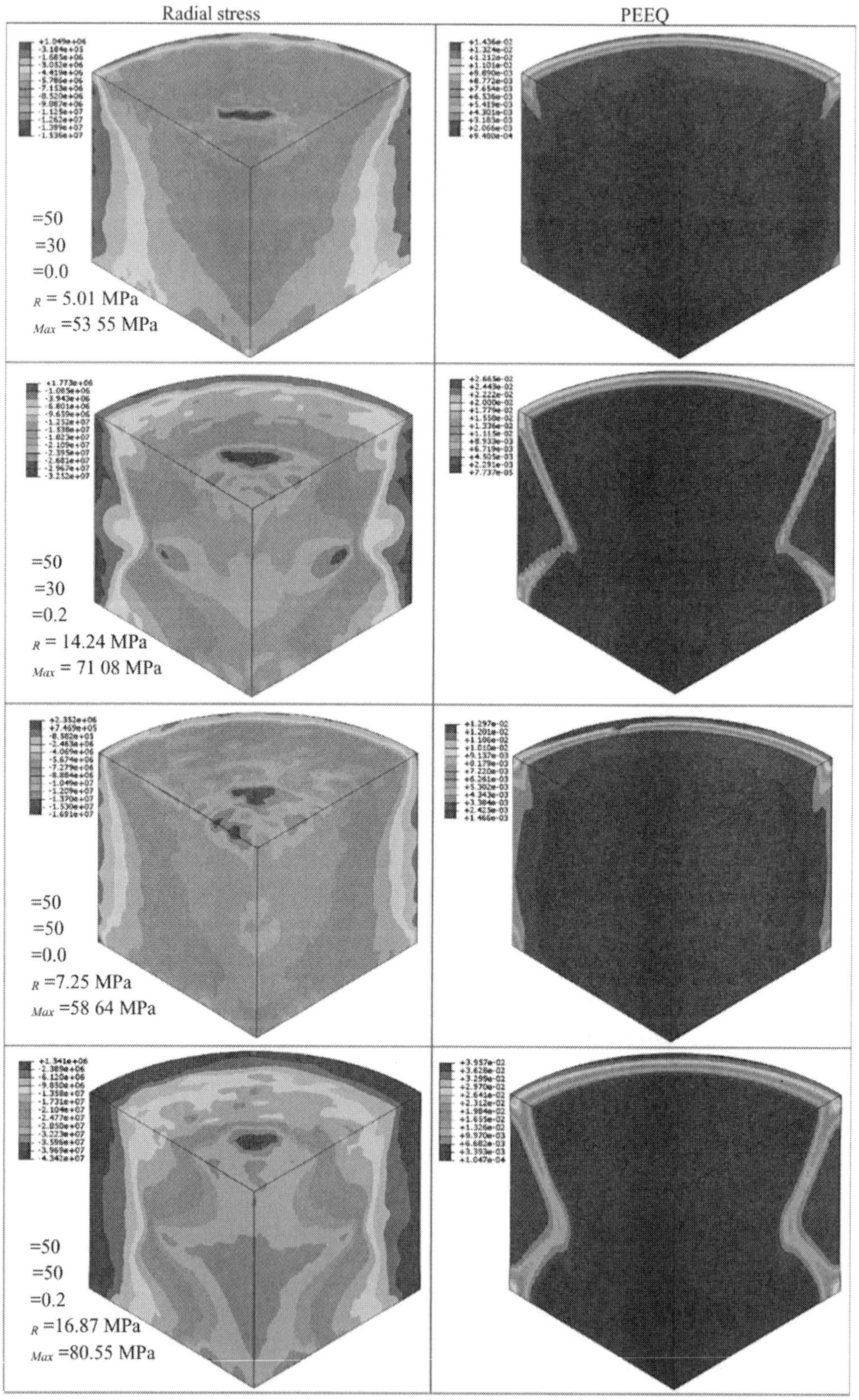

Figure 4. Contour plots of radial stress and PEEQ for a log(strain-rate)=3 for various material parameters sets.

compressive strength, which remains constant at 40 MPa. However, it can be seen in Fig. 3(a) (SHPB simulations) that parameters β and ψ largely influence the dynamic response of mortar as indicated by the increase of DIF when the strain-rate increases, particularly for strain-rates higher than $10^2\,s^{-1}$. This results indicate that a good calibration of parameters β and ψ should be performed when strain-rates higher than $10^2\,s^{-1}$ are simulated.

The effect of friction on the dynamic response in SHPB tests is well-known, which leads to errors in the DIF [4,25,26], particularly in concrete-like materials due to their coarser surface when compared to metals [27] and therefore friction should not be completely neglected even though lubricants are used. Figure 3(a) shows the results of the effect of the friction coefficient μ on the dynamic response of mortar for various β and ψ. It can be seen that when $\beta = \psi = 0$, the increase of μ does affect the dynamic response of the material; however when β and ψ increase, the DIF increases with the increase of μ. This can be explained by the fact that when β increases the material becomes more pressure dependant, and therefore, more sensitive to the induced lateral confinement due to the friction and dilation. Results from Grote et al. [3] are included in Fig. 3 for comparison. It can be seen that when $\beta = \psi = 50°$ and $\mu = 0.2$ the numerical predictions are closer to the experimental results; however, this value is very large because $\mu \approx 0.05$–0.1 when lubricants are used in SHPB tests. Unfortunately, the friction coefficient is not given in Grote et al. [3] neither it is mentioned if lubricants were used. It is believed that the increase of DIF shown in Grote et al. [3] is partially due to genuine strain-rate effects and partially due to structural effects. Figure 4 shows contour plots of radial stress and equivalent plastic strain (PEEQ) for specimens at high strain-rates of approximately $10^3\,s^{-1}$ for various sets of β, ψ and μ. It can be observed that the lateral confinement, which can be expressed by the compressive radial stresses, increases when the friction coefficient is increased as expected. It can also be seen that for specimens with same interface friction coefficient, the lateral confinement increases when the dilation parameter is increased, and the rate of the increase is higher for higher strain-rates.

From Fig. 2, it can be seen that the peak stresses due to the lateral confinement P are somewhat comparable with the peak stresses in Fig. 4 for the levels of average radial stresses that act as lateral confinement in the samples. It is observed that PEEQ also increases when the dilation parameter is increased (for example, in Fig. 4, compare the PEEQ contours when $\psi = 30°$ and $\mu = 0.2$, with the PEEQ contours when $\psi = 50°$ and $\mu = 0.2$). This confinement effect is produced in the hourglass-shaped zone showed in Fig. 5, which is delimited by the fault regions with high PEEQ and volumetric expansion or dilation. This phenomenon has been observed by Elmer VII et al. [17] in numerical simulations using the K&C model in LS-DYNA. It is worth mentioning that the dilation parameter is crucial in the numerical prediction of concrete strength particularly for strain-rates higher than $10^2\,s^{-1}$ and a good calibration of this parameter should be performed to avoid an over prediction of strength when strain-rate effect is incorporated into the model.

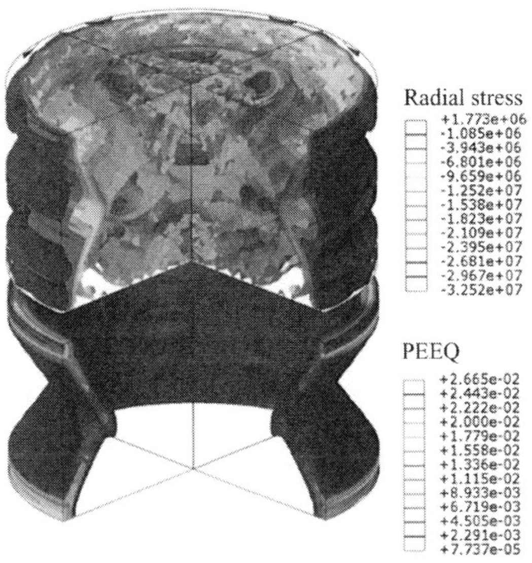

Figure 5. Contour plots of radial stress and PEEQ for a log(strain-rate)= 3 and $\beta = 50°$, $\psi = 30°$ and $\mu = 0.2$.

Figure 3(b) shows numerical simulations for specimens with diameters of 12 mm and 20 mm. It can be seen that an increase in the specimen diameter while keeping the same length (6 mm) leads to an increase of DIF which agrees with previous research [8,27]. This clearly demonstrates that inertia has an effect on the dynamic response, which should not be considered as genuine strain-rate effect.

4. Conclusions

The numerical study conducted in this paper shows that the increase of the dynamic increase factor (DIF) with strain-rate for concrete-like materials in SHPB tests is a complex phenomenon related not only to material strain-rate effects but also to structural effects. It is demonstrated that several factors including dilation, friction coefficient and lateral inertia may contribute to lateral confinement, which is the main mechanism in the increase of DIF, particularly in the strain-rate range of 10^2–$10^3\,s^{-1}$. Although, some genuine strain-rate effects may exist as suggested by meso-scale simulations in previous investigations, the findings in this study show that structural effects have a significant contribution to the increase of DIF. Therefore, an effort to correctly calibrate existing phenomenological models and interpret the results obtained from SHPB tests should be done to include genuine strain-rate effect on DIF in continuum macroscopic models.

References

[1] P. Bischoff, S. Perry, Mater. Struct. **24**, 425 (1991).
[2] C.A. Ross, J.W. Tedesco, S.T. Kuennen, ACI Mater. J. **92** (1995).
[3] D.L. Grote, S.W. Park, M. Zhou, Int. J. Impact Eng. **25**, 869 (2001).

[4] Q.M. Li, H. Meng, Int. J. Solids Struct. **40**, 343 (2003).
[5] J.F. Georgin, J.M. Reynouard, Cem. Concr. Compos. **25** (2003).
[6] P. Rossi, F. Toutlemonde, Mater. Struct. **29** (1996).
[7] M.M. Al-Mousawi, S.R. Reid, W.F. Deans, Proc. Inst. Mech. Eng. Part C: J. Mech. Eng. Sci. **211** (1997).
[8] M. Zhang, H.J. Wu, Q.M. Li, F.L. Huang, Int. J. Impact Eng. **36**, 1327 (2009).
[9] X.Q. Zhou, H. Hao, Int. J. Solids Struct. **45** (2008).
[10] G. Cusatis, Int. J. Impact Eng. **38**, 162 (2011).
[11] Z.C. Mu, A.N. Dancygier, W. Zhang, D.Z. Yankelevsky, Int. J. Impact Eng. **49**, 91 (2012).
[12] L. Snozzi, A. Caballero, J.F. Molinari, Cem. Concr. Res. **41** (2011).
[13] C. Qin, C. Zhang, Int. J. Impact Eng. **38** (2011).
[14] C.A. Ross, *Crack patters resulting from high strain-rate tests on concrete*, 1992, Air Force Engineering and Services Center: Tyndall AFB, Florida, USA.
[15] Z. Song, Y. Lu, Int. J. Impact Eng. **46** (2012).
[16] C. Noble, E. Kokko, I. Darnell, T. Dunn, L. Hagler, L. Leininger, *Concrete model descriptions and summary of benchmark studies for blast effects simulations*, 2005, Lawrence Livermore National Laboratory (LLNL), Livermore, CA.
[17] W. Elmer VII, E. Taciroglu, L. McMichael, Int. J. Impact Eng. **45**, 1 (2012).
[18] Simulia, *Abaqus Analysis User's Manual (Version 6.11)*, 2011, SIMULIA.
[19] S.W. Park, Q. Xia, M. Zhou, Int. J. Impact Eng. **25**, 887 (2001).
[20] J.-F. Jiang, Y.-F. Wu, Int. J. Solids Struc. **49**, 445 (2012).
[21] B.M. Luccioni, M. Luege, Int. J. Impact Eng. **32**, 1248 (2006).
[22] Y.B. Lu, Q.M. Li, Int. J. Protective Struct. **1**, 363 (2010).
[23] G. Ravichandran, G. Subhash, J. Am. Ceram. Soc. **77**, 263 (1994).
[24] B. Song, W. Chen, Lat. Am. J. Solids Struct. **2**, 113 (2005).
[25] R.S. Hartley, T.J. Cloete, G.N. Nurick, Int. J. Impact Eng. **34**, 1705 (2007).
[26] D.J. Kim, K. Sirijaroonchai, S. El-Tawil, A.E. Naaman, Int. J. Impact Eng. **37**, 141 (2010).
[27] Q.M. Li, Y.B. Lu, H. Meng, Int. J. Impact Eng. **36**, 1335 (2009).

EPJ Web of Conferences **94**, 04032 (2015)
DOI: 10.1051/epjconf/20159404032

Indentation of aluminium foam at low velocity

Xiaopeng Shi[1], Yinggang Miao[1], Shuangyan Liu[1], Yulong Li[1,a], and Guoxing Lu[2]

[1] School of Aeronautics, Northwestern Polytechnical University, Xi'an 710072, China
[2] Faculty of Science, Engineering and Technology, Swinburne University of Technology, Hawthorn, Vic 3122, Australia

Abstract. The indentation behaviour of aluminium foams at low velocity (10 m/s ~ 30 m/s) was investigated both in experiments and numerical simulation in this paper. A flat-ended indenter was used and the force-displacement history was recorded. The Split Hopkinson Pressure bar was used to obtain the indentation velocity and forces in the dynamic experiments. Because of the low strength of the aluminium foam, PMMA bar was used, and the experimental data were corrected using Bacon's method. The energy absorption characteristics varying with impact velocity were then obtained. It was found that the energy absorption ability of aluminium foam gradually increases in the quasi-static regime and shows a significant increase at ~10 m/s velocity. Numerical simulation was also conducted to investigate this process. A 3D Voronoi model was used and models with different relative densities were investigated as well as those with different failure strain. The indentation energy increases with both the relative density and failure strain. The analysis of the FE model implies that the significant change in energy absorption ability of aluminium foam in indentation at ~10 m/s velocity may be caused by plastic wave effect.

1. Introduction

As a lightweight energy-absorbing material, aluminium foam has been widely used in automotive industry and aerospace engineering [1] Researchers showed more and more interests in mechanical properties of this material, especially the energy-absorbing mechanism. When this material is frequently used in blast-protected devices [2], indentation becomes a potential mode of failure. Ramachandra et al. [3,4] experimentally investigated the indentation behaviour of an aluminium foam under dynamic loads. In their research, drop-tower experiments were used to investigate the energy-absorption ability of aluminium foams under a velocity of 3~30 m/s. It was found that there is significant change in energy absorption ability of aluminium foam in indentation at ~10 m/s. But the experimental data at velocity of 3~30 m/s were limited, and the indentation velocity was not constant in drop-tower tests. Besides, the role of the tearing energy in indentation at low velocity is still unknown. In this paper, the SHPB (Split Hopkinson Pressure bar) was used in indentation tests and an FEM model was established to investigate the energy absorption ability at the velocity of 10~30 m/s. Empirical formulas were obtained to describe the relationship between the indentation energy and the indentation velocities.

2. Materials and method

2.1. Materials and specimens

Commercial closed-cell aluminium foam, ALPORAS, was used in this study. This foam was made from aluminium with 1.5% of Ca and 1.5% of Ti as well. The average pore size is ~3.5 mm and the nominal relative density of this material is ~0.09 as provided by the manufacturer. A typical macroscopic view of this material is shown in Fig. 1.

The dynamic indentation tests were conducted on blocks of 20 × 35 × 50 mm. A steel plate with thickness of 2 mm was used as indenter and was attacthed to the end of incident bar of SHPB system. To eliminate friction between the specimen and the indenter, the shank of indenter was chamfered at a small angel. The experimental set-up is shown in Fig. 2.

The projected area of the indenter is 20 mm × 20 mm, which covers ~30 cells to ensure the measured data represent the average response of this material. For comparison, the dynamic compression specimen is designed as a 20 × 20 × 20 mm block. Both indentation and compression specimen are shown in Fig. 3.

2.2. Experimental method

SHPB system was used in both the indentation and compression experiments. The conventional SHPB analysis is based on the one-dimensional elastic wave propagation assumption. The incident pulse $\varepsilon_i(t)$, the reflected pulse $\varepsilon_r(t)$ and the transmission pulse $\varepsilon_t(t)$ can be recorded by the strain gauges attached to the incident bar and the transmission bar. According to one-dimensional stress wave theory, these signals can be shifted to bar-specimen interfaces. Thus, the velocities and the forces at both faces of the sample can be obtained from the following equations:

$$V_{input} = C_0[\varepsilon_i(t) - \varepsilon_r(t)] \quad (1)$$

$$V_{output} = C_0\varepsilon_t(t) \quad (2)$$

[a] Corresponding author: liyulong@nwpu.edu.cn

Article available at http://www.epj-conferences.org or http://dx.doi.org/10.1051/epjconf/20159404032

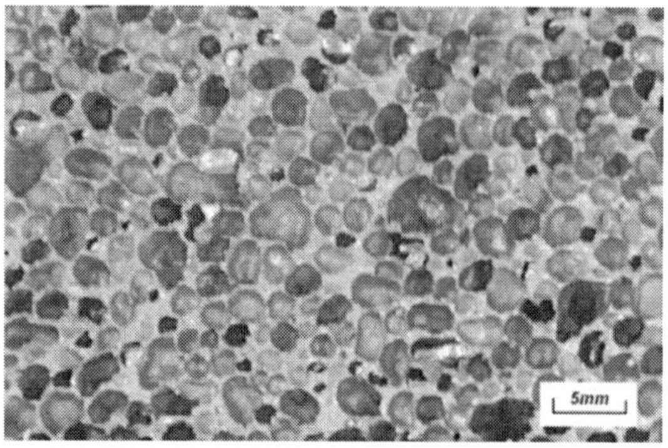

Figure 1. Macroscopic view of ALPORAS foam.

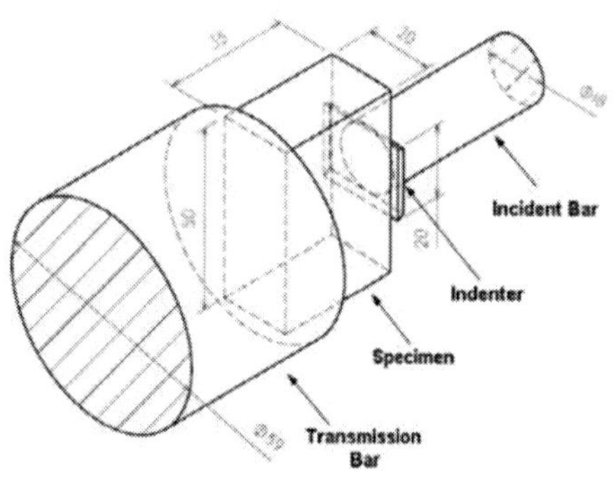

Figure 2. Sketch of the set-up for dynamic indentation test. The dimensions are in mm.

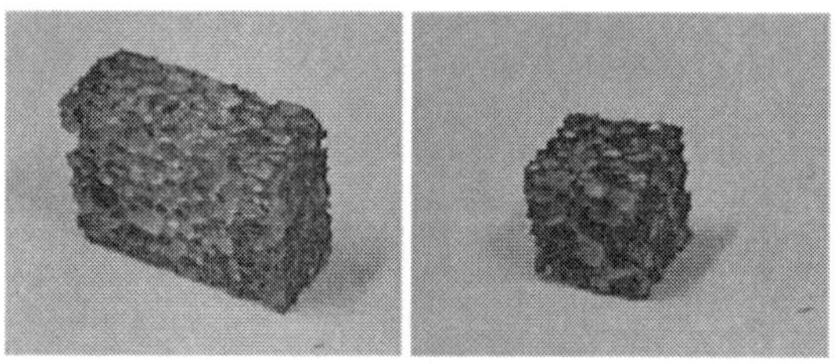

Figure 3. The specimen for indentation test (left) and the specimen for compression test (right).

$$F_{input} = ES_i[\varepsilon_i(t) + \varepsilon_r(t)] \tag{3}$$

$$F_{output} = ES_t\varepsilon_t(t) \tag{4}$$

where V_{input}, V_{output}, F_{input}, F_{output} are particle velocities and forces at the interfaces, C_0, E, S_i and S_t are, respectively, the longitudinal wave speed in the pressure bars, Young's modulus, the cross-sectional area of incident bar and the cross-sectional area of the transmission bar.

Due to the low strength of cellular materials, the signals from strain gauges on metallic bars would be weak, which leads to a low signal-noise ratio. PMMA bars were used in this study to avoid this problem. The incident bar is 2 m in length with a diameter of 18 mm and the transmission bar is 1 m in length with a diameter of 59 mm. However, as PMMA is a viscoelastic material and the wave dispersion effect cannot be neglected when stress wave propagates in the bar, the signals need correction before being used. Bacon [5] developed an experimental method to correct the dispersion effect. If the lateral motion can be neglected in a straight cylindrical bar, the normal stress $\sigma(x, t)$ and the longitudinal strain $\varepsilon(x, t)$ are related to the axial displacement $u(x, t)$ at cross-section x and at time t by

$$\frac{\partial\sigma(x,t)}{\partial x} = \rho\frac{\partial^2 u(x,t)}{\partial t^2} \tag{5}$$

$$\varepsilon(x,t) = \frac{\partial u(x,t)}{\partial x}. \tag{6}$$

Equations (5) and (6) can be written in one equation in the Fourier domain:

$$\frac{\partial^2\tilde{\sigma}(x,\omega)}{\partial x^2} = -\rho\omega^2\tilde{\varepsilon}(x,\omega) \tag{7}$$

where ρ, ω, $\tilde{\sigma}(x, \omega)$ and $\tilde{\varepsilon}(x, \omega)$ denote the mass density of the bar, the angular frequency, the Fourier transforms of the stress and strain, respectively.

Then, the linear viscoelastic behavior of the material can be expressed as follows:

$$\tilde{\sigma}(x,\omega) = E^*(\omega)\tilde{\varepsilon}(x,\omega) \tag{8}$$

where $E^*(\omega)$ is the complex Young's modulus. From Eqs. (7) and (8), the one-dimension equation of axial motion of a viscoelatic bar can be written as follows:

$$\left(\frac{\partial^2}{\partial x^2} - \gamma^2\right)\tilde{\varepsilon}(x,\omega) = 0 \tag{9}$$

where the propagation coefficient γ can be expressed as follows:

$$\gamma^2 = -\frac{\rho\omega^2}{E^*(\omega)}. \tag{10}$$

The general solution for equation (9) is:

$$\tilde{\varepsilon}(x,t) = \tilde{P}(\omega)e^{-\gamma x} + \tilde{N}(\omega)e^{\gamma x} \tag{11}$$

where the $\tilde{P}(\omega)$ and $\tilde{N}(\omega)$ are the Fourier transforms of the strains at $x = 0$ due to the waves traveling in the directions of increasing and decreasing x, respectively.

To determine the propagation coefficient γ, an experiment was conducted. A strain guage was attached in the middle of the incident bar where x was set 0. The incident bar was impacted by a very short projectile at its left end with its right end ($x = d$) free. Two strain signals without superpostion was recorded as $\varepsilon_1(t)$ and $\varepsilon_2(t)$, which refers to the incident wave and the reflected wave, respectively. According to Eq. (11):

$$\tilde{\varepsilon}_1 = \tilde{P} \text{ and } \tilde{\varepsilon}_2 = \tilde{N}. \tag{12}$$

Since the right end of the incident bar is free, the strain at the right end is 0:

$$\tilde{P}e^{-\gamma d} + \tilde{N}e^{\gamma d} = 0. \tag{13}$$

Then the propagation coefficient γ can be solved as:

$$e^{-\gamma 2d} = -\frac{\tilde{\varepsilon}_1(\omega)}{\tilde{\varepsilon}_2(\omega)}. \tag{14}$$

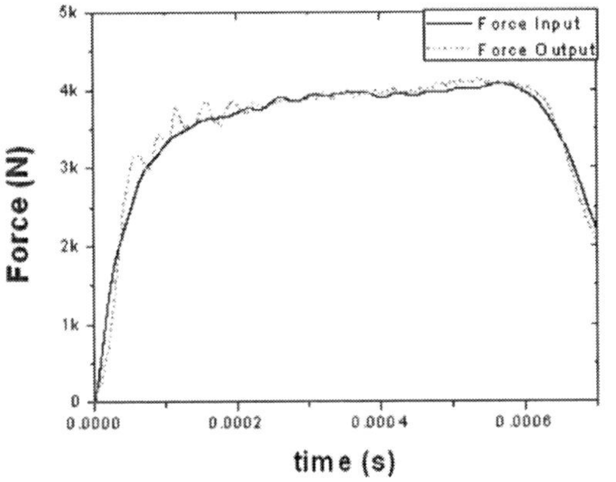

Figure 4. Comparision of input and output forces.

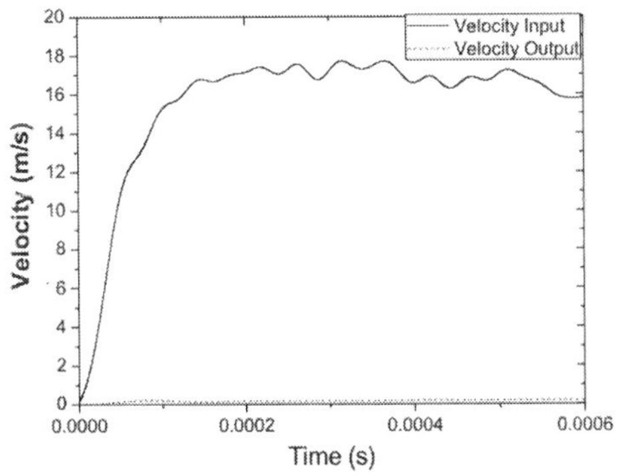

Figure 5. Velocities in a dynamic indentation test using SHPB.

A MATLAB program using the above method was written to correct the experiment result. To verify the correction result, a SHPB test without specimen was performed. Using Eqs. (3) and (4), the input and output forces can be obtained and these forces should be equal. Figure 4 shows the comparition bewteen these two forces and the result indicates that the correction method can be used in our study.

3. Experimental result and discussion

3.1. Compression and indentation forces

A typical pair of velocity histories of the dynamic indentation test is shown in Fig. 5. The particle velocity at the end of the transmission bar is nearly zero in the test. For velocity of the incident bar, there is a plateau of a duration of ~0.0005 s, which means the impact velocity is constant.

Both indentation and compression tests were conducted and typical curves of normalized force versus normalized indentation depth (compression depth for compression tests) are shown in Fig. 6 and Fig. 7 Note that the force F was normalized with respect to the contact area between the indenter and specimen. The normalized indentation depth was the indentation depth or compression depth d divided by the thickness of the specimen L.

Deshpande and Fleck [6] tested Alulight and Duocel foams but no velocity sensitivity was observed. While in

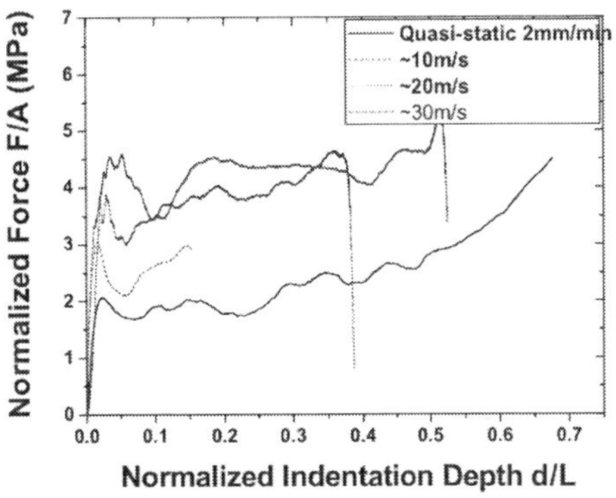

Figure 6. Normalized force vs. Normalized indentation depth for dynamic and quasi-static tests.

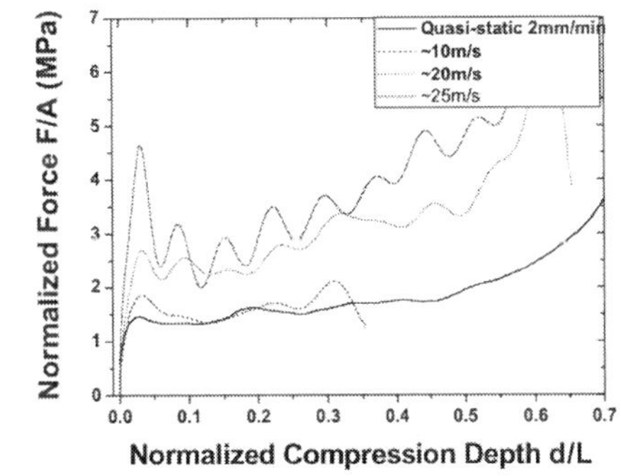

Figure 7. Normalized force vs. Normalized compression depth for dynamic and quasi-static tests.

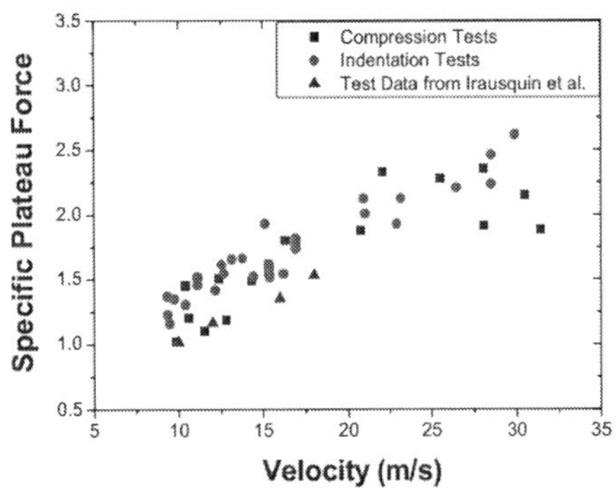

Figure 8. Specific force vs. Loading velocity in the tests.

this research, in both indentation tests and conpression tests, the plateau force shows great velocity sensitivity. Compared with the case of quasi-static loading, the plateau force in indentation at ~30 m/s velocity is nearly doubled. Note that the plateau forces in indentation tests are higher than in compression tests due to the forces needed to tear the materials beside the indenter. To show the relationship between the normalized plateau force and the loading velocity, the dynamic plateau forces diveded by quasi-stiac value of all specimens are plotted in Fig. 8, as well as the data from Irausquin et al. [7] They evaluated the dynamic behaviour of closed-cell aluminum foam ALPORAS 10% using SHPB.

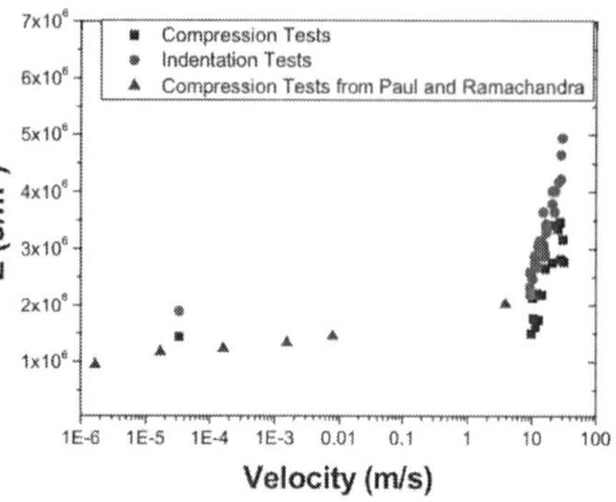

Figure 9. Energy absorbed per unit displaced volume, E vs. the displacement velocity.

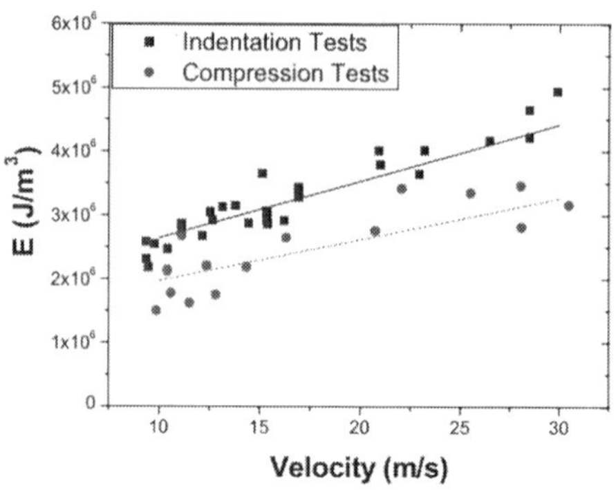

Figure 10. Energy absorbed per unit displaced volume, E vs. the displacement velocity at 10 m/s–30 m/s.

In both our tests and Irausquin's tests, there is a significant increase in the specific plateau forces with the increasing velocity between 10 m/s~30 m/s. As the the cell wall material is mainly made up of aluminum, which has no obvious rate sensitivity, the cell wall material cannot lead to this increase. On the other hand, according to Gibson and Ashby [8], Dashpande and Fleck's [6] research, the contribution of gas trapped in cells is so small as to be negligible for this increase. Microinertia effects of the cell walls [7,9] or the plastic wave [10] may cause this phenomenon. Further discussion will be conducted in FEM analysis.

3.2. Variation of Energy

The energy absorbed per unit displaced volume, E, is plotted as a function of the indenter velocity in Fig. 9. Quasi-static experiment result is also added. For comparison, Paul and Ramachandra's [4,11] compression tests results are plotted. Note that the size of indenter which Ramachandra used is different from this research, their indentation tests results are not presented here. For a given velocity within the range of 10^{-6}m/s ~30 m/s, E for indentation is significantly larger than the corresponding value for compression due to the mechanisms of tear and shear involved in indentation. And it can be observed that E varies gradually with the velocity of up to ~10 m/s and it increases dramatically above a velocity of ~10 m/s. This phenomenon implies different mechanism domainates in these two ranges. As researched by Paul [11], the gradual increase of E or plateau stress is due to the microinertia effects of the cell wall, another mechanism shoul be added for the dramatically increase above a velocity of ~10 m/s.

To observe the variation of energy absorbed above the velocity of 10 m/s clearly, Fig. 10 is presented. The energy absorbed per unit volume by indentation can be written as follows:

$$E_{ind} = E_{com} + E_{tear} \quad (15)$$

where E_{com} and E_{tear} are energy absorbed per unit volume by compression tests and by indenter tearing the material beside itself. Equation (15) is correct only if the plastic deformation area near the tearing surface is so small that the plastic energy in this area can be neglected. Figure 11 shows the typical specimen after indentation.

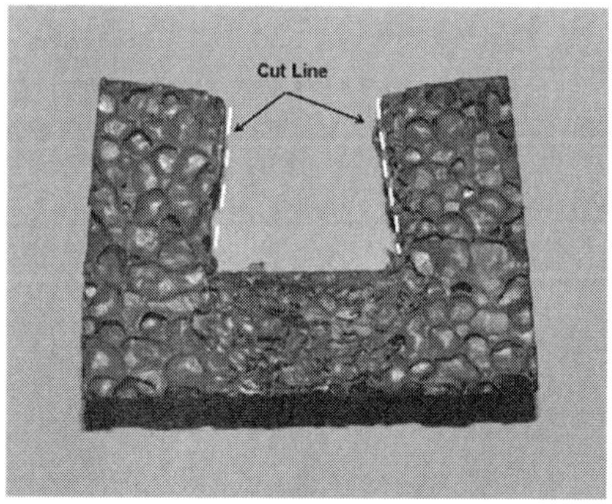

Figure 11. Dynamic indentation specimen after test.

The cells beside the cut lines fractured neatly which means little plasitc energy was dissipated near the cut line. On the other hand, in our previous work [12], no obvious plastic deformation was observed beside the indenter in quasi-static indentaion. As to dynamic indentation, this area should be smaller. So Eq. (15) is used to calculate the tearing energy γ in dynamic loading. Then

$$E_{ind} = E_{com} + \frac{2\gamma}{w} \quad (16)$$

where w is the width of the indenter. From Fig. 10, it can be observed that the difference between E_{ind} and E_{com} does not change much for velocity of 10m/s–30 m/s, which means the tearing energy γ does not change significantly in this velocity range. Tearing energy γcalculated here is ~15 kJ/m^3, which is a little higher than quasi-static value ~12 kJ/m^3 from our previous study [12].

4. FEM analysis

To investigate the reason for the increase of the energy absorbed by specimen at velocity of 10 m/s-30 m/s, an FEM analysis using ABAQUS Explict was conducted. A 3D Voronoi model was used to simulate the compression and indentation process. The geometric information of the voronoi model was obtained from a C++ based program,

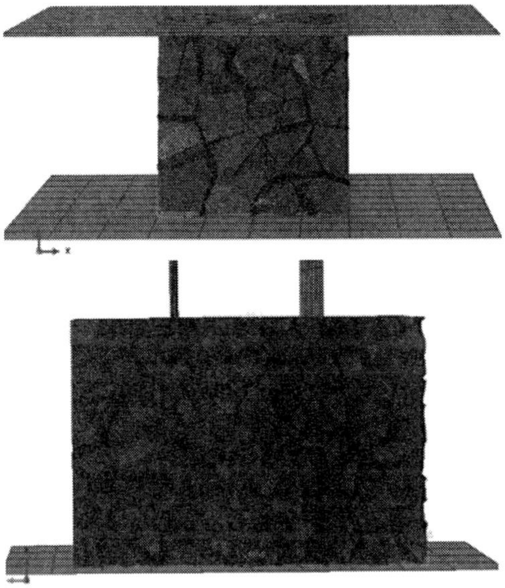

Figure 12. FEM model for compression (above) and indentation (below).

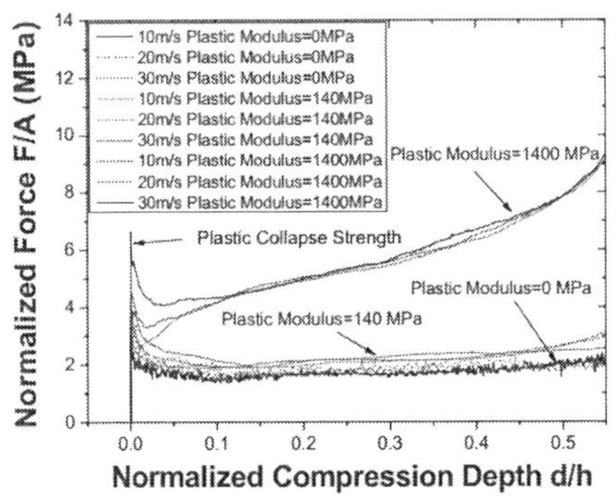

Figure 13. FEM result for compression.

Voro++ [13], and then imported to CAE program to establish an FEM model. The FEM model is shown in Fig. 12, where the indenter and the support was modelled as rigid body. The cell walls were made up of S4R and S3R elements and the thickness of the cell walls was ajusted to fit the mass of the foam specimen. Hard contact was used in the model. No friction and no slide between the foam and the support was considered in our simulation. The compression simulation was firstly conducted. An elastic-perfect plastic model and elastic-plastic hardening model were used for cell wall material respectively. The elastic modulus and the yield stress of the material are 69 GPa and 50 MPa respectively. For elastic-plastic hardening model, the two different plastic modulus was set as 140 MPa and 1400 MPa. The normalized force as the function of normalized compression depth is shown in Fig. 13. Note the the plastic modulus for elastic-perfect plastic model is 0.

Neither the plastic collapse strength nor the plateau force of the model with elastic-perfect plastic material increases with increasing velocity, which means that

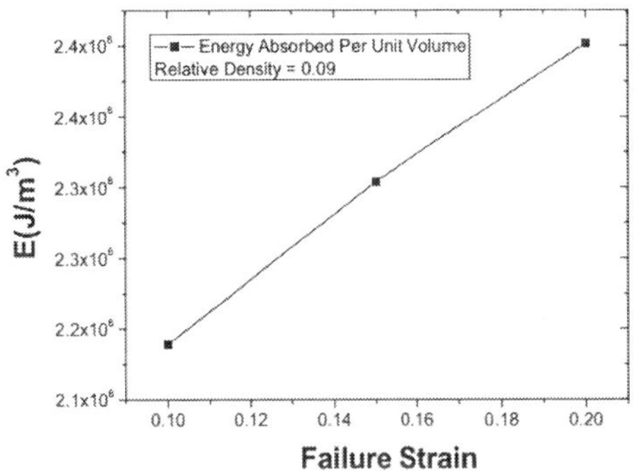

Figure 14. The energy absorbed per unit volume in indentation changing with failure strain.

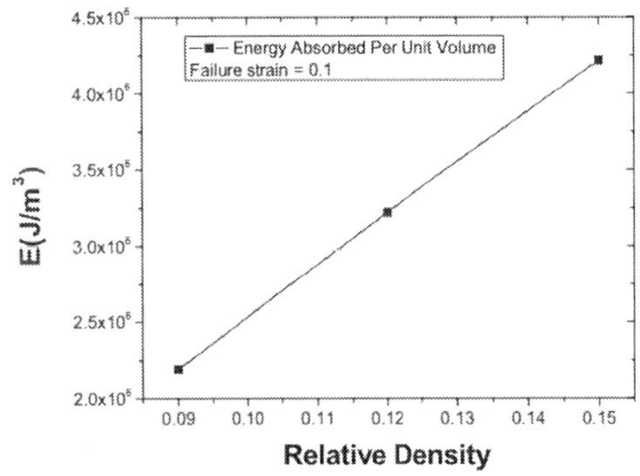

Figure 15. The energy absorbed per unit volume in indentation changing with relative density.

the inertia effects of the cell wall does not dominate in this velocity range. While the plastic hardening model shows great velocity sensitivity about the collaspe strength especially for the model with plastic modulus of 1400 MPa. The reason for this is the effect of plastic wave propagated in cell walls. The pasitc wave velocity C_p can be expressed as

$$C_p = \sqrt{E_p/\rho}. \tag{17}$$

Where E_p is the plastic modulus and ρ is the density of the material. Considering the plastic wave effect, the dynamic plastic collapse stress can be written as

$$\sigma_{pl} = \sigma_{qs} + \rho C_p V_i. \tag{18}$$

Where σ_{qs} and V_i are quasi-static plastic collapse stress and node velocity of the cell wall. Equation (18) explained the increase of velocity sensitivity with increasing plastic modulus of the material. It should be noted that the plateau strengh of all the models did not show the velocity sensitivity, the reason for this difference from the tests result needs further investigation.

To simulate the indentation process, the elements near the tearing surface were eliminated when the plastic strain reached ε_f. The indentation at 10 m/s was simulated and different ε_f was used to investigate the relationship between the failure strain and the indentation energy. With failure strain 0.1, models with relative density of 0.09, 0.12, 0.15 were also analysed. To simplify the analysis, elastic-perfectly plastic material model was used. The results are shown in Figs. 14 and 15. The indentation energy increases linearly with both failure strain of the elements and relative density of the foam. This increase may due to the linear increasing plastic work needed to tear the element with increasing failure strain and cell wall thichness.

5. Conclusions

In this paper, the indentation behaviour of aluminum foam at velocity of 10 m/s–30 m/s was investigated. SHPB was used to conduct the tests and 3D Voronoi foam model was used to simulate the compression and the indentation process. It was found that the compression and indentation energy absorbed per unit volume change slowly for the loading velocity below ~10 m/s and increases significantly at velocity of 10 m/s–30 m/s. FEM analysis implies different mechanism in these two ranges. The inertia effect of the cell wall dominates at first and then the plastic wave plays an important role. The relationship bewteen the energy aborption ability and the failure strain of the element as well as the relative density of the model was also investiged.Tthere is a linear relationship between the energy absorbed per unit volume in indentaion and both the failure strain of the element and the relative density of the Voronoi model.

The authors would like to thank the financial support by the National Natural Science Foundation of China (Contract no. 11372256) and the 111 Project (Contract no. B07050). Part of this work was conducted when the first author (X.Shi) was attached for two years in Nanyang Technological University, Singapore, under a scholarship from the Chinese Scholarship Council. This paper was also supported by Center for High Performance Computing of Northwestern Polytechnical University, China.

References

[1] J.A. Reglero, M.A. Rodriguez-Perez, E. Solorzano, J.A. de Saja, Materials & Design, **32**, 2 (2011).
[2] W.H. Hou, F. Zhu, G.X. Lu, D.N. Fang, International Journal of Impact Engineering, **37**, 10 (2010).
[3] P.S. Kumar, S. Ramachandra, U. Ramamurty, Materials Science and Engineering a-Structural Materials Properties Microstructure and Processing, **347**, 1–2 (2003).
[4] S. Ramachandra, P.S. Kumar, U. Ramamurty, Scripta Mater, **49**, 8 (2003).
[5] C. Bacon, Experimental Mechanics, **38**, 4 (1998).
[6] V.S. Deshpande, N.A. Fleck, International Journal of Impact Engineering, **24**, 3 (2000).
[7] I. Irausquin, J.L. Perez-Castellanos, V. Miranda, F. Teixeira-Dias, Materials & Design, **47** (2013).
[8] L.J. Gibson, M.F. Ashby, Cellular solids: structure and properties, 2nd ed., Cambridge University Press, Cambridge; New York, 1997.
[9] H. Zhao, I. Elnasri, S. Abdennadher, Int J Mech Sci, **47**, 4–5 (2005).
[10] P.J. Tan, S.R. Reid, J.J. Harrigan, Z. Zou, S. Li, Journal of the Mechanics and Physics of Solids, **53**, 10 (2005).
[11] A. Paul, U. Ramamurty, Materials Science and Engineering a-Structural Materials Properties Microstructure and Processing, **281**, 1–2 (2000).
[12] X.P. Shi, Y.L. Li, G.X. Lu, C.J. Shen, Materials Science and Engineering a-Structural Materials Properties Microstructure and Processing, 614, (2014).
[13] `http://math.lbl.gov/voro++/about.html`.

EPJ Web of Conferences **94**, 04033 (2015)
DOI: 10.1051/epjconf/20159404033

Finite element simulation of impact response of wire mesh screens

Caizheng Wang[a], Krishna Shankar, and Alan Fien

School of Engineering and Information Technology, UNSW Canberra, ACT 2600, Australia

Abstract. In this paper, the response of wire mesh screens to low velocity impact with blunt objects is investigated using finite element (FE) simulation. The woven wire mesh is modelled with homogeneous shell elements with equivalent smeared mechanical properties. The mechanical behaviour of the woven wire mesh was determined experimentally with tensile tests on steel wire mesh coupons to generate the data for the smeared shell material used in the FE. The effects of impacts with a low mass (4 kg) and a large mass (40 kg) providing the same impact energy are studied. The joint between the wire mesh screen and the aluminium frame surrounding it is modelled using contact elements with friction between the corresponding elements. Damage to the screen of different types compromising its structural integrity, such as mesh separation and pulling out from the surrounding frame is modelled. The FE simulation is validated with results of impact tests conducted on woven steel wire screen meshes.

1. Introduction

Security screen doors and window grilles (Fig. 1) are increasingly employed in a number of domestic, commercial and industrial buildings in Australia, for allowing free circulation of air in hot weather, keeping insects from entering the premises, and more importantly providing substantial resistance to forced entry [1]. Security screens are also increasingly being employed to protect against damage from natural sources such as cyclone debris. Hence, the determination of impact resistance of wire mesh security screens is essential and necessary. Traditionally, the design of these security screens has been done using trial and error methods and experimental impact testing. Extensive experimental testing is impractical and expensive and can be avoided by using numerical models that predict the response of the security screen to impact.

However, there is hardly any literature directly related to the modelling of wire mesh security screens. There have been some studies conducted on modelling of steel meshes for rock fall protection [2,3] and simulation of impact and ballistic resistance of woven fabrics [4–8]. There are several approached employed to represent woven structures. Single scale modelling techniques include representing the entire fabric as homogenized membranes or plates [4,5], and full detailed modelling with 3D solid finite elements [6]. There are also some meaningful explorations of multi-scale modelling methods, such as G. Nilakantan [7,8] who used a hybrid element analysis method (with both solid and shell elements) to model impact on textile fabrics. Modelling individual wires of a woven wire mesh screen can be very complex and will considerably increase the simulation time (there are over 300 wires in a typical screen running in each direction and criss-crossing over each other); hence, it was considered expedient to investigate the approach using homogeneous sheet structure with equivalent properties to simulate the discrete mesh first, as employed by some previous researchers [4,5,8].

In this paper, explicit FE software, LS-Dyna is used to model security screens subjected to impact loading. The impact conditions that are simulated correspond to the Australian Standards for Security Screens (AS 5039), which require that the security screen withstand five impacts at 100J energy without developing a breach large enough to allow intruder access [9]. FE model of same energy level impact of security screens are developed for impactor mass of 4 kg and 40 kg in order to study the influence of velocity on the impact response on the wire mesh security screen. The steel woven wire mesh is modelled as a homogeneous continuous sheet using fully integrated Belytschko-Tsay Shell elements and the equivalent smeared were determined using macroscopic material properties of the wire mesh obtained from tensile testing on a Shimadzu Hydraulic machine. A bilinear material model available in the material library as MAT24 was used for the wire mesh. The aluminium frame is modelled using brick elements and elastic-linear material law. The clamping of the wire mesh to the frame is modelled using Tie break contact elements.

2. Tensile testing woven wire mesh samples

2.1. Tensile test setup and results

To determine the tensile properties of the woven wire mesh, tensile testing was conducted on coupons using the Shimadzu Hydraulic Tensile Test machine shown in Fig. 2. The load was applied under displacement control at the rate of 1 mm/min. A laser extensometer was used to measure the strain. Two reflective tabs were put on each sample marking the gauge length over which extension

[a] Corresponding author: caizheng.wang@student.adfa.edu.au

Figure 1. Stainless steel wire mesh security screen.

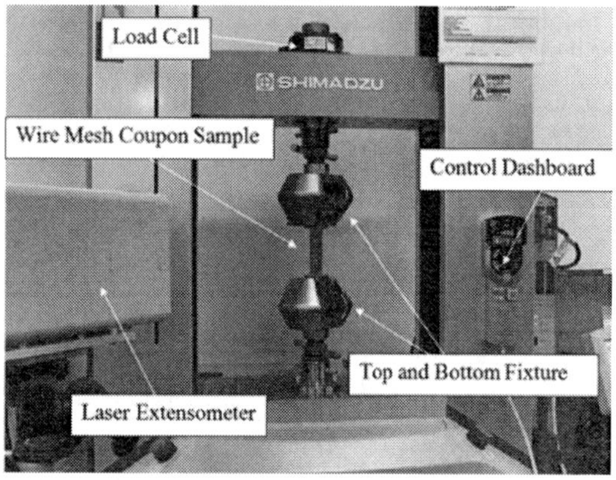

Figure 2. Tensile test setup on Shimadzu test machine.

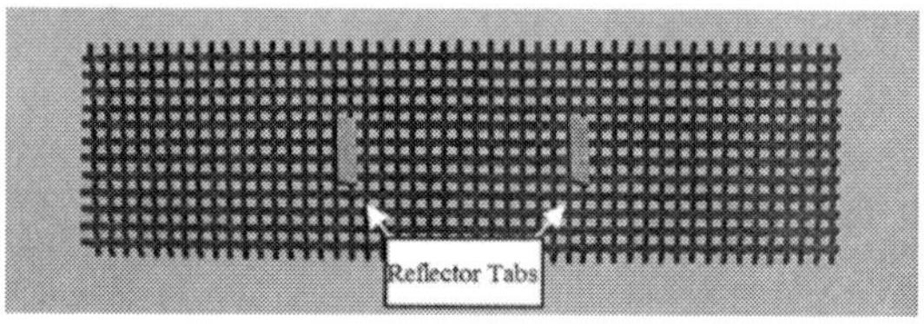

Figure 3. Stainless steel wire mesh tensile test sample.

was measured to determine strain. During the testing, the sample was gripped in pin-mounted fixtures to ensure that only tensile force is applied on the coupon.

The wire mesh coupons employed in this study were cut from an commercially marketed security screen in Australia, made of 304 stainless steel wires (Nominal Young's Modulus E = 193 GPa) with a diameter of 0.9 mm and spacing of 2.5 mm. A total of 5 coupons were tested. Each coupon had a width of 30 mm containing 12 wires in the loading direction, and a nominal length of 30–40 mm (see Fig. 3).

Figure 4 shows all the samples after testing. It can be seen that the last sample in the picture failed outside the gauge length, so the data from this sample was not included. The plots of load vs engineering strain for the four valid samples are shown in Fig. 5.

2.2. Equivalent smeared material properties

Several methods can be employed to model a periodic inhomogeneous structure with a homogeneous continuous

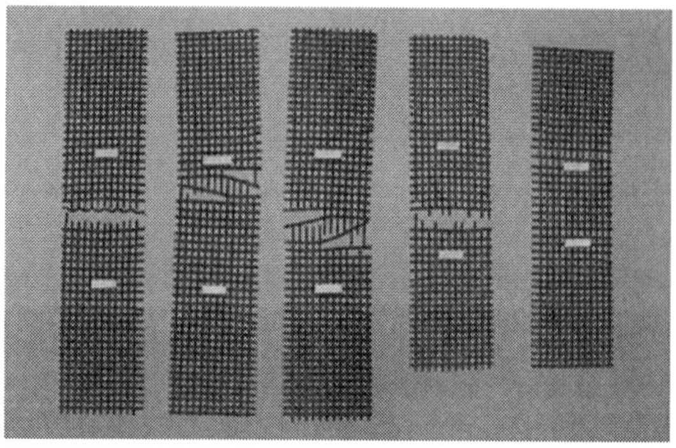

Figure 4. Samples after tensile testing.

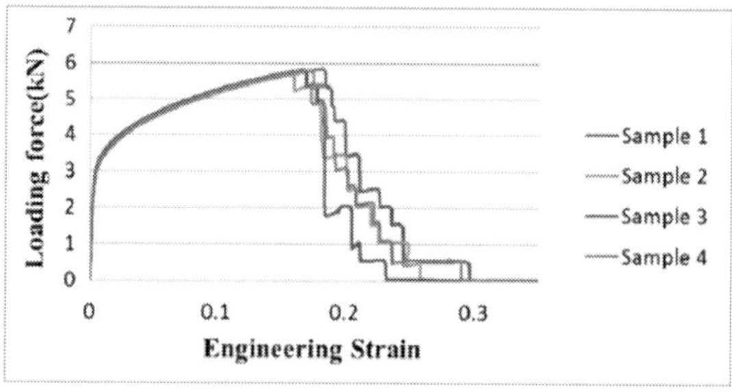

Figure 5. Force vs. engineering strain for tensile test samples.

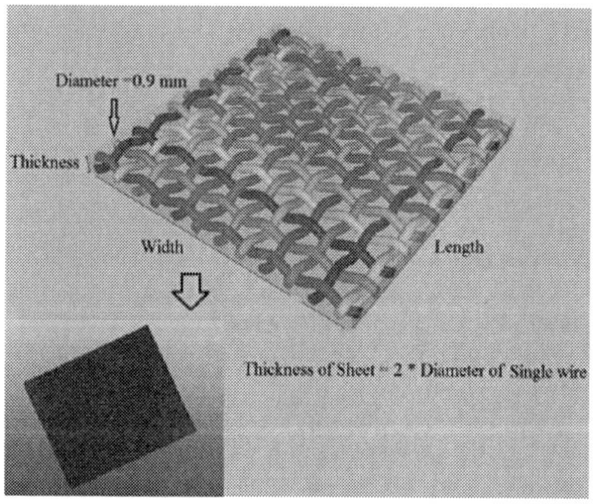

Figure 6. Woven wire mesh and equivalent homogeneous sheet.

structure. C.T. Lim et al. [5] used a mathematical manipulation of the three-element spring-dashpot model to obtain viscoelastic nature material properties of a woven fabric. B.J.E. van Rens [10] determined a perforated plate's smeared properties by calculating finite element model of a representative volume unit (RVU) of the plate. A. Shahkarami and R. Vaziri [6] created a 3D finite element mesh of the unit cell from a woven fabric using the measured fabric cross-sectional micro-images to investigate the smeared material properties.

In this study, the RVU approach of B.J.E. van Rens [10] is adopted. In this approach, by equating the volume of the representative homogeneous sheet (see Fig. 6) to the enclosed rectangular volume of the perforated mesh, the

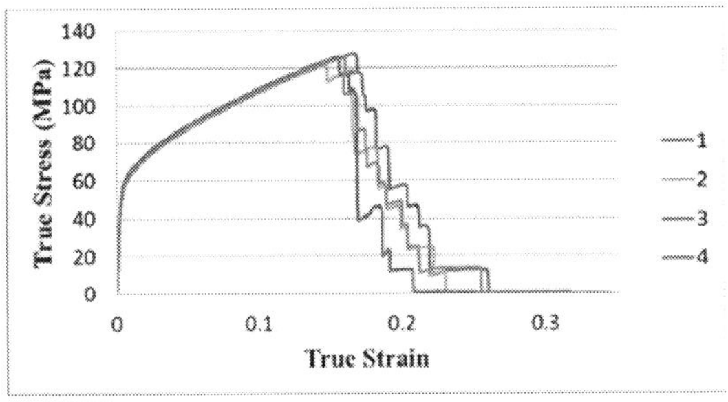

Figure 7. Stress-strain curve of equivalent homogeneous sheet.

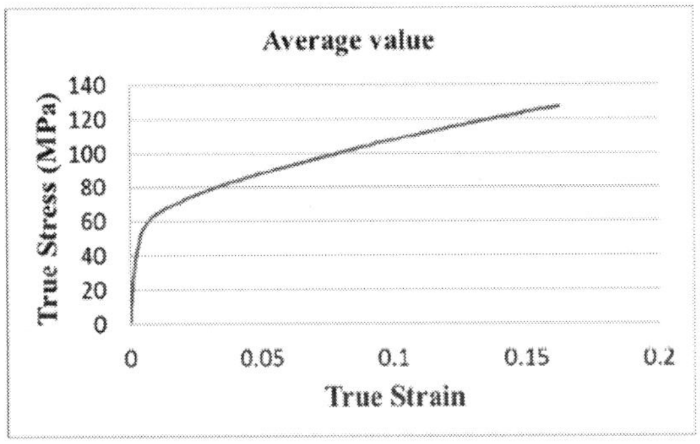

Figure 8. Average equivalent true stress-strain curve.

thickness of the homogenous sheet is obtained as twice the diameter of the wire mesh [8]. The modulus of the sheet material is then determined so as to provide the same effective in-plane stiffness, i.e., gradient of the load strain curve as that measured from coupon tests. Since the response of the wire mesh under tensile loading is non-linear, the equivalent modulus of the representative sheet has to be determined in this manner at all points along the curve, to obtain the stress strain curve of the representative sheet. It is to be noted that in the RVU approach, since the equivalent sheet has a thickness of twice the diameter of the wires; hence although its in-plane stiffness would be the same as the over-all in-plane stiffness of the wire mesh, the bending stiffness would be much higher. Figure 7 shows the true stress vs true strain curve of the representative homogenous sheet determined in this manner for the four individual coupons tested. As can be seen the four curves lie very close to each other, except in the post failure region, and the averaged stress strain data (see Fig. 8) is employed for generating the input data for bilinear material model in the FE simulation. In the initial linear region, where the stiffness is constant, the modulus of the equivalent sheet is obtained as 27.28 GPa.

3. Numerical models of wire mesh screens

The most common failure mode observed in impact testing of wire mesh screens is the pull out of the sheet from the frame. This occurs due to the tensile forces on the individual wires exceeding the friction provided by clamping in the frame, and is accompanied by the wires behind screws getting cut as the wire mesh pulls out. The focus of this study was to simulate this mode of failure.

In the LS-Dyna simulation, the homogeneous sheet was modelled with Shell 163 elements with equivalent bi-linear properties listed in Table 1. The frame surrounding the mesh (see Fig. 9) was modelled using Solid 164 elements with standard aluminium alloy elastic properties (Young's Modulus 73 GPa and Poissson's ratio 0.33). The impactor was modelled as rigid with Solid 164 elements. The element type and material models are listed in Table 2.

To investigate the influence of velocity two impactor masses, 4 kg and 40 kg were employed with the same input impact energy (Figs. 9(a) and 9(b)).

The 'AUTOMATIC_SURFACE_TO_SURFACE' contact was used to define the contact between the impactor and screen, while the 'CONTACT' type between frames and, shell screen was 'TIEBREAK_SURFACE_TO_SURFACE, which is typically used to represent bolted or spot- welded connections. Fully fixed (zero displacements and rotations) boundary conditions were applied on the underside surface of the bottom frame and at the outer edges of the top frame.

The screen is 800 mm × 800 mm in geometry, while the frames are 25 mm in width and 5 mm in thickness. The location of impact loading point is also the same, 250 mm away from frame two adjacent frame edges. The hemispherical impactors of 4 kg and 40 kg mass had radii of 50 mm and 100 mm, respectively.

4. Simulation results

4.1. Effect of impact velocity

A comparison of the simulations of a single impact of 100 J using the two impactors masses of 4 kg (input

Table 1. Equivalent smeared material properties of the representative sheet.

Young's Modulus (GPa)	Poisson's ratio	Yield Stress (MPa)	Yield Strain	Tangent Modulus (GPa)	Failure Strain
27.28	0.3	36	0.20%	0.57	16%

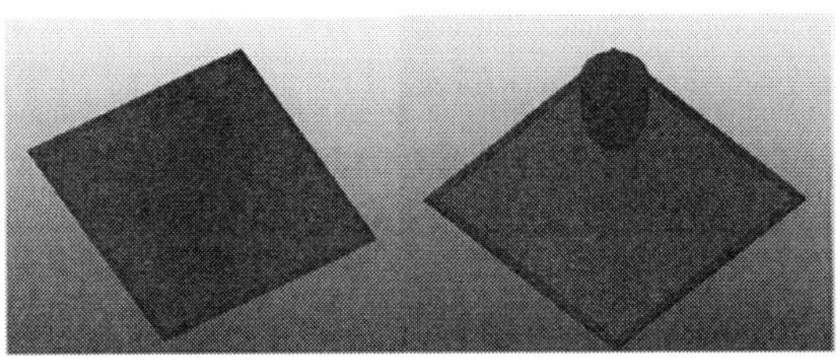

(a) (b)

Figure 9. Finite Element (FE) models (a) 4 kg and (b) 40 kg impactors.

Table 2. Element and Material types.

Components	Element type	No. of elements	Material type
Impactor	Solid 164	5000	Rigid
Screen	Shell 163	6400	MAT24
Frame	Solid 164	12400	Elastic

Table 3. Equivalent material properties of the smeared sheet.

	Maximum Von Mises Stress in screen	Maximum Von Mises Stress in frame	Maximum Screen Deflection
40 kg impactor	84.20 MPa	239.3 MPa	57.25 mm
4 kg impactor	103.29 MPa	231.9 MPa	54.11 mm

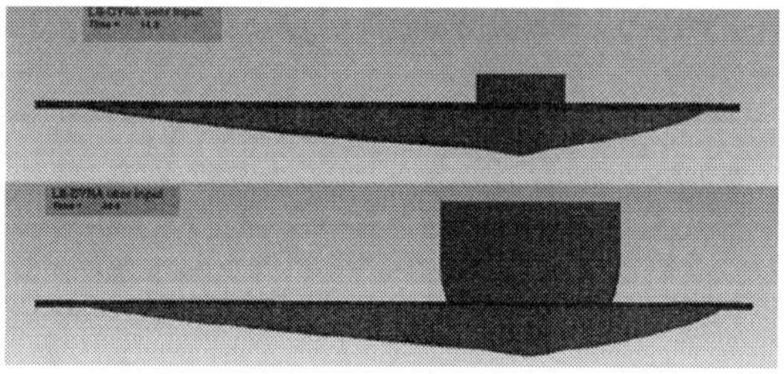

Figure 10. Deformed shape of screen impacted with (a) 4 kg and (b) 40 kg mass.

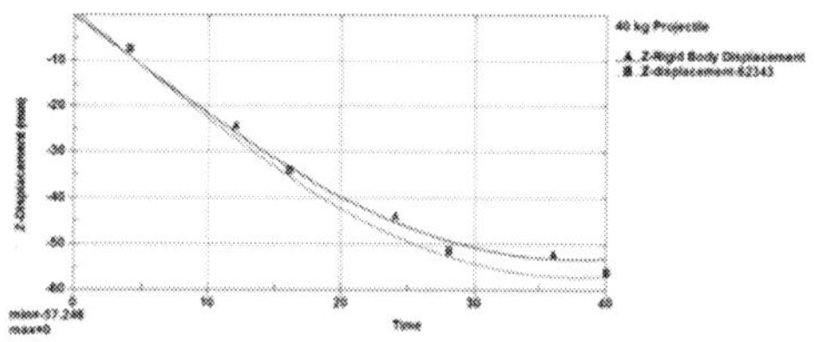

Figure 11. Z–displacements of 4 kg impactor and screen at contact point.

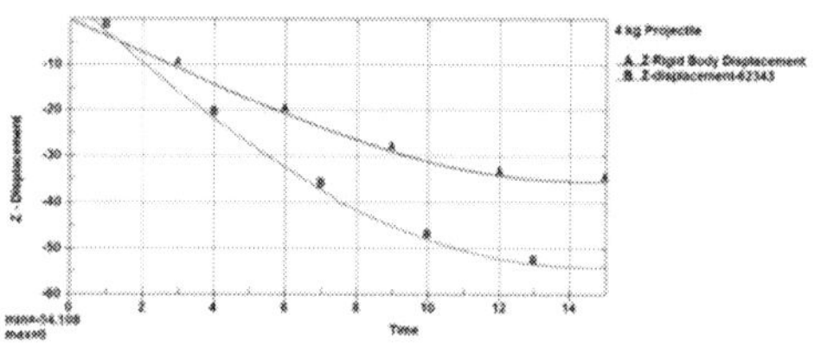

Figure 12. Z–displacements of 40 kg impactor and screen at contact point.

velocity of 7.07 m/s) and 40 kg (input velocity of 2.24 m/s) was first performed. Figure 10 shows the deformed shapes of the screens impacted with the 4 kg and the 40 kg masses at the instant of maximum contact force. Figures 11 and 12 respectively show the displacement time histories of the impactor and the contact point of the screen, with the 4 kg mass and the 40 kg mass. It is evident from these figures that the screen deflects more when impacted with the 40 kg mass than with the 4 kg mass (maximum deflection of 57.25 mm compared to 54.11 mm of the latter), even the kinetic energy of both are impactors before contact are the same. The maximum stress experienced by the screen are also higher when impacted with the lower mass compared to those for the higher mass (see Table 3), which may be attributed to the higher velocity of the impactor with the lower mass providing less time for the stresses and energy to be transmitted to the surrounding frame. Figures 13 and 14

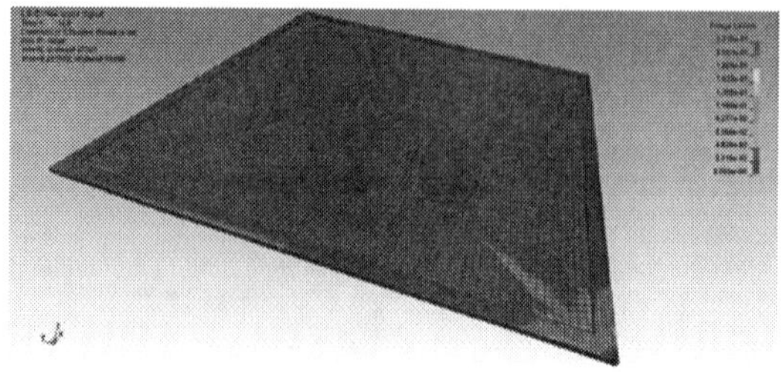

Figure 13. Contour map of Von Mises Stress of screen on impact with 4 kg mass at 100 J at 14.8 s.

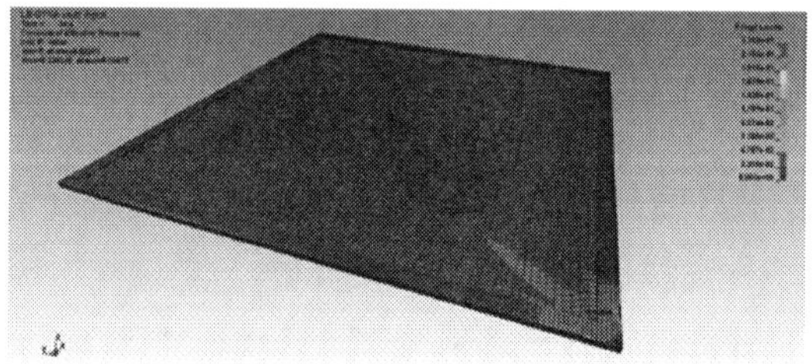

Figure 14. Contour map of Von Mises Stress of screen on impact with 40 kg mass at 100 J at 34.4 s.

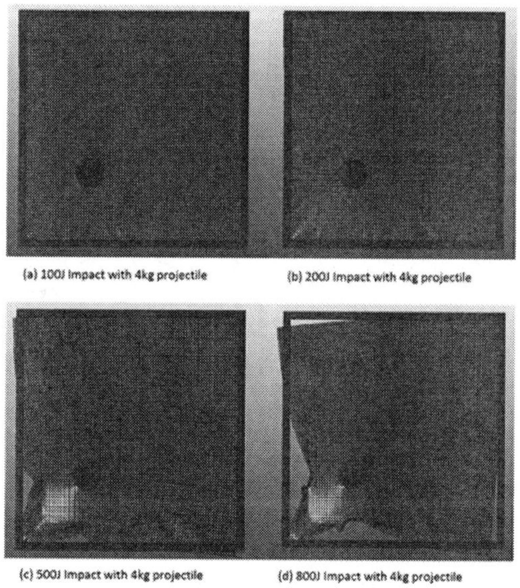

Figure 15. Response of wire mesh security screen on impact with 4 kg mass at different energy levels.

show the contour maps of Von Mises stress distributions in the screen and the frame, 14.8 seconds after impact with 4 kg mass and 34.4 seconds after impact with 40 kg mass, respectively. The development of high stresses in the scree and the frame at the corner near the impact point are evident in both cases.

4.2. Impacts at increasing energy levels

Simulations were performed at four impact energy levels, viz., 100 J, 200 J, 500 J and 800 J, with the 4 kg mass and the 40 kg mass to investigate the development of

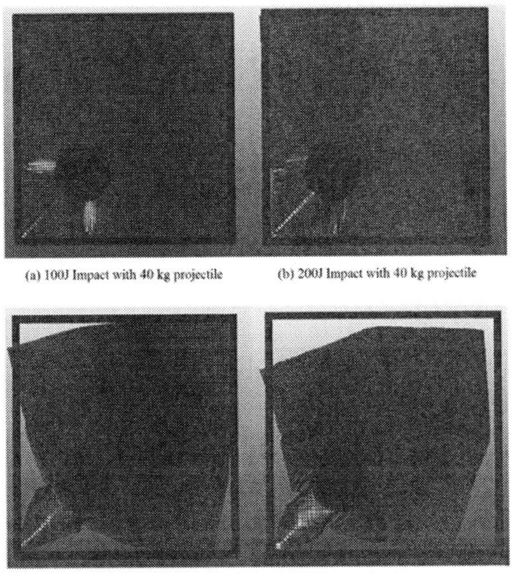

Figure 16. Response of wire mesh security screen on impact with 40 kg mass at different energy levels.

Figure 17. Woven wire mesh security screen after a singular impact of 500 J.

damage. The deformed shapes of the screen impacted with the 4 kg mass at these four energy levels are shown in Figs. 15(a) to 15(d) and those for impacts with the 40 kg mass are shown in Figs. 16(a) to 16(d). It can be seen that while the screens are undamaged when impacted with both masses at 100 J, the wire mesh stats pulling out from the corner on impact with 200 J with the 40 kg mass, but is still fairly undamaged when impacted at 200 J with the 4 kg mass. At 500 J both screens exhibit considerable damage in the form of pull out of the mesh from the frame at the impact corner, developing into pull out at other corners and along the sides of the frame when the energy is increased to 800 J.

For comparison, Fig. 17 shows the photograph of a wire mesh screen impacted with 500 J of energy with a 4 kg mass during impact testing. As can be seen, the real screen also failed by mesh pulling out from the frame edges at the impact corner. The breaches between the frame and the mesh in the experimental screen are 510 mm and 420 mm long, while those obtained in the simulation for impact at 500 J with the 4 kg mass are 511 mm and 494 mm, indicating good agreement between the simulation and the testing.

5. Conclusion

This paper looked at the impact behaviour of woven wire mesh security screen subjected to low velocity impact loading using a smeared homogeneous model for the perforated screen. It is seen the responses of screen are different for impacts at the same energy with a low mass and relatively high mass with the same impact energy. With a 4 kg mass the maximum deflection is lower but the stresses are higher in the mesh, while the higher mass of 40 kg produces higher screen deflection and higher stresses in the frame. The latter may account for the fact damage in the form of separation between the mesh and the frame appears to occur at lower impact energies when the 40 kg mass is employed compared to damage initiation when impacted with the lower mass.

The damage mode and failure observed in experimental testing was very similar to those predicted by the simulation, providing validation for the finite element model developed. To predict other modes of failure, the model has to be developed further to include details of the geometry of the frame and individual woven wires, which will be undertaken in future work.

References

[1] D. Smith, S. Grove, Window security screen and latch assembly, U.S. Patent, Appl. No.: 13/410,379 (2012)

[2] A. Volkwein, Numerical simulation of flexible rockfall protection systems, Computing in Civil Engineering (2005)

[3] E.C. Morton, A.G. Thompson & E. Villaescusa, Testing and analysis of steel wire mesh for mining applications of rock surface support, 11th congress of the International Society for Rock Mechanics, Vol. **2**, 1061–1064 (2007)

[4] A. Shahkarami, R. Vaziri, A continuum shell finite element model for impact simulation of woven fabrics. International Journal of Impact Engineering 2007, **34**:104e19

[5] C.T. Lim, V.P.W. Shim, Y.H. Ng, Finite-element modeling of the ballistic impact of fabric armor, International Journal of Impact Engineering 2003, **28**:13e31

[6] E. Glatt, S. Rief, A. Wiegmann, Structure and pressure drop of real and virtual metal wire meshes, Berichte des Fraunhofer ITWM, ISSN 1434-9973, Nr. 157 (2009)

[7] G. Nilakantan, M. Keefe, T. A. Bogetti, J. W. Gillespie, Multiscale modeling of the impact of textile fabrics based on hybrid element analysis,

International Journal of Impact Engineering, Vol. **37**, 1056–1071(2010)

[8] G. Nilakantan, M. Keefe, T. A. Bogetti, R. Adkinson, J. W. Gillespie, On the finite element analysis of woven fabric impact using multiscale modeling techniques, International Journal of Solids and Structures, Vol. **47**, Issue 17, 2300–2315 (2010)

[9] S. Australia, AS 5039-2008 Security screen doors and security window grilles, in, Standards Australia, Sydney, NSW (2008)

[10] B. J. E. van Rens, W. A. M. Brekelmans, F. P. T. Baaijens, Homogenization of the elasto plastic behavior of perforated plates, Computers and Structures, 537–545, **69** (1998)

EPJ Web of Conferences **94**, 04034 (2015)
DOI: 10.1051/epjconf/20159404034

Effect of dynamic strain rate on micro-indentation properties of pure aluminum

Hiroyuki Yamada[1,a], Midori Hotta[2], Tsuyoshi Kami[3], Nagahisa Ogasawara[1], and Xi Chen[4]

[1]National Defense Academy, School of Systems Engineering, 1-10-20 Hashirimizu Yokosuka Kanagawa, Japan
[2]National Defense Academy, Graduate Student, School of Systems Engineering, 1-10-20 Hashirimizu Yokosuka Kanagawa, Japan
[3]Columbia University, Graduate Student, 500 West 120th Street, New York, USA
[4]Columbia University, 500 West 120th Street, New York, USA

Abstract. Indentation is widely used to investigate the elastic and plastic properties of mechanical materials, which includes the strain rate sensitivity. The indentation exhibits an inhomogeneous strain distribution in contrast to compression and tensile tests with homogeneous deformation. Thus, the strain rate of the indentation may form the inhomogeneous distribution. Therefore, the effect of strain rate distribution of the indentation on pure aluminum with respect to the strain rate dependence of strength in order to clarify the effect of the strain rate on the indentation technique. First, the numerical simulation was established using the Cowper-Symonds equation as the dynamic constitutive equation. Secondary, the strain rate distribution was calculated from the equivalent plastic strain distribution. The strain rate distribution was quite different from the strain distribution, which showed that the strain rate at the crater rim was higher than that beneath the indenter. Finally, we try to perform the averaging of strain rate distribution in order to make an index of strain rate in the indentation. The average of strain rate distribution was calculated using the equivalent plastic strain above a boundary value that is the critical strain and the representative strain. There is correlation between the average strain rate and the loading curvature, which shows that the average strain rate can express as the representative of strain rate for the indentation technique.

1. Introduction

Instrumented indentation tests are widely used to investigate the elastic and plastic properties of mechanical materials [1–3]. A rigid indenter penetrates normally into a homogeneous solid, where the indentation load (P) and displacement (h) are continuously recorded during loading and unloading.

In the micro-indentation test, the hardness has been reported to increase with decreasing indentation depth [4,5]. Doerner and Nix [4] pointed out that the material strength was affected by the strain rate in the indentation test. When the indentation test is performed at a constant displacement rate, $\dot{h}$, the strain rate can be defined as follows:

$$\dot{\varepsilon}_i = c(\dot{h}/h). \tag{1}$$

where c is the constant and $\dot{\varepsilon}_i$ is called indentation strain rate. Thus, the Eq. (1) indicates that the strain rate becomes very high as the indentation depth decreases.

While the pure aluminum with strain rate dependence of strength has been evaluated for investigating the effect of strain rate. In a previous study, the authors [3] have investigated properties of 99.999 mass% pure aluminum (hereinafter referred to as 5N aluminum). Figure 1 shows the typical true stress-strain curves in the quasi-static test and the impact test for 5N aluminum. It was confirmed that the flow stress was increased with increasing the strain rate. In addition, the typical load-displacement curves for 5N aluminum in the indentation test are shown in Fig. 2. During the loading process, the load increased with increasing loading rate. Thus, it was revealed that the load of the micro-indentation test was affected by the loading rate. As shown in Figs. 1 and 2, the effect of loading rate on load during indentation was in good agreement with the trend of the strain rate dependence of flow stress.

In addition, the authors [6] have also investigated the effect of high strain rate on the load-displacement relation by comparing the results of a micro-indentation test at a loading rate of 70 mN/s and the results obtained by the finite element method (FEM) analysis using the annealed pure copper having strain rate dependence of strength. As a result of the FEM analysis, the equivalent plastic strain rate at the early stage of testing reached the dynamic strain rate, approximately $10^1\ s^{-1}$, underneath the sharp indenter in the micro-indentation test. This result has shown the effect of strain rate on the indentation.

However, the indentation exhibits an inhomogeneous strain distribution in contrast to the compression and the tensile tests with homogeneous deformation. Thus, the strain rate of the indentation may form the inhomogeneous distribution.

The purpose of this study is to investigate the effect of strain rate distribution of the indentation on pure aluminum with respect to strain rate dependence of strength in order to clarify in detail the effect of high strain rate on the indentation test.

[a] Corresponding author: `ymda@nda.ac.jp`

Article available at http://www.epj-conferences.org or http://dx.doi.org/10.1051/epjconf/20159404034

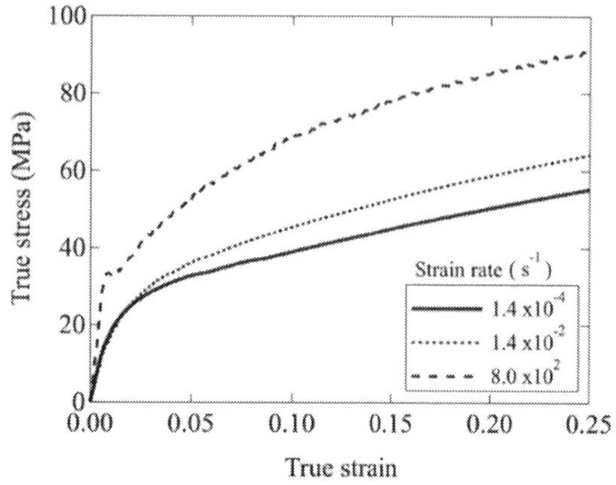

Figure 1. True stress-strain curves for 5N aluminum in the quasi-static test and the impact test.

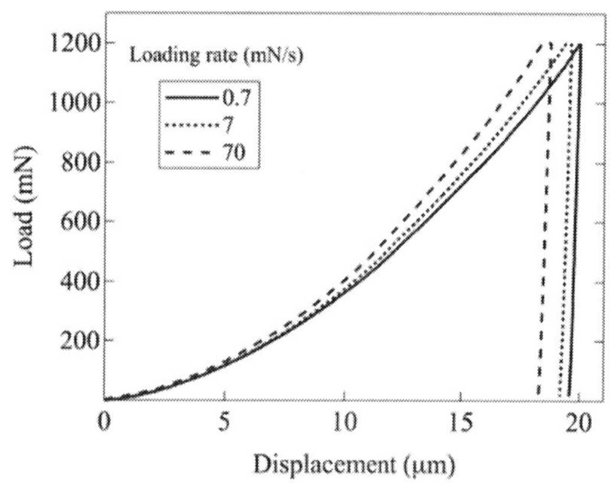

Figure 2. Load-displacement curves for 5N aluminum in the indentation test.

2. The effect of the indentation strain rate

The flow stress with respect to the strain rate could be expressed with exponential function defined by following equations.

$$\sigma = \sigma_0 \left[1 + \left(\frac{\dot{\varepsilon}}{\gamma} \right)^m \right] \quad (2)$$

where σ_0 is the static stress, m and γ are the strain rate parameters. The above equation is called the Cowper-Symonds constitutive equation. This constitutive equation is widely used to represent the strain rate sensitivity of metal materials in a simple way.

In general, the FEM analysis has the implicit and the explicit methods. The latter, such as LS-DYNA, is designed for dynamic analysis, which means the problem of the stress wave propagation. In a previous study [6], the authors used the explicit methods in order to use the Cowper-Symonds model, because the high strain rate phenomenon was confirmed in the indentation test. However, the typical indentation test using commercially apparatus has longer test duration than the high strain rate test to evaluate the dynamic and the impact phenomena. Therefore, the mass scaling technique has been used in the previous study [6], which can shorten the analysis time using the explicit method. Thereby, it is necessary to make the density of indenter 10000 times to shorten the analysis time. However, this leads to the inertia force in specimen due to the acceleration. The actual indenter was moved at a few μm/s, which did not take place the inertia force in specimen. Thus, we try to use the implicit code for the analysis of the indentation affected by the strain rate.

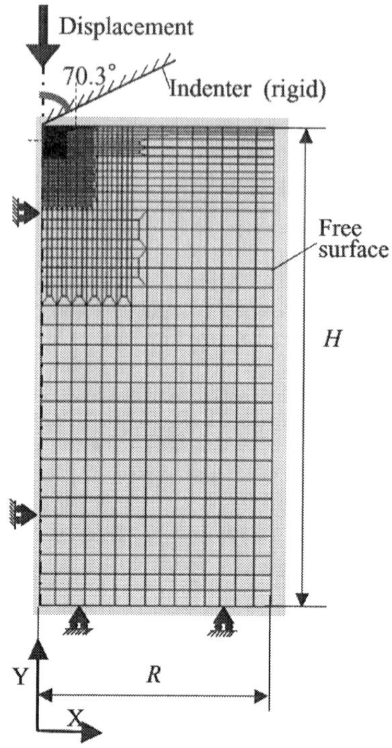

Figure 3. Simulation model.

In this study, the numerical simulations were performed using the commercial finite element package ANSYS, the implicit code. Figure 3 shows the simulation model. For simplicity, the indentation simulation was modeled as an axisymmetric and elasto-plastic analysis. The conical indenter with included with half apex angle of 70.3° was modeled as a rigid body. This gives approximately the same ratio of indentation depth to contact area as the Berkovich indenter. A typical mesh for the axisymmetric indentation model comprises about 4-node elements. The mesh has approximately 29,600 elements. As shown in Fig. 3, the mesh of the model is designed to become coarser with distance from the indenter tip in order to reproduce the strain distribution during the indentation. The minimum element size was $0.075\,\mu$m $\times$ $0.125\,\mu$m underneath the indenter. Thus, more accurate results and convergences can be obtained.

In the indentation test, there are two indenter control methods: the load control (loading rate, $\dot{P}$, is constant) and the displacement control (displacement rate, $\dot{h}$, is constant). When used the load control, the strain rate changes depending on materials during indentation. Therefore, the displacement control method is used in this study. To evaluate the role of indenter velocity, the analysis

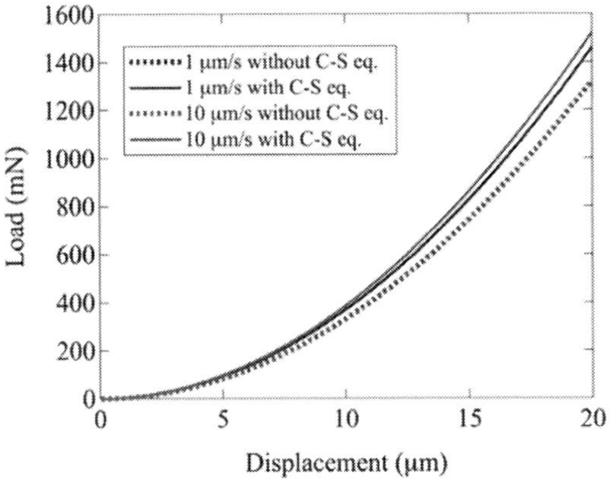

Figure 4. Load-displacement curves obtained through the FEM analysis at displacement rate of 1 μm/s and 10 μm/s.

was performed at $\dot{h} = 1$ and 10 μm/s. The friction between indenter and specimen surface is 0.15, which assumes the friction between diamond and metal. We chose to the high-pure aluminum with 99.999 mass% purity for the material model, which is proved to have the strain rate dependence as shown in Fig. 1. The values of strain rate parameters of the Cowper-Symonds model are $m = 0.16$ and $\gamma = 5000$.

3. The effect of strain rate distribution on indentation

3.1. Loading curvature-displacement relationship

Under quasi-static conditions, the loading segment of the load-displacement relation obeys a parabolic relation, known as Kick's law:

$$P = Ch^2 \tag{3}$$

where C is the loading curvature, which is a constant value. The loading curvature depends on both the elasto-plastic properties of materials and the indenter angle. It was known [4] that the loading curvature decreased with increasing displacement immediately after indentation, and then this value was convergent, regardless of the loading rate. This is because the indentation was affected by the strain rate at the early stage of testing.

The load-displacement curves were obtained through the FEM analysis at $\dot{h} = 1\,\mu$m/s and 10 μm/s (Fig. 4). And then the loading curvature-displacement relations obtained through the load-displacement curves were calculated using Eq. (3), as shown in Fig. 5. In the FEM analysis without the Cowper-Symonds model, the loading curvature kept constant, and the load-displacement curves at $\dot{h} = 1\,\mu$m/s and 10 μm/s coincided. On the other hand, the decrease of the loading curvature immediately after initial indentation and the convergence were observed when used the Cowper-Symonds model, regardless of the $\dot{h}$. In addition, the loading curvature was increased when the displacement rate was increased. Thus, we can replicate the effect of strain rate on the indentation during the loading using the implicit method, as is the case in the explicit method [6].

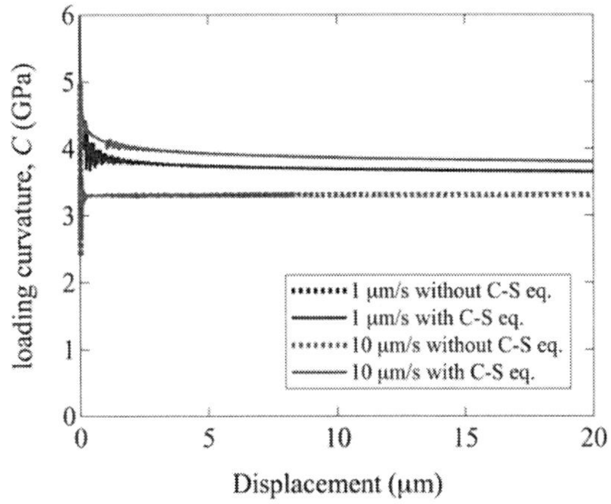

Figure 5. Loading curvature displacement curves obtained from Fig. 4 at displacement rate of 1 μm/s and 10 μm/s.

3.2. Relationship between strain and strain rate distribution

Figure 6 shows the equivalent plastic strain distributions at $h = 0.2, 2, 20\,\mu$m and $\dot{h} = 1\,\mu$m/s. It was clear that the strain was concentrated beneath the indenter. Furthermore, the relationship between the equivalent plastic strain distribution and the displacement indicated the correlation.

In this study, we cannot obtain directly the strain rate distribution using ANSYS software. Thus, the strain rate at each node point was calculated from the equivalent plastic strain, $\varepsilon_{plastic}$, as follows:

$$\dot{\varepsilon}_{node} = \frac{d\varepsilon_{plastic}}{dt} \tag{4}$$

where, dt is the output time step of the simulation, which is 2 ms in this study. The strain rate distribution was shown using the numerical software, MATLAB. Figure 7 shows the strain rate distributions at $h = 0.2, 2, 20\,\mu$m and $\dot{h} = 1\,\mu$m/s. It was revealed that the value for the strain rate distribution is decreased with increasing the displacement.

The strain rate distribution was quite different from the strain distribution at each displacement. As shown in Fig. 6, the strain beneath the indenter keeps high when the displacement is increased, which indicate that there is little change in the amount of strain around the indenter tip. On the other hand, it was confirmed that the strain rate at the crater rim was higher than that beneath the indenter. The strain of the crater rim changes newly with increasing the displacement, which means that the amount of the strain rate change was high. We infer that the high strain rate area does not express as a representative of strain rate for the indentation, since that area is very small in comparison with overall strain rate distribution.

The strain rate distributions for the same indentation strain rate calculated from Eq. (1) were compared. Figure 8 shows the strain rate distribution at $\dot{h} = 10\,\mu$m/s and $h = 2\,\mu$m, which is the condition of $\dot{\varepsilon}_i = 5c$. The value of the strain rate distribution at $\dot{h} = 1\,\mu$m/s and $h = 0.2\,\mu$m almost coincided with that at $\dot{h} = 10\,\mu$m/s

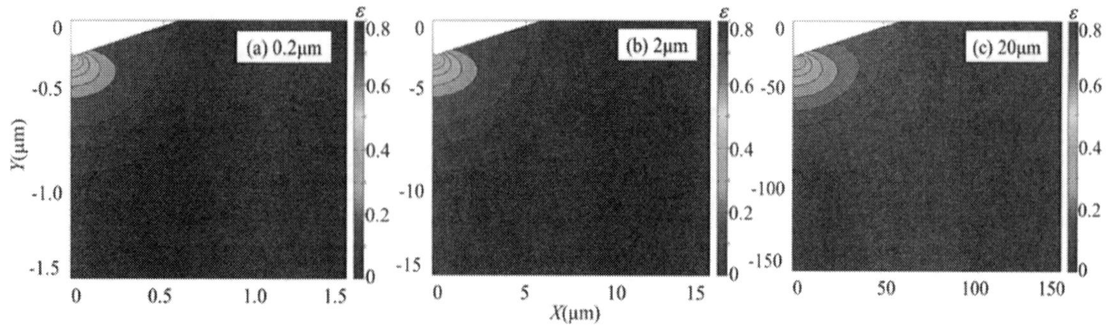

Figure 6. Equivalent plastic strain distributions at displacement rate of 1 μm/s.

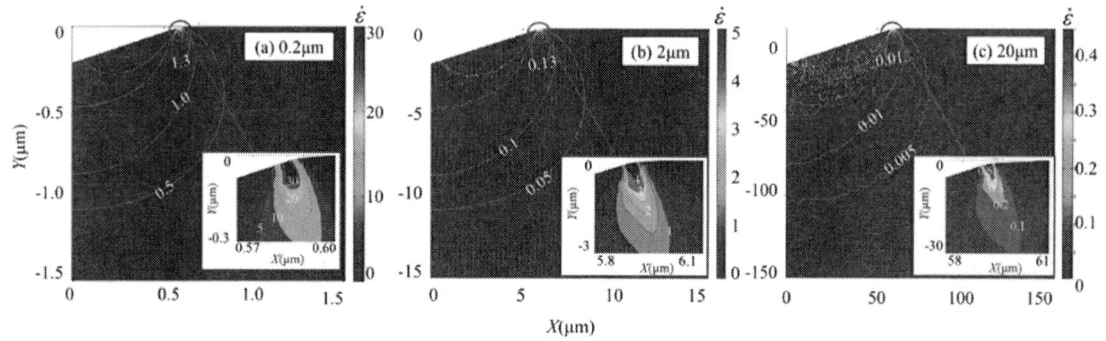

Figure 7. Strain rate distributions at displacement rate of 1 μm/s.

and $h = 2\,\mu$m. This result shows that these distributions of the same indentation strain rate have the similarity rule.

4. Introduction of average strain rate

4.1. Definition of average strain rate

As discussed in the previous chapter, the effect of the strain rate distribution on indentation is not apparent in the results of FEM analysis. Therefore, we try to perform the averaging of strain rate distribution in order to make an index of strain rate in the indentation. In this study, the average of strain rate distribution is calculated using the equivalent plastic strain above a boundary value, which defines as average strain rate, $\dot{\varepsilon}_{av}$. The candidates for this boundary value are the critical strain, ε_c, and the representative strain, ε_R.

Liu *et al.* [7] pointed out that the critical strain was identified as the upper boundary of the detectable range of indentation deformation, and moderate tailoring of the constitutive behavior beyond this range cannot be effectively detected by the reverse analysis of the *P-h* curve. The authors [8] proposed that the critical strain as follows:

$$\varepsilon_c = \left(2.3n^2 - 0.3n - 1.0\right)(\alpha - 88.4). \qquad (5)$$

where n is the work-hardening exponent and α is a cone with half apex angle. When using the conical indenter with half apex angle of 70.3°, ε_c is 0.15.

Dao *et al.* [9] introduced the concept of the representative strain, which is used for the indentation analysis to normalize the load-displacement curves of the indentation. The authors [10] proposed more exact the representative strain as follows:

$$\varepsilon_R = 0.0638 \cot\alpha. \qquad (6)$$

In the present study, ε_R is 0.03.

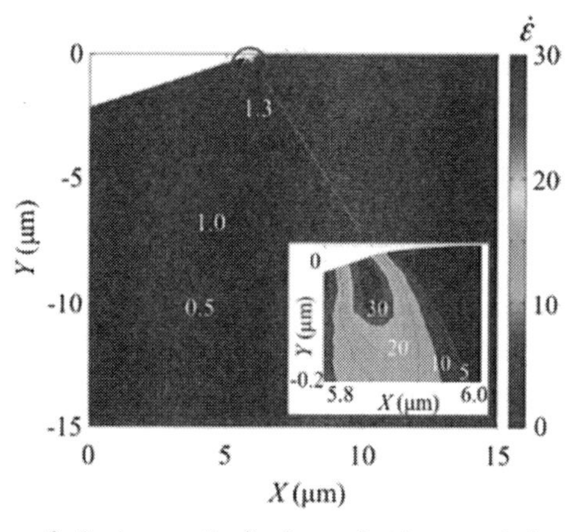

Figure 8. Strain rate distribution under the same indentation strain rate with Fig. 7 (a) ($\dot{h} = 10\,\mu$m/s, $h = 2\,\mu$m).

4.2. Calculation of average strain rate

A process of calculation for the average strain rate is shown below.

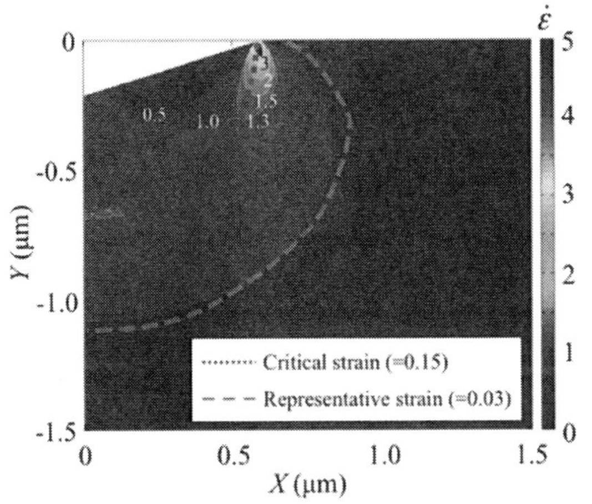

Figure 9. Relationship between the boundary value and the strain rate distribution at $\dot{h} = 1\,\mu$m/s and $h = 0.2\,\mu$m, which shows the miniature of the strain rate range of Fig. 7(a).

(a) The strain rate of each element, $\dot{\varepsilon}_{elm}$, is calculated as follows:

$$\dot{\varepsilon}_{elm} = \frac{1}{4}\sum_{k=1}^{4}\dot{\varepsilon}_k \tag{7}$$

where, $\dot{\varepsilon}_k$ is the strain rate at each node point making up the element.

(b) An area filling $\varepsilon_{elm} \geq \varepsilon_c$ or ε_R is extracted in the equivalent plastic strain distributions.

(c) A volume size at each element in the area noted above, V_{elm}, is calculated, since the elements are different from each other in this study. The area at each element, S_{elm}, is calculated using the shoelace formula as follows:

$$S_{elm} = \frac{1}{2}\sum_{k=1}^{4}\left(x_k y_{k+1} - x_{k+1} y_k\right) \tag{8}$$

where x_k and y_k are coordinate data of each node making up the element. V_{elm} is calculated using Pappus-Guldinus theorem as follows:

$$V_{elm} = S_{elm} \times 2\pi \times \frac{1}{4}\sum_{k=1}^{4} x_k. \tag{9}$$

(d) The average strain rate is calculated as follows:

$$\dot{\varepsilon}_{av,X} = \frac{\sum(\dot{\varepsilon}_{elm} \times V_{elm})}{\sum V_{elm}} \tag{10}$$

where, $\dot{\varepsilon}_{av,X}$ is the average strain rate at each boundary value.

4.3. Relationship between average strain rate and loading curvature

Figure 9 shows the relationships between the boundary value and the strain rate distribution at $\dot{h} = 1\,\mu$m/s and $h = 0.2\,\mu$m. The representative strain encompassed a broad strain rate range area compared to the critical strain.

Table 1. The relationship between $\dot{\varepsilon}_{av,R}$ and the loading curvature.

Data / h (μm)	$\dot{\varepsilon}_{av,R}$ (s^{-1})		C (GPa)	
	1 μm/s	10 μm/s	1 μm/s	10 μm/s
0.2	7.90×10^{-1}	7.63×10^{0}	4.03	4.33
2	7.86×10^{-2}	7.88×10^{-1}	3.80	4.01
20	6.76×10^{-3}	7.95×10^{-2}	3.65	3.81

When calculated using the equation (10) at $\dot{h} = 1\,\mu$m/s and $h = 0.2\,\mu$m, the average strain rates of the critical strain and the representative strain were 7.25×10^{-1} s^{-1} and 7.90×10^{-1} s^{-1}, respectively. In both the boundary strains, the average strain rates reached approximately 10^{-1} s^{-1}. In this condition, the highest strain rate is approximately 3×10^{1} s^{-1}, as shown in Fig. 7(a). It was thus clear that the average strain rate was not affected by the highest strain rate around the crater rim.

The relationship between $\dot{\varepsilon}_{av,X}$ and the loading curvature were investigated. As an example, the terms of $\dot{\varepsilon}_{av,R}$ is listed in Table 1. The average strain rate was decreased with increasing the displacement. This is because the increase in volume at low strain rate area becomes marked when the displacement was increased. In addition, it was clarified that the loading curvature was decreased with increasing the displacement, which indicates the similar trend of the value of $\dot{\varepsilon}_{av,R}$.

Therefore, the average strain rates and the loading curvature under the same indentation strain rates were compared. The same indentation strain rates are indicated by the arrows in Table 1. The average strain rate and the loading curvature for the same indentation strain rates showed almost the same values. Thus, there is correlation between the average strain rate and the loading curvature, which show that the average strain rate can express as the representative of strain rate for the indentation technique.

5. Summary

In order to clarify the effect of high strain rate on the indentation technique, the effect of strain rate distribution of the indentation on pure aluminum with respect to the strain rate dependence of strength was investigated. The obtained results are summarized as follows.

(1) The effect of strain rate on the indentation during the loading was replicated using the implicit method, as is the case in the explicit method.

(2) The strain rate distribution was quite different from the strain distribution, which showed that the strain rate at the crater rim was higher than that beneath the indenter.

(3) The strain rate distributions of the same indentation strain rate have the similarity rule.

(4) We established the calculation method of the average strain rate, which can express as the representative of strain rate for the indentation technique.

This study was supported in part by the Grant-in-Aid for Young Scientists (A) (Grant No. 25709004) from the Ministry of Education, Culture, Sports, Science and Technology, Japan. The authors also wish to thank the Light Metal Educational Foundation, Inc., Osaka in Japan for financial support.

References

[1] Y.T. Cheng, C.M. Cheng, Mater. Sci. Eng. **R44**, 91 (2004).
[2] A. Gouldstone, N. Chollacoop, M. Dao, J. Li, A.M. Minor, Y.-L. Shen, Acta Mater. **55**, 4015 (2007).
[3] N. Ogasawara, X. Chen, N. Chiba, J. Mater. Res. **21**, 947 (2006).
[4] M.F. Doerner, W.D. Nix, J. Mater. Res. **1**, 601 (1986).
[5] E. Harsono, S. Swaddiwudhipong, Z.S. Li, L. Shen, Int. J. Solids Structures **48**, 972 (2011).
[6] H. Yamada, Y. Shimizu, N. Ogasawara, X. Chen, J. JSEM **12**, 88 (2012).
[7] L. Liu, N. Ogasawara, N. Chiba, X. Chen, J. Mater. Res. **3**, 784 (2009).
[8] N. Ogasawara, H. Yamada, Proc. JSEM 2010 Annu. Conf. Exp. Mech., 247 (2010).
[9] M. Dao, N. Chollacoop, K.J. Van Vliet, T.A. Venkatesh, S. Suresh, Acta mater. **49**, 3899 (2001).
[10] N. Ogasawara, N. Chiba, X. Chen, J. Mater. Res. **20**, 2225 (2005).

EPJ Web of Conferences **94**, 04035 (2015)
DOI: 10.1051/epjconf/20159404035

Multiaxial behavior of foams – Experiments and modeling

Laurent Maheo[1,2,a], Sandra Guérard[3,b], Gérard Rio[2,c], Adrien Donnard[3,d], and Philippe Viot[3,e]

[1] Ecoles de Saint-Cyr Coëtquidan, CREC, 56381 Guer, France
[2] Université de Bretagne Sud, LIMATB, 56100 Lorient, France
[3] Arts et Metiers ParisTech, Campus de Bordeaux-Talence, I2M, 33405 Talence, France

Abstract. Cellular materials are strongly related to pressure level inside the material. It is therefore important to use experiments which can highlight (i) the pressure-volume behavior, (ii) the shear-shape behavior for different pressure level. Authors propose to use hydrostatic compressive, shear and combined pressure-shear tests to determine cellular materials behavior. Finite Element Modeling must take into account these behavior specificities. Authors chose to use a behavior law with a Hyperelastic, a Viscous and a Hysteretic contributions. Specific developments has been performed on the Hyperelastic one by separating the spherical and the deviatoric part to take into account *volume change* and *shape change* characteristics of cellular materials.

1. Introduction

Foams and more generally cellular materials are specifically used as filling material in sandwich structures with a function of energy dissipation. Their ability to resist to impact and shock, combined with their low density, now make them ideal candidates for passive safety and energy dissipation [1].

Foam behavior has been studied by authors in numerous publications according to simple characterization tests that allow obtaining input parameters for the models proposed by industrial codes [2]. Classically, only uniaxial compressive experiments are needed to feed the numerical model. Even if the numerical models succeed in describing foam behavior under uniaxial loading conditions, they are not well adapted to represent multiaxial behavior occurring under complex loadings, such as punch impacts. To improve mechanical description of foams, Hanssen *et al.* [3] proposed several experiments. One of this test which is very important to describe foam behavior is the hydrostatic compressive test [4] and some numerical models take into account the behavior under hydrostatic pressure conditions [2].

Aware of the importance of hydrostatic pressure on foams behavior, other authors have developed hydrostatic compressive [4] and multi-axial tests [5]. These tests are performed to achieve a representation of a load surface (or behavior transition) of foams in the equivalent stress plane as a function of hydrostatic stress. Other representation that takes into account, independently, loading direction characterized by the Lode angle, hydrostatic pressure and stress intensity has also been used recently and seems to be a good way to describe cellular materials [5].

In this scientific framework, authors decided to investigate experimentally, and numerically, cellular materials behavior. Foam behavior will be described numerically using a generalist yet powerful phenomenological model, namely HVH (Hyperelastic, Viscous, Hysteretic) model [6]. The HVH model has been enhanced to take into account foam specificities such as *volume change* and *shape change* sensitivities, and the coupling behavior between the *volume change* and the *shape change* behaviors. Experimental study has been led to get mechanical characteristics of the numerical model and to guide the choice for the numerical developments. Shear tests have therefore been performed to get *shape change* behavior. Shear tests under different level of pressure can be possible to identify the coupling between pressure and shear by using an original hexapod device that allows combined loadings, such as uniaxial compression/tension and shear tests. Digital Image Correlation technique has been used to analyze the area of the specimen that is subjected to shear behavior and obtain the shear strain field. The stress is calculated thanks to a 3-dimensional sensor. Experimental results clearly show that shear behavior is related to the level of hydrostatic pressure. Numerical model has therefore been improved accordingly and shows good agreement to describe foam behavior under multiaxial loading conditions.

2. Experiments

Every tests presented is this paper have been performed on a polypropylene (PP) foam, named Arpro© and produced by JSP society. The density of the foam is between 90 and 100 kg.m^3.

2.1. Uniaxial compressive test

Uniaxial compressive tests can be performed in quasi-static conditions using an electromechanical device or in

[a] e-mail: laurent.maheo@univ-ubs.fr
[b] e-mail: sandra.guerard@ensam.eu
[c] e-mail: gerard.rio@univ-ubs.fr
[d] e-mail: adrien.donnard@ensam.eu
[e] e-mail: philippe.viot@ensam.eu

Article available at http://www.epj-conferences.org or http://dx.doi.org/10.1051/epjconf/20159404035

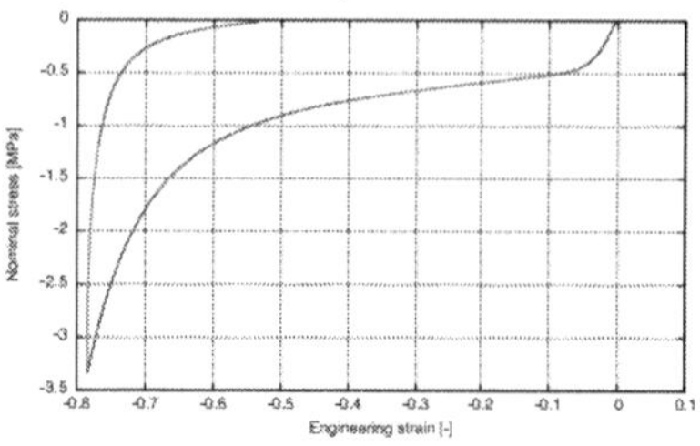

Figure 1. Classical compressive test on a PP foam. Evolution of the nominal compressive stress versus the engineering compressive strain.

Figure 2. Classical compressive test on a PP foam. Foam specimen for engineering strain values of 0.00, 0.35 and 0.70.

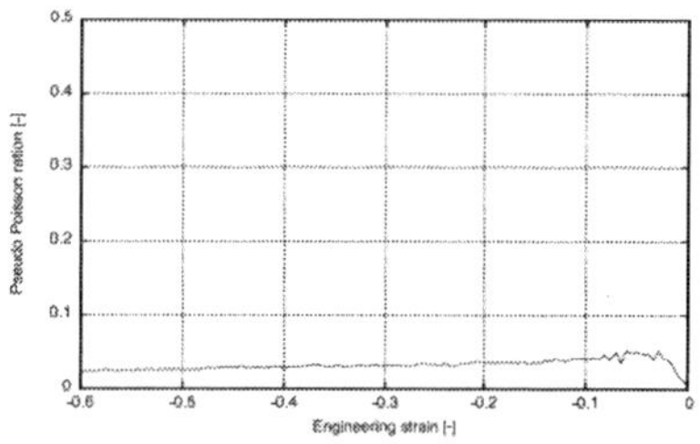

Figure 3. Classical compressive test on a PP foam. Evolution of the pseudo Poisson's ratio versus the engineering compressive strain.

dynamic conditions using a flywheel device or a Split-Hopkinson Pressure Bar device as in [7–9]. Classical foam behavior is observed in Fig. 1 with a first elastic stage, a second stress plateau stage and a third densification stage. Nominal stress is classical calculated: $\sigma_{nom} = \frac{F_z}{S_0}$ where F_z is the force along the z-axis and S_0 is the cross section area.The engineering strain is calculated as: $\varepsilon_{eng} = \frac{h-h_0}{h_0}$ with h and h_0 the current and the initial height of the sample, respectively.

It can also be noticed in Fig. 2 that the transversal displacement of the foam under uniaxial compressive loading conditions is quite small even for large transformations.

In this context of large transformations a pseudo Poisson's ratio ν_p can be calculated using image analysis. Every images obtained during the test is analyzed with an ImageJ© procedure. (i) An area is defined at the center of the foam specimen and related to its height. (ii) A threshold and an outline macro give the outlines of the specimen. (iii) The specimen width over the initial width ratio is calculated for every image using a mean of the position of each specimen side. The pseudo Poisson's ratio can be calculated as: $\nu_p = -\frac{l-l_0}{l_0}\frac{h_0}{h-h_0}$ with l and l_0 the current and the initial width, respectively. Its evolution can be displayed versus the engineering compressive strain in Fig. 3. It can be observed that during the compressive test, the pseudo Poisson's ratio is about 0.03.

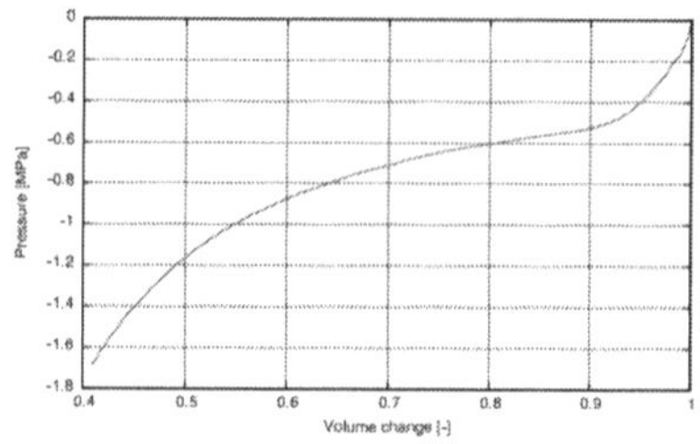

Figure 4. Hydrostatic compressive test on a PP foam. Evolution of the pressure versus the *volume change*.

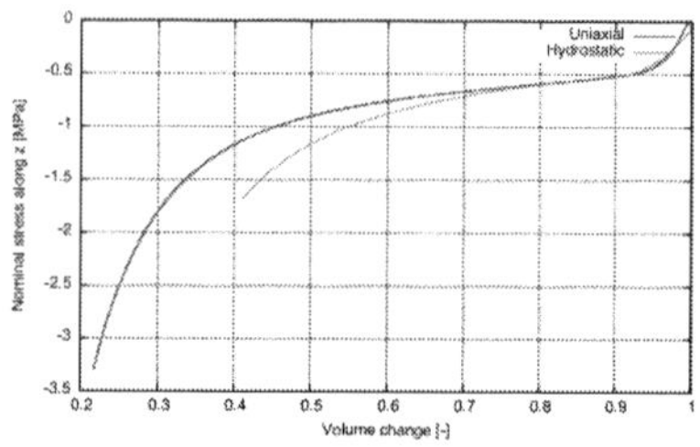

Figure 5. Comparison of uniaxial and hydrostatic compressive tests on a PP foam. Evolution of the σ_z versus the *volume change*.

2.2. Hydrostatic compressive test

Hydrostatic compressive tests on PP foam in quasi-static and dynamic conditions have already been performed by Viot and published in [4]. The results are presented Fig. 4.

The shape of the hydrostatic curve looks like the one of the uniaxial compressive test. For this reason, it seems interesting to plot both curves of the uniaxial and hydrostatic compressive tests on the same figure. With the assumption of a null-pseudo Poisson's ratio, the relation between the engineering strain ε_{eng} and the volume variation V is: $V = 1 - \varepsilon_{eng}$. If the results are presented with σ_z vs. V, Fig. 5, it can be noticed that the hydrostatic behavior is quite the same than the uniaxial compressive behavior. The level of the stress plate is almost the same and is approximately equal to 0.5–1.0 MPa. At the beginning of both tests, before a volume change of $\varepsilon_z = 0.7$, the difference between two tests does note exceed 10%. This shows the importance of the *volume change* contribution on the foam behavior.

2.3. Shear test

2.3.1. *Discussion about specimen dimensions*

In order to evaluate the *shape change* contribution, shear tests are needed. However, if we want to exhibit the influence of the pressure on the shear behavior, combined compressive-shear experiments should be used. With these considerations, the shape of foam specimen must allow to perform combined compressive-shear tests using an hexapod device, see Figs. 6 and 7.

The Hexapod device allows to perform three translations and three rotations. An additional seventh electromechanical jack can be independently used to improve compressive velocity but has been locked for this study. A three-force sensor is located at the top of the foam specimen in order to record the force vector during

Figure 6. Hexapod device using for combined shear-compressive tests.

Figure 7. Foam specimen before a combined shear-compressive test.

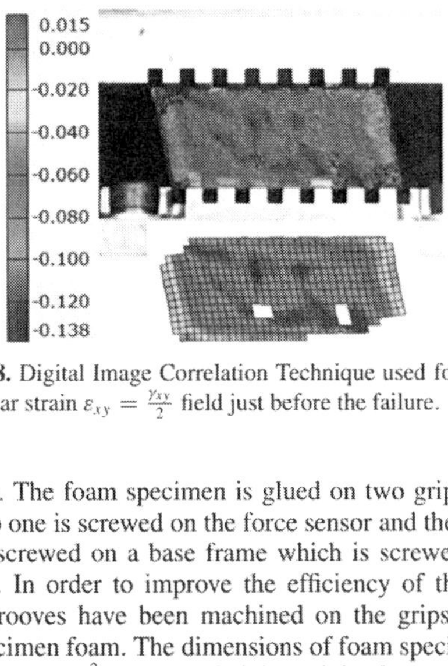

Figure 8. Digital Image Correlation Technique used for a shear test. Shear strain $\varepsilon_{xy} = \frac{\gamma_{xy}}{2}$ field just before the failure.

the test. The foam specimen is glued on two grip pieces. The top one is screwed on the force sensor and the bottom one is screwed on a base frame which is screwed in the ground. In order to improve the efficiency of the grips, 5mm-grooves have been machined on the grips and on the specimen foam. The dimensions of foam specimen are $75\times30\times40$ mm^3 in length, height and depth, respectively. These dimensions do not guarantee pure shear test as mentioned by Bouvier *et al.* [10]. Indeed, some flexural phenomena occur at both side of the specimen and can not be neglected. However this test can give some first needed informations.

2.3.2. Shear test results

Three quasi-static shear tests have been performed at a strain rate of $\dot{\gamma}_{xy} = 10^{-2}$ s^{-1} until a strain level of $\gamma_{yx}^{max} = 1$. Digital Image Correlation technique has been used to measure strain on the specimen using a zone of interest of 128 pixels (see Fig. 8). It can be seen two areas

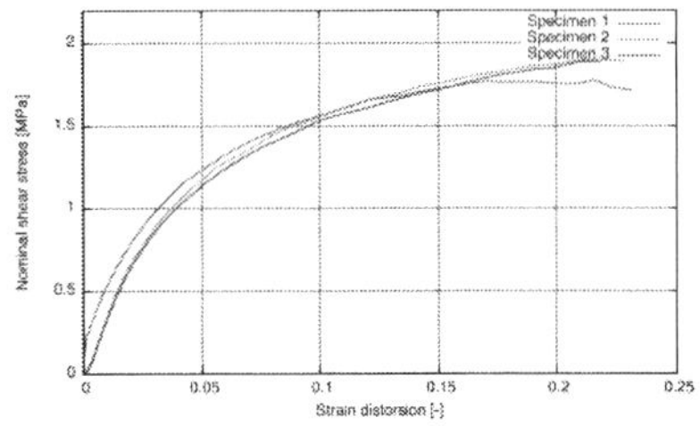

Figure 9. Shear behavior with the evolution of the shear stress τ_{xy} vs. the strain distorsion γ_{xy} until the failure.

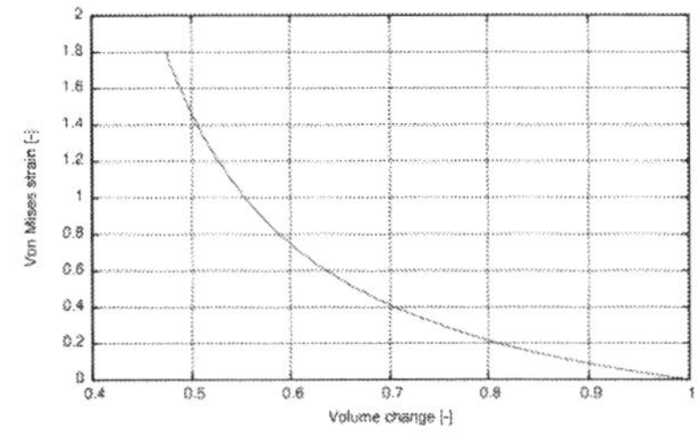

Figure 10. Evolution of the *shape change* vs. the *volume change* during an uniaxial compressive test with a pseudo Poisson's ration of $\nu = 0.03$.

on this figure. The first one is the area which is located at both side of the specimen and is subjected to flexion behavior. The size of this area decreases with the height of the specimen. The second one is a non-homogeneous strain field area which is located at the center of the specimen. This non-homogeneity is due to foam beads with different densities which involve localization of deformation [9]. A average of strain field can be performed in this area to get the mean strain distorsion. The failure occurs for the three tested specimen at a strain level of $\gamma_{yx}^{mean} = 0.224 \pm 0.06$.

Using the measurements of the three-axis force sensor, it can be possible to plot the evolution of the shear stress vs. the strain distorsion (see Fig. 9).

It can be shown that the results are quite repeatable and the shear stress reach a value of 2 MPa, higher than the stress value plateau for hydrostatic and uniaxial compressive test. Aware of the presence of flexion behavior on both side of the specimen, future tests on specimen with an adapted geometry need to be performed.

2.4. Experiments summary

In order to sum up the three previous experiments using only one diagram, authors decided to plot the evolution of the principal stress σ_I vs. two independent variables:

- the volume change V,
- the shape change using the Von Mises strain $\varepsilon_{VM} = \sqrt{\frac{3}{2}\bar{\bar{\varepsilon}}_D : \bar{\bar{\varepsilon}}_D}$ where $\bar{\bar{\varepsilon}}_D$ is the deviatoric strain tensor.

Hydrostatic compressive test and shear test assume that no *shape change* and no *volume change* occur during each test, respectively. Thanks to the measurement of transversal strain during the uniaxial compressive test, i.e. $\nu = 0.03$, it is possible to plot the evolution of the *shape change* vs. the *volume change*, see Fig. 10.

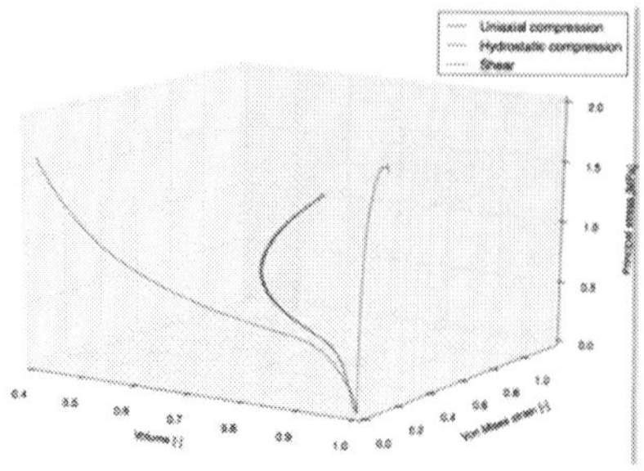

Figure 11. Three dimensional representation of the three experiments. Evolution of the principal stress vs. the *shape change* (Von Mises strain ε_{VM}) and vs. the *volume change* (Vol. V).

It is therefore possible to plot in the three axis (Prinicpal stress, Volume and Von Mises strain) the evolutions of each experiment, see Fig. 11. The hydrostatic compressive test and the shear test do not depend of the *shape change* and the *volume change*, respectively. Concerning the evolution of the uniaxial compressive test, it can be noted that at the beginning of the test, the foam is mainly subjected to *volume change* and progressively, the contribution of *shape change* increases. With these considerations, we can say that the *shape change* is related to the pressure level inside the foam and the modeling development must take into account this key point.

3. Modeling

3.1. HVH model

The numerical model used to describe foam behavior is the Hyperelastic – Viscous – Hysteretic (HVH) model. It has been used successfully to describe the behavior of materials such as shape memory materials NiTi [11], elastomeric materials [6] or polypropylene materials [12]. This model is based on the addition of three powers $\mathcal{P}_{hyp}$, $\mathcal{P}_{vis}$ and $\mathcal{P}_{hys}$ describing the three behavior contributions:

$$\begin{aligned}\mathcal{P}_{int} &= \mathcal{P}_{hyp} + \mathcal{P}_{vis} + \mathcal{P}_{hys}\\ \int_{\mathcal{D}} \bar{\bar{\sigma}} : \bar{\bar{D}}\, dV &= \int_{\mathcal{D}} \left(\bar{\bar{\sigma}}_{hyp} + \bar{\bar{\sigma}}_{vis} + \bar{\bar{\sigma}}_{hys}\right) : \bar{\bar{D}}\, dV\end{aligned} \quad (1)$$

where $\mathcal{P}_{int}$ is the internal power, $\bar{\bar{\sigma}}_{hyp}$, $\bar{\bar{\sigma}}_{vis}$ and $\bar{\bar{\sigma}}_{hys}$ are the hyperelastic, the viscous and the hysteretic stress tensor, $\bar{\bar{D}}$ is the strain rate tensor and $\mathcal{D}$ the studied domain.

The viscous model chosen for this study is the classical generalized Maxwell's model composed of n spring-dashpot in parallel. The parameters of the model can be identified by relaxation tests. The hysteretic model is the one used in the research works of Rio, Favier *et al.*, for example in [11]. The hysteretic stress σ_{hys} is assumed to be purely deviatoric and irreversible. These two previous contributions do not concern the topic of this article and readers can find more explanations about the hysteretic model in [11, 13, 14].

The hyperelastic contribution is developed by treating individually the spherical and the deviatoric behaviors in order to take into account the differences between the *shape change* and the *volume change* behaviors in cellular materials. The hyperelastic potential ω_{hyp} can be written as the contribution of a spherical ω_{hyp}^{sph} and a deviatoric ω_{hyp}^{dev} contribution:

$$\omega_{hyp} = \omega_{hyp}^{sph} + \omega_{hyp}^{dev}. \quad (2)$$

3.2. Hyperelastic model – spherical behavior

The spherical contribution of the hyper elastic model is only related to the *volume change* V. With a phenomenological approach, it is possible to describe hydrostatic behavior using a polynomial form of potential equation which check (i) the positivity of the potential over the volume change and (ii) the null-value of the potential for null-*volume change*, i.e. $V = 1$. With these considerations, a hyperelastic potential is proposed:

$$\omega_{hyp}^{sph}(V) = a_i V^i\, (V-1)^2 \quad (3)$$

where a_i are the parameters of the i-order polynom. The pressure can be obtained easily from the potential. The free energy E can be obtained by:

$$E = \int_{\mathcal{D}} \omega dV = \int_{\mathcal{D}_{ref}} \omega \sqrt{g}\, d\theta^i \quad (4)$$

where g is the determinant of the metric tensor and θ_i the material coordinates. The power $\dot{E}$ can be calculated using classical equations of continuum mechanics with the Cauchy stress tensor $\bar{\bar{\sigma}}$ and the strain rate tensor $\bar{\bar{D}}$ based on the Almansi's strain:

$$\dot{E} = \int_{\mathcal{D}_{ref}} \frac{d(\omega\sqrt{g})}{dt}\, d\theta^i = \int_{\mathcal{D}_{ref}} \bar{\bar{\sigma}} : \bar{\bar{D}}\, \sqrt{g}\, d\theta^i. \quad (5)$$

This involves that:

$$\bar{\bar{\sigma}} : \bar{\bar{D}} = \frac{1}{\sqrt{g}} \frac{\partial(\omega\sqrt{g})}{\partial t}. \quad (6)$$

In the case of an hydrostatic compressive test,

$$\bar{\bar{\sigma}} : \bar{\bar{D}} = -p\, \mathbf{I}_{\bar{\bar{D}}} = -p\, \frac{\dot{V}}{V} \quad (7)$$

where $\mathbf{I}_{\bar{\bar{D}}}$ is the first invariant, i.e. the trace, of the tensor $\bar{\bar{D}}$. By developing equation 6 with $V = \sqrt{g}/\sqrt{g_0}$ and using Eq. (7), it finally comes:

$$-p = V\, \frac{\partial \omega}{\partial V} + \omega. \quad (8)$$

It is therefore possible to express the pressure $p(V)$ with the a_i parameters of the model and for a 4th-order polynom, it gives for example:

$$\begin{aligned}\omega(V) &= a_0 + (-2a_0 + a_1)V^1 + (a_0 - 2a_1 + a_2)V^2\\ &+ (a_1 - 2a_2 + a_3)V^3 + (a_2 - 2a_3 + a_4)V^4\\ &+ (a_3 - 2a_4)V^5 + (a_4)V^6\end{aligned} \quad (9)$$

$$\begin{aligned}-p(V) &= a_0 + 2(-2a_0 + a_1)V^1 + 3(a_0 - 2a_1 + a_2)V^2\\ &+ 4(a_1 - 2a_2 + a_3)V^3 + 5(a_2 - 2a_3 + a_4)V^4\\ &+ 6(a_3 - 2a_4)V^5 + 7(a_4)V^6\end{aligned} \quad (10)$$

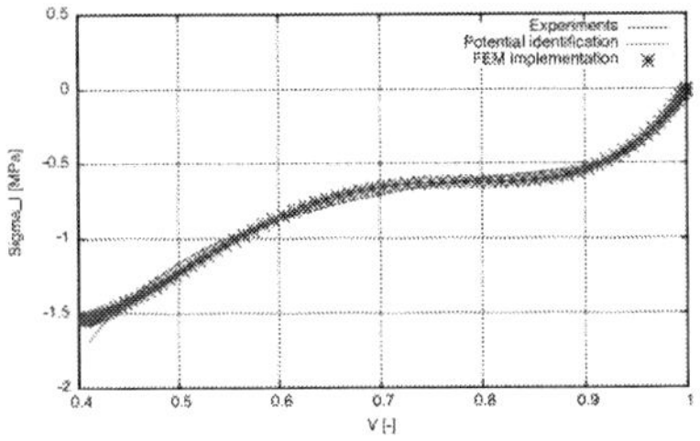

Figure 12. Spherical potential ω_{hyp}^{sph} identification thanks to equation 11 and FEM implementation validation in Herezh++ [15].

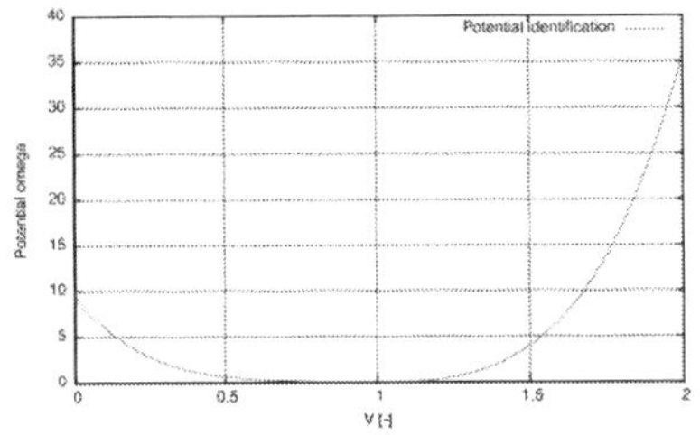

Figure 13. Spherical potential ω_{hyp}^{sph} shape.

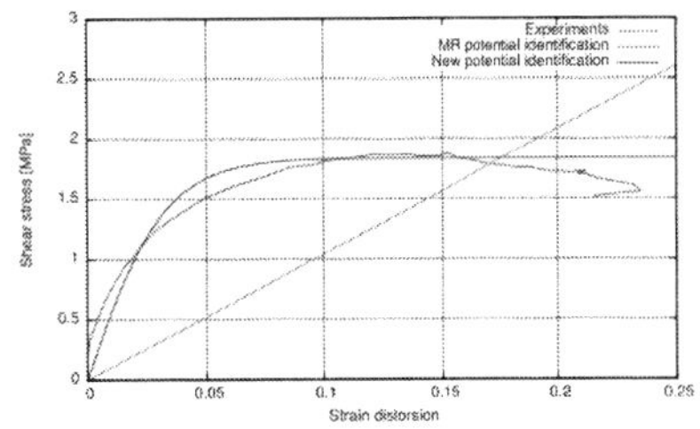

Figure 14. Deviatoric potential ω_{hyp}^{sph} identification thanks to Eq. (18).

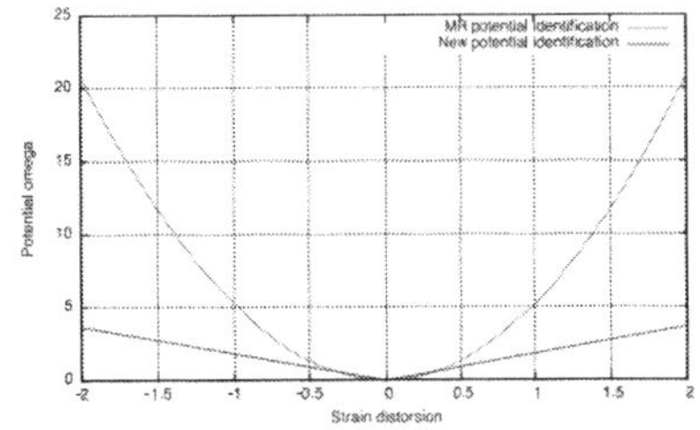

Figure 15. Deviatoric potential ω_{hyp}^{dev} shape.

The determination of the spherical model parameters becomes easy with the results of an hydrostatic compressive test. The Figs. 12 and 13 show the identified model compared to experimental results and the shape of the spherical potential.

The use of a polynomial function for the spherical hyperelastic potential is quite accurate to model spherical behavior (see Fig. 12). This potential check the thermodynamic conditions with a positive value of $\omega_{hyp}^{sph} > 0\ \forall V$ and a null-value of $\omega_{hyp}^{sph} = 0 \Leftrightarrow V = 1$ (see Fig. 13).

3.3. Hyperelastic model – deviatoric behavior

Many deviatoric potentials exist and can be used for hyperelastic behavior. The Money-Rivlin [16,17] and Ogden [18] models are the famous ones. These potentials are based on the invariants of the Cauchy-Green stress tensor $\bar{\bar{B}}$ as the Mooney-Rivlin (MR) one:

$$\omega_{MR}(\mathbf{I}_{\bar{\bar{B}}}, \mathbf{II}_{\bar{\bar{B}}}, \mathbf{III}_{\bar{\bar{B}}}) = C_{10}(\bar{\mathbf{I}}_{\bar{\bar{B}}} - 3) + C_{01}(\bar{\mathbf{II}}_{\bar{\bar{B}}} - 3) \quad (11)$$

with:

$$\begin{aligned} \mathbf{I}_{\bar{\bar{B}}} &= \lambda_1^2 + \lambda_2^2 + \lambda_3^2 \\ \mathbf{II}_{\bar{\bar{B}}} &= \lambda_1^2\lambda_2^2 + \lambda_1^2\lambda_3^2 + \lambda_2^2\lambda_3^2 \\ \mathbf{III}_{\bar{\bar{B}}} &= \lambda_1^2\lambda_2^2\lambda_3^2 \end{aligned} \quad (12)$$

where λ_i are the stretches of the Cauchy-Green tensor and:

$$\bar{\mathbf{I}}_{\bar{\bar{B}}} = \mathbf{I}_{\bar{\bar{B}}} \cdot \mathbf{III}_{\bar{\bar{B}}}^{-\frac{1}{3}} \quad \text{and} \quad \bar{\mathbf{II}}_{\bar{\bar{B}}} = \mathbf{II}_{\bar{\bar{B}}} \cdot \mathbf{III}_{\bar{\bar{B}}}^{-\frac{2}{3}}. \quad (13)$$

The use of the Cauchy-Green stress tensor is a good way to separate *shape change* and *volume change*. However, the parameters of the MR potential are difficult to identify directly and moreover in a case of shear test. Indeed, the shear test involves these three principal stretches:

$$\lambda_{1,2}^2 = \left(1 + \frac{\gamma^2}{2}\right) \pm \left(\gamma\sqrt{1 + \frac{\gamma^2}{4}}\right) \quad \text{and} \quad \lambda_3^2 = 1 \quad (14)$$

with γ is the strain distorsion and the invariants can be calculated as:

$$\bar{\mathbf{I}}_{\bar{\bar{B}}} = \bar{\mathbf{II}}_{\bar{\bar{B}}} = \lambda^2 + \frac{1}{\lambda^2} + 1 \quad (15)$$

which leads to an indetermination to identify separately the C_{10} and the C_{01} parameters. With these considerations, a new model based on the potential of Favier used in [6] and using the Cauchy-Green stress tensor invariants has been proposed.

$$\omega(\bar{\mathbf{I}}_{\bar{\bar{B}}}) = \frac{Q_\sigma^2}{\mu_1} \ln\left[\cosh\left(\frac{\mu_1}{Q_\sigma}\sqrt{(\bar{\mathbf{I}}_{\bar{\bar{B}}} - 3)}\right)\right] + \mu_2\,(\bar{\mathbf{I}}_{\bar{\bar{B}}} - 3) \quad (16)$$

with Q_σ, μ_1 and μ_2 the three parameters of the deviatoric potential.

It is therefore possible to calculate the stress from the expression of the deviatoric tensor:

$$\sigma^{ij} = \frac{\partial\omega}{\partial\varepsilon_{Aij}} = \frac{\partial\omega}{\partial\bar{\mathbf{I}}_{\bar{\bar{B}}}}\frac{\partial\bar{\mathbf{I}}_{\bar{\bar{B}}}}{\partial\gamma}\frac{\partial\gamma}{\partial\varepsilon_{Aij}} \quad (17)$$

and gives:

$$\sigma^{12} = Q_\sigma \tanh\left(\frac{\mu_1}{Q_\sigma}\gamma\right) + 2\mu_2\,\gamma. \quad (18)$$

The three parameters of the deviatoric potential has been identified with $\mu_1 = 56.2$ MPa, $Q_\sigma = 1.84$ MPa and μ_2 has been fixed to 0 MPa in order to prevent the potential from being negative. Results are presented in Figs. 14 and 15. New potential identification can be performed easily compared with MR potential because the value of $\mu_1 + \mu_2$ is the slope of the first part of the identified curve, μ_2 is the slope of the second part of the identified curve and Q_σ is the intersection of the second slope and the ordinate axis.

4. Conclusion

This research work proposed to study a PP foam by performing some experiments to exhibit its *volume change* and its *shape change* by using hydrostatic compressive test and shear test, respectively. The dependance of pressure level for cellular materials leads the authors to develop an original combined pressure-shear test using an hexapod device.

Finite Element modeling must take into account spherical and deviatoric contributions separately. With these considerations, two Hyperelastic models have been proposed and studied. The first spherical one is based on a polynomial function and can be easy identified from a hydrostatic compressive test. The second deviatoric one is based on the first invariant of the Cauchy-Green stress tensor and can also be easy to identify using a shear test.

Further works will focus on the combined compressive-shear test to get the *shape change* under different pressure level. These experiments will give informations about the coupling between *shape change* and *volume change*. Hyperelastic model will be therefore adapted.

References

[1] N.J. Mills, C. Fitzgerald, A. Gilchrist, R. Verdejo. Compos Sci Technol **63**(16), 2389–2400 (2003)

[2] O. Hallquist. LS-Dyna3D theoritical manual, Livermore Software Technology (1998)

[3] A.G. Hanssen, O.S. Hopperstad, M. Langseth, H. Ilstad. Int J Mech Sci **44**(2), 359–406 (2002)

[4] P. Viot. International Journal Impact Engineering **36**, 975–989 (2009)

[5] E. Combaz, C. Bacciarini, R. Charvet, W. Dufour, F. Dauphin, A. Mortensen. Acta Materialia **58**, 5168–5183 (2010)

[6] A. Vandenbroucke, H. Laurent, N. Ait Hocine, G. Rio. Computational Materials Science **48**, 495–503 (2010)

[7] P. Viot, F. Beani, J.-L. Lataillade. J Mater Sci **40**, 5829–5837 (2005)

[8] R. Bouix, P. Viot, J.-L. Lataillade. Int J Imp Eng **36**, 329–342 (2009)

[9] L. Maheo, P. Viot. Int J Imp Eng **53**, 84–93 (2013)

[10] S. Bouvier, H. Haddadi, P. Levée, C. Teodosiu. J Mater Proc Technol **172**(1), 96–103 (2006)

[11] B.S. Shariat, Y. Liu, G. Rio. J Intel Mater Syst Struct **25**(12), 1445–1455 (2014)

[12] M. Zrida, H. Laurent, G. Rio, S. Pimbert, V. Grolleau, N. Masmoudi, C. Bradai. Comput Mater Sci **45**(2), 516–527 (2009)

[13] P. Guélin. J Mécanique **19**(2), 217–247 (1980)

[14] P. Pegon. PhD thesis, Institut National Polytechnique, Université Joseph Fourier, Grenoble (1988)

[15] G. Rio. Herezh++: FEM software for large transformations in solids. Dépôt APP - Certification IDDN.FR.010.0106078.000.R.P.2006.035.20600, LIMATB-UBS (2015)

[16] M. Monney. J Appl Phys **11**(9), 582–592 (1940)

[17] R.S. Rivlin. Philos Trans Roy Soc London Ser A? Math Phys Eng Sci A **241**(835), 379–397 (1948)

[18] R.W. Ogden. Mineola (NY): Dover Publication Inc. (1984)

EPJ Web of Conferences **94**, 04036 (2015)
DOI: 10.1051/epjconf/20159404036

Investigation into stress wave propagation in metal foams

Lang Li, Pu Xue[a], and Yue Chen

School of Aeronautics, Northwestern Polytechnical University, Xi'an 710072, PR China

Abstract. The aim of this study is to investigate stress wave propagation in metal foams under high-speed impact loading. Three-dimensional Voronoi model is established to represent real closed-cell foam. Based on the one-dimensional stress wave theory and Voronoi model, a numerical model is developed to calculate the velocity of elastic wave and shock wave in metal foam. The effects of impact velocity and relative density of metal foam on the stress wave propagation in metal foams are explored respectively. The results show that both elastic wave and shock wave propagate faster in metal foams with larger relative density; with increasing the impact velocity, the shock wave propagation velocity increase, but the elastic wave propagation is not sensitive to the impact velocity.

1. Introduction

Metal foam is one kind of ideal lightweight energy-absorbing material, which can both bear great deformation and maintain relatively constant stress level under compression loading. It has been widely applied in various engineering fields, especially under impact environment [1,2]. Therefore, it is particularly important to study the dynamic characteristics under high-speed impact.

Research in dynamic mechanical behaviour of metal foams mainly includes three aspects, namely, the strain rate effect of base material, the inertia effect and the stress wave propagation in foams. The existing study pay much attention to the former two aspects. Some researchers point that the dynamic stress enhancement in metal foams under dynamic impact is mainly caused by inertia effect, other than strain rate effect of base material [3,4], however, some other researchers hold the opposite point [5,6]. As for the stress wave propagation, some idealized stress strain models, such as RPPL, RLHPL, EPPL etc. [7–9] are proposed to describe the stress wave propagation in metal foams. However, the existing research is limited in the study of stress wave propagation velocity, due to the limitation of test methods. In this paper, a numerical model will be developed to calculate the propagation velocity of stress wave.

2. Numerical modelling

3D-Voronoi model with uniform cell wall thickness is constructed by Voronoi tessellations to represent aluminum alloy closed-cell foam. The bilinear strain-hardening relationship is used to represent the true stress-strain relationship of the base material. The material properties are listed in Table1. The different relative densities are obtained by changing the thickness of cell walls, as follows

$$\rho^*/\rho_s = \frac{\sum_{i=1}^{n} A_i \cdot t}{V} \quad (1)$$

in which ρ^* is the density of foam, ρ_s is the density of base material, A_i is the area of the cell wall and V is the total volume of foam model. n is the number of cell wall and t is the cell wall thickness.

The foam model is set between two rigid plates, and the top plate moving towards the supporting plate at a prescribe velocity, as shown in Fig. 1. The dynamic compression process is implemented using commercialized software LS-DYNA.

3. Results and discussion

3.1. Elastic wave propagation

Under high-speed impact, elastic wave firstly forms in the metal foam and propagates from the impact side to the support side making the entire mental foam at the state of elastic deformation. Since the propagation of elastic wave costs some time to get through the metal foam, there exist time delay (Δt_e) between the stress curves at the impact side and the support side, as is shown in Fig. 3 where the prescribed impact velocity is 200m/s and the relative density is 0.1. Through dividing the length (l) of the metal foam by the time delay, the propagation velocity of elastic wave in metal foams with a given relative density under certain impact velocity can be obtained as

$$v_e = {}^{l}/_{\Delta t_e}. \quad (2)$$

Metal foams with different relative densities are compressed under four different impact velocities, 125, 150, 175 and 200(m/s) respectively, and the elastic wave propagation velocities are obtained through Eq. (2), as

[a] Corresponding author: p.xue@nwpu.edu.cn

Table 1. Material parameters of the base material.

Density (Kg/m3)	Young's modulus (GPa)	Tangent modulus (GPa)	Yield stress (MPa)	Poisson's ratio
2700	69	0.47	76	0.35

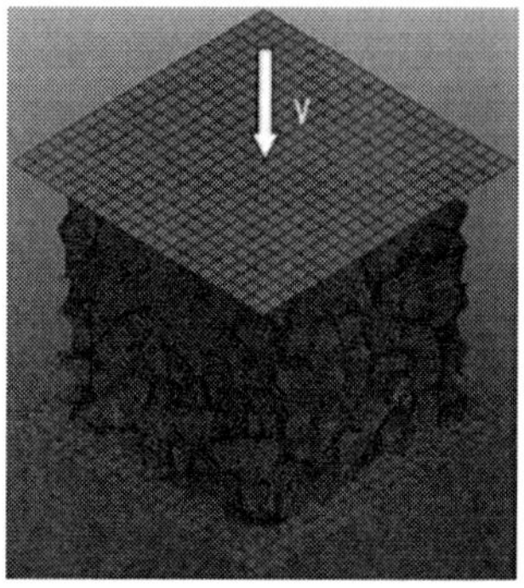

Figure 1. Finite element model of 3D-Voronoi foam.

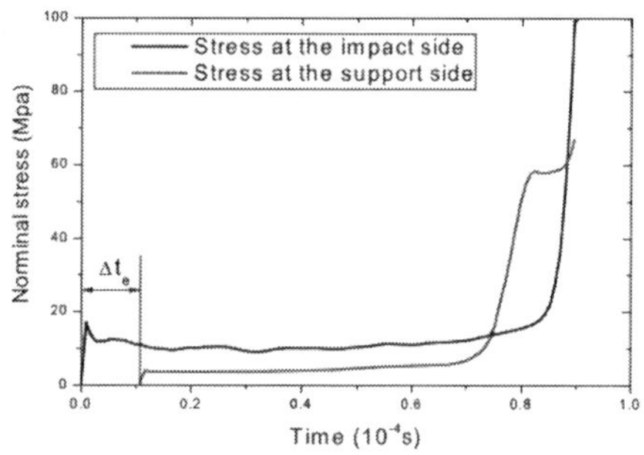

Figure 2. Stress vs. time curve for metal foam with the relative density of 0.1.

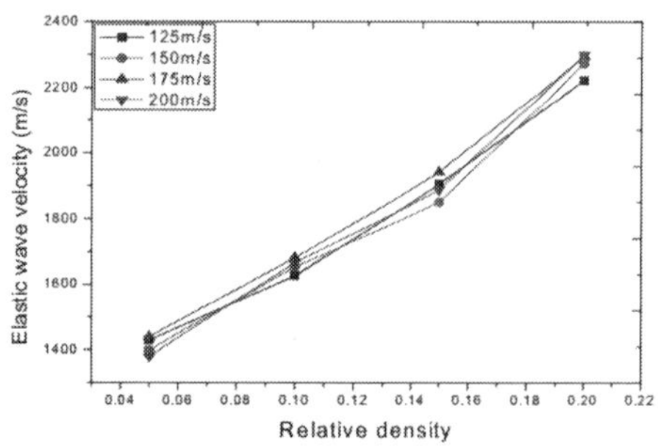

Figure 3. Elastic wave propagation velocity vs. relative density under various impact velocities.

is shown in Fig. 3. It is found that the elastic wave propagation velocity increase with the relative density become larger, but the effect of impact velocity on the propagation velocity of elastic wave is not obvious.

3.2. Shock wave propagation

When the impact load is large enough, the plastic collapse occurred , the deformation is localized and forms a narrow band, called a "shock front". The shock wave propagates

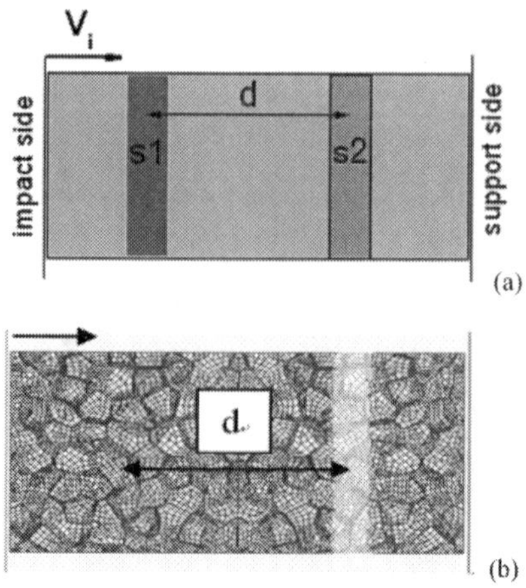

Figure 4. (a) Theoretical model, (b) Finite element model.

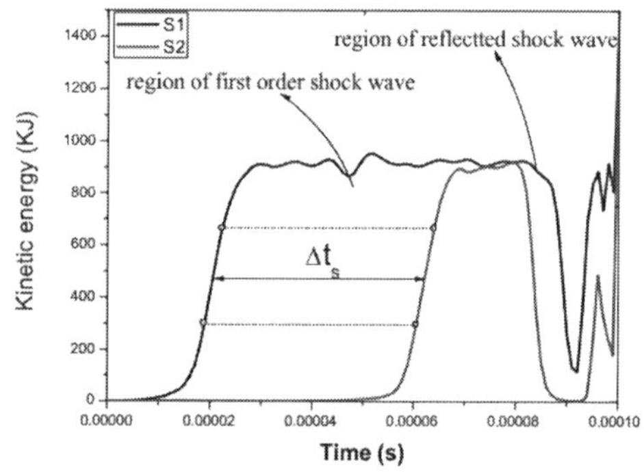

Figure 5. Kinetic energy vs. time curve for foam layer S1 and S2.

from the impact side to the support side. According to one-dimensional stress wave theory, ahead of the shock front, metal foam is assumed to be at rest, and behind the shock front, metal foam is fully compacted. When the shock wave reaches to the support side, the shock wave will be reflected and the metal foam is compressed again.

3.2.1. Calculation model of shock wave propagation velocity

When the metal foam is compressed at a constant impact velocity, two layers (S1 and S2) of metal foam with the same layer width are chosen to be the object of study, as is shown in Fig. 4. Figure 5 shows the total kinetic energy vs. time curve of S1 and S2 respectively. According to the analysis above, during the shock front pass through the foam layer S1 and S2, the material particles in S1 and S2 will experience three stages: before the shock front reaches, the layer of foam will be at rest and its total kinetic energy is zero; when the shock front is passing through, the particles are speed up from zero to a certain value and their total kinetic energy will experience a rise; when the shock leaves the layer, all particles in the layer have gained a certain velocity and their total kinetic energy will keep platform. By recording the time delay between S1 and S2, the shock wave propagation velocity in metal foams can be

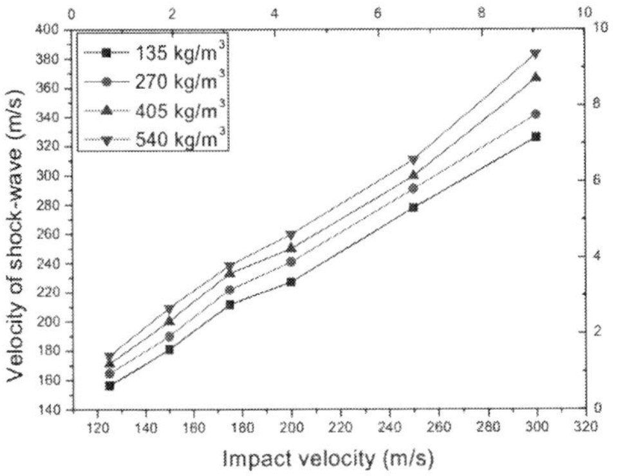

Figure 6. The velocity of shock wave vs. impact velocity for different density metal foams.

obtained as:

$$v_p = {}^{d}/_{\Delta t_s}. \tag{3}$$

Where d is the distance between S1 and S2, and Δt_s is the time for shock wave propagates from S1 to S2 which can be obtained from Fig. 5.

It can be seen from Fig. 5 that the kinetic energy curve alternately rising and falling, this is because when the shock wave arrives at the support side, surface reflection occurs and the metal foam undergo secondary compression process, in this paper, only the first order shock wave is discussed.

3.2.2. Investigation into shock wave propagation velocity

Figure 6 shows the shock wave propagation velocity vs. impact velocity curves for four different relative density metal foams, where the shock wave propagation velocity is calculated by Eq. (3). The diagram results show that the shock wave is dependent on the impact velocity and the density of the metal foam; with the increase of impact velocity the shock wave propagates faster in metal foam; for a given impact velocity, the shock wave propagates faster in metal foams with larger relative density.

4. Summary

Based on the one-dimensional stress wave theory and Voronoi model, a numerical model is developed investigate the stress wave propagation in metal foams. It is found that both elastic wave and shock wave propagate faster in metal foams with larger relative density; with increasing the impact velocity, the shock wave propagation velocity increase, but the elastic wave propagation is not sensitive to the impact velocity.

References

[1] G.X. Lu, T.X. Yu. Energy absorption of structures and materials(Cambridge University, Cambridge 2003)

[2] X.M. Qiu, J. Zhang, T.X. Yu, Int. J. Impact. Eng., **36**, 1231–1241 (2009)

[3] G.W. Ma, Z.Q.Ye, Z.S. Shao, Int. J. Impact. Eng., **36**, 6 (2009)

[4] Y.D. Liu, J.L. Yu, Z.J. Zheng, J.R. Li, Int. J. Solids. Struct., **46**, 3988–3998 (2009)

[5] A. Paul, U. Ramamurty, Mater. Scleng. A, **281**, 1–7 (2000)

[6] H. Zhao, Elnasri and H.J. Li. Adv. Eng. Mater, **8**, 877–883 (2006)

[7] S.R. Reid, C. Peng, Int. J. Impact. Eng., **19**, 531–570 (1997)

[8] P.J. Tan, S.R. Reid, J.J. Harrigan, Z. Zou, S. Li, J. Mech. Phys. Solids, **53**, 2206–2230 (2005)

[9] Z.J. Zheng, Y.D. Liu, J.L. Yu, S.R. Reid, Int. J. Impact. Eng., **42**, 66–79 (2012)

EPJ Web of Conferences **94**, 04037 (2015)
DOI: 10.1051/epjconf/20159404037

Theoretical study of the porosity effects on the shock response of graphitic materials

Nicolas Pineau[1,a], Emeric Bourasseau[1], Jean-Bernard Maillet[1], Laurent Soulard[1], and David Hébert[2]

[1] CEA/DAM/DIF, 91297 Arpajon, France
[2] CEA/DAM/CESTA, 33116 Le Barp, France

Abstract. In this paper we present a theoretical study of the shock compression of porous graphite by means of combined Monte Carlo and molecular dynamics simulations using the LCBOPII potential. The results show that the Hugoniostat methods can be used with "pole" properties calculated from porous models to reproduce the experimental Hugoniot of pure graphite and diamond with good accuracy. The computed shock temperatures show a sharp increase for weak shocks which we analyze as the heating associated with the closure of the initial porosity. After this initial phase, the temperature increases with shock intensity at a rate comparable to monocrystalline graphite and diamond. These simulations data can be exploited in view to build a full equation of state for use in hydrodynamic simulations.

1. Introduction

The responses of carbon materials under high pressure and temperature constraints have motivated a large number of experimental and theoretical studies in the academic and applied physics fields [1–9]. However building accurate equations of state (EOS) remains a very challenging problem due to a lack of reliable experimental data on the thermal effects associated to its diverse forms and transformations: indeed temperature measurements can be particularly difficult under extreme, highly dynamic constraints. This difficulty increases in the case of porous systems due to the differences between the initial samples used in the reference experiments. For instance, experimentally shock-compressed graphite can span porosity rates ranging up to more than 20%, yielding graphite to diamond (G/D) transition pressures from 18 to 45 GPa. These values lie substantially above the estimated G/D coexistence line (in the 5–8 GPa range according to experiments and simulations [2,10,11]), which signs a possible metastability of shocked graphite in the diamond-predominant regime (reversely, diamond is well-known to be metastable in the low temperature/pressure regime, where graphite is the most stable phase). This effect results in a wide spread of experimental data in the G/D transition regime [12–17].

Theoretical studies of the shock compression of graphite are scarce. Recently, a DFT-based molecular dynamics study yielded one point of the Hugoniot curve beyond the G/D transition using the Multi-Scale Shock Technique [18] (MSST-DFT). Then the full Hugoniot curve of monocrystalline graphite was calculated [9] using the LCBOPII empirical potential [19,20], in excellent agreement with MSST-DFT, but with a large over-estimation of the G/D transition pressure (G/D).

In this paper we address the shock compression of porous graphite using atomistic (Monte Carlo and molecular dynamics) simulations with the LCBOPII potential using specifically designed porosity models. We compare our Hugoniot data to experimental results and discuss the role of the sample porosity on the resulting equation of state.

2. Numerical setup

This study is based on a combination of Monte Carlo (MC) and molecular dynamics (MD) simulations. We ran the MC simulations with the GIBBS code [21] and the MD simulations with the STAMP code [22]. The MC simulations used samples of limited sizes (up to 5.000 atoms) of bulk graphite and diamond with 3D periodic conditions. The MC Hugoniot curves were calculated using the SCA method [23,24]. We used MD to calculate the specific energies of the various porosity models at 300 K, in the NVT ensemble.

We designed two porosity models for the porous Hugoniot calculations: a so-called "mosaic" model to treat the intergranular component of the porosity, and a so-called "intragranular" model to treat the porosity inside graphite grains. We chose a pole density of 1.81 g/cm^3 corresponding to a 20% porosity: this density is close to the densities of the experimental samples presented for comparison. In the mosaic model (Fig. 1-left) the system consists of a $2 \times 2 \times 2$ stack of graphite cubes of size $6.1 \times 6.0 \times 6.0\,\text{nm}^3$ with alternating orientations: the spacing between the cubes is adjusted to obtain the desired density. In the intragranular model (Fig. 1-right) a paralleliped cavity of size $7.8 \times 7.5 \times 7.6\,\text{nm}^3$ is cut into a bulk graphite cube of size $13.6 \times 13.2 \times 13.3\,\text{nm}^3$: the fraction of removed atoms is adjusted to reach a density within 0.1% of the mosaic model. These two models may be considered as characteristic of the different types of

[a] Corresponding author: nicolas.pineau@cea.fr

Article available at http://www.epj-conferences.org or http://dx.doi.org/10.1051/epjconf/20159404037

Figure 1. Structure of the mosaic (left), sc-intragranular (right). In order to highlight the geometry of the pores, we colored the atoms according to their coordination: blue for the fully coordinated graphite atoms (sp^2), green for under-coordinated atoms.

porosity that can be found in a compacted powder (inside a grain for the intragranular case and between grains for the mosaic case); however the defect sizes probably lie at the lower boundary of the actual experimental ranges, thus maximizing the "porous energy" with respect to real samples. We obtained specific energies of −56.9 MJ/kg and −57.8 MJ/kg for the mosaic and intragranular case (to be compared to a specific energy of −58.6 MJ/kg for bulk graphite at 300 K).

3. Results and discussion

In Fig. 2 we present the results of the MC hugoniostat (MC-SCA) calculations for bulk and porous graphite and diamond with a comparison to a set of experimental data for graphite powders compacted to 1.76 g/cm^3 and 1.88 g/cm^3.

The results show that the Hugoniot of bulk graphite is close to the experimental data for pressures lower than 20 GPa, whereas the diamond shock pressure is substantially underestimated on the full range. These discrepancies are corrected by the use of the properties of porous graphite at the "pole": in particular the increment in shock pressure is sufficient to fill the gap between bulk diamond and experiments, yielding an excellent agreement for pressures superior to 40 GPa. This effect is also visible, although less marked, for porous graphite.

For the temperature EOS, the agreement between the SESAME data and our calculations for bulk graphite is excellent. However the impact of porosity on the shock temperature is substantial and not accounted for in the EDM3 porous EOS. In particular, a sharp temperature increase is found for low shock intensities (+700 K for the intragranular case, +1200 K for the mosaic case) which may result from the closure of the porosity induced by the shock: similar trends are found starting from graphite or diamond. After this initial heating phase, the shock temperatures increase at distinct, quasi constant rates for graphite and diamond (~37.4 K/GPa vs. ~63.3 K/GPa respectively) reflecting the difference in heat capacity between the two bulk solids (0.71 kJ/kg/K [26] vs. 0.52 kJ/kg/K [27] at 300 K). Note that the change in magnitude of the computed slopes agree qualitatively with the ones estimated by the bulk EOS (~12.7 K/GPa vs. ~42.8 K/GPa).

The hugoniostat simulations of graphite and diamond show strong metastabilities of the bulk phases (at low and high pressure for diamond and graphite respectively). Accordingly these simulations fail to reproduce the intermediate states found experimentally between 20 and 40 GPa, which likely correspond to the transition of graphite toward diamond. This result is not fully surprising since i) our samples have limited sizes which may hinder the nucleation of diamond in shocked graphite (and reversely), and ii) graphite and diamond are known to be strongly metastable beyond their respective thermodynamic domains. This draws a limit to the validity of the calculated shock temperatures in view to build a full EOS for shocked carbon, and calls for a new strategy in order to explore this intermediate regime. Note that the large dispersion of the experimental data in the 20–40 GPa range could result from the partial shock-conversion of the initial graphite to diamond, with a probable influence of the initial sample density and structural properties on the G/D transition rate.

The results at low and high pressure can be exploited in view to construct a partial EOS for shocked porous graphite, including the temperature which is very difficult to access experimentally. First a domain of confidence must be drawn for the pure graphite and diamond phases. The large number of available shock experiments tells

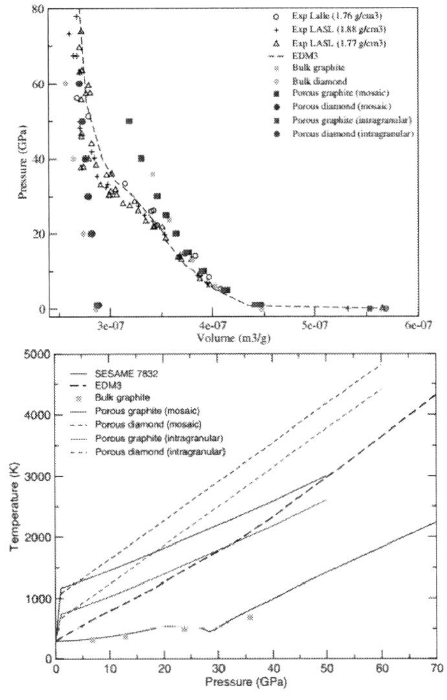

Figure 2. $P(V)$ and $T(P)$ Hugoniot diagrams of the bulk and porous systems calculated from MC simulations. The orange symbols are for the bulk graphite pole, blue lines for the mosaic porous pole, red lines for the intragranular porous pole. Full lines are for graphite simulations and dashed lines for diamond. The black symbols are a few set of experimental measurements (the corresponding pole densities are given in the legend). The corresponding SESAME (bulk graphite, $T(P)$ diagram) and EDM3 [25] (porous carbon, both diagrams) equations of state are provided (full and dashed black lines respectively).

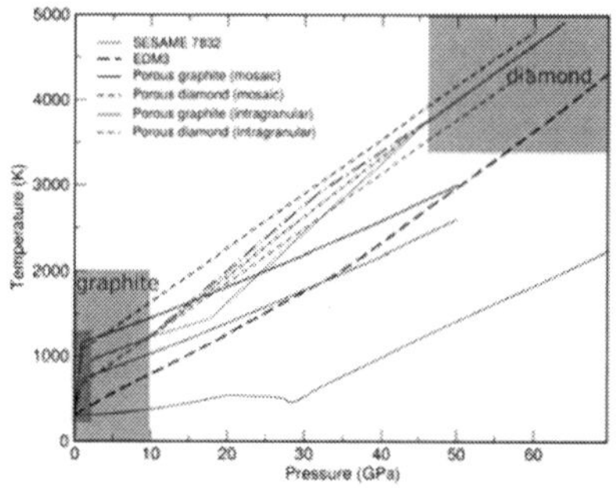

Figure 3. Same as Fig. 2: domains of confidence for graphite and diamond (grey regions) and schematics of potential EOS models are proposed (green lines).

us that graphite is always preserved for shock pressures lower than 15 GPa, and diamond is always found above 40 GPa. Therefore we will rely on the LCBOPII Hugoniot for pure graphite and pure diamond in these two respective low and high pressure domains, either in the $P(V)$ or the $T(P)$ diagrams. Examples of potential equations of state in the $T(P)$ diagram are given in Fig. 3. Note that new experimental measurements of the shock temperature appear necessary to validate the final EOS, in particular in the low shock intensity regime, where the sharp temperature increase is predicted.

4. Conclusion

We used MC and MD simulations to unravel the behaviour and EOS of polycrystalline graphite under shock compression. In the MC-SCA framework, the complex structure of the polycrystalline samples was accounted for by using a pole energy estimated from several model of porous structures, corresponding the variety of porosity types in granular materials.

Our results show that a reasonable quantitative estimation of the porous energy yields shock compression results in good agreement with the available $P(V)$ experimental data in the P<15 GPa and P>40 GPa regimes. The simulations predict a sharp heating in the low compression regime which may be assigned to the closure of the porosity. These data can be used to build an EOS with reasonable confidence in the low compression and high compression regimes, where respectively graphite and diamond are experimentally, and unambiguously, known to be the only stable form of shocked carbon.

Due to the strong metastability of graphite and diamond beyond their respective domains of predominance, the intermediate regime cannot be investigated using bulk graphite or diamond as the initial material. We are currently working on a new procedure using hybrid G/D samples in the MC-SCA framework: we believe that the use of pre-nucleated diamond in shocked graphite is a necessary step to explore this regime with small-scale atomistic simulations.

Besides, we use large-scale MD simulations to unravel the shock-compression mechanisms at play in porous carbon: in particular we investigate the dependance of the G/D transition threshold on porosity, which may explain partially the major discrepancy between experimental and simulation Hugoniot curves.

This work shows that a set of simple but well designed atomistic simulations can be used to obtain reliable information on the thermodynamic response of complex materials. From these data, complete EOS can be proposed and used for finite element methods, in a multiscale approach. The key ingredients to a full model are an accurate potential energy function, a set of approximate structural models which, taken altogether, are representative of the complexity of the system, and experimental data in the full targeted thermodynamic range.

Further developments, or generalization, of temperature measurement setups applied to the shock compression of solids is indispensable in order to validate the thermal component of the proposed EOS.

References

[1] F.P. Bundy, J. Chem. Phys. **38**, 618 (1963)
[2] F.P. Bundy, J. Chem. Phys. **38**, 631 (1963)
[3] S.V. Zybin, M.L. Elert, C.T. White, Phys. Rev. B **66**, 220102 (2002)
[4] H. Nagao, K.G. Nakamura, K. Kondo, N. Ozaki, K. Takamatsu, T. Ono, T. Shiota, D. Ichinose, K.A. Tanaka, K. Wakabayashi et al., Phys. Plasmas **13**, 052705 (2006)
[5] J.H. Eggert, D.G. Hicks, P.M. Celliers, D.K. Bradley, R.S. McWilliams, R. Jeanloz, J.E. Miller, T.R. Boehly, G.W. Collins, Nature Physics **6**, 40 (2010)
[6] R.S. McWilliams, J.H. Eggert, D.G. Hicks, D.K. Bradley, P.M. Celliers, D.K. Spaulding, T.R. Boehly, G.W. Collins, R. Jeanloz, Phys. Rev. B **81**, 014111 (2010)
[7] R.Z. Khaliullin, H. Eshet, T.D. Kuhne, J. Behler, M. Parrinello, Nature Materials **10**, 693 (2011)
[8] N.A. Marks, M. Lattemann, D.R. McKenzie, Phys. Rev. Lett. **108**, 075503 (2012)
[9] N. Pineau, J. Phys. Chem. C **117**, 12778 (2013)
[10] R. Berman, F. Simon, Z. Elektrochem. **59**, 333 (1955)
[11] L.M. Ghiringhelli, J.H. Los, E.J. Meijer, A. Fasolino, D. Frenkel, Phys. Rev. Lett. **94**, 145701 (2005)
[12] B.J. Alder, R.H. Christian, Phys. Rev. Lett. **7**, 367 (1961)
[13] W.H. Gust, Phys. Rev. B **22**, 4744 (1980)
[14] D.J. Erskine, W.J. Nellis, Nature **349**, 317 (1991)
[15] W.J. Nellis, A.C. Mitchell, A.K. McMahan, J. Appl. Phys. **90**, 696 (2001)
[16] K. Yamada, Y. Tanabe, Carbon **40**, 261 (2002)
[17] G.I. Kanel', G.S. Bezruchko, A.S. Savinykh, S.V. Razorenov, V.V. Milyavskii, K.V. Khishchenko, High Temperature **48**, 806 (2010)
[18] C.J. Mundy, A. Curioni, N. Goldman, I.F.W. Kuo, E.J. Reed, L.E. Fried, M. Ianuzzi, J. Chem. Phys. **128**, 184701 (2008)
[19] J.H. Los, L.M. Ghiringhelli, E.J. Meijer, A. Fasolino, Phys. Rev. B **72**, 214102 (2005)

[20] J.H. Los, L.M. Ghiringhelli, E.J. Meijer, A. Fasolino, Phys. Rev. B **73**, 229901(E) (2006)
[21] P. Ungerer, B. Tavitian, A. Boutin, *Applications of molecular simulation in the oil and gas industry* (IFP Publications, Paris, France, 2005)
[22] STAMP is a massively parallel MD code developped at CEA/DAM
[23] J.B. Maillet, G. Stoltz, Applied Math. Research eXpress **2008**, abn004 (2009)
[24] E. Bourasseau, J.B. Maillet, N. Desbiens, G. Stoltz, J. Phys. Chem. A **115**, 10729 (2011)
[25] D. Hébert, G. Seisson, I. Bertron, J. Chevalier, C. Thessieux, J. Quessada, S. Tastet, J. Phys. Conf. Ser. **500**, 112057 (2014)
[26] S. Picard, D. Burns, P. Roger, Metrologia **44**, 294 (2007)
[27] A.C. Victor, J. Chem. Phys. **36**, 1903 (1962)

EPJ Web of Conferences **94**, 04038 (2015)
DOI: 10.1051/epjconf/20159404038

A 3-D perspective of dynamic behaviour of heterogeneous solids

Yong Lu[a] and Rongxin Zhou

Institute for Infrastructure and Environment, School of Engineering, The University of Edinburgh, The King's Buildings, Edinburgh EH9 3JL, UK

Abstract. The dynamic behaviour of concrete-like materials under high strain rates has been a subject of continuous scrutiny over the years. A prevailing explanation attributes much of the dynamic increase of strength, especially under compression, to the macroscopic inertia confinement. Studies conducted by the authors' group using meso-scale computational models suggest that the heterogeneity of the material composition, in particular the involvement of the aggregates, also plays a sensible part in the process of damage evolution and the increase of the bulk strength under high strain rates, and a detailed investigation into this effect would benefit if a realistic representation of the heterogeneity in 3D can be achieved. This paper presents some recent progress in the development of a 3-D meso-scale computational model incorporating randomly-shaped 3-D aggregate particles, including the general validation of the model, and application in the simulation of the dynamic response of concrete under high strain rate compression.

1. Introduction

Concrete is highly non-homogenous material with large heterogeneities. Generally speaking, the description of concrete may be classified into three distinctive scales, namely macroscale, mesoscale and microscale. On the macroscale concrete is considered as homogeneous material and treated as such in classical finite element models. On the mesoscale concrete is considered as a composite material comprising of coarse aggregate, mortar matrix and the interfacial transition zone (ITZ). A mesoscale model permits a direct description of the material heterogeneity and hence allows for a realistic prediction of the development of damage within the muti-phase material. At a microscopic level, the mortar matrix is further divided into fine aggregates and hardened cement paste with pores embedded inside. As far as the composite mechanical behaviour is concerned, sub-division of the individual phases does not appear to introduce significant effect [1], and hence modelling of concrete on a mesoscale is deemed to be an effective choice when material heterogeneity needs to be taken into account.

A number of studies on the 2D mesoscale modelling of concrete can be found in the existing literature (e.g. [1,2]. However, 2D mesoscale modelling of concrete-like material has inherent limitation in representing a realistic stress and strain field in concrete specimen, particularly with regard to the lateral inertial confining effect in dynamic loading. Thus a pseudo 3D model has been developed and applied in the analysis of the strain rate effect under dynamic compression in [3]. Understandably, a true 3D mesoscale model is a more desirable solution in representing the material heterogeneous structures and giving more reliable results. However, modelling the random mesoscale structure of concrete in 3D is considerably more complex.

The primary difficulty arises from the representation of a random aggregate structure. To circumvent this difficulty, simple shapes of aggregate particles like spheres or mixed spheres and ellipsoids are mostly used in previous research. Further proposal of a modified version of the ellipsoid function [4] made it possible to better approximate real aggregates. Recently a more accurate approximation of particle shapes represented by polyhedrons has been adopted by some researchers. This method can approximate the real aggregate shape better than simple spheres and ellipsoids but it cannot satisfy the size grading curve. Direct mapping of the aggregate particles from physical samples has been explored using computer image analysis and computed tomography (CT-scan).

In an effort to develop a comprehensive mesoscale model for concrete, a series of development work has been conducted in this research group. The model is aimed at achieving a realistic representation of the actual shapes and sizes of aggregate particles and at the same time allowing for high volumetric ratios of aggregates to be attained. In the proposed approach, 3D polyhedral-shaped aggregates are represented by convex hull in R^3, which is the minimum convex set containing a series of points generated randomly. Flaky and elongated shapes of polytopes are also considered to enable greater flexibility in the simulation of real shapes of aggregate particles. The standard take-and-place procedure is used for generating the main 3D meso-structure. Extending from the basic framework [5], the present study focuses on improving the efficiency in the generation of the 3D mesoscale geometry and the robustness of finite element meshing for the highly unstructured mesoscale model, as well as enabling the realisation of high density packing of aggregates. The mesoscale model generated from the enhanced procedure is verified against standard experimental observations under quasi-static compression.

[a] Corresponding author: `yong.lu@ed.ac.uk`

Article available at http://www.epj-conferences.org or http://dx.doi.org/10.1051/epjconf/20159404038

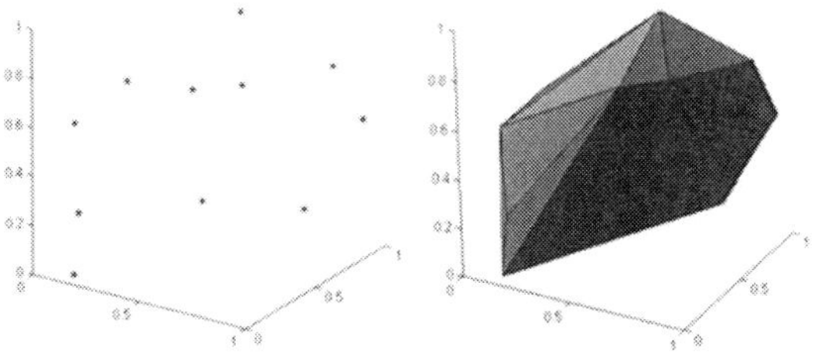

Figure 1. An example convex Hull.

The model is then applied to simulate the dynamic compressive behaviour of concrete under high strain rate compression.

2. Generation of the 3D mesoscale geometry with random aggregates

2.1. Basic operations

The generation of the 3D mesoscale model involves first of all the creation of 3D aggregates in random shapes and sizes following prescribed distributions and packed to a desired density, or volumetric ratio. The procedure for the creation of the 3D mesoscale model geometry follows the commonly used "take and place" routine in a sequential manner, starting with the largest aggregate size group, and carry on until the smallest size group is fulfilled.

In the present work, we create the aggregate particles by bounded polyhedrons. The convex hull for a set of points is the minimum convex polyhedron containing all the predefined points. Thus it is possible to generate a random shaped aggregate particle from a set of 3D random points. The number of the points which are used to generate a polytope can be varied. A larger number of random points are used to produce gravel shapes while a relatively small number of points are used for angular shapes of aggregates. The convexity of the polyhedron can be automatically fulfilled without separate checking at every step. Furthermore, the take and place procedure is used to guarantee a prescribed aggregate size grading according to the Fuller curve. Figure 1 gives an example of a convex hull generated by Matlab with a computational geometry function named Qhull.

The shapes of aggregate particles created above are in-polyhedrons of spheres, which are relatively regular. Both flaky and elongated aggregates can be realised from the regular polytopes by shrinking or elongation. The Euler angles α β and γ, which represent the coordinate system rotation in R^3 are utilised to ensure the shrinkage or the elongation to take place in a random direction, not just limited along x, y and z axes.

After generating an individual aggregate particle, the 'place' process is carried out. In this process, the aggregate generated from the "take" process is positioned into the predefined concrete space in a random manner, subjected to prescribed physical constraints. The most important and time consuming step in the placing process is the intersection checking between two polytopes. Clearly it is neither economical nor necessary to check each newly placed aggregates against all existing aggregates for intersection. Thus a pre-selection algorithm is used to identify the existing polytopes that may have a chance to intersect with the one being placed. A local checking space is defined as a bounding sphere which shares the same centre with convex hull and covers all vertices of the polytope. Subsequent intersection check only needs to be carried out for polytopes whose bounding spheres intersect with that of the current particle.

A linear time algorithm is then employed to determine the distance between two convex hulls. The algorithm is based on a hierarchical representation of polyhedrons. Once the hierarchical representation is constructed, it is possible to detect the intersection condition by iteratively calculating the minimum distance which is not necessarily on the vertices between two particles.

2.2. Enhancement on the placing of aggregates

In the existing "place" procedure, once an aggregate being placed is found to intersect with any existing aggregates, this aggregate will be abandoned and a new polytope is regenerated and placed into a new position, and the checking process is carried out all over again. This cycle continues until no intersection occurs. In order to increase the success rate of placing the aggregate and improve the placing efficiency, a translate-and-rotate procedure is employed on the aggregate which is being placed The detail of this procedure is described as follows:

For the random translating procedure, point P_i (X_i, Y_i, Z_i) is firstly defined as a coordinate of any one vertex of the particle which will be placed in the concrete specimen. Then a small distance is translated on the particle. Thus the coordinate of the vertex after moving should be $Q_i(X_m, Y_m, Z_m)$ where

$$0 < X_i, Y_i, Z_i < 50; 0 < X_m, Y_m, Z_m < 50; \quad (1)$$

and

$$-1.5 < X_m - X_i, Y_m - Y_i, Z_m - Z_i < 1.5, \quad (2)$$

Here a value of 1.5 is chosen for limiting the translating procedure in a certain space.

For the random rotating procedure, according to Euler's rotation theorem, any rotation in 3-dimenisonal Euclidean space can be achieved by three elemental rotations. Thus three random degrees α, β, γ which are defined as the random rotating degrees about x, y and z axis, respectively, are introduced to locate the aggregate particle after rotating. The coordinate of a vertex after rotating can be expressed as

$$R_i = Z(\gamma)Y(\beta)X(\alpha)P_i. \quad (3)$$

And the final expression of the coordinate is

$$R_i = \begin{bmatrix} cos\gamma cos\alpha - sin\gamma cos\beta sin\alpha & cos\gamma sin\alpha + sin\gamma cos\beta cos\alpha & sin\gamma sin\beta \\ sin\gamma cos\alpha \; cos\gamma cos\beta sin\alpha & cos\gamma cos\beta cos\alpha \; sin\gamma sin\alpha & cos\gamma sin\beta \\ sin\alpha sin\beta & -sin\beta cos\alpha & cos\beta \end{bmatrix} P_i \quad (4)$$

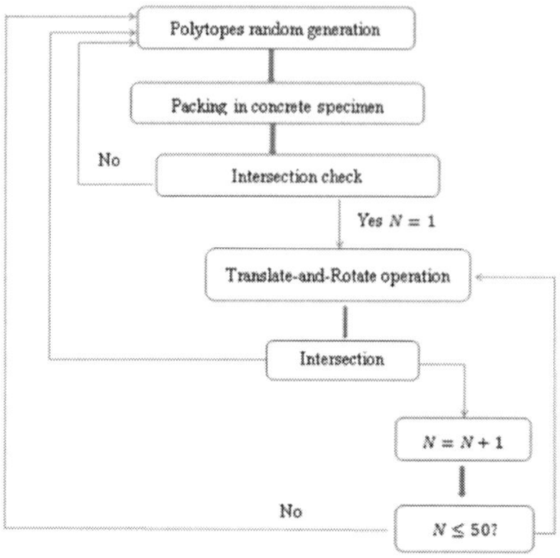

Figure 2. Flow chart of the placing procedure including translate-and-rotate operation.

Figure 3. Example of 3D mesoscale structure and meshing results.

It should be noted here the translate-and-rotate procedure does not guarantee the condition of no intersection between two particles. A limited number of translate-and-rotate operations can be set to avoid being trapped into useless attempts. From the trial analysis in the present study, a limit of 50 times appears to be adequate. Figure 2 presents a flow chart for the "take and place" procedure incorporating the adjustment operation.

When compared with the procedure without consideration of the translate-and-rotate method it is found that the translate-and-rotate procedure not only reduces the computing time but also effects to increase the aggregate volume percentage significantly. When a high volume ratio of aggregates is desired, however, the procedure would become extremely time-consuming or even fail. Thus a alternative approach would be needed to further increase the aggregate percentage, and this will be touched upon later.

3. Finite element meshing and generation of supplementary aggregates

Another challenge for 3D mesoscale modelling of concrete is to discretise the specimen into elements strictly according to different phase boundaries. Due to the existence of randomly shaped aggregate particles, the meso-structure is highly unstructured. For meshing unstructured domain, triangle and tetrahedral meshing are mostly used in the gird refinement. Specific smoothing algorithms for meshing unstructured domain include Octree, advancing front and Delaunay refinement. A typical way to work around the difficulty arising from meshing directly the highly unstructured mesoscale geometry is by firstly performing a background meshing and then bundling groups of the meshed elements into aggregates of targeted (or mapped) shapes. The obvious drawback of such an approach is that the actual surfaces of the aggregates cannot be preserved and instead saw-tooth shaped boundaries will result. Consequently the interaction between aggregates and mortar matrix becomes unrealistic which can in turn affect the representation of the mesoscopic failure mechanisms. In the present mesoscale model we endeavour to mesh the mesoscale structure of concrete directly. An advanced meshing code, called TetGen [6], is employed to do this task. The advantage of using this mesh method is that it can ultimately retain realistic boundaries between aggregate and mortar. This provides an essential basis for the simulation of stress concentration, crack initiation and damage accumulation in concrete from a mesoscopic perspective. Figure 3 shows an example of the 3D structure and meshing results of a cubic concrete specimen. Note that up to this point only aggregate and mortar elements have been generated; the polytope particles represent the aggregates and the remaining domain belongs to the mortar material.

The generation of the third phase, i.e. the ITZ can be made by different methods depending upon the way the ITZ is to be modelled, for example by "shrinking" the already formed aggregates to leave an interface layer between the aggregates and the mortar matrix which is subsequently meshed as ITZ. Using cohesive elements is another option; however, due to complex stress conditions, typical cohesive element formulation is found to be hardly suitable, so this option is still subject to the availability of enhanced cohesive element. In the present study we use the equivalent solid layer approach to represent the effect of ITZ. To simplify the meshing process which is already very complex as discussed above, in the current 3D model the equivalent ITZ layer is generated by choosing elements that immediately come into contact with the aggregates. An in-house selection program is developed using MATLAB to identify the layer of elements surrounding aggregate particles. Figure 4 shows the identification results and the ITZ elements thus defined.

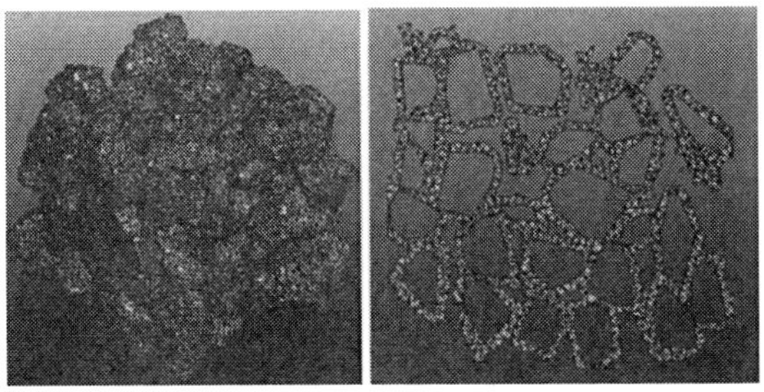

Figure 4. ITZ layer; left = 3D view, right = a plane cut view.

A practical restriction to the take-and-place method in 3D meso-scale modelling is that the packing density is low and it is difficult to achieve a realistic aggregate volume ratio of 40–50%. From our experience the maximum aggregate volume ratio that may be achieved from the standard take-and-place procedure described earlier is around 30%. Therefore, additional aggregates need to be created to augment the aggregates and increase the total volume ratio. In the present work we propose to create additional aggregates from the mortar domain by grouping mortar elements to form secondary aggregates in a quasi-random manner. The detail of this procedure is reported in a separate paper. Through this approach, high aggregate volume ratio to the order of 50–60% may be achieved.

4. FE analysis using the mesoscale model

In the present analysis 3D mesoscale model for a 50-mm cubic specimen of concrete is developed. This sample size is typical in dynamic testing of concrete for high strain rate loading. The specimen is subjected to simulated quasi-static test before it is used for the simulation under high strain rate loading. Considering computational stability (especially in large deformation range) and that the mesh size in the 3D mesoscale model is usually considerably fine, 4-node tetrahedron element is used in the simulations.

4.1. Material models

At the mesoscopic level, damage and the nonlinear behaviour occur primarily in the mortar matrix and along ITZ. Therefore a suitable material model for these two phases is important for the final simulation results. The material model employed for the two parts in this study is the K&C Concrete Damage Model, also known as material #72R in LS-DYNA. The material model is capable of describing the material failure due to tension, shear, as well as compression under various stress conditions, and it also includes pressure and strain rate dependent features. The detail information of this material model can be found in [7]. The material model has been tested extensively and is found to be a suitable candidate for quasi-static as well as dynamic applications.

The coarse aggregates usually have significantly higher strength than mortar. It is deemed to be reasonable to use a linear elastic material model for aggregates under low rate loadings. However for high dynamic loading such as shock and blast, the rapid propagation of stress wave could result in high stresses being developed in aggregates in a very different way as compared to low rate loading conditions, and this could subject the aggregates to potential failure [8]. Under this circumstance, a nonlinear material model becomes necessary. Herein we also use the concrete damage model for aggregates with however failure strength matching that of the chosen aggregate type.

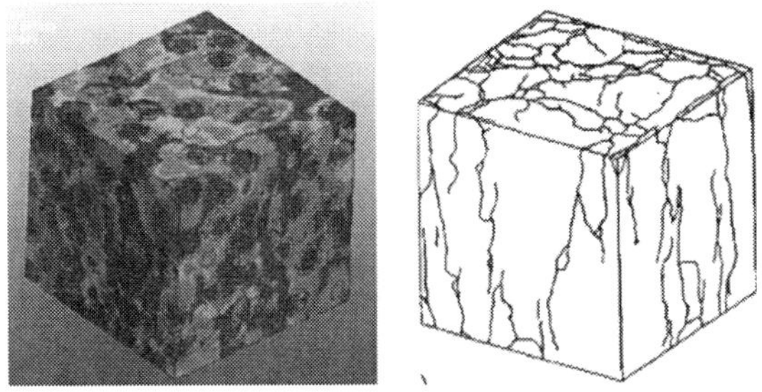

Figure 5. Damage patterns under lower loading-face friction: left = simulated, right = experimental (after [9]).

4.2. Model configurations

To simulate the uniaxial loading condition, the specimen is fixed at the bottom face and loaded from the top edge in the vertical (z) direction. In order to be able to produce the full range of the concrete response including the softening stage, the loading is applied in a displacement-controlled mode by imposing a velocity boundary condition. The transient analysis code LS-DYNA is employed to perform the simulation using an explicit time integration procedure, which enables an easy realisation of dynamic loading.

5. Model verification and application in analysis of concrete under high strain rate compression

5.1. Verification under quasi-static compression

Experimental results show that the compressive behaviour of concrete specimen can be strongly influenced by frictional restraint between specimen and loading platen which is known as loading-face friction effect. The present 3D mesoscale model is also examined by this effect as a basic model validation against relevant experimental observations.

As can be seen in Figure, the damage pattern from the simulation using the mesoscale model agrees well with the experimental result. Under the uniaxial compression with a lower friction boundary condition, a series of column-type of fractured blocks occurred in the concrete specimen because of the formation of the -cracks almost parallel to the applied load. Similar agreement between the simulated and experimental results for higher friction cases has also been achieved (not shown here).

5.2. Dynamic compression

Classical experimental results have shown that the "apparent" dynamic compressive strength increases with the increase of the strain rate. A Dynamic Increase Factor (DIF), which is defined as the ratio of the apparent dynamic strength to the static strength of the sample specimens, is commonly used to account for the strength enhancement due to high strain rates. However, the true mechanism underlying the occurrence of the DIF is still a subject of continued debate.

As far as dynamic compression is concerned, various analytical and numerical studies in more recent years suggest that that the lateral inertial confinement plays a key role in the enhancement of the dynamic compressive strength (e.g. [10,11]). More recent numerical studies using 2D mesoscale models provide further support to this argument (e.g. [3]). However, 2D mesoscale model has inherent limitation in the representation of the lateral inertial confinement. A pseudo 3D model proposed in (Lu and Song), in which a slice of 2D mesoscale model is sandwiched between two halves of homogeneous models, made it possible to retain the mesoscale features within the plane of interest while lateral inertia can develop naturally in a 3D environment. However, clearly a true 3D mesoscale model is a more desirable solution in representing more robustly the material heterogeneity and enabling more reliable results.

The present 3D mesoscale model is therefore employed to simulate the dynamic behaviour of a concrete specimen under dynamic compression. For a comparison, a 2D homogeneous model, a 2D mesoscale model and a 3D homogeneous model are also analysed for the same variation range of the strain rates. For an un-altered observation of the contribution of the structural inertial effect, the constituent materials are considered to be rate insensitive, i.e., no embedded strain rate enhancement factor is adopted in the material properties in all the models. Thus any increase in the nominal (apparent) compressive strength of the simulated test specimen is attributable only to the sample-wide dynamic effect, as well as the material heterogeneity. The nominal dynamic compressive strength is evaluated from the average stress on both the loading and supporting faces, and the stress on these two faces is in turn calculated as the total nodal force in the axial direction divided by the cross-section area. This effectively resembles the "three-wave" approach in evaluating the stresses from the standard SHPB tests.

The dynamic loading is simulated by applying a velocity boundary condition in a similar way as in the quasi-static analysis, but with a high velocity and a different time history in order to achieve a desirable strain rate and stress distribution. Note that in the dynamic simulation herein, no friction induced lateral confinement is considered at the loading face.

Figure 6 shows the variation of the predicted DIF with the strain rate from the simulations using the models mentioned above. Comparison also includes four empirical equations on DIF of concrete, namely a) CEB-FIP DIF model, b) and c) two empirical models, and d) a semi-empirical model.

It is worth mentioning here that the strain rate analysed in SHPB-like compression tests should be limited by the dimension (length) of the specimen. In the current numerical experiment, the dimension of concrete specimen

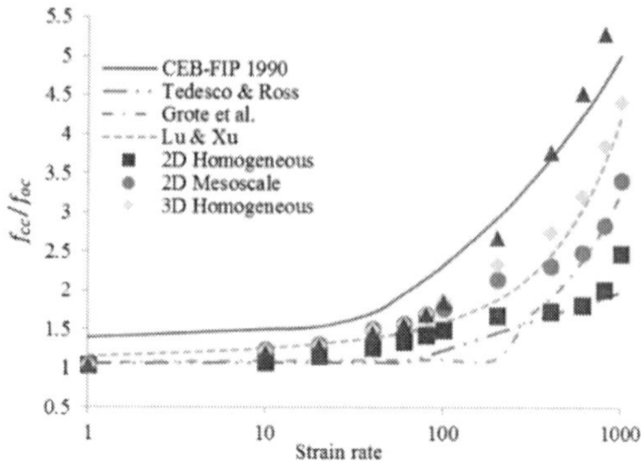

Figure 6. Predicted DIF with the strain rate.

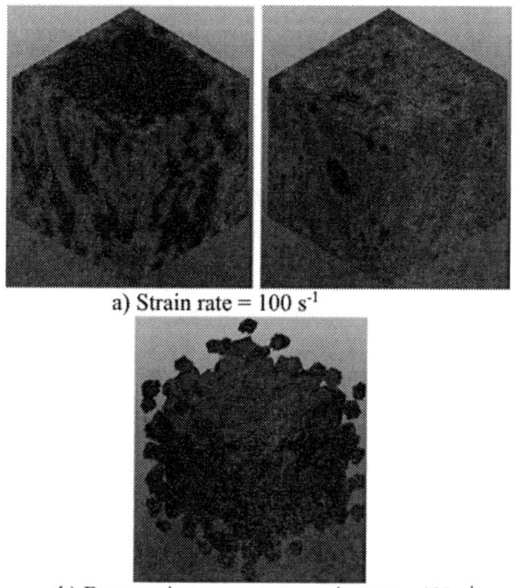

a) Strain rate = 100 s^{-1}

b) Damage in aggregates, strain rate = 400 s^{1}

Figure 7. Damage contours from 3D mesoscale model.

is 50 mm for which a strain rate up to about 100 s^{-1} is considered as acceptable The simulations in the higher strain rate range do not satisfy the stress equilibrium and strain uniformity requirements for the size of the specimen. Therefore the DIF results in this range may be considered for a reference purpose only.

As can be observed from Fig. 6, all models exhibit a significant increase in the nominal compressive strength as the strain rate increases, despite that no strain rate enhancement has been incorporated in the material constitutive model. The general trend of the DIF curves in numerical models resemble well with the curves given by the empirical formulas. The 3D mesoscale tends to predict the upper bound DIF among all numerical results, and this is deemed to be attributable to the enhanced contribution from the aggregates.

Figure 7 shows selected damage contours from the 3D mesoscale simulation results, including distribution of damage within the aggregate particles. Thus, very detailed data can be extracted from the 3D mesoscale model which will permit scrutiny of the damage mechanisms from various mesoscopic perspectives.

6. Conclusions

This paper reports the latest progress in the development of a comprehensive 3D mesoscale model framework for modelling of concrete allowing the realisation of practically any shapes and packing density of aggregates. The enhancements include improved packing efficiency by incorporating a translate-and-rotate algorithm in the placing operation, and increased packing density by creating secondary aggregates from the meshed mortar domain in a quasi-random manner. Furthermore, an efficient meshing program is adopted to perform direct meshing of the highly unstructured mesoscale geometric domains.

3D mesoscale model is then generated for the analysis of a representative 50-mm cubic specimen. The model is subjected to quasi-static compression, as well as tension (not shown in this paper), and results are found to agree favourably with typical experimental observations. The model is subsequently applied to simulate dynamic compression under high strain rates. The simulation results provide a full 3D perspective of the stress wave phenomenon and evolution of damage in a realistic 3D perspective. It further demonstrates that the mesoscopic heterogeneity can make a significant contribution towards the overall dynamic increase factor (DIF) under high strain rate compression.

The 3D mesoscale model provides a robust modelling platform for investigation into micro-mesoscopic mechanisms for concrete as well as general heterogeneous composites under dynamic loads.

References

[1] Z. Tu, and Y. Lu, Mesoscale modelling of concrete for static and dynamic response analysis Part 1: model development and implementation. Structural Engineering and Mechanics, **37**(2), 197–213 (2011)

[2] Z.M. Wang, A.K.H. Kwan, and H.C. Chan, Mesoscopic study of concrete I: generation of random aggregate structure and finite element mesh. Computers & Structures, **70**(5), 533–544 (1999)

[3] Y. Lu, Z. Song and Z. Tu, Analysis of dynamic response of concrete using a mesoscale model incorporating 3D effects, International Journal of Protective Structures, **1**(2), 197–218 (2010)

[4] S. Häfner, et al., Mesoscale modeling of concrete: Geometry and numeric, Computers & structures, **84**(7), 450–461 (2006)

[5] Z. Song, and Y. Lu, An algorithm for generation of 3D mesostructure of concrete for finite element analysis, in 19th UK ACME Conference, Edinburgh, UK (2011)

[6] H. Si, A Quality Tetrahedral mesh generator and 3D Delaunay triangulator Version 1.5, *User's Manual* (2013)

[7] L.J. Malvar, J.E. Crawford and K.B. Morrill, K&C concrete material model release III - automated generation of material model input, Technical Report TR-99-24.3, Karagozian & Case (2000)

[8] Z. Song, and Y. Lu, Mesoscopic analysis of concrete under excessively high strain rate compression and implications on interpretation of test data, International Journal of Impact Engineering, **46**, 41–55 (2012)

[9] van Vliet, M.R. and J.G. van Mier, Experimental investigation of concrete fracture under uniaxial compression, Mechanics of Cohesive-frictional Materials, **1**(1), 115–127 (1996)

[10] F.V. Donze, S.-A. Magnier, L. Daudeville, C. Mariotti and L. Davenne. Numerical study of compressive behavior of concrete at high strain rates. J Eng Mech – ASCE, **125**, 1154–63 (1999)

[11] Q.M. Li and H. Meng, About the dynamic strength enhancement of concrete-like materials in a split Hopkinson pressure bar test, International Journal of solids and structures, **40**(2), 343–360 (2003)

EPJ Web of Conferences **94**, 04039 (2015)
DOI: 10.1051/epjconf/20159404039

Relaxation model for dynamic plastic deformation of materials

Yuri Petrov[1,2], Elijah Borodin[1,3,a], Ezio Cadoni[4], and Nina Selyutina[2]

[1]IPME RAS, Extreme States Dynamics Department, V.O., 199178 Bolshoj pr., 61 St. Petersburg, Russia
[2] Saint Petersburg State University, 7/9, 199034, Universitetskaya nab., St. Petersburg, Russia
[3] Chelyabinsk State University, 454001 Br. Kashirinykh, 129, Russia
[4] University of Applied Sciences of Southern Switzerland, DynaMatLab, 6952 Canobbio, Switzerland

Abstract. Mechanical model for plastic relaxation of stresses in metals has been proposed. Our approach based on a characteristic plastic relaxation time, which is assumed as a material parameter. In a number of tasks values of its parameter become similarly high and it leads to dynamic effect appearing even at low strain rates deformation condition. The phenomenon of yield drop is one of these manifestations. We also discuss a physically based assumption (in the frames of dislocation theory) for analytical form of relaxation time. We find that there are two different characteristic relaxation time parameters, which is playing a main role for different stages of plastic deformation in low defect crystals.

1. Introduction

Several cases of unexpected mechanical behavior of metals in deformation can readily be explained by taking their dynamic characteristics into account. One such well-known paradox is the yield drop [1–3] and sharp yield point phenomenon in metal whiskers [4] when, in the case of quasi-static loading rates 10^{-5}–$10^{-2}s^{-1}$, the stresses attained in the material are dozens times greater than the value of its quasi-static yield point [4].

In pure metals this effect can be explained in terms of the material "dislocation starvation" [5], when the amount of mobile defects is insufficient for ensuring the required plastic strain rate and the elastic stresses continue to grow. From the mechanical standpoint, this phenomenon expresses the fact that the material already experiences dynamic deformations for such "quasi-static" strain rates. Namely, for a material with such properties, these strain rates are already sufficient for exciting the dynamic deformation mode, which cannot be described on the basis of quasi-static concepts that there are critical flow stresses in the material. To describe the dynamic characteristics of the material, it is necessary to explicitly take into account the fact that the material cannot unboundedly experience plastic relaxations of the arising stresses and a certain time is required for the development of relaxation process inside the material.

This can be attained by introducing an additional dynamic characteristic of the material itself, i.e. the characteristic plastic relaxation time parameter. Obviously, this parameter should be independent of the deformation process characteristics but should express the mechanical properties of the material itself in a wide range of loading rates.

[a] Corresponding author: elbor7@gmail.com

2. Models for description of plastic flow of metals

The elasto-viscoplastic deformation of a solid can be described in the framework of the Voigt model or the Maxwell model of a very viscous fluid [6]. In the Maxwell model, the equation for change in the shear stress σ_τ can be written as

$$\frac{d\sigma_\tau}{dt} = 2G\dot{\varepsilon} - \frac{\sigma_{eff}}{\tau} H(\sigma_{eff}) \qquad (1)$$

where $\dot{\varepsilon}$ is the strain rate; G is the shear modulus; τ is the stress relaxation time such that, under deformation, the material behaves elastically at significantly shorter times and visco-plastically at significantly longer times; $\sigma_{eff} = \sigma_\tau - y_b$ is the effective stress such that, at its positive value, the relaxation of shear stresses becomes possible; y_b is the barrier stress corresponding to the given mechanism of plasticity; and $H(x)$ is the Heaviside function. The integration of Eq. (1) at a constant strain rate $\dot{\varepsilon} = const$ and the initial condition $\sigma_\tau(0) = \sigma_y^0$ gives the following dependence of the stress on the time:

$$\sigma_\tau(t) = \sigma_y^0 + 2G\tau\dot{\varepsilon}(1 - \exp(-t/\tau)). \qquad (2)$$

For $\dot{\varepsilon} = 0$, we have $\sigma_\tau(t) = \sigma_y^0$. At a fixed strain rate, the maximum shear stresses achieved in the material at $t \to \infty$ can be represented in the form

$$\sigma_\tau^{\max} = \sigma_y^0 + 2G\tau\dot{\varepsilon}. \qquad (3)$$

These stresses determine the dynamic yield strength of the material and consist of two parts: the barrier stress σ_y^0 and the "dynamic addition" $2G\tau\dot{\varepsilon}$, which depends linearly on the strain rate, the characteristic relaxation time, and the elastic constants of the metal.

In terms of the aforementioned models of an elasto-viscoplastic body, there are no fundamental differences between the shear strengths under the quasi-static and dynamic deformation conditions. Everything is determined by the parameters accounting for the internal structure and defect substructure of the material. The transition between the static and dynamic regimes, according to formula (3), occurs at a strain rate $\dot{\varepsilon}_{tr} \sim \sigma_y^0/2G\tau \equiv \sigma_y^0/\mu$, where the coefficient $\mu = 2G\tau$, which characterizes the dynamic viscosity, depends on the elastic properties and defect substructure of the metal. For different concentrations of defects in the material, the strain rate can take almost any value. As an example, we consider the dislocation glide in which the plastic strain rate is determined by the Orowan equation [7–9]

$$\dot{\varepsilon}_{pl} = \sum_{\beta} \rho_D^{\beta} b V_D^{\beta}$$

where ρ_D^{β} is the scalar dislocation density, which is numerically equal to the length of dislocation lines per unit volume; b is the Burgers vector; and is the V_D^{β} magnitude of the dislocation velocity in the slip system β. According to this equation, for a specified plastic strain rate, it is necessary to use a particular dislocation density $\rho_D^{(cr)} = \dot{\varepsilon}_{pl}/bV_D^{\beta}$ for the effective relaxation of shear stresses generated in the material (for the fulfillment of the condition $\dot{\varepsilon}_{pl} \sim \dot{\varepsilon}$). The dislocation velocity cannot exceed the transverse velocity of sound in the metal [10]. If the density of dislocations is less than $\rho_D^{(cr)}$, their amount is insufficient to ensure the required plastic strain rate, which will lead to a sharp increase in the yield strength of the metal and to the manifestation of its dynamic features even at low strain rates. The phenomenon of the lack of dislocations is well known in the physics of nano-objects, where it is referred to as the dislocation starvation of the material [11]. As will be shown below, the characteristic time of plastic relaxation is inversely proportional to the dislocation velocity and increases without limit when $\rho_D \to 0$. Therefore, even at a low barrier resistance y_b under quasi-static deformation conditions, the yield strength can increase up to values of the theoretical tensile strength, which is observed in pure single crystal whiskers. The transition strain rate $\dot{\varepsilon}_{tr}$, at which the dynamic effects become significant, in this case, can be arbitrarily small. Such an increase in the yield strength proves to be natural for extremely high strain rates, when normal dislocation densities $\rho_D \sim 10^{12}\mathrm{m}^{-2}$ are no longer sufficient to ensure the required plastic strain rate. In the dislocation starvation relaxation time is determined by the rate of generation of new defects in the crystal lattice [12].

The similar formalism is used in the yield criterion proposed in [13,14], where it was also shown that the explicit introduction of the characteristic relaxation time makes it possible to determine the yield strength for an arbitrary law of loading of a material and to consider, within a unified model, both the fast and slow loading. For a Voigt body, the strain dependence of the stress is described by the formula

$$\sigma = 2G\varepsilon + \mu\dot{\varepsilon}, \tag{4}$$

which can also be rewritten as $\dot{\varepsilon} + 2G\varepsilon/\mu = \sigma/\mu$. It is assumed that, apart from the elastic strains, under dynamic conditions there is a viscous addition proportional to the strain rate. The integration of expression (4) with respect to time by multiplying both sides by $2G\exp(-t/\tau)$ gives

$$2G\varepsilon = \frac{1}{\tau}\int_0^t \sigma(s)exp\left(-\frac{t+s}{\tau}\right)ds, \tag{5}$$

where we introduced the characteristic time $\tau = \mu/2G$.

The left hand side of this equation represents the elastic stress that we would have in the quasi-statics by setting $\dot{\varepsilon} = 0$ in formula (4). Equation (5) can be interpreted in the framework of the concept of fading memory [15]: the load acting earlier at instants of time $s \ll t$ made a significantly smaller contribution to the current state of the defect structure as compared to the recently applied load. The current stress should be replaced by its "relaxed" value. Moreover, it should be noted that the stresses acting in the material at previous instants of time are included in the equation with smaller and smaller weighting coefficients. In the general case, this leads to the inequality in the integral form

$$\int_0^t \sigma(s)K(t-s)ds \le \sigma_y^0 \tag{6}$$

where the kernel of the integral operator $K(t)$ is the fading memory function. By comparing expressions (6) and (7), we find that the Voigt model corresponds to the exponential memory decay:

$$K(t) = \frac{1}{\tau}exp\left(-\frac{t}{\tau}\right) \tag{7}$$

where, as in expression (6), the $\tau = \mu/2G$ is the characteristic time. As a simpler approximation of the exponential law (7), we can propose the step decay function

$$K(t) = \begin{cases} \frac{1}{\tau}, 0 \le t \le \tau \\ 0, t > \tau \end{cases} \tag{8}$$

Eq. (7) then gives the integral yield criterion for a metal in the form [7,8]

$$\frac{1}{\tau}\int_{t-\tau}^{t}\left(\frac{\sigma(s)}{\sigma_y^0}\right)^{\alpha} ds \le 1 \tag{9}$$

where $\sigma(s)$ is the function specifying the time dependence of the applied stresses, which are assumed to be zero at $t < 0$ [14]; τ is the characteristic time of stress relaxation ("incubation time") [13,14]; σ_y^0 is the yield strength under quasi-static loading and α is the dimensionless parameter describing the sensitivity of the material to the level of stresses, usually $\alpha \ge 1$ [14].

We shall assume that the deformation in the specimen increases linearly as $\varepsilon(t) = \dot{\varepsilon}tH(t)$, where $H(t)$ is the Heaviside function. We introduce the dimensionless relaxation function $0 < \gamma(t) \le 1$ by the condition:

$$\gamma(t) = \begin{cases} 1 & , \frac{1}{\tau}\int\limits_{t-\tau}^{t}\left[\frac{\Sigma(s)}{\sigma_y^0}\right]^{\alpha} ds \le 1 \\ \left\{\frac{1}{\tau}\int\limits_{t-\tau}^{t}\left[\frac{\Sigma(s)}{\sigma_y^0}\right]^{\alpha} ds\right\}^{-1/\alpha} & , \frac{1}{\tau}\int\limits_{t-\tau}^{t}\left[\frac{\Sigma(s)}{\sigma_y^0}\right]^{\alpha} ds \le 1 \end{cases} \tag{10}$$

Here $\Sigma(t) = 2G\dot{\varepsilon}tH(t)$is the function coinciding with the stress in the specimen at the stage of elastic deformation, i.e., before the time t_* of appearance of the macroscopic plastic flow; this time is calculated by criterion (9). The actual stresses in the deformed specimen for $t \geq t_*$ is determined by the relation

$$\sigma(t) = 2Gg(t)\varepsilon(t), \qquad (11)$$

where $g(t) = G\gamma^{1-\beta}(t)$ and β is a scalar parameter $(0 \leq \beta \leq 1)$ which controls the strengthening degree. (The case $\beta = 0$ corresponds to the absence of strengthening.) In the case where the stress relaxation is caused by the simultaneous action of different mechanisms of plasticity, it is advisable to introduce into the model several characteristic relaxation times [16] or, in a more general case, to specify a function describing the contribution from processes with different characteristic times.

3. Numerical simulation and discussion

We shall investigate well-known yield drop phenomenon in metal with low dislocation density. A large amount of tensile experiments demonstrate that even in the case of quasi-static loading rates 10^{-4}–$10^{-2}s^{-1}$, the stresses attained in the material can be dozens times greater than the value of its quasi-static yield point [1–5]. When the amount of mobile defects is insufficient for ensuring the required plastic strain rate and the elastic stresses continue to grow. From the mechanical standpoint, this phenomenon expresses the fact that the material already experiences dynamic deformations for such "quasistatic" strain rates. Namely, for a material with such properties, these strain rates are already sufficient for exciting the dynamic deformation mode, which can not be described on the basis of quasistatic concepts that there are critical flow stresses in the material. To describe the dynamic characteristics of the material, it is necessary to explicitly take into account the fact that the material cannot unboundedly experience plastic relaxations of the arising stresses and a certain time is required for the development of relaxation process inside the material. This can be attained by introducing an additional dynamic characteristic of the material itself, i.e., the characteristic plastic relaxation time parameter. Obviously, this parameter should be independent of the deformation process characteristics but should express the mechanical properties of the material itself in a wide range of loading rates.

3.1. Numerical simulations of yield drop phenomenon

Figure 1 shows an example of the calculation of the stress–strain curve for a linear-in-time increase of the strain according to formulas (9)–(11) for a hypothetical material with strain hardening, which roughly corresponds to a soft steel ($G = 100$ GPa, $\sigma_y^0 = 200$ MPa, $\beta = 0.18$, $\dot{\varepsilon} = 8.5 \times 10^{-3}\,s^{-1}$). Fitting relaxation time is 0.4 s that two orders of magnitude smaller then for whiskers samples of metals [18] and many orders of magnitude more than in usual metal samples.

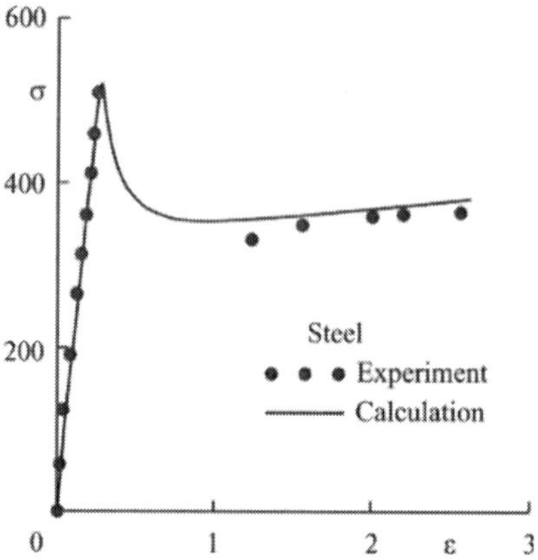

Figure 1. Sharp yield point effect with the subsequent strain hardening in the soft steel at a strain rate of $8.5 \times 10^{-3}\,s^{-1}$. The calculation is performed according to model (9)–(11). Dots are the experimental data taken from [1].

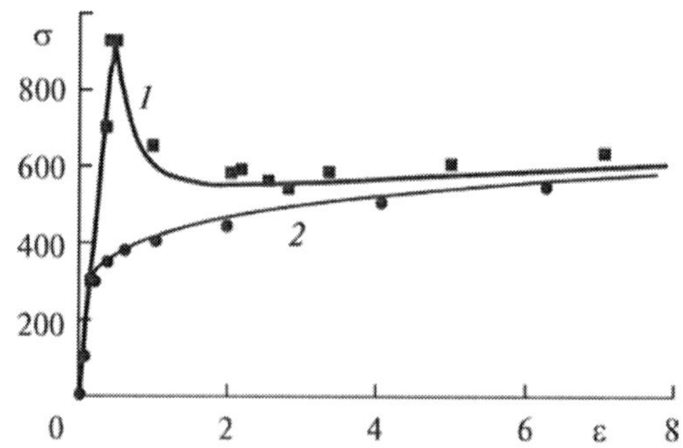

Figure 2. Steel specimens ($G = 78$ GPa, $\sigma_y^0 = 310$ MPa, $\alpha = 1$). The points correspond to the experimental data given in [17].

Figure 2 illustrates the simulation of the sharp yield point effect with the subsequent deformation strengthening according to the model (9)–(11) in the case of deformation of steel specimens ($G = 78$ GPa, $\sigma_y^0 = 310$ MPa, $\alpha = 1$). The points correspond to the experimental data given in [17]. In this material, the sharp yield point effect is already not observed at quasi-static strain rates $5 \cdot 10^{-2}\,s^{-1}$ (curve 2), but is clearly observed in the case of dynamic deformation at the rates $10^3\,s^{-1}$ (curve 1). The points correspond to the recently obtained experimental data [17]. It is remarkable that in this case the obtained value of relaxation time scale is already equal only to 14 microseconds that are still too many.

Figures 3 and 4 show dependences stress of iron on deformation degree under strain rate in the order 10^{-3} based on (9)–(11). The constant characteristic time in Fig. 3 is increased upon condition decrease of the dislocations density (curve 1 – $10\,cm^{-2}$, curve 2–$10^3\,cm^2$, curve 3–$10^5\,cm^{-2}$, curve 4–$10^7\,cm^{-2}$) and growth of the "yield drop" peak.

As shown in Fig. 4, the behavior of iron under deformations characterized by dependence with static yield stress 145 MPa and less constant characteristic time 0.46 versus parameters in Fig. 3.

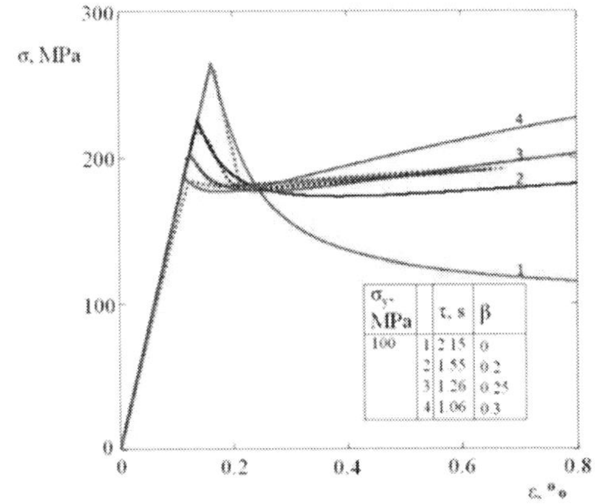

Figure 3. The stress-strain curves for small deformations of iron with the single relaxation time parameter. Initial dislocation densities for curves are (1) – $10\,cm^{-2}$, (1) – $10^3\,cm^{-2}$, (3) – $10^5\,cm^{-2}$, (4) – $10^7\,cm^{-2}$. Our simulation and data of G.T. Hahn (dots) [5].

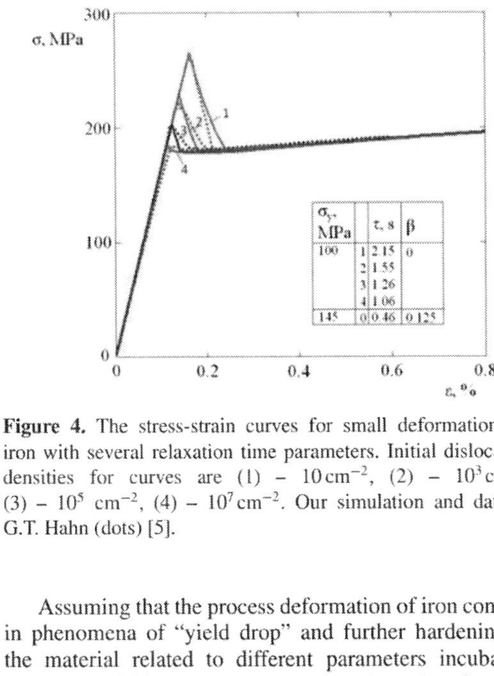

Figure 4. The stress-strain curves for small deformations of iron with several relaxation time parameters. Initial dislocation densities for curves are (1) - $10\,cm^{-2}$, (2) - $10^3\,cm^{-2}$, (3) - $10^5\,cm^{-2}$, (4) - $10^7\,cm^{-2}$. Our simulation and data of G.T. Hahn (dots) [5].

Assuming that the process deformation of iron consists in phenomena of "yield drop" and further hardening of the material related to different parameters incubation time and yield stress, stress-strain piecewise function is plotted in Fig. 4. The parameter of the hardening β is equal to zero in "yield drop" part. The relative transition point between two processes for each curve is calculated on incubation time of "yield drop" part and time beginning plastic deformation of "hardening" part.

As shown above, decrease of the dislocations density result in increase incubation time on "yield drop" part. Thus, the characteristic time of "yield drop" part can be involved with kinetic of defects, introducing hardening parameter β in relaxation function $\gamma(t)$ on second part curve in Fig. 3. On account of the macroscopic parameters of the yield stress and incubation time can be characterized stress-strain curve in view of micro-defects.

3.2. Physical based interpretations for relaxation time parameters

In the case of a steady-state flow, the expression (2) determines the level of maximal stresses attained in the material, and it can be rewritten as $\sigma_\tau^{\max}$ (3). The maximum shear stresses (3) are proportional to the yield point of the material. It follows from (3) that, in the case of small characteristic times and strain rates, we have a constant value of the yield point, which can vary only due to the strain strengthening [19]. One can determined static yield strength σ_y^0 by the Taylor and Hall-Petch lows [16,19]. On the other hand, at high strain rates or at large characteristic relaxation times, the second term in (3) becomes significant and the dependence of the yield point on the strain rate is rather complicated.

We use Eq. (3) to estimate the characteristic time of plastic relaxation due to the dislocation sliding. Consider the dislocation model of crystal plasticity, which states that the plastic strain is caused by displacements of separate defects of the crystal structure [19]. At times exceeding the acceleration time of dislocations (of the order of 10^{-10}–10^{-9}s), we can assume that their motion is steady-state and $\dot{\sigma} = 0$. Then, under the condition that the high external stresses are significantly greater than the barrier stresses $\sigma_\tau \gg \sigma_y$, it follows from (3) that

$$\sigma_\tau = 2G\tau\dot{\varepsilon}. \tag{12}$$

In [13], it was shown that, by substituting (3) into the Orovan relation (10) and by using the dislocation dynamic equation [16,18], we obtain the following expression for the characteristic stress relaxation time:

$$\tau_D = \frac{B_f}{Gb^2\rho_D} \sim \frac{B_f}{E_D} \tag{13}$$

where E_D is the total elastic energy of dislocation lines per unit volume and $B_f = (\nu_D/2)\,\rho b^2$ characterizes the rate of scattering of the dislocation kinetic energy [20]; $\nu_D \sim 10^{13}\,s^{-1}$ is the nearly Debye frequency whose inverse value gives the time of scattering of the kinetic energy of mobile dislocations and ρ is the material density. The time (13) is inversely proportional to the dislocation density and can vary in sufficiently broad ranges. At the initial deformation stages, which are important for the sharp yield point effect, the relaxation time is completely determined by the density of dislocations which are present in the material before the deformation. That is true for metal wishkers with extremely low initial dislocation density [18]. For yield drop phenomenon one can see from Fig. 3, Fig. 4 that relations of characteristic time to dislocation density has logarithmic form. Similar relation has the time needs for increasing of dislocation density [8,9].

$$\tau_w \sim \frac{\varepsilon_D}{0.2} \frac{\ln\left(\rho_D^{cr}/\rho_D^0\right)}{BV_D^2}, \tag{14}$$

here V_D velocity of dislocations relative to the substance, $\varepsilon_D \sim 8\,eV/b$ is the energy of the dislocations formation per unit length [9,20].

From the standpoint of the above-mentions models of an elastoviscoplastic body, there are no fundamental differences between the shear strength in the case of "quasi-static" strain and in the case of "dynamic" strain. Everything is determined by the values of the parameters describing the internal structure and the defect substructure of the material. For various defect concentrations in the material, this strain rate can take practically any values. The defect structure and the mechanical properties of the material vary in the course of plastic deformation.

4. Conclusion

Thus, the dynamic deformation mode is determined not by external factors but by the relation between the strain rate and the material ability to relaxation of elastic stresses arising in it. The value of the characteristic plastic relaxation time determines the relaxation properties of the material and is one of its basic dynamic characteristics. This time is independent of the strain rate and is determined at the micro level by the dynamic properties of the basic carriers of plastic deformation and by their amount per unit volume. The coincidence of the values of characteristic time obtained by various approaches confirms the claim that its nature is universal. At relatively small strains, the mechanical properties are determined by a single value of the characteristic time corresponding to the initial dislocation density. For large deformation nature of the characteristic relaxation time can may change. Also at strong plastic deformations, it is necessary to take into account the irreversible changes in the defect structure of the material, which can be described by simple kinetic equations for the dislocation density variations. Macroscopically, this can be expressed as a decrease in the characteristic time of plastic deformation with time. In the case of deformation of pure metals with low dislocation density the characteristic relaxation time turns out to be very large; i.e., it is several orders of magnitude greater than the relaxation times typical of usual specimens of metals with large dislocation densities [14].

The research was supported by the Program of the President of the Russian Federation (Grant No. MD-286.2014.1), by the Russian Foundation for Basic Research (Grants Nos. 14-01-31454 and 14-01-00814), by Program No. 25 of Presidium of the Russian Academy of Sciences, by Marie Curie Foundation TAMER No. 610547, and the Ministry of Education and Science of Russian Federation (competitive part of State Task of NIR CSU No. 3.1334.2014/K).

A special acknowledgement goes to the Scientific & Technological Cooperation Programme Switzerland-Russia for the financial support of the Utilisation of Specific Infrastructure Project between DynaMat Laboratory of the University of Applied Sciences of Southern Switzerland and the St. Petersburg State University.

References

[1] M.M. Hutchinson, J. Iron Steel Inst. **186** (1957)
[2] S.S. Brenner, J. Appl. Phys. **28**, 9 (1957)
[3] B.V. Petukhov, Technical Physics **71**, 11 (2001)
[4] R.E. Smallman, R.J. Bishop *Metals and Materials: Science, Processes, Applications* (Butterworth-Heinemann Ltd, 1995)
[5] G.T. Hahn, Acta Met. **10**, 17 (1962)
[6] L.D. Landau, E.M. Lifshitz, *Course of theoretical physics, Vol. 7: Theory of elasticity* (Nauka, Moscow, 2003; Butterworth–Heinemann, Oxford, 2005)
[7] E.N. Borodin, A.E. Mayer, Phys. Solid State. **54**, 4 (2012)
[8] V.S. Krasnikov, A.E. Mayer, A.P. Yalovets, Int. J. Plast. **27** (2011)
[9] A.E. Mayer, K.V. Khishchenko, P.R. Levashov, P.N. Mayer, J. Appl. Phys. **113** (2013)
[10] *Shock waves and high-strain-rate phenomena in metals*, Ed. by M. A. Meyers and L. E. Murr (Plenum, New York, 1981; Metallurgiya, Moscow, 1984)
[11] J.R. Greer, J.Th.M. De Hosson, Prog. Mater. Sci. **56** (2011)
[12] R.W. Armstrong, F.J. Zerilli, J. Phys. D: Appl. Phys. **43** (2010)
[13] A.A. Gruzdkov, Yu.V. Petrov, V.I. Smirnov, Phys. Solid State **44**, 11 (2002)
[14] A.A. Gruzdkov, E.V. Sitnikova, N.F. Morozov, Y.V. Petrov, Math. Mech. Solids **14**, 72 (2009)
[15] Yu.N. Robotnov, *Elements of hereditary solid mechanics* (Nauka, Moscow, 1977; Mir, Moscow, 1980)
[16] E.N. Borodin, A.E. Mayer, Yu.V. Petrov, A.A. Gruzdkov, Physics of the Solid State, **56**, 12 (2014)
[17] E. Cadoni, F. D'Aiuto, and C. Albertini, " Dynamic Behavior of Advanced High Strength Steel Used in the Automobile Structures," DYMAT 1, 135–141 (2009)
[18] Yu.V. Petrov, E.N. Borodin, Physics of the Solid State, **57**, 2 (2015)
[19] M.A. Meyers, K.K. Chawla, *Mechanical behavior of materials* (Cambridge Univ. Press, New York, 2009)
[20] A.E. Dudorov, A.E. Mayer, Vestnik Chelyabinsk Gos. Univ. **39**, 12 (2011)

EPJ Web of Conferences **94**, 04040 (2015)
DOI: 10.1051/epjconf/20159404040

Simulation of dynamic response of projectile and granular target

Kinya Ogawa[1,a], Shin-nosuke Takeda[2], Hidetoshi Kobayashi[2], and Kenichi Tanigaki[2]

[1] Institute of Space Dynamics, Ondoyama, Narutaki, Ukyo-KU, Kyoto 616-8255, Japan
[2] School of Engineering Science, Osaka University 1-3 Machikaneyama-cho, Toyonaka, Osaka 560-8531, Japan

Abstract. In the present study, the three-dimensional elastic discrete element method (DEM) and the elastic finite element method (FEM) analysis were, respectively, applied to the randomly distributed identical spheres as a granular target and to the cylindrical projectiles in order to clarify dynamic response during penetration. In the target, it was found that highly densified region was formed just ahead of the projectile and began to propagate spherically at much higher velocity than that of projectile leaving relatively rarefied region. It was also found that the peak resistance during penetration was closely related to the initial formation of densified region and was expressed in terms of momentum change of target particles accelerated by the projectile. Stress wave propagation in the projectiles with various body lengths was investigated during penetration and was discussed in connection with dynamic response of target particles ahead of the projectile. Effect of mechanical properties of projectile on the peak resistance was also investigated and was understood in connection with dynamic behavior of target particles.

1. Introduction

Collision impact between continuous medium and granular materials such as sand, soil, and heterogeneous brittle aggregates is interesting subject and has been investigated over several decades. Resistance during penetration has been identified in experiments and a various formulations for the resistance versus penetration relations have been proposed [1,2]. Direct measurement of resistance has been also tried [3,4]. It would be the first and most fundamental approach to clarify dynamic characteristics in the problem based on elastic wave propagation theory. It is also necessary to find out effect of mechanical impedance of materials upon the dynamic behavior of subject.

In the present study the discrete element method together with the finite element method were applied to simulate the dynamic behaviors of penetration into granular target and the characteristics of the problem were fully discussed in connection with dynamic movement of particles together with dynamic response of projectile.

2. Computational setup

2.1. Projectile

Figure 1 shows a projectile and a target of granular medium. A projectile was a cylinder of 42 mm in diameter and three kinds of length with a flat end. Three dimensional elastic FEM model was applied to the projectile having Young' modulus of 69 GPa and density of 3.8×10^3 Kg/m^3 with Poisson's ratio of 0.3. Initial velocity of the projectile was in the range of 60 m/s to 240 m/s.

2.2. Target

The target was formed by randomly distributing identical spheres of 2.18 mm in diameter and of 3.76 g/cm^3 in density with Young's modulus of 310 GPa into a spheroidal container of 218 mm in diameter and of 160 mm in depth as shown in the figure. Average packing density was changed from about 60% to 48%. Elastic Discrete Element Method (DEM) was applied for the target.

Reactions between spheres were expressed by springs both in normal and tangential directions, together with a slider to represent static and kinetic frictions as shown in Fig. 2. Numerical simulations were carried out with time increment of 2.94×10^{-8} s.

3. Results

3.1. Behavior of target

To clarify deformation state in the particulate target we have focus our attention to the change of apparent packing density which may reflects to stress and strain in the sense of continuous mechanics. Figure 3 represents distribution of packing density caused by penetration of projectile at initial impact velocity of 120 m/s and 240 m/s. Just after the impact, higher density region appears just ahead of the projectile, and then, at almost the peak resistance which will be explained in detail, the region spreads spherically and begins to propagate downward at much higher velocity than that of projectile leaving relatively rarefied region. This rarefied region is essentially caused by difference of velocities of the projectile and the high density region and such velocity difference would be decreased with the increase of initial impact velocity.

With increase of initial impact velocity, the high density region grows much clearly and tails up to just ahead of the projectile as seen in the figure.

[a] Corresponding author: paddy-og@khaki.plala.or.jp

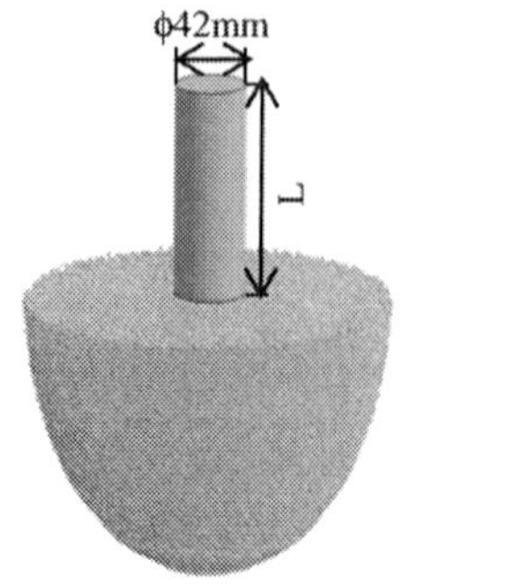

Figure 1. Target and flat-ended cylindrical projectile.

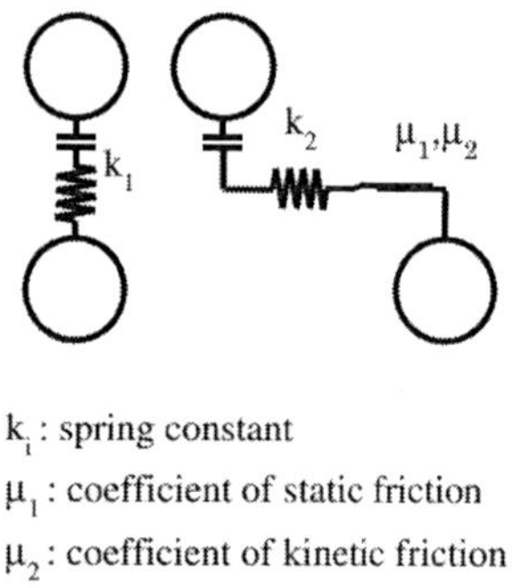

Figure 2. Mechanical model of granular medium with elastic springs and sliders to represent static and kinetic friction.

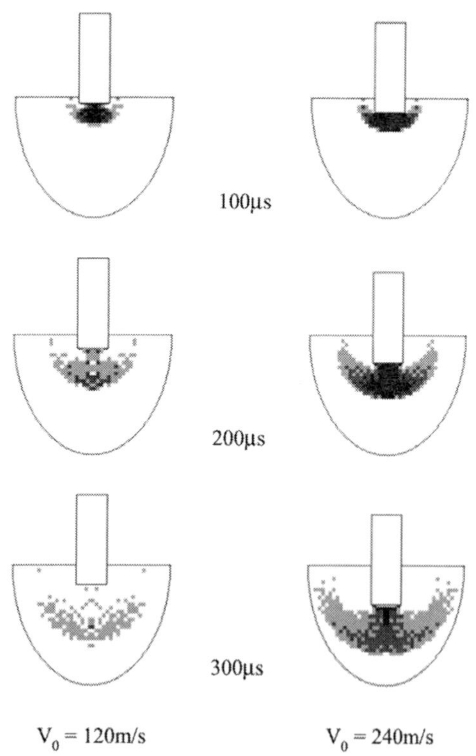

Figure 3. Distribution of density in target during penetration at different impact velocity.

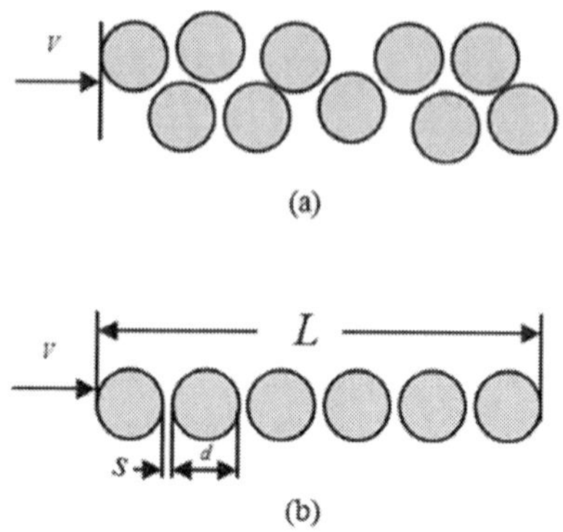

Figure 4. a), b) Illustration of arrangement of spheres.

The propagation velocity of such high density region was determined as follows. The location of the most distant particle which began to move at time t in vertical direction is expressed by (Y, r) and that at time $t + \Delta t$ is as $(Y + \Delta Y, r + \Delta r)$, and then the propagation velocity, C, can be evaluated as

$$C = (\Delta Y / \Delta t). \tag{1}$$

In the case of granular medium the elastic wave velocity depends on the density. Elastic wave propagation in randomly arranged particles with separations as illustrated in Fig. 4a can be simplified by that in aligned particles with separations as shown in Fig. 4b. When the velocity, V, is applied to the left end particle, then it moves and hits its next neighbor particle and so on. Then we have the following relation for the elastic wave propagation;

$$\frac{nd + (n-1)s}{C} = \frac{nd}{C_0} + \frac{(n-1)s}{V} \tag{2}$$

where C_0 is the elastic wave velocity in the case of $s = 0$, and n is a number of particles in the length L. Equation (2) can be rewritten as follows.

$$\frac{1}{C} = \alpha \frac{1}{C_0} + \beta \frac{1}{V}, \alpha = \frac{1}{(1 + s/d)}, \beta = \frac{s/d}{(1 + s/d)}. \tag{3}$$

In Fig. 5, the propagation velocities at the peak resistance of projectile are plotted against the velocity of projectile following the Eq. (3) for different packing densities. The values of s/d can be obtained from the slope for respective lines. Since the close packing density is given by $\pi\sqrt{2}/6\,(1 + s/d)^{-3}$, the estimated values are of 63%, 59%, 54% and 44% which are slightly different from the used packing density of 60%, 57%, 53% and 48%, respectively in the present simulation. The values of C_0 can be obtained by using the respective intercepts of lines together with the values of s/d, and are ranged from 500 m/s to 530 m/s almost corresponding to the value of 530 m/s for rectangular lattice arrangement without separation [5].

3.2. Wave propagation in a projectile

Elastic wave propagation in the projectile was investigated and was exampled in Fig. 6. At the impact onto the

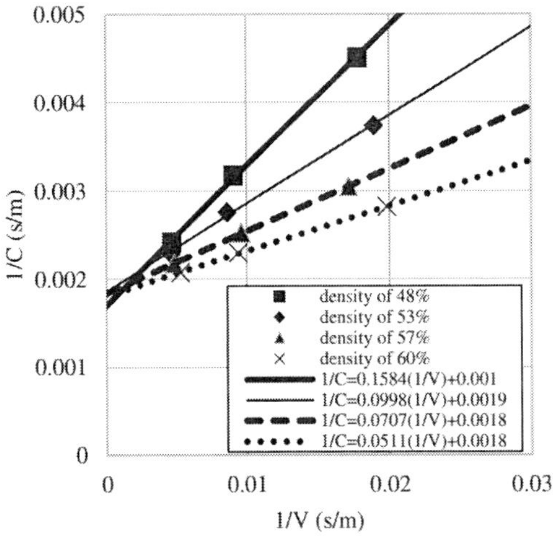

Figure 5. Relations (a), (b), (c), and (d) for packing density of 60%, 57%, 53%, and 48%, respectively.

target, the stress state near the projectile head region is not uniform due to the individual collision to thc target particles as shown in (a), while such elastic state quickly spreads laterally and propagates upward almost alike unidirectional compression wave propagation as shown in (b). After the elastic wave reaches to the upper free end of the projectile, it reflects back as the tensile elastic wave and propagates down ward as shown in (c). Averaging the stress over the sectional area at the locations, A, and B, as indicated in the figure, the relations of the stresses at the A, and the B to time are shown in Fig. 7. It is apparently observed that the elastic wave reflections occur during penetration of the projectile. Assuming the elastic wave propagation without retardation and free end reflection on the upper end of the projectile, the force at the impact head can be evaluated by superposing the elastic waves detected at the locations A, and B, and is given in Fig. 8(a,b) for the projectiles of 120 mm and 200 mm in length. It should be noted that the force decreases after the peak value with periodical change corresponding to the reflection of stress wave as indicated by an arrow, since the periods of 48 and 80 microseconds for the projectiles of 120 mm and 200 mm in length, respectively. It is also apparent that the force evaluated from acceleration of the projectile well corresponds to the force obtained by stress wave analysis, even though significant undulations appear at around the peak value. The relations between reaction force and time are strongly influenced by impact velocity as shown in Fig. 9. Significant increase of peak value can be understood by change of momentum of particles in target as discussed before [6].

3.3. Effect of mechanical properties of projectile

Effect of mechanical properties of projectile on the resistance force during penetration was also investigated.

Based on the one-dimensional elastic wave propagation theory, the reaction force of the projectile can be evaluated as follows. Let's denote ρ, C, and ρ_t, C_t, are

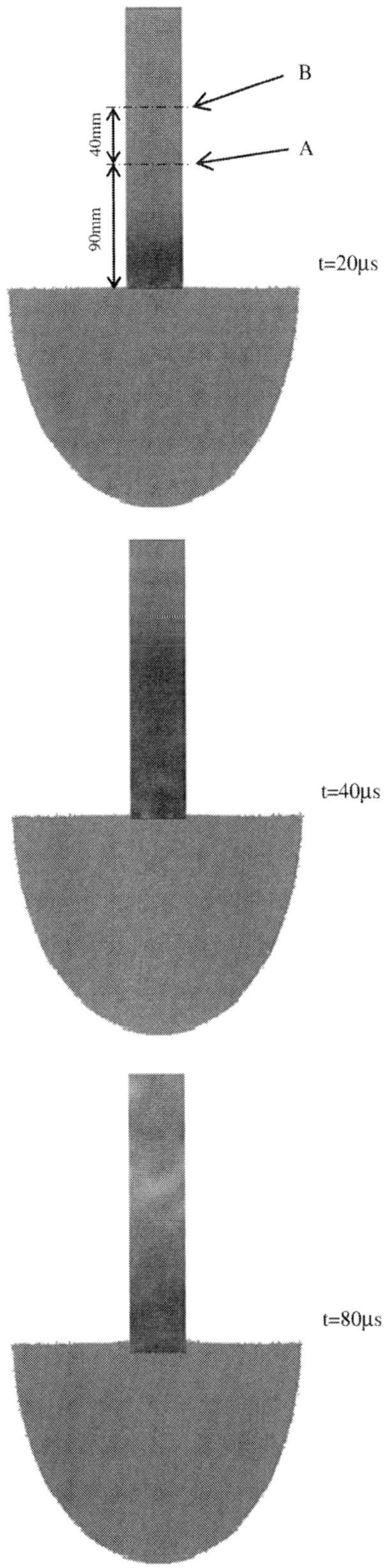

Figure 6. Distribution of axial stress in projectile in time.

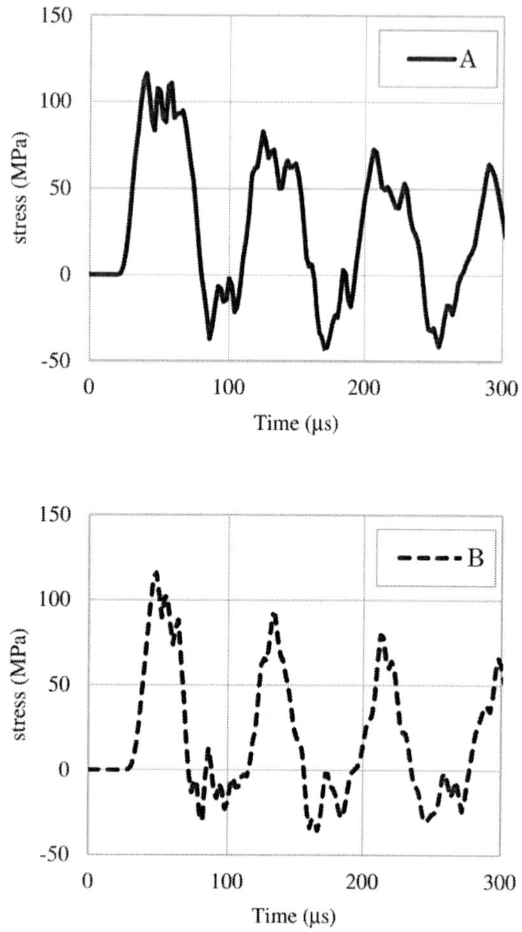

Figure 7. Stress waves detected at locations A, and B in a projectile.

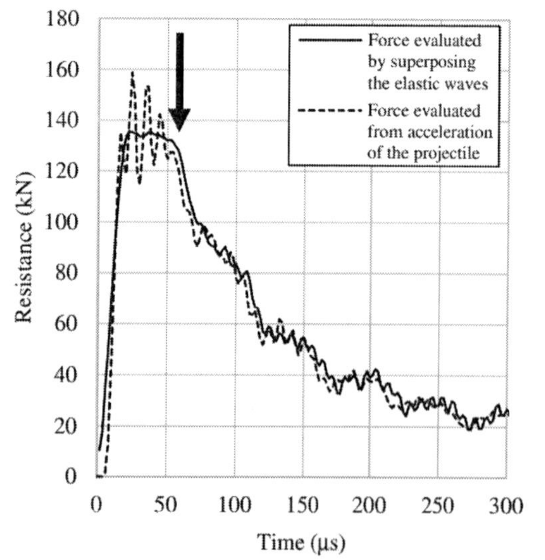

Figure 8. a) Force-time relations derived from stress wave detected in projectile of 120 mm in length.

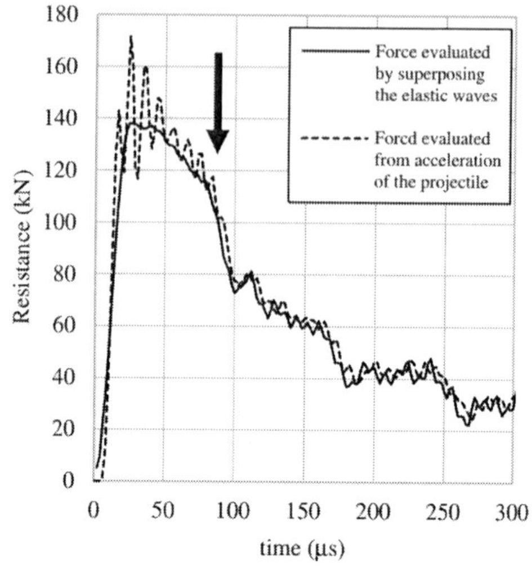

Figure 8. b) Force-time relations derived from stress wave detected in projectile of 200 mm in length.

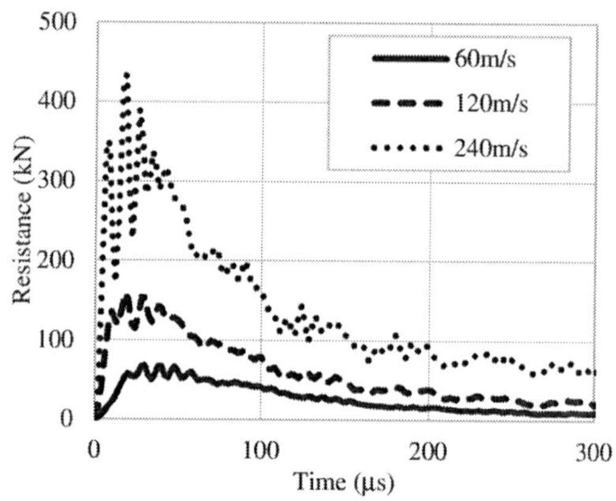

Figure 9. Force-time relations at various impact velocity.

the respective density and wave velocity of the cylindrical projectile and the target.

The reaction force F at initial impact velocity of V_0 can be expressed as follows

$$F = A_0\sigma_T = A\sigma_R = \frac{V_0}{\left(\frac{1}{\rho CA} + \frac{1}{\rho_t C_t A_t}\right)}. \tag{4}$$

Where, A and A_t are the sectional area of projectile and target, and σ_R and σ_T are the reflected and the transmitted stress wave amplitude in the projectile and the target, respectively. The effective sectional area of target, A, can be estimated as before [6], and is almost constant independent of the impact velocity. On the other hand, elastic wave velocity of the target depends on the impact velocity as shown in Fig. 5.

Now, we can see the effect of mechanical properties of projectile upon the resistance force as shown in Fig. 10. For the projectile of initial values of Young's modulus of 69 GPa and of density of 2.8×103 Kg/m^3, the

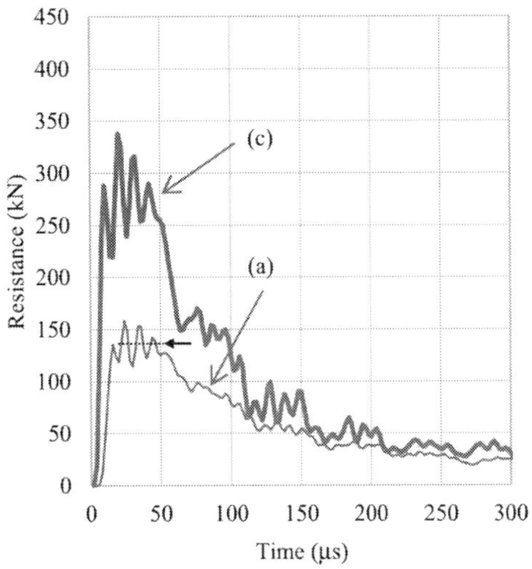

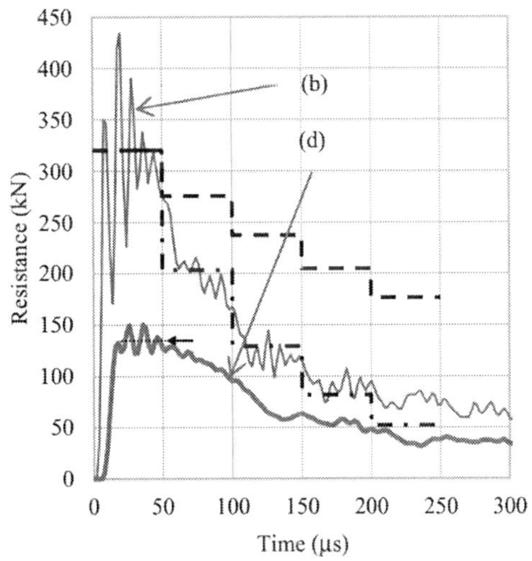

Figure 10. Effect of mechanical impedance and momentum of projectile on force-time curve.

force-time relations (a) and (b) are respectively obtained at initial impact velocities of 120 m/s and 240 m/s. By reducing the Young's modulus and the density of the projectile to the half of their initial values the force –time relation shown by (c) is obtained at initial impact velocity of 240 m/s. The relation denoted by (d) is obtained at initial impact velocity of 120 m/s for the projectile of the Young's modulus of half to the initial value and the density of twice the initial value. It is apparent from Eq. (4) that the value of reaction force F is the same, when the value of ρCA is kept constant. Actually, almost identical values of initial reaction force F indicated by the arrows on the curves (a) and (d) are obtained at the initial impact velocity of 120 m/s. It is very interesting that the momentum of projectile for curves (d) is twice of the momentum of projectile for curve (a), since the density for the former projectile is double of the latter. It is also found that the peak reaction forces for the curves (b) and (c) are different so far, even though the momentum is kept constant. From the peak reaction force, F, for the curves (a) and (b), the value of $\rho_t C_t A_t$ in the Eq. (4) can be evaluated as 1.19 KN/(m/s) and 1.43 KN/(m/s), respectively. Then, the peak reaction forces for the curves (d) and (c) can be estimated as 135 KN and 299 KN, respectively. As shown in the figure, fairly good predictions are obtained. With taking account of dependence of the wave velocity C_t in the target on the impact velocity as was mentioned before [6], evaluated values of $\rho_t C_t A_t$ at impact velocity of 120 m/s and at 240 m/s are still slightly different, but this can be understood by considering a possible change of A_t by impact velocity already reported [6].

After the initial reaction, the stress wave in the projectile goes up and down and re-impact is continued to the target. Therefore, the same kind of equation like Eq. (4) is obtained for respective reverberations as follows;

$$F_i = \frac{\left(1 - \frac{\rho_t C_t A_t}{\rho C A}\right)}{\left(1 + \frac{\rho_t C_t A_t}{\rho C A}\right)} F_{i-1}. \tag{5}$$

Here, the reaction force after i-th reflection is represented as F_i.

In the figure, the relations obtained by Eq. (5) are represented by broken lines for the case of constant $\rho_t C_t A_t$, and by chained lines for the case of variable $\rho_t C_t A_t$. Judging from the results shown in Fig. 3, it would be reasonable to taking account of increase of A_t, since highly compressed region grows with penetration.

4. Concluding remarks

Three dimensional dynamic FEM and DEM simulations on the projectile penetration into granular target were performed and the followings were concluded.

(1) Propagation of elastic compression wave generated by penetration was observed and the dependence of wave velocity on density of particulate medium was clarified by means of simple expression.
(2) Periodical undulations on the reaction force curve were well understood in connection with propagation of elastic wave in the projectile.
(3) Mechanical impedance of projectile plays a predominant role on collision phenomena as well as the mechanical properties of target materials.

References

[1] Allen, W.A., Mayfield, E.B. and Morrison H.L., Jour. App. Phy., **28**, 370 (1957).
[2] Backman, M.E. and Goldsmith, W., Int. Jour. of Eng. Sci., **16**, 1 (1978).
[3] Bazhenov, V.G., Bragov, A.M. and Kotov, V.L., Jour. App. Mech. Tech. Phy., **50**, 1011 (2009).
[4] Hauver, G.E., Inter. Jour. of Eng. Sci., **16**, 871 (1978).
[5] Zhang, Y., Master's thesis for Osaka University, Japan(2013).
[6] Ogawa, K., Takeda, S. and Kobayashi, H., Mech. Eng. Jour. Jap., **02**, (2015), (to be published).

EPJ Web of Conferences **94**, 04041 (2015)
DOI: 10.1051/epjconf/20159404041

Kinetic model for mechanical twinning and its application for intensive loading of metals

Alexander Mayer[1,a] and Elijah Borodin[2,1]

[1] Chelyabinsk State University, Department of General and Applied Physics, 454001 Br. Kashirinykh str. 129, Chelyabinsk, Russia
[2] IPME RAS, Extreme States Dynamics Department, 199178 Bolshoj pr. V.O., 61 St. Petersburg, Russia

Abstract. In this report, we present our twinning model intended for simulation of the dynamic deformation of metals with low values of the stacking fault energy, as well as the results of application of the model to numerical simulation of intensive loading of metals. Generation of a twin is described as an appearance of a stacking fault with size more than some critical value, while growth of a twin is considered as a cooperative movement of partial dislocation along the stacking fault. The twin nucleation rate is expressed through the energy released due to annihilation of dislocations. Movement of partial dislocations in the course of twin growth passes under the action of elastic stress field and phonon drag. The surface energy of the growing twin continuously increases which leads to the appearance of an additional force. Application of this model allows us to investigate plastic response of metals at various dynamic loading conditions and initial defect structures. Influence of twinning at Taylor rod compaction experiments is analyzed including formation of the shape of the lateral surface.

1. Introduction

Dislocation plasticity [1–5] is the main mechanism of the plastic deformation of metals in most cases. Twinning becomes an alternative plasticity mechanism at low temperatures and high strain rates [1,6,7]; it occurs if value of the stacking fault energy is less than 100 mJ/m^2, which is the case for copper, steel and for many alloys. Mechanical twinning reveals itself in many dynamical processes including the Taylor impact tests [8,9] applied to a metal with a low value of the stacking fault energy.

There are a lot of dislocation plasticity models including applicable for the dynamic conditions of deformation [3–5]. At the same time, the majority of the existing twinning models are designed for quasi-static conditions; most of the models are empirical and parameterized by a number of adjustable parameters.

In this report, we present our twinning model intended for simulation of the dynamic deformation of metals with low values of the stacking fault energy, as well as the results of application of the model to numerical simulation of intensive loading of metals. The model is based on the kinetics of twinning and takes into account generation of new and growth of existing twins.

2. Mathematical model

Total plastic strain is represented as a result of the combined action of two competing processes: the dislocation motion and the mechanical twinning. The plastic deformation tensor w_{ik} is represented by a sum: $w_{ik} = w_{ik}^{\mathrm{D}} + w_{ik}^{\mathrm{T}}$, where w_{ik}^{D} is caused by the dislocation motion and w_{ik}^{T} is caused by the mechanical twinning. These tensors are determined through evolution of corresponding defect subsystems. For dislocation plasticity w_{ik}^{D} we use the model proposed in [4,5].

2.1. Continuum mechanics equations

The common part of the mathematical model consists of three conservation laws. The first one of them is the continuity equation:

$$\frac{1}{\rho}\frac{d\rho}{dt} = -\sum_{k=1}^{N}\frac{\partial\, v_k}{\partial x_k}, \tag{1}$$

where ρ is the substance density; v_k is the velocity vector; x_k is the Cartesian coordinates; N is the number of dimensions of the considered problem, index k numerates spatial directions. The total time derivatives are used, which is valid for Lagrangian particles moving with substance. The equation of substance motion:

$$\rho\frac{dv_i}{dt} = -\frac{\partial P}{\partial x_i} + \sum_{k=1}^{N}\frac{\partial S_{ik}}{\partial x_k}, \quad i = 1,\ldots,N, \tag{2}$$

where P is the pressure, which is determined from a wide-range equation of state $P = P(\rho, U)$ [10]; S_{ik} is the tensor of stress deviators, which characterizes the shear stresses; U is the internal energy determined from the energy conservation law in the following form:

$$\rho\frac{dU}{dt} = -\frac{P}{\rho}\frac{d\rho}{dt} + \sum_{i=1}^{N}\sum_{k=1}^{N} S_{ik}\frac{dw_{ik}}{dt}, \tag{3}$$

where the second term in the right-hand part is the heat release due to the plastic deformation.

[a] Corresponding author: mayer@csu.ru

Article available at http://www.epj-conferences.org or http://dx.doi.org/10.1051/epjconf/20159404041

The generalized Hooke law [11] with accounting of the plastic strain w_{ik} is used for determination of the stress deviators:

$$S_{ik} = 2G\left[u_{ik} - \frac{1}{3}\delta_{ik}\sum_{l=1}^{N} u_{ll} - w_{ik}\right], \quad i, k = 1, \ldots, N, \tag{4}$$

where G is the shear modulus, δ_{ik} is the bivalent mixed tensor, u_{ik} is the tensor of macroscopic deformations:

$$\frac{du_{ik}}{dt} = \frac{1}{2}\left[\frac{\partial \upsilon_i}{\partial x_k} + \frac{\partial \upsilon_k}{\partial x_i}\right] + + \frac{1}{2}\sum_{l=1}^{N}\left\{u_{il}\left(\frac{\partial \upsilon_k}{\partial x_l} - \frac{\partial \upsilon_l}{\partial x_k}\right) + u_{lk}\left(\frac{\partial \upsilon_i}{\partial x_l} - \frac{\partial \upsilon_l}{\partial x_i}\right)\right\} \tag{5}$$

where the first term in the right-hand part is the infinitesimal strain rate tensor; the second term accounts for the change of the tensor components in the laboratory coordinate system due to the substance rotation.

2.2. Model of twinning

Plastic deformation caused by twinning can be expressed through the volume fraction of twins:

$$\frac{dw_{ik}^{\mathrm{T}}}{dt} = \sum_{\gamma}\frac{d\alpha^{\gamma}}{dt}\left(\tau_i^{\gamma} n_k^{\gamma} + \tau_k^{\gamma} n_i^{\gamma}\right)\varepsilon_{\mathrm{T}} + + \frac{1}{2}\sum_{l=1}^{N}\left\{w_{il}^{\mathrm{T}}\left(\frac{\partial \upsilon_k}{\partial x_l} - \frac{\partial \upsilon_l}{\partial x_k}\right) + w_{lk}^{\mathrm{T}}\left(\frac{\partial \upsilon_i}{\partial x_l} - \frac{\partial \upsilon_l}{\partial x_i}\right)\right\} \tag{6}$$

where α^{γ} is the volume fraction of twinned material; ε_{T} is the deformation of twinned material with respect to the initial one, for example $\varepsilon_{\mathrm{T}} = 1/\sqrt{2}$ in FCC metals [6]; unit vectors τ_k^{γ} and n_k^{γ} describe the crystallographic orientations of twins; index γ numerates all possible orientations. The second term in the right-hand part plays the same role as in Eq. (5). We suppose that twins are cylindrical and have the same radius and thickness in a physically small substance element, than the volume fraction is equal to

$$\alpha^{\gamma} = N_{\mathrm{T}}^{\gamma}\cdot\left(\pi\left(R_{\mathrm{T}}^{\gamma}\right)^2\cdot h_{\mathrm{T}}^{\gamma}\right) + N_{\mathrm{I}}^{\gamma}\cdot\left(\pi\left(R_{\mathrm{I}}^{\gamma}\right)^2\cdot h_{\mathrm{I}}^{\gamma}\right), \tag{7}$$

where N_{T}^{γ} and N_{I}^{γ} are the local concentrations of twins, mobile and immobilized correspondently; R_{T}^{γ} and R_{I}^{γ} are radiuses of mobile and immobilized twins; h_{T}^{γ} and h_{I}^{γ} are their thicknesses. All these values are determined by the kinetics of twins.

Energy of a twin consists of the surface part determined by the stacking fault energy γ_{T}, the energy in external stresses and the energy of elastic deformations of the surrounding matrix [12]. Differentiation of this energy over radius or thickness gives the forces tending to change the corresponding size of the twin:

$$F_R^{\gamma} = 4\pi\varepsilon_{\mathrm{T}}\cdot R_{\mathrm{T}}^{\gamma}h_{\mathrm{T}}^{\gamma}\cdot\sum_{k=1}^{N}\sum_{i=1}^{N} S_{ik}n_i^{\gamma}\tau_k^{\gamma} - -2\pi\gamma_{\mathrm{T}}\cdot\left(2R_{\mathrm{T}}^{\gamma} + h_{\mathrm{T}}^{\gamma}\right) - \Phi\cdot\left(h_{\mathrm{T}}^{\gamma}\right)^2, \tag{8}$$

$$F_h^{\gamma} = 2\pi\varepsilon_{\mathrm{T}}\cdot\left(R_{\mathrm{T}}^{\gamma}\right)^2\cdot\sum_{k=1}^{N}\sum_{i=1}^{N} S_{ik}n_i^{\gamma}\tau_k^{\gamma} - -2\pi\gamma_{\mathrm{T}}\cdot R_{\mathrm{T}}^{\gamma} - 2\Phi\cdot R_{\mathrm{T}}^{\gamma}h_{\mathrm{T}}^{\gamma}, \tag{9}$$

where $\Phi = 2\pi^3(2-\nu)/[3(1-\nu)]\,G\varepsilon_{\mathrm{T}}^2$ and ν is the Poisson ratio. Growth of the twin radius or thickness is connected with motion of twinning dislocations at its edges of the twin. The balance between the forces (8), (9) and the drag forces acting on these dislocations allows one to find the growth equations:

$$\frac{dR_{\mathrm{T}}^{\gamma}}{dt} = F_R^{\gamma}/B, \quad \frac{dh_{\mathrm{T}}^{\gamma}}{dt} = F_h^{\gamma}/B, \tag{10}$$

where B is the drag coefficient.

Expressions (8) and (9) give the critical radius R_{c}^{γ} and thickness h_{c}^{γ} of twin at which forces are equal to zero–a twin larger than the critical one will grow up. Immobilized twins can not grow in radius but they can become thicker similar to Eqs. (9), (10).

Kinetics equations are written from the energy-wise consideration. The twinning becomes an active channel of plastic deformation when the dislocation plasticity becomes ineffective. Effectiveness of the dislocation channel is restricted by annihilation of dislocations. A certain part of the plastically dissipated energy is spent on the formation of new defects [13,14]. In the conditions of active annihilation, the dissipated energy can not be stored in the dislocation subsystem and should be transmitted to anther type of defects, twins for example. Thus, the generation rate of twins is as follows:

$$\dot{N}_{\mathrm{TW}}^{\gamma+} = \varepsilon_{\mathrm{D}}Q_{\mathrm{a}}^{\beta}/\left(4\pi\left(R_{\mathrm{c}}^{\gamma}\right)^2\gamma_{\mathrm{T}}\cdot\gamma_0\right), \tag{11}$$

where Q_a is the total annihilation rate of dislocations; γ_0 is the total number of possible orientations of twins. A twin becomes immobilized then it reaches a grain boundary or another twin; the immobilization rate can be written as follows:

$$\dot{N}_{\mathrm{T}}^{\gamma-} = N_{\mathrm{T}}^{\gamma}\cdot\left|dR_{\mathrm{T}}^{\gamma}/dt\right|\cdot\left(\Delta^{-1} + d^{-1}\right), \tag{12}$$

where d is the grain diameter. The average distance between the twins Δ is connected with their volume fraction [6] $\Delta^{-1} = \sum_{\gamma}\left[\alpha^{\gamma}/\left\{(1-\alpha^{\gamma})h_{\mathrm{T}}^{\gamma}\right\}\right]$. Finally, the kinetics equations are the following

$$\frac{dN_{\mathrm{T}}^{\gamma}}{dt} = \dot{N}_{\mathrm{T}}^{\gamma+} - \dot{N}_{\mathrm{T}}^{\gamma-}\frac{dN_{\mathrm{I}}^{\gamma}}{dt} = \dot{N}_{\mathrm{T}}^{\gamma-}. \tag{13}$$

Implementation of this twinning model to the plate impact tests gives results corresponding to the experimental observations, as it is shown in [15].

3. Numerical implementation

Equations of the twinning model, the dislocation plasticity model [4,5] and the continuum mechanics equations with supplementary relations are numerically solved within 2D cylindrical version of the finite-deference computer code CRS, which 1D realization is briefly

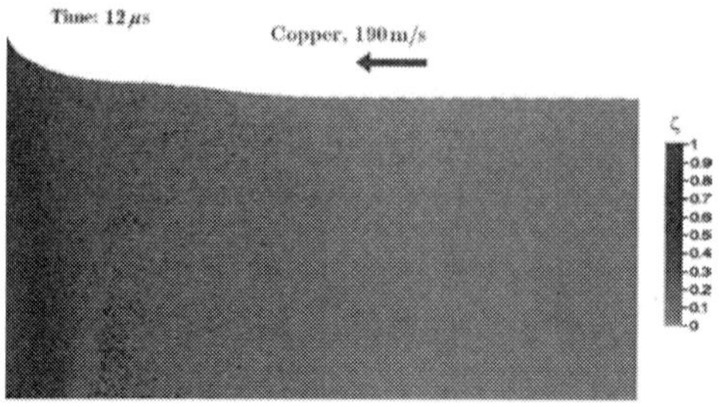

Figure 1. Shape of copper rod in the process of compaction with initial velocity of 190 m/s (12 μs from onset of impact) and spatial distribution of the volume fraction of twins.

described in [16]. Method of division on physical processes is used: each time step the continuum mechanics Eqs. (1)–(3) are solved at first and the macroscopic deformation (5) is defined, then the evolution of dislocations [4,5] is calculated, then the evolution of twins (6)–(13), and, finally, the new stress field is determined with the use of the wide range equation of state [10] for pressure and Hook law (4) for deviators. The continuum mechanics equations are solved with using the finite-difference numerical method [17]. Equations of kinetics of dislocations and twins are solved predominantly by explicit Euler method.

4. Results of calculations

Modelling of the Taylor anvil on rod compaction of copper samples at various impact velocities (up to 500 m/s) was performed. Calculations shows, that the dislocation plasticity is predominant mechanism at low impact velocities ($<$ 100 m/s), while twinning begins substantially influence at higher velocities.

Figure 1 presents an example of calculation results: the spatial distribution of the total volume fraction of twins at a moderate compaction velocity. Shock wave already passes through the sample. The twinning predominates in the part of the rod, which adjoins to the anvil; dislocation plasticity here is suppressed due to the hardening at the expense of twins. Dislocation plasticity is active in that part of the rod, in which the volume fraction of twins is low. Superposition of twinning and dislocation plasticity results in a complex form of the cross-sections of the deformed samples.

5. Conclusions

A kinetics-based model of the mechanical winning in metals at the high rate deformation is proposed. Growth of twins is considered as a collective slipping of the twinning dislocations; the finite rate of nucleation of twins is taken into account. The model of twinning is a complement to the dislocation plasticity model [4,5] and requires only one additional parameter, which is the sacking fault energy.

The twinning model is applied together with the dislocation plasticity model to description of the plastic deformation in the dynamic Taylor anvil on rod tests. The obtained results have a qualitative correspondence with the experimental data about the form of the compacted samples. Calculations let us to reveal the contributions of the dislocation plasticity and twinning in different stages of the deformation process. Influence of twinning increases with the increase of the impact velocity. On the one hand, the twinning suppresses the dislocation plasticity due to the intense hardening of material in this area. On the other hand, the twinning provides an additional plastic deformation and extends the area of large deformations.

This study was supported by grants of the President of the Russian Federation (MD-286.2014.1) and the Ministry of Education and Science of the Russian Federation (State Task No. 3.1334.2014/K, Chelyabinsk State University).

References

[1] J.P. Hirth, J. Lothe, *Theory of dislocations* (Wiley & Sons, New York, 1982)

[2] M.A. Meyers, K.K. Chawla, *Mechanical behavior of materials* (Cambridge University Press, New York, 2009)

[3] M.A. Meyers, D.J. Benson, O. Vohringer, B.K. Kad, Q. Xue, H.-H. Fu, Mater. Sci. Eng. A **322**, 194 (2002)

[4] V.S. Krasnikov, A.E. Mayer, A.P. Yalovets, Int. J. Plast. **27**, 1294 (2011)

[5] A.E. Mayer, K.V. Khishchenko, P.R. Levashov, P.N. Mayer, J. Appl. Phys. **113**, 193508 (2013)

[6] J.W. Christian, S. Mahajan, Prog. Mat. Sci. **39**, 1 (1995)

[7] E.N. Borodin, S.A. Atroshenko, A.E. Mayer, Tech. Phys. **59**, 1163 (2014)

[8] G. Taylor, Proc. Royal Soc. London A, 194 (1948)

[9] M.L. Wilkins, M.W. Guinan, J. Appl. Phys. **44**, 1200 (1973)

[10] V.E. Fortov, K.V. Khishchenko, P.R. Levashov, I.V. Lomonosov, Nucl. Instrum. Meth. Phys. Res. A **415**, 604 (1998)

[11] L.D. Landau, E.M. Lifshitz, *Course of Theoretical Physics Vol. VII Theory of Elasticity* (Pergamon, New York, 1986)

[12] J.D. Eshelby, *Collected works of J.D. Eshelby* (2006)

[13] C. Kittel, *Introduction to Solid State Physics* (Wiley, New York, 2004)

[14] O.A. Plekhov, O.B. Naimark, N. Saintier, T. Palin-Luc, Tech. Phys. **54**, 1141 (2009)

[15] E.N. Borodin, A.E. Mayer, IOP Conf. Series: Mater. Sci. Eng. **63**, 012034 (2014)

[16] A.E. Mayer, E.N. Borodin, V.S. Krasnikov, P.N. Mayer, J. Phys.: Conf. Ser. **552**, 012002 (2014)

[17] A.P. Yalovets, J. Appl. Mech. Tech. Phys. **38**, 137 (1997)

EPJ Web of Conferences **94**, 04042 (2015)
DOI: 10.1051/epjconf/20159404042

Numerical investigations on pressurized AL-composite vessel response to hypervelocity impacts: Comparison between experimental works and a numerical code

Jérôme Mespoulet[a], Fabien Plassard, and Pierre-Louis Hereil

THIOT INGENIERIE, Route Nationale, 46130 Puybrun, France

Abstract. Response of pressurized composite-Al vessels to hypervelocity impact of aluminum spheres have been numerically investigated to evaluate the influence of initial pressure on the vulnerability of these vessels. Investigated tanks are carbon-fiber overwrapped prestressed Al vessels. Explored internal air pressure ranges from 1 bar to 300 bar and impact velocity are around 4400 m/s. Data obtained from experiments (Xray radiographies, particle velocity measurement and post-mortem vessels) have been compared to numerical results given from LS-DYNA ALE-Lagrange-SPH full coupling models. Simulations exhibit an under estimation in term of debris cloud evolution and shock wave propagation in pressurized air but main modes of damage/rupture on the vessels given by simulations are coherent with post-mortem recovered vessels from experiments. First results of this numerical work are promising and further simulation investigations with additional experimental data will be done to increase the reliability of the simulation model. The final aim of this crossed work is to numerically explore a wide range of impact conditions (impact angle, projectile weight, impact velocity, initial pressure) that cannot be explore experimentally. Those whole results will define a rule of thumbs for the definition of a vulnerability analytical model for a given pressurized vessel.

1. Introduction

The context of this paper is the vulnerability of high pressure vessels subjected to high velocity impact of space debris. The increase number of these debris leads to a new regard on these phenomena. From an operational point of view, it is essential to know the limit between the perforation regime and the bursting regime of the vessel in function of the internal pressure and of the projectile impact characteristics (velocity, mass, impact angle ...).

One of the first study on this topic has been performed by NASA in 1963 [1]. Experimental investigations consist in the impact into liquid-filled tanks. The goal was to study the effects of some of the primary variables contributing to bursting or catastrophic fracturing of liquid-filled tanks. A lot of studies have been done by NASA and SwRI on this subject [2–5].

Since 1995, EMI and ESA group has presented a lot of experimental, numerical and analytic works on the vulnerability of high pressure vessels [6–10]. Interesting experimental results are the visualization of debris cloud in high pressure gas and the deceleration of this debris by this gas. Majority of tests have been realized at normal incidence, at velocity around 7 km/s with aluminum spheres as projectiles. The impact tests were performed on cylindrical pressure vessels which were unshielded or shielded and made of AL5754, Al2219, unalloyed Ti or Ti6Al4V. Most of the cylindrical vessels had a wall thickness of 1 mm, a diameter of 150 mm and a length of 350 mm.

[a] Corresponding author: mespoulet@thiot-ingenierie.com

In 2001, a French group (EADS, INSA, CNRS and CNES) has presented some numerical simulation with SPH algorithm on the behavior of high pressure composite tanks under an hypervelocity impact. This numerical study has been used to prepare a following experimental campaign on composite tanks [11,12].

The aim of the work presented in this paper, is to numerically analyze the different regimes of a pressurized composite/metal vessel subjected to an hypervelocity impact. These regimes are the penetration mode, the critical rupture and fragmentation mode, and finally the bursting mode. Driving parameters are impact velocity, debris characteristics and internal pressure level. The effects of internal gas pressure vessel on the state of the vessel will be analyzed in the impact regime of an 8 mm diameter aluminum sphere at an impact velocity of 4400 m/s. The first difference with previous published studies is the tested vessel which is an overwrapped carbon fiber on an aluminum liner. The second difference concerns the value of the internal pressure explored, which will be 300 bar, the maximum pressure admitted in this vessel of a volume of 1.1 liter.

The results presented in this paper are based on a more important program supported by CNES and which includes hypervelocity impact experiments performed with some specific measurements in order to have enough data to discriminate the calculations. All these results are compared to numerical simulations in order to evaluate the reliability of such calculations in this specific field of hypervelocity impacts against high pressure vessels.

The stake of numerical simulations of high pressure vessels vulnerability is to be able to extrapolate the

Article available at http://www.epj-conferences.org or http://dx.doi.org/10.1051/epjconf/20159404042

Table 1. Parameters of the hypervelocity impact tests campaign.

Test n°	Projectile velocity (m/s)	Vessel pressure (bar)
#HE0183	4334 ± 60	1
#HE0184	4325 ± 60	200
#HE0188	4310 ± 60	300

experimental results at velocities up to 16 km/s which are not achievable in laboratory on Earth.

The challenge of theses simulations is to combine several complex effects with a full 3D model. The aim of this paper is to evaluate the capability of commercial codes to reproduce the different events that occur during experimental tests.

2. Experimental work

The hypervelocity impact tests were performed with the HERMES two-stage light gas gun at Thiot Ingenierie Laboratory [13]. Vessels used in this study are used for paintball games. No details on the manufacturing and material characteristics are given by the supplier but the technology of these vessels seems to be closed to the technology of composite tanks that are used in satellites. Part of hypervelocity impact test campaign at 90° impact angle that have been numerically investigated are presented in Table 1.

Analysis of the data and the results obtained during the four tests of hypervelocity impact on composite-Aluminum vessel confirms the results already presented on pressurized vessel: the velocity of the debris cloud generated by the impact is reduced by the effect of the gas. Failure of the pressurized vessel increases with the inflated pressure, but for all the pressure investigated (200–300 bar) failure mode is the same: it leads to 2, 3 or 4 main debris of the aluminum layer of the vessel and a highly damaged state for the composite layers. Failure mode is directly dependent of the composite texture. The impact of the aluminum sphere leads to a hole in the aluminum liner and to the rupture of the composite fibers. The confining stress induced by these overwrapped fibers is then released in this circumferential portion which results in the rupture of the aluminum portion in this same location.

3. Description of numerical simulations

3.1. Context

Simulations have been done using the nonlinear explicit code LS-DYNA. This study combines several complex problems with strong interferences between them:

- Multi layered vessel made of aluminum liner pre-stressed by external tension fibers winding composite materials,
- Rupture modes of these materials at hypervelocity impact,
- Initialization of the gas pressure inside the vessel with replication of shock waves generated by impact,
- Complete interactions between projectile/vessel, projectile/gas and vessel/gas.

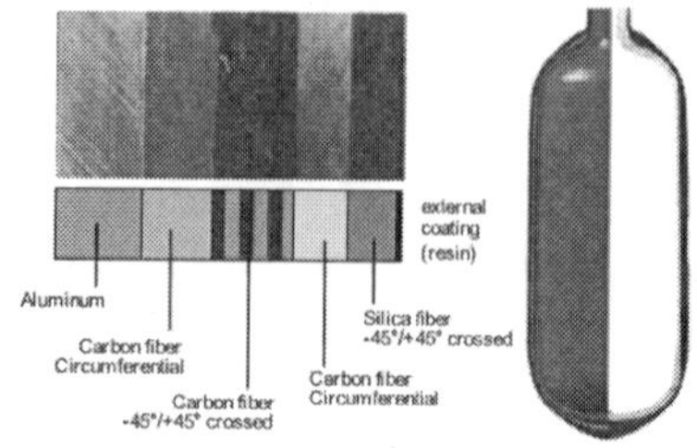

Figure 1. Cut view with description of the Aluminum-Composite vessel (left) and comparison of cut-view with model (right).

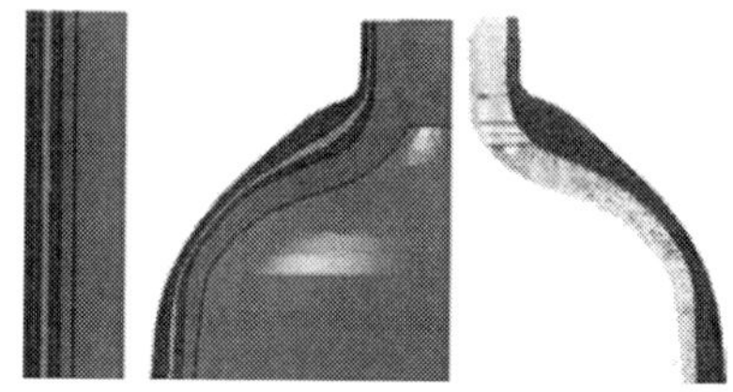

Figure 2. Comparison of vessel end cut-view with model numerical (left) and detail of the model impact area (right).

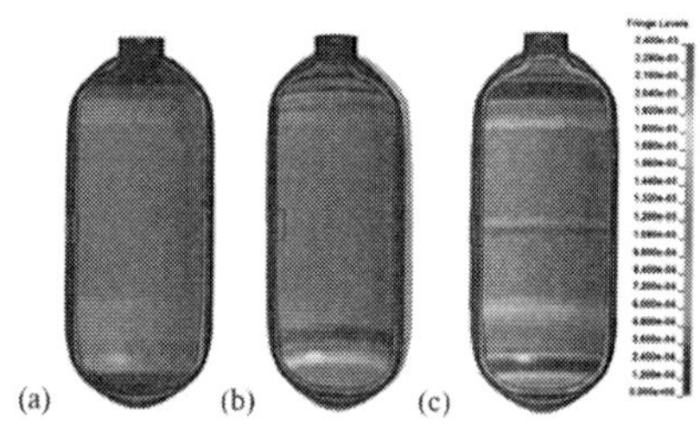

Figure 3. Comparison of initial Von Mises stress (0-240 MPa) in aluminum layer (a) at 0 bar (pres-stress at rest), (b) 200 bar and (c) and 300 bar.

3.2. Solvers, material laws and interactions

The vessel is modeled with a lagrangian formulation, using the real thickness variations taken from the half cross section of the vessel (Fig. 1). The aluminum liner thickness varies between 2 and 4 mm. The composite external material is made of 4 plies layered with: carbon fiber circumferential, carbon fiber −45°/ + 45 °C rossed, carbon fiber circumferential and silica fiber −45°/ + 45 °C rossed. Meshing are done with hexagonal solid elements with 0.5 to 2 mm segment characteristic length.

Comparison of tested vessel cut-views with model are presented on Fig. 1 (large view) and in details (Fig. 2).

A static nonlinear step is first performed to take into account the prestresses induced by the composite fibers winding. Without this step, calculations show that the vessel is not able to withstand an internal pressure of 300 bar. A mean consistent value of 100 MPa Von Mises stresses is arbitrarely chosen for the aluminum liner as initial compressive stress at rest. A calculation with the desired values of gas internal pressure is then performed according to the initial pressure in the experimental test (Fig. 3).

Air inside and outside the vessel is modeled with an ALE formulation. The vessel is totally immersed in the

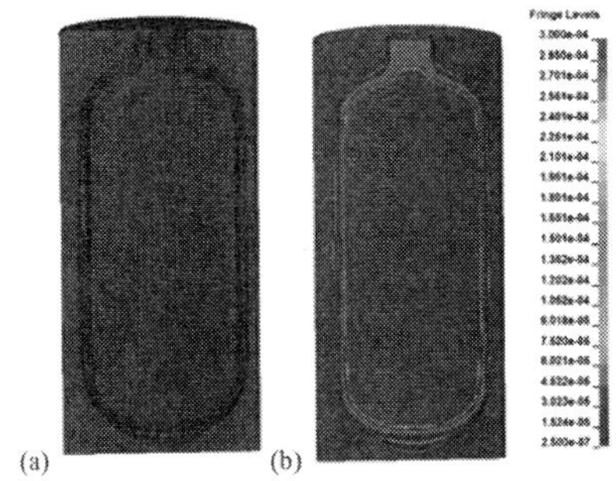

Figure 4. (a) Meshing view of Lagrangian parts immerged in ALE fluid part, (b) Fluid/structure interaction model with 300 bar initial pressure.

ALE mesh. Air inside the vessel is initialized with the same pressure as in experiment (Fig. 4). Air outside the vessel is initialized with a partial vacuum of 250 mbar. A perfect gas equation of state is used for Air in ALE grid. The stress/strain state (prestresses + internal pressure) is taken as initial state of explicit impact calculations.

The 8 mm diameter aluminum sphere is modeled with SPH formulation. The SPH solver was first introduced in the seventies for astrophysics. It is a Lagrangian method with a variable nodal connectivity. It is a gridless method which can handle severe distortions without grid tangling and so does not need the use of erosion algorithms. This technique is then well adapted for hypervelocity phenomena simulations [10–12, 14–16].

The aluminum parts are modeled with a classic elasto-plastic model with an erosion criterion based on strain. The composite layers are modeled with a composite material model with 9 elastic constitutive parameters and a Chang-Chang failure model criterion based on the brittle limit for each direction.

The main difficulty is the correct treatment of the interfaces between each formulation:

- Interaction between SPH sphere and vessel modeled with 3D Lagrangian elements is treated with a node/surface interface,
- Interaction between vessel and air is managed with a fluid/structure interface,
- Interaction between the ball and the gas is managed by a recently developed interface between SPH and ALE formulation. At this time, this interaction capability is only available with a beta version of Ls-Dyna program [16, 17].

4. Results and comparison with experiments

4.1. #HE0183

The first test simulated is #HE0183 which is characterized by an internal pressure of 1 bar inside the composite vessel. For this test, the shock wave generated in the gas is not decisive and so calculation is performed without ALE grid.

The shape of aluminum rupture (Fig. 5) does not reproduce exactly the experiment but it can be explained

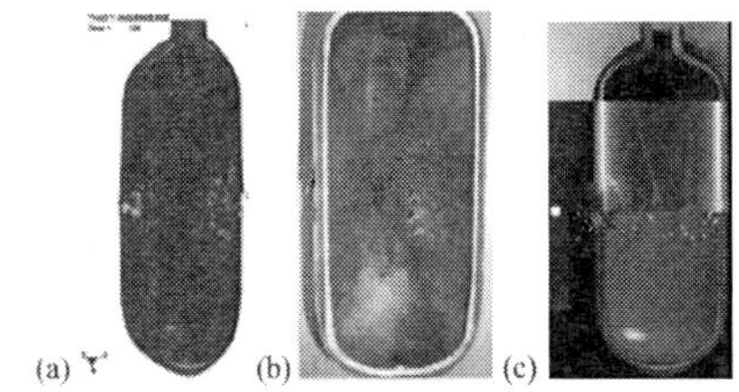

Figure 5. HE#183: Illustration of damaged zones at the inside rear face obtained (a) in simulation, (b) in experiment.; (c) Comparison between calculation and Xray radiography 41 μs after impact.

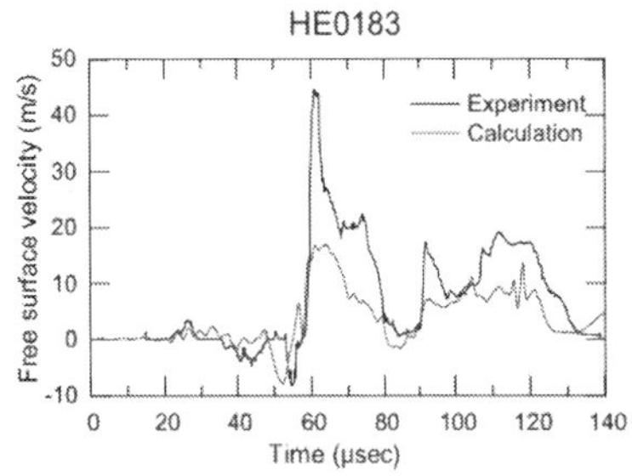

Figure 6. HE#183: Experiment and calculation comparison of the free surface velocity.

by the fact that the mesh is too coarse to simulate this phenomenon and the rupture model would need to be improved. The direction and the shape of composite damages are consistent with the experimental test. Fragments do not pass through the vessel skin but damage on external side is consistent with experiment. The last picture on this figure shows a comparison with the Xray radiography taken 41 μs after impact. We can see that this time corresponds to the arrival time of the debris cloud on the opposite side of the vessel.

Figure 6 shows a comparison of the free surface particle velocity. From a chronometry point of view, calculation is consistent with experiment.

4.2. #HE0184

Test HE0184 has been performed with an internal air pressure of 200 bar. In this configuration, taking into account fluid is essential to obtain the global behavior. Indeed, pressure build up acting on inside face increases the initial stress in the aluminum liner and amplifies tension in composite external skin.

Figure 7 shows the evolution of shock waves in the pressurized air and the motion of the SPH nodes after impact. The fragmentation of the projectile is not well reproduced. Some numerical tests have shown that the interface between SPH and ALE is so high that behavior of SPH part is misrepresented. No more expansion of SPH nodes occurs when nodes interact with ALE fluid material. The projectile is almost stopped by the gas and fragment impacts the opposite of the aluminum liner at low velocity (<200 m/s). This phenomenon is consistent with the experimental observation.

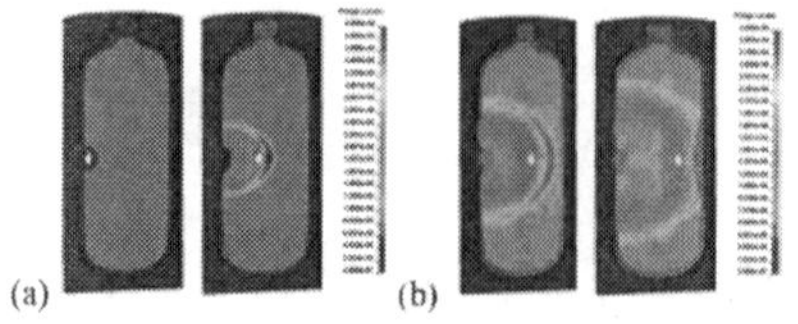

Figure 7. HE#184: Evolution of shock wave pressure in pressurized air (a) 30 and 85 μs (0–400 bar) and (b) 153 and 208 μs after impact (0-300 bar).

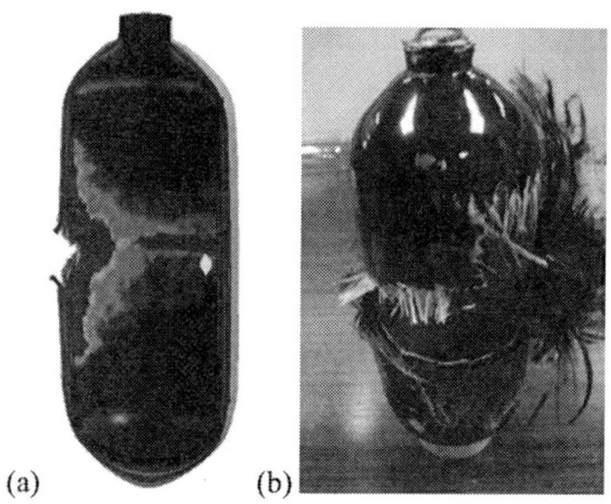

Figure 8. HE#184: (a) Inside view of simulation damaged zones in Al liner, (b) Postmortem reservoir after #HE0184.

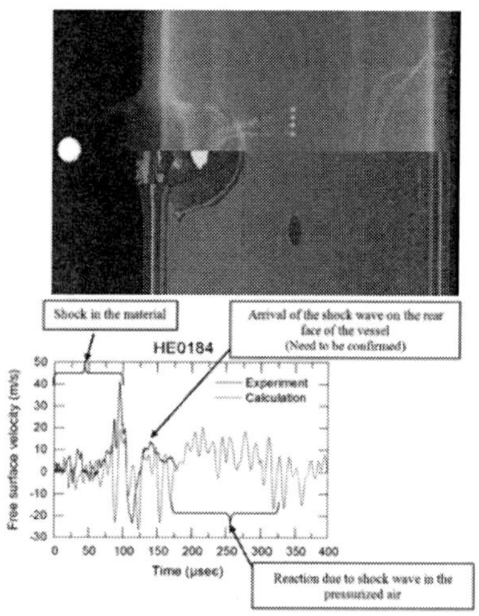

Figure 9. HE#184: Comparison between calculation and Xray radiography 41 μs after impact (top), Test and calculation comparison of the free surface velocity (bottom).

The shock wave in the gas reaches the opposite side 140μs after impact. Calculations indicate that the vessel rupture is caused by initial pressure that acts on internal side of the vessel and opens it. This rupture is not total which is consistent with the real case but the damaged zone is not reproduced (Fig. 8).

Figure 9 shows the comparison with the Xray radiography. Shock wave in the vessel is reproduced, but its propagation velocity is slower than in the numerical model (shock front is symbolized by red axis lines). Hypothesis of this discrepancy have been listed below:

- aluminum liner deformation is less than in simulation, so a smaller amount of gas is initially pushed by the liner deformation,
- an ideal gas is probably not sufficient to reproduce the real physic, especially shock velocity in this high pressure state,
- Interaction between the ball and the gas is managed by a meshing size should be reduced,
- strain rate dependent model should be used.

Figure 9 shows also the free surface particle velocity. Initial oscillations can be explained by penalty algorithm implementation for dynamic treatment of the fluid/structure. Nevertheless, the first peak corresponding to shock waves propagation in vessel material at 95 μs and the effect of the shock wave in the air with duration of around 125 μs can be distinguished.

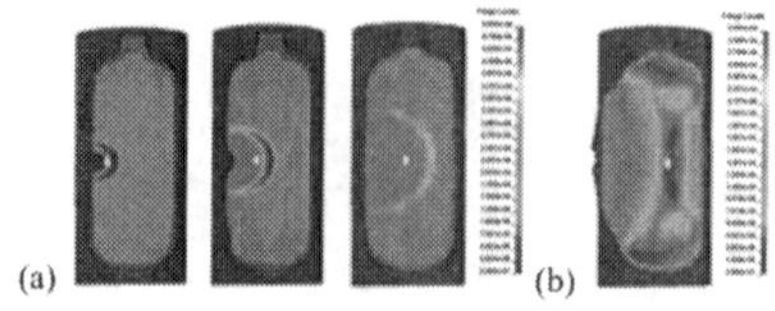

Figure 10. HE#188: Evolution of shock wave pressure in pressurized air (a) 48, 89 and 144 μs (0–500 bar) and (b) 300 μs after impact (0–48 bar).

4.3. #HE0188

Test HE0188 is performed with an internal air pressure of 300 bar. Figure 7 shows the evolution of shock waves in the pressurized air and the motion of the SPH nodes after impact. Once again, the fragmentation of the projectile is not well reproduced: no expansion of SPH nodes after ALE interaction. The projectile is once again stopped by the gas as in experiment. The shock wave in the gas reaches the opposite side at 160 μs. We can see that rupture of the vessel is caused by initial pressure. The pressure on the ruptured zone is responsible for the opening of the vessel. Damage on the vessel occurs before reflection of the shock wave on the opposite side of the vessel.

Figure 11 shows the rupture of the vessel with the same large parts as in the experiment. Simulation give two caps and the fractured body but sizes are not exactly the same than the test. Once again a finer mesh and a better knowledge of the individual behavior of material that compose the vessel (especially composites) needs to be investigated.

Figure 12 shows the comparison with the Xray radiographies for two different times. We can see that the shock wave in the vessel is reproduced. Once again, the shock wave propagation velocity is slower in the numerical model due probably to hypothesis explained in the previous paragraph

Figure 13 shows the free surface particle velocity at the rear face of the vessel. The oscillations at the beginning can be explained by the dynamic treatment of the fluid/structure with implementation of the penalty

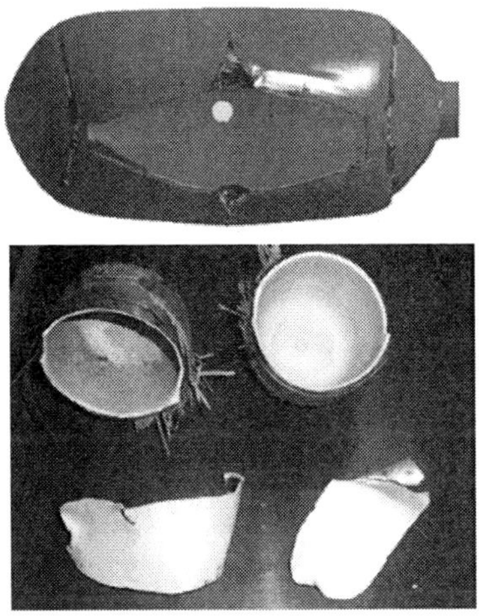

Figure 11. HE#188: Inside view of damaged zones in Al liner, (top), Postmortem reservoir after #HE0188 (bottom).

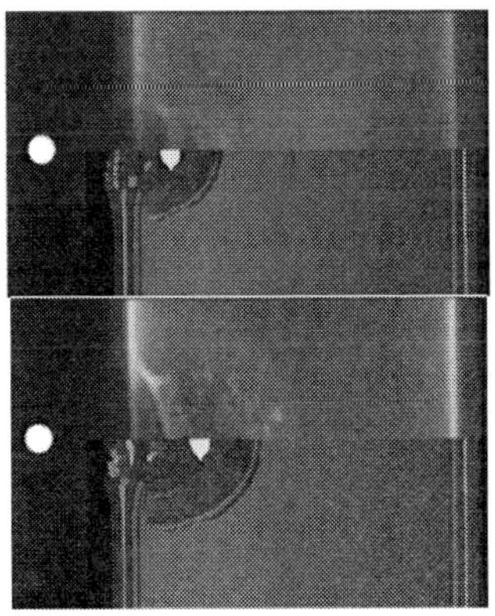

Figure 12. HE#184: Comparison between calculation and Xray radiography 13 μs after impact (top), 28 μs after impact (bottom).

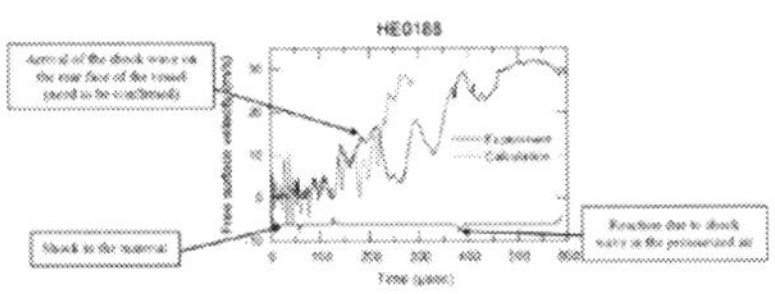

Figure 13. HE#188: Experiment and calculation comparison of the free surface velocity).

algorithm. Nevertheless we are able to distinguish some peaks corresponding to the propagation of the shock wave in the vessel material around 100 μs. Then the velocity of the rear side of the exploded vessel increases.

5. Analysis

Figure 15 presents the evolution of the projectile velocity versus time for the three studied cases: HE00183 not pressurized, HE0184 pressurized at 200 bar and HE00188 pressurized at 300 bar. This graph shows the effect of the high pressure gas on the velocity decrease of the projectile.

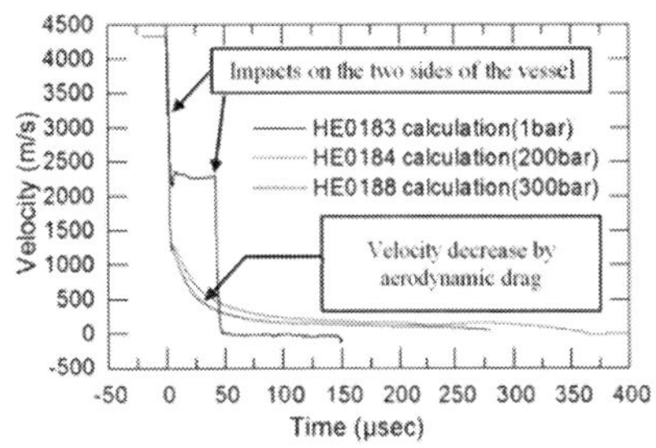

Figure 14. Evolution of the projectile versus time for the three simulations #HE0183, #HE0184 and #HE0188.

At the impact on the first side, the non-pressurized vessel decreases the projectile velocity at 2250 m/s, while pressurized vessels decrease the projectile at 1250 m/s. The fragments are stopped due to aerodynamic drag in the pressurized gas.

6. Conclusions and future works

The responses of gas-filled pressure composite-aluminum vessels subjected to hypervelocity impact are numerically analyzed from selected experimental results. It has been found during experimental phase that failure of the pressurized vessel increases with the inflated pressure and that failure mode depends directly on the stacking sequence of composite.

The numerical study developed in this paper has shown that a commercial code as Ls-Dyna has tools to reproduce global phenomenon from experimental setup. A crossed study between experiments and calculations allows us to understand the storyline of the hypervelocity impacts on pressurized composite vessels. The main difficulty is the management of interactions between SPH formulations, Lagrangian finite elements and ALE element for gas. However, an important work is still needed to improve the consistency and the robustness of the code. Especially for the new SPH/ALE interaction which shows limits in term of physics. Materials knowledge of the studied vessel has also to be improved in order to fill the code with right parameters in material models.

The authors thank CNES and Thiot Ingenierie for funding. In addition, the authors appreciate the support provided by Thiot-Ingenierie Facility and personnel including Valérie Berger, Ghislain Darnis and Fabrice Vigié.

References

[1] F.S. Stepka, R.P. Dengler, C.R. Morse, *Preliminary Investigation of Catastrophic Fracture of Liquid-Filled Tanks Impacted by High-Velocity Particles*, NASA Technical Note D-1537, LeRC, OH (1963)

[2] J.P. Whitney and P.D. White, *Designing hypervelocity impact testing for thin-walled pressure vessels*, Proceeding of the 43rd ARA meeting, Columbus, OH (1992)

[3] J.P. Whitney, *Hypervelocity impact tests of shielded and unshielded pressure vessels*, NASA Johnson Space center,Report JSC-32294 (1993)
[4] E. Christiansen, J.H. Kerr and J.P. Whitney, *Debris cloud ablation in gas-filled pressure vessels*, Int. J. of Impact Engng **20**, 7 (1997)
[5] S.A. Mullin, H. Couque, B.G. Cous-Palais, B.G. Grosch and J.D. Walker, *Bursting of shielded pressure vessels subject to HVI*, Int. J. of Impact Engng **20**, 7 (1997)
[6] M. Lambert and E. Schneider, *Hypervelocity impacts on gas filled pressure vessels*, Int. J. of Impact Engng **20**, 7 (1997)
[7] F.E. Schäfer, E. Schneider and M. Lambert, *An experimental study to investigate HVI on pressure vessels*, Second European Conference on Space Debris, Darmstadt, Germany (1997)
[8] F.E. Schäfer, E. Schneider and M. Lambert, *Impact fragment cloud propagating in a pressure vessel*, Act. Astr. **39**, 1 (1997)
[9] I.Y. Telitchev, F.E. Schäfer and E. Schneider, *Hypervelocity impacts on gas filled pressure vessels*, Int. J. of Impact Engng **23**, 7 (1999)
[10] M. Palmieri, F.E. Schäfer, S. Hiermaier and M. Lambert, *Numerical simulation of non-perforating impacts on shielded gas-filled pressure vessels*, Int. J. of Impact Engng **26**, 7 (2001)
[11] C. Maveyraud, J.P. Vila, D. Sornette, C. Le Floch, J.M. Dupilier and R. Salomé, *Numerical modelling of the behavior of high pressure vessel tank under an hypervelocity impact*, Proceeding of the 3rd Int. Conf. on Spacecraft Propulsion, ESA (2000)
[12] R. Salomé V. Albouys, C. Le Floch, D. Sornette and J.P. Vila, *High pressure composite tank behavior under an hypervelocity impact*, ESA (2001)
[13] P. Thiot, P.L. Héreil, J. Mespoulet, G. Darnis, V. Gasquet, W. Chan and K. Taylor, *Technical description and capabilities of THIOT INGENIERIE's Two-Stage Light-Gas-Gun*, Proceeding of the 60th ARA meeting, Baltimore (2009)
[14] L.D. Libersky and A.G. Petscheck, *High Strain Lagragian Hydrodynamics*, J. of Comp. Phys. **109** (1993)
[15] F. Plassard, J. Mespoulet and P.L. Héreil, *Hypervelocity impact of aluminium sphere against aluminium plate: experiment and LSDYNA correlation*, Proceeding of the 10th Inter. LS-Dyna Users conf., Starsbourg (2011)
[16] H. Chen and J. Wang, *LS-Dyna ALE nodal coupling*, Proceeding of the 12th Inter. LS-Dyna Users conf. (2012)
[17] *Recent developments in LSs-Dyna*, LS-Dyna User forum (2013)

EPJ Web of Conferences **94**, 04043 (2015)
DOI: 10.1051/epjconf/20159404043

Numerical simulation of large deformations of amorphous polymer with finite element method: Application to normal impact test

C.A. Bernard[1,a], J.P.M. Correia[1], N. Bahlouli[1], and S. Ahzi[1,2]

[1] University of Strasbourg/CNRS, ICUBE Laboratory, 2 rue Boussingault, 67000 Strasbourg, France
[2] Qatar Environment and Energy Research Institute, Qatar Foundation, PO Box 5825, Doha, Qatar

Abstract. During the last decades, the part of polymeric materials considerably increased in automotive and packaging applications. However, their mechanical behaviour is difficult to predict due to a strong sensitivity to the strain rate and the temperature. Numerous theories and models were developed in order to understand and model their complex mechanical behaviour. The one proposed by Richeton et al. [Int. J. Solids Struct. 44, 7938 (2007)] seems particularly suitable since several material parameters possess a strain rate and temperature sensitivity. The aim of this study is to implement the proposed constitutive model in a commercial finite element software by writing a user material subroutine. The implementation of the model was verified on a compressive test. Next a normal impact test was simulated in order to validate the predictive capabilities of the model. A good agreement is found between the FE predictions and the experimental results taken from the literature.

1. Introduction

Amorphous polymeric materials exhibit interesting mechanical and physical properties. Thus, polymers are widely used for large-strain applications in the automotive or packaging industries. Below the glass transition temperature, amorphous polymers exhibit a complex behaviour which is strain rate and temperature dependent [1,2]. According to the literature, the initial elastic linear response is followed by a viscoelastic response which ends at the yield point. Then, a viscoplastic response takes place characterized by a strain softening followed by a large strain hardening.

Nowadays, finite element (FE) simulations are widely used to improve geometry of structures. However, the mechanical behaviour of polymeric materials needs to be correctly reproduced. During the last decades, numerous models based on physical considerations were developed [3–5] in order to simulate the thermo-mechanical behaviour of amorphous polymer in the glassy region. A model particularly suitable is the one developed by Richeton et al. [4]. In the model, the elastic modulus and yield strength are strain rate and temperature dependent. They also proposed a thermally activated flow rule based on the cooperative model initially presented in the work of Fotheringham and Cherry [6]. These constitutive equations [4] allow reproducing the stress-strain curves of amorphous polymers over a wide range of strain rate and of temperature.

The aim of this study is to implement the constitutive equations proposed by Richeton et al. [4] in the commercial FE code ABAQUS/Explicit through a VUMAT subroutine. Firstly, the model formulation is detailed. Then, the integration algorithm is presented and validation tests at constant strain rate are performed in order to verify the correct implementation of the constitutive equation. Finally, FE simulations of an impact test were carried out. A good agreement is observed between the FE predictions and the experimental results [7].

2. Constitutive equations

The three-dimensional constitutive model used to describe the mechanical behaviour of amorphous polymers under finite strain is based on the work of Richeton et al. [4]. According to the kinematics of finite strain, the deformation gradient tensor **F** is multiplicatively decomposed into an elastic, thermal and plastic components:

$$\mathbf{F} = \mathbf{F}^e\mathbf{F}^{thp} \text{ with } \mathbf{F}^{thp} = \mathbf{F}^{th}\mathbf{F}^p \qquad (1)$$

with $\mathbf{F}^p$ the plastic deformation gradient defined from the reference configuration to the first relaxed configuration, the plastic deformation gradient defined from the reference configuration to the first relaxed configuration, $\mathbf{F}^{th}$ the thermal deformation gradient defined from the first configuration to the second relaxed configuration and $\mathbf{F}^e$ the elastic deformation gradient defined from the second relaxed configuration to the current configuration. The polar decomposition of **F**, $\mathbf{F}^e$ and $\mathbf{F}^p$ are given by:

$$\mathbf{F}^i = \mathbf{R}^i\mathbf{U}^i = \mathbf{V}^i\mathbf{R}^i \text{ with } i = e, p \qquad (2)$$

with **R** the rotation tensor, **U** the right stretch tensor and **V** the left stretch tensor. The velocity gradient tensor **L** is given by:

$$\mathbf{L} = \dot{\mathbf{F}}\mathbf{F}^{-1}. \qquad (3)$$

Incorporating Eq. (1) into Eq. (3), one obtains:

$$\mathbf{L} = \mathbf{L}^e + \mathbf{F}^e\left(\mathbf{L}^{th} + \mathbf{F}^{th}\mathbf{L}^p\mathbf{F}^{th-1}\right)\mathbf{F}^{e^{-1}} \qquad (4)$$

[a] Corresponding author: chrystelle.bernard@etu.unistra.fr

with

$$\mathbf{L}^e = \dot{\mathbf{F}}^e\mathbf{F}^{e^{-1}}, \mathbf{L}^{th} = \dot{\mathbf{F}}^{th}\mathbf{F}^{th^{-1}}, \mathbf{L}^p = \dot{\mathbf{F}}^p\mathbf{F}^{p^{-1}}. \quad (5)$$

The velocity gradient tensor $\mathbf{L}$ can be decomposed as follows:

$$\mathbf{L} = \mathbf{D} + \mathbf{W} \quad (6)$$

where $\mathbf{D}$ is the rate of deformation tensor (symmetric part) and $\mathbf{W}$ the spin tensor (skew-symmetric part). According to Eq. (6), the elastic, thermal and plastic velocity gradients are given by:

$$\mathbf{L}^e = \mathbf{D}^e + \mathbf{W}^e, \mathbf{L}^{th} = \mathbf{D}^{th} + \mathbf{W}^{th}, \mathbf{L}^p = \mathbf{D}^p + \mathbf{W}^p. \quad (7)$$

In the following, two basic assumptions are made. Firstly, we assumed that the plastic flow is incompressible. Thus,

$$\det\mathbf{F}^p = 1, \det\mathbf{F}^{th} = 1, \det\mathbf{F}^{thp} = 1. \quad (8)$$

We also assumed that the thermoplastic flow is irrotational. By consequence,

$$\mathbf{W}^p = 0, \mathbf{W}^{th} = 0, \mathbf{W}^{thp} = 0. \quad (9)$$

In this work, the elastic behaviour of amorphous polymers is assumed to be linear isotropic and can be modelled using the classical Hooke's law. The Cauchy stress tensor $\mathbf{T}$ is defined by:

$$\mathbf{T} = \frac{1}{\det\mathbf{F}^e}\mathbf{C}^e \ln\mathbf{V}^e \quad (10)$$

with $\ln\mathbf{V}^e$ the Hencky strain and $\mathbf{C}^e$ the fourth order stiffness tensor defined by:

$$\mathbf{C}^e_{ijkl} = \frac{E}{1+\nu}\left[\frac{1}{2}\delta_{ik}\delta_{jl} + \delta_{il}\delta_{jk} + \frac{\nu}{1-2\nu}\delta_{ij}\delta_{kl}\right] \quad (11)$$

with E the Young's modulus, ν the Poisson's ratio and δ the Kronecker-delta symbol. The strain rate and temperature dependence of the Young's modulus is given by Richeton et al. [8]:

$$\begin{aligned} E(\theta,\dot{\varepsilon}) = \; & (E_1(\dot{\varepsilon}) - E_2(\dot{\varepsilon}))\exp\left[-\left(\frac{\theta}{T_\beta(\dot{\varepsilon})}\right)^{m_1}\right] \\ + \; & (E_2(\dot{\varepsilon}) - E_3(\dot{\varepsilon}))\exp\left[-\left(\frac{\theta}{T_g(\dot{\varepsilon})}\right)^{m_2}\right] \\ + \; & E_3(\dot{\varepsilon})\exp\left[-\left(\frac{\theta}{T_f(\dot{\varepsilon})}\right)^{m_3}\right] \end{aligned} \quad (12)$$

where θ is the absolute temperature, $\dot{\varepsilon}$ is the effective strain rate, T_β is the β-transition temperature, T_g is the glass transition temperature, T_f is the flow temperature, E_i are the instantaneous stiffness of the material measured by DMA at the beginning of each transition region (glassy, glass transition and rubbery region) and m_i the Weibull moduli corresponding to the statistics of bond breakage. The different instantaneous moduli and transition temperatures are strain rate dependent and follow the expressions:

$$\begin{cases} E_i(\dot{\varepsilon}) = E_i^{ref}\left(1 + s_i\log\left(\dot{\varepsilon}/\dot{\varepsilon}^{ref}\right)\right) \\ \frac{1}{T_\beta(\dot{\varepsilon})} = \frac{1}{T_\beta^{ref}} + \frac{R}{\Delta H_\beta}\ln\left(\dot{\varepsilon}^{ref}/\dot{\varepsilon}\right) \\ T_g(\dot{\varepsilon}) = T_g^{ref} - \frac{c_2\log(\dot{\varepsilon}^{ref}/\dot{\varepsilon})}{c_1+\log(\dot{\varepsilon}^{ref}/\dot{\varepsilon})} \\ T_f(\dot{\varepsilon}) = T_f^{ref}\left(1 + 0.01\log\left(\dot{\varepsilon}/\dot{\varepsilon}^{ref}\right)\right) \end{cases} \quad (13)$$

with E_i^{ref}, T_β^{ref}, T_g^{ref}, T_f^{ref} the instantaneous moduli and the transition temperatures at the chosen reference strain rate $\dot{\varepsilon}^{ref}$, s_i the strain rate sensitivity parameters, ΔH_β the β-activation energy, R the ideal gas constant and c_1, c_2 the WLF parameters [9]. According to literature [10–12], the Poisson's ratio of amorphous polymers is strain rate and temperature dependent in the glass transition region. In this region, the Poisson's ratio sharply increases and achieves a value close to the Poisson's ratio of a rubbery material. In order to take into account this phenomenon, the following phenomenological model is proposed by Bernard et al. [13]:

$$\nu(\theta,\dot{\varepsilon}) = \begin{cases} \nu_0 + (\nu_c - \nu_0)\exp\left[-\frac{\theta-(T_g(\dot{\varepsilon})+\Delta\theta)^2}{2\omega^2}\right] \\ \text{if } \theta < T_g + \Delta\theta \\ \nu_c \\ \text{if } \theta \geq T_g + \Delta\theta \end{cases} \quad (14)$$

where ν_0 and ν_c are respectively the Poisson's ratio values at the beginning and at the end of the glass transition region, $\Delta\theta$ is related to the half of the temperature range across which the glass transition occurs and where $2\omega = \omega_1/\sqrt{\ln 4}$ with ω_1 the full width at half maximum of the Gaussian function.

The end of the elastic behaviour is characterized by the yield point. The yield strength is modelled with the equations proposed by Richeton et al. [14]. Below the glass transition temperature T_g, the yield strength τ_y is given by:

$$\tau_y = \tau_i(\theta) + \frac{2k\theta}{V}\sinh^{-1}\left(\frac{\dot{\gamma}\,(\theta < T_g)}{\dot{\gamma}_0\exp\left(-\frac{\Delta H_\beta}{R\theta}\right)}\right)^{1/n} \quad (15)$$

with $\tau_i(\theta)$ the evolving internal shear stress at the absolute temperature θ, k the Bolzmann constant, V the activation volume, $\dot{\gamma}$ the effective shear strain rate, $\dot{\gamma}_0$ a pre-exponential shear strain rate constant and n describes the cooperative character of the yield strength. At the yield point, the internal stress is expressed as:

$$\tau_{i_y}(\theta) = \tau_i(0) - m\theta + \alpha_p \text{ with } P = -\frac{1}{3}\mathrm{tr}(\mathbf{T}) \quad (16)$$

with $\tau_i(0)$ the internal stress at 0 K, m a material constant, α_p a pressure sensitivity parameter and P the hydrostatic pressure.

After the yield point, the mechanical behaviour of amorphous polymer is characterized by a strain softening followed by an important strain hardening due to a reorientation and rearrangement of polymer chain segments. In order to describe the plastic flow, we introduced the expression of the thermoplastic strain rate tensor $\mathbf{D}^{thp}$ as:

$$\mathbf{D}^{thp} = \mathbf{D}^{p} + \beta(\theta)\dot{\theta}\mathbf{I} \tag{17}$$

with $\beta(\theta)$ the thermal expansion coefficient equal to $\beta(298\,K)$ if $\theta < T_g$ (in agreement with [15,16]) and $\mathbf{I}$ the identity tensor. The plastic strain rate tensor $\mathbf{D}^{p}$ is given by:

$$\mathbf{D}^{p} = \frac{\dot{\gamma}^{p}}{\sqrt{2}\tau}\overline{\mathbf{T}}^{*'} \tag{18}$$

where $\overline{\mathbf{T}}^{*'}$ is the deviatoric part of the driving stress tensor $\overline{\mathbf{T}}^{*}$ expressed in the isothermally unloading configuration, τ is the effective shear stress defined by:

$$\tau = \left[\frac{1}{2}\overline{\mathbf{T}}^{*'} : \overline{\mathbf{T}}^{*'}\right]^{1/2} \tag{19}$$

and $\dot{\varepsilon}^{p}$ is the effective plastic shear rate. Below T_g, the effective plastic shear rate is expressed as:

$$\dot{\gamma}^{p} = \dot{\gamma}_0 \exp\left(-\frac{\Delta H_\beta}{R\theta}\right)\sinh^{n}\left(\frac{(\tau - \tau_i)V}{2k\theta}\right) \tag{20}$$

where the material constants are the same as in Eq. (15). The strain softening is modelled by the evolution of the internal stress τ_i. This stress evolves significantly from its initial value τ_{i_y}. In agreement with Boyce et al. [17], the evolution of τ_i follows the phenomenological law given by:

$$\dot{\tau}_i = h\left(1 - \frac{\tau_i}{\tau_{ps}}\right)\dot{\gamma}^{p} \tag{21}$$

with h the softening slope and τ_{ps} the stress referring to the preferred structural state of the material. In Eqs. (18) and (19), the deviatoric part of the driving stress tensor appears. It is expressed in the isothermally unloading configuration. In the intermediate configuration, the driving stress tensor is given by:

$$\overline{\mathbf{T}}^{*} = \mathbf{R}^{e^T}\mathbf{T}\mathbf{R}^{e} - \mathbf{B} \tag{22}$$

with $\mathbf{B}$ the backstress tensor which allow to take into account the important strain hardening of the material. In [4], the backstress tensor $\mathbf{B}$ is assumed to be coaxial to the plastic stretch tensor $\mathbf{V}^{p}$. In accordance with Arruda and Boyce [18], the principal components of $\mathbf{B}$ are given by:

$$\mathbf{B}_i = C_R(\theta)\frac{\sqrt{N(\theta)}}{3}\mathcal{L}^{-1}\left(\frac{\lambda_{chain}}{\sqrt{N(\theta)}}\right)\frac{\lambda_i^2 - \lambda_{chain}^2}{\lambda_{chain}} \quad \text{for } i = 1, 2, 3 \tag{23}$$

with C_R the rubbery modulus, N the number of rigid chain links between entanglements, $\mathcal{L}^{-1}$ the inverse Langevin function that is estimated with the Padé approximation proposed by Cohen [19], λ_i are the principal components of $\mathbf{V}^{p}$, and λ_{chain} is the plastic stretch on a chain (or called network locking stretch):

$$\lambda_{chain} = \sqrt{\frac{\lambda_1^2 + \lambda_2^2 + \lambda_3^2}{3}}. \tag{24}$$

According to Richeton et al. [4], C_R and N are temperature dependent parameters. More details on the model can be found in [4].

3. Computational procedure

The constitutive model presented above was implemented into the FE code ABAQUS/Explicit through a user material subroutine VUMAT [20].

At each step time and for each integration point of the elements, the FE code provides the Cauchy stress tensor $\mathbf{T}$ and the user-defined state variables defined at time t, and the time increment Δt. In addition, the FE code estimates the strain increment tensor $\Delta\varepsilon$, the deformation gradient tensor $\mathbf{F}$, the temperature θ and the right stretch tensor $\mathbf{U}$ at time $t + \Delta t$. At the end of each time increment, the user needs to return to ABAQUS/Explicit the updated values of Cauchy stress tensor $\mathbf{T}$ and of state variables.

At the beginning of the computation, at time $t = 0$, the material is assumed to have an elastic behaviour ($\mathbf{F}^{p} = \mathbf{F}^{th} = \mathbf{F}^{thp} = \mathbf{I}$) and have an initial temperature θ. In small strain, amorphous polymers exhibit an elastic behaviour. Thus, the elastic deformation gradient tensor $\mathbf{F}^{e}$ equals the deformation gradient tensor $\mathbf{F}$. After performing the polar decomposition on $\mathbf{F}^{e}$, the Cauchy stress tensor $\mathbf{T}$ is computed at time $t + \Delta t$ using Eq. (10). When the effective stress τ reaches the yield strength τ_y, the plastic and thermal deformation gradient tensors deviate from the identity tensor. These tensors have to be derived from the viscoplastic flow rules, Eqs. (17)–(24). In order to update the Cauchy stress tensor $\mathbf{T}_{t+\Delta t}$, an implicit time computation scheme was used. More details about the computational procedure could be found in [13].

4. Computational test

To check the implementation of the constitutive model, preliminary FE simulations were performed. The material is a polycarbonate (PC) studied by Richeton et al. [4]. The material was tested in uniaxial compression at 25°C and at a constant strain rate of $10\,\mathrm{s}^{-1}$. The material parameters can be found in the work of Richeton et al. [4].

The validation tests were performed on a square firstly meshed with one element CPE4R (continuum plane strain element with reduced integration) and next divided into 10×10 elements. The meshed square was subjected to a compression test with a constant strain rate under isothermal and adiabatic conditions.

A good agreement is obtained between the numerical predictions found in [4] and the FE predictions with one element. However, for a large number of elements, a distortion of the mesh is noticed as illustrated by Fig. 1(a). This phenomenon is due to an instability of the effective strain rate, $\dot{\varepsilon}_{t+\Delta t}$ that leads to a non-convergence of the problem as shown in Fig. 1(b) for three different elements. Thus, two averaging techniques are investigated in order to stabilize the effective strain rate: geometry homogenization and time homogenization.

4.1. Geometry homogenization

In the performed compressive FE tests, the predicted effective strain rate oscillates around a mean value before diverging. From the previous remark, we assume that the effective strain rate values should be equal for all the elements with a same abscissa. Thus, at each time

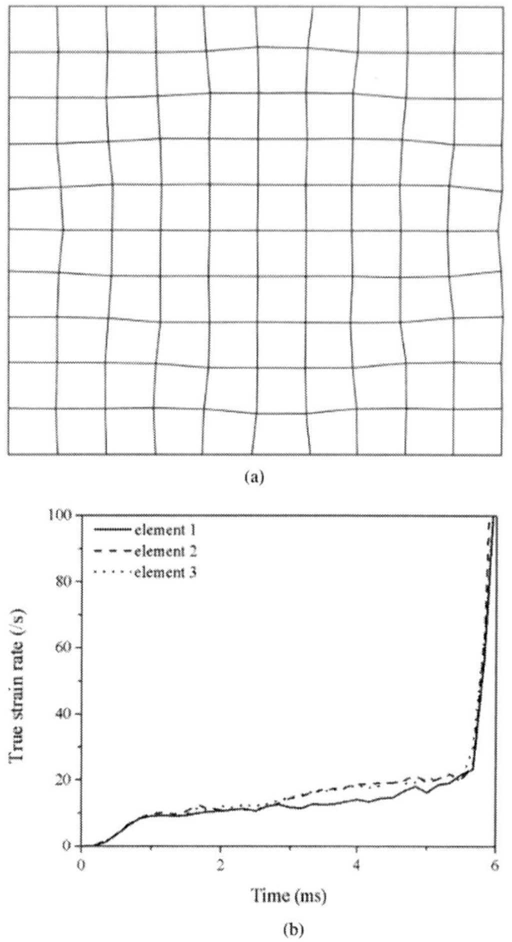

Figure 1. (a) Deformed shape of a square geometry meshed with 10×10 elements tested in uniaxial compression at 25°C and at $10\,s^{-1}$. (b) Strain rate-time curve for three elements of the mesh.

increment and for all the elements with the same abscissa, an average effective strain rate is computed. Thus, the geometry averaged effective strain rate $\dot{\varepsilon}_{HG}$ is defined by:

$$\dot{\varepsilon}_{HG} = \frac{1}{n_e}\sum_{i=1}^{n_e} \dot{\varepsilon}_i \tag{25}$$

with n_e the number of element with the same abscissa and $\dot{\varepsilon}_i$ the effective strain rate for the element i.

4.2. Time homogenization

The time increment averaging method is based on the assumption that the effective strain rate evolves smoothly. The time increment homogenization method consists in averaging the effective strain rate on a given interval of time for each element. Thus, the time averaged effective

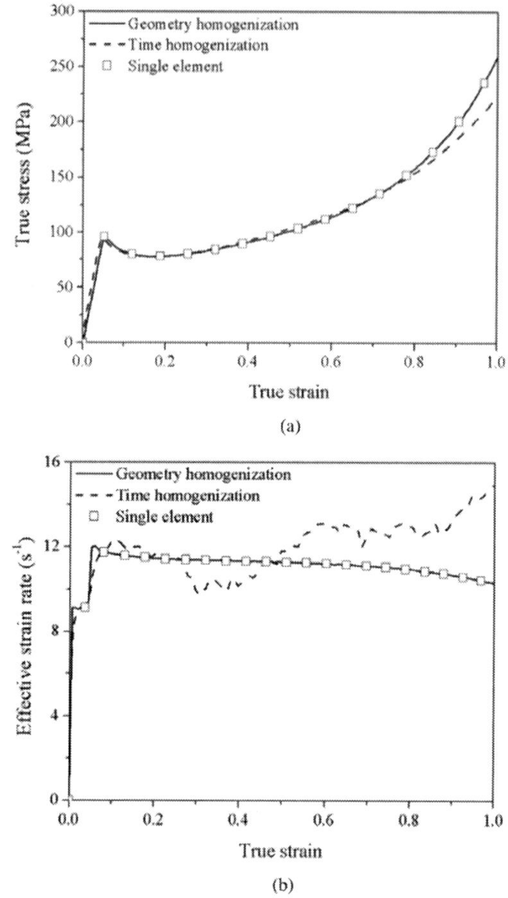

Figure 2. FE predictions of uniaxial compressive test at a constant strain of $10\,s^{-1}$ for an unique element and 10×10 elements: (a) stress-strain curve and (b) effective strain rate.

strain rate $\dot{\varepsilon}_{HT}$ is defined by:

$$\dot{\varepsilon}_{HT}^{t_N} = \frac{\sum_{j=i}^{N-1}\left(\dot{\varepsilon}_{HT}^{t_j} * \Delta t_j\right) + \dot{\varepsilon}^{t_N} * \Delta t_N}{\sum_{j=i}^{N} \Delta t_j} \tag{26}$$

where $t_N - t_i$ represents the interval of time where the time homogenization is done.

4.3. Results and discussion

In Fig. 2, the mechanical response and the strain rate response obtained with 10×10 elements using geometry homogenization and time homogenization are compared with the ones obtained with one element. For both homogenization method used, a good agreement is found. We can note a discrepancy, in the stress-strain response (Fig. 2(a)), occurring at 80% strain in the case of the time homogenization method. The analysis of the strain rate response (Fig. 2(b)) shows that the strain rate predicted with the time homogenization geometry is slightly instable

Table 1. Material parameters of the proposed model for PC identified on the experiments given in Ames et al. [7].

Parameters	PC
$\dot{\varepsilon}^{ref}$ (s^{-1})	1
E_i^{ref} (MPa)	3400/920/20
T_i^{ref} (K)	195/421/436
m_i	1/125/5
s	0.011
n	6.5
V_a M^3	5.28×10^{-28}
$\tau_i(0)$ (MPa)	76.59
m (MPa K^{-1})	0.133
$\dot{\gamma}_0$ (s^{-1})	3.32×10^{-28}
c_1	16.19
c_2 (○C)	55.6
τ_{ps}/τ_i	0.45
h (MPa K^{-1})	120
$C_R(0)$ (MPa)	44.208
a (MPa K^{-1})	0.0968
$N(0)$	1.949
b (K^{-1})	2.51×10^{-3}
ω	11
$\Delta\theta$ (K)	10

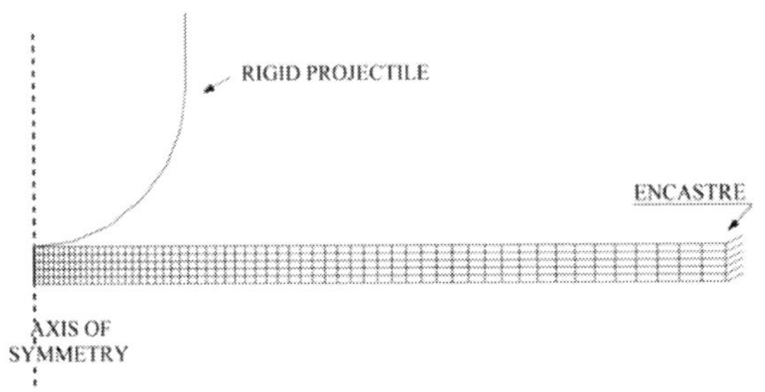

Figure 3. FE mesh used in the thermo-mechanically coupled analysis of normal impact test.

contrary to the strain rate predicted by the geometry homogenization. The two homogenization techniques gave good results. In spite of some instabilities of the effective strain rate for the time homogenization method, this technique is more suitable since it is easier to use.

5. Normal impact test

A normal impact test was modelled and simulated with ABAQUS/Explicit. This impact test and the experimental results are taken from the work of Ames et al. [7]. In this test a circular plate of PC is clamped and is impacted by a spherical-tipped cylindrical projectile. The mechanical behaviour of the PC was described with the constitutive model presented above. Some material properties were determined by Ames et al. [7]. The other parameters were identified from stress-strain curves presented in [7] and are listed in Table 1.

ABAQUS/Explicit code was used for the FE simulation using the VUMAT subroutine previously validated. In order to reduce the computation time, the axi-symmetry of the problem was used and thus only one slice of the

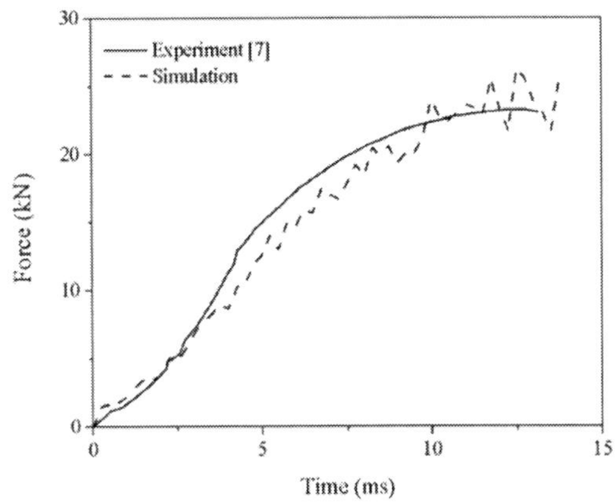

Figure 4. Reaction force versus time response for the projectile.

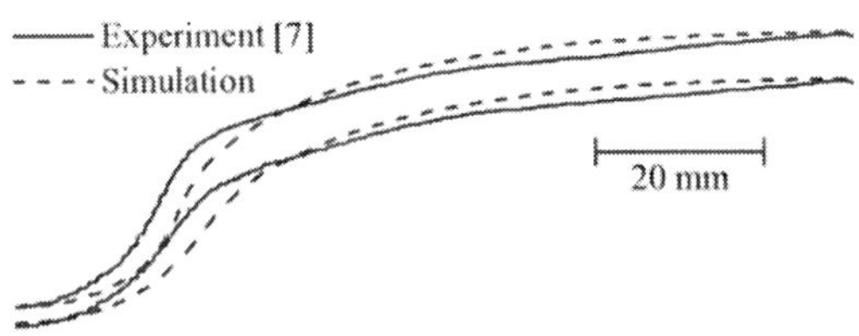

Figure 5. Comparison between experimental and numerically predicted section profiles.

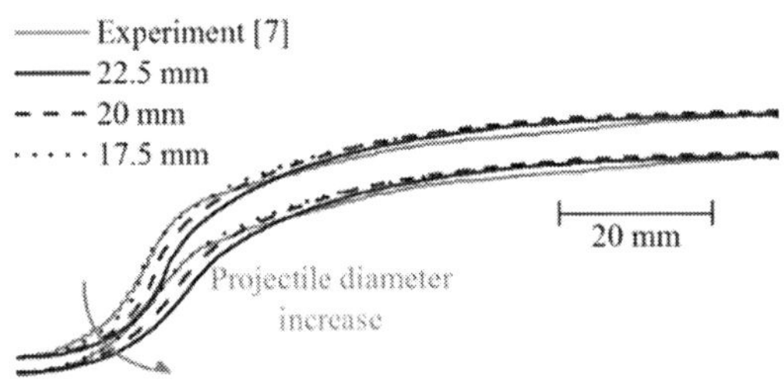

Figure 6. Influence of the projectile diameter size in the section profiles.

geometry was meshed. The PC plate has an initial radius of 101.6 mm and a thickness of 5.334 mm. The projectile, with a mass of 80 kg and a radius of 22.5 mm, impacts the plate at a velocity of 3.6 ms^{-1}.

The PC plate was meshed with 300 elements CAX4RT (4-node bilinear displacement and temperature, reduced integration with hourglass control element). All the nodal displacements were constrained on the right side of PC plate. The geometry of the test and of the mesh are presented in Fig. 3. The contact between the projectile and the plate is assumed frictionless.

At the end of the impact test, the reaction force on the projectile (Fig. 4) and the numerically predicted profile surface (Fig. 5) are compared with the experimental results [7]. For both figures, a good agreement is found between the predicted results and the experimental ones. Moreover, a temperature increase of 44 K has been observed under the projectile. This value is closed with FE predictions found in [7].

However, in Fig. 5, a discrepancy between the profile shapes is observed under the projectile. In order to analyse the effect of projectile geometry, FE simulations with three different projectile diameters were carried out. The predicted results are compared with the experimental ones measured by Ames et al. [7] in Fig. 6. We can note that the best fit is obtained with a projectile diameter of 17.5 mm.

6. Conclusion

An elastic-viscoplastic model has been implemented in the commercial FE code ABAQUS/Explicit, via a VUMAT subroutine, in order to simulate the mechanical response of amorphous polymers. The numerical implementation of the constitutive model was first validated for a unique and several elements in plane strain conditions. In order to stabilize the mechanical response, homogenization techniques for averaging the strain rate have been proposed. Both homogenization methods gave good results. A normal impact test on a PC plate was modelled and simulated. A good agreement was found between the numerical prediction and the experimental results taken from the literature [7].

References

[1] M.R. Boumbimba, E. Hablot, K. Wang, N. Bahlouli, S. Ahzi, L. Avérous, Compos. Sci. Technol. **71**, 674 (2011)

[2] K. Wang, F. Addiego, N. Bahlouli, S. Ahzi, Y. Rémond, V. Toniazzo, Compos. Sci. Technol. **95**, 89 (2014)

[3] E.M. Arruda, M.C. Boyce, R. Jayachandran, Mech. Mater. **19**, 193 (1995)

[4] J. Richeton, S. Ahzi, K.S. Vecchio, F.C. Jiang, A. Makradi, Int. J. Solids Struct. **44**, 7938 (2007)

[5] L. Anand, N.M. Ames, V. Srivastava, S.A. Chester, Int. J. Plasticity **25**, 1474 (2009)

[6] D.G. Fotheringham, B.W. Cherry, J. Mater. Sci. **13**, 951 (1978)

[7] N.M. Ames, V. Srivastava, S.A. Chester, L. Anand, Int. J. Plasticity **25**, 1495 (2009)

[8] J. Richeton, G. Schlatter, K.S. Vecchio, Y. Rémond, S. Ahzi, Polymer **46**, 8194 (2005)

[9] M.L. Williams, R.F. Landel, J.D. Ferry, J. Am. Chem.l Soc. **77**, 3701 (1955)

[10] G.N. Greaves, A.L. Greer, R.S. Lakes, T. Rouxel, Nat. Mater. **10**, 823 (2011)

[11] P. H. Mott, J. R. Dorgan, C. M. Roland, J. Sound Vib. **312**, 572 (2008)

[12] S. Pandini, A. Pegoretti, eXPRESS Polym. Letters **5**, 685 (2011)

[13] C.A. Bernard, J.P.M. Correia, S. Ahzi, N. Bahlouli, (in preparation) (2015)

[14] J. Richeton, S. Ahzi, L. Daridon, Y. Rémond, Polymer **46**, 6035 (2005)

[15] J. Bicerano, *Prediction of polymer properties*, (CRC Press, 2002)

[16] D.W. van Krevelen, K. te Nijenhuis, *Properties of polymers : their correlation with chemical structure; their numerical estimation and prediction from additive group contributions*, (Elsevier Science & Technology Books, 2009)

[17] M.C. Boyce, D.M. Parks, A.S. Argon, Mech. Mater. **7**, 15 (1988)

[18] E. M. Arruda, M.C. Boyce, J. Mech. Phys. Solids **41**, 389 (1993)

[19] A. Cohen, Rheol. Acta **30**, 270 (1991)

[20] *ABAQUS/Explicit Users Manual* (2007)

EPJ Web of Conferences **94**, 04044 (2015)
DOI: 10.1051/epjconf/20159404044

Experimental and numerical analysis concerning the behaviour of OL50 steel grade specimens coated with polyurea layer under dynamics loadings

Florina Bucur[1], Eugen Trana[1], Adrian Rotariu[1], Adinel Gavrus[2,a], Cristian Barbu[1], and Dominique Guines[2]

[1] Military Technical Academy, 31-39 Av. G. Cosbuc, 050141 Bucharest, Romania
[2] LGCGM (EA 3913)-INSA Rennes, UEB, 20 Av. des Buttes de Coësmes, 35708 Rennes, France

Abstract. This study refers to an experimental and numerical evaluation of a polyurea coating layer influence on the dynamic behaviour of OL50 specimens. Mechanical quasi-static and dynamic tensile tests were performed in axial loading conditions, for 2 mm steel plate specimens. Several metallic specimens have been previously coated with 1.5 mm and 3 mm respectively thickness polyurea layer and tested in traction. The findings results indicate that the presence of polyurea changes the loading pattern of metallic material in the necking area. In terms of polyurea coated metal specimens fracture, there was clearly observed a change of fracture limit. One possible explanation of this phenomenon is the modification of triaxiality state in the necking zone, fact proven by the numerical simulations. Test results indicate that the presence of polyurea layer delays the necking onset phenomenon which precedes the OL50 metallic specimen fracture.

1. Introduction

Defining constitutive law for all kind of materials has always been a challenging issue addressed by the international scientific community in a continuous manner.

Investigating multilayer materials proves to be even a much more difficult process. However, addressing this problem is a need since numerous studies [1–8] underlined the advantages of using this kind of materials in all kind of real life applications.

The present paper focus on the behavior of OL50 steel grade coated with a polyurea layer under both static and dynamic loadings.

A detailed investigation of polyurea had been previously performed in several papers [1,3–5,7]. As a conclusion of the earliest mentioned paper one can say that the polyuria is a recent polymer with a hyper-elastic behavior and good mechanical properties. Due to its good and rapid adherence and short drying period polyurea can easily be used on walls, fabrics and metals coating.

Other interesting characteristics of polyurea refer to its viscoelastic behaviour (strain rate dependent), hysteresis and softening response to cyclic loading and also stoichiometry dependent mechanical properties.

If polyurea investigation is of a more recent interest, metals were thoroughly studied over the last century. Due to this effort, metals behavior dependency on strain, strain rate, temperature, triaxiality state and Lode angle is well known in nowadays. These aspects were addressed in numerous papers such [8,9,12–15].

A previous work on metallic-elastomer layered plates response under static, dynamic and impulsive loading had been conducted by Amini [5,6].

[a] Corresponding author: adinel.gavrus@insa-rennes.fr

Table 1. Mechanical characteristics of steel OL50.

Quality class	The thickness or diameter of a, in mm $a \leq 16$		
1a	Rp0,2 [N/mm^2] Min	Rm [N/mm^2]	A5 (%)
	290	490–610	21

The objective of the current study is twofold:

1. experimental investigations on diffuse and localized necking and fracturing of OL50 steel grade dog-bone and biaxial specimens, both in the absence and the presence of polyurea coating;
2. numerical investigations of triaxiality state in the necking area for the tested configurations.

For this study, based on the large scale applications, OL50 steel grade has been selected for investigations. In order to evaluate the change of its mechanical behavior in the presence of a polyurea layer, several dog-bone specimens were manufactured according to ISO 6892-1 standard [16].

2. Experimental part

2.1. Testing set-up

OL50 steel grade is a structural steel characterized by: $20.1 \cdot 10^{10}\,\text{N/m}^2$ Young modulus, $8.25 \cdot 10^{10}\,\text{N/m}^2$ shear modulus, 160–200 HB hardness and 24–29 kgf/mm^2 fatigue strength. According to the manufacturing data sheet at room temperature other mechanical characteristics are according to the ones presented in Table 1.

Figure 1. a – Test assembly, b – Type of specimens.

The extra material is part of the polymer composite that is remarkable for being sticky and hyperplastic, namely polyurea brand EUROPOL®.

Laboratory tests consisted in performing several tensile trials in both quasi-static and dynamic regime. Three different geometric configurations had been selected for testing:

1. G1 - short simple specimens (50 mm active area),
2. G2 - intermediary specimens (50 mm active area): bilayer - polyurea-steel (PS) and multilayer - polyurea-steel-polyurea (PSP),
3. G3 - long simple specimens (80 mm active area).

In all these three configurations a 2 mm thick steel specimen was used. For G2 and G3 configurations the final thickness for the tested specimens in the active area after spraying the polyurea was about 5 mm as follows: for bilayer specimens, polyurea layer thickness had about 3 mm and for multilayer specimens 1.5 mm on each side. On the specimen length, the polyurea has been sprayed centrally on a length of 70 mm with 50 mm active area according to Fig. 1b.

Tests were performed on a universal testing machine belonging to the Laboratory of Civil and Mechanical Engineering & Technological Platform of INSA Rennes (France). The tests were performed at 1 mm/s, 0.5 m/s, 1 m/s and 2 m/s regime.

Figure 1 presents the specimens configurations and the used clamping assembly in the experimental device.

During the tests, displacement of the machine arm and the force had been recorded via machine transducers.

Experimental biaxial tests have been also realized in static and dynamic conditions using special cruciform specimens. Unfortunately the data recorded in those tests have poor quality being extremely affected by oscillatory phenomena, which limited their classical analysis, but necessary to understand the qualitative material behavior for high strain rates.

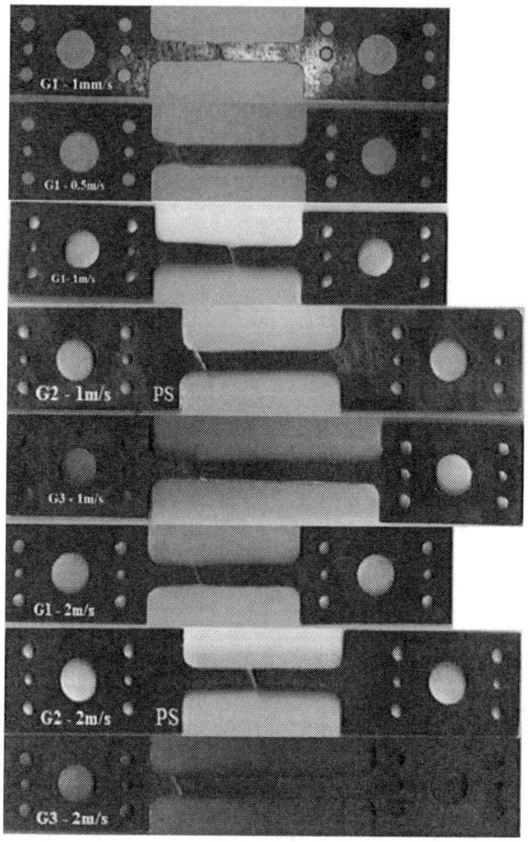

Figure 2. Necking and fracture area of specimens obtained from uniaxial tensile tests.

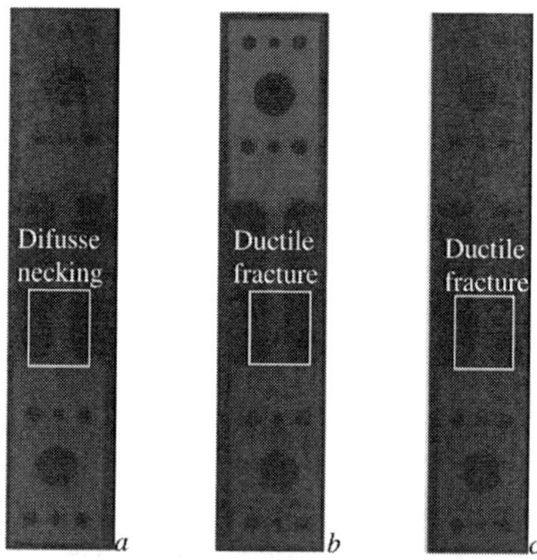

Figure 3. PSP Multilayer specimen radiography: a, b - 1 m/s, c- 2 m/s.

2.2. Experimental observation

All experimental tests indicate the specific behavior of ductile metals, namely the occurrence of a necking zone before fracture moment as shown in Fig. 2.

Necking and fracture patterns for tested specimen are presented in Fig. 2 and Fig. 3. For the G2 PSP configuration X-ray analysis was performed.

Relevant data regarding post-test measurements of OL50 steel grade dog-bone specimens are detailed in Table 2. Firstly as can be seen also in Fig. 3 one of the

Table 2. Post data measured dimension on test samples (*l0-active area before tests).

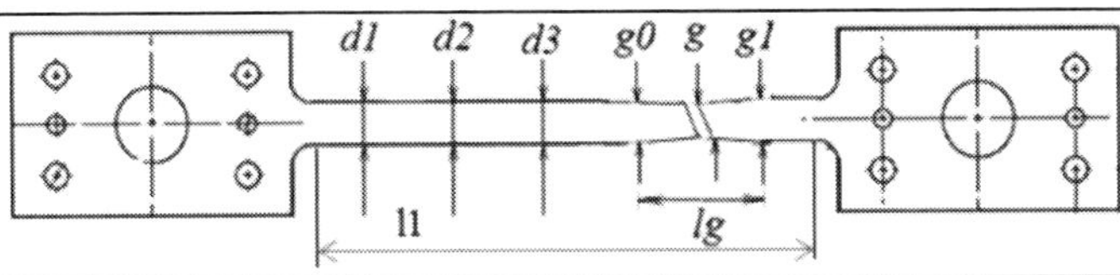

Geometry	Loading Speed	Measured dimensions (mm)						Diffuse necking length(mm)	Specimen thickness in the necking area (mm)	Maximum load (N)
		l0	l1	dm	g0	g1	g	lg		
G1	1m/s	50	64,2	9,65	9,2	9,24	7,66	24	0,67	9591.75
		50	65,2	9,53	9,2	9,31	7,58	28,5	0,67	9319.59
		50	65,1	9,35	9,22	9,06	7,54	21	0,66	Unrec. data
	2m/s	50	63,1	9,51	9,2	9,1	7,94	35,9	0,58	9686.25
		50	63,6	9,45	9,25	9,18	7,46	28	0,59	9553.95
		50	62,1	9,42	9,25	9,2	7,44	22	0,58	9333.765
G3	1m/s	80	97	8,96	8,87	8.88		13,1		9372
		80	100	8,94	8,63	8,57	7,28	18	0,7	9261
		80	99,3	9,15	8,72	8,63	7,88	15	0,7	9380
	2m/s	80	96,6	9,28	8,9	8,92	8,36	13,2		9545
		80	101,2	8,95	8,9	8,87	7,9	14	0,64	9404
		80	101,9	8,96	8,73	8,8	7,8	13,5	0,66	9396
G2-PS	1m/s	50	62,02	8,95	8,55	8,73	7,54	15,5	0,8	9444.935
		50	66,25	8,71	8,48	8,52	7,92	13,31	0,82	9601.8
	2m/s	50	66,01	8,71	8,46	8,6	7,04	19,3	0,75	9444.95

specimens hasn't exhibit fracture during testing. This is due to the machine functioning, which requires an imposed travel distance before the test.

In a general way the post-testing analysis prove that polyurea strongly sticks to the metallic layer and the decohesion being restricted only to the fracture area (for unfractured specimen decohesion in the necking area is not present).

For the G2-PS configuration a decrease of diffuse necking zone and a localized necking less pronounced was observed. Furthermore the elongation increase compared with G1 configuration was noted (Fig. 2).

During the test, due to used transducers presence, force vs. displacement data were recorded. An example of such data for G1 configuration and 1 m/s loading speed is plotted in Fig. 4.

As one can observe in Fig. 4 that due to the dynamic behavior of the sensor and elastic deformations of the machine, the curve profiles present oscillations more or less pronounced, which required a mathematical smoothing process.

Force vs displacement data recorded during experimental tests were processed to true stress vs. true strain data, by the use of classical analytical model based on the assumption of tension and deformation homogeneity in the central section of the specimen. The results are presented in Fig. 5.

A comparative analysis between G1 and G2 configuration test specimens in terms of force vs. displacement agrees with post-test measurements which highlighted an increase elongation in G2 configuration test specimens.

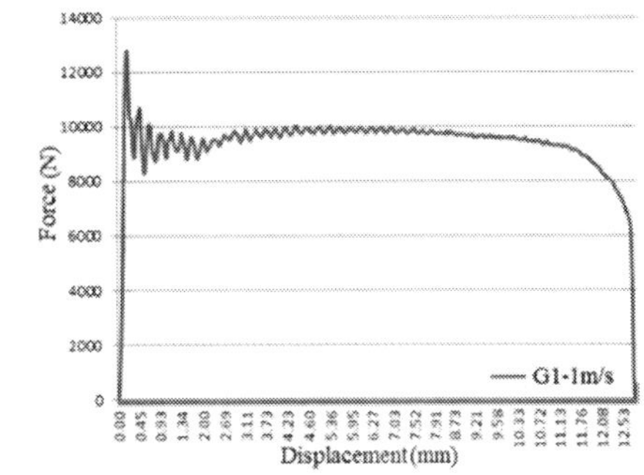

Figure 4. The experimental curve force vs. displacement evolution for test at 1 m/s.

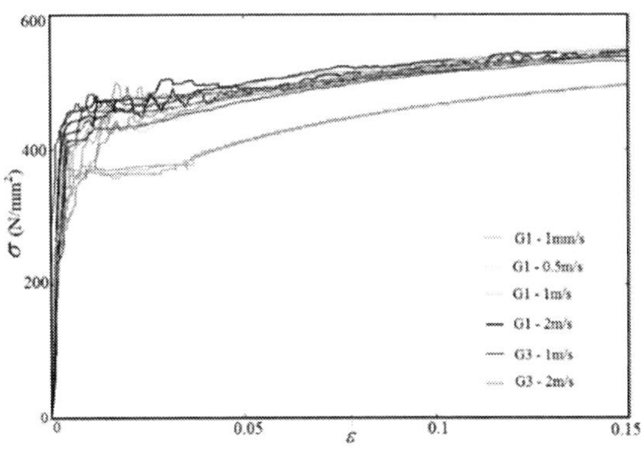

Figure 5. The experimental true stress vs. true strain curve evolution for all test range.

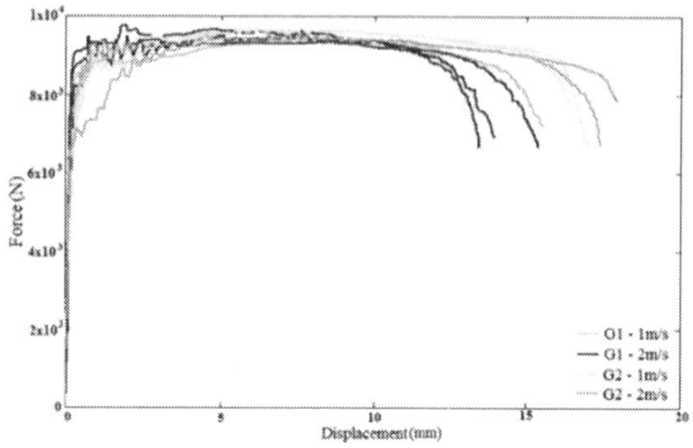

Figure 6. The experimental force vs. displacement curve evolution for specimens G1 and G2.

Table 3. Initial values of strain rates.

Geometry	Loading Speed	Initial value of strain rates
G1,G2	1 m/s	$20\,s^{-1}$
	2 m/s	$40\,s^{-1}$
G3	1m/s	$12.5\,s^{-1}$
	2 m/s	$25\,s^{-1}$

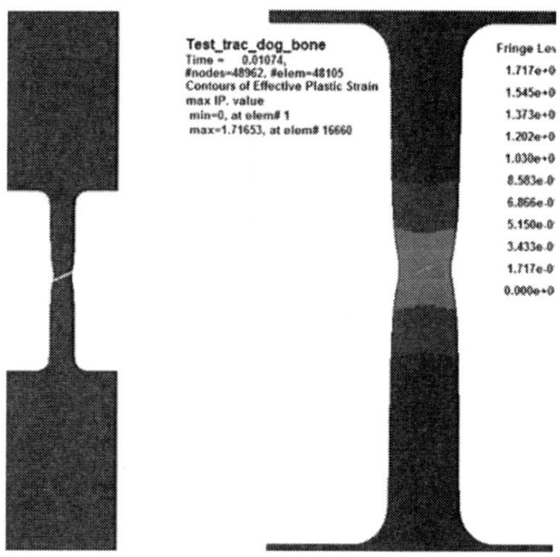

Figure 7. Ductile fracture for uniaxial specimen with detail on the area subjected to dynamic load.

3. Numerical simulation of uniaxial tensile tests

In order to investigate by numerical means the physical phenomena occurred during uniaxial tensile tests, several simulations had been performed using LsDyna commercial software.

Numerical simulations were performed for both G1 and G2 bi-layered configurations.

For G1 configuration case the 2D numerical model involved the use of 15600 type 2 shell elements. The mesh length for active areas was 0.25 mm and 1 mm for the rest of the specimen. However, knowing that an accurate result involves smaller mesh dimensions an Adaptive_Mesh triggered by shell thickness had been implemented in the

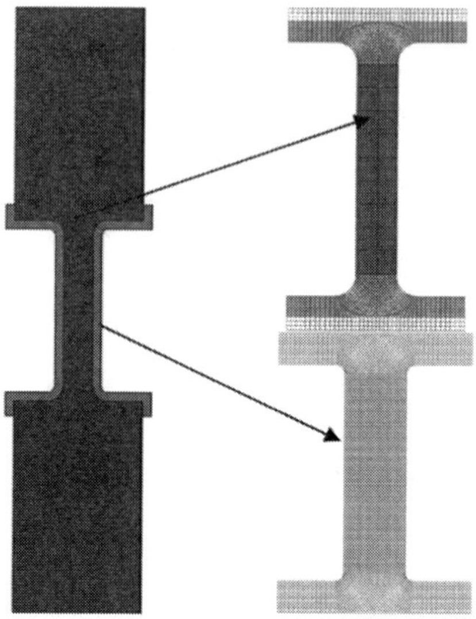

Figure 8. Dog-bone specimen mesh and details on the area subjected to dynamic load.

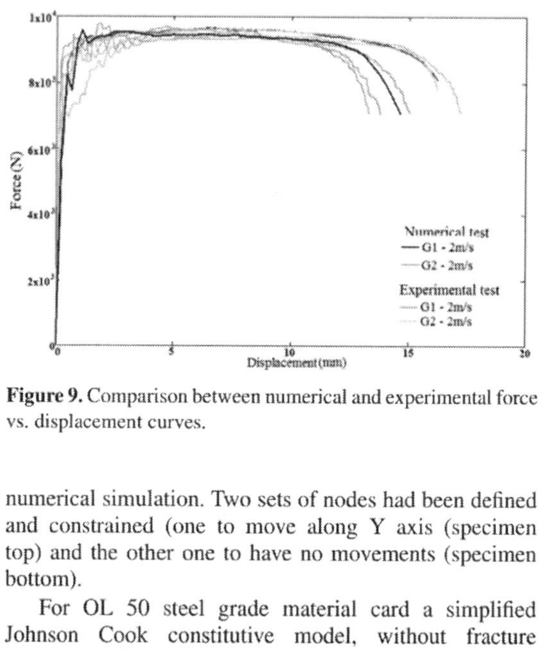

Figure 9. Comparison between numerical and experimental force vs. displacement curves.

numerical simulation. Two sets of nodes had been defined and constrained (one to move along Y axis (specimen top) and the other one to have no movements (specimen bottom).

For OL 50 steel grade material card a simplified Johnson Cook constitutive model, without fracture parameters, has been chosen. The material parameters were obtained by classical data processing of the quasi-static and dynamic experimental test results. The identified values for material parameters were: $A = 382.4\,\text{MPa}$, $B = 226.5\,\text{MPa}$, $C = 0.019$, $n = 0.201$.

Starting from experimental dynamic tests, analysis tensile loads at constant velocity are considered. Initial values of strain rates are defined by $\dot{\varepsilon}_0 = v_0/l_0$ where (v_0) is the loading speed and (l_0) is the active area. The initial values of strain rates are presented in Table 3.

According to a steel reference value characterizing the start of dynamics phenomena close to $100\,s^{-1}$ [17], these values correspond to an intermediate strain rate range. Concerning the material failure, during numerical

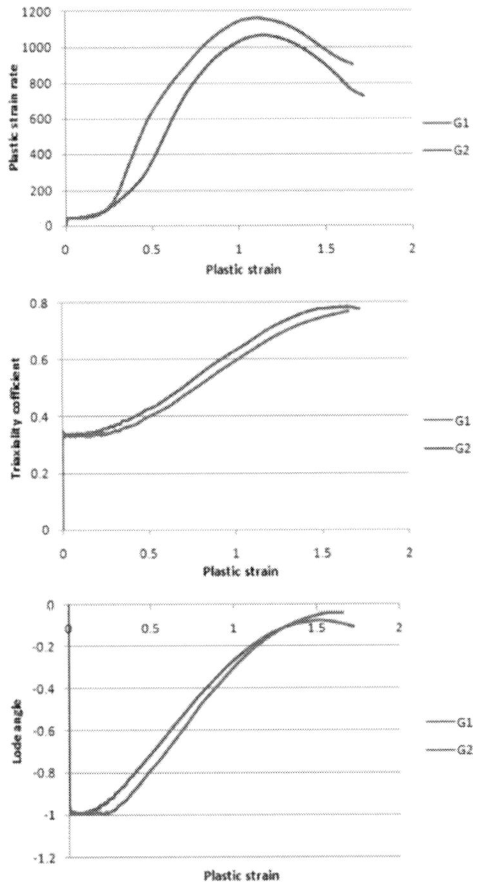

Figure 10. Parameters evolution depending on the effective plastic strain for G1 and G2.

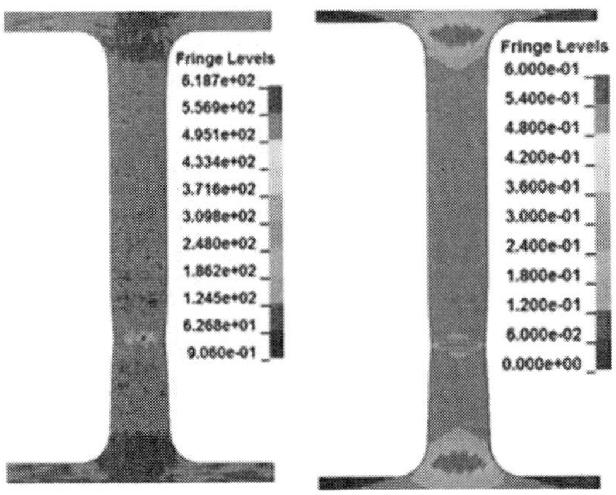

Figure 11. Iso-values of effective strain rate and triaxiality obtained for a plastic strain value of 0.4.

simulations total length, thickness in the necking area and specimen width in the same area had been observed and compared with the measured ones in order to identify the values for the fracture strain. Also, a supplementary check was performed in order to ensure that the numerical simulation is in agreement with the real test. For this purpose force vs. displacement for the specimen top

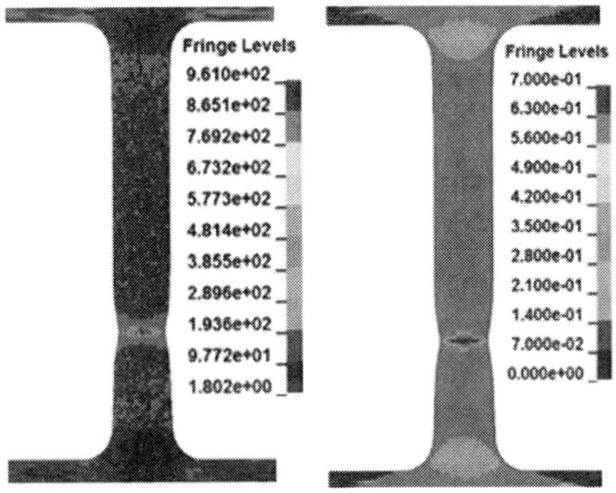

Figure 12. Iso-values of effective strain rate and triaxiality obtained for a plastic strain value of 0.8.

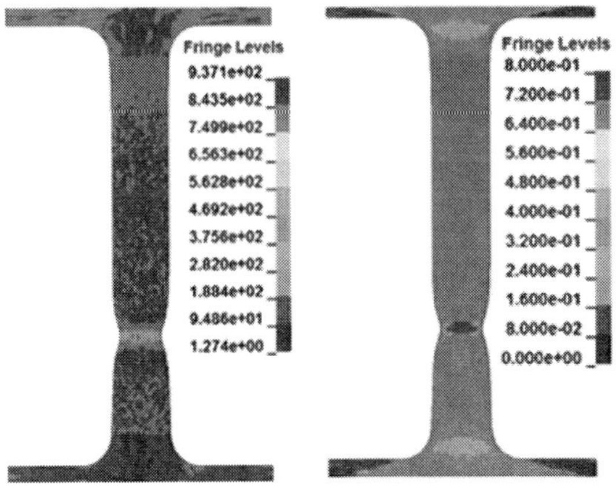

Figure 13. Iso-values of effective strain rate and triaxiality obtained for a plastic strain value of 1.2.

was plotted against the numerical one showing good agreement. In Fig. 7 is highlighted the fracture pattern for G1 specimen during the dynamic traction.

For G2 configuration case, since polyuria was added not only on the front of the specimen but also on the sides, a 3D modeling approach seems necessary (Fig. 8). For the 3D simulation the same steps as in 2D case were imposed. The only main difference was related to Adaptive_Mesh card. Due to the impossibility to define such a card for hexahedron elements this line was skipped. The 3D numerical simulation involved the use of type 1 hexahedron elements. Between the two layers (steel and polyurea) a Contact Tiebreak type without considering debonding was defined since post-test observations have indicated that up to fracture polyurea remains in contact with the metallic layer.

For polyurea a hyperelastic_rubber constitutive model was used. The material parameters were adopted from available scientific literature [7].

The 3D simulation (Fig. 9) shows that when a polyurea layer is added, an additional elongation of the steel specimen is observed as in the experimental tests.

The numerical simulations allowed also performing for the necking center area a comparative analysis (between G1 and G2 test configurations) regarding strain rate, triaxiality and Lode angle evolution (Fig. 10).

In Figs. 11–13 are presented the iso-values of effective strain rate and triaxiality for different values of plastic strain during the necking process: 0.4, 0.8 and 1.2. The information are presented in order to highlight the heterogeneity of the three studied phases: diffuse necking, localized necking and fracture. The images corespond to a 3D numerical simulation of G1 test configuration.

4. Conclusions

The experimental results show imperceptible increases of the maximum load for the coated specimens. In terms of elongation until fracture a 20% increase is observed and an 18% increase of consumed energy is computed.

Geometric analysis showed distinct changes concerning the necking area which was reduced by approximately 45% for the coated specimens. Concerning the average width of the specimens in the area unaffected by the necking this one was reduced by 7%, indicating clearly that the presence of a polyurea layer change the evolution of necking phenomenon, information that is consistent with the data presented in other scientific works as [8,9].

To improve the analysis of dynamic tensile loadings, numerical investigations were performed using the Ls-Dyna solver.

Numerical results indicate that the presence of the polyurea layer changes the loading pattern of metallic material in the necking area. In fact the polyurea takes an important role during the tensile loading and increases the material ductility. It is also observed that the stress triaxality it is influenced by the use of the polymeric layer, due to its strain hardening visco-elastic behavior. Furthermore due to strain rate increase phenomenon characterizing the necking area this behavior is even more highlighted.

In terms of fracture limit the coated specimens exhibit a clear decrease due to a different evolution of strain rate, triaxiality and Lode angle.

As a general conclusion the experimental test results indicates that the presence of a polyurea layer increases the ductility and delays the necking phenomenon due to elasto-viscoplastic properties of the polymer.

This paper has been supported in part by UEFISCDI GRANT 278/2014.

References

[1] K. Ravi-Chandar, *Experimental Characterization of Nonlinear Viscoelastic and Adhesive Properties of Elastomers* - Report, The University of Texas (2006)

[2] M. Feaga, *The effect of projectile strike velocity on the performance of polyurea coated RHA plates under ballistic impact* – Disertation, Lehigh University (2007)

[3] L. Xue, W. Mock Jr. T. BELYTSCHKO, Mechanics of Materials No. **42**, pp. 981–1003 (2010)

[4] J. Shim, D. Mohr, Mechanics of Materials No. **43**, pp. 349–360 (2011)

[5] M.R. Amini, J. Isaacs, S. Nemat-Nasser, Mechanics of Materials, **42, 6**, pp. 628–639 (2010)

[6] M.R. Amini, *Effect of polyurea on dynamic response and fracture resistance of steel plates under impulsive loads* - Disertation, University of California (2007)

[7] C. Gamonpilas, R. McCuiston, Polymer, **53**, pp. 3655–3658 (2012)

[8] Z. Xue, A. Vaziri, J.W. Hutchinson, Journal of the Mechanics and Physics of Solids, **56,** 93–113 (2008)

[9] Z. Xue, John W. Hutchinson, Mechanics of Materials, **39**, 473–487 (2007)

[10] Ling Y., AMP Journal of Technology, **5**, (1996)

[11] Lee Y.-W., Wierzbicki T., International Journal of Impact Engineering, **31**, pp. 1253–1276 (2005)

[12] G.R. Johnson, W.H. Cook, Engineering Fracture Mechanics, **21**, 31–48 (1985)

[13] A. Gavrus, Key Eng. Mat. **504-506,** pp. 697–702 (2012)

[14] T. Wierzbicki, Y. Bao, T.-W. Lee, Y. Bail, International Journal of Mechanical Science **47** 719–743 (2005)

[15] J. Barsoum, S. Faleskog, Procedia Engineering, **10**, 69–75 (2011)

[16] International standard, **ISO 6892-1**, (2009)

[17] A. Gavrus, P. Caestecker, E. Ragneau, B. Davoodi, J. Phys. IV **110**, pp. 353–358 (2003)

EPJ Web of Conferences **94**, 04045 (2015)
DOI: 10.1051/epjconf/20159404045

Sensitivity of high strain rate of structural elements in relation to dynamics properties of material

Leopold Kruszka[1,a], Iurii S. Vorobiov[2], and Nataliia Iu. Ovcharova[2]

[1]The Gen. Jaroslaw Dabrowski Military University of Technology, 2 Gen. S. Kaliski St., Warsaw 00-908, Poland
[2]A.N. Podgorny Institute of Mechanical Engineering Problems, National Ukrainian Academy of Sciences, 2/10 D. Pozharskoho St., Kharkov 61046, Ukraine

Abstract. Protective structures such as process chambers, protective boxes, facing elements of vehicles, personal protection equipment, motors cases etc. widely used in modern technology has been tested in the following project. Under the influence of impact loads, the three-dimensional dynamic stress-strain state with finite dynamic displacements and deformations has arised. The deformations occur in the elastic- plastic stage. In the analysis of the high-rate deformations, the dynamic properties of the materials, which are determined based on experimental data should be taken into account. The dynamic stress state of structural elements depends essentially on the dynamic properties of the materials used. The problem is solved using the finite element method, which takes into account the specifics of the process. Boundary conditions in the element nodes must satisfy the equality movement as well as derivatives. The formed function allows to describe continuous and smooth stress changes. The numerical analysis of the dynamic stress-strain state of structural elements under impact loads, takes into account different dynamic properties of the materials. A series of numerical calculations allows to reveal the features of high-rate deformation elements of protective structures and makes recommendations to improve their dynamic strength under different loading conditions.

1. Introduction and problem statement

Analysis of the dynamic stress-strain state of a number of modern structures, such as input and output devices, cases of gas turbine engines, safety boxes, process chambers, lining elements of vehicles, personal protective elements of energy systems, aerospace, transport and military equipment, under the influence of local shock and impulse loads is an actual and complex problem [1–6]. Identification of danger zones localization of stresses in impact places allows to find ways to reduce their level and increase the dynamic stability of elements of critical structures. This requires the construction of refined mathematical models. Together with intense impact loads, three dimensional dynamic stress-strain state develops in the elastic-plastic stage. Strength properties of most materials change depending on the value and rate of deformation. Therefore three-dimensional models with the dynamic properties of the material, which are determined on the basis of experimental studies are used [1–7]. In different papers different approaches and methods of accounting of the dynamic properties of materials are used. Known constitutive model described in [1]. A method for determining the dynamic properties of the material by a Hopkinson bar is presented in [2]. Common approaches to the modeling of the dynamic properties of the materials are included in the monographs [3,5]. Large elastic-plastic deformations are investigated in [4]. The physical nature of the behavior of materials under the influence of different types of loads is considered in [6]. Influence of dynamic properties of materials in the process of high-rate deformation of structural elements is an actual and vital problem. If the equation of state reflects the elastic-plastic character of deformation and dynamic properties of the material, then the problem is physically nonlinear. Under intense loads finite displacement and deformation arises resulting in geometrical nonlinearity of problems [8].

A mathematical model for the analysis of the problem must take into account the heterogeneity of the material's structural elements as originally given as well as arisen during high-rate deformation. Therefore, all the material characteristics are variable in the spatial coordinates and time. Dependencies for stress from strains are determined based on a dynamic version of the theory of plastic deformation. Dynamic hardening of materials reflects the dependencies of the stress intensity σ_i from intensities εi of the strain and of the strain rate $\sigma_i = \sigma_i(\varepsilon_i, \dot{\varepsilon}_i)$ [7,9].

Figure 1 shows some of the dependencies of the relative dynamic elastic limit σ_{sd}/σ_s for a number of structural materials obtained on the basis of experimental data [7]. Those curves correspond to structural materials: 1 – ЭИ 878 (Х15Н8А18) steel, 2 – Х18Н10Т steel, 3 – ОТ41 alloy, 4 – ЭП410 (Х15Н5Д2Т) steel, 5 – АМr2М alloy.

It can be seen that the rate of deformation provides the greatest influence on steel alloys and less influence on aluminium.

In region I the experiments were carried out in a range of strain rate $8 \cdot 10^1 - 6 \cdot 10^2$ 1/s, and in region II in the range of $1,3 \cdot 10^2 - 3,5 \cdot 10^3$ 1/s. It is established that the equation $\sigma_{sd}/\sigma_s = \left[1 - \left(\frac{\dot{\varepsilon}}{D}\right)^{\frac{1}{n}}\right]$ quite accurately describes the dependence of the dynamic elastic limit the strain rate.

[a] Corresponding author: lkruszka@wat.edu.pl

Article available at http://www.epj-conferences.org or http://dx.doi.org/10.1051/epjconf/20159404045

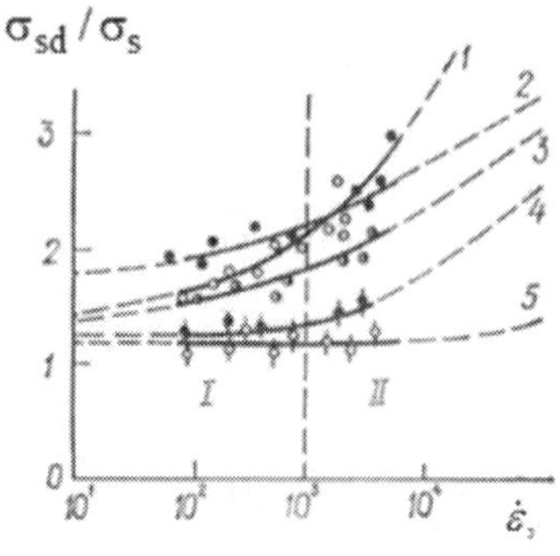

Figure 1. Experimental dependence of the dynamic elastic limit in relation to the strain rate [7].

The value of the coefficients D and n for various materials are given in [7]. In the general case strain-rate hardening of the material may be described as

$$\sigma = E\varepsilon_s\left[1+\left(\frac{\dot{\varepsilon}_{eq}}{D}\right)^{1/n}\right]+\sigma_s\left\{\frac{\varepsilon_{eq}}{\varepsilon_s}-\left[1+\left(\frac{\dot{\varepsilon}_{eq}}{D}\right)^{1/n}\right]\right\}^m \quad (1)$$

where, ε_s, D, m, n – characteristics of strain hardening and rate hardening of the material; ε_{eq}, $\dot{\varepsilon}_{eq}$ – the elastic-plastic deformation and rate deformation of specimens measured during the tests.

The first term reflects an increase of the dynamic yield stress, and the second - large plastic deformation [7].

However, in numerical studies different dependencies for describing the dynamic properties of the material are used. Their form is determined by the way of the experimental data processing used by different authors [1–5]. It is interesting to compare the results of calculations using different, the most common dependencies.

The problem of the variant of finite element method is solved takes into account the specifics of the process. An important requirement for the choice of the type of finite elements is a accounting of all specified factors. The boundary conditions at the nodes of the elements must satisfy the equality displacements as well as derivatives. The functions formed allow to describe the continuous and smooth variation of stresses.

2. Analysis of the numerical results

Numerical analysis based on the finite element method allows to determine the distribution, magnitude and time variation of the dynamic displacements, strains and stresses in different types of loads [8–10]. However, the numerical results depend strongly from the method of accounting of the dynamic properties of the material. For comparison, consider the impact on the rectangular element of the steel structure with thickness $h = 8$ mm, impactor mass $m = 0,1$ kg with a velocity 200 m/s for different accounting methods of dynamic properties of the material.

Trying to solve the problem in the elastic formulation, the increase of intensity of the load leads, as expected, to overestimation of maximum stress (to 800 MPa) when the undervaluation of displacements equals about 0.25 mm.

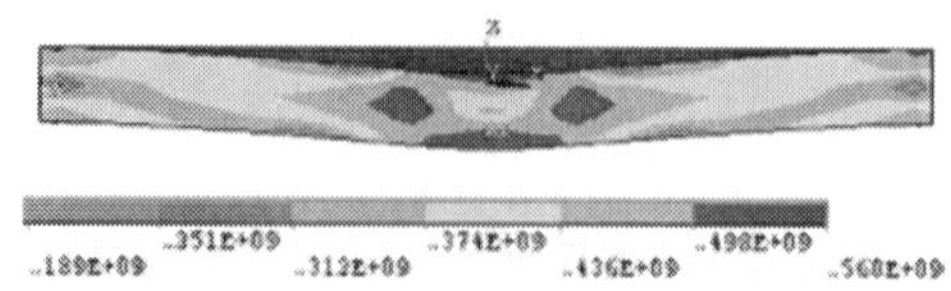

Figure 2. Distribution of equivalent stress σ_{eq} under the influence of the impactor with the dynamic properties of the material according to the relationship (2).

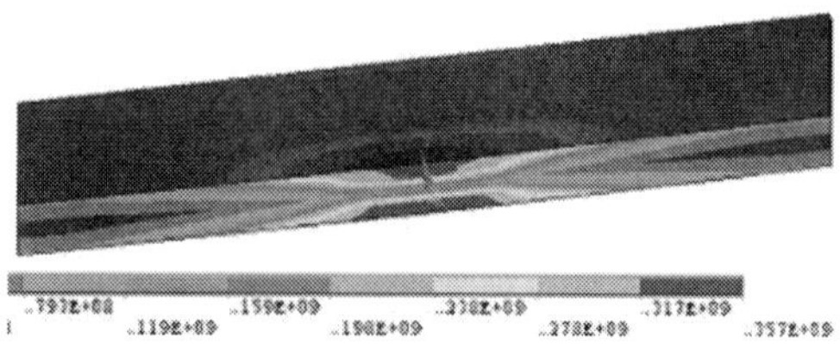

Figure 3. Distribution of equivalent stress under the influence of the impactor in a refined setting.

In [8–10] it is shown that it is appropriate to use the account of the dynamic properties of material by the Perzyna dependency [5], which is a special case of the relation (1)

$$\sigma_i = \left[1+\left(\frac{\dot{\varepsilon}_i^{pl}}{\gamma}\right)^m\right]E\varepsilon_i \quad (2)$$

where E – elastic modulus;
m and r – strain rate sensitivity coefficients;
$\dot{\varepsilon}_i^{pl}$ – deformation rate at the plastic stage.
In this case a decrease in the largest equivalent stress (500 MPa) and the increase in displacements (2.8 mm), besides changing the character of distribution of equivalent stresses σ_{eq} (Fig. 2) can be noticed.

This model can be refined by taking into account finite deformations and increasing the number of degrees of freedom at the nodes. In this case, the maximum equivalent stress is further reduced (350 MPa), and the amount of displacement is increased (Fig. 3).

Dynamic properties of the material can be addressed in some other form using Peirce's dependency [4], which uses the same notation as in the Perzyna dependency (2).

$$\sigma_i = \left[1+\left(\frac{\dot{\varepsilon}_i^{pl}}{\gamma}\right)\right]^m E\varepsilon_i. \quad (3)$$

In this case finite displacement and a large number of degrees of freedom at the nodes is also taken into account. (Fig. 4).

In this case there has been some reduction of the maximum equivalent stresses (312 MPa) and the value of maximum displacements does not change. When unloading there is a sharp reduction of stresses and redistribution of dynamic stress-strain state.

It can be noticed that during refining of the elastic-plastic and dynamic properties of the material, as well as taking into account finite deformations, increasingly evident nonlinear character of deformation with increasing dynamic changes of the stress-strain state in the pipe of

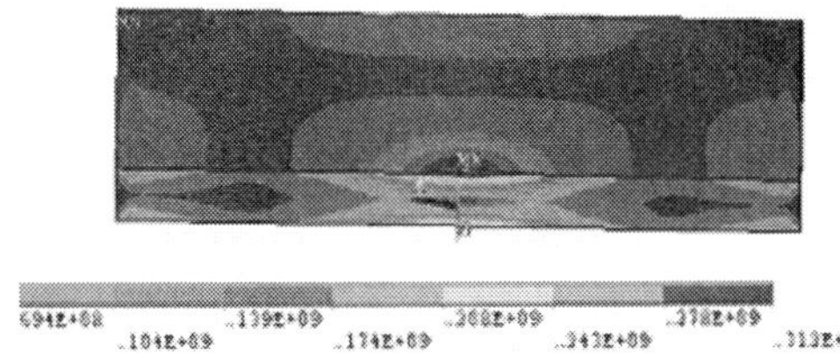

Figure 4. Distribution of equivalent stresses under the influence of a drummer in relation to the dynamic properties of the material according to the relationship (3).

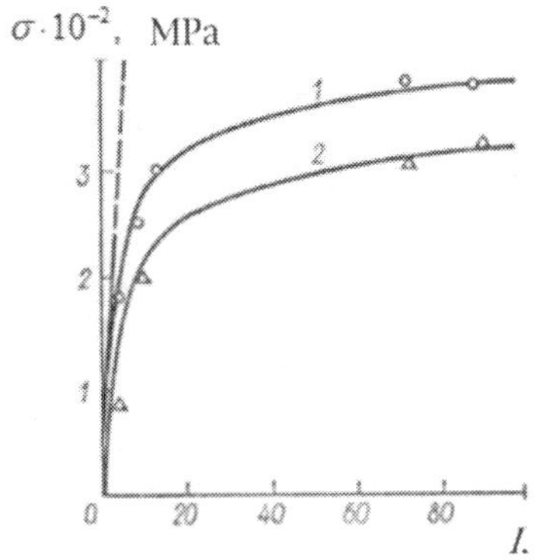

Figure 5. Change of the axial (1) and radial (2) stresses in the pipe with the increase of impact load [7].

gas pipeline with an increase in the local shock load by total momentum I are presented [7].

The radius of the pipe equals as follows: 0,5 m, thickness 0,01 m, length 5,5 m. In the analysis of large plastic deformations the dependence which is a special case of (1) was used. displacements and strains at lower values of stresses can be noticed. To further illustrate this phenomenon the results of numerical studies of the

$$\sigma_i = A \cdot \varepsilon_i^m \left[1 - \left(\frac{\dot{\varepsilon}_i}{D}\right)^{\frac{1}{n}}\right] \tag{4}$$

where $A = 1323$ MPa - hardening modulus in the plastic stage; $m = 0,14$; $n = 3,8$; $D = 520$ 1/s.

Figure 5 shows a comparison of the numerical (curves) and experimental (points) results for the axial (curve 1) and radial (curve 2) stress [7]. It shows the influence of the plastic and dynamic properties of the material and the non-linear character of deformation as the load increases. The dotted line shows the limit of the elastic.

3. Conclusions

Dynamic properties of the material have a significant influence on the development of high-rate process of elastic-plastic deformation of structural elements under intense impact loads. Solution of the problem in the elastic formulation is possible only at low speeds and low loads, when $\left(\frac{\dot{\varepsilon}_i}{D}\right)^{1/n} \ll 1$. When unloading, differences in the character of the stress-strain state can be observed, the unloading process starts at different stages of the stress-strain state.

When choosing a sufficiently appropriate elastic-plastic and dynamic characteristics of the material reliable results can be achieved, which can properly evaluate the allowable impact loading or dynamic strength of structural elements. It should strive to comparison of the results obtained when using different accounting methods of dynamic properties of the material. It allows to increase the accuracy of numerical calculations. In any case, multivariate numerical studies allow to identify character of elastic-plastic high-rate deformation of structural elements and secure their dynamic strength.

References

[1] G.R. Johnson, W.H. Cook, *A constitutive model and data for metals subjected to large strains high strain rates and high temperatures*, Proceedings of the 7th International Symposium on Ballistics, The Hague, The Netherlands, April 1983.

[2] L. Kruszka, W.K. Nowacki, *New application of the Hopkinson pressure bar technique to determining dynamic behavior of materials*, J. of Theoretical and Applied Mechanics, **2** (1996).

[3] M.A. Meyers, *Dynamics behavior of materials*, Wiley, New York 1994.

[4] D. Peirce and D.R.J. Owen, *A model for large deformations of elastic-viscoplastic solids at finite strains: computational issues*, in: *Finite Inelastic Deformations: Theory and Applications*, Springer-Verlag, Berlin 1992.

[5] P. Perzyna, *Fundamental problems in viscoplasticity*, Advances in Applied Mechanics, **9** (1968).

[6] V.T. Troshchenko, A.A. Lebedev, V.A. Strizhalo. *Mechanical behaviour of materials under different types of loading* (in Russian), Logos, Kiev, 2000.

[7] Iu.S. Vorobiov, A.V. Kolodyazhny, V.I. Sevryukov and E.G. Yanyutin, *High-speed deformation of structure elements* (in Russian), Naukova Dumka, Kiev 1989.

[8] Iu.S. Vorobiov, N.Iu. Ovcharova, L. Kruszka, *Nonlinear deformations of structures cylindrical element under local shock*, Proceedings of the 4th, International Conference on Nonlinear Dynamics ND-KhPI2013 - Ukraine, pp. 351–357, 2013.

[9] Iu.S. Vorobiov, L. Kruszka, N.Iu. Ovcharova, *Finite Element Analysis of Local Shock Loading on Structures Cylindrical Elements,* Proceedings of The 8th International Symposium on Impact Engineering (ISIE2013), Osaka University, Japan, pp. 499–504, 2013.

[10] Iu.S. Vorobiov, N.Iu. Ovcharova, *Dynamics of structural elements under impact loads* (in Russian), Vibration in Engineering and Technologies, **74**, (2014).

EPJ Web of Conferences **94**, 04046 (2015)
DOI: 10.1051/epjconf/20159404046

Deformation of compound shells under action of internal shock wave loading

Marina Chernobryvko[1], Leopold Kruszka[2,a], and Konstantin Avramov[1]

[1] A. N. Podgorny Institute for Mechanical Engineering Problems, National Academy of Sciences of Ukraine, 2/10 Dm. Pozharskoho St., 61046 Kharkiv, Ukraine
[2] The General Jaroslaw Dabrowski Military University of Technology, 2 Gen. Sylwester Kaliski St., 00-908 Warsaw, Poland

Abstract. The compound shells under the action of internal shock wave loading are considered. The compound shell consists of a thin cylindrical shell and two thin parabolic shells at the edges. The boundary conditions in the shells joints satisfy the equality of displacements. The internal shock wave loading is modelled as the surplus pressure surface. This pressure is a function of the shell coordinates and time. The strain rate deformation of compound shell takes place in both the elastic and in plastic stages. In the elastic stage the equations of the structure motions are obtained by the assumed-modes method, which uses the kinetic and potential energies of the cylindrical and two parabolic shells. The dynamic behaviour of compound shells is treated. In local plastic zones the 3-D thermo-elastic-plastic model is used. The deformations are described by nonlinear model. The stress tensor elements are determined using dynamic deformation theory. The deformation properties of materials are influenced by the strain rate behaviour, the influence of temperature parameters, and the elastic-plastic properties of materials. The dynamic yield point of materials and Pisarenko-Lebedev's criterion of destruction are used. The modified adaptive finite differences method of numerical analysis is suggested for those simulations. The accuracy of the numerical simulation is verified on each temporal step of calculation and in the case of large deformation gradients.

1. Introduction

The compound shells are widely used in rocket production and aircraft construction. The compound shell consists of a cylindrical shell and two parabolic shells at the edges. These shells are by the action of internal shock wave loading. The interaction of the thin-walled structures with shock wave can lead to the self-sustained vibrations with large amplitudes and large deformations with destruction.

In spite of wide use of parabolic shells, few efforts were made to analyze dynamics of such structures. Free vibrations of parabolic shells of revolution are analyzed in [1]. The first order shear deformation theory is used to analyze these shells. The discretization of the system by means of the differential quadrature technique leads to the eigenvalue problem. Basing on the first-order shear deformation theory, the paper [2] focuses on the dynamic behaviour of moderately thick functionally graded parabolic panels and shells of revolution. The generalized differential quadrature method has been used to discretize the system equations.

The article [3] presents the results of an experimental and numerical investigation into the response of partially confined right-circular stainless steel cylinders to air-blast loading. The blast loading was generated by detonating spheres of plastic explosive at two axial positions along the centre line of the cylinders.

The calculation of material properties following from change of deformation speed, temperature parameters, and elastic-plastic behaviour of materials are analyzed in [4]. Rate-independent finite elastic-plastic equations with thermo-coupled effects are proposed to bypass the yield condition and loading–unloading conditions in [5].

In paper [6] a thermo-elastic-plastic model for the fiber-metal laminated (FML) beams is established. The elastic-plastic behaviour of the FML beam is modelled by a macromechanical plasticity theory, where the composite layers in the FML are assumed to be linearly elastic and the aluminium layers are taken as elastic-plastic. Additionally, the thermal stresses in the FML beam are studied by introducing heat conduction equation. Considering the mismatch in properties of different layers, the interfacial damage is investigated based on cohesive zone model and shear-lag model, and the evolution of the interfacial damage is addressed as well. The incremental thermo-elastic-plastic governing equations for the FML beams are solved by finite difference method and iteration method.

In this paper a thermo-elastic-plastic model for the compound shells is investigated. The deformations are described by nonlinear model. The stress tensor elements are determined using dynamic deformation theory. The deformation properties of materials are influenced by the strain rate behaviour, the influence of temperature parameters, and the elastic-plastic properties of materials. The dynamic yield point of materials and Pisarenko-Lebedev's criterion of destruction are used.

[a] Corresponding author: lkruszka@wat.edu.pl

Article available at http://www.epj-conferences.org or http://dx.doi.org/10.1051/epjconf/20159404046

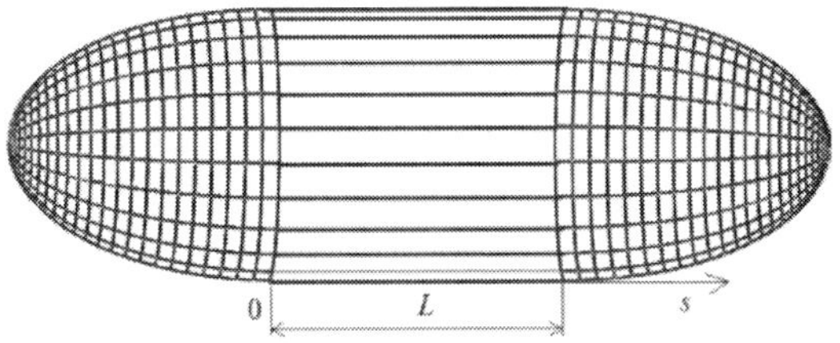

Figure 1. The scheme of the compound shell.

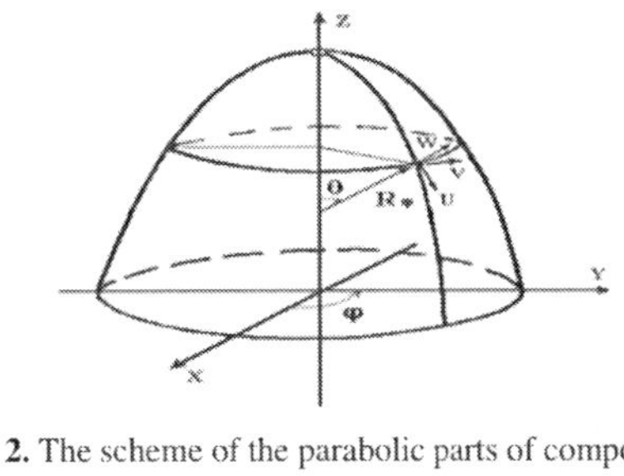

Figure 2. The scheme of the parabolic parts of compound shell.

2. Problem formulation and equations of motions

We study the construction as the shell consisting of three parts. The compound shell consists of a thin cylindrical shell and two thin parabolic shells at the edges, Fig. 1. The boundary conditions in the shells' joints satisfy the equality of displacements.

The dynamic equations for components of the displacement vector $U_c = U(u, w, \beta)$ can be written within the linear Timoshenko theory of shells. For the cylindrical part of the compound shell, the equations of motion are [7]:

$$\frac{\partial^2 u}{\partial s^2} + \frac{\nu}{R_c}\frac{\partial w}{\partial s} = c_1\frac{\partial^2 u}{\partial t^2},$$

$$\frac{\partial^2 w}{\partial s^2} + \frac{\partial \beta}{\partial s} - b_1\left(\frac{w}{R_c} + \nu\frac{\partial u}{\partial s}\right) = c_2\frac{\partial^2 w}{\partial t^2} - P_c(s, \phi, t),$$

$$\frac{\partial^2 \beta}{\partial s^2} - b_2\left(\frac{\partial w}{\partial s} + \beta\right) = c_1\frac{\partial^2 \beta}{\partial t^2}, \qquad (1)$$

where $s \in [0, L]$, ν – Poisson's ratio, $c_1 = \frac{\rho(1-\nu^2)}{E}$, $c_2 = \frac{\rho}{kG}$, $b_1 = \frac{2}{(1-\nu)kR_c}$, $b_2 = \frac{6(1-\nu)k}{h^2}$, R_c – radius of the cylindrical middle surface, h – thickness of the cylindrical shell, k – Timoshenko's shear coefficient, $E = E(\dot{\varepsilon}, T)$, $G = G(\dot{\varepsilon}, T)$ – modules of elasticity for tension and compression, which are functions of the strain rate $\dot{\varepsilon}$ and temperature T, ρ – material density, $P_c(s, \varphi, t)$ – internal shock wave loading acting on the cylindrical part.

It is necessary to define loading acting on the cylindrical part of the shell. Generally, these fields are determined for each type of explosion in an experimental way. It is possible to take a general form of the loading for numerical experiment:

$$P_c(s, \varphi, t) = Q_{\max}(s, \varphi)e^{-\frac{t-t_0(s,\varphi)}{\theta(s,\varphi)}}F_1(s, \varphi, t), \qquad (2)$$

where $Q_{\max}(s, \varphi)$ – the characteristic pressure wave amplitude, $\theta(s, \varphi)$ – speed of the pressure wave attenuation, $t_0(s, \varphi)$ – detention period of the load act, $F_1(s, \varphi, t)$ – distribution function of pressure along cylindrical shell.

The displacements $u(s, \varphi, t)$, $w(s, \varphi, t)$, $\beta(s, \varphi, t)$ are expanded using the eigenmodes of free vibrations as:

$$u(s, \varphi, t) = \sum_{n=1}^{N_u} q_n^{(u)}(t)U_n(s, \varphi);$$

$$w(s, \varphi, t) = \sum_{n=1}^{N_w} q_n^{(w)}(t)W_n(s, \varphi), \qquad (3)$$

$$v(s, \varphi, t) = \sum_{n=1}^{N_B} q_n^{(B)}(t)B_n(s, \varphi);$$

where $q^{(u)} = [q_1^{(u)}, \ldots, q_{N_u}^{(u)}]$; $q^{(w)} = [q_1^{(w)}, \ldots, q_{N_w}^{(w)}]$; $q^{(B)} = [q_1^{(B)}, \ldots, q_{N_B}^{(B)}]$; are vectors of the generalized coordinates; $U_n(s, \varphi)$, $W_n(s, \varphi)$, $B_n(s, \varphi)$ are eigenmodes of free vibrations.

The scheme of the parabolic parts of compound shell is shown in Fig. 2.

For the parabolic part of the compound shell the potential energy for components of the displacement vector $u(\theta, \varphi, t)$, $v(\theta, \varphi, t)$, $w(\theta, \varphi, t)$ is [7]:

$$\Pi = H_1 \times \int_0^{2\pi}\int_{\theta_1}^{\theta_2} \times \left[(E_1 + E_2)^2 - 2(1-\nu)\left(E_1E_2 - \frac{1}{4}\Omega_1^2\right)\right]A_1 d\theta d\varphi$$
$$+ H_2\int_0^{2\pi}\int_{\theta_1}^{\theta_2}\left[\{(K_1 + K_2)^2 - 2(1-\nu)(K_1K_2 - \Omega_2^2)\}\right.$$
$$\times A_1 + 2\{(E_1K_1 + E_2K_2) + \nu(E_1K_2 + E_2K_1)$$
$$\left. + (1-\nu)\Omega_1^2\Omega_2^2\} AB_1\right] d\theta d\varphi, \qquad (4)$$

where R_0 – curvature radius at the shell critical point $\theta = 0$, ν – Poisson's ratio, h – the shell thickness; ρ – density of the shell material, $H_1 = \frac{Eh}{2(1-\nu^2)}$, $H_2 = \frac{h^2}{12}H_1$,

$A_1 = R_0^2 \frac{\sin\theta}{\cos^4\theta}$, $AB_1 = R_0 \frac{\sin\theta(1+\cos^2\theta)}{\cos^3\theta}$,

$$E_1 = \frac{1}{R_0}\left(\frac{\partial u}{\partial\theta} + w\right)\cos^3\theta,$$

$$K_1 = \frac{1}{R_0^2}\left[\left(\frac{\partial u}{\partial\theta} + \frac{\partial^2 w}{\partial\theta^2}\right)\cos^6\theta - \left(\frac{\partial w}{\partial\theta} + 3u\right)\cos^5\theta\sin\theta\right],$$

$$E_2 = \frac{1}{R_0}\left(\frac{\partial v}{\partial\varphi} + u\cdot\cos\theta + w\cdot\sin\theta\right)ctg\theta,$$

$$K_2 = \frac{1}{R_0^2}\left[\frac{\partial v}{\partial\varphi}ctg\theta\cos\theta - \frac{\partial^2 w}{\partial\varphi^2}ctg^2\theta + \left(\frac{\partial w}{\partial\theta} - u\right)ctg\theta\cos^4\theta\right],$$

$$\Omega_1 = \frac{1}{R_0}\left(\frac{\partial v}{\partial\theta}\cdot\cos^3\theta - v\cdot\cos^2\theta\sin\theta + w\cdot\frac{\cos^4\theta}{\sin\theta} + \frac{\partial u}{\partial\varphi}ctg\theta\right),$$

$$\Omega_2 = \frac{1}{R_0^2}\left[\frac{\partial w}{\partial\varphi}\frac{\cos^3\theta}{\sin\theta} - \frac{\cos^4\theta}{\sin\theta} \times \left(\frac{\partial^2 w}{\partial\theta\partial\varphi} - \frac{\partial u}{\partial\varphi} - \frac{\partial v}{\partial\theta} + v\right)\frac{\cos^4\theta}{\sin\theta}\right].$$

The shell kinetic energy takes the following form:

$$\mathrm{T} = \frac{\rho h R_0^2}{2}\int_0^{2\pi}\int_{\theta_1}^{\theta_2}\left(\left(\frac{\partial u}{\partial t}\right)^2 + \left(\frac{\partial v}{\partial t}\right)^2 + \left(\frac{\partial w}{\partial t}\right)^2\right)B_2 d\theta\, d\varphi, \tag{5}$$

where $B_2 = \frac{\sin\theta}{\cos^4\theta}$.

The virtual work of the internal shock wave loading acting on the parabolic part takes the following form:

$$\delta\mathrm{A} = R_0^2\int_0^{2\pi}\int_{\theta_1}^{\theta_2}P_{pr}(\theta,\varphi,t)\,\delta w\frac{\sin\theta}{\cos^4\theta}d\theta\, d\varphi, \tag{6}$$

where $P_{pr}(\theta,\varphi,t)$ –the pressure of the supersonic gas stream acting on the shell; δw – virtual displacement of the shell.

In an analogous (2) we can write down $P_{pr}(\theta,\varphi,t)$ for a parabolic shell:

$$P_{pr}(\theta,\varphi,t) = Q_{\max}(\theta,\varphi)e^{-\frac{t-t_0(\theta,\varphi)}{\theta(\theta,\varphi)}}F_2(\theta,\varphi,t), \tag{7}$$

where $F_2(\theta,\varphi,t)$ - distribution function of pressure along the parabolic shell.

The displacements $u(\theta,\varphi,t)$, $v(\theta,\varphi,t)$, $w(\theta,\varphi,t)$ are:

$$\begin{aligned} u(\theta,\varphi,t) &= \sum_{n=1}^{N_u} q_n^{(u)}(t)U_n(\theta,\varphi);\\ v(\theta,\varphi,t) &= \sum_{n=1}^{N_v} q_n^{(v)}(t)V_n(\theta,\varphi);\\ w(\theta,\varphi,t) &= \sum_{n=1}^{N_w} q_n^{(w)}(t)W_n(\theta,\varphi), \end{aligned} \tag{8}$$

where $q^{(u)} = [q_1^{(u)},\ldots,q_{N_u}^{(u)}]$; $q^{(v)} = [q_1^{(v)},\ldots,q_{N_v}^{(v)}]$; $q^{(w)} = [q_1^{(w)},\ldots,q_{N_w}^{(w)}]$ are vectors of the generalized coordinates; $U_n(\theta,\varphi)$, $V_n(\theta,\varphi)$, $W_n(\theta,\varphi)$ are eigenmodes of free vibrations [8].

Along the junction of the parabolic shell with the cylinder:

$$\begin{aligned} u(\theta,\varphi,t) &= u(s,\varphi,t),\\ v(\theta,\varphi,t) &= v(s,\varphi,t),\\ w(\theta,\varphi,t) &= w(s,\varphi,t). \end{aligned} \tag{9}$$

Let us note that the assumed-modes method determines the displacements for the deformation process.

3. Thermo-elastic-plastic model

The strain tensor ε_{ij} is nonlinear as follows [9]:

$$\varepsilon_{ij} = \frac{\partial u_i}{\partial x_j} + \frac{\partial u_i}{\partial x_j}\frac{\partial u_j}{\partial x_i}. \tag{10}$$

The Eq. (10) was used for the parabolic parts.

It should be noted that the system of Eqs. (1) determines the displacements for the deformation process in the cylindrical part. Therefore, the strain-displacement relations can be taken both in the nonlinear as well as in the linearized form. Analysis of the cylindrical shell linearization of the strain-displacement relations does not lead to essential errors. So, the strains are given by:

$$\varepsilon_1 = \frac{\partial u}{\partial z},\quad \varepsilon_2 = \frac{1}{R_c}w,\quad \varepsilon_3 = \frac{\partial\beta}{\partial z},\quad \varepsilon_{12} = \frac{\partial w}{\partial z} - \beta, \tag{11}$$

where $\varepsilon 1$ – tensile strain of the medial surface compression in the longitudinal direction, $\varepsilon 2$ – tensile strain of the middle surface in the circumferential direction, $\varepsilon 3$ – bending of the surface in the longitudinal direction, $\varepsilon 12$ – shear strain of the middle surface.

In further mathematical modelling of the problem we also take into account that some zones of plastic strain can develop thus forming adiabatic shear bands under detonation loading in the material. Therefore, further solution is reduced by introducing definition of the equivalent strain:

$$\varepsilon_{eq} = \varepsilon_{eq}(\varepsilon_1(t), \varepsilon_2(t), \varepsilon_3(t), \varepsilon_{12}(t)), \tag{12}$$

The stress tensor elements σ_{ii} and σ_{ij} are determined in a dynamic deformation theory

$$\begin{aligned} \sigma_{ii} - \sigma_0 &= \frac{4\sqrt{3}\cdot\sigma_{eff}(\varepsilon_{eq},\dot\varepsilon_{eq},T)}{9\varepsilon_{eq}}(\varepsilon_{ii} - \varepsilon_0),\\ \sigma_{ij} &= \frac{2\sqrt{3}\cdot\sigma_{eff}(\varepsilon_{eq},\dot\varepsilon_{eq},T)}{3\varepsilon_{eq}}\varepsilon_{ij} \end{aligned} \tag{13}$$

where $\sigma_0 = \sigma_{ii}/3$ – mean value of diagonal element of stress tensor, $\varepsilon_0 = K\sigma_0 + \alpha T$ – mean value of diagonal element of strain tensor, ε_{eq} – equivalent strain, $\dot\varepsilon_{eq}$ – equivalent strain rate σ_{eff} – effective stress.

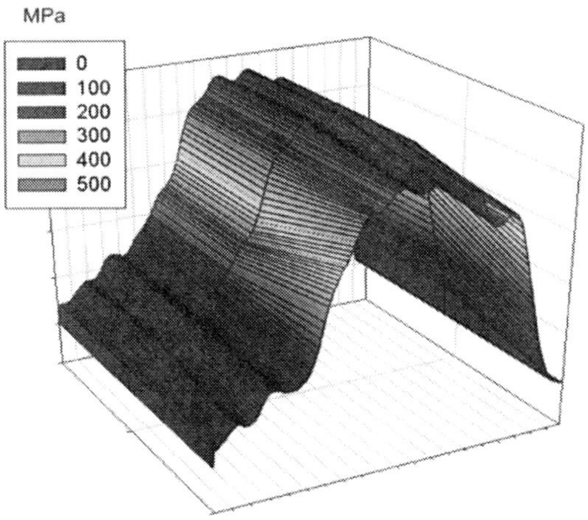

Figure 3. The field of stress obtained along of compound shell.

A multiplier $\sigma_{eff}(\varepsilon_{eq}, \dot{\varepsilon}_{eq}, T)/\varepsilon_{eq}$ contains dynamic material properties $\sigma_{eff}(\varepsilon_{eq}, \dot{\varepsilon}_{eq}, T)$:

$$\sigma_{eff} = \sigma_y^{st}\left\{\left[1+\left(\frac{\dot{\varepsilon}_{eq}}{D}\right)^{\frac{1}{n}}\right]-\left(\frac{\varepsilon_{eq}}{\varepsilon_y^{st}}-\left[1+\left(\frac{\dot{\varepsilon}_{eq}}{D}\right)^{\frac{1}{n}}\right]\right)^m\right\}\times \times\left(1-\frac{T-T_k}{T_p-T_k}\right)^r, \quad (14)$$

where σ_y^{st} – the static yield point of materials, ε_y^{st} – the strain at static yield point, D, n, m, r – experimental parameters of dynamic material properties, T_k – temperature which determines material properties, T_p – melting temperature of material.

The dynamic yield point of materials is as follows:

$$\sigma_y^d = \sigma_y^{st}\left[1+\left(\frac{\dot{\varepsilon}_{eq}}{D}\right)^{\frac{1}{n}}\right]. \quad (15)$$

The verification of destruction is conducted with the use of Pisarenko-Lebedev's criterion [10]:

$$\sigma_{dstr} = \frac{\sigma_+}{\sigma_-}\cdot\sigma_{eff} + \sigma_{11}\cdot\left(1-\frac{\sigma_+}{\sigma_-}\right), \quad (16)$$

where σ_+ – yield point of materials in tension, σ_- – yield point of materials in compression.

This dynamic problem has a nonlinear character. The finite difference method [9] allows taking an adequate decision on knots of mesh. The choice of small step allows conducting a linear task.

4. Numerical simulation

The followings parameters of material were used: $E = 1,3\cdot10^{11}$ Pa; $E_1 = 10,07\cdot10^8$ Pa; $\sigma_y^{st} = 4,9\cdot10^8$ Pa; $\sigma_{dstr} = 6,9\cdot10^8$ Pa; $\rho = 4,5\cdot10^3$ kg/m^3; $D = 4,45\cdot10^3$ 1/s; $T_p = 1680\,°C$; $T_k = 27\,°C$; $\alpha = 8,3\cdot10^8 1/°C$, $\nu = 0.3 \div 0.5$; $n = 4,84$; $m = 0,21$; $r = 1$.

The results of numerical calculations are presented in Fig. 3.

5. Summary

This work presents the strength analysis of the compound shell under the action of internal shock wave loading. This dynamic thermo-elastic-plastic problem is non-linear. Its solution requires to connect mathematical modelling with experimental data of material dynamic characteristics. The main feature of the method is the calculation of influence of the strain hardening and temperature on the dynamic plastic deformation.

This research is partially supported by the grant of National Ukrainian Academy of Sciences in space research for 2012-2016 (contract No. II-67-14).

References

[1] C. Dahlberg, J. Faleskog, Eur. J. of Mechanics A/Solids **44** (2010), 1–16.
[2] F. Tornabene, E. Viola, Eur. J. of Mechanics A/Solids **28** (2009), 991–1013.
[3] G.S. Langdon, A. Ozinsky, S. Chung Kim, Eur. J. of of Impact Engineering, **73** (2014), 1–14.
[4] M.A. Meyers, Dynamics Behavior of Materials (New York: Wiley, 1994).
[5] L. Kruszka, W.K. Nowacki, J. of Thermal Stresses, **18** (1995), 313–334.
[6] Heng Xiao, Yingli Lia, Yiming Fua, Int. J. of Plasticity, **63** (2014), 211–228.
[7] M. Amabili, *Nonlinear Vibrations and Stability of Shells and Plates*, 2008, (Cambridge University Press, Cambridge).
[8] M.V. Chernobryvko, K.V. Avramov, V.N. Romanenko, T.J. Batutina, A.M. Tonkonogenko, Int. J. of Theor. and Appl. Mech., **49** (2014), 27–32.
[9] M.V. Chernobryvko, L. Kruszka, Yu.S. Vorobiev Applied Mechanics and Materials, **566**, 493–498.
[10] G.S. Pisarenko, A.A. Lebedev, *Deformation and strength of materials under complex stress state*, 1976, (Naukova Dumka, Kiev).

EPJ Web of Conferences **94**, 04047 (2015)
DOI: 10.1051/epjconf/20159404047

Numerical study on dynamic compressive deformation and elasto-plastic wave propagation of foam materials

Kenichi Tanigaki[1,a], Toru Idouji[1], Keitaro Horikawa[1], Hidetoshi Kobayashi[1], and Kinya Ogawa[2]

[1] Grad. Sch. of Eng. Sci., Osaka University, Toyonaka, Osaka, Japan
[2] Space Dynamics Lab. Kyoto, Kyoto, Japan

Abstract. Finite element models of closed-cell foam structures were created using the three-dimensional Voronoi tessellation method coupled with the random sequential addition algorithm. The dynamic compressive deformation behaviors of the models were numerically studied using LS-DYNA code. The deformation mode of the models changed gradually as the deformation rate increases. Also, the generation and the propagation of plastic wave was clearly observed with the rate of 100 m/s. The longitudinal elastic wave velocity showed a weak negative dependency on the deformation rate although the strain rate dependence of material properties was not considered. Furthermore, a prediction method for the dynamic stress state on the impact side based on the static stress-strain relationship was presented.

1. Introduction

A stress-strain curve of foam materials generally consists of three stages; elastic region, plateau region, and densification region. Since the plateau region denotes ability of large deformation under low stress, foam structures show excellent energy absorption. However, polymer foam, which is widely used commercially as impact absorbing materials, could not absorb much energy because of polymer's low yield strength. For this reason, metal forms such as aluminum forms have been developed, which are expected to be used as a shock absorber suitable for high-speed collisions.

The mechanical properties of aluminum foams have been studied experimentally. For instance, from our previous work of high-speed compressions of aluminum foams with a velocity of about 30 m/s, it is shown that plastic waves are generated clearly and the stress state observed at the impact side is quite different from that of the other side. However, the relationship among the nonuniform stress-strain state, the generation of plastic wave, and the deformation speed is still unclear.

In this study, we created FEM models of closed-cell foam structures and carried out compression test simulation with various velocities from 1 to 100 m/s, using LS-DYNA code [1].

2. Calculation detail

2.1. FEM modelling of closed cell foam structures

The three-dimensional Voronoi tessellation method [2] was employed to create closed-cell structures to simulate dynamic deformation behaviors of aluminum foams. First, about 200 nuclei were randomly generated in a $10 \times 10 \times 10\,\text{mm}^3$ cube using the random sequential addition (RSA) algorithm [3], which requests nuclei have a minimum distance and excludes exceedingly close ones. Then, Voronoi cells enclosed by perpendicular bisectors of nuclei were generated. Thus the three-dimensional Voronoi tessellation coupled with the RSA algorithm gave us uniform and randomly distributed polyhedral cells with the aspect ratio of approximately one. Next, the cells were slightly shrunken by moving all vertexes inward to their nuclear. The downsized cells and the space between them were considered as bubbles and cell-walls, respectively. We adjusted the degree of shrinkage of the cells to make the relative density of the foam structures 10%. These geometries of structures were produced using MATLAB and the FEM model was meshed by ANSYS workbench. The characteristic length of element size was set to be 0.5 mm.

Figure 1 shows a $10 \times 10 \times 10\,\text{mm}^3$ model made of approximately 200,000 first order tetrahedral solid elements, which was used for compression test simulations to investigate the deformation rate dependence of mechanical properties of foam structures. In addition, an extended $10 \times 10 \times 50\,\text{mm}^3$ model (approximately 1,000,000 elements) was also created in the same manner for the purpose of observing elastic and plastic wave propagation.

2.2. Dynamic simulations using LS-DYNA

The cell-wall material was assumed to be bilinear elastic-plastic with Young's modulus $E = 69\,\text{GPa}$, Tangent modulus $E' = 12.5\,\text{MPa}$, Poisson's ratio $\nu = 0.33$, Yielding stress $\sigma_y = 35\,\text{MPa}$, and mass density $\rho = 2700\,\text{kg/m}^3$, which are the values of mechanical properties of polycrystalline pure aluminum.

The models were arranged between two rigid walls as shown in Fig. 2. One of the walls was fixed (Fixed side),

[a] Corresponding author: `tanigaki@me.es.osaka-u.ac.jp`

Article available at http://www.epj-conferences.org or http://dx.doi.org/10.1051/epjconf/20159404047

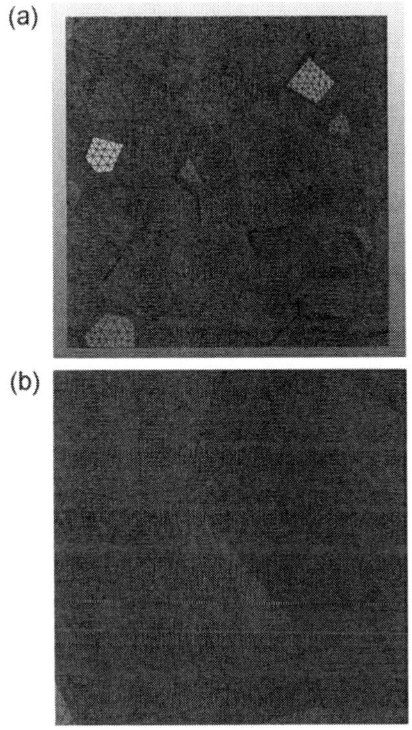

Figure 1. (a) 10 × 10 × 10 mm³ FEM model consists of first-order tetrahedral solid elements and (b) its close-up view.

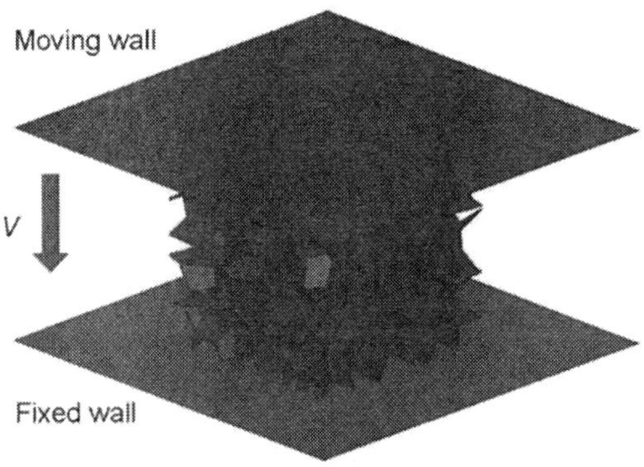

Figure 2. Schematic diagram of compression test simulation of the 10 × 10 × 10 mm³ FEM model with two rigid walls.

and the other moved to compress the model with a constant velocity (Impact side). The velocity $v = 1$–150 m/s were used for the dynamic compression test simulations. All calculations were performed using LS-DYNA code [1].

To obtain the stress strain relationship of the models, the nominal stresses of the models were calculated from the contact reaction forces of both rigid walls and the initial apparent cross sectional area of the models ($A = 100\,\text{mm}^2$). The nominal strains of the models were given from a velocity of the moving rigid wall v and a length of the models L.

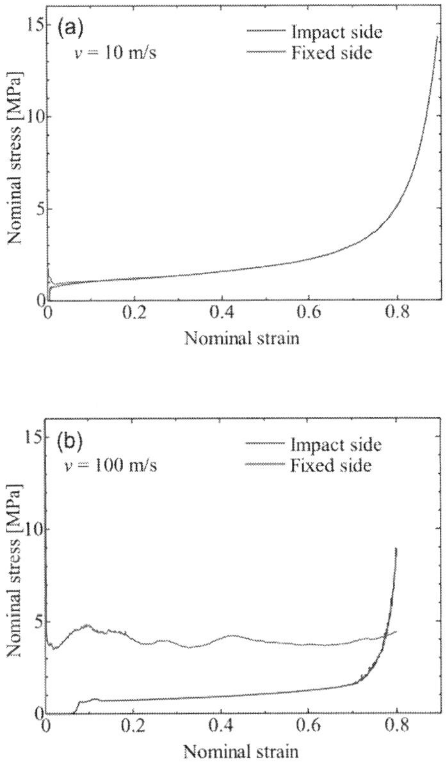

Figure 3. The stress-strain curves of the 10 × 10 × 10 mm³ FEM model obtained from impact side (red lines) and fixed side (blue lines). (a) $v = 10$ m/s. (b) $v = 100$ m/s.

3. Rate dependence of stress-strain curve and deformation mode

Figure 3 shows the stress-strain curves obtained from the impact side and the fixed side of the 10 × 10 × 10 mm³ model with the deformation rate of 10 m/s and 100 m/s. The stress state of the model can be considered uniform except the very early stage of the deformation when the deformation rate is $v = 10\,\text{m/s} = (\dot{\varepsilon} = 10^3\,\text{s}^{-1})$. On the other hand, with the deformation rate of 100 m/s ($\dot{\varepsilon} = 10^4\,\text{s}^{-1}$), the stress states on the impact and fixed sides of the model were quite deferent.

Deformation behaviors of the model with various rates ($v = 10$, 30, 50, 70, 90 and 100 m/s) are shown in Fig. 4. The deformation mode of the 10 × 10 × 10 mm³ model changed gradually as the deformation rate increases and a plastic wave was clearly observed with the rate of $v = 100$ m/s.

The barreling effect is generally not negligible in compressive deformations of dense materials with such dimensions. In contrast, the plastic strain of the model was distributed almost in one-dimension and the plastic wave had a plane wavefront because voids in the cells allow the apparent volume of foam structures to decrease significantly in the plateau region.

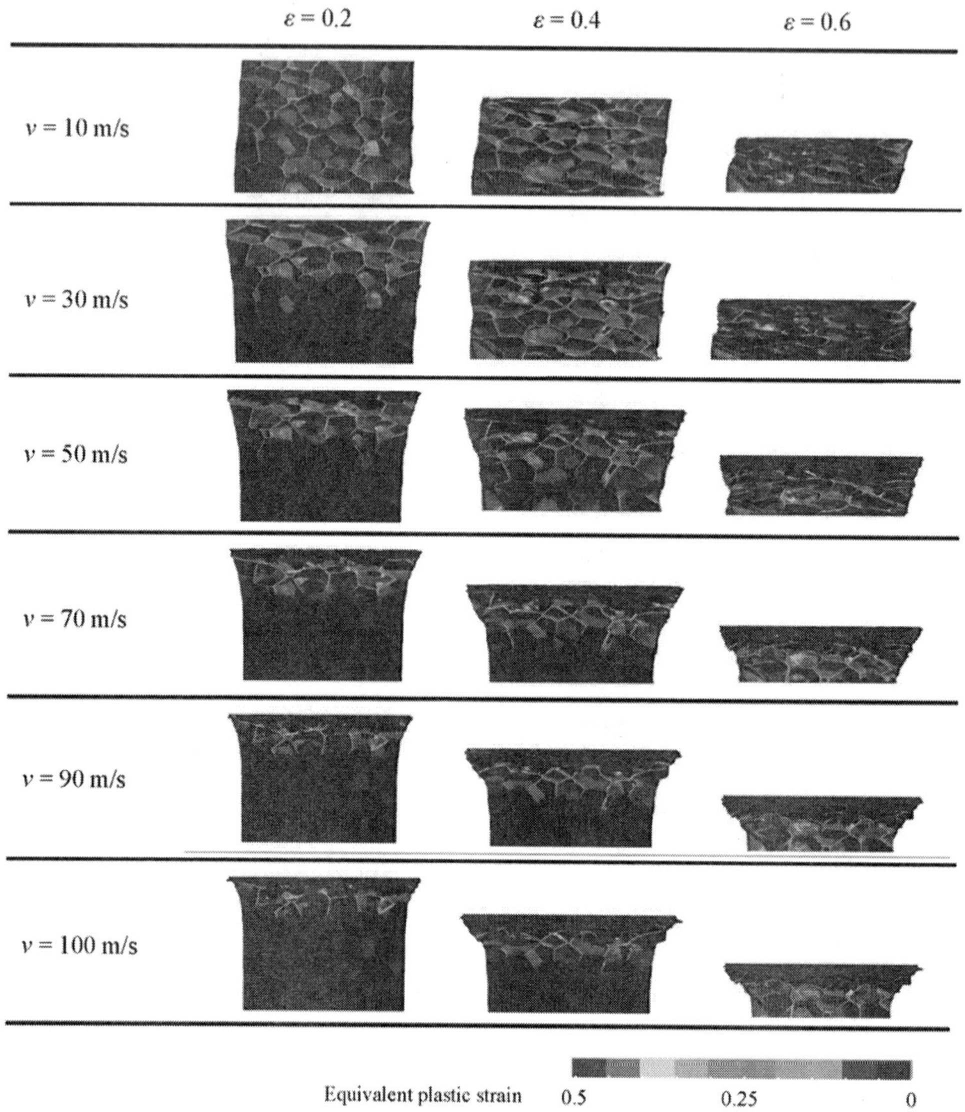

Figure 4. The deformation behaviour of the $10 \times 10 \times 10\,\text{mm}^3$ model for each deformation rate ($v = 10$–100 m/s).

4. Generation and propagation of plastic wave

Figure 4 shows the equivalent plastic strain distribution of each element. To compare and discuss the behaviour of the plastic wave and the macroscopic stress-train relationship (Fig. 3), a macroscopic one-dimensional strain distribution was calculated as follows. First, the model was sliced into slabs with a thickness of 0.5 mm. The z-coordinate of the slabs were determined by averaging those of all elements included in each slab.

Then, the generation and the propagation of macroscopic strain waves in the $10 \times 10 \times 10\,\text{mm}^3$ model were obtained by a time development of intervals of the slabs as shown in Fig. 5. A rapid increase of the strain caused by impact collision ($v > 30$ m/s) was observed on the impact side. When the deformation rates were relatively slow, the increase becomes uniform because of the effect of the reflected wave from the fixed side. By contrast, with higher rates, it maintained its shape and propagated toward the fixed side. However, a small influence of the reflected wave was still observed even with the maximum rate of $v = 100$ m/s since the length of the $10 \times 10 \times 10\,\text{mm}^3$ model was too short.

Here, The extended $10 \times 10 \times 50\,\text{mm}^3$ model was employed to avoid the effect of the reflected wave. Figure 6 shows the plastic wave propagation behaviour of the $10 \times 10 \times 50\,\text{mm}^3$ model with the deformation rate of $v = 100$ m/s. The constant strain region of $\varepsilon = 0.8$ following the wave front was clearly observed, and hence, the model can be assumed to be separated into two regions, the plastic-deformed and the non-deformed regions.

The strain ε of the plastic deformed region is given by

$$\varepsilon = v/C_p, \tag{1}$$

where C_p denotes the velocity of the plastic wave.

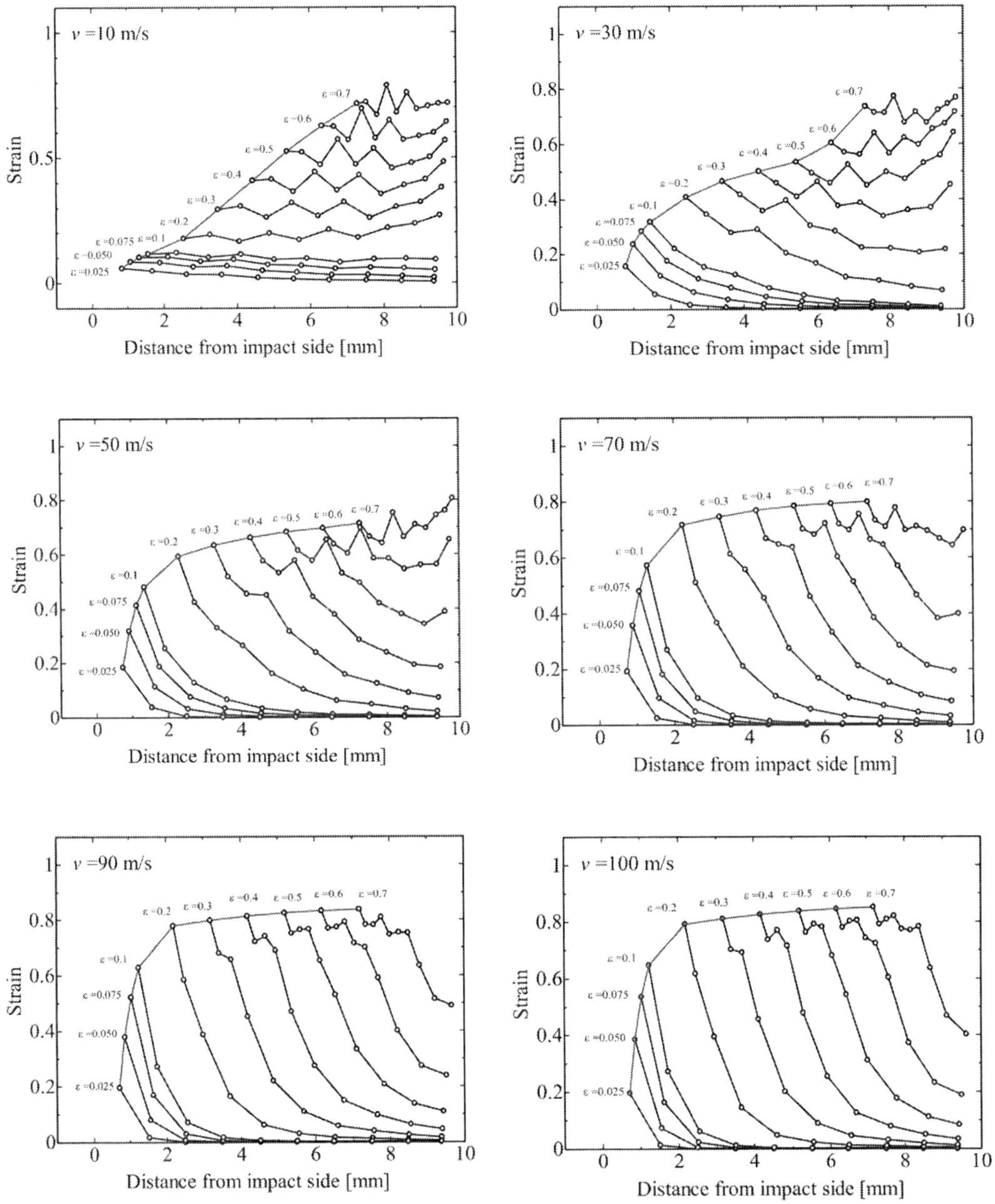

Figure 5. The deformation rate dependence (v = 10–100 m/s) of the macroscopic one-dimensional strain distribution in the 10 × 10 × 10 mm^3 model. The red lines indicate the strain observed on the impact side.

The change in momentum mv and the impulse $F\Delta T$ of the plastic-deformed region can be written as

$$mv = (C_p \Delta T A \rho^*)v, \quad (2)$$

$$F\Delta T = \sigma A \Delta T, \quad (3)$$

where A and ρ^* are the apparent cross sectional area and the apparent mass density of the model, respectively. Combining Eqs. (1), (2) and (3) leads to the stress on the impact side

$$\sigma = \rho^* v^2 / \varepsilon. \quad (4)$$

Using Eq. (4) and the static stress-strain curve of the model [Fig. 3(a)], the stress on the impact side σ was calculated to be approximately 3.8 MPa, which was in good agreement with the value obtained from the raction forces on the fixed side with the deformation rate of v = 100 m/s [Fig. 3(b)].

5. Propagation velocity of elastic wave

A longitudinal elastic wave velocity propagating in foam structures is given by

$$C_e = \sqrt{E^*/\rho^*}, \quad (5)$$

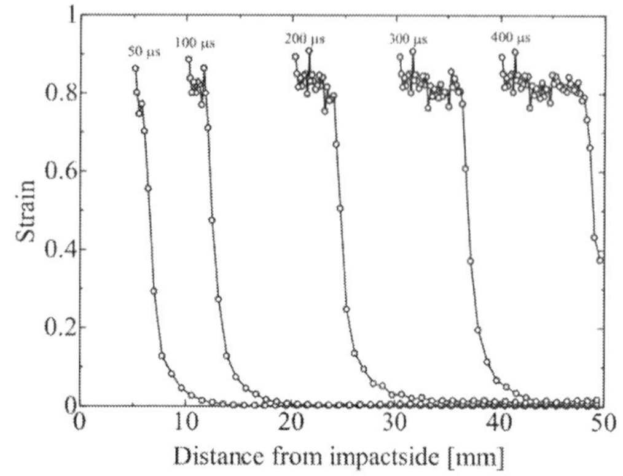

Figure 6. The macroscopic plastic wave propagation behaviour in the $10 \times 10 \times 50\,\text{mm}^3$ model with the deformation rate of $v = 100\,\text{m/s}$.

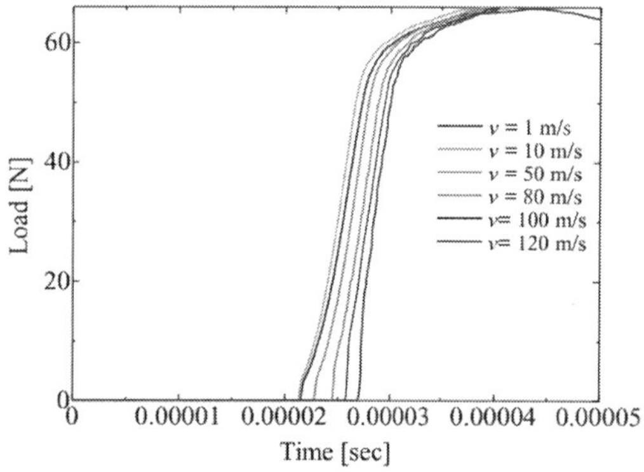

Figure 7. The rising times of the reaction forces observed on the fixed rigid wall using the $10 \times 10 \times 50\,\text{mm}^3$ extend model.

where E^* denotes the apparent Young's modulus of foam structures. To determine the apparent Young's modulus accurately, an additional quasi-static compression test simulation of the $10 \times 10 \times 10\,\text{mm}^3$ model with a velocity of $v = 0.01\,\text{m/s}$ ($\dot{\varepsilon} = 10^0\,\text{s}^{-1}$) was performed. The result showed an entirely uniform deformation and the apparent Young's modulus of $E^* = 1.07\,\text{GPa}$ was obtained. Using this value and the apparent mass density $\rho^* = 270\,\text{kg/m}^3$ (the relative density $\rho^*/\rho = 10\%$), the longitudinal elastic wave velocity was calculated to be $C_e = 1987\,\text{m/s}$.

Figure 7 shows the rising times of the reaction forces observed on the fixed rigid wall using the $10 \times 10 \times 50\,\text{mm}^3$ extended model. Elastic wave velocities were calculated from them for each deformation rate.

The resultant elastic wave velocities of the model were about 2000 m/s. In particular, with the deformation rate of less than 10 m/s, they were in good agreement with the value estimated theoretically. However, the longitudinal elastic wave velocity showed a weak negative dependency on the deformation rate as shown in Fig. 8.

In this study, the strain rate dependence of material properties was not considered. Also, the effect of air compression in the cells was ignored. Therefore, the apparent Young's modulus and the elastic wave velocity should be independent of the deformation rate.

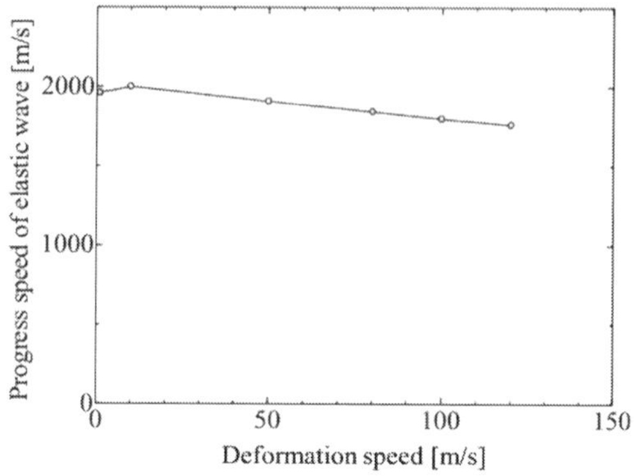

Figure 8. The deformation rate dependence of the longitudinal elastic wave velocity of the $10 \times 10 \times 50\,\text{mm}^3$ extend model.

A frequency dispersion due to their discontinuity in structure is a possible cause of that. Because of a long wavelength of low frequency components, the models behave as a homogeneous medium. However, high frequency components of the stress wave caused by high speed collisions, which have short wavelengths, propagate in the cell walls and their phase velocity is assumed to decrease.

6. Conclusion

In this study, the $10 \times 10 \times 10\,\text{mm}^3$ and the $10 \times 10 \times 50\,\text{mm}^3$ finite element models of closed-cell foam structures were created using the three-dimensional Voronoi tessellation method coupled with the random sequential addition algorithm. The dynamic compressive deformation behaviors of the models ($v = 10$–$100\,\text{m/s}$) of the models were numerically studied using LS-DYNA code.

The plastic deformation mode of the models changed gradually as the deformation rate increases. The generation and the propagation of the plastic wave was clearly observed with the rate of $v = 100\,\text{m/s}$. Also, the longitudinal elastic wave velocity showed a weak negative dependency on the deformation rate although the strain rate dependence of material properties was not considered.

Furthermore, a prediction method for the dynamic stress state on the impact side based on the static stress-strain relationship was presented.

References

[1] Livermore Software Technology Corporation, `http://www.lstc.com/products/ls-dyna`, (1987).
[2] A. Okabe, B. Boots, K. Sugihara, and S. N. Chiu. *Spatial tessellations: Concepts and applications of Voronoi diagrams*, Wiley, second edition (2000).
[3] G. Tarjus, P. Schaaf, and J. Talbot, J. Stat. Phys., **63**, 1-2, 167 (1990).

EPJ Web of Conferences **94**, 04048 (2015)
DOI: 10.1051/epjconf/20159404048

Transitioning a unidirectional composite computer model from mesoscale to continuum

Sidney Chocron[1,a], Ramón Zaera[2], James Walker[1], Alon Brill[3], Roman Kositski[3], Doron Havazelet[4], Ulrich Heisserer[5], and Harm van der Werff[5]

[1] Southwest Research Institute, San Antonio, TX, USA
[2] Universidad Carlos III de Madrid, Madrid, Spain
[3] Rafael, Haifa, Israel
[4] Ministry of Defense, Government of Israel
[5] DSM Dyneema, The Netherlands

Abstract. Ballistic impact on composites has been a challenging problem as seen in the abundant literature about the subject. Continuum models usually cannot properly predict deflection history on the back of the target while at the same time giving reasonable ballistic limits. According to the authors the main reason is that, while continuum models are very good at reproducing the elastic characteristics of the laminate, the models do not capture the behaviour of the "failed" material. A "failed" composite can still be very effective in stopping a projectile, because it can behave very similar to a dry woven fabric. The failure aspect is much easier to capture realistically with a mesoscale model. These models explicitly contain yarns and matrix allowing the matrix to fail while the yarns stay intact and continue to offer resistance to the projectile. This paper summarizes the work performed by the authors on the computationally expensive mesoscale models and, using them as benchmark computations, describes the first steps towards obtaining more computationally effective models that still keep the right physics of the impact.

1. Introduction

Composite materials have become essential in applications like aerospace, defense, and biomedical devices. Because of their intrinsic inhomogeneities and the fact that their properties depend on the direction in the material, it has been a challenge to understand how these materials deform under applied forces, in particular dynamic forces. The laminate theory is able to predict the properties of a laminate from the properties and orientation of the single laminae that form the composite. This theory has been very successful in predicting the elastic deformation of composite materials where no slip occurs between fibers. But once the laminate starts to slip, yield or fail the theory is not applicable and continuum computational models are not accurate enough for many applications like impact.

Southwest Research Institute® (SwRI) has been working for a few years on the behavior of these materials under ballistic impact. The studies performed seem to show that, because of the lack of good strength and failure models, it is (so far) not possible to generate a reasonably accurate first principles computational continuum model of the impact on a composite. By "reasonably accurate" we mean a model capable of predicting qualitatively delamination and failure patterns and quantitavely determine deflection and ballistic limit of various laminates impacted by different threats.

A pretty unique mesoscale yarn-level model was developed by SwRI during the last few years using a commercial code (LS-DYNA), see [1]. The model was carefully built starting from impact on a single yarn and checking with experiments and theory that wave propagation and failure were being properly predicted by the computations. From these validated yarns, fabric layers were meshed to build a fabric system. The prediction of deflections vs. time and ballistic limits was shown to be very good for materials like Kevlar®, Dyneema®, and PBO®, although the size of panels was limited to squares 8 to 12 inches and a maximum number of layers around 30. These computations take 2 to 3 days in 24 cores when using SwRI's computer cluster.

The logical next step was to add matrix to the yarn- level fabric computation to obtain the mesoscale model of a composite. This was done both for woven composites (Kevlar/Epoxy [2], S-2 glass/Epoxy) and for unidirectional systems (Polyethylene/Polyurethane) [3]. Again the models were able to predict qualitatively deformation and failure patterns, and quantitatively deflection and ballistic limits. Computation times with composites were even longer than with fabrics sometimes taking up to a week of computation on 16 to 24 cores in the cluster.

The focus of the last part of the project was to explore a continuum approach to the problem as well as different ways to speed-up the computations, for example using shell elements, while trying to keep the main physical phenomena in the computation. The mesoscale simulations were used as a benchmark or reference because it is thought that they contain the fundamental physics of the problems. The work is still in progress so this paper should

[a] Corresponding author: schocron@swri.edu

be considered only an initial exploration of the problem and, by no means, the final answer on how an UHMWPE composite should be simulated.

2. Materials

All the materials tested in this project were received from DSM-Dyneema through a joint-research project. All the materials (yarn, single layer composite and multi-layer composites) are based on the same fiber, SK76, for consistency.

2.1. Filament and yarns

Although no testing or impact simulation is performed at the fiber level, the properties of the fibers are what makes UHMWPE an outstanding material. The following properties were provided by Van der Werff (personal communication): modulus of 1350 cN/Tex (132 GPa), strength of 38.4 cN/dtex (3.8 GPa), linear density of 2 dtex (0.0002 g/m), strain to failure of 3.5%, diameter 16.1 μm, density of 980 kg/m^3, which provide a sound speed of 11.6 km/s. These come from single filament tests at a speed of 25 mm/min in specimen with a 50 mm gage length and agree with properties found elsewhere in the literature.

2.2. Single layers

Single layers of HB80 were also provided by DSM Dyneema with an approximate size of 20 in $\times$ 20 in. These are SK76 yarns in a polyurethane matrix. The single layers consist of 4 "plies" with $[0/90]_2$ fiber orientation. Each "ply" actually has approximately 2.5 filaments through the thickness. The HB80 single layer (i.e. four plies) has an areal density of 0.145 kg/m^2.

The matrix material is polyurethane, which represents a 17% matrix content by weight for the composite. The reader is encouraged to refer to [3] for a thorough description of the architecture.

2.3. HB80 laminates

Laminates were manufactured by DSM Dyneema in Greenville, North Carolina, USA from the single layers described in the previous section. Nominal thicknesses were 6.1 mm, 11.7 mm, and 23 mm, with, respectively, 37, 74, and 149 layers. Material tests were performed on specimens cut from the 11.7 mm panels. Panels labeled as A were hot-pressed at 20.7 MPa while panels labeled as B were hot-pressed at 27.6 MPa.

3. Material tests performed

The material tests performed focused on delamination properties of the laminate because the tensile properties of the composite are well characterized from the work by Heisserer [4]. The material tests consisted of: lap shear, torsion, and in-plane and out-of-plane compression tests. The lap-shear and compression tests were performed at atmospheric and high (500 MPa) confinement pressures and were presented in [3]. Characterization of the material at high pressures is thought to be essential as the projectile

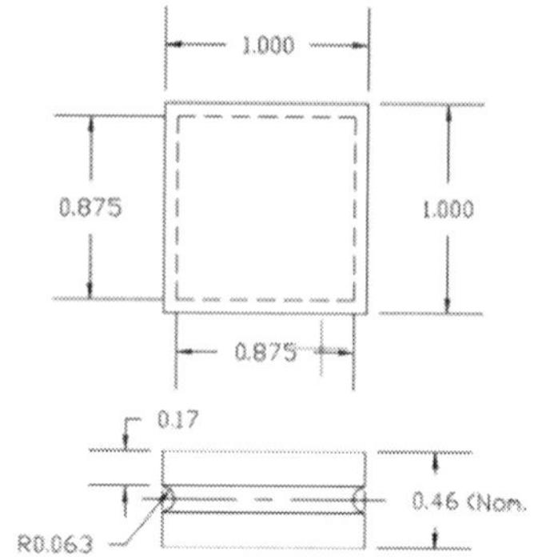

Figure 1. Drawing of the torsion specimens (dimensions in inches).

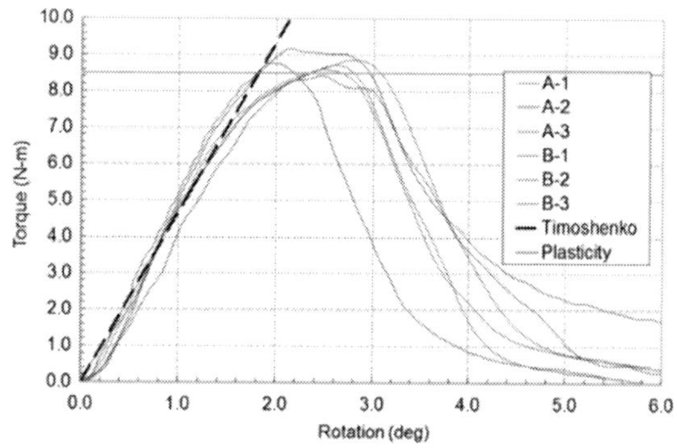

Figure 2. Torque curves for the torsion specimens.

penetrating the laminate is exerting high pressure through the thickness. Only the torsion results are reported in this paper. These provide a shear strength than can later be used in the numerical models.

Figure 1 shows the torsion specimens. A torque is applied on the top and bottom of the specimen which fails in shear in the center of the notch. The torque applied is in the z-direction, i.e. perpendicular to the plane of the laminate. These tests should provide similar information to the lap-shear tests and can be used as an independent validation. The angular velocity of the piston during the test was 4.25×10^{-3} rad/s.

The torsion tests show that curing pressure does not affect the delamination stress. Figure 2 shows the torque curves obtained for A and B specimens as a function of the rotation of the head. The rotation was corrected for compliance by performing a test on a steel specimen. This allows estimating the shear modulus and strength in shear.

From elasticity theory (see, for example, Timoshenko [5]), it is possible to estimate the slope expected in Fig. 2. The torque T, shear modulus G, angle turned per meter θ, and side length $2a$ are related by:

$$T = 0.1406\, G\theta\, (2a)^4. \qquad (1)$$

The elastic modulus that provided the match shown in Fig. 2 (dashed black line) was 70 MPa. Poisson's ratio is unknown so a value of 0.4 was assumed as typical for a polymer. The shear modulus is related to the elastic

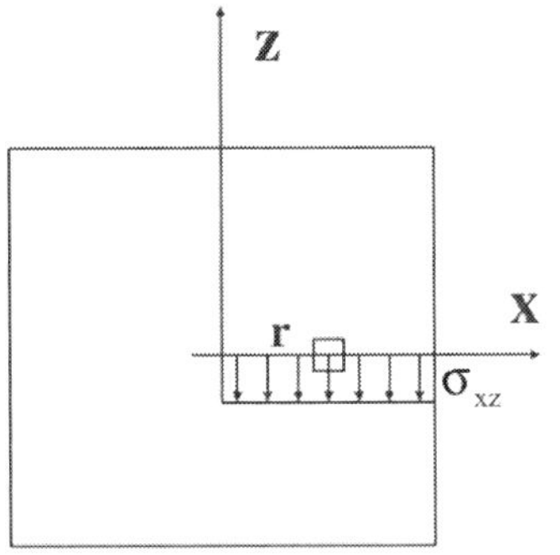

Figure 3. Stresses on gage section of the torsion specimen.

modulus E and Poisson ratio ν by:

$$G = E/(2(1+\nu)) = 25\,\text{MPa}. \quad (2)$$

The gage length of the gage section was $l = 3.2$ mm (twice the radius of the gage section 2×0.063 in $= 0.126$ in $=$ 3.2 mm) and was used to calculate the angle turned per unit of length. Note that "rotation" in Fig. 2 is an absolute rotation while θ in Eq. (1) is rotation per meter; i.e., if the absolute rotation is ϕ then $\theta = \phi/l$. The gage section is not uniform so l is only an approximate number. With the semi-width of the specimen being $a = 11.1$ mm, the theoretical slope of the torque versus rotation line is 266.8 Nm/rad (shown as "Timoshenko" line in Fig. 2).

The maximum shear stress in the cross section is given by (see again Timoshenko)

$$\tau_{max} = 1.351\ G\theta_{max}a. \quad (3)$$

Where θ_{max} is the rotation angle per unit of length when the maximum torque happens, i.e., according to Fig. 2 when the rotation of the piston is 2.5 degs or 3.5e-2 rad. So $\tau_{\max} = 4.1$ MPa on the cross section when failure happens.

An alternative way to estimate the torque is by using plasticity theory. Assuming the whole gage section is flowing plastically when yielding during the test it is straightforward to estimate the torque analytically. In view of Fig. 2 this assumption might not be actually realistic since the behavior of the specimen is more brittle than ductile, i.e. the torque reaches a maximum and then fails suddenly instead of staying at a plateau for a long time. Nevertheless it is thought it is an interesting exercise to find the estimated strength from plasticity theory.

Figure 3 shows the stresses on the gage section of the specimen for pure torsion. Only shear stresses are applied during this test and the figure shows, specifically, the stress on the x-axis. Note that for this derivation x and z axes are in-plane while y-axis is out-of-plane, i.e. through the thickness of the laminate. Since the assumption is that the whole section is yielding then the equivalent stress at any point in the section is equal to the yield stress Y (for a Von Mises flow surface):

$$\sigma_{eq} = \sqrt{3(\sigma_{xz}^2 + \sigma_{yz}^2)} = Y \rightarrow \sigma_{xz} = \frac{Y}{\sqrt{3}}. \quad (4)$$

Since there is 1/8 symmetry on the cross section we can calculate the torque needed on 1/8 of the section and multiply by 8:

$$T = 8\int_{\theta=0}^{\theta=\pi/4}\int_{r=0}^{r=a/\cos\theta} r\frac{Y}{\sqrt{3}}r\,dr\,d\theta$$
$$= \frac{8Y}{3\sqrt{3}}\int_0^{\pi/4}\frac{a^3}{\cos^3\theta}d\theta = 1.77\,Ya^3. \quad (5)$$

So, for a measured torque of approximately 8.5 Nm we have $Y = T/(1.77a^3) = 3.5$ MPa. This implies $\sigma_{xz} =$ 2.0 MPa.

The shear strength derived from elastic theory and plasticity are quite different (4.1 MPa vs. 2.0 MPa). The fact that lap-shear test also provides a shear strength of approximately 2.0 MPa might just be a coincidence since the authors think that it can be hardly justified that the whole section is yielding before failure. It is more likely that the torsion test is providing a larger shear strength because of some unintended out-of-plane compression (possibly warping) force or other artifact of the test. This could be better understood by performing detailed numerical simulations.

4. Summary of mesoscale model and results

The mesoscale model was carefully validated in a way similar to the one described in [1] for woven fabrics. The main point of the validation is making sure that both wave propagation in the material and failure of the fibers are properly captured.

4.1. Numerical model of the strip

The first step to create the UD model was to reproduce the ballistic impact results presented in Chocron et al. [6] for single-layer/single-ply and for single-layer/three-plies (90/0/90, where the 0 direction is on the length direction of the strip.)

The fibers in the real material have a diameter of 17 μm and, for numerical purposes, were bundled into strips 2-mm wide. This would correspond to approximately bundling 300 fibers in a single solid element.

The bundle is only one solid element through the thickness so it does not have any bending stiffness, as expected from a bundle of fibers. The thickness of this element is the same than the thickness of the single Dyneema HB80 UD layers: 38.59 μm. The element is orthotropic with the stiff direction in the same direction than the UD fibers and the elastic and failure properties used are the ones found in the literature, as reported in the materials section.

The polyurethane matrix material was placed side by side with the fibers "bundle" to account for the "dead weight" and cohesion between the fibers. This is only a rough approximation to the real material where the matrix is homogenously distributed among the fibers. The resin was assumed to be a homogenous isotropic material with a density of 0.97 g/cm^3 and elastic perfectly plastic (Young's modulus 70 MPa, Poisson's ratio 0.45, strength

3.5 MPa, and failure at an effective plastic strain of 10%). Transverse impact simulations on the strip were performed at different velocities to check the strain, the transverse wave velocity, and the critical velocity (impact velocity that fails immediately the fibers). These were compared to the predictions from theory and the tests performed in [6]. The results were very satisfactory as shown in [3].

4.2. Numerical model of the single layer

The same strips were used to form a single layer (4 plies, i.e. $[0/90]_2$) of size 30 cm × 30 cm. This single layer model has ~260,000 nodes (131,000 solid elements) with the same thickness than the real material (4 × 38.59 μm). The areal density of the numerical layer obtained was 150 g/m^2, very close to the 145 g/ m^2 of the real material.

The layer to layer interaction, i.e. matrix between layers, was described by a very simplistic "tiebreak" contact model available in the LS-DYNA code. This model destroys the bond between the layers when a combination of normal and shear stresses reaches a threshold. The normal stress to failure selected was 1 GPa and the shear stress 10 MPa. These numbers on their own might seem completely unrealistic but, the shear and delamination tests under high pressure confinement provide similar numbers [3].

The numbers used in the numerical simulation are, of course, a very rough approximation of the complex process that is happening during impact. Right below the projectile the pressures are enormous, certainly higher than 500 MPa, so the strength of the material is probably very large. Far from the projectile the pressures seen by the polyethylene are much smaller, i.e. the strength of the matrix is probably the one measured in atmospheric conditions. Additionally, the out-of-plane strength in tension and compression are probably very different. Unfortunately the code used does not allow right now such a complex behavior for contact between layers so a crude average strength needs to be used for the simulations.

Transverse wave speeds measured in [6] agree reasonably well with the ones measured in the numerical simulations for single layers.

4.3. Numerical model of the multiple-layer targets

Targets for numerical simulations were produced using the same mesh geometry presented in the single-layer section. In real targets the number of layers is very large, for example, see the materials section, a 6-mm-thick target has 37 layers or 37 × 4 = 148 plies.

Depending on the thickness of the laminate, the number of plies and, consequently, the computation time, can be enormous. The plies were consolidated by a factor of four, i.e. the consolidated ply was four times thicker than the original one. The effect on the computation was a significant speed-up while the results were not greatly affected [3].

Figure 4 shows an example of the impact of a fragment on a laminate. The mesoscale model allows capturing realistically the following features: 1) the fibers are being pulled towards the impact point, 2) the fibers from the first layers impacted are pulled less than the ones in the last layers because they fail faster, 3) this creates a characteristic delamination pattern on the edges that is similar to the one observed in the real targets.

Figure 4. Example of a laminate impacted by a fragment. Note the delamination on the sides of the target.

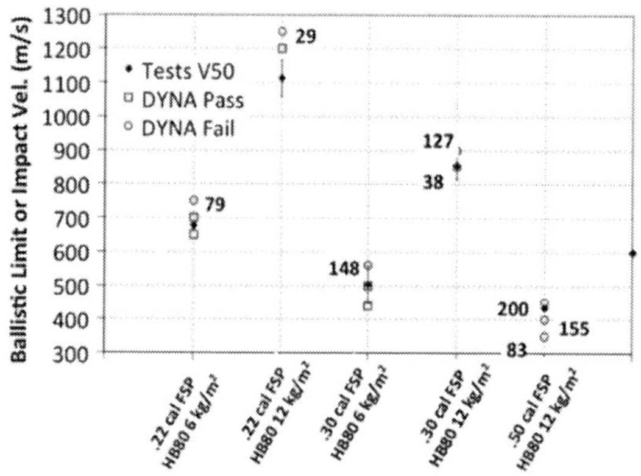

Figure 5. Comparison of the experimental ballistic limits (black diamond) with the numerical results (open squares for stops and open circles for target failure).

Unfortunately, when comparing the transverse wave (i.e. the pyramid) from the tests and the simulations for thick laminates, it was found that the numerical model is "stiffer", i.e. the transverse wave in the full laminate does not propagate as fast as in the test. This is probably due to the very simplistic layer-to-layer contact condition.

Nevertheless the model was able to predict well the ballistic limits. Figure 5 compares the experimental and numerical ballistic limits for two different targets (Dyneema HB80, 6 and 12 kg/m^2, respectively, 6.1 and 11.7 mm thick) and three fragment sizes: 0.22 inch, 0.30 inch, and 0.5 inch caliber. The black diamond shows the experimental ballistic limit (usually a 4-shot V50) with error bars. The open squares represent computations with LS-DYNA where the projectile stopped. The open circles are also computations where the projectile went through the target. The number next to the open circles is the residual velocity of the projectile after going through the target.

The figure shows that the numerical prediction for the 6 kg/m^2 targets is very good. The model slightly overpredicted for the .22 cal FSP vs. the 12 kg/m^2. For the .30 and .50 cal FSPs against the 12 kg/m^2 the projectile did not stop in the computations but the residual velocities, as shown in the figure, were small enough.

5. Effort with membrane elements

During the second part of this effort, the objective was to obtain results similar to the mesoscale model with considerably less computational effort. Also, a simple material description like the one used in the mesoscale model is desirable. The key of this new model has been using two-dimensional plane stress finite elements (membrane), instead of the 3D brick elements, to represent the laminate plies. Compared to the brick elements, the potential advantages of using two-dimensional elements are: i) reduced number of nodes, ii) larger characteristic dimension, resulting in a larger critical (stable) time increment, iii) only two contact surfaces instead of six in the brick element, decreasing the burden of tracking the contact surfaces.

The commercial finite element code ABAQUS/Explicit was selected for this approach. The 4-node membrane element with reduced integration (M3D4R in ABAQUS notation) was found to be stable even at large displacements. The orthotropic elastic model with failure condition based on maximum stress was implemented in the user subroutine VFABRIC, specific for plane stress elements, thus replicating the constitutive behavior assumed in the mesoscale model.

The matrix adhesion and ulterior delamination between each pair of plies i and $i+1$ was reproduced through a cohesive interaction. The nominal traction stress vector across the contact consists of three components that represent the normal t_n, and the two shear tractions t_s and t_t. The corresponding separations are denoted by δ_n, δ_s and δ_t. The tensile behavior, valid for $t_n > 0$, is driven by a linear law relating each stress with the corresponding separation; according to Utomo [7], the same values for the three stiffnesses have been adopted $t_j = K \cdot \delta_j$, all of them equal to the unconfined compressive out-of plane stiffness of the laminate $K = E_3 = 4.3$ GPa. Failure of the layer-to-layer interaction consists of two parts: a damage initiation criterion and a damage evolution law. For initiation, a quadratic stress criterion has been considered. On the other side, the damage evolution law describes the rate at which the cohesive stiffness is degraded once the corresponding initiation criterion is reached. In our case, a brittle failure was considered by setting to zero the energy dissipated as a result of the damage process.

Regarding the compressive behavior, ABAQUS/Explicit offers the possibility of using a pressure-overclosure relation through a user-defined curve, which was formerly identified as a key feature in order to avoid a large out-of-plane compressibility of the laminate due to low penalty stiffness in the contact between plies. This curve has been defined considering a state of uniaxial-strain deformation, which is deemed to be representative of the strong confinement developed in the region directly affected by impact, and elastic behavior. A non-linear (logarithmic) measure of the deformation has been adopted. Leading to an asymptotic increase of pressure as the overclosure between plies approaches the ply thickness. Additionally, a Coulomb frictional behavior was considered, with a low value of the friction coefficient $\mu = 0.02$, according to Chocron et al. [2]. The compressive contact condition is stated for the interaction between each ply i and all the subsequent $i+1$, $i+2, \ldots, N$; N being the total number of plies in the laminate.

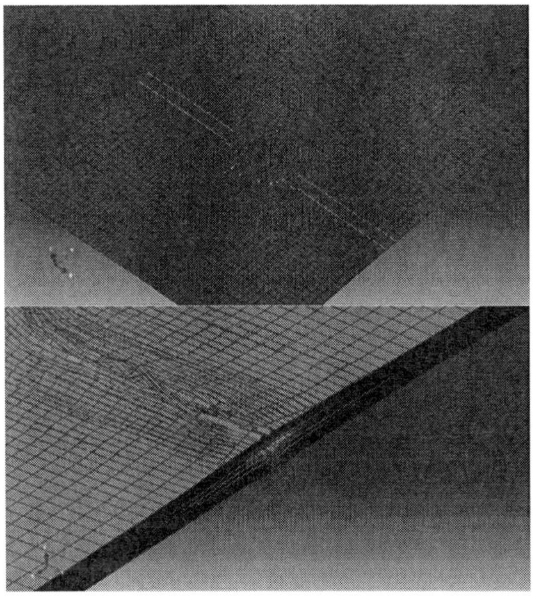

Figure 6. Normal impact of a .30 cal FSP against 148-plies laminate at 400 m/s. Top: interply damage contours at 32 μs. Bottom: detail at the mid-point of one of the edges of the laminate at 70 μs.

The cohesive forces drop to zero once the failure (delamination) criterion is satisfied. If the surfaces are in contact again after failure of the cohesive bond, penetrations will give rise to compressive contact stresses following the above mentioned pressure-overclosure law, but not further cohesive constraints are enforced.

Figure 6 (top) shows the damage produced by the impact of a .30 cal FSP at 400 m/s on the laminate. Damage in this context means that the matrix between the layers has failed, i.e. the red indicates zones that are delaminated. As expected the principal yarns (the ones directly under the projectile) are the ones that delaminate first during the impact. Additionally, the model predicts an extensive delamination of the upper half of the target, an effect that is also observed experimentally. This constitutes an evidence of the suitability of the approach to capture the interaction between the different plies in the laminate upon impact.

Beyond the previous qualitative descriptions of the performance of the model, a comparison of the time histories of the FSP velocity obtained with the mesoscale model and with the membrane element model is presented in Fig. 7. As can be seen, the prediction of both models is very similar. The mesoscale model consolidates 4 plies with the same fiber orientation into a single layer to reduce the computational effort. On the other side, the membrane model considers one layer of membrane elements for each

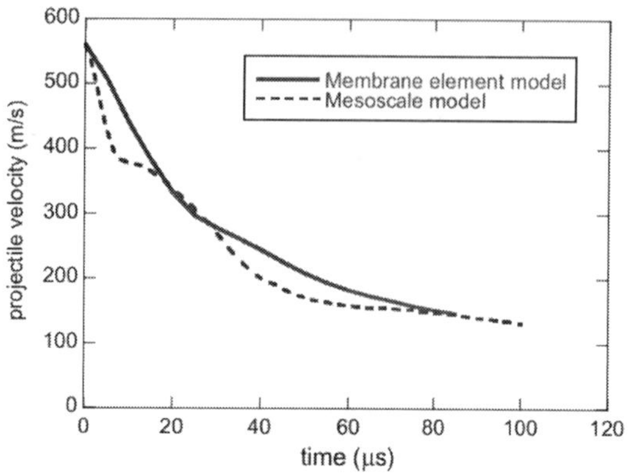

Figure 7. Comparison of the projectile velocity history for the membrane model and the mesoscale model. Normal impact of a .30 cal FSP against 148-plies laminate at 599 m/s.

layer. The membrane model takes approximately ten times less computation time than the mesoscale models.

6. Brief comment on the continuum description

A limited part of the effort was dedicated to computations with the whole panel being an orthotropic continuum. Interestingly, by choosing the in-plane elastic moduli the same than the modulus of the fiber, it was possible to reproduce the strain expected from the theory for a single layer. Also, it was possible to qualitatively reproduce some of the features like fiber pulling in the cross fibers under the projectile impact for thicker laminates.

By tuning the shear moduli for the 6.1-mm-thick target, it was possible to find moduli that matched the history of the velocity of the projectile for different impact velocities, in a similar fashion to what is shown in Fig. 7. The velocity of the projectile going through the laminate is extremely sensitive to the shear moduli, i.e., a small change in the moduli produced a large effect in the velocity history.

Unfortunately the moduli values that worked for the thin 6.1-mm laminate did not work for thicker laminates. The authors believe that the reason is that damage happens very fast to the first few layers impacted decreasing the shear modulus dramatically, probably close to zero. But the layers that are not damaged still provide significant stiffness to the laminate. This stiffness degradation is still not being captured by the models implemented and is the subject of future work on impact in composites.

7. Conclusions

A mesoscale computational model was described for unidirectional composites. The model is very successful at reproducing delamination and yarn pull as observed in the experiments. It also predicts properly ballistic limits for the limited set of tests performed. But mesoscale models take 1 to 7 days of computation time, depending on the projectile, and the size and thickness of the target.

To decrease the computation time a numerical model based on a 2D membrane element has been developed. The model has been validated in a very limited way by comparing it to the mesoscale model. It looks promising but is still work in progress. A continuum model has also been explored. From this preliminary exploration it looks like the model will need to degrade the shear elastic moduli to produce realistic results.

References

[1] S. Chocron, E. Figueroa, N. King, T. Kirchdoerfer, A. Nicholls, E. Sagebiel, C. Weiss, and C. Freitas, "Modeling and validation of full fabric targets under ballistic impact," *Composites Science and Technology*, vol. **70**, no. 13, pp. 2012–2022, 2010.

[2] S. Chocron, A. Carpenter, Z. Asaf, and L. Levin, "Impact simulations of kevlar® composite at the yarn level," in *Proceedings of the 28th International Symposium on Ballistics, Atlanta, Georgia, USA, 22–26 September*, 2014.

[3] S. Chocron, A.E. Nicholls, A. Brill, A. Malka, T. Namir, D. Havazelet, H. van der Werff, U. Heisserer, and J.D. Walker, "Modeling unidirectional composites by bundling fibers into strips with experimental determination of shear and compression properties at high pressures," *Composites Science and Technology*, vol. **101**, no. 0, pp. 32–40, 2014. Online June 25, `http://dx.doi.org/10.1016/j.compscitech.2014.06.016`.

[4] U. Heisserer, "Tensile properties of Dyneema HB plies (HB2, HB26, HB50, HB80)," tech. rep., DSM-Dyneema, 2011.

[5] S. Timoshenko and J.N. Goodier, *Theory of Elasticity*. New York: Mc.Graw-Hill, 1951.

[6] S. Chocron, N. King, R. Bigger, J. Walker, U. Heisserer, and H. van der Werff, "Impacts and waves in Dyneema® HB80 strips and laminates," *J. Appl. Mech.*, vol. **80**, no. 3, pp. 031806–10, 2013.

[7] B.D. Heru-Utomo, *High-speed Impact Modeling and Testing of Dyneema Composite*. PhD thesis, Technical University Delft, The Netherlands, 2011.

EPJ Web of Conferences **94**, 04049 (2015)
DOI: 10.1051/epjconf/20159404049

Numerical simulation of the effect of regular and sub-caliber projectiles on military bunkers

Pavel Jiricek[a] and Marek Foglar

CTU in Prague, Fac. Of Civil Engineering, Thákurova 7, 166 29 Praha 6, Czech Republic

Abstract. One of the most demanding topics in blast and impact engineering is the modelling of projectile impact. To introduce this topic, a set of numerical simulations was undertaken. The simulations study the impact of regular and sub-calibre projectile on Czech pre-WW2 military bunkers. The penetrations of the military objects are well documented and can be used for comparison. The numerical model composes of a part from a wall of a military object. The concrete block is subjected to an impact of a regular and sub-calibre projectile. The model is divided into layers to simplify the evaluation of the results. The simulations are processed within ANSYS AUTODYN software. A nonlinear material model of with damage and incorporated strain-rate effect was used. The results of the numerical simulations are evaluated in means of the damage of the concrete block. Progress of the damage is described versus time. The numerical simulation provides good agreement with the documented penetrations.

1. Introduction

In the end of 1930s, Röchling company used an earlier patent of Škoda Plzeň to design projectile with diameter smaller than diameter of the barrel. This sub-calibre projectile was lead within a barrel by guiding elements, which detached after the projectile left the cannon. To ensure, that a "thinner" projectile could carry the same amount of the explosives as a regular caliber projectile, the sub-calibre projectile has to be extended in length. The sub-calibre projectiles were 1.7–3.6 m long arrow shaped projectiles.

Sub-calibre projectiles made by Röchling were tested on Czech pre-WWII military bunkers (earlier occupied by Nazi-Wehrmacht) in October 1939 and in winter of 1940 in the Králíky area and three years later in Náchod area. Total sum of all sub-calibre projectiles fired on the Czech fortification were 139. Shots were fired from imminent distance (66.3–200 m), i.e. from unrealistic positions in the means of battlefield situation. These tests showed better efficiency of sub-calibre projectiles in comparison to regular ones.

The presented impact and penetration data comes from military archives and [1].

Shots fired from 21 cm howitzer with the same angle of impact lead to different penetration depths for regular and sub-calibre projectiles. For example sub-calibre projectile Röchling 129 mm (weight 158 kg) impact velocity 373.9 m/s penetrated the reinforced concrete of the bunker 200 cm deep. On the other hand regular projectile of calibre 210 mm (weight 121.4 kg) penetrated the concrete 45 cm deep with impact velocity 227.8 m/s. This represents a nearly four times greater efficiency. A 15 cm howitzer calibre tests lead to similar results for regular and sub-calibre projectiles.

[a] Corresponding author: pavel.jiricek@fsv.cvut.cz

Figure 1. Front wall of the infantry bunker N-S-91 damaged by sub-calibre projectiles Röchling (the protective embankment was removed before the probe shootings).

The test results obtained from 21 cm calibre howitzer were used for numerical simulations. Both groups of the results are summed in Table 1 [1]. The projectiles were tested without the explosive content.

2. Numerical simulations

The evaluation of the effect of different calibre types projectiles requires complex computational models using nonlinear material models capable of handling high velocity impact behaviour. Due to these conditions ANSYS Autodyn was chosen.

2.1. Computational model

Computational model composes of bunker front wall section, dimensions 2 × 2 × 3.5 m. Using Lagrangean

Article available at http://www.epj-conferences.org or http://dx.doi.org/10.1051/epjconf/20159404049

Table 1. Depth of penetration of regular and sub-calibre projectiles shot on bunker K-Bg-S-11 [1]. The analysed data sets are emphasized.

Type of the projectile	Impact angle	Impact velocity	Depth of penetration
Röchling 15 cm Rö, Gr., calibre 105 mm, weight 75 kg	70°	507.0 m/s	216 cm
	71°	460.9 m/s	199 cm
	85°	297.5 m/s	122 cm
Anti-fortification grenade 15 cm Gr. 19 Be., calibre 150 mm, weight 75 kg	71°	736.0 m/s	114 cm
	84°	520.5 m/s	67 cm
	76°	320.6 m/s	40 cm
Röchling 21 cm Rö, Gr. 44 Be., calibre 129 mm, weight 158 kg	85°	399.5 m/s	200 cm
	86°	**373.9 m/s**	**200 cm**
	88°	294.1 m/s	160 cm
Anti-fortification grenade 21 cm Gr. 18 Be., calibre 210 mm, weight 121,4 kg	75°	257.7 m/s	42 m
	87°	**227.8 m/s**	**45 cm**
	89°	289.9 m/s	77 cm

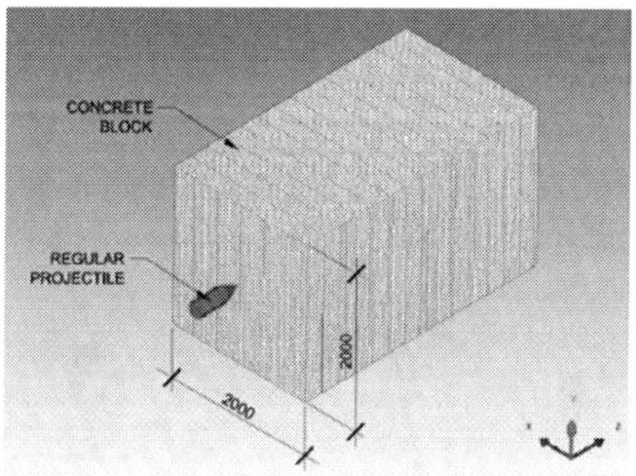

Figure 2. Computational model: regular projectile.

mesh with size of the elements 50 × 50 × 50 mm, the part of the wall is divided into the layers with thickness 0.25 m to simplify the evaluation of the results. On the sides, top and the bottom of the wall displacement of the nodes are restricted. The boundary conditions are restraints corresponding to the dimensions of the surrounding structure. The dimension of the model was optimized to provide the best possible results with minimized computational time.

Material model of the military object's concrete wall is chosen to describe the damage of the wall during the impact. It is essential to take into account strain-rate effect of the concrete in case of high velocity impact loading.

Accordingly, the RHT concrete model was chosen. RHT Concrete is part of ANSYS Autodyn material library. Material model was set to match concrete grade C30/37; mean compressive strength of the concrete was 45 MPa (the compressive strength of the concrete of the bunker as found in the military archives).

Geometry of both considered projectiles were modelled as accurate as possible. The material model of regular projectile was considered bilinear steel. Bilinear material model is chosen for sub-calibre projectile as well, but

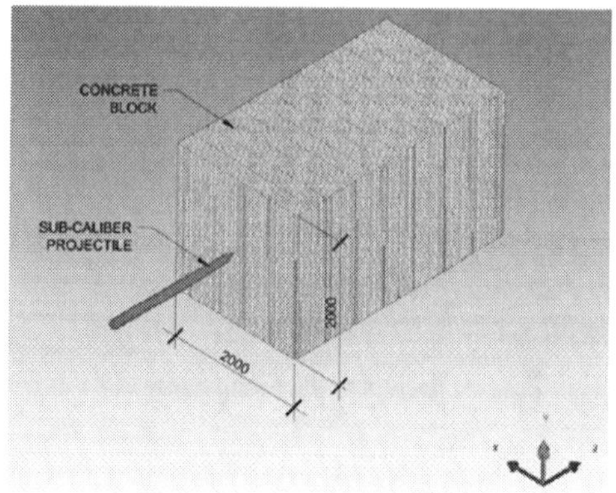

Figure 3. Computational model: sub-calibre projectile.

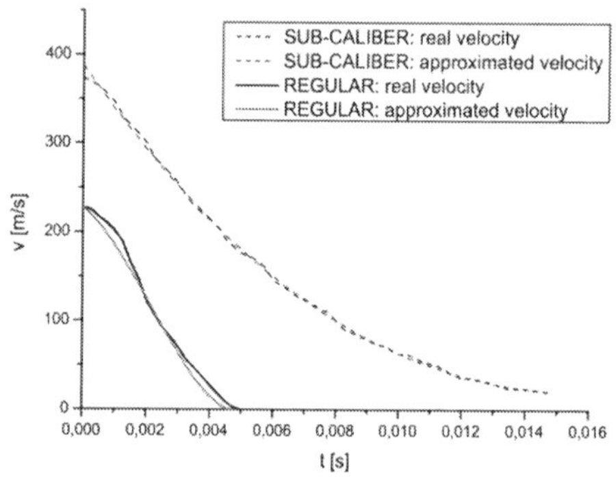

Figure 4. Graph plotting the peed of the projectile during the penetration of the concrete block.

steel is according to archive documents considered as chrome-vanadium with corresponding material properties, including the tip of the shot. Initial conditions are taken as presented in Table 1; velocity of sub-calibre projectile is set to 373.9 m/s and 227.8 m/s for regular calibre projectile.

The computational model was simplified in the mean of the impact angle, which was set to 90°. Guiding elements for sub-calibre projectile were neglected as well as the fletching, which has no effect on the impact itself and increase the computational time.

2.2. Results

2.2.1. Regular projectile

The numerical simulations resulted into a graph plotting the speed of the projectile during the penetration of the concrete block, see Fig. 4. Impact duration of the regular projectile is about 5 ms and the speed of the projectile can be approximated with a 3rd degree curve.

The damage of concrete during penetration is an important parameter of impact modelling. Maximal damage on the front of the wall occurs during first 2.5 ms of the impact (Fig. 5).

The volume of crushed concrete is, in the case of regular projectile, cone shaped with depth 0.6 m. The projectile itself penetrated wall 50 cm deep and crushed concrete volume is 0.14 m^3 (Fig. 6). The results show

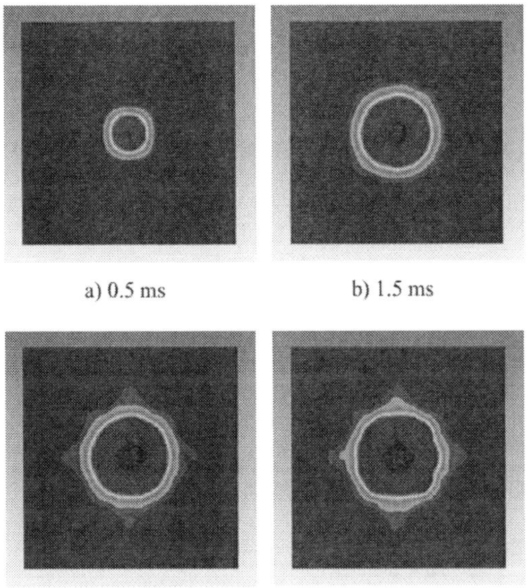

a) 0.5 ms b) 1.5 ms

c) 2.5 ms d) 4.5 ms

Figure 5. Damage of the front of the wall in mean of time.

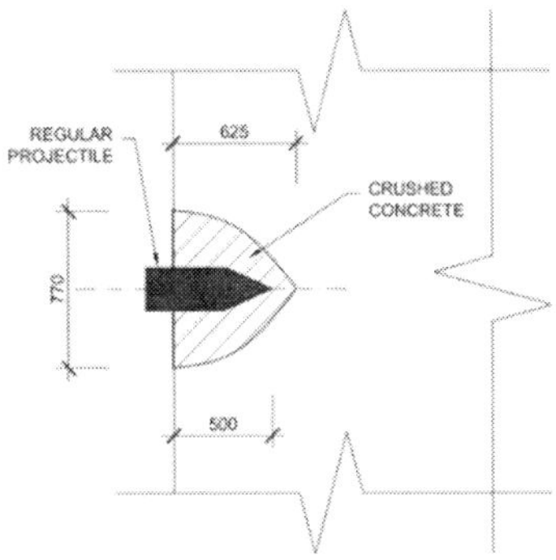

Figure 6. Shape of the resulting shape of damaged concrete: regular projectile.

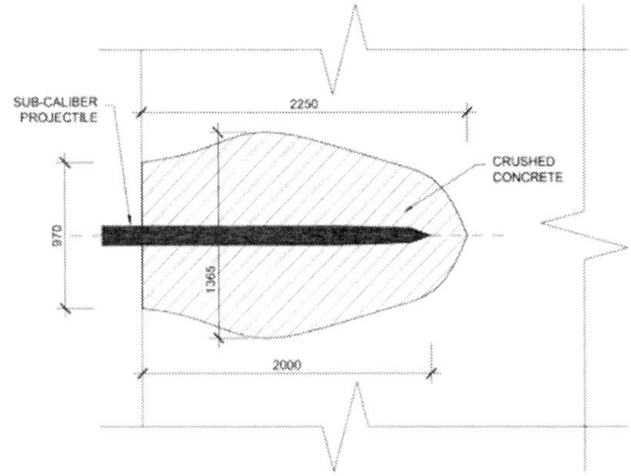

Figure 7. Shape of the resulting shape of damaged concrete: sub-calibre projectile.

Figure 8. Sub-calibre projectile Röchling stuck in infantry bunker N-S-91.

great agreement with the documented penetrations (90%) (Table 1).

2.2.2. Sub – calibre projectile

Unlike for the regular projectile, impact duration of sub-calibre projectile is 3 times longer, 15 ms. The curve of the velocity of the projectile during impact is quite different (Fig. 4) in comparison to the regular projectile but it can also be approximated with 3rd degree curve.

According to the shape and speed of the sub-calibre projectile, maximal extent of damage of the concrete wall is approximately in the depth of 0.75 m and total damage permeates to the depth 2.2 m. The volume of crushed concrete is 2.07 m^3 and projectile itself is stuck 2.0 m deep in the wall. The agreement with documented penetrations is very good (over 90%, Table 1).

3. Comparison

From the graph plotting the speed of the projectile during the penetration of the concrete block, a higher deceleration of regular projectile compared to sub-calibre projectile can be seen. It is mostly caused by the shape of the projectiles. Regular projectile is approximately barrel shaped with cone in the front; the ratio of diameter to length is 1:3. On the other hand sub-calibre projectile is barrel shaped tapered from the middle of the length with cone in the front; the ratio of diameter to length is 1:18.

According to the shape and larger cross-sectional area, than sub-calibre projectile, the impact force of a regular calibre dissipates on the surface of the wall and the penetration is therefore shallow (Fig. 6). On the other hand sub-calibre projectile acts as a needle and the penetration is much deeper (Figs. 7 and 10). Similar results were obtained by numerical simulation of projectile with and without spike impacting concrete target with different volumes of fiber reinforcement [13].

As seen in Figs. (2 and 3), the shape and velocity of sub-calibre and regular projectiles during impact differ significantly. In the case of a regular projectile, the maximal damage of the concrete is on the front wall of the infantry bunker and subsequent slowdown of the projectile processes in short time lapse. Due to these circumstances, the whole kinetic energy is dissipated on the front of the wall. On the other hand, sub-calibre projectile penetrates

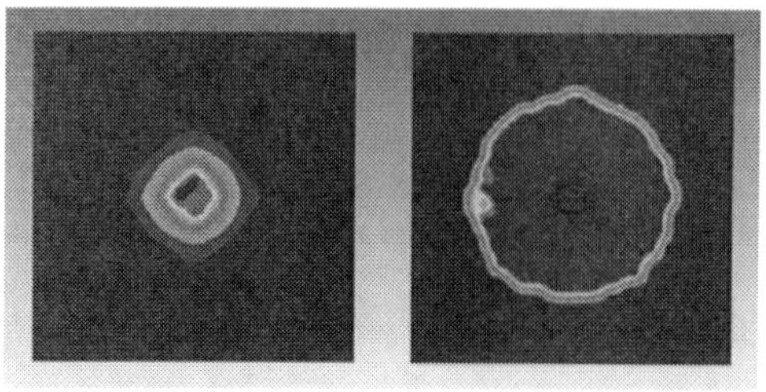

Figure 9. Damage of the front wall 0.5 m deep of the infantry bunker; left: regular projectile; right: sub-calibre projectile.

the concrete of the bunker much easier due to smaller diameter and higher speed, which can be seen at Figs. 4 and 7. Thus, sub-calibre projectile causes more damage acts deeper than regular projectile.

The maximum damage is concentrated the surface of the wall, in the case of regular projectile and in the depth of 0.75 m in case of sub-calibre projectile. Marginal difference can be observed in volume of the crushed concrete, where regular projectile damages 0.14 m^3 volume of the concrete and sub-calibre projectile 2.07 m^3. Simultaneously sub-calibre projectile penetrates concrete wall 2.0 m deep in contrast to 0.5 m deep penetration of regular projectile.

Sub-calibre projectile weights 158 kg and its initial velocity is 373.9 m/s, that is 65% higher than regular projectile with speed 227.8 m/s and weight 121.4 kg. Kinetic energy can be basically expressed as:

$$E_k = {}^1/_2 m\, v^2. \tag{1}$$

The kinetic energy is 11 MJ for sub-calibre projectile and 3 MJ in case of regular projectile. Nearly 4 times lower energy of regular projectile than sub-calibre one matches lower volume of crushed concrete and damage on the front of the wall which should be larger than in case of sub-calibre projectile.

Sub-calibre projectile penetrate concrete of the bunker much easily than regular projectile. Kinetic energy of the regular projectile is much lower than sub-calibre projectile, but decelerates three times shorter. Deceleration could be simply expressed as:

$$a = \Delta v / \Delta t. \tag{2}$$

Using elementary physics (2) and simplified results, deceleration could be approximated to 45,000 m/s^2 for regular projectile and 25,000 m/s^2 in case of sub-calibre projectile. These simplified decelerations are used to evaluate impact force from the change of the momentum, 3.9 MN for sub-calibre projectile and 5.5 MN for regular projectile.

According to high velocities of impact loading on the front of the infantry bunker, significant strain-rate effect would occur. Dynamic increase factor (DIF), is well known phenomenon [2,3] based on increase of the concrete strength (both tensile and compressive) with higher impact loading and described as the ratio of dynamic strength to static strength. Strain-rates are approximately 90-/s for sub-calibre projectile and 200-/s in case of regular

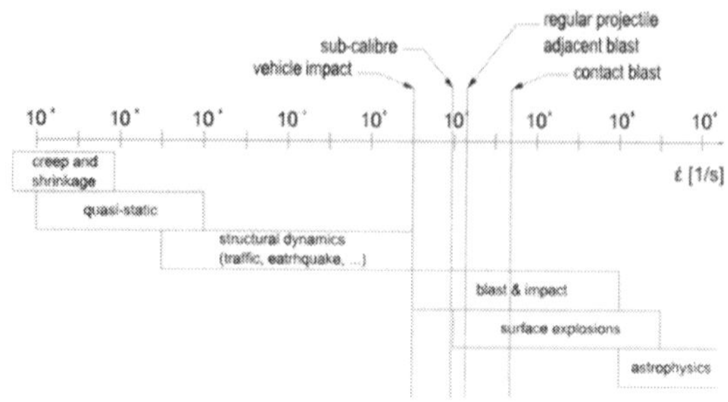

Figure 10. Strain rates of different dynamic loading cases, according to [3].

Table 2. Comparison of the results obtained from numerical simulation.

Type of the projectile	Weight Weight	Kinetic energy	Deceleration	Impact force
Regular	121 kg	3 MJ	45,000 m/s^2	5.5 MN
Sub-calibre	158 kg	11 MJ	25,000 m/s^2	3.9 MN

Table 3. Comparison of the results of strain rates according to different types of extreme loading.

Extreme loading	$\dot{\varepsilon}$ [-/s]
Vehicle impact [7]	10
Sub-calibre	90
Regular-calibre	200
Adjacent blast [11,14]	200
Contact blast [6]	2,400

projectile. According to [3], dynamic increase factor in the means of compressive strength reaches the value of 2.1 for sub-calibre projectile and 2.7 in case of regular projectile. The difference in strain-rate between sub-calibre and regular projectile is mainly the shape of the projectiles and the distribution of the energy during the impact loading and penetration.

Concrete as a semi-brittle material is very strain-rate sensitive. Strain rates obtained from several numerical simulation are plotted in Fig. 10, which presents values of strain rate for different dynamic loadings. For better understanding, findings from previous studies are presented. Vehicle impact loading [7] strain rates are approximately 10-/s which represents the top of the range of structural dynamics (Fig. 10). Adjacent blast [11], [14] strain rates come up to same values as regular projectile and corresponds with range for blast and impact range given by [3]. The highest strain rates are obtained during contact blast [6].

For the in Table 3 presented strain rates, the value of dynamic increase factor (DIF) can reach values from 1.15 to 37.25 according to the approach used; for details see [12].

The total effectiveness of the used ammunition is much considering the explosives fillings of the projectiles, because the sub-calibre projectile detonates as a recessed charge [4], which is more effective than contact explosion [5,6] of charge of the regular projectile.

4. Conclusions

The presented simulation was performed to evaluate penetration of two types of the projectiles to the front wall of a military bunker. The sub-calibre projectile penetrates the concrete 4 times deeper than regular projectile and is therefore more effective.

According to depth of penetration of the projectiles into wall of the bunker, substantial compliance with archive data and was achieved [1]. The shape and depth of damage caused by the projectiles in numerical simulations is similar to the one observed on tested bunkers. The obtained results show that the described numerical simulations can be used to predict similar situations.

The financial support of the Czech Science Foundation grant project 13-30441S is gratefully acknowledged.

References

[1] M. Ráboň, O. Gregar, B. Kachlík, et. al., *Val na obranu republiky – Československé opevnění z let 1935-1938 na Králicku* (Spolek přátel československého opevnění Brno, s.r.o., Brno, 2005)

[2] M. Drahorád, et. al., Beton TKS, **1/2012**, 74–79 (2012)

[3] CEB–FIP Model Code 2010, Comité Euro–International du Béton, (2010)

[4] J. Henrych, *Dynamika výbuchu a její užití* (Academia, Praha, 1973)

[5] M. Foglar, et. al., Beton TKS, **4/2011**, 90–93 (2011)

[6] M. Foglar, M. Kovář, International Journal of Impact Engineering, **59**, 18–28, (2013)

[7] P. Jiříček, M. Foglar, Bridge Maintenance, Safety, manegement and Life Extension 2014, 762–768, (2014)

[8] CEB-FIP Model Code 1990, Comité Euro-International du Béton, (1993)

[9] L. J. Malvar, C. A. Ross, ACI Material Journal, **95(6)**, 735–739, (1998)

[10] J. W. Tedesco, C. A. Ross, Journal of Pressure Vessel Technology, **120(4)**, 398–405, (1998)

[11] E. Sochorová, M. Foglar V. Křístek, Reliability, Risk and Safety. Back to the Future, 1–8 (2010)

[12] P. Jiříček, M. Foglar, Proceeding of 23rd Military Aspects of Blast and Shock 2014, (2014)

[13] U. Nyström, K. Gylloft, International Journal of Impact Engineering, **38**, 95–105, (2011)

[14] M. Foglar, J. Pachman, V. Pelikán, R. Hájek, M. Künzel, WIT Transactions on The Built Enviroment, **141**, 171–179, (2014)

[15] M. Kovář, M. Foglar, R. Hájek, WIT Transactions on The Built Enviroment, **141**, 159–169, (2014)

EPJ Web of Conferences **94**, 04050 (2015)
DOI: 10.1051/epjconf/20159404050

Modeling fragmentation with new high order finite element technology and node splitting

Lars Olovsson[1], Jérôme Limido[2,a], Jean-Luc Lacome[2], Arve Grønsund Hanssen[3], and Jacques Petit[4]

[1] IMPETUS Afea AB, Sördalavägen 22, 14160 Huddinge, Sweden
[2] IMPETUS Afea SAS, 6 rue du Cers, 31330 Grenade, France
[3] IMPETUS Afea AS, Strandgaten 32, 4400 Flekkefjord, Norway
[4] CEA, DAM, GRAMAT, 46500 Gramat, France

Abstract. The modeling of fragmentation has historically been linked to the weapons industry where the main goal is to optimize a bomb or to design effective blast shields. Numerical modeling of fragmentation from dynamic loading has traditionally been modeled by legacy finite element solvers that rely on element erosion to model material failure. However this method results in the removal of too much material. This is not realistic as retaining the mass of the structure is critical to modeling the event correctly. We propose a new approach implemented in the IMPETUS AFEA SOLVER® based on the following: New High Order Finite Elements that can easily deal with very large deformations; Stochastic distribution of initial damage that allows for a non homogeneous distribution of fragments; and a Node Splitting Algorithm that allows for material fracture without element erosion that is mesh independent. The approach is evaluated for various materials and scenarios: -Titanium ring electromagnetic compression; Hard steel Taylor bar impact, Fused silica Taylor bar impact, Steel cylinder explosion, The results obtained from the simulations are representative of the failure mechanisms observed experimentally. The main benefit of this approach is good energy conservation (no loss of mass) and numerical robustness even in complex situations.

1. Introduction

Random ductile and brittle fragmentation modelling is still a challenging task. Indeed, such fragmentation generally occurs in a complex multi-physics framework, as in classic tube explosion or implosion [1].

From a mechanical point of view a reliable material model is essential. At first, it is necessary to capture the strain localizations at the right time in order to deal with fragment size. Classic models for ductile materials used for this purpose are: Steinberg-Cochran-Guinan [2], Johnson-Cook [3], Zerilli-Armstrong [4], Mechanical Threshold Stress [5] and their evolutions. Most of these models take into account strain, strain rate, temperature, pressure, saturation stress and very high strain rates viscosities. They seem accurate enough to be used for the numerical simulations of the strain localizations. Moreover, random localizations shouldn't be naturally initiated in accurate numerical simulations. It is necessary to introduce a stochastic aspect in material models like random elastoplastic properties or a random initial damage distribution.

From a numerical point of view it is necessary to have a very accurate formalism in order to avoid activating fragmentation due to numerical errors. For example, Eulerian and ALE approaches must cope with interface reconstruction approximation errors. This is a very difficult task in the case of strong 3D fragmentation [6]. In a pure Lagrangian approach strain localization should only occur due to initial singularities (cross section step…) or wave crossing (spalling…). This is not the case for example with first order tetrahedron finite element approximations that can produce random fragmentation without any initial material heterogeneities [7,8]. Accuracy is thus essential for a reliable random fragmentation numerical model.

Of course, random strain localization is only the first step in modelling a tube explosion. Is it necessary to apprehend 3D domain opening and multi-species interaction (high explosive and metals for example). In most cases, Eulerian or ALE approaches are chosen but here again complex 3D interface reconstruction and contacts are very challenging and leads to energy conservation problems. A pure Lagrangian approach is generally not possible. Actually, solutions to treat domain opening are often limited to element erosion (3D X-FEM or meshless approaches for multiple cracks are still not mature for industrial applications). It means that a very fine mesh is needed to limit mass loss during computation. Furthermore, very large deformations are most of the time impossible to treat with classic FE, specially the HE part.

We propose to describe and evaluate a new monolithic purely Lagrangian approach implemented in the IMPETUS AFEA SOLVER®. This approach is based on the following:

- New High Order Finite Elements (Advanced Solid Element Technology ASET) that offers very high accuracy compared to classic first order elements
- Stochastic distribution of initial damage or elastoplatic properties that allows for a realistic strain localisation

[a] Corresponding author: jerome@impetus-afea.com

– Node splitting algorithm that allows for material fracture without element erosion that is mesh "independent".
– New Lagrangian Particle method for HE modeling.

The approach is evaluated for various materials and scenarios: -Titanium ring electromagnetic compression; Hard steel Taylor bar impact, Fused silica Taylor bar impact, Steel cylinder explosion,

2. High order finite elements for transient dynamics

2.1. Description

As highlighted in the introduction numerical accuracy is primordial to capture strain localization as an initiation mechanism for fragmentation. Legacy explicit solver use reduced integration first order finite elements (1 Gauss point) + hourglass control. These elements have a very low accuracy but are very popular because of low computation time. In practice, a large amount of elements is needed to compensate for insufficient accuracy. This is consistent with the very fine mesh needed to limit mass loss for an element erosion strategy for crack modeling. Nevertheless, real 3D applications like warhead design often lead to an unrealistic number of elements.

We propose ASET, an Advanced Solid Element Technology based on fully integrated third order approximations (64 Gauss points). These elements bypass the need for hourglass control and give much superior accuracy and deformability. For sure, it is not realistic to use this type of element with erosion. Thus, we developed a node splitting algorithm (see Sect. 4.1) that allows exact mass conservation and realistic computation time. We illustrate ASET accuracy with a classic pinched cylinder test.

2.2. Wriggers's pinched cylinder test

We evaluate the ability of 3rd order hexahedron finite elements in the framework of elastoplastic bending of thin structures. This popular benchmark proposed by Wriggers et al. [9] consists of a pinched tube in the middle. This tube has an inner radius dimension r = 300 mm, a thickness t = 3 mm and a length L = 600 mm. The material is treated as an elastoplastic law, with a Young's modulus E = 3 e^3 MPa, Poisson's ratio $\nu = 0.3$ and isotropic hardening that can be expressed as $\sigma = 24.3 + 300\varepsilon_p$ MPa.

In order to compare the solution described in Wriggers et al. [9] with IMPETUS we consider the plastic strain for various displacements (one-half model see Fig. 1). The observed results are very close. Unlike Wriggers who used 1600 shell elements, we used 50 3rd order elements. We highlight the fact we only used 1 element through thickness. This leads to a very high aspect ratio, about 20. That means we ignored all classic meshing rules imposed by legacy solvers (at least 4 elements through thickness and an aspect ratio smaller than 4). Nevertheless, accuracy is preserved as proved in Fig. 2 by the Force/Displacement comparison.

It is apparent that ASET 3rd order finite elements can deal with a strongly non-linear problem even with a very bad mesh quality.

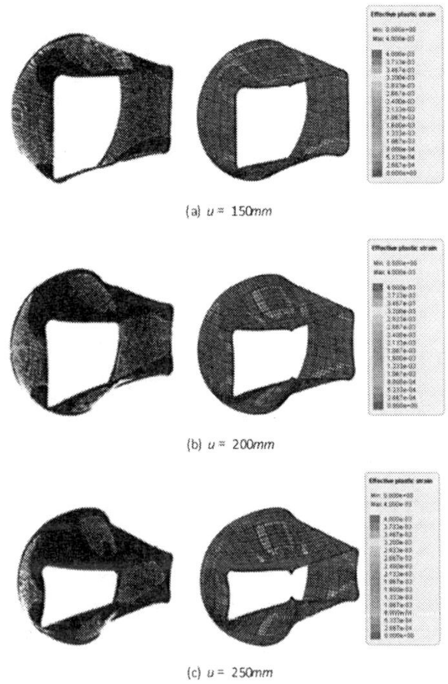

Figure 1. Plastic deformation visualization at different displacement. Right: IMPETUS Solver, left: Wriggers et al. [9].

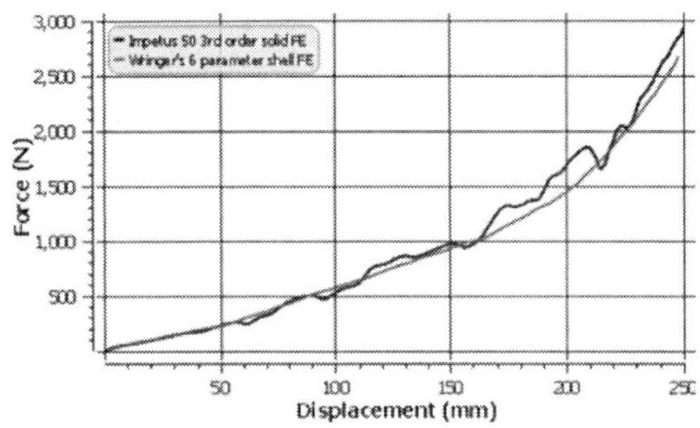

Figure 2. Force/Displacement at center (red: [9] black: Impetus).

3. Constitutive relations and random initialisation

As depicted in Sect. 1 the introduction of a stochastic distribution of initial damage or initial variation of elastoplastic properties is very important. The combination with a reliable constitutive relation gives the foundation for a fragmentation model.

3.1. Stochastic "Defects" distribution

A distribution function f(D) describes the number of "defects" per unit volume of matter.

$$f(D) = \begin{cases} a\mathrm{e}^{-bD} & \forall\, D \leq D_{max} \\ 0 & \forall\, D > D_{max}. \end{cases}$$

Note that the maximum initial damage cannot be larger than D_{max}. The number of defects N per unit volume of

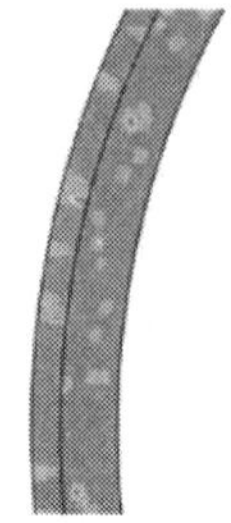

Figure 3. Initial defect distribution example.

matter in the range D_0 to Dmax can be calculated by integrating f(D) from D_0 to Dmax:

$$N = \int_{D_0}^{D_{max}} f(D) dD.$$

Based on the assumed distribution f(D) one can show that the probability p of having at least one initial defect larger than or equal to D_0 in a volume v is:

$$p = 1 - \mathbf{e}^{-N_v}.$$

This probability expression can be used to assign an initial defect level to each integration point in the model. The defect level is obtained by solving the expression for D_0 (given a random number p and an integration point volume v).

3.2. Ductile material model

We chose to use a slightly modified version of the popular Johnson-Cook model [3]. This is a constitutive model for ductile metals with thermal softening and strain rate hardening. The yield stress is defined as:

$$\sigma_y = f\left(\varepsilon_{eff}^p\right) g(D) \left(1 - \left(\frac{T - T_0}{T_m - T_0}\right)^m\right) \left(1 + \frac{\dot{\varepsilon}_{eff}^p}{\varepsilon_0}\right)^c$$

where g(D) is a damage softening and D is the damage level, defined as:

$$g(D) = \begin{cases} 1 & \forall\, D \leq s_0 \\ 1 + \frac{D - S_0}{1 - S_0}(S_1 - 1) & \forall\, D > S_0. \end{cases}$$

That is, g(D) drops linearly from 1 at D = s_0 to s_1 at D = 1 (full damage).

That is the way the yield stress can be randomly distributed at the initial state (see Fig. 3).

3.3. Brittle material model

We chose to develop a slightly modified version of the popular Johnson-Holmquist material model [10]. This is a constitutive relation for ceramic materials with different failure mechanisms in compression and tension. The material is assumed to have a pressure dependent shear resistance. At positive pressures, plastic flow is a combination of shearing and dilatation. Inelastic dilatation is interpreted as crushing that gradually reduces the shear resistance of the material. A brittle fracture criterion is used on the tensile side (see [11] for more details).

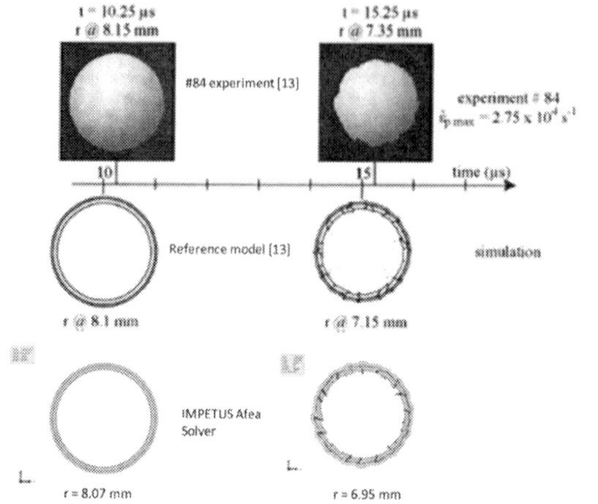

Figure 4. Electromagnetic cylinder compression (Top) Experiment [13], (Middle) 2D Ouranos model [13], (Bottom) 2.5D Impetus Afea model.

4. Random ductile fragmentation

In this section, we analyse 3 reference tests for random ductile fragmentation.

4.1. Petit's electromagnetic compressions test

The experiment of electromagnetic cylindrical compression was developed at the French Atomic Agency (CEA) to study the behaviour of ductile materials at large strain and high strain rate [12]. This section will focus on Ti6Al4V. Clearly this particular titanium alloy is very sensitive to adiabatic shearing. One test carried out at the CEA is used as the example.

The tube inner section is observed with a frame camera. Figure 4 (axial view) shows that adiabatic shear bands appear. Experimental observations proved that he adiabatic shearing threshold does not depend only on the strain but on the strain rate history too.

The CEA numerical result is based on the Ouranos 2D Lagrangian solver and taken as reference (details can be found in [13]). We developed an approach based on third order finite elements (see Sect. 2) and a ductile material model (see Sect. 3.2). Both solvers used an equivalent loading pressure to electromagnetic load (neglecting overheating).

We obtained good correlation between the experiment and our numerical approach (see Fig. 4). One can notice that Petit's model is largely more physically advanced than ours (modified Zerilli-Armstrong). Nevertheless, our goal here is to demonstrate we can capture strain localization with relatively few high order elements (Impetus model : 2400 elements, Ouranos 2D model ~180,000 elements). This allows extending our approach in 3D (see Fig. 5) with largely reduced computation time compared to the reference model.

4.2. Borvik's taylor bar impact

SimLab NTNU conducted very interesting Taylor tests on ARNE tool steel, hardened to nominally Rockwell

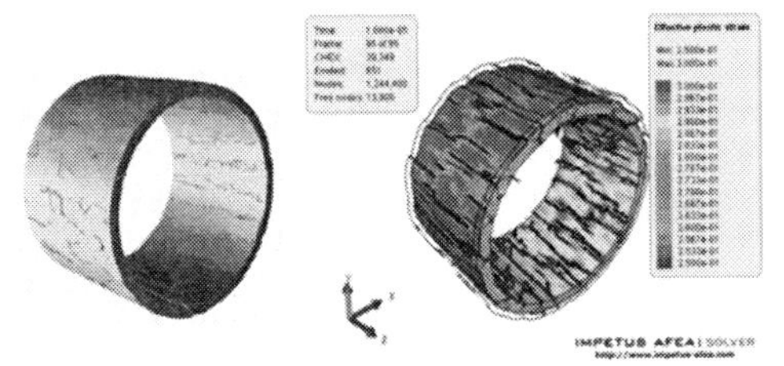

Figure 5. Electromagnetic cylinder compression (Left) 3D Ouranos model, (Right) 3D Impetus Afea Solver model.

Figure 6. Taylor bar impact at 250.5 m/s [14] (Top) experiment (Bottom) Impetus numerical model.

C values of 52 [14], see Fig. 6. An impact velocity about 250 m/s leads to quasi brittle failure. Proposed numerical model relies on cubic hexahedron elements (see Sect. 2) and a modified Johnson-Cook constitutive model (see Sect. 3.1 and 3.2). An additional numerical aspect is introduced here: Node Splitting. Indeed, the Fig. 6 model uses a very coarse mesh that prohibits the element erosion approach to generate fragments. We thus developed a node splitting technique based on the following 3 steps:

- Loop over all integration points surrounding the node that is to be split. Mark the two integration points with largest damage.
- Define a vector v between these two integration points.
- Try to split the node such that the propagating crack plane has its normal in a direction as close to v as possible.

This model exhibits very good energy balance as no element is eroded during computation. Moreover, fragment mass distribution is in good agreement with experimental data.

4.3. Goto's cylinder explosion

4.3.1. Experiment description

This case is inspired by the experiments presented in [15] corresponding to the fragmentation of a cylinder due to HE detonation.

An AISI 1018 steel cylinder height H = 203.2 mm, outside diameter De = 50.8 mm and a thickness e = 3 mm is fragmented by the detonation of an explosive LX-17. The explosive LX-10 allows initiating the detonation in order to obtain a quasi-plane wave in the cylinder. In addition to the distribution of the fragments, their thickness and microstructure are studied in [15] in order to assess the corresponding fracture deformation.

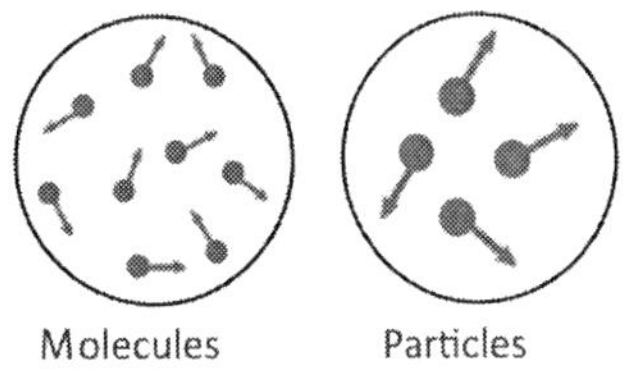

Figure 7. Particles vs. Molecules.

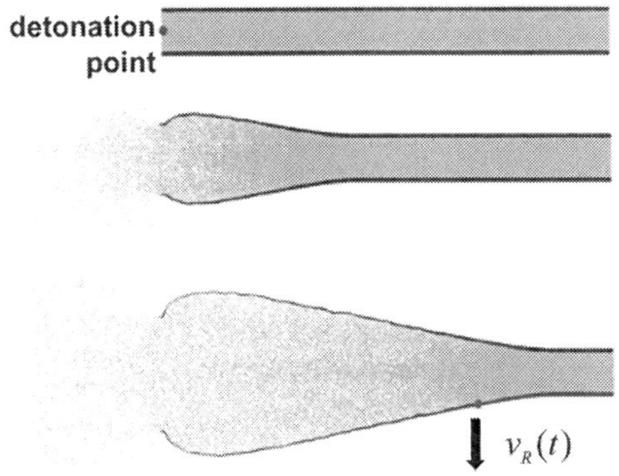

Figure 8. Cylinder Test used for HE Calibration.

4.3.2. Blast particle method

High explosive modeling is a challenging task especially in the case of a strong interaction with a strongly deformed/fractured domain. Most of the referenced studies used an Euler/Lagrange coupled approach. As noticed in Sect. 1, this classic approach is very complex and leads to energy conservation problems. We developed a purely Lagrangian Meshless Method applied to modeling the complete Blast Event [16]. The implementation of the Blast Particle Method (BPM) in the IMPETUS Afea Solver® is described in detail in [17] and compared with experimental results in [17, 18].

As a short description, the HE model is based upon the Kinetic Molecular Theory for gas. The basic assumptions are presented below, but the first two are not valid for HE so modifications to the theory are necessary.

The basic assumptions for Molecular theory are:

- The average distance between particles is large compared to the particle size.
- Equilibrium exists.
- Molecules obey Newton's Law.
- Molecular collision is perfectly elastic.

One Particle represents typically 10^{15}–10^{20} molecules (see Fig. 7).

Calibration for specific types of explosives is accomplished by using a traditional cylinder test, as shown in Fig. 8.

Parameters that are used to characterize the HE particle model:

ρ Density

E0 Internal energy

D Detonation velocity

ν Co-volume (Optimized in cylinder test simulations).

γ Ratio of heat capacities(Optimized in cylinder test simulations).

Air is also modeled with the same approach as the HE.

4.3.3. *Numerical model*

Here again the proposed numerical model for ductile steel cylinder is based on 3rd order hexahedron elements (see Sect. 2), modified Johnson-Cook constitutive model (see Sect. 3.1 and 3.2) and Node Splitting Algorithm (see Sect. 4.1). The High explosive is modelled using the innovative Discrete Particle Method (DPM) [16]. Coupling between DPM and FEM is based on contact algorithm that transfers particles impulses to the structure. DPM is independent of the structures complexity thanks to its meshless nature. This means that treatment is the same for a simple cylinder or for a complex warhead. The main advantage of this monolithic Lagrangian approach is exact mass conservation and very good energy conservation. Moreover, a close range cylinder explosion impacting a structure like a plate is trivial. This means that both fragment creation and their effect on the plate would be taken into account.

A visualization of the numerical model is presented in Fig. 9. We can distinguish red dots that are related to DPM and FEM in blue. The dynamics in the simulation is in good agreement with experiments [15]. The mesh size used is quite coarse and limits the analysis of fragment morphology. Indeed, experimentally obtained fragments exhibit inclined shear fractures parallel to the fragment length, whereas the fragment length is defined by flat fracture surfaces. These inclined shear fractures are not reproduced in this model due to the coarse mesh and the Node Splitting Algorithm itself. This aspect requires a specific study. Nevertheless, the obtained fragment mass distribution is acceptable as shown in Fig. 10.

Regarding computational considerations, the GPU parallelization of the solver is very efficient. In fact, the model ran in less than 1h on a standard workstation equipped with an Nvidia K20 GPU processor. This aspect is very important as it allows for real industrial applications and future developments.

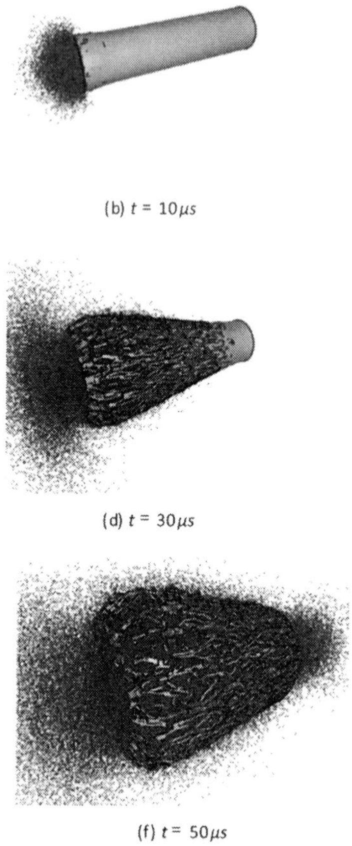

Figure 9. Tube explosion IMPETUS Afea Solver model.

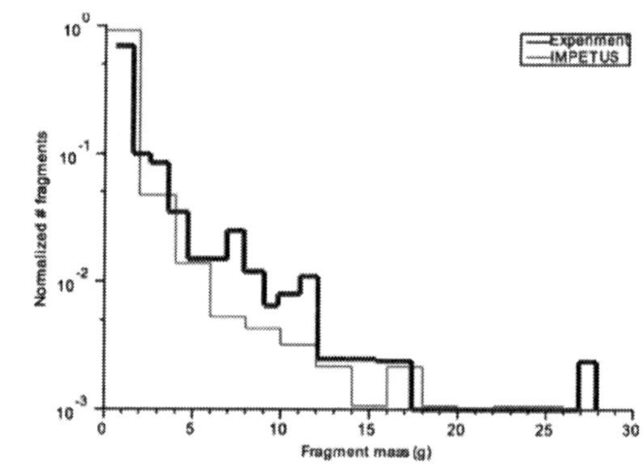

Figure 10. Fragment mass distribution (black: experiment [15], red: Impetus Afea Solver).

5. Brittle random fragmentation

We now focus on evaluating the approach on a brittle material: fused silica. Fused silica is a high purity synthetic amorphous silicon dioxide widely used in the military industry as window material. It may be subjected to high-energy ballistic impacts. Under such dynamic conditions, post-yield response of the ceramic as well as the strain rate related effects become significant and should be accounted for in the constitutive modelling. In this study, we use a modified Johnson-Holmquist (J-H) model (see Sect. 3.2). Parameters have been identified by Ruggiero et al. [11] by an inverse calibration technique, on selected validation test configurations.

Numerical simulations were performed with 3rd order hexahedron elements (see Sect. 2), a modified Johnson-Holmquist constitutive model (see Sects. 3.1 and 3.3) and a Node Splitting Algorithm (see Sect. 4.1). This approach overcomes classic numerical drawbacks associated with element erosion and numerical inaccuracy of fragmentation activation.

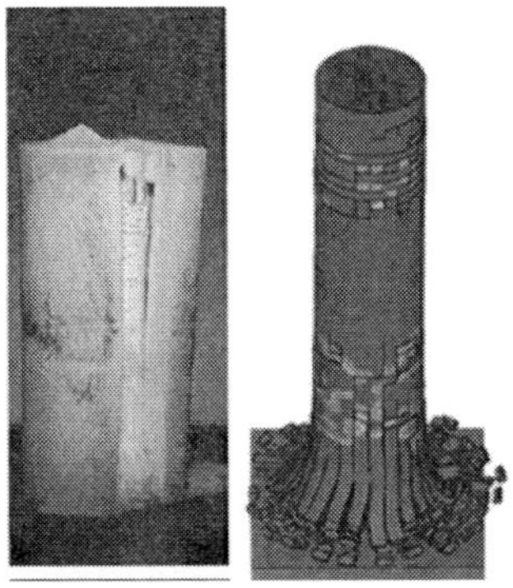

Figure 11. Fused Silica Taylor impact (left: experiment, right: Impetus Afea Solver) [11].

Taylor impact at 89 m/s is shown in Fig. 11. Spall fracture development in the rear portion of the cylinder as well as separation in the front section crushed region is exhibited. Qualitative comparison with a high speed camera selected frame is in a fairly good agreement.

6. Conclusions

It was demonstrated on several challenging ductile and brittle random fragmentation cases relevance of this new approach. The ASET Element Technology coupled to the Discrete Particle Method (DPM) demonstrated a very good alternative to classic approaches as it proposes a fully Lagrangian framework with exact mass conservation with reduced computation time. Industrial applications targeted are directly linked to warhead design.

Material models used in this study were quite simple as the focus was on evaluating this very innovative approach. This means that in order to be really predictive a strong effort is required to introduce more physics into the material models.

References

[1] D. Grady, Shock Wave and High Pressure Phenomena, XVI, 376 p. (2006)

[2] D.J. Steinberg, S.G. Cochran, M.W. Guinan, J. Appl. Phys., vol. **51**, no. 3, pp. 1498–1504 (1980)

[3] G.R. Johnson, W.H. Cook, Proceedings of the 7th Int. Symp. on Ballistics, pp. 541–547 (1983)

[4] F.J. Zerilli, R.W. Armstrong, J. Appl. Phys., vol. **61**, no.5, pp. 1816–1825 (1987)

[5] P.S. Follansbee, U.F. Kocks, Acta metall., vol. **36**, no. 1, pp. 81–93 (1988)

[6] R. Bell and E. Hertel, SAND92-1716, Sandia National Laboratories (1992)

[7] Q.B. Diep, J.F. Moxnes, G. Nevstad, Proceedings of the 21st Int. Symp. on Ballistics, pp. 752–758 (2004)

[8] A. Rusinek, R. Zaera, Int. J. Impact Engng, vol. **34**, no. 4, pp. 799–822 (2007)

[9] P. Wriggers, R. Eberlein, S. Reese, Int. J. of Solids and Structures, **33**:3309–3326. (1996)

[10] G.R. Johnson, T.J. Holmquist, AIP Conf. Proc., **309**, 981–4 (1994)

[11] A. Ruggiero, G. Iannitti, G. Testa, J. Limido, J.L. Lacome, L. Olovsson, M. Ferraro, N Bonora, J. Phys.: Conf. Ser. **500**, 182036 (2014)

[12] J. Petit, Y.A. Alexeev, S.P. Ananiev, M.N. Kazeev, J. Phys. IV France **7**, Colloque C3, (1997)

[13] J. Petit, Ch.14 in Materials under Extreme Loadings: Application to Penetration and Impact (2010)

[14] T. Karlsen; A.B. Kjølseth, Master thesis, Institutt for konstruksjonsteknikk (2012)

[15] D.M. Goto, R. Becker, T.J. Orzechowski, H.K. Springer, A.J. Sunwoo, C.K. Syn, Int. J. Impact Engn, **35**, 1547–1556 (2008)

[16] L. Olovsson, A.G. Hanssen, T. Børvik, M. Langseth, European J. of Mechanics A/Solids, Vol **29**, pp. 1–6 (2010)

[17] T. Børvik, L. Olovsson, A.G. Hanssen, K.P. Dharmasena, Wadley H.N.G. Hansson, J. of the Mech. and Phys. of Solids Vol **59**, pp. 940–958 (2011)

EPJ Web of Conferences **94**, 04051 (2015)
DOI: 10.1051/epjconf/20159404051

Finite element analysis of hypervelocity impact behaviour of CFRP-Al/HC sandwich panel

Vaibhav A. Phadnis[1] and Vadim V. Silberschmidt[2,a]

[1] AMRC with Boeing, The University of Sheffield, UK
[2] Wolfson School of Mechanical and Manufacturing Engineering, Loughborough University, Loughborough, UK

Abstract. The mechanical response of CFRP-Al/HC (carbon fibre-reinforced/epoxy composite face sheets with Al honeycomb core) sandwich panels to hyper-velocity impact (up to 1 km/s) is studied using a finite-element model developed in ABAQUS/Explicit. The intraply damage of CFRP face sheets is analysed by mean of a user-defined material model (VUMAT) employing a combination of Hashin and Puck criteria, delamination modelled using cohesive-zone elements. The damaged Al/HC core is assessed on the basis of a Johnson Cook dynamic failure model while its hydrodynamic response is captured using the Mie-Gruneisen equation of state. The results obtained with the developed finite-element model showed a reasonable correlation to experimental damage patterns. The surface peeling of both face sheets was evident, with a significant delamination around the impact location accompanied by crushing HC core.

1. Introduction

Advanced composite materials such as CFRP/Al-HC (carbon fibre-reinforced plastic face-sheets/aluminium honeycomb core) sandwich panels typically find their application in satellite structures due to their excellent mechanical properties such as high specific strength, stiffness and ability to provide tailored properties. In the absence of proper shielding, these structures can be often exposed to impacts by space debris. Many satellites already employ honeycomb sandwich panels with CFRP face-sheets covered by multi-layer thermal insulation as external walls [1]. Furthermore, FRPs are often used in the construction of spacecraft components such as antenna struts, low distortion frames and lightweight robotic arms [2–5].

Still, many concerns remain about the possibility of employing polymer-matrix composites as the primary satellite structure for long-duration missions because of severe environmental conditions encountered in space. One of the major problems is related to the vulnerability to impact damage from meteoroids and man-made space debris that populate the near-Earth orbital environment. Such particles travel at speeds of several km/s, thus presenting a serious hazard for existing satellites. Previous researchers found that damage due to high velocity impacts (HVIs) on composite materials consists of an impact crater, surface peeling of the outer laminae on both the front and rear side of the plate and a significant internal delamination around the impact point, that can lead to severe degradation of the material's mechanical strength. Moreover, if a hypervelocity particle has enough energy to fully perforate the impacted plate, a debris cloud emerging from the rear face of the target can significantly damage adjacent structures and internal components of the spacecraft.

This study focuses on the development of a finite element (FE) model to simulate a HVI behaviour (with velocities up to 1 km/s) of a CFRP-Al/HC sandwich panel using general-purpose FE software Abaqus/Explicit [6]. A user-defined material model (VUMAT) developed to study damage of CFRP face sheets employing combined Hashin and Puck criteria was used, while the interply damage was modelled by means of cohesive-zone elements. A hydrodynamic response of Al/HC was captured using a Mie-Gruneisen equation of state and its damage was modelled using a Johnson Cook dynamic failure model. The developed FE model was validated with experimental data from Wicklein et al. [7] and showed a reasonable correlation between damage patterns experimentally and numerically.

2. Finite-element model

A quarter of the full geometry of the sandwich plate was modelled owing to its symmetry about relevant planes; this resulted in significantly reduced computational efforts. A steel projectile of 3 mm diameter was impacted on the sandwich structure with an impact velocity ranging from 700 m/s to 1 km/s. The general contact algorithm in ABAQUS/explicit[6] was used to simulate contact conditions between the projectile and the composite face-sheets and Al/HC, and between the layers of CFRP laminate by defining appropriate contact-pair properties. The setup of FE model is shown in Fig. 1.

The elastic response of CFRP laminate was accounted for; it is discussed in Sect. 2.1. An element-deletion approach was used to model a material-erosion process based on initiation and evolution of damage in the meshed elements. Criteria for damage initiation and evolution are discussed in Sect. 2.2.

[a] Corresponding author: v.silberschmidt@lboro.ac.uk

Article available at http://www.epj-conferences.org or http://dx.doi.org/10.1051/epjconf/20159404051

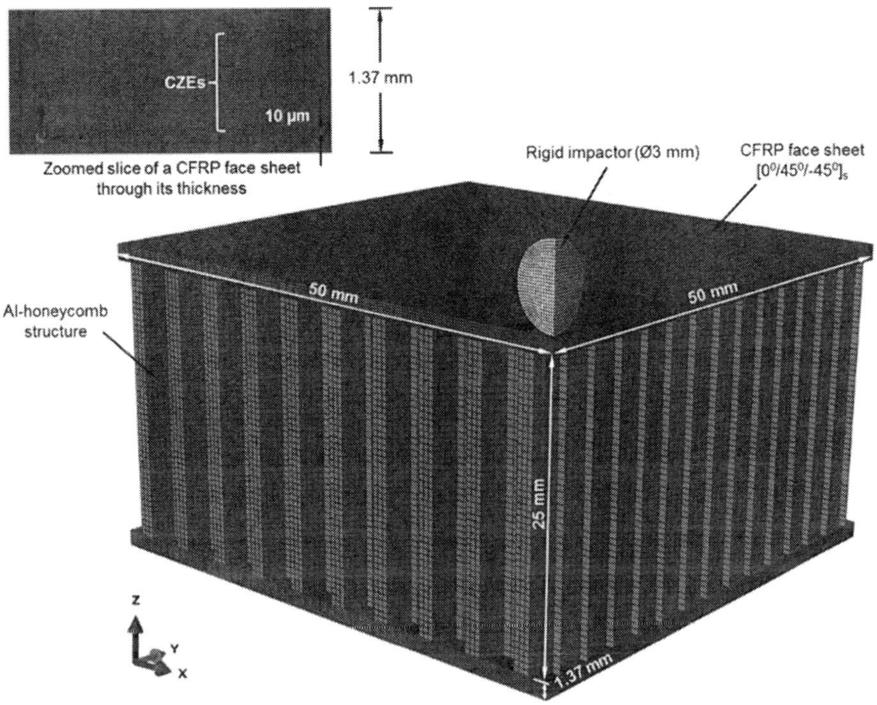

Figure 1. Setup of FE model.

2.1. Response of UD composite laminate

An elastic stress-strain relationship for a UD composite ply was introduced assuming the orthotropic damaged elasticity response. Its constitutive behaviour before damage is expressed using a generalised Hooke's law in the tensorial form:

$$\{\sigma\} = \mathbf{C}\,\{\epsilon\}. \tag{1}$$

The effect of strain-rate on elastic moduli and strength was defined in order to accurately account for a high strain-rate response of the laminated composite material. It should be noted that for carbon/epoxy-like systems, where carbon fibres are brittle and show no strain-rate dependence, the strain-hardening at higher loading rates is entirely due to epoxy matrix; while in glass/epoxy like systems both glass fibres and epoxy matrix may contribute towards this.

The strain-rate effects in the elastic and shear moduli are defined by the following expressions:

$$\begin{aligned} E(\dot{\epsilon}) &= E(\dot{\epsilon}_0) + \left[m_e \cdot \log\left(\frac{\dot{\epsilon}}{\dot{\epsilon}_0}\right) + 1\right] \\ G(\dot{\gamma}) &= G(\dot{\gamma}_0) + \left[m_s \cdot \log\left(\frac{\dot{\gamma}}{\dot{\gamma}_0}\right) + 1\right], \end{aligned} \tag{2}$$

where $E(\dot{\epsilon})$ and $G(\dot{\gamma})$ are elastic and shear moduli for a given strain rate, $E(\dot{\epsilon}_0)$ and $G(\dot{\gamma}_0)$ are elastic and shear moduli at the reference strain rate $\dot{\epsilon}_0$ and $\dot{\gamma}_0$, respectively, in three directions and three planes of a 3D co-ordinate system. m_e and m_s are the material parameters obtained by fitting a curve to the experimentally obtained data. The effects of strain rate on ply strength are discussed in the following sections where initiation of different damage modes is considered.

2.1.1. *Damage initiation*

Modelling of composites and their damage at a laminate level typically requires input of several parameters, including homogenised ply properties, interply strength and information about the laminate lay-up. Here, a layer-by-layer modelling strategy was adopted to capture failure in each ply. This offers several advantages. Firstly, full 3D stress states can be analysed. Typically, FE models of composite deformation involve the use of 2D shell elements to simulate composite plies, which do not allow for accurate representation of stress through the composite thickness.

Secondly, intraply and interply damage can be introduced discretely along with phenomenological models that account for complex interaction between them. Here, a material model offering a combination of Hashin's [7] and Puck's failure criteria [8] was used to implement the advantages of both. The Hashin's criteria were employed to estimate damage in carbon fibres while damage in epoxy matrix was modelled using the Puck's criteria. The empirical forms of these criteria are discussed below.

(a) Fibre damage

– in tension.

Here, tensile failure of fibres is defined as a function of tensile strength of lamina along the fibre direction as well as shear strength in transverse directions (in and out-of plane). The empirical relation is expressed in the following form:

$$\left(\frac{\sigma_{11}}{X_{1t}}\right)^2 + \left(\frac{\sigma_{12}}{S_{12}}\right)^2 + \left(\frac{\sigma_{13}}{S_{13}}\right)^2 \geq 1, \quad d_{ft} = 1 \quad (3)$$

where X_{1t} is the tensile strength of a composite in fibre direction, while S_{12} and S_{13} are its shear strengths in 1–2 and 1–3 planes, respectively. d_{ft} is a scalar damage variable indicating fibre damage in tension such that $0 \leq d_{ft} \leq 1$, 0 represents no damage, while 1 indicates fully-damaged state; this nomenclature applies to all damage modes discussed below. Here, X_{1_t} is assumed equal to its quasi-static magnitude since fibres show negligible or no strain-rate dependence at high loading rates.

– in compression.

Longitudinal compressive failure of composites is driven by shear mechanisms in an intricate way. A first crack may be started by shear fracture of fibres followed by rotation and in-plane shearing of the matrix at the crack tip, which in turn promotes kink-band development. The empirical relation for this damage mode is expressed in the following form

$$\left(\frac{\sigma_{11}}{K(\dot{\epsilon})X_{1c}^{qs}}\right) + \left(\frac{\sigma_{12}}{K(\dot{\gamma})X_{12}^{qs}}\right) + \left(\frac{\sigma_{13}}{K(\dot{\gamma})X_{13}^{qs}}\right) \geq 1, \quad d_{fc} = 1. \quad (4)$$

The scaling parameter K is defined as:

$$K = \frac{S_{12}^{dyn}(\dot{\gamma}_{12})}{S_{12}^{qs}} = \frac{S_{13}^{dyn}(\dot{\gamma}_{13})}{S_{13}^{qs}}. \quad (5)$$

Here, $S_{12}^{dyn}(\dot{\gamma})$ and $S_{13}^{dyn}(\dot{\gamma})$ are in-plane and out-of-plane shear strengths of a ply in dynamic loading condition at a given strain rate and S_{12}^{qs} and S_{23}^{qs} are its shear strengths in quasi-static loading condition.

(b) Matrix damage.

The phenomenological matrix-failure criteria for 3D QS failure analysis of UD composites were previously formulated by Puck and Schürmann [8]. These criteria forms a basis of the criterion presented here. The applicability of the originally proposed tensile and compressive failure criteria for quasi-static loading conditions was extended to the model material's response under dynamic loading.

The polymer matrix material in a CFRP composite demonstrates strain-rate sensitivity at high strain rates ($\sim 10^3/s$) which are typical for high-velocity impact events. This effect becomes significant, particularly in transverse directions, where the polymer matrix is a primary load bearing. An approach, similar to that employed in modeling compressive failure of fibres was adopted here to account for strain-rate sensitivity. The following criteria were proposed to model failure in the

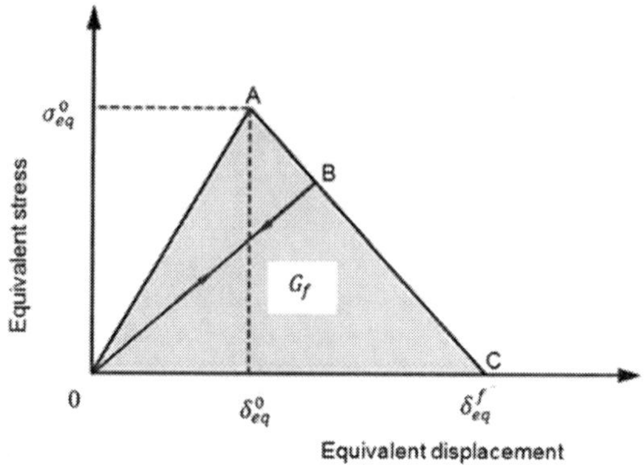

Figure 2. Concept of equivalent stress vs. equivalent displacement in evolution of damage variable.

matrix material in compression and tension:

$$\left(\frac{\sigma_{22}}{K(\dot{\epsilon})X_{2t}^{qs}}\right)^2 + \left(\frac{\sigma_{12}}{K(\dot{\gamma})S_{12}^{qs}}\right)^2 + \left(\frac{\sigma_{23}}{K(\dot{\gamma})S_{23}^{qs}}\right)^2 \geq 1;$$

$$\sigma_{22} + \sigma_{33} > 0, \quad d_{mt} = 1 \ldots \text{(tensile failure)}, \quad (6)$$

$$\sigma_{22} + \sigma_{33} < 0, \quad d_{mc} = 1 \ldots \text{(compressive failure)}.$$

2.1.2. Damage evolution

To alleviate mesh dependency during strain softening, a characteristic length (L_c) of a finite element was introduced into the formulation so that the constitutive law was expressed as a stress-displacement relation as shown in Fig. 2. The positive slope of the stress-displacement curve prior to damage initiation corresponds to a linear-elastic material behaviour; while the negative slope after damage initiation is achieved by evolution of damage parameter d as following:

$$d_i = \frac{\epsilon_{eq}^{fi}\left(\epsilon_{eq} - \epsilon_{eq}^0\right)}{\epsilon_{eq}\left(\epsilon_{eq}^{fi} - \epsilon_{eq}^0\right)}. \quad (7)$$

where ϵ_{eq}^0 is the strain at damage initiation, and ϵ_{eq}^f is the strain at which the material is completely damaged. $i = f_t, f_c, m_t, m_c$ corresponds to damage variables associated with failure modes of fibre and matrix material in tension (t) and compression (c). The strain at failure was expressed as,

$$\epsilon_{eq}^f = \frac{2G_f}{\sigma_{eq}^f.L_c}, \quad (8)$$

where G_f is the fracture energy, and L_c represents characteristic element length (for a typical hexahedral mesh, it represents the shortest distance between diagonally opposite nodes of a smallest element). The elastic and strength properties for CFRP composite laminates were taken from [7] and [11]. The mechanical and shock-related properties for Al-HC core were obtained from the manufacturer's catalogue, and the Abaqus technical brief [10], respectively, owing to identical material system. These are listed in Table 1.

Table 1. Material properties for Al 6061-T6 core.

For Mie-Grüneisen EOS model	
Reference density (kg/m^3)	2700
Grüneisen coefficient	1.97
Wave speed (m/s)	5240
Parameter, s	1.40
Reference temperature (K)	293.2
Specific heat (J/kg K)	885
For Johnson Cook plasticity and failure model	
A	324.1
B	113.8
n	0.42
$\theta_{melt}\,(K)$	925
$\theta_{transition}\,(K)$	293.2
m	1.34
c	0.002
$\dot{\epsilon}_0$ (1/s)	1

2.1.3. Element deletion

An element-deletion approach [11] used to remove the failed elements from the mesh was based on the magnitude of damage variables, $d_i = f_t, f_c, m_t, m_c$ as calculated using Eq. (7) applied to discrete damage modes in the composite laminate. The element was removed when the magnitude of any of the damage variables associated with the fibre failure reached $d_{max}(= 1)$, at an integration point of the element. When this condition was satisfied, the element was removed from the mesh, and it offered no subsequent resistance to deformation.

2.2. Response of Al honeycomb

A hydrodynamic response of Al honeycomb was accounted for using the Mie-Gruneisen equation of state, while its deviatoric response was modelled using the linear elastic and Johnson-Cook plasticity model available in ABAQUS/Explicit. An identical element deletion strategy was used to remove excessively deformed elements from the simulation as discussed in Sect. 2.1.3.

3. Results and discussion

Delamination patterns at the entry and exit sides of CFRP face sheets were studied using the developed FE model. The extent of delamination between adjacent plies after the projectile perforated the CFRP/Al-HC structure is shown in Figs. 3 and 4. It should be noted that though the FE model was quarter-symmetric, full delamination planes were obtained during the post-processing offering better observation of these results. The ply-level damage was also studied: here fibre damage/spallation was the dominant mode of damage. The fibre damage on the front and back faces of entry and exit side face sheets is shown in Figs. 5a and b. Shear-driven crushing of Al core directly below the point of impact was also evident from simulations. This

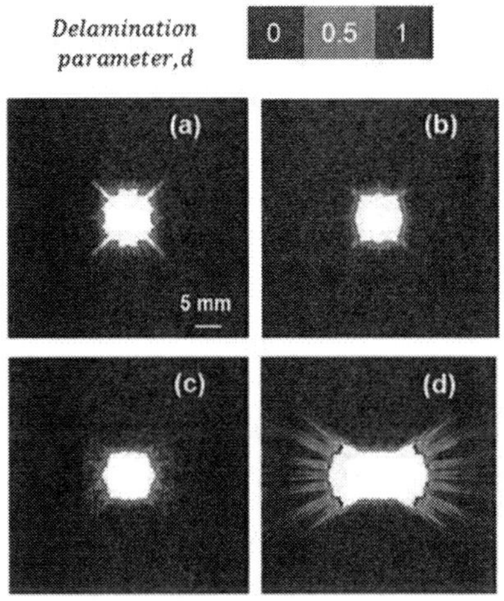

Figure 3. Entry-side delamination of CFRP face sheet between 0° and 45°(a), −45° and 45°(b), 45° and 0°(c), 45° and −45° (d) plies. v_i = 1 km/s, projectile radius 3 mm.

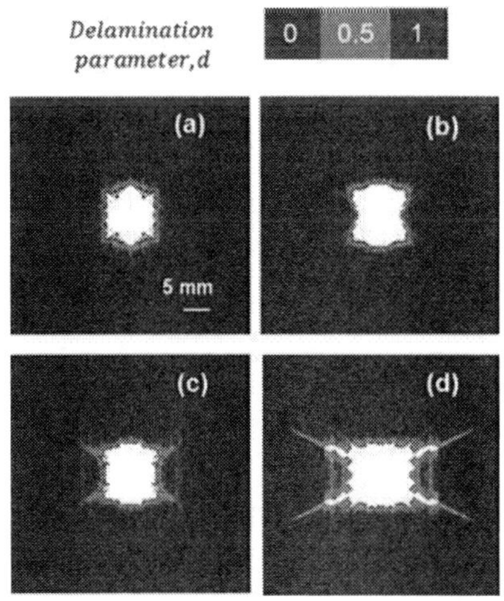

Figure 4. Exit-side delamination of CFRP face sheet between 0° and 45°(a), −45° and 45°(b), 45° and 0°(c), 45° and −45°(d) plies. v_i = 1 km/s, projectile radius 3 mm.

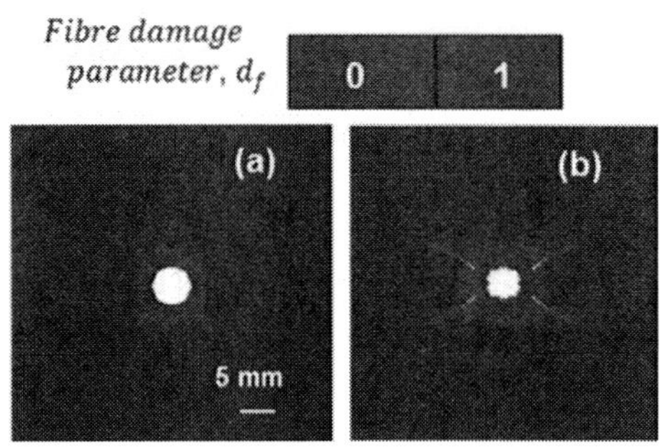

Figure 5. Fibre damage on surface of CFRP sheets at entry(a) and exit(b) sides. v_i = 1 km/s, projectile radius 3 mm.

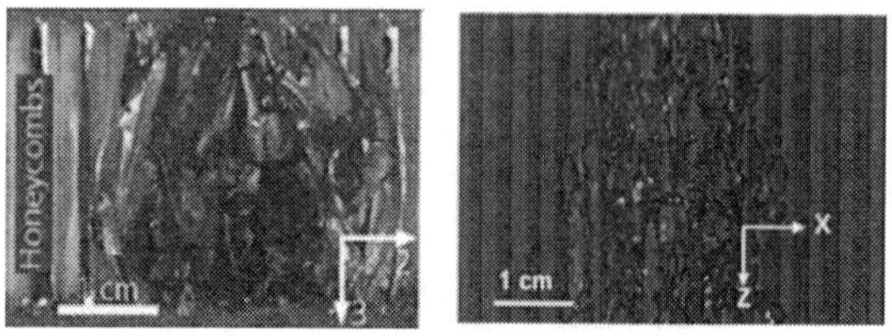

Figure 6. Damage of Al-HC core:comparison of experiment and FE analysis. $v_i = 1$ km/s, projectile radius 3 mm.

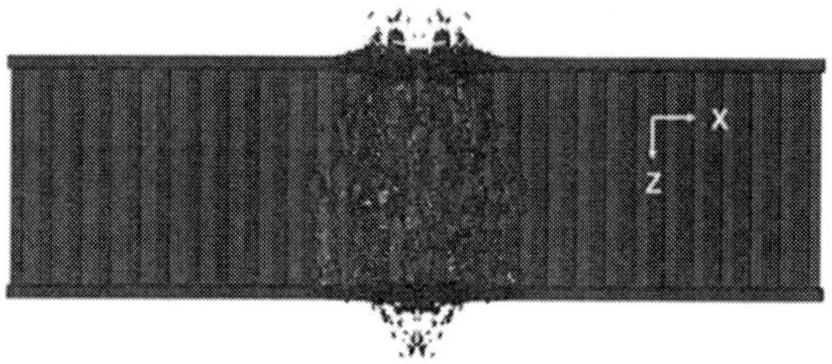

Figure 7. Cut view of final state of studied composite laminate showing surface peel up of CFRP face sheets. $v_i = 1$ km/s, projectile radius 3 mm.

was compared with the experimental results of [7] and shown in Fig. 6 (X–Z plane of simulation corresponds to 2–3 plane of experimental results). The cut-view showing final configuration of CFRP/Al-HC composites structure is presented in Fig. 7.

4. Conclusions

In this work, the hypervelocity ($v_i = 1$ km/s, projectile radius 3 mm) impact response of the CFRP/Al-HC composite structure was studied using a developed 3D finite-element model. Damage to CFRP face sheets was observed in terms of dynamic spall and surface peeling, while Al-HC core failed due to core cracking driven by shear. The simulation results offer a reasonable agreement with experimental observations.

References

[1] Riedel, W., Harwick, W., White, D.M. & Clegg, R.A. (2003) ADAMMO – advanced material damage models for numerical simulation codes. ESA CR(P) 4397, EMI report I 75/03, Freiburg.

[2] Riedel, W., Nahme, H., White, D.M. & Clegg R.A. (2006) Hypervelocity impact damage prediction in composites: part II – experimental investigations and simulations. Int. J. Impact Engng **33**, 670–80.

[3] Hiermaier. S., Riedel, W., Hayhurst, C., Clegg, R.A. & Wentzel, C. (1999) AMMHIS – advanced material models for hypervelocity impact simulations. Final report, EMI report E 43/98, ESA CR(P) 4305, Freiburg.

[4] White, D.M., Wicklein, M., Clegg, R.A. & Nahme, H. (2008) Multi-layer insulation material models suitable for hypervelocity impact simulations. Int. J. Impact Engng **35**(12), 1853-1860.

[5] White, D.M., Taylor, E.A. & Clegg, R.A. (2003) Numerical simulation and experimental characterization of direct hypervelocity impact on a spacecraft hybrid carbon fibre/Kevlar composite structure. Int. J. Impact Eng **29**(1–6), 779–90.

[6] ABAQUS 6.10. Theory manual (2010) Dassault System.

[7] Wicklein, M., Ryan, S., White, D.M. & Clegg, R.A. (2008) Hypervelocity impact on CFRP: Testing, material modelling, and numerical simulation. Int. J. Impact Engng **35**: 1861–1869.

[8] Hashin, Z. (1980) Failure criteria for unidirectional fibre composites. J. Appl. Mech. **47**: 321–329.

[9] Puck, A. & Schürmann, H. (1998) Failure analysis of FRP laminates by means of physically based phenomenological models. Comp. Sci. Technol. **58**(7): 1045–1067.

[10] Simulia Abaqus technical brief (2012) Simulation of the ballistic performance of aluminium plates with Abaqus/Explicit, USA.

[11] Phadnis, V.A., Makhdum, F., Roy, A. & Silberschmidt, V.V. (2013) Drilling in carbon/epoxy composites: Experimental investigations and finite element implementation. Comp. Pt A **47**: 41–51.

EPJ Web of Conferences **94**, 04052 (2015)
DOI: 10.1051/epjconf/20159404052

Penetration analysis of projectile with inclined concrete target

S.B. Kim[a], H.W. Kim, and Y.H. Yoo

Agency for Defence Development, Yuseong, PO Box 35, Daejeon, South Korea

Abstract. This paper presents numerical analysis result of projectile penetration with concrete target. We applied dynamic material properties of 4340 steels, aluminium and explosive for projectile body. Dynamic material properties were measured with static tensile testing machine and Hopkinson pressure bar tests. Moreover, we used three concrete damage models included in LS-DYNA 3D, such as SOIL_CONCRETE, CSCM (cap model with smooth interaction) and CONCRETE_DAMAGE (K&C concrete) models. Strain rate effect for concrete material is important to predict the fracture deformation and shape of concrete, and penetration depth for projectiles. CONCRETE_DAMAGE model with strain rate effect also applied to penetration analysis. Analysis result with CSCM model shows good agreement with penetration experimental data. The projectile trace and fracture shapes of concrete target were compared with experimental data.

1. Introduction

Generally, the concrete penetration experiment evaluates survivability, deceleration, penetration depth and residual stress of projectile [1–5]. In this study, we performed a penetration test and analysis of warhead with concrete target. The main purpose of test is to evaluate the survivability of warhead against to inclined concrete target.

Analysis results were compared with the test result of sled equipment. Concrete damage models that were provided by the LS-DYNA commercial codes [6] were applied for penetration analysis of concrete target.

2. Penetration analysis

2.1. Modelling and analysis condition

Projectile and concrete target were modelled as a half shapes in consideration of symmetry, as shown in Fig. 1. Steel supporters were modelled as a rigid body (rigid wall) to support the concrete target. Supporters were located at front and rear side and a lower part of two concrete targets.

Figure 2 shows the detailed projectile shapes. Projectile is composed of 4340 steel warhead case, explosive and fuze and aluminium shroud head and body. Figure 3 shows concrete target with strength of 5000 psi. Two layers of concrete are inclined with 30 degrees. Initial impact conditions are the same with the sled experimental conditions. Projectile of 232 kg impacts with initial velocity of 333 m/s.

Target shape after sled test is shown in Fig. 4. Warhead penetrated the first layer of concrete, but did not pass through the second layer of concrete. Finally, the warhead has been fractured at the middle region of body.

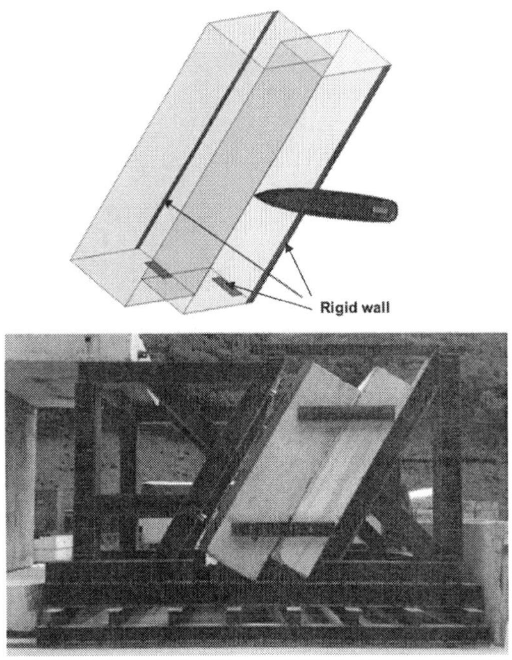

Figure 1. Concrete target.

2.2. Mechanical properties

Material of warhead body is 4340 steel with heat treatment. Simplified Johnson-Cook model for steel is utilized for penetration analysis. Figure 5 shows the yield stress distribution with strain rate obtained from the static tensile testing machine and Hopkinson pressure bar testing machine. Table 1 show the coefficients for J-C model of 4340 steel. Figure 6 shows concrete strength distribution with respect to the test age. Sled test was performed after

[a] Corresponding author: ksb@add.re.kr

Article available at http://www.epj-conferences.org or http://dx.doi.org/10.1051/epjconf/20159404052

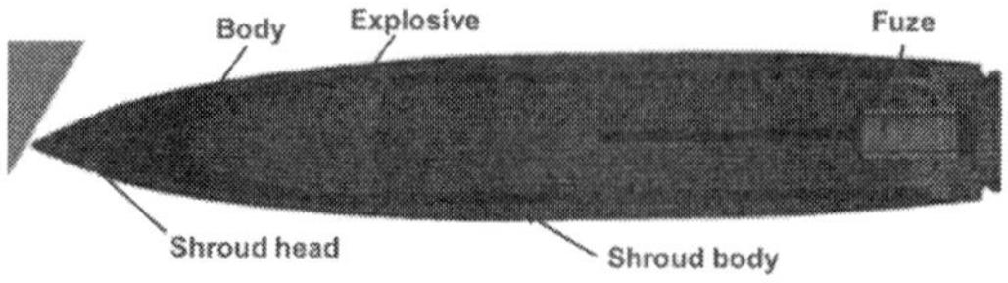

Figure 2. Projectile modelling.

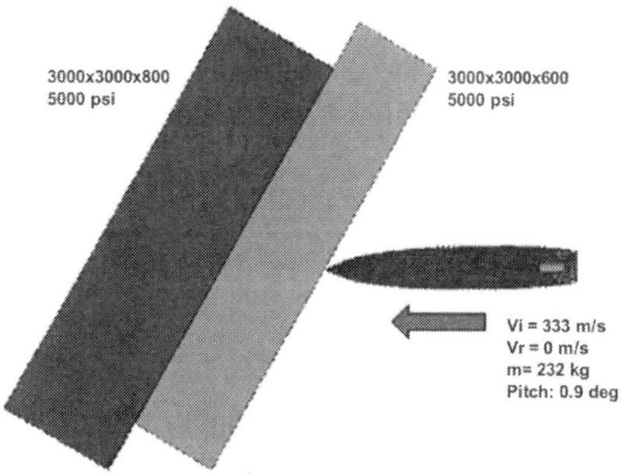

Figure 3. Projectile and target modelling.

Figure 4. Concrete shape after sled penetration test.

Table 1. Coefficients of J-C model.

Coefficients of J-C model, $\sigma = (A+B\varepsilon^n)(1+C\ln(\varepsilon/\varepsilon_0))$	
A	1330 MPa
B	1004 MPa
n	0.5
C	0.021

4 weeks. The average concrete strength is about 5000 psi based on cube test.

2.3. Analysis result

Commercial F.E. code, LS-DYNA 3D v.971 was used with several included concrete models. Three concrete material models such as SOIL_CONCRETE, CSCM and CONCRETE_DAMAGE were utilized.

Analysis result with SOIL_CONCRETE model shows that the warhead has been failed to penetrate the target, as shown in Fig. 7. Aluminium shroud was fractured and remained at the tail part of warhead. Generally, shroud of projectile should be designed to be well fractured during

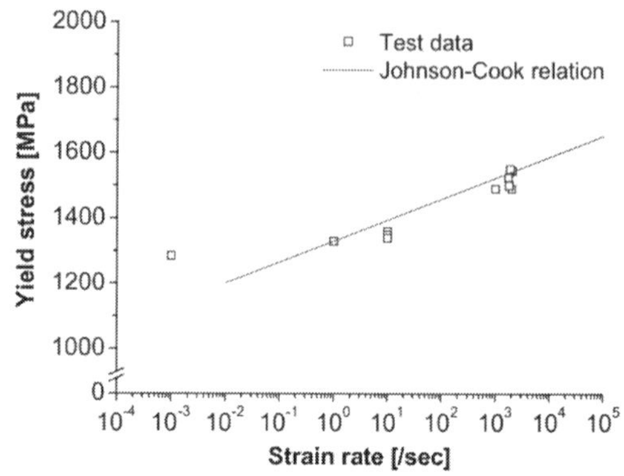

Figure 5. Yield stress distribution with strain rate for 4340 steel.

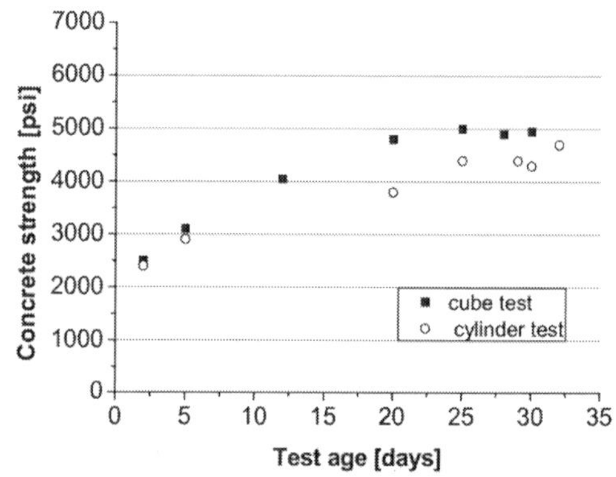

Figure 6. Concrete strength with test age.

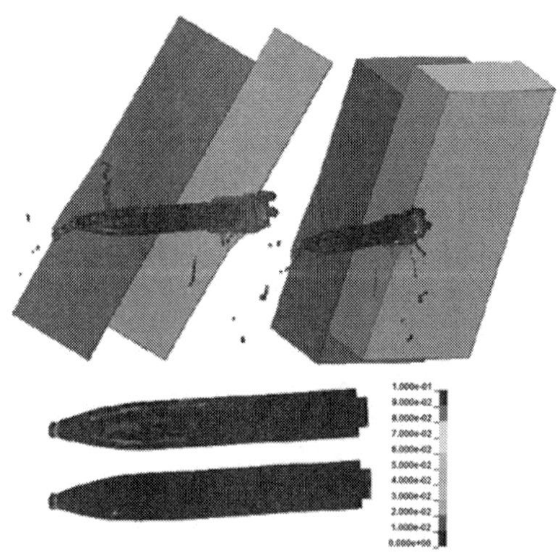

Figure 7. SOIL_CONCRETE model.

the target penetration. Warhead trajectory with inclined target is different from the experimental result. Warhead was stopped with a little degree of descent direction compared with initial impact direction. Therefore, this model may be suitable to evaluate the penetration depth and residual velocity with normal direction target. However it may not be appropriate to estimate the exact

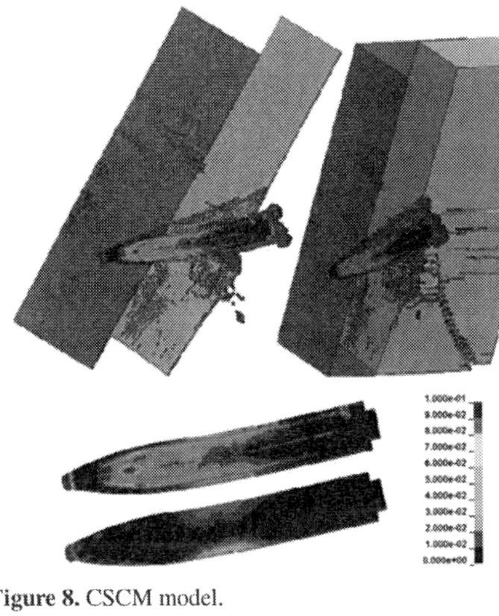

Figure 8. CSCM model.

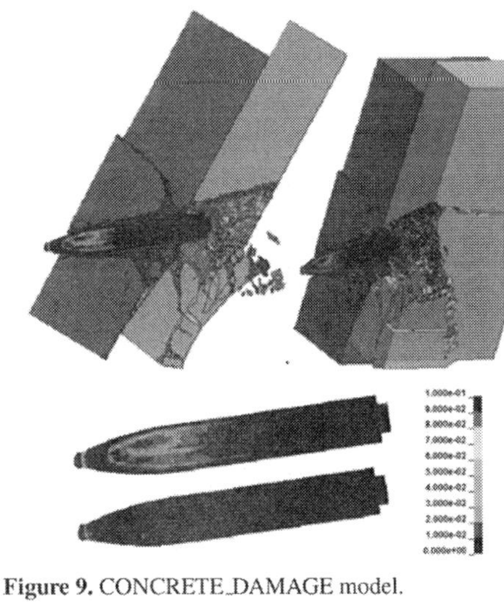

Figure 9. CONCRETE_DAMAGE model.

projectile trajectory or concrete target fracture shapes, especially for inclined target.

Figure 8 shows the analysis result with the CSCM model. Lower part of the first layer concrete cracked down toward to the edge. And warhead has stopped toward the bottom of the concrete target. In particular, the plastic deformation amount of warhead is higher than 7% at final state. This plastic strain concentrates in the central portion of the warhead, and it is likely to be fractured at this part. For 4340 steel, the fracture elongation for 4340 steel is about 7 ~ 8% in static tensile test. During the dynamic impact penetration, fracture elongation could be decrease compared to static state.

CONCRETE_DAMAGE model shows the most severe concrete fracture, as shown in Fig. 9. This means that the concrete strength is smaller than the other concrete models. The trajectory of the warhead side down similar to the test results, but the penetration depth of the warhead in this model is the larger than the other two models. The other two models did not consider the strain rate effect of concrete. Standard CONCRETE_DAMAGE model in LS-DYNA does not use the strain rate effect of concrete. However, dynamic properties of concrete was applied to this analysis for the consideration of strain rate effect [7,8]. Generally, as strain rate increases, the strength of concrete also increases. Though considering the strain rate effect with this model, concrete strength is weak in comparison to other two models.

Thus, three analysis results with different concrete models show obviously different penetration depth, trajectory and fracture shapes in spite of the same strength for concrete (5000 psi) and the same eroding values for numerical element deletion. Compared to the penetration depth, the concrete is considered weakly in the order of CSCM, SOIL_CONCRETE and CONCRETE_DAMAGE model.

3. Conclusions

This paper presents the numerical analysis with various concrete models compared with the experimental results. Analysis result with CSCM model shows good agreement with penetration experimental data. The traces of projectile and failure shapes of concrete target were compared with experimental data.

References

[1] D.J. Frew, S.J. Hanchak, M.L. Green, M.J. Forrestal, Int. J. Impact Eng. **21**, 489 (1998)

[2] D.J. Frew, M.J. Forrestal, S.J. Hanchak, ASME J. Appl. Mech. **67**, 841 (2000)

[3] M.J. Forrestal, S.J. Hanchak, ASME J. Appl. Mech. **69**, 853 (2000)

[4] M.J. Forrestal, D.J. Frew, J.P. Hickerson, T.A. Rohwer, Int. J. Impact Eng. **28**, 479 (2003)

[5] T.L. Warren, A.F. Fossum, D.J. Frew, Int. J. Impact Eng. **30**, 477 (2004)

[6] LS-DYNA Keyword User's Manual ver. 971, Livermore Software Technology Corporation (2007)

[7] L.J. Malvar, J.E. Crawford, J.W. Wesevich, D. Simons, Int. J. Impact Eng. **19**, 847 (1997)

[8] L.J. Malvar, C.A. Ross, ACI Materials J. **95**, 735 (1998)

EPJ Web of Conferences **94**, 04053 (2015)
DOI: 10.1051/epjconf/20159404053

Using plastic instability to validate and test the strength law of a material under pressure

Cyril Bolis[a], Denis Counilh, and Brice Savale

CEA, DAM, DIF, 91297 Arpajon, France

Abstract. In dynamical experiments (pressures higher than 10 GPa, strain rate around 104–106 s^{-1}), metals are classically described using an equation of state and a strength law which is usually set using data from compression or traction tests at low pressure (few MPa) and low strain rates (less than 103 s^{-1}). In consequence, it needs to be extrapolated during dynamical experiments. Classical shock experiments do not allow a fine validation of the stress law due to the interaction with the equation of state. To achieve this aim, we propose to use a dedicated experiment. We started from the works of Barnes et al. (1974 and 1980) where plastic instabilities initiated by a sinusoidal perturbation at the surface of the metal develop with the pressure. We adapted this principle to a new shape of initial perturbation and realized several experiments. We will present the setup and its use on a simple material: gold. We will detail how the interpretation of the experiments, coupled with previous characterization experiments helps us to test the strength lax of this material at high pressure and high strain rate.

1. Introduction

Testing and validating strength law at high pressure and high strain rate is a challenging problem. Classically, these laws are set using characterization experiments such as Hopkinson's bars or compression/traction machines at low pressure and low strain rates (lower than $10^3\,s^{-1}$). Even with these limitations, they are used to simulate shock experiments where the pressure is driven by explosives or by laser shots and where the strain rate can reach 10^4 to $10^7\,s^{-1}$. Without any direct data at these regimes, the material behavior has to be extrapolated.

To validate this extrapolation, we looked at experiments which are directly sensitive to the strength law. Usual shock experiments are often dedicated to equation of state measurements and are only sensitive to the strength law in a second order. To have direct measurements, we were inspired by the works of Barnes et al. [1,2] who made the link between the growth of Rayleigh-Taylor instability and the strength law. Since the last years, this technic was upgraded using laser driven plasmas to generate pressures over 1 Mbar [3] of proton radiography at Prad [4] to achieve multiple views of the perturbation growth.

In this paper, we will show how we were inspired from Prad experiments to design our Rayleigh-Taylor instability experiment. We will present the principle and the setup that was tested with two gold targets. The results of the two experiments are analyzed and compared to numerical simulations. Lastly, we will discuss how to fit the perturbation growth by changing the parameters of the strength law.

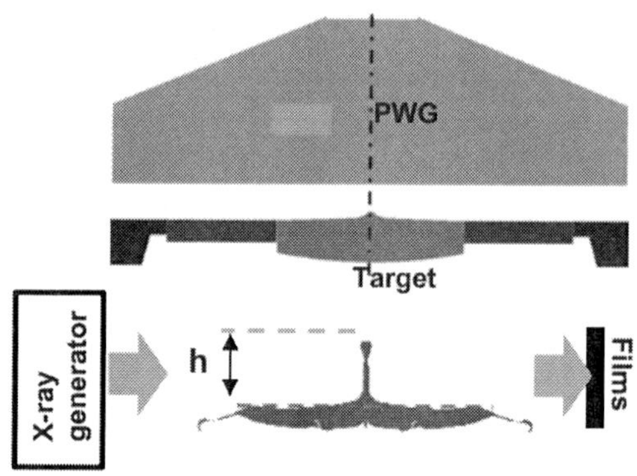

Figure 1. Principle of the experiment.

2. Experimental setup

2.1. Principle

Similarly to the technic developed by Barnes, we used a high explosive plane wave generator (PWG) to generate the pressure (Fig. 1). The target is located a few millimeters from the surface of the explosive (HMW based). Therefor, the pressure is generated by the release of detonation products. It can be assimilated to an isentropic compression which produces the growth of an axisymmetric perturbation at the center of the target in front of the plane wave generator.

During the experiment, the shape of the target is acquired using low energy X-ray generator (mean spectra 80 keV) coupled to radiographic films on which we measure the position of the target and the perturbation length (h in Fig. 1). As the perturbation is axisymmetric, three couples generator/film on different axis

[a] Corresponding author: cyril.bolis@cea.fr

Article available at http://www.epj-conferences.org or http://dx.doi.org/10.1051/epjconf/20159404053

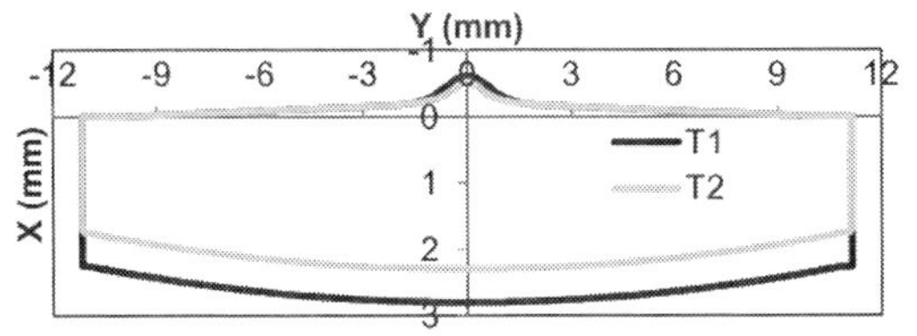

Figure 2. Targets shapes.

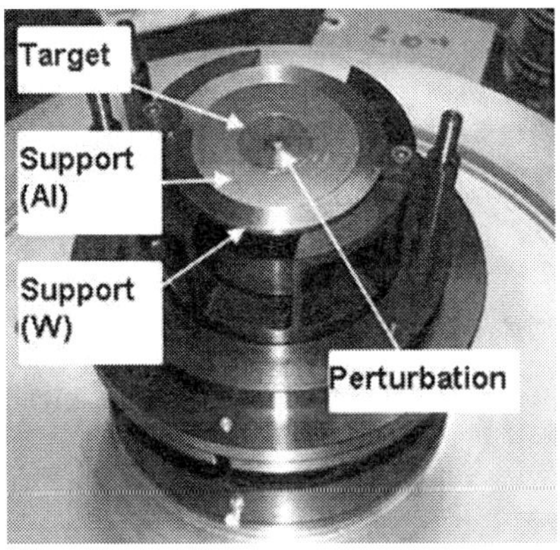

Figure 3. Target support: upper view.

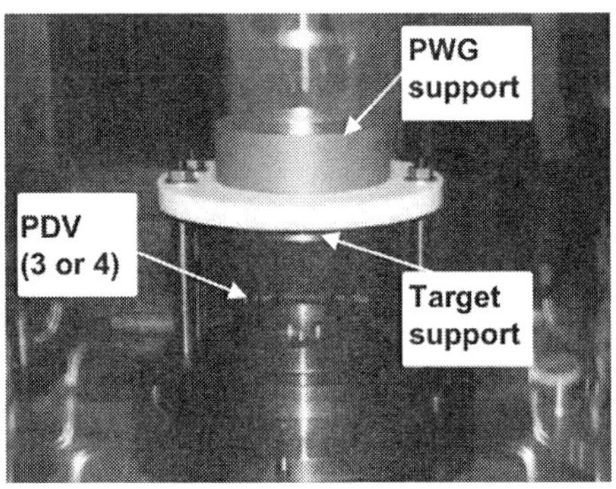

Figure 4. Target support: side view.

(around every 120°) are used to image the target at three different times.

To test our setup, two experiments were realized (T1 and T2) using two different target shapes (Fig. 2). Main differences arise from the target thickness and the perturbation shapes (Gaussian for T1, design by simulation for T2) and amplitude (900 μm for T1, 300 μm for T2). For both targets, pure gold was chosen as a well-known and simple material.

2.2. Setup

The setup of the experiment is presented on the next figures. The target is supported by two metallic pieces, in light metal (aluminum for T1 and copper for T2) and in heavy metal (tungsten in both cases) (Fig. 4). The first one will flight faster than the target whereas the second one will

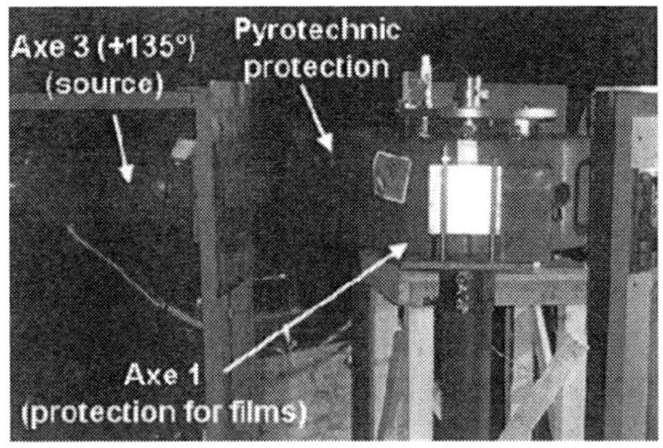

Figure 5. Pyrotechnic protections and radiographic chain.

almost not move. This setup allows us to avoid some side effects and to ensure the visibility of the target during the experiment.

The plane wave generator is mounted in front of the perturbation at a fixed distance (Fig. 4) at the opposite side, under the radiographic windows, three (four for T2) PDV probes are used to follow the first moment of the acceleration of the target.

The setup is placed inside a closed glass cylinder which is used to make primary void around the target. It is also used as a first pyrotechnic protection for the three radiographic chains. The second one is assured by a steel cylinder with steel plates in the radiographic axes. The holes in the cylinder are used as collimators for the X-rays. In front of each source, a box containing sensitive films is placed (Fig. 5).

For each experiment, all data were acquired. The target flight is recorded on the films and the perturbation length is measured (Fig. 6). Its evolution is clearly visible on radiography at the three times. One can noticed on the bottom of the image at 20.4 μs the trace of support in copper, whereas the one in tungsten stay at the top of the image.

3. Results and first simulations

Before discussing of the perturbation growth, the pressure law applied to the target has to be checked first. For this, the velocities signals for the first moments and the target position for the later ones were used.

Simulations were realized using a hydrocode in an eulerian scheme. Plane wave generator was simulated using a JWL equation of state and an Arrhenius's type kinetic of detonation. The mesh size was 50 μm square.

3.1. Velocities

Three PDV probes were used to measure the beginning of the acceleration of the target at r = 7 mm (on T2 a fourth PDV probe was used to follow the center of the target but did not give relevant results). These probes record only a few microseconds of the flight due to the detonation products which shadow the target.

For each shot, the three probes recorded the same signal which confirms the flat flight of the target. As can been seen in Fig. 7, simulations well fit the signals. It

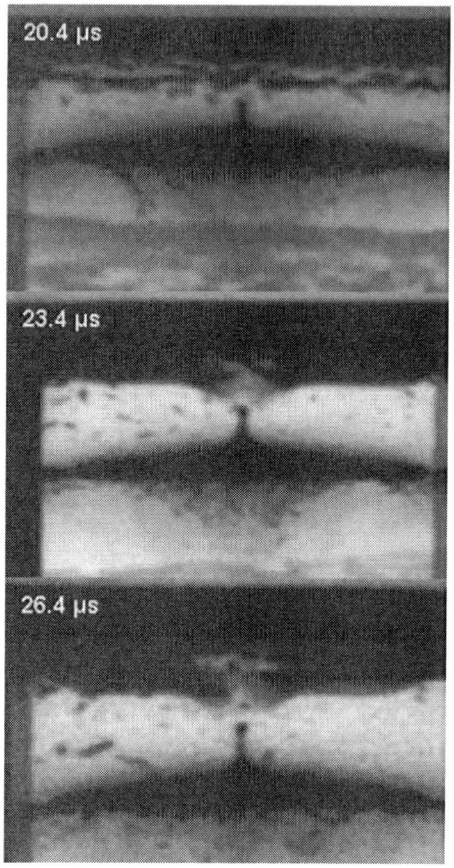

Figure 6. Radiographies from the second experiment.

can be noticed that velocities on T2 were recorded on a longer duration. This can be explained by the differences on the choice of the light support (aluminum for T1, copper for T2). The copper piece is heavier and moves slower than the aluminum one. In T2, the detonation products had to wait longer before the support let them leak in direction of PDVs. This allows us to check that the second acceleration of the target (around 12 μs) is also well reproduced.

3.2. Targets position

For longer times, we focused on the position of the target (r = 10 mm in front of PWG) at radiographic times. This position was measured on films and on simulations using the same technique. The comparison between experiments and numerical simulations is presented in Fig. 8 and shows a very good agreement for the two shots.

Having the first part of the acceleration with PDVs and the later part with the position of the target, we can conclude that the simulation well describes the pressure law generated by the plane wave generator and applied to the target during the experiment. The simulation of the growth of the perturbation can now be discussed in detail.

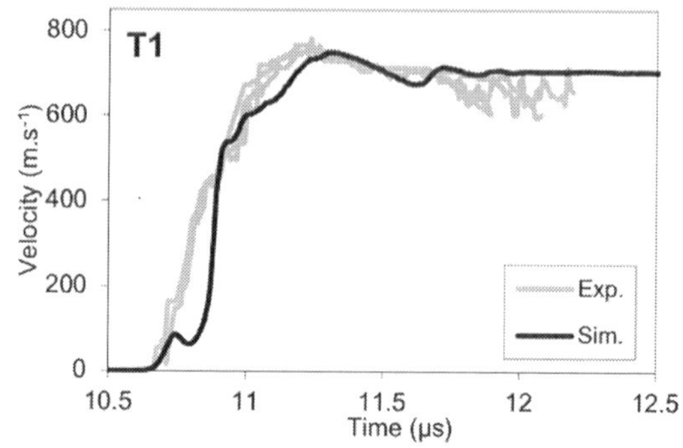

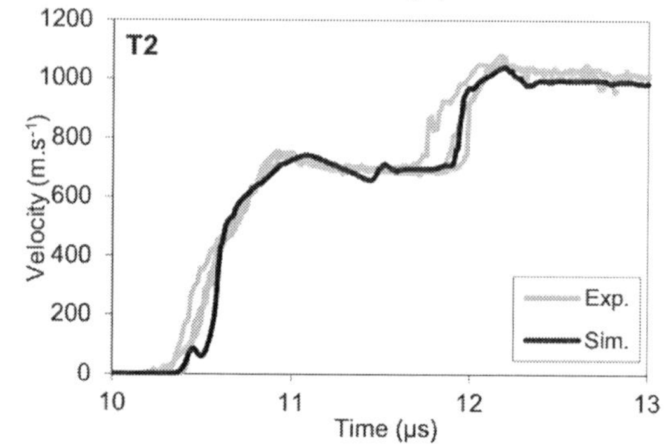

Figure 7. Velocities at r = 7 mm: Comparison between experiments and simulations.

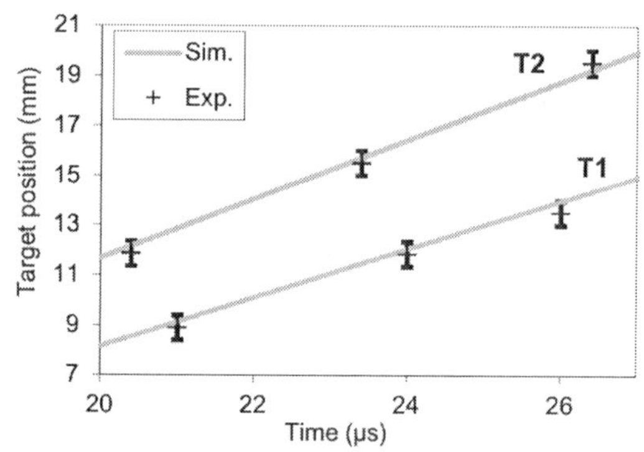

Figure 8. Target position: comparison between experiments and simulations.

4. Perturbation length

4.1. Strength law

The method used to adjust the parameters of the strength law of gold on the two experiments is discussed here. The material used for the targets was previously characterized and a Steinberg Cochran Guinan's law (SCG [5]) was set. The formulation of the law is reminded in formulas (1) for the shear modulus (G) and (2) for the elastic limit (Y). As gold is a fcc metal, its dependency to strain rate should be weak and the SCG model should be sufficient for a first approach.

$$G = G_0 \left[1 + \frac{G'_P}{G_0} P \left(\frac{\rho}{\rho_0}\right)^{-\frac{1}{3}} + \frac{G'_T}{G_0} (T - T_0) \right] \quad (1)$$

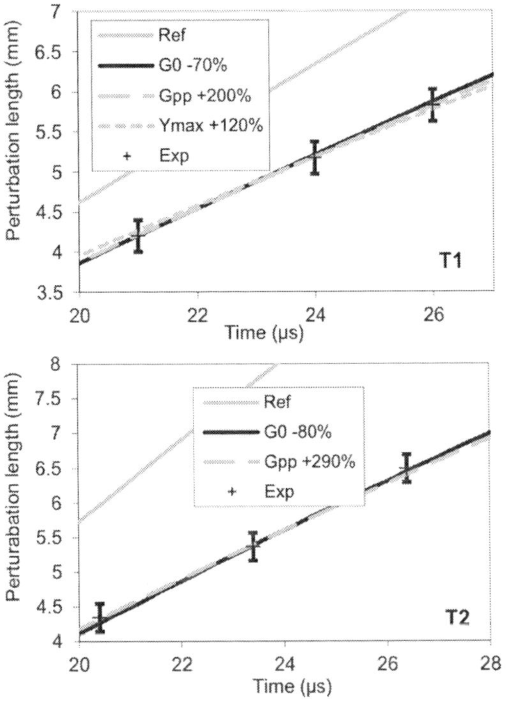

Figure 9. Perturbation length for the two shots and comparison between fitted simulations.

$$Y = \frac{G(P,T)}{G_0} \min\left[Y_0(1+\beta\,\varepsilon_P)^n\,;\,Y_{max}\right]. \quad (2)$$

Various simulations have been led to test the growth rate sensibility to the different part of the material description (equation of state, damage, mesh…). Only the strength law exhibits a real effect on the growth rate.

4.2. Uncoupled simulations

In this part, the parameters of the strength law were adjusted separately for the two experiments. The initial shear modulus (G_0), its pressure derivative (G'_P = Gpp in figures) and the maximal elastic limit (Y_{max}) were fitted separately for the two shots. Results are shown in Fig. 9. In both cases, the simulations using the reference parameters (ref in the following figures) overestimate the perturbation length. By modifying the parameters, simulations can be adjusted at the three experimental times. But, the changes of the parameters are not consistent between the two experiments. Thus, the pressure derivative of the shear modulus has to be increased at +200% for T1 and +290% for T2 whereas no adjustment can be achieve using Y_{max} for T2. Other tests have been carried out with other sensitive parameters without more success.

One can notice that the behavior is opposite between the initial shear modulus (G_0 − 70% for T1) and its pressure derivative (G'_P + 200% for T1) whereas they both seem to play in the same way in formulas. In fact, the difference comes from one hypothesis of the SCG model where the pressure derivatives of the shear modulus and of the elastic limit are equal (term $G(P,T)/G_0$ in Eq. (2)). If the two derivatives (G'_P and Y'_P) are separated, it can be observed on simulations that the dependency raises from the pressure derivative of the elastic limit (Y'_P) which has to be increased by +200% for T1.

This first test evidences that the growth rate of the two experiments using different initial perturbation shapes cannot be adjusted using only one parameter. A coupled research using more parameters will be discussed in the next paragraph.

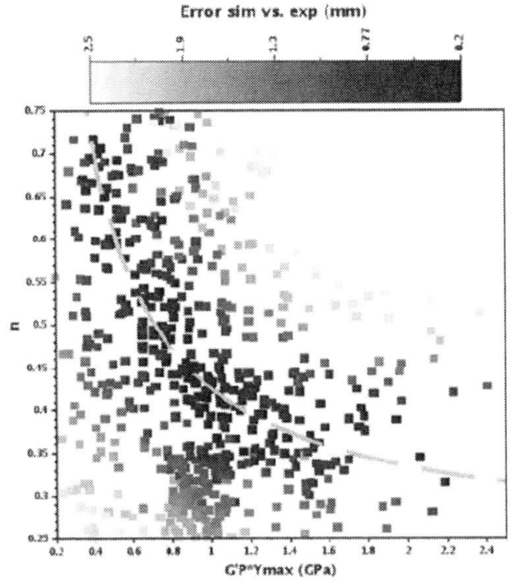

Figure 10. Perturbation length: Distance from simulations to both experiments (dash line = minimum error line).

4.3. Coupled simulations

In this part, the two experiments were adjusted simultaneously using more than one parameter at a time. Due to the number of parameters to test and the time required by each simulation, a Monte-Carlo's like approach where several sets of parameters are computed using a random law was used. For each set, the simulations for the two shots were computed and the distance between the experimental perturbation lengths and the numerical ones was measured.

The first run of simulations did not show any consistency until a dependency of several parameters was exhibit. Thus, Fig. 10 shows the distance between simulations and experiments as a function of the hardening parameter n and the product $G'_P Y_{max}$. In this graph, the existence of a valley of minimum error (symbolized by a dash line) which emphasis a correlation between these parameters can be observed.

This correlation between n and Y_{max} in the SCG model can be understood using an energetic approach. Indeed, for a given experiment (energy given by the pressure law ≈ mechanical energy) and a given perturbation growth (≈ strain), different sets of parameters can achieve the same final mechanical energy (Fig. 11).

If the iso values of the mechanical energy are plot as a function of the two parameters (Fig. 12), the same shapes

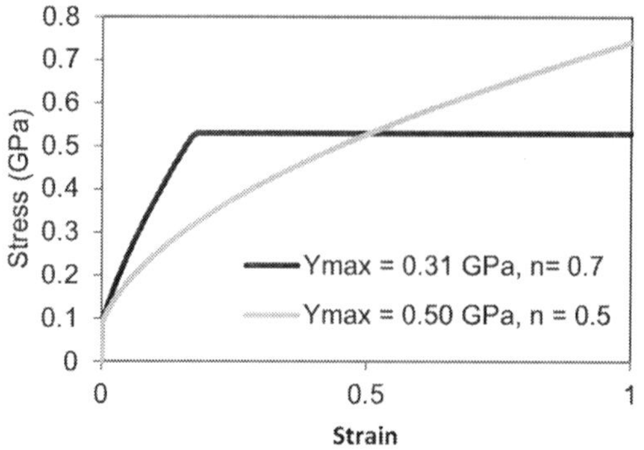

Figure 11. Stress/strain curves for two sets of parameters having the same mechanical energy at strain = 1.

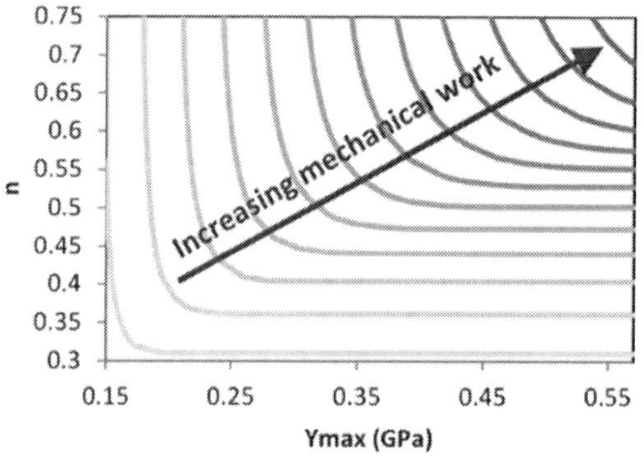

Figure 12. Iso-mechanical energy for a given strain as a function of n and Y_{max}.

as observed in Fig. 10 is found. The two experiments having the same characteristics (same explosives, same distance), the pressure is similar on both experiments and the influence of G'_P is the same.

From the research of parameters, the five best sets can be extracted. They all show various values (1st set: $G'_P = 4.98$, $n = 0.35$, $Y_{max} = 295$ MPa, 5th set: $G'_P = 3.97$, $n = 0.41$, $Y_{max} = 264$ MPa) but with an error to experiment lower than the experimental uncertainty. Target shapes at latest time are well reproduced as shown in Fig. 13.

Using several parameter modifications, we succeeded to simulate the two experiments with the same strength law. From this study, we have also shown that these sets are probably not unique and that other solutions can be achieved.

4.4. Comparison to characterization experiments

To try to reduce the possibilities, the best sets were compared to characterization experiments. None of them allow reproducing the behavior identified in the past and well fitted by the reference set (Fig. 14). At the opposite, the new parameters almost fit data from Suzuki et al. [6].

In their article, Suzuki et al emphasizes the sensitivity of pure gold to its microstructure and the way samples are prepared. A first hypothesis to explain the difference with previous characterization can be an effect of the machining

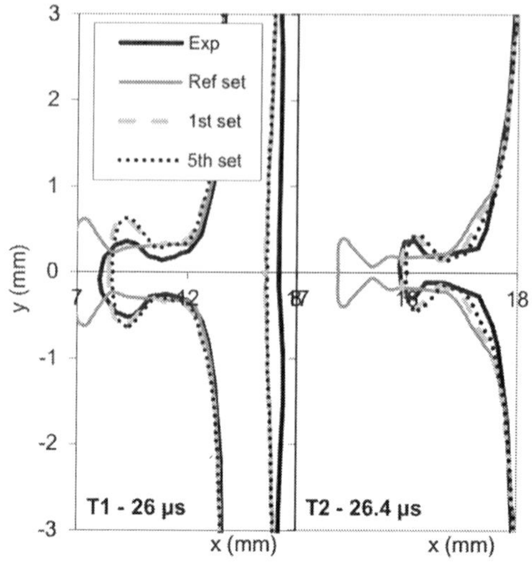

Figure 13. Comparison between experiment and simulated shapes for the last radiography of the two experiments.

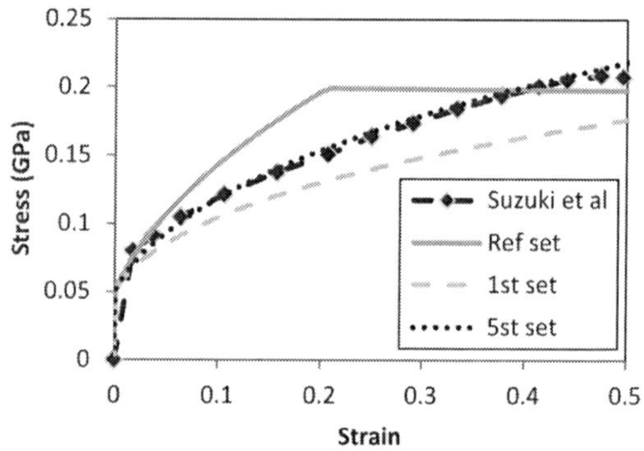

Figure 14. Stress/Strain curves for different sets of parameters.

of the target on its microstructure and then on the strength law.

We can also wonder about the validity of the simulations (several tests have been performed to limit some numerical effects) or of the SCG law for such experiment. As the difference of solicitations is important between our experiments and the characterization ones, we can wonder if the same deformation mechanisms are activated in both cases.

5. Conclusion

From the initial works of Barnes et al, we proposed a new setup using an axisymmetric perturbation which allows realizing easily multiple radiographies during the same experiment.

We test our setup on a well-known material, pure gold, with two different initial perturbation shapes. From PDV and target positions, we validated the pressure law generated by the release of detonation products and started discussing about the perturbation growth.

After several tries, we succeeded in fitting some parameters of the SCG law to the two experiments.

We have shown that these sets are not unique and present some notable differences with the reference issued from the characterization experiments. This disparity can be explained by the sensitivity of pure gold to its microstructure. We also wonder about the validity of our simulations and of the SCG law to compute such experiments.

Several works will be done in the future to try to answer this question. In particular, new experiments using other materials will be realized and interpreted in the same way.

References

[1] J.F. Barnes et al., "Taylor instability in solids", J. Appl. Phys. **45**, 727 (1974)

[2] J.F. Barnes et al., "Further experimentation on Taylor instability in solids", J. Appl. Phys. **51**, 4678 (1980)

[3] H.S. Parks et al., "Strong stabilization of the Rayleigh-Taylor instability by material strength at megabar pressures", Phys. Plasmas **17**, 056314 (2010)

[4] R.T. Olson et al., "The effect of microstructure on Rayleigh-Taylor instability growth in solids", J. Phys: Conf. Ser. **500** 112048 (2014)

[5] D.J. Steinberg, S.G. Cochran, M.W. Guinan, "A constitutive model for metals applicable at high-strain rate", J. of Applied Physics **51** (3) , march 1980

[6] T. Suzuki et al., "*Strength Enhancement and Deformation Behavior of Gold after Equal Channel Angular Pressing*", Materials Transactions, Vol **45**, n°7 (2004), pp. 2200–2208

EPJ Web of Conferences **94**, 04054 (2015)
DOI: 10.1051/epjconf/20159404054

On what controls the spacing of spontaneous adiabatic shear bands in collapsing thick-walled cylinders

Zev Lovinger[1,2,a], Zvi Rosenberg[1,2], and Daniel Rittel[2]

[1] Rafael – Advanced defence systems, Haifa, Israel
[2] Faculty of Mechanical Engineering, Technion – Israel Institute of Technology

Abstract. Shear bands formation in collapsing thick walled cylinders occurs in a spontaneous manner. The advantage of examining spontaneous, as opposed to forced shear localization, is that it highlights the *inherent susceptibility* of the material to adiabatic shear banding without prescribed geometrical constraints. The Thick-Walled Cylinder technique (TWC) provides a controllable and repeatable technique to create and study multiple adiabatic shear bands. The technique, reported in the literature uses an explosive cylinder to create the driving force, collapsing the cylindrical sample. Recently, we developed an electro-magnetic set-up using a pulsed current generator to provide the collapsing force, replacing the use of explosives. Using this platform we examined the shear band evolution at different stages of formation in 7 metallic alloys, spanning a wide range of strength and failure properties. We examined the number of shear bands and spacing between them for the different materials to try and figure out what controls these parameters. The examination of the different materials enabled us to better comprehend the mechanisms which control the spatial distribution of multiple shear bands in this geometry. The results of these tests are discussed and compared to explosively driven collapsing TWC results in the literature and to existing analytical models for spontaneous adiabatic shear localization.

1. Introduction

The initiation and growth of shear bands have been extensively studied since the pioneering work of Zener and Hollomon [1], using various experimental techniques with which the formation of a shear band is well defined. The Kolsky - bar technique is widely used for the study of shear localization with specially designed shear specimens, such as the "Hat" [2] and "Punch" [3] geometries, or the Shear Compression Specimen (SCS) [4,5]. The torsion bar is used to induce ASB in tubular specimens. This technique was used in the seminal work of Marchand and Duffy [6], who identified the successive stages in the evolution of the adiabatic shear bands. Other experiments used pre-notched plates with a symmetric double notched specimen, as in Kalthoff [7], or an eccentric single notched specimen, as in Mason [8] and Zhou [9]. The common basis for all of these well-used techniques is that they all refer to a state of *forced* shear localization. In such cases, shear bands are formed in *well-defined regions*, dictated by the specimen geometry, parallel to the direction of the maximum shearing load. A controlled shear band has many advantages, enabling one to study the evolution of the shear localization process. Yet, the process of shear initiation is in fact dictated geometrically, and therefore externally forced

In order to examine the initiation stage of shear localization, a different set-up is used, referring to *spontaneous* shear localization (as referred to by Chen et al. [10]). In this case, the loading characteristics do not constrain the specific locations or directions of shear banding and multiple shear bands can evolve in regions and in directions other than the direction of the prescribed load. The major advantage of examining spontaneous shear localization is that it highlights the inherent susceptibility of the material to adiabatic shear banding.

The evolution of shear bands in the collapse of a thick-walled cylinder is an excellent example for spontaneous shear localization. In this case, an external pressure induces an inward collapse of the cylinder. The cylindrical symmetry lacks a transverse boundary, and shear bands which evolve from the inner surface of the collapsing cylinder, do not form on a clear boundary or by stress concentration. Free of transverse boundary constraints, the initiation of the shear bands occurs spontaneously. Nesterenko et al. [11,12] introduced a well-controlled and repeatable technique to create multiple shear bands in collapsing thick walled cylinders driven by an explosive loading.

Stokes et al. [13] conducted several TWC experiments on the Pegasus-II facility using electro-magnetic forces as the driving collapsing force. A pronounced advantage of this loading technique is due to the more uniform collapse of the specimen, as compared with the explosively driven ones, which are also driven by an axial force due to the directional ignition of the explosive driver. More recently, we introduced an electromagnetically driven version of the TWC technique [14,15], using a pulsed current generator. This method has the advantage that it is highly controllable and somewhat simpler than the use of explosives or alternatively, the use of a large scale Mega-Joule energy facility. In a TWC experiment, one can measure the number, spacing and length of the spontaneously initiated

[a] Corresponding author: cvloving@gmail.com

ASBs, as in [11] and [14]. Consequently, the TWC techniques, either explosive or electromagnetic, offer an ideal experimental platform to validate existing models for ASB formation, and use the data to calibrate material models for shear band formation.

In this work, we conducted on the EM platform tests on 7 mettalic alloys, examining the number of shear bands and spacing at different stages of collapse. For some of the materials we compared the results with those obtained in the explosively driven tests reported in the literature. We further compared the results with theoretical models in the literature for evaluating the spacing between shear bands.

2. Experimental work

2.1. Experimental set-up

We are using an Electro-Magnetic (EM) set-up to provide the collapsing force on the cylindrical specimen. The EM set up is based on a pulsed current generator (PCG) using a capacitor bank system. The capacitors are charged with voltages of ~30 kV which are released using an array of low inductance rail spark-gaps, allowing a fast discharge into the conductor system. The currents created are of the order of 1 MA, with a rise-time of ~1 μsec. The conductor system is made of two plates, each attached to an opposite pole, separated by an isolating layer. The specimen is an assembly of coaxial cylinders: an inner Cu cylinder, the metallic specimen, an outer Cu cylinder (attached to the upper conductor plate), an insulating plastic cylinder and an outer brass cylinder (attached to the lower conductor plate). When the voltage is discharged, a current flows on both sides of the plastic isolator: through the outer Brass cylinder and the outer Cu cylinder, but in *opposite* directions, creating repulsive magnetic forces between them. The controllable parameters in the set-up are the voltage to which we charge the capacitors, resulting in different magnetic pressures, and the specimen's geometry, controlling the extent of collapse and the stage of shear band evolution. The main diagnostics is post-mortem: the collapsing cylinders, which come to a stop at the end of the experiment, are cut and polished to reveal the spatial distribution of shear bands and the spacing and lengths of the bands are measured.

2.2. Experimental results

Using the EM set-up described above, we conducted tests on 7 different materials: Stainless Steel – ss304L, Titanium (CP grade2), Ti6Al4V – Wrought material, Ti6Al4V – LMD (Laser metal deposition – "printed" material) was chose to examine the influence of microstructure and grain size in comparison with the wrought material, cast aluminum Al-A356, cast Magnesium Mg-AM50 and OFHC Copper: heat treated copper to vary grain size and examine the effect of grain-size on shear band distribution. The studied effects of grain size on shear band evolution are not discussed here.

The polished and etched specimens reveal the multiple shear bands which evolved in the specimen at different stages of collapse. For each specimen we characterized: the final dimensions (these measurements determined

Table 1. Results summary for all tested materials.

Material	Avg. Spacing [mm]	Avg. Number of SB	Grain size [mm]	Compression failure strain
CP-Ti	0.205	32	150	0.50
Ti6Al4V wrought	0.46	20	2	0.20
Ti6Al4V LMD	0.43	21	100	0.20
ss304L	0.14	52	100	0.45
Mg-AM50	0.24	35	10	0.30
Al-A356	–	No SB	100	>0.70
OFHC-Cu With different Grain sizes	–	No SB	20–300	>0.85

the final effective strain which were reached at each experiment), the number of shear bands (counting all shear bands at all lengths) and the average spacing between shear bands. We found that the number of shear bands remained constant during the collapse, meaning no new shear bands initiate during the later stages of propagation. Thus, the numbers of shear bands in specimens at an early stage or late stage of evolution are comparable. The number of shear bands and spacing in all tested materials are summarized in Table 1.

It can be seen that we reached with the different materials very different behaviors: materials with a large number of shear bands and small spacing (ss304L), materials with a small number of shear bands and large spacing (Ti6Al4V materials), an intermediate behavior for CP-Ti and Mg and materials which showed no shear band evolution up to plastic strains of 85% (Cu and Al).

3. Results and discussion

In this study we examined our experimental results to answer two questions:

(1) How do our results compare with the larger specimens tested in the explosively driven thick wall cylinder (ED-TWC) experiments in the literature?
(2) How do our results compare with the theoretical approaches predicting spacing, for different materials?

3.1. Comparison with ED-TWC experiments

The geometry of the specimens we tested in the research, using the EM platform, is scaled down by a factor of 4 in comparison with the geometry tested in the explosively driven experiments [16,17]. As we examined specimens at different stages of collapse, the spacing between shear bands change (become smaller) with the extent of collapse. Thus, we normalized the results in each specimen to the geometry upon initiation and used this definition for the ED-TWC specimens reported in the literature as well. The data in the literature enabled us to compare our results for ss304L, CP-Ti and Ti6Al4V. The comparison is given in Table 2.

Table 2. Comparison of shear bands' spacing between the EM and Explosively driven tests.

Material	Electro-magnetically driven TWC-tests (this work) [mm]	Explosively-driven TWC tests [mm]
Ti	0.21	0.24
Ti6Al4V	0.46	0.55
ss304	0.14	0.14

The comparison between the two cases shows an overall good agreement with differences of only 10%–20%, despite the large scale-factor. The slightly larger spacing in the explosively driven tests could be possibly explained by the smaller strain rate (strain rates in the explosively driven tests are $\sim$6–8 $\cdot 10^4$ and in the EM driven tests are $\sim$2–3 $\cdot 10^5$).

We concluded from this comparison that the spacing between shear bands does not depend on geometrical dimensions of the specimen (e.g radius, wall thickness) but rather on material properties.

3.2. Comparison with theoretical spacing models

There are several theoretical models in the literature for evaluating the spacing between shear bands in TWC tests. The two main models, referring to the spacing at the initiation stage, are those of Wright and Ockendon [18] and of Molinari [19]. The published experimental results are compared with these models. These models are developed following a perturbation analysis on the set of conservation equations including thermal conductivity and a constitutive equation. The basis for these analyses is the classical approach for shear band evolution by thermal softening overcoming strengthening effects of strain and strain rate. The spacing is calculated as the wavelength which shows the fastest growth-rate. The expressions derived by Wright & Ockendon [18] and Molinari [19] are:

$$L_{WO} = 2\pi \left[\frac{k \cdot C}{\dot{\gamma}_0^3 a^2 \tau_0}\right]^{1/4} \cdot m^{3/4},$$

$$L_{Mo} = 2\pi \left[\frac{k \cdot C}{\dot{\gamma}_0^3 a^2 \tau_0}\right]^{1/4} \times \left[\frac{m^3(1-aT_0)^2}{(1+m)}\right]^{1/4} \quad (1)$$

$$\tau = \tau_0 \left[1 - a(T - T_0)\right] \left(\frac{\dot{\gamma}}{\dot{\gamma}_0}\right)^m$$

where τ_0 is the static shear stress, a is the thermal sensitivity coefficient, $\dot{\gamma}$ is the strain-rate, m is the strain-rate sensitivity, $\dot{\gamma}_0$ is the reference strain-rate set to be 1e4 $\sec^{-1}$, k is thermal conductivity and C is the heat capacity. Strain rate in our tests were taken from the numerical simulations and we used here a value of $\dot{\gamma}$ =2–3 $\cdot 10^5 \sec^{-1}$. For the explosively driven tests, strain rate was $\dot{\gamma} = 6$–$8 \cdot 10^4 \sec^{-1}$. The calculated spacings suing Eq. (1) in comparisson with the measured ones are presented in Table 3.

Table 3. Calculated spacing using Eq. (1).

Material	Model Model	Calculated [mm]	Measured [mm]
CP-Ti	WO	0.075	0.205
	Molinari	0.07	
Ti6Al4V (Wrought)	WO	0.11	0.46
	Molinari	0.10	
Ti6Al4V (LMD)	WO	0.11	0.43
	Molinari	0.10	
ss304L	WO	0.24	0.14
	Molinari	0.22	
Mg-AM50	WO	0.31	0.24
	Molinari	0.29	
Al-A356	WO	0.21	— (>0.70)
	Molinari	0.19	
OFHC- Cu	WO	0.42	— (>0.85)
	Molinari	0.39	

From the comparison between the calculated and the experimental spacings, we drew the following conclusions:

(1) The theoretical expressions do not match the experimental results, as far as spacing is considered.

(2) The high strain rate in our tests has a very pronounced effect on the calculated spacing values. While for the explosively driven tests, the spacing values for CP-Ti and ss304L showed a limited match, the higher strain rate (by a factor of 5) result in much smaller spacing values for which the matching is lost. The effect of the strain rate overrides all other differences in the coefficients, according to a variety of values reported in the literature.

(3) The models predict small differences in the spacing of CP-Ti and Ti6Al4V, indicated also in the work of Xue et al. [17]. Yet, the experimental results show **a factor of 2–3** between the two materials.

(4) The models predict small spacing values for Aluminum and Copper which do not agree with our experimental results, in which no shear bands evolved.

4. Summary and conclusions

In this work, we examined the shear band evolution in TWC tests on 7 metallic alloys, using an EM platform for driving the collapsing cylinders. We measured and compared the shear band spacing in the different materials and compared them with results obtained in larger scaled, explosively driven tests reported in the literature and to theoreical predictive models for spacing.

From the examination of the results obtained on the EM platform using scaled down geometry in comparison with the explosively driven specimens reported in the literature, we found that spacing is not characterized by a geometrical scale of the specimen size (e.g radius, wall thickness) but rather by material properties. The examination of the different materials enabled us to better comprehend the mechanisms which control the spatial distribution of multiple shear bands in this geometry. The incompatibility of the 1D models (W&O, Molinari) to predict spacing in the experiments strengthens the thesis that the thermal softening approach does not capture the full physics of shear localization. Further examination of these conclusions is currently examined through numerical simulations using energy based failure criteria which are indicating that plasticity mechanisms of the material are dominating the spatial evolution of shear bands in this geometry.

The financial support of the Israel Science Foundation – ISF (grants 2017688 and 2109912) and of the Pazi Foundation are kindly acknowledged.

References

[1] C. Zener, J. H. Hollomon, J. of Appl. Phys, pp. 22–32 (1943).
[2] L.W. Meyer and S. Manwaring, *Metallurgical Applications of Shock-Wave and High Strain Rate Phenomena* (Marcel Dekker, NY, 1986).
[3] W. Dabboussi, J.A. Nemes, Int J. Mech. Sci., Vol **47**, pp. 1282–1299 (2005).
[4] A. Dorogoy and D. Rittel, Exp. Mech., Vol. **45** (2), 167–177 (2005).
[5] A. Dorogoy and D. Rittel, Exp. Mech., Vol. **45** (2), pp. 178–185 (2005).
[6] A. Marchand, J. Duffy, J. Mech. Phys. Solids Vol. **36** No. 3 (1988).
[7] J.F. Kalthoff, S. Wrinkler, *Impact Loading and Dynamic Behavior of Materials* (EDS), Vol. **1**, pp. 185–195 (1987).
[8] J.J. Mason, A.J. Rosakis, G. Ravichandran, J. Mech. Phys. Solids **42**, 1679–1697 (1994).
[9] M. Zhou et al., J. Mech. Phys. Solids Vol **44**, No. 6, pp. 981–1006 (1996).
[10] Y.J. Chen et al., Mat. Sci. and Engineering A **286**, pp. 70–82 (1999).
[11] V.F. Nesterenko, M.P. Bondar, Expl. and Shock Waves, Vol **30** (1994).
[12] V.F. Nesterenko et al., Fizika Goreniya i Vzryva, vol. **25**, no. 4, pp. 154–155 (in Russian) (1989).
[13] J.L. Stokes et al., *Proceedings of International Conference on Fundamental Issues and Applications of Shock-Wave and High-Strain-Rate Phenomena* (Elsevier, 2001).
[14] Z. Lovinger et al., IJIE, Vol. **38**, 918–929 (2011).
[15] Z. Lovinger et al., JMPS (to be published) (2015).
[16] Q. Xue et al., Mat. Sci. and Engineering A **384**: 35–46 (2004).
[17] Q. Xue et al., Acta Materialia **50**, pp. 576–596 (2002).
[18] T.W. Wright, H. Okendon, Int. J. of Placticity, Vol. **12**, No. 7 pp. 927–934 (1996).
[19] A. Molinari, J. of the Mech. Phys. of Solids, Vol. **45**, No. 9 pp. 1551–1575 (1997).

EPJ Web of Conferences **94**, 04055 (2015)
DOI: 10.1051/epjconf/20159404055

Flow and failure of an aluminium alloy from low to high temperature and strain rate

Rafael Sancho, David Cendón, and Francisco Gálvez[a]

Technical University of Madrid, Department of Materials Science, c/ Profesor Aranguren, 28040 Madrid, Spain

Abstract. The mechanical behaviour of an aluminium alloy is presented in this paper. The study has been carried out to analyse the flow and failure of the aluminium alloy 7075-T73. An experimental study has been planned performing tests of un-notched and notched tensile specimens at low strain rates using a servo-hydraulic machine. High strain rate tests have been carried out using the same geometry in a Hopkinson Split Tensile Bar. The dynamic experiments at low temperature were performed using a cryogenic chamber, and the high temperature ones with a furnace, both incorporated to the Hopkinson bar. Testing temperatures ranged from $-50\,^{\circ}$C to $100\,^{\circ}$C and the strain rates from $10^{-4}\,s^{-1}$ to $600\,s^{-1}$. The material behaviour was modelled using the Modified Johnson-Cook model and simulated using LS-DYNA. The results show that the Voce type of strain hardening is the most accurate for this material, while the traditional Johnson-Cook is not enough accurate to reproduce the necking of un-notched specimens. The failure criterion was obtained by means of the numerical simulations using the analysis of the stress triaxiality versus the strain to failure. The diameters at the failure time were measured using the images taken with an image camera, and the strain to failure was computed for un-notched and notched specimens. The numerical simulations show that the analysis of the evolution of the stress triaxiality is crucial to achieve accurate results. A material model using the Modified Johnson-Cook for flow and failure is proposed.

1. Introduction

As a consequence of the structural integrity degradation caused by impact loads, coupled with cost savings, impact simulations are becoming more common during design process.

Simulating the impact behaviour of a material involves the need to reproduce the flow and fracture behaviour of such material accurately, taking into account large strains, strain rate effects and thermal softening. Usually, the constitutive equations used for that purpose define the equivalent stress as a function of the equivalent plastic strain, equivalent plastic strain rate and temperature (Johnson-Cook constitutive equation, 1983 [1] or Zerilli-Armstrong, 1987 [2]). Furthermore, experimental tests have demonstrated that fracture behaviour of ductile materials strongly depends on stress triaxiality (Johnson-Cook, 1985 [3]; Børvik et al., 2003 [4], Xue-Wierzbicki, 2004 [5]); being smaller the equivalent fracture strain for high stress triaxiality values.

In the present work, authors study the flow and failure of the alloy AA 7075-T73 at different temperatures and strain rates, trying to describe its mechanical behaviour using the modified Johnson-Cook model proposed by Børvik et al. [6] and implemented in the LS-DYNA numerical code.

Numerical simulations, using LS-DYNA finite element code, were carried out to determine a more suitable fracture envelope [7,8] than that using Bridgman's stress triaxiality formulation [9].

[a] Corresponding author: fgalvez@mater.upm.es

2. Material model

Due to the wide spread of the Johnson-Cook model for describing the flow and fracture behaviour of ductile metals under impact loads, a Johnson-Cook type model was chosen.

2.1. Constitutive model

Based on the hypothesis of strain equivalence (Lemaitre, 1992 [10]), Børvik et al. [6] reformulated the Johnson-Cook model [1,3] in order to take into account damage evolution in the constitutive equation, replacing the Cauchy stress tensor σ by an effective stress tensor $\tilde{\sigma}$:

$$\tilde{\sigma} = \frac{\sigma}{1-\beta D} \tag{1}$$

where D is the damage parameter and β is a flag to control the coupling of the model. Note that β can only take the values 0 and 1, switching the modified Johnson-Cook model from the uncoupled to the coupled material model.

The constitutive relation, using Voce type hardening [12,13], reads:

$$\tilde{\bar{\sigma}} = \left\{A + B\bar{r}^n = \sum_{i=1}^{2} Q_i\left[1-\exp(-C_i\bar{r})\right]\right\} \times \left(1+\dot{\bar{r}}^*\right)^c\left(1-T^{*m}\right) \tag{2}$$

$A, B, n, Q_1, Q_2, C_1, C_2, c$ and m are material constants, $\dot{\bar{r}}^* = \dot{\bar{r}}/\dot{r}_0$ is a dimensionless strain rate, being $\dot{r}_0$ a user-defined strain rate and $T^* = (T-T_r)/(T_m-T_r)$

the homologous temperature. T_r is the reference temperature (usually the room temperature), T_m the melting temperature and T the current temperature.

Assuming adiabatic conditions, the rate of temperature increase is defined as:

$$\dot{T} = \frac{\chi}{\rho C_p} \boldsymbol{\sigma} : \dot{\boldsymbol{\varepsilon}}_p \tag{3}$$

where ρ is the density, C_p is the specific heat at constant pressure and χ is the Taylor-Quinney coefficient, which is often assigned the value 0.9 for the majority of metals [14].

2.2. Failure criterion

Failure criterion is based on damage evolution; but unlike the Johnson-Cook model (1985) [3], damage does not start to accumulate until a threshold equivalent plastic strain $\bar{\varepsilon}_p^d$ is reached. The material fails when damage gets to a critical value D_c. The damage evolution law is then defined as:

$$\dot{D} = 0 \quad \text{for } \bar{\varepsilon}_p \leq \bar{\varepsilon}_p^d \tag{4}$$

$$\dot{D} = D_c \frac{\dot{\bar{\varepsilon}}_p}{\bar{\varepsilon}_p^f - \bar{\varepsilon}_p^d} \quad \text{for } \bar{\varepsilon}_p > \bar{\varepsilon}_p^d. \tag{5}$$

The equivalent plastic strain to failure $\bar{\varepsilon}_p^f$ is homologous to the modified strain hardening function:

$$\bar{\varepsilon}_p^f = \left[D_1 + D_2 \exp\left(D_3 \sigma^*\right)\right] \left[1 + \dot{\bar{\varepsilon}}_p^*\right]^{D_4} \left[1 + D_5 T^*\right] \tag{6}$$

where D_1, D_2, D_3, D_4 and D_5 are material constants and $\sigma^* = \sigma_H / \bar{\sigma}$ is the stress triaxiality.

3. Material description

The material studied was the wrought high-strength aerospace aluminium alloy AA 7075 T73. Its principal alloying elements are Zn, Mg and Cu, being Zn and Mg elements which improve the mechanical properties of aluminium significantly. Cu alloying element improves the coupling mechanical behaviour-stress corrosion cracking resistance. The T73 temper was applied to achieve the best stress corrosion resistance.

4. Experimental procedure

In order to calibrate the modified Johnson-Cook model, three groups of experiments (quasi-static, dynamic and dynamic at high and low temperature tensile tests) were carried out.

Different initial stress triaxialities were obtained by means of testing axysimmetric specimens with different notch radii (un-notched, R = 4 mm, R = 2 mm and R = 1 mm) [7]:

$$\sigma^* = \frac{1}{3} + \ln\left(1 + \frac{r}{2R}\right) \tag{7}$$

being r and R the initial cross section and notch radii.

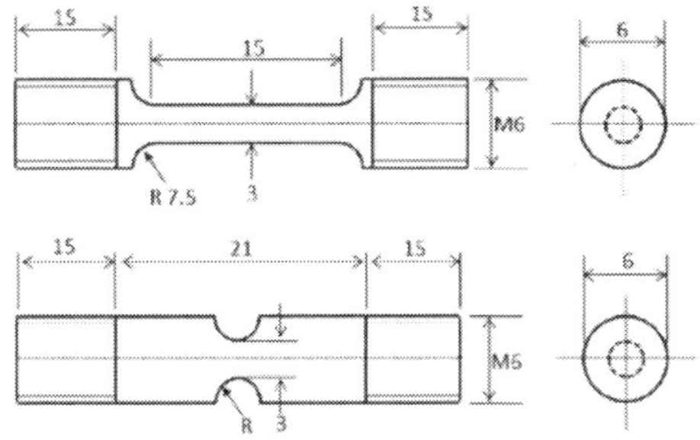

Figure 1. Dimensions [mm] and geometry of the axisymmetric smooth (upper) and notched (lower) specimens used for the experimental procedure.

Geometry and dimensions of the tested specimens are shown in Fig. 1.

4.1. Quasi-static tensile tests

Quasi-static tensile tests were carried out at room temperature and under a strain rate of $2.6 \times 10^{-4}\,\text{s}^{-1}$. The tests were filmed (see Fig. 2) and the initial d_0 and fracture d_f diameters of the samples were determined and utilized for the calculation of the equivalent plastic strain to failure (see Table 1) using the equation:

$$\bar{\varepsilon}_p^f = 2 \ln\left(\frac{d_0}{d_f}\right). \tag{8}$$

The axisymmetric smooth specimens were instrumented with an optical extensometer in order to obtain the axial strain. Assuming volume conservation, nominal true stress σ_N and nominal true strain ε_N from the corresponding engineering values were obtained applying the expression:

$$\varepsilon_N = \ln(1 + e); \quad \sigma_N = s(1 + e) \tag{9}$$

where s and e are the engineering stress and strain respectively. Taking into account the additive decomposition of the strains, the true nominal plastic strain is calculated as:

$$\varepsilon_{PN} = \varepsilon_N - \varepsilon_e = \varepsilon_N - \frac{\sigma_N}{E} \tag{10}$$

where ε_e is the elastic strain and E is the Young's modulus.

4.2. Dynamic tensile tests

Dynamic tensile tests at room temperature were carried out with a Split Hopkinson Tension Bar (SHTB). The tests were conducted at an approximate strain rate of $600\,\text{s}^{-1}$. Considering one-dimensional wave propagation inside the input and output bars [14, 17], the specimen engineering stress is given by:

$$s = \frac{F}{A_0} = \frac{E_b A_b}{A_0} \varepsilon_t \tag{11}$$

where F is the force applied over the specimen by the SHTB, E_b is the elastic modulus of the input and output bars, A_b is the input and output bar cross-section area

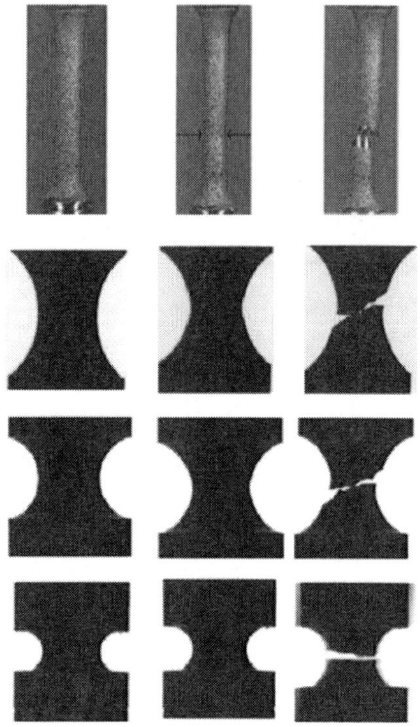

Figure 2. Video frames for axysimmetric smooth and notched specimens. Figure shows one representative specimen for each one of the different geometry tested under quasi-static conditions. From left to right, the initial, just-before-fracture and just-after-fracture frames.

and ε_t is the strain of the transmitted wave. The specimen engineering strain e and strain rate $\dot{e}$ are:

$$e = -\frac{2c_0}{l_s}\int_0^t \varepsilon_r dt \tag{12}$$

$$\dot{e} = -\frac{2c_0}{l_s}\varepsilon_r \tag{13}$$

where $c_0 = \sqrt{E_b/\rho_b}$ is the elastic wave propagation velocity inside the input and output bars, ρ_b is the mass density of the bars and l_s is the specimen initial length.

The diameters of the fractured specimens were measured and the equivalent plastic strain to failure values were calculated (see Table 1).

It should be stressed that dynamic tensile tests, unlike quasi-static tensile tests, are under adiabatic conditions due to its short duration (about 1 ms), causing an increase of temperature in the material.

4.3. Dynamic tensile tests at various temperatures

Dynamic tensile tests (strain rate ~$600\,s^{-1}$) were also carried out at −50 °C, −10 °C and 100 °C with the SHTB.

The specimens at 100 °C were tested inside a furnace and the specimens at −10 °C and −50 °C were tested inside a cryogenic chamber using dry ice.

Table 1. Equivalent plastic strain to failure data from the experimental tests.

Specimen	σ^*	$\bar{\varepsilon}_p^f$	T [°C]
Quasi-static regime ($2.6 \times 10^{-4}\,s^{-1}$)			
Smooth	0.333	0.407	25
Notched R4	0.505	0.231	25
Notched R2	0.652	0.165	25
Notched R1	0.893	0.129	25
Dynamic regime ($600\,s^{-1}$)			
Smooth	0.333	0.469	25
Notched R4	0.505	0.281	25
Notched R2	0.652	0.212	25
Notched R1	0.893	0.176	25
Smooth	0.333	0.472	100
Notched R4	0.505	0.284	100
Notched R2	0.652	0.212	100
Notched R1	0.893	0.172	100
Smooth	0.333	0.460	−10
Notched R4	0.505	0.279	−10
Notched R2	0.653	0.213	−10
Notched R1	0.893	0.169	−10
Smooth	0.333	0.416	−50
Notched R4	0.505	0.250	−50
Notched R2	0.652	0.209	−50
Notched R1	0.893	0.163	−50

The equivalent plastic strain to failure values were also calculated and are recorded in Table 1.

5. Calibration of the modified Johnson-Cook model

Calibrating the modified Johnson-Cook model means obtaining the material constants that better fit the flow and fracture behaviour of the material. Concerning the fracture behaviour (see Eq. (6)), some authors [7,8,18] have demonstrated that stress triaxiality in the centre of axysimmetric specimens is not constant during tensile test, so numerical simulations were used for studying the stress triaxiality histories.

Numerical simulations of the experimental tests were performed using LS-DYNA finite element code with explicit time integration. The used material model was the modified Johnson-Cook model implemented in LS-DYNA [19]. It is important to point that the material model was considered an uncoupled model setting $\beta = 0$ (see Eq. (1)). Moreover, the threshold equivalent plastic strain $\bar{\varepsilon}_p^d$ and the critical damage value D_c took values of 0 and 1, respectively.

5.1. Mesh size study

Finite element analyses involving strain localization and failure tend to be sensitive to the node density [6], so numerical simulations on meshes with different element densities were carried out. All specimens were modelled using 2D axysimmetric meshes. Smooth axysimmetric specimens were modelled with 5, 10, 15 and 20 elements across the radius; while notched specimens

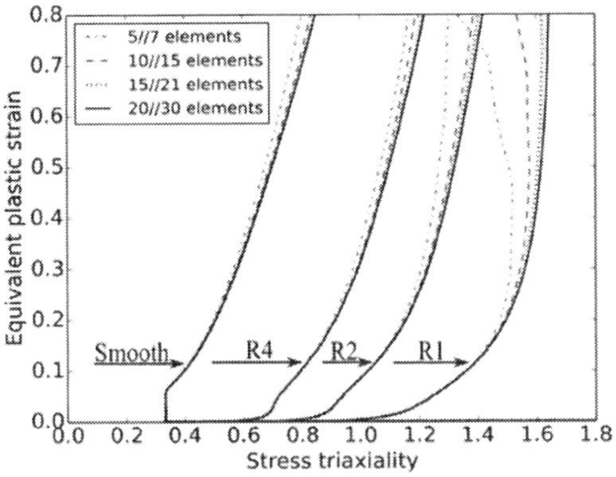

Figure 3. Stress triaxiality evolution of the critical elements for specimens with different mesh densities.

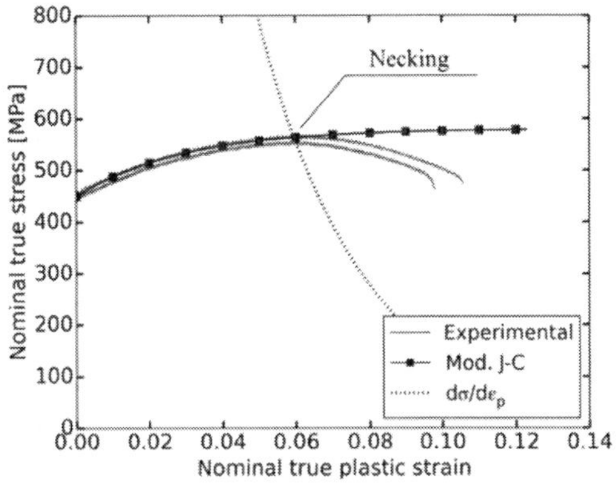

Figure 4. Nominal true stress vs. nominal true plastic strain curves, fitting the material behaviour with the modified Johnson-Cook hardening law with Voce expression.

with notch radii $R = 4$ mm, $R = 2$ mm and $R = 1$ mm, were modelled with 7, 15, 21, 30 elements across the minimum cross section radius.

Once the simulations were performed, stress triaxiality evolution vs the equivalent plastic strain of the critical elements were plotted (see Fig. 3), concluding that 15 element mesh and 21 element mesh are the critical mesh density for smooth and notched specimens.

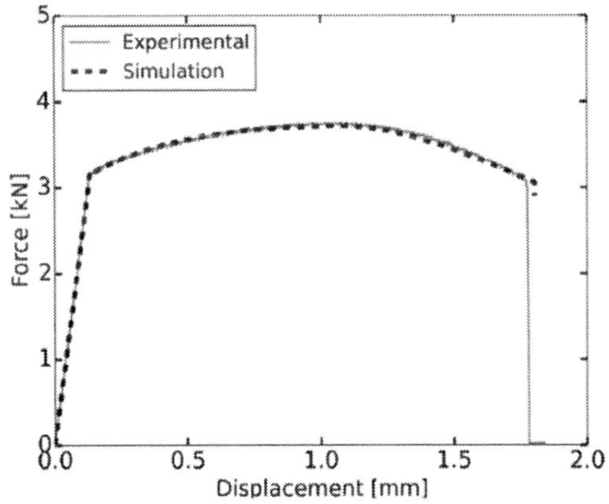

Figure 5. Load-displacement curve of smooth axisymmetric specimen under quasi-static loading condition compared with the numerical simulation.

5.2. Calibration of quasi-static constants

The first group of constants were calibrated using the quasi-static tensile tests of axisymmetric smooth and notched specimens. The specimens were modelled by using four-node 2D axisymmetric elements with reduced integration and a stiffness-based hourglass control. The quasi-static tensile tests were simulated prescribing a displacement to the nodes corresponding to one end of the specimen and fixing the nodes corresponding to the opposite end. Quasi-static loading conditions were controlled [19].

The values of the material constants A, B, n, Q_1, Q_2, C_1, C_2 were determined with the experimental nominal true stress-plastic strain curves. Figure 4 shows that Voce hardening constitutive relation fits quite well with the experimental values. Moreover, numerical simulations for the axisymmetric smooth specimen showed good agreement with the experimental force-displacement curve (Fig. 5).

To determine the fracture parameters D_1, D_2 and D_3, numerical simulations of the quasi-static tensile tests using no failure criterion were performed in order to check which are the most critical elements in the specimens and collect the triaxiality and the equivalent plastic strain histories of such elements until the time step in which the equivalent plastic strain to failure of the simulated specimens were equal to the experimental one. Figure 6 shows clearly that critical elements are those situated in the centre of specimens (axis of symmetry). The stress triaxiality and equivalent plastic strain histories of the critical elements were used to determine, using the Eq. (6), the fracture parameters D_1, D_2 and D_3 (see Fig. 7).

5.3. Calibration of *c*, *m*, D_4 and D_5

The constant c was obtained analysing the quasi-static and dynamic tensile tests of smooth axisymmetric specimens performed at 25°C, being c the slope of a line that relates the natural logarithm of the flow stress at a certain equivalent plastic strain with the natural logarithm of the dimensionless plastic strain rate. The thermal softening exponent m was calibrated using the dynamic tensile tests of smooth specimens performed at −50°C, −10°C, 25 °C and 100 °C (Fig. 8). Since the flow stress at 0.6% equivalent plastic strain of the smooth specimen tested at −50 °C was lower than that at −10 °C and 25 °C, the material constitutive response at that temperature could not be modelled.

Calibration of D_4 was similar to that carried out during the adjustment of D_1, D_2 and D_3, running simulations of the dynamic tensile tests at room temperature without failure criterion. Stress triaxiality and equivalent plastic strain histories from critical elements were collected and used for adjusting D_4 parameter. Figure 9, showed a good agreement between the computed equivalent fracture strain to failure and the measured one experimentally.

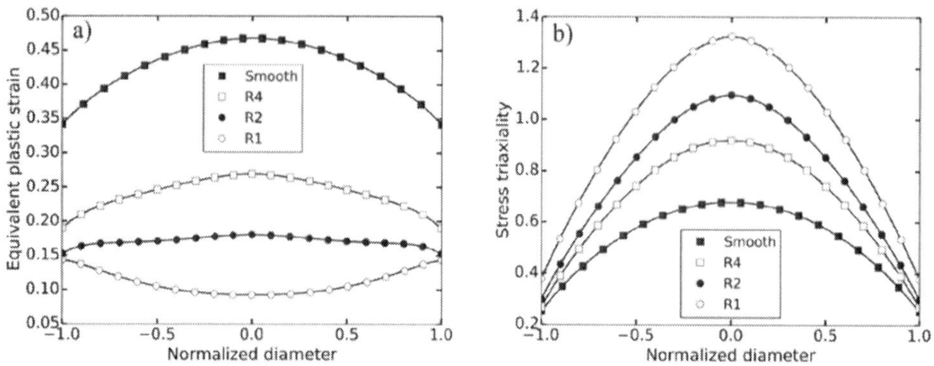

Figure 6. Variation of the equivalent plastic strain (a) and stress triaxiality (b) along the elements of the minimum cross section for the specimens simulated under quasi-static loading conditions. The values from −1 to 0 of normalized diameter were plotted assuming stress triaxiality and equivalent plastic strain symmetry.

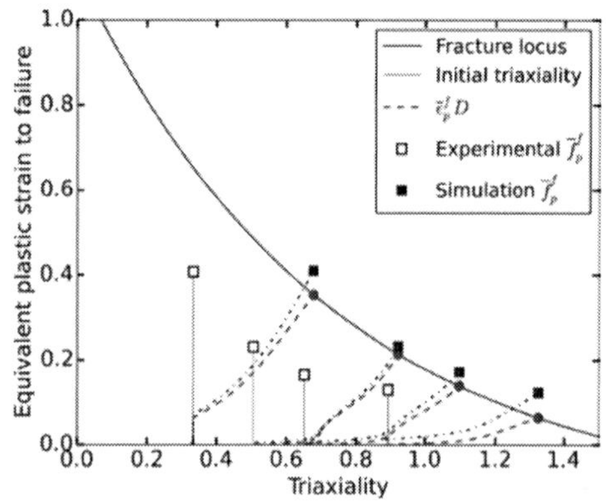

Figure 7. Fracture locus in the equivalent plastic strain to failure vs. triaxiality for the quasi-static regime.

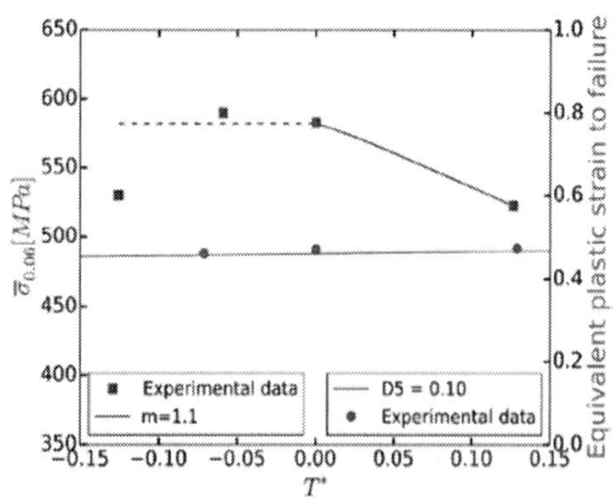

Figure 8. Flow stress at an equivalent plastic strain of 0.6% and equivalent plastic strain to failure against the homologous temperature.

The equivalent plastic strain to failure of the dynamic tensile tests of smooth axysimmetric specimens tested at temperatures of −10°C, 25°C, and 100°C were used for the calibration of the thermal softening constant D_5. Since the constitutive behaviour of the aluminium alloy at −50°C could not be modelled, its fracture behaviour was not taken into account. Equivalent plastic strain to failure vs homologous temperature space, which was calculated taking into account rising temperature in critical elements, was plotted (Fig. 8) to determine the value of D_5 with a linear regression.

Table 2. Modified Johnson-Cook material constants for AA 7075 T3.

Constitutive relation						
A[GPa]	Q_1 [GPa]	Q_2[GPa]	C_1	C_2	c	m
0.45	0.13	0	31	0	0.0026	1.1
Failure criterion						
D_1	D_2	D_3	D_4	D_5		
−0.14	1.25	−1.37	0.02	0.1		

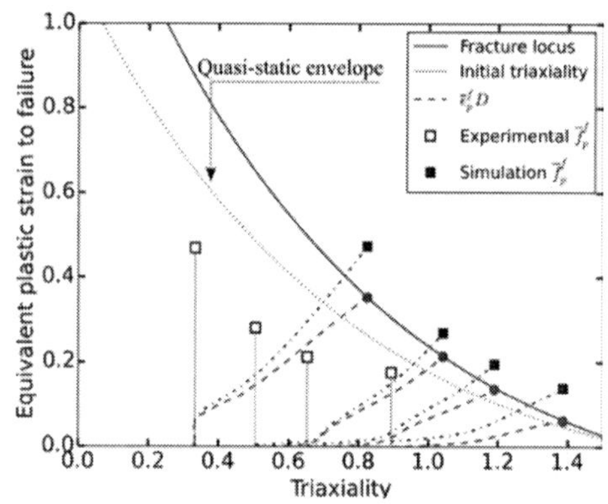

Figure 9. Fracture locus in the equivalent plastic strain to failure vs. triaxiality for dynamic tensile tests at room temperature.

Finally, Fig. 10 shows the good agreement between the experimental nominal true stress-strain curves and those obtained from the simulation.

To conclude, all the constants of the model are showed in Table 2.

6. Concluding remarks

The modified Johnson-Cook material model was calibrated for the aluminium alloy 7075-T73. In order to calibrate the model three groups of mechanical tensile tests were performed: quasi-static at room temperature, dynamic at room temperature and dynamic at low and high temperature.

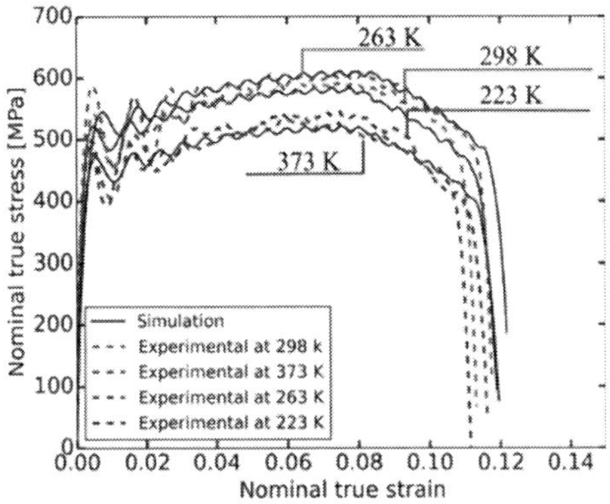

Figure 10. Nominal true stress-strain curves for the experimental and simulated dynamic tensile tests at various temperatures. The curves correspond to the smooth specimens.

To reproduce the constitutive behaviour of the material, Voce hardening law was necessary. Experimental results showed a weakening of the material at −50°C that prevented material modelling at that temperature.

Numerical simulations were needed in order to calibrate the failure criterion since stress triaxiality changes during the loading process. Damage in the material is driven by the accumulated plastic strain and amplified by the stress triaxiality, so the competition between them determines where fracture initiation will occur. In our case, fracture initiates at the centre of the specimen where the stress triaxiality is higher.

Finally, numerical study showed good agreement between the experimental and numerical data; so, a successful calibration was achieved.

They would also like to thank the Spanish Ministry of Science and Innovation for financial support through project BIA2011-24445, and ITP (Industria de Turbopropulsores) for the funding through CENIT-Openaer project.

References

[1] G.R. Johnson, W.H. Cook, in: Proc 7th Int Sympo. Ballistics, 541–547 (1983)

[2] F.J. Zerilli, R.W. Armstrong, J. Appl. Phys. **61**, 1816–1825 (1987)

[3] G.R. Johnson, W.H. Cook, Eng. Fract. Mech **21**, 31–48 (1985)

[4] T. Børvik, O.S. Hopperstad, M. Langseth, K. Labibes, C. Albertini, Eur. J. Mech. A/Solids **22**, 1–13 (2003)

[5] T. Wierzbicki, Y. Bao, Y-W. Lee, Y. Bai, Int. J. Appl. Mech **47**, 719–743 (2009)

[6] T. Børvik, O.S. Hopperstad, T. Berstad, M. Langeseth, Eur. J. Mech. A/Solids **20**, 685–712 (2001)

[7] B. Erice, F. Gálvez, D.A. Cendón, V. Sánchez-Gálvez, Eng. Fract. Mech **79**, 1–17 (2012)

[8] T. Børvik, O.S. Hopperstad, T. Berstad, Eur. J. Mech. A/Solids **22**, 15–32 (2003)

[9] P.W. Bridgman, *Studies in large plastic flow and fracture* (McGraw-Hill, 1952)

[10] J. Lemaitre, *A course on Damage Mechanics* (Springer-Verlag, 1996)

[11] G.T. Camacho, M. Ortiz, Int. J. Comp. Meth. Appl. Mech. Eng. **142**, 269–301 (1997)

[12] E. Voce, J. Int. Met. **74**, 537 (1948)

[13] E. Voce, Metallurgia **51**, 219 (1955)

[14] M.A. Meyers, *Dynamic Behavior of Materials* (John Wiley & Sons, INC, 1994)

[15] F.A. McClintock, J. Appl. Mech. **35**, 363–371 (1968)

[16] J.R. Rice, D.M. Tracey, J. Mech. Phys. Solids. **17**, 201–217 (1969)

[17] H. Kolsy, *Stress Waves in Solids* (Dover, 1963)

[18] S. Dey, T. Børvik, O.S. Hopperstad, M. Langseth, Comput. Mater. Sci. **38**, 176–191 (2006)

[19] *LS-DYNA KEYWORD USER'S MANUAL* (Version 971 R6, 2012)

EPJ Web of Conferences **94**, 04056 (2015)
DOI: 10.1051/epjconf/20159404056

Numerical simulation of ZrO2(Y2O3) ceramic plate penetration by cylindrical plunger

V. Bratov[1,2,a], N. Kazarinov[2,b], and Y. Petrov[1,2,c]

[1] Institute of Problems of Mechanical Engineering, 1199178 Saint Petersburg, Russia
[2] Saint Petersburg State University, 198504 Saint Petersburg, Russia

Abstract. In this paper dynamic fracture process due to high-speed impact of steel plunger into ceramic sample is simulated. The developed numerical model is based on finite element method and a concept of incubation time criterion, which is proven to be applicable in order to predict brittle fracture under high-rate deformation. Simulations were performed for ZrO2(Y2O3) ceramic plates. To characterize fracture process quantitatively fracture surface area parameter is introduced and controlled. This parameter gives area of new surface created during dynamic fracture of a sample and is essentially connected to energetic peculiarities of fracture process. Multiple simulations with various parameters made it possible to explore dependencies of fracture area on plunger velocity and material properties. Energy required to create unit of fracture area at fracture initiation (dynamic analogue of Griffith's surface energy) was evaluated and was found to be an order of magnitude higher as comparing to its static value.

1. Introduction

Dynamic fracture properties of ceramic materials are primarily studied due to applications where ceramic/multi-layered ceramics or ceramic composites serve as a protection against mechanical and thermal impacts on an object [1,2] however theoretical problems primarily include understanding the mechanisms driving static and dynamic fracture of this special class of materials (extremely brittle, very high strength and low toughness [3]).

In this paper an approach that was previously successfully applied to simulate dynamic fracture in other classes of quasi-brittle materials [4,5] was applied to study penetration of ceramic plate by a cylindrical plunger. The focus is primarily on energetic peculiarities of the process, including investigations of histories of fracture surface evolution in fractured ceramic material, calculations of specific fracture energy at dynamic rupture initiation and studies of correlation between new fracture surface within ceramic material created as a result of a plunger impact and initial plunger energy (velocity) as well as the properties of the ceramic material.

This study is deliberately restricted to the simplest possible geometry and contact problem formulation. Thermal effects are not considered in this work. In the considered case, an elastic cylindrical plunger with properties typical for steel normally hits a circular homogeneous ceramic plate, behaving as linear elastic material up to the moment of fracture initiation.

2. Problem formulation

Both the plunger and the target are supposed to be linear elastic with stress-strain state given by the Lame equations and the Hook's law:

$$\rho\frac{\partial^2 U_i}{\partial t^2} = (\lambda+\mu)\nabla_i\left(\nabla\cdot\bar{U}\right) + \mu\Delta U_i \qquad (1)$$

$$\sigma_{i,j} = \delta_{i,j}\lambda\nabla\cdot\bar{U} + \mu\left(\frac{\partial U_i}{\partial x_j} + \frac{\partial U_j}{\partial x_i}\right). \qquad (2)$$

The ceramic plate is supposed to be fixed on its outer radius. The plunger initial velocity V is in the direction normal to the plate surface. Due to the symmetry of the problem, a 2D problem with axial symmetry is solved.

Material properties for the target are typical for ZrO2(Y2O3) ceramics [6,7]. The plunger is supposed to have properties of steel. Tables 1 and 2 give the material properties utilized for the simulation.

3. Fracture criterion

Nowadays it is known and generally recognized that classical fracture criteria (critical stress criterion, critical stress intensity factor criterion, etc.) are inapplicable in order to predict fracture caused by dynamic high-rate loads. Incubation time fracture criterion [8] can be utilized for correct and robust prediction of critical conditions leading to fracture of material loaded by impact loads. In [9,10] this approach was successfully used to simulate dynamic fracture in other classes of quasi-brittle materials. It is supposed that a similar approach can be used to predict fracture initiation, evolution and arrest in ceramic materials [11] for the studied class of problems. The criterion for

[a] e-mail: vladimir@bratov.com
[b] e-mail: nkazarinov@gmail.com
[c] e-mail: yp@yp1004.spb.edu

Article available at http://www.epj-conferences.org or http://dx.doi.org/10.1051/epjconf/20159404056

Table 1. Target material properties.

Material property	Value
Elastic module E, GPa	200
Poisson ratio	0.25
Ultimate stress σ_c, MPa	750
Fracture toughness K_{Ic}, MPa$\sqrt{m}$	13.3
Density ρ, kg/m^2	6000

Table 2. Plunger material properties.

Material property	Value
Elastic module E, GPa	200
Poisson ratio	0.25
Density ρ, kg/m^2	7860

fracture at a point x, at time t, reads as:

$$\frac{1}{\tau}\int_{t^*-\tau}^{t^*}\frac{1}{d}\int_{x^*-d}^{x^*}\sigma(x,t)\,dx\,dt \geq \sigma_c, \tag{3}$$

where τ is the microstructural time of a fracture process (or fracture incubation time) – a parameter characterizing the response of the material to applied dynamical loads (i.e. τ is constant for a given material and does not depend on problem geometry, the way a load is applied, the shape of a load pulse or its amplitude). d is the characteristic size of a fracture process zone and is constant for the given material and chosen scale. $\sigma(x,t)$ is stress at a point, changing with time, and σ_c is its critical value (ultimate stress or critical tensile stress found in quasi-static conditions).

Assuming:

$$d = \frac{2}{\pi}\frac{K_{Ic}^2}{\sigma_c^2}, \tag{4}$$

where K_{Ic} is the critical stress intensity factor for mode I loading (mode I fracture toughness), measured in quasi-static experimental conditions.

4. Simulation technique

Fracture criterion (3) was integrated into finite element computational scheme. By the choice of the element size to be equal to d – characteristic size of the fracture process zone, the linear increment of a single fracture event will also be equal to d, as implied by the incubation time approach [12].

Once fracture condition is executed somewhere in the plate, rupture should take place and this is simulated by splitting the nodes in the point where the fracture criterion was executed. Regular mesh with rectangular elements allows for fracture in two perpendicular directions. Each element of the target mesh has its own set of nodes – neighbouring elements do not have common nodes. However, nodes coincident in location have their degrees of freedom (DOF's) coupled, which is giving the same numerical formulation as a common node. Once the fracture condition is executed in one of the perpendicular directions, the restriction on DOF's in the same direction is removed – a new surface is created.

The solution output provides a possibility to monitor the change of fracture surface area after every solution step. After the solution completion time dependence of the created fracture area can be provided for analysis. Since the problem is solved in two-dimensional formulation, the fracture surface is calculated considering axial symmetry of the problem. In first case the total length of microcrack faces is calculated – once elements are separated the length of their newly appeared free edges is added to the total fracture length. On another hand, if the axial symmetry is considered one should multiply the length of microcrack faces by, where is the distance from the microcrack to the symmetry axis and thus, the fracture surface area created in the bulk of the plate is estimated. Both approaches do not describe real three-dimensional fracture pattern, but the second one seems to be more realistic. Typical fracture surface area – time dependence is shown in Fig. 1.

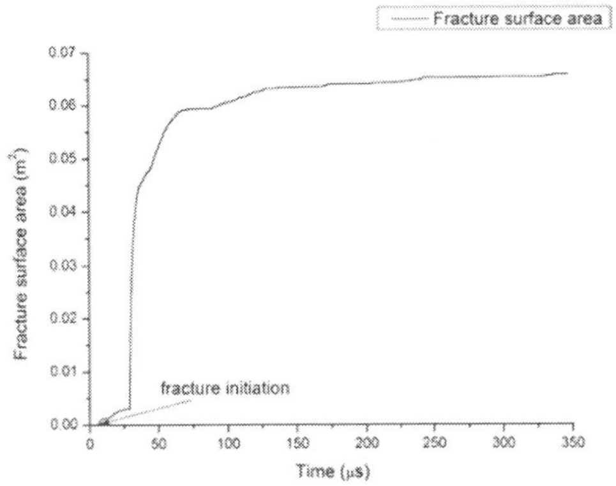

Figure 1. Fracture surface area as a function of time.

It should be noticed here, that parts of the ceramic target separated in course of the fracture process do not interact with each other, which is, of course, a significant simplification of the model.

5. Results and discussion

Utilized fracture criterion (3) requires a material strength parameter to be evaluated in an independent test (ex. [13]). Such an experimental study was not yet performed for ceramic materials and the incubation time is unknown for the materials used in the simulation. This opens a possibility to investigate the influence of the incubation time on behaviour of the fractured media. Reasonable range for possible incubation time variation was chosen keeping in mind values typical for other brittle materials [10]. Initial velocity of the plunger was another parameter that was alternated in course of simulations. As a result, dependencies of fracture process on the incubation time value and the plunger initial velocity were received.

The main parameter characterizing fracture in this study is the induced damage – the new surface created in a result of the interaction between the ceramic sample and the plunger. Induced damage was calculated according to the procedure described in the previous section. Figure 2 gives the received final damage as the plunger velocity V is alternated. Residual velocity of the plunger was also registered as another parameter of the process providing opportunity to find threshold initial velocity

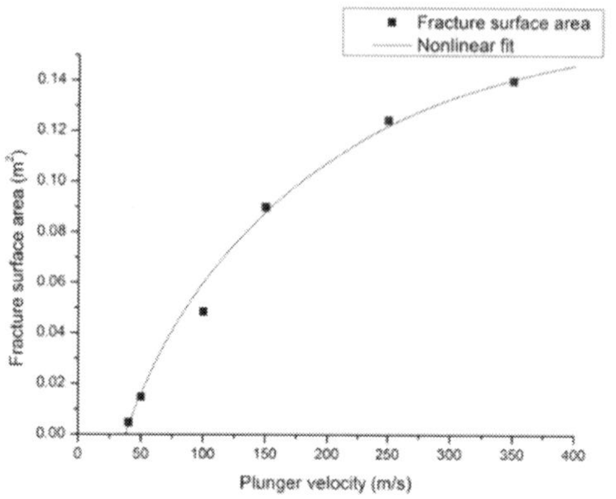

Figure 2. Fracture surface area as a function of plunger initial velocity.

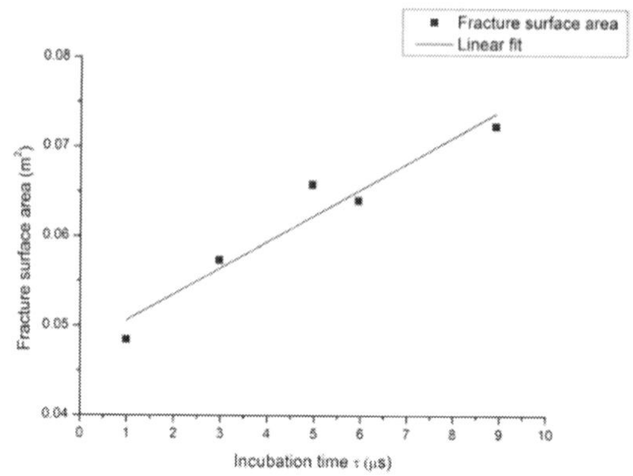

Figure 3. Final fracture surface area versus incubation time dependence.

corresponding to the situation when the plunger is completely stopped by the target.

Having the initial velocity of the plunger, its kinetic energy can be easily calculated as $E = mV^2/2$, where m is the plunger mass. Kinetic energy of the plunger after the interaction with the target can be calculated analogously using its residual velocity. Difference between these energies is transferred to the ceramic plate and can be used for fracture. Knowledge of this energy and the corresponding area of fracture created in the result of the interaction provide a possibility to calculate a value of specific dynamic fracture surface energy γ_d analogous to specific fracture surface energy introduced by Griffith [14].

This energy can be calculated as:

$$\gamma_d = \left.\frac{dE}{dS}\right|_{S=0}, \qquad (5)$$

where E stands for energy transferred to the ceramic plate and S is the area of fracture surface created in a result of interaction. Calculation of γ_d using (6) gives $\gamma_d = 1343\frac{J}{m^2}$ for the studied case. This value is considerably higher (approximately an order of magnitude) than fracture surface energy evaluated in quasi-static loading conditions [15]. The similar behaviour of dynamic analogue of Griffith's surface energy was previously observed for PMMA in [16].

Variation of the incubation time used in fracture criterion (3) in fact, means the variation of the material as the incubation time is a material property responsible for material reaction on dynamic loading. Figure 3 gives the final fracture surface area S as a function of incubation time τ. Presented data was obtained for the same projectile velocity – 100 m/s. Greater incubation times lead to higher values of final fracture surface area. This is explained by the fact that for lower values of incubation time, fracture condition is executed earlier, resulting in fracture concentrated in the areas close to the contact between the ceramic plate and the plunger while for bigger values of the incubation time fracture is spread further from the contact area.

6. Conclusions

The presented investigation is the first attempt to simulate penetration of ceramics utilizing incubation time fracture criterion controlling fracture of ceramics. Relatively simple model geometry and constitutive laws gave a possibility to concentrate the study on the most important fracture process characteristics such as dependency of final fracture surface area on plunger velocity and calculation of dynamic analogue of Griffith's specific fracture energy. The use of incubation time fracture criterion provided an opportunity to examine fracture process parameters with consideration to microstructural processes preceding macroscopic fracture of the target. Moreover, the dependence of fracture surface area (and hence the amount of damage induced) on incubation time was studied.

References

[1] G. Ben-Dor, A. Dubinsky and T. Elperin, Journal of Mechanics and Structures **1**(7), (2006), 1161–1177

[2] J.A. Zukas, Transactions on the Built Environment **22** (1996) 103–115

[3] Robert O. Ritchie, Nature Materials **10** (2011), 817–822

[4] V. Bratov, Acta Mechanica Sinica **27**(4) (2011), 541–549

[5] V. Bratov, Y. Petrov, International Journal of Fracture **146** (2012), 53–60

[6] T. Masaki, Journal of American Mechanical Society **69**(8) (1986), 638–640

[7] A.J.A. Winnubst, K. Keizer, A.J. Burggraaf, Journal of Materials Science **18** (1983), 1958–1966

[8] V.A. Bratov, N.F. Morozov, Yu.V. Petrov, *Dynamic Strength of Continuum* (St. Petersburg University Press, Saint Petersburg 2009) 224

[9] N.A. Kazarinov, V.A. Bratov, Yu.V. Petrov, Doklady Physics **454**(6) (2014), 557–560

[10] V. Smirnov, Y.V. Petrov, V. Bratov, Science China Physics, Mechanics and Astronomy **55**(1), 78–85

[11] Y.V. Petrov, V.V. Taraban, Vestnik of Saint Petersburg University **1**, 101–106

[12] Y.V. Petrov, B.L. Karihaloo, V.A. Bratov, International Journal of Engineering Science **61**, 3–9
[13] S.I. Krivosheev, N.F. Morozov, Yu.V. Petrov, Mechanics of Solids **165** (1999), 78–85
[14] A. Griffith, Philosophical transactions of the Royal Society of London, A**221** (1921), 163–198
[15] B. Lawn, *Fracture of Brittle Solids Second Edition* (Press Syndicate of the University of Cambridge, Melbourne 1993), 378
[16] V.A. Bratov, A.A. Gruzdkov, S.I. Krivosheev, Yu.V. Petrov, Doklady Physics **49**(5) (2004), 338–341

EPJ Web of Conferences **94**, 04057 (2015)
DOI: 10.1051/epjconf/20159404057

Analysis on formulas of concrete plate under contact explosion

Zheng Wang[a], Longhe Liang, and Jianfeng Lou

Institute of Applied Physics and Computational Mathematics, Beijing 100094, China

Abstract. Based on many test data of concrete plates of several thicknesses, this paper presents some new simple engineering formulas for computing critical thickness of scabbing and perforation, the front explosion diameter and the rear scabbing diameter of reinforced concrete plates under contact explosion. For the damage problem of a 20 cm thick concrete target under contact explosion of 1.25 kg TNT charge, the paper gives the numerical simulation results of LS-DYNA software for comparison. The damage zone of concrete plate in numerical simulation is determined by analysis of tension pressure resulted from shock wave reflection on the free boundary. And the numerical simulation results are in basic agreement on the results of these engineering formulas.

1. Introduction

Explosion damage for structures is an important problem and focuses the attention of everyone. For concrete plates under explosion, the front impact crater damage and the rear scabbing damage may occur. For a certain charge of contact explosion, it is called as critical scabbing when the rear scabbing cracks occur. It is called as critical perforation when the front impact crater is collected through out to the rear scabbing hole.

Based on the experimental data, the engineering formula for computing critical thickness of scabbing and perforation of concrete plates under explosion has been studied for many years. In literature [1,2], Quanping Zheng etc. have reported much experimental results and had detailed analysis on early engineering formula. They reviewed that PCDM equation and the amended PCDM equation which shown in Eqs. (1) and Eq. (2) are used mostly.

$$h_s = 0.048\left(\frac{d}{W^{1/3}}\right)^{-0.3}\left(\frac{W}{W+C}\right)^{-0.3} W^{1/3} \quad (1)$$

$$h_s = 0.0115\left(\frac{d}{W^{1/3}}\right)^{-0.7}\left(\frac{W}{W+C}\right)^{-1.31} W^{1/3}. \quad (2)$$

In which, h_s is the critical scabbing thickness, h_p is critical perforation thickness, d is the distance from the charge to the plate, W is the charge mass and C is the confined case mass. But as shown in Eq. (1) and Eq. (2), d is too sensitive for small value, so PCDM equations can not be used to analysis problems under contact explosion.

In the first part of the paper, some new engineering formula for analysis on damage of reinforced concrete plates under contact explosion are presented based on conclusion of tests data in literature [1]. Then some numerical simulation results of LS-DYNA software are given for comparison.

[a] Corresponding author: gly_w@sina.com

2. Fitting formulas

The research is based on the experimental data in literature [1].

Five kinds of thickness of reinforced concrete plate are used in experiments shown in Table 1. The compression strength of concrete are 37.5–118.6 MPa, and the charge mass are 0.064–16kg equivalent TNT.

Table 1 is the analysis and conclusion on critical scabbing and perforation thickness of the reinforced concrete under contact explosion, and the experimental data are quoted from the literature [1].

For the reinforced concrete plate under contact explosion, fitting experimental data (see Fig. 1) in Table 1, h_s and h_p can be received as Eq. (3) and Eq. (4):

$$h_s = -0.06114 + 0.38682\ W^{1/3} \quad (3)$$

$$h_p = 0.000876713 + 0.18345\ W^{1/3} \approx 0.18345\ W^{1/3}. \quad (4)$$

Figure 1 shows that linearity fitting as relation of $W^{1/3}$ has good results.

Figure 2 is the comparison of fitting results with data reported in literature [2]. Figure 2 shows that the scabbing thickness curve computed by Eq. (3) is consist with results in literature [2]. The curve for $0.5h_s$ is also shown in Fig. 2. It is reported by BRL Lab in the United state that h_p equals $0.5h_s$ for high velocity impact condition. The perforation thickness curve computed by Eq. (4) is different from results in literature [2] but it is consist with the curve for $0.5h_s$. As analysis on experimental data in Table 1, there are much tests data for perforation condition, and the damage either from explosion or high velocity impact are all rooted from shock wave propagation and reflection in the target, so Eq. (4) should be credible.

Using the similar approach, the front explosion diameter d_e and the rear scabbing diameter d_s can be received as Eq. (5) and Eq. (6) and the comparison of

Table 1. Experimental data from literature [1] (h is the thickness of the reinforced concrete, W_s and W_p are charge mass for critical scabbing and perforation.)

h (m)	W_s (kg)	W_p (kg)	Comment
0.1	≤ 0.064	0.2	For 0.064 kg charge, scabbing height is 1.1cm. 5 tests for 0.2–0.25 kg charge, plates are perforated or critical perforated.
0.2	/	1.2	1 test for 1.2 kg charge, the plate is perforated, 2 tests for charge mass of 1.4 kg, one plate is perforated and the other which has 2% reinforced fibers is critical perforated.
0.25	≤ 0.6	/	For 0.2 kg charge, no scabbing. For 0.6 kg and 1.0 kg charge, scabbing height are 12 cm and 14 cm.
0.3	≥ 0.6	3.8	For 0.6 kg charge, scabbing occur; For 3.8 kg charge, and the one plate is perforated other is critical perforated.
0.4	1.6	11	3 tests 1.6 kg charge, scabbing block occur;1 test for 10 kg charge and 2 tests for 12 kg charge, plates are critical perforated.

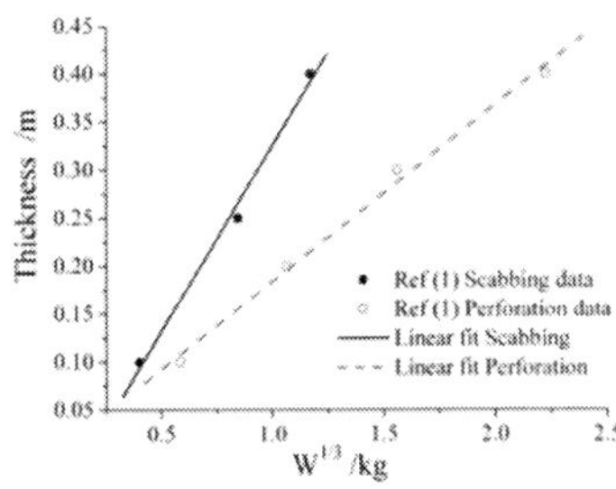

Figure 1. Fitting critical scabbing and perforation thickness.

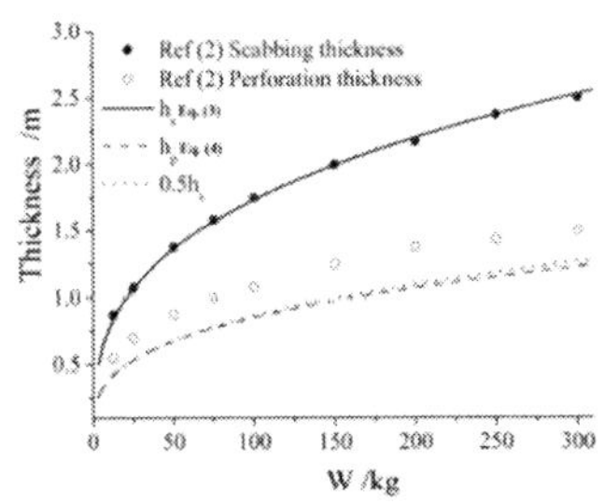

Figure 2. Comparison with data in literature [2].

fitting results with experimental data are shown in Fig. 3 and Fig. 4.

$$d_e = 0.32\ W^{1/3} \tag{5}$$

$$d_s = -0.1 + 0.58\ W^{1/3}. \tag{6}$$

In Fig. 3 and Fig. 4, there are much experimental data which are quoted from the literature [1]. It shows that data in Fig. 3 and Fig. 4 are more widely dispersed, so

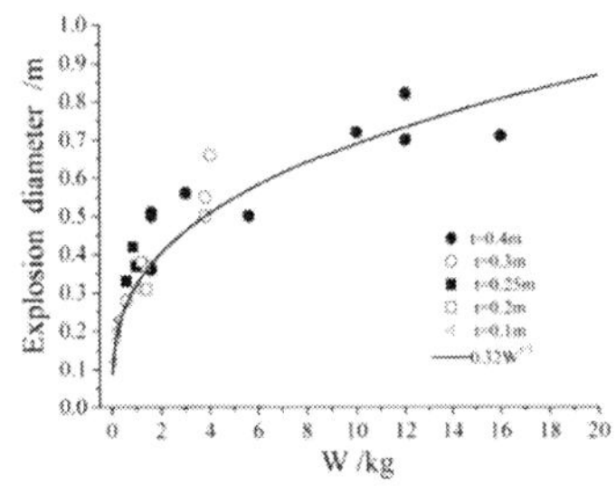

Figure 3. Fitting front explosion diameter d_e.

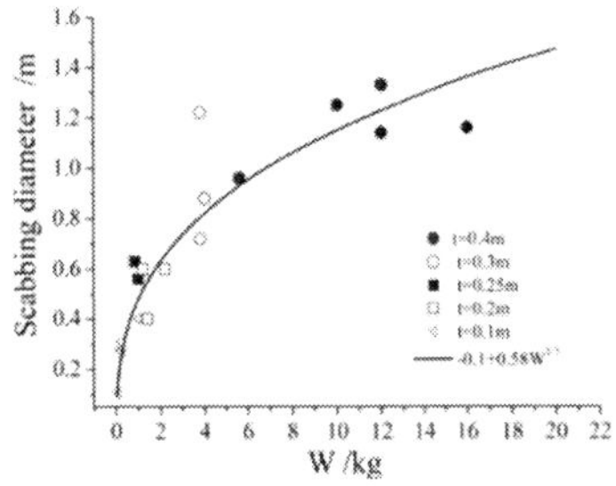

Figure 4. Fitting rear scabbing diameter d_s.

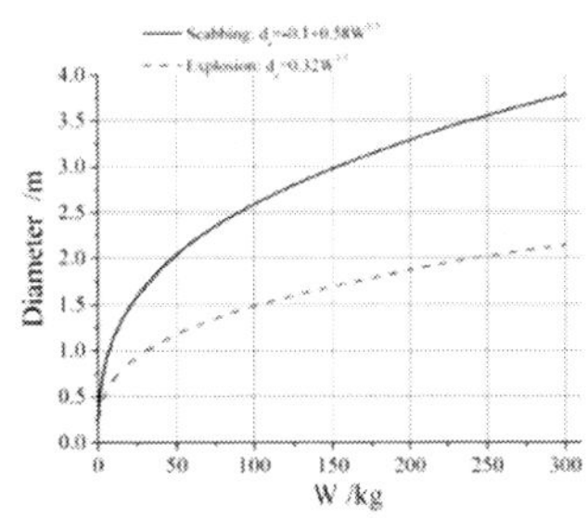

Figure 5. Extended results of de and d_s.

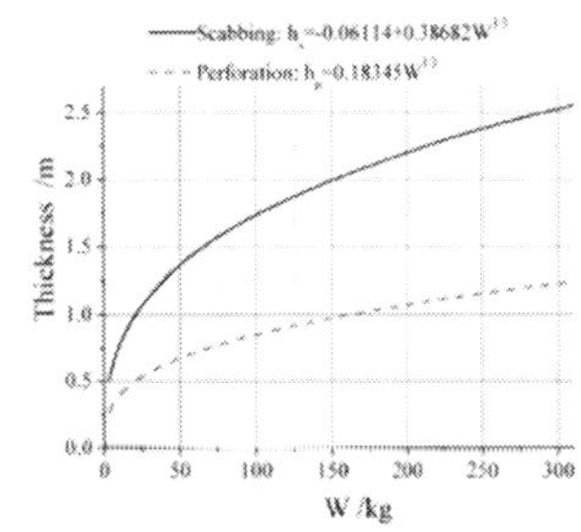

Figure 6. Extended results of hs and h_p.

there is a certain deviation by using the fitting formula as Eq. (5) and Eq. (6). And it is noted that these data include 5 kinds of thickness for concrete plate, but the thickness have little effect on the front explosion diameter d_e or the rear scabbing diameter d_s.

These results of above simple fitting formula can be extended to large charge mass such as 300 kg which is shown in Fig. 5 and in Fig. 6. It shows that the rear scabbing diameter d_s is larger than the front explosion diameter d_e in Fig. 5 and $h_p = 0.5h_s$ is approximately right in Fig. 6.

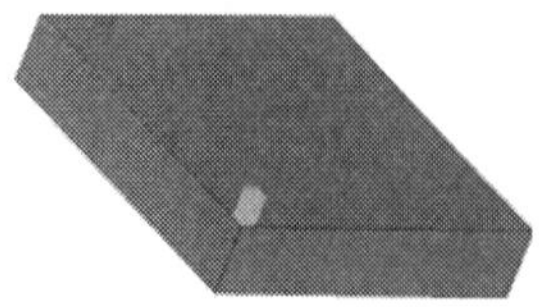

Figure 7. 1/4 model of numerical simulation.

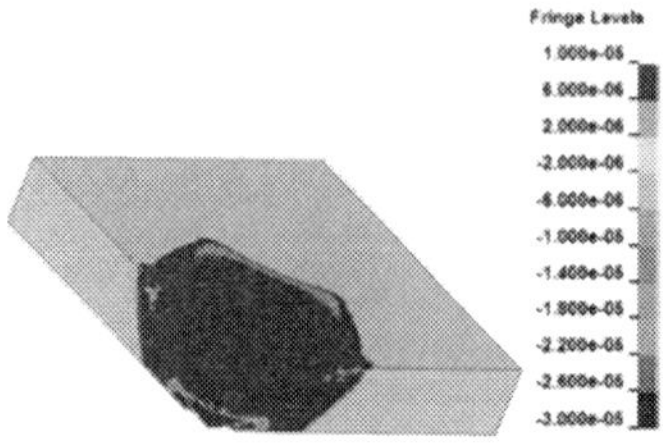

Figure 8. Pressure distribution at 100 μs.

Figure 9. Pressure distribution at 120 μs.

3. Numerical simulation

According to above analysis, for 0.2 m thick concrete plate under contact explosion of 1.25 kg TNT charge, through out perforation will occur. The paper gives the numerical simulation results of LS-DYNA software for comparison. Figure 7 is the 1/4 model of simulation. ALE method is used. The yellow part in Fig. 7 is explosive and computed using fluid model, the concrete is simulated using solid elements.

The equation of state for detonation products is Jones-Wilkins-Lee (JWL) as

$$P = \mathrm{A}\left(1 - \frac{\omega}{\mathrm{R}_1\bar{v}}\right)\exp(-\mathrm{R}_1\bar{v}) + \mathrm{B}\left(1 - \frac{\omega}{\mathrm{R}_2\bar{v}}\right)\exp(-\mathrm{R}_2\bar{v}) + \omega\frac{E}{\bar{v}} \tag{7}$$

where $\bar{v} = v/v_0$ is relative volume, E is internal energy per unit reference volume. The others are parameters such as A = 373.77 GPa, B = 3.7471 GPa, $\mathrm{R}_1 = 4.5$, $\mathrm{R}_2 = 0.9$, $\omega = 0.35$. And the density of explisive is 1.63 g/cm^3, the detonation speed is $D_{CJ} = 6.93$ km/s, the detonation pressure is $p_{CJ} = 21.0$ GPa, and the initial internal energy per unit reference volume is 6.08 GPa.

The HJC model and the corresponding parameters from the literature [3] are used only setting the compression strength as 35 MPa and the tension strength as 3.5 MPa for concrete plate.

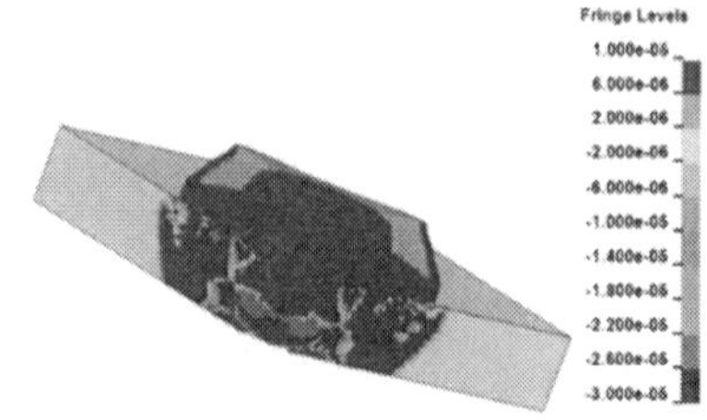

Figure 10. Pressure distribution at 125 μs.

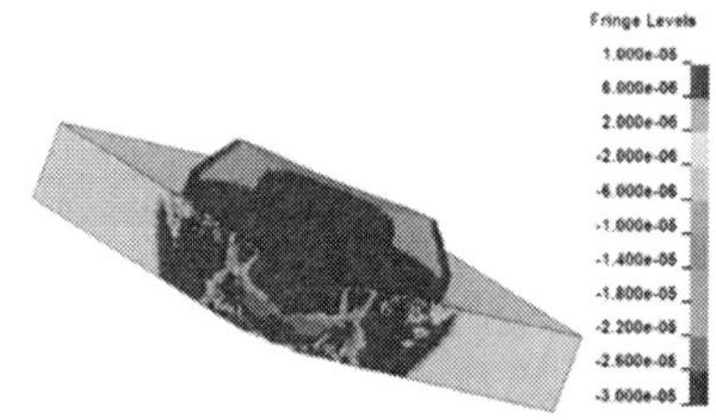

Figure 11. Pressure distribution at 130 μs.

Figures 8–11 show the pressure distribution results of the explosive production and the concrete plate for different times. The units in these figures are 100 GPa. The blue sign means the tension pressure is below the tension strength of concrete. For lack of accuracy damage model of concrete under explosion loading, the damage zone of concrete plate in numerical simulation is determined by the value of tension pressure resulted from shock wave reflection on the free boundary. According to analysis on tension pressure distribution, as shown in Fig. 11, the through out perforation will occur in the concrete plate. The front explosion diameter d_e is about 0.52 m, and the rear scabbing diameter d_s is about 0.74 m. These results are in consistent to the results of Eq. (5) and Eq. (6).

4. Conclusions

Based on analysis on experimental data, the paper presents new simple engineering formula for computing critical thickness of scabbing and perforation, the front explosion diameter and the rear scabbing diameter of reinforced concrete plates under contact explosion. And the numerical simulation results are in basic agreement on the results of these formulas.

References

[1] X.J. Li, Q.P. Zheng, Y. Yang, Explosion and shock wave [Chin.], **29**, 385 (2009)

[2] Q.P. Zheng, Q.H. Qian, Z.S. Zhou, Eng. Mech. [Chin.], **20**, 47 (2003)

[3] T.J. Holmquist, G.R. Johnson, W.H. Cook, *14th Int. Symp. Ballistics*, 591 (1993)

EPJ Web of Conferences **94**, 04058 (2015)
DOI: 10.1051/epjconf/20159404058

The theoretical analysis and numerical simulation of the metal jet incoherence

Jun Liu[a] and Zheng Wang

Institute of Applied Physics and Computational Mathematics, Beijing 100094, China

Abstract. By the analysis of the classic jet theory and literature, the regions of no-jet, convergent jet and incoherent jet are given in convergent point coordinate system. The theoretical analysis of metal jet incoherence formation explains Walker's jet incoherence theory which is based on experiment, and it's proved that in the supersonic portion of the metal jet, both the incoherence jet and the convergence jet can be formed. The jet incoherence estimation equation is given by our theoretical analysis. At last the numerical simulation of the copper jet formation process at different speeds and angles is done. The typical images of bifurcated jet, hollow jet and dense rarefaction jet are given by calculations to confirm the theoretical analysis results.

1. Introduction

Metal collision to form a shaped jet in low speed or supersonic conditions may have many complex physical phenomena, it may not produce metal jet or generate unsteady jet. In the study of metal shaped jet, Birkhoff put forward the classic jet theory. Pugh develop this theory, and puts forward the quasi-steady jet theory. But neither of the constant jet or the quasi-steady jet theory can correctly predict the conditions in which it can't form metal jet or generate unsteady jet. Because these theories are based on the incompressible fluid assumption.

In the process of metal shaped jet formation, the convergent metal jet is formed generally. Due to the shock wave blockade and strength blockade the metal jet may be not formed. In the incoherence jet conditions it can be divided to bifurcated jet and rarefaction jet. And the rarefaction jet can be divided into hollow jet and dense rarefaction jet. It is generally believed that the hollow jet will evolve into bifurcated jet in a long distance, but the dense rarefaction jet is different from the bifurcation jet.

Because the radial expansion of the metal jet will greatly reduce the penetration ability, the research of the incoherence jet formation is becoming a hot spot. Although the supersonic collision to form stable metal jet is possible, the incoherence jet phenomenon have been observed from experiments in the supersonic region. The assumption is proposed by Zhou [1] (θ_c is the limit angle to produce the attached shock wave):

(1) In the subsonic flow collision, the convergent jet can be formed;
(2) In the supersonic flow collision, if $\theta < \theta_c$, that will do not produce jet;
(3) In the supersonic flow collision, if $\theta > \theta_c$, the incoherence jet is formed.

[a] Corresponding author: caepcfd@126.com

Experiments show that the incoherence jet is effected by the material sound velocity. Walker [2,3] put forward the opposite view from Zhou: The incoherence of metal jet is not decided by the shock wave in supersonic collision, but the state in high pressure collision area is important for the convergent jet formation.

In this paper, the areas of no-jet, convergent jet and incoherent jet are given by the analysis of the classic jet theory in convergent point coordinate. And then, the Walker's incoherence jet theory is introduced and analyzed. As the supplement of Walker's theory, the metal jet incoherence is analysed in different supersonic speed ranges. At the end, the multi-material elastic-plastic hydrodynamic Euler method is used to simulate the metal jet incoherence phenomena. The simulation of the copper jet formation process at different speeds and angles is given to confirm our theoretical analysis results.

2. Theory of metal jet formation

In the metal shock blockade theory, the relationship between pressure p, density and velocity deflection angle θ_1 can be given by analysis of the physical characteristics between wave front and wave after. Combined with the Hugoniot relation and the equation of state, the $p-\theta$ polar curve of Cu is given in Fig. 3.

In the polar curve, $\theta_{\max}$ is the max deflection angle to produce the attached shock wave at different initial velocity q_0. And then the metal shock blockade curve can be computed (Fig. 4-L_1), the shock wave blockage area is the region below this curve in where there is no jet formation.

The strength blockade theory is based on ideal elastoplastic assumption, it's considered that The normal velocity $q_\perp$ must be greater than the material strength Y_0 to form the metal jet. So the critical angle of the strength

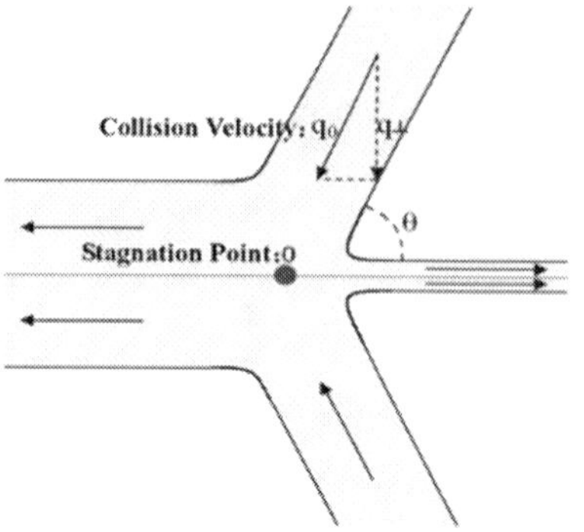

Figure 1. The metal jet formation in convergent point coordinate system (q_0 is collision velocity, θ is collision angle, O is stagnation point.)

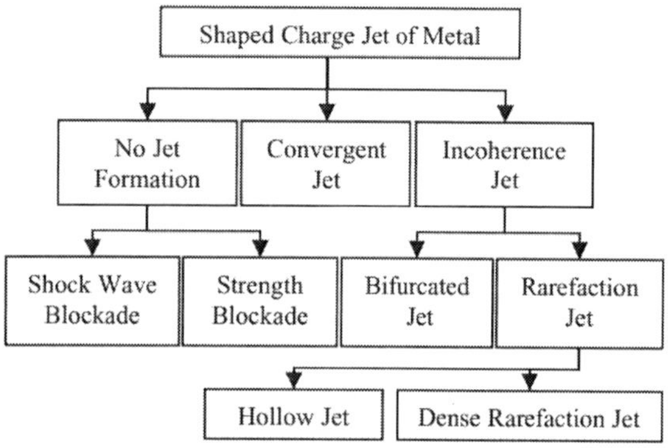

Figure 2. The classification of the shaped charge jet.

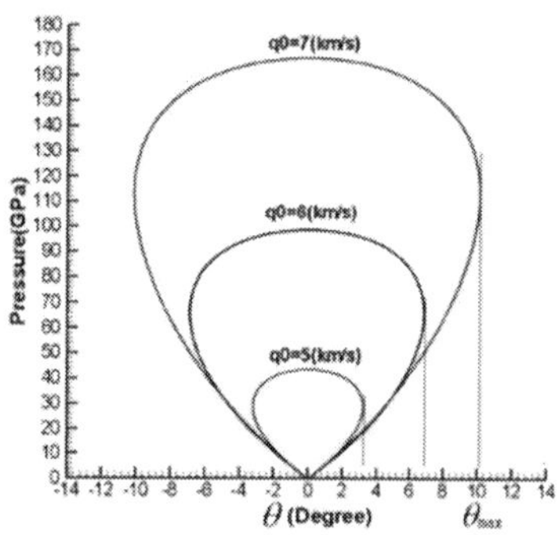

Figure 3. The pressure-angle polar curve of copper using Mie-Gruneisen equation of state.

blockade θ_2 is defined as

$$\sin\theta_2 = \frac{1}{q_0}\sqrt{\frac{2Y_0}{\rho}} \tag{1}$$

the metal strength blockade curve is computed (Fig. 4-L_2), and if $\theta < \theta_2$, there is no jet formation.

The metal jet incoherence experiment indicates that the convergent jet can be formed when the collapse velocity is less than 1.23 times the body sound speed in copper

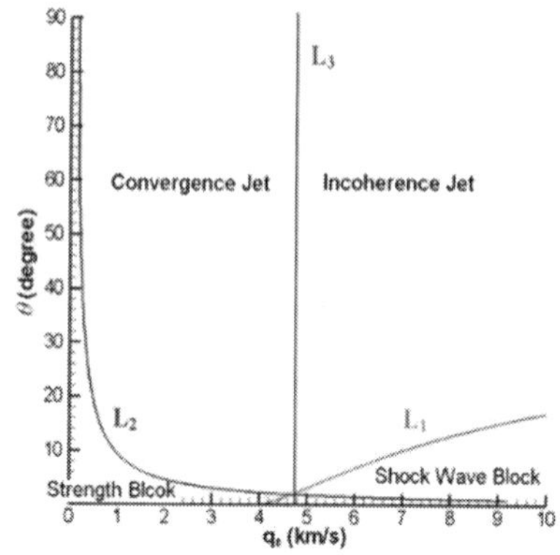

Figure 4. The region of no-jet, convergent jet and incoherent jet of copper in convergent point coordinate.

shaped cover. By extending this criterion to other shaped cover materials, Walker [2] give out the criterion below:

(1) In the convergent point coordinate, if $q_0 > C_L$, the bifurcated jet or discontinuous jet is formed;
(2) In the convergent point coordinate, if $q_0 < C_L$, the convergent jet is formed;
(3) In the convergent point coordinate, if $q_0 \approx C_L$, the hollow jet is formed;

C_L is the low longitudinal wave velocity, it's defined as:

$$C_L = \sqrt{3\frac{1-\gamma}{1+\gamma}}C_0, \quad C_0 = \sqrt{\frac{K}{\rho}} \tag{2}$$

γ is the poisson coefficient, K is the bulk modulus. So the metal jet incoherence curve is given by Fig. 4-L_3 which identify the jet incoherence region.

Cu: $\gamma = 0.35, C_L = 1.20C_0$; Fe: $\gamma = 0.29, C_L = 1.28C_0$; Ni: $\gamma = 0.30$, $C_L = 1.27C_0$. These data are very close to the experimental results of copper: $q_\perp = 1.23C_0$. Therefore, using the 1.23 multiplier for other materials as the jet incoherence criterion seems to be reasonable. However, the jet formation process is very complex. In the shaped cover collapse before flying to convergent point, it has experienced a shock loading and unloading expansion. Therefore, in the dynamic convergence collision it is not the original state, especially the collision area has a higher pressure. In 1996, Walker [3] reiterated that using the volume velocity is more reasonable, and the experimental factor 1.23 may not equal to C_L/C_0. Note that Walker's analysis of jet incoherence theory is not perfect, we give a demonstration of the jet incoherence theory in the convergent point coordinates, to verify the Walker's experimental criterion.

3. Theoretical analysis of the metal jet incoherence

Two metal rod with infinite length and limited width collide at a certain angle and speed. We can only analyse the upper half part because of the symmetry. Define the upstream density, stream tube width and velocity is $\rho_0, A_0, \vec{q}_0$ in a stream tube l_1 which do not close to

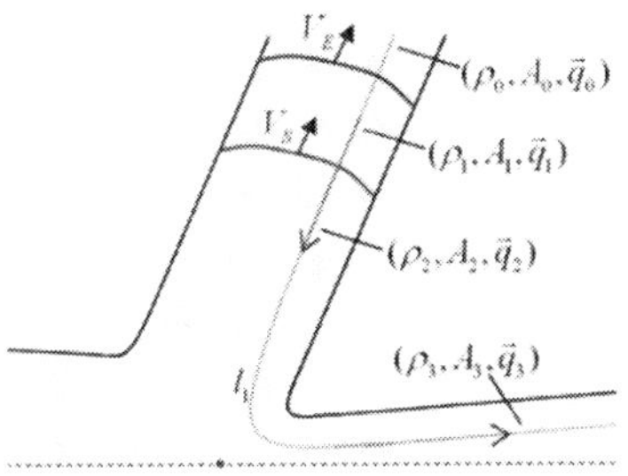

Figure 5. The formation of metal-collision jet and the position of shock wave.

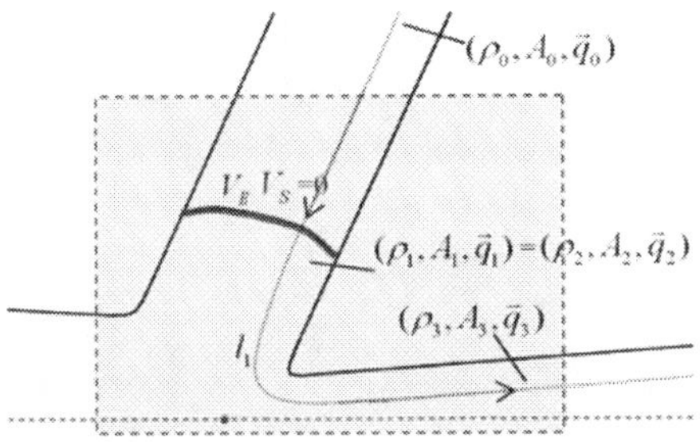

Figure 6. The region of steady flow when $|\vec{q}_0| > C_L$.

the outer surface; behind the elastic longitudinal wave, the density, stream tube width and velocity is $\rho_1, A_1, \vec{q}_1$; behind the body wave, the density, stream tube width and velocity is $\rho_2, A_2, \vec{q}_2$; after crossing through the high pressure area and unloading to zero pressure condition, the density, stream tube width and velocity is $\rho_3, A_3, \vec{q}_3$. The velocity of the elastic longitudinal wave is V_E, and the velocity of the body wave is V_S.

It's intuitively considered that the judgment of the jet incoherence is: if $A_3 > A_0$, the metal jet is incoherence; if $A_3 \leq A_0$, the metal jet is convergent. The initial velocity range is divided into two kinds of situations: $|\vec{q}_0| > C_L$ and $C_0 < |\vec{q}_0| < C_L$.

(1) $|\vec{q}_0| > C_L$.

In this condition, the position of the elastic longitudinal wave and body wave is not moved ($|V_E| = |V_S| = 0$). The process of upstream area unloading to zero pressure area can be seemed as a steady flow.

In an ideal without heat conduction, Bernoulli integration can be done in steady region along the streamline:

$$\frac{\vec{q}^2}{2} + e + \frac{p}{\rho} + \tilde{V} = C(\psi) \tag{3}$$

$C(\psi)$ is the function of streamline number ψ (in uniform stream, $C(\psi)$ is a constant). In this model $\tilde{V} = 0$, along the streamline we can get:

$$\frac{\vec{q}_0^2}{2} + e_0 + \frac{p_0}{\rho_0} = \frac{\vec{q}_3^2}{2} + e_3 + \frac{p_3}{\rho_3}. \tag{4}$$

The upstream state have $p_0 = 0$, $e_0 = 0$. If the density does not change with temperature, the state unloading to zero

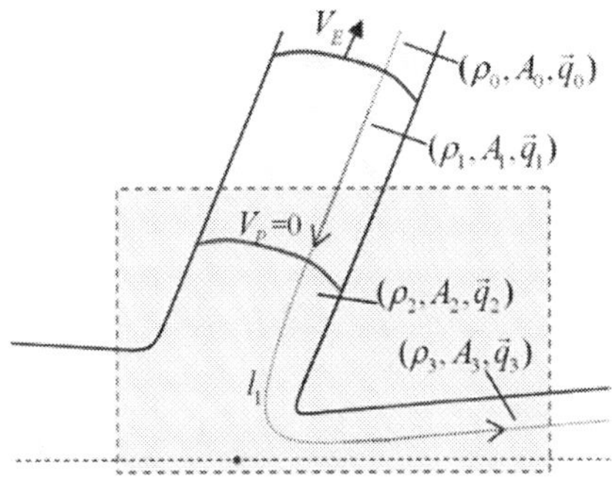

Figure 7. The region of steady flow at $C_0 < |\vec{q}_0| < C_L$.

pressure area have $p_3 = 0$, $\rho_3 = \rho_0$, $e_3 > 0$. Get into (4):

$$|\vec{q}_0| > |\vec{q}_3|. \tag{5}$$

Combine with the conservation of mass ($\rho_0 |\vec{q}_0| A_0 = \rho_3 |\vec{q}_3| A_3$), we can get:

$$A_0 < A_3. \tag{6}$$

This analysis result show that the metal jet is incoherence when $|\vec{q}_0| > C_L$, and it's the same as Walker's experimental results.

(2) $C_0 < |\vec{q}_0| < C_L$.

In this condition, the position of body wave is not moved ($|V_S| = 0$), but the elastic longitudinal wave velocity is not zero ($|V_E| > 0$). The process from behind elastic longitudinal wave area to unloading to zero pressure area can be seemed as a steady flow.

The elastic wave velocity is $D = C_L - |\vec{q}_0|$, the shock wave relations is:

$$\begin{aligned}
&\rho_1(D - |\vec{q}_1|) = \rho_0(D - |\vec{q}_0|)\\
&\rho_1(D - |\vec{q}_1|)^2 + p_1 = \rho_0(D - |\vec{q}_0|)^2 + p_0\\
&e_1 + \frac{p_1}{\rho_1} + \frac{(D - |\vec{q}_1|)^2}{2} = e_0 + \frac{p_0}{\rho_0} + \frac{(D - |\vec{q}_0|)^2}{2}
\end{aligned} \tag{7}$$

We use the state equation below to simplify analyzing:

$$p = (\gamma - 1)\rho e + C_0^2(\rho - \rho_0). \tag{8}$$

By Bernoulli integration (3), along the streamline we can get:

$$\frac{\vec{q}_1^2}{2} + e_1 + \frac{p_1}{\rho_1} = \frac{\vec{q}_3^2}{2} + e_3 + \frac{p_3}{\rho_3}. \tag{9}$$

In this place, we also have $p_3 = 0$, $\rho_3 = \rho_0$. On the simultaneous (7), (8) and (9) :

$$\frac{\vec{q}_3^2}{2} = \frac{\vec{q}_0^2}{2} + \left(\frac{2(C_L - \vec{q}_0)(C_L^2 - C_0^2)}{C_L(\gamma + 1)} - e_3\right) \tag{10}$$

By the conservation of mass ($\rho_0 |\vec{q}_0| A_0 = \rho_3 |\vec{q}_3| A_3$), we can get:

$$\begin{aligned}
&\frac{A_0^2}{A_3^2} = 1 + \frac{2}{\vec{q}_0^2} Q,\\
&Q = \frac{2(C_L - \vec{q}_0)(C_L^2 - C_0^2)}{C_L(\gamma + 1)} - e_3.
\end{aligned} \tag{11}$$

Table 1. The parameters of constitutive equations and state equation of copper.

Constitutive Equation		Equation of State	
Y_0(GPa)	0.12	ρ_0(g/cm^3)	8.93
Y_{max}(GPa)	0.64	γ_0	1.99
β	36	C_0(km/s)	3.94
n	0.45	S_1	1.489
b(GPa^{-1})	0.028	S_2	0
h(K^{-1})	3.8e-4	S_3	0
T_{m0}(K)	1790	a	0.47
ε_{00}	−0.1178	$P_{\min}(GPa)$	1.8
ε_{01}	−0.2344		
ε_{02}	7.529		
ε_{03}	15.26		
ε_{04}	21.90		
R(kJ/g/K)	1.3d-4		

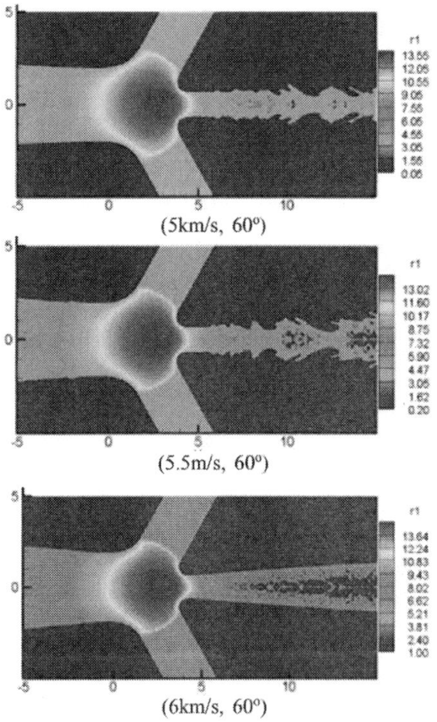

Figure 8. The density image at 60° with different collision velocity.

So, the positive or negative of Q determine whether the metal jet is incoherence or not in $C_0 < |\vec{q}_0| < C_L$:

(1) if $Q < 0$, the metal jet is convergence;
(2) if $Q > 0$, the metal jet is incoherence.

In the Eq. (11), if $|\vec{q}_0| \geq C_L$, $Q < 0$, the incoherence jet is formed, this result is just the same as analysis in previous section.

When $C_0 < |\vec{q}_0| < C_L$, the metal jet may be complex, it does not follow Walker's analysis. According to our theoretical analysis, the incoherence situation can be evaluated by Eq. (14).

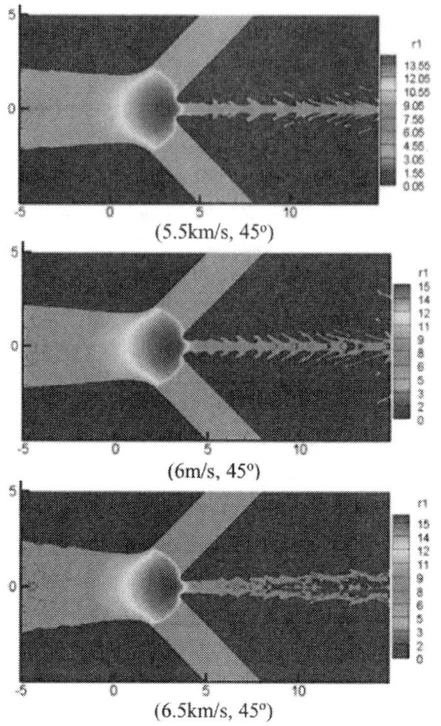

Figure 9. The density image at 45° with different collision velocity.

4. Numerical simulation of metal jet incoherence

The multi-material elastic-plastic hydrodynamic Euler program "MEPH" [4,5] is used to simulate the metal jet incoherence phenomena. The Steinberg-Guinan constitutive model is used:

$$G = G_0\left[1 + bp\cdot\left(\frac{\rho}{\rho_0}\right)^{1/3} - h\,(T - 300)\right]$$

$$Y = Y_0\left(1 + \beta\varepsilon^p\right)^n\left[1 + bp\cdot\left(\frac{\rho}{\rho_0}\right)^{1/3} - h\,(T - 300)\right]$$

$$Y_0\left(1 + \beta\varepsilon^p\right)^n \leq Y_{\max} \qquad (12)$$

$$T = (E - \varepsilon_0)/3R,$$

$$\varepsilon_0 = \varepsilon_{00} + \varepsilon_{01}x + \varepsilon_{02}x^2 + \varepsilon_{03}x^3 + \varepsilon_{04}x^4, x = 1 - \rho_0/\rho$$

$$T_m = T_{m0}\left(\frac{\rho_0}{\rho}\right)^{2/3}\exp\left[2\gamma_0\,(1 - V)\right].$$

Mie-Gruneisen state equation is used:

$$p = \rho_0 c^2\mu\frac{\left[1 + (1 - \frac{\gamma_0}{2})\mu - \frac{a}{2}\mu^2\right]}{\left[1 - (S_1 - 1)\mu - \frac{S_2\mu^2}{(\mu+1)} - \frac{S_3\mu^3}{(\mu+1)^2}\right]^2} + (\gamma_0 + a\mu)\,E, \qquad (13)$$

$$\mu = \frac{\rho}{\rho_0 - 1}.$$

The simulations with different collision velocity is done under θ=60°, 45°.

In the Fig. 8 and Fig. 9, the collision velocity is bigger than 1.23 times the speed of the body sound, the simulation results show that copper jet is incoherence. And with the increase of the stream velocity, the copper jet incoherence forms is evolved from hollow jet to bifurcation jet gradually. With the decrease of collision angle, the hollow jet and bifurcation jet is formed more difficultly.

5. Summary

In this paper, the classical jet theory is reviewed, and the regions of no-jet, convergent jet and incoherent jet are given in convergent point coordinate. The metal jet incoherence is analyzed in different supersonic speed ranges: when $|\vec{q}_0| > C_L$ the metal jet must be incoherence; but when $C_0 < |\vec{q}_0| < C_L$ the metal jet forms are complex, either incoherence jet or convergence jet may be formed. And the jet incoherence judgment equation is given by our theoretical analysis In $C_0 < |\vec{q}_0| < C_L$. At the end, both the dense rarefaction jet, hollow jet and bifurcation jet is simulated to verify the correctness of the theoretical analysis. Jet incoherence theory research and its numerical simulation can be widely applied in shaped charge jet, high-velocity impact and micro-jet formation.

The first author was supported by NNSF(11371069, 11371065, 11372052).

References

[1] Pei Chi Chou, Joseph Carleone, Robert R. Karpp, J. Appl. Phys., **47**, 7: 2975–2981 (1976)

[2] Walker J.D., *14th International Symposium on Ballistics*, 165–172 (1993)

[3] Walker J.D., *16th International symposium on Ballistics*, 457–462 (1996)

[4] He C J, Yu Z L, Feng Q J, Chinese Journal of Explosion and Shock Waves, **19**, 3: 216–221 (1999)

[5] Liu Jun, He C J, Liang X H. Chinese Journal of High Pressure Physics, **22**: 72–78 (2008)

EPJ Web of Conferences **94**, 04059 (2015)
DOI: 10.1051/epjconf/20159404059

Elastic-plastic effect study in hypervelocity impact by SPH method

Xiangfei Gong[1,2,a], Shudao Zhang[2], and Jiming Yang[1]

[1] University of Science and Technology of China, 230022 Hefei, China
[2] Institute of Applied Physics and Computational Mathematics,100094 Beijing, China

Abstract. A new distributed parallel SPH programming scheme using AMR (adaptive mesh refinement) background grid is proposed, in which the size of a grid is decided based on maximal smoothed length of local particles. In detonation simulation, computational complexity of the new scheme is better than O(NlgN) which is the best order of magnitude in the previous methods. Several hypervelocity impact problems are simulated using the new SPH scheme with state equations or strength model. The simulation results are discussed comparison to experimental and computed results in other literature.

1. Introduction

Numerical simulation is increasingly becoming a significant means for engineering application and science research. In spacecraft protection problems the velocity of impact objects is generally higher than ten kilometers per second. In the computer simulation of this kind of hypervelocity impact issue Lagrangian methods seem to be more favored than Euler methods because of the development of meshfree Lagrangian algorithms which aren't puzzled by large deformations [1].

For centuries, numerical methods based on grid or mesh play the essential roll in computer simulation, despite the great success, numerical method based on grid suffer from severe difficulties in some aspects including free surface, deformable boundary, moving interface, large deformation, etc. Over the past years meshfree methods, e.g. smoothed particle hydrodynamics (SPH), have been an allimportant research focus of more effective numerical methods for complicated problems [2]. In SPH method, the materials in computational domain are represented by discrete particles, which possess of material properties and have generally unalterable mass. The particles possessing different material properties naturally trace the moving interface along with the movement of the particle. There are no fixed links between particles in SPH that allows a straightforward handling of large deformations, on the other hand that makes it necessary to find new neighboring particles at every step, which is very time-consuming so that SPH is commonly highly inefficient compared with grid based methods. The inefficiency is a crucial matter that restricts extensive application of SPH, so it's very important to improve the neighbor particles search efficiency in SPH method [3].

The algorithm in this paper references to tree structure of link-list method [4] and variable size of bucket method [5]. Background grid of adaptive mesh refinement (AMR) [6] is implemented as buckets that the new search algorithm can be regarded as a modified bucket search algorithm with tree-structure buckets. Size of background grids of the new algorithm is variable that it can be applied when smoothing length is alterable, while higher search efficiency can be gained by the ordering between buckets when finding neighbor buckets.

In the paper hypervelocity impact examples are calculated using the new SPH algorithm with materials described by state equations or by strength model. And the simulation results are discussed comparison to experimental data.

2. SPH algorithm with AMR background grid

In this paper, SPH method is used for the numerical simulation and the extensive presentation can be found in [7]. Here some noticeable contents of SPH method are addressed.

2.1. Governing equations in SPH

After kernel and particle approximation, for Euler equations controlling the ideal fluid dynamic problems, we have

$$\frac{d\rho_i}{dt} = \sum_{j=1}^{N} m_j v_{ij}^{\beta} \cdot \frac{\partial W_{ij}}{\partial x_i^{\beta}} \tag{1}$$

$$\frac{dv_i^{\alpha}}{dt} = -\sum_{j=1}^{N} m_j \left(\frac{p_i}{\rho_i^2} + \frac{p_j}{\rho_j^2}\right) \frac{\partial W_{ij}}{\partial x_i^{\alpha}} \tag{2}$$

$$\frac{de_i}{dt} = \frac{1}{2}\sum_{j=1}^{N} m_j \left(\frac{p_i}{\rho_i^2} + \frac{p_j}{\rho_j^2}\right) v_{ij}^{\beta} \frac{\partial W_{ij}}{\partial x_i^{\beta}}. \tag{3}$$

Where ρ denotes the scale density, m is mass of a particle, v is velocity, W is kernel function, p is static pressure and e is internal energy pre unit mass.

[a] Corresponding author: gong_xiangfei@iapcm.ac.cn

Article available at http://www.epj-conferences.org or http://dx.doi.org/10.1051/epjconf/20159404059

For Navier-Stokes(N-S) equations controlling the general fluid dynamics problems, we have

$$\frac{d\rho_i}{dt} = \sum_{j=1}^{N} m_j v_{ij}^{\beta} \cdot \frac{\partial W_{ij}}{\partial x_i^{\beta}} \tag{4}$$

$$\frac{dv_i^{\alpha}}{dt} = -\sum_{j=1}^{N} m_j \left(\frac{\sigma_i^{\alpha\beta}}{\rho_i^2} + \frac{\sigma_j^{\alpha\beta}}{\rho_j^2} \right) \frac{\partial W_{ij}}{\partial x_i^{\alpha}} \tag{5}$$

$$\frac{de_i}{dt} = \frac{1}{2}\sum_{j=1}^{N} m_j \left(\frac{p_i}{\rho_i^2} + \frac{p_j}{\rho_j^2} \right) v_{ij}^{\beta} \frac{\partial W_{ij}}{\partial x_i^{\beta}} + \frac{\mu_i}{2\rho_i} \varepsilon_i^{\alpha\beta} \varepsilon_j^{\alpha\beta}. \tag{6}$$

Where σ denotes the total stress tensor and ε denotes the strain tensor.

2.2. Equations of state (EOS) and constitution model

In computational simulation, the projectile and target is provided by Mie-Gruneisen equation of state. Elasticity is implemented incrementally and the Johnson-Cook strength model [8] describes von Mises plasticity as a function of strain, strain rate and temperature,

$$\sigma = [A + B\varepsilon^n][1 + C \ln \dot{\varepsilon}^*][1 - T^{*m}]. \tag{7}$$

Where σ is the von Mises tensile flow stress, ε is the equivalent plastic strain, $\dot{\varepsilon}^* = \dot{\varepsilon}/\dot{\varepsilon}_0^*$ is the dimensionless plastic strain rate for $\dot{\varepsilon}_0 = 1.0\,s^{-1}$, and T^* is the homologous temperature.

2.3. SPH smoothing function

Cubic B-spline smoothing function is adopted in simulation which is maybe the most frequently used smoothing function:

$$W(R,h) = \alpha_d \times \begin{cases} \frac{2}{3} - R^2 + \frac{1}{2}R^3, & 0 \le R < 1, \\ \frac{1}{6}(2-R)^3, & 1 \le R < 2, \\ 0, & R \ge 2. \end{cases} \tag{8}$$

In one-, two- and three-dimensional space, $\alpha_d = 1/h$, $15/7\pi h^2$ and $3/2\pi h^3$, respectively. Gauss kernel and quintic B-spline is also tried, while the results are almost the same.

2.4. AMR background grid

2.4.1. Mesh form

In background grid search algorithm, the uniform size of background grid is restricted by the maximal smoothing length of particles $l > \max(\kappa h_i)$ $\quad i = 1, \ldots, N$, where N is the number of all particles.

In AMR background grids, size of the jth grid can be restricted as $\max(\kappa h_i) < l_j < 2\alpha \max(\kappa h_i)$ $\quad i = 1, N_j$, $\alpha > 1.0$ and N_j is the number of particles in the jth grid.

What should be paid attention to is that grids whose coarser father grid are just the same in 2^d tree mesh structure can be combined to one grid. In the AMR grid configurations, neighbor particles of a given particle can be found in its situated grid or in grids of the same size adjoining to the aforementioned grid.

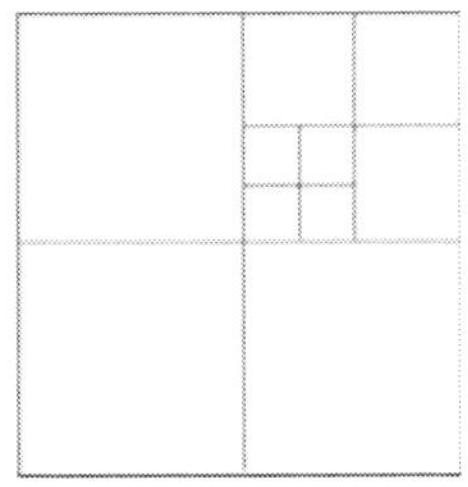

Figure 1. Background grid of adaptive mesh refinement.

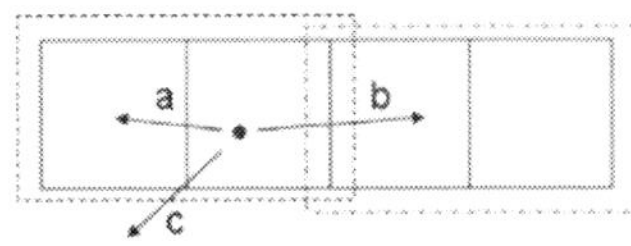

Figure 2. Particle movement.

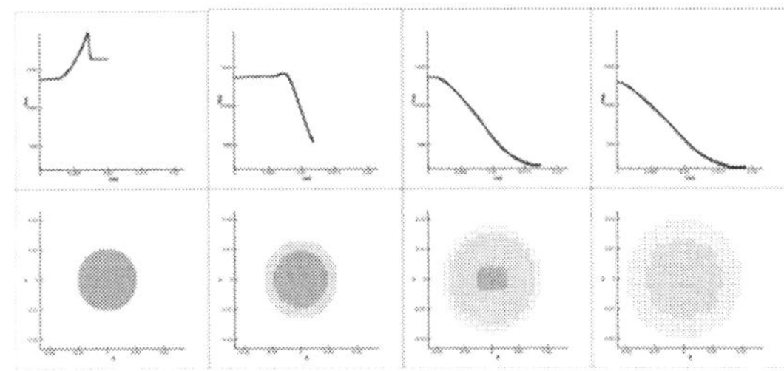

Figure 3. Density and background grids at different computational step.

2.4.2. Particle movement and creating new grid

A particle must belong to a new grid when its new location isn't in the primary situated grid. There are three cases, (a) the new grid is already exist, and it is situated in the same course as the primary grid, (b) the new grid is situated in a different course, (c) the new grid is maybe not exist then it should be created. Correspondingly, (a) the particle is transferred to the link-list of the new grid, (b) the particle is transferred to the new course, (c) a new grid is created in the course where the primary grid is situated.

3. Simulation results

3.1. 2D explosive detonation

Two-dimensional explosive detonation is simulated by AMR background grid SPH. TNT explosive is presented by JWL equations of state, detonation velocity $D_{CJ} = 6930\,\text{m/s}$and detonation pressure $P_{CJ} = 2.1 \times 10^{10}\,\text{Pa}$.

Figure 3 shows the density distributing and grids at different computational moment, which demonstrates the background grid evolvement.

Evolvement of C/N is showed in Fig. 4, in which C denotes the searching amount and N denotes the particle amount, so the C/N denotes the average number of searching needed by each particle. It shows that the evolvement rule is the same in spite of different particle amount.

The comparison of 5N/M and C/N is showed in Fig. 5, where M denotes the background grid amount and

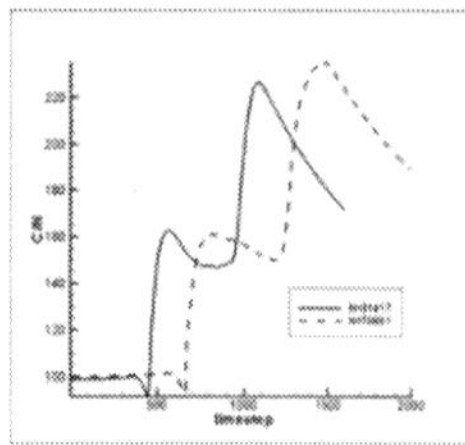

Figure 4. Evolvement of C/N.

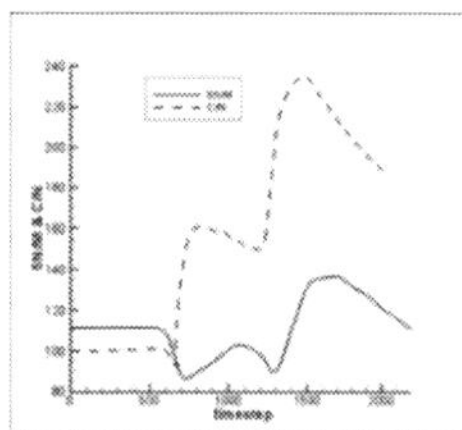

Figure 5. Comparison of 5N/M and C/N.

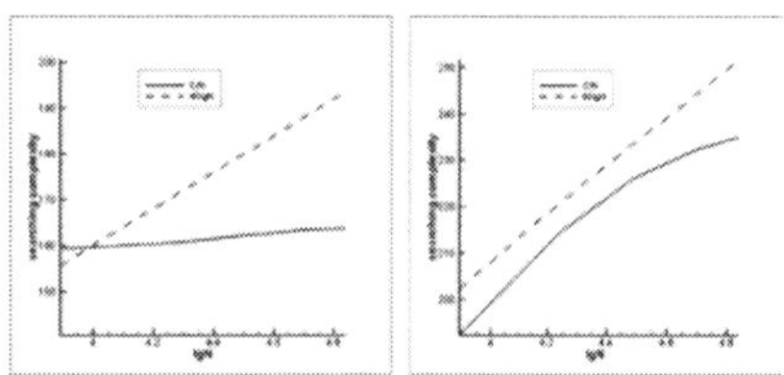

Figure 6. Vertex of C/N and lg(N) in level −1 and level −2.

N/M denotes the average number of particles in each grid. The same trends show the influence of N/M to C/N that C/N increases and reduces along with the N/M.

In Fig. 6 the dashed line, mlg(N), increases faster than the real line, C/N, while N increases. Here m is a constant. Then we get $C/N < m\lg(N)$, i.e. $C < mN\lg(N)$, in the order of magnitude, $C < O(N\lg N)$.

3.2. Sphere impacting a single plate

The aforementioned computational approach is used to simulate the impact of a sphere onto a plate at 6.15 km/s. The diameter of the sphere is 15.88 mm, and the thickness of the plate is 0.772 mm. The sphere material is 2017-T4 aluminum and the plate material is 6061-T6 in experimentation, they are both substituted as 2024-T351 aluminum in computational simulation because the two model parameters are not available while are similar in composition and response as 2024-T351 aluminum.

In Fig. 7 the colored images show the computed results at 2.5, 4.0 and 8.1 μs after impact. Computed results with just state equations are almost the same as that with constitution model. The achromatic images are from experimental radiograph at 0.0, 8.1 and 23.2 μs in literature [1]. The computed sphere shape are similar to the experimental shape but are obviously flattened in the impact direction and there are more neat corners which show the insufficiency of strength in local area. There are two possible causes of this disagreement, the constitution model is not enough or the SPH method is too smoothed to keep local rigidity.

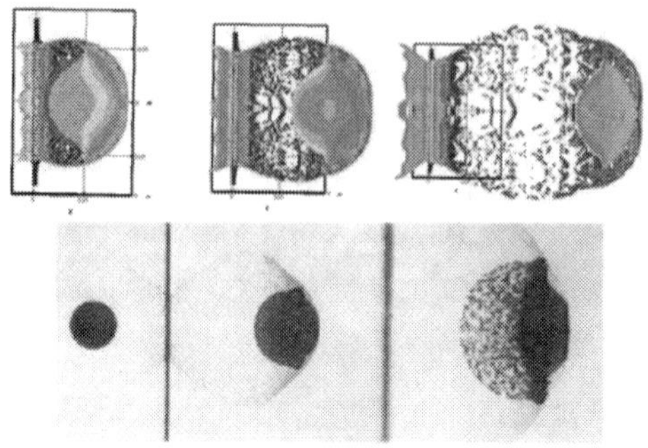

Figure 7. A sphere impact a single plate.

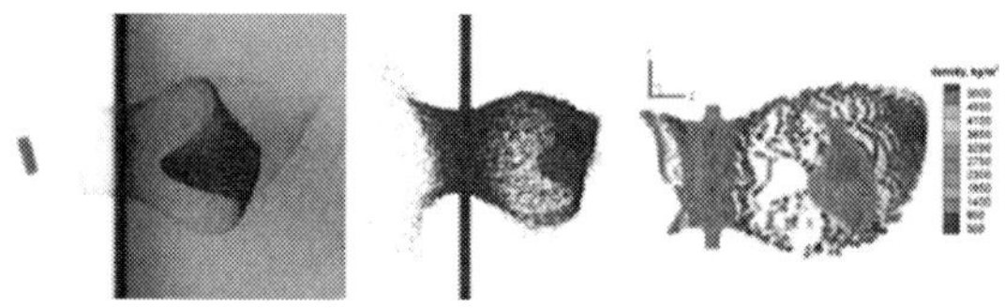

Figure 8. Oblique cylinder impact a single plate.

3.3. Oblique cylinder impacting a single plate

An OFHC copper cylinder impacts onto a 6061-T6 aluminum plate. The impact velocity is 6.47 km/s, and the obliquity is 20.6°. The cylinder diameter is 7.72 mm, and its length is 2.36 mm. The plate thickness is 2.03 mm.

In Fig. 8 the left is radiograph, the middle is computed result in [1] and the right is computed result in the aforementioned computational approach. In the result of [1] projectile material is colored red, and target material blue, while projectile and target material are displayed according to density grading in results of this paper. The computed debris-cloud results are similar to the computed results in [1]. The size of backward cone is underpredicted in the same way while the forward cone is more close to the test data. Computed results with state equations are almost the same as results with constitution model in the same way.

4. Conclusions

A new SPH algorithm with AMR background grid is proposed referencing to link-list and bucket method, and the neighbor particle searching efficiency is improved. By comparison to the experimental data, the computational approach is showed to be feasible to hypervelocity impact simulations. From the computed results the following conclusions can be gotten that impact materials can be described well enough by state equations in hypervelocity impact problems and that the constitution model in this paper is maybe not enough or the SPH method is not suitable for local details description of elastic-plastic effect in hypervelocity impact.

References

[1] S.R. Beissel, C.A. Gerlach, G.R. Johnson. Int. J. Impact Eng., **33** (2006)

[2] M.B. Liu, G.R. Liu, Arch. Comput. Method E., **17** (2010)

[3] W.S. Yang, X.Y. Meng, Z.H. Wang, Ship Science and Technology, **34**, 12 (2012)
[4] J.J. Monaghan, J.C. Lattanzio, Astron. Astr., 149 (1985)
[5] G.R. Liu, Z.H. Tu, Comput. Method Appl. M., **191** (2002)
[6] X.F. Gong, S.D. Zhang, S. Jiang, Comput. Phys., **4** (2006) (in Chinese)
[7] G.R. Liu, M.B. Liu, *Smoothed particle hydrodynamics: a meshfree method.* (Hunan University, Changsha, 2005)
[8] G.R. Johnson, W.H. Cook, Eng. Fract. Mech., **21**, 1 (1985)

EPJ Web of Conferences **94**, 04060 (2015)
DOI: 10.1051/epjconf/20159404060

Analysis of forming characteristics of Ta EFP according to material model

H.J. Kim[a], Y.S. Yi, and L.J. Park

Agency for Defence Development, Republic of Korea

Abstract. This paper presents numerical analysis result of forming characteristics of Ta explosively formed penetrator (EFP) according to various material models and their values. Dynamic material properties of Ta were measured with static tensile testing machine and Hopkinson pressure bar tests. We used AUTODYN hydrodynamic code to simulate these phenomena. We used three material models, such as Von-Mises model, linear hardening model and Johnson-Cook model. We also compared the numerical results with the EFP forming test data. The numerical results show that material model and its parameter are so important to predict the shape of formed penetrator and Von-Mises model predicts the shape of the formed liner most well. We also analysed the influence of liner thickness on EFP formation using the verified numerical model.

1. Introduction

EFP, an explosively formed penetrator, is used to penetrate a target effectively at standoff distances. When detonated, the metal plate is deformed by the explosion and formed into a penetrator.

EFP's performance is closely related to the density of the liner metal, velocity and length of the formed penetrator. Thus, Tantalum which is dense and ductile metal is widely used in liner metal and many researchers have studied the formation of Ta EFP [1–3].

S. Pappu and L. E. Murr [1] have studied microstructural features of liner in Ta, iron and copper. Johnson-Cook and Zerilli-Armstrong material model were compared in numerical simulation. They suggested that ZA model was better for Ta EFP.

G. Hussain et al.[4] have studied material model of liner in Cu and Mild steel. Johnson-Cook and modified Johnson-Cook model were compared in numerical simulation with semi-empirical equations. They showed modified Johnson-Cook was in good agreement with experimental results.

As mentioned above, Many researchers have suggested the different suitable material models one another although the liner material they used was the same, Ta. Therefore, to analyze the reason for this difference, we conducted numerical analysis on the forming characteristics of Ta EFP according to material models. In numerical simulation, we used the material parameters applied to the material model gotten from the MTS static tensile tests and Hopkinson pressure bar tests about the Ta material used to make the liner of EFP for the forming tests. We evaluated three material models, such as Von-Mises model, linear hardening model and Johnson-Cook model. By comparing these numerical results with the FXR images gotten from the EFP forming test, we analyzed the forming characteristics of Ta EFP according to material models, estimated some reasons for the difference between former results and suggested the suitable material model about the Ta liner material used to our tests. Also, using the verified numerical model, we analysed the influence of liner thickness on EFP formation.

2. Numerical modelling of EFP

2.1. EFP numerical simulations model setup

In this study, Ta EFP was numerically simulated by the Autodyn 2D axisymmetric hydrodynamic code using eulerian model. To get parameters of the material model for numerical hydrodynamic code, static and dynamic tests were conducted with a MTS and a Split Hopkinson Pressure Bar (SHPB) on tantalum. Obtained results are shown in Fig. 1.

The charge diameter of the EFP used to numerical simulation is 90 mm. The explosive used is LX14. The material of the case is SS304. We used the shock Hugoniot equation of state for Ta liner and SS304 case, and JWL equation of state for LX14 explosive. We used three material models, Von-Mises model, linear hardening model and Johnson-Cook model for Ta liner, and Von-Mises model for the case. The schematic image of EFP and simulation model is shown in Fig. 2. The detonation was chosen as a single point in AUTODYN code in the location shown in Fig. 2. The detailed models for each material are listed in Table 1.

Figure 3 shows the numerical results of the formed penetrator using Von-Mises model, linear hardening model and Johnson-Cook model for Ta penetrator at 0.3 ms, respectively. The penetrator feature of JC model was not folded into the center, and that of linear hardening model had several throats.

[a] Corresponding author: hjkim@add.re.kr

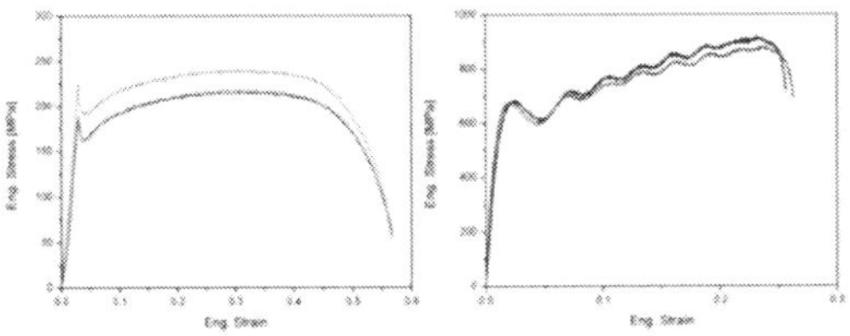

Figure 1. Stress-strain curves of (Left) Tantalum at static state and (Right) Tantalum at dynamic state.

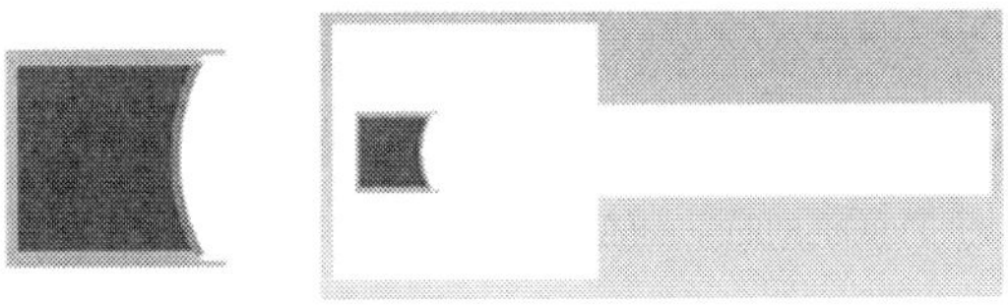

Figure 2. Schematic image of (Left) Ta EFP and (Right) Eulerian domain.

Table 1. Material model used.

Part	Material	EOS	Material model
Liner	Tantalum	Shock Hugoniot	Von-Mises, JC, linear-hardening
Case	SS 304	Shock Hugoniot	Von-Mises
Explosive	LX14	JWL	None
Domain	Air	Ideal gas	None

	Von-Mises	Linear-hardening	Johnson-Cook
Penetrator feature			

Figure 3. Penetrator features according to the material strength model (a) Von-Mises, (b) linear-hardening and (c) Johnson-cook model.

2.2. Experiment result

To validate numerical simulation results, Ta EFP forming test was performed. Figure 4 shows the picture of experimental setup. To acquire picture of penetrator feature, Flash X-ray (FXR) system was used at 0.3 ms after detonated. Figure 5 shows the FXR image.

Figure 6 shows the picture of penetrator features. The Von-Mises model is in well agreement with the experimental data. The simulated tip velocity and total length of Ta were 1.46 km/s and 71.2 mm with 19% and 6% error respectively, and the maximum diameter and minimum diameter defined in Fig. 6 were 19 mm and 14 mm with 16% and 2% error respectively. Also, L/D was 3.7 with 25% error. Although L/D error is about 25%, the errors of other values are less than 20%. Therefore, Von-Mises was the suitable model for Ta EFP in our case.

Also through this study, we estimated that the difference of suitable model between former researchers and our result or between former researchers was caused by the difference of the shape of EFP or that of Ta material properties. In fact, Ta has wide range of its static and dynamic properties.

Figure 4. Picture of the experimental setup.

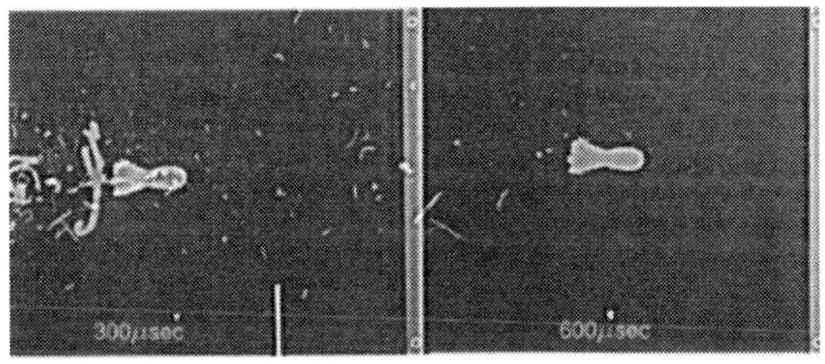

Figure 5. Result image of FXR.

	FXR image	Autodyn 2D
Penetrator feature		
Velocity [km/s]	1.23	1.46 (119%)
Length [mm]	67.4	71.2 (94%)
Max. Diameter [mm]	22.7	19 (84%)
Min. Diameter [mm]	13.7	14 (102%)
L/D (Length/Max. Dia.)	3	3.7 (125%)

Figure 6. Penetrator features of (a) FXR image and (b) numerical result.

3. Influence of liner thickness on EFP formation

Using the verified numerical model above, we analysed the influence of liner thickness on EFP formation. We conducted numerical simulations to change the liner thickness from 3 mm to 4 mm by the Autodyn 2D code. Figure 7 shows the numerical results of penetrator characteristics according to the thickness of the liner. When the thickness of the liner was decreased, the penetrator was more elongated, that is, the length of penetrator was longer, the L/D was larger and the

Liner thickness	4t	3.8t	3.5t	3t
Penetrator feature				
Velocity [km/s]	1.46	1.49 (102%)	1.62 (111%)	1.83 (125%)
Length [mm]	71.2	77.5 (109%)	86.5 (121%)	104.7 (147%)
Max. Diameter [mm]	19	18.6 (98%)	17.2 (91%)	15.8 (83%)
Min. Diameter [mm]	14	5.4 (39%)	-	-
L/D (Length/Max. Dia.)	3.7	4.1 (111%)	5.0 (135%)	6.6 (178%)

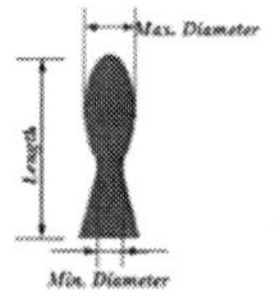

Figure 7. Penetrator features according to the liner thickness.

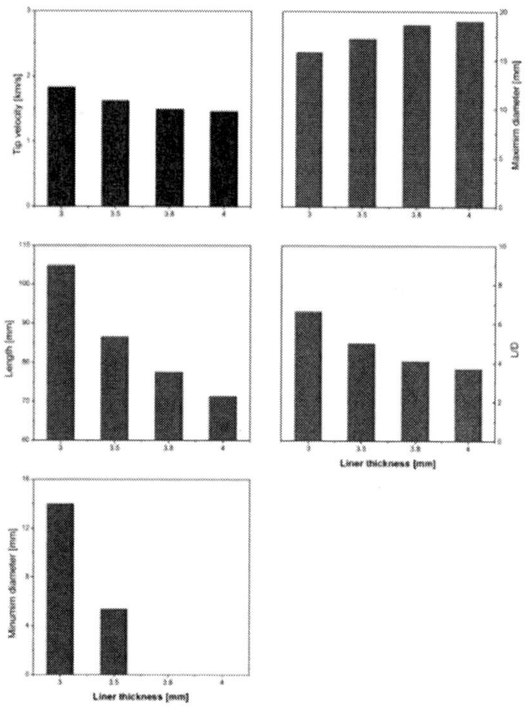

Figure 8. Numerical results according to (a) tip velocity, (b) maximum diameter, (c) length, (d) L/D and (e) minimum diameter.

diameters were smaller. Also, the velocity was faster. From our results, we can estimated the EFP of 3.8 mm thickness liner had the best penetration performance among them.

When the thickness of the liner was thinner than 3.8 mm, the penetrators were divided into several fragments like the 3.5 mm and 3 mm thickness liner shown in Fig. 7.

Figure 8 shows the numerical results variation of EFP characteristics according to the variation of the thickness of the liner. As shown in Fig. 8, the sensitivity of length, L/D and minimum diameter of EFP by the thickness variation was larger than that of velocity and maximum diameter. The minimum diameter was the most sensitive among them and it effected negative influence on the penetration performance of EFP unlike other parameters. So, we estimated that when we would design EFP varying the thickness of liner, we should consider that the threshold value occurs in minimum diameter of EFP.

4. Conclusions

Ta EFPs were numerically studied to investigate forming characteristics of Ta liner comparing the EFP forming test data. Penetrator features were numerically simulated by Autodyn 2D code and three types of strength model were used such as Von-Mises, linear hardening and Johnson-Cook model. To obtain parameters of the material model, material tests were conducted with a MTS and SHPB on Ta. The result of Von-Mises model was in good agreement with our test data. From this study, we estimated that the difference of suitable model between former researchers and our result or between former researchers was caused by the difference of the shape of EFP or that of Ta material properties.

Also we conducted the numerical simulation about the influence of liner thickness on EFP formation. From these results, we concluded that the thinner liner of EFP was, the better the penetration performance of it was. But also we concluded that the minimum diameter was the most sensitive among them and it effects negative influence on the penetration performance of EFP unlike other parameters. So, we estimated that when we would design EFP varying the thickness of liner, we should consider that the threshold value occurs in minimum diameter of EFP.

References

[1] S. Pappu and L. E. Murr, Journal of Material Science **37**, 233–248 (2002)

[2] G. Hussain, A. Q. Malik and K. Sanaullah, Journal of Engineering & Technology **30**, 417–426 (2011)

[3] R. A. Requeiro and M. F. Horstemeyer Advances in Computational Engineering, 384–389 (2000)

[4] G. Hussain et al. Journal of Energetic Materials **31**, 143–155 (2013)

EPJ Web of Conferences **94**, 04061 (2015)
DOI: 10.1051/epjconf/20159404061

Numerical analysis of high strain rate failure of electro-magnetically loaded steel sheets

Borja Erice[1,a] and Dirk Mohr[1,2,3]

[1] Solid Mechanics Laboratory (CNRS-UMR 7649), Department of Mechanics, École Polytechnique, Palaiseau, France
[2] Impact and Crashworthiness Lab, Department of Mechanical Engineering, Massachusetts Institute of Technology, Cambridge MA, USA
[3] Department of Mechanical and Process Engineering, ETH Zurich, Switzerland

Abstract. Electro-magnetic forces provide a potentially power full means in designing dynamic experiments with active control of the loading conditions. This article deals with the development of computational models to simulate the thermo-mechanical response of electro-magnetically loaded metallic structures. The model assumes linear electromagnetic constitutive equations and time-independent electric induction to estimate the Joule heating and the Lorentz forces. The latter are then taken into account when evaluating stress equilibrium. A thermo-visco-plastic model with Johnson-Cook type of temperature and strain rate dependence and combined Swift-Voce hardening is used to evaluate the material's thermo-mechanical response. As a first application, the model is used to analyse the effect of electro-magnetic loading on the ductility of advanced high strength steels.

1. Introduction

Modern equipment is able to discharge large capacitor energies in specifically designed coils generating electro-magnetic (EM) forces in conductive workpieces in periods of time in the order of 0.1 ms. Far from being spatially concentrated forces, their area of influence extends to that of the EM field generated in the coil. This might give an unprecedented level of tailoring and control on dynamic experiments. Additionally, a significant increase in ductility on conductive metals is reported with respect to conventional mechanical forming processes.

The objective of this numerical study is twofold. First, understand how the EM loading can affect to a dynamic deformation process and second dilucidate which are the underlying mechanisms behind the increase on ductility and delayed fracture. To do so, numerical simulations of electro-magnetically vs. mechanically loaded sheet metal plates were compared.

2. Problem description

Our analysis is concerned with the introduction of a hemispherical dome (of 50 mm radius) into an initially flat 1.4 mm thick steel sheet (see Fig. 1). Two different high speed loading scenarios are considered to achieve this: (1) electro-magnetic loading, and (2) mechanical impact loading. The goal is to compare these two techniques and to come up with an explanation of the well-known failure retardation in forming processes through electro-magnetic load application.

[a] Corresponding author: borja@lms.polytechnique.fr

2.1. Electro-magnetically applied load

The magnetic field is created by circulating a pulsed electric current through a coil that is positioned next to a conductor material. Figure 2 shows a serial RLC circuit that provides the required current, with L_{RLC} and R_{RLC} denoting the inductance and resistivity of the circuit, respectively. The current of intensity $i_c(t)$ circulating through the coil with a resistance R_{coil} and an inductance L_{coil} is generated by the sudden discharge of the energy stored in capacitor C. Assuming an equivalent $\bar{R}$, $\bar{L}$, C circuit, the expression for the current intensity of the underdamped response circuit [1] reads

$$i_c(t) = \frac{V_0}{\omega \bar{L}} \exp\left(-\frac{\bar{R}}{2\bar{L}}t\right) \sin(\omega t), \omega = \sqrt{\frac{1}{\bar{L}C} - \left(\frac{\bar{R}}{2\bar{L}}\right)^2} \quad (1)$$

with ω denoting the natural frequency.

Spirally wounded coils are typically employed to apply the Lorentz forces as uniformly as possible to the workpiece [2]. The configuration adopted for the present numerical study is detailed in Fig. 1(a). A $5 \times 5\,\mathrm{mm}^2$ square cross-section coil with four turns is located underneath the steel specimen. The overall coil diameter is chosen such as that the coil covers the entire 100 mm diameter die. The current intensity $i_c(t)$ is imposed through the coil as indicated in red colour in Fig. 1(a). The direction of the prescribed current intensity is such that the Lorentz forces generated in the workpiece are oriented upwards, in the positive direction of the z-axis. As a result, the workpiece is pushed into the receiving die. The die is assumed to be non–conductive, fully rigid and fixed in space. A blank holder is positioned underneath the plate to clamp the specimen. The current intensity is

Article available at http://www.epj-conferences.org or http://dx.doi.org/10.1051/epjconf/20159404061

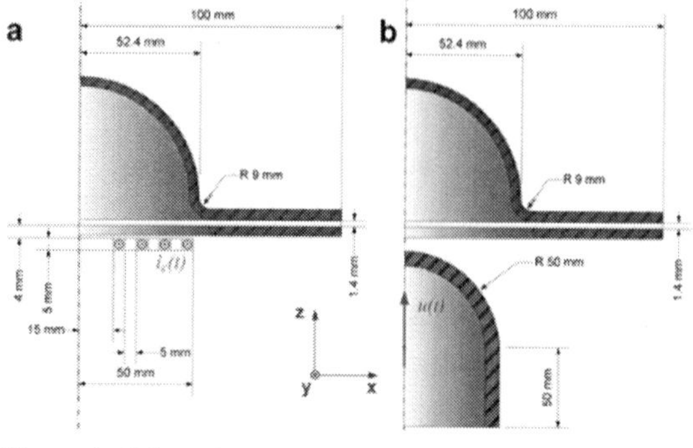

Figure 1. Schematic view of the EM (a) and mechanical (b) problems.

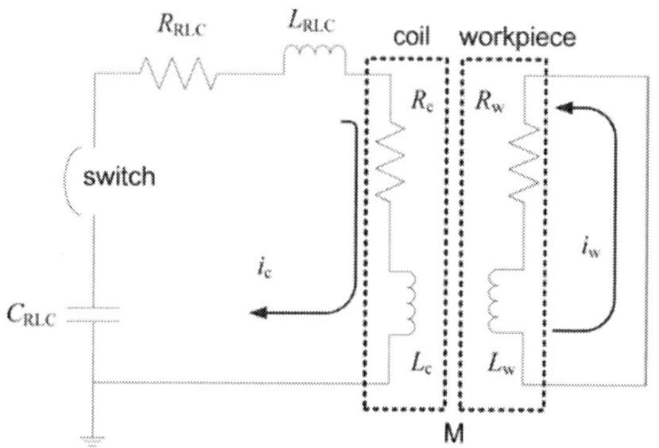

Figure 2. Simplified RLC circuit used in the electro-magnetic loading processes. M represents the mutual inductance.

calculated based Eq. (1) with an initial voltage of $V_0 = 11\,\text{kV}$, a capacitance of $C = 900\,\mu\text{F}$ and an inductance of $L = 1\,\mu\text{H}$. The capacitor energy is released rapidly, generating a damped current intensity that is attenuated within about 0.3 ms. The entire forming process will also be completed within this duration.

2.2. Mechanically applied load

In the mechanical process, the sheet is loaded through a hemispherical punch. The duration of a conventional mechanical sheet metal forming process is of the order of seconds, while the typical duration of the electro-magnetic process is about four orders of magnitude shorter. In our comparative study, we consider two mechanical loading scenarios: a slow case that is complete within 1 s, and a fast case that completes the forming within 0.3 ms. Moreover, to ensure the direct comparability, the punch velocity history in the fast process is chosen such that it matched the velocity of the dome apex in the electro-magnetic process.

3. Thermo-mechanical constitutive model

The mechanical constitutive response of the material presented here is based on the work by Mohr et al. [3], which subsequently was modified by Roth and Mohr [4] to take into account strain rate and thermal softening effects.

The model assumes the additive decomposition the strain tensor $\boldsymbol{\varepsilon}$ as:

$$\boldsymbol{\varepsilon} = \boldsymbol{\varepsilon}^e + \boldsymbol{\varepsilon}^p \tag{2}$$

with the elastic and plastic parts $\boldsymbol{\varepsilon}^e$ and $\boldsymbol{\varepsilon}^p$. The elastic strain and Cauchy stress tensors are related through Hooke's law as follows:

$$\boldsymbol{\sigma} = \mathbf{C} : \left(\boldsymbol{\varepsilon} - \boldsymbol{\varepsilon}^p\right) \tag{3}$$

where $\mathbf{C}$ is the positive-definite fourth-order symmetric tensor that contains the elastic modulus E and the Poisson's ratio ν,

$$\mathbf{C} = \frac{\nu E}{(1+\nu)(1-2\nu)} \boldsymbol{I} \otimes \boldsymbol{I} + \frac{E}{(1+\nu)} \mathbf{I} \tag{4}$$

being $\boldsymbol{I}$ and $\mathbf{I}$ the identity second-order and fourth-order tensors, respectively.

The Hill 1948 orthotropic yield function [5] is adopted:

$$\phi(\boldsymbol{\sigma}, \boldsymbol{\alpha}) = \bar{\sigma}_\phi(\boldsymbol{\sigma}) - Y(\boldsymbol{\alpha}, \dot{\boldsymbol{\alpha}}) = 0 \tag{5}$$

where Y is a deformation resistance, $\boldsymbol{\alpha}$ is a set of internal hardening variables, and $\dot{\boldsymbol{\alpha}} = d\boldsymbol{\alpha}/dt$ their time derivative. $\bar{\sigma}_\phi$ is the yield function equivalent stress that can be expressed conveniently with a linear transformation of the stress tensor as follows [6]:

$$\bar{\sigma}_\phi(\boldsymbol{\sigma}) = \sqrt{\boldsymbol{\sigma} : \mathbf{P} : \boldsymbol{\sigma}} \tag{6}$$

with the positive-definite fourth-order symmetric tensor,

$$\mathbf{P} = \begin{bmatrix} 1 & P_{12} & -(1+P_{12}) & 0 & 0 & 0 \\ & P_{22} & -(P_{12}+P_{22}) & 0 & 0 & 0 \\ & & 1+2P_{12}+P_{22} & 0 & 0 & 0 \\ & & & P_{33} & 0 & 0 \\ & & \text{sym} & & 3 & 0 \\ & & & & & 3 \end{bmatrix}. \tag{7}$$

The three independent components P_{12}, P_{22} and P_{33} can be expressed as a function of the uniaxial tension yield stresses in three directions with respect to the rolling direction, typically at 0, 45 and 90 degrees and the equi-biaxial tension yield stress [3]. The yield function defines a von Mises yield surface in the case of isotropic yielding being $P_{12} = -0.5$, $P_{22} = 1.0$ and $P_{33} = 3.0$. A non-associated flow potential of the form,

$$\bar{\sigma}_\psi(\boldsymbol{\sigma}) = \sqrt{\boldsymbol{\sigma} : \mathbf{G} : \boldsymbol{\sigma}} \tag{8}$$

has been chosen to represent the flow rule $\dot{\boldsymbol{\varepsilon}}^p = \dot{\lambda}\,(\partial\psi/\partial\boldsymbol{\sigma})$. The choice of a non-associated flow rule (NAFR) for sheet metal forming is based on experimental observations. $\mathbf{G}$ is a positive-definite fourth-order symmetric tensor that has the same form as the $\mathbf{P}$. The three independent components G_{12}, G_{22} and G_{33} are defined as a function of the Lankford ratios corresponding to the same three directions used in the yield function [3].

The set of internal variables are the equivalent plastic strain, the equivalent plastic strain rate and the temperature; their evolution equations read:

$$\dot{\bar{\varepsilon}}_p = \dot{\lambda}\frac{\bar{\sigma}_\psi}{\bar{\sigma}_\phi}, \ddot{\bar{\varepsilon}}_p = \frac{d\dot{\bar{\varepsilon}}_p}{dt} = \frac{d}{dt}\left(\dot{\lambda}\frac{\bar{\sigma}_\psi}{\bar{\sigma}_\phi}\right), \dot{T} = \dot{\lambda}\frac{\chi\left(\dot{\bar{\varepsilon}}_p\right)}{\rho C_p}\bar{\sigma}_\psi \tag{9}$$

where ρ is the mass density andC_p is the specific heat. Typically, the temperature increase is assumed to be only due to adiabatic effects and the Taylor-Quinney coefficient χ is assumed to be constant and equal to 0.9. In an isothermal situation such a coefficient would be null. A smooth transition between isothermal and adiabatic conditions is proposed as [4]:

$$\chi\left(\dot{\bar{\varepsilon}}_p\right) = \chi_0\left(\dot{\bar{\varepsilon}}_p - \dot{\varepsilon}_0\right)^2\left(3\dot{\varepsilon}_a - 2\dot{\bar{\varepsilon}}_p - \dot{\varepsilon}_0\right)\Big/\left(\dot{\varepsilon}_a - \dot{\varepsilon}_0\right)^3 \tag{10}$$

being $\dot{\varepsilon}_0$ and $\dot{\varepsilon}_a$ the limit strain rates for the isothermal and adiabatic domains, respectively.

Isotropic hardening is assumed accounting for strain hardening, strain rate hardening and the thermal softening has been considered independent [7–9]. Therefore, the constitutive response can be conveniently expressed in three separate terms that multiply each other as [4]:

$$Y\left(\bar{\varepsilon}_p, \dot{\bar{\varepsilon}}_p, T\right) = Y_\varepsilon\left(\bar{\varepsilon}_p\right) Y_{\dot{\varepsilon}}\left(\dot{\bar{\varepsilon}}_p\right) Y_T(T). \tag{11}$$

The strain hardening is defined as a weighted linear combination of a Swift and Voce laws. It reads:

$$Y_\varepsilon = \alpha\left\{A\left(\varepsilon_0 + \bar{\varepsilon}_p\right)^n\right\} + (1-\alpha)\{\sigma_0 + Q_1\left[1 - \exp\left(-C_1\bar{\varepsilon}_p\right)\right]\} \tag{12}$$

where A, ε_0, n, σ_0, Q_1, C_1 are material constants and α is the weighting coefficient. The strain rate hardening and the thermal softening have the same form as those defined by Johnson and Cook [9],

$$Y_{\dot{\varepsilon}} = 1 + C\ln\left(\frac{\dot{\bar{\varepsilon}}_p}{\dot{\varepsilon}_0}\right), Y_T = 1 - \left(\frac{T - T_r}{T_m - T_r}\right)^m \tag{13}$$

where C and m are material constants, T_r and T_m are the reference and melting temperatures respectively.

The Hosford-Coulomb failure criterion is a stress state dependent failure criterion. The stress state is characterized by the stress triaxiality η and the Lode angle parameter $\bar{\theta}$. The expression of the equivalent plastic strain to failure is composed of three independent terms. The first term represents the stress-state-dependent failure criterion, while the second and the third are the expressions that introduce the strain rate and temperature effects defined as in [10]. The equivalent plastic strain to failure of the failure criterion is:

$$\bar{\varepsilon}_p^f = \bar{\varepsilon}_{\eta,\bar{\theta}}^f\left(\eta, \bar{\theta}\right)\bar{\varepsilon}_{\dot{\varepsilon}}^f\left(\dot{\bar{\varepsilon}}_p\right)\bar{\varepsilon}_T^f(T). \tag{14}$$

The independent terms are defined as follows:

$$\bar{\varepsilon}_{\eta,\bar{\theta}}^f = b(1+c)^{\frac{1}{n}}\left(\left\{\frac{\left(F_1^a + F_2^a + F_3^a\right)}{2}\right\}^{\frac{1}{a}} + c\left(2\eta + f_1 + f_3\right)\right)^{-\frac{1}{n}} \tag{15}$$

$$F_1 = f_1 - f_2,\quad F_2 = f_2 - f_3,\quad F_3 = f_1 - f_3$$

$$f_1 = \frac{2}{3}\cos\left[\frac{\pi}{6}\left(1-\bar{\theta}\right)\right],\ f_2 = \frac{2}{3}\cos\left[\frac{\pi}{6}\left(3+\bar{\theta}\right)\right] \tag{16}$$

$$f_3 = -\frac{2}{3}\cos\left[\frac{\pi}{6}\left(1+\bar{\theta}\right)\right]$$

$$\bar{\varepsilon}_{\dot{\varepsilon}}^f = 1 + D_4\ln\left(\frac{\dot{\bar{\varepsilon}}_p}{\dot{\varepsilon}_0}\right), \bar{\varepsilon}_T^f = 1 + D_5\frac{T - T_r}{T_m - T_r} \tag{17}$$

where a, b, c, D_4 and D_5 are material constants. The damage indicator evolution law (failure for $D = 1$) is defined as

$$\dot{D} = \frac{1}{\bar{\varepsilon}_p^f\left(\eta, \bar{\theta}, \dot{\bar{\varepsilon}}_p, T\right)}\dot{\bar{\varepsilon}}_p. \tag{18}$$

It should be noted that since the model is uncoupled, i.e. no softening due to damage evolution; the damage indicator is not an internal variable of the plasticity model. The time discretized equations were integrated in an elastic predictor-plastic corrector-type set of equations and the return to the yield surface was executed with a cutting plane algorithm. The calibration procedure for the DP590 steel can be found in [4] (see Table 1).

4. Numerical set-up

Due to its symmetry conditions only one fourth of the geometry was modeled. The discretization was performed using eight-node reduced-integration solid elements with stiffness-based hourglass control. The die and the punch were modelled as non-conductive rigid bodies. The die was fixed in space, while a displacement $u(t)$ was prescribed to the punch. Six elements over the thickness were used for the work piece, whereas for the die and punch, since they were modelled as rigid bodies, only one element was used through the thickness. The in-plane element size of the plate and dies was 1 mm. The end of the plate was considered to be free of constrains. The only constraint comes from the blanket holder that prevents any specimen motion in z-direction (see Table 1).

A Boundary Element Method (BEM) is employed to circumvent the meshing of the surrounding air [11,12]. The electromagnetic and the mechanical solvers are coupled through the addition of Lorentz forces to the stress equilibrium equations in form of nodal forces. Following the developer's recommendation [11], the chosen time step of the electromagnetic solver was about 100 times the mechanical time step. Thus, the electromagnetic solver will be active every 100 mechanical increments. The wires were assumed to be made of pure electrical quality copper with $E = 117$ GPa, $\nu = 0.3$ and a mass $\rho = 8920\,\mathrm{kg/m^3}$. The wire turns were modelled as individual parts. The same current intensity was prescribed in all coils. Therefore, the current displacement effect between turns was considered negligible. In the electromagnetic forming all the materials involved were considered to exhibit their free-space magnetic permeability. The electrical conductivity of the die and coils were considered to be zero and 5.96×10^7 S/m respectively. The electrical conductivity for the DP 590 steel is contained in Table 1.

Table 1. Material constants for the DP590 steel.

$\rho\left(\frac{kg}{m^3}\right)$	$C_p\left(\frac{J}{kg°C}\right)$	E (GPa)	ν	χ_0	$\dot{\varepsilon}_a\left(s^{-1}\right)$	$\Sigma_0\left(\frac{S}{m}\right)$
7850.0	420.0	210.0	0.3	0.9	1.379	1.5×10^6
P_{12}	P_{22}	P_{33}	G_{12}	G_{22}	G_{33}	
−0.5000	1.0000	3.0000	−0.4946	0.9318	2.4653	
A (MPa)	ε_0	n	σ_0 (MPa)	Q_1 (MPa)	C_1	α
1031.00	0.00128	0.199	350.00	324.25	25.92	0.64
C	$\dot{\varepsilon}_0\left(s^{-1}\right)$	m	T_0 (°C)	T_r (°C)	T_m (°C)	
0.0136	0.001164	0.921	25.0	25.0	1400.6	
a	b	c	n	D_4	D_5	
1.970	0.820	0.000	0.199	0.025	0.000	

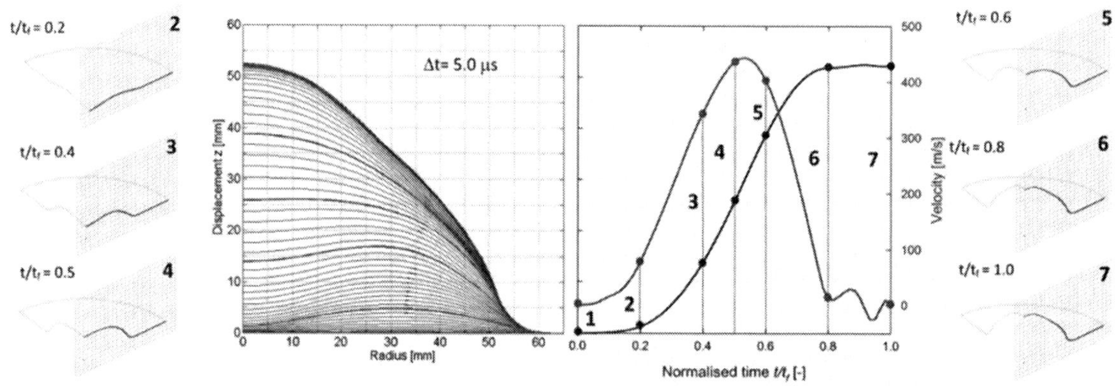

Figure 3. Evolution of the cross-section upper surface depicted in blue in the lateral pictures (left). Displacement and velocity vs. normalized time of the point located on the center of the plate (right).

5. Results and discussion

Firstly, we analyze the EM forming process and the characteristic steps leading to the final sheet shape. The right hand side plot in Fig. 3 shows the displacement (black) and velocity (blue) histories of the point located in the center of the upper surface of the sheet. The time has been normalized by the total duration of the simulations (t_f). The plot is marked with seven solid points that correspond to $t/t_f = [0.0, 0.2, 0.4, 0.5, 0.6, 0.8, 1.0]$. For each of these positions, a snapshot of the deformed geometry was taken (see left and right of Fig. 3). To enhance the visualization of the progressive deformation of the sheet, a cross-section cut along one of the specimen symmetry planes is illustrated in blue over a grid formed by 5 × 5 mm squares. Additionally, on the left hand side of Fig. 3 the evolution with time of such cross-section upper line is plotted every 5 μs time intervals. The red curves correspond to the initial and final states, $t/t_f = [0.0, 1.0]$, whereas the blue curves correspond to the intermediate steps, $t/t_f = [0.2, 0.4, 0.5, 0.6, 0.8]$.

In the initial stage of the plate forming, i.e. from steps one to two, the Lorentz forces generated on the sheet metal plate were concentrated in the area of influence of the coils. As a result, there was a lack of magnetic pressure on the center of the specimen. Its effect on the deformed shape vanished quickly after about 10 μs (beyond point two in Fig. 3).

From there onwards, the velocity was then transferred from the part driven by Lorentz forces towards the center of the plate. Moreover, the inertia of the workpiece appeared to be driving the deformation process beyond point four where the velocity was about to reach its maximum (see Fig. 3 (right)). In the next stage, from point four to five, the displacement of the area close to the center kept displacing at the same pace as it did it the stage before. Conversely, there was a significant loss of velocity that indicated the commencing of the progressive slow-down of the plate (see Fig. 3 (right)). During the subsequent stage, from step five to six, it suffered a dramatic drop on velocity completing the slow-down process and attaining virtually its maximum displacement. In the last simulated stage, between points six and seven, there were nearly null forces that could affect the deformation of the workpiece. Hence, the center point practically maintained its displacement, while the velocity oscillated around zero.

Figure 4 shows the upper dome profile evolution of the two mechanically-loaded cases (slow and fast) compared to the EM-loaded case. Note that the time increment for each one of the black solid curves plotted in the slow case is the 20 ms and green curves coincide with the time the first fracture appears. In the slow case, displayed in Fig. 4(a), the plate deformation was driven by the force applied with the punch. That means that the parts of the plate that were not in contact with it were dragged immediately by the zones that were actually in contact. The first zone of the workpiece to be contacted by the punch was the center of it. Consequently, it can be seen how it was the first zone to be lifted and then the rest of the plate followed. The way in which the plate was being

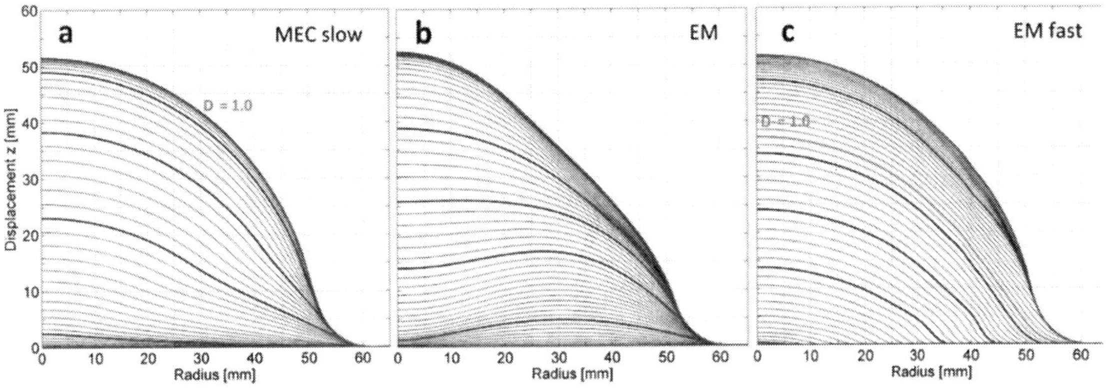

Figure 4. Evolution with time of the slow (a), EM (b) and fast (c) cases.

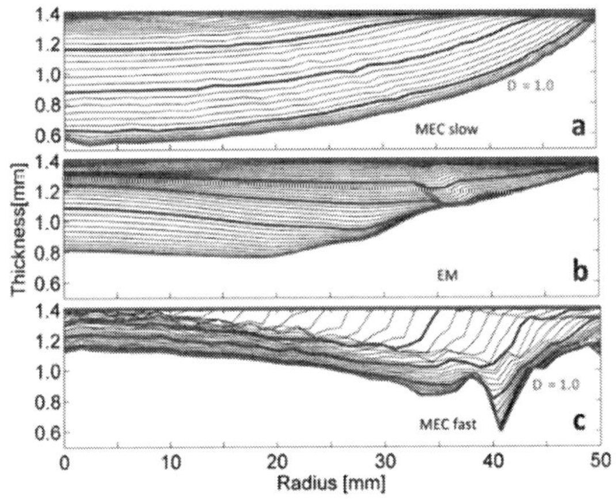

Figure 5. Evolution of the thickness with time of the slow (a), EM (b) and fast (c) cases.

formed made the thinning to be localized within a narrow plastic zone. Therefore, the workpiece necked rapidly in the center area causing its fracture before it reached the pre-established maximum displacement. The progressive thinning and necking on the center area can be seen in Fig. 5(a).

The fast case resembles an impact problem, where the punch acts like a guided projectile and the sheet as the target. The only difference here is that the projectile has a prescribed motion in contrast with a typical impact problem. In Fig. 4(c) we observe that even though the principle applied to move the plate was the same as that in the slow case, the plate did not deform in the same way. The center of the workpiece "saw" the contact but the rest of it still did not noticed the impact yet. Figure 5(c) depicts perhaps more clearly such phenomenon. There was a stress wave that travelled from the first point where the punch impacted, the center, towards the end of the plate. This relieved such part from strains. There was a moment when the plate did not move by the effect of the punch but by the effect of the inertia. Recall that a similar effect occurred in the in the EM case. The inertia then dragged the central part of the plate, which was not in contact with the punch anymore, necking severely the zone that was highly strained shortly before.

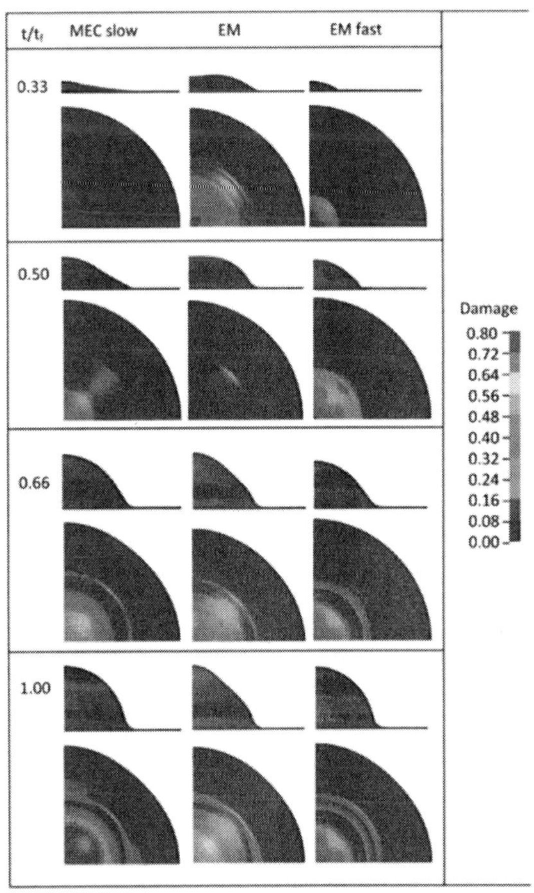

Figure 6. Damage contours of the numerical simulations carried out for the slow (a), EM (b) and fast (c) cases.

The differences on the damage accumulation of the three study cases, slow, EM and fast, are represented in Fig. 6. The differences between the slow and fast cases are clearly distinguishable here. On one hand, the progressive plastic strain accumulation, and therefore damage accumulation, in the central zone, can be observed in the slow case. On the other hand, the absence of such accumulation in the fast case central zone can be clearly seen. In fact, the plastic strain accumulation, as explained before, occurs away from the specimen center. In any case, both slow and fast cases presented a severe thinning and subsequent necking as can be seen in Fig. 5. Conversely to these cases, the more uniform force distribution of the EM case, allowed the plate to thin relatively uniformly and without necking, revealing EM loaded process as the less harmful for the material in terms of fracture. The manner in which the load was applied in this case allowed the strains to distribute more uniformly over the plate and extend the load-applying area not solely to the center, but to far more distance than the mechanically-driven forces allowed. On top of that, the load-applying period was small enough to benefit from strain rate and thermal softening effects which also increase the material's ductility. Indeed, Fig. 6 shows how the damage accumulation in the plate is far from the levels of the mechanically loaded cases.

6. Conclusions

This paper presents a computational model predicting the high strain rate failure of electro-magnetically loaded structures. In addition to the electro-magnetic and mechanical solvers, it makes use of a newly-developed rate- and temperature dependent user material subroutine. In particular, the Hill yield surface with a non-associated flow rule was used in conjunction with a Johnson-Cook type of isotropic rate-dependent hardening law. The stress-state, strain-rate and temperature dependent Hosford-Coulomb fracture initiation model was used to evaluate the damage accumulation on the deformation processes.

In a first application, the modeling approach was used to analyze the apparent ductility increase in electromagnetic forming through comparison with simulations with conventional mechanical loading. The analysis showed that the load application in the mechanical cases caused pronounced strain localization within the deforming sheet, whereas significantly more uniform strain distributions are observed for electromagnetic loading. The delayed failure in electromagnetic forming is thus mainly attributed to the increased strain field uniformity, whereas the increase in ductility due to strain rate and temperature effects is seen as secondary.

The first author is grateful for the partial financial support through the French National Research Agency (Grant ANR-11-BS09-0008, LOTERIE).

References

[1] A. Alonso Rodriguez and A. Valli, *Eddy Current Approximation of Maxwell Equations*. MS&A, ed. T.Y. Hou, et al. Vol. 4. (2010): Springer.

[2] V. Psyk, D. Risch, B.L. Kinsey, A.E. Tekkaya, and M. Kleiner, *Electromagnetic forming—A review*. Journal of Materials Processing Technology, (2011). **211**(5): p. 787–829.

[3] D. Mohr, M. Dunand, and K.-H. Kim, *Evaluation of associated and non-associated quadratic plasticity models for advanced high strength steel sheets under multi-axial loading*. International Journal of Plasticity, (2010). **26**(7): p. 939–956.

[4] C.C. Roth and D. Mohr, *Effect of Strain Rate on Ductile Fracture Initiation in Advanced High Strength Steel Sheets: Experiments and Modeling*. International Journal of Plasticity (2014) (In press).

[5] R. Hill, *A theory of the yielding and plastic flow of anisotropic metals*. roceedings of the Royal Society of London. Series A, Mathematical and Physical Sciences, (1948). **193**(1033): p. 281–297.

[6] M.A. Crisfield, *Non-linear Finite Element Analysis of Solids and Structures, Volume 2: Advanced Topics*. (1997), London: John Wiley & Sons.

[7] T. Børvik, Hopperstad, O.S., Berstad, T., Langseth, M., *A computational model of viscoplasticity and ductile damage for impact and penetration*. European Journal of Mechanics – A/Solids, (2001). **20**: p. 685–712.

[8] S. Chocron, B. Erice, and C.E. Anderson, *A new plasticity and failure model for ballistic application*. International Journal of Impact Engineering, (2011). **38**(8-9): p. 755–764.

[9] G.R. Johnson and W.H. Cook. *A Constitutive Model and Data for Metals Subjected to Large Strains, High Strain Rates and High Temperatures*. in *7th International Symposium on Ballistics*. (1983). The Hague.

[10] G.R. Johnson and W.H. Cook, *Fracture characteristics of three metals subjected to various strains, strain rates, temperatures and pressures*. Engineering Fracture Mechanics, (1985). **21**: p. 31–48.

[11] P. L'Eplanettenier and I. Çaldichoury, *EM Theory Manual, Livermore Software Technology Corporation (LSTC)*. (2012). 40.

[12] LSTC, *LS-DYNA KEYWORD USER'S MANUAL Version 971 R7.0*. (2013), Livermore, California: Livermore Software Technology Corporation.

EPJ Web of Conferences **94**, 05001 (2015)
DOI: 10.1051/epjconf/20159405001

Industrial application of Instrumented DWTT in evaluating material resistance to ductile fracture for modern pipeline steels

Jian Fang[1,a], Jian-wei Zhang[1], and Xing Huang[2]

[1] Baosteel Research Institute, Baoshan Iron & Steel Co., Ltd., Shanghai 201900, China
[2] Shenzhen WANCE Testing Machine Co., Ltd., Shenzhen 518055, China

Abstract. $CTOA_C$ of typical X70, X80 and X90 pipeline steels plates were evaluated by using energy based regression method upon their single specimen. Clear difference in the material resistance to ductile cracking could be distinguished by the values of $CTOA_C$ and their characteristic fracture surface. The region of stable cracking on the fracture surface could be identified from the dynamic crack extension curve, with the help of key-curve method. In the meanwhile, the crucial material based parameter required by many other $CTOA_C$ algorithm, ($A^*\sigma_f$) of present X70, X80 and X90 plates were experimentally determined, also proving the validity of theoretical geometric factor A^* to be 0.30–0.34, close to the commonly cited constant of 1/3.

1. Introduction

Drop-Weight-Tear-Test (DWTT) has been widely applied as an alternative mill trial to characterize the material resistance to ductile facture for modern high strength and toughness pipeline steels, like X70, X80. Considering the larger ligament of DWTT specimen for crack propagating, the results from DWTT including fracture appearance, Ductile Brittle Transition Temperature value are usually consistent with the full-scale tests, more than Charpy test using sub-size specimens. Most significantly, the critical Crack Tip Opening Angle, $CTOA_C$ over the steady-state stage of crack propagation derived from DWTT [1] has been also ascertained as a measure of crack arrest capacity, conveniently and economically conducted in a mill test. In the recent works, a single-specimen energy based regression method [2] to estimate $CTOA_C$ has been proposed on the basis of well-known Martinelli-Venzi rigid-plastic ductile fracture model [3]. By using the regression of absorbed energy against the impact load as input, the regression method shall give more reliable consequences than conventional simple energy input approaches [4]. However, how to define the regression range especially confined in the steady-state stage is still a technical difficulty and influence the quality of $CTOA_C$ results.

In the present study, key-curve method [5] has been introduced to link the microscopic ductile cracking behaviour from the obtained DWTT force-displacement curve to the captured macroscopic fracture surface of broken DWTT specimen, helpful to the accurate definition of stable region for ductile fracture.

[a] Corresponding author: fangjian@baosteel.com

2. Principles

2.1. Energy based regression method

As above, V-notch pressed DWTT specimens are commonly used for a mill test. The single-specimen energy based regression method for a DWTT-type specimen is described as Eq. (1).

$$CTOA_C = 2\arctan\left[\frac{-4r^*}{S}\cdot\frac{\Delta E}{\Delta P}\right]_{a_1\to a_2}. \quad (1)$$

Where, S is the span between two support rollers (= 254 mm), r^* is a rotation factor, for DWTT specimens with larger ligaments, it has been determined to be 0.57 for high-strength steels including API X70–X100 pipeline steel appropriately [6].

Thus, the regression value of impact energy vs. load within the stable cracking range that is defined by the crack length from a_1 to a_2 may be used to evaluate instantaneous $CTOA_C$, only requiring values of r^* and S.

2.2. Key-curve method

Key-curve (KC) method developed by Kobayashi [5] has been proved satisfactorily to estimate the dynamic crack extension based on the instrumented impact. The procedure of KC method is described in Eqs. (2) and (3). (more detailed in Ref. [2]) Firstly, the key curve is analytically established by fitting the original relationship between the force (P) and displacement (y) on the pre-peak part of the instrument curve, obtaining the parameters k and n.

$$\frac{PW}{b_0^2} = k\left(\frac{y}{W}\right)^n. \quad (2)$$

Where, P is impact load, b_0 (= W − a_0) is the initial ligament width, y is load line displacement by the tup as

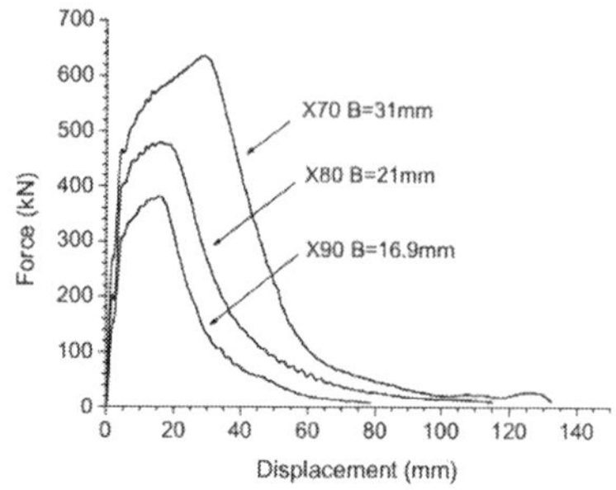

Figure 1. Curves for different instrumented DWTT samples.

above. n and k are material based constants, which should be determined on each force-displacement relationship by curve fitting from the beginning to the maximum force point.

Secondly, an amount of the crack extension could be estimated at any displacement on the post-peak part of the load-displacement curve, as Eq. (2).

$$\Delta a = W - \left(\left(\frac{P W^{n+1}}{k y^n} \right)^{1/2} + a_0 \right). \quad (3)$$

As a result, an amount of the crack extension is to be estimated at any displacement on the post-peak part of force-displacement curve.

3. Experimental

An instrumented pendulum impact machine with 40 kJ capacity and 8 m/s velocity was used to apply DWTT tests at room temperature on the standard V-notch pressed specimens sampled from three typical API X70, X80 and X90 pipeline steel plates. Given the machine's physical parameters like hammer mass, velocity, etc., the instrument load-displacement relation could be calculated by double integration of recorded load-time trace from the force measuring device (including the instrumented tup and AD/DA units), based on the Newton's law. Consequently, the total DWTT energy could be subdivided into the crack initiation and propagation parts, as illustrated in Fig. 1.

After preparation of the instrumented tup and being statically calibrated linearly over the load scale up to 800kN (accuracy better than ±1%, relative error), the DWTT energy calculated by instrument method is nearly the same as the measured from the pendulum dial, less than ±2% (full measurement scale, relative error).

4. Results and discussions

4.1. Instrumented force-displacement curves

The force-displacement curves for specimens with their individual thickness and different strength & toughness are shown in Fig. 1. Except for the location and size of curves due to the initial ligaments of specimens that experience the different plastic bending work hardening before crack onset and the total impact energy absorbed, all the curves have the similar features as followed.

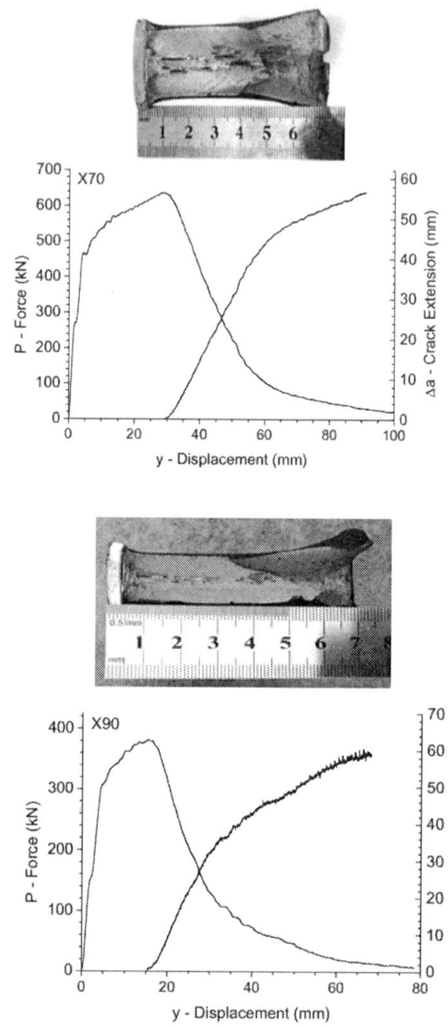

Figure 2. Dynamic crack extension Δa-y, from instrumented force-displacement relation by KC method, and its validation by measuring the region of stable cracking on the fracture surface.

It is clearly shown that after the initial elastic response, there is a significant plastic deformation before crack initiation. For convenience, here in the paper, the point at the maximum force is identified as the crack initiation nominally. Somewhere after the peak on the curve, the force starts to drop approximately linear with increasing tup displacement. Following the linear force-dropping period, the curve begins to tail off, due to the formation of shear lips, as shown in Fig. 2.

In some other works [7], the linear segment on the post-peak force-displacement curve was defined as the region of stable cracking. However $CTOA_C$ calculated within the range on the directly obtained force-displacement curve without any information or check about the real crack extension will confront the unforeseeable measurement deviation. For example, the influence of the initial cleavage zone nearby the notch root.

Table 1. The results of ($A^*\sigma_f$) from three typical high strength pipeline steel plates, and verification of the constant A^*.

No.	B mm	YS MPa	TS MPA	($A^*\sigma_f$) GPa	σ_f MPa	A^*
X70	31	506	641	0.261	746	0.35
X80	21	610	710	0.292	858	0.34
X90	16.9	680	800	0.293	962	0.30

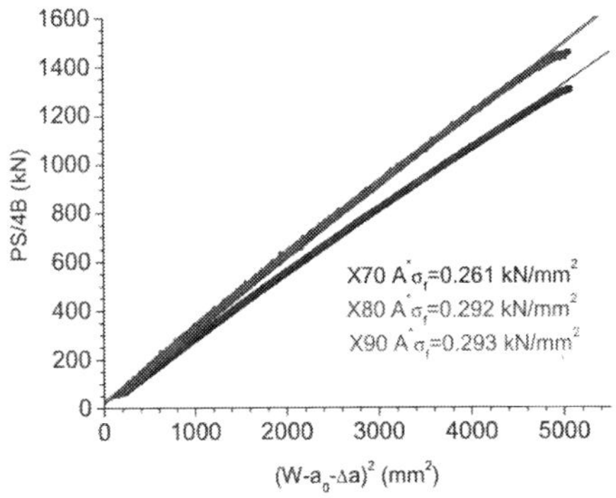

Figure 3. Plot of (PS/4B) vs. $[(W - a_0 - \Delta a)^2]$ relations for determining ($A^*\sigma_f$) as slope of linear regression.

4.2. Dynamic crack extension and constant of A*

Key-curve method deduced from the force-displacement curve of DWTT testing provides a way to study the entire cracking history, which is also recorded by the fracture surface. The calculated dynamic crack extension, e.g. the $\Delta a - y$, could be validated and confirmed by the direct measurement on the fracture surface, illustrated by Fig. 2.

Clearly on the fracture surface, the stable propagation featured by the region of silky and dull grey in appearance could be identified out of the initial cleavage fracture in bright-white & crystallite triangle area nearby the notch root, and the final formation of shear lips. As a result, the stable cracking of sample X70 is obtained nearly between the crack lengths of $10 \sim 40$ mm, and for sample X80 and X90, the regions are shortened to $10 \sim 35$ mm, in coincidence with the length of the linear segments indicated on the dynamic crack extension curve $\Delta a - v$.

In some works [3,6], the crack extension was estimated by limit loading hypothesis based on the change of the force over the propagation, compared with the maximum force at crack initiation, as Eq. (4).

$$P = \frac{4A^*\sigma_f B(W-a)^2}{S}. \tag{4}$$

Where, A^* is a material related constant (assumed to be 1/3 [8]), σ_f is a dynamic flow stress accurately determined from a high strain rate tensile. B is the thickness. $(W - a)$ is the remaining ligament. Derived from the limit loading hypothesis, a nearly linear correlation between (PS/4B) and $(W - a)^2$ might exist over the ductile fracture, which also offers an experimental way to obtain the parameter product of ($A^*\sigma_f$) that is required by some $CTOA_C$ algorithms [4].

In Table 1, the material based parameter product ($A^*\sigma_f$) was obtained as the slope of linear regression (PS/4B) vs. $(W\text{-}a_0\text{-}\Delta a)^2$ over the crack propagation, in Fig. 3. Here, dynamic crack extension Δa, was estimated by using KC method independently as above.

From Fig. 3, values of ($A^*\sigma_f$) for three typical high strength and toughness pipeline steels are 0.261 GPa for X70, 0.292 GPa for X80 and 0.293 GPa for X90 respectively. Moreover, once given the material properties, σ_f could be approximately evaluated as 1.3 times the average of yield and ultimate tensile strengths practically [6,9]. Therefore the unknown geometrical constant A^* for these three steels were experimentally calculated about 0.30–0.35, verified very close to the commonly cited constant 1/3.

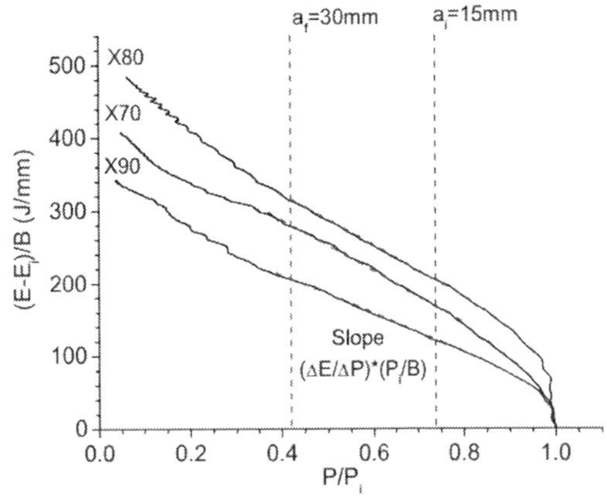

Figure 4. Plots of $[(E - E_i)/B]$ vs. (P/P_i) relations and linear regression for $CTOA_C$ as slope.

4.3. Estimation of $CTOA_C$

Towards the single-specimen energy based regression method, CTOAC could be determined with the input of regression slope ($\Delta E/\Delta P$) over the steady-state stage. The influence of specimen thickness was removed by normalizing Eq. (1) to obtain an equivalent expression shown as Eq. (5), for a convenient comparison of three sample all together.

$$CTOA_C = 2\arctan\left[\frac{-4Br^*}{P_i S}\cdot\frac{\Delta[(E-E_i)/B]}{\Delta(P/P_i)}\right]_{a_i\to a_f}. \tag{5}$$

The regression range in Eq. (5) is defined as the crack propagation from a_i to a_f, strictly confined in the region of stead-state stage that could be easily indicated from $\Delta a - y$ curve as introduced above. In the present paper, same regression range between $a_i = 15$ mm and $a_f = 30$ mm are selected for all three sample for a clear comparison, as shown in Fig. 4.

In the figure, from the right to left along the curve, energy is consumed for crack propagation with the load decreasing to zero. After crack initiation (i.e. maximum load point, $P/P_i = 1$), the slope of the curve is approximately constant in the portion illustrated by parallel vertical dashed lines, indicating the crack propagation from a_i to a_f.

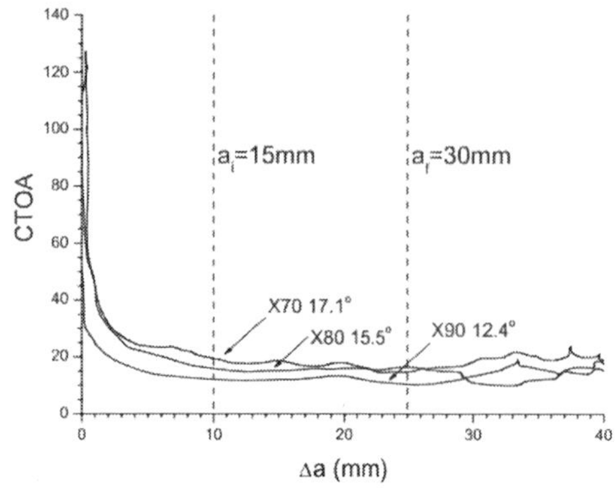

Figure 5. Plots of CTOA evolution for samples X70, X80 and X90 during crack propagation, indicating the stability of $CTOA_C$ over steady-state stage from a_i to a_f.

The x-axis of Fig. 4 is the inverse of crack extension and the slope of tangent at any point on the curve is related to the instantaneous CTOA value by Eq. (5).

(1) All samples show a similar evolution of CTOA value from crack initiation to thoroughly fracture, that is, the specimen experiences an high CTOA level, relative to the large slope of the curve, then followed by a constant $CTOA_C$ over the steady-state stage, and ends up with a rising period of CTOA due to shear lips, as shown in Fig. 5. This phenomenon has also been observed in other works by using in-situ high-speed video measurement for DWTT dynamically [9] and compact tension under quasi-static loading [10].

(2) Over the period of stable cracking, $CTOA_C$ is proportional to the absolute value of fitted slope. Samples X70 and X80 have the close values of $CTOA_C$ about 17.1° and 15.5° respectively, considerably higher than that of X90 about 12.4°. Similar values of CTOAC were also reported by Xu [11] in his study on the typical high grade pipeline steels X52-X100.

5. Conclusions

(1) Force-displacement curves of the standard V-notch pressed specimens sampled from three typical high grade X70, X80 and X90 pipeline steels were obtained by a 40kJ instrumented pendulum DWTT machine.

(2) Key-curve method was applied to evaluate the dynamic crack extension, Δa-y, helpful to define the region of stable ductile cracking that was also confirmed on the fracture surface. In the meanwhile, the validation of limit loading hypothesis was ascertained by linear fitting (PS/4B) vs. $(W - a_0 - \Delta a)^2$ relation, also providing an experimental way to the values of $(A^*\sigma_f)$. The value of A^* was acquired about 0.30–0.35, close to the commonly cited constant 1/3.

(3) The values of $CTOA_C$ indicates that X70 and X80 have nearly the same resistance to ductile fracture, higher than that of X90.

References

[1] D.J. Horsley, Engng. Fract. Mech. **70**, 547–552 (2003)

[2] J. Fang, J.W. Zhang, L. Wang, Int. J. Fract. **187**, 123–131 (2014)

[3] A. Martinelli, S. Venzi, Engng. Fract. Mech. **53**, 263–277 (1996)

[4] M. Di Biagio, A. Fonzo, G. Mannucci, et al. *Ductile fracture propagation resistance for advanced pipeline design* (GRI Report No. GRI-04/0129, Gas Research Institute, 2005)

[5] T. Kobayashi, I. Yamamoto, M. Niinomi, Engng. Fract. Mech. **24**, 773–782 (1986)

[6] S. Xu, R. Bouchard, W.R. Tyson, Engng. Fract. Mech. **74**, 2459–2464 (2007)

[7] G.M. Wilkowski, D.L. Rudland, Y.Y. Wang, D. Horsley, et al., *Proc. 4th International Pipeline Conference (IPC 2002)*, Paper IPC 2002-27132 (2002)

[8] J.F. Knott, *Fundamentals of Fracture Mechanics*, (Butterworths, London, 1973)

[9] D.L. Rudland, G.M. Wilkowski, Z. Feng, et al., Engng. Fract. Mech. **70**, 567–577 (2003)

[10] J.Q. Wang, J. Shuai, Engng. Fract. Mech. **79**, 36–49 (2012)

[11] S. Xu, W.R. Tyson, *Proc. 7th International Pipeline Conference (IPC 2008)*, Paper IPC 2008-64060 (2008)

EPJ Web of Conferences **94**, 05002 (2015)
DOI: 10.1051/epjconf/20159405002

Application of the dynamic characterization of metals in automotive industry

Fabio D'Aiuto[1,a], Daniele De Caro[1], Claudio Federici[1], Michele M. Tedesco[1], Alessandro Ziggiotti[1], and Ezio Cadoni[2]

[1] Centro Ricerche FIAT, 10135 Torino, Italy
[2] DynaMat Laboratory, University of Applied Sciences of Southern Switzerland, 6952 Canobbio, Switzerland

Abstract. This paper presents the experimental methodology used by R&D EMEA – Global Materials Labs Department to test metals at high strain rate of $500\,s^{-1}$. The implementation of dynamic results in commercial FEM Software LS – DYNA for crash simulation are presented. The effects of the strain rate on the tensile properties of metals, used in automotive field, are evaluated using results obtained from a direct tension split Hopkinson bar, built in collaboration with the University of Applied Sciences of Southern Switzerland DynaMat Lab. Finally the complete mechanical characterization of the Magnesium alloy AZ31B is presented, from static up to dynamic tests, showing its applications in FCA (Fiat Chrysler Automobiles), problems and future developments.

1. Introduction

Mechanical behaviour of materials can change at high strain rates. Generally, by increasing the strain rate many materials show hardening effect, a decrease in failure strain and an increase in strength. This is a very important topic in the automotive field, where there are a lot of mechanical components subject to impulsive loads; automotive crashworthiness is one such example. Several vehicle components are designed to function as energy absorbers in the event of crash, such that the deceleration seen by the driver is not so harsh to cause severe bodily injury. In order to design structures able to absorb properly the energy, the material must be characterized in a way that reproduces the actual working conditions [1]. Therefore dynamic material testing method assuring results of high precision must be designed, in order to understand the strain rate sensitivity of materials.

It is scientifically well recognized that the most satisfactory testing method for accurate measurement of the dynamic mechanical properties of materials is the Hopkinson bar technique, that allows the generation of a loading pulse well controlled in rise time, amplitude and duration, giving rise to the propagation of an uniaxial elastic plane stress wave [2]. Different versions of the Hopkinson bar technique were developed to investigate dynamic tensile properties of materials. R&D EMEA – Global Materials Labs Department, in collaboration with the University of Applied Sciences of Southern Switzerland DynaMat Lab, built a direct tension split Hopkinson bar, able to carry out high strain rate tension tests on metals in the range of $500–1500\,s^{-1}$; it is the tension version of the Modified Hopkinson Bar developed by Albertini and Montagnani in seventies [3,4] and nowadays widely used [5–10].

[a] Corresponding author: Fabio.DAIUTO@crf.it

Experimental dynamic results are used in commercial FEM software to perform crash simulations. Indeed regarding the optimization of materials and structures in the body component, the finite element method has been considered a powerful tool to estimate crash performance [11]. Using FEM software the most important choice is the material model used to describe the flow behaviour of materials under various strain rates. There are several phenomenological and physics based constitutive models available in literature. Physic based model can provide more accurate representation of material behaviour over a wide range of temperature and strain rates, however they are not preferred because they require more data from some controlled experiments. For this reason phenomenological models are preferred in practical applications such as numerical simulations of components subjected to impulsive loads [12–14]. However, the researchers are continually studying and improving the material models in the numerical simulation due to the ever-increasing demands for more accurate predictions of dynamic material behaviour. The purpose of this work is to show the methodology used in FCA lead the dynamic characterization of metals and the implementation of these results in commercial FEM software for the crash simulation. Moreover, the complete characterization of the Magnesium alloy AZ31B will be presented, from static up to dynamic tests, showing its applications, problems and future developments.

2. Mechanical testing at high strain rate

The tensile tests at high strain rates are performed on the Modified Hopkinson Bar [3–10], installed at the GML laboratory. In Fig. 1 a sketch of the apparatus is shown.

The tensile load pulse is generated by releasing a certain amount of elastic mechanical energy stored in a portion of the input bar (pre-tensioned bar) through static

Article available at http://www.epj-conferences.org or http://dx.doi.org/10.1051/epjconf/20159405002

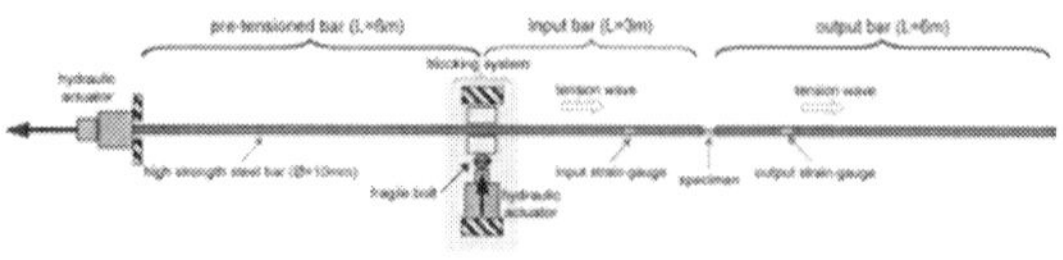

Figure 1. Scheme of the Modified Hopkinson Bar.

tensioning, up to a maximum stress value lower than its yield stress. At this scope a mechanical clamp [15] is used to grip the incident bar. The section of the incident bar between the clamp and the hydraulic actuator carries a static tensile load. The remainder of the input bar (to the right of the clamp) is unloaded. When the elastic mechanical energy is stored in the pre-tensioned bar and the specimen is inserted between the input and output bar, a notched bolt in the clamp is broken using a second hydraulic actuator, and an elastic tensile pulse is generated, propagating down the incident bar from the clamp toward the specimen. This pulse has a duration that is the double of the elastic wave travel time to move from the clamped section toward the free end of the input bar, and an amplitude that is half of the initial stored tensile pulse. As the wave reaches the input bar-specimen interface, a part of the pulse is reflected as a compression wave into the input bar and the remaining part is transmitted through the specimen. When the wave transmitted into the specimen arrives at the specimen-output bar interface, it is partially transmitted into the output bar and partially reflected into the specimen. If the specimen length is short so that the time taken by the wave to propagate through the specimen is short compared to the total time of the test, many reflections inside the specimen are created, allowing an homogenous stress and strain distribution along the specimen gauge length until fracture.

In this case the total length of the input bar is 9 m, the portion of it that is pre-tensioned during the test is 6 m, and the output bar length is 6 m; both bars have a diameter of 10 mm and are made in high strength steel. This configuration permits to achieve tensile pulse duration of 2.4 ms, allowing the deformation at constant high strain rate until fracture of high ductility specimens.

The semi-conductor strain-gage station is glued on the input bar at 750 mm from the specimen in order to record the deformation ε_I of the bar generated by the incident tension pulse during the propagation toward the specimen and the deformation ε_R caused by the part of the incident tension pulse reflected at the interface incident bar-specimen, reflection which is correlated with the deformation of the specimen. Another strain-gage station is glued on the output bar at the same distance from the specimen as the strain-gauge station on the incident bar; this second strain-gauge is used to record the deformation ε_T provoked on the bar by the part of the incident pulse which has been sustained by the specimen and has been therefore transmitted in the output bar; the transmitted pulse is proportional to the engineering stress in the specimen. A third strain-gauge station is bonded in the pre-tensioned bar to check the preload.

On the basis of the recorded signals ε_I, ε_R and ε_T, and applying the one-dimensional elastic plane stress wave propagation theory to the input bar-specimen-output bar system, it is possible to calculate the stress, strain and strain-rate in the specimen.

Figure 2. Modified Hopkinson Bar installed at the GML Laboratory.

With the described apparatus it is possible to test a wide range of materials at high strain rate, in the range of 500–1500 s^{-1}. In Fig. 2 the Modified Hopkinson Bar installed at the GML Laboratory is shown.

3. Crash simulation

The main reason that pushes the automotive industries to investigate on the dynamic behaviour of materials is the need to perform crash test simulations. The objective of crashworthiness related design is to guarantee a prescribed safety level for the occupants during the impact events. In order to achieve it, it is necessary to design a vehicle structure collapsing in a controlled manner during the collisions to adsorb properly the impact energy created by the crash event. The underestimation or overestimation of strain rates in crashworthiness design may reduce the structure energy absorption capability, increasing occupant's body accelerations, resulting in more injuries.

3.1. Numerical model

In FCA the vehicle frontal crash simulations are conducted by LS-DYNA. To implement the mechanical behaviours of materials in commercial FEM software, the true stress plastic true strain data, from static and dynamic experimental results, are fitted by the following expression:

$$\sigma_{TRUE} = \sigma_Y(\dot{\varepsilon})\,[1 + C(\dot{\varepsilon})\varepsilon_{TRUE\ PLASTIC}]^{n(\dot{\varepsilon})}. \quad (1)$$

Where σ_Y is the yield strength of materials. Only experimental results until the onset of necking are used. The constants C and n are calculated for each curve at different strain rate, respecting the Considère criterion in the necking point [16]:

$$\frac{d\sigma}{d\varepsilon} = \sigma. \quad (2)$$

The equation is also extended in the field of large deformation. An example of the constitutive model at different strain rate is reported in Fig. 3.

This set of curves is built for each material used in FCA, and implemented in LS-DYNA to describe completely the mechanical behaviour of materials in a

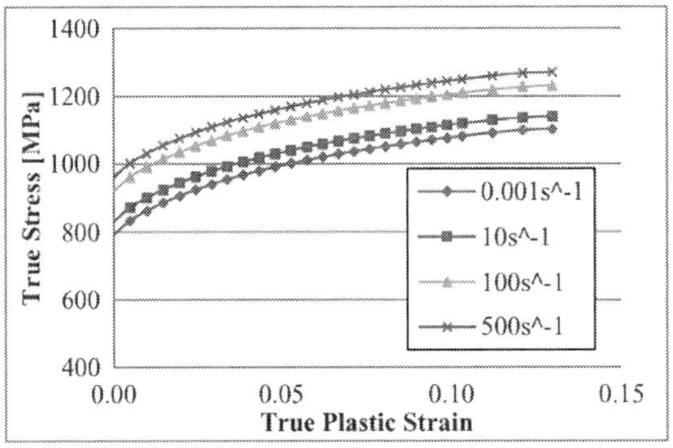

Figure 3. Constitutive model at different strain rate.

wide range of strain rate. This method is largely used in automotive impact engineering and is known as Piecewise Linear Plasticity. In LS-DYNA, a table is used to define for each strain rate value a load curve ID that gives the stress versus effective plastic strain. The lowest strain rate given in the table is applied if the strain rate falls below that minimum value. Likewise, the highest strain rate of the experimental data is used as a saturation plateau for the strain rate effects. In the simulation, the strain rate for each element is calculated and a linear interpolation between the experimentally determined strain rates is utilized to calculate the resulting stress in the plastic region. Moreover the simulations are conducted through dynamic explicit non-linear finite element model [17].

3.2. Crash analysis

Generally different types of crash analysis on BIW (body in white) are performed to test the safety of the vehicle:

- EURONCAP front crash, 40% overlap offset frontal crash at 64 km/h into deformable barrier.
- USNCAP, 100% frontal crash at 56 km/h into rigid barrier.
- Side impact test with moving deformable barrier, at 50km/h, at 90 degree angle.

4. Magnesium alloy characterization

In an effort to improve the fuel efficiency of automobiles, car designers are investigating new material to reduce the overall vehicle weight. Magnesium alloys are good candidates to achieve that weight reduction due in part to their low density and high specific strength. Magnesium is the lightest structural metals with a density of approximately one-fourth that of steel and two-thirds that of aluminium, and as the specific strength and stiffness of magnesium exceeds that of most commonly used metals and some plastic-based metals, magnesium based materials are actively pursued by companies for weight-critical applications. Currently more than 60% of vehicle weight is due to use of steel or cast iron in the body structure, on the contrary, for example in America, the contribution to the overall vehicle weight of Mg alloys is only of 0.3% [18]. But bringing Mg parts to the market requires a several important study of its

Table 1. Chemical composition of AZ31B in wt.%.

Constituents	Al	Mn	Zn	Mg
wt.%	2.8	0.31	0.5	Balanced

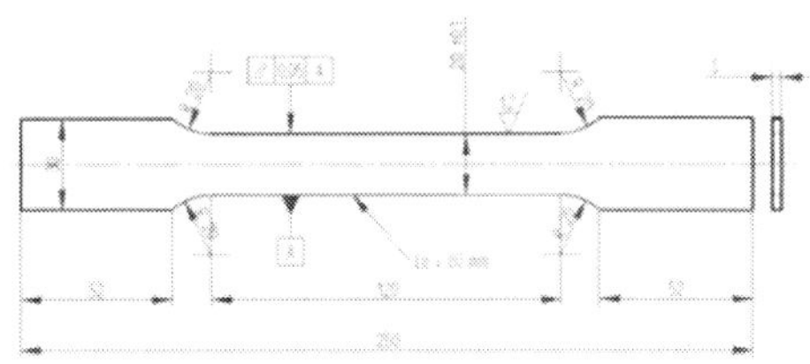

Figure 4. Static specimen geometry.

mechanical behaviour, its anisotropy, and it is also crucial to understand Mg alloys behaviour at high strain rates, especially in the automotive field where the components are often subjected to crash events.

Currently in literature there are few information about the tensile properties of Mg alloys at high strain rates, and the studies available are mostly on extruded and cast material, and predominantly in compression [19–22]. Ulacia et al. [19] performed an exhaustive testing campaign on AZ31-O sheet at dynamic ($\dot{\varepsilon} \sim 10^3\,S^{-1}$) and low ($\dot{\varepsilon} \sim 10^3\,S^{-1}$) strain rates, in tension and compression. He showed the different microstructural evolution at high strain rates and high temperatures. J. Xiao et al. [20] carried out impact test on AZ31 Mg alloy reinforced by 1% vol silicon carbide nanoparticles, showing strong rate dependence, increasing in flow stress as the strain rate increases. Hasenpouth [22] performed tensile test on AZ31B magnesium alloy sheets, at low, medium and high strain rates. He found a clear increase in tensile strength as the strain rate increases, showing a stress rise of approximately 60 MPa.

The complete characterization of the Magnesium alloy AZ31B is presented, from static up to dynamic tests, realized by the GML Laboratory in collaboration with SUPSI.

4.1. Material

Nowadays AZ31B is the most common commercial magnesium alloy available in sheet form. It was obtained from Elektron, whose chemical composition in wt% is given in Table 1.

4.2. Quasi Static Test

The quasi static test ($\dot{\varepsilon} \sim 10^3\,S^{-1}$) were performed by means of an electromechanical universal testing machine, with 200 kN maximum load bearing capacity, installed at GML Laboratory. In order to measure the strain on the specimen gauge length a video extensometer was used. The sample geometry is reported in Fig. 4. The tests were carried out on specimens oriented in longitudinal, transversal and diagonal directions respect to the rolling direction.

In Figs. 5 and 6 the engineering and true curves along the three orientations considered are reported.

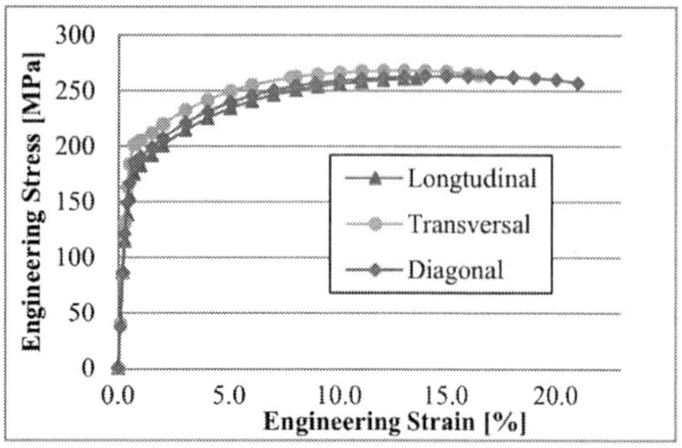

Figure 5. Static engineering stress vs. strain curves for three orientations.

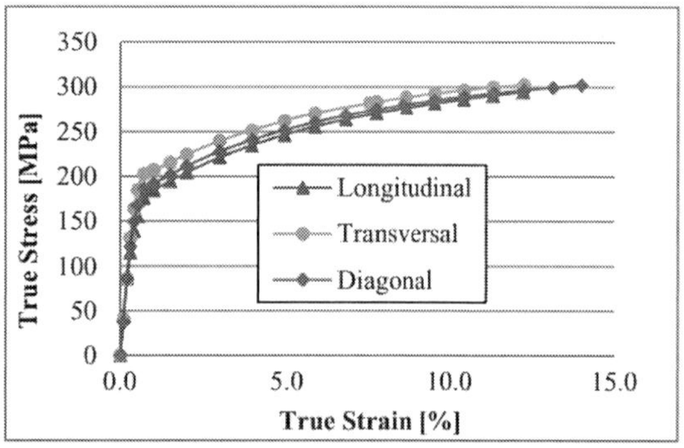

Figure 6. Static true stress vs. strain curves for three orientations.

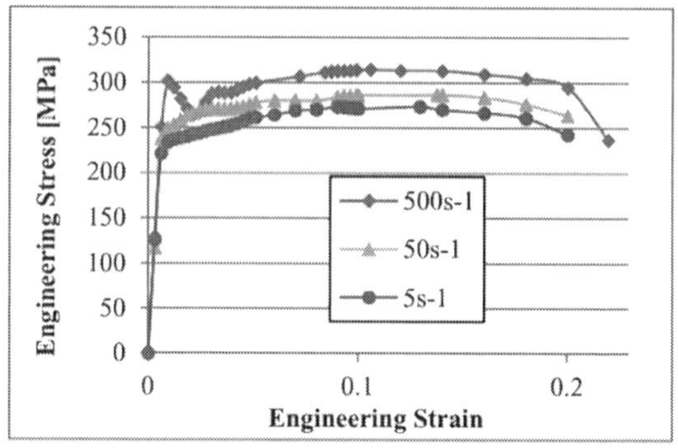

Figure 7. Dynamic engineering stress vs. strain curves for three strain rates.

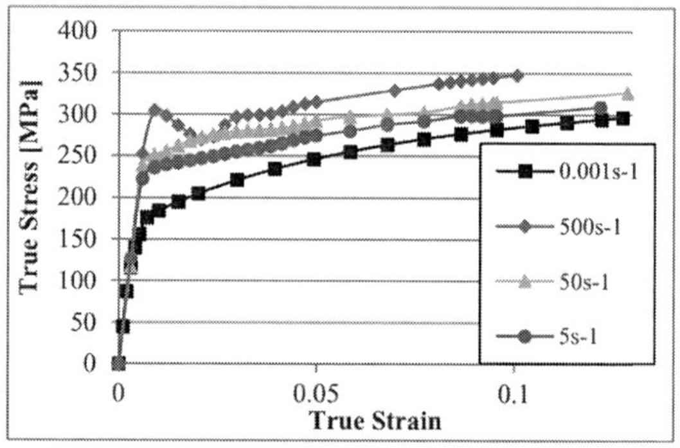

Figure 8. Dynamic true stress vs. strain curves for three strain rates.

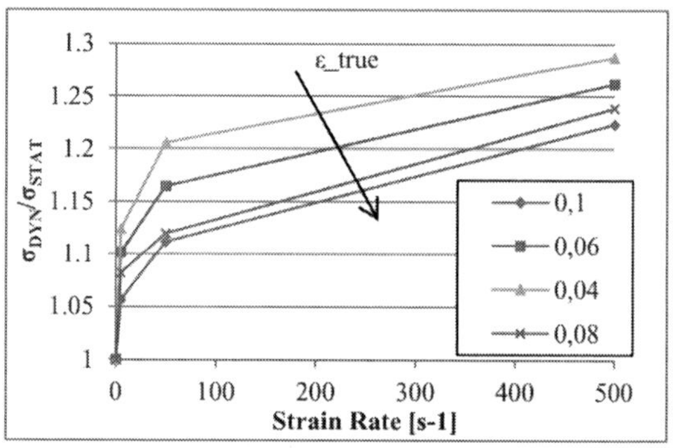

Figure 9. DIF vs. strain rate curves, at four different true strains.

As shown in the figures above, the mechanical properties of material along the three directions are slightly different, both in elongation and flow plastic stress, attesting the anisotropy of magnesium alloys. However the dynamic tests were carried out only on specimens oriented along the rolling direction.

4.3. Dynamic test

Dynamic tensile tests on AZ31B alloy were conducted at medium (5–50 s^{-1}) and high (500 s^{-1}) strain rate.

The medium strain rate tests were performed by SUPSI on Hydro-pneumatic machine (HPM) [6,7,7,10].

The high strain rate tests, at 500 s^{-1}, were carried out on the Modified Hopkinson Bar installed at GML Laboratory.

The specimens tested were flat, having 10 mm gauge length, 4 mm width and 1.5 mm thickness; the geometry is the same reported in [6,10]. The experimental data were analysed to obtain the stress versus strain curves at three different strain rate regimes. The engineering stress versus engineering strain curves and true stress versus true strain curves of Mg alloy are compared in Fig. 7 and Fig. 8, respectively. The true curves were plotted until the neck point for each strain rate regime. Related to the quasi-static results, for comparison only the true stress versus true strain curve, due to the different geometry used in quasi-static and dynamic tests, is reported.

The results clearly show that the AZ31B alloy is very sensitive to the strain rate; in fact the flow stress increases as the strain rate rises from 0.001 up to 500 s^{-1}. Moreover the fracture engineering strain is almost the same as strain rate increases, so the area under stress-strain curve is more at high strain rate, increasing the fracture energy and toughness of material.

To better understand the effect of the strain rate on mechanical behaviour of Mg alloy, in Fig. 9 the dynamic stress-static stress ratio versus strain rate for four different true strains is plotted. The figure shows an increasing in strain rate effect as the strain rate increases, but it is also possible notice that the strain rate sensitivity on Mg alloy seems decrease as the true strain increases. This particular behaviour of Mg alloy is better represented in Fig. 10

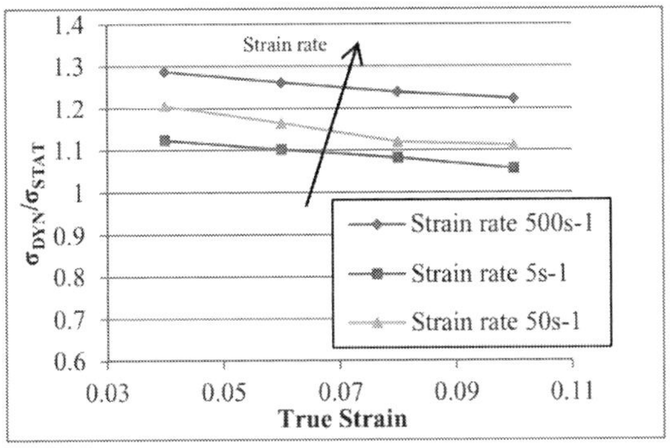

Figure 10. DIF vs. true strain curves, at three different strain rates.

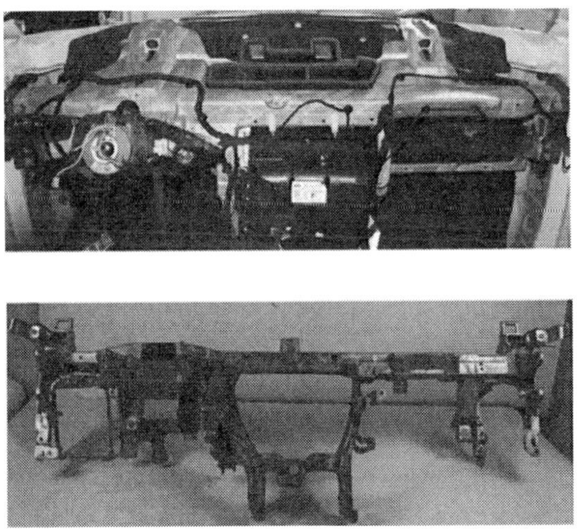

Figure 11. Instrument Panel Beam in Mg alloy.

that plots the dynamic stress-static stress ratio (known as Dynamic Increase Factor – DIF) versus true strain for three different strain rates.

4.4. Applications

FCA in the last years has increased the usage of Magnesium alloys in its products in order to perform a significant weight reduction of the vehicle. The main applications of Mg alloys are currently related to the instrument panel beam, the seat structure, gearbox and steering wheel, as depicted in Fig. 11.

At the moment the Mg alloys are used only as high pressure die casting (HPDC) and not in sheet form, for their actual manufacturing difficulties. Unfortunately, due to their crystallographic structure and texture, magnesium alloys exhibit low ductility at room temperature and strong anisotropy in their constitutive behaviour. Therefore, elevated temperature stamping is needed to produce magnesium alloy parts, which increases their production cost. Improvements in magnesium alloy sheet are thus needed and research interest on this activity has greatly increased in the past few years.

5. Conclusions

The experimental methodology used by GML to test materials at high strain rates is presented in this paper. Moreover the dynamic results implementation for crash simulation is described. The complete mechanical characterization of AZ31B Magnesium alloy is investigated. The quasi-static tests are performed along three different orientations, attesting the anisotropy of magnesium alloys. The results obtained from dynamic tests show the strain rate effect on material, that exhibits an increasing in strength as the strain rate rises from 0.001 up to $500\,s^{-1}$; on the contrary the engineering fracture strain is almost the same at different strain rates, resulting in enhancement of energy adsorbed at fracture.

In conclusion, due to the increasing demand for cutting down energy consumption and greenhouse gas emissions, the Mg alloys are a good choice in order to reduce the overall weight of the vehicle for more economic use of fuel. However nowadays the good mechanical behaviour of Mg alloys in sheet form, and its advantages for weight-critical applications, are obstructed by their actual manufacturing difficulties.

References

[1] A.T. Owens, *Strain response of particulate composites under high rates of loading*, MS Thesis, Auburn University, Alabama (2007).

[2] H. Kolsky, Proc. Phys. Soc. Sect.B62, 676–700 (1949).

[3] C. Albertini, M. Montagnani, Nucl. Eng. Des. **37**, 115–124 (1976).

[4] C. Albertini, M. Montagnani, Inst. of physics conf. series **21**, 22–32 (1974).

[5] E. Cadoni, M. Dotta, D. Forni, P. Spätig, J. Nucl. Mat. **414**(3), 360–366 (2011).

[6] N.K. Singh, E. Cadoni, M.K. Singha, N.K. Gupta, Mat. Des. **32**, 5091–5098 (2011).

[7] E. Cadoni, L. Fenu, D. Forni, Constr. Build. Mat. **35**, 399–407 (2012).

[8] E. Cadoni, M. Dotta, D. Forni, N. Tesio, C. Albertini, Mat. Des. **49**, 657–666 (2013).

[9] C. Albertini, E. Cadoni, G. Solomos, Phil. Trans. R. Soc. A 2014 **372**, 20130197.

[10] E. Cadoni, F. D'Aiuto, C. Albertini, Proc. DYMAT2009, EDP Sciences, **1**, 135–141 (2009).

[11] A.K. Pickett, T. Pyttel, F. Payen, F. Lauro, N. Petrinic, H. Werner, Int. J. Impact Eng. **30**, 853–872 (2004).

[12] F.J. Zerilli, R.W. Armstrong, J. Appl. Phys. **61**, (1987).

[13] G.R. Cowper, P.S. Symonds, Brown University Applied Mathematics Report **28**, 1–46 (1958).

[14] G.R. Johnson, W.H. Cook, Eng. Fract. Mech., **21**, 31 (1985).

[15] M. Montagnani, C. Albertini, U. Buzzi, M. Forlani, 1973 Dispositif de contrainte a accumulation mecanique pour essais dinamique de traction. Institut national de la propriete industrielle: Paris Brevet No. 74.17085. The Patent Office of London, Patent No. 1.473.683, 1974. Italian Patent No. 50008A.

[16] K.S. Havner, Int. J. Plast. **20**, 965–978 (2004).
[17] Z. Shen, X. Qiao, H. Chen, Proceedings of the FISITA 2012 World Automotive Congress, Vol. 9, Automotive Safety Technology.
[18] G.S. Cole, A. North, American Automotive Strategic Vision of Magnesium, USAMP report 2007.
[19] I. Ulacia, N.V. Dudamell, F. Galvez, S. Yi, M.T. Perez-Prado, I. Hurtado, Act. Mat. **58**, 2988–2998 (2010).
[20] J. Xiao. D.W. Shu, X.J. Wang, Int. J. Mech. Scie. **89**, 381–390 (2014).
[21] M.T. Tucker, M.F. Horstemyer, P.M. Gullett, H. El Kadiri, W.R. Whittington, Scr. Mater **60**, 182–185 (2009).
[22] D. Hasenpouth, Tensile high strain rate behaviour of AZ31B Magnesium Alloy Sheet, (2010).
[23] E.D.H. Davies, S.C. Hunter, J. Mech. Phys. Solids **11**, 155–179 (1963).

EPJ Web of Conferences **94**, 05003 (2015)
DOI: 10.1051/epjconf/20159405003

High-rate behaviour of iron ore pellet

Gustaf Gustafsson[1,a], Hans-Åke Häggblad[1], Pär Jonsén[1], and Masahiro Nishida[2]

[1] Luleå University of Technology, Division of Mechanics of Solid Materials, 97187 Luleå, Sweden
[2] Nagoya Institute of Technology, Gokiso-cho, Showa-ku, Aichi 466-8555, Japan

Abstract. Iron ore pellets are sintered, centimetre-sized spheres of ore with high iron content. Together with carbonized coal, iron ore pellets are used in the production of steel. In the transportation from the pelletizing plants to the customers, the iron ore pellets are exposed to different loading situations, resulting in degradation of strength and in some cases fragmentation. For future reliable numerical simulations of the handling and transportation of iron ore pellets, knowledge about their mechanical properties is needed. This paper describes the experimental work to investigate the dynamic mechanical properties of blast furnace iron ore pellets. To study the dynamic fracture of iron ore pellets a number of split Hopkinson pressure bar tests are carried out and analysed.

1. Introduction

Handling of iron ore pellets is an important part in the production chain for many producers of iron ore pellets. Knowledge about this sub process is very important for further improved efficiency and increased product quality. After production in the pelletizing plants, the iron ore pellets pass through a number of transportation and handling systems like conveyor belts, silo filling, silo discharging and transport by rail and ship. During these treatments, the pellets are exposed to different loadings, resulting in degradation of their strength and generation of fines. To study and optimize processes of transportation and handling of bulk material, traditionally half- or full-scale experiments have been used [1,2]. The focus of these studies is often the pressures on the surrounding structures and not the stresses in the bulk material. Numerical simulations of these processes give a possibility to study the processes in more detail. Most of the work so far in simulation and constitutive modelling of granular materials like soils [3] and metal powders [4,5] is carried out with continuum based methods like the finite element (FE) method. For iron ore pellets flow particle based methods like the smoothed particle hydrodynamics (SPH) method [6] can be used. Such models give the flow, stresses and strains on a continuum length scale but not the stress state inside the individual granules. To study the local contact behaviour of compressed granular material, detailed models of the individual particles are necessary. Such models are available for some granular materials in 2D, see for example [7] and for static loading of pellets, see [8].

This paper describes the experimental work to investigate dynamic mechanical properties for blast furnace iron ore pellets. The data for single iron ore pellets worked out in this paper could be used to model the individual particles in an assembly.

[a] Corresponding author: gustaf.gustafsson@ltu.se

2. Theoretical considerations

Iron ore pellets are a hard and brittle porous material and can be compared with materials like rocks and ceramics. Theories developed for these materials have been used in this study.

2.1. Compression of an irregular sphere

A theoretical and practical study of the stress state in an irregular sphere-like test piece subjected to the point load test is given in [9]. The results show that the stress state in an irregular test piece subjected to concentrated loads may be, in the vicinity of the axis of loading, much the same as that in a perfectly spherical test piece compressed diametrically. According to their results the maximum tensile strength, S_t for an irregular test piece occur near the centre and is approximately given by

$$S_t = c\frac{2F}{\pi D^2} \quad (1)$$

where F is the applied force and D is the distance between the points of loading. The constant c is experimentally determined and takes a value between 1.21 and 1.36 in the centre of the test piece for rock materials. For more details, see [9].

As a measure of the irregularity of the specimens, a shape index, η, is derived. The distance, D, is divided by the average diameter, $\overline{D}$, of the specimens according

$$\eta = \frac{D}{\overline{D}}. \quad (2)$$

The average diameter is calculated as the diameter of a sphere with the same mass, M, and density, ρ, as the specimen.

$$\overline{D} = \sqrt[3]{\frac{6M}{\pi\rho}}. \quad (3)$$

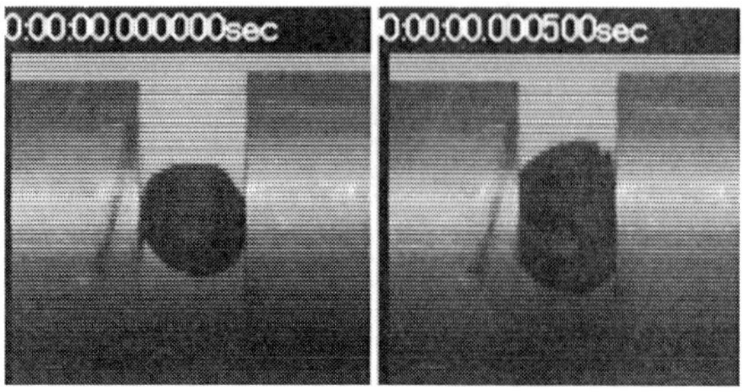

Figure 1. Iron ore pellet before loading (left) and after breakage (right).

2.2. Statistical measures

Some of the data analysed will be expressed in terms of the statistical expectation values μ, standard deviations S, confidence intervals and prediction intervals. Assuming a normal distribution, the 95% confidence interval and 95% prediction interval are given by Eqs. (4) and (5), respectively

$$\hat{\mu} - 1.96\frac{S}{\sqrt{N}} < \mu < \hat{\mu} + 1.96\frac{S}{\sqrt{N}} \tag{4}$$

$$\hat{\mu} - 1.96S < x < \hat{\mu} + 1.96S. \tag{5}$$

Where $\hat{\mu}$ is an estimate of the expectation value, x is the measured data, N is the number of tests and S is the standard deviation, according to

$$\hat{\mu} = \frac{1}{N}\sum_{i=1}^{N} X_i \tag{6}$$

$$S = \sqrt{\frac{1}{N}\sum_{i=1}^{N}(x_i - \overline{x})^2}. \tag{7}$$

3. Materials and methods

The material tested is iron ore pellets from LKAB (Luossavaara-Kiirunavaara AB) in Malmberget, Sweden. In the blast furnace, iron ore is reduced and smelted to liquid crude iron. Fine-grained iron ore is too fine material for burdening in the blast furnace and must therefore be sintered into larger entities. Iron ore pellets are sintered, centimetre-sized spheres of grained ore with high iron content ($\approx$ 67 wt% Fe).

3.1. Tested iron ore pellets

To obtain a more consistent quality and material characteristics in the iron ore pellets test samples, laboratory produced pellets are used in this study. In the laboratory, the iron ore pellets are produced in a small-scale pelletizing process including all the steps in the pelletizing plants. In the last step the iron ore pellets are sintered in batches of about 50 kg. In the laboratory production, the iron ore pellets are handled more careful and the production parameters are better controlled. The sizes of the iron ore pellets are screened out to a range of 9 mm to 12 mm before testing. See Fig. 1 for an iron ore pellet from one of the dynamic tests. The average density for the tested pellets batch was measured to $\rho = 3.69\,\mathrm{kg/m^3}$.

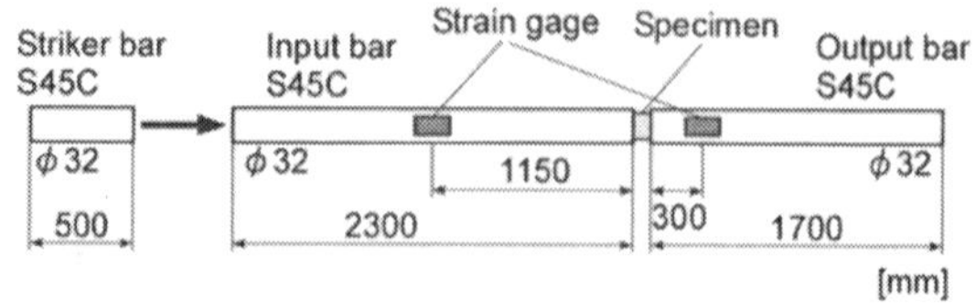

Figure 2. Experimental set-up for split Hopkinson pressure bar (SHPB) tests.

Table 1. Material constants of SHPB input and output bars.

Density	Elastic wave velocity of the input/output bar, c_b	Young's modulus, E
7.85 g/cm^3	5110 m/s	208 GPa

Table 2. Test descriptions and specimen data.

Test serie	Number of specimens	Mass, M, [g] average $\pm S$	Diameter, D, [mm] average $\pm S$	Shape index, η average $\pm S$
1	10	3.22 ± 0.51	11.88 ± 0.69	1.01 ± 0.054
2	8	3.17 ± 0.46	11.24 ± 0.76	0.95 ± 0.044
3	10	3.29 ± 0.32	11.72 ± 0.30	0.98 ± 0.039
4	13	3.05 ± 0.41	11.50 ± 0.60	0.99 ± 0.045
5	30	2.95 ± 0.52	11.67 ± 0.71	1.02 ± 0.045

3.2. Split Hopkinson pressure bar load tests

Dynamic testing of iron ore pellet is carried out using the split Hopkinson pressure bar (SHPB) assembly. The input and output bars for the SHPB experiments were made of steel, and their diameter and length were 32 mm and 2300 mm/1700 mm respectively. Strain gages were put on both sides of input bar and output bar 1150 mm and 300 mm away from the specimen, see Fig. 2. The compressive force histories of specimens were calculated from the strain of bars measured by the strain gages based on the following equation [10]:

$$F(t) = AE\left[\varepsilon_i(t) + \varepsilon_r(t) + \varepsilon_t(t)\right] \tag{8}$$

where ε_i, ε_r and ε_t are the axial strains in the input bar induced by the incident wave and reflected wave from the specimen and the axial strain in the output bar induced by the transmitted wave. E is Young's modulus and A is the areas of the input and output bars. Material constants of the steel bars used in the calculation are shown in Table 1. A steel striker with a diameter of 32 mm and a length of 500 mm was used in the tests. A summation of the sizes and number of specimens for the test series are shown in Table 2.

4. Experimental results and discussion

The shape of the iron ore pellets is not perfectly spherical; they have an irregular smooth shape. Nor do all iron ore pellets have the same size as shown in Table 1. The

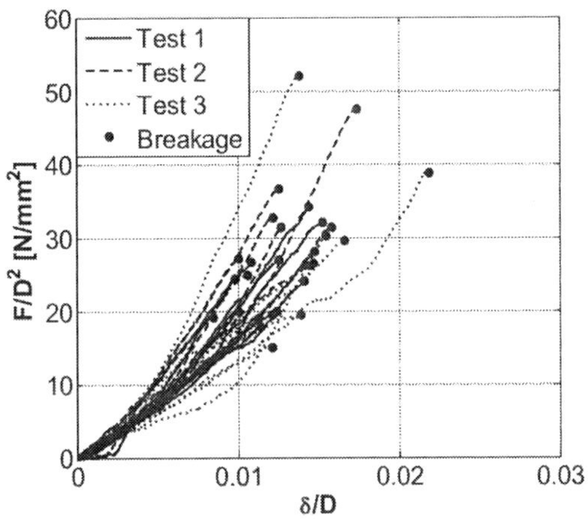

Figure 3. Load versus displacement including fracture. Test series 1–3.

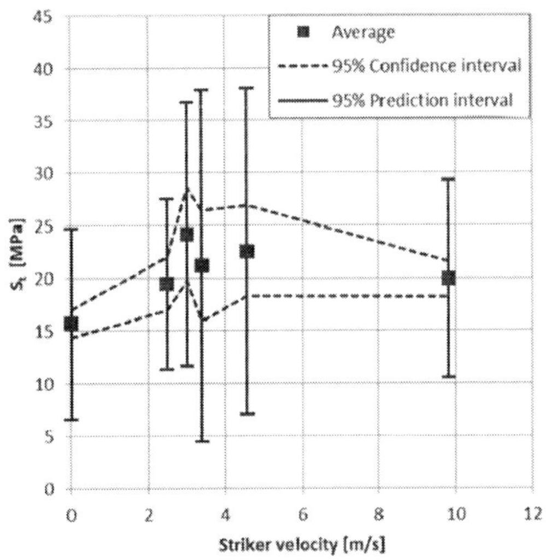

Figure 4. Fracture strength (S_t) as function of striker velocity. Average breakage level with 95% confidence interval and 95% prediction interval. Static data from reference [11].

diameter, D, is defined as the distance between the loading points at each test.

Dynamic tests of *71* iron ore pellets with five different striker velocities are carried out. The load displacement curves for some of the experiments are shown in Fig. 3. To compensate for size variations the load, F, is divided by D^2 resulting in a stress-like measure and the displacement, δ, is divided by D resulting a strain measure. The fractures are indicated with dots.

The stress state in the centre of the specimen is assumed be compressive in the loading direction and tension perpendicular to this direction. Using c = 1.21 in Eq. (1) gives the tension stress values. The average fracture

Table 3. Results of dynamic test series.

Test serie	Striker velocity, v [m/s] average $\pm S$	Deformation velocity, $\dot{\delta}/D$, [1/s] average	Fracture load, F/D^2 [N/mm^2] average $\pm S$	Fracture strength, S_t [MPa] average $\pm S$
1	2.49 ± 0.017	172	25.3 ± 5.36	19.5 ± 4.13
2	3.01 ± 0.007	213	31.5 ± 8.34	24.2 ± 6.42
3	3.38 ± 0.070	223	27.5 ± 11.1	21.2 ± 8.54
4	4.56 ± 0.044	277	29.4 ± 10.3	22.6 ± 7.92
5	9.82 ± 0.076	723	25.8 ± 6.20	19.9 ± 4.78

load and fracture strength are shown in Table 3, see also Fig. 4. In Table 3 is also the deformation velocity given, it can be interpreted as a measure of the strain rate.

5. Conclusions

This investigation has provided some new knowledge about the mechanical properties and behaviour of iron ore pellets. From the experimental measurements, fracture strength are evaluated for different deformation velocities and determined in terms of statistical means and confidence intervals. The results of this study show that the stiffness seems to be approximately the same for the tested range of deformation velocities and the fracture strength show a small decrease with increased deformation velocity. Compared to data from quasi-static testing the fracture strength is increased with approximate 35%.

References

[1] C.F. Chen, J.M. Rotter, J.Y. Ooi, Z. Zhong, Chemical Engineering Science, **60**, 3029–3041 (2005)

[2] J.Y. Ooi, J.F. Chen, J.M. Rotter, Powder Technology, **99**, 272–284 (1998)

[3] W.F. Chen, E. Mizuno, *Nonlinear analysis in soil mechanics* (Elsevier, Netherlands, 1990)

[4] P.R. Brewin, O. Coube, P. Doremus, J.H. Tweed, *Modelling of powder die compaction* (Springer, London, 2008)

[5] P. Jonsén, H.-Å. Häggblad, Journal of Solids and Structures. **44**, 6398–6411 (2007)

[6] G. Gustafsson, H. Häggblad, M. Oldenburg, *NUMIFORM 2007*, 1483–1488 (2007)

[7] D.T. Gethin, R.S. Ransing, R.W. Lewis, M. Dutko, A.J.L. Crook, Computers and Structures, **79**, 1287–1294 (2001)

[8] G. Gustafsson, H.-Å. Häggblad, P. Jonsén, Powder Technology, **239**, 231–238 (2013)

[9] Y. Hiramatsu, Y. Oka, International Journal of Rock Mechanics and Mining Science, **3**, 89–99 (1966)

[10] W. Chen, and B. Song, *Split Hopkinson (Kolsky) bar: design, testing and applications* (Springer 2010)

[11] G. Gustafsson, H.-Å. Häggblad, P. Jonsén, Powder Technology, **235**, 126–135 (2013)

EPJ Web of Conferences **94**, 05004 (2015)
DOI: 10.1051/epjconf/20159405004

Experimental and numerical analysis of the dynamic behaviour in tension of an armour steel for applications in defence industry

Ezio Cadoni[1,a], Matteo Dotta[2], Daniele Forni[1,2], Gianmario Riganti[1], and Hanspeter Kaufmann[3]

[1] DynaMat Laboratory, University of Applied Sciences of Southern Switzerland, 6952 Canobbio, Switzerland
[2] Department of Structural, Geotechnical and Building Engineering – Politecnico di Torino, 10129 Torino, Italy
[3] RUAG Defence AG, 3602 Thun, Switzerland

Abstract. The dynamic behaviour of armour steel in tension was investigated over a wide range of strain-rates on round specimens. The experiments were carried out by means of a Split Hopkinson Tensile Bar device and by a Hydro Pneumatic Machine. The target strain rate were set at the following six levels: 10^{-3}, 5, 25, 100, 500 and 1000 s^{-1}. Two material models were calibrated and used to replicate the experiments and to simulate blasting event on steel plate. Finally, the two responses are compared.

1. Introduction

The armour steels are widely used in protection of particular structures (i.e. vehicles, buildings, personnel etc.). Those structures are designed to withstand severe dynamic events such as impact or blasting loads. In order to improve the degree of protection and to provide optimized solution for safety is fundamental the knowledge of the armour steel under the same conditions generated in real dynamic events. These processes cannot be analysed without numerical simulations. However, accuracy and realistic structural assessment are highly influenced by the material modelling. Hence, it is recommended to investigate the material responses in a wide range of strain rates.

This work investigates the dynamic behaviour of high hardness armour steel (HBW 570–640) in tension at different strain rate such as 10^{-3}, 5, 25, 100, 500 and 1000 s^{-1}. The test data were analysed with the purpose to identify material parameters. Two techniques were applied: the calibration of well-known model and the reverse simulation of the Split Hopkinson Tensile Bar (SHTB) test [1]. Material models developed were assessed and discussed by means of a numerical example of blast on steel plate.

2. Material

The high hardness armour steel here analysed is a commercial material product used in the construction of defence vehicles or in temporary structures when high level of protection is required. The Ultimate Tensile Strength is higher than 2 GPa while the yield strength is typically higher than 1.5 GPa. The elastic modulus is 207 GPa, density is 7850 kg/m^3 and the melting temperature is 1800 K. The chemical compositions of the armour steel is shown in Table 1.

Table 1. Chemical composition according to the manufacture.

C max %	0.470
Si %	0.700
Mn max %	1.000
P max %	0.010
S max %	0.005
Cr max %	1.500
Ni max %	3.000
Mo max %	0.700
B max %	0.005

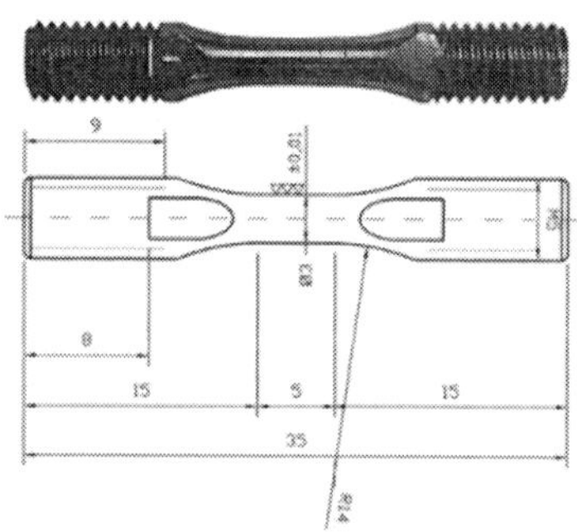

Figure 1. Specimen geometry.

2.1. Specimens

Round samples having 3 mm in diameter and 5 mm of gauge length have been used (see Fig. 1) [2–8].

3. Experimental set-up

Tensile testing at several strain-rates has been performed using different experimental set-ups. All tests were performed at room temperature, and the target strain-rates were set at the following six levels: 10^{-3}, 5, 25, 100, 500

[a] Corresponding author: ezio.cadoni@supsi.ch

Article available at http://www.epj-conferences.org or http://dx.doi.org/10.1051/epjconf/20159405004

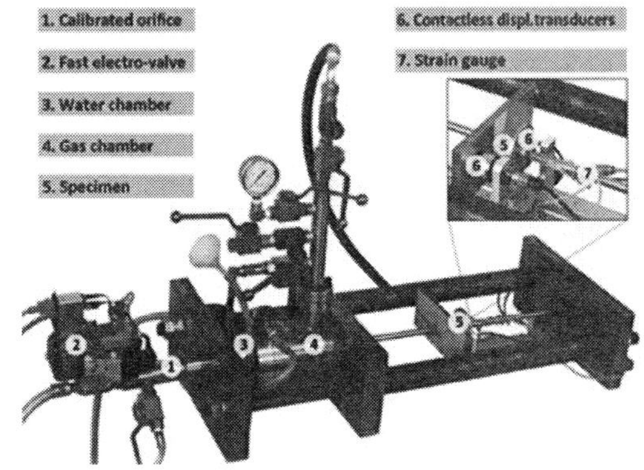

Figure 2. Hydro-Pneumatic Machine.

and $1000\,s^{-1}$. At least three tensile tests were carried out for each strain-rate.

3.1. Quasi-static tests

Quasi-static tests have been performed by means of an universal electromechanical testing machine Zwick/Roell–Z50 which has the maximum load bearing capacity of 50 kN.

3.2. Intermediate strain-rate

Intermediate strain-rate behaviour in tension has been investigated by means of a hydro-pneumatic machine (HPM), whose scheme is reported in Fig. 2.

The functioning of the machine is described in [4,6].

3.3. High strain rate

The high strain-rate tests have been performed using a Split Hopkinson Tensile Bar [6] that consists into two cylindrical high strength steel bars, having a diameter of 10 mm, with a length of respectively 9 and 6 meters for input and output bar, as represented in Fig. 3.

The armour steel specimen is assembled between the two bars. Input and output bars were instrumented with strain gauges which measure incident, reflected and transmitted pulses acting on the specimen cross section.

The present set-up has the following condition: i) the bar diameter (10 mm) is small in comparison with the pulse length (12 m); ii) the specimen length is short so that the time taken by the wave to propagate through the specimen is short compared to the total time of the test. This condition allows many reflections inside the specimen necessary for reaching a homogeneous stress and strain distribution along the specimen gauge length, what means also equilibrium of the forces acting on both ends of the specimen.

Having fulfilled the conditions i) and ii) and being the two bars elastically loaded the one-dimensional elastic plane stress wave propagation theory can be applied to the input bar-specimen-output bar system as it is extensively shown in [4–8], where the three relationships are proofed which allow to calculate the stress, the strain and the strain rate in the specimen material versus time from the records

Figure 3. Split Hopkinson Tensile Bar device.

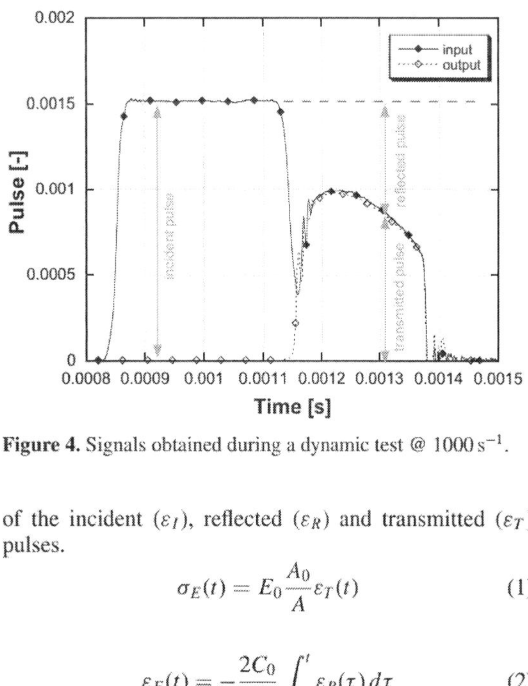

Figure 4. Signals obtained during a dynamic test @ $1000\,s^{-1}$.

of the incident (ε_I), reflected (ε_R) and transmitted (ε_T) pulses.

$$\sigma_E(t) = E_0 \frac{A_0}{A} \varepsilon_T(t) \tag{1}$$

$$\varepsilon_E(t) = -\frac{2C_0}{L} \int_0^t \varepsilon_R(\tau)\, d\tau \tag{2}$$

$$\dot{\varepsilon}_E(t) = \frac{2C_0}{L} \varepsilon_R(t) \tag{3}$$

where, E_0 is the bars elastic modulus, A_0 is the bars cross-sectional area, A is the specimens cross-sectional area, L is the specimen gauge length and C_0 is the sound velocity of the bar material.

In Fig. 4 the signals for a SHTB test @ $1000\,s^{-1}$ are shown. By using these signals and the equations (1) and (3) it is possible describe the history of the strain rate and of the stress (see Fig. 5). Thanks to the length of the SHTB

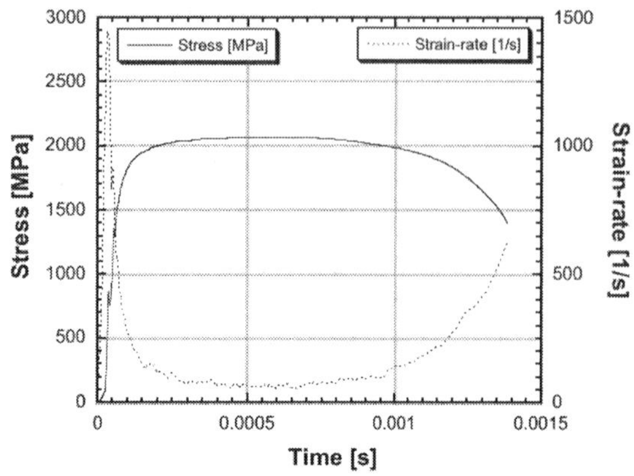

Figure 5. Engineering stress and strain-rate vs. time $100\,s^{-1}$.

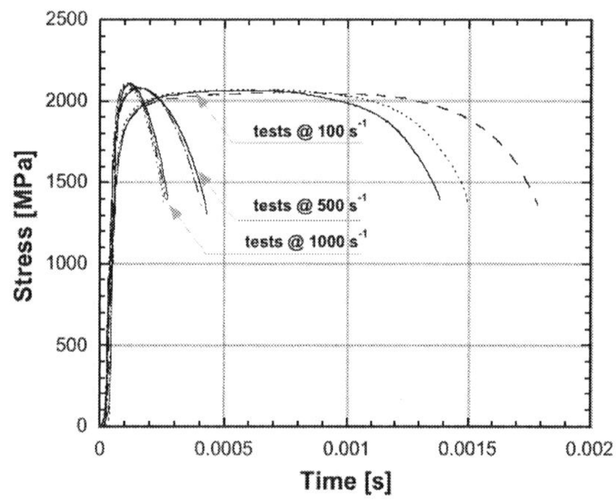

Figure 6. Engineering stress vs. time curves.

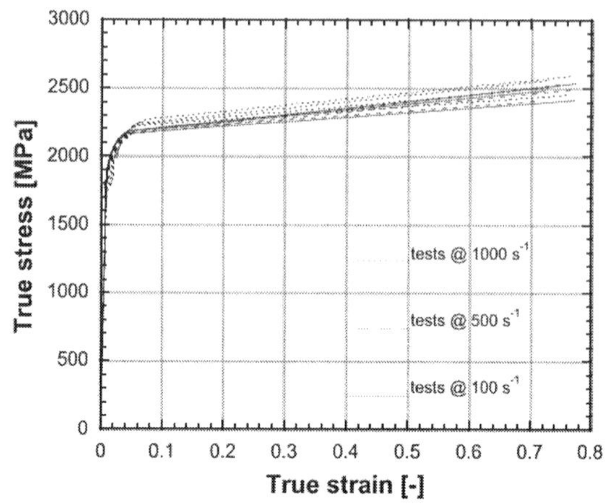

Figure 7. True stress vs. true strain at different strain-rates.

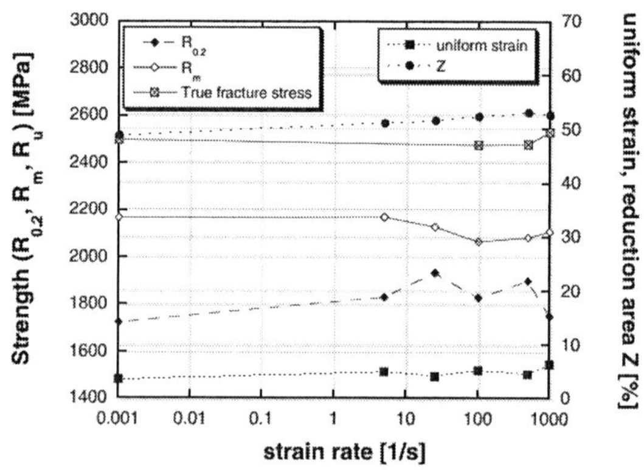

Figure 8. Mechanical properties in function of the strain-rate.

device and the strength of this armour steel, strain rates a little bit lower than $100\,s^{-1}$ were obtained, as shown in Fig. 5. This fact is guaranteed by the constancy of the strain rate during the deformation in the plastic region.

4. Results

By using different pre-loads in the pre-tensioned bar is possible to obtain different strain rates such as 100, 500 and $1000\,s^{-1}$ (see Fig. 6).

The armour steel analysed shows quite low strain rate sensitivity. This can be observed in Fig. 7 where the true stress versus true strain curves at different high strain-rates are depicted.

Analysing Fig. 8, a better and brief representation of the strain-rate sensitivity can be highlighted. Here the trends of the yield stress ($R_{0.2}$), the ultimate tensile strength (R_m) and the fracture stress (R_u) are reported as function of the strain-rate. In the same plots also the uniform and the reduction in cross-sectional area (Z) are shown. For each of these trends, a visual trend line is included to help distinguish the data sets.

The characteristics of fracture as the reduced area of the specimen cross section after failure in the necking zone as well as the fracture strain, have been obtained by means of two images taken before and after the test of each specimen, as shown in Fig. 9. In order to do this, the specimen is at best reconstructed by bringing together the two broken parts and both the diameter and the meridional radius of curvature at the reduced section have been measured. Post-mortem examination of the specimens has also been made after fracture obtaining measurement of the diameter and the meridional radius of curvature at the reduced section that are used for the last point of the true stress-strain diagram by the Bridgman formula.

In Fig. 9 can be observed the differences between the specimen after quasi-static and high strain rate tests. In the quasi-static the necking zone is larger than the dynamic one, this means that in dynamic the deformation is more localized.

5. Numerical analysis

The numerical analysis of the dynamic behaviour of armour steel in tension was conducted using the obtained experimental results. The aim of this study was to verify the capacity of the material constitutive models to replicate the material response in case of dynamic events. In particular, two procedures were chosen. The first was the use of the Johnson-Cook model [9] calibrated with the experimental results, the second was the use of

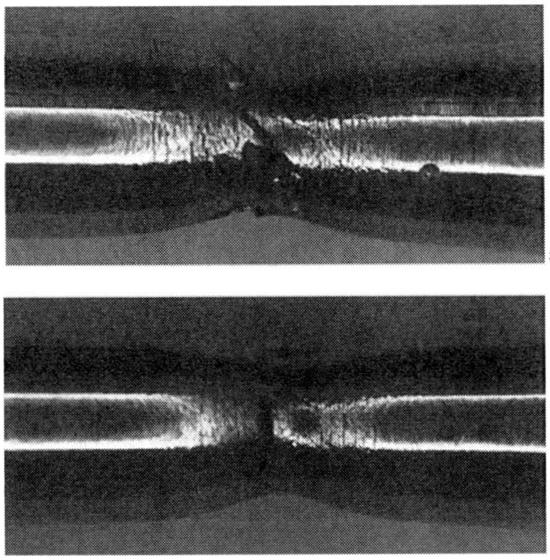

Figure 9. Photos of specimens after test: a) static; b) $1000\,s^{-1}$.

the *MAT PIECEWISE_LINEAR_PLASTICITY model, present in LS-Dyna code, calibrated on the experimental signals. These two procedures were used in the FEM simulation of the dynamic tensile test carried out with the SHTB. A second assessment was performed simulating numerically a blasting event on an armour steel plate.

5.1. Calibration of the material model

Dynamic response of materials is usually expressed through models based on the viscoplasticity theory.

They are usually expressed as the product of functions of key variables such as temperature, strain rate, stress and strain, as in Eq. (4).

$$\sigma_y(\varepsilon, \dot{\varepsilon}, T) = f(\sigma_y, \varepsilon) \cdot g(\dot{\varepsilon}) \cdot h(T). \quad (4)$$

Equation of state can be further added to this basic formulation in order to represent pressure-volume-energy/temperature relation.

Those material models are capable to represent the strain rate sensitivity, plasticity and are widely used in impact, ballistic, blast, and crash event simulations.

Among the advantages of those models can be highlighted: the simplicity in use; limited number of dynamic tests to identify dynamic sensitivity; the use of quasi-static stress strain to identify material response; their availability into the FEM codes.

As example the widely used Johnson-Cook (JC) relationship [9] has been calibrated by using the experimental results obtained at several strain rates. The parameters are shown in Table 2. Only isotropic hardening and strain-rates hardening have been considered, due to the unavailability of temperature data.

5.2. Material model and Finite Element simulation of the SHTB

In order to verify the material model correctness the identity between the experimental and the simulated output

Table 2. Johnson-Cook parameters for the analysed materials.

A [MPa]	B [MPa]	n [–]	C_{mean} [–]
1763	3367	0.4873	−0.01566

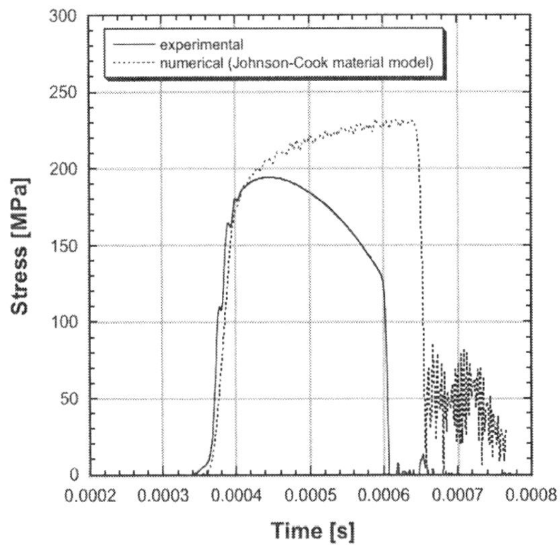

Figure 10. Comparison between experimental and simulated, by Johnson-Cook model, of the signal in output bar.

bar gauge record in the dynamic tensile test using JC model and *MAT_PIECEWISE_LINEAR_PLASTICITY model, available in the LsDyna code, was checked.

The use of the JC model, with the parameter shown in Table 2, into the FE simulation of the SHTB generates the output bar stress shown in Fig. 10.

By comparing the FE simulations of the SHTB using the *MAT_PIECEWISE_LINEAR_PLASTICITY model in correspondence of the output strain gauge and the experimental response of the armour steel it is possible to generate the material stress-strain data. Figure 11 shows the output gauge signals comparison in that case.

The constraints due to the fixed interpolation rule in the JC model origins difference respect the real response. In particular the use into the JC interpolation of the minimal stress difference criteria moves the necking point leading to increased deformation energy. The stress/strain/strain-rate data point obtained by the SHTB simulation are expected to be more realistic because the breakage time and the consequent plastic work are controlled by the experimental output signal.

5.3. Material model and FE simulation of blasting

The differences between the two models can be focused by mean of a numerical example. A shell structure of size $1300 \times 1300 \times 0.13$ m, with 10 mm thickness is subject to a blast loading due to the explosion of 20 kg of TNT 200 mm below the centroid. The event was modelled with LS-Dyna solver.

Two material models were assigned to the structure, by means of the JC form and the data point. Finally, the two responses are compared. The two cases internal energy

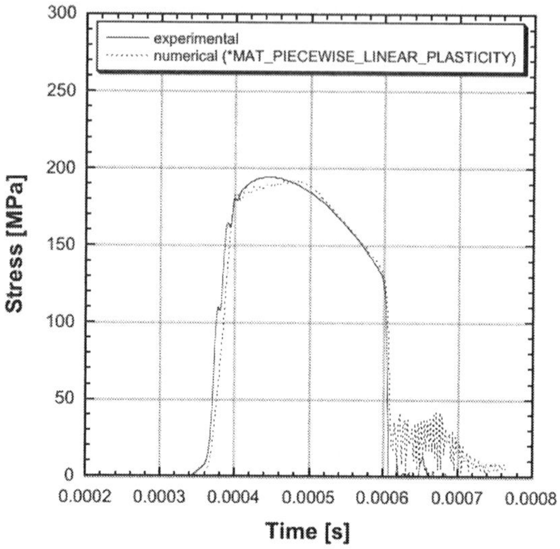

Figure 11. Comparison between experimental and simulated, by *MAT_PIECEWISE_LINEAR_PLASTICITY, of the signal in output bar.

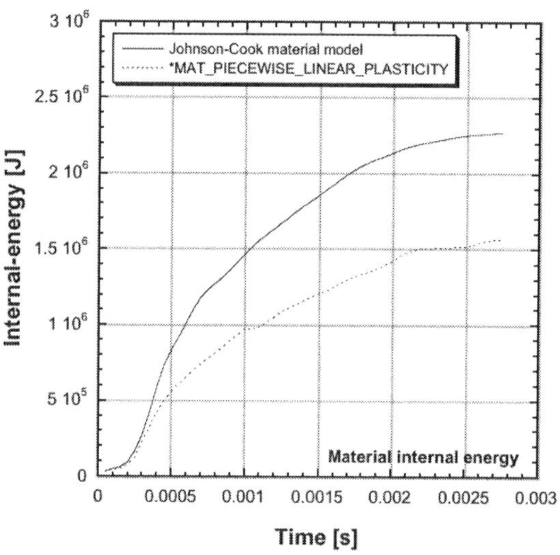

Figure 12. Material internal energy time history comparison.

(Fig. 12) such as the final deformation (Fig. 13) shows relevant differences.

5.4. Remarks

A drawback of the JC material model can be found in the respect of the necking point that affects the deformation energy of the material. Indeed the interpolation based on the real stress-strain does not constrain the necking point, while an interpolation based on the necking point does not respect the yield point or the ultimate tensile strength.

The JC material model can be tuned to single cases to fit the experimental evidences. This does not ensure the applicability of the JC material model to other geometrical and loading conditions. For material with high ultimate deformation, the problem increases in evidence.

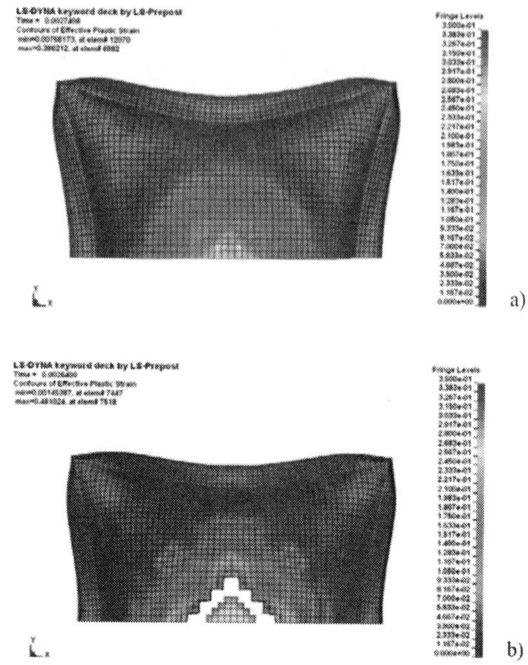

Figure 13. Contour of equivalent plastic strain during blasting: a) JC material model b) *MAT_PIECEWISE_LINEAR_PLASTICITY.

The previous inconvenient can be avoided using a material definition by using a tabular form into the space of stress, strain and strain rate. This simple form better reproduces the experimental response of the material.

When this table is produced, by using the finite element simulation of the SHTB, the local strain rates are considered with significant advantages in the material model reliability.

To obtain univocal definition of material parameters, an increased of the tests number and the use of automatic optimizators are required.

With reference to the previous numerical example, differences of 40% in deformation energy due to different material characterization were found.

6. Conclusion

The present work reveals that the armour steel investigated is low strain rate sensitive in the strain rate range from $0.001\,s^{-1}$ to $1000\,s^{-1}$ under uniaxial tension. In the true stress versus true strain curves a moderate increase of strain hardening in the material with the increasing strain rate was observed. The true fracture stress as well as the reduction of cross section area at fracture, increases with the increasing strain rate. Thanks to the length of the SHTB a quite low strain rate were obtained.

The Johnson-Cook model was calibrated using the experimental results. By reverse analysis a *MAT_PIECEWISE_LINEAR_PLASTICITY material model was also obtained. The reliability of the two models was verified by mean of FEM simulation of the SHTB test.

Finally, a numerical assessment was performed simulating a blast event on a steel plate by using the two material models. This comparison showed differences of 40% in deformation energy.

References

[1] G. Riganti, E. Cadoni, Mat. Des. **57**, 156–167 (2014).
[2] E. Cadoni, M. Dotta, D. Forni, S. Bianchi, H. Kaufmann, EPJ WoC **26**, 05004 (2012).
[3] E. Cadoni, M. Dotta, D. Forni, S. Bianchi, Appl. Mech. Mat. **82** 124–129 (2011).
[4] E. Cadoni, L. Fenu, D. Forni, Constr. Build. Mat. **35**, 399–407 (2012).
[5] D. Asprone, E. Cadoni, A. Prota, , ACI Struct. J. **106**, 523–529 (2009).
[6] E. Cadoni, M. Dotta, D. Forni, P. Spätig, J. Nucl. Mat. **414**(3), 360–366 (2011)
[7] E. Cadoni, M. Dotta, D. Forni, N. Tesio, C. Albertini, Mat. Des. **49**, 657 – 666 (2013).
[8] C. Albertini, E. Cadoni, G. Solomos, Phil. Trans. Roy. Soc. A: **372**, 20130197, (2014).
[9] Johnson, G.J., Cook, W.H. Proceedings of the Seventh International Symposium on Ballistics, The Hague, 1983, pp. 541–547.

EPJ Web of Conferences **94**, 05005 (2015)
DOI: 10.1051/epjconf/20159405005

Rate dependence of electrical and mechanical properties of conductive polymer nanocomposites

J.R. Foley[1], C.L. Stilson[1], K.K.G. Smith[1,2], C.M. McKinion[1,3], C. Chen[4], S. Ganguli[4], and A.K. Roy[4]

[1] Air Force Research Laboratory, Eglin AFB, FL, USA
[2] National Research Council Postdoctoral Research Associate, Eglin AFB, FL, USA
[3] Doolittle Institute, Fort Walton Beach, FL, USA
[4] Air Force Research Laboratory, Wright-Patterson AFB, OH, USA

Abstract. Conductive polymer nanocomposites with enhanced electrical and thermal properties show promise as an alternative solution for electronic materials. For example, electronic interconnect materials will have comparable electrical and thermal conductivity to solder with an increased operating range of strain and temperature. This paper documents the fabrication and experimental evaluation of a prototype conductive polymer nanocomposite. Material selection, fabrication processes, and initial characterization of a low Tg polymer with a high fill ratio of carbon nanotubes is presented. The electrical and thermal properties of the composite are measured and compared with predictions. The mechanical properties are measured using dynamic mechanical analysis (DMA) over a wide temperature range. The mechanical and electrical responses of the conductive polymer composite are simultaneously measured at higher strain rates using a modified split Hopkinson pressure bar (SHPB) apparatus. The dynamic stress-strain response is obtained using traditional analytic methods (e.g., two- and three-wave analysis). The electrical response is observed using constant current excitation with high bandwidth (>500 kHz) instrumentation. The dynamic compression data implies the change in electrical resistance is solely a function of the material deformation, i.e., the material exhibits constant electrical conductivity and is insensitive to the applied loads. DMA and SHPB dynamic data are used to estimate the parameters in a Mulliken-Boyce constitutive model, and the resulting behavior is critically evaluated. Finally, progress towards improving the polymer composite's mechanical, electrical, and thermal properties are discussed.

1. Introduction

The global electronics industry is dominated by components designed for commercial electronics [1]. Accordingly, the design architecture, fabrication process, materials, component design, and testing [2] all reflect the mass-production, commercial-off-the-shelf (COTS) nature of consumer electronics. However, these same processes, materials, etc., are often used in electronics subjected to much harsher operating environments. The corresponding failure modes are diverse and depend on the geometry, materials, and physics of operation of the devices [3]. For example, Fig. 1 shows an SEM image of the cross-section of a surface mount resistor subjected to severe mechanical shock environments [4,5]. The mechanical loading initiated a crack that has propagated along the interface between the fiber-reinforced polymer printed circuit board and the copper trace material. The particular damage mechanism, known as pad lift-off [6] or pad cratering, occurs at the interconnect between the assembly and the component. Other electronics failure modes at device interconnects include solder joint fracture, component termination failures, or component stress cracking.

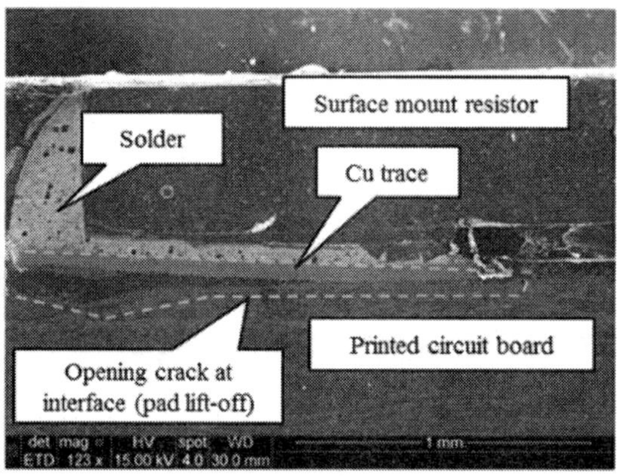

Figure 1. SEM image of a surface-mount electronic component cross-section after high shock loading. The dashed line indicates the onset of pad lift-off at the printed circuit board-copper trace interface.

Developing alternatives for the packaging and assembly of traditional electronics requires understanding of the multifunctional response of candidate materials. This paper documents the fabrication and initial characterization of the combined thermal, electrical, and rate-dependent mechanical response of a new electrically-conductive polymer composite.

2. Material development

A conductive polymer nanocomposite (CPnC) was proposed as an alternative material for electronic

Article available at http://www.epj-conferences.org or http://dx.doi.org/10.1051/epjconf/20159405005

Table 1. Design objectives for the conductive polymer nanocomposite in comparison to Sn-37Pb and SAC305 solders.

Design Metric	Sym.	Sn-37Pb	SAC-305	CPnC Goal	Units
Electrical resistivity	ρ	0.146	0.132	≤ 0.5	$\mu\Omega \cdot m$
Thermal conductivity	κ	50.9	58	≥ 58	$W/m \cdot K$
Tensile Elongation	$\Delta L/L_0$	~50	35	>100	%
Glass Trans. Temp.	T_g	N/A*	N/A*	<273	K
Viscosity	η	N/A*	N/A*	<1e6	poise

*Note: glass transition temperature and viscosity are not meaningful for solder alloys except during the reflow process and are therefore not applicable.

interconnects. This section reviews the design objectives, composition and fabrication process of the initial composite, and characterization of the material.

2.1. Material performance objectives

The primary performance objective for the CPnC is to reproduce the thermal and electrical properties of solder while exhibiting strain-tolerance over a wide range of operating conditions. A eutectic lead-based solder alloy, Sn-37Pb (63% Sn, 37% Pb), and a lead-free solder alloy, SAC305 (96.5% Sn, 3% Ag, 0.5% Cu), were chosen as references [7]. The strain tolerance was given indicated by the strain to failure (i.e., ultimate strain). Viscosity, particularly shear-thinning behavior, was sought to aid in mixing, processing, and ultimately enable manufacturing concepts. Finally, the electrical properties should be insensitive to strain (i.e., not change resistivity as a function of deformation). Table 1 below lists the design objectives for the study.

2.2. Fabrication process

A low T_g polymer matrix was chosen instead such that the material would remain compliant during dynamic events and extend the ultimate strain. The polymer, Epon 828 with Jeffamine D2000 hardener, has relatively poor electrical and thermal performance (i.e., conductivity) in the cured state. In order to improve the electrical and thermal conductivity in a manner similar to commercially available conductive epoxies (e.g., the Ag-epoxy examined in [8]), conductive nanoparticles were considered as the filler. Candidate fillers included Ag nanoparticles and multiwalled carbon nanotubes (MWCNT).

Initial compositions investigated included two heavily filled CPnC's. The first consisted of 20% (by weight) Ag nanoparticles, 10% MWCNT and the Epon 828/D2000 matrix. The second was 20% MWCNT with the Epon 828/D2000 matrix. The processing procedure involves three mixing cycles of 2000 rpm for 3 minutes each in a Thinky planetary centrifugal mixer. After each mix cycle, the composite material is degassed under vacuum (20 kPa) while continuing mixing for 2 minutes. After three mixing cycles, the mixture is cast in either silicone or Teflon molds, placed in vacuum bags, and subsequently cured in an autoclave. The curing temperature profile involves (1) a ramp from room temperature to 120 °C at a rate of 1 °C/min., (2) hold at 120 °C for 3 hours, then (3) a cool down to room temperature; pressure is maintained at 690 kPa and a vacuum of −100 kPa (gauge pressure) is applied to the parts during curing. The vacuum and pressure are released after the parts cooled and the samples then separated from the molds.

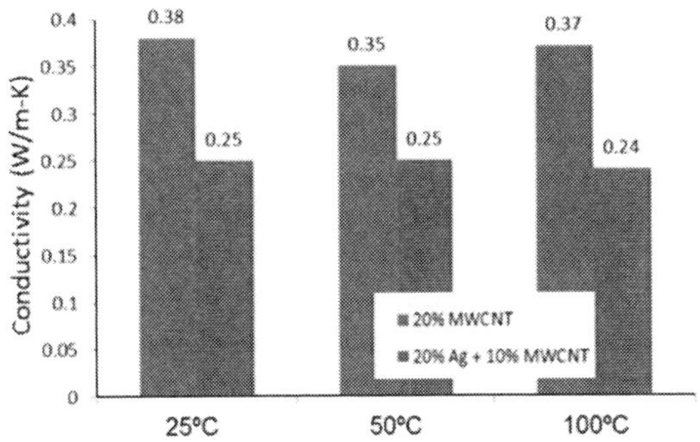

Figure 2. Thermal conductivity at three different temperatures.

2.3. Characterization

After fabrication of the samples, basic thermal, electrical, and mechanical charaterization of the candidate CPnC's was performed. Electrical resistivity was estimated using the van der Pauw technique. The 20% MWCNT/Epon 828/D-2000 prototype material was measured to have an electrical resistivity of $0.5\,\mu\Omega \cdot m$, reaching the design goal. The 20% Ag/10% MWCNT/Epon 828/D2000 was observed to have a much higher resistivity,

The thermal conductivity was measured using laser flash analysis. The 20% MWCNT/Epon 828/D2000 and the 20% Ag/10% MWCNT/Epon 828/D2000 were measured at 25 °C, 50 °C, and 100 °C. The results are shown in Fig. 2 below. While both conductivities were significantly below the design goal, the 20% MWCNT material showed higher average thermal conductivity ($0.37\,W/m \cdot K$) than the 20% Ag/10% MWCNT composite ($0.25\,W/m \cdot K$).

One of the key unanswered questions revolves around the microstructure, specifically whether the nanoparticles were uniformly distributed throughout the matrix and the nature the interfaces between nanoparticles. SEM and TEM analyses were performed; images from the SEM analysis on the 20% Ag/10% MWCNT/Epon 828/D2000 and the 20% MWCNT/Epon 828/D2000 are shown in Fig. 3 and Fig. 4, respectively. The 20% mass loading of the silver nanoparticle flakes was initially expected to contribute strongly to the enhanced transport properties within the material. However, Fig. 3 illustrates that the very dense Ag is sparsely distributed and does not enhance the overall network due to the isolated location of each particle, explaining the lower thermal conductivity and higher electrical resistivity. Commercial conductive epoxies achieve enhanced transport through much higher mass loadings, typically 65% or higher [9]. Additionally, there were few observable interfaces between the two types

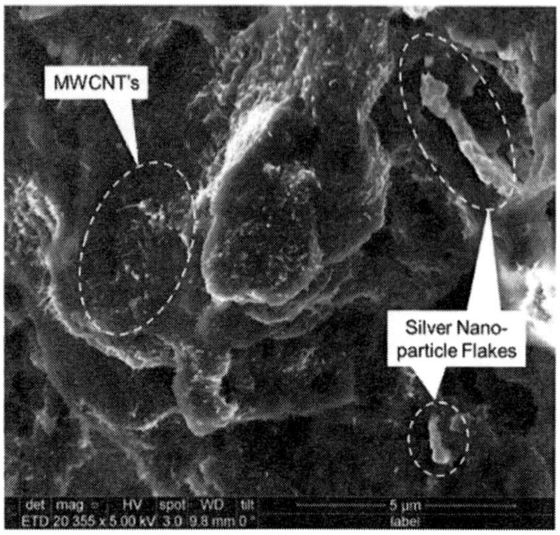

Figure 3. SEM image of the 20% Ag/10% MWCNT CPnC showing carbon nanotubes and the flakes of silver nanoparticles (matrix has been removed).

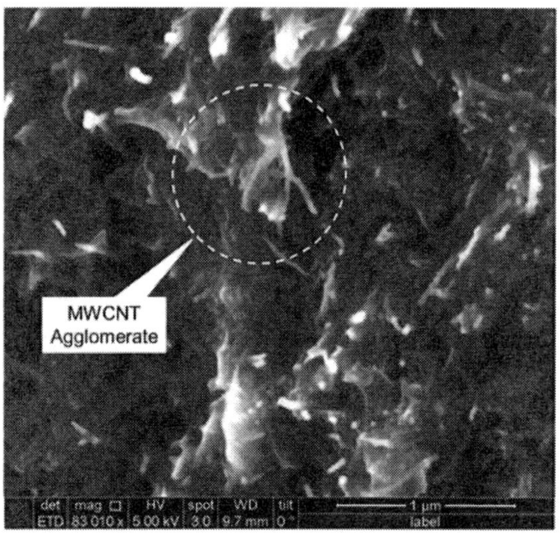

Figure 4. SEM image of the 20% MWCNT CPnC showing multiwalled carbon nanotubes with significant agglomeration (matrix material has been removed).

of nanoparticle. The MWCNT-only CPnC, shown in Fig. 4, exhibits slightly enhanced transport relative to the silver-loaded CPnC, presumably through conduction within the agglomerated nanotubes.

The overall better performance of the 20% MWCNT composite–with a lower electrical resistivity and higher thermal conductivity–led to its selection as the best candidate for more detailed analysis. Due to the importance of processability for electronics materials, the viscosity of the CPnC was measured as a function of curing temperature profile, shown in Fig. 5. The melt-thinning

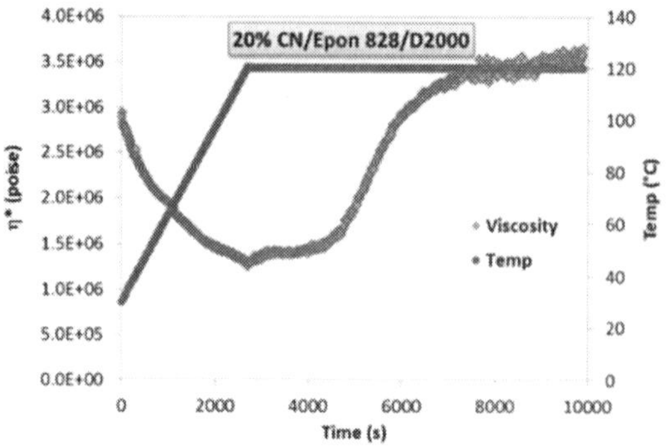

Figure 5. Viscosity and temperature vs. time for the 20% MWCNT CPnC.

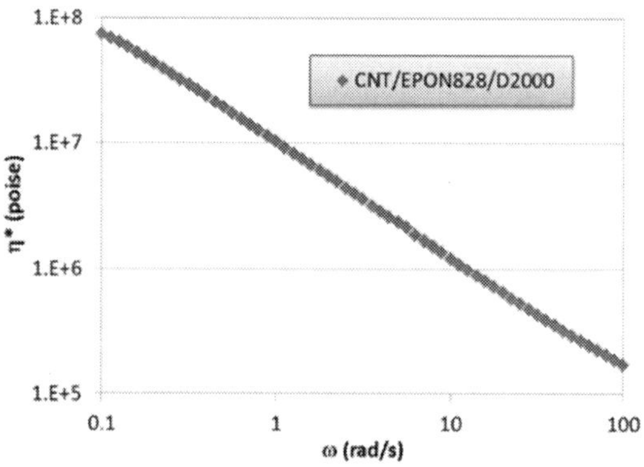

Figure 6. Shear-thinning behavior of the 20% MWCNT CPnC.

of the matrix contributes to the reduced viscosity through the temperature, reaching a minimum of approximately 1.3e6 poise at the cure temperature. The shear-thinning behavior of the composite is shown in Fig. 6; the 20% MWCNT material exhibits significant shear-thinning, which is desirable for many processing techniques.

2.4. Dynamic mechanical analysis

Dynamic mechanical analysis (DMA) [10] was then performed on both candidate CPnC's. DMA samples of 17.9 mm long, 11 mm wide, and 3.1 mm thick were cast for each of the candidate materials and the matrix material. The samples were tested in a single cantilever configuration in a TA Instruments Q800 [11,12]. The excitation frequency was set at 1 Hz and the samples were ramped through the temperature range of $-120\,°C$ to $+\,120\,°C$ at a rate of 3 °C/minute. The displacement was held constant at 20 μm for this analysis.

Typical DMA data for the CPnC's is shown in Fig. 7 at the measured frequency of 1 Hz. The low T_g is readily apparent, and the room temperature (298 K) modulus of both candidates are approximately 70 MPa. The frequency can be converted to an equivalent strain rate ($\dot{\varepsilon}$) using the equivalence relationship [13],

$$\dot{\varepsilon} = 4\omega d_0 / l_g, \tag{1}$$

where ω is the angular frequency (in rad/s), d_0 is the amplitude of the displacement, and l_g is the specimen gage

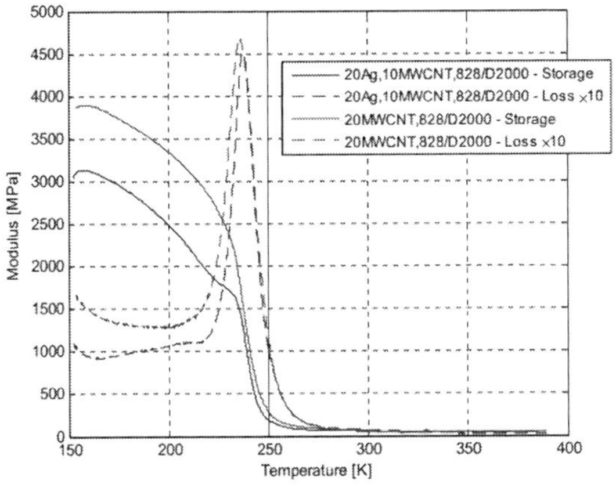

Figure 7. DMA storage and loss modulus data for two candidate CPnC's.

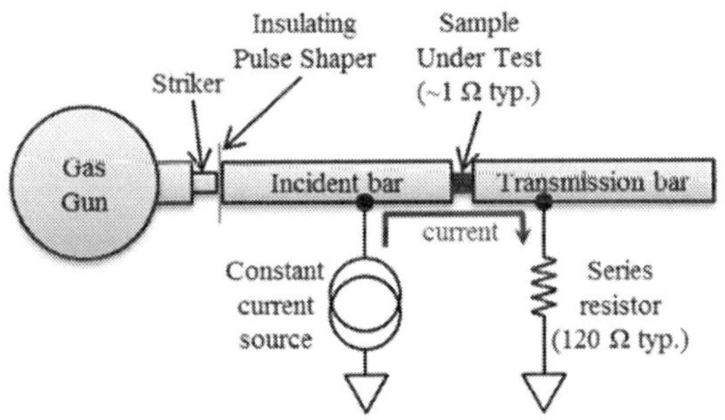

Figure 8. Schematic diagram of modified SHPB apparatus used to simultaneously monitor mechanical and electrical response of the sample under test.

length. For this sample geometry, the equivalent strain rate is $\dot{\varepsilon} \approx 0.03\,\mathrm{s}^{-1}$.

3. Dynamic compression experiments

3.1. Apparatus

A split Hopkinson pressure bar (SHPB) apparatus [14] was modified to allow simultaneous measurement of the electrical resistance of the sample by using the incident and transmission bars as electrodes. A schematic is shown in Fig. 8. 6 foot (1.81 m) long aluminum alloy (6061) bars with 1 in. (25.4 mm) diameter were used for both the incident and transmission bar.

The technique to simultaneously measure the electrical and mechanical response of the material is most similar to previous work by Heeder et al. [15]. The primary distinction is the much higher mass loading of the 20% MWCNT CPnC investigated in this work. Other recent studies have attempted similar measurements but on much higher equivalent resistance composites [16, 17].

A Precision Filters 28144 constant current signal conditioner [18] is used to provide the excitation current. Electrical connection to the incident and transmission bars is made using hose clamps compressing a copper mesh (to which the wires are soldered) as shown in Figure. Due to the low resistivity of the nanocomposite (the sample

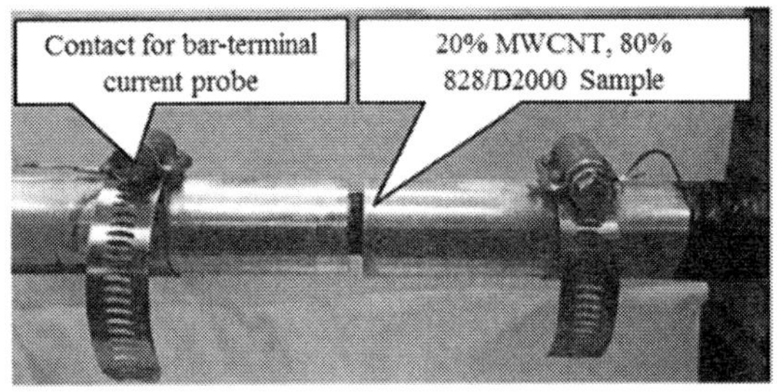

Figure 9. Photographic detail of SHPB experiment showing sample and contact configuration.

resistance is typically less than 1 Ω), a series resistor is necessary to raise the total resistance to the expected range of a strain gage (in this case, approximately 120 Ω). Uncured conductive epoxy is used in lieu of conductive grease as a multifunctional surface finish: it reduces friction at the sample-bar interfaces and also reduces the contact resistance.

Dynamic strain is measured on the incident and transmission bars using Kyowa semiconductor strain gages [19]; the same Precision Filters system used to measure the sample resistance provides the strain gage conditioning. Both sensor and sample resistance signals are digitized by a National Instruments data acquisition system with PXI-6133 cards [20].

3.2. Sample material, preparation, and geometry

The 20% MWCNT nanocomposite samples were prepared as noted in Sect. 2.2 and cast into 0.625 in. (16 mm) diameter rods. SHPB sample disks with variable lengths between 0.125 in. (3 mm) and 0.188 in. (4.8 mm) were cut from these rod.

3.3. Results and discussion

As expected with a low T_g matrix, the 20% MWCNT CPnC exhibited a compliant, rubbery response as shown in the stress-strain curve shown in Fig. 10 with a measured strain rate of $522\,\mathrm{s}^{-1}$. The material exhibited excellent recovery, exhibiting no measurable change in thickness after dynamic compression and no discernible damage.

The electrical response of the material in various tests is shown as a function of time is shown in Fig. 11 and Fig. 12. The bulk resistivity of the sample was initially hypothesized to follow the compressive stress (or pressure) due to the MWCNT networks acquiring more contact points, decreasing the overall composite material's resistance. However, as Fig. 11 shows, there was no direct correlation to support this hypothesis. The opposite scenario, where the bulk resistivity (ρ_0) is constant, would yield a scenario where the change in resistance (ΔR) follows from the geometric distortion of the sample,

$$\Delta R = \rho_0 \Delta(L/A), \tag{2}$$

where L and A are the instantaneous length and area of the conductive path in the sample. Assuming full participation and one-dimensional compression, this result would be proportional to the net compression of the sample. This

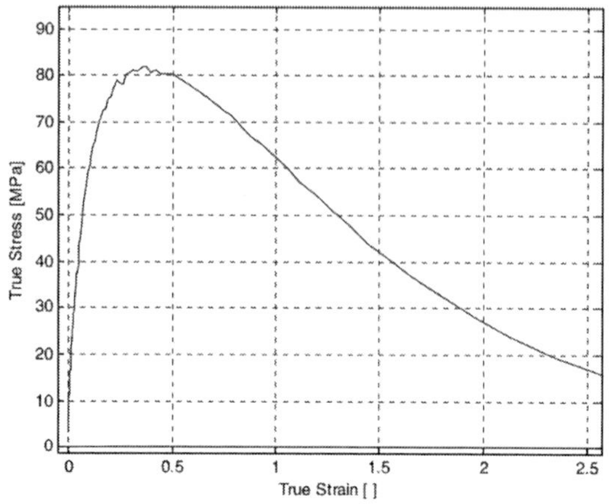

Figure 10. Typical stress-strain diagram for the 20% MWCNT CPnC.

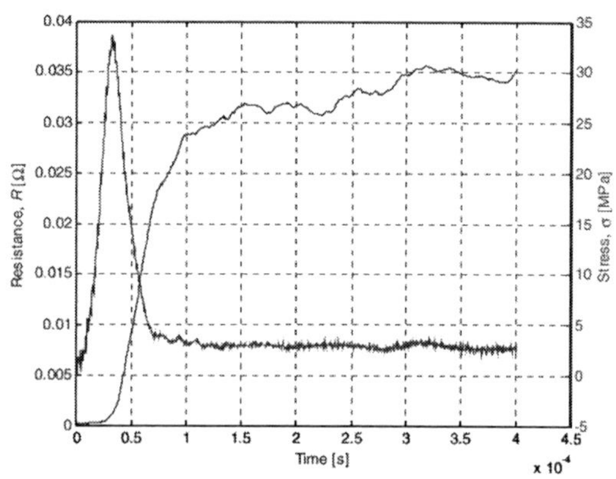

Figure 11. Change in resistance and stress vs. time for the nanocomposite.

proportionality is evident in Fig. 12, supporting this conclusion.

3.4. Constitutive analysis

The Ogden model [21] was initially considered due to the rubbery nature of the nanocomposite, however the calcualted compression of the CPnC with minimal radial distortion suggested this was inappropriate.

The Mulliken-Boyce (M-B) constitutive model was then chosen for analyzing the data based on previous work [22,23]. The M-B model [24,25] is a two phase (α and β) viscoelastic-viscoplastic model for glassy polymers that incorporates both polymer network stress and the chain stresses. The model has been shown to accurately reproduce rate-dependent behavior of polymers. DMA data [26] is used to estimate the activation energies, mean temperatures, and shift factors associated with different phases of a polymeric material associated in the M-B model. However, the CPnC's considered are well above the glass transition temperature. Nonetheless, the model was implemented with the assumption that the nanotube networks can be modeled by the chain stresses

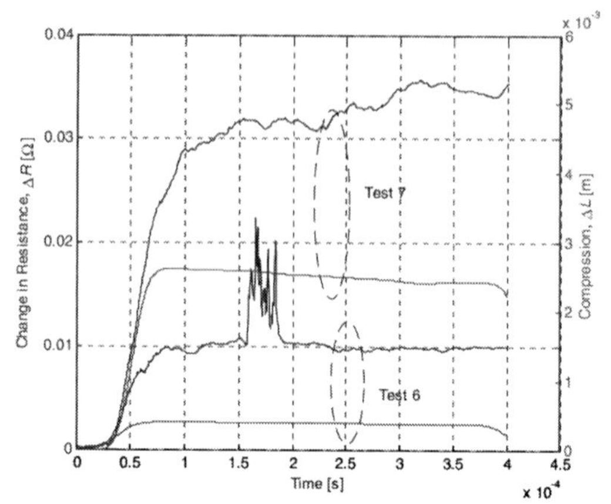

Figure 12. Resistance change (left axis) and total deformation (right axis) for two tests. The roughly proportional response implies little change in the bulk resistivity as a function of pressure. (An intermittent electrical contact is evident in the resistance during test 6.)

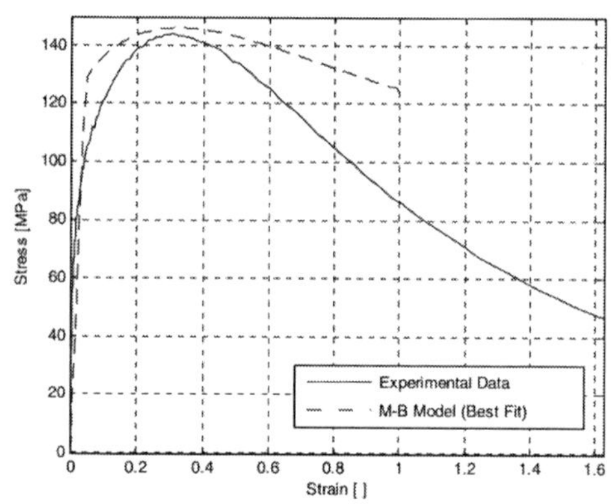

Figure 13. Overlay of expermental and Mulliken-Boyce model stress-strain diagrams for a 20% MWCNT CPnC sample.

analagous to the polymer (quantified as a Langevin spring element in the viscoelastic model).

A genetic algorithm was implemented in Matlab [27] to estimate the M-B model parameters. The DMA data was used for the modulus as a function of temperature with the β phase assumed to be negligible. A resulting fit is shown in Fig. 13. The rubbery modulus was estimated to be 20 MPa with a stretch factor (N) of 2.9, indicating significant participation in the network stress term. The estimates for the phase activation energies ($\sim 10^{-19}$ J) nor attempt frequencies ($\sim 10^{19}\,s^{-1}$) are not expected to be accurate due to the system being above the glass transition temperature. Overall, the parameter estimates were limited by the lack of shift factors associated with the DMA (i.e., no frequency-dependent information), variability between dynamic stress-strain data sets, and lack of dynamic data at multiple strain rates. A more appropriate constitutive model should also be sought, given the inappropriate use of the M-B model in a solely rubbery phase.

4. Future work

The development of the CPnC is at a very immature stage. While the initial results are promising, the thermal conductivity and electrical resistivity need to be significantly improved. CNT interface functionalization (metallic approach) is being pursued as a possible means to further improve the properties of the next iteration of the nanocomposite. More thorough examination of the rate-dependent effects, including a wider range of frequencies of DMA data and adding both quasi-static and dynamic tensile and compressive testing at different strain rates, are also underway. Improvements to the modified SHPB are also underway, such as using bars with lower impedance (e.g., hollow Al bars) and utilizing longer duration loading pulses. Detailed examination of the coupling between the mechanical and electrical response of the material is also in progress. Lastly, analysis of the stress-strain data using different constitutive models is planned.

5. Conclusion

The development of a conductive polymer nanocomposites with enhanced electrical and thermal properties was pursued as an alternative material for electronic interconnects. The material synthesis method and electrical/thermal characterization of two prototype CPnC's compositions–20% Ag/10% MWCNT and 20% MWCNT in Epon 828/D2000 matrix–was presented. Mechanical properties of the resulting rubbery material was measured over a wide temperature range using dynamic mechanical analysis (DMA). The mechanical and electrical responses of the conductive polymer composite were simultaneously measured at higher strain rates using a modified split Hopkinson pressure bar (SHPB) apparatus. The change in electrical resistance of the sample during dynamic deformation is shown to be largely a function of the material deformation. If the bulk resistivity is assumed to be constant, the resulting change in resistance implies the material is largely insensitive to the applied loads. The stress-strain curves were estiamted using a Mulliken-Boyce model with marginal results. Finally, ongoing research towards improving the polymer composite's mechanical, electrical, and thermal properties was discussed.

The authors wish to thank the U.S. Air Force Office of Scientific Research (Project Officer: Dr. Byung-Lip "Les" Lee) for their support of this project. Opinions, interpretations, conclusions, and recommendations are those of the authors and are not necessarily endorsed by the United States Air Force.

References

[1] W. Custer, IPC Outlook (downloaded from www.ipc.org) (2011)
[2] JEDEC, Board level drop test method of components for handheld electronic products (2003)
[3] R.A. Amy, G.S. Aglietti, and G. Richardson, Shock & Vibration. **16**, 45 (2009)
[4] R.D. Lowe, et al., Proceedings of the 85th Shock and Vibration Symposium (2014)
[5] R.D. Lowe, et al., Proceedings of the 85th Shock and Vibration Symposium (2014)
[6] J.W. Dally, P. Lall, and J.C. Suhling, Mechanical Design of Electronic Systems (College House Enterprises Knoxville, TN, 2008)
[7] T. Siewert, et al., Database for Solder Properties with Emphasis on New Lead-free Solders (National Institute of Standards and Technology Colorado, 2002)
[8] X. Luo and D.D.L. Chung, Journal of Materials Science. **34**, 273 (1999)
[9] Tra-Con, Tra-Duct 2902 Room Temperature Conductive Silver Epoxy Adhesive (Technical Product Data and MSDS) (2002)
[10] Introduction to Dynamic Mechanical Analysis (DMA): A Beginner's Guide (PerkinElmer, Inc. Waltham, MA, 2008)
[11] DMA Q800 Specifications (TA Instruments, 2010)
[12] TA Instruments Dynamic Mechanical Analyzer (TA284) (TA Instruments, 2011)
[13] A.D. Mulliken and M.C. Boyce, International Journal of Solids and Structures. **43**, 1331 (2006)
[14] G.T. Gray III, in ASM Handbook. Vol 8: Mechanical Testing and Evaluation, edited by H. Kuhn and D. Medlin (ASM International, Materials Park, 2002), pp. 462
[15] N. Heeder, et al., Journal of Materials Science. **47**, 3808 (2012)
[16] A.S. Lim, et al., Composites Science and Technology. **71**, 616 (2011)
[17] L. Jin-tao, C. Jian-kang, and Z. Ming-hua, Polymers & Polymer Composites. **20**, 45 (2012)
[18] Precision 28144 Quad-Channel Wideband Transducer Conditioner with Voltage and Current Excitation (Datasheet) (Precision Filters, Inc. Ithaca, 2009)
[19] Kyowa KSP-2-120-E4 Strain Gage Datasheet (Kyowa Electronic Instruments Co., 2014)
[20] NI PXI-6133 Specifications. (Available from ni.com) edn. (National Instruments Austin, 2003)
[21] O.A. Shergold, N.A. Fleck, and D. Radford, Int. J. Impact Engng. **32**, 1384 (2006)
[22] J. Jordan, J. Foley, and C. Siviour, in Mechanics of Time Dependent Materials (Monterey, CA, 2008)
[23] J. Jordan, J. Foley, and C. Siviour, Mechanics of Time-Dependent Materials. **12**, 249 (2008)
[24] A.D. Mulliken. Massachusetts Institute of Technology, 2004
[25] A.D. Mulliken. Massachusetts Institute of Technology, 2006
[26] J.R. Foley, J.L. Jordan, and C.R. Siviour, Proceedings of DYMAT 2012 (2012)
[27] Matlab User's Manual (version 2011b) (Mathworks, 2011)

EPJ Web of Conferences **94**, 05006 (2015)
DOI: 10.1051/epjconf/20159405006

Blast impact behaviour of concrete with different fibre reinforcement

Martina Drdlová[1, a], René Čechmánek[1], and Radek Řídký[2]

[1] Research institute for Building Materials, Hněvkovského 65, 61700 Brno, Czech Republic
[2] SVS FEM, Škrochova 42, 61500 Brno, Czech Republic

Abstract. The paper summarizes the results of the development of special concrete intended for the explosion resistance applications, with the emphasis on minimal secondary fragments formation at the explosion. The fine-grained concrete matrix has been reinforced by various types of short dispersed fibers (metallic, mineral and polymer) of different sizes and by their combination and the effect of the fibre reinforcement on the physico-mechanical properties and blast resistance was observed. The concrete prism specimens have been subjected to the determination of mechanical parameters (compressive and flexural strength at quasi-static load). The blast tests were conducted on the slab specimens prepared from selected mixtures. The material characteristics and explosion test data have been used for numerical investigation, which defined the optimal wall composition and dimensions of the concrete element which should resist the explosion defined by type, size, weight and placement of the blast. In the next step the test elements resistance was verified by real explosion test.

1. Introduction

Efforts to eliminate possible threats of terrorist attacks in areas with high concentrations of people is mostly realized in a passive way. For example, trash bins and baggage lockers were removed from the potentially critical public spaces, isolation of large areas and long time forced downtimes are required while the suspect objects are investigated. Hence the alternative approach, which results in the reduction of citizen discomfort and in increase of safety at the same time, is on demand. For lower explosion load levels the reinforced high performance concrete is a good and mainly inexpensive solution compared to other systems recently applied.

The high performance concrete itself as a quasi-brittle material does not conform to the resistance requirements, as at the explosion many secondary fragments are formed while the concrete breaks. The additional reinforcement in form of random dispersed fibres can increase the blast resistance of the concrete. By the addition of steel fibres to the conventional concrete, the impact strength value has increased 2–15 times depending on the amount of the fibres [1]. Better performance of fibre reinforced concrete (FRC) under impact load compared to plain concrete without reinforcement was also confirmed by [2–4]. Maalej et al. [5] found that hybrid fibre-reinforced concrete (using combinations of two or more types of fibres in the same concrete mixture) offered increased shatter resistance with reduced scabbing, spalling and fragmentation, whilst exhibiting better energy absorption.

Fibre reinforcement in combination with high performance matrix is a way to improve blast resistance of concrete. Optimal composition with emphasis on the compatibility of the matrix and fibre is required, especially in the case of relatively thin wall elements. The aim of the experiments presented in this paper is to develop a high-grade concrete with increased resistance to blast, which will be used for elements of urban infrastructure, for example litter bins with blast resistance up to 1000 g of TNT.

[a] Corresponding author: drdlova@vustah.cz

2. Materials and test procedures

Concrete mixtures with various fibre reinforcement were prepared. For the mechanical performance investigation, the prism $100 \times 100 \times 400$ mm specimens were made and the bulk density according to ČSN ISO 6275, compressive strength and flexural strength at 28 days of age in accordance with ČSN EN 12390-3, were measured. Four types of fibres have been incorporated into the matrix to determine their influence on physico-mechanical properties: Glass fibres (G) – length 12 mm, diameter 0,14 μm, basalt fibres (B) – length 6 mm, diameter 12 μm, steel fibres (S) – length 6 mm, diameter 200 μm, polypropylene (P) – length 12 mm, diameter 18 μm. Mix proportions of fibres are listed in Table 1. Fine-grain high performance concrete (consisting of fine SiO_2 sand, cement 52.5R, microsilica, and superplasticizer) was selected as a standard. Total volume of fibres was 3 vol.%. For the explosion tests the concrete boards $40 \times 500 \times 500$ mm were produced from selected mixtures and test procedure according to certified methodology MT0-B-VTUO 10/9 developed by Military research institute of CR [6] was applied.

The methodology consists of measuring vibrations and further acceleration of a test stand holding the sample, which is subjected to the explosion load. The concrete specimen is fixed in the steel frame of the stand and the stand is placed on a sand foundation. The cylinders of Semtex C4 plastic high explosive weighting 77.0 g and

Table 1. Mix proportion of fibres in the mixtures.

Designation	Fibre content (kg/m^3)			
	Glass	Steel	Basalt	Polypropylene
REF	0	0	0	0
G100	53.7	0	0	0
B50	0	0	26.9	0
S400	0	234.9	0	0
P4	0	0	0	2.3
G50B37	26.9	0	20.5	0
G50P10	27.7	0	0	5.5
G50B37P2	27.7	0	20.3	1.1
G40B20PP4	22.2	0	10.3	2.2
S200G50	28.7	114.8	0	0
S200B30	0	113.6	15.9	0
S130G22B17P1	12.6	75.7	9.5	0.6

Figure 1. Test stand with a charge.

with dimensions d = 45 mm, h = 45 mm are used as testing charge. The test stand with the charge is shown in Fig. 1.

The acceleration detector is attached to the steel frame of the stand – the data from the acceleration sensors is collected for at least 500 milliseconds using sampling frequency of 200 kHz. The acoustic detector of explosion is used for triggering the data collection. The results of the measurements are data characterizing the material capability of absorbing the blast energy. The obtained data are processed and the absolute value of acceleration is gained which is then integrated. The integrated value enables to compare the energy absorbed by particular materials – the lower the integral value, the more blast energy was absorbed by the specimen. The energy absorption is one of the parameters – however the integrity of the sample must also be taken into consideration when evaluating the overall sample blast resistance performance. Hence the comparison of the samples is done using the following equation: 1/(Relative attenuation*Integrity), where Integrity value reflects the integrity of the sample after subjected to the blast load – the integrity was marked using numbers from 1 (the sample is without visually noticeable changes after blast load) to 5 (the sample is broken to small pieces after blast load). The outputs of the equation given above are shown in Fig. 6 and were used to compare the overall performance of samples when

Figure 2. Blast test layout.

subjected to the blast load – the higher value, the better overall blast resistance of the sample (combination of the absorbed energy and preservation of the sample integrity) [7]. The obtained data provided interesting information about the materials response on blast load and was used for adjustment of the material model in subsequent numerical simulations. The aim of numerical investigation was to find optimal dimensions and wall composition of the litter bin, which will be able to resist 1000 g of TNT charge detonated inside the vessel. Numerical investigation was performed using implicit/explicit solver LS-Dyna.

Based on the findings obtained during the numerical investigation, real litter bin was manufactured and subjected to the blast load. Test procedure according to the methodology M-VTÚO 11/11 [8] was applied to evaluate the blast resistance of the litter bin. The test layout is depicted in Fig. 2. At the distance of 1,5 m from the axis of the tested object, the check panels are deployed and fixed to the compacted sand ground. Test panels consist of steel frame, which holds hardboard-polystyrene-aluminium plate sandwich. The dimensions of the check panel are 1 × 2 m. Semtex C4 plastic high explosive weighting 770 g (equivalent of 1000 g of TNT) was used as a testing charge. The integrity of the tested object, its global displacement, creation of secondary fragments and condition of the check panels are observed and evaluated. To declare the test as satisfactory, the global displacement of the specimen must not exceed 1 m, the overall integrity of the object must be preserved and the check panels must not be penetrated by secondary fragments.

3. Test results and discussion

3.1. Physico-mechanical parameters

Bulk density, flexural and compressive strengths at 28 days of age were determined. The average values of at least 5 specimens of the same mixture are presented in Table 2.

Both flexural and compressive strength are influenced by fibre incorporation. Almost all mixtures with fibres showed reduced compressive strength values in comparison with REF specimen without fibres. The opposite effect of fibre incorporation was observed in the case of flexural

Table 2. Physico-mechanical parameters of the specimens.

Designation	Compressive strength (MPa)	Flexural strength (MPa)	Bulk density (MPa)
REF	**110.3**	**6.3**	**2,244**
G100	**98.1**	**18.8**	**2,136**
B50	88.4	6.4	2,144
S400	115.0	18.1	2,424
P4	99.8	8.7	2,171
G50B37	86.9	11.3	2,136
G50P10	93.8	15.4	2,150
G50B37P2	82.2	13.2	2,134
G40B20PP4	94.5	11.5	2,152
S200G50	92.2	22.1	2,300
S200B30	109.6	22.3	2,276
S130G22B17P1	**117.9**	**23.3**	**2,255**

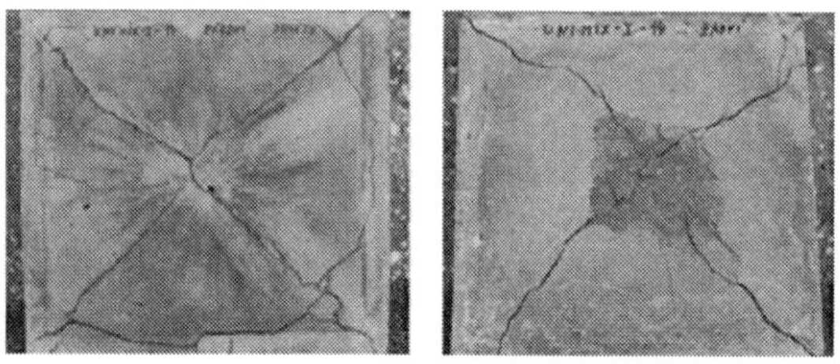

Figure 3. Specimen S130G22B17P after blast test. Front side (left), back side (right).

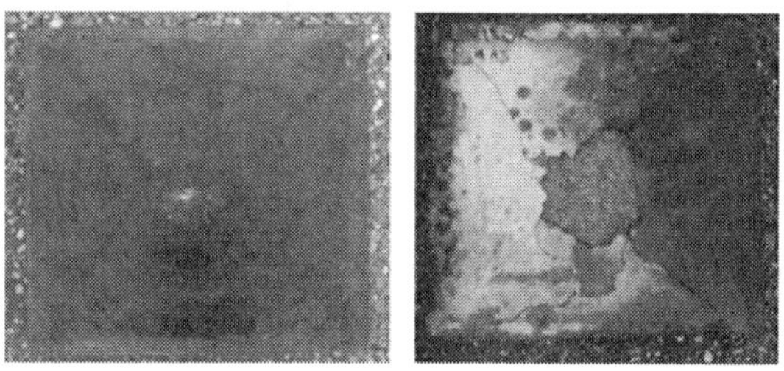

Figure 4. Specimen G100 after blast test. Front side (left), back side (right).

strength – any fibre brought enhancement of the flexural strength, with high scatter of values dependent on fibre type. The highest flexural strength values were achieved using fibre blends containing steel fibres.

3.2. Blast tests conducted on slab specimens

Blast test according to the methodology MT0-B-VTUO 10/9 described above was conducted with three selected mixtures signed bold in Table 2. The mixtures were selected due to their extraordinary mechanical properties, in particular flexural strength. Two slabs of each mixture were examined.

The average values of overall blast resistance of the materials evaluated and calculated according to the methodology MT0-B-VTUO 10/9 are graphically depicted in Fig. 6. The photographs of the slabs are shown in Figs. 3, 4 and 5.

The best blast resistance was achieved using glass fibres as a reinforcement (Specimen G100). The test

Figure 5. Specimen REF after blast test.

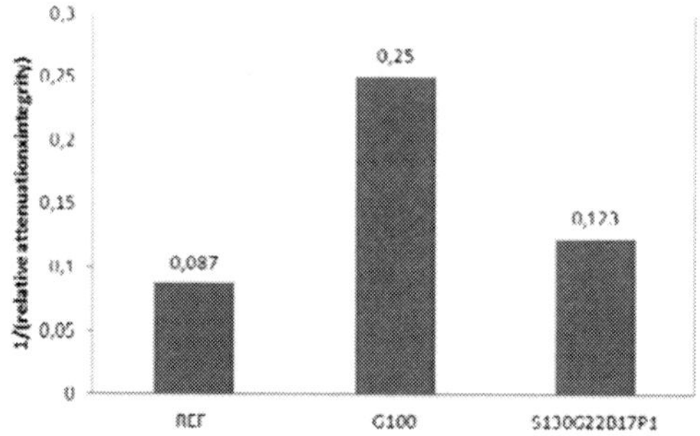

Figure 6. Explosion resistance rate of three selected specimens.

specimen remained integral after the blast test, on the front side only hair cracks appeared, from the bottom side one flat fragment was ejected (weight 85 g). The specimen with hybrid reinforcement (glass, steel, polypropylene and basalt) showed lower blast resistance, probably due to demanding production procedure connected with difficulties in achieving the good homogeneity of the material. Large cross cracks were created on both sides of the test slab and one spall (weight 180 g) was separated from the specimen, without overall disintegration of the slab. The blast test performed on the REF specimen without reinforcement results in total damage of the specimen (see Fig. 5).

3.3. Numerical investigation and real element blast test

In this step, the aim of the research work was to find dimensional parameters and structural composition of the litter bin, which will be able to resist the blast wave created by detonation of 770 g C4 explosive (equivalent of 1000 g of TNT) without fragmentation. Material G100 fine-grained high performance concrete reinforced with glass fibres was selected for the numerical investigation and subsequent blast tests, as it exhibited excellent workability and mechanical properties, very good production technology and explosion resistance.

RHT model was selected for simulations of the concrete material. All data obtained from tests described above was used to adjust the material model. Bilinear plastic kinematic model was used for laminate layer. The design of the vessel geometry and dimensions was based on previous research for lower level 250 g of TNT, which indicates the minimum required wall thickness of 200 mm. The litter bin was dimensioned in accordance with these requirements and due to high level of load, the layer of

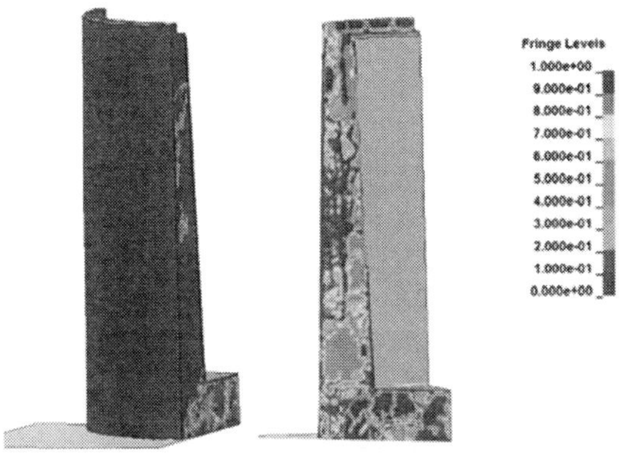

Figure 7. Litter bin after explosion detonated in the middle of the vessel (concrete with inner laminate layer 3.5 mm and outer laminate layer 1 mm).

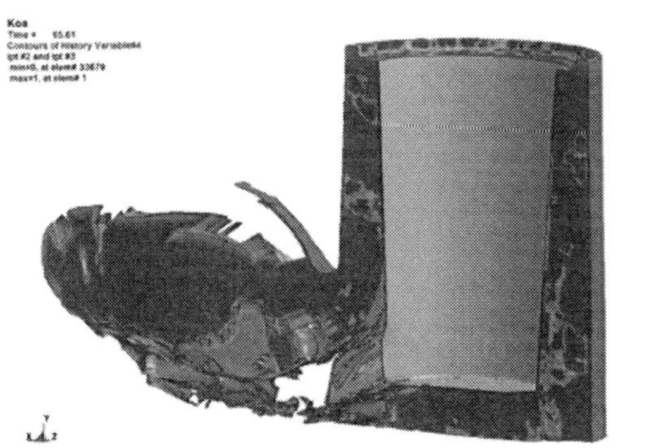

Figure 8. Litter bin with outer laminate layer 1 mm after explosion detonated at the bottom of the vessel.

laminate (both inside and outside) was added. Inner layer thickness was 3.5 mm, outer layer 1 mm. At the first step, the charge was placed in the middle of the vessel's free space. The results of numerical simulations showed, that even though the partial failure of concrete occurred, the laminate shell transmitted created tensile strength and the litter bin remained integral (see Fig. 7).

In the next step, the charge was placed at the bottom of the vessel and it was investigated if the wall would be able to transmit the strain created by detonation of such placed explosive.

As seen in Fig. 8, the litter bin resistance was insufficient, the outer layer failure accompanied with massive fragmentation occurred. The outer laminate layer thickness was increased to 5 mm and the same test was carried out (see Fig. 9). According to results of numerical simulations, the litter bin with given dimensions, geometry and composition should resist the blast load of 1000 g TNT charge. Subsequent real blast test of manufactured concrete litter bin (carried out according to the methodology M-VTUO 11/11 described above) confirmed the results of numerical simulations. The inner layer of the vessel was destroyed, while the outer layer showed neither damage nor cracks. No global

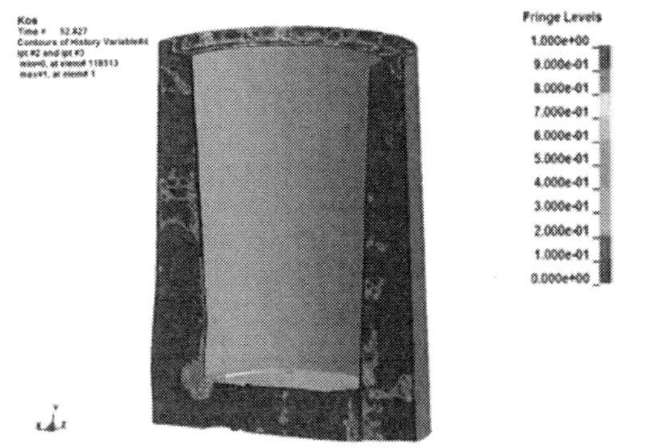

Figure 9. Litter bin with outer laminate layer 3.5 mm after explosion detonated at the bottom of the vessel.

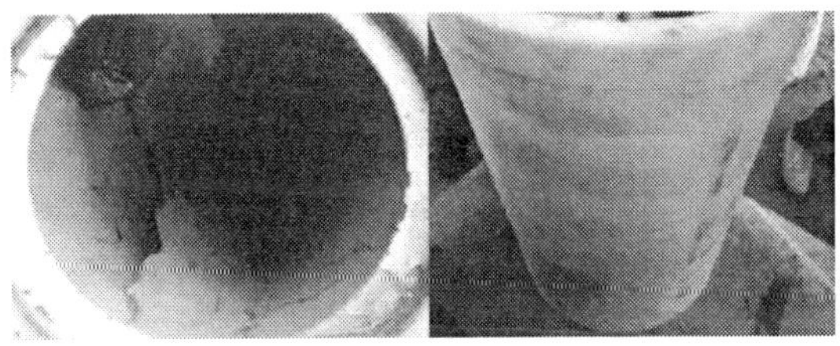

Figure 10. Litter bin after explosion.

displacement of the bin was observed, the check panels were clean without any damage caused by penetration of the fragments.

4. Conclusions and summary

Based on the results of this experimental investigation, the following conclusions are drawn:

To ensure the same volume fraction of the fibres in the matrix, very different mass fraction of fibres was required. Despite the volume fraction of fibres was constant for all mixtures, the consistency varied considerably. Some of the fibres at a dosage of 3 vol.% of fresh mixture were almost unworkable (basalt), other fibres provided good workability even when added in higher amounts (steel). It was observed that the type of incorporated fibres affects flexural strength significantly.

With the addition of any fibres and fibre blends used, the flexural strength was enhanced. Very different contribution to the strength enhancement was observed for each fibre type. The improved mechanical performance (particularly flexural strength) together with improved fracture performance caused by the reinforcement brought about enhanced explosion resistance as well. The level of the enhancement depends on the incorporated fibre type. Extensive further research will be focused on investigation of the relations between the fibre properties (length, tensile strength, modulus of elasticity) and blast resistance of concrete reinforced with them.

Good agreement was concluded between the results of the blast experiments and simulations. Adjusted RHT model can be successfully used for simulations of the fibre-concrete material. Numerical simulation can be successfully applied when reduction of the costs connected with development of blast resistance materials and products is required.

The authors wish to express their gratitude and sincere appreciation to the authority of The Technology Agency of the Czech Republic, project No. TA02000162 for financial support.

References

[1] E.K. Schrader, Impact resistance and test procedure for concrete, ACI J **78** (1981)

[2] Z.S. Tabatabaei, J.S. Volz, J. Baird, B.P. Gliha, D.I.Keener, Int J Impact Eng **57** (2013)

[3] H. Su, J. Xu, Constr Build Mat **45** (2013)

[4] E. Musselman, Blast and impact resistance of carbon fibre reinforced concrete. Ph.D dissertation. (advisor Andres Schokker), The Pennsylvania State University. University park, Pennsylvania, USA, 2007

[5] M. Maalej, S.T. Qeck, J. Zhang, J Mat Civil Eng **17** (2005)

[6] J. Křesťan, J. Krátký, A. Dvořák, Zkušební metoda MT0-B-VTUO 10/9 Hodnocení protivýbuchové odolnosti a míry absorpce rázové vlny ve vzorku na zkušebním stavu T0-B 10/09 (2009)

[7] H. Szklorzova, R. Holešinský, J. Krátký, J. Křesťan, S. Rolc, L. Smetanová. In Proceedings Recent advances in cement and concrete (2012)

[8] Belko, J., Čalkovský, P., Křesťan, J. Testování bezpečnostních objektů M-VTÚO 11/11, Metodika VTÚO Brno (2011)

EPJ Web of Conferences **94**, 05007 (2015)
DOI: 10.1051/epjconf/20159405007

Experimental analysis and constitutive modelling of steel of A-IIIN strength class

Leopold Kruszka[1,a] and Jacek Janiszewski[2]

[1] Military University of Technology, Faculty of Civil Engineering and Geodesy, 2 Gen. Sylwester Kaliski St., 00-908 Warsaw, Poland
[2] Military University of Technology, Faculty of Mechatronics and Aviation, 2 Gen. Sylwester Kaliski St., 00-908 Warsaw, Poland

Abstract. Fundamentally important is the better understanding of behaviour of new building steels under impact loadings, including plastic deformations. Results of the experimental analysis in wide range of strain rates in compression at room temperature, as well as constitutive modelling for and B500SP structural steels of new A-IIIN Polish strength class, examined dynamically by split Hopkinson pressure bar technique at high strain rates, are presented in table and graphic forms. Dynamic mechanical characteristics of compressive strength for tested building structural steel are determined as well as dynamic mechanical properties of this material are compared with 18G2-b steel of A-II strength class, including effects of the shape of tested specimens, i.e. their slenderness. The paper focuses the attention on those experimental tests, their interpretation, and constitutive semi-empirical modelling of the behaviour of tested steels based on Johnson-Cook's model. Obtained results of analyses presented here are used for designing and numerical simulations of reinforced concrete protective structures.

1. Introduction

Examining the behaviour of structural construction steels in a wide range of deformation rates is important for solving the problems of ballistic impacts and penetrations at high-speeds, the dynamics of destruction and the safety of reinforced concrete and steel structures of protective and defensive buildings exposed to exceptional impacts. Nowadays, contemporary protection structures constitute neuralgic elements of critical infrastructure, in particular of public buildings and important engineering structures. These structures are designed to reduce the effects of a terrorist attack using both explosive materials and projectiles on the protected building structures and prevent the objects from construction disaster. The calculation of the ultimate limit state and usability of these supporting structures of buildings to dynamic loads using computer methods require knowledge of the constitutive equations describing the behaviour of the used construction materials. Constitutive equations even in a simplified form as a rule contain a number of model and material parameters to be determined experimentally.

Nowadays, for the design and construction of modern defensive and protective structures, new constants B500SP and St3SY-b-500 in the highest A-IIIN strength class are used. Their basic static properties by standards are given in the table No 1.

Since these steels have been recently introduced to the practice of engineering in civil engineering, in the available literature, there is no data on both the dynamic mechanical properties of these materials and their material and model constants for any constitutive equation in a wide range of strain rates. As shown by previous studies [1,2], for most metals and their alloys, differences in their behaviour during static and dynamic deformations are so clear that the results of static tests cannot be used to analyse and solve dynamic problems. Hence, there is an increase in demand for comprehensive experimental studies, especially for the dynamic properties of modern structural construction steels.

Table 1. Static mechanical properties of structural steels B500SP and St3SY-b-500.

The initial yield point [N/mm^2]	500
Tensile strength [N/mm^2]	575
Modulus of longitudinal elasticity [GPa]	200
Standards	PN-H 93220:2006, PN-EN 0080:2005(U)

Experimental studies of the behaviour of the structural steel of B500SP and, for comparison purpose, 18G2-b steel of A-II strength class during the uniaxial stress state compression at high speeds of deformation were carried out using the compression Hopkinson bar test technique. The end result of this research work, having the aspect of application, is the proposed Johnson-Cook model for the given steels, which is widely used in the computer codes for the numerical analysis of engineering structures dynamically loaded. It should be noted that the design issues of protective structures and strengthening of elements of supporting structures of public buildings sensitive to the terrorist attacks, belong to "the engineering combat against terrorism" [3].

[a] Corresponding author: lkruszka@wat.edu.pl

Figure 1. Compression split Hopkinson pressure bar used in investigations.

2. Static and dynamic strength testing procedure

To provide base-line material properties, materials compression strength tests under quasi-static loading were carried out. Quasi-static compression tests were performed on MTS Criterion 45 testing machine using specimens with 9 mm in diameter and 3, 6 and 9 mm length. The strain-rate was maintained at $10^{-3}\,s^{-1}$. Material samples for both quasi-static and dynamic compression strength tests were machined from the same bars with diameter of 12 mm.

A classical compression split Hopkinson pressure bar was applied to studied high-strain-rate mechanical response of given steels. The apparatus presented in Fig. 1 consists mainly of a striker launching system (air pressure gun), a striker, an input bar, an output bar (bar system), a velocity measuring device and a computer-controlled high-frequency data acquisition system.

The input bar and the output bar were $1218 \pm 0,25$ mm each long, while the striker length was $200 \pm 0,05$ mm. Both the bars and the striker had a common diameter of 12.05 mm and were made of commercial maraging steel grade 350, which was heat treated to guarantee high strength property of bars (nominal quasi-static yield strength $R_{0.2} = 2320$ MPa, Young modulus $E = 190,6$ GPa, sound speed $C_o = 4866$ m/s). Each bars were supported by 4 linear bearing stands, which were mounted on optical bench allowing precise alignment of the bars system.

The elastic deformation signals in the input and output, transmitted bars were captured using pair of strain gages attached symmetrically on the opposite surfaces of the bars and in theirs middle length. The strain gages were connected to the opposite legs of the Wheatstone bridge, which was a typical full bridge configuration. In the other legs of the tensometer bridge, the dummy resistors were mounted, which resistance matched with the strain gages resistance. The typical electrical strain gages of 1.6 mm gage length were used (CEA-13-062UW-350, Vishay Micro Measurements). The amplified signals of the strain gages were recorded at a frequency of 1 MHz with the use of a signal conditioning unit (SGA-0B V5 Wheatstone bridge with signal conditioning amplifiers, ESA Messtechnik) and LeCroy WJ354A high-speed digital oscilloscope (data acquisition system).

In order to achieve an appropriate loading condition for high strain rate tests (e.g. stress equilibrium) a 4 mm diameter wave pulse shapers made of Cu-ETP copper sheet were usually used. The wave pulse shapers were manufactured from copper sheet with the use of *sheet metal punching technique*. In dependence on impact velocity of striker bar, the thickness of pulse shaper was 0.1, 0.2, 0.3 and 0.4 mm, respectively. For the example, waves profiles obtained for dynamic experiments without and with pulse shaper were presented in Fig. 2.

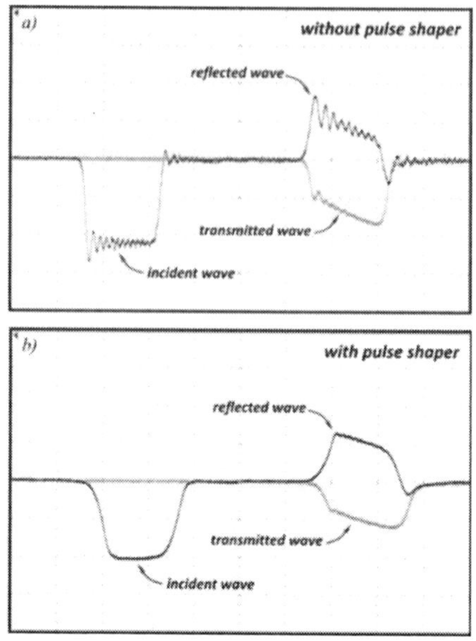

Figure 2. Comparison waves profiles obtained during Hopkinson experiments without (a) and with 0.3 mm thick wave pulse shaper (b).

The high-strain-rate investigations were carried out on specimens with different slenderness (L/Dratio, where L and D – length and diameter of specimen), which was 0.25, 0.5, 0.75 and 1, respectively. As a consequence, influence of specimen slenderness on plastic flow stress was also estimated. At least three dynamic tests were performed for each specimen geometry.

3. Results and discussion

3.1. Mechanical response of tested structural steels

Compilation of the quasi-static and dynamic strength tests for different slenderness of specimen is presented in Fig. 3. It can be seen clearly a minor impact of specimen slenderness on true stresses determined at 0.05 strain. Only in the case of specimens with slenderness of 0.25 it was found higher level of plastic flow stresses for both type of steels in comparison to other specimen slendernesses. Therefore further considerations on obtained results were made without taking into account of experimental data for specimens with the 0.25 slenderness.

On the basis of results presented in Fig. 3 and also in Fig. 4, it is observed a significant difference in mechanical strength for both tested materials. The B500SP steel is characterized by higher plastic flow stress than 18G2-b

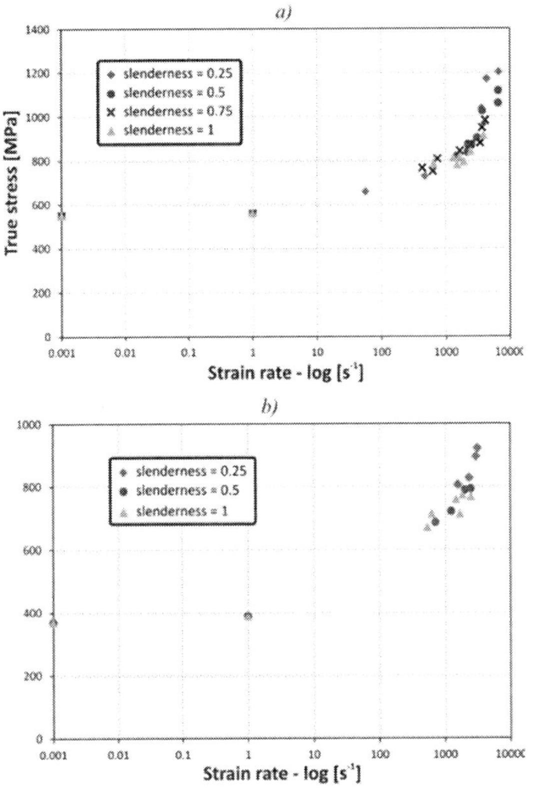

Figure 3. Specimen slenderness influence on plastic stress flow of the B500SP steel (a) and the 18G2-b steel (b) under different loading conditions.

steel both under static and dynamic loading conditions. The true stresses at small strains for B500SP steel are about 200 MPa higher than ones for 18G2-b steel. However, for strains above 0.15, differences in stresses decreases to about 100 MPa. Figure 4 shows also that character of stress changes is slightly variant: the original peak of stress for B500SP steel has oscillating nature, whereas in case of second structural steel the beginning of plastic flow seems to be smooth. Moreover, it was found the B500SP steel has got similar strain hardening sensitivity to the 18G2-b steel. In turn, strain rate sensitivity of the B500SP steel seems to be lower in comparison to 18G2-b steel. It is proved by results presented both in Fig. 4 and especially in Fig. 5. For quasi-static and moderate range of strain rate, B500SP steel is insensitive to strain rate, whereas 18G2-b steel demonstrates modest strain rate sensitivity. Above strain rate level of $10^3\ s^{-1}$, flow stress increase for B500SP steel seems to be also relatively smaller than for 18G2-b steel.

3.2. Determination of Johnson-Cook constitutive model parameters

In order to assess the Johnson-Cook materials parameters for tested steels, the experimental data was fit to the

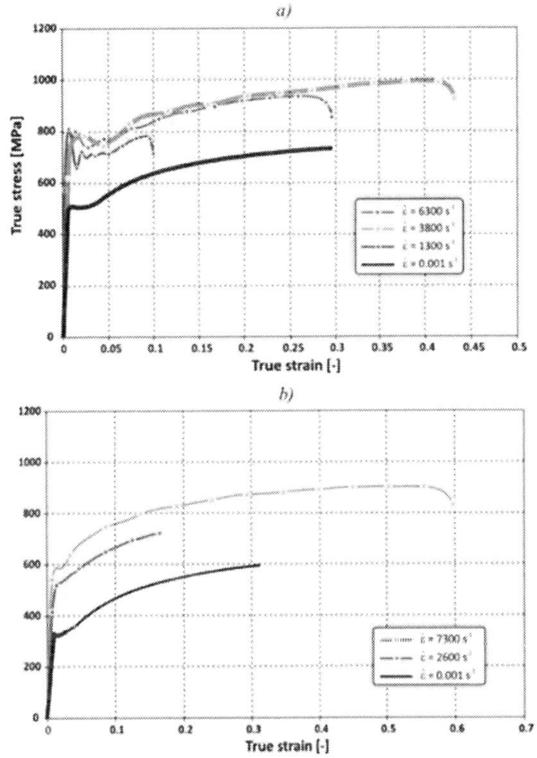

Figure 4. Comparison of stress-strain curves of the B500SP (a) and the 18G2-b steel (b) under different loading conditions.

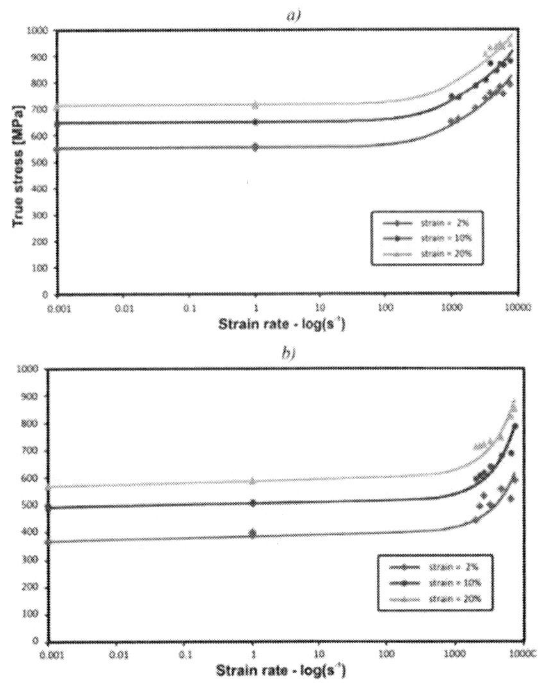

Figure 5. Flow stress strain rate sensitivity at 2/10/20% true plastic strain for B500SP (a) and 18G2-b (b) steels.

Table 2. Johnson-Cook constitutive model parameters for B500SP and 18G2-b steels.

Steel	*A*[MPa]	*B* [MPa]	*n*	*C*
18G2-b	340	625	0.40	0.024
B500SP	530	820	0.49	0.013

simplified Johnson-Cook constitutive model (Eq. (1)) using a least-squares nonlinear regression procedure (*genfit function*) in Mathcad software.

$$\sigma = (A + B\varepsilon^n)(1 + C\ln(\dot{\varepsilon})) \tag{1}$$

where: A, B, n and C are material constants. A is the yield stress for $\dot{\varepsilon}_c = 1\,s^{-1}$, B and n represents the strain hardening, C is the strain rate constant [4].

It should be noted here that the model parameters identification was made for B, n and C constants, whereas a value of A was estimated on the base of quasi-static compression test. Moreover, the procedure of finding parameters was performed on three experimental stress-strain curves, which were obtained for strain rate range of $1200 \div 3000\,s^{-1}$. The best sets of J-C parameters were the result of the compromise between fitting errors determined for all selected experimental curves. Values of J-C material constants for B500SP and 18G2-b steels are presented in Table 2.

In order to illustrate the convergence of obtained results, the experimental flow stress curves for both steels at various strain rates was compared in Fig. 6 with flow stress data generated using the J-C constitutive equation with identified constants. It can be seen from this figure that flow stress computed using J-C model is in a satisfactory agreement with experimental data both in the case of B500SP steel and 18G2-b.

4. Summary

Dynamic stress – strain curves in compression of tested both steels were obtained by split Hopkinson pressure bar using a pulse shaper technique. The effects of strain rate on mechanical properties of those materials were also briefly addressed. The parameters of material Johnson-Cook model was identified by these results. Those two steels exhibit similar strain hardening as well as strain rate hardening.

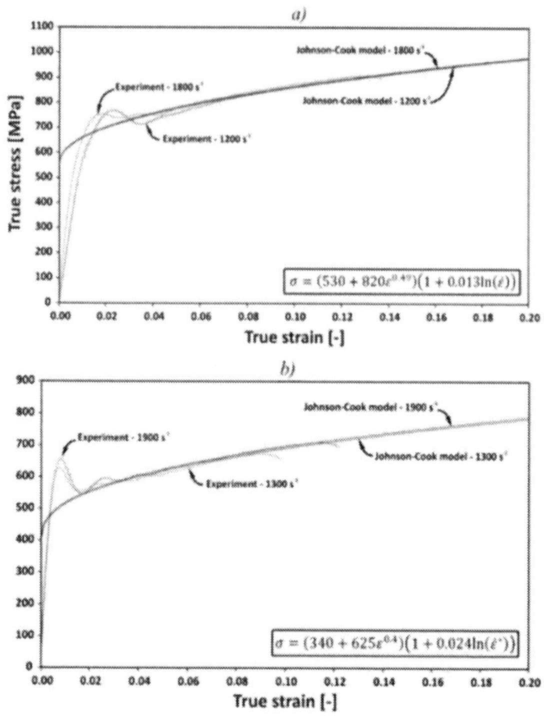

Figure 6. Comparison of Johnson-Cook model predictions with selected experimental results for B500SP (a) and 18G2-b steel (b).

The authors thank to Mr Jakub Bolanowski for his assistance in this work.

References

[1] M.A. Meyers, Dynamic Behaviour of Materials (Johns Wiley and Sons, INC, New York-Chichester-Brisbane-Toronto-*Singapoure*, 1994).

[2] J.A. Zukas, ed., High-Velocity Impact Dynamics *(Wiley, New York*, 1990).

[3] L. Kruszka, R. Rekucki, Applied Mechanics and Materials, **82** (2011).

[4] G.R. Johnson, W.H. Cook, Proceedings of the 7th International Symposium on Ballistics, The Hauge, The Netherlands, April (1983).

EPJ Web of Conferences **94**, 05008 (2015)
DOI: 10.1051/epjconf/20159405008

Comparative experimental study of dynamic compressive strength of mortar with glass and basalt fibres

Leopold Kruszka[1,a], Wojciech Moćko[2], Luigi Fenu[3], and Ezio Cadoni[4]

[1] Gen. Jaroslaw Dabrowski Military University of Technology, 2 Gen. Sylwester Kaliski St., 00-908 Warsaw, Poland
[2] Motor Transport Institute, 80, Jagiellonska St., 03-301 Warsaw, Poland
[3] University of Cagliari, 09123 Cagliari, Italy
[4] University of Applied Sciences of Southern Switzerland, DynaMat Laboratory, 6952 Canobbio, Switzerland

Abstract. Specimen reinforced with glass and basalt fibers were prepared using Standard Portland cement (CEM I, 52.5 R as prescribed by EN 197-1) and standard sand, in accordance with EN 196-1. From this cementitious mixture, a reference cement mortar without fibers was first prepared. Compressive strength, modulus of elasticity, and mod of fracture were determined for all specimens. Static and dynamic properties were investigated using Instron testing machine and split Hopkinson pressure bar, respectively. Content of the glass fibers in the mortar does not influence the fracture stress at static loading conditions in a clearly observed way. Moreover at dynamic range 5% content of the fiber results in a significant drop of fracture stress. Analysis of the basalt fibers influence on the fracture stress shows that optimal content of this reinforcement is equal to 3% for both static and dynamic loading conditions. Further increase of the fiber share gives the opposite effect, i.e. drop of the fracture stress.

1. Introduction

Concrete is considered as a brittle material. Its tensile strength is relatively low in comparison with compressive strength. In order to increase tensile properties of mortar reinforcement in a form of fibers may be included. As a consequence mortar matrix composite reinforced with fibers is created. Various fibres may be applied for the purposes of matrix strengthening, i.e. steel, glass, synthetic, basalt, carbon, organic. Diameter of reinforcement is usually within the range from 0.2 mm to 2 mm, whereas length is between 25 mm and 60 mm. Fibers are homogenously distributed in random directions inside the matrix, however there are some special solutions where fibers are placed along major directions.

Optimal content of reinforcement may substantially increase tensile strength of mortar composite, and slightly increase compressive behaviour. Moreover fiber presence may influence stress-strain behaviour of material. Since matrix material without reinforcement shows brittle behaviour, therefore fracture is rapid without capability to take a load after crack initiation. Addition of fiber may change this behaviour because after fracture of matrix reinforcement is responsible for carrying the load.

Basalt fibers has very good physical and mechanical properties therefore mortar based composites reinforced with this material are often used in special structures, i.e. nuclear reactors covers or high buildings elevations. Since high costs of basalt fibers application of this types of composites is limited. The second discussed material for reinforcement, i.e. glass fibre has lower tensile strength, higher elongation and three times lower costs in comparison with basalt one. However it is not corrosive resistant what substantially reduces its practical application.

Dynamic mechanical properties of fibre reinforced composites are described by compressive dynamic increase factor (CDIF) and tensile dynamic increase factor (TDIF). In order to determine high strain rate properties split Hopkinson pressure bar (SHPB) methodology is usually applied. This technique was applied to estimate compressive and bending strength of steel fibers reinforced mortar [1]. It was found that increase of steel fibers content up to 1.5% increases both compressive (from 51 MPa to 75 MPa) and bending strength (from 6 MPa to 10 MPa). It was also confirmed, that fiber addition prevents brittle fracture of composite. Dynamic tests [2] done with use of pre-tension Hopkinson bar show that high performance fibre reinforced concrete properties may be strongly affected by high temperature. It is extremely important in many civil engineering and military applications where structure resistance against dynamic loadings may be lowered by fire.

This work presents analysis of compressive strength of concrete reinforced by basalt and glass fibres under quasi-static and dynamic loading conditions.

2. Experimental method

Both types of specimen (reinforced with glass and basalt fibers) were prepared using Standard Portland cement (CEM I, 52.5 R as prescribed by EN 197-1) and standard sand, in accordance with EN 196-1. From this cementitious mixture, a reference cement mortar (water/cement ratio: 0,5) without fibers was first prepared. The fiber reinforced specimens were then prepared by adding a given fiber

[a] Corresponding author: lkruszka@wat.edu.pl

Article available at http://www.epj-conferences.org or http://dx.doi.org/10.1051/epjconf/20159405008

Table 1. Flexural tensile and compressive strength of reference and fiber reinforced mortars obtained through static tests.

Material	Flexural tensile strength [MPa]	Compressive strength [MPa]
Reference mortar	$6{,}57 \pm 0{,}13$	$70{,}20 \pm 1{,}65$
Glass fiber 3%	$7{,}78 \pm 0{,}31$	$57{,}58 \pm 2{,}44$
Glass fiber 5%	$8{,}37 \pm 0{,}69$	$50{,}46 \pm 2{,}13$
Basalt fiber 3%	$8{,}72 \pm 0{,}97$	$57{,}15 \pm 0{,}90$
Basalt fiber 5%	$8{,}53 \pm 0{,}63$	$52{,}10 \pm 1{,}42$

content (in weight) of glass and basalt fibers to the cementitious mixture. Glass fibers and basalt fibers with same size (12 mm in length and 14 μm in diameter) were used, thus allowing to better compare the respective performance in fibre reinforced concrete under dynamic loading conditions, taking into account applications of those materials to protective structures. For the same reason, both types of specimens were prepared with same fiber content (3% and 5% in weight). Four different types of fiber reinforced mortar were hence obtained. Compressive strength, modulus of elasticity, and mod of fracture were determined for all specimens. Basic quasi-static parameters of tested materials are shown in Table 1. Static and dynamic properties of tested specimens were investigated experimentally using universal Instron 8802 testing machine equipped with Aramis measure system, and split Hopkinson pressure bar equipped in Phanton v1210 high speed camera, respectively. Quasi-static tests were carried out at strain rate equal to 0.005 1/s whereas dynamic tests were performed at strain rate equal to 800/s.

Split Hopkinson Pressure Bar [3–9], presented in Fig. 1, was equipped with incident and transmitter bars 20 mm in diameter and 2000 mm in length, which were made of high strength maraging steel, $\sigma_y = 2100$ MPa. The signals acquired from the strain gauges were amplified by the wideband bridge Vishay amplifier and digitized by an oscilloscope. The initial velocity of the striker, which was accelerated in a pressure gas launcher was measured by two sets of diodes and photo detectors coupled to a digital counter. Based on the waveforms recorded by a digital oscilloscope for transmitted ε_T (t) and reflected ε_R (t) waves and the known cross sectional area of the bars A and the specimen A_S, the speed of the elastic wave propagation in the material of the bars C_0 and the test-piece length L, it is possible to determine stress $\sigma(t)$, strain $\varepsilon(t)$ and strain rate $\dot{\varepsilon}(t)$ in the specimen using the following formulas:

$$\delta(t) = E\left(\frac{A}{A_S}\right)\varepsilon_T(t) \qquad (1)$$

$$\varepsilon(t) = -\frac{2C_0}{L}\int \varepsilon_R(t)dt \qquad (2)$$

$$\dot{\varepsilon}(t) = \frac{d\varepsilon(t)}{dt} = \frac{-2C}{L}\varepsilon_R(t). \qquad (3)$$

3. Results

The course of compression tests at quasi-static and dynamic loading conditions are shown if Figs. 2–6.

Figure 1. Split Hopkinson bar arrangement applied in the dynamic tests. 1 – digital oscilloscope, 2 – base, 3 – steel bars, 4 – pressure tank, 5 – tensometer bridge.

a)

b)

Figure 2. Deformation of reference mortar at (a) quasi-static loading conditions, (b) dynamic loading conditions.

Additionally quasi-static results were analysed using digital image correlation method. It may be observed that damage mode changes with strain rate, i.e. at quasi-static loadings number of leading cracks is lower than at dynamic loading conditions. This behaviour is independent on tested material.

Influence of reinforcement content on the compressive strength is shown in Fig. 7 and Fig. 8. Compressive strength of matrix determined at quasi-static conditions

a)

b)

Figure 3. Deformation of specimen reinforced with 3% of glass fiber at (a) quasi-static loading conditions, (b) dynamic loading conditions.

a)

b)

Figure 4. Deformation of specimen reinforced with 5% of glass fiber at (a) quasi-static loading conditions, (b) dynamic loading conditions.

a)

b)

Figure 5. Deformation of specimen reinforced with 3% of basalt fiber at (a) quasi-static loading conditions, (b) dynamic loading conditions.

a)

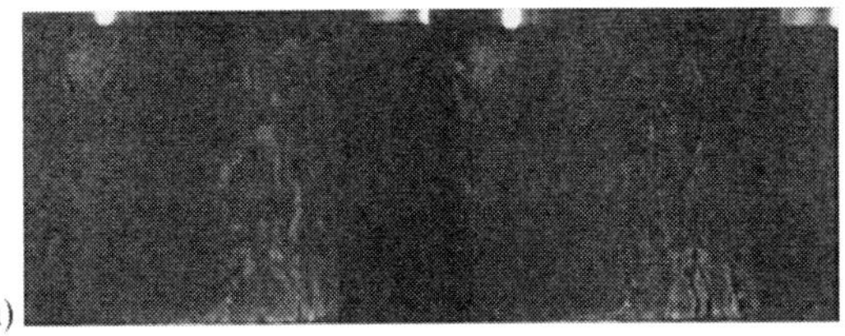

Figure 6. Deformation of specimen reinforced with 5% of basalt fiber at (a) quasi-static loading conditions.

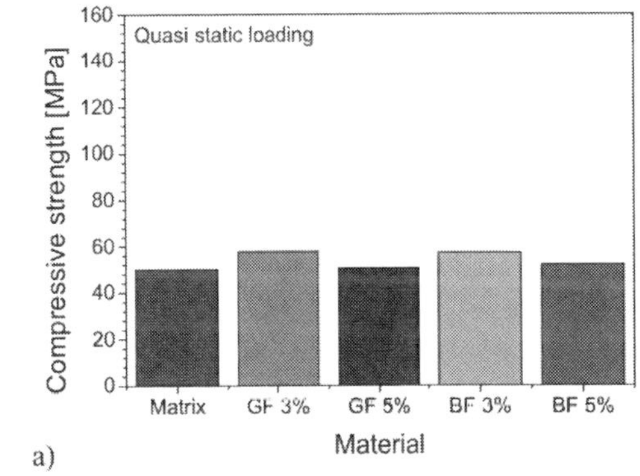

a)

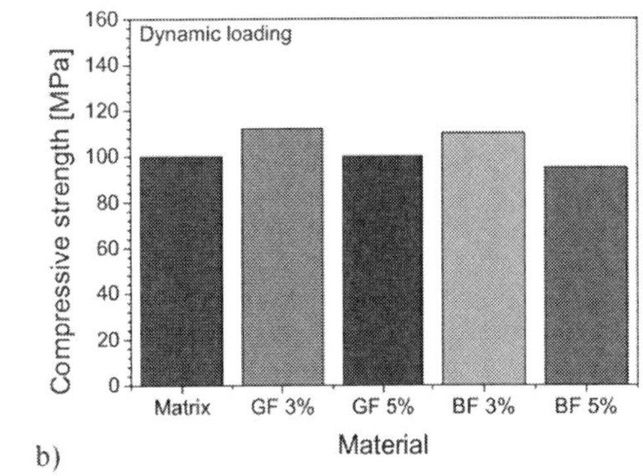

b)

Figure 7. Compressive strength of tested materials at (a) quasi-static loading conditions, (b) dynamic loading conditions.

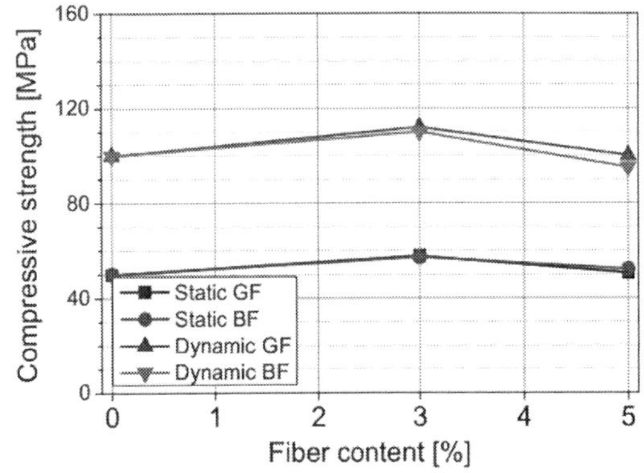

Figure 8. Influence of fiber content on the compressive strength of tested materials.

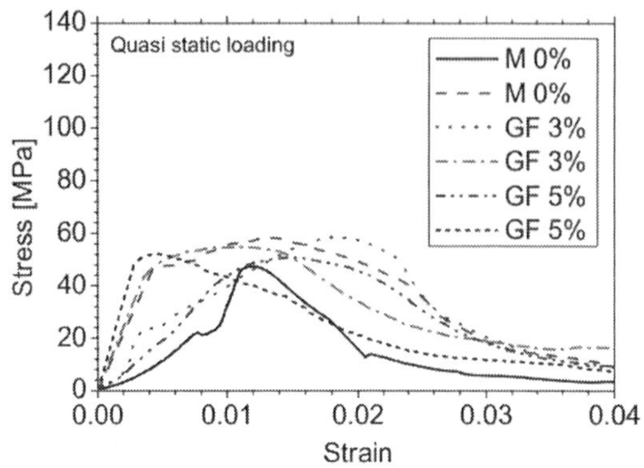

Figure 9. Stress-strain curves of reference and glass fibre reinforced mortar at quasi-static loading conditions.

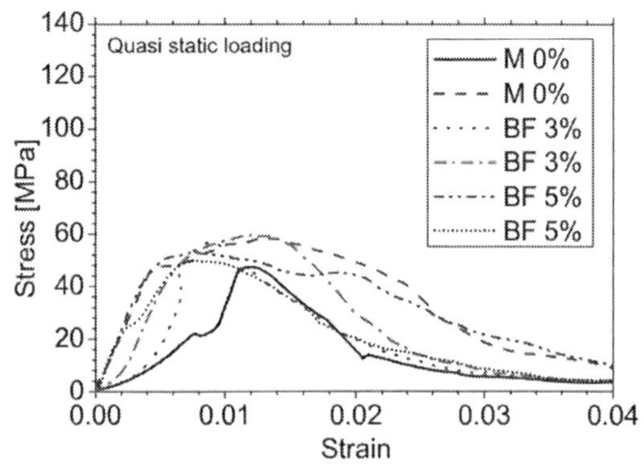

Figure 10. Stress-strain curves of reference and basalt fibre reinforced mortar at quasi-static loading conditions.

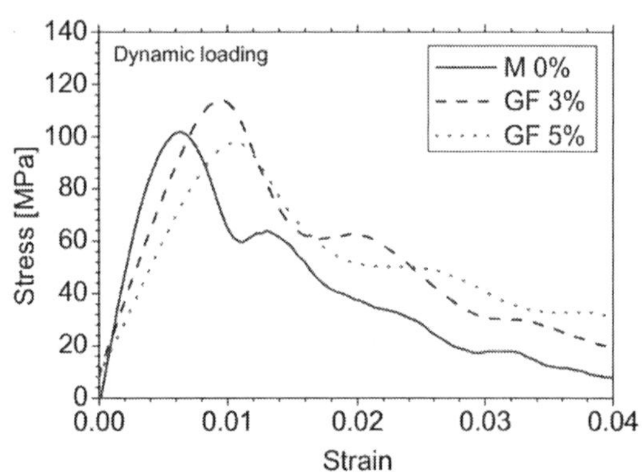

Figure 11. Stress-strain curves of reference and glass fibre reinforced mortar at dynamic loading conditions.

is equal to 50 MPa. Increase of strain rate up to 800/s induces increase of strength up to 100 MPa. Therefore dynamic strength is two time higher than quasi-static one. Fiber content equal to 3% gives strengthening effect at low at high strain rate deformation independently on fiber material. Further increase of fibre content results in opposite effect, i.e. drop of compressive strength.

Stress-strain curves of matrix and composites are presented in Figs. 9–12. It may be observed that in

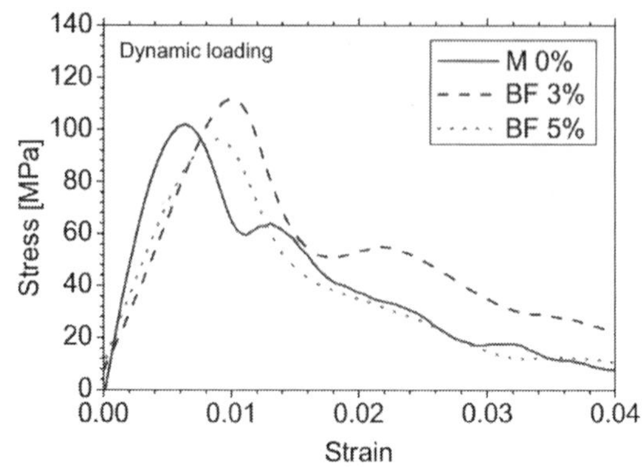

Figure 12. Stress-strain curves of reference and basalt fibre reinforced mortar at dynamic loading conditions.

the dynamic loading condition damage has more rapid character than at quasi-static one. As a consequence stress value after reaching maximum value quickly drops to 0. Effect of load carrying by reinforcement fibers after crack initiation was not observed.

4. Summary

Analysis of the basalt and glass fibers influence on the fracture stress shows that optimal content of this reinforcement is equal to 3% for both static and dynamic loading conditions. Further increase of the fiber share gives the opposite effect, i.e. drop of the fracture stress.

References

[1] T. Ch. Yet, R. Hamid, M. Kasamuri, Dynamic Stress-Strain Behaviour of Steel Fiber Reinforced High-Performance Concrete with Fly Ash, Advances in Civil Engineering **2012**, 1–6 (2012)

[2] A. Caverzan, E. Cadoni, M. Prisco, Behaviour of advanced cementitious composites under dynamic loading and fire, In: Proceedings of the International Workshop on Structures Response to Impact and Blast. Haifa. 2009.

[3] H. Kolsky, An Investigation of the mechanical properties of materials at very hign rates of deformation of loading. Proc. Phys. Soc. **62B**, 647–700 (1949)

[4] W. Moćko, J. Janiszewski, M. Grązka, Application of an extended Rusinek-Klepaczko constitutive model to predict the mechanical behavior of 6082-T6 aluminum under Taylor impact test conditions. J. Strain Analysis **48**, 364–375 (2013)

[5] W. Moćko, J. Janiszewski, J. Radziejewska, M. Grązka, Analysis of deformation history and damage initiation for 6082-T6 aluminium alloy loaded at classic and symmetric Taylor impact test conditions. Int. J. Impact Eng. **75**, 203–213 (2015)

[6] W. Moćko, Analysis of the impact of the frequency range of the tensometer bridge and projectile geometry

on the results of the measurement by the split Hopkinson pressure bar method. Metrol. Meas. Syst. **20**, 555–564 (2013)
[7] W. Moćko, J.A. Rodríguez-Martínez, Z.L. Kowalewski, A. Rusinek, Compressive viscoplastic response of 6082-T6 and 7075-T6 aluminium alloys under wide range of strain rate at room temperature: Experiments and modelling. Strain **48**, 498–509 (2012)
[8] W. Sumelka, Thermoelasticity in the Framework of the Fractional Continuum Mechanics, J. Therm. Stresses **37,** 678-706 (2014)
[9] W. Sumelka, Fractional viscoplasticity, Mech. Res. Commun. **56**, 31–36 (2014)

EPJ Web of Conferences **94**, 05009 (2015)
DOI: 10.1051/epjconf/20159405009

Effect of strain rate and temperature on mechanical properties of selected building Polish steels

Wojciech Moćko[1] and Leopold Kruszka[2,a]

[1] Motor Transport Institute, 80, Jagiellońska st., 03-301 Warsaw, Poland
[2] Gen. Jaroslaw Dabrowski Military University of Technology, 2, Gen. Sylwester Kaliski st., 00-908 Warsaw, Poland

Abstract. Currently, the computer programs of CAD type are basic tool for designing of various structures under impact loading. Application of the numerical calculations allows to substantially reduce amount of time required for the design stage of such projects. However, the proper use of computer aided designing technique requires input data for numerical software including elastic-plastic models of structural materials. This work deals with the constitutive model developed by Rusinek and Klepaczko (RK) applied for the modelling of mechanical behaviour of selected grades structural St0S, St3SX, 18GS and 34GS steels and presents here results of experimental and empirical analyses to describe dynamic elastic-plastic behaviours of tested materials at wide range of temperature. In order to calibrate the RK constitutive model, series of compression tests at wide range of strain rates, including static, quasi-static and dynamic investigations at lowered, room and elevated temperatures, were carried out using two testing stands: servo-hydraulic machine and split Hopkinson bar. The results were analysed to determine influence of temperature and strain rate on visco-plastic response of tested steels, and show good correlation with experimental data.

1. Introduction

In practical applications the cases of extreme loadings of building structure may be also considered, i.e. fire, earthquake or threats related to terrorism [1]. The last one are appears in recent years and may take a form of detonations and blast loadings or impact of large objects into building [2,3]. Since the mechanical properties of structural materials strongly depends on temperature and strain rate, therefore it is justifiable to take into account those relations during calculations [4]. Moreover, knowing precise material characteristics the FEM simulation of building behaviour under sudden impulse loadings of may by conducted.

This work deal with the constitutive model developed by Rusinek and Klepaczko (RK) applied for the modelling of mechanical behaviour of selected grades structural St0S, St3SX, 18GS and 34GS building steels and presents here results of experimental and empirical analyses to describe elastic-plastic behaviours of tested materials. In order to calibrate the RK constitutive model, series of compression tests at wide range of strain rates, including static, quasi-static and dynamic investigations at lowered, room and elevated temperatures, were carried out using two testing stands: servo-hydraulic machine and split Hopkinson bar. The results were analysed to determine influence of temperature and strain rate on visco-plastic response of tested steels, and show good correlation with experimental data.

Chemical composition of steels being analysed is presented in Table 1.

Table 1. Chemical composition of selected structural steels [5].

Steel class Grade	C max [%]	Si max [%]	P max [%]	S max [%]	M max [%]	Cr max [%]	Ni max [%]	Cu max [%]	Mo max [%]
A–0 St0S-b	0.23	0.40	0.07	0.06	1.0	–	–	–	–
A–I St3SX	0.22	0.07	0.05	0.05	–	–	–	–	–
A–II 18G2	0.22	0.55	0.05	0.05	1.5	0.3	0.3	0.35	0.1
A–III 34GS	0.22	0.70	0.07	0.05	1.2	0.3	0.3	0.35	0.1

2. Analysis of visco-plastic response of structural steels

In order to obtain stress-strain curves of selected steels at various temperatures, i.e. lowered and elevated ones, two various experimental methods were applied using thermal chambers. At low strain rate compression tests were conducted using a servo-hydraulic testing machine. An electro-mechanical extensometer was applied for the axial strain measurements. The specimens used in all of the tests performed were machined from round bars using the machining. Interfaces were lubricated using MoS_2 in order to reduce friction effects between the anvil and the specimen under compression.

At high strain rates well-known Split Hopkinson Pressure Bar methodology was applied. The signals acquired from the strain gauges were amplified by the wideband bridge circuit and digitized by an oscilloscope. The initial velocity of the striker, which was accelerated in

[a] Corresponding author: lkruszka@wat.edu.pl

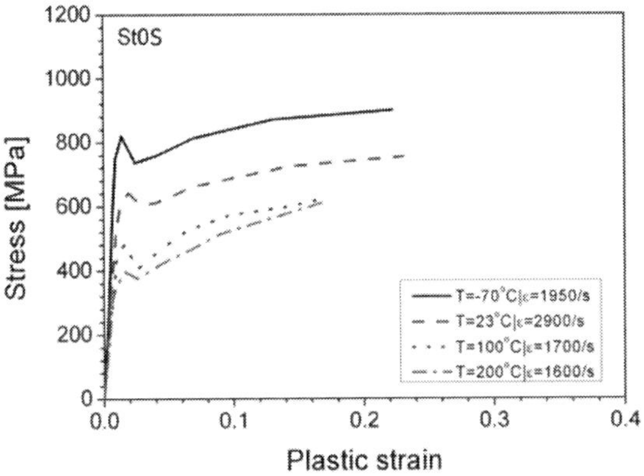

Figure 1. Dynamic stress-strain curves of St0S steel at wide range of temperatures.

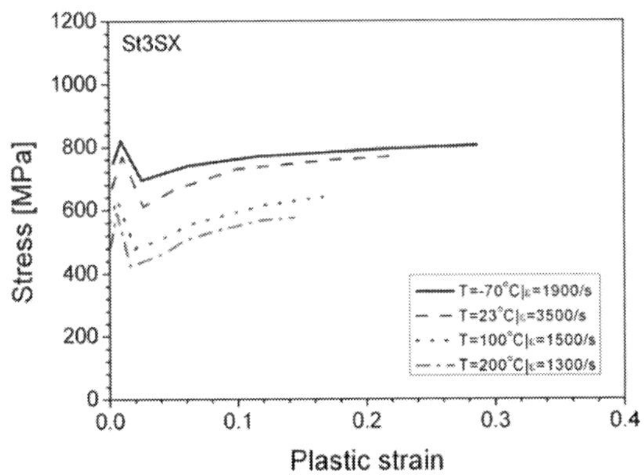

Figure 2. Dynamic stress-strain curves of St3SX steel at wide range of temperatures.

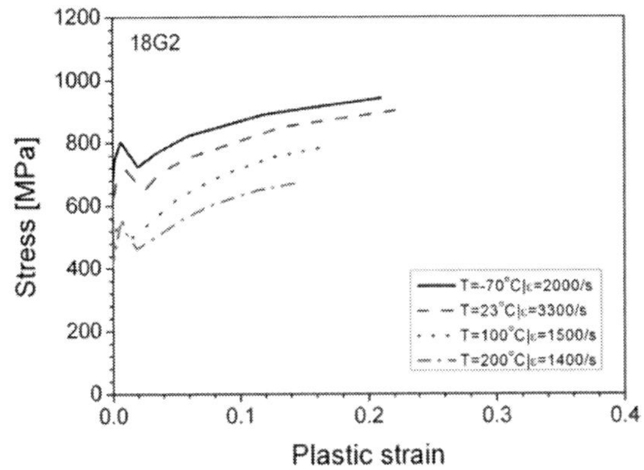

Figure 3. Dynamic stress-strain curves of 18G2 steel at wide range of temperatures.

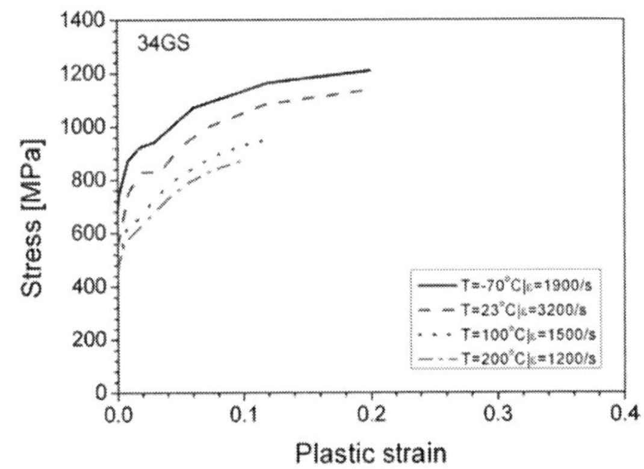

Figure 4. Dynamic stress-strain curves of 34GS steel at wide range of temperatures.

a pressure gas launcher was measured by two sets of diodes and photo detectors coupled to a digital counter. Based on the waveforms recorded by a digital oscilloscope for transmitted $\varepsilon_T(t)$ and reflected $\varepsilon_R(t)$ waves and the known cross sectional area of the bars A and the specimen A_S, the speed of the elastic wave propagation in the material of the bars C_0 and the test-piece length L, it is possible to determine stress $\sigma(t)$, strain $\varepsilon(t)$ and strain rate $\dot{\varepsilon}(t)$ in the specimen using the following formulas:

$$\sigma(t) = E\left(\frac{A}{A_S}\right)\varepsilon_T(t) \tag{1}$$

$$\varepsilon(t) = -\frac{2C_0}{L}\int \varepsilon_R(t)dt \tag{2}$$

$$\dot{\varepsilon}(t) = \frac{d\varepsilon(t)}{dt} = \frac{-2C_0}{L}\varepsilon_R(t). \tag{3}$$

The dynamic stress-strain characteristics of St0S, St3SX, 18GS and 34GS steels obtained at wide range of temperatures are presented in Figs. 1–4. The shape of curves for different steel grades is similar, however clearly visible differences in flow stress magnitude may be found. All tested grades of steel shows strong strain hardening effect, which may be observed as an increase of flow stress in a function of strain. The second mechanism which may be observed is strain rate hardening. It may be found that nature of flow stress increase in a function of strain rate is similar for all steel grades.

3. The Rusinek-Klepaczko constitutive model

The overall flow stress is decomposed into three stress components [6]: the internal stress $\bar{\sigma}_\mu(\bar{\varepsilon}^p, \dot{\bar{\varepsilon}}^p, T)$, the effective stress $\bar{\sigma}^*(\dot{\bar{\varepsilon}}^p, T)$ and the drag stress $\bar{\sigma}_d(\dot{\bar{\varepsilon}}^p)$, as shown in Eq. (4).

$$\begin{aligned}&\overline{\sigma}(\overline{\varepsilon}_p, \dot{\overline{\varepsilon}}_p, T) = \\ &\frac{E(T)}{E_0}\left[\sigma_\mu(\overline{\varepsilon}_p, \dot{\overline{\varepsilon}}_p, T) + \sigma^*(\dot{\overline{\varepsilon}}_p, T)\right] + \bar{\sigma}_d(\dot{\bar{\varepsilon}}_p).\end{aligned} \tag{4}$$

Because the Young's modulus is temperature dependent, it has the form shown in Eq. (5).

$$\mathrm{E(T)} = \mathrm{E_0}\left\{1 - \frac{\mathrm{T}}{\mathrm{T_m}}\exp\left[\theta^*\left(1 - \frac{\mathrm{T}}{\mathrm{T_m}}\right)\right]\right\} \tag{5}$$

where E_0, T_m and θ^* denote the Young's modulus at $T = 0\,\mathrm{K}$, the melting temperature and the characteristic homologous temperature, respectively. This expression allows the definition of the material thermal softening as a function of the crystal lattice [7], and empirically demonstrates the influence of the material structure on the temperature sensitivity.

The internal stress $\sigma_\mu(\bar{\varepsilon}_p, \dot{\bar{\varepsilon}}_p, T)$ is defined by a power-law type phenomenological function shown in Eq. (6).

$$\bar{\sigma}_\mu\left(\bar{\varepsilon}^p, \dot{\bar{\varepsilon}}^p, T\right) = B(\dot{\bar{\varepsilon}}^p, T)(\varepsilon_0 + \bar{\varepsilon}^p)^{n(\dot{\bar{\varepsilon}}^p, T)} \qquad (6)$$

where $B(\dot{\bar{\varepsilon}}^p, T)$ and $n(\dot{\bar{\varepsilon}}^p, T)$ are the modulus of plasticity and the strain hardening exponent, respectively, both of which are rate- and temperature-dependent, and is the strain level that defines the material yield stress at a specific strain rate and temperature.

The explicit formulations describing the modulus of plasticity $B(\dot{\bar{\varepsilon}}^p, T)$ and the strain hardening exponent $n(\dot{\bar{\varepsilon}}^p, T)$ are given in Eqs. (7)–(8).

$$B\left(\dot{\bar{\varepsilon}}^p, T\right) = B_0\left(\left(\frac{T}{T_m}\right)\log\left(\frac{\dot{\varepsilon}_{\max}}{\dot{\bar{\varepsilon}}^p}\right)\right)^{-\upsilon} \qquad (7)$$

$$n\left(\dot{\bar{\varepsilon}}^p, T\right) = n_0\left\langle 1 - D_2\left(\frac{T}{T_m}\right)\log\left(\frac{\dot{\bar{\varepsilon}}^p}{\dot{\varepsilon}_{\min}}\right)\right\rangle \qquad (8)$$

where B_0 is a material parameter, υ defines the temperature sensitivity, n_0 is the strain hardening exponent at T = 0 K, D_2 is a material constant and $\dot{\varepsilon}_{\min}$ is the lower limit of the model. The McCauley operator has the following properties: $\langle\bullet\rangle = \bullet$ if $\langle\bullet\rangle \geq 0$, or $\langle\bullet\rangle = 0$ if $\langle\bullet\rangle \leq 0$.

The effective stress $\sigma * (\bar{\varepsilon}_p, T)$, defined in Eq. (4), is derived from the concept of a thermal activation analysis. It describes the combined effects of the strain rate and temperature on plastic flow, and it is defined to be independent of strain [8].

$$\bar{\sigma}^*\left(\dot{\bar{\varepsilon}}^p, T\right) = \sigma_0^*\left\langle 1 - D_1\left(\frac{T}{T_m}\right)\log\left(\frac{\dot{\varepsilon}_{\max}}{\dot{\bar{\varepsilon}}^p}\right)\right\rangle^{m^*} \qquad (9)$$

where $\bar{\sigma}^*$ is a material parameter related to the threshold thermal stress, D_1 is a material constant, $\dot{\varepsilon}_{\max}$ is the maximum strain rate accepted for a particular material and m^* is a constant that determines the rate and temperature dependences of the material flow stress.

The drag stress $\bar{\sigma}_d(\dot{\bar{\varepsilon}}_p)$ is defined by Eq. (10). This formulation, originally developed by Nemat-Nasser [9], provides a phenomenological description of the rate-dependent strain within the viscous drag regime [7].

$$\bar{\sigma}_d\left(\dot{\bar{\varepsilon}}^p\right) = \chi \cdot \left[1 - \exp\left(-\alpha \cdot \dot{\bar{\varepsilon}}^p\right)\right] \qquad (10)$$

where χ and α are constants that represent the material rate sensitivity caused by the drag resistance to dislocation motion.

In the case of adiabatic conditions of deformation, the constitutive relation is combined with the energy balance principle, as shown in Eq. (11). This relation allows for an approximation of the thermal softening of the material via adiabatic heating.

$$\Delta T\left(\bar{\varepsilon}^p, \bar{\sigma}\right) = \frac{\beta}{\rho C_p}\int_0^{\bar{\varepsilon}^p} \bar{\sigma}\left(\bar{\varepsilon}^p, \dot{\bar{\varepsilon}}^p, T\right) d\bar{\varepsilon}^p \qquad (11)$$

Table 2. RK model coefficient for analysed structural steels.

Steel class Grade	θ	B_0	ν	σ_0	D_1	m	n_0	D_2
A–0 St0S-b	0.9	845	0.01	1200	0.50	3.4	0.28	0.10
A–I St3SX	0.9	797	0.02	1100	0.53	2.8	0.28	0.05
A–II 18G2	0.9	927	0.01	1000	0.53	2.7	0.24	0.01
A–III 34GS	0.9	1200	0.05	1500	0.53	3.4	0.21	0.01

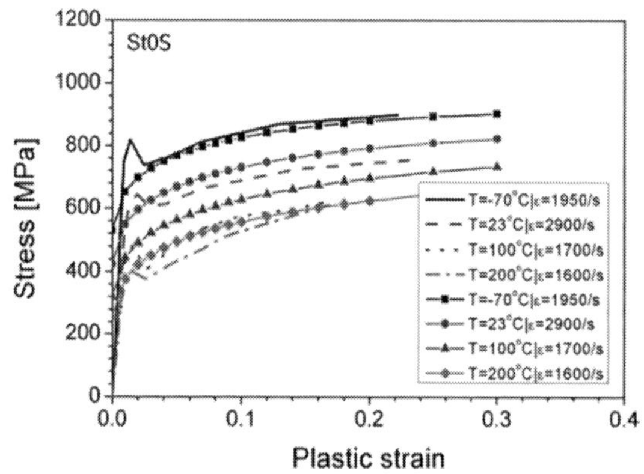

Figure 5. Comparison of experimental and RK model dynamic stress-strain curves at wide range of temperatures based data of St0S steel.

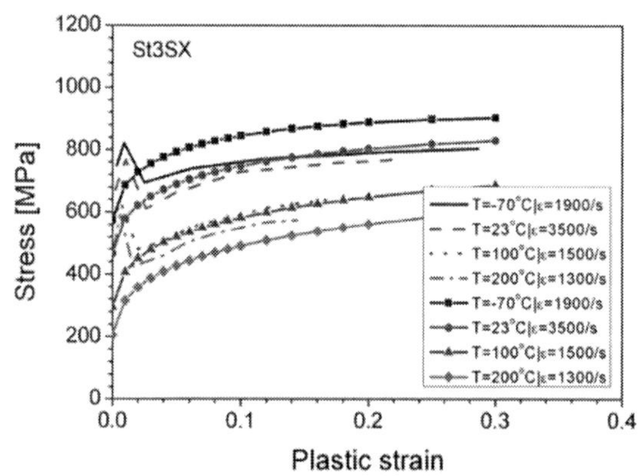

Figure 6. Comparison of experimental and RK model dynamic stress-strain curves at wide range of temperatures based data of St3SX steel.

where β is the Quinney-Taylor coefficient, which is assumed to be constant, ρ is the material density and C_p is the specific heat at a constant pressure. The transition from isothermal to adiabatic conditions is assumed to occur at $\dot{\bar{\varepsilon}}_p = 10\,\mathrm{s}^{-1}$, in agreement with experimental observations and numerical estimations reported in the literature [10]. The RK model coefficients obtained according to calibration procedure described in details previously [11] are presented in Table 2.

The dynamic stress-strain curves obtained experimentally and calculated on the basis of RK equation at wide

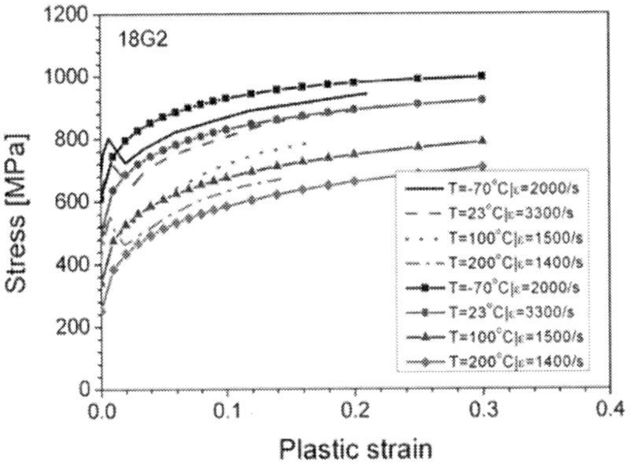

Figure 7. Comparison of experimental and RK model dynamic stress-strain curves at wide range of temperatures based data of 18G2 steel.

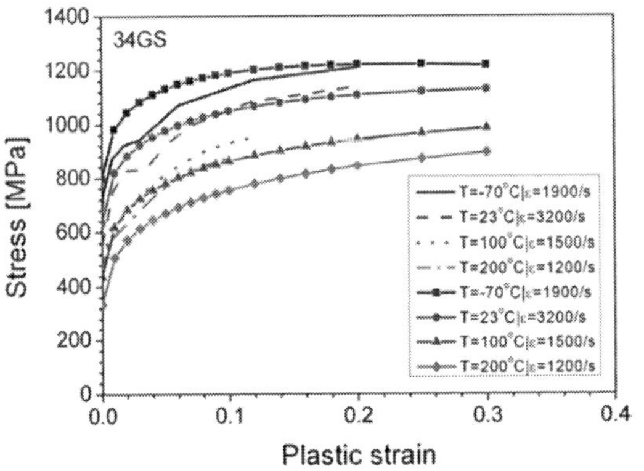

Figure 8. Comparison of experimental and RK model dynamic stress-strain curves at wide range of temperatures based data of 34 GS steel.

range of temperatures are presented in Figs. 5–8. For all analyzed cases model based data shows reasonable agreement with experimental ones at lowered, room and elevated temperatures as well.

4. Summary

The viscous-plastic characteristics of St0S, St3SX, 18 GS and 34 GS steels depends on temperature and strain rate. Therefore it is strongly recommended to take those relations into consideration in the case of numerical analysis of building behavior under extreme loading conditions i.e. fire, earthquake, blast or impact of building objects. The elasto-plastic behavior of structural steel may be expressed in a form of Rusinek-Klepaczko constitutive equation, which shows good correlation with experimental data.

References

[1] L. Kruszka, R. Rekucki, Applied Mechanics and Materials **82**, 422 (2011)

[2] V. G. Bazhenov, A I. Kibetz, L. Kruszka, International Journal of Protective Structures **3**, 449 (2012)

[3] M. V. Chernobryvko, Y. S. Vorobyov, L. Kruszka, International Journal of Protective Structures **3**, 141 (2012)

[4] W. Mocko, J.A. Rodriguez-Martinez, Z.L. Kowalewski, A. Rusinek. Strain **48**, 498 (2012)

[5] W. Moćko, L. Kruszka, Procedia Engineering **57**, 789 (2013)

[6] A. Rusinek, J. R. Klepaczko, International Journal of Plasticity **17**, 87 (2001)

[7] A. Rusinek, J.A. Rodriguez-Martinez, J.R. Klepaczko, R.B. Pecherski, Materials & Design **30**, 1748 (2009)

[8] S. Nemat-Nasser, W.G. Guo, V.F. Nesterenko, Mechanics of Materials **33**, 425 (2001)

[9] R. Kapoor, S. Nemat-Nasser, Metallurgical and Materials Transactions **A31**, 815 (1999)

[10] O. Oussouaddi, J.R. Klepaczko, Journal de Physique IV **1**, 323 (1991).

[11] A. Rusinek, J.A. Rodriguez-Martinez, A. Arias, International Journal of Mechanical Sciences **52**, 120 (2010)

EPJ Web of Conferences **94**, 05010 (2015)
DOI: 10.1051/epjconf/20159405010

Behaviour of a birch plywood under various experimental conditions

Lydie Caetano[1], Bertrand Galpin[2], Vincent Grolleau[1,a], and Jean-Damien Capdeville[3]

[1] Univ. Bretagne-Sud, EA 4250, LIMATB, 56100 Lorient, France
[2] Ecoles Militaires de Saint- Cyr Coëtquidan, Laboratoire de Mécanique et Matériaux, 56381 Guer, France
[3] GTT, Gaztransport et Technigaz, 78470 Saint Rémy-Lès-Chevreuses, France

Abstract. Liquid Natural Gaz -LNG- carriers must provide cryogenic reliability and safety for at least 40 years. Most of the insulation systems belong to the so-called membrane type, in which the "insulation complex" is composed of a metal membrane as a barrier, a plywood panel, insulation foam and a plywood panel glued to the inner hull. During construction and exploitation at cryogenic temperature, an impact of the membrane may occur and could lead to a LNG leakage. In order to evaluate this risk, and the role of the plywood in the perforation process of the membrane, a series of original impact tests are carried out at room and cryogenic temperatures on metal-plywood samples thanks to a specially designed catapult. In order to obtain a reliable numerical simulation of the impact, a series of impact and out of plane compression tests at room and cryogenic temperatures are carried out. Thanks to a design experiment, we identified the influence of the moisture content, from 2% up to 30%, and of the temperature, from −173 °C up to +20 °C, on the strain rate sensitivity of the plywood.

1. Introduction

Liquefied Natural Gas shipping transport is receiving more and more attention. LNG carriers must provide cryogenic reliability and safety for at least 40 years. Most of the insulation systems belong to the so-called membrane type, in which the "insulation complex" is composed of a metal membrane as a barrier (1.2 mm thick, 304 grade stainless steel), a plywood panel (12 mm thick, cross-ply plywood made from 9 birch veneers birch wood), insulation foam (Reinforced PolyUrethane Foam), and a plywood panel glued to the inner hull. During construction and exploitation, an impact of the membrane may occur and could lead to a LNG leakage. In order to evaluate this risk, and the role of the plywood in the perforation process of the membrane, a series of original impact tests are carried out on metal-plywood samples. A catapult and an instrumented projectile are used to perform impact tests. The results of the first tests lead us to conclude that the plywood plays a great role in the absorption of the impact energy. In a second phase, the out of plane behaviour of the plywood is studied under various experimental conditions of moisture content (MC), strain rate and temperature. As some other works revealed it, temperature [1], MC [2] and strain rate [3] have a great effect on the wood behaviour. But a huge interaction between temperature, strain rate and moisture content influences the stress plateau level [4,5]. Our work focuses on these interactions during the out of plane compression of the tested birch plywood.

[a] Corresponding author: vincent.grolleau@univ-ubs.fr

2. Impact tests

This section focus on the experimental set-up and the experimental evidence of the role the plywood plays on the perforation process of the membrane.

2.1. Experimental set-up

The instrumented projectile is propelled thanks to a specially designed pneumatic catapult, Fig. 1. The pneumatic catapult consists of a pneumatic piston propelling the projectile with an available kinetic energy of 2200 J (5kg at 30 m/s). The projectile is guided thanks to guiding rods.

The 2.5 kg projectile is composed of two parts, (i) the base block, which is instrumented with an accelerometer, (ii) the sensing projection, instrumented using strain gages in order to measure the impact force applied to the sensing projection end, Fig. 1, [6,7]. We use a pair of tensile half-bridge-embedded, opposite-mounted gauges so as to filter possible flexural strains. The signal is amplified using a quick A2 Vishay bay and recorded with an acquisition adapter (1-MHz and 16-bit digitizing). The sensing projection is covered by a so-called cap-impacting end, which protects the gages. 5 geometries of the impacting end, i.e. flat, dihedral, trihedral, hemispherical and plate edge are available, Fig. 1. Here, the presented tests are limited to the 6mm thick, 90° dihedral shape.

The membrane and the plywood-foam sandwich are simply put-up on the catapult bed and a blank-holder applies a 2 MPa pressure load on the periphery of the membrane, along a 50 mm wide strip. The dimension of the membrane is 400 × 400 mm, 270 mm thick.

For cryogenic temperature tests, the plywood-foam sandwich is soaked onto liquid nitrogen prior to the test and liquid nitrogen is continuously sprinkled on

Article available at http://www.epj-conferences.org or http://dx.doi.org/10.1051/epjconf/20159405010

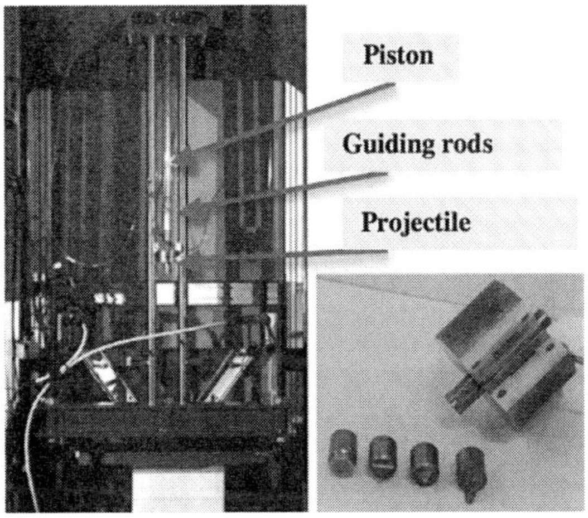

Figure 1. Pneumatic catapult (left side) and instrumented projectile along with some impacting ends (right side).

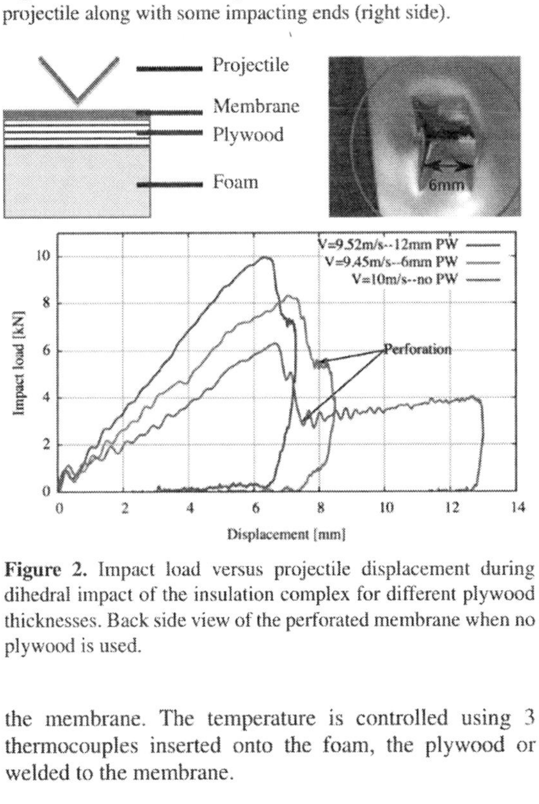

Figure 2. Impact load versus projectile displacement during dihedral impact of the insulation complex for different plywood thicknesses. Back side view of the perforated membrane when no plywood is used.

the membrane. The temperature is controlled using 3 thermocouples inserted onto the foam, the plywood or welded to the membrane.

2.2. Impact test results

In order to evaluate the role of the plywood in the perforation process, a series of impact test is performed with 12 mm thick, 6 mm thick or without plywood between the membrane and the foam. The presented results are obtained at ambient temperature, and the initial velocity of the 2.5 kg, dihedral shape instrumented projectile is 9.7 ± 0.5 m/s. Figure 2 shows the evolution of the impact load, namely the contact force between the membrane and the projectile, as a function of the solid displacement of the projectile. Thus, the area under the curve represents the amount of initial kinetic energy consumed during the impact. It is clear from Fig. 2 that the plywood influences drastically the perforation process of the membrane: when a 12 mm thick plywood is used, the perforation is not complete and the membrane deforms plastically until the projectile rebounds. Under the same conditions, a perforation occurs for 6 mm thick plywood and without plywood. The maximum impact load and the total lost in kinetic energy when the maximum impact load is reached are 40% higher in the case of 12 mm thick plywood than without plywood.

During the impacts, the plywood is mainly loaded under out of plane compression conditions. The lack of published results on this subject for the cryogenic conditions leads us to study more precisely the plywood behaviour under cryogenic and high strain rate conditions.

3. Out of plane compression of the plywood

The 12 mm thick cross-ply plywood is made from 9 birch veneers, which are peeled from logs. Thus the out of plane axis of the board refers to the radial direction of the wood log. The compressive behaviour of the wood along this direction is known to be temperature, rate and moisture content sensitive, and well predicted from the theory of cellular solids [8].

3.1. Experimental protocol

A series of quasi-static out of plane compression tests have been performed with various diameters of specimens, from 12 up to 50 mm, and various numbers of stacked samples, from 1 to 3, in order to ensure that no size-effect would affect our results. In the following, a single 28 mm diameter circular sample is used.

The amount of moisture in wood,

$$MC = \frac{m_{wet} - m_{dry}}{m_{dry}} \tag{1}$$

where m_{wet} is the mass of the specimen at a given moisture content and m_{dry} is the mass of the ovendry specimen, is expressed as a percentage of wood mass when oven-dried [9].

In the following, the moistures contents are 2, 4, 8, 15% and fiber saturation point (FSP), about 28%.

Quasi-static tests refer to the 4.e-4, 4.e-3, 4.e-2 and 4.e-1 [/s] strain rate tests, while high strain rate tests refer to a 700/s strain rate.

In order to consider the variability due to the natural origin of the wood, each experimental condition is repeated at least five times. The curves presented Figs. 5 and 6 are the most representative ones of each series.

The temperatures of the tests refers to the initial temperature of the sample at the beginning of each test, and are in the $[-173°; +25°]$°C range. Every sample is radially drilled at mid span; the hole is 0.63 mm diameter and 9 mm long. These dimensions allow us to insert a T-type thermocouple within the sample. The area of each sample is about 615 mm^2 while the total area of this hole is less than 6 mm^2, thus the behaviour of the sample is not affected by the hole.

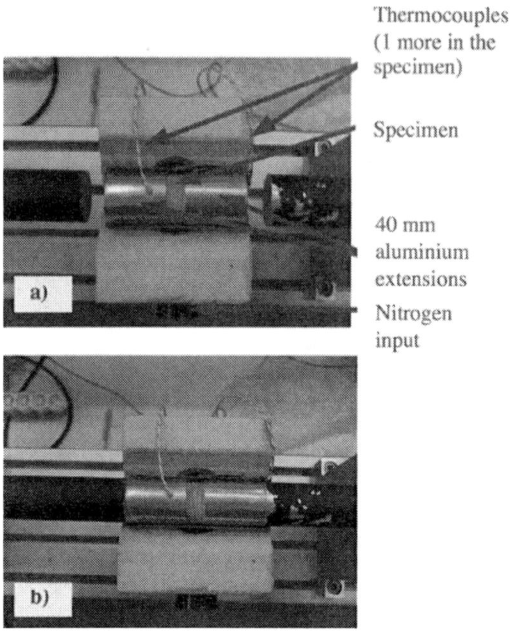

Figure 3. Extension rods – a) before contact – b) in contact.

Figure 4. Cryogenic chamber completely close.

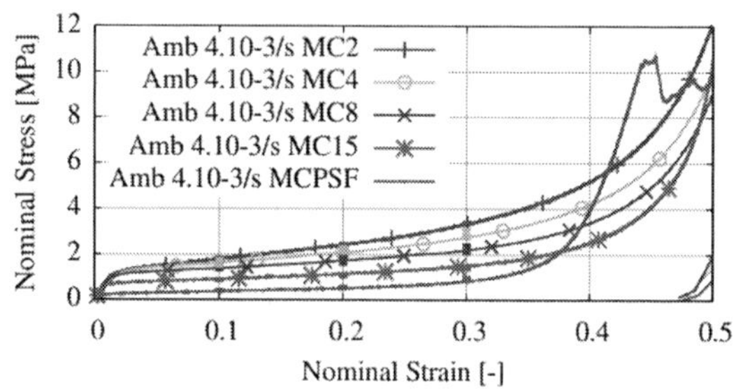

Figure 5. Engineering nominal stress versus nominal strain evolution for 12 mm thick birch plywood at 4e-3s-1 and room temperature for various moisture contents.

3.2. Quasi-static tests

For quasi-static tests, a cryogenic chamber was conceived. Temperature is controlled thanks to 3 thermocouples located on the compression plateau, the chamber and the sample. A control loop regulates the temperature thanks to a remote valve and a liquid nitrogen sprinkler located on the top of the chamber. During the tests, the amplitude of variation of the initial temperature was found to be less than $\pm 10\,^{\circ}$C.

3.3. High strain rate SHPB tests

For SHPB compression tests, 30 mm diameter aluminium bars and projectile are used. The 1 m long projectile impact generates a 400 μs incident signal duration and the initial speed, about 10 m/s, leads to a strain rate about 700/s.

For cryogenic tests, we decided to keep the bars at room temperature, and in order to avoid any warm-up of the sample when in contact with the bars, the sample is cool-down along with two 40 mm long 30 mm diameter aluminium extensions rods placed on the two faces of the sample, Figs. 3 and 4. Preliminary tests showed that the extensions do not affect the strain wave propagation even at cryogenic temperatures, and the only consequence is the modification of the strain-gage to bar end distances. Prior to the test, the sample and the two extension rods are placed in a climatic chamber. This specially designed chamber is made of polyurethane foam V-holder, and a foam cap. The axis of the chamber is coaxial to the bars axis and can move freely along this axis. A sprinkler cool-down the sample and the extension rods with a temperature regulated nitrogen gaz. During the tests, the temperatures of the injected nitrogen, the sample and the extension rods are recorded, the amplitude of variation of the initial temperature was found to be less than ± 5°C.

4. Results

Under quasi-static and ambient temperature conditions, the compressive behaviour of the 12 mm thick birch plywood follows the classic behaviour of cellular materials. The MC affects drastically the stress level: the higher the MC, the lower the stress plateau value. The plywood behaviour can be described in three states: the elastic part, Fig. 5 until about 0.015 nominal strain at 8%MC, a stress plateau, Fig. 5 until about 0.45 strain at 8%MC, and the densification part, Fig. 5 over 0.45 strain at 8%MC.

The strain rate sensitivity of the plywood behaviour is huge. For example, for a 8%MC and ambient temperature test, the engineering stress is 100% higher at 0.1 nominal strain and 120% higher at 0.2 strain at 700/s strain rate than for 4.e-3/s. This difference exists under cryogenic temperature too and for each moisture contents, greater or less.

The cryogenic condition plays also a major role. For example, when the temperature falls from room to cryogenic conditions, for a 8%MC and 4.e-3/s strain rate test, the stress is 140% higher at 0.1 strain, 180% higher at 0.2 strain and 230% higher at 0.3 strain. It can be noticed that the evolution is not linear, as it is not linear for the effect of the strain rate too.

Under cryogenic and high strain rate conditions, it is clear from Fig. 6 that the interaction between temperature drop, MC and strain rate sensitivity reverse the MC effect on the stress plateau value: the higher the MC, the higher the stress plateau value under cryogenic, $-173\,^{\circ}$C, and high strain rate conditions about 700/s.

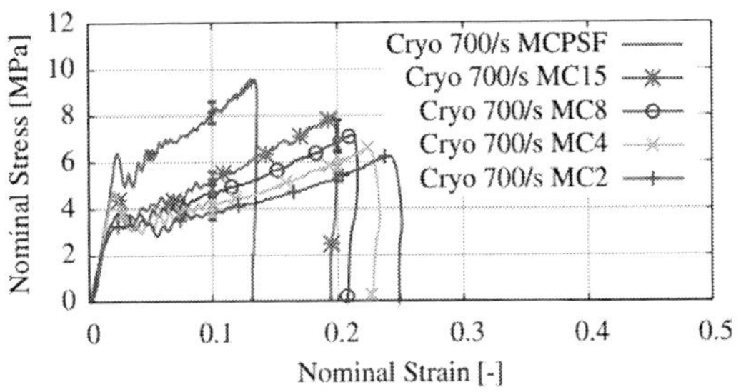

Figure 6. Engineering nominal stress versus nominal strain evolution for a 12 mm thick birch plywood at 700s-1 and cryogenic temperature for various moisture contents.

5. Conclusion

Impact tests have been performed on metal-plywood-foam complex. It has been proven that the plywood influences drastically the perforation process of the metal membrane. Some original compression tests have been done and different devices to impose moisture content, temperature and/or strain rate were developed. The main experimental result is the significant level of the strain rate, moisture content and temperature interactions on the stress plateau level during out of plane compression of this 12 mm thick birch plywood.

The results lead us to several conclusions on the effect of these three parameters on birch plywood behaviour and bring us experimental data to develop a numerical model.

References

[1] K.H. Boller, *Strength of wood at low temperatures, Low temperature test methods and standards for containers,* Committee on packing, packaging and preservation (December 1953)

[2] K. Siim, R. Kask, H. Lille and E. Täkker, *8th International DAAAM Baltic Conference, Study of physical and mechanical properties of birch plywood depending on moisture content,* Ind. Eng. (2012)

[3] S.R. Reid and C. Peng, *Dynamic uniaxial crushing of wood*, IJIE, **19**: 531–570 (1997)

[4] C.C. Gerhards, *Effects of moisture content and temperature on the mechanical properties of wood : an analysis of immediate effects,* W&F, **14**: 4–36 (1982)

[5] S. Widehammar, *Stress-Strain Relationships for Spruce Wood: Influence of Strain Rate, Moisture Content and Loading Direction*, SEM, **44**: 44–48 (February 2004)

[6] B. Galpin, V. Grolleau, S. Umiastowski, G. Rio, and L. Mahéo. *Design and application of an instrumented projectile for load measurements during impact*, IJC, **13**(2): 139–148 (2008)

[7] S. Tanimura. *A new method for measuring impulsive force at contact parts*, Expe. Mec., **24**: 271–276 (1984)

[8] L.J. Gibson and M.F. Ashby. *Cellular Solids – Structure and Properties*, Cambridge University Press, second edition (August 1999)

[9] Forest Service. *Wood handbook, wood as an engineering material. Technical report, United States Department of Agriculture*, Forest Service (2010)

EPJ Web of Conferences **94**, 05011 (2015)
DOI: 10.1051/epjconf/20159405011

Dynamic behavior and failure of the base and heat affected materials of a HSS fillet welded joint

Julien Carrier[1,2,a], Eric Markiewicz[1], Grégory Haugou[1], David Lebaillif[2], Nicolas Leconte[1], and Hakim Naceur[1]

[1] Université de Valenciennes, LAMIH CNRS 8201, Le Mont-Houy, CISIT, 59313 Valenciennes Cedex 9, France
[2] Nexter Systems, 11 allée des Marronniers, 78022 Versailles, France

Abstract. Welded joints, due to their manufacturing process, are commonly weakened areas. This study analyses the dynamic behavior of the Base Metal (BM) and the Heat-Affected Zone (HAZ) materials of a HSS (High Strength Steel) fillet welded joint. First, a specific approach is developed to generate the HAZ material using a thermal treatment. Hardness and grain size are used to validate the replicated HAZ. This approach appears efficient and repeatable. Secondly, the true stress-strain quasi-static and dynamic behaviors up to failure of the BM and the HAZ are determined. This characterization is performed thanks to video tracking procedure and Bridgman-LeRoy correction. The comparison between these two materials shows that the thermal field of the welding process increases the HAZ yield stress and hardening while decreasing the strain at failure. It appears that the base metal is not rate sensitive from quasi-static up to $1350\,s^{-1}$. On the contrary, the heat affected material appears to be rate sensitive but by softening. This unexpected dynamic material softening requires further analyses.

1. Introduction

Finite element modeling is currently used to model the dynamic behavior of structures submitted to blast events. It allows to design armoured vehicles by evaluating the strength and integrity of the structure against battlefield threats. Welded joints, due to their manufacturing process, are weakened areas. Since it is difficult to model the failure of welded joints in a coarse model, a meso-scale model is proposed. So, mechanical properties up to failure of the constitutive materials: Base Metal (BM), Heat-Affected Zone (HAZ) and Weld Metal (WM) have to be supplied. In this paper, only the BM and the HAZ are presented. Multi-scale approaches have been already successfully applied on a mild-steel spot-welded assembly ([1] & [2]) or on aluminum welded joint [3]. A similar method is investigated for a HSS fillet welds assembly subjected to dynamic loadings. Several studies about the mechanical strength of welded assemblies published in the open literature cover a large field of loadings: fatigue [4], quasi-static [5] and low velocity loadings [6]. However, finite element models of military applications are not well expanded in literature. Gaudreault [7] is almost the only one who presented a study dedicated to a welded HSS joint subjected to an explosion. The welded joint is modeled by a shell element with the HAZ materials properties. The weld geometry is not considered so the stress concentration (and the fracture behavior) could not be really predictive. Few papers reported the HSS material behavior at high strain rate used for armoured vehicles ([8] & [9]), even less under a tensile loading as well as the dynamic behavior of welded assembly manufactured with these specific steels. Besides, open literature does not cover mechanical strength of the associated HAZ material. In this paper, a specific approach is presented to replicate the HAZ material using a thermal treatment. Then, quasi-static and dynamic behavior up to failure of the BM and the HAZ are compared and analyzed.

2. Experimental program

The manufacturing process and tensile specimen geometries, as well as experimental tests, measurement facilities and analysis procedure are presented in this section.

2.1. Materials

The BM is a hot-rolled HSS that offers high strength, ductility and good weldability. Due to the small size of the HAZ, dumbbell tensile specimens cannot be extracted directly from the welded joint to perform quasi-static and dynamic tests. As a consequence, a specific approach is developed to generate the HAZ material using a thermal treatment. Both hardness and grain size are controlled to validate the replicated HAZ with respect to the real one. Grain size is measured by the grain size number (*G which is an integer)* through the following relation (defined by the norm):

$$M = 8.2^G \quad (1)$$

where M is the number of grain per mm^2 on the observed sample surface. The Vickers hardness is measured along 6 paths across the BM, HAZ and WM. As seen in Fig. 1, the hardness between BM and HAZ increase significantly.

The hardness in the heat affected material doesn't remain constant along the path. The question is: how to take into account the gradient of properties? Our choice is to determine the extreme characteristics of the HAZ: on one hand it's given by the BM and, on the other hand, by

[a] Corresponding author: j.carrier2@nexter-group.fr

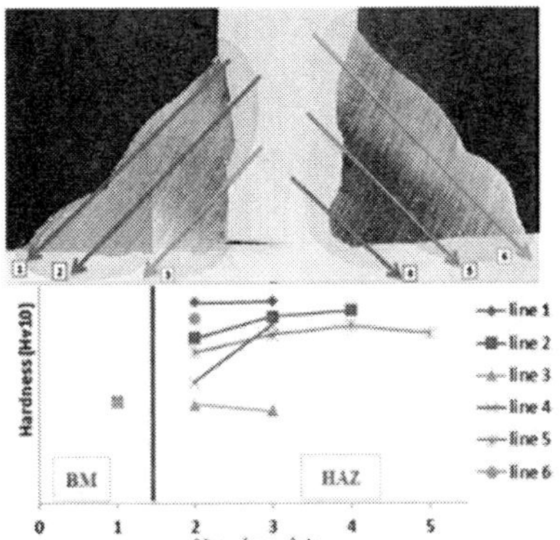

Figure 1. Hardness specimen and hardness measurement in the BM and HAZ materials.

the maximum hardness zone in the HAZ. This hardness is then considered for the heat treatment. Furthermore, this maximum hardness is located where the maximum deformation will occur under dynamic loadings due to high geometrical stress concentration. The aim of this heat treatment is not to re-create the real and complex thermal cycle induced by the welding process (i.e. by using a Gleeble® thermomechanical simulator), but only to obtain specimens with mechanical properties close to the real HAZ.

The heat treatment is detailed below. The sample is put in a preheat furnace. The temperature is risen to a specific temperature (austenitic temperature – Tc) and remains constant during several minutes. The sample is then tempered to fix the microstructure. To adjust the final hardness of the sample a low temperature stress-relieving treatment is done.

A preliminary study was carried out to adjust the temperature (Tc) to obtain the correct austenite grain size in the material before the tempering (Fig. 2a). The first temperature tested was too low to completely remove ferrite and pearlite in the material. As the austenitic temperature raises, the hardness and the grain size number decrease.

The HAZ microstructure and the hardness are finally validated by comparison with the real one (obtained in the welded joint) (Fig. 2).

To avoid excessive distortions of the specimen, the thermal cycle is applied on the rough sample, and then the final geometry is machined.

The heat treatment process is applied on 5 samples per batch: 4 samples will be machined at the tensile specimen geometry; the last one is the batch control specimen. This one is cut in the middle to make hardness measurements on the position corresponding to the gage cross section (Fig. 3). 6 batches of 5 samples are made. Hardness measurements remain constant for all batches: less than 2% of dispersion is measured.

2.2. Specimen geometries

Quasi-static and dynamic tests are performed on smooth axisymmetric specimens at room temperature. As shown in Fig. 4, specimen geometries with different gage lengths

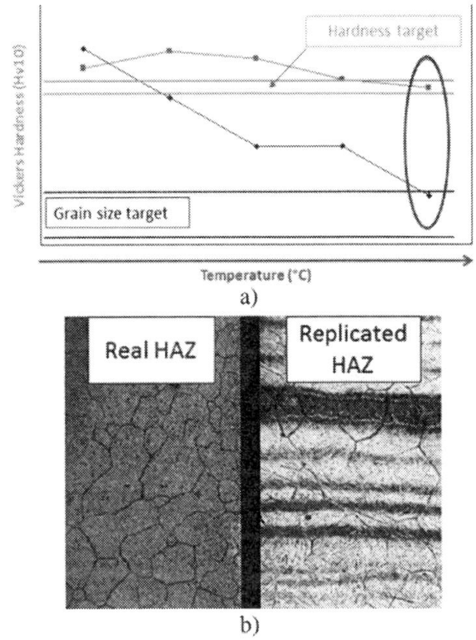

Figure 2. a) Description of the optimized heat treatment; b) comparison of real and replicated HAZ microstructure.

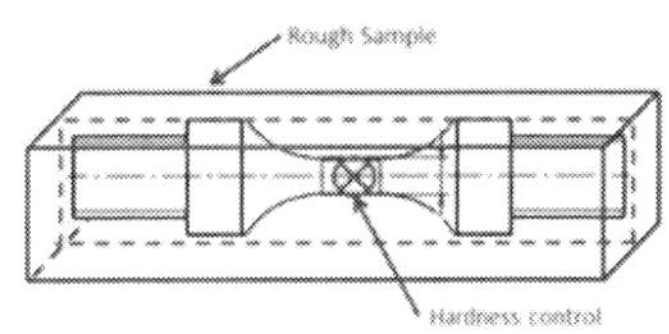

Figure 3. Rough sample.

and diameters are required to cover a large range of strain rates from quasi-static to $1350\,s^{-1}$. Quasi-static geometry complies the standards (Fig. 4a). Different dynamic geometries are considered in order to compare BM and HAZ materials properties at the same strain rate. Gage diameters have been adapted to perform dynamic tests with the same pre-loading force on a pre-stretched tension bar. The strain rates are measured in the gage length of the sample during plasticity before necking and after the wave stabilization. All tests are performed 3 times for quasi-static tests and 5 times for dynamic tests.

2.3. Pre-stretched tension bar

Experimental program based on dynamic loadings is carried out using Hopkinson bars technique. As the high ductility properties of the base material and heat affected zones have been observed under quasi static conditions, a high duration time is wished. For that, as shown by Fig. 5, a pre-stretched bar device, based on the original concept proposed by Albertini [10], is used to determine the set of flow curves for moderate strain rates up to failure. Both bars have a common diameter equal to 11 mm and the specimen is inserted between positions D and E. The pre-loading force F – stabilized between A and B – is

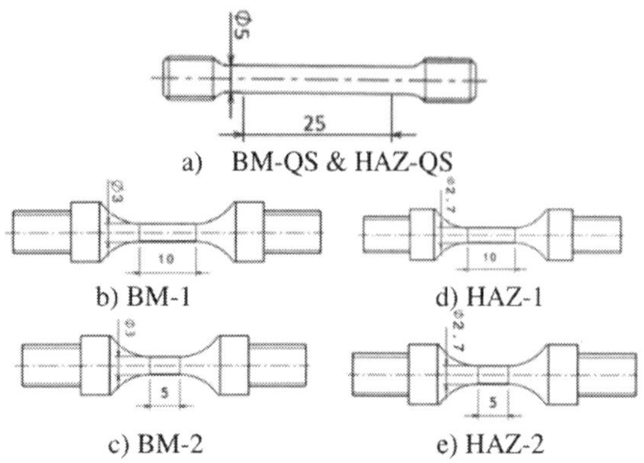

Figure 4. Geometries of tensile specimens: a) quasi-static for BM and HAZ; b) c) dynamic for BM; d) e) dynamic for HAZ.

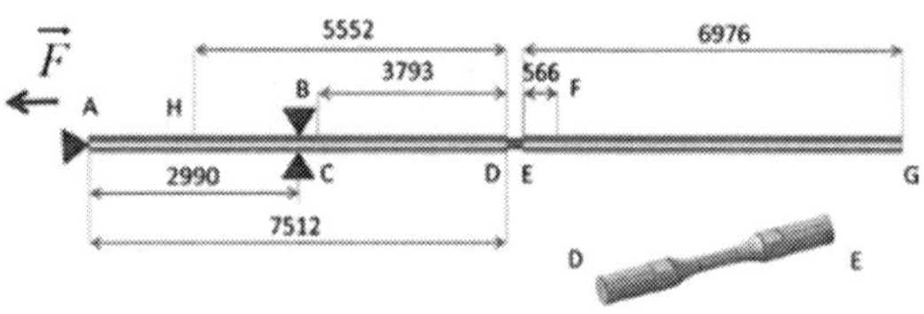

Figure 5. Sketch of the pre-stretched tension bar – Dimensions in mm.

provided by two quasi-static loadings applied on the input bar (Fig. 5).

This ensures the storage of an elastic energy brutally released when the brittle fracture of the fuse occurs at the position B. The elastic wave propagates with a constant duration time depending on the length of the pre-stretched part (AB) along the input bar and loads the specimen up to failure. A set of three full strain bridges HBM LY 41 350Ω – 1.5 mm in length) is strategically positioned along the experimental set-up to detect the elastic waves system using a high speed recorder (Yokogawa DL 750 – 4 channels). The data are finally recorded to compute the stress/strain curves [11]. Here, the position in B is chosen so that no superposition of the reflected wave with the incident one can occur at the position of the strain gage bonded in C. The strain rate value is changed by varying the pre-loading force F of the pre-stretched device.

2.4. Measurement of specimen contour evolution by video tracking

Based on previously published works ([12] & [13]), a methodology is set up to measure the specimen contour evolution after necking to obtain the true stress-strain curve of the material. A CMOS high speed camera (Photron APX RS 3000) is added to record the shape evolution of the specimen during the test. As sharp edges between the background and the specimen are needed, a flashlight is placed behind the specimen so as to highlight its shadow.

Then, the elongation and reduction of the specimen can be quantified using a Matlab® routine. Figure 6 provides details on the optical measurement device and the image analysis progress. The speed of the camera is set to 75000

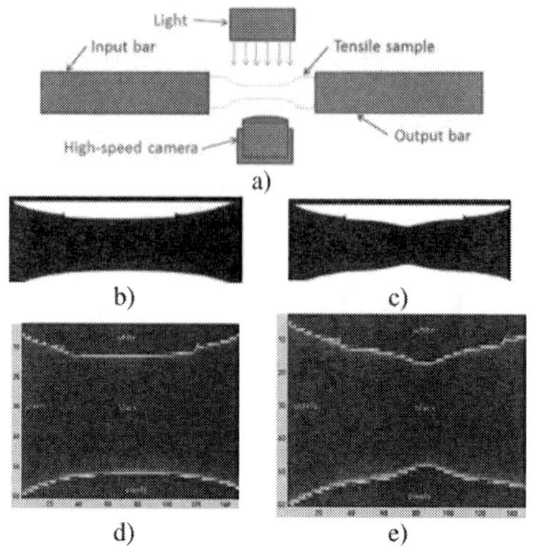

Figure 6. a) Optical measurement device; b) c) Initial images at the start of test and during test; d) e) Images after filtering procedure.

and 90000 f/s depending on the range of strain rates so as to ensure over 20 images during the test.

Diameter reduction measurement, in particular the smallest diameter after necking, is obtained by a tracking algorithm with a filter on the contrast gradient. The minimum diameter is the minimum distance between the two filtered edges of the specimen. The first image, just before the test starts, is used to calibrate the original gage diameter of the specimen (in pixels) (Fig. 6b & d). The final diameter measured by video tracking (last diameter at fracture initiation) is compared to the post-mortem diameter to validate the digital calibration. The true stress-strain curve is obtained using the following relations:

$$\varepsilon_{true} = 2\ln(D_0) - 2\ln(D) \tag{2}$$

where D_0 is the initial gage diameter and D is the minimum diameter in the necked area.

$$\sigma_{true} = \sigma_{zz} = \frac{4F}{\pi \cdot D^2} \tag{3}$$

where F is the force measured during the test.

The high speed camera is limited to a picture resolution of 64 × 250 pixels on the specimen gage section. This low resolution and the limited number of recorded images (about 20) lead to an error of about 0.05 mm on the measured diameter. Therefore, after the onset of necking, an error of about 50MPa on the true stress and 0.03 on the true strain is estimated.

2.5. Bridgman-LeRoy correction

True stress-true strain diagram cannot be computed beyond necking, unless the effects of necking on the geometry of the tensile specimen and the stress state are accurately quantified. Necking produces a triaxial stress state that does not reflect the true uniaxial flow stress of the material. Therefore, the true uniaxial stress (σ_{zz}) must be multiplied by a correction factor to correct the effect of the triaxial

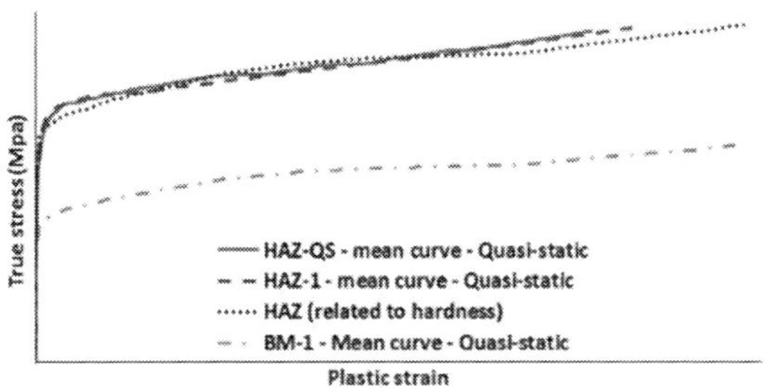

Figure 7. BM and HAZ quasi-static true stress-strain curves.

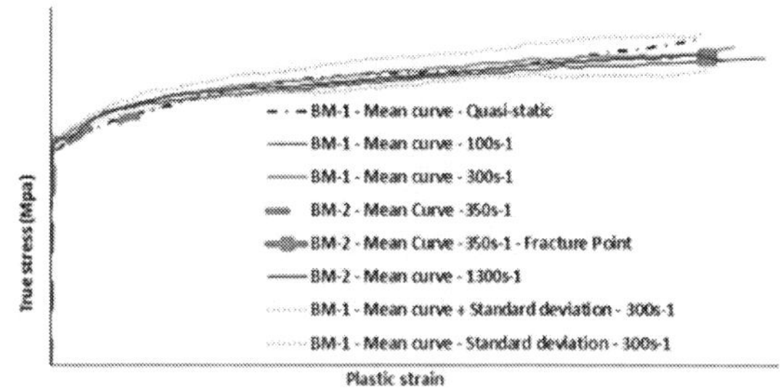

Figure 8. BM – True stress-strain curve – Strain rate dependency.

stresses and obtain the equivalent stress (σ_{eq}). Necking effects are easily quantified for specimens with circular cross-sections using the Bridgman correction approach. In the center of the sample the triaxiality stress ratio is calculated by the relation:

$$\sigma^* = \frac{1}{3} + \ln\left(1 + \frac{a}{2R}\right). \tag{4}$$

Where a is the current radius of the neck and R is the radius of curvature of the neck surface in the longitudinal plane at the minimum section. The Bridgman and LeRoy correction ([14] & [15]) is defined by:

$$\overline{\sigma_{eq}} = \frac{\sigma_{zz}}{\left(1 + \frac{2R}{a}\right)\ln\left(1 + \frac{a}{2R}\right)}. \tag{5}$$

LeRoy [15] found that the neck geometry could be calculated with the following approximation and gives good results for steel with $\kappa = 1.1$:

$$\frac{a}{R} = \kappa\left(\overline{\varepsilon_p} - \varepsilon_u\right) \text{ for } \overline{\varepsilon_p} > \varepsilon_u. \tag{6}$$

3. Experimental results

In the first part of this section, the BM and HAZ mechanical behaviors are compared thanks to their true stress-strain behavior laws. The strain-rate dependency of each material is investigated in the second part. Those true stress-strain behavior laws are obtained up to failure by using specimen contour evolution by video tracking associated with the Bridgman-LeRoy correction. It can be noted that, across all further analyses, for each test the strain rate is considered to be constant until failure occurs and is equal to the value measured in the gage length before necking. In the neck, the local strain rate increases significantly with the diameter reduction. To access to this strain rate evolution the measurement of strain fields by DIC would be necessary.

3.1. Quasi-static BM and HAZ tests results

The quasi-static true stress-strain behaviors for both materials are presented in Fig. 7. The specimen geometries are referred by BM-1, HAZ-QS and HAZ-1 (Fig. 4). Except the experimental scatter at failure, the true stress-strain behavior related to the HAZ with the two different specimen geometries are similar (curves HAZ-1 & HAZ-QS in Fig. 7). The Bridgman-LeRoy correction allows to determine the local material behavior. The microstructure of the MB is a martensitic lath structure which provided ductility and toughness. The heat treatment has created a new martensitic structure with a smaller grain size number. So, the yield stress and ultimate tensile stress have increased on the HAZ material but strain at failure has decreased. In fact, the applied thermal cycle is known to act as a heat treatment which hardens the material.

An alternative way, based on an engineering approximation, allows to calibrate the HAZ behavior law by scaling with a constant factor, the BM behavior law related to the hardness of HAZ material. The obtained HAZ behavior is quite close to the real measured HAZ behavior (Fig. 7).

3.2. BM strain rate dependency

In this section, the BM true stress-strain behavior is analyzed for strain rates from the quasi-static range up to 1300s^{-1} (Fig. 8). The specimen geometries are referenced by BM-1 and BM-2 which only differ by the gage length (Fig. 4). Only the experimental scatter at $300\,\text{s}^{-1}$ with the BM-1 specimen geometry is presented here. To compare the impact of the specimen geometry, additional tests at $300\,\text{s}^{-1}$ (and $350\,\text{s}^{-1}$) for both BM specimen geometries (BM-1 and BM-2) are led. As a video problem appeared during test at $350\,\text{s}^{-1}$ on the BM-2 specimen geometry, only the pre-necking behavior law and the stress-strain point at failure calculated post-mortem are presented. We observe that BM-2 hardening law is contained in the BM-1 experimental scatter at $300\,\text{s}^{-1}$. All mean curves, from quasi-static up to $1300\,\text{s}^{-1}$, are contained in the BM-1 experimental scatter at $300\,\text{s}^{-1}$, so the BM hardening is not strain rate sensitive.

We can conclude that, thanks to the specimen contour tracking and the Bridgman-LeRoy correction, the geometry effects after necking do not seem to impact the BM material behavior or it is drowned in the experimental scatter. Regarding the strain at failure, a reverse engineering approach, based on the engineering stress and strain behavior laws up to failure, is however needed to identify a geometry insensitive failure criterion.

Triaxiality factor for quasi-static and dynamic tests (with the different geometries) are in the order of 0.8 at failure in the center of the specimen.

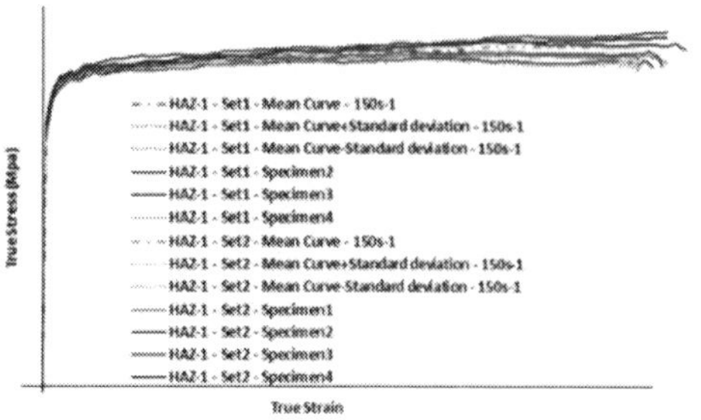

Figure 9. HAZ – True stress-strain curve – Experimental scatter.

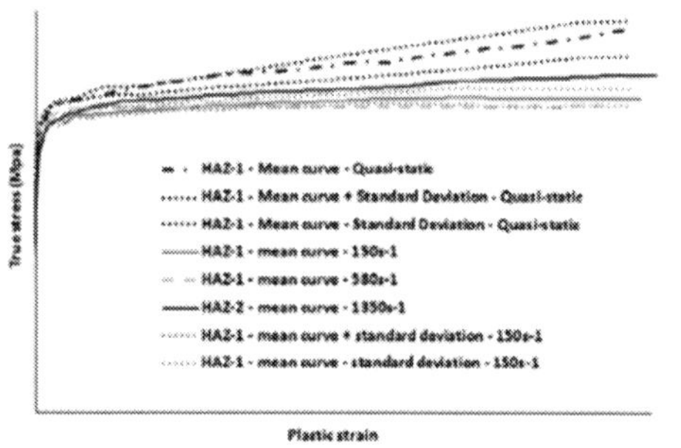

Figure 10. HAZ – True stress-strain curve – Strain rate dependency.

3.3. HAZ strain rate dependency

In this last part, the HAZ behavior is studied at different strain rates from quasi-static up to $1350\,s^{-1}$ (Fig. 10). The specimen geometries are referenced by HAZ-1 and HAZ-2 (Fig. 10). To study and validate the repeatability of the applied thermal cycle, two sets of 4 samples are tested with the same strain rate ($150\,s^{-1}$). One set shows more scatter (a relative difference of 7% on the strain at failure) against 1% for the other. Nevertheless, the both two mean curves are included in the experimental scatter of the other one (Fig. 9). It can be concluded that thermal cycle is a repeatable operation.

Figure 10 shows, firstly, a softening between quasi-static and dynamic tests results, both on yield stress and stress at failure. Secondly, for dynamic tests, the strain rate dependency between $150\,s^{-1}$ and $1350\,s^{-1}$ is not explicit. Indeed, with the HAZ-1 geometry, there is a softening with the increase of the strain rate ($150\,s^{-1}$ and $580\,s^{-1}$). Tests performed with the HAZ-2 geometry at $1350\,s^{-1}$ shows a slight hardening by comparison to HAZ-1 dynamic tests results. Experimental scatters could not explain this gap. Stress triaxiality factor are quite similar between all those tests.

At this time we are not able to explain those unexpected results. Regarding the softening, the microstructure and the chemical composition generated by the heat treatment will be deeply analyzed across the tested specimens. Hardness measurements, across a longitudinal section in the post-mortem sample, will be also made. Regarding potential geometry effects, quasi-static and dynamic ($580\,s^{-1}$) tests with the HAZ-2 specimen geometry will be led to quantify if geometrical effects impact the HAZ true stress-strain curve.

4. Conclusion and prospects

This paper presents the evolution of the behavior between BM and HAZ materials of a HSS fillet welded joint assembly from quasi-static to dynamic loadings. First, a specific approach is developed to generate the HAZ material using a thermal treatment. Hardness and grain size are used to validate the replicated HAZ material. This approach appears efficient and repeatable. To cover a large range of strain rates, different specimen geometries are needed. A direct method, using video tracking of the specimen geometry and Bridgman-LeRoy correction is used to identify the true stress-strain curve. Secondly, quasi-static and dynamic behaviors from BM and HAZ are compared. The thermal cycle affects significantly the material behavior. Indeed, on the HAZ the hardening is increased whereas strain at failure is decreased. Regarding the strain rate effects, it appears that the BM is insensitive while the HAZ presents a softening between quasi-static and dynamic tests. Further analyses are needed on HAZ sample (hardness, chemical composition and microstructure) to understand this dynamic material softening.

References

[1] E. Markiewicz, P. Ducrocq, P. Drazetic, G. Haugou, T. Fourmentraux, J.Y. Berard, Int. J. of Materials and Product Technology, **16**, 484-509 (2001).
[2] B. Langrand, E. Markiewicz, Int. J. of Impact Engineering, **37-7**, 792–805 (2010).
[3] T. Wang, Ph.D. thesis, NTNU, 2006.
[4] F. Labesse-Jied, B. Lebrun, E. Petitpas, J.L. Robert, Proceedings of the 6th Int. Conference on Biaxial/Multiaxial Fatigue and Fracture., Lisbon, 2001, Editor: M. De Freitas, 81–89.
[5] T.K. Chan, R.F.D. Porter Goff, Thin-Walled Structures, **36**, 265–287 (2000).
[6] H.G. Hildrum, Ph.D. thesis, NTNU, 2002.
[7] P. Gaudreault, A. Bouamoul, R. Durocher, B. St-Jean, Proceedings of the 22nd Int. Symposium on Ballistics, Vancouver, BC, Canada, 2005.
[8] B. Gailly, Ph.D. thesis, Mines de Paris, 1996.
[9] P. Chevrier, Ph.D. thesis, Université de Metz, 1998.
[10] C. Albertini, M. Montagnani, Mechanical Properties at High Strain Rates, **21**, 22–32 (1974).
[11] H. Kolsky, *Stress waves in solids*, Dover Publications Inc., New York, 1963.
[12] O.S. Hopperstad, T. Borvik, M. Langseth, K. Labibes, C. Albertini, Eur. J. of Mech. A/Solids, **22**, 1–13 (2003).
[13] T. Borvik, O.S. Hopperstad, T. Berstad, Eur. J. of Mech. A/Solids, **22**, 15–32 (2003).
[14] P.W. Bridgman, *Studies in Large Plastic Flow and Fracture*, McGraw-Hill, 1952.
[15] G. Le Roy, J.D. Embury, G. Edward, M.F. Ashby, Acta Metall., **29**, 1509–1522 (1981).

AUTHOR INDEX

AUTHOR INDEX

AUTHOR INDEX

AUTHOR INDEX

AUTHOR INDEX

EDP Sciences – Web of Conferences
17, Avenue du Hoggar
Parc d'Activité de Courtabouf
BP 112
F-91944 Les Ulis Cedex A
France

ISBN 978-1-5108-2330-3

9 781510 823303

CURRAN ASSOCIATES INC.
proceedings.com